**Oral Bioavailability and Drug Delivery**

## Wiley Series in Drug Discovery and Development

**Binghe Wang**, Series Editor

*Drug Delivery: Principles and Applications*
Edited by Binghe Wang, Teruna Siahaan, and Richard A. Soltero

*Computer Applications in Pharmaceutical Research and Development*
Edited by Sean Ekins

*Glycogen Synthase Kinase-3 (GSK-3) and Its Inhibitors: Drug Discovery and Development*
Edited by Ana Martinez, Ana Castro, and Miguel Medina

*Aminoglycoside Antibiotics: From Chemical Biology to Drug Discovery*
Edited by Dev P. Arya

*Drug Transporters: Molecular Characterization and Role in Drug Disposition*
Edited by Guofeng You and Marilyn E. Morris

*Drug-Drug Interactions in Pharmaceutical Development*
Edited by Albert P. Li

*Dopamine Transporters: Chemistry, Biology, and Pharmacology*
Edited by Mark L. Trudell and Sari Izenwasser

*Carbohydrate-Based Vaccines and Immunotherapies*
Edited by Zhongwu Guo and Geert-Jan Boons

*ABC Transporters and Multidrug Resistance*
Edited by Ahcène Boumendjel, Jean Boutonnat, and Jacques Robert

*Drug Design of Zinc-Enzyme Inhibitors: Functional, Structural, and Disease Applications*
Edited by Claudiu T. Supuran and Jean-Yves Winum

*Kinase Inhibitor Drugs*
Edited by Rongshi Li and Jeffrey A. Stafford

*Evaluation of Drug Candidates for Preclinical Development: Pharmacokinetics, Metabolism, Pharmaceutics, and Toxicology*
Edited by Chao Han, Charles B. Davis and Binghe Wang

*HIV-1 Integrase: Mechanism and Inhibitor Design*
Edited by Nouri Neamati

*Carbohydrate Recognition: Biological Problems, Methods, and Applications*
Edited by Binghe Wang and Geert-Jan Boons

*Chemosensors: Principles, Strategies, and Applications*
Edited by Binghe Wang and Eric V. Anslyn

*Medicinal Chemistry of Nucleic Acids*
Edited by Li He Zhang, Zhen Xi, and Jyoti Chattopadhyaya

*Plant Bioactives and Drug Discovery: Principles, Practice, and Perspectives*
Edited by Valdir Cechinel Filho

*Dendrimer-Based Drug Delivery Systems: From Theory to Practice*
Edited by Yiyun Cheng

*Cyclic-Nucleotide Phosphodiesterases in the Central Nervous System: From Biology to Drug Discovery*
Edited by Nicholas J. Brandon and Anthony R. West

*Drug Transporters: Molecular Characterization and Role in Drug Disposition, 2nd Edition*
Edited by Guofeng You and Marilyn E. Morris

*Drug Delivery: Principles and Applications, 2nd Edition*
Edited by Binghe Wang, Longqin Hu, and Teruna J. Siahaan

*Carbon Monoxide in Drug Discovery: Basics, Pharmacology, and Therapeutic Potential*
Edited by Binghe Wang and Leo E. Otterbein

*Drug Transporters: Molecular Characterization and Role in Drug Disposition, Third Edition*
Edited by Guofeng You and Marilyn E. Morris

*Hydrogen Sulfide: Chemical Biology Basics, Detection Methods, Therapeutic Applications, and Case Studies, First Edition*
Edited by Michael D. Pluth

*Nucleic Acids in Medicinal Chemistry and Chemical Biology: Drug Development and Clinical Applications, First Edition*
Edited by Lihe Zhang, Xinjing Tang, Zhen Xi, and Jyoti Chattopadhyaya

*Oral Bioavailability and Drug Delivery: From Basics to Advanced Concepts and Applications, First Edition*
Edited by Ming Hu and Xiaoling Li

# Oral Bioavailability and Drug Delivery

## From Basics to Advanced Concepts and Applications

*Edited by*

*Ming Hu*
University of Houston, Houston, TX, USA

*Xiaoling Li*
University of the Pacific, TJ Long School of Pharmacy, Stockton, CA, USA

WILEY

Published by John Wiley & Sons, Inc., Hoboken, New Jersey.
Published simultaneously in Canada.



***Library of Congress Cataloging-in-Publication Data:***

Names: Hu, Ming, Ph. D., editor. | Li, Xiaoling, Ph. D., editor.
Title: Oral Bioavailability and Drug Delivery : from basics to advanced concepts and applications / edited by Ming Hu and Xiaoling Li.
Other titles: Wiley series in drug discovery and development.
Description: Hoboken, New Jersey : John Wiley & Sons, Inc., [2024] | Series: Wiley series in drug discovery and development | Includes bibliographical references and index.
Identifiers: LCCN 2023018561 (print) | LCCN 2023018562 (ebook) | ISBN 9781119660651 (cloth) | ISBN 9781119660668 (adobe pdf) | ISBN 9781119660682 (epub)
Subjects: MESH: Biological Availability | Administration, Oral | Drug Development | Pharmacokinetics
Classification: LCC RS420 (print) | LCC RS420 (ebook) | NLM QV 38 | DDC 615.1/9–dc23/eng/20230829
LC record available at https://lccn.loc.gov/2023018561
LC ebook record available at https://lccn.loc.gov/2023018562

Cover design: Wiley
Cover Images: © La Gorda/Shutterstock (Human digestive system); © Tridsanu Thopet/Shutterstock (falling capsules medication); © Dan Thornberg/Shutterstock (Basketball sitting on a rim with net); © Jenner Images/Getty Images (Falling white pills with one pink pill in mid air)

Set in 9.5/12.5pt STIXTwoText by Straive, Chennai, India

*Dedicated to*

*My parents, Bailing Li and Jie Hu, for giving me life*
*My grandmother, Yunzhi Su, for raising me*
*My wife, Xinghang Ma, and sons, Richard and Louis, for their unconditional love, encouragement, and understanding.*

*Xiaoling Li*

*Dedicated to*

*My parents, Zhengye Hu and Qihua Chang, for Their Encouragement and Praise*
*My wife, Yanping Wang, for Her Love and Companionship*
*My daughter and son, Vivian and William, for Becoming Great Adults*
*My grandchildren, Stella and Arthur, for Bringing Joy and Continuing the Legacy*

*Ming Hu*

# Contents

# List of Contributors

***Khondoker Alam***
Department of Pharmaceutical Sciences
University of Oklahoma Health Sciences Center College of Pharmacy
Oklahoma, OK
USA

***Mamoun Alhamadsheh***
Department of Pharmaceutics & Medicinal Chemistry
Thomas J. Long School of Pharmacy
University of the Pacific
Stockton, CA
USA

***Jessica T. Babic***
Department of Pharmacy
Memorial Hermann-Texas Medical Center
Houston, TX
USA

***Matthew Behymer***
Department of Industrial and Physical Pharmacy
College of Pharmacy
Purdue University
West Lafayette, IN
USA

***Marival Bermejo***
Department of Engineering: Pharmacy and Pharmaceutical Technology Section
School of Pharmacy
Universidad Miguel Hernández de Elche
Elche
Alicante
Spain

***Chandan Bhugra***
GlaxoSmithKline
Collegeville, PA
USA

***Dion R. Brocks***
Faculty of Pharmacy and Pharmaceutical Sciences
University of Alberta
Alberta
Canada

***Valentina Bryant***
Department of Obstetrics and Gynecology
University of Texas Medical Branch
Galveston, TX
USA

***Dinh Bui***
Department of Pharmacological and Pharmaceutical Sciences
College of Pharmacy
University of Houston
Houston, TX
USA

***Stephen M. Carl***
Department of Industrial and Physical Pharmacy
College of Pharmacy
Purdue University
West Lafayette, IN
USA

and

Bristol-Myers Squibb Research Institute
Discovery Pharmaceutics
Princeton, NJ
USA

***Sergio C. Chai***
Department of Chemical Biology and Therapeutics
St. Jude Children's Research Hospital
Memphis, TN
USA

***Jae H. Chang***
Exelixis
Alameda, CA
USA

***Rajan Chaudhari***
Intelligent Molecular Discovery Laboratory
Department of Experimental Therapeutics
The University of Texas MD Anderson Cancer Center
Houston, TX
USA

***Min Chen***
Research Center for Biopharmaceutics and Pharmacokinetics
College of Pharmacy
Jinan University
Guangzhou
China

***Shun Chen***
Shenzhen Pharmacin Co., Ltd
Shenzhen
China

***Taosheng Chen***
Department of Chemical Biology and Therapeutics
St. Jude Children's Research Hospital
Memphis, TN
USA

***Senping Cheng***
Triastek, Inc.
Nanjing
Jiangsu
China

***Paresh P. Chothe***
Global Drug Metabolism and Pharmacokinetics
Takeda Development Center Americas, Inc. (TDCA)
Lexington, MA
USA

***Xin Y. Chu***
Department of Pharmacy
Faculty of Science
National University of Singapore
Singapore
Singapore

***Jack Cook***
Clinical Pharmacology
Pfizer Inc.
East Lyme, CT
USA

***Alexandra Crowe***
Department of Pharmaceutical Sciences
University of Oklahoma Health Sciences Center College of Pharmacy
Oklahoma, OK
USA

***Neal M. Davies***
Faculty of Pharmacy and Pharmaceutical Sciences
University of Alberta
Alberta
Canada

***Jin Dong***
Clinical Pharmacology and Quantitative Pharmacology
Clinical Pharmacology and Safety Sciences, R&D
AstraZeneca
Gaithersburg, MD
USA

***Ting Du***
Department of Pharmaceutical Sciences
College of Pharmacy and Health Sciences
Texas Southern University
Houston, TX
USA

***Christabel Ebuzoeme***
Department of Pharmaceutical Sciences
College of Pharmacy and Health Sciences
Texas Southern University
Houston, TX
USA

***Ayman El-Kattan***
IFM Therapeutics
Boston, MA
USA

***Imoh Etim***
Department of Pharmaceutical Sciences
College of Pharmacy and Health Sciences
Texas Southern University
Houston, TX
USA

***Taleah Farasyn***
Department of Pharmaceutical Sciences
University of Oklahoma Health Sciences Center College of Pharmacy
Oklahoma, OK
USA

***Melanie A. Felmlee***
Department of Pharmaceutics and Medicinal Chemistry
University of the Pacific
Thomas J. Long School of Pharmacy
Stockton, CA
USA

***Lon W.R. Fong***
Intelligent Molecular Discovery Laboratory
Department of Experimental Therapeutics
The University of Texas MD Anderson Cancer Center
Houston, TX
USA

***Marcus Laird Forrest***
Department of Pharmaceutical Chemistry
School of Pharmacy
University of Kansas
Lawrence, KS
USA

***Song Gao***
Department of Pharmaceutical Sciences
College of Pharmacy and Health Sciences
Texas Southern University
Houston, TX
USA

***Yi Gao***
AbbVie Inc.
Formulation Sciences, Development Sciences, Research and Development
North Chicago, IL
USA

***Lucila Garcia-Contreras***
Department of Pharmaceutical Sciences
University of Oklahoma Health Sciences Center College of Pharmacy
Oklahoma, OK
USA

***Jonathan M.E. Goole***
Pharmacotherapy and Galenic Pharmacy Research Department
Institute of Pharmacy
Universite Libre de Bruxelles
Brussels
Belgium

***Olafur S. Gudmundsson***
Bristol-Myers Squibb Research Institute
Discovery Pharmaceutics
Princeton, NJ
USA

***Shuchi Gupta***
Department of Pharmaceutics & Medicinal Chemistry
Thomas J. Long School of Pharmacy
University of the Pacific
Stockton, CA
USA

***Paul W.S. Heng***
Department of Pharmacy
Faculty of Science
National University of Singapore
Singapore

***Dea Herrera-Ruiz***
Universidad Autónoma del Estado de Morelos
Facultad de Farmacia
Cuernavaca
Mexico

***Tze Ning Hiew***
Department of Pharmaceutical Sciences and Experimental Therapeutics
College of Pharmacy
University of Iowa
Iowa City, IA
USA

***Paul C.L. Ho***
Department of Pharmacy
Faculty of Science
National University of Singapore
Singapore
Singapore

and

School of Pharmacy
Monash University Malaysia
Kuala Lumpur
Malaysia

**Ming Hu**
Joint Laboratory for Translational Cancer Research of Chinese Medicine of the Ministry of Education of the People's Republic of China
International Institute for Translational Chinese Medicine
Guangzhou University of Chinese Medicine
Guangzhou
Guangdong
China

and

Department of Pharmacological & Pharmaceutical Sciences
College of Pharmacy
University of Houston
Houston, TX
USA

**Beibei Huang**
Intelligent Molecular Discovery Laboratory
Department of Experimental Therapeutics
The University of Texas MD Anderson Cancer Center
Houston, TX
USA

**Bhaskara R. Jasti**
Department of Pharmaceutics and Medicinal Chemistry
Thomas J Long School of Pharmacy
University of the Pacific
Stockton, CA
USA

**Eun-Jung (Zenobia) Jeong**
Korean Institute of Toxicology
Daejeon
South Korea

**Yankang Jing**
Department of Pharmaceutical Sciences
Computational Chemical Genomics Screening Center
University of Pittsburgh
Pittsburgh, PA
USA

and

Pharmacometrics and System Pharmacology (PSP)
PharmacoAnalytics Program
School of Pharmacy
University of Pittsburgh
Pittsburgh, PA
USA

National Center of Excellence for Computational Drug Abuse Research
University of Pittsburgh
Pittsburgh, PA
USA

and

Drug Discovery Institute
University of Pittsburgh
Pittsburgh, PA
USA

and

Departments of Computational Biology and Structural Biology
School of Medicine
University of Pittsburgh
Pittsburgh, PA
USA

**Gregory T. Knipp**
Department of Industrial and Physical Pharmacy
College of Pharmacy
Purdue University
West Lafayette, IN
USA

**Amit Kokate**
Department of Pharmaceutics and Medicinal Chemistry
University of the Pacific
Thomas J. Long School of Pharmacy
Stockton, CA
USA

and

Office of Pharmaceutical Quality
Center for Drug Evaluation and Research
Food and Drug Administration
Silver Spring, MD
USA

**Kaustubh Kulkarni**
Novartis Institutes for BioMedical Research
Cambridge, MA
USA

**Ritika Kurian**
Department of Pharmaceutical Sciences
University of Maryland School of Pharmacy
Baltimore, MD
USA

***Annie Lee***
Department of Pharmaceutics and Medicinal Chemistry
University of the Pacific
Thomas J. Long School of Pharmacy
Stockton, CA
USA

***Sue-Chih Lee***
Food and Drug Administration
Offices of Research and Standards
Office of Generic Drugs
Center for Drug Evaluation and Research
Silver Spring, MD
USA

***Heran Li***
Department of pharmaceutical analysis
School of Pharmacy
China Medical University
Liaoning
Shenyang
China

***Li Li***
Department of Pharmacological and Pharmaceutical Sciences
College of Pharmacy
University of Houston College of Pharmacy
Houston, TX
USA

***Sanming Li***
Department of physical chemistry
School of Pharmacy
Shenyang Pharmaceutical University
Shenyang
China

***Xiaoling Li***
Triastek, Inc.
Nanjing
Jiangsu
China

and

Department of Pharmaceutics and Medicinal Chemistry
Thomas J Long School of Pharmacy
University of the Pacific
Stockton, CA
USA

***Xiaoyan Li***
Joint Laboratory for Translational Cancer Research of Chinese Medicine of the Ministry of Education of the People's Republic of China
International Institute for Translational Chinese Medicine
Guangzhou University of Chinese Medicine
Guangzhou
Guangdong
China

***Qing Lin***
Department of Pharmaceutics
West China School of Pharmacy
College of Polymer Science and Engineering
Sichuan University
Chengdu
China

***David J. Lindley***
Department of Industrial and Physical Pharmacy
College of Pharmacy
Purdue University
West Lafayette, IN
USA

***Puchun Liu***
PL Consulting – Pharmaceutical Leverage
New York, NY
USA

***Xin Liu***
Triastek, Inc.
Nanjing
Jiangsu
China

***Zhongqiu Liu***
Joint Laboratory for Translational Cancer Research of Chinese Medicine of the Ministry of Education of the People's Republic of China
International Institute for Translational Chinese Medicine
Guangzhou University of Chinese Medicine
Guangzhou
Guangdong
China

***Chuang Lu***
Drug Metabolism and Pharmacokinetics
Sanofi-Genzyme
Waltham, MA
USA

***Yuanfeng Lyu***
School of Pharmacy
Faculty of Medicine
The Chinese University of Hong Kong
Hong Kong SAR

***Venugopal P. Marasanapalle***
Novartis Institutes for BioMedical Research
Telangana
India

***Mehul Mehta***
Center for Drug Evaluation and Research
Food and Drug Administration
Silver Spring, MD
USA

***Mikolaj Milewski***
Merck & Co., Inc.
Rahway, NJ
USA

***Jonathan Moreno***
Department of Pharmaceutical Sciences
College of Pharmacy
Western University of Health Sciences
Pomona, CA
USA

***Marilyn E. Morris***
University at Buffalo (State University of New York)
School of Pharmacy and Pharmaceutical Sciences
Department of Pharmaceutical Sciences
Buffalo, NY
USA

***Marcel Musteata***
Department of Pharmaceutical Sciences
Albany College of Pharmacy and Health Sciences
Albany, NY
USA

***Ajit S. Narang***
Department of Pharmaceutical Sciences
ORIC Pharmaceuticals, Inc.
South San Francisco, CA
USA

***Christopher Nayar***
Department of Pharmaceutical Sciences
College of Pharmacy
Western University of Health Sciences
Pomona, CA
USA

***Hamdah Al Nebaihi***
Faculty of Pharmacy and Pharmaceutical Sciences
University of Alberta
Alberta
Canada

***Michael Ng***
Department of Pharmaceutics and Medicinal Chemistry
University of the Pacific
Thomas J. Long School of Pharmacy
Stockton, CA
USA

***Li F. Pan***
Department of Industrial and Physical Pharmacy
College of Pharmacy
Purdue University
West Lafayette, IN
USA

***Miki Susanto Park***
Department of Pharmaceutics and Medicinal Chemistry
Thomas J. Long School of Pharmacy
University of the Pacific
Stockton, CA
USA

***Arjun D. Patel***
Department of Pharmaceutics & Medicinal Chemistry
Thomas J. Long School of Pharmacy
University of the Pacific
Stockton, CA
USA

***Sandeepraj Pusalkar***
Global Drug Metabolism and Pharmacokinetics
Takeda Development Center Americas
Inc. (TDCA)
Lexington, MA
USA

***Yihong Qiu***
QPD Solutions LLC
Lincolnshire, IL
USA

***Tianjing Ren***
University at Buffalo (State University of New York)
School of Pharmacy and Pharmaceutical Sciences
Department of Pharmaceutical Sciences
Buffalo, NY
USA

***Yi Rong***
Department of Pharmacological and Pharmaceutical Sciences
College of Pharmacy
University of Houston
Houston, TX
USA

***Ana Ruiz-Garcia***
Metrum Research Group
San Diego, CA
USA

***Erik Rytting***
Department of Obstetrics and Gynecology
University of Texas Medical Branch
Galveston, TX
USA

***Mansi Shah***
Department of Obstetrics and Gynecology
University of Texas Medical Branch
Galveston, TX
USA

***Pramila Sharma***
Department of Pharmaceutics and Medicinal Chemistry
Thomas J Long School of Pharmacy
University of the Pacific
Stockton, CA
USA

***Rashim Singh***
Department of Pharmacological and Pharmaceutical Sciences
College of Pharmacy
University of Houston
Houston, TX
USA

***Leslie T. Steen***
Department of Pharmaceutical Sciences
University of Maryland School of Pharmacy
Baltimore, MD
USA

***Robin Sunsong***
Department of Pharmaceutical Sciences
College of Pharmacy and Health Sciences
Texas Southern University
Houston, TX
USA

***Vincent H. Tam***
Department of Pharmacy Practice and Translational Research
University of Houston
Houston, TX
USA

***Timothy Tracy***
Triastek, Inc.
Nanjing
Jiangsu
China

and

Tracy Consultants
LLC
Huntsville, AL
USA

***Yifan Tu***
Department of Pharmaceutical Science
Roche Pharma Research and Early Development
China Innovation Center of Roche
Pudong
Shanghai
China

and

Department of Pharmacological and Pharmaceutical Sciences
College of Pharmacy
University of Houston
Houston, TX
USA

***Jayabharathi Vaidyanathan***
Center for Drug Evaluation and Research
Food and Drug Administration
Silver Spring, MD
USA

***Pushkaraj Wagh***
Department of Pharmaceutical Sciences
College of Pharmacy
Western University of Health Sciences
Pomona, CA
USA

***Vrushali Waknis***
AbbVie, Inc.
Irvine, CA
USA

***Jennifer Waltz***
Department of Obstetrics and Gynecology
University of Texas Medical Branch
Galveston, TX
USA

***Hongbing Wang***
Department of Pharmaceutical Sciences
University of Maryland School of Pharmacy
Baltimore, MD
USA

***Jeffrey Wang***
Department of Pharmaceutical Sciences
College of Pharmacy
Western University of Health Sciences
Pomona, CA
USA

***Liping Wang***
Joint Laboratory for Translational Cancer Research of Chinese Medicine of the Ministry of Education of the People's Republic of China
International Institute for Translational Chinese Medicine
Guangzhou University of Chinese Medicine
Guangzhou
Guangdong
China

***Lu Wang***
Department of Pharmacological and Pharmaceutical Sciences
College of Pharmacy
University of Houston
Houston, TX
USA

***Stephen Wang***
Novartis Institutes for Biomedical Research
Cambridge, MA
USA

***Zeren Wang***
Shenzhen Pharmacin Co., Ltd
Shenzhen
China

***Lori B. Ward***
Department of Industrial and Physical Pharmacy
College of Pharmacy
Purdue University
West Lafayette, IN
USA

***Baojian Wu***
Research Center for Biopharmaceutics and Pharmacokinetics
College of Pharmacy
Jinan University
Guangzhou
China

and

School of Pharmaceutical Sciences
Guangzhou University of Chinese Medicine
Guangzhou
China

***Lei Wu***
Triastek, Inc.
Nanjing
Jiangsu
China

***Cindy Xia***
Global Drug Metabolism and Pharmacokinetics
Takeda Development Center Americas, Inc. (TDCA)
Lexington, MA
USA

***Wen Xie***
Center for Pharmacogenetics and Department of Pharmaceutical Sciences
University of Pittsburgh
Pittsburgh, PA
USA

***Xiang-Qun Xie***
Department of Pharmaceutical Sciences
Computational Chemical Genomics Screening Center
University of Pittsburgh
Pittsburgh, PA
USA

and

Pharmacometrics and System Pharmacology (PSP) PharmacoAnalytics Program
School of Pharmacy
University of Pittsburgh
Pittsburgh, PA
USA

and

National Center of Excellence for Computational Drug Abuse Research
University of Pittsburgh
Pittsburgh, PA
USA

and

Drug Discovery Institute
University of Pittsburgh
Pittsburgh, PA
USA

and

Departments of Computational Biology and Structural Biology
School of Medicine
University of Pittsburgh
Pittsburgh, PA
USA

***Zuoxu Xie***
Department of Pharmacological and Pharmaceutical Sciences
College of Pharmacy
University of Houston
Houston, TX
USA

***Jaime A. Yáñez***
Escuela Profesional de Medicina Humana
Universidad Privada San Juan Bautista
Filial de Ica
Ica
Peru

***Mengbi Yang***
School of Pharmacy
Faculty of Medicine
The Chinese University of Hong Kong
Hong Kong SAR

***Xinning Yang***
Food and Drug Administration
Offices of Clinical Pharmacology
Office of Translational Sciences
Center for Drug Evaluation and Research
Silver Spring, MD
USA

***Mengdi Ying***
Joint Laboratory for Translational Cancer Research of Chinese Medicine of the Ministry of Education of the People's Republic of China
International Institute for Translational Chinese Medicine
Guangzhou University of Chinese Medicine
Guangzhou
Guangdong
China

***Lawrence Yu***
Center for Drug Evaluation and Research
Food and Drug Administration
Silver Spring, MD
USA

***Wei Yue***
Department of Pharmaceutical Sciences
University of Oklahoma Health Sciences Center College of Pharmacy
Oklahoma, OK
USA

***Hani Zaher***
Corning Gentest$^{SM}$ Contract Research Services
Corning Life Sciences
Woburn, MA
USA

***George Zhang***
Corning Gentest$^{SM}$ Contract Research Services
Corning Life Sciences
Woburn, MA
USA

***Lei Zhang***
Food and Drug Administration
Offices of Research and Standards
Office of Generic Drugs
Center for Drug Evaluation and Research
Silver Spring, MD
USA

***Ling Zhang***
Department of Pharmaceutics
West China School of Pharmacy
College of Polymer Science and Engineering
Sichuan University
Chengdu
China

***Shuxing Zhang***
Intelligent Molecular Discovery Laboratory
Department of Experimental Therapeutics
The University of Texas MD Anderson Cancer Center
Houston, TX
USA

***Xinyuan Zhang***
Food and Drug Administration
Offices of Clinical Pharmacology
Office of Translational Sciences
Center for Drug Evaluation and Research
Silver Spring, MD
USA

***Zhirong Zhang***
Department of Pharmaceutics
West China School of Pharmacy
College of Polymer Science and Engineering
Sichuan University
Chengdu
China

***HaiAn Zheng***
Department of Pharmaceutical Sciences
Albany College of Pharmacy and Health Sciences
Albany, NY
USA

***Yujuan Zheng***
Life Sciences Department
Simulations Plus, Inc.
Lancaster, CA
USA

and

Current affiliation: Amador Bioscience
Quantitative Clinical Pharmacology Department
Germantown, MD
USA

***Zicong Zheng***
Department of Pharmacological and Pharmaceutical Sciences
College of Pharmacy
University of Houston
Houston, TX
USA

***Deliang Zhou***
Small Molecules CMC
BeiGene Ltd
Beijing
China

***Zhu Zhou***
Department of Chemistry
York College
The City University of New York
Jamaica, NY
USA

***Wei Zhu***
Clinical Pharmacology and Pharmacometrics
Janssen Research & Development
Raritan, NJ
USA

***Xiaochun Zhu***
Global Drug Metabolism and Pharmacokinetics
Takeda Development Center Americas, Inc. (TDCA)
Lexington, MA
USA

***Zhong Zuo***
School of Pharmacy
Faculty of Medicine
The Chinese University of Hong Kong
Hong Kong SAR

# Foreword

In Spring of 1983, I took a position at The University of Michigan. There I met my first Chinese student from mainland China, Ming Hu, and began a personal and professional relationship that has lasted for more than 40 years. His initial project was to develop permeability techniques in an animal model that eventually were adapted to be a human methodology and led to important advances, such as the Biopharmaceutics Classification System (BCS), now recognized and accepted worldwide. He followed this up with studies in Professor Borchardt's laboratory developing the early tissue culture techniques, so widely used today. Thus Professor Hu has been in the mechanistic oral biopharmaceutics field, since the very beginning. He is now a Professor at University of Houston and one of the two editors of this book. I am very pleased to have observed his contributions to science and his success as a scientist over 40 years I have known him and followed his career. It is a sincere pleasure to write this foreword for this monograph on mechanistic oral biopharmaceutics, a broad and diverse field, with a great deal of science, lying below the surface. The monograph is co-edited by Ming and his long-time friend and former classmate at Shanghai Medical University, Professor Xiaoling Li at University of the Pacific.

This book has two purposes, to give readers a contemporary understanding of the science of oral bioavailability and to present the state-of-the-art tools that can be used to advance the science of oral bioavailability and solve problems in the development of drug products for orally administration. It presents the significant advances in the science of oral bioavailability over the last five decades. This multidisciplinary scientific field has steadily progressed from an emphasis on physical sciences such as solubility and solid-state properties, to incorporating the significant recent advances in the biological sciences that emphasize transporters, enzymes, and the biological and physiological processes that influence their expression and function.

I'll note some of the evolutionary and perhaps revolutionary steps this field of oral bioavailability has taken over last five decades. In the 1960s and 1970s, application of the physical sciences to the problem of oral drug delivery produced the first wave of major advances that shaped the development of the modern commercial oral dosage form and the science of oral bioavailability. Important physicochemical principles and strategies such as manipulation of dissolution via physical manipulation of the drug and drug product and chemical modification using prodrugs were developed. These approaches are routinely considered and applied in the drug product development process today. The principles governing sustained and controlled release formulations were developed in those "early" years (e.g. Higuchi equation), and have become widely applied in the later decades of the twentieth century. In the 1980s, important progress in science of oral bioavailability was led by the development of two critical absorption models, rat intestinal segment perfusion model (developed in my laboratory) and Caco-2 cell mono-layer culture model (developed in Dr. Ronald T. Borchardt's lab). Professor Hu studied in both laboratories and was an early contributor to the development of both of these systems for the study of oral absorption. These methods have since become widely adopted by the pharmaceutical industries. This set the basis for predicting oral absorption and partitioning bioavailability into its component process, dissolution/release, transport/permeation and metabolism, notability distinguishing absorption and systemic availability. During the 1980s, major advances were also made in the study of metabolism in the intestine as well as the liver, particularly the cytochrome P450s and resultant potential drug–drug interaction mechanisms. In addition to predicting oral absorption, my laboratory also pioneered the concept of exploiting the intestinal mucosal cell peptide transporter (hPEPT1) to improve the oral absorption of polar drugs by making a prodrug, chemically combining the drug and an amino acid with a peptide-bond like structure.

This mechanistic concept is the basis for the absorption of many polar drugs and prodrugs. The development of several approved prodrugs including valacyclovir and valganciclovir, while originally empirical, is based on these transport mechanisms. In the 1990s, I established the concept of the BCS, partitioning drugs into classes for drug development and drug product regulation. This BCS approach has found wide use in drug discovery, development as well as regulation. It has been adopted by regulatory authorities and governments around the world as a basis for the regulation of drug product quality. More recent development in dosage form development suggests that 3D printing technology may ultimately be used in making controlled delivery formulation with unique release characteristics.

During this same period, the US Food and Drug Administration began the mandate of requiring studies that predict drug–drug interactions based on the sciences that were developed during the past three decades. Study of efflux transporters began in the 1990s, exploded in 2000s, and continued for the last 22 years in the twenty-first century. While efforts to make an inhibitor of *p*-glycoprotein for anticancer application have not produced an approved drug, a compound called ritonavir has found its way into many oral drug formulations to improve their oral bioavailability. In addition to p-glycoprotein, other efflux transporters including multidrug resistance-related proteins (MRPs), breast-cancer-resistant protein (BCRP) are also extensively studied. More recently, hepatic and intestinal uptake transporters such as organic anion transporting peptides (OATP), organic anion transporters (OATs), organic cation transporters (OCTs), and carboxylic acid transporter (CAT) are being shown to be involved in the uptake of drugs previously thought was only taken up passively. Such advances in our mechanistic understanding of oral bioavailability will most certainly lead to future advances in personalized medicine, since genomic data about polymorphism and epigenetics are no longer too expensive to obtain for individuals.

The advances in the science of oral bioavailability are driven by the need to develop orally administered drugs, which remains the most acceptable patient-compliant means of administering drugs to patients across the globe today. Although the scientific basis was most often the pursuit of industrial scientists, a lack of rapid advancement in the science of oral bioavailability became recognized as a hurdle in the drug development process in the early 1990s as many highly potent compounds (high-affinity ligands) against for example HIV in vitro were inactive in humans. In a timely or even a watershed event, the National Institute of Health in 1994 organized a conference on "Oral Bioavailability," where scientists of various backgrounds addressed the complex problem facing potent yet poorly bioavailable drug candidates, particularly anti-HIV candidates. Senior managements in many of the major pharmaceutical companies became aware of and recognized the importance of "bioavailability" as the pharmaceutical industry was working hard to fast track the development of anti-HIV drugs. This led to investment by the pharmaceutical industries in the technology and scientists to tackle this oral delivery problem. While actual numbers can be hard to obtain and interpret, my impression is that the attention to bioavailability has led to major decrease in the percentage of clinical trial failures due to oral bioavailability problems. Looking even further into the future, I believe the science of oral bioavailability will be driven by the needs for personalized medicine, the treatment of individual patients, both drugs, delivery system and dose and by the commercial need to increase the efficiency and efficacy of oral drug product development. This book provides a comprehensive survey of the modern study of the science of oral bioavailability in the twenty-first century.

*Gordon L. Amidon, Ph.D.*
The University of Michigan
Ann Arbor, MI

# Preface

Since the concept of bioavailability was introduced, significant progress has been made in understanding the science of oral bioavailability and in improving the oral delivery of drugs. Yet, we also find that there is still much to be discovered to have a good handle on oral bioavailability. As a subject, bioavailability encompasses the knowledge and technologies from various disciplines. A pharmaceutical scientist in a specific research area will benefit from a treatise on the topic. Hence, the objective of this book is to provide the framework for fundamental concepts and contemporary practice of bioavailability in pharmaceutical research and drug development.

It is our belief that this book provides both the basic concepts to a novice and the advanced knowledge to veteran pharmaceutical scientists and graduate students in related research fields. Chapter 1 gives a high-level summary of this book. The basic concepts of bioavailability are covered in Chapters 2–16. From Chapters 16–27, the advanced concepts of bioavailability are discussed in greater depth. Various approaches and methods for improving and studying bioavailability are highlighted in Chapters 27–42. Lastly, the computation approach to oral bioavailability is discussed in Chapters 43–45. The comprehensive coverage of topics on bioavailability in this book offers readers a choice of logically building their knowledge on bioavailability from basic concepts to advanced applications or *à la carte* based on their specific needs.

A book with such diverse contents requires a multidisciplinary effort. Without the efforts of contributors from different areas, this book would not be a reality. We would like to personally thank all authors for their contributions and patience during the completion of this book project. Sincere thanks are gratefully extended to Mr. Jonathan Rose and Ms. Daniela Bez at John Wiley and Sons, Inc. for their patience, understanding, support, and confidence in us. We would also like to express our appreciation to Mrs. Sonya North for her invaluable secretarial assistance. Finally, we would like to thank the world-renowned scientist and leading expert in bioavailability, Professor Gordon L. Amidon, for writing an insightful and inspiring foreword for this book.

23rd November 2023

*Ming Hu, Ph.D.*
Department of Pharmacological
and Pharmaceutical Sciences
College of Pharmacy
University of Houston
Houston, Texas

*Xiaoling Li, Ph.D.*
Department of Pharmaceutics
and Medicinal Chemistry
Thomas J. Long School of Pharmacy
University of the Pacific
Stockton, California

# 1

# Barriers to Oral Bioavailability – An Overview

*Ming Hu*[1] *and Xiaoling Li*[2]

[1] *Department of Pharmacological and Pharmaceutical Sciences, College of Pharmacy, University of Houston, Houston, TX, USA*
[2] *Department of Pharmaceutics and Medicinal Chemistry, Thomas J Long School of Pharmacy, University of the Pacific, Stockton, CA, USA*

## 1.1 Introduction

Oral bioavailability is a critical factor that determines the efficacy and adverse effects of drugs. Therefore, study of oral bioavailability has received considerable attention in scientific arena. Unfortunately, we are unable to predict bioavailability as a priori to this date, although we have made significant progress in understanding various components of this complex puzzle including solubility (e.g. aqueous solubility), partition coefficients (e.g. octanol/water), absorption (e.g. permeability across the Caco-2 cell membrane), metabolism (e.g. microsome-mediated phase I metabolism), and excretion (e.g. efflux via p-glycoprotein). However, few of these understanding would allow us to accurately predict a drug candidate's bioavailability in humans. Therefore, oral bioavailability remains to be a highly experimental parameter that eludes prediction from modern computational or experimental approaches, although some preliminary progress has been made in recent years. Lack of oral bioavailability is also one of the leading causes of drug failures in product development and clinical trials, although the problem appears to be alleviating somewhat in the last decade [1].

The reasons why oral bioavailability is such a challenge for development of drugs or drug-like substances (e.g. nutraceuticals) are several folds: first, many physicochemical and biological factors contribute to the bioavailability of a compound; second, many scientific disciplines are involved but few if any scientists could comprehend all areas; third, reliable scaling from animal models to humans is often absent; and fourth, oral bioavailability is often seriously affected by the diet and drug–drug interactions, none of which can be adequately controlled in a large clinical trial. Another limitation is that we are normally only able to gain access to limited body fluids such as blood and urine, and fluids surrounding the target tissues/cells are often not accessible. Therefore, bioavailability, a measure of the extent and rate of absorption and the elimination processes, really represents only systemic blood exposure to drugs (Figure 1.1). Therefore, it is not surprising that this is sometimes unsatisfactory as bioavailability would not correlate with efficacy. However, scientists are still making all efforts to unveil the unknowns and to understand bioavailability.

Oral bioavailability remains a major challenge to the development of nutraceuticals and naturally derived chemopreventive agents using polyphenols. Polyphenols (e.g. flavonoids and isoflavonoids) are derived from plants, and are consumed in the forms of fruits, vegetables, species, and herbs. In Asian countries, soy isoflavones and curcumin spices are excellent sources of polyphenols [2–4]. In European countries, lignans are a source of polyphenols in cereal bran or whole grain breads or flex seed oils [3, 4]. However, many people do not take sufficient quantities of naturally occurring polyphenols [5], a pill that will mimic the effects of ingesting fruits and vegetables may bring us better health one day. However, challenges remain due to the poor bioavailability of polyphenols, and many of the principles outlined in this book can be used to improve their oral bioavailability.

Taken together, we can say that developing dosage form with an appropriate drug, drug candidate, or other pharmacologically active agents is going to be challenge, but reward is equally great as the vast majority of top revenue and prescription leaders are orally administered drugs. Therefore, we will devote this chapter to briefly introduce each of the factors that influence bioavailability and provide the link to the appropriate

*Oral Bioavailability and Drug Delivery: From Basics to Advanced Concepts and Applications*, First Edition. Edited by Ming Hu and Xiaoling Li.

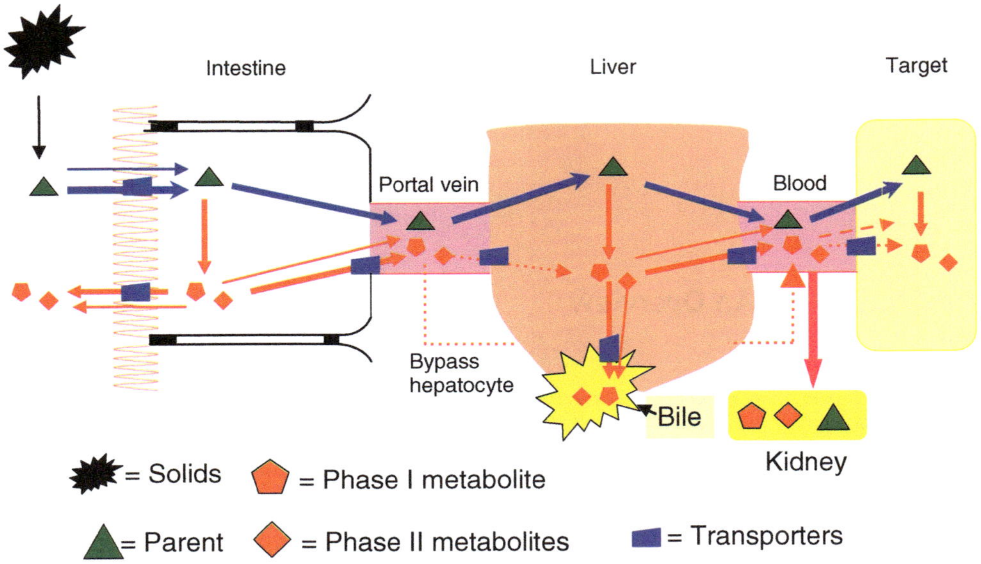


**Figure 1.1** Organ Bioavailability Barriers to Drugs. We depicted a diagram that shows the processes that include dissolution from the solids to molecules, transport of the dissolved molecules via passive and carrier-mediated uptake transporters into the cells, phase I and phase II metabolism inside the enterocytes and beyond. Drug metabolism mostly occurs in the liver. Drug elimination is mainly via bile and kidney, so other elimination route (e.g. exhalation) is not shown.

chapters in this book where you can have more details on special topics.

### 1.1.1 Physicochemical Factors

A vast array of physicochemical factors will affect a drug's oral bioavailability, and chief among these is the solubility. Solubility can be as simple as the drug concentration in water, but predicting how much will dissolve and how fast a solid will dissolve (i.e. the dissolution rate) in vivo is not trivial given the complexity of contents in gastrointestinal (GI) tract.

Solubility is affected by multiple factors include types of solvent or media (e.g. aqueous versus organic), composition of the media (e.g. pH, ionic strength, surfactants etc.), and temperature. On the other hand, the rate of dissolution is affected by factors such as particle size, salt form, and crystalline form. Nanoparticles have been attracting interest to be a part of modern drug dosage form design since the nanoparticle approach can often alleviate challenges associated with poor solubility and low dissolution rate [6]. Other approaches for enhancement of solubility and dissolution rate such as self-emulsifying and amorphous solid dispersion have also been playing an important role in product development of poor water-soluble drugs or drug candidates. Although often these approaches can only enhance the solubility of drugs for short period of time in GI tract, the result for enhancement of drug absorption has been proven meaningful in improving oral bioavailability.

The physicochemical properties of the dosage form will also impact drug bioavailability. Dosage form designs that can affect the appearance of drug in dissolved form such as disintegration time and release would have significant effect on oral bioavailability. For example, an enteric-coated dosage form will not dissolve until it reaches the intestine, where the pH value is much higher than stomach. Some dosage forms are coated with rate-limiting membranes that can control when and how fast a drug is released from the dosage form. The controlled release of drugs available for absorption creates an effective dissolution rate that is the drug release rate from the dosage form. In other words, the apparent dissolution rate can be altered by the use of dosage forms, which determines the availability of drug for absorption.

Varieties of pharmaceutical technologies and drug delivery approaches have been used to improve the physicochemical properties of the drugs, and many of the chapters in this book have presented important basic principles and technologies.

### 1.1.2 Biological Factors

Oral administration is a preferred route of delivery for small molecular drugs because the intestine has a surface area. An intestine surface area of 100–200 $m^2$ has been widely used in textbooks, various discussions and simulations for long time. However, Helander and

Fändriks [7] reported an estimation of small intestine surface area of 30 $m^2$ based on the data from microscopy of villi and microvilli biopsies. The surface area is much smaller than previously used figure. Nevertheless, it is still a sizable surface area considering other surface areas related to human body. Since oral absorption route can comfortably accommodate the drugs with molecular weight up to 600, it makes the oral route a viable and preferred option for chronical administration of small molecular drugs. Intestinal tract is a very accommodating organ for drug absorption because this organ is built to absorb nutrients. If drugs are good for us, should not the intestinal tract be built to do what benefits us? It was not until 1990s that the myth was found to be untrue if medicinal chemists can develop active compounds in vitro, formulation scientists can deliver them in vivo. The development of two classes of compounds in the 1990s, rennin inhibitors and HIV protease inhibitors, proves to the future drug development scientists that oral bioavailability matters and intestine is not just an absorption organ.

For orally administered drugs, GI tract is not always a hospitable place. There are multiple barriers to the absorption of drugs in vivo. The first barrier is variable pH of the GI. For an acidic drug, this could mean slow dissolution in the stomach, where pH is low. For a basic drug, it could mean poor solubility (solids precipitating out) in the intestinal fluids where pH is high. The second barrier is the digestive enzymes that are secreted into the GI tract, which amounts to about 45 g/day in adult humans. Although the primary functions of these enzymes are to digest nutrients such as protein, carbohydrates, and nucleotides, their presence is one of the primary reasons why protein and genetic materials (for gene therapy) cannot be delivered orally. The third barrier is the intestinal epithelium, a complex tissue with advanced cellular structures and metabolic functionality. The presence of cellular junctions, especially tight junctions, makes it impassable via the paracellular route for molecules with molecular weight greater than 200. Therefore, the vast majority of the drug molecules must use the transcellular route. Transcellular route is marred in complication. Although it was always believed that lipophilic molecules will have an easy access to the transcellular route, the presence of various efflux transporters that bind with lipophilic molecules could seriously limit the absorption of lipophilic molecules. In addition, if a molecule is too lipophilic (e.g. $\log P > 5$), it may stick to the cellular membrane. Because intestinal epithelial cells only have a functional existence of three to four days, molecules that bind too tightly will be eventually lost when the epithelial cells are sloughed off. For drugs that are hydrophilic or otherwise cannot penetrate the intestinal epithelium by passive diffusion, they must have special structural motifs that make them attractive for the nutrient transporters such as the small peptide transporter 1 (or PepT1), amino acid transporters, and nucleobases transporters. Assuming drug molecules get into the epithelial cells, there are intestinal first-pass metabolisms capable of further degrading their chance to reach the systemic circulation. These metabolisms are primary phase II metabolism although CYP3A4 is thought to be decently active in the enterocytes. The last barrier to oral bioavailability, perhaps the most well known, is the first-pass metabolism in the liver. Since all drugs absorbed via the GI (except last few centimeters of the rectum) have to enter the portal vein and encounter hepatocytes (each of which can be called metabolic "superstar"), escaping liver metabolism is the last step in the oral bioavailability processes. Therefore, a drug may need to overcome chemical barrier formed by extreme acidic or basic environment, diffusional barrier formed by the epithelia, and biochemical barrier formed by enzymes. Each of the barriers would reduce the chance of drug molecules to reach the systemic circulation leading to a lower oral bioavailability.

Due to the complexity of biological system, the difficulty of controlling the large number of inherent factors that can affect bioavailability makes predicting oral bioavailability a very challenging task. However, the rapid advances in the human genomics and proteomics hold the promise to better predict factors determining human responses to drugs. More and more new drugs are now marketed with a companion diagnostic test to better achieve the goal of a personalized medicine.

### 1.1.3 Diet and Food Effects

Development of drugs is becoming more global and multidimensional. The day when a standard diet is appropriate across the globe is probably over. Traditionally diet and food effects have focused on the protein contents, caloric intake, and fat amounts, and few if any have carefully examined the effects of other more exotic dietary components such as spices. More recently, consumers are taking ever large quantities of dietary supplements with increasing frequency and variety. How the changes in the modern diet impact drug development is not yet well addressed. Therefore, diet and food effects on drug absorption as well as bioavailability are an important topic in drug development and treatment.

### 1.1.4 Drug–Drug Interactions

Drug–drug interactions remain a serious concern for the development of new drugs and treatment. Based on the target patient population, certain types of drug interactions are not tolerated by the manufacturer, FDA, or both. Traditionally, drug–drug interactions are classified into pharmacodynamic interactions and pharmacokinetic interactions, and the chapters in this book will mainly deal with the latter since this is a book focused on oral bioavailability.

Classical pharmacokinetic drug–drug interactions typically involve phase I metabolic enzymes and clinical examples of this type of interactions are well documented in the literature. From a pharmacokinetic point of view, drug–drug interaction may cause a rise or a fall in body exposure of drug, i.e. change in $C_{max}$ and AUC values. From a mechanistic point of view, a rise in exposure is typically related to inhibition of enzyme activities or downregulation of relevant metabolic enzymes, whereas a fall in exposure is typically related to activation of dormant enzymes or induction of relevant metabolic enzymes.

More recently, FDA included uptake and efflux transporters into the drug–drug interaction universe with relevant guidance. Unlike enzyme-mediated drug–drug interactions, transporter-mediated drug–drug interactions often lead to more complex pharmacokinetic changes especially when the drug is a substrate of both transporters and enzymes, which is often the case. When drug–drug interactions are mediated by the interplay between enzymes and efflux transporters, the pharmacokinetic behaviors are more difficult to sort out as more than one component may be involved in more than an organ. For example, when CYP3A4 interplays with P-gp in the intestine versus in the liver, the consequence of drug–drug interaction may be very different. When both are involved, it is nearly impossible to delineate the contributions from each of the relevant components. Altogether, taking into consideration of transporter-mediated and interplay dominant drug–drug interactions have complicated the drug development process and increased the complexity and cost of development, it has also decreased the late-stage failure of drugs due to pharmacokinetic problems.

### 1.1.5 Scientific Disciplines Involved

A vast array of scientists is involved in the development of orally administered drugs. Aside from classical biologists and medicinal chemists that are involved in the drug discovery phase, more preclinical ADME works are now integrated into the drug discovery area. Once a candidate is selected, additional ADME work plus toxicology will be needed to further advance the candidate into clinical trials. Afterward, physician researchers, nurse practitioners, biostatisticians, marketing professionals, and pharmaceutical economists became involved in the clinical studies. At the same time, a different group of scientists, many with engineering background, is making decisions on the manufacturing processing parameters. Therefore, it is not entirely unexpected that scientists in different groups do not always have the proper background to fully understand each other. A purpose of this book is to provide topics relevant to oral bioavailability in different chapters for scientists in different disciplines to gain an understanding of their colleagues' terminology and basic methodologies, and for graduate students and postdocs to gain basic and advanced knowledge in their discipline and in related areas of oral bioavailability research.

## References


1 Ho, R.J. and Chien, J.Y. (2009). Drug delivery trends in clinical trials and translational medicine: updated analysis of ClinicalTrials.gov database. *J. Pharm. Sci.* 98: 1928–1934.

2 Aggarwal, B.B., Sundaram, C., Malani, N., and Ichikawa, H. (2007). Curcumin: the Indian solid gold. *Adv. Exp. Med. Biol.* 595: 1–75.

3 Fletcher, R.J. (2003). Food sources of phyto-oestrogens and their precursors in Europe. *Br. J. Nutr.* 89 (Suppl 1): S39–S43.

4 Slavin, J. (2003). Why whole grains are protective: biological mechanisms. *Proc. Nutr. Soc.* 62: 129–134.

5 Adhami, V.M. and Mukhtar, H. (2006). Polyphenols from green tea and pomegranate for prevention of prostate cancer. *Free Radical Res.* 40: 1095–1104.

6 Chaubal, M.V. (2004). Application of drug delivery technologies in lead candidate selection and optimization. *Drug Discovory Today* 9: 603–609.

7 Helander, H.F. and Fandriks, L. (2014). Surface area of the digestive tract - revisited. *Scand. J. Gastroenterol.* 49: 681–689.

# 2

# Solubility of Pharmaceutical Solids

*Pramila Sharma[1], Yi Gao[2], Heran Li[3], Bhaskara R. Jasti[1], Sanming Li[4], and Xiaoling Li[1]*

[1] Department of Pharmaceutics and Medicinal Chemistry, Thomas J Long School of Pharmacy, University of the Pacific, Stockton, CA, USA
[2] AbbVie Inc., Formulation Sciences, Development Sciences, Research and Development, North Chicago, IL, USA
[3] Department of pharmaceutical analysis, School of Pharmacy, China Medical University, Liaoning, Shenyang, China
[4] Department of physical chemistry, School of Pharmacy, Shenyang Pharmaceutical University, Shenyang, China

## 2.1 Introduction

Aqueous solubility is a key physicochemical attribute for an active pharmaceutical ingredient (API). Adequate drug solubility in intestinal fluid is essential to provide a concentration gradient necessary to drive the drug molecule across the gastrointestinal (GI) membrane and get absorbed (Tiong and Elkordy). Drugs with poor aqueous solubility show inadequate oral absorption and subsequently a poor bioavailability [1, 2]. Essentially, solubility, dissolution, and permeability are among the most important parameters that control the rate and extent of drug absorption.

Poor solubility has become a prevalent problem in recent decades. The advent of combinatorial chemistry, automation, and artificial intelligence in drug discovery has brought new opportunities in the area of pharmaceutical industry and resulted in an exponential increase in the number of drug-like molecules. High-throughput screening has led to compounds with high pharmacological activity. However, many of these compounds have low water solubility as they are more lipophilic, with higher molecular weights due to the increased molecular complexity. Solubility of these compounds tends to be far from optimal for developing into a product [3, 4]. It is estimated that the proportion of the poorly water-soluble marketed oral drugs is more than 40%, and about 90% of new chemical entities (NCE) are practically insoluble [5–8]. These NCEs are difficult to be formulated using conventional pharmaceutical formulation approaches and often fail early in the product development process.

It is increasingly challenging for formulation scientists to develop suitable dosage forms for poorly soluble drugs during their preclinical and clinical assessment and as commercial drug products. The effect of drug solubility on bioavailability has been studied since the 1960s, and the correlations between the factors affecting solubility, dissolution, and bioavailability were widely published including pH [9–16], salts [17–22], API particle size [23–25], impurities, polymorphs [26–29], lipids [30, 31], and amorphous solid dispersions (ASDs) [32]. In particular, the solid-state properties of different polymorphs are related to their solubility, thus affecting drug absorption and bioavailability. An early example of impact of polymorphs was reported in 1967, where the effect of polymorphs on the bioavailability of chloramphenicol palmitate was investigated and crystal form B had more bioavailability than form A [33]. Evidently, pharmaceutical development of drugs with poor water solubility requires more than conventional formulations to achieve sufficient absorption and exposure. Various solubility enhancement techniques have been developed and applied to improve the aqueous solubility and dissolution of poorly water-soluble drugs such as pH adjustment, cosolvents and/or surfactants, micronization or nanosizing, complexations, salt formation, and amorphous formation.

This chapter focuses on the fundamentals and pharmaceutical-relevant aspects of *in vitro* solubility and briefly discusses the *in vivo* solubility of solid pharmaceutical compounds in three sections: fundamentals of solubility and dissolution including factors affecting solubility and methods to determine solubility, relationship between solubility and oral bioavailability, and strategies to improve the solubility of poorly water-soluble compounds.

*Oral Bioavailability and Drug Delivery: From Basics to Advanced Concepts and Applications,* First Edition. Edited by Ming Hu and Xiaoling Li.

## 2.2 Fundamentals of Solubility

### 2.2.1 Definition

A solution is a homogeneous mixture of two or more components on a molecular level. In a two-component system, the component present in the larger proportion is generally referred to as the solvent, and the other as the solute which is often a solid when pharmaceutical compounds are concerned. When a solute is placed in contact with a solvent, a mixing process occurs due to the propensity of all molecules toward randomization, resulting in an increase in overall entropy of the system. The solute molecules start to break away from the surface and pass into the solvent system. The detached solute molecules are free to move randomly throughout the solvent and eventually form a uniform solution. Solubility is defined as the concentration of the solute in a saturated solution at a certain temperature and pressure. It is important to realize that "saturated" implies phase equilibrium in the processes of solute dissolution from solid phase and crystallization from liquid phase for a solid in a liquid [34]. As per the USP <1236>, solubility is defined as the concentration limit, at thermodynamic equilibrium, to which a solute may be uniformly mixed into a solvent. This is referred to as equilibrium or saturated solubility. Because of the aqueous environment in the biological system and most of the drugs being solid at room temperature, pharmaceutically relevant solubility typically deals with solids in an aqueous system.

Solubility can be expressed by a qualitative, semiquantitative, or quantitative relationship between the solute and the solvent or solution. Quantitative expression of solubility gives an accurate ratio of the amount of solute to solvent or solution at a certain temperature. In qualitative terms, solubility can be divided into either soluble or insoluble based on a visual inspection of the dissolution of a compound in a given solvent at given temperature. The semiquantitative expression of solubility adopts descriptive terms to express the relationship between the solute and the solvent. The *U.S. Pharmacopeia and National formulary* describes the solubility of a compound in seven different categories (e.g. very soluble, freely soluble, etc.), as shown in Table 2.1.

In pharmaceutical sciences, the solubility values have traditionally been expressed quantitatively by using the concentration of solute in solution or the ratio of solute in solvent at saturation. Solubility is stated in units of concentration such as molarity and molality, mole fraction, mole ratio, weight/volume, and weight/weight.

**Table 2.1** Terms of Approximate Solubility.

| Term | Parts of solvent required for one part of solute | Solubility range (mg/ml) |
|---|---|---|
| Very soluble | Less than 1 | > 1000 |
| Freely soluble | From 1 to 10 | 100–1000 |
| Soluble | From 10 to 30 | 33–100 |
| Sparingly soluble | From 30 to 100 | 10–33 |
| Slightly soluble | From 100 to 1000 | 1–10 |
| Very slightly soluble | From 1000 to 10 000 | 0.1–1 |
| Practically insoluble | 10 000 and over | < 0.1 |

Commonly used units are microgram or milligram or gram per milliliter. Solubility is expressed in units of concentration, including percentage on a weight or volume basis, molarity, molality, mole fraction, and parts. It follows from the previous discussion that the equilibrium solubility of a solute will depend on its relative affinity toward solvent molecules and fellow solute molecules. Therefore, the strength of molecular interactions, both inter and intra, affects solubility.

### 2.2.2 Thermodynamics of Solubility

Solubility is a thermodynamic property of a substance. Thermodynamics deals with a system. In chemical terms, system is made up of components/composition variables. In physical terms, a system can contain a number of phases. Phase is a homogeneous part of a system, which indicates all variables of the system such as temperature, pressure, and composition are uniform within it. The relationship between the phases, compositions, and the number of independent variables in a phase equilibrium system can be described in a concise manner by the Gibbs phase rule, which was developed by Josiah Willard Gibbs during the years 1875–1878. The phase rule is expressed as follows:

$$F = C - P + 2 \tag{2.1}$$

where $P$ is the number of phases in equilibrium; $C$ is the number of components; and $F$ is the number of degrees of freedom, that is, the number of independent variables in the system (usually pressure, temperature, and concentration). $F$ is the least number of intensive variables that have to be fixed to define a system. For example, for sodium chloride in an unsaturated condition, the value of $F$ is $2-1+2 = 3$, which includes pressure, temperature, and concentration. For a saturated solution with excess amount of sodium chloride at phase equilibrium, the number of phases is 2. Hence, the number of degrees of freedom is equal to 2, which means, as the temperature and pressure are set, the concentration of sodium chloride is constant. When pressure remains constant at 1 atm, the measured solubility is a constant at a given temperature. Therefore, once the temperature of a saturated solution is defined, the solubility is also defined. The term "saturated" in definition of solubility implies equilibrium with respect to the processes of dissolution and crystallization for solubility of a solid in a liquid in most of pharmaceutical-related context [35].

As the solvent bulk is saturated with the solute molecules, the redeposition process starts to occur. Once solute molecules have reached the capacity of solvent in the system, the rate of molecules leaving becomes equal to the rate of redeposition (dynamic equilibrium). The concentration of the solute in the solvent at this equilibrium state is defined as the thermodynamic solubility. The system related to the solubility is at equilibrium and the equilibrium is established for a given solute between two phases: that is, in most of pharmaceutical-relevant cases a pure drug in solid phase and its dissolved form in liquid phase. At equilibrium, the chemical potentials of both phases are equal. Therefore, solubility is an equilibrium concept, while dissolution is a kinetic phenomenon. In addition to the equal chemical potentials, each phase in the equilibrium system must also be in thermal equilibrium (uniform temperature, $T$) and mechanical equilibrium (uniform pressure, p). A proper understanding of the thermodynamics of solubility provides the basis to identify the factors that affect solubility and how to modulate the solubility.

When the equilibrium is established through dissolution and mixing, the state of the system is governed by Gibbs free energy [36] at constants $T$ and $P$ according to the equation:

$$\Delta G_{\text{soln}} = \Delta H_{\text{soln}} - T\Delta S_{\text{soln}} \tag{2.2}$$

where $G$ is free energy, $H$ is enthalpy, $S$ is entropy, and $T$ is temperature. $\Delta G$, $\Delta H$, and $\Delta S$ refer to the changes in free energy, enthalpy, and entropy of the solution during the dissolution process, respectively. Negative values of $\Delta G_{\text{soln}}$ favor the dissolution process. An appreciation of how numerous factors affect the free energy of solution, or, more specifically, how they affect the enthalpy and entropy of a solution would enhance the understanding of solubility. Entropy is the tendency of a system to attain maximum disorder. As molecules of a solute and solvent in a mixture are more randomly distributed than when they are in two separate phases, entropy increases after mixing. Enthalpy of mixing is the net effect of the attractive (or repulsive) intermolecular forces among the solvent and solute molecules. The dissolution process may result in a positive enthalpy change especially when the size of the solute is much larger than that of the solvent.

For an ideal solution, it is assumed that the mixing is ideal, the solute and the solvent are of similar size and shape, and there is no self-association or solvation between solvent and solute. Hence, the enthalpy change is zero. The free energy change is determined by the entropy change. For a solid in liquid solution, the partial free energy change involved in the transfer of 1 mol of solute from the solute phase to a saturated solution [35] is expressed as:

$$\Delta G_{\text{solute}} = T\Delta S_{\text{solute}} = -\Delta H_{\text{f}}\left(\frac{T_0 - T}{T_0}\right) = RT\ \ln X^i_{\text{solute}} \tag{2.3}$$

where $X^i_{\text{solute}}$ is the ideal solubility of the solute expressed in mole fraction. The superscript $i$ refers to an ideal solution. $T_0$ is the melting point of the solid solute in absolute degrees, $T$ is the absolute temperature of the solution, $R$ is the gas constant, and $\Delta H_{\text{f}}$ is the heat absorbed when the solid melts. Therefore, in an ideal solution, the solubility of the solute can be derived from the following equation:

$$-\log X^i_{\text{solute}} = \frac{\Delta H_f}{2.303\,RT}\left(\frac{T_0 - T}{T_0}\right) \tag{2.4}$$

Notably, nearly all solutions encountered in the pharmaceutical arena are not ideal. In real solutions, which are formed when the affinity of solute molecules for each other is different than that toward the solvent molecules, or vice versa, the size and shape of the solvents and solutes are always different and self-association and coassociation among the solvents and solutes always exist [37]. In this case, the solubility of a solute can be derived from its activity ($a_{\text{solute}}$), which represents "effective concentration" of a solute in the solution and is expressed as concentration ($X_{\text{solute}}$) multiplied by its activity coefficient ($\gamma_{\text{solute}}$):

$$a_{\text{solute}} = X_{\text{solute}}\gamma_{\text{solute}} \tag{2.5}$$

Taking logarithms of Eq. (2.5),

$$\log a_{\text{solute}} = \log X_{\text{solute}} + \log \gamma_{\text{solute}} \tag{2.6}$$

In an ideal solution $\gamma_{\text{solute}}$=1, so

$$\log a_{\text{solute}} = \log X^i{}_{\text{solute}} = \frac{\Delta H_{\text{f}}}{2.303\,RT}\left(\frac{T_0 - T}{T_0}\right) \tag{2.7}$$

In real solution, the activity coefficient depends on the nature of both the solute and the solvent as well as the temperature of the solution and can be obtained by considering the intermolecular forces of attraction that must be overcome in the dissolution process. The activity coefficient can be expressed as:

$$\log \gamma_{\text{solute}} = \frac{V_{\text{solute}}\ \Phi^2_{\text{solute}}}{2.303\ RT}\left(\delta_{\text{solvent}} - \delta_{\text{solute}}\right)^2 \qquad (2.8)$$

where $V_{\text{solute}}$ is the molar volume of solute, $\Phi_{\text{solute}}$ is the volume fraction of solvent, and $\delta_{\text{solvent}}$ and $\delta_{\text{solute}}$ are the solubility parameters of solvent and solute, respectively [38]. The solubility parameters account for the associations between the solute itself as well as the solute and the solvent. Therefore, the solubility of the solute in real solution expressed in mole fraction can be derived from Eqs. (2.6) and (2.8):

$$-\log \gamma_{\text{solute}} = \frac{\Delta H_f}{2.303\ RT}\left(\frac{T_0 - T}{T_0}\right) + \frac{V_{\text{solute}}\ \Phi^2_{\text{solute}}}{2.303\ RT}\left(\delta_{\text{solvent}} - \delta_{\text{solute}}\right)^2 \qquad (2.9)$$

### 2.2.3 Solubility and Dissolution

As described above, solubility is a thermodynamic property and considered as a constant when the number of degrees of freedom is defined. However, reaching the solubility is a kinetic process which is best described by the dissolution. The rate of dissolution reflects the amount of drug dissolved over a given time period. When drug is presented in GI for absorption, the solubility represents the maximum concentration that is available for absorption and the dissolution rate determines the amount that is available to be absorbed before reaching the solubility of drug in the GI track. Dissolution rate plays an important role in drug absorption, especially for those drugs with dissolution being the rate-limiting step in the absorption process, especially for the high permeability but low solubility drug or drug candidates such as biopharmaceutical classification system (BCS) IIa compounds. The dissolution rate is described by the well-known Noyes and Whitney Equation (2.10), which was established based on the diffusion layer model [39].

$$\frac{dM}{dt} = \frac{DA}{h}\left(C_S - C\right) \approx \frac{DAC_S}{h} \qquad (2.10)$$

where $M$ stands for amount of drug dissolved, $t$ is the time duration for the amount of the drug dissolved, $D$ is diffusivity of the drug molecule in solution which is an intrinsic property of the drug, $A$ is the surface area of the drug solid that faces the dissolution medium, $h$ is the diffusion layer thickness, $C_s$ is the thermodynamic solubility of the drug in the dissolution medium, and $C$ represents the drug concentration in the bulk medium away from the diffusion layer.

Equation (2.10) indicates that dissolution rate not only depends on the drug solubility in the dissolution medium, but also the surface area of the drug solid particles. Therefore, these two parameters play important role in oral absorption of drugs with limited solubility and dissolution rate (see Figure 2.1).

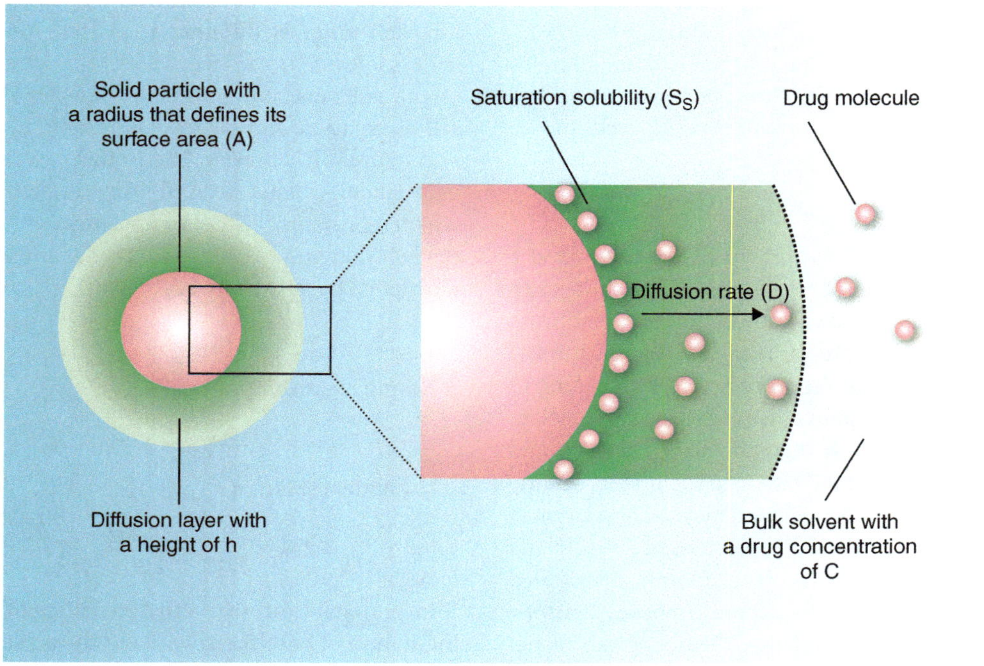


Source: Dahlgren and Lennernäs [40].

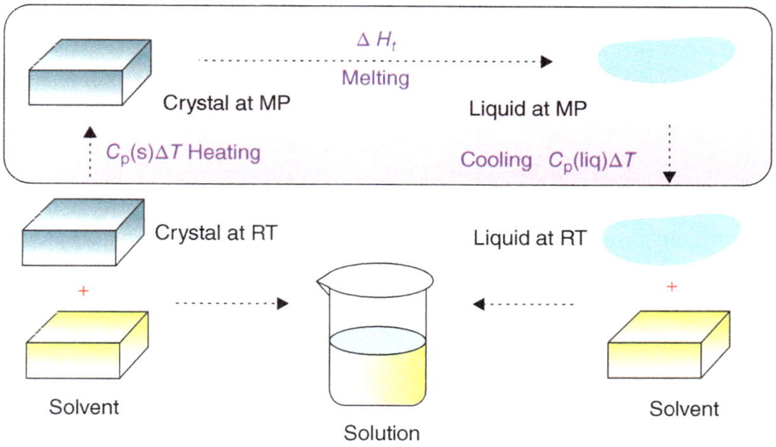


**Figure 2.1** The dissolution process of crystal drug.

### 2.2.4 Factors that Affect Solubility

As shown in Table 2.2, under constant temperature and pressure, the solubility is affected by physicochemical properties of the solute (drug molecule) including innate chemical structure of the solute (molecular weight, hydrogen bonding capabilities, and p$K_a$), the solution-state properties (lipophilicity), and solid-state properties (melting point, polymorphs, and crystallinity). The surrounding factors imposed on to the solute also impact the solubility, such as physicochemical properties of the solvent (pH, ionic strength, temperature, and so forth). The approaches that can change these solubility factors for the purpose of increasing solubility are also shown in Table 2.2.

#### 2.2.4.1 Solute Factors

***2.2.4.1.1 Molecular Structure of Solute*** The nature of solute is of paramount importance in determining its solubility in any solvent system because the chemical structure governs the physicochemical properties of a drug or drug candidate. Minor changes in the chemical structure of a compound can have significant effect on its solubility. For example, incorporation of a hydrophilic hydroxyl group can create a substantial increase in aqueous solubility, as seen in the 100-fold difference of water solubility between phenol and benzene.

As the drug candidate is selected on the basis of the biological and pharmacological effects, the solubility of a drug or drug candidate is usually a given property. Although there is little or no extraordinary flexibility for radical alteration of the chemical structure once the pharmacological effects of a specific chemical structure are identified, the chemical structure may be modified by converting to different salts or forming a complex while keeping the pharmacophore intact for its

**Table 2.2** Effects of various factors on solubility.

| Type of properties | Factors | Solubility-enhancing strategies |
|---|---|---|
| Intrinsic API molecular properties- | Molecular size, shape, Lipophilicity, H-bonding capability, Polarity, | increase entropy and/or reduce enthalpy of mixing: adding ionizable groups, hydrogen bonding, polar groups, out-of-plane substitutions, reducing molecular weight, and constructing a prodrug |
| API solid-state properties | Amorphous, Polymorph, Solvates, Melting point | Reduce enthalpy of mixing: converting to amorphous form, forming high energy solid, and crystal engineering |
| Intrinsic API solution properties | Osmotic properties, drug ionization in solution, drug diffusion in solution | Reduce enthalpy of mixing: salt formation |
| Dissolution media properties | pH, polarity, surface tension, added surfactants, cosolvents | Reduce enthalpy of mixing: complexations, micellar solubilization, and lipid-based formulations |
| Solubility measurement parameters | Temperature, agitation method and time, filtration and analytical method for concentration determination | |

activity. If the consideration for solubility modifications is incorporated into drug design, the alteration of the chemical structure becomes feasible.

The key attributes of the chemical structure related to solubility include physicochemical properties of a molecule such as lipophilicity and melting point [34, 41]. Modification of molecular structure to change lipophilicity, size, p$K_a$, and melting point will alter the water solubility. Although the chemical structure alternation is not a usual means after the candidate compounds are selected, the following structural modifications techniques such as adding ionizable groups [42, 43], hydrogen bonding [44], polar groups [45, 46], out-of-plane substitutions [47, 48], and reducing molecular weight have been used for enhancing aqueous solubility. Although the chemical structure change has been used in drug discovery stage for solubility improvement, it should also be noted that chemical structure change after a drug approval would be considered as a new chemical entity and subjected to extensive testing for approval. Solid-state properties of a solute, such as crystal structure, affect molecular interaction between solute molecules themselves. The lattice energy difference between amorphous and crystalline material has significant impact on solubility at a molecular level [49, 50].

***2.2.4.1.2 Particle Size*** Most often drug product is formulated using crystalline APIs. The crystalline solid particles can be viewed as molecular aggregates of a drug compound. The size of the crystalline solid can affect the dissolution rate because as the particle becomes smaller, the surface-area-to-volume ratio increases the surface area, which allows a greater interaction with the solvent, as depicted in Eq. (2.11). When the particle size is reduced below certain threshold, the solubility can also increase the relationship between thermodynamic solubility and particle size as described by the Thomson–Freundlich equation (also known as the *Kelvin equation*) for spherical particles [51]:

$$\frac{S_d}{S_0} = \exp\left(\frac{\gamma V_m}{R T d}\right) \tag{2.11}$$

where $d$ refers to the particle size or diameter of the particles, $S_d$ is the solubility of the drug particles, $S_0$ is the solubility of the bulk drug at temperature $T$, $V_m$ is the molar volume, $\gamma$ is the surface free energy (surface tension), and $R$ is the gas constant. As the particle radius decreases, the solubility of the drug increases. For the solubility $S_d$ to differ significantly from the solubility $S_0$ of the bulk material, the exponential term needs to be much greater than 1. A substantial increase in solubility would not occur until the particle size was significantly below 200 nm.

As an example, the relationship between particle size and ration of solubility can be simulated according to Kelvin equation as shown in Figure 2.2. The increase in solubility would not be significant for the particle size greater than 10 nm. Hence, relying on particle size reduction only for solubility enhancement may not be

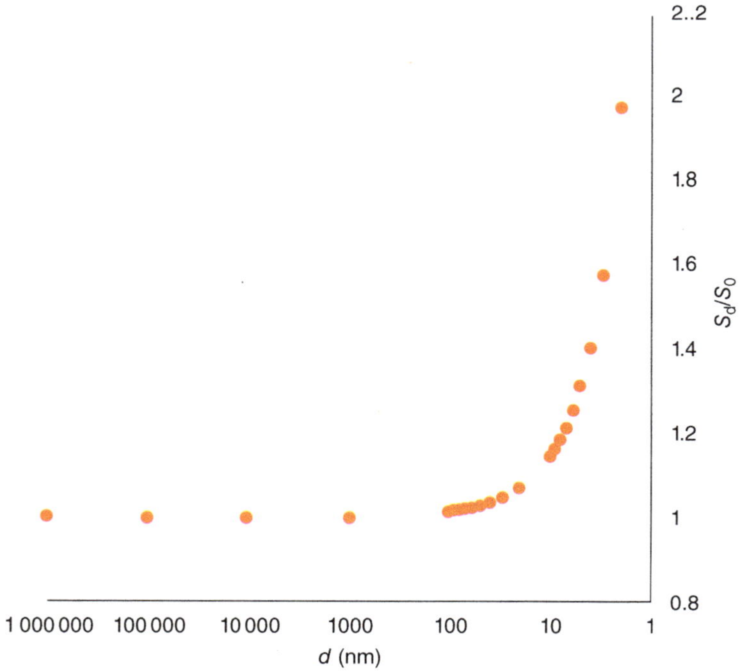


**Figure 2.2** Relationship between particle size and solubility.

pharmaceutically meaningful, and the gain in reducing particle size is mainly in dissolution rate increase.

***2.2.4.1.3 Lipophilicity*** Lipophilicity or hydrophilicity is one of the key attributes that is determined by the chemical structure. Lipophilicity represents the affinity of a molecule or a moiety for a lipophilic environment. It is commonly measured by its distribution behavior in a biphasic system (e.g. partition coefficient in octan-1-ol/water, expressed as log *P*). Lipophilicity is determined by van der Waals, dipolar, hydrogen bond, and ionic interactions. Compounds with higher lipophilicity usually have lower water solubility (Figure 2.3). In many cases, log *P* can be used as a predictor of solubility [52, 53], particularly when the drug is in amorphous form.

***2.2.4.1.4 Crystalline Solid and Polymorphs*** Low solubility can be also due to high melting point of the crystalline form. Higher melting point indicates higher lattice energy. It takes more energy for the molecules in a higher melting point crystalline form to dissolve; therefore, it leads to lower solubility. To make it more complex, a substance may crystallize to form more than one crystalline form. The phenomenon is called polymorphism. Polymorphs are different physical states of solutes in the solid state that often exhibit different melting point, although they have the same chemical structure in melt or in solution. Since the melting point of the solid is related to its solubility, the polymorphs will most likely have different solubilities. The most stable compound has the lowest free energy, thus the lowest solubility. By combining the effect of lipophilicity and crystal lattice energy, the solubility may be predicted as demonstrated by an general solubility equation (GSE) developed by Yalkowsky and Banerjee [54].

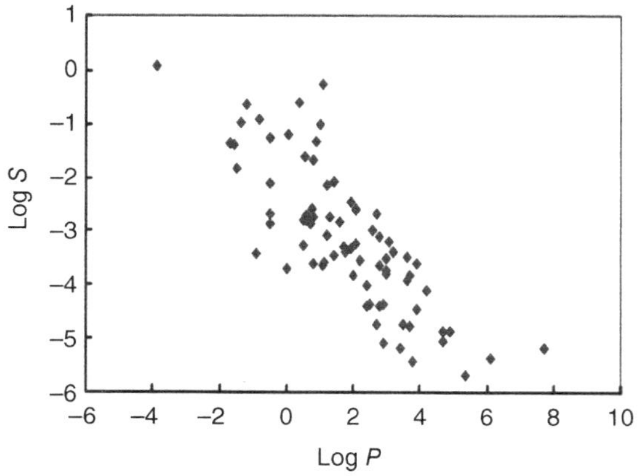


**Figure 2.3** Relationship between solubility and lipophilicity of 80 drug or drug-like molecules.

$$\log (S_0) = -0.01\ (\text{MP} - 25°\text{C}) - \log P + 0.5 \quad (2.12)$$

where $S_0$ is the intrinsic solubility, log *P* is the octanol/water partition coefficient (a measure of lipophilicity), and MP is the melting point (a measure of crystal lattice strength).

A typical marketed drug has a calculated log*P* (clog*P*) of 2.5 as it corresponds to the upper limit of good solubility predicted by the GSE. GSE constraint of clog$P > 2.5$ does not reflect ionization within the molecule and its impact on solubility enhancement. A modified GSE was derived by Hill and Young [55] replacing log *P* with $\log D_{\text{pH7.4}}$

$$\log S_{0\text{pH7.4}} = -0.01\ (\text{MP} - 25°\text{C}) - \log D_{\text{pH7.4}} \quad (2.13)$$

where $\log D_{\text{pH7.4}}$ is the octanol/water distribution coefficient at pH 7.4. Hill and Young further found significant corollary between the number of aromatic ring systems and $c\log D_{\text{pH7.4}}$ and proposed a solubility forecast index (SFI):

$$\text{SFI} = c\log D_{\text{pH7.4}} + \text{number of aromatic rings} \quad (2.14)$$

However, the predicted solubility values are rough and can be quite different from the real values. More accurate solubility values are necessary in all drug product development. Therefore, solubility of drug substances is usually measured experimentally.

#### 2.2.4.2 Solvent Factors

Solubility is not only determined by the chemical structure and its related properties of the solute but also depends upon the physicochemical properties of the solvent.

***2.2.4.2.1 pH and Ionic Strength*** In aqueous environments, weak electrolytes do not completely ionize. An equilibrium is established between ionized species and unionized species. The pH of the solution plays a role in the ionization degree of weak electrolytes which in turn affects the solubility of weak electrolyte solutes. A solute and solvent system can be described by using a phase equilibrium and an ionization equilibrium. For a monoprotic weak acid, HA (or base, B), when a solute molecule, HA (or B), is in equilibrium with its precipitated form, HA(s) (or B(s)), the process is denoted by the following equilibrium:

$$\text{HA (s)} \rightleftharpoons \text{HA} \quad \text{or} \quad \text{B (s)} \rightleftharpoons \text{B} \quad (2.15)$$

and the corresponding equilibrium constant is defined as:

$$S_0 = \frac{[\text{HA}]}{[\text{HA (s)}]} = [\text{HA}] \quad \text{or} \quad S_0 = \frac{[\text{B}]}{[\text{B (s)}]} = [\text{B}] \quad (2.16)$$

where $S_0$ is defined as the intrinsic solubility, the unionized species. In a saturated solution, the total solubility $S$, at a particular pH, is the sum of the concentrations of all the compound species (ionized and unionized) dissolved in the aqueous solution:

$$S = [A^-] + [HA] \quad \text{or} \quad S = [B] + [BH^+] \tag{2.17}$$

where [HA] is a constant (intrinsic solubility), while [A−] is a variable depending on pH (ionized species). According to Henderson–Hasselbalch equation:

$$K_a = \frac{[H^+][A^-]}{[HA]} \quad \text{or} \quad K_a = \frac{[H^+][B]}{[BH^+]} \tag{2.18}$$

$$S = [A^-] + [HA] = \frac{[HA]K_a}{[H^+]} + [HA]$$
$$= [HA]\left(\frac{K_a}{[H^+]} + 1\right) = S_0\left(10^{pH-pK_a} + 1\right) \tag{2.19}$$

or

$$S = [B] + [BH^+] = [B] + \frac{[H^+][B]}{K_a}$$
$$= [B]\left(1 + \frac{[H^+]}{B}\right) = S_0\left(10^{pK_a-pH} + 1\right) \tag{2.20}$$

For diprotic zwitterion, $XH_{2+}$ [56]:

$$S = S_0\left(1 + 10^{pH-pK_{a2}} + 10^{pK_{a1}-pH}\right) \tag{2.21}$$

For a weak acid with two ionizable acid groups, $XH_2^{2+}$ [56]

$$S = S_0\left(1 + 10^{pH-pK_{a2}} + 10^{2pH-pK_{a1}-pK_{a2}}\right) \tag{2.22}$$

However, the ideal relationship between solubility and pH does not always hold true [57]. As pH change raises solubility, at some point salts will reach saturation and precipitate, limiting further increases in solubility. Solubility of a salt is limited by its solubility products ($K_{SP}$).

For a weak acid, as pH is raised, its solubility increases with more free acid ionized and some solid HA dissolved. If the pH keeps increasing, a salt of the free acid may form at a particular pH. At the turning point of this pH, the salt solubility reaches the solubility product and starts to precipitate. According to Gibbs phase rule, during this process, the pH is a constant, and this special pH point has been designated as the Gibbs' p$K_a$ (p$Ka^{GIBBS}$) (also $pH_{max}$) [58, 59]. Therefore, below the $pH_{max}$ the crystalline free acid is the thermodynamically stable form, and above the $pH_{max}$ the crystalline salt is the thermodynamically stable form. If the solubility measurement is conducted at a pH below the $pH_{max}$ the recovered residual solid will be the crystalline free acid. If the solubility measurement is conducted at a pH above the $pH_{max}$ the recovered residual solid will be the crystalline salt. The equilibrium equation associated with this phenomenon is

$$HA\ (s) \rightleftharpoons A^-\ (s) + H^+ \tag{2.23}$$

$$K_a^{GIBBS} = \frac{[H^+][A^-]_s}{[HA]_S} = [H^+] \tag{2.24}$$

Therefore, when the aqueous medium at a given pH is saturated with the free acid ($pH < pK_a^{GIBBS}$), the effective solubility at that pH is expressed as (Figure 2.4):

$$S_{acid}\left(pH < pK_a^{GIBBS}\right) = [A^-] + [HA]_S$$
$$= [HA]_S\left(1 + \frac{K_a}{[H^+]}\right)$$
$$= S_0\left(1 + 10^{pH-pK_a}\right) \tag{2.25}$$

As the pH increases to higher than $pK_a^{GIBBS}$, and the salt becomes the saturation species, the equilibrium solubility at this particular pH can be expressed as:

$$S_{acid}\left(pH > pK_a^{GIBBS}\right) = [A^-]_S + [HA]$$
$$= [A^-]_S\left(1 + \frac{[H^+]}{K_a}\right)$$
$$= S_i\left(1 + 10^{pK_a-pH}\right) \tag{2.26}$$

where $S_i$ refers to the solubility of the salt under a certain concentration of the counterion.

Similarly, for a weak base, when the aqueous medium at a given pH is saturated with the free base ($pH > pK_a^{GIBBS}$), the total solubility at this pH may be expressed as follows:

$$S_{base}\left(pH > pK_a^{GIBBS}\right) = [B]_S + [BH^+]$$
$$= [B^+]_S\left(1 + \frac{[H^+]}{K_a}\right)$$
$$= S_i\left(1 + 10^{pK_a-pH}\right) \tag{2.27}$$

As the pH reduces to lower than $pK_a^{GIBBS}$ and the salt is in the saturated form, the effective solubility can be

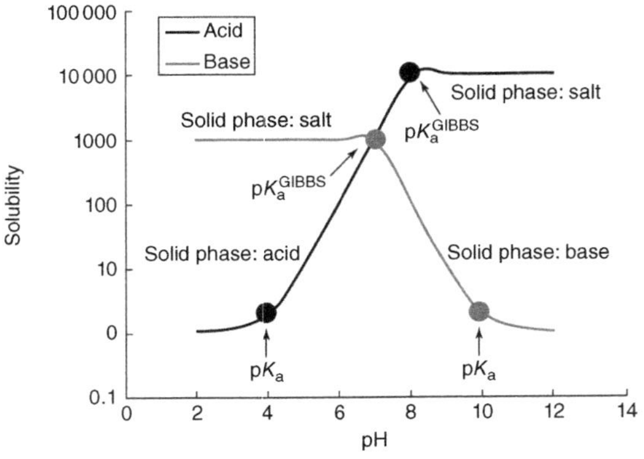


**Figure 2.4** Solubility versus pH.

**Table 2.3** Effect of pH and ionic strength on the solubility of weak electrolytes.

| | | Solubility |
|---|---|---|
| Weak acid | $pH < pK_a^{GIBBS}$ | $S_0\left(10^{pH-pK_a}+1\right)$ |
| | $pH > pK_a^{GIBBS}$ | $S_i\left(10^{pK_a-pH}+1\right)$ |
| | | In the absence of excess counterion: $S_i = \sqrt{K_{SP}}$ |
| | | In the presence of excess counterion: $S_i = K_{SP}/[X^+]$, $[X^+]$: the concentration of Counterion |
| Weak base | $pH < pK_a^{GIBBS}$ | $S_0\left(10^{pK_a-pH}+1\right)$ |
| | $pH > pK_a^{GIBBS}$ | $S_i\left(10^{pH-pK_a}+1\right)$ |
| | | In the absence of excess counterion: $S_i = \sqrt{K_{SP}}$ |
| | | $S_i = K_{SP}/[Y^-]$, $[Y^-]$: the concentration of counterion |

expressed as:

$$S_{base}\left(pH < pK_a^{GIBBS}\right) = [BH^+]_S + [B] = [BH^+]_S\left(1+\frac{K_a}{[H^+]}\right) = S_0\left(1+10^{pH-pK_a}\right) \quad (2.28)$$

where $S_0$ refers to the intrinsic solubility. Generally, the salt solubility plateau is reached when the difference between the solubility of the unionized and the ionized species is about 4 log units for acids and 3 log units for bases [60, 61]. However, there are several exceptions [11, 62, 63]. In the presence of different counterions, the salt solubility of a certain acid or base can vary. Salt forms formed by relatively stronger counterions present higher solubility [64–66]. Aqueous solubilities of haloperidol salts differed depending on the salt-forming agents used [67]. The salt of methanesulfonic acid showed the highest solubility.

As the salt becomes saturated in the solution, the equilibrium that exists when pH is above $pK_a^{GIBBS}$ for the salt of an acidic drug or below $pK_a^{GIBBS}$ for the salt of a basic drug can be expressed as:

$$X^+A^-\ (s) \rightleftharpoons A^- + X^+ \quad \text{or} \quad BH^+Y^-\ (s) \rightleftharpoons BH^+ + Y^- \quad (2.29)$$

$$K_{SP} = [A^-][X^+] \quad \text{or} \quad K_{SP} = [BH^+][Y^-] \quad (2.30)$$

where $[X^+]$, $[Y^-]$ represent the counterion concentration of acid and base, respectively.

In the absence of the counterion from other substances,

$$S_i = [A^-]_s = [X^+] = \sqrt{K_{SP}} \text{ or } S_i = [BH^+]_s = [Y^-] = \sqrt{K_{SP}} \quad (2.31)$$

The total solubility of a salt can be expressed as:

$$S_{salt}\left(pH > pK_a^{GIBBS}\right) = [A^-]_S + [HA] = [A^-]_S\left(1+\frac{K_a}{[H^+]}\right) = \sqrt{K_{SP}}\left(1+10^{pH-pK_a}\right) \quad (2.32)$$

$$S_{salt}\left(pH < pK_a^{GIBBS}\right) = [BH^+]_S + [B] = [BH^+]_S\left(1+\frac{K_a}{[H^+]}\right) = \sqrt{K_{SP}}\left(1+10^{pH-pK_a}\right) \quad (2.33)$$

The solubility decreases as common ion effect takes place. In the presence of an excess amount of the counterion, a decrease in solubility is be observed, which can be expressed as follows:

$$S_{salt}\left(pH > pK_a^{GIBBS}\right) = [A^-]_S + [HA] = [A^-]_S\left(1+\frac{[H^+]}{K_a}\right) = \frac{K_{SP}}{[X^+]}\left(1+10^{pK_a-pH}\right) \quad (2.34)$$

$$S_{salt}\left(pH < pK_a^{GIBBS}\right) = [BH^+]_S + [B] = [BH^+]_S\left(1+\frac{K_a}{[H^+]}\right) = \frac{K_{SP}}{[Y^-]}\left(1+10^{pH-pK_a}\right) \quad (2.35)$$

Table 2.3 summarizes the total solubility of weak electrolytes at different pH.

For nonelectrolytes, the effect of ionic strength on solubility depends on its influence on the structure of the water or its ability to compete with solvent water molecules. Both effects are described by the Setschenow equation:

$$\log\frac{S_0}{S} = kM \quad (2.36)$$

where $S_0$ is the solubility of the nonelectrolyte in pure water, $S$ is the solubility of the nonelectrolyte in the salt solution, $M$ is the concentration of the salt, and $k$ is the salting out constant (Setschenow constant).

*Cosolvents:* The ability of a liquid to act as a solvent for a particular solute can also be defined by solubility parameter. This approach was introduced by Hildebrand and Scott in 1962, based on the concept that strength of intermolecular cohesive forces of a liquid can be expressed in terms of solubility parameter. If the solubility parameter of solute is similar to that of solvent/cosolvent compositions, it will offer the higher solubility/miscibility of solute in the solvent. Influence of factors such as hydrogen bond formation and ionization is also considered to calculate solubility parameters. Furthermore, electrostatic properties, e.g. dielectric constant and dipole moment, are often linked by empirical or semiempirical relationships to these solubility parameters [39].

An example of prediction of Hansen solubility parameters (HSP) is given below:

$$\delta^2 = \delta_d^2 + \delta_p^2 + \delta_h^2 \tag{2.37}$$

where $\delta$ = total HSP, $\delta_d$ = dispersion HSP, $\delta_p$ = polar HSP, $\delta_h$ = hydrogen-bonded HSP.

Values of $\delta$, $\delta_d$, $\delta_p$, and $\delta_h$ for a solute and different pure solvent can be predicted using HSPiP software by entering the simplified molecular-input line-entry system (SMILES) into the HSPiP software. SMILES for each component can be obtained from PubChem database of each component [68].

In formulation studies, the cosolvents are usually less polar than water. Although less polar solvent would suppress the ionization of weak electrolyte organic compound drugs or drug candidates, the increase in intrinsic solubility in a less polar cosolvent system plays a more significant role in total solubility and resulted in an increase of total solubility. Besides common ions and cosolvents, total aqueous drug solubility is also influenced by other additives such as complexing agents, and surfactants.

### 2.2.5 Pharmaceutically Relevant Solubility

Drug dissolving in the biological system is the absolute necessity for absorption since drug molecules are required to be present in a dissolved form to be transported across biological membranes. In drug discovery, a commonly asked question by medicinal chemists is, "what is the minimum solubility required for a compound?" The minimum required solubility depends on the permeability and the dose of a compound [5, 50, 69]. The bioavailability of an orally administered drug depends primarily on its solubility in the GI tract, permeability across cell membranes, and metabolism. The correlation among solubility, permeability, and dose can be described with the following plot (Figure 2.5); this forms the basis for the BCS [70, 71]. Please see Chapter 9 for Biopharmaceutics Classification System.

BCS divides drugs and drug candidates into four classes based on two important parameters: aqueous solubility and permeability. The BCS has aided the characterization of drug products and is often used as a decision-making tool in drug formulation and development. In this system, compounds are defined as a combination of solubility and permeability (Figure 2.6), for example, high solubility, high permeability or low solubility, and high permeability. High solubility is defined as the highest dose strength that is soluble in 250 ml or less of aqueous media across the physiological pH range (pH 1–6.8), at 37 ± 1 °C, while the drugs that are not completely soluble under these conditions are considered to have low solubility. As a guideline, the following solubility classification ranges are suggested for medicinal chemists in drug discovery project teams [72].

| | |
|---|---|
| <10 μg/ml | Low solubility |
| 10–60 μg/ml | Moderate solubility |
| >60 μg/ml | High solubility |

When a compound is identified as a drug candidate, its solubility is one of the crucial parameters that may decide the fate of the candidate in the drug development process. Although BCS has been used in drug development for qualitative to semiquantitative projection of drug absorption, the *in vitro/in vivo* correlations (IVIVC) have not been, which limits the use of dissolution tests as a surrogate for human testing in BA/BE studies [73]. However, there are some limitations of BCS in pharmaceutical development. BCS mentions the solubility of drug molecule is defined as the lowest solubility in buffer solutions with a pH range from 1 to 6.8. This pH range usually covers stomach and small

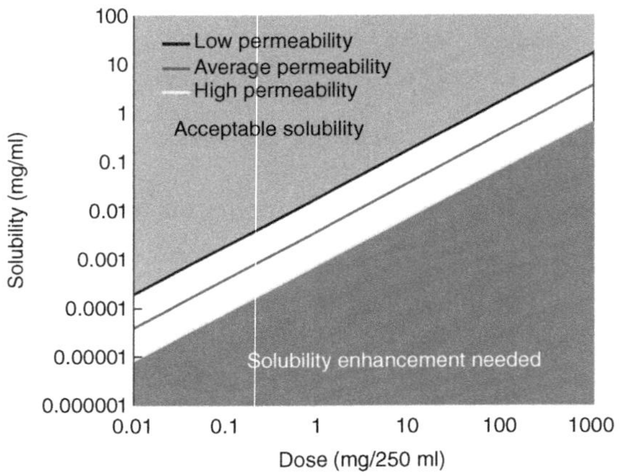


**Figure 2.5** Correlation between solubility, permeability, and dose.

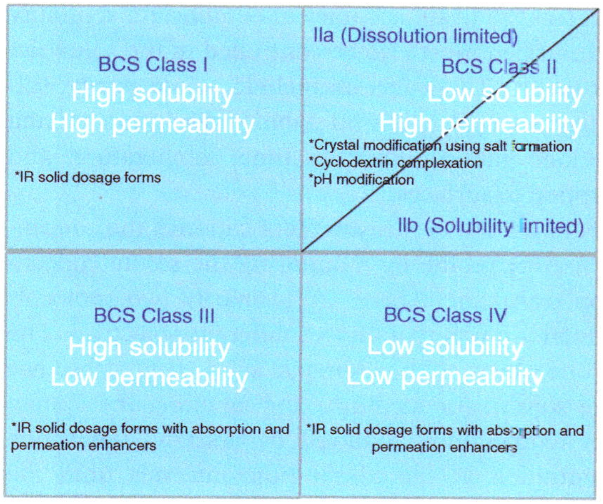


**Figure 2.6** Biopharmaceutics classification system and various approaches for solubility enhancement.

intestine regions. For neutral compounds, it would be unnecessary to determine their solubility across the whole pH range. Similarly, lipophilic compounds may show poor solubility in these buffers but will be solubilized by bile components in vivo. Considering the intestinal solubility is different from buffer solubility, a revised classification system from BCS for oral drugs was proposed [74]. The revised system is termed the developability classification system (DCS). The assignment of the DCS is based on intestinal solubility (such as solubility in simulated small intestine fluid), the permeability in the small intestine, and an estimate of the particle size needed to overcome dissolution rate limited absorption. The DSC was further refined as rDCS in 2018. The DCS offers a guideline in assessing drug developability and to define formulation strategies for poorly water-soluble drug candidates. It should be noted that an accurate estimation of the dose: solubility (D/S) ratio is not possible in early stages of development since only a rough estimate of the expected dose range of the compound is available [75].

Classification of drugs and drug candidates using BCS helps the pharmaceutical scientists to identify the approach or strategy in drug discovery and development. Solubility in this context is a thermodynamic parameter. When a drug enters a biological system, for example, after oral administration, drug dissolution and transit are all kinetic processes. Most of drug absorption is driven by passive diffusion and driving force of diffusion is the concentration gradient between the GI tract and blood circulation. Therefore, the concentration of drug in GI plays a key role in drug absorption. Solubility is the concentration of solute at saturation, which serves as the upper limit of drug concentration in GI. If the drug is in supersaturation state, the concentration is higher than its solubility. The supersaturation state is not thermodynamically stable; however, it does exhibit a higher concentration than solubility. This high concentration results in a higher concentration gradient, which leads to the higher absorption. Due to the instability of the concentration at supersaturation state and also representing a limit of drug dissolved in the solvent system, this concentration at supersaturation state has been called kinetic solubility. Although the term solubility is used, the kinetic solubility is not a thermodynamic parameter and is not established based on the phase equilibrium and chemical potential equilibrium between solid-state solute and dissolved solute. Nevertheless, kinetic solubility is a pharmaceutically relevant parameter [76, 77].

Since the contents of GI track are more complicated than buffer solutions with controlled pH and ionic strength, the pharmaceutically relevant solubility is determined experimentally *in vitro* using the conditions that mimic the *in vivo* biological system, such as simulated GI fluids.

#### 2.2.5.1 Medium of Solubility Determination

The solubility measured in aqueous media and in simulated gastric and small intestine fluids provides essential data to help estimation of the biopharmaceutical properties of a drug and for developing the pharmaceutical products. The solubility of a compound is often determined in water, buffer solutions, or simulated biological fluids. The drug absorption is not only dependent on the solubility, but also on the physiological conditions in the GI tract, including intestinal pH, GI transit time, bile salts, lipid content, the food components, and the presence of enzymes [10], as well as the formulation design and manufacturing processes of the dosage forms. Simulating the biological environment becomes important in order to understand the solubility of a compound *in vivo*. Various biorelevant media have been developed to estimate the solubility within the GI tract [78, 79]. The solubility measurements in biorelevant media have also been reported for IVIVC [70, 80]. Although various media have been proposed for simulating the fasting physicochemical characteristics and/or intragastric composition, literature data on how the simulated media compares to intragastric contents are limited [81]. Compared with data in other frequently used media, solubility data in fasted simulated gastric fluid (FaSSGF) provide a better assessment of intragastric solubility during a bioavailability study in the fasted state. However, it is more difficult to simulate the fed state due to the complexity of the fed state *in vitro* [82].

#### 2.2.5.2 *In Vitro* Solubility

A number of different terms have been used in the literature to describe solubility, such as apparent solubility, apparent equilibrium solubility, kinetic solubility, precipitation solubility, thermodynamic solubility, and equilibrium solubility. These terms are confusing and sometimes contradictory to each other. There should be limited terms to describe the solubility. The equilibrium solubility (sometimes called as saturation solubility) is the same as thermodynamic solubility, which are all refer to the drug concentration obtained when the solute and solution have reached the same chemical potential, or at the equilibrium state. Adding "apparent" before "solubility" simply implies the system has not yet reached equilibrium; for example, apparent solubility is usually used to describe the solubility of an amorphous API, since amorphous material is metastable and will eventually crystallize at equilibrium. Kinetic solubility is not a term of solubility but often refers to the procedure of solubility measurement, which generates a dissolution profile with time or an apparent solubility. Different terms have different protocols to assess solubility. Most widely used methods in the pharmaceutical industry for solubility determination are equilibrium solubility and kinetic solubility [55]. Before designing any solubility experiments, it is necessary to identify whether the purpose is to determine the equilibrium solubility or a kinetic dissolution rate [83]. Many techniques for determining aqueous solubility have been developed and evaluated [83, 84]. The most used and gold standard method is the shake-flask method. Other methods include precipitation method, potentiometric acid–base titration method, and high-throughput methods [85–87]. Each solubility determination method has its own fit for purpose, with considerations of drug amount, test time, etc. All equilibrium methods primarily consist of two parts: saturating the solvent with the solute and measuring the amount of solute present per unit of the solvent at that saturation state [88]. With increasing number of compounds moving into drug development from drug discovery, high-throughput methods that determine kinetic and/or equilibrium solubility have been developed to meet the needs of drug discovery research.

Equilibrium solubility (or thermodynamic solubility) represents the saturation solubility of a compound in equilibrium with an excess of undissolved compound at the end of the dissolution process [34]. The residual solid at the end of the experiment is analyzed, often by powder x-ray diffraction to ensure that there have been no solid form changes caused by polymorphism, hydration, co-crystal formation, etc. Other factors that may affect the result accuracy of equilibrium solubility determination must also be controlled in the solubility experiments. These factors include solute purity, solvent type, and pH, chemical stability in solution, mixing conditions and equilibration time, temperature, and adsorption to surfaces.

Kinetic solubility measurement captures the concentration–time profile by monitoring the solute concentration over time. There are primarily two protocols for kinetic solubility with two different objectives. The first one starts with adding a solid drug substance to the solvent (or medium) and agitating the system at set temperature, followed by assaying the solute concentration at spaced-out time intervals until the concentration is plateaued. A plot of the concentration with time shows the solute concentration gradually increases until a maximum, at which point the equilibrium solution is reached. This concentration–time profile is essentially a dissolution profile of the sample in a dissolution medium. The second kinetic solubility protocol is referred to as solvent shift method, starting with a high-concentration solution of the solute dissolved in a good solvent, which is then spiked into the solvent (or solubility medium), and then continuously monitors the solute concentration. In this procedure, the solute dissolution process is forfeited, and the maximum solute concentration is achieved momentarily followed by precipitation (if the added solute amount exceeds the solubility). The solubility by the solvent shift usually represents a kinetic solubility value since the initial maximum concentration is often equivalent to the amorphous solubility, and the solubility obtained changes depending on the rate of precipitation and the nature of the precipitated solid form. Nevertheless, the solvent shift method is fast and easy to automate, and thus is widely applied during early drug discovery stage. In practice, the compound is dissolved in DMSO which is then added to aqueous media to precipitate the compound. In this quick assessment, the physical character of the precipitated species is not usually determined. Therefore, the solvent shift type of kinetic solubility often overestimates the true solubility [89], and may or may not be equal to the equilibrium solubility of crystalline forms.

***2.2.5.2.1 Shake-Flask Method*** The shake-flask method is one of the most traditional solubility methods introduced more than 50 years ago and is still considered as the most reliable technique in establishing solubility. An excess amount of drug is added to a medium to form a saturated solution. The flasks are shaken at defined temperature. Aliquots of solution are withdrawn and filtered. The amount of dissolved drug is determined by assaying the filtered supernatant. However, one should be aware of possible surface adsorption to either the filter or centrifuge tube, which could result in underestimation of true solubility values. The equilibrium is determined when the concentration of the compound reaches a constant value using nephelometry, UV absorbance, or potentiometric measurements. Therefore, the sampling intervals are important in determination of solubility. For drugs with a low dissolution rate, days or even weeks may be required to achieve equilibrium; a two-day equilibration time is commonly used. The pH plays an important role in the determination of solubility for electrolyte compound. If solubility–pH profiles are necessary, the measurements need to be performed in several different pH buffers, as defined by the BCS scheme. The shake-flask method is considered as the "gold standard" by the industry and the US Food and Drug Administration (FDA). However, the shake-flask method is time consuming and needs certain amount of material; thus, it may not be an efficient method feasible for early stages in drug discovery with large number of compounds to be screened [90, 91].

To increase the efficiency and reduce the sample amount, a semi-high-throughput, reliable, and miniaturized form of the shake-flask solubility assay has been developed. A few milliliters of the medium are added to a vial and closed with a plunger containing a filter [83, 92, 93]. The sample is shaken at a controlled speed and temperature until it reaches the phase equilibrium. The samples are filtered, and the filtrates are analyzed by HPLC (high-performance liquid chromatography). This method requires a smaller amount of material but is still time consuming in gathering a pH–solubility profile for multiple compounds. Various considerations are required while determining the solubility using this approach. Bergstrom et al. [57] discovered another modified version of the shake-flask method where the solvent volume was reduced to 1000, 500, 200, 100, and even 50 μl. No statistically considerable discrepancies were observed in solubility values in this volume range. Thus, it was established that volume reduction does not affect the precision of the method. Furthermore, many other attempts were made to miniaturize the classical shake-flask method, minimize the necessary solvent volume and by doing so, decrease the consumption of the tested compound, and increase the method output [94–98].

***2.2.5.2.2 Precipitation Method*** In order to meet the high-throughput needs of drug discovery research, kinetic-based methods, which are less time-consuming, have been introduced. These methods use drug solutions prepared from DMSO stocks and induce precipitation by diluting with buffer solutions or biorelevant medium. The detection system is usually based on light-scattering methods using laser light or single-wavelength emissions, which detect precipitated particles in solution [69, 89, 99]. Kinetic solubility values obtained by these methods are time dependent. Supersaturation may occur during precipitation. Solubility values are likely to be higher than the thermodynamic solubility and may not be reproducible among different kinetic methods. Therefore, the value of kinetic solubility may not be compared with that from other solubility determination methods.

***2.2.5.2.3 Potentiometric Acid–Base Titration Method*** Stuart and Box used a method called chasing equilibrium to measure the intrinsic solubility of ionizable compounds [100]. Chasing equilibrium provides results by monitoring pH changes induced by precipitation. The equilibrium pH at which the sample is neither further precipitating nor redissolving is recorded. This method estimates intrinsic solubility values of organic acids, bases, and ampholytes within 20–80 minutes. The principle is based on dissolving the substance at a certain pH where it exists in ionized form and then precipitate of its unionized form is obtained by changing the pH using either strong acid or strong base titrant. The rate of change of pH due to precipitation or dissolution is monitored and intrinsic solubility of the neutral form of the compound is determined.

Potentiometric-based approaches were first introduced into solubility profiling by Avdeef [58] and Avdeef and Berger [101] to provide intrinsic solubility and comprehensive profiling of solubility within a pH range. This method involves titration of a basic compound from high pH to low pH or an acidic compound from low pH to high pH. Apparent ionization constant is calculated from the pH of each point in full titration curve. Intrinsic solubility is calculated from the shift compared to the aqueous ionization constant.

$$\log S_0 = \log \frac{C}{2} - \left|\mathrm{p}K_\mathrm{a}^{\mathrm{app}} - \mathrm{p}K_\mathrm{a}\right| \tag{2.38}$$

where $C$ is the sample concentration, $S_0$ is the intrinsic solubility of the unchanged species weak electrolyte, $\mathrm{p}K_\mathrm{a}{}^{\mathrm{app}}$ is the apparent $\mathrm{p}K_\mathrm{a}$ determined under conditions of precipitation, and $\mathrm{p}K_\mathrm{a}$ is the true $\mathrm{p}K_\mathrm{a}$ measured in the absence of the solid phase. Once $S_0$ and $\mathrm{p}K_\mathrm{a}$ are determined, it becomes straightforward to derive the solubility–pH profile. The properties of the state where an uncharged species (e.g. HX) coprecipitates with its ionized form (e.g. $X^-$ salt) can be described by:

$$pK_a^{(S)} - pK_a = \log S_{HX} - \log S_X \quad (2.39)$$

where $pK_a^{(s)}$ is the ionization constant of substance in the solid state and $S_{HX}$ and $S_X$ refer to the solubility constant of the unionized and ionized species, respectively. However, this method takes 3–10 hours for the equilibrium to be established.

To overcome this limitation, Avdeef and Berger used a method, called *dissolution template titration* (DTT), to reach the phase equilibrium in a shorter time [101]. This method is further described in computational approaches section.

Both, the DTT method and the chasing equilibrium method, provide a faster approach in obtaining solubility–pH profiles from a single titration compared to the standard shake-flask method, but both lack the high-throughput speed for drug discovery screening compared to the kinetic solubility methods.

***2.2.5.2.4 Computational Approaches*** Computational methods are beneficial in early drug discovery as usually only limited amounts of drug compounds are available for various preformulation or characterization studies. Computational methods for solubility determination are based on fragments of drug or drug candidate chemical structure, log *P*, solvation properties, or a combination of models. Fragment-based models attempt to predict compound solubility as a sum of compound fragment contributions, including atoms, bonds, or larger substructures, where $n_i$ are values of different molecular descriptors $i$ for the given molecule and $a_i$ are the corresponding coefficients determined by regression analysis from the following equation:

$$\log S = \sum_i a_i n_i + a_0 \quad (2.40)$$

When applying fragment contribution methods for prediction of compound solubility, a large number of fragments are necessary to get reliable results. Numerous methods exist for solubility prediction based on fragment contributions. Some of the more popular approaches include models published by Chen [102], Ghasemi [103], Gao [104].

A method, called DTT, to reach the phase equilibrium in a shorter time was developed by [101]. With this method, $pK_a$ and log *P* of the solute are taken as input parameters. The log *P* in octanol is used to estimate the intrinsic solubility ($S_0$) using the Hansch–Yalkowsky equation.

$$\log S_0 = 1.17 - \log P \quad (2.41)$$

or an improved version for ionizable molecules of moderate lipophilicity [105]:

$$\log S_0 = -2.17 - 0.0082 \log P - 0.134 (\log P)^2 \quad (2.42)$$

Using this information, a simulated titration curve is created, which acts as a template for the experiment before titration is initiated.

The solubility calculation can also be performed by using empirical model such as Eqs. (2.12), (2.13), and (2.14).

#### 2.2.5.3 Estimation of Solubility in the GI Tract

Fluids in the GI tract have a complex composition, and their composition varies with food intake. Although it is not practical to determine the solubility of the drug *in vivo*, the solubility of drugs in fluids aspirated from the GI has been used as an approach for determining *in vivo* solubility. Utilizing human aspirates in the estimation of intragastric solubility study is limited by high costs and availability of human subjects. Also, the reproducibility of solubility data is highly dependent on the aspiration protocol and meals. Accurate estimation of intragastric solubility remains problematic [106]. In one study, human intestinal luminal fluid (HIF) was extracted to study the solubility and its dependence on pH and buffering capacity in these fluids [107]. Mesalamine and prednisolone were selected as the model compounds. In the different GI regions, the solubility of the non-ionizable compound, prednisolone, did not change; however, the solubility of mesalamine with a $pK_a$ of 2.3 and 5.7 significantly changed from the jejunal to the colonic fluids. The solubility of mesalamine was shown to be dependent on pH; in addition, a direct correlation was shown between the buffer capacity and mesalamine solubility. Another study [108] indicated that postprandial conditions tended to increase the solubilizing capacity of HIF. A total of 28 compounds exhibit enhanced solubilities; only four compounds classified as weak acids (glibenclamide, glipizide, indomethacin, and sulfasalazine) had lower solubilities in fed versus fasted state conditions. As direct measurement of luminal concentrations of the drug is awkward and not reproducible, more effort has been directed at determination of solubility in simulated human GI fluids.

A rapid solubility-screening assay was developed with a focus on BCS class II drug solubility in animal and simulated human GI fluids. The assay enables biologically promising drug leads to be evaluated for solubility limitations earlier in the drug development process, minimizes GI fluid needs, and produces *in vitro* solubility information with potential *in vivo* implications [109]. The solubility of the most thermodynamically stable solid form is determined using physiologically relevant media, e.g. simulated gastric fluid (SGF), fasted-state simulated intestinal fluid (FaSSIF), and fed-state simulated intestinal fluid (FeSSIF). However, they might not truly mimic *in vivo* behavior. For example, a weakly basic compound might completely dissolve in SGF but can precipitate with increase in pH as it moves toward small intestine. The precipitate obtained could develop physical character such as amorphous or crystalline and change in particle size. Therefore, the results of precipitation in GI track may either enhance or negatively impact the oral absorption [110]. Although several simulated intestinal fluids were investigated by including bile salts and lecithin to simulate fasted and fed conditions for determining the potential food effects *in vivo*, the aspirates tend to provide a higher solubilization capacity compared to *in vitro* simulation media due to the higher amounts and variety of bile salts and lecithin. Simulated intestinal fluids use sodium taurocholate as a representative of bile salt, whereas detailed characterization of human intestinal fluids showed it is a complex mixture of different bile salts such as sodium glycocholate, sodium glycodeoxycholate, and sodium taurodeoxycholate including sodium taurocholate. A study showed that individual bile salts differ in their efficiency as crystallization inhibitors and was related to hydrophobicity of the bile salts [111].

## 2.3 Solubility and Oral Bioavailability

Administration via the oral route has been, and still is, the most popular and convenient route of drug delivery due to its ease of administration, high patient compliance, costeffectiveness, and dosage form design flexibility. However, challenges for oral delivery still exist, especially for BCS class II and class IV drugs with their poor bioavailability. The oral bioavailability depends on several factors such as aqueous solubility, dissolution rate, first pass metabolism, and efflux by transporters. The orally administered drug would not have any therapeutic effect if it did not enter the systemic circulation. The extent of the drug that becomes available for therapeutic effect in the systemic circulation depends on the drug available for absorption in GI, while the drug available for GI absorption is decided by the rate of dissolution and the aqueous solubility of the drug. For a compound to be absorbed in the intestine after oral dosing, the solid dosage form or suspension must disintegrate, dissolve,

**Table 2.4** Absorption limiting steps and conditions.

| Absorption limiting steps | Conditions |
|---|---|
| Dissolution limiting | $T_{diss} > 199$ min<br>$P_{eff} > 2 \times 10^{-4}$ cm/s<br>$D_{abs} \gg$ dose |
| Permeability limiting | $T_{diss} < 50$ min<br>$P_{eff} < 2 \times 10^{-4}$ cm/s<br>$D_{abs} \gg$ dose |
| Solubility limiting | $T_{diss} < 50$ min<br>$P_{eff} > 2 \times 10^{-4}$ cm/s<br>$D_{abs} <$ dose |

Source: Yu [112].

and diffuse across the intestinal epithelium to enter the systemic circulation (Figure 2.7). The dissolution rate of a solute in a solvent is directly proportional to its solubility, as described by the Noyes–Whitney equation [39]:

$$\frac{\mathrm{d}C}{\mathrm{d}t} = \frac{kDA}{hV}\left(C_{\mathrm{s}} - C\right) \tag{2.43}$$

where $k$, $D$, and $A$ are the dissolution rate constant, diffusion coefficient, and solute surface area, respectively; $C_s$ and $C$ are concentrations of the solute at the particle surface and the bulk medium, respectively; $V$ is the volume of the dissolution medium; and $h$ is the thickness of the stagnant diffusion layer. If the concentration of the solute at the particle surface reaches its maximum value or solubility, and the concentration in the bulk solvent is absorbed instantaneously, then

$$\frac{\mathrm{d}C}{\mathrm{d}t} = \frac{kDA}{hV}\left(C_{\mathrm{s}}\right) \tag{2.44}$$

Although solubility is an important factor for bioavailability, it is not the only factor that affects the bioavailability. An integrated model has been established for determining the causes of poor oral drug absorption, where physicochemical characteristics of

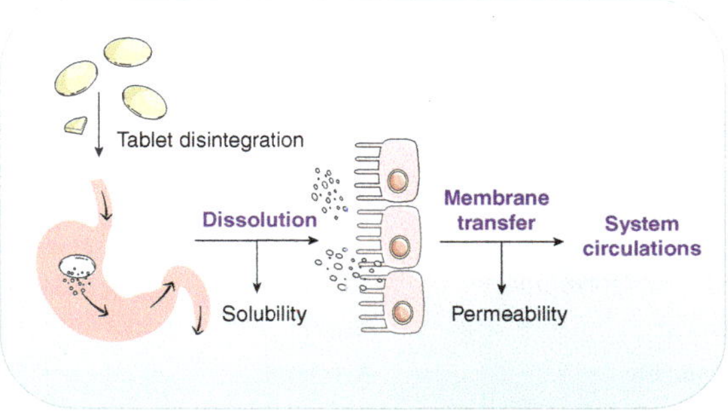


**Figure 2.7** Simplified drug absorption schematic.

the drug, formulation factors, as well as physiological properties of the biological system were taken into consideration [112]. Important parameters such as dissolution time ($T_{\text{diss}}$), absorbable dose ($D_{\text{abs}}$), and effective permeability ($P_{\text{eff}}$) were used to determine the cause of poor oral absorption (Table 2.4).

$$T_{\text{diss}} = \frac{\rho h r_0}{3DC_s} \tag{2.45}$$

$$D_{\text{abs}} = P_{\text{eff}} C_s A \left(T_{\text{si}}\right) \tag{2.46}$$

where $A$ is the effective absorption area, $T_{\text{si}}$ is the mean small intestinal transit time, $C_s$ is the aqueous solubility, $D$ is the diffusion coefficient, $h$ is the diffusion layer thickness, $r_0$ is the initial radius of particles, and $\rho$ is the density of drug. As shown in Table 2.4, solubility becomes an issue for drug absorption or bioavailability when short transit time and low permeability are involved. The successful rate of drug product development would increase significantly if a computational method was used to quantitatively predict the effect of solubility on absorption prior to extensive formulation efforts for the solubility enhancement of lead drug candidates. Investigation in computation methods to predict solubility and absorption is ongoing, but the methods have not been validated due to the complex and dynamic processes involved in an *in vivo* environment. As the scientific understanding of the biological system and molecular aspects of drug candidates advances,

**Table 2.5** Marketed products utilizing solubility-enhancement technologies.

| Compound | Marketed product | Technology utilized |
|---|---|---|
| *Nanosuspensions* | | |
| Sirolimus | Rapamune | Nanocrystal (Elan) |
| Aprepitant | Emend | Nanocrystal |
| Fenofibrate | TriCor | Nanocrystal |
| Fenofibrate | Triglide | SkyePharma |
| Megastrol acetate | Megase ES | Nanocrystal |
| *Solid solution/dispersions* | | |
| Nabilone | Cesamet | Solid dispersion in Polyvinylpyrrolidone (PVP) |
| Lopinavir/ritonavir | Kaletra | Solid solution in Polyvinylpyrrolidone-vinyl acetate (PVP-VA) |
| Itraconazole | Sporanox | Solid dispersion in Hydroxypropyl methylcellulose (HPMC) |
| Etravirine | Intelence | Solid dispersion in HPMC |
| Griseofulvin | Gris-PEG | Solid dispersion in PEG |
| *Cyclodextrin* | | |
| Itraconazole | Sporanox (oral solution) | Hydroxypropyl-beta-cyclodextrin (HP-$\beta$-CD) |
| Mitomycin | MitoExtra(IV infusion) | Hydroxypropyl-beta-cyclodextrin (HP-$\beta$-CD) |
| Chlordiazepoxide | Transillium | $\beta$-cyclodextrin |
| Omiprazole | Omebeta(tablet) | $\beta$-cyclodextrin |
| Aripiprazole | Abilify (i.m. solution) | Sulfobutylether-β-Cyclodextrin (SBE-$\beta$-CD) |
| Nitroglycerin | Nitropen(Sublingual tablet) | $\beta$-cyclodextrin |
| *Surfactant/Cosolvent* | | |
| Cyclosporine A | Sandimmune | 65% Cremophor EL, 35% ethanol |
| Paclitaxel | Taxol | Cremophor EL |
| *Microemulsions* | | |
| Cyclosporine A | Restasis | — |
| Ibuprofen | Solufen | — |
| Saquinavir | Fortovase | — |
| *Lipid-based formulation* | | |
| Ritonavir | Norvir | Oleic acid |
| Dutasteride | Avodart | Mono-di-glycerides of caprylic/capric acid |
| Isotretinoin | Claravis | Polysorbate 80, hydrogenated vegetable oil, and soybean oil |

the quantitative description of bioavailability based on physicochemical properties of drugs and physiological processes as well as interactions between the compounds, formulation, and intragastric fluids will eventually be established.

## 2.4 Strategies to Improve Solubility

Several approaches are used to improve solubility and dissolution of poorly water-soluble compounds. The fundamental principle of solubility enhancement is focused on increasing concentration of drug in a solution and surface area of a drug particle. To increase the concentration of drug in a solution, methods like solubilization or attempts to increase apparent solubility are used. To increase solute–solvent interactions, methods such as pH modification, complexation using complexing agents such as cyclodextrins, cosolvents, and surfactants/or lipids are used. Furthermore, techniques such as salt formation, cocrystals/metastable crystal forms, ASDs, and lipid-based drug delivery systems are used to modify solid state of a drug molecule. In addition, micronizing and nanosizing are used to reduce the particle size and increase the surface area.

Although several factors may be responsible for the low bioavailability, a particularly widespread problem is poor absorption due to low solubility and slow and/or incomplete drug dissolution in the lumen of the GI tract. Formulation efforts have been shown to improve the solubility and dissolution rate limited absorptions. A summary of some commercial products that utilize solubility and dissolution enhancement techniques is listed in Table 2.5.

The most frequently applied formulation approaches to deliver poorly soluble drugs in solid oral dosage forms include salt formation, ASD, lipid-based drug delivery, and increasing surface area (reducing particle size).

### 2.4.1 Formulation through Salt Formation

Salt dissolves faster than the free acid or free base crystalline form to overcome dissolution rate obstacles. An acceptable salt is identified through a rigorous salt screen and characterization process, often a result of close collaborations from multiple disciplines. Salt selection criteria should consider dissolution rate, stability, viable crystallization process (manufacturability), and ease of formulating with the salt. The advantage of salt is obvious. The salt can be formulated by conventional formulation processes of mixing with excipients, blending, and tableting or encapsulation. Due to the simpler and lower cost formulation process, salt formation is a preferred formulation approach over other more complexed supersaturation generation techniques such as ASDs or lipid-based drug delivery systems.

However, for salts of poorly water-soluble compounds, the major risk is the physical instability because of salt disproportionation phenomenon. When the salt disproportionation occurs, the salt converts back to the parent form (free acid or free base) which is poorly water soluble. The conversion can potentially occur when in contact with aqueous solutions. This is because the salt is a metastable solid form when the solution pH is away from the $pH_{max}$. For an example of an acidic drug, as illustrated in Figure 2.8, the salt of flurbiprofen would dissolve to generate a supersaturation (concentration above the solubility of the free acid flurbiprofen) when the pH is below the $pH_{max}$ of 7.3. The supersaturation is a driving force for the free acid form to crystallize. The precipitation of the less soluble free acid crystalline form will reduce the solubility and potentially less absorption *in vivo*. Therefore, stabilizing the salt during dissolution is the key for success of a salt formulation. Examples of salt approach include drug Diclofenac where the sodium or potassium salt of diclofenac increased solubility more than 400 times more than diclofenac free acid (which is <0.02 mg/ml), and drug where Enalapril maleate salt increased solubility from 0.21 to 25 mg/ml (see Figure 2.9).

### 2.4.2 Formulation Through Amorphous Solid Dispersions

ASD formulations have been utilized to deliver many life-saving drugs commercially. ASD is formed when the API is turned into amorphous form which is molecularly dispersed into a polymer matrix or a polymer–surfactant matrix. ASD is distinct from other enabling formulation approaches by the two combined performance-enhancing features. The first feature is to create the amorphous API which gives rise to the amorphous advantage: the higher apparent solubility with supersaturation potential. The second feature is to release the API as nanodroplets during dissolution if the drug loading is below the limit of congruency (LoC). Appropriately selected excipients in ASD formulations can drive the amorphous API to quickly dissolve and reach amorphous solubility in solution, along the way creating nano-sized amorphous particles. ASD essentially realizes both formulation goals of (i) solubility increase and (ii) particle size reduction [114]. Most of other techniques can only realize either goal. For

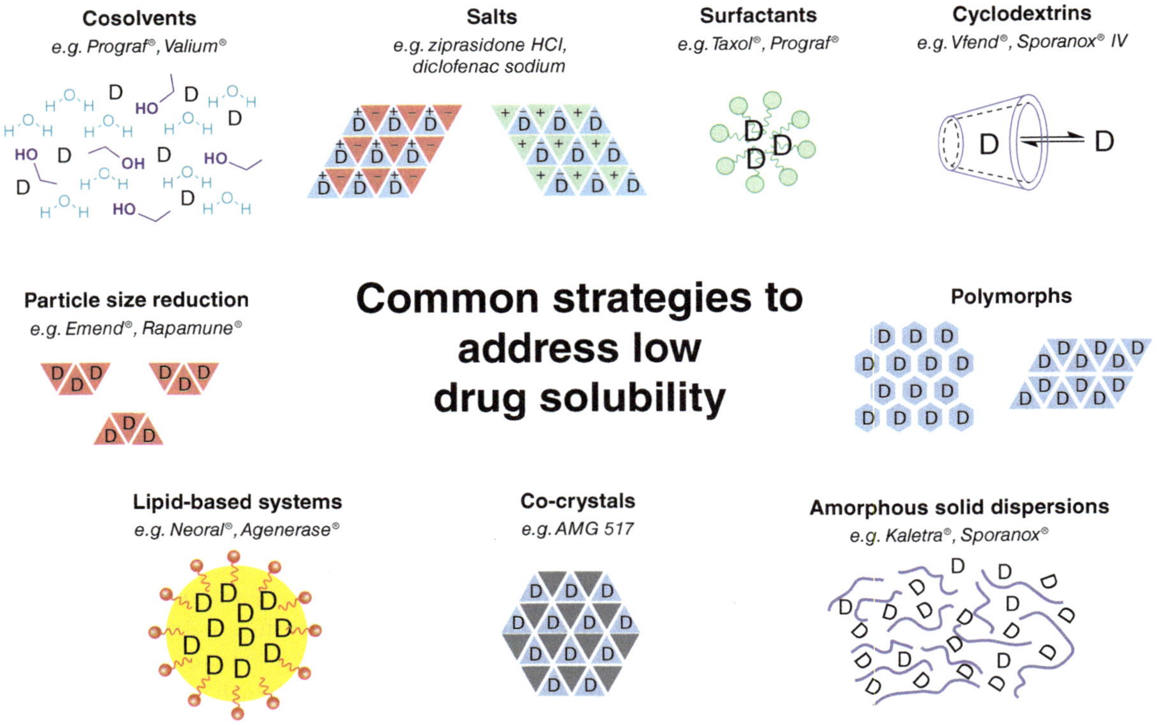


**Figure 2.8** Strategies used to enhance solubility of poorly water-soluble drugs. Source: Image adapted from [113].

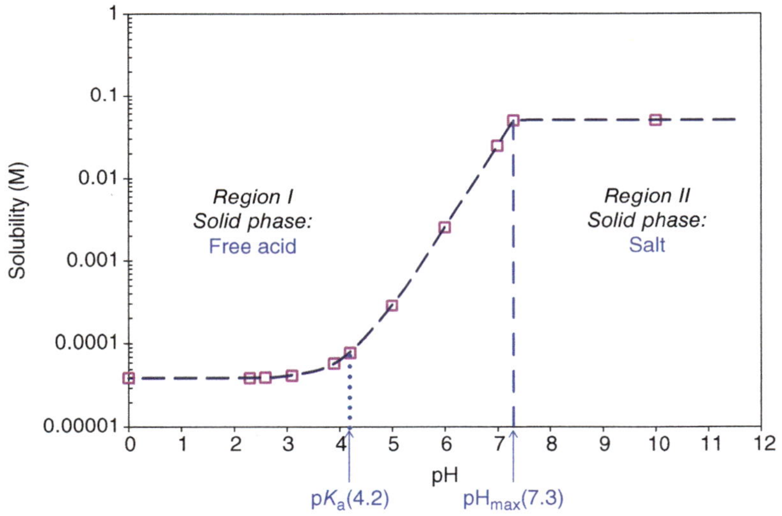


**Figure 2.9** pH-Solubility Profile of a Weakly Acidic Drug, Flurbiprofen, with a p$K_a$ of 4.2 [114]

example, adding cyclodextrin only improves solubility, and particle size-reducing technique only can reduce the API particle size. Therefore, ASD has become one of the most powerful ammunitions to tackle poorly water-soluble drugs.

Maintaining the API to be stable in amorphous state in the ASD formulations is not without significant challenges. The amorphous form is not thermodynamically stable, and crystallization may occur during drug product storage. The unwanted crystallization, even presented at small amount, may be detrimental. The crystalline seeds often trigger more or sometimes massive crystallization during dissolution of the ASD formulation that could eliminate the solubility advantage of the amorphous formulation. Another challenge is to obtain a high drug loading for ASDs to meet pill burden requirement. It has been found that drug release from ASD suffers a sudden decline at drug loadings

**Table 2.6** Brief introduction to formulation through other approaches.

| Solubility enhancement technologies | Basic principle | Examples | References |
|---|---|---|---|
| Nanosizing | Increase in surface area will increase in rate of dissolution, change in particle curvature, and introduction of defects into crystal lattice. | a. Improved oral bioavailability of cilostazol, fenofibrate, progesterone, proquazone.<br>b. 14-fold increase in in vitro dissolution of itraconazole suspension.<br>c. Mesoporous silica nanospheres with high drug loading of itraconazole resulted in 1.5 times enhancement of oral bioavailability | [2, 113, 117] |
| Incorporation of complex agents | | Cyclodextrins are found it at least 35 pharmaceutical products especially for improvement of drug solubility. Some of the products are mentioned in Table 2.5. | [118, 119] |
| Surfactants/Micelles and polymeric micelles | | Polymeric micelles increased the bioavailability of fenofibrate compared to the coarse suspension. | [120–123] |
| Prodrugs | Structural modifications techniques, such as adding ionizable groups hydrogen bonding, polar groups, out-of-plane substitutions, and reducing molecular weight, have been used for enhancing the aqueous solubility of a drug or drug candidate. Chapter 32 discusses the prodrug strategies in details. | | [42–44, 124] |
| Liquisolid compacts to deliver poorly water-soluble liquid state APIs | Liquid medications are blended with suitable selected powder excipients to yield dry-looking, nonadherent, free-flowing, and readily compressible powders. | This technique improved in vitro release of poorly water-soluble drugs such as carbamazepine, hydrocortisone, and naproxen. | [125, 126] |

exceeding the "limit of congruency" (LoC). The LoC is low for many poorly water-soluble drugs. Therefore, ASD formulation requires more effort. Careful selection of appropriate polymer(s) is necessary to ensure amorphous API stability in the drug product. Optimizing drug loading by choosing the right polymer(s) and surfactant(s) is also important to achieve sufficient release and supersaturation with no or minimal crystallization during dissolution. Extensive literatures and books have been published on how to develop and manufacture ASDs. The readers are recommended to refer to these literatures.

### 2.4.3 Formulation Through Lipid and Surfactant-Based Formulations

The lipid and surfactant-based drug delivery systems (LSBDDSs) are among the oldest formulation approaches for enhancing bioavailability of poorly water-soluble hydrophobic compounds. The formulation is made by dissolving the drug substance in the lipid and/or surfactant excipients. Since a LSBDDS is often a solution at ambient temperature, the release of the drug from LSBDDS after ingested *in vivo* is not a dissolution per se, but a miscibility process to make it more accurate. LSBDDS upon mixing with aqueous medium generates new structural elements, small lipid droplets, and/or emulsion droplets. Such a process is called emulsification. These small droplets keep the drug in solution state and prevent them from precipitation. Based on the type of lipidic excipients and formulation composition, Pouton and Porter [115] divided lipid-based drug delivery systems into IV Types:

- Type I: Oils (tri-, di-, and monoglycerides) only
- Type II: Oils and less water-soluble surfactants
- Type III: Oils, water-soluble surfactants, and cosolvents
- Type IV: Water-soluble surfactants and cosolvents (no oils)

As in any formulation approaches, which type of lipid-based drug delivery systems to select is dependent on the drug properties and the target drug product profiles. Types III and IV have gained more interest since the surfactants and cosolvents in these formulations help to emulsify easily, resulting in small or micro-sized droplets. Thus, Types III and IV are also called self-micro-emulsifying drug delivery systems (SMEDDSs) and self-nano-emulsifying drug delivery systems (SNEDDSs).

Despite the potential advantage of the LSBDDSs to increase solubility, the total market share in the oral formulation space still ranged merely between 2% and 4% [116]. There are several limitations in LSBDDSs. One issue is drug loading as a result of limited solubility of API in lipids. Another is the chemical and physical stability of the API in solution state; chemical reaction with impurities in the lipidic excipients may occur, and adduction formation between the API and lipidic excipients is also possible which is a polymorphic conversion leading to precipitation. The third is the requirement of compatibility with capsule shells and a more complex manufacturing process. Due to these limitations of LSBDDS, ASD has emerged as a better technology to deliver poorly soluble drugs. However, LSBDDS is much easier to prepare and therefore widely used in preclinical studies.

Formulation of Type III and Type IV LSBDDSs involves a logical selection of excipients starting with a solubility screen in the excipients. In general, lipid, surfactant, and cosolvent excipients that provide the highest solubility are then mixed in miscible proportions to produce the formulation prototypes. The prototypes are screened by dilution or dissolution tests, stability evaluations, and compatibility with capsule materials.

### 2.4.4 Formulation Through Other Approaches

Other formulation approaches to increase solubility and dissolution are briefly summarized in Table 2.6. In case all approaches fail to deliver the compound, the last resort is to change the chemical structure to make it more soluble, such as making it into a prodrug. It should be mindful that a prodrug is a new chemical entity and requires to be redeveloped from the beginning of the toxicology evaluations to the clinical trials, which is highly costly. A better way is for the formulation scientist to get involved early and work closely with the chemists to evaluate the prodrug approach early in the discovery programs.

## 2.5 Summary

In this chapter, the importance of the fundamentals of solubility of pharmaceutical active ingredients has been discussed and emphasized. As the solubility of a compound is determined by the properties of both the solute and the solvent, understanding the interplay between their physicochemical properties, including crystal lattice, polymorphs, types of salt, and solution pH, will lead to a successful formulation design with the available solubility enhancement technologies.

As the solubility is a crucial factor in dissolution and drug absorption, modulating the solubility becomes an effective means of improving the bioavailability of a drug. With increasing number of low water-soluble compounds entering drug development pipelines, solubility has become a significant physicochemical property in product development.

## Abbreviations

| | |
|---|---|
| API | active pharmaceutical ingredient |
| ABCs | amphiphilic block copolymers |
| BCS | Biopharmaceutics classification system |
| CMC | critical micelle concentration |
| DDIs | drug–drug interactions |
| DMEs | drug metabolizing enzymes |
| D/S | dose: Solubility |
| FaSSGF | fasted state simulated gastric fluid |
| FDA | Food and Drug Administration |
| GI | gastrointestinal |
| GSE | general solubility equation |
| HPC | hydroxypropylcellulose |
| HPMC | hydroxypropylmethylcellulose |
| IVIVC | *in vitro*/*in vivo* correlation |
| NCE | new chemical entity |
| PEG | polyethylene glycol |
| PEG-*b*-P(ALA-*co*-MAA) | poly(ethylene glycol)-block-poly(alkyl acrylate-*co*-methacrylic acid) |
| PVP | polyvinylpyrrolidone |
| PVPVA | polyvinylpyrrolidone/Vinyl acetate |
| SEF | self-emulsifying formulation |
| SMEDDS | self-micro emulsifying drug delivery systems |
| SNEDDS | self-nano emulsifying drug delivery systems |
| SMILES | simplified molecular-input line-entry system |

## References


1 Dolinina, E.S., Vorobyeva, E.V., and Parfenyuk, E.V. (2015). Development of novel delivery system for warfarin based on mesoporous silica: adsorption characteristics of silica materials for the anticoagulant: pharmaceutical development and technology. *Pharm Dev Technol* 21: 1546.

2 Ren, X.N., Cheng, S., Liang, Y. et al. (2020). Mesoporous silica nanospheres as nanocarriers for poorly soluble drug itraconazole with high loading capacity and enhanced bioavailability. *Micro Meso Mat.* 305: 110389.

3 Xie, X.M., Li, Y., and Zhao, D.Z. (2020). Oral administration of natural polyphenol-loaded natural polysaccharide-cloaked lipidic nanocarriers to improve efficacy against small-cell lung cancer. *Nanomed. Nanotechnol.* 29: 102261.

4 Yamashita, S. and Furubayashi, T. (1998). In vitro-in vivo correlations: application to water-insoluble drugs. *Gattefosse* 91: 25–31.

5 Lipinski, C. (2000). Drug-like properties and the causes of poor solubility and poor permeability. *J Pharmacol Toxicol Methods* 44: 235–249.

6 Lipinski, C. (2002). Poor aqueous solubility-an industry wide problem in drug discovery. *Am Pharm Rev* 5: 82–85.

7 Venkatesh, R. and Lipper, A. (2000). Role of the development scientist in compound lead selection and optimization. *J Pharm Sci* 89: 145–154.

8 Zhang, X.W., Xing, H.J., Zhao, Y., and Ma, Z.G. (2018). Pharmaceutical dispersion techniques for dissolution and bioavailability enhancement of poorly water-soluble drugs. *Pharmaceutics* 10 (3): 74–106.

9 Li, S. (2005). Investigation of solubility and dissolution of a free base and two different salt forms as a function of pH. *Pharm Res* 22: 628–635.

10 Meshali, H. (1984). Simultaneous solubility and dissolution rate of sulfamethoxazole and trimethoprim in binary mixtures. *Pharmazie* 39: 407–408.

11 Serajuddin, A. and Jarowski, C. (1985b). Effect of diffusion layer pH and solubility on the dissolution rate of pharmaceutical bases and their hydrochloride salts I: phenazopyridine. *J Pharm Sci* 74: 142–147.

12 Serajuddin, A., Sheen, P., Mufson, D. et al. (1986). Preformulation study of a poorly water-soluble drug alpha-pentyl-3-(2-quinolinylmethoxy)benzenemethanol: selection of the base for dosage form design. *J Pharm Sci* 75: 492–496.

13 Serajuddin, A. and Jarowski, C. (2010). Effect of diffusion layer ph and solubility on the dissolution

rate of pharmaceutical acids and their sodium salts ii: salicylic acid, theophylline, and benzoic acid. *J Pharm Sci* 74 (2): 148–154.

14 Sheng, J., Kasim, N., Chandrasekharan, R., and Amidon, G. (2006). Solubilization and dissolution of insoluble weak acid, ketoprofen: effect of pH combined with surfactant. *Eur J Pharm Sci* 29: 306–314.

15 Streng, W., Hsi, S., Helms, P., and Tan, B. (1984). General treatment of pH-solubility profiles of weak acids and bases and the effects of different acids on the solubility of a weak base. *J Pharm Sci* 73: 1679–1684.

16 Wang, Z. (2002). Solubility of E2050 at various pH: a case in which apparent solubility is affected by the amount of excess solid. *J Pharm Sci* 91: 1445–1455.

17 Bighley, L., Berge, S., and Monkhouse, D. (1995). *Salt Forms of Drugs and Absorptions*. New York: Marcel Dekker.

18 Black, S., Collier, E., Davey, R., and Roberts, R. (2007). Structure, solubility, screening, and synthesis of molecular salts. *J Pharm Sci* 96: 1053–1068.

19 Bogardus, J. (1982). Common ion equilibria of hydrochloride salts and the setschenow equation. *J Pharm Sci* 71: 588–590.

20 Bogardus, J. and Blackwood, R. (1979). Solubility of doxycycline in aqueous solution. *J Pharm Sci* 68: 188–194.

21 David, S.E., Timmins, P., and Conway, B.R. (2012). Impact of the counterion on the solubility and physicochemical properties of salts of carboxylic acid drugs. *Drug Dev Ind Pharm* 38 (1): 93–103.

22 Miyazaki, S., Oshiba, M., and Nadai, T. (1981). Precaution on use of hydrochloride salts in pharmaceutical formulation. *J Pharm Sci* 70: 594–596.

23 Glsn, T., Akda, Y., Izat, N. et al. (2019). Effect of particle size and surfactant on the solubility, permeability and dissolution characteristics of deferasirox. *Journal of Research in Pharmacy* 23 (5): 851–859.

24 Johnson, K.C. (2012). Comparison of methods for predicting dissolution and the theoretical implications of particle-size-dependent solubility. *J Pharm Sci* 101 (2): 681–689.

25 Wu, W. and Nancollas, G. (1998). A new understanding of the relationship between solubility and particle size. *J Solution Chem* 27: 521–531.

26 Gong, N., Yang, D., Jin, G. et al. (2019). Structure, characterization, solubility and stability of podophyllotoxin polymorphs. *J Mol Struct* 1195: 323–330.

27 Kato, Y. and Kohketsu, M. (1981). Relationship between polymor- phism and bioavailability of amobarbitol in the rabbit. *Chem Pharm Bull* 28: 268–272.

28 Kobayashi, Y., Ito, S., Itai, S., and Yamamoto, K. (2000). Physicochemical properties and bioavailability of carbamazepine polymorphs and dihydrate. *Int J Pharm* 193: 137–146.

29 Pandit, J., Gupta, S., Gode, K., and Mishra, B. (1984). Effect of crystal form on the oral absorption of phenylbutazone. *Int J Pharm* 21: 129–132.

30 Nguyen, T.-H., Hanley, T., Porter, C.J.H. et al. (2010). Phytantriol and glyceryl monooleate cubic liquid crystalline phases as sustained-release oral drug delivery systems for poorly water-soluble drugs I. Phase behaviour in physiologically-relevant media. *Journal of Pharmacy and Pharmacology* 62 (7): 844–855.

31 Warren, D.B., Anby, M.U., Hawley, A., and Boyd, B.J. (2011). Real time evolution of liquid crystalline nanostructure during the digestion of formulation lipids using synchrotron small-angle X-ray scattering. *Langmuir* 27 (15): 9528–9534.

32 Schittny, A., Huwyler, J., and Puchkov, M. (2020). Mechanisms of increased bioavailability through amorphous solid dispersions: a review. *Drug Delivery* 27 (1): 110–127.

33 Aguiar, A., Krc, J., Kinkel, A., and Samyn, J. (1967). Effect of polymorphism on the absorption of chloramphenicol from chloramphenicol palmitate. *J Pharm Sci* 56: 847–853.

34 Mantri, R.V., Sanghvi, R., and Zhu, H.J. (2017). Solubility of pharmaceutical solids. In: *Developing Solid Oral Dosage Forms*, 2e, 3–22. Academic press.

35 Lorimer, J.W. and Cohen-Adad, R. (2003). Thermodynamics of solubility. *The Experimental Determination of Solubilities* 6: 17–76.

36 Gibbs, J. W. (1957). A method of geometrical representation of the thermodynamic properties by means of surfaces. The Collected Works of J. Willard Gibbs, Ph. D., LL. D, 33–54. Yale University Press.

37 Pitzer, K.S. (1973). Thermodynamics of electrolytes. I. Theoretical basis and general equations. *The Journal of Physical Chemistry* 77 (2): 268–277.

38 Sinko, P. and Martin, A.N. (2006). *Martin's Physical Pharmacy and Pharmaceutical Sciences*. Philadelphia: Lippincott Williams & Wilkins.

39 Aulton, M.E. (2017). Dissolution and solubility. In: *Aulton's Pharmaceutics E-Book: The Design and Manufacture of Medicines*, vol. 18, 21–22. Elsevier Health Sciences.

40 Dahlgren, D. and Lennernäs, H. (2022). 1.17 - Oral Drug Delivery, Absorption and Bioavailability. In: *Comprehensive Pharmacology* (ed. T. Kenakin), 406–437. Elsevier https://doi.org/https://doi.org/10.1016/B978-0-12-820472-6.00022-0.

41 Faller, B. and Ertl, P. (2007). Computational approaches to determine drug solubility. *Adv Drug Deliv Rev* 59: 533–545.

42 Cojocaru, O.A., Bica, K., Gurau, G. et al. (2013). Prodrug ionic liquids: functionalizing neutral active pharmaceutical ingredients to take advantage of the ionic liquid form. *Med Chem Comm* 4 (3): 559–563.

43 Li, Y., Zhu, Y.M., Jiang, H.J. et al. (2000). Synthesis and antimalarial activity of artemisin in derivatives containing an amino group. *J Med Chem* 43: 1635–1640.

44 Xie, L., Yu, D., Wild, C. et al. (2004). Anti-AIDS agents. 52. Synthesis and anti-HIV activity of hydroxymethyl (3′R,4′R)-3′,4′-di-*O*-(*S*)-camphanoyl-(+)-cis-khellactone derivatives. *J Med Chem* 47: 756–760.

45 Kim, I., Morisseau, C., Watanabe, T., and Hammock, B.D. (2004). Design, synthesis, and biological activity of 1,3-disubstituted ureas as potent inhibitors of the soluble epoxide hydrolase of increased water solubility. *J Med Chem* 47: 2110–2122.

46 Marinelli, L., Fornasari, E., Eusepi, P. et al. (2019). Carvacrol prodrugs as novel antimicrobial agents. *European Journal of Medicinal Chemistry* 178: 515–529.

47 Fujita, Y., Yonehara, M., Tetsuhashi, M. et al. (2010). β-Naphthoflavone analogs as potent and soluble aryl hydrocarbon receptor agonists: improvement of solubility by disruption of molecular planarity. *Bioorganic & Medicinal Chemistry* 18 (3): 1194–1203.

48 Fray, M., Bull, D.J., Carr, C.L. et al. (2001). Structure-activity relationships of 1,4-dihydro-(1*H*,4*H*)-quinoxaline-2,3-diones as *N*-methyl-D-aspartate (glycine site) receptor antagonists. 1. Heterocyclic substituted 5-alkyl derivatives. *J Med Chem* 44: 1951–1962.

49 Hancock, B. and Parks, M. (2000). What is the true solubility advantage for amorphous pharmaceuticals? *Pharm Res* 17: 397–404.

50 Parks, C., Koswara, A., Devilbiss, F. et al. (2017). Solubility curves and nucleation rates from molecular dynamics for polymorph prediction-moving beyond lattice energy minimization. *Phy Chem Chem Physics* 19 (7): 5285–5295.

51 Swalin, R. (1962). *Thermodynamics of Solids*. Hoboken, NJ: Wiley.

52 Cammarata, A. (1972). Interrelationship of the regression models used for structure-activity analyses. *J Med Chem* 15: 573–577.

53 Cammarata, A., Yau, S., and Rogers, K. (1971). Structure activity relationships having a basis in regular solution theory. *J Med Chem* 14: 1211–1212.

54 Ran, Y., Jain, N., and Yalkowski, S. (2001). Prediction of aqueous solubility of organic compounds by the general solubility equation (GSE). *J Chem Inf Model* 41: 1208–1217.

55 Elder, D. and Holm, R. (2013). Aqueous solubility: simple predictive methods (in silico, in vitro and bio-relevant approaches). *International Journal of Pharmaceutics* 453 (1): 3–11.

56 Avdeef, A. (2007). Solubility-pH equations for mono-, di-, and triprotic molecules. *Adv Drug Deliv Rev* 59: 568–590.

57 Bergstrom, C., Luthman, K., and Artursson, P. (2004). Accuracy of calculated pH dependent aqueous drug solubility. *Eur J Pharm Sci* 22: 387–398.

58 Avdeef, A. (1998). pH-Metric solubility. 1. Solubility-pH profiles from Bjerrum plots. Gibbs buffer and p$K_a$ in the solid state. *Pharm Pharmacol Commun* 4: 165–178.

59 Avdeef, A., Berger, C., and Brownell, C. (2001). pH-metric solubility 2: correlation between the acid-base titration and the saturation shake-flask solubility pH methods. *Pharm Res* 17: 85–89.

60 Avdeef, A. (2003). *Absorption and Drug Development – Solubility, Permeability and Charge State*. Hoboken, NJ: Wiley-Interscience.

61 Avdeef, A. (2004). Physicochemical profiling. (Solubility, permeability and charge state). *Current Topics in Medicinal Chemistry* 1 (4): 277–351.

62 Serajuddin, A. and Jarowski, C. (1985a). Effect of diffusion layer pH and solubility on the dissolution rate of pharmaceutical acids and their sodium salts. II: Salicylic acid, theophylline, and benzoic acid. *J Pharm Sci* 74: 148–154.

63 Serajuddin, A., Sheen, P.C., and Augustine, M. (1987). Common ion effect on solubility and dissolution rate of the sodium salt of an organic acid. *J Pharm Pharmacol* 39: 587–591.

64 Cavanagh, K.L., Maheshwari, C., and Rodríguez-Hornedo, N. (2018). Understanding the differences between cocrystal and salt aqueous solubilities. *J Pharm Sci* 1: 113–120.

65 Davies, G. (2001). Changing the salt, changing the drug. *Pharm J* 266: 322–323.

66 Serajuddin, A. (2007). Salt formation to improve drug solubility. *Adv Drug Deli Rev* 59 (7): 603–616.

67 Li, S., Doyle, P., Metz, S. et al. (2005). Effect of chloride ion on dissolution of different salt forms of haloperidol, a model basic drug. *J Pharm Sci* 94: 2224–2231.

68 Alqarni, M.H., Haq, N., Alam, P. et al. (2021). Solubility data, Hansen solubility parameters and thermodynamic behavior of pterostilbene in some pure solvents and different (PEG-400+ water) cosolvent compositions. *Journal of Molecular Liquids* 331: 115700.

69 Lipinski, C., Lombardo, F., Dominy, B., and Feeney, P. (1997). Experimental and computational approaches to estimate solubility and permeability in drug discovery and development settings. *Adv Drug Deliv Rev* 23: 3–25.

70 Amidon, G., Lennernaes, H., Shah, V., and Crison, J. (1995). A theoretical basis for a biopharmaceutic drug classification: the correlation of in vitro drug product dissolution and in vivo bioavailability. *Pharm Res* 12: 413–420.

71 Varma, M.V., Gardner, I., Steyn, S.J. et al. (2012). Ph-dependent solubility and permeability criteria for provisional biopharmaceutics classification (BCS and BDDCS) in early drug discovery. *Mol Pharm* 9 (5): 1199–1212.

72 Kerns, E. and Di, L. (2008). *Drug-like Properties: Concepts, Structure Design and Methods: from ADME to Toxicity Optimization*. Burlington, MA: Academic Press.

73 Gesenberg, C., Mathias, N.R., Xu, Y. et al. (2019). Utilization of in vitro, in vivo and in silico tools to evaluate the pH-dependent absorption of a BCS class ii compound and identify a pH-effect mitigating strategy. *Pharm Res* 36 (12): 164.

74 Butler, J.M. and Dressman, J.B. (2010). The developability classification system: application of biopharmaceutics concepts to formulation development. *Journal of Pharmaceutical Sciences* 99 (12): 4940–4954.

75 Rosenberger, J., Butler, J., and Dressman, J. (2018). A refined developability classification system. *Journal of Pharmaceutical Sciences* 107 (8): 2020–2032.

76 Smith, B.T. (2015). *Remington Education: Physical Pharmacy*. Pharmaceutical Press.

77 Singh 2006 Y. Martin's physical pharmacy and pharmaceutical sciences. New Jersey: Department of Pharmaceutics Ernest Mario School of Pharmacy Rutgers, The State University of New Jersey. 2006.

78 Goss, S.L., Lemons, K.A., Kerstetter, J.E., and Bogner, R.H. (2010). Determination of calcium salt solubility with changes in ph and p(co(2)), simulating varying gastrointestinal environments. *J Pharm Pharmacol* 59 (11): 1485–1492.

79 Grove, M., Pedersen, G., Nielsen, J., and Mullertz, A. (2005). Bioavail- ability of seocalcitol 1: relating solubility in biorelevant media with oral bioavailability in rats - effect of medium and long chain triglycerides. *J Pharm Sci* 94: 1830–1838.

80 Van de Waterbeemd, H. (1998). The fundamental variables of the biopharmaceutics classification system (BCS): a commentary. *Eur J Pharm Sci* 7: 1–3.

81 Jones, H. (2006). Predicting pharmacokinetic food effects using biorelevant solubility media and physiologically based modeling. *Clin Pharmacokinet* 45: 1213–1226.

82 Bhattachar, S., Deschenes, L., and Wesley, J. (2006). Solubility: it's not just for physical chemists. *Drug Discov Today* 11: 1012–1018.

83 Veseli, A., Simon, Ž., and Kristl, A. (2019). A review of methods for solubility determination in biopharmaceutical drug characterization. *Drug Dev Ind Pharm* 45: 1–25.

84 Etzweiler, F., Senn, E., and Schmid, H.W.H. (1995). Method for measuring aqueous solubilities of organic compounds. *Anal Chem* 67: 655–658.

85 Jorgensen, W. and Duffy, E. (2002). Prediction of drug solubility from structure. *Adv Drug Deliv Rev* 54: 355–366.

86 Kerns, E. (2001). High throughput physicochemical profiling for drug discovery. *J Pharm Sci* 90: 1838–1858.

87 Pan, L., Ho, Q., Tsutsui, K., and Takahashi, L. (2001). Comparison of chromatographic and spectroscopic methods used to rank compounds for aqueous solubility. *J Pharm Sci* 90: 521–529.

88 Peltonen, L. and Hirvonen, J. (2018). Drug nanocrystals-versatile option for formulation of poorly soluble materials. *Int J Pharma.* 537 (1–2): 73–83.

89 Alsenz, J. and Kansy, M. (2007). High throughput solubility measurement in drug discovery and development. *Advanced Drug Delivery Reviews* 59 (7): 546–567.

90 Grant, D. and Higuchi, T. (1990). *Solubility Behavior of Organic Compounds*. New York: Wiley.

91 Krisztina, T., Dóra, T., Dóra, C. et al. (2019). Solubility analysis of venlafaxine hydrochloride polymorphs by shake-flask method and real time monitoring. *Acta Pharm* 89 (2): 88–96.

92 Chen, X. and Venkatesh, S. (2004). Miniature device for aqueous and non-aqueous solubility

measurements during drug discovery. *Pharm Res* 21: 1758–1761.

93 Glomme, A., Marz, J., and Dressman, J. (2005). Comparison of a miniaturized shake-flask solubility method with automated potentiometric acid/base titrations and calculated solubilities. *J Pharm Sci* 94: 1–16.

94 Chen, J., Wang, J., and Schwab, L.P. (2014). Benzimidazole analogs as potent hypoxia inducible factor inhibitors: synthesis, biological evaluation, and profiling drug-like properties. *Anticancer Res* 34: 3891–3904.

95 Gopi, S.P., Banik, M., and Desiraju, G.R. (2017). New cocrystals of hydrochlorothiazide: optimizing solubility and membrane diffusivity. *Cryst Growth Des* 17: 308–316.

96 Rajput, L. (2014). Stable crystalline salts of haloperidol: a highly water-soluble mesylate salt. *Cryst Growth Des* 14: 5196–5205.

97 Ruff, A., Fiolka, T., and Kostewicz, E.S. (2017). Prediction of Ketoconazole absorption using an updated in vitro transfer model coupled to physiologically based pharmacokinetic modelling. *Eur J Pharm Sci* 100: 42–55.

98 Silva, C.C., Cirqueira, M., and Martins, F.T. (2013). Lamivudine salts with 1,2-dicarboxylic acids: a new and a rare synthon with double pairing motif fine-tuning their solubility. *CrystEngComm* 15: 6311–6317.

99 Bevan, C. and Lloyd, R. (2000). A high-throughput screening method for the determination of aqueous drug solubility using laser nephelometry in microtiter plates. *Anal Chem* 72: 1781–1787.

100 Stuart, M. and Box, K. (2005). Chasing equilibrium: measuring the intrinsic solubility of weak acids and bases. *Anal Chem* 77: 983–990.

101 Avdeef, A. and Berger, C. (2001). pH-metric solubility. 3. Dissolution titration template method for solubility determination. *Eur J Pharm Sci* 14: 281–291.

102 Chen, X. (2002). Prediction of aqueous solubility of organic com- pounds using a quantitative structure - property relationship. *J Pharm Sci* 91: 1838–1852.

103 Ghasemi, J. (2007). QSPR prediction of aqueous solubility of drug- like organic compounds. *Chem Pharm Bull* 55: 669–674.

104 Gao, H. (2002). Estimation of aqueous solubility of organic compounds with QSPR approach. *Pharm Res* 19: 497–503.

105 Yalkowsky, S. and Banerjee, S. (2001). *Methods of Estimation for Organic Compounds*. New York: Marcel Dekker.

106 Vertzoni, M. (2007). Estimation of intragastric solubility of drugs: in what medium? *Pharm Res* 24: 909–917.

107 Fadda, H., Sousa, T., Carlsson, A. et al. (2010). Drug solubility in luminal fluids from different regions of the small and large intestine of humans. *Mol Pharm* 7: 1527–1532.

108 Augustijns, P., Wuyts, B., Hens, B. et al. (2014). A review of drug solubility in human intestinal fluids: implications for the prediction of oral absorption. *Eur J Pharm Sci* 57: 322–332.

109 Guo, J., Elzinga, P.A., Hagerman, M.J., and Herron, J.N. (2008). Rapid throughput solubility screening method for BCS class II drugs in animal GI fluids and simulated human GI fluids using a 96-well format. *J Pharm Sci* 97: 1427–1442.

110 Skolnik, S., Barrett, J., Yang, W. et al. (2022). Discovery solubility measurement and assessment with drug development in mind. *Drug Discovery Today* 1 (October): 31–50.

111 Indulkar, A.S., Gao, Y., Raina, S.A. et al. (2018). Crystallization from supersaturated solutions: role of lecithin and composite simulated intestinal fluid. *Pharmaceutical Research* 35 (8): 1–14.

112 Yu, L. (1999). An integrated model for determining causes of poor oral drug absorption. *Pharm Res* 16: 1883–1887.

113 Williams, H.D., Trevaskis, N.L., Charman, S.A. et al. (2013). Strategies to address low drug solubility in discovery and development. *Pharmacological Reviews* 65 (1): 315–499.

114 Gao, Y., Gesenberg, C., and Zheng, W. (2017). Oral formulations for preclinical studies: principle, design, and development considerations. In: *Developing Solid Oral Dosage Forms*, 455–495. Elsevier.

115 Pouton, C.W. and Porter, C.J.H. (2008). Formulation of lipid-based delivery systems for oral administration: materials, methods and strategies. *Advanced Drug Delivery Reviews* 60 (6): 625–637.

116 Strickley, R.G. (2007). Currently marketed oral lipid-based dosage forms: drug products and excipients. In: *Oral Lipid-Based Formulations*, 23–54. CRC Press.

117 Sun, W., Mao, S., Shi, Y. et al. (2011). Nanonization of itraconazole by high pressure homogenization: stabilizer optimization and effect of particle size on oral absorption. *Journal of Pharmaceutical Sciences* 100 (8): 3365–3373.

118 Rajewski, R. (1996). Pharmaceutical applications of cyclodextrins.2. In vivo drug delivery. *J Pharm Sci* 85: 1142–1169.

**119** Zhong, Z., Zhong, W., Jiong, Z. et al. (2016). The pharmaceutical applications of cyclodextrins. *Progress in Modern Biomedicine* 2016: 19.

**120** Jones, M. and Leroux, J. (1999). Polymeric micelles— a new generation of colloidal drug carriers. *Eur J Pharm Biopharm* 48: 101–111.

**121** Yang, H., Zhao, X., Zhang, X. et al. (2018). Optimization of bioreducible micelles self-assembled from amphiphilic hyperbranched block copolymers for drug delivery. *J Polym Sci A* 56: 1383–1390.

**122** Rosler, A., Vandermeulen, G.W., and Klok, H.A. (2012). Advanced drug delivery devices via self-assembly of amphiphilic block copolymers. *Adv Drug Deliv Rev* 64 (suppS): 270–279.

**123** Kataoka, K., Harada, A., and Nagasaki, Y. (2012). Block copolymer micelles for drug delivery: design, characterization and biological significance. *Adv Drug Deliv Rev* 64: 37–48.

**124** Vierling, P. and Greiner, J. (2003). Prodrugs of HIV protease inhibitors. *Current Pharmaceutical Design* 9 (22): 1755–1770.

**125** Tayel, S.A., Soliman, I.I., and Louis, D. (2008). Improvement of dissolution properties of carbamazepine through application of the liquisolid tablet technique. *European Journal of Pharmaceutics and Biopharmaceutics* 69 (1): 342–347.

**126** Tiong, N. and Elkordy, A.A. (2009). Effects of liquisolid formulations on dissolution of naproxen. *European Journal of Pharmaceutics and Biopharmaceutics* 73 (3): 373–384.

# 3

# *In Vitro* Dissolution of Pharmaceutical Solids

*Tze Ning Hiew*[1] *and Paul W.S. Heng*[2]

[1] *Department of Pharmaceutical Sciences and Experimental Therapeutics, College of Pharmacy, University of Iowa, Iowa City, IA, USA*
[2] *Department of Pharmacy, Faculty of Science, National University of Singapore, Singapore*

## 3.1 Dissolution Theory and Fundamentals

### 3.1.1 Background and Rationale

Dissolution can be defined as the process of a solid going into solution. In pharmaceutical solids, dissolution of the active pharmaceutical ingredient (API) is a critical rate-limiting step to its absorption and is often preceded by disintegration of the solid dosage form.

The early understanding of dissolution commenced with a publication by Noyes and Whitney [1], which defined the relationship between dissolution rate and concentration gradient, assuming constant surface area. This formed the basis of the well-known Noyes–Whitney equation. This equation was modified by Nernst [2] and Brunner [3] to incorporate Fick's law of diffusion, which established the relationship between dissolution rate and dissolution coefficient.

Historically, disintegration was thought to be more critical than dissolution because it was deemed to be directly related to bioavailability as smaller particles were reported to be better absorbed. Hence, it was not surprising that pharmacopeial tests for disintegration testing existed for a few decades (1934 in the Swiss Pharmacopeial Helvetica, 1950 in the United States Pharmacopeia) before those for dissolution testing were available. It was not until 1970 when the first official dissolution test for solid oral dosage forms was incorporated in the United States Pharmacopeia (USP) 18. Due to wide variations in the results obtained using different dissolution apparatuses, there was a strong push by the U.S. Food and Drug Administration (FDA) for standardization of dissolution testing in the subsequent years. Today, the USP 42 has over 950 monographs with dissolution tests. The USP currently has seven official methods described in two general chapters: USP Apparatus 1 – rotating basket, Apparatus 2 – paddle, Apparatus 3 – reciprocating cylinder, Apparatus 4 – flow-through cell, Apparatus 5 – paddle over disk, Apparatus 6 – cylinder, and Apparatus 7 – reciprocating cylinder. The first four apparatus are described in the general chapter ⟨711⟩ Dissolution [4], while the latter three apparatus are described in the general chapter ⟨724⟩ Drug Release [5]. The Rufinamide Tablets monograph was the first USP monograph with a dissolution test using the USP Apparatus 4, which became official in the USP 36–NF 31 in 2013 [6]. The USP Apparatus 5 – paddle over disk and Apparatus 6 – cylinder were developed for the evaluation of transdermal delivery systems. The USP Apparatus 7 – reciprocating cylinder is by far the most versatile, as it can be used to evaluate a variety of dosage forms including transdermal and oral dosage forms. Interestingly, the USP Apparatus 5, 6, and 7 were officially incorporated into the USP before Apparatus 3 and 4. The paddle over disk, cylinder, and reciprocating cylinder were made official in 1988 together with a section on drug release testing of transdermal systems [6]. The applications and limitations of each apparatus will be elaborated in Sections 3.3.1, 3.4.1, and 3.4.2.

This chapter will not elaborate in detail on well-established concepts that have been comprehensively covered and are readily available in standard references, but aim to provide an overview of fundamental concepts as well as more recent developments in this field.

### 3.1.2 Why Dissolution Testing

The oral bioavailability of a drug invariably depends on its dissolution upon ingestion, absorption in the small intestine, and finally transport to its target site

*Oral Bioavailability and Drug Delivery: From Basics to Advanced Concepts and Applications*, First Edition. Edited by Ming Hu and Xiaoling Li.

of action (see also Chapters 1 and 4). Consequently, *in vitro* dissolution testing, particularly using simulated biological fluids, is widely regarded as a rapid and simple surrogate of *in vivo* performance, which may not be available in the early discovery and developmental phases. *In vitro* dissolution testing has later established itself as one of the standard tests in the drug development process and quality control of commercial products. Dissolution testing as part of quality control is important as dissolution failure can result in drug product recalls. Modified-release products are especially at greater risk of dissolution failure, with 96% of recalls associated with dissolution failure made between 2011 and 2013 affecting modified-release oral dosage forms [7].

The purpose of *in vitro* dissolution testing is to predict how a dosage form will perform *in vivo*. Immediate-release dosage forms have different rate-determining release mechanism compared to modified-release dosage forms. Therefore, the dissolution method should be dosage form specific and clinically relevant. For the former, a dissolution test should be sufficiently robust to capture variables such as wetting and disintegration time, while pH-dependent drug release and onset of dissolution should be taken into account for the latter. As *in vivo* data become increasingly available in the later stages of product development, *in vivo–in vitro* correlation (IVIVC) can be established (see also Chapter 16). Adherence to *in vitro* dissolution specifications can demonstrate batch-to-batch consistency of a manufactured product and provide a form of quality assurance for its oral bioavailability. Where modifications to the formulations are required, IVIVC data can help to set more realistic specifications for *in vitro* dissolution testing in order to facilitate approvals in post-filing changes [8–12]. IVIVC is especially relevant to dosage forms where *in vivo* dissolution is the rate-limiting step for bioavailability [13]. Therefore, it is mostly applicable to extended-release dosage forms, as the dissolution of this dosage form is intentionally designed to be the rate-limiting step.

### 3.1.3 Theoretical Concepts

The first concept of dissolution was published by Noyes and Whitney [1] and the original equation is expressed as in Eq. (3.1). Their experimental setup at that time ensured that the surface area of materials remained constant during dissolution because the materials were present in excess of the amount needed to form a saturated solution.

$$\frac{\mathrm{d}C}{\mathrm{d}t} = K(C_\mathrm{s} - C) \tag{3.1}$$

where $\frac{\mathrm{d}C}{\mathrm{d}t}$ is dissolution rate, $C$ is instantaneous concentration at time $t$, $C_\mathrm{s}$ is saturation concentration and $K$ is a constant.

The original equation was modified in 1904 by Nernst [2] and Brunner [3] to incorporate the diffusion layer concept and Fick's second law of diffusion to derive Eq. (3.2) where the constant $K$ is explicated further by the diffusion coefficient of the drug $D$, surface area $A$, volume of the dissolution medium $V$, and the thickness of diffusion layer $h$.

$$\frac{\mathrm{d}C}{\mathrm{d}t} = \frac{DA(C_\mathrm{s} - C)}{Vh} \tag{3.2}$$

By integrating Eq. (3.2), Hixson and Crowell [14] were able to relate time to the cubic root of weight, and under sink conditions, the term $(C_\mathrm{s} - C)$ can be considered a constant. This gave rise to Eq. (3.3) where $W_\mathrm{o}$ is the initial weight and $k_2$ is a constant. All three approaches listed regarded diffusion of the drug across a stagnant film of liquid around the solid as the rate-limiting step.

$$\sqrt[3]{W_\mathrm{o}} - \sqrt[3]{W} = k_2 t \tag{3.3}$$

### 3.1.4 Properties of the API

Based on Eq. (3.2), dissolution rate is influenced by the diffusion coefficient, $D$ of an API, where a higher diffusion coefficient would translate into faster dissolution rate. The diffusion coefficient of an API is related to its molecular weight. Larger molecules have a lower diffusivity and transverse slowly through the diffusion layer.

Dissolution rate is also related to the solubility term, $C_\mathrm{s}$ in Eq. (3.2) (see also Chapter 2). Highly soluble compounds tend to dissolve faster. In addition, since most drugs are weak acids or bases, their solubility is affected by p$K_\mathrm{a}$. These APIs exhibit pH-dependent solubility, as the pH of the medium will affect the degree of ionization. Thus, aqueous solubilities of the APIs are often determined over the physiologically relevant pH range of 1.2–7.5 to predict the effect of pH on dissolution and sometimes even oral absorption.

Other properties affecting the aqueous solubility of a drug include its melting point, log $P$, and solid-state form. A high melting point API is more poorly soluble as its higher lattice energy makes it more difficult to dissolve. The log $P$ value of an API is a measure of its lipophilicity. APIs with high log $P$ values are more poorly water soluble, as they do not interact favorably with water molecules. The solid-state form of the API also plays a considerable part in its solubility, especially for larger molecules. Many APIs can exist in different solid-state forms notably: crystalline, amorphous, and solvate. Due to their orderliness, crystalline solids often dissolve more slowly than the corresponding amorphous forms, which do not possess any fixed-ordered molecular arrangements. For crystalline APIs that exhibit polymorphism, polymorphs can have different solubility profiles due to differences in their lattice energies. A polymorph with lower lattice energy will be more soluble. When formulated as a salt of a weak acid or weak base, solubility will improve due to favorable thermodynamics involving aqueous solvation of ions and the ability of the salt to alter the resultant pH on the solid surface or in the diffusion layer. The latter is especially important, as evident by findings that the microenvironmental conditions in the diffusion layer on the drug particles can play a considerable role in enhancing dissolution [15].

Since dissolution rate is directly proportional to the surface area of the drug particle, the drug can be further processed to achieve a smaller particle size range. Smaller particles have larger surface areas, which when exposed to the dissolution medium will lead to faster dissolution rates. In fact, micronization is one of the most common approaches to improve the dissolution performance of poorly soluble compounds [16]. However, micronized particles have large surface energies and a tendency to agglomerate, which reduce the effective surface area for dissolution. In certain cases, very rapid dissolution may induce localized supersaturation and recrystallization into larger, slowly dissolving crystals.

### 3.1.5 Examples of Techniques to Alter Dissolution Rate Based on the Dissolution Theory

Drug discovery pipelines in pharmaceutical research and development have been increasingly filled with poorly soluble compounds, with about 90% of the drugs in development being characterized as poorly soluble [17]. There are many strategies available to maximize dissolution rate and enhance drug bioavailability, such as particle size reduction, formation of solid dispersions, and crystal engineering. To illustrate strategies used for dissolution rate improvement based on Eq. (3.2), the use of nanotechnology and amorphous solid dispersion (ASD) as formulation strategies will be discussed in Sections 3.1.5.1 and 3.1.5.2.

#### 3.1.5.1 Nanotechnology – Changing the Surface Area Term in the Equation

Nanotechnology involves the study of materials at the nanoscale, which ranges from about 1–100 nm. Drug particles in the nano-size range are otherwise referred to as nanoparticles. These particles have high specific surface areas, which lead to an increase in dissolution rate. Nanocrystalline-based drug products in the market include Emend® and Rapamune®, and a more comprehensive list can be found in the publication by Malamatari et al. [18].

Nanoparticles can be produced through a top-down or bottom-up approach [19, 20]. Examples of top-down approaches include wet milling and high-pressure homogenization. As its name implies, wet milling involves dispersing the drug in aqueous solution to form a suspension, which is wet-milled using rotor–stator mixers or media mills [18]. High-pressure homogenization can be achieved through microfluidization or piston-gap homogenization. Microfluidization involves the use of two jet streams, which are collided together at high velocity under high pressure. The shear force from the collision results in particle size reduction. Piston-gap homogenization comminutes particles by forcing them through a narrow gap at high pressure [21]. As the static pressure in the gap is less than the vapor pressure of water at room temperature, water starts to boil and bubbles are formed in the gap. When the bubbles exit the gap, they collapse and generate cavitation energy that breaks down the particles.

The generation of nanoparticles with the bottom-up approach can be achieved through precipitation or controlled droplet evaporation. Precipitation usually involves mixing a drug solution with an anti-solvent to induce precipitation of the drug [19]. Upon evaporation, the solute in droplets will form solid particles. The key to obtaining solid particles in the nano-size range lies in controlling the size of the droplets. This can be achieved through the use of nano-spray dryers [22]. Discussion of how nanoparticles can be used to improve oral delivery is also found in Chapter 30.

#### 3.1.5.2 Solid Dispersions – Changing the Solubility Term in the Equation

Solid dispersions can be simply defined as molecular mixtures of poorly soluble drugs in a hydrophilic polymer matrix. They were first described in the 1960s as eutectic mixtures or molecular mixtures dispersed in crystalline carriers such as urea and mannitol. However, with this approach, dissolution rate was not markedly improved because crystalline dispersions were more thermodynamically stable than amorphous systems [23]. Consequently, the latter approaches focused on ASDs where the drug (in its amorphous state) and polymeric carrier (e.g. copovidone, polymethacrylates, and hypromellose) are molecularly mixed. ASD formulations may also contain surfactants such as docusate sodium, poloxamer 188, polyoxyethylene 40 monostearate, sodium dodecyl sulfate, or polysorbate 80. Ghebremeskel et al. [24] reported that surfactants caused solvation/plasticization, manifesting in the reduction of (i) the melting point of API, (ii) glass transition temperatures of the polymers, and (iii) the combined glass transition temperature of the solid dispersion formed from quench cooling. Surfactants are also added to increase the drug dissolution rate. Thermal analysis using modulated differential scanning calorimetry, powder X-ray diffraction, Raman spectroscopy, and polarized light microscopy can be used to evaluate the physicochemical stability of the ASDs, which can have shelf lives of more than 12 months [23].

Compared to micronization, ASD formulation is a more efficient way of particle size reduction because the former can only achieve 1–5 μm ranges compared to molecular scales in ASDs. In the past decade, solid dispersion-based formulations have become more prevalent in the market, both as new chemical entities (e.g. Kaletra®) and reformulated older drugs (e.g. Noxafil® Delayed-Release Tablet) [25]. Compared to its crystalline counterpart, dissolution of the higher energy amorphous form of the drug in the gastrointestinal fluid enables the formation of a supersaturated solution, which is thermodynamically favorable to promote rapid absorption. In the absence of recrystallization, a metastable equilibrium can be achieved between the amorphous material and the solution phase, where the maximum achievable concentration of the amorphous drug in solution (bulk aqueous phase) is known as its amorphous solubility [26]. In some cases, the concentration of drug released from ASD can exceed its amorphous solubility. Since the drug molecules can no longer dissolve in the bulk aqueous phase, phase separation occurs to yield colloidal drug-rich droplets via liquid–liquid phase separation [27]. Liquid–liquid phase separation is desirable as the drug-rich droplets serve as a drug reservoir to replenish the drug removed through transport across the membrane to ensure that maximum drug concentration in the bulk aqueous phase is always maintained [28].

The issue of recrystallization has been examined by the use of polymeric carrier to increase the overall glass transition temperature of the ASD and thus, decrease molecular mobility. Konno and Taylor [29] reported that hypromellose, povidone, or hypromellose acetate succinate can inhibit the nucleation of amorphous felodipine and prevent recrystallization by increasing the kinetic barrier to nucleation. Indeed, studies have reported that with kinetic stabilization, recrystallization in ASDs can be prevented for as long as 25 years [30, 31]. Recent studies with copovidone have shown that, in addition to its role as a crystallization inhibitor, the polymer in the ASD also plays a role in controlling the rate of drug dissolution in congruently releasing formulations. In these formulations, dissolution is polymer-controlled, and the drug and polymer have similar normalized dissolution rates, which are largely dictated by the polymer [32–34].

ASDs are often produced from melting or melt processes such as melt extrusion and high-shear melt granulation. However, the high temperatures associated with these techniques exclude their use for thermolabile drugs. An alternative for such drugs would be to employ solvent evaporation techniques such as spray drying [35], freeze drying [36], spray freeze drying [37], and the use of supercritical fluids [38].

The major advantage of ASDs over micronization in enhancing drug dissolution is the provision of a hospitable adjacent medium for the dissolved drug. With the polymer present, the risk of is recrystallization reduced. While the colloidal drug-rich droplets are expected to subsequently crystallize as the system is in metastable equilibrium, liquid–liquid phase separation can persist for drugs which crystallize more slowly, therefore, reaping the benefits of the colloidal drug-rich droplets to ensure that supersaturation is always maintained.

## 3.2 Dissolution of Drug Products

Dissolution of a solid dosage form commences when it comes into contact with the dissolution medium (wetting), followed by penetration of the medium into the solid, which generally assists in the disintegration/deaggregation into primary particles. The last step involves the API going into solution within the dissolution medium. The dissolution process has been schematically presented by Wagner [39] and reproduced in Figure 3.1 below.

### 3.2.1 Type of Dosage Forms and Common Dissolution Mechanisms

#### 3.2.1.1 Immediate-Release Dosage Forms

Immediate-release dosage forms are designed to dissolve without modifications to delay or prolong drug release. Most immediate-release dosage forms contain disintegrants to facilitate disintegration and deaggregation of the dosage form upon contact with the dissolution medium. Thus, it is important that disintegrants are

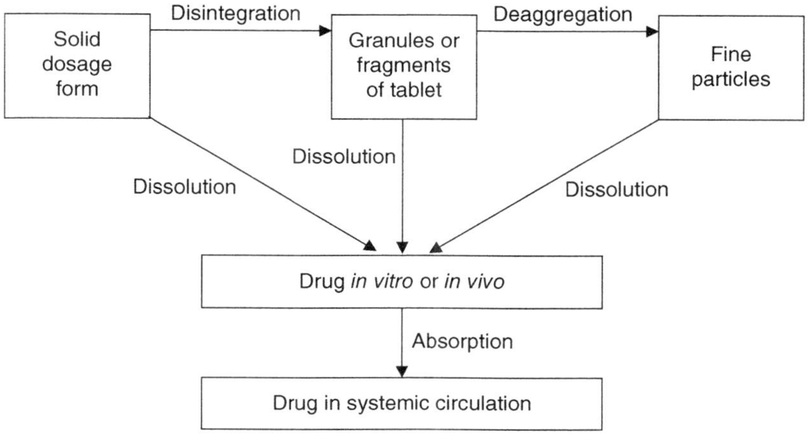


**Figure 3.1** Schematic representation of the dissolution process of a solid dosage form. Source: Adapted from Wagner [39].

efficient as disintegration precedes dissolution [40]. Upon disintegration, the primary drug particles come into contact with the dissolution medium and undergo dissolution, as governed by Eq. (3.2).

#### 3.2.1.2 Modified-Release Dosage Forms

Modified-release dosage forms can be further subdivided into extended-release and delayed-release dosage forms. Extended-release dosage forms are designed to release APIs in a controlled manner over an extended period. Examples of extended-release dosage forms include matrix tablets and film-coated tablets/pellets. Matrix tablets are made by blending the API with a swellable hydrophilic excipient (e.g. hypromellose, guar gum, and sodium alginate) prior to compression. Upon contact with the dissolution medium, the hydrated excipient swells to form a gel layer around the tablet surface. The gel layer retards the penetration of water to the tablet core and acts as a diffusion barrier for drug release [41, 42]. With time, water will penetrate deeper into the tablet core, which results in a thicker gel formed [43]. At the same time, the fully hydrated gel layer at the surface erodes through polymer chain disentanglement. Eventually, an equilibrium is achieved between the rate of hydration and erosion of the polymer. Highly water-soluble drugs are released from matrix tablets through diffusion across the gel layer, while poorly water-soluble drugs enter the dissolution medium primarily through erosion of the tablet matrix with time. Matrix tablets can also be formulated with water-insoluble excipients such as polyvinyl acetate and ethyl cellulose. Ethyl cellulose undergoes plastic deformation during compression to form a dense compact [44], especially when a smaller particle size fraction is used [45]. The high tablet density retards the rate of water penetration into the tablet matrix, which in turn controls the drug dissolution rate. The water-insoluble excipient like polyvinyl acetate can also be pre-blended with a water-soluble excipient such as povidone, which acts as a pore former. One example of a polyvinyl acetate/povidone blend is Kollidon® SR. Since povidone is readily soluble in the aqueous medium, it will leach out from the tablet upon wetting to form pores, which enable the drug to diffuse outward into the dissolution medium.

In addition to matrix tablet formulations, tablets and pellets can be coated with a water-insoluble or poorly water-soluble polymer to modify their drug release profiles. Polymers that are commonly used to film-coat tablets and pellets to achieve a controlled-release profile include ethyl cellulose and methacrylic acid copolymer. The coated pellets are subsequently filled into hard-shell capsules or compressed with excipients to form multiple-unit pellet system tablets. With ethyl cellulose films, drug release mechanism involves a combination of osmotically driven release and diffusion through the polymer coat [46]. The methacrylic acid copolymer coating dispersion consists of copolymers of ethyl acrylate, methyl methacrylate, and a low content of a methacrylic acid ester with quaternary ammonium groups in the form of trimethylammonioethyl methacrylate chloride. The ammonium groups are present as salts and because they are soluble in water, they subsequently produce permeable polymer by the formation of pores. The pores enable water permeation into the pellet core to dissolve the drug and elicit its release.

Delayed-release dosage forms are designed to achieve slow release of the API over a prescribed amount of time after administration. Delayed-release dosage forms are usually coated with an enteric polymer (e.g. hypromellose acetate succinate or phthalate) to prevent drug release in the acidic gastric environment, and drug release will commence with the coat dissolving in the intestinal environment.

### 3.2.2 Factors Affecting Dissolution

#### 3.2.2.1 Drug and Excipient Properties

As mentioned in Section 3.1.4, the properties of an API affect its dissolution rate. Besides the properties of the API, the drug product formulation can affect its dissolution behavior, depending on the target release profiles. If modulating the rate and extent of drug release is necessary, a highly soluble drug can be formulated to dissolve more slowly, for example, with polymers (e.g. polymethacrylates, polyethylene glycol, polyvinyl acetate, polyvinyl alcohol, polylactide-*co*-glycolides, hypromellose, and ethyl cellulose). Formulations where the active is dispersed within a matrix or encapsulated by a film of such materials, form the basis of modified-release dosage forms. Contrary to modified-release dosage forms, immediate-release dosage forms are designed to release the drug quickly. For table dosage forms, the quick release can be achieved by adding disintegrants such as cross-linked povidone or sodium starch glycolate to promote the disintegration and deaggregation of tablet dosage forms. Also, pH-modifying agents, such as citric acid, can be added to improve the dissolution of poorly soluble, weakly basic drugs. Surfactants such as sodium dodecyl sulfate as well as wetting agents may also be added to improve surface wetting and in some cases for solubilization by micelle formation.

#### 3.2.2.2 Drug Product Formulation and Processing Conditions

The choice of processing technique may also influence drug release from a product. Spray drying forms highly porous particles that offer themselves very well for dissolution. Fluid bed granulation enables even distribution of hydrophilic binder and loose agglomerates formed can potentially be easier to dissolve. Denser granules by high shear granulation and roller compaction will generally result in slower dissolving particles. Often, dissolution properties are more affected by the formulation than the granulation techniques. Well-formulated granules will produce tablets with good disintegration properties. Poorly soluble drugs or poorly wetting drugs can be made hydrophilic using hydrophilic binders or surface-active additives. The optimal blending of hydrophobic APIs with hydrophilic excipients can help initiate wetting and drug dissolution upon disintegration. This is better achieved by the use of a liquid binder containing hydrophilic additives than by dry granulation methods such as roller compaction.

## 3.3 *In Vitro* Dissolution Methods for Ensuring Quality of Commercial Drug Products

### 3.3.1 Compendial Dissolution Apparatus

The basket and paddle methods are the two most commonly used dissolution methods and are applicable to a wide range of solid dosage forms including capsules, tablets (immediate- and modified-release), beads, and even floating dosage forms. They consist simply of a covered glass vessel fitted with a basket or paddle at the end of a motorized shaft. The vessel is filled with dissolution medium, which in the simplest form, is simply deionized water, hydrochloric acid, or buffers at various physiological pHs. The entire setup is partially immersed in a water bath maintained at 37 °C. The solid dosage form is dropped into the basket or the vessel directly and aliquot samples of the dissolution medium are collected at fixed time intervals throughout the test. These samples are then analyzed using ultraviolet (UV) spectrophotometry or high-performance liquid chromatography, with calibration using reference standards.

While both methods are simple to setup and use, there are significant concerns about the lack of reproducibility and accuracy in the data obtained using the basket and paddle methods. This is attributed to the variability in the position and wobble of the shaft, the position of the dosage form, sampling points, calibration errors, and differences in the type of equipment [47, 48]. Uneven hydrodynamic distribution within the vessel leads to variability in the shear forces experienced by the tablet, mass transfer of the solid to the liquid, and mixing of the tablet fragments [49, 50]. These variabilities are well studied and controlled during dissolution method development and validation.

Using particle tracking velocimetry and planar-induced laser fluorescence techniques, researchers were able to map velocity fields within the mixing vessel and provide information about flow patterns and recirculation zones [49, 50]. Computational fluid dynamic models demonstrated considerable spatial heterogeneity in the strain rate and shear forces present within the vessel under normal operating conditions. The highest shear forces were experienced at the walls of the vessels whereas the lowest were between the agitator shaft and the vessel wall above the paddles. Such heterogeneous mixing patterns have also led to poor mixing of the buffer and acid media during the pH transition step of the USP method A for enteric-coated tablets [51].

Due to the limitations of the basket and paddle methods, the USP Apparatus 3 was incorporated in 1991 for the testing of extended-release products. Based on the disintegration tester, it consists of a set of cylindrical flat-bottomed glass outer vessels and a set of glass reciprocating inner vessels, which are fitted with meshes at the bottom. By adjusting the agitation rate, it was possible to produce similar dissolution profiles as those produced by the USP Apparatus 2 without the issues associated with the latter [52]. This made it possible to use the USP Apparatus 3 for testing of immediate-release tablets containing highly soluble or poorly soluble drugs in addition to extended-release formulations.

With the introduction of more novel dosage forms such as suppositories, implants as well as controlled-release formulations or multi-particulates, the USP Apparatus 4 (flow-through cell) may be a more suitable dissolution setup. The flow-through cell is made of transparent and inert material, mounted vertically with a filter system that prevents the escape of undissolved particles from the top of the cell; standard cell diameters are 12 and 22.6 mm; the bottom cone is usually filled with small glass beads of about 1 mm diameter with one bead of about 5 mm positioned at the apex to protect the fluid entry tube; and a tablet holder is available for positioning of special dosage forms, for example, inlay tablets [4]. The assembly consists of a reservoir and a pump for the dissolution medium, a flow-through cell, and a water bath that maintains the dissolution medium at $37 \pm 0.5\,^{\circ}C$. The pump forces the medium upward through the flow-through cell at standard flow rates of 4, 8, and 16 ml/min. It must deliver a constant flow ($\pm 5\%$ of the nominal flow rate); the flow profile is sinusoidal with a pulsation of $120 \pm 10$ pulses/min [4].

As the USP Apparatuses 5 and 6 are not used for the evaluation of oral dosage forms, they will not be further elaborated here. While the Apparatus 7 was initially introduced alongside the inclusion of a section on transdermal systems, it can also be used to evaluate solid dosage forms such as extended-release tablets [5] as well as long-acting implants. Examples of USP monographs with the Apparatus 7 listed for dissolution test include nifedipine [53] and verapamil hydrochloride [54] extended-release tablets. The Apparatus 7 was also used to evaluate the individual pellet coat quality of sustained-release pellets [55].

The reciprocating holder assembly in Apparatus 7 consists of a set of containers of known volume, usually made of glass or other suitable inert material, a motor and drive assembly with the ability to reciprocate the system vertically and in automated systems, to index the system horizontally to different rows of vessels, as well as a set of sample holders. The sample holder for an oral extended-release tablet can be a rod or spring holder. Using the Apparatus 7 will require attaching the dosage form onto a suitable holder. The solution containers are partially immersed in a water bath to maintain the dissolution medium at $37 \pm 0.5\,^{\circ}C$. During the test, the sample holder is suspended from a vertically reciprocating shaker such that the dosage form is continuously immersed in the dissolution medium. The immersion frequency is about 30 cycles/min with an amplitude of about 2 cm. At each sampling interval, the solution containers are removed from the water bath, cooled to room temperature, and topped up with a suitable solvent (e.g. water) to correct for evaporative losses. The dissolution medium is then analyzed spectrophotometrically for drug content with reference to a calibration standard.

In general, apparatus selection is made based on knowledge of the formulation design. The dissolution test should ideally evaluate the practical aspects of dosage form performance. A compendial apparatus should be selected, but non-compendial apparatus may be used with proper justification.

### 3.3.2 General Compendial Dissolution Conditions

When designing a dissolution test, in addition to selecting an appropriate apparatus, other factors that should be considered include medium volume, composition, and pH. Depending on the apparatus selected, rotational speed of the basket/paddle, dip rate, or flow rate should also be optimized. The duration of the test and sampling times selected should be practical based on knowledge of the dosage form. The dissolution conditions for each drug product are usually specified in the individual monograph.

One goal in dissolution method development is to have sink conditions. Sink conditions are defined as having a medium volume of at least three times the volume required to form a saturated solution of the API. Appropriate selection of dissolution medium and volume to satisfy sink conditions is important to discriminate between an acceptable and unacceptable batch. The media commonly used in dissolution testing include water, diluted hydrochloric acid, pH-adjusted buffer, and simulated gastric or intestinal fluid (with or without enzyme). Surfactants may also be added to enhance the solubility of the API. Simulated gastric fluid generally contains a dilute solution of hydrochloric acid, sodium chloride to adjust for osmolarity, and purified pepsin derived from porcine stomach mucosa, which is the main enzyme found in the stomach. Compendial simulated intestinal fluid, on the other hand, is made up of a pH 6.8 phosphate buffer and pancreatin, which is an extract of the pancreas and rich in amylases, proteases, and lipases. More information on factors to be considered when developing and validating dissolution procedures can be found in general chapter ⟨1092⟩ The Dissolution Procedure: Development and Validation [56].

## 3.4 *In Vitro* Dissolution Methods in Product Development

### 3.4.1 Intrinsic Dissolution Apparatus

First described by Wood et al. [57], the rotating disk intrinsic dissolution apparatus is used for the determination of the intrinsic dissolution rate of an API [58]. This dissolution apparatus consists of a punch and die. A surface plate is screwed onto the base of the die, which has three threaded holes. The material whose intrinsic dissolution rate is to be measured is filled into the die cavity. The punch is then inserted into the die cavity and the material is compressed with a hydraulic press [58]. The surface plate is then removed, and the die is fixed onto the shaft, which provides support as the die is lowered into the dissolution medium.

The stationary disk assembly consists of a steel punch, die, and base plate. Preparation of the sample for compression is carried out by adding material into the die affixed onto the base plate and fastened with three washers and nuts. The test material in the die cavity is compressed by the insertion of the punch and compaction force applied. The base plate is subsequently removed, a gasket placed around the threaded shoulder of the die and a polypropylene cap screwed on as support for the assembly.

While the rotating disk assembly can be paired with a round-bottom dissolution vessel, a specially designed dissolution vessel with a flat bottom is required for the stationary disk assembly. In both the rotating and stationary disks, a compacted pellet of the material is formed upon compression. As only the material on the bottom of the die is exposed, the area exposed to the dissolution medium is clearly defined. This allows for the calculation of the intrinsic dissolution rate for a pure substance, typically normalized for surface area and expressed as the rate per unit area (mass/area/time). Since the area exposed during dissolution should remain constant during the dissolution experiment, it is important that bubbles do not form on the compact surface. Therefore, deaeration of the dissolution medium to remove dissolved gases is strongly recommended.

One limitation of the rotating and stationary disks is the need for a reasonably large quantity of test material, which may not always be readily available. Information on the dissolution rate of a material is especially important during the early developmental phase for new chemical entities because it could predict potential bioavailability problems [59]. The intrinsic solid-state properties of a material such as its crystallinity and the presence of any other polymorphs also affect its dissolution rate. This information is crucial for product development, as techniques to improve dissolution may have to be employed during the formulation stage to ensure good bioavailability if the material has poor aqueous solubility.

### 3.4.2 Use of Non-Compendial Apparatus

#### 3.4.2.1 Surface Dissolution Imager (SDI)

The need for a large quantity of test material, especially if it is expensive or scarce, during the early formulation stages may be circumvented by miniaturizing the dissolution tests [60–62]. One such example is the commercially available Sirius SDI UV imaging system. Requiring only 3–10 mg of samples, the SDI system consists of a flow-through cell equipped with a xenon lamp source and a charge-coupled device camera chip. A detailed schematic of the SDI can be found in the publication by Østergaard et al. [63]. Initially, the SDI was used primarily to monitor the release of pure APIs [64–66], but it has since been used as an analytical tool to understand the dissolution behavior of polymers [67], other dosage forms [63, 68–70], as well as drug–counterion [71] and drug–excipient [72] interactions.

#### 3.4.2.2 Small-Volume Apparatus

The small-volume apparatus is based on the compendial USP one-liter vessel setup, with modifications to fit a smaller vessel designed to be used with 50–200 ml of dissolution medium, and accordingly, the paddle dimensions are also reduced. Small-volume apparatus is useful when the dosage form has very low drug content. It may be useful in the dissolution testing of extended-release dosage forms, where small amounts of API are released and may otherwise be difficult to quantify using common spectrophotometric methods.

#### 3.4.2.3 Other Non-Compendial Apparatus

Other non-compendial apparatus used in dissolution testing include apex vessels and modified flow-through cells. Apex vessels may be used to eliminate coning (mounding of undissolved material in the stagnant zone below the paddle), especially if increasing the paddle rotational speed does not address this problem. Modified flow-through cells may be used for special dosage forms such as powders.

### 3.4.3 Use of Non-Compendial Dissolution Media

As the main objective of dissolution testing is to forecast the *in vivo* performance of a drug product, it is important that the *in vitro* test mimics the *in vivo* conditions as closely as possible. In the past two decades, several biorelevant gastrointestinal media that simulate the fasted and fed states have been developed. A fasted state simulated gastric fluid (FaSSGF) was proposed by Vertzoni et al. [73] to mimic the fasting gastric content, which was found to be a more representative assessment of the solubility and dissolution behavior of poorly soluble, weakly basic drugs in the fasted state [73, 74]. Adjusted to pH 1.6, the FaSSGF comprises sodium taurocholate to represent bile salts, lecithin, pepsin, and sodium chloride.

In the postprandial state, the luminal composition of the stomach becomes more complicated and unpredictable, as it is highly dependent on the composition of the meal ingested. The FDA recommends that a high-fat (approximately 50% of total caloric content of the meal) and high-calorie (approximately 800–1000 calories) meal be used as a test meal for bioavailability and bioequivalence studies [75]. Examples of dissolution media that closely resemble the FDA's specifications include milk and Ensure® Plus [76]. Compared to milk, Ensure® Plus was reported to have the closest physicochemical behavior and composition to the standard breakfast recommended by the FDA, while milk better simulates a breakfast with lower fat content [77]. As the postprandial gastric pH decreases continuously from pH 6.4 to 2.7 upon meal ingestion [78], three "snapshot" media with pH values of 6.4, 5.0, and 3.0 were developed by Jantratid et al. [79] to represent the "early", "middle", and "late" phases of gastric digestion, respectively. Among the three media, the middle one, which comprises sodium chloride, acetic acid, sodium acetate, ortho-phosphoric acid, sodium dihydrogen phosphate, and milk/buffer in a 1 : 1 ratio was designated as fed state simulated gastric fluid (FeSSGF). In 2019, a new fed state gastric fluid (FEDGAS) was developed by Biorelevant.com to meet the full-fat content of an FDA high-fat meal and can be prepared at three different pH values.

To simulate the fasted and fed states in the intestine, two biorelevant media – fasted state simulated intestinal fluid (FaSSIF) and fed state simulated intestinal fluid (FeSSIF) were proposed by the International Pharmaceutical Federation in 1997 [80]. Both media can significantly improve the dissolution rates of poorly soluble, lipophilic drugs (e.g. BCS Class II compounds). FaSSIF contains sodium taurocholate, lecithin, sodium hydroxide, sodium phosphate monobasic as well as sodium chloride, and is adjusted to pH 6.5. On the other hand, FeSSIF has a lower pH, higher buffer capacity, and osmolarity [81]. FeSSIF formulations differ from FaSSIF in terms of the concentrations of taurocholate and lecithin because the presence of food is known to stimulate bile and phospholipid secretion in humans. Acetate or phosphate buffers are also included in FeSSIF formulations to provide the buffering capacity.

Based on the *in vivo* data by Porter et al. [82], modifications to the biorelevant media were proposed to better represent the fasted and fed states in the intestine. In the updated FaSSIF, otherwise known as FaSSIF-V2, the amount of lecithin was decreased from 0.75 to 0.2 mM. In addition, maleate buffer was used in place of phosphate buffer as its $pK_{a_2}$ of 6.27 allows for sufficient buffer capacity to cover the required pH range of 5.4–6.5, thereby enabling the preparation of both the fasted and fed state media without exceeding the physiologically relevant osmolality. Besides, maleic acid also has the advantage of being able to retard the rancidity of fats and oils [79].

In the development of an updated FeSSIF, a "snapshot" approach similar to that of FeSSGF was used. The updated FeSSIF, otherwise known as FeSSIF-V2, is composed of sodium taurocholate, lecithin, glyceryl monooleate, sodium oleate, maleic acid, sodium hydroxide, and sodium chloride. Glyceryl monooleate and sodium oleate were added to reflect lipolysis products in the media [79].

Since the colon is mainly responsible for the digestion of polysaccharides, simulated colonic fluids will need to contain the respective enzymes present, e.g. galactomannase for a guar gum-based formulation [83]. There are currently two human SCoFs, SCoF1, and SCoF2, which simulate the distal and proximal colons, respectively. SCoF1 comprises potassium and sodium chloride, monobasic potassium phosphate, and dibasic sodium phosphate, while SCoF2 contains sodium hydroxide and acetic acid [84]. To simulate the fasted and fed states of the ascending colon, Vertzoni et al. [85] developed the fasted state simulated colonic fluid (FaSSCoF) and fed state simulated colonic fluid (FeSSCoF). Based on tris/maleate buffer, FaSSCoF and

FeSSCoF contain bile salt extract, phosphatidylcholine, palmitic acid, and bovine serum albumin. Additionally, FeSSCoF also contains sodium chloride and glucose.

To test drug release from sublingual, buccal, and rapidly disintegrating tablets, simulated salivary fluids containing a variety of electrolytes, mucin, and α-amylase can be used [86–90].

### 3.4.4 Use of Non-Sink Conditions

While sink conditions should typically be used in dissolution testing, non-sink conditions are sometimes used for dissolution testing of enabling formulations for poorly soluble drugs (e.g. ASDs) in the research setting. The benefit of using ASD as a formulation approach is that supersaturation with respect to crystalline solubility can be achieved and maintained in the presence of polymers. It is critical that supersaturation is maintained during transit in the gastrointestinal tract, as crystallization of the amorphous drug would result in the loss of its superiority as an enabling formulation. Amorphous solids can crystallize via two routes – solid-mediated (matrix) and solution-mediated (nucleation and growth) crystallization. It is only by using non-sink conditions that supersaturation can be generated during dissolution so that any risk of drug crystallization can be detected.

## 3.5 Automation in Dissolution Testing and Prediction

### 3.5.1 Automated Sampling

Manual sampling of the dissolution medium during a dissolution run is prone to human error and variability. Thus, automation of the sampling process is highly desirable for good quality control of dissolution data. Fully automated dissolution, sometimes accompanied by an analytical assay, enables measurements to be carried out with high precision and reproducibility. For prolonged dissolution test runs, issues like clogged filters and bubbles can be potential problems, contributing to inaccuracies. Bubbles are of particular nuisance when online circulation and measurement systems are used. Bubbles, if present in the measurement cell, can cause reading errors. Therefore, the importance of deaerating the medium prior to dissolution tests cannot be overstated. Arrangement of sampling to a fraction collector, then into an analytical system can certainly avoid the problems associated with bubble formation during the dissolution run. In some cases, the positioning of sampling probes can be problematic. The development of a hollow shaft sampling system by Sotax Corporation is an example of an interesting adaptation.

### 3.5.2 *In Situ* Analysis

Analysis of the amount of drug released into the dissolution medium is usually determined by taking samples from the dissolution vessel, either by continuous flow or using a fraction collector by an automated dissolution system. The sampling process not only affects steady-state environment of a dosage form undergoing dissolution but also is prone to various occasional errors such as air bubbles, filter failures, missed timings, tubing leaks, and others. A UV probe system may be developed as immersion and measure followed by withdrawal and wash systems. The major concern of a UV probe system is the filtering of medium and constant refreshing of the medium at the measurement site without affecting the bulk flow stream of the dissolution medium in the vessel.

A possible solution is to use fiber optic UV probes for *in situ* absorbance measurements. This will allow very fast sampling times, even continuous, and would be of great value to rapidly dissolving dosage forms, especially during dosage form development. However, dissolution test is not of great necessity for many rapidly dissolving dosage forms. Additionally, the use of *in situ* UV probes is not without drawbacks. During the dissolution of solid dosage forms, air bubbles or insoluble components such as excipients could potentially interfere with the measurement by either absorbing or scattering light [91]. With ASD formulations, the presence of colloidal drug-rich droplets could also present a challenge in drug quantification. The impact of air bubbles on the collected spectral data was shown in a study by Lu et al. [92]. A filter can be placed over the opening of the fiber optic probe, but blockage of the filter would result in a stagnant volume within the probe, which is not reflective of the drug concentration in the bulk solution. In assessing the impact of particle absorption effects on quantification, Mie theory calculations were shown to be able to model the effect of particle size and suspension concentration on the resultant spectra [93].

The challenges associated with the use of *in situ* UV probes could be resolved by applying mathematical filters to the spectral data to eliminate the contribution of non-API components [91]. The first method that can be used is baseline correction. Baseline correction is useful for data affected by wavelength-independent scattering, which is incidentally the more common form of scattering observed [92]. In most cases, a baseline offset based on a single wavelength or two-point correction is sufficient, usually using wavelength(s) where the API has no absorbance. If needed, a broader wavelength range can also be used to perform the baseline correction. However, if wavelength-dependent scattering is observed, a polynomial equation may be required to model the wavelength dependences of light scattering from particles [91].

The second method of spectral correction involves the use of derivative spectroscopy. As the absorbance spectrum is usually expressed as absorbance as a function of wavelength, its first-order derivative is the rate of change of absorbance with respect to wavelength. While the first-order derivative spectrum is usually sufficient to eliminate the effects from a baseline shift, one shortcoming of using this transformed spectrum is that the absorbance corresponding to the wavelength

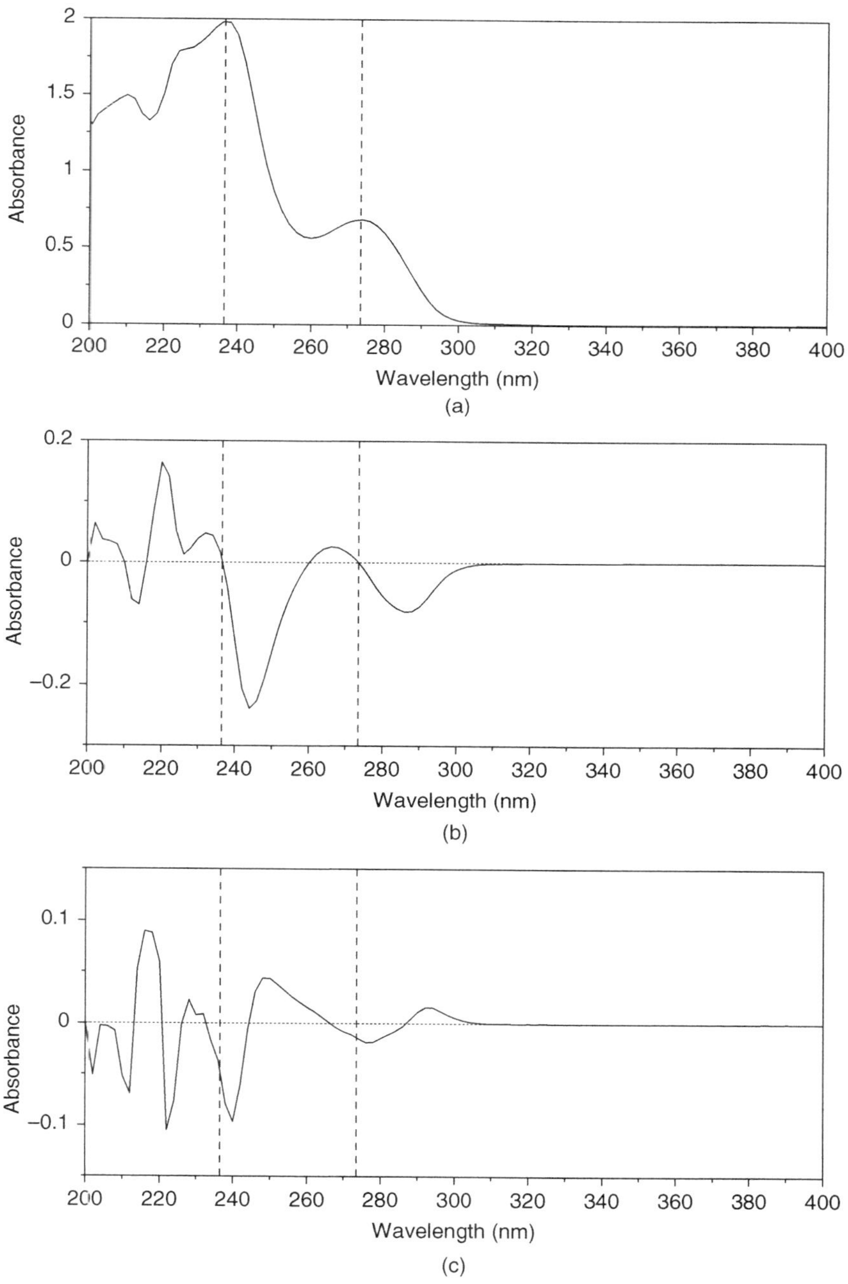


**Figure 3.2** (a) The spectrum of acetylsalicylic acid obtained using a UV probe, and its (b) first and (c) second derivative spectra.

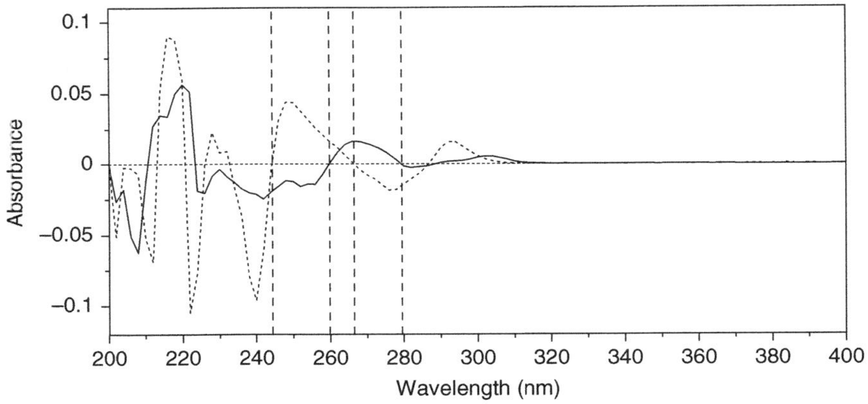


**Figure 3.3** The second-derivative spectra of acetylsalicylic acid (dotted line) and acetaminophen (solid line), with their respective unique zero-intercept points.

where maximum absorbance is observed for the original spectrum falls on the abscissa (Figure 3.2b). Therefore, the second-order derivative spectrum is more commonly used, as the wavelength corresponding to the maximum absorbance presents as a nonzero absorbance value in this spectrum (Figure 3.2c).

One additional benefit of using derivative spectroscopy is the ability to quantify the concentration of more than one component (usually API) in the dissolution media. Traditionally, liquid chromatography has been the mainstay for the separation and quantification of two or more components, but derivative spectroscopy opens up new opportunities for faster quantification. The main idea behind the use of derivative spectroscopy for quantification is that the second-derivative spectra of different UV-absorbing components cross the abscissa at different wavelengths. These crossing points, henceforth referred to as zero-intercept points, are unique for each component. As the UV probe is able to collect absorbance data across a range of wavelengths, the wavelength(s) corresponding to the zero-intercept point(s) of one component can be used to quantify the concentration of the other component, which has a nonzero absorbance value at the same wavelength (Figure 3.3). However, if the two components have very similar spectra, then this method may not be so suitable as the absorbance value of the first component at the zero-intercept point of the second component may be too low for accurate quantification. In this case, other chemometrics methods such as partial least square regression can be used.

## 3.6 Conclusions

Dissolution testing of pharmaceutical solids and solid dosage forms is based on a simple theoretical concept but plays a key role in product development and quality control of commercial drug products. While the existing standard pharmacopeial tests are well established, testing methods and equipment are constantly evolving and becoming increasingly sophisticated to cope with novel products that are developed in the pharmaceutical industry. More importantly, with new findings that improve our understanding of the dissolution behavior of various solubility-enabling formulations and systems, the design and selection of dissolution test must take into consideration its accuracy in quantifying the amount of drug released over time. Accurate and representative *in vitro* dissolution tests are especially important as more APIs developed today are categorized as poorly water-soluble, which poses a greater challenge in establishing *in vivo–in vitro* relationship or IVIVC. Therefore, apart from the consideration of which apparatus to use, careful selection of other dissolution conditions such as the dissolution medium, medium volume, rotational speed, and pH also affect the test outcomes. The desired dissolution method for pharmaceutical commercial product tests should be universally acceptable and easy to implement yet sufficiently discriminating at assessing product quality, eventually assuring biopharmaceutical performance.

## References

1 Noyes, A.A. and Whitney, W.R. (1897). The rate of solution of solid substances in their own solutions. *J. Am. Chem. Soc.* 19 (12): 930–934.

2 Nernst, W. (1904). Theorie der Reaktionsgeschwindigkeit in heterogenen Systemen. *Z. Phys. Chem.* 47U (1): 52–55.

3 Brunner, E. (1904). Reaktionsgeschwindigkeit in heterogenen Systemen. *Z. Phys. Chem.* 47U (1): 56–102.
4 USP 43–NF 38 (2020). ⟨711⟩ Dissolution, United States Pharmacopeial Convention, GUID-AC788D41-90A2-4F36-A6E7-769954A9ED09_1_en-US
5 USP 43–NF 38 (2020). ⟨724⟩ Drug release, United States Pharmacopeial Convention, GUID-1BF488DB-86ED-4A33-AF13-71C2F1EDA2F2_5_en-US.
6 Brown, W.E. and Marques, M.R. (2014). USP and dissolution—20 years of progress. *Dissolution Technol.* 21 (3): 24–27.
7 Xu, W., Boehm, G., and Zheng, Q. (2016). Factors to consider in developing individual pharmaceutical product quality risk profiles useful to government procurement agencies. *Acta Pharm. Sin. B* 6 (1): 79–92.
8 U.S. Food and Drug Administration, Center for Drug Evaluation and Research (1997). Dissolution testing of immediate release solid oral dosage forms. Guidance for Industry, Office of Training and Communications, Division of Communications Management, The Drug Information Branch, Rockville, MD.
9 U.S. Food and Drug Administration, Center for Drug Evaluation and Research (1997). Extended release oral dosage forms: development, evaluation, and application of in vitro/in vivo correlations. Guidance for Industry, Office of Training and Communications, Division of Communications Management, The Drug Information Branch, Rockville, MD.
10 U.S. Food and Drug Administration, Center for Drug Evaluation and Research (2014). Bioavailability and bioequivalence studies submitted in NDAs or INDs—general considerations. Guidance for Industry, Office of Communications, Division of Drug Information, Center for Drug Evaluation and Research, Food and Drug Administration, Silver Spring, MD.
11 U.S. Food and Drug Administration, Center for Drug Evaluation and Research (2017). Waiver of in vivo bioavailability and bioequivalence studies for immediate-release solid oral dosage forms based on a biopharmaceutics classification system. Guidance for Industry, Office of Communications, Division of Drug Information, Center for Drug Evaluation and Research, Food and Drug Administration, Silver Spring, MD.
12 U.S. Food and Drug Administration, Center for Drug Evaluation and Research (2018). Dissolution testing and acceptance criteria for immediate-release solid oral dosage form drug products containing high solubility drug substances. Guidance for Industry, Office of Communications, Division of Drug Information, Center for Drug Evaluation and Research, Food and Drug Administration, Silver Spring, MD.
13 USP 43–NF 38 (2020). ⟨1088⟩ In vitro and in vivo evaluation of dosage forms, United States Pharmacopeial Convention, GUID-0818140F-BBE0-4124-96DB-4A37CBF25659_3_en-US.
14 Hixson, A.W. and Crowell, J.H. (1931). Dependence of reaction velocity upon surface and agitation. *Ind. Eng. Chem.* 23 (8): 923–931.
15 Stavchansky, S.A. and McGinity, J.W. (1989). Bioavailability in tablet technology. In: *Pharmaceutical Dosage Forms: Tablets*, vol. 2 (ed. H.A. Lieberman, L. Lachman, and J.B. Schwartz), 349–570. New York: Marcel Dekker, Inc.
16 Loh, Z.H., Samanta, A.K., and Sia Heng, P.W. (2015). Overview of milling techniques for improving the solubility of poorly water-soluble drugs. *Asian J. Pharm. Sci.* 10 (4): 255–274.
17 Loftsson, T. and Brewster, M.E. (2010). Pharmaceutical applications of cyclodextrins: basic science and product development. *J. Pharm. Pharmacol.* 62 (11): 1607–1621.
18 Malamatari, M., Taylor, K.M.G., Malamataris, S. et al. (2018). Pharmaceutical nanocrystals: production by wet milling and applications. *Drug Discov. Today* 23 (3): 534–547.
19 Kwok, P.C. and Chan, H.K. (2014). Nanotechnology versus other techniques in improving drug dissolution. *Curr. Pharm. Des.* 20 (3): 474–482.
20 Li, W., Ng, K.-Y., and Heng, P.W.S. (2014). Development and evaluation of optimized sucrose ester stabilized oleanolic acid nanosuspensions prepared by wet ball milling with design of experiments. *Biol. Pharm. Bull.* 37 (6): 926–937.
21 Müller, R.H., Jacobs, C., and Kayser, O. (2003). DissoCubes—a novel formulation for poorly soluble and poorly bioavailable drugs. In: *Modified-Release Drug Delivery Technology* (ed. M.J. Rathbone, J. Hadgraft, and M.S. Roberts), 135–149. New York: Marcel Dekker.
22 Arpagaus, C., Collenberg, A., Rütti, D. et al. (2018). Nano spray drying for encapsulation of pharmaceuticals. *Int. J. Pharm.* 546 (1): 194–214.
23 Vasconcelos, T., Sarmento, B., and Costa, P. (2007). Solid dispersions as strategy to improve oral bioavailability of poor water soluble drugs. *Drug Discov. Today* 12 (23): 1068–1075.
24 Ghebremeskel, A.N., Vemavarapu, C., and Lodaya, M. (2007). Use of surfactants as plasticizers in preparing solid dispersions of poorly soluble API: selection of polymer–surfactant combinations using

solubility parameters and testing the processability. *Int. J. Pharm.* 328 (2): 119–129.

25 Jermain, S.V., Brough, C., and Williams, R.O. (2018). Amorphous solid dispersions and nanocrystal technologies for poorly water-soluble drug delivery – an update. *Int. J. Pharm.* 535 (1): 379–392.

26 Taylor, L.S. and Zhang, G.G.Z. (2016). Physical chemistry of supersaturated solutions and implications for oral absorption. *Adv. Drug Deliv. Rev.* 101: 122–142.

27 Ilevbare, G.A. and Taylor, L.S. (2013). Liquid–liquid phase separation in highly supersaturated aqueous solutions of poorly water-soluble drugs: implications for solubility enhancing formulations. *Cryst. Growth Des.* 13 (4): 1497–1509.

28 Indulkar, A.S., Gao, Y., Raina, S.A. et al. (2016). Exploiting the phenomenon of liquid–liquid phase separation for enhanced and sustained membrane transport of a poorly water-soluble drug. *Mol. Pharm.* 13 (6): 2059–2069.

29 Konno, H. and Taylor, L.S. (2006). Influence of different polymers on the crystallization tendency of molecularly dispersed amorphous felodipine. *J. Pharm. Sci.* 95 (12): 2692–2705.

30 Theil, F., Anantharaman, S., Kyeremateng, S.O. et al. (2017). Frozen in time: kinetically stabilized amorphous solid dispersions of nifedipine stable after a quarter century of storage. *Mol. Pharm.* 14 (1): 183–192.

31 Theil, F., Milsmann, J., Kyeremateng, S.O. et al. (2017). Extraordinary long-term-stability in kinetically stabilized amorphous solid dispersions of fenofibrate. *Mol. Pharm.* 14 (12): 4636–4647.

32 Indulkar, A.S., Lou, X., Zhang, G.G.Z., and Taylor, L.S. (2019). Insights into the dissolution mechanism of ritonavir-copovidone amorphous solid dispersions: importance of congruent release for enhanced performance. *Mol. Pharm.* 16 (3): 1327–1339.

33 Que, C., Lou, X., Zemlyanov, D.Y. et al. (2019). Insights into the dissolution behavior of ledipasvir–copovidone amorphous solid dispersions: role of drug loading and intermolecular interactions. *Mol. Pharm.* 16 (12): 5054–5067.

34 Saboo, S., Mugheirbi, N.A., Zemlyanov, D.Y. et al. (2019). Congruent release of drug and polymer: a "sweet spot" in the dissolution of amorphous solid dispersions. *J. Control. Release* 298: 68–82.

35 Chauhan, B., Shimpi, S., and Paradkar, A. (2005). Preparation and evaluation of glibenclamide-polyglycolized glycerides solid dispersions with silicon dioxide by spray drying technique. *Eur. J. Pharm. Sci.* 26 (2): 219–230.

36 Srinarong, P., Kouwen, S., Visser, M.R. et al. (2010). Effect of drug-carrier interaction on the dissolution behavior of solid dispersion tablets. *Pharm. Dev. Technol.* 15 (5): 460–468.

37 Srinarong, P., Hämäläinen, S., Visser, M.R. et al. (2011). Surface-active derivative of inulin (Inutec® SP1) is a superior carrier for solid dispersions with a high drug load. *J. Pharm. Sci.* 100 (6): 2333–2342.

38 Rodier, E., Lochard, H., Sauceau, M. et al. (2005). A three step supercritical process to improve the dissolution rate of Eflucimibe. *Eur. J. Pharm. Sci.* 26 (2): 184–193.

39 Wagner, J.G. (1970). Biopharmaceutics: 17. Rate of dissolution *in vitro* and *in vivo* part II. Introduction and historical highlights. *Drug Intell. Clin. Pharm.* 4 (2): 32–37.

40 Desai, P.M., Liew, C.V., and Heng, P.W.S. (2016). Review of disintegrants and the disintegration phenomena. *J. Pharm. Sci.* 105 (9): 2545–2555.

41 Mastropietro, D., Park, K., and Omidian, H. (2017). Polymers in oral drug delivery. In: *Comprehensive Biomaterials II* (ed. P. Ducheyne), 430–444. Oxford: Elsevier.

42 Wan, L.S.C., Heng, P.W.S., and Wong, L.F. (1991). The effect of hydroxypropylmethylcellulose on water penetration into a matrix system. *Int. J. Pharm.* 73 (2): 111–116.

43 Wan, L.S.C., Heng, P.W.S., and Wong, L.F. (1995). Matrix swelling: a simple model describing extent of swelling of HPMC matrices. *Int. J. Pharm.* 116 (2): 159–168.

44 Katikaneni, P.R., Upadrashta, S.M., Rowlings, C.E. et al. (1995). Consolidation of ethylcellulose: effect of particle size, press speed, and lubricants. *Int. J. Pharm.* 117 (1): 13–21.

45 Crowley, M.M., Schroeder, B., Fredersdorf, A. et al. (2004). Physicochemical properties and mechanism of drug release from ethyl cellulose matrix tablets prepared by direct compression and hot-melt extrusion. *Int. J. Pharm.* 269 (2): 509–522.

46 Ozturk, A.G., Ozturk, S.S., Palsson, B.O. et al. (1990). Mechanism of release from pellets coated with an ethylcellulose-based film. *J. Control. Release* 14 (3): 203–213.

47 Borst, I., Ugwu, S., and Beckett, A.H. (1997). New and extended applications for USP drug release apparatus 3. *Dissolution Technol.* 4 (1): 11–18.

48 Healy, A.M., McCarthy, L.G., Gallagher, K.M., and Corrigan, O.I. (2002). Sensitivity of dissolution rate to location in the paddle dissolution apparatus. *J. Pharm. Pharmacol.* 54 (3): 441–444.

49 Baxter, J.L., Kukura, J., and Muzzio, F.J. (2005). Hydrodynamics-induced variability in the USP apparatus II dissolution test. *Int. J. Pharm.* 292 (1): 17–28.

50 Kukura, J., Baxter, J.L., and Muzzio, F.J. (2004). Shear distribution and variability in the USP Apparatus 2 under turbulent conditions. *Int. J. Pharm.* 279 (1): 9–17.

51 Miller, D.A., Gamba, M., Sauer, D. et al. (2007). Evaluation of the USP dissolution test method A for enteric-coated articles by planar laser-induced fluorescence. *Int. J. Pharm.* 330 (1): 61–72.

52 Yu, L.X., Wang, J.T., and Hussain, A.S. (2002). Evaluation of USP apparatus 3 for dissolution testing of immediate-release products. *AAPS PharmSci* 4 (1): 1–5.

53 USP 43–NF 38 (2020). Nifedipine extended-release tablets, United States Pharmacopeial Convention, GUID-82A28F33-BDA8-4040-8294-864692EAE4A7_7_en-US.

54 USP 43–NF 38 (2020). Verapamil hydrochloride extended-release tablets, United States Pharmacopeial Convention, GUID-42ECE4EC-BB14-49CB-A2A5-A813E0B42364_5_en-US.

55 Xu, M., Liew, C.V., and Heng, P.W.S. (2015). Evaluation of the coat quality of sustained release pellets by individual pellet dissolution methodology. *Int. J. Pharm.* 478 (1): 318–327.

56 USP 43–NF 38 (2020). ⟨1092⟩ The dissolution procedure: development and validation, United States Pharmacopeial Convention, GUID-CE0902BA-77AC-422D-8BF0-A221B5DE6012_5_en-US.

57 Wood, J.H., Syarto, J.E., and Letterman, H. (1965). Improved holder for intrinsic dissolution rate studies. *J. Pharm. Sci.* 54 (7): 1068–1068.

58 USP 43–NF 38 (2020). ⟨1087⟩ Intrinsic dissolution—dissolution testing procedures for rotating disk and stationary disk, United States Pharmacopeial Convention, GUID-A257EA8C-9E25-48B2-A77C-4E5D3B98945A_2_en-US.

59 Yu, L.X., Carlin, A.S., Amidon, G.L., and Hussain, A.S. (2004). Feasibility studies of utilizing disk intrinsic dissolution rate to classify drugs. *Int. J. Pharm.* 270 (1): 221–227.

60 Alsenz, J., Haenel, E., Anedda, A. et al. (2016). Miniaturized INtrinsic DISsolution Screening (MINDISS) assay for preformulation. *Eur. J. Pharm. Sci.* 87: 3–13.

61 Avdeef, A. and Tsinman, O. (2008). Miniaturized rotating disk intrinsic dissolution rate measurement: effects of buffer capacity in comparisons to traditional Wood's apparatus. *Pharm. Res.* 25 (11): 2613–2627.

62 Berger, C.M., Tsinman, O., Voloboy, D. et al. (2007). Technical note: miniaturized intrinsic dissolution rate (Mini-IDR™) measurement of griseofulvin and carbamazepine. *Dissolution Technol.* 14 (4): 39–41.

63 Østergaard, J., Meng-Lund, E., Larsen, S.W. et al. (2010). Real-time UV imaging of nicotine release from transdermal patch. *Pharm. Res.* 27 (12): 2614–2623.

64 Boetker, J.P., Savolainen, M., Koradia, V. et al. (2011). Insights into the early dissolution events of amlodipine using UV imaging and Raman spectroscopy. *Mol. Pharm.* 8 (4): 1372–1380.

65 Gordon, S., Naelapää, K., Rantanen, J. et al. (2013). Real-time dissolution behavior of furosemide in biorelevant media as determined by UV imaging. *Pharm. Dev. Technol.* 18 (6): 1407–1416.

66 Hulse, W.L., Gray, J., and Forbes, R.T. (2012). A discriminatory intrinsic dissolution study using UV area imaging analysis to gain additional insights into the dissolution behaviour of active pharmaceutical ingredients. *Int. J. Pharm.* 434 (1): 133–139.

67 Pajander, J., Baldursdottir, S., Rantanen, J., and Østergaard, J. (2012). Behaviour of HPMC compacts investigated using UV-imaging. *Int. J. Pharm.* 427 (2): 345–353.

68 Hiew, T.N., Tan, D.L.H., Tiang, Y.L., and Heng, P.W.S. (2019). Understanding the release performance of pellets with hydrophobic inclusions in sustained-release coating. *Int. J. Pharm.* 557: 229–237.

69 Jensen, S.S., Jensen, H., Cornett, C. et al. (2015). Real-time UV imaging identifies the role of pH in insulin dissolution behavior in hydrogel-based subcutaneous tissue surrogate. *Eur. J. Pharm. Sci.* 69: 26–36.

70 Ye, F., Yaghmur, A., Jensen, H. et al. (2011). Real-time UV imaging of drug diffusion and release from Pluronic F127 hydrogels. *Eur. J. Pharm. Sci.* 43 (4): 236–243.

71 Asare-Addo, K., Walton, K., Ward, A. et al. (2018). Direct imaging of the dissolution of salt forms of a carboxylic acid drug. *Int. J. Pharm.* 551 (1): 290–299.

72 Hiew, T.N., Alaudin, M.I.B., Chua, S.M., and Heng, P.W.S. (2018). A study of the impact of excipient shielding on initial drug release using UV imaging. *Int. J. Pharm.* 553 (1): 229–237.

73 Vertzoni, M., Dressman, J., Butler, J. et al. (2005). Simulation of fasting gastric conditions and its importance for the in vivo dissolution of lipophilic compounds. *Eur. J. Pharm. Biopharm.* 60 (3): 413–417.

**74** Dressman, J.B., Vertzoni, M., Goumas, K., and Reppas, C. (2007). Estimating drug solubility in the gastrointestinal tract. *Adv. Drug Deliv. Rev.* 59 (7): 591–602.

**75** U.S. Food and Drug Administration, Center for Drug Evaluation and Research (2002). Food-effect bioavailability and fed bioequivalence studies. Guidance for Industry, Office of Training and Communications, Division of Drug Information, Rockville, MD.

**76** Klein, S., Dressman, J.B., Butler, J. et al. (2004). Media to simulate the postprandial stomach I. Matching the physicochemical characteristics of standard breakfasts. *J. Pharm. Pharmacol.* 56 (5): 605–610.

**77** Klein, S. (2010). The use of biorelevant dissolution media to forecast the *in vivo* performance of a drug. *AAPS J.* 12 (3): 397–406.

**78** Kalantzi, L., Goumas, K., Kalioras, V. et al. (2006). Characterization of the human upper gastrointestinal contents under conditions simulating bioavailability/bioequivalence studies. *Pharm. Res.* 23 (1): 165–176.

**79** Jantratid, E., Janssen, N., Reppas, C., and Dressman, J.B. (2008). Dissolution media simulating conditions in the proximal human gastrointestinal tract: an update. *Pharm. Res.* 25 (7): 1663.

**80** Aiache, J.M., Aoyagi, N., Blume, H. et al. (1997). FIP guidelines for dissolution testing of solid oral products. *Dissolution Technol.* 4 (4): 5–14.

**81** Greenwood, D.E. (1994). Small intestinal pH and buffer capacity: implications for dissolution of ionizable compounds. Doctoral dissertation. University of Michigan, Ann Arbor, MI. Retrieved from ProQuest Dissertations Publishing (9423193).

**82** Porter, C.J.H., Trevaskis, N.L., and Charman, W.N. (2007). Lipids and lipid-based formulations: optimizing the oral delivery of lipophilic drugs. *Nat. Rev. Drug Discov.* 6 (3): 231–248.

**83** Yang, L. (2008). Biorelevant dissolution testing of colon-specific delivery systems activated by colonic microflora. *J. Control. Release* 125 (2): 77–86.

**84** USP 43–NF 38 (2020). ⟨1236⟩ Solubility measurements, United States Pharmacopeial Convention, GUID-B21E4B4E-B5EF-4EA6-879E-2D03A2476ABB_2_en-US.

**85** Vertzoni, M., Diakidou, A., Chatzilias, M. et al. (2010). Biorelevant media to simulate fluids in the ascending colon of humans and their usefulness in predicting intracolonic drug solubility. *Pharm. Res.* 27: 2187–2196. https://doi.org/10.1007/s11095-010-0223-6.

**86** Davis, R.E., Hartman, C.W., and Fincher, J.H. (1971). Dialysis of ephedrine and pentobarbital from whole human saliva and simulated saliva. *J. Pharm. Sci.* 60 (3): 429–432.

**87** Duffó, G.S. and Castillo, E.Q. (2004). Development of an artificial saliva solution for studying the corrosion behavior of dental alloys. *Corrosion* 60 (6): 594–602.

**88** Kartal, A., Marvola, J., Matheka, J. et al. (2010). Computational prediction of local drug effect on carcinogenic acetaldehyde in the mouth based on in vitro/in vivo results of freely soluble L-cysteine. *Drug Dev. Ind. Pharm.* 36 (6): 715–723.

**89** Mashru, R.C., Sutariya, V.B., Sankalia, M.G., and Parikh, P.P. (2005). Development and evaluation of fast-dissolving film of salbutamol sulphate. *Drug Dev. Ind. Pharm.* 31 (1): 25–34.

**90** Tavss, E.A., Gaffar, A., and King, W.J. (1984). Studies on the formation of electrostatic complexes between benzethonium chloride and anionic polymers. *J. Pharm. Sci.* 73 (8): 1148–1152.

**91** Nir, I. and Lu, X. (2018). *In situ* UV fiber optics for dissolution testing – what, why, and where we are after 30 years. *Dissolution Technol.* 25 (3): 70–77.

**92** Lu, X., Lozano, R., and Shah, P. (2003). *In-situ* dissolution testing using different UV fiber optic probes and instruments. *Dissolution Technol.* 10 (4): 10–15.

**93** Van Eerdenbrugh, B., Alonzo, D.E., and Taylor, L.S. (2011). Influence of particle size on the ultraviolet spectrum of particulate-containing solutions: implications for *in-situ* concentration monitoring using UV/Vis fiber-optic probes. *Pharm. Res.* 28 (7): 1643–1652.

# 4

# Biological and Physiological Features of the Gastrointestinal Tract Relevant to Oral Drug Absorption

*Paul C.L. Ho*[1,2]

[1] *Department of Pharmacy, Faculty of Science, National University of Singapore, Singapore, Singapore*
[2] *School of Pharmacy, Monash University Malaysia, Kuala Lumpur, Malaysia*

## 4.1 Introduction

In clinical practice, oral delivery remains the preferred route of drug administration, due to convenience and ease of administration and greater stability of the dosage forms. These lead to improved patient compliance, lower production cost, and access to larger markets. However, the gastrointestinal tract possesses a range of biological and physiological barriers that limit intestinal absorption. Biological barriers include the mucus layer, villi and microvilli, the unstirred water layer, and the epithelial cell layer, which are composed of villus and crypt cells, goblet cells, and M cells of the Peyer's patch. Physiological factors include the range of luminal and microclimate pH, enzymatic activities found in the lumen and within the enterocytes, specific transport mechanisms that limit absorption, intestinal transit time, agents such as bile salts secreted into the intestinal lumen, age, and gender. This chapter will focus on some biological features, such as the gastrointestinal anatomical structures and cell types, and physiological factors such as the gastrointestinal pH, age, and gender that would affect or regulate the absorption of drugs through the gastrointestinal tract, whereas the other topics, such as the drug metabolism enzymes and transporters will be discussed in more details in other chapters in this book.

## 4.2 Biological Features of Gastrointestinal Tract

The gastrointestinal tract (Figure 4.1) is essentially an open-end, epithelium-lined tube that extends from the mouth to the anus, the surface of which is covered with a mucous membrane or mucosa, that serves as the first barrier to entry of materials into the body. The single layer of columnar epithelial cells is interspersed with a variety of specialized cells, such as goblet cells, endocrine cells, and M cells. Stomach, though not the major absorption organ, can absorb some non-ionized, lipophilic molecules of moderate size, especially the weak acidic drugs. However, absorption there is limited by the relatively small epithelial surface; and its epithelium is dominated by surface mucosal cells rather than absorptive cells. Absorption of drugs mainly takes place in the small intestine which is divided into three portions; the first 20 to 30 cm is the duodenum, the next 2.5 m is the jejunum, and the final 3.5 m is the ileum. The duodenum has a thick wall with a deeply folded mucous membrane embedded with duodenal digestive glands and Brunner's glands. Brunner's glands are found only in the submucosa of the duodenum. The function of Brunner's gland is to produce a protective, nonenzymatic secretion to neutralize gastric acid. The surface of jejunum is thicker and more vascular than the duodenum and has more villi than the ileum. Ileum, on the other hand, has more lymphatic follicles (Peyer's patches) than any other parts of the intestine.

The epithelial layer of the small intestine is folded into villi (Figure 4.2) and the surface of villi is covered with microvilli (estimated at 3000–7000 per cell in the small intestine; fewer in large intestine) (Figure 4.3), providing a large surface area for absorption [2–4]. Table 4.1 shows the relative anatomical lengths and surface areas of various regions of the gastrointestinal tract in humans. Next to the epithelium is a thin layer of loose connective tissue (lamina propria), which is infiltrated with blood and lymphatic capillaries. The epithelial cells are tightly bound to each other. The tight junction between the adjacent epithelial cells prevents intercellular passage of drug molecules from

*Oral Bioavailability and Drug Delivery: From Basics to Advanced Concepts and Applications,* First Edition. Edited by Ming Hu and Xiaoling Li.

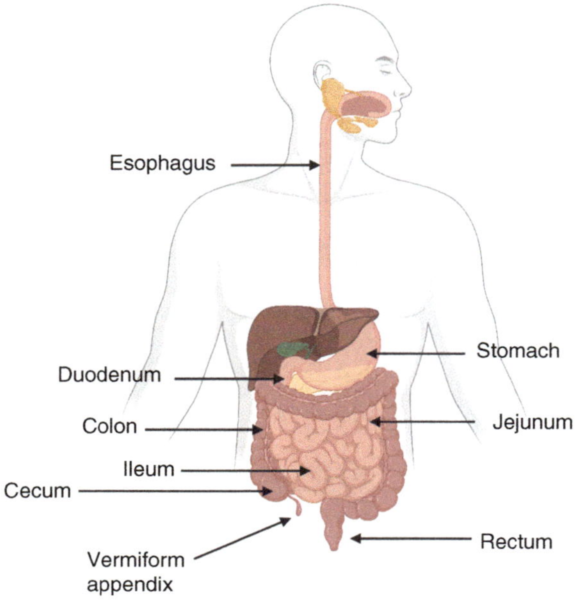


**Figure 4.1** The gastrointestinal tract of humans.

the intestinal lumen to the lamina propria (Figure 4.3). Absorption of drugs from the lumen in the intestinal tract requires the drug molecule to pass through the unstirred aqueous layer, the mucous layer, glycocalyx, in order to reach the epithelium, the lamina propria, and the walls of the blood or lymph capillaries there. In the blood capillaries, drug molecules would then be carried away to the systemic circulation through the liver. For highly lipid-soluble drugs, particularly if administered in an oily vehicle, they could be absorbed into the lymphatic circulation via the central lacteals in the villi (Figure 4.4). Lower down in the gastrointestinal tract is the colon, which is also not the major site for drug absorption, due to its small surface area as a result of lack of villi and microvilli, sluggish movement, low volume of available dissolution fluid, and the reduced permeability of the colonic epithelium to polar compounds.

Drug absorption process takes place during the transit down the gastrointestinal tract and mixing within intestinal lumens, which is accomplished by contractions of the separate muscular layers, the circular and the longitudinal muscle layers under the lamina propria, the muscularis mucosa. The intestinal motility will help dissolution of solid drugs in the intestinal fluids that facilitate their absorption in the intestinal lumens. Some drugs display region-specific absorption that can be caused by differential drug solubility and stability in different regions of the intestine as a result of changes in environmental pH, degradation by enzymes present in the lumen of the intestine, or interaction with endogenous components such as bile [5]. Active transport mechanisms for drugs involving carriers and pump systems at specific regions of the intestine also contribute to this phenomenon of absorption windows [6].

### 4.2.1 Absorption Pathways

Most drugs are absorbed via passive diffusion and a small portion of drugs are absorbed through facilitated and carrier-mediated pathways. The latter absorption pathways have been discussed in greater details in other chapters in this book. Briefly, passive diffusion can be subdivided into transcellular and paracellular absorption pathways.

#### 4.2.1.1 Transcellular and Paracellular Absorption Pathways

Paracellular and transcellular pathways are the two pathways by which drugs may cross the intestinal epithelium. A number of small hydrophilic, ionized drugs are absorbed via the paracellular pathway [7]. However, absorption via this route is generally low

**Table 4.1** The relative anatomical lengths and surface areas of various regions of the gastrointestinal tract in humans [3].

| Region of gastrointestinal tract | Average length (cm) | Average diameter (cm) | Absorbing surface area ($m^2$) |
|---|---|---|---|
| Stomach | 20 | 15 | ~0.11 |
| Duodenum | 25 | 5 | ~0.09 |
| Jejunum | 300 | 5 | ~60 |
| Ileum | 300 | 5 | ~60 |
| Cecum | 10–30 | 7 | ~0.05 |
| Colon | 150 | 5 | ~0.25 |
| Rectum | 15–19 | 2.5 | ~0.015 |

Source: Ritschel [3] / Europe PMC.

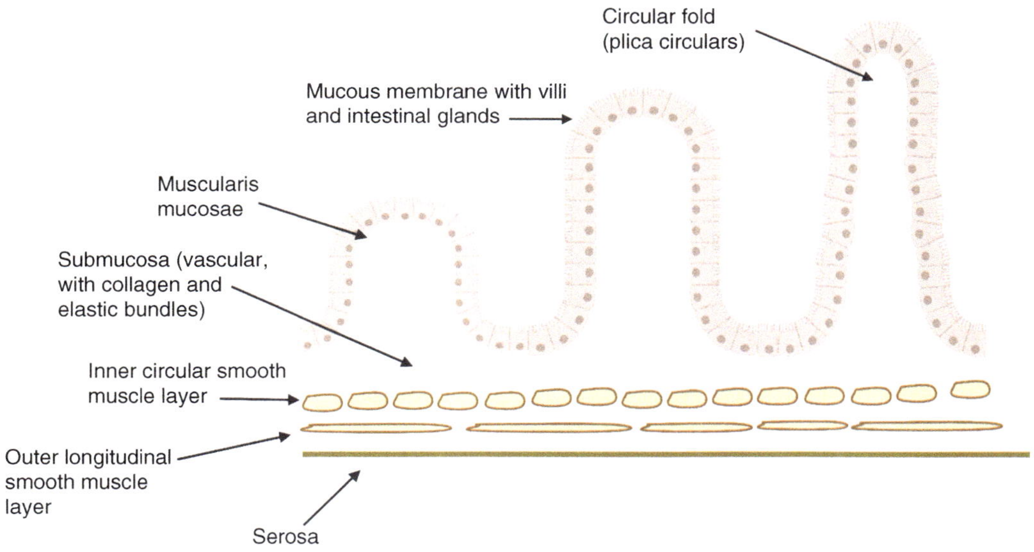


**Figure 4.2** Cross-sections of the proximal small intestine of humans. The surface areas for absorption from the lumina of the small intestine of humans are greatly increased by the presence of millions of tiny projections of villi.

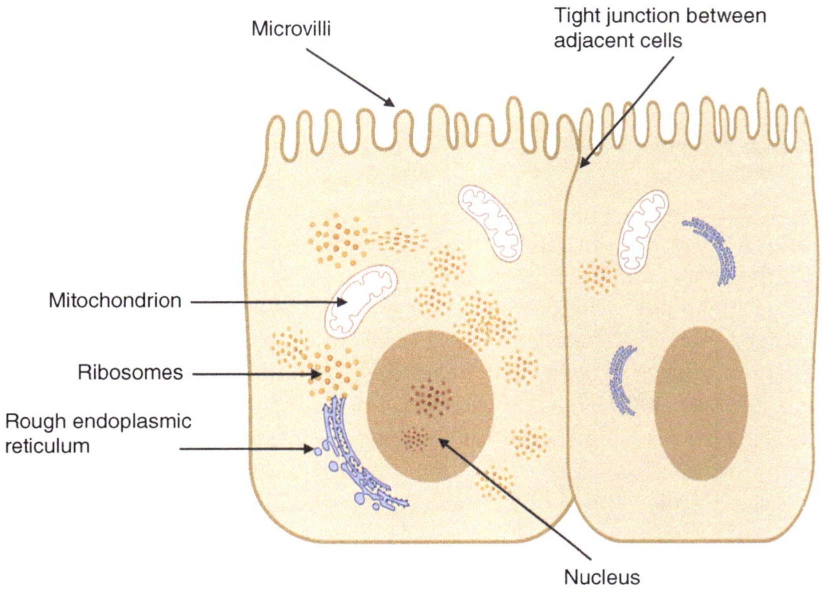


**Figure 4.3** Diagram of a typical intestinal epithelial cell (enterocyte). The cells are joined tightly to neighboring cells by tight junctions. The apical surface of cell has numerous microscopic microvilli that provide a large surface for absorption.

since intercellular tight junctions with a pore diameter of 4–8 Å restrict free transepithelial movement of most drug molecules between epithelial cells [8]. Many orally administered drugs are lipophilic and undergo passive transcellular absorption [9]. The absorption is simply due to diffusion down a concentration gradient. The absorption rate depends on the lipid solubility of the molecule in the lipid bilayer. A drug molecule must cross the apical cell membrane into the interior of the epithelial columnar cells. On emerging from the columnar cell, the drug molecule must cross the underlying basement membrane into the lamina propria.

#### 4.2.1.2 Facilitated and Carrier-Mediated Absorption Pathways

In contrast to passive diffusion, carrier-mediated absorption requires the reversible interaction of the drug

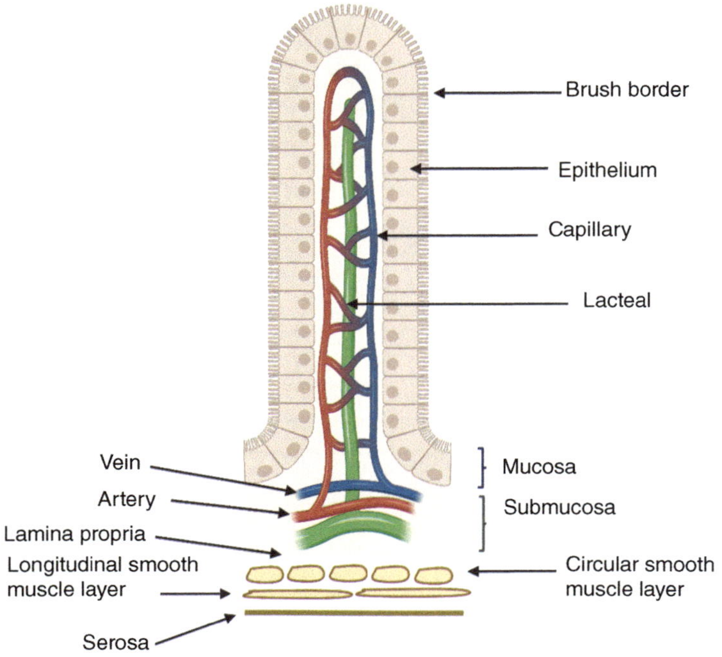


**Figure 4.4** Diagrammatic representation of an intestinal villus. The large blind-ending lymphatic vessel, the lacteal is responsible for absorption of higher molecular weight substances, including fats.

molecule with a carrier, usually a protein in the apical membrane of the epithelial columnar cell. The absorption process can occur up a concentration gradient, i.e. from a region of lower concentration to a region of higher concentration; and the transport is an energy consumption process, e.g. the ATP-binding cassette (ABC) dependent proton-coupled folate transporter/haem carrier protein 1 (PCFT/HCP1) is involved in the intestinal absorption of methotrexate [10]. In comparison, facilitated absorption process also involves a carrier, but does not require energy. Therefore, concentration gradient, as in passive diffusion, is the driving force for this absorption process. The classical example of facilitated absorption is the intestinal absorption of glucose [11]. A comprehensive review on carrier-mediated transport of pharmaceutical drugs was reported [12]. Now, it is well established that intestinal transporters such as P-gp, BCRP, PEPT1, and OATP2B1 can affect oral absorption of many drugs substantially either by facilitating their intestinal uptake or by pumping them back to intestinal lumen, limiting their oral bioavailability [13]. Some of these intestinal transporters such as P-gp, ASBT, and PEPT1 display considerable differences in their abundance along the human intestine, forming the "absorption window" to some drugs [13, 14]. Drug transporters are also involved in relevant clinical drug–drug interaction (DDI). Efflux transporters such as P-gp and BCRP pump drugs from inside the enterocytes back into the gut lumen. Inhibition of such intestinal efflux transporters would decrease absorption time and even increase oral bioavailability. In contrast, inhibition of intestinal uptake transporter such as (OATP) 2B1 would prolong absorption [15]. Sodhi and Benet then proposed applying parameters of mean absorption time (MAT) and time to maximum concentration ($t_{max}$) to evaluate the involvement of intestinal transporters in oral DDIs. To illustrate the application, the authors evaluated the DDI of many drugs in the presence of rifampin, an inducer of intestinal transporters. The exposure of rosuvastatin, a substrate of OATP1B1 and OATP1B3 was increased by threefolds in both White and Asian subjects under the influence of rifampin. A significant decrease in MAT and $t_{max}$ indicated that the bioavailability was enhanced due to induction of the intestinal transporters. In another example of apixaban, a substrate of both CYP3A4 and P-gp, multiple dosing of rifampin resulted in an approximately twofold reduction in its exposure. However, there was no change in MAT and $t_{max}$, suggesting that the *in vitro* susceptibility of apixaban to P-gp is not clinically significant. An observed decrease in MRT indicated an increase in clearance causing the reduction in exposure. Therefore, by comparing MAT, $t_{max}$, and MRT, one could discern the contribution of intestinal transporters on drug exposure.

### 4.2.2 Peyer's Patches

Peyer's patches are aggregations of lymphoid tissues that are mostly found in ileum, located at the antimesenteric border of the intestine. They appear as elongated thickenings of the intestinal epithelium measuring a few centimeters in length. Microscopically, Peyer's patches comprise of lymphoid follicles, which are separated from the intestinal lumen by a single layer of specialized epithelium (FAE, follicle-associated epithelium) containing M cells (the membranous epithelial cells) and enterocytes. This epithelium is different from the villus epithelium in that the enterocytes are more cuboidal, and it contains fewer goblet cells with no secretory component [16]. The M cells do not have fully developed microvilli, are pinocytic, and contain numerous vesicles in the basolateral cytoplasm enclosing lymphocytes and occasional macrophages [17]. M cells play a central role in the intestinal immune response since they transport macromolecules and therefore have a specific role in antigen uptake. Recently, Peyer's patch M cells have been considered as a potential target for drug and vaccine delivery [16, 18–21]. Uptake of nanoparticles through Peyer's patches following oral administration has been explored as an approach to deliver drugs to lymphatic organs like lung through the lymphatic pathways [18].

## 4.3 Physiological Features of Gastrointestinal Tract

### 4.3.1 pH of Gastrointestinal Fluids

The source of hydrogen ions in the gastrointestinal tract is from the parietal cells in the stomach. Secretion is controlled by both neural and hormonal feedback mechanisms. When the acidic content of the stomach reaches the duodenum, it stimulates the secretion of an alkaline fluid rich in bicarbonate anions (pH 8) from the pancreas. This fluid together with the bile and alkaline fluids secreted by the Brunner's glands in the submucosa of the duodenum, neutralizes the acidic contents reaching the duodenum.

The pH of the fluids varies considerably along the length of the gastrointestinal tract, and there are also considerable variations within the normal pH range among individuals. Gastric fluid is highly acidic with a pH range of 1–3.5; whereas the pH of the fluids in the small intestine is relatively more alkaline with a pH range of 5–6 in the duodenum, and a pH of 8 in the lower ileum. The fluid in the large intestine generally has a pH of about 8 [22–24]. There appears to be a diurnal cycle of gastric acidity, the fluids are more acidic at night and the pH values fluctuate during the day primarily in response to food ingestion. Gastric fluid pH generally increases after meals, and then slowly decreases over the following few hours. A more recent study indicated that besides elevation in gastric pH, significant increases in buffer capacity, and bile salt levels, but decreases in surface tension were detected in luminal fluids in the upper gastrointestinal tract after meals [25]. These changes induced by food may influence oral absorption of some drugs.

Gastrointestinal pH may influence drug absorption in a number of ways. The degree to which a weak electrolyte drug ionizes in the gastrointestinal fluids is a function of pH of the fluid and $pK_a$ of the drug. In general, the solubility and dissolution rate of an ionized form of a drug is higher than the unionized form of the same drug. On the other hand, the unionized form of a drug will be absorbed faster than the ionized counterpart at any particular site along the gastrointestinal tract. However, most drugs in solution are absorbed much faster from the small intestine regardless of differences in pH between the stomach, small intestine, and large intestine. This is a reflection of the overwhelmingly much larger absorptive surface area available in the small intestine (Table 4.1).

Some acidic drugs, such as salicylate are supposed to be fully ionized in the alkaline luminal fluids. Despite this, salicylate is still well absorbed from the small intestine. To account for the absorption of these compounds, a more acidic microclimate close to the brush order has been proposed [26]. This was supported by microelectrode measurements, which suggested a surface pH of between 4.5 and 6.0, depending on the luminal glucose concentration.

The aqueous solubility of a weak electrolyte drug is influenced by pH, and its rate of dissolution from a solid dosage form will be pH dependent. For a poorly soluble weak electrolyte drug administered in a solid dosage form, its rate of dissolution in the gastrointestinal fluids may be the rate-limiting step in the absorption process. Hence, the *in situ* gastrointestinal pH would exert a major influence on the dissolution rate and even the overall absorption rate and extent of such a drug administered in a solid dosage form. As a result, a poorly soluble weak acidic drug, which only exhibits a high dissolution rate in an alkaline solution, would be expected to be rapidly absorbed from a solid dosage form when the drug passes through the acidic environment of the stomach and reaches the more alkaline intestinal fluids where rapid drug dissolution can occur. For a poorly soluble weak basic drug, which only exhibits a high dissolution rate in an acidic aqueous solution, it must first dissolve in the acidic gastric fluid before it reaches the small intestine where the drug may reprecipitate, but in very fine particles with large surface area readily available for redissolution. Any undissolved weak basic drug in the alkaline small intestinal fluids would be expected to have relatively low absorption rate.

Absorption of a drug can also be affected by its chemical stability in the gastrointestinal tract. Any drugs, which exhibit low chemical stability in either acidic or alkaline environments, would be expected to have impaired absorption if they are not properly protected from the hostile physiological environment in their dosage form designs.

### 4.3.2 Biliary Secretion

Bile is produced by the hepatic cells in the liver and secreted through the hepatic duct into the gallbladder where bile is concentrated by removing sodium ions, chloride, and water from the bile fluid. Bile is a complex mixture and the major components include cholesterol, bicarbonate ions, phospholipids mainly as lecithin, bilirubin, and bile salts such as sodium glycocholate and sodium taurocholate. Bile salts are water-soluble derivatives of cholesterol. The proportions of sodium and potassium ions in bile are similar to that found in plasma, while the concentrations of chloride and bicarbonate are often lower, and the bile acids make up the difference in the ion balance.

During food intake, following stimulation by secretin and cholecystokinin, bile fluid is secreted into the duodenum. Bile acids are poorly absorbed in the proximal small intestine but are reabsorbed by an active process in the terminal ileum. After absorption, bile acids are substantially cleared by the liver but are resecreted in the bile in a process known as enterohepatic recirculation. Many drugs such as morphine, warfarin, indomethacin, clindamycin, and erythromycin all undergo enterohepatic recirculation to different extents clinically [27]. These compounds are metabolized in the liver to produce metabolites generally larger and more polar than the parent compounds. The process favors excretion of the metabolites into bile, which is eventually emptied via the bile duct into the duodenum. These metabolites may be reabsorbed from the intestinal lumen, either after biotransformation back to parent compounds by the intestinal microflora or even without biotransformation. The process of enterohepatic recirculation can increase the half-life of these compounds in the blood, leading to prolonged therapeutic or even toxic effects. It has been found that biliary excretion is the major route of elimination of drugs with molecular weights greater than 400 (e.g. valsartan, with a molecular weight of 435 and 88% being excreted into bile in humans) [28, 29].

Bile fluid is important in the absorption of lipid-soluble vitamins, steroids, and dietary lipids. Bile salts assist in the intestinal absorption of lipophilic compounds by enhancing the dissolution rate and solubility [30]. They are also critical for the lymphatic uptake of lipids and lipid-soluble drugs. The improved dissolution rate in the intestine may be due to the improvement of the wetting properties of the solid drug or the increased solubility of drug in micellar solutions [31, 32]. Bile acids, on the other hand, have also been found to reduce the bioavailability of some hydrophilic drugs, such as atenolol, possibly due to strong interaction between these hydrophilic drugs with bile acid micelles [33].

### 4.3.3 Gastrointestinal Microflora

There is almost nil or very little flora in the stomach, but there is a significant increase in the number of microflora toward the terminal part of the small intestine. In the colon, 60% of the fecal mass is made up by the anaerobic bacteria, the most common of which are *Bifidobacterium*, *Lactobacillus*, and *Bacteroides* [34]. This gastrointestinal microflora can metabolize a variety of chemicals including drugs, which can lead to reduced absorption in the intestine. In some cases, these metabolic activities can also lead to toxicological consequences [28], e.g. beta-glucuronidase in intestinal microflora converts irinotecan hydrochloride (CPT-11), an antitumor camptothecin derivative, to its active form (7-ethyl-10-hydroxycamptothecin), which causes severe forms of diarrhea clinically [35]. Inhibition of the beta-glucuronidase activity in the intestinal microflora by antibiotics markedly ameliorated diarrhea and reduced cecal damage [35, 36].

The intestinal microflora possesses a wide range of metabolic activities including hydrolysis of glucosides, glucuronides, sulfate esters, amides, esters, dehydroxylation, deamidation, decarboxylation, dealkylation, deamination, reduction of double bonds, nitro and diazo groups, acetylation and esterification [28, 37, 38]. Table 4.2 shows some metabolic reactions of the intestinal flora and examples of drugs involved [37].

**Table 4.2** Some metabolic reactions of the intestinal microflora and examples of drugs are involved [37].

| Metabolic reactions | Examples of drug involved |
|---|---|
| *Reduction* | |
| Nitro compounds | Clonazepam |
| Sulfoxides | Sulindac |
| 21-Hydroxycorticoids | Aldosterone |
| Double bonds | Digoxin |
| Azo compounds | Prontosil |
| *Hydrolysis* | |
| Nitrate esters | Glyceryl trinitrate |
| Sulfate esters | Sodium picosulfate |
| Succinate esters | Carbenoxolone |
| Amides | Methotrexate |
| Glucuronides | Morphine glucuronide |
| Glucosides | Sennosides |
| *Removal of functional groups* | |
| N-Dealkylation | Methamphetamine |
| Deamination | Flucytosine |
| *Other reactions* | |
| Heterocyclic ring fission | Levamisole |
| Side-chain cleavage | Steroids |

Source: Adapted from Abu Shamat [37].

Intestinal microflora plays a major role in enterohepatic circulation of a number of drugs such as digoxin, oral contraceptives, promazine, and chloramphenicol [39]. Metabolites of some drugs are excreted through the bile duct into intestinal lumen, where the metabolites, especially the glucuronide conjugates are hydrolyzed by the microflora there and converted back to parent compounds for reabsorption. The process will extend drug levels in the body.

Interethnic variation in intestinal microflora affects the metabolic reduction of the cardiac drug digoxin by the flora. It was found that the production of the digoxin reduction products was 36% in North Americans compared to 13.7% in South Indians [40, 41]. However, the number and species of organisms capable of reducing digoxin were found similar in stool cultures from Indians and Americans. The main difference was found in the metabolic activity of the anaerobic gut flora presumably mediated by environmentally determined factors [41]. Thus, the interethnic variation in microflora may affect the delivery of pharmacologically active drugs required for efficacy. *p*-cresol is produced by intestine-dwelling bacteria. A recent study indicated that individuals having high predose urinary levels of *p*-cresol sulfate had low postdose urinary ratios of acetaminophen sulfate to acetaminophen glucuronide [42]. It was suggested that, in individuals with high bacterially mediated *p*-cresol generation, competitive O-sulfonation of *p*-cresol reduces the effective systemic capacity to sulfonate acetaminophen. Thus, many other sulfonation reactions could also be potentially affected by competition with *p*-cresol, leading to variable drug metabolism and response. Up to date, it is well established that intestinal microflora is capable of enzymatically transforming drugs into active, inactive, and even toxic metabolites, thus potentially affecting the pharmacokinetics and bioavailability of orally administered medications [43]. However, its impact on clinical outcomes is unknown yet, though interindividual variability of gut microbial composition would be able to explain partially on the differences in drug efficacy and toxicity observed in certain patient populations. A latest study showed that enterococci with tyrosine decarboxylases (TDC) can decarboxylate levodopa to dopamine in small intestine and the study also indicated that the natural variation of the *TDC*-gene positively correlated with the daily dose requirement of levodopa in PD patients [44]. The finding illustrated the potential involvement of bacteria to the bioavailability of levodopa. A recent study developed a quantitative experimental framework for mapping the ability of the human gut microbiome in the context of microbial communities to metabolize small-molecule drugs [45]. The screen was able to identify microbiome-encoded genes responsible for specific metabolic events, and how drug–microbiome interactions vary between individuals. Such an approach could be applied in drug development and personalized medicine.

### 4.3.4 Gastrointestinal Transit and Motility

#### 4.3.4.1 Gastric Emptying Rate

Stomach is not the major organ for drug absorption due to its small epithelial surface. Most drugs are best absorbed from the small intestine after oral administration. Thus, the gastric emptying rate, i.e. the rate at which the drug leaves the stomach and enters the duodenum, will affect the overall rate of drug absorption, and thereby the onset of the therapeutic action. A classic example showed a good correlation between stomach emptying time and peak plasma concentration for acetaminophen [46], i.e. the faster gastric emptying rate (shorter gastric emptying time) gave the higher peak drug plasma concentration. In contrast, absorption of a poorly soluble weak basic drug, e.g. nitrofurantoin could be enhanced by delayed gastric emptying, as the drug is allowed to dissolve in gastric acid prior to being absorbed rapidly from the small intestine [47]. Drug absorption can be increased using gastroretentive strategy to overcome the short transit time of drug in primarily absorption site of small intestine. This topic is elaborated in Chapter 33.

Gastric emptying rate will also affect the absorption of drug that is susceptible to chemical or enzyme degradation in the stomach. Drug in enteric dosage form is formulated to release drug in the small intestine but not in the stomach for the purpose of either to minimize the side effect of gastric irritation of the drug or to protect the drug from chemical or enzyme degradation in the stomach. Any delay in the gastric emptying of this enteric dosage form into the small intestine will delay the onset of the therapeutic action.

In general, the gastric emptying of solution or suspension is faster than non-disintegrating solid dosage form; and the presence of food in the stomach reduces the gastric emptying rate and can potentially delay the absorption of drugs. However, for most drugs, the influence of food on absorption, if any, is clinically insignificant [48]. Food intake may influence absorption of drugs in different directions – increasing, decreasing, or having no consistent effect on the amount absorbed. The outcomes depend on the physicochemical properties of the drug itself, and also the types of the drug formulation. Gu and coworkers attempted to predict effect of food on extent of drug absorption based on physicochemical properties [49]. They noticed that solubilization effect of food primarily accounted for the positive food effect on absorption, while interference of food with absorption caused negative effect on absorption of compounds that are highly hydrophilic and probably with narrow window of absorption.

#### 4.3.4.2 Intestinal Motility

Since the small intestine is the primary site of drug absorption, the longer the residence time in the small intestine the greater the potential for efficient drug absorption. There are two types of intestinal movements, propulsive and mixing. The propulsive movements primarily determine the intestinal transit rate and thus the residence time of a drug in the small intestine. Chyme, the intestinal content passes through the human small intestine at a rate of 1–4 cm per min [50]. The velocity of transport is faster in the proximal portions of the small intestine, i.e. duodenum and proximal jejunum, and decreases as chyme reaches the ileum. On average, chyme passes through the entire small intestine in 3–4 hours. Transit time for chyme in the large intestine is much slower, depending on the amount of fiber in the diet. Transit time through the large intestine in healthy humans is 2–4 days [2].

Intestinal transit or residence time will be important for drugs, which are absorbed by intestinal carrier-mediated system, as these drugs are preferentially absorbed in the location of highest carrier density, e.g. riboflavin is absorbed mainly in the proximal small intestine through a sodium-independent carrier-mediated system [50]. Factors that influence intestinal motility will influence the bioavailability of riboflavin. Oral absorption of riboflavin was found higher in hypothyroid children, but lower in hyperthyroid children when compared to the normal children [51]. Riboflavin absorption became normal in both groups after successful treatment of the thyroid disorder.

Intestinal residence time will also affect the absorption of drugs, which dissolve slowly in the intestinal fluids; dosage forms, which release drugs slowly, such as the controlled or sustained release dosage forms, as they pass through the gastrointestinal tract, and the enteric-coated dosage forms, which release drug only when they reach the small intestine.

## 4.4 Other Physiological Factors

### 4.4.1 Age Effects

A number of physiological changes with aging might impair drug absorption from the gastrointestinal tract. Gastric fluid pH increases dramatically with age, due to decreased acid and pepsin secretion and the incidence of achlorhydria is quite prevalent in the aging population [52, 53]. Elevated gastric pH may affect drug ionization, dissolution, degradation, and gastric emptying time. A reduction in the number of absorbing cells and splanchnic blood flow with a smaller intestinal surface area might also impair or delay drug absorption [54, 55]. Gastric emptying is slightly delayed in the elderly with decreased peristalsis, and colonic transit is slowed secondary to region-specific loss of neurons [56, 57]. Despite these physiological changes in the gastrointestinal tract, to date, impaired absorption has been demonstrated in the aged for only some compounds (e.g. galactose, 3-methylglucose, calcium, thiamine, and iron) absorbed by active transport mechanisms [58]. Absorption of most drugs, which is accomplished by passive diffusion, does not appear to be adversely affected in the aged.

The integrity of drug absorption in the aged versus young is usually assessed by comparing the rate of absorption; time to peak plasma concentration; peak plasma concentration; and area under the plasma drug concentration-time curve. Based on these pharmacokinetic parameters, paracetamol, acetylsalicylic acid, phenylbutazone, practolol, and sulfamethizole were found to be absorbed normally in the elderly [59, 60]. The absorption of indomethacin, propoxyphene [61], digoxin, and theophylline [62], may also be unchanged in the elderly.

For drugs undergoing significant first-pass metabolism in the gastrointestinal tract and/or liver, there may be a significant increase in the extent of drug absorption (bioavailability) in the elderly, most probably due to decreased efficiency of these first-pass processes. Such increases have been noted in patients taking levodopa [63], chlormethiazole [64], propranolol [65], and labetalol [66].

Gastrointestinal absorption of prazosin has been found to be decreased in old age [67]. The pharmacokinetic behavior of prazosin was studied in the elderly following oral and intravenous administration, and its absolute bioavailability was found to be about 30% lower in the elderly as compared to the young. No reduction in drug clearance was noted in the age, indicating that drug disposition was unaffected by age. The mechanism for the observed decrease in absorption of prazosin was believed to be a result of decreased gastric motility and splanchnic blood flow in the aged. Absorption of clorazepate dipotassium has also been found to be adversely affected by age [68]. The most likely explanation is that of decreased acid secretion by the stomach since the absorption of this drug is dependent on the dissolution of the compound in the gastric acid prior to emptying into the small intestine.

### 4.4.2 Gender Effects

Oral absorption of drugs can be influenced by a number of factors including gastric acid secretion, gastric emptying rate, gastrointestinal blood flow, and surface area, in addition to presystemic intestinal and hepatic metabolism and transport. Some of these factors have been investigated for the gender effects. No difference was observed in the gastrointestinal pH between men and women under the respective fasted and fed states [22]. No gender effect was also observed in the characteristics of stomach and proximal jejunal fluids, including pH, osmolality, electrolyte concentrations, and levels of bile acids and proteins, though all these measured parameters with the exception of the total concentration of proteins varied significantly among the individuals [69]. However, gastrointestinal motility has been shown to be influenced by sex hormones [70, 71], indicating gastric emptying and intestinal transit times may vary during pregnancy and the menstrual cycles. With a new developed radiological procedure using radiopaque rings as markers and fluoroscopy for detection, gastric emptying, small-bowel transit, and colonic transit were measured simultaneously in a cohort of healthy subjects [72]. It was found that gastric emptying, small-bowel transit, and colonic transit were significantly slower in female healthy subjects compared to males. Small-bowel transit was significantly faster in women with overweight and in postmenopausal women compared to lean and premenopausal women, respectively. Changes in gastric emptying time have been observed in pregnancy [71], but the changes were variable during the menstrual cycle [73].

Only a few clinical studies have shown gender differences in drug absorption and bioavailability. Verapamil, after oral administration, was cleared faster in men than in women, but the difference was not observed after intravenous administration of the drug [74]. It was postulated to be caused by gender differences in the intestinal process. Mizolastine, an oral antihistamine was found to have a longer duration for absorption in men than in women [75]. In an earlier study, plasma levels of aspirin and salicylate after low oral doses of aspirin were found significantly higher in females than in males, suggesting to be due to intrinsically lower metabolic activity in that gender [76]. A later study also demonstrated that following oral administration, aspirin was absorbed more quickly in females than in males, although the overall bioavailability was the same in both gender groups [77]. More recently, it was illustrated that there was gender difference in the oral absorption of refampicin [78]. It was postulated to be caused by the gender-related differences in the extent of intestinal and hepatic cytochrome P450 (CYP) 3A induction by rifampicin.

Despite gastrointestinal motility has been shown to be affected by sex hormones, only few studies showed differences in drug absorption between males and females, and the findings could vary among studies. As the absorption often determined from the drug plasma concentration after oral administration could be influenced by a number of confounding factors including distribution, metabolism, and excretion; and also the health status of the test subjects, it would be difficult to discern gender effect on drug absorption. The current available evidence indicates that there is little gender effect on drug absorption.

## 4.5 Conclusion

The biological and physiological features of gastrointestinal tract together with the physicochemical properties of drugs are important parameters that determine their absorption from the intestine. Based on this information, it is possible to predict oral absorption of drugs by silico modeling, especially for those that are absorbed via passive diffusion [79]. An analysis of human pharmacokinetic parameters of antituberculosis drugs revealed associations between lipophilicity and volume of distribution, clearance, plasma protein binding, and oral bioavailability [80]. Most of these drugs with favorable pharmacokinetic properties complied with Lipinski's rule of five. In recent years, there is exploding information on intestinal transporters and carrier-mediated drug absorption [12, 13, 81, 82]. The current evidence indicates that carrier-mediated and active uptake of drugs may be more common than is usually assumed [12]. It was suggested that targeting drug substrates for absorptive intestinal membrane transporters could be a feasible strategy for optimizing drug absorption [81, 83]. Intestinal transporters might also mediate potential DDI [84]. Intestinal microflora has been identified to play an important role in drug metabolism and oral absorption [85]. Intestinal microbiome has been found to modulate the fate of more than 30 approved drugs. Its clinical impact has yet to be established. It envisions that modulation of intestinal microflora could potentially be applied to achieve individualized drug response, i.e. personalized medicine.

## References


1 DeSesso, J.M. and Jacobson, C.F. (2001). Anatomical and physiological parameters affecting gastrointestinal absorption in humans and rats. *Food Chem. Toxicol.* 39: 209–228.

2 Granger, D.N., Barrowman, J.A., and Kvietys, P.R. (1985). *Clinical Gastrointestinal Physiology*. Philadelphia: W.B. Saunders.

3 Ritschel, W.A. (1991). Targeting in the gastrointestinal tract – new approaches. *Methods Find Exp. Clin. Pharmacol.* 13: 313–336.

4 Synder, W.S., Cook, M.J., Nasset, E.S. et al. (1975). *Report of the Task Group on Reference Man*. New York: Pergamon.

5 Macheras, P., Reppas, C., and Dressman, J.B. (1995). *Biopharmaceutics of Orally Administered Drugs*. Herfordshire: Ellis Horwood.

6 Bardelmeijer, H.A., van Tellingen, O., Schellens, J.H.M., and Beijnen, J.H. (2000). The oral route for the administration of cytotoxic drugs: strategies to increase the efficiency and consistency of drug delivery. *Invest. New Drugs* 18: 231–241.

7 Hayashi, M., Tomita, M., and Awzu, S. (1997). Transcellular and paracellular contribution to transport processes in the colorectal route. *Adv. Drug Delivery Rev.* 28: 191–204.

8 Nellans, H.N. (1991). Paracellular intestinal transport: Modulation of absorption. *Adv. Drug Delivery Rev.* 7: 339–364.

9 Hunter, J. and Hirst, B.H. (1997). Intestinal secretion of drugs. The role of P-glycoprotein and related drug efflux systems in limiting oral drug absorption. *Adv. Drug Delivery Rev.* 25: 129–157.

10 Yokooji, T., Mori, N., and Murakami, T. (2009). Site-specific contribution of proton-coupled folate transporter/haem carrier protein 1 in the intestinal absorption of methotrexate in rats. *J. Pharm. Pharmacol.* 61: 911–918.

11 Kellett, G.L. (2001). The facilitated component of intestinal glucose absorption. *J. Phys.* 531: 585–595.

12 Dobson, P.D. and Kell, D.B. (2008). Carrier-mediated cellular uptake of pharmaceutical drugs: an exception or the rule? *Nat. Rev. Drug Discov.* 7: 205–220.

13 Müller, J., Keiser, M., Drozdzik, M., and Oswald, S. (2017). Expression, regulation and function of intestinal drug transporters: an update. *Biol. Chem.* 398: 175–192.

14 Ashmawy, S.M., El-Gizawy, S.A., El Maghraby, G.M., and Osman, M.A. (2019). Regional difference in intestinal drug absorption as a measure for the potential effect of P-glycoprotein efflux transporters. *J. Pharm. Pharmacol.* 71: 362–370.

15 Sodhi, J.K. and Benet, L.Z. (2020). The necessity of using changes in absorption time to implicate intestinal transporter involvement in oral drug-drug interactions. *AAPS J.* 22: 111.

16 Yeh, P.Y., Ellens, H., and Smith, P.L. (1998). Physiological considerations in the design of particulate dosage for oral vaccine delivery. *Adv. Drug Delivery Rev.* 34: 123–133.

17 Kraehenbuhl, J.P. and Neutra, M.R. (2000). Epithelial M cells: differentiation and function. *Annu. Rev. Cell Dev. Biol.* 16: 301–332.

18 Bachhav, S.S., Dighe, V.D., and Devarajan, P.V. (2018). Exploring Peyer's patch uptake as a strategy

for targeted lung delivery of polymeric rifampicin nanoparticles. *Mol. Pharm.* 15: 4434–4445.

19 Brayden, D.J. and Baird, A.W. (2004). Apical membrane receptors on intestinal M cells: potential targets for vaccine delivery. *Adv. Drug Delivery Rev.* 56: 721–726.

20 Jepson, M.A., Clark, M.A., and Hirst, B.H. (2004). M cell targeting by lectins: a strategy for mucosal vaccination and delivery. *Adv. Drug Delivery Rev.* 56: 511–525.

21 Managuli, R.S., Raut, S.Y., Reddy, M.S., and Mutalik, S. (2018). Targeting the intestinal lymphatic system: a versatile path for enhanced oral bioavailability of drugs. *Expert Opin. Drug. Deliv.* 15: 787–804.

22 Dressman, J.B., Berardi, R.R., Dermentzoglou, L.C. et al. (1990). Upper gastrointestinal (GI) pH in young, healthy men and women. *Pharm. Res.* 7: 756–761.

23 Evans, D.F., Pye, G., Bramley, R. et al. (1988). Measurement of gastrointestinal pH profiles in normal ambulant human subjects. *Guts* 29: 1035–1041.

24 Russell, T.L., Beradi, R.R., Barnett, J.L. et al. (1993). Upper gastrointestinal pH in 79 healthy, elderly North American men and women. *Pharm. Res.* 10: 187–196.

25 Kalantzi, L., Goumas, K., Kalioras, V. et al. (2006). Characterization of the human upper gastrointestinal contents under conditions stimulating bioavailability/bioequivalence studies. *Pharm. Res.* 23: 165–176.

26 Lucas, M. (1984). The surface pH of the intestinal mucosa and its significance in the permeability of organic anions. In: *Pharmacology of Intestinal Permeation II* (ed. T.Z. Csáky), 119–163. Berlin: Springer-Verlag.

27 Roberts, M.S., Magnusson, B.M., Burczynski, F.J., and Weiss, M. (2002). Enterohepatic circulation: physiological, pharmacokinetic and clinical implications. *Clin. Pharmacokinet.* 41: 751–790.

28 Calabrese, E.J. (1983). Biliary excretion. In: *Principles of Animal Extrapolation* (ed. E.J. Calabrese), 176–192. New York: Wiley.

29 Yang, X.N., Gandhi, Y.A., Duignan, D.B., and Morris, M.E. (2009). Prediction of biliary excretion in rats and humans using molecular weight and quantitative structure-pharmacokinetic relationships. *AAPS J.* 11: 511–525.

30 Porter, C.J.H., Pouton, C.W., Cuine, J.F., and Charman, W.N. (2008). Enhancing intestinal drug solubility using lipid-based delivery system. *Adv. Drug Delivery Rev.* 60: 673–691.

31 Miyazaki, S., Inoue, H., Yamahira, T., and Nadai, T. (1979). Interaction of drugs with bile components, effect of bile salts on the dissolution behavior of indomethacin and phenylbutazone. *Chem. Pharm. Bull.* 27: 2468.

32 Miyazaki, S., Yamahira, T., Inoue, H., and Nadai, T. (1980). Interaction of drugs with bile components, 2. Effect of bile on the absorption of indomethacin and phenylbutazone in rats. *Chem. Pharm. Bull.* 28: 323–326.

33 Barnwell, S.G., Laudanski, T., Dwyer, M. et al. (1993). Reduced bioavailability of atenolol in man: the role of bile acids. *Int. J. Pharm.* 89: 245–250.

34 Guarner, F. and Malagelada, J.R. (2003). Gut flora in health and disease. *Lancet* 361: 512–519.

35 Takasuna, K., Hagiwara, T., Hirohashi, M. et al. (1996). Involvement of beta-glucuronidase in intestinal microflora in the intestinal toxicity of the antitumor camptothecin derivative irinotecan hydrochloride (CPT-11) in rats. *Cancer Res.* 56: 3752–3757.

36 Brandi, G., Dabard, J., Raibaud, P. et al. (2006). Intestinal microflora and digestive toxicity of irinotecan in mice. *Clin. Cancer Res.* 12: 1299–1307.

37 Abu Shamat, M. (1993). The role of the gastrointestinal microflora in the metabolism of drugs. *Int. J. Pharm.* 97: 1–13.

38 Wilson, I.D. and Nicholson, J.K. (2009). The role of gut microbiota in drug response. *Curr. Pharm. Res.* 15: 1519–1523.

39 Ilett, K.F., Tee, L.B.G., Reeves, P.T., and Minchin, R.F. (1990). Metabolism of drugs and other xenobiotics in the gut lumen and wall. *Pharmacol. Ther.* 46: 67–93.

40 Alam, A.N., Saha, J.R., Dobkin, J.F., and Lindenbaum, J. (1988). Interethnic variation in the metabolic inactivation of digoxin by the gut flora. *Gastroenterology* 95: 117–123.

41 Mathan, V.I., Wiederman, J., Dobkin, J.F., and Lindenbaum, J. (1989). Geographic differences in digoxin inactivation, a metabolic activity of the human anaerobic gut flora. *Gut* 30: 971–977.

42 Clayton, T.A., Baker, D., Lindon, J.C. et al. (2009). Pharmacometabonomic identification of a significant host-microbiome metabolic interaction affecting human drug metabolism. *Proc. Natl. Acad. Sci. U S.A.* 106: 14728–14733.

43 Flowers, S.A., Bhat, S., and Lee, J.C. (2020). Potentially implications of gut microbiota in drug pharmacokinetics and Bioavailability. *Pharmacotherapy* 40: 704–712.

44 Van Kessel, S.P., Frye, A.K., El-Gendy, A.O. et al. (2019). Gut bacterial tyrosine decarboxylases restrict

levels of levodopa in the treatment of Parkinson's disease. *Nat. Commun.* 10: 310.
45 Javdan, B., Lopez, J.G., Chankhamjon, P. et al. (2020). Personalized mapping of drug metabolism by the human gut microbiome. *Cell* 181: 1661–1679.
46 Heading, R.C., Nimmo, J., Prescott, L.F., and Tothill, P. (1973). The dependence of paracetamol absorption on the rate of gastric emptying. *Br. J. Pharmacol.* 47: 415–421.
47 Rosenberg, H.A. and Bates, T.R. (1976). The influence of food on nitrofurantoin bioavailability. *Clin. Pharmacol. Ther.* 20: 227–232.
48 Sjöqvist, F., Borgå, O., Dahl, M.L., and ML'E, O. (1996). Fundamentals of clinical pharmacology. In: *Avery's Drug Treatment*, 4e (ed. T.M. Speight and N.G. Holford), 1–73. Auckland: ADIS press.
49 Gu, C.H., Li, H., Levons, J. et al. (2007). Predicting effect of food on extent of drug absorption based on physicochemical properties. *Pharm. Res.* 24: 1118–1130.
50 Said, H.M. and Mohammed, Z.M. (2006). Intestinal absorption of water-soluble vitamins: an update. *Curr. Opin. Gastroenterol.* 22: 140–146.
51 Levy, G., MacGillivray, M.H., and Procknal, J.A. (1972). Riboflavin absorption in children with thyroid disorders. *Pediatrics* 50: 896.
52 Feldman, M. (1997). The mature stomach. Still pumping out acid? *JAMA* 278: 681–682.
53 Vanzant, F.R. (1932). The normal range of gastric acidity from young to old age: an analysis of 3746 records. *Arch. Med.* 49: 345–359.
54 Bender, A.D. (1968). Effect of age on intestinal absorption: implication for drug absorption in the elderly. *J. Am. Geriatr. Soc.* 16: 1331–1339.
55 Ciccocioppo, R., Di Sabatino, A., Luinetti, O. et al. (2002). Small bowel enterocyte apoptosis and proliferation are increased in the elderly. *Gerontology* 48: 204–208.
56 Orr, W.C. and Chen, C.L. (2002). Aging and neural control of the GI tract: IV. Clinical and physiological aspects of gastrointestinal motility and aging. *Am. J. Physiol. Gastrointest. Liver Physiol.* 283: G1226–G1231.
57 Wiley, J.W. (2002). Aging and neural control of the GI tract: III. Senescent enteric nervous system: lessons from extraintestinal sites and nonmammalian species. *Am. J. Physiol. Gastrointest. Liver Physiol.* 283: G1020–G1026.
58 Kinirons, M.T. and Crome, P. (1997). Clinical pharmacokinetic consideration in the elderly. An update. *Clin. Pharmacokinet.* 33: 302–312.
59 Triggs, E.J., Nation, R.L., Long, A., and Ashley, J.J. (1975). Pharmacokinetics in the elderly. *Eur. J. Clin. Pharmacol* 8: 55–62.
60 Castleden, C.M., Volans, C.N., and Raymond, K. (1977). The effect of ageing on drug absorption from the gut. *Age Ageing* 6: 138–143.
61 Melander, A., Bodin, N.O., Danielson, K. et al. (1978). Absorption and elimination of d-propoxyphene, acetylsalicylic acid and phenazone in a combination tablet (DOLERON): comparison between young and elderly subjects. *Acta Med. Scand.* 203: 121–124.
62 Stevenson, I.H., Salem, S.A.M., O'Malley, K. et al. (1981). Age and drug absorption. In: *Drug absorption* (ed. L.F. Prescott and W.S. Nimmo), 253. Auckland: ADIS Press.
63 Evans, M.A., Triggs, E.J., Broe, G.A., and Saines, N. (1980). Systemic availability of orally administered L-dopa in the elderly parkinsonian patients. *Eur. J. Clin. Pharmacol.* 17: 215–221.
64 Nation, R.L., Vine, J., Triggs, E.J., and Learoyd, B. (1977). Plasma level of chlormethiazole and two metabolites after oral administration to young and aged human subjects. *Eur. J. Clin. Pharmacol* 12: 137–145.
65 Castleden, C.M. and George, C.F. (1979). The effect of ageing on the hepatic clearance of propranolol. *Br. J. Clin. Pharmacol.* 7: 49–54.
66 Kelly, J.G., McGarry, K., O'Malley, K., and O'Brien, E.T. (1982). Bioavailability of labetalol increases with age. *Br. J. Clin. Pharmacol.* 14: 304–305.
67 Rubin, P.C., Scott, P.J.W., and Reid, J.L. (1981). Prazosin disposition in young and elderly subjects. *Br. J. Clin. Pharmacol.* 12: 401–404.
68 Shader, R.I., Georgotas, A., Greenblatt, D.J. et al. (1978). Impaired absorption of desmethyldiazepam from clorazepate by magnesium aluminium hydroxide. *Clin. Pharmacol. Ther.* 24: 308–315.
69 Lindahl, A., Ungell, A.L., Knutson, L., and Lennernäs, H. (1997). Characterization of fluids from the stomach and proximal jejunum in men and women. *Pharm. Res.* 14: 497–502.
70 Hutson, W.R., Roehrkasse, R.L., and Wald, A. (1989). Influence of gender and menopause on gastric emptying and motility. *Gastroenterology* 96: 11–17.
71 Singer, A.J. and Brandt, L.J. (1991). Pathophysiology of the gastrointestinal tract during pregnancy. *Am. J. Gastroenterol.* 86: 1695–1712.
72 Sadik, R., Abrahamsson, H., and Stotzer, P.O. (2003). Gender differences in gut transit shown with a newly developed radiological procedure. *Scand. J. Gastroenterol.* 38: 36–42.

73 Caballero-Plasencia, A.M., Valenzuela-Barrano, M., Martin-Ruiz, J.L. et al. (1999). Are there changes in gastric emptying during the menstrual cycle? *Scand. J. Gastroenterol.* 34: 772–776.

74 Krecic-Shepard, M.E., Barnas, C.R., Slimko, J., and Schwartz, J.B. (2000). Faster clearance of sustained release verapamil in men versus women: continuing observations on sex-specific differences after oral administration of verapamil. *Clin. Pharmacol. Ther.* 68: 286–292.

75 Mesnil, F., Mentre, F., Dubruc, C. et al. (1998). Population pharmacokinetic analysis of mizolastine and validation from sparse data on patients using nonparametric maximum likelihood method. *J. Pharmacokinet. Biopharm.* 26: 133–161.

76 Ho, P.C., Triggs, E.J., Bourne, D.W.A., and Heazlewood, V.J. (1985). The effects of age and sex on the disposition of acetylsalicylic acid and its metabolites. *Br. J. Clin. Pharmacol.* 19: 675–684.

77 Aarons, L., Hopkins, K., Rowland, M. et al. (1989). Route of administration and sex differences in the pharmacokinetics of aspirin, administered as its lysine salt. *Pharm. Res.* 6: 660–666.

78 Gorski, J.C., Vannaprasaht, S., Hamman, M.A. et al. (2003). The effect of age, sex, and rifampin administration on intestinal and hepatic cytochrome P450 3A activity. *Clin. Pharmacol. Ther.* 74: 275–287.

79 Zhao, Y.H., Le, J., Abraham, M.H. et al. (2001). Evaluation of human intestinal absorption data and subsequent derivation of a quantitative structure-activity relationship (QSAR) with the Abraham descriptors. *J. Pharm. Sci.* 90: 749–784.

80 Lakshminarayana, S.B., Huat, T.B., Ho, P.C. et al. (2015). Comprehensive physicochemical, pharmacokinetic and activity profiling of anti-TB agents. *J. Antimicrob. Chemother.* 70: 857–867.

81 Steffansen, B., Nielsen, C.U., Brodin, B. et al. (2004). Intestinal solute carriers: an overview of trends and strategies for improving oral drug absorption. *Eur. J. Pharm. Sci.* 21: 3–16.

82 Zakeri-Milani, P. and Valizadeh, H. (2014). Intestinal transporters: enhanced absorption through P-glycoprotein-related drug interactions. *Expert Opin. Drug Metab. Toxicol.* 10: 859–871.

83 Hetal, P.T. and Jagruti, L.D. (2015). Influence of excipients on drug absorption via modulation of intestinal transporters activity. *Asian J. Pharm.* 9: 69–82.

84 Reznicek, J., Ceckova, M., Ptackova, Z. et al. (2017). MDR1 and BCRP transporter-mediated drug-drug interaction between rilpivirine and abacavir and effect on intestinal absorption. *Antimicrob. Agents Chemother.* 61: e00837–e00817.

85 Sun, C., Chen, L., and Zhen, Z. (2019). Mechanisms of gastrointestinal microflora on drug metabolism in clinical practice. *Saudi Pharm. J.* 27: 1146–1156.

# 5

# Absorption of Drugs Via Passive Diffusion and Carrier-Mediated Pathways

*Amit Kokate[1,2], Jae H. Chang[3], and Miki S. Park[1]*

[1] *Department of Pharmaceutics and Medicinal Chemistry, University of the Pacific, Thomas J. Long School of Pharmacy, Stockton, CA, USA*
[2] *Office of Pharmaceutical Quality, Center for Drug Evaluation and Research, Food and Drug Administration, Silver Spring, MD, USA*
[3] *Exelixis, Alameda, CA, USA*

## Disclaimer

The chapter reflects the views of the authors and should not be construed to represent FDA's views or policies.

## 5.1 Introduction

The oral route is a noninvasive means of drug administration that is preferred over other methods because it is both convenient and economical. Physicians do not require special skills nor need particular supplies such as syringes or needles to administer the medicine. Meanwhile, patients do not need complicated instructions to take the medications and therefore can minimize their visits to their physicians as they can administer the drugs themselves in the comfort of their homes. Additionally, oral delivery offers a versatile route for the administration of both immediate-release and modified-release oral dosage forms. Such advantages benefit both the physicians and the patients, which can lead to increased compliance and ultimately to higher probability of a successful drug therapy. Although some drugs are specifically targeted to the gastrointestinal (GI) tract, such as bismuth subsalicylate for heartburn and ezetimibe for the reduction of cholesterol absorption, most drug treatments are intended to act on targets that are not localized to the intestine, such as morphine to the central nervous system and cyclosporine to the lymphocytes. Therefore, it is necessary for orally administered drugs to enter the systemic circulation in order to reach their intended target to elicit a particular pharmacodynamic effect, i.e. drugs need to get absorbed.

The term absorption is often used to refer to both systemic absorption and intestinal absorption and occasionally, it is not clear, which absorption process is being studied. Therefore, a distinction needs to be made between the two absorption terminologies to avoid possible confusion. Systemic absorption is defined as "the process by which unchanged drug proceeds from site of administration to site of measurement within the body, usually plasma in an arm vein" [1]. It is not a parameter specific to oral administration but is a prerequisite step for any drug that is intended for systemic action and is administered extravascularly (e.g. subcutaneous and intramuscular injection, inhalation, transdermal, and rectal administration). The rate and extent of systemic oral drug absorption will depend on the ability of the drug to:

1. Dissociate, disintegrate, and dissolve (if not administered in solution form).
2. Withstand chemical and enzymatic degradation in the GI lumen (e.g. bacterial esterases and peptidases).
3. Diffuse across unstirred layer and mucous coat to reach GI cells.
4. Traverse across the GI membranes.
5. Escape metabolism in the gut wall (e.g. gut first-pass effect of CYP3A4).
6. Withstand degradation in portal blood.
7. Escape liver metabolism and/or excretion into the bile (liver first-pass effect).
8. Escape lung metabolism and/or exhalation (lung first-pass effect).

The intestinal epithelium comprising chiefly of a single layer of absorptive, columnar epithelial cells (enterocytes) forms the major barrier to oral drug absorption. Intestinal absorption describes a subset of systemic absorption and is defined as the movement of unchanged drugs across enterocytes. It is dependent on the first five criteria described above. This chapter will

*Oral Bioavailability and Drug Delivery: From Basics to Advanced Concepts and Applications*, First Edition. Edited by Ming Hu and Xiaoling Li.

focus on intestinal absorption, and specifically on how the molecules navigate across the enterocytes. The two mechanisms that will be discussed are *passive diffusion* and *carrier-mediated* pathways.

## 5.2 Passive Diffusion

Diffusion is defined as "the process by which solute molecules are transported from one part of a system to another as a result of random molecular motion" [2]. In passive diffusion, the movement of molecules across the lipid bilayer does not require the expenditure of energy nor does the lipid bilayer play an active role in the diffusion process. Instead, the driving force is the electrochemical potential gradient where molecules move from high electrochemical potential gradient region (i.e. in the GI lumen) to low electrochemical potential gradient region (i.e. the portal blood).

Two mechanisms associated with passive diffusion are *paracellular* and *transcellular* pathways (Figure 5.1). In the paracellular pathway, molecules move across the membranes through water-filled pores of tight junction located *between* the cells; whereas in the transcellular pathway, diffusion is *through* the lipid bilayer membranes.

Passive transcellular diffusion is considered to be the main intestinal absorption mechanism for drugs due to its very high capacity and a large surface area [3]. Lipophilic drugs are generally believed to permeate extremely well because of greater partitioning into the phospholipid bilayer of the cell membrane. One commonly used method of improving transcellular drug diffusion is via chemical modification (e.g. ester bond) of the active drug to form a prodrug with increased lipophilicity [4]. Hydrolysis of the ester bond catalyzed by esterases distributed throughout the body releases the active drug.

Between the two pathways, the paracellular route is generally thought to be a minor pathway for most drugs, especially since the tight junctions account for only ~0.01% to 0.1% of the total surface area of the intestinal membrane [5]. Moreover, tight junctions become more compact down the GI tract (i.e. along the jejunum to the colon) thereby, further reducing the contributions from the paracellular route [6]. Due to the nature of the *aqueous* pores that are bound by tight junctions of the enterocytes, flux via the paracellular route is strongly dictated by the physicochemical properties of the drug. These molecules are usually small and hydrophilic ($\log P < 0$) with poor partitioning into the cell membrane, and with a molecular weight of less than 250 Da [7]. Furthermore, it has been demonstrated that positively charged molecules can move more readily between the cells than negatively charged molecules since the cellular membrane is negatively charged [8]. Examples of drugs that utilize paracellular pathway as a mechanism of absorption are cimetidine [9], ranitidine [10], and sotalol [11]. The limited absorption efficiency of the paracellular pathway and an increased understanding of the physiology of the tight junctions have stimulated research in the development of non-toxic compounds that can enhance drug absorption by reversibly opening the tight junctions (tight junction modulators) [12].

For both pathways, dependence on the intrinsic properties of the drug suggests that dissolution (or the *rate* at which the drug dissolves into solution) may have a significant effect on intestinal absorption. Consequently, dosage formulation can be a sensitive parameter as

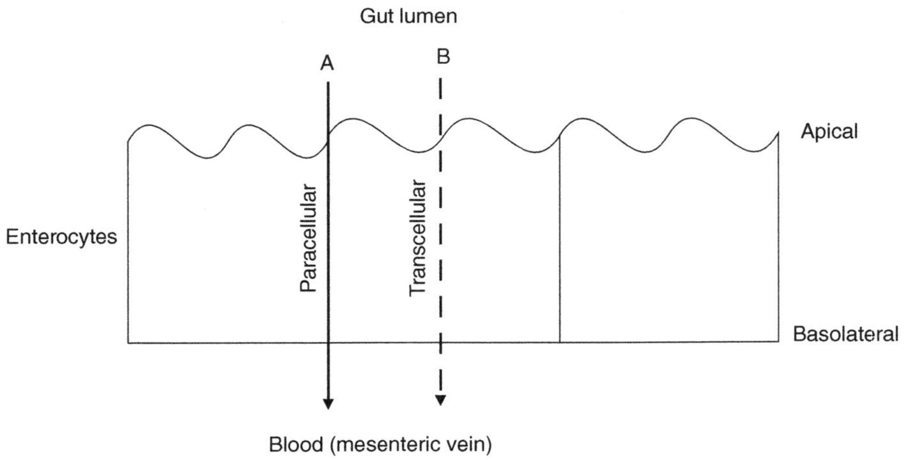


**Figure 5.1** Schematic illustrating the paracellular and transcellular pathways of intestinal drug absorption. (A) Paracellular absorption of drug that utilizes the tight junction between the cells to traverse across the intestinal epithelia. (B) Transcellular absorption that allows drugs to traverse across the intestinal epithelial via passive diffusion.

exemplified by cyclosporine's Neoral®. In addition to the drug characteristics, the anatomy and physiological state of the GI environment at the site of absorption such as the pH and intestinal transit/residence time regulated by variables such as the gastric emptying time and disease states (to alter the integrity of the GI) may also be sensitive parameters. However, unlike the paracellular route, there is a higher likelihood that drugs passing through the intestine by the transcellular route may interact with cellular proteins such as drug transporters and metabolizing enzymes. Therefore, for many drugs, transcellular pathway oftentimes entails both diffusion and active component(s).

### 5.2.1 Passive Diffusion Described by Fick's First Law

In order for molecules to move across a single bilayer to get into the enterocytes, they must undergo three sequential processes: partition from the aqueous phase into the lipid phase, diffusion across the lipid bilayer, and repartition into the cytosolic aqueous phase. These steps need to be repeated if the molecule is to exit the cells to reach the portal blood. The kinetics of passive diffusion for nonionized molecules is described by Fick's law of diffusion. Fick's first law is a quantitative description of diffusion through a given surface area. It states that the observed rate of diffusion is proportional to the concentration (chemical) gradient. In one dimension, the mathematical form of Fick's first law can be described as follows:

$$J = Aj = -AD\frac{dC}{dX} \tag{5.1}$$

where

| | | |
|---|---|---|
| $J$ | = | total flux or rate of passive diffusion for nonionized molecules (unit: amount/time) |
| $A$ | = | the area through which diffusion occurs (unit: square of length) |
| $j$ | = | total flux per unit area (unit: amount/[time·area]) |
| $D$ | = | diffusion coefficient or diffusivity (unit: area/time) |
| $C$ | = | concentration (unit: amount/volume) |
| $X$ | = | distance (unit: length) |
| $dC/dX$ | = | change in concentration (chemical) gradient (unit: amount/volume) |

The negative sign in the equation connotes that the direction of flux is opposite to the increase in the concentration of the diffusants (molecules that move from one location to another).

Fick's first law can be modified to add another dimension as illustrated in Figure 5.2. Here, $h$ is the thickness

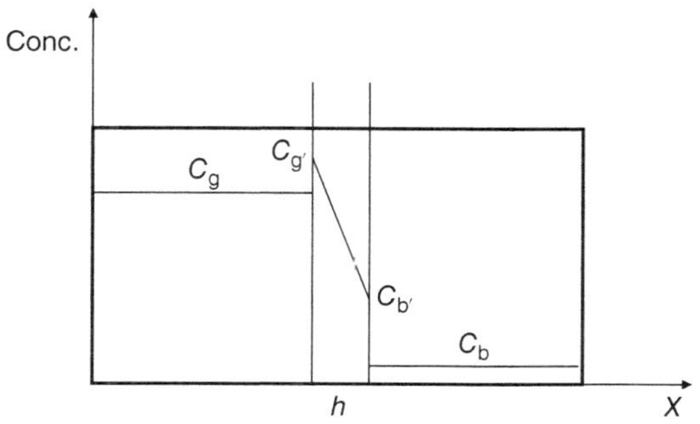


**Figure 5.2** A simplified model of diffusion across gastrointestinal membranes.

of the diffusional barrier, $C_g$ is the drug concentration in the gut lumen, $C_b$ is the drug concentration in the blood compartment, $C_{g'}$ is the drug concentration in the apical compartment (epithelial membrane facing the gut side), and $C_{b'}$ is the drug concentration in the basolateral compartment (epithelial membrane facing the blood side). The relationship between $C_g$ and $C_{g'}$ and $C_b$ and $C_{b'}$ are as follows [13]:

$$K = \frac{C_{g'}}{C_g} = \frac{C_{b'}}{C_b} \tag{5.2}$$

where $K$ = membrane: water partition coefficient (no unit; often estimated experimentally as the partition between octanol and water). Based on the relationships defined above, the rate of passive diffusion across GI membranes can now be described as follows:

$$j = \frac{D^*K}{h}(C_g - C_b) \tag{5.3}$$

Since the values of $D$, $K$, and $h$ are constant for a given molecule and diffusional barrier, Eq. (5.3) can be further simplified by defining a new parameter, permeability coefficient ($P$, unit: length/time) where:

$$P = \frac{D^*K}{h} \tag{5.4}$$

Substituting Eq. (5.4) into Eq. (5.3) yields,

$$j = P^*(C_g - C_b) \tag{5.5}$$

or

$$J = P^*A^*(C_g - C_b) \tag{5.6}$$

Usually, the drug concentration in the blood compartment, $C_b$, is extremely small compared to $C_g$, the drug concentration in the gut lumen.

Equation (5.5) then becomes,

$$j = P^*C_g \tag{5.7}$$

It is noteworthy to mention here that Fick's first law simplifies intestinal absorption by treating the

GI membranes and cytosol as one homogeneous or isotropic diffusional barrier [13]. In actuality, however, the site of absorption is complex, which can be segregated into multiple heterogeneous layers with varying properties [14, 15].

### 5.2.2 Rate-Limiting Step in Intestinal Drug Absorption

As described above, once a drug is given orally, it faces many obstacles before it is able to reach the systemic circulation. The precondition that must be met for intestinal drug absorption is not only that the drug is in solution within the specified formulation, but that the rate of dissolution is sufficient to allow the drug to become homogenous with the intestinal fluids as it arrives at the site of absorption. However, the rate of dissolution is not the only obstacle as the drug must also be able to cross the GI tract. As the drug travels through the mouth, esophagus, stomach, duodenum, jejunum, ileum, and colon, it has the potential to traverse these membranes in each of these places. If it is not able to cross these membranes by the time it reaches the colon, it will end up in the feces and the extent of intestinal absorption would not be complete. With each of these processes, there are rates associated with every step and accordingly, the slowest process will be the rate-limiting step that will determine the overall rate of intestinal drug absorption. Moreover, because the GI tract is finite, the GI transit time is limited, which will have an overall impact on the extent of absorption. Specifically, as defined by Fick's First Law, intestinal absorption of a drug may be *dissolution rate-limited, perfusion rate-limited, diffusion/permeability rate-limited*, or *gastric-emptying rate limited*.

#### 5.2.2.1 Gastric-Emptying Rate-Limited Absorption

Equation (5.6) indicates that the passive diffusion is directly proportional to the surface area but indirectly proportional to the thickness of the lipid bilayer. Indeed, the surface area of the small intestine is immense due to the presence of villi and microvilli (200 $m^2$ for small intestine versus 1 $m^2$ for stomach), while the small intestinal membrane is much thinner when compared to that of the stomach [16]. According to a recent study based on radiological techniques, the surface area of the small intestine was a much conservative 30 $m^2$, which was still significantly greater than the surface area of the stomach (0.5 $m^2$) [17]. These physiological parameters make the small intestine an ideal site for drug absorption and thus, the availability of intact drug in the intestine controlled by the rate at which a drug is being delivered to the small intestine (via the stomach) could be the rate-limiting step. This is known as the gastric-emptying limited absorption. Gastric emptying may be slowed by external factors such as intense exercise or food intake. Delay in gastric emptying may be beneficial or detrimental depending on the situation. For example, if a drug has poor permeability, delay in gastric emptying may be beneficial as it gives more time for the drug to become absorbed. However, if the drug is not stable in the acidic environment of the stomach such as with penicillin, a delay in the rate of gastric emptying may result in reduced intestinal drug absorption since less intact drug becomes available in the small intestine.

#### 5.2.2.2 Perfusion Rate-Limited Absorption

The rate at which the drug is carried away from the site of absorption once it reaches the systemic circulation to the active site will determine the drug concentration in the blood ($C_b$). GI tract is richly supplied with both lymphatic and blood vessels. Once drugs traverse the GI membranes, they can potentially enter either the lymphatic or the blood capillaries. Therefore, the rate of blood flow (or lymphatic flow for those molecules that are being carried away via the lymph system) to the intestine will affect $C_b$. The majority of intestinally absorbed drugs enter the portal blood because the rate of fluid flow in the portal blood is approximately 500-fold higher than that of the intestinal lymph [18]. However, if a drug has $\log P > 5$ and/or a long-chain triglyceride with solubility >50 mg/g, it is more likely to be delivered to the systemic circulation via the intestinal lymph rather than the portal blood. For drugs that possess high dissolution rate and can easily diffuse through membranes, intestinal perfusion will be the rate-limiting step for these drugs and it is called perfusion-limited absorption.

#### 5.2.2.3 Dissolution Rate-Limited Absorption

Equation (5.3) states that the larger the concentration gradient across the intestinal membranes ($C_g - C_b$), faster will be the rate of passive diffusion. Most orally administered drugs are not in solution but are given as solid dosage forms (tablet, capsule, or suspension) to be taken with water. As a result, before a drug can diffuse across the enterocytes, it needs to dissociate, disintegrate, and dissolve. The efficiency of these steps is dependent on various parameters including formulation, aqueous solubility, and dissolution rate, which in turn will affect drug concentration in the gut lumen. Formulation and aqueous solubility are constants associated with the physicochemical properties of the drug.

Dissolution rate can be described with Nernst–Brunner equation [16]:

$$\frac{dQ}{dt} = \frac{DS}{h}(C_h - C_g) \tag{5.8}$$

where:

- $Q$ = the amount of drug already in solution (unit: amount)
- $dQ/dt$ = the rate of increase of the amount of drug in solution dissolving from a solid (unit: amount/time)
- $D$ = diffusion coefficient of the drug molecule (unit: area/time)
- $S$ = effective surface area of the drug particles (unit: area)
- $h$ = effective diffusion boundary layer thickness adjacent to the dissolving surface (unit: length)
- $C_h$ = saturation solubility of the drug in the boundary layer at any particular pH (unit: amount/volume)
- $C_g$ = concentration of drug in the bulk fluids (unit: amount/volume)

For drugs that can permeate the intestinal membrane easily, absorption has been observed to be dissolution-limited for drugs that exhibit low aqueous solubility and/or poor dissolution rate. An example of a drug exhibiting dissolution-limited absorption is griseofulvin [19]. Under fasting conditions, the extent of griseofulvin intestinal absorption was poor. However, when griseofulvin was taken alongside a high-fat meal, its absorption increased. The most likely reason for the change has been attributed to the content of the high-fat meal that resulted in an increase in the rate of dissolution.

#### 5.2.2.4 Diffusion/Permeability Rate-Limited Absorption

Diffusion coefficient and partition coefficient are two variables that affect drug membrane permeability as described by Eq. (5.4). Diffusion coefficient or diffusivity (represented by $D$) represents the mobility of drug molecules in a given diffusional barrier [13]. According to Eq. (5.3), it is directly proportional to the rate of passive diffusion. Its value is dependent on several parameters including the diffusional barrier, temperature, and the physicochemical properties of the drug. Stokes–Einstein equation quantitatively captures this relationship as follows [13]:

$$D = \frac{kT}{6\pi\eta r} \tag{5.9}$$

where,

- $D$ = diffusion coefficient in a given diffusional barrier (unit: area/time)
- $k$ = Boltzmann constant ($1.3\,806\,503 \times 10^{-23}$ $m^2$ $kg/s^2$ K)
- $T$ = absolute temperature (unit: degree)
- $\pi$ = pi (no unit)
- $\eta$ = viscosity (unit: amount/[time·length])
- $r$ = radius of the diffusant (unit: length)

As defined by the Stokes–Einstein equation, diffusion coefficient is inversely proportional to the radius of the diffusant. Hence, for a given diffusional barrier existing at a particular temperature, the diffusion coefficient becomes smaller as the size of the molecules increases and vice-versa as the size of the molecules decreases. Since the smaller value of the diffusion coefficient denotes slower rate of passive diffusion, these equations suggest that the size of the molecules may limit passive diffusion.

In addition to the size, the lipophilicity of the molecule may also affect intestinal absorption. Partition coefficient is defined as the ratio of drug concentration in the membrane compared to water. Equations (5.3) and (5.4) show that the partition coefficient is directly proportional to the rate of passive diffusion and is, therefore, an important descriptor that can govern drug permeability across a biological membrane. As the plasma membrane is made up of phospholipids, a lipophilic molecule will have a higher partition coefficient value compared to a hydrophilic molecule, suggesting that the lipophilic molecule will have a higher rate of passive diffusion than the hydrophilic molecule (with everything else being equal). As the membrane-water partition coefficient is difficult to obtain, an organic solvent like *n*-octanol, which has been suggested to mimic the lipophilicity of the cell membrane, is used in place of the membrane phase [20].

Another factor to consider in intestinal absorption is the charge state of the drug molecules. For weakly acidic or basic compounds, their $pK_a$ and the pH of the GI tract will determine the extent of deprotonation and protonation, respectively. Because the pH varies greatly along the GI tract (from about 1–3 in the stomach to 7–8 in the ileum), the extent of ionization will vary depending on the region of GI tract. The Henderson–Hasselbalch relationship shown in Eq. (5.10) and Eq. (5.11) equates the degree of ionization to pH and $pK_a$. For a weak acid, it predicts that the extent of deprotonation will be higher as the pH increases; whereas for a weak base, protonation will be higher as the pH decreases. As a result, because it is more difficult for charged compounds to pass through the lipid bilayer compared to neutral compounds, intestinal absorption for weakly acidic compounds will be lower at the higher pH (small intestines) compared to the lower pH (stomach), while the opposite is true for weakly basic compounds.

$$\mathrm{pH} = \mathrm{p}K_\mathrm{a} + \log\frac{[\mathrm{A}^-]}{[\mathrm{HA}]} \tag{5.10}$$

$$\mathrm{pH} = \mathrm{p}K_\mathrm{a} + \log\frac{[\mathrm{B}]}{[\mathrm{BH}^+]} \tag{5.11}$$

Drug lipophilicity is generally expressed quantitatively as partitioning of a neutral drug species between *n*-octanol and water ($\log P$). However, $\log P$ does not encompass the extent of ionization of ionizable molecules. As 95% of all drugs have ionizable groups, distribution coefficient ($\log D$) that considers the extent of ionization as well as intrinsic lipophilicity, has been demonstrated to be a better descriptor for oral drug permeability [21]. In addition, other studies have demonstrated better correlation of Caco-2 permeability with $\log D$ in comparison to $\log P$ [22]. The correlation between these lipophilicity descriptors ($\log P$ versus $\log D$) and permeability has also been explored in other types of biological membranes. For example, $\log D$ at salivary pH 6.8 demonstrated better correlation with permeability across the buccal mucosa compared to $\log P$ [23].

Although according to the classical pH-partition hypothesis, only the neutral species of an ionizable drug can permeate across a lipophilic membrane barrier, deviations from the pH-partition hypothesis have been reported, possibly due to a substantial permeation of the charged species of the drug through the membrane [24].

Assuming that both the ionized and unionized species contribute independently to the total steady-state flux of an ionizable drug at a given pH, the total drug flux across biological membranes can be described by expanding Eq. (5.7) [25, 26]:

$$j_\mathrm{T} = j_\mathrm{u} + j_\mathrm{i} = P_\mathrm{u}{}^*C_\mathrm{u} + P_\mathrm{i}{}^*C_\mathrm{i} \tag{5.12}$$

where $j$, $P$, and $C$ stand for flux, permeability, and concentration, respectively; the subscripts T, u, and i refer to total, unionized, and ionized species of a drug.

Upon dividing $j_\mathrm{T}$, $P_\mathrm{u}{}^*C_\mathrm{u}$, and $P_\mathrm{i}{}^*C_\mathrm{i}$ with the total drug concentration ($C_\mathrm{T}$), Eq. (5.12) can be modified to describe the total permeability ($P_\mathrm{T}$) as:

$$P_\mathrm{T} = P_\mathrm{u}{}^*f_\mathrm{u} + P_\mathrm{i}{}^*f_\mathrm{i} = P_\mathrm{i} + f_\mathrm{u}{}^*(P_\mathrm{u} - P_\mathrm{i}) \tag{5.13}$$

$f_\mathrm{u}$ and $f_\mathrm{i}$ refer to fraction of drug unionized ($C_\mathrm{u}/C_\mathrm{T}$) and ionized ($C_\mathrm{i}/C_\mathrm{T}$), respectively. The concentrations of the unionized and ionized species ($C_\mathrm{u}$ and $C_\mathrm{i}$) depend on pH and can be calculated using the Henderson–Hasselbalch equation.

The readers are also referred to a novel modeling approach by Pade et al. to predict the passive intestinal permeability in different regions of the intestinal tract that considers the luminal pH, bile salt concentrations, surface area for drug permeation (based on intestinal dimensions, and villous morphology), and the unstirred mucus layer thickness and its microenvironment pH, in addition to physicochemical properties of the drug (octanol–water partition coefficient, molecular weight, $\mathrm{p}K_\mathrm{a}$) [27].

We have discussed many parameters that can affect intestinal drug absorption by passive diffusion. Among them, intestinal drug absorption is recognized to be dependent on two major factors: solubility and intestinal permeability. In fact, a majority of the drugs were found to undergo either dissolution or diffusion/permeability rate-limited absorption [28]. Ideally, a drug molecule needs to possess certain properties such that it has the right balance of solubility and intestinal permeability. That is, in order to be able to reach the systemic circulation from the GI lumen, a drug cannot be too lipid-soluble that it will not dissolve in the aqueous fluids of the GI tract, nor be so water-soluble that it cannot permeate the lipid bilayer of the GI membranes.

Lipinski et al. [29] attempted to characterize the intestinal absorption potential of compounds with the "rule of 5" (Ro5). According to the rule, poor intestinal absorption or permeation is most likely when a particular compound meets at least two of the outlined parameters:

1. More than five hydrogen bond donors (OH + NH count).
2. Molecular weight greater than 500.
3. $c\mathrm{Log}\,P > 5$ or $M\log P > 4.15$ and,
4. More than 10 hydrogen bond acceptors (O + N atom count).
5. Compound classes that are substrates for transporters are exceptions to the rule

Ro5 was developed as a quick screen to eliminate compounds that lack intestinal absorption potential. However, it is important to keep in mind that for compounds that pass the Ro5 screen, the extent of systemic absorption (bioavailability) could still be low for those compounds due to extensive first-pass metabolisms in the gut and liver (and perhaps lung). Furthermore, Lipinski's Ro5 is not applicable when the permeability of a compound is carrier-mediated.

Lately, certain potential drug targets have been reported to be "difficult-to-drug" with conventional Ro5-compliant compounds, and this has resulted in

**Table 5.1** Various forms and the characteristics of solute movement in biological matrix [35].

| Type of solute movement | Protein? | Saturable? | Against concentration gradient? | Energy dependent? |
|---|---|---|---|---|
| Passive diffusion | N | N | N | N |
| Facilitated diffusion-channel protein | Y | N | N | N |
| Facilitated diffusion-carrier-protein | Y | Y | N | N |
| Active-primary | Y | Y | Y | Y |
| Active-secondary | Y | Y | Y | Y |

Source: Adapted from Amidon and Sadâee [35].

substantial interest in beyond rule of 5 (bRo5) molecules that have demonstrated the potential to modulate such drug targets [30]. Certain bRo5 compounds have been reported to exhibit significant absorption after oral administration. These include tetracycline, which violates one rule of the Ro5, and rifampicin, which violates three of the rules. Both compounds resulted in high passive lipoidal diffusion in liposomal permeation assay regardless of their unfavorable properties [31]. According to the authors, the possible reasons for some bRo5 compounds to be absorbed by passive lipoidal diffusion include the ability to neutralize hydrogen bond donors and acceptors in the apolar membrane environment by intramolecular hydrogen bonding that can mask polar structures and thus increase membrane transport, the adoption of an elongated shape so that the molecule can penetrate the sterically hindered lipid bilayer, and the presence of a significant fraction of neutral species.

Certain cyclic peptides are known to violate several Ro5 criteria and still exhibit appreciable oral bioavailability [32]. Cyclosporin A is an example of a cyclic peptide that undergoes passive diffusion and demonstrates reasonable oral bioavailability. The unusual passive membrane permeability of Cyclosporin A has been attributed to (i) a combination of N-methylation that results in a reduction in the number of NH protons (HBD) and (ii) a flexible structure that permits conformational transition depending on the solvent, i.e. an open conformation under aqueous conditions that permits interaction with the solvent by hydrogen bonding versus a closed conformation leading to intramolecular hydrogen bonding and a lowering of the polar surface area in the lipid bilayer [33].

## 5.3 Carrier-Mediated Transport

In addition to passive diffusion, drugs can also traverse membranes via carrier-mediated pathways. Carrier-mediated transport involves a direct, physical interaction between the drug and proteins that are embedded within the lipid bilayer. Carrier-mediated drug flux can either be a passive or an active process. Facilitated diffusion is the term used to describe a passive carrier-mediated transport. As in passive diffusion, its driving force is the electrochemical potential gradient across the membranes. In contrast, active carrier-mediated transport requires energy as the driving force for transport. The source of energy can come from various sources such as light, oxidation reactions, ATP hydrolysis (or phospho-enolpyruvate), or cotransport with other ions [34]. Based on the source of energy, active carrier-mediated transport can be further categorized as *primary* or *secondary* active transport. The source of energy for primary active transport is usually attained via ATP hydrolysis. For secondary active transport, the energy source is oftentimes harnessed by ion gradients (e.g. $Na^+$, $H^+$) generated through ATP hydrolysis or voltage gradient. Table 5.1 summarizes properties that are characteristic of various transport processes.

Proteins involved in the carrier-mediated pathway can be classified as either channels or drug transporters. Channels are aqueous-filled pores and/or tunnels formed by the arrangement of their transmembrane domain within the lipid bilayer. Also known as ion channels, they often exist as an assembly of multiple proteins. Because the composition of the transmembrane domains is unique depending on the type of channels, it may exhibit certain selectivity toward compounds that have specific physicochemical properties. The most common type of channels is sodium and potassium channels that move metal ions; however, they can recognize and move a variety of nonmetal substrates. For example, the pore-forming antifungal agent alamethicin has been shown to move multiple substrates for UDP-glucuronosyltransferases in microsomal vesicles [36, 37].

While channel proteins partake only in facilitated diffusion, drug transporters can be either facilitated or active. The mechanism of transport can be characterized by four thermodynamic cycles as illustrated

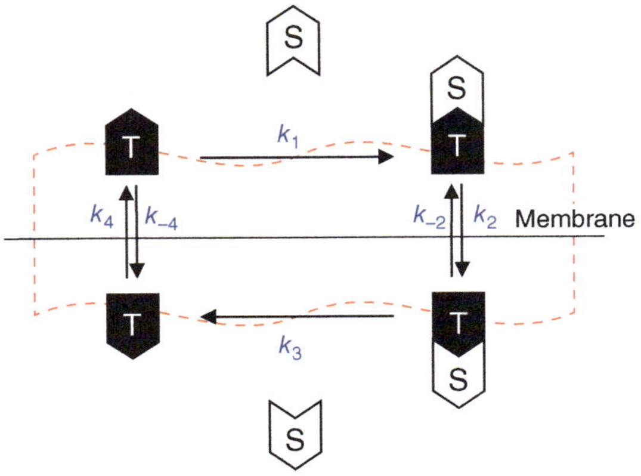


**Figure 5.3** Schematic diagram of drug transport across a membrane. *k*'s are kinetic constants. T is Transporter and S is Solute.

in Figure 5.3. In the first cycle, the molecule binds to the drug transporter ($k_1$). It has been demonstrated that while the substrate binding site of a transporter is accessible on only one side of the membrane at any given time, substrates are able to access both sides of the membrane for ion channels [34]. A transporter may recognize only one substrate or many different substrates. For example, P-glycoprotein (P-gp) is a well-known efflux transporter, which can transport a diverse type of compounds [38]. A substrate can also be transported by more than one transporter. For example, fexofenadine is a substrate of both P-glycoprotein and organic anion-transporting protein, an uptake transporter (OATP) [39]. In the second cycle, the drug transporter moves the molecule across the membrane barrier ($k_2/k_{-2}$). In the third cycle, the drug transporter unloads its payload ($k_3$). In the final cycle, the drug transporter resets to its original conformation where it is ready to accept a new substrate ($k_4/k_{-4}$). It is important to note that only unbound drug would be available to bind to the transporters. In this manner, drug transporters behave like enzymes. Consequently, their kinetics can be defined by $J_{max}$ and $K_t$ [35]. According to Michaelis–Menten kinetics, rate of transport is defined as:

$$\text{rate of transport} = \frac{J_{max} \cdot C}{K_t + C} \tag{5.14}$$

where,

$J_{max}$ = maximum rate of transport (unit: amount/time)
$K_t$ = Michaelis constant (unit: amount/volume)
$C$ = concentration of the substrate (unit: amount/volume)

$J_{max}$ refers to the capacity of the transporter to move a particular substrate. It depends on the amount of the transporter expressed in the membrane and how quickly it translocates a substrate (turnover rate). $K_t$ represents the concentration of substrate at which the rate of drug transport is half of $J_{max}$. Specifically, it refers to the affinity of the transporter for a particular substrate. According to the above relationship, when $C \ll K_t$, the rate of transport increases proportionately to increases in the substrate concentration. However, when $C > K_t$, the rate of transport approaches a constant value that is equal to $J_{max}$.

Generally, because there are multiple steps required for transporters to shuttle substrates across the lipid bilayer, its rate of transport is usually slower than rates observed with ion channels. In fact, the rate of transport through channels has been shown to be several orders of magnitude larger than with drug transporters [40]. In addition, while the substrate must directly interact with transporters, the gating mechanism of ion channels does not require physical interaction with the substrate. As a result, the rate of transport through ion channels generally does not show concentration dependence while transporters demonstrate saturable kinetics. Instead, the relationship between concentration and rate of transport for ion channels resembles passive diffusion. Assuming a sink condition, where drug concentration in the blood is negligible (the drug is removed from the site of absorption as soon as it is absorbed), the rate of transport by passive diffusion is linearly correlated to drug concentration in the gut lumen. That is, the rate of transport is doubled if gut lumen drug concentration is doubled. In contrast, the rate for transporters increases linearly with increasing substrate concentrations in the gut lumen until the concentration exceeds a certain threshold value where all the available binding sites are occupied and the rate of transport is equal to $J_{max}$ [35]. Figure 5.4 compares the rate of transport between passive diffusion and transporter-mediated transport.

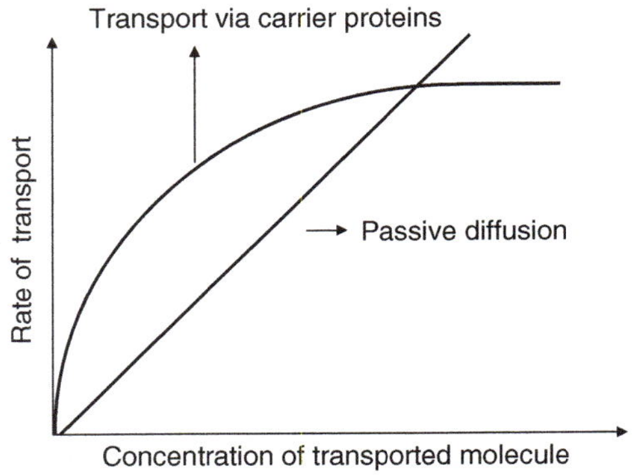


**Figure 5.4** Kinetics of passive diffusion compared to carrier-protein mediated transport.

Depending on the mechanism of transport, drug transporters can be further classified as *uniporters*, *symporters*, and *antiporters*. As the name implies, uniporters bind and transport only one type of substrate at a time. There are both passive and active uniporters such as glucose transporters (GLUTs) [41] and equilibrative nucleoside transporters (ENTs) [42], which are driven by electrochemical gradients; whereas P-gp, breast cancer resistance protein (BCRP), multidrug resistance-associated proteins (MRPs) and $Na^+K^+$-ATPase, which are driven by ATP. In contrast, symporters and antiporters are active transporters, which can move more than one type of substrates; usually a drug molecule along with a metal ion. Specifically, symporters (or cotransporters) transport both ions and substrates simultaneously in the same direction while antiporters (or countertransporters) simultaneously transport ion in one direction and substrate in the opposite direction. Because the driving force for symporters and antiporters are voltage or ion gradient (usually $Na^+$ or $H^+$), they are also called ion-coupled solute transporters. Furthermore, since these transporters rely on ion gradients generated by ATP hydrolysis, they are also known as secondary active transporter. Some examples of symporters are $Na^+$-glucose cotransporter (SGLT1) [43] and $H^+$-dipeptide cotransporter (PEPT1) [44]. Examples of antiporters are $Na^+/H^+$ antiporter and $Na^+/Ca^{2+}$ antiporter.

The role of transporters varies depending on their function and localization within the gut lumen. They can be classified as either *efflux* or *uptake* (or *influx*) depending on the direction of transport [44]. An efflux transporter shuttles compounds out of the cells, usually against the concentration gradient, a process known as secretion. In contrast, an uptake transporter is absorptive as it directs compounds into the cells, usually along the concentration gradient. The epithelial lining of the gut lumen that houses these transporters are polarized, where the "apical" side faces the intestinal lumen while the "basolateral" side faces the portal blood. The polarized enterocytes asymmetrically express a variety of drug transporters and because of the presence of tight junctions, which prevent the mixing of proteins on the apical and basolateral sides, asymmetry is maintained. Consequently, the asymmetric expression results in vectorial drug transport [44]. Figure 5.5 illustrates select drug transporters that are localized to the apical and basolateral membranes of the human intestine. The actual number of transporters present on intestinal membrane is much larger than presented in Figure 5.5. For simplicity, we only list a few examples to illustrate the point. Here, the function of uptake transporters at the apical membrane such as PEPT1 will be absorptive while that of an efflux transporter such as P-gp will be secretory.

### 5.3.1 Competitive, Noncompetitive, and Un-competitive Inhibition

Many transporters can recognize more than one substrate and hence, different drugs can compete for

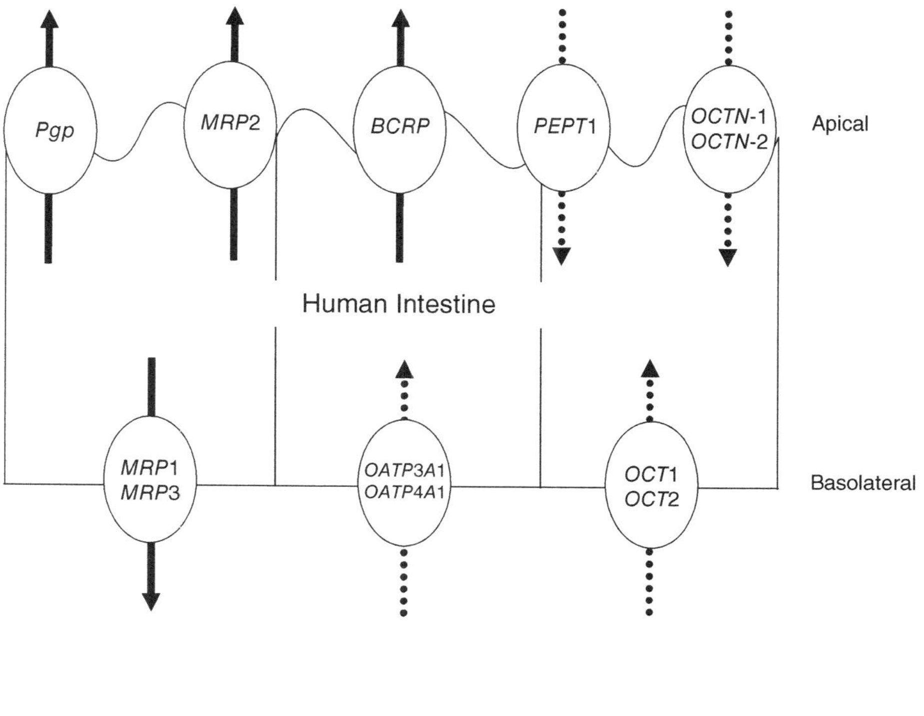


**Figure 5.5** Schematic showing select drug transporters expressed in the human intestine.

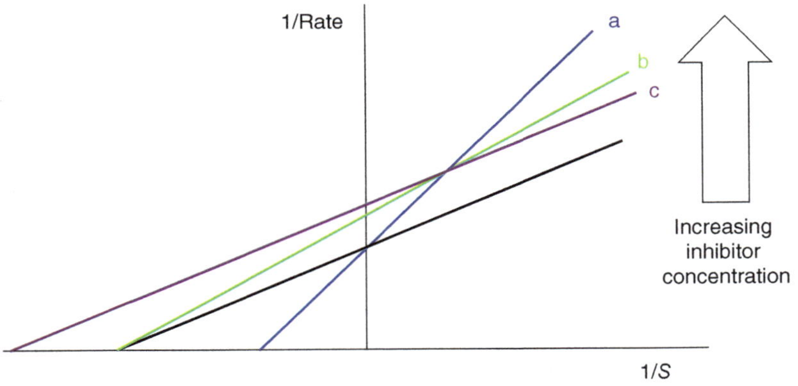


**Figure 5.6** Schematic of inverse rate versus inverse substrate concentration plot. Black line = without inhibitor; line a = with a competitive inhibitor; line b = with a noncompetitive inhibitor, and line c = with an uncompetitive inhibition. As the concentration of competitive and noncompetitive inhibitors increases, the slopes of lines a and b will increase accordingly. As the concentration of un-competitive inhibitor increases, the *y*-intercept increases without affecting the slope.

the same binding site(s). As a result, the function of transporters can be inhibited. There are three types of inhibition: competitive, noncompetitive, and un-competitive [45]. Competitive inhibition occurs when both the substrate and the inhibitor compete for the same binding site. The rate of transport of a drug in the presence of a competitive inhibitor is given by a modification of Eq. (5.14):

$$J = \frac{J_{max} C}{K_t \left(1 + \frac{I}{K_i}\right) + C} \tag{5.15}$$

where,

$I$ = inhibitor concentration (unit: amount/volume)
$K_i$ = Inhibition constant (unit: amount/volume)

Noncompetitive inhibition occurs when the inhibitor does not bind to the transporter active site but to an allosteric site. Although the inhibitor does not occupy the transporter active binding sites, binding to an allosteric site lowers the affinity of the transporter for the substrate by inducing conformation changes within the tertiary structure of the transporter. The rate of transport of a drug in the presence of a noncompetitive inhibitor is defined by Eq. (5.16):

$$J = \frac{J_{max} C}{(K_t + C)\left(1 + \frac{I}{K_i}\right)} \tag{5.16}$$

Finally, un-competitive inhibition occurs when the inhibitor binds to the intermediate of the substrate-transporter complex to terminate the translocation step. The rate of transport of a drug in the presence of an uncompetitive inhibitor is defined by Eq. (5.17):

$$J = \frac{J_{max} C}{\left(\frac{K_t}{\left(1+\frac{I}{K_i}\right)} + C\right)\left(1 + \frac{I}{K_i}\right)} \tag{5.17}$$

The different types of inhibition can be experimentally differentiated by investigating the kinetics across various substrate and inhibitor concentrations [46]. By plotting the inverse rate against the inverse substrate concentration with increasing inhibitor concentrations, inhibition can be classified as competitive, noncompetitive, or uncompetitive if the lines at increasing inhibitor concentration intersect at the *y*-axis, intersect at the *x*-axis, or if the lines do not intersect, respectively (Figure 5.6).

As shown above, drug transporters are selective, saturable, and inhibitable. In addition, many transporters demonstrate genetic polymorphisms, which may affect their function. Because of these properties, intestinal drug absorption via carrier-mediated systems may show dose-dependent absorption and could have potential drug–drug interactions during coadministration with various medications. Although much emphasis has been given to drug interactions occurring in the liver, intestinal drug interaction may also occur. For example, both the AUC and $C_{max}$ for fexofenadine, a substrate for P-gp and OATP, decreased threefold when administered with grapefruit juice, an inhibitor of OATP [47]. Moreover, drug interactions can also be site-specific since some transporters are expressed at specific regions of the intestine, such as the bile acid transporters in the ileum [48].

Transporters were once thought to play an insignificant role in intestinal drug absorption because it was thought that they were few in number and that they affected only few drugs [49]. However, as our knowledge and understanding grow, it is increasingly being recognized that drug transporters play a significant role in the intestinal absorption of many drugs. It has been suggested that drugs with a lower lipophilicity (e.g. log $D < 0$), a relatively high polar surface area/hydrogen-bonding capacity (e.g. PSA > 75 $Å^2$), and an affinity for transporter(s) have the potential to experience a significant contribution to the overall permeability by carrier-mediated transport. Conversely, in case of drugs with a positive log $D$ (e.g. log $D > 0$)

**Table 5.2** SLC transporter family members with known expression in GI tract.

| Human gene name | Protein name | Aliases |
|---|---|---|
| **SLC1: The high-affinity glutamate and neutral amino acid transporter family** | | |
| SLC1A1 | EAAC1, EAAT3 | System $X\text{-}_{AG}$ |
| SLC1A4 | ASCT1, SATT | System ASC |
| SLC1A5 | ASCT2, AAAT | System ASC |
| **SLC2: The facilitative GLUT transporter family** | | |
| SLC2A2 | GLUT2 | |
| SLC2A5 | GLUT5 | |
| SLC2A9 | GLUT9 | GLUTX |
| **SLC3: The heavy subunits of the heteromeric amino acid transporters** | | |
| SLC3A1 | rBAT | NBAT, D2 |
| SLC3A2 | 4F2hc | CD98hc |
| **SLC4: The bicarbonate transporter family** | | |
| SLC4A2 | AE2 | |
| SLC4A3 | AE3 | |
| SLC4A4 | NBCe1 | NBC, NBC1 |
| SLC4A7 | NBCn1 | NBC2, NBC3 |
| **SLC5: The sodium–glucose cotransporter family** | | |
| SLC5A1 | SGLT1 | |
| SLC5A4 | SGLT3 | SAAT1 |
| SLC5A8 | SMCT1 | AIT |
| SLC5A9 | SGLT4 | |
| SLC5A11 | SGLT6 | KSTRabbit, ST1RabbitSMIT2 |
| SLC5A12 | SMCT2 | |
| **SLC6: The sodium-and chloride-dependent neurotransmitter transporter family** | | |
| SLC6A8 | CT1 | CRTR |
| SLC6A9 | GLYT1 | |
| SLC6A14 | ATB(0+) | |
| SLC6A19 | $B^0AT1$ | HND |
| SLC6A20 | SIT1 | XT3, Xtrp3 |
| **SLC7: The cationic amino acid transporter/glycoprotein-associated family** | | |
| SLC7A1 | CAT-1 | ATRC1 |
| SLC7A5 | LAT1 | [assoc. with 4F2hc] |
| SLC7A6 | $y^+LAT2$ | [assoc. with 4F2hc] |
| SLC7A7 | $y^+LAT1$ | [assoc. with 4F2hc] |
| SLC7A8 | LAT2 | [assoc. with 4F2hc] |
| SLC7A91 | $B^{0,+}AT$ | [assoc. with rBAT] |
| SLC7A10 | Asc-1 | [assoc. with 4F2hc] |
| **SLC8: The $Na^+/Ca^{2+}$ exchanger family** | | |
| SLC8A1 | NCX1 | NACA, NCE |

*(continued)*

**Table 5.2** (Continued)

| Human gene name | Protein name | Aliases |
|---|---|---|
| **SLC9: The $Na^+/H^+$ exchanger family** | | |
| SLC9A1 | NHE1 | APNH |
| SLC9A2 | NHE2 | |
| SLC9A3 | NHE3 | |
| SLC9A4 | NHE4 | |
| SLC9A6 | NHE6 | KIAA0267 |
| SLC9A7 | NHE7 | |
| SLC9A8 | NHE8 | KIAA0939 |
| **SLC10: The sodium bile salt cotransport family** | | |
| SLC10A2 | ASBT | IBAT, ISBT |
| SLC10A3 | P3 | |
| SLC10A6 | P6 | |
| **SLC11: The proton-coupled metal ion transporter family** | | |
| SLC11A2 | DMT1 | NRAMP2, DCT |
| **SLC12: The electroneutral cation-Cl cotransporter family** | | |
| SLC12A2 | NKCC1 | BSC2 |
| SLC12A4 | KCC1 | |
| SLC12A6 | KCC3 | |
| SLC12A7 | KCC4 | |
| SLC12A8 | CCC9 | |
| SLC12A9 | CIP | |
| **SLC13: The human $Na^+$-sulfate/carboxylate cotransporter family** | | |
| SLC13A2 | NaC1 | |
| **SLC15: The proton oligopeptide cotransporter family** | | |
| SLC15A1 | PEPT1 | Oligopeptide transporter 1, $H^+$/peptide transporter1 |
| **SLC16: The monocarboxylate transporter family** | | |
| SLC16A1 | MCT1 | |
| SLC16A2 | MCT8 | XPCT (was MCT7) |
| SLC16A10 | TAT1 | |
| **SLC17: The vesicular glutamate transporter family** | | |
| SLC17A3 | NPT4 | |
| SLC17A4 | $Na^+/PO_4^-$ cotransporter homolog | |
| **SLC18: The vesicular amine transporter family** | | |
| SLC18A2 | VMAT2 | SVAT, SVMT, VAT2 (MAT) |
| SLC18A3 | VAChT | |
| **SLC19: The folate/thiamine transporter family** | | |
| SLC19A1 | RFT | RFC, FOLT |
| SLC19A2 | ThTr1 | ThT1 |
| SLC19A3 | ThTr2 | |
| **SLC20: The type III $Na^+$-phosphate cotransporter family** | | |
| SLC20A1 | PiT-1 | |
| SLC20A2 | PiT-2 | Amphotropic murine leukemia virus receptor, Gibbon ape leukemia virus receptor 2, GLVR2, Glvr-2, MLVAR |

*(continued)*

Table 5.2 (Continued)

| Human gene name | Protein name | Aliases |
|---|---|---|
| **SLC21/SLCO: The organic anion-transporting family** | | |
| SLCO2A1 | OATP2A1 | SLC21A2 |
| SLCO3A1 | OATP3A1 | SLC21A11 |
| SLCO4A1 | OATP4A1 | SLC21A12 |
| **SLC22: The organic cation/anion/zwitterion transporter family** | | |
| SLC22A5 | hOCTN2 | CT1 |
| SLC22A13 | OCTL1 | hORCTL3 |
| SLC22A14 | OCTL2 | hORCTL4 |
| **SLC23: The $Na^+$-dependent ascorbic acid transporter family** | | |
| SLC23A1 | SVCT1 | YSPL3, NCBT2 |
| SLC23A2 | SVCT2 | YSPL2, KIAA0238, NBTL1, NCBT1 |
| **SLC24: The $Na^+/(Ca^{2+}$-$K^+)$-exchanger family** | | |
| SLC24A3 | NCKX3 | |
| **SLC25: The mitochondrial carrier family** | | |
| SLC25A2 | ORC2 (ornithine carrier 2) | ORNT2 |
| SLC25A14 | UCP5 (uncoupling protein 5) | BMCP1, brain mitochondrial carrier protein |
| SLC25A15 | ORC1 (ornithine carrier 1) | ORNT1 |
| SLC25A22 | GC1 (glutamate carrier 1) | |
| SLC25A23 | APC2 | ScaMC-2 |
| SLC25A25 | APC3 | ScaMC-3 |
| SLC25A26 | SAMC | |
| **SLC26: The multifunctional anion exchanger family** | | |
| SLC26A2 | | DTDST |
| SLC26A3 | | DRA, CLD |
| SLC26A6 | | CFEX, PAT-1 |
| SLC26A11 | | |
| **SLC27: The fatty acid transport protein family** | | |
| SLC27A4 | FATP4 | |
| **SLC28: The $Na^+$-coupled nucleoside transport family** | | |
| SLC28A1 | CNT1 | |
| SLC28A2 | CNT2 | SPNT |
| SLC28A3 | CNT3 | |
| **SLC29: The fatty acid transport protein family** | | |
| SLC29A1 | ENT1 | |
| SLC29A2 | ENT2 | |
| SLC29A3 | ENT3 | |
| SLC29A4 | ENT4 | |
| **SLC30: The zinc efflux family** | | |
| SLC30A1 | ZNT1 | |
| SLC30A2 | ZNT2 | |
| SLC30A4 | ZNT4 | Dri27 |
| SLC30A5 | ZNT5 | ZTL1 |
| SLC30A6 | ZNT6 | |
| SLC30A7 | ZNT7 | ZNTL2 |

(continued)

**Table 5.2** (Continued)

| Human gene name | Protein name | Aliases |
|---|---|---|
| **SLC31: The copper transporter family** | | |
| SLC31A1 | hCtr1 | hCtr1, Ctr1 |
| SLC31A2 | hCtr2 | hCtr2, Ctr2 |
| **SLC34: The type II $Na^+$-phosphate cotransporter family** | | |
| SLC34A2 | NaPi-IIb | |
| **SLC36: The proton-coupled amino acid transporter family** | | |
| SLC36A1 | LYAAT1 | |
| **SLC37: The sugar-phosphate/phosphate exchanger family** | | |
| SLC37A1 | SPX1 | G3PP |
| SLC37A3 | SPX3 | |
| **SLC38: The system A&N, sodium-coupled neutral amino acid transporter family** | | |
| SLC38A2 | SNAT2 | ATA2, KIAA1382, SA1, SAT2 |
| SLC38A5 | SNAT5 | JM24, SN2 |
| **SLC39: The metal ion transporter family** | | |
| SLC39A1 | ZIP1 | ZIRTL |
| SLC39A3 | ZIP3 | |
| SLC39A4 | ZIP4 | |
| SLC39A5 | ZIP5 | |
| SLC39A6 | LIV-1 | |
| SLC39A7 | KE4 | |
| SLC39A8 | BIGM103 | |
| SLC39A9 | | |
| SLC39A10 | | |
| SLC39A11 | | |
| SLC39A13 | | |
| SLC39A14 | ZIP14 | |
| **SLC40: The basolateral iron transporter family** | | |
| SLC40A1 | IREG1 | Ferroportin, MTP1 |
| **SLC43: The $Na^+$-independent, system L-like amino acid transporter family** | | |
| SLC43A2 | LAT4 | |
| **SLC46: Heme transporter family** | | |
| SLC46A1 | HCP1 | MGC9564 |

Source: Adapted from SLC tables.

and low polar surface area/hydrogen-bonding capacity (e.g. PSA < 75 $A^2$), diffusion through the passive lipoidal membrane might contribute predominantly to the overall permeability, irrespective of affinity for a transporter(s), owing to the large surface area of the lipoidal membrane [50].

There are many different transporters in the human intestine. Most drug transporters belong to two major superfamilies: ABC (ATP binding cassette) and SLC (solute carrier) superfamilies. Many transporters in ABC superfamily are primary active efflux transporters while many transporters in the SLC superfamily are uptake transporters that can be passive (uniporters), coupled (symporters), or exchangers (antiporters).

Intestinal ABC transporters such as ABCB1 (P-glycoprotein, P-gp), ABCG2 (breast cancer resistance protein, BCRP), and ABCC2 (multidrug resistance-associated protein, MRP2) that are expressed in the apical (luminal) membrane of the enterocytes can limit the intestinal absorption of many drugs by effluxing the compound back into gut lumen [51].

At present, SLC superfamily is divided into 65 subfamilies and new transporter genes continue to be discovered and added to the SLC superfamily ([52] and slc.bioparadigms.org). Members of the subfamilies with known expression in intestines are listed in Table 5.2. The substrate specificity of these transporters may not be as diverse as the ABC transporters, but together, they are able to transport a wide range of compounds including xenobiotics and endogenous compounds such as vitamins, sugars, and amino acids [53]. SLC transporters that have been reported to participate in carrier-mediated transport in the intestine include oligopeptide transporter (PEPT1), organic anion polypeptide transporters (OATP1A2 and OATP2B1), monocarboxylate transporters (MCT1), carnitine/organic cation transporters (OCTN1 and OCTN2) and the plasma membrane monoamine transporter (PMAT) [51]. One of the more important and most studied uptake transporters is PEPT1. Also known as $H^+$/peptide transporter 1, it is a member of the SLC15 family (SLC15A1). PEPT1 is a high-capacity transporter with a broad substrate specificity that is highly expressed on the apical membrane of enterocytes. It is a symporter whose transport of di- and tripeptides is coupled to the transport of protons. In addition to di- and tripeptides as substrates, a variety of xenobiotics are recognized by PEPT1 including amoxicillin, ampicillin, bestatin, cefaclor, cefalexin, cefixime, cefadroxil, enalapril, midodrine, temocapril, temocaprilate, valganciclovir, and valacyclovir [54].

The ability of PEPT1 to uptake drugs into enterocytes has been used to improve the extent of absorption of drugs with poor intestinal permeability. Intestinal PEPT1 has evolved as an attractive target for prodrugs owing to its high capacity and multispecificity [55]. For example, acyclovir is a polar molecule with low extent of systemic absorption resulting from poor passive diffusion. When L-valine was coupled to acyclovir via an ester linkage to produce the prodrug valacyclovir, the extent of systemic absorption significantly increased. Valacyclovir demonstrates an increased oral bioavailability (>50%) compared to 15% for acyclovir [56]. The reason for the improved absorption has been attributed to the addition of L-valine, which allowed PEPT1 to become involved in the absorption process [57]. This technique was also used to increase the extent of systemic absorption of ganciclovir.

Although the expression and activity of intestinal efflux ABC transporters have been extensively studied, much remains unknown about the expression, regulation, and function of the intestinal uptake SLC transporters, and the area is still evolving [58]. One such example is OATP2B1, an OATP transporter with a broader substrate specificity compared to PEPT1, which was earlier reported to be localized in the apical membrane of enterocytes [59]. However, recent studies have indicated that OATP2B1 is likely expressed in the basolateral membrane of enterocytes, and its overall contribution to intestinal absorption of its substrates is unclear [60]. Further research is required to evaluate the expression, functionality, and molecular mechanisms of intestinal drug uptake by SLC transporters, considering the high variability in the oral absorption of substrate drugs and significant potential for transporter-mediated drug–drug interactions.

## 5.4 Summary

In this chapter, we have examined the absorption of drugs via passive diffusion and carrier-mediated pathways. Drugs undergo passive diffusion by paracellular and/or transcellular pathways, with the latter being the more dominant route. A drug can cross the GI membranes by one pathway or by a combination of these pathways. The relative contribution of each pathway depends on the physicochemical properties of the drug, the formulation, and the anatomy and physiology at the site of absorption.

Mathematical treatment of passive diffusion and carrier-mediated transport was briefly discussed. The various rate-limiting factors for intestinal absorption were identified. Finally, the different types of carrier-mediated transport mechanisms were reviewed, and the readers were introduced to uptake/efflux transporters that are currently understood to play a significant role in drug absorption.

## References

1 Derendorf, H. and Schmidt, S. (2019). *Rowland and Tozer's Clinical Pharmacokinetics and Pharmacodynamics: Concepts and Applications*. Wolters Kluwer.

2 Yu, L.X. and Amidon, G.L. (2000). Analytical solutions to mass transfer. In: *Transport Processes in Pharmaceutical Systems* (ed. G.L. Amidon, P.I. Lee, and E.M. Topp), 23–54. New York: Marcel Dekker.

3 Lennernas, H. (2007). Intestinal permeability and its relevance for absorption and elimination. *Xenobiotica* 37: 1015–1051.

4 Imai, T. and Ohura, K. (2010). The role of intestinal carboxylesterase in the oral absorption of prodrugs. *Curr. Drug Metab.* 11: 793–805.

5 Stenberg, P., Bergstrom, C.A., Luthman, K., and Artursson, P. (2002). Theoretical predictions of drug absorption in drug discovery and development. *Clin. Pharmacokinet.* 41: 877–899.

6 Ungell, A.L., Nylander, S., Bergstrand, S. et al. (1998). Membrane transport of drugs in different regions of the intestinal tract of the rat. *J. Pharm. Sci.* 87: 360–366.

7 Thomas, V.H., Bhattachar, S., Hitchingham, L. et al. (2006). The road map to oral bioavailability: an industrial perspective. *Expert Opin. Drug Metab. Toxicol.* 2: 591–608.

8 Karlsson, J., Ungell, A., Grasjo, J., and Artursson, P. (1999). Paracellular drug transport across intestinal epithelia: influence of charge and induced water flux. *Eur. J. Pharm. Sci.* 9: 47–56.

9 Piyapolrungroj, N., Zhou, Y.S., Li, C. et al. (2000). Cimetidine absorption and elimination in rat small intestine. *Drug Metab. Dispos.* 28: 65–72.

10 Bourdet, D.L., Pollack, G.M., and Thakker, D.R. (2006). Intestinal absorptive transport of the hydrophilic cation ranitidine: a kinetic modeling approach to elucidate the role of uptake and efflux transporters and paracellular vs. transcellular transport in Caco-2 cells. *Pharm. Res.* 23: 1178–1187.

11 Yang, Y., Faustino, P.J., Volpe, D.A. et al. (2007). Biopharmaceutics classification of selected beta-blockers: solubility and permeability class membership. *Mol. Pharm.* 4: 608–614.

12 Lemmer, H.J. and Hamman, J.H. (2013). Paracellular drug absorption enhancement through tight junction modulation. *Expert Opin. Drug Deliv.* 10: 103–114.

13 Li, X. and Jasti, B. (2005). Theory and applications of diffusion and dissolution. In: *Theory and Practices of Contemporary Pharmaceutics* (ed. T.K. Ghosh and B.R. Jasti), 197–216. Boca Rotan: CRC Press.

14 Cong, D., Doherty, M., and Pang, K.S. (2000). A new physiologically based, segregated model to explain route-dependent intestinal metabolism. *Drug Metab. Dispos.* 28: 224–235.

15 Tam, D., Tirona, R.G., and Pang, K.S. (2003). Segmental intestinal transporters and metabolic enzymes on intestinal drug absorption. *Drug Metab. Dispos.* 31: 373–383.

16 Hoener, B.-a. and Benet, L.Z. (2002). Factors influencing drug absorption and drug availability. In: *Modern Pharmaceutics* (ed. G.S. Banker and C.T. Rhodes), 145–186. New York: Marcel Dekker.

17 Helander, H.F. and Fandriks, L. (2014). Surface area of the digestive tract – revisited. *Scand. J. Gastroenterol.* 49: 681–689.

18 Trevaskis, N.L., Charman, W.N., and Porter, C.J. (2008). Lipid-based delivery systems and intestinal lymphatic drug transport: a mechanistic update. *Adv. Drug Deliv. Rev.* 60: 702–716.

19 Fujioka, Y., Metsugi, Y., Ogawara, K. et al. (2008). Evaluation of in vivo dissolution behavior and GI transit of griseofulvin, a BCS class II drug. *Int. J. Pharm.* 352: 36–43.

20 Artursson, P., Palm, K., and Luthman, K. (2001). Caco-2 monolayers in experimental and theoretical predictions of drug transport. *Adv. Drug Deliv. Rev.* 46: 27–43.

21 Bhal, S.K., Kassam, K., Peirson, I.G., and Pearl, G.M. (2007). The rule of five revisited: applying $\log D$ in place of $\log P$ in drug-likeness filters. *Mol. Pharm.* 4: 556–560.

22 Hou, T.J., Zhang, W., Xia, K. et al. (2004). ADME evaluation in drug discovery. 5. Correlation of Caco-2 permeation with simple molecular properties. *J. Chem. Inf. Comput. Sci.* 44: 1585–1600.

23 Kokate, A., Li, X., and Jasti, B. (2008). Effect of drug lipophilicity and ionization on permeability across the buccal mucosa: a technical note. *AAPS PharmSciTech* 9: 501–504.

24 Dahlgren, D. and Lennernas, H. (2019). Intestinal permeability and drug absorption: predictive experimental, computational and in vivo approaches. *Pharmaceutics* 11: 411.

25 Palm, K., Luthman, K., Ros, J. et al. (1999). Effect of molecular charge on intestinal epithelial drug transport: pH-dependent transport of cationic drugs. *J. Pharmacol. Exp. Ther.* 291: 435–443.

26 Kokate, A., Li, X., Singh, P., and Jasti, B. (2008). Effect of thermodynamic activities of the unionized and ionized species on drug flux across buccal mucosa. *J. Pharm. Sci.* 97: 4294–4306.

27 Pade, D., Jamei, M., Rostami-Hodjegan, A., and Turner, D.B. (2017). Application of the MechPeff model to predict passive effective intestinal permeability in the different regions of the rodent small intestine and colon. *Biopharm. Drug Dispos.* 38: 94–114.

28 Zhao, Y.H., Le, J., Abraham, M.H. et al. (2001). Evaluation of human intestinal absorption data and subsequent derivation of a quantitative structure-activity relationship (QSAR) with the Abraham descriptors. *J. Pharm. Sci.* 90: 749–784.

29 Lipinski, C.A., Lombardo, F., Dominy, B.W., and Feeney, P.J. (2001). Experimental and computational approaches to estimate solubility and permeability in drug discovery and development settings. *Adv. Drug Deliv. Rev.* 46: 3–26.

30 Poongavanam, V., Doak, B.C., and Kihlberg, J. (2018). Opportunities and guidelines for discovery of orally absorbed drugs in beyond rule of 5 space. *Curr. Opin. Chem. Biol.* 44: 23–29.

31 Kramer, S.D., Aschmann, H.E., Hatibovic, M. et al. (2016). When barriers ignore the "rule-of-five". *Adv. Drug Deliv. Rev.* 101: 62–74.

32 Nielsen, D.S., Shepherd, N.E., Xu, W. et al. (2017). Orally absorbed cyclic peptides. *Chem. Rev.* 117: 8094–8128.

33 Dougherty, P.G., Sahni, A., and Pei, D. (2019). Understanding cell penetration of cyclic peptides. *Chem. Rev.* 119: 10241–10287.

34 Oh, D.M. and Amidon, G.L. (1999). Overview of membrane transport. *Pharm. Biotechnol.* 12: 1–27.

35 Amidon, G.L. and Sadâee, W. (1999). *Membrane Transporters as Drug Targets*. New York: Kluwer Academic/Plenum Publishers.

36 Chang, J.H., Yoo, P., Lee, T. et al. (2009). The role of pH in the glucuronidation of raloxifene, mycophenolic acid and ezetimibe. *Mol. Pharm.* 6: 1216–1227.

37 Fisher, M.B., Campanale, K., Ackermann, B.L. et al. (2000). In vitro glucuronidation using human liver microsomes and the pore-forming peptide alamethicin. *Drug Metab. Dispos.* 28: 560–566.

38 Wu, C.Y. and Benet, L.Z. (2005). Predicting drug disposition via application of BCS: transport/absorption/elimination interplay and development of a biopharmaceutics drug disposition classification system. *Pharm. Res.* 22: 11–23.

39 Cvetkovic, M., Leake, B., Fromm, M.F. et al. (1999). OATP and P-glycoprotein transporters mediate the cellular uptake and excretion of fexofenadine. *Drug Metab. Dispos.* 27: 866–871.

40 Oh, D.M., Han, H.K., and Amidon, G.L. (1999). Drug transport and targeting. Intestinal transport. *Pharm. Biotechnol.* 12: 59–88.

41 Seatter, M.J. and Gould, G.W. (1999). The mammalian facilitative glucose transporter (GLUT) family. *Pharm. Biotechnol.* 12: 201–228.

42 King, A.E., Ackley, M.A., Cass, C.E. et al. (2006). Nucleoside transporters: from scavengers to novel therapeutic targets. *Trends Pharmacol. Sci.* 27: 416–425.

43 Scheepers, A., Joost, H.G., and Schurmann, A. (2004). The glucose transporter families SGLT and GLUT: molecular basis of normal and aberrant function. *JPEN J. Parenter. Enteral Nutr.* 28: 364–371.

44 Ito, K., Suzuki, H., Horie, T., and Sugiyama, Y. (2005). Apical/basolateral surface expression of drug transporters and its role in vectorial drug transport. *Pharm. Res.* 22: 1559–1577.

45 Giacomini, K.M. and Sugiyama, Y. (2006). Membrane transporters and drug response. In: *Goodman & Gilman's the Pharmacological Basis of Therapeutics* (ed. L.S. Goodman, A. Gilman, L.L. Brunton, et al.), 41–70. New York: McGraw-Hill.

46 Garrett, R.H. and Grisham, C.M. (2004). *Biochemistry*. Saunders College Publishing.

47 Dresser, G.K., Bailey, D.G., Leake, B.F. et al. (2002). Fruit juices inhibit organic anion transporting polypeptide-mediated drug uptake to decrease the oral availability of fexofenadine. *Clin. Pharmacol. Ther.* 71: 11–20.

48 Smith, P.L. (2000). The role of permeability studies in preclinical evaluation. In: *Oral Drug Absorption* (ed. J.B.D.H. Lennernas), 119–136. Marcel Dekker.

49 Dressman, J.B., Amidon, G.L., and Fleisher, D. (1985). Absorption potential: estimating the fraction absorbed for orally administered compounds. *J. Pharm. Sci.* 74: 588–589.

50 Smith, D., Artursson, P., Avdeef, A. et al. (2014). Passive lipoidal diffusion and carrier-mediated cell uptake are both important mechanisms of membrane permeation in drug disposition. *Mol. Pharm.* 11: 1727–1738.

51 Estudante, M., Morais, J.G., Soveral, G., and Benet, L.Z. (2013). Intestinal drug transporters: an overview. *Adv. Drug Deliv. Rev.* 65: 1340–1356.

52 Schumann, T., Konig, J., Henke, C. et al. (2020). Solute carrier transporters as potential targets for the treatment of metabolic disease. *Pharmacol. Rev.* 72: 343–379.

53 Hagenbuch, B. and Meier, P.J. (2003). The superfamily of organic anion transporting polypeptides. *Biochim. Biophys. Acta* 1609: 1–18.

54 Dobson, P.D. and Kell, D.B. (2008). Carrier-mediated cellular uptake of pharmaceutical drugs: an exception or the rule? *Nat. Rev. Drug Discov.* 7: 205–220.

55 Zhang, Y., Sun, J., Sun, Y. et al. (2013). Prodrug design targeting intestinal PepT1 for improved oral absorption: design and performance. *Curr. Drug Metab.* 14: 675–687.

56 Minhas, G.S. and Newstead, S. (2019). Structural basis for prodrug recognition by the SLC15 family of proton-coupled peptide transporters. *Proc. Natl. Acad. Sci. U.S.A.* 116: 804–809.

**57** Sugawara, M., Huang, W., Fei, Y.J. et al. (2000). Transport of valganciclovir, a ganciclovir prodrug, via peptide transporters PEPT1 and PEPT2. *J. Pharm. Sci.* 89: 781–789.

**58** Oswald, S. (2019). Organic anion transporting polypeptide (OATP) transporter expression, localization and function in the human intestine. *Pharmacol. Ther.* 195: 39–53.

**59** Tamai, I. (2012). Oral drug delivery utilizing intestinal OATP transporters. *Adv. Drug Deliv. Rev.* 64: 508–514.

**60** Keiser, M., Kaltheuner, L., Wildberg, C. et al. (2017). The organic anion-transporting peptide 2B1 is localized in the basolateral membrane of the human jejunum and Caco-2 monolayers. *J. Pharm. Sci.* 106: 2657–2663.

# 6

# Determinant Factors for Passive Absorption of Drugs

*Wei Zhu*[1] *and Mikolaj Milewski*[2]

[1] *Clinical Pharmacology and Pharmacometrics, Janssen Research & Development, Raritan, NJ, USA*
[2] *Merck & Co., Inc., Rahway, NJ, USA*

## 6.1 Introduction

Oral drug delivery remains the most common route by which small-molecule therapeutic agents are delivered into systemic circulation. Some undeniable benefits of utilizing this route as compared to parenteral alternatives include ease of self-administration, simplicity of dosing instructions, painless application, noninvasiveness, long history of use, and cost-effectiveness [1]. Familiarity with this mode of drug delivery by medical doctors, nurse practitioners, pharmacists, and patients translates into high expertise in the medical field as well as high patient preference and compliance [2, 3]. Additionally, this route of drug administration is fully aligned with the primary function of the gastrointestinal (GI) tract of food ingestion and nutrient absorption.

## 6.2 Fundamentals of Drug Absorption

In most general sense the drug absorption process in the human GI tract is heavily dependent on the physicochemical properties of the drug and its formulation (i.e. biopharmaceutics properties) and the nature of the anatomical barrier – the gut wall (i.e. physiological properties). The process of drug transport from the dosage form into the systemic blood circulation can be simplistically thought of as two consecutive steps: drug release from the dosage form into GI fluids (disintegration and dissolution) and drug transport across the intestinal mucosa (permeation). Here, we will first briefly review human GI tract anatomy and physiology before focusing on the permeation across the intestinal wall.

Wei Zhu and Mikolaj Milewski made equal contributions to this book chapter.

### 6.2.1 Gut Wall

The overall structure of the GI tract can be divided into stomach, small intestine (duodenum, jejunum, and ileum), and large intestine (cecum, colon, and rectum). It is widely recognized that the small intestine is the major site for drug absorption. In humans the average length of the small intestine is around 7 m while the large intestine is 1.5 m while the surface area of the stomach is around 0.050 $m^2$, small intestine is 200 $m^2$, and large intestine is 0.35 $m^2$ [4]. On the macroscopic level, the intestinal mucosa of the small intestine is characterized by several features that facilitate the physiological function of food processing and digestion of nutrients such as circular folds (*plica circularis*) and villi. These macroscopic features were described in Chapter 4 and will not be revisited here. Instead, we will review on the microscopic structure of the intestinal barrier and how it impacts the drug absorption process. On the microscopic level, each villus is composed of epithelium formed by enterocytes with protruding microvilli. On top of the microvilli, there is a water-rich mucus layer. The microvilli collectively form a so-called brush border which effectively increases absorptive surface area of the intestinal wall. Adjacent enterocytes are connected by tight junctions. Underneath enterocytes are richly perfused blood capillaries, removing permeating molecules from the intestine and carrying them to portal vein, liver, and finally the systemic circulation. The driving force for the diffusive absorption is the permeant concentration gradient across the intestinal barrier, i.e. concentration gradient that exists between the lumen of the intestine and blood. The first step in the absorption process is the diffusion across the mucus layer followed by either passive transcellular transport (across an enterocyte), passive paracellular transport (between enterocytes through tight junctions), or an

*Oral Bioavailability and Drug Delivery: From Basics to Advanced Concepts and Applications*, First Edition. Edited by Ming Hu and Xiaoling Li.

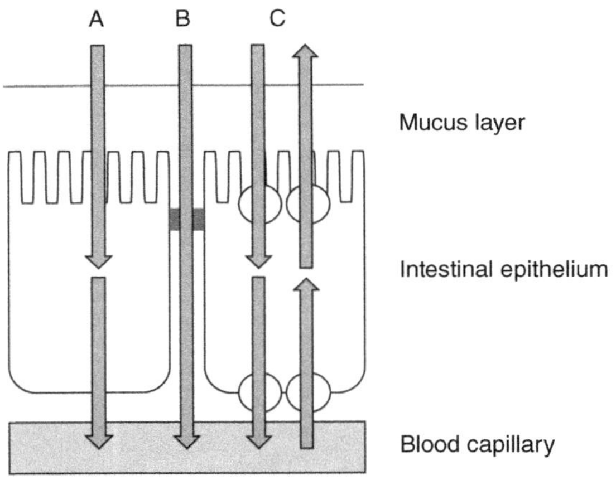


**Figure 6.1** A schematic representation of permeation pathways across the intestinal epithelium: (A) transcellular; (B) paracellular; (C) transporter-mediated.

active transport (transporter-mediated passage across an enterocyte) (see Figure 6.1).

## 6.2.2 Pathways of Absorption

### 6.2.2.1 Transcellular Pathway

The passive transcellular absorption can be thought of as involving transport across a 2-ply membrane: the unstirred water layer (UWL, corresponding physiologically to the mucus later) and the lipoidal layer (corresponding to the bilayer lipid membrane of the enterocyte). Since these two steps are sequential the total diffusive resistance is equal to the sum of the individual diffusive resistances from individual layers:

$$R_{\mathrm{trans}} = R_{\mathrm{UWL}} + R_{\mathrm{m}} \tag{6.1}$$

where $R_{\mathrm{trans}}$ is the diffusional resistance of the transcellular pathway, $R_{\mathrm{UWL}}$ is the diffusional resistance of the UWL, and $R_{\mathrm{m}}$ is the diffusional resistance of the epithelial membrane.

Alternatively, it can be expressed in terms of individual permeation coefficients as:

$$\frac{1}{P_{\mathrm{trans}}} = \frac{1}{P_{\mathrm{UWL}}} + \frac{1}{P_{\mathrm{m}}} \tag{6.2}$$

where $P_{\mathrm{trans}}$ is the permeability coefficient of the transcellular pathway, $P_{\mathrm{UWL}}$ is the permeability coefficient of the UWL, and $P_{\mathrm{m}}$ is the permeability coefficient of the epithelial membrane. Hence, the physicochemical properties of a drug molecule will impact the rate of permeation across individual layers and the overall rate of absorption.

The mucus layer is produced by goblet cells and is a complex glycoprotein colloid gel which coats the lining of the epithelial brush border in the small intestine [5]. Physiologically, it enables passage of the nutrients (and from the purpose of this discussion also drug molecules) from the intestinal lumen to the epithelium while protecting epithelium from direct contact with a bacterial flora [6]. Although the actual composition and structure of the native intestinal mucus is complex involving at least two distinctive groups of mucins and showing heterogenicity dependent of the location in the intestinal tract for the purpose of the drug transport studies it is usually modeled as a simple UWL [7].

The molecular diffusion across a UWL depends chiefly on the size of the permeant. For modeling purposes molecular size is usually approximated with molecular weight, MW, given high correlation between these two properties for organic molecules. Additionally, the UWL permeability to drugs is often simply modeled as reciprocal to the cubic root of the MW:

$$P_{\mathrm{UWL}} \sim \frac{1}{\mathrm{MW}^{1/3}} \tag{6.3}$$

On the other hand, enterocyte cell membrane consists largely of phospholipid molecules forming a bilayer structure. The phospholipids are positioned with their hydrophilic polar heads facing toward the outer surface of the membrane while hydrophobic lipid chains extend toward the interior. Beyond phospholipids, other key components of the cell membrane include cholesterol and glycolipids which impart the membrane with its functional properties [8]. A permeating drug needs to first partition out of the aqueous environment of the mucus layer and into the lipid bilayers. Consequently, the molecule diffuses through the lipoidal pathway within the lipid bilayer down the concentration gradient. The ability of a drug to access bilayer lipids depends heavily on its ionization state, lipophilicity, and molecular size. In simple permeation models, the emphasis is usually put on the importance of lipophilicity which is traditionally measured by permeant's octanol/water partition coefficient $K_{\mathrm{o/w}}$. This measure of lipophilicity is readily available for many established drug products and investigational molecules. Even in absence of experimental values, there are multiple easily accessible software packages offering computational predictive estimates of $K_{\mathrm{o/w}}$ values [9]. Additionally, it is typically assumed that the ionized species cannot partition into the lipid bilayer and the fraction of drug ionized is excluded from the pool of permeating molecules. Hence:

$$P_{\mathrm{trans}} = a \cdot K_{\mathrm{o/w}}^{\alpha} \tag{6.4}$$

where "$a$" is a preexponential term and $\alpha$ is the exponent capturing preferential difference in drug molecules' ability to partition into enterocyte lipid membrane bilayer *versus* octanol solvent. More sophisticated treatments of permeation through bilayer have also been developed with explicit treatment of permeant speciation, recognition of partial ability of ionized species to permeate through lipoidal pathways, and impact of molecular size on the ability of the lipid bilayer to accommodate larger permeants [10, 11]. Additionally, it is important to recognize $P_{trans}$ will only correlate linearly with $K_{o/w}$, as suggested above, for moderately lipophilic permeants. For highly lipophilic compounds a permeability plateau in $P_{trans}$ is typically observed due to a limitation coming from diffusion through UWL layers on the apical side of the plasma membrane.

#### 6.2.2.2 Paracellular Pathway

The passive paracellular absorption can be thought of as involving transport across a 2-ply membrane: the UWL (corresponding physiologically to the mucus later) followed by a narrow aqueous channel (corresponding physiologically to a tight junction between adjacent enterocytes). Since these two steps are sequential the total diffusive resistance is equal to the sum of the individual diffusive resistances from individual layers:

$$R_{\text{para}} = R_{\text{UWL}} + R_{\text{channel}} \tag{6.5}$$

where $R_{para}$ is the diffusional resistance of the paracellular pathway, $R_{UWL}$ is the diffusional resistance of the UWL, and $R_{channel}$ is the diffusional resistance of aqueous channel between enterocytes.

Alternatively, it can be expressed in terms of individual permeation coefficients as:

$$\frac{1}{P_{\text{para}}} = \frac{1}{P_{\text{UWL}}} + \frac{1}{P_{\text{channel}}} \tag{6.6}$$

where $P_{para}$ is the permeability coefficient of the paracellular pathway, $P_{UWL}$ is the permeability coefficient of the UWL, and $P_{channel}$ is the permeability coefficient of aqueous channel between enterocytes. Hence, the physicochemical properties of a drug molecule will impact the rate of permeation across individual layers and the overall rate of absorption.

The cell junctions are contact points between adjacent cells or cells and cellular matrix. Tight junctions specifically represent only one of several types of cell junctions present in epithelial membranes including tight junctions, desmosomes, adherens junctions, and gap junctions. Built of multiprotein complexes and encircling the perimeter of enterocytes tight junctions ensure cell-to-cell adhesion and play a critical role in the control of molecule permeation through extracellular matrix, i.e. provide a barrier function. Both localization in the GI tract and chemical signaling pathways affect the level of permeability through this permeation pathway leading to tissues in varying degree of "tightness" (low extracellular permeability) or "looseness" (high extracellular permeability) [12–15]. From the absorption perspective, the passage of a permeant through the tight junction between intestinal epithelial cells was modeled as a diffusion through a narrow negatively charged channel using a molecular sieve function and a field-of-force function [11, 16]. Small and positively charged species can permeate with relative ease while large and negatively charged permeants are largely excluded from this pathway.

#### 6.2.2.3 Transporter-Mediated Pathway

It should be pointed out that an intestinal membrane is host to a variety of active carriers which may play a role in drug absorption. Transporters can be functionally categorized as uptake-type (facilitating drug transport from apical to basolateral slide of the membrane) or efflux-type (transporting permeants from basolateral to apical side) and belong to several structural subgroups. Active transport requires energy and, unlike passive transport, is highly selective. The active transport generally applies to permeants structurally similar to endogenous substances [17, 18]. The review of active transport across intestinal membranes is outside of the scope of this chapter.

#### 6.2.2.4 Total Passive Permeation (Paracellular and Transcellular Pathways)

Overall, paracellular diffusion is of practical importance only for small and hydrophilic compounds. The relative contribution of paracellular pathway to total drug transport is generally recognized as minor since the fractional area of tight junctions (the paracellular route) as compared to the permeation area of the apical slide of enterocytes (the transcellular route) is only 0.01–0.1% [19]. Hence,

$$P_{\text{eff}} = (1 - f_{\text{para}}) * P_{\text{trans}} + f_{\text{para}} * P_{\text{para}} \tag{6.7}$$

where $P_{eff}$ is the total passive permeability coefficient of epithelial membrane, $f_{para}$ is the fractional area of apical size of epithelial membrane available for paracellular pathway, $P_{trans}$ is the permeability coefficient of the transcellular pathway, and $P_{para}$ is the permeability coefficient of the paracellular pathway. The general permeability–lipophilicity relationship for a small-molecule total passive intestinal transport is sigmoidal with plateau on the low-lipophilicity end corresponding to the paracellular transport and plateau on the high-lipophilicity end corresponding to the UWL diffusion limitation (see Figure 6.2).

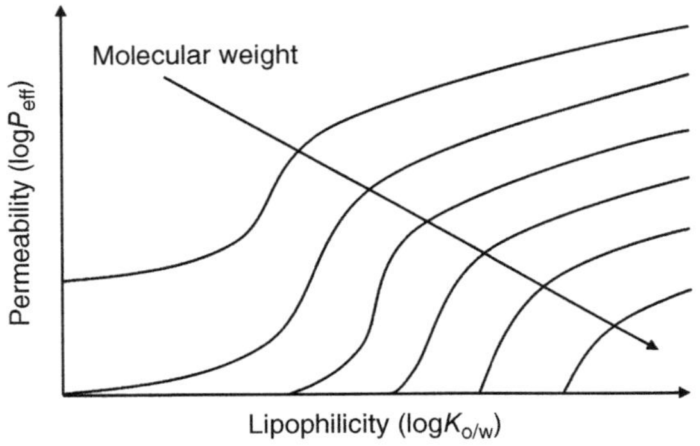


**Figure 6.2** General permeability–lipophilicity relationship in intestinal drug transport as a function of molecular weight. Source: Gian Camenisch et al. [11] / with permission of Elsevier.

### 6.2.3 Driving Forces for Passive Absorption

#### 6.2.3.1 Concentration Gradient

The primary driving force for passive drug absorption in the GI tract is the activity gradient of the permeating drug across the intestinal mucosa often approximated by a concentration gradient. This approach works well for dilute aqueous solutions. Following oral drug ingestion and dissolution of the dosage form the permeant concentration in the GI fluids exceeds that of drug concentration in the intestinal blood capillaries, hence establishing a concentration gradient across intestinal membrane. This gradient provides the thermodynamic driving force for transcellular and paracellular diffusion of drug molecules from intestinal lumen into blood circulation.

The steady-state form of Fick's first law of diffusion describes one-dimensional diffusion across a simple membrane:

$$J = P_{\text{eff}} \cdot \Delta C = \left(\left(1 - f_{\text{para}}\right) * P_{\text{trans}} + f_{\text{para}} * P_{\text{para}}\right) \cdot \Delta C \quad (6.8)$$

where $J$ is drug flux across intestinal mucosa, $P_{\text{eff}}$ is the total passive permeability coefficient of epithelial membrane, $f_{\text{para}}$ is the fractional area of apical size of epithelial membrane available for paracellular pathway, $P_{\text{trans}}$ is the permeability coefficient of the transcellular pathway, $P_{\text{para}}$ is the permeability coefficient of the paracellular pathway, and $\Delta C$ is the drug concentration difference established across epithelial membrane.

This equation can be further simplified to:

$$J = P_{\text{eff}} \cdot \Delta C \approx P_{\text{trans}} \cdot \Delta C \quad (6.9)$$

for nonhydrophilic and larger permeants where transcellular transport dominates over paracellular one according to the discussion in Section 6.2.2.4. Equations (6.8) and (6.9) demonstrate the interplay between the thermodynamic driving force for absorption, $\Delta C$, and kinetic factors related to nature of drug and barrier properties, $P_{\text{eff}}$, which were described in the former sections.

#### 6.2.3.2 Pressure Gradient

Beyond diffusion, a convective drug transport mode is possible across the paracellular route. The thermodynamic driving force for convective fluid movement is the pressure gradient established across intestinal mucosa. It was postulated that sodium-coupled transport of organic solutes from lumen to intracellular spaces creates the osmotic force for fluid absorption and causes opening of tight junctions thereby promoting absorption of nutrients by solvent drag [20, 21]. This phenomenon can be described by Kedem–Katchalsky equation which accounts for diffusive and convective (i.e. solvent-drag coupled) solute transport across porous membranes:

$$J_{\text{s}} = P_{\text{eff}} \cdot \Delta C + (1 - \sigma_{\text{f}}) \cdot J_{\text{v}} \cdot \overline{C} \quad (6.10)$$

where $J_{\text{s}}$ is the solute (drug) flux across intestinal mucosa, $P_{\text{eff}}$ is the total passive permeability coefficient of epithelial membrane, $\Delta C$ is the drug concentration difference established across epithelial membrane, $\sigma_{\text{f}}$ is the solvent drag reflection coefficient (ultrafiltration coefficient) for drug caused by membrane restriction, $J_{\text{v}}$ is the volumetric water flux, and $\overline{C}$ is intramembrane drug concentration for convection. The practical importance of this mode of transport for drug absorption was, however, challenged by a study which demonstrated lack of effect of fluid absorption on *in vivo* permeability of terbutaline in human jejunum [22].

## 6.3 Absorption Determining Factors

### 6.3.1 Physiological Factors

It is commonly accepted that following an oral administration, with few exceptions, drug absorption occurs in the intestine. Drug release from the dosage form as well as drug availability to permeate across intestinal mucosa are dependent on physiological factors that can have a significant impact on the rate and extent of drug absorption. Below, we briefly discuss key physiological factors that are known to play an important role in the absorption process:

#### 6.3.1.1 Gastric Emptying Time

Stomach anatomy features relatively thick mucosa and small surface area (around $0.05\,\text{m}^2$) as compared to the small intestine ($200\,\text{m}^2$). For these reasons, stomach does not contribute to drug absorption for most molecules. However, the residence time a drug spends in stomach limits the rate with which it appears in the small intestine, the major absorption site for most molecules. Consequently, for a drug formulated in an immediate-release dosage form the gastric emptying can become the rate-determining step in the absorption process. In fasted state 15–30 minutes residence time is common. This fact has been recognized in FDA regulatory guidance [23] in assertion that for BCS class 1 molecules (high solubility, high permeability) dissolution of >85% in 0.1N HCl media within 15 minutes implies that the dosage form "behaves like a solution and generally should not have any bioavailability problems." Gastric emptying time has also been studied extensively with respect to the impact of meal ingestion. A caloric content of the meal was shown to correlate positively with prolonged gastric emptying. Under fed conditions, the stomach pylorus gradually releases chyme at a rate of approximately 4 kcal/min across an 2 mm opening into the duodenum [24]. Consequently, depending on caloric content of the meal one to three hours gastric emptying times are typical. Additionally, detailed kinetics of gastric emptying vary depending on fed conditions. While gastric emptying time profile often follow a first-order process in fasted state a zero-order process was found to be more adequate to describe fed state conditions. Beyond presence of food, the gastric emptying time is also influenced by a host of factors including psychological stress, exercise, or pathological conditions. These changes can facilitate or hinder drug absorption process depending on the characteristics of the drug substance and drug formulation.

#### 6.3.1.2 Gastric pH

In addition to the gastric residence time stomach pH can play an important role in drug solubilization and stability. The fasted-state stomach pH oscillates around values of 1.5–3.0. This implies that for weak-base drugs, especially from BCS classes II and IV (low solubility–high permeability and low solubility–low permeability, respectively), beneficial conditions exist for dissolution and solubilization in the stomach. On the other hand, weak-acid drugs may demonstrate limited ability to solubilize under such pH conditions. The pH-stability profiles of individual drugs will be dependent on their chemical structure. For drugs undergoing degradation under acidic conditions enteric coating dosage forms can provide a mitigation plan. Hence, any changes in physiological values of the stomach fluid may have a profound impact on the absorption of ionizable molecules and drug stability. Additionally, under fed conditions, the pH of the stomach is elevated with typical values around pH 4.5–6.0. Multiple other physiological conditions beyond food intake such as concomitant medication use (e.g. antacids ingestion) and pathophysiological conditions such as achlorhydria contribute to changes in the gastric pH.

#### 6.3.1.3 Intestinal Transit Times

The time it takes to move from one segment of GI tract to the next is referred to as the transit time. Mechanistically, transport in small intestine depends on gastric emptying and relies on propulsion, mixing, and segmentation of chyme. A migrating motor complex (MMC) propels ingested content from stomach across the small intestine into the colon [25]. The physiological function of large intestine is water and electrolytes absorption as well as food waste storage. The control over colonic motor function is largely exerted by the nervous system. Intermittent transport across colon is based on high-amplitude contractions leading to stool formation [26]. The absolute intestinal transit times depend to some extent on a method of measurement and a kinetic model adopted to report the data. Commonly used methods are magnetic resonance imaging [27], scintigraphic analysis following ingestion of probes labeled with a small amount of radionuclide, radiography followed by a barium-labeled probe, SmartPill™ motility capsule, or breath hydrogen analysis. Studies have demonstrated that ingested foods and drugs do not necessarily move synchronously through the digestive system and they may leave given segments of the GI tract in order different from how they arrived. Nonetheless, some transit average values are commonly reported and helpful in approximating the time a drug dosage form spends in respective sections of the absorptive tract. From the perspective of drug absorption, the critical residence times are related to small intestine, and for some compounds, large intestine. The small intestine transit (SIT) time is typically in the range of three to four hours although it is subject to intersubject variability with values as low as one hour and as high as seven hours reported. The large intestine transit is highly variable and usually reported to be 12–72 hours range [Variability of GI transit in healthy women and men [28, 29]].

### 6.3.2 Physicochemical Factors

The physicochemical properties of a drug have a profound effect on the absorption and oral bioavailability [30, 31]. These properties are largely inherent factors of drug molecules and are often taken into consideration in pharmaceutical industry as part of the developability screening of a drug candidate in the discovery stage. The impact of various physicochemical factors on absorption of drugs is detailed in this section.

#### 6.3.2.1 Solubility

Solubility is one of the major factors impacting the drug absorption. Any drug to be absorbed must be presented in the form of an aqueous solution at the site of absorption. Low aqueous solubility is the major problem encountered with formulation development of new chemical entities (NCEs) as well as for the generic development, as more than 40% NCEs developed in pharmaceutical industry are practically insoluble in water [32]. For compounds with limited solubility, the supersaturation of the intestinal fluid plays a major part in their absorption [33]. Various solubility enhancement techniques have been used for the drugs with poor aqueous solubility which include physical and chemical modifications of drug such as particle size reduction, crystal engineering, salt formation, as well as formulation technologies such as amorphous solid dispersion, use of surfactant, and nanoparticle technology [34–36].

#### 6.3.2.2 Permeability

Permeability is another major factor governing the absorption of the drugs. Permeability may be defined as the ability of any drug molecule or any substance to permeate a cell membrane, endothelium, or epithelium. The intestinal permeability of an orally administered drug (given the value of effective permeability, or $P_{eff}$) has been widely used to determine the rate and extent of the absorption of the drug through the GI tract in humans. Section 6.2.2 provides a cursory overview of permeability of the small intestine. Additionally, Dahlgren and Lennernas provided a comprehensive summary of predictive experimental, computational, and *in vivo* approaches on intestinal permeability and drug absorption [37].

The biopharmaceutical classification system (BCS) was originally developed by Amidon et al. on the basis of solubility and permeability characteristics of a molecule [38]. According to this classification system, drugs are classified into four categories, namely, class I, II, III, and IV. Class I is highly soluble and highly permeable, class II is poorly soluble and highly permeable, class III is highly soluble and poorly permeable, and class IV is poorly soluble and poorly permeable. To maximize the oral absorption, the solubility/permeability interplay between these two parameters must be taken into consideration [39]. The US Food and Drug Administration (FDA), European Medicines Agency (EMA), and other jurisdictions use such system to evaluate the BCS-based biowaiver requests for BCS I and/or III drug products from the sponsors [40].

#### 6.3.2.3 "pH Partition Theory" and p$K_a$

Most of the drugs can be classified as either weak acids or weak bases. The "pH-partition theory," originally proposed by Shore et al. in 1957 [41], illustrated one of the basic concepts that govern the oral drug absorption process. When a drug is given orally, the extent of the ionization of these drugs is driven by the p$K_a$ of the molecule and pH inherent to the GI tract. The oral absorption of a drug takes place mainly by a passive diffusion of the unionized form of the molecule through the lipophilic intestinal epithelial membrane [42, Section 2.2]. The weakly acidic and weakly basic drugs generally undergo ionization in the GI tract, and the fraction of the ionized versus unionized form of an acidic or a basic drug can be calculated by a modified Henderson–Hasselbalch equation. For example, for monoacidic or monobasic compounds, it can be expressed as Eq. (6.11) or Eq. (6.12) below, respectively:

$$\log\left[\frac{\text{ionized}}{\text{unionized}}\right] = \text{pH} - \text{p}K_a \tag{6.11}$$

$$\log\left[\frac{\text{unionized}}{\text{ionized}}\right] = \text{pH} - \text{p}K_a \tag{6.12}$$

Since high aqueous solubility is required for fast dissolution of a drug, while the partitioning behavior of a drug molecule between aqueous and nonaqueous solvents is also one of the essential components to facilitate the permeation across the epithelial membrane, the interplay between these two interrelated parameters has a major impact on the rate and extent of oral drug absorption across the GI tract [43]. It is worth noting that an acidic drug with a low p$K_a$ or a basic drug with a high p$K_a$ is not necessarily guaranteed to have a high solubility in the GI tract unless the drug has a high intrinsic solubility (solubility of the unionized form) [44].

#### 6.3.2.4 Particle Size and Surface Area

The dissolution rate of the active pharmaceutical ingredient (API) is the rate at which drug dissolves in the GI fluids. It can be the rate-limiting step of the absorption (to be further discussed in Section 6.4.4), and it is fundamentally associated with the particle size of the drug when administered orally. Larger particle size of the drug leads to a small surface area and vice versa [45]. API particles need to be solubilized before the drug can get absorbed through the intestinal epithelial membrane. Thus, the particle size can play an important role in determining the absorption of the drug. Reduction in particle size of the drug has been often used in the pharmaceutical industry which leads to an increased surface area available for solubilization process and an enhanced bioavailability of poorly soluble drugs [34, 46]. The role of particle size in drug dissolution can be explained by the Noyes–Whitey equation which was originally proposed in 1897 (Eq. (6.13)) as following:

$$\frac{dC}{dt} = \frac{D \cdot S}{h \cdot V}(C_s - C) \tag{6.13}$$

where $dC/dt$ is dissolution rate, $D$ is drug diffusion coefficient, $S$ is surface area of drug particles, $V$ is volume of dissolution medium, $h$ is diffusional layer thickness, $C_s$ is solubility of the drug, and $C$ is drug concentration in solution. As shown in Eq. (6.13), the dissolution rate is directly proportional to the surface area of the drug particle which in turn is increased with decrease of the particle size. Chaumeil summarized that reduction of particle size via different milling processes was applied to griseofulvin, progesterone, spironolactone and diosmin, etc. to improve dissolution rate and oral absorption, and consequently their bioavailability and clinical efficacy [47].

#### 6.3.2.5 Salt Form

As previously mentioned, most of the drug molecules are either weak acids or weak bases. These molecules tend to form strong ionic interactions with an oppositely charged counter ion during crystallization process resulting in a salt form. The use of API salt form in the pharmaceutical industry is well documented in the literature, particularly for parent compounds with low aqueous solubility [34, 48, 49].

In general, the dissolution process of salt form is usually associated with a change of the surface pH of the diffusion layer of the drug particle, resulting in a significant increase in the dissolution rates and higher apparent solubility ($C_{app}$) of the drug molecules in the diffusion layer over its inherent solubility at the pH of the dissolution medium ($C_s$). Nelson first reported such observation in 1957 that the salts of acidic theophylline with high diffusion layer pHs had greater *in vitro* dissolution rates than those exhibiting a lower diffusion layer pH [50]. Similar observations were reported with respect to alteration of the drug dissolution rate by modifying the diffusion layer pH at the surface of the dissolving solid if other liabilities of the salt form can be overcome or minimized [51–53].

Overall, scientists in the pharmaceutical industry should leverage the understanding of solubility/dissolution, chemical stability, manufacturability, clinical target, etc. to determine the best path forward for API form selection when multiple options are available.

#### 6.3.2.6 Polymorphic Form and Amorphous API

Polymorphism was defined by McCrone in 1965 who states that "polymorph is a solid crystalline phase of a given compound resulting from the possibility of at least two crystalline arrangements of the molecules of that compound in the solid state" [54]. Therefore, polymorphs are chemically identical with differences in molecular packing, resulting in different physical properties such as molecular density, crystal shape, and melting temperature. Despite these physical differences, the solubility improvement for different polymorphs is usually within 2-fold, and such difference in the solubility value is not expected to have a significant effect on absorption and bioperformance [55].

On the other hand, there are significant differences between crystalline polymorphs and their amorphous state. Amorphous drugs are solid forms which have no crystal structure. Due to the lack of crystalline order in amorphous solids, significant improvement in solubility over that of crystalline forms was observed ranging from less than 2- to up to 100-fold. Therefore, the amorphous API tends to have significantly higher dissolution rate as compared to their crystalline polymorphs which may significantly increase the rate and extent of oral absorption and bioavailability especially for compounds with a low crystalline solubility [56]. However, it should be noted that the amorphous solids are less thermodynamically stable due to a lack of a 3-dimensional crystalline structure and higher free energy. The physical and chemical stability of the amorphous systems as well as the mitigation approach have been extensively discussed in the literature [57–60].

#### 6.3.2.7 Lipinski Rule of 5

With the consideration of various physicochemical properties playing a significant role in absorption and bioperformance of a drug, Lipinski et al. attempted to characterize the intestinal absorption potential (AP) with the "Rule of 5 (Ro5)," which is one of the most widely used concepts to qualitatively predict oral drug absorption [61, 62]. According to Lipinski et al., well-absorbing molecules do not violate more than 2 of the following rules:

1. More than 5 hydrogen bond donors (defined as –NH or –OH groups)
2. More than 10 hydrogen bond acceptors (defined as oxygen or nitrogen atoms, including those that are part of hydrogen-bond donors)
3. Molecular weight (MW) > 500
4. Calculated lipophilicity ($c\text{Log}K_{o/w}$) > 5

Although it was subsequently reported that the Ro5 does not serve to discriminate drugs from "nondrugs"

[63] and it was recently argued that an up-to-date molecular mass and hydrogen bond donor cutoff criteria are needed according to the physicochemical properties of recently FDA-approved drugs [64], the Ro5 is a readily useful and quick screening tool to eliminate compounds which lack intestinal AP at the discovery stage. It should be noted that a compound passing Lipinski's Ro5 criteria may still have low bioavailability due to extensive metabolism in gut and/or liver, and the Ro5 does not apply to the compounds when the permeation is carrier-mediated.

### 6.3.3 Formulation Factors Affecting Oral Absorption

In addition to the physicochemical factors, various formulation factors may also impact the oral absorption of a drug. In a simple solution formulation, drug is fully solubilized and readily absorbable. There is no process preceding absorption with possible exception precipitation and re-solubilization processes in the GI tract for certain drugs. However, for a solid dosage formulation, disintegration of the dosage form is the first step so that the granules and primary drug particles can be released from the dosage form to dissolve and absorb into the system circulation. This is mostly applicable to tablets and to some extent capsules disintegrated from a capsule shell. For formulations intended for immediate release, a faster disintegration usually results in rapid absorption [65]. The disintegration time is typically affected by the amount of binder and compression force [66].

The next step after disintegration of a solid dosage form is the dissolution of the drug particles in the GI fluids. Drug dissolution is a prerequisite for drug absorption which influences the rate and extent of the absorption of the drug. As discussed in Section 6.3.2.4, the dissolution rate can be the rate-limiting step of the absorption, and it can be described by the Noyes–Whitney equation as previously illustrated in Eq. (6.13). The dissolution rate depends on the hydrodynamics of the GI fluid, drug diffusivity through the diffusion layer, solubility of the drug in the GI fluid, surface area of the drug particles, and so forth. There are various formulation factors which impact at least one or more aforementioned parameters, such as solubilization enhancement technologies, usage of the API wetting agent, and usage of excipients such as diluents and/or lubricants in the dosage forms [67].

From formulation/manufacturing perspective, a high-shear wet granulation process usually yields a tablet that dissolves faster than those made by other granulating methods [68, 69]. However, wet granulation has several limitations on physical/chemical stability such as formation of crystal bridge or chemical degradation [69, 70]. A higher compression force yields a tablet with greater hardness and reduced wettability resulting in longer disintegration time. However, a higher compression force may cause crushing of drug particles (especially needle-like particles) into smaller ones with higher effective surface area which in turn increases the dissolution rate. Therefore, the effect of compression force should be thoroughly studied on a case-by-case basis [71].

It is estimated that about 40% of drugs with market approval and nearly 90% of molecules in the discovery pipeline are poorly water-soluble [72]. As the number of poorly soluble drugs has increased, efforts to enhance solubility have evolved so that the dissolution rate can be directly enhanced with increased apparent solubility according to the Noyes–Whitney equation. Solubility enhancement technology has been widely used and reported in the literature, including solid dispersions, nanoparticles, self-emulsifying drug delivery systems (SEDDS), co-solvency, co-crystallization, and so forth [72–78].

As discussed previously, the formulations intended for immediate release generally result in relatively rapid drug absorption. On the other hand, the drug release from modified-release (MR) or controlled-release (CR) dosage forms is deliberately changed thereby altering the timing and/or the rate of release of the drug substance in the GI tract. The MR products have been widely used in pharmaceutical industry to achieve a desired therapeutic target, reduce the intensity of adverse effects, and/or achieve better patient compliance. Examples of the MR formulations include sustained-release formulation, enteric-coated formulation, and gastric-retentive formulation [79–82].

## 6.4 Rate Limiting Steps in Absorption and Prediction of Dosing Amount Absorbed

The BCS proposed by Amidon et al. suggested that the elementary key constraints for the oral drug absorption are the solubility/dissolution of the drug and the permeability through the GI epithelial membrane. To maximize the oral absorption of a compound, the solubility/dissolution/permeability relationship and interplay must be taken into consideration. On the other hand, it is essential to know the oral absorption process and understand predictive tools/models on *in vivo* absorption across the drug discovery and development stages in the pharmaceutical industry.

In this section, theoretical biopharmaceutical concepts such as the maximum absorbable dose (MAD) will be introduced. The absorption-limiting step classification (including permeability limited, dissolution limited, and solubility–permeability limited absorption) will be discussed along with a theoretical approximation of fraction of dose absorbed (Fa) in general cases.

### 6.4.1 The Maximum Absorbable Dose (MAD)

The MAD concept, widely used in drug discovery and early stages of drug development, was initially introduced to guide the settings of particle size specifications to reduce the variability on absorption for poorly soluble drugs [83]. The MAD was defined as the maximum amount of drug that could be absorbed in the GI tract if the drug remains at saturation solubility for the entire duration of the absorption time (GI transit time). Sun et al. [84] modified the original version of the MAD equation by replacing absorption rate constant and fluid volume with effective human intestinal permeability coefficient ($P_{\text{eff}}$) and effective intestinal surface area to calculate the MAD in Eq. (6.14) below:

$$\text{MAD} = P_{\text{eff}} \cdot C_s \cdot A \cdot T_{\text{si}} \tag{6.14}$$

where $C_s$ is the solubility, $T_{\text{si}}$ is the mean intestinal residence time of the drug (three to four hours for the small intestine), and A is the absorption surface area.

The MAD was designed to give a rough approximation of how much drug could be absorbed orally. If the projected clinical dose was less than the MAD, then conventional formulations should be able to provide adequate absorption to achieve therapeutic plasma concentration. On the other hand, if the projected dose was greater than the MAD, then solubility-enhancing formulations might be needed to achieve the therapeutic target. One of the important implications of the MAD equation is that permeability and solubility are compensatory to each other, which means that a high permeability may offset low solubility when a low solubility in the GI tract becomes restrictive to absorption [85].

### 6.4.2 Dose Number (Do), Absorption Number (An), and Dissolution Number (Dn)

Other concepts which are widely used in theoretical biopharmaceutics are dose number (Do), absorption number (An), and dissolution number (Dn). These three dimensionless numbers were originally proposed by Oh et al. when estimating the Fa from suspensions of poorly soluble compounds [86]. These parameters can be defined as Eqs. (6.15)–(6.17) below [44]:

$$\text{Dose number (Do)} = \text{Dose}/(C_s \cdot V) \tag{6.15}$$

$$\text{Absorption number (An)} = (P_{\text{eff}}/R) \cdot T_{\text{si}} \tag{6.16}$$

$$\text{Dissolution number (Dn)} = (3D \cdot C_s)/(\rho \cdot r^2) \cdot T_{\text{si}} \tag{6.17}$$

where $C_s$ is solubility of the drug; $V$ is the volume of the GI fluid (typically 250 ml for stomach at fasted state or 500 ml for small intestine); $P_{\text{eff}}$ is the effective permeability coefficient of the drug which is a function of permeability of epithelial membrane and permeability of the UWL (Section 6.2.2.1); $R$ is the radius of the gut; $T_{\text{si}}$ is the mean intestinal residence time of the drug; $D$ is diffusivity of the dissolved drug, $\rho$ is the density of the drug particle, and $r$ is the initial radius of the drug particle. These parameters have been utilized for assessment of whether dose/solubility ratio, drug permeability, and/or dissolution rate of the drug substance/product are likely to be the rate-limiting step(s) of oral absorption in the GI tract as the dose increases. The detailed absorption limiting steps are discussed in Sections 6.4.3–6.4.5.

### 6.4.3 Permeability Limited Absorption

The oral absorption of a drug will be permeability limited if there is an inadequate rate of transfer of drug mass across the intestinal membrane. Permeability-limited absorption occurs for low permeability and highly soluble drugs or drugs dosed in solution formulation, assuming no precipitation occurs. The absolute amount of drug absorbed increases proportionally as the dosing amount increases because the drug concentration increases in the intestinal fluid when dose increases, although Fa remains constant. In this case, the oral absorption of the drug becomes permeability limited, and the amount absorbed can be estimated as Eq. (6.18) below [87, 88] assuming the small intestine is a cylindric tube:

$$\text{Fa} = 1 - e^{-2\text{An}} \tag{6.18}$$

It should be noted that permeability is one of the key drug dependent variable as it does not vary across different API forms or polymorphs.

### 6.4.4 Dissolution Rate Limited Absorption

The oral absorption of a drug will be dissolution rate limited if the rate of dissolution is slow compared to the time of transit in the GI tract and permeation across the intestinal membrane. Dissolution rate limited absorption occurs when the permeation rate is much larger than the dissolution rate, thereby "sink condition" is created and any dissolved drug instantly disappears from the intestinal lumen. In this case, the absorbed amount increases proportionally as the dose increases, because the total solid surface area is in proportion to the dose. Particle size reduction increases the surface area of drug particles resulting in an increase in the absorbed amount, although Fa does not change when dose increases. Assuming a "sink condition" is created throughout the absorption site, the amount absorbed can be approximated as Eq. (6.19) below [87, 88]:

$$\mathrm{Fa} = 1 - e^{-\mathrm{Dn}} \tag{6.19}$$

In this case, particle size and intestinal solubility are the two key drug-dependent variables to govern the absorption of the drug.

### 6.4.5 Solubility–Permeability Limited Absorption

When the dissolution rate of a drug is much faster than the permeation rate and the dose-to-solubility ratio exceeds the intestinal fluid volume, the concentration of a dissolved drug in the GI tract reaches close to the equilibrium solubility of the drug [87]. Although a poorly soluble compound is usually lipophilic and may have a high membrane permeability, the effective permeability value has an upper limit due to the existence of UWL (see Section 6.2.2.1). Therefore, as the dose increases the undissolved drug may accumulate in the intestinal lumen resulting in a solubility–permeability limited absorption (originally referred as solubility limited absorption in the literature [44, 89]). In this case, the absorbed amount does not increase proportionally with the dose, and the Fa decreases when dose increases. The amount absorbed can be calculated as in Eq. (6.20) below, under assumption that the dissolved amount is always equal to the solubility of the drug multiplied by the intestinal volume [87, 88]:

$$\mathrm{Fa} = \frac{\mathrm{MAD}}{\mathrm{Dose}} = \frac{2\mathrm{An}}{\mathrm{Do}} \tag{6.20}$$

### 6.4.6 Estimation of the Fraction Absorbed (Fa) in General Cases

By applying the steady-state approximation, an analytical solution for Fa estimation was proposed [88, 90]. This equation, which was called "Fa equation" by Sugano and Terada [87], captures the essential processes of oral drug absorption (except for the case of supersaturation). The amount absorbed can be approximated in Eq. (6.21) below:

$$\mathrm{Fa} = 1 - e^{-\frac{1}{\frac{1}{\mathrm{Dn}} + \frac{\mathrm{Do}}{2\mathrm{An}}}} \tag{6.21}$$

The Fa equation suggested that Fa can be estimated by three dimensionless parameters: Dn, Do, and An, and either Dn or An/Do can be the dominant factor on Fa estimation when one largely exceeds the other.

## 6.5 Overview of In Silico Prediction of Absorption and Pharmacokinetics for Oral Dosage Forms

Oral drug absorption is a complex process involving a large number of interdependent events occurring simultaneously in the GI tract. These events include disintegration of the formulation to release drug particles; dissolution of the drug particles into the GI fluid; transition of the formulation, drug particles, and dissolved drug along down the GI tract; precipitation of the drug during the transition in GI tract and redissolution of the precipitates; and permeation of the dissolved drug through the gut wall and reaching of the system circulation. These events can be described via various differential equations which need to be simultaneously solved to obtain some of the key outcomes, such as Fa and bioavailability (%F).

As discussed before, the absorption and pharmacokinetic behavior of a drug are governed by various physiological, physicochemical, and formulation factors, while low oral bioavailability is a common issue faced by scientists in drug discovery and development stages. On the other hand, *in vivo* evaluation in preclinical species can be time consuming, while the cross-species translation of the bioperformance of a drug may be questionable and unreliable [91]. The *in silico* prediction of *in vivo* absorption and pharmacokinetics of a drug is therefore of great importance in the pharmaceutical industry. In order to facilitate such prediction, numerous mathematical models have been investigated and developed to integrate the experimental data of a compound to calculate the oral absorption of a drug over the last three decades. These mathematical frameworks started from the "static" concept such as AP, the MAD, the aforementioned dimensionless numbers on solubility, permeability, and dissolution (Do, An, and Dn) [83, 86, 92]. It then expanded to the earlier stage of compartmental model such as the mixing tank model or the multiple mixing tank model, which modeled the GI tract as a single or multiple well-stirred compartment with an assumption on mixing and movement of the drug across the compartment to allow the integration of dissolution and permeation processes simultaneously [93, 94].

Following the mixing tank model, the compartment absorption transit model was then introduced by Yu et al. to represent the distribution of a drug in the GI tract [89, 95], which later on led to the Advanced Compartmental Absorption and Transit (ACAT) model that is implemented in the software GastroPlus®. Another multicompartment model with similar concept such as the Advanced Dissolution Absorption and Metabolism (ADAM) model was later introduced and was implemented in the software simCYP®. In most of these multicompartmental oral absorption and pharmacokinetic models, the stomach, the small intestine, and the large intestine are expressed as one, seven, and one compartments, respectively. Overall, the importance of *in silico* biopharmaceutical modeling has been recognized through decades of research, development, and investigation on this field, and the terminology of Physiologically Based Biopharmaceutics Modeling (PBBM) has been recently invented to distinguish it from the traditional terminology of Physiologically Based Pharmacokinetic Modeling (PBPK modeling) with the emphasis on the prediction of absorption process of the drug.

The applications of PBBM in the pharmaceutical industry have been well documented in the literature. In the discovery stage, PBBM can be used to assess the potential of *in vivo* absorption of drug candidates with limited physicochemical data. If the absorption of the drug is of critical importance PBBM can be used for candidate screening and early decision-making. Prior to the clinical Phase 1 study, the PBBM-based strategies for prediction of human pharmacokinetics are often employed and *in vivo* data from preclinical species may be leveraged together with the available *in vitro* data. The models should be continuously refined at later stages of drug development to inform CMC decisions, such as API form selection and formulation optimization. Applications of PBBM in evaluation of formulation performance with solubilization enhancement technology such as amorphous solid dispersion have been reported [96]. In addition, as more information is obtained in later stage of drug development, a thorough parameter sensitivity analysis (PSA) via PBBM to evaluate the impact of critical material attributes (CMA) and critical processing parameters (CPPs) on *in vivo* bioperformance creates the opportunity of establishing the margin of a product design "safe space." The safe space herein represents the boundaries defined by *in vitro* specifications (i.e. *in vitro* dissolution/release), within which two formulation batches are expected to be bioequivalent to the other. Pepin et al. in 2016 reported the application of PBBM in combination with clinical PK data to define the particle size specifications and dissolution safe space of Lesinurad IR tablets in the US market registration [97].

In addition, PBBM can serve as a useful tool in the investigation of gastric pH effect and food effect on oral bioavailability of a drug. The clinically relevant drug–drug interactions (DDIs) for drugs with steep pH-solubility profile when coadministered with gastric acid-reducing agents is a potential issue during drug development. Significant efforts on formulation optimization may be required to mitigate such DDI effect, along with a potential clinical assessment. It was reported that PBBM can play an important role in mechanistic understanding of such effects and guiding the formulation optimization process [98, 99]. Evaluation of food effect via PBBM is more complicated due to physiological changes in GI transit time, pH along the GI tract, bile secretion resulting in different solubilization and permeation of the drug, potential food–drug interaction, and so forth. However, numerous examples using PBBM to estimate food effects were investigated and reported including the assessment of different meal types (low, medium, and high fat/calorie meals) [100, 101].

As "bottom up" PBBM models are relatively new research tools for pharmaceutical scientists, the integration of interplay between GI physiology and physicochemical/formulation factors across different PBBM software/applications remains a challenge with outstanding discrepancies between *in vitro*, *in vivo* preclinical, and clinical conditions. To address this challenge an OrBiTo (Oral Biopharmaceutical Tools) initiative was formed in 2012 connecting European universities, regulators, and multiple industrial partners with a goal of integrating physiologically based *in silico* biopharmaceutics models capturing the full complexity of GI drug absorption. Recent publications by the *in silico* task subteam of the OrBiTo initiative identified the gaps in system parameters by analyzing *in silico* performance across different compound classes, investigated the prediction accuracy, and evaluated the relative performance of three of the major PBBM software packages in the market [102–105].

## 6.6 Summary

In this chapter, we have reviewed the passive absorption of drugs from the GI tract. First, we examined the physiology of the human stomach and intestines and discussed the passive permeation from gut lumen through mucosal pathways leading into blood circulation. Drug concentration gradient across the mucosal membrane was identified as the major driving force for diffusive transport. We then embarked on more detailed discussion of key physiological and physicochemical factors which determine the rate and extent of *in vivo* absorption of therapeutics. Multitude of factors including GI pH, transit times, drug solubility, permeability, ionization, solid-state forms, and particle size were identified and described. Additionally, formulation dissolution and solubilization techniques applicable to low-solubility drugs were briefly acknowledged and reviewed. Then, we moved on to develop a practical understanding on how permeability, dissolution, or solubility can each become a factor limiting the amount of drug absorbed depending on characteristics of the therapeutic of interest and its formulation. Finally, the utility and challenges related to the use of physiologically based computational absorption models were described. Of course, this chapter only highlights key topics and is not meant to be treated as an exhaustive review of the whole field of oral drug absorption. The authors hope that this overview can provide an appropriate introduction for a scientist new to this field or become a handy source of references for a more inquisitive reader.

## References


1 Lennernas, H. and Abrahamsson, B. (2005). *The use of biopharmaceutic classification of drugs in drug discovery and development: current status and future extension. J. Pharm. Pharmacol.* 57 (3): 273–285.

2 Pfeiffer, P. et al. (2006). *Patient preference for oral or intravenous chemotherapy: a randomised cross-over trial comparing capecitabine and Nordic fluorouracil/leucovorin in patients with colorectal cancer. Eur. J. Cancer* 42 (16): 2738–2743.

3 Liu, G. et al. (1997). *Patient preferences for oral versus intravenous palliative chemotherapy. J. Clin. Oncol.* 15 (1): 110–115.

4 DeSesso, J.M. and Jacobson, C.F. (2001). *Anatomical and physiological parameters affecting gastrointestinal absorption in humans and rats. Food Chem. Toxicol.* 39 (3): 209–228.

5 Schroeder, B.O. (2019). *Fight them or feed them: how the intestinal mucus layer manages the gut microbiota. Gastroenterol. Rep. (Oxford)* 7 (1): 3–12.

6 Macierzanka, A., Mackie, A.R., and Krupa, L. (2019). *Permeability of the small intestinal mucus for physiologically relevant studies: Impact of mucus location and ex vivo treatment. Sci. Rep.* 9 (1): 17516.

7 Obata, K. et al. (2005). *Prediction of oral drug absorption in humans by theoretical passive absorption model. Int. J. Pharm.* 293 (1–2): 183–192.

8 Alberts, B. (2002). *Molecular Biology of the Cell*, 4e. New York: Garland Science, 1548 p.

9 van de Waterbeemd, H. and Gifford, E. (2003). *ADMET in silico modelling: towards prediction paradise? Nat. Rev. Drug Discovery* 2 (3): 192–204.

10 Sugano, K. et al. (2001). *High throughput prediction of oral absorption: improvement of the composition of the lipid solution used in parallel artificial membrane permeation assay. J. Biomol. Screening* 6 (3): 189–196.

11 Camenisch, G., Folkers, G., and van de Waterbeemd, H. (1998). *Shapes of membrane permeability-lipophilicity curves: extension of theoretical models with an aqueous pore pathway. Eur. J. Pharm. Sci.* 6 (4): 325–329.

12 Yu, Q.H. and Yang, Q. (2009). *Diversity of tight junctions (TJs) between gastrointestinal epithelial cells and their function in maintaining the mucosal barrier. Cell Biol. Int.* 33 (1): 78–82.

13 Shin, K., Fogg, V.C., and Margolis, B. (2006). *Tight junctions and cell polarity. Annu. Rev. Cell Dev. Biol.* 22: 207–235.

14 Fanning, A.S., Mitic, L.L., and Anderson, J.M. (1999). *Transmembrane proteins in the tight junction barrier. J. Am. Soc. Nephrol.* 10 (6): 1337–1345.

15 Balda, M.S. et al. (1992). *Structure, regulation, and pathophysiology of tight junctions in the gastrointestinal tract. Yale J. Biol. Med.* 65 (6): 725–735; discussion 737–740.

16 Camenisch, G., Folkers, G., and van de Waterbeemd, H. (1996). *Review of theoretical passive drug absorption models: historical background, recent developments and limitations. Pharm. Acta Helv.* 71 (5): 309–327.

17 Bush, K.T. et al. (2017). *The drug transporter OAT3 (SLC22A8) and endogenous metabolite communication via the gut-liver-kidney axis. J. Biolumin. Chemilumin.* 292 (38): 15789–15803.

18 Hagenbuch, B. and Meier, P.J. (2004). *Organic anion transporting polypeptides of the OATP/SLC21 family: phylogenetic classification as OATP/SLCO superfamily, new nomenclature and*

*molecular/functional properties. Pflugers Arch.* 447 (5): 653–665.

19 Stenberg, P. et al. (2002). *Theoretical predictions of drug absorption in drug discovery and development. Clin. Pharmacokinet.* 41 (11): 877–899.

20 Pappenheimer, J.R. and Reiss, K.Z. (1987). *Contribution of solvent drag through intercellular junctions to absorption of nutrients by the small intestine of the rat. J. Membr. Biol.* 100 (2): 123–136.

21 Kedem, O. and Katchalsky, A. (1958). *Thermodynamic analysis of the permeability of biological membranes to non-electrolytes. Biochim. Biophys. Acta* 27 (2): 229–246.

22 Fagerholm, U. et al. (1995). *The lack of effect of induced net fluid absorption on the in vivo permeability of terbutaline in the human jejunum. J. Drug Targeting* 3 (3): 191–200.

23 Guidance for Industry: Dissolution Testing of Immediate Release Solid Oral Dosage Forms. 1997. https://www.fda.gov/media/70936/download.

24 Hellstrom, P.M., Gryback, P., and Jacobsson, H. (2006). *The physiology of gastric emptying. Best Pract. Res. Clin. Anaesthesiol.* 20 (3): 397–407.

25 Muller, M., Canfora, E.E., and Blaak, E.E. (2018). Gastrointestinal transit time, glucose homeostasis and metabolic health: modulation by dietary fibers. *Nutrients* 10 (3).

26 Bharucha, A.E. (2012). *High amplitude propagated contractions. Neurogastroenterol. Motil.* 24 (11): 977–982.

27 Schiller, C. et al. (2005). *Intestinal fluid volumes and transit of dosage forms as assessed by magnetic resonance imaging. Aliment Pharmacol. Ther.* 22 (10): 971–979.

28 Degen, L.P. and Phillips, S.F. (1996). *Variability of gastrointestinal transit in healthy women and men. Gut* 39 (2): 299–305.

29 Read, N.W. et al. (1980). *Transit of a meal through the stomach, small intestine, and colon in normal subjects and its role in the pathogenesis of diarrhea. Gastroenterology* 79 (6): 1276–1282.

30 Pouton, C.W. (2006). *Formulation of poorly water-soluble drugs for oral administration: physicochemical and physiological issues and the lipid formulation classification system. Eur. J. Pharm. Sci.* 29 (3-4): 278–287.

31 Hurst, S. et al. (2007). *Impact of physiological, physicochemical and biopharmaceutical factors in absorption and metabolism mechanisms on the drug oral bioavailability of rats and humans. Expert Opin. Drug Metab. Toxicol.* 3 (4): 469–489.

32 Savjani, K.T., Gajjar, A.K., and Savjani, J.K. (2012). *Drug solubility: importance and enhancement techniques. ISRN Pharm.* 2012: 195727.

33 Grohganz, H. et al. (2014). *Refining stability and dissolution rate of amorphous drug formulations. Expert Opin. Drug. Deliv.* 11 (6): 977–989.

34 Williams, H.D. et al. (2013). *Strategies to address low drug solubility in discovery and development. Pharmacol. Rev.* 65 (1): 315–499.

35 Kalepu, S. and Nekkanti, V. (2016). *Improved delivery of poorly soluble compounds using nanoparticle technology: a review. Drug Delivery Transl. Res.* 6 (3): 319–332.

36 Kumar, S. et al. (2013). *Drug carrier systems for solubility enhancement of BCS class II drugs: a critical review. Crit. Rev. Ther. Drug. Carrier Syst.* 30 (3): 217–256.

37 Dahlgren, D. and Lennernas, H. (2019). Intestinal permeability and drug absorption: predictive experimental, computational and in vivo approaches. *Pharmaceutics* 11 (8).

38 Amidon, G.L. et al. (1995). *A theoretical basis for a biopharmaceutic drug classification: the correlation of in vitro drug product dissolution and in vivo bioavailability. Pharm. Res.* 12 (3): 413–420.

39 Dahan, A. and Miller, J.M. (2012). *The solubility-permeability interplay and its implications in formulation design and development for poorly soluble drugs. AAPS J.* 14 (2): 244–251.

40 Davit, B.M. et al. (2016). *BCS biowaivers: similarities and differences among EMA, FDA, and WHO requirements. AAPS J.* 18 (3): 612–618.

41 Shore, P.A., Brodie, B.B., and Hogben, C.A. (1957). *The gastric secretion of drugs: a pH partition hypothesis. J. Pharmacol. Exp. Ther.* 119 (3): 361–369.

42 Kubinyi, H. (1979). *Lipophilicity and biological acitivity. Drug transport and drug distribution in model systems and in biological systems. Arzneim.-Forsch.* 29 (8): 1067–1080.

43 Yang, Y. et al. (2012). *Beyond size, ionization state, and lipophilicity: influence of molecular topology on absorption, distribution, metabolism, excretion, and toxicity for druglike compounds. J. Med. Chem.* 55 (8): 3667–3677.

44 Martinez, M.N. and Amidon, G.L. (2002). *A mechanistic approach to understanding the factors affecting drug absorption: a review of fundamentals. J. Clin. Pharmacol.* 42 (6): 620–643.

45 Sharma, P.A. et al. (2015). *Nanomaterial based approaches for the diagnosis and therapy of cardiovascular diseases. Curr. Pharm. Des.* 21 (30): 4465–4478.

46 Wu, Y. et al. (2004). *The role of biopharmaceutics in the development of a clinical nanoparticle formulation of MK-0869: a Beagle dog model predicts improved bioavailability and diminished food effect on absorption in human. Int. J. Pharm.* 285 (1–2): 135–146.

47 Chaumeil, J.C. (1998). *Micronization: a method of improving the bioavailability of poorly soluble drugs. Methods Find Exp. Clin. Pharmacol.* 20 (3): 211–215.

48 Chowhan, Z.T. (1978). *pH-solubility profiles or organic carboxylic acids and their salts. J. Pharm. Sci.* 67 (9): 1257–1260.

49 Nelson, E. (1958). *Comparative dissolution rates of weak acids and their sodium salts. J. Am. Pharm. Assoc.* 47 (4): 297–299.

50 Nelson, E. (1957). *Solution rate of theophylline salts and effects from oral administration. J. Am. Pharm. Assoc.* 46 (10): 607–614.

51 Li, S. et al. (2005). *IV-IVC considerations in the development of immediate-release oral dosage form. J. Pharm. Sci.* 94 (7): 1396–1417.

52 Hendriksen, B.A., Felix, M.V., and Bolger, M.B. (2003). *The composite solubility versus pH profile and its role in intestinal absorption prediction. AAPS PharmSci* 5 (1): E4.

53 Strickley, R.G. (2004). *Solubilizing excipients in oral and injectable formulations. Pharm. Res.* 21 (2): 201–230.

54 Bernstein, J. (1989). *Polymorphism in drug design and delivery. Prog. Clin. Biol. Res.* 289: 203–215.

55 Pudipeddi, M. and Serajuddin, A.T. (2005). *Trends in solubility of polymorphs. J. Pharm. Sci.* 94 (5): 929–939.

56 Schittny, A., Huwyler, J., and Puchkov, M. (2020). *Mechanisms of increased bioavailability through amorphous solid dispersions: a review. Drug. Deliv.* 27 (1): 110–127.

57 Singhal, D. and Curatolo, W. (2004). *Drug polymorphism and dosage form design: a practical perspective. Adv. Drug Delivery Rev.* 56 (3): 335–347.

58 Kaushal, A.M., Gupta, P., and Bansal, A.K. (2004). *Amorphous drug delivery systems: molecular aspects, design, and performance. Crit. Rev. Ther. Drug. Carrier Syst.* 21 (3): 133–193.

59 Lin, X. et al. (2018). *Physical stability of amorphous solid dispersions: a physicochemical perspective with thermodynamic. Kinet. Environ. Aspects. Pharm. Res.* 35 (6): 125.

60 Lehmkemper, K. et al. (2018). *Physical stability of API/polymer-blend amorphous solid dispersions. Eur. J. Pharm. Biopharm.* 124: 147–157.

61 Lipinski, C.A. et al. (2001). *Experimental and computational approaches to estimate solubility and permeability in drug discovery and development settings. Adv. Drug Delivery Rev.* 46 (1–3): 3–26.

62 Lipinski, C.A. (2000). *Drug-like properties and the causes of poor solubility and poor permeability. J. Pharmacol. Toxicol. Methods* 44 (1): 235–249.

63 Oprea, T.I. (2000). *Property distribution of drug-related chemical databases. J. Comput.-Aided Mol. Des.* 14 (3): 251–264.

64 Shultz, M.D. (2019). *Two decades under the influence of the rule of five and the changing properties of approved oral drugs. J. Med. Chem.* 62 (4): 1701–1714.

65 Kelly, K. et al. (2003). *Comparison of the rates of disintegration, gastric emptying, and drug absorption following administration of a new and a conventional paracetamol formulation, using gamma scintigraphy. Pharm. Res.* 20 (10): 1668–1673.

66 Chowhan, Z.T. (1980). *Role of binders in moisture-induced hardness increase in compressed tablets and its effect on in vitro disintegration and dissolution. J. Pharm. Sci.* 69 (1): 1–4.

67 Jambhekar, S.S. and Breen, P.J. (2013). *Drug dissolution: significance of physicochemical properties and physiological conditions. Drug Discovery Today* 18 (23–24): 1173–1184.

68 Badawy, S.I. et al. (2000). *Effect of process parameters on compressibility of granulation manufactured in a high-shear mixer. Int. J. Pharm.* 198 (1): 51–61.

69 Thapa, P. et al. (2019). *Effects of granulation process variables on the physical properties of dosage forms by combination of experimental design and principal component analysis. Asian J. Pharm. Sci.* 14 (3): 287–304.

70 Shanmugam, S. (2015). *Granulation techniques and technologies: recent progresses. Bioimpacts* 5 (1): 55–63.

71 Akseli, I. et al. (2017). *A practical framework toward prediction of breaking force and disintegration of tablet formulations using machine learning tools. J. Pharm. Sci.* 106 (1): 234–247.

72 Kalepu, S. and Nekkanti, V. (2015). *Insoluble drug delivery strategies: review of recent advances and business prospects. Acta Pharm. Sin. B* 5 (5): 442–453.

73 Mahmood, A. and Bernkop-Schnurch, A. (2019). *SEDDS: a game changing approach for the oral administration of hydrophilic macromolecular drugs. Adv. Drug Delivery Rev.* 142: 91–101.

74 Jouyban, A. (2019). *Review of the cosolvency models for predicting solubility in solvent mixtures: an update. J. Pharm. Pharm. Sci.* 22 (1): 466–485.

75 Rodrigues, M. et al. (2018). *Pharmaceutical cocrystallization techniques.* Advances and challenges. *Int. J. Pharm.* 547 (1–2): 404–420.

76 Edgar, J.Y.C. and Wang, H. (2017). *Introduction for design of nanoparticle based drug delivery systems. Curr. Pharm. Des.* 23 (14): 2108–2112.

77 Singh, A., Worku, Z.A., and Van den Mooter, G. (2011). *Oral formulation strategies to improve solubility of poorly water-soluble drugs. Expert Opin. Drug. Deliv.* 8 (10): 1361–1378.

78 Cid, A.G. et al. (2019). *Solid dispersion technology as a strategy to improve the bioavailability of poorly soluble drugs. Ther Deliv* 10 (6): 363–382.

79 Maincent, J. and Williams, R.O. 3rd, (2018). *Sustained-release amorphous solid dispersions. Drug Delivery Transl. Res.* 8 (6): 1714–1725.

80 Sugihara, H. et al. (2014). *Development of a gastric retentive system as a sustained-release formulation of pranlukast hydrate and its subsequent in vivo verification in human studies. Eur. J. Pharm. Sci.* 53: 62–68.

81 Nair, A.B., Gupta, R., Kumria, R. et al. (2010). *Formulation and evaluation of enteric coated tablets of proton pump inhibitor. J. Basic Clin. Pharm.* 1 (4): 215–221.

82 Mohamed, M.F. et al. (2019). *Development of in vitro-in vivo correlation for upadacitinib extended-release tablet formulation. AAPS J.* 21 (6): 108.

83 Johnson, K.C. and Swindell, A.C. (1996). *Guidance in the setting of drug particle size specifications to minimize variability in absorption. Pharm. Res.* 13 (12): 1795–1798.

84 Sun, D. et al. (2004). *In vitro testing of drug absorption for drug 'developability' assessment: forming an interface between in vitro preclinical data and clinical outcome. Curr. Opin. Drug Discovery Dev.* 7 (1): 75–85.

85 Butler, J.M. and Dressman, J.B. (2010). *The developability classification system: application of biopharmaceutics concepts to formulation development. J. Pharm. Sci.* 99 (12): 4940–4954.

86 Oh, D.M., Curl, R.L., and Amidon, G.L. (1993). *Estimating the fraction dose absorbed from suspensions of poorly soluble compounds in humans: a mathematical model. Pharm. Res.* 10 (2): 264–270.

87 Sugano, K. and Terada, K. (2015). *Rate- and extent-limiting factors of oral drug absorption: theory and applications. J. Pharm. Sci.* 104 (9): 2777–2788.

88 Sugano, K. (2009). *Introduction to computational oral absorption simulation. Expert Opin. Drug Metab. Toxicol.* 5 (3): 259–293.

89 Yu, L.X. (1999). *An integrated model for determining causes of poor oral drug absorption. Pharm. Res.* 16 (12): 1883–1887.

90 Sugano, K. (2009). *Computational oral absorption simulation for low-solubility compounds. Chem. Biodivers.* 6 (11): 2014–2029.

91 Chanteux, H. et al. (2015). *Cross-species differences in the preclinical pharmacokinetics of CT7758, an alpha4beta1/alpha4beta7 integrin antagonist. Drug Metab. Dispos.* 43 (9): 1381–1391.

92 Dressman, J.B., Amidon, G.L., and Fleisher, D. (1985). *Absorption potential: estimating the fraction absorbed for orally administered compounds. J. Pharm. Sci.* 74 (5): 588–589.

93 Dressman, J.B. and Fleisher, D. (1986). *Mixing-tank model for predicting dissolution rate control or oral absorption. J. Pharm. Sci.* 75 (2): 109–116.

94 Luner, P.E. and Amidon, G.L. (1993). *Description and simulation of a multiple mixing tank model to predict the effect of bile sequestrants on bile salt excretion. J. Pharm. Sci.* 82 (3): 311–318.

95 Yu, L.X. et al. (1996). *Transport approaches to the biopharmaceutical design of oral drug delivery systems: prediction of intestinal absorption. Adv. Drug Delivery Rev.* 19 (3): 359–376.

96 Mitra, A., Zhu, W., and Kesisoglou, F. (2016). *Physiologically based absorption modeling for amorphous solid dispersion formulations. Mol. Pharmaceutics* 13 (9): 3206–3215.

97 Pepin, X.J. et al. (2016). *Justification of drug product dissolution rate and drug substance particle size specifications based on absorption PBPK modeling for Lesinurad immediate release tablets. Mol. Pharmaceutics* 13 (9): 3256–3269.

98 Mitra, A. et al. (2011). *Using absorption simulation and gastric pH modulated dog model for formulation development to overcome achlorhydria effect. Mol. Pharmaceutics* 8 (6): 2216–2223.

99 Mitra, A. et al. (2020). *Prediction of pH-dependent drug-drug interactions for basic drugs using physiologically based biopharmaceutics modeling: industry case studies. J. Pharm. Sci.* 109 (3): 1380–1394.

100 Tistaert, C. et al. (2019). *Food effect projections via physiologically based pharmacokinetic modeling: predictive case studies. J. Pharm. Sci.* 108 (1): 592–602.

101 Li, M. et al. (2018). *Predictive performance of physiologically based pharmacokinetic models for the effect*

*of food on oral drug absorption: current status. CPT Pharmacometrics Syst .Pharmacol.* 7 (2): 82–89.

**102** Ahmad, A. et al. (2020). *IMI - oral biopharmaceutics tools project - evaluation of bottom-up PBPK prediction success part 4: prediction accuracy and software comparisons with improved data and modelling strategies. Eur. J. Pharm. Biopharm.* 156: 50–63.

**103** Margolskee, A. et al. (2017). *IMI - oral biopharmaceutics tools project - evaluation of bottom-up PBPK prediction success part 1: characterisation of the OrBiTo database of compounds. Eur. J. Pharm. Sci.* 96: 598–609.

**104** Margolskee, A. et al. (2017). *IMI - oral biopharmaceutics tools project - evaluation of bottom-up PBPK prediction success part 2: an introduction to the simulation exercise and overview of results. Eur. J. Pharm. Sci.* 96: 610–625.

**105** Darwich, A.S. et al. (2017). *IMI - oral biopharmaceutics tools project - evaluation of bottom-up PBPK prediction success part 3: identifying gaps in system parameters by analysing in silico performance across different compound classes. Eur. J. Pharm. Sci.* 96: 626–642.

# 7

# Protein Binding and Drug Distribution

*HaiAn Zheng and Marcel Musteata*

*Department of Pharmaceutical Sciences, Albany College of Pharmacy and Health Sciences, Albany, NY, USA*

## 7.1 Introduction

Once entered human body, most drug molecules bind to proteins and tissues to certain extents, before reaching their therapeutic targets, which are also mostly proteins. These drug–protein binding form drug–protein complex that is usually reversible. Most drug molecules bind to binding site(s) on proteins through noncovalent interactions. A drug–protein binding equilibrium can be characterized to evaluate its impact on drug distribution before the drug is metabolized, excreted, or reaches the target.

Drug–protein binding may significantly impact drug distribution and bioavailability. Firstly, most drug–protein complex cannot cross physiological barriers, thus bound drug molecules are restricted within physiological or pharmacokinetics compartment(s). Secondly, the bound drug in a protein complex cannot be exposed to metabolism enzymes or other degradation mechanisms, thus reduces drug clearance. Also, drug–protein complex is more difficult to be excreted by kidney. Lastly and importantly, the bound drug molecules are usually inactive, as only free drug molecules can bind to their biomolecular target to exert pharmacological effects. Therefore, understanding protein binding is fundamental to study drug distribution for every candidate for drug design and development.

Protein–drug binding equilibrium can be studied *in vitro* by dialysis, ultrafiltration (UF), or other methods that distinguish and quantify the free drug from drug–protein complex. For example, dialysis membrane is used to separate drug–protein complex from free drugs, so their binding behavior can be investigated kinetically, or after reaching equilibrium. These methods and recent advancements are discussed in this chapter. It is also critical to ensure an *in vitro* method to be created, correlated, and/or calibrated to *in vivo* profile. Choosing representative protein(s), contraction ratios, and experimental conditions are important for method development and optimization to study drug–protein binding.

Drug binds to various proteins in many tissues and organs, wherever reachable. Most drugs enter the blood and interact with plasma protein extensively during circulation. This is the most important protein binding process, which impacts the distribution and bioavailability of most drugs. Therefore, this chapter focuses on drug binding with plasma proteins, while other drug–protein binding mechanisms also can be deliberated by using the same principles.

## 7.2 Protein–Drug Binding in Plasma

After a drug is absorbed or is directly injected into the blood, it interacts with complex blood components, as listed in Table 7.1. Human blood contains plasma (~55% v/v) and blood cells (~45% v/v), which can be separated by centrifugation. The plasma contains about 92% of water with dissolved glucose, inorganic ions, and other small molecules, and the other 8% of contents are proteins. Plasma proteins, especially albumin and alpha-1-acid glycoprotein (AGP) often bind to drugs, as well as endogenous small molecules such as free fatty acids and steroids [1].

Human serum albumin (HSA) is the most abundant protein in plasma (~4% w/v of plasma, ~0.6 mM in plasma, or 32~53 g/l of blood). It constitutes more than half of the protein contents in plasma and maintains important functions. The albumin stabilized the extracellular volume and osmotic pressure of the blood. It also serves as carrier protein for many endogenous molecules, such as free fatty acids, steroids, and thyroid

*Oral Bioavailability and Drug Delivery: From Basics to Advanced Concepts and Applications,* First Edition. Edited by Ming Hu and Xiaoling Li.

**Table 7.1** Complex components of the blood plasma

| Blood fractions | Contents | Molecule | Molecular weight (kDa) | Plasma concentration (g/L) | Commonly binds to drug |
|---|---|---|---|---|---|
| **Plasma** (55%v/v) | *Protein colloids* | Albumin | ~69 | 35~50 | Acidic |
| | (8% w/v of plasma) | AGP | ~44 | 0.4~1 | Basic |
| | | Lipoprotein (particles) | 200~3400 | Varies | Basic |
| | | Immunoglobulins | ~150 | 2.0~2.5 | |
| | | Blood-clotting proteins | | | |
| | | Other protein and factors | | | |
| | Water & solutes | with inorganic ions, glucose, | | | |
| | (92% w/v of plasma) | and other solutes | | | |
| **Blood cells** (45%v/v) | | | | | |

hormones. As drugs enter blood, albumin binds most of them to some extent.

HSA is a water-soluble globular protein of a molecular size of about 67 kDa. Its monomeric polypeptide chain is composed of 585 amino acids. More than 100 crystal structures have been published, and more than half of them are complexed with specific ligands (PDB .org). The protein has three structurally homologous alpha-helical domains (I–III), each containing subdomains A and B (Figure 7.1, [3]). HSA has several distinct binding sites that can accommodate molecules with diverse structures, but most ligands are hydrophobic molecules with negatively charged function groups, such as fatty acids, thyroxine, hemin, and bilirubin. Drugs bind to albumin mainly by hydrophobic interactions, although some electrostatic attractions were also found [4]. It has up to seven binding sites for endogenous long-chain fatty acids (FA1-FA7). Drugs and drug-like molecules often bind to primary sites 1 and 2 that are located on subdomain IIA and IIIA (Figure 7.1), while others could be the secondary for drug–protein binding [2]. The binding capacity of HSA varies depending on ligand properties.

Many drugs can bind at Binding Site 1, such as phenytoin, phenylbutazone, sulfonamides, and valproic acid. The Binding Site 2 is shared by semisynthetic penicillin, benzodiazepine, probenecid and medium-chain fatty acids, and many others. When more than one drug that can bind to the same binding site is administrated together, they will compete and displace each other and increase their free drug concentrations overall. These are examples of drug–drug interactions that are caused by competitive drug–protein binding.

AGP is the second important serum protein in blood for drug–protein binding, which is also called orosomucoid (ORM). It accounts for about 1~3% of total plasma protein contents, about 0.4–1 g/l in plasma. But its expression level is variable, responding to injury, inflammation, or infections. AGP is negatively charged at physiological pH and binds to basic and neutrally charges lipophilic endogenous molecules. It also found binding to several positively charged basic drugs.

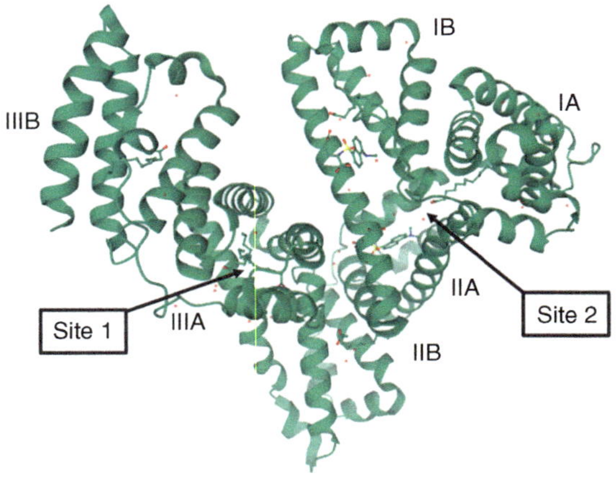


**Figure 7.1** A crystal structure of human serum albumin complexed with dansyl-L-glutamate and myristic acid (PDB, 2XSI, [2]). Source: Adapted from Ryan et al. [2].

Human AGP is a single chain of 183 amino acids, and is heavily glycosylated to a molecular weight of around 44 kDa. Its crystal structure represented in Figure 7.2 (PDB 3KQ0) reveals a typical folding of eight-stranded beta-barrel. The branched, partly hydrophobic, and partly acidic cavity, together with the loop 1 and the two sugar side chains at its entrance, explains the diverse ligand binding behavior of AGP, which is known to vary with changes in glycosylation pattern [5].

Lipoprotein particles are complexes of lipids and proteins, with a central hydrophobic core of cholesterol and triglycerides that are surrounded by phospholipids

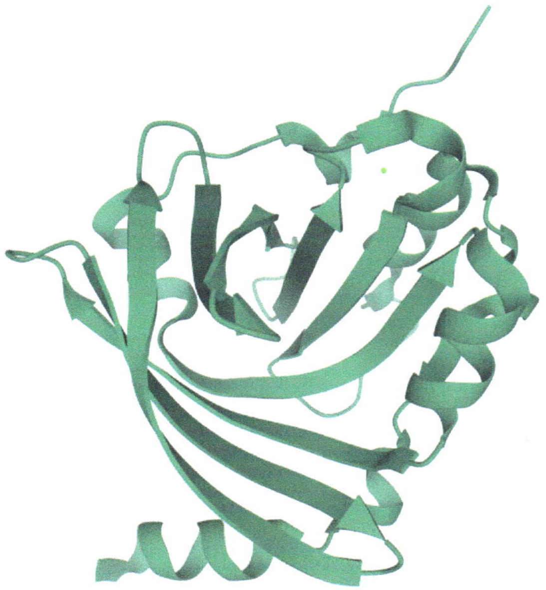

**Figure 7.2** Crystal structure of human alpha1-acid glycoprotein (PDB 3KQ0). Source: Adapted from Schönfeld et al. [5].

membrane and apolipoproteins. These particles can be classified according to their densities as very-low-density lipoproteins (VLDL), low-density lipoproteins (LDL), intermediate-density lipoproteins (IDL), and high-density lipoproteins (HDL). They are responsible for transporting lipids to the liver. Their bindings to drug usually are not significant, but when albumin or other specific binding sites are saturated, they also impact some drug's biodistribution.

Some drugs can also bind to blood cells and their components such as hemoglobin, as well as many tissues. But drug binding to plasma proteins is the most important and dominating factor, therefore, is the focus of this chapter.

## 7.3 Modeling of Binding Equilibria

When drug molecules bind to protein molecules, numerous equilibria are possible. These can be expressed by different kinds of equilibrium constants, which reveal diverse viewpoints in envisioning the equilibria.

The binding of a drug $C$ by a protein $M$ with $n$ binding sites may be expressed in terms of a *stoichiometric* analysis or on the basis of a *site-oriented* model. The reliance of binding on drug level can always be connected in terms of $n$ stoichiometric binding constants, $k_i$, even if there are connections between sites that heighten or decrease binding affinities. A matching correlation in terms of site binding constants, $K_j$, under the broadest conditions rests on the definition of $n \cdot 2^{n-1}$ different constants of which $2^n - 1$ are independent. The binding constants resulting from the two approaches are not the same in magnitude or in the binding step to which they must be allocated [6–8]. Both simulations are presently encountered in technical writings, at times short of openly recognizing, which model is employed.

### 7.3.1 Stoichiometric Model

For the equilibrium between a compound (drug) $C$ and a protein molecule $M$ with several binding sites, the stepwise stoichiometric equilibria can be represented briefly as follows:

$$M + C \Longleftrightarrow MC_1$$
$$MC_1 + C \Longleftrightarrow MC_2$$
$$\ldots$$
$$MC_{b-1} + C \rightleftarrows MC_b$$

After writing the related equations for the stoichiometric equilibrium constants $k_i$ and species concentrations, the concentration of each complex $[MC_i]$ may be conveyed as a function of free drug concentration $[C]$, free protein concentration $[M]$, and stoichiometric binding constants:

$$k_1 = [MC_1]/([M]\cdot[C]) \Rightarrow [MC_1] = k_1 \cdot [M]\cdot[C] \quad (7.1)$$

$$k_2 = [MC_2]/([MC_1]\cdot[C]) \Rightarrow [MC_2] = k_1 \cdot k_2 \cdot [M]\cdot[C]^2 \quad (7.2)$$

$$\ldots$$

$$k_b = [MC_b]/([MC_{b-1}]\cdot[C]) \Rightarrow [MC_b] = k_1 \cdot k_2 \cdot \ldots \cdot k_b \cdot [M]\cdot[C]^b \quad (7.3)$$

The number of molecules of drug bound per molecule of protein ($N$) can be expressed either as a function of free drug concentration ($C_f = [C]$), total drug concentration ($C_t$), and protein concentration ($C_m$), or as a function of the concentration of the individual species:

$$N = \frac{C_t - C_f}{C_m} = \frac{\text{moles bound drug}}{\text{moles total protein}} = \frac{[MC_1] + 2\cdot[MC_2] + \ldots + n[MC_n]}{[M] + [MC_1] + [MC_2] + \ldots + [MC_n]} \quad (7.4)$$

With the necessary substitutions, obtained from Eq. (7.1) to (7.3), a general equation describing $N$ as a function of stoichiometric binding constants ($k_i$) and free drug concentration ($C_f$) can now be obtained: [9]

$$N = \frac{C_t - C_f}{C_m} = \frac{k_1 \cdot C_f + 2\cdot k_1 \cdot k_2 \cdot C_f^2 + \ldots + b\cdot(k_1 \cdot k_2 \cdot \ldots \cdot k_n)\cdot C_f^n}{1 + k_1 \cdot C_f + k_1 \cdot k_2 \cdot C_f^2 + \ldots + (k_1 \cdot k_2 \cdot \ldots \cdot k_n)\cdot C_f^n} \quad (7.5)$$

### 7.3.2 Site-Oriented Model

In the case of the site-oriented approach, each site is considered to have a fixed, invariant affinity, and its binding of drug is independent of events at other sites. Accordingly, stoichiometric binding constants ($k_i$) are replaced with site binding constants ($K_i$), and the binding equilibrium may be written in the same way for all binding sites ($M_i$ represents the $i$th binding site on the protein)

$$\mathrm{M_i + C \rightleftarrows M_iC}$$

The equation for the site-oriented equilibrium constant and species concentrations may be written as a function of drug–protein complex, free ligand, and free receptor concentration:

$$K_i = [M_iC]/([M_i]\cdot[C]) \Rightarrow [M_iC] = K_i\cdot[M_i]\cdot[C] \tag{7.6}$$

For each binding site, the number of moles of ligand bound per moles of receptor ($i$th site) is:

$$N_i = \frac{[M_iC]}{[M_i]+[M_iC]} = \frac{K_i[M_i][C]}{[M_i]+K_i[M_i][C]} = \frac{K_iC_f}{1+K_iC_f} \tag{7.7}$$

Then, for the receptor as a whole:

$$N = \frac{C_t - C_f}{C_m} = \sum_{i=1}^{n} N_i = \sum_{i=1}^{n} \frac{K_iC_f}{1+K_iC_f} \tag{7.8}$$

Equation (7.8), even though commonly utilized and described in technical documents because of its minimalism and suitability, often results in nonrealistic values for binding constants and free levels. This frequently occurs when the binding constants rise with bigger occupancy of the protein molecule, as is the case with numerous enzymes or carrier proteins (like hemoglobin). Recently, the phenomenon was also described for drug binding to plasma proteins and attributed to self-induction of binding [10]. In such cases, the stoichiometric binding model should be used.

Equation (7.8), when used for one binding site and one protein, can be used to calculate the popular free drug fraction:

$$\frac{C_t - C_f}{C_m} = \frac{KC_f}{1+KC_f} \tag{7.9}$$

### 7.3.3 Free Drug Fraction

When the protein concentration is much higher than the drug concentration, and there is only one binding site, the free concentration of protein $[M]$ is almost the same as the total concentration of protein ($C_m$), and Eq. (7.6) can be rewritten as:

$$K = \frac{C_t - C_f}{C_m\cdot C_f} \tag{7.10}$$

which can be rewritten as:

$$K = \frac{\frac{C_t}{C_f} - 1}{C_m} \tag{7.11}$$

$$K\cdot C_m = \frac{C_t}{C_f} - 1 \tag{7.12}$$

$$1 + K\cdot C_m = \frac{C_t}{C_f} \tag{7.13}$$

$$\text{free drug fraction} = \frac{C_f}{C_t} = \frac{1}{1+K\cdot C_m} \tag{7.14}$$

Equation (7.14) is typically used to calculate the free drug fraction as a function of the total concentration of protein and the drug–protein binding constant. The equation shows that the free drug fraction is independent of the drug concentration, as long as the protein concentration is in large excess.

## 7.4 Bioanalytical Methods for Studying Drug–Protein Binding

Drug distribution in a living system is mostly affected by the movement of blood, which includes two main drug-binding proteins: AGP and albumin (HSA). Drugs that are not attached to blood proteins are immediately available for elimination and transformation (unidirectional processes) as well as for moving into tissues and attaching to receptors (reversible processes). Therefore, unbound concentrations are understood in pharmaceutical sciences to be better correlated to drug effect than total values. Nevertheless, most medicinal drug scrutinizing judgments continue to be centered on the quantity of unbound plus bound drug levels [11]. This divergence is largely initiated by past bias in the direction of determining total drug levels, but also by the greater practical obstacle of precisely quantifying unbound drug concentrations. Some scientists claim that unbound concentrations do not need to be measured when there is a constant unbound drug fraction within and between individuals [12]. However, this is seldom the case: while the unbound fraction is reasonably constant when drug levels are much lower than binding protein levels, the unbound fraction hinges on the level of binding protein, which is infrequently the same among people [10, 13, 14]. Moreover, the binding traits of proteins can be altered radically in genetic or metabolic diseases, as well as in people who have medical conditions such as burns or starvation. Likewise, several experiments described nonlinear protein binding even at low drug levels [10, 13, 15–17]. In such instances of atypical protein binding, the total drug level is not precisely related to drug activity and the unbound drug level should be examined instead.

Numerous methods have been used for analysis of unbound levels of analytes, and most of them are centered on separating the sample into unbound and bound parts followed by direct chemical analysis of unbound molecules [18]. Such methods based on fractionating include affinity chromatography, equilibrium dialysis (ED), UF, ultracentrifugation, and electrophoresis. These approaches are usually extensive, the examined drugs can be permanently separated by the sample preparation device and may be erroneous if the binding equilibrium shifts while the sample is separated. The liquid phase used by column-based approaches is very distinct from the makeup of a live system and can cause substantial biases [19–21]. Likewise, this happens for methods using mass spectrometers that are relevant only when the samples are prepared in compatible solutions, and generate valuable results only if the proportion between binding protein and drug is 1 : 1. If the binding ratio between protein and drug is more than 1, the mass spectra of the drug–protein adducts become too complex to be used.

While methods established using proteins attached to a surface are based on a rapid conversion between free and attached states, UF and ultracentrifugation can be used when the change between states is slow, so the unattached molecules can be separated without changing the molecules bound to the protein. Flow-dialysis approaches are dependable but were applied when the binding ratio is 1 : 1. Capillary electrophoresis is very flexible and is applicable for any binding ratio between drug and binding protein, but the heat inside the capillary can vary substantially during a test and the approach only works with a limited number of compatible solution compositions.

A new sample processing technique that has been applied for monitoring unbound drug levels is solid phase microextraction (SPME) [14, 22, 23]. Even though this method based on partial fractionation has been used for quite a while, it is seldom cited in review papers about unbound level measurement, possibly since it is largely used in fundamental studies and has been used only in recent times for unbound level measurement. The strong points of SPME consist of small size and the equilibrium-based approach, which allow the un-intrusive determination of drug–protein binding. Approaches based on SPME are appropriate for samples from biological organisms and at a broad array of levels if a suitable extraction phase is selected.

Approaches that can determine unbound drug levels without dissociating the drug from the protein consist of spectroscopy, calorimetry, and surface plasmon resonance. These approaches hinder the least the drug–protein association equilibrium but are not appropriate for intricate biological samples.

### 7.4.1 Outline of Bioanalytical Approaches

Examination of the association of drugs with proteins is an essential early investigation in drug discovery, whether it is dedicated to the measurement of overall protein binding or the association between drugs and receptors. While the value of overall protein binding in plasma is a key factor when it comes to therapeutic applications, the quantitative description of drug binding to specific proteins is much more essential for exploring the interaction of drugs with their receptors. In order to attain truthful results, the binding constants are generally measured with isolated proteins that are either bound to a solid support or prepared in solutions that match physiological environments. The equilibrium between ligands and receptors is usually examined based on variations in the characteristics of the unbound or bound ligand molecules. Still, a few approaches such as calorimetry and surface plasmon resonance hinge on variations in the sequence of the binding receptor.

Determination of unbound drug levels in complicated biological samples is done either to establish the degree of total plasma protein binding (without essentially categorizing a precise binding protein) or to guarantee therapeutic effectiveness for drugs with narrow therapeutic windows and increased attachment to plasma proteins.

Measurement of the general extent of drug binding to macromolecules is a critical part of drug discovery, both in the preclinical and clinical stages. While the key drug-binding proteins are HSA and AGP, plasma includes many other proteins; therefore, most pharmaceuticals will have a definite level of binding with plasma constituents. To examine these relations, a drug–plasma association determination is needed.

### 7.4.2 Affinity Chromatography

The majority of studies regarding binding of drugs to particular plasmatic proteins are using affinity chromatography with the protein-bound on column. With this method, the examined compound is introduced as a small amount (zonal elution) or very large amount (frontal analysis) onto a column with the protein, and the profile of the eluate is observed. The benefit of zonal elution is that it uses a low amount of sample but accuracy and precision can be low occasionally [24]. The high volume of sample used in frontal analysis allows saturation of the protein attached to the separation column, which produces superior accuracy and precision at the detriment of having to use a larger amount of compound; although this is not an issue with widely available compounds, it can thwart usage of this approach for newly discovered drugs. Once, an affinity chromatography column is manufactured, it can be applied in many instances for any drugs that need to be investigated. The biggest trials for this method are to reduce unspecific attachment to the stationary phase and to make reproducible columns minus changing the binding affinity and natural structure of the receptor. This is typically done by making use of end-capped stationary phases customized for specific drugs and by careful development of the protein attachment approach.

A very active study group in affinity chromatography from the University of Nebraska issued numerous publications depicting the equilibrium between separated drugs and proteins. The unbound concentration of carbamazepine in a mixture with AGP was examined using an immobilized AGP column under monitored conditions. The scientists found low-affinity exchanges with the chromatographic support and high-affinity ones with the protein. By adjusting for the interaction with the support, the equilibrium constant between carbamazepine and AGP at 37 °C and pH 7.4 was precisely measured. Competition investigations revealed that these exchanges were happening at the same site that attaches propranolol on AGP. Controlled-conditions studies revealed that the shift in enthalpy was the main operating force for the binding of carbamazepine to AGP. Their findings offered a much more comprehensive image of how carbamazepine attaches to AGP in human plasma [20]. In another study, the same study group researched binding of lidocaine to two plasma proteins, HSA and AGP, using both frontal analysis and zonal elution. It was demonstrated that lidocaine has potent binding to AGP and weak to moderate binding to HSA. Competitive trials with site discriminatory probes demonstrated that lidocaine interacts with Sudlow site II of HSA and the propranolol site of AGP. Their analysis reveals how affinity chromatography can be applied to investigate the binding of a drug with several serum proteins and deliver comprehensive data on the interaction sites and equilibrium constants that are implicated in such processes [25]. A detailed study of the binding of antimalarial compounds to plasma proteins was done by Zsila et al. [21]. The researchers used multiple approaches, such as induced circular dichroism and affinity chromatography to research particular connections between six antimalarial agents of quinoline and acridine types to AGP. Induced circular dichroism spectra showed binding of amodiaquine, primaquine, tafenoquine, and quinacrine to AGP, the serum level of which greatly increases in Plasmodium infections. Association constant values of about $10^5$–$10^6$ $M^{-1}$ could be established. Analysis of the UV spectra of the drug–AGP complexes indicated the presence of the ligands into the central hydrophobic space of the protein. Results of fluorescence tests also validated the AGP binding of these drugs and offered further insights into binding specifics. Fluorescence and circular dichroism displacement tests revealed the high-affinity AGP binding of mefloquine (association constant of about $10^6$ $M^{-1}$). HSA association constants assessed from affinity chromatography ($10^3$–$10^5$ $M^{-1}$) were observed to lag behind those for AGP. In the case of chloroquine, no substantial binding interaction was observed either with AGP or HSA. The scientists also presented the pharmacological facets of their findings in great detail [21].

The preceding findings revealed that chromatographic columns with bound AGP keep the initial binding affinity in solution and can be manufactured reproducibly – which have been the major difficulties when attaching AGP to solid supports.

An extremely carefully performed research about the binding of phenytoin to attached HSA revealed that the drug can bind with the protein at the warfarin-azapropazone, indole-benzodiazepine, tamoxifen, and digitoxin sites. This was achieved through frontal analysis and competitive binding zonal elution experiments, the latter of which used four probe drugs for the major and minor binding sites of HSA infused into the presence of liquid phases containing known levels of phenytoin. Both allosteric interactions and immediate binding for phenytoin seem to happen at the warfarin-azapropazone and tamoxifen sites. It was determined that this rather complex binding system shows the significance of discovering the binding regions on HSA for particular drugs as a means of identifying their transport in plasma and describing their ability for drug–drug interactions [24].

Anesthetics are a category of drugs that have to be especially well scrutinized, given the high risk they pose. Because numerous studies indicated that their effect is better related to unbound levels, numerous researchers are examining drug–drug interactions by determining the binding constants between anesthetics and HSA. For example, propofol and halothane are clinically used broad anesthetics, which are carried primarily by HSA in the plasma. Scientists from the University of Pennsylvania Medical Center described anesthetic–HSA interactions in solution using elution chromatography, isothermal titration calorimetry, hydrogen-exchange experiments, and geometric analyses of high-resolution structures. They discovered that propofol has a much stronger binding affinity for HSA than halothane. The binding stoichiometry of the two compounds was also significantly distinct, with propofol binding to albumin in a 2 : 1 ratio as contrasted with 7 : 1 for halothane. Hydrogen-exchange experiments in sequestered recombinant domains of HSA indicated that propofol-binding sites are mainly found in domain III, while halothane sites are more broadly scattered. Besides the pharmacokinetic consequences of propofol dislocating halothane from some HSA sites, their data advocates that HSA might be an appropriate platform for additional characterization of the general anesthetic structure–activity relationship [26]. The group has issued an interesting commentary about the lack of competition between bilirubin and propofol for binding sites on HSA, founded on the fact that association of these two compounds happens at different locations [27].

### 7.4.3 Solid Phase Microextraction

SPME is a quick sample preparation approach in which a minor quantity of extracting phase is put in interaction with a sample for a measured period of time. Because of the fractional extraction method, SPME can be used to determine unbound drug concentrations in either negligible or non-negligible mode [28]. The extraction phase has to be sensibly selected so that it does not extract the binding protein. Yet, although method development must be carefully done, SPME allows for an important variety of extraction phases and analytical methods, which is demonstrated by the large number of research papers published on the topic. A researcher from Aristotle University applied this technique to study the binding of seven molecules to HSA. The preferred extraction mode was "negligible depletion" when the level of analyte in the sample remains nearly the same after extraction. This way, the bound drug–unbound drug equilibria are not shifted. Calibration curves were built for each drug by HPLC-fluorescence and HPLC-UV analysis. Binding of each compound to HSA was studied individually [19]. As matrix effects are critical when quantifying unbound levels, Heringa et al. examined the impact of proteins on the rate of microextraction. It was found that there is a significant effect from the protein being there on the rate of octylphenol uptake and blockage of the microextraction phase seems to happen. Nevertheless, the quantity of protein adsorbed on the probe was low and did not reduce or enhance the determined uptake of octylphenol. Based on measurement of unbound octylphenol, the authors effectively determined the apparent affinity constant for serum albumin [29].

### 7.4.4 Ultrafiltration

Possibly the most widespread fractionation technique for determining unbound drug levels in various biological mixtures is UF. UF is frequently the technique of choice particularly in clinical labs because of its convenience and rapidity. The biological sample is usually centrifuged in devices with semipermeable filters and the unbound compound is determined in the ultrafiltrate.

Some of the main hurdles with using UF in clinical laboratories are the temperature at which the procedure is done and the specificity and sensitivity of the determination procedure. Even though it is now well acknowledged that variations in temperature considerably change the degree of drug binding to proteins, most clinical labs carry on measuring unbound levels by UF at room temperature (25 °C). In a comprehensive investigation, Kodama et al. determined the unbound phenytoin concentration by UF at body temperature and room temperature and revealed a change of 44% in binding affinity, which obviously shows the importance of controlling the temperature for such determinations [30].

Another problem is the fact that many clinical methods are based on antibodies that have cross-reactivity to drug metabolites and sometimes even to endogenous compounds. While this may not be an issue when total drug levels are determined, it can lead to reduced accuracy when the metabolites are less bound to serum proteins than the unchanged drug. In such situations, the proportion of drug to metabolite level in the ultrafiltrate can surge in an erratic mode, depending on the metabolic rate and time of sample retrieval. Subject to the degree of metabolite cross-reactivity, errors can be as high as 100%. In these circumstances, quantitative techniques using chromatography should be employed. Regrettably, due to rapidity and accessibility, immunoassays remain the quantitative approach of choice in many clinical labs.

Furthermore, some UF-based methods have as disadvantage of nonspecific binding of drugs to the UF apparatus. Nevertheless, this can simply be verified by filtering buffer samples without macromolecules but with known ligand levels and determining the amount in the ultrafiltrate. Nonspecific binding can then be measured by relating the determined concentration with the prepared one. Lastly, there is a lot of inconsistency between studies concerning the molecular mass cutoff of the membrane used for UF, ranging from 3 to 50 kDa. Presently, there is no agreement concerning the ideal experimental method and more study is desirable.

Some of the most scrutinized unbound compound levels in medical practice are from the class of antiepileptics. Of these, phenytoin and valproic acid are well-known medicines that were comprehensively investigated. Iwamoto et al. assessed the association between unbound phenytoin concentrations and clinical responses, as well as the factors swaying protein binding of phenytoin. Their outcomes show that the unbound phenytoin level is more useful than the total concentration for monitoring antiepileptic effects in patients receiving phenytoin monotherapy. Also, they found that the unbound phenytoin fraction was considerably influenced by aging, creatinine clearance, and plasma albumin concentrations [31]. Other scientists used the same analytical approach to validate a new equation for computing normalized phenytoin levels for patients with atypical serum albumin concentrations [32]. A team from the Miyazaki Medical College Hospital has issued several papers regarding the unbound concentration of phenytoin in pediatric patients. Binding parameters of phenytoin to pediatric serum proteins were associated with *in vivo* and *in vitro* binding parameters in adult subjects reported by other investigators. Their results advocate that although the number of binding sites is the same, there are some changes in binding constants between pediatric and adult subjects, with adult albumin having an affinity approximately 1.2 times higher [33]. Mamiya et al. investigated drug–drug interactions between phenytoin and valproic acid in patients with severe motor and cerebral disabilities with epilepsy and found that hypoalbuminemia and valproate coadministration with phenytoin increased the unbound fraction [34].

Valproic acid is a widely prescribed anticonvulsant and mood stabilizer. Although unbound concentration monitoring for valproic acid is not as well recognized as for phenytoin, many studies indicate its usefulness. All these studies were completed by using UF and immunoassay. Alvarez et al. confirmed that measurement of unbound fractions is difficult and established a model for forecasting unbound valproic acid level as a function of total amount and albumin level. The new model was confirmed against investigational data attained by UF. Their investigation revealed that there are substantial dissimilarities between total valproate and unbound valproate which further endorses the need to determine or calculate the unbound portion [35]. Ueshima et al. from the Okayama University Hospital studied the rapport between unbound valproic acid levels and effectiveness in the case of intractable epileptic youngsters. An augmented unbound plasma level of valproic acid detected in such high-dose therapy is expected to regularly cause toxicity, but the unbound level is rarely determined in therapeutic drug monitoring activity, and the total valproic acid level is commonly measured in its place. The researchers revealed that the free level nonlinearly increased as the total level increased, and that free valproic acid concentrations in infants are usually higher and vary more widely than those in grown-up patients. They finished by showing that free levels in neonates and infants should be carefully checked and used to personalize dosage regimens of valproic acid in intractable epileptic children [36]. The unbound level of valproic acid can also change unpredictably subsequent to rapid intravenous administration. Researchers from Abbott Laboratories and the University of Alabama characterized valproate protein binding in patients with epilepsy who achieve transient high (>150 mg/l) total serum levels following rapid infusion at very large amounts given. This was performed by determining both total and free valproic acid levels. One- and two-binding site representations were modeled in a nonlinear mixed effects population analysis framework. Because of the rapid administration of high doses, the authors found free valproate levels much higher than in previous studies, further highlighting the importance of monitoring unbound concentrations [37]. Sproule et al. also recognize that understanding total valproate levels can be complicated. The researchers endorse monitoring free valproate levels in order to streamline the understanding of drug concentrations, particularly with dosage changes at higher levels and in elderly patients, even if albumin levels are within the normal window [38].

The measurement of unbound levels for phenytoin and valproic acid are some of the best-known applications of monitoring unbound concentrations in clinical practice.

Several reports that challenge the present paradigm about the calculation and measurement of unbound drug levels were published by a scientist from Genentech [6, 10, 13, 39, 40]. Many articles show that the free-drug ratio is concentration dependent, which should be considered in the understanding of drug pharmacokinetics as well as in modeling. Naproxen binding to proteins in human serum is considered as an illustration of the method. It is shown that the supposition that all binding sites of the protein have the same affinity yields the slowest possible concentration increase of the unbound drug fraction, while the assumption that a drug binds to a single binding site yields the highest possible value of the unbound fraction for a given drug level. Unbound drug levels were obtained by UF after the drug was incubated with human serum for one hour. The ultrafiltrate was analyzed by LC-MS. No nonspecific binding of naproxen to the centrifugal filter devices was observed by comparison of the naproxen concentrations in plasma water and in the ultrafiltrate obtained after its centrifugation [6, 40].

A systematic study of atazanavir pharmacokinetics was performed by Barrail-Tran et al. The researchers determined the in vitro binding features of atazanavir and estimated whether plasma protein binding to HSA and AGP influences the pharmacokinetics of atazanavir in HIV-infected patients. For the in vitro study, atazanavir protein binding features were analyzed in AGP- and HSA-containing purified solutions. Atazanavir was found to bind to AGP on a high-affinity saturable site and to HSA on a low-affinity nonsaturable site. For the in vivo study, biological samples from 51 patients were collected prior to drug intake at week 6. Atazanavir levels were determined by LC-MS/MS, with unbound levels being fractionated by UF. HSA concentrations, AGP concentrations, and phenotypes were also measured in these patients. Their results indicate that atazanavir pharmacokinetics varies reasonably with its protein binding, especially to AAG, without predictable clinical concerns [41]. This is not surprising, since atazanavir is bound less than 90% to serum proteins.

Dawidowicz et al. published two papers about the abnormal association between the unbound fraction of a drug and its total level in drug–protein systems. Their studies are surprising, indicating that the unbound fraction of some drugs go up with the decrease in total levels. They used both purified HSA and human plasma to investigate the binding of propofol, fatty acids, indomethacin, and lidocaine to proteins. Unbound drug compounds were fractionated by UF through a 10 kDa molecular mass cutoff membrane and measured by LC. The tests carried out in this research show that ligand hydrophobicity changes the relationship between the unbound ligand level and its total concentration and that similar abnormal variations of the unbound drug level are observed not only for drugs interacting with different binding sites on HSA but also for basic local anesthetics that bind to AGP. Their outcomes extended the present understanding about drug–protein binding and are supporting further studies [15, 16]. UF has also been combined with capillary electrophoresis for measuring unbound levels and to study the pharmacological effect of anthracyclines and taxanes. The kinetics of nonspecific binding of anthracyclines in the presence of taxane and vice-versa were examined. For determination, a novel sweeping method was developed and showed improved limits of detection for both drugs versus previously reported approaches. The proposed technique was successfully utilized for serum samples. The main benefits were short analysis time, high peak efficiencies, and applicability for the fractionation of different mixtures of anthracyclines and taxanes [42].

### 7.4.5 Equilibrium Dialysis

While UF is the most prevalent technique, particularly in clinical sceneries, ED is still taken as the golden standard for determining unbound drug levels and continues to be used in many research environments. The method is based on drug diffusion across a semipermeable membrane that separates the sample to be investigated from a buffer solution. The membrane should be permeable to the drug and not to the protein. The device is incubated until equilibrium is reached and the unbound drug is determined in the buffer solution. The main difficulties with this approach are extended equilibration periods (up to two days), problems for drugs with low solubility in water, volume changes due to variations in osmotic pressure, artifacts generated by electrically charged drugs or proteins, and nonspecific adsorption to the ED apparatus. Although quick ED devices have been introduced, the equilibration time is still long, up to six hours. Moreover, this quick equilibration technique requires stirring and mixing which may result in sample loss and/or leakage through the membrane. Since ED is mostly used in research settings, determination of the unbound ratio is usually done by chromatography and mass spectrometry, and not immunoassays as is the case for UF.

Many articles report using both ED and UF to measure unbound drug levels, for cross-validation reasons. For example, Vita et al. used both approaches to determine the serum protein binding of roscovitine at 25 and 37 °C. Binding of roscovitine to HSA was constant (about 90%) within the concentration window investigated while binding to AGP decreased with increasing drug level, showing that albumin is more important in clinical settings. Protein binding was higher at 25 °C compared to 37 °C. The outcomes found by ED were established to be a decent match to those achieved by UF [43]. In an extremely cited paper, Nowak et al. employed ED to endorse results for unbound mycophenolic acid determinations obtained by UF. HSA, high concentrations of the primary glucuronide metabolite, and sodium salicylate were established to considerably influence mycophenolic acid binding to proteins. The inference of the project was that the pharmacological activity of the drug is a function of the free level [44].

A very hands-on approach for determining unbound drug levels in whole blood using rapid ED was developed by Chen et al. Chlorpromazine, chloroquine, propranolol, and a proprietary compound of different erythrocyte partitioning properties were evaluated to measure the unbound level in whole blood. Extra provisions were taken in each phase to circumvent the hemolysis of erythrocytes. The drug levels in blood and isotonic buffer were measured by LC-MS/MS. Their data prove that equilibrium can be obtained inside six hours with good mixing and stirring. Initial data established technique acceptability for drugs with low, medium, and high erythrocyte partitioning. Additionally, these investigational outcomes can be applied to elucidate nonlinear pharmacokinetic profiles and to forecast effect and doses across species [45].

It is a generally recognized theory that CNS activity of a compound is determined by the free brain drug level. Nevertheless, little experimental data are accessible in databases to provision this theory. Liu et al. methodically investigated this theory by exploring the relationship between in vitro binding affinity and in vivo activity determined as the drug level occupying 50% of the transporters for eighteen serotonin and dopamine transporter inhibitors. The free brain levels were determined from total brain levels and the free brain fraction, which was determined by the brain homogenate technique. The in vitro free ratio in brain homogenate and plasma for each drug was measured by a 48-well rapid ED device. The results of this study were very remarkable, showing that forecast of the biophase drug level by using the free brain level rather than the total brain levels results in an approximately 100-fold enhancement in exactness. A 10-fold enhancement was also noticed by using the free serum levels rather than the total plasma levels to predict the biophase level in the brain. These projects strongly support the theory that drug action in the brain is much better related to the free brain drug levels [46, 47].

As in the above projects, numerous scientists are focusing on determining unbound drug levels in the target organs and not in blood. The unbound drug in the tissue is the most significant parameter for pharmacokinetic and pharmacodynamic modeling. Nevertheless, this is not a trivial undertaking, and many projects are still done by obtaining tissue samples and determining total levels.

Hsueh et al. estimated the in vitro and in vivo connection between unbound drug levels in serum and in fatty tissue for six commercial drugs with varied properties and an eclectic range of plasma protein binding values. The biological samples (plasma and homogenized diluted fatty tissue) were dialyzed against buffer for 4 hours in a 96-well ED apparatus to determine the unbound drug level. Of specific concern was the technique for determining the free level in the fatty tissue. Since the undiluted homogenized tissue could not be processed directly, a wide range of dilutions were made. The unbound drug ratio was measured in these diluted samples. Finally, the fraction free in undiluted fatty tissue was measured by extrapolating from the free fraction in diluted tissue versus dilution factor. The overall deduction of the project was that the unbound drug concentration in fatty tissue is comparable to the unbound level in plasma within a 3-fold error window [48]. Even though this inaccuracy can be too great when accurate results are needed, the scientists propose that unbound drug levels in serum could be employed as a substitute for unbound levels in tissues, particularly in a resource-limited setting.

Gefitinib, an inhibitor of epidermal growth factor receptor tyrosine kinase, shows wide intersubject pharmacokinetic variability, which may cause alterations in treatment consequences. Since free drug levels are more pertinent to pharmacological and toxicological reactions, gefitinib binding in serum and factors influencing this process were considered both in vitro and in cancer patients. An ED method using 96-well microdialysis plates was enhanced and certified for measuring the free ratio of gefitinib in serum. It was discovered that gefitinib was broadly bound in human, rat, mouse, and dog serum with mean fraction unbound values of 3.4%, 3.8%, 5.1%, and 6.0% correspondingly. Therefore, the variable plasma protein levels detected in cancer patients will disturb gefitinib-free levels with consequences for intersubject variation in drug toxicity and response, meriting the requirement to determine the concentrations of unbound compounds [49]. Comparable studies were done for anticoagulants such as warfarin and phenprocoumon [50, 51].

Important trials in determining unbound levels are faced for drugs with great binding that is active at low levels. Moreover, some drugs have significantly different binding as a function of patient gender, age, and disease state. For these complex circumstances, the gold standard technique for measuring unbound levels is isotope dilution ED. This method was employed by Hackbarth et al. to explore the accuracy of calculated unbound testosterone concentrations. Even though total testosterone is usually thought to be appropriate for diagnosing major androgen excess or deficiency, unbound (bioactive) testosterone is of greater diagnostic importance. Such cases are met in hyper or hypothyroidism, liver cirrhosis, obesity, or exogenous sex hormone use, especially estrogen supplementation. Serum levels of the corresponding binding globulin also rise with age, often shifting unbound testosterone concentrations disproportionately to total testosterone levels. The researchers of the study contrasted three recognized and two new equations that varied only by their testosterone association constants with isotope dilution ED in two patient clusters with not the same gender allocations. Irrespective of the equation, ≥25% of samples presented objectionable deviation from results obtained by isotope dilution ED. The researchers concluded that the application of many unbound testosterone equations to broader populations will often yield outcomes that vary substantially from isotope dilution ED [52]. Clearly, the best method in this situation would be to continually determine the unbound testosterone level.

### 7.4.6 Other Approaches

Many other approaches have been developed to measure unbound drug levels and drug–protein binding, but their applicability is presently rather narrow. A remarkable study that extends the concept of unbound levels to larger molecules such as proteins was done by Speeckaert et al. to study the possible association of vitamin D binding protein with lipoproteins. The availability of vitamin D binding protein in lipoprotein fractions was inspected using precipitation, gel permeation chromatography, and ultracentrifugation. Total and actin-unbound protein levels were evaluated by immunonephelometry and enzyme-linked immunosorbent assay. The project revealed that the lipid-bound vitamin D binding protein fraction is of greater clinical significance than originally believed [53]. In order to test "the unbound bilirubin hypothesis," Calligaris et al. verified in vitro cytotoxicity in four cell lines exposed to different unbound bilirubin levels acquired by changing total bilirubin/albumin fraction, using serum albumins with different binding affinities, and/or displacing unconjugated bilirubin from albumin with a sulfonamide. Unbound bilirubin was measured by an adapted, negligibly diluted peroxidase approach. These studies clearly showed that bilirubin-induced cytotoxicity in each cell line can be predicted very well by unbound bilirubin regardless of the production site and level of albumin, or total bilirubin concentration [54].

One of the approaches that is progressively more being applied for unbound level determination in development laboratories is SPME. Although the approach has been mainly used for projects with isolated proteins, some articles present its usefulness for intricate biological matrices. The capability of SPME to determine protein binding for drugs with very high binding constants is above all significant as this is challenging to be done with standard approaches such as ED and UF. To investigate such compounds, extraction phases with high extraction constants for the studied chemical should be used. The applicability of this approach for determining drug binding to proteins in human plasma was established when it was validated on five common medicines: ibuprofen, warfarin, verapamil, propranolol, and caffeine, with high, intermediate, and low binding properties [14]. In another article, non-fouling microextraction coatings that can be applied in direct contact with bio-fluids were established and used for quick drug quantification and determination of drug plasma protein binding [55].

### 7.4.7 Alternatives to Determining Plasma Protein Binding

When unbound drug levels are too problematic to determine or the technique is too costly, the total drug level can be normalized by using different procedures, such as those published by Sheiner, Winter, and Tozer for phenytoin [11, 32, 56] or those suggested by Hackbarth for testosterone [52]. However, it was shown numerous times that equations are typically not as good as actually determining unbound levels [32, 35, 57, 58].

Consequently, as bioanalytical techniques come to be more sensitive, accurate, and precise, we will surely observe an increase in monitoring of unbound drug levels.

## 7.5 Impact of Drug–Protein Binding on Pharmacokinetic Parameters

The extent of protein binding in plasma is one of the important factors affecting the drug distribution and action. The extent of distribution is of clinical importance because it determines the amount of the drug that must be administered to produce desirable therapeutic concentration. The impact of protein binding on distribution and pharmacokinetics can be discussed in terms of volume of distribution ($V_d$), clearance, and half-life:

### 7.5.1 Volume of Distribution

Apparent $V_d$ describes the extent of distribution as:

$$V_d = \frac{\text{total amount of drug in the body}}{\text{plasma concentration}}$$

$V_d$ is expressed as the fluid volume that would be required to contain the total drug amount, at the same concentration as detected in the plasma. It can be considered as a virtual parameter of mathematical ratio, rather than a real physiological volume. $V_d$ also represents the volume of plasma that must be cleared by the elimination organs (liver and kidneys), which extract the drug and remove it from the body.

Although $V_d$ seems a virtual and abstract parameter, it is important because it relates the total body drug load to plasma concentration and together with clearance determines the elimination rate and half-life. The degree of protein binding in both plasma and tissues can significantly affect the extent of $V_d$. A high intracellular, as well as tissue protein binding, can lead to a high value of $V_d$. A high plasma protein binding leads to a low value of $V_d$. So $V_d$ somewhat represents the relative drug affinity between the plasma and tissues.

### 7.5.2 Clearance

Clearance is the volume of plasma cleared of drug per unit time:

$$\text{CL} = \frac{\text{Rate of elimination from the body}}{\text{Plasma concentration}}.$$

It is an important pharmacokinetic parameter that determines the duration of drug action, the time interval between doses, and the time to reach equilibrium during repeated dosing. It indicates the average plasma concentrations at steady state, which determines the therapeutic activity. It is the best measurement of the ability of the body to remove the drug away through the elimination organs, mainly liver and kidneys. Because clearance is constant, the greater CL value, the greater will be the rate of drug removal from the body. Clearance of a drug can significantly be affected by the extent of its protein binding. Higher drug–protein bound tends to have a lower systemic clearance. Higher volume of distribution will typically result in slower drug elimination rates. Also, extensive protein binding in blood can result in low hepatic clearance. It was confirmed by studies of the relationship between plasma protein binding and hepatic clearance under equilibrium conditions.

In renal clearance, protein binding is very important in the urinary excretion of unchanged drugs, since it can affect its glomerular filtration, active secretion, and renal metabolism. Glomerular filtration is especially influenced by the present of unbound drug in blood since it is a physical filtration process with limits in molecular weight. Clearance of a lower plasma protein binding drug is more extensive.

### 7.5.3 Half-life

Half-life measures the rate of drug elimination, which ranges from a few minutes to many days for different drugs. Half-life is determined by drug metabolism and excretion, while the drug is circulating between blood and tissue. Half-life is the best indicator of changes in drug concentration with time and changes in therapeutic activity for many drugs after a single dose:

$$t_{1/2} = 0.693\, V_d/\text{CL}$$

As half-life is dependent on both the volume of distribution and the clearance, it is more difficult to directly predict the effects of protein binding on half-life.

## 7.6 Physicochemical Factors that Affect Protein–Drug Binding and Drug Distribution

Drug–protein binding equilibrium is sensitive to environmental changes, such as temperature, pH, and ion strength. *In vivo*, these physiological conditions are precisely maintained. For *in vitro* protein binding assays, these parameters should be carefully controlled to ensure the method is relevant for clinical applications.

Under standard physiological conditions, the physicochemical properties of drug molecule determine drug–protein binding profile. The lipophilicity of the drug molecule is important, as hydrophobic interaction is the common mechanism of protein binding, The molecular weight, conformation, and ionization also impact their interactions at the binding site. The structure and affinity relationship should be discussed for each drug. Also, the protein binding of a drug is affected by the presence of endogenous compounds or other drugs. If these endogenous factors vary under a disease stage, they will impact the protein binding profile as discussed below.

## 7.7 Physiological and Pathological Factors that Affect Protein–Drug Binding and Drug Distribution

Interindividual differences of drug–protein binding are common because protein bind can be impacted by different physiological and/or pathological conditions of different patients. They are usually caused by increased or decreased levels of albumin, AGP, and other proteins that are involved in protein binding equilibrium. They may also be caused by changing concentrations of endogenous ligands that competing binding sites, as well as drug–drug interactions. These variabilities may lead to significant differences in drug pharmacokinetics and bioavailability, which bring safety and efficacy concerns.

Plasma protein level is related to age, gender, and disease conditions. For example, Neonate and geriatric patients have lower albumin levels than normal adults. The plasma protein levels are controlled by process of protein synthesis, catabolism, distribution, and elimination. Burns, injuries, and surgeries may cause increased extracellular albumin and temperately reduced plasma albumin concentration. Liver diseases reduced protein synthesis, so also reduce albumin levels. The albumin concentration can also be reduced by renal failure, trauma, and many other disease conditions [59].

Disorders on serum protein profile and an inversion of the albumin/globulins ratio are characteristics of rheumatoid disease. The arthritic patients present a lower intravascular and body content of albumin than normal subjects. It has been suggested that plasma albumin levels at the acute rheumatoid disease phase are reduced to a third of their normal levels, whereas the levels of AGP increase up to twenty-fold. These changes are reflected by similar changes in their synthesis rate by the liver [60, 61].

Increasing free fatty acid level in plasma may be caused by physical, pathological, or psychological stress, such as trauma, surgery, or UV exposure that stimulate sympathetic and endocrine systems. Increased fatty acids plasma concentration can interfere drug–protein binding and increase free drug concentration, because of their strong binding affinity to albumin. For example, the role of stress on lidocaine concentrations as well as its protein binding was investigated. Under stress, lidocaine concentrations in serum showed a marked elevation in mouse models [62, 63]. Similarly, elevated lipid levels by high-lipid diet can also affect lidocaine level due to drug displacement from its binding site of proteins [64].

Other disease types, including cardiovascular, hepatic, renal, and diabetic diseases, also have significant changes in plasma protein and endogenous ligand levels. As these patients usually require multiple drug treatment, drug–protein binding often need to be considered for medication management and dosing calculation for optimal outcomes. There are been intensive clinical reviews and pharmacotherapeutics discussions focus on these clinical applications. Overall, drug–protein binding can significantly impact drug distribution and bioavailability, and is an important reason of inter-patient variations. Therefore, each drug's protein binding is characterized to control and ensure their clinical outcomes.

## References


1 Shargel, L. and Yu, A.B.C. (ed.) (2015). *Applied Biopharmaceutics & Pharmacokinetics*, 7e. McGraw-Hill.

2 Ryan, A.J., Ghuman, J., Zunszain, P.A. et al. (2011). Structural basis of binding of fluorescent, site-specific dansylated amino acids to human serum albumin. *J. Struct. Biol.* 174 (1): 84–91.

3 He, X.M. and Carter, D.C. (1992). Atomic structure and chemistry of human serum albumin. *Nature* 358: 209–215.

4 Zunszain, P.A., Ghuman, J., Komatsu, T. et al. (2003 Jul). Crystal structural analysis of human serum albumin complexed with hemin and fatty acid. *BMC Struct. Biol.* 7 (3): 6.

5 Schönfeld, D.L., Ravelli, R.B.G., Mueller, U., and Skerra, A. (2008). The 1.8-Å crystal structure of α1-Acid glycoprotein (orosomucoid) solved by UV RIP reveals the broad drug-binding activity of this human plasma lipocalin. *J. Mol. Biol.* 384 (2): 393–405.

6 Berezhkovskiy, L.M. (2007). On the calculation of the concentration dependence of drug binding to plasma proteins with multiple binding sites of different affinities: determination of the possible variation of the unbound drug fraction and calculation of the number of binding sites of the protein. *J. Pharm. Sci.* 96 (2): 249–257.

7 Niri, V.H., Mathers, J.B., Musteata, M.F. et al. (2009). Monitoring BTEX and aldehydes in car exhaust from a gasoline engine during the use of different chemical cleaners by solid phase microextraction-gas

chromatography. *Water Air Soil Pollut.* 204 (1-4): 205–213.

8 Trainor, G.L. (2007). The importance of plasma protein binding in drug discovery. *Expert Opin. Drug Discovery* 2 (1): 51–64.

9 Klotz, I.M. (1997). *Ligand-Receptor Energetics: A Guide for the Perplexed.* New York: Wiley.

10 Berezhkovskiy, L.M. (2010). On the influence of protein binding on pharmacological activity of drugs. *J. Pharm. Sci.* 99 (4): 2153–2165.

11 Lee, M. (2009). *Basic Skills in Interpreting Laboratory Data.* Bethesda, MD: American Society of Health-System Pharmacists.

12 Benet, L.Z. and Hoener, B.A. (2002). Changes in plasma protein binding have little clinical relevance. *Clin. Pharmacol. Ther.* 71 (3): 115–121.

13 Berezhkovskiy, L.M. (2010). On the possibility of self-induction of drug protein binding. *J. Pharm. Sci.* 99 (10): 4400–4405.

14 Musteata, F.M., Pawliszyn, J., Qian, M.G. et al. (2006). Determination of drug plasma protein binding by solid phase microextraction. *J. Pharm. Sci.* 95 (8): 1712–1722.

15 Dawidowicz, A.L., Kobielski, M., and Pieniadz, J. (2008). Anomalous relationship between free drug fraction and its total concentration in drug-protein systems - I. Investigation of propofol binding in model HSA solution. *Eur. J. Pharm. Sci.* 34 (1): 30–36.

16 Dawidowicz, A.L., Kobielski, M., and Pieniadz, J. (2008). Anomalous relationship between free drug fraction and its total concentration in drug-protein systems - II. Binding of different ligands to plasma proteins. *Eur. J. Pharm. Sci.* 35 (1-2): 136–141.

17 Vuignier, K., Schappler, J., Veuthey, J.L. et al. (2010). Drug-protein binding: a critical review of analytical tools. *Anal. Bioanal.Chem.* 398 (1): 53–66.

18 Lambrinidis, G., Vallianatou, T., and Tsantili-Kakoulidou, A. (2015 Jun). In vitro, in silico and integrated strategies for the estimation of plasma protein binding. A review. *Adv. Drug Deliv. Rev.* 23 (86): 27–45. https://doi.org/10.1016/j.addr.2015.03.011. . PMID: 25819487.

19 Theodoridis, G. (2006). Application of solid-phase microextraction in the investigation of protein binding of pharmaceuticals. *J. Chromatogr. B* 830 (2): 238–244.

20 Xuan, H., Joseph, K.S., Wa, C.L., and Hage, D.S. (2010). Biointeraction analysis of carbamazepine binding to alpha (1)-acid glycoprotein by high-performance affinity chromatography. *J. Sep. Sci.* 33 (15): 2294–2301.

21 Zsila, F., Visy, J., Mady, G., and Fitos, I. (2008). Selective plasma protein binding of antimalarial drugs to alpha(1)-acid glycoprotein. *Bioorg. Med. Chem.* 16 (7): 3759–3772.

22 Musteata, M.L. and Musteata, F.M. (2009). Analytical methods used in conjunction with SPME: a review of recent bioanalytical applications. *Bioanalysis* 1 (6): 1081–1102.

23 Vuckovic, D., Cudjoe, E., Musteata, F.M., and Pawliszyn, J. (2010). Automated solid-phase microextraction and thin-film microextraction for high-throughput analysis of biological fluids and ligand–receptor binding studies. *Nat. Protoc.* 5 (1): 140–161.

24 Chen, J.Z., Ohnmacht, C., and Hage, D.S. (2004). Studies of phenytoin binding to human serum albumin by high-performance affinity chromatography. *J. Chromatogr. B* 809 (1): 137–145.

25 Soman, S., Yoo, M.J., Jang, Y.J., and Hage, D.S. (2010). Analysis of lidocaine interactions with serum proteins using high-performance affinity chromatography. *J. Chromatogr. B* 878 (7-8): 705–708.

26 Liu, R.Y., Meng, Q.C., Xi, J. et al. (2004). Comparative binding character of two general anaesthetics for sites on human serum albumin. *Biochem. J* 380: 147–152.

27 Zhou, R. and Liu, R. (2010). Does bilirubin change the free concentration of propofol? *Acta Anaesthesiol. Scand.* 54 (5): 653–654.

28 Vaes, W.H.J., Urrestarazu Ramos, E., Verhaar, H.J.M. et al. (1996). Measurement of the free concentration using solid-phase microextraction: binding to protein. *Anal. Chem.* 68 (24): 4463–4467.

29 Heringa, M.B., Hogevonder, C., Busser, F., and Hermens, J.L.M. (2006). Measurement of the free concentration of octylphenol in biological samples with negligible depletion-solid phase microextraction (nd-SPME): analysis of matrix effects. *J. Chromatogr. B* 834 (1-2): 35–41.

30 Kodama, H., Kodama, Y., Itokazu, N. et al. (2001). Effect of temperature on serum protein binding characteristics of phenytoin in monotherapy paediatric patients with epilepsy. *J. Clin. Pharm. Therap.* 26 (3): 175–179.

31 Iwamoto, T., Kagawa, Y., Naito, Y. et al. (2006). Clinical evaluation of plasma free phenytoin measurement and factors influencing its protein binding. *Biopharm. Drug Dispos.* 27 (2): 77–84.

32 Anderson, G.D., Pak, C., Doane, K.W. et al. (1997). Revised Winter-Tozer equation for normalized phenytoin concentrations in trauma and elderly patients

with hypoalbuminemia. *Ann. Pharmacother.* 31 (3): 279–284.

33 Kodama, H., Kodama, Y., Shinozawa, S. et al. (2000). In vivo binding characteristics of phenytoin to serum proteins in monotherapy pediatric patients with epilepsy. *Int. J. Clin. Pharmacol. Therap.* 38 (1): 25–29.

34 Mamiya, K., Yukawa, E., Matsumoto, T. et al. (2002). Synergistic effect of valproate coadministration and hypoalbuminemia on the serum-free phenytoin concentration in patients with severe motor and intellectual disabilities. *Clin. Neuropharmacol.* 25 (4): 230–233.

35 Alvarez, M.T.R., Valino, L.E., Gabriel, E.F. et al. (2010). Relation between total and unbound valproate concentration and seric albumin. *Pharm. World Sci.* 32 (2): 289.

36 Ueshima, S., Aiba, T., Sato, T. et al. (2008). Individualized dosage adjustment of valproic acid based on unbound serum concentration in intractable epileptic children. *Yakugaku Zasshi-J. Pharm. Soc. Jpn.* 128: 92–92.

37 Dutta, S., Faught, E., and Limdi, N.A. (2007). Valproate protein binding following rapid intravenous administration of high doses of valproic acid in patients with epilepsy. *J. Clin. Pharm. Therap.* 32 (4): 365–371.

38 Sproule, B., Nava-Ocampo, A.A., and Kapur, B. (2006). Measuring unbound versus total valproate concentrations for therapeutic drug monitoring. *Therap. Drug Monit.* 28 (5): 714–715.

39 Berezhkovskiy, L.M. (2008). Some features of the kinetics and equilibrium of drug binding to plasma proteins. *Expert Opin. Drug Metab. Toxicol.* 4 (12): 1479–1498.

40 Berezhkovskiy, L.M. (2009). Consideration of the linear concentration increase of the unbound drug fraction in plasma. *J. Pharm. Sci.* 98 (2): 383–393.

41 Barrail-Tran, A., Mentre, F., Cosson, C. et al. (2009). Influence of alpha-1 glycoprotein acid concentrations and variants on atazanavir pharmacokinetics in HIV-infected patients included in the ANRS 107 trial. *Antimicrob. Agents Chemother.* 54 (2): 614–619.

42 Shakalisava, Y., Regan, F., and McKeon, L. (2011). *Feasibility of the Ultrafiltration Procedure when Studying Pharmacological Effect of Anthracyclines and Taxanes.* Atlanta, GA: Pittcon.

43 Vita, M., Abdel-Rehim, M., Nilsson, C. et al. (2005). Stability, pKa and plasma protein binding of roscovitine. *J. Chromatogr. B* 821 (1): 75–80.

44 Nowak, I. and Shaw, L.M. (1995). Mycophenolic-acid binding to human serum-albumin - characterization and relation to pharmacodynamics. *Clin. Chem.* 41 (7): 1011–1017.

45 Chen, S., Zhang, J., Huang, T.N. et al. (2007). A practical method for measuring free drug concentration in whole blood using rapid equilibrium dialysis (RED) device. *Drug Metab. Rev.* 39: 290.

46 Liu, X.R. (2009). Unbound brain concentration determines brain receptor occupancy. *Drug Metab. Rev.* 41: 27.

47 Liu, X.R., Vilenski, O., Kwan, J. et al. (2009). Unbound brain concentration determines receptor occupancy: a correlation of drug concentration and brain serotonin and dopamine reuptake transporter occupancy for eighteen compounds in rats. *Drug Metab. Dispos.* 37 (7): 1548–1556.

48 Hsueh, M.M., Wang, M.M., Sleckza, B. et al. (2009). Theory vs. realty: is free drug concentration in adipose equal to free concentration in plasma? *Drug Metab. Rev.* 41: 130.

49 Li, J., Brahmer, J., Messersmith, W. et al. (2006). Binding of gefitinib, an inhibitor of epidermal growth factor receptor-tyrosine kinase, to plasma proteins and blood cells: in vitro and in cancer patients. *Invest. New Drugs* 24 (4): 291–297.

50 Vorum, H. and Honore, B. (1996). Influence of fatty acids on the binding of warfarin and phenprocoumon to human serum albumin with relation to anticoagulant therapy. *J. Pharm. Pharmacol.* 48 (8): 870–875.

51 Vorum, H., Jorgensen, H.R.I., and Brodersen, R. (1993). Variation in the Binding-Affinity of Warfarin and Phenprocoumon to Human Serum-Albumin in Relation to Surgery. *Eur. J. Clin. Pharmacol.* 44 (2): 157–162.

52 Hackbarth, J.S., Hoyne, J.B., Grebe, S.K., and Singh, R.J. (2011). Accuracy of calculated free testosterone differs between equations and depends on gender and SHBG concentration. *Steroids* 76 (1-2): 48–55.

53 Speeckaert, M.M., Taes, Y.E., De Buyzere, M.L. et al. (2010). Investigation of the potential association of vitamin D binding protein with lipoproteins. *Ann. Clin. Biochem.* 47 (2): 143–150.

54 Calligaris, S.D., Bellarosa, C., Giraudi, P. et al. (2007). Cytotoxicity is predicted by unbound and not total billrubin concentration. *Pediatr. Res.* 62 (5): 576–580.

55 Musteata, M.L., Musteata, F.M., and Pawliszyn, J. (2007). Biocompatible solid phase microextraction coatings based on polyacrylonitrile and SPE phases. *Anal. Chem.* 79 (18): 6903–6911.

56 Bauer, L.A. (2008). *Phenytoin. Applied Clinical Pharmacokinetics*, 485–490. USA: McGraw-Hill Medical.

57 Mauro, L.S., Mauro, V.F., Bachmann, K.A., and Higgins, J.T. (1989). Accuracy of two equations in determining normalized phenytoin concentrations. *DICP* 23 (1): 64–68.

58 Tandon, M., Pandhi, R., Garg, S.K., and Prabhakar, S.K. (2004). Serum albumin-adjusted phenytoin levels: an approach for predicting drug efficacy in patients with epilepsy, suitable for developing countries. *Int. J. Clin. Pharmacol. Therap.* 42 (10): 550–555.

59 Jusko, W.J. and Gretch, M. (1976). Plasma and tissue protein binding of drugs in pharmacokinetics. *Drug Metab. Rev.* 5 (1): 43–140. https://doi.org/10.3109/03602537608995839. PMID: 829788.

60 Ballantyne, F.C., Fleck, A., and Dick, W.C. (1971). Albumin metabolism in rheumatoid arthritis. *Ann. Rheum. Dis.* 30 (3): 265–270.

61 Majithia, V. and Geraci, S.A. (2007). Rheumatoid arthritis: diagnosis and management. *Am. J. Med.* 120 (11): 936–939.

62 Saranteas, T., Tesseromatis, C., Potamianou, A. et al. (2002). Stress-induced lidocaine modification in serum and tissues. *Eur. J. Drug Metab. Pharmacokinet.* 27 (4): 229–232.

63 Saranteas, T., Mourouzis, C., Dannis, C. et al. (2004 Jul). Effect of various stress models on lidocaine pharmacokinetic properties in the mandible after masseter injection. *J. Oral Maxillofac. Surg.* 62 (7): 858–862.

64 Karakitsos, D., Karabinis, A., and Kotsiou, A. (2009). Changes of lidocaine levels in rat serum and tissue under stress and hyperlipidemic diets. *Eur. J. Drug Metab. Pharmacokinet.* 34 (3-4): 143–146.

# 8

# Drug Transport Across the Placental Barrier

*Valentina Bryant, Mansi Shah, Jennifer Waltz, and Erik Rytting*

*Department of Obstetrics and Gynecology, University of Texas Medical Branch, Galveston, TX, USA*

## 8.1 Introduction

Texas history elicits the call: "Remember the Alamo!" The importance of understanding maternal-to-fetal drug transport could be highlighted by the call: "Remember Thalidomide!" Only a few years after the sedative and antiemetic medication had entered the market, it was discovered that thalidomide use during pregnancy was the cause of a number of severe birth effects, including undeveloped limbs [1, 2].

In many instances, the use of medication during pregnancy cannot simply be avoided. Treatments for certain chronic conditions and other illnesses that arise are necessary because efforts to improve the health of the mother will promote the development of her unborn child (or children) [3]. In fact, a report from the French Pregnancy Cohort indicates that in 36 065 pregnancies followed between 2010 and 2013, 89.9% of pregnant women used at least one medication. The average (± s.d.) number of prescriptions filled was 8.1 (± 6.5) [4]. Accordingly, there is significant clinical relevance to the question of fetal exposure to maternal medication. In addition, although less common, there are instances where the placental transfer of a maternal dose facilitates fetal drug therapy [5].

This chapter will provide an overview of the pharmacokinetics of drugs administered during pregnancy, placental development and structure, functions of the human placenta, mechanisms of drug transport across the placenta, mechanisms of drug metabolism within the placenta, strategies to alter drug transport across the placenta, and finally, experimental models of the human placenta.

## 8.2 Pharmacokinetics of Drugs Administered During Pregnancy

Most clinical trials exclude pregnant women; therefore, dosing regimens for drug prescribed to pregnant women are based on data from clinical trials performed in men and nonpregnant women [6]. Anatomical and physiological changes in multiple organ systems compared to the nonpregnant state – and even changes associated with gestational age – can affect drug pharmacokinetics and pharmacodynamics during pregnancy [7, 8]. Therefore, it is imperative to understand the basic changes that occur during pregnancy which affect the absorption, distribution, metabolism, and elimination of drugs in order to minimize the risks associated with either insufficient therapeutic response or toxic effects in either the mother or the developing fetus. While additional details can be found in more thorough reviews [6, 9], some important aspects are summarized below:

### 8.2.1 Absorption

Variable absorption of drugs administered orally during pregnancy can be caused by numerous factors, including increased stomach pH, reduced gastric transit time and motility, altered uptake and efflux transporter activity, and changes in hepatic or intestinal metabolism [7, 10, 11].

### 8.2.2 Distribution

Pregnancy-induced changes such as tissue perfusion, increased cardiac output, stroke volume, heart rate, plasma volume, and total body water can affect the

*Oral Bioavailability and Drug Delivery: From Basics to Advanced Concepts and Applications,* First Edition. Edited by Ming Hu and Xiaoling Li.

volume of distribution ($V_d$) of drugs [7, 11, 12]. An increase in total body water leads to lower plasma concentrations for hydrophilic drugs, and thus, a higher $V_d$. On the other hand, increased total body water causes a dilution of plasma protein concentrations, and this can affect free and total plasma drug concentrations, especially for drugs with high protein binding [13]. As the gestation progresses and the placenta and the fetus grow, blood flow to the uterus is increased, which may affect the rate at which some drugs cross the placenta. Increased drug distribution to the placenta and fetus would lead to an increased $V_d$ [14].

### 8.2.3 Metabolism

The activity of some drug-metabolizing enzymes is altered during pregnancy, which leads to changes in plasma drug and metabolite concentrations [15]. The hepatic enzymes CYP3A4 and CYP2D6 account for the biotransformation of a vast number of drugs [16]. Both of these enzymes have shown increased activity during pregnancy compared to the nonpregnant state. Increases in drug metabolism can lead to lower plasma drug concentrations, which may lead to reduced therapeutic effects. However, in the case of prodrug administration, increased enzymatic activity can lead to higher plasma concentrations of the active metabolite, with potential concerns for toxicity. On the other hand, the reduced activity of CYP2C19 could lead to higher plasma concentrations of drugs metabolized by this enzyme [17, 18]. Glucuronidation by UGT1A4 is also increased during pregnancy, likely due to upregulation caused by higher levels of estrogen [9, 19]. It is not yet fully understood how pregnancy might enhance or diminish interindividual differences in drug-metabolizing enzyme activity attributed to genetic variability and polymorphisms [20, 21].

### 8.2.4 Excretion

Renal blood flow and the glomerular filtration rate increase during pregnancy, which can lead to increased excretion for some drugs [22]. For example, since lithium is eliminated primarily via the kidney, reduced plasma concentrations of this drug have been observed in pregnant women [23, 24].

A drug's ability to cross the placenta may also change as a function of gestational age, which can affect the amount of drug remaining in the maternal circulation and the ratio of drug reaching the fetal circulation. Ongoing and future pharmacokinetic studies of drugs prescribed during pregnancy will help to identify situations where dose adjustments may be necessary to improve therapeutic efficacy and reduce risks to the pregnant patient and her developing offspring.

## 8.3 Placental Development and Structure

The placenta is an organ formed in early pregnancy as the interface between the maternal and fetal circulations in order to support and sustain fetal development. This section will provide a brief overview of the gross anatomy of the placenta and the process by which the placenta develops.

The placenta grows proportionally with the fetus. The average placenta at birth weighs 470 g, is 2.5 cm thick and 22 cm in diameter [25]. The fetal portion of the placenta is the chorionic plate (see Figure 8.1). The umbilical cord is attached to the placenta and the umbilical vessels radiate out across the chorionic plate. A properly formed umbilical cord will contain two umbilical arteries to bring deoxygenated blood away from the fetus and one umbilical vein to bring oxygenated blood to the fetus [25]. The maternal side of the placenta is referred to as the basal plate or decidua basalis. It should be noted that the basal plate and myometrium contain fetal cells that have penetrated and expanded into maternal tissue [26]. The maternal surface – which faces the uterus during gestation – is divided into convex areas called lobes, or cotyledons [27]. The fetomaternal zone is composed of the intervillous space, which contains maternal blood, and the villous trees within the lobes, which contain fetal tissue and fetal blood vessels. This is the location of nutrient and waste exchange [28].

The components of the villous trees are as follows: the stem villus is the bottom of the tree containing fetal arteries and veins, while the intermediate villi contain smaller vessels, including arterioles, venules, and capillaries. The terminal villi are directly involved in nutrient and gas exchange. Lastly, anchoring villi attach the placenta to the uterine wall [28]. The umbilical cord connects the placental circulation to the fetal circulation [29].

Before implantation, the endometrium is remodeled to create a suitable environment for pregnancy through the process of decidualization, which begins six days after ovulation and is achieved through high levels of progesterone [30, 31]. Seven days after fertilization, the morula, a ball of approximately 16 cells, develops into the fluid-filled blastocyst which implants into the decidua. The outer cell mass of the blastocyst, or trophectoderm, will develop into the placenta. By the

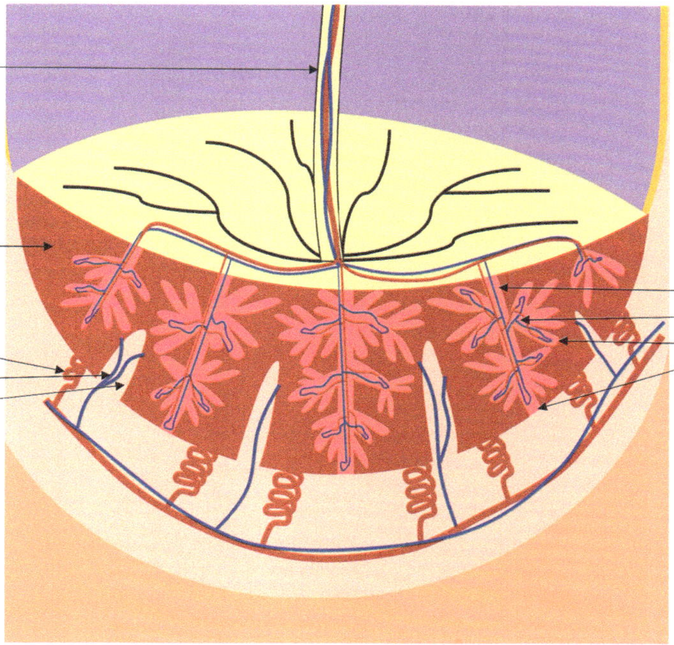


**Figure 8.1** Cross-sectional illustration of the human placenta near term.

eighth day, the invading trophoblast can be divided into two primitive populations of cells: the cytotrophoblasts and the syncytiotrophoblasts, each having their own functions in the development of the placenta [32].

The cytotrophoblasts are the regenerative stem cells of the placenta. The syncytiotrophoblasts, on the other hand, do not have the capacity to proliferate and must be replenished by cytotrophoblasts [33]. Primitive syncytiotrophoblasts may aid in the degradation of the extracellular matrix by releasing proteolytic enzymes [34]. Cytotrophoblasts and syncytiotrophoblasts can then infiltrate by forming finger-like projections into the decidua. The syncytiotrophoblasts will serve as the interface between the mother and fetus for the entire length of gestation. The syncytiotrophoblast layer prevents the migration of maternal white blood cells because it lacks intracellular gaps. Immune tolerance created by the syncytiotrophoblasts and other measures toward fetal antigens is essential to allow the invasion of cytotrophoblasts into the decidua basalis. Otherwise, the mother's immune system might recognize the fetal antigens as foreign and attack [35]. Proteolytic enzymes also erode maternal capillaries to fill lacunae – spaces formed between projections of trophoblast – with maternal blood. The trophoblastic lacunae eventually evolve into the intervillous space. Pregnancy starts in a relatively low oxygen environment for the first 10–12 weeks, until the trophoblast plugs no longer block the maternal spiral arteries and maternal blood fully fills the intervillous space [32]. Oxygenated maternal blood flows through the spiral arteries into the intervillous space, where it pushes more deoxygenated blood out through maternal veins [29].

In early pregnancy, the chorionic villi consist of a layer of syncytiotrophoblasts, a layer of villous cytotrophoblasts, trophoblast basal lamina, connective tissue, basal lamina of the endothelium, and capillary endothelium. The core contains connective tissue, fetal macrophages called Hofbauer cells, and fetal vessels. The barrier adapts as the fetus continues to grow. Eventually, in some locations the barrier will have become thin, consisting only of a syncytiotrophoblast layer, a fused basal lamina, and the capillary endothelium of the fetal vessels [36, 37]. Throughout the second to the fourth month of gestation, the villi branch out to increase the surface area to maximize gas and nutrient transport. Then, starting in the fourth month, septa form from the basal plate, growing into the intervillous space to divide the villi into cotyledons [38]. The placenta continues to grow until delivery. At term, the surface area of the syncytiotrophoblast layer is 12 $m^2$, and maternal blood flow reaches 600–700 ml/min [29].

## 8.4 Functions of the Human Placenta

The placenta is essential throughout pregnancy for endocrine functions, transport and metabolism,

**Table 8.1** Examples of hormones produced by the placenta and their main functions.

| Hormone | Main Function |
|---|---|
| *Steroid hormones* | |
| Progesterone | • Support the endometrium<br>• Prevent uterine contraction |
| Estrogen | • Stimulate growth of myometrium<br>• Stimulate mammary gland development |
| *Peptide hormones* | |
| Human Chorionic Gonadotropin (hCG) | • Support corpus luteum graviditatis<br>• Maternal recognition of pregnancy |
| Human Placental Lactogen (HPL) | • Alter maternal metabolism to increase nutrient availability and promote fetal growth and development |
| Corticotropin-releasing hormone (CRH) | • Support fetal lung maturation and surfactant production |
| Placental Growth Hormone (GH) | • Replace pituitary GH from mother<br>• Alter maternal metabolism<br>• Promote fetal growth |
| Insulin-Like Growth Factor (IGF) | • Stimulate placental growth |
| Vascular Endothelial Growth Factor (VEGF) | • Support growth and maintenance of the fetoplacental unit<br>• Assist delivery of oxygen and nutrients to fetus |
| Relaxin | • Prepare for implantation<br>• Relax pelvic ligaments for childbirth |

Source: Adapted from Refs. [43–48].

and protection [39]. Details regarding transport and metabolism follow in the sections below. It should be noted that when considering these categories of placental function, there is some overlap, as some of the placenta's protective capabilities come through its transport and metabolism functions. For example, efflux transporter proteins can limit the transplacental transfer of some xenobiotic compounds, and metabolizing enzymes may convert certain drugs to less active metabolites [40, 41].

In terms of endocrine functions, steroid and peptide hormones are produced by the placenta, primarily by the syncytiotrophoblasts [42]. These hormones are generally important for the development of the feto-placental unit, and some specific functions are listed in Table 8.1. For example, a hormone produced by the placenta from the onset of pregnancy is hCG, which signals to the mother that she is pregnant [43]. The continuation of pregnancy requires the spiral arteries to continue to supply the uterus instead of undergoing degradation as occurs during menstruation. hCG rapidly increases in early pregnancy and binds to LH receptors, which maintains the corpus luteum graviditatis, which can then secrete progesterone and estrogen needed to support the pregnancy until the placenta matures [44].

The placenta can also help to protect the fetus from pathogens during pregnancy. The syncytium is a multinucleated cellular mass, so it lacks the intercellular gaps that bacteria, protozoa, and viruses usually utilize to infect the host. However, the immune-tolerant environmental factors necessary for fetal invasion may allow some pathogens to overcome the placental barrier [35]. Vertical transmission was initially discovered with rubella in 1941 [49]. Other pathogens that can be transmitted vertically include toxoplasmosis, syphilis, cytomegalovirus, human immunodeficiency virus, herpes simplex virus, and varicella-zoster virus [35, 50–55].

The placenta takes on many roles of the undeveloped fetal liver. For example, trophoblast cells contain enzymes which handle bile acids, biliary pigments, and xenobiotics [40]. Although some transport and metabolism functions highlighted below may reduce fetal exposure to some compounds, other drugs may not be so affected. Even low concentrations of certain substances could be detrimental. Thalidomide is not the only example of drug-induced teratogenicity. From weeks three to seven, the embryo is especially susceptible to morphological abnormalities because of organogenesis and body patterning [38]. Toxic substances that cross the placenta – such as alcohol, anticonvulsants, and warfarin – can impair cell migration and differentiation in this critical period.

## 8.5 Mechanisms of Drug Transport Across the Placenta

The rate and extent of placental drug transfer depend upon placental-related factors such as surface area, thickness, pH, metabolism, blood flow, drug transporters,

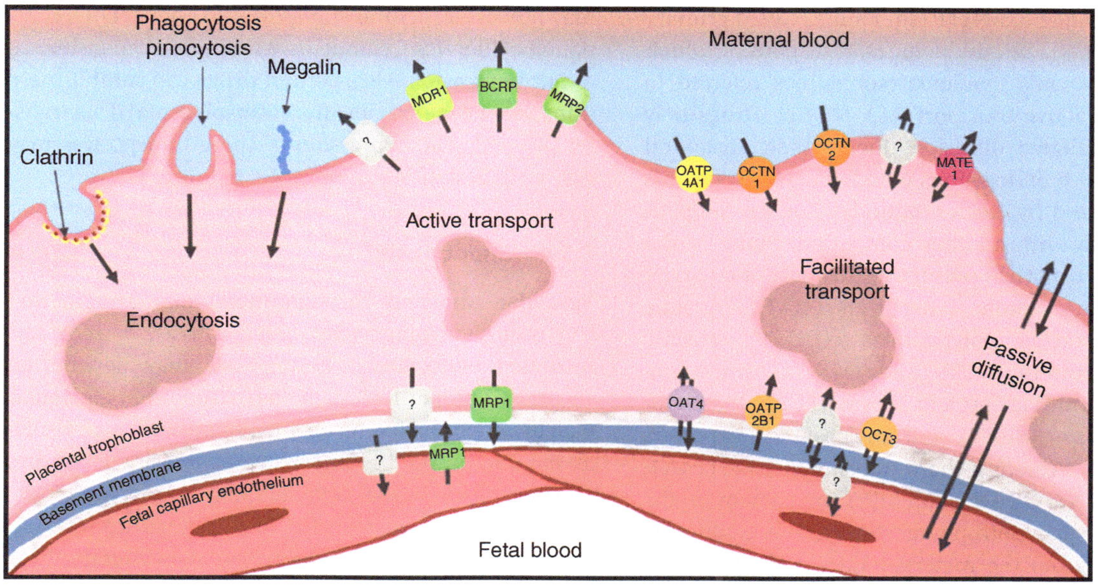


**Figure 8.2** Transport processes within the human placental barrier. It should be noted that this does not represent an exhaustive list of transporters present in the placental trophoblast or fetal capillary endothelium. Trophoblast cells represent the rate-limiting barrier separating maternal and fetal circulations within the placenta. MDR1, multidrug resistance protein 1 (P-glycoprotein); BCRP, breast cancer resistance protein; MRP2, multidrug resistance-associated protein 2; OATP4A1, organic anion transporting polypeptide 4A1; OCTN1, novel organic cation transporter 1; OCTN2, novel organic cation transporter 2; MATE1, multidrug and toxin extruding protein 1; MRP1, multidrug resistance-associated protein 1; OAT4, organic anion transporter 4; OATP2B1, organic anion transporting polypeptide 2B1; OCT1, organic cation transporter 1; OCT3, organic cation transporter 3. Question marks indicate transport processes requiring further characterization. Source: Reproduced from Al-Enazy et al. [13].

and drug-specific factors such as molecular weight, partition coefficient, p$K_a$, protein binding, and concentration [56]. As the gestation progresses, the ratio of cytotrophoblast cells to syncytiotrophoblast cells changes, with the proportion of cytotrophoblast cells gradually decreasing [57]. This leads to a thinner barrier [39, 58]. A number of transporters are present on both sides of the trophoblast layer and in the fetal capillary endothelium which are involved in active and facilitative transport mechanisms [13, 59]. Some of these same transporters are important for the maternal-to-fetal delivery of carbohydrates, amino acids, and other nutrients, or for the removal of waste products from the fetal circulation [60]. Figure 8.2 highlights some of the major transport mechanisms of the placental barrier, which are discussed in more detail below:

### 8.5.1 Passive Diffusion

This energy-independent process can occur via transcellular or paracellular pathways. Depending upon the drug's concentration gradient, a drug can diffuse through trophoblast cells or water channels, governed by Fick's law of diffusion. As the pregnancy progresses, several factors can increase the passive diffusion of drugs across the placenta, including increased blood flow to placenta, increased villous surface area, and a thinner trophoblast layer. Passive diffusion is the mechanism by which a number of small, lipophilic drugs cross the placenta [56, 61].

Passive diffusion is also important in gas and lipid exchange. The placental membrane is highly permeable to respiratory gases. Productive gas exchange between the maternal to fetal circulation is made possible by the relatively higher affinity for oxygen and lower affinity for carbon dioxide found in fetal hemoglobin compared to adult hemoglobin. This ensures the correct directionality during exchange, i.e. oxygen to the fetus and carbon dioxide to the mother [39]. In terms of lipids, triglycerides are unable to cross the placenta, but once lipoprotein lipases on the maternal side cleave them into fatty acids and glycerol, these can cross the placental membrane [60].

### 8.5.2 Facilitated Diffusion

In this process, a drug diffuses from a higher concentration gradient to a lower concentration gradient (similar to the gradient driving passive diffusion), but using a carrier/transporter. This process is distinct from active transport dependent on ATP (sometimes called primary active transport), but it includes transporters which are driven by ionic gradients (sometimes referred to as secondary active transport) [13, 60, 62]. Drugs may cross via facilitated diffusion due to their structural similarities with endogenous compounds. The process can be hindered by carrier saturation or by competitive inhibition with another compound. Some of the facilitative transporters present in the placental barrier include organic anion transporters (OATs), organic anion transporting polypeptides (OATPs), organic cation transporters (OCTs and OCTNs), multidrug and toxin-extruding protein 1 (MATE1), the serotonin transporter (SERT), and nucleoside transporters (NTs) [13, 63]. Amongst these transporters, OATP4A1, MATE1, OCTN1, OCTN2, and SERT are present on the apical side of syncytiotrophoblast cells, thus facilitating the diffusion from the maternal side to the fetal side [13, 59, 63]. OAT4, OCT3, and OATP2B1 are present on the basolateral side, facilitating the uptake of drugs from the fetal side into the trophoblast cells [63].

Glucose is a major source of energy for the fetus, and its primary mechanism of transport is facilitated diffusion. Glucose carriers, such as GLUT1, are expressed on both the maternal and fetal membranes of the placenta [64]. The transport of glucose into the placenta is also important to form a source of glycogen storage in this organ. The fetal liver is unable to perform gluconeogenesis, so this source can ensure the fetus has a continual supply of energy [60]. The transfer of amino acids across the placenta is mediated by accumulative, exchange, and facilitated transporters [65]. These amino acid transporters have specificity to a certain amino acid or group of amino acids, based on size and charge [60]. A considerable number of amino acid transporters cotransport sodium, reflecting the more complex and integrated nature of ion transport across the placenta [39, 60].

### 8.5.3 Active Transport

This carrier-mediated and energy-dependent drug transport can occur against a concentration gradient. Active ATP-binding cassette transporters in the placenta include P-glycoprotein (P-gp, MDR1), breast cancer resistance protein (BCRP), and multidrug resistance-associated proteins (MRPs). P-gp, BCRP, and MRP2 are present on the apical membrane of the syncytiotrophoblast cells and are responsible for the efflux of several compounds, including many drugs. MRP1 and MRP5 are present on the basolateral side [13, 63, 66]. Drugs which are substrates for these transporters may possess structural similarities to endogenous compounds. Therefore, competitive inhibition can influence the drug transfer. Also, certain drugs can inhibit these transporters, which can affect transplacental drug transfer [41, 63]. The expression of these transporters may vary with gestational age [16, 63, 66].

### 8.5.4 Endocytosis

Vesicular transport through membrane invagination can occur via pinocytosis, phagocytosis, or receptor-mediated endocytosis. Endocytosis may depend on substrate properties such as structure, molecular weight, lipophilicity, or on membrane properties such as membrane fluidity and receptor density. Clathrin-, megalin-, and cubilin-mediated endocytosis across the placenta have been reported in the literature [67, 68]. Drug delivery via nanoparticles, dendrimers, liposomes, or exosomes can also take place via endocytosis [67, 69–72].

Another aspect of the placenta's role in protecting the developing fetus is the maternal-to-fetal transfer of IgG, which is mediated by endocytosis. Trophoblast cells endocytose IgG at the maternal membrane at neutral pH, and then the Fc portion of IgG binds to FcRn in the newly acidic endosome. The bound IgG can undergo transcytosis to the fetal membrane and is released at neutral pH. The IgG transported to the fetus will help to provide passive immunity to the fetus while its immune system develops [73].

## 8.6 Mechanisms of Drug Metabolism Within the Placenta

The biotransformation of drugs administered during pregnancy can occur in multiple locations, with the maternal liver being the most prominent for most drugs [14]. Pregnancy-related changes in the activity of hepatic enzymes were mentioned in the pharmacokinetics section above and are reviewed in detail elsewhere [14, 16, 74]. This section will instead focus on drug-metabolizing enzymes within the placenta. The developing fetus can be exposed to parent drug molecules that cross the placenta, drug metabolites formed in maternal organs, and drug metabolites formed within the placenta. Depending on the chemical nature of the metabolites in the maternal circulation and the extent to which those metabolites cross the placenta, the biotransformation of drugs within the placenta may be an important factor affecting the levels of certain drug metabolites in the fetal circulation [75].

Both Phase I and Phase II enzymes are expressed in the placenta. Phase I enzymes include cytochrome P450 enzymes (CYPs), hydroxysteroid dehydrogenases (HSDs), and epoxide hydrolase [13, 76]. Phase II enzymes in the placenta include uridine 5′-diphosphate glucuronosyltransferase enzymes (UGTs), sulfotransferases (SULTs), glutathione-S-transferase-π (GST-π), and *N*-acetyltransferase 1 (NAT1) [13, 76]. The expression of placental drug metabolizing enzymes can vary with maternal age [77], gestational age [78], disease state [79], and smoking status [80].

CYP enzymes that have been detected in the placenta at least at the mRNA level include CYP1A1, CYP1A2, CYP1B1, CYP2B6, CYP2C8, CYP2C9, CYP2D6, CYP2E1, CYP2F1, CYP2J2, CYP3A4, CYP3A5, CYP3A7, CYP4B1, CYP11A1, CYP17, CYP19A1, CYP19A11, CYP24A1, and CYP27B1 [13, 76, 81]. One example of placental drug biotransformation by a CYP is the N-dealkylation of buprenorphine to norbuprenorphine by CYP19 [82]. HSD enzymes in the placenta include 3β-HSD, 11β-HSD, and 17β-HSD [13, 81]. An example of drug metabolism by an HSD is the conversion of bupropion to both erythrohydrobupropion and threohydrobupropion by 11β-HSD [81].

UGTs present in the placenta include UGT1A1, UGT1A4, UGT1A6, UGT1A9, UGT2B4, UGT2B7, UGT2B10, UGT2B11, UGT2B15, and UGT2B15 [13, 76]. An example of UGT-catalyzed placental drug biotransformation is the glucuronidation of 3′-azido-3′-deoxythymidine [83]. SULTs in the placenta include SULT1A1, SULT1A3, SULT1B1, SULT1E1, SULT2A1, SULT2B1a, and SULT2B1b [13, 76]. As an example, SULT1A3 is responsible for the placental sulfation of dopamine [84].

## 8.7 Strategies to Alter Drug Transport Across the Placenta

Taking medications to treat maternal complications during pregnancy presents a risk of adverse off-target effects on the placenta and the fetus; therefore, limited placental transfer and disposition are desirable for these drugs. On the other hand, treating fetal complications *in utero* through maternal drug administration implies sufficient drug delivery across the placental barrier for better treatment efficacy, while limiting nonspecific effects on the mother. Furthermore, targeting the placenta itself for pharmacological treatment would require efficient uptake of the drug by the trophoblasts, while limiting transplacental passage for fetal safety.

As transporters and metabolizing enzymes play important roles in the transfer of many drugs across the placenta, understanding the processes by which the expression or activity of these transporters or enzymes may be altered may lead to strategies to alter the distribution of drugs into and across the placenta. Still, it is important to consider that any factors that may alter the transport or biotransformation of drugs within the placenta could also affect the transport or metabolism of endogenous substrates, such as essential nutrients. The activities of a number of placental enzymes are known to be affected by drug-induced or environmental factors [63, 81, 85, 86]. Coadministration of P-gp inhibitors can increase the transplacental transfer of drugs which are substrates of this efflux transporter [63, 87–90], but this could lead to potential off-target effects since this transporter is not exclusively expressed in the placenta. Such an approach could increase the tissue distribution of other endogenous and exogenous substrates.

Nanoparticle-based drug delivery platforms represent a promising tool to control drug delivery across the placental barrier. A variety of delivery systems have been developed as drug carriers, including liposomes (phospholipid-based nanovesicles), micelles, emulsions, nanogels, carbon nanotubes, polymeric nanocapsules, quantum dots, dendrimers, and metallic nanoparticles [91, 92]. For transplacental passage, a nanoparticle would need to cross the trophoblast layer and the fetal endothelium to enter the fetal capillary lumen. Nanoparticles typically enter the placenta via pinocytosis, caveolae- or clathrin-mediated endocytosis, or phagocytosis [93]. The exocytosis of nanoparticles could occur via lysosomal secretion, endosomal escape, and the secretion of multivesicular bodies. The uptake of nanoparticles by the fetal capillary endothelium could happen via similar mechanisms [93]. Nanoparticles with a size below 25 nm might cross the placenta paracellularly through placental pores (transtrophoblastic channels) [93, 94].

Cumulative evidence from *in vivo* animal models and *ex vivo* placental perfusion studies suggests that maternal-to-fetal transfer is more likely for smaller nanocarriers (<240 nm) [92, 95]. However, other characteristics such as shape, core structure material (such as lipid, protein, polymer, or carbon), surface composition affecting pH, zeta potential/charge, and hydrophilicity could affect the passage of some nanomaterials across the placenta and thus could be tuned for particular clinical applications [92, 96–98]. Thus, incorporation of a drug into the nanoparticle (via either nanocarrier surface coating, entrapment, or encapsulation) could be used to modify a drug's properties and consequently circumvent its conventional route of placental disposition.

Nanoparticle-based drug delivery systems for the treatment of maternal conditions in pregnancy would aim to restrict therapy to maternal tissue while preventing placental accumulation and passage of the drug by minimizing its circulation in a free form. For example, the treatment of preterm birth could benefit from carrier-mediated drug delivery to improve outcomes for the newborn [99]. Indomethacin, a first-line tocolytic agent used to suppress preterm labor, is a nonsteroidal anti-inflammatory drug that acts by decreasing prostaglandin production in the maternal uterus [100]. Indomethacin crosses the placental barrier [101], and its usefulness is limited by fetal and neonatal side effects such as oligohydramnios [102] and antenatal closure of the ductus arteriosus [103]. Liposomes are biodegradable and have been clinically used to improve drug delivery in chemotherapeutic and antimicrobial therapies [104]. Liposome-based systems have been suggested as a safe option for both the mother and her fetus [105]. Liposomes targeted to the oxytocin receptor have been proposed for directing tocolytic therapy to the uterus, and data suggest that these approaches could also reduce fetal drug exposure [99, 106, 107].

Fetal conditions in pregnancy are often treated via maternal administration of the drugs due to risks associated with direct fetal injections [5]. Due to short half-lives and increased clearance of some drugs in pregnancy (e.g. digoxin), higher doses are required to maintain adequate steady-state levels in the maternal circulation for sufficient drug transfer across the placenta, thus exacerbating unintentional maternal effects [108]. In this context, an optimal nanocarrier-based drug formulation for prenatal fetal therapy would have enhanced selectivity toward the placenta with efficient transplacental transfer. Encapsulation of digoxin into PEGylated poly(lactic-*co*-glycolic acid) nanoparticles enhanced its transfer across BeWo b30 cell trophoblast monolayers (this *in vitro* experimental model is described in Section 8.8.5) from the maternal compartment to the fetal compartment by twofold, which is similar to the effect of P-gp inhibition on the transfer of free digoxin. Thus, it appears that this nanoparticle formulation shields digoxin from P-gp-mediated efflux [108]. Encapsulation of dexamethasone (a drug for prenatal treatment of congenital adrenal hyperplasia in females) into polymeric nanoparticles increased the apparent permeability of the drug by more than 10-fold in the same *in vitro* model [109]. Further optimization of nanocarriers for targeted placental delivery could include surface functionalization with ligands targeting receptors such as Fc receptors in the placenta [5], placental chondroitin sulfate A binding peptide [93], or with the tumor homing peptides CGKRK and iRGD, which selectively recognize integrins in the placenta and uterine spiral arteries [110]. It should be noted that the placental biodisposition of any nanocarrier drug delivery system is likely dependent on gestational age, as the placenta undergoes changes with the progression of pregnancy [5]. There is still much work to do to develop and optimize nanoparticle-based drug formulations specific for pregnancy and evaluate their biodisposition, safety, and efficacy using *in vitro*, *ex vivo*, and *in vivo* experimental models before moving onto clinical trials with pregnant human subjects.

## 8.8 Experimental Models of the Human Placenta

Assessing the extent and mechanisms of placental drug biodisposition is necessary to evaluate potential effects of these drugs on placental function and fetal development, and can help clinicians, scientists, and pregnant women to make informed decisions in health care. Paired samples of maternal and umbilical cord blood collected at the time of delivery provide fetal-to-maternal drug concentration ratios, which are very informative to evaluate the transplacental transfer of the drugs at term or at near-term stages of pregnancy. However, these ratios are highly variable due to differences of timing between maternal drug administration and sampling as well as difficulties in collecting maternal and cord blood samples simultaneously. Due to several factors affecting drug distribution that may change as a function of gestational age, these ratios may differ from fetal-to-maternal drug ratios in the first and second trimesters of pregnancy, but it would not be considered ethical to obtain such data from early pregnancy, as the risks of cordocentesis limit its use to diagnose rare disorders such as fetal anemia and neonatal alloimmune thrombocytopenia [111].

A practical experimental model for studying placental drug disposition should recapitulate properties of the human placental barrier and mimic its functions. Advances in the development of such models have facilitated feasible and systemic approaches for the preclinical assessment of drugs and to predict their biodisposition across the maternal–fetal barrier. Such models can be especially useful for drugs that are routinely used by pregnant women or that are proposed for use during pregnancy. Below, a number of experimental approaches used to study placental function and drug biodisposition are summarized.

### 8.8.1 Animal Models

Pregnant animals represent a whole-body system and thus have an undeniable advantage in studying placental drug disposition. Mammalian animals, including rats [112], mice [113, 114], rabbits [115], camels [116], pigs [117], sheep [118], and nonhuman primates [119, 120] have served as *in vivo* models portraying the functions and structure of human placenta.

The rapid development and high reproductive rate of mice, along with some similarities in the placental structures of mice and humans, make the mouse model a popular choice for studying placental drug disposition. Indeed, both mouse and human placentas are hemochorial, meaning that trophoblast tissue is in direct contact with maternal blood [113]. Although the murine placenta has three trophoblast layers between the maternal and fetal circulation as opposed to a single layer in the human placenta, there are some similarities between the specific placental plasma membranes and the diffusional exchange and transporter protein-mediated transfer of nutrients in the placenta are comparable in both species [113]. Additional details regarding the similarities and differences of placental transfer mechanisms in mice and humans are reported elsewhere [113, 121, 122]. Knockout and humanized mice may also facilitate studies of compromised pregnancies – such as those with fetal growth restriction, preeclampsia, or other related conditions – and their responses to therapeutics [123–127].

Pregnant sheep have been used to study the placental biodisposition of drugs because of similarities in fetal–placental vasculature to that of human pregnancy [128]. Moreover, the pregnant sheep model provides an option to conduct repetitive sampling from both the maternal and fetal sides of the placenta, which is not feasible in a mouse model. Despite anatomical differences in placental structures (sheep have a cotyledonary, epitheliochorial placenta, whereas the human placenta is discoidal and hemochorial), the villous tree of the sheep cotyledon shares similar physiological traits with the villous tree of human placenta. However, human and ovine placentas have differences in the placental expression and localization of the glucose transporter isoforms GLUT1 and GLUT3. Nevertheless, pregnant sheep have been used to model fetal growth restriction [128] and to study the placental biodisposition of drugs and other substances [118, 129, 130].

Despite the use of animal models in various pregnancy-related studies, marked differences between species in terms of duration of gestation, placental anatomy, and expression of placental metabolizing enzymes and transporters do not allow for direct extrapolation of the data from animal models to humans. Moreover, ethical, logistical, and financial constraints may hamper the routine use of some animal models to mimic human pregnancy.

### 8.8.2 *Ex Vivo* Placental Perfusion

The *ex vivo* perfusion of the human placental lobule currently represents a gold standard model in the field of placental transport [131]. A recent meta-analysis showed a good correlation between *ex vivo* placental perfusion data and *in vivo* transplacental drug transfer data. However, there were several outliers, which may be related to differences in the timing of collecting umbilical cord and maternal blood samples [132].

*Ex vivo* placental perfusion is noninvasive and is the only experimental model allowing for the study of placental biodisposition of various compounds in organized human tissue. In this model, the maternal and fetal blood vessels of a single placental cotyledon are cannulated to reestablish the physiological circulation that takes place *in vivo*. Depending on the experimental setup, *ex vivo* placental perfusion may be configured either as an open–open perfusion, or closed–closed. The methodology on the composition of the maternal and fetal perfusates, perfusion flow rates, oxygenation, and control measures for tissue integrity and viability vary among laboratories [133]. Reference compounds such as antipyrine and creatinine are used to confirm appropriate overlap of the maternal and fetal circulations and are also utilized to normalize the experimental data to account for interindividual variability. The placental perfusion model can be used to determine both the kinetics and the mechanisms of placental transfer [134, 135]. The involvement of specific enzymes and transporters can be examined through the use of specific inhibitors during perfusion, the quantification of mRNA and protein expression, as well as the evaluation of polymorphisms [136–138].

Because placental tissue is obtained at the time of delivery, the perfusion method routinely utilizes term or near-term placentas. Due to dynamic changes in the expression of transport proteins, serum albumin concentrations, and maternal–fetal diffusion distance, the results obtained from placental perfusion experiments cannot be directly extrapolated to earlier stages of pregnancy, which is a shortcoming of this model

[139, 140]. A limited number of reports have compared *ex vivo* placental transfer in term placentas to late second-trimester placentas [82, 141]. The transfer of methadone, a known P-gp substrate, was 30% higher in term placentas (38–41 weeks of gestation) than in preterm placentas (27–34 weeks), which is in agreement with the reported decline in P-gp protein expression toward term [141].

Although proteins can be added and the pH can be adjusted, the experimental perfusion media will not perfectly match maternal and fetal blood. The duration of the placental perfusion experiments is typically limited to two to six hours [142]. Since the placental maternal vasculature is damaged due to tissue separation from the uterine wall at the time of delivery, there are no standard criteria for basal cannulation and maternal perfusion. Accordingly, there may be differences when comparing study outcomes from different laboratories. This model is labor-intensive and technically challenging, with success rates ranging from 15% to 50%, due to failures such as fetal media leaking [133, 142, 143]. Although the experimental flow rates of the maternal and fetal perfusates do not match the speed of *in vivo* placental blood flow [29, 144], *ex vivo* placental perfusion still serves as an effective model to study placental drug disposition.

### 8.8.3 *In Vitro* Trophoblast Tissue Preparations

Human placental explants and villous explants contain multiple cell types similar to the whole organ, but the lack of a continuous membrane prevents the use of this model to study transplacental transfer. Nevertheless, explants are still suitable to investigate mechanisms of transfer and regulation of drug-metabolizing enzymes and transporters, as well as endocrine functions, cellular proliferation, and differentiation [145, 146]. Similar to *ex vivo* placental perfusion, freshly collected placental tissue is used for this experimental setup.

Subcellular fractions of human placental tissue are prepared via differential centrifugation and have been extensively used to study placental drug metabolism and transport. Microsomal preparations from endoplasmic reticulum containing membrane-bound CYP enzymes have been used extensively to study the metabolism of xenobiotics [75, 147]. Enzymes localized in mitochondrial membranes also contribute to placental drug metabolism [148]. The cytosolic fraction of subcellular placental preparations has been utilized to study the role of non-membrane-bound enzymes in placental drug metabolism as well [86, 149].

Vesicles prepared from either the apical brush border membrane or from the basal membrane of trophoblast tissue are useful to study the expression and function of placental uptake and efflux transporters [150, 151]. The membrane vesicle preparation is a mixture of right-side-out-oriented vesicles (ROVs) and those oriented inside–out (IOVs). Accordingly, the transporter substrate binding site is situated either on the outside of the vesicle membrane bilayer or on the inside. Thus, the enriched population of IOVs is desirable to study efflux transporters, while uptake transport studies rely on the operation of ROVs [151, 152]. The purity of the isolated membrane vesicles is assessed using specific markers [153]. Technical difficulties such as extensive substrate diffusion into the vesicles represent a significant limitation in the use of placental membrane vesicles to study placental transport [154]. Additional shortcomings of subcellular fractions and membrane vesicles include cross-contamination during preparation as well as the potential loss of enzyme or transporter activity during purification or storage [155].

### 8.8.4 Primary Cells from Human Placenta

Primary villous cytotrophoblasts can be isolated and frozen. They provide a good model system to study mechanisms affecting transplacental drug transfer, such as uptake, efflux, and metabolism [146]. Isolated primary villous cytotrophoblast cells represent normal cells, thereby having some advantages over immortalized or malignant cancer cell lines. These primary cells can be isolated from different stages of pregnancy, enabling the comparison of experimental data from early pregnancy and term. However, some limitations of cultured primary trophoblast cells are that they do not proliferate, transporter expression may change after several days of *in vitro* incubation, and they do not form monolayers with tight junctions [146, 156, 157]. Instead, the cells syncytialize spontaneously, forming aggregates with intercellular gaps, which precludes transcellular transport studies on semipermeable membranes, although there have been attempts to seed multiple overlapping cell layers [157].

### 8.8.5 Continuous Cell Lines

Several continuous cell lines from human placental trophoblast cells have been developed which can overcome the lack of proliferation of cultured primary trophoblast cells. Ideally, continuous trophoblastic cell lines should possess phenotypical and functional characteristics consistent with that of the parent cells to serve as a good surrogate for the primary cells [158]. Some cell lines have been derived from first-trimester trophoblasts transfected with SV40 viruses, such as SGHPL cells and HTR-8/SVneo. Both of these cell lines express markers of extravillous trophoblast cells and the HTR-8/SVneo cells may also contain a mixture of mesenchymal cells [158, 159].

Choriocarcinoma-derived cell lines include the BeWo, Jar, and JEG-3 cell lines [160–162]. BeWo cells express many of the same transporters and enzymes as primary cultured cytotrophoblast cells, and they secrete hormones such as human placental lactogen, human chorionic gonadotropin, progesterone, and estrogen; therefore, they are used extensively as a model of villous trophoblast cells [163, 164]. The BeWo b24 and b30 clones can form a confluent polarized monolayer when grown on a semipermeable support, thereby providing a suitable *in vitro* model of the placental barrier [13, 165]. JEG-3 cells can also form a monolayer under specific culture conditions [166]. These *in vitro* model systems have been employed to study asymmetrical transcellular distribution of multiple compounds across placental trophoblast cells [166, 167]. In addition, the BeWo, Jar, and JEG-3 cell lines have been used to study xenobiotic metabolism within the human placenta [142]. Still, these cell lines have their shortcomings. There are some differences in the expression of some transporters and enzymes, which could affect predictions of *in vivo* conditions [156, 163]. It should be noted that a placental perfusion study with $n = 4$ would represent the average of results from four genetically distinct placentas with variable transporter and enzyme expression profiles, whereas, in the case of cell line experiments, all study replicates are from cells derived from a single individual [167]. Cell culture passage number may affect reproducibility, and it should also be mentioned that these trophoblast cell lines are a simplified representation of the placental barrier, as they lack the full complement of stromal cells and the fetal endothelium present *in vivo* [168]. Nevertheless, excellent correlation has been observed between the transfer rates of several compounds studied by both the BeWo b30 cell line and *ex vivo* placental perfusion [169].

### 8.8.6 Placenta-on-a-Chip

The placenta-on-a-chip model is configured on the basis of a microfluidic device in which trophoblast cells and endothelial cells are cultured in confluent monolayers opposite each other on a porous polymeric membrane representing the maternal and fetal interfaces of the placenta (see Figure 8.3) [170].

BeWo or JEG-3 cells are commonly used to represent the trophoblast epithelial layer, while human umbilical vein endothelial cells (HUVECs) or human placental villous endothelial cells (HPVECs) can be used to represent the endothelial cell layer [114, 171–173]. The model configuration allows independent control of the fluid flows in the microchannels representing the fetal capillary compartment and the intervillous space. The placenta-on-a-chip represents an *in vitro* model recapitulating components of the physiology and functionality of the human placenta for simulation and quantitative analysis of placental drug biodisposition.

Transcellular distribution of glucose was tested using placenta-on-a-chip dynamic model systems with JEG-3/HUVEC [173] and BeWo/HPVEC [171] coculture and yielded similar transfer rates of 17–39.1% and 34.8%, respectively. These values were similar to those reported from *ex vivo* placental perfusion (26.5–38.3%). The function of efflux transporters has also been observed in this model [172]. Key factors for establishing apical–basolateral polarity and microvilli on the cell surface include membrane thickness and composition as well as the presence of the physiologically relevant sheer stress (0.001–0.1 dyn $cm^2$) [174–177]. Some of the limitations regarding the utility of placenta-on-a-chip model systems include the inability to account for gestational age-related changes in the placenta, nonspecific interactions of compounds with the chip's polymeric matrix, and potential issues with detection sensitivity when operating with small volumes [114].

Other unconventional approaches utilized to model the placental barrier include a custom-built perfusion chamber with the amniotic membrane as a matrix to coculture trophoblast and endothelial cells on each side. Similar to the placenta-on-a-chip, this model incorporates physiologically relevant sheer stress for better approximation of the *in vivo* situation [176]. In another approach, slices of human placental tissue were mounted on and pressed between the half-chambers of an Ussing chamber, but the data quality is dependent on the manner of tissue slicing [174, 178].

### 8.8.7 *In Silico* Models

Physiologically based pharmacokinetic models (PBPK) use mathematics to describe an organism as an interconnected system of compartments representing predefined organs or tissues [168]. For each compartment, a system of differential equations can be built to reflect the mass balance relying on the anatomical and physiological structure of the body, but it is also dependent on the model's assumptions. The model can incorporate

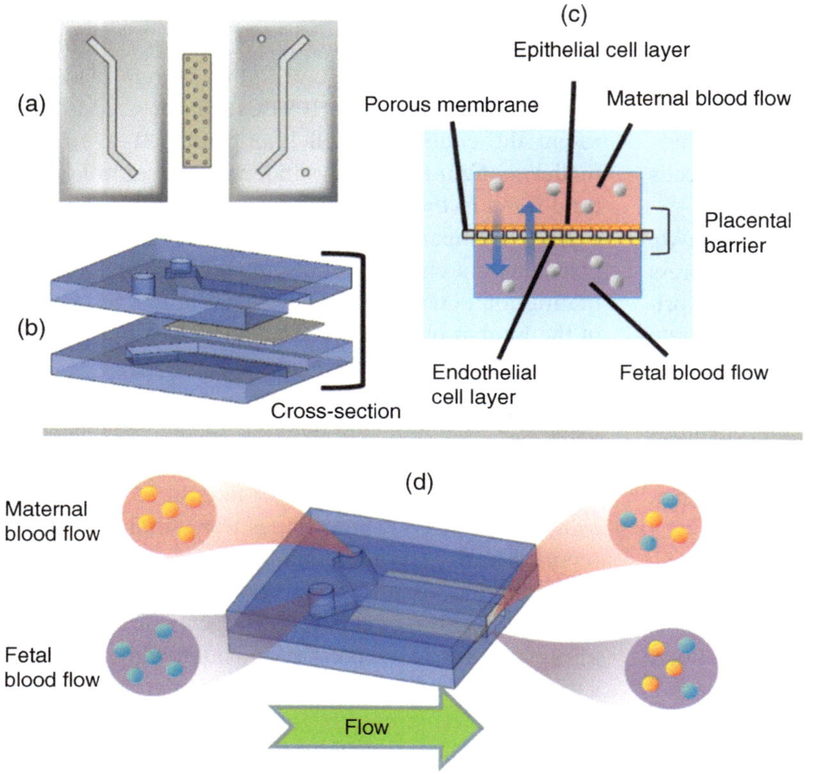


**Figure 8.3** The placenta-on-a-chip consists of two layers of poly(dimethylsiloxane) separated by a porous membrane and channel on each side. (a) Top and bottom layers with a porous membrane separating the channels before being attached. (b) Channels aligned and ready to be attached and the porous membrane placed in between the layers to separate the midsections of each channel (where two cell layers interact). (c) Cross-sectional view of the channels. (d) Experimental concept that shows the maternal and fetal bloodstreams perfused through the channels (not to scale). Source: Pemathilaka et al. [170]/Elsevier/Public Domain CC BY 4.0.

parameters such as a drug's physicochemical properties as well as data derived from *in vitro* and *ex vivo* experiments.

PBPK models have been used to predict the maternofetal pharmacokinetics of medications based on data from *ex vivo* perfusion experiments or *in vitro* cell line studies to estimate placental transfer constants and partition coefficients. Upon integration of these parameters into a pregnancy whole-body PBPK model, predicted values can be compared to *in vivo* data for validation [179–181]. It should be noted that models built using experimental data obtained at the time of delivery may not provide accurate extrapolation to earlier stages of pregnancy. The integration of parameters reflecting the activity of placental metabolizing enzymes and transporters would be expected to further optimize the performance of PBPK models in pregnancy [182].

## References


1 McBride, W.G. (1961). Thalidomide and congenital abnormalities. *Lancet* 2: 90927–90928.

2 Vargesson, N. (2019). The teratogenic effects of thalidomide on limbs. *J. Hand Surg. Eur. Vol.* 44: 88–95.

3 Dathe, K. and Schaefer, C. (2019). Arzneimittelbehandlung der schwangeren. *Dtsch. Arztebl. Int.* 116: 783–790.

4 Bérard, A., Abbas-Chorfa, F., Kassai, B. et al. (2019). The French pregnancy cohort: medication use during pregnancy in the French population. *PLoS One* 14: 19095.

5 Rytting, E. and Ahmed, M.S. (2013). Fetal drug therapy. In: *Clinical Pharmacology During Pregnancy* (ed. D.R. Mattison), 55–72. Academic Press Inc.

6 Ansari, J., Carvalho, B., Shafer, S.L., and Flood, P. (2016). Pharmacokinetics and pharmacodynamics of drugs commonly used in pregnancy and parturition. *Anesth. Analg.* 122: 786–804.

7 Costantine, M.M. (2014). Physiologic and pharmacokinetic changes in pregnancy. *Front. Pharmacol.* 5: 00065.

8 Tasnif, Y., Morado, J., and Hebert, M.F. (2016). Pregnancy-related pharmacokinetic changes. *Clin. Pharmacol. Ther.* 100: 53–62.

9 Pariente, G., Leibson, T., Carls, A. et al. (2016). Pregnancy-associated changes in pharmacokinetics: a systematic review. *PLoS Med.* 13: 02160.

10 Loebstein, R., Lalkin, A., and Koren, G. (1997). Pharmacokinetic changes during pregnancy and their clinical relevance. *Clin. Pharmacokinet.* 33: 328–343.

11 Ward, R.M. and Varner, M.W. (2019). Principles of pharmacokinetics in the pregnant woman and fetus. *Clin. Perinatol.* 46: 383–398.

12 Dawes, M. and Chowienczyk, P.J. (2001). Pharmacokinetics in pregnancy. *Best Pract. Res. Clin. Obstet. Gynaecol.* 15: 819–826.

13 Al-Enazy, S., Ali, S., Albekairi, N. et al. (2017). Placental control of drug delivery. *Adv. Drug Deliv. Rev.* 116: 63–72.

14 Feghali, M., Venkataramanan, R., and Caritis, S. (2015). Pharmacokinetics of drugs in pregnancy. *Semin. Perinatol.* 39: 512–519.

15 Isoherranen, N. and Thummel, K.E. (2013). Drug metabolism and transport during pregnancy: how does drug disposition change during pregnancy and what are the mechanisms that cause such changes? *Drug Metab. Dispos.* 41: 256–262.

16 Pinheiro, E.A. and Stika, C.S. (2020). Drugs in pregnancy: pharmacologic and physiologic changes that affect clinical care. *Semin. Perinatol.* 44: 151221.

17 Ke, A.B., Nallani, S.C., Zhao, P. et al. (2014). Expansion of a PBPK model to predict disposition in pregnant women of drugs cleared via multiple CYP enzymes, including CYP2B6, CYP2C9 and CYP2C19. *Br. J. Clin. Pharmacol.* 77: 554–570.

18 McGready, R., Stepniewska, K., Seaton, E. et al. (2003). Pregnancy and use of oral contraceptives reduces the biotransformation of proguanil to cycloguanil. *Eur. J. Clin. Pharmacol.* 59: 553–557.

19 Chen, H., Yang, K., Choi, S. et al. (2009). Up-regulation of UDP-glucuronosyltransferase (UGT) 1A4 by 17β-estradiol: a potential mechanism of increased lamotrigine elimination in pregnancy. *Drug Metab. Dispos.* 37: 1841–1847.

20 Rytting, E., Nanovskaya, T.N., Wang, X. et al. (2014). Pharmacokinetics of indomethacin in pregnancy. *Clin. Pharmacokinet.* 53: 545–551.

21 Shah, M., Xu, M., Shah, P. et al. (2019). Effect of $CYP_2C_9$ polymorphisms on the pharmacokinetics of indomethacin during pregnancy. *Eur. J. Drug Metab. Pharmacokinet.* 44: 83–89.

22 Anderson, G.D. (2005). Pregnancy-induced changes in pharmacokinetics: a mechanistic-based approach. *Clin. Pharmacokinet.* 44: 989–1008.

23 Westin, A.A., Brekke, M., Molden, E. et al. (2017). Changes in drug disposition of lithium during pregnancy: a retrospective observational study of patient data from two routine therapeutic drug monitoring services in Norway. *BMJ Open* 7: e015738.

24 Deligiannidis, K.M. (2010). Therapeutic drug monitoring in pregnant and postpartum women: recommendations for SSRIs, lamotrigine, and lithium. *J. Clin. Psychiatry* 71: 649–650.

25 Cunningham, F.G., Leveno, K.J., Bloom, S.L. et al. (2018). Placental abnormalities. In: *Williams Obstet*, 25e (ed. A. Moyer and R.Y. Brow), 111. New York, NY: McGraw-Hill Education.

26 Burton, G.J. and Jauniaux, E. (2017). The cytotrophoblastic shell and complications of pregnancy. *Placenta* 60: 134–139.

27 Cunningham, F.G., Leveno, K.J., Bloom, S.L. et al. (2018). Implantation and placental development. In: *Williams Obstet*, 25e (ed. A. Moyer and R.Y. Brow), 80. New York, NY: McGraw-Hill Education.

28 Rampersad, R., Cervar-Zivkovic, M., and Nelson, D.M. (2011). Development and anatomy of the human placenta. In: *The Placenta: From Development to Disease* (ed. H.H. Kay, D.M. Nelson, and Y. Wang), 17–26. Wiley.

29 Wang, Y. and Zhao, S. (2010). Placental blood circulation. In: *Vascular Biology of the Placenta* (ed. D.N. Granger and J. Granger), 3. Morgan & Claypool Publishers.

30 Gellersen, B. and Brosens, J.J. (2014). Cyclic decidualization of the human endometrium in reproductive health and failure. *Endocr. Rev.* 35: 851–905.

31 Okada, H., Tsuzuki, T., and Murata, H. (2018). Decidualization of the human endometrium. *Reprod. Med. Biol.* 17: 220–227.

32 James, J.L., Carter, A.M., and Chamley, L.W. (2012). Human placentation from nidation to 5 weeks of gestation. Part I: what do we know about formative placental development following implantation? *Placenta* 33: 327–334.

33 Huppertz, B. and Gauster, M. (2011). Trophoblast fusion. In: *Cell Fusion in Health and Disease*, Advances in Experimental Medicine and Biology (ed. T. Dittmar and K.S. Zänker), 81–95. Springer Dordrecht.
34 Cohen, M., Meisser, A., and Bischof, P. (2006). Metalloproteinases and human placental invasiveness. *Placenta* 27: 783–793.
35 Pereira, L. (2018). Congenital viral infection: traversing the uterine-placental interface. *Annu. Rev. Virol.* 5: 273–299.
36 Herrick, E.J. and Bordoni, B. (2019). *Embryology, Placenta*. StatPearls Publishing.
37 James, J.L. and Chamley, L.W. (2016). Structure and development of the human placenta. In: *Placenta: The Tree of Life* (ed. O. Parolini), 1–12. CRC Press.
38 Pansky, B. (1982). *Review of Medical Embryology*. Macmillan.
39 Gude, N.M., Roberts, C.T., Kalionis, B., and King, R.G. (2004). Growth and function of the normal human placenta. *Thromb. Res.* 114: 397–407.
40 Marin, J.J.G., Macias, R.I.R., and Serrano, M.A. (2003). The hepatobiliary-like excretory function of the placenta. A review. *Placenta* 24: 431–438.
41 Syme, M.R., Paxton, J.W., and Keelan, J.A. (2004). Drug transfer and metabolism by the human placenta. *Clin. Pharmacokinet.* 43: 487–514.
42 Oratz, S. (2014). The hormones of the placenta. *Sci. J. Lander. Coll. Arts Sci.* 8: 7.
43 Napso, T., Yong, H.E.J., Lopez-Tello, J., and Sferruzzi-Perri, A.N. (2018). The role of placental hormones in mediating maternal adaptations to support pregnancy and lactation. *Front. Physiol.* 9: 01091.
44 Cole, L.A. (2010). Biological functions of hCG and hCG-related molecules. *Reprod. Biol. Endocrinol.* 8: 102.
45 Vuorela, P., Hatva, E., Lymboussaki, A. et al. (1997). Expression of vascular endothelial growth factor and placenta growth factor in human placenta[1]. *Biol. Reprod.* 56: 489–494.
46 Hiden, U., Glitzner, E., Hartmann, M., and Desoye, G. (2009). Insulin and the IGF system in the human placenta of normal and diabetic pregnancies. *J. Anat.* 215: 60–68.
47 Hansell, D.J., Bryant-Greenwood, G.D., and Greenwood, F.C. (1991). Expression of the human relaxin H1 gene in the decidua, trophoblast, and prostate. *J. Clin. Endocrinol. Metab.* 72: 899–904.
48 Prouillac, C. and Lecoeur, S. (2010). The role of the placenta in fetal exposure to xenobiotics: importance of membrane transporters and human models for transfer studies. *Drug Metab. Dispos.* 38: 1623–1635.
49 Gregg, N.M. (1941). Congenital cataract following German measles in the mother. In: *Problems of Birth Defects* (ed. T.V.N. Persaud), 170–180. Springer.
50 Dunn, D., Wallon, M., Peyron, F. et al. (1999). Mother-to-child transmission of toxoplasmosis: risk estimates for clinical counselling. *Lancet* 353: 1829–1833.
51 Thiébaut, R., Leproust, S., Chêne, G., and Gilbert, R. (2007). Effectiveness of prenatal treatment for congenital toxoplasmosis: a meta-analysis of individual patients' data. *Lancet* 369: 115–122.
52 Fiumara, N.J., Fleming, W.L., Downing, J.G., and Good, F.L. (1952). The incidence of prenatal syphilis at the Boston City Hospital. *N. Engl. J. Med.* 247: 48–52.
53 Soilleux, E.J. and Coleman, N. (2003). Transplacental transmission of HIV: a potential role for HIV binding lectins. *Int. J. Biochem. Cell Biol.* 35: 283–287.
54 Siegfried, N., van der Merwe, L., Brocklehurst, P., and Sint, T.T. (2011). Antiretrovirals for reducing the risk of mother-to-child transmission of HIV infection. *Cochrane Database Syst. Rev.* 7: CD003510.
55 James, S.H., Sheffield, J.S., and Kimberlin, D.W. (2014). Mother-to-child transmission of herpes simplex virus. *J. Pediatr. Infect. Dis. Soc.* 3: S19–S23.
56 Griffiths, S.K. and Campbell, J.P. (2015). Placental structure, function and drug transfer. *Contin. Educ. Anaesth. Crit. Care Pain* 15: 84–89.
57 Baergen, R.N. (2005). *Manual of Benirschke and Kaufmann's Pathology of the Human Placenta*. Springer Science & Business Media.
58 Donnelly, L. and Campling, G. (2014). Functions of the placenta. *Anaesth. Intensive Care Med.* 15: 136–139.
59 Vähäkangas, K. and Myllynen, P. (2009). Drug transporters in the human blood-placental barrier. *Br. J. Pharmacol.* 158: 665–678.
60 Lager, S. and Powell, T.L. (2012). Regulation of nutrient transport across the placenta. *J. Pregnancy* 2012: 179827.
61 Van Der Aa, E.M., Copius Peereboom-Stegeman, J.H.J., Noordhoek, J. et al. (1998). Mechanisms of drug transfer across the human placenta. *Pharm. World Sci.* 20: 139–148.
62 Staud, F., Cerveny, L., and Ceckova, M. (2012). Pharmacotherapy in pregnancy; effect of ABC and SLC transporters on drug transport across the

placenta and fetal drug exposure. *J. Drug Targeting* 20: 736–763.

63 Rubinchik-Stern, M. and Eyal, S. (2012). Drug interactions at the human placenta: what is the evidence? *Front. Pharmacol.* 3: 00126.

64 Barta, E. and Drugan, A. (2010). Glucose transport from mother to fetus – a theoretical study. *J. Theor. Biol.* 263: 295–302.

65 Cleal, J.K., Lofthouse, E.M., Sengers, B.G., and Lewis, R.M. (2018). A systems perspective on placental amino acid transport. *J. Physiol.* 596: 5511–5522.

66 Zu Schwabedissen, H.E.U.M., Grube, M., Heydrich, B. et al. (2005). Expression, localization, and function of MRP5 (ABCC5), a transporter for cyclic nucleotides, in human placenta and cultured human trophoblasts: effects of gestational age and cellular differentiation. *Am. J. Pathol.* 166: 39–48.

67 Akour, A.A., Kennedy, M.J., and Gerk, P. (2013). Receptor-mediated endocytosis across human placenta: emphasis on megalin. *Mol. Pharm.* 10: 1269–1278.

68 Akour, A.A., Kennedy, M.J., and Gerk, P.M. (2015). The role of megalin in the transport of gentamicin across BeWo cells, an *in vitro* model of the human placenta. *AAPS J.* 17: 1193–1199.

69 Mitchell, M.D., Peiris, H.N., Kobayashi, M. et al. (2015). Placental exosomes in normal and complicated pregnancy. *Am. J. Obstet. Gynecol.* 213: S173–S181.

70 Cui, C., Xue, Y.N., Wu, M. et al. (2013). Cellular uptake, intracellular trafficking, and antitumor efficacy of doxorubicin-loaded reduction-sensitive micelles. *Biomaterials* 34: 3858–3869.

71 Bajoria, R. and Sooranna, S.R. (1998). Liposome as a drug carrier system: prospects for safer prescribing during pregnancy: a review. *Placenta* 19: 265–287.

72 Lambot, N., Lybaert, P., Boom, A. et al. (2006). Evidence for a clathrin-mediated recycling of albumin in human term placenta[1]. *Biol. Reprod.* 75: 90–97.

73 Palmeira, P., Quinello, C., Silveira-Lessa, A.L. et al. (2012). IgG placental transfer in healthy and pathological pregnancies. *Clin. Dev. Immunol.* 2012: 985646.

74 Jeong, H. (2010). Altered drug metabolism during pregnancy: hormonal regulation of drug-metabolizing enzymes. *Expert Opin. Drug Metab. Toxicol.* 6: 689–699.

75 Zharikova, O.L., Fokina, V.M., Nanovskaya, T.N. et al. (2009). Identification of the major human hepatic and placental enzymes responsible for the biotransformation of glyburide. *Biochem. Pharmacol.* 78: 1483–1490.

76 Blanco-Castañeda, R., Galaviz-Hernández, C., Souto, P.C.S. et al. (2020). The role of xenobiotic-metabolizing enzymes in the placenta: a growing research field. *Expert. Rev. Clin. Pharmacol.* 13: 247–263.

77 Collier, A.C., Tingle, M.D., Paxton, J.W. et al. (2002). Metabolizing enzyme localization and activities in the first trimester human placenta: the effect of maternal and gestational age, smoking and alcohol consumption. *Hum. Reprod.* 17: 2564–2572.

78 Pavek, P. and Smutny, T. (2014). Nuclear receptors in regulation of biotransformation enzymes and drug transporters in the placental barrier. *Drug Metab. Rev.* 46: 19–32.

79 Perez-Sepulveda, A., Monteiro, L.J., Dobierzewska, A. et al. (2015). Placental aromatase is deficient in placental ischemia and preeclampsia. *PLoS One* 10: 0139682.

80 Storvik, M., Huuskonen, P., Pehkonen, P., and Pasanen, M. (2014). The unique characteristics of the placental transcriptome and the hormonal metabolism enzymes in placenta. *Reprod. Toxicol.* 47: 9–14.

81 Wang, X., Abdelrahman, D.R., Zharikova, O.L. et al. (2010). Bupropion metabolism by human placenta. *Biochem. Pharmacol.* 79: 1684–1690.

82 Fokina, V.M., Patrikeeva, S.L., Zharikova, O.L. et al. (2011). Transplacental transfer and metabolism of buprenorphine in preterm human placenta. *Am. J. Perinatol.* 28: 25–32.

83 Collier, A.C., Keelan, J.A., Van Zijl, P.E. et al. (2004). Human placental glucuronidation and transport of 3′azido-3′-deoxythymidine and uridine diphosphate glucuronic acid. *Drug Metab. Dispos.* 32: 813–820.

84 Mitra, P. and Audus, K.L. (2009). Expression and functional activities of selected sulfotransferase isoforms in BeWo cells and primary cytotrophoblast cells. *Biochem. Pharmacol.* 78: 1475–1482.

85 Pasanen, M. (1999). The expression and regulation of drug metabolism in human placenta. *Adv. Drug Deliv. Rev.* 38: 81–97.

86 Fokina, V.M., Patrikeeva, S.L., Oncken, C. et al. (2017). The role of placental carbonyl reducing enzymes in biotransformation of bupropion and 4-methylnitrosamino-1-(3-pyridyl)-1-butanone. *Drug Metab. Lett.* 11: 29–34.

87 Smit, J.W., Huisman, M.T., Van Tellingen, O. et al. (1999). Absence or pharmacological blocking of

placental P-glycoprotein profoundly increases fetal drug exposure. *J. Clin. Invest.* 104: 1441–1447.

88 Nanovskaya, T., Nekhayeva, I., Karunaratne, N. et al. (2005). Role of P-glycoprotein in transplacental transfer of methadone. *Biochem. Pharmacol.* 69: 1869–1878.

89 Mölsä, M., Heikkinen, T., Hakkola, J. et al. (2005). Functional role of P-glycoprotein in the human blood-placental barrier. *Clin. Pharmacol. Ther.* 78: 123–131.

90 Ito, S. (2001). Transplacental treatment of fetal tachycardia: implications of drug transporting proteins in placenta. *Semin. Perinatol.* 25: 196–201.

91 Bhatia, S. (2016). Nanoparticles types, classification, characterization, fabrication methods and drug delivery applications. In: *Natural Polymer Drug Delivery Systems*, 33–93. Springer Cham.

92 Menezes, V., Malek, A., and Keelan, J.A. (2012). Nanoparticulate drug delivery in pregnancy: placental passage and fetal exposure. *Curr. Pharm. Biotechnol.* 12: 731–742.

93 Zhang, B., Liang, R., Zheng, M. et al. (2019). Surface-functionalized nanoparticles as efficient tools in targeted therapy of pregnancy complications. *Int. J. Mol. Sci.* 20: 3642.

94 Kertschanska, S., Schröder, H., and Kaufmann, P. (1997). The ultrastructure of the trophoblastic layer of the degu (*Octodon degus*) placenta: a re-evaluation of the "channel problem.". *Placenta* 18: 219–225.

95 Wick, P., Malek, A., Manser, P. et al. (2010). Barrier capacity of human placenta for nanosized materials. *Environ. Health Perspect.* 118: 01200.

96 Keelan, J.A., Leong, J.W., Ho, D., and Iyer, K.S. (2015). Therapeutic and safety considerations of nanoparticle-mediated drug delivery in pregnancy. *Nanomedicine* 10: 2229–2247.

97 Bajoria, R., Fisk, N.M., and Contractor, S.F. (1997). Liposomal thyroxine: a noninvasive model for transplacental fetal therapy. *J. Clin. Endocrinol. Metab.* 82: 3271–3277.

98 Menjoge, A.R., Rinderknecht, A.L., Navath, R.S. et al. (2011). Transfer of PAMAM dendrimers across human placenta: prospects of its use as drug carrier during pregnancy. *J. Control. Release* 150: 326–338.

99 Paul, J.W. and Smith, R. (2018). Preventing preterm birth: new approaches to labour therapeutics using nanoparticles. *Best Pract. Res. Clin. Obstet. Gynaecol.* 52: 48–59.

100 Abou-Ghannam, G., Usta, I.M., and Nassar, A.H. (2012). Indomethacin in pregnancy: applications and safety. *Am. J. Perinatol.* 29: 175–186.

101 Moise, K.J., Ou, C.N., Kirshon, B. et al. (1990). Placental transfer of indomethacin in the human pregnancy. *Am. J. Obstet. Gynecol.* 162: 549–554.

102 Kirshon, B., Moise, K.J., Mari, G., and Willis, R. (1991). Long-term indomethacin therapy decreases fetal urine output and results in oligohydramnios. *Am. J. Perinatol.* 8: 86–88.

103 Vermillion, S.T., Scardo, J.A., Lashus, A.G. et al. (1997). The effect of indomethacin tocolysis on fetal ductus arteriosus constriction with advancing gestational age. *Am. J. Obstet. Gynecol.* 177: 256–261.

104 Riehemann, K., Schneider, S.W., Luger, T.A. et al. (2009). Nanomedicine—challenge and perspectives. *Angew. Chem. Int. Ed.* 48: 872–897.

105 Kaga, M., Li, H., Ohta, H. et al. (2012). Liposome-encapsulated hemoglobin (hemoglobin-vesicle) is not transferred from mother to fetus at the late stage of pregnancy in the rat model. *Life Sci.* 91: 420–428.

106 Refuerzo, J.S., Alexander, J.F., Leonard, F. et al. (2015). Liposomes: a nanoscale drug carrying system to prevent indomethacin passage to the fetus in a pregnant mouse model. *Am. J. Obstet. Gynecol.* 212: 508.e1–508.e7.

107 Refuerzo, J.S., Leonard, F., Bulayeva, N. et al. (2016). Uterus-targeted liposomes for preterm labor management: studies in pregnant mice. *Sci. Rep.* 6: 34710.

108 Albekairi, N.A., Al-Enazy, S., Ali, S., and Rytting, E. (2015). Transport of digoxin-loaded polymeric nanoparticles across BeWo cells, an *in vitro* model of human placental trophoblast. *Ther. Deliv.* 12: 1325–1334.

109 Ali, H., Kalashnikova, I., White, M.A. et al. (2013). Preparation, characterization, and transport of dexamethasone-loaded polymeric nanoparticles across a human placental *in vitro* model. *Int. J. Pharm.* 454: 149–157.

110 King, A., Ndifon, C., Lui, S. et al. (2016). Tumor-homing peptides as tools for targeted delivery of payloads to the placenta. *Sci. Adv.* 2: 1600349.

111 Berry, S.M., Stone, J., Norton, M.E. et al. (2013). Fetal blood sampling. *Am. J. Obstet. Gynecol.* 209: 170–180.

112 Stragierowicz, J., Sitarek, K., Grobelski, B., and Kilanowicz, A. (2018). The effect of prenatal exposure on disposition of hexachloronaphthalene in female Wistar rats and fetal compartment. *Int. J. Occup. Med. Environ. Health* 31: 685–695.

113 Dilworth, M.R. and Sibley, C.P. (2013). Review: transport across the placenta of mice and women. *Placenta* S34–S39.

114 Pemathilaka, R.L., Reynolds, D.E., and Hashemi, N.N. (2019). Drug transport across the human placenta: review of placenta-on-a-chip and previous approaches. *Interface Focus* 9: 0031.

115 Skoda, G., Hoffmann, O.I., Gócza, E. et al. (2017). Placenta-specific gene manipulation in rabbits. *J. Biotechnol.* 259: 86–90.

116 Saadeldin, I.M., Swelum, A.A.A., Elsafadi, M. et al. (2017). Isolation and characterization of the trophectoderm from the Arabian camel (*Camelus dromedarius*). *Placenta* 57: 113–122.

117 Czaja, E., Wieciech-Rojkiewicz, I., Knapczyk-Stwora, K. et al. (2017). The impact of antiandrogen flutamide on the hypoxia inducible factor 1α and vascular endothelial growth factor a gene and protein expression in the pig placenta during late pregnancy. *Folia Histochem. Cytobiol.* 55: 159–167.

118 Bazargan, M., Foster, D.J.R., Muhlhausler, B.S. et al. (2016). Limited fetal metabolism of rosiglitazone: elimination via the maternal compartment in the pregnant ewe. *Reprod. Toxicol.* 61: 162–168.

119 Pentšuk, N. and Van Der Laan, J.W. (2009). An interspecies comparison of placental antibody transfer: new insights into developmental toxicity testing of monoclonal antibodies. *Birth Defects Res. B Dev. Reprod. Toxicol.* 86: 328–344.

120 Benveniste, H., Fowler, J.S., Rooney, W.D. et al. (2003). Maternal-fetal in vivo imaging: a combined PET and MRI study. *J. Nucl. Med.* 44: 1522–1530.

121 Irvin-Choy, N.S., Nelson, K.M., Gleghorn, J.P., and Day, E.S. (2020). Design of nanomaterials for applications in maternal/fetal medicine. *J. Mater. Chem. B* 8: 6548–6561.

122 Maltepe, E., Bakardjiev, A.I., and Fisher, S.J. (2010). The placenta: transcriptional, epigenetic, and physiological integration during development. *J. Clin. Invest.* 120: 1016–1025.

123 Stanley, J.L., Andersson, I.J., Hirt, C.J. et al. (2012). Effect of the anti-oxidant tempol on fetal growth in a mouse model of fetal growth restriction. *Biol. Reprod.* 87: 1–8.

124 Kazuki, Y., Akita, M., Kobayashi, K. et al. (2016). Thalidomide-induced limb abnormalities in a humanized CYP3A mouse model. *Sci. Rep.* 6: 21419.

125 Kalabis, G.M., Petropoulos, S., Gibb, W., and Matthews, S.G. (2007). Breast cancer resistance protein (Bcrp1/Abcg2) in mouse placenta and yolk sac: ontogeny and its regulation by progesterone. *Placenta* 28: 1073–1081.

126 Saad, A.F., Kechichian, T., Yin, H. et al. (2014). Effects of pravastatin on angiogenic and placental hypoxic imbalance in a mouse model of preeclampsia. *Reprod. Sci.* 21: 138–145.

127 Carver, A.R., Andrikopoulou, M., Lei, J. et al. (2014). Maternal pravastatin prevents altered fetal brain development in a preeclamptic CD-1 mouse model. *PLoS One* 9: e100873.

128 Barry, J.S. and Anthony, R.V. (2008). The pregnant sheep as a model for human pregnancy. *Theriogenology* 69: 55–67.

129 James, M.O., Li, W., Summerlot, D.P. et al. (2010). Triclosan is a potent inhibitor of estradiol and estrone sulfonation in sheep placenta. *Environ. Int.* 36: 942–949.

130 Gauderat, G., Picard-Hagen, N., Toutain, P.L. et al. (2016). Bisphenol A glucuronide deconjugation is a determining factor of fetal exposure to bisphenol A. *Environ. Int.* 86: 52–59.

131 Schneider, H., Panigel, M., and Dancis, J. (1972). Transfer across the perfused human placenta of antipyrine, sodium, and leucine. *Am. J. Obstet. Gynecol.* 114: 822–828.

132 Hutson, J.R., Garcia-Bournissen, F., Davis, A., and Koren, G. (2011). The human placental perfusion model: a systematic review and development of a model to predict *in vivo* transfer of therapeutic drugs. *Clin. Pharmacol. Ther.* 90: 67–76.

133 Mathiesen, L., Mose, T., Mørck, T.J. et al. (2010). Quality assessment of a placental perfusion protocol. *Reprod. Toxicol.* 30: 138–146.

134 Nanovskaya, T.N., Patrikeeva, S.L., Paul, J. et al. (2013). Transplacental transfer and distribution of pravastatin. *Am. J. Obstet. Gynecol.* 209: 373.e1–373.e5.

135 Hemauer, S.J., Yan, R., Patrikeeva, S.L. et al. (2008). Transplacental transfer and metabolism of 17-α-hydroxyprogesterone caproate. *Am. J. Obstet. Gynecol.* 199: 169.e1–169.e5.

136 Vinot, C., Gavard, L., Tréluyer, J.M. et al. (2013). Placental transfer of maraviroc in an *ex vivo* human cotyledon perfusion model and influence of ABC transporter expression. *Antimicrob. Agents Chemother.* 57: 1415–1420.

137 Tertti, K., Petsalo, A., Niemi, M. et al. (2011). Transfer of repaglinide in the dually perfused human placenta and the role of organic anion transporting polypeptides (OATPs). *Eur. J. Pharm. Sci.* 44: 181–186.

138 Nekhayeva, I.A., Nanovskaya, T.N., Hankins, G.D.V., and Ahmed, M.S. (2006). Role of human placental efflux transporter P-glycoprotein in the transfer of buprenorphine, levo-α-acetylmethadol, and paclitaxel. *Am. J. Perinatol.* 23: 423–430.

139 Mathias, A.A., Hitti, J., and Unadkat, J.D. (2005). P-glycoprotein and breast cancer resistance protein expression in human placentae of various gestational ages. *Am. J. Physiol. Regul. Integr. Comp. Physiol.* 289: R963–R969.

140 Sun, M., Kingdom, J., Baczyk, D. et al. (2006). Expression of the multidrug resistance P-glycoprotein, (ABCB1 glycoprotein) in the human placenta decreases with advancing gestation. *Placenta* 27: 602–609.

141 Nanovskaya, T.N., Nekhayeva, I.A., Hankins, G.D.V., and Ahmed, M.S. (2008). Transfer of methadone across the dually perfused preterm human placental lobule. *Am. J. Obstet. Gynecol.* 198: 126.e1–126.e4.

142 Myllynen, P. and Vähäkangas, K. (2013). Placental transfer and metabolism: an overview of the experimental models utilizing human placental tissue. *Toxicol. In Vitro* 27: 507–512.

143 Conings, S., Amant, F., Annaert, P., and Van Calsteren, K. (2017). Integration and validation of the *ex vivo* human placenta perfusion model. *J. Pharmacol. Toxicol. Methods* 88: 25–31.

144 Myllynen, P., Mathiesen, L., Weimer, M. et al. (2010). Preliminary interlaboratory comparison of the ex vivo dual human placental perfusion system. *Reprod. Toxicol.* 30: 94–102.

145 Miller, R.K., Genbacev, O., Turner, M.A. et al. (2005). Human placental explants in culture: approaches and assessments. *Placenta* 26: 439–448.

146 Orendi, K., Kivity, V., Sammar, M. et al. (2011). Placental and trophoblastic in vitro models to study preventive and therapeutic agents for preeclampsia. *Placenta* 32: S49–S54.

147 Vähäkangas, K., Raunio, H., Pasamen, M. et al. (1989). Comparison of the formation of benzo[*a*]pyrene diolepoxide-dna adducts *in vitro* by rat and human microsomes: evidence for the involvement of *p*-450IA1 and *p*-450IA2. *J. Biochem. Toxicol.* 4: 79–86.

148 Fokina, V.M., Zharikova, O.L., Hankins, G.D.V. et al. (2012). Metabolism of 17-alpha-hydroxyprogesterone caproate by human placental mitochondria. *Reprod. Sci.* 19: 19248.

149 Partanen, H.A., El-Nezami, H.S., Leppänen, J.M. et al. (2010). Aflatoxin B1 transfer and metabolism in human placenta. *Toxicol. Sci.* 113: 216–225.

150 Hemauer, S.J., Patrikeeva, S.L., Nanovskaya, T.N. et al. (2010). Role of human placental apical membrane transporters in the efflux of glyburide, rosiglitazone, and metformin. *Am. J. Obstet. Gynecol.* 202: 383.e1–383.e7.

151 Afrouzian, M., Al-Lahham, R., Patrikeeva, S. et al. (2018). Role of the efflux transporters BCRP and MRP1 in human placental bio-disposition of pravastatin. *Biochem. Pharmacol.* 156: 467–478.

152 Noguchi, S., Nishimura, T., Fujibayashi, A. et al. (2015). Organic anion transporter 4-mediated transport of olmesartan at basal plasma membrane of human placental barrier. *J. Pharm. Sci.* 104: 3128–3135.

153 Glazier, J.D. and Sibley, C.P. (2006). In vitro methods for studying human placental amino acid transport: placental plasma membrane vesicles. *Methods Mol. Med.* 122: 241–252.

154 Gilibili, R.R., Chatterjee, S., Bagul, P. et al. (2017). Coproporphyrin-I: a fluorescent, endogenous optimal probe substrate for $ABCC_2$ ($MRP_2$) suitable for vesicle-based $MRP_2$ inhibition assay. *Drug Metab. Dispos.* 45: 604–611.

155 Pasanen, M., Raunio, H., and Pelkonen, O. (1985). Freezing affects the activity and subcellular distribution profile of human placental xenobiotic- and steroid-metabolizing enzymes. *Placenta* 6: 527–535.

156 Evseenko, D.A., Paxton, J.W., and Keelan, J.A. (2006). ABC drug transporter expression and functional activity in trophoblast-like cell lines and differentiating primary trophoblast. *Am. J. Physiol. Regul. Integr. Comp. Physiol.* 290: R1357–R1365.

157 Hemmings, D.G., Lowen, B., Sherburne, R. et al. (2001). Villous trophoblasts cultured on semi-permeable membranes form an effective barrier to the passage of high and low molecular weight particles. *Placenta* 22: 70–79.

158 Shiverick, K.T., King, A., Frank, H.G. et al. (2001). Cell culture models of human trophoblast II: trophoblast cell lines – a workshop report. *Placenta* 22: S104–S106.

159 Abou-Kheir, W., Barrak, J., Hadadeh, O., and Daoud, G. (2017). HTR-8/SVneo cell line contains a mixed population of cells. *Placenta* 50: 1–7.

160 Pattillo, R.A., Gey, G.O., Delfs, E. et al. (1971). The hormone-synthesizing trophoblastic cell *in vitro*: a model for cancer research and placental hormone synthesis. *Ann. N. Y. Acad. Sci.* 172: 288–298.

161 Pattillo, R.A. and Gey, G.O. (1968). The establishment of a cell line of human hormone-synthesizing trophoblastic cells *in vitro*. *Cancer Res.* 28: 1231–1236.

162 Kohler, P.O. and Bridson, W.E. (1971). Isolation of hormone-producing clonal lines of human choriocarcinoma. *J. Clin. Endocrinol. Metab.* 32: 683–687.

163 Serrano, M.A., Macias, R.I.R., Briz, O. et al. (2007). Expression in human trophoblast and choriocarcinoma cell lines, BeWo, Jeg-3 and JAr of genes involved in the hepatobiliary-like excretory function of the placenta. *Placenta* 28: 107–117.

164 Wolfe, M.W. (2006). Culture and transfection of human choriocarcinoma cells. *Methods Mol. Med.* 121: 229–239.

165 Bode, C.J., Jin, H., Rytting, E. et al. (2006). In vitro models for studying trophoblast transcellular transport. *Methods Mol. Med.* 122: 225–239.

166 Ikeda, K., Utoguchi, N., Tsutsui, H. et al. (2011). *In vitro* approaches to evaluate placental drug transport by using differentiating JEG-3 human choriocarcinoma cells. *Basic Clin. Pharmacol. Toxicol.* 108: 138–145.

167 Poulsen, M.S., Rytting, E., Mose, T., and Knudsen, L.E. (2009). Modeling placental transport: correlation of *in vitro* BeWo cell permeability and *ex vivo* human placental perfusion. *Toxicol. In Vitro* 23: 1380–1386.

168 Bouazza, N., Foissac, F., Hirt, D. et al. (2019). Methodological approaches to evaluate fetal drug exposure. *Curr. Pharm. Des.* 25: 496–504.

169 Li, H., Van Ravenzwaay, B., Rietjens, I.M.C.M., and Louisse, J. (2013). Assessment of an in vitro transport model using BeWo $b_{30}$ cells to predict placental transfer of compounds. *Arch. Toxicol.* 87: 1661–1669.

170 Pemathilaka, R.L., Caplin, J.D., Aykar, S.S. et al. (2019). Placenta-on-a-chip: in vitro study of caffeine transport across placental barrier using liquid chromatography mass spectrometry. *Global Challenges* 3: 00112.

171 Blundell, C., Tess, E.R., Schanzer, A.S.R. et al. (2016). A microphysiological model of the human placental barrier. *Lab Chip* 16: 3065–3073.

172 Blundell, C., Yi, Y.S., Ma, L. et al. (2018). Placental drug transport-on-a-chip: a microengineered in vitro model of transporter-mediated drug efflux in the human placental barrier. *Adv. Healthc. Mater.* 7: 00786.

173 Lee, J.S., Romero, R., Han, Y.M. et al. (2016). Placenta-on-a-chip: a novel platform to study the biology of the human placenta. *J. Matern. Fetal Neonatal Med.* 29: 1046–1054.

174 Arumugasaamy, N., Navarro, J., Kent Leach, J. et al. (2019). *In vitro* models for studying transport across epithelial tissue barriers. *Ann. Biomed. Eng.* 47: 1–21.

175 Miura, S., Sato, K., Kato-Negishi, M. et al. (2015). Fluid shear triggers microvilli formation via mechanosensitive activation of TRPV6. *Nat. Commun.* 6: 8871.

176 Levkovitz, R., Zaretsky, U., Gordon, Z. et al. (2013). *In vitro* simulation of placental transport: part I. Biological model of the placental barrier. *Placenta* 34: 699–707.

177 Levkovitz, R., Zaretsky, U., Jaffa, A.J. et al. (2013). *In vitro* simulation of placental transport: part II. Glucose transfer across the placental barrier model. *Placenta* 34: 708–715.

178 Song, D., Guo, J., Han, F. et al. (2013). Establishment of an *in vitro* model of the human placental barrier by placenta slice culture and ussing chamber. *Biosci. Biotechnol. Biochem.* 77: 1030–1034.

179 De Sousa Mendes, M., Hirt, D., Vinot, C. et al. (2016). Prediction of human fetal pharmacokinetics using *ex vivo* human placenta perfusion studies and physiologically based models. *Br. J. Clin. Pharmacol.* 81: 646–657.

180 Zhang, Z. and Unadkat, J.D. (2017). Development of a novel maternal-fetal physiologically based pharmacokinetic model II: verification of the model for passive placental permeability drugs. *Drug Metab. Dispos.* 45: 939–946.

181 Zhang, Z., Imperial, M.Z., Patilea-Vrana, G.I. et al. (2017). Development of a novel maternal-fetal physiologically based pharmacokinetic model I: insights into factors that determine fetal drug exposure through simulations and sensitivity analyses. *Drug Metab. Dispos.* 45: 920–938.

182 Dallmann, A., Liu, X.I., Burckart, G.J., and van den Anker, J. (2019). Drug transporters expressed in the human placenta and models for studying maternal-fetal drug transfer. *J. Clin. Pharmacol.* 59: S70–S81.

# 9

# Biopharmaceutics Classification System: Theory and Practice

*Mehul Mehta, Jayabharathi Vaidyanathan, and Lawrence Yu*

*Center for Drug Evaluation and Research, Food and Drug Administration, Silver Spring, MD, USA*

## 9.1 Introduction

Since Prof. Gordon L. Amidon introduced the Biopharmaceutics Classification System (BCS) in 1995 [1], its validity and applicability have become the subject of extensive research and discussion [2]. The BCS is a framework that is intended to classify a drug substance based on its aqueous solubility and intestinal permeability. Considering gastrointestinal (GI) transit and environment could impact the rate and extent of drug absorption in addition to dissolution and permeation, the BCS met with much skepticism in the literature initially.

The main skepticism was mainly from the definition of permeability and its measurement. In 1988, Prof. Gordon L. Amidon discovered that human oral fraction dose absorption can be estimated by rat intestinal membrane permeability for passive and carrier-mediated compounds [3]. This relationship was further verified with human permeability data from Lennernas et al. [4]. However, the equations to calculate the permeability or predict the fraction of dose absorbed have been criticized as the small intestine was considered as fully extended cylinder [5]. As we now know, this assumption does not impact the equation to link the fraction dose absorbed and permeability. Over the last 25 years, we have witnessed its soundness and value in BCS-based biowaiver and drug development [6–8].

Another issue with the BCS is the solubility as it is not only factor controlling the dissolution. In 2004, nevertheless, the FDA investigated the intrinsic dissolution rate of 15 compounds and found a good relationship between intrinsic dissolution and solubility [9]. In addition, it was observed that the BCS solubility classification is overly conservative [10]. On the other hand, BCS was originally developed for regulatory utility. Therefore, its conservativism is understandable as the regulatory use needs to minimize the false positive, meaning the compound that is not highly soluble is classified as highly soluble.

The FDA issued the first regulatory guidance for BCS-based biowaiver in 2000 [11]. The European Medicines Agency (EMA) published the BCS-based biowaiver in 2010 [12]. Health Canada and the World Health Organization (WHO) published the BCS biowaiver guidance in 2014 and 2015, respectively [13, 14]. In 2020, the International Council for Harmonisation of Technical Requirements for Pharmaceuticals for Human Use (ICH) issued the M9 guideline on BCS-based biowaivers [15, 16]. In addition to contributing to biowaiver, the BCS has been used in the FDA's dissolution guidance [17] and regulatory assessment of drug applications [18].

This chapter will describe the theory and practice of BCS and its utility in biowaiver and other regulatory guidance and decision-making with practical case studies.

## 9.2 Theory

### 9.2.1 Bioavailability and Bioequivalence

Studies to measure bioavailability (BA) and/or establish bioequivalence (BE) of a product are important elements in support of regulatory drug applications. BA means the rate and extent to which the active ingredient or active moiety is absorbed from a drug product and becomes available at the site of action (21 eCFR 320.1(a)). BA data provide an estimate of the fraction of the drug absorbed, as well as provide information related to the pharmacokinetics of the drug. BE means the absence of a significant difference in the rate and extent to which the active ingredient or active moiety in pharmaceutical equivalents or pharmaceutical

*Oral Bioavailability and Drug Delivery: From Basics to Advanced Concepts and Applications*, First Edition. Edited by Ming Hu and Xiaoling Li.

alternatives becomes available at the site of drug action when administered at the same molar dose under similar conditions in an appropriately designed study. Studies to establish BE between two products are important for major formulation or manufacturing changes as well as generic drug development. In practice, BE for oral drug products is demonstrated by no significant differences in the rate and extent of drug absorption of test and reference drug products.

BA and BE problems were observed in the early 1970s when patients who took digoxin had variable or poor responses to the medication [19]. The FDA recognized the existence of BA/BE problems and established the BE standards in the 1980s. The FDA recommended the following methods for BE documentation [20]:

(a) Comparative pharmacokinetic studies
(b) Comparative pharmacodynamic studies;
(c) Comparative clinical trials;
(d) Comparative *in vitro* tests;
(e) Any other approach deemed adequate by FDA.

Experiences in 1990s revealed that comparative pharmacokinetic studies are mostly used for BE demonstration of systemically absorbed drug products while pharmacodynamic studies and clinical trials were generally employed for locally acting drug products. However, in 1990s, *in vitro* tests alone were rarely utilized for the purpose of BE establishment, and unnecessary human studies were performed for a certain class of drug products.

Dissolution tests were adopted in 1960s to replace disintegration tests that were adopted in 1950s. However, the relationship between dissolution and BA was generally uncertain [21]. There was a growing interest in the subject of dissolution, absorption, distribution, and metabolism. The correlation was established for digoxin between *in vitro* dissolution and *in vivo* performance, providing incentive that dissolution may be used for prediction of *in vivo* BA in addition to serving as quality control. There was a pressing need to establish a scientific basis so that scientists understand when the relation of *in vitro* to *in vivo* could be established, and when it could not.

### 9.2.2 Oral Drug Absorption Prediction

#### 9.2.2.1 Plug Flow Model

Prof. Gordon L. Amidon's research in 1970s and 1980s focused on the mechanistic understanding and prediction of oral drug absorption [22]. By 1990, Prof. Amidon's lab had developed methods for predicting absorption in humans using animal intestinal jejunal permeability for soluble drugs and separated the absorption and metabolism components of systemic availability [3, 23]):

$$F_{abs} = 1 - \exp\left(-2P^*_{eff}\right)$$

where $P^*_{eff}$ is the dimensionless effective permeability. This discovery suggests the complex process of permeability across intestinal membrane can be described by a simple parameter, effective permeability, greatly facilitating the development of the BCS.

The fundamental starting point for the BCS is Fick's First Law applied to intestinal membrane:

$$J = P_{eff} \times C$$

where $J$, $P_{eff}$, and $C$ are the drug flux (mass/area/time) through the intestinal membrane, effective permeability, and drug concentration. This equation assumes that the drug concentration at the other side of membrane is zero, which often is the case. The small intestine is assumed to be a cylindrical tube with the surface area of $2\pi RL$, where $R$ is the radius and $L$ is the length of the tube. From mass balance, we have:

$$-\frac{dM}{dt} = Q\left(C_0 - C_{out}\right) = 2\pi RP_{eff} \int_0^L C\,dz$$

where $M$ is the amount of a drug, $Q$ is the volumetric flow rate, $C_0$ and $C_{out}$ are the inlet and outlet concentrations, respectively. Under the steady-state assumption, the fraction of dose absorbed, Fabs is $1 - C_{out}/C_0$ and transit time $<Tsi> = \pi R^2 L/Q$. Therefore, with plug flow model [23], we have:

$$F_{abs} = 1 - \frac{C_{out}}{C_0} = \frac{2 <Tsi> P_{eff}}{R} \int_0^1 C^* dz^*$$

where $C^*$ and $z^*$ are dimensionless variables, $C^* = C/C_0$, $z^* = z/L$. $<Tsi>$ is the mean human small intestinal transit time that is 199 minutes [24]. If inlet and outlet concentrations are below the solubility, then the equation above can be integrated and result in:

$$F_{abs} = 1 - \mathrm{Exp}\left(-\frac{2 <Tsi> P_{eff}}{R}\right)$$

Chiou [5] argued that the experimental conditions of perfusion studies do not reflect the reality. This is likely true as the small intestine is flat with many pockets of water. However, as the prediction and perfusion assume the intestine is a cylindrical tube, the net effect is that the absorption prediction is in good agreement with the experimental data.

#### 9.2.2.2 Compartmental Absorption and Transit Model

The issue with the plug flow model is that it assumes steady state and the small intestine is considered as fully extended cylinder. In reality, any absorption in humans is not steady state as it will vary with time. Second, the small intestine is likely flat with pockets of water. Therefore, Yu et al. [24, 25] developed a compartmental absorption transit (CAT) model. The basic equation for the CAT model for nondegradable and highly soluble drugs is described as follows:

$$\frac{dM_n}{dt} = K_t M_{n-1} - K_t M_n, n = 1,2, \ldots, 7$$

$$\frac{dM_a}{dt} = K_a \sum_{n=1}^{7} M_n$$

where $M_n$ is the amount of drug at the nth compartment and $K_a$ is the amount of drug absorbed, $t$ is time, $n$ is the number of total compartments, $K_t$ is the transit rate constant and $K_a$ is the absorption rate constant. $K_a$ is proportional to the permeability. The final fraction dose absorbed from the CAT is:

$$F_{abs} = 1 - (1 + 0.54P_{eff})^{-7}$$

Figure 9.1 shows the predicted values for the fraction of dose absorbed. It suggests that both the CAT and plug flow model give an excellent prediction, laying the foundation of the BCS.

### 9.2.3 Biopharmaceutics Classification System

The above analysis shows two key parameters, which determine the rate and extent of absorption of a drug substance: solubility and permeability. Thus, drugs can be divided into high/low solubility-permeability classes, and the expectations regarding *in vitro–in vivo* correlations are more clearly stated. Table 9.1 summarizes these expectations that are discussed in more detail below:

Class I: High Solubility–High Permeability Drugs. This is the class where the drug is well absorbed, though its systemic availability may be low due to first-pass metabolism. The rate-limiting step to oral drug absorption is drug dissolution or gastric emptying if dissolution is relatively rapid. For immediate-release dosage forms that dissolve rapidly, the absorption rate is controlled by gastric emptying, and no correlation with dissolution is expected. In

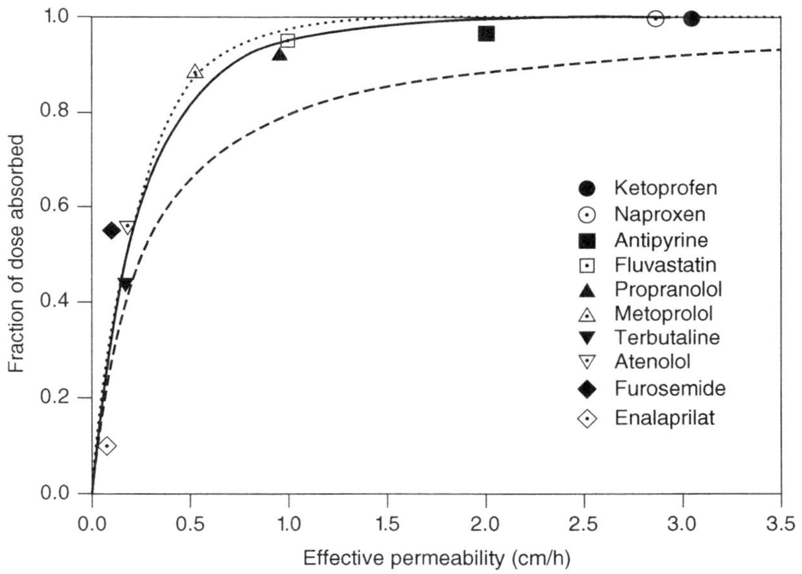


**Figure 9.1** The fraction of dose absorbed as function of the human effective permeability, where (—) represents the prediction of the compartmental absorption and transit model, (---) represents the single compartment model, and (...) represents the plug flow model. Source: Yu and Amidon [26]/with permission of Elsevier.

**Table 9.1** *In vitro–in vivo* (IVIV) correlation expectations for immediate release drug products based on BCS.

| Class | Solubility | Permeability | IVIVC expectations |
|---|---|---|---|
| I | High | High | IVIVC is highly unlikely as gastric emptying is likely the rate-limiting step |
| II | Low | High | IVIVC is possible unless dose is extremely high. It could still be challenging as *in vitro* dissolution is too much different from the vivo in practice |
| III | High | Low | Limited or no IVIVC is expected as the absorption is limited by permeability. |
| IV | Low | Low | Limited or no IVIVC is expected |

practice, the dissolution of 85% in 30 minutes ensures BA/BE.

Class II: Low Solubility–High Permeability Drugs. This is the class where *in vivo* dissolution is the rate-limiting step in oral drug absorption unless that dose is extremely high and small intestine is saturated with drugs where absorption is limited by solubility. Therefore, *in vitro–in vivo* relationships are possible. *In vitro* dissolution should reflect *in vivo* conditions as much as possible.

Class III: High Solubility–Low Permeability. For this class of drugs, permeability is the rate-controlling step in drug absorption. This class of drugs typically exhibits high variability. No *in vitro* and *in vivo* correlations are expected.

Class IV. Low Solubility–Low Permeability Drugs. This class of drugs presents significant issues for effective drug delivery and establishing *in vitro* and *in vivo* relationships is really challenging.

The FDA's 2000 BCS guidance provides the first definition of high/low solubility and permeability. The solubility classification of a drug in the BCS is based on the highest dose strength in an IR product. A drug substance is considered *highly soluble* when the highest strength is soluble in 250 ml or less of aqueous media over the pH range of 1.0–7.5. Otherwise, the drug substance is considered poorly soluble. The volume estimate of 250 ml is derived from typical BE study protocols that prescribe administration of a drug product to fasting human volunteers with a glass (about 8 oz) of water.

The permeability classification is based directly on the extent of intestinal absorption of a drug substance in humans or indirectly on measurements of the rate of mass transfer across the human intestinal membrane. Animal or *in vitro* models capable of predicting the extent of intestinal absorption in humans may be used as alternatives, e.g. *in situ* rat perfusion models and *in vitro* epithelial cell culture models. A drug substance is considered highly permeable when the extent of intestinal absorption is determined to be 90% or higher. Otherwise, the drug substance is lowly permeable.

Since its issuing of the 2000 BCS Biowaiver guidance, the biowaiver extensions have been our focus [27]. The suggested changes include redefining BCS solubility and permeability class boundaries and biowaiver extension to BCS Class II and III drugs. In 2017, the FDA issued the revised BCS guidance and revised definition of high/low solubility and permeability. A drug substance is considered highly soluble when the highest strength is soluble in 250 ml or less of aqueous media within the pH range of 1–6.8 at $37 \pm 1\,°C$. A drug substance is considered to be highly permeable when the systemic BA or the extent of absorption in humans is determined to be 85% or more of an administered dose based on a mass balance determination (along with evidence showing stability of the drug in the GI tract) or in comparison to an intravenous reference dose.

ICH recently developed and finalized the BCS-based Biowaiver guidance [15]. ICH's definitions of solubility and permeability are very similar to the FDA 2017 guidance [28]. However, ICH uses the highest single therapeutic dose instead of the FDA's highest dose strength. The topic of the highest single therapeutic dose is not new. The FDA had extensive debates and discussions before the issuing of the 2000 BCS guidance. One viewpoint discussed was to use the highest dose strength as this was used in the FDA BE studies. As BCS is used mainly for waivers of BE studies, the highest dose strength should be used in the FDA's biowaiver guidance. The other viewpoint discussed was to use the highest therapeutic dose as BCS was developed based on the understanding of the absorption rate-limiting steps and the use of the highest therapeutic dose more closely reflects the reality.

## 9.3 BCS-based Biowaiver

The origin of the term "biowaiver" can be traced to the US eCFR title 21, part 320 [29]. This part defines the BA and BE requirements for drug applications and within that, the section 320.22 lays out the criteria for waiver of evidence of *in vivo* BA or BE. As stated in that section, an applicant may request FDA to waive the requirement for the submission of evidence measuring the *in vivo* BA or demonstrating the *in vivo* BE of the drug product that is the subject of the application; it further states that an applicant shall submit a request for waiver with the application. Thus, "biowaiver" is the waiver of the *in vivo* BA or BE study that is required as per eCFR 320.

Furthermore, as conveyed in eCFR 320, several types of drug products like parenterals, gases, and oral solutions are eligible for biowaiver provided required conditions are met. Biowaivers can also be granted to different (typically lower) strengths within an immediate-release oral drug product line if the required conditions are met. For complete details, the reader is referred to the FDA guidance [30].

As discussed earlier in this chapter, BCS-based biowaivers form a distinct category where the biowaiver is based on the fundamental properties of the drug substance – solubility and permeability – and the drug product, i.e. rapid and comparable dissolution. Over last several years, the BCS-based biowaiver approach has gained tremendous interest amongst the regulatory, academic, and industry scientists worldwide, which ultimately led to the recognition of this as a topic for international harmonization by the ICH by unanimous support. The M9 BCS Based Biowaivers Expert Working Group (EWG) was thus formed in November 2016 and over the next three years and six meetings, the EWG created and cleared the M9 guideline through all the ICH clearance steps; the final M9 guideline, along with the training slides, was published in November 2019. The rest of this section summarizes key aspects from the M9 guideline essential for BCS-based biowaivers. The reader is advised to refer to the M9 guideline for additional important details and is also strongly urged to refer to the companion document entitled M9 Questions and Answers, which provides detailed answers to several important issues that were brought up by the external stakeholders during the comment period [16].

As stated in the guideline, BCS-based biowaiver can be granted to products that are classified as BCS Class I (high solubility and high permeability) or BCS Class III (high solubility and low permeability). In order for the biowaiver to be applicable, the drug substance in test and reference products has to be identical. A biowaiver may also be applicable if test and reference products contain different salts provided that both belong to BCS Class I. A test product containing a different ester, ether, isomer, mixture of isomers, complex, or derivative of a drug substance from that of the reference product is not eligible for the BCS-based biowaiver. Prodrugs are eligible for a BCS-based biowaiver when absorbed as the prodrug.

BCS-based biowaiver is limited to immediate release, solid oral dosage forms or suspensions that deliver the drug to the systemic circulation; narrow therapeutic index drugs are not eligible. Drug products with buccal or sublingual absorption are not eligible for a BCS-based biowaiver application. Additionally, the BCS-based biowaiver is applicable only when the mode of product administration includes water.

Fixed-dose combination (FDC) products are eligible for a BCS-based biowaiver provided all drug substances contained in the combination drug product meet the required criteria outlined in the guideline. BCS-based biowaivers are applicable to FDCs, which are the same dosage form and strength. FDC formulations containing only BCS Class I drugs should meet criteria regarding excipients for a BCS Class I drug. FDC formulations containing only BCS Class III drugs, or BCS Class I and BCS Class III drugs, should meet criteria regarding excipients for a BCS Class III drug.

For a drug substance to be classified as highly soluble, the highest single therapeutic dose should be completely soluble in 250 ml or less of aqueous media over the pH range of 1.2–6.8 at $37 \pm 1\,°C$. If the highest single therapeutic dose does not meet this criterion but the highest strength of the reference product is soluble under the stated conditions, additional data should be submitted to justify the BCS-based biowaiver approach as further detailed in the guideline.

If the drug substance has an absolute BA of ≥85%, the drug substance is considered to have high permeability (HP). The same conclusion of HP can also be reached if ≥85% of the administered dose is recovered in urine as unchanged (parent drug), or as the sum of parent drug, Phase 1 oxidative, and Phase 2 conjugative metabolites. Permeability can be also assessed by validated and standardized *in vitro* methods using Caco-2 cells; this does require demonstration of permeability independent of active transport. If mass balance studies are used to demonstrate HP, additional data to document the drug's stability in the GI tract should be provided unless ≥85% of the dose is recovered as unchanged drug in urine. Demonstration of stability in the GI tract is required if *in vitro* Caco-2 studies are used to support HP.

Qualitative and quantitative differences in excipients are permitted for BCS Class I drugs, except for excipients that may affect absorption; such excipients should be qualitatively the same and quantitatively similar, i.e. within ± 10% of the amount of excipient in the reference product. Additionally, the cumulative difference should be within ± 10% for excipients that may affect absorption. BCS Class III drug substances are more susceptible to the effects of excipients since they may have site-specific absorption and therefore there can be a greater number of mechanisms through, which excipients can affect their absorption than for BCS Class I drugs. Hence, for BCS Class III drugs, all excipients should be qualitatively the same and quantitatively similar (except for film coating or capsule shell excipients), i.e. within ± 10% of the amount of excipient in the reference product, and the cumulative difference for these excipients should be within ± 10%.

For the BCS-based biowaiver consideration, comparative *in vitro* dissolution tests should be conducted using one batch representative of the proposed commercial manufacturing process for the test product relative to the reference product. These experiments should use compendial apparatus and suitably validated analytical method(s). To qualify for a BCS-based biowaiver for BCS Class I, drug substance in both the test product and reference product should display either very rapid (≥85%

for the mean percent dissolved in ≤15 minutes) *in vitro* dissolution characteristics, or rapid (≥85% for the mean percent dissolved in ≤30 minutes) and similar *in vitro* dissolution characteristics based on the similarity factor *f*2 under all of the defined conditions [15].

When considering the BCS-based biowaiver for FDC formulations, dissolution profiles should meet the criteria for all drug substances in the FDC product. For FDCs containing both BCS Class I and BCS Class III drugs, the dissolution criteria for the applicable BCS class for each component should be applied.

## 9.4 BCS Waiver Case Studies

The following examples will highlight how the principles of BCS are being used by the FDA to make biowaiver decisions. The examples also provide the type of data that is submitted to demonstrate high solubility, HP, and rapid dissolution.

### 9.4.1 Case Study 1

#### 9.4.1.1 Solubility

The applicant provided solubility data using different media (water, 0.1 N HCl, 0.01 N HCl, pH 4.5 acetate buffer, pH 6.8 acetate buffer, pH 7.7 phosphate buffer, and pH 8.7 alkaline borate buffer). Equilibrium solubility testing was performed at the abovementioned buffers at 37 °C in triplicates in each medium.

As shown in Table 9.2, the mean solubility of the drug was in the range of 39.54 to 105.47 mg/ml. The lowest solubility of the test product (39.54 mg/ml), exceeded the solubility requirement of 0.4 mg/ml (100 mg/250 ml). The drug was also found to be stable up to 25 hours at 37 °C in respective buffers. It was concluded that the solubility experiments were conducted as described in the BCS guidance and that the drug demonstrated high solubility.

**Table 9.2** Solubility data for example 1.

| Medium | Solubility (mg/ml) |
|---|---|
| Water | 69.95 |
| 0.1 HCl | 105.47 |
| 0.01 HCl | 74.16 |
| pH 4.5 Acetate buffer | 82.4 |
| pH 6.8 Phosphate buffer | 52.41 |
| pH 7.7 Phosphate buffer | 39.54 |
| pH 8.7 Alkaline Borate buffer | 51.38 |

#### 9.4.1.2 Permeability

The absolute BA of the drug was approximately 87%. Data were also available from a mass balance study for this product. The mean total recovery following the oral dose was ~100% with 87% recovered in urine and approximately 79% excreted intact in urine. When mass balance studies are used to demonstrate HP, additional data to document the drug's stability in the GI tract is required, unless 85% or more of the drug is excreted unchanged in the urine as per the FDA BCS guidance. There was no information regarding drug's stability in the GI tract. As per the guidance, a single method may be sufficient (i) when the absolute BA is 85% or more, (ii) when 85% or more of the administered drug is excreted unchanged in urine, or (iii) when 85% or more of the administered drug is recovered in urine as parent and metabolites with evidence indicating stability in GI tract. Since in this case, the absolute BA data showed conclusive demonstration of HP, mass balance information was used as supportive data to make the determination of HP.

#### 9.4.1.3 Dissolution

The dissolution testing was performed using 500 ml of three different mediums (0.1 N HCl, pH 4.5 acetate buffer, and pH 6.8 phosphate buffer) using Apparatus I at 100 rpm. Dissolution testing was performed for all available strengths. The data showed that both the test and reference products dissolved ≥85% in 15 minutes and met the criteria for very rapidly dissolving (≥85% in 15 minutes) in all three media and it was concluded that the dissolution profiles for the test and reference products are similar (*f*2 testing was not needed).

The overall conclusion was that the drug product was classified as BCS class 1 and eligible for BCS-based biowaiver.

### 9.4.2 Case Study 2

#### 9.4.2.1 Solubility

The applicant submitted solubility data over the pH range of 1.0, 3.0, 5.0, and 6.8. The results showed that the solubility of drug across the pH range 1.0 to 6.8 was higher than 170 mg/ml (equilibrium solubility). The aqueous solubility of >170 mg/ml in all pH conditions exceeded that needed for classifying drug as highly soluble using either 500 mg (highest dose strength) or 900 mg (highest dose per administration). Adequate solubility measurements demonstrated high solubility based on maximum allowable dose and highest strength at equilibrium conditions. The provided solubility information over the pH range of 1.0, 3.0, 5.0, and 6.8, clearly indicated that the drug is a highly soluble drug substance, as per the BCS guidance.

#### 9.4.2.2 Permeability

The applicant conducted *in vitro* permeability study using in situ rat intestine perfusion method for this drug and provided acceptable data as required per the BCS guidance. The method was validated for BCS permeability classification, using twenty model compounds ranging from highly to moderately to poorly permeable compounds. In the study, metoprolol was used as a model compound to mark the boundary for HP and atenolol represented low permeability (LP). The applicant's model was able to distinguish between highly permeable and poorly permeable compounds. According to the BCS guidance attachment A, high, moderate, and low permeability are the drugs, which have fraction absorbed (Fa) in humans >85%, 50–84%, or <50%, respectively. The rat *in situ* perfusion had demonstrated method suitability with delineation between HP and LP model drugs. The results showed that the drug is a highly permeable drug substance because its permeability is higher than that of metoprolol (>85% Fa in humans). The mean apparent permeability value for drug was similar across test concentrations spanning two orders of magnitude (1%, 10%, and 100% of the 500 mg dose strength in 250 ml); and the lack of dependence of the measured permeability on initial drug concentrations, 20, 200, and 2 mg/ml indicates an apparent passive transport. In addition, GI stability data show that drug was stable in simulated gastric fluid (SGF) for 1 hour and up to three hours in simulated intestinal fluid (SIF) with <5% degradation. This indicates that the drug is stable in GI tract. For this case, since Caco-2 *in vitro* permeability may be limited by drug cytotoxicity; the selection of an *in situ* method was justifiable. Thus, the results concluded the HP of the drug substance.

#### 9.4.2.3 Dissolution

The applicant conducted acceptable comparative dissolution testing on the 500 mg strength using the USP apparatus II at 50 rpm in a volume of 500 ml of the following media: (i) 0.1 N HCl or SGF USP without enzymes; (ii) pH 4.5 buffer; and (iii) pH 6.8 buffer or SIF USP without enzymes and the strength of the drug product exhibit very rapid *in vitro* dissolution. The multimedia pH dissolution testing demonstrated that the test product meets the BCS guidance definition of very rapidly dissolving (≥85% in 15 minutes).

The overall conclusion was that the drug product was classified as BCS class 1 and eligible for BCS-based biowaiver.

### 9.4.3 Case Study 3

#### 9.4.3.1 Formulation

This drug was submitted for a BCS class 3 waiver request. The inactive ingredients in the test product were qualitatively (*Q*1) the same and quantitatively (*Q*2) very similar as those in the formulation of the reference product. There was some difference (0.0242 mmol) in one of the excipients, mannitol between the test formulation and the reference listed drug (RLD) formulation. Per the current FDA BCS guidance, "*excessive quantities of certain excipients, such as surfactants (e.g., polysorbate 80) and sweeteners (e.g., mannitol or sorbitol) may be problematic.*" Based on the data obtained from ranitidine and cimetidine studies in literature [31], there was no evidence suggesting that 0.0242 mmol difference of mannitol would raise a concern on the small intestine transit time and/or absorption difference of the drug of interest between the test and reference products.

The extrapolation from ranitidine/cimetidine data from the literature to this drug may be feasible because both ranitidine and cimetidine are BCS class 3 compounds. Furthermore, sorbitol/mannitol has been shown to reduce ranitidine/cimetidine transit time in the small intestine [32]. Thus, the test formulation was determined to be adequate in supporting the BCS class 3 waiver request.

#### 9.4.3.2 Solubility

Equilibrium solubility data were provided using the following testing media: 0.1 N HCl, pH 4.5 acetate buffer, pH 6.8 phosphate buffer, pH 7.7 phosphate buffer, pH 8.7 alkali borate buffer, and pH 9.7 alkali borate buffer. In each case, the solubility of the drug exceeded the minimum required solubility of 0.004 mg/ml (1 mg/250 ml). It was also noted that per RLD label, the maximum daily dose of this drug is also the maximum single dose and was the same as the highest tablet strength. Data were also provided that demonstrated the equilibrium solubility conditions were reached after 240 minutes of incubation time, in each of above 6 testing media. The applicant conducted the equilibrium solubility study with adjusting the solution final pHs matching to the target pHs after being filtered as per the FDA BCS guidance. Based on the abovementioned results, the drug was characterized as highly soluble.

#### 9.4.3.3 Dissolution

Quality control dissolution testing on all strengths using the FDA-recommended dissolution method: USP Apparatus 2 (paddle) at 50 rpm in 0.1 N HCl (500 ml) and was found adequate. Multimedia dissolution testing was conducted using USP Apparatus 2 (paddle) at 50 rpm in 0.01 N HCl, pH 4.5 acetate buffer, and pH 6.8 phosphate buffer. The data showed that the dissolution profiles were comparable between the test and reference products in each of the BCS waiver guidance requested media (0.1 N HCl, pH 4.5 acetate buffer, and pH 6.8 phosphate buffer). The data also met the BCS dissolution criteria in that the drug product was considered very rapidly dissolving as no less than (NLT) 85% of drug was dissolved within 15 minutes. Overall, this product was classified as BCS class 3 drug product and determined to be eligible for BCS-based biowaiver.

## 9.5 BCS: Additional Regulatory Applications

Besides BCS-based biowaiver, another significant application of BCS is dissolution testing. In 2018, the FDA issued a guidance entitled "Dissolution Testing and Acceptance Criteria for Immediate-Release Solid Oral Dosage Form Drug Products Containing High Solubility Drug Substances." This guidance provides recommendations applicable to IR solid oral drug products for standard dissolution release methods and setting of acceptance criteria, that can be used in lieu of extensive dissolution method development and validation studies and dissolution acceptance criterion/criteria proposals.

This guidance is a part of the FDA overall effort to develop and implement patient-centric quality standards (PCQS). PCQS are defined as the set of criteria and acceptance ranges to which a drug product should conform to in order to deliver the therapeutic benefit indicated in the label. PCQS can increase flexibility within the pharmaceutical manufacturing sector while maintaining quality by establishing acceptance criterion/criteria limits based on clinical performance instead of process capability or manufacturing process control. PCQS avoid under- or over-discriminating specifications.

The guidance recommends using the following testing conditions for the dissolution test of IR oral drug products containing high-solubility drug substances:

A. Basket Method (USP Apparatus 1)
   - Stirring rate = 100 rpm
   - 500 ml of 0.1 N HCl in aqueous medium
   - No surfactant in medium
   - $37 \pm 0.5\,°C$

B. Paddle Method (USP Apparatus 2)
   - Stirring rate = 50 rpm
   - 500 ml of 0.1 N HCl in aqueous medium
   - No surfactant in medium
   - $37 \pm 0.5\,°C$

When either dissolution method A or B is used, the guidance recommends setting the dissolution acceptance criterion as "Q = 80% in 30 minutes."

The BCS is also used in the FDA's polymorphism guidance to determine if drug substance polymorphism will impact the BA/BE of the drug product [33]. For a drug whose absorption is only limited by its dissolution, large differences in the apparent solubilities of the various polymorphic forms are likely to affect BA/BE. On the other hand, for a drug whose absorption is only limited by its intestinal permeability, differences in the apparent solubilities of the various polymorphic forms are less likely to affect BA/BE. Furthermore, when the apparent solubilities of the polymorphic forms are sufficiently high and drug dissolution is rapid in relation to gastric emptying, differences in the solubilities of the polymorphic forms are unlikely to affect BA/BE.

Finally, the BCS has been used extensively in regulatory reviews and applications. Specific example is the use of BCS in the FDA's new initiative: Knowledge-aided Assessment and Structured Application (KASA) [18]. The KASA system is designed to: (i) capture and manage knowledge – such as established conditions – during the lifecycle of a drug product; (ii) establish rules and algorithms for risk identification, mitigation, and communication; (iii) perform computer-aided analyses of applications to compare regulatory standards and quality risks across approved applications and facilities; and (iv) provide a structured assessment that radically eliminates text-based narratives and summarization of provided information. When fully developed and implemented, KASA will enrich regulatory quality oversight through lifecycle management of products and facilities. Ultimately, KASA will advance FDA's focus on pharmaceutical *quality*, the foundation for ensuring the safety and efficacy of drugs.

## 9.6 Summary

BCS is a concept that was developed to meet the regulatory needs. It resulted in the first FDA guidance that was entirely based on the results of the regulatory-sponsored research. The FDA issued its first

BCS guidance in 2000. However, it took 20 years for its acceptance worldwide, as witnessed by the issuing of M9, the ICH BCS Guidance. Over the years, we have witnessed how a scientifically fundamental concept involving basic properties of the drug substance and drug product evolved into regulatory guidance, resulting in approvals of hundreds of drug applications. By providing appropriate *in vitro* methods to replace expensive, time-consuming, and avoidable *in vivo* BE studies, the BCS has facilitated availability of good quality drug products at reduced cost and at a faster pace for benefit of patients globally.

## References


1 Amidon, G.L., Lennernäs, H., Shah, V.P. et al. (1995). A theoretical basis for a biopharmaceutic drug classification: the correlation of in vitro drug product dissolution and in vivo bioavailability. *Pharm. Res.* 12: 413–420.

2 Capsugel Symposia Series (1995–1997). Biopharmaceutics drug classification and international drug regulation. seminars and open forums, Tokyo, Japan, July 15, 1997; Geneva, Switzerland, May 14, 1996; and Princeton, NJ, USA, May 17, 1995.

3 Amidon, G.L., Sinko, P.J., and Fleisher, D. (1998). Estimating human oral fraction dose absorbed: a correlation using rat intestinal membrane permeability for passive and carrier-mediated compounds. *Pharm. Res.* 5: 651–654.

4 Lennernäs, H., Ahrenstedt, O., Hällgren, R. et al. (1992). Regional jejunal perfusion, a new in vivo approach to study oral drug absorption in man. *Pharm. Res.* 9: 1243–1251.

5 Chiou, W.L. (1994). Determination of drug permeability in a flat or distended stirred intestine. Prediction of fraction dose absorbed in humans after oral administration. *Int. J. Clin. Pharmacol. Therap.* 32: 474–482.

6 Cook, J., Addicks, W., and Wu, Y.H. (2008). Application of the biopharmaceutical classification system in clinical drug development—an industrial view. *AAPS J.* 10: 306–310. https://doi.org/10.1208/s12248-008-9036-5.

7 Ku, M.S. (2008). Use of the biopharmaceutical classification system in early drug development. *AAPS J.* 10: 208–212.

8 Mehta, M.U., Uppoor, R.S., Conner, D.P. et al. (2017). Impact of the US FDA "Biopharmaceutics Classification System" (BCS) guidance on global drug development. *Mol. Pharm.* 14: 4334–4338.

9 Yu, L.X., Carlin, A.S., Amidon, G.L. et al. (2004). Feasibility studies of utilizing disk intrinsic dissolution rate to classify drugs. *Int. J. Pharm.* 270: 221–227.

10 Yazdanian, M., Briggs, K., Jankovsky, C. et al. (2004). The "high solubility" definition of the current FDA guidance on biopharmaceutical classification system may be too strict for acidic drugs. *Pharm. Res.* 21: 293–299.

11 Food and Drug Administration Guidance for Industry (August 2000), Waiver of in vivo bioavailability and bioequivalence studies for immediate release solid oral dosage forms based on a biopharmaceutics classification system, CDER/FDA.

12 European Medicines Agency (2010). Committee for Medicinal Products for Human Use (CHMP), guideline on the investigation of bioequivalence.

13 Health Canada Guidance Document (2014). Biopharmaceutics classification system based biowaiver.

14 World Health Organization (2015). WHO technical report series, No. 992 annex 7. Multisource (Generic) pharmaceutical products: guidelines on registration requirements to establish interchangeability.

15 European Medicines Agency (2020, February). ICH M9 guideline on biopharmaceutics classification system-based biowaivers.

16 European Medicines Agency (2020, February). ICH M9 guideline on biopharmaceutics classification system-based biowaivers – questions and answers.

17 Food and Drug Administration Guidance for Industry (August 2018). Dissolution testing and acceptance criteria for immediate-release solid oral dosage form drug products containing high solubility drug substances guidance for industry. CDER/FDA.

18 Yu, L.X., Raw, A., Wu, L. et al. (2019). FDA's new pharmaceutical quality initiative: knowledge-aided assessment & structured applications. *Int. J. Pharm.* https://doi.org/10.1016/j.ijpx.2019.100010.

19 Yu, A., Sun, D., Li, V.L. et al. (2014). Bioequivalence history. In: *FDA Bioequivalence Standards. AAPS Advances in the Pharmaceutical Sciences Series* (ed. L.X. Yu and B.V. Li), 1–28. AAPS press.

20 Food and Drug Administration Guidance for Industry (2003). Bioavailability and bioequivalence studies for orally administered drug products – general considerations.

21 Cohen, J.L., Hubert, B.B., Leeson, L.J. et al. (1990). The development of USP dissolution

and drug release standards. *Pharm. Res.* 7: 983–987.

22 Shah, V.P. and Amidon, G.L. (2014). A theoretical basis for a biopharmaceutic drug classification: the correlation of in vitro drug product dissolution and in vivo bioavailability, Pharm. Res. 12: 413–420, 1995—Backstory of BCS. *AAPS J.* 16: 894–898.

23 Sinko, P.J., Leesman, G.D., and Amidon, G.L. (1991). Predicting fraction dose absorbed in humans using a macroscopic mass balance approach. *Pharm. Res.* 8: 979–988.

24 Yu, L.X., Lipka, E., Crison, J.R. et al. (1996a). Transport approaches to the biopharmaceutical design of oral drug delivery systems: prediction of intestinal absorption. *Adv. Drug Delivery Rev.* 19: 359–376.

25 Yu, L.X., Crison, J.R., and Amidon, G.L. (1996b). Compartmental transit and dispersion model analysis of small intestinal transit flow in humans. *Int. J. Pharm.* 140: 111–118.

26 Yu, L.X. and Amidon, G.L. (1999). A compartmental absorption and transit model for estimating oral drug absorption. *Int. J. Pharm.* 186: 119–125.

27 Yu, L.X., Amidon, G.L., Polli, J.E. et al. (2002). Biopharmaceutics classification system: the scientific basis for biowaiver extension. *Pharm. Res.* 19: 921–925.

28 Food and Drug Administration Guidance for Industry (December 2017). Waiver of in vivo bioavailability and bioequivalence studies for immediate release solid oral dosage forms based on a biopharmaceutics classification system. CDER/FDA.

29 eCFR. (n.d.). Title 21, Chapter 1, Subchapter D, Part 320. https://www.ecfr.gov/cgi-bin/text-idx?SID=f52df72070eace0cd59944c96641b3aa&mc=true&node=pt21.5.320&rgn=div5. (accessed 12 September 2020).

30 Food and Drug Administration Guidance for Industry (2013). Bioequivalence studies with pharmacokinetic endpoints for drugs submitted under an ANDA.

31 Chen, M.L., Sadrieh, N., and Yu, L. (2013). Impact of osmotically active excipients on bioavailability and bioequivalence of BCS class III drugs. *AAPS J.* 15: 1043–1050.

32 Chen, M.L., Straughn, A.B., Sadrieh, N. et al. (2007). A modern view of excipient effects on bioequivalence: case study of sorbitol. *Pharm. Res.* 24: 73–80.

33 Food and Drug Administration Guidance for Industry (July 2007). ANDAs: Pharmaceutical Solid Polymorphism, CDER/FDA.

# 10

# Effects of Food on Drug Absorption

*Zhu Zhou[1], Venugopal P. Marasanapalle[3], Xiaoling Li[2], and Bhaskara R. Jasti[2]*

[1]Department of Chemistry, York College, The City University of New York, Jamaica, NY, USA
[2]Department of Pharmaceutics and Medicinal Chemistry, TJ Long School of Pharmacy, University of the Pacific, Stockton, CA, USA
[3]Novartis Institutes for BioMedical Research, Telangana, India

## 10.1 Introduction

Coadministration of food with drugs may have a significant impact on drug absorption and impacting its safety and efficacy. Therefore, it is important to evaluate the food effects on drug absorption during drug development to optimize the instructions of food intake relative to drug administration. The simultaneous intake of food and drugs could increase or reduce the extent of bioavailability, delay or accelerate the rate of absorption, or cause no change in the absorption of drugs. For immediate-release dosage forms, food effects are classified as positive, negative, and no food effects based on the change in Area Under the Curve (AUC) of a drug.

Food can interfere with the absorption of drugs by causing physicochemical or physiological changes in the gastrointestinal (GI) tract and its associated systems. Additionally, some food such as grapefruit juice can cause specific food–drug interactions at the level of the intestinal monolayer as various transporters and enzymes are located in the enterocytes. The focus of this chapter is the various physicochemical and physiological effects caused by diet that results in food effects. The role of food intake in causing changes in solubility, permeability, and physiological changes in metabolism and eventually food effects, are included. Specific examples of food–drug interactions through different mechanisms including intestinal enzymes and transporters are stated in the chapter. Additionally, methods to predict the food effects from physicochemical characteristics are briefly summarized.

### 10.1.1 Food–Drug Interactions (Food Effects) in Clinical Therapy and Drug Development

Of all the current drugs in the market, about 62% are administered by the oral route [1]. When compared with other routes of drug administration such as transdermal, parenteral, nasal, and pulmonary; oral administration is preferred since it is more convenient and inexpensive. Oral drug administration is most commonly accompanied by food in clinical conditions. Food is a complex mixture of chemicals, each potentially can interact with drugs. Additionally, the physicochemical and physiological changes caused by food intake in the gastrointestinal tract (GIT) and its associated organ systems can result in pharmacokinetic changes of coadministered drugs [2, 3]. Depending upon a drug's onset and duration of action and its therapeutic window, changes in its pharmacokinetics could affect its therapeutic outcome [4]. Kang and Ratain indicated 34 of the 99 (34.3%) orally administered drugs approved by the US Food and Drug Administration (FDA) between January 2000 and May 2009 should be taken with a specific dietary protocol to avoid significant food–drug interactions [5]. In recent years, despite the development of improved drug delivery technologies, food effects still cause significant problems. In a recent publication, O'Shea et al. found significant food effects were present in 67 of 157 (42.68%) orally administered drugs licensed by the European Medicines Agency (EMA) and FDA since 2010 until 2017 [6]. Particularly, oncology drugs often show changes in oral bioavailability when taking with food due to many of them being administered in a relatively high-doses albeit their aqueous solubility is low [7].

*Oral Bioavailability and Drug Delivery: From Basics to Advanced Concepts and Applications*, First Edition. Edited by Ming Hu and Xiaoling Li.

#### 10.1.1.1 Regulatory Considerations of Food Effects

Evaluation impact of food intake on the rate and extent of absorption is an important part of the registration process for a new drug. According to the FDA, all new chemical entities should conduct food effect studies except in certain scenarios that FDA may waive such studies for drugs that are designated as Biopharmaceutical Classification System (BCS) class 1 immediate-release drug. Administration of drug with 240 ml of water on an overnight (at least 10 hours) fasted stomach is defined as *fasted state* (preprandial) administration whereas administration of drug 30 minutes after starting and completing a meal on an overnight (at least 10 hours) fasted stomach is defined a *fed state* (postprandial) administration [8]. During clinical trials, blood plasma concentrations are analyzed with fed and fasted healthy participants. A clinically significant food effect is considered to be present if the 90% confidence interval for the ratio of the population geometric means between fed and fasted treatments, based on log-transformed data, fails to meet 0.8 to 1.25 (80–125%) for either AUC or Maximum Concentration ($C_{max}$) unless other criteria for the drug are more appropriate [8].

To trigger the highest effects on GIT physiology and systemic availability of the drug, both FDA and EMA recommend the administration of a high-calorie (800–1000 kcal), and high-fat (500–600 kcal of total calories from fat) food as a test meal to evaluate the food effects on oral drug bioavailability, and the drug should be administered 30 minutes after the start of the meal with a glass of water (240 ml) [8, 9]. According to FDA, if administered with a high-fat meal would cause loss of drug efficacy or unacceptable toxicity, then the administration of a drug with a low-fat meal may be more appropriate [8]. Both FDA and EMA provide examples for the composition of a typical high-fat breakfast; two eggs fried in butter, two strips of bacon, two slices of toast with butter, four ounces of hash brown potatoes, and eight ounces of whole milk or equivalent calorie content meal [8, 9]. In the past 15 years, the majority of pharmacokinetic data published on food effects have used this standard meal while in the European Union regulatory setting, studies of the effects of a "moderate" meal are permitted for the drug recommended to be taken with food, and sometimes different food compositions may be conducted [10].

#### 10.1.1.2 Food–Drug Interactions in Clinical Therapy

The purpose of the food effect study is to determine whether there is a clinically significant food effect on drug availability. Depending upon its therapeutic area and cost-of-goods, increased drug absorption may be advantageous for therapy, e.g. chloroquine [11]. On the other hand, decreased bioavailability when coadministered with food may be disadvantageous, e.g. alendronate [12]. Drug administration along with or after food reduces gastric discomfort such as nausea or vomiting for many drugs, e.g. metformin and corticosteroids [12]. If the patient is advised to take the drug after food, then there is a relatively higher compliance, because drug administration is "tied-up" with food intake such as roxithromycin [12, 13]. Depending on the therapeutic significance of food effects on drug absorption, US FDA label recommendations are made to the patients about their usage; some examples are shown in Table 10.1.

A dramatic impact of food on drug absorption can be illustrated with the tyrosine kinase inhibitor, lapatinib.

**Table 10.1** U.S. FDA label recommendations for usage of drugs based on known food effects.

| Drugs | Type of food | Food effect | U.S. FDA label recommendation[a)] |
|---|---|---|---|
| Flibanserin, sildenafil citrate | High-fat meal | Delayed the rate | None |
| Ropinirole | High-fat meal | Accelerated the rate | |
| Flibanserin, imbruvica, vemurafenib | High-fat meal | Increased the extent | |
| Daclatasvir, ropinirole | High-fat meal | Reduced the extent | |
| Boceprevir, telaprevir | Low/high-fat meal | Increased the extent | Administer with food |
| Eluxadoline | High-fat meal | Reduced the extent | |
| Clodronate | Standard breakfast | Reduced the extent | On empty stomach |
| Migalastat hydrochloride | High fat/light meal/one hour after a light meal | | |
| Naloxegol, sonidegib | High-fat meal | Increased the extent | |

a) Data obtained from FDA Drug Label (from Drugs@FDA database) on 20 June 2020.

After a standard high-caloric meal, the oral bioavailability of a single dose of 1500 mg of lapatinib increased on average by 325% [14]. Thus FDA recommends taking lapatinib either one hour before or one hour after eating. This food intake advice raised an intense debate. Ratain and Cohen suggested to enable a smaller dose with food to reduce the cost of this expensive treatment and also reduce the severe GI side effects [15]. On the other hand, increasing the extent of absorption will not necessarily lead to a label impact. For vemurafenib, when coadministered with food the AUC increased about fivefold, however, no lower dose in the presence of food is advised in both US FDA and European Union product labels [16].

If a drug is intended for chronic use, then a restriction is imposed on the patient to take it consistently in either a fasted state or fed state [17]. Even though food can cause clinically significant changes in single-dose pharmacokinetics, this effect is minimized in chronic treatment. Food is also noted to reduce the fluctuations of peak and trough plasma concentrations of some drugs, e.g. atenolol [18]. For patients' compliance, drug administration without restriction on food intake is ideal, especially in chronic use. Gemifloxacin for bacterial infections and stavudine therapy for human immunodeficiency virus (HIV) infected patients are good examples of drugs administered without food restriction [19, 20]. However, there are exceptions, rufinamide exhibits higher fluctuations and enhanced absorption in the fed state when compared with the fasted state, and according to the FDA label, it is recommended to be administered with food [21] (FDA BANZEL® product label).

#### 10.1.1.3 Positive, Negative, and No Food Effects

The pharmacokinetic changes or food–drug interactions have been classified as an increase or a reduced extent of bioavailability, delay or accelerated rate of absorption, or no change in the absorption of drugs by Welling [3, 22]. A drug is considered to exhibit delayed or accelerated rate of absorption when its time of peak plasma concentration ($T_{max}$) is increased or decreased, respectively, in the fed state when compared with the fasted state [3]. There may or may not be any statistical difference in its $C_{max}$ and its extent of absorption ($AUC_{0-\infty}$ or bioavailability) [3]. A drug is considered to exhibit increased or reduced extent of bioavailability, when its $AUC_{0-\infty}$ is increased or decreased, respectively, in the fed state when compared with the fasted state [3]. If coadministration of food results in no significant differences in $C_{max}$, $T_{max}$, or $AUC_{0-\infty}$, then food is considered to not affect the absorption of such drugs [3].

Fed state drug administration can affect one or more pharmacokinetic parameters ($T_{max}$, $C_{max}$, and $AUC_{0-\infty}$) of a drug when compared with fasted state drug administration [2]. For example, food was shown to delay the rate but increase the extent of absorption of flibanserin [6]; accelerate the rate and increase the extent of absorption of rufinamide [21]; delay the rate and decrease the extent of absorption of avitriptan [23, 24]. Therefore, for the sake of simplicity drugs may be classified to exhibit *positive*, *negative*, or *no food effects* based on the change of only one pharmacokinetic parameter, the $AUC_{0-\infty}$; since $AUC_{0-\infty}$ reflects the extent of absorption of a drug, and defines the effectiveness of therapy. $C_{max}$ is dependent on both rate and extent of drug absorption and is used to understand the toxicity of the drug [25]. A food effect is defined as *positive* when coadministration of food causes an increase in the extent of drug absorption; *negative* when food causes decreased extent of drug absorption; and lastly, *no effect* when coadministration of food causes no change in the $AUC_{0-\infty}$ of a drug when compared with fasted state. Thus, changes in $T_{max}$ and $C_{max}$ of a drug on coadministration of food are not considered in this classification. Examples of pharmacokinetic profiles of how food intake can influence drug absorption and drugs exhibiting positive, negative, and no food effects are summarized in Figure 10.1 and Table 10.2.

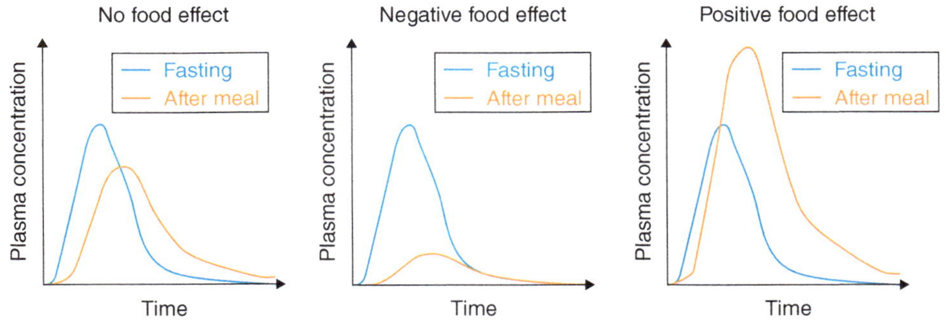


**Figure 10.1** Pharmacokinetic classification of food–drug interactions (food effects). Source: Adapted from Koziolek et al. [26].

**Table 10.2** Examples of drugs known to exhibit food effects.

| Positive food effects | No food effects | Negative food effects |
|---|---|---|
| Alectinbib, chloroquine, deramciclane, dolutegravir, imbruvica, lurasidone hydrochloride, naloxegol, nintedanib, propranolol, sonidegib, rufinamide etc. | Abacavir, captopril, ibuprofen, lamivudine, ledipasvir, perindopril, prednisone, sparfloxacin, suvorexant, terbinafine, etc. | Betrixaban, daclatasvir, droxidopa. eluxadoline, ixazomib, migalastat hydrochloride, tipiracil etc. |

#### 10.1.1.4 Absorption of Drugs

Orally administered drugs undergo minimal absorption from the conduit regions of the GIT, namely, mouth and esophagus [27]. The stomach acts as a reservoir for temporary storage of food and fluids, and only a few drugs are absorbed from the stomach, one of the reasons being its limited surface area for drug absorption [28]. Orally ingested food and administered drugs are predominantly absorbed in the upper small intestine (duodenum and jejunum) [29]. The different mechanisms of drug transport across the intestinal wall are passive diffusion, endocytosis, and active/carrier-mediated transport [30].

#### 10.1.1.5 Metabolism and Transport in the Gastrointestinal Tract (see also Chapter 11)

***10.1.1.5.1 Metabolism*** Degradation by luminal enzymes and metal ion complexation are the other important parameters that are influenced by the presence of food, affecting the absorption of drugs [3, 31, 32]. For drugs that do not undergo gastric or enzymatic degradation, and do not interact with metal ions/$Ca^{2+}$ or bile salts, another barrier encountered during intestinal permeation is the presence of metabolic enzymes in the intestinal enterocytes and liver. The metabolic reactions are classified into phase I and phase II metabolic reactions. Phase I metabolism involves functional group changes, such as a ring or side-chain hydroxylation, nitro group reduction, aldehyde oxidation, dealkylation, and deamination. Phase II metabolism involves conjugation reactions in which the drug molecule is combined with solubilizing groups such as glucuronic acid and glycine to form polar excretable conjugates [33].

The major phase-I metabolizing enzyme system, the cytochrome P-450 (CYP), is expressed in the liver and the gut wall [34–37], although the CYP abundance in the small intestine various greatly from CYP abundance in the liver [38]. CYP3A and CYP2C9 are the most importnt enzymes in the small intestine, accounting in total for more than 95% of the entire CYP content [16]. Food constituents such as cruciferous vegetables can cause metabolic enzyme induction and affect drug absorption [35, 39].

In the gut lumen, microflora-mediated hydrolysis and reduction can be important but are often neglected [40]. More than 30 marketed drugs are substrates for bacterial enzymes and many others are likely to be given renewed importance and research of the gut microbiome. Additionally, some drug delivery systems and prodrugs that are poorly absorbed in the small intestine and stomach targeting the colon are designed to release the pharmacologically active substance by microbial activity [16]. For drugs with a low solubility and/or permeability that cause prolonged GI residence times, there might be microbial interactions involved. Therefore, metabolic processes coordinated by the microbiome need to be considered as the gut microbiota can affect drug bioavailability [41].

The fraction of drug that bypasses the enterocytic metabolism passes into the splanchnic hepatic (SH) blood flow and travels through the portal vein into the liver, which is the main metabolic organ and is responsible for extensive transformation of some drugs. The liver is a large organ consisting of a huge chemical pool (approximately 25 μmol of CYP) and is capable of metabolizing diverse substances [29, 42]. It prevents the entry of many drug molecules into the systemic circulation (first-pass metabolism or presystemic metabolism). Blood from the venules flows into the branches of hepatic sinusoids (hepatic acinus), which lie between the hepatic plates and form a complex capillary bed [29, 42]. Thus, the hepatic cells are continuously exposed to portal venous blood [42].

***10.1.1.5.2 Efflux Transporters*** Efflux transporters present on the intestinal mucosal membrane may reduce the systemic absorption of drugs by limiting their permeation across the intestinal wall [39, 43]. Of particular importance in efflux transporters is, P-glycoprotein (P-gp), first identified in resistant cancer cells [39, 44], which belongs to a large family of efflux proteins called ATP-binding cassette (ABC) proteins comprising of multidrug resistance-associated protein (MDR) and multidrug resistance-associated protein transporters (MRP) [45]. P-gp is a membrane-bound ATP-dependent transporter responsible for the efflux of structurally diverse chemotherapeutic agents from the cell's interior to its exterior (in the intestine, from the serosal side to the mucosal side, the lumen). These efflux proteins along with the phase I metabolic enzyme CYPs are present at high concentrations on the villus tip of the enterocytes [39, 46]. The concentration levels of P-gp mRNA increase longitudinally along the GIT; the stomach has the lowest levels and the colon has the highest levels [39, 47]. The main dietary inhibitors of P-gp are furanocoumarins and flavonoids from fruit and vegetables. Other than P-gp, efflux transporters such as breast cancer resistance protein (BCRP) and MRPs are also located in the apical membrane of the intestine [16]. Efflux proteins and CYP enzymes work synergistically in the enterocytes and share many common substrates and prevent their entry into the systemic circulation, thus forming the first line of barriers for oral drug absorption [39, 45].

***10.1.1.5.3 Uptake Transporters*** Organic anion-transporting polypeptides (OATPs) are transmembrane transport proteins categorized as members of the solute carrier (SLC) family (detailed discussion also in Chapter 21). Among 11 individual human OATP transporters that comprise 6 families (OATP1-6), OATP1A2 and OATP2B1 are reportedly expressed in the human intestine, although the expression level of OATP1A2 in the intestine remains controversial [48]. Considerable progress has been made in understanding the role of intestinal efflux transporters in drug absorption, and some of them (P-gp and BCRP) have been routinely studied. However, much less attention has been paid to intestinal influx transporters, although increasing number of food–drug interactions have been attributed to the inhibition of influx transporters, particularly OATPs. Specific recommendations on how these transporters should be evaluated are currently lacking, indicating a gap between research and clinical recommendations. It should be noted the effect of an uptake transporter is opposite to the efflux transporters, it facilitates absorption and will decrease the AUC ratio. When substrates of intestinal OATPs, e.g. atenolol and fexofenadine, are coadministered with apple juice, grapefruit juice, or orange juice, ~85% reduction in their exposure was observed. However, some of these results were not reflected in their FDA product labels. Additional research is needed to understand the overall clinical relevance of these intestinal uptake transporters including identifying well-characterized clinical probe substrates and inhibitors [48, 49]. Besides OATPs, oligopeptide transporter (PepT1) expressed in the apical side of the intestine has been reported to involve in the absorption of di and tripeptides [50].

#### 10.1.1.6 Permeability in the Gastrointestinal Tract

Orally administered drugs and ingested food are mainly absorbed in the upper small intestine (duodenum and jejunum) [29]. The mucosa of the large intestine, consisting of the colon and the rectum resembles the small intestine, except for the presence of villi. This reduces the surface area of the large intestine and affecting the passage of drugs [28]. The tight junctions between the cells of the large intestine are narrower when compared with the small intestine, thus there is restricted paracellular movement of many molecules across large intestine [28]. However, the residence time in the large intestine is longer than in the small intestine enabling slow absorption of drugs [28]. The physical barriers for permeability in the colon are formed by the viscosity of the contents, the presence of gas bubbles, and dietary fiber [51]. These factors may also contribute to increasing the thickness of the unstirred water layer [51]. Additional discussion on absorption of drugs by passive diffusion is found in Chapters 5 and 6.

#### 10.1.1.7 Food-Induced Changes on Drug Bioavailability

The extent of food–drug interactions is dependent on factors such as the physicochemical properties of the drug, formulation, type, and size of the meal, the order in which food and drug are ingested, and the time interval between the administration of food and drug [52, 53]. The physiological changes following food intake in the stomach, intestine, and circulation affect drug absorption. The physiological changes that occur in the stomach include but are not limited to, prolonged gastric retention, increased pH, fluid volume, secretion of gastric acid and pepsin activity, and slower gastric emptying rate. The physiological changes that occur in the intestine are increased buffer capacity, motility-propulsive (peristalsis), intestinal enzymes, bile salts, osmolality, mixing and other local constrictive movements, and a potential decrease in permeability and surface tension [3, 4, 6, 42, 54]. Food intake also results in increased enzyme secretion in the pancreas and release of bile salts in the gall bladder [6, 54]. The important physicochemical and physiological differences in the fasted state and fed state GIT are summarized in Table 10.3 below, although the values might mask considerable inter and intraindividual variation. Food, besides causing physiological changes in the GIT, also interacts directly or indirectly with drugs and influences their bioavailability.

**Table 10.3** Physicochemical and physiological differences in fed and fasted states in the GIT.

| Physiological state | Stomach | | Small intestine | |
|---|---|---|---|---|
| | Fasted | Fed | Fasted | Fed |
| Fluid volume | 10–50 ml [10] | 1500 ml [55]<br>FDA breakfast: 580 ± 38 ml [56] | 10–177 ml [57, 58] | 13–95 ml [58] |
| pH | 1–3 [11, 55] | 2–5 [55, 59] | 6–8 [10] | 5 [60] |
| Osmolality | Median: 98–140 mOsmol/kg [61] | Median: 217–559 mOsmol/kg [61] | Duodenum: 124–266 mOsmol/kg [62] | Duodenum: 250–367 mOsmol/kg [62] |
| Buffer capacity | Median: 7–18 mmol/l/ΔpH [61] | Median: 14–30 mmol/l/ΔpH [61] | 2.4–13 mmol/l/pH [62] | 13.2–30 mmol/l/pH [62] |
| Surface tension | 35–45 mN/m [61] | 30–31 mN/m [62] | Duodenum: 33.3–46 mN/m [62] | Duodenum: 32.2–36.7 mN/m [62] |
| Residence time | 25 min [62] | 40 min [62] | 3–4 h [63] | |
| Bile salts | Median: 0.1 mM [62] | 0.06 mM [62] | 0–17 mM [62] | 0.2–34 mM [62] |

The different changes caused by food can affect drug bioavailability by altering one or more of the following drug parameters: stability, solubility, and permeability. The solubility and permeability of drugs can be affected by complexation with ions such as $Ca^{2+}$ in food, pH change, and viscosity. If solubility is the limiting factor for oral bioavailability, food-induced fluid volume can impact drug absorption by changing the luminal drug concentration. For orally administered drugs, a specific volume of luminal fluid is needed for the drug to promote the active ingredient release and its absorption into the body. Patients fed under FDA's standard high-caloric and high-fat meals are shown to have an increase of gastric content volume in the stomach. This increase in gastric volume is maintained for approximately six hours due to the low rate of caloricchyme elimination. Therefore, the intake of a drug two hours after a moderate or large meal will not produce a fasted state administration of the drug. Alteration in the volume of luminal fluids, particularly in the stomach and small intestine, and their kinetics, are likely to play a critical role in the food effects [16].

The pH of the environment in the GIT is significantly altered by the presence of food which will impact the solubility and potentially change oral bioavailability. The characteristics of the stomach and upper intestine are most important for drug absorption of immediate-release dosage forms [6, 16]. In a fasted state, the pH of the stomach is between 1 and 3 and increases temporarily when food is ingested that gradually decreases as gastric acid is secreted in response to food intake [55]. Absorption of drugs occurs to a certain extent in the stomach for drugs that are acidic or neutral in the stomach. Increased pH in the stomach might change the solubility and dissolution of ionizable compounds, which would increase AUC and $C_{max}$ for weak acids such as cefuroxime but decrease AUC and $C_{max}$ for weak bases such as indinavir and dipyridamole [6]. Milk or milk products contain calcium ions and can increase the pH in the stomach. This may result in precipitation of some drugs in the intestine upon chelation with calcium, resulting in the dissolution of enteric-coated tablets and drug release [64]. The luminal environment of the upper intestinal tract has a lower pH (pH 5) in the postprandial state when compared with the preprandial state pH (pH 6.5) [60]. Though this luminal pH change might impact drugs with a $pK_a$ value between 5 and 7, the relevance of this impact on food–drug interactions has not yet been demonstrated [16].

The presence of food stimulates motility which causes changes in dissolution and absorption of the drugs. The mixing results in higher absorption of some drugs owing to effective contact of drugs with the intestine [3, 65]. The interplay between transit of dosage forms and GI motility can be complex and is impacted by different factors. Emptying of solids can be delayed up to two hours while emptying of liquid formulations would occur quite rapidly in the fasted state. In the fed state, particles smaller than 3–4 mm and liquids will empty with food at a rate invariably slower than the fasted state, whereas particles greater than 7 mm can be retained which displays a significant lag time. However, small intestine transit time appears to be similar in the fasted and fed state. Delayed emptying leads to an increase in stomach exposure and can decrease bioavailability due to increased degradation of unstable drug or drug interaction with the secreted gastric acid, e.g. didanosine [3, 6, 56, 63, 66–68].

The presence of food can cause the physicochemical properties of the luminal contents changes. The composition of the luminal fluids is impacted by buffer capacity and buffer species that can determine the dissolution rate of ionizable drugs and enteric coatings [54]. Buffer capacity and osmolality are higher in the fed state in both the stomach and intestine [61, 62, 69]. The intake of a meal can change the fluid components such as bile salts that can change drug dissolution and solubilization that potentially alters bioavailability [16]. In the fasted state, the intestinal lumen consists of residual undigested food and fluids, and low bile and enzyme concentrations [70]. In the fed state, the intestinal lumen becomes a more complex environment, consisting of gastric acid from the chyme, enzyme systems, higher concentration of bile salts, heavy metals or calcium from ingested food, inorganic salts, food particulate matter, dietary fiber, fat globules, partially digested carbohydrates, and proteins, all of which can interact physically or chemically with drugs [3, 4, 23, 24, 31, 32, 65, 71]. Fiber, heavy metals, and calcium present in food result in adsorption or complexation or precipitation of drugs [3, 31, 32, 65, 71, 72]. Chelation of drugs with polyvalent metal cations such as calcium and magnesium in food and also complex formation with proteins can also impede drug absorption, e.g. alendronate forms insoluble complexes with multivalent cations such as calcium and magnesium ions [3, 32]. Ciprofloxacin and other fluoroquinolones are known to bind to heavy metals and calcium ions to form insoluble chelates [3].

The rate of drug absorption is also dependent on water flux from the intestinal lumen to the systemic circulation, which is dependent on the hyper, hypo, or isotonicity of luminal contents [32, 73]. Some drugs can be affected by the convective flow of fluids across the GI wall caused as a result of the osmotic nature of the food (hyper/hypotonicity) [32, 73–75]. Water and other fluids move across the intestinal wall (solvent drag) and interfere with paracellular passive diffusion of drugs when the intestinal contents in the fed state cause hypertonicity [32]. The drugs phenobarbital sodium and salicylate were effective and exhibited higher plasma concentrations, respectively, when administered in dilute solutions when compared with concentrated solutions [32, 74]. The presence of fats in food increases the secretion of bile and absorption of some drugs, which affect the partitioning of drugs between the lumen and the intestinal wall [75, 76].

Drugs that have low permeability are sometimes given in the form of prodrugs and the presence of food can decrease the rate of hydrolysis of the prodrugs, enhancing their absorption [77]. The amount of drug that enters the systemic circulation is a function of presystemic clearance [3]. The increased splanchnic blood flow after a meal may influence the absorption of extensively metabolized drugs as a result of changes in the clearance of drugs during the first pass through the liver [32, 53]. After a meal, a larger fraction of drugs bypass the liver because of the increased splanchnic blood flow; therefore, resulting in increased bioavailability, e.g. propranolol [3, 78].

Modified (e.g. sustained/controlled/delayed) release products are designed to deliver drugs over a long period of time and contain drugs in a higher quantity than immediate-release dosage forms [3, 79]. These products have been designed to overcome the negative food effects by releasing the drug further down the GIT to reduce the interaction with the ingested meal. However, an interaction with food may cause "dose dumping" for some sustained release or controlled release formulations such as theo-24 which is known as during gastric transit high shear forces of up to 500 mbar can arise and lead to dose dumping. These events can cause serious and prolonged high circulating drug levels [3, 75, 80, 81]. Coadministration of controlled-release formulations with food results in a slower absorption rate and a higher fraction of drugs undergoing presystemic metabolism [3, 16].

In summary, food-induced changes can increase the bioavailability of drugs by bile solubilization [32, 72, 76], higher GI motility [3, 23, 24, 65]. On the other hand, adsorption of the drug to dietary constituents reduced contact time with GIT, increase in viscosity of the intestinal fluid [4, 6], higher degradation by gastric acid [82], complex formation and/or precipitation of the drug by bile salts, heavy metal ions and/or calcium [31, 32, 71, 83], could decrease their absorption. The other factors that alter drug absorption are: hyper or hypotonicity [32, 73, 74], formulation [75, 84–88], contact angle or wetting [55], and modification of pH of the urine [83]. However, due to the inter and intraindividual variability of the physiology of the GI, the drug absorption can vary greatly. Adding food intake as another impacting factor would increase the variability and complexity of drug absorption [6, 16].

## 10.2 Mechanisms of Food Effects

### 10.2.1 Factors Causing Positive Food Effects

The major factor for positive food effects, which refers to increase in systemic exposure, is the increase in dissolution and solubilization of poorly aqueous soluble drugs. The increased GI motility along with the increased concentration of bile and lipid digestion products in the fed state is believed to improve dissolution and solubilization of poorly water-soluble drugs in the lumen [6, 72, 76, 83]. For example, food intake would increase bile production, enhanced dissolution thus increasing the absorption of carbamazepine. The temporal rise in the pH of the stomach in the fed state due to the ingestion of food is attributed to reduce ionization of the basic drug chloroquine and enhanced absorption [11]. Dilution (higher volume), efficient mixing, and slow gastric emptying rate are believed to cause positive food effects for drugs that are predominantly absorbed in the upper GIT. For example, gastric emptying delay can provide advantages for drugs such as dicumarol and spironolactone to promote more dissolution and absorption which increases bioavailability [89]. In the fasted state, very little intestinal contractions occur, and the luminal contents are essentially stagnant [55]. This could cause a thick aqueous diffusional boundary layer on the surface of the small intestine and prolonged residence time [55]. In the fed state, the increased intestinal activity increases dissolution and transit rate, and thinner boundary layer (30–100 μm) [55, 90]. Therefore, the absorption of lipophilic drugs will be more efficient in the fed state [55]. And, in the fed state, the lipophilic drugs are possibly taken up by micelles and ferried across the diffusional boundary layer, increasing their contact with the intestinal wall and resulting in positive food effects [3, 22, 91]. For example, abiraterone acetate is a lipophilic drug used to treat metastatic prostate cancer known for its limited bioavailability. AUC and $C_{max}$ of abiraterone acetate have increased by about 4.5- and 6.5-fold, respectively, in fed state compared with fasted conditions [92].

Reduction in first-pass metabolism by different mechanisms such as increasing lymphatic uptake, changing blood flow, and decreasing enteric metabolism can result in increased bioavailability [6]. Food constituents may also compete with prodrugs (e.g. cefpodoxime proxetil) as substrates to enzymes, and decrease their degradation in the intestinal lumen, resulting in a positive food effect [77]. The transient increase in SH blood flow is believed to be the reason for food-induced reduction of metabolism of some drugs and increased systemic availability [3, 93]. For high permeability drugs, which undergo extensive presystemic metabolism, the increased SH blood flow in the fed state, is presumed to result in a larger fraction to bypass liver metabolism during absorption [94].

Bioavailability can be impacted by food with specific composition acting like inhibitors, binding to drug metabolic enzymes such as CYP3A4. Additionally, increased bioavailability of certain drugs might be due to the inhibition of intestinal efflux transporters. The most classic example of specific food–drug interaction is grapefruit juice's inhibition of intestinal CYP3A4. The intestinal CPY3A enzymes play a significant role in the first-pass mechanisms for drugs like felodipine and simvastatin. As a result, concomitant administration of grapefruit juice with CYP3A substrate drugs can result in significant changes in drug exposure, leading to an increase in bioavailability [16, 37]. Won et al. indicated that fruit juices, phenolic and polyphenolic compounds present in food are major inhibitors for intestinal CYPs, phase II conjugation enzymes, uptake, and efflux transporters [95].

### 10.2.2 Factors Causing Negative Food Effects

The most common cause of negative food effects, which decrease systemic drug exposure, is direct physicochemical interactions between food and drug in the fed state. In the fed state, drug concentration and diffusivity change due to increased viscosity [4, 6, 22, 55, 65, 96]. The changes in tonicity cause convective movement of fluids across the intestinal wall that affects drug permeability [32, 74, 97]. The elevation in media viscosity can cause significant delay in tablet disintegration, hindering the diffusion of a drug to the absorptive membranes and preventing drug release, particularly causing problems to poorly permeable drugs [6, 98]. Negative food effects exhibited by certain drugs are due to complexation to metal ions/$Ca^{2+}$ [99, 100]. Alendronic acid, ciprofloxacin, clodronate Na, didanosine, estramustine, etidronic acid, penicillamine, tetracycline, and other fluoroquinolones are known to bind to heavy metals and $Ca^{2+}$ to form insoluble chelates. For example, low fat and high carbohydrate and protein diet caused the most significant decrease in the AUC of metformin when compared with the fasted state administration due to solvent drag created by the high osmotic pressure from carbohydrate [89, 101–103].

The physiological changes caused by the presence of food negatively affect the absorption, especially for drugs displaying instability and potentially acid lability in the GIT. Hydrophilic, low-permeability drugs which are ionizable in the intestinal pH conditions are prone to negative food effects [6, 104]. Such drugs have less than 90% absorption from the intestine in the fasted state, this may not only be due to limited solubility but may also be due to limited membrane permeability

[4, 13, 105, 106]. The lower pH in the upper intestinal tract, despite the presence of bile micelles, causes precipitation, and slower dissolution rate for drugs such as furosemide [104]. Whereas, for hydrophilic basic drugs, such as atenolol, the decreased pH may result in higher ionization, which in turn, may result in lower permeability and lower absorption in the fed state when compared with the fasted state [104]. Changes in solubility and permeability due to pH changes in the upper intestinal tract in fed and fasted states were found to contribute to negative food effects of hydrophilic, ionizable, low permeability drugs [104]. Low permeability, non-ionizable, and acidic/basic drugs with relatively higher p$K_a$s/p$K_b$s (not ionizable between pH 5 and 6.7), will not be affected by pH in their solubility and permeability properties and do not exhibit negative food effects [106].

Inhibition of uptake transporters may also play a role in negative food effects, especially for poorly permeable drugs that often rely on the uptake transporters. Yu, Zhou et al. summarized intestinal uptake transporter OATP2B1/1A2-mediated food–drug interactions. Common juices such as apple juice, grapefruit juice, and orange juice are decreasing the exposure of some of the coadministered OATP 2B1/1A2 substrates, e.g. atenolol, celiprolol, and fexofenadine. This may be avoided by limiting juices consumption with known affected drugs. Furthermore, the interactions might be prevented by staggering the administration of OATPs substrates and juices. For example, delaying administration of fexofenadine by two hours or four hours after drinking grapefruit juice would lead to a lower level of inhibition or no inhibition, respectively. Because many statin drugs are substrates for both intestinal and hepatic OATPs and inhibition at these two sites will have the opposite effect, transporters at both sites need to be considered together [48, 49, 107].

## 10.3 Prediction of Food Effects

There are several *in vitro/in vivo/in silico* approaches to predict and assess the effects of food on drug pharmacokinetics in drug discovery and early drug development [108]. Empirical observations have been made in literature correlating the BCS and the effects of food on drugs [4, 13, 109]. The effects of food were correlated with the physicochemical properties of drug molecules such as solubility, partition coefficient, and p$K_a$. using logistic regression analysis and BCS, respectively [110, 111]. A study of structurally dissimilar drug candidates correlated percent food effects to acidic/basic dissociation constant ($K_a$ or $K_b$), and to Caco-2 permeability [106]. Recent FDA guidance suggests the possible consideration of the BCS I to waive food effect studies as more than 80% of BCS class 1 immediate-release products showed negligible effects by high-fat meals. [8]. Several *in vitro* dissolution studies in biorelevant media to predict pharmacokinetic profiles for lipophilic compounds under fed conditions have been reported. Koziolek et al. suggested an *in vitro* fed stomach model that was set up specifically for modeling release and dissolution in the fed stomach [56, 69, 108]. Furthermore, an *in situ* rat jejunal permeability model (single-pass intestinal perfusion technique) was used to simulate negative food effects, by using the pH-conditions of fed and fasted states (pH 5 versus 6.7) of the human intestine [106]. Commonly used animals to predict the food effect in humans are dogs, although the anatomic and physiological species differences may lead to inaccurate prediction. Physiologically based pharmacokinetic (PBPK) modeling involves mathematical models that incorporate information about biochemical processes as well as the body's anatomical and physiological components into the model structure. In the past years, it has been widely used to predict drug absorption including food effects. However, there are knowledge gaps about how to integrate the food-induced changes in human physiology, formulation, and active pharmaceutical ingredients which prevent the effective use of PBPK models [108].

## 10.4 Summary

In this chapter, the importance of studying food effects in clinical studies and the different types of food effects were introduced. The different physicochemical and physiological mechanisms that could alter drug absorption when taken with food were explained. The food-induced changes in physicochemical and physiological properties such as permeability, efflux, influx, and metabolism were emphasized with specific examples of drugs. It is important to note that the GI functionality is very complex and variable, the role of inter and intraindividual differences, the timing of taking the drugs, and meal compositions are also needed to be considered when evaluating the food impact on absorption.

## Abbreviations

| | |
|---|---|
| GIT | gastrointestinal tract |
| FDA | Food and Drug Administration |
| EMA | European Medicines Agency |
| BCS | Biopharmaceutical Classification System |
| AUC | area under the curve |
| $C_{max}$ | maximum concentration |
| HIV | human immunodeficiency virus |
| $T_{max}$ | time to reach maximum concentration |
| CYP | cytochrome P-450 |
| SH | splanchnic hepatic |
| ABC | ATP-binding cassette |
| MDR | multidrug resistance |
| MRP | multidrug resistance associated protein |
| BCRP | breast cancer resistance protein |
| P-gp | P-glycoprotein |
| OATP | organic anion-transporting polypeptide |
| SLC | solute carrier |
| PepT1 | oligopeptide transporter |
| PBPK | physiologically based pharmacokinetic |

## References


1 Zhong, H., Chan, G., Hu, Y. et al. (2018). A comprehensive map of FDA-approved pharmaceutical products. *Pharmaceutics* 10 (4): 263.

2 Singh, B.N. (1999). Effects of food on clinical pharmacokinetics. *Clin. Pharmacokinet.* 37 (3): 213–255.

3 Welling, P.G. (1996). Effects of food on drug absorption. *Annu. Rev. Nutr.* 16: 383–415.

4 Fleisher, D., Li, C., Zhou, Y. et al. (1999). Drug, meal and formulation interactions influencing drug absorption after oral administration. Clinical implications. *Clin. Pharmacokinet.* 36 (3): 233–254.

5 Kang, S. and Ratain, M. (2010). Inconsistent labeling of food effect for oral agents across therapeutic areas: differences between oncology and non-oncology products. *Clin. Cancer Res.* 16 (17): 4446–4451.

6 O'Shea, J., Holm, R., O'Driscoll, C.M., and Griffin, B.T. (2019). Food for thought: formulating away the food effect – a PEARRL review. *J. Pharm. Pharmacol.* 71 (4): 510–535.

7 Willemsen, A., Lubberman, F.J.E., Tol, J. et al. (2016). Effect of food and acid-reducing agents on the absorption of oral targeted therapies in solid tumors. *Drug Discov. Today* 21 (6): 962–976.

8 CDER (2019). Guidance for industry: assessing the effects of food on drugs in INDs and NDAs – clinical pharmacology considerations, FDA.

9 European Medicines Agency (EMA) (2012). Guideline on the investigation of drug.

10 Koziolek, M., Alcaro, S., Augustijns, P. et al. (2019). The mechanisms of pharmacokinetic food-drug interactions – a perspective from the UNGAP group. *Eur. J. Pharm. Sci.* 134 (15): 31–59.

11 Tulpule, A. and Krishnaswamy, K. (1982). Effect of food on bioavailability of chloroquine. *Eur. J. Clin. Pharmacol.* 23 (3): 271–273.

12 Grannell, L. (2019). When should I take my medicines? *Aust. Prescr.* 42 (3): 86–89.

13 Li, Z., Vachharajani, N.N., and Krishna, R. (2002). On the assessment of effects of food on the pharmacokinetics of drugs in early development. *Biopharm. Drug Dispos.* 23 (4): 165–171.

14 Koch, K.M., Reddy, N.J., Cohen, R.B. et al. (2009). Effects of food on the relative bioavailability of lapatinib in cancer patients. *J. Clin. Oncol.* 27 (8): 1191–1196.

15 Ratain, M.J. and Cohen, E.E. (2007). The value meal: how to save $1,700 per month or more on lapatinib. *J. Clin. Oncol.* 25 (23): 3397–3398.

16 Koziolek, M., Görke, K., Neumann, M. et al. (2019). Development of a bio-relevant dissolution test device simulating mechanical aspects present in the fed stomach. *Eur. J. Pharm. Sci.* 2014 (57): 250–256.

17 Kindler, J., Ruegg, P.C., Neuray, M., and Pacha, W. (1987). Effect of food intake on plasma levels and antihypertensive response during maintenance therapy with endralazine. *Eur. J. Clin. Pharmacol.* 32 (4): 367–372.

18 Melander, A., Stenberg, P., Liedholm, H. et al. (1979). Food-induced reduction in bioavailability of atenolol. *Eur. J. Clin. Pharmacol.* 16 (5): 327–330.

19 Allen, A., Bygate, E., Clark, D. et al. (2000). The effect of food on the bioavailability of oral gemifloxacin in healthy volunteers. *Int. J. Antimicrob. Agents* 16 (1): 45–50.

20 Kaul, S., Christofalo, B., Raymond, R.H. et al. (1998). Effect of food on the bioavailability of stavudine in subjects with human immunodeficiency virus infection. *Antimicrob. Agents Chemother.* 42 (9): 2295–2298.

21 Cardot, J.M., Lecaillon, J.B., Czendlik, C., and Godbillon, J. (1998). The influence of food on the disposition of the antiepileptic rufinamide in

healthy volunteers. *Biopharm. Drug Dispos.* 19 (4): 259–262.
22 Welling, P.G. (1989). Effects of food on drug absorption. *Pharmacol. Ther.* 43 (3): 425–441.
23 Marathe, H.P., Greene, S.D., Kollia, G.D., and Barbhaiya, R.H. (1998). Evaluation of the effect of food on the pharmacokinetics of avitriptan. *Biopharm. Drug Dispos.* 19: 381–394.
24 Marathe, P.H., Greene, D.S., Lee, J.S., and Barbhaiya, R.H. (1998). Assessment of effect of food, gender, and intra-subject variability in the pharmacokinetics of avitriptan. *Biopharm. Drug Dispos.* 19 (3): 153–157.
25 Midha, K.K., McKay, G., Rawson, M.J. et al. (2001). Effects of food on the pharmacokinetics of methylphenidate. *Pharm. Res.* 18 (8): 1185–1189.
26 Koziolek, M., Grimm, M., Schneider, F. et al. (2016). Navigating the human gastrointestinal tract for oral drug delivery: uncharted waters and new frontiers. *Adv. Drug Deliv. Rev.* 101: 75–88.
27 DeSesso, J.M. and Jacobson, C.F. (2001). Anatomical and physiological parameters affecting gastrointestinal absorption in humans and rats. *Food Chem. Toxicol.* 39 (3): 209–228.
28 Dressman, J.B., Amidon, G.L., Reppas, C., and Shah, V.P. (1998). Dissolution testing as a prognostic tool for oral drug absorption: immediate release dosage forms. *Pharm. Res.* 15 (1): 11–22.
29 Thummel, K.E., Shen, D.D., and Shen, D.D. (1997). Enzyme-catalyzed processes of first-pass hepatic and intestinal drug extraction. *Adv. Drug Deliv. Rev.* 27 (2–3): 99–127.
30 Barthe, L., Woodley, J., and Houin, G. (1999). Gastrointestinal absorption of drugs: methods and studies. *Fundam. Clin. Pharmacol.* 13 (2): 154–168.
31 Neuvonen, P.J. (1976). Interactions with the absorption of tetracyclines. *Drugs* 11 (1): 45–54.
32 Welling, P.G. (1977). Influence of food and diet on gastrointestinal drug absorption: a review. *J. Pharmacokinet. Biopharm.* 5 (4): 291–334.
33 Gennaro, R.A. (1995). *Remington: The Science and Practice of Pharmacy*. Easton, PA: Mack Publishing Company.
34 McKinnon, R.A., Burgess, W.M., Hall, P.D.L.M. et al. (1995). Characterisation of CYP3A gene subfamily expression in human gastrointestinal tissues. *Gut* 36 (2): 259–267.
35 Wacher, V.J., Salphati, L., and Benet, L.Z. (2001). Active secretion and enterocytic drug metabolism barriers to drug absorption. *Adv. Drug Deliv. Rev.* 46 (1–3): 89–102.
36 Wacher, V.J., Wu, C.Y., and Benet, L.Z. (1995). Overlapping substrate specificities and tissue distribution of cytochrome P450 3A and P-glycoprotein: implications for drug delivery and activity in cancer chemotherapy. *Mol. Carcinog.* 13 (3): 129–134.
37 Yu, J., Zhou, Z., Owens, K.H. et al. (2017). What can be learned from recent new drug applications? A systematic review of drug interaction data for drugs approved by the US FDA in 2015. *Drug Metab. Dispos.* 45 (1): 86–108.
38 Drozdzik, M., Busch, D., Lapczuk, J. et al. (2018). Protein abundance of clinically relevant drug-metabolizing enzymes in the human liver and intestine: a comparative analysis in paired tissue specimens. *Clin. Pharmacol. Ther.* 104 (3): 515–524.
39 Benet, L.Z., Izumi, T., Zhang, Y. et al. (1999). Intestinal MDR transport proteins and P-450 enzymes as barriers to oral drug delivery. *J. Control. Release* 62 (1–2): 25–31.
40 Peters, S., Jones, C.R., Ungell, A.L., and Hatley, O.J. (2016). Predicting drug extraction in the human gut wall: assessing contributions from drug metabolizing enzymes and transporter proteins using preclinical models. *Clin. Pharmacokinet.* 55 (6): 673–696.
41 Enright, E.F., Gahan, C.G.M., Joyce, S.A., and Griffin, B.T. (2016). The impact of the gut microbiota on drug metabolism and clinical outcome. *Yale J. Biol. Med.* 89 (3): 375–382.
42 Guyton, A.C. and Hall, J.E. (1996). *Textbook of Medical Physiology*. Philadelphia: Saunders Company.
43 Benet, L.Z., Wu, C.-Y., Hebert, M.F., and Wacher, V.J. (1996). Intestinal drug metabolism and antitransport processes: a potential paradigm shift in oral drug delivery. *J. Control. Release* 39 (2–3): 139–143.
44 Gottesman, M.M. and Pastan, I. (1993). Biochemistry of multidrug resistance mediated by the multidrug transporter. *Annu. Rev. Biochem.* 62: 385–427.
45 Chan, L.M., Lowes, S., and Hirst, B.H. (2004). The ABCs of drug transport in intestine and liver: efflux proteins limiting drug absorption and bioavailability. *Eur. J. Pharm. Sci.* 21 (1): 25–51.
46 Thiebaut, F., Tsuruo, T., Hamada, H. et al. (1987). Cellular localization of the multidrug-resistance gene product P-glycoprotein in normal human tissues. *Proc. Natl. Acad. Sci. U.S.A.* 84 (21): 7735–7738.
47 Fojo, A.T., Ueda, K., Slamon, D.J. et al. (1987). Expression of a multidrug-resistance gene in human

tumors and tissues. *Proc. Natl. Acad. Sci. U.S.A.* 84 (1): 265–269.

48 Yu, J., Zhou, Z., Tay-Sontheimer, J. et al. (2017). Intestinal drug interactions mediated by OATPs: a systematic review of preclinical and clinical findings. *J. Pharm. Sci.* 106 (9): 2312–2325.

49 Zhou, Z. (2019). Intestinal OATP-mediated drug interactions: where are we? American Society of Clinical Pharmacology and Therapeutics Webinar (29 July 2019).

50 Smith, D.E., Clémençon, B., and Hediger, M.A. (2013). Proton-coupled oligopeptide transporter family SLC15: physiological, pharmacological and pathological implications. *Mol. Asp. Med.* 34 (2–3): 323–326.

51 Norris, D.A., Puri, N., and Sinko, P.J. (1998). The effect of physical barriers and properties on the oral absorption of particulates. *Adv. Drug Deliv. Rev.* 34 (2–3): 135–154.

52 Welling, P.G. (1984). Interactions affecting drug absorption. *Clin. Pharmacokinet.* 9 (5): 404–434.

53 Williams, L., Hill, D.P. Jr., Davis, J.A., and Lowenthal, D.T. (1996). The influence of food on the absorption and metabolism of drugs: an update. *Eur. J. Drug Metab. Pharmacokinet.* 21 (3): 201–211.

54 Varum, F.J., Htton, G.B., and Basit, A.W. (2013). Food, physiology and drug delivery. *Int. J. Pharm.* 457 (2): 446–460.

55 Horter, D. and Dressman, J.B. (2001). Influence of physicochemical properties on dissolution of drugs in the gastrointestinal tract. *Adv. Drug Deliv. Rev.* 46 (1-3): 75–87.

56 Koziolek, M., Grimm, M., Garbacz, G. et al. (2014). Intragastric volume changes after intake of a high-caloric, high-fat standard breakfast in healthy human subjects investigated by MRI. *Mol. Pharm.* 11 (5): 1632–1639.

57 Grimm, M., Koziolek, M., Saleh, M. et al. (2018). Gastric emptying and small bowel water content after administration of grapefruit juice compared to water and isocaloric solutions of glucose and fructose: a four-way crossover MRI pilot study in healthy subjects. *Mol. Pharm.* 15 (2): 548–559.

58 Schiller, C., Fröhlich, C.P., Giessmann, T. et al. (2005). Intestinal fluid volumes and transit of dosage forms as assessed by magnetic resonance imaging. *Aliment. Pharmacol. Ther.* 22 (10): 971–979.

59 Malagelada, J.R., Longstreth, G.F., Summerskill, W.H.J., and Go, V.L.W. (1976). Measurement of gastric functions during digestion of ordinary solid meals in man. *Gastroenterology* 70 (2): 203–210.

60 Avdeef, A. (2001). Physicochemical profiling (solubility, permeability and charge state). *Curr. Top. Med. Chem.* 1 (4): 277–351.

61 Koziolek, M., Garbacz, G., Neumann, M., and Weitschies, W. (2013). Simulating the postprandial stomach: physiological considerations for dissolution and release testing. *Mol. Pharm.* 10 (5): 1610–1622.

62 Mudie, D.M., Amidon, G.L., and Amidon, G.E. (2010). Physiological parameters for oral delivery and *in vitro* testing. *Mol. Pharm.* 7 (5): 1388–1405.

63 Fadda, H.M., McConnell, E.L., Short, M.D. et al. (2009). Meal-induced acceleration of tablet transit through the human small intestine. *Pharm. Res.* 26: 356–360.

64 Williams, L., Davis, J.A., and Lowenthal, D.T. (1993). The influence of food on the absorption and metabolism of drugs. *Med. Clin. North Am.* 77 (4): 815–829.

65 Toothaker, R.D. and Welling, P.G. (1980). The effect of food on drug bioavailability. *Annu. Rev. Pharmacol. Toxicol.* 20: 173–199.

66 Abuhelwa, A.Y., Williams, D.B., Upton, R.N., and Foster, D.J. (2017). Food, gastrointestinal pH, and models of oral drug absorption. *Eur. J. Pharm. Biopharm.* 112: 234–248.

67 Custodio, J.M., Wu, C.Y., and Benet, L.Z. (2008). Predicting drug disposition, absorption/elimination/transporter interplay and the role of food on drug absorption. *Adv. Drug Deliv. Rev.* 60 (6): 717–733.

68 Shyu, W.C., Knupp, C.A., Pittman, K.A. et al. (1991). Food-induced reduction in bioavailability of didanosine. *Clin. Pharmacol. Ther.* 50 (5 Pt 1): 503–507.

69 Pentafragka, C., Symillides, M., McAllister, M. et al. (2019). The impact of food intake on the luminal environment and performance of oral drug products with a view to *in vitro* and *in silico* simulations: a PEARRL review. *J. Pharm. Pharmacol.* 71 (4): 557–580.

70 Galia, E., Nicolaides, E., Hörter, D. et al. (1998). Evaluation of various dissolution media for predicting *in vivo* performance of class I and II drugs. *Pharm. Res.* 15 (5): 698–705.

71 Deppermann, K.M. and Lode, H. (1993). Fluoroquinolones: interaction profile during enteral absorption. *Drugs* 45 (Suppl 3): 65–72.

72 Humberstone, A.J., Porter, C.J., and Charman, W.N. (1996). A physicochemical basis for the effect of food on the absolute oral bioavailability of halofantrine. *J. Pharm. Sci.* 85 (5): 525–529.

73 Ochsenfahrt, H. and Winne, D. (1971). The contribution of solvent drag to the intestinal absorption of the acidic drugs benzoic acid and salicylic acid from the jejunum of the rat. *Naunyn Schmiedeberg's Arch. Pharmacol.* 281: 197–217.

74 Borowitz, J.L., Moore, P.F., Yim, G.K.W., and Miya, T.S. (1971). Mechanism of enhanced drug effects produced by dilution of the oral dose. *Toxicol. Appl. Pharmacol.* 19: 164–168.

75 Zhi, J., Rakhit, A., Patel, I.H. et al. (1995). Effects of dietary fat on drug absorption. *Clin. Pharmacol. Ther.* 58 (5): 487–491.

76 Rolan, P.E., Mercer, A.J., Weatherley, B.C. et al. (1994). Examination of some factors responsible for a food-induced increase in absorption of atovaquone. *Br. J. Clin. Pharmacol.* 37 (1): 13–20.

77 Crauste-Manciet, S., Decroix, M., Farinotti, R., and Chaumeil, J.C. (1997). Cefpodoxime-proxetil hydrolysis and food effects in the intestinal lumen before absorption: *in vitro* comparison of rabbit and human material. *Int. J. Pharm.* 157 (2): 153–161.

78 Melander, A. and McLean, A. (1983). Influence of food intake on presystemic clearance of drugs. *Clin. Pharmacokinet.* 8 (4): 286–296.

79 Darwish, M., Hellriegel, E.T., and Xie, F. (2008). Single-dose pharmacokinetics of once-daily cyclobenzaprine extended release 30 mg versus cyclobenzaprine immediate release 10 mg three times daily in healthy young adults : a randomized, open-label, two-period crossover, single-centre study. *Clin. Drug Investig.* 28 (12): 793–801.

80 Hendeles, L.H. and Weiberger, M. (1985). Theophylline product and dosing interval selection for chronic asthma. *J. Allergy Clin. Immunol.* 76: 285–291.

81 Kziolek, M., Kostewicz, E., and Vertzoni, M. (2018). Physiological considerations and *in vitro* strategies for evaluating the influence of food on drug release from extended-release formulations. *AAPS PharmSciTech* 19: 2885–2897.

82 Lentz, K.A. (2008). Current methods for predicting human food effect. *AAPS J.* 10 (2): 282–288.

83 Drabant, S., Nemes, K.B., Horváth, V. et al. (2004). Influence of food on the oral bioavailability of deramciclane from film-coated tablet in healthy male volunteers. *Eur. J. Pharm. Biopharm.* 58 (3): 689–695.

84 Benziger, D.P., Kaiko, R.F., Miotto, J.B. et al. (1996). Differential effects of food on the bioavailability of controlled-release oxycodone tablets and immediate-release oxycodone solution. *J. Pharm. Sci.* 85 (4): 407–410.

85 Karim, A., Burns, T., Janky, D., and Hurwitz, A. (1985). Food-induced changes in theophylline absorption from controlled-release formulations. Part II: Importance of meal composition and dosing time relative to meal intake in assessing changes in absorption. *Clin. Pharmacol. Ther.* 38: 642–647.

86 Karim, A., Burns, T., Wearley, L. et al. (1985). Food-induced changes in theophylline absorption from controlled-release formulations. Part I: Substantial increased and decreased absorption with Uniphyl tablets and Theo-Dur Sprinkle. *Clin. Pharmacol. Ther.* 38: 77–83.

87 Schug, B.S., Brendel, E., Wolf, D. et al. (2002). Formulation-dependent food effects demonstrated for nifedipine modified-release preparations marketed in the European Union. *Eur. J. Pharm. Sci.* 15: 279–285.

88 Waldman, S.A. and Morganroth, J. (1995). Effect of food on the bioequivalence of different verapamil sustained-release formulations. *J. Clin. Pharmacol.* 35: 163–169.

89 Yaheya, M. and Ismail, M. (2009). Drug-food interactions and role of pharmacist. *Asian J. Pharm. Clin. Res.* 2 (4): 1–10.

90 Lennernas, H. (1998). Human intestinal permeability. *J. Pharm. Sci.* 87 (4): 403–410.

91 Schmidt, L.E. and Dalhoff, K. (2002). Food-drug interactions. *Drugs* 62 (10): 1481–1502.

92 Papangelou, A., Olszanski, A.J., Stein, C.A. et al. (2017). The effect of food on the absorption of abiraterone acetate from a fine particle dosage form: a randomized crossover trial in healthy volunteers. *Oncol. Ther.* 5: 161–170.

93 McLean, A.J., McNamara, P.J., DuSouich, P. et al. (1978). Food, splanchnic blood flow, and bioavailability of drugs subject to first-pass metabolism. *Clin. Pharmacol. Ther.* 24 (1): 5–10.

94 Liedholm, H. and Melander, A. (1986). Concomitant food intake can increase the bioavailability of propranolol by transient inhibition of its presystemic primary conjugation. *Clin. Pharmacol. Ther.* 40 (1): 29–36.

95 Won, C.S., Oberlies, N.H., and Paine, M.F. (2012). Mechanisms underlying food-drug interactions: inhibition of intestinal metabolism and transport. *Pharmacol. Ther.* 136 (2): 186–201.

96 Reppas, C., Eleftheriou, G., Macheras, P. et al. (1998). Effect of elevated viscosity in the upper gastrointestinal tract on drug absorption in dogs. *Eur. J. Pharm. Sci.* 6 (2): 131–139.

97 Saffar, F., Aiache, J.M., and Andre, P. (1995). Influence of food on the disposition of the antidiabetic

drug metformin in diabetic patients at steady-state. *Methods Find. Exp. Clin. Pharmacol.* 17 (7): 483–487.

**98** Radwan, A., Wagner, M. et al. (2014). Bio-predictive tablet disintegration: effect of water diffusivity, fluid flow, food composition and test conditions. *Eur. J. Pharm. Sci.* 57: 273–279.

**99** Massarella, J.W., DeFeo, T.M., Brown, A.N. et al. (1989). The influence of food on the pharmacokinetics and ACE inhibition of cilazapril. *Br. J. Clin. Pharmacol.* 27 (Suppl 2): 205S–209S.

**100** Melander, A., Danielson, K., Hanson, A. et al. (1976). Reduction of isoniazid bioavailability in normal men by concomitant intake of food. *Acta Med. Scand.* 200 (1–2): 93–97.

**101** Gunnarsson, P.O., Davidsson, T., Andersson, S.B. et al. (1990). Impairment of estramustine phosphate absorption by concurrent intake of milk and food. *Eur. J. Clin. Pharmacol.* 38 (2): 189–193.

**102** Holmes, B., Brogden, R.N., and Richards, D.M. (1985). Norfloxacin. A review of its antibacterial activity, pharmacokinetic properties and therapeutic use. *Drugs* 30 (6): 482–513.

**103** Monk, J.P. and Campoli-Richards, D.M. (1987). Ofloxacin. A review of its antibacterial activity, pharmacokinetic properties and therapeutic use. *Drugs* 33 (4): 346–391.

**104** Marasanapalle, V.P., Crison, J.R., Ma, J. et al. (2009). Investigation of some factors contributing to negative food effects. *Biopharm. Drug Dispos.* 30 (2): 71–80.

**105** Blume, H.H. and Schug, B.S. (1999). The biopharmaceutics classification system (BCS): class III drugs – better candidates for BA/BE waiver? *Eur. J. Pharm. Sci.* 9 (2): 117–121.

**106** Marasanapalle, P. V. (2007). Factors contributing to and predictive models for drugs exhibiting negative food effects of unknown mechanisms. Doctorate thesis. Pharmaceutics and Medicinal Chemistry. Stockton, CA, University of the Pacific. p. 212.

**107** Savannah, J.M., Wu, L., Ritchie, T.K., and Unadkat, J. (2019). Organic anion transporting polypeptide 2B1 – more than a glass-full of drug interactions. *Pharmacol. Ther.* 196: 204–215.

**108** Li, M., Zhao, P., Pan, Y., and Wagner, C. (2018). Predictive performance of physiologically based pharmacokinetic models for the effect of food on oral drug absorption: current status. *CPT Pharmacometrics Syst. Pharmacol.* 7 (2): 82–89.

**109** Amidon, G.L., Lennernas, H., Shah, V.P., and Crison, J.R. (1995). A theoretical basis for a biopharmaceutic drug classification: the correlation of *in vitro* drug product dissolution and *in vivo* bioavailability. *Pharm. Res.* 12 (3): 413–420.

**110** Gu, C.H., Li, H., Levons, J. et al. (2007). Predicting effect of food on extent of drug absorption based on physicochemical properties. *Pharm. Res.* 24 (6): 1118–1130.

**111** Singh, B.N. (2005). A quantitative approach to probe the dependence and correlation of food-effect with aqueous solubility, dose/solubility ratio, and partition coefficient (Log *P*) for orally active drugs administered as immediate-release formulations. *Drug Dev. Res.* 65 (2): 55–75.

# 11

# Drug Metabolism in Gastrointestinal Tract

*Rashim Singh, Dinh Bui, and Ming Hu*

*Department of Pharmacological and Pharmaceutical Sciences, College of Pharmacy, University of Houston, Houston, TX, USA*

## 11.1 Introduction

Metabolism is defined as the process of biochemical modification (i.e. biotransformation) of an endogenous substance or xenobiotics, usually catalyzed by an enzyme system. The primary physiological role of metabolism in the human body is to maintain the homeostasis of a variety of endogenous substrates, such as hormones. All the major enzymes reported in the human body are shown to have endogenous substrates. Besides the primary function of maintaining a hormonal balance during the lifetime of human beings, human metabolizing enzymes also have a secondary function of metabolizing xenobiotics. The main purpose of this function is to aid in the detoxification and excretion of chemicals, including drugs, from the human body.

Once the xenobiotics are absorbed from the intestine after oral administration, the drug metabolizing enzymes and transporters in various metabolic organs plays an important role in their disposition in humans. The main organ of drug metabolism is the liver, but other organs such as the intestine, kidney, lung, and placenta can also be involved in this process. Most drugs administered through oral routes are hydrophobic enough to get absorbed through intestinal cells through passive diffusion. Once they reach the intestinal or liver cells, most are converted by various Phase I and/or Phase II enzymes to a more hydrophilic metabolite which then gets excreted out of the body via kidney and other excretion routes. The present chapter will discuss the metabolism of xenobiotics in the gastrointestinal tract (GIT) with a focus on the intestine.

Here, we will discuss the importance and contribution of the GIT as a metabolic organ for xenobiotics. We will touch briefly on the main metabolic enzyme families in the GIT, along with organ expression, regulation by nuclear receptors (NRs), and pathophysiological conditions associated with the modulation and genetic polymorphism of these enzymes, as well as the effects of age, pharmacogenetics, diet, and environment on the drug metabolism in the GIT. We will briefly discuss the coupling processes that exist between the enzyme systems and efflux transporters, which play a critical role in the local, enteric, and enterohepatic recycling that prolongs the exposure of drugs. In this 2nd edition, we will discuss some new topics of interest related to the metabolism of drugs in the GIT. We will explore the application of the newly proposed biopharmaceutics drug disposition classification system (BDDCS) that can be successfully used to classify drugs based on enzyme–transporter interaction to the intestinal metabolism of drugs. Furthermore, we will discuss the various important phenomena that relate to intestinal drug metabolism such as drug–drug/natural product interactions, drug-induced toxicities, and bidirectional interactions with the human gut microbiome (GM). Lastly, we will discuss the potential role of intestinal metabolism-based drug designing and lead optimization to develop drugs for gastrointestinal diseases and disorders.

### 11.1.1 Gastrointestinal Tract (GIT) as Drug Disposition Organ

The gross structures of the luminal surfaces of the GIT components are essentially similar, covered with a layer of columnar epithelial cells, goblet cells, and endocrine cells. However, the cell types and functions vary along the length of GIT. The main function of the stomach cell is to secrete acid, pepsinogen, gastrin, and intrinsic factors to aid in the process of digestion of food. The stomach also expresses some metabolic enzymes. The epithelial cells in the small intestine have villi and microvilli on their surface which provide a large area of

*Oral Bioavailability and Drug Delivery: From Basics to Advanced Concepts and Applications,* First Edition. Edited by Ming Hu and Xiaoling Li.

absorption for the nutrients and water, and also contain various enzymes [1–3] and transporters [4, 5]. The epithelial cells in the colon primarily function to absorb fluid and electrolytes and express various metabolic enzymes [1, 6].

The average lengths and absorbing surface areas of the GIT organs determined postmortem are the esophagus, 25 cm, and 0.02 $m^2$; stomach, 20 cm, and 0.11 $m^2$; duodenum, 25 cm, and 0.09 $m^2$; jejunum, 300 cm, and 60 $m^2$; ileum, 300 cm, and 60 $m^2$; and colon, 150 cm, and 0.25 $m^2$, respectively. The bacterial flora in the GIT are also capable of metabolizing many drugs and their counts (reported as $\log_{10}$ number of viable organisms per g wet weight) vary considerably in different parts of GIT: stomach (0–5); proximal small intestine (0–5); distal small intestine (6–7); and large intestine (7–10) [1, 6].

The primary function of the GIT is to absorb nutrients and water, and as a result, may also absorb potentially beneficial and harmful xenobiotics. In addition to this, the GIT also acts as the first line of defense against foreign particles and xenobiotics. Upon oral administration, the bioavailability of drugs can be limited by their absorption from gastric and intestinal epithelium and/or their first-pass metabolism (FPM). The metabolism of drugs before entering the systemic circulation is referred to as FPM [7, 8].

In classic literature, the liver has been accepted as the major site of FPM, but recent studies have indicated that the intestine also contributes substantially to the overall FPM of certain drugs such as cyclosporine, nifedipine, and verapamil. The small intestine has also been shown to contribute significantly to the first-pass oxidative metabolism of midazolam catalyzed by mucosal CYP3A4 [9–13]. In some studies, intestinal metabolism has even been shown to play a greater role than the liver [9, 11, 14–18].

The important role of the intestine in the FPM of orally ingested xenobiotics is supported by the fact that numerous metabolic enzymes are expressed in various parts of the intestine; it is the first site of exposure for xenobiotics and has a large surface area available for both absorption and metabolism. Also, it is important to note that the intestinal cells are exposed to a higher concentration of drug as compared to liver cells as the intestine is the site of drug release from the dosage form. The latter can lead to a higher contribution by the intestine in metabolizing the drugs. However, successful assessments of the relative contributions of the liver and intestine to the FPM of xenobiotics are difficult to make, due to limitations of available techniques to decouple the two components [1, 15, 19, 20]. However, recent attempts have been made to compare the contribution of liver and intestine to the overall bioavailability and drug–drug interactions (DDIs) [9, 21–23].

Lin et al. [8] and Doherty and Charman [24] have suggested the overall greater metabolic capability and contribution of the liver in the FPM, based on the greater weight of the human liver (~1.5 kg) than the small intestine (~0.7 kg), and greater cytochrome P450 (CYP) expression levels and total microsomal protein content in the liver than intestine [8, 24]. However, this fact does not necessarily diminish the fact that the intestine is significantly capable of metabolizing the orally ingested drugs before they are transported to the liver and therefore can play an important contribution to the overall metabolism and bioavailability of drugs and other natural compounds. Though all components of GIT show the presence of metabolic enzymes, significant metabolic activities are seen in the small intestine [2, 8, 25–27].

### 11.1.2 Drug-Metabolizing Enzymes in GIT

Drug metabolism in the intestine can be broadly divided into phase I and phase II metabolism. Phase I contributes to more than 70% of the total drug metabolism, while phase II metabolism contributes to less than 30% of the overall drug metabolism in humans (Figure 11.1a) [3, 5, 28–32]. Phase I reactions are usually oxidation, hydrolysis, and reduction, while phase II reactions are primarily the conjugation of drugs and phase I metabolites with substances such as glucuronic acid, sulfuric acid, or amino acids.

#### 11.1.2.1 Enzymes for Phase I Metabolism

In phase I reactions, the molecular weight of the parent drug either decreases, such as in hydrolysis and dealkylation reactions, or increases, such as in oxidation and hydroxylation reactions. Metabolites of phase I are usually biologically less active or inactive, but occasionally create compounds with increased biological activity. Another important function of the phase I enzyme is to make the compound more hydrophilic and/or create an intermediate that can further be a substrate of phase II metabolism [33].

In normal physiology, phase I metabolism of endogenous substances or xenobiotics can generate free radicals which increase with aging and are responsible for various oxidative stress-related conditions in the body. In the case of various environmental carcinogens, these free radicals can be very active and can bind to DNA or proteins, thereby causing mutations and cancers [34]. The most important family of enzymes in

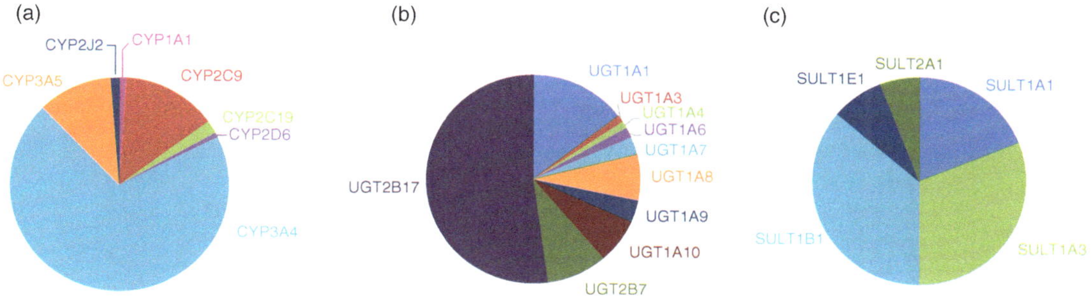


**Figure 11.1 Relative expression levels of major CYPs (a), UGT (b), and SULT (c) isoforms expressed in the human intestine.** UGT, uridine dinucleotide phosphate (UDP) glucuronosyltransferase; CYP, cytochrome P450; SULT, sulfotransferases.

phase I metabolism is CYP which contributes toward more than 80% of total phase I metabolism [29].

***11.1.2.1.1 Cytochrome P450*** CYPs comprise a superfamily of hemeproteins that contain a single-iron protoporphyrin IX prosthetic group. CYPs catalyze various oxidation, reduction, and hydroxylation reactions, such as hydroxylation of saturated hydrocarbons, epoxidation of double bonds, oxidation, and dealkylation reactions. In the oxidation reaction, CYP uses molecular oxygen, inserts one of its oxygen atoms into a substrate, and reduces the second oxygen to a water molecule, utilizing two electrons that are provided by NADPH (nicotinamide adenine dinucleotide phosphate) via a reductase protein or a CYP reductase [33].

CYP superfamily is subdivided into families and subfamilies that are classified based on amino acid sequence homology. 18 families of CYP genes and 43 subfamilies have been identified in humans, out of which only three families, CYP1, CYP2, and CYP3, are responsible for drug metabolism in humans [35, 36]. Most of the drug metabolism by CYPs is accounted for by the CYP1A2 (8.9%), 2A6 (3.4%), 2B6 (7.2%), 2C8 (4.7%), 2C9 (12.8%), 2C19 (6.8%), 2D6 (20%), 2E1 (3%), CYP2J2 (3%), and 3A4/5 (30.2%) isoforms [37]. Among these, CYP3A (CYP3A4 and 3A5) and CYP2C (CYP2C8, 2C9, 2C18, and 2C19) are the most abundant subfamilies, accounting for ~40% and ~30% of CYP metabolism, respectively [29, 34] (Figure 11.1a). Among endogenous substrates of CYPs are cholesterol and bile acids, fatty acids, steroids, prostaglandins, and vitamins A and D.

CYPs are involved in some important endogenous functions in humans. CYP 1, 2, 3, and 4 families are involved in the metabolism of arachidonic acid to form eicosanoids (leukotrienes and prostaglandins), which are further metabolized by CYP5A1 and 8A1 to form compounds which play an important role in platelet aggregation. Other functions are metabolism and synthesis of cholesterol and bile acids (CYP7A1, 27A1, 39A1, 46A1, 51A1, 7B1, 8B1, 3A4, 2D6); steroids (CYP11A1, 11B1, 11B2, 17A1, 19A1, 21A2); and vitamin D (CYP24A1, 27A1, 27B1, 2A4, 2D6) [34, 38–41].

CYPs are expressed throughout the body, mainly in the liver, small intestine, stomach, colon, and lungs. Apart from those organs with intimate exposure to the environment, CYPs are also present in the kidney and placenta. There are 18 functional CYPs in humans. The quantitative expressions of the major drug-metabolizing CYP isoforms, except CYP3A5 and CYP1A1, in the human intestine are shown in Table 11.1 [28]. CYP3A4 has been shown to be the predominant CYP isoform expressed in the small intestine [2, 5, 28, 32].

Total microsomal protein content decreases markedly along the length of the small intestine, from the duodenum to the ileum [5, 28]. Total CYP content increased slightly from the duodenum to the jejunum and then decreased sharply toward the ileum. Human small intestine expression levels of CYP isoforms show greater interindividual variability as compared to human liver expression levels of CYP isoforms. Such interindividual variations were seen at the levels of gene expression, protein abundance, and catalytic activity of the enzyme [2, 5, 28, 30, 32, 43].

#### 11.1.2.2 Enzymes for Phase II Metabolism

Phase II metabolism, also known as conjugation reactions, is catalyzed by various transferases leading to an increase in molecular weight of the parent molecule. It usually increases the hydrophilicity of the drug so that it can be eliminated from the body, except for methylation (e.g. methylation of catechols), which creates metabolites that are either similarly nonpolar or sometimes even less polar. The inherent function of phase II metabolism is to protect or detoxify, so phase II metabolites, in general, are less bioactive (or toxic) than phase I metabolites, except in some cases such as morphine, ezetimibe, and PhIP (2-Amino-1-methyl-6-phenylimidazol[4,5-β]pyridine)

**Table 11.1** Quantitative protein expression of CYPs and UGTs in human jejunum and ileum [28].

| Enzyme isoforms | Jejunum ($n = 4$) | | Ileum ($n = 12$) | |
|---|---|---|---|---|
| | Mean ± S.D. | Range (min–max) | Mean ± S.D. | Range (min–max) |
| CYP2C9 | 1.52 ± 0.61 | (0.81–2.17) | 1.42 ± 0.60 | (0.92–3.04) |
| CYP2C19 | 1.08 ± 0.11 | (0.99–1.23) | 1.25 ± 0.60 | (0.79–3.04) |
| CYP2D6 | 1.91 ± 0.46 | (1.31–2.30) | 1.64 ± 0.55 | (1.00–2.63) |
| CYP2J2 | 1.04 ± 0.30 | (0.72–1.43) | 0.90 ± 0.16 | (0.62–1.11) |
| CYP3A4 | 33.33 ± 5.70 | (24.80–36.75) | 22.04 ± 10.60 | (7.58–34.70) |
| UGT1A1 | 2.93 ± 2.17 | (0.83–5.35) | 1.52 ± 0.55 | (0.72–2.58) |
| UGT1A3 | 1.26 ± 0.29 | (0.87–1.54) | 1.10 ± 0.23 | (0.90–1.69) |
| UGT1A6 | 0.75 ± 0.02 | (0.74–0.77) | 0.75 ± 0.02 | (0.71–0.79) |
| UGT2B7 | 1.84 ± 0.97 | (1.03–3.03) | 1.17 ± 0.23 | (0.88–1.65) |
| UGT2B15 | 0.39 ± 0.03 | (0.34–0.41) | 0.37 ± 0.03 | (0.32–0.44) |
| UGT2B17* | 6.8 ± 6.0 | (0.06–20.8) | 10.8 ± 10.8 | (0.06–28.7) |

* n=6 human intestinal microsomes samples [42]

Source: Adapted from Couto et al. [28].

glucuronides. Usually, C-glucuronides are more reactive and toxic metabolites than O-glucuronides [7, 44–46].

The most important phase II enzymes involved in the metabolism of drugs are UGTs (UDP-glucuronosyl transferases, or Uridine diphosphate glucuronosyl transferases) and SULTs (sulfotransferases). Both enzyme systems are distributed widely in many tissues including the intestine. UGTs are membrane-bound enzymes and are located in the endoplasmic reticulum in cells, while SULTs are present in the cytosol [7, 47–49].

Another important class of phase II enzymes involved in the metabolism of toxins and relieving oxidative stress is glutathione S-transferases (GSTs). GST is particularly important in the detoxification of toxicants or oxidants because it can reduce super or regular oxidants, thereby reducing these species' ability to cause molecular or cellular damage. GSTs play a critical role in our ability to defend against environmental assaults or ingested toxins. GSTs do not appear to be as important for the metabolism of drugs, and the interested reader should refer to the review articles by various authors [50–54].

***11.1.2.2.1 UDP-Glucuronosyl Transferases (UGTs)*** UGTs utilize UDP-glucuronic acid as a cofactor and transfer glucuronic acid to the substrates, forming hydrophilic conjugates called glucuronides (or β-D-glucopyranosiduronic acids, and these conjugates are usually β-glucuronidase sensitive). Substrates of UGTs include endogenous compounds such as steroids, bile acids, bilirubin, hormones, dietary constituents like flavonoids, xenobiotics such as morphine and valproic acid, and environmental toxins and carcinogens such as PhIP [44, 46, 55–57]. There is increasing evidence that for most phenolic compounds, whether they are endogenous molecules or drugs or polyphenols, intestinal UGT is the main organ for glucuronidation.

For the glucuronidation reaction, the substrates typically possess hydroxyl (alcoholic, phenolic), carboxyl, sulfuryl, carbonyl, or amino (primary, secondary, or tertiary) groups in their structure. O-glucuronidations are more common than N- and C-glucuronidation. The normal function of glucuronidation in mammals is to maintain the homeostasis of various endogenous substrates, and to transport and detoxify xenobiotics [26, 44, 46, 56, 57].

Most UGT substrates are lipophilic and therefore may be toxic to cells. For these compounds to be eliminated through renal and biliary routes, they need to be converted to more hydrophilic metabolites such as glucuronides, and hence glucuronidation by UGTs is considered as probably the most important detoxification pathway for a lot of xenobiotics and/or their phase-I metabolites. Hydrophilic phase II metabolites usually need a transporter to get excreted out of intestinal/liver cells. Various efflux transporters such as MRP2 and BCRP are shown to be involved in the efflux of phase II metabolites [57–65].

UGTs are mainly expressed in the liver and various extrahepatic organs such as the small intestine, bile ducts, stomach, colon, kidney, pancreas, brain, and prostate. Also, some UGTs are expressed (albeit not particularly abundantly) in the lungs, esophagus, mammary, and adrenal glands. 22 UGT isoforms have been identified so far belonging to 1A, 2A, 2B, 3A, and

8A subfamilies [66]. The quantitative expressions of major drug-metabolizing UGT isoforms in the human intestine are shown in Table 11.1 [28, 42].

The most abundant isoform expressed in the small intestine is UGT1A1, which is equally well expressed in the liver too. The most abundant UGT isoform gene expressions are reported for UGT1A1, UGT1A6, UGT1A10, UGT2A3, and UGT2B7 in human jejunum [5]. Quantitatively, UGT1A8 and 1A10 are major isoforms expressed in the small intestine, while UGT1A9 is a major isoform expressed in the liver. The relative protein expression of major UGT isoforms in the intestine is shown in Figure 11.1b [2, 28, 30, 67]; however, there are some discrepancies in reporting. Similar to CYPs, human small intestine UGTs also show greater inter-subject variability in their gene expression, protein abundance, and catalytic activity as compared to human liver UGTs. The human colon has shown markedly lower expression levels of UGTs as compared to the small intestine, making it more susceptible to carcinogens [2, 28, 45, 46, 49, 68, 69].

***11.1.2.2.2 Sulfotransferases (SULTs)*** SULTs are phase II enzymes that catalyze the transfer of a sulfonate group ($SO_3^-$) from the coenzyme 3′-phosphoadenosine-5′-phosphosulfate (PAPS) to a substrate. SULTs can either be cytosolic or membrane-bound which are located in the Golgi apparatus. Membrane-bound SULTs, which are not the focus of our discussion, sulfonate macromolecules such as peptides, proteins, lipids, and glycosaminoglycans can affect their structural and functional characteristics and even functions. Cytosolic SULTs, which are usually the ones involved in drug metabolism, are responsible for the sulfonation of xenobiotics and endogenous substrates such as hormones, steroids, bile acids, and neurotransmitters [70–76].

In most cases, the addition of a sulfonate moiety to a compound increases its water solubility and decreases its biological activity. The sulfation reaction in living cells is reversed by sulfatase, which hydrolyzes the sulfonated conjugates. SULTs maintain and regulate the endocrine status of an individual by modulating the activity of steroid hormones and their biosynthesis. Another important function of SULT is the defense against xenobiotics. Although the primary goal of sulfate-conjugation of xenobiotics is to detoxify the drugs, it may also result in the activation of certain chemical procarcinogens, such as turning certain dietary and environmental agents into carcinogens [70, 71, 73, 74, 76–79].

In humans, SULT superfamily consists of three families, SULT1, SULT2, and SULT4, with at least thirteen distinct members: SULT1 (1A1, 1A2, 1A3, 1A4, 1B1, 1C2, 1C4, 1E1); SULT2 (2A1 and 2B1_v1, 2B1_v2); and SULT4 (4A1_v1 and 4A1_v2). The relative protein expression in the human intestine of 5 major SULT isoforms is shown in Figure 11.1c. Quantitatively, the small intestine mainly expresses SULT 1A1 and 1A3, which increases from jejunum to ileum, followed by 1B1 and 2A1, whereas the liver mainly expresses 1A1 and 2A1, followed by 1E1 [2, 3, 77, 78, 80–82]. SULT family of enzymes has not been extensively studied in the last decade with smaller advancement in its understanding compared to other phase II enzyme families.

### 11.1.3 Regulation of Drug-Metabolizing Enzymes by Nuclear Receptors

NRs are the largest known family of transcription factors that function as modulators of gene expression in tissues and they appear to be particularly important for several conjugation enzymes such as UGTs and SULTs. Forty-nine known members of the NR superfamily have been reported which share key structural features. Specifically, ligand-activated transcription factors within the NR superfamily appear to be critical in controlling the expression of drug-metabolizing enzymes in organs such as the intestine and liver. Therefore, NRs are proposed to have a significant impact on drug disposition [83–90].

Based on the linear amino acid positions along the polypeptide chain and the function, NR can be divided into 3 major protein domains: activation function 1 (AF-1), DNA-binding, and ligand-binding domains. AF-1 domain is located at the N-terminus and is responsible for ligand-independent activation of the NR via the recognition of other transcription factors and coactivators. DBD directly interacts with DNA response elements on target genes to modulate the transcription and is located at the center. LBD is located at the C-terminus and is responsible for binding ligands such as hormones, bile acids, lipids, eicosanoids, drugs, and other xenobiotics, which then activate NR [84, 91–93].

The unliganded NRs bind to transcriptional corepressors in the cytoplasm, which recruit specific histone deacetylase-containing complexes. Once a drug is transported through the cell membrane, the drug ligand binds to the LBD in the cytoplasm and causes the dissociation of the histone deacetylase-containing complex and translocation of the ligand-bound NR to the nucleus. Once in the nucleus, the ligand–receptor

**Table 11.2** Summary of the effect of nuclear receptor activation on target gene expression of the drug-metabolizing enzymes in human intestine [90–92, 94–97].

| Target gene | Nuclear receptor | Nuclear receptor ligands | Effect on target genes |
|---|---|---|---|
| | | **Phase II** | |
| UGT1A1 | PXR | Rifampin | ↑ |
| | CAR | Phenobarbital | ↑ |
| | AhR | Napthaflavone, TCDD | ↑ |
| GST | PXR | Dexamethasone | ↑ |
| SULT2A1 | PXR | Dexamethasone | ↑ |
| | VDR | Vitamin D3 | ↑ |
| | FXR | Chenodeoxycholic acid | ↑ |
| | | **Phase I** | |
| CYP3A4 | PXR | Rifampin, hyperforin, clotrimazole, phenytoin, phenobarbital, dexamethasone, zearalenone, | ↑ |
| | | flucloxacillin, artemisinin, nicotine, nifedipine, nicardipine, omeprazole, paclitaxel, etoposide, carbamazepine, topotecan, mevastatin, | |
| | | Ritonavir, ketoconazole | ↓ |
| | CAR | Artemisinin, phenobarbital, CITCO | ↑ |
| | GR | Dexamethasone | ↑ |
| | HNF 4α | | ↑ |
| | VDR | 1α,25 hydroxyvitamin D3, lithocholic acid, 19-nor-1α,25 dihydroxyvitamin D2 | ↑ |
| | FXR | Chenodeoxycholic acid,21 GW 406421 | ↑ |
| CYP2C9 | PXR | Hyperforin, rifampin, nifedipine, nicardipine, isradipine, dexamethasone | ↑ |
| | CAR | Phenobarbital | ↑ |
| | GR | Dexamethasone | ↑ |
| CYP2C19 | CAR | Phenobarbital | ↑ |
| | GR | Dexamethasone | ↑ |

PXR, pregnane X receptor; CAR, constitutive androstane receptor; GR, glucocorticoid receptor; HNF4α, hepatocyte nuclear factor 4α; FXR, farnesoid X receptor; VDR, vitamin D receptor; AhR, aryl hydrocarbon receptor; UGT, UDP-glucuronosyltransferase; CYP, cytochrome; GST, glutathione S-transferase; SULT, sulfotransferase; CITCO, 6-(4-chloropheny-l)imidazo[2,1-*b*][1,3]thiazole-5-carbaldehyde-*O*-(3,4-dichlorobenzyl)oxime; TCDD, 2,3,7,8-tetrachlorodibenzo-*p*-dioxin.

complex recruits coactivators and forms either homodimers or, in many cases, forms a heterodimer with the retinoid X receptor (RXR). These homodimer or heterodimer complexes then bind to the response elements in the regulatory (promoter and enhancer) regions of target genes and increase transcription of the target genes. Table 11.2 shows the NRs activated by certain ligands which can mostly upregulate and sometimes downregulate the gene expression of the major phase I and phase II enzymes [39, 83, 85, 87, 91, 98–102].

### 11.1.4 Diseases Associated with Drug-Metabolizing Enzymes

Most of the diseases and/or disorders associated with CYP and UGT enzymes are related to their endogenous functions in the body. The imbalance of CYP expression or activity can disturb various endogenous functions. The defect in metabolic regulation and homeostasis of sex hormones can also be responsible for causing cancer in sexual organs.

One of the major diseases is autoimmune hepatitis, which is a chronic inflammatory disease of the liver where there is a loss of self-tolerance toward the liver. In this disease, drug-metabolizing enzymes have been identified as human hepatocellular autoantigens. In response to these autoantigens, our immune system expresses auto-antibodies in the liver and kidney called liver/kidney microsomal (LKM) antibodies. LKM auto-antibodies target CYP2D6 (LMK1), CYP2C9 (LMK2), and CYP1A2 (LM, found only in liver tissue), and LKM3 auto-antibodies target UGT1A1, 1A4, and 1A6 and also show some reactivity with UGT2B isoforms. [41, 46, 49, 103–105].

Apart from the diseases based on the endogenous function of CYPs and UGTs, modulation of expression of enzymes involved in detoxification of potential carcinogens can cause chemical carcinogenesis. CYP 1A1 and 1A2 are involved in the metabolism of aryl hydrocarbons and environmental mutagens and toxicants. UGTs (1A and 2B) transform various potential environmental carcinogens such as benzopyrenes and heterocyclic amines into reactive metabolites.

Chemical carcinogenesis due to CYPs and UGTs modulation can happen in two different ways: either the enzymes responsible for detoxification/bioactivation have undergone mutation, upregulation, or downregulation in the diseased tissue, rendering the tissue more prone to carcinogenic and/or toxicogenic effect of the toxic compounds; or the metabolizing enzymes itself bioactivate the carcinogens into reactive metabolites, which have the potential of forming DNA/protein adducts and thus cause cancers in tissues like mammary and prostate tissues [41, 46, 49, 56, 103–106].

UGT1A1 mutation/polymorphism can also cause unconjugated hyperbilirubinemia in humans. Three types are known: Crigler–Najjar syndrome type I (complete lack of bilirubin glucuronidation), Crigler–Najjar syndrome type II (low amount of mono- and di-bilirubin glucuronides), and Gilbert syndrome (fluctuating unconjugated hyperbilirubinemia in response to psychological stress, infection, fasting, or physical activity) [56, 104, 106–108].

Under unconjugated hyperbilirubinemia conditions, serum bilirubin levels are very high. Bilirubin is a breakdown product of catabolism of heme-containing proteins such as CYP450 and hemoglobins. Bilirubin due to its hydrophobicity remains bound to albumin in serum. Elimination of bilirubin occurs by its conversion into a hydrophilic diglucuronide by UGT1A1, and subsequent transport of the diglucuronide by ABC efflux proteins. Failure of either metabolism or transport of bilirubin glucuronides to the bile can increase serum levels of bilirubin, causing albumin to saturate and bilirubin to accumulate in tissues. This situation can manifest itself into clinically relevant jaundice and cause further toxicity in the liver and brain [56, 104, 106, 108, 109].

## 11.2 Role of Intestinal Efflux Transporters in the Drug Disposition

The efflux transporters for the parent drug and/or its metabolites play an important role in deciding the actual intracellular drug disposition. It involves efflux transporters at apical, basolateral, or both membranes. So, it is more complex to sort out the mechanism of DDIs due to drugs being a substrate of enzymes and drugs and metabolites being a substrate of transporters [63, 110–116].

The role of an efflux transporter in drug disposition in the intestine can be prominently seen where the parent drug molecule, once taken up by the intestinal cell, is effluxed back into the intestinal lumen by the transporters. This process increases the chances of the drug molecules encountering the various enzyme systems multiple times during their absorption and transport from the intestine to the systemic circulation. Also, the role of efflux transporters is particularly seen in cases where the conjugation reaction results in a highly hydrophilic metabolite that is incapable of passively diffusing across the cellular membrane. Hence, the excretion of conjugated metabolites from cells is dependent upon their transport by the efflux transporters on the apical and/or the basolateral membranes of enterocytes [63, 110, 112–115].

### 11.2.1 Efflux Transporters in Intestine

Intestinal epithelial cells express a variety of uptake transporters (such as PEPT1 and OATPs) and efflux transporters (P-gp, MRPs, and BCRP) on the apical and basolateral membranes, for the influx and efflux of the endogenous and xenobiotic substrates, respectively. Figure 11.2 shows the intestinal expression of the major efflux and influx transporters in the intestinal cells. The efflux transporters have a critical role in limiting influx and facilitating efflux to prevent the intracellular accumulation of the parent drug and its phase II metabolites. In tumor cells, such functions confer multidrug resistance for various anticancer agents [4, 117–121].

The enterocytes show localized expression of the efflux transporters on the apical and basolateral membranes. The apical membrane expresses known transporters such as P-gp, MRP2, and BCRP, while the basolateral membrane expresses MRP1, MRP3, MRP4, and MRP5. The small intestine shows a high expression of MRP3, P-gp, and BCRP and relatively lower expression of other MRPs, whereas the colon expresses P-gp highly but shows a lower expression of MRP1, MRP2, and MRP3 but no BCRP [2, 4, 5, 117, 119–122].

The expression of the efflux transporter also varies between different segments of the small intestine. In the human intestine, the mRNA and protein levels of the efflux transporters were reported in the rank

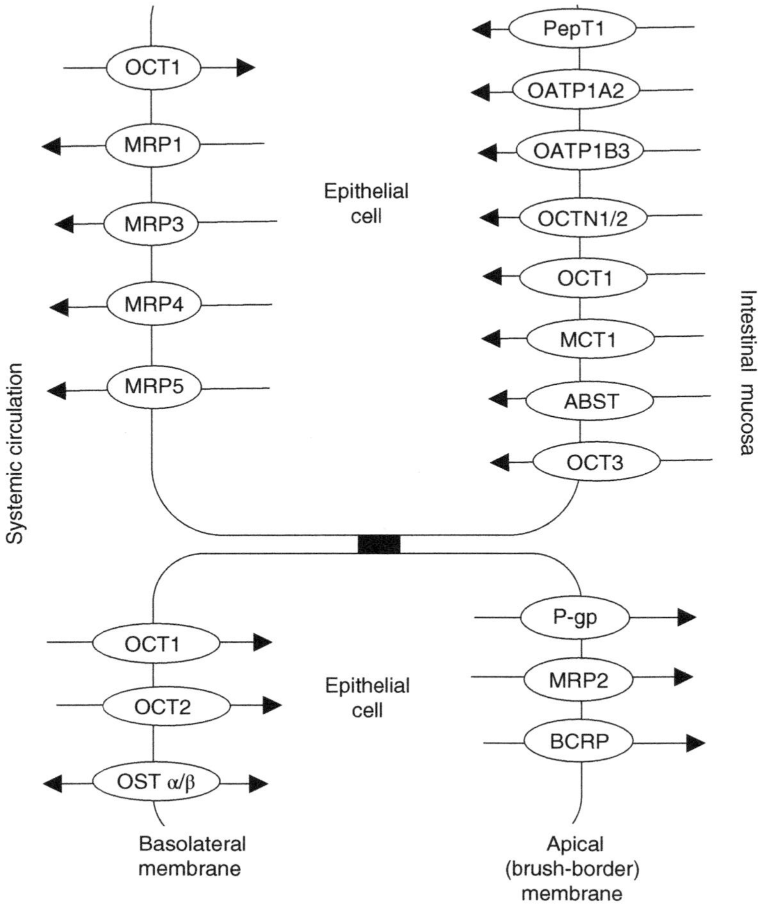


**Figure 11.2 Expression of efflux transporters on the apical and basolateral membranes of the intestine.** P-gp, P-glycoprotein; MRP1/2/3/4/5, multidrug resistance protein; BCRP, breast cancer resistance protein 1/2/3/4/5; OATP 1A2/1B1, organic anion transporting polypeptides 1A2/1B1; OCT1/2/3, organic cation transporter 1/2/3; MCT1, monocarboxylate transporters 1; ABST, apical sodium-dependent bile acid transporter; PepT1, Peptide transporter 1; and OCTN1/2, organic cation transporters novel ½; OSTα/β, organic solute transporter alpha/beta.

order in the various segments of the intestine. The rank order of efflux transporter gene expression was MRP3 ≫ P-gp > MRP2 > MRP5 > MRP4 > MRP1 in the duodenum; BCRP ~ MRP2 > P-gp ~ MRP3 ~ MRP6 ~ MRP5 ~ MRP1 > MRP4 > MDR3 in the jejunum; P-gp > MRP3 ≫ MRP1 ~ MRP5 ~ MRP4 > MRP2 in the terminal ileum; and MRP3 ≫p-gp > MRP4 ~ MRP5 > MRP1 ≫ MRP2 in the colon [120, 121]. However, the reported protein expression levels of the efflux transporters in the duodenum (P-gp ~ MRP3 ≫BCRP > MRP2 > MRP1), jejunum (P-gP ≫ BCRP ~ MRP3 > MRP1 > MRP2), and ileum (P-gP ~ MRP3 > BCRP >MRP1 > MRP2) showed that mRNA levels do not necessarily correlate with the protein expression level [5].

The P-gp, MRPs, and BCRP also show substrate overlapping in the transport of parent drugs and their metabolites, which can result in possible DDI in the liver and intestine. It also explains the possible compensation process in the efflux of the drug and conjugated metabolites in $TR^-$ rats (MRP2-deficient rats), $MRP2^-$ /$MRP3^-$ /$BCRP^-$ mice, wherein the absence of one efflux transporter, expression of other efflux transporters on apical and/or basolateral membrane of liver and intestine increases to compensate for the deficiency [110, 113, 123–127].

#### 11.2.1.1 P-gp

P-gp or MDR1 (*ABCB1*) is an ATP-binding cassette (ABC) efflux transporter. In the human intestine, P-gp is highly expressed on the apical surface of superficial columnar epithelial cells of the ileum and colon, and expression gradually decreases proximally into the jejunum, duodenum, and stomach [4, 128]. The relative P-gp levels progressively increase from the proximal to the distal regions by approximately 2-fold. Interindividual variation in P-gp expression can range from 1.5- to

**Table 11.3** Summary of the effect of nuclear receptor activation on target gene expression of the drug transporters in the intestine [83, 89, 130, 131].

| Target gene | Nuclear receptor | Nuclear receptor ligands | Effect on target genes |
|---|---|---|---|
| MDR1/ABCB1 | PXR | Flucloxacillin, artemisinin, rifampin, hyperforin | ↑ |
| | CAR | Artemisinin | ↑ |
| MRP2/ABCC2 | PXR | Rifampin, hyperforin | ↑ |
| | CAR | Phenobarbital | ↑ |
| | FXR | Chenodeoxycholic acid | ↑ |
| MRP4/ABCC4 | CAR | Phenobarbital | ↑ |
| BSEP/ABCB11 | FXR | Cholic acid, chenodeoxycholic acid | ↑ |
| | | Lithocholic acid | ↓ |
| BCRP/ABCG2 | PPARγ | Rosiglitazone | ↑ |
| | | GW9662 | ↓ |
| NTCP/SLC10A1 | | Dexamethasone | ↑ |
| SLCO1B3/OATP1B3 | PXR | Chenodeoxycholic acid | ↑ |
| SLCO1A2/OATP1A2 | PXR | | ↑ |

PXR, pregnane X receptor; CAR, constitutive androstane receptor; GR, glucocorticoid receptor; FXR, farnesoid X receptor; PPARγ, peroxisome proliferator-activated receptor; MDR, multidrug resistance; MRP, multidrug resistance-associated protein; BSEP, bile salt export pump; BCRP, breast cancer resistance protein; NTCP, $Na^+$ taurocholate-transporting polypeptide; OATP, organic anion-transporting polypeptide.

3.0-fold. It has also been shown that P-gp expression is about seven times higher in the intestine than in the liver and CYP3A4 expression is about 3 times higher in the intestine than in the liver, suggesting the critical participation of intestinal P-gp in coupling in the intestine and limiting oral drug bioavailability [129].

#### 11.2.1.2 MRP2

MRP2 (*ABCC2*) is also an ATP-binding cassette (ABC) efflux transporter. MRP2 protein is highly expressed on the apical membrane in the intestine. Unlike P-gp expression, MRP2 expression is higher in the proximal and very low in the distal region of the human intestine. The MRP2 protein expression increases from the duodenum to the jejunum and then decreases from the jejunum to the distal ileum, while MRP2 mRNA expression is somewhat similar to all segments of the small intestine. MRP2 has been shown to be involved in the intestinal and hepatobiliary excretion of the parent drug and/or its phase II conjugated metabolites. This suggests their involvement in the enteric- and entero-hepatic recycling and the enzyme–transporter coupling of the drug-disposition process [4, 120, 121].

#### 11.2.1.3 BCRP

Unlike P-gp and MRP2 ABC transporters, which have 2 ATP-binding domains and 2 sets of transmembrane domains, BCRP (*ABCG2*) is a half ABC transporter and has a single ATP-binding domain and a single set of transmembrane domains. BCRP expression in the small intestine and colon is localized on the apical membrane. BCRP expression is highest in the duodenum and decreased somewhat toward the distal region of the human small intestine. BCRP expression also gradually decreases from the small intestine to the colon; the end of the colon expressing only about half of the expression in the duodenum [4, 117].

### 11.2.2 Regulation of Efflux Transporters by Nuclear Receptors (See Also Chapter 41)

Similar to drug-metabolizing enzymes, certain NRs are also reported to induce or downregulate the expression levels of the intestinal efflux transporters (Table 11.3). Some of these NRs, such as CAR, PXR, and FXR, can regulate the expression of both enzymes and efflux transporters. In some cases, with a particular ligand, both enzyme and efflux transporters are induced by a particular NR, while in others, one of them can get induced while the other can be downregulated [83, 87, 89, 94, 102, 131].

## 11.3 Drug Metabolism–Transporter Coupling in Drug Disposition in GIT

The coupling of the drug-metabolizing enzyme with the drug transporters (especially efflux) in drug disposition can be defined as a phenomenon where the process of

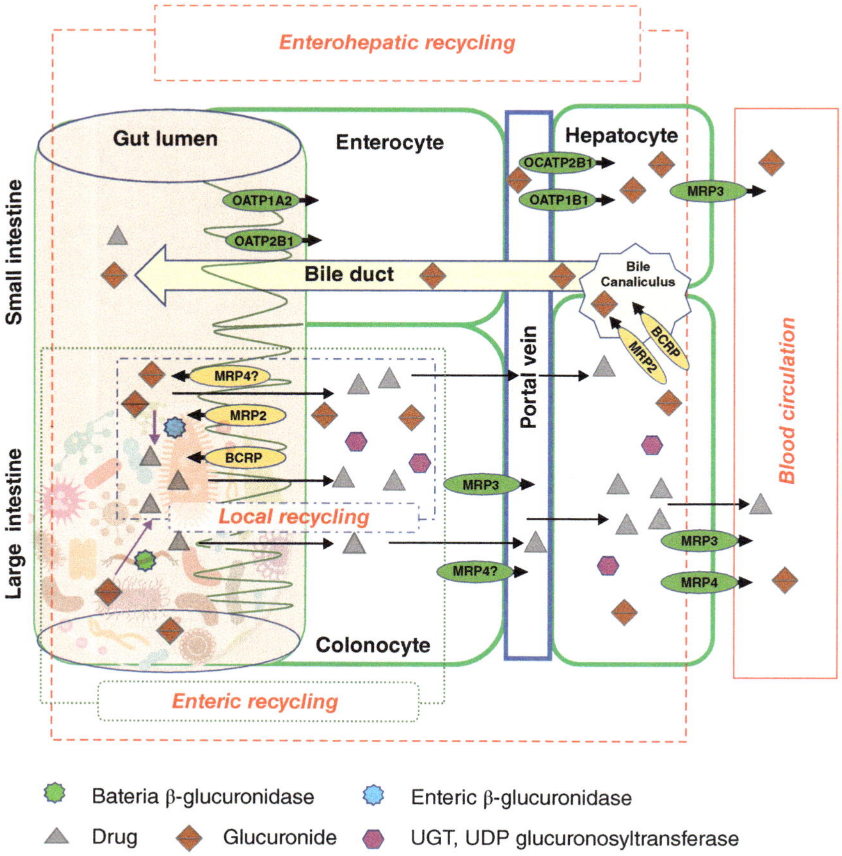


**Figure 11.3 Triple Recycling.** Parent drugs get absorbed and some percentage of them get metabolized by the enzymes in the intestinal cells. The rest of the parent drug is either effluxed back into the intestinal lumen or get transported to the blood. Metabolites can get excreted either on the apical or basolateral side. In the gut, the metabolites are either converted back into parent drugs, which then get reabsorbed into the intestinal cell again. The conversion of metabolite to parent drug can be either carried by the beta glucuronidases of the enterocytes (local recycling) or by the microbial beta glucuronidases, primarily in the ileum and colon (enteric recycling). From the mesenteric vein, drugs and metabolites are taken up by the liver cells. In the liver, the drug can get metabolized again. The parent drug and the metabolites are then excreted into bile, which gets emptied in the duodenum after a meal. In the gut, the drug is then reabsorbed again (entero-hepatic recycling).

drug metabolism and excretion/elimination of a drug and its metabolites can influence each other and change the dynamics substantially [62, 132]. The coupling of enzymes and transporters is found in the biological systems to maximize the absorption of nutrients, such as proteins and carbohydrates. It also often plays an important role in the cellular excretion of hydrophilic phase II conjugated metabolites and thus has a profound impact on the recycling of drugs, systemic and local bioavailability, drug metabolism and excretion, and drug–drug and drug–herb interactions (as shown in Figure 11.3).

### 11.3.1 Triple Recycling of Drugs

The efficient coupling of conjugate enzymes and efflux transporters in the liver and intestine enables the triple recycling, enterohepatic, enteric, and/or local recycling of the drug undergoing glucuronidation and sulfonation (as shown in Figure 11.3). These recycling processes prolong the exposure of a drug to systemic circulation because of the presence of mammalian and microbial hydrolases. Many compounds undergo only enterohepatic recycling but some compounds such as polyphenols undergo duo or triple recycling schemes involving enteric, local, and/or enterohepatic recycling [62, 77, 133–136]. Triple recycling plays a pivotal role in governing the disposition of orally administered phenolics such as flavonoids, resulting in their low systemic bioavailability but higher local bioavailability in the gut and longer-than-expected apparent half-life [137, 138].

#### 11.3.1.1 Enterohepatic Recycling

This is a classical recycling process where the action of the liver and intestine is necessary to complete the recycling process. Here, the parent drug and/or its metabolites when excreted from the basolateral side of the enterocytes into the mesenteric vein may be taken up by the liver cells from the blood through passive diffusion and/or various uptake transport mechanisms. In hepatocytes, drugs can get metabolized again. The parent drug and the phase II metabolites are then excreted into bile by the efflux transporters on the apical side. The bile, containing the drug and its metabolites, gets emptied into the duodenum after a meal. In the intestine, the parent drug excreted or released from phase II metabolites by the action of gut microflora is then reabsorbed again by the intestinal cells. The process constitutes "enterohepatic recycling" [6, 62, 133–136, 139]. The hepatic enzyme(s)–transporter(s) coupling has high capacity and can enable efficient enterohepatic recycling and display associated "double peak" phenomenon, as a result of bile emptying from the gall bladder after a meal.

Recently, a new recycling phenomenon of hepatoenteric recycling has been published, where both liver and intestines are involved in the recycling of drugs, similar to enterohepatic recycling. However, there are two major differences between these recycling mechanisms. First, the liver plays the role of the metabolic organ in the enterohepatic recycling, whereas it only plays the role of recycling organ for the intestinally metabolized drugs. Second, hepatic uptake transporter (OATPs) plays a very critical rate-limiting role in hepatoenteric recycling as metabolites produced by the intestine require an uptake transporter for metabolites to enter the liver for recycling [140]. Hepatoenteric recycling is one of the most complex drug disposition phenomena involving intestinal drug metabolic enzymes, liver apical and intestinal basolateral efflux transporters, and liver uptake transporters. A drug may undergo either enterohepatic or hepatoenteric recycling or both efficiently based on the relationship between drug and metabolite structure with both efflux and uptake transporters.

#### 11.3.1.2 Enteric Recycling

When a drug is orally administered, it is absorbed by the intestinal cells from the intestinal lumen (apical side), using one or more transport pathways (passive diffusion/active transport/facilitated transport). Once inside the intestinal cell, certain percentages of parent drugs are acted upon by various phase II enzymes and are metabolized into phase II conjugate(s). The rest of the drug can either be effluxed back into the intestinal lumen by efflux transporter(s) or transported from the basolateral side into the mesenteric blood system by passive diffusion or by efflux transporters on the basolateral side, many of which have an apical counterpart (e.g. MRP2 vs. MRP3) [59, 133, 134, 136, 139, 141, 142].

Metabolites formed can be excreted either on the apical or on the basolateral side by passive diffusion (phase I metabolites) or using efflux transporters (hydrophilic phase II metabolites). The hydrophilic phase II metabolites can be excreted by similar efflux transporters on the apical or basolateral side or completely different transporters on two sides. In the gut lumen, certain phase II metabolites such as glucuronides and sulfates are acted upon by the gut microflora, mostly in the distal intestine, and converted back into the parent drug. The drug is then reabsorbed again by the intestinal cells in the later segments of the intestine. The process constitutes "Enteric recycling" and is repeated again and again, till the drug is eventually eliminated from the body [59, 133, 134, 136, 139, 141, 142].

Contrary to enterohepatic recycling, for drugs that mainly undergo enteric recycling, excretion of conjugates by enterocytes, which shows intestinal enzyme(s)–transporter(s) coupling, is less likely to have a double peak phenomenon, mainly because metabolites are gradually and continuously excreted and emptied into the large intestine, where they are expected to be hydrolyzed and reabsorbed [6, 62, 133–136, 139].

#### 11.3.1.3 Local Recycling

In both enterohepatic and enteric recycling, glucuronides are converted to aglycone by the action of microbial β-glucuronidases in the large intestine. However, the β-glucuronidase enzyme is also located in the lysosome of various human cells [143]. These catalyze the hydrolysis of β-D-glucuronic acid residues from mucopolysaccharides such as heparan sulfate [144]. In the brush border membrane of the small intestine of neonates, mucosal cells β-glucuronidases convert conjugated bilirubin to its unconjugated form for reabsorption.

Xia et al. recently defined a third and equally important recycling phenomenon called local recycling of drugs [138], where the action of microbial β-glucuronidases is not essential for the conversion of glucuronides to the parent compound, so the recycling could complete in the small intestine alone using human β-glucuronidases. Flavonoids, wogonin [138], and tilianin [137] were reported to undergo a local recycling process due to hydrolysis by enterocyte-derived β-glucuronidase. Though both wogonin and tilianin undergo the triple recycling processes, local recycling contribution to overall recycling was found to be appreciable [137, 138].

Local recycling of flavonoid glucuronides may significantly prolong the residence time and local exposure of flavonoids in the intestine, thereby positively impacting their beneficial activities in the gut, irrespective of their poor systemic bioavailabilities. Local recycling of xenobiotics and their active or inactive metabolites can result in either enhanced gut protection or damage. Drugs that are extensively metabolized by intestinal UGTs, e.g. raloxifene [25, 26, 145], ezetimibe [146, 147], and resveratrol [148, 149], are expected to participate significantly in the local recycling process, resulting in prolonged systemic half-life, gut residence time, and exposure, and thereby enhanced gut-specific biological activities. So far, the colonic microbial β-glucuronidase is considered essential for the conversion of SN-38 glucuronide to the toxic SN-38 (the active metabolite of irinotecan) to enter enteric and enterohepatic recycling [150–152], which has been shown as the main cause of SN-38-induced colonic toxicity. However, that does not fully explain the damage in the upper small intestine. The novel phenomenon of local recycling on the other hand could be used to explain the upper intestinal toxicity of chemotherapeutic agents.

## 11.3.2 Theories of Drug Metabolism–Transporter Coupling

Two important and distinct theories have been proposed in the literature to explain the coupling of metabolic enzymes and efflux transporters for the disposition of xenobiotics. One deals with the drug being substrate of both phases I enzymes and the efflux transporters (double jeopardy) and the other deals with parent drug being a substrate of phase II enzymes and conjugated metabolites (and sometimes parent drug too) being a substrate of the efflux transporter (revolving door).

### 11.3.2.1 "Double Jeopardy" Theory

The "double jeopardy" theory suggests that when a drug is a substrate of efflux transporter and a phase I enzyme, it can result in multiple or repeated exposures of the substrate to the same enzyme(s). This, in turn, can result in a higher extent of metabolism of the drug as the chances or probability of the substrate getting metabolized increases with the increase in the number of times the drug gets exposed. Hence, the systemic bioavailability of drugs might decrease if the drug is a substrate of both enzymes and efflux transporters. Usually, this phenomenon is seen in drugs that undergo phase I metabolism (Figure 11.4) [132, 153, 154].

Due to the repeated efflux of the parent drug, there are lesser chances of cellular exposure and hence less potential for interacting with intracellular targets, including those for regulation. On the other hand, the potential for DDI increases for drugs that are substrates of the same enzyme and transporter systems. There exists an inverse relationship between the enzyme system and efflux transporters in the intestine and liver. In the intestine, during the absorption process, the drug encounters the efflux transporters such as P-gp first, before coming into contact with enzymes. Whereas, in the liver drugs come first in contact with the enzyme system in hepatocytes and then the efflux transporter on the bile canicular membrane [132, 153, 154].

For substrates of "double jeopardy," efflux only occurs at the apical membrane. This is different from the "revolving door theory," where efflux can happen at both apical and basolateral membranes. The latter is more complicated to sort out than the first one.

### 11.3.2.2 "Revolving Door" Theory

According to the "revolving door" theory, the hydrophilic conjugated metabolites cannot freely move out of the cells by passive diffusion. Rather, they depend on the action of the efflux transporters for their excretion. Hence, the disposition fate of the drug is decided by the fact one of these two processes may act as the rate-limiting step. In other words, either the formation of the metabolite by enzymes or the excretion of metabolites by efflux transporters can be the rate-limiting step. The consequences of the coupling according to this theory can be an imbalance between formation and excretion of hydrophilic conjugated metabolites (Figure 11.5) [59, 62, 133, 141, 142].

If the excretion of metabolites by efflux transporters is the rate-limiting step, then the rate of metabolite excretion will be lesser than the maximal rate of metabolite formation. It can increase the accumulation of metabolites inside the cell which might have one of two consequences or both. First, the reverse-enzymatic reaction can cause the conjugated metabolite to be converted back to the parent drug compound. This can decrease the net forward rate of metabolite formation. Or the metabolite can cause inhibition of the forward enzyme reaction (negative feedback mechanism via classical product inhibition) which in turn can decrease the net rate of metabolite formation [59, 62, 133, 141, 142].

On the other hand, if the formation rate is the rate-limiting step, then the efflux transporter can readily remove the formed metabolites which could have inhibitory effects via product inhibition. Therefore, the action of efflux transporters can favor the forward enzymatic reaction of conversion of parent drug into

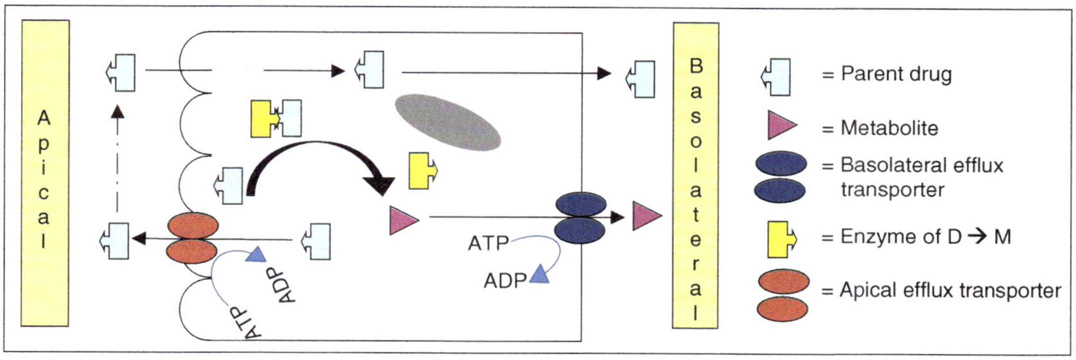


**Figure 11.4 "Double jeopardy" Theory.** The figure depicts the transport of drug molecules by passive diffusion from the apical to the basolateral membrane of the cell. The drug is a substrate of efflux transporter on the apical side and also a substrate of one or more enzymes that convert the parent drug into metabolite(s). The drug is absorbed from and effluxed to the luminal side multiple times, and this increases the number of times a single drug molecule encounters the enzyme(s) and increases the chances of its getting converted to metabolites.

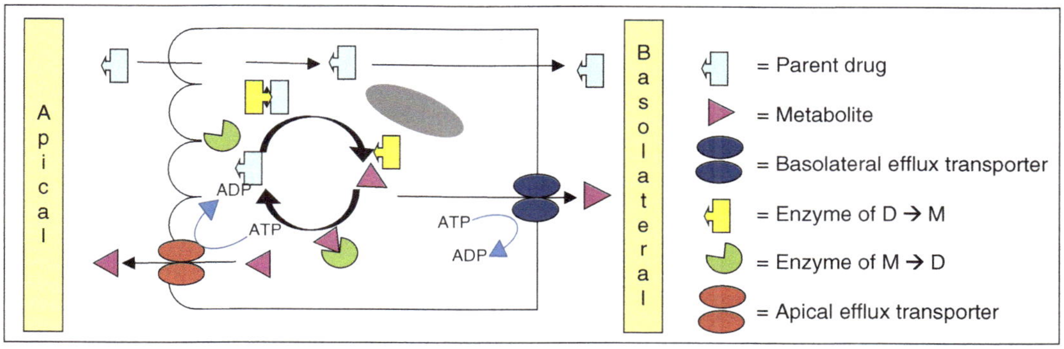


**Figure 11.5 "Revolving door" Theory.** The figure depicts the transport of drug molecules by passive diffusion from the apical to the basolateral membrane of the cell. One enzyme converts the parent drug into conjugated metabolite (forward reaction), and the other enzyme converts the conjugated metabolite(s) back into the parent drug (reverse reaction). The conjugated metabolite(s) can be excreted either on the apical membrane or the basolateral membrane using the same or different efflux transporters. The total rate and extent of excretion of metabolites on the apical and basolateral sides control the rate of formation of metabolites and hence the bioavailability of drugs.

a metabolite. This can result in higher-than-expected excretion rates of metabolites based on cellular enzyme activities measured using subcellular fractions and thereby impact the bioavailability of parent drug [59, 62, 133, 141, 142].

### 11.3.3 Features of Drug Metabolism–Transporter Coupling Process

The process of drug metabolism–transporter coupling can be simple, involving one enzyme isoform and one specific transporter. However, it is often more complex due to the involvement of multiple individual processes that can affect and influence each other toward the overall drug disposition and pharmacokinetic profile. For example, glutathione is used as a co-substrate by efflux transporters such as MRPs (Multidrug-resistance-associated proteins). It is also a cofactor of the enzyme GSTs. The over-activity of GST can diminish glutathione pool in cells, and hence in turn can reduce efflux of conjugated metabolites and therefore decrease the overall metabolism and bioavailability of parent drugs [62, 133, 142].

The various factors or individual processes that contribute to the overall coupling phenomenon in drug disposition are: (i) efflux of the parent drug at the apical or basolateral side or both; (ii) transport of metabolites by some of the efflux transporters from the cytosol to intestinal lumen or blood; (iii) involvement of multiple transporters in the efflux of parent drug and/or metabolite(s); (iv) involvement of multiple enzyme systems and isoforms in metabolizing the parent drug; (v) formation of multiple metabolites each using similar yet different sets of transporters; and (vi) effect of the overall contribution of one enzyme or transporter system to the drug disposition due to the changes in their activities or expression levels on the contribution of others enzyme and transporters due to compensation.

### 11.3.4 Challenges in Delineating Drug Metabolism–Transporter Coupling Process

The abovementioned individual processes can lead to the formation of a complex network. The advantage of this network is that it enhances the body's defense against ingestion of toxic chemicals, as the failure of any of the individual elements of the network, be it a transporter or an enzyme, is highly unlikely to cause the failure of the entire network. The importance of particular transport or enzymatic metabolic processes in the network is hard to know, so it is difficult to pinpoint and target the enzyme and transporter whose inhibition can collapse or disrupt the network and hence can promote the transport of higher amounts of the parent compound and thus improve the bioavailability of various pharmacological and/or chemopreventive agents [24, 132, 133, 153, 154].

The scientific community has been facing a lot of challenges in sorting out this complex process. The major reason has been the lack of specific inhibitors for enzymes or transporters involved in drug disposition. To overcome this, various approaches, such as gene-silencing and transfection, have been used in various cell lines to under or overexpress certain enzymes and/or transporters, respectively. But it is difficult to extrapolate the *in vitro* data obtained by using these approaches to *in vivo* conditions [24, 113, 132, 133, 153, 154].

Another major problem is that the compensation process is seen in animals that are deficient in one transporter or enzyme because of knockout or transgenic manipulation or natural breeding. There is a higher expression of enzymes and transporters of the same family or subfamily, but the mechanism of compensation is not known when gene(s) of interest is silenced or induced for a prolonged period [24, 113, 132, 133, 153, 154].

### 11.3.5 Drug–Drug Interactions due to Drug Metabolism–Transporter Coupling Process

Besides increasing drug metabolism, the coupling process also increases the potential of drug–drug and drug–herbal interactions. However, the coupling makes it difficult to sort out the mechanisms of drug interactions as it is much more difficult to sort out a complex network of barriers as discussed above. More than one agent and/or multiple mechanisms of modulation of action (inhibition or induction) might affect the output from the network, i.e. clinically significant drug interaction. Even if the mechanism of drug interaction is somehow established, it is difficult to develop drugs that will not modulate the enzyme and/or the efflux transporter involved in the interaction. The drugs that interact with enzymes and/or transporters can cause modulation of the cellular excretion process, resulting in more severe drug interactions [24, 113, 132].

## 11.4 Factors Affecting Intestinal Drug Metabolism

Interindividual variability in the rate of metabolism of a drug in the intestine can be as high as 50-fold. The various factors responsible for variability among the population are genetic polymorphism [38, 39, 56, 80, 98, 104, 155–157], age [158, 159], gender [117, 160], species [38], strain [60, 142, 161], diet and environmental factors [44, 124, 159, 162], induction and inhibition of drug-metabolizing enzymes [15, 40, 84, 102, 163], and induction and inhibition of drug influx and efflux transporters [63, 110, 112, 113, 154]. Also, physiological conditions such as pregnancy [164, 165], pathological conditions or disease states [159], route of administration [166–168], dosing strength, and the dosage form can impact DDIs. Induction and inhibition of the drug-metabolizing enzymes and efflux transporters by ligand-based NR activation are discussed above. Some of the major factors affecting drug metabolism, such as pharmacogenetics, diet, and environment, are discussed below.

### 11.4.1 Pharmacogenetics

Genetic polymorphism in the genes encoding the drug-metabolizing enzymes and the efflux transporters in the intestine responsible for the transport of conjugated metabolites can cause interindividual variation in human drug biotransformation. Additionally, drug-encoding transcriptional factors that regulate the expression of drug-metabolizing enzymes and efflux transporters in the intestine also can contribute to interindividual variability. The genetic polymorphism can either cause a decrease or an increase in the expression of drug-metabolizing enzymes and efflux transporters. It can also affect the inherent enzyme and transport activity and the response of enzymes and transporters to their respective inducers [24, 56, 68, 98, 103, 124, 157, 169, 170].

An increase in expression levels of drug-metabolizing enzymes can cause serious side effects and toxicity if the production of active metabolites with narrow therapeutic windows and carcinogens is increased. In the case of inactive metabolites, it may also have lower therapeutic effects due to inadequate plasma concentrations of the parent drug. The reverse can happen if the expression levels of drug-metabolizing enzymes are lowered. There can be incidences of toxic plasma concentration of drugs or lack of efficacy if the metabolite is the active entity as in the case of prodrugs [24, 56, 68, 98, 103, 124, 157, 169, 170].

In principle, a change, whether it be an increase or decrease, in the expression level of efflux transporters can cause changes in drug metabolism, for example, if efflux of metabolites from the intestinal cells is the rate-limiting step in the overall disposition of the drug, then a change in transporter expression will affect the drug's disposition. In practice, there is a lot of substrate overlap among the various efflux transporters, which can limit any significant impact of transporter expression level changes on the overall extent of drug metabolism [24, 56, 68, 98, 103, 124, 157, 169, 170].

The extent of the effect of the genetic polymorphism is dependent on the reaction involved, for example, phase I and phase II reactions; the chemical properties of the drug, such as polarity and isoform/enzyme selectivity; and the overall contribution of enzyme/isoform in the process of active biotransformation and/or detoxification. Genetic polymorphism can also impact the susceptibility of individuals to cancer and other diseases associated with higher exposure to drugs or their reactive metabolites [24, 56, 68, 98, 103, 124, 157, 169, 170].

Due to genetic polymorphism, based on the level of activity of a particular enzyme, the population can be clinically categorized into poor, intermediate, fast, and ultrafast metabolizers for a particular drug, which can have a decisive effect on which drug can be prescribed to an individual and at what dose.

#### 11.4.1.1 Phase I Enzymes

Among phase I enzymes, CYP has shown the most common and clinically significant genetic polymorphisms. CYP2D6 metabolizes various important classes of prescription drugs, such as antidepressants, antipsychotics, β-adrenergic blockers, and antiarrhythmics. About 17 allelic variants of CYP2D6 have been identified, characterized, and reported in the literature. Approximately 15% of whites in the US and Europe have CYP2D6 alleles with low metabolic activity. Incidences of polymorphism are also reported for CYP1A1, CYP1A2, CYP2A6, CYP2C9, CYP2C19, CYP2E1, and CYP3A4 [39, 98, 157].

#### 11.4.1.2 Phase II Enzymes

Among phase II, polymorphism most commonly occurs in N-acetyltransferase 2 (NAT2) and glutathione S-transferase M1 (GSTMl) which have little contribution to the metabolism of xenobiotics. Major phase-II drug-metabolizing enzymes are UGTs and SULTs, which are still under investigation for the presence of polymorphism. Due to large substrate overlap among the isoforms of UGTs in metabolizing xenobiotics, the genetic polymorphism might not translate into any significant clinical phenotype for most cases [46, 56, 80, 104, 106, 108, 157, 169, 171].

The most common and significant polymorphism reported in about 7% of population for phase II enzyme is Gilbert's syndrome. This syndrome is caused by the deficiency of the UGT isoform UGT1A1, which is responsible for the metabolism of bilirubin, an endogenous substrate. Apart from UGT 1A1, polymorphism has also been reported in other UGT1As (1A3, 1A4, 1A6, 1A7, 1A8, 1A9, and 1A10) and 2Bs (2B4, 2B7, and 2B15). So far, 33 polymorphic alleles of the human UGT1A1 gene have been reported in association with unconjugated hyperbilirubinemia. Out of which, 4 alleles (UGT1A1*6, UGT1A1*27, UGT1A1*28, UGT1A1*29) are shown to cause Gilbert's syndrome, 19 polymorphic alleles cause Crigler–Najjar syndrome type I, and 10 alleles cause Crigler–Najjar syndrome type II [46, 56, 104, 106, 108].

In SULTS, 1A1 and 1A2 have shown genetic polymorphism. A SULT1A1 allele has been reported with a frequency of 31–37% in Caucasians and is associated with low enzyme activity and a reduced level of SULT1A1 protein. The rare SULT1A1*3 allelic variant shows increased affinity for the substrates *p*-nitrophenol and PAPS. Six different allozymes of SULTT1A2 have been detected, some of these variants have reduced affinity for the substrate *p*-nitrophenol and the cofactor PAPS, whereas others only have minor effects [78, 80, 169, 171].

### 11.4.2 Diet and Environment

Various dietary and environmental components can affect the drug metabolism in humans by inducing the drug-metabolizing phase I and phase II enzymes. They can be monofunctional inducers (inducing either phase I or phase II enzymes) or bifunctional inducers (inducing both phase I and phase II enzymes) [172]. The inducers activate the transcriptional factor complexes of the phase II enzyme genes [173]. Induction of enzymes can either decrease the bioavailability of parent drugs or increase the bioavailability of active metabolites.

Various dietary components that have shown to be the potent inducers of phase II enzymes are isoflavonoids (such as genistein) [174]; flavonols (such as kaempferol found in Kale) [175, 176]; polyphenols (such as ellagic acid found in berries) [177, 178]; epicatechin (such as ECGC found in green tea) [179, 180]; and curcumin, a major component of turmeric [181]. Two widely used food additives butylated hydroxyanisole (BHA) and butylated hydroxytoluene (BHT) have also been shown to increase tissue phase II enzyme activities [182–185].

Upregulation of phase I enzymes generally is not desirable as they can promote the conversion of drugs into reactive metabolites which can aggravate various pathological conditions including aging, cardiac problems, and cancers. On the other hand, the upregulation of phase II enzymes can aid in the detoxification of reactive and electrophilic metabolites of phase I reactions and other dietary and environmental carcinogens, thereby reducing the incidence of cancers [172].

PhIP (2-amino-1-methyl-6-phenylimidazo[4,5-*b*]pyridine) is the most abundant heterocyclic amine found in fried and grilled meat and is formed partly to heat-dependent condensation of creatinine and phenylalanine during the high-temperature cooking process. PhIP is shown to be a potential carcinogen and can cause mutagenesis in the large and small intestines of mice and dose-dependent tumor formation in rats and mice; most commonly in the colon, prostate, and mammary gland. The metabolic activation and detoxification of PhIP play an important role in its carcinogenic and mutagenic nature. PhIP along with other heterocyclic aromatic amines requires bioconversion into active metabolites to become carcinogenic [186, 187].

One of the most important carcinogenic metabolites of PhIP is *N*-hydroxy PhIP, which can cause the breaking of DNA strands and the exchange of sister chromatids, thus forming DNA adducts. The biological potency of PhIP-induced carcinogenicity is strongly dependent on its metabolism into *N*-hydroxy PhIP. The conversion of PhIP into *N*-hydroxy PhIP can be reduced by the induction of detoxifying phase II enzymes [44].

Environmental insecticides (such as Organophosphorus) [188] and naturally occurring pyrethrin [189] are also shown to affect the drug metabolism in rodents. Polluted breathing air contains polycyclic aromatic hydrocarbons (PAHs) from *cigarette* smoke, wood smoke, vehicle exhausts, asphalt roads, or agricultural burn smoke. PAHs are inducers of CYP1A2, which metabolizes PAHs to various reactive carcinogenic metabolites and probably can cause lung and colon cancers [190].

Various adverse incidences are also reported by the enzyme induction or inhibition by drug/herbal components. Herbal supplement of St. John Wart induces CYP3A4, which increases the metabolism of cyclosporin (immunosuppressant) into inactive metabolites, causing rejection of liver transplant in patients [191–193]. Literature data suggest that grapefruit juice inhibits intestinal CYP3A and other enzymes, which causes a percentage increase in AUC ranging from 60% (cyclosporine), 230% (atorvastatin), to 1250% to 1400% (lovastatin/simvastatin) [194].

## 11.5 Biopharmaceutics Drug Disposition Classification System

It is widely recognized that solubility and permeability are the fundamental properties determining the bioavailability of an orally administered drug. In 1995, Amidon and coworkers proposed a biopharmaceutic classification system (BCS) used in the pharmaceutical industry and drug regulatory [195]. The BCS takes into account three major factors of oral administrated drug absorption: solubility, dissolution, and permeability. The BCS Class I (high solubility, high permeability, and rapid dissolution) drugs are authorized to use the *in vitro* solubility and permeability to waive the bioequivalence studies by FDA. [196].

However, BCS failed to consider the impact of intestinal enzymes and transporters on the bioavailability of orally administered drugs. Benet and Wu proposed a BDDCS, suggesting that the drug metabolism may be used for characterizing high intestinal permeability drugs (Table 11.4) [197, 198]. BDDCS could be used to predict drug disposition and DDIs in the intestine and the liver, taking into account the effect of the enzyme and transporter [199]. BDDCS allows scientists to consider formulation and drug-designing strategies for achieving the desired pharmacokinetic profile and efficacy/toxicity outcome for new drug entities under development.

BDDCS mainly uses intestinal permeability and the fraction of dose metabolized (more or less than 70%) for predicting the importance of hepatic and intestinal transporters in determining drug disposition and predicting DDIs [199]. Recent studies in our lab have shown that compounds metabolized extensively in the intestine may be more disposed toward different recycling processes, thereby affecting their systemic

**Table 11.4** Transporters effects predicted by BDDCS following oral dosing.

| | High solubility | Low solubility |
|---|---|---|
| High permeability/metabolism | **Class 1**<br>Transporter effects minimal in the gut and liver. For example, verapamil, diazepam, acetaminophen, midazolam, lidocaine, and quinidine. | **Class 2**<br>Efflux transporter effects predominate in the gut, but both uptake and efflux transporters can affect the liver. For example, atorvastatin, carbamazepine, ketoconazole, raloxifene, and warfarin. |
| Low permeability/metabolism | **Class 3**<br>Absorptive transporter effects predominate (but can be modulated by efflux transporters). For example, erythromycin, metformin, amoxicillin, and cetirizine. | **Class 4**<br>Absorptive and efflux transporter effects could be important. For example, neomycin and amphotericin B. |

exposure, half-life, and intestinal efficacy and toxicity (unpublished data). However, BDDCS uses overall drug metabolism (liver and intestine together); hence, the role of only intestinal drug metabolism contribution to the drug disposition classification system is not well understood yet. In conclusion, further research is needed to measure or predict the contribution of hepatic vs intestinal metabolic enzymes and uptake/efflux transporters to the overall disposition for further extending the BDDCS application for the drugs targeted toward treating intestinal diseases/disorders and for predicting the drug-induced intestinal toxicity potential of novel drugs.

## 11.6 Metabolism-Based Drug–Drug and Drug–Natural Product Interactions

### 11.6.1 Drug–Drug Interaction

The evaluation of DDIs is essential to access drug safety and effectiveness. In the past, DDIs were among the most frequent reasons for drug development failure in the market, change of labels, or withdrawal from markets. FDA issued the guidance for *in vitro* drug metabolism studies and the guidance for *in vivo* drug interaction studies in 1997 and 1999, respectively, which played a crucial role in reducing the failure of metabolism-based DDIs in drug development. Nowadays, drug metabolism and the potential for DDIs are addressed in the early phase of drug development in the pharmaceutical industry.

Based on FDA guidance, more than 20% change in the systemic exposure of drug or active metabolite due to competitive inhibition or induction of drug-metabolizing enzymes or drug efflux/uptake transporters involved in the disposition of phase II metabolites could result in clinically relevant DDIs and therefore warrant DDI clinical trials. Since the detoxification of drugs by phase II enzymes constitutes the major contribution of intestinal metabolism, we will focus on the DDI mediated by phase II enzymes (especially UGTs) and transporters involved in glucuronide transport in this section.

Several TKIs (erlotinib, gefitinib, and sorafenib) have been shown to inhibit UGTs [200–202]. The DDI potential of TKIs' inhibitory effects on the UGT activities has been reported. [200, 202]. Erlotinib showed the potency to competitively inhibit glucuronidation by UGT1A1. Gefitinib has been found to inhibit different UGT enzymes, such as UGT1A1, UGT1A7, and UGT2B7. Lapatinib exhibited competitive inhibition against UGT1A1, and imatinib demonstrated broad inhibition on several UGTs, especially UGT2B17. Lapatinib and imatinib may cause clinically significant DDIs when coadministered with UGT1A1 or 2B17 substrates [202].

The other phase II metabolism enzymes, SULTs, also play an essential role in drug metabolism and detoxification of many compounds such as estrogens and flavonoids [203]. A recent study found that tumor suppressor p53 could regulate the expression of SULTs [204]. Many drugs have been shown to inhibit SULT activities, such as mefenamic acid, salicylic acid, clomiphene, and danazol [205], which can impact the bioavailability and clearance of drugs cleared by sulfate conjugation pathways. For example, morphine is excreted in the urine of neonates and children majorly as morphine-3-sulfate, and SULT inhibition can significantly impact the clearance of morphine [206]. Similarly, many glucocorticoids, such as dexamethasone, can induce SULT activity, resulting in faster clearance of drugs from the body [207].

### 11.6.2 Drug–Natural Product Interactions

Over the last decade, there has been a significant worldwide rise in the use of herbal products including dietary supplements. In the United States alone, the sale of dietary supplements has almost doubled from (USD 4.8 to USD 8.8 billion) from 2008–2018 [208]. Although there is a widespread public perception that herbs, herbal mixtures, or their pure active components used as dietary supplements are relatively safe for human consumption, research has demonstrated that these products may impact the efficacy and safety of therapeutic drugs, due to undesirable drug–natural products interactions via pharmacokinetic and pharmacodynamic interactions. Pharmacokinetic-based drug–natural product interactions can either be beneficial or harmful due to changes in drug exposure and disposition.

The possible pharmacokinetic mechanism may involve the inhibition or induction of drug-metabolizing enzymes, such as cytochrome P-450 or UGT isoenzymes, and impairment of hepatic or renal function via modulation of efflux and uptake transporters. Several natural products have been reported to interact with drugs in humans for either beneficial or harmful effects, such as coenzyme Q10 [209, 210], Dong Quai [211], ephedra [212], Ginko Biloba [213, 214], ginseng [215, 216], glucosamine sulfate [35, 217], ipriflavone [218], melatonin [219, 220], St. John's wort [221, 222], XCHT [223, 224], grapefruit juice [194, 225], and green tea polyphenols [226]. However, most of the interactions were case reports and systemic studies are needed to investigate interaction mechanisms for many reported interactions.

Many herb–drug interaction articles in the literature are often not based on human studies. Most reports are inferred from *in vivo* animal studies, *in vitro* cellular assays, or hypothesized from other sources. However, it is essential to pay attention to the interaction with narrow therapeutic index drugs, such as warfarin and digoxin. For example, St. John's wort – a potent CYP3A4 inducer – has been reported to cause significant decreases in the drug levels of midazolam, amitriptyline, cyclosporine, digoxin, indinavir, irinotecan, warfarin, alprazolam, simvastatin, and ethinylestradiol [191, 192, 221, 222, 227–229].

On the other hand, an increase in felodipine bioavailability after oral administration with grapefruit juice has been associated with CYP3A4 inhibition [194]. This could lead to lower-than-desired blood pressure or a faster-than-desired heartbeat. Similarly, *Schisandra sphenanthera* extract (commonly known as magnolia vine) used as a hepatoprotective, anti-inflammatory and detoxifying agent, has been shown to reduce the clearance of drugs like midazolam and tacrolimus by 50% (AUC ratio $\geq 2.0$) in a clinical interaction study. Schisantherin A and schisandrol B have been reported to inhibit CYP3A4 and CYP2C8, thereby increasing bosutinib exposure (AUC ratio 3.0) [230]. Paine's lab has done significant work in understanding and modeling CYP-based drug–herb interactions [231, 232]. Valerian and Ginkgo biloba were reported to increase the activity of CYP3A4 and 2D6, and CYP1A2 and 2D6, respectively [233]. However, it needs to be noted that change in intestinal CYP activities due to natural products is more important for drugs, with a higher contribution to overall FPM by the intestinal CYP metabolism, such as buspirone (78%) and nisoldipine (56%) [43].

Natural products especially herbal mixtures can both induce and inhibit UGT activities. The changes in the UGT activities can be tissue-specific, such that intestinal but not hepatic UGTs are inhibited or upregulated, resulting in changes in efficacy and/or toxicity of drugs extensively metabolized by intestinal UGTs. *Glycyrrhiza glabra* roots extract (1 g/kg, p.o. for 6 days) and its active constituent glycyrrhizin (23 mg/kg, p.o.) were shown to increase the UGT1A activities by about 2-fold in Wistar rats [234]. Cao et al. reported that icariin, a bioactive constituent isolated from Chinese herbal mixture Ying Yang Huo, and its intestinal metabolites were potential inhibitors against several UGT isoforms. Icariin was found to exhibit potent inhibition against UGT1A3, whereas icariside II inhibited UGT1A4, UGT1A7, UGT1A9, and UGT2B7, and icaritin inhibited UGT1A7 and UGT1A9. The quantitative prediction of risks indicated that *in vivo* inhibition against intestinal UGT1A3, UGT1A4, and UGT1A7 would likely occur after oral administration of icariin products (*Epimedium pubescens* decoction) [235]. Similarly, green tea constituents, catechins, (-)-epicatechin gallate, and (-)-epigallocatechin gallate, are also shown to inhibit several UGTs that can impact the metabolism of raloxifene, an FDA-approved drug for postmenopausal osteoporosis and the reduction of invasive breast cancers [236].

Similarly, several natural products have been reported to modulate SULT activities. Green, black, and oolong teas were reported to inhibit SULT1A3, potentially increasing the bioavailability of orally administered substrates of SULT1A3, such as salbutamol [237]. The wine anthocyanins and flavonols (e.g. quercetin), apple and grape juice, curcumin in turmeric, several flavonoids, and certain phytoestrogens (daidzein, genistein) are also shown to inhibit SULT activities, whereas caffeic acid and retinoic acid are reported as SULT inducers [238]. The change in SULT activities can have a major impact on the pharmacokinetics of drugs with significant SULT metabolism such as resveratrol metabolism by intestinal SULT1A1 [239].

As discussed in Section 11.4, the intestinal and hepatic apical efflux transporters (MRP2 and BCRP) and hepatic uptake (OATP, OAT) transporters play an important role in the disposition of glucuronides and conjugate metabolism via the recycling pathway. Therefore, any changes (inhibition or induction) in the activities of these transporters due to drugs and natural products can impact the overall drug metabolism and disposition. Details of transporter-mediated DDIs are discussed somewhere else in this book.

## 11.7 Metabolic Interactions Between Gut Microbiome and Drugs in GIT

In recent years, there has been increasing interest in understanding the metabolism of drugs by the community of microbes residing in the human gut, commonly known as the gut microbiome or GM, due to its impact on the clinical efficacy and safety of drugs [240–243]. As discussed in Sections 11.5 and 11.6, the GM plays a significant role in enteric, enterohepatic, and hepatoenteric (a newly described phenomenon) recycling through the hydrolysis of biliary-excreted phase II metabolites, affecting the local and systemic bioavailabilities and half-lives of various drugs and their metabolites [137, 138, 140, 244]. This interaction of the GM with the host metabolism of drugs is shown to influence the safety and efficacy of drugs both in healthy and diseased microbiota in humans [242]. Also, healthy human GM is known to break down dietary components and drugs to generate beneficial microbial metabolites that are responsible for maintaining the homeostasis of different organs and gut health. The disbalance of microbiota structure or dysbiosis can lead to various GI-related diseases and disorders and has been implicated in several metabolic and neurological diseases [245–247].

The GM can act on drugs either through direct or indirect mechanisms. Direct mechanisms could be the activation of prodrugs or the deactivation of drugs via microbial detoxification pathways. Some examples of direct biotransformation of drugs by the GM are the reduction of digoxin by *E. lentum*, reduction of the prodrug sulfasalazine to the active 5-aminosalicylate, hydroxylation of lovastatin and flurbiprofen, omeprazole, etc. [248]. This is a fast-developing area, but there has been limited knowledge of direct biotransformation of drugs by the GM with clinical implications.

GM-mediated biotransformation of a drug is generally different from that performed by the host liver. The GM mainly produces nonpolar low-molecular-weight metabolites through hydrolytic and reductive metabolism, as opposed to the liver, which primarily generates polar and high-molecular-weight metabolites via oxidative and conjugative metabolism [248]. Interestingly, however, the microbial CYPs are soluble enzymes in contrast to membrane-bound mammalian CYPs and can be similarly induced by CYP inducers, such as phenobarbital [249]. Moreover, similar microbial enzymes can be secreted by different bacterial species. For example, ginsenoside Rb1 from the herb ginseng is metabolized into bioactive 20-O-*b*-D-glucopyranosyl-20(*S*)-protopanaxadiol (compound K) by *Ruminococcus* spp., *Bacteroides* spp., and *Bifidobacterium* spp. [250].

As discussed above, microbe-generated drug metabolites can either be like or distinct from the host-generated drug metabolites depending upon the mode of biotransformation [251]. The microbial metabolites can be either derived from the direct transformation of drugs or dietary components by the GM (e.g. compound K from ginsenoside) [252]; from host metabolites modified by the GM (e.g. biliary-secreted phase II metabolites of raloxifene, morphine, and ezetimibe) [253–255]; and de novo synthesis by the GM (e.g. short-chain fatty acids and polysaccharide A) [256].

Since various disease states and drug (e.g. antibiotics) treatments can change the overall microbiota structure and metabolic potential of GM, the microbial drug metabolism could significantly affect the absorption, bioavailability, disposition, and efficacy (systemic and local) as well as the drug-induced intestinal and/or systemic toxicity of drugs. For example, the GM-mediated hydrolysis of glycyrrhizin into 18β-glycyrrhetic acid and ginsenoside into compound K, which are more bioavailable compared to parent compounds [252, 257]. Similarly, SN-38 glucuronide hydrolysis to SN-38 may result in increased intestinal toxicity in patients undergoing irinotecan chemotherapy [152]. GM can change the disposition of drugs directly by competing with host drug-metabolism or indirectly by changes in GM structure that can result in altered expression levels of multiple drug-metabolizing enzymes and transporters in the liver and kidney, as shown in studies using germ-free mice and antibiotics-treated mice [248]. For example, changes in BCRP protein levels changed the pharmacokinetics of anticancer drugs, such as mitoxantrone and topotecan [122].

There exists a clear relationship between GM and host-metabolism as the GM-mediated release of certain metabolites can act as signaling molecules for mucosal enteroendocrine cells, resulting in the release of hormones including CCK, PYY, GLP-1, GIP, and 5-HT which can regulate the host-metabolism. Therefore, drug or dietary interventions that alter GM can also be used to treat metabolic disorders [258].

## 11.8 Metabolism-Based Xenobiotic-Induced Toxicity

The gut epithelial layer's integrity and function are essential for the absorption, intestinal metabolism and disposition of oral drugs. The gut epithelial layer's integrity and function are adversely impacted by pathological conditions due to gut-related diseases or disorders such as inflammatory bowel diseases, irritable bowel syndrome, intestinal cancer, and *C. difficile* infection. Additionally, drug-induced GIT toxicity plays an important role in negatively impacting the gut metabolic function, both host and microbial metabolism. Due to the significant contribution of intestinal metabolism of several orally administered drugs [12, 259] to the FPM and enterohepatic recycling, any changes in intestinal drug-metabolizing enzyme and transporter function or the GM can significantly alter the intestinal exposure of drugs, which in many instances leads to the drug-induced intestinal toxicity. The changes in intestinal drug-metabolizing enzyme and transporter function can be either due to changes in protein expression or DDI.

The intestinal exposure to the toxic substances, such as SN-38 (the active metabolite of pro-drug irinotecan) and *N*-hydroxy-PhIP (DNA adduct forming, mutagenic metabolite of dietary carcinogen PhIP), that causes the intestinal toxicity also involves the action of microbial beta-glucuronidases on the O-glucuronide of SN-38 and $N^2/N^3$-glucuronides of PhIP secreted in bile [244, 248]. In the case of irinotecan toxicity, the reduction in UGT1A1 activities due to genetic polymorphisms [260] or SN-38-induced downregulation of intestinal UGT protein expression and activities [224], can result in changes in systemic and intestinal exposure of SN-38, respectively. A case-controlled study of Japanese cancer patients revealed that those with variant UGT1A1 alleles were at significantly higher risk of severe adverse reactions to irinotecan [106, 108, 261].

Published research done in our lab has shown that certain drug treatments, such as irinotecan treatment, can downregulate the intestinal UGT expression, leading to increased colonic exposure of SN-38, disturbance of homeostasis of endobiotic and normal gut physiology, as well as the detoxification of dietary carcinogens and therapeutic drugs by glucuronidation in the intestine, leading to drug-induced severe gut toxicity [150–152]. The loss of UGT-mediated detoxification capability in the inflamed gut, either due to drug toxicity or pathological conditions, has been correlated with dysregulation of several NRs responsible for UGT regulation, due to imbalance of endobiotics acting as transcription factors for these NRs. This loss of UGT, which precedes severe diarrhea in the rodent model, was successfully attenuated by traditional Chinese medicine, Xiao Chai Hu Tang that can upregulate intestinal UGT activities [224].

Another example of intestinal drug metabolism-induced toxicity is morphine-induced constipation, which is determined by the colonic exposure to morphine and its 3-O-glucuronide (M3G) and 6-O-glucuronide (M6G). Morphine treatment-induced constipation in about 40–80% of patients by acting on the μ-opioid receptor in the colon and M6G was reported to be 100-fold more potent agonist, whereas M3G has 300-fold lower affinity toward μ receptors than morphine [262–264]. Since M6G is primarily formed by UGT2B7, whereas M3G can be formed by most UGT1A isoforms, the intestinal ratio and levels of morphine, M6G, and M3G are dependent on UGT1A and 2B activities in enterocytes. The increase in UGT1A activities by natural products (such as Xiao Chai Hu Tang) could lead to an increase in morphine intestinal detoxification and reduced constipation, whereas an increase in UGT2B gut activities can exacerbate the side effect.

## 11.9 GIT Metabolism-Based Drug-Designing and Lead Optimization in Drug Development

As detailed in this chapter, the intestine, due to the metabolic capability of enterocytes and gut microbes and the triple recycling of drugs and metabolites, and the long length of the GIT can extensively metabolize several orally administered drugs. Due to this capability, intestinal drug metabolism has the potential to be successfully used as a metabolism-based drug-designing and lead optimization tool in the drug development process. Especially for drugs targeted toward intestinal disease and disorders, higher drug exposure in the GIT than systemic circulation is highly desired. To achieve that, lead molecules can be structurally designed to either have low intestinal permeability or to enter one or more recycling pathways for prolonged residence time and higher local exposure in the intestine. Since the enteric and local recycling require the drug to be in the conjugated form (phase II metabolites), compounds need to be good candidates for UGTs and/or SULTs. This phenomenon also requires the hydrolysis of conjugates back to the parent molecule for reabsorption, usually dependent on the activities of β-glucuronidases in enterocytes or microbial β-glucuronidases and sulfatases. In general, UGT-mediated metabolism shows better recycling efficiency via the triple recycling mechanisms and therefore could act as one of the main lead-optimizing parameters.

On the other hand, in cases of drug-induced intestinal injury, such as in irinotecan chemotherapy, where the hydrolysis of glucuronides (SN-38 glucuronide) back to the parent form (SN-38, the active phase I metabolite of irinotecan) by the gut microbes can lead to severe gastrointestinal toxicity. The irinotecan-induced intestinal injury is an interplay of various NR regulation, inflammation signaling, metabolism, and disposition processes, such as downregulation of UGT-mediated detoxification of SN-38 by the enterocytes, biliary excretion of SN-38, and SN-38 glucuronides by the hepatic efflux transporters, and the hydrolysis of SN-38 glucuronide to SN-38 by the gut microbes [150–152, 170]. To prevent or reduce irinotecan-induced intestinal injury, theoretically, future irinotecan analogs with reduced UGT-mediated metabolic clearance and enterohepatic recycling need to be designed. However, it needs to be noted that oral administration of topotecan, an analog of irinotecan, which is not a UGT substrate, still causes significant diarrhea in cancer patients [265], suggesting that drug-designing of analogs to achieve lower UGT metabolism alone may not lead to the desired outcome in this case.

Alternatively, new drug treatments can inhibit the hepatic apical efflux transporters to reduce the biliary excretion of SN-38 and its glucuronide, induce the intestinal UGTs for increased detoxification of SN-38 in enterocytes [152], or inhibit the microbial β-glucuronidases [266] to reduce the overall SN-38 exposure to enterocytes needs to be designed. In summary, metabolism-based drug-designing is an understudied and underutilized tool for lead optimization in drug development, and more focused research is needed to explore its utilization in improving local efficacy and reducing intestinal toxicity mediated by the intestinal drug metabolism.

## 11.10 Summary

The present chapter has given an overview of various components of drug metabolism in the human GIT. Drug metabolism in the GIT contributes significantly to the FPM of many drugs and hence plays an important role in deciding the oral bioavailability and disposition of drugs, especially those with a susceptible phenolic group. Overall intestinal disposition of drugs is usually the interplay of metabolism by phase I and phase II enzymes, transport of parent drug and/or conjugated metabolites by efflux and uptake transporters, and/or metabolism by the gut microbes. The coupling of multiple enzymes and transporters presents a complex network of bioavailability barriers, which are difficult to overcome for certain compounds. Interindividual variability in the expressions and activities of drug-metabolizing enzymes and transporters and the GM composition due to the genetic variation, polymorphism, and dysbiosis is an important determinant of the drug-induced toxicity, drug efficacy, and clinical drug–drug/drug–herb interactions in humans.

Since the liver and intestine share various drug-metabolizing enzymes, it is very difficult to elucidate the actual contribution of hepatic versus intestinal metabolism in the overall FPM of xenobiotics. Moreover, the clinical impact of modulation of intestinal-metabolizing enzymes and transporters on the drug disposition is still not well understood. More effective tools need to be developed to further understand the interactions between various components of the coupling barrier network, which can help to overcome the various DDI possibilities and oral bioavailability challenges. Finally, the mechanisms of regulation of intestinal phase II-metabolizing enzymes and efflux/uptake transporters are still under investigation, though we have made some progress in understanding the various mechanisms of drug-induced gastrointestinal toxicities in the last decade. Improved preclinical models with good clinical translation potential are needed to further exploit our understanding of intestinal drug metabolism by host and microbes to prevent and/or treat the gastrointestinal disorders/diseases and drug-induced toxicities, as well as better modulate the local (gut) and systemic exposure of drugs by metabolism-directed drug-designing.

## Abbreviations

| | |
|---|---|
| BCRP | breast cancer resistance protein |
| BDDCS | biopharmaceutics Drug Disposition Classification System |
| CYP | cytochrome P450 |
| DDI | drug–drug interaction |
| DNPI | drug natural product interaction |
| FPM | first-pass metabolism |
| GIT | gastrointestinal tract |
| GM | gut microbiome |
| MRP2 | multidrug resistance protein 2 |
| P-gp | P-glycoprotein |
| PhIP | 2-amino-1-methyl-6-phenylimidazol ([4,5-β]-)pyridine |
| mRNA | messenger ribonucleic acid |
| NR | nuclear receptor |
| SULT | sulfotransferases |
| UGT | uridine diphosphate glucuronosyl transferases |

## References


1 Ding, X. and Kaminsky, L.S. (2003). Human extrahepatic cytochromes P450: function in xenobiotic metabolism and tissue-selective chemical toxicity in the respiratory and gastrointestinal tracts. *Annu. Rev. Pharmacol. Toxicol.* 43: 149–173. https://doi.org/10.1146/annurev.pharmtox.43.100901.140251.

2 Fritz, A., Busch, D., Lapczuk, J. et al. (2019). Expression of clinically relevant drug-metabolizing enzymes along the human intestine and their correlation to drug transporters and nuclear receptors: An intra-subject analysis. *Basic Clin. Pharmacol. Toxicol.* 124: 245–255. https://doi.org/10.1111/bcpt.13137.

3 Peters, S.A., Jones, C.R., Ungell, A.L., and Hatley, O.J. (2016). Predicting drug extraction in the human gut wall: assessing contributions from drug metabolizing enzymes and transporter proteins using preclinical models. *Clin. Pharmacokinet.* 55: 673–696. https://doi.org/10.1007/s40262-015-0351-6.

4 Takano, M., Yumoto, R., and Murakami, T. (2006). Expression and function of efflux drug transporters in the intestine. *Pharmacol. Ther.* 109: 137–161. https://doi.org/10.1016/j.pharmthera.2005.06.005.

5 Vaessen, S.F., van Lipzig, M.M., Pieters, R.H. et al. (2017). Regional expression levels of drug transporters and metabolizing enzymes along the pig and human intestinal tract and comparison with Caco-2 cells. *Drug Metab. Dispos.* 45: 353–360. https://doi.org/10.1124/dmd.116.072231.

6 Kararli, T.T. (1995). Comparison of the gastrointestinal anatomy, physiology, and biochemistry of humans and commonly used laboratory animals. *Biopharm. Drug Dispos.* 16: 351–380. https://doi.org/10.1002/bdd.2510160502.

7 Krishna, D.R. and Klotz, U. (1994). Extrahepatic metabolism of drugs in humans. *Clin. Pharmacokinet.* 26: 144–160. https://doi.org/10.2165/00003088-199426020-00007.

8 Lin, J.H., Chiba, M., and Baillie, T.A. (1999). Is the role of the small intestine in first-pass metabolism overemphasized? *Pharmacol. Rev.* 51: 135–158.

9 Gao, Y., Yang, C., Wang, L. et al. (2020). Comparable intestinal and hepatic first-pass effect of YL-IPA08 on the bioavailability and effective brain exposure, a rapid anti-PTSD and anti-depression compound. *Front. Pharmacol.* 11: 588127. https://doi.org/10.3389/fphar.2020.588127.

10 Gertz, M., Harrison, A., Houston, J.B., and Galetin, A. (2010). Prediction of human intestinal first-pass metabolism of 25 CYP3A substrates from in vitro clearance and permeability data. *Drug Metab. Dispos.* 38: 1147–1158. https://doi.org/10.1124/dmd.110.032649.

11 Liu, Y.T., Hao, H.P., Xie, H.G. et al. (2010b). Extensive intestinal first-pass elimination and predominant hepatic distribution of berberine explain its low plasma levels in rats. *Drug Metab. Dispos.* 38: 1779–1784. https://doi.org/10.1124/dmd.110.033936.

12 Paine, M.F., Shen, D.D., Kunze, K.L. et al. (1996). First-pass metabolism of midazolam by the human intestine. *Clin. Pharmacol. Ther.* 60: 14–24. https://doi.org/10.1016/S0009-9236(96)90162-9.

13 Thummel, K.E., O'Shea, D., Paine, M.F. et al. (1996). Oral first-pass elimination of midazolam involves both gastrointestinal and hepatic CYP3A-mediated metabolism. *Clin. Pharmacol. Ther.* 59: 491–502. https://doi.org/10.1016/S0009-9236(96)90177-0.

14 Bui, D., Li, L., Yin, T. et al. (2020). Pharmacokinetic and metabolic profiling of key active components of dietary supplement magnolia officinalis extract for prevention against oral carcinoma. *J. Agric. Food.*

*Chem.* 68: 6576–6587. https://doi.org/10.1021/acs.jafc.0c01475.

15 Fromm, M.F., Busse, D., Kroemer, H.K., and Eichelbaum, M. (1996). Differential induction of prehepatic and hepatic metabolism of verapamil by rifampin. *Hepatology* 24: 796–801. https://doi.org/10.1002/hep.510240407.

16 Gavhane, Y.N. and Yadav, A.V. (2012). Loss of orally administered drugs in GI tract. *Saudi Pharm. J.* 20: 331–344. https://doi.org/10.1016/j.jsps.2012.03.005.

17 Holtbecker, N., Fromm, M.F., Kroemer, H.K. et al. (1996). The nifedipine-rifampin interaction. Evidence for induction of gut wall metabolism. *Drug Metab. Dispos.* 24: 1121–1123.

18 Wu, C.Y., Benet, L.Z., Hebert, M.F. et al. (1995). Differentiation of absorption and first-pass gut and hepatic metabolism in humans: studies with cyclosporine. *Clin. Pharmacol. Ther.* 58: 492–497. https://doi.org/10.1016/0009-9236(95)90168-X.

19 Kunta, J.R., Lee, S.H., Perry, B.A. et al. (2004). Differentiation of gut and hepatic first-pass loss of verapamil in intestinal and vascular access-ported (IVAP) rabbits. *Drug Metab. Dispos.* 32: 1293–1298. https://doi.org/10.1124/dmd.104.000752.

20 Lee, Y.H., Perry, B.A., Lee, H.S. et al. (2001). Differentiation of gut and hepatic first-pass effect of drugs: 1. Studies of verapamil in ported dogs. *Pharm. Res.* 18: 1721–1728. https://doi.org/10.1023/a:1013374630274.

21 Alqahtani, S., Bukhari, I., Albassam, A., and Alenazi, M. (2018). An update on the potential role of intestinal first-pass metabolism for the prediction of drug-drug interactions: the role of PBPK modeling. *Expert Opin. Drug Metab. Toxicol.* 14: 625–634. https://doi.org/10.1080/17425255.2018.1482277.

22 Kirby, B.J. and Unadkat, J.D. (2010). Impact of ignoring extraction ratio when predicting drug-drug interactions, fraction metabolized, and intestinal first-pass contribution. *Drug Metab. Dispos.* 38: 1926–1933. https://doi.org/10.1124/dmd.110.034736.

23 Yoon, I.S., Choi, M.K., Kim, J.S. et al. (2011). Pharmacokinetics and first-pass elimination of metoprolol in rats: contribution of intestinal first-pass extraction to low bioavailability of metoprolol. *Xenobiotica* 41: 243–251. https://doi.org/10.3109/00498254.2010.538090.

24 Doherty, M.M. and Charman, W.N. (2002). The mucosa of the small intestine: how clinically relevant as an organ of drug metabolism? *Clin. Pharmacokinet.* 41: 235–253. https://doi.org/10.2165/00003088-200241040-00001.

25 Kosaka, K., Sakai, N., Endo, Y. et al. (2011). Impact of intestinal glucuronidation on the pharmacokinetics of raloxifene. *Drug Metab. Dispos.* 39: 1495–1502. https://doi.org/10.1124/dmd.111.040030.

26 Mizuma, T. (2009). Intestinal glucuronidation metabolism may have a greater impact on oral bioavailability than hepatic glucuronidation metabolism in humans: a study with raloxifene, substrate for UGT1A1, 1A8, 1A9, and 1A10. *Int. J. Pharm.* 378: 140–141. https://doi.org/10.1016/j.ijpharm.2009.05.044.

27 Obach, R.S., Zhang, Q.Y., Dunbar, D., and Kaminsky, L.S. (2001). Metabolic characterization of the major human small intestinal cytochrome p450s. *Drug Metab. Dispos.* 29: 347–352.

28 Couto, N., Al-Majdoub, Z.M., Gibson, S. et al. (2020). Quantitative proteomics of clinically relevant drug-metabolizing enzymes and drug transporters and their intercorrelations in the human small intestine. *Drug Metab. Dispos.* 48: 245–254. https://doi.org/10.1124/dmd.119.089656.

29 Guengerich, F.P. (2006). Cytochrome P450s and other enzymes in drug metabolism and toxicity. *AAPS J.* 8: E101–E111. https://doi.org/10.1208/aapsj080112.

30 Gundert-Remy, U., Bernauer, U., Blomeke, B. et al. (2014). Extrahepatic metabolism at the body's internal-external interfaces. *Drug Metab. Rev.* 46: 291–324. https://doi.org/10.3109/03602532.2014.900565.

31 Klomp, F., Wenzel, C., Drozdzik, M., and Oswald, S. (2020). Drug-drug interactions involving intestinal and hepatic CYP1A enzymes. *Pharmaceutics* 12: https://doi.org/10.3390/pharmaceutics12121201.

32 Schelstraete, W., Clerck, L., Govaert, E. et al. (2019). Characterization of porcine hepatic and intestinal drug metabolizing CYP450: comparison with human orthologues from a quantitative, activity and selectivity perspective. *Sci. Rep.* 9: 9233. https://doi.org/10.1038/s41598-019-45212-0.

33 Williams, D.A., Lemke, T.L., and Foye, W.O. (2002). *Foye's Principles of Medicinal*. Lippincott Williams & Wilkins.

34 Guengerich, F.P. (2003). Cytochromes P450, drugs, and diseases. *Mol. Interventions* 3: 194–204. https://doi.org/10.1124/mi.3.4.194.

35 Nelson, B.A., Robinson, K.A., and Buse, M.G. (2000). High glucose and glucosamine induce insulin resistance via different mechanisms in 3T3-L1 adipocytes. *Diabetes* 49: 981–991. https://doi.org/10.2337/diabetes.49.6.981.

36 Nelson, D.R., Koymans, L., Kamataki, T. et al. (1996). P450 superfamily: update on new sequences, gene mapping, accession numbers and nomenclature. *Pharmacogenetics* 6: 1–42. https://doi.org/10.1097/00008571-199602000-00002.

37 Zanger, U.M. and Schwab, M. (2013). Cytochrome P450 enzymes in drug metabolism: regulation of gene expression, enzyme activities, and impact of genetic variation. *Pharmacol. Ther.* 138: 103–141. https://doi.org/10.1016/j.pharmthera.2012.12.007.

38 Hruby, K. and Anzenbacher, P. (1997). Cytochrome P450--its significance in the biotransformation of xenobiotics and interspecies comparisons. *Cesk Fysiol* 46: 34–39.

39 Lamba, J.K., Lin, Y.S., Schuetz, E.G., and Thummel, K.E. (2002). Genetic contribution to variable human CYP3A-mediated metabolism. *Adv. Drug Delivery Rev.* 54: 1271–1294. https://doi.org/10.1016/s0169-409x(02)00066-2.

40 Lin, J.H. and Lu, A.Y. (1998). Inhibition and induction of cytochrome P450 and the clinical implications. *Clin. Pharmacokinet.* 35: 361–390. https://doi.org/10.2165/00003088-199835050-00003.

41 Nebert, D.W. and Russell, D.W. (2002). Clinical importance of the cytochromes P450. *Lancet* 360: 1155–1162. https://doi.org/10.1016/S0140-6736(02)11203-7.

42 Zhang, H., Basit, A., Busch, D. et al. (2018). Quantitative characterization of UDP-glucuronosyltransferase 2B17 in human liver and intestine and its role in testosterone first-pass metabolism. *Biochem. Pharmacol.* 156: 32–42. https://doi.org/10.1016/j.bcp.2018.08.003.

43 Xie, F., Ding, X., and Zhang, Q.Y. (2016). An update on the role of intestinal cytochrome P450 enzymes in drug disposition. *Acta Pharm. Sin. B* 6: 374–383. https://doi.org/10.1016/j.apsb.2016.07.012.

44 Nowell, S.A., Massengill, J.S., Williams, S. et al. (1999). Glucuronidation of 2-hydroxyamino-1-methyl-6-phenylimidazo[4,5-*b*]pyridine by human microsomal UDP-glucuronosyltransferases: identification of specific UGT1A family isoforms involved. *Carcinogenesis* 20: 1107–1114. https://doi.org/10.1093/carcin/20.6.1107.

45 Strassburg, C.P., Nguyen, N., Manns, M.P., and Tukey, R.H. (1999). UDP-glucuronosyltransferase activity in human liver and colon. *Gastroenterology* 116: 149–160. https://doi.org/10.1016/s0016-5085(99)70239-8.

46 Tukey, R.H. and Strassburg, C.P. (2000). Human UDP-glucuronosyltransferases: metabolism, expression, and disease. *Annu. Rev. Pharmacol. Toxicol.* 40: 581–616. https://doi.org/10.1146/annurev.pharmtox.40.1.581.

47 Cappiello, M., Giuliani, L., and Pacifici, G.M. (1991a). Distribution of UDP-glucuronosyltransferase and its endogenous substrate uridine 5'-diphosphoglucuronic acid in human tissues. *Eur. J. Clin. Pharmacol.* 41: 345–350. https://doi.org/10.1007/BF00314965.

48 Cappiello, M., Giuliani, L., Rane, A., and Pacifici, G.M. (1991b). Dopamine sulphotransferase is better developed than p-nitrophenol sulphotransferase in the human fetus. *Dev. Pharmacol. Ther.* 16: 83–88.

49 Fisher, M.B., Paine, M.F., Strelevitz, T.J., and Wrighton, S.A. (2001). The role of hepatic and extrahepatic UDP-glucuronosyltransferases in human drug metabolism. *Drug Metab. Rev.* 33: 273–297. https://doi.org/10.1081/dmr-120000653.

50 Hayes, J.D., Flanagan, J.U., and Jowsey, I.R. (2005). Glutathione transferases. *Annu. Rev. Pharmacol. Toxicol.* 45: 51–88. https://doi.org/10.1146/annurev.pharmtox.45.120403.095857.

51 Lo, H.W. and Ali-Osman, F. (2007). Genetic polymorphism and function of glutathione S-transferases in tumor drug resistance. *Curr. Opin. Pharmacol.* 7: 367–374. https://doi.org/10.1016/j.coph.2007.06.009.

52 McIlwain, C.C., Townsend, D.M., and Tew, K.D. (2006). Glutathione S-transferase polymorphisms: cancer incidence and therapy. *Oncogene* 25: 1639–1648. https://doi.org/10.1038/sj.onc.1209373.

53 Salinas, A.E. and Wong, M.G. (1999). Glutathione S-transferases – a review. *Curr. Med. Chem.* 6: 279–309.

54 Townsend, D.M. and Tew, K.D. (2003). The role of glutathione-S-transferase in anti-cancer drug resistance. *Oncogene* 22: 7369–7375. https://doi.org/10.1038/sj.onc.1206940.

55 Kutsuno, Y., Itoh, T., Tukey, R.H., and Fujiwara, R. (2014). Glucuronidation of drugs and drug-induced toxicity in humanized UDP-glucuronosyltransferase 1 mice. *Drug Metab. Dispos.* 42: 1146–1152. https://doi.org/10.1124/dmd.114.057083.

56 Nagar, S. and Remmel, R.P. (2006). Uridine diphosphoglucuronosyltransferase pharmacogenetics and cancer. *Oncogene* 25: 1659–1672. https://doi.org/10.1038/sj.onc.1209375.

57 Yang, G., Ge, S., Singh, R. et al. (2017). Glucuronidation: driving factors and their impact on glucuronide disposition. *Drug Metab. Rev.* 49: 105–138. https://doi.org/10.1080/03602532.2017.1293682.

58 Ge, S., Tu, Y., and Hu, M. (2016a). Challenges and opportunities with predicting in vivo phase II metabolism via glucuronidation from in vitro data. *Curr. Pharmacol. Rep.* 2: 326–338. https://doi.org/10.1007/s40495-016-0076-8.
59 Hu, M., Chen, J., and Lin, H. (2003). Metabolism of flavonoids via enteric recycling: mechanistic studies of disposition of apigenin in the Caco-2 cell culture model. *J. Pharmacol. Exp. Ther.* 307: 314–321.
60 Joseph, T.B., Wang, S.W., Liu, X. et al. (2007). Disposition of flavonoids via enteric recycling: enzyme stability affects characterization of prunetin glucuronidation across species, organs, and UGT isoforms. *Mol. Pharmaceutics* 4: 883–894. https://doi.org/10.1021/mp700135a.
61 Liu, X., Tam, V.H., and Hu, M. (2007). Disposition of flavonoids via enteric recycling: determination of the UDP-glucuronosyltransferase isoforms responsible for the metabolism of flavonoids in intact Caco-2 TC7 cells using siRNA. *Mol. Pharmaceutics* 4: 873–882.
62 Liu, Z. and Hu, M. (2007). Natural polyphenol disposition via coupled metabolic pathways. *Expert Opin. Drug Metab. Toxicol.* 3: 389–406. https://doi.org/10.1517/17425255.3.3.389.
63 Suzuki, H. and Sugiyama, Y. (2000). Role of metabolic enzymes and efflux transporters in the absorption of drugs from the small intestine. *Eur. J. Pharm. Sci.* 12: 3–12. https://doi.org/10.1016/s0928-0987(00)00178-0.
64 Xu, H., Kulkarni, K.H., Singh, R. et al. (2009). Disposition of naringenin via glucuronidation pathway is affected by compensating efflux transporters of hydrophilic glucuronides. *Mol. Pharmaceutics* 6: 1703–1715. https://doi.org/10.1021/mp900013d.
65 Zhu, W., Xu, H., Wang, S.W., and Hu, M. (2010). Breast cancer resistance protein (BCRP) and sulfotransferases contribute significantly to the disposition of genistein in mouse intestine. *AAPS J.* 12: 525–536. https://doi.org/10.1208/s12248-010-9209-x.
66 Ouzzine, M., Gulberti, S., Ramalanjaona, N. et al. (2014). The UDP-glucuronosyltransferases of the blood-brain barrier: their role in drug metabolism and detoxication. *Front. Cell. Neurosci.* 8: 349. https://doi.org/10.3389/fncel.2014.00349.
67 Lv, X., Zhang, J.B., Hou, J. et al. (2019). Chemical probes for human UDP-glucuronosyltransferases: a comprehensive review. *Biotechnol. J.* 14: e1800002. https://doi.org/10.1002/biot.201800002.
68 Fisher, M.B., Vandenbranden, M., Findlay, K. et al. (2000). Tissue distribution and interindividual variation in human UDP-glucuronosyltransferase activity: relationship between UGT1A1 promoter genotype and variability in a liver bank. *Pharmacogenetics* 10: 727–739. https://doi.org/10.1097/00008571-200011000-00007.
69 Strassburg, C.P., Manns, M.P., and Tukey, R.H. (1998). Expression of the UDP-glucuronosyltransferase 1A locus in human colon. Identification and characterization of the novel extrahepatic UGT1A8. *J. Biolumin. Chemilumin.* 273: 8719–8726. https://doi.org/10.1074/jbc.273.15.8719.
70 Falany, C.N. (1997a). Enzymology of human cytosolic sulfotransferases. *FASEB J.* 11: 206–216. https://doi.org/10.1096/fasebj.11.4.9068609.
71 Falany, C.N. (1997b). Sulfation and sulfotransferases. Introduction: changing view of sulfation and the cytosolic sulfotransferases. *FASEB J.* 11: 1–2. https://doi.org/10.1096/fasebj.11.1.9034159.
72 Marto, N., Morello, J., Monteiro, E.C., and Pereira, S.A. (2017). Implications of sulfotransferase activity in interindividual variability in drug response: clinical perspective on current knowledge. *Drug Metab. Rev.* 49: 357–371.
73 Negishi, M., Pedersen, L.G., Petrotchenko, E. et al. (2001). Structure and function of sulfotransferases. *Arch. Biochem. Biophys.* 390: 149–157. https://doi.org/10.1006/abbi.2001.2368.
74 Petrotchenko, E.V., Pedersen, L.C., Borchers, C.H. et al. (2001). The dimerization motif of cytosolic sulfotransferases. *FEBS Lett.* 490: 39–43. https://doi.org/10.1016/s0014-5793(01)02129-9.
75 Suiko, M., Kurogi, K., Hashiguchi, T. et al. (2017). Updated perspectives on the cytosolic sulfotransferases (SULTs) and SULT-mediated sulfation. *Biosci. Biotechnol., Biochem.* 81: 63–72. https://doi.org/10.1080/09168451.2016.1222266.
76 Yoshinari, K., Petrotchenko, E.V., Pedersen, L.C., and Negishi, M. (2001). Crystal structure-based studies of cytosolic sulfotransferase. *J. Biochem. Mol. Toxicol.* 15: 67–75. https://doi.org/10.1002/jbt.1.
77 Chen, G., Zhang, D., Jing, N. et al. (2003a). Human gastrointestinal sulfotransferases: identification and distribution. *Toxicol. Appl. Pharmacol.* 187: 186–197. https://doi.org/10.1016/s0041-008x(02)00073-x.
78 Gamage, N., Barnett, A., Hempel, N. et al. (2006). Human sulfotransferases and their role in chemical metabolism. *Toxicol. Sci.* 90: 5–22. https://doi.org/10.1093/toxsci/kfj061.
79 Xie, Y. and Xie, W. (2020). The role of sulfotransferases in liver diseases. *Drug Metab. Dispos.* 48: 742–749. https://doi.org/10.1124/dmd.120.000074.

80 Hildebrandt, M.A., Carrington, D.P., Thomae, B.A. et al. (2007). Genetic diversity and function in the human cytosolic sulfotransferases. *Pharmacogenomics J.* 7: 133–143. https://doi.org/10.1038/sj.tpj.6500404.
81 Riches, Z., Stanley, E.L., Bloomer, J.C., and Coughtrie, M.W. (2009). Quantitative evaluation of the expression and activity of five major sulfotransferases (SULTs) in human tissues: the SULT "pie". *Drug Metab. Dispos.* 37: 2255–2261. https://doi.org/10.1124/dmd.109.028399.
82 Teubner, W., Meinl, W., Florian, S. et al. (2007). Identification and localization of soluble sulfotransferases in the human gastrointestinal tract. *Biochem. J* 404: 207–215. https://doi.org/10.1042/BJ20061431.
83 Chen, Y., Tang, Y., Guo, C. et al. (2012). Nuclear receptors in the multidrug resistance through the regulation of drug-metabolizing enzymes and drug transporters. *Biochem. Pharmacol.* 83: 1112–1126. https://doi.org/10.1016/j.bcp.2012.01.030.
84 Handschin, C. and Meyer, U.A. (2003). Induction of drug metabolism: the role of nuclear receptors. *Pharmacol. Rev.* 55: 649–673. https://doi.org/10.1124/pr.55.4.2.
85 Hu, D.G., Meech, R., McKinnon, R.A., and Mackenzie, P.I. (2014). Transcriptional regulation of human UDP-glucuronosyltransferase genes. *Drug Metab. Rev.* 46: 421–458. https://doi.org/10.3109/03602532.2014.973037.
86 Kliewer, S.A. (2003). The nuclear pregnane X receptor regulates xenobiotic detoxification. *J. Nutr.* 133: 2444S–2447S. https://doi.org/10.1093/jn/133.7.2444S.
87 Prakash, C., Zuniga, B., Song, C.S. et al. (2015). Nuclear receptors in drug metabolism, drug response and drug interactions. *Nucl. Recept. Res.* 2: https://doi.org/10.11131/2015/101178.
88 Tanner, N., Kubik, L., Luckert, C. et al. (2018). Regulation of drug metabolism by the interplay of inflammatory signaling, steatosis, and xeno-sensing receptors in HepaRG cells. *Drug Metab. Dispos.* 46: 326–335. https://doi.org/10.1124/dmd.117.078675.
89 Urquhart, B.L., Tirona, R.G., and Kim, R.B. (2007). Nuclear receptors and the regulation of drug-metabolizing enzymes and drug transporters: implications for interindividual variability in response to drugs. *J Clin Pharmacol* 47: 566–578. https://doi.org/10.1177/0091270007299930.
90 Willson, T.M. and Kliewer, S.A. (2002). PXR, CAR and drug metabolism. *Nat. Rev. Drug Discovery* 1: 259–266. https://doi.org/10.1038/nrd753.
91 Stanley, L.A., Horsburgh, B.C., Ross, J. et al. (2006). PXR and CAR: nuclear receptors which play a pivotal role in drug disposition and chemical toxicity. *Drug Metab. Rev.* 38: 515–597. https://doi.org/10.1080/03602530600786232.
92 Timsit, Y.E. and Negishi, M. (2007). CAR and PXR: the xenobiotic-sensing receptors. *Steroids* 72: 231–246. https://doi.org/10.1016/j.steroids.2006.12.006.
93 Tirona, R.G. and Kim, R.B. (2005). Nuclear receptors and drug disposition gene regulation. *J. Pharm. Sci.* 94: 1169–1186. https://doi.org/10.1002/jps.20324.
94 Chai, X., Zeng, S., and Xie, W. (2013). Nuclear receptors PXR and CAR: implications for drug metabolism regulation, pharmacogenomics and beyond. *Expert Opin. Drug Metab. Toxicol.* 9: 253–266. https://doi.org/10.1517/17425255.2013.754010.
95 Gregory, P.A., Lewinsky, R.H., Gardner-Stephen, D.A., and Mackenzie, P.I. (2004). Regulation of UDP glucuronosyltransferases in the gastrointestinal tract. *Toxicol. Appl. Pharmacol.* 199: 354–363. https://doi.org/10.1016/j.taap.2004.01.008.
96 Runge-Morris, M., Kocarek, T.A., and Falany, C.N. (2013). Regulation of the cytosolic sulfotransferases by nuclear receptors. *Drug Metab. Rev.* 45: 15–33. https://doi.org/10.3109/03602532.2012.748794.
97 Wang, Y.M., Ong, S.S., Chai, S.C., and Chen, T. (2012). Role of CAR and PXR in xenobiotic sensing and metabolism. *Expert Opin. Drug Metab. Toxicol.* 8: 803–817. https://doi.org/10.1517/17425255.2012.685237.
98 Nebert, D.W., McKinnon, R.A., and Puga, A. (1996). Human drug-metabolizing enzyme polymorphisms: effects on risk of toxicity and cancer. *DNA Cell Biol.* 15: 273–280. https://doi.org/10.1089/dna.1996.15.273.
99 Quattrochi, L.C. and Guzelian, P.S. (2001). Cyp3A regulation: from pharmacology to nuclear receptors. *Drug Metab. Dispos.* 29: 615–622.
100 Rushmore, T.H. and Kong, A.N. (2002). Pharmacogenomics, regulation and signaling pathways of phase I and II drug metabolizing enzymes. *Curr. Drug. Metab* 3: 481–490. https://doi.org/10.2174/1389200023337171.
101 Wu, K.C. and Lin, C.J. (2019). The regulation of drug-metabolizing enzymes and membrane transporters by inflammation: evidences in inflammatory diseases and age-related disorders. *J. Food Drug Anal.* 27: 48–59. https://doi.org/10.1016/j.jfda.2018.11.005.
102 Xu, C., Li, C.Y., and Kong, A.N. (2005). Induction of phase I, II and III drug metabolism/transport

by xenobiotics. *Arch. Pharmacal Res.* 28: 249–268. https://doi.org/10.1007/BF02977789.

103 Anzenbacher, P. and Anzenbacherova, E. (2001). Cytochromes P450 and metabolism of xenobiotics. *Cell. Mol. Life Sci.* 58: 737–747. https://doi.org/10.1007/pl00000897.

104 Miners, J.O., McKinnon, R.A., and Mackenzie, P.I. (2002). Genetic polymorphisms of UDP-glucuronosyltransferases and their functional significance. *Forensic Toxicol.* 181–182: 453–456. https://doi.org/10.1016/s0300-483x(02)00449-3.

105 Thummel, K.E. (2007). Gut instincts: CYP3A4 and intestinal drug metabolism. *J. Clin. Invest.* 117: 3173–3176. https://doi.org/10.1172/JCI34007.

106 Mackenzie, P.I., Miners, J.O., and McKinnon, R.A. (2000). Polymorphisms in UDP glucuronosyltransferase genes: functional consequences and clinical relevance. *Clin. Chem. Lab. Med.* 38: 889–892. https://doi.org/10.1515/CCLM.2000.129.

107 Mi, X.X., Yan, J., Ma, X.J. et al. (2019). Analysis of the UGT1A1 genotype in hyperbilirubinemia patients: differences in allele frequency and distribution. *Biomed. Res. Int.* 2019: 6272174. https://doi.org/10.1155/2019/6272174.

108 Strassburg, C.P., Kneip, S., Topp, J. et al. (2000). Polymorphic gene regulation and interindividual variation of UDP-glucuronosyltransferase activity in human small intestine. *J. Biolumin. Chemilumin.* 275: 36164–36171. https://doi.org/10.1074/jbc.M002180200.

109 Fujiwara, R., Nguyen, N., Chen, S., and Tukey, R.H. (2010). Developmental hyperbilirubinemia and CNS toxicity in mice humanized with the UDP glucuronosyltransferase 1 (UGT1) locus. *Proc. Natl. Acad. Sci. U.S.A.* 107: 5024–5029. https://doi.org/10.1073/pnas.0913290107.

110 Endres, C.J., Hsiao, P., Chung, F.S., and Unadkat, J.D. (2006). The role of transporters in drug interactions. *Eur. J. Pharm. Sci.* 27: 501–517. https://doi.org/10.1016/j.ejps.2005.11.002.

111 Estudante, M., Morais, J.G., Soveral, G., and Benet, L.Z. (2013). Intestinal drug transporters: an overview. *Adv. Drug Delivery Rev.* 65: 1340–1356. https://doi.org/10.1016/j.addr.2012.09.042.

112 Kim, R.B. (2002). Transporters and xenobiotic disposition. *Forensic Toxicol.* 181–182: 291–297. https://doi.org/10.1016/s0300-483x(02)00296-2.

113 Lin, J.H. (2007). Transporter-mediated drug interactions: clinical implications and in vitro assessment. *Expert Opin. Drug Metab. Toxicol.* 3: 81–92. https://doi.org/10.1517/17425255.3.1.81.

114 Liu, X. (2019). Overview: role of drug transporters in drug disposition and its clinical significance. *Adv. Exp. Med. Biol.* 1141: 1–12. https://doi.org/10.1007/978-981-13-7647-4_1.

115 Mao, Q., Lai, Y., and Wang, J. (2018). Drug transporters in xenobiotic disposition and pharmacokinetic prediction. *Drug Metab. Dispos.* 46: 561–566. https://doi.org/10.1124/dmd.118.081356.

116 Shugarts, S. and Benet, L.Z. (2009). The role of transporters in the pharmacokinetics of orally administered drugs. *Pharm. Res.* 26: 2039–2054. https://doi.org/10.1007/s11095-009-9924-0.

117 Gutmann, H., Hruz, P., Zimmermann, C. et al. (2005). Distribution of breast cancer resistance protein (BCRP/ABCG2) mRNA expression along the human GI tract. *Biochem. Pharmacol.* 70: 695–699. https://doi.org/10.1016/j.bcp.2005.05.031.

118 Murakami, T. and Takano, M. (2008). Intestinal efflux transporters and drug absorption. *Expert Opin. Drug Metab. Toxicol.* 4: 923–939. https://doi.org/10.1517/17425255.4.7.923.

119 Schinkel, A.H. and Jonker, J.W. (2003). Mammalian drug efflux transporters of the ATP binding cassette (ABC) family: an overview. *Adv. Drug Delivery Rev.* 55: 3–29. https://doi.org/10.1016/s0169-409x(02)00169-2.

120 Taipalensuu, J., Tornblom, H., Lindberg, G. et al. (2001). Correlation of gene expression of ten drug efflux proteins of the ATP-binding cassette transporter family in normal human jejunum and in human intestinal epithelial Caco-2 cell monolayers. *J. Pharmacol. Exp. Ther.* 299: 164–170.

121 Zimmermann, C., Gutmann, H., Hruz, P. et al. (2005). Mapping of multidrug resistance gene 1 and multidrug resistance-associated protein isoform 1 to 5 mRNA expression along the human intestinal tract. *Drug Metab. Dispos.* 33: 219–224. https://doi.org/10.1124/dmd.104.001354.

122 Kuno, T., Hirayama-Kurogi, M., Ito, S., and Ohtsuki, S. (2016). Effect of intestinal flora on protein expression of drug-metabolizing enzymes and transporters in the liver and kidney of germ-free and antibiotics-treated mice. *Mol. Pharmaceutics* 13: 2691–2701. https://doi.org/10.1021/acs.molpharmaceut.6b00259.

123 Chu, X.Y., Strauss, J.R., Mariano, M.A. et al. (2006). Characterization of mice lacking the multidrug resistance protein MRP2 (ABCC2). *J. Pharmacol. Exp. Ther.* 317: 579–589. https://doi.org/10.1124/jpet.105.098665.

124 Huang, S.M. and Lesko, L.J. (2004). Drug-drug, drug-dietary supplement, and drug-citrus fruit and

other food interactions: what have we learned? *J Clin Pharmacol* 44: 559–569. https://doi.org/10.1177/0091270004265367.

125 Laramy, J.K., Kim, M., Parrish, K.E. et al. (2018). Pharmacokinetic assessment of cooperative efflux of the multitargeted kinase inhibitor ponatinib across the blood-brain barrier. *J. Pharmacol. Exp. Ther.* 365: 249–261. https://doi.org/10.1124/jpet.117.246116.

126 Lin, J.H. (2003). Drug–drug interaction mediated by inhibition and induction of P-glycoprotein. *Adv. Drug Delivery Rev.* 55: 53–81. https://doi.org/10.1016/s0169-409x(02)00171-0.

127 Vlaming, M.L., van Esch, A., Pala, Z. et al. (2009). Abcc2 (Mrp2), Abcc3 (Mrp3), and Abcg2 (Bcrp1) are the main determinants for rapid elimination of methotrexate and its toxic metabolite 7-hydroxymethotrexate in vivo. *Mol. Cancer Ther.* 8: 3350–3359. https://doi.org/10.1158/1535-7163.MCT-09-0668.

128 Ho, Y.J., Jeng, L.B., Yang, M.D. et al. (2003). A trial of single photon emission computed tomography of the liver with technetium-99m tetrofosmin to detect hepatocellular carcinoma. *Anticancer Res.* 23: 1743–1746.

129 Richter, M., Gyemant, N., Molnar, J., and Hilgeroth, A. (2004). Comparative effects on intestinal absorption in situ by P-glycoprotein-modifying HIV protease inhibitors. *Pharm. Res.* 21: 1862–1866. https://doi.org/10.1023/b:pham.0000045240.81664.be.

130 Rigalli, J.P., Tocchetti, G.N., and Weiss, J. (2019). Modulation of ABC transporters by nuclear receptors: physiological, pathological and pharmacological aspects. *Curr. Med. Chem.* 26: 1079–1112. https://doi.org/10.2174/0929867324666170920141707.

131 Staudinger, J.L., Woody, S., Sun, M., and Cui, W. (2013). Nuclear-receptor-mediated regulation of drug- and bile-acid-transporter proteins in gut and liver. *Drug Metab. Rev.* 45: 48–59. https://doi.org/10.3109/03602532.2012.748793.

132 Benet, L.Z., Cummins, C.L., and Wu, C.Y. (2004). Unmasking the dynamic interplay between efflux transporters and metabolic enzymes. *Int. J. Pharm.* 277: 3–9. https://doi.org/10.1016/j.ijpharm.2002.12.002.

133 Jia, X., Chen, J., Lin, H., and Hu, M. (2004). Disposition of flavonoids via enteric recycling: enzyme-transporter coupling affects metabolism of biochanin A and formononetin and excretion of their phase II conjugates. *J. Pharmacol. Exp. Ther.* 310: 1103–1113.

134 Liu, Y., Liu, Y., Dai, Y. et al. (2003). Enteric disposition and recycling of flavonoids and ginkgo flavonoids. *J. Altern. Complement. Med.* 9: 631–640. https://doi.org/10.1089/107555303322524481.

135 Roberts, M.S., Magnusson, B.M., Burczynski, F.J., and Weiss, M. (2002). Enterohepatic circulation: physiological, pharmacokinetic and clinical implications. *Clin. Pharmacokinet.* 41: 751–790. https://doi.org/10.2165/00003088-200241100-00005.

136 Silberberg, M., Morand, C., Mathevon, T. et al. (2006). The bioavailability of polyphenols is highly governed by the capacity of the intestine and of the liver to secrete conjugated metabolites. *Eur. J. Nutr.* 45: 88–96. https://doi.org/10.1007/s00394-005-0568-5.

137 Dai, P., Zhu, L., Luo, F. et al. (2015). Triple recycling processes impact systemic and local bioavailability of orally administered flavonoids. *AAPS J.* 17: 723–736. https://doi.org/10.1208/s12248-015-9732-x.

138 Xia, B., Zhou, Q., Zheng, Z. et al. (2012). A novel local recycling mechanism that enhances enteric bioavailability of flavonoids and prolongs their residence time in the gut. *Mol. Pharmaceutics* 9: 3246–3258. https://doi.org/10.1021/mp300315d.

139 Chen, J., Lin, H., and Hu, M. (2003b). Metabolism of flavonoids via enteric recycling: role of intestinal disposition. *J. Pharmacol. Exp. Ther.* 304: 1228–1235. https://doi.org/10.1124/jpet.102.046409.

140 Tu, Y., Wang, L., Rong, Y. et al. (2021). Hepatoenteric recycling is a new disposition mechanism for orally administered phenolic drugs and phytochemicals in rats. *Elife* 10: https://doi.org/10.7554/eLife.58820.

141 Jeong, E.J., Jia, X., and Hu, M. (2005a). Disposition of formononetin via enteric recycling: metabolism and excretion in mouse intestinal perfusion and Caco-2 cell models. *Mol. Pharmaceutics* 2: 319–328.

142 Wang, S.W., Chen, J., Jia, X. et al. (2006). Disposition of flavonoids via enteric recycling: structural effects and lack of correlations between in vitro and in situ metabolic properties. *Drug Metab. Dispos.* 34: 1837–1848.

143 Naz, H., Islam, A., Waheed, A. et al. (2013). Human beta-glucuronidase: structure, function, and application in enzyme replacement therapy. *Rejuvenation Res.* 16: 352–363. https://doi.org/10.1089/rej.2013.1407.

144 McCarter, J.D. and Withers, S.G. (1994). Mechanisms of enzymatic glycoside hydrolysis. *Curr. Opin.*

*Struct. Biol.* 4: 885–892. https://doi.org/10.1016/0959-440x(94)90271-2.

**145** Jeong, E.J., Liu, Y., Lin, H., and Hu, M. (2005b). Species- and disposition model-dependent metabolism of raloxifene in gut and liver: role of UGT1A10. *Drug Metab. Dispos.* 33: 785–794. https://doi.org/10.1124/dmd.104.001883.

**146** Ghosal, A., Hapangama, N., Yuan, Y. et al. (2004). Identification of human UDP-glucuronosyltransferase enzyme(s) responsible for the glucuronidation of ezetimibe (Zetia). *Drug Metab. Dispos.* 32: 314–320. https://doi.org/10.1124/dmd.32.3.314.

**147** Kosoglou, T., Statkevich, P., Johnson-Levonas, A.O. et al. (2005). Ezetimibe: a review of its metabolism, pharmacokinetics and drug interactions. *Clin. Pharmacokinet.* 44: 467–494. https://doi.org/10.2165/00003088-200544050-00002.

**148** Brill, S.S., Furimsky, A.M., Ho, M.N. et al. (2006). Glucuronidation of trans-resveratrol by human liver and intestinal microsomes and UGT isoforms. *J. Pharm. Pharmacol.* 58: 469–479. https://doi.org/10.1211/jpp.58.4.0006.

**149** Ge, S., Yin, T., Xu, B. et al. (2016b). Curcumin affects phase II disposition of resveratrol through inhibiting efflux transporters MRP2 and BCRP. *Pharm. Res.* 33: 590–602. https://doi.org/10.1007/s11095-015-1812-1.

**150** Hecht, J.R. (1998). Gastrointestinal toxicity or irinotecan. *Oncology (Williston Park)* 12: 72–78.

**151** Horikawa, M., Kato, Y., and Sugiyama, Y. (2002). Reduced gastrointestinal toxicity following inhibition of the biliary excretion of irinotecan and its metabolites by probenecid in rats. *Pharm. Res.* 19: 1345–1353. https://doi.org/10.1023/a:1020358910490.

**152** Sun, R., Zhu, L., Li, L. et al. (2020). Irinotecan-mediated diarrhea is mainly correlated with intestinal exposure to SN-38: critical role of gut Ugt. *Toxicol. Appl. Pharmacol.* 398: 115032. https://doi.org/10.1016/j.taap.2020.115032.

**153** Knight, B., Troutman, M., and Thakker, D.R. (2006). Deconvoluting the effects of P-glycoprotein on intestinal CYP3A: a major challenge. *Curr. Opin. Pharmacol.* 6: 528–532. https://doi.org/10.1016/j.coph.2006.04.009.

**154** Lau, Y.Y., Wu, C.Y., Okochi, H., and Benet, L.Z. (2004). Ex situ inhibition of hepatic uptake and efflux significantly changes metabolism: hepatic enzyme-transporter interplay. *J. Pharmacol. Exp. Ther.* 308: 1040–1045. https://doi.org/10.1124/jpet.103.061770.

**155** Board, P., Blackburn, A., Jermiin, L.S., and Chelvanayagam, G. (1998). Polymorphism of phase II enzymes: identification of new enzymes and polymorphic variants by database analysis. *Toxicol. Lett.* 102–103: 149–154. https://doi.org/10.1016/s0378-4274(98)00300-2.

**156** Neber, D.W. and Roe, A.L. (2001). Ethnic and genetic differences in metabolism genes and risk of toxicity and cancer. *Sci. Total Environ.* 274: 93–102. https://doi.org/10.1016/s0048-9697(01)00732-x.

**157** Ozawa, S. (1998). Interindividual differences in efficacy and toxicity induced by therapeutic drugs and xenobiotics in relation to genetic polymorphisms in xenobiotic metabolizing enzymes. *Kokuritsu Iyakuhin Shokuhin Eisei Kenkyusho Hokoku* 116: 69–81.

**158** Benedetti, M.S., Whomsley, R., and Canning, M. (2007). Drug metabolism in the paediatric population and in the elderly. *Drug Discovery Today* 12: 599–610. https://doi.org/10.1016/j.drudis.2007.06.011.

**159** Wilkinson, G.R. (1997). The effects of diet, aging and disease-states on presystemic elimination and oral drug bioavailability in humans. *Adv. Drug Delivery Rev.* 27: 129–159. https://doi.org/10.1016/s0169-409x(97)00040-9.

**160** Shapiro, B.H., Agrawal, A.K., and Pampori, N.A. (1995). Gender differences in drug metabolism regulated by growth hormone. *Int. J. Biochem. Cell Biol.* 27: 9–20. https://doi.org/10.1016/1357-2725(94)00056-5.

**161** Chen, J., Wang, S., Jia, X. et al. (2005). Disposition of flavonoids via recycling: comparison of intestinal versus hepatic disposition. *Drug Metab. Dispos.* 33: 1777–1784. https://doi.org/10.1124/dmd.105.003673.

**162** Felton, J.S. and Malfatti, M.A. (2006). What do diet-induced changes in phase I and II enzymes tell us about prevention from exposure to heterocyclic amines? *J. Nutr.* 136: 2683S–2684S. https://doi.org/10.1093/jn/136.10.2683S.

**163** Lu, A.Y. (1998). Drug-metabolism research challenges in the new millennium: individual variability in drug therapy and drug safety. *Drug Metab. Dispos.* 26: 1217–1222.

**164** Larrey, D., Letteron, P., Foliot, A. et al. (1986). Effects of pregnancy on the toxicity and metabolism of acetaminophen in mice. *J. Pharmacol. Exp. Ther.* 237: 283–291.

**165** Tsutsumi, K., Kotegawa, T., Matsuki, S. et al. (2001). The effect of pregnancy on cytochrome P4501A2, xanthine oxidase, and *N*-acetyltransferase activities in humans. *Clin. Pharmacol. Ther.* 70: 121–125. https://doi.org/10.1067/mcp.2001.116495.

166 Assaf, R.A., Dundee, J.W., and Gamble, J.A. (1975). The influence of the route of administration on the clinical action of diazepam. *Anaesthesia* 30: 152–158. https://doi.org/10.1111/j.1365-2044.1975.tb00826.x.

167 Houston, J.B. and Levy, G. (1976). Effect of route of administration on competitive drug biotransformation interaction: salicylamide-ascorbic acid interaction in rats. *J. Pharmacol. Exp. Ther.* 198: 284–294.

168 Shand, D.G., Kornhauser, D.M., and Wilkinson, G.R. (1975). Effects of route of administration and blood flow on hepatic drug elimination. *J. Pharmacol. Exp. Ther.* 195: 424–432.

169 Glatt, H. and Meinl, W. (2004). Pharmacogenetics of soluble sulfotransferases (SULTs). *Naunyn-Schmiedeberg's Arch. Pharmacol.* 369: 55–68. https://doi.org/10.1007/s00210-003-0826-0.

170 Tukey, R.H., Strassburg, C.P., and Mackenzie, P.I. (2002). Pharmacogenomics of human UDP-glucuronosyltransferases and irinotecan toxicity. *Mol. Pharmacol.* 62: 446–450. https://doi.org/10.1124/mol.62.3.446.

171 Nowell, S. and Falany, C.N. (2006). Pharmacogenetics of human cytosolic sulfotransferases. *Oncogene* 25: 1673–1678. https://doi.org/10.1038/sj.onc.1209376.

172 Prestera, T., Holtzclaw, W.D., Zhang, Y., and Talalay, P. (1993). Chemical and molecular regulation of enzymes that detoxify carcinogens. *Proc. Natl. Acad. Sci. U.S.A.* 90: 2965–2969. https://doi.org/10.1073/pnas.90.7.2965.

173 Moinova, H.R. and Mulcahy, R.T. (1999). Up-regulation of the human gamma-glutamylcysteine synthetase regulatory subunit gene involves binding of Nrf-2 to an electrophile responsive element. *Biochem. Biophys. Res. Commun.* 261: 661–668. https://doi.org/10.1006/bbrc.1999.1109.

174 Wang, W., Liu, L.Q., Higuchi, C.M., and Chen, H. (1998). Induction of NADPH:quinone reductase by dietary phytoestrogens in colonic Colo205 cells. *Biochem. Pharmacol.* 56: 189–195. https://doi.org/10.1016/s0006-2952(98)00141-5.

175 Uda, Y., Price, K.R., Williamson, G., and Rhodes, M.J. (1997). Induction of the anticarcinogenic marker enzyme, quinone reductase, in murine hepatoma cells in vitro by flavonoids. *Cancer Lett.* 120: 213–216. https://doi.org/10.1016/s0304-3835(97)00311-x.

176 Yannai, S., Day, A.J., Williamson, G., and Rhodes, M.J. (1998). Characterization of flavonoids as monofunctional or bifunctional inducers of quinone reductase in murine hepatoma cell lines. *Food Chem. Toxicol.* 36: 623–630. https://doi.org/10.1016/s0278-6915(98)00022-2.

177 Barch, D.H., Rundhaugen, L.M., and Pillay, N.S. (1995). Ellagic acid induces transcription of the rat glutathione S-transferase-Ya gene. *Carcinogenesis* 16: 665–668. https://doi.org/10.1093/carcin/16.3.665.

178 Barch, D.H., Rundhaugen, L.M., Thomas, P.E. et al. (1994). Dietary ellagic acid inhibits the enzymatic activity of CYP1A1 without altering hepatic concentrations of CYP1A1 or CYP1A1 mRNA. *Biochem. Biophys. Res. Commun.* 201: 1477–1482. https://doi.org/10.1006/bbrc.1994.1870.

179 Khan, N., Afaq, F., Saleem, M. et al. (2006). Targeting multiple signaling pathways by green tea polyphenol (-)-epigallocatechin-3-gallate. *Cancer Res.* 66: 2500–2505. https://doi.org/10.1158/0008-5472.CAN-05-3636.

180 Stoner, G.D. and Mukhtar, H. (1995). Polyphenols as cancer chemopreventive agents. *J. Cell Biochem. Suppl.* 22: 169–180. https://doi.org/10.1002/jcb.240590822.

181 Dinkova-Kostova, A.T. and Talalay, P. (1999). Relation of structure of curcumin analogs to their potencies as inducers of Phase 2 detoxification enzymes. *Carcinogenesis* 20: 911–914. https://doi.org/10.1093/carcin/20.5.911.

182 McLellan, L.I., Harrison, D.J., and Hayes, J.D. (1992). Modulation of glutathione S-transferases and glutathione peroxidase by the anticarcinogen butylated hydroxyanisole in murine extrahepatic organs. *Carcinogenesis* 13: 2255–2261. https://doi.org/10.1093/carcin/13.12.2255.

183 McLellan, L.I. and Hayes, J.D. (1989). Differential induction of class alpha glutathione S-transferases in mouse liver by the anticarcinogenic antioxidant butylated hydroxyanisole. Purification and characterization of glutathione S-transferase Ya1Ya1. *Biochem. J* 263: 393–402. https://doi.org/10.1042/bj2630393.

184 McLellan, L.I., Judah, D.J., Neal, G.E., and Hayes, J.D. (1994). Regulation of aflatoxin B1-metabolizing aldehyde reductase and glutathione S-transferase by chemoprotectors. *Biochem. J* 300 (Pt 1): 117–124. https://doi.org/10.1042/bj3000117.

185 Sharma, R., Ahmad, H., Singhal, S.S. et al. (1993). Comparative studies on the effect of butylated hydroxyanisole on glutathione and glutathione S-transferases in the tissues of male and female

CD-1 mice. *Comp. Biochem. Physiol. C: Pharmacol. Toxicol.* 105: 31–37. https://doi.org/10.1016/0742-8413(93)90053-n.

186 Nakagama, H., Nakanishi, M., and Ochiai, M. (2005). Modeling human colon cancer in rodents using a food-borne carcinogen, PhIP. *Cancer Sci.* 96: 627–636. https://doi.org/10.1111/j.1349-7006.2005.00107.x.

187 Ochiai, M., Watanabe, M., Kushida, H. et al. (1996). DNA adduct formation, cell proliferation and aberrant crypt focus formation induced by PhIP in male and female rat colon with relevance to carcinogenesis. *Carcinogenesis* 17: 95–98. https://doi.org/10.1093/carcin/17.1.95.

188 Calvin, G., Long, P.H., Stitzel, K.A. et al. (1988). Ethylenediaminetetra(methylenephosphonic acid): genotoxicity, biodistribution, and subchronic and chronic toxicity in rats. *Food Chem. Toxicol.* 26: 601–610. https://doi.org/10.1016/0278-6915(88)90231-1.

189 Graham, P.S., Hellyer, R.O., and Ryan, A.J. (1970). The kinetics of inhibition of drug metabolism in vitro by some naturally occurring compounds. *Biochem. Pharmacol.* 19: 769–775. https://doi.org/10.1016/0006-2952(70)90240-6.

190 Vakharia, D.D., Liu, N., Pause, R. et al. (2001). Effect of metals on polycyclic aromatic hydrocarbon induction of CYP1A1 and CYP1A2 in human hepatocyte cultures. *Toxicol. Appl. Pharmacol.* 170: 93–103. https://doi.org/10.1006/taap.2000.9087.

191 Wang, Z., Gorski, J.C., Hamman, M.A. et al. (2001). The effects of St John's wort (*Hypericum perforatum*) on human cytochrome P450 activity. *Clin. Pharmacol. Ther.* 70: 317–326.

192 Wang, Z., Hamman, M.A., Huang, S.M. et al. (2002). Effect of St John's wort on the pharmacokinetics of fexofenadine. *Clin. Pharmacol. Ther.* 71: 414–420. https://doi.org/10.1067/mcp.2002.124080.

193 Zhou, S.F., Zhou, Z.W., Li, C.G. et al. (2007). Identification of drugs that interact with herbs in drug development. *Drug Discovery Today* 12: 664–673. https://doi.org/10.1016/j.drudis.2007.06.004.

194 Bressler, R. (2006). Grapefruit juice and drug interactions. Exploring mechanisms of this interaction and potential toxicity for certain drugs. *Geriatrics* 61: 12–18.

195 Amidon, G.L., Lennernas, H., Shah, V.P., and Crison, J.R. (1995). A theoretical basis for a biopharmaceutic drug classification: the correlation of in vitro drug product dissolution and in vivo bioavailability. *Pharm. Res.* 12: 413–420. https://doi.org/10.1023/a:1016212804288.

196 Niazi, S.K.N. (2017). *Handbook of Pharmaceutical Manufacturing Formulations*, 3e, 27–35.

197 Benet, L.Z. and Larregieu, C.A. (2010). The FDA should eliminate the ambiguities in the current BCS biowaiver guidance and make public the drugs for which BCS biowaivers have been granted. *Clin. Pharmacol. Ther.* 88: 405–407. https://doi.org/10.1038/clpt.2010.149.

198 Wu, C.Y. and Benet, L.Z. (2005). Predicting drug disposition via application of BCS: transport/absorption/ elimination interplay and development of a biopharmaceutics drug disposition classification system. *Pharm. Res.* 22: 11–23. https://doi.org/10.1007/s11095-004-9004-4.

199 Benet, L.Z. (2013). The role of BCS (biopharmaceutics classification system) and BDDCS (biopharmaceutics drug disposition classification system) in drug development. *J. Pharm. Sci.* 102: 34–42. https://doi.org/10.1002/jps.23359.

200 Liu, Y., Ramirez, J., House, L., and Ratain, M.J. (2010a). Comparison of the drug-drug interactions potential of erlotinib and gefitinib via inhibition of UDP-glucuronosyltransferases. *Drug Metab. Dispos.* 38: 32–39. https://doi.org/10.1124/dmd.109.029660.

201 Peer, C.J., Sissung, T.M., Kim, A. et al. (2012). Sorafenib is an inhibitor of UGT1A1 but is metabolized by UGT1A9: implications of genetic variants on pharmacokinetics and hyperbilirubinemia. *Clin. Cancer Res.* 18: 2099–2107. https://doi.org/10.1158/1078-0432.CCR-11-2484.

202 Zhang, N., Liu, Y., and Jeong, H. (2015). Drug–drug interaction potentials of tyrosine kinase inhibitors via inhibition of UDP-glucuronosyltransferases. *Sci. Rep.* 5: 17778. https://doi.org/10.1038/srep17778.

203 Li, Y., Meng, Q., Yang, M. et al. (2019). Current trends in drug metabolism and pharmacokinetics. *Acta Pharm. Sin. B* 9: 1113–1144. https://doi.org/10.1016/j.apsb.2019.10.001.

204 Sun, J., Wen, Y., Zhou, Y. et al. (2018). p53 attenuates acetaminophen-induced hepatotoxicity by regulating drug-metabolizing enzymes and transporter expression. *Cell Death Dis.* 9: 536. https://doi.org/10.1038/s41419-018-0507-z.

205 Wang, L.Q. and James, M.O. (2006). Inhibition of sulfotransferases by xenobiotics. *Curr. Drug. Metab* 7: 83–104. https://doi.org/10.2174/138920006774832596.

206 Pacifici, G.M. (2016). Metabolism and pharmacokinetics of morphine in neonates: a review. *Clinics (Sao Paulo)* 71: 474–480. https://doi.org/10.6061/clinics/2016(08)11.

207 Bian, H.S., Ngo, S.Y., Tan, W. et al. (2007). Induction of human sulfotransferase 1A3 (SULT1A3) by glucocorticoids. *Life Sci.* 81: 1659–1667. https://doi.org/10.1016/j.lfs.2007.09.029.

208 Gaston, T.E., Mendrick, D.L., Paine, M.F. et al. (2020). "Natural" is not synonymous with "Safe": toxicity of natural products alone and in combination with pharmaceutical agents. *Regul. Toxicol. Pharm.* 113: 104642. https://doi.org/10.1016/j.yrtph.2020.104642.

209 Landbo, C. and Almdal, T.P. (1998). Interaction between warfarin and coenzyme Q10. *Ugeskr Laeger* 160: 3226–3227.

210 Spigset, O. (1994). Reduced effect of warfarin caused by ubidecarenone. *Lancet* 344: 1372–1373. https://doi.org/10.1016/s0140-6736(94)90736-6.

211 Page, R.L. and Lawrence, J.D. (1999). Potentiation of warfarin by dong quai. *Pharmacotherapy* 19: 870–876. https://doi.org/10.1592/phco.19.10.870.31558.

212 Theoharides, T.C. (1997). Sudden death of a healthy college student related to ephedrine toxicity from a ma huang-containing drink. *J. Clin. Psychopharmacol.* 17: 437–439. https://doi.org/10.1097/00004714-199710000-00025.

213 Chung, K.F., Dent, G., McCusker, M. et al. (1987). Effect of a ginkgolide mixture (BN 52063) in antagonising skin and platelet responses to platelet activating factor in man. *Lancet* 1: 248–251. https://doi.org/10.1016/s0140-6736(87)90066-3.

214 Galluzzi, S., Zanetti, O., Binetti, G. et al. (2000). Coma in a patient with Alzheimer's disease taking low dose trazodone and gingko biloba. *J. Neurol. Neurosurg. Psychiatry* 68: 679–680. https://doi.org/10.1136/jnnp.68.5.679a.

215 Vuksan, V., Sievenpiper, J.L., Koo, V.Y. et al. (2000). American ginseng (*Panax quinquefolius* L) reduces postprandial glycemia in nondiabetic subjects and subjects with type 2 diabetes mellitus. *Arch. Intern. Med.* 160: 1009–1013. https://doi.org/10.1001/archinte.160.7.1009.

216 Zhu, M., Chan, K.W., Ng, L.S. et al. (1999). Possible influences of ginseng on the pharmacokinetics and pharmacodynamics of warfarin in rats. *J. Pharm. Pharmacol.* 51: 175–180. https://doi.org/10.1211/0022357991772105.

217 Monauni, T., Zenti, M.G., Cretti, A. et al. (2000). Effects of glucosamine infusion on insulin secretion and insulin action in humans. *Diabetes* 49: 926–935. https://doi.org/10.2337/diabetes.49.6.926.

218 Monostory, K., Vereczkey, L., Levai, F., and Szatmari, I. (1998). Ipriflavone as an inhibitor of human cytochrome P450 enzymes. *Br. J. Pharmacol.* 123: 605–610. https://doi.org/10.1038/sj.bjp.0701633.

219 Hartter, S., Grozinger, M., Weigmann, H. et al. (2000). Increased bioavailability of oral melatonin after fluvoxamine coadministration. *Clin. Pharmacol. Ther.* 67: 1–6. https://doi.org/10.1067/mcp.2000.104071.

220 von Bahr, C., Ursing, C., Yasui, N. et al. (2000). Fluvoxamine but not citalopram increases serum melatonin in healthy subjects – an indication that cytochrome P450 CYP1A2 and CYP2C19 hydroxylate melatonin. *Eur. J. Clin. Pharmacol.* 56: 123–127. https://doi.org/10.1007/s002280050729.

221 Johne, A., Schmider, J., Brockmoller, J. et al. (2002). Decreased plasma levels of amitriptyline and its metabolites on comedication with an extract from St. John's wort (Hypericum perforatum). *J. Clin. Psychopharmacol.* 22: 46–54. https://doi.org/10.1097/00004714-200202000-00008.

222 Obach, R.S. (2000). Inhibition of human cytochrome P450 enzymes by constituents of St. John's Wort, an herbal preparation used in the treatment of depression. *J. Pharmacol. Exp. Ther.* 294: 88–95.

223 Ohnishi, N., Okada, K., Yoshioka, M. et al. (2002). Studies on interactions between traditional herbal and western medicines. V. Effects of Sho-saiko-to (Xiao-Cai-hu-Tang) on the pharmacokinetics of carbamazepine in rats. *Biol. Pharm. Bull.* 25: 1461–1466. https://doi.org/10.1248/bpb.25.1461.

224 Sun, R., Basu, S., Zeng, M. et al. (2019b). Xiao-Chai-Hu-Tang (XCHT) intervening irinotecan's disposition: the potential of XCHT in alleviating irinotecan-induced diarrhea. *Curr. Cancer Drug Targets* 19: 551–560. https://doi.org/10.2174/1568009618666181029153255.

225 Bailey, D.G., Malcolm, J., Arnold, O., and Spence, J.D. (1998). Grapefruit juice-drug interactions. *Br. J. Clin. Pharmacol.* 46: 101–110. https://doi.org/10.1046/j.1365-2125.1998.00764.x.

226 Werba, J.P., Misaka, S., Giroli, M.G. et al. (2018). Update of green tea interactions with cardiovascular drugs and putative mechanisms. *J. Food Drug Anal.* 26: S72–S77. https://doi.org/10.1016/j.jfda.2018.01.008.

227 Durr, D., Stieger, B., Kullak-Ublick, G.A. et al. (2000). St John's wort induces intestinal P-glycoprotein/MDR1 and intestinal and hepatic CYP3A4. *Clin. Pharmacol. Ther.* 68: 598–604. https://doi.org/10.1067/mcp.2000.112240.

228 Piscitelli, S.C., Burstein, A.H., Chaitt, D. et al. (2000). Indinavir concentrations and St John's wort.

*Lancet* 355: 547–548. https://doi.org/10.1016/S0140-6736(99)05712-8.

229 Roby, C.A., Anderson, G.D., Kantor, E. et al. (2000). St John's Wort: effect on CYP3A4 activity. *Clin. Pharmacol. Ther.* 67: 451–457. https://doi.org/10.1067/mcp.2000.106793.

230 Adiwidjaja, J., Boddy, A.V., and McLachlan, A.J. (2020). Potential for pharmacokinetic interactions between *Schisandra sphenanthera* and bosutinib, but not imatinib: in vitro metabolism study combined with a physiologically-based pharmacokinetic modelling approach. *Br. J. Clin. Pharmacol.* 86: 2080–2094. https://doi.org/10.1111/bcp.14303.

231 Brantley, S.J., Gufford, B.T., Dua, R. et al. (2014). Physiologically based pharmacokinetic modeling framework for quantitative prediction of an herb-drug interaction. *CPT: Pharmacometrics Syst. Pharmacol.* 3: e107. https://doi.org/10.1038/psp.2013.69.

232 Cox, E.J., Maharao, N., Patilea-Vrana, G. et al. (2019). A marijuana-drug interaction primer: precipitants, pharmacology, and pharmacokinetics. *Pharmacol. Ther.* 201: 25–38. https://doi.org/10.1016/j.pharmthera.2019.05.001.

233 Hellum, B.H., Hu, Z., and Nilsen, O.G. (2007). The induction of CYP1A2, CYP2D6 and CYP3A4 by six trade herbal products in cultured primary human hepatocytes. *Basic Clin. Pharmacol. Toxicol.* 100: 23–30. https://doi.org/10.1111/j.1742-7843.2007.00011.x.

234 Moon, A. and Kim, S.H. (1997). Effect of *Glycyrrhiza glabra* roots and glycyrrhizin on the glucuronidation in rats. *Planta Med.* 63: 115–119. https://doi.org/10.1055/s-2006-957625.

235 Cao, Y.F., He, R.R., Cao, J. et al. (2012). Drug-drug interactions potential of icariin and its intestinal metabolites via inhibition of intestinal UDP-glucuronosyltransferases. *Evid. Based Complement Altern. Med.* 2012: 395912. https://doi.org/10.1155/2012/395912.

236 Tian, D.D., Kellogg, J.J., Okut, N. et al. (2018). Identification of intestinal UDP-glucuronosyltransferase inhibitors in green tea (*Camellia sinensis*) using a biochemometric approach: application to raloxifene as a test drug via in vitro to in vivo extrapolation. *Drug Metab. Dispos.* 46: 552–560. https://doi.org/10.1124/dmd.117.079491.

237 Nishimuta, H., Ohtani, H., Tsujimoto, M. et al. (2007). Inhibitory effects of various beverages on human recombinant sulfotransferase isoforms SULT1A1 and SULT1A3. *Biopharm. Drug Dispos.* 28: 491–500. https://doi.org/10.1002/bdd.579.

238 Hodges, R.E. and Minich, D.M. (2015). Modulation of metabolic detoxification pathways using foods and food-derived components: a scientific review with clinical application. *J. Nutr. Metab* 2015: 760689. https://doi.org/10.1155/2015/760689.

239 Springer, M. and Moco, S. (2019). Resveratrol and its human metabolites-effects on metabolic health and obesity. *Nutrients* 11: https://doi.org/10.3390/nu11010143.

240 Clarke, G., Sandhu, K.V., Griffin, B.T. et al. (2019). Gut reactions: breaking down xenobiotic-microbiome interactions. *Pharmacol. Rev.* 71: 198–224. https://doi.org/10.1124/pr.118.015768.

241 Kang, M.J., Kim, H.G., Kim, J.S. et al. (2013). The effect of gut microbiota on drug metabolism. *Expert Opin. Drug Metab. Toxicol.* 9: 1295–1308. https://doi.org/10.1517/17425255.2013.807798.

242 Li, H., He, J., and Jia, W. (2016). The influence of gut microbiota on drug metabolism and toxicity. *Expert Opin. Drug Metab. Toxicol.* 12: 31–40. https://doi.org/10.1517/17425255.2016.1121234.

243 Tarasiuk, A. and Fichna, J. (2019). Gut microbiota: what is its place in pharmacology? *Expert Rev. Clin. Pharmacol.* 12: 921–930. https://doi.org/10.1080/17512433.2019.1670058.

244 Zhang, J., Lacroix, C., Wortmann, E. et al. (2019). Gut microbial beta-glucuronidase and glycerol/diol dehydratase activity contribute to dietary heterocyclic amine biotransformation. *BMC Microbiol.* 19: 99. https://doi.org/10.1186/s12866-019-1483-x.

245 Oliphant, K. and Allen-Vercoe, E. (2019). Macronutrient metabolism by the human gut microbiome: major fermentation by-products and their impact on host health. *Microbiome* 7: 91. https://doi.org/10.1186/s40168-019-0704-8.

246 Rastelli, M., Knauf, C., and Cani, P.D. (2018). Gut microbes and health: a focus on the mechanisms linking microbes, obesity, and related disorders. *Obesity (Silver Spring)* 26: 792–800. https://doi.org/10.1002/oby.22175.

247 Tarashi, S., Siadat, S.D., Ahmadi Badi, S. et al. (2019). Gut bacteria and their metabolites: which one is the defendant for colorectal cancer? *Microorganisms* 7: https://doi.org/10.3390/microorganisms7110561.

248 Sun, C., Chen, L., and Shen, Z. (2019a). Mechanisms of gastrointestinal microflora on drug metabolism in clinical practice. *Saudi Pharm. J.* 27: 1146–1156. https://doi.org/10.1016/j.jsps.2019.09.011.

249 Narhi, L.O. and Fulco, A.J. (1986). Characterization of a catalytically self-sufficient 119,000-dalton cytochrome P-450 monooxygenase induced by barbiturates in Bacillus megaterium. *J. Biolumin. Chemilumin.* 261: 7160–7169.

250 Kim, K.A., Jung, I.H., Park, S.H. et al. (2013). Comparative analysis of the gut microbiota in people with different levels of ginsenoside Rb1 degradation to compound K. *PLoS One* 8: e62409. https://doi.org/10.1371/journal.pone.0062409.

251 Swanson HI. Drug metabolism by the host and gut microbiota: a partnership or rivalry? *Drug Metab Dispos.* 2015;43(10):1499-1504. https://doi.org/10.1124/dmd.115.065714.

252 Sharma, A. and Lee, H.J. (2020). Ginsenoside compound K: insights into recent studies on pharmacokinetics and health-promoting activities. *Biomolecules* 10: https://doi.org/10.3390/biom10071028.

253 Hardwick, R.N., Fisher, C.D., Street, S.M. et al. (2012). Molecular mechanism of altered ezetimibe disposition in nonalcoholic steatohepatitis. *Drug Metab. Dispos.* 40: 450–460. https://doi.org/10.1124/dmd.111.041095.

254 Kemp, D.C., Fan, P.W., and Stevens, J.C. (2002). Characterization of raloxifene glucuronidation in vitro: contribution of intestinal metabolism to presystemic clearance. *Drug Metab. Dispos.* 30: 694–700. https://doi.org/10.1124/dmd.30.6.694.

255 Wang, F., Meng, J., Zhang, L. et al. (2018). Morphine induces changes in the gut microbiome and metabolome in a morphine dependence model. *Sci. Rep.* 8: 3596. https://doi.org/10.1038/s41598-018-21915-8.

256 Postler, T.S. and Ghosh, S. (2017). Understanding the holobiont: how microbial metabolites affect human health and shape the immune system. *Cell Metab.* 26: 110–130. https://doi.org/10.1016/j.cmet.2017.05.008.

257 Kim, D.H., Hong, S.W., Kim, B.T. et al. (2000). Biotransformation of glycyrrhizin by human intestinal bacteria and its relation to biological activities. *Arch. Pharmacal Res.* 23: 172–177. https://doi.org/10.1007/BF02975509.

258 Martin, A.M., Sun, E.W., Rogers, G.B., and Keating, D.J. (2019). The influence of the gut microbiome on host metabolism through the regulation of gut hormone release. *Front. Physiol.* 10: 428. https://doi.org/10.3389/fphys.2019.00428.

259 Kolars, J.C., Awni, W.M., Merion, R.M., and Watkins, P.B. (1991). First-pass metabolism of cyclosporin by the gut. *Lancet* 338: 1488–1490. https://doi.org/10.1016/0140-6736(91)92302-i.

260 Soepenberg, O., Dumez, H., Verweij, J. et al. (2005). Phase I and pharmacokinetic study of oral irinotecan given once daily for 5 days every 3 weeks in combination with capecitabine in patients with solid tumors. *J. Clin. Oncol.* 23: 889–898. https://doi.org/10.1200/JCO.2005.01.008.

261 Ando, M., Hasegawa, Y., and Ando, Y. (2005). Pharmacogenetics of irinotecan: a promoter polymorphism of UGT1A1 gene and severe adverse reactions to irinotecan. *Invest. New Drugs* 23: 539–545. https://doi.org/10.1007/s10637-005-4022-6.

262 Galligan, J.J. and Akbarali, H.I. (2014). Molecular physiology of enteric opioid receptors. *Am. J. Gastroenterol. Suppl.* 2: 17–21.

263 Osborne, P.B., Chieng, B., and Christie, M.J. (2000). Morphine-6 beta-glucuronide has a higher efficacy than morphine as a mu-opioid receptor agonist in the rat locus coeruleus. *Br. J. Pharmacol.* 131: 1422–1428. https://doi.org/10.1038/sj.bjp.0703697.

264 Pergolizzi, J.V. Jr., Raffa, R.B., Pappagallo, M. et al. (2017). Peripherally acting mu-opioid receptor antagonists as treatment options for constipation in noncancer pain patients on chronic opioid therapy. *Patient Prefer Adherence* 11: 107–119. https://doi.org/10.2147/PPA.S78042.

265 Gerrits, C.J., Schellens, J.H., Burris, H. et al. (1999). A comparison of clinical pharmacodynamics of different administration schedules of oral topotecan (Hycamtin). *Clin. Cancer Res.* 5: 69–75.

266 Bhatt, A.P., Pellock, S.J., Biernat, K.A. et al. (2020). Targeted inhibition of gut bacterial beta-glucuronidase activity enhances anticancer drug efficacy. *Proc. Natl. Acad. Sci. U.S.A.* 117: 7374–7381. https://doi.org/10.1073/pnas.1918095117.

# 12

# Liver Drug Metabolism

*Ritika Kurian, Leslie T. Steen, and Hongbing Wang*

*Department of Pharmaceutical Sciences, University of Maryland School of Pharmacy, Baltimore, MD, USA*

## 12.1 Introduction

Following absorption across the gut lumen, drugs and nutrients enter via the portal blood to the liver, where they encounter a myriad of enzymes, which are collectively responsible for preparing all substances, endogenous or exogenous, for further tissue distribution and elimination. The liver has evolved a pathway of metabolism and elimination for xenobiotics and endobiotics that can be easily divided into three phases. Phase I consists of a mixed-function oxidase system responsible for oxidation, reduction, or hydrolysis of substrates to make them increasingly polar and better suited for phase II metabolism. In phase II, conjugating enzymes introduce endogenous hydrophilic moieties to substrate compounds, leading to a significant increase in their polarity and hydrophilicity. Lastly, phase III involves the removal of these conjugated and oxidized substrates by efflux transporters into the bloodstream or bile for excretion in urine or faeces, respectively. The bioavailability of a particular drug depends on which enzymes recognize the drug and their metabolism processes, the abundance of these enzymes and the inducibility and/or inhibition of these enzymes. This chapter introduces the major organ of metabolism, the liver and concentrates on the basics of phase I and phase II metabolism and their overall impact on drug bioavailability.

## 12.2 Hepatic Structure and Function

As the central hub for metabolism of pharmaceutical agents as well as all other ingested compounds, the liver plays a pivotal role in the fields of pharmacology and toxicology. Located below the diaphragm and to the right of the stomach, both the location and anatomy of the liver support its multiple and diverse functions. The liver is composed mainly of hepatocytes that are arranged in linear chords separated by spaces or sinusoids. The hepatocytes are organized into the basic functional units termed *lobules*. The lobule is an almost hexagonal-shaped cluster of hepatocytes divided and supported by a connective tissue scaffold, anchored by portal triads at the outside edges (which contain small branches of a hepatic artery, a portal vein, and a bile duct) and centralized on the central vein (Figure 12.1). Oxygenated blood from the heart reaches the liver through the hepatic artery, while nutrient-enriched venous blood from the intestinal tract arrives via the portal vein. In contrast with other organs, delivery of arterial blood from the heart via the hepatic artery makes up only 25% of the total liver blood supply. The remaining 75% is composed of venous blood that comes directly from the gut and intestine, further reinforcing the prominence of the liver in the process of metabolism. Smaller branches of the hepatic artery and the portal vein within the portal triad deliver blood to the lobule, which fills the sinusoids surrounding the hepatocytes and empties through the central vein. The sinusoids, lined with endothelial cells, do not come into direct contact with blood, but a "leaky" endothelial barrier created by these endothelial cells permits entry of oxygen, nutrients, and drugs into the apical surface of the hepatocyte. At the basal-lateral surface of the hepatocyte, a small intercellular space creates a canaliculus where bile is collected. Bile, produced by the hepatocytes, is transported to the small intestine through canaliculi and bile ducts and aids in digestion through emulsification of lipids. Bile flows at the basal-lateral surface of the hepatocyte in a direction opposite to the blood flow at the apical surface, from central vein to portal triad and finally emptys into the bile duct.

*Oral Bioavailability and Drug Delivery: From Basics to Advanced Concepts and Applications,* First Edition. Edited by Ming Hu and Xiaoling Li.

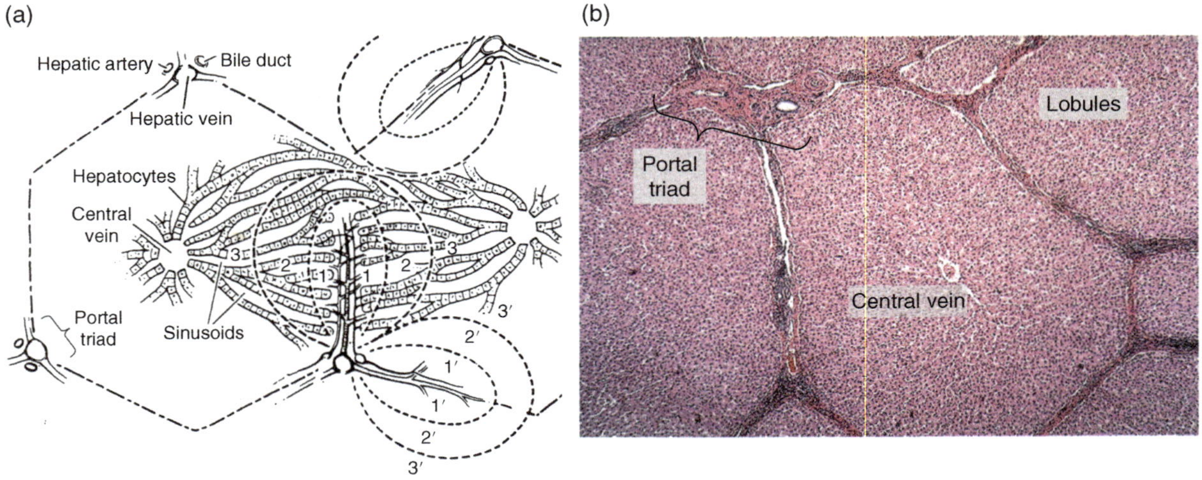


**Figure 12.1** Diagram of liver microstructure. (a) The architecture of the liver lobule with individual structures labeled. Hepatocytes are divided into zones based on proximity to the portal blood flow: (1) periportal, (2) midzonal, and (3) centrilobular. (b) Histological section of liver.

From the portal triad to the central vein, the lobule is divided into three zones based on a series of gradient functional changes: periportal (zone 1), midzonal (zone 2), and centrilobular (zone 3) (Figure 12.1). The function of hepatocytes in each of these zones is unique and discrete due to the differing environments and gradient changes among these zones. Hepatocytes in the periportal region (zone 1) receive oxygenated and nutrient-rich blood, whereas centrilobular hepatocytes (zone 3) encounter blood depleted of both nutrients and oxygen, which has been processed during its transit through the hepatic lobule. The midzonal region differs from both zone 1 and zone 3 and is believed to have an intermediate environment. Midzonal necrosis has been observed in patients with yellow fever and is also associated with patients who have suffered episodes of shock before their deaths [1]. Hepatocytes also display a maturation process from zone 1 to zone 3, where smaller hepatocytes are located in zone 1 and the larger, more mature hepatocytes are found in zone 3. Notably, this maturation phenomenon coincides with the regional distribution of several major cytochrome P450 (CYP) enzymes and metabolism-mediated liver toxicity. For instance, the analgesic acetaminophen (also known as Tylenol®) is metabolized to the more toxic intermediate, *N*-acetyl-quinoneimine (NAPQI), by CYP2E1 and CYP3A4, causing the primary lesion in acetaminophen toxicity to be located in the centrilobular region (zone 3), where both CYP2E1 and CYP3A4 are highly expressed [2].

The liver, classified as a digestive gland, is involved in several important functions such as energy homeostasis and cholesterol metabolism, filtering blood and processing nutrients, producing bile, and processing haemoglobin. In addition, the liver plays a major role in detoxification of endogenous compounds, such as ammonia and bile acids, and exogenous compounds, such as environmental contaminants and pharmaceutical drugs. Strategically positioned to receive blood directly from the gut and small intestine, the liver is the first organ to encounter all orally absorbed compounds and is therefore endowed with enzymes and proteins capable of neutralizing potentially harmful insults. To be therapeutically effective, orally delivered drugs, specifically, must endure the harsh environment of the stomach as well as the enzymatic attack by the liver before reaching the systemic circulation or a therapeutic target. This process is referred to as first-pass metabolism and is the cause of a significant loss of available drug fraction after oral administration. Propranolol, a beta channel blocker, is highly prone to first pass metabolism leaving <30% of the dose available for absorption into systemic circulation, but the reduced drug fraction does not impair its antihypertensive effect. However, nitroglycerin, which is extensively used as a vasodilator for treatment of anginal attacks is also affected by first pass metabolism and fails to attain therapeutic plasma concentrations after oral administration, restricting its administration to sublingual, intravenous or transdermal exposure routes.

The remainder of this chapter will focus on the enzymes and mechanisms that contribute to pharmaceutical drug metabolism and their impact on the bioavailability of orally administered drugs.

## 12.3 Phase I Drug Metabolism

The basic need for a system of metabolism or biotransformation is driven by the general lipophilic nature of xenobiotics and endobiotics, including drugs, industrial chemicals, environmental pollutants, pesticides, dietary constituents, naturally occurring toxins, steroid hormones, vitamins, bile acids, fatty acids, and eicosanoids. These compounds freely cross lipid membranes and travel through the blood via lipoproteins, making it difficult for the body to control the biological or pharmacological activity of such lipophilic molecules. Thus, the goal of phase I metabolism is to make lipophilic substrates more water soluble by modifying their chemical structure, and therefore more excretable. Phase I reactions, including oxidation, reduction, and hydrolysis, involve introduction of functional groups such as –OH, $-NH_2$, –SH, or –COOH, and this usually results in modest increases in polarity and hydrophilicity relative to the parent compounds. Phase I enzymes include the CYPs, flavin monooxygenases (FMOs), dehydrogenases, reductases, and esterases [3–5]. The CYP family of metabolizing enzymes makes the largest contribution to phase I metabolism of xenobiotics, is responsible for the metabolism of nearly 75% of the clinically approved drugs [6], and thus will be the focus of this section.

### 12.3.1 Cytochrome P450

#### 12.3.1.1 General Characteristics

Cytochrome P450 proteins (CYPs) are heme-containing pigments found in the microsomes of hepatocytes. Located in the lipid-rich environment of the smooth endoplasmic reticulum, these monooxygenases are appropriately positioned to encounter hydrophobic xenobiotics seeking refuge from the more aqueous environment of the cytoplasm. CYPs were first characterized by their ability to bind oxygen and carbon monoxide, thus, were named for their maximum wavelength absorption at 450 nm when bound to carbon monoxide. All other known carbon monoxide-binding hemoproteins have a characteristic maximum absorption at 420 nm, making this characteristic of CYPs unique.

Although the liver represents the major site of CYP expression, extrahepatic tissues such as the small intestine, colon, kidney, lung, skin, placenta, heart, and the brain also display P450 expression. Individual organs do not necessarily express the variety or abundance of CYP enzymes found in the liver, but localized expression of P450 enzymes allows for organ-specific metabolism and localized detoxification. This is an important consideration for compounds or metabolites that have a tendency to accumulate in a particular organ or tissue, or when evaluating pharmaceutical drugs administered transdermally or by inhalation.

Measurement of CYP enzymes across the human population has led to the discovery of significant interindividual variation in expression and activity. This variation was observed in nearly all CYP enzymes and results from complex interactions occurring among genetic, environmental, and pathophysiological factors. Historical evidence demonstrated that CYP expression was known to be inducible, but modern technology identified xenobiotic receptors, namely, pregnane X receptor (PXR), constitutive active receptor (CAR), and aryl hydrocarbon receptor (AhR) as the mechanisms of xenobiotic-mediated induction of most CYP enzymes. Inducer compounds act by binding to the receptor and interacting with response elements located in the promoter region of individual CYP enzymes to initiate transcription, enhance expression, and accelerate enzymatic activity. Alternatively, repression of CYP expression has also been observed, but the mechanisms of this change are less well understood. Under conditions of infection and inflammation, the release of cytokines and other inflammatory by-products can suppress gene transcription and leads to a decrease in CYP enzyme expression and activity [7]. Within the CYP gene sequence itself, single nucleotide polymorphisms (SNPs) have been found in the promoter and coding regions affecting the expression and/or activity of the enzyme. Pharmaceutical drugs metabolized by polymorphic CYP enzymes can have significantly altered efficacy and toxicity. Toxic side effects seen in patients taking debrisoquine were helpful in identifying the polymorphisms in CYP2D6 [8]. Polymorphisms of CYP2C9 and CYP2B6 with decreased activity lead to reduced metabolism of substrate drugs like losartan and efavirenz, respectively [9, 10].

#### 12.3.1.2 Biochemistry

Structurally, the CYP enzyme is a macrocyclic porphyrin molecule consisting of a centrally bound iron (Fe) ion (Figure 12.2). The porphyrin molecule is composed of four pyrrole rings connected by carbon bonds, and the pyrrolic nitrogen atoms are oriented towards the center of the ring, creating a region for binding metal ions. The properties of an individual porphyrin molecule are solely determined by the metal ion bound in the center. Porphyrin molecules are found throughout biology with a variety of metal ions bound to the center. For example, a porphyrin molecule with a centrally bound magnesium (Mg) ion, also known as *chlorophyll*, helps plants absorb the sun's energy, while a porphyrin molecule

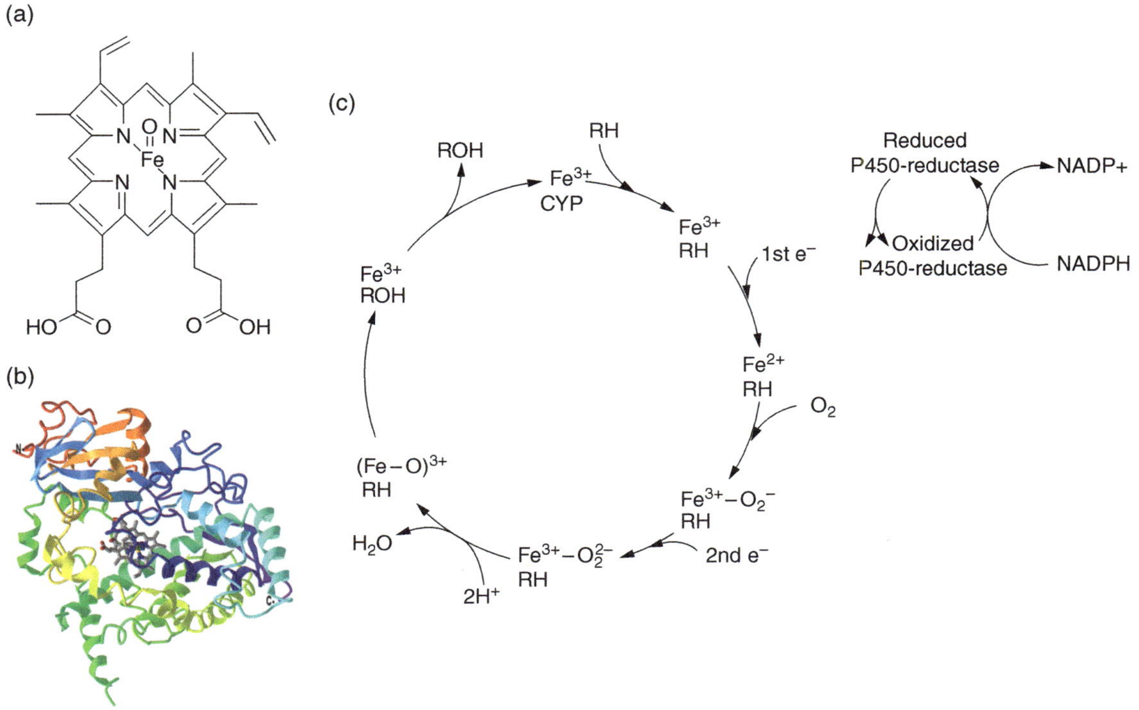


**Figure 12.2** Cytochrome P450s and the catalytic cycle. (a) The basic structure of the prophyrin molecule, P450 with a centrally bound iron (Fe). (b) 3D ribbon structure of CYP3A4 is colored from N-terminus to C-terminus in a progression from blue to green to yellow to red. The ball-and-stick model in the center represents the heme group. (c) The catalytic cycle of P450 starts with the addition of substrate (RH) to a resting heme ($Fe^{3+}$). Source: Adapted from Williams et al. [11].

bound with cobalt (Co) in the center represents the essential vitamin B12. Porphyrin molecules, such as hemoglobin and CYP, have iron (Fe) bound in the center that enables their interactions with molecular oxygen. The iron or heme center present in hemoglobin supports the efficient transport and storage of oxygen, but the heme center of the CYP enzymes allows for activation and utilization of oxygen. CYPs catalyze a monooxygenation reaction, which is characterized by using one molecule of diatomic molecular oxygen ($O_2$) to oxidize a substrate while the other molecule is reduced to water. The basic monooxygenation reaction catalyzed by P450s is shown here:

$$R + O_2 + NADPH + H^+ = ROH + H_2O + NADP^+$$

where, "R" = the substrate, and "ROH" = the product.

In this reaction, NADPH is the electron donor providing electrons required for the reaction and the flavoprotein NADPH-P450 reductase is responsible for the transfer of electrons (Figure 12.2). Here, molecular oxygen ($O_2$) is reduced to water, the substrate (R) is transformed into an oxidized product (ROH), and the P450 enzyme is regenerated, ready to encounter another molecule and the cycle begins again. The catalytic cycle of P450 starts with the binding of a substrate followed by addition of the first electron resulting in a change to the charged state of iron within the heme moiety of the enzyme. Then, oxygen binds the complex and sequential addition of the second electron and protons helps release water, oxidized substrate, and CYP in its original state (Figure 12.2). The power in this system comes from (i) a multiplicity of P450 isozymes with variable substrate profiles; (ii) the redundancy of available electron donors; and (iii) the versatile iron core of the cytochrome protein itself.

Despite the efficiency of the CYP catalytic cycle, the process of reducing molecular oxygen to water is inherently dangerous due to the formation of highly reactive intermediates (Figure 12.3). Interruption of the P450 catalytic cycle between substrate and oxidized product can cause the release of reactive oxygen species, superoxide anion ($O_2^{\cdot-}$), hydrogen peroxide ($H_2O_2$), or other free radical intermediates. Free radicals, defined by possession of a single unpaired electron, may be charged or neutral molecules that initiate a radical cascade by interacting with molecular oxygen and generating reactive oxygen species (Figure 12.3). These reactive byproducts are prone to damaging membranes (via lipid peroxidation) and DNA (via strand breaks), but they are also capable

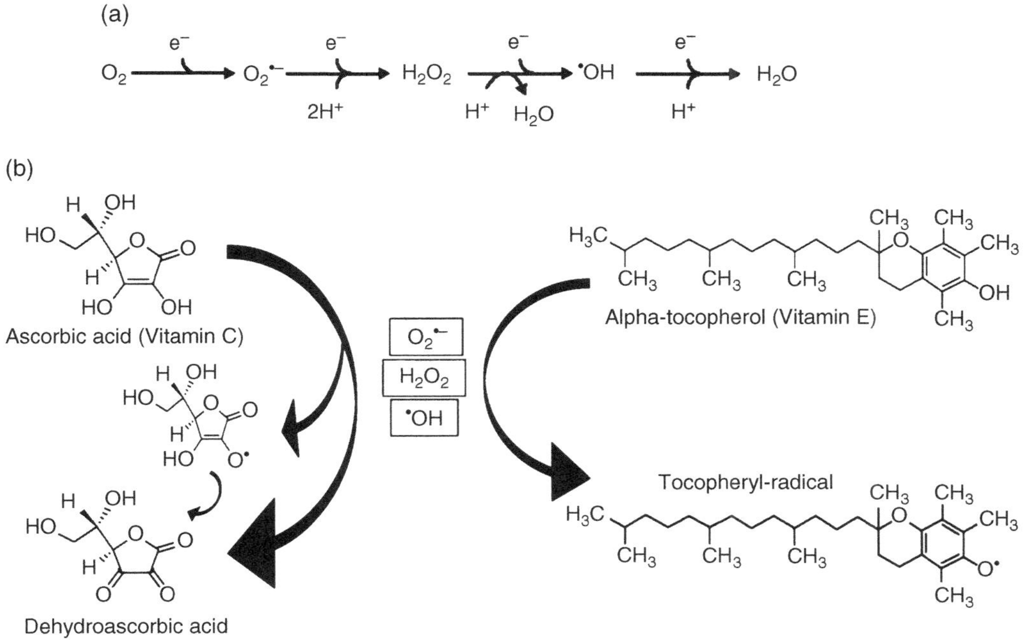


**Figure 12.3** Reactive oxygen species. (a) The conversion of molecular oxygen to water has the potential to generate reactive intermediates by addition of individual electrons ($e^-$) or protons ($H^+$). (b) Antioxidants, vitamins C and E donate a proton(s) to the reactive oxygen radicals ($O_2^{\cdot}$, $H_2O_2$, $^{\cdot}OH$) and prevent propagation of reactive species and subsequent cellular damage. In response to reactive oxygen species, Vitamin C (ascorbic acid, AA) is oxidized to dehydroascorbic acid (DAA) via a semidehydro-radical intermediate (SDAA), which interrupts oxygen radical propagation. Vitamin C is regenerated by (1) reaction of 2 SDAA → 1AA + 1 DAA; and (2) oxidation of glutathione (2 GSH → GSSG). Vitamin E (α-tocopherol) terminates the chain of reactive propagation reactions via a tocopheryl-radical, which uses its chromanol ring structure to stabilize the unpaired electron and is subsequently reduced by ascorbic acid to its original form. Source: Groff et al. [12] / West Publishing.

of activating other molecules and propagating a state of oxidative stress within the cell. Oxidative stress occurs from an imbalance between the collective action of all reactive oxygen species and the mechanisms to detoxify them, resulting in damage to cellular constituents and overall cell function.

As discussed above, detoxification is the primary objective of P450-mediated metabolism, but some parent compounds are metabolized to active metabolites or toxic byproducts under a process called bioactivation. Bioactivation of a parent compound may be an undesirable outcome in most cases, yet some pharmaceutical drugs have been designed to take advantage of this process in order to exert their therapeutic effect. For example, tamoxifen (TAM), long used for the treatment of women with metastatic breast cancer, is referred to as a prodrug because it requires bioactivation to its secondary metabolite, 4-hydroxy-*N*-desmethyl-TAM (endoxifen), which displays strong binding affinity towards estrogen receptors and causes suppression of estrogen-dependent cell growth [13–15]. TAM metabolites demonstrate almost 30-100-fold greater potency than the parent compound in decreasing estrogen-dependent cell proliferation. Other chemotherapeutics, such as cyclophosphamide and ifosfamide, require metabolic activation to the reactive metabolite, phosphoramide mustard, which displays the cytotoxic qualities [8, 16, 17]. In addition to activating pharmaceutical prodrugs, CYP-mediated metabolism is the predominant process generating carcinogenic metabolites from exogenous compounds. Ironically, bioactivation can be responsible for the initiation of the carcinogenic process and for activating the drug necessary for its cure.

Reactive products generated due to bioactivation (or metabolic activation) can be broadly classified into two categories: electrophiles or free radicals. Electrophilic metabolites (containing a positive center) elicit their toxic effect directly by binding nonenzymatically to nucleophilic molecules such as RNA, DNA, peptides, and proteins. For example, acetaminophen is metabolized by P450s to a quinonimine intermediate (NAPQI), which actively binds hepatic proteins, resulting in liver necrosis (Figure 12.4). Hepatocytes have developed mechanisms to neutralize the harmful effects of these bioactivated metabolites – antioxidants, such

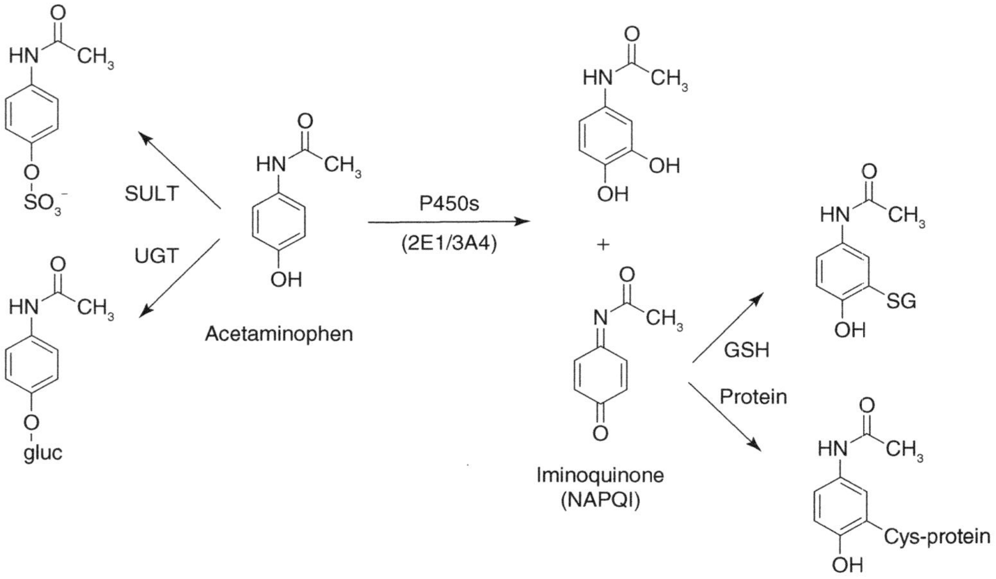


**Figure 12.4** Acetaminophen metabolism. The metabolism of acetaminophen illustrates a number of mechanisms from phase I and phase II reactions.

as vitamins C and E, interact with reactive oxygen species, ending the cycle of radical propagation [18] (Figure 12.3), and the tripeptide, glutathione (GSH), is an abundant and effective quencher of electrophilic molecules. While bioactivation reactions make up a very small portion of metabolic oxidations, their products are a primary cause of pathology and disease, and, thus, a primary concern in toxicity and drug development. Fortunately, the vast majority of phase I reactions successfully oxidize their substrate and the product proceeds to phase II metabolism or efflux excretion.

#### 12.3.1.3 Nomenclature

CYPs are a superfamily of hemoproteins with hundreds of members found in all species, from plants and bacteria to vertebrates and mammals. Historically, P450s were named based on their location and function, but rapidly increasing number of identified P450s and the observed overlap of location and function of characterized P450s led to confusion when these trivial names used. In 1993, a P450 nomenclature committee devised a plan, individual enzymes that share ≥40% identity are grouped into a family (denoted by an Arabic numeral, CYP1), those sharing ≥55% identity are classified into subfamilies (denoted by a letter, CYP2B, CYP2C), and individual enzymes within the same subfamily are identified by the second Arabic numeral (CYP2C8, CYP2C9) [19, 20].

Among the more than 100 identified mammalian families of CYP proteins, 18 CYP families are expressed in human tissues, and 57 human genes have been identified that encode for CYP proteins [21, 22]. The vast majority of CYP expression is dedicated to homeostatic functions, including the metabolism of vitamins, fatty acids, bile acids, prostaglandins, hormones, and cholesterol [23]. Members of the CYP families 11B, 17A, 19A, and 21A all coordinately participate in enzymatic alterations of a cholesterol precursor to synthesize endogenous hormones such as cortisol, testosterone, and estradiol. Families CYP7A, 8B, and 27A are involved in bile acid biosynthesis, and CYP51A enzymatically generates cholesterol from a precursor molecule [24, 25]. In addition to the CYPs responsible for homeostasis, members of CYP families 1–4 are exclusively dedicated to the metabolism of all exogenous compounds, including the food we eat and the drugs we take. The human CYP isozymes responsible for exogenous or xenobiotic metabolism are CYP1A1, 1A2, 2A6, 2B6, 2C8, 2C9, 2C19, 2D6, 2E1, and 3A4. An up-to-date list of clinically relevant drugs metabolized by each isozyme can be found at http://medicine.iupui.edu/clinpharm/ddis/table.asp.

All CYP enzymes share a basic structure but, subtle differences in the active substrate-binding site result in CYP isoforms having distinct but often overlapping substate profiles. The geometry, hydrophobic character, and size of the CYP-binding pocket determine the variable substrate profiles of CYP enzymes. CYP2C8 and 2C9 share 80% amino acid identity, yet differ in their enzymatic efficiency as well as the reactions catalyzed.

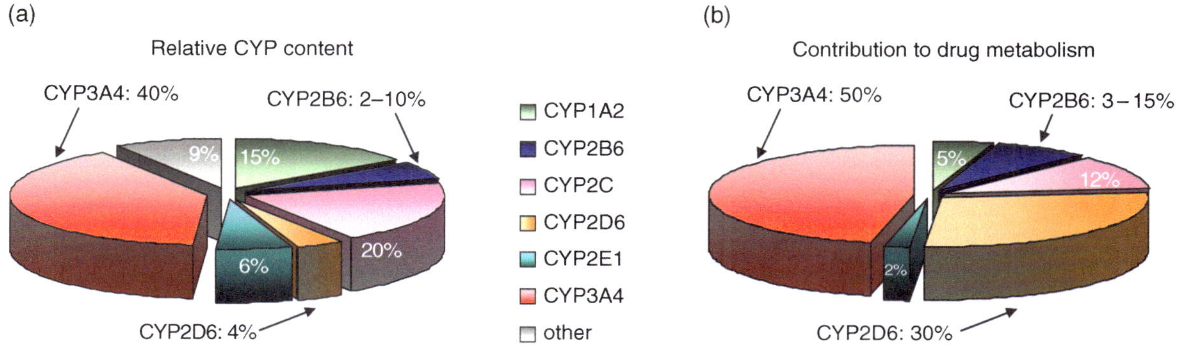


**Figure 12.5** Human liver CYP isoforms. The major hepatic CYP isoforms involved in xenobiotic metabolism are presented based on relative content (a) and contribution to metabolism (b). CYP3A4 is the most abundantly expressed CYP at 40% and is responsible for at least 50% of drug metabolism; meanwhile, CYP2D6 only represents 4% of CYP content but contributes 30% of CYP-mediated drug metabolism. CYP2B6, once thought to be of negligible importance, is expressed at a highly variable 2–10% of CYP content, and its contribution to drug metabolism is growing (3–15%) as researchers identify new substrate drugs. Source: Wang et al. [27] / BENTHAM SCIENCE PUBLISHERS LTD.

Using the substrate warfarin for comparison, CYP2C8 specifically generates the 5-hydroxy metabolite while CYP2C9 generates the 4-hydroxy metabolite [5, 26]. Among the CYPs that share a common substrate, there is usually a significant difference in affinity allowing for a specific interaction of enzyme and substrate, but providing a secondary, redundant mechanism should substrate concentrations exceed the capacity of the initial CYP enzyme. Differences in substrate affinities are also accompanied by significantly variable expression levels of individual isozymes. CYP3A4 is notoriously known as the most abundantly expressed CYP protein in the human liver, accounting for as much as 40% of total hepatic CYP content (Figure 12.5). On the contrary, CYP2D6 and CYP2E1 account for 4% and 6% of hepatic CYP content, respectively. It is important to understand the *in vivo* realities of metabolism, especially in the context of choosing an appropriate drug therapy.

#### 12.3.1.4 Induction

The induction of CYP enzymes responsible for xenobiotic metabolism provides a simple, adaptive mechanism for responding to xenobiotic challenges. Prior to an identified mechanism, it was known that a variety of inducer compounds displayed only a few known patterns of isozyme induction; this observation is now known to be mediated by nuclear receptors [28]. Currently, the most common molecular mechanism for P450 induction in response to xenobiotic compounds is receptor-mediated transcriptional activation. Additionally, mechanisms such as enhanced mRNA stabilization have been shown to specifically contribute to the xenobiotic-induced upregulation of CYP2E1 [29]. The transcriptional upregulation of P450s, such as CYP3A4, CYP2B6, CYP2Cs, and CYP1As, is mediated by a variety of structurally unrelated chemical inducers and has therapeutic consequences for pharmaceutical drugs.

The inducible nature of monooxygenase activity in the liver was discovered in the early 1970s followed by the identification and characterization of P450 enzymes in the mid-late 1980s [30]. The advent of DNA and cloning technologies in the 1990s led to the discovery of more than 2500 genes for P450 enzymes [22]. However, the mechanism of inducible P450 expression remained a mystery until the discovery and characterization of two orphan nuclear receptors, PXR and CAR, in the late 1990s [31–36]. Both nuclear receptors, categorized as "orphans" without a known endogenous ligand, contained a large and flexible ligand-binding pocket able to accommodate a variety of structures found in known inducer compounds [37, 38]. Similar to other nuclear receptor superfamily members, PXR and CAR execute their transcription effects by (i) binding ligand; (ii) heterodimerizing with a common partner, retinoid X receptor (RXR); and (iii) binding to conserved hexameric sequences, or response elements in the promoter region of regulated gene products. Further characterization of PXR and CAR, has determined that they also share a similar structure, overlapping ligand profiles, response elements, and target genes. Several metabolic genes belonging to phases I, II, and III, have been identified as target genes regulated by PXR, or CAR, or both. A summary of metabolic genes regulated by PXR and CAR is shown in Figure 12.6.

Discovery of PXR in the late 1990's was a major breakthrough in our understanding of the transcriptional regulation of major drug metabolizing enzymes. PXR was identified as the primary mechanism for

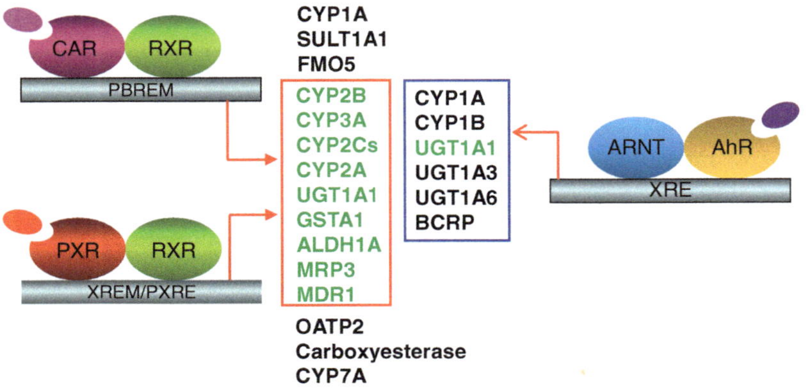


**Figure 12.6** Target genes of PXR, CAR, and AhR. On the left, PXR and CAR are shown dimerized to their common partner, RXR and sitting response elements XREM (xenobiotic response enhancer module) and PBREM (phenobarbital response enhancer module), respectively. Overlapping target genes are boxed in the center with CAR-specific targets shown above and PXR-specific targets shown below. On the right, AhR is shown bound to partner, ARNT and activating its target genes after binding to the XRE (xenobiotic response element; often called the dioxin response element). UGT1A1 represents a common target gene of all three receptors.

regulating inductive expression of CYP3A4, the isoform responsible for metabolism of nearly 50–60% of all pharmaceutical drugs and a critical determinant of bioavailability [39–42]. Several common and clinically relevant drugs, such as rifampicin, artemisinin, simvastatin, troglitazone, and spironolactone were identified as PXR ligands, and many of those were also known to be CYP3A4 substrates [28, 43, 44]. Based on this discovery, drugs, and other shared substrates of PXR and CYP3A4, have the ability to increase the expression of the enzyme responsible for their own metabolism, thereby accelerating their own rate of catalysis. However, this regulatory mechanism of shared substrates holds the very real possibility of unintended, yet devastating drug–drug interactions, and as such, early drug development has now included screening for PXR activation.

As the mechanisms of transcriptional regulations were uncovered, PXR and CAR were found to regulate CYP3A4, CYP2B6, and CYP2Cs, to a greater or lesser degree, but the CYP1As were already known to be primarily regulated by the AhR. Historically, AhR was known to mediate the deleterious effects of dioxins and related industrial compounds through transcriptional regulation of CYP1A1 and CYP1A2 [45, 46]. Today, CYP1A2 is a clinically relevant enzyme with common and pharmaceutical drug substrates such as caffeine, naproxen, haloperidol, and verapamil, and inducer compounds such as omeprazole (Prilosec®). Recent investigations of AhR have discovered its role in the receptor-mediated upregulation of other metabolic genes, shown in Figure 12.6. Different from nuclear receptors PXR and CAR, AhR belongs to the basic helix-loop-helix (bHLH) Per-ARNT-Sim (PAS) family of proteins where AhR is the only ligand-activated member. AhR recognizes specific response elements in the promoter of target genes as a heterodimer with a distinct partner ARNT (Ah receptor nuclear translocator) [47, 48]. Despite their basic differences, PXR, CAR, and AhR are often grouped as xenobiotic receptors; receptors responsible for binding exogenous compounds or drugs and mediating the upregulation of metabolic genes.

Interestingly, several of the identified inducers of CYP expression are also substrates of CYP-mediated metabolism suggesting that these compounds are effectively responsible for accelerating their own metabolism [30]. Under a circumstance where the parent inducer compound is toxic, then such an increase in metabolic activity would be beneficial and lower the potential for toxicity. However, if the parent inducer compound is metabolized to toxic intermediates, upregulating CYP expression would increase the potential for toxicity. Unfortunately, circumstances of xenobiotic-induced metabolism rarely involve only one compound with one enzyme. In reality where individuals are exposed to inducer and substrate compounds simultaneously, inducers affect not only their own metabolism but also that of any other co-administered drug as substrates of the induced enzyme. For example, cotreatment of the dietary supplement, St. John's Wort (SJW), used as an antidepressant, with oral contraceptives results in the SJW-mediated induction of CYP3A4 expression, accelerated metabolism of contraceptives by the induced CYP3A4 expression, and increased likelihood of unexpected pregnancy based on decrease therapeutic efficacy of the oral contraceptives; the same reaction of SJW was seen with co-administered immunosuppressants, HIV protease inhibitors and anticonvulsants [49–51]. Multidrug therapy regimens are widespread for the treatment of cancer, HIV, and aging, but with this regulatory mechanism in mind, the components of the drug "cocktail" are selected based on their ability to inhibit or upregulate enzymes responsible for affecting other components of the multidrug therapy. Physicians and pharmacists are aware of these drug behaviors and interactions, but need to be diligent, especially when evaluating the intended and unintended consequences of affecting the activity of CYP3A4 and other clinically relevant CYP isoforms.

#### 12.3.1.5 Inhibition

Contrary to the previously unknown mechanism of P450 enzyme induction, enzyme inhibition is already well characterized. In fact, enzyme inhibition is often the primary mechanism of action of a drug––the monoamine oxidase inhibitor (MAOI) class of antidepressants, angiotensin-converting enzyme (ACE) inhibitor family of antihypertensives, and antibiotics such as ciprofloxacin that inhibit topoisomerase II and topoisomerase IV. However, many drugs demonstrate the ability to inhibit P450-mediated enzymatic activity in addition to their intended inhibitory therapeutic effect. The consequences of CYP inhibition can vary from a minor increase in plasma concentrations to a sustained decrease in systemic clearance of substrate drugs, which could lead to harmful drug–drug interactions. Enzymatic inhibition occurs by multiple mechanisms: (i) the inhibitor interferes with the binding of substrate, (ii) the inhibitor reduces catalytic activity of the enzyme, or (iii) a combination of the two. The mechanisms of CYP inhibition can be divided into two basic classes: reversible and irreversible, the difference being the magnitude of the inhibitor's affinity for the enzyme.

Reversible inhibition occurs when the inhibitor binds to the enzyme noncovalently and establishes an equilibrium of binding and dissociation with the enzyme. Inhibitors can be broadly categorized as competitive, noncompetitive, or uncompetitive based on where they bind and in what order they bind with respect to substrate, and each category of inhibitor is predominantly defined by their effect on kinetic parameters. In competitive inhibition, the substrate and the inhibitor compete for the active site of the enzyme and the kinetic signature displays an increase in $K_m$ but leaves $V_{max}$ unaffected. In the case of most CYP enzymes, competitive inhibitors are often also substrates for the enzyme. Noncompetitive and uncompetitive inhibitions occur when the inhibitor binds to the allosteric site adjacent to the active site of the enzyme, resulting in altered enzymatic activity without competing with substrate binding. Noncompetitive inhibitors have an equal affinity for free enzyme (E) and the enzyme–substrate (ES) complex and therefore only decreases $V_{max}$. In contrast, uncompetitive inhibitors bind only the ES complex and display a decrease in both $K_m$ and $V_{max}$ values. More often, inhibitors fall into an intermediate or mixed inhibitor category that displays characteristics of competitive and noncompetitive inhibition. Mixed inhibitors can interfere with substrate binding, affect enzyme conformation, and alter catalytic activity by binding to both E and ES enzyme forms, but their affinity for E and ES are not the same. Similar to noncompetitive inhibitors, mixed inhibitors decrease $V_{max}$, but unequal affinities for E and ES leads to an increase in $K_m$.

Regardless of category or kinetic signature, reversible inhibition is common occurrence in multidrug therapies. For example, the antifungal itraconazole is a substrate and potent inhibitor of CYP3A4 and can considerably decrease the metabolism of a synthetic corticosteroid, methylprednisolone. The increase in the half-life and $C_{max}$ of methylprednisolone requires a 50% reduction in the dose of the corticosteroid to avoid unintended side effects [52]. The antihistamine terfenadine and the antifungal ketoconazole compete for binding to CYP3A4 [53]. Ketoconazole inhibits the CYP3A4-dependent metabolism of terfenadine, causing increased plasma concentrations of the parent drug, blocking cardiac potassium channels, and causing ventricular arrhythmias [53]. Alternatively, coadministration of CYP3A4 substrates and HIV protease inhibitors, ritonavir and saquinavir results in an improved pharmacokinetic profile of saquinavir following ritonavir-mediated inhibition of CYP3A4 activity [54]. Simply based on comparative affinities, the outcome of competitive inhibition, whether detrimental or beneficial, will depend on the characteristics of the individual substrate drugs.

Irreversible inhibition occurs when an enzyme is covalently modified by the inhibitor causing permanent alteration of the enzyme and preventing catalytic activity. Irreversible inhibitors usually contain reactive functional groups such as nitrogen mustards, aldehydes, and alkenes, which can react with nucleophilic amino acid side chains and form covalent adducts [55]. Despite differences in the mechanism of inhibition, all irreversible inhibitors share (i) a kinetic signature in which $K_m$ is unaffected, but $V_{max}$ is decreased and

(ii) the necessity of *de novo* protein synthesis to restore enzymatic activity. Synthesis of effective amounts of functional new CYP proteins takes time, during which enzyme inhibition continues and drug–drug interactions become increasingly evident. Irreversible inhibitors can directly alter the enzyme's active site by reacting with electrophilic functional groups and amino acid residues within the binding pocket. Following active site binding, irreversible inhibitors can also cause a conformational change in the active site preventing the enzyme from catalyzing or releasing the inhibitor, thereby, permanently inactivating the enzyme. While some irreversible inhibitors exhibit their inhibitory effect as parent drugs, most require metabolic conversion to a reactive metabolite responsible for covalent binding of the heme moiety or CYP apoprotein and inactivation of the enzyme [55]. Mechanism-based inactivation of CYP, also known as *suicide inhibition* occurs within the enzyme active site as the enzyme catalytically generates its own inhibitory molecule. For example, a calcium channel blocker, diltiazem, forms potent metabolites such as *N*-desmethyl diltiazem and *N*,*N*-didesmethyl diltiazem via CYP3A and the resultant P-450-iron (II)-metabolite complex can cause inhibition of CYP3A4 [56]. The broad-spectrum antibiotic chloramphenicol gets oxidized to oxamyl moiety that can acylate a lysine residue in the CYP active center, preventing catalytic turnover of the enzyme [57]. However, chloramphenicol inactivation of CYP enzymes is not uniform, as studies in rat liver microsomes have revealed that CYPA1 and CYP1A2 families are resistant to this inactivation [58].

Reversible and irreversible inhibition represent another critical consequence of co-administered drug treatments of which physicians and pharmacists need to be aware. Pharmaceutical drugs, however, are not the only factors participating in enzyme inhibition; other factors such as diet, vitamins, and herbal supplements can also contribute to the metabolism and efficacy of drug treatments. Most notable of the diet–drug interactions is the effect of grapefruit juice on efficacy and safety of co-administered drugs. Discovered accidentally, grapefruit juice specifically inhibits intestinal CYP3A4 activity, causing plasma concentrations of CYP3A4 substrate drugs such as statins and opioids to be greater than expected. Patients who consumed grapefruit juice while taking statin drugs experienced an unexpected increase in statin plasma concentrations and suffered serious, even lethal consequences due to this specific diet–drug interaction. Today, grapefruit juice consumption is strictly contraindicated for a variety of prescription drugs, such as statins, opioids, antihistamines, and others [59]. However, based on its localized site of action, grapefruit juice can be considered beneficial to drugs that suffer significant first pass metabolism and limited plasma concentrations. The calcium channel blocker felodipine suffers extensive CYP3A4-mediated catalysis during first pass metabolism, but consumption of grapefruit juice leads to greater than five-fold increase in plasma concentrations [59–61]. This interaction could be used to adjust the therapeutic index of felodipine, making it an effective therapy in an appropriate patient population. Grapefruit juice is not alone in its ability to cause unexpected drug interactions. Cranberry juice has been found to inhibit CYP2C9, known to be responsible for the metabolism of the anticoagulant warfarin, and has the potential to increase the international normalized ratio (INR) of patients taking warfarin, greatly increasing the likelihood of toxic consequences among such patients [62]. Although enzyme inhibition has been previously adopted as a strategy to exert therapeutic effects, understanding the various mechanisms of CYP inhibition can help circumvent potential drug–drug interactions resulting in serious consequences.

### 12.3.2 Additional Phase I Enzymes

In addition to CYPs, a number of other enzymes can catalyze phase I metabolism by catalyzing oxidation, hydrolysis, and reduction reactions. Compared to CYP, these enzymes are generally more substrate specific, recognizing a particular residue or functional group with greater affinity, and limited to executing a single enzymatic reaction. This feature complements the broad substrate, low-affinity characteristics of P450 metabolism, which executes a variety of reactions by several CYP isozymes to obtain its oxidized products.

Flavin-containing monooxygenases (FMOs), alcohol dehydrogenase, and aldehyde dehydrogenase contribute to the non-P450 oxidation reactions. Hydrolysis and reduction reactions are accomplished by carboxylesterase (CES), epoxide hydrolase (EH), and reductases for azo-, nitro-, and carbonyl-groups, to name a few. Several of these enzymes are specific to an organ or tissue – true acetylcholinesterase and pseudocholinesterase, which catalyze the hydrolysis of carboxylic acid esters, amides, and thioesters, are found in the blood (erythrocyte membranes and serum, respectively) while absent in other tissues of the body. These additional enzymes catalyzing hydrolysis, reduction, and oxidation reactions enable each organ or tissue to have an individualized metabolic response, in addition to the primary metabolic response from the liver.

#### 12.3.2.1 Flavin Monooxygenases

Like P450 enzymes, flavin- (or FAD-) containing monooxygenases (FMO) are microsomal enzymes that utilize NADPH and oxygen to oxidize xenobiotic substrates. FMOs attack nucleophilic heteroatoms (N, S, and P) present in an array of compounds, and many of these compounds are also substrates for P450-mediated oxidations. P450s and FMOs differ in their oxidation processes, in which P450s use the oxygenated heme prosthetic group to carry or transfer electrons while FMOs use FAD [63]. Despite the shared substrates and a similar catalytic cycle, FMOs lack the variety and abundance of P450 expression. The abundance of FMO enzymes depends on multiple factors, including cofactor supply, physiological and environmental factors, as well as diet. The expression of FMOs is limited to the lung, kidney, and liver, and the entire FMO gene family is comprised of five genes (FMO1–FMO5) [64]. FMO3 is the predominant isoform in the human liver and plays a crucial role in the biotransformation of cimetidine, tamoxifen, methamphetamine, olanzapine, and nicotine [65].

#### 12.3.2.2 Carboxylesterases Hydrolysis

Carboxylesterases (CES's) are glycoproteins that belong to the family of esterases and mediate the hydrolysis of CES into corresponding alcohols and carboxylic acids. CESs are predominantly expressed in tissues that have a barrier function, such as epithelia of lung, intestine, liver, kidney, and spleen. These enzymes generate pharmacologically active metabolites from the hydrolysis of parent compounds containing ester or amide functional groups. Ester derivatives of therapeutic agents are often used as prodrugs, meaning that CESs represent an important class of enzymes involved in prodrug activation. There are five potential CES genes identified, but only two isoforms have been thoroughly investigated [66]. CES1 is primarily involved with the metabolism of substrates that are small and planar, for example tamiflu, ritalin, and plavix. The second isoform, CES2 preferentially targets larger and bulkier molecules such as the chemotherapeutic, CPT-11 (7-ethyl-10-[4-(1-piperidino)-1-piperidino] carbonyloxycamptothecin), also known as irinotecan, and the narcotic, cocaine [67]. CES2 metabolically converts CPT-11 to ethyl camptothecin (SN-38) which therapeutically targets colorectal and cervical cancers [68]. CES substrates also include endogenous compounds such as platelet-activating factor, diacylglycerol, and other esterified lipids that can undergo CES mediated hydrolysis.

Studies have suggested that certain CES isoenzymes for example RL2 might play a role in early stage hepatocarcinogenesis in rats because RL2 mediates the hydrolysis of diacylglycerol, an endogenous bioactivator of protein kinase C, which is involved with cell proliferation and carcinogenesis [69].

#### 12.3.2.3 Epoxide Hydrolase

EH is another enzyme participating in hydrolysis reactions of phase I metabolism, yet with a very specialized function. The primary substrates of EH catalytic hydrolysis are compounds that contain alkene epoxides and arene oxides which are the products of P450-mediated oxidation of aliphatic alkenes and aromatic hydrocarbons, respectively. These EH substrates start as sterically strained and highly reactive compounds, but the EH catalyzed dihydroxylation reaction converts them to innocuous dihydrodiol derivatives. Most epoxide intermediates are reactive, destructive, and potentially carcinogenic, for example, soluble EH catalyzes the conversion of leukotoxin (epoxide derived from linoleic acid) to the toxic metabolite leukotoxin diol, resulting in multiple organ failure and respiratory distress syndrome after severe body burns [70, 71]. However, not all epoxides are toxic. The major metabolite of the antiepileptic drug, carbamazepine is the 10,11-epoxide, which is stable enough to be measured as the predominant metabolite in the blood and therefore not a substrate of EH. In fact, unaltered carbamazepine-10,11-epoxide is found in the urine of exposed patients. The potential for epoxide metabolites to react with endogenous molecules, namely DNA and proteins requires immediate and local detoxification to prevent toxic effects. That is why EH is ubiquitously expressed in virtually all tissues of the body. In the liver, specifically, the expression of EH parallels that of the CYPs, such that both are predominantly expressed in centrilobular hepatocytes (zone 3). The coexpression of EH and CYPs within close proximity of each other helps prevent the escape of reactive epoxide intermediate compounds and the occurrence of epoxide-mediated toxicities.

## 12.4 Phase II Drug Metabolism

Conventionally, phase I and phase II processes are presented as the linear progression of metabolism for compounds such as phenytoin, but every single compound does not necessarily have to undergo phase II metabolism followed by phase I reactions. A few compounds can effectively be excreted from the body as phase I oxidized metabolites and are not subjected to phase II metabolism. On certain occasions, some compounds might require two rounds of phase I metabolism before being conjugated by phase II enzymes. P450-mediated metabolism can convert aromatic hydrocarbons to an epoxide intermediate, which is then recognized by EH (mentioned above). Sulfur-containing xenobiotics can undergo sulfoxidation from a sulfide (S) to a sulfoxide (SO), and ultimately to a sulfone ($SO_2$) all mediated by P450s or FMOs, such as albendazole [5]. Lastly, the parent form of some compounds can act as a substrate for phase II conjugating enzymes without phase I oxidation. Acetaminophen, for example, is competitively metabolized by phase I (CYP2E1/CYP3A4) and phase II (UGT (uridine diphosphate-glucuronosyltransferase), ST) enzymes, where P450-mediated metabolism leads to the generation of the reactive NAPQI intermediate and phase II conjugations result in detoxification (Figure 12.4).

After phase I reactions, the soluble oxidized products are minimally soluble, and are primed for attack by phase II conjugating enzymes. Phase II metabolism consists of conjugating enzymes and related cofactors that form covalent bonds with polar substituents, such as sugars, sulfates, and amino acids donated from related cofactors to exogenous and endogenous compounds. The addition of bulky, polar substituents renders these metabolites increasingly water-soluble facilitating their elimination. The three major primary mechanisms of Phase II metabolism are glucuronidation, sulfation, and GSH conjugation. For reference, other mechanisms are listed in Table 12.1. All three mechanisms share two commonalities – a necessary cofactor, whose concentration impacts the capacity of the pathway, and a catalyzing enzyme represented in a number of gene products, which provides variety and redundancy to a critical metabolic process, as summarized in Table 12.1. These enzymes and cofactors play a vital role in the modification of endogenous steroids, dietary constituents, and pharmaceutical drugs, preparing them for excretion. The pathways of both glucuronidation and sulfation are balanced by specificity and capacity, while GSH conjugation participates in metabolism as part of a larger role in cellular protection. Collectively, all three mechanisms of phase II metabolism are essential to neutralizing and eliminating a variety of undesired compounds from the body.

### 12.4.1 Glucuronidation

Glucuronidation of a compound occurs by the addition of a glucuronic acid (GA) moiety donated from the cofactor uridine diphosphate-glucuronic acid (UDP-GA) and catalyzed by UGT (uridine diphosphate-glucuronosyltransferase) (Table 12.1). UDP-GA is a high-energy cofactor and is the product of addition of glucose-1-phosphate to a uridine triphosphate molecule. The GA is an "activated" glucose molecule that easily leaves the cofactor and is donated to the substrate compound. Upon conjugation, the GA moiety increases the molecular weight as well as water solubility of the substrate. The carboxylic acid moiety of the GA is ionized at physiological pH, increasing its solubility and facilitating recognition by transporter proteins. Glucuronidation is considered a high-capacity phase II conjugation pathway in the human body as well as in other mammals. Based on the abundance of UDP-GA and the variety of UGT isoforms, our glucuronidation system has the capacity and flexibility to metabolize a multitude of endogenous and exogenous compounds. Endogenous compounds, such as bilirubin, thyroid hormones, testosterone, and progesterone, are substrates for glucuronidation. In addition, a variety of drugs such as morphine, acetaminophen, buprenorphine, and SN-38, the active metabolite of anticancer prodrug irinotecan, are also glucuronidated [72, 73]. In 2002, it was reported that 1 in 10 of the top 200 prescribed drugs undergo glucuronidation as a clearance mechanism. For example, raloxifene, used extensively for osteoporosis, is the target of extensive glucuronidation, which results in 2% bioavailability [74, 75].

UGTs are equipped with the ability to recognize a variety of substrate compounds and therefore, could be regarded as the P450s of phase II metabolism. Similar to P450 enzymes, UGTs are membrane-bound enzymes localized to the endoplasmic reticulum of the cell that execute their major catalytic feats with a number of enzyme isoforms. Four human UGT families have been identified, namely, UGT1, UGT2, UGT3, and UGT8 with UGT1 and UGT2 family members being primarily responsible for the glucuronidation of both endogenous and exogenous substrates. Families UGT1 and UGT2 comprise more than 16 isoforms (more information can be found at the UGT homepage, (http://som.flinders.edu.au/FUSA/ClinPharm/UGT). The UGT1 family consists of nine enzymes produced by the alternative splicing of exons in a single gene locus. For gene UGT1A, there are 12 DNA regions identified as exon 1 that can be selected to be spliced with exons 2 through 5, which are constant for all UGT1A enzymes; there are

**Table 12.1** Summary of phase II enzymes and their cofactors

| Mechanism | Enzyme | Co-factors |
|---|---|---|
| Glucuronidation | UGTs:<br>UGT1 and UGT2 | Uridine diphospho-glucuronic acid (UDP-GA)<br>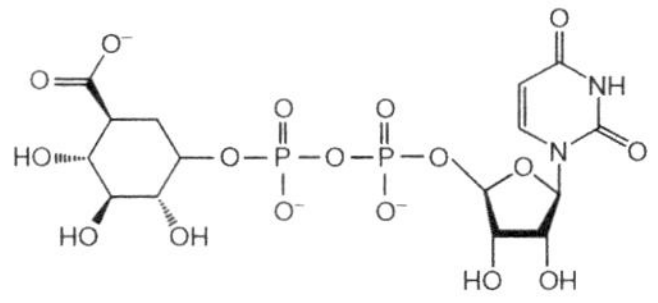 |
| Sulfation | SULTs:<br>SULT1 and SULT2 | 3′-phosphoadenosine 5′-phosphosulfate (PAPS)<br>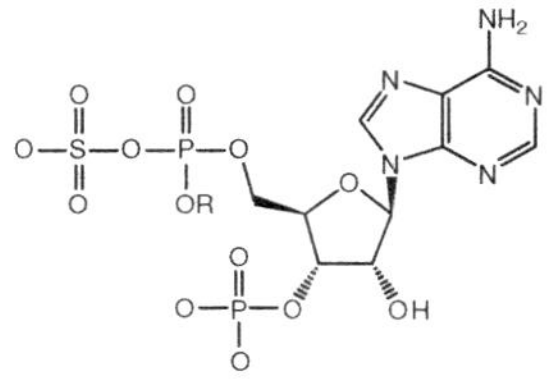 |
| Glutathion conjugation | GSTs:<br>alpha, mu, pi, omega, theta and zeta | Glutathione<br>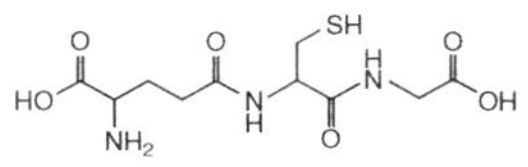<br>[Glu-Cys-Gly] |
| Methylation | Methyl transferases:<br>*O*-methyl transferases<br>*N*-methyl transferases<br>*S*-methyl transferases | *S*-adenosyl methionine (SAM)<br>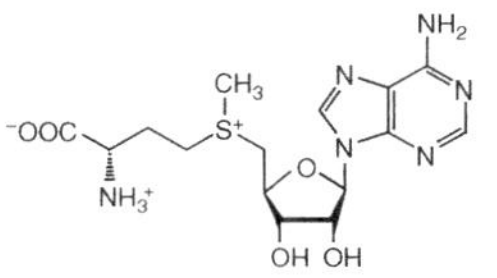 |
| Acetylation | *N*-acetyl transferase:<br>NAT-1 and NAT-2 | Acetyl CoA<br>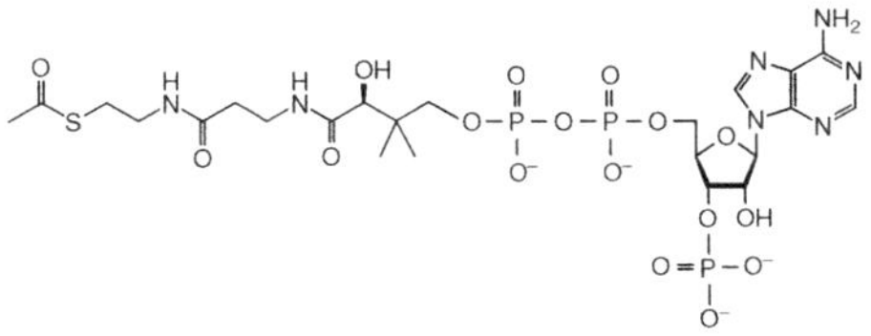 |
| Amino acid conjugaton | Acyl-CoA synthetase and *N*-acyl transferase | Glycine OR glutamate<br>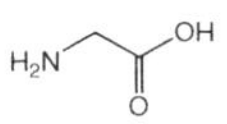 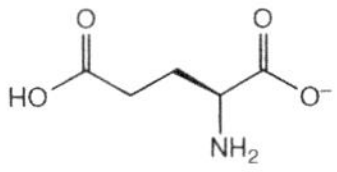 |

also 3 exon 1 regions identified as pseudogenes that do not generate functional RNA or protein products. Such a variety present in the N terminal region provides a molecular basis to explain the diversity of substrates conjugated via the glucuronidation pathway. Members of the UGT2 family (UGT2B4, 7, 10, 11, 15, 17) also contribute to the capacity of the pathway, but they do so as individual gene products in the traditional way. However, the unique genomic organization of the UGT1A family can have some clinical and/or pharmacological consequences. Individual mutations in any of the 12 possible exon 1 regions will affect one enzyme and alter the conjugation of a limited number of compounds, while defects in exons 2 through 5 would affect all UGT1A enzymes. Crigler–Najjar syndrome is a congenital defect characterized by a decrease or loss of bilirubin glucuronidation resulting from mutations in the UGT1A gene [76]. The location of the mutation is directly correlated to the severity of the disease:; mutations in exons 2–5 leads to a complete loss of conjugation; mutations in exon 1 that are specific for bilirubin also result in complete loss of conjugation; whereas mutations in other exon 1 sequences may reduce or only slightly affect conjugation activity [5]. This syndrome is characterized by inefficient bilirubin conjugation, but the effects of altered UGT activity can extend to other substrates for conjugation. Crigler–Najjar syndrome is a rare but often fatal condition caused by the inactivation of a single enzyme resulting in toxic effects of bilirubin accumulation on the central nervous system. The severity of a disease state resulting from the dysregulation of this conjugating pathway only underscores its importance to the metabolic process.

Although glucuronide metabolites can be present both in urine and bile, the preferred route of elimination is largely dependent on the size of the parent compound with minor impact from the glucuronide addition. Regardless of the selection criteria, each elimination pathway has associated complications. Endogenously, bile acids are recycled and reabsorbed from the gut after being glucuronidated in the liver by a process called enterohepatic recirculation. Glucuronidases in the intestinal microflora hydrolyze the glucuronide conjugate and release the bile acid and GA moiety. However, the glucuronidase activity and the enterohepatic recirculation are not specific or exclusive to bile acid and its glucuronide conjugates. The pathway intended to recycle bile acids by recirculation and hydrolysis can result in reabsorption of potential glucuronide metabolites of many pharmaceutical drugs. Drugs susceptible to enterohepatic recirculation tend to display an increased half-life and decreased clearance, increased bioavailability, an increase in potential therapeutic toxicities, and unpredictable plasma concentrations that fall and spike as glucuronide metabolites are conjugated and hydrolyzed, respectively. Some environmental compounds vulnerable to enterohepatic recirculation are known to cause pathologic effects and initiate carcinomas.

Glucuronides excreted in the urine undergo complications similar to those of biliary excretion. Acidic conditions in the bladder result in the hydrolysis of glucuronide conjugate and release of a reactive unconjugated substrate and glucuronic acid. The nature of the β-glucuronide bond makes glucuronide metabolites particularly susceptible to pH-dependent hydrolysis compared to other conjugated metabolites. There is no risk of reabsorption of the compound across the bladder epithelium, but the compound released from its glucuronide conjugate may be reactive and potentially harmful to the bladder. Occupational exposure to benzidine dyes caused bladder cancer in exposed workers as a direct consequence of the pH-dependent hydrolysis of benzidine-glucuronides in the bladder. The consequences of glucuronide hydrolysis can influence drug choice, therapeutic efficacy, and nontarget-related side effects, regardless of the size of the drug.

### 12.4.2 Sulfation

Sulfate conjugation, in addition to glucuronidation, is another phase II pathway that facilitates the neutralization and elimination of undesirable compounds from the body. Sulfation of a compound is accomplished by a transfer of $SO^{3-}$(sulfate) from the universal sulfate donor 3′-phosphoadenosine-5′-phosphosulfate (PAPS), and the reaction is catalyzed by sulfotransferases (SULTs) (Table 12.1). Phenolic compounds and alcohols are favorable substrates for SULTs, and the resulting sulfuric acid esters are highly water-soluble and readily excretable. The high-energy sulfate donor PAPS is synthesized by a two-step reaction that combines inorganic sulfate and two molecules of ATP. ATP is an abundant product of oxidative phosphorylation, but inorganic sulfate must be derived from the amino acid cysteine by a complex oxidative process. The limited availability of free cysteine in the cell restricts the cellular concentration of PAPS to low levels and, therefore, limits the capacity of the sulfate conjugation pathway. In order to compensate for limited PAPS resources, SULTs share a majority of substrate compounds with UGTs, which results in a cooperative, redundant pathway of conjugation. Based on the available concentrations of their respective cofactors, the sulfation pathway, with limited PAPS resources functions in a high-specificity, low-capacity fashion, while the glucuronidation pathway, with ample UGT-GA resources, functions in a low-specificity, high-capacity fashion. A classic example of a shared substrate for UGT and SULT is acetaminophen (Figure 12.4) and the relative amount of each conjugated metabolite is affected in a dose-dependent manner. At low doses, acetaminophen sulfate is the predominant metabolite, but as the dose of the drug increases, there is a proportionate decrease in the formation of sulfate conjugates and an increase in glucuronide conjugates.

In comparison with UGTs, there is considerably less information available about the functions of SULTs, and this is especially true with respect to SULT genes. There are 47 identified mammalian SULT isoforms, 13 of which represent human SULT enzymes and are divided into 3 families: SULT1, SULT2, and SULT4 [77, 78]. For the purposes of drug metabolism, SULT1 and SULT2 are the two most important enzyme families. While SULT1E1, SULT2A1, and SULT2B1 are actively involved in conjugation of steroid hormones, SULT1A1 is primarily involved in sulfate conjugation of most drugs and other exogenous compounds. In general, sulfation is considered an effective detoxification pathway, and the resulting sulfate conjugates are highly water-soluble and readily excreted. However, some sulfate conjugates are chemically unstable, causing them to degrade into reactive, carcinogenic compounds. This metabolic bioactivation of compounds by sulfate conjugation is uncommon and can be influenced by an individual's SULT expression profile. Extrinsic factors such as diet may alter the SULT's expression profile in an individual, thus affecting the individual's cancer risk. Antioxidants and known inhibitors of SULT1A1 activity, namely quercetin and curcumin, have demonstrated anticancer effects.

### 12.4.3 Glutathione

In phase II conjugation reactions, glucuronidation and sulfation are primarily responsible for increasing the solubility and excretability of relatively stable phase I oxidized products. However, conjugation with glutathione (GSH) is for protection against reactive metabolites produced by phase I oxidations and reactive oxygen species. As mentioned previously, metabolism is not the only source of reactive oxygen species, and as such, GSH is not confined to participating in phase II metabolism, but is part of a global cellular mechanism of protecting against reactive molecules.

GSH is a tripeptide with a gamma peptide linkage of cysteine, glycine, and glutamate synthesized by sequential addition of amino acids. In contrast to the bulky adducts used by UGT and SULT conjugation, GSH is a 'soft' nucleophile due to the polarizability of sulfur and is promptly deprotonated to a highly reactive GS- ion that is free to quench the electrophilic centers of reactive molecules (predominately carbon atoms, but also heteroatoms O, N, and S). This nucleophilic attack can occur spontaneously, but the rate of conjugation is significantly increased by glutathione *S*-transferases (GSTs) (Table 12.1). In the case of carbon electrophiles, conjugation with GSH can replace a good leaving group (–Cl,–Br,–$NO_2$,–CHO, etc.), or be added across a carbon–carbon double bond [5]. Unlike carbon electrophiles, conjugation of electrophilic heteroatoms requires two molecules of GSH. The first reaction is similar to the conjugation reactions with electrophilic carbon atoms, but then a second GSH molecule attacks the conjugate bond, nonenzymatically, generating a reduced product and oxidized glutathione (GSSG). GSSG is nonreactive, can be easily recycled into GSH, and is part of a feedback loop that actively regulates intracellular GSH levels. Cellular concentrations of GSH are kept extremely high (at ~10 mM), and GSTs account for as much as 10% of total cellular protein [5]. By comparison, GSH cellular levels are approximately 30-fold greater than UDP-GA levels and 130-fold greater than PAPS levels. High GSH and GST levels attest to the importance of glutathione conjugation for the protection of intracellular mechanisms against reactive molecules. GSTs are monomeric enzymes that are catalytically active as homo- or heterodimers. GSTs consists of two sites, the hydrophobic H-site, where a substrate binds and an adjacent hydrophilic G-site for GSH binding. These sites together constitute the active site of the enzyme and accelerate the interaction between GSH and electrophilic centers by stabilizing the GS-ion. This reaction leads to the formation of a stable, nonreactive thioether bond between the cysteine residue of GSH and an electrophilic substrate [79]. Human GSTs are divided into three main families: cytosolic, mitochondrial, and membrane-bound microsomal [80]. To accomplish the global neutralization of electrophilic compounds through GSH conjugation, various types of GST genes are necessary. The genes belonging to the family of cytosolic GSTs are predominately involved in xenobiotic metabolism. Within the cytosolic GST family, there are seven classes of GST enzymes: alpha, mu, omega, pi, sigma, theta, and zeta [81]. Enzyme classes differ slightly depending on sequence and substrate profile. The omega class of

GSTs (GSTO1 and GSTO2) displays a unique 19-residue N-terminal extension resulting in an inability to recognize substrates of other GST enzymes. The theta class displays only 7% overall sequence identity with alpha, mu, and pi classes [82]. In theta-class enzymes, GSTT1 and GSTT2, substituting the stabilizing tyrosine residue in the GSH binding site with a serine residue enables structural rearrangements unique to the class of theta enzymes [81]. Due to these distinct features among various classes of GST enzymes, heterodimer formation is restricted to pairing with enzymes within the same class. Representative enzymes from the classes of alpha, mu, omega, theta, and zeta are expressed in the liver; expression of alpha and theta enzymes being the most abundant. The alpha GST class has five members, GSTA1–A5. Hepatic expression of GSTA1, A2, and A4 is well documented, while GSTA3 is found exclusively in steroidogenic tissue and GSTA5 has yet to be detected [81]. Patterns of expression are not restricted by GST class since individual members of each class display distinct expression patterns dependent more on function rather than classification.

In addition to conjugation reactions and neutralization of reactive molecules, there is evidence to suggest that the GSTs participate in MAP kinase and ERK signaling cascades, the catabolism of tyrosine and phenylalanine, and chemotherapeutic resistance in tumor cells. GST pi, specifically, is commonly found in tumor cells, while its expression in normal tissue is widespread, expression in the liver is regarded as insignificant, being found only in the bile duct epithelium (Figure 12.1), Overexpression of GST pi (GSTP1) encourages chemotherapeutic resistance through increased conjugation with GSH and direct interaction with the MAP kinase cascade causing inhibition of apoptosis [83, 84]. Currently, drug development research is focusing on developing GST pi inhibitors to prevent this acquired resistance mechanism.

## 12.5 Novel Platforms for Drug Metabolism Studies

As the relevance of drug metabolism has gained significant traction in the field of drug discovery and development, there is a constant need to develop improved *in vitro* techniques to study drug metabolism and potential drug-drug interactions of lead candidates. Conventionally, drug metabolism studies involved measuring the disappearance of the drug candidate in different liver cell fractions such as microsomes, cytosol and S9. Human primary hepatocytes (HPH) have also been utilized to determine metabolic stability. There is an extensive amount of information available which elaborately discussed these standard metabolic studies in detail and hence, will not be included in this book chapter [85–87]. Moreover, these traditional approaches do not accurately mimic the physiological conditions which are necessary to evaluate the metabolic profile of a lead candidate, and this can result in discrepancies between *in vitro* and *in vivo* data. Quite often drug safety issues have been identified at a later stage of drug development due to the inefficient nonclinical toxicological prediction, leading to a high cost of drug development. To help accelerate the identification and development of safer drugs, physiologically relevant *in vitro* tools with high-throughput potentials are often demanded. *In vitro* cell-based microphysiological (3D) systems that can accurately emulate human *in vivo* response and the cellular microenvironment hold great potential for improving the assessment of drug metabolism and toxicity.

### 12.5.1 Micropatterned Liver Cell Coculture

The nonparenchymal cells (NPC) in liver consists of sinusoidal endothelial cells, Kupffer cells, hepatic stellate cells, and cholangiocytes. Heterotypic contact/interactions between hepatocytes and NPC, help regulate cell growth, migration, and differentiation. In the liver, cell–cell interactions are crucial to maintain organ function and hence *in vitro* experimental conditions should ideally mimic these interactions [88]. However, randomly distributed liver cocultures exhibit suboptimal cell–cell interactions, resulting in unstable and lower liver function. In contrast, micropatterned cocultures (MPCCs), created by use of semiconductor-based lithographic techniques enable optimum cell–cell interactions, which stabilize and enhance liver function for ≥4 weeks [89] (Figure 12.7). These MPCCs can be adopted to standard 24-,96-, and 384- well plate formats to evaluate drug clearance, DDI, metabolic profiling and drug induced liver injury [89, 91–93]. Cell viability of HPH suspension cultures usually declines within four to six hours and hence, cannot be utilized for drug incubation studies beyond this time frame. In contrast, MPCCs of HPH can maintain optimum P450 and phase II conjugation activity for ~4 weeks [91]. Additionally, certain transporters also have optimum activity in these systems compared to standard hepatocyte monocultures. Thus, these MPCC's enable to mimic cell–cell interactions that are critical to the native liver and thereby, offer the advantage over standard monocultures by ensuring an accurate and

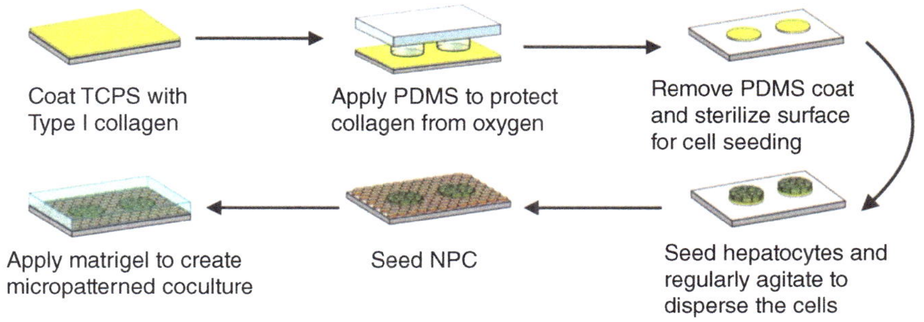


**Figure 12.7** Micropatterned liver cell coculture model. Procedure for the development of hepatocyte-based micropatterned coculture using polydimethylsiloxane (PDMS). Primary human hepatocytes (PHH) will selectively attach to collagen, while nonparenchymal cells (NPCs) will surround PHH. Source: Berger et al. [90].

"long-term" prediction of drug metabolism and toxicity. Additionally, proper incorporation of NPC for the investigation of immune-related and inflammation-mediated hepatotoxicity or drug–drug interactions have shown promise to elucidate behavior of large molecules such as cytokines and therapeutic proteins [94]. Appropriate microphysiological models that recapitulate hepatic architecture and function are key to accurate predictions of *in vivo* behavior for drug development in an efficient and cost-effective manner.

### 12.5.2 Spheroids

HPH when maintained in conventional 2D monolayer, de-differentiate and loose hepatocyte specific functions typically within one week. Hence, the use of this system for long-term studies is limited. Various other complex culture systems such as scaffold-based systems and bioreactors have been previously utilized to mimic hepatic microenvironment, but these models lack scalability and are usually complex for experimental setup. To overcome these challenges, hepatocytes can be cultured as spheroids, which are 3D spherical cellular aggregates. In a spherical platform, hepatocytes can survive for a longer period of time (up to 35 days) and maintain robust physiologically relevant metabolic capacity. Secretion of essential liver function biomarkers such as albumin and urea are not significantly impacted during this period. Moreover, cellular polarity and formation of functional bile ducts has also been demonstrated [95]. Limitations of the using of the spheroid model include but not limited to: (i) difficulty in medium change and liquid handling; (ii) larger spheroids can affect oxygen and nutrition distribution, while smaller spheroids only provide limited biomass for subsequent analysis; and (iii) difficulty in capturing clear morphology of particular cells in spheroids. Overcoming these challenges would further expand the application of this simplified 3D microphysiological system in studying various aspects of liver function, liver disease and drug induced liver injuries.

### 12.5.3 Liver on a Chip

Liver-on-chips represent a promising technology because of their capability to model functional aspects of organ-level physiology by incorporating properly localized hepatocytes and NPC under continuous microfluidic medium perfusion. This platform recreates the epithelial-endothelial interface, provide mechano-actuation to emulate unique hepatic structure and utilizes microfluidics to maintain nutrient-waste homeostasis. Together, these features offer more accurate recreation of *in vivo* relevant phenotypes and have become attractive tools for preclinical drug efficacy and safety assessment [96] (Figure 12.8). Dual-cell chips that consist of hepatocyte and liver sinusoidal endothelial cells have displayed production of higher amounts of albumin than conventional sandwich monoculture plates. Similarly, the drug metabolizing capacity of these microchips over a 14 day period was greater than those exhibited by freshly isolated primary hepatocytes. Furthermore, liver-on-chips can be highly beneficial to model microscopic end points such as steatosis and fibrosis to identify compounds that are potentially toxic. Therefore, these microfluidic devices can be potentially utilized for safety testing, disease modeling, biomarker identification and predicting compounds that might cause hepatotoxicity [97, 98].

## 12.6 Drug Metabolism and Its Impact on Adverse Drug Reactions

While the actions of phase I and phase II enzymes are intricate and well-studied, an observation of significant interindividual variability inherent in metabolism can affect the response to any given pharmaceutical drug. The variability manifests as altered clearance kinetics and plasma concentrations, decreased efficacy, or increased toxicity-related side effects. Investigations into the mechanisms of interindividual variability have identified individual P450s, UGTs, or GSTs as critical components of variability, each in a substrate-specific

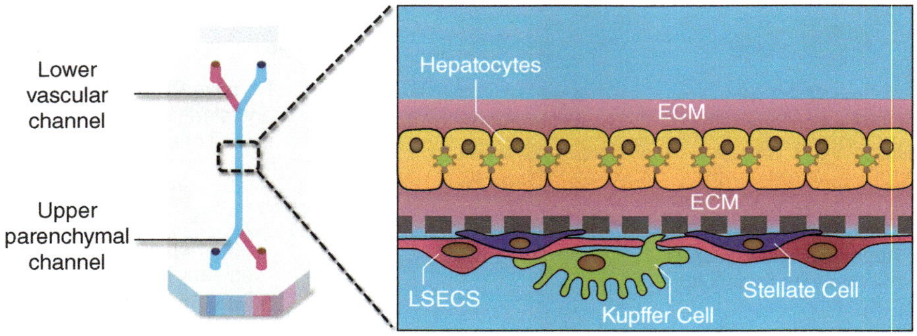


**Figure 12.8** Representation of a Liver Chip that recapitulates the cytoarchitecture of a liver. Primary hepatocytes were grown in extra cellular matrix (ECM) sandwich format and placed in the upper parenchymal channel. Nonparenchymal cells (NPCs) grown in the lower vascular channel on the opposite side. Source: Jang et al. [97] / American Association for the Advancement of Science.

context. In this section, the consequences of variability in drug metabolism, namely, adverse drug reactions (ADRs), and the causes of variabilities, specifically genetic polymorphisms will be discussed.

The World Health Organization (WHO) defines ADR as a noxious response to a medication that is unintended at doses usually administered for diagnosis, prophylaxis or treatment. ADRs have become a constant concern for the medical and pharmaceutical communities, with the growing use of prescription drugs among the general population and the prevalence of multidrug therapies for chronic conditions such as aging, HIV infection and cancer. Trend analysis of cross-sectional data on ADR-related deaths over an 8-year period between 1999 and 2006 revealed that ADR deaths increased with time. The classes of drugs commonly associated with ADR deaths included anticoagulants, opioids and immunosuppressants [99]. A review of patients admitted to hospitals due to ADRs determined that 57% of these ADRs were unrecognized by the physician at the time of admission and 18.6% of the prescribed drugs were contraindicated [100]. According to the Center for Disease Control and Prevention (CDC)'s Ambulatory Medical Care Utilization Estimates for 2005, 1% of all medical encounters were a result of adverse effects to medical treatment, accounting for 10.8 million visits to the doctor or hospital. According to the data obtained from National Ambulatory Medical Care Survey (NAMCS) and the National Hospital Ambulatory Medical Care Survey (NHAMCS) out of the 1.8 million visits to physician offices, outpatient clinics and emergency departments, ADRs accounted for 41.1%, or ~740,000 effective patients [101]. Although clinicians can help reduce the occurrence of some of the reported ADRs by avoiding administrations of contraindicated drugs, the risk-benefit paradigm of pharmaceutical treatments indicates that certain ADRs might be unavoidable and might be an unfortunate cost for pharmaceutical drug treatment and the reality of polypharmacy.

Drug behaviors, such as induction and inhibition of metabolizing enzymes (discussed above), are responsible for the occurrence of ADRs. Inhibition of phase I enzymes has various outcomes, including therapeutically beneficial results, but the consequences of enzyme inhibition are defined by the nature of coadministered drug, the specific enzyme involved, and the individual's genetic background. Often, inhibition of CYP enzymes leads to toxic side effects. Induction of P450 enzymes, such as CYP3A4 and CYP2B6, as a result of PXR- and CAR-mediated upregulation has consequences ranging from reduced half-life and increased clearance to decreased therapeutic index and increased toxicity; these responses are to some extent specific to the characteristics of both the drug and the individual. The induction and inhibition of non-CYP phase I and phase II enzymes have been studied less extensively and are believed to be only minor contributors to the development of ADRs. Based on the number of substrates, inducers, and inhibitor compounds, CYP3A4 along with nuclear receptors PXR and CAR are recognized as the major contributors to the occurrence of ADRs.

Several genetic polymorphisms of metabolic genes have been discovered as the primary mechanism for the observation of significant interindividual variations in drug metabolism and have subsequently, given rise to the research area of pharmacogenetics. Pharmacogenetics is the science of understanding how an individual's genetic make-up can be utilized to personalize therapeutic treatment to achieve maximal efficacy with minimal toxic side effects. Several types of polymorphic changes can occur in an individual's genetic sequence including, copy number variants, insertions and deletions, and SNPs. SNPs have been found in genes that encode for phase I, phase II enzymes as well as transporter proteins, and are identified as a major mechanism of individual variability in drug response. A current list of all identified SNPs for CYP enzymes is available at www.cypalleles.ki.se.

SNPs have been identified in phase I, phase II, and phase III transporter genes, but the largest impact on drug metabolism comes from CYP genes. CYP polymorphisms have been estimated to have an adverse impact on the therapeutic effect of around 20–25% drug treatments [102]. Among the other CYP enzymes, CYP2D6 is the most clinically important polymorphic enzyme involved in drug metabolism. As the only CYP lacking inducible expression, the more than 100 gene variants of CYP2D6 are the largest contributing factor to interindividual variation of enzymatic activity. The abundance and frequency of CYP2D6 alleles affect drugs that cause nonresponse to therapy as well as ADRs [103]. Although CYP3A4 is significantly more abundant and plays a greater role in drug metabolism, polymorphic alleles of CYP3A4 contribute far less to interindividual variation than polymorphic alleles of CYP2D6. While a number of SNPs have been identified in the CYP3A4 gene locus, none of them are responsible for any specific observed variation, leaving the mechanism of CYP3A4 interindividual variability yet to be determined.

The polymorphic behavior of CYP2D6 and the importance of CYP3A4 were known long before genetic polymorphisms were realized, but today, following the identification of SNPs, a previously overlooked isozyme, CYP2B6, has been attracting attention. Early investigations into the contribution of CYP2B6 to metabolism were largely unsuccessful due to a lack of specific molecular tools. However, new substrate drugs, such as the chemotherapeutic, cyclophosphamide and the antiretroviral, efavirenz are highlighting the importance of CY2B6 polymorphisms with regards to the efficacy of therapeutic treatments [9, 104].

The clinical impact of SNPs is also found in phase II gene, especially UGTs. A popular chemotherapy treatment for colorectal cancers, irinotecan, has demonstrated significant interindividual variation and ADRs. Irinotecan is a prodrug and requires metabolic activation of its active metabolite, SN-38, which is a substrate for glucuronidation. The toxic side effects of irinotecan treatment have been directly connected to the SNPs identified in the UGT1A1 gene [83]. Recently, there has been a steady increase in the practice of polypharmacy, further emphasizing the importance of understanding the metabolism of the coadminstered drugs to help avoid DDI's that could potentially be harmful.

## 12.7 Conclusion

Metabolism in the liver is an intricate operation involving multiple enzymatic mechanisms to address the body's interactions with various foreign substances. Orally administered drugs must endure a battery of enzymatic attacks by phase I and phase II enzymes following absorption from the intestinal lumen. Unfortunately, the objective of metabolism is to get rid of "toxic" compounds from the body, regardless of endogenous substance or drug compound, while the objective of pharmaceutical treatment is to ensure that drug compounds are retained in the body for a sufficient duration to execute a therapeutic effect. A delicate balance between these two goals needs to be maintained; but various factors, such as a redundant enzyme system, enterohepatic circulation, and unique drug behaviors, make it difficult to establish this balance. The growing use of multidrug treatment (or polypharmacy) has increased awareness about drug behavior such as induction and inhibition, along with the contribution of genetics, diet, environment, and lifestyle the response of a drug. Therefore, the success of a pharmaceutical drug treatment is achieved with respect for the patient as a whole and not just the hallmarks of each disease or condition.

## References


1 de la Monte, S.M., Arcidi, J.M., Moore, G.W., and Hutchins, G.M. (1984). Midzonal necrosis as a pattern of hepatocellular injury after shock. *Gastroenterology* 86 (4): 627–631.

2 Anundi, I., Lahteenmaki, T., Rundgren, M. et al. (1993). Zonation of acetaminophen metabolism and cytochrome P450 2E1-mediated toxicity studied in isolated periportal and perivenous hepatocytes. *Biochem. Pharmacol.* 45 (6): 1251–1259.

3 Guengerich, F.P. (2001). Common and uncommon cytochrome P450 reactions related to metabolism and chemical toxicity. *Chem. Res. Toxicol.* 14 (6): 611–650.

4 Nelson, S.D. and Gordon, W.P. (1983). Mammalian drug metabolism. *J. Nat. Prod.* 46 (1): 71–78.

5 Parkinson, A. (2001). Biotransformation of xenobiotics. In: *Casarett and Doull's Toxicology: The*

*Basic Science of Poisons* (ed. C.D. Klassen), 213. McGraw-Hill Inc.
6 Preissner, S.C., Hoffmann, M.F., Preissner, R. et al. (2013). Polymorphic cytochrome P450 enzymes (CYPs) and their role in personalized therapy. *PLoS One* 8 (12): e82562.
7 Morgan, E.T. (2009). Impact of infectious and inflammatory disease on cytochrome P450-mediated drug metabolism and pharmacokinetics. *Clin. Pharmacol. Ther.* 85 (4): 434–438.
8 Kimura, S., Umeno, M., Skoda, R.C. et al. (1989). The human debrisoquine 4-hydroxylase (CYP2D) locus: sequence and identification of the polymorphic CYP2D6 gene, a related gene, and a pseudogene. *Am. J. Hum. Genet.* 45 (6): 889–904.
9 Haas, D.W., Ribaudo, H.J., Kim, R.B. et al. (2004). Pharmacogenetics of efavirenz and central nervous system side effects: an Adult AIDS Clinical Trials Group study. *AIDS* 18 (18): 2391–2400.
10 Maekawa, K., Adachi, M., Matsuzawa, Y. et al. (2017). Structural basis of single-nucleotide polymorphisms in cytochrome P450 2C9. *Biochemistry* 56 (41): 5476–5480.
11 Williams, P.A., Cosme, J., Vinkovic, D.M. et al. (2004). Crystal structures of human cytochrome P450 3A4 bound to metyrapone and progesterone. *Science* 305 (5684): 683–686.
12 Groff, J.L., Gropper, S.S., and Hunt, S.M. (1995). *Advanced Nutrition and Human Metabolism*. New York: West Publishing.
13 Desta, Z., Ward, B.A., Soukhova, N.V., and Flockhart, D.A. (2004). Comprehensive evaluation of tamoxifen sequential biotransformation by the human cytochrome P450 system in vitro: prominent roles for CYP3A and CYP2D6. *J. Pharmacol. Exp. Ther.* 310 (3): 1062–1075.
14 Osborne, C.K. (1998). Tamoxifen in the treatment of breast cancer. *Drug Ther.* 339.
15 Stearns, V., Johnson, M.D., Rae, J.M. et al. (2003). Active tamoxifen metabolite plasma concentrations after coadministration of tamoxifen and the selective serotonin reuptake inhibitor paroxetine. *J. Natl. Cancer Inst.* 95 (23): 1758–1764.
16 Wang, D. and Wang, H. (2012). Oxazaphosphorine bioactivation and detoxification The role of xenobiotic receptors. *Acta Pharm. Sin. B* 2 (2).
17 Xie, H.J., Yasar, U., Lundgren, S. et al. (2003). Role of polymorphic human CYP2B6 in cyclophosphamide bioactivation. *Pharmacogenomics J.* 3 (1): 53–61.
18 Gropper, S.S. and Smith, J.L. (1995). *Advanced Nutrition and Human Metabolism*. New York: West Publishing.
19 Nebert, D.W., Nelson, D.R., Adesnik, M. et al. (1989). The P450 superfamily: updated listing of all genes and recommended nomenclature for the chromosomal loci. *DNA* 8 (1): 1–13.
20 Nelson, D.R., Kamataki, T., Waxman, D.J. et al. (1993). The P450 superfamily: update on new sequences, gene mapping, accession numbers, early trivial names of enzymes, and nomenclature. *DNA Cell Biol.* 12 (1): 1–51.
21 Guengerich, F.P. (2006). Cytochrome P450s and other enzymes in drug metabolism and toxicity. *AAPS*.
22 Nelson, D.R. (2009). The cytochrome P450 homepage. *Human Genomics* 4: 59–65. https://drnelson.uthsc.edu/CytochromeP450.html.
23 Nebert, D.W. and Russell, D.W. (2002). Clinical importance of the cytochromes P450. *Lancet* 360 (9340): 1155–1162.
24 Hafner, M., Rezen, T., and Rozman, D. (2011). Regulation of hepatic cytochromes p450 by lipids and cholesterol. *Curr. Drug Metab.* 12 (2): 173–185.
25 Lorbek, G., Lewinska, M., and Rozman, D. (2012). Cytochrome P450s in the synthesis of cholesterol and bile acids-from mouse models to human diseases. *FEBS J.* 279 (9): 1516–1533.
26 Klassen, C.D. (2001). *Casarett and Doull's Toxicology: The Basic Science of Poisons*. New York: McGraw-Hill.
27 Wang, H. and Tompkins, L.M. (2008). CYP2B6: new insights into a historically overlooked cytochrome P450 isozyme. *Curr. Drug Metab.* 9 (7): 598–610.
28 Wang, H. and LeCluyse, E.L. (2003). Role of orphan nuclear receptors in the regulation of drug-metabolising enzymes. *Clin. Pharmacokinet.* 42 (15): 1331–1357.
29 Kessova, I. and Cederbaum, A.I. (2003). CYP2E1: biochemistry, toxicology, regulation and function in ethanol-induced liver injury. *Curr. Mol. Med.* 3 (6): 509–518.
30 Kliewer, S.A., Goodwin, B., and Willson, T.M. (2002). The nuclear pregnane X receptor: a key regulator of xenobiotic metabolism. *Endocr. Rev.* 23 (5): 687–702.
31 Baes, M., Gulick, T., Choi, H.S. et al. (1994). A new orphan member of the nuclear hormone receptor superfamily that interacts with a subset of retinoic acid response elements. *Mol. Cell. Biol.* 14 (3): 1544–1552.

32 Bertilsson, G., Heidrich, J., Svensson, K. et al. (1998). Identification of a human nuclear receptor defines a new signaling pathway for CYP3A induction. *Proc. Natl. Acad. Sci. U.S.A.* 95 (21): 12208–12213.

33 Blumberg, B., Sabbagh, W. Jr., Juguilon, H. et al. (1998). SXR, a novel steroid and xenobiotic-sensing nuclear receptor. *Genes Dev.* 12 (20): 3195–3205.

34 Honkakoski, P., Zelko, I., Sueyoshi, T., and Negishi, M. (1998). The nuclear orphan receptor CAR-retinoid X receptor heterodimer activates the phenobarbital-responsive enhancer module of the CYP2B gene. *Mol. Cell. Biol.* 18 (10): 5652–5658.

35 Kliewer, S.A., Moore, J.T., Wade, L. et al. (1998). An orphan nuclear receptor activated by pregnanes defines a novel steroid signaling pathway. *Cell* 92 (1): 73–82.

36 Lehmann, J.M., McKee, D.D., Watson, M.A. et al. (1998). The human orphan nuclear receptor PXR is activated by compounds that regulate CYP3A4 gene expression and cause drug interactions. *J. Clin. Invest.* 102 (5): 1016–1023.

37 Watkins, R.E., Wisely, G.B., Moore, L.B. et al. (2001). The human nuclear xenobiotic receptor PXR: structural determinants of directed promiscuity. *Science* 292 (5525): 2329–2333.

38 Xu, R.X., Lambert, M.H., Wisely, B.B. et al. (2004). A structural basis for constitutive activity in the human CAR/RXRalpha heterodimer. *Mol. Cell* 16 (6): 919–928.

39 Barwick, J.L., Quattrochi, L.C., Mills, A.S. et al. (1996). Trans-species gene transfer for analysis of glucocorticoid-inducible transcriptional activation of transiently expressed human CYP3A4 and rabbit CYP3A6 in primary cultures of adult rat and rabbit hepatocytes. *Mol. Pharmacol.* 50 (1): 10–16.

40 Guengerich, F.P. (1999, 1999). Cytochrome P-450 3A4: regulation and role in drug metabolism. *Pharmacol. Toxicology.* 39 (1): 1–17.

41 Li, A.P., Kaminski, D.L., and Rasmussen, A. (1995). Substrates of human hepatic cytochrome P450 3A4. *Toxicology* 104 (1-3): 1–8.

42 Zhou, S.-F. (2008). Drugs behave as substrates, inhibitors and inducers of human cytochrome P450 3A4. *Curr. Drug Metab.*

43 Faucette, S.R., Zhang, T.C., Moore, R. et al. (2007). Relative activation of human pregnane X receptor versus constitutive androstane receptor defines distinct classes of CYP2B6 and CYP3A4 inducers. *J. Pharmacol. Exp. Ther.* 320 (1): 72–80.

44 Honkakoski, P., Sueyoshi, T., and Negishi, M. (2003). Drug-activated nuclear receptors CAR and PXR. *Ann. Med.* 35 (3): 172–182.

45 Poland, A., Glover, E., and Kende, A.S. (1976). Stereospecific, high affinity binding of 2,3,7,8-tetrachlorodibenzo-p-dioxin by hepatic cytosol. Evidence that the binding species is receptor for induction of aryl hydrocarbon hydroxylase. *J. Biol. Chem.* 251 (16): 4936–4946.

46 Quattrochi, L.C. and Tukey, R.H. (1989). The human cytochrome Cyp1A2 gene contains regulatory elements responsive to 3-methylcholanthrene. *Mol. Pharmacol.* 36 (1): 66–71.

47 Denison, M.S., Fisher, J.M., and Whitlock, J.P. Jr., (1988). The DNA recognition site for the dioxin-Ah receptor complex. Nucleotide sequence and functional analysis. *J. Biol. Chem.* 263 (33): 17221–17224.

48 Hoffman, E.C., Reyes, H., Chu, F.F. et al. (1991). Cloning of a factor required for activity of the Ah (dioxin) receptor. *Science* 252 (5008): 954–958.

49 Izzo, A.A. and Ernst, E. (2001). Interactions between herbal medicines and prescribed drugs: a systematic review. *Drugs* 61 (15): 2163–2175.

50 Piscitelli, S.C., Burstein, A.H., Chaitt, D. et al. (2000). Indinavir concentrations and St John's wort. *Lancet* 355 (9203): 547–548.

51 Ruschitzka, F., Meier, P.J., Turina, M. et al. (2000). Acute heart transplant rejection due to Saint John's wort. *Lancet* 355 (9203): 548–549.

52 Varis, T., Kaukonen, K.M., Kivistö, K.T., and Neuvonen, P.J. (1998). Plasma concentrations and effects of oral methylprednisolone are considerably increased by itraconazole. *Clin. Pharmacol. Ther.* 64 (4): 363–368.

53 KivistO, K.T., Neuvonen, P.J., and Klotz, U. (1994). *Inhibition of Terfenadine Metabolism*, 27. Pharmacokinet: Clin.

54 Schmitt, C., Hofmann, C., Riek, M. et al. (2009). Effect of saquinavir-ritonavir on cytochrome P450 3A4 activity in healthy volunteers using midazolam as a probe. *Pharmacotherapy* 29 (10): 1175–1181.

55 Lin, J.H. and Lu, A.Y. (1998). Inhibition and induction of cytochrome P450 and the clinical implications. *Clin. Pharmacokinet.* 35 (5): 361–390.

56 Ma, B., Prueksaritanont, T., and Lin, J.H. (1999). Drug interactions with calcium channel blockers possible involvement of metabolite-intermediate complexation with CYP3A. *Am. Soc. Pharmacol. Exp. Therap.* 28: 2.

57 Halpert, J. (1981). Covalent modification of lysine during the suicide inactivation of rat liver

cytochrome P-450 by chloramphenicol. *Biochem. Pharmacol.* 30 (8): 875–881.

58 Halpert, J.R., Miller, N.E., and Gorsky, L.D. (1985). On the mechanism of the inactivation of the major phenobarbital-inducible isozyme of rat liver cytochrome P-450 by chloramphenicol. *J. Biol. Chem.* 260 (14): 8397–8403.

59 Bailey, D.G., Malcolm, J., Arnold, O., and Spence, J.D. (1998). Grapefruit juice-drug interactions. *Br. J. Clin. Pharmacol.* 46 (2): 101–110.

60 Lown, K.S., Bailey, D.G., Fontana, R.J. et al. (1997). Grapefruit juice increases felodipine oral availability in humans by decreasing intestinal CYP3A protein expression. *J. Clin. Invest.* 99 (10): 2545–2553.

61 Mazi-Kotwal, N. and Seshadri, M. (2012). Drug interactions with grapefruit juice. *Br. J. Med. Pract.* 5.

62 Yamreudeewong, W., Henann, N.E., Fazio, A. et al. (1995). Drug-food interactions in clinical practice. *J. Fam. Pract.* 40 (4): 376–384.

63 Basaran, R. and Can Eke, B. (2017). Flavin containing monooxygenases and metabolism of xenobiotics. *Turk. J. Pharm. Sci.* 14 (1): 90–94.

64 Hines, R.N. (2006). Developmental and tissue-specific expression of human flavin-containing monooxygenases 1 and 3. *Expert Opin. Drug Metab. Toxicol.* 2 (1): 41–49.

65 Phillips, I.R. and Shephard, E.A. (2017). Drug metabolism by flavin-containing monooxygenases of human and mouse. *Expert Opin. Drug Metab. Toxicol.* 13 (2): 167–181.

66 Hatfield, M.J., Umans, R.A., Hyatt, J.L. et al. (2016). Carboxylesterases: general detoxifying enzymes. *Chem. Biol. Interact.* 259 (Pt B): 327–331.

67 Hatfield, M.J., Tsurkan, L., Garrett, M. et al. (2011). Organ-specific carboxylesterase profiling identifies the small intestine and kidney as major contributors of activation of the anticancer prodrug CPT-11. *Biochem. Pharmacol.* 81 (1): 24–31.

68 Satoh, T., Hosokawa, M., Atsumi, R. et al. (1994). Metabolic activation of CPT-11, 7-ethyl-10-[4-(1-piperidino)-1- piperidino]carbonyloxycamptothecin, a novel antitumor agent, by carboxylesterase. *Biol. Pharm. Bull.* 17 (5): 662–664.

69 Maki, T., Hosokawa, M., Satoh, T., and Sato, K. (1991). Changes in carboxylesterase isoenzymes of rat liver microsomes during hepatocarcinogenesis. *Jpn. J. Cancer Res.* 82 (7): 800–806.

70 Decker, M., Arand, M., and Cronin, A. (2009). Mammalian epoxide hydrolases in xenobiotic metabolism and signalling. *Arch. Toxicol.* 83 (4): 297–318.

71 Moghaddam, M.F., Motoba, K., Borhan, B. et al. (1996). Novel metabolic pathways for linoleic and arachidonic acid metabolism. *Biochim. Biophys. Acta*.

72 Fisher, M.B., Paine, M.F., Strelevitz, T.J., and Wrighton, S.A. (2001). The role of hepatic and extrahepatic UDP-glucuronosyltransferases in human drug metabolism. *Drug Metab. Rev.* 33 (3-4): 273–297.

73 Tukey, R.H. and Strassburg, C.P. (2000). Human UDP-glucuronosyltransferases: metabolism, expression, and disease. *Annu. Rev. Pharmacol. Toxicol.* 40: 581–616.

74 Kemp, D.C., Fan, P.W., and Stevens, J.C. (2002). Characterization of raloxifene glucuronidation in vitro: contribution of intestinal metabolism to presystemic clearance. *Drug Metab. Dispos.* 30 (6): 694–700.

75 Williams, J.A., Hyland, R., Jones, B.C. et al. (2004). Drug-drug interactions for UDP-glucuronosyltransferase substrates: a pharmacokinetic explanation for typically observed low exposure (AUCi/AUC) ratios. *Drug Metab. Dispos.* 32 (11): 1201–1208.

76 Ritter, J.K., Yeatman, M.T., Ferreira, P., and Owens, I.S. (1992). Identification of a genetic alteration in the code for bilirubin UDP-glucuronosyltransferase in the UGT1 gene complex of a Crigler-Najjar type I patient. *J. Clin. Invest.* 90 (1): 150–155.

77 Gamage, N., Barnett, A., Hempel, N. et al. (2006). Human sulfotransferases and their role in chemical metabolism. *Toxicol. Sci.* 90 (1): 5–22.

78 Negishi, M., Pedersen, L.G., Petrotchenko, E. et al. (2001). Structure and function of sulfotransferases. *Arch. Biochem. Biophys.* 390 (2): 149–157.

79 Eaton, D.L. and Bammler, T.K. (1999). Concise review of the glutathione S-transferases and their significance to toxicology. *Toxicol. Sci.* 49.

80 Hayes, J.D., Flanagan, J.U., and Jowsey, I.R. (2005). Glutathione transferases. *Annu. Rev. Pharmacol. Toxicol.* 45: 51–88.

81 McIlwain, C.C., Townsend, D.M., and Tew, K.D. (2006). Glutathione S-transferase polymorphisms: cancer incidence and therapy. *Oncogene* 25 (11): 1639–1648.

82 Sheehan, D., Meade, G., Foley, V.M., and Dowd, C.A. (2001). Structure, function and evolution of glutathione transferases: implications for classification of non-mammalian members of an ancient enzyme superfamily. *Biochem. J.* 360 (Pt 1): 1–16.

83 Coleman, M.D. (2010). Conjugation and transport processes. *Human Drug Metab.* 125–157.

84 Dang, D.T., Chen, F., Kohli, M. et al. (2005). Glutathione S-transferase pi1 promotes tumorigenicity in HCT116 human colon cancer cells. *Cancer Res.* 65 (20): 9485–9494.

85 Ackley, D.C., Rockich, K.T., and Baker, T.R. (2004). *Metabolic Stability Assessed by Liver Microsomes and Hepatocytes. Optimization in Drug Discovery* (ed. I.V. Methods, Z. Yan, and G.W. Caldwell), 151–162. Totowa, NJ: Humana Press.

86 Kassel, D.B. (2004). Applications of high-throughput ADME in drug discovery. *Curr. Opin. Chem. Biol.* 8 (3): 339–345.

87 Obach, R.S. (1999). Prediction of human clearance of twenty-nine drugs from hepatic microsomal intrinsic clearance data: an examination of in vitro half-life approach and nonspecific binding to microsomes. *Drug Metab. Dispos.* 27 (11): 1350–1359.

88 Bhatia, S.N., Balis, U.J., Yarmush, M.L., and Toner, M. (1999). Effect of cell–cell interactions in preservation of cellular phenotype: cocultivation of hepatocytes and nonparenchymal cells. *FASEB J.* 13 (14): 1883–1900.

89 Khetani, S.R. and Bhatia, S.N. (2008). Microscale culture of human liver cells for drug development. *Nat. Biotechnol.* 26 (1): 120–126.

90 Berger, D.R., Ware, B.R., Davidson, M.D. et al. (2015). Enhancing the functional maturity of induced pluripotent stem cell-derived human hepatocytes by controlled presentation of cell-cell interactions in vitro. *Hepatology* 61 (4): 1370–1381.

91 Bell, C.C., Hendriks, D.F., Moro, S.M. et al. (2016). Characterization of primary human hepatocyte spheroids as a model system for drug-induced liver injury, liver function and disease. *Sci. Rep.* 6: 25187.

92 Khetani, S.R., Kanchagar, C., Ukairo, O. et al. (2013). Use of micropatterned cocultures to detect compounds that cause drug-induced liver injury in humans. *Toxicol. Sci.* 132 (1): 107–117.

93 Wang, W.W., Khetani, S.R., Krzyzewski, S. et al. (2010). Assessment of a micropatterned hepatocyte coculture system to generate major human excretory and circulating drug metabolites. *Drug Metab. Dispos.* 38 (10): 1900–1905.

94 Nguyen, T.V., Ukairo, O., Khetani, S.R. et al. (2015). Establishment of a hepatocyte-kupffer cell coculture model for assessment of proinflammatory cytokine effects on metabolizing enzymes and drug transporters. *Drug Metab. Dispos.* 43 (5): 774–785.

95 Tostoes, R.M., Leite, S.B., Serra, M. et al. (2012). Human liver cell spheroids in extended perfusion bioreactor culture for repeated-dose drug testing. *Hepatology* 55 (4): 1227–1236.

96 Low, L.A., Mummery, C., Berridge, B.R. et al. (2020). Organs-on-chips: into the next decade. *Nat. Rev. Drug Discovery.*

97 Jang, K.-J., Otieno, M.A., Ronxhi, J. et al. (2019). Reproducing human and cross-species drug toxicities using a Liver-Chip. *Sci. Transl. Med.* 11 (517): eaax5516.

98 Underhill, G.H. and Khetani, S.R. (2017). Bioengineered liver models for drug testing and cell differentiation studies. *Cell. Mol. Gastroenterol. Hepatol.* 5 (3): 426–439.e421.

99 Shepherd, G., Mohom, P., Yacoub, K., and May, D.W. (2012). Adverse drug reaction deaths reported in United States vital statistics 1999-2006. *Ann. Pharmacother.* 46.

100 Dormann, H., Criegee-Rieck, M., Neubert, A. et al. (2003). Lack of awareness of community-acquired adverse drug reactions upon hospital admission: dimensions and consequences of a dilemma. *Drug Saf.* 26 (5): 353–362.

101 Burt, C.W., McCaig, L.F., and Rechtsteiner, E.A. (2007). Ambulatory medical care utilization estimates for 2005. *Adv. Data* (388): 1–15.

102 Ingelman-Sundberg, M. (2004). Human drug metabolising cytochrome P450 enzymes: properties and polymorphisms. *Naunyn-Schmiedeberg's Arch. Pharmacol.* 369 (1): 89–104.

103 Ingelman-Sundberg, M. (2005). Genetic polymorphisms of cytochrome P450 2D6 (CYP2D6): clinical consequences, evolutionary aspects and functional diversity. *Pharmacogenomics J.* 5 (1): 6–13.

104 Nakajima, M., Komagata, S., Fujiki, Y. et al. (2007). Genetic polymorphisms of CYP2B6 affect the pharmacokinetics/pharmacodynamics of cyclophosphamide in Japanese cancer patients. *Pharmacogenet. Genomics* 17 (6): 431–445.

# 13

# Urinary Excretion of Drugs and Drug Reabsorption

*Jessica T. Babic[1], Jack Cook[2], and Vincent H. Tam[3]*

[1]Department of Pharmacy, Memorial Hermann-Texas Medical Center, Houston, TX, USA
[2]Clinical Pharmacology, Pfizer Inc., East Lyme, CT, USA
[3]Department of Pharmacy Practice and Translational Research, University of Houston, Houston, TX, USA

## 13.1 Introduction

The kidneys are important organs that remove endogenous and exogenous waste from the body. These highly perfused organs are also critical toward maintaining the acid–base balance, fluid volume, and regulating the fluid osmolality as well as solute/electrolyte concentrations [1]. Depending on their physiochemical properties, drugs may be excreted unchanged in the urine or may be metabolized to more polar derivatives prior to elimination via the kidneys.

The following sections will present pharmacokinetic principles pertinent to the understanding of renal handling of drugs. It is intended to provide the underlying physiological and mechanistic concepts along with the relevant mathematical treatment. Several examples from the literature are cited to illustrate the application of these pharmacokinetic principles in clinical settings.

## 13.2 Kidney as an Eliminating Organ

The functional unit of the kidney is a nephron (Figure 13.1). Each kidney is composed of about 1 million nephrons, which first filter the fluid from the blood and subsequently convert it to urine. Filtration occurs at the glomerular capillaries, which are enclosed in the Bowman's capsule. Following filtration, the fluid flows into the proximal tubule. These kidney components are part of the renal cortex. Subsequently, the filtrate passes into the descending and ascending limbs of the Loop of Henle, both of which lie in the renal medulla. The ascending limb brings the fluid to the distal tubule in the renal cortex. Thereafter, it flows through the connecting tubule into the collecting duct before being excreted as urine.

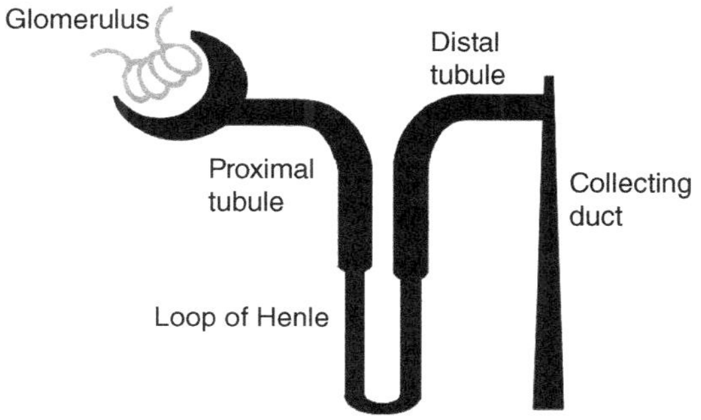


**Figure 13.1** Structure of a nephron.

As the filtered fluid traverses the nephron, it allows the reabsorption of certain substances into the blood (ions like sodium and chloride, organic solutes like urea and glucose) and secretion of some others in the reverse direction (organic acids and bases like bile salts and catacholamines). These processes coupled with the differential fluid exchange (described below) alter the concentration of solutes in the urine along the different tubular segments. For instance, poor reabsorption of urea makes it progressively more concentrated as the fluid moves from proximal tubule to the collecting duct; whereas, high degree of reabsorption of glucose lowers its concentration to an almost undetectable level in the urine.

Even though the kidneys receive 20–25% of the cardiac output and filter approximately 125 ml of fluid per minute, the final urine output is only around 1 ml/min. This difference arises since the renal tubules reabsorb over 99% of the filtered fluid. Reabsorption of filtered water mainly occurs in the proximal tubule by virtue of an osmotic gradient arising from sodium reabsorption.

*Oral Bioavailability and Drug Delivery: From Basics to Advanced Concepts and Applications,* First Edition. Edited by Ming Hu and Xiaoling Li.

Ions like sodium, potassium, and chloride are differentially reabsorbed along the nephron. Sodium reabsorption in the proximal tubule is mediated by the active sodium–potassium ATPase pump. In addition, protein carrier molecules also mediate sodium transport, which in-turn triggers the cotransport of glucose and amino acids. Chloride reabsorption is also coupled to sodium reabsorption via a similar cotransport mechanism.

The sodium–potassium ATPase pump and cotransport protein carriers are also present in the Loop of Henle. This segment allows the reabsorption of ions like sodium, chloride, potassium, calcium, and bicarbonate. Further down in the distal tubule, sodium is reabsorbed and potassium is secreted, again by the action of the sodium–potassium ATPase pump. This part (like the proximal tubule and collecting duct) plays an important role in the acid–base regulation by virtue of the secretion of hydrogen ions, which have the ability to remove bicarbonate ions.

Organic acids and bases that are the end products of metabolism are eliminated from the body by secretion into the tubular fluid. Exogenous compounds like *p*-aminohippuric acid (PAH) can also be secreted by transporters present in the tubular cells. These transporters are saturable and can cause drug–drug interactions as they have the ability to act upon a variety of substrates. The efflux transport pumps like P-glycoprotein (P-gp) are found on the brush border membrane of the proximal tubular cells and may cause drug interactions with renally secreted agents. Besides drug transport, metabolism can also occur in the kidneys. Glucuronidation and sulfation are two main pathways for renal drug biotransformation. Renal metabolism can also be mediated by peptidases, which act on peptides and proteins.

Drug elimination from the kidney is the net outcome of the three aforementioned processes – glomerular filtration, tubular secretion, and tubular reabsorption. Therefore, the renal clearance ($CL_{ren}$) of a drug can be expressed as:

$$CL_{ren} = CL_f + CL_{sec} - F[CL_f + CL_{sec}] \quad (13.1)$$

$CL_f$ and $CL_{sec}$ represent filtration and secretion components of $CL_{ren}$ and F denotes the fraction of drug that is ultimately reabsorbed from the renal tubules into the bloodstream [2]. Each of these processes is discussed in detail in the next section.

$CL_{ren}$ can be viewed as the constant that relates the instantaneous rate of urinary excretion to the corresponding plasma concentration (Cp) [2].

$$\frac{dA}{dt} = CL_{ren} \cdot Cp \quad (13.2)$$

This equation can be rearranged and integrated over a time interval ($\Delta t$).

$$A_{\Delta t} = CL_{ren} \cdot AUC_{\Delta t} \quad (13.3)$$

$AUC_{\Delta t}$ is the area under the plasma-concentration curve over the duration $\Delta t$, and $A_{\Delta t}$ is the amount of drug excreted in this period. If the entire plasma concentration–time curve is considered for AUC calculation, then the above equation yields:

$$Ae = CL_{ren} \cdot AUC_{inf} \quad (13.4)$$

Ae is the portion of the administered dose that is excreted in the urine (Ae = fe · Dose; where, fe is the fraction of dose that is excreted) and $AUC_{inf}$ is the total area under the plasma-concentration curve. This equation can be used to estimate the time-averaged $CL_{ren}$ for an intravenously or extravascularly administered agent but suffers from the limitation that it requires a short to intermediate half-life, for the complete urine recovery of the drug.

Another important relationship between the renal clearance and the total clearance ($CL_{tot} = Dose/AUC_{inf}$) can be derived from Eq. (13.4).

$$CL_{ren} = fe \cdot CL_{tot} \quad (13.5)$$

Accordingly, the nonrenal component of total clearance can be expressed as:

$$CL_{nonren} = (1 - fe) \cdot CL_{tot} \quad (13.6)$$

$CL_{ren}$ can also be estimated from the excretion rate of the drug and the plasma concentration as described by Eq. (13.2). Rate of urinary excretion is measured over a finite time interval ($\Delta t$) and plotted against the plasma concentration recorded at the midpoint of the urine collection interval (a measure of the average concentration over the collection period). The slope of the straight line relationship (passing through origin) yields an estimate of $CL_{ren}$ (Figure 13.2). It should be noted that this method assumes linear kinetics. The slope at any point in the curve represents the average clearance for that average concentration.

If the drug is administered intravenously and the renal pathway accounts for the elimination of the entire dose, and the renal clearance is independent of time and concentration, then the cumulative amount of drug excreted in urine at different time points can be used to calculate the elimination rate constant (or half-life) by means of the "sigma-minus method" [2].

$$\ln(Ae - Ae_t) = \ln(Ae) - K \cdot t \quad (13.7)$$

As seen from Eq. (13.7), a semilogarithmic plot of the fraction remaining to be excreted, i.e. difference

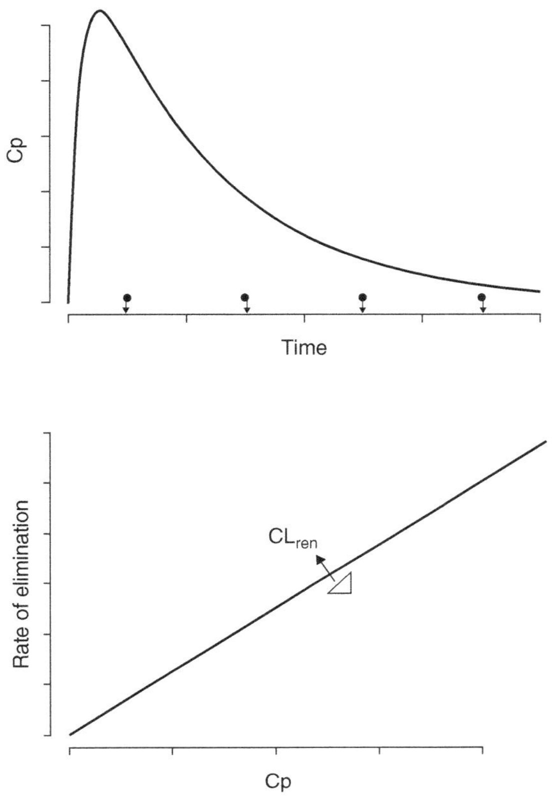


**Figure 13.2** Estimation of renal clearance from plasma and urinary data. The top panel shows the plasma concentration–time profile from an extravascular dose. Plasma samples are collected at the midpoints of urine collection intervals (filled circles with arrows). The bottom panel shows the linear relationship between the rate of urinary excretion and the plasma concentration.

between the dose (Ae) and the cumulative amount excreted till time "$t$" ($Ae_t$) against time, will yield a straight line with the negative slope representing the renal elimination rate constant (Figure 13.3). This method can be extended to drugs, which are not solely eliminated by renal excretion. Ae in such a case will represent the cumulative amount of drug excreted unchanged in the urine at time infinity, i.e. fe · Dose. However, the estimated $K$ will correspond to the rate constant for overall elimination and not any individual pathway (Figure 13.3).

### 13.2.1 Glomerular Filtration

Glomerular capillaries filter small molecules and low molecular weight proteins via a passive process that depends on the unbound solute concentration and factors like size, electrical charge, and net filtration pressure. The filtrate is normally free of plasma proteins and blood cells due to their large sizes. However, it

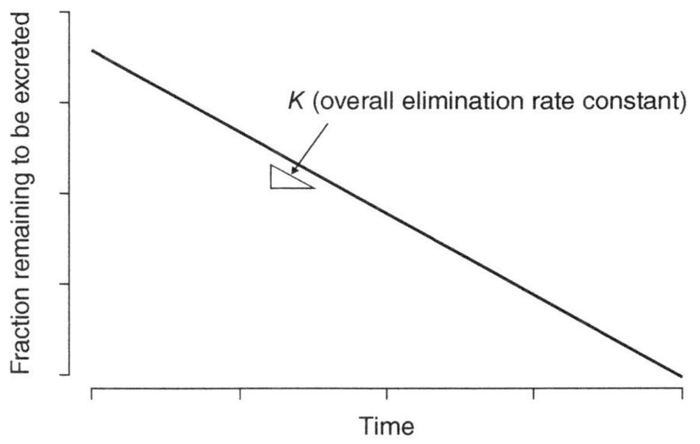


**Figure 13.3** Semilogarithmic plot of the amount of the drug remaining to be excreted versus time-sigma-minus method for the calculation of overall elimination rate constant from urinary excretion data.

contains small molecules like sodium, glucose, and creatinine in similar concentrations as in plasma. Intermediate-sized proteins (molecular weight less than 20 000 g/mol) have limited permeability and therefore low filterability [3].

The lack of blood components in the glomerular filtrate makes the luminal drug concentrations similar to the unbound concentrations in plasma (Cu). The rate of filtration can be related to Cu by a clearance term that represents the volume of plasma water filtered per unit time and is termed as the Glomerular Filtration Rate (GFR).

$$\mathrm{GFR} = \frac{\mathrm{Filtration\ Rate}}{\mathrm{Cp} \cdot \mathrm{fu}} \tag{13.8}$$

Cp denotes the total plasma concentration, fu is the unbound fraction and their product is the unbound concentration Cu.

The blood flow to the kidney averages 1200 ml/min in a healthy individual of which only around 10% is filtered at the glomerulus. Thus the typical value of GFR under normal conditions is 120–125 ml/min. This value can be compared to the estimate of $CL_{ren}$ to elucidate the renal handling mechanism of the drug. If $CL_{ren}$ exceeds fu · GFR, it suggests that drug undergoes net secretion in addition to filtration. Similarly, a value of $CL_{ren}$ lower than fu · GFR indicates net tubular reabsorption.

GFR estimation is fundamental to the assessment of kidney function. In fact, the National Kidney Foundation uses it as one of the main diagnostic criteria for Chronic Kidney Disease (CKD) [4]. Historically, the most widely used method for estimating GFR has been the Cockcroft–Gault formula, which calculates creatinine clearance ($CL_{cr}$, an index of GFR) based on serum creatinine levels, age, gender, and body weight.

$$CL_{cr}(\text{ml/min}) = \frac{[\{140 - \text{age (years)}\} \cdot \text{weight (kg)}]}{72 \cdot \text{serum creatinine (mg/dl)}} \cdot 0.85 \text{ (if female)} \quad (13.9)$$

This method is simple and easy to use since it only requires the measurement of endogenous creatinine, which is produced as a by-product of skeletal muscle metabolism. However, it tends to overestimate creatinine clearance owing to a small fraction of creatinine, which is secreted in the urine. A more reliable method for GFR determination involves administering an exogenous marker like inulin and measuring its urinary excretion rate. The rate of renal excretion of inulin reflects the filtration process exclusively as this compound does not undergo any tubular secretion or reabsorption. However, the difficulty in implementation of this method with respect to intravenous administration of inulin, controlled urine sampling, and a difficult analytical method, limits its routine use [5]. Some of the other filtered markers that have been used clinically or experimentally for estimating GFR include iothalamate, iohexol, and Cystatin C [6].

Other serum creatinine-based equations for the estimation of GFR have also been reported. One is the "Modification of Diet in Renal Disease (MDRD)" equation, which uses serum creatinine, age, race, and gender for the calculation of GFR [4]. This relationship between serum creatinine concentration and GFR (measured from the renal clearance of 125I-iothalamate) was developed by Levey et al. from the MDRD study [7].

$$\begin{aligned} &\text{GFR (ml/min/1.73 m}^2) \\ &\quad = 175 \cdot \text{serum creatinine (mg/dl)}^{-1.154} \\ &\quad\quad \cdot \text{age (years)}^{-0.203} \cdot 0.742 \text{ (if female)} \\ &\text{GFR (ml/min/1.73 m}^2) \\ &\quad = 175 \cdot \text{serum creatinine (mg/dl)}^{-1.154} \\ &\quad\quad \cdot \text{age (years)}^{-0.203} \cdot 1.21 \text{ (if black)} \quad (13.10) \end{aligned}$$

However, this equation was developed based on individuals with CKD and has been shown to be less accurate in patients with values >60 ml/min/1.73 m$^2$ [8, 9]. In order to overcome this limitation, the "CKD-Epidemiology Collaboration (CKD-EPI)" equation was recently proposed, using the same variables. Studies have found this equation to allow for more accurate GFR estimation and fewer false positive diagnoses of CKD, when compared to the MDRD equation [10, 11].

$$\begin{aligned} &\text{GFR}_\alpha \text{ (ml/min/1.73 m}^2) \\ &\quad = 141 \cdot \min \text{ (serum creatinine}/\kappa, 1)^\alpha \\ &\quad\quad \cdot \max \text{ (serum creatinine}/\kappa, 1)^{-1.209} \cdot 0.993^{\text{Age}} \\ &\quad\quad \cdot 1.159 \text{ (if black)} \cdot 1.1018 \text{ (if female)} \quad (13.11) \end{aligned}$$

$\kappa$ is 0.7 for females and 0.9 for males, $\alpha$ is −0.329 for females and −0.411 for males, min indicates the minimum of Scr/$\kappa$ or 1, and max indicates the maximum of Scr/$\kappa$ or 1, and age is measured in years.

In children, the Schwartz formula is employed, which yields an estimate of GFR from the product of the height of the child and a numerical constant ($X$) normalized with respect to the serum creatinine levels [12]. $X$ depends on the age of the child and can take a value of 0.45 for infants less than 1 year of age or 0.55 for children up to 12 years.

$$\text{GFR (ml/min/1.73 m}^2) = \frac{X \cdot \text{height (cm)}}{\text{serum creatinine (mg/dl)}} \quad (13.12)$$

Creatinine clearance should be used with caution when estimating renal function since it may not be a robust measure in certain disease states and physiological conditions. For instance, serum creatinine does not capture the altered renal kinetics of fluconazole in HIV patients [6]. Similarly, a lack of concordance between GFR and creatinine clearance is observed in liver cirrhosis [13] and renal transplantation [14]. Also, since serum creatinine is generated from muscle, patients with very high or low muscle mass (e.g. weight lifters or bed-bound individuals) could have creatinine values not representative of their renal function, resulting in GFR estimations lower or higher than the true GFR [15]. Another factor to consider is the variability in serum creatinine assays among laboratories and even within the same laboratory. In 2005, a standardized creatinine assay was introduced, and most laboratories now use this method. However, variable serum creatinine values still affect the current applicability of PK/PD studies conducted prior to the use of this assay.

The discrepancy between the two measures warrants the inspection of other pathways for quantification of renal function. A "cocktail approach" has been evaluated, which involves simultaneous monitoring of markers for different kidney functions (glomerular filtration, active secretion, and tubular reabsorption). Gross et al. [16], administered a cocktail of sinistrin (for glomerular filtration), PAH (for organic anion secretion), pindolol (for organic cation secretion), and fluconazole (for tubular reabsorption) and successfully demonstrated that the cocktail approach could be used to characterize the individual kidney pathways. This approach can prove extremely beneficial when compensatory mechanisms in the kidney may cause an increase in contribution of one pathway in response to a decline in performance of another.

Serum creatinine may also be an insensitive measure of renal function in patients with acute kidney injury (AKI); an extensive degree of tissue injury must have occurred before a substantial increase in serum creatinine is observed. Therefore, more sensitive markers of kidney injury are needed in this population. Recently, novel renal biomarkers have been investigated for this purpose, including neutrophil gelatinase-associated lipocalin (NGAL), kidney injury molecule-1 (KIM-1), Cystatin C, interleukin-6 (IL-6), interleukin-8 (IL-8), interleukin-18 (IL-18), *N*-acetyl-glucosaminidase (NAG), glutathione transferases (GST) and liver fatty acid binding protein (LFABP) [17]. KIM-1, a biomarker upregulated in the proximal tubule of kidney cells in response to kidney injury, has shown promise in predicting AKI in several studies. Li et al. demonstrated that urinary KIM-1 has a high predictive value for contrast-induced AKI in patients after cardiac catheterization, while Pang et al. showed its utility in predicting vancomycin-associated AKI [18, 19]. Several additional biomarkers have shown promise in predicting kidney injury in animal models, but additional data is needed to determine their clinical utility.

For drugs primarily eliminated via the kidneys, their disposition kinetics are determined by the patient's renal function. A decline in renal function, therefore, leads to a decrease in drug elimination. A dose adjustment may be required if the renal pathway accounts for more than 30% of total elimination [20]. The slowed elimination manifests as an increase in half-life. Cephalosporins such as ceftazidime and cefepime exhibit this behavior and require a dosing regimen modification in patients with renal insufficiency [21, 22]. Particular attention is needed for drugs that have a narrow therapeutic index like gentamicin and vancomycin [23].

When it comes to selecting the most appropriate estimation of GFR to use when adjusting medication doses, there has been much debate. For many years, the US Food and Drug Administration (FDA) recommended calculating creatinine clearance using the Cockcroft–Gault equation during drug development [24]. Therefore most published guidance on dose adjustment in patients with decreased kidney function is based on this parameter. As of 2010, FDA guidance for industry included both the Cockcroft–Gault and MDRD equations as acceptable methods for stratifying subjects by renal function and recommended reporting pharmacokinetic results using both methods. Medications to be approved are expected to have renal dose adjustments by eGFR using MDRD in the package insert, such as meropenem–vaborbactam and delafloxacin.

The dependence of the overall elimination rate constant ($K$) on renal function (e.g. $CL_{cr}$), for drugs exhibiting the aforementioned characteristics, can be described by a linear relationship where the elimination rate constants for the nonrenal and the renal pathways (knr and ke, respectively) are given by the intercept and the product of the slope ($m$) and renal function (RF), respectively (Figure 13.4).

$$K = \text{knr} + m \cdot \text{RF} \tag{13.13}$$

Such plots can offer insight into the relative contribution of these pathways and the velocity of renal elimination, i.e. shallow versus steep slope.

### 13.2.2 Active Secretion

Renal secretion is a carrier-mediated process that transports endogenous and exogenous substances from the blood circulation into the tubular lumen. The transport process is a function of the substrate concentration, competition from other substrates, and perfusion [23, 25].

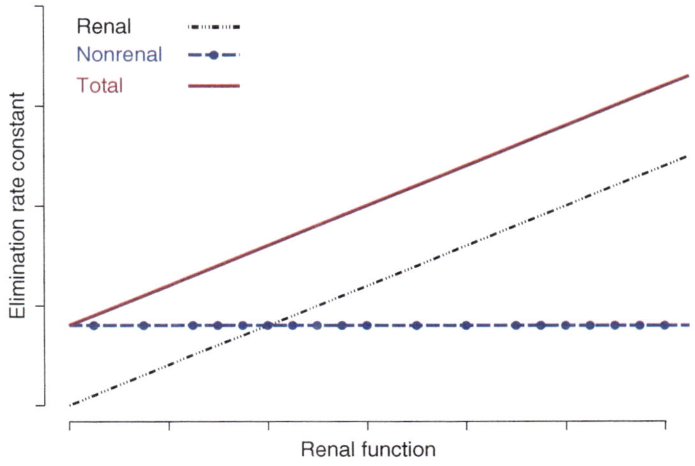


**Figure 13.4** Relationship between the elimination rate constant (renal, nonrenal, and overall) and the renal function.

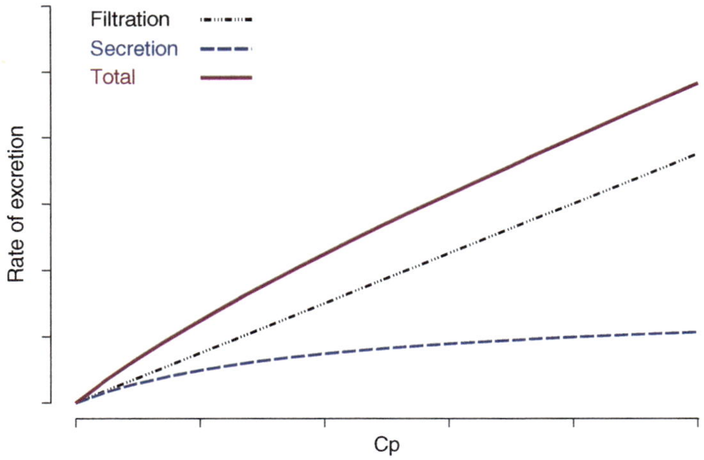


**Figure 13.5** Relationship between excretion rate and plasma concentration for a drug that undergoes active tubular secretion. The total excretion rate arises from the secretion (saturable) and filtration (linear) components.

The kidney has separate systems in the proximal tubule, which are responsible for secretion of organic cationic and anionic drugs. A detailed discussion of these transport systems is presented later in the chapter.

These carrier-mediated systems are capacity-limited indicating that they can get saturated at a high concentration of the substrate. The hyperbolic relationship between the rate of secretion in this case and the corresponding plasma concentration can be expressed by the Michaelis–Menten equation:

$$\frac{dS}{dt} = \frac{S_{max}\,Cp}{S_{50} + Cp} \tag{13.14}$$

where, $S_{max}$ is the maximal rate of secretion (capacity of the secretory system, e.g. 16 mg/min for creatinine), Cp is the plasma concentration of the substrate at the secretory site and $S_{50}$ is an affinity constant indicating the concentration at which the transport system is operating at half the maximal capacity. If the substrate concentration is well below the $S_{50}$, the rate of secretion will follow apparent first-order kinetics as shown in Eq. (13.15).

$$\frac{dS}{dt} = \left(\frac{S_{max}}{S_{50}}\right) Cp \tag{13.15}$$

The ratio $S_{max}/S_{50}$ represents the maximum value of intrinsic clearance ($CL_{int}$) of the transport system in the absence of any flow limitations.

However, when Cp far exceeds $S_{50}$, the transport system is completely saturated and the rate of secretion asymptotes to $S_{max}$ (Eq. (13.16)). This suggests that any further increase in concentration will not translate to an increase in the secretion rate.

$$\frac{dS}{dt} = S_{max} \tag{13.16}$$

The aforementioned Michaelis–Menten behavior can be deciphered from a plot of the excretion rate of the drug versus the plasma concentration, which will reflect the summation of a linear (filtration) and nonlinear (secretion) process (Figure 13.5).

Equation (13.14) can be modified to incorporate the effect of a competitive inhibitor, which serves to reduce the affinity of the substrate for the transport system in a concentration-dependent manner. The rate of secretion can now be expressed in terms of the inhibitor constant $K_i$ and its instantaneous plasma concentration $C_i$.

$$\frac{dS}{dt} = \frac{S_{max}\,Cp}{S^*_{50} + Cp} \tag{13.17}$$

where, $S^*_{50} = S_{50}\left(1 + \frac{C_i}{K_i}\right)$

Renal tubular secretion can also be described by the "well-stirred model" to study the effects of perfusion and plasma protein binding on the secretory process (see also Chapter 7).

$$CL_{sec} = \frac{Q \cdot CL_{int} \cdot fu}{Q + CL_{int} fu} \tag{13.18}$$

$Q$ is the renal blood flow, fu is the fraction of drug unbound in plasma, and $CL_{int}$ is the intrinsic clearance of the transport system as described earlier. The total renal clearance can now be expressed as the following equation:

$$CL_{ren} = fu \cdot GFR + \frac{Q \cdot CL_{int} \cdot fu}{Q + CL_{int} fu} - F\left[fu \cdot GFR + \frac{Q \cdot CL_{int} \cdot fu}{Q + CL_{int} fu}\right] \tag{13.19}$$

where, $F$ is the fraction reabsorbed. As can be seen from Eq. (13.19), when the estimated $CL_{ren}$ exceeds fu · GFR, it is indicative of tubular secretion.

Equation (13.19) can be further simplified depending on the efficiency of the transporter relative to the delivery of the drug to the transport site. In the "perfusion-limited" scenario, the intrinsic clearance of the transport system is much higher compared to the

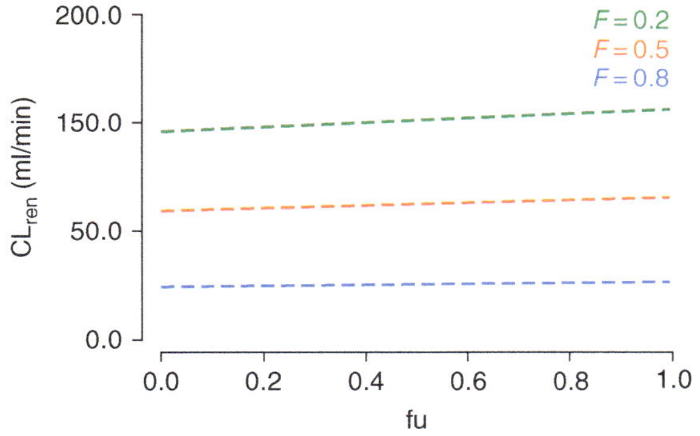


**Figure 13.6** Relationship between the free fraction of a drug in plasma and the change in renal clearance for a drug, which exhibits perfusion-limited renal elimination.

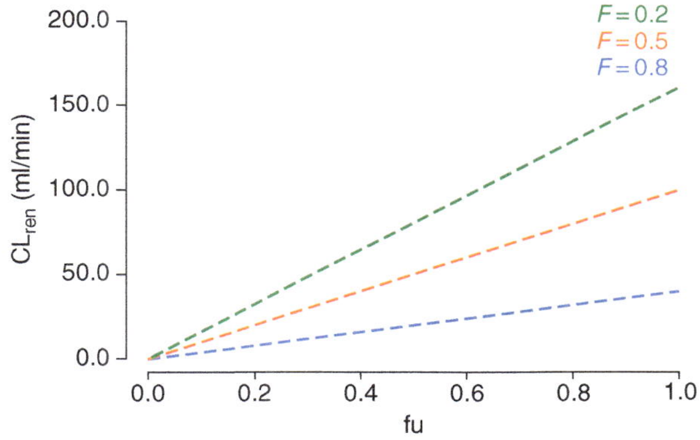


**Figure 13.7** Relationship between the free fraction of a drug in plasma and the change in renal clearance for a drug, which exhibits secretion-limited renal elimination.

blood flow ($fu \cdot CL_{int} \gg Q$, e.g. PAH). Consequently, the above equation reduces to:

$$CL_{ren} = fu \cdot GFR(1 - F) + Q(1 - F) \qquad (13.20)$$

A plot of $CL_{ren}$ versus fu will yield a straight line with slope $GFR(1 - F)$ and intercept $Q(1 - F)$ (Figure 13.6). Cook and Smith [26], demonstrated using simulations that changing the fraction unbound will have little impact on the change in renal clearance for such a drug. PAH, the prototypical drug for this scenario has an additional property that it undergoes no tubular reabsorption. These disposition features make it a useful marker for characterizing active secretion and measuring the renal blood flow.

In the opposite scenario, when the $Q \gg fu \cdot CL_{int}$, i.e. the secretion process is rate limiting, Eq. (13.19) simplifies to:

$$CL_{ren} = fu[(GFR + CL_{int}) \cdot (1 - F)] \qquad (13.21)$$

Again, the $CL_{ren}$ can be expressed as a linear function of fu (e.g. furosemide), with the term in square brackets representing the slope of the straight-line relationship (intercept is 0) (Figure 13.7). In this case, change in the free fraction will lead to a pronounced consequent change in the renal clearance; however, the rate of change (slope) will be a function of the fraction reabsorbed ($F$). The slope will be shallow for highly reabsorbed drugs and steep for drugs, which are primarily eliminated by filtration and secretion [26].

### 13.2.3 Tubular Reabsorption

Tubular reabsorption can occur via passive diffusion across the tubular membrane or by a carrier-mediated process. The former is the primary mechanism of reabsorption for most drugs, whereas the latter has been associated with vitamins and amino acids. The movement of the filtered drug from the tubules back into the circulation makes the estimated renal clearance value lower than the filtration clearance, i.e. $CL_{ren} < fu \cdot GFR$. This does not overrule the presence of a secretion component; however, it implies that if secretion is present it contributes to a lesser extent toward renal elimination. In other words, $CL_{ren}$ reflects filtration and net reabsorption in the presence of tubular secretion.

Similar concentrations in the plasma and luminal fluid can occur for a compound whose reabsorption has reached an equilibrium state. The renal clearance of such a compound (e.g. ethanol) is dependent on the urine flow and increases linearly with an increase in the filtered fraction [27]. However if the equilibrium is not yet established, an "ascending convex relationship" between renal clearance and urine flow is observed as in the case of propylene glycol [28].

The pH of urine has a typical value of around 6.2 but it can vary from 4.5 to 7.5. The change in urinary pH can influence the ionization state of a variety of drugs, which in turn can restrict or facilitate their movement across biological membranes as explained by the "pH-partition hypothesis" (unionized form is favored) [3]. At a certain pH, the Henderson–Hasselbalch equation can yield the fraction unionized for a weak acid or a weak base depending on the $pK_a$ value (Eqs. (13.22) and (13.23), respectively).

$$\text{uniounionized (weak acid)} = \frac{1}{\left(1 + 10^{(pH - pK_a)}\right)} \qquad (13.22)$$

$$\text{uniounionized (weak base)} = \frac{1}{\left(1 + 10^{(pK_a - pH)}\right)} \qquad (13.23)$$

An enhanced acidic environment can suppress the ionization of weak acids and facilitate the ionization of weak bases. The opposite trend is observed in the presence of basic conditions. As a consequence of decreased ionization, reabsorption is increased, which concomitantly reduces the renal clearance. In clinical practice, urine can be made alkaline (up to a pH of 7.5) by administering sodium bicarbonate and acidified (up to a pH of 4.5) by ammonium chloride.

Weak bases like ephedrine (p$K_a$ 9.6) and phenylpropanolamine (p$K_a$ 9.4) exhibit pH sensitivity by registering an increase in tubular reabsorption under alkaline urinary conditions [29, 30]. Similarly, acidification enhances the renal excretion of weak bases as seen in case of methadone (p$K_a$ 9.2) [31]. On the other end, alkalinization of urine increases the excretion rate of weak acids like salicylate (p$K_a$ 3) and chlorpropamide (p$K_a$ 4.8) [32] on account of increased ionization.

Very weak acids like acetaminophen (p$K_a$ 9.5) do not demonstrate any pH-dependence as they remain predominantly unionized at all values of urine pH [33]. Similarly, very weak bases like propoxyphene (p$K_a$ 6.3) are insensitive to changes in pH, primarily because of their excellent reabsorption across the entire urine pH range [3]. Strong acids (p$K_a > 2$) and strong bases (p$K_a > 12$) are expected to remain ionized between the urine pH extremes of 4.5 and 7.5. Thus their reabsorption characteristics should not be a function of changes in urinary pH.

The renal handling of compounds such as water-soluble vitamins involves active tubular reabsorption, which is a saturable process. Riboflavin is an example of a water-soluble vitamin, which shows a decrease in renal clearance with declining serum levels [34]. This phenomenon is attributed to saturable tubular reabsorption, which can be characterized by a capacity ($T_{max}$) and an affinity term ($T_{50}$), respectively. The mathematical treatment of such a capacity-limited system has been described in detail in Section 13.2.3. As mentioned earlier, the presence of net saturable reabsorption can be detected by plotting the excretion rate against the plasma drug concentration (Figure 13.8). A less than proportional increase in the rate of excretion with the concentration is suggestive of an asymptotic increase in the renal clearance due to saturable tubular reabsorption.

Active tubular reabsorption may also lead to drug interactions with other substrates of the transport system. A few published studies have reported the interaction between drugs and endogenous compounds. Ampicillin is preferentially reabsorbed compared to uric acid and therefore increases the urinary excretion of urate [35]. Drug interactions may also occur due to change in the urine pH thereby affecting the rate of reabsorption. Carbonic-anhydrase inhibitors alkalinize urine by reducing bicarbonate reabsorption

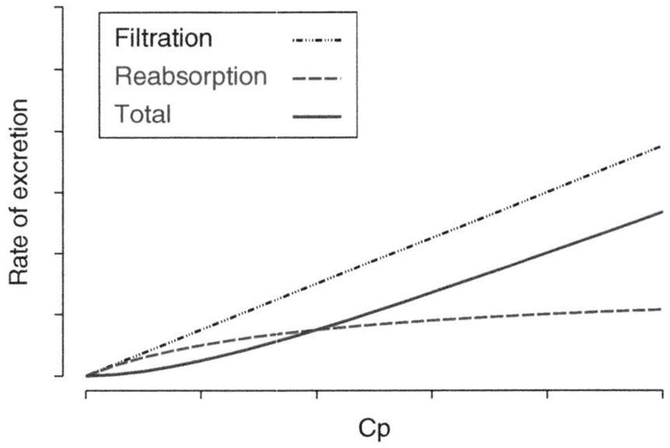


**Figure 13.8** Relationship between excretion rate and plasma concentration for a drug that undergoes active tubular reabsorption. The total excretion rate arises from the reabsorption (saturable) and filtration (linear) components.

and may increase the reabsorption of weak bases and the elimination of weak acids. Therapeutic interventions have taken advantage of such drug-endogenous compound interactions. Glucose is reabsorbed via sodium-dependent glucose cotransporters (SGLT) in the kidney. SGLT2 inhibitors such as canagliflozin, dapagliflozin, and empagliflozin have been approved by the FDA for use with diet and exercise to lower blood sugar in adults with type 2 diabetes. URAT1 is a transporter that is responsible for the reabsorption of urate. URAT1 inhibitors such as lesinurad have been shown to be effective in the treatment of gout [36]. Finally, diuretic agents like mannitol that have the ability to induce diuresis and increase urine flow can significantly increase the elimination of drugs like digoxin that undergo net tubular reabsorption [37]. This occurs because the increased urine flow causes digoxin to move past the reabsorptive sites before equilibrium can be established.

## 13.3 Drug Transporters and Their Role in Renal Elimination

In the past decade, significant progress has been made in the understanding of the role of drug transporters in renal secretion and reabsorption of endogenous as well as exogenous compounds. Figure 13.9 illustrates the major transporters important for secretion and reabsorption in the kidneys.

### 13.3.1 Transporters Involved in Active Tubular Secretion

The organic ion transporter family (SLC22A) plays an important role in the renal secretion process, which

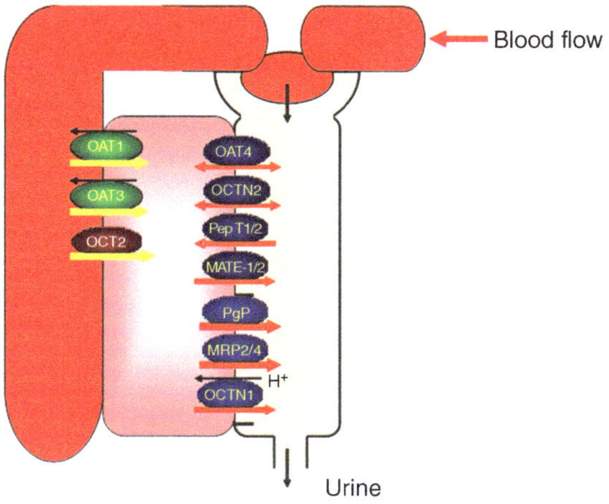


**Figure 13.9** Major renal transporters involved in the renal secretion of drugs. Source: Adapted from Sahi et al. [38].

comprises the organic cation transporters (OCTs and OCTNs) and organic anion transporters (OATs) [39]. In both the cation and anion systems, the carrier-mediated process is responsible for basolateral uptake followed by efflux across the brush border membrane in the renal proximal tubule cells [25]. Among the organic cation transporters, OCT2 is highly expressed in the human kidney [40]. It interacts with a variety of structurally diverse organic cations such as drugs, toxins, endogenous amines, and metabolism products with relatively low molecular weights [41]. For example, tetraethylammonium (TEA), cimetidine, metformin, procainamide, quinidine, acyclovir, zidovudine, etc. are known to interact with OCT2 [42]. In the OCTN family, OCTN1 and OCTN2 are the two major isoforms, expressed on the brush border membrane of human kidney. OCTN1 transports drugs including quinidine, doxorubicin, verapamil, and TEA [43]. OCTN2 is secretory for drugs, such as verapamil, quinidine, pyrilamine, and reabsorptive for carnitine, a cofactor essential for β-oxidation of fatty acids [44]. The OATs are involved in the secretion of low molecular weight organic anions. OAT1 and OAT3, the two major isoforms of this family, are expressed on the basolateral membrane, whereas OAT4 is localized on the brush border membrane [45]. OAT1 and OAT3 are believed to mediate the uptake of anionic substrates into the renal proximal tubule. It is speculated that OAT4 may be involved in effluxing organic anions from the renal tubules into the urine [46]. These transporters are involved in the renal elimination of many drugs, such as PAH, anticancer agents, nonsteroidal anti-inflammatory drugs (NSAIDs), antiviral drugs and β-lactam antibiotics [47–49].

In the kidney, P-gp is a major efflux transporter, which interacts with structurally diverse compounds like anticancer drugs (e.g. doxorubicin, etoposide, and paclitaxel), immunosuppressants (e.g. cyclosporin A), and steroids (e.g. aldosterone and cortisol) [50, 51]. Besides P-gp, other efflux transporters, such as multidrug resistance associated protein 2 and 4 (MRP2 and MRP4), have also been identified on the brush border of renal proximal tubules [45]. MRP2 mainly interacts with the conjugated organic anions, and may play a role in the renal secretion of conjugates to glutathione and glucuronide. MRP4 can interact with methotrexate (MTX) and antiviral drugs, such as adefovir and cidofovir. Recently, human multidrug and toxin extrusion/H(+)-organic cation antiporters, MATE1 and MATE2-K, have been identified as two new organic cation transporters expressed on the apical side of kidney cells [52, 53]. Like OCT2, the MATE system can transport TEA and metformin and is inhibited by cimetidine [53]. Based on limited studies, these transporters appear to be responsible for the tubular secretion of cationic drugs across the brush border membrane into the urine.

### 13.3.2 Transporters Involved in Reabsorption

Peptide transporters (PEPTs, SLC15A) are important for the reabsorption of compounds in the kidney. Two isoforms of peptide transporters, PEPT1 and PEPT2 are localized on the brush border membrane of the renal proximal tubule cells [54]. These transporters have affinity to di- or tri-peptides, but not single amino acids or tetra-peptides. They are involved in interacting with several therapeutic drugs, including β-lactam antibiotics (cephalexin), angiotensin-converting enzyme (ACE) inhibitors (enalapril and quinapril), and antiviral drugs (valacyclovir and valganciclovir). The expression of hPEPT2 in kidney is higher than that of hPEPT1, although fewer therapeutic drugs have been reported to be substrates of PEPT2. Besides PEPTs, nucleoside transporters expressed on the brush border membrane are also known to interact with nucleosides and nucleoside analogs. However, more investigation is required for better understanding of the role of transporters in the reabsorption of compounds in the kidneys.

### 13.3.3 Transporter-Mediated Drug-Drug Interactions (DDI)

Renal transport systems have broad substrate specificity. Therefore, concomitant use of drugs that depend on renal transporters for their urinary elimination may increase the risk of interactions. The degree of change in renal clearance caused by a DDI, however, depends on the contribution of the respective pathway(s) toward overall renal elimination.

Numerous interactions among renally cleared anionic drugs have been reported in human PK studies, and a summary of these is presented by Masereew and Russel [25] (Table 13.1). Table 13.2 gives an overview of known organic anion drug interactions at the level of renal secretory clearance. Concomitant use of OAT substrate drugs can cause a reduction in the intrinsic clearance for renal secretion, possibly leading to an increase in plasma concentration, which in turn can cause systemic toxicities. For example, the peak concentration and half-life of cefadroxil were increased 1.4- and 1.3-fold, respectively, following coadministration of probenecid, and its urinary excretion rate constant fell by 58% [55]. Similarly, severe methotrexate toxicity was observed, due to an increase in serum concentrations, after coadministration with NSAIDs, probenecid, or penicillin [56, 57]. In some cases, this interaction may be considered favorable, and concomitant dosing is used to extend the drug bioavailability. Probenecid coadministration with oseltamivir has been shown to decrease oseltamivir clearance by approximately 50% and subsequently double the plasma levels of oseltamivir, a strategy that could potentially be intentionally employed to extend the supply of oseltamivir in the event of a shortage [58]. However, some inhibitors of the OAT family may not cause a significant interaction, because of relatively low plasma concentrations. Hence, it is important to take the systemic concentration and the potency of inhibitors into consideration in order to predict the clinically relevant transporter-mediated DDIs.

Besides organic anionic drugs, renal interactions between organic cationic drugs have also been reported (Table 13.3). Cimetidine is a potent inhibitor of the organic cation system and inhibits the renal secretion of amiloride, metformin, ranitidine, and triamterene [25]. This property of cimetidine is also utilized in clinical practice to improve the accuracy of GFR estimation through creatinine clearance. Cimetidine administration causes creatinine to be exclusively eliminated by filtration on account of an inhibition of the secretion pathway [59]. Trimethoprim is another drug that has a high affinity for the organic cation transport system [60]. As in the case of anionic drugs, systemic concentrations should be monitored for concomitantly administered organic cationic drugs, which have the potential to compete for the cationic transport system.

Drug interactions arising because of inhibition of the efflux transporters are not common; however, serious adverse events have been associated with them. One key example is severe digoxin toxicity when coadministrated with P-gp inhibitors, such as quinidine, clarithromycin, and ritonavir [61–63]. It was found out that tubular secretion of digoxin was diminished, which resulted in an increase in plasma digoxin levels. Digoxin interactions caused by P-gp inhibition also occur at the gut where concentrations of the inhibitor are much higher than those in the kidney. DDIs involving some renal transporters can occur at multiple sites because of their nonspecific tissue expression. Thus, the observed effects reflect an overall summation across these multiple sites. Specific effects of such renal efflux transporters may require more detailed studies to characterize these interactions.

At the level of reabsorption, only a limited number of clinical studies have reported interactions with endogenous compounds. As discussed previously, ampicillin increases uric acid's urinary excretion rate [35]. The metabolite of allopurinol and oxipurinol increases the renal clearance of xanthine due to inhibition of tubular xanthine reabsorption [64]. Similarly, cisplatin treatment results in a strong increase in renal carnitine excretion, possibly due to inhibition of carnitine reabsorption by OCTN2 in renal proximal tubules [65]. However, as active transport constitutes only a small component of tubular reabsorption, which otherwise is primarily a passive diffusion process, the clinical relevance of transporter-mediated DDIs has not been fully established.

## 13.4 Renal Elimination and Bioavailability

Urinary excretion data can be used to assess the extent of bioavailability of drugs administered extravascularly or to assess relative bioavailability. The following section provides three methods for estimating the bioavailable fraction, when urine data are used alone or when urine data are used in conjunction with plasma data.

### 13.4.1 Urine Data Alone

Bioavailability is most often assessed by measuring drug concentrations in the blood, plasma, or serum. In some instances where less invasive measures are desired, it is possible to assess the extent of absorption solely through urine data. Theoretically, urine excretion data can also be used to measure the rate of absorption. However, with many drugs, absorption is nearly complete within a few hours. This requires relatively short urine collection

**Table 13.1** Anionic drug interactions and non-interactions for tubular secretion observed in human pharmacokinetic studies.

| Inhibitor drug and dose | Inhibited drug | Dose | Inh. $CL_{ren}$ (%) |
|---|---|---|---|
| *Diatrizoate* | | | |
| 13 000 mg iv | *p*-Aminohippurate | 8–12 mg/kg iv | 6[a)] |
| *Didanosine* | | | |
| 67–250 mg | Zidovudine | 50–200 mg | — |
| | Zidovudine-gluc. | — | 50 |
| *Diflunisal* | | | |
| 500 mg po bid | Triamterene | 100 mg po | NS |
| | pOH-triamterene-sulf. | — | 32 |
| *Ibuprofen* | | | |
| 10–25 mg po | Methotrexate | 7.5 mg po | 20 |
| *Indomethacin* | | | |
| 25 mg po q12h × 6 | Zidovudine | 200 mg po | 11 (NS) |
| | Zidovudine gluc. | — | 18 |
| *Naproxen* | | | |
| 250 mg po q12h × 6 | Zidovudine | 200 mg po | 5 (NS) |
| | Zidovudine gluc. | — | 21 |
| *Probenecid* | | | |
| 1000 mg iv | *p*-Aminohippurate | 10 mg/kg | 85 |
| | Amoxicillin | | |
| 2000 mg po | Azlocillin | 2000 mg iv | 47 |
| 2000 mg po | *R*-Carbenicillin | 1100 mg iv | 32 |
| | *S*-Carbenicillin | 900 mg iv | 21 |
| 1500–3000 mg po | Cidofovir | 5 mg/kg iv | 32 |
| 500 mg po q6h × 13 | Cimetidine | 300 mg iv | 25 |
| 2000–4000 mg/d po | Cisplatin | 60–100 mg/m²/d iv | 41 |
| 25 mg/kg iv | Chlorothiazide | 2.5 mg/kg iv | 53 |
| 500 mg po q6h × 6 | Ciprofloxacin | 200 mg iv | 64 |
| | Diflunisal | | |
| | Diflunisal-acylgluc. | — | 70 |
| | Diflunisal-phenolgluc. | — | 70 |
| | Diflunisal-sulf. | | — |
| 1500 mg po | Famotidine | 20 mg po | 89 |
| 1000 mg po q12h × 2 | Furosemide | 40 mg iv | 72 |
| 500 mg po q6 × 16 | Ganciclovir | 1000 mg po q8hx12 | 19 |
| 25 mg/kg iv | Hydrochlorothiazide | 2.5 mg/kg iv | 83 |
| 2000 mg po | Nafcillin | 500 iv | 72 |
| 1000 mg po | Paracetamol | 1500 mg | +13 (NS) |
| | Paracetamol-sulf. | — | 21 |
| | Paracetamol-gluc. | — | 79 |
| 2000 mg po | Procainamide | 750 mg iv | 3 (NS) |
| 1000 mg po q11 × 2 | *R*-Sulbenicillin | 1470 mg iv | 67 |
| | *S*-Sulbenicillin | 530 mg iv | 69 |
| 500 mg iv | Temocillin | 3000 mg iv | 34 |

**Table 13.1** (Continued)

| Inhibitor drug and dose | Inhibited drug | Dose | Inh. $CL_{ren}$ (%) |
|---|---|---|---|
| 2000 mg po | Ticarcillin | 5000 g iv | 22 |
| 25 mg/kg iv | Trichlorthiazide | 2.5 mg/kg iv | 89 |
| 500 mg po q6h × 12 | Zidovudine | 200 mg po q6h × 24 | 37 |
| | Zidovudine-gluc. | — | 77 |
| 500 mg po q6h × 4 | Zidovudine | 2 mg/kg po bid | 25 |
| | Zidovudine-gluc. | — | 95 |
| *Salicylate* | | | |
| 2000 mg + 33 mg/min iv | Methotrexate | 5–15 mg iv | 35 |
| 12.5–20 po | Methotrexate | 7.5 mg po | 20 |
| 45–50 mg/kg q24h × 16 | Methotrexate | 10 mg/m² iv | 8 (NS) |
| *Zidovudine* | | | |
| 100 mg po q4 × 35 | Ganciclovir | 1000 mg po | +14.6 (NS) |

a) Calculated from renal extraction values.
inh, inhibited; iv, intravenous; po, per os; NS, not significant (generally $p > 0.05$), if not indicated, $p$, 0.05; sulf., sulfate conjugate; gluc., glucuronide.
Source: Masereeuw and Russel [25] / Taylor & Francis.

intervals that are impractical due to limitations on subjects' ability to produce urine and completely empty their bladder.

When renal ($CL_{ren}$) and systemic ($CL_{tot}$) clearances are constant, the amount of drug excreted unchanged in the urine (Ae) can be determined using Eqs. (13.24) and (13.25).

$$\mathrm{Ae} = F \cdot \mathrm{Dose} \cdot \frac{\mathrm{CL_{ren}}}{\mathrm{CL_{tot}}} \tag{13.24}$$

where $F$ is the absolute bioavailability. Equation (13.24) can be rearranged to:

$$F = \frac{\mathrm{Ae}}{\mathrm{Dose}} \cdot \frac{\mathrm{CL_{tot}}}{\mathrm{CL_{ren}}} \tag{13.25}$$

If there are two dosage forms (Form1 and Form2), the relative bioavailability ($F_{rel}$) of Form1 compared to Form2 is the ratio of their absolute bioavailabilities ($F_{Form1}$ and $F_{Form2}$, respectively).

$$F_{rel} = \frac{F_{Form1}}{F_{Form2}} = \frac{\frac{\mathrm{Ae_{Form1}}}{\mathrm{Dose_{Form1}}} \cdot \frac{\mathrm{CL_{tot}}}{\mathrm{CL_{ren}}}}{\frac{\mathrm{Ae_{Form2}}}{\mathrm{Dose_{Form2}}} \cdot \frac{\mathrm{CL_{tot}}}{\mathrm{CL_{ren}}}} = \frac{\frac{\mathrm{Ae_{Form1}}}{\mathrm{Dose_{Form1}}}}{\frac{\mathrm{Ae_{Form2}}}{\mathrm{Dose_{Form2}}}} = \frac{\mathrm{Ae_{Form1}} \cdot \mathrm{Dose_{Form2}}}{\mathrm{Ae_{Form2}} \cdot \mathrm{Dose_{Form1}}} \tag{13.26}$$

*Example:* Table 13.3 shows urinary excretion data from a furosemide bioequivalence study [66]. In that study, the bioavailability of five tablet products was compared to furosemide solution. The last column of Table 13.3 displays the bioavailability of each product relative to the solution. Since the doses were the same, this value was calculated as the ratio of Ae values (Test/Reference).

### 13.4.2 Plasma and Urine Data Including Data from Intravenous Administration

Absolute bioavailability calculations presented elsewhere in this book have assumed that renal clearance is constant. However, there are many instances where renal clearance is nonlinear and such estimates will be biased. In cases where nonrenal clearance is constant, Øie and Jung [67] have proposed the following formula for determining absolute bioavailability in cases of changing or nonlinear renal clearance:

$$F = \frac{(\mathrm{Dose_{IV}} - \mathrm{Ae_{IV}}) \cdot \frac{\mathrm{AUC_{inf}}^{\mathrm{test}}}{\mathrm{AUC_{inf}}^{\mathrm{IV}}} + \mathrm{Ae_{test}}}{\mathrm{Dose_{test}}} \tag{13.27}$$

$AUC_{inf}$ is the area under the plasma-concentration time curve from time zero to infinity and the subscripts IV and test refer to parameters obtained after intravenous and test formulations, respectively.

*Example:* The differences in methodology to determine absolute bioavailability and the associated assumptions are illustrated in Table 13.4. Grahnén et al. [68] administered single doses of cimetidine to healthy subjects in study employing a crossover design. It was subsequently noted that renal clearance was not constant following each treatment and bioavailability was calculated using AUC ratios, Ae ratios, and by the method of Øie and Jung [67].

**Table 13.2** Cationic drug interactions and non-interactions for tubular secretion observed in human pharmacokinetic studies.

| Inhibitor drug and dose | Inhibited drug | Dose | Inh. $CL_{ren}$ (%) |
|---|---|---|---|
| *Amiloride* | | | |
| 5 mg po q12h × 24 | Cimetidine | 400 mg po q12h × 24 | 5 (NS) |
| *Cimetidine* | | | |
| 400 mg po q 12h × 24 | Amiloride | 5 mg po q12h × 24 | 16 |
| 400 mg po q12h × 14 | Bisoprolol | 10 mg po q24h × 7 | 45 |
| 400 mg po q12h × 3 | Cephalexin | 500 mg po | 21 |
| 1.2–2.0 g po/d SS | Creatinine | Endogenous | 20 |
| 5 mg/kg iv | Creatinine | Endogenous | 16 |
| | $^{14}$C-creatinine | 0.04 pCi/min iv | 23 |
| 400 mg po q6h × 4 | Creatinine | Endogenous | 37 |
| 400 mg po q6h × 8 | Creatinine | Endogenous | 13 |
| 400 mg po q6h × 8 | Creatinine | Endogenous | 26 |
| 400 mg po q 12h × 28 | Disopyramide | 150 mg iv | NS |
| 400 mg po | Disopyramide | 300 mg po | +20 (Dur) |
| 400 mg po ql2h × l 0 | Metformin | 250 mgq24h × 5 | 28 |
| 600 mg po q12h × 3 | Nicotine | 30 mg/kg iv | 47 |
| | Cotinine | — | +7 (NS) |
| 400 mg po q 12 h × 8 | *R*,*S*-pindolol | 15 mg po | 39 (*R*) |
| | | | 30 (*S*) |
| 300 mg po q6h × 16 | Procainamide | 500 mg po q6 × 13 | 36 |
| | *N*-Ac. procainamide | — | 27 |
| 1000 mg po in 3 h | Procainamide | 1000 mg po | 43 |
| | *N*-Ac. procainamide | — | 24 |
| 200 mg po | Procainamide | 500 mg po | 31 |
| | *N*-Ac. procainamide | — | 21 (NS) |
| 400 mg po | Procainamide | 500 mg po | 40 |
| | *N*-Ac. procainamide | — | 16 |
| 300 mg po q6h × 14 | Procainamide | 1000 mg po | 36 |
| | *N*-Ac. procainamide | — | 16 |
| 300 mg po q6h × 28 | Quinidine | 330 mg po | 32 |
| 400 mg po q12h × 3 | Ranitidine | 150 mg po | 25 |
| 400 mg po q12h × 8 | Triamterene | 100 mg q24h × 4 | 62 |
| | pOH-triamterene-sulf. | — | 16 (NS) |
| 400 mg po q12h × 12 | *R*,*S*-verapamil[a)] | 160 mg po | — |
| | *R*-D-617[a)] | — | 20 (NS) |
| | *S*-D-617[a)] | — | 30 |
| 300 mg po q8h × 4 | Zidovudine | 600 mg po/d | 56 |
| | Zidovudine gluc. | — | 23 |
| *Cisplatin* | | | |
| 90 mg/m$^2$ | Bleomycin | 30 mg/m$^2$/d iv | 22[b)] |
| *Famotidine* | | | |
| 10 mg iv | Creatinine | Endogenous | NS |
| 40 mg po q24h × 6 | Procainamide | 5 mg/kg iv + 8 (NS) (269) | |
| | *N*-Ac. procainamide | — | +9 (NS) |

**Table 13.2** (Continued)

| Inhibitor drug and dose | Inhibited drug | Dose | Inh. $CL_{ren}$ (%) |
|---|---|---|---|
| *Metformin* | | | |
| 250 mg po q24h × 5 | Cimetidine | 400 mg q12h × 10 | NS |
| *Methadone* | | | |
| 50 mg/d | Zidovudine | 150 mg iv | 27 |
| *Ranitidine* | | | |
| 150 mg po | Cimetidine | 400 mg po q12h × 13 | 1 (NS) |
| 150 mg po | Disopyramide | 300 mg po | 7 (NS) |
| 300 mg po q12h × 3 | Nicotine | 30 mg/kg iv | 41 (NS) |
| | Cotinine | — | 43 (NS) |
| 150 mg po q12h × 8 | Procainamide | 500 mg po q6h × 13 | 3 (NS) |
| | *N*-Ac. procainamide | — | 4 (NS) |
| 150 mg po q12h × 2 | Procainamide | 1000 mg po | 18 |
| | *N*-Ac. procainamide | — | 10 |
| 750 mg po in 12 h | Procainamide | 1000 mg po | 36 |
| | *N*-Ac. procainamide | | 38 |
| 150 mg po q12h × 4 | Triamterene | 100 mg po q24h × 4 | 51 |
| | pOH-triamterene-sulf. | — | 47 |
| 150 mg po q12h × 2 | Zidovudine | 600 mg po/d | 5 (NS) |
| | Zidovudine-gluc. | — | 9 (NS) |
| *Triamterene* | | | |
| 100 mg po q24h × 4 | Cimetidine | 400 mg po q12h × 8 | 14 (NS) |
| 100 mg po q24h × 4 | Ranitidine | 150 mg po q12h × 4 | 14 |
| *Trimethoprim* | | | |
| 100 mg po q12h × 20 | Creatinine | Endogenous | 35[c)] |
| 5 mg/kg po q l 2h × 20 | Creatinine | Endogenous | 21 |
| 5 mg/kg po q6h × 40 | Creatinine | Endogenous | 16 |
| 200 mg po q12h × 19 | Digoxin | 1 mg iv | 17 |
| 200 mg po q24h × 4 | *R,S*-pindolol | 10 mg po ql 2h × 7 | 37 (*S*;y)[d)] |
| | | | 25 (*S*;e) |
| | | | 37 (*R*;y) |
| | | | 34 (*R*;e) |
| 200 mg po q24h × 4 | Procainamide | 500 mg po q6h × 12 | 47 |
| | *N*-Ac. procainamide | — | 13 |
| 100 mg po q12h × 8 | Procainamide | 1000 mg po | 45 |
| | *N*-Ac. procainamide | — | 26 |
| 150 mg po | Zidovudine | 3 mg/kg iv | 50 |
| | Zidovudine gluc. | — | 20 |
| *Verapamil* | | | |
| 120 mg po q12h × 7 | Almotriptan | 12.5 po | |

inh, inhibited; iv, intravenous; NS, not significant (generally $p > 0.05$), if not indicated, $p < 0.05$; po, per os; *N*-Ac., *N*-acetyl metabolite; gluc., glucuronide.

a) *S*- and *R*-verapamil were administered simultaneously on two occasions with one of enantiomers deuterated. The verapamil enantiomers were cleared by GFR only. D-617 is *N*-dealkylated verapamil.

b) Mean of two patients corrected for creatinine clearance.

c) Increase in serum creatinine.

d) Kinetics in young (y) and elderly (e) subjects.

Source: Masereeuw and Russel [25] / Taylor & Francis.

**Table 13.3** Amount of furosemide excreted following single oral administration of 40 mg tablets to 12 healthy volunteers.

| Formulation | Ae (mg) | $F_{rel}$ (%) |
|---|---|---|
| Tablet A | 18.3 | 76.3 |
| Tablet B | 17.9 | 74.6 |
| Tablet C | 18.7 | 77.9 |
| Tablet D | 20.5 | 85.4 |
| Tablet E | 16.7 | 69.6 |
| Solution F (Reference) | 24.0 | 100.0 |

**Table 13.4** Renal clearance and bioavailability of cimetidine after intravenous and oral administration.

| Dose (mg) | $CL_{ren}$ (ml/min) | $F_{AUC}$ (%) | $F_{Ae}$ (%) | $F_{AUC+Ae}$ (%) |
|---|---|---|---|---|
| 100 i.v. | 597 | 100 | 100 | 100 |
| 100 p.o. | 533 | 85 | 79 | 80 |
| 400 p.o. | 256 | 123 | 51 | 73 |
| 800 p.o | 469 | 99 | 64 | 71 |

$F_{AUC}$ = Absolute bioavailability calculated as the ratio of AUC values.
$F_{Ae}$ = Absolute bioavailability calculated as the ratio of Ae values.
$F_{AUC}$ = Absolute bioavailability calculated by the method of Øie and Jung [67].

It is interesting to note that with the data presented, one cannot be sure which estimate is most accurate. The method that uses the ratio of AUC values assumes a constant total clearance and thus in this case assumes that the change in renal clearance is offset by an equal and opposite change in nonrenal clearance. The method that uses the ratio of Ae values assumes that nonrenal clearance changes in proportion to renal clearance. The method of Øie and Jung [67] assumes that nonrenal clearance is constant.

### 13.4.3 Plasma and Urine Data Following Extravascular Administration

Hinderling and Shi [69] have proposed a method that under certain conditions allows the computation of mean absolute bioavailability following only extravascular (e.g. oral) administration. The method requires that the drug be studied in a population with varying renal function. It assumes linear kinetics, that nonrenal clearance and bioavailability are not dependant on renal clearance. The method works best for drugs that are predominately renally cleared.

The method was derived by first considering that nonrenal clearance ($CL_{nonren}$) can be estimated from total systemic clearance ($CL_{tot}$) and renal clearance:

$$CL_{tot} = CL_{nonren} + CL_{ren} \tag{13.28}$$

Dividing both sides of the equation by $F$ yields:

$$\frac{CL_{tot}}{F} = \frac{CL_{nonren}}{F} + \frac{CL_{ren}}{F} \tag{13.29}$$

Note that $CL_{tot}/F$ = Dose/$AUC_{inf}$ and that $CL_{ren}$ = Ae/$AUC_{inf}$. A regression of $CL_{ren}$ values on $CL_{tot}/F$ values will yield a straight line with a slope of $1/F$. Thus the absolute bioavailability can be estimated as the reciprocal of the slope. Further, the nonrenal clearance can be estimated as the ratio: intercept/slope.

*Example:* Data from a study [70] investigating the pharmacokinetics of pregabalin in subjects with varying renal function are presented in Figure 13.10. The estimate of absolute bioavailability (reciprocal of the slope) was 113% (95% confidence interval 97.7–134%) and is consistent with the ≥90% reported in the product label [71].

## 13.5 Augmented Renal Clearance

Augmented renal clearance (ARC) is a relatively new concept where increased creatinine clearance and enhanced clearance of renally eliminated medications are seen in critically ill patients [72]. Patients at risk for ARC include those with sepsis, trauma, burns, and severe neurologic injury. Identification of patients at risk for ARC is important as this phenomenon may lead to inadequate exposure to medications, including beta-lactams and vancomycin. Mechanisms of renal clearance changes in ARC are not fully understood but appear to be multifactorial. Commonalities among this patient population include aggressive fluid resuscitation, potential capillary leak, and the following expansion of drug volume of distribution. In addition, endogenous inflammatory responses may contribute to alterations in renal perfusion and glomerular filtration.

The challenge in managing patients with ARC comes in predicting the degree of enhanced renal clearance and adjusting drug doses appropriately. While therapeutic drug monitoring can be especially useful in this situation for medications where it is routinely employed (e.g. vancomycin and aminoglycosides), it may not be practical for other agents where availability of assays and lengthy turnaround times are barriers to implementation (e.g. beta-lactams). However, this is something that should be considered in particularly at-risk populations

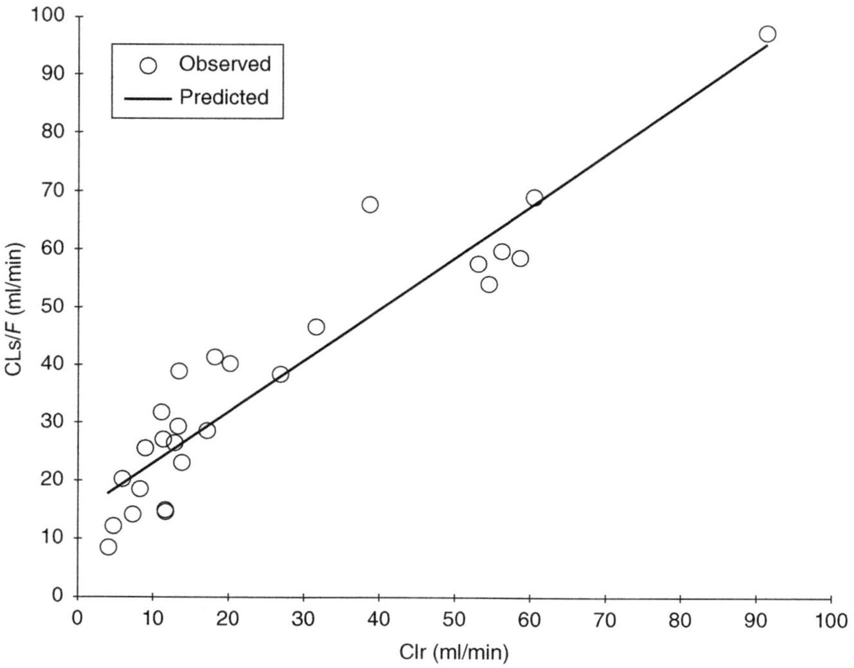


**Figure 13.10** Scatter plot of individual subjects' pregabalin renal clearance ($CL_{ren}$) and oral clearance ($CL/F$) in subjects with various degrees of renal function (and not on hemodialysis). Regression line: $CL_{tot}/F = 0.886CL_{ren} + 14.1$ ml/min. Source: Adapted from Randinitis et al. [70].

to avoid consequences of antimicrobial underdosing, including treatment failure and development of drug resistance, while balancing patient safety concerns with overdosing. The impact and management of ARC in nonantimicrobials are less described.

## References


1 Guyton, A.C. and Hall, J.E. (1996). *Textbook of Medical Physiology*. Philadelphia: W.B. Saunders Company.

2 Gibaldi, M. and Perrier, D. (1982). *Pharmacokinetics*. New York: Marcel Dekker.

3 Rowland, M. and Tozer, T.N. (1995). *Clinical Pharmacokinetics Concepts and Applications*. Lippincott Williams & Wilkins Philadelphia.

4 Coresh, J. and Stevens, L.A. (2006). Kidney function estimating equations: where do we stand? *Curr. Opin. Nephrol. Hypertens.* 15: 276–284.

5 Prigent, A. (2008). Monitoring renal function and limitations of renal function tests. *Semin. Nucl. Med.* 38: 32–46.

6 Tett, S.E., Kirkpatrick, C.M., Gross, A.S., and McLachlan, A.J. (2003). Principles and clinical application of assessing alterations in renal elimination pathways. *Clin. Pharmacokinet.* 42: 1193–1211.

7 Levey, A.S., Bosch, J.P., Lewis, J.B. et al. (1999). A more accurate method to estimate glomerular filtration rate from serum creatinine: a new prediction equation. Modification of diet in renal disease study group. *Ann. Intern. Med.* 130: 461–470.

8 Levey, A.S. and Stevens, L.A. (2010). Estimating GFR using the CKD Epidemiology Collaboration (CKDEPI) creatinine equation: more accurate GFR estimates, lower CKD prevalence estimates, and better risk predictions. *Am. J. Kidney Dis.* 55: 622–627.

9 Levey, A.S., Stevens, L.A., Schmid, C.H. et al. (2009). A new equation to estimate glomerular filtration rate. *Ann. Intern. Med.* 150: 604–612.

10 Matsushita, K., Selvin, E., Bash, L. et al. (2010). Risk implications of the new CKD-EPI equations as compared to the MDRD study equation for estimated GFR: the ARIC study. *Am. J. Kidney Dis.* 55: 648–659.

11 Shafi, T., Matsushita, K., Selvin, E. et al. (2012). Comparing the association of GFR estimated by the CKD-EPI and MDRD study equations and mortality: the third national health and nutrition examination survey (NHANES III). *BMC Nephrol.* 13: 42.

12 Schwartz, G.J., Brion, L.P., and Spitzer, A. (1987). The use of plasma creatinine concentration for estimating glomerular filtration rate in infants, children, and adolescents. *Pediatr. Clin. North Am.* 34: 571–590.

13 Proulx, N.L., Akbari, A., Garg, A.X. et al. (2005). Measured creatinine clearance from timed urine collections substantially overestimates glomerular filtration rate in patients with liver cirrhosis: a systematic review and individual patient meta-analysis. *Nephrol. Dial. Transplant.* 20: 1617–1622.

14 Nankivell, B.J., Gruenewald, S.M., Allen, R.D., and Chapman, J.R. (1995). Predicting glomerular filtration rate after kidney transplantation. *Transplantation* 59: 1683–1689.

15 Matzke, G.R., Aronoff, G.R., Atkinson, A.J. et al. (2011). Drug dosing consideration in patients with acute and chronic kidney disease – a clinical update from Kidney Disease: Improving Global Outcomes (KDIGO). *Kidney Int.* 80: 1122–1137.

16 Gross, A.S., McLachlan, A.J., Minns, I. et al. (2001). Simultaneous administration of a cocktail of markers to measure renal drug elimination pathways: absence of a pharmacokinetic interaction between fluconazole and sinistrin, *p*-aminohippuric acid and pindolol. *Br. J. Clin. Pharmacol.* 51: 547–555.

17 Vanmassenhove, J., Vanholder, R., Nagler, E., and Van Biesen, W. (2013). Urinary and serum biomarkers for the diagnosis of acute kidney injury: an in-depth review of the literature. *Nephrol. Dial. Transplant.* 28: 254–273.

18 Li, Q., Huang, Y., Shang, W. et al. (2020). The predictive value of urinary kidney injury molecular 1 for the diagnosis of contrast-induced acute kidney injury after cardiac catheterization: a meta-analysis. *J. Interv. Cardiol.* 2020: 4982987.

19 Pang, H.M., Qin, X.L., Liu, T.T. et al. (2017). Urinary kidney injury molecule-1 and neutrophil gelatinase-associated lipocalin as early biomarkers for predicting vancomycin-associated acute kidney injury: a prospective study. *Eur. Rev. Med. Pharmacol. Sci.* 21: 4203–4213.

20 Brater, D.C. (2002). Measurement of renal function during drug development. *Br. J. Clin. Pharmacol.* 54: 87–95.

21 Tam, V.H., McKinnon, P.S., Akins, R.L. et al. (2003). Pharmacokinetics and pharmacodynamics of cefepime in patients with various degrees of renal function. *Antimicrob. Agents Chemother.* 47: 1853–1861.

22 Welage, L.S., Schultz, R.W., and Schentag, J.J. (1984). Pharmacokinetics of ceftazidime in patients with renal insufficiency. *Antimicrob. Agents Chemother.* 25: 201–204.

23 Bonate, P.L., Reith, K., and Weir, S. (1998). Drug interactions at the renal level. Implications for drug development. *Clin. Pharmacokinet.* 34: 375–404.

24 U.S. Food and Drug Administration (2010). Guidance for industry: pharmacokinetics in patients with impaired renal function – Study design, data analysis, and impact on dosing and labeling.

25 Masereeuw, R. and Russel, F.G. (2001). Mechanisms and clinical implications of renal drug excretion. *Drug Metab. Rev.* 33: 299–351.

26 Cook, J.A. and Smith, D.E. (1985). Theoretical limits of changes in plasma protein binding on renal drug clearance. *J. Pharm. Sci.* 74: 108–109.

27 Tang-Liu, D.D., Tozer, T.N., and Riegelman, S. (1983). Dependence of renal clearance on urine flow: a mathematical model and its application. *J. Pharm. Sci.* 72: 154–158.

28 Yu, D.K. and Sawchuk, R.J. (1987). Pharmacokinetics of propylene glycol in the rabbit. *J. Pharmacokinet. Biopharm.* 15: 453–471.

29 Wilkinson, G.R. and Beckett, A.H. (1968). Absorption metabolism and excretion of the ephedrines in man. I. The influence of urinary pH and urine volume output. *J. Pharmacol. Exp. Ther.* 162: 139–147.

30 Zimmerman, C.L., O'Connell, M.B., and Soria, I. (1990). The effects of urine pH modification on the pharmacokinetics and pharmacodynamics of phenylpropanolamine. *Pharm. Res.* 7: 96–102.

31 Nilsson, M.I., Widerlov, E., Meresaar, U., and Anggard, E. (1982). Effect of urinary pH on the disposition of methadone in man. *Eur. J. Clin. Pharmacol.* 22: 337–342.

32 Neuvonen, P.J. and Karkkainen, S. (1983). Effects of charcoal, sodium bicarbonate, and ammonium chloride on chlorpropamide kinetics. *Clin. Pharmacol. Ther.* 33: 386–393.

33 Kietzmann, D., Bock, K.W., Krahmer, B. et al. (1990). Paracetamol test: modification by renal function, urine flow and pH. *Eur. J. Clin. Pharmacol.* 39: 245–251.

34 Jusko, W.J. and Levy, G. (1970). Pharmacokinetic evidence for saturable renal tubular reabsorption of riboflavin. *J. Pharm. Sci.* 59: 765–772.

35 Sommers, D.K. and Schoeman, H.S. (1987). Drug interactions with urate excretion in man. *Eur. J. Clin. Pharmacol.* 32: 499–502.

36 Pérez-Ruiz, F., Jansen, T., Tausche, A.K. et al. (2019). Efficacy and safety of lesinurad for the treatment of hyperuricemia in gout. *Drugs Context* 8: 212581.

37 Koren, G., Klein, J., Bentur, Y., and Giesbrecht, E. (1989). The effects of mannitol diuresis on digoxin and phenobarbital handling by the kidney: implications for tubular reabsorption and secretion of the cardiac glycoside. *Clin. Invest. Med.* 12: 279–284.

38 Sahi, J. (2005). Use of in vitro transporter assays to understand hepatic and renal disposition of new drug candidates. *Expert Opin. Drug Metab. Toxicol.* 1: 409–427.

39 Inui, K.-I., Masuda, S., and Saito, H. (2000). Cellular and molecular aspects of drug transport in the kidney. *Kidney Int.* 58: 944–958.

40 Koepsell, H., Gorboulev, V., and Arndt, P. (1999). Molecular pharmacology of organic cation transporters in kidney. *J. Membr. Biol.* 167: 103–117.

41 van Montfoort, J.E., Hagenbuch, B., Groothuis, G.M. et al. (2003). Drug uptake systems in liver and kidney. *Curr. Drug Metab.* 4: 185–211.

42 Dresser, M.J., Leabman, M.K., and Giacomini, K.M. (2001). Transporters involved in the elimination of drugs in the kidney: organic anion transporters and organic cation transporters. *J. Pharm. Sci.* 90: 397–421.

43 Yabuuchi, H., Tamai, I., Nezu, J.-I. et al. (1999). Novel membrane transporter OCTN1 mediates multispecific, bidirectional, and pH-dependent transport of organic cations. *J. Pharmacol. Exp. Ther.* 289: 768–773.

44 Wu, X., Huang, W., Prasad, P.D. et al. (1999). Functional characteristics and tissue distribution pattern of organic cation transporter 2 (OCTN2), an organic cation/carnitine transporter. *J. Pharmacol. Exp. Ther.* 290: 1482–1492.

45 Lee, W. and Kim, R.B. (2004). Transporters and renal drug elimination. *Annu. Rev. Pharmacol. Toxicol.* 44: 137–166.

46 Cha, S.H., Sekine, T., Kusuhara, H. et al. (2000). Molecular cloning and characterization of multispecific organic anion transporter 4 expressed in the placenta. *J. Biol. Chem.* 275: 4507–4512.

47 Cha, S.H., Sekine, T., Fukushima, J.-i. et al. (2001). Identification and characterization of human organic anion transporter 3 expressing predominantly in the kidney. *Mol. Pharmacol.* 59: 1277–1286.

48 Takeda, M., Babu, E., Narikawa, S., and Endou, H. (2002a). Interaction of human organic anion transporters with various cephalosporin antibiotics. *Eur. J. Pharmacol.* 438: 137–142.

49 Takeda, M., Khamdang, S., Narikawa, S. et al. (2002b). Human organic anion transporters and human organic cation transporters mediate renal antiviral transport. *J. Pharmacol. Exp. Ther.* 300: 918–924.

50 Lecureur, V., Sun, D., Hargrove, P. et al. (2000). Cloning and expression of murine sister of P-glycoprotein reveals a more discriminating transporter than MDR1/P-glycoprotein. *Mol. Pharmacol.* 57: 24–35.

51 Yamazaki, M., Neway, W.E., Ohe, T. et al. (2001). In vitro substrate identification studies for P-glycoprotein-mediated transport: species difference and predictability of in vivo results. *J. Pharmacol. Exp. Ther.* 296: 723–735.

52 Otsuka, M., Matsumoto, T., Morimoto, R. et al. (2005). A human transporter protein that mediates the final excretion step for toxic organic cations. *Proc. Natl. Acad. Sci. U.S.A.* 102: 17923–17928.

53 Tanihara, Y., Masuda, S., Sato, T. et al. (2007). Substrate specificity of MATE1 and MATE2-K, human multidrug and toxin extrusions/$H^+$-organic cation antiporters. *Biochem. Pharmacol.* 74: 359–371.

54 Ganapathy, M.E., Brandsch, M., Prasad, P.D. et al. (1995). Differential recognition of beta-lactam antibiotics by intestinal and renal peptide transporters, PEPT 1 and PEPT 2. *J. Biol. Chem.* 270: 5672–25677.

55 Mariño, E.L. and Dominguez-Gil, A. (1981). The pharmacokinetics of cefadroxil associated with probenecid. *Int. J. Clin. Pharmacol. Ther. Toxicol.* 19: 506–508.

56 Thyss, A., Kubar, J., Milano, G. et al. (1986). Clinical and pharmacokinetic evidence of a life-threatening interaction between methotrexate and ketoprofen. *Lancet* 327: 256–258.

57 Tracy, T.S., Krohn, K., Jones, D.R. et al. (1992). The effects of a salicylate, ibuprofen, and naproxen on the disposition of methotrexate in patients with rheumatoid arthritis. *Eur. J. Clin. Pharmacol.* 42: 121–125.

58 Holodiny, M., Penzak, S.R., Straight, T.M. et al. (2008). Pharmacokinetics and tolerability of oseltamivir combined with probenecid. *Antimicrob. Agents Chemother.* 52: 3013–3021.

59 van Acker, B.A.C., Koopman, M.G., Arisz, L. et al. (1992). Creatinine clearance during cimetidine administration for measurement of glomerular filtration rate. *Lancet* 340: 1326–1329.

60 Boom, S.P., Gribnau, F.W., and Russel, F.G. (1992). Organic cation transport and cationic drug interactions in freshly isolated proximal tubular cells of the rat. *J. Pharmacol. Exp. Ther.* 263: 445–450.

61 Hedman, A., Angelin, B., Arvidsson, A. et al. (1990). Interactions in the renal and biliary elimination of digoxin: stereoselective difference between quinine and quinidine. *Clin. Pharmacol. Ther.* 47: 20–26.

62 de Maat, M.M., Ekhart, G.C., Huitema, A.D. et al. (2003). Drug interactions between antiretroviral

drugs and comedicated agents. *Clin. Pharmacokinet.* 42: 223–282.

63 Wakasugi, H., Yano, I., Ito, T. et al. (1998). Effect of clarithromycin on renal excretion of digoxin: interaction with P-glycoprotein[ast]. *Clin. Pharmacol. Ther.* 64: 123–128.

64 Turnheim, K., Krivanek, P., and Oberbauer, R. (1999). Pharmacokinetics and pharmacodynamics of allopurinol in elderly and young subjects. *Br. J. Clin. Pharmacol.* 48: 501–509.

65 Heuberger, W., Beradi, S., Jacky, E. et al. (1998). Increased urinary excretion of carnitine in patients treated with cisplatin. *Eur. J. Clin. Pharmacol.* 54: 503–508.

66 McNamara, P.J., Foster, T.S., Digenis, G.A. et al. (1987). Influence of tablet dissolution on furosemide bioavailability: a bioequivalence study. *Pharm. Res.* 4: 150–153.

67 Oie, S. and Jung, D. (1979). Bioavailability under variable renal clearance conditions. *J. Pharm. Sci.* 68: 128–129.

68 Grahnen, A. (1984). Cimetidine bioavailability and variable renal clearance. *Eur. J. Clin. Pharmacol.* 27: 623–624.

69 Hinderling, P.H. and Shi, J. (1995). Absolute bioavailability estimated from oral data. *J. Pharm. Sci.* 84: 385–386.

70 Randinitis, E.J., Posvar, E.L., Alvey, C.W. et al. (2003). Pharmacokinetics of pregabalin in subjects with various degrees of renal function. *J. Clin. Pharmacol.* 43: 277–283.

71 Pfizer (2007). *Lyrica Product Insert*. New York, NY: Pfizer Inc.

72 Cook, A.M. and Hatton-Kolpek, J. (2019). Augmented renal clearance. *Pharmacotherapy* 39: 346–354.

# 14

# Excretion of Drugs and Their Metabolites into the Bile

*Song Gao[1], Imoh Etim[1], Robin Sunsong[1], Christabel Ebuzoeme[1], Ting Du[1], and Dinh Bui[2]*

[1] *Department of Pharmaceutical Sciences, College of Pharmacy and Health Sciences, Texas Southern University, Houston, Texas, USA*
[2] *Department of Pharmacological and Pharmaceutical Sciences, College of Pharmacy, University of Houston, Houston, Texas, USA*

## 14.1 Introduction

Biliary secretion can be understood as bile excretion from the liver into the small intestine through the bile duct, which is one of the major physiological phenomena of the hepatobiliary system. The components of bile mainly contain water, bile acids, cholesterol, lipids, and other organic components [1, 2]. Drugs or dietary components can also be excreted into the bile as parent or metabolite forms and eliminated through feces. Certain secreted drugs and metabolites can be reabsorbed in the intestine with or without the facilitation of intestinal microflora to form enterohepatic recycling (EHR).

Drug excretion through bile is affected by physicochemical, physiological, and pathological factors, including the hydrophilicity and structural properties (e.g., molecular weight and number of charges) of the compounds, hepatic metabolism (e.g., glucuronidation and oxidation), hepatic uptake and efflux, and disease conditions. Many *in vitro* (e.g., cell culture models), *in situ* (e.g., intestinal perfusion or portal vein infusion model with bile duct cannulation), and *in vivo* (e.g., transporter or enzyme deficient or overexpressing animals) models have been established to study the kinetics and mechanism of biliary excretion. In humans, biliary excretion studies have been performed in both patients and healthy volunteers using different techniques for bile collection. Hundreds of research papers on biliary excretion have been published every year.

Biliary secretion eliminates drugs from the systemic circulation and decreases plasma exposure. Therefore, it is important to evaluate biliary excretion in new molecular identities. For the purpose of enhancing bioavailability, biliary excretion should be avoided. However, if the parent drug undergoes EHR, or the phase II conjugates of the drug undergo EHR facilitated by microflora hydrolysis, drug molecules can be reabsorbed into the systemic circulation to form multiple absorption peaks in the PK profiles and prolong the half-lives. For example, it was reported that SN-38, a chemotherapeutic drug that undergoes efficient EHR, was detected in the plasma 21 days after a single dose in a clinical trial [3]. Whether biliary excretion should be avoided or utilized in designing new drugs depends on the diseases and the pharmaceutical purposes.

Other than drug development, evaluation of biliary excretion is also very important in clinics to determine liver functions or dose adjustments. Usually, the enzyme functions or biliary efflux transporters are disrupted in chronic diseases, which consequentially affects systemic drug exposure. Therefore, the dose should be adjusted accordingly if it is a substrate of that particular enzyme or transporter. For example, it was reported that in patients with cirrhosis or cholangitis, the biliary excretion (i.e., biliary clearance) of mezlocillin was dramatically decreased (<10-fold), and the dose of mezlocillin for treating biliary tract infections is suggested to be increased [4].

This chapter discusses the anatomy of the liver and biliary system, physiological functions of bile excretion, drugs that undergo biliary secretion, the impact of biliary excretion on ADME and PK, transporters involved in biliary excretion, physiological and pathological factors affecting biliary excretion, and preclinical and clinical models for biliary excretion study.

*Oral Bioavailability and Drug Delivery: From Basics to Advanced Concepts and Applications,* First Edition. Edited by Ming Hu and Xiaoling Li.

## 14.2 Anatomy and Physiology of the Liver and Biliary System

### 14.2.1 Anatomy of Liver and Biliary System

The liver plays an essential role in processing endogenous and xenobiotic compounds, and also in the synthesis and turnover of most plasma proteins. It is well-known that the liver is a detoxification organ that "cleans" blood from the portal vein before entering the systemic circulation. Biliary excretion is the drainage of "waste" generated by the liver [5]. From the anatomic point of view, the liver, the largest internal organ in the body weighing approximately 2–3% of body weight, is located in the right upper quadrant of the abdominal cavity protected by the rib cage. The liver lies to the right of the stomach and overlies the gall bladder, which is maintained by a series of avascular ligamentous attachments [6]. Two types of blood from two separate sources are flowing through the liver: one is oxygenated blood from the hepatic artery (20–25%), and the other is nutrient-rich blood from the hepatic portal vein (75–80%) [7]. The portal vein is joined by the hepatic artery and the common bile duct, forming the portal triad enclosed in the hepatoduodenal ligament.

The liver is mainly (~70–80% of liver mass) constituted by hepatocytes, in which metabolic enzymes and drug transporters are highly expressed. Hepatocytes are a class of polarized cells having two surface domains: the sinusoidal domain and the canalicular domain. The sinusoidal domain faces the blood in charge of exchanging compounds between hepatocytes with the blood and the canalicular domain is an intercellular space between two neighbor cells that are enclosed by tight junctions and is the beginning of the bile drainage system [8] (Figure 14.1a). Additional functional polarity is mainly provided by the action of canalicular ATP-binding cassette (ABC) transporters that facilitate biliary excretion. Other than hepatocytes, another type of cell is cholangiocytes, which are cuboidal epithelial cells responsible for the secretion, reabsorption, and regulation of bile components. The biliary system is a very complex three-dimensional network draining bile into the small intestine. The unique functions of hepatocytes require that they maintain two flow systems that are countercurrent to each other, the secretion of bile and the processing and secretion of sinusoidal blood components. This flow system is sustained due to the hepatocyte's highly polarized state. Hepatocytes secrete the bile into the bile canaliculi which are enclosed by the tight junction between adjacent hepatocytes. Then the bile flows from the bile canaliculi to the bile duct networks which are lined with cholangiocytes and eventually converge into large bile ducts which enter the gall bladder and then to the small intestine [9, 10] (Figure 14.1b).

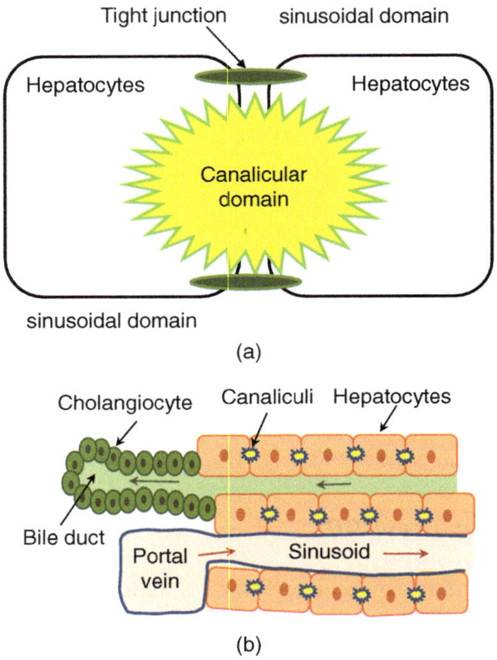


**Figure 14.1 Structures of hepatocytes and biliary system.** Two domains of hepatocytes (a). Biliary system (b).

### 14.2.2 Physiological Function of Biliary Excretion

Biliary secretion, defined as bile excretion from the liver into the small intestine through the bile duct, is one of the major physiological phenomena of the liver and biliary system. Hepatocytes secrete a yellow–brown or yellow–green juice called bile into bile canaliculi. The secreted bile contains large quantities of bile acids, cholesterol, and other organic molecules. As bile flows through the bile ducts from canaliculi into the small intestine, its composition is modified by the addition of a watery, bicarbonate-rich section from ductal epithelial cells [11]. In most vertebrates, bile secreted by the liver is stored in the gallbladder, a reservoir between the liver and the small intestine, and will be secreted into the duodenum after eating. Several species of mammals, such as rats, horses, and deer, lack a gallbladder, and concentrations of the biliary composition of these species are usually slightly lower than those from bladder animals [12]. It is estimated that an adult human produces 400–800 ml of bile per day [13] and bile production in other animals are various depending on their body weights. For example, an adult rat (250–300 g) can produce approximately 25–30 ml of bile per day while a rabbit (2.0–2.5 kg) can excrete approximately 200–260 ml of bile per day [13].

One of the major physiological functions of biliary excretion is to secrete signaling and facilitating molecules (e.g., bile acids, bilirubin) to solubilize hydrophobic compounds for enhanced absorption or digestion of foodstuffs and also to interact with the microbiome in the intestine [14, 15]. Another major function of biliary secretion is to eliminate hepatic metabolites. The liver is an important organ that can metabolize xenobiotic compounds, especially drugs. After administration, drug molecules enter the liver through blood, portal vein (oral administration), or systemic circulation (other administration routes). Since the liver is an organ that highly expresses metabolic enzymes (e.g., CYPs, UGTs, and SULTs), drugs can be efficiently metabolized in the liver to form metabolites. Usually, the majority of drug molecules and metabolites are excreted into the bile as "wastes" following different transport mechanisms to achieve detoxification purposes [16], which forms one of the major drug elimination pathways.

## 14.3 Biliary Excreted Drugs and Metabolites

Many drugs and or their metabolites are excreted through the bile. Currently, the structure–excretion relationship is not quite clear. But it is known that the chemical structure of a compound plays a very integral role in biliary excretion. A change of a single chemical group in the structure of a compound can affect biliary excretion. For example, biliary excretion is increased by the addition of a second –OH group in position 41 of the 4-hydroxybiphenylglucuronide compounds and sulfonamides, the substitution of various chemical groups alters their biliary excretion. Some drugs can be excreted through biliary excretion as parent compounds (Table 14.1), while others are excreted only after being metabolized in the hepatocytes (Table 14.2). Hepatic transporters play a dominant role in excreting different compounds.

## 14.4 Impact of Biliary Excretion on ADME and Pharmacokinetics

Biliary excretion has a significant impact on drug ADME and pharmacokinetics. For drugs administered through the oral route, the drug molecules are absorbed in the small intestine and enter the liver through the portal vein, where part of them could be metabolized by the hepatic enzymes (e.g., CYPs and UGTs). For some drugs, part of the parent drugs can be excreted into the bile and then enter the small intestine together with the bile juice, where the drugs can be reabsorbed and enter the liver again to form enterohepatic recycling. This recycling can occur multiple times until most of the drugs are eliminated through feces. For example, it is estimated that a bile acid molecule can be recycled >20 times between the liver and the intestine before it is eliminated [69]. For some other drugs, their phase II conjugates (e.g., glucuronides and sulfates) can be efficiently excreted into the bile through efflux transporters (e.g., P-gp, BCRP, and $MRP_2$). These conjugates are usually hydrophilic and difficult to be reabsorbed in the intestine. However, intestinal microflora can hydrolyze these phase II conjugates to release the parent drugs, which can be reabsorbed to participate in enterohepatic recycling [70]. Phase I metabolites, such as reduced or oxidized drug metabolites, usually cannot be converted back to the parent drug in the intestine. Therefore, even if the metabolites are reabsorbed in the intestine, the drug molecules cannot be recycled. For drugs administered through other routes, drug molecules reach the liver through blood and can also be excreted into the bile. If the intestinal permeability of the drug is good enough, the excreted drug can be reabsorbed in the intestine, where part of the drug can be recovered back to the systemic circulation and the other part participates in EHR, otherwise, the drug will be eliminated through feces.

When the excreted drugs are reabsorbed through either parent or phase II metabolites pathway to form EHR, multiple absorption peaks usually appear in the PK profile [71]. In addition, the terminal half-life will be prolonged, and AUC will be increased due to recycling (Figure 14.2). For example, genistein, a dietary component, participates efficiently in EHR, resulting in a long half-life ($t_{1/2}$, 5–9 h) in animals and humans [72]. However, drug concentrations at the elimination phase are usually very low. There is no evidence showing that pharmacological efficacy will be prolonged due to EHR. On the contrary, low concentration and long-term exposure to certain drugs (e.g., antibiotics) may induce drug resistance [73]. Therefore, biliary excretion of these drugs must be paid attention to in drug development and clinical practice.

The direct impact of biliary excretion on biodistribution is not experimental clear. Theoretically, when drugs are excreted to the bile, the systemic exposure

**Table 14.1** Drugs secreted into bile as parent compounds.

| S/N | Therapeutic class | Parent compound | Chemical structure | Transporters | References |
|---|---|---|---|---|---|
| 1. | Antibacterial agents | Mezlocillin | 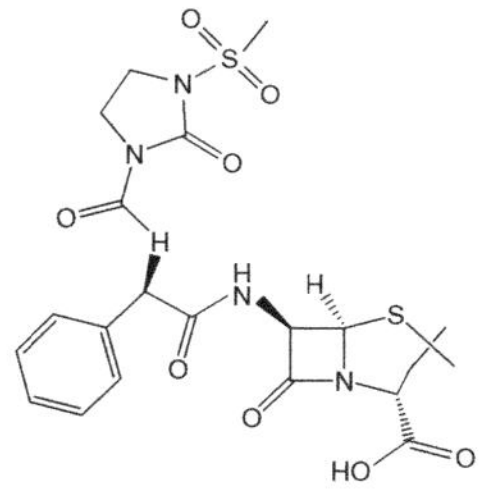 | Unidentified | [17] |
| 2. | | Azlocillin | 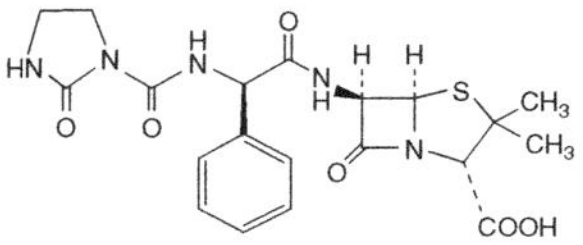 | Unidentified | [17] |
| 3. | | Trovafloxacin | 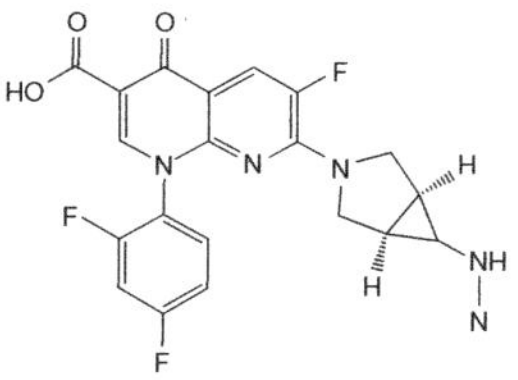 | Unidentified | [18] |
| 4. | | Dalfopristin/Quinupristin | 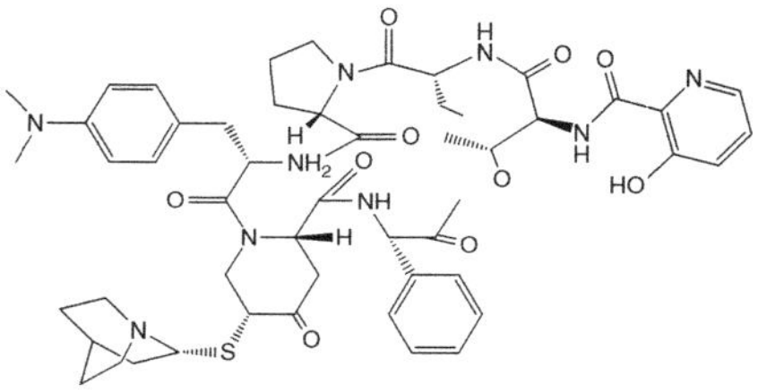 | Unidentified | [19] |
| 5. | | Sulfacetamide | 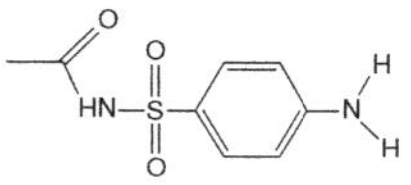 | OAT | [20, 21] |
| 6. | | Sulfadimethoxine | 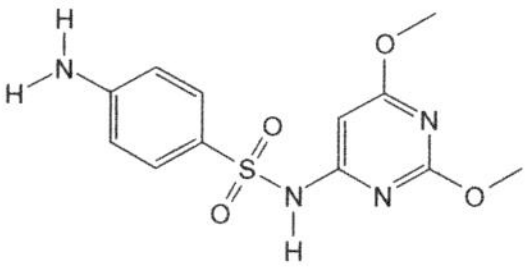 | OAT | [20, 21] |
| 7. | | Ampicillin | 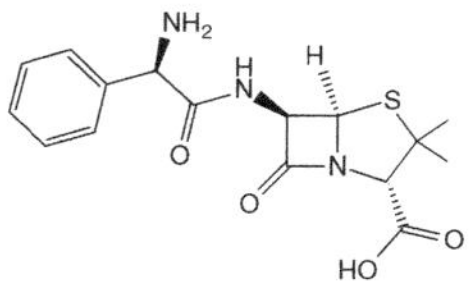 | P-gp | [22, 23] |
| 8. | | Amoxicillin | 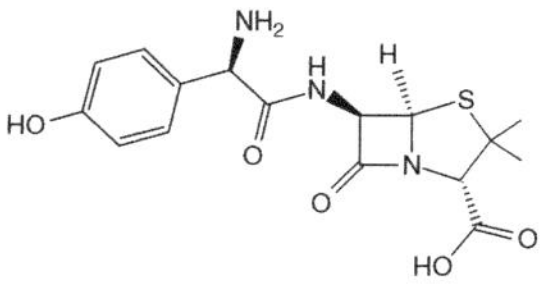 | hPepT1, hPepT2 | [24, 25] |

*(continued)*

**Table 14.1** (Continued)

| S/N | Therapeutic class | Parent compound | Chemical structure | Transporters | References |
|---|---|---|---|---|---|
| 9. | | Nafcillin | 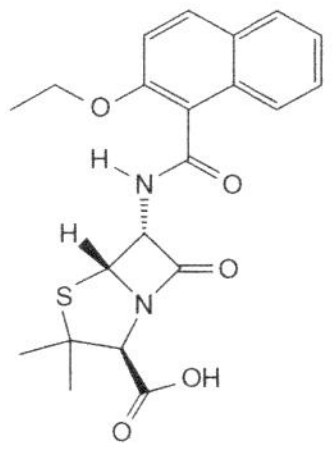 | OATP1B1, OATP1B3 | [26] |
| 10. | | Cefamandole | 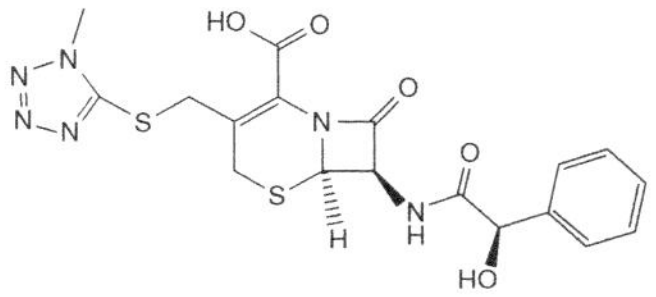 | OAT3 | [27, 28] |
| 11. | | Cefoxitin | 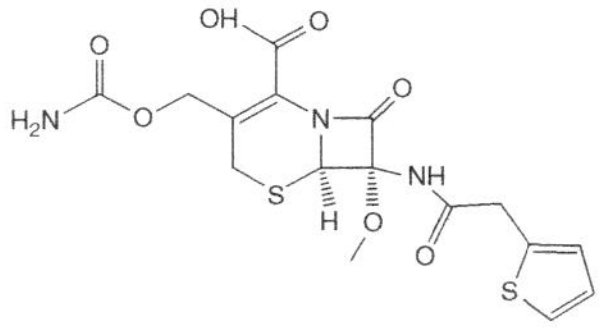 | OAT1 | [29] |
| 12. | | Ceftriaxone | 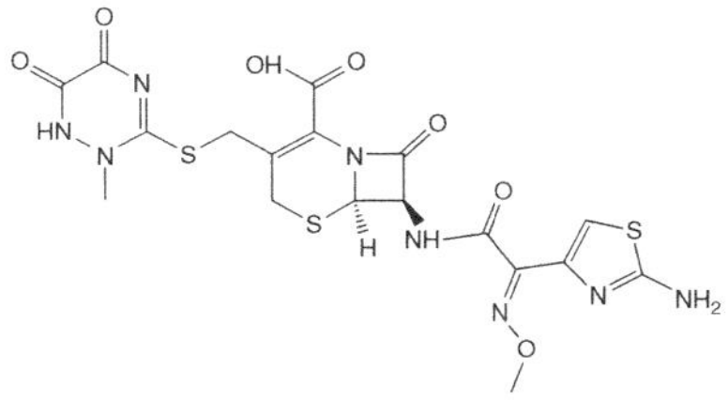 | OAT3 | [28, 30] |
| 13. | | Cefoperazone | 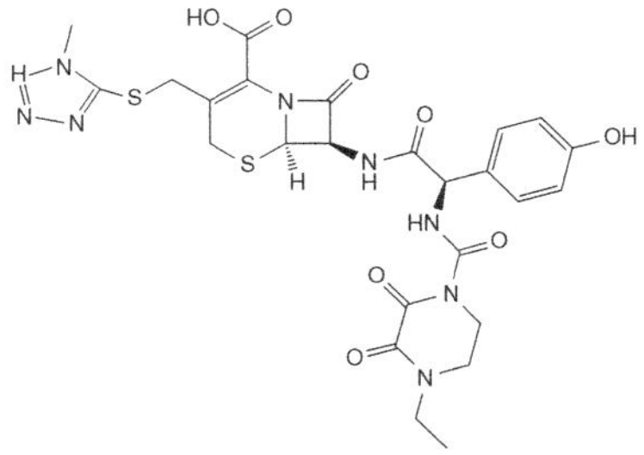 | OAT3 | [28] |
| 14. | | Aztreonam | 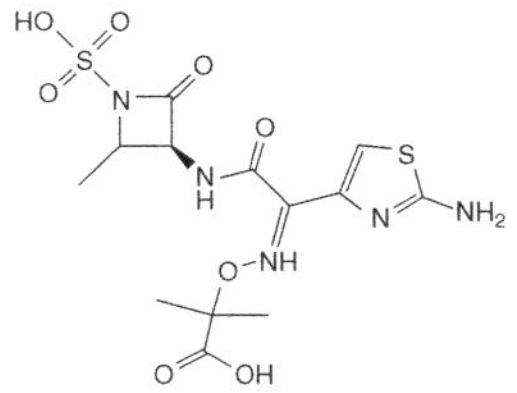 | Unidentified | [31] |

(*continued*)

**Table 14.1** (Continued)

| S/N | Therapeutic class | Parent compound | Chemical structure | Transporters | References |
|---|---|---|---|---|---|
| 15. | | Rosaramicin | 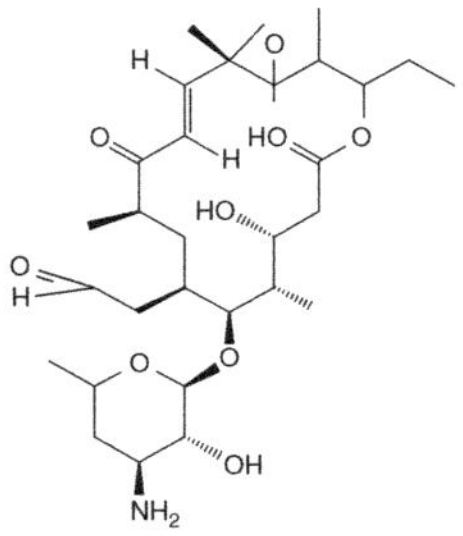 | Unidentified | [32] |
| 16. | | Thiamphenicol | 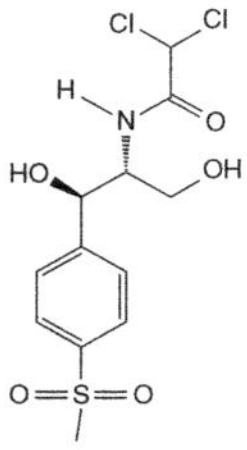 | Unidentified | [33] |
| 17. | | Rifampicin | 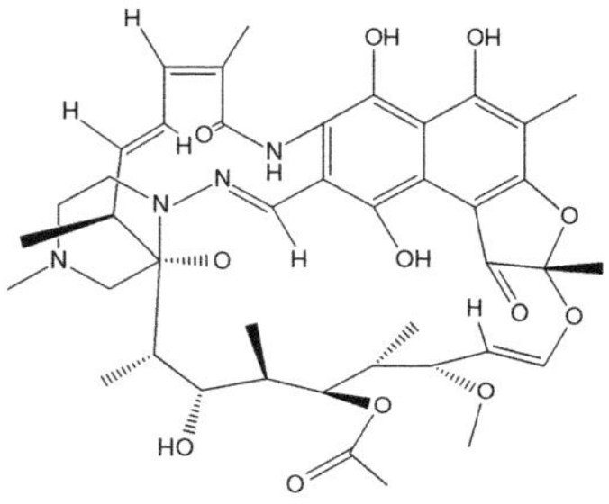 | P-gp, OATP1B1 | [34] |
| 18. | | Pefloxacin | 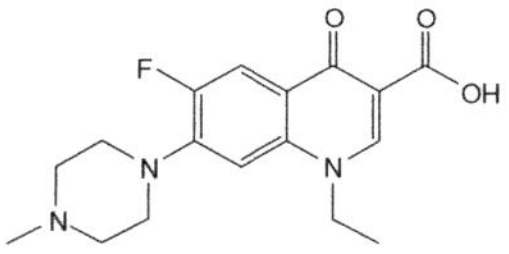 | ABCG2 | [35] |
| 19. | | Norfloxacin | 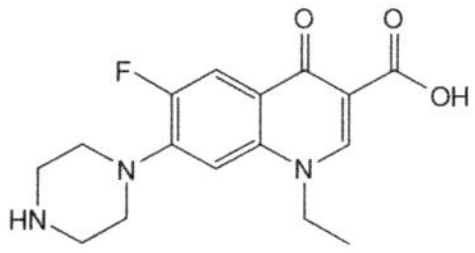 | ABCG2 | [36, 37] |
| 20. | | Novobiocin | 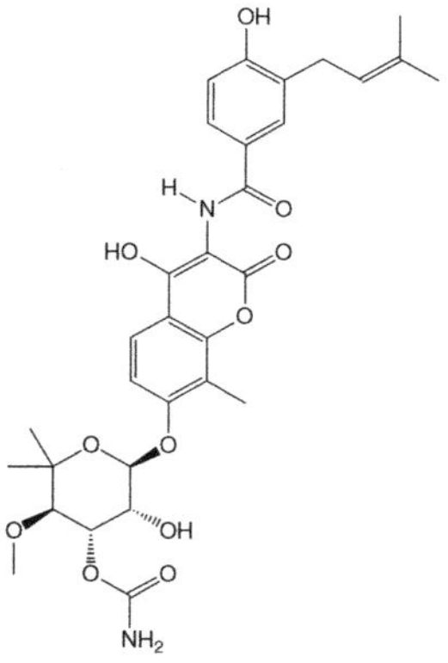 | Unidentified | [38] |

(*continued*)

**Table 14.1** (Continued)

| S/N | Therapeutic class | Parent compound | Chemical structure | Transporters | References |
|---|---|---|---|---|---|
| 21. | Anticancer agents | Irinotecan | 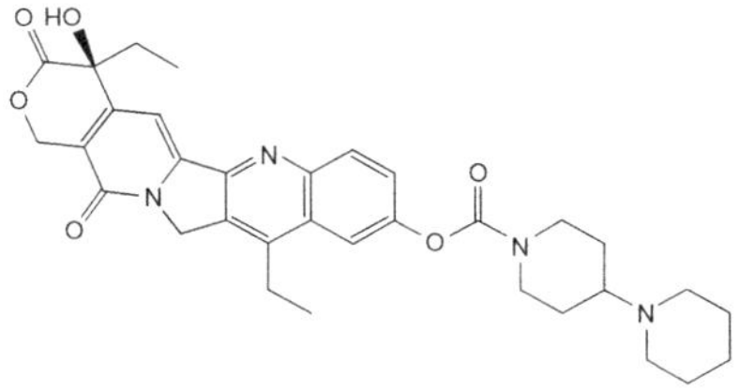 | P-gp | [39] |
| 22. | | Paclitaxel | 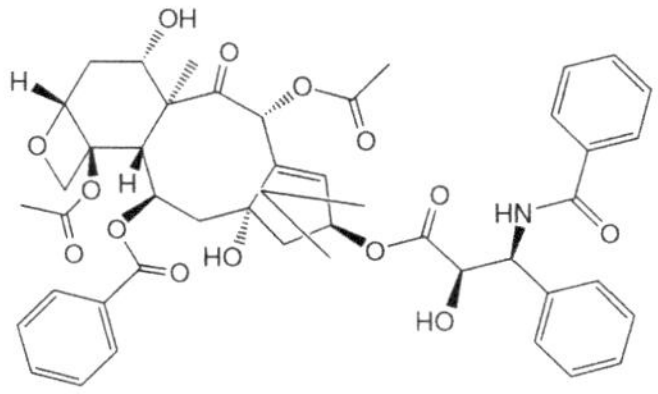 | P-gp, MRP2 | [40, 41] |
| 23. | | Doxorubicin | 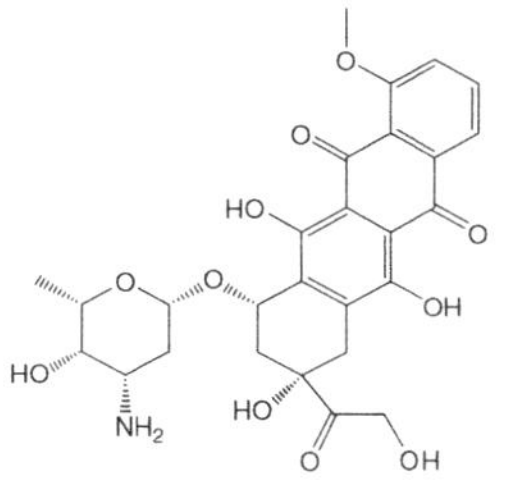 | P-gp | [42, 43] |
| 24. | Antihistamine agent | Levocetirizine | 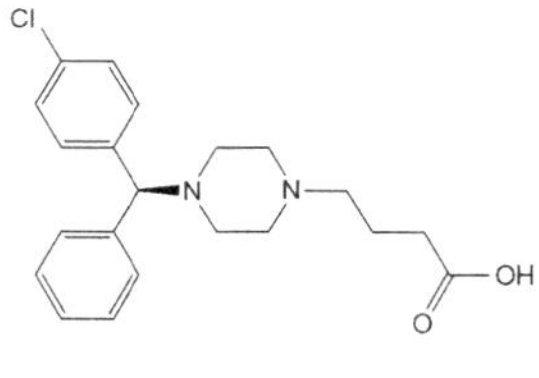 | OAT4 | [44, 45] |
| 25. | Antihypertensive agent | Valsartan | 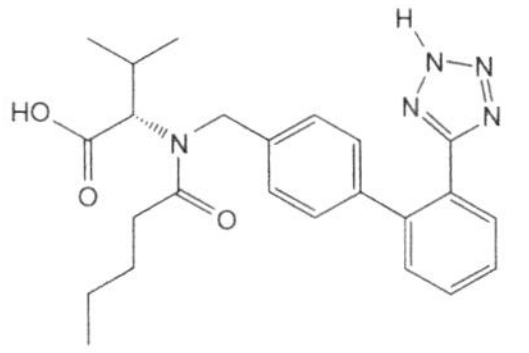 | OATP1B1/Mrp2 | [46, 47] |
| 26. | Antilipidemic agent | Ezetimibe | 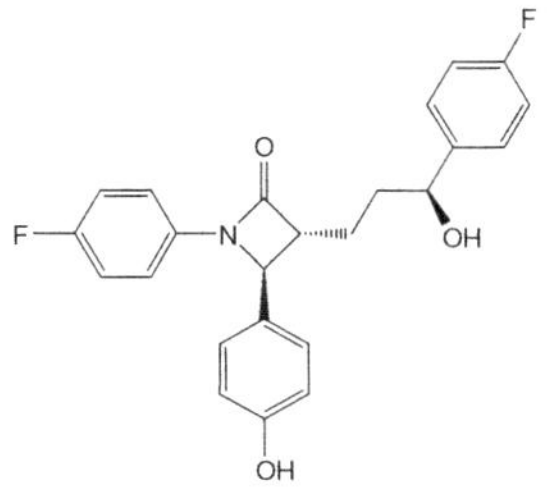 | MRP2, BCRP | [48] |

(continued)

**Table 14.1** (Continued)

| S/N | Therapeutic class | Parent compound | Chemical structure | Transporters | References |
|---|---|---|---|---|---|
| 27. | Antipsychotic agent | Quetiapine | 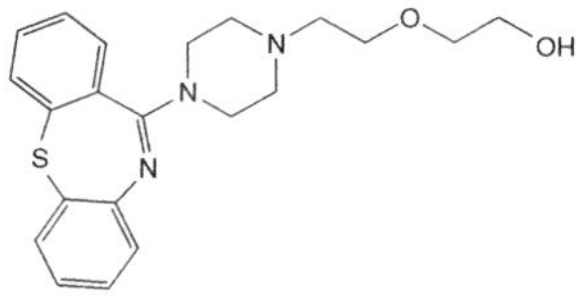 | P-gp | [49, 50] |
| 28. | Antiretroviral agent | Lopinavir/Ritonavir | 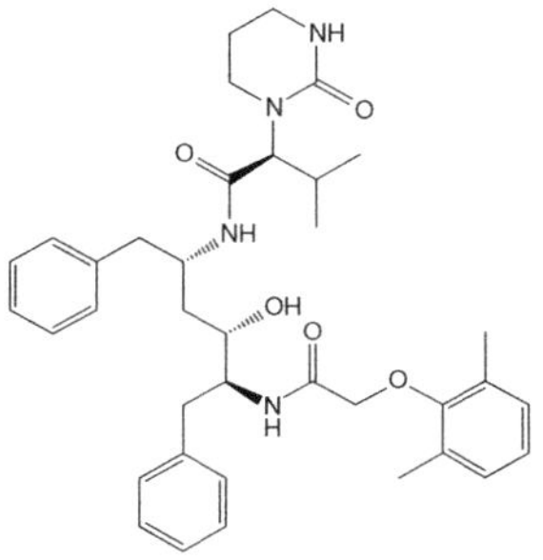 | P-gp, MRP2 | [51, 52] |
| 29. | Hormonal therapy agent | Stilbesterol | 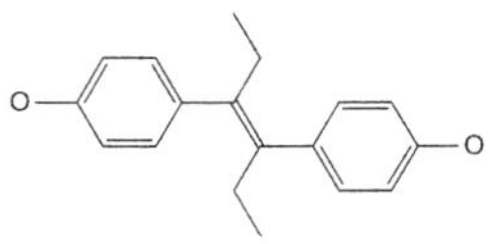 | Unidentified | [53] |
| 30. | Non-opioid analgesic agent | Acetaminophen | 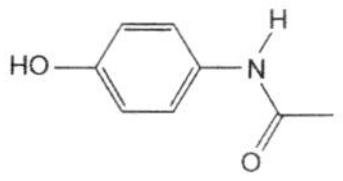 | MRP2, MRP3, MRP4, BCRP | [54, 55] |
| 31. | NSAIDs | Diclofenac | 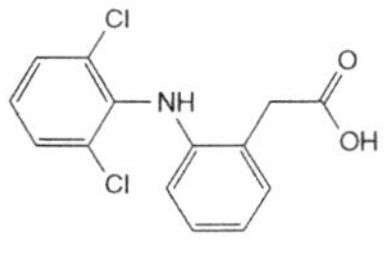 | MRP2, BCRP | [56, 57] |
| 32. | | Ibuprofen | 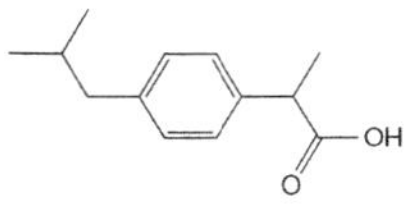 | MRP2 | [56] |
| 33. | | Meloxicam | 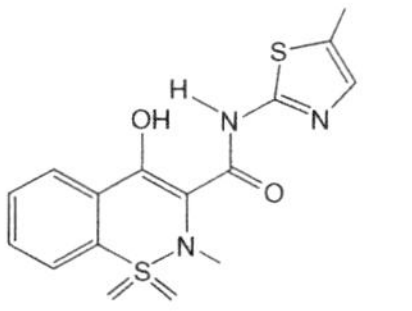 | ABCG2 | [58] |
| 34. | Vitamin | 1,25-Ddihydroxycholecalciferol | 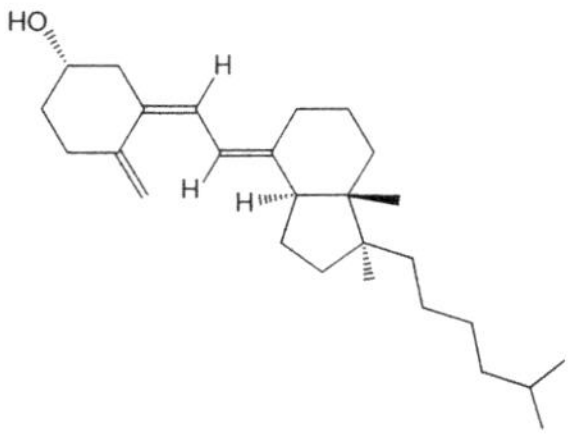 | P-gp | [59, 60] |

**Table 14.2** Drugs secreted into the bile as metabolites.

| S/N | Therapeutic class | Compound | Metabolite | Chemical structure | Transporters | References |
|---|---|---|---|---|---|---|
| 1. | Antibacterial agents | Sulfaguanidine | $N^4$-acetylsulfaguanidine | 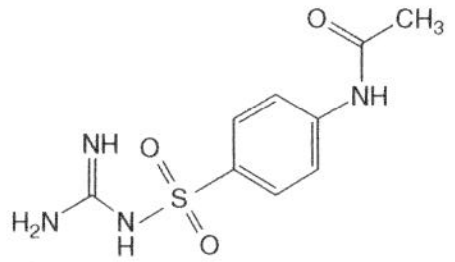 | OAT | [20, 21] |
| 2. | | Sulfanilamide | $N^4$-acetylsulfanilamide | 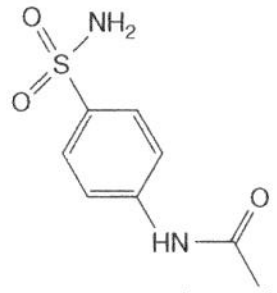 | OAT | [20, 21] |
| 3. | | Nafcillin | 2-Ethoxy-1-naphthamido penicillioc acid | 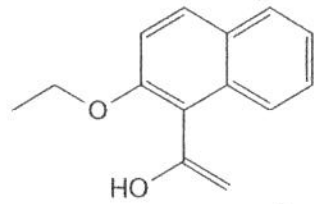 | OATP1B1, OATP1B3 | [26] |
| 4. | | Rosaramicin | 20-Bis-ureidorosaramicin | 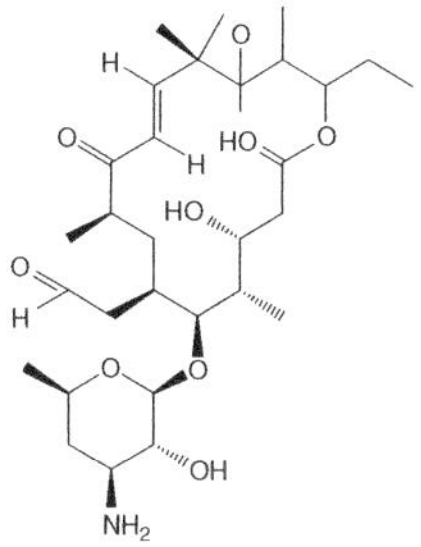 | Unidentified | [32] |
| 5. | | Rifampicin | 25-Desacetyl-rifampicin | 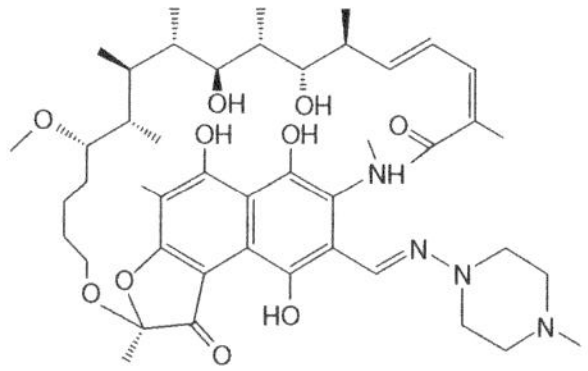 | P-gp, OATP1B1 | [34, 61] |
| 6. | Anticancer agents | Irinotecan | SN-38 and SN-38 glucuronide | 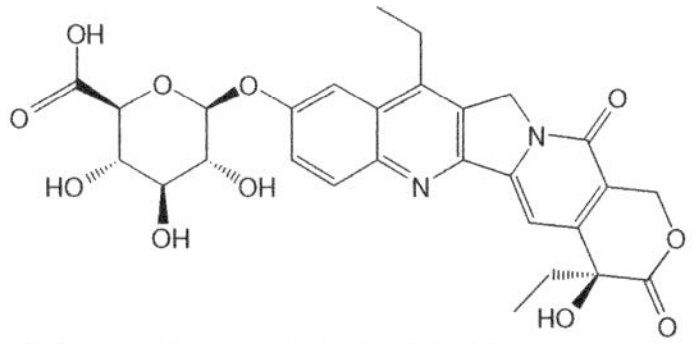 | P-gp | [39] |

*(continued)*

**Table 14.2** (Continued)

| S/N | Therapeutic class | Compound | Metabolite | Chemical structure | Transporters | References |
|---|---|---|---|---|---|---|
| 7. | | Paclitaxel | 6α-Hydroxyl paclitaxel | 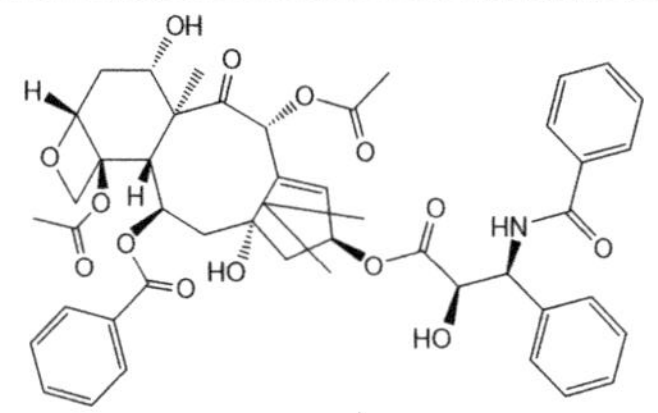 | P-gp | [40, 62] |
| 8. | | Doxorubicin | $4^1$-Epidoxorubicin | 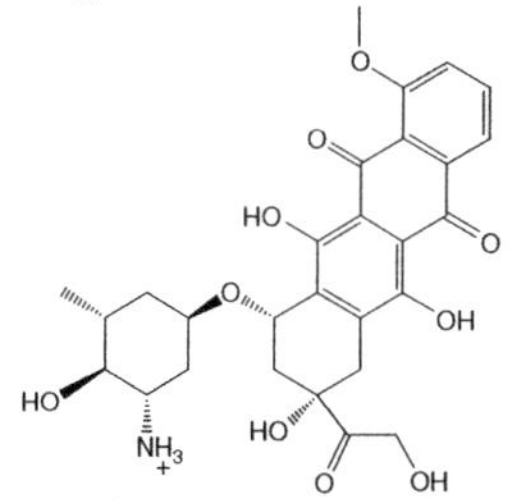 | P-gp | [63] |
| 9. | Antidiabetic agent | Rosiglitazone | *N*-desmethyl-*p*-hydroxy rosiglitazone | 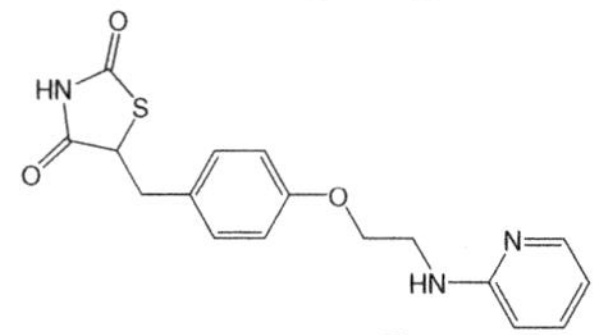 | P-gp | [64] |
| 10. | Antihistamine agent | Desloratadine | 3-Hydroxydesloratadine-*O*-glucuronide | 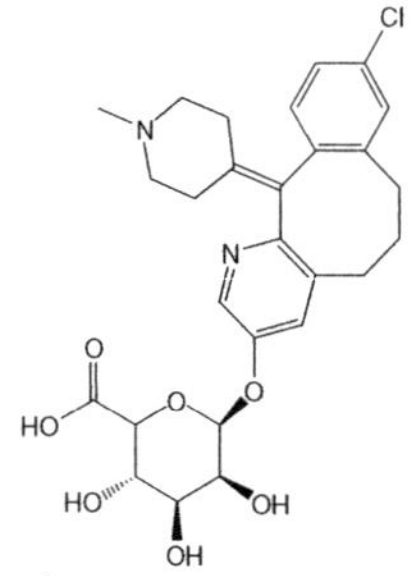 | Unidentified | [44] |
| 11. | Antiprogestational steroids | Mifepristone | Metapristone | 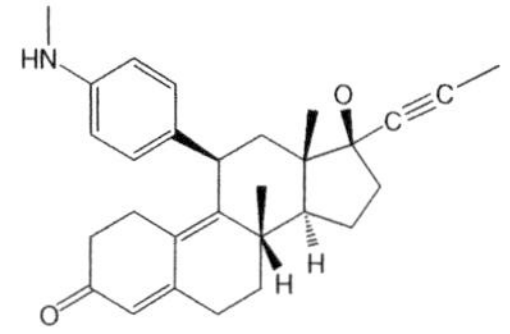 | P-gp | [65] |

| | | | | | | |
|---|---|---|---|---|---|---|
| 12. | Anxiolytic | Lorazepam | Lorazepam glucuronide | 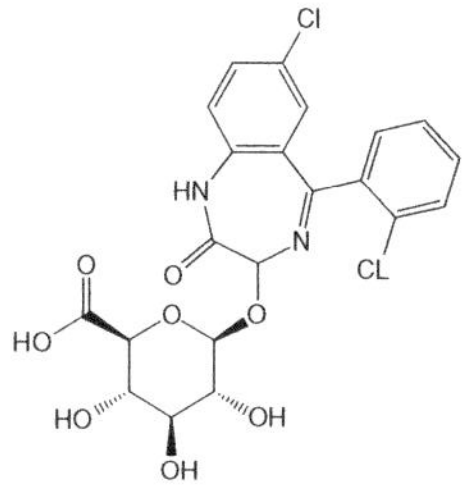 | Unidentified | [66] |
| 13. | Immunosuppressant agent | Mycophenolic acid | 7-*O*-mycophenolic acid glucuronide | 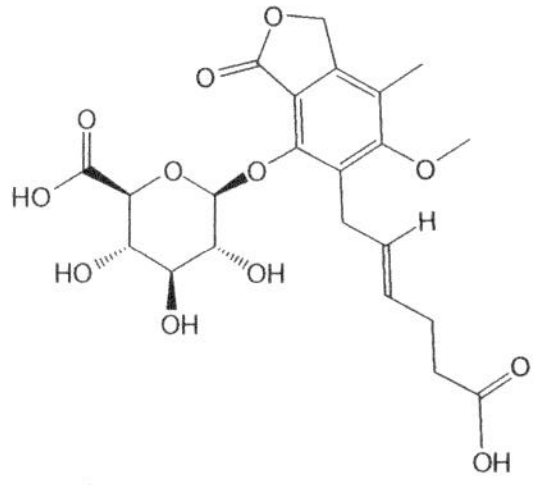 | MRP2 | [67] |
| 14. | Non-opioid analgesic agent | Acetaminophen | *N*-acetyl-*p*-benzoquinone imine and acetaminophen glucuronide | 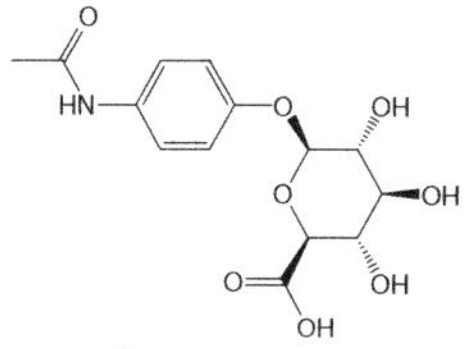 | MRP2, BCRP | [55] |
| 15. | NSAIDs | Aspirin | Salicylic acid | 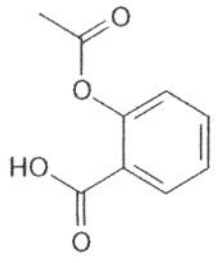 | P-gp | [68] |
| 16. | | Diclofenac | Diclofenac acyl glucuronide | 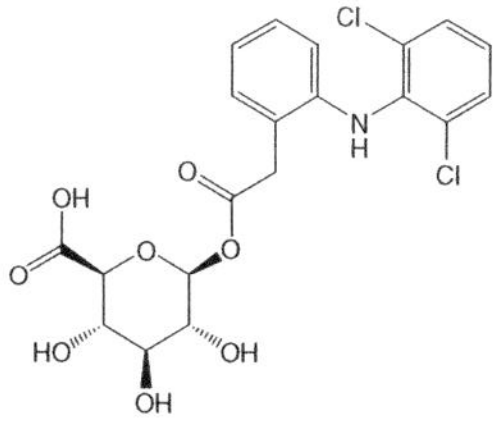 | MRP2, BCRP | [56, 57] |

(*continued*)

Table 14.2 (Continued)

| S/N | Therapeutic class | Compound | Metabolite | Chemical structure | Transporters | References |
|---|---|---|---|---|---|---|
| 17. | | Ibuprofen | Ibuprofen acyl glucuronide | 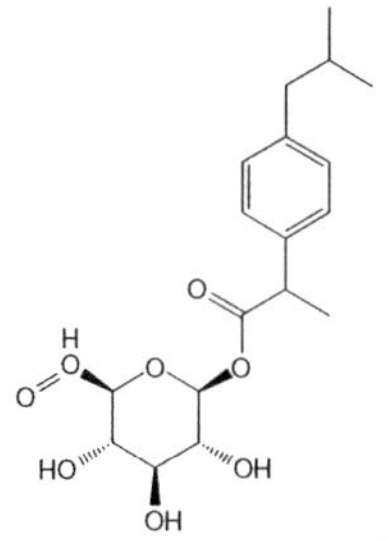 | MRP2 | [56] |
| 18. | | Meloxicam | $5^1$-Carboxymeloxicam | 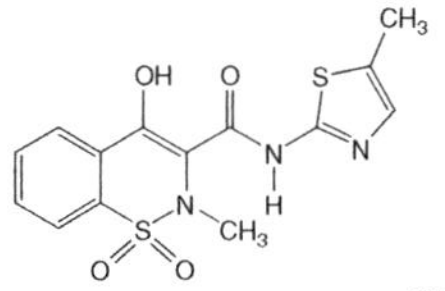 | BCRP | [58] |
| 19. | Vitamin | Vitamin D | 25-Hydroxycholecalciferol | 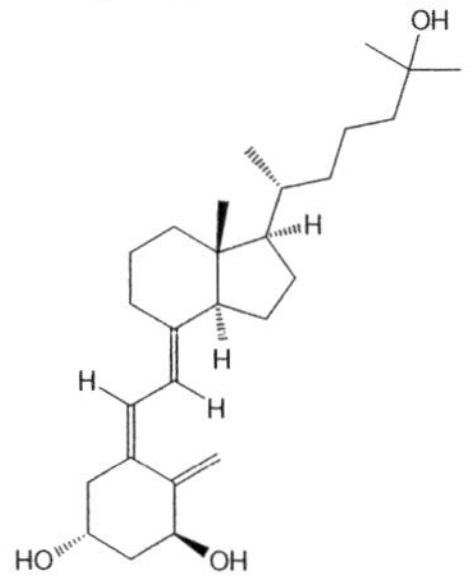 | P-gp | [60] |

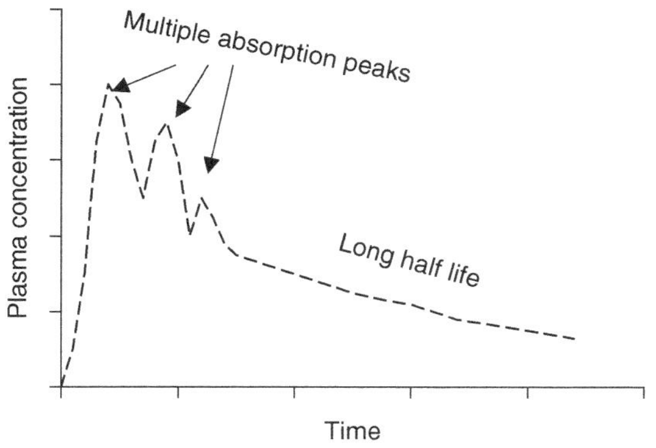


**Figure 14.2 PK profile of enterohepatic recycling compounds.** Multiple absorption peaks and long half-lives are typical features in the PK profiles of drugs undergo enterohepatic recycling.

will be decreased and drug distributions in other organs should be reduced. In case the drugs undergo EHR, the systemic drug exposure (i.e. AUC) will be increased due to prolonged half-life. Tissue concentrations may be increased due to increased plasma AUC. For the liver, bile duct, and intestine, drug molecules will be exposed to these organs repeatedly and pharmacological effects in these organs may be enhanced due to repeated drug exposure. We have attempted to utilize this feature of biliary excretion to design drugs that are only available in the colon to avoid high drug exposure in the other organs and off-target side effects [74].

Biliary excretion is an elimination process. However, the excretion facilitated by efflux transporters can always work competitively or cooperatively with the metabolism catalyzed by different enzymes. In hepatocytes, drug-related efflux transporters (e.g., P-gp, BCRP, and $MRP_2$) and metabolic enzymes (e.g., CYPS, UGTs, and SULTs) are highly expressed [75, 76] and the interplay between biliary excretion and hepatic metabolism is not rare. The well-known interplay is probably between CYP3A and P-gp in the liver and the observation of interplay extends to other enzymes and transporters. It was reported that the increment of efflux transporter activity could decrease the metabolic rate under linear kinetic conditions in the liver [77, 78]. The underlying rationale is that when efflux transporter activity is induced, more drugs will be excreted out of the hepatocytes and less of the substrate is available for the metabolic enzymes. On the contrary, inhibition of efflux transporters will lead to increased metabolism rates. In addition to the interplay between efflux transporters and metabolic enzymes, uptake transporters can also be involved in the interplay, where the uptake transporter (e.g., OATP), metabolic enzymes (e.g., UGTs), and efflux transporters (e.g., MRP2 and BCRP) work together to govern the intracellular drug concentrations [79]. Eventually, drug and metabolite levels or the ratio of drug/metabolites exposed in the plasma will be affected. The interplay between biliary excretion and hepatic metabolism could be quantitatively described using PBPK modeling [80, 81].

Biliary and renal excretion are two major elimination pathways, and these two pathways can compensate with each other. Some drugs are mainly eliminated through biliary excretion while others are primarily cleared from the kidney. When biliary excretion is changed, drug exposure in the plasma should be affected which will consequentially affect renal clearance. On the contrary, when renal elimination is reduced, compensatory biliary excretion will occur [82]. For example, it was reported that in the bile duct occlusion rats, the urine concentration of gadobenate ion, a diagnostic agent that can be eliminated through both renal and biliary excretion, was significantly higher than that of normal rats. Interestingly, in the renal vessel occlusion rats, the biliary excretion of gadobenate ion was significantly higher than normal rats [83]. The compensatory effect between biliary and renal elimination was also found in humans. For example, it was reported that biliary elimination of ceftriaxone increased in patients with renal dysfunction when compared to normal renal function [84].

## 14.5 Hepatic Transporters Involved in Biliary Excretion

Biliary excretion is a sequence of events involving the uptake of drugs from the sinusoidal (i.e. blood side or basolateral side) side into the hepatocytes followed by metabolism inside the cells and then drug and/or metabolites efflux out to the canalicular side (i.e. biliary excretion). In some cases, the drug molecules can also be effluxed into the blood side (i.e. basolateral efflux). Some compounds (e.g., lipophilic molecules) influx from the plasma to hepatic cytosol or from the cytoplasm into the bile by passive diffusion, but numerous drugs and metabolites are facilitated by transporters. There are many transport proteins expressed in the hepatocytes, which are polarized cells with different sides as shown in Figure 14.3. The function of uptake transporters is to pump drug molecules from the blood into the hepatocytes, and the function of efflux transporters is to pump drugs or metabolites from the hepatocytes to the canalicular. During drug development, some drugs (e.g., compounds for the treatment of bile duct

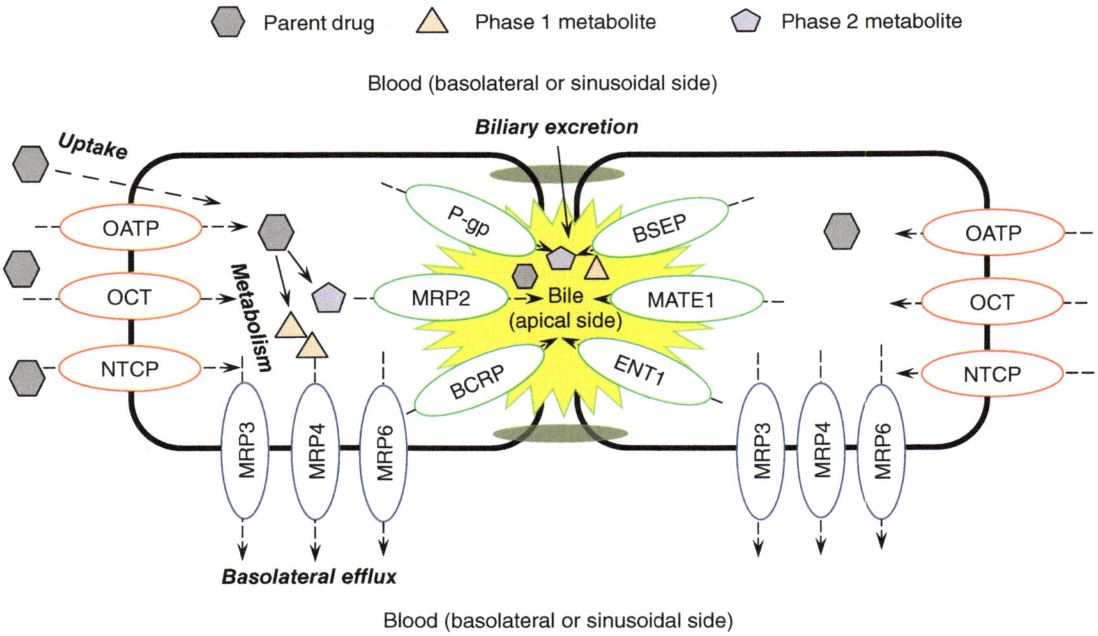


**Figure 14.3 Hepatic transporters.** Uptake and basolateral and apical efflux transporters are highly expressed in hepatocytes.

diseases) may need high biliary excretion, while others require low biliary excretion for enhanced systemic bioavailability. The transporter expression levels in the liver tissues in humans and animals (e.g., rats, mice, and dogs) have been well studied using LC-MS/MS [85–87], however, the knowledge of structure–transport activity relationships for these transporters is not quite clear and more efforts on this research area are appreciated in designing desired drugs.

### 14.5.1 Uptake Transporters

The hepatic uptake transporters located in the basolateral (e.g., sinusoidal) side as shown in Figure 14.3 pump substrates from the blood into the hepatocytes. Although uptake transporters do not participate in biliary excretion, the high drug concentration inside the hepatocytes will facilitate drug efflux into the bile. The hepatocytes uptake was probably first reported in a liver perfusion model showing an uptake of bile acid from the basolateral side by a carrier-mediated, sodium-dependent process that followed the Michaelis–Menten kinetics [88]. Later, a few uptake transporters were identified and characterized. The predominant transporters involved in human hepatic uptake include OATPs, NTCP, OCTs, and OATs [89]. The uptake transporters do not directly transport any substrates into the bile, but they can work corporately with hepatic enzymes and efflux transporters to facilitate biliary excretion. In drug development, design molecules targeting uptake transporters could increase intracellular drug concentrations. For example, medicinal chemists designed drugs targeting OATP uptake transporters to increase liver tissue concentrations for the treatment of liver diseases [90].

#### 14.5.1.1 Organic Anion Transporting Polypeptides (OATPs)

The OATPs, a class of transmembrane proteins that belongs to the solute carrier transmembrane proteins family, transport a variety of organic anions into the hepatocytes from blood and play an important role in drug disposition in the liver [91]. OATPs are sodium-independent transporters but usually function as bidirectional transporters: they influx substrates into the cells by exchanging glutathione from cytoplasm to outside [92]. At least 11 human OATP isoforms have been found including OATP1A2, 1B1, 1B3, 1C1, 2A1, 2B1, 3A1, 4A1, 4C1, 5A1, and 6A1. In human liver tissue, OATP1A2, 1B1, 1B3, and 2B1 are predominantly expressed [93]. According to computer-based hydropathy analysis, all OATP isoforms share a similar transmembrane domain organization, with 12 predicted transmembrane domains and a large fifth extracellular loop [94]. At the function level, OATPs are involved in the hepatic uptake of many drugs, and sometimes it could be the rate-limiting step of drug clarence from the liver. Typical substrates are organic anions such as glucuronide, sulfates, glutathione conjugates of drugs and dietary components, and genetic variability in genes encoding OATP transporters can result in marked inter–individual differences in biliary excretion [95].

#### 14.5.1.2 Organic Anion Transporters (OATs) and Organic Cation Transporters (OCTs)

OATs and OCTs are two subfamily of the solute carrier 22 (SCL22) membrane protein. There are at least 6 OAT isoforms, including OAT1, 2, 3, 4, 6, 7, and 10, and at least 3 OCT isoforms, including OCT1, 2, and 3, found in humans, but only OAT2 and OCT1 are highly expressed in the liver [96]. The OAT2 is a sodium-independent multi-specific organic anion/dimethyl-dicarboxylate exchanger [97]. The intriguing substrates of OAT2 are cyclic nucleotides, but they also transport some drugs such as bumetanide, pravastatin, benzylpenicillin, cimetidine, erythromycin, 2′,3′-dideoxycytidine, methotrexate, and zidovudine[98]. OCT1 expressed in the liver facilitates drug diffusion and allows the entry of organic cations into cells. Recent findings revealed that OCT1 can mediate the uptake of drugs for treating various diseases such as imatinib, biguanides, proguanil, and morphine [99–102].

#### 14.5.1.3 Sodium/Taurocholate Cotransporting Polypeptide (NTCP)

NTCP is a family member of the solute carrier 10 (SCL10) and is highly expressed on the basolateral membrane of hepatocytes. As a sodium-dependent uptake transporter, the primary responsibility of NTCP is the uptake of bile acids from the sinusoids [103]. It was reported that NTCP necessitated the binding of two sodium-ions together with a bile salt molecule for transporting and its function depends on the charge of the substrate, the driving force of the uptake may vary and may consist of the energy available from the chemical and/or the electrochemical gradient of sodium [104]. NTCP is one of the main transporters in the enterohepatic circulation of bile acids and as such, when inhibited by drugs, hepatotoxicity may occur. In the liver, although NTCP is distributed equally along all liver lobules, uptake of conjugated bile acids occurs predominantly in periportal cells, probably due to high bile acid concentrations in that area [105]. Compounds with bile acid-like structures would most likely be a good substrate of NCTP [106].

### 14.5.2 Efflux Transporters

There are two types of efflux transporters expressed in the hepatocytes (Figure 14.3). One is canalicular efflux transporters pumping substrates from the cell into the bile canalicular networks and the other is basolateral efflux transporters (e.g., MRP3, MRP4, and MRP6) pumping substrates from the cell into the blood [107]. These transporters belong to the ABC protein family, suggesting that drugs can be transported against concentration gradients and ATP is required [108]. The major efflux transporters that participate in drug and metabolite biliary excretion are P-gp, MRP2, and BCRP. Other efflux transporters such as BSEP (bile salt export), ENT1 (equilibrative nucleoside transporter 1), and MATE1 (multidrug and toxin extrusion protein 1), mainly modulate the homeostasis of important endogenous compounds (e.g., bile acids), but may also be direct or indirectly involved in biliary excretion of certain drugs and metabolites [109]. In drug development, many efforts have been inputted to design new entities to avoid these efflux transporters or using inhibitors for lower biliary excretion to enhance systemic bioavailability [110, 111].

#### 14.5.2.1 P-glycoprotein (P-gp)

P-gp, encoded by the ABCB gene, also called Multidrug Resistance Protein 1 (MDR1), is probably the most well-studied transporter. P-gp was discovered in the 1970s and the first mammalian crystal structure was elucidated about 30 years later in 2009 [112, 113]. P-gp is involved in drug resistance and acts by removing cytotoxic drugs from cells. Now, countless high-quality papers have been published on P-gp, most of which are on the impact of P-gp on drugs' bioavailability, and how to avoid efflux by P-gp, especially in cancer cells that overexpressed P-gp to cause drug resistance. P-gp transports numerous hydrophobic drugs and metabolites, such as daunorubicin, doxorubicin, etoposide, paclitaxel, SN-38, digoxin, cyclosporin A, and quinine from the cytoplasm to outside of the cells. In the liver, P-gp facilitates biliary excretion [114]. Typical P-gp substrates possess one or more planar aromatic rings that enable interaction with a hydrophobic region of the drug-binding domain of the protein and substrates with a cationic charge at physiological pH, a bulky structure (e.g., molecular weight >400), and a log partition coefficient >1 are more favorited [115, 116]. Three-dimensional quantitative structure–activity relationship (QSAR) models for P-gp substrates and inhibitors have been developed to quantitatively describe the capability of efflux and inhibition, and predict the binding sites [117, 118]. Fluorescent probes, such as rhodamine 123, are often used as probe substrates to analyze P-gp efflux activity [119, 120].

In the liver, P-gp is highly expressed in the apical (i.e. canalicular) side mediating biliary excretion. Reducing biliary excretion by targeting hepatic P-gp has been reported at least in animal models. For example, it was reported that oral administration of verapamil or quercetin, which are P-gp inhibitors, can significantly reduce biliary excretion of intravenously administrated irinotecan, a substrate of P-gp, into the bile, thereby, decreasing irinotecan and its major metabolite SN-38's concentration in the gut and potentially mitigating irinotecan-induced gut toxicity [121, 122].

#### 14.5.2.2 Multidrug Resistance-associated Protein 2 (MRP2)

MRP2, encoded by the ABCC2 gene and previously named as cMOAT, is the most widely studied apical member of the MRP family, which contains at least 9 isoforms in mammalians expressed on the basolateral (i.e. sinusoidal) side such as MRP3, MRP4, or the apical (i.e. canalicular) side such as MRP2. The MPR family, a class of ABC-transporters pumping endogenous compounds and xenobiotics outside of the cell, was first discovered in the study explaining mechanisms of resistance of tumor cells against antineoplastic agents [123]. That is probably where their names come from. Based on the structures, MRPs can be divided into two groups: a "short" group (i.e. MPR4, 5, 8, and 12) with four domains comprising two membrane-spanning domains, and a "long" group (i.e. MRP1, 2, 3, 6, and 7) with five domains [124]. MPR2 was first cloned from rat liver in 1996 [125], but the function of MRP2 was recognized long before its cloning by studies on the hepatobiliary elimination of organic anions in normal and transport-deficient mutant rats [126, 127]. The spectrum of MRP2 substrates is mainly anionic conjugates such as bilirubin, bilirubin conjugates, leukotriene C4, estrone-sulfate, SN-38, SN-38 conjugates, phenolic-glucuronides, and phenolic-sulfates[125, 128].

In the liver, MRP2 is also highly expressed in the apical (i.e. canalicular) side and primarily mediates the efflux of bile acids, glutathione, and their conjugates. Interestingly, the interplay between efflux transporters MRP2 and uptake transporter OATP1 mediates intracellular homeostasis of glutathione [129]. Many phase II conjugates (e.g., glucuronides and sulfates) are good substrates of MRP2 [130]. The biliary excretion of these compounds is highly influenced by the function and expression of MRP2.

#### 14.5.2.3 Breast Cancer Resistance Protein (BCRP)

BCRP, encoded by the ABCG2 gene, is also an ABC efflux transporter highly expressed in hepatocytes. The BCRP was first cloned in 1998 from a multidrug-resistant breast cancer cell line that confers resistance to chemotherapeutic drugs such as mitoxantrone and topotecan [131]. That is probably where the BCRP name comes from. Structurally, BCRP protein contains six transmembrane domains and functions as a homodimer or homotetramer [132]. The major functions of BCRP are to restrict the distribution of its substrates into organs and their elimination from excretory organs, mediating both biliary and renal excretion. Substrates of BCRP include different wide ranges of chemotherapeutic drugs such as mitoxantrone, camptothecin derivates, flavopiridol, methotrexate, imatinib, gefitinib, vinblastine, cisplatin, and paclitaxel. Other than anticancer drugs, BCRP also efflux conjugates of certain organic anions, particularly sulfated and glucuronide conjugates, such as estrone-3-sulfate, SN-38-sulfate, resveratrol-glucuronides, and genistein-glucuronide [133, 134]. Fluorescent probes for BCRP are available, such as BODIPY-prazosin, Hoechst 33342, and pheophorbide A which can be used to determine the function of BCRP [135]. BCRP is generally co-expressed with P-gp, and these two efflux transporters may share some substrates, inhibitors, and inducers [136].

In the liver, BCRP is also highly expressed in the apical (i.e. canalicular) side. Although biliary excretion of drugs through BCRP is usually not the major elimination pathway, biliary excretion of drug metabolites, particularly glucuronides and sulfates, through BCRP is very common. However, inhibition of drug biliary excretion usually could not significantly increase systemic drug exposure but can dramatically increase liver drug concentrations [137].

## 14.6 Factors Affecting Biliary Secretion

### 14.6.1 Gender Variation in Biliary Excretion

The expression of certain drug transporters (e.g., BCRP) is influenced by sex hormones and therefore is gender dependent [138]. It is understandable that biliary excretion of drugs or metabolites is significantly different in different genders if the drug or metabolites are good substrates of the hormone-related transporters. Preclinical models using animals have demonstrated the variation in biliary excretion in different sex [139]. However, whether these findings are translatable to humans is not clear and variation of biliary excretion in different sex in humans is an understudied research area.

In the animal model, the hepatic expression of BCRP at the protein level is significantly higher in males compared to that in females in both mice and rats.

Therefore, the favorite substrates (e.g., glucuronides and sulfates of phenolic compounds) usually have a higher biliary excretion rate in males compared to those in females. For example, it was reported that biliary excretion of acetaminophen-glucuronide and sulfate was about twofolds higher in male mice when compared to those in females in an intestinal perfusion model with bile duct cannulation. Western blot experiment showed that the hepatic BCRP expression in males were significantly higher than that in females [140], suggesting that more glucuronides and sulfates were pumped out by BCRP. In addition, the biliary excretion of acetaminophen- glucuronide and sulfate were similar in BCRP-deficient mice. Similarly, biliary excretion differences in different genders were also observed for other BCRP substrates such as nitrofurantoin and 2-amino-1-methyl-6-phenylimidazo[4,5-*b*]pyridine (PhIP) using a gallbladder cannulated in-vivo mouse model. Also, the difference was eliminated in BCRP-deficient mice [141]. These findings support the assumption that BCRP expression is gender-dependent, which could lead to different biliary excretion.

However, in humans, evidence to support different BCRP expressions in different sex is not found. On the contrary, a recent study showed that expression of hepatic BCRP in humans aged from 7-days to 87-years old is not sex-dependent at both mRNA and protein levels [142]. The conclusion is consistent with that from another study using a more accurate LC-MS method to quantify hepatic BCRP expression in humans [143]. The variation in function level for BCRP substrates is not clear. More effort is needed to demonstrate the variation of biliary excretion in different sex in humans due to BCRP expression to verify the findings in animals.

Efflux transporter MRP2 is also regulated by sex hormones and the pattern of growth hormone secretion, therefore, its expression level is different in different genders [144]. In adult rats, the mRNA level of MRP2 in females is about fourfold higher than that in males [144]. However, in mice, the result is contradictory. It was reported that in C57BL6 mice, the hepatic MRP2 mRNA level is similar in both males and females [145]. Although different methods have been developed to relatively quantify hepatic MRP2 in human liver tissues [86], the difference of MRP2 expression in different sex in humans is understudied and biliary excretion of drugs and metabolites facilitated by MRP2 in different sex is unclear.

Other major hepatic efflux transporters, including P-gp and BSEP, are not hormone-related transporters, and their expression levels are found to be gender independent in human tissues [146, 147]. Therefore, these transporters are not a contributor to variation of biliary excretion in different genders.

The expressions of hepatic uptake transporters (e.g., OATP1A1, OATP1A2, OATP1A4, and MRP4) are also different at mRNA levels in mice and rats [145, 148], which may contribute to variation in biliary excretion.

### 14.6.2 Age Variation in Biliary Excretion

Age-dependent biliary excretion was observed in animal models. For example, it was reported that biliary excretion of glutathione-related sulfur increased ninefold in 10 weeks of rats compared to that of 2 weeks [149]. Another study showed that in the biliary excretion of ouabain, a natural product that can be used for the treatment of hypotension and some arrhythmias, was only 1/3 in 24-week old rats when compared to that in 4-month old rats with intravenously drug administration [150]. Biliary excretion of digitoxin also showed a significant difference (only 43%) in old rats (e.g., 30-month) when compared to juvenile rats (3-month) [151]. Variation of biliary excretion in different ages is compound dependent as it was reported that for i.v. injected allyl isothiocyanate, biliary excretion is similar in three-month and two-month rats [152]. This is probably because different compounds are different substrates of hepatic efflux transporters. However, data on biliary excretion in different ages in humans are sparse and controversial. It is assumed that the expression of certain hepatic transporters is lower in infants when compared to those in adults based on the clinical observation that infants cannot dispose of some of the endogenous and exogenous compounds. However, there is no strong evidence to support this assumption. Other pieces of evidence suggested biliary excretion was age-dependent where bile acid level in the gut is significantly lower when compared to infants with adults. However, when normalized by body surface, the bile acids section rate was similar in different ages [153].

Mechanism studies showed that expression levels of some transporters are significantly lower in infants when compared to in adults at mRNA level. For example, it was reported that hepatic expression of MRP2, OATP1B1, and OATP1B3 were significantly lower in pediatric age than those in adults and the mRNA levels of P-gp in the liver of fetuses and neonates were about 20- to 30-fold lower than those of adults [154]. The other two major efflux transporters (i.e. MRP2 and BCRP) involved in biliary excretion have similar results. These results were consistent with those reported by other researchers [155, 156]. However, at the protein level, it was reported that expressions of hepatic transporters (e.g., P-gp and BCRP) were age-independent [143, 157]. Although some of the results at the protein level are somewhat controversial because there are some studies showing that hepatic transporters (e.g., MRP2) protein expressions were age-dependent at least in the first three months of life [158, 159]. At the functional level, more evidence is needed to demonstrate age-dependent in biliary excretion as a meta-analysis result showed that for azithromycin, ceftriaxone, digoxin, and buprenorphine, which are highly excreted through bile, the biliary excretion appears to be similar from the first few months of postnatal age to be adults [160].

### 14.6.3 Other Factors Affect Biliary Excretion

Other than the above physiological and pathological factors, biliary excretion of drugs and metabolites can also be affected by certain drugs. At normal conditions, the average bile amount secreted by the liver is relatively consistent. However, some drugs can influence the total bile flow. Consequentially, concentrations of drugs and metabolites in the bile will be affected when the total bile amount secreted from the liver is changed. For example, it was reported that cefmetazole, a broad-spectrum cephalosporin, can increase bile flow up to twofold in rats following a dose-dependent manner from 40–400 μmol/kg through i.v. injection in the first hour after drug administration [161]. Another example also showed that methimazole, an antithyroid medicine, increased bile flow in rats up to twofold [162]. The effect of different drugs on bile flow summarized in an early review showed that many drugs can affect bile flow by increasing the total volume of the secreted bile and changing the biliary composition. Therefore, to study biliary excretion of drugs and metabolites, especially in new drug candidate's evaluations, the total bile volume and the biliary composition should not be neglected. The total excreted amount is a more accurate evaluation when compared to concentration.

Biliary excretion is also influenced by lifestyle, diet habitats, and disease conditions. For example, high-fat dietary habit leads to an increase in lipid intake and is thought to be responsible for increased risk of dyslipidemia and liver diseases that may consequentially affect hepatic transporter expression and biliary secretion [163]. In some diseases, such as obstruction, cholangitis, gallstone, and cholangiocarcinoma, the bile duct is blocked, and bile flow is affected. Therefore, the biliary excretion of drugs and metabolites will be affected.

## 14.7 Biliary Excretion Research Models

Measuring biliary excretion in humans is difficult. Therefore, many *in vitro*, *in situ*, and *in vivo* animal models are developed to study hepatic metabolism and hepatobiliary excretion. New techniques are reported to optimize the culture conditions of these models and novel models are continuously invented. Specific chemical inhibitors, siRNA, genetic-modified animals, can also be used in these models to make them suitable and powerful for screening and mechanism studies. In addition, mathematic modeling is also attempted based on drug classes in the Biopharmaceutics Drug Disposition Classification System (BDDCS) to predict biliary secretion [164]. In humans, biliary excretion can be estimated using fecal recovery or duodenal aspirates. However, these are not direct measurements and results may not be accurate. For example, drug recovery from feces may be under-evaluated in the case of enterohepatic recycling. Besides, microflora could metabolize the drug into unusual metabolites which are difficult to measure unless radiolabel drugs are used. In patients who have diseases in the digestive system (e.g., bile-duct blocked), bile-duct is cannulated to bypass the block and a bile shunt (i.e., T-tube) can be introduced out of the body for bile collection. In these cases, biliary excretion can be accurately determined using bile samples.

### 14.7.1 *In Vitro* Models

Different *in vitro* models using hepatocytes, such as suspended and plated hepatocytes, cocultures, and a three-dimensional model, have been established to study drug metabolism and excretion, among which the sandwich-cultured hepatocytes (SCH) model is probably the best-known one [165]. Hepatocytes are polarized cells expressing different metabolic enzymes (e.g., CYPs and UGTs) and both uptake and efflux transporters (e.g., OATP and P-gp), which enable us to evaluate drug disposition and study drug disposition mechanisms. However, under conventional cultured conditions, hepatocytes rapidly differentiate losing the distinct phenotypes, which means that expression of transporters could partially or totally be lost and it is unable to differentiate between sinusoidal efflux and canalicular excretion [166, 167]. In the SCH model, hepatocytes are cultured between two layers of gelled collagen to form a "sandwich" (Figure 14.4). This structure allows hepatocytes to interact with each other to form *in vivo*-like properties with proper localization of

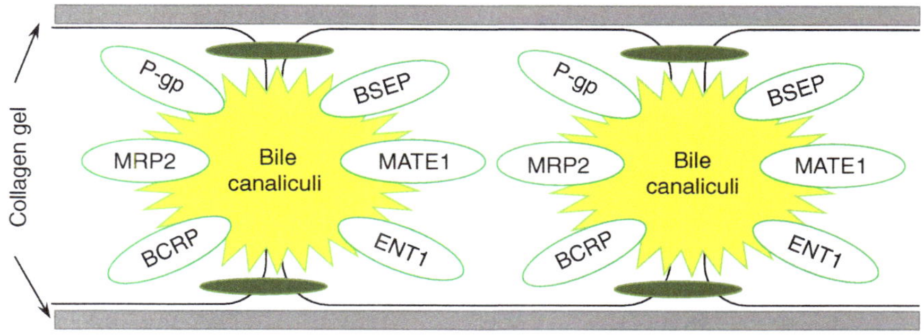


**Figure 14.4 SCH model.** Hepatocytes are clamped into two layers of collagen gels and bile canalicular networks can form.

basolateral and canalicular transporters, including P-gp, MRPs, and BCRP, as well as the formation of functional bile networks. [168, 169]. It is also reported that the bile canalicular networks can be disrupted by calcium and magnesium depletion in culture medium [170]. Therefore, absorption and efflux functions can be studied using this model by modulating the concentration of calcium or magnesium levels in the buffer.

Sandwich-cultured hepatocytes were first introduced in the late 1980s and early 1990s using rat hepatocytes [171, 172]. Both animal and human hepatocytes can be used in this model. Hepatocytes from transporter-deficient animals can also be used to demonstrate the impact of that particular transporter on drug disposition [173]. In addition, vectors expressing siRNA can be used to selectively knockdown a transporter in the hepatocytes [174], which makes this model more powerful in studying hepatic uptake and efflux. To culture the sandwich hepatocytes, the first step is to isolate primary hepatocytes from the liver. Then the fresh primary hepatocytes are seeded in a plate pretreated with collagen gel solution at 30 μl/cm$^2$ for at least one hour. Then, the second layer of collagen gel is spread over the cells one day after culturing the system in a normal medium with 5% of $CO_2$, and the medium is allowed to be replaced at least 30 minutes after coating the second collagen layer. Type I collagen gel prepared from rat tails is usually used as the coating material in the SCH model. Other coating materials (e.g., heparin gel) were also successfully used in this model [175]. For rat hepatocytes, the bile canaliculi can be formed after 24–48 hours and a more uniform network will be established three to seven days after seeding and can be maintained for at least two to three weeks [176]. Allowing SCH to have a relatively wide experimental window. However, it is much more difficult to isolate fresh human primary liver cells due to lack of fresh liver tissue. Cryopreserved human hepatocytes are usually used instead. A comprehensive review was published recently describing experimental details using human hepatocytes in SCH model [177].

It was reported that to optimize bile canalicular network formation for hepatobiliary excretion studies, culture conditions, such as cell density, volume of collagen, media type, media additives (e.g., epidermal growth factor), culture time, and concentrations of certain ions (e.g., $Ca^{2+}$ and $Mg^{2+}$), have to be tightly controlled [178, 179]. A conventional SCH is prepared as follows: primary hepatocytes are maintained in DMEM medium supplemented with 10% FBS, 0.5 U/ml insulin, 7 ng/ml glucagon, 200 U/ml penicillin, 200 mg/ml streptomycin, 7.5 mg/ml hydrocortisone, and 20 ng/ml TGF at 37 °C in a humidified 10% $CO_2$ atmosphere. The medium was changed every day. To grow a sandwich, a cell culture dish (e.g., 60 mm dish) is coated with collagen (e.g., 1 mg in 1 ml solution for a 60 mm dish), which is incubated at 37 °C for 45 minutes. Then, hepatocytes are seeded onto the dish (e.g., 2 million in 2 ml for a 60 mm dish) and cultured for 24 hours at 37 °C with 5% of $CO_2$. After removing the medium, the second layer of collagen (e.g., 1 mg in 1 ml solution for a 60 mm dish) is coated on top of the cell layer. Medium is then added after the system is incubated at 37 °C for 45 minutes. In the following days, the medium is changed daily [180]. The culture duration for a cell plate used to study hepatobiliary excretion is based on optimal canalicular network development in each species. For example, for rat and human SCH, the cells are usually used after 3 days, typically 5 days, after seeding [181, 182]. A detailed protocol of SCH culturing is described in recent publications [183, 184].

Although the SCH model was established in the late 1980s, it was started to be used for hepatic transporter evaluation a decade later when fluorescein-labeled taurocholate was used to demonstrate the function of $MRP_2$ in a rat SCH model [181]. In SCH model, hepatocytes can be cultured in single cell culture dishes/flasks or multi-well plates (e.g., 6 or 24 well plates), providing high-throughput screening capability. To evaluate hepatic uptake and hepatobiliary excretion using the SCH model, regular Hanks' Balanced Salt solution (HBSS) buffer and calcium-free buffers are used parallelly or

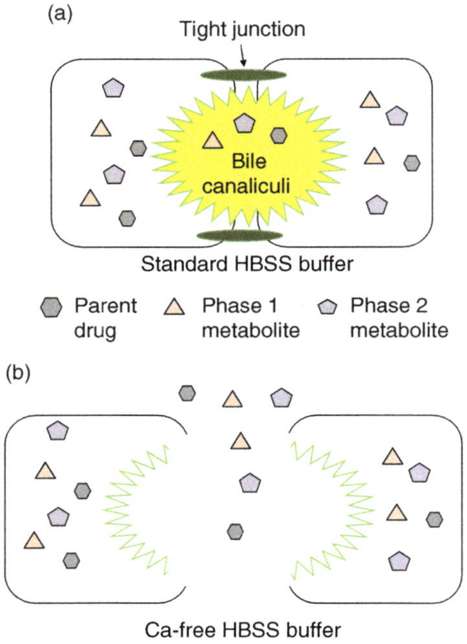


**Figure 14.5 SCH model with standard HBSS and Ca-free HBSS buffers.** Tight junction is formed in standard HBSS buffer and substrates are captured in the bile canalicular network (a). Tight junction could not form in Ca-free HBSS buffer and substrates are released to the cell culture medium (b).

sequentially in the incubation system. Because calcium maintains tight junctions between cells, the bile canalicular network will be disrupted in calcium-free buffer. Therefore, drugs and drug metabolites excreted from the hepatocytes will be released into the incubation buffer, if there is any, in calcium-free buffer but will be captive in standard buffer containing calcium (Figure 14.5). Then both incubation buffer and cell lysates will be collected for drug quantification. By comparing the intercellular and medium drug and metabolite concentrations, hepatic drug accumulation and biliary excretion can be calculated. To conduct an experiment, drugs are incubated with the hepatocytes cultured in a sandwich configuration in standard HBSS buffer and calcium-free HBSS buffer parallelly for a predetermined period and analytes in the medium and the cells are quantified.

To determine the intracellular accumulation, hepatocytes cultured in the sandwich configuration are incubated with standard or calcium-free HBSS for 10 minutes at 37 °C. Then the buffers are removed, and the cell layers are washed with ice-cold buffer gently for three times, after which a standard HBSS solution containing the tested drug is added to the preincubated cells to further incubate for a predetermined period of time based on the drugs. Then, the solution is removed, and the cells are washed again with ice-cold blank standard HBSS buffer, after which the cells are collected to prepare cell lysates to determine the intracellular levels of the analytes using analytical tools such as HPLC or LC-MS. Cell accumulation is calculated by measuring the intracellular drug concentrations using cell lysates. It is estimated that the hepatocellular volume of rats SCH in 6 or 24 well plates is $6.2\times10^{-6}$ μl/cell [185]. Therefore, the intracellular amount of the test drug can be calculated accordingly. Other than the intracellular accumulation, hepatic uptake clearance can also be determined using the following equation:

$$\mathrm{Cl_{uptake}} = \frac{\mathrm{Accumulation_{standard\ buffer}}}{T \times \mathrm{concentration\ in\ buffer}}$$

where $T$ is the incubation time and concentration in buffer is the initial drug concentration in the incubation buffer assuming the concentration at time $T$ is less than 10% of difference when compared to the initial concentration.

Biliary excretion can be determined using Biliary Excretion Index (BEI) following the equation:

$$\mathrm{BEI} = \frac{\mathrm{Accumulation_{standard\ buffer}} - \mathrm{Accumulation_{calcium\text{-}free\ buffer}}}{\mathrm{Accumulation_{standard\ buffer}}} \times 100$$

where the $\mathrm{Accumulation_{standard\ buffer}}$ and $\mathrm{Accumulation_{calcium\text{-}free\ buffer}}$ are obtained by quantitating the analytes preincubated in standard buffer and calcium-free buffer, respectively. It is understandable as the $\mathrm{Accumulation_{standard\ buffer}}$ is the amount/concentration inside the cell and the bile canalicular network and the $\mathrm{Accumulation_{calcium\text{-}free\ buffer}}$ is those inside the cells. Their difference is that excreted into the bile canalicular network, revealing hepatobiliary excretion.

The SCH model can also be used to study sinusoidal (i.e. basolateral) and bile efflux [186–188]. First, the cells are preloaded with the tested drug by incubating it with the working solution in a standard HBSS buffer for a certain period of time. Then, the working solution is removed, and the cells are washed with standard HBSS buffer. To determine efflux, cells are incubated with blank standard HBSS buffer and calcium-free HBSS buffer in different wells, parallelly. Samples are taken from the incubation buffers periodically (no longer than 30 minutes) and drug/metabolites concentrations in the standard and calcium-free buffers are determined. Basolateral and bile efflux is determined as follows:

$$\text{Sinusoidal efflux} = \text{Accumulated Amount in supernatant}_{\text{Standard buffer}}$$

$$\text{Bile efflux} = \text{Accumulated Amount in supernatant}_{\text{Calcium-free buffer}} - \text{Accumulated Amount in supernatant}_{\text{Standard buffer}}$$

Inhibitors for efflux transporters (e.g., verapamil for P-gp, Ko-143 for BCRP, and MK-571 for $MPR_2$) can also be used after preloading to determine the impact of a particular transporter on efflux to basolateral and bile.

SCH model is relatively simple, controllable, and suitable for screening and mechanism studies when compared to animal models or human studies. However, the disadvantages of SCH are not neglectable. Firstly, primary hepatocytes are required, and culture conditions and procedures are complex, which may be difficult for some labs. Additionally, for human SCH, it is not easy to get fresh human liver tissue to prepare primary hepatocytes. Although cryopreserved human hepatocytes can be used instead of fresh primary cells, whether protein expression changes or not well-studied. Secondly, cell properties can be affected by culture conditions and it is difficult to control batch-to-batch variations. Thirdly, although SCH has bile canaliculi formation, it is reported that drug metabolizing enzymes (e.g., CYPs) may not fully reflect those in the *in vivo* situation, and therefore SCH is not a suitable model for hepatic metabolism studies [189].

Other *in vitro* models are also available for hepatic metabolism and hepatobiliary excretion studies. For example, canalicular plasma membrane vesicles [190, 191], hepatocyte couplets [192, 193], 3D-sandwich cultured hepatocytes [194, 195], spheroid cultured hepatocytes [196, 197], and microphysiological system [198, 199]. All these models have their advantages. The most challenging disadvantage for these *in vitro* models is probably the *in vivo*-like properties such as the formation of biliary canaliculi network.

To overcome this major disadvantage, an isolated liver perfusion model is established [200]. In this model, livers are isolated and stationed in a chamber supplied with oxygenated Krebs-Ringer bicarbonate buffer and maintained at 37 °C. The portal vein is cannulated for drug perfusion and the bile duct is cannulated for bile collection. In the isolated liver perfusion model, hepatocytes are in an intact liver with *in vivo*-like properties and the biliary excretion results should be very similar to that *in vivo* when compared to cell culture-based models. Recently, a fancier model was introduced called precision-cut tissue slices (PCTS). This is an *ex vivo* model that can be treated as a mini organ containing all cells of the liver tissues in their natural environment, leaving intercellular and cell-matrix interactions intact. Therefore, the PCTS model is highly suitable to study hepatic drug disposition [201]. However, the model preparation is complex. More studies are needed to optimize and simplify this model to make it a routine model for most pharmaceutical labs.

### 14.7.2 *In Situ* Perfusion with Bile Duct Cannulation in Rodents

Intestinal perfusion model using rodents (i.e. rats and mice), which was initiated in the 1980s [202], is recognized by the FDA as a preclinical model to study intestinal membrane permeabilities and drug absorption in the intestine [203, 204]. With bile duct cannulation, bile can be collected to determine metabolism in the liver and biliary excretion of drugs and metabolites. In this model, animals are under anesthesia with the abdomen open. The tested drug is perfused constantly through one or multiple cannulated segment(s) of the intestine in an osmotic pressure-controlled buffer or can be infused through the jugular vein (e.g., HBSS buffer) using syringe pumps to mimic oral or intravenous administration. Drug absorption can be determined by comparing drug concentrations in the perfusates collected from the outlet canular with the initial concentrations. Periodically, samples can be collected through bile duct cannulation to determine drug and metabolite concentrations for drug metabolism in the liver and biliary excretion studies (Figure 14.6). Experimental detail is described in a separate chapter in this book [205].

Both rats and mice can be used as experimental animals in intestinal perfusion studies, but rats are more popular than mice because bile duct cannulation is much easier. For mice, it is difficult to cannulate the bile duct as it is too tiny, and the bile sample is usually

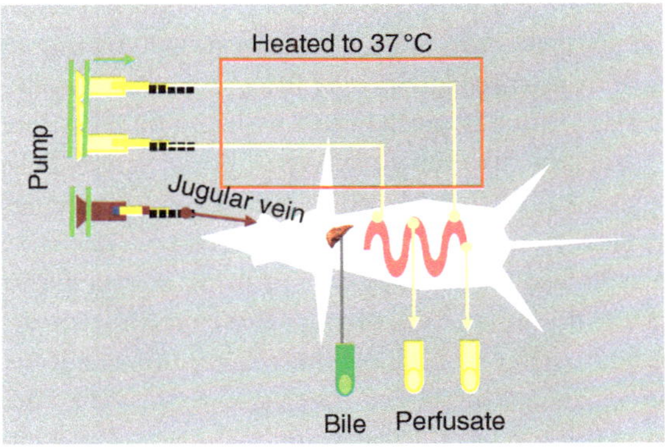


**Figure 14.6 Intestinal perfusion model.** Drug solution is perfused through intestine or infused through jugular vein and bile is collected through bile duct canular for biliary excretion study.

collected at the end of the experiment by collecting the gallbladder. Therefore, only a one-time point of biliary excretion can be afforded when a mouse is used. The tested drug is usually perfused through the intestine to mimic oral administration [206, 207]. In addition, the drug also can be perfused through the portal vein (i.e. liver perfusion) to avoid metabolism in the intestine [208] or through the jugular vein to mimic i.v.f injection in humans [209], depending on the purpose of the study.

The major advantage of this model is that animals are alive, and hepatocytes are in the intact liver supplied with blood flow, which is almost identical to the *in vivo* status. Additionally, chemical inhibitors and genetic-modified animals (e.g., MRP2-deficient rats, BCRP-deficient, and P-gp-deficient mice) are available to be used in this model to verify the impact of a transporter on drug biliary excretion. Disadvantages of the perfusion model are: (i) complex surgery must be conducted; (ii) animals can not survival after the experiment; and (iii) rats do not have gallbladder and the bile flow is directly from the liver consistently during the experiment, which is different from that in humans, and results for some drugs in rats perfusion studies may not be translatable.

### 14.7.3 *In Vivo* Animal Models

The above *in vitro* and *in situ* models are suitable for biliary excretion screening and mechanism studies and results from these models can be used to predict what will happen *in vivo*. Eventually, *in vivo* models must be used to verify the results obtained from other models. The bile-duct cannulated animal model is used as an *in vivo* model to study drug biliary excretion. In this model, the bile-duct is cannulated using a plastic tube with the other end of the tube in the duodenum to form a cycle allowing bile to return to the small intestine to maintain normal physiological functions [210, 211]. A segment of the plastic tube is placed outside of the body with a joint, where the tube can be open for bile collection. Different surgical techniques and approaches to increase the surgical success rates are available [210, 212]. Bile can be collected periodically or continuously over a few days from the jointed opening in the cannula outside of the body. After bile collection, the cannular will be reconnected allowing bile to flow back to the body. The bile samples will be processed and analyzed. Bile-duct cannulation is intricate and technically complex, and a skilled person is desired. The surgery should be aseptic because animals need to recover and survive. In addition, cannulas are at risk of twisting and blocking and animals need to be housed separately in metabolic cages. Fortunately, bile-duct cannulated animals (e.g., rats and mice) are commercially available from different vendors (e.g., Charles River Laboratories and Envigo Corporation), which promote this model to be used in biliary excretion studies.

### 14.7.4 Models for Human Studies

Although biliary excretion is an important elimination pathway, it is very rare to conduct clinical trials to determine biliary excretion in drug development because it is difficult to collect bile in humans using a noninvasive method due to the anatomy of the hepatobiliary tract. However, collecting bile using bile drainage catheters for treatment purposes is available in patients who have diseases that result in bile duct blocking (e.g., cholangitis, biliary obstruction) [213]. This technique usually needs surgery to put a transhepatic catheter (i.e. a plastic tube) into the bile duct and bypass the blocked part. A bile shunt (T-tube) can be used to divert part of the bile out of the body for external collection to determine the biliary excretion of drugs, drug metabolites, and endogenous compounds [214–216]. For example, it was reported the biliary excretion of bile acids, bile acids metabolites (i.e., bile acid sulfates), and bilirubin was determined in pancreatic carcinoma and bile duct obstruction patients by collecting bile using a bile drainage catheter [214]. Another example is that biliary excretion of mezlocillin in hepatic disease patients was determined using bile drainage catheters and doses of mezlocillin were suggested to be adjusted based on biliary and renal clearances [4].

Another more invasive method is cholecystectomy. In the case of certain diseases, the gallbladder needs to be removed. Before surgery, drugs can be administered to patients and bile can be collected from the removed bladder to determine biliary excretion. This method is more invasive and only allows one-time point bile collection, but it is still useful in patients who have certain diseases (e.g., choledocholithiasis, cholecystitis, stones, and gallstones) when the gallbladder must be removed. For example, the concentrations of piperacillin and tazobactam in the bile and gallbladder wall tissues were determined in patients after cholecystectomy and the results can be used for treatment referral in clinics [217].

Endoscopic Nasobiliary Drainage (END) can also be used to collect bile samples in patients. END was first introduced in the late 1970s in pancreatocholedochal catheterization [218] and now it is an established technique that is being used in the surgical management of acute cholangitis [219]. To collect bile from END, a tube is placed into the bile duct through the nasal under endoscopy. Different types of nasobiliary catheters with various diameters, shapes, and materials are currently available [220, 221]. Multiple time points are available using END and drug and metabolite biliary excretion can be assessed using the bile collected from the drainage. For example, biliary excretion of aspirin was determined in patients with ENDs [68].

For healthy volunteers, the invasive methods above are not applicable. Fecal recovery is an alternative method, in which feces are collected and fecal drug concentrations are determined for biliary excretion evaluation. In this approach, drugs and metabolites in the feces may be different from those concentrations in the bile because of reabsorption in the intestines or the metabolism by the intestine and the microflora, resulting in underestimated results. Usually, radioactive drugs are used to determine fecal recovery because microflora may create unknown and unusual metabolites [222, 223]. This method is noninvasive and can be used in drug development. However, this method could not distinguish between biliary excretion and intestinal secretion processes. In addition, for an orally administered drug, if the drug were not absorbed totally in the intestine due to solubility or permeability challenges, the unabsorbed drug will affect the evaluation [224].

Another alternative method in healthy volunteers is using the aspiration of duodenal fluids. Duodenal fluids somehow can represent the bile composition and therefore they can be used to determine biliary excretion of drugs and metabolites [225]. In this method, an oroenteric tube is placed into the duodenum, and the duodenal fluid for biliary excretion studies. This method has been used in biliary excretion studies previously [226–228].

## 14.8 Concluding Remarks

Biliary excretion is an important elimination pathway for many drugs and influences drug disposition and systemic exposure. Certain drugs are excreted through the bile as parent compounds, while others are excreted as metabolites. Many factors, such as uptake and efflux transporters, age, gender, and disease conditions, can affect biliary excretion, which makes the evaluation of biliary excretion difficult. There are *in vitro*, *in situ*, and *in vivo* models available to study biliary excretion, among which SCH model is a very useful *in vitro* model for biliary excretion screening and mechanism studies. For humans, different techniques have been developed for bile collection to study biliary excretion in patients and even in healthy volunteers. In drug development, an approach to utilize or avoid biliary excretion in designing new drugs is currently not available because the structure–excretion relationship is not well understood. More effects are expected to be conducted in this area.

## Abbreviations

| | |
|---|---|
| AUC | area under the curve |
| BCRP | breast cancer resistance protein |
| BSEP | bile salt export |
| CYP | cytochrome P450 |
| EHR | enterohepatic recycling |
| ENT1 | equilibrative nucleoside transporter 1 |
| HBSS | hanks' balanced salt solution |
| MRP2 | multidrug resistance protein 2 |
| mRNA | messenger-ribonucleic acid |
| MATE | multidrug and toxin extrusion protein 1 |
| NTCP | sodium/taurocholate cotransporting polypeptide |
| OATPs | organic anion transporting polypeptides |
| OATs | organic anion transporters |
| OCTs | organic cation transporters |
| P-gp | Pp-glycoprotein |
| SAR | structure–activity relationship |
| SCH | sandwich cultured hepatocytes |
| SULT | sulfotransferases |
| UGT | uridine diphosphate glucuronosyl transferases |

## References

1 Birru, W.A., Warren, D.B., Ibrahim, A. et al. (2014). Digestion of phospholipids after secretion of bile into the duodenum changes the phase behavior of bile components. *Mol. Pharmaceutics* 11: 2825–2834.

2 Gowda, G.A., Somashekar, B.S., Ijare, O.B. et al. (2006). One-step analysis of major bile components in human bile using 1H NMR spectroscopy. *Lipids* 41: 577–589.

3 Mathijssen, R.H., Verweij, J., Loos, W.J. et al. (2002). Irinotecan pharmacokinetics-pharmacodynamics: the clinical relevance of prolonged exposure to SN-38. *Br. J. Cancer* 87: 144–150.

4 Hildebrandt, R., Gundert-Remy, U., Raedsch, R. et al. (1995). Biliary excretion of mezlocillin in patients with hepatic disease. *Int. J. Clin. Pharmacol. Ther.* 33: 384–387.

5 Grant, D.M. (1991). Detoxification pathways in the liver. *J. Inherit. Metab. Dis.* 14: 421–430.

6 Abdel-Misih, S.R. and Bloomston, M. (2010). Liver anatomy. *Surg. Clin. North Am.* 90: 643–653.

7 Corless, J.K. and Middleton, H.M. 3rd, (1983). Normal liver function. a basis for understanding hepatic disease. *Arch. Intern. Med.* 143: 2291–2294.

8 Kanamura, S., Kanai, K., and Watanabe, J. (1990). Fine structure and function of hepatocytes during development. *J. Electron Microsc. Tech.* 14: 92–105.

9 Sarnova, L. and Gregor, M. (2017). Biliary system architecture: experimental models and visualization techniques. *Physiol. Res.* 66: 383–390.

10 Vakili, K. and Pomfret, E.A. (2008). Biliary anatomy and embryology. *Surg. Clin. North Am.* 88: 1159–1174. vii.

11 Jones, R.S. and Meyers, W.C. (1979). Regulation of hepatic biliary secretion. *Annu. Rev. Physiol.* 41: 67–82.

12 Mcmaster, P.D. (1922). Do species lacking a gall bladder possess its functional equivalent? *J. Exp. Med.* 35: 127–140.

13 Boyer, J.L. (2013). Bile formation and secretion. *Compr. Physiol.* 3: 1035–1078.

14 Di Ciaula, A., Garruti, G., Lunardi Baccetto, R. et al. (2017). Bile acid physiology. *Ann. Hepatol.* 16: s4–s14.

15 Ridlon, J.M., Kang, D.J., Hylemon, P.B., and Bajaj, J.S. (2014). Bile acids and the gut microbiome. *Curr. Opin. Gastroenterol.* 30: 332–338.

16 Rollins, D.E. and Klaassen, C.D. (1979). Biliary excretion of drugs in man. *Clin. Pharmacokinet.* 4: 368–379.

17 Gundert-Remy, U., Frohnapfel, F., Jourdan, W. et al. (1982). Estimation of biliary excretion of ureidopenicillins in healthy volunteers using marker dilution technique. *Br. J. Clin. Pharmacol.* 13: 795–801.

18 Melnik, G., Schwesinger, W.H., Teng, R. et al. (1998). Hepatobiliary elimination of trovafloxacin and metabolites following single oral doses in healthy volunteers. *Eur. J. Clin. Microbiol. Infect. Dis.* 17: 424–426.

19 Delgado, G. Jr., Neuhauser, M.M., Bearden, D.T., and Danziger, L.H. (2000). Quinupristin-dalfopristin: an overview. *Pharmacotherapy* 20: 1469–1485.

20 Millburn, P., Smith, R.L., and Williams, R.T. (1967). Biliary excretion in foreign compounds. Sulphonamide drugs in the rat. *Biochem. J.* 105: 1283–1287.

21 Ahn, S.Y. and Bhatnagar, V. (2008). Update on the molecular physiology of organic anion transporters. *Curr. Opin. Nephrol. Hypertension* 17: 499–505.

22 Kampmann, J., Lindahl, F., Hansen, J.M., and Siersbaek-Nielsen, K. (1973). Effect of probenecid on the excretion of ampicillin in human bile. *Br. J. Pharmacol.* 47: 782–786.

23 Siarheyeva, A., Lopez, J.J., and Glaubitz, C. (2006). Localization of multidrug transporter substrates within model membranes. *Biochemistry* 45: 6203–6211.

24 Maudgal, D.P., Maxwell, J.D., Lees, L.J., and Wild, R.N. (1982). Biliary excretion of amoxycillin and ceftriaxone after intravenous administration in man. *Br. J. Clin. Pharmacol.* 14: 213–217.

25 Li, M., Anderson, G.D., Phillips, B.R. et al. (2006). Interactions of amoxicillin and cefaclor with human renal organic anion and peptide transporters. *Drug Metab. Dispos.* 34: 547–555.

26 Nakakariya, M., Shimada, T., Irokawa, M. et al. (2008). Identification and species similarity of OATP transporters responsible for hepatic uptake of beta-lactam antibiotics. *Drug Metabol. Pharmacokinet.* 23: 347–355.

27 Waterman, N.G., Eickenberg, H.U., and Scharfenberger, L. (1976). Concentration of cefamandole in serum interstitial fluid, bile, and urine. *Antimicrob. Agents Chemother.* 10: 733–735.

28 Shitara, Y., Horie, T., and Sugiyama, Y. (2006). Transporters as a determinant of drug clearance and tissue distribution. *Eur. J. Pharm. Sci.* 27: 425–446.

29 Humphries, N.L., Tasker, D.G., and Karran, S.J. (1982). Comparative biliary excretion of cephalosporins in normal and partially hepatectomized rats. *Antimicrob. Agents Chemother.* 22: 170–171.

30 Patel, I.H. and Kaplan, S.A. (1984). Pharmacokinetic profile of ceftriaxone in man. *Am. J. Med.* 77: 17–25.

31 Brogden, R.N. and Heel, R.C. (1986). Aztreonam. a review of its antibacterial activity, pharmacokinetic properties and therapeutic use. *Drugs* 31: 96–130.

32 Lin, C.C., Chung, M., Gural, R. et al. (1984). Pharmacokinetics and metabolism of rosaramicin

in humans. *Antimicrob. Agents Chemother.* 26: 522–526.

33 Fuglesang, J. and Bergan, T. (1982). Chloramphenicol and thiamphenicol. *Antibiot Chemother* 31: 1–21.

34 Schuetz, E.G., Schinkel, A.H., Relling, M.V., and Schuetz, J.D. (1996). P-glycoprotein: a major determinant of rifampicin-inducible expression of cytochrome P4503A in mice and humans. *Proc. Natl. Acad. Sci. U.S.A.* 93: 4001–4005.

35 Montay, G., Goueffon, Y., and Roquet, F. (1984). Absorption, distribution, metabolic fate, and elimination of pefloxacin mesylate in mice, rats, dogs, monkeys, and humans. *Antimicrob. Agents Chemother.* 25: 463–472.

36 Dan, M., Serour, F., Gorea, A. et al. (1987). Concentration of norfloxacin in human gallbladder tissue and bile after single-dose oral administration. *Antimicrob. Agents Chemother.* 31: 352–353.

37 Merino, G., Alvarez, A.I., Pulido, M.M. et al. (2006). Breast cancer resistance protein (BCRP/ABCG2) transports fluoroquinolone antibiotics and affects their oral availability, pharmacokinetics, and milk secretion. *Drug Metab. Dispos.* 34: 690–695.

38 May, J.M., Owens, T.W., Mandler, M.D. et al. (2017). The antibiotic novobiocin binds and activates the ATPase that powers lipopolysaccharide transport. *J. Am. Chem. Soc.* 139: 17221–17224.

39 Sugiyama, Y., Kato, Y., and Chu, X. (1998). Multiplicity of biliary excretion mechanisms for the camptothecin derivative irinotecan (CPT-11), its metabolite SN-38, and its glucuronide: role of canalicular multispecific organic anion transporter and P-glycoprotein. *Cancer Chemother. Pharmacol.* 42 (Suppl): S44–S49.

40 Monsarrat, B., Mariel, E., Cros, S. et al. (1990). Taxol metabolism. Isolation and identification of three major metabolites of taxol in rat bile. *Drug Metab. Dispos.* 18: 895–901.

41 Lagas, J.S., Vlaming, M.L., Van Tellingen, O. et al. (2006). Multidrug resistance protein 2 is an important determinant of paclitaxel pharmacokinetics. *Clin. Cancer Res.* 12: 6125–6132.

42 Bao, L., Hazari, S., Mehra, S. et al. (2012). Increased expression of P-glycoprotein and doxorubicin chemoresistance of metastatic breast cancer is regulated by miR-298. *Am. J. Pathol.* 180: 2490–2503.

43 Krishna, R., Mcintosh, N., Riggs, K.W., and Mayer, L.D. (1999). Doxorubicin encapsulated in sterically stabilized liposomes exhibits renal and biliary clearance properties that are independent of valspodar (PSC 833) under conditions that significantly inhibit nonencapsulated drug excretion. *Clin. Cancer Res.* 5: 2939–2947.

44 Molimard, M., Diquet, B., and Benedetti, M.S. (2004). Comparison of pharmacokinetics and metabolism of desloratadine, fexofenadine, levocetirizine and mizolastine in humans. *Fundam. Clin. Pharmacol.* 18: 399–411.

45 Noguchi, S., Nishimura, T., Mukaida, S. et al. (2017). Cellular uptake of levocetirizine by organic anion transporter 4. *J. Pharm. Sci.* 106: 2895–2898.

46 Brookman, L.J., Rolan, P.E., Benjamin, I.S. et al. (1997). Pharmacokinetics of valsartan in patients with liver disease. *Clin. Pharmacol. Ther.* 62: 272–278.

47 Yamashiro, W., Maeda, K., Hirouchi, M. et al. (2006). Involvement of transporters in the hepatic uptake and biliary excretion of valsartan, a selective antagonist of the angiotensin II AT1-receptor, in humans. *Drug Metab. Dispos.* 34: 1247–1254.

48 Kosoglou, T., Statkevich, P., Johnson-Levonas, A.O. et al. (2005). Ezetimibe: a review of its metabolism, pharmacokinetics and drug interactions. *Clin. Pharmacokinet.* 44: 467–494.

49 Devane, C.L. and Nemeroff, C.B. (2001). Clinical pharmacokinetics of quetiapine: an atypical antipsychotic. *Clin. Pharmacokinet.* 40: 509–522.

50 Wang, J.S., Zhu, H.J., Markowitz, J.S. et al. (2006). Evaluation of antipsychotic drugs as inhibitors of multidrug resistance transporter P-glycoprotein. *Psychopharmacology* 187: 415–423.

51 Kumar, G.N., Jayanti, V.K., Johnson, M.K. et al. (2004). Metabolism and disposition of the HIV-1 protease inhibitor lopinavir (ABT-378) given in combination with ritonavir in rats, dogs, and humans. *Pharm. Res.* 21: 1622–1630.

52 Agarwal, S., Pal, D., and Mitra, A.K. (2007). Both P-gp and MRP2 mediate transport of Lopinavir, a protease inhibitor. *Int. J. Pharm.* 339: 139–147.

53 Millburn, P., Smith, R.L., and Williams, R.T. (1967). Biliary excretion of foreign compounds. Biphenyl, stilboestrol and phenolphthalein in the rat: molecular weight, polarity and metabolism as factors in biliary excretion. *Biochem. J.* 105: 1275–1281.

54 Hjelle, J.J. and Klaassen, C.D. (1984). Glucuronidation and biliary excretion of acetaminophen in rats. *J. Pharmacol. Exp. Ther.* 228: 407–413.

55 Mcgill, M.R. and Jaeschke, H. (2013). Metabolism and disposition of acetaminophen: recent advances in relation to hepatotoxicity and diagnosis. *Pharm. Res.* 30: 2174–2187.

56 Schneider, H.T., Nuernberg, B., Dietzel, K., and Brune, K. (1990). Biliary elimination of

non-steroidal anti-inflammatory drugs in patients. *Br. J. Clin. Pharmacol.* 29: 127–131.

57 Zhang, Y., Han, Y.H., Putluru, S.P. et al. (2016). Diclofenac and its Acyl glucuronide: determination of in vivo exposure in human subjects and characterization as human drug transporter substrates in vitro. *Drug Metab. Dispos.* 44: 320–328.

58 Garcia-Lino, A.M., Blanco-Paniagua, E., Astorga-Simon, E.N. et al. (2020). Abcg2 transporter affects plasma, milk and tissue levels of meloxicam. *Biochem. Pharmacol.* 175: 113924.

59 Mawer, E.B., Backhouse, J., Holman, C.A. et al. (1972). The distribution and storage of vitamin D and its metabolites in human tissues. *Clin. Sci.* 43: 413–431.

60 Fan, J., Liu, S., Du, Y. et al. (2009). Up-regulation of transporters and enzymes by the vitamin D receptor ligands, 1alpha,25-dihydroxyvitamin D3 and vitamin D analogs, in the Caco-2 cell monolayer. *J. Pharmacol. Exp. Ther.* 330: 389–402.

61 Tirona, R.G., Leake, B.F., Wolkoff, A.W., and Kim, R.B. (2003). Human organic anion transporting polypeptide-C (SLC21A6) is a major determinant of rifampin-mediated pregnane X receptor activation. *J. Pharmacol. Exp. Ther.* 304: 223–228.

62 Bender, J., Fang, J., and Simon, R. (2017). A computational study of the inhibition mechanisms of P-glycoprotein mediated paclitaxel efflux by kinase inhibitors. *BMC Syst. Biol.* 11: 108.

63 Shen, F., Chu, S., Bence, A.K. et al. (2008). Quantitation of doxorubicin uptake, efflux, and modulation of multidrug resistance (MDR) in MDR human cancer cells. *J. Pharmacol. Exp. Ther.* 324: 95–102.

64 Hemauer, S.J., Patrikeeva, S.L., Nanovskaya, T.N. et al. (2010). Role of human placental apical membrane transporters in the efflux of glyburide, rosiglitazone, and metformin. *Am. J. Obstet. Gynecol.* 202 (383): e1–e7.

65 Sarkar, N.N. (2002). Mifepristone: bioavailability, pharmacokinetics and use-effectiveness. *Eur. J. Obstet. Gynecol. Reprod. Biol.* 101: 113–120.

66 Greenblatt, D.J. and Engelking, L.R. (1988). Enterohepatic circulation of lorazepam and acetaminophen conjugates in ponies. *J. Pharmacol. Exp. Ther.* 244: 674–679.

67 Patel, C.G., Ogasawara, K., and Akhlaghi, F. (2013). Mycophenolic acid glucuronide is transported by multidrug resistance-associated protein 2 and this transport is not inhibited by cyclosporine, tacrolimus or sirolimus. *Xenobiotica* 43: 229–235.

68 Brune, K., Nuernberg, B., and Schneider, H.T. (1993). Biliary elimination of aspirin after oral and intravenous administration in patients. *Agents Actions Suppl.* 44: 51–57.

69 Larusso, N.F., Korman, M.G., Hoffman, N.E., and Hofmann, A.F. (1974). Dynamics of the enterohepatic circulation of bile acids. Postprandial serum concentrations of conjugates of cholic acid in health, cholecystectomized patients, and patients with bile acid malabsorption. *New Engl. J. Med.* 291: 689–692.

70 Yang, G., Ge, S., Singh, R. et al. (2017). Glucuronidation: driving factors and their impact on glucuronide disposition. *Drug Metabol. Rev.* 49: 105–138.

71 Roberts, M.S., Magnusson, B.M., Burczynski, F.J., and Weiss, M. (2002). Enterohepatic circulation: physiological, pharmacokinetic and clinical implications. *Clin. Pharmacokinet.* 41: 751–790.

72 Yang, Z., Kulkarni, K., Zhu, W., and Hu, M. (2012). Bioavailability and pharmacokinetics of genistein: mechanistic studies on its ADME. *Anti-Cancer Agents Med. Chem.* 12: 1264–1280.

73 Sandegren, L. (2014). Selection of antibiotic resistance at very low antibiotic concentrations. *Ups. J. Med. Sci.* 119: 103–107.

74 Ming Hu YM, Song Gao, 2015. Locally bioavailable drugs. In. International patent WO2016172159A1.

75 Almazroo, O.A., Miah, M.K., and Venkataramanan, R. (2017). Drug Metabolism in the Liver. *Clin. Liver Disease* 21: 1–20.

76 Giacomini, K.M., Huang, S.M., Tweedie, D.J. et al. (2010). Membrane transporters in drug development. *Nat. Rev. Drug Discovery* 9: 215–236.

77 Liu, L. and Pang, K.S. (2005). The roles of transporters and enzymes in hepatic drug processing. *Drug Metab. Dispos.* 33: 1–9.

78 Liu, L. and Pang, K.S. (2006). An integrated approach to model hepatic drug clearance. *Eur. J. Pharm. Sci.* 29: 215–230.

79 Nies, A.T., Schwab, M., and Keppler, D. (2008). Interplay of conjugating enzymes with OATP uptake transporters and ABCC/MRP efflux pumps in the elimination of drugs. *Expert Opin. Drug Metab. Toxicol.* 4: 545–568.

80 Pang, K.S., Maeng, H.J., and Fan, J. (2009). Interplay of transporters and enzymes in drug and metabolite processing. *Mol. Pharmaceutics* 6: 1734–1755.

81 Wu, B. (2012). Pharmacokinetic interplay of phase II metabolism and transport: a theoretical study. *J. Pharm. Sci.* 101: 381–393.

82 Srinivas, N.R. (2010). Altered disposition of drugs in acute renal failure rat models: drug development strategies and perspectives. *Arzneim.-Forsch.* 60: 731–748.

83 Kirchin, M.A., Lorusso, V., and Pirovano, G. (2015). Compensatory biliary and urinary excretion of gadobenate ion after administration of gadobenate dimeglumine (MultiHance((R))) in cases of impaired hepatic or renal function: a mechanism that may aid in the prevention of nephrogenic systemic fibrosis? *Br. J. Radiol.* 88: 20140526.

84 Imafuku, A., Sawa, N., Sekine, A. et al. (2018). Risk factors of ceftriaxone-associated biliary pseudolithiasis in adults: influence of renal dysfunction. *Clin. Exp. Nephrol.* 22: 613–619.

85 Khatri, R., Fallon, J.K., Rementer, R.J.B. et al. (2019). Targeted quantitative proteomic analysis of drug metabolizing enzymes and transporters by nano LC-MS/MS in the sandwich cultured human hepatocyte model. *J. Pharmacol. Toxicol. Methods* 98: 106590.

86 Ohtsuki, S., Schaefer, O., Kawakami, H. et al. (2012). Simultaneous absolute protein quantification of transporters, cytochromes P450, and UDP-glucuronosyltransferases as a novel approach for the characterization of individual human liver: comparison with mRNA levels and activities. *Drug Metab. Dispos.* 40: 83–92.

87 Wang, L., Prasad, B., Salphati, L. et al. (2015). Interspecies variability in expression of hepatobiliary transporters across human, dog, monkey, and rat as determined by quantitative proteomics. *Drug Metab. Dispos.* 43: 367–374.

88 Reichen, J. and Paumgartner, G. (1976). Uptake of bile acids by perfused rat liver. *Am. J. Physiol.* 231: 734–742.

89 Riley, R.J., Foley, S.A., Barton, P. et al. (2016). Hepatic drug transporters: the journey so far. *Expert Opin. Drug Metab. Toxicol.* 12: 201–216.

90 Tu, M., Mathiowetz, A.M., Pfefferkorn, J.A. et al. (2013). Medicinal chemistry design principles for liver targeting through OATP transporters. *Curr. Top. Med. Chem.* 13: 857–866.

91 Hagenbuch, B. and Meier, P.J. (2004). Organic anion transporting polypeptides of the OATP/SLC21 family: phylogenetic classification as OATP/SLCO superfamily, new nomenclature and molecular/functional properties. *Pflügers Arch.* 447: 653–665.

92 Ballatori, N., Hammond, C.L., Cunningham, J.B. et al. (2005). Molecular mechanisms of reduced glutathione transport: role of the MRP/CFTR/ABCC and OATP/SLC21A families of membrane proteins. *Toxicol. Appl. Pharmacol.* 204: 238–255.

93 Fahrmayr, C., Fromm, M.F., and Konig, J. (2010). Hepatic OATP and OCT uptake transporters: their role for drug-drug interactions and pharmacogenetic aspects. *Drug Metabol. Rev.* 42: 380–401.

94 Hagenbuch, B. and Meier, P.J. (2003). The superfamily of organic anion transporting polypeptides. *Biochim. Biophys. Acta* 1609: 1–18.

95 Kalliokoski, A. and Niemi, M. (2009). Impact of OATP transporters on pharmacokinetics. *Br. J. Pharmacol.* 158: 693–705.

96 Nigam, S.K., Bush, K.T., Martovetsky, G. et al. (2015). The organic anion transporter (OAT) family: a systems biology perspective. *Physiol. Rev.* 95: 83–123.

97 Kobayashi, Y., Ohshiro, N., Sakai, R. et al. (2005). Transport mechanism and substrate specificity of human organic anion transporter 2 (hOat2 [SLC22A7]). *J. Pharm. Pharmacol.* 57: 573–578.

98 Burckhardt, G. (2012). Drug transport by Organic Anion Transporters (OATs). *Pharmacol. Ther.* 136: 106–130.

99 Brosseau, N. and Ramotar, D. (2019). The human organic cation transporter OCT1 and its role as a target for drug responses. *Drug Metabol. Rev.* 51: 389–407.

100 Emoto, C., Fukuda, T., Johnson, T.N. et al. (2017). Characterization of Contributing factors to variability in morphine clearance through PBPK modeling implemented with OCT1 transporter. *CPT: Pharmacometrics Syst. Pharmacol.* 6: 110–119.

101 Segal, E.D., Yasmeen, A., Beauchamp, M.C. et al. (2011). Relevance of the OCT1 transporter to the antineoplastic effect of biguanides. *Biochem. Biophys. Res. Commun.* 414: 694–699.

102 Watkins, D.B., Hughes, T.P., and White, D.L. (2015). OCT1 and imatinib transport in CML: is it clinically relevant? *Leukemia* 29: 1960–1969.

103 Anwer, M.S. and Stieger, B. (2014). Sodium-dependent bile salt transporters of the SLC10A transporter family: more than solute transporters. *Pflügers Arch.* 466: 77–89.

104 Weinman, S.A. (1997). Electrogenicity of Na(+)-coupled bile acid transporters. *Yale J. Biol. Med.* 70: 331–340.

105 Groothuis, G.M., Hardonk, M.J., Keulemans, K.P. et al. (1982). Autoradiographic and kinetic demonstration of acinar heterogeneity of taurocholate transport. *Am. J. Physiol.* 243: G455–G462.

106 Dong, Z., Ekins, S., and Polli, J.E. (2015). A substrate pharmacophore for the human sodium

taurocholate co-transporting polypeptide. *Int. J. Pharm.* 478: 88–95.

107 Klaassen, C.D. and Aleksunes, L.M. (2010). Xenobiotic, bile acid, and cholesterol transporters: function and regulation. *Pharmacol. Rev.* 62: 1–96.

108 Liu, X. (2019). ABC family transporters. *Adv. Exp. Med. Biol.* 1141: 13–100.

109 Chandra, P. and Brouwer, K.L. (2004). The complexities of hepatic drug transport: current knowledge and emerging concepts. *Pharm. Res.* 21: 719–735.

110 Choi, Y.H. and Yu, A.M. (2014). ABC transporters in multidrug resistance and pharmacokinetics, and strategies for drug development. *Curr. Pharm. Des.* 20: 793–807.

111 Yuan, Z.W., Li, Y.Z., Liu, Z.Q. et al. (2018). Role of tangeretin as a potential bioavailability enhancer for silybin: pharmacokinetic and pharmacological studies. *Pharmacol. Res.* 128: 153–166.

112 Li, Y., Yuan, H., Yang, K. et al. (2010). The structure and functions of P-glycoprotein. *Curr. Med. Chem.* 17: 786–800.

113 Zhou, S.F. (2008). Structure, function and regulation of P-glycoprotein and its clinical relevance in drug disposition. *Xenobiotica* 38: 802–832.

114 Oude Elferink, R.P., Meijer, D.K., Kuipers, F. et al. (1995). Hepatobiliary secretion of organic compounds; molecular mechanisms of membrane transport. *Biochim. Biophys. Acta* 1241: 215–268.

115 Schmid, D., Ecker, G., Kopp, S. et al. (1999). Structure-activity relationship studies of propafenone analogs based on P-glycoprotein ATPase activity measurements. *Biochem. Pharmacol.* 58: 1447–1456.

116 Smit, J.W., Weert, B., Schinkel, A.H., and Meijer, D.K. (1998). Heterologous expression of various P-glycoproteins in polarized epithelial cells induces directional transport of small (type 1) and bulky (type 2) cationic drugs. *J. Pharmacol. Exp. Ther.* 286: 321–327.

117 Ekins, S., Kim, R.B., Leake, B.F. et al. (2002). Application of three-dimensional quantitative structure-activity relationships of P-glycoprotein inhibitors and substrates. *Mol. Pharmacol.* 61: 974–981.

118 Wang, R.B., Kuo, C.L., Lien, L.L., and Lien, E.J. (2003). Structure-activity relationship: analyses of p-glycoprotein substrates and inhibitors. *J. Clin. Pharm. Ther.* 28: 203–228.

119 Molnar, J., Engi, H., Hohmann, J. et al. (2010). Reversal of multidrug resitance by natural substances from plants. *Curr. Top. Med. Chem.* 10: 1757–1768.

120 Takeshita, A., Shigeno, K., Shinjo, K. et al. (2001). All-trans retinoic acid (ATRA) differentiates acute promyelocytic leukemia cells independently of P-glycoprotein (P-gp) related multidrug resistance. *Leuk. Lymphoma* 42: 739–746.

121 Bansal, T., Awasthi, A., Jaggi, M. et al. (2008). Pre-clinical evidence for altered absorption and biliary excretion of irinotecan (CPT-11) in combination with quercetin: possible contribution of P-glycoprotein. *Life Sci.* 83: 250–259.

122 Bansal, T., Mishra, G., Jaggi, M. et al. (2009). Effect of P-glycoprotein inhibitor, verapamil, on oral bioavailability and pharmacokinetics of irinotecan in rats. *Eur. J. Pharm. Sci.* 36: 580–590.

123 Stavrovskaya, A.A. (2000). Cellular mechanisms of multidrug resistance of tumor cells. *Biochemistry* 65: 95–106.

124 Jedlitschky, G., Hoffmann, U., and Kroemer, H.K. (2006). Structure and function of the MRP2 (ABCC2) protein and its role in drug disposition. *Expert Opin. Drug Metab. Toxicol.* 2: 351–366.

125 Nies, A.T. and Keppler, D. (2007). The apical conjugate efflux pump ABCC2 (MRP2). *Pflügers Arch.* 453: 643–659.

126 Hosokawa, S., Tagaya, O., Mikami, T. et al. (1992). A new rat mutant with chronic conjugated hyperbilirubinemia and renal glomerular lesions. *Lab. Anim. Sci.* 42: 27–34.

127 Huber, M., Guhlmann, A., Jansen, P.L., and Keppler, D. (1987). Hereditary defect of hepatobiliary cysteinyl leukotriene elimination in mutant rats with defective hepatic anion excretion. *Hepatology* 7: 224–228.

128 Gao, S. and Hu, M. (2010). Bioavailability challenges associated with development of anti-cancer phenolics. *Mini Rev. Med. Chem.* 10: 550–567.

129 Ballatori, N. and Rebbeor, J.F. (1998). Roles of MRP2 and oatp1 in hepatocellular export of reduced glutathione. *Semin. Liver Dis.* 18: 377–387.

130 Zhou, S.F., Wang, L.L., Di, Y.M. et al. (2008). Substrates and inhibitors of human multidrug resistance associated proteins and the implications in drug development. *Curr. Med. Chem.* 15: 1981–2039.

131 Robey, R.W., Polgar, O., Deeken, J. et al. (2007). ABCG2: determining its relevance in clinical drug resistance. *Cancer Metastasis Rev.* 26: 39–57.

132 Kage, K., Tsukahara, S., Sugiyama, T. et al. (2002). Dominant-negative inhibition of breast cancer resistance protein as drug efflux pump through the

inhibition of S-S dependent homodimerization. *Int. J. Cancer* 97: 626–630.

133 Ge, S., Wei, Y., Yin, T. et al. (2017). Transport-glucuronidation classification system and PBPK modeling: new approach to predict the impact of transporters on disposition of glucuronides. *Mol. Pharmaceutics* 14: 2884–2898.

134 Mao, Q. and Unadkat, J.D. (2015). Role of the breast cancer resistance protein (BCRP/ABCG2) in drug transport--an update. *AAPS J.* 17: 65–82.

135 Prochazkova, J., Kubala, L., Kotasova, H. et al. (2011). ABC transporters affect the detection of intracellular oxidants by fluorescent probes. *Free Radical Res.* 45: 779–787.

136 Kodaira, H., Kusuhara, H., Ushiki, J. et al. (2010). Kinetic analysis of the cooperation of P-glycoprotein (P-gp/Abcb1) and breast cancer resistance protein (Bcrp/Abcg2) in limiting the brain and testis penetration of erlotinib, flavopiridol, and mitoxantrone. *J. Pharmacol. Exp. Ther.* 333: 788–796.

137 Lee, C.A., O'connor, M.A., Ritchie, T.K. et al. (2015). Breast cancer resistance protein (ABCG2) in clinical pharmacokinetics and drug interactions: practical recommendations for clinical victim and perpetrator drug-drug interaction study design. *Drug Metab. Dispos.* 43: 490–509.

138 Tanaka, Y., Slitt, A.L., Leazer, T.M. et al. (2005). Tissue distribution and hormonal regulation of the breast cancer resistance protein (Bcrp/Abcg2) in rats and mice. *Biochem. Biophys. Res. Commun.* 326: 181–187.

139 Liu, W., Kulkarni, K., and Hu, M. (2013). Gender-dependent differences in uridine 5'-diphospho-glucuronosyltransferase have implications in metabolism and clearance of xenobiotics. *Expert Opin. Drug Metab. Toxicol.* 9: 1555–1569.

140 Lee, J.K., Abe, K., Bridges, A.S. et al. (2009). Sex-dependent disposition of acetaminophen sulfate and glucuronide in the in situ perfused mouse liver. *Drug Metab. Dispos.* 37: 1916–1921.

141 Merino, G., Van Herwaarden, A.E., Wagenaar, E. et al. (2005). Sex-dependent expression and activity of the ATP-binding cassette transporter breast cancer resistance protein (BCRP/ABCG2) in liver. *Mol. Pharmacol.* 67: 1765–1771.

142 Riches, Z., Abanda, N., and Collier, A.C. (2015). BCRP protein levels do not differ regionally in adult human livers, but decline in the elderly. *Chem. Biol. Interact.* 242: 203–210.

143 Prasad, B., Lai, Y., Lin, Y., and Unadkat, J.D. (2013). Interindividual variability in the hepatic expression of the human breast cancer resistance protein (BCRP/ABCG2): effect of age, sex, and genotype. *J. Pharm. Sci.* 102: 787–793.

144 Simon, F.R., Iwahashi, M., Hu, L.J. et al. (2006). Hormonal regulation of hepatic multidrug resistance-associated protein 2 (Abcc2) primarily involves the pattern of growth hormone secretion. *Am. J. Physiol. Gastrointest. Liver Physiol.* 290: G595–G608.

145 Maher, J.M., Slitt, A.L., Cherrington, N.J. et al. (2005). Tissue distribution and hepatic and renal ontogeny of the multidrug resistance-associated protein (Mrp) family in mice. *Drug Metab. Dispos.* 33: 947–955.

146 Meier, Y., Pauli-Magnus, C., Zanger, U.M. et al. (2006). Interindividual variability of canalicular ATP-binding-cassette (ABC)-transporter expression in human liver. *Hepatology* 44: 62–74.

147 Prasad, B., Evers, R., Gupta, A. et al. (2014). Interindividual variability in hepatic organic anion-transporting polypeptides and P-glycoprotein (ABCB1) protein expression: quantification by liquid chromatography tandem mass spectroscopy and influence of genotype, age, and sex. *Drug Metab. Dispos.* 42: 78–88.

148 Fu, Z.D., Csanaky, I.L., and Klaassen, C.D. (2012). Effects of aging on mRNA profiles for drug-metabolizing enzymes and transporters in livers of male and female mice. *Drug Metab. Dispos.* 40: 1216–1225.

149 Gregus, Z., Stein, A.F., and Klaassen, C.D. (1987). Age-dependent biliary excretion of glutathione-related thiols in rats: role of gamma-glutamyltransferase. *Am. J. Physiol.* 253: G86–G92.

150 Kitani, K., Zsolnai-Nagy, I., Kanai, S. et al. (1988). Correlation between the biliary excretion of ouabain and the lateral mobility of hepatocyte plasma membrane proteins in the rat--the effects of age and spironolactone pretreatment. *Hepatology* 8: 125–131.

151 Kitani, K., Sato, Y., and Van Bezooijen, K. (1982). The effect of age on the biliary excretion of digitoxin and its metabolites in female BN/Bi rats. *Arch. Gerontol. Geriatrics* 1: 43–54.

152 Borghoff, S.J. and Birnbaum, L.S. (1986). Age-related changes in the metabolism and excretion of allyl isothiocyanate. a model compound for glutathione conjugation. *Drug Metab. Dispos.* 14: 417–422.

153 Balistreri, W.F. (1983). Immaturity of hepatic excretory function and the ontogeny of bile acid metabolism. *J. Pediatr. Gastroenterol. Nutr.* 2 (Suppl 1): S207–S214.

154 Mooij, M.G., Schwarz, U.I., De Koning, B.A. et al. (2014). Ontogeny of human hepatic and intestinal transporter gene expression during childhood: age matters. *Drug Metab. Dispos.* 42: 1268–1274.

155 Miki, Y., Suzuki, T., Tazawa, C. et al. (2005). Steroid and xenobiotic receptor (SXR), cytochrome P450 3A4 and multidrug resistance gene 1 in human adult and fetal tissues. *Mol. Cell. Endocrinol.* 231: 75–85.

156 Van Kalken, C.K., Giaccone, G., Van Der Valk, P. et al. (1992). Multidrug resistance gene (P-glycoprotein) expression in the human fetus. *Am. J. Pathol.* 141: 1063–1072.

157 Tang, L., Hines, R.N., Eg, S., and Meibohm, B. (2007). Age associated protein expression of P-glycoprotein (MDR1/P-gp) and MRP2 in human pediatric liver. *Clin. Pharmacol. Ther.* 81: S101.

158 Li, N., Zhang, Y., Hua, F., and Lai, Y. (2009). Absolute difference of hepatobiliary transporter multidrug resistance-associated protein (MRP2/Mrp2) in liver tissues and isolated hepatocytes from rat, dog, monkey, and human. *Drug Metab. Dispos.* 37: 66–73.

159 Mooij, M.G., Van De Steeg, E., Van Rosmalen, J. et al. (2016). Proteomic analysis of the developmental Trajectory of human Hepatic membrane transporter Proteins in the first three months of life. *Drug Metab. Dispos.* 44: 1005–1013.

160 Johnson, T.N., Jamei, M., and Rowland-Yeo, K. (2016). How does in vivo biliary elimination of drugs change with age? evidence from in vitro and clinical data using a systems pharmacology approach. *Drug Metab. Dispos.* 44: 1090–1098.

161 Gonzalez, J., Fernandez, C., Marino, E. et al. (1989). Biliary excretion and choleretic effect of cefmetazole in rats. *Antimicrob. Agents Chemother.* 33: 1970–1974.

162 Ruiz, J., Munoz, M.E., Esteller, A., and Gonzalez, J. (1988). The effect of methimazole on bile flow and bile acid secretion in the rat. *Biochem. Pharmacol.* 37: 583–589.

163 Yamanashi, Y., Takada, T., and Suzuki, H. (2018). Associations between lifestyle-related diseases and transporters involved in intestinal absorption and biliary excretion of cholesterol. *Biol. Pharm. Bull.* 41: 1–10.

164 Hosey, C.M., Broccatelli, F., and Benet, L.Z. (2014). Predicting when biliary excretion of parent drug is a major route of elimination in humans. *AAPS J.* 16: 1085–1096.

165 Tetsuka, K., Ohbuchi, M., and Tabata, K. (2017). Recent progress in hepatocyte culture models and their application to the assessment of drug metabolism, transport, and toxicity in drug discovery: the value of tissue engineering for the successful development of a microphysiological system. *J. Pharm. Sci.* 106: 2302–2311.

166 Borlak, J. and Klutcka, T. (2004). Expression of basolateral and canalicular transporters in rat liver and cultures of primary hepatocytes. *Xenobiotica* 34: 935–947.

167 Oude Elferink, R.P., Ottenhoff, R., Liefting, W.G. et al. (1990). ATP-dependent efflux of GSSG and GS-conjugate from isolated rat hepatocytes. *Am. J. Physiol.* 258: G699–G706.

168 Swift, B., Pfeifer, N.D., and Brouwer, K.L. (2010). Sandwich-cultured hepatocytes: an in vitro model to evaluate hepatobiliary transporter-based drug interactions and hepatotoxicity. *Drug Metabol. Rev.* 42: 446–471.

169 Yang, K., Guo, C., Woodhead, J.L. et al. (2016). Sandwich-cultured hepatocytes as a tool to study drug disposition and drug-induced liver injury. *J. Pharm. Sci.* 105: 443–459.

170 Brouwer, K.L., Keppler, D., Hoffmaster, K.A. et al. (2013). In vitro methods to support transporter evaluation in drug discovery and development. *Clin. Pharmacol. Ther.* 94: 95–112.

171 Dunn, J.C., Tompkins, R.G., and Yarmush, M.L. (1991). Long-term in vitro function of adult hepatocytes in a collagen sandwich configuration. *Biotechnol. Progr.* 7: 237–245.

172 Dunn, J.C., Yarmush, M.L., Koebe, H.G., and Tompkins, R.G. (1989). Hepatocyte function and extracellular matrix geometry: long-term culture in a sandwich configuration. *FASEB J.* 3: 174–177.

173 Yang, K., Pfeifer, N.D., Hardwick, R.N. et al. (2014). An experimental approach to evaluate the impact of impaired transport function on hepatobiliary drug disposition using Mrp2-deficient TR- rat sandwich-cultured hepatocytes in combination with Bcrp knockdown. *Mol. Pharmaceutics* 11: 766–775.

174 Yue, W., Abe, K., and Brouwer, K.L. (2009). Knocking down breast cancer resistance protein (Bcrp) by adenoviral vector-mediated RNA interference (RNAi) in sandwich-cultured rat hepatocytes: a novel tool to assess the contribution of Bcrp to drug biliary excretion. *Mol. Pharmaceutics* 6: 134–143.

175 Revzin, E. (2015). Heparin hydrogel sandwich cultures of primary hepatocytes. *Eur. Polym. J.* 72: 726–735.

176 Lecluyse, E.L., Audus, K.L., and Hochman, J.H. (1994). Formation of extensive canalicular networks by rat hepatocytes cultured in collagen-sandwich configuration. *Am. J. Physiol.* 266: C1764–C1774.

177 Fardel, O., Moreau, A., Le Vee, M. et al. (2019). Evaluation of drug biliary excretion using sandwich-cultured human hepatocytes. *Eur. J. Drug Metabol. Pharm.* 44: 13–30.

178 Chandra, P., Lecluyse, E.L., and Brouwer, K.L. (2001). Optimization of culture conditions for determining hepatobiliary disposition of taurocholate in sandwich-cultured rat hepatocytes. *In Vitro Cell. Dev. Biol. Anim.* 37: 380–385.

179 Hamilton, G.A., Jolley, S.L., Gilbert, D. et al. (2001). Regulation of cell morphology and cytochrome P450 expression in human hepatocytes by extracellular matrix and cell-cell interactions. *Cell Tissue Res.* 306: 85–99.

180 Berthiaume, F., Moghe, P.V., Toner, M., and Yarmush, M.L. (1996). Effect of extracellular matrix topology on cell structure, function, and physiological responsiveness: hepatocytes cultured in a sandwich configuration. *FASEB J.* 10: 1471–1484.

181 Liu, X., Lecluyse, E.L., Brouwer, K.R. et al. (1999). Biliary excretion in primary rat hepatocytes cultured in a collagen-sandwich configuration. *Am. J. Physiol.* 277: G12–G21.

182 Lu, Y., Slizgi, J.R., Brouwer, K.R. et al. (2016). Hepatocellular disposition and transporter interactions with tolvaptan and metabolites in sandwich-cultured human hepatocytes. *Drug Metab. Dispos.* 44: 867–870.

183 Gijbels, E., Vanhaecke, T., and Vinken, M. (2019). Establishment of sandwich cultures of primary human hepatocytes. *Methods Mol. Biol.* 1981: 325–333.

184 Vinken, M., Elaut, G., Henkens, T. et al. (2006). Rat hepatocyte cultures: collagen gel sandwich and immobilization cultures. *Methods Mol. Biol.* 320: 247–254.

185 Uhal, B.D. and Roehrig, K.L. (1982). Effect of dietary state on hepatocyte size. *Biosci. Rep.* 2: 1003–1007.

186 Feng, B., Xu, J.J., Bi, Y.A. et al. (2009). Role of hepatic transporters in the disposition and hepatotoxicity of a HER2 tyrosine kinase inhibitor CP-724,714. *Toxicol. Sci.* 108: 492–500.

187 Pfeifer, N.D., Yang, K., and Brouwer, K.L. (2013). Hepatic basolateral efflux contributes significantly to rosuvastatin disposition I: characterization of basolateral versus biliary clearance using a novel protocol in sandwich-cultured hepatocytes. *J. Pharmacol. Exp. Ther.* 347: 727–736.

188 Yang, K., Pfeifer, N.D., Kock, K., and Brouwer, K.L. (2015). Species differences in hepatobiliary disposition of taurocholic acid in human and rat sandwich-cultured hepatocytes: implications for drug-induced liver injury. *J. Pharmacol. Exp. Ther.* 353: 415–423.

189 Kimoto, E., Walsky, R., Zhang, H. et al. (2012). Differential modulation of cytochrome P450 activity and the effect of 1-aminobenzotriazole on hepatic transport in sandwich-cultured human hepatocytes. *Drug Metab. Dispos.* 40: 407–411.

190 Inoue, M., Kinne, R., Tran, T. et al. (1983). Rat liver canalicular membrane vesicles. isolation and topological characterization. *J. Biol. Chem.* 258: 5183–5188.

191 Suchy, F.J., Courchene, S.M., and Blitzer, B.L. (1985). Taurocholate transport by basolateral plasma membrane vesicles isolated from developing rat liver. *Am. J. Physiol.* 248: G648–G654.

192 Coleman, R. and Roma, M.G. (2000). Hepatocyte couplets. *Biochem. Soc. Trans.* 28: 136–140.

193 Coleman, R., Wilton, J.C., Stone, V., and Chipman, J.K. (1995). Hepatobiliary function and toxicity in vitro using isolated hepatocyte couplets. *Gen. Pharmacol.* 26: 1445–1453.

194 Deharde, D., Schneider, C., Hiller, T. et al. (2016). Bile canaliculi formation and biliary transport in 3D sandwich-cultured hepatocytes in dependence of the extracellular matrix composition. *Arch. Toxicol.* 90: 2497–2511.

195 Murray, J.W., Han, D., and Wolkoff, A.W. (2014). Hepatocytes maintain greater fluorescent bile acid accumulation and greater sensitivity to drug-induced cell death in three-dimensional matrix culture. *Phys. Rep.* 2, e12198.

196 Bell, C.C., Hendriks, D.F., Moro, S.M. et al. (2016). Characterization of primary human hepatocyte spheroids as a model system for drug-induced liver injury, liver function and disease. *Sci. Rep.* 6: 25187.

197 Lauschke, V.M., Shafagh, R.Z., Hendriks, D.F.G., and Ingelman-Sundberg, M. (2019). 3D primary hepatocyte culture systems for analyses of liver diseases, drug metabolism, and toxicity: emerging culture paradigms and applications. *Biotechnol. J.* 14: e1800347.

198 Bhatia, S.N. and Ingber, D.E. (2014). Microfluidic organs-on-chips. *Nat. Biotechnol.* 32: 760–772.

199 Sano, E., Mori, C., Matsuoka, N. et al. (2019). Tetrafluoroethylene-propylene elastomer for fabrication of microfluidic organs-on-chips resistant to drug absorption. *Micromachines (Basel)* 10: 793.

200 Rg, B.K.T. (1996). *Isolated Perfused Liver, in Models for Assessing Drug Absorption and Metabolism*. New York: Plenum Press.

201 De Graaf, I.A., Olinga, P., De Jager, M.H. et al. (2010). Preparation and incubation of precision-cut liver and intestinal slices for application in drug metabolism and toxicity studies. *Nat. Protoc.* 5: 1540–1551.

202 Sinko, P.J. and Amidon, G.L. (1988). Characterization of the oral absorption of beta-lactam antibiotics. I. Cephalosporins: determination of intrinsic membrane absorption parameters in the rat intestine in situ. *Pharm. Res.* 5: 645–650.

203 Roig, T. and Vinardell, M.P. (1991). Intestinal perfusion in vivo for the study of absorptive processes. *Comp. Biochem. Physiol. A: Physiol.* 98: 3–7.

204 Stappaerts, J., Brouwers, J., Annaert, P., and Augustijns, P. (2015). In situ perfusion in rodents to explore intestinal drug absorption: challenges and opportunities. *Int. J. Pharm.* 478: 665–681.

205 Hu, M. and Li, X. (eds) (2011). Chapter 29 Intestinal perfusion methods for oral drug absorption. In: *Oral Bioavailability*. https://www.google.com/books/edition/Oral_Bioavailability/DmMsqDO1IW0C?hl=en&gbpv=1&dq=oral+bioavailability,+ming+hu+and+xiaolin+Li&printsec=frontcover.

206 Kovacs, N.P., Almasi, A., Garai, K. et al. (2019). Investigation of intestinal elimination and biliary excretion of ibuprofen in hyperglycemic rats. *Can. J. Physiol. Pharmacol.* 97: 1080–1089.

207 Zhang, Y., Wang, C., Liu, Z. et al. (2018). P-gp is involved in the intestinal absorption and biliary excretion of afatinib in vitro and in rats. *Pharmacol. Rep.* 70: 243–250.

208 Zeng, M., Sun, R., Basu, S. et al. (2016). Disposition of flavonoids via recycling: direct biliary excretion of enterically or extrahepatically derived flavonoid glucuronides. *Mol. Nutr. Food Res.* 60: 1006–1019.

209 Sun, R., Basu, S., Zeng, M. et al. (2019). Xiao-Chai-Hu-Tang (XCHT) intervening irinotecan's disposition: the potential of XCHT in alleviating irinotecan-induced diarrhea. *Curr. Cancer Drug Targets* 19: 551–560.

210 Burden, N., Kendrick, J., Knight, L. et al. (2017). Maximizing the success of bile duct cannulation studies in rats: recommendations for best practice. *Lab. Anim.* 51: 457–464.

211 Rolf, L.L. Jr., Bartels, K.E., Nelson, E.C., and Berlin, K.D. (1991). Chronic bile duct cannulation in laboratory rats. *Lab. Anim. Sci.* 41: 486–492.

212 Tonsberg, H., Holm, R., Bjerregaard, T.G. et al. (2010). An updated and simplified method for bile duct cannulation of rats. *Lab. Anim.* 44: 373–376.

213 Van Den Hazel, S.J., De Vries, X.H., Speelman, P. et al. (1996). Biliary excretion of ciprofloxacin and piperacillin in the obstructed biliary tract. *Antimicrob. Agents Chemother.* 40: 2658–2660.

214 Dooley, J.S., Bartholomew, C., Summerfield, J.A., and Billing, B.H. (1984). The biliary excretion of sulphated and non-sulphated bile acids and bilirubin in patients with external bile drainage. *Clin. Sci.* 67: 61–68.

215 Sellman, R., Kanto, J., and Pekkarinen, J. (1975). Biliary excretion of diazepam and its metabolites in man. *Acta Pharmacol. Toxicol.* 37: 242–249.

216 Slatter, J.G., Schaaf, L.J., Sams, J.P. et al. (2000). Pharmacokinetics, metabolism, and excretion of irinotecan (CPT-11) following I.V. infusion of [(14)C]CPT-11 in cancer patients. *Drug Metab. Dispos.* 28: 423–433.

217 Westphal, J.F., Brogard, J.M., Caro-Sampara, F. et al. (1997). Assessment of biliary excretion of piperacillin-tazobactam in humans. *Antimicrob. Agents Chemother.* 41: 1636–1640.

218 Nagai, N., Toli, F., Oi, I. et al. (1976). Continuous endoscopic pancreatocholedochal catheterization. *Gastrointes. Endoscopy* 23: 78–81.

219 Itoi, T., Kawai, T., Sofuni, A. et al. (2008). Efficacy and safety of 1-step transnasal endoscopic nasobiliary drainage for the treatment of acute cholangitis in patients with previous endoscopic sphincterotomy (with videos). *Gastrointes. Endoscopy* 68: 84–90.

220 Mukai, S., Itoi, T., Baron, T.H. et al. (2017). Indications and techniques of biliary drainage for acute cholangitis in updated Tokyo Guidelines 2018. *J. Hepato-Biliary-Pancreat. Sci.* 24: 537–549.

221 Tsuyuguchi, T., Takada, T., Kawarada, Y. et al. (2007). Techniques of biliary drainage for acute cholangitis: Tokyo guidelines. *J. Hepato-Biliary-Pancreat. Surg.* 14: 35–45.

222 Anggard, E., Gunne, L.M., Homstrand, J. et al. (1975). Disposition of methadone in methadone maintenance. *Clin. Pharmacol. Ther.* 17: 258–266.

223 Beumer, J.H., Garner, R.C., Cohen, M.B. et al. (2007). Human mass balance study of the novel anticancer agent ixabepilone using accelerator mass spectrometry. *Invest. New Drugs* 25: 327–334.

224 Chandrasekaran, A., Mckeand, W.E., Sullivan, P. et al. (2009). Metabolic disposition of [14C]bazedoxifene in healthy postmenopausal women. *Drug Metab. Dispos.* 37: 1219–1225.

225 Janowitz, P., Swobodnik, W., Wechsler, J.G. et al. (1990). Comparison of gall bladder bile and endoscopically obtained duodenal bile. *Gut* 31: 1407–1410.

226 Ghibellini, G., Johnson, B.M., Kowalsky, R.J. et al. (2004). A novel method for the determination of biliary clearance in humans. *AAPS J.* 6: e33.

227 Matsson, E.M., Eriksson, U.G., Knutson, L. et al. (2011). Biliary excretion of ximelagatran and its metabolites and the influence of erythromycin following intraintestinal administration to healthy volunteers. *J. Clin. Pharmacol.* 51: 770–783.

228 Maudgal, D.P., Lanzini, A., Northfield, T.C. et al. (1985). Quantification of temocillin biliary excretion and gallbladder bile concentration in healthy subjects. *Drugs* 29 (Suppl 5): 146–150.

# 15

# Pharmacokinetic Behaviors of Orally Administered Drugs

*Hamdah Al Nebaihi[1], Dion R. Brocks[1], Jaime A. Yáñez[2], Marcus Laird Forrest[3], and Neal M. Davies[1]*

[1] *Faculty of Pharmacy and Pharmaceutical Sciences, University of Alberta, Alberta, Canada*
[2] *Escuela Profesional de Medicina Humana, Universidad Privada San Juan Bautista, Filial de Ica, Ica, 11004, Peru*
[3] *Department of Pharmaceutical Chemistry, School of Pharmacy, University of Kansas, Lawrence, KS, USA*

## Objectives

- To introduce the reader to some basic principles of oral pharmacokinetics
- To provide background knowledge of the pharmacokinetic parameters involved after oral administration.
- To explain the different physicochemical and physiological factors that affect oral disposition of drugs.

## 15.1 Introduction

Pharmacokinetics is a discipline that studies the temporal and spatial disposition of a drug in a system. The discipline mainly involves the movement of drugs throughout the body and can also be applied to vitamins, nutraceuticals, hormones, toxins, and phytonutrients. However, throughout the chapter, the term drug will be utilized for consistency. Pharmacokinetics examines the rate and extent of the four basic processes involved in the time and spatial disposition of a drug in the body namely: Absorption, Distribution, Metabolism, and Elimination (collectively referred to as ADME).

Absorption (see also Chapters 4–6) is the process of a substance entering the blood stream; distribution (see also Chapter 7) is the dispersion or dissemination of substances throughout the fluids and tissues of the body. Metabolism (see also Chapters 11 and 12) is the irreversible transformation of substances into parent-derived metabolites, and excretion (see also Chapter 13) is the elimination of the substances from the body. Pharmacokinetics (PK) thus provides the quantitative relationship between administered doses (single dose or multiple dosing regimens) and the observed blood (or plasma or serum), urine, and/or tissue concentrations of the drug over time thus, both delineating and describing quantitative and qualitative information on the absorption, distribution, metabolism, and elimination of a particular drug (Figure 15.1).

An interrelated parallel discipline to pharmacokinetics is pharmacodynamics (PD), which can be defined as the quantitative relationship between the observed blood (or plasma or serum), urine, and/or tissue concentration of the active available drug at the effect site and the magnitude of the observed pharmacological effect(s) over time. Thus, pharmacodynamics deals with the time course of drug action, and it is intimately related to the pharmacokinetics of the drug. In summation, the discipline of pharmacokinetics explores what the body does to the drug (ADME), while pharmacodynamics explores what a drug does to the body (pharmacological effect).

### 15.1.1 Orders of Reaction

As described, pharmacokinetics deals with the temporal and spatial disposition of a drug in a system, which is equivalent to movement of a drug throughout the body. As a consequence of this movement, there is an initial concentration of drug entering the body and a certain concentration exiting the body over a discrete amount of time. Therefore, the amount of time it takes for that process to occur is dependent on the velocity and the initial concentration of drug that the body is exposed to. The velocity at which the process occurs is termed *rate*, and the way in which a drug concentration affects the overall rate of a pharmacokinetic process is called the *order of reaction*.

*Oral Bioavailability and Drug Delivery: From Basics to Advanced Concepts and Applications*, First Edition. Edited by Ming Hu and Xiaoling Li.

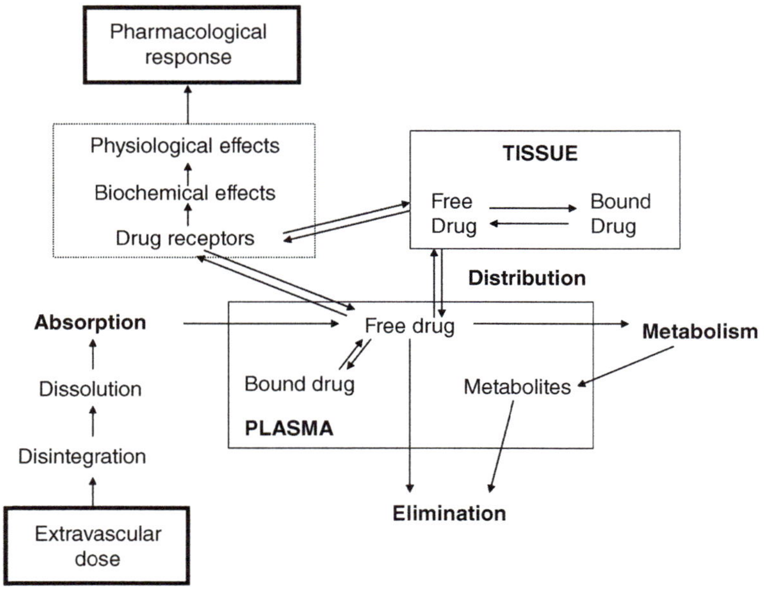


**Figure 15.1** Schematic representation showing the interrelation between absorption, distribution, metabolism, and elimination (ADME) kinetics as well as pharmacodynamics (pharmacological response).

There are two types of orders of reaction in pharmacokinetic (zero and first order), which are controlled by their dependency or absence of dependency on concentration.

#### 15.1.1.1 Zero-Order Reaction

In zero-order reactions, the elimination of drug proceeds at a constant rate and is independent of the concentration of drug. This is because the kinetic order of a reaction is determined by the exponent ($n$) of the rate equation: $\mathrm{d}C/\mathrm{d}t = k{\cdot}C^n$. Since in zero-order reaction, $n$ is equal to zero, the concentration term ($C^n = C^0$) becomes 1, and Eq. (15.1) can be derived.

$$\frac{\mathrm{d}C}{\mathrm{d}t} = -k_0 \tag{15.1}$$

*where*, $\mathrm{d}C/\mathrm{d}t$ represents the change in concentration over time. The rate (velocity) of the reaction is equal to $-k_0$ which is the zero-order rate constant.

It can be observed from Eq. (15.1) that the rate ($\mathrm{d}C/\mathrm{d}t$) is equal to a rate constant ($k_0$); thus, a zero-order reaction is independent of concentration, and the rate is constant throughout the reaction. Furthermore, the zero-order reaction can be integrated to yield a linear equation of the form: $y = m{\cdot}x + b$; in which $x$ and $y$ represent the variable for their corresponding axis, $m$ is the slope of the line and $b$ is the $y$-intercept. The integrated version of the equation is shown in Eq. (15.2).

$$C_t = C_0 - k_0 \cdot t \tag{15.2}$$

*where*, $C_t$ represents the concentration at any determined time, $C_0$ is the initial concentration at time zero, $k_0$ is the zero-order rate constant, and $t$ is time. Note that the units for $k_0$ are units of concentration divided by units of time (e.g. (μg/ml)/h, (ng/ml)/min or equivalent).

Based on Eq. (15.2), zero-order reactions will correspond to a straight line in rectilinear axis graph as shown in Figure 15.2, which indicates that concentration decreases linearly with time.

The necessary time to eliminate half of the initial concentration (half-life, $t^{1/2}$) for zero-order reactions can be calculated from Eq. (15.3), which is derived from the linear Eq. (15.2).

$$t^{1/2} = 0.5 \cdot \frac{C_0}{k_0} \tag{15.3}$$

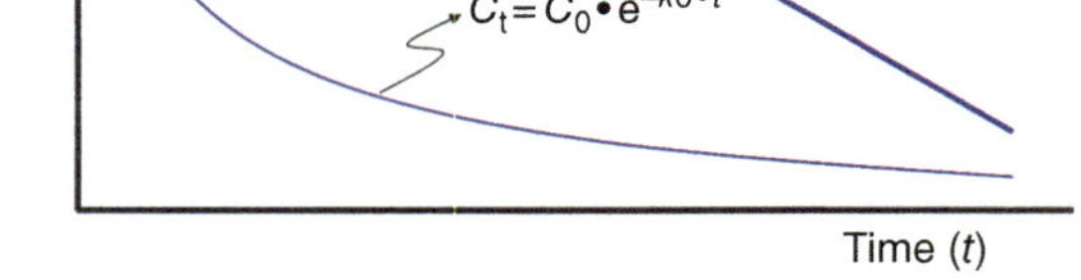


**Figure 15.2** Concentration–time profile of zero-order and first-order reactions in linear scale. Note that zero-order reaction can be mathematically represented as a linear line in the rectilinear graph and it is the reaction that follows the equation of the line.

*where*, $t^{1/2}$ represents half-life, $C_0$ is initial concentration, and $k_0$ is the zero-order rate constant.

From Eq. (15.3), it can be observed that half-life depends on initial concentration and increases with higher initial concentrations. Very few drugs such as ethanol and some drugs with modified release profiles undergo zero-order kinetics. Some processes that undergo zero-order kinetics are intravenous (IV) infusion since there is constant drug input into the body at a constant rate as an infusion (Eq. (15.1)). Another example of zero-order kinetics can be the disposition of drugs taken in overdose situations. An excessive amount of drug taken can overload metabolic processes because there is either too much drug present for the metabolic systems to handle, or the active transport mechanisms are overloaded since there is more drug than the carriers can process.

#### 15.1.1.2 First-Order Reaction

Contrary to zero-order kinetics, in first-order reactions the rate of elimination of drug depends on concentration. As described above, the kinetic order of a reaction is determined by the exponent ($n$) of the rate equation: $dC/dt = k{\cdot}C^n$. Since in first-order reaction, $n$ is equal to 1, the concentration term is present as part of the equations as shown in Eq. (15.4).

$$\frac{dC}{dt} = -k \cdot C \tag{15.4}$$

*where*, $k$ is the first order rate constant.

Based on Eq. (15.4), it can be observed that the rate of a first-order kinetic reaction depends on concentration and the rate changes throughout the reaction depending on concentration. Furthermore, drug concentration changes with respect to time as a function of drug concentration remaining and a first-order rate constant ($k$).

As shown in Figure 15.2, once the equation for first-order reaction is plotted in rectilinear coordinates it exhibits a curvilinear behavior and does not fit the equation of a linear line: $y = m{\cdot}x + b$ (Eq. (15.5)). Thus, in first-order reaction the concentration decreases exponentially with time.

$$C_t = C_0 \cdot e^{-k \cdot t} \tag{15.5}$$

*where*, $C_t$ represents the concentration at a determined time, $C_0$ is the initial concentration at time zero, $-k$ is the first-order rate constant, and $t$ is time. Note that the units for $k$ are the units of reciprocal time ($h^{-1}$, $min^{-1}$, or equivalent).

In order to fit Eq. (15.5) to the equation of the line, the exponential term needs to be transformed by applying the natural logarithm function (Eq. (15.6)) or the regular logarithm function (Eq. (15.7)).

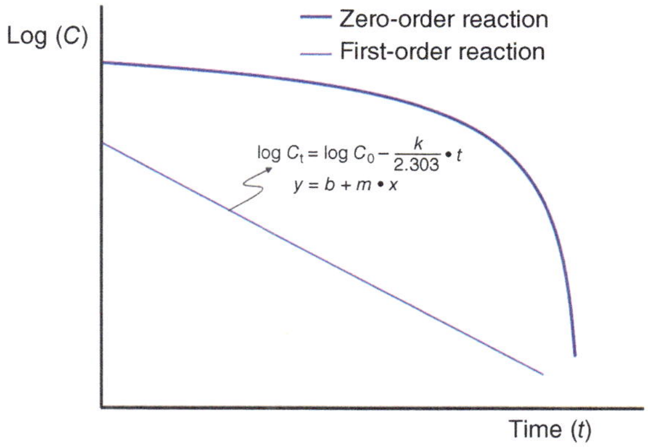


**Figure 15.3** Concentration–time profile of zero-order and first-order reactions in semi-logarithm graph. Note that first-order reaction can be mathematically explained as a line in the log-linear scale and it is the only process that follows the equation of linear line.

$$\ln C_t = \ln C_0 - k \cdot t \tag{15.6}$$

$$\log C_t = \log C_0 - \frac{k \cdot t}{2.303} \tag{15.7}$$

*where*, ln represents natural logarithm, log is logarithm to the base 10, $C_t$ is the concentration at a determined time, $C_0$ is the initial concentration at time zero, $-k$ is the first order rate constant, and $t$ is time.

The logarithm base 10 equation (Eq. (15.7)) will be used from now on for simplicity of further calculations. Based on the equation, first-order reactions will correspond to a straight line in a semi-log (linear-log) graph as shown in Figure 15.3.

Half-life ($t^{1/2}$) for first-order reaction can be calculated from Eq. (15.8), which is derived from the linear Eq. (15.7).

$$t^{1/2} = \frac{0.693}{k} \tag{15.8}$$

*where*, $t^{1/2}$ represents half-life and $k$ is the first-order rate constant.

From Eq. (15.8), it can be observed that half-life does not depend on initial concentration ($C_0$). The vast majority of drugs follow simple first-order kinetics, and first-order absorption is the most common process after oral administration, in which the absorption rate constant ($k_a$) can be determined by simple compartmental or non-compartmental modeling and analysis.

### 15.1.2 Pharmacokinetic Models

Pharmacokinetics (PK) utilizes mathematical models and equations to describe, understand, and predict the rate processes of absorption, distribution, metabolism, and elimination using concentration–time data obtained by experimentation. A PK model can be described as a mathematical approach to characterize the concentration–time profile of drugs; thus, characterizing the rate of the processes and the fate of a drug in a biological system after a single dose or after multiple dosing regimens. Various PK models can be applied to concentration–time data depending on the route of administration and on the dosage form. The different mathematical equations that characterize the PK of a drug are developed based on the drug's concentration–time profiles through specific pharmacokinetic parameters. These parameters can be estimated either by compartmental (model-dependent) or non-compartmental (model-independent) models.

#### 15.1.2.1 Compartmental Models

A pharmacokinetic compartment is not a real physiological or anatomic entity rather it is a tissue, organ, or set of any of anatomical compartment that have similar drug affinity, blood flow, rate of uptake or distribution, and subsequent clearance. Physiologically a biological system such as the human body is composed of a vast number of compartments and an administered drug will have access to multiple organs, tissues, cells, and organelles. Most commonly the compartments are described as the central and peripheral compartments; the central compartment refers to blood and rapidly perfused organs with blood, while the peripheral compartment refers to other organs and tissues such as muscle and fat.

Generally, the behavior of drug can be readily described using a one-compartment open model or a two-compartment open model. However, rarely depending on the drug's concentration–time profile a multi-compartment model might need to be employed. A rule of thumb is to first employ a one-compartment model to fit the experimental data and if adequate fitting is not obtained, then try fitting the data to a two-compartment model or a more sophisticated model. The concentration–time profiles of a drug are usually characterized from the two most readily accessible sampled body fluids: blood (or plasma or serum) and urine. Therefore, the concentration–time profile in blood (or plasma or serum) or the urinary cumulative excretion versus time curve can be utilized in compartmental models. Extravascular administration (e.g. oral administration) plasma concentration–time profile can be usually fitted to a one-compartment model. There are two main pharmacokinetic models: mammillary models and physiological based pharmacokinetic (PBPK) models.

***15.1.2.1.1 Mammillary Model*** A mammillary is the most common PK model in which one or more peripheral (tissue) compartments are joined to a central (sampling) compartment. In the case of extravascular administration, there is an extra process of absorption as described by the input absorption rate constant ($k_a$) (Figure 15.4, Model 1), the same situation applies for two-compartment models (Figure 15.4, Model 2) where intercompartmental rate constants are also present.

***15.1.2.1.2 Physiologically Based Pharmacokinetic (PBPK) Model*** PBPK is a special type of model where the physiology and anatomy of the animal or human body and the metabolic information of the drug are incorporated. It is also known as blood flow or perfusion model, and it is based on the physiological parameter that blood flow distributes drug to various organs as shown in Figure 15.5. PBPK modeling has become widely used in computer simulation programs as the conceptual model to allow dose, route, and interspecies

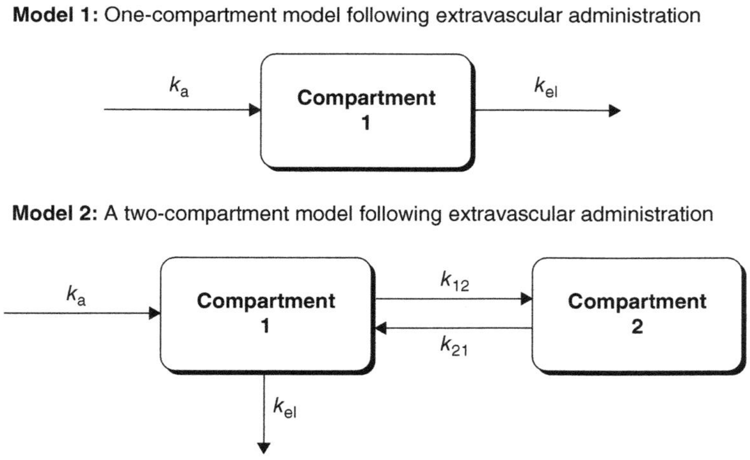


**Figure 15.4** Mammillary pharmacokinetic compartment models; *where*, compartment 1 represents the central (sampling) compartment, compartment 2 is the peripheral (tissue) compartment, $k_{el}$ is the elimination rate constant, $k_a$ is the absorption rate constant, $k_{12}$ and $k_{21}$ are the intercompartmental rate constants.

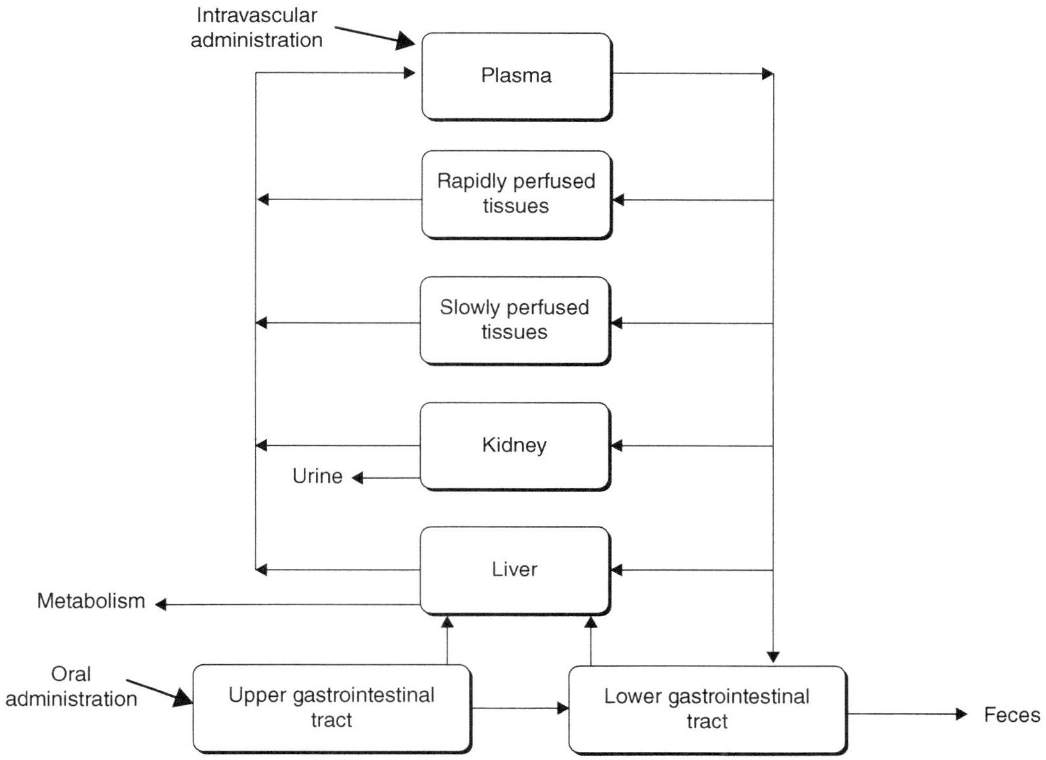


**Figure 15.5** Physiologically based pharmacokinetic (PBPK) model incorporating physiological compartments depending on the drug's distribution.

extrapolations. Since PBPK also allows for the incorporation of metabolic properties of the drug, metabolism can also be predicted. Therefore, this model allows not only for the estimation and prediction of absorption, distribution, and elimination (as the mammary model), but also for metabolism.

#### 15.1.2.2 Non-compartmental Model

As described above compartmental models require the assumption of a specific compartmental model and sometimes the modeling can be complex and time consuming. Therefore, non-compartmental analysis was developed based on the statistical model theory and is independent on the route of administration or compartmental theory restrictions. Furthermore, non-compartmental analysis can be applied to datasets that otherwise would be modeled using a compartmental model as long as linear pharmacokinetics is assumed (generally in the elimination phase).

### 15.1.3 Pharmacokinetics of Oral Administration

Oral administration requires that the drug enters into the bloodstream (systemic circulation) across the gastrointestinal barrier. This process is called absorption and it is only present in extravascular routes of administration such as oral administration. From solid oral dosage forms absorption consists of a number of sequential phases. First, the drug needs to be released from the dosage form and undergo dissolution, which will depend on physicochemical factors of the drug (Section 15.2) (e.g. dosage form, lipophilicity, and solubility), and subsequently on the environment of the body (physiological factors, Section 15.3).

The blood/plasma concentration–time profile after oral administration is more complex than intravenous administration, since the rate of change of drug concentration will be a function of absorption rate ($k_a$) and elimination rate ($k_{el}$) constants. These constants are generally first-order rate constants and provide the "speed" at which a drug enters the body (absorption rate constant) and is eliminated (elimination rate constant). A typical plasma concentration–time plot after oral administration is shown in Figure 15.6, it can be observed that the plasma concentration at time zero ($C_0$) is zero, followed by a first-order increase in plasma concentration (absorption phase) reaching a maximum concentration ($C_{max}$) at a determined time ($t_{max}$). A distribution phase ($V_d$) occurs but is not discernable in an oral concentration–time plot as it is

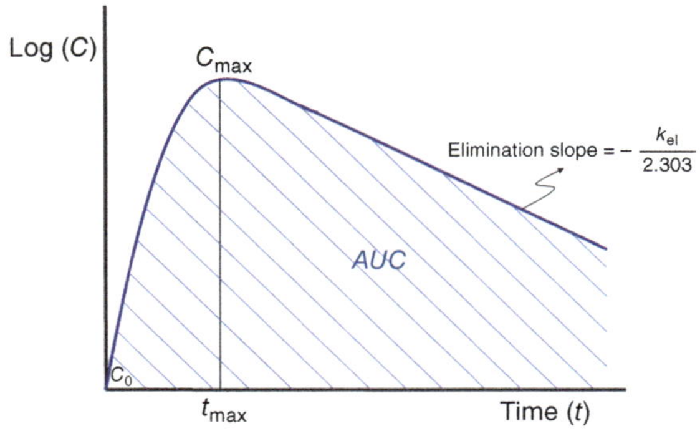


**Figure 15.6** Plasma concentration–time profile following oral administration.

**Table 15.1** Influence of changes in absorption ($k_a$) or elimination rate constants ($k_{el}$) on area under the curve ($AUC_{0-\infty}$), maximum concentration ($C_{max}$), and time to achieve maximum concentration ($t_{max}$).

| Change in rate constant | $AUC_{0-\infty}$ | $C_{max}$ | $t_{max}$ |
|---|---|---|---|
| Faster $k_a$<br>No change in $k_{el}$ | No change | Higher | Shorter |
| Slower $k_a$<br>No change in $k_{el}$ | No change | Lower | Longer |
| Faster $k_{el}$<br>No change in $k_a$ | Lower | Lower | Shorter |
| Slower $k_{el}$<br>No change in $k_a$ | Higher | Higher | Longer |

Source: Adapted from Ritschel and Kearns [1].

masked by bioavailability ($V_d/F$), while a first-order elimination phase is usually evident in the terminal concentration–time points. The area under the concentration–time curve (AUC) is a pharmacokinetic parameter that gives information on extent of absorption over time.

The calculation of AUC from time zero to infinity is a two-step process. The AUC from time zero to the time of the last measurable concentration ($AUC_{0-t}$) is often calculated by the trapezoidal rule or an integration program. Then, the AUC from the time of last measurable concentration to infinity ($AUC_{t-\infty}$) can be extrapolated following Eq. (15.9).

$$AUC_{t-\infty} = \frac{C_{last}\ ((\mu g/ml)\ h)}{K_{el}} \quad (15.9)$$

*where*, $C_{last}$ is the last measurable concentration and $k_{el}$ is the elimination rate constant. Thus, $AUC_{0-\infty}$ equals the sum of $AUC_{0-t}$ plus $AUC_{t-\infty}$.

However, it has to be noted that the concentrations of drug reaching the central compartment are dependent on the absorption from the gut. Once a drug enters the oral cavity, it passes through the gastrointestinal (GI) tract, and it may be absorbed in different segments of the GI tract depending on the surface area of the particular segments (i.e. small intestine has the largest surface area), residence time of the drug in a particular GI segment (i.e. gastric emptying and intestinal motility), which can be influenced by the input rate dosage form of the drug (e.g. enteric coating and sustained release), and the physiochemical properties of the drug ($pK_a$, solubility). Depending on the formulation of the drug (e.g. controlled release) significant changes in $C_{max}$ and $t_{max}$ can be observed (compared to an immediate release formulation) and the appearance of a lag time in plasma concentrations ($t_{lag}$) can also be observed. Therefore, it is important to note that different oral dosage forms (e.g. solution, capsule, and sustained released tablet), disease states (e.g. intestinal atrophy and Crohn's disease), and gastrointestinal contents (e.g. fasting, administration before, and during or after a meal) may alter the absorption rate constant ($k_a$) of a drug. The elimination rate constant ($k_{el}$) may also be altered due to a variety of disease states in particular renal or hepatic impairment. If the dose remains constant parameters $t_{max}$ and $C_{max}$ are dependent on the absorption and elimination rate constants ($k_a$ and $k_{el}$, respectively), while $AUC_{0-\infty}$ is dependent on the elimination rate constant ($k_{el}$) only as can be observed in Table 15.1.

Furthermore, in the case of an instant release formulation the elimination rate constant ($k_{el}$) can be estimated from the terminal slope of the extravascular concentration–time profile; however, in the case of certain drugs with slow rate of input and certain controlled-released formulation, the absorption process can be a rate-limiting step for entry into the systemic circulation. A phenomenon called "flip-flop kinetics" can occur, and it is characterized when the rate of absorption ($k_a$) is slower than the rate of elimination ($k_{el}$) (i.e. $k_a \ggg k_{el}$). Flip-flop kinetics can be easily characterized by examining the pharmacokinetics after intravenous administration, where a $k_a$ does not occur. After IV administration, the ascending portion of the pharmacokinetic plasma concentration–time profile is the elimination phase ($k_{el}$). In "flip-flop" kinetics the ascending portion of the plasma concentration–time profile is the absorption phase ($k_a$). An example of "flip-flop" kinetics is shown in Figure 15.7. It can be observed that the slope of the terminal phase of the extravascular administration is not parallel to the slope of the terminal phase of the intravenous administration. In order to identify the true elimination phase and absorption phase, the slope of the terminal phase of the intravenous administration will be used to calculate

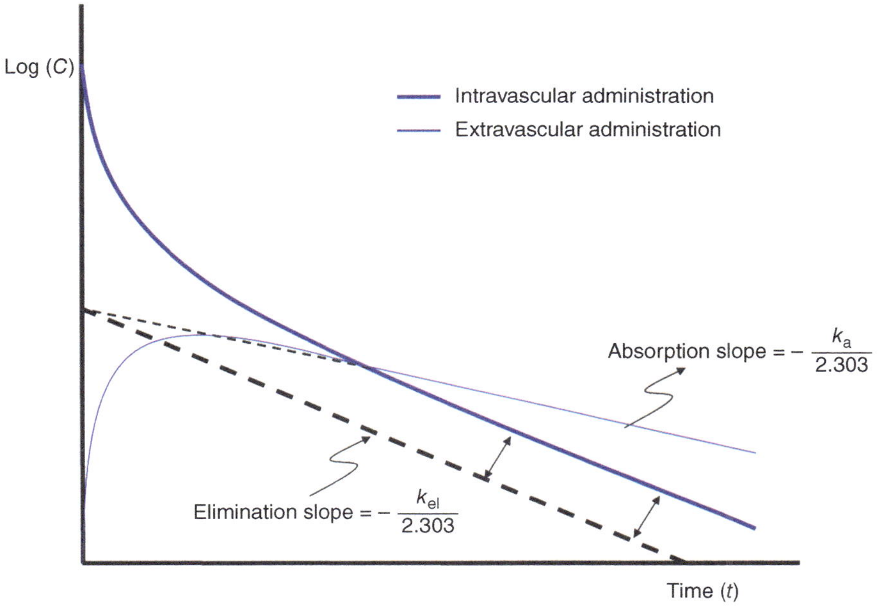


**Figure 15.7** Plasma concentration–time profile following an extravascular dose (oral administration) and an intravascular dose exhibiting flip-flop kinetics exhibiting absorption rate-limited elimination. It can be observed that the terminal portion of the extravascular curve is not parallel with the intravenous curve. The latter reflects the true elimination of the drug.

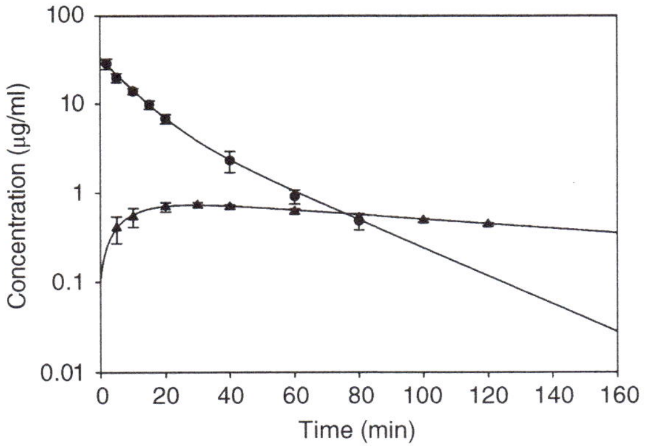


**Figure 15.8** Mean plasma concentrations (± SD, $n = 6$) of acamprosate after intravenous bolus and oral administration of 9.3 mg/kg. Source: Zornoza et al. [2]/with permission of John Wiley & Sons.

the elimination rate constant ($k_{el}$). The absorption rate constant ($k_a$) will be obtained from the slope of the terminal phase of the extravascular administration.

This "flip-flop" phenomenon has been observed with multiple drugs; for instance, Zornoza et al. [2] observed it with acamprosate. Acamprosate (calcium bis acetyl-homotaurine) is a drug used to prevent relapse in patients with alcohol (ethanol)-dependence. It can be observed in Figure 15.8 that acamprosate after intravenous bolus and oral administration of 9.3 mg/kg follows flip-flop kinetics since the terminal slope in the oral curve is lower than the slope observed in the intravenous curve [2]. This indicates that the real rate-limiting step is related to acamprosate absorption.

Another important factor in oral concentration–time profiles is bioavailability ($F$), which is defined as the fraction of the dose that reaches the systemic circulation intact. This value ranges from 0% to 1% or 0% to 100%. Bioavailability can be calculated as described in Eq. (15.10).

$$F = \frac{\mathrm{AUC}_{\mathrm{po}}}{\mathrm{AUC}_{iv}} \times \frac{\mathrm{Dose}_{iv}}{\mathrm{Dose}_{\mathrm{po}}} \tag{15.10}$$

Bioavailability of a drug is influenced by multiple factors. First of all, a drug administered orally has to be released from the dosage form and undergo dissolution and then be absorbed into the systemic circulation having to pass through various membranes. It is possible that a drug is metabolized by first-pass organs (i.e. gut, liver, and lung), which can reduce the concentration of the parent compound reaching the systemic circulation, hence reducing bioavailability of the parent compound. Secondly, depending on the different dosage forms administered varying amounts of the parent drug can be delivered and absorbed; thus, differences in onset, duration, and intensity of the pharmacological effect are possible.

Therefore, it can be observed that absorption plays an extremely important role in the oral administration of drugs and that multiple factors can affect oral drug absorption. These factors will contribute to variability in oral plasma concentration–time profiles, absorption and elimination rates, bioavailability, and ultimately pharmacological effect. The salient factors affecting oral concentration–time profiles can be subdivided into two groups: physicochemical and physiological. The intent of this chapter is to provide a comprehensive, rather than exhaustive, appraisal of pharmacokinetic behaviors of orally administered drugs. Illustrative examples are selected from drugs currently in clinical use, as well as drugs currently being developed. An effort has been made to concentrate on the biomedical literature since 1980, although reference is made to earlier publications when appropriate to be illustrative or for historical significance to the development of the field.

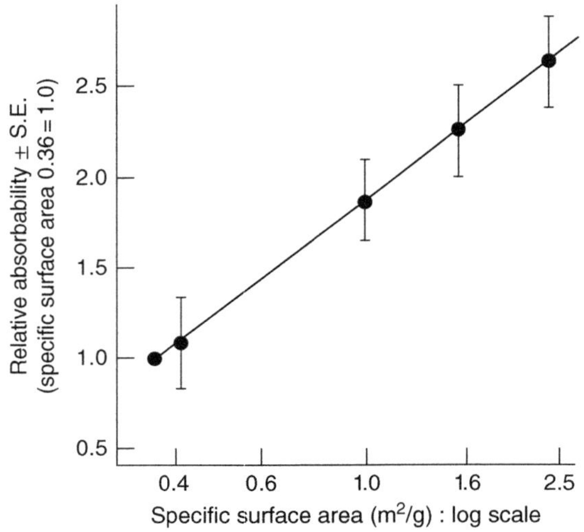


**Figure 15.9** Regression of relative absorbability of griseofulvin on log-specific surface area. Source: Atkinson et al. [4]/with permission of Springer Nature.

## 15.2 Physicochemical Factors Affecting Oral Concentration Time Profiles

The Biopharmaceutics Classification System (BCS) classifies drugs absorption characteristics according to solubility and permeability. According to the BCS, drug substances are classified into four groups as follows:

Class I – High Permeability, High Solubility
Class II – High Permeability, Low Solubility
Class III – Low Permeability, High Solubility
Class IV – Low Permeability, Low Solubility

Class I compounds are generally very well absorbed, while Class II compounds exhibit dissolution-rate limited absorption, Class III compounds exhibit permeability-rate limited absorption, while Class IV compounds have very poor oral bioavailability [3].

### 15.2.1 Dissolution Rate Limited Absorption

Various xenobiotics, which possess unique pharmacological properties including very poorly water-soluble drugs, have been made bioavailable in a manner that has made it possible to turn them from scientifically interesting molecules into commercially successful formulated drugs. The pharmaceutical scientific foundation of our understanding of the effect of reducing particle size on bioavailability essentially began with griseofulvin in 1962 [4] (Figure 15.9).

In this classical example of the Class II compound griseofulvin, bioavailability is clearly related in a linear manner with reduced particle size and thereby a corresponding increase in the effective surface area of the particles. It has been known in the pharmaceutical art for over 40 years that particle size may significantly affect the oral absorption of poorly soluble compounds. Small particles with large surface area provide a higher peak plasma concentration and bioavailability and reduced time to peak plasma concentration. Subsequently, dissolution studies developed as an *in vitro* prognostic tool for *in vivo* drug absorption. In particular, the extent of dissolution may often be the rate-limiting step to absorption, particularly for poorly water-soluble compounds [5].

In modern pharmaceutical formulation development, it is clear that the pharmaceutical industry considers many different options and invests considerable effort and expense in order to develop solutions to the issues surrounding formulation and particularly the problem of limited bioavailability of poorly soluble drugs. There are significant "real-life" difficulties resulting from attempts to make poorly soluble pharmaceutical compound bioavailable.

In cases of poorly soluble pharmaceutical compounds, particle size reduction increases surface area and a change in dissolution and bioavailability is expected.

However, for the vast majority of important poorly water-soluble compounds marketed throughout the world, alternative approaches and strategies have been considered and formulations have been developed without incurring to particle size changes to solve the oral bioavailability issue [6].

The bioavailability of a drug formulation generally depends on two main factors: solubility in water and absorption/permeability. Reducing particle size increases the particles' surface area, which will increase the extent of dissolution according to the Noyes Whitney Eq. (15.11). The modified Noyes–Whitney equation relates the dissolution of particles to the properties of the solid and the media of dissolution:

$$\frac{dW}{dt} = \frac{DA(Cs - C)}{L} \tag{15.11}$$

*where,*

$dW/dt$ is the rate of dissolution.
$A$ is the surface area of the solid.
$C$ is the concentration of the solid in the bulk dissolution medium.
$C_s$ is the concentration of the solid in the diffusion layer surrounding the solid.
$D$ is the diffusion coefficient.
$L$ is the diffusion layer thickness.

The particle size of raw material to be used in drug formulation has been shown to be an important determinant of the dissolution rate and bioavailability of several drugs. When the particle size is made smaller, the surface area of the drug, which comes into contact with solvent is increased, and this leads to faster dissolution rate; and examples of such a drug is griseofulvin. Another classical example of this type of drug is digoxin. Digoxin bioavailability intimately depends on the particle surface area. Table 15.2 of Jounela et al. [7] demonstrated differences in dissolution disintegration and particle size of three varieties of digoxin tablets. Tablets of different particle sizes were used between 7 and 102 μm with the rate and extent of dissolution is different for the three formulations and product B failed the dissolution test at 18% dissolution after 60 minutes, and there was a 13% difference in extent of dissolution of A and C.

These *in vitro* dissolution findings are reflected in the oral concentration time pharmacokinetic profiles in Figure 15.10 with significant differences apparent in oral bioavailability occurring with formulations that have large particle size and poor dissolution.

### 15.2.2 Solubility-Limited Absorption

Oral bioavailability is influenced by both solubility and stability of drug, which can be pH dependent. Variation in pH within the gastrointestinal tract can have multiple etiologies. Changes in gastric acidity can result in alterations in the oral concentration–time profiles of drugs with pH-sensitive dissolution. Drugs may demonstrate pH-dependent solubility characteristics. As such drugs may behave as several discrete fractions with the GI because of the variation in pH values found in the acidic stomach to the distal more alkaline intestine.

Discontinuous absorption of drugs has frequently been demonstrated in the early absorption phase following oral administration to human and animals. For example, the solubility of allopurinol is known to be low at acidic pH and high at an alkaline pH. The slow increase in plasma concentration during the early phase of the absorption process, observed in some studies, appears to be due to the slow solubility of the drug in the stomach. Therefore, different dissolution characteristics may account for the two-phase absorption profile of allopurinol and need to be considered to be for the analysis of pharmacokinetic analysis of plasma allopurinol concentration data. After a period of first-order gastric emptying, almost all of the remaining drug in the stomach empties by gastric peristalsis at the same time to the small intestine absorption site where the pH value is more optimal for dissolution of the drug. These two sequential processes can result in a rapid increase in plasma concentration in the latter phase of the absorption. This observation suggests that the low solubility of a drug and its gastric emptying behavior can be the major factors governing its discontinuous absorption process and resultant oral concentration–time profile behavior.

Typical plasma-concentration data from dogs after oral administration of a 100 mg tablet of allopurinol could not be fitted to a conventional compartment model but could be well-fitted to a two-fraction absorption model. In situations of pH-dependent solubility,

**Table 15.2** Particle size, disintegration time and dissolution rate of three varieties of digoxin tablets.

| Tablet | Particle size, mean diameter μ (90% confidence limits) | Disintegration time, min | Dissolution rate, percentage in solution at 15 min | 30 min | 60 min |
|---|---|---|---|---|---|
| A | 13 (5–30) | 5.5 | 58 | 59 | 71 |
| B | 102 (80–150) | 4.5 | 12 | 18 | 18 |
| C | 7 (3–10) | 5.5 | 49 | 55 | 62 |

Source: Adapted from Jounela et al. [7]/with permission of Springer Nature.

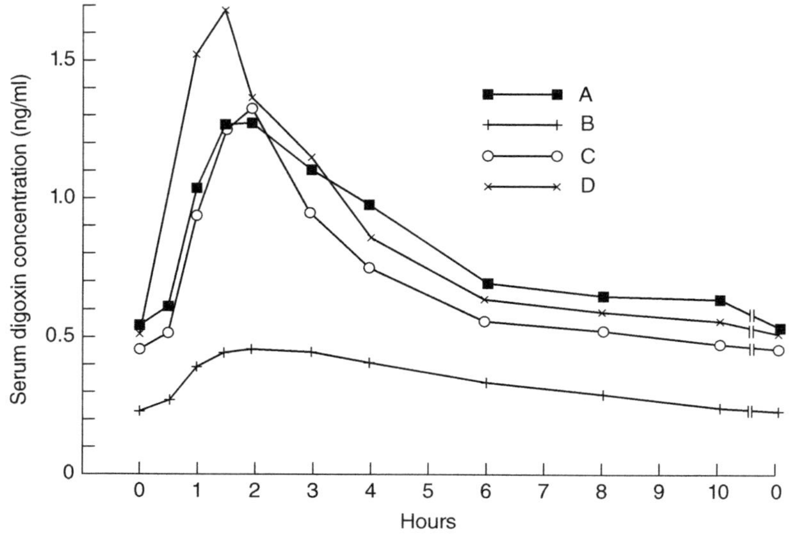


**Figure 15.10** Steady-state serum digoxin concentration after administration of 0.25 mg of four varieties of digoxin products; mean values from seven subjects. Source: Jounela et al. [7]/with permission of Springer Nature.

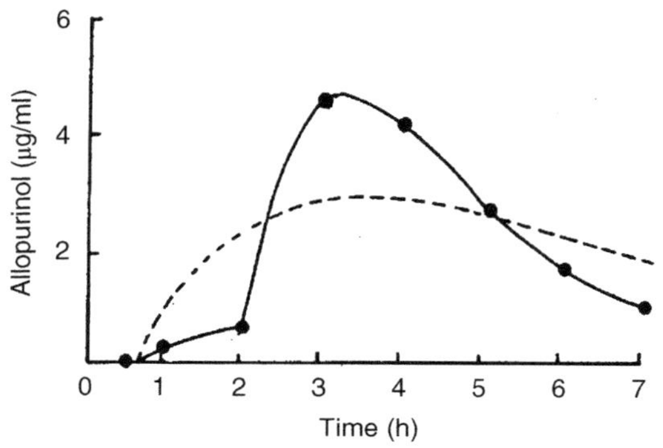


**Figure 15.11** Nonlinear least squares regression curve for plasma allopurinol concentration data (regular tablet) in dogs (100-mg dose). Key: (- - -) conventional one-compartment model (–) two-fraction absorption model. Source: Murata et al. [8]/with permission of Elsevier.

drug can be subdivided into two fractions in the GI tract. These discrete fractions may have very different rate constants with different input rates hence the appearance of small concentrations of drug giving the appearance of a lag time and then subsequent large increase in plasma concentrations, respectively [8] (Figure 15.11).

Weakly basic drugs in particular may have a slower dissolution at a higher pH as drugs will be primarily dissociated into the unionized form. Other weak bases ketoconazole and dipyridamole have been used as model drugs to study pH-dependent absorption. In a canine model of drug absorption, gastric pH was controlled in order to predict formulation performance. Adequate gastric acidity is a prerequisite for sufficient dissolution and thus oral absorption of basic compounds. Studies were performed including no treatment (pH 0.9–8.3), pentagastrin treatment to stimulate gastric acid secretion (pH 2–3), and famotidine treatment to block gastric acid secretion (pH 5–7.5). Marketed differences in systemic ketoconazole and dipyridamole exposure with altered pH were apparent. An increase in $C_{max}$ and AUC in the pentagastrin treatment as compared to control was apparent. Famotidine treatment that reduced gastric acidity did not significantly affect $C_{max}$ and AUC [9] (Figure 15.12).

#### 15.2.2.1 Modified Release Formulation

A conventional regular-release dosage form is designed to rapidly release the dose of drug following administration and drug concentrations rise up to a peak and decline. Conventional dosage forms do not maintain a constant plasma concentration of drug in systemic circulation over time. Modified release preparations such as delayed release, repeat action prolonged release sustained release and extended release, controlled release, and modified release dosage forms have been developed in attempts to maintain more constant plasma concentrations, avoid excessive fluctuations in maximum and minimum concentrations, and increase patient compliance through less frequent administration and ultimately increase effectiveness of the drug (Figure 15.13). These delivery systems impart some degree of input rate control into the dosage form. The input rate of absorption is controlled by formulation as long as a release from the formulation is the rate-limiting step in the absorption process. Zero-order absorption can occur when a zero-order controlled release delivery system is used. Pharmacokinetic models assuming zero-order absorption assume that drug residing in the GI tract is absorbed at a constant rate $k_0$

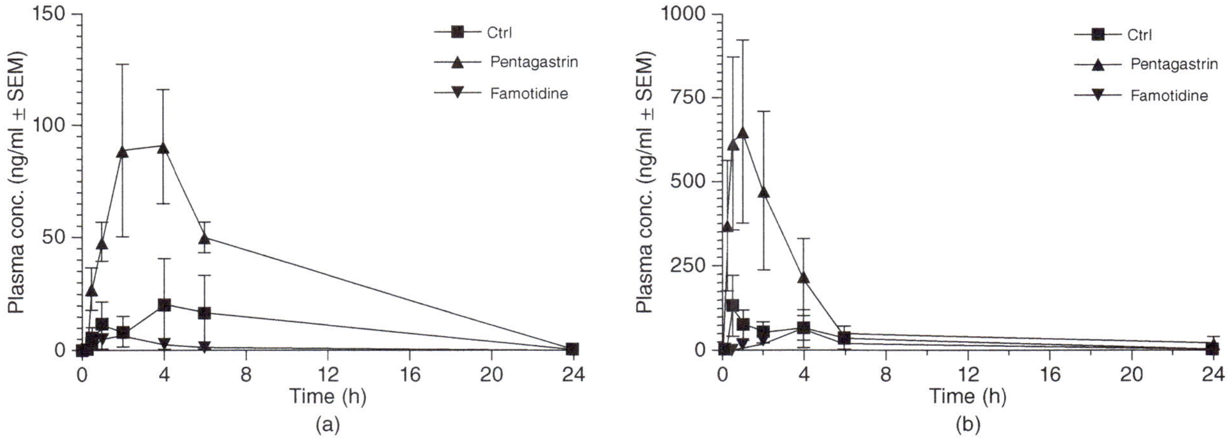


**Figure 15.12** Ketoconazole (a) and dipyridamole (b) plasma profile in dogs, pH-dependent absorption. Results are control (no treatment), pentagastrin, and famotidine-treated dogs shown as mean concentration (ng/ml, ± SEM, $N = 4$). Source: Zhou et al. [9]/with permission of Springer Nature.

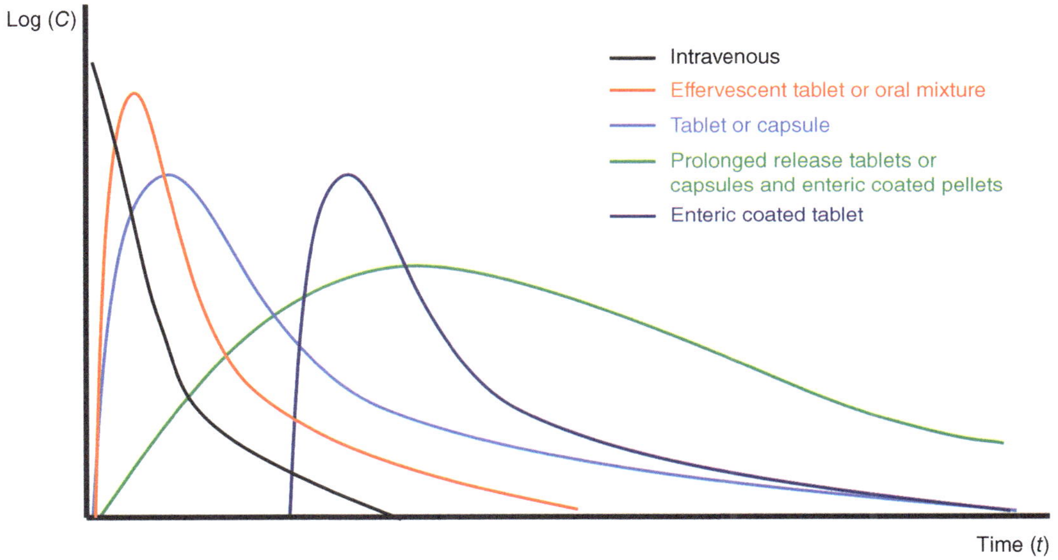


**Figure 15.13** Theoretical plasma concentration–time curves of model formulations.

and is constant until the amount remaining in the gut is depleted.

There are five oral formulations of valproic acid (VPA)/divalproex sodium currently approved in the United States. These formulations have unique formulation characteristics and input rates that provide distinct oral concentration–time profiles due to marked absorption differences. The comparable area under the curve across the formulations demonstrates that the formulation affects drug release and thus *in vivo* absorption. The VPA syrup demonstrated rapid absorption with the highest $C_{max}$ and shortest $t_{max}$. The extended-release tablet was designed to deliver drug at a zero-order rate over more than 18 hours and had the slowest absorption with the lowest $C_{max}$ and longest $t_{max}$ [10] (Figure 15.14).

Sustained-release preparations can be designed to have both a fast and slow-release component. These components may exhibit different behavior in the GI tract after oral absorption and their input into the GI tract and subsequent absorption vary. The mean plasma concentration of pindolol after oral administration of a sustained release dosage form, to eight healthy subjects was analyzed using both a two and a three-fraction absorption model. These models assume one compartment with two or three first-order absorption

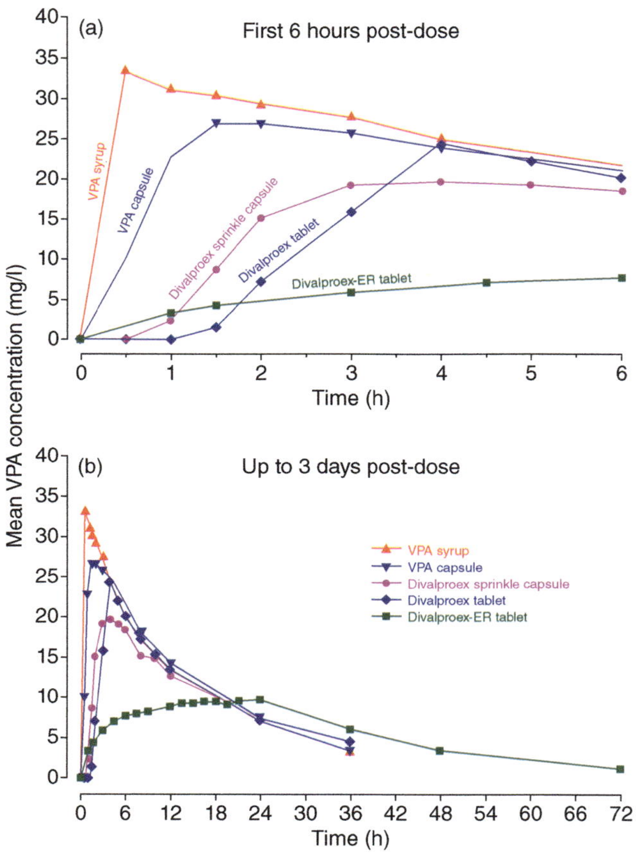


**Figure 15.14** Comparison of mean plasma VPA concentration–time profiles during absorption phase (panel a) and over 72 hours (panel b) for five VPA/divalproex formulations following a 250 mg single dose. Source: Reproduced from Dutta and Reed [10]/with permission of Elsevier.

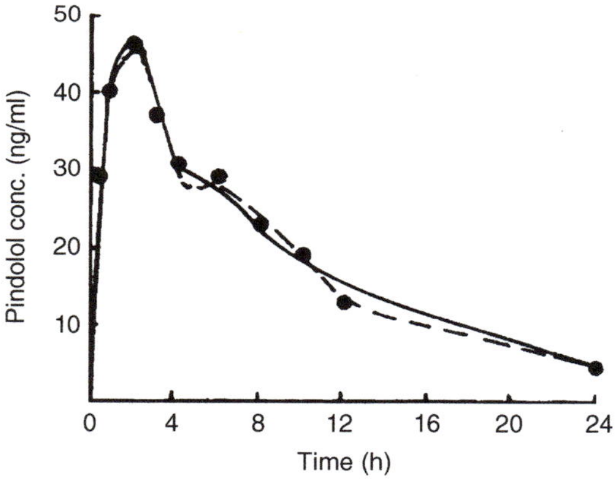


**Figure 15.15** Nonlinear least squares regression curve for mean pindolol concentration (sustained release tablet) in humans (30 mg-dose). Key: (–) two fraction absorption model (– – –) three fraction absorption model. Source: Bresolle et al. [11]/with permission of Elsevier.

processes from two or three discrete fractions of drug, one first-order metabolic processes, and one first-order urinary excretion process. Good simulation curves were obtained with both models (Figure 15.15) [11].

The absorption of sustained-release oral sodium valproate demonstrates a double-peaking phenomenon in concentration–time profiles. The complexity of the input process lead to the data subsequently fitted to a double Weibull function a continuous probability distribution with a probability density function and then first-order elimination was used. In this case, the Weibull model was used as an approximation of the overall input process and fitted the data more successfully than a single $k_a$ or $k_0$ input [11].

#### 15.2.2.2 Nanocrystals

Oral bioavailability of poorly water-soluble drugs may be severely limited by slow dissolution due to extremely high crystal energy and hydrophobicity. For compounds with large oil-water partitions and high melting points, nanocrystals can overcome these limitations by providing the compound in a crystalline size that is pharmaceutically acceptable, typically the mean particle size is <1 μm and the 99% of particles are <5 μm [12]. Numerous methods exist for the production of nanocrystals including homogenization [13, 14], milling [15], solvent evaporation [16], and supercritical $CO_2$ precipitation [17]. Several nanocrystal drugs are

**Table 15.3** Pharmacokinetic parameters following oral administration of danazol nanocrystals, cyclodextrin inclusion complexes, and a convention formulation to beagles ($n$ = 5, 20 mg/kg dose).

| Formulation | $C_{max}$ (μg/ml) | $t_{max}$ (h) | AUC (μg h/ml) | Absolute bioavailability |
|---|---|---|---|---|
| Nanocrystals | 3.01 ± 0.80 | 1.5 ± 0.3 | 16.5 ± 3.2 | 82.3 ± 10.1 |
| Cyclodextrin-complexes | 3.94 ± 0.14 | 1.2 ± 0.2 | 20.4 ± 1.9 | 106.7 ± 12.3 |
| Conventional suspension | 0.2 ± 0.06 | 1.7 ± 0.4 | 1.0 ± 0.4 | 5.1 ± 1.9 |
| Cyclodextrin (IV) | | | 19.8 ± 0.6 | 100 |

Source: Adapted from Liversidge and Cundy [18].

currently in the market including Wyeth Pharmaceuticals' Rapamune® formulation of sirolimus and Merck's Emend® formulation of aprepitant.

A report by Liversidge and Cundy [18] demonstrates that a nanocrystal suspension can dramatically increase the oral bioavailability of danazol compared to conventional formulation in polyvinylpyrrolidone K-15 (Table 15.3). Pharmacokinetic parameters of the nanocrystalline danazol were statistically the same as cyclodextrin formulations, suggesting that nanocrystalization overcame rate limitations on absorption due to dissolution.

Nanocrystalline tablet sirolimus was compared to an oral solution in 477 patients with acute renal allograph rejection [19]. The oral solution exhibited a greater whole blood $C_{max}$ and lower $t_{max}$ than the nanocrystal formulation, but the $AUC_{0-24h}$ and CL/F were identical (Table 15.4). Overall, the nanocrystalline sirolimus was therapeutically equivalent to the oral solution, with the added benefits of improved stability at room temperature and better palatability.

#### 15.2.2.3 Cyclodextrins

Cyclodextrins can improve the solubility, stability, and absorption of drug formulations, and currently worldwide over 30 marketed pharmaceutics contain cyclodextrins [20]. Cyclodextrins are composed of cyclic α-D-glucopyranosides (6, 7, and 8 units in α, β, and γ-cyclodextrin, respectively) that form a hydrophilic shell with a hydrophobic interior in which small hydrophobic drug molecules (typically <500 Da and $\log P_{o/w} > 2$) can form a host-guest inclusion complex. The ability of cyclodextrins to improve oral bioavailability has been reviewed numerous times [20–27]. Table 15.5 provides a representative list of drug oral bioavailability in oral formulations with and without cyclodextrins.

#### 15.2.2.4 Complexation: Formation of Poorly Absorbable Bile Salt Micelles

There are several examples of drugs that form poorly absorbable bile salt micelles in the small intestine that lead to multiple peaking phenomena in oral concentration–time profile including several beta-blockers. Pafenolol is a beta-blocker that demonstrated discontinuous oral absorption properties in animals and men. The plasma concentration–time profile of pafenolol exhibits two distinct peaks (Figure 15.16), with a truncated and dose-dependent bioavailability suggesting incomplete intestinal uptake and absorption [47].

One mechanistic explanation for the appearance of multiple absorption peaks involves the formation of a poorly absorbed micelle complex of the drug with bile salts. Complex formation between pafenolol and the bile acids in the lumen thereby

**Table 15.4** Pharmacokinetics parameters of oral solution and nanocrystalline tablet sirolimus in patients following 2 mg/day dosing.

| Formulation | Study day | $AUC_{0-24h}$ (ng h/ml) | $C_{max}$ (ng/ml) | $t_{max}$ (h) |
|---|---|---|---|---|
| Nanocrystalline tablet | 90 ($n$ = 10) | 232 ± 99 | 14.2 ± 5.0 | 4.60 ± 4.20 |
| | 30 ($n$ = 13) | 208 ± 73 | 13.6 ± 5.9 | 2.92 ± 1.71 |
| | 1 ($n$ = 12) | 130 ± 50 | 12.6 ± 6.2 | 5.67 ± 6.37 |
| Solution | 90 ($n$ = 12) | 178 ± 61 | 12.9 ± 4.3 | 2.67 ± 1.23 |
| | 30 ($n$ = 16) | 207 ± 110 | 16.0 ± 7.9 | 1.75 ± 1.00 |
| | 1 ($n$ = 13) | 197 ± 127 | 29.3 ± 13.1 | 2.15 ± 1.41 |

Source: Adapted from Mathew et al. [19].

**Table 15.5** Comparison of oral formulations with and without cyclodextrin.

| Drugs (species) | Water solubility (mg/ml) | Dosed CD: drug (molar) | CD type | Comparison formulation | AUC increase (× AUC control) | References |
|---|---|---|---|---|---|---|
| Acyclovir (rat) | 1.3 | 1 | β-CD | Suspension | 1.2 (total) | [28] |
| Albendazole (sheep) | 0.000 16 | 35.4 | HP-β-CD | Suspension | 1.4 (0–∞) | [29] |
| Artemisinin (human) | 0.084 | 1 | β-CD | Tablet | 1.7 (0–∞) | [30] |
| Carbamazepine (rat) | 0.5 | 1 | HP-β-CD | Powder | 2.2 (0–12 h) | [31] |
| Carmofur (rabbit) | 0.18 (pH 2.5) | 1 | DM-β-CD | Powder | 7.5 | [32] |
| Cinnarizine (dog) | 0.0023 (pH 4.5) | 2 | SBE-4-CD | Capsule | 7.8 (0–∞) | [28] |
| Diazepam (human) | 0.0057 | 1.5 | β-CD | Powder | 1.4 (0–4 h) | [33] |
| Digoxin (dog) | 0.0034 | 4 | γ-CD | Tablet | 5.4 (0–24 h) | [34] |
| Dipyridamole (dog) | 0.008 | 1 | β-CD | Tablet | 1.5 | [33] |
| Fluoxetine (human) | 3.46 (pH 1–10) | — | γ-CD | Water solution | 2.5 (0–∞) | [35] |
| Flurbiprofen (human) | 0.0313 | 1 | HP-β-CD | Drug | Increase | [36] |
| Glibenclamide (dog) | 0.00618 | 44 | β-CD | Capsule | 4.9 (absolute bioavailability) | [37] |
| Ibuprofen (rabbit) | 0.47 | 1 | β-CD | Powder | 1.9 (% recovery in urine) | [38] |
| Ketoprofen (dog) | 0.133 | 1 | β-CD | Powder | 1.5 (% recovery in urine) | [38] |
| Nifedipine (rabbit) | 0.0054 | 1 | β-CD | Capsule | 2.6 (0–10 h) | [39] |
| Piroxicam (rabbit) | 0.0036 | 2.5 | DM-β-CD | Powder | 2.2 (0–∞) | [40] |
| Prednisolone (human) | 0.25 | 2 | β-CD | Tablet | 1.2 (0–8 h) | [41] |
| Rutin (dog) | 0.045 | 1 | HP-β-CD | Tablet | 2.9 (0–∞) | [42] |
| Salbutamol (dog) | High | — | TB-β-CD | Gelatin capsule | 4.6 (0–24 h) | [43] |
| Spironolactone (human) | 0.0242 | 1.6 | HP-β-CD | capsule | 3.6 (0–24 h) | [44] |
| Tacrolimus (rat) | <0.005 | 68.8 | DM-β-CD | Capsule | 4.5 (0–12 h) | [45] |
| Tolbutamide (rabbit) | 0.1053 | 979.6 | β-CD | suspension | 1.4 (0–∞) | [46] |

Source: Carrier et al. [21]/with permission of Elsevier.

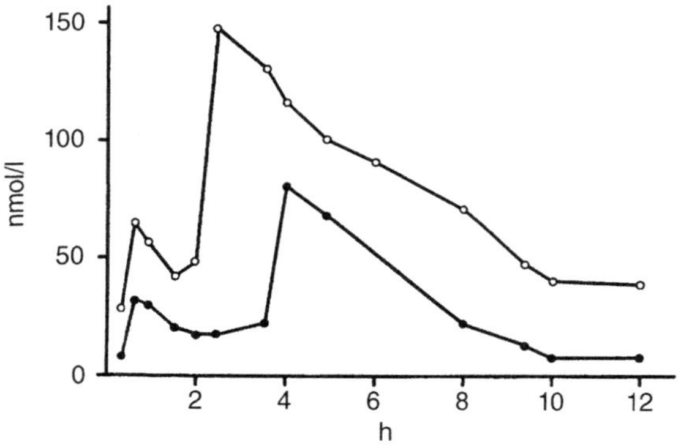


**Figure 15.16** Plasma concentration–time profile in two subjects following single oral administration of 25 mg (filled circle) or 50 mg (open circle) of pafenolol. Source: Reproduced from Lennernas and Regardh [47]/with permission of Springer Nature.

prevents intestinal uptake and accounts for the low, dose-dependent, and variable absorption and the double peaks that occur in the plasma concentration–time profile of pafenolol. Evidence in rats suggesting this double-peaking phenomenon of pafenolol was not present in bile-depleted rats [48–52]. Erratic fluctuation in plasma pharmacokinetics of acebutolol and nadolol observed after oral administration in human subjects is also apparent. The first peak concentration of the drug in plasma is suggested to be due to rapid micelle formation in the small intestine with cessation of absorption. Subsequently, dissociation of the micelle in the ileum due to active reabsorption of bile salts results in a secondary peak in the plasma concentration profile of the drug. Food-related reduced absorption of drugs that form bile micelle drug complexes may result in increased formation due to the stimulation of bile acid secretion by a meal [53–57].

In summary, there are several examples of drugs that can make unabsorbable micelles with bile salts. The cessation of absorption after oral administration occurs rapidly and is consistent with micelle formation in the proximal part of the small intestine, due to high concentrations of bile acids in the ileum, In the distal ileum dissociation of micelles occurs (no uptake of intact micelles takes place) and the bile acids are actively reabsorbed, followed by enhanced intestinal absorption of free drug and a second peak in the plasma concentration profile occurs.

## 15.3 Physiological Factors Affecting Oral Concentration Time Profiles

### 15.3.1 Drug Metabolism

After oral administered drugs are often converted to metabolites. In addition, sometimes the drug is a prodrug and is dependent on metabolism for activation. A pharmacokinetic model involving metabolism of a parent compound to a metabolite is presented in Figure 15.17.

The rate of change of the amount of metabolite in the body is the sum of the rate of formation KF and the rate of elimination $K(m)$. The amount of a metabolite is dependent on formation and elimination. $dA(m)/dt = (KF * A) - (K(m) * A(m))$. There are two scenarios either formation or elimination is the rate-limiting step. The rate-limiting step will have a smaller rate constant. In the case of a formation-rate-limited drug. $KF < K(m)$, the slowest step is the formation of the metabolite (i.e. elimination of drug). This is the usual case for most polar metabolites. The metabolite declines in parallel and with the same half-life as the parent drug.

In the case of an elimination-rate limited drug $KF > K(m)$, the slowest step is elimination of metabolite once it is formed. This is the unusual case and metabolite accumulates and metabolite elimination is rate-limiting step and decline is dependent on its $K(m)$. The half-life of a metabolite is therefore greater than the parent drug ($t½$ M > $t½$ drug) (Figures 15.18 and 15.19). In any metabolic sequence, substances formed beyond the rate-limiting step (*smallest rate constant*) decline with the $t½$ of the slowest step.

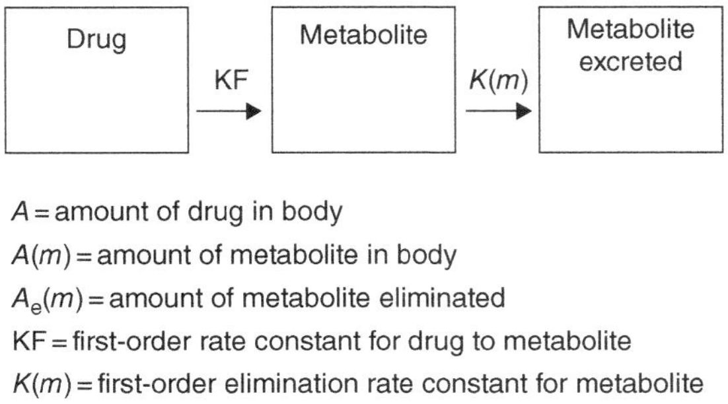


**Figure 15.17** Compartmental model of drug metabolism.

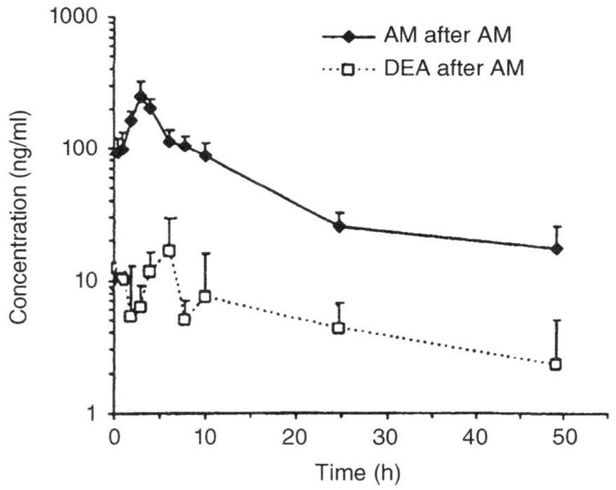


**Figure 15.18** Plasma concentration versus time curves of amiodarone (AM) and desmethylamiodoarone (DEA) (mean ± SD) following the oral administrations of 10 mg/kg amiodarone HCl to Sprague–Dawley rats. Source: Reproduced from Shayeganpour et al. [58]/with permission of John Wiley & Sons.

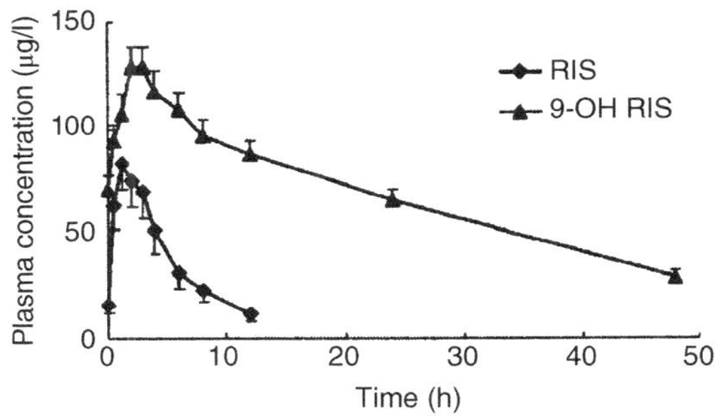


**Figure 15.19** Mean steady-state plasma-concentration–time curves of RIS (risperidone) and 9-OH RIS. $N = 23$ Mean ± SEM. Source: Zhou et al. [220]/with permission of Springer Nature.

### 15.3.2 Enterohepatic Recycling

Enterohepatic recycling is a physiological process involving the circulation of bile and bile acids produced in the liver to the small intestine where they are subsequently reabsorbed back into the liver. Bile acids and some drugs are transported to the gall bladder for temporary storage and subsequently move to the duodenum via the common bile duct after a meal during the process of digestion. Bile acids that are delivered to the duodenum are extensively recycled by the enterohepatic circulation as are certain drugs.

In the lower part of the small intestine (ileum), conjugated drugs and bile salts are reabsorbed into the hepatic portal circulation. Bacteria present in the small intestine can deconjugate certain xenobiotics back to lipid-soluble parent compounds. Conjugates of glucuronic acid and sulfates are hydrolyzed by bacterial flora and the aglycone form may be passively absorbed into the hepatic portal circulation in the distal intestine. The effect of enterohepatic recirculation is that both bile salts and certain drugs remain in the enterohepatic circulation for a prolonged period of time as a consequence of this recycling process [59].

There are a variety of drugs such as many nonsteroidal anti-inflammatory drugs (i.e. indomethacin), cardiac glycosides, antibiotics, and opioids that undergo varying amounts of enterohepatic recirculation. First-order absorption processes are commonly used to characterize drug absorption processes; however, for xenobiotics that undergo enterohepatic recirculation first-order absorption processes are unable to depict correctly the disposition of the drug into the systemic circulation. For these drugs, the reabsorption process leads to characteristic double or multiple peaks in the oral plasma concentration curves (Figure 15.20).

The time course of morphine concentration after a 7.6 mg/kg oral administration in rats showed a relatively rapid absorption phase with maximum concentrations obtained about 40 minutes. After that, the levels decline exponentially up to 240 minutes when the second peak begins to appear (Figure 15.20). It has been established that morphine undergoes enterohepatic recirculation and the second peak is due to this process. In the liver, morphine is mainly metabolized to glucuronides, which are eliminated via the bile. In the small intestine, morphine glucuronides are hydrolyzed by bacteria. The liberated morphine can be reabsorbed, and thus the enterohepatic recirculation circle is complete. The release of the drug may occur spontaneously but can also be triggered by circadian variations in gallbladder emptying and by the intake of food. Enterohepatic recirculation is unlikely if the second peak is observed after oral administration but not observed often to less extent after intravenous administration [60].

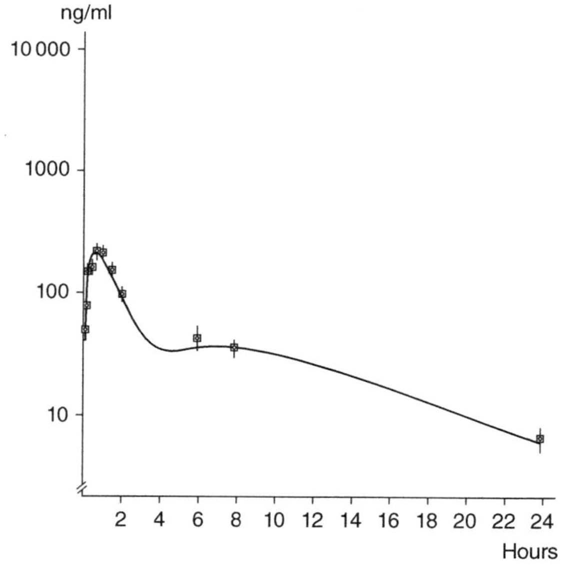


**Figure 15.20** Serum concentrations of morphine after oral administration of 7.6 mg/kg to 10 rats. Source: Dahlstrom and Paalzow [60]/with permission of Springer Nature.

### 15.3.3 Gastric Emptying and Intestinal Transit Time

The rate and extent of drug absorption after oral administration can be affected by the residence time of the drug in the stomach and along the intestine; this can be particularly important for Class I drugs with high solubility and high permeability and Class III drugs with high solubility and low permeability. In turn, the residence time is dictated to a large extent by gastric emptying and gastrointestinal motility. The duration of gastrointestinal motility cycle is highly variable ranging from 15 minutes to greater than 3 hours and is characterized by four phases. Phase I is a quiescent period and last 20–90 minutes, phase II consists of intermittent and irregular contractions and its duration is ~2 hours. These contractions gradually increase in strength culminating in a short period of intense contractions called phase III, the activity phase or the housekeeping wave lasts for 3–25 minutes. The phase IV is a brief transition period from phase III back to phase I. As very little absorption of most drugs occurs from the stomach relative to the small intestine, drug is retained in the stomach until it is delivered to the intestine and subsequently absorbed. For drugs with a high water solubility, dissolution is fast and is most likely not a rate limiting for absorption. Under these circumstances, gastric emptying may be a critical determining factor in drug absorption. Since gastrointestinal motility, gastric emptying, and intestinal transit rates are variable and since most drugs are studied in the fasted state, it seems reasonable to suggest that plasma level curves can exhibit double peaks due to this gastrointestinal physiological variability.

Pharmacokinetic simulations of drug administration during four situations provide distinct shapes in oral concentration plasma profiles. Type A (first $C_{max}$ < second $C_{max}$) occurs when the drug is administered during the first 15 minutes of phase I activity in the stomach. Type B (single $C_{max}$ peak) occurs when the drug is administered during the first 15 minutes of phase I to the beginning of phase II, 60 minutes

(15–60 minutes). Type C (first $C_{max}$ > second $C_{max}$) occurs when the drug is administered from the beginning of phase II to 30 minutes into Phase II (60–90 minutes). Type D (first $C_{max}$ = second $C_{max}$) occurs when the drug is administered close to, or in conjunction with, gastrointestinal motility cycle in the stomach, and includes the remainder of the motility cycle (90–105 minutes). Double peaks in the plasma concentration–time curve frequently occur after administration of cimetidine, as a liquid or tablet in the fasted state. Double peaks do not occur if cimetidine is administered intravenously or concomitantly with food [61].

The formulation of a drug with a variety of excipients including sugars and polyethylene glycol is regularly undertaken. Recent studies have demonstrated that the incorporation of certain excipients in a formulation can alter the transit behavior of a dosage form and may not simply be "inert" substances [6]. The incorporation of mannitol as an excipient into a pharmaceutical formulation of cimetidine leads to a reduced bioavailability for this drug, which is primarily absorbed from the small intestine compared to the same drug formulated with sucrose. The small intestinal transit times were significantly shortened after administration of the mannitol solution and tablet with the transit time of the mannitol solution containing cimetidine 23% of those of the sucrose solution (Figure 15.21) [62].

Interestingly, the sucrose formulations exhibited a double-peak phenomenon to a greater degree than the mannitol preparations, which may further suggest the

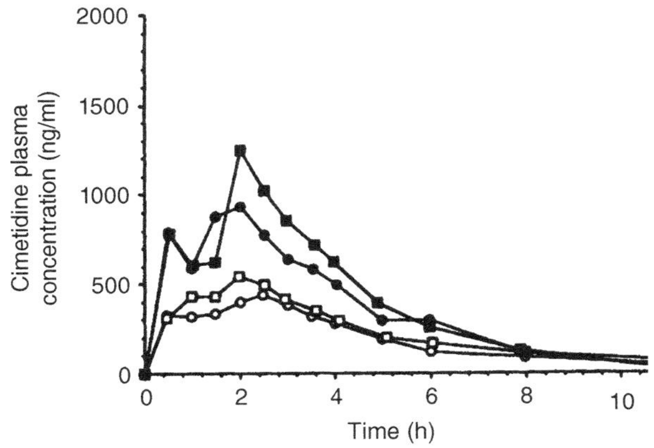


**Figure 15.21** Individual cimetidine plasma concentration (ng/ml) time profiles for subjects following administration of test formulations: (open circles) chewable mannitol tablet, (closed circles) sucrose tablet (open squares) mannitol solution, (closed squares) sucrose solution Source: Reproduced from Adkin et al. [62]/with permission of Springer Nature.

residence time available for absorption is affected by the excipient. Mannitol may also decrease bioavailability because of solvent drag because it osmotically holds water in the small intestine, which leads to a net flux of water into the lumen of the small intestine rather than normal flux into plasma.

For the commonly used pain reliever acetaminophen (paracetamol), it has been suggested that gastric emptying is rate limiting in the absorption of the drug. Variability in stomach emptying rates and intestinal blood flow rates over the course of the entire absorption process after a single dose would likely result in variable rates of appearance in the plasma. When administered with the anticholinergic muscarinic antagonist propantheline, which delays gastric emptying the absorption of paracetamol was slowed (Figure 15.22). Conversely, the absorption of acetaminophen was accelerated by the gastrointestinal motility modifier metoclopramide, which stimulates gastric emptying by increasing gastric peristalsis and relaxes the pyloric sphincter shortening gastric emptying time. There was no change in the extent of absorption as measured by urinary excretion [63, 64].

An interesting double-peak phenomenon has been reported for the pharmacokinetics of alprazolam and its metabolites after oral administration in rats. Alprazolam is a benzodiazepine anxiolytic and demonstrates double peaks in serum concentration data as well as double peaks in its metabolites (Figure 15.23). It is hypothesized that this is due to reduction in gastric motility caused by the muscle relaxant effect of alprazolam leading to bi-phasic gastric emptying [65]. A similar oral pharmacokinetic profile is observed for epinastine and nonsedating histamine 1 antagonist with this double-peaking pattern suggested to be a consequence of reduction in gastric motility delaying transit time [66].

Class III BCS drugs may be sensitive to changes in GI transit and permeability. For instance, a lipoidic excipient caprylocaproyl macroglyceride was utilized in an oral capsule formulation of a BCS III compound UK081252. The caprylocaproyl macroglycerides enhanced the absorption of the compound after oral administration due to a strong permeability-enhancing effect; however, the liquid-filled formulation consistently also produced a double-peak phenomena in the oral concentration–time profile (Figure 15.24). This lipid-based surfactant may inhibit gastric emptying for a variable period (up to two hours) to temporarily limit the supply of compound available for absorption through an indirect physiological effect with a pharmacokinetic consequence [67].

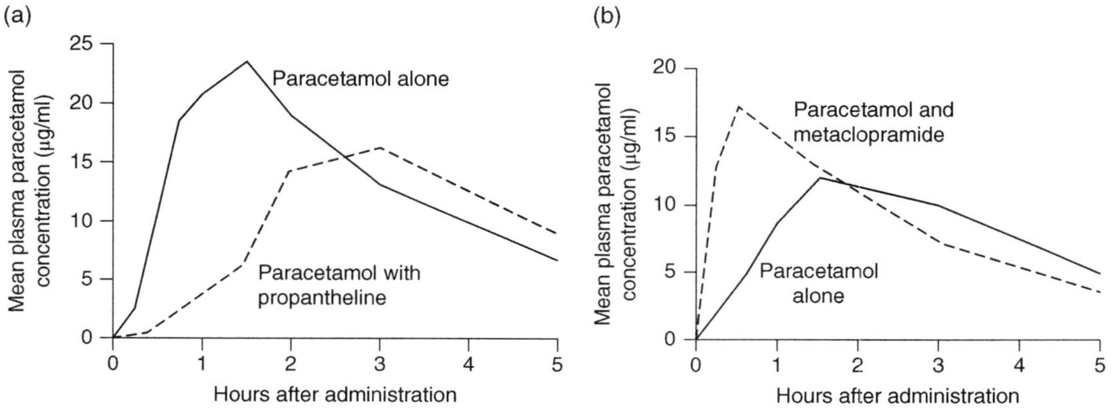


**Figure 15.22** Inhibitory effect of propantheline on paracetamol absorption in six patients (a) Increased rate of paracetamol absorption after metoclopramide in five healthy volunteers (b). Source: Nimmo et al. [63]/with permission of BMJ Publishing.

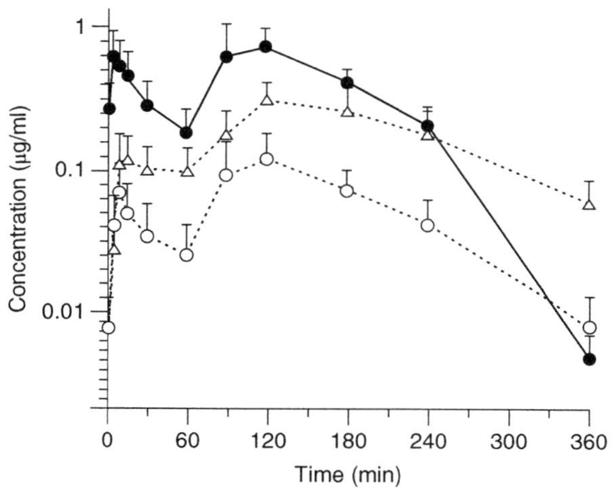


**Figure 15.23** Mean serum alprazolam and its metabolite (4-OH and alpha OHALP) concentration time profiles after oral 12.5 mg/kg alprazolam administration. Source: Wang et al. [65]/with permission of American Society for Pharmacology and Experimental Therapeutics.

Another interesting drug absorption case is that of the administration of danazol. After oral administration, danazol was found to have double-peaking phenomena whereas only single peaks were present after intra-intestinal administration to the proximal jejunum or proximal ileum. The double peaking appeared to be related to a combination of inhibition of gastric emptying regulated absorption due to the presence of lipid monoglyceride in the emulsifying formulation; however, as danazol solubility was dependent on bile salt solubilization within the upper intestine the secondary peaks are characteristically higher than the initial peaks. In cases such as cimetidine and penicillamine when plasma profiles were subsequently studied after administration to fed subjects only single peaks were observed however, for danazol double peaks were present regardless of administration with or without food. Suggesting that danazol absorption phases depend on bile salt solubilization with similar $C_{max}$ values after jejunal or oral administration [68–70].

### 15.3.4 Chronopharmacokinetics

Chronopharmacokinetics can be defined as the study of the temporal changes in ADME influenced by the time of drug administration and the phase of the circadian rhythm in which the species is in. A circadian rhythm is about a 24-hour cycle in which the different physiological processes of living beings occur, determining their sleeping and feeding patterns. Physiological functions change throughout the day (a circadian rhythm). It has been observed that temporal variations exist for drug absorption from the GI tract (circadian variation in gastric acid, pH, motility, blood flow), plasma protein binding, metabolism (variations in enzyme activity, hepatic blood flow), and elimination (renal blood flow, glomerular filtration, urinary pH). Therefore, it is important to consider the possibility of circadian rhythms at the moment of designing pharmacokinetic experiments in clinical trials. The first reported chronopharmacokinetics studies were performed by Reinberg et al. [71, 72] and assessed ethanol and sodium salicylate in adult healthy volunteers. A dose of 0.67 g/kg or 1 g solution/24 hours of ethanol or sodium salicylate, respectively, was ingested at 07 : 00, 11 : 00, 19 : 00, and 23 : 00 hours, with a week between tests. It was observed that higher $C_{max}$ and higher ethanol plasma concentrations were observed at 07 : 00 hours than the other times [71], while sodium salicylate was observed to have higher renal clearance, lower $t_{max}$, and lower $t^{1/2}$ when it ingested between 19 : 00 and

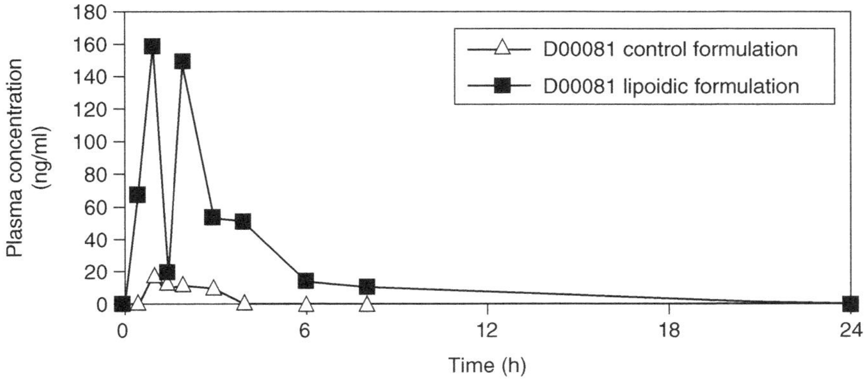


**Figure 15.24** Plasma concentration–time profile of UK-81252 after oral administration to an individual dog. Source: Chang and Shojaei [67]/with permission.

23:00 hours [72]. Since these seminal investigations that drugs can be influenced by circadian changes, chronopharmacokinetics has been widely reported and appreciated. In general, it has been observed that most lipophilic drugs are absorbed faster when taken in the morning compared to the evening; while little difference has been reported for water-soluble compounds [73]. Drug distribution has also been observed to be affected because circadian rhythms affect the blood flow to different organs, and affect the protein binding of different drugs (acidic and basic). Drug metabolism can also be affected because hepatic blood flow has been reported to be greater at 8 a.m., while metabolism slows during the night with subjects resting in the supine position. This is critical because in general, the metabolism of drugs with high extraction ratio depends on hepatic blood flow, while drugs with low extraction ratio depend on liver enzymatic activity. Therefore, since drugs with high extraction ratio are blood-flow dependent their clearance will be affected by circadian rhythms [73].

Pranlukast is a novel, potent, and selective orally administered leukotriene ($LTD_4$) receptor antagonist approved in Japan for the treatment of asthma, and currently under clinical trials in USA and Europe. When a single dose (300 mg) was administered to healthy subjects at two different occasions: 08:00–09:00 or 20:00–21:00 hours after 30 minutes of standard meal, higher $t_{max}$, and $C_{max}$ values were observed after evening administration (20:00–21:00 hours) (Figure 15.25). This temporal dosing difference translated into significantly higher AUC (56% increase), $C_{max}$ (14% increase), and $t_{max}$ (2.5 hours later) values after evening administration [74].

On the other hand, when sumatriptan succinate (100 mg, immediate release capsule), used for migraine treatment, was administered to healthy volunteers

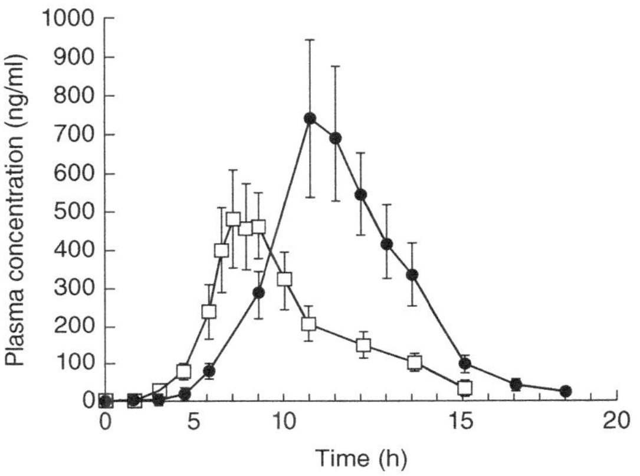


**Figure 15.25** Arithmetic mean ± SEM plasma concentration versus time profiles of healthy male volunteers given 200 mg pranlukast orally after morning (08–09 : 00 hours open squares) or evening (20–21 : 00 hours closed circles) administration. Source: Brocks et al. [74]/with permission of John Wiley & Sons.

at four different times (07:00, 13:00, 19:00, and 01:00 hours), the morning dosing (07:00 hours) exhibited higher $C_{max}$ values (59.09 ± 10.53 ng/ml) than after the 19:00 hours administration (41.88 ± 12.21 ng/ml) (Figure 15.26). It was also observed that AUC was significantly higher following the 07:00 hours than the 19:00 hours administration, while the oral clearance and volume of distribution were significantly lower after morning administration (07:00 hours) [75].

### 15.3.5 P-Glycoprotein and Other Transporters

Drug absorption was originally believed to be primarily a passive process but is now known to be influenced by major metabolizing enzymes, influx, and efflux membrane transporters found in the enterocytes. Different families of transporters have been identified

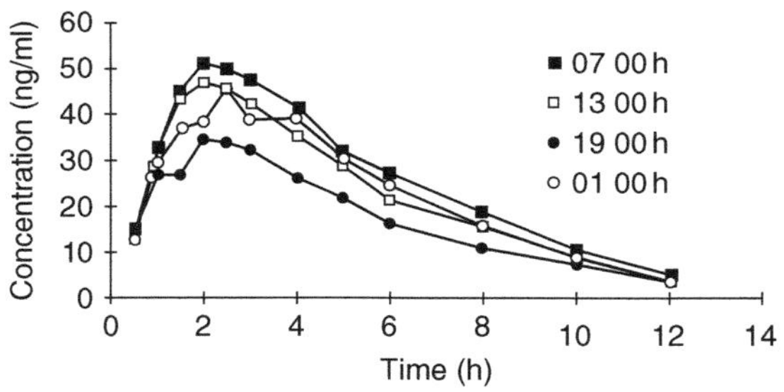


**Figure 15.26** Mean ($n = 12$) serum concentrations of sumatriptan after oral administration of 100 mg at four different time points. Source: Poondru et al. [75]/with permission Oxford University Press.

to play a role in carrier-mediated transport to facilitate intestinal absorption. Some of these families of transporters include lipid transporters/bile acid transporters, monocarboxylate transporters (organic anionic transporters), organic cation transporters, nucleoside transporter, intestinal dipeptide transporter, amino acid transporters, vitamin transporters, phosphate transporters, and P-glycoprotein [76]. Besides the diversity of transporters, they are often also site-specific, and some drugs such as β-lactam antibiotics are substrates for multiple transporter systems [77]. The most studied transporters are the amino acid transporter PepT1 and P-glycoprotein (P-gp). A brief description of PepT1 and PepT2 will follow; however, this section will concentrate on P-gp since its effect on the pharmacokinetics of multiple drugs has been studied in great part due to the availability of P-gp deficient mice [78].

The absorption of peptoid, dipeptides, and tripeptides drugs into the apical gastrointestinal membrane depends on the apical $H^+$/dipeptide transporter (PepT1), followed by transport across the basolateral membrane by PepT2 (a $H^+$ independent transporter) [79]. PepT1 density is site-specific since it increases from the duodenum to the ileum, and it is more abundant at the villus tip.

P-glycoprotein (P-gp) is the product of the multidrug resistance 1 (MDR1) gene, (also known as ABCB1). It was originally described in chemotherapeutic-resistant tumor cells, but it is also widely expressed in blood-tissue barriers (blood–brain barrier, placenta, and blood-testis barrier), and normal tissues with excretory function (small intestine and kidneys) [80]. P-gp has multiple substrates that are shared with a major metabolizing cytochrome P-450 (CYP3A4); however, some drugs such as digoxin, digitoxin, bunitrolol, levofloxacin, and fexofenadine are transported by P-gp but are not metabolized by CYP3A4, while other drugs such as midazolam are metabolized by CYP3A4 but are not transported by P-gp [80]. Furthermore, an opposing efflux-metabolism gradient can be observed in the small intestine, in which the level of P-gp expression increases from proximal to distal regions, while CYP3A4 levels decrease from proximal to distal regions [81].

The importance of P-gp on the bioavailability of certain drugs such as HIV protease inhibitors, chemotherapeutic agents, and analgesics has been reported. For instance, Taxol® (paclitaxel supplemented with Cremophor EL and ethanol) was orally administered (10 mg/kg) to female FVB *mdr1a*(−/−) (P-gp deficient) and *mdr1a*(+/+) (wild-type, P-gp present) mice. Significantly higher plasma concentrations were observed in the P-gp deficient mice (Figure 15.27) indicating that the disruption of the *mdr1a* P-glycoprotein gene plays resulted in dramatic changes in the pharmacokinetics of paclitaxel and that P-glycoprotein plays an important role in the efflux of this chemotherapeutic agent causing its poor uptake. Significantly higher AUC (6.4-fold), $C_{max}$ (11.4-fold), and oral bioavailability (3.14-fold) values were observed for the P-gp knockout mice compared to the wild-type [82].

The importance of P-glycoprotein in paclitaxel update has also been confirmed after concomitant administration of valspodar (also known as SDZ-PSC 833), a

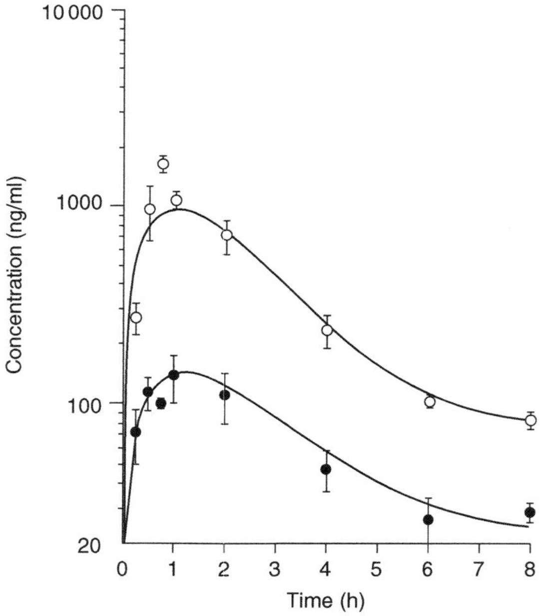


**Figure 15.27** Plasma concentration–time curves of paclitaxel in female FVB wt (closed circle) and mdr1a (−/−) (open circle) mice after oral administration of paclitaxel (10 mg/kg). The data are shown as the mean concentration, and error bars represent the SEM (n=2-10 per time point). Source: Reproduced from Sparreboom et al. [82]/with permission of National Academy of Science.

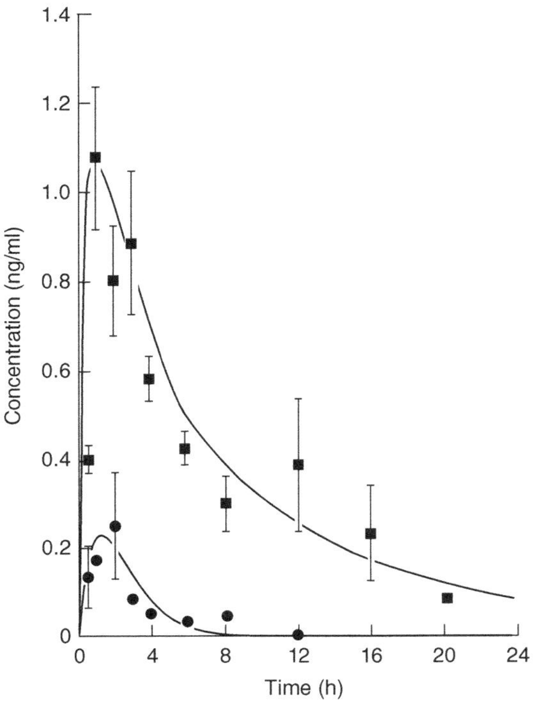


**Figure 15.28** Plasma concentration–time curves after oral administration of paclitaxel to mice. Per kg of body weight the mice received either 10 mg of paclitaxel alone (closed circle control group) or in combination with 50 mg of SDZ PSC 833 (closed squares). Symbols and error bars represent mean ± standard error. The plasma concentration of paclitaxel in the control group at $t$ - 12 hours was below 25 ng/ml. Source: van Asperen et al. [83]/with permission of Springer Nature.

P-glycoprotein blocker. It was observed that when Taxol (10 mg/kg, paclitaxel supplemented with Cremophor EL and ethanol) was concomitantly administered with 50 mg/kg of SDZ-PSC 833 dramatic increases in paclitaxel plasma concentrations were achieved compared to Taxol alone (Figure 15.28). This inhibition of P-gp translated into a significant increase in oral AUC (11-fold) from 735 ± 134 ng h/ml (Taxol alone) to 8066 ± 819 ng h/ml (coadministration with SDZ-PSC 833) [83].

Similar pharmacokinetic differences have been correlated to observed changes in pharmacodynamics of different drugs such as fentanyl, morphine, and methadone for which an increased and prolonged antinociception was observed in *mdr1a*(−/−) (P-gp deficient) mice compared to *mdr1a*(+/+) (wild-type, P-gp present) mice [84]. Similar results were observed with alfentanil, a synthetic opioid used as an anesthetic and as postsurgical analgesic; in this study, it was observed that a three times lower dose of alfentanil was sufficient in *mdr1a*(−/−) (P-gp deficient) mice to achieve the same nociception levels than *mdr1a*(+/+) (wild-type, P-gp present) mice [85].

As mentioned above, there are also other transport systems that have been recently characterized and that might play important roles in further research in current and future drugs. For instance, the organic anions and cation transporters are amphiphilic solute facilitators and the lipid and cholesterol transporters play a role in the absorption of lipophilic compounds such as cholesterol and long-chain fatty acids (LCFA). The lipid and cholesterol transporters disprove the previous assumption that LCFAs could diffuse freely through the enterocyte membrane since LCFAs actually can undergo saturable absorption [76].

As it can be observed, the membrane transporters play a crucial role in the way and amount that drugs can be absorbed from the GI tract into the central compartment. This is a growing field of research and there are very comprehensive and complete reviews of membrane transporters for further reading [76, 79, 80, 86, 87].

### 15.3.6 Site-Specific Absorption

As drugs administered orally are absorbed into the systemic circulation from mainly the small intestine. Some drugs are known to be absorbed from other portions of the GI tract including the stomach and colon. Distinct double peaking in oral plasma-concentration–time profiles is a phenomena that can occur and site-dependent absorption of drugs may be the underlying mechanisms. Ranitidine produced an oral concentration–time curve with a pronounced secondary peak. The absorption characteristics of ranitidine were studied after delivery to three locations in the GI tract in healthy volunteers (Figure 15.29). Subjects received ranitidine HCl 150 for injection via a nasoenteric tube directly into their stomach, or jejunum or cecum sequentially in three periods [88]. Concentrations following cecal dosing were much lower than concentrations following gastric and jejunal dosing. Double plasma concentration peaks were observed in several subjects after gastric and jejunal dosing. This effect seems to be most pronounced in the fasting state and is blunted by food. Results suggest that the rate of ranitidine absorption was much slower following cecal administration than after gastric or jejunal dosing suggesting the possibility of site-specific absorption phenomena for ranitidine with two areas of efficient absorption separated by an area of poor absorption. Double peaking after administration of ranitidine to the jejunum suggests that this phenomenon is not related to variability in gastric emptying.

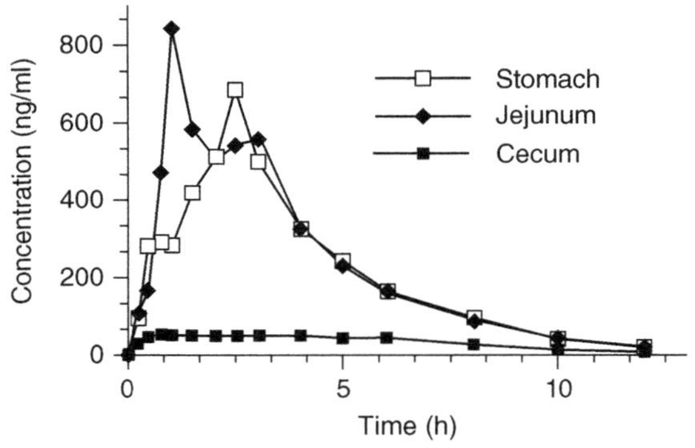


**Figure 15.29** Typical subject's plasma ranitidine concentration–time profiles following bolus administration of ranitidine HCL 1560 mg into the stomach, jejunal and cecum. Note the double concentration peaks after both jejunal and stomach administration. Source: Reproduced from Williams et al. [88]/with permission of Springer Nature.

It is widely thought that pharmaceutical excipients are inert and their role in bioavailability is linked to their role in solubilization. The presence of some excipients such as PEG 400 can increase small intestinal transit and may affect drugs such as ranitidine that have an absorption window while other excipients such as propylene glycol, vitamin E TPGS, and Capmul MCM do not. Volunteers who received an immediate-release pellet formulation of ranitidine (150 mg) encapsulated within a hard gelatin capsule and a liquid preparation of either 150 ml orange juice (control) or 150 ml orange juice containing 10 g PEG 400 (test). The absolute bioavailability from the pellet formulation was reduced by 31% and the small intestine transit time was shortened by 37% with the administration of PEG 400 in the test preparation. Interestingly, the appearance of double peaks was less evident in the presence of PEG 400 (Figure 15.30). It has also been shown that sodium acid pyrophosphate, a pharmaceutical excipient used in effervescent formulation reduced small intestinal transit time and bioavailability of ranitidine. It is possible that PEG 400 may have other direct effects on the GI tract such as osmotic effects, which alter oral concentration–time profiles [89–91].

In addition, there is evidence for a second site of absorption of veralipride in the human small intestine using a new drug delivery telemetric shuttle. A telemetric shuttle was used to coadminister veralipride and veralipride D3 in order to examine the possibility of two sites of absorption in the small intestine. The telemetric capsule delivered drug in solution into the small intestine at specific time periods after oral administration near the connection between the jejunum and ileum. It appears that a second site in the small intestine absorbs drug (Figure 15.31). The absorption rate constants and the intestinal transit overt this site may be a limiting factor to absorption. A telemetric capsule allows for the site-specific administration of drug and is a noninvasive tool for the investigation of drug absorption in the GI tract [92].

### 15.3.7 Genetic Polymorphism

Variability in oral pharmacokinetics in patients has been related to different factors such as age, sex, disease, drug and food interaction, and also to genetic factors. Inter-patient variability refers to variability observed between different people and requires an individualized dosing based on each patient's characteristics, while intra-patient variability refers to changes in PK within the same patient on a given day (see Section 15.3.4). However, the amount of research in the last years has increased dramatically toward an improved drug and dosage selection based on the comprehensive understanding of a patient's genetics. The discipline of pharmacogenetics is progressing at a rapid pace. Genetic polymorphisms in the various genes coding for metabolizing enzymes, receptors, and transporters, structural, and signaling proteins that modulate ADME parameters of a drug are being delineated. Pharmacogenetics was generated from the initial observations in 1958 about the existence of multiple forms of metabolizing enzymes [93]. However, the term pharmacogenetics (also known as pharmacogenomics) as such was first reported in 1964 in a seminal paper in which the patient variability issue already raises as a major concern for adequate drug regimens [94]. Since this time research has identified that not only drug metabolism is affected by genetic polymorphism, but also transport and distribution are greatly affected inducing extensive inherited variability [95, 96]. Since genetic polymorphisms can affect both pharmacokinetics and pharmacodynamics, it needs to be emphasized that genetic variants in transporters, plasma proteins, and metabolizing enzymes induce PK variability, while genetic variants in receptors, ion channels, enzymes, and immune molecules tend to induce PD differences. Since the scope of this chapter is on oral concentration–time profiles as they relate to pharmacokinetics, this section will center on the effects of genetic variants in transporters, plasma proteins, and metabolizing enzymes over the pharmacokinetics of different drugs.

An example of genetic variant in transporters is the case of the MDR1 genotype, and as mentioned above one product of this gene is P-glycoprotein (P-gp). It was

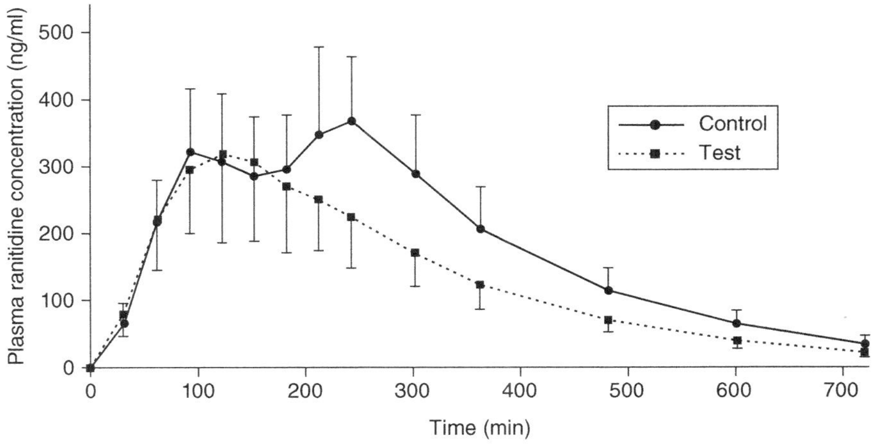


**Figure 15.30** (Mean ± SD) Ranitidine profiles in the absence (Control) and presence (test) of PEG 400. Source: Koch et al. [89]/with permission of Springer Nature.

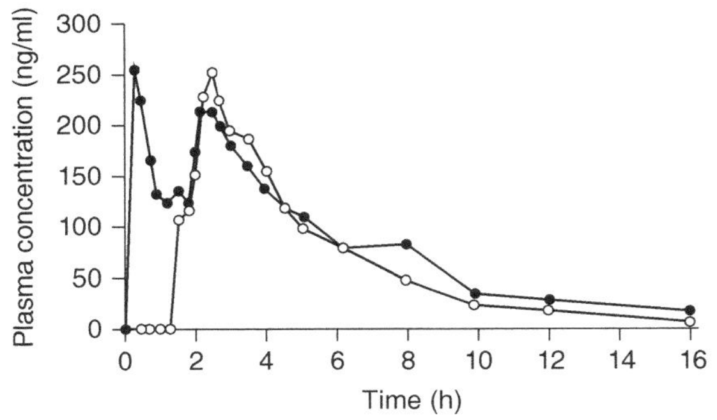


**Figure 15.31** Mean plasma concentration time profile of veralipride (Closed circles) and veralipride-D3 (open circles) from five subjects. Delivery of veralipride-D3 was done 1.5 hours after T0 using a telemetric capsule. Source: Reproduced from Staveris et al. [92]/with permission of Europe PMC.

observed in a study conducted on 114 healthy Japanese subjects (65 males and 49 females, 19–76 years old) that 35.1% were homozygous for wild-type allele (C/C), 52.6% were compound heterozygotes (C/T), and 12.3% were homozygous for the mutant allele (T/T). Furthermore, when these subjects were given a single oral dose of digoxin (0.25 mg) and Digosin® tablet, and dramatic changes in serum concentrations were observed between the different genotypes (Figure 15.32). These changes translated into significant differences in $AUC_{0\text{-}4h}$ between C/C (4.11 ± 0.57 ng h/ml), C/T (3.20 ± 0.49 ng h/ml), and T/T (3.27 ± 0.58 ng h/ml), while similar $AUC_{0\text{-}24h}$ were observed (8.84 ± 3.28, 5.80 ± 1.94, and 5.74 ± 3.04 ng h/ml) for C/C, C/T, and T/T, respectively [97].

Genetic variants in metabolizing enzymes have been observed with different enzymes, such as the human arylamine *N*-acetyltransferases 1 and 2 (NAT1 and NAT2) are polymorphic phase II metabolizing enzymes that acetylate arylamine compounds. They play an important role in the detoxification of different occupational chemicals and carcinogens (i.e. benzidine, β-napththylamine, and 4-aminobiphenyl) and in the metabolism and/or activation of different drugs such as hydrazines (i.e. isoniazid, hydrazaline, and phenylzine), arylamines (i.e. dapsone, procainamide, and sulfamethazine), and drugs that are metabolized to amines such as sulfasalazine, clonazepam, nitrazepam, and caffeine. Polymorphism at the NAT1 locus has been reported to produce at least 26 allelic variants, which generate phenotypic variations in terms of NAT1 catalytic activity (also known as acetylator of an individual), leading to inter-patient variability and/or predisposition to disease in humans [98]. For instance, different acetylator phenotypes have been observed in subjects with different ethnicities (Table 15.6) [99].

The different acetylator variants have been reported to affect pharmacokinetics of different drugs. For instance, when isoniazid (employed for tuberculosis treatment) was administered to children (0.1–196 months age) as a single oral dose (10 mg/kg), dramatic changes in plasma concentrations of the parent drug isoniazid (Figure 15.33) and the metabolite acetyl isoniazid were observed.

**Table 15.6** Acetylation phenotypes and ethnicity.

| Population | %Slow | %Hetero fast | %Homo fast |
|---|---|---|---|
| South Indians | 59 | 35.6 | 5.4 |
| Caucasians | 58.6 | 35.9 | 5.5 |
| Blacks | 54.6 | 38.6 | 6.8 |
| Eskimos | 10.5 | 43.8 | 45.7 |
| Japanese | 12 | 45.3 | 42.7 |
| Chinese | 22 | 49.8 | 28.2 |

Source: Adapted from Kalow et al. [99]/with permission.

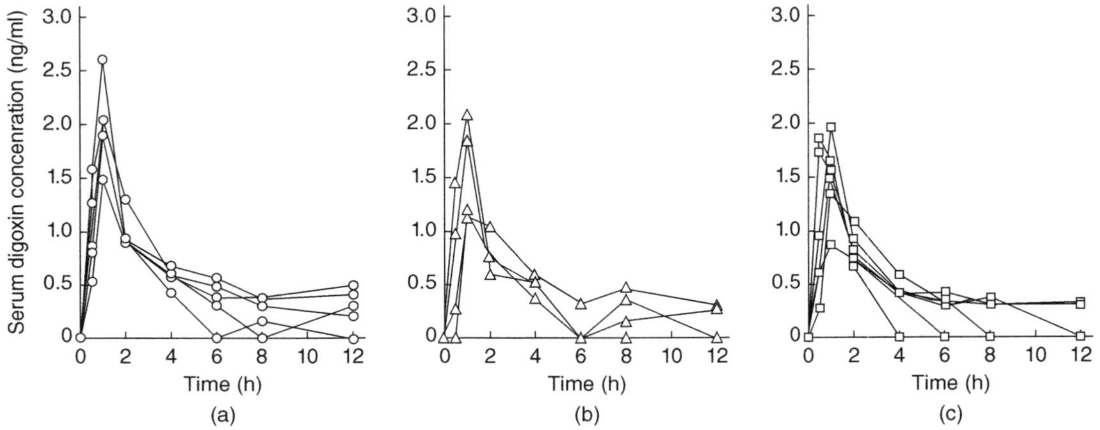


**Figure 15.32** Serum concentration–time profiles of digoxin for three genotype groups: (a) open circle, C/C, $n = 5$; (b) open triangle, C/T, $n = 4$; (c) open square, T/T, $n = 6$ after a single oral administration of digoxin. Source: Sakaeda et al. [97]/with permission.

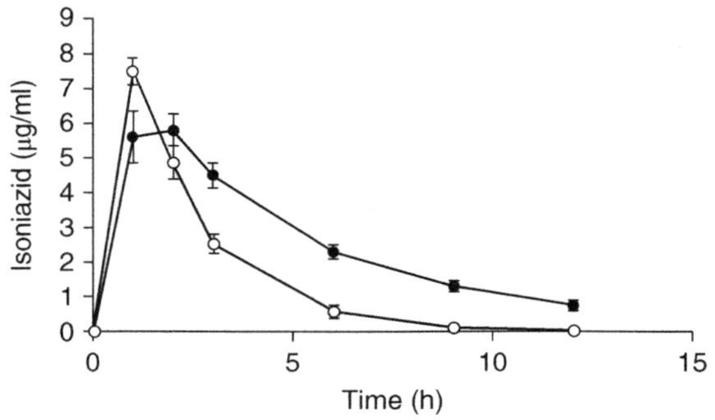


**Figure 15.33** Mean plasma concentration–time (± SEM) profile for isoniazid in slow (closed circle) and in fast acetylators (open circle). Source: Rey et al. [100]/with permission of Springer Nature.

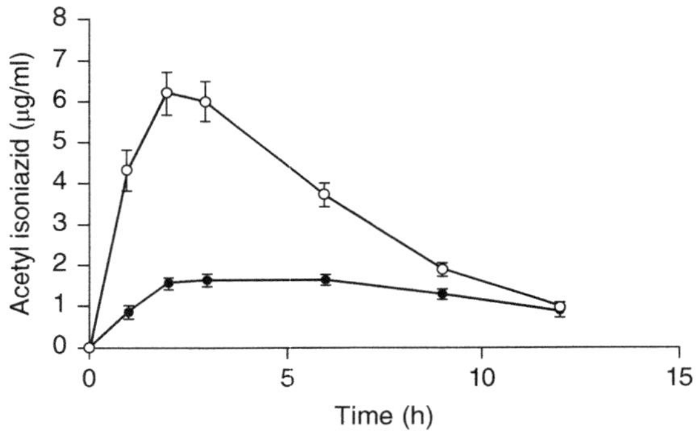


**Figure 15.34** Mean plasma concentration–time (± SEM) profile for acetyl isoniazid in slow (closed circle) and in fast acetylators (open circle). Source: Rey et al. [100]/with permission of John Wiley & Sons.

It can be observed that slow acetylators will present overall higher parent (isoniazid) and lower metabolite (acetyl isoniazid) plasma concentrations (Figure 15.34). These differences also translated into PK differences, it was observed that in the case of parent (isoniazid) no significant changes in $C_{max}$ were observed. However, the mean apparent oral clearance was significantly lower in the slow acetylator (0.298 ± 0.099 l/h/kg) than in the fast acetylator group (0.528 ± 0.234 l/h/kg). The mean apparent $V_d$ and half-life were higher in the slow acetylators (1.56 ± 0.65 l/kg and 3.88 ± 1.89 hours, respectively) than in the fast acetylators (1.06 ± 0.45 l/kg and 1.64 ± 1.1 hours, respectively). In the case of the metabolite (acetyl isoniazid) it was observed that slow acetylators exhibited significantly lower $C_{max}$ values (1.89 ± 0.48 mg/l) and higher half-life (6.47 ± 2.84) compared to the fast acetylators (6.28 ± 2.05 mg/l and 3.36 ± 0.92 hours for $C_{max}$ and half-life respectively) [100].

There are multiple examples reporting genetic polymorphism in different cytochrome P450s (CYP450s) [101–105]. For instance, genetic polymorphisms in a major metabolic enzyme (CYP3A4, present in intestine and liver) have been shown to play an important role in the metabolism of different drugs such as cyclosporine. In a study, 26 healthy Chinese volunteers (11 women and 15 men, taking no other medicine including oral contraceptives, free of grapefruit juice intake for at least two weeks, and no intake of caffeine or alcohol before administration) were administered with a single oral dose 4 mg/kg cyclosporine capsule (Neoral®). Among these subjects, 12 were wild homozygotes (*1/*1), 12 heterozygotes (*1/*18B), and 2 mutant homozygotes (*18B/*18B), and significant changes in whole blood cyclosporine concentrations (lower $C_{max}$ for *18B/*18B) were observed especially for the mutant homozygotes (*18B/*18B) (Figure 15.35). This change

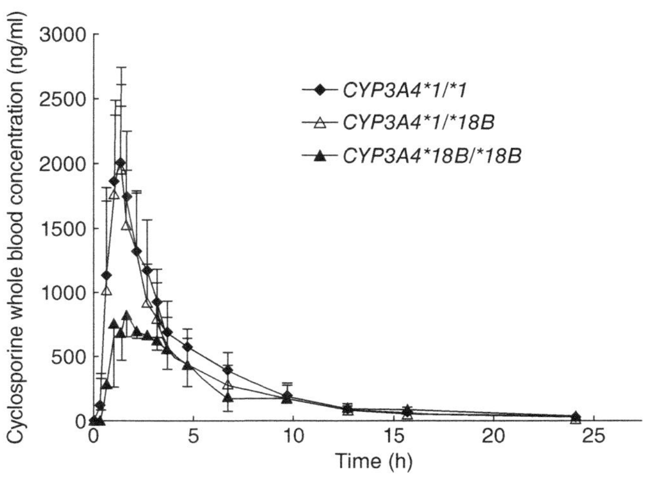


**Figure 15.35** Whole blood cyclosporine concentration (mean values ± SE of the mean)-time profiles among CYP3A4*1/*1 group, the CYP3A4*1/*18B group, and the CYP3A4*18B/*18B group after oral administration of 4 mg/kg of cyclosporine. Source: Hu et al. [101]/with permission.

in whole blood concentrations translated into lower $AUC_{0-4h}$ (2561 ng h/ml, $p$-value = 0.021) and $AUC_{0-12h}$ (4154 ng h/ml, $p$-value = 0.056) values for subjects with CYP3A4 *18B/*18B compared to *1/*1 (5026 and 7184 ng h/ml, respectively) and *1/*18B (4434 and 6215 ng h/ml, respectively) [101].

Therefore, it can be observed that genetic polymorphism plays an important role in the ADME parameters of drugs and that the field of pharmacogenetics is blooming due to the necessity to better predict pharmacokinetics disposition of drugs in order to design individualized drug regimens, which can better predict therapeutic effects and compliance. There are extensive reviews on this topic dealing with genetic polymorphisms of different molecules and their effect on pharmacokinetics and pharmacodynamics [106–110].

### 15.3.8 Disease

Disease plays an important role in the pharmacokinetic and pharmacodynamic variability of drugs in patients. For instance, gastrointestinal motility is increased by colitis, diarrhea, and flu, thus decreasing drug absorption. Inflammatory bowel disease and Crohn's disease can have decreased drug absorption as scar tissue develops in the GI tract altering motility and available surface area. Inflammatory diseases in general decrease protein binding to alpha1-acid glycoprotein (AAG) increasing the systemic concentrations of unbound basic drugs. Renal insufficiency has also been shown to induce the accumulation of endogenous substances that compete with drugs for protein binding sites (specially albumin), affecting plasma concentrations of highly bound drugs. Liver diseases may influence first-pass metabolism, and affect bile production altering absorption. Migraines and acute myocardial infarction also reduce gastric emptying. The most well-characterized diseases that influence the pharmacokinetics of orally administered drugs include inflammation (rheumatoid arthritis and Crohn's disease) and hepatic disorders (liver cirrhosis).

#### 15.3.8.1 Inflammation

Inflammation is a complex process that involves a series of events encompassing numerous stimuli such as infection, ischemia, antigen–antibody interactions, chemical, thermal, or mechanical injury. The inflammatory responses have been characterized to occur in three distinct phases, each apparently mediated by different mechanisms: an acute phase characterized by local vasodilatation and increased capillary permeability, a subacute phase characterized by infiltration of leukocyte and phagocyte cells, and a chronic proliferative phase, in which tissue degeneration and fibrosis occur [111]. Disease states can involve acute and/or chronic inflammatory processes, and during the distinct phases of inflammation, there are different inflammatory mediators involved that play key pathophysiological roles.

For instance, rheumatoid arthritis (RA) is a painful autoimmune disease that targets joints resulting in inflammation and affecting locomotor activity. Furthermore, it has been reported that patients with RA present decreased glomerular filtration rate (GFR), elevated tubular enzyme, and serum proteins (proteinuria) in urine [112]. Inflammation has also been reported to increase protein binding, reducing the amount of unbound drug, which has been reported to reduce the clearance of multiple drugs [113, 114].

It has been reported that inflammation is unlikely to alter the pharmacokinetics of drugs used to treat inflammatory conditions if they exhibit low hepatic

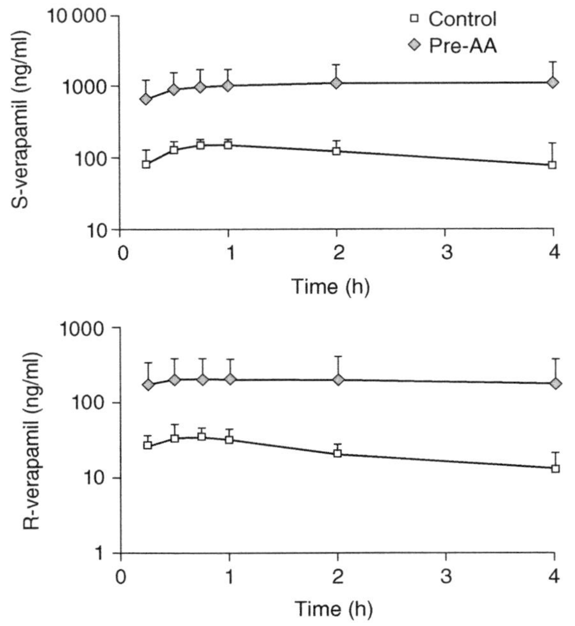


**Figure 15.36** Mean plasma veparamil concentrations of control ($n = 6$) and pre-AA ($n - 4$) rats following administration of single oral 20 mg/kg racemic veparamil doses. Error bars represent standard deviation of the means. Lines connect experimental data points. Source: Ling and Jamali [116]/with permission of The American Society for Pharmacology and Experimental Therapeutics.

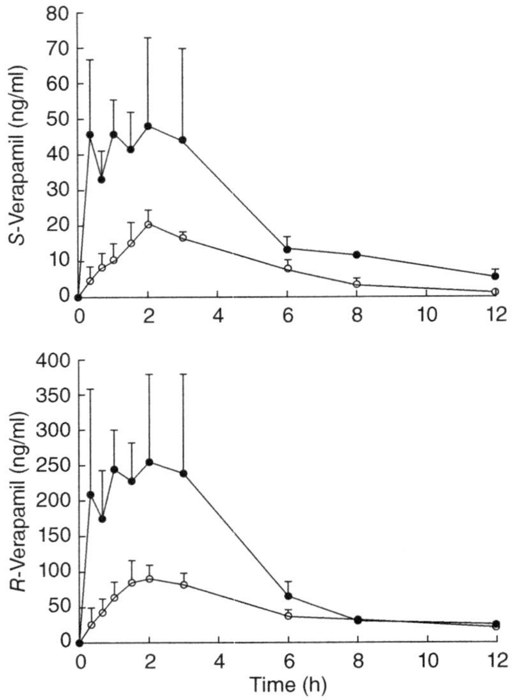


**Figure 15.37** Serum veparamil enantiomers concentration–time profile in healthy control subjects (open circle) and patients with rheumatoid arthritis (closed circle) after receiving a single 80 mg oral dose of veparamil. Error bars represent SEM ($n = 8$/group). Source: Mayo et al. [117]/with permission of John Wiley & Sons.

extraction ratios. However, patients with rheumatoid arthritis (about 40%) concomitantly take other drugs that are highly dependent on hepatic function (i.e. cardiovascular drugs) [115]. For instance, verapamil, a phenylalkylamine calcium channel blocker, used in the treatment of hypertension was administered orally (20 mg/kg) to pre-adjuvant arthritic (pre-AA) rats. Pre-AA rats exhibited significantly higher plasma concentrations of both enantiomers compared to the control rats (Figure 15.36). This translated into 6.11-fold and 3.57-fold increase in $C_{max}$ for *S*- and *R*-verapamil, respectively compared to control rats and 9.34-fold and 9.51-fold increases in $AUC_{0\text{–}4h}$ for *S*- and *R*-verapamil, respectively compared to control rats. It was also observed that pre-AA rats presented a significant reduction in total microsomal cytochrome P450 (49%), CYP3A1/2 (58%), and CYP1A1/2 (42%) isoforms microsomal content [116].

Similar results were observed when 80 mg racemic verapamil (Isoptin®) was orally administered to RA patients compared to healthy volunteers. It was observed that RA induces a significant increase in serum concentration of both enantiomers (Figure 15.37), exhibiting fourfold and threefold increases in $AUC_{0\text{–}\infty}$ of *S*- and *R*-verapamil, respectively. Significant differences in apparent oral clearance (CL/F) were also observed, in which RA patients exhibited 2.23-fold and 1.56-fold lower CL/F values (*S*- and *R*-veparamil, respectively) than healthy volunteers. The higher variability observed in RA patients produced no significant differences in $C_{max}$ and terminal half-life. However, the authors also observed that RA resulted in sixfold and fivefold decreases in protein binding for *R*- and *S*-verapamil, respectively [117].

Crohn's disease is an inflammatory bowel disease that has unknown etiology implicated by T-cell activation involved in the underlying pathophysiology [118]. The conventional therapy includes corticosteroids and aminosalicylates [119] that target the autoimmune and inflammatory processes present in the distal intestinal tract. Main targets include interleukin-2 (IL-2), IL-4, and lymphotoxins such as tumor necrosis factor-α (TNF-α), which are increased in Crohn's disease [120]. For instance, budesonide is an effective glucocorticosteroid in the treatment of mild to moderate Crohn's disease. It was observed that when 18 mg budesonide (Entocort® controlled-release capsules) were administered to Crohn's disease patients

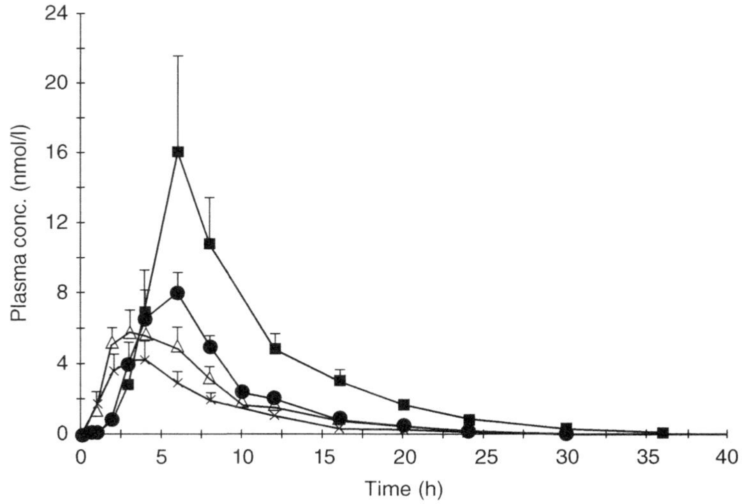


**Figure 15.38** Mean (SEM) plasma concentrations of budesonide after Entocort controlled release-capsules (closed square, 18 mg single dose in fed patients: closed circle 18 mg single dose in fed health subjects: open triangle 18 mg single dose in fasted healthy subjects) and standard immediate-release ($^{2}H_8$)-budesonide capsules (×, 4 mg single dose in fed patients). Source: Edsbacker et al. [121]/with permission of John Wiley & Sons.

and healthy volunteers higher plasma concentrations (1.57-fold higher $C_{max}$) were achieved in patients compared to controls (Figure 15.38). Furthermore, also higher mean absorption times (MAT) and systemic availability (1.41-fold and 1.78-fold, respectively) were observed for patients compared to controls. In addition, controlled-release capsules exhibited higher $t_{max}$, $C_{max}$, and MAT values compared to the immediate-release standard capsules [121].

#### 15.3.8.2 Hepatic Disorders

The liver is the primary site responsible for the metabolism and metabolic clearance of many drugs. Many drugs are metabolized by the cytochrome P450 (CYP450) enzyme system present in the liver. It has been estimated that more than 50% of the widely-used drugs are metabolized by CYP450 isozymes [122]. Hepatic impairment will result in physiological changes in the liver such as parenchymal cell injury, cell death, and changes in function and circulation that can result in intrahepatic and extrahepatic recirculation [123]. For instance, cirrhosis is a common hepatic disorder, which is a consequence of chronic liver disease characterized by the presence of fibrous scar tissue that replaces liver tissue leading to a progressive loss of liver function. Therefore, alterations in hepatic function may result in the necessity to adjust the dosage regimen especially in drugs highly metabolized and cleared by the liver.

Roflumilast is an isozyme-selective inhibitor of phosphodiesterase 4 ($PDE_4$) prescribed for the treatment of chronic obstructive pulmonary disease (COPD) and asthma. It is metabolized by CYP3A4 and 1A2 isozymes to a primary active metabolite, roflumilast *N*-oxide, which has a $PDE_4$ selectivity profile and potency similar to the parent roflumilast [124]. When roflumilast (250 μg) was administered once daily for 14 days to healthy control subjects, patients with mild liver cirrhosis (Child-Pugh A), and patients with moderate liver cirrhosis (Child-Pugh B), higher plasma concentrations of roflumilast and roflumilast-*N*-oxide were observed as the severity of cirrhosis increased (Figure 15.39). It was observed that apparent oral clearance (CL/F) and systemic exposure ($AUC_{24}$) gradually increase with the severity of the liver disease. In patients with mild liver cirrhosis (Child-Pugh A), the apparent oral clearance (CL/F) and total exposure ($AUC_{24}$) of parent roflumilast were 37% lower and 51% higher, respectively, compared to the healthy control subjects, whereas no differences in $C_{max}$ between both groups were observed. In patients with moderate liver cirrhosis (Child-Pugh B), the apparent oral clearance (CL/F), total exposure ($AUC_{24}$), and $C_{max}$ of the metabolite roflumilast N-oxide were 44% lower, 92% higher, and 27% higher, respectively, compared to the healthy control subjects [125].

Another example of the impact of hepatic disorders on the oral pharmacokinetic profile of a drug is evident after oral administration of solifenacin succinate, a once-daily anti-muscarinic agent prescribed for overactive bladder. Metabolism also plays an important role in the elimination of solifenacin whereas less than 15% is excreted as intact parent. Main four metabolites: M2 (solifenacin *N*-oxide), M3 (4*R*-hydroxy solifenacin), M4 (4*R*-hydroxy solifenacin *N*-oxide), and M5 (solifenacin *N*-glucuronide). The first three metabolites are formed in the liver by CYP3A4, while the last metabolite is formed by direct glucuronidation of solifenacin. It was observed that after administration of a single oral dose (10 mg) to healthy subjects and moderate hepatic impaired patients exhibited a 60% increase in $AUC_{0-\infty}$ was observed in subjects with moderate

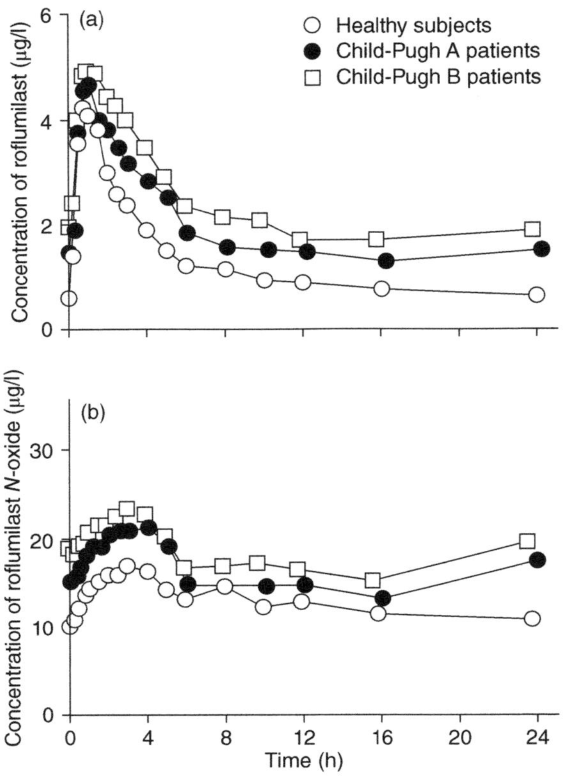


**Figure 15.39** Geometric mean steady-state plasma concentration–time courses of (a) roflumilast and (b) roflumilast *N*-oxide in healthy subjects and patients with mild (Child-Pugh A) and moderate (Child-Pugh B) liver cirrhosis after repeated oral once-daily administration of roflumilast 250 µg for 14 days. Source: Hermann et al. [125]/with permission of Springer Nature.

hepatic impairment (Figure 15.40). Hepatic impaired subjects exhibited a twofold longer half-life and slightly shorter $t_{max}$ than healthy subjects. However, no significant changes in $C_{max}$ were observed between both groups. A smaller fraction of solifenacin was excreted in urine (7.9 ± 3.13%) for hepatically impaired subjects compared to healthy subjects (11.3 ± 3.90%). The M2 and M4 metabolites exhibited similar patterns in pharmacokinetic parameters. The $C_{max}$ was about 1.5- to 1.6-fold lower and about half of the fraction was excreted in hepatic impaired compared to healthy subjects. The half-life for both metabolites was 1.5-fold higher in hepatic impaired compared to healthy subjects, while similar renal clearance was observed in both groups [126].

Overall, it can be observed that in hepatic impaired patients metabolism may be affected leading to higher concentrations of parent compound circulating in plasma. Alterations in circulating plasma concentrations might lead to potential clinical toxicities. Furthermore, a decrease in metabolite(s) may be therapeutically detrimental in cases of metabolically active metabolites.

#### 15.3.8.3 Pain

It is widely recognized that inter-subject pharmacokinetic variability can be linked to different factors such as genetics, food intake, gender, and age, but also to cognitive and emotional factors such as pain. This is a critical factor since a good understating of pain-related effects on the pharmacokinetics of different drugs is important to be able to treat algesia induced by different diseases such as cancer and autoimmune diseases. It has been observed that acute pain alters absorption by causing delays in $t_{max}$, and lower exposure [127]. In this study, racemic ibuprofen (200 or 600 mg) was administered one week before (control, no pain) and again after surgical removal of wisdom teeth (test, acute pain), and the serum concentrations were quantified over six hours. It can be observed in Figure 15.41 that there are lower serum concentrations (lower $C_{max}$ values) of *R*- and *S*-ibuprofen and a delay in $t_{max}$ at the 600 mg dose (similar results were observed with the lower dose of 200 mg).

The $AUC_{0-2h}$ was lower after surgery for both enantiomers, but $AUC_{0-6h}$ was lower only for *S*-ibuprofen. It is notable that there is unidirectional metabolic chiral inversion of *R*- to *S*-ibuprofen normally and the serum ibuprofen enantiomer concentrations are reversed by surgery. The oral concentration–time profiles with ibuprofen suggest that with ibuprofen that acute pain increases the onset of analgesic effect, while reducing the rate and extent of exposure to drug [127].

### 15.3.9 Pregnancy

Pregnant women are susceptible to illnesses and/or pregnancy-induced complications that might require a drug regime. Furthermore, women with chronic diseases (diabetes mellitus, epilepsy, etc.) will continue taking their medications during pregnancy. An understanding of oral concentration–time profiles is important because pregnancy induces physiological changes that may modify drug disposition. For instance, during pregnancy changes in concentration of serum proteins, increases in body water, body fat, liver metabolism, and renal plasma flow (almost double) have been reported [128]. Changes in plasma volume and concentration of serum proteins (available for protein binding) may alter the apparent volume of distribution ($V_d$) and bioavailability of different drugs. Furthermore, changes in $V_d$ and clearance can cause

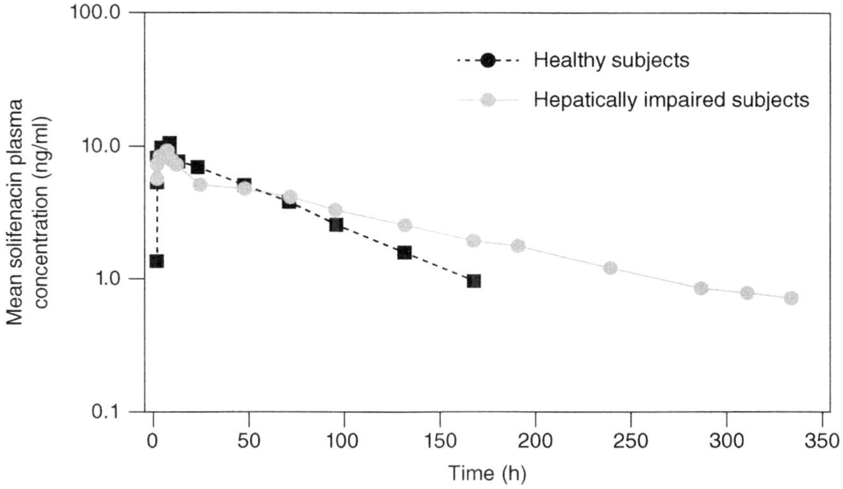


**Figure 15.40** Mean plasma concentration–time profile of solifenacin in healthy subjects ($n = 8$) and in subjects with hepatic impairment ($n = 8$) after a single oral 10 mg dose of solifenacin. Source: Kuipers et al. [126]/with permission of The Japanese Pharmacological Society.

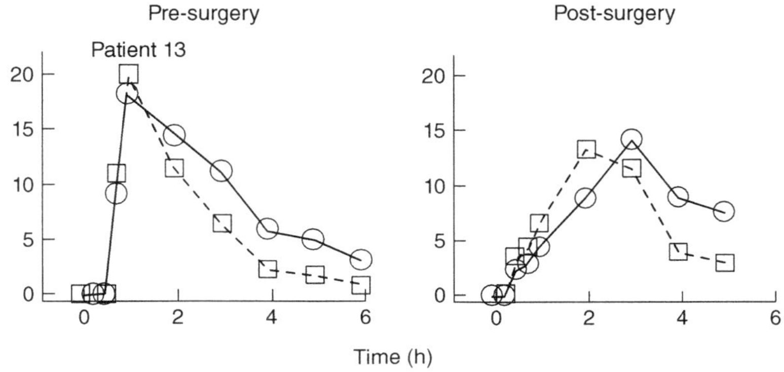


**Figure 15.41** An individual patient's serum ibuprofen enantiomer concentration–time curve following oral administration of 600 mg racemic ibuprofen one week before (Pre-surgery, control) and at the time of first experience moderate to severe pain (Post-surgery). *S*-ibuprofen (open circles), *R*-ibuprofen (closed squares). Source: Jamali and Kunz-Dober [127]/with permission of John Wiley & Sons.

increases or decreases in the terminal elimination half-life of drugs during pregnancy. Depending on the extent of drug metabolism (isozyme(s) involved) and excretion patterns (amount of unchanged drug excreted by the kidneys), changes in pharmacotherapy may be needed during pregnancy. Renal excretion of unchanged drugs and metabolism of drugs catalyzed by select cytochrome P450 (CYP) isoenzymes (i.e. CYP3A4, CYP2D6, and CYP2C9) and uridine diphosphate glucuronosyltransferase (UGT) isoenzymes (i.e. UGT1A4 and UGT2B7) are increased during pregnancy. Therefore, the dose of drugs predominantly excreted unchanged by the kidneys and/or metabolized by these isozymes may need to be increased during pregnancy in order to achieve the desired therapeutic effect. On the other hand, the activity of CYP1A2 and CYP2C19 activity is reduced during pregnancy; thus, the doses of drugs metabolized by these isozymes may need to be reduced in order to avoid potential clinical toxicities [129].

There are excellent comprehensive reviews in the pregnancy-related effects on pharmacokinetics of different types of drugs are available in the literature [129–134]. There are a plethora of drugs that exhibit pharmacokinetic changes during pregnancy, including antiretrovirals, antiarrhythmic agents, antiepileptic drugs, antibiotics, and among others. The examples described below include zidovudine (a retroviral), and cefatrizine (a β-lactam antibiotic).

Zidovudine (Retrovir) was the first agent approved for the treatment of HIV disease, it has also been reported to reduce the risk of maternal-fetal HIV transmission by two-thirds [135]. A crossover study was conducted with three HIV-infected women receiving zidovudine (200 mg orally every four hours) from 19 to 39 weeks of pregnancy and postpartum. Significant pharmacokinetic differences between ante- and post-partum profiles were observed, $AUC_{0\text{–}4h}$ changed from 1.19 to 1.81 mg/l h ($p = 0.02$), and CL/F (2.26 versus 1.67 l/h/kg, $p = 0.05$) [136, 137].

Another example is cefatrizine (a β-lactam antibiotic), which is a second-generation oral cephalosporin that is substantially active against strains of streptococci, staphylococci, enterobacteria, and other common Gram-positive and Gram-negative bacteria. In a

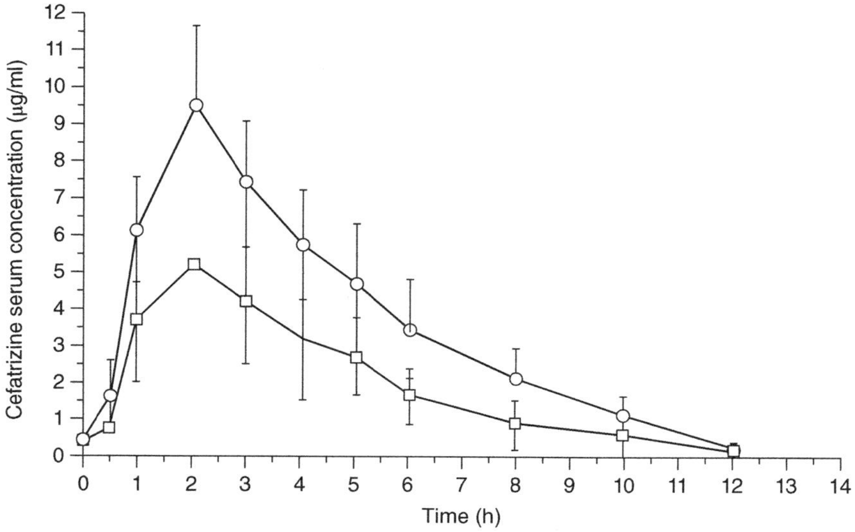


**Figure 15.42** Cefatrizine serum concentrations (mean ± SD, µg/ml) in 20 female volunteers (circles) and 20 women on the 23rd week of gestation (squares) after the administration of the third oral dose of 1 g cefatrizine, every 12 hours. Source: Papantoniou et al. [138]/with permission of Karger.

comparative study, healthy female volunteers and pregnant women on the 23-week of gestation received orally three doses of cefatrizine (1 g) every 12 hours. Lower serum concentrations were observed in the pregnant women group (Figure 15.42). A substantial decrease in $AUC_{0-12h}$ and $C_{max}$ (43 and 44%, respectively) was observed in pregnant women compared to nonpregnant. Furthermore, a 1.63-fold (2.48 versus 1.52 hours) increase in half-life of the drug was observed in pregnant-women [138].

## 15.4 Food-Effects and Oral Concentration Time Profiles

### 15.4.1 Effects of Lipids

For orally administered drugs, data showing the effect of food are a requirement for successful new drug applications. In many such evaluations, the standard FDA breakfast is used. This meal consists of two eggs fried in butter, two strips of bacon, two slices of toast with butter, 4 oz of hash brown potatoes, and 8 oz of whole milk. Nutritionally this breaks down into approximately 150 protein calories, 250 carbohydrate calories, and 500–600 fat calories. High-fat meals can affect drug solubilization in the intestinal tract in several ways. In humans, high-fat meals induce increased concentrations of the hormone cholecystokinin, which stimulates gall bladder contraction and increases in concentrations of bile within the intestinal tract. In animals without a gall bladder such as the rat, it has been observed that concentrations of bile salts within the freely-flowing bile, and the flow rate of bile, is increased [139–141]. The net effect is formation of mixed micelles, which can aid in drug solubilization within and absorption of drug from the lumen of the intestinal tract. For more water-soluble drugs, it is possible that the presence of the lipids within the intestines actually decreases solubility of the drugs, leading to decreases in absorption. The presence of the high-fat meal is also associated with a decrease in gastrointestinal motility, which can further influence the rate or extent of drug absorption.

The coadministration of oral lipids with drug may have variable effects on drug bioavailability and clearance. In some cases, drug plasma concentrations may be decreased, whereas for others either no change or an increase will be observed. The net effect is dependent on several factors including the physicochemical properties of the drug, presence of metabolism and/or transport, and whether or not the drug is subject to binding by lipoproteins. Changes in the maximum concentrations ($C_{max}$) and time to reach $C_{max}$ ($t_{max}$) are common, with actual changes in the AUC being less common.

In many cases due to their routine conduct during drug development, publication of results of food-effect studies are not available, although they do form the basis for administration recommendations. There are some good examples from the published literature that can be used to illustrate the effects of oral lipids on systemic drug concentrations. Increases in plasma concentrations with high-fat meals have been observed with drugs such as amiodarone, halofantrine, cyclosporine, and nifedipine. Each of these drugs shares the property of being highly lipophilic compound.

Amiodarone is an important antiarrhythmic drug. With the coadministration of a tablet formulation of amiodarone given with a high-fat meal, it was observed that the mean plasma AUC more than doubled, from

14 to 33 mg h/l. The $C_{max}$ increased even more in the presence of food (3.7-fold), but there was no change in $t_{max}$, indicating that a greater proportion of the dose had likely made it intact to the systemic circulation [142]. The increase in the concentrations of the active metabolite of amiodarone, desethylamiodarone did not increase as much as those of the parent drug. In the rat, the coadministration of 1% cholesterol in peanut oil with amiodarone as methylcellulose suspension yielded similar results, with increases in AUC and $C_{max}$ of 1.8- and 2.7-fold, respectively [143].

When halofantrine was coadministered with a high-fat meal to human volunteers, its AUC and $C_{max}$ were been reported to increase in humans by nearly 3- and 6.5-fold, respectively [144]. For this drug, unlike the case for amiodarone, there seemed to be an increase in the rate of absorption after the high-fat meal ($t_{max}$, 3.3 hours fed, 6 hours fasted). Similar findings regarding the effects of oral lipids on halofantrine plasma concentrations were noted in the rat, where the AUC and $C_{max}$ were increased by at least twofold (Figure 15.43) [145]. In dogs, there was an astounding increase in the oral AUC of halofantrine when a high-fat meal was coadministered with halofantrine [146]. Plasma AUC rose by over 12-fold for this drug in the dog. There was a much lower effect of the oral lipid on the main metabolite of halofantrine and desbutylhalofantrine, which rose by only 2.1-fold.

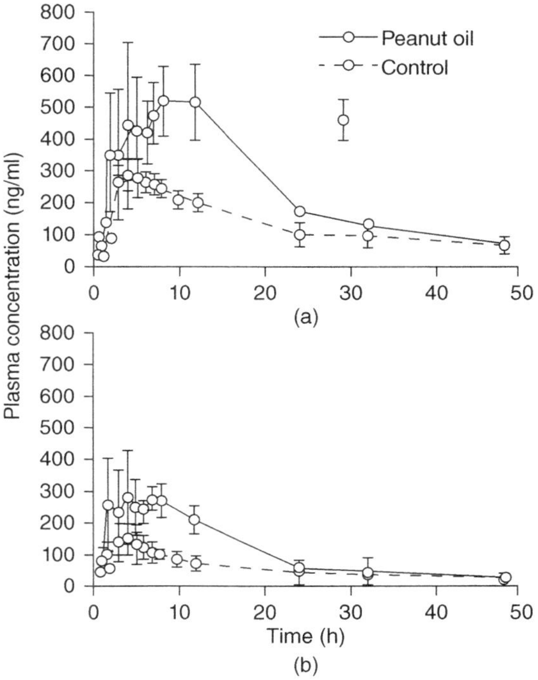


**Figure 15.43** Mean ± SD plasma concentration versus time profiles of halofantrine enantiomers after 7 mg/kg oral dose of (±)-halofantrine HCl given to fasted ($n = 5$) and peanut-oil fed rats ($n = 5$). (a) (+)-Halofantrine; (b) (−)-Halofantrine Source: Reproduced from Brocks and Wasan [145]/with permission of Elsevier.

Some of the increase in plasma concentrations of amiodarone and halofantrine seem to be attributable to more than a simple increase in solubilization of drug within the intestinal tract [143, 145, 147]. Both of these drugs bind to lipoproteins, which are increased in plasma after ingestion of a high-fat meal. In line with this, there is a decrease in the unbound fraction of the drug in the plasma in the postprandial state, which can cause a decrease in the clearance and the volume of distribution of the drug. Consequently, part of the increase in the plasma concentrations is not really due to an increase in drug bioavailability, but rather to reduced clearance of the drug. In rat and dog, this has been demonstrated to be the case. The effect of such increased lipoprotein binding can only be verified by intravenous administration of the drug in the fasted and fed states, which is not typically done as part of food-effect studies.

Cyclosporine is a unique drug in that after a high-fat meal its clearance actually increases, presumably due to an increase in uptake of lipoprotein-bound drug by the liver [148]. Despite the increase in clearance, however, after its oral administration blood cyclosporine AUC in humans is increased. In this case, the effect of increase in bioavailability outweighs the counteracting increase in clearance.

Sometimes formulation effects can play a role in the extent of food effects. A good example of this was recently published for nifedipine, where the bioequivalence of a reference formulation [149], Adalat OROS by Bayer SpA, was compared to that of a test formulation, Nifedipine (by Merck), in the presence of a high-fat meal. After the high-fat meal, the Nifedipine formulation was observed to have a significantly higher AUC and $C_{max}$, indicating that under these conditions the formulations were not bioequivalent. This was an apparent case of "dose-dumping" of the drug from the test formulation caused by the ingestion of the meal.

There are a number of studies indicating that high-fat meals can cause a change in measures of drug absorption, but not bioavailability. One example of this is misoprostol, where the high-fat meal caused a reduction in $C_{max}$ and increase in $t_{max}$, but without a change in overall drug AUC [150]. Ketoconazole displayed a decrease rate of absorption after administration of a high-fat meal. The bioavailability was not significantly altered by the meal, although the authors stated that a trend toward an increase was present [151]. In contrast, for another azole antifungal drug, itraconazole, a significant increase in bioavailability was observed [152].

A good example of a drug where bioavailability and rate of absorption are decreased is provided by a polymer-coated erythromycin particle tablet formulation. In this case, the drug bioavailability was drastically reduced by over 70% when given with a high-fat meal, with prolonged lag times for absorption being apparent in a number of subjects [153]. A high-fat meal had no effect on the rate or extent of absorption of a controlled-release morphine formulation [154].

### 15.4.2 Effect of Citrus Juices

The coingestion of certain fruit juices are known to be associated with either changes in metabolism or transport of drugs, thus impeding or enhancing the extent of drug absorption. Grapefruit juice is the most studied to date, although other evaluations have been made including sweet and sour (Seville) varieties of orange juices, tangerine, lime, and even apple juice [155]. Mechanistically from the perspective of metabolism, inhibition of CYP3A4 (particularly in the intestinal tract) has been implicated to be the primary focus of inhibition by these juices. There is some evidence that P-glycoprotein and organic anion transporters are also inhibited by components in some of these juices (Figure 15.44). The causative ingredients of the effects are thought to include flavonoids and furanocoumarins. Interestingly some of the flavonoids, including hesperetin, naringenin, and eriodictyol are chiral and exhibit juice brand-to-brand differences in content and in stereoselectivity in plasma concentrations after ingestion [156].

Some of the more significant pharmacokinetic effects of the juices on plasma concentrations of orally administered drugs include some drugs with important concentration versus effect relationships for effect/toxicity, such as halofantrine [157], verapamil, [158] and amiodarone [159], although there is at least one report showing a minimal interaction between verapamil with grapefruit juice [160].

For instance, a single oral dose of amiodarone (17 mg/kg) coadministered with 300 ml of grapefruit juice was given to healthy subjects; while the same volume of water was employed as control. It was observed that the group that received grapefruit juice exhibited an increase in $C_{max}$ (1.8-fold) and AUC (1.5-fold) compared to the control group (Figure 15.45) [159].

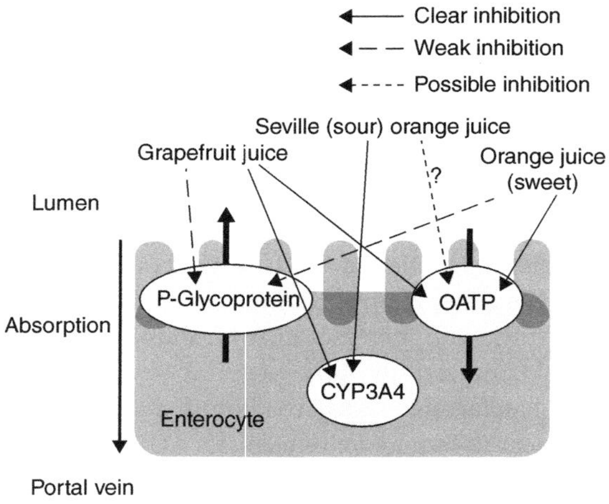


**Figure 15.44** The target sites in enterocytes of the small intestine for various citrus juices. CYP3A4 = cytochrome P450 3A4; OATP = organic anion transporting polypeptide. Source: Saito et al. [155]/with permission of Springer Nature.

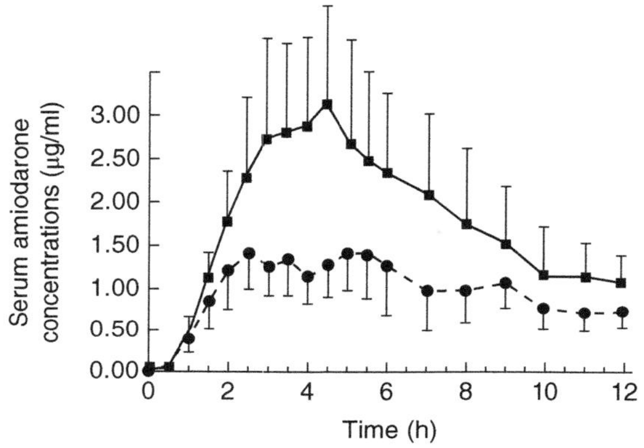


**Figure 15.45** Mean serum concentrations versus time graph for amiodarone in 11 subjects receiving amiodarone with grapefruit juice (solid line and full squares) or with water (broken line and full circles). Each point is mean ± SEM. Source: Libersa et al. [159]/with permission of John Wiley & Sons.

## 15.5 The Impact of the Lymphatic System on Oral Bioavailability

### 15.5.1 Overview of the Lymphatic System

The lymphatic system is a vital component of the mammalian immune system [161] that is ubiquitous except in the central nervous system [162]. It also serves as a secondary circulatory system to drain endogenous materials and fluids from extracellular space toward the cardiovascular system. The lymphatic system has been explored as a potential means of optimizing drug delivery to the systemic circulation [161]. The majority of orally administered drugs traverse the intestinal enterocytes and enter the adjoining blood capillaries; this blood supply eventually merges into the hepatic portal vein. The intestinal mesenteric lymphatics provide an alternate avenue for achieving systemic availability of digested dietary lipids and xenobiotics including fat-soluble vitamins and fat-soluble drugs [163, 164].

The lymphatic path of absorption is promoted by the presence of dietary lipids in the intestinal tract or by the coadministration with formulation-derived lipids. Lipids including cholesterol esters [165], and those possessing a glycerol backbone such as triglycerides [166] and phospholipids [167, 168], are de-esterified within the intestinal lumen by the action of esterases and lipases into products such as cholesterol and di- and mono-glycerides. These products are then able to be absorbed into the enterocytes where they are again re-esterified [166]. These esterified lipids are either stored within the cells or assembled with apoproteins into intestinal lipoprotein, primarily as chylomicrons and very low-density lipoproteins (VLDL). This process may incidentally allow the incorporation of other lipophilic constituents present in the enterocytes (drugs and vitamins). Lipoprotein-associated drug is then secreted from the enterocyte into surrounding intestinal lacteals, which drain into the thoracic duct. Ultimately, the thoracic lymph duct drains the lymph into the venous system. From the perspective of drug absorption, this process, unlike that of blood capillary entry into the mesenteric circulation, circumvents any possible hepatic first-pass metabolism [164].

Thus, blood and lymph are two potential absorptive pathways for the systemic availability of orally administered drugs [163]. Small lipophilic molecules may be predominantly absorbed into mesenteric blood due to the significantly higher blood compared to lymph flow rates [169]. In general drugs with triglyceride solubility exceeding 50 mg/ml, log partition coefficient above 5, or molecular weight more than 500 exhibits a favored absorption via the lymphatic capillaries [169]. However, penclomedine ($\log P = 5.48$ and triglyceride solubility = 175 mg/ml) [170] and CI-976, a lipophilic lipid regulator ($\log P = 5.83$ and triglyceride solubility = 100 mg/ml) were reported to have low lymphatic transport [171].

### 15.5.2 Intestinal Lipoprotein Assembly

The major lipoproteins secreted by the intestine are chylomicrons and VLDL [172]. The production of chylomicrons increases at higher post-prandial lipid loads, whereas VLDL appears predominantly under fasting conditions (low intestinal lipid loads) [172]. The assembly of each form of lipoprotein occurs by two separate independent pathways. That was determined using a hydrophobic surfactant (Pluronic L81) to inhibit the secretion of chylomicrons but not VLDL from enterocytes [173, 174]. The initial step in the lipoprotein assembly is the synthesis of apo-B in the rough endoplasmic reticulum, which associates with phospholipid present in the endoplasmic reticulum membrane. The addition of neutral triglycerides to the apo-B phospholipid complex leads to the production of a primordial lipoprotein that gets released into the lumen of the endoplasmic reticulum [175]. The second step is formation of large triglyceride-rich lipid droplets in the smooth endoplasmic reticulum. The primary source for droplet formation is the absorbed luminal biliary phospholipids, mostly from a high-fat meal [176]. Ultimately, the primordial lipoprotein formed in the rough endoplasmic reticulum fuses with the triglyceride-rich lipid droplets formed in the smooth endoplasmic reticulum [177]. This last step is called core expansion and most likely occurs in the junction between the rough and smooth endoplasmic reticulum [177].

The amount of lipid added to the primordial lipoprotein during core expansion will distinguish between chylomicrons and VLDL production [177]. The lipoproteins rich in lipophilic-core lipids are moved to the Golgi apparatus [176]. Golgi vesicles fuse with the basolateral membrane of the enterocyte and release the lipoprotein particles into the lamina propria where they are favorably absorbed by the mesenteric lymphatic vessels, then drained into blood circulation via thoracic duct [176].

The pre- and post-prandial lymphatic transport has been discussed elsewhere [162]. Briefly, under fasting conditions, the main contribution of endogenous fatty acids comes from bile [178, 179] and with a minor contribution from enterocyte desquamation [178] and from de novo synthesis within the enterocyte [178, 180]. Bile is important to the solubilization and intestinal absorption of food-derived lipids and lipophilic drugs. It allows for the formation of transient mixed micelles of bile salts and phospholipids [181, 182] that once in the enterocyte allows for the self-assemble into lipoproteins [162]. In the case of post-prandial conditions, there is an increase in lymphatic flow, particularly after the consumption of high-fat meals [183]. Of note, the use of conventional dosage forms for hydrophobic drugs where low water solubility is problematic, lipid-based microemulsion, or self-micro-emulsifying formulations are useful approaches to enhance solubility and lymphatic absorption [184]. Because of the wide variety of lipid-based formulations that can be used, the

Lipid Formulation Classification System (LFCS) [185] details their characteristics and potential limitations in achieving small particle sizes and how prone the drug is to precipitate with intestinal fluid dilution and formulation-derived lipid digestion [162]. This may also have the effect of increasing the proportion of the dose that enters the blood via the lymphatic route [183].

### 15.5.3 Integration of Drug with Intestinal Lymph

Drug accessibility to either intestinal lymph vessels or blood vessels after oral dosing depends on endothelial cell architecture. The lymphatic endothelial cells are characterized by the relatively wide cell-to-cell junction and discontinuous basal lamina [186]. Contrarily, the vascular endothelial layer possesses tight junctions and a continuous basal lamina [186]. Another factor to consider is the rate of fluid flow, which is 500-fold higher than that of the mesenteric lymph. Most small lipophilic molecules access the blood more extensively than the lymph even when the drugs' physicochemical properties ($\log P > 5$ and a triglyceride solubility >50 mg/ml) may be amenable to the lymphatic route [187]. Delivery strategies that work to move the drug to the lymphatic pathway are essential considerations.

A correlation between the appearance of triglyceride and administered drug in post-prandial lymph has been observed [188]. Moxidectin, a parasiticide widely utilized in veterinary, is a potential candidate for intestinal lymph absorption [188]. It has a *n*-octanol to water partition coefficient $\log P$ of 6 and poor water solubility (0.006 mg/l). It was observed that a considerable proportion (22%) of orally administered moxidectin to dogs gained access to the intestinal lymph flow [188]. The antimalarial drug halofantrine has been well used by others to study lymphatic drug uptake [189–192]. It has a $\log P > 8$ with triglycerides solubility >50 mg/ml and has been shown to undergo absorption by the lymph pathway. In rats, the proportion of absorbed halofantrine to lymphatic system was shown to be different when three structurally-different triglycerides varying in proportion and position of medium and long-chain fatty acids on the glycerol backbone were coadministered. This highlights the possibility that the formulation design might have on promoting the lymph rather than the blood capillary pathway of systemic availability [193]. An apparent difference in transport rate for halofantrine has been observed because it gets metabolized to desbutylhalofantrine in the enterocyte, which causes it to be removed from the lymph precursor pool [187].

Lymphatic absorption of antineoplastic and antiviral agents has drawn attention from many as tumor cells and viruses can propagate through the lymphatic system [194, 195]. This realization has prompted interest toward the study and development of lymph-promoting drug shunting beyond the improvement of increased systemic bioavailability, to directed therapeutic advantages [196, 197]. As example, the anticancer drug doxorubicin was encapsulated into liposomes, which were preferentially absorbed into the lymphatic circulation after it was administered intraperitoneally to rats [161]. In the thoracic duct, the concentration of doxorubicin from liposomal doxorubicin was >fivefold higher than when non-liposomal doxorubicin was given. This suggests the drugs encapsulated into liposomes and administered intraperitoneally are absorbed and passed into lymphatic vessels, and then remain in lymph nodes [161]. Xie et al. reported that using liposomes as a lymphatic drug delivery system may have a role in improving the therapeutic index of antineoplastic drugs and reduce their side effects.

### 15.5.4 Assessment Tools for Intestinal Lymph Drug Transport

Given the expanding interest in the contribution of lymphatics to the transport of xenobiotics [162, 184, 187], it becomes important to identify models that allow for the ability to recognize when a drug may follow this pathway. This is crucial for designing formulations that promote the pathway. Several techniques and models have been implemented to assess the intestinal lymph absorption and transportation by collection of mesenteric lymph [164]. Mesenteric and intestinal lymph-cannulated rat models are the most commonly used and described methods [198]. After highly invasive surgical procedures for cannulation of the lymphatic vessels, lymph may be collected when the rat is anesthetized [199] or when the rat is conscious but restrained or given free movement while tethered to a harness [189, 200] or the conscious unrestrained rat model where intravenous and oral administration and continuous lymph collection over 12 hours and systemic blood collection over 30 hours can be performed [201, 202]. Such experiments involve the administration of drugs orally by gavage or duodenal infusion with rehydration solutions being administered to compensate for the fluid losses caused by the diversion of the lymph [201].

Lymph cannulation models have also been studied in other species including mice, sheep, and dogs [190, 203, 204]. However, lymph collection from thoracic duct might not accurately estimates the intestinal lymph since it also receives lymph fluid from other regions (directed from the liver and lumbar regions) [199]. Implementing lymph cannulation may pose changes in lymph flow and transport because lymph vessels and flows normally work against a pressure gradient force, which is obviated once the vessel is cannulated [205]. Such cannulations are challenging for routine screening for lymph uptake as it requires highly adept technical surgical skills [164]. Because this type of cannulation requires an irreversible surgery, intestinal lymphatic drug transport is not studied directly in humans.

Utilizing chemical inhibitors is an alternative model, which indirectly study the intestinal lymph [206]. It relies on measuring the level of administered drugs in the blood with and without present of the inhibitors [206]. Those chemicals blockers (colchicine, cycloheximide, Pluronic L81, vinca alkaloids) have potential to inhibit the production of intestinal lipoprotein. They may inhibit the formation of intestinal lymph and transport [206], which may pose an alternative to the use of invasive surgery [207]. Unfortunately, some of these agents are toxic and have been used at doses that are detrimental to the well-being of the animals. More study of their use is required to maximize their potential benefit, which would be worthwhile given that they might have represented a more desirable procedure for screening for lymphatic uptake and formulation development.

Most understanding of membrane transporter and intracellular drug transport has been driven by the application of the human colonic adenocarcinoma Caco-2 cell line [175]. While it has utility for some understanding of these processes intracellularly, it is also colonic in origin, a site where intrinsic enzyme and transporter protein populations differ from those where most drug absorption will occur (in the small intestine). It also lacks blood and lymph flows making it impossible to use as a stand-alone tool to understand lymphatic versus blood capillary routes of absorption. More importantly, Caco-2 cells synthesize apo B-48 and apo B-100, while enterocytes only synthesize apo B-48 [182]. Another critical difference is that Caco-2 cells employ the glycerol-3-phosphate pathway [208] and produce various lipoproteins, but chylomicrons not being the major lipoprotein produced like in the case of enterocytes [182]. The lipolysis predictive model has been described to assess the release of drug from a lipid-based drug delivery system and whether the drug precipitates following lipolysis [209], which simulates *in vivo* conditions. However, because a successful lipid-delivery system needs to keep the drug solubilized following lipolysis, *in vivo* studies are required to fully understand the cumulative and expansive role of all of the factors involved in systemic bioavailability.

### 15.5.5 Interpretation of Oral Concentration-Time Profiles Attributable to Lymphatic Absorption

As discussed above, because of the reliance of chylomicron formation on fat absorption [200], the diversion of drugs via the lymphatic rather than blood capillary pathway will inevitably increase when the drug is taken with fatty meals. The appearance of drug in the blood would be expected to parallel the appearance of lipids in the blood after a high-fat meal. Kassis et al. examined the correlation between fat intake and post-prandial appearance of triglycerides in rats. After overnight fasting, the rats were given a lipid bolus (2.2 ml Liposyn II and 0.8 ml saline) through the duodenum, while they were fully anesthetized with isoflurane. After recovery, lymph was collected every 15 minutes for 4 hours in addition to the fasting lymph that was collected before lipid bolus. It was noted the triglyceride concentration surged steeply up to one hour then it fell gradually, yet it did not return to the baseline fasting values within four hours [183]. Similar observation was denoted by Holm et al. after oral administration of halofantrine to the rats in a mixture consisting of free fatty acids with different levels of unsaturation. There was a linear correlation between the level of halofantrine and triglycerides up to nine hours post-dosing [193].

Because of the link between lymphatic absorption and lipoprotein appearance in the blood, some consideration is warranted for the relationship between lipoprotein binding of drugs and estimation of drug bioavailability. If a high-fat meal was administered with a drug with properties amenable to lymphatic absorption, higher blood levels compared to the fasting state might be attributed to lymphatic diversion. However, this will only occur if the drug was rapidly extracted by the liver and hence possessed a high hepatic extraction ratio ($E$). A low hepatic $E$ diverted to the lymph would not affect the overall blood levels of the drug. Such an event of increased blood concentrations could more likely be explained by other factors. This may include an increased solubility in the gastrointestinal lumen. Another possibility would be an increased level of binding of the drug in the blood. Such a binding in the case of a low hepatic extraction would cause an increase in the area under the concentration versus time curve due to a decrease in clearance, which for a low $E$ drug is directly dependent on the degree of binding in the blood [163].

Only when the lymphatically diverted drug possesses higher levels of hepatic extraction could the possibility of an increase in systemic exposure be attributed to lymphatic diversion. Drugs that are of low extent of hepatic extraction (ratio of blood clearance by hepatic blood flow <0.3) are not expected to have their measures of systemic availability significantly increased by diversion of absorption from the mesenteric blood capillary to the lymphatic mode of absorption. Some increases will be expected for drugs that have moderate levels of

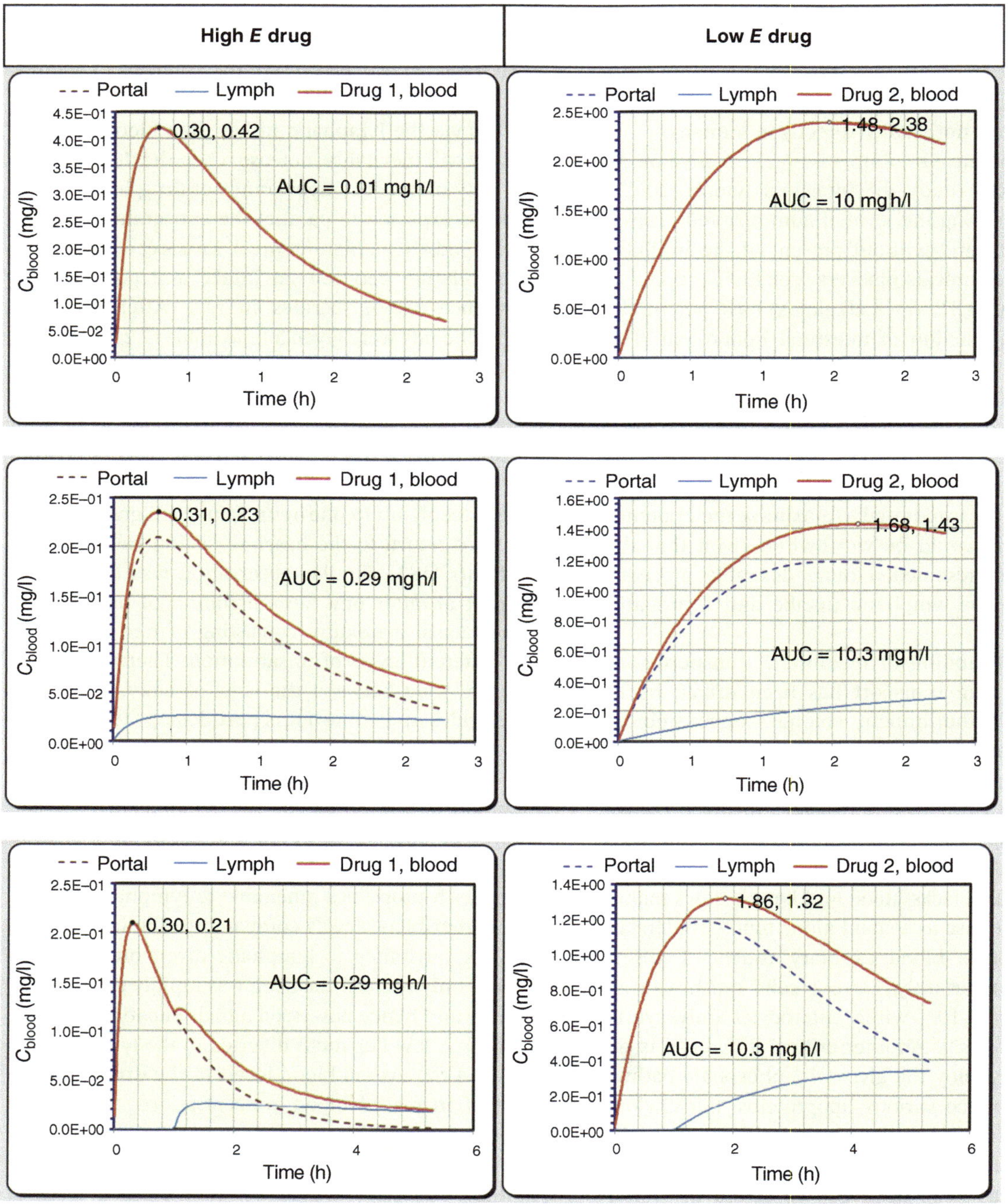


**Figure 15.46** Simulated effects of lymphatic absorption on the concentration versus time curves of drugs when both blood and lymphatic absorption pathways exist. Left column shows simulation for a highly extracted drug ($E = 0.98$); right column shows a low $E$ drug ($E = 0.053$). Top row shows 100% portal pathway; center row shows 50% portal and 50% lymphatic pathway with no lag time; bottom row shows 50% lymphatic pathway with a 1-hour lag time. Values in each panel shows the $t_{max}$ and $C_{max}$.

extraction, but the largest impact of lymphatic diversion of absorption will occur for drugs that possess high hepatic extraction (>0.7). This is an important factor to consider when undertaking efforts to increase systemic availability by increasing lymphatic absorption [163]. We have developed an in silico model for lymphatic absorption with an added delay compared to mesenteric blood capillary absorption using the instructional program uSIMPK25 [163]. As observed in Figure 15.46, there is the potential to lead to secondary peaks in the concentration versus time profile [163], which correlates to actual concentration–time profiles for xenobiotics. Puerarin and berberine chloride undergo lymphatic absorption and their concentration versus time curves do appear to show double peaks [210, 211].

Another important factor to consider in looking at data from drugs that are potentially diverted to the lymphatics is plasma protein binding. Drugs that serve as a good candidate for lymphatic absorption are apt to also be bound to lipoproteins in the blood [212–218]. For low hepatic extraction drugs, any such binding may cause a decrease in the area under the blood concentration versus time curves due to decreasing the clearance of the total drug concentrations. This could lead to the conclusion that the increase was due to lymphatic diversion, which would be an incorrect conclusion. First, diversion from the liver would not affect bioavailability by itself because for low extraction ratio drugs, most would get past the liver intact anyways. The increase could only be explained by an increase in binding. But even with an increase in binding and AUC, the unbound concentrations would be unchanged.

For drugs that are of moderate to high extraction, however, it might appear by looking at the total (bound + unbound) drug concentrations that the fed condition has little effect on the area under the curve whether or not the drug enters by lymph or blood capillaries from the intestinal tract. However, if one considers the unbound concentrations, which are the therapeutically relevant drug component in the blood, one would see that the fed condition leads to an increase in systemic exposure when the drug is subject to lymphatic compared to blood capillary absorption [163].

The coadministration of an appreciable amount of lipid (food-derived or direct lipid dose) diminishes enterocyte metabolism because of the sequestration of drugs into larger lipid droplets that reduce the accessibility of drugs to the enzymes located in the smooth endoplasmic reticulum [162]. This has been observed for benzo(α)pyrene [219] and halofantrine [187]. It is important to mention that ATP-binding cassette (ABC) transporters such as P-gp and MRP-2 are believed to influence intestinal lipoprotein formation and intracellular cholesterol trafficking and that ABCA1 appears to enhance the formation of nascent high-density lipoprotein [162]. Furthermore, P-gp has been shown to have coupled with CYP3A affecting the disposition of drugs such as cyclosporine, ketoconazole, lovastatin, and ritonavir [162].

To summarize, drug passage into the lymphatic system represents an important aspect of the absorption process. It can lead to complexity in the interpretation of the concentration versus time profiles of drugs subject to this mode of systemic availability. Consideration of not only the drug physicochemical properties, but just as importantly, the basic pharmacokinetic properties of drugs is essential in making decisions as to whether lipid formulation design should be attempted to enhance therapeutic benefit by increasing drug diversion to the intestinal lymph.

## 15.6 Summation

For the pharmacokinetic analysis of oral drug concentration–time data simple first-order processes are not always involved and many irregular absorption processes are apparent in the literature. The absorption and alterations in absorption of drugs over time is a very complex process that is intimately affected by both physicochemical and physiological factors. Pharmaceutical chemists and formulation scientists can alter drug physicochemical properties such as particle size, solubility, and permeability in attempts to enhance bioavailability oral concentration–time profiles. However, the understanding of physiology in terms of enterohepatic recirculation, intestinal lymphatics, gastrointestinal motility, disease states, food consumption, and genetics are often able to describe the salient factors that impact drug concentration during absorption. The problem with drug absorption depends largely on its physicochemical and physiological properties within humans or animal models. Therefore, it is desired to design a good strategy to understand in detail and investigate the mechanism of absorption for drugs, as this is a parameter of variability, which ultimately may affect the pharmacodynamics and patient outcome.

## Abbreviations

| | |
|---|---|
| ADME | absorption, distribution, metabolism and elimination |
| BCS | biopharmaceutics classification system |
| CYP | cytochrome P450 |
| GFR | glomerular filtration rate |
| IV | intravenous |
| MAT | mean absorption time |
| MDR1 | multidrug resistance 1 gene |
| NAT | *N*-acetyl transferase |
| PBPK | physiological based pharmacokinetic model |
| PD | pharmacodynamics |
| P-gp | P-glycoprotein |
| PK | pharmacokinetics |

## References


1 Ritschel, W.A. and Kearns, G.L. (2004). *Handbook of Basic Pharmacokinetics: Including Clinical Applications*. APhA Publications.
2 Zornoza, T., Cano-Cebrian, M.J., Hipolito, L. et al. (2006). Evidence of a flip-flop phenomenon in acamprosate pharmacokinetics: an in vivo study in rats. *Biopharm. Drug Dispos.* 27: 305–311.
3 Amidon, G.L., Lennernas, H., Shah, V.P., and Crison, J.R. (1995). A theoretical basis for a biopharmaceutic drug classification: the correlation of in vitro drug product dissolution and in vivo bioavailability. *Pharm. Res.* 12: 413–420.
4 Atkinson, R.M., Bedford, C., Child, K.J., and Tomich, E.G. (1962). Effect of particle size on blood griseofulvin-levels in man. *Nature* 193: 588–589.
5 Skelly, J.P., Van Buskirk, G.A., Savello, D.R. et al. (1993). Scaleup of immediate release oral solid dosage forms. *Pharm. Res.* 10: 313–316.
6 Strickley, R.G. (2004). Solubilizing excipients in oral and injectable formulations. *Pharm. Res.* 21: 201–230.
7 Jounela, A.J., Pentikainen, P.J., and Sothmann, A. (1975). Effect of particle size on the bioavailability of digoxin. *Eur. J. Clin. Pharmacol.* 8: 365–370.
8 Murata, K., Noda, K., Kohno, K., and Samejima, M. (1987). Pharmacokinetic analysis of concentration data of drugs with irregular absorption profiles using multi-fraction absorption models. *J. Pharm. Sci.* 76: 109–113.
9 Zhou, R., Moench, P., Heran, C. et al. (2005). pH-dependent dissolution in vitro and absorption in vivo of weakly basic drugs: development of a canine model. *Pharm. Res.* 22: 188–192.
10 Dutta, S. and Reed, R.C. (2007). Distinct absorption characteristics of oral formulations of valproic acid/divalproex available in the United States. *Epilepsy Res.* 73: 275–283.
11 Bressolle, F., Gomeni, R., Alric, R. et al. (1994). A double Weibull input function describes the complex absorption of sustained-release oral sodium valproate. *J. Pharm. Sci.* 83: 1461–1464.
12 Wong, J., Brugger, A., Khare, A. et al. (2008). Suspensions for intravenous (IV) injection: a review of development, preclinical and clinical aspects. *Adv. Drug Delivery Rev.* 60 (8): 939–954.
13 Dearn, A.R. (1995). Atovaquone pharmaceutical compositions. US Patent US6649659B1, filed 04 October 1999, issued 18 November 2003.
14 Westesen, K. and Siekmann, B. (1998). Solid lipid particles, particles of bioactive agents, and methods for the manufacture and use there of. WO1994020072A1.
15 Liversidge, G.G., Cundy, K.C., Bishop, J.F., and Czekal, D.A. (1992). Surface-modified drug nanoparticles. European Union. US Patent 5,145,684, filed 25 January 1991, issued September 8, 1992.
16 Sarkari, M., Brown, J., Chen, X. et al. (2002). Enhanced drug dissolution using evaporative precipitation into aqueous solution. *Int. J. Pharm.* 243: 17–31.
17 Niu, F., Roby, K.F., Rajewski, R.A. et al. (2006). Paclitaxel nanoparticles: production using compressed $CO_2$ as antisolvent, characterization and animal model studies. In: *Polymeric Drug Delivery Volume II – Polymeric Matrices and Drug Particle Engineering* (ed. S. Svenson), 262–277. Washington, DC: American Chemical Society.
18 Liversidge, G.G. and Cundy, K.C. (1995). Particle-size reduction for improvement of oral bioavailability of hydrophobic drugs: 1. Absolute oral bioavailability of nanocrystalline danazol in beagle dogs. *Int. J. Pharm.* 125: 91–97.
19 Mathew, T.H., Van Buren, C., Kahan, B.D. et al. (2006). A comparative study of sirolimus tablet versus oral solution for prophylaxis of acute renal allograft rejection. *J. Clin. Pharmacol.* 46: 76–87.
20 Davis, M.E. and Brewster, M.E. (2004). Cyclodextrin-based pharmaceutics: past, present and future. *Nat. Rev. Drug Discovery* 3: 1023–1035.

21 Carrier, R.L., Miller, L.A., and Ahmed, I. (2007). The utility of cyclodextrins for enhancing oral bioavailability. *J. Controlled Release* 123: 78–99.

22 Challa, R., Ahuja, A., Ali, J., and Khar, R.K. (2005). Cyclodextrins in drug delivery: an updated review. *AAPS PharmSciTech* 6: E329–E357.

23 Rajewski, R.A. and Stella, V.J. (1996). Pharmaceutical applications of cyclodextrins. 2. In vivo drug delivery. *J. Pharm. Sci.* 85: 1142–1169.

24 Szejtli, J. (1994). Medicinal applications of cyclodextrins. *Med. Res. Rev.* 14: 353–386.

25 Thompson, D.O. (1997). Cyclodextrins – enabling excipients: their present and future use in pharmaceuticals. *Crit. Rev. Ther. Drug Carrier Syst.* 14: 1–104.

26 Uekama, K. (2004). Design and evaluation of cyclodextrin-based drug formulation. *Chem. Pharm. Bull.* 52: 900–915.

27 Uekama, K. and Otagiri, M. (1987). Cyclodextrins in drug carrier systems. *Crit. Rev. Ther. Drug Carrier Syst.* 3: 1–40.

28 Luengo, J., Aranguiz, T., Sepulveda, J. et al. (2002). Preliminary pharmacokinetic study of different preparations of acyclovir with beta-cyclodextrin. *J. Pharm. Sci.* 91: 2593–2598.

29 Evrard, B., Chiap, P., DeTullio, P. et al. (2002). Oral bioavailability in sheep of albendazole from a suspension and from a solution containing hydroxypropyl-β-cyclodextrin. *J. Controlled Release* 85: 45–50.

30 Wong, J.W. and Yuen, K.H. (2001). Improved oral bioavailability of artemisinin through inclusion complexation with β- and γ-cyclodextrins. *Int. J. Pharm.* 227: 177–185.

31 Choudhury, S. and Nelson, K.,.F. (1992). Improvement of oral bioavailability of carbamazepine by inclusion in 2-hydroxypropyl-β-cyclodextrin. *Int. J. Pharm.* 85: 175–180.

32 Kikuchi, M., Hirayama, F., and Uekama, K. (1987). Improvement of oral and rectal bioavailabilities of carmofur by methylated β-cyclodextrin complexations. *Int. J. Pharm.* 38: 191–198.

33 Stracciari, G.L., Malvisi, J., Anfossi, P., and Fregnan, G.B. (1989). Pharmacokinetics of dipyridamole-beta-cyclodextrin complex in dogs. *Arch. Int. Pharmacodyn. Ther.* 300: 7–13.

34 Uekama, K., Fujinaga, T., Otagiri, M. et al. (1981). Enhanced bioavailability of digoxin by γ-cyclodextrin complexation. *J. Pharmacobiodyn.* 4: 735–737.

35 Geczy, J., Bruhwyler, J., Scuvee-Moreau, J. et al. (2000). The inclusion of fluoxetine into gamma-cyclodextrin increases its bioavailability: behavioral, electrophysiological and pharmacokinetic studies. *Psychopharmacology* 151: 328–334.

36 Ramprakash, G. and Nagarsenker, M.S. (2001). Cyclodextrin based oral delivery of flurbiprofen: effect of method of preparation on drug release and studies on in-vivo performance. In: *28th International Symposium on Controlled Release of Bioactive Materials and 4th Consumer & Diversified Products Conference*, 772–773. San Diego, CA: Controlled Release Society, Inc.

37 Savolainen, J., Jarvinen, K., Taipale, H. et al. (1998). Co-administration of a water-soluble polymer increases the usefulness of cyclodextrins in solid oral dosage forms. *Pharm. Res.* 15: 1696–1701.

38 Nambu, N., Shimoda, M., Takahashi, Y. et al. (1978). Bioavailability of powdered inclusion compounds of nonsteroidal antiinflammatory drugs with β-cyclodextrin in rabbits and dogs. *Chem. Pharm. Bull.* 26: 2952–2956.

39 Emara, L.H., Badr, R.M., and Elbary, A.A. (2002). Improving the dissolution and bioavailability of nifedipine using solid dispersions and solubilizers. *Drug Dev. Ind. Pharm.* 28: 795–807.

40 Elkheshen, S.A., Ahmed, S.M., and Al-Quadeib, B.T. (2002). Inclusion complexes of piroxicam with β-cyclodextrin derivatives in comparison with the natural β-cyclodextrin. *Pharm. Ind.* 64: 708–715.

41 Uekama, K., Otagiri, M., Uemura, Y. et al. (1983). Improvement of oral bioavailability of prednisolone by β-cyclodextrin complexation in humans. *J. Pharmacobiodyn.* 6: 124–127.

42 Miyake, K., Arima, H., Hirayama, F. et al. (2000). Improvement of solubility and oral bioavailability of rutin by complexation with 2-hydroxypropyl-β-cyclodextrin. *Pharm. Dev. Technol.* 5: 399–407.

43 Hirayama, F., Horikawa, T., Yamanaka, M., and Uekama, K. (1995). Enhanced bioavailability and reduced metabolism of salbutamol by perbutanoyl-β-cyclodextrin after oral administration in dogs. *Pharm. Sci.* 1: 517–520.

44 Vila-Jato, J.L., Blanco, J., and Vilar, A. (1986). Spironolactone/β-cyclodextrin complex: oral bioavailability in humans. *Pharm. Technol.* 32: 82–85.

45 Arima, H., Yunomae, K., Miyake, K. et al. (2001). Comparative studies of the enhancing effects of cyclodextrins on the solubility and oral bioavailability of tacrolimus in rats. *J. Pharm. Sci.* 90: 690–701.

46 Veiga, F., Fernandes, C., and Teixeira, F. (2000). Oral bioavailability and hypoglycaemic activity of tolbutamide/cyclodextrin inclusion complexes. *Int. J. Pharm.* 202: 165–171.

47 Lennernas, H. and Regardh, C.G. (1993b). Evidence for an interaction between the beta-blocker pafenolol and bile salts in the intestinal lumen of the rat leading to dose-dependent oral absorption and double peaks in the plasma concentration-time profile. *Pharm. Res.* 10: 879–883.

48 Lenneras, H. and Regardh, C.G. (1990). Pharmacokinetics of pafenolol in the rat: a suitable model for studying absorption mechanisms of a drug exhibiting unusual absorption properties in man. *Biopharm. Drug Dispos.* 11: 619–631.

49 Lennernas, H. and Regardh, C.G. (1993a). Dose-dependent intestinal absorption and significant intestinal excretion (exsorption) of the beta-blocker pafenolol in the rat. *Pharm. Res.* 10: 727–731.

50 Lennernas, H. and Regardh, C.G. (1993c). Regional gastrointestinal absorption of the beta-blocker pafenolol in the rat and intestinal transit rate determined by movement of 14C-polyethylene glycol (PEG) 4000. *Pharm. Res.* 10: 130–135.

51 Lennernas, H., Renberg, L., Hoffmann, K.J., and Regardh, C.G. (1993). Presystemic elimination of the beta-blocker pafenolol in the rat after oral and intraperitoneal administration and identification of a main metabolite in both rats and humans. *Drug Metab. Dispos.* 21: 435–440.

52 Regardh, C.G., Lundborg, P., Gabrielsson, M. et al. (1990). Pharmacokinetics of a single intravenous and oral dose of pafenolol – a $beta_1$-adrenoceptor antagonist with atypical absorption and disposition properties – in man. *Pharm. Res.* 7: 1222–1227.

53 Grosvenor, M.P. and Lofroth, J.E. (1995). Interaction between bile salts and beta-adrenoceptor antagonists. *Pharm. Res.* 12: 682–686.

54 Mostafavi, S.A. and Foster, R.T. (2003). Influence of cimetidine co-administration on the pharmacokinetics of acebutolol enantiomers and its metabolite diacetolol in a rat model: the effect of gastric pH on double-peak phenomena. *Int. J. Pharm.* 255: 81–86.

55 Piquette-Miller, M., Foster, R.T., Kappagoda, C.T., and Jamali, F. (1991). Pharmacokinetics of acebutolol enantiomers in humans. *J. Pharm. Sci.* 80: 313–316.

56 Piquette-Miller, M. and Jamali, F. (1997). Pharmacokinetics and multiple peaking of acebutolol enantiomers in rats. *Biopharm. Drug Dispos.* 18: 543–556.

57 Yamaguchi, T., Oida, T., Ikeda, C., and Sekine, Y. (1986). Intestinal absorption of a beta-adrenergic blocking agent nadolol. III. Nuclear magnetic resonance spectroscopic study on nadolol-sodium cholate micellar complex and intestinal absorption of nadolol derivatives in rats. *Chem. Pharm. Bull.* 34: 4259–4264.

58 Shayeganpour, A., Hamdy, D.A., and Brocks, D.R. (2007). Pharmacokinetics of desethylamiodarone in the rat after its administration as the preformed metabolite, and after administration of amiodarone. *Biopharm. Drug Dispos.* 29: 159–166.

59 Roberts, M.S., Magnusson, B.M., Burczynski, F.J., and Weiss, M. (2002). Enterohepatic circulation: physiological, pharmacokinetic and clinical implications. *Clin. Pharmacokinet.* 41: 751–790.

60 Dahlstrom, B.E. and Paalzow, L.K. (1978). Pharmacokinetic interpretation of the enterohepatic recirculation and first-pass elimination of morphine in the rat. *J. Pharmacokinet. Biopharm.* 6: 505–519.

61 Oberle, R.L. and Amidon, G.L. (1987). The influence of variable gastric emptying and intestinal transit rates on the plasma level curve of cimetidine; an explanation for the double peak phenomenon. *J. Pharmacokinet. Biopharm.* 15: 529–544.

62 Adkin, D.A., Davis, S.S., Sparrow, R.A. et al. (1995a). The effect of different concentrations of mannitol in solution on small intestinal transit: implications for drug absorption. *Pharm. Res.* 12: 393–396.

63 Nimmo, J., Heading, R.C., Tothill, P., and Prescott, L.F. (1973). Pharmacological modification of gastric emptying: effects of propantheline and metoclopramide on paracetamol absorption. *Br. Med. J.* 1: 587–589.

64 Clements, J.A., Heading, R.C., Nimmo, W.S., and Prescott, L.F. (1978). Kinetics of acetaminophen absorption and gastric emptying in man. *Clin. Pharmacol. Ther.* 24: 420–431.

65 Wang, Y., Roy, A., Sun, L., and Lau, C.E. (1999). A double-peak phenomenon in the pharmacokinetics of alprazolam after oral administration. *Drug Metab. Dispos.* 27: 855–859.

66 Ogiso, T., Kasutani, M., Tanaka, H. et al. (2001). Pharmacokinetics of epinastine and a possible mechanism for double peaks in oral plasma concentration profiles. *Biol. Pharm. Bull.* 24: 790–794.

67 Chang, R.K. and Shojaei, A.H. (2004). Effect of a lipoidic excipient on the absorption profile of compound UK 81252 in dogs after oral administration. *J. Pharm. Pharm. Sci.* 7: 8–12.

68 Adkin, D.A., Davis, S.S., Sparrow, R.A. et al. (1995b). The effects of pharmaceutical excipients on small intestinal transit. *Br. J. Clin. Pharmacol.* 39: 381–387.
69 Charman, W.N., Rogge, M.C., Boddy, A.W. et al. (1993a). Absorption of danazol after administration to different sites of the gastrointestinal tract and the relationship to single- and double-peak phenomena in the plasma profiles. *J. Clin. Pharmacol.* 33: 1207–1213.
70 Charman, W.N., Rogge, M.C., Boddy, A.W., and Berger, B.M. (1993b). Effect of food and a monoglyceride emulsion formulation on danazol bioavailability. *J. Clin. Pharmacol.* 33: 381–386.
71 Reinberg, A., Clench, J., Aymard, N. et al. (1975a). Circadian variations of the effects of ethanol and of blood ethanol values in the healthy adult man. Chronopharmacological study. *J. Physiol.* 70: 435–456.
72 Reinberg, A., Clench, J., Ghata, J. et al. (1975b). Circadian rhythms in the urinary excretion of salicylate (chronopharmacokinetics) in healthy adults. *C. R. Acad. Sci. Hebd. Seances Acad. Sci. D.* 280: 1697–1699.
73 Baraldo, M. (2008). The influence of circadian rhythms on the kinetics of drugs in humans. *Expert Opin. Drug Metab. Toxicol.* 4: 175–192.
74 Brocks, D.R., Upward, J., Davy, M. et al. (1997). Evening dosing is associated with higher plasma concentrations of pranlukast, a leukotriene receptor antagonist, in healthy male volunteers. *Br. J. Clin. Pharmacol.* 44: 289–291.
75 Poondru, S., Devaraj, R., Boinpally, R.R., and Yamasani, M.R. (2000). Chronopharmacokinetics of sumatriptan in healthy human subjects. *J. Pharm. Pharmacol.* 52: 1085–1090.
76 Martinez, M.N. and Amidon, G.L. (2002). A mechanistic approach to understanding the factors affecting drug absorption: a review of fundamentals. *J. Clin. Pharmacol.* 42: 620–643.
77 Iseki, K., Sugawara, M., Sato, K. et al. (1999). Multiplicity of the $H^+$-dependent transport mechanism of dipeptide and anionic beta-lactam antibiotic ceftibuten in rat intestinal brush-border membrane. *J. Pharmacol. Exp. Ther.* 289: 66–71.
78 Schinkel, A.H., Smit, J.J., van Tellingen, O. et al. (1994). Disruption of the mouse *mdr1a* P-glycoprotein gene leads to a deficiency in the blood-brain barrier and to increased sensitivity to drugs. *Cell* 77: 491–502.
79 Yang, C.Y., Dantzig, A.H., and Pidgeon, C. (1999). Intestinal peptide transport systems and oral drug availability. *Pharm. Res.* 16: 1331–1343.
80 Fromm, M.F. (2004). Importance of P-glycoprotein at blood-tissue barriers. *Trends Pharmacol. Sci.* 25: 423–429.
81 Mouly, S. and Paine, M.F. (2003). P-glycoprotein increases from proximal to distal regions of human small intestine. *Pharm. Res.* 20: 1595–1599.
82 Sparreboom, A., van Asperen, J., Mayer, U. et al. (1997). Limited oral bioavailability and active epithelial excretion of paclitaxel (Taxol) caused by P-glycoprotein in the intestine. *Proc. Natl. Acad. Sci. U.S.A.* 94: 2031–2035.
83 van Asperen, J., van Tellingen, O., Sparreboom, A. et al. (1997). Enhanced oral bioavailability of paclitaxel in mice treated with the P-glycoprotein blocker SDZ PSC 833. *Br. J. Cancer* 76: 1181–1183.
84 Thompson, S.J., Koszdin, K., and Bernards, C.M. (2000). Opiate-induced analgesia is increased and prolonged in mice lacking P-glycoprotein. *Anesthesiology* 92: 1392–1399.
85 Kalvass, J.C., Olson, E.R., and Pollack, G.M. (2007). Pharmacokinetics and pharmacodynamics of alfentanil in P-glycoprotein-competent and P-glycoprotein-deficient mice: P-glycoprotein efflux alters alfentanil brain disposition and antinociception. *Drug Metab. Dispos.* 35: 455–459.
86 Ayrton, A. and Morgan, P. (2001). Role of transport proteins in drug absorption, distribution and excretion. *Xenobiotica* 31: 469–497.
87 Suzuki, H. and Sugiyama, Y. (2000). Role of metabolic enzymes and efflux transporters in the absorption of drugs from the small intestine. *Eur. J. Pharm. Sci.* 12: 3–12.
88 Williams, M.F., Dukes, G.E., Heizer, W. et al. (1992). Influence of gastrointestinal site of drug delivery on the absorption characteristics of ranitidine. *Pharm. Res.* 9: 1190–1194.
89 Koch, K.M., Parr, A.F., Tomlinson, J.J. et al. (1993). Effect of sodium acid pyrophosphate on ranitidine bioavailability and gastrointestinal transit time. *Pharm. Res.* 10: 1027–1030.
90 Basit, A.W., Newton, J.M., Short, M.D. et al. (2001). The effect of polyethylene glycol 400 on gastrointestinal transit: implications for the formulation of poorly-water soluble drugs. *Pharm. Res.* 18: 1146–1150.
91 Schulze, J.D., Ashiru, D.A., Khela, M.K. et al. (2006). Impact of formulation excipients on human intestinal transit. *J. Pharm. Pharmacol.* 58: 821–825.

92 Staveris, S., Houin, G., Dufour, A. et al. (1994). Evidence for a second site of absorption of veralipride in the human small intestine. Use of a new drug delivery telemetric shuttle. *Arzneim.-Forsch.* 44: 1068–1072.

93 Hsu, T.C. and Klatt, O. (1958). Mammalian chromosomes in vitro. IX. On genetic polymorphism in cell populations. *J. Natl. Cancer Inst.* 21: 437–473.

94 Helmut, B. and Werner, G.H. (1964). Pharmacogenetics – results and problems. *Arch. Julius Klaus Stift. Vererbungsforsch. Sozialanthropol Rassenhyg* 39: 9–24.

95 Evans, W.E. and McLeod, H.L. (2003). Pharmacogenomics – drug disposition, drug targets, and side effects. *N. Engl. J. Med.* 348: 538–549.

96 Weinshilboum, R. (2003). Inheritance and drug response. *N. Engl. J. Med.* 348: 529–537.

97 Sakaeda, T., Nakamura, T., Horinouchi, M. et al. (2001). MDR1 genotype-related pharmacokinetics of digoxin after single oral administration in healthy Japanese subjects. *Pharm. Res.* 18: 1400–1404.

98 Rodrigues-Lima, F. and Dupret, J.M. (2004). Regulation of the activity of the human drug metabolizing enzyme arylamine *N*-acetyltransferase 1: role of genetic and non genetic factors. *Curr. Pharm. Des.* 10: 2519–2524.

99 Kalow, W. (1982). Ethnic differences in drug metabolism. *Clin. Pharmacokinet.* 7: 373–400.

100 Rey, E., Gendrel, D., Treluyer, J.M. et al. (2001). Isoniazid pharmacokinetics in children according to acetylator phenotype. *Fundam. Clin. Pharmacol.* 15: 355–359.

101 Hu, Y.F., Tu, J.H., Tan, Z.R. et al. (2007). Association of CYP3A4*18B polymorphisms with the pharmacokinetics of cyclosporine in healthy subjects. *Xenobiotica* 37: 315–327.

102 Kim, R.B., Yamazaki, H., Chiba, K. et al. (1996). In vivo and in vitro characterization of CYP2E1 activity in Japanese and Caucasians. *J. Pharmacol. Exp. Ther.* 279: 4–11.

103 Kirchheiner, J., Sasse, J., Meineke, I. et al. (2003). Trimipramine pharmacokinetics after intravenous and oral administration in carriers of CYP2D6 genotypes predicting poor, extensive and ultrahigh activity. *Pharmacogenetics* 13: 721–728.

104 Kubo, M., Koue, T., Maune, H. et al. (2007). Pharmacokinetics of aripiprazole, a new antipsychotic, following oral dosing in healthy adult Japanese volunteers: influence of CYP2D6 polymorphism. *Drug Metab. Pharmacokinet.* 22: 358–366.

105 Lennard, M.S., Silas, J.H., Freestone, S. et al. (1982). Oxidation phenotype – a major determinant of metoprolol metabolism and response. *N. Engl. J. Med.* 307: 1558–1560.

106 Bolt, H.M., Roos, P.H., and Thier, R. (2003). The cytochrome P-450 isoenzyme CYP2E1 in the biological processing of industrial chemicals: consequences for occupational and environmental medicine. *Int. Arch. Occup. Environ. Health* 76: 174–185.

107 Chen, M.L. (2006). Ethnic or racial differences revisited: impact of dosage regimen and dosage form on pharmacokinetics and pharmacodynamics. *Clin. Pharmacokinet.* 45: 957–964.

108 Kirchheiner, J., Roots, I., Goldammer, M. et al. (2005). Effect of genetic polymorphisms in cytochrome p450 (CYP) 2C9 and CYP2C8 on the pharmacokinetics of oral antidiabetic drugs: clinical relevance. *Clin. Pharmacokinet.* 44: 1209–1225.

109 Meletiadis, J., Chanock, S., and Walsh, T.J. (2006). Human pharmacogenomic variations and their implications for antifungal efficacy. *Clin. Microbiol. Rev.* 19: 763–787.

110 Sakaeda, T. (2005). MDR1 genotype-related pharmacokinetics: fact or fiction? *Drug Metab. Pharmacokinet.* 20: 391–414.

111 Rotelli, A.E., Guardia, T., Juarez, A.O. et al. (2003). Comparative study of flavonoids in experimental models of inflammation. *Pharmacol. Res.* 48: 601–606.

112 Boers, M., Dijkmans, B.A., Breedveld, F.C. et al. (1990). Subclinical renal dysfunction in rheumatoid arthritis. *Arthritis Rheum.* 33: 95–101.

113 Laethem, M.E., Belpaire, F.M., Wijnant, P. et al. (1994). Influence of endotoxin on the stereoselective pharmacokinetics of oxprenolol, propranolol, and verapamil in the rat. *Chirality* 6: 405–410.

114 Piafsky, K.M., Borga, O., Odar-Cederlof, I. et al. (1978). Increased plasma protein binding of propranolol and chlorpromazine mediated by disease-induced elevations of plasma $\alpha_1$ acid glycoprotein. *N. Engl. J. Med.* 299: 1435–1439.

115 Houston, M.C. (1991). Nonsteroidal anti-inflammatory drugs and antihypertensives. *Am. J. Med.* 90: 42S–47S.

116 Ling, S. and Jamali, F. (2005). Effect of early phase adjuvant arthritis on hepatic P450 enzymes and pharmacokinetics of verapamil: an alternative approach to the use of an animal model of inflammation for pharmacokinetic studies. *Drug Metab. Dispos.* 33: 579–586.

117 Mayo, P.R., Skeith, K., Russell, A.S., and Jamali, F. (2000). Decreased dromotropic response to verapamil despite pronounced increased drug

concentration in rheumatoid arthritis. *Br. J. Clin. Pharmacol.* 50: 605–613.

118 Mueller, C., Knoflach, P., and Zielinski, C.C. (1990). T-cell activation in Crohn's disease. Increased levels of soluble interleukin-2 receptor in serum and in supernatants of stimulated peripheral blood mononuclear cells. *Gastroenterology* 98: 639–646.

119 Latteri, M., Angeloni, G., Silveri, N.G. et al. (2001). Pharmacokinetics of cyclosporin microemulsion in patients with inflammatory bowel disease. *Clin. Pharmacokinet.* 40: 473–483.

120 Fluckiger, S.S., Schmidt, C., Meyer, A. et al. (1995). Pharmacokinetics of orally administered cyclosporine in patients with Crohn's disease. *J. Clin. Pharmacol.* 35: 681–687.

121 Edsbacker, S., Bengtsson, B., Larsson, P. et al. (2003). A pharmacoscintigraphic evaluation of oral budesonide given as controlled-release (Entocort) capsules. *Aliment Pharmacol. Ther.* 17: 525–536.

122 Wrighton, S.A., VandenBranden, M., and Ring, B.J. (1996). The human drug metabolizing cytochromes P450. *J. Pharmacokinet. Biopharm.* 24: 461–473.

123 Morgan, D.J. and McLean, A.J. (1995). Clinical pharmacokinetic and pharmacodynamic considerations in patients with liver disease. An update. *Clin. Pharmacokinet.* 29: 370–391.

124 Bundschuh, D.S., Eltze, M., Barsig, J. et al. (2001). In vivo efficacy in airway disease models of roflumilast, a novel orally active PDE4 inhibitor. *J. Pharmacol. Exp. Ther.* 297: 280–290.

125 Hermann, R., Nassr, N., Lahu, G. et al. (2007). Steady-state pharmacokinetics of roflumilast and roflumilast N-oxide in patients with mild and moderate liver cirrhosis. *Clin. Pharmacokinet.* 46: 403–416.

126 Kuipers, M., Smulders, R., Krauwinkel, W., and Hoon, T. (2006). Open-label study of the safety and pharmacokinetics of solifenacin in subjects with hepatic impairment. *J. Pharm. Sci.* 102: 405–412.

127 Jamali, F. and Kunz-Dober, C.M. (1999). Pain-mediated altered absorption and metabolism of ibuprofen: an explanation for decreased serum enantiomer concentration after dental surgery. *Br. J. Clin. Pharmacol.* 47: 391–396.

128 Davis, M., Simmons, C.J., Dordoni, B. et al. (1973). Induction of hepatic enzymes during normal human pregnancy. *J. Obstet. Gynaecol. Br. Commonw.* 80: 690–694.

129 Anderson, G.D. (2005). Pregnancy-induced changes in pharmacokinetics: a mechanistic-based approach. *Clin. Pharmacokinet.* 44: 989–1008.

130 Loebstein, R., Lalkin, A., and Koren, G. (1997). Pharmacokinetic changes during pregnancy and their clinical relevance. *Clin. Pharmacokinet.* 33: 328–343.

131 Mirochnick, M. and Capparelli, E. (2004). Pharmacokinetics of antiretrovirals in pregnant women. *Clin. Pharmacokinet.* 43: 1071–1087.

132 Mitani, G.M., Steinberg, I., Lien, E.J. et al. (1987). The pharmacokinetics of antiarrhythmic agents in pregnancy and lactation. *Clin. Pharmacokinet.* 12: 253–291.

133 Pennell, P.B. (2003). Antiepileptic drug pharmacokinetics during pregnancy and lactation. *Neurology* 61: S35–S42.

134 Philipson, A. (1979). Pharmacokinetics of antibiotics in pregnancy and labour. *Clin. Pharmacokinet.* 4: 297–309.

135 Connor, E.M., Sperling, R.S., Gelber, R. et al. (1994). Reduction of maternal-infant transmission of human immunodeficiency virus type 1 with zidovudine treatment. Pediatric AIDS Clinical Trials Group Protocol 076 Study Group. *N. Engl. J. Med.* 331: 1173–1180.

136 Acosta, E.P., Page, L.M., and Fletcher, C.V. (1996). Clinical pharmacokinetics of zidovudine. An update. *Clin. Pharmacokinet.* 30: 251–262.

137 Watts, D.H., Brown, Z.A., Tartaglione, T. et al. (1991). Pharmacokinetic disposition of zidovudine during pregnancy. *J. Infect. Dis.* 163: 226–232.

138 Papantoniou, N., Ismailos, G., Daskalakis, G. et al. (2007). Pharmacokinetics of oral cefatrizine in pregnant and non-pregnant women with reference to fetal distribution. *Fetal Diagn. Ther.* 22: 100–106.

139 Dumaswala, R., Berkowitz, D., and Heubi, J.E. (1996). Adaptive response of the enterohepatic circulation of bile acids to extrahepatic cholestasis. *Hepatology* 23: 623–629.

140 Klaassen, C.D. and Watkins, J.B. 3rd, (1984). Mechanisms of bile formation, hepatic uptake, and biliary excretion. *Pharmacol. Rev.* 36: 1–67.

141 Knox, R., Stein, I., Levinson, D. et al. (1991). Effect of fat pre-feeding on bile flow and composition in the rat. *Biochim. Biophys. Acta* 1083: 65–70.

142 Meng, X., Mojaverian, P., Doedee, M. et al. (2001). Bioavailability of amiodarone tablets administered with and without food in healthy subjects. *Am. J. Cardiol.* 87: 432–435.

143 Shayeganpour, A., Jun, A.S., and Brocks, D.R. (2005). Pharmacokinetics of amiodarone in hyperlipidemic and simulated high fat-meal rat models. *Biopharm. Drug Dispos.* 26: 249–257.

144 Milton, K.A., Edwards, G., Ward, S.A. et al. (1989). Pharmacokinetics of halofantrine in man: effects of food and dose size. *Br. J. Clin. Pharmacol.* 28: 71–77.

145 Brocks, D.R. and Wasan, K.M. (2002). The influence of lipids on stereoselective pharmacokinetics of halofantrine: important implications in food-effect studies involving drugs that bind to lipoproteins. *J. Pharm. Sci.* 91: 1817–1826.

146 Humberstone, A.J., Porter, C.J., and Charman, W.N. (1996). A physicochemical basis for the effect of food on the absolute oral bioavailability of halofantrine. *J. Pharm. Sci.* 85: 525–529.

147 Humberstone, A.J., Porter, C.J., Edwards, G.A., and Charman, W.N. (1998). Association of halofantrine with postprandially derived plasma lipoproteins decreases its clearance relative to administration in the fasted state. *J. Pharm. Sci.* 87: 936–942.

148 Gupta, S.K., Manfro, R.C., Tomlanovich, S.J. et al. (1990). Effect of food on the pharmacokinetics of cyclosporine in healthy subjects following oral and intravenous administration. *J. Clin. Pharmacol.* 30: 643–653.

149 Wonnemann, M., Schug, B., Anschutz, M. et al. (2008). Comparison of two marketed nifedipine modified-release formulations: an exploratory clinical food interaction study. *Clin. Ther.* 30: 48–58.

150 Karim, A., Rozek, L.F., Smith, M.E., and Kowalski, K.G. (1989). Effects of food and antacid on oral absorption of misoprostol, a synthetic prostaglandin E1 analog. *J. Clin. Pharmacol.* 29: 439–443.

151 Lelawongs, P., Barone, J.A., Colaizzi, J.L. et al. (1988). Effect of food and gastric acidity on absorption of orally administered ketoconazole. *Clin. Pharm.* 7: 228–235.

152 Barone, J.A., Koh, J.G., Bierman, R.H. et al. (1993). Food interaction and steady-state pharmacokinetics of itraconazole capsules in healthy male volunteers. *Antimicrob. Agents Chemother.* 37: 778–784.

153 Randinitis, E.J., Sedman, A.J., Welling, P.G., and Kinkel, A.W. (1989). Effect of a high-fat meal on the bioavailability of a polymer-coated erythromycin particle tablet formulation. *J. Clin. Pharmacol.* 29: 79–84.

154 Kaiko, R.F., Lazarus, H., Cronin, C. et al. (1990). Controlled-release morphine bioavailability (MS Contin tablets) in the presence and absence of food. *Hosp. J.* 6: 17–30.

155 Saito, M., Hirata-Koizumi, M., Matsumoto, M. et al. (2005). Undesirable effects of citrus juice on the pharmacokinetics of drugs: focus on recent studies. *Drug Saf.* 28: 677–694.

156 Yanez, J.A., Remsberg, C.M., Miranda, N.D. et al. (2008). Pharmacokinetics of selected chiral flavonoids: hesperetin, naringenin and eriodictyol in rats and their content in fruit juices. *Biopharm. Drug Dispos.* 29: 63–82.

157 Charbit, B., Becquemont, L., Lepere, B. et al. (2002). Pharmacokinetic and pharmacodynamic interaction between grapefruit juice and halofantrine. *Clin. Pharmacol. Ther.* 72: 514–523.

158 Ho, P.C., Ghose, K., Saville, D., and Wanwimolruk, S. (2000). Effect of grapefruit juice on pharmacokinetics and pharmacodynamics of verapamil enantiomers in healthy volunteers. *Eur. J. Clin. Pharmacol.* 56: 693–698.

159 Libersa, C.C., Brique, S.A., Motte, K.B. et al. (2000). Dramatic inhibition of amiodarone metabolism induced by grapefruit juice. *Br. J. Clin. Pharmacol.* 49: 373–378.

160 Zaidenstein, R., Dishi, V., Gips, M. et al. (1998). The effect of grapefruit juice on the pharmacokinetics of orally administered verapamil. *Eur. J. Clin. Pharmacol.* 54: 337–340.

161 Xie, Y., Bagby, T.R., Cohen, M.S., and Forrest, M.L. (2009). Drug delivery to the lymphatic system: importance in future cancer diagnosis and therapies. *Expert Opin. Drug Delivery* 6: 785–792.

162 Yanez, J.A., Wang, S.W., Knemeyer, I.W. et al. (2011). Intestinal lymphatic transport for drug delivery. *Adv. Drug Delivery Rev.* 63: 923–942.

163 Brocks, D.R. and Davies, N.M. (2018). Lymphatic drug absorption via the enterocytes: pharmacokinetic simulation, modeling, and considerations for optimal drug development. *J. Pharm. Pharm. Sci.* 21: 254s–270s.

164 Trevaskis, N.L., Hu, L., Caliph, S.M. et al. (2015). The mesenteric lymph duct cannulated rat model: application to the assessment of intestinal lymphatic drug transport. *J. Vis. Exp.* 97: 52389.

165 Weng, W., Li, L., van Bennekum, A.M. et al. (1999). Intestinal absorption of dietary cholesteryl ester is decreased but retinyl ester absorption is normal in carboxyl ester lipase knockout mice. *Biochemistry* 38: 4143–4149.

166 Patel, J.P. and Brocks, D.R. (2009). The effect of oral lipids and circulating lipoproteins on the metabolism of drugs. *Expert Opin. Drug Metab. Toxicol.* 5 (11): 1385–1398.

167 Blomstrand, R. (1955a). The intestinal absorption of phospholipids in the rat. *Acta Physiol. Scand.* 34: 147–157.

168 Blomstrand, R. (1955b). On the intestinal absorption of phospholipids in the rat in the absence of bile. *Acta Physiol. Scand.* 34: 158–161.

169 Porter, C.J. and Charman, W.N. (2001). Intestinal lymphatic drug transport: an update. *Adv. Drug Delivery Rev.* 50: 61–80.

170 Myers, R.A. and Stella, V.J. (1992). Factors affecting the lymphatic transport of penclomedine (NSC-338720), a lipophilic cytotoxic drug – comparison to DDT and hexachlorobenzene. *Int. J. Pharm.* 80: 51–62.

171 Hauss, D.J., Mehta, S.C., and Radebaugh, G.W. (1994). Targeted lymphatic transport and modified systemic distribution of Ci-976, a lipophilic lipid-regulator drug, via a formulation approach. *Int. J. Pharm.* 108: 85–93.

172 Tso, P., Lindstrom, M.B., and Borgstrom, B. (1987). Factors regulating the formation of chylomicrons and very-low-density lipoproteins by the rat small intestine. *Biochim. Biophys. Acta* 922: 304–313.

173 Tso, P., Balint, J.A., and Rodgers, J.B. (1980). Effect of hydrophobic surfactant (Pluronic L-81) on lymphatic lipid transport in the rat. *Am. J. Phys* 239: G348–G353.

174 Tso, P., Drake, D.S., Black, D.D., and Sabesin, S.M. (1984). Evidence for separate pathways of chylomicron and very low-density lipoprotein assembly and transport by rat small intestine. *Am. J. Physiol.* 247: G599–G610.

175 van Greevenbroek, M.M. and de Bruin, T.W. (1998). Chylomicron synthesis by intestinal cells in vitro and in vivo. *Atherosclerosis* 141 (Suppl 1): S9–S16.

176 Thomson, A.B., Keelan, M., Garg, M.L., and Clandinin, M.T. (1989). Intestinal aspects of lipid absorption: in review. *Can. J. Physiol. Pharmacol.* 67: 179–191.

177 Hussain, M.M. (2000). A proposed model for the assembly of chylomicrons. *Atherosclerosis* 148: 1–15.

178 Trevaskis, N.L., Porter, C.J., and Charman, W.N. (2005). Bile increases intestinal lymphatic drug transport in the fasted rat. *Pharm. Res.* 22: 1863–1870.

179 Tso, P., Balint, J.A., and Simmonds, W.J. (1977). Role of biliary lecithin in lymphatic transport of fat. *Gastroenterology* 73: 1362–1367.

180 Shiau, Y.F., Popper, D.A., Reed, M. et al. (1985). Intestinal triglycerides are derived from both endogenous and exogenous sources. *Am. J. Physiol.* 248: G164–G169.

181 Hofmann, A.F. and Borgstrom, B. (1962). Physico-chemical state of lipids in intestinal content during their digestion and absorption. *Fed. Proc.* 21: 43–50.

182 Nordskog, B.K., Phan, C.T., Nutting, D.F., and Tso, P. (2001). An examination of the factors affecting intestinal lymphatic transport of dietary lipids. *Adv. Drug Delivery Rev.* 50: 21–44.

183 Kassis, T., Yarlagadda, S.C., Kohan, A.B. et al. (2016). Postprandial lymphatic pump function after a high-fat meal: a characterization of contractility, flow, and viscosity. *Am. J. Physiol. Gastrointest. Liver Physiol.* 310: G776–G789.

184 Porter, C.J.H., Trevaskis, N.L., and Charman, W.N. (2007). Lipids and lipid-based formulations: optimizing the oral delivery of lipophilic drugs. *Nat. Rev. Drug Discovery* 6: 231–248.

185 Pouton, C.W. and Porter, C.J. (2008). Formulation of lipid-based delivery systems for oral administration: materials, methods and strategies. *Adv. Drug Delivery Rev.* 60: 625–637.

186 Pepper, M.S. and Skobe, M. (2003). Lymphatic endothelium: morphological, molecular and functional properties. *J. Cell Biol.* 163: 209–213.

187 Trevaskis, N.L., Charman, W.N., and Porter, C.J. (2008). Lipid-based delivery systems and intestinal lymphatic drug transport: a mechanistic update. *Adv. Drug Delivery Rev.* 60: 702–716.

188 Lespine, A., Chanoit, G., Bousquet-Melou, A. et al. (2006). Contribution of lymphatic transport to the systemic exposure of orally administered moxidectin in conscious lymph duct-cannulated dogs. *Eur. J. Pharm. Sci.* 27: 37–43.

189 Caliph, S.M., Charman, W.N., and Porter, C.J. (2000). Effect of short-, medium-, and long-chain fatty acid-based vehicles on the absolute oral bioavailability and intestinal lymphatic transport of halofantrine and assessment of mass balance in lymph-cannulated and non-cannulated rats. *J. Pharm. Sci.* 89: 1073–1084.

190 Khoo, S.M., Edwards, G.A., Porter, C.J., and Charman, W.N. (2001). A conscious dog model for assessing the absorption, enterocyte-based metabolism, and intestinal lymphatic transport of halofantrine. *J. Pharm. Sci.* 90: 1599–1607.

191 Porter, C.J., Charman, S.A., and Charman, W.N. (1996a). Lymphatic transport of halofantrine in the triple-cannulated anesthetized rat model: effect of lipid vehicle dispersion. *J. Pharm. Sci.* 85: 351–356.

192 Porter, C.J., Charman, S.A., Humberstone, A.J., and Charman, W.N. (1996b). Lymphatic transport of halofantrine in the conscious rat when administered as either the free base or the hydrochloride salt:

effect of lipid class and lipid vehicle dispersion. *J. Pharm. Sci.* 85: 357–361.

193 Holm, R., Porter, C.J., Mullertz, A. et al. (2002). Structured triglyceride vehicles for oral delivery of halofantrine: examination of intestinal lymphatic transport and bioavailability in conscious rats. *Pharm. Res.* 19: 1354–1361.

194 Karaman, S. and Detmar, M. (2014). Mechanisms of lymphatic metastasis. *J. Clin. Invest.* 124: 922–928.

195 Pantaleo, G., Graziosi, C., Demarest, J.F. et al. (1993). HIV infection is active and progressive in lymphoid tissue during the clinically latent stage of disease. *Nature* 362: 355–358.

196 Makwana, V., Jain, R., Patel, K. et al. (2015). Solid lipid nanoparticles (SLN) of Efavirenz as lymph targeting drug delivery system: elucidation of mechanism of uptake using chylomicron flow blocking approach. *Int. J. Pharm.* 495: 439–446.

197 Ravi, P.R., Aditya, N., Kathuria, H. et al. (2014). Lipid nanoparticles for oral delivery of raloxifene: optimization, stability, in vivo evaluation and uptake mechanism. *Eur. J. Pharm. Biopharm.* 87: 114–124.

198 Bollman, J.L., Cain, J.C., and Grindlay, J.H. (1948). Techniques for the collection of lymph from the liver, small intestine, or thoracic duct of the rat. *J. Lab Clin. Med.* 33: 1349–1352.

199 Boyd, M., Risovic, V., Jull, P. et al. (2004). A stepwise surgical procedure to investigate the lymphatic transport of lipid-based oral drug formulations: cannulation of the mesenteric and thoracic lymph ducts within the rat. *J. Pharmacol. Toxicol. Methods* 49: 115–120.

200 Kohan, A.B., Howles, P.N., and Tso, P. (2012). Methods for studying rodent intestinal lipoprotein production and metabolism. *Curr. Protoc. Mouse Biol.* 2: 219–230.

201 Edwards, G.A., Porter, C.J., Caliph, S.M. et al. (2001). Animal models for the study of intestinal lymphatic drug transport. *Adv. Drug Delivery Rev.* 50: 45–60.

202 Hauss, D., Fogal, S., and Ficorilli, J. (1998). Chronic collection of mesenteric lymph from conscious, tethered rats. *Contemp. Top. Lab. Anim. Sci.* 37: 56–58.

203 Kota, J., Machavaram, K.K., McLennan, D.N. et al. (2007). Lymphatic absorption of subcutaneously administered proteins: influence of different injection sites on the absorption of darbepoetin alfa using a sheep model. *Drug Metab. Dispos.* 35: 2211–2217.

204 Trevaskis, N.L., Caliph, S.M., Nguyen, G. et al. (2013). A mouse model to evaluate the impact of species, sex, and lipid load on lymphatic drug transport. *Pharm. Res.* 30: 3254–3270.

205 Kassis, T., Kohan, A.B., Weiler, M.J. et al. (2012). Dual-channel in-situ optical imaging system for quantifying lipid uptake and lymphatic pump function. *J. Biomed. Opt.* 17: 086005.

206 Dahan, A. and Hoffman, A. (2005). Evaluation of a chylomicron flow blocking approach to investigate the intestinal lymphatic transport of lipophilic drugs. *Eur. J. Pharm. Sci.* 24: 381–388.

207 Xiao, C. and Lewis, G.F. (2012). Regulation of chylomicron production in humans. *Biochim. Biophys. Acta* 1821: 736–746.

208 Trotter, P.J. and Storch, J. (1993). Fatty acid esterification during differentiation of the human intestinal cell line Caco-2. *J. Biol. Chem.* 268: 10017–10023.

209 Dahan, A. and Hoffman, A. (2008). Rationalizing the selection of oral lipid based drug delivery systems by an in vitro dynamic lipolysis model for improved oral bioavailability of poorly water soluble drugs. *J. Controlled Release* 129: 1–10.

210 Elsheikh, M.A., Elnaggar, Y.S.R., Hamdy, D.A., and Abdallah, O.Y. (2018). Novel cremochylomicrons for improved oral bioavailability of the antineoplastic phytomedicine berberine chloride: optimization and pharmacokinetics. *Int. J. Pharm.* 535: 316–324.

211 Tang, T.T., Hu, X.B., Liao, D.H. et al. (2013). Mechanisms of microemulsion enhancing the oral bioavailability of puerarin: comparison between oil-in-water and water-in-oil microemulsions using the single-pass intestinal perfusion method and a chylomicron flow blocking approach. *Int. J. Nanomed.* 8: 4415–4426.

212 Attili-Qadri, S., Karra, N., Nemirovski, A. et al. (2013). Oral delivery system prolongs blood circulation of docetaxel nanocapsules via lymphatic absorption. *Proc. Natl. Acad. Sci. U.S.A.* 110: 17498–17503.

213 Blaner, W.S., Li, Y., Brun, P.J. et al. (2016). Vitamin A absorption, storage and mobilization. *Subcell Biochem.* 81: 95–125.

214 Dahan, A., Mendelman, A., Amsili, S. et al. (2007). The effect of general anesthesia on the intestinal lymphatic transport of lipophilic drugs: comparison between anesthetized and freely moving conscious rat models. *Eur. J. Pharm. Sci.* 32: 367–374.

215 Fidge, N.H., Shiratori, T., Ganguly, J., and Goodman, D.S. (1968). Pathways of absorption

of retinal and retinoic acid in the rat. *J. Lipid Res.* 9: 103–109.

216 Patel, J.P., Hamdy, D.A., El-Kadi, A.O., and Brocks, D.R. (2012). Effect of serum lipoproteins on stereoselective halofantrine metabolism by rat hepatocytes. *Chirality* 24: 558–565.

217 Speeckaert, M.M., Taes, Y.E., De Buyzere, M.L. et al. (2010). Investigation of the potential association of vitamin D binding protein with lipoproteins. *Ann. Clin. Biochem.* 47: 143–150.

218 Urien, S., Barre, J., Morin, C. et al. (1996). Docetaxel serum protein binding with high affinity to alpha$_1$-acid glycoprotein. *Invest. New Drugs* 14: 147–151.

219 Van Veld, P.A., Vetter, R.D., Lee, R.F., and Patton, J.S. (1987). Dietary fat inhibits the intestinal metabolism of the carcinogen benzo[*a*]pyrene in fish. *J. Lipid Res.* 28: 810–817.

220 Zhou, Z.L., Li, X., Peng, H.Y. et al. (2006). Multiple dose pharmacokinetics of risperidone and 9-hydroxyrisperidone in Chinese female patients with schizophrenia. *Acta Pharmacol. Sin.* 27: 381–386.

# 16

# *In Vitro-In Vivo* Correlations of Pharmaceutical Dosage Forms

*Deliang Zhou*[1] *and Yihong Qiu*[2]

[1] *Small Molecules CMC, BeiGene Ltd, Beijing, PR China*
[2] *QPD Solutions LLC, Lincolnshire, IL, USA*

## 16.1 Introduction

Linking *in vitro* tests to the *in vivo* performance of pharmaceutical products has been a continuing effort of pharmaceutical scientists. The therapeutic efficacy of a dosage form is determined by the concentration of its active(s) in the body or at the site(s) of action, which depends on the rate and extent of drug absorption and disposition in the body. Solid oral drug absorption has two critical elements in general: the *in vivo* dissolution in the gastrointestinal (GI) lumen, and the permeation of drug molecules by crossing the GI epithelial membrane into the bloodstream. While the permeation and disposition are inherent properties of a particular molecule under normal circumstances, the *in vivo* dissolution of a drug depends greatly on the properties of the drug, dosage form design, and drug release environment in the GI tract. This has stimulated extensive research to understand, model, and predict the *in vivo* performance based on *in vitro* dissolution/release tests.

These efforts in developing a relationship between dissolution and bioavailability may be traced back to as earlier as 1950s [1]. A number of studies subsequently revealed that differences in dissolution rate could cause differences in clinical effects, including onset time, efficacy, and toxicity [2, 6]. Development and evolution of various dissolution apparatus, methodologies, test media and models is a result of these study findings in an attempt to control drug product quality.

The general notion of using *in vitro* test to predict *in vivo* performance should not, however, be limited to the aspect of drug dissolution or release, which is by far the most common application of *in vitro-in vivo* correlations. For example, permeation studies with various *in vitro* epithelial cell cultures (e.g. Caco-2), using cultured human tissues or excised animal intestinal membranes, or even synthetic membranes (e.g. parallel artificial membrane permeability assay, or PAMPA) have increased our understanding on the permeation properties of many drug molecules. Attempts to design a "complete" GI absorption model have also emerged, such as the TNO *in vitro* gastro-intestinal models to simulate the successive dynamic processes in GI tract from the stomach to the small and large intestines despite its limited applicability. Nevertheless, steady progresses have been made. In a nutshell, the biopharmaceutics classification system (BCS) [7] has highlighted the important aspects of oral drug absorption based on the simple concepts of solubility and permeability.

Linking *in vitro* testing to the *in vivo* performance of a pharmaceutical formulation can be generally categorized as the *in vitro-in vivo* relationship (IVIVR) or association (IVIVA). These relationships may be quantitative, semiquantitative, or even merely rank ordering. Once a predictive model is established and validated appropriately, an *in vitro-in vivo* correlation (IVIVC) is then achieved. This chapter discusses the basic concepts and applications of IVIVC based on the connection between *in vitro* dissolution/release and *in vivo* drug absorption, with a focus on the conventional convolution/deconvolution-based IVIVC for oral dosage forms. More in-depth treatments of this subject can be found in textbooks and review articles [8, 9]. The same principles hold true for other long-acting dosage forms, such as extended-release injectables, transdermal patches, or ocular implants (https://doi.org/10.1016/j.xphs.2021.10.031, https://doi.org/10.1016/j.ijpharm.2021.120738, https://doi.org/10.1016/j.jconrel.2020.11.055, https://doi.org/10.2174/1567201814666170516113406, https://doi.org/10.1016/j.addr.2021.113829). However, they are not within the scope of this chapter. We also introduce

*Oral Bioavailability and Drug Delivery: From Basics to Advanced Concepts and Applications*, First Edition. Edited by Ming Hu and Xiaoling Li.

the concept and application of the physiologically based biopharmaceutical modeling (PBBM), which has advanced significantly in developing IVIVC for more complex situations.

Historically, the *in vitro* dissolution test has been used primarily as a quality control tool for product and process and evaluation of batch-to-batch variability. However, the dissolution rate of a specific dosage form could be an arbitrary parameter, depending on various factors such as type of apparatus, medium, and agitation. Unless it is demonstrated that the *in vitro* dissolution behavior reflects the *in vivo* performance in humans, the data cannot be used to predict or judge the clinical effectiveness of a drug product. Although the dissolution specifications are typically based on clinical and/or bioavailability batches, passing these specifications does not necessarily guarantee the bioequivalence of future production batches. On the other hand, a failure may not necessarily indicate a lack of bioequivalence, because these specification limits are not quantitatively linked to in vivo absorption. Once an IVIVC is established, the dissolution test can be used as a meaningful quality control tool to provide maximum assurance and predictability to its *in vivo* performance [10].

## 16.2 Categories of *In Vitro-In Vivo* Correlations

Due to the significant role that an IVIVC can play in assuring product performance, regulatory agencies in the US and EU have been actively involved in understanding and developing IVIVC, which results in science-based regulatory guidance and directives in the late 1990's. USP [11] defines IVIVC as the establishment of a relationship between a biological property, or a parameter derived from a biological property produced by a dosage form, and a physicochemical characteristic of the same dosage form. It has also laid out a framework for defining different levels of correlation, which are still in use today. FDA has put more emphasis on the quantitative aspect, as it defines [10] IVIVC as "A predictive mathematical model describing the relationship between an *in vitro* property of a dosage form (usually the rate or extent of drug dissolution or release) and a relevant *in vivo* response (e.g. plasma drug concentration or amount of drug absorbed)." In its 1997 guidance [10], FDA presented its comprehensive perspective on methods of developing and evaluating an IVIVC, setting dissolution specifications using an IVIVC, and using dissolution as a surrogate for bioequivalence study useful for certain manufacturing and postapproval changes.

Currently, IVIVC is categorized into levels of A, B, C, and multiple C as explained as below [10]:

Level A correlation represents a point-to-point relationship between the *in vitro* dissolution and the *in vivo* input rate (e.g. the *in vivo* dissolution from the dosage form), or the *in vivo* response time course (e.g. plasma concentration). Level A correlation is usually established by a two-stage procedure: deconvolution of the drug plasma concentration profile followed by comparison of the fraction of drug absorbed to the fraction of drug dissolved. Alternatively, it can be developed by a single-stage procedure, i.e. directly modeling the drug plasma concentration using the *in vitro* dissolution as the input based on the convolution principle.

Level B IVIVC correlates the summary parameters that characterize the *in vitro* and *in vivo* time courses based on statistical moment analysis, e.g. correlations between the mean *in vitro* dissolution time to the mean *in vivo* dissolution time, or to mean *in vivo* residence time. This is not considered as a point-to-point mapping because a number of profiles may produce similar mean statistical parameters.

Level C correlation establishes a single point relationship between a dissolution parameter, for example, amount dissolved *in vitro* at a particular time (e.g. $Q_{60}$) or the time required for dissolution of a fixed amount (e.g. $t_{50\%}$), and a summary parameter that characterizes the *in vivo* time course (e.g. AUC, $C_{max}$, or $T_{max}$).

A multiple Level C correlation establishes the relationship between *in vitro* dissolution and one or several pharmacokinetic parameters of interest, at multiple time points.

In regulatory applications as well as pharmaceutical development, Level A correlation is the most informative as it describes a "full" understanding (model) that is valid within the boundary of the correlation (e.g. types of dosage form, mechanisms of drug release, dose strength, etc.). It can be used to predict the entire *in vivo* plasma profile based on the *in vitro* release data, and is thus useful in setting meaningful dissolution specifications, applying quality control measures, and obtaining biowaivers in postapproval changes (e.g. formulation, equipment, process, and manufacturing sites). Level C correlation can provide useful guidance in early product development. While multiple Level C correlation, when validated, can be as useful in the same context where Level A is used. The establishment of a multiple Level C correlation often warrants further exploration of Level A correlation. Level B correlation lacks the uniqueness and is therefore least useful for regulatory applications.

Convolution and deconvolution techniques play an important role in understanding as well as establishing an IVIVC, particularly, the Level A correlation. This is discussed in Section 16.3.

## 16.3 Convolution and Deconvolution

### 16.3.1 Linear System, Convolution, and Deconvolution

A linear system is a mathematical model of a system based on the use of a linear operator. A general deterministic system can be described by operator, $G$, which maps an input, $f(t)$, as a function of time to an output or response, $r(t)$, a type of black box description. Because linear systems satisfy the properties of superposition and scaling, the behavior of the system subjected to a complex input can be described as a sum of responses from simpler inputs. This mathematical property makes the solution of modeling equations simpler than many nonlinear systems. For time-invariant systems, this is the basis of the impulse response or the frequency response methods, which describe a general input function $f(t)$ in terms of unit impulses or frequency components.

Linear systems find important applications in the pharmaceutical field because the body's response to a drug input can be considered as a linear system [12] in most cases. Nonlinear response, which may be necessary to describe systems close to their saturation capacity (such as enzymatic metabolism at high concentration), is beyond the scope of this discussion. In developing IVIVC, one of the critical steps is to understand and obtain drug input for correlation with the *in vitro* drug release from the dosage form. In general, the plasma concentration after a unit dose of *i.v.* bolus contains all information with regard to drug disposition in the body. Thus, it can be defined as a unit impulse function, $g(t)$, that is characteristic of the body system with an instantaneous input. The unit impulse function can often be represented as monoexponential or polyexponential decays (i.e., PK profiles after *i.v.* administration). When a drug molecule is introduced into the systemic circulation by routes other than *i.v.* bolus (such as oral) with a rate described by an input function, $f(t)$, the system response or plasma concentration, $r(t)$, can be calculated via convolution of the input function, $f(t)$, and the unit impulse function, $g(t)$,

$$r(t) = \int_0^t f(\tau)g(t-\tau)\mathrm{d}\tau \tag{16.1}$$

or, in simpler notion:

$$r(t) = f(t) * g(t) \tag{16.2}$$

Various notions have been used in the literature for the unit impulse function, such as the system function, the weighting function, the characteristic function, the disposition function, or the kernel. In IVIVC application, the unit impulse function is not limited to the body response after *i.v.* bolus dose, but also solution or immediate release (IR) dosage forms, depending on the situations and how system is defined.

Often we know or can measure the response after a unit impulse (a reference dose), $g(t)$, and the response after certain input, $r(t)$ (e.g., PK data of oral tablets). The question is how to find the particular input function, $f(t)$ (e.g., in vivo absorption of oral tablets). Deconvolution is the inverse operation of convolution that is used to address this question. Alternatively, deconvolution can also be applied to search the system function, $g(t)$ when $f(t)$ and $r(t)$ are known. Deconvolution may be written as the following notions:

$$f(t) = r(t)//g(t) \text{ or } g(t) = r(t)//f(t) \tag{16.3}$$

The convolution integral may appear too abstract. However, it is evident from the superposition principle: the total response of a system after various inputs at different times is the sum of all partial responses, taking into account the time-shifting. To illustrate the point, let's take a look at the body response after a number of discrete *i.v.* bolus ($D_i$, $i = 0, 1, 2, \ldots, n$) are introduced into a system at time $t_i$ ($i = 0, 1, 2, \ldots, n$). It is further assumed that the drug disposition follows a one-compartment model, with an elimination rate constant of $k_e$, and $t_0 = 0$. The amount of drug in the body at any time is then:

$$r(t) = \sum_{i=0}^{n} D_i e^{-k_e(t-t_i)} \tag{16.4}$$

For example, Figure 16.1 shows the response after *i.v.* bolus injections at $t = 0, 1, 2, 3.5$, in the amount of 1,

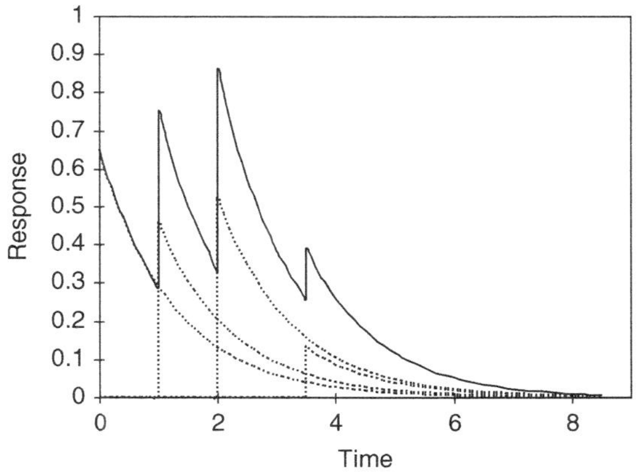


**Figure 16.1** Illustration of the Superposition Principle.

0.7, 0.8, and 0.2, respectively, for a system characterized by $g(t) = 0.65e^{-0.8t}$. The solid line represents the overall system response while the dash lines are the response for each individual dose.

When the input rate is continuous instead of discrete doses as above, the summation then becomes integration, which leads to the convolution integral.

It might be interesting to see how pharmacokinetic models can be derived in a different, but more straightforward manner with the convolution concept. For example, amount of drug in the body during *i.v.* infusion, a zero-order input of drug molecules in the blood, assuming one-compartment model, can be obtained as:

$$D_B(t) = \int_0^t k_{in} \exp[-k_e(t-\tau)]d\tau = \frac{k_{in}}{k_e}[1 - \exp(-k_e t)] \tag{16.5}$$

where $k_{in}$ and $k_e$ are the rate of infusion and the elimination rate constant, respectively.

Similarly, the amount of drug in the body during first-order absorption can be readily obtained by convolution:

$$D_B(t) = \int_0^t F k_a D_0 \exp(-k_a t) \cdot \exp[-k_e(t-\tau)]d\tau = \frac{FD_0 k_a}{k_a - k_e}[\exp(-k_e t) - \exp(-k_a t)] \tag{16.6}$$

where $F$ is the total bioavailable fraction, $D_0$ is the dose administered, and $k_a$ is the first-order absorption rate constant.

Convolution and deconvolution have been applied extensively to delineate the various rate processes in biopharmaceutics and pharmacokinetics, such as *in vivo* dissolution, absorption, metabolism, elimination, and pharmacodynamics. Typical examples of their applications in IVIVC are shown in Table 16.1 and illustrated in Figure 16.2.

**Table 16.1** Typical system definitions in linear system analysis for oral delivery.

| Case | Unit impulse response, $g(t)$ | Input response, $r(t)$ | Input function $f(t)$ |
|---|---|---|---|
| I | Plasma levels from *i.v.* bolus | Plasma levels from oral solution | Absorption in GI |
| II | Plasma levels from *i.v.* bolus | Plasma levels from oral solid dosage form | Dissolution and absorption in GI |
| III | Plasma levels from oral solution (or IR dosage form) | Plasma levels from oral solid dosage form | Dissolution in GI |

### 16.3.2 Convolution Methods

Performing convolution as defined by the integral is straightforward when each function is relatively simple and well-defined. However, in many cases, the mathematical expression can become cumbersome. Thus, other methods may be preferred, such as the Laplace transform (Section 16.3.3.1). Numerical convolution plays an important role when $g(t)$ and $r(t)$ do not assume a specific analytical form. It is very useful in IVIVC applications since drug release or plasma concentration data are discrete and usually do not conform to exact functional forms except for those of *i.v.* bolus administration.

In mathematics, continuous function may be approximated by a series of discrete numbers, at certain values of the independent variable. The convolution signal can also be evaluated similarly. Let's divide the time domain into $n+1$ nodes, $t_i$, where $i = 0, 1, 2, \ldots, n$. The value of each function, $r(t)$, $f(t)$, and $g(t)$, at the nodes $t_i$, is written as $r_i$, $f_i$, and $g_i$, respectively. Then the general numerical convolution scheme is shown below:

$$r_0 = 0$$
$$r_i = \int_0^{t_i} f(\tau)g(t_i - \tau)d\tau = \sum_{j=1}^{i} \int_{t_{j-1}}^{t_j} f(\tau)g(t_i - \tau)d\tau \tag{16.7}$$

The numerical computation is then carried out in each interval $[t_{i-1}, t_i]$, using various approximations, such as the staircases and trapezes.

#### 16.3.2.1 Point-Area Method

In point-area method [13, 14], the input function is assumed to be constant (i.e. staircase) in each interval $[t_{i-1}, t_i]$, $\tilde{f}_i$, which may be approximated conveniently by its mean or midpoint value, or simply as $\tilde{f}_j = (f_j + f_{j-1})/2$. Let's further define $\Delta t_j = t_j - t_{j-1}$, $\Delta f_j = f_j - f_{j-1}$. The mean of a function in the interval $[t_{i-1}, t_i]$, can be calculated as:

$$\tilde{f}_j = \frac{1}{t_j - t_{j-1}} \int_{t_{j-1}}^{t_j} f(t)dt \tag{16.8}$$

This leads to the point-area numerical scheme:

$$r_i = \sum_{j=1}^{i} \tilde{f}_j \int_{t_{j-1}}^{t_j} g(t_i - \tau)d\tau = \sum_{j=1}^{i} \tilde{f}_j \tilde{g}_{i-j} \Delta t_j \tag{16.9}$$

Because of the assumption of constant input rate in each interval, the convolution integral can be assessed by the mean value of each of the input and weighting functions in a simple manner. One may notice that

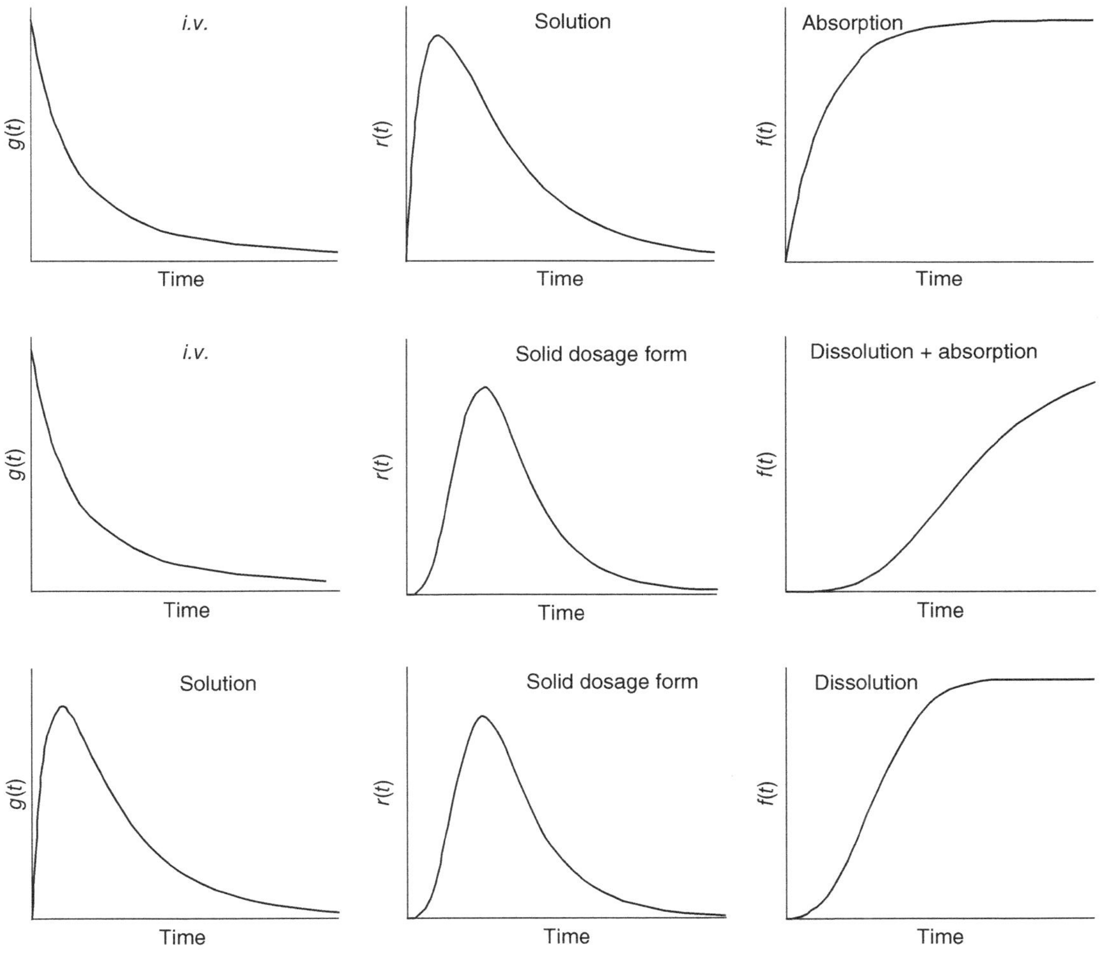


**Figure 16.2** Illustration of some system definitions in linear system analysis.

$\Delta F_j = \tilde{f}_j \Delta t_j$ is the cumulative input in the interval $[t_{j-1}, t_j]$, therefore, the point-area formula can be rewritten as:

$$r_i = \sum_{j=1}^{i} \Delta F_j \cdot \tilde{g}_{i-j} \tag{16.10}$$

which is quite useful, because the input functions in most IVIVC applications are usually in the cumulative formats, e.g. amount dissolved. Chiou [15] has provided an equivalent method-based midpoint value approximation.

#### 16.3.2.2 Trapezoidal Method

A slight improvement to the point-area scheme is the trapezoidal [16] method, where the integral $\int_0^{t_i} f(\tau)g(t_i - \tau)\mathrm{d}\tau$ is approximated by the trapezes formed by the function product $f(\tau)g(t_i - \tau)$ in each of the integration interval, i.e.

$$r_i = \frac{1}{2} \sum_{j=1}^{i} (f_{j-1} \cdot g_{i-j+1} + f_j \cdot g_{i-j}) \Delta t_j \tag{16.11}$$

One challenge with either point-area or trapezoidal numerical scheme is the handling of unequal time intervals. With equally spaced time steps, implementation of the above numerical schemes is straightforward. However, unequal time interval is almost the norms in pharmacokinetic sampling. Because of the exponential decaying nature of most pharmacokinetics, sampling with equal time steps is not a wise use of resources. To deal with unequally spaced time steps, one usually needs to resort to numerical interpolation. This can be done before the convolution by generating profiles based on equal time intervals, or during the convolution by interpolating function values at certain required time points that are not in the original experimental dataset. Figure 16.3 shows an example of convolution with unequal time steps based on the point-area scheme. All these numerical schemes can be easily implemented in a computer program, such as Microsoft Excel [17].

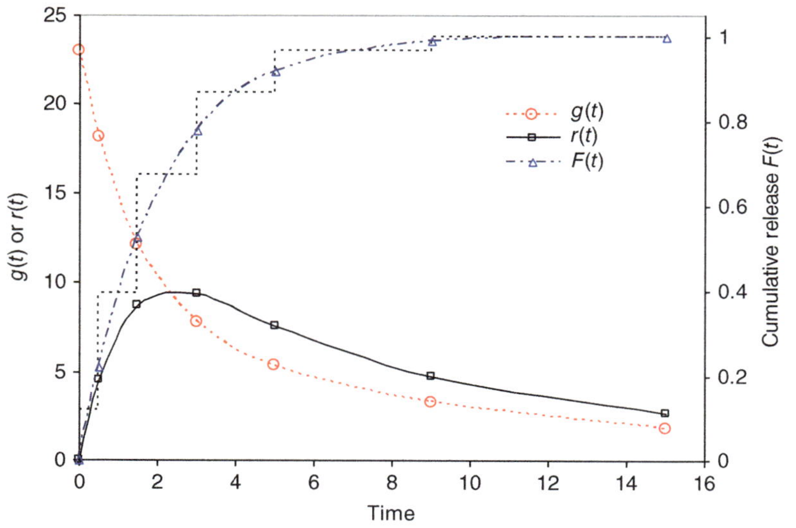


**Figure 16.3** Illustration of point-area convolution with unequal time steps.

### 16.3.3 Deconvolution Methods

Deconvolution is more frequently needed in IVIVC applications to estimate the *in vivo* dissolution/absorption of a formulation when the body responds following administrating a reference solution or *i.v.* formulation is available. Deconvolution has also been applied to determine the unit impulse response, albeit less frequently. Here we briefly discuss some basic deconvolution methods.

#### 16.3.3.1 Algebraic Approach

When the algebraic forms of the input and unit impulse functions are known, convolution and deconvolution can be conveniently conducted using the analytical forms. One of the techniques most useful in this case is the Laplace transform, which is defined as:

$$L\{f(t)\}c = \int_0^{\infty} e^{-st} f(t)dt \tag{16.12}$$

The Laplace integral transforms a function from the time domain to the Laplacian domain (denoted by "*s*"). For example, the Laplace transform of the monoexponential $ae^{-bt}$ is $a/(s+b)$. Most operations, such as differentiation and integration, are much simplified in the Laplace domain. For example, the convolution integral becomes simple multiplication:

$$L\{r(t)\} = L\left\{\int_0^{t_i} f(\tau)g(t_i - \tau)d\tau\right\} = L\{f(t)\} \cdot L\{g(t)\} \tag{16.13}$$

Therefore, the deconvolution is simply a function division in the Laplacian domain:

$$L\{f(t)\} = L\{r(t)\}/L\{g(t)\} \tag{16.14}$$

$$f(t) = L^{-1}[L\{r(t)\}/L\{g(t)\}] \tag{16.15}$$

The $L^{-1}$ represents the inverse Laplace transform. In some cases, the mathematic form in the Laplacian domain may be simple enough and the inverse transform is intuitive. For example, when the unit impulse function is a monoexponential, i.e. $g(t) = Ae^{-bt}$,

$$L\{f(t)\} = \frac{sL\{r(t)\} + bL\{r(t)\}}{A} \tag{16.16}$$

$$f(t) = \frac{r'(t) + br(t)}{A} \text{ (assuming } r_0 = 0) \tag{16.17}$$

where $r(t)$ represents drug plasma concentration, and $r'(t)$ is the derivative. The amount of drug absorbed from time 0 to $t$, is then obtained by integration:

$$X_a(t) = \frac{r(t) + b\int_0^t r(t)dt}{A} \tag{16.18}$$

where no assumption on the form of the response function has been made. Veng-Pedersen has provided the general deconvolution solutions when the unit impulse function can be described by the polyexponential [18].

In most cases, however, the inverse transform may have to resort to a technique called partial fraction because the expressions are cumbersome. Numerical partial fractions can also be aided by computers, which are based on various techniques utilized in solving problems of linear algebra.

#### 16.3.3.2 Mass Balance Approaches

Wagner-Nelson [19] and Loo-Riegelman [20, 21] deconvolution methods are based on mass balance principle and compartmental analysis. The former is derived based on analysis of the one-compartment model, where the total amount of drug absorbed equals to the amount in body, plus the amount that has been eliminated at any time point:

$$X_a(t) = VC(t) + k_e V \int_0^t C(t)\mathrm{d}t \qquad (16.19)$$

where $X_a$(t), $C(t)$, $V$, and $k_e$ are amounts of drug absorbed, and the drug concentration at time $t$, the apparent volume of distribution, and the elimination rate constant, respectively. The following equation is used when expressed in fraction absorbed:

$$F_a(t) = \frac{C(t) + k_e \int_0^t C(t)\mathrm{d}t}{k_e \int_0^\infty C(t)\mathrm{d}t} \qquad (16.20)$$

where $F_a(t)$ is the fraction of bioavailable drug absorbed ($FD$) at time $t$, $D$ is the administered dose, and $F$ is absorbed fractional coefficient of the dose. It can be seen that Wagner-Nelson method is a special case of Eq. (16.18).

In 1983, Wagner [22] published his exact Loo-Rieglman method for absorption analysis of one to three compartment drugs. For drugs following biexponential disposition, the amount absorbed at any time is the sum of the amounts in the central and peripheral compartments, plus the amount that has already been cleared, i.e.

$$X_a(t) = V\left[C(t) + k_{1E} \int_0^t C(t)\mathrm{d}t + k_{12}\mathrm{e}^{-k_{21}t} \int_0^t C(t)\mathrm{e}^{k_{21}t}\mathrm{d}t\right] \qquad (16.21)$$

where $k_{1E}$ is the elimination rate constant from the central compartment, $k_{12}$, and $k_{21}$ is the microconstants between compartments, with subscript 1 and 2 representing the central and peripheral compartments, respectively. Similar equation can be derived for triexponential disposition. All these equations can be considered as special cases in Veng-Pedersen's polyexponential treatments [18].

#### 16.3.3.3 Numerical Approaches

The numerical convolution schemes, either the point-area method or the trapezoidal method, can be rearranged such that the input function is evaluated numerically provided the unit impulse function and the response function are known. It should be noted that special treatment on the first point of input function, i.e. $f_0$, is required during deconvolution when $g_0 > 0$ (i.e. *i.v.* bolus input) [23].

Direct numerical deconvolution methods suffer from instability [24] in the presence of errors in the experimental data, due to the ill-conditioned deconvolution matrix in many instances. Improvements over the direct deconvolution methods have been proposed. One technique is called the constrained deconvolution [25], where the value of a deconvoluted node is only allowed to be nonnegative, nonincreasing, or nondecreasing. Least-square [26] method is another improvement, where the input and/or system functions can be assumed to be polyexponential [27, 28], polynomial [29, 30], spline [31], prescribed functions [32], etc. Constrained least square [33, 34] has also been used for the more advanced numerical deconvolution programs.

## 16.4 Development and Assessments of an IVIVC

In the 1997 guidance [10], FDA provided a comprehensive scientific framework and regulatory guideline for IVIVC model development, evaluation, and applications. In general, the key components in establishing an IVIVC include (i) study design, (ii) model building, and (iii) model validation based on appropriate statistical assessments.

### 16.4.1 Study Design and General Considerations

To develop an IVIVC, generally two or more, preferably three or more, formulations with different *in vitro* release rates and different *in vivo* absorption profiles are required. Use of only one formulation could still be possible, according to the guidance, when the *in vitro* release profile is shown to be independent of the dissolution test conditions. The osmotic pump may be considered as an example delivery technology that is least sensitive to test conditions.

All formulations should be tested under the same dissolution conditions and the dissolution profiles should differ sufficiently (e.g. by 10%) to allow for a meaningful correlation. In principle, any *in vitro* test method may be used to obtain the dissolution characteristics of the dosage forms as long as it is proven to be predictive of *in vivo* performance. However, The IVIVC guidance prefers *in vitro* data generated in an aqueous medium using USP apparatus I (basket) or II (paddle), operating within an appropriate range of rotation speeds (e.g. 50–100 rpm), though USP apparatus III (reciprocating cylinder) or IV (flow through cell) may also be used. Other methodologies may also be used when justified. A minimum of 12 individual dosage units should be tested, with coefficient of variation less than 10%. The key is to develop a dissolution methodology to adequately discriminate among formulations. The dissolution methods may need to be revisited based on the *in vivo* performance of the formulations, as shown in the case of divalproex sodium [35–37].

Bioavailability studies for IVIVC development should be performed in humans with sufficient sample size, in order to adequately characterize the absorption profiles of the drug products. The number of subjects may depend on the inherent variability of a particular drug molecule in human. The Guidance prefers crossover studies, but parallel studies or cross-study analyses are also acceptable, provided there is a common reference (*i.v.*, oral solution, or IR dosage form) treatment. In addition, IVIVCs are usually developed in the fasted state. When a drug is not tolerated in the fasted state, studies may be conducted in the fed state.

### 16.4.2 IVIVC Model Building

#### 16.4.2.1 Level A Correlation

This level of correlation defines a direct relationship between *in vitro* dissolution and *in vivo* responses. The corresponding *in vivo* response can be plasma concentrations or the amount of drug released and/or absorbed *in vivo*. The latter is obtained by deconvoluting the observed plasma concentration–time profile from the reference treatment (*i.v.*, solution formulation, or IR formulation). Level A correlation is usually established by a two-stage procedure, or alternatively by the single-stage approach, as briefly described below.

***16.4.2.1.1 Two-Stage Approach*** The two-stage approach is the most frequently used in building IVIVC models. It involves estimation of the *in vivo* release/absorption profile from plasma concentration–time data using an appropriate deconvolution technique for each formulation, followed by comparison with the percent released *in vitro*, preferably using a basic linear model with intercept, *a*, and slope, *b*:

$$(\%\ \text{absorbed})_{in\ vivo} \text{ or } (\%\ \text{release})_{in\ vivo} = a + b\,(\%\ \text{released})_{in\ vitro} \quad (16.22)$$

Ideally, the slope is close to unity, indicating 1 : 1 correlation between *in vitro* dissolution and *in vivo* absorption/dissolution. A negative intercept implies that the *in vivo* process lags behind the *in vitro* dissolution. A positive intercept has no clear physiological meaning. It can be a result of relatively high variability or curvature at the early time points. A scaling factor, such as time-shifting and time-scaling parameters, may be used. However, whatever the data treatment, it should be the same for each of the formulations. Nonlinear models, while uncommon, may also be appropriate. Figure 16.4 provides a typical model-building process based on two-stage approaches.

***16.4.2.1.2 Single-Stage Approach*** Alternatively, IVIVC can be estimated with a single-stage approach, where the time course of plasma concentrations is directly

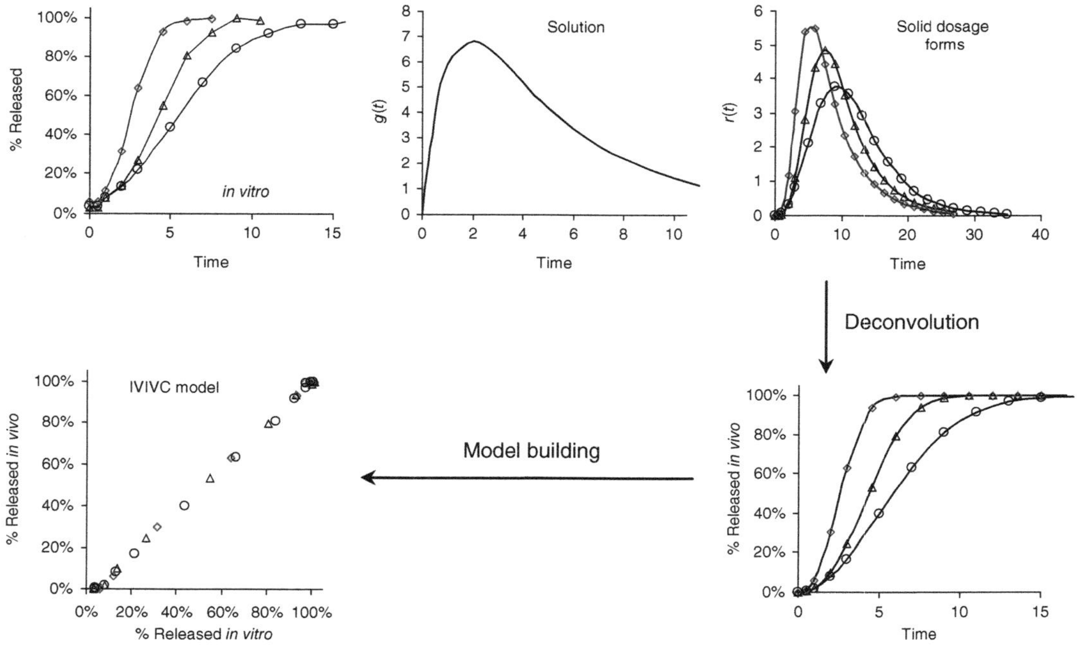


**Figure 16.4** Illustration of the two-stage approach for IVIVC.

predicted by convoluting the dissolution data based on appropriate model. The plasma concentration predicted from the model is then compared directly to the observed. This convolution-based modeling focuses on the ability to predict measured quantities, rather than indirectly estimating "*in vivo*" fraction absorbed and/or released. Veng-Pedersen et al. have elaborated this approach based on four tablet formulations of carbamazepine [38].

One may have to pay attention to the absolute bioavailability when the convolution approach is applied. For drugs showing different regional absorptions, overestimation of plasma concentrations can occur for the slower-releasing formulations because the convolution predicts the same dose-normalized AUC as the reference dose. This situation may be dealt with by truncating the absorption profiles, or treating the absorption differently in different regions, based on experimental evidence and/or prior knowledge.

One advantage of the convolution method is its capability of implementation of nonlinear pharmacokinetics. The convolution-based approach was also suggested as preferred over the deconvolution-based approach for assessing time-scaled IVIVC [39].

#### 16.4.2.2 Level B Correlation

Level B IVIVC compares meantimes that molecules spend in *in vitro* and *in vivo* spaces. It also requires *in vivo* testing of at least three dosage forms. Based on statistical moment analysis [40, 41], mean residence time (MRT), mean absorption time (MAT), or mean *in vivo* dissolution time ($\text{MDT}_{\textit{in vivo}}$) can be calculated and related to mean *in vitro* dissolution time ($\text{MDT}_{\textit{in vitro}}$). The analysis is very general without an assumption on kinetics of the drug transfer in a system. The correlations are reflective of the entire dissolution or plasma concentration curves by using all the *in vitro* and *in vivo* data. One major issue is that several different curves can produce similar or same meantime values. All parameters can be model-independent.

The time course of plasma concentrations or dissolution can be regarded as a statistical distribution curve of drug molecules in the body system. Meantime parameters can be defined as the first moment of a particular distribution, whereas the second moment of the distribution corresponds to the variance or spread of the distribution.

According to the above notion, the following can be defined: the mean residence time, MRT, from the plasma time profile; mean absorption time, MAT, from the absorption profile; and the mean dissolution time, MDT, from the concentration–time profile. The general equation is:

$$\text{MRT (MAT or MDT)} = \frac{\int_0^\infty tC(t)\mathrm{d}t}{\int_0^\infty C(t)\mathrm{d}t} \tag{16.23}$$

It can further be shown that the mean residence times are additive for successive processes, e.g.:

$$\text{MAT}_{\text{p.o}} = \text{MRT}_{\text{p.o.}} - \text{MRT}_{i.v.} \tag{16.24}$$

$$\text{MDT}_{\text{solid, }\textit{in vivo}} = \text{MRT}_{\text{solid}} - \text{MRT}_{\text{soln}} \tag{16.25}$$

*In vitro* dissolution is often expressed as fraction cumulative amount, $F(t)$, dissolved over time. A dissolution profile may be regarded as distribution function of the residence times of drug molecules in the formulation. The mean dissolution time can be calculated as the AUC defined by the amount released axis and the release profile, i.e. the AUC of the profile where the time and the amount released axes are switched:

$$\text{MDT}_{\textit{in vitro}} = \frac{\int_0^\infty \frac{\mathrm{d}F(t)}{\mathrm{d}t} t\mathrm{d}t}{F_\infty} = \frac{\int_0^\infty t\mathrm{d}F}{F_\infty} \tag{16.26}$$

where $F_\infty$ is the maximum amount released. The integral in the above equation is actually the area between the drug dissolution curve and its asymptote.

#### 16.4.2.3 Level C Correlation

Level C correlation requires *in vivo* testing of three or more formulations having different release rates for establishing a linear or nonlinear relationship between *in vitro* and *in vivo* parameters because each data point of the correlation plot corresponds to one formulation. The summary parameters for *in vivo* profile may be AUC, $C_{\text{max}}$, and $T_{\text{max}}$, while the *in vitro* parameters are often percent released at time $t$ ($Q_t$). A single-point Level C correlation may be helpful for formulation development. It can also be used to justify setting of a dissolution specification, at a particular time point. The information is generally insufficient for justifying a biowaiver. A multiple Level C correlation establishes correlation at several time points with one or more bioavailability parameters of interest. If the correlation is over the entire dissolution profile, it may be useful to support a biowaiver. A multiple Level C correlation should be based on at least three dissolution time points covering the early, middle, and late (preferably >80% released or asymptote reached) stages of the dissolution profile.

A multiple Level C correlation often triggers further exploration of a Level A correlation because the development of the Level A correlation becomes more likely according to the FDA guidance.

### 16.4.3 Model Validation

An IVIVC, when developed, should be demonstrated consistently with two or more, preferably three or more formulations with different release rates. The same relationship should hold for all formulations in the investigation. However, if one or more formulations (usually with the highest or lowest release rate) do not show the same relationship, the correlation may still hold within the range of release rates defined by rest formulations.

One of the statistical assessments is to compare the fit of a reduced model where all tested formulations are fitted to a single correlation line with that of a full model where each formulation is fitted to a different correlation line. If both models fit well, the IVIVC is considered validated. If the full model fits well, but the reduced model does not, or if the full model is statistically different from the reduced model at significance level of 0.05, the IVIVC becomes formulation dependent and is therefore invalid. A correlation is drug-specific and often found applicable only to a specific type of formulation.

The FDA guidance [10] has laid out detailed criteria to assess the predictive performance of a proposed IVIVC through the estimation of prediction error, PE, defined as: %PE = [(observed value – predicted value)/observed value] × 100. Determination of PE involves calculation of the *in vivo* absorption profiles from the *in vitro* data using the established IVIVC model, followed by prediction of the corresponding plasma concentration profiles via convolution. The guidance further elaborates approaches to validate the model internally and externally.

The internal PE evaluates how tightly the model describes the data used to develop the IVIVC. The internal validation may suffice when the IVIVC has been derived using three or more formulations with different release rates for a non-narrow therapeutic index drug. Internal validation requires that the mean absolute %PE should be within 10% for $C_{max}$ and AUC and that the %PE for each individual formulation should not exceed 15%.

The external validation requires dataset that was not used in the original development of the IVIVC, such as formulations with different release rates, formulations with minor manufacturing process changes, or less preferable, formulations from a different manufacturing batch. The external validation affords greater "confidence" in the model and is recommended by FDA for the following scenarios: (i) narrow therapeutic index drug; (ii) only two formulations were used to develop the correlation; (iii) calculation of the internal PE is inconclusive. A limit of 10% PE is also required to establish an external validation. PE of 10–20% indicates inconclusive predictability and illustrates the need for further study using additional datasets. PE of greater than 20% indicates lack of predictability.

Multiple Level C correlations should be based on at least three dissolution time points covering the early, middle, and late stages of the dissolution profile. The assessment of the predictability will depend on the type of application for which the correlation is to be used. These same methods and criteria hold as in the case of Level A correlation.

## 16.5 Applications of an IVIVC

The objective of developing an IVIVC is to identify an *in vitro* test that can serve as a predictor of its *in vivo* performance. When this is achieved, IVIVC can be used as a surrogate for human bioavailability studies, which is useful to (i) set meaningful dissolution specifications to ensure product quality, (ii) obtain biowaivers in cases where bioequivalence would be required, such as certain postapproval changes, (iii) guide formulation and process development and scale-up.

### 16.5.1 Setting Dissolution Specifications

Generally, *in vitro* dissolution specifications are established to ensure batch-to-batch consistency and are important tools to provide process control and quality assurance. Dissolution specifications are usually derived from the data for the bioavailability batches based on relevant regulatory guidance. For extended-release dosage forms, the range at any dissolution time point specification has to be within ±10% of the mean profile obtained from the bioavailability batch in the absence of an IVIVC.

A validated Level A correlation allows for setting dissolution specifications such that all lots within the specification limits will be bioequivalent. The procedure involves predicting the *in vivo* plasma time course from dissolution profiles via convolution based on the IVIVC model. The dissolution specification is determined such that the dissolution profiles defined by the upper and lower limits will produce *in vivo* profiles with a maximum difference of 20% as predicted in $C_{max}$ and AUC. By doing so, a more meaningful quality control tool can be established to assure product quality. In addition, more flexible and wider dissolution specification (e.g. >±10%) is acceptable so long as the predicted $C_{max}$ and AUC *are within 20%.*

Similarly, in the case of Level C and multiple Level C IVIVC models, specification ranges should be set at the correlation time point such that there is a maximum of 20% difference in the predicted AUC or $C_{max}$. If the correlation involves more than one parameter, the one resulting in tighter limits should always be used. In addition, drug release at the last time point should be at least 80%, which should be taken into consideration during IVIVC development.

### 16.5.2 Waiver of *In Vivo* Bioavailability Studies

In addition to providing enhanced control of product quality, a validated IVIVC can be used as a surrogate for human bioequivalence study. As a result, bioequivalence studies may be waivered based on an established Level A or multiple Level C correlation in many scenarios where an *in vivo* study is required. Examples include postapproval changes of manufacture sites or certain types of process, formulation, and equipment, as well as approval of new strengths of modified release products. For immediate release solid dosage forms, the biowaiver request for INDs, NDAs, ANDAs, and postapproval changes is usually based on the consideration of the drug's BCS class, therapeutic index, IVIVR and potential effect of excipients on bioavailability. Comprehensive regulation, scientific rationales, considerations, approaches, and evaluation criteria associated with these biowaivers have been clearly laid out in various regulatory guidelines [10, 42–44].

## 16.6 Challenges

Challenges associated with a successful IVIVC development often reside in the physicochemical, biological, and pharmacokinetic properties of a drug substance, formulation design, *in vitro* and *in vivo* study design, and modeling methodology.

First, no correlation is expected when dissolution is not a rate-limiting step for drug absorption. For example, as the rate of permeation becomes slower, the *in vivo* performance of different formulations usually cannot be differentiated by drug release rate. Therefore, correlation is often lacking for IR dosage forms for most BCS class I and III drugs. On the other hand, IVIVC is possible for both IR and ER dosage forms of BCS class II, and ER dosage forms of BCS class I drugs since drug release is typically slower than the permeation through the intestinal membrane.

Second, the convolution and deconvolution approaches underlying the IVIVC development are based on linear system analysis [12], which is valid only under certain conditions and assumptions. Two phenomena have been observed for real systems: saturation and time-variant responses. Saturation can occur with protein- and/or enzyme-mediated pathways in the body, including absorption, metabolism, and distribution. Time variant response may be caused by biological and physiological rhythms, enzyme induction, sensitization or desensitization. Absorption characteristics of a drug may vary depending on regions in the GI tract whereas this is not differentiated in the treatment of convolution or deconvolution. A number of drugs have been shown to exhibit enterohepatic recycling. All of the above can lead to deviation from the linearity assumption, directly or indirectly, and complicate the data analysis and its reliability. Particularly when a drug substance is not well understood, one would be left in the dark without knowing the exact cause of failure; otherwise, a modified model may still be possible. In that sense, we should avoid treating IVIVC as purely a black box of convolution and deconvolution. Understanding drug properties and their interplay with GI tract is essential.

Third, the mean *in vivo* responses used in IVIVC development ignores the inter- and intra-subject variability, which might be an important consideration for drugs that are inherently more variable than others. The better-controlled fasting conditions used to characterize biological response in IVIVC study may also contribute to the issue because physiological and biological factors impacting drug absorption could change significantly and more variably under fed conditions.

Our understanding of properties of a dosage form, particularly the interplays of various factors during *in vivo* release, is still very limited. Totally different *in vivo* release behavior may result from a similar dosage form design for "slightly" different drugs. In many cases, we have to rely on the *in vivo* profiles to invent our "mimicking" *in vitro* dissolution test methods. For example, the performances of three acetaminophen matrix tablets were better mimicked by dissolution testing with the incorporation of glass beads to mimic the increased shear during *in vivo* dissolution in order to account for the observed differences in oral absorption [45]. The release of divalproex sodium HPMC matrix tablets was shown to be erosion-control in the gut lumen, contrary to the conventional wisdom of diffusion-controlled release for a water-soluble molecule. The faster than *in vitro* and erosion-controlled *in vivo* release of divalproex sodium is probably due to the increased GI motility and attrition and has been successfully "mimicked" by incorporation of surfactant for the purpose of IVIVC [35–37].

Our lack of understanding has certainly prevented us from developing more universal *in vitro* methods applicable to different drug molecules or to different types of dosage forms. Establishing an IVIVC still remains a daunting task with an element of trial and error. Even if successfully established, it is often valid only for a specific type of dosage form, with the same type of rate-controlling excipient, the same release mechanism and within a certain defined range.

A review [46] in 2016 indicated that only 54 IVIVCs were submitted to FDA from 2008 to 2015 with a low acceptance rate of 40% and that the annual number of submissions was not increasing, which caused some concern as it might dwindle the quality by design (QbD) efforts advocated by regulatory agencies. The authors called for the development of biopredictive dissolution methods and the use of nonconventional approaches, including mechanistic/physiologically based approaches to increase the likelihood of IVIVC success.

## 16.7 Physiologically Based Biopharmaceutics Models (PBBM)

*In vivo* dissolution of drug from a solid dosage form and the subsequent oral absorption are complex and involve a multitude array of concurrent physical, chemical, and biological processes, which typically involve disintegration, dissolution/ionization, precipitation/redissolution, micellization, permeation across the GI membrane via passive diffusion, or by an uptake transporter, or being shuttled back by an efflux transporter (such as Pgps or BCRPs), enzymatic degradation in the gut lumen or metabolism in the gut wall, etc. The impact or lack of impact of the above processes to the overall oral absorption of a drug depends on the properties of the drug molecule, formulation design, and dose. In addition, significant differences exist in various segments of the GI tracts in terms of surface area, motility, expression levels of enzymes and transporters, amount and pH of the GI fluid, and residence time, not to mention the intra-subject and inter-subject variability. Furthermore, these parameters may be altered by many physiological, pathological, and pharmaceutical factors.

The conventional IVIVC modeling tends to ignore differences in the *in vivo* dissolution/absorption indifferent regions of the GI tract and lump them into a gross profile. Thus, it certainly has limitations. In general, it is applicable to many controlled-release dosage forms of BCS I drug molecules with high solubility and high permeability in different sections of the GI tract. The high solubility of BCS I drugs implies that the dissolved molecules are unlikely to precipitate and cause solubility-limited absorption. The high permeability in all the GI regions relevant for drug absorption makes it possible not to differentiate duodenum, jejunum, ileum, and colon and treat absorption throughout the GI in a similar way. For controlled release of low solubility/high permeability BCS II drugs, as long as the amount released at each GI location is relatively small and can be mostly absorbed at the site of release, establishment of an IVIVC is also feasible. Examples include ER of nifedipine and other BCS drugs. However, this approach may be insufficient for other situations.

A drug molecule may have different permeability along the GI tract, usually higher in the upper GI and lower in the lower GI such as in the colon. When permeation becomes the rate-limiting step, or even partially, during the absorption process, it is inadequate not to differentiate the rate of absorption at different GI segments, particularly in the colon where the absorption may be drastically reduced. This certainly has implications for BCS III and BCS IV molecules. For poorly soluble BCS II and BCS IV drug molecules, dose, and solubility in the GI tract play a critical role in the oral absorption. Drug concentration in the GI tract may be modeled by particle dissolution based on the Nernst–Brunner equation, and the GI drug concentration may reach solubility, depending on the rate of depletion (absorption). For supersaturable formulations, such as lipids, amorphous solid dispersions, or appropriate salt forms, the drug concentration in the GI is often modeled using the amorphous solubility. While the supersaturation may drive a higher oral absorption, it may also cause drug molecules to crystallize and precipitate in the GI tract. Therefore, the scenarios in the GI dissolution are very dynamic, which is further complicated by the ionization process for frequently encountered weakly acidic/basic drug molecules. While we don't in general know what are the exact *in vivo* scenarios for a specific dosage form without devoting efforts in probing its behavior, to the least, the simplistic *in vitro* dissolution profile used in describing the dosage form is unlikely realistic for BCS II and IV drugs in most cases. Therefore, building a quantitative *in vitro-in vivo* correlation is often challenging.

Physiologically based pharmacokinetic (PBPK) models describe drug movement and disposition in the body based on physiological, anatomical, biochemical and physicochemical parameters of the drug molecule, which was developed even before the compartmental

PK models. Physiologically based absorption model (PBAM) may be considered as the extension of the PBPK model to include the GI physiology that is relevant to oral drug absorption. Physiologically based biopharmaceutics models (PBBM) are based on the same principles as PBAM but with an emphasis on biopharmaceutics [47]. PBBM may provide advantages for the oral absorption modeling of BCS II/III/IV drugs where the complex dynamics such as those mentioned above exist in different regions of the GI tract and impact the validity of the assumptions implicit in conventional IVIVC treatment via convolution/deconvolution.

PBPK modeling can be considered as an extension of the conventional IVIVC or a more complex form of IVIVC. These models incorporate not just the solubility and its pH-dependency, dissolution but also various biological and physiological variables such as permeation, transit time, transporters, enzymes, distribution, fluid types and volumes in different segments of the GI tract. Parameterization of these models usually starts with *in vitro* assays, ex-vivo studies and knowledge from prior studies or literature, and may be optimized through means of model fitting. To ensure the reliability and accuracy, model validation and verification are essential in developing IVIVC based on PBPK models. Given the limited data, over-parameterization can be a drawback and should be treated cautiously.

Commercial software packages such as GastroPlus (Simulations Plus Inc.), Simcyp Simulator (Simcyp Ltd), PK-Sim (Bayer Technology Services), MATLAB (MathWorks Inc.), and STELLA (isee systems) are available for PBAM modeling. Attempts have been made to account for all relevant processes during the GI absorption of drugs, including release from the dosage form; degradation, complexation, micellization in the GI tract; and the various mechanisms of drug uptake, efflux, and first-pass metabolism in either gut wall or liver; and to describe the interplay of these factors in determining the rate and extent of drug absorption from the GI tract. Most of these computational modeling software can be traced back to the compartmental absorption and transit (CAT) [48] model developed by Yu et al., which later evolved into the advanced compartmental absorption and transit (ACAT) [49] model (GastroPlus), and the advanced dissolution, absorption, and metabolism (ADAM) [50] model (Simcyp).

The CAT model divides the GI tract into seven compartments [48], considering gastric emptying, small intestinal transit, *in vivo* dissolution, and absorption. Dissolution is assumed to be immediate and essentially behaves like solution. The unabsorbed drug transits from one compartment to the next in a first-order fashion and is conveniently described by the following equation:

$$\frac{\mathrm{d}D_i}{\mathrm{d}t} = k_\mathrm{t}D_{i-1} - k_{\mathrm{t},i}D_i - k_\mathrm{a}D_i \tag{16.27}$$

where $D_i$ is the fractional dose, $k_{\mathrm{t},i}$ is the first-order transit rate constant, and $k_\mathrm{a}$ is the first-order absorption rate constant, all in the $i$th compartment.

The ACAT model [49] adds gut metabolism and modeling capability of modified-release drug products to the CAT model. Nonlinear metabolism/transport kinetics are also adopted in this model. The model incorporates drug physicochemical factors ($\mathrm{p}K_\mathrm{a}$, solubility, and permeability), dosage formulation factors (dose, particle size, and density), and physiological factors (gastric emptying, intestinal transit rate, first-pass metabolism, luminal transport, GI blood flow, and food effects). Drug exists in six states, i.e. unreleased, undissolved, dissolved, degraded, metabolized, and absorbed. Recent development has also incorporated particle dissolution.

The ADAM model [50], while still based on the CAT model, also considers drug dissolution, GI fluid transit, drug degradation, intestinal metabolism, gut wall permeation, and active transport. Similarly, differential equations are used to describe the dynamics of unreleased drug trapped in the formulation, those available for dissolution, the amount of dissolved drug, and the amount in the enterocyte. The ADAM model considers drug physicochemical properties, dosage formulation, and physiological factors to achieve similar results.

PBBM can be deemed as the application of the PBPK modeling in understanding the role of biopharmaceutics to facilitate the establishment of the in vitro link to *in vivo* pharmacokinetic profile [47]. These models are built on mechanistically sound representations of *in vivo* human absorption with plausible parameters representing the physiology. When in vitro dissolution reflects the *in vivo* dissolution, the model can predict the *in vivo* plasma drug concentration profile. As the model is validated appropriately, PBBM can be used to establish IVIVC, dissolution specification, dissolution safe space, and control of critical material attributes (CMAs) such as API particle size or critical process parameters (CPPs) in the framework of QbD. PBPM also enables virtual BE simulations by incorporating appropriate intra- and inter-subject physiology variability (such as GI pH, transit time, permeability, and clearance), which is lacking by the conventional convolution/deconvolution approach. Interested readers may refer to other literature for additional details [51–55].

Figure 16.5 provides an example of dissolution safe space for lesinurad IR tablets developed based on a ACAT model [56]. The pivotal batch has the fastest

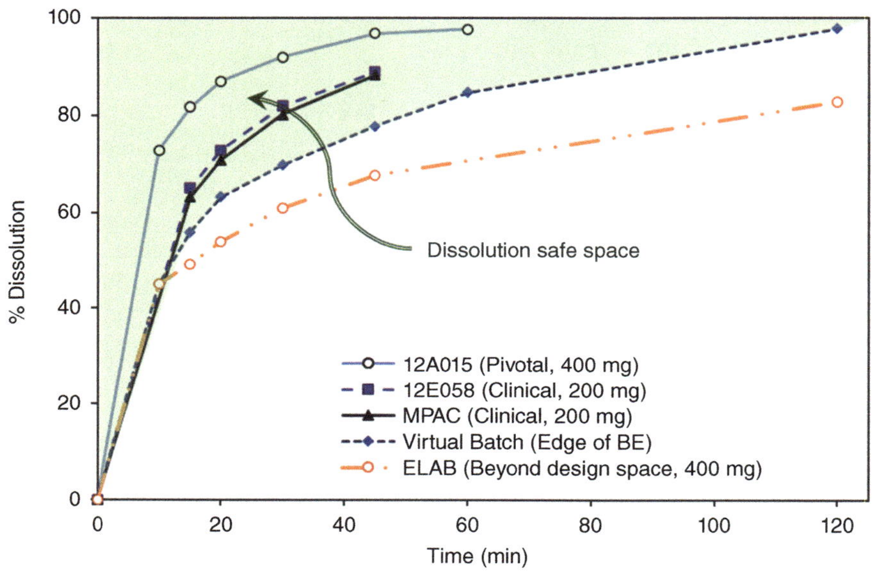


**Figure 16.5** Dissolution space for anticipated bioequivalence to lesinurad IR tablets using the QC dissolution test. ELAB 400 mg was a lab batch manufactured outside the intended commercial granulation operating ranges and failed BE to the pivotal biobatch. Source: Adapted from Pepkin et al. [56].

dissolution rate and is typical of the commercial and clinical batches. Batch MPAC and 12E058 show slower dissolution rate but are clinically acceptable. The batch ELAB was a development lot, manufactured outside of the intended commercial operating ranges for granulation and showed slow and incomplete dissolution. A relative bioavailability study indicated that batch ELAB had only 80% (90% confidence interval: 68.2–93.8%) and 88.1% (90% confidence interval: 79.9–97.2%) of the Cmax and AUC of the pivotal biobatch.

A PBPK model was constructed incorporating the properties of the drug molecule and physiological factors, based on two human PK studies (e.g. an absolute bioavailability study with concurrent oral and *i.v.* micro-dosing $^{14}C$ technique and a human ADME study), and successfully reproduced the pharmacokinetic profiles in the relative BA study. A virtual trial was then conducted, with introduced random intra- and inter-subject variability in stomach pH, transit time, and permeability, to define the boundary of bioequivalence in terms of the *in vitro* dissolution profile and the results are shown in Figure 16.5. Therefore, a clinically meaningful dissolution safe space (and specification limit) is derived for this product.

The US Food and Drug Administration has issued a draft guidance on this subject and recommended PBPK modeling tool to support postapproval manufacturing or formulation changes for solid oral drugs [57].

## 16.8 Summary

Development of *in vitro-in vivo* correlation is an important objective during product development and should be proactively and systematically pursued on a product-by-product basis. An established IVIVC is useful in providing maximum quality assurance to product release, guiding product and process development and supporting biowaivers for certain manufacturing and postapproval changes. It is also an important aspect underpinning the QbD framework for modernized drug product development, manufacturing, and control. A single *in vitro-in vivo* correlation for different products of the same drug can not be accomplished in most cases due to the complex dynamic GI environments and their interplay with API and formulation. Often, an *in vitro-in vivo* correlation is applicable only for a specific type of drug product. Development of *in vitro-in vivo* correlations requires integrated knowledge of physicochemical and biological properties of drug substance, dosage form, and their interplay with the biology and physiology of the GI tract as well as in vitro tests. While the conventional convolution/deconvolution-based approach is applicable to ER dosage forms when dissolution is rate-limiting, it may be challenging for more complex situations. Steady progresses have been made on mechanistically based PBBM modeling to link the *in vivo* drug performance to their *in vitro* dissolution.

## References


1 Dokoumetzidis, A. and Macheras, P. (2006). A century of dissolution research: from Noyes and Whitney to the biopharmaceutics classification system. *Int. J. Pharm.* 321: 1–11.

2 Martin, C.M., Rubin, M., O'Malley, W.G. et al. (1968). Brand, generic drugs differ in man. *JAMA* 205: 23–36.

3 MacLeod, C., Rabin, H., Ruedy, J. et al. (1972). Comparative bioavailability of three brands of ampicillin. *Can. Med. Assoc. J.* 107: 203–209.

4 Levy, G. (1964). Effect of dosage form properties on therapeutic efficacy of tolbutamide tablets. *Can. Med. Assoc. J.* 90: 978–979.

5 Varley, A.B. (2020). The generic inequivalence of drugs. *JAMA* 206: 1745–1748.

6 Lindenbaum, J., Mellow, M.H., Blackstone, M.O., and Butler, V.P. Jr. (2010). Variation in biologic availability of digoxin from four preparations. *N. Engl. J. Med.* https://doi.org/10.1056/NEJM197112092852403 285 (24): 1344–1347.

7 Amidon, G.L., Lennernäs, H., Shah, V.P., and Crison, J.R. (1995). A theoretical basis for a biopharmaceutic drug classification: the correlation of in vitro drug product dissolution and in vivo bioavailability. *Pharm. Res.* 12: 413–420.

8 Chilukuri, D.M., Sunkara, G., and Young, D. (ed.) (2007). *Pharmaceutical Product Development: In Vitro-In Vivo Correlation*. New York: Informa Healthcare USA Inc.

9 Qiu, Y. (2009). In vitro-in vivo correlations: Fundamentals, development considerations, and applications. In: *Developing Solid Oral Dosage Forms: Pharmaceutical Theory and Practice* (ed. Y. Qiu, Y. Chen, and G. Zhang), 365–377. Burlington, MA: Academic Press.

10 U.S. Department of Health, Food and Drug Administration, Center for Drug Evaluation and Research (September 1997). Guidance for Industry: Extended release oral dosage forms: development, evaluation, and application of in vitro/in vivo correlations.

11 USP 32–NF 27 (2009). Chapter 1088. In vitro and in vivo evaluation of dosage forms. The United States Pharmacopeial Convention.

12 Cutler, D.J. (1978). Linear systems analysis in pharmacokinetics. *J. Pharmacokinet. Biopharm.* 6: 265–282.

13 Vaughan, D.P. and Dennis, M. (1978). Mathematical basis of point-area deconvolution method for determining in vivo input functions. *J. Pharm. Sci.* 67: 663–665.

14 Hanano, M. (1967). Studies on absorption and excretion of drug. VII. A new estimation method for the release of drugs from dosage forms and the availability in vivo. *Chem. Pharm. Bull.* 15: 994–1001.

15 Chiou, W.L. (1980). New calculation method of mean total body clearance of drugs and its application to dosage regimens. *J. Pharm. Sci.* 69: 90–91.

16 Langenbucher, F. (1982). Numerical convolution/deconvolution as a tool for correlating in vitro with in vivo drug availability. *Pharm. Ind.* 44: 1166–1172.

17 Langenbucher, F. (2003). Handling of computational in vitro/in vivo correlation problems by Microsoft Excel: III. Convolution and deconvolution. *Eur. J. Pharm. Biopharm.* 56: 429–437.

18 Veng Pedersen, P. (1980). Model-independent method of analyzing input in linear pharmacokinetic systems having polyexponential impulse response. I: Theoretical analysis. *J. Pharm. Sci.* 69: 298–305.

19 Wagner, J.G. and Nelson, E. (1964). Kinetic analysis of blood levels and urinary excretion in the absorptive phase after single doses of drug. *J. Pharm. Sci.* 53: 1392–1403.

20 Loo, J.C.K. and Riegelman, S. (1968). New method for calculating the intrinsic absorption rate of drugs. *J. Pharm. Sci.* 57: 918–928.

21 Wagner, J.G. (1975). Application of the Loo-Riegelman absorption method. *J. Pharmacokinet. Biopharm.* 3: 51–67.

22 Wagner, J.G. (1983). Pharmacokinetic absorption plots from oral data alone or oral/intravenous data and an exact Loo-Riegelman equation. *J. Pharm. Sci.* 72: 838–842.

23 Langenbucher, F. and Moeller, H. (1983). Correlation of in vitro drug release with in vivo response kinetics. Part I: Mathematical treatment of time functions. *Pharm. Ind.* 45: 623–628.

24 Gamel, J., Rousseau, W.F., Katholi, C.R., and Mesel, E. (1973). Pitfalls in digital computation of the impulse response of vascular beds from indicator-dilution curves. *Circ. Res.* 32: 516–523.

25 Kiwada, H., Morita, K., Hayashi, M. et al. (1977). A new numerical calculation method for deconvolution in linear compartment analysis of pharmacokinetics. *Chem. Pharm. Bull.* 25: 1312–1318.

26 Veng Pedersen, P. (1980). Novel approach to bioavailability testing: statistical method for comparing drug input calculated by a least-squares deconvolution technique. *J. Pharm. Sci.* 69: 318–324.

27 Veng Pedersen, P. (1980). Novel deconvolution method for linear pharmacokinetic systems with polyexponential impulse response. *J. Pharm. Sci.* 69: 312–318.

28 Gillespie, W.R. and Veng-Pedersen, P. (1985). A polyexponential deconvolution method. Evaluation of the "gastrointestinal bioavailability" and mean in vivo dissolution time of some ibuprofen dosage forms. *J. Pharmacokinet. Biopharm.* 13: 289–307.

29 Cutler, D.J. (1978). Numerical deconvolution by least squares: use of polynomials to represent the input function. *J. Pharmacokinet. Biopharm.* 6: 243–263.

30 Yeh, K.C. and Small, R.D. (1989). Pharmacokinetic evaluation of stable piecewise cubic polynomials as numerical integration functions. *J. Pharmacokinet. Biopharm.* 17: 721–740.

31 Verotta, D. (1993). Two constrained deconvolution methods using spline functions. *J. Pharmacokinet. Biopharm.* 21: 609–636.

32 Cutler, D.J. (1978). Numerical deconvolution by least squares: use of prescribed input functions. *J. Pharmacokinet. Biopharm.* 6: 227–241.

33 Verotta, D. (1993). Estimation and model selection in constrained deconvolution. *Ann. Biomed. Eng.* 21: 605–620.

34 Verotta, D. (1989). An inequality-constrained least-squares deconvolution method. *J. Pharmacokinet. Biopharm.* 17: 269–289.

35 Qiu, Y., Cheskin, H.S., Engh, K.R., and Poska, R.P. (2003). Once-a-day controlled-release dosage form of divalproex sodium I: formulation design and in vitro/in vivo investigations. *J. Pharm. Sci.* 92: 1166–1173.

36 Qiu, Y., Garren, J., Samara, E. et al. (2003). Once-a-day controlled-release dosage form of divalproex sodium II: development of a predictive in vitro drug release method. *J. Pharm. Sci.* 92: 2317–2325.

37 Dutta, S., Qiu, Y., Samara, E. et al. (2005). Once-a-day extended-release dosage form of divalproex sodium III: development and validation of a level A in vitro-in vivo correlation (IVIVC). *J. Pharm. Sci.* 94: 1949–1956.

38 Veng-Pedersen, P., Gobburu, J.V.S., Mayer, M.C., and Straughn, A.B. (2000). Carbamazepine level-A *in vivo-in vitro* correlation (IVIVC): a scaled convolution based predictive approach. *Biopharm. Drug Dispos.* 21: 1–6.

39 Bressolle-Gomeni, R.G. and Françoise. (2020). Comparison of alternative population modeling approaches for implementing a Level A IVIVC and for assessing the time-scaling factor using deconvolution and convolution-based methods. *AAPS J.* 22: 1–13.

40 Podczeck, F. (1993). Comparison of in vitro dissolution profiles by calculating mean dissolution time (MDT) or mean residence time (MRT). *Int. J. Pharm.* 97: 93–100.

41 Veng-Pedersen, P. (1989). Mean time parameters in pharmacokinetics, definition, computation and clinical implications. *Part I. Clin. Pharmacokinet.* 17: 345–366.

42 U.S. Department of Health, Food and Drug Administration, Center for Drug Evaluation and Research (August 1997). Guidance for Industry: Dissolution testing of immediate release solid oral dosage forms.

43 U.S. Department of Health, Food and Drug Administration, Center for Drug Evaluation and Research (August 2000). Guidance for Industry: Waiver of in vivo bioavailability and bioequivalence studies for immediate-release solid oral dosage forms based on a biopharmaceutics classification system.

44 U.S. Department of Health, Food and Drug Administration, Center for Drug Evaluation and Research (September 1997). Guidance for Industry: SUPAC-MR: modified release solid oral dosage forms scale-up and postapproval changes: chemistry, manufacturing, and controls; In vitro dissolution testing and in vivo bioequivalence documentation.

45 Sako, K., Sawada, T., Nakashima, H. et al. (2002). Influence of water soluble fillers in hydroxypropylmethylcellulose matrices on in vitro and in vivo drug release. *J. Controlled Release* 81: 165–172.

46 Seo, S.S.-S., Min, L., John, D. et al. (2016). Regulatory experience with in vivo in vitro correlations (IVIVC) in new drug applications. *AAPS J.* 18: 1379–1390.

47 Heimbach, T., Suarez-Sharp, S., Kakhi, M. et al. (2019). Dissolution and translational modeling strategies toward establishing an in vitro-in vivo link—a workshop summary report. *AAPS J.* 21: 29.

48 Yu, L.X. and Amidon, G.L. (1999). A compartmental absorption and transit model for estimating oral drug absorption. *Int. J. Pharm.* 186: 119–125.

49 Yu, W.H., Sau Lawrence, L., and Lawrence, X. (2009). Mechanistic approaches to predicting oral drug absorption. *AAPS J.* 11: 217–224.

50 Tucker, M.J., David, T., Jiansong, Y. et al. (2009). Population-based mechanistic prediction of oral drug absorption. *AAPS J.* 11: 225–237.

51 Chen, M.J.D., Po-Chang, C., Fang, M. et al. (2020). A physiologically based pharmacokinetic model of vismodegib: deconvoluting the impact of saturable plasma protein binding, pH-dependent solubility and nonsink permeation. *AAPS J.* 22: 1–10.
52 Amidon, M.B., Bart, H., Joseph, D. et al. (2020). A mechanistic physiologically-based biopharmaceutics modeling (PBBM) approach to assess the in vivo performance of an orally administered drug product: from IVIVC to IVIVP. *Pharmaceutics* 12: 74.
53 Comfort, H.Z., Binfeng, X., Jennifer, S. et al. (2014). Application of physiologically based absorption modeling to formulation development of a low solubility, low permeability weak base: mechanistic investigation of food effect. *AAPS PharmSciTech* 15: 400–406.
54 Wu, L.M., Youssef, M.M., Liang, Z. et al. (2020). Using a physiologically based pharmacokinetic absorption model to establish dissolution bioequivalence safe space for oseltamivir in adult and pediatric populations. *AAPS J.* 22: 1–10.
55 Grimstein, M., Yang, Y., Zhang, X. et al. (2019). Physiologically based pharmacokinetic modeling in regulatory science: an update from the U.S. Food and Drug Administration's Office of Clinical Pharmacology. *J. Pharm. Sci.* 108: 21–25.
56 Pepin, X.J.H., Flanagan, T.R., Holt, D.J. et al. (2016). Justification of drug product dissolution rate and drug substance particle size specifications based on absorption PBPK modeling for lesinurad immediate release tablets. *Mol. Pharmaceutics* 13: 3256–3269.
57 US FDA (2020). Draft Guidance: The use of physiologically based pharmacokinetic analyses – biopharmaceutics applications for oral drug product development, manufacturing changes, and controls.

# 17

# Advanced Concepts in Oral Bioavailability Research – An Overview

*Baojian Wu*[1,2], *Min Chen*[1], *and Ming Hu*[3]

[1] *Research Center for Biopharmaceutics and Pharmacokinetics, College of Pharmacy, Jinan University, Guangzhou, China*
[2] *School of Pharmaceutical Sciences, Guangzhou University of Chinese Medicine, Guangzhou, China*
[3] *Department of Pharmacological and Pharmaceutical Sciences, College of Pharmacy, University of Houston, Houston, Texas*

Oral route is the most common means of drug administration. Bioavailability of oral drugs is of particular interest to the drug development community as it determines drug concentrations (exposure) in target organs and therefore pharmacological effects and sometimes adverse effects. Much progress has been made in understanding the factors affecting oral bioavailability and advanced concepts have emerged in the field. Here, we provide an overview of these advanced concepts, most of which are related to drug-metabolized enzymes (also referred as enzymes in this chapter) and transporters. Other important areas are physiological and physicochemical characteristics that affect dosage form performance, and various formulation approaches that improve oral bioavailability.

Gut microbiota is depicted as a "virtual or hidden organ" because it has impact on human health and diseases by regulating many physiological functions [1]. It is of no surprise that gut microbiota also modulates drug disposition and pharmacokinetic behaviors. The best evidence for involvement of gut microbiota in drug disposition refers to the phenomenon of enterohepatic circulation in which enzymes (e.g. glucosidases and glucuronidases) secreted by gut microbiota hydrolyze the metabolites back to parent drug and enable drug reabsorption [2, 3]. In fact, gut microbiota produces the enzymes catalyzing non-hydrolytic reactions such as reduction and dehydroxylation (e.g. azoreductase, nitroreductase, and nitrate reductase) in addition to the hydrolytic enzymes [4, 5]. These microbial enzymatic reactions have been shown to affect oral bioavailability of many drugs [6–8]. For instance, reductive metabolism by gut microbiota reduces the bioavailability of nabumetone, a nonsteroidal anti-inflammatory drug [7]. It has become a new concept that the gut microbiota is another metabolically active compartment together with classical drug metabolism in the body that influences the bioavailability of oral drugs (Figure 17.1) [1].

Scientists have recognized that drug metabolism and transport are two interdependent processes. It is thus inappropriate to interpret the metabolic or transport activity alone without taking the other one into account. The interdependence of metabolism and transport is described as "metabolism-transport interplay" or "enzyme–transporter interplay" (Figure 17.1) [9, 10]. In general, there are two types of metabolism-efflux transport interplay, termed as parallel and concatenated interplay [11]. No matter which type, the interplay enables the enzymes and transporters to work synergistically to remove drug molecules from the cells for elimination, thereby impacting drug disposition and bioavailability [11]. For instance, the interplay between phase II enzymes and efflux transporters accounts for low oral bioavailability of flavonoids [12–14]. To be specific, flavonoids are metabolized by phase II enzymes in the enterocytes and hepatocytes to form hydrophilic metabolites, which are subsequently pumped out of cells by efflux transporters [13]. The metabolite formation and excretion are so efficient that the compounds can be rapidly eliminated, and their exposure levels are very low in the body.

Considering the role of transporters in drug disposition and the synergistic interplay of transporters with enzymes, it is essential to have a tool that predicts the importance of transporters and the occurrence of the interplay. To this end, Benet and coworkers have developed a Biopharmaceutics Drug Disposition Classification System (BDDCS), which characterizes drugs into four classes according to their solubility and metabolism (Class I: high solubility, extensive metabolism; Class II:

*Oral Bioavailability and Drug Delivery: From Basics to Advanced Concepts and Applications,* First Edition. Edited by Ming Hu and Xiaoling Li.

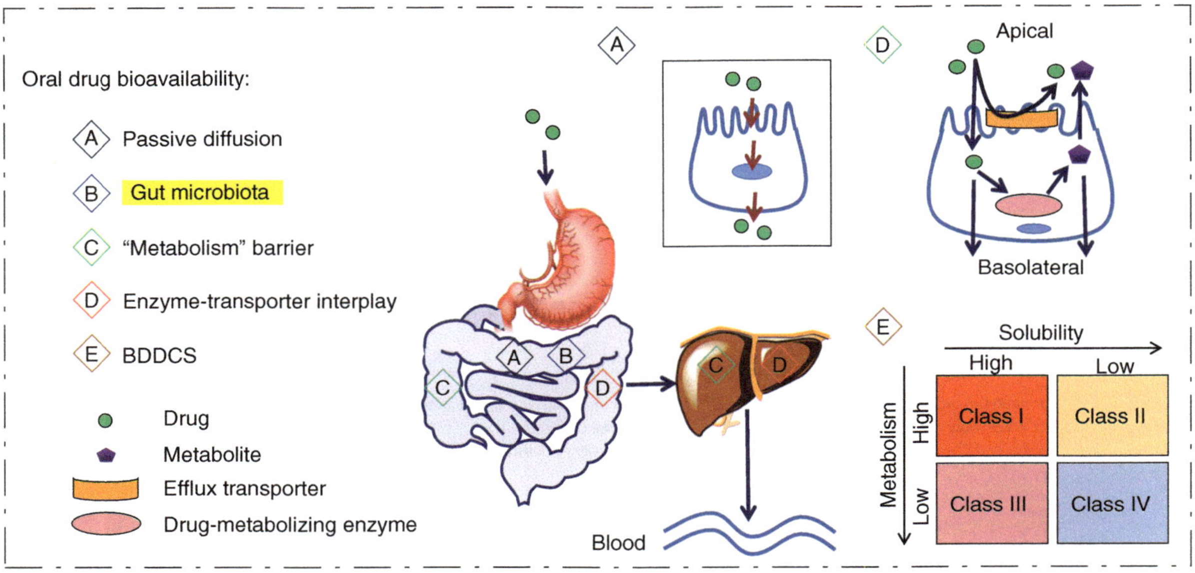


**Figure 17.1 Examples of advanced concepts in oral bioavailability research.** A. Passive diffusion is the main mechanism of drug transport for a major fraction of clinical drugs. B. Gut microbiota is a metabolically active compartment that influences the bioavailability of oral drugs. C. Extensive first-pass metabolism is a predominant barrier in oral drug delivery. D. Enzyme-transporter interplay in the intestine and liver. E. Biopharmaceutics Drug Disposition Classification System (BDDCS) can predict the importance of transporters and the occurrence of the interplay.

low solubility, extensive metabolism; Class III: high solubility, poor metabolism; and Class IV: low solubility, poor metabolism) [15]. BDDCS can be used to predict drug disposition and potential drug–drug interactions with an emphasis on defining which drugs would be amenable to enzyme-only and transporter-only disposition and drug–drug interactions, as well as where enzyme–transporter interplay may be important (Figure 17.1) [16]. For example, BDDCS predicts that atorvastatin, a Class II drug, may potentially exhibit a drug–drug interaction with the inhibitors of hepatic uptake transporters. Experiments with animals and humans confirm that atorvastatin is a substrate of organic anion transporting polypeptides (OATPs, a class of uptake transporters) and that the OATP inhibitor rifampin significantly increases the total area under the curve (AUC) of atorvastatin by 6.8-fold [17]. BDDCS predicts potential effects of uremic toxins (inhibitors of hepatic uptake transporters) on hepatic disposition of erythromycin, a class III drug [18]. This prediction is validated by the experimental findings that the hepatic clearance of erythromycin in end-stage renal disease patients is decreased by 31% and that bioavailability is increased by 36% [18]. It is believed that BDDCS is a seminal concept in the field of drug disposition and bioavailability and will have a profound impact on drug discovery and development.

Computational (*in silico*) models for predicting absorption, distribution, metabolism, and excretion (ADME) properties and oral bioavailability are important tools that permit rapid identification of compounds that may exhibit good systemic exposure and efficacy. The Lipinski's "rule-of-five" is the first and rather useful approach for *in silico* prediction of oral absorption. It states that if two of the five following factors are out of the range, the compounds would have poor absorption and permeability: molecular weight >500, >5 H-bond donors, log$P$ > 5, and >10 H-bond acceptors [19]. However, quantitative prediction of oral bioavailability by quantitative structure–activity relationship (QSAR) models is limited by the lack of molecular descriptors for effective characterization of first-pass metabolism. To overcome this problem, scientists have introduced new rules (e.g. pharmacophore-based active conformations and multiple active conformations for single substrate with multiple metabolic sites) or novel molecular descriptors (e.g. descriptors from 3D molecular interaction fields such as CoMFA and VolSurf, and topological descriptors with fragment/bond formation such as TOPS-MODE) to improve the *in silico* metabolism models [20–23]. The advances in modeling drug metabolism would facilitate the development of reliable and predictive *in silico* tools for oral bioavailability. In addition to QSAR models, physiologically based *in silico* models for the prediction of oral bioavailability have been developing rapidly. Two examples of such models are used by software packages developed by GastroPlus and Simcyp [24, 25]. These models can help identify molecular parameters that have the greatest effects on the absorption for a given class of screened compounds.

Oral bioavailability is known to be influenced by a number of factors including physicochemical property, permeability, and first-pass (presystemic) metabolism. For a long time, poor solubility and permeability were regarded as two major limiting factors to oral drug absorption. The Biopharmaceutical Classification Systems or BCSs was used to separate drugs into four classes, with Class I as highly soluble and highly permeable, Class II as poorly soluble but highly permeable, Class III as highly soluble but poorly permeable, and Class IV as poorly soluble and poorly permeable [26]. Depending on the drugs, formulation strategies have largely focused on solving the solubility and/or permeability problems [27]. The formulations used to tackle the solubility- and/or permeability-limited bioavailability include polymeric micelles, lipid nanocapsules, solid lipid nanoparticles, and self-microemulsifying delivery systems [28].

Oral bioavailability of drugs may also be limited by severe first-pass metabolism. Some have argued that the majority (>70%) of marketed drug and new drug entities are subjected to extensive metabolism *in vivo* [16]. Hence, extensive metabolism is a predominant barrier in oral drug delivery (Figure 17.1). Drug delivery techniques for drugs that undergo extensive metabolism should be directed to effectively overcome the "metabolism" barrier. Of note, nanoformulations have been shown to improve oral bioavailability of presystemically metabolized drugs via targeted inhibition of metabolism. For example, sodium oleate-based nanoemulsions greatly enhance oral absorption of chrysin [29]. Labrasol containing nanoemulsions improve oral bioavailability of resveratrol by inhibiting intestinal glucuronidation [30].

It is a general view that both passive diffusion and carrier-mediated processes play crucial roles in drug transport, disposition, and bioavailability. However, passive diffusion is the main mechanism of drug transport for a major fraction of clinical drugs [31, 32]. The efficiency of passive diffusion is dependent on membrane lipids and the lipophilicity of transported molecules [32]. Currently, multiple methods are used to determine the role of passive diffusion in drug transport [32]. These methods are mainly based on various artificial membranes (e.g. black lipid membrane and liposomes) consisting of a bilayer made of zwitterionic phospholipids [32]. There is a good correlation between artificial membrane permeability and cellular membrane permeability [32]. Notably, parallel artificial membrane permeation assay (PAMPA), a method that measures drug permeability from a donor compartment to an acceptor compartment through a lipid-infused artificial membrane, is widely used as an *in vitro* model of passive diffusion [33]. Simplicity, low cost, high throughput, and wide pH range render this assay a highly attractive one in modern drug discovery.

In some cases, the rate-limiting step in drug absorption from the gastrointestinal tract is drug dissolution and release from the dosage forms. Drug dissolution is a process by which a solid substrate dissolves, and is controlled by the affinity between the solid and the liquid medium that surrounds it. The scientists have recognized that physicochemical properties and physiological conditions play an important role in drug dissolution and absorption processes and therefore oral bioavailability. The physicochemical properties showing a strong effect on drug dissolution include diffusivity, particle size, equilibrium solubility, polymorphism, and dissociation constant, whereas the important physiological conditions include gastric fluid and its composition, surface tension of the fluids, and surface active agents. Of note, co-ingestion of food can contribute to the solubility enhancement of drugs (e.g. dicumarol and griseofulvin) in the gastrointestinal tract ([34, 35]). Current dissolution medium does not accurately reflect the physiological conditions of the upper gastrointestinal tract. This is probably the main reason for the difficulty in predicting *in vivo* dissolution from *in vitro* dissolution. Therefore, the ability to predict the importance of dissolution in oral drug absorption and bioavailability would depend on improved design of test equipment and the medium used in dissolution tests.

The current book seeks to address some of these concepts with Chapters 18–35 that are written by experts in their field of research to give the reader an up-to-date overview on the status of various advanced concepts in oral drug delivery and bioavailability. After Chapter 35, a series of method chapters (Chapters 36–45) presented some methodological approaches to the study of oral bioavailability.

## Abbreviations

| | |
|---|---|
| ADME | absorption, distribution, metabolism, and excretion |
| AUC | area under the curve |
| BCS | Biopharmaceutical Classification System |
| BDDCS | Biopharmaceutics Drug Disposition Classification System |
| CoMFA | comparative molecular field analysis |
| OATPs | organic anion transporting polypeptides |
| PAMPA | parallel artificial membrane permeation assay |
| QSAR | quantitative structure–activity relationship |

## References


1 Noh, K., Kang, Y.R., Nepal, M.R. et al. (2017). Impact of gut microbiota on drug metabolism: an update for safe and effective use of drugs. *Arch. Pharmacal Res.* 40 (12): 1345–1355.

2 Sousa, T., Paterson, R., Moore, V. et al. (2008). The gastrointestinal microbiota as a site for the biotransformation of drugs. *Int. J. Pharm.* 363 (1-2): 1–25.

3 Zhang, S.H., Wang, Y.Z., Meng, F.Y. et al. (2015). Studies of the microbial metabolism of flavonoids extracted from the leaves of *Diospyros kaki* by intestinal bacteria. *Arch. Pharmacal Res.* 38 (5): 614–619.

4 Kim, D.H. and Kobashi, K. (1986). The role of intestinal flora in metabolism of phenolic sulfate esters. *Biochem. Pharmacol.* 35 (20): 3507–3510.

5 Rowland, I.R., Mallett, A.K., Bearne, C.A., and Farthing, M.J. (1986). Enzyme activities of the hindgut microflora of laboratory animals and man. *Xenobiotica* 16 (6): 519–523.

6 Haiser, H.J. and Turnbaugh, P.J. (2013). Developing a metagenomic view of xenobiotic metabolism. *Pharmacol. Res.* 69 (1): 21–31.

7 Jourova, L., Anzenbacher, P., and Anzenbacherova, E. (2016). Human gut microbiota plays a role in the metabolism of drugs. *Biomed. Pap. Med. Fac. Univ. Palacky Olomouc Czech Repub.* 160 (3): 317–326.

8 Wilson, I.D. and Nicholson, J.K. (2017). Gut microbiome interactions with drug metabolism, efficacy, and toxicity. *Transl. Res.* 179: 204–222.

9 Pang, K.S., Maeng, H.J., and Fan, J. (2009). Interplay of transporters and enzymes in drug and metabolite processing. *Mol. Pharmaceutics* 6 (6): 1734–1755.

10 Wang, S., Xing, H., Zhao, M. et al. (2016). Recent advances in understanding of kinetic interplay between phase II metabolism and efflux transport. *Curr. Drug Metab.* 17 (10): 922–929.

11 Shi, S. and Li, Y. (2014). Interplay of drug-metabolizing enzymes and transporters in drug absorption and disposition. *Curr. Drug Metab.* 15 (10): 915–941.

12 Crozier, A., Del Rio, D., and Clifford, M.N. (2010). Bioavailability of dietary flavonoids and phenolic compounds. *Mol. Aspects Med.* 31 (6): 446–467.

13 Jiang, W. and Hu, M. (2012). Mutual interactions between flavonoids and enzymatic and transporter elements responsible for flavonoid disposition via phase II metabolic pathways. *RSC Adv.* 2 (21): 7948–7963.

14 Setchell, K.D., Brown, N.M., Desai, P. et al. (2001). Bioavailability of pure isoflavones in healthy humans and analysis of commercial soy isoflavone supplements. *J. Nutr.* 131 (4): 1362S–1375S.

15 Wu, C.Y. and Benet, L.Z. (2005). Predicting drug disposition via application of BCS: transport/absorption/elimination interplay and development of a biopharmaceutics drug disposition classification system. *Pharm. Res.* 22 (1): 11–23.

16 Benet, L.Z. (2013). The role of BCS (biopharmaceutics classification system) and BDDCS (biopharmaceutics drug disposition classification system) in drug development. *J. Pharm. Sci.* 102 (1): 34–42.

17 Lau, Y.Y., Huang, Y., Frassetto, L., and Benet, L.Z. (2007). Effect of OATP1B transporter inhibition on the pharmacokinetics of atorvastatin in healthy volunteers. *Clin. Pharmacol. Ther.* 81: 194–204.

18 Sun, H., Huang, L., Frassetto, L., and Benet, L.Z. (2004). Effects of uremic toxins on hepatic uptake and metabolism of erythromycin. *Drug Metab. Dispos.* 32: 1239–1246.

19 Lipinski, C.A., Lombardo, F., Dominy, B.W., and Feeney, P.J. (2001). Experimental and computational approaches to estimate solubility and permeability in drug discovery and development settings. *Adv. Drug Delivery Rev.* 46 (1–3): 3–26.

20 Dong, D. and Wu, B. (2012). In silico modeling of UDP-glucuronosyltransferase 1A10 substrates using the VolSurf approach. *J. Pharm. Sci.* 101 (9): 3531–3539.

21 Hou, T., Wang, J., Zhang, W., and Xu, X. (2007a). ADME evaluation in drug discovery. 6. Can oral

bioavailability in humans be effectively predicted by simple molecular property-based rules? *J. Chem. Inf. Model.* 47 (2): 460–463.

22 Hou, T., Wang, J., Zhang, W., and Xu, X. (2007b). ADME evaluation in drug discovery. 7. Prediction of oral absorption by correlation and classification. *J. Chem. Inf. Model.* 47 (1): 208–218.

23 Wu, B., Morrow, J.K., Singh, R. et al. (2011). Three-dimensional quantitative structure-activity relationship studies on UGT1A9-mediated 3-O-glucuronidation of natural flavonols using a pharmacophore-based comparative molecular field analysis model. *J. Pharmacol. Exp Ther.* 336 (2): 403–413.

24 Jamei, M., Turner, D., Yang, J. et al. (2009). Population-based mechanistic prediction of oral drug absorption. *AAPS J.* 11 (2): 225–237.

25 Yu, L.X. and Amidon, G.L. (1999). A compartmental absorption and transit model for estimating oral drug absorption. *Int. J. Pharm.* 186 (2): 119–125.

26 FDA Guidance (2017). Guidance for Industry: Waiver of In Vivo Bioavailability and Bioequivalence Studies for Immediate-Release Solid Oral Dosage Forms Based on a Biopharmaceutics Classification System. https://www.fda.gov/media/70963/download.

27 Agrawal, U., Sharma, R., Gupta, M., and Vyas, S.P. (2014). Is nanotechnology a boon for oral drug delivery? *Drug Discovery Today* 19 (10): 1530–1546.

28 Alam, M.A., Al-Jenoobi, F.I., Al-Mohizea, A.M., and Ali, R. (2015). Understanding and managing oral bioavailability: physiological concepts and patents. *Recent Pat. Anti-cancer Drug Discovory* 10 (1): 87–96.

29 Dong, D., Quan, E., Yuan, X. et al. (2017). Sodium oleate-based nanoemulsion enhances oral absorption of chrysin through inhibition of UGT-mediated metabolism. *Mol. Pharmaceutics* 14 (9): 2864–2874.

30 Zhou, J., Zhou, M., Yang, F.F. et al. (2015). Involvement of the inhibition of intestinal glucuronidation in enhancing the oral bioavailability of resveratrol by labrasol containing nanoemulsions. *Mol. Pharmaceutics* 12 (4): 1084–1095.

31 Di, L., Artursson, P., Avdeef, A. et al. (2012). Evidence-based approach to assess passive diffusion and carrier-mediated drug transport. *Drug Discovery Today* 17 (15–16): 905–912.

32 Sugano, K., Kansy, M., Artursson, P. et al. (2010). Coexistence of passive and carrier-mediated processes in drug transport. *Nat. Rev. Drug Discovery* 9 (8): 597–614.

33 Ottaviani, G., Martel, S., and Carrupt, P.A. (2006). Parallel artificial membrane permeability assay: a new membrane for the fast prediction of passive human skin permeability. *J. Med. Chem.* 49 (13): 3948–3954.

34 Macheras, P.E., Koupparis, M.A., and Antimisiaris, S.G. (1990). Drug binding and solubility in milk. *Pharm. Res.* 7 (5): 537–541.

35 Welling, P.G. (1977). Influence of food and diet on gastrointestinal drug absorption: a review. *J. Pharmacokinet. Biopharm.* 5 (4): 291–334.

# 18

# Expression and Pharmaceutical Relevance of Intestinal Transporters

*Melanie A. Felmlee, Michael Ng, and Annie Lee*

*Department of Pharmaceutics and Medicinal Chemistry, University of the Pacific, Thomas J. Long School of Pharmacy, Stockton, CA, USA*

## 18.1 Introduction

The plasma membrane forms the permeability barrier in all cells, allowing for the influx or uptake of essential nutrients and xenobiotics for biological processes, the encapsulation and sequestering of metabolic intermediates within cells, and the extrusion of metabolites and waste products. Gases and small, uncharged polar molecules are readily permeable and pass through the phospholipid bilayer unimpeded via simple diffusion down a chemical concentration gradient. However, most chemical substances essential for life such as ions, amino acids, and sugars are largely impermeable. Selective permeability for polar and hydrophilic molecules is governed by a class of integral membrane proteins called transporters, which are embedded in the cell surface and in organelle membranes and can act in a primary or secondary activated transport capacity.

Primary transporters exclusively use the hydrolysis of ATP as an energy source to mediate active transport and include the ion-driven P-, V-, and F-type ATPases, along with ATP-binding cassette (ABC). ABC transporters are one of the largest gene superfamilies of transporter proteins found in organisms ranging from bacteria to humans, and ABC transporters are responsible for a wide spectrum of substrate trafficking across cell membranes. All members contain two transmembrane domains, which form the substrate translocation pathway, and two nucleotide-binding domains, which bind and hydrolyze ATP. Though structurally conserved, each ABC protein binds specifically to a single substrate or group of related substrates, including ions, amino acids, sugars, nucleosides, cholesterol, peptides, phospholipids, oligonucleotides, polysaccharides, and proteins.

In contrast to ABCs, secondary active transporters mediate trafficking by coupling the energetically unfavorable movement of one substrate to the favorable shuttling of a different substrate independent of ATP-hydrolysis. These reactions utilize biological conditions such as extracellular ion gradients and membrane potentials as a driving force, coupled to the physiochemical properties of the trafficked substrates. The solute carrier family (SLC) is the second largest superfamily of membrane proteins in the human genome next to the G protein-coupled receptors. They are comprised predominantly of equilibrative, facilitative, and secondary active symporters and antiporters, with over 400 unique members identified. These members are further categorized within 52 families based on sequence homology. Facilitative SLCs, such as the SLC2 family, rely on an electrochemical gradient to catalyze the movement of substrates across membranes, whereas secondary active SLC, like the sodium/glucose cotransporters (SLC5 family) utilize ion gradients generated by ATP-dependent pumps to transport substrates against the concentration gradient. SLC members exhibit diverse substrate binding and translocation of polar and hydrophilic solutes, such as inorganic ions (SLC11, SLC30, SLC 39, and SLC40), amino acids (SLC1, SLC3, SLC6, SLC7, SLC25, and SLC36), neurotransmitters (SLC1 and SLC6), and sugars (SLC2, SLC5, and SLC50), and are essential regulatory components for cell processes and physiological homeostasis.

ABCs and SLCs are ubiquitously expressed in the intestines, heart, liver, kidney, lungs, blood–brain barrier, and tumors and thus, are emerging candidates for pharmacological and clinical targeting for disease treatment. SLCs are involved in more than 80 monogenetic disorders, and 190 different SLCs have been found to be mutated in human diseases [1, 2]. ABCs function as efflux transporters involved in the extrusion of chemotherapeutic drugs in tumor models. The early molecular characterization of ABC1, or

*Oral Bioavailability and Drug Delivery: From Basics to Advanced Concepts and Applications,* First Edition. Edited by Ming Hu and Xiaoling Li.

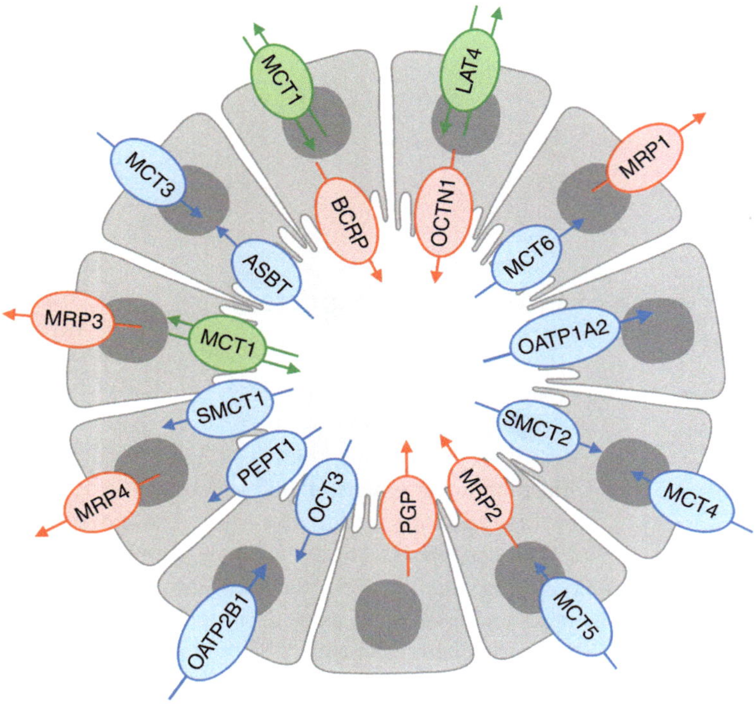


**Figure 18.1** Intestinal localization of ABC and SLC transporters. Uptake transporters are blue; efflux transporters are red; and bidirectional transporters are green. Source: Created with BioRender.com.

P-glycoprotein as a multidrug resistance protein in cancer cells signified the clinical importance of the ABC superfamily in biopharmaceutical drug transport. SLCs' natural role in transporting nutrients and vitamins has made them increasingly researched in drug–drug and drug–nutrient interaction studies.

Intestinal SLCs and ABCs are determinants of oral bioavailability of drugs that are their substrates, and their localization is detailed in Figure 18.1. Thus, knowledge of transporter expression and substrate characteristics can be utilized to modify compounds into prodrugs or mimic known substrates [3]. Drugs can be conjugated with bile acid analogs for transport with apical sodium-dependent bile acid transporter (ASBT), with fatty acids for MCT1 uptake, with an amino acid for PEPT1, and with L-carnitine for uptake by OCTN2 [3–5]. Along with synthesis, a multitude of factors needs to be considered as the effects on bioavailability are nuanced, exemplified by the pharmacokinetics of trans-resveratrol and cyclosporine [6, 7]. The polyphenol trans-resveratrol demonstrates that metabolism can work in concert with efflux transporters, a contributor to its poor bioavailability, with resulting metabolites exhibiting kinetic differences [6]. The compounds undergo phase II metabolism in the intestines and the resulting sulfate and glucuronide conjugates are extruded via MRP2 and breast cancer resistance protein (BCRP) with 3-glucuronide and sulfate conjugates exhibiting low and high affinity and capacity for BCRP [6]. Cyclosporine demonstrates the role of drug–drug interactions in significantly impacting efficacy [7]. Cyclosporine studies have demonstrated that coadministration with St. John's Wort induces CYP3A4 and P-glycoprotein (Pgp) expression, leading to subtherapeutic cyclosporine levels and cases of transplant rejections [7, 8].

## 18.2 Intestinal Drug Transport

The gastrointestinal tract is the major organ for the disposition of oral drugs. The intestine is the major site of nutrient and mineral absorption in the body during digestion, and orally administered drugs are dependent on intestinal absorption to enter systemic circulation. The small intestine is divided into three sections beginning proximally with the duodenum near the pylorus of the stomach, and proceeding to the jejunum, ileum and, ileocecal valve that connects to the large intestine. The intestinal tract is lined with simple columnar epithelial sheets with villi folds exposed to the intestinal lumen that enhances substrate absorptive efficiency. Individual enterocytes are packed in close proximity to provide this absorptive layer, as well as forming tight junctions that restrict intercellular and transepithelial movement of drug molecules. Like all epithelial cells, intestinal enterocytes are classified as polarized with their plasma membrane organized into two discrete regions: the blood-facing basolateral and lumen-directed apical membranes. Enterocytes contain microscopic microvilli along the apical membrane that form the brush border, which increases surface area and provides additional cell surface occupancy for drug transporters. The apical and basolateral membranes contain distinct expressional patterns of transporters that mediate coordinated transcellular absorption and transport of nutrients and drug substrates from the intestinal lumen to blood circulation via the portal vein.

Multiple drug transporter families and isoforms are present along the length of the intestine, with differences in transporter expression and localization along the different segments of the small intestine. This review focuses on transporter protein expression and localization in the intestinal tract, which functions as a natural barrier to xenobiotics and influences the influx and efflux of drugs and their metabolites. Passive diffusion of drugs and xenobiotics across intestinal enterocytes will not be covered. The following sections will detail human intestinal expression, localization, and activity of SLC and ABC transporters. Transporter substrates and inhibitors are summarized in Tables 18.1 and 18.2.

## 18.3 Uptake Transporters

### 18.3.1 Apical Sodium-Dependent Bile Acid Transporter (ASBT)

ASBT (SLC10A2) is a bile acid transporter that functions as an electrogenic sodium–solute cotransporter that transports sodium and substrates in a ratio of 2 : 1 or greater [4]. ASBT transports unconjugated bile acids, and their glycine and taurine conjugates [4, 74]. ASBT is expressed in the terminal ileum, renal proximal tubules, and biliary epithelium [75]. While ASBT has no known drug substrates, it is a potential target for the development of prodrugs with improved oral bioavailability due to its transport of glycine and taurine conjugates [76].

ASBT plays a major role in the intestines. Passive absorption, facilitated transport, and active transport of bile acids at the apical and basolateral membrane by ASBT and OSTα/β (organic solute transporter), respectively, enable more than 90% of bile acids to be recycled [4, 77]. While passive absorption is feasible throughout the small intestine and colon, uptake by ASBT is limited by its strategic residence in the ileum, promoting maximal bile acid-assisted absorption of fats in the more proximal regions [4]. ASBT expression in the duodenum is comparatively modest, and its presence in the jejunum and colon is negligible, the ASBT protein expression ratio of the terminal ileum to that of proximal jejunum is about 100 [78]. Groer et al. [79] determined ASBT to be quantitatively absent in a patient's jejunum tissue sample and another study by Brück et al. [80] determined that the average ASBT abundance in sampled jejunal tissues ($n = 6$) was a mere $0.2 \pm 0.3$ pmol/mg. ASBT was undetectable in three Caco-2 lines that underwent a 21-day culture period, with the caveat that these cell lines are not identical and that such results may not be congruent depending on the culture period and procedure [80]. To decipher species differences, Akazawa et al. [81] utilized LC–MS/MS and found human ASBT protein levels ($n = 2$) were roughly at least 1.28 times greater in the ileum than in cynomolgus monkey ($n = 1$).

Besides facilitating fat transport, bile acids regulate their own production. Bile acids bind to the farnesoid X receptor (FXR), subsequently promoting production of fibroblast growth factors, FGF19 (human) and FGF15 (mouse) [82]. These fibroblast growth factors travel via the hepatic portal vein to the liver, where downstream effects ultimately repress CYP7A1 expression [82]. When ABST was inactivated in mice there were decreased bile acid concentrations and greater production via the CYP7A1-assisted classical pathway as opposed to the alternative pathway [82]. This rationale can be exploited for the treatment of hyperlipidemia involving a two-pronged approach: ASBT inhibition will reduce bile acid levels, which lowers cholesterol absorption and necessitates cholesterol conversion for bile acid synthesis [9]. Other than hyperlipidemia, ASBT inhibitors are potential therapeutic strategies for diseases like nonalcoholic fatty liver disease, type 2 diabetes, chronic idiopathic constipation, inflammatory bowel disease, hepatic encephalopathy, and cholestasis [9]. Such inhibitors are benzothiazepine, naphthol, steroid, benzamide, and quinoline derivatives [9]. In addition to acting as a target for these disease states, ASBT can be utilized to enhance bioavailability of pharmaceutical agents [4]. Prototype prodrugs via conjugation to bile acid analogs at the side chain or C-3 have been synthesized [4].

### 18.3.2 L-alpha Amino Acid Transporters (LATs) (see also Chapter 19)

LATs belong to the SLC7 and SLC43 families with four members currently identified: LAT1 (SLC7A5), LAT2 (SLC7A8), LAT3 (SLC43A1), and LAT4 (SLC43A2). LAT1 and LAT2 complex with SLC3A2 to form system L [10], and mediate sodium-independent amino exchange of large neutral amino acids (LAT1 and LAT2) and small

**Table 18.1** Uptake transporter substrates and inhibitors.

| Transporter | Substrate | Inhibitor |
|---|---|---|
| ASBT | Cholic acid (and taurine and glycine conjugates) [9] | Benzothiapine, naphthol, steroid, benzamide, and quinoline derivatives [9] |
| | Chenodeoxycholic acid (and taurine and glycine conjugates) [9] | |
| | Lithocholic acid [9] | |
| | Deoxycholic acid [9] | |
| LAT1/LAT2 | LAT1 and/or LAT2: | |
| | L-Dopa [10] | |
| | Gabapentin [10] | |
| | Dopamine [10] | |
| | Melphalan [10] | |
| | T3 (triiodothyronine) [10] | |
| | T4 (thyroxine) [10] | |
| | LAT1 | |
| | Large neutral amino acids [10, 11] | |
| | With less affinity: | |
| | L-asparagine [12] | |
| | L-glutamine [12] | |
| | LAT2: | |
| | Large neutral amino acids [10, 11] | |
| | Small neutral amino acids (e.g. Glycine, cysteine) [10, 13] | |
| MCT | MCT1 | MCT1 |
| | Pyruvate [14] | CHC [14] |
| | Butyrate [14] | pCMBS [14] |
| | Lactate [14] | Phloretin [14] |
| | Acetoacetate [14] | Quercetin [14] |
| | α-Oxoisovalerate [14] | XP13512 [14] |
| | α-Oxoisohexanoate [14] | AZD3965 [14] |
| | α-Ketoisovalerate [14] | |
| | XP13512 [14] | |
| | Penicillins [3] | |
| | Valproic Acid [3] | |
| | Nateglinide [3] | |
| | Atorvastatin [3] | |
| | β-lactam antibiotics [3] | |
| | MCT6 | |
| | Bumetanide [14] | |
| | Nateglinide [14] | |
| | Probenecid [14] | |
| | MCT7 | |
| | Ketone bodies [14] | |

Table 18.1 (Continued)

| Transporter | Substrate | Inhibitor |
|---|---|---|
| | MCT8 | |
| | Thyroid hormones [14] | |
| | MCT10 | |
| | Thyroid hormones [14] | |
| | MCT9 | |
| | Carnitine [14] | |
| | MCT12 | |
| | Creatinine [14] | |
| SMCT1 | γ-Hydroxybutyrate [15] | Ibuprofen [16] |
| | β-Hydroxybutyrate [16] | Ketoprofen [16] |
| | Acetoacetate [16] | |
| | Acetate [16] | |
| | Nicotinate [16] | |
| | Butyrate [16] | |
| | Propionate [16] | |
| | L-lactate [16] | |
| | Pyruvate [16] | |
| | 5-Aminosalicylate [16] | |
| | Salicylate [16] | |
| | Benzoate [16] | |
| OATP2B1 | Baicalin [17] | Atorvastatin [20] |
| | Irbesartan [18] | Fluvastatin [23] |
| | Saquinavir [19] | Celecoxib [24] |
| | Ketoconazole [18] | Sulfasalazine [20] |
| | Triiodothyronine [18] | Resperine [23] |
| | 5-Carboxyfluorescin (5-CF) [20] | Cyclosporine [23] |
| | Astemizole [18] | Glibenclamide [20] |
| | Methotrexate [19] | Fluticasone [23] |
| | 4′,5′-Dibromofluorescin (DBF) [20] | Ezetimibe [23] |
| | Atorvastatin [21] | Irbesartan [24] |
| | Pravastatin [22] | Narirutin [25] |
| | Etoposide [18] | Estrone-3-sulfate [20] |
| | Tenoposide [18] | Itraconazole [23] |
| | Sulfasalazine [22] | Erlotinib [24] |
| | Glyburide [19] | Zafirlukast [23] |
| | | Talinolol [20] |
| | | Curcumin [26] |
| | | Allura Red AC [26] |
| | | Green tea catechins [19] |
| | | Grapefruit juice [19] |

*(continued)*

**Table 18.1** (Continued)

| Transporter | Substrate | Inhibitor |
|---|---|---|
| PEPT1 | Di- and tri-peptides [27, 28] | Lys[Z($NO_2$)]-Pro [31] |
| | Zidovudine [3] | Conjugates of polyrotaxane-dipeptide [32] |
| | Didanosine [3] | |
| | NP-647 [29] | |
| | Acyclovir [3] | |
| | Ganciclovir [3] | |
| | Fosinopril [30] | |
| | L-Dopa-Phe [30] | |
| OCT1 | Metformin [33] | Irinotecan [34] |
| | Lamotrigine [33] | Berberine [39] |
| | Morphine [34] | Mitoxantrone [39] |
| | Ranitidine [35] | Prazosin [36] |
| | Tetraethylammonium [36] | Lamotrigine [39] |
| | Methylnaltrexone [37] | Itraconazole [39] |
| | Ondansetron [38] | Moxifloxacin [39, 40] |
| | Sumatriptan [38] | |
| | Meptazinol [37] | |
| | Ethambutol [39, 40] | |
| | Hydromorphone [37] | |
| | Fenoterol [38] | |
| | Oxymorphone [37] | |

neutral amino acids (LAT2) [11]. LAT3 and LAT4 do not require SLC3A2 for functional activity, and demonstrate sodium-, chloride- and pH-independent transport, functioning as facilitated diffusion transporters. In contrast to LAT1 and LAT2, LAT3 and LAT4 have narrower substrate specificity transporting leucine, isoleucine, phenylalanine, and methionine (LAT4 only) [11]. LAT1 is ubiquitously expressed with the greatest expression observed in the brain, placenta, testis, and bone marrow [83]. LAT2 is primarily expression in the intestine, colon, and kidney [10]. LAT3 expression is observed in the intestine, liver, and pancreas [11]. LAT4 expression is in the intestine, colon, and kidney [84].

The intestinal presence of LATs is inconclusive. In a northern blot study, Yanagida et al. [12] were unable to detect human LAT1 expression in the small intestine and colonic tissue of humans. Similarly, in the small intestine of mice, Lat1 did not reach the threshold needed to be detected [85]. Dave et al. [86] found Lat1 to be absent from NMRI mouse intestines with the protein focally present in the brain. Alternatively, LAT2 may have a greater role in the intestine than the former isoform, albeit levels are still low or undetectable in humans [10]. In mice, Lat2 (Slc7a8) abundance was relatively steady with an abundance of $17.1 \pm 0.5$, $21.1 \pm 0.4$, and $19.9 \pm 0.4$ fmol/μg plasma membrane protein in the duodenum, jejunum, and ileum, respectively [85]. Western blot and immunohistochemistry techniques revealed Lat2 expression in the stomach and the basolateral membrane of mouse villus enterocytes, and absent from the colon [86].

Immunofluorescence techniques in human duodenum, distal ileum, and ascending colon revealed that LAT4 was present on the basolateral membrane, with greater aggregation in the villus in contrast to the crypts [84]. However, the GI tract is not a predominant expression site relative to the other tissue compartments studied, including the kidney, placenta, and peripheral blood leukocytes [84]. Nevertheless, to clarify species differences, Bodoy et al. [11] conducted

**Table 18.2** Efflux transporter substrates and inhibitors.

| Transporter | Substrate | Inhibitor |
|---|---|---|
| **P-glycoprotein (P-gp)** | Loperamide [41] | Carvedilol [50] |
| | Digoxin [42] | Green tea [51] |
| | Sumatriptan [43] | Venetoclax [52] |
| | Fexofenadine [44] | Ritanovir [45] |
| | Dabigatran [45] | Amiodarone [53] |
| | Talinolol [46] | TPGS [46] |
| | Mitoxantrone [47] | Itraconazole [44] |
| | Atenolol [47] | HM30181 [41] |
| | Lopinavir [48] | Haloperidol [54] |
| | Atorvastatin [49] | PEG-300 [55] |
| | | Tween-80 [55] |
| | | Grapefruit juice [56] |
| | | Verapamil [57] |
| | | Pluronic-85 [3] |
| | | Cyclosporin [49] |
| | | Atorvastatin [49] |
| **MRP2** | Ritonavir [49] | Daunorubicin [49] |
| | Methotrexate [58] | MK571 [61] |
| | Glutathione conjugates [59] | Probenecid [61] |
| | Taurolithocholate 3-sulfate [49] | Leflunomide [61] |
| | Taurochenodexycholate 3-sulfate [49] | Cyclosporin [49] |
| | Lopinavir [48] | Bilirubin mono- and bis-glucuronide [49] |
| | Fosinopril [58] | |
| | Azithromycin [60] | |
| | irinotecan [49] | |
| | Bilirubin mono- and bis-glucuronide [49] | |
| **BCRP** | Topotecan [62] | WK-X-34 [68] |
| | Mitoxantrone [63] | Cremphor EL [69] |
| | Methotrexate [63] | Tween 20 [69] |
| | Sulfasalazine [64] | Pluronic P85 [69] |
| | Irinotecan [65] | Cyclosporin A [65] |
| | Rosuvastatin [64] | Lapatinib [64] |
| | Abacavir [66] | Ketoconazole [70] |
| | Prazosin [49] | Itraconazole [70] |
| | Estrone 3-sulfate [67] | Sirolimus [65] |
| | Cimetidine [63] | Tacrolimus [65] |
| | SN-38-glucoronide [67] | Curcumin [64] |
| | | Nicardipine [63] |
| | | Saquinavir [49] |

*(continued)*

**Table 18.2** (Continued)

| Transporter | Substrate | Inhibitor |
|---|---|---|
| | | Ritonavir [49] |
| | | Rilpivirine [66] |
| | | Omeprazole [63] |
| | | Flavinoids [71] |
| **OCTN1** | Tetraethylammonium [72] | Cediranib [72] |
| | Choline [72] | Camptothecin [72] |
| | Ergothioneine [72] | |
| | Acetylcholine [72] | |
| **OCTN2** | Colistin [73] | |
| | TEA [3] | |
| | L-carnitine [3] | |
| | γ-butyrobetaine [72] | |
| | Acetyl-carnitine [72] | |
| | Acyl-carnitine [72] | |
| | Valproic acid [3] | |
| | Ipratropium [3] | |

a northern blotting study with mouse GI tissue; it was discovered that Lat4 mRNA had the greatest presence in the small intestine and kidney compared to the other tissues studied. *In situ* hybridization studies narrowed expression primarily to the crypt with sparse expression in the villus [11]. Maric et al. [84] advocated for citrulline to be possible biomarker for the transporter. Citrulline is synthesized from other amino acids in the epithelial enterocytes, and thus, plasma concentrations may be reflective of amino acid efflux transporters [84]. Indeed, LAT4 mRNA expression was observed to correlate with plasma citrulline [84]; however, the correlation was not retained when type 2 diabetic patients and obese patients were removed from statistical analysis [84].

The relevance of LATs to drug development is due to their utilization for prodrug transport [87]. LAT1 and/or LAT2 substrates include melphalan, gabapentin, and L-dopa, which employ the carriers to promote uptake to the target site [10, 87]. For instance, L-dopa and gabapentin utilize LAT1 to cross the blood–brain barrier [87]. Although the carrier facilitates uptake at the effect site, its implications in intestinal bioavailability may not be clinically relevant as other transporters or passive absorption are more pertinent to intestinal absorption [10].

### 18.3.3 Proton-Dependent Monocarboxylate Transporters (MCTs)

Proton-dependent MCTs belong to the SLC16A family, with 14 isoforms identified based on sequence homology and conserved motifs. Of these 14 members, MCT1 (SLC16A1), MCT2 (SLC16A7), MCT3 (SLC16A8), and MCT4 (SLC16A3) have been the most extensively studied and characterized in terms of expression, function, and substrates. These substrates include short-chain fatty acids L-lactate, butyrate, pyruvate, and ketone bodies, such as acetoacetate, that are essential for carbohydrate, lipid, and amino acid metabolism for energy supply to tissues throughout the human body. Only MCT1-4 and MCT7 (SLC16A6) have been experimentally demonstrated to transport monocarboxylates, with MCT1-4 exhibiting electroneutral proton-coupled symporter activity in a stoichiometric ratio of 1 : 1. MCT8 and MCT10 (SLC16A2/SLC16A10) have been shown to shuttle thyroid hormones T2, rT3, T3, and T4, with MCT10 also responsible for transporting aromatic amines [88, 89]. MCT9 (SLC16A9) has demonstrated carnitine specificity [90], and MCT12 (SLC16A12) has been shown to be a creatinine transporter [91]. The endogenous binding substrates and regulatory mechanisms of MCT5, MCT6, MCT11, MCT13, and MCT14 (SLC16A4/SLC16A5/SLC16A11/

SLC16A13/SLC16A14) in humans are not yet fully characterized.

MCT 1, 3, 4, 5, 6, 8, 10, 12, and 13 are expressed in the human GI tract [14, 92]. With respect to the intestine, MCT1 is the most widely characterized with limited data available for other isoforms [14]. MCT1 has greater expression in the colon, consistent with the uptake of the bacterial fermentation byproduct butyrate as an energy source by colonic enterocytes [93]. The meta-analysis by Hardwood et al. [78] determined relative MCT1 expression along the small intestine to be comparably consistent with levels of the duodenum and ileum 1.25 and 1.29 times that of the jejunum. The highest expression was observed in the colon with relative expression 4.72 times that of the jejunum [78]. Individuals with irritable bowel disease have downregulated MCT1, depressing the uptake of butyrate [3].

Although studies concurred on the general trend of relative MCT1 expression, there were contradictory results about membrane localization [92, 94–97]. Human tissue immunohistochemistry studies demonstrated localization at the basolateral side of the surface epithelia of the rectum and proximal and sigmoid colon, and the upper crypts of the proximal colon [95]. Conversely, according to immunoblotting and immunohistochemical studies conducted by Gill et al. [92], the apical membrane is the primary localization site of MCT1. MCT3 was discovered primarily on the basolateral side, with descending order of expression as follows: ileum, proximal colon, distal colon, and jejunum [92]. Immunoblotting studies demonstrated the basolateral localization of both MCT4 and MCT5, with absence in the jejunum and an ascending trend in carrier levels from the ileum to the distal colon [92]. Kohyama et al. [98] identified MCT6 at the apical membrane of small intestine epithelium, while Gill et al. [92] claimed MCT6 to be absent in the small intestine and colon. The chaperone protein CD147 (basigin) mediates localization of MCT1, 3, and 4 to the membrane [14]. In CD147 deficiency, embigin is used in lieu to modulate membrane trafficking of MCT1 [14].

There have been MCT localization studies in animals, with the focus on MCT1. Akazawa et al. [81] found human ($n$ = 2) MCT1 expression in the ileum was roughly at least 1.34 times greater than the monkey sample ($n$ = 1). In mice and rats, Mct1 mRNA was located using *in situ* hybridization and X-ray visualization and protein was examined using immunohistochemistry; in both studies, Mct1 gene and protein expression was most evident in the caecum followed by the colon and had a qualitatively reduced presence in the small intestine and stomach [95]. Specifically in the small intestine of mice, there was the greatest abundance of Mct1 in the ileum (37.7 ± 2.5 mol/μg plasma membrane protein) compared to that of the duodenum (27.6 ± 1.2 mol/μg plasma membrane protein) and jejunum (25.1 ± 1.1 mol/μg plasma membrane protein) [85]. Furthermore, immunostaining and immunoelectron microscopy identified Mct1 at the basolateral membrane, but indiscernible at the junctional complexes, enteroendocrine cells, intraepithelial lymphocytes, goblet cells, and the apical membrane [95]. Utilizing qPCR, mouse MCT2, and 5 were predominantly in the stomach, while MCT4 mRNA was predominantly in the duodenum [99]. Utilizing immunoblotting, MCT 1, and 4 were spotted in porcine small intestine and colon, while MCT2 was present in the small intestine [100].

In addition to the aforementioned endogenous compounds, there are many confirmed and speculated xenobiotic substrates of MCT1, including atorvastatin, valproic acid, penicillin, and the drug of abuse γ-hydroxybutyrate, and inhibitors, including phloretin and α-cyano-hydroxy-cinnamate [3, 14]. In MCT1-transfected MDA-MB231 cells, ketoprofen, probenecid, and ibuprofen reduced GHB (γ-hydroxybutyrate) transport into cells [15]. Prototype prodrugs and nanoparticles targeting MCT1 have also been synthesized [3]. Wang et al. [3] designed and evaluated gabapentin, 5-FU, and gemcitabine prodrugs. The gabapentin prodrug, XP1352, enhanced bioavailability from 25 to 84% in monkey studies [3]. 5-Fluorouracil was linked with octanedioic acid or hexanedioic acid, enhancing bioavailability by one to four times the *in vivo* baseline [3]. The gemcitabine prodrug was synthesized with an amide linkage and increased both bioavailability (8.8-fold) and drug release [3]. Xenobiotic substrates have also been identified for MCT6 [14, 98]. Nateglinide transport in MCT6-expressing Xenopus laevis oocytes was demonstrated by Kohyama et al. [98] and thus, the transporter was proposed to aid in the intestinal uptake of the antidiabetic agent. Additional substrates are the loop diuretic bumetanide and the gout treatment probenecid [14].

### 18.3.4 Sodium-Dependent Moncarboxylate Transporters (SMCT)

Sodium-coupled monocarboxylate transporters belong to the SLC5A family and include SMCT1 (SLC5A8) and SMCT2 (SLC5A12). SMCT1 is an electrogenic transporter that binds sodium and monocarboxylates in a heterodimeric ratio ranging from 4 : 1 to 2 : 1 dependent on substrate [101]. SMCT2 is predicted to have electroneutral binding of sodium and monocarboxylates in a 1 : 1 ratio, with a lower affinity and higher capacity for substrates as compared to SMCT1. Endogenous substrates trafficked by SMCT1 and SMCT2 are similar to those of MCTs, including short-chain fatty acids, lactate, pyruvate, and ketone bodies [16, 88]. SMCT expression has been demonstrated in the kidney, intestine, brain, and thyroid gland (SMCT1 only) [102–104].

In humans, both isoforms are purported to be at the apical membrane of the intestine [16]. In humans, SMCT2 protein expression was detected in the jejunum and was absent in the colon and duodenum [99]. SMCT2 was found mainly at the luminal side of central crypt cells with reduced presence at the more proximal crypt cells [99]. Teramae et al. [99] studied smct1/2 localization: smct1 was primarily at the later regions of the GI and smct2 was chiefly located in the jejunum with both carriers identified at the apical membrane. Utilizing *in situ* hybridization and qPCR, smct1 mRNA was identified in the late ileum (I-2) and large intestine [99]. Immunostaining revealed smct1 protein in the apical membrane of the large intestine crypt cells, and the mid-crypts of the caecum [95]. smct2 mRNA was identified in the jejunum (J-1 and J-2) and to a lesser extent, the proximal ileum (I-1) [99]. There was no remarkable presence in the large intestine, duodenum, or distal ileum [99].

There is more abundant information regarding SMCT1 xenobiotic transport than its counterpart [16]. SMCT1 is downregulated in colorectal tumors as butyrate uptake by this transporter can impede histone deacetylase functionality [16]. The SNP at position 251 substitutes a phenylalanine for valine, diminishing transporter functionality and consequently, likely depresses substrate absorption [16]. GHB has been identified as a substrate of SMCT1 when expressed in FRTL-5 cells, but its clinical implications in oral administration of GHB has not been fully elucidated [15]. SMCT1 may be involved in the uptake of 5-aminosalicylate, which is indicated for ulcerative colitis [16]. On the other hand, ibuprofen and other NSAIDs are known to bind to SMCT1, obstructing the uptake of substrates [16].

### 18.3.5 Organic Anion Transporting Polypeptides (OATPs)

OATPs belong to the SLCO superfamily (previously SLC21A) with 11 OATPs identified in humans, belonging to six families [105]. OATPs are bidirectional transporters that transport a diverse range of compounds in an ATP and sodium-independent manner; however, the driving force for transport has not been elucidated [105]. Substrates include bile salts, thyroid hormones, polypeptides, and a wide range of pharmaceutical agents [105, 106]. OATP2A1, OATP3A1, and OATP4A1 have ubiquitous tissue expression, while other family members have restricted distribution. OATPs demonstrate species differences with respect to tissue distribution and specific isoforms [106]. In humans, two OATPs are expressed in the intestine, OATP1A2 and OATP2B1 [105].

OATP2B1 is more abundant than OATP1A2, with the latter exhibiting minimal or undetectable expression depending on the study [79, 80, 107, 108]. In fact, the presence of OATP1A2 in the intestine is dubious despite its proposed implication in enhancing apical xenobiotic transport [109]. Groer et al. [79] determined OATP1A2 to be absent in the jejunum and ileum ($n = 1$) and Brück et al. were [80] unable to measure OATP1A2 in jejunal tissues ($n = 6$). In parallel, OATP1A2 was undetectable along the small intestine and colon of organ donors ($n = 6$) [107]. OATP1A2 was undetectable in Caco-2 lines ($n = 3$) that underwent a 21-day culture period [80]. Conversely, mass spectrometry detected a mean OATP1A2 abundance of $1.50 \pm 0.04$ and $1.59 \pm 0.31$ pmol/mg of total mucosal protein in the jejunum ($n = 4$) and ileum ($n = 12$), respectively [108]. There was no significant difference in protein expression in these two regions with the stipulation that the lack of variability may be attributed to the use of separate cohorts for the ileum and jejunum [108].

There are discrepancies among sources with regard to apical or basolateral residence of OATP2B1 [22, 110, 111]. Keiser et al. [22] executed membrane isolation tangentially with mass spectroscopy in human jejunum samples ($n = 6$ surgery patients), immunostaining using Caco-2 cells, and efflux ratio determination of the substrate pravastatin, all of which supported its possible basolateral localization. QconCAT in tangent with mass spectrometry was used to detect a mean OATP2B1 expression of $2.65 \pm 1.95$ and $3.71 \pm 2.50$ pmol/mg of total mucosal protein in the jejunum and ileum, respectively [108]. OAT2B1 had greater expression as compared to OATP1A2, OST-$\alpha$, and OST-$\beta$ [108]. Brück et al [80] found the average OATP2B1 protein abundance in the jejunum tissues ($n = 6$) was $0.5 \pm 0.3$ pmol/mg and the average OATP2B1 expression in Caco-2 lines ($n = 3$) was $3.3 \pm 0.5$ pmol/mg. It should be noted that in all studies the sample number evaluated is small, indicative of the difficulty in obtained human intestinal tissue.

OATPs have a spectrum of substrates, with some overlap among isoforms [112]. Potential OATP1A2 substrates include saquinavir, methotrexate, and levofloxacin [113]. OATP2B1 substrates include atorvastatin, astemizole, etoposide, triiodothyronine, ketoconazole, baicalin, and glyburide [17–19, 21]. Green tea catechins, grapefruit juice, and xenobiotics, such as ezetimibe and irbesartan, have inhibitory effects on OATP2B1 [19, 23, 24]. In a randomized and crossover study ($n = 14$), patients were administered either enantiomer of fexofenadine during the treatment periods [114]. Coadministration with grapefruit juice reduced AUC of the S-enantiomer by 52% and the R-enantiomer by 39%, illuminating possible stereoselectivity [114]. In addition to drug and food interactions, there are pharmacogenetic considerations, including OATP2B1 SNPs 1467C>T and 935G>A, that have been implicated in reduced drug absorption profiles for fexofenadine and montelukast, respectively [19].

### 18.3.6 Peptides Transporters (PepTs)

PEPT1 and PEPT2 belong to the SLC15A family [30]. These proton-dependent di- and tripeptide carriers are hypothesized to have intracellularly located termini and an expansive extracellular loop linking 9 and 10 of the 12-alpha-helical transmembrane domains [27, 28]. Pre-miR-27a-3p, pre-miR-193a-3p, and pre-miR-193b-3p were observed to interact with 3′-UTR of the PepT1 gene, eluding to posttranscriptional regulation by these microRNA [115]. PEPT1 has a high capacity, but low affinity [3]. It is a significant transporter in the small intestine where it can aid in the oral absorption of certain drugs such as L-dopa-Phe, beta-lactam antibiotics, antivirals, and ACE inhibitors [30]. mRNA expression is additionally present in the liver, kidneys, heart, lungs, pancreas, uterus, and placenta [27]. PEPT2 is notable in the kidney, but mRNA expression is ubiquitous with expression in, but not limited to the liver, lungs, testes, and ovaries [27].

PepT1 is a major protein in the small intestine; in fact, a study by Drodzik et al. [107] found that when examining nine major transporters (4 ABC and 5 SLCs), PepT1 made up 50% of the total protein amount. PepT1 is amply expressed in the upper region of the jejunum with a mean of 3.69 pmol/mg of total membrane protein in Caucasian adults [78]. The highest protein expression was in the terminal ileum, 1.36 times that of the proximal jejunum [78]. Similarly, Groer et al. [79] measured PEPT1 levels to be approximately 2.5 pmol/mg in the jejunum and nearly double this in the ileum. Moreover, Brück et al. [80] found the average PEPT1 protein expression in jejunum sampled tissues ($n = 6$) to be $4.0 \pm 0.8$ pmol/mg, which was greater than that of ABCC2, ABCC3, ABCB1, ABCG2, OCT1, OCT3, OATP2B1, and ASBT while PEPT1 was found to be liberally distributed in the small intestine, there was negligible expression in the colon [78]. Drozdzik et al. [107] demonstrated a significantly greater ($p < 0.001$) presence of PEPT1 in the small intestine (2.63–4.89 pmol/mg) than in the colon (0.19–0.31 pmol/mg).

In mice, pept1(slc15a1) abundance was relatively steady; transporter expression was $4.95 \pm 0.77$, $6.92 \pm 0.69$, and $6.71 \pm 1.54$ fmol/μg plasma membrane protein in the duodenum, jejunum, and ileum, respectively [85]. In order to facilitate extrapolation to human outcomes, Hu et al. [116] successfully created a mouse model to express humanized PEPT1. To investigate functionality, an *in situ* perfusion experiment with the dipeptide glySar was conducted, demonstrating a greater capacity yet lower potency in the non-transgenic mouse than its transgenic counterparts [116]. Oral dosing of glySar revealed similar $C_{max}$ and $T_{max}$ values in the transgenic (huPEPT1) and non-transgenic (pept1) mouse [116].

To enhance bioavailability, prodrugs of the antiviral agents oseltamivir and zanamivir have been synthesized targeting PEPT1 [28]. In addition, Sun et al. [5] designed an ester prodrug, Peramivir-(CH2)2-l-Val, to improve poor absorption ($F = 0.03$ for parent compound). In rat studies, the prodrug bioavailability ($F$) was determined to be 0.653 and due to the rapid cleavage of the ester, parent drug release was expeditious [5].

### 18.3.7 Organic Cation Transporters (OCTs)

OCTs belong to the SLC22 family with three members (OCT1 (SLC22A1), OCT2 (SLC22A2), and OCT3 (SLC22A3)) identified in humans, and three corresponding orthologs in rats and mice [117]. All OCTs are predicted to have similar membrane topology with 12 transmembrane domains, an intracellular loop and an extracellular loop, and transporter substrates in a sodium-independent electrogenic manner [118, 119]. OCT substrates are typically organic cations and weak bases carrying a positive charge at physiological pH; however, uncharged compounds, including cimetidine, are substrates [119]. Transport may be bidirectional depending on the electrochemical gradient of the substrate [117]. OCTs demonstrate a relatively wide tissue distribution with all three isoforms expressed in the human small intestine [118].

The membrane localization of OCT1 is inconclusive [120]. OCT1 is detectable at low levels along the small intestine [78, 120]. Absolute expression in the distal jejunum, duodenum, proximal ileum, and distal ileum was 0.87, 1.03, 1.07, and 1.11 times that of the proximal jejunum [78]. The lowest protein abundance was in the colon, which was 0.73 times that of the proximal jejunum [78]. Brück et al [80] found the average OCT1 abundance of six jejunum tissues was $0.7 \pm 0.5$ pmol/mg. Contrasting the previous studies, Gröer et al. [79] found OCT1 was measurable in the ileum but not in the jejunum; however, the study was limited to a single intestinal sample. The localization of OCT1 has yet to be conclusively deduced, and is complicated by its bidirectional modality [120]. A review by Wenzel et al. [120] postulated that OCT1 resides on the basolateral side, enabling uptake of cationic substances in the bloodstream for the purpose of elimination. However, Han et al. [121] carried out immunostaining on intestinal samples of human and Caco-2 cells, with each of the mediums exhibiting the concurring presence of OCT1 at the luminal side.

OCT2 expression is neither evident in the small intestine nor colon [122]. OCT3 was expressed along the GI tract, with greatest absolute expression in the colon, which was 2.25 times that of the proximal jejunum [78]. Of note, protein levels in the terminal ileum were a bit higher than that of the proximal ileum with respective expressions 1.26 and 1.07 times that of the proximal jejunum [78]. Utilizing immunofluorescence techniques on jejunum tissue, Müller et al. [123] located OCT3 on the apical membrane. The average OCT3 abundance in the jejunum tissues ($n = 6$) was $0.1 \pm 0.03$ pmol/mg, but OCT3 was undetectable in Caco-2 lines ($n = 3$, culture time = 21 days) [80]. Drozdzik et al. [107] measured transporter levels in the GI tissue of organ donors ($n = 6$). In the colon (0.11–0.14 pmol/mg) there was a significantly greater presence of OCT3 than in the small intestine (0.05–0.06 pmol/mg) [107].

OCTs are extensively involved in intestinal drug uptake and exhibit a wide substrate specificity [124]. OCT1 substrates include antinociceptive agents (morphine and its prodrug codeine, oxymorphone, and hydromorphone) and the μ-opioid antagonist methylnaltrexone indicated for opioid-induced constipation [34, 37]. Other substrates include ethambutol, imatinib, acyclovir, lamotrigine, and metformin [33, 35, 38–40]. The lattermost, a first-line type 2 diabetes agent, relies on PMAT (plasma membrane monoamine transporter), SERT (serotonin transporter), and OCT1 for intestinal transport [125]. A systemic review by Mato et al. [126] included a study population of 10 different ethnicities and discerned 34 OCT1 polymorphisms, with Met408Val implicated in decreasing gene expression, consequently decreasing intestinal uptake via the transporter. However, the effect of the SNP is nuanced, as Han Chinese with this polymorphism exhibited a significant decrease in hemoglobin A1C levels, an indicator of the drug's efficacy [126]. OCT3 substrates include $MPP^+$ (methyl 4-phenylpyridinium), tetraethylammonium, and neurotransmitters such as serotonin, dopamine, and epinephrine [127, 128].

## 18.4 Efflux Transporters

### 18.4.1 P-glycoprotein (Pgp)

Pgp (ABCB1) is an ABC transporter that is encoded by the MDR1 gene in humans. In contrast, Pgp is encoded by three genes in rodents, mdr1a (Abcb1a), mdr1b (Abcb1b), and mdr2. Pgp mediates the active efflux of a wide range of compounds against a concentration gradient. Substrates include antiretrovirals, antibiotics, chemotherapeutic agents, and analgesics [129]. Pgp is widely expressed in epithelial cells making up biological barriers in the intestine, brain, and placenta, and contributes to multidrug resistance in cancer therapy [130, 131].

Pgp is expressed at the apical membrane of enterocytes [132]. Protein levels in descending order are as follows: distal ileum, distal jejunum, proximal ileum, duodenum, and colon [78]. Levels in the distal jejunum, duodenum, and colon were 1.86, 0.57, and 0.36 that of the proximal jejunum [78]. The ileum, which was delineated into four subsections (ileum I-IV) by Harwood et al [78]; expression in the proximal (I-II) and distal ileum (III-IV) was 1.78 and 2.06–2.18 times greater than that of the proximal jejunum. Another study observed the same trend of ascending ABCB1 protein expression from the duodenum to the ileum [107]. Additionally, there was a significantly greater presence of ABCB1 in the small intestine (0.29–1.06 pmol/mg) compared to the colon (0.15–0.37 pmol/mg) [107]. QconCAT in tangent with mass spectrometry was used to detect a mean Pgp amount of $0.74 \pm 0.22$ and $0.96 \pm 0.56$ pmol/mg of total mucosal protein in the jejunum and ileum, respectively [108]. Al-Majdoub et al. [133] obtained jejunal ($n = 4$) and ileal ($n = 12$) biopsy samples, which had a mean Pgp expression of $1.02 \pm 0.75$ pmol/mg of mucosal protein. The average ABCB1 expression of six human jejunum sampled tissues was $0.5 \pm 0.2$ pmol/mg and the average ABCB1 expression of three Caco-2 lines was $1.0 \pm 0.1$ pmol/mg [80]. In addition, Pgp expression in Caco-2 cells was quantified at different culture times with progressive increases in expression between weeks 1 and 4 [80].

In mice, Mdra1 saw an ascending trend from the duodenum ($5.22 \pm 0.38$ fmol/μg plasma membrane protein) to the ileum (9.48 ± 0.45 fmol/μg plasma membrane protein) [85]. In the small intestine of mouse, both mdr1b and mdr2 were beneath the quantification threshold [85]. In a second study, LC tangent MS was used to quantify the impact of sex and food on Pgp abundance in the different sections of the GI tract [134]. Jejunal and ileal Pgp levels were 83% and 58% greater in fasted male rats relative to fasted female rats [134]. The inverse was observed in the postprandial state with female rats exhibiting greater Pgp levels in these regions than the fed males [134]. In the ileum and rectum of rhesus macaque tissue, Pgp mRNA was double that of the humanized mouse model [135].

PGP has an extensive catalog of xenobiotic substrates, inducers, and inhibitors. The transporter offers protection against potentially harmful substances but presents challenges, especially in oncology as tumor cells overexpress PGP, promoting the efflux of chemotherapeutic agents [136]. PGP substrates include digoxin, sumatriptan, mitoxantrone, atenolol, fexofenadine, and dabigatran, with certain agents like atorvastatin acting both as a substrate and inhibitor [42–45, 47, 49]. For substrates with a narrow therapeutic window, it is imperative to investigate the influence of MDR1 on the pharmacokinetic profile and pharmacodynamic endpoints. Examples of inducers include St. Johns' Wort, rifampicin, carbamazepine, phenytoin, and curcumin [8, 137, 138]. Inhibitors of the transporter tend to have molecular axis of at least 18, a minimum log $P$ of 2.92, and tertiary basic nitrogen(s) [49]. Inhibitors include haloperidol, amiodarone, itraconazole, and carvedilol [44, 50, 53, 54]. The focus should not only be on active ingredients as excipients like nonionic surfactants (i.e. Tween 80) and PEG-300 and compounds in juices and green tea also have been demonstrated to inhibit PGP [49, 55].

### 18.4.2 Multidrug Resistance-associated Protein (MRP)

MRPs are members of the ABCC family of transporters with nine members (MRP1–MRP9; ABCC1–ABCC6 and ABCC10–ABCC12) [139]. The family has three additional members, the cystic fibrosis transmembrane conductance regulator protein, and the SUR proteins (SUR1 and SUR2). Similar to PGP, MRPs transport a diverse range of compounds including antiretrovirals, anticancer agents, and antibiotics [139]. MRP1–MRP6 have differing tissue expression with one or more isoforms expressed in the intestine, colon, brain, liver, kidney, and placenta [139].

According to the designations by Harwood et al. [78], MRP2 is on the luminal side, while MRP1, 3, and 4 are on the basolateral membrane. Relative MRP1 expression was most prevalent in the proximal jejunum and lowest in the duodenum, in which expression in the latter was almost half (0.45) of the former [78]. In between was the colon, distal ileum, distal jejunum, and proximal ileum, with relative expression levels 0.93, 0.89, 0.88, and 0.86 of the proximal duodenum, respectively [78]. In the duodenum, MRP1 exhibited lower gene expression than PGP and MRP2-5 [140]. Villin-normalized MRP1 mRNA levels in the duodenum were one-half to one-third of the terminal ileum and four regions of the colon [140]. MRP1 mRNA levels in the terminal ileum were lower than that of PGP and MRP3, but comparable to that of MRP4 and MRP5 [140].

With the exclusion of the jejunum, which was not included in the study, there was a descending trend of MRP2 from the duodenum to colon, with sparse levels in the colon and expression in the terminal ileum roughly half that of the duodenum [100]. MRP2 expression was 3.33 ± 0.76 and 2.66 ± 0.65 pmol/mg of total mucosal protein in the jejunum and ileum, respectively [108]. There was more MRP2 than PGP and BCRP in these regions [108]. Similarly, in Caucasian adults, MRP2 exhibited greater protein expression in the proximal jejunum than PGP, BCRP, and MRP3 [78]. The lowest abundance was in the duodenum, and the greatest abundance was in the colon, with levels 0.30 and 1.65 that of the proximal jejunum [78]. There were relatively steady levels from the proximal jejunum to distal ileum, with expression in the proximal and distal ileum 0.89 and 0.92 times that of the proximal jejunum [78]. The average MRP2 levels in the jejunum ($n$ = 6) was 0.9 ± 0.5 pmol/mg and expression in Caco-2 lines was 0.8 ± 0.2 pmol/mg [80].

In the duodenum and four regions of the colon (ascending, descending, transverse, and sigmoid), MRP3 had greater mRNA levels than that of PGP and MRP1, 2, 4, and 5 [140]. MRP3-5 had roughly similar levels in the duodenum and ileum, but levels were two to three times greater in the colon [140]. However, caution should be given to gene expression data as miR-192-5p was proposed to regulate translation of ABCC3 mRNA [115]. Such microRNA regulation along with post-translation regulation may lead to discrepancies between protein and gene expression, a relatively unanimous theme observed among transporters [115]. In the colon (1.53–2.11 pmol/mg), there was a significantly greater presence of ABCC3 than in the small intestine (0.51–0.70 pmol/mg) [107]. ABCC3 protein abundance in the jejunum was 0.5 ± 0.2 pmol/mg [80].

Human MRP1 expression in the ileal tissue was roughly 1.56 times greater than that of the monkey tissue [81]. Monkey MRP2 expression was roughly at least 1.68 times greater in the jejunum than the two human samples [81]. Immunohistochemistry staining of two humanized mouse models and monkey tissue revealed MRP4 was localized in the lamina propria and heavily concentrated in basolateral membrane of mucosal cells [135]. Rost et al. [141] conducted studies with rat intestine tissue and identified Mrp2 and Mrp3 at the brush border and basolateral membrane, respectively [141]. Monkey MRP6 expression in the jejunum and ileum was roughly 4.34 and 3.88 times greater than the human samples [81].

Like PGP, MRPs are involved in the extrusion of pharmaceutical agents. MRP2 substrates include glucuronide, sulfate, and glutathione conjugates, including taurolithocholate-3-sulfate, glucuronide conjugates of bilirubin [49, 142]. Other substrates include antivirals (ritonavir, lopinavir) and chemotherapeutic agents (antimetabolite methotrexate, irinotecan) [48, 49, 58]. Inhibitors include antineoplastic agents, such as daunorubicin, immunosuppressant cyclosporin, and the gout drug probenecid [49, 61].

### 18.4.3 Breast Cancer Resistance Protein (BCRP)

BCRP (ABCG2) is an ABC transporter with no sequence similarity to PGP or MRP1. In contrast to other ABC transporters, BCRP is a "half-transporter" with six transmembrane domains and one nucleotide binding domain, which forms homodimers in the membrane [67]. BCRP functions as an efflux transporter transporting sulfo-conjugated organic anions, hydrophobic and amphipathic drugs, demonstrating substantial overlap in substrate specificity with PGP. BCRP is expressed in a wide range of tissues, including the intestine, colon, placenta, brain, liver, and kidney [49, 143].

BCRP levels were greatest in the distal jejunum, which had a 2.47 times greater abundance than the proximal jejunum [78]. BCRP protein expression in the proximal ileum, duodenum, and colon were, respectively 0.94, 0.63, and 0.44 that of the proximal jejunum [78]. Another study supported this trend: there was significantly greater expression in the small intestine (0.19–0.41 pmol/mg) than in the colon (0.04–0.16 pmol/mg) [107]. Couto et al. [108] determined the BCRP abundance to be $0.59 \pm 0.27$ and $0.45 \pm 0.13$ pmol/mg of total mucosal protein in the jejunum ($n$ = 4) and ileum ($n$ = 12), respectively. The average BCRP amount in six jejunum tissues was $0.3 \pm 0.2$ pmol/mg [80]. Additionally, the mean BCRP ($n$ = 10) expression in the jejunum and ileum was $0.37 \pm 0.29$ pmol/mg of mucosal protein [133]. In mice, Bcrp exhibited an ascending trend from the duodenum ($4.79 \pm 0.18$ fmol/μg plasma membrane protein) to the ileum ($6.97 \pm 0.22$ fmol/μg plasma membrane protein) [85]. The average BCRP expression of three Caco-2 lines was $0.5 \pm 0.1$ pmol/mg, and culture time did not influence expression [80].

BCRP was originally identified in MCF-7 breast cancer cells and is implicated in efflux of xenobiotics, posing an obstacle for tumor delivery [144]. BCRP substrates include chemotherapeutic agents (camptothecins, TKIs, methotrexate, mitoxantrone, and nucleoside analogs), select toxins, glucuronide and sulfated compounds (estrone-3-sulfate), rosuvastatin, cimetidine, and sulfasalazine [63, 64, 67]. BCRP inhibitors are nicardipine, azole antifungals, flavonoids, tacrolimus, omeprazole, HIV protease inhibitors, and antineoplastic agents (tyrosine kinase inhibitors) [63, 71]. Substrates and inhibitors of both PGP and BCRP are difficult to distinguish in human clinical trials. BCRP inhibition can occur by allosteric means, impediment of ATPase functionality, and competitive inhibition [63]. Because of the possibility of more than one binding site, competitive inhibition has been postulated to occur at differing sites depending on the inhibitor [63]. Aiding this theory, an inhibitor and protein mutation at a specific site may only impact certain substrates [63].

### 18.4.4 Novel Organic Cation Transporters (OCTN)

OCTNs are transporters in the SLC22A family with two isoforms in humans, OCTN1 (SLC22A4) and OCTN2 (SLC22A5). Rodents express three Octn transporters, Octn1, Octn2, and Octn3. OCTN1 is predicted to have 11 transmembrane domains and one nucleotide binding domain, while OCTN2 is predicated to have 12 transmembrane domains [27]. OCTN1 functions as an electroneutral proton gradient-dependent antiporter resulting in substrate efflux, while OCTN2 functions as a sodium-independent carnitine cotransporter and a sodium-dependent transporter of organic cations [117]. Expression of OCTN1 and OCTN2 is observed in a wide range of tissues, including the intestine, colon, liver, kidney, and heart [118].

Harwood et al. [78] determined that the relative OCTN1 expression was highest in the proximal jejunum and the distal jejunum was 0.63 of its more proximal counterpart. Levels in the colon, duodenum, proximal ileum, and terminal ileum were 0.24, 0.46, 0.78, and 0.80 that of the proximal jejunum [78]. Another study concluded that there was more OCTN1 in the ileum versus the colon [145]. Sugiuara et al. [146]

identified OCTN1 at the luminal membrane of human small intestinal tissue. Meier et al. [147] presented villin-normalized OCTN1 mRNA, which was present in the duodenum, ileum, and colon at relatively steady levels. OCTN2 gene expression was significantly greater in the colon compared to the duodenum [147]. Koepsell [122] theorized that the apical residence of OCTN2 in rats is possibly generalizable to humans.

OCTN1 substrates often possess an amine, carbonyl, or carboxyl group [72]. Example substrates include choline and ergothioneine, while inhibitors are cediranib and camptothecin [72]. There have been attempts to synthesize OCTN2-targeted prodrugs to promote uptake into enterocytes [3]. OCTN2 prodrug potential lies in the fact that it has a wider substrate capacity than PEPT1 and ASBT, which are also frequently targeted to enhance bioavailability [3]. OCTN2 has a high capacity and affinity for carnitine, evident by the 50-fold intracellular presence of carnitine in contrast to the extracellular [3]. Xenobiotics can be linked at the hydroxy position of carnitine; when this strategy was used for prednisolone, transport rose 1.79-fold [3]. Another strategy is to mimic carnitine structure, which processes amine and carboxylic ends, with greater affinity the more substituents on the amine [3].

## 18.5 Summary

Select members of the ABC and SLC transporters are the main proteins implicated in the active transport of nutrients and drugs in the intestine. ABCs are powered by ATP hydrolysis and in the intestine, act predominantly as efflux transporters (Pgp, MRPs, and BCRP). In contrast, SLCs are usually powered by an ion concentration gradient and can act as influx, efflux, or bidirectional transporters. Intestinal SLC subfamilies/members include ASBT, LATs, MCTs, SMCTs, PepTs, OCTs, OATPs, and OCTN. The major themes surrounding these proteins include structuring prodrugs that are substrates for these transporters, directly targeting the transporters for disease treatment, pinpointing species differences for more accurate *in vivo* to *in vivo* extrapolation and PBPK modeling, and identifying clinically relevant inhibitors, substrates, and inducers [148].

There are study differences in culture timing (i.e. Caco-2 cell lines), quantification techniques (LC/MS–MS, immunoblotting), protein digestion strategies, protein references, and fraction used (total membrane or cytoplasmic) [80, 149]. Besides laboratory protocol differences, interindividual variances exist and are further exacerbated by food and drug interactions [49, 134]. Also important to note is that mRNA is usually not indicative of protein abundance due to posttranscriptional and translational regulation [115]. Thus, as disparate protein quantification results are inevitable, there should be a focus on trends, with a call for elucidation of the nuances and contradictory results.

## References


1 Cesar-Razquin, A., Snijder, B., Frappier-Brinton, T. et al. (2015). A call for systematic research on solute carriers. *Cell* 162 (3): 478–487.

2 Lin, L., Yee, S.W., Kim, R.B., and Giacomini, K.M. (2015). SLC transporters as therapeutic targets: emerging opportunities. *Nat. Rev. Drug Discovery* 14 (8): 543–560.

3 Wang, G., Zhao, L., Jiang, Q. et al. (2020). Intestinal OCTN2- and MCT1-targeted drug delivery to improve oral bioavailability. *Asian J. Pharm. Sci.* 15 (2): 158–172.

4 Dawson, P.A. (2011). Role of the intestinal bile acid transporters in bile acid and drug disposition. *Handb. Exp. Pharmacol.* 201: 169–203.

5 Sun, Y., Gan, W., Lei, M. et al. (2018). PEPT1-mediated prodrug strategy for oral delivery of peramivir. *Asian J. Pharm. Sci.* 13 (6): 555–565.

6 Planas, J.M., Alfaras, I., Colom, H., and Juan, M.E. (2012). The bioavailability and distribution of trans-resveratrol are constrained by ABC transporters. *Arch. Biochem. Biophys.* 527 (2): 67–73.

7 Nicolussi, S., Drewe, J., Butterweck, V., and Meyer Zu Schwabedissen, H.E. (2020). Clinical relevance of St. John's wort drug interactions revisited. *Br. J. Pharmacol.* 177 (6): 1212–1226.

8 Hennessy, M., Kelleher, D., Spiers, J.P. et al. (2002). St Johns wort increases expression of P-glycoprotein: implications for drug interactions. *Br. J. Clin. Pharmacol.* 53 (1): 75–82.

9 Yang, N., Dong, Y.-Q., Jia, G.-X. et al. (2020). ASBT(SLC10A2): a promising target for treatment of diseases and drug discovery. *Biomed. Pharmacother.* 132: 110835.

10 del Amo, E.M., Urtti, A., and Yliperttula, M. (2008). Pharmacokinetic role of L-type amino acid transporters LAT1 and LAT2. *Eur. J. Pharm. Sci.* 35 (3): 161–174.

11 Bodoy, S., Martin, L., Zorzano, A. et al. (2005). Identification of LAT4, a novel amino acid transporter with system L activity. *J. Biol. Chem.* 280 (12): 12002–12011.

12 Yanagida, O., Kanai, Y., Chairoungdua, A. et al. (2001). Human L-type amino acid transporter 1 (LAT1): characterization of function and expression in tumor cell lines. *Biochim. Biophys. Acta (BBA)* 1514 (2): 291–302.

13 Segawa, H., Fukasawa, Y., Miyamoto, K.-I. et al. (1999). Identification and functional characterization of a $Na^+$-independent neutral amino acid transporter with broad substrate selectivity*. *J. Biol. Chem.* 274 (28): 19745–19751.

14 Felmlee, M.A., Jones, R.S., Rodriguez-Cruz, V. et al. (2020). Monocarboxylate transporters (SLC16): function, regulation, and role in health and disease. *Pharmacol. Rev.* 72 (2): 466–485.

15 Cui, D. and Morris, M.E. (2009). The drug of abuse gamma-hydroxybutyrate is a substrate for sodium-coupled monocarboxylate transporter (SMCT) 1 (SLC5A8): characterization of SMCT-mediated uptake and inhibition. *Drug Metab. Dispos.* 37 (7): 1404–1410.

16 Ganapathy, V., Thangaraju, M., Gopal, E. et al. (2008). Sodium-coupled monocarboxylate transporters in normal tissues and in cancer. *AAPS J.* 10 (1): 193–199.

17 Kalapos-Kovács, B., Juhász, V., Temesszentandrási-Ambrus, C. et al. (2018). Baicalin is a substrate of OATP2B1 and OATP1B3. *Phytother. Res.* 32 (8): 1647–1650.

18 Schäfer, A.M., Bock, T., and Meyer zu Schwabedissen, H.E. (2018). Establishment and validation of competitive counterflow as a method to detect substrates of the organic anion transporting polypeptide 2B1. *Mol. Pharmaceutics* 15 (12): 5501–5513.

19 Yu, J., Zhou, Z., Tay-Sontheimer, J. et al. (2017). Intestinal drug interactions mediated by OATPs: a systematic review of preclinical and clinical findings. *J. Pharm. Sci.* 106 (9): 2312–2325.

20 Kawasaki, T., Shiozaki, Y., Nomura, N. et al. (2020). Investigation of fluorescent substrates and substrate-dependent interactions of a drug transporter organic anion transporting polypeptide 2B1 (OATP2B1). *Pharm. Res.* 37 (6): 115.

21 Grube, M., Köck, K., Oswald, S. et al. (2006). Organic anion transporting polypeptide 2B1 is a high-affinity transporter for atorvastatin and is expressed in the human heart. *Clin. Pharmacol. Ther.* 80 (6): 607–620.

22 Keiser, M., Kaltheuner, L., Wildberg, C. et al. (2017). The organic anion-transporting peptide 2B1 is localized in the basolateral membrane of the human jejunum and Caco-2 monolayers. *J. Pharm. Sci.* 106 (9): 2657–2663.

23 Unger, M.S., Schumacher, L., Enzlein, T. et al. (2020). Direct automated MALDI mass spectrometry analysis of cellular transporter function: inhibition of OATP2B1 uptake by 294 drugs. *Anal. Chem.* 92 (17): 11851–11859.

24 Unger, M.S., Mudunuru, J., Schwab, M. et al. (2020). Clinically relevant OATP2B1 inhibitors in marketed drug space. *Mol. Pharmaceutics* 17 (2): 488–498.

25 Morita, T., Akiyoshi, T., Sato, R. et al. (2020). Citrus fruit-derived flavanone glycoside narirutin is a novel potent inhibitor of organic anion-transporting polypeptides. *J. Agric. Food. Chem.* 68 (48): 14182–14191.

26 Tikkanen, A., Pierrot, E., Deng, F. et al. (2020). Food additives as inhibitors of intestinal drug transporter OATP2B1. *Mol. Pharmaceutics* 17 (10): 3748–3758.

27 Klaassen, C.D. and Aleksunes, L.M. (2010). Xenobiotic, bile acid, and cholesterol transporters: function and regulation. *Pharmacol. Rev.* 62 (1): 1–96.

28 Brandsch, M. (2013). Drug transport via the intestinal peptide transporter PepT1. *Curr. Opin. Pharmacol.* 13 (6): 881–887.

29 Khomane, K.S., Nandekar, P.P., Wahlang, B. et al. (2012). Mechanistic insights into PEPT1-mediated transport of a novel antiepileptic, NP-647. *Mol. Pharmaceutics* 9 (9): 2458–2468.

30 Brandsch, M., Knutter, I., and Bosse-Doenecke, E. (2008). Pharmaceutical and pharmacological importance of peptide transporters. *J. Pharm. Pharmacol.* 60 (5): 543–585.

31 Knütter, I., Theis, S., Hartrodt, B. et al. (2001). A novel inhibitor of the mammalian peptide transporter PEPT1. *Biochemistry* 40 (14): 4454–4458.

32 Yui, N., Ooya, T., Kawashima, T. et al. (2002). Inhibitory effect of supramolecular polyrotaxane–dipeptide conjugates on digested peptide uptake via intestinal human peptide transporter. *Bioconjugate Chem.* 13 (3): 582–587.

33 Shu, Y., Sheardown, S.A., Brown, C. et al. (2007). Effect of genetic variation in the organic cation

transporter 1 (OCT1) on metformin action. *J. Clin. Invest.* 117 (5): 1422–1431.

34 Zhu, P., Ye, Z., Guo, D. et al. (2018). Irinotecan alters the disposition of morphine via inhibition of organic cation transporter 1 (OCT1) and 2 (OCT2). *Pharm. Res.* 35 (12): 243.

35 Meyer, M.J., Seitz, T., Brockmöller, J., and Tzvetkov, M.V. (2017). Effects of genetic polymorphisms on the OCT1 and OCT2-mediated uptake of ranitidine. *PLoS One* 12 (12): e0189521.

36 Moss, D.M., Liptrott, N.J., Siccardi, M., and Owen, A. (2015). Interactions of antiretroviral drugs with the SLC22A1 (OCT1) drug transporter. *Front. Pharmacol.* 6: 78.

37 Meyer, M.J., Neumann, V.E., Friesacher, H.R. et al. (2019). Opioids as substrates and inhibitors of the genetically highly variable organic cation transporter OCT1. *J. Med. Chem.* 62 (21): 9890–9905.

38 Morse, B.L., Kolur, A., Hudson, L.R. et al. (2020). Pharmacokinetics of organic cation transporter 1 (OCT1) substrates in Oct1/2 knockout mice and species difference in hepatic OCT1-mediated uptake. *Drug Metab. Dispos.* 48 (2): 93–105.

39 Zhou, S., Zeng, S., and Shu, Y. (2021). Drug-drug interactions at organic cation transporter 1. *Front. Pharmacol.* 12 (97).

40 Te Brake, L.H., van den Heuvel, J.J., Buaben, A.O. et al. (2016). Moxifloxacin is a potent in vitro inhibitor of OCT- and MATE-mediated transport of metformin and ethambutol. *Antimicrob. Agents Chemother.* 60 (12): 7105–7114.

41 Kim, T.E., Lee, H., Lim, K.S. et al. (2014). Effects of HM30181, a P-glycoprotein inhibitor, on the pharmacokinetics and pharmacodynamics of loperamide in healthy volunteers. *Br. J. Clin. Pharmacol.* 78 (3): 556–564.

42 Verstuyft, C., Strabach, S., El-Morabet, H. et al. (2003). Dipyridamole enhances digoxin bioavailability via P-glycoprotein inhibition. *Clin. Pharmacol. Ther.* 73 (1): 51–60.

43 Wilt, L.A., Nguyen, D., and Roberts, A.G. (2017). Insights into the molecular mechanism of triptan transport by P-glycoprotein. *J. Pharm. Sci.* 106 (6): 1670–1679.

44 Shimizu, M., Uno, T., Sugawara, K., and Tateishi, T. (2006). Effects of itraconazole and diltiazem on the pharmacokinetics of fexofenadine, a substrate of P-glycoprotein. *Br. J. Clin. Pharmacol.* 61 (5): 538–544.

45 Kumar, P., Gordon, L.A., Brooks, K.M. et al. (2017). Differential influence of the antiretroviral pharmacokinetic enhancers ritonavir and cobicistat on intestinal P-glycoprotein transport and the pharmacokinetic/pharmacodynamic disposition of dabigatran. *Antimicrob. Agents Chemother.* 61 (11).

46 Bogman, K., Zysset, Y., Degen, L. et al. (2005). P-glycoprotein and surfactants: effect on intestinal talinolol absorption. *Clin. Pharmacol. Ther.* 77 (1): 24–32.

47 Saaby, L., Helms, H.C., and Brodin, B. (2016). IPEC-J2 MDR1, a novel high-resistance cell line with functional expression of human P-glycoprotein (ABCB1) for drug screening studies. *Mol. Pharmaceutics* 13 (2): 640–652.

48 Agarwal, S., Pal, D., and Mitra, A.K. (2007). Both P-gp and MRP2 mediate transport of Lopinavir, a protease inhibitor. *Int. J. Pharm.* 339 (1-2): 139–147.

49 Takano, M., Yumoto, R., and Murakami, T. (2006). Expression and function of efflux drug transporters in the intestine. *Pharmacol. Ther.* 109 (1-2): 137–161.

50 Baris, N., Kalkan, S., Güneri, S. et al. (2006). Influence of carvedilol on serum digoxin levels in heart failure: is there any gender difference? *Eur. J. Clin. Pharmacol.* 62 (7): 535–538.

51 Albassam, A.A. and Markowitz, J.S. (2017). An appraisal of drug-drug interactions with green tea (*Camellia sinensis*). *Planta Med.* 83 (6): 496–508.

52 Chiney, M.S., Menon, R.M., Bueno, O.F. et al. (2018). Clinical evaluation of P-glycoprotein inhibition by venetoclax: a drug interaction study with digoxin. *Xenobiotica* 48 (9): 904–910.

53 Ledwitch, K.V. and Roberts, A.G. (2017). Cardiovascular ion channel inhibitor drug-drug interactions with P-glycoprotein. *AAPS J.* 19 (2): 409–420.

54 Iwaki, K., Sakaeda, T., Kakumoto, M. et al. (2006). Haloperidol is an inhibitor but not substrate for MDR1/P-glycoprotein. *J. Pharm. Pharmacol.* 58 (12): 1617–1622.

55 Hugger, E.D., Novak, B.L., Burton, P.S. et al. (2002). A comparison of commonly used polyethoxylated pharmaceutical excipients on their ability to inhibit P-glycoprotein activity in vitro. *J. Pharm. Sci.* 91 (9): 1991–2002.

56 Chen, M., Zhou, S.Y., Fabriaga, E. et al. (2018). Food-drug interactions precipitated by fruit juices other than grapefruit juice: an update review. *J. Food Drug Anal.* 26 (2s): S61–s71.

57 Zhu, T., Howieson, C., Wojtkowski, T. et al. (2017). The effect of verapamil, a P-glycoprotein inhibitor, on the pharmacokinetics of peficitinib, an orally administered, once-daily JAK inhibitor. *Clin. Pharmacol. Drug Dev.* 6 (6): 548–555.

58 Green, B.R. and Bain, L.J. (2013). Mrp2 is involved in the efflux and disposition of fosinopril. *J. Appl. Toxicol.* 33 (6): 458–465.

59 Keppler, D. (1999). Export pumps for glutathione S-conjugates. *Free Radical Biol. Med.* 27 (9-10): 985–991.

60 Sugie, M., Asakura, E., Zhao, Y.L. et al. (2004). Possible involvement of the drug transporters P glycoprotein and multidrug resistance-associated protein Mrp2 in disposition of azithromycin. *Antimicrob. Agents Chemother.* 48 (3): 809–814.

61 Liao, X.Y., Deng, Q.Q., Han, L. et al. (2020). Leflunomide increased the renal exposure of acyclovir by inhibiting OAT1/3 and MRP2. *Acta Pharmacol. Sin.* 41 (1): 129–137.

62 Sawangrat, K., Yamashita, S., Tanaka, A. et al. (2019). Modulation of intestinal transport and absorption of topotecan, a BCRP substrate, by various pharmaceutical excipients and their inhibitory mechanisms of BCRP transporter. *J. Pharm. Sci.* 108 (3): 1315–1325.

63 Mao, Q. and Unadkat, J.D. (2015). Role of the breast cancer resistance protein (BCRP/ABCG2) in drug transport--an update. *AAPS J.* 17 (1): 65–82.

64 Lee, C.A., O'Connor, M.A., Ritchie, T.K. et al. (2015). Breast cancer resistance protein (ABCG2) in clinical pharmacokinetics and drug interactions: practical recommendations for clinical victim and perpetrator drug-drug interaction study design. *Drug Metab. Dispos.* 43 (4): 490–509.

65 Gupta, A., Dai, Y., Vethanayagam, R.R. et al. (2006). Cyclosporin A, tacrolimus and sirolimus are potent inhibitors of the human breast cancer resistance protein (ABCG2) and reverse resistance to mitoxantrone and topotecan. *Cancer Chemother. Pharmacol.* 58 (3): 374–383.

66 Reznicek, J., Ceckova, M., Ptackova, Z. et al. (2017). MDR1 and BCRP transporter-mediated drug-drug interaction between rilpivirine and abacavir and effect on intestinal absorption. *Antimicrob. Agents Chemother.* 61 (9).

67 Ni, Z., Bikadi, Z., Rosenberg, M.F., and Mao, Q. (2010). Structure and function of the human breast cancer resistance protein (BCRP/ABCG2). *Curr. Drug Metab.* 11 (7): 603–617.

68 Kawahara, I., Nishikawa, S., Yamamoto, A. et al. (2020). The impact of breast cancer resistance protein (BCRP/ABCG2) on drug transport across Caco-2 cell monolayers. *Drug Metab. Dispos.* 48 (6): 491–498.

69 Yamagata, T., Kusuhara, H., Morishita, M. et al. (2007). Effect of excipients on breast cancer resistance protein substrate uptake activity. *J. Controlled Release* 124 (1-2): 1–5.

70 Gupta, A., Unadkat, J.D., and Mao, Q. (2007). Interactions of azole antifungal agents with the human breast cancer resistance protein (BCRP). *J. Pharm. Sci.* 96 (12): 3226–3235.

71 Pick, A., Müller, H., Mayer, R. et al. (2011). Structure-activity relationships of flavonoids as inhibitors of breast cancer resistance protein (BCRP). *Bioorg. Med. Chem.* 19 (6): 2090–2102.

72 Pochini, L., Galluccio, M., Scalise, M. et al. (2019). OCTN: a small transporter subfamily with great relevance to human pathophysiology, drug discovery, and diagnostics. *SLAS Discov.* 24 (2): 89–110.

73 Visentin, M., Gai, Z., Torozi, A. et al. (2017). Colistin is substrate of the carnitine/organic cation transporter 2 (OCTN2, SLC22A5). *Drug Metab. Dispos.* 45 (12): 1240–1244.

74 Dawson, P.A. (2017). Roles of ileal ASBT and OSTalpha-OSTbeta in regulating bile acid signaling. *Dig. Dis.* 35 (3): 261–266.

75 Balakrishnan, A. and Polli, J.E. (2006). Apical sodium dependent bile acid transporter (ASBT, SLC10A2): a potential prodrug target. *Mol. Pharmaceutics* 3 (3): 223–230.

76 Zheng, X. and Polli, J.E. (2010). Synthesis and in vitro evaluation of potential sustained release prodrugs via targeting ASBT. *Int. J. Pharm.* 396 (1-2): 111–118.

77 Beaudoin, J.J., Brouwer, K.L.R., and Malinen, M.M. (2020). Novel insights into the organic solute transporter alpha/beta, OSTα/β: From the bench to the bedside. *Pharmacol. Therapeut.* 211: 107542.

78 Harwood, M.D., Zhang, M., Pathak, S.M., and Neuhoff, S. (2019). The regional-specific relative and absolute expression of gut transporters in adult caucasians: a meta-analysis. *Drug Metab. Dispos.* 47 (8): 854–864.

79 Gröer, C., Brück, S., Lai, Y. et al. (2013). LC–MS/MS-based quantification of clinically relevant intestinal uptake and efflux transporter proteins. *J. Pharm. Biomed. Anal.* 85: 253–261.

80 Brück, S., Strohmeier, J., Busch, D. et al. (2017). Caco-2 cells - expression, regulation and function of drug transporters compared with human jejunal tissue. *Biopharm. Drug Dispos.* 38 (2): 115–126.

81 Akazawa, T., Uchida, Y., Miyauchi, E. et al. (2018). High expression of UGT1A1/1A6 in monkey small intestine: comparison of protein expression levels of cytochromes P450, UDP-glucuronosyltransferases,

and transporters in small intestine of cynomolgus monkey and human. *Mol. Pharmaceutics* 15 (1): 127–140.

82 Dawson, P.A. (2017). Roles of ileal ASBT and OSTα-OSTβ in regulating bile acid signaling. *Dig. Dis.* 35 (3): 261–266.

83 Scalise, M., Galluccio, M., Console, L. et al. (2018). The human SLC7A5 (LAT1): the intriguing histidine/large neutral amino acid transporter and its relevance to human health. *Front. Chem.* 6: 243.

84 Maric, S., Fluchter, P., Guglielmetti, L.C. et al. (2020). Plasma citrulline correlates with basolateral amino acid transporter LAT4 expression in human small intestine. *Clin. Nutr.*

85 Akazawa, T., Uchida, Y., Tachikawa, M. et al. (2016). Quantitative targeted absolute proteomics of transporters and pharmacoproteomics-based reconstruction of P-glycoprotein function in mouse small intestine. *Mol. Pharmaceutics* 13 (7): 2443–2456.

86 Dave, M.H., Schulz, N., Zecevic, M. et al. (2004). Expression of heteromeric amino acid transporters along the murine intestine. *J. Physiol.* 558 (Pt 2): 597–610.

87 Puris, E., Gynther, M., Auriola, S., and Huttunen, K.M. (2020). L-Type amino acid transporter 1 as a target for drug delivery. *Pharm. Res.* 37 (5): 88.

88 Morris, M.E. and Felmlee, M.A. (2008). Overview of the proton-coupled MCT (SLC16A) family of transporters: characterization, function and role in the transport of the drug of abuse gamma-hydroxybutyric acid. *AAPS J.* 10 (2): 311–321.

89 Friesema, E.C., Ganguly, S., Abdalla, A. et al. (2003). Identification of monocarboxylate transporter 8 as a specific thyroid hormone transporter. *J. Biol. Chem.* 278 (41): 40128–40135.

90 Suhre, K., Shin, S.Y., Petersen, A.K. et al. (2011). Human metabolic individuality in biomedical and pharmaceutical research. *Nature* 477 (7362): 54–60.

91 Abplanalp, J., Laczko, E., Philp, N.J. et al. (2013). The cataract and glucosuria associated monocarboxylate transporter MCT12 is a new creatine transporter. *Hum. Mol. Genet.* 22 (16): 3218–3226.

92 Gill, R.K., Saksena, S., Alrefai, W.A. et al. (2005). Expression and membrane localization of MCT isoforms along the length of the human intestine. *Am. J. Physiol. Cell Physiol.* 289 (4): C846–C852.

93 Canani, R.B., Costanzo, M.D., Leone, L. et al. (2011). Potential beneficial effects of butyrate in intestinal and extraintestinal diseases. *World J. Gastroenterol.* 17 (12): 1519–1528.

94 Al-Mosauwi, H., Ryan, E., McGrane, A. et al. (2016). Differential protein abundance of a basolateral MCT1 transporter in the human gastrointestinal tract. *Cell Biol. Int.* 40 (12): 1303–1312.

95 Iwanaga, T., Takebe, K., Kato, I. et al. (2006). Cellular expression of monocarboxylate transporters (MCT) in the digestive tract of the mouse, rat, and humans, with special reference to slc5a8. *Biomed. Res.* 27 (5): 243–254.

96 Buyse, M., Sitaraman, S.V., Liu, X. et al. (2002). Luminal leptin enhances CD147/MCT-1-mediated uptake of butyrate in the human intestinal cell line Caco2-BBE. *J. Biol. Chem.* 277 (31): 28182–28190.

97 Sivaprakasam, S., Bhutia, Y.D., Yang, S., and Ganapathy, V. (2017). Short-chain fatty acid transporters: role in colonic homeostasis. *Compr. Physiol.* 8 (1): 299–314.

98 Kohyama, N., Shiokawa, H., Ohbayashi, M. et al. (2013). Characterization of monocarboxylate transporter 6: expression in human intestine and transport of the antidiabetic drug nateglinide. *Drug Metab. Dispos.* 41 (11): 1883–1887.

99 Teramae, H., Yoshikawa, T., Inoue, R. et al. (2010). The cellular expression of SMCT2 and its comparison with other transporters for monocarboxylates in the mouse digestive tract. *Biomed. Res.* 31 (4): 239–249.

100 Sepponen, K., Ruusunen, M., Pakkanen, J.A., and Poso, A.R. (2007). Expression of CD147 and monocarboxylate transporters MCT1, MCT2 and MCT4 in porcine small intestine and colon. *Vet. J.* 174 (1): 122–128.

101 Coady, M.J., Chang, M.H., Charron, F.M. et al. (2004). The human tumour suppressor gene SLC5A8 expresses a $Na^+$-monocarboxylate cotransporter. *J. Physiol.* 557 (Pt 3): 719–731.

102 Gopal, E., Fei, Y.J., Sugawara, M. et al. (2004). Expression of slc5a8 in kidney and its role in Na(+)-coupled transport of lactate. *J. Biol. Chem.* 279 (43): 44522–44532.

103 Barac-Nieto, M., Murer, H., and Kinne, R. (1980). Lactate-sodium cotransport in rat renal brush border membranes. *Am. J. Phys.* 239 (5): F496–F506.

104 Iwanaga, T. and Kishimoto, A. (2015). Cellular distributions of monocarboxylate transporters: a review. *Biomed. Res.* 36 (5): 279–301.

105 Roth, M., Obaidat, A., and Hagenbuch, B. (2012). OATPs, OATs and OCTs: the organic anion and cation transporters of the SLCO and SLC22A gene superfamilies. *Br. J. Pharmacol.* 165 (5): 1260–1287.

106 Hagenbuch, B. and Meier, P.J. (2003). The superfamily of organic anion transporting polypeptides. *Biochim. Biophys. Acta* 1609 (1): 1–18.

107 Drozdzik, M., Gröer, C., Penski, J. et al. (2014). Protein abundance of clinically relevant multidrug transporters along the entire length of the human intestine. *Mol. Pharmaceutics* 11 (10): 3547–3555.

108 Couto, N., Al-Majdoub, Z.M., Gibson, S. et al. (2020). Quantitative proteomics of clinically relevant drug-metabolizing enzymes and drug transporters and their intercorrelations in the human small intestine. *Drug Metab. Dispos.* 48 (4): 245–254.

109 Glaeser, H., Bailey, D.G., Dresser, G.K. et al. (2007). Intestinal drug transporter expression and the impact of grapefruit juice in humans. *Clin. Pharmacol. Ther.* 81 (3): 362–370.

110 Sai, Y., Kaneko, Y., Ito, S. et al. (2006). Predominant contribution of organic anion transporting polypeptide OATP-B (OATP2B1) to apical uptake of estrone-3-sulfate by human intestinal Caco-2 cells. *Drug Metab. Dispos.* 34 (8): 1423–1431.

111 Kobayashi, D., Nozawa, T., Imai, K. et al. (2003). Involvement of human organic anion transporting polypeptide OATP-B (SLC21A9) in pH-dependent transport across intestinal apical membrane. *J. Pharmacol. Exp. Ther.* 306 (2): 703–708.

112 Thakkar, N., Lockhart, A.C., and Lee, W. (2015). Role of organic anion-transporting polypeptides (OATPs) in cancer therapy. *AAPS J.* 17 (3): 535–545.

113 Kalliokoski, A. and Niemi, M. (2009). Impact of OATP transporters on pharmacokinetics. *Br. J. Pharmacol.* 158 (3): 693–705.

114 Akamine, Y., Miura, M., Komori, H. et al. (2015). The change of pharmacokinetics of fexofenadine enantiomers through the single and simultaneous grapefruit juice ingestion. *Drug Metab. Pharmacokinet.* 30 (5): 352–357.

115 Bruckmueller, H., Martin, P., Kähler, M. et al. (2017). Clinically relevant multidrug transporters are regulated by microRNAs along the human intestine. *Mol. Pharmaceutics* 14 (7): 2245–2253.

116 Hu, Y., Xie, Y., Wang, Y. et al. (2014). Development and characterization of a novel mouse line humanized for the intestinal peptide transporter PEPT1. *Mol. Pharmaceutics* 11 (10): 3737–3746.

117 Koepsell, H. and Endou, H. (2004). The SLC22 drug transporter family. *Pflugers Arch.* 447 (5): 666–676.

118 Koepsell, H. (2013). The SLC22 family with transporters of organic cations, anions and zwitterions. *Mol. Aspects Med.* 34 (2-3): 413–435.

119 Koepsell, H., Lips, K., and Volk, C. (2007). Polyspecific organic cation transporters: structure, function, physiological roles, and biopharmaceutical implications. *Pharm. Res.* 24 (7): 1227–1251.

120 Wenzel, C., Drozdzik, M., and Oswald, S. (2021). Organic cation transporter 1 an intestinal uptake transporter: fact or fiction? *Front. Pharmacol.* 12: 648388.

121 Han, T.K., Everett, R.S., Proctor, W.R. et al. (2013). Organic cation transporter 1 (OCT1/mOct1) is localized in the apical membrane of Caco-2 cell monolayers and enterocytes. *Mol. Pharmacol.* 84 (2): 182–189.

122 Koepsell, H. (2020). Organic cation transporters in health and disease. *Pharmacol. Rev.* 72 (1): 253–319.

123 Müller, J., Lips, K.S., Metzner, L. et al. (2005). Drug specificity and intestinal membrane localization of human organic cation transporters (OCT). *Biochem. Pharmacol.* 70 (12): 1851–1860.

124 Boxberger, K.H., Hagenbuch, B., and Lampe, J.N. (2014). Common drugs inhibit human organic cation transporter 1 (OCT1)-mediated neurotransmitter uptake. *Drug Metab. Dispos.* 42 (6): 990–995.

125 McCreight, L.J., Bailey, C.J., and Pearson, E.R. (2016). Metformin and the gastrointestinal tract. *Diabetologia* 59 (3): 426–435.

126 Mofo Mato, E.P., Guewo-Fokeng, M., Essop, M.F., and Owira, P.M.O. (2018). Genetic polymorphisms of organic cation transporter 1 (OCT1) and responses to metformin therapy in individuals with type 2 diabetes: a systematic review. *Medicine* 97 (27): e11349.

127 Li, D.C., Nichols, C.G., and Sala-Rabanal, M. (2015). Role of a hydrophobic pocket in polyamine interactions with the polyspecific organic cation transporter OCT3. *J. Biol. Chem.* 290 (46): 27633–27643.

128 Song, W., Luo, Q., Zhang, Y. et al. (2019). Organic cation transporter 3 (Oct3) is a distinct catecholamines clearance route in adipocytes mediating the beiging of white adipose tissue. *PLoS Biol.* 17 (1): e2006571.

129 Ashraf, R., Ronaldson, P.T., and Bendayan, R. (2014). Drug transport in the brain. In: *Drug Transporters: Molecular Characterization and Role in Drug Disposition* (ed. G. You and M.E. Morris), 273–301. Hoboken, NJ, USA: Wiley.

130 Cordon-Cardo, C., O'Brien, J.P., Boccia, J. et al. (1990). Expression of the multidrug resistance gene product (P-glycoprotein) in human normal and tumor tissues. *J. Histochem. Cytochem.* 38 (9): 1277–1287.

131 Tamaki, A., Ierano, C., Szakacs, G. et al. (2011). The controversial role of ABC transporters in clinical oncology. *Essays Biochem.* 50 (1): 209–232.

132 Berggren, S., Gall, C., Wollnitz, N. et al. (2007). Gene and protein expression of P-glycoprotein, MRP1, MRP2, and CYP3A4 in the small and large human intestine. *Mol. Pharmaceutics* 4 (2): 252–257.

133 Al-Majdoub, Z.M., Couto, N., Achour, B. et al. (2021). Quantification of proteins involved in intestinal epithelial handling of xenobiotics. *Clin. Pharmacol. Ther.* 109 (4): 1136–1146.

134 Dou, L., Gavins, F.K.H., Mai, Y. et al. (2020). Effect of food and an animal's sex on P-glycoprotein expression and luminal fluids in the gastrointestinal tract of wistar rats. *Pharmaceutics* 12 (4).

135 Thompson, C.G., Fallon, J.K., Mathews, M. et al. (2017). Multimodal analysis of drug transporter expression in gastrointestinal tissue. *Aids* 31 (12): 1669–1678.

136 Aouali, N., Eddabra, L., Macadré, J., and Morjani, H. (2005). Immunosuppressors and reversion of multidrug-resistance. *Crit. Rev. Oncol. Hematol.* 56 (1): 61–70.

137 Zurth, C., Koskinen, M., Fricke, R. et al. (2019). Drug-drug interaction potential of darolutamide: in vitro and clinical studies. *Eur. J. Drug Metab. Pharmacokinet.* 44 (6): 747–759.

138 Elmeliegy, M., Vourvahis, M., Guo, C., and Wang, D.D. (2020). Effect of P-glycoprotein (P-gp) inducers on exposure of P-gp substrates: review of clinical drug-drug interaction studies. *Clin. Pharmacokinet.* 59 (6): 699–714.

139 Nies, A.T. and Lang, T. (2014). Multidrug resistance proteins of the ABCC subfamily. In: *Drug Transporters: Molecular Characterization and Role in Drug Disposition* (ed. G. You and M.E. Morris), 161–185. Hoboken, NJ: Wiley.

140 Zimmermann, C., Gutmann, H., Hruz, P. et al. (2005). Mapping of multidrug resistance gene 1 and multidrug resistance-associated protein isoform 1 to 5 mRNA expression along the human intestinal tract. *Drug Metab. Dispos.* 33 (2): 219–224.

141 Rost, D., Mahner, S., Sugiyama, Y., and Stremmel, W. (2002). Expression and localization of the multidrug resistance-associated protein 3 in rat small and large intestine. *Am. J. Physiol. Gastrointest. Liver Physiol.* 282 (4): G720–G726.

142 König, J., Nies, A.T., Cui, Y. et al. (1999). Conjugate export pumps of the multidrug resistance protein (MRP) family: localization, substrate specificity, and MRP2-mediated drug resistance. *Biochim. Biophys. Acta* 1461 (2): 377–394.

143 Maliepaard, M., Scheffer, G.L., Faneyte, I.F. et al. (2001). Subcellular localization and distribution of the breast cancer resistance protein transporter in normal human tissues. *Cancer Res.* 61 (8): 3458–3464.

144 Doyle, L.A., Yang, W., Abruzzo, L.V. et al. (1998). A multidrug resistance transporter from human MCF-7 breast cancer cells. *Proc. Natl. Acad. Sci. U.S.A.* 95 (26): 15665–15670.

145 Girardin, M., Dionne, S., Goyette, P. et al. (2012). Expression and functional analysis of intestinal organic cation/L-carnitine transporter (OCTN) in Crohn's disease. *J. Crohns. Colitis.* 6 (2): 189–197.

146 Sugiura, T., Kato, S., Shimizu, T. et al. (2010). Functional expression of carnitine/organic cation transporter OCTN1/SLC22A4 in mouse small intestine and liver. *Drug Metab. Dispos.* 38 (10): 1665–1672.

147 Meier, Y., Eloranta, J.J., Darimont, J. et al. (2007). Regional distribution of solute carrier mRNA expression along the human intestinal tract. *Drug Metab. Dispos.* 35 (4): 590–594.

148 Harwood, M.D., Neuhoff, S., Carlson, G.L. et al. (2013). Absolute abundance and function of intestinal drug transporters: a prerequisite for fully mechanistic in vitro-in vivo extrapolation of oral drug absorption. *Biopharm. Drug Dispos.* 34 (1): 2–28.

149 Harwood, M.D., Achour, B., Neuhoff, S. et al. (2016). In vitro-in vivo extrapolation scaling factors for intestinal P-glycoprotein and breast cancer resistance protein: Part I: A cross-laboratory comparison of transporter-protein abundances and relative expression factors in human intestine and Caco-2 cells. *Drug Metab. Dispos.* 44 (3): 297–307.

# 19

# Amino Acid Transporters

*Liping Wang[1], Xiaoyan Li[1], Mengdi Ying[1], Ming Hu[1,2], and Zhongqiu Liu[1]*

[1]Joint Laboratory for Translational Cancer Research of Chinese Medicine of the Ministry of Education of the People's Republic of China, International Institute for Translational Chinese Medicine, Guangzhou University of Chinese Medicine, Guangzhou, Guangdong, China
[2]Department of Pharmacological & Pharmaceutical Sciences, College of Pharmacy, University of Houston, Houston, TX, USA

## 19.1 Introduction

The absorption, distribution, and excretions of many (if not most) of nutrients, xenobiotics, drugs, environmental toxins, and their metabolites are controlled (i.e. facilitated or limited) in the body by membrane transporters. The transporters can be classified into influx transporters and efflux transporters. Based on their energy requirements, they can be further classified as active transporters and facilitated transporters [1]. These transporters usually belong to either the ATP-binding cassette (ABC) family or the solute carriers (SLCs) family. ABC transporters are widely distributed in all organisms, from bacteria to mammals, with about 600 predicted or reported transporters. However, only 49 ABC genes have been identified in humans and approximately 9 ABCs were shown to affect drug pharmacokinetics and pharmacodynamics [2, 3]. The known ABC transporters include P-glycoprotein (P-gp/ABCB1), multidrug resistance proteins (MRPs/ABCCs), and breast cancer resistance proteins (BCRP/ABCG2). The SLC family, the second-largest family of membrane proteins in the human genome, appears to be more significant than ABC transporters in humans with over 400 proteins classified into 65 subfamilies [4]. Irrespective of the difference in numbers, these transporters are membrane-bond proteins with one or more binding sites and 6–12 (occasionally 14) transmembrane domains connected by intra and extracellular loops.

Despite their possible structural differences, these transporters are often classified into five major groups on the basis of structures, substrate specificities, and transport mechanisms [5]. The five groups are as follows: (i) organic ion transporter superfamily (OAT, OCT, and OCTN-SLC 22A subfamily); (ii) ATP-dependent transporter superfamily (ABC); (iii) peptide transporter family (SLC15A subfamily); (iv) organic anion transporting polypeptide family (OATP-SLCO subfamily); and (v) amino acid transporter superfamilies.

Amino acids support the normal physiological functions such as cell growth and homeostasis (maintenance of normal cell functions). Amino acids are essential not only for protein synthesis but also as a carbon and nitrogen source in the synthesis of purine and pyrimidine nucleotides, amino sugars, and glutathione (GSH). After the digestion of proteins (e.g. those derived from food), the resulting small peptides and amino acids are efficiently absorbed by the enterocytes of the small intestine [6]. The peptides absorbed inside into the enterocytes are hydrolyzed into amino acids and subsequently pooled with those absorbed by amino acid transporters for release into the blood and transport to all tissues for protein synthesis. Unused amino acids are filtered and subsequently reabsorbed in the kidney. The activities of amino acid transporter are frequently referred to as system, which implies a functionally identified transport activity that appears to be similar in a variety of cell types. It was recognized that amino acid transport systems accept groups of amino acids rather than individual amino acids [7, 8].

Amino acid transporters belong to a class of the specialized transporter, which are plasma membrane proteins mediating the transport of amino acids across the plasma membrane [8]. The specific transporters are located in apical and basolateral membranes of the intestinal and kidney epithelial cells to ensure the efficient vectorial transport of amino acids through the epithelia to the systemic circulation [9, 10]. Substrate recognition of amino acid transporters is generally strict

*Oral Bioavailability and Drug Delivery: From Basics to Advanced Concepts and Applications*, First Edition. Edited by Ming Hu and Xiaoling Li.

**Table 19.1** The currently known families of amino acid transporters in epithelial and endothelial cells.

| Group | Acronym | Full family name | $Na^+$-dependent | Transporter type | Expression |
|---|---|---|---|---|---|
| Glycoprotein-associated amino acid transporters | LAT2 | System L transporter (4F2 heteromer) | – | L | Small intestine, renal epithelial cells of the proximal tubule, heart, liver, muscle |
| | $y^+$LAT1,2 | System $y^+$L transporter (4F2 heteromer) | –/+[a] | $y^+$L | Brain (astrocytes), small intestine, testis, kidney |
| | xCT | Cystine glutamate exchanger (4F2 heteromer) | – | $X_c^-$ | Astrocytes, macrophages, retinal cells |
| | Asc-1 | System asc transporter (4F2 heteromer) | – | Asc | Strocytes of the caudal brain, spinal cord, lung |
| | $b^0$AT | $b^{0+}$ amino acid transporter (rBAT heteromer) | – | $b^{0+}$ | Apical membrane of epithelial cells of the small intestine, the renal proximal tubule |
| Cationic amino acid transporters | CAT1-4 | Cationic amino acid transporter 1,2A/B,3 | –/+[a] | $y^+$ | Ubiquitous |
| Excitatory amino acid transporters | EAAT1/4/5 | Excitatory amino acid transporter 1-5(GLAST,GLT-1) | + ($K^+$ exchange) | $X^-{}_{AG}$ | Brain, retina |
| | EAAT2 | Excitatory amino acid transporter 2 | + ($K^+$ exchange) | $X^-{}_{AG}$ | Brain (astrocytes, Bergmann glia, motor neurons), liver, pancreas |
| | EAAT3 | Excitatory amino acid transporter 3 | + ($K^+$ exchange) | $X^-{}_{AG}$ | Brain (neurons), intestine, kidney, liver, heart, placenta |
| | ASCT1 | ASC transporter | + | ASC | Ubiquitous |
| | ASCT2 | ASC transporter | + | ASC | Lung, skeletal muscle, large intestine |
| | $ATB^0$ | $B^{0+}$ amino acid transporter | + | $B^{0,+}$ | Lung, mammary gland |
| Neutral amino acid transporters | SN1 | System N transporter | +($H^+$, exchange) | N | Pancreatic β-cells, liver, brain, skeletal muscle, kidney |
| | SN2 | System N transporter | +($H^+$, exchange) | N | Pancreatic, α-cells, stomach, brain, liver |
| | SAT1 | System A transporter 1 | + | A | Neural progenitor cells |
| | SAT2 | System A transporter 2 | + | A | Pancreatic, β-cells, brain, liver, kidney |

| | | | | | |
|---|---|---|---|---|---|
| | SAT3 | System A transporter 3 | + | A | Pancreatic α-cells, hepatocytes, presynaptic neurons |
| | $B^0AT1$ | $Na^+$-dependent neutral amino acid transporter B(0)AT1 | + | $B^0$ | S1 segment of the proximal tubule, duodenum, jejunum, ileum |
| | $B^0AT2$ | $Na^+$-dependent neutral amino acid transporter B(0)AT2 | + | $B^0$ | Brain |
| | PAT1 | Proton-coupled amino acid transporter 1 | $H^+$ | Imino-glycine | Small intestine |
| | PAT2 | Proton-coupled amino acid transporter 2 | $H^+$ | Imino-glycine | S1 segment of the proximal tubule |
| Monocarboxylate transporters | TAT1 | System T transporter | – | T | Basolateral membrane of intestinal epithelium, renal proximal tubule cells, perivenous hepatocytes |
| Neurotransmitter transporters | GAT1-3 | GABA transporter | +(+Cl) | GABA | Brain |
| | GLYT1 | Glycine transporter 1 | +(+Cl) | Gly | Lungs, brain, retina, liver, kidney |
| | GLYT2 | Glycine transporter 2 | +(+Cl) | Gly | Glycinergic neurons |
| | PROT etc. | Proline/betaine transporters | +(+Cl) | Pro | Glutamatergic neurons |
| | $ATB^0$ | $B^{0+}$ amino acid transporter | +(+Cl) | $B^{0,+}$ | Lung, mammary gland |
| | LAT3 | LNAA transporter small subunit 3 | – | L | Pancreas, skin, muscle, liver, kidney podocytes, prostate |
| | LAT4 | LNAA transporter small subunit 3 | – | L | Placenta, basolateral membrane of small intestine and kidney, proximal tubule, distal convoluted tubule |

a) Transports neutral amino acid substrates $Na^{\lessgtr}$-dependently and cationic amino acids $Na^{\lessgtr}$-independently.
Source: [12], [13], [14], [15].

because amino acid transporters have to reply on three features of substrate molecules for their recognition: a positive charge conferred by the α-amino group, a negative charge conferred by the α-carboxyl group, and the spatial and/or electric characteristic of the substrate amino acid side chain. Among amino acid transporters, there is one particular member (see below) that exhibits broad substrate selectivity so that it can participate in the transport of variety of exogenous compounds including toxic compounds occurring in the environment [11].

The classical-system-based classification of amino acid transporter is based on functional characteristics. This system classification is now being superseded by the classification based on similarities in transporter protein structures, even though the system can often be readily identified with a particular transporter gene family. Structurally related transporters have been grouped into one of six distinct gene families, which are summarized alongside their functional correlates in Table 19.1. In this chapter, we will first introduce the molecular features, transporter properties, and mechanisms of the multispecific nature of amino acid drug transporters. We will then discuss the implication related to absorption of drugs and regulation of the amino acid transporters.

## 19.2 Classification of Amino Acid Transporters and their Functions

Molecular cloning approaches have led to the identification of $Na^+$-dependent and $Na^+$-independent amino acid transporters. $Na^+$-dependent transporters that possess the ability to concentrate amino acids inside the cells [8] use free energy stored as $Na^+$-electrochemical potential gradient across the plasma membrane to transport amino acid substrates. Table 19.2 summarizes the nomenclature and the classical nomenclature used to designate these different amino acid transporter systems. The ionic dependency and $K_m$ values corresponding to the different transport systems were complied from publications and do not necessarily reflect transport properties of all cell types. Nevertheless, most amino acid transporters are classified into transporters that can transport cationic, neutral, imino, and anionic amino acids.

Cationic amino acids, such as L-arginine, L-lysine, and L-ornithine, are transported by the $Na^+$-independent systems $y^+$ (SLC7A1), $y^+L$ (SLC3A2/SLC7A7 and SLC7A6), $b^{0,+}$ (SLC3A1/SLC7A9), and $Na^+$-dependent system $B^{0,+}$ (SLC6A14). System $y^+$ is the major cationic amino acid transport system overexpressed in NO-producing cells and is thus likely to play a key role in regulating L-arginine supply for NO synthesis. Available evidence indicates that bovine aortic endothelial cells express a taurine transporter sharing a high degree of sequence homology with that of the mTAUT cDNA isolated from the brain.

Neural amino acids are transported by the $Na^+$-independent amino acid transporters such as the (i) neutral amino acid transport system L (SLC3A2/SLC7A8, SLC43A1, and SLC43A2) [25], (ii) small neutral amino acid preferring system ASC (SLC1A4 and SLC1A5) [26], and (iii) aromatic amino acid selective system T (SLC16A10) [27]. These neutral amino acids are also transported by multispecific $Na^+$-independent amino acid transporters including the basic amino acid selective system $y^+$ [28], and basic and neutral amino acid transport system $b^{0,+}$, and $y^+L$ [29, 30], but not cysteine and glutamate-preferring system $x_c^-$ (SLC3A2/SLC7A11) [31]. Neutral amino acid transporters such as leucine and other large hydrophobic neutral amino acids are transported by system L [25], amino acid transporters for alanine and other small and polar neutral amino acids are classified as system A and system ASC.

Anionic amino acid transporters are mainly responsible for the transport of amino acids with two carboxylic acid groups including aspartic acid and glutamic acid. System $x_c^-$ is a $Na^+$-independent anionic amino acid transport system, and is critical in amino acid exchanger function in various cell types [32, 33]. The system $X^-_{AG}$ amino acid transporters are high affinity, $Na^+$-dependent transporters for aspartate and glutamate, and they are often grouped into the EAAT family, much of it expressed by the cells in CNS.

Another way to classify amino acid transporters is based on role of $Na^+$: $Na^+$-dependent and $Na^+$-independent transporters, although such a classification method may be controversial. In the kidney, the reabsorption of cationic amino acids (Figure 19.1) is well established, but the roles of $Na^+$ dependence and electrogenicity remain controversial. Thus, using membrane vesicles, investigators have demonstrated $Na^+$-dependent secondary active transport [35], electrogenic but $Na^+$-independent transport, or electroneutral transport in the absence of $Na^+$ [36]. In the intestine, enterocytes express several classes of amino acid transporters on their apical membrane. These ensure the efficient removal of amino acids, such as NHE3/SLC9A3 and $B^0$AT1/SLC6A19, from the lumen of the intestine. A different set of transporters ($y^+$LAT1/SLC7A7/SLC3A2, TaT1/SLC6A10,

**Table 19.2** Amino acid transport systems and identified transporters in mammalian.

| Transporter system | Associated protein or cDNA | Amino acid selectivity | Ionic dependency | $K_m$ (μM) | Tissue/cellular expression | Example of physiological |
|---|---|---|---|---|---|---|
| $y^+$ | CAT-1 | Cationic | No | 70–250 | Ubiquitous | L-Arginine transport into primary human T lymphocytes and T-cell activation |
| | CAT-2A or CAT-2/2α | Cationic | No | 2150–5200 | Liver, pancreatic islets | L-arginine transport, NO synthesis, insulin secretion, macrophage activation, regulation of inflammatory processes of the GI tract |
| | CAT-2B or CAT-2/2β | Cationic | No | 38–380 | Liver, pancreatic islets | L-arginine transport, NO synthesis, insulin secretion, macrophage activation, regulation of inflammatory processes of the GI tract |
| | CAT-3 | Cationic | No | 40–120 | Brain, thymus, ovary, testis | L-arginine uptake in primary cortical neurons, NMDA receptor activation on the mTOR pathway, neuronal process formation |
| | CAT-4 | Cationic (Arg) | NO | 450–910 | Placenta, testis | Transport of the cationic amino acids (arginine, lysine, and ornithine) |
| $y^+L$ | $y^+$LAT1 | Leu, Lys, Arg, Orn, Hys, Met, Ala, Cys | $Na^+$ | 16.2–31.7 | Basolateral membrane of the small intestine and kidney | Nitric oxide synthesis in human umbilical vein endothelial cells via transport of L-arginine |
| | $y^+$LAT2 | Leu, Arg, glu, Lys, Orn, Hys, Met | $Na^+$ | 120–295 | Brain (astrocytes), small intestine, testis | L-arginine transport in astrocytes, NO synthesis, podocyte development and function |
| L | LAT1 | Large neutral | No | 30–300 | Monocytes, macrophages testis, pancreatic β-cells | L-Leu transport, mTOR activation, pro-inflammatory cytokine production, pancreatic β-cell signaling and function |
| | LAT2 | Large neutral | No | 30–300 | Small intestine, renal epithelial cells of the proximal tubule, heart, liver, muscle | Flux of amino acids across cell, barriers, amino acid renal reabsorption, optimal hearing function, platelet reactivity, clopidogrel response |
| | LAT3 | Neutral | No | 8–1000 | Ubiquitous | mTOR activation, Regulation of cellular thyroid hormone levels |
| | LAT4 | Neutral | No | 180–5000 | Ubiquitous | mTOR activation, Regulation of cellular thyroid hormone levels |
| $x_c^-$ | xCT | Cystine/glutamate | No | 40–92 | Astrocytes, macrophages, retinal cells | L-cystine uptake, glutathione synthesis |
| Asc | Asc-1 | Small neutral | No | 9–23 | Astrocytes of the caudal brain and spinal cord, lung, small intestine | L-Ala, L-Gly, L-Ser, L-Cys transport, glycinergic inhibitory function in the central nervous system |
| | Asc-2 | Small neutral | No | 2.9 | n.r. | n.r. |

*(continued)*

**Table 19.2** (Continued)

| Transporter system | Associated protein or cDNA | Amino acid selectivity | Ionic dependency | $K_m$ (μM) | Tissue/cellular expression | Example of physiological |
|---|---|---|---|---|---|---|
| $b^{0,+}$ | $b^{0,+}$AT | Neutral/cationic cystine | No | 88 | Apical membrane of epithelial cells of the small intestine and the renal proximal tubule | Transport of L-cystine, maintenance of glomerular and renal tubular function |
| A | ATA1 | Neutral/N-Me group | $Na^+$ | 890 | Neural progenitor cells, placenta, heart, lung, skeletal muscle, spleen, stomach, and testis | Insulin-regulated glucose metabolism, mTOR activation |
| | ATA2 | Neutral/N-Me group | $Na^+$ | 390 | Ubiquitously | mTOR activation, embryonic development, glutaminolysis |
| | ATA3 | Neutral/N-Me group | $Na^+$ | >200 | Pancreatic B-cells, brain, retina, liver, kidney | Insulin secretion, glucose recycling, nitric oxide synthesis |
| N | SN 1 | Gly, Pro, Ala, Ser, Cys, Gln, Met, Hys, Lys, Arg | $Na^+$/ Coupled; $H^+$/ Antiporter | 150–1600 | Pancreatic β-cells, liver, brain, skeletal muscle, kidney | L-Gln release from astrocytes and pancreatic β-cells, glucose |
| | SN 2 | Gln, Asn, Hys, Ala | $Na^+$/ Coupled; $H^+$/ Antiporter | 150–1600 | Stomach, brain, liver, lung, and intestinal tract | Glutaminolysis, glutamine- glutamate cycle |
| | SNAT7 | Gln, Ala, Hys, Asn, Ser | $Na^+$ | n.r. | GABAergic and other neurons in brain, spinal cord liver, skeletal muscle | Energy metabolism, cell growth, maintenance of L-Glu neurotransmitter pool in the brain |
| ASC | ASCT1[a] | Neutral | $Na^+$ | 9–464 | Ubiquitous | L-Ser transport from astrocytes to neurons, neuronal development and function |
| | ASCT2[a] | Neutral | $Na^+$ | 9–464 | Lung, skeletal muscle, large intestine | L-Gln transport, mTOR activation, supply of carbon source for TCA cycle through L-Gln transport |
| $B^{0,+}$ | ATB$^{0,+}$ | Neutral/cationic | $Na^+/Cl^-$ | 23–140 | Lung, mammary gland | Regulation of tryptophan availability for serotonin synthesis, appetite regulation, development of early embryos |
| Pro | PROT[a] | Proline | $Na^+/Cl^-$ | 6 | Glutamatergic neurons | Presynaptic regulatory role in excitatory synaptic transmission |
| Gly | GLYT 1[a] | Glycine | $Na^+/Cl^-$ | 17 | Lung, brain, retina, liver, kidney | Development of basic respiratory pattern and tone |
| | GLYT 2[a] | Glycine | $Na^+/Cl^-$ | 17 | Glycinergic neurons | Clearance of extracellular L-Gly |

| | | | | | | |
|---|---|---|---|---|---|---|
| Beta-like | mTAUT[a] | β-Amino-acids | $Na^+/Cl^-$ | 3–13 | Ubiquitous | Taurine and β-alanine transport |
| | GAT-1-3<br>BGT-1 | GABA and/or β-amino acids | $Na^+/Cl^-$ | 1–20 | Brain | Terminates the action of GABA by its high-affinity $Na^+$-dependent reuptake into presynaptic terminals |
| $X^-_{AG}$ | EAAT1[a] | L-Glu, D/L-Asp | $Na^+$,<br>$H^+$, $K^+$ | 18–97 | Brain (astrocytes, oligodentrites), interfollicular epidermis and sebaceous gland | L-Glu clearance from synaptic cleft, stem/progenitor cell activation during skin growth |
| | EAAT2[a] | L-Glu, D/L-Asp | $Na^+$,<br>$H^+$, $K^+$ | 18–97 | Brain (astrocytes, Bergmann glia, motor neurons), liver, pancreas | L-Glu clearance from synaptic cleft, brain development |
| | EAAT3[a] | L-Glu, D/L-Asp, L-Cys | $Na^+$,<br>$H^+$, $K^+$ | 18–97 | Brain (neurons), intestine, kidney, liver, heart, placenta | Learning, memory, motor neuron behavior, GABA and glutathione synthesis, protection from oxidative stress |
| | EAAT4[a] | L-Glu, D/L-Asp | $Na^+$,<br>$H^+$, $K^+$ | 18–97 | Cerebellum (Purkinje cells) | L-Glu clearance and synaptic activity of cerebellar Purkinje cells |
| | EAAT5[a] | L-Glu, D/L-Asp | $Na^+$,<br>$H^+$, $K^+$ | 18–97 | Retina (axon terminals of bipolar cells) | Visual processing |
| System IMINO | SIT1/IMINO | L-Pro, L-Hyp | $Na^+/Cl^+$ | ~200 | Brain, intestine, choroid plexus<br>kidney/apical membrane of the S2 and S3 segment of the proximal tubule | L-Pro reabsorption |

a) Cloned high-affinity amino acid transporters derived from the brain or retina that may not exhibit kinetic properties identified on other cell types. Lowercase trems for the classical transport systems indicates $Na^+$ independence. n.r., not reported; Ala, alanine; Ser, serine; Cys, cysteine; Gln, glutamine; Asn, asparagine; His, histidine; Glu, glutamate; Asp, aspartate; Leu, leucine; Lys, lysine; Arg, arginine; Orn, ornithine; Met, methionine; Gly, glycine; Thr, threonine; Pro, proline; Tyr, tyrosine; Val, valine; L-Hyp, L-hydroxyproline, respectively.

*Note*: The substrate specificity, ionic dependency, and Michaelis constant ($K_m$) values have been demonstrated from the cited references and reviews.

Source: [16]; [17]; [13]; [18]; [19]; [20]; [21]; [15]; [22]; [23]; [24]; [14].

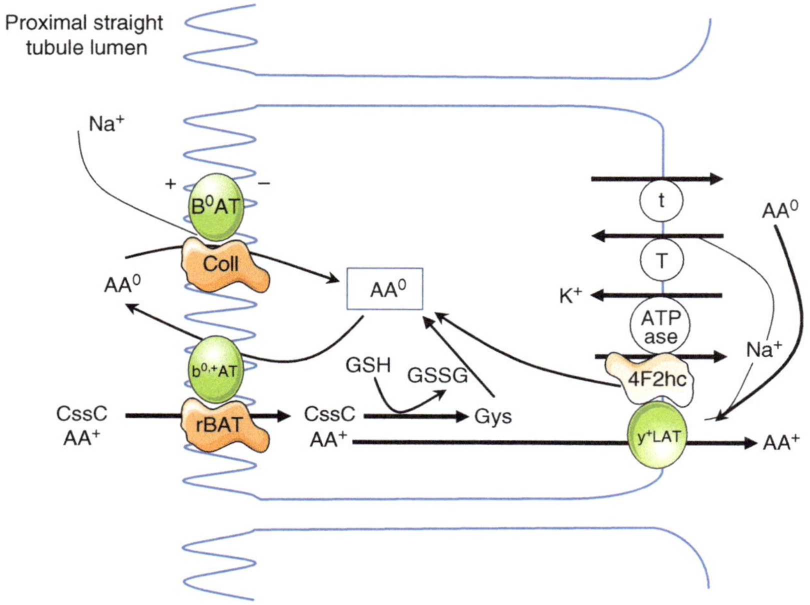


**Figure 19.1** The illustration depicts renal reabsorption of cationic amino acids ($AA^+$) and cysteine (CssC). The various broad-scope systems ($b^{0,+}$, $y^+L$, and $B^{0,+}$) and interactions with neutral amino acids ($AA^0$) are shown. GSH, glutathione; GSSG, oxidized glutathione. Source: Adapted from Verrey et al. [34].

LAT1/SLC7A5/SLC3A2, LAT2/SLC7A8/SLC3A2, LAT4/SLC43A2, etc.) is found in the basolateral membrane, allowing amino acids to be released into the blood stream after nutrient intake (Figure 19.2, [13]).

Amino acids are major nutrients obligatory for the survival of cells and several have been shown to play an important role in supporting cancer cell growth. Amino acid transporters in tumor cells are assumed to be upregulated to satisfy these cells' demand for amino acids, and SLC1A5, SLC7A5, SLC7A11, and/or SLC6A14 were found to be highly expressed in cancer cells. Their high expression occurs in certain cancer types with a direct or indirect involvement of the oncogene c-Myc. Together these factors are functionally coupled to promote cancer growth (Figure 19.3, [37]). In recent years, amino acid transporters have been mentioned as "hot" targets for cancer drug discovery.

The above discussion was primarily focused on general characteristics of amino acid transporters that may affect their functions in different tissue and cell types. We focus on their roles in the intestine and the kidney as these are organs that are likely to impact the bioavailability of amino-acid-like drugs such as L-DOPA (3,4-dihydroxy-l-phenylalanine), L-methyldopa, melphalan, and gabapentin. Sections 19.2.1–19.2.12 describe several important amino acid transporter systems and individual transporters when appropriate.

### 19.2.1 System y+

System $y^+$ is a $Na^+$-independent transporter for cationic amino acids [7, 38]. System $y^+$ also binds some neutral amino acids with lower affinity in the presence of $Na^+$. Cationic amino acid transporter (CAT) proteins were among the first amino acid carriers identified in mammalian cells and are classified as members of the SLC family 7 [39]. For example, L-arginine, L-lysine, and L-ornithine are all transported by CAT-1 (SLC7A1), which is pH-independent, sensitive to trans-stimulation, and saturated at circulating plasma concentrations (~0.1–0.2 mM) [40].

Four other related cationic transport proteins, CAT-2A/SLC7A2A, CAT-2B/SLC7A2B, CAT-3/SLC7A3, and CAT-4/SLC7A4, have now been demonstrated to function in different mammalian species [40, 41]. CAT-1, -2A, and -2B are glycosylated and located in the plasma membrane. CAT-2 is usually expressed in the liver and pancreatic islets, and CAT-2B is always induced under inflammatory conditions in various cells such as T cells, macrophages, lung, and testis [42]. CAT-2A is a low-affinity carrier for cationic amino acids, which is unlike CAT-1. On the other hand, CAT-3 mediates $Na^+$-independent transport of cationic amino acids but with substrate specificities and $K_m$ values different from those of other CAT isoforms [43, 44]. Lastly, human CAT-3 is not neuron-specific with a high selectivity for cationic amino acids, such that it does not transport L-citrulline, L-methionine, L-cysteine, or L-glutamate [43, 44].

Heterodimeric amino acid transporters, a subfamily of SLC7, are composed of a heavy chain (rBAT or 4F2hc) and another relative light chain linked by a disulfide bridge [45, 46]. The heavy subunit may be necessary to bind the complexity to the membrane, whereas the

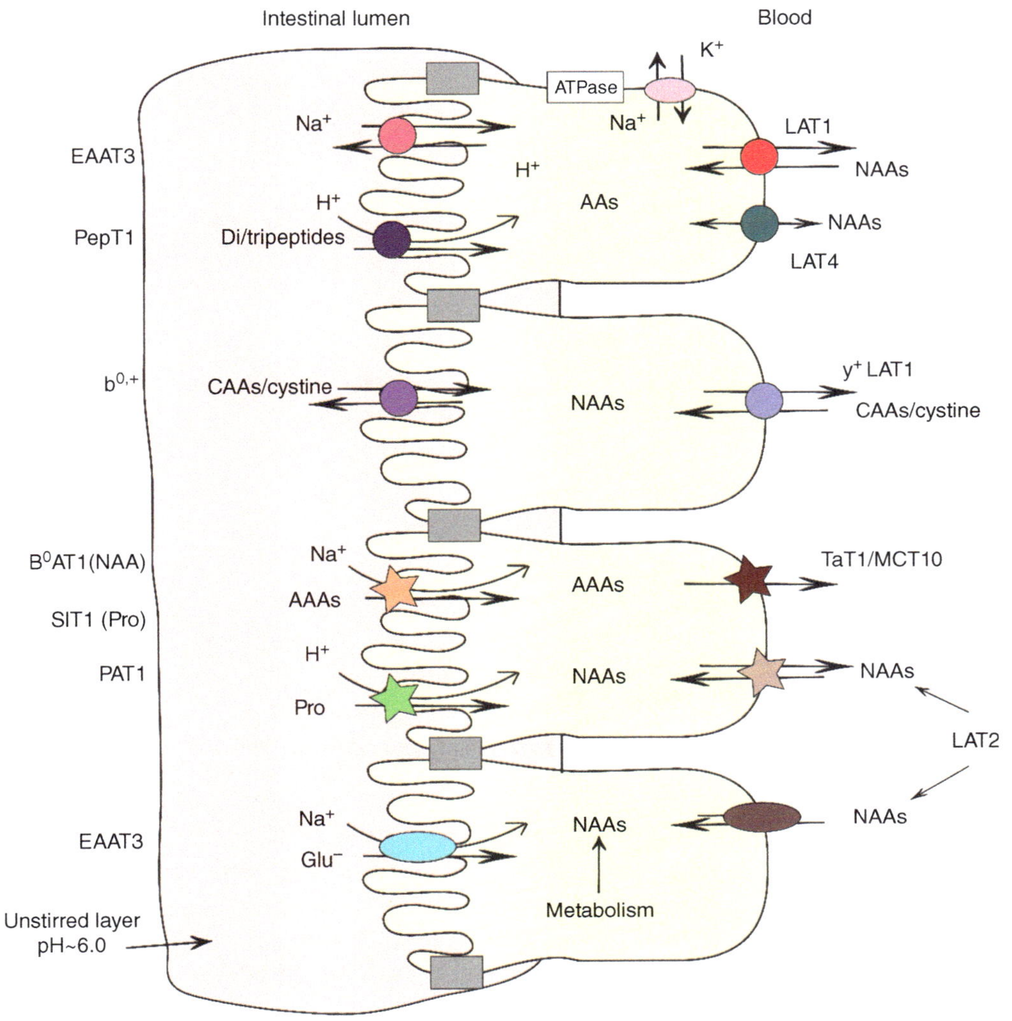


**Figure 19.2** Intestinal absorption of amino acids. The inwardly directed $Na^+$ electrochemical gradient that is used as a driving force for the uptake of $Na^+$-coupled amino acids transport is maintained by the basolateral $Na^+/K^+$-ATPase. The unstirred mucin layer at the brush-border membrane surface maintains a slightly acidic pH of ~6.0, which could support the $H^+$-coupled uptake mechanism. AA, amino acid; CAA, cationic amino acid; NAA, neutral amino acid. Source: Adapted from Kandasamy et al. [15].

light chain may act as a conduit to transport [47–49]. Heterodimeric amino acid transporters involving LAT1, LAT2, $y^+$LAT-1, and $y^+$LAT-2. xCT, Asc-1 and $b^{0,+}$AT have also been identified [45, 46].

### 19.2.2 System $y^+$L

System $y^+$L has a broad specificity among amino acid transporter [29]. System $y^+$L-like functionality has also been found in the intestine, placenta, lymphocytes, and platelets [45]. The 4F2hc/$y^+$LAT1 and 4F2hc/$y^+$LAT2 are the molecular correlates of system $y^+$L. 4F2hc form a complex with CD147 and monocarboxylate transporters (MCTs), suggests their joint role in functional response [50]. System $y^+$L can bind cationic and neutral amino acids, and its specificity varies depending on the ionic ingredients. However, system $y^+$L is stereoselective, electroneutral, and sensitive to trans-stimulation; and it can also mediate high affinity, $Na^+$-independent cationic amino acid transport [29, 45].

### 19.2.3 System L

System L is a classical $Na^+$-independent transporter and is most active with branch-chained and aromatic neutral amino acids with different affinity to a range of large and small neutral amino acids. Transport by the system L is trans-stimulated by intracellular substrates of this carrier and in some cases may be increased by low extracellular pH. The cDNAs encoding

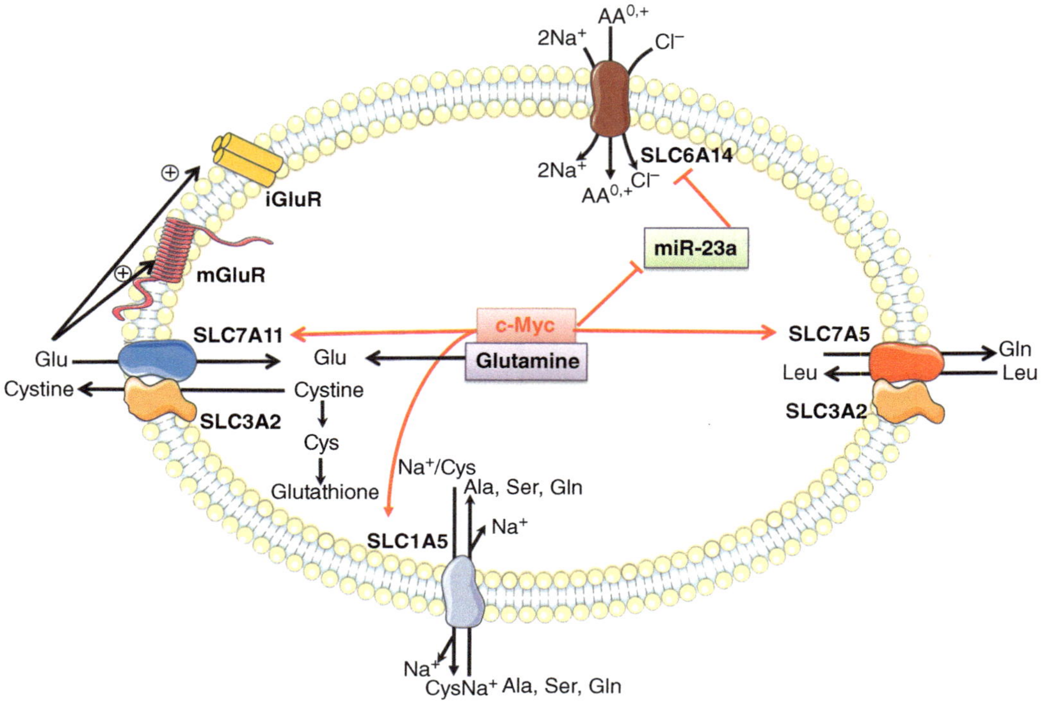


**Figure 19.3** The amino acid transporters that are upregulated in cancer and their functional interrelationship. Source: Adapted from Bhutia et al. [37].

system L-like transporters shown in Table 19.1 have been identified. Co-expression of 4F2hc and LAT-1 or LAT-2 mRNAs in oocytes can induce $Na^+$-independent transport and can have an extensive specificity for small and large zwitterionic amino acids. The transport is also sensitive to trans-stimulation [51, 52]. The expression of a tumor variant of LAT-1, a tumor-associated gene-1/L-amino acid transporter-1 (TA1/LAT-1) is substantially upregulated in hepatocytes deprived of L-arginine, indicating that LAT-1 may act as a sensor of amino acid deprivation [53].

### 19.2.4 System $x_c^-$

System $x_c^-$, a $Na^+$-independent anionic amino acid transport system, is not sensitive to inhibition by dibasic and neutral amino acids. System $x_c^-$ could also enable exchange of L-cystine for intracellular L-glutamate with a 1 : 1 stoichiometry. Cysteine can be oxidized to L-cystine in the extracellular fluid. The L-cystine in the cells is rapidly reduced to L-cysteine, which is the rate-limiting precursor for GSH synthesis [31, 54]. L-Cystine transport by system $x_c^-$ is inhibited by L-glutamate and homocysteine [54]. System $x_c^-$ activity in *Xenopus oocytes* required two cDNA transcripts. One is the 4F2hc protein and the other is a novel protein of 502 amino acids with 12 putative transmembrane domains [54]. Even though the oxidant-sensing transcription factor Nrf2 can bind to the antioxidant response/electrophile response element (ARE/EpRE) in the 5′-flanking region of stress response genes, system $x_c^-$ activity is not induced in Nrf2-deficient macrophages [55].

### 19.2.5 System $b^{0,+}$

Identification of system $b^{0,+}$ in the intestine and renal tubules was accomplished on the basis that a defect in system $b^{0,+}$ in the human kidney can cause hereditary hyperaminoaciduria cystinuria [46, 56]. A potential subunit of system $b^{0,+}$ is the cloned sequence rBAT/$b^{0,+}$AT [46, 57], which has a molecular mass of ~54 kDa and has also been identified in the heart, liver, placenta, and lung [46]. System $b^{0,+}$ in humans may transport L-cystine and neutral and dibasic amino acids. The affinities of dibasic and L-cystine are usually higher than those of neutral amino acids [57]. The rBAT/$b^{0,+}$AT heterodimer mediates the electrogenic exchange of extracellular cationic amino acids and cystine for intracellular neutral amino acids, except imino acids [58].

### 19.2.6 System $B^{0,+}$

System $B^{0,+}$ has been identified as a $Na^+$-dependent transport system for cationic and neutral amino acids [59]. The substrate specificity of system $B^{0,+}$ is similar to that of system $b^{0,+}$, but system $B^{0,+}$ also can accept L-alanine, L-serine, and 2-aminobicyclo-(2,2,1)-heptane-2-carboxylic acid [56]. A cDNA encoding a $Na^+/Cl^-$-dependent carrier called $ATB^{0,+}$ has been reported to transport cationic and neutral amino acids [60]. System $B^{0,+}$ shows an extensive specificity for neutral amino acids and a high-affinity for cationic and neutral amino acids. The $ATB^{0,+}$ transporter may play an important role in regulating L-arginine transport for NO synthesis [60].

### 19.2.7 System A

System A is a classical $Na^+$-dependent system and a key target for hormonal regulation, mediating the uptake of amino acids with small side chains including alanine, serine, proline, and glutamine. Its transport activity can be upregulated in response to amino acid deprivation or hypertonic stress [8]. Neutral amino acids are the most reactive with system A because they have short, polar, linear, and N-methylated side chains. The transport of system A is reduced or inhibited markedly by lower intracellular pH and competitive substrates or inhibitors for this carrier. Three isoforms of system A were identified, and all three belong to the SLC38 family and exhibit differential tissue expression patterns [61]. The SLC38 family of transporters has a total of 11 members in humans and they encode amino acid transporters called sodium-coupled neutral amino acid transporters (SNATs). To date, five SNATs have been characterized and functionally subdivided into systems A (ATA1/SNAT1/SLC38A1, ATA2/SNAT2/SLC38A2, and ATA3/SNAT4/SLC38A4) and system N (SNAT3/SLC38A3 and SNAT5/SLC38A5) showing the highest transport for glutamine and alanine. Among the three isoforms, ATA1 and ATA2 have similar functions in oocyte saturation, voltage, $Na^+$ dependence, and pH sensitivity [62]. As to tissue distribution, ATA1 is mainly expressed in placenta and the brain [63], ATA2 in endothelial cells [64, 65], and ATA3 only in the liver [64]. ATA3 has a great affinity for cationic amino acids compared to that for neutral amino acids [66].

### 19.2.8 System *asc*

System *asc* is a $Na^+$-independent, high-affinity transport system that mediates the entry of small neutral amino acids like L-alanine, L-serine, L-cysteine, glycine, L-threonine, and 2-aminoisobutyric acid. 4F2hc/Asc-1 and Asc-2 are the two isoforms of system asc (alanine–serine–cysteine) [67, 68]. cDNAs from mouse and human brains encoding asc-type amino acid transport have been isolated [67, 68]. The encoding proteins were designated Asc-1 and hAsc-1 and are structurally related to the family of amino acid transporters linked by disulfide bonds to the type II membrane glycoproteins such as 4F2hc and rBAT. These Asc-1 transporters require 4F2hc for functional expression but do, not require $Na^+$ or $Cl^-$ for functions. These Asc-1 transporters could transport many small amino acids including alanine, serine, and cysteine [67, 68]. Co-expression of Asc-2 with either 4F2hc or rBAT in oocytes or COS-7 cells did not induce transport activity, whereas a fusion protein consisting of the C-terminus of Asc-2 and the N-terminus of either 4F2hc or rBAT fragment mediated L-serine uptake. AGT1, another transporter, is structurally related to Asc-2, and shows a high-affinity for $Na^+$-independent transport of L-aspartate and L-glutamate when employed as a fusion protein with either rBAT or 4F2hc [69], suggesting that Asc-2 and AGT1 is a new subfamily of the heterodimeric amino acid transporters.

### 19.2.9 System *ASC*

System ASC is a ubiquitously expressed $Na^+$-dependent system. It transports small amino acids like L-alanine, L-serine, and L-cysteine with a high stereoselectivity and sensitivity to trans-stimulation. Its transport is unaffected by amino acid starvation [70, 71]. Anionic amino acids, such as L-glutamate may inhibit or activate the system ASC at lower extracellular pH [72]. The cDNA (ASCT1, SLC1A4) encoding system ASC has been identified in human. Approximately 40% of the sequence is associated with the excitatory amino acid transporter (EAAT) family of glutamate transporters [73, 74]. Another identified cDNA (ASCT2, SLC1A5) encoding system ASC could play an important role in facilitating the efflux of L-glutamine from glial cells [75]. Although the ASC transporters may enable exchange of neutral amino acids, the net uptake, and the transport is a substrate-dependent anion transmembrane movement [76].

### 19.2.10 System N

System N shows a preference for the transport of L-glutamine, L-asparagine, and L-histidine, but also helps in the transport of N-methylated substrates. This system is insensitive to regulation by glucagon and insulin. System N has a reduced rate of transport at lower extracelluar pH [8, 76]. System N may tolerate the substitution of $Li^+$ for $Na^+$ as the cotransported ion. The subtypes of System N in skeletal muscle were named system $N^m$ and have a much weaker $Li^+$ tolerance and higher pH sensitivity than the hepatic system N [77, 78]. The neuronal system $N^b$ shows a similar $Li^+$ tolerance and pH sensitivity to liver system N but is inhibited by L-glutamate [79].

A subtype of system N was identified in the rat brain, named SN1 (SNAT3), with highest expression in the kidney, heart, and the brain [8]. System N1 mediates $Na^+$-coupled transport of L-glutamine and other neutral amino acids. The mechanism involves the efflux of $H^+$ by the transporter, resulting in intracellular alkalinization. Transport of system SN1 is mediated with inward transport of two $Na^+$s and one amino acid, with coupling efflux to one $H^+$ [80]. L-glutamine transport in system N1 is regulated by changes in the cell membrane with depolarization, resulting in switching from uptake to efflux as well as by transmembrane L-glutamine gradients [79]. Another member of system N, namely, SN2 (SNAT5), was cloned with extensive expression in the lung and the stomach. Recently, a novel glutamine transporter SNAT7 encoded by the SLC38A7 gene was identified, and this transporter was previously assumed to be a member of system N [81].

### 19.2.11 System $X_{AG}^-$

The high affinity, $Na^+$-dependent transporters for aspartate and glutamate are grouped into the EAAT family. Some of these transporters are predominantly expressed in the brain, whereas others are also expressed in other cells. In particular, human fibroblasts are endowed with a system for high-affinity glutamate transport, named years ago as system $X^-_{AG}$, which is characterized by the operational features of EAAT transporters, such as strict sodium- and potential-dependence and reactivity toward D-aspartate but not D-glutamate [82]. High-affinity L-glutamine transporters of the SLC1 family (corresponding to system $X^-_{AG}$) mediate L-glutamine uptake into epithelial cells and neurons (e.g. SLC1A1/EACC1) as well as into glial cells (e.g. SLC1A2/GLT1 and SLC1A3/GLAST) [15].

### 19.2.12 System IMINO

System IMINO is moderately stereospecific for L-proline and reacts with neutral amino acids at lower affinities than with their N-methylated derivatives. It mediates the $Na^+$-dependent transport of imino acids such as proline and pipecolate in a variety of intact tissue [23]. Sodium/imino acid transporter 1 (SIT1, SLC6A20) is a $Na^+$-dependent imino acid transporter with properties of the classical system IMINO.

## 19.3 Epithelial Amino Acid Transporters

### 19.3.1 Neutral Amino Acid Transporters in the Apical Membrane of Renal Epithelial Cells

The accumulation of neutral amino acids in kidney is enabled by the $Na^+$-electrochemical gradient. One $Na^+$: one amino acid ratio was needed for the transport of phenylalanine and alanine [83]. Increased extracellular $Na^+$ may change the $K_m$ of alanine rather than the $V_{max}$ of the transportation, further, the $K_m$ value for $Na^+$-dependent transport was increased when the membrane was hyperpolarized [84]. Neutral amino acids represent >80% of the free plasma amino acids and are all transported by the luminal $B^0AT1$ transporter, albeit at variable apparent affinities. A new finding is that the luminal $Na^+$-dependent neutral amino acid transporters of the SLC6 family require an associated protein for their surface expression (e.g. $B^0AT1$), and implied for the L-proline transporter SIT1 and for $B^0AT3$ (XT2, SLC6A18) [34].

Transporters of neutral amino acids have overlapping substrate specificity in that infusion of a single neutral amino acid into the blood markedly affects secretion and clearance of the other neutral amino acids [9]. For example, the reabsorption of histidine can be decreased by almost all neutral amino acids, anionic amino acids, and cationic amino acids [85], whereas alanine can inhibit reabsorption of other neutral amino acids and cationic amino acids [86]. In contrast, anionic amino acids do not increase the excretion of neutral amino acids [86]. The "nonspecific" inhibition occurs because of the competition by several transporters for the same electrochemical gradient (e.g. sodium). For example, the uptake of amino acids can be inhibited by D-glucose [87]. The substrate specificity for the renal proximal neutral amino acid transporter was demonstrated by kinetic studies, in that there was strong inhibition of the L-alanine uptake by L-valine and/or L-hydroxyproline. However, cationic or anionic acids were not effective in inhibiting the uptake of L-alanine [88].

In the kidney, low-affinity uptake of serine [89], phenylalanine [83], and the branch-chained amino acids [90] occur in the proximal convoluted tubule.

A high-affinity uptake of the same amino acids is present in the proximal straight tubule, suggesting that a high-affinity variant of system $B^0$ is expressed in the proximal straight tubule. The transport of tryptophan is mediated by system $B^0$, and both $Na^+$-dependent (low affinity) and $H^+$-dependent transporters contribute to its transport [91]. Additionally, the transport of neutral amino acids is also partially inhibited by lysine [87].

Taken together, a common, low-affinity, $Na^+$-dependent neutral amino acid transporter (system $B^0$) is present in the proximal convoluted tubule. System $B^0$ is also different from the $Na^+$-dependent transporters of neutral amino acids (system A and system ASC) in nonepithelial cells [8]. System $B^0$ is also different from system A (SNAT isoforms) and ASC (ASCT2) in epithelial cells in that $B^0$ is a high-affinity transporter in the proximal straight tubule. On the other hand, ASCT2 mediates the system ASC-like transport activity in the apical membrane (Section 23.2).

### 19.3.2 Neutral Amino Acid Transporters in the Apical Membrane of Intestinal Epithelial Cells

The transport of neutral amino acids is entirely $Na^+$-dependent in the gut, and these amino acids are not accumulated in the intestinal epithelium. Neutral amino acid uptake is complemented by the apical peptide transporter 1 (PepT1) and system $B^{0,+}$, which are found in distal parts of the intestine. Similarly, system ASC activity (ASCT2) is also found in the distal part of the intestine. The major neutral amino acid transporter at the apical membrane of intestine epithelium is $B^0$AT1 [13].

Phenylalanine, methionine, and leucine display high affinity (or low $K_m$ values) for the neutral amino acid transporters, while alanine and particularly glycine display much lower affinity [92]. In the rat intestine, a single transporter for neutral amino acids has a broad specificity. On the other hand, two transporter systems were found in the rabbit ileum [93]: one that transports most of neutral amino acids and the other that transports large neutral amino acids (LNAAs). On the basis of these findings, glycine, alanine, and serine are almost exclusively transported by system 1, but LNAA are transported by both systems. It was reported that system 1 is likely to be equivalent to system $B^0$ in the proximal convoluted tubule of the kidney, whereas system 2 seems to be a $Na^+$-dependent transporter for most neutral amino acids [94]. In cultured Caco-2 cultured cells, a $Na^+$-dependent transporter of phenylalanine was found to be shared by methionine, but not by other amino acids [10]. The uptake of phenylalanine is inhibited partially by most of the tested amino acids and completed inhibited by methionine and phenylalanine, indicating that $B^0$ is responsible for the transport of phenylalanine [95]. Therefore, both system $B^0$ and system ASC play a key role in amino acid absorption in the intestine. System $B^{0,+}$ is localized to the ileum but also seems to be more extensively expressed in the colon. These transporters were also described in Section 23.2.

### 19.3.3 Neutral Amino Acid Transporters in the Basolateral Membrane of Epithelial Cells

A $Na^+$-independent transporter with a broad substrate specificity was found in rat small intestine [87], which was similar to the transporter of system L of nonepithelial cells. The uptake of neutral amino acids into the membrane was trans-stimulated by intravesicular substrates. The transporter is inhibited by *p*-chloromercuribenzenesulfonate (pCMBS) and *N*-ethylmaleimide (NEM). A system A-like activity, a system ASC-like activity, and a novel activity were identified to response in a $Na^+$-dependent manner [96] to the nutritional amino acids when the intestine was starved or in disorder. System ASC activity was identified at the brush-border membrane in both the kidney and the intestine. The intestine in particular sequesters large amounts of glutamine from the circulation [97]. Approximately one-third of the arterial glutamine is removed in a single pass of the small intestine [98]. The uptake of glutamine through the basolateral membrane in the intestine is via a $Na^+$-dependent transporter that has the properties of system A [99]. The uptake of glutamine may be inhibited by other small neutral amino acids and certain amino acid analogs.

It is unclear whether the $Na^+$-dependent transporter of tryptophan in brush border may involve additional transporters [100]. Interestingly, uptake of serine may be inhibited by phenylalanine but not by D-serine. On the other hand, uptake of cysteine is $Na^+$-independent manner in the rat intestine [101]. Two transporters for cysteine were detected. One was a low-affinity system, which showed an affinity for alanine, serine, and cysteine, and was similar to system ASC [102]. The other was a high-affinity system for aromatic amino acids, which was similar to system L because of the inhibition by an amino acid analog 2-amino-2-norbornane-carboxylic acid (BCH) [103]. The uptake of glutamine in rat kidney is different compared to that of other neutral amino acids because it proceeds almost entirely in a $Na^+$-dependent manner [104]. The transporter has a low affinity to glutamine, but it is upregulated by metabolic acidosis.

The transporter, system L-like of and highly specific system for neutral amino acid has been detected as the heterodimeric transporter involving 4F2hc/LAT2 [18]. At the molecular level, the close relationship between this transporter and that of a $Na^+$-dependent glutamine transporter located at the basolateral membrane of the kidney suggests that the transporter is most likely to be SNAT3, although this transporter displays system A-like functions in the intestine. System ASC has been identified in the basolateral membrane and the apical membrane [18]. The function of system ASC is controlled by the heteromeric amino acid transporter 4F2hc/asc-1. These transporters are described in Section 23.2.

### 19.3.4 Transporters of Cationic Amino Acids and Cystine in the Apical Membrane of Renal Epithelial Cells

Cationic amino acids are transported against an electrochemical gradient because they have a positive charge. Their transportation is controlled by systems $b^{0,+}$ and $y^+$, which has also been identified in a variety of nonepithelial cells [34, 105]. The transport of cationic amino acids does not involve transmembrane $Na^+$ gradient. For example, the initial rate of lysine uptake did not change in the presence or absence of $Na^+$ [106, 107]. In general, the transport of cationic amino acid is $Na^+$-independent [36], and is activated by an inside negative membrane potential, which suggests a mechanism of electrogenic transport.

Uptake of cationic amino acid is strongly activated by intracellular and intravesicular cationic amino acids [107, 108]. The transport of arginine, ornithine, and lysine occurs through a common special transport system in the kidney [109], but the transport of cystine has been controversial. Infusion of lysine or arginine resulted in increased excretion of both cationic amino acids and cystine but did not increase their plasma concentration [110]. In contrast, mutual inhibition [111] and mutual exchange or diffusion were reported for cationic amino acids [108], but not in the presence of cystine in the kidney cortex slices. There are two transporters for lysine: one is sensitive to alanine, phenylalanine, proline, and β-alanine, and the other is sensitive to alanine and phenylalanine [87]. This is similar to system $b^{0,+}$, which is inhibited by phenylalanine and alanine [30]. The transport of lysine was inhibited by arginine, cystine, phenylalanine, and methionine with or without $Na^+$ in the incubation medium, and also by glucose or glutamate, which may exclude nonspecific depolarization as the mechanism of inhibition.

Kinetic studies demonstrated the presence of only one system for lysine transport as the two transporters have similar affinity in the presence or absence of $Na^+$ [107]. Moreover, the system was responsible for cystine but not for other neutral amino acids [35].

Cystine interacts with system $B^0$, and, consequently, cystine transport could be inhibited by neutral amino acids in the presence of $Na^+$. The mechanism for the uptake of cystine is complicated because of the following factors: (i) it has limited solubility; (ii) cystine may be changed into cysteine when inside the cells of cortical slices [112]; (iii) cystine may bind to proteins [113]; and (iv) cystine shows $Na^+$-dependent uptake [113]. Cysteine can be transported by the neutral amino acid transport system $B^0$ although it does not affect the transport of cationic amino acids [114].

The transport of cationic amino acids is partially dependent on $Na^+$ because a $Na^+$-independent transporter has a high affinity for cationic amino acids. The transporter is shared with cystine and is inhibited by neutral amino acids. The transporter is functionally linked to system $b^{0,+}$ [30]. The heteromeric transporter rBAT/$b^{0,+}$AT of cationic amino acid has been identified as the mediator of this activity. However, the molecular identity of the $Na^+$-stimulated CAT remains to be identified.

### 19.3.5 Transporters of Cationic Amino Acids and Cystine in the Apical Membrane of Intestinal Epithelial Cells

There exists mutual inhibition between arginine, cystine, lysine, ornithine, and methionine in the accumulation of cationic amino acids by the intestinal epithelium [115, 116]. The transport of lysine may be classified into a $Na^+$-dependent and a $Na^+$-independent mode [117]. Intracellular leucine can accelerate lysine uptake in the intestine. All neutral amino acids trans-stimulate the uptake of lysine with the exception of isoleucine. The interaction between cationic and neutral amino acids in the intestine seems to differ in that there is a lack of inhibition of neutral amino acid transport by cationic amino acids, which is most likely caused by markedly lower affinity of the cationic amino acids to amino acid transporters responsible for neutral amino acid absorption [117]. Therefore, system $B^{0+}$ has the potential to act as a small component of neutral amino acid transport. The transporter rBAT/$b^{0,+}$AT is the major intestinal apical membrane transporter for cystine and cationic amino acids. The mediator of epithelial cell cationic amino acid efflux is the $y^+$LAT1 transporter. However, a cationic amino acid transport activity, resembling that of system $y^+$ appears to be present in nonepithelial cells. The CAT family is responsible for nonepithelial cell cationic amino acid transport [13].

The transport of cationic amino acids in the small intestine is less complex than in the proximal tubule because no stimulation of uptake for lysine in the rat intestine was detected in the presence of a $Na^+$-gradient [118]. However, a $Na^+$-dependent CAT in rat intestine can be induced by a protein-rich diet [119]. The activity of transport was completely inhibited by cationic amino acids and in part by neutral amino acids [120]. System $B^{0,+}$, a $Na^+$-dependent transport activity for neutral and cationic amino acids, has been demonstrated in the rabbit terminal ileum; this has been described in Section 23.3. System $B^{0,+}$ has properties similar to the high-protein diet, which can induce CAT. System $b^{0,+}$, a $Na^+$-independent transporter for cationic amino acids, has also been demonstrated to be the heteromeric transporter rBAT/$b^{0,+}$AT (Section 23.2).

### 19.3.6 Transporters of Cationic Amino Acids and Cystine in the Basolateral Membrane of Epithelial Cells

The basolateral heterodimeric obligatory amino acid exchanger $y^+$LAT1-4F2hc plays an important role for the transepithelial transport of its cationic substrates, and it transports its substrates out of the cells in exchange for extracellular neutral amino acids and $Na^+$ [34].

The absorption of lysine in intestinal epithelial cells is strongly influenced by leucine when added to the luminal side. Two different mechanisms are proposed to explain this effect. In the first mechanism, the intracellular leucine acts as an allosteric modulator to increase the basolateral efflux of lysine [117]. For the second mechanism, the apical heterologous exchange is responsible for the effect. However, using basolateral vesicles from the rat jejunum, it was demonstrated that an electrogenic uniporter for cationic amino acids was strongly influenced by leucine at micromole concentration level [121]. Therefore, the exchange mechanism was not likely to be involved. An amino acid exchanger with high affinity for neutral and cationic amino acids was identified in erythrocytes [29]. The transporter was also identified in vesicles isolated from placenta [122]. System $y^+$L was defined as the transporter responsible for the interaction with neutral and cationic amino acids. Neutral amino acids need $Na^+$ for interaction with the system, while cationic amino acids can be transported in the absence of $Na^+$. Cloning of system $y^+$L is expressed in the basolateral membrane (Section 23.2) in the kidney and intestine. System $y^+$L may be influenced by lysinuric protein intolerance (Section 23.2).

### 19.3.7 Transporters of Anionic Amino Acids in the Apical Membrane of Renal Epithelial Cells

The kidney tubular luminal influx of anionic amino acids requires EAAT3 (EAAC1/SLC1A1), which has been shown to cotransport its substrates with three $Na^+$ ions and one proton in exchange for one $K^+$ ion [34]. The uptake of glutamate and aspartate increases greatly in the presence of a $Na^+$ gradient [123, 124]. The initial rates of the uptake can be increased by increasing extracellular $Na^+$ concentration. The transporter for its substrates has a high affinity. The transporter is stereospecific for glutamate but not for aspartate, and it cannot be inhibited by neutral or cationic amino acids. Other inhibitors are L-cysteate and DL-threo-3-hydroxyaspartate. The uptake of glutamate can be increased by an inside negative membrane potential in the presence of intravesicular $K^+$ [125], which was proposed to involve the cotransport of three $Na^+$ and an antiport of one $K^+$. The transporter may work in the absence of intravesicular $K^+$ as a slow electroneutral transporter. Diffusion may not play a key role in glutamate transport because the uptake is very slow at absence of $Na^+$ [125]. System-ASC-like activity or a low affinity of system $B^0$-like activity for glutamate transport at low pH has been identified [124]. On the other hand, glutamate and aspartate can be rapidly metabolized in the renal parenchyma and can be transaminated to form 2-oxoglutarate and oxaloacetate, respectively [126]. The properties of the glutamate transporter in the kidney are consistent with those of the glutamate transporter in the brain [127].

### 19.3.8 Transporters of Anionic Amino Acids in the Apical Membrane of Intestinal Epithelial Cells

Glutamate uptake in the intestine is strictly $Na^+$-dependent [128]. A glutamate-specific system is not shared by D-aspartate because glutamate uptake is around five times higher than that of D-aspartate at equal concentrations with only one transport system for D-aspartate, but two for glutamate [128, 129]. The capacity of transport is inhibited by many of neutral amino acids including phenylalanine [128, 129]. The low-affinity glutamate transport was controlled by system $B^0$ [130] and system ASC at low pH [7]. Moreover, the inhibition of glutamate transport by neutral amino acids did not coincide with the relative affinity for system $B^0$. In the acidic condition of the intestine, the low-affinity system seems to make a marked contribution to the uptake of glutamate. Fifty percent of the intestinal glutamate uptake is regulated by system ASC [131]. It has been suggested that ASCT2 expression is limited to the ileum and colon, the dominant role of system ASC in jejunum is a rabbit-specific function [13].

### 19.3.9 Transporters of Anionic Amino Acids in the Basolateral Membrane of Epithelial Cells

For the transport of both glutamate and aspartate, a $Na^+$-dependent transporter has been identified in the basolateral membrane of renal and intestinal epithelial cells [132, 133]. The transporter can be activated by intracellular $K^+$, which is similar to the apical system [132]. EAAT2 (GLT-1/SLC1A2) is expressed in the basolateral membrane of the kidney [126]. Glutamate is rapidly metabolized in both the kidney and the intestine. For example, in the intestine, more than 98% of glutamate is changed into α-ketoglutarate [126], and consequently metabolized. As a result, most of the glutamate and aspartate in the nutrition cannot enter the portal circulation [98]. However, a specific efflux transporter for glutamate in the basolateral membrane has not been reported.

### 19.3.10 Transporters of Imino Acids and Glycine in the Apical Membrane of Renal Epithelial Cells

Proline is absorbed well in the proximal tubule of the kidney, whereas the absorption of glycine is limited [9]. A common transport system is responsible for the transport of praline, glycine, and prolinuria [134]. The absorption of glycine, proline, and hydroxyproline in the renal tubular is incomplete in newborn mammals compared to that in adults [135]. The deficient absorption of praline and glycine is almost due to a lack of the two independent high-affinity transporters for praline and glycine; in contrast, the common low-affinity system is active in newborns [18]. In the rat, the high affinity, proline-specific system appears in the second week, and the high affinity, glycine system enhances its activity in the third week [136]. The transport of glycine was mediated by a system that is most likely $B^0$, which would be shared with alanine, and by a system shared by praline [137]. The two systems are $Na^+$ and energy-dependent, although the high-affinity system is responsible for praline because change in $Na^+$ concentration can largely influence the $V_{max}$ rather than the $K_m$ of the glycine transporter [138]. Glycine and proline are partially transported by separate systems according to their pH dependence [138]. In addition, a high-affinity system for glycine is shared by neutral amino acids in the proximal straight tubule.

Proline transport is entirely $Na^+$- or $H^+$-dependent [139]. The $H^+$-dependent proline/glycine transporter exists together with more than one $Na^+$-dependent proline uptake system. The $H^+$-coupled system was decreased in the proximal convoluted tubule; whereas transport in the proximal straight tubule was strictly $Na^+$ coupled [140]. In summary, at least three different transporters are responsible for proline and glycine: (i) a $Na^+$-independent but $H^+$-dependent transporter for glycine and proline; (ii) a specific transporter for proline and hydroxyproline; and (iii) a $B^0$-like transporter for glycine and proline.

### 19.3.11 Transporters of Imino Acids and Glycine in the Apical Membrane of Intestinal Epithelial Cells

Three different mechanisms are responsible for transport of proline and glycine in the intestine [115, 141]: (i) the general system $B^0$ of neutral amino acids; (ii) a proline, hydroxyproline transporter, a sarcosine, *N*-dimethylglycine, and betaine transporter; and (iii) a mediating system for the uptake of glycine and proline. It appears that the intestinal proline uptake in rabbit is mediated by the $B^0$ and IMINO transporter systems. The imino acid carrier (PAT1) and system $B^0$ are the principal proline transport components in rat intestine. In humans, sa1, PAT1, and SIT1 are mediators of glycine and proline intestinal transport [13].

### 19.3.12 Transporters of Imino Acids and Glycine in the Basolateral Membrane of Epithelial Cells

The transport of glycine across the basolateral membrane is $Na^+$-independent in the proximal convoluted tubule but is $Na^+$-dependent in the proximal straight tubule [142]. Because few amino acids remain in the lumen of the proximal straight tubule, the epithelial cells may be responsible for the supply of amino acids from the blood. The $Na^+$-dependent glycine transporter is similar to that of the apical system. Proline transport is entirely $Na^+$-independent in the kidney [143]. A $Na^+$-independent transporter for proline was also found in the small intestine [144]. The uptake of proline may be inhibited by L-proline, leucine, alanine, sarcosine, glycine, lysine, OH-proline, taurine, β-alanine, serine, and phenylalanine, while other amino acids such as threonine or methionine did not affect the uptake of proline. The glycine transporter GlyT1A seems to mediate basolateral membrane glycine transport [145].

### 19.3.13 Transporters of β-Amino Acids and Taurine in the Apical Membrane of Renal Epithelial Cells

There is a common system for taurine, β-alanine, and GABA (γ-aminobutyric acid B-subtype receptor) in the kidney proximal tubule [146]. Uptake of β-alanine is via a $Na^+$-coupled transporter in the kidney [147, 148]. β-Alanine transport is also sensitive to $Cl^-$ anions [147]. β-alanine transport has a stoichiometry of >2 $Na^+$:1 $Cl^-$:β-alanine [149]. Uptake of β-alanine is strongly inhibited by taurine and hypotaurine, but only weakly by GABA, and not by any other amino acids [149]. A low-affinity transporter is responsible for taurine in the proximal convoluted tubule and appears to be $H^+$-coupled [91, 147]. Another low-affinity $Na^+$-dependent taurine transporter is not $Cl^-$-dependent in the proximal straight tubule [91]. The $H^+$-dependent transport was inhibited by β-alanine, proline, glycine, and alanine, which is similar to that of the imino acid carrier. The upregulation of the high-affinity taurine transporter can occur when the diet is rich in sulfur-containing amino acids or in taurine content [150, 151]. In summary, a high-affinity $Na^+$- and $Cl^-$-dependent transporter for taurine and β-alanine and a low-affinity $H^+$-dependent transporter displayed characteristics of the imino acid transporter.

### 19.3.14 Transporters of β-amino Acids and Taurine in the Apical Membrane of Intestinal Epithelial Cells

Two major transporters are responsible for β-alanine and taurine in the intestine [116]: (i) a high-affinity low-capacity transporter that is $Na^+$- and $Cl^-$-dependent [152, 153] and (ii) a $H^+$-dependent, low affinity, high capacity transporter, which is transported by β-alanine, taurine, GABA, proline, glycine, and betaine. However, there are some species differences. For example, in the rat, taurine transport is $Na^+$- and $Cl^-$-dependent but is not shared by β-alanine, whereas in rabbit a high-affinity $Na^+$ and $Cl^-$-dependent transporter is shared by β-alanine and taurine [152, 154]. Similar to that in the kidney, a high-affinity $Na^+$- and $Cl^-$-dependent transporter for taurine and β-alanine and a low-affinity $H^+$-dependent transporter displayed characteristics typical of the imino acid carrier in the intestine.

### 19.3.15 Transporters of β-amino Acids and Taurine in the Basolateral Membrane

Taurine transport in the basolateral membrane may be mediated by a very similar system to the $Na^+$- and $Cl^-$-dependent apical system for β-amino acids and taurine [155, 156]. An osmolyte plays an important role in taurine transport in the proximal straight tubule, which is indicated by the presence of $Na^+$-coupled transporters in both the apical and basolateral membranes in the nephron. The absorption of taurine appears to occur in the proximal convoluted tubule [157]. Additionally, the GABA/betaine transporter known as BGT1 (SLC6A12), which is expressed at the basolateral membrane of epithelial cells, transports β-alanine to a smaller extent than that of taurine [13].

### 19.3.16 Transporters of Amino-Acid-Like Drugs

Amino-acid-like drugs including L-DOPA, α-methyldopa, melphalan, and gabapentin are substrates of L-type amino acid transporters [158]. Among the amino acid transport systems, system L is of importance in providing the cell uptake of large, neutral, branch-chained, or aromatic amino acids, especially some essential amino acids. In system L, there are two important transporters, namely, a cloned transporter named LAT1, and LAT2, named on the basis of its homology [51, 52]. LAT1 and LAT2 are $Na^+$-independent transporters, whose transports are driven by a second amino acid transporter that carriers an amino acid that is a common substrate for both $Na^+$-dependent and $Na^+$-independent systems. LAT1 was found to be highly expressed in the blood–brain barrier (BBB), blood–retinal barrier (BRB), testis, bone marrow, placenta, and tumor [159]. LAT2 is distributed in the kidney, colon, and intestine [51]. LAT1 and LAT2 have been demonstrated to be responsible for the transport of several amino-acid-like drugs such as L-DOPA, melphalan, baclofen, 3-*O*-methyldopa, α-methyltyrosine, gabapentin, α-methyldopa, and thyroid hormones, thus affecting their pharmacokinetic characteristics [158]. The transport of these amino-acid-like drugs is described in Sections 19.3.16.1–19.3.16.4.

#### 19.3.16.1 Transporters for L-DOPA

Levodopa or L-DOPA, an intermediate involved in the biosynthesis of dopamine, is extensively used for the treatment of Parkinson's disease. Many amino acids contribute to the absorption of L-DOPA. The uptake of L-DOPA occurs through LAT1 and LAT2 in the intestinal epithelial cells [160–163], the brain capillary endothelium [164, 165], and renal epithelial cells [166]. The $K_m$ values of L-DOPA to LAT1 are 28–102 μM in different cultured brain capillary endothelial cells of animals [164, 165]. L-DOPA is poorly absorbed by passive diffusion across the cell membranes because of its hydrophilic molecule. However, it has reasonable bioavailability because it can be transported by LAT2 in the intestine [162, 167]. The concentrations of L-DOPA in plasma were increased when low-protein meals are taken with LNAA in Parkinson's patients. The mechanism for the above phenomenon may involve competition for the LAT system and the resulting changes in the pharmacokinetics of L-DOPA [168, 169]. In contrast, the concentrations of L-DOPA in brain were decreased after it is taken with high-protein meals. This may be caused by a large increase in the plasma amino acid concentrations, which could saturate the L-amino acid transport system [170]. L-DOPA is also transported by the amino acid transporter rBAT (SLC3A1), which is expressed in the apical membrane of the proximal tubule cell and other epithelia of the renal cortex [166]. Furthermore, a $Na^+$-independent transporter TAT1 was demonstrated to transport aromatic amino acids L-DOPA and 3-*O*-methyldopa. The transporter is expressed in the intestine, placenta, liver, and kidney proximal tubules [171].

#### 19.3.16.2 Transporters for Melphalan

Melphalan, used for the treatment of multiple myeloma and carcinoma of the breast and ovary, is a nitrogen mustard derivative of the neutral amino acid L-phenylalanine. Melphalan at higher concentrations is transported by the system L demonstrated in tumor cells [172], LPC1 plasmacytoma cells [173], and in the BBB cells *in vitro* and *in vivo* [164, 174]. Melphalan uses an active transporter because it is hydrophilic and it is a charged molecule at neutral pH. There may be certain contributions of LAT-mediated transport as well [175]. The $K_m$ of melphalan is 150 μM for LAT transporters in the rat BBB [174]. Melphalan's affinity for the carrier of system L is low and significant competition by the endogenous LNAA may slow down the transport [174]. The polymorphisms of genomic DNA in LAT1 and LAT2 have been demonstrated in cancer patients. Investigation into their influence on the pharmacokinetic variability of melphalan, however, produced no correlation between the polymorphisms and melphalan clearance [176]. However, it is difficult to show a direct correlation with clearance because LAT1 and LAT2 may simultaneously affect the absorption, distribution, and elimination of melphalan.

#### 19.3.16.3 Transporters for Baclofen and Gabapentin

The chemical structures of both baclofen and gabapentin are similar to that of GABA. They may be transported across the BBB to be used as agents for the treatment of epilepsy and postherpetic neuralgia patients. Baclofen is extensively absorbed by rapid passive diffusion [177]. Baclofen is transported by a saturable neutral amino acid carrier of system L [178], and it weakly inhibited leucine uptake [164]. The distribution of baclofen may be restricted by probenecid-sensitive OAT (organic anion transport) system as an efflux transporter [179, 180]. Gabapentin is a hydrophilic analogue of baclofen. It has been proven that gabapentin is a substrate of LAT1 through the *in vitro* model of BBB [181]. The relative bioavailability of gabapentin varied from 80% to 40% [182]. Active transport of gabapentin plays a significant role in intestinal absorption and competitive inhibition by L-leucine suggests that system LAT2 is involved in its absorption [182]. Food does not have any interaction with gabapentin during intestinal absorption [183].

#### 19.3.16.4 Transporters for Thyroxine and Triiodothyronine

Thyroxine and triiodothyronine are lipophilic molecules that indicate high passive transcellular permeability. Several transporters for thyroid hormone have been identified that can be classified into organic amino acid transporters and amino acids transporters [184]. System L, such as LAT1 and LAT2, might be the contributors to thyroid hormone transport [185]. LAT1 is overexpressed in many types of tumor cells, continuously supplying amino acids for tumor cell growth. Therefore, LAT1 is a potential drug target in cancer. Although there are more than 100 inhibitors of LAT1 reported, only few of them are potent and selective, including the most advanced compound JPH203, a triiodothyronine-derived amino acid [186]. Moreover, monocarboxylate transporter 8 (MCT8) is also demonstrated to be responsible for the transport of cellular thyroid hormone uptake and efflux, which cooperates with LAT1 and LAT2 to participate in transferring thyroid hormones from astrocytes to neurons [187].

## 19.4 Endothelial Amino Acid Transporters

### 19.4.1 Transporters in the Blood–Brain Barrier

Endothelial cells on the BBB are linked by zonulae occludens. They limit the intercellular diffusion of circulating solutes such as monoamines. Free intracellular amino acid levels in the brain are interrelated to rates of influx across the BBB, and the limited synthesis of neurotransmitters such as serotonin, dopamine, and histamine [188, 189]. Cerebrovascular endothelial cells have a specific carrier, which can mediate the influx and efflux of amino acids across the luminal and abluminal (basal) membranes of the BBB. Therefore, multiple amino acid transporters function at the BBB to maintain brain homeostasis. Among these, the glycoprotein-associated amino acid transporters play an important role in amino acid homeostasis in brain parenchyma.

There are two $Na^+$-independent amino acid transport systems in the brain capillary endothelium: one is the transport system for LNAA (system L or a high-affinity isoform L1) and the other is for cationic (system $y^+$) amino acids. $Na^+$-dependent amino acid transport systems A, $B^{0,+}$, ASC, $X^-_{AG}$, and one for β-amino acids have also been characterized in the cerebral vasculature. Generally, the high affinity, $Na^+$-dependent transporters are mainly located in the abluminal membrane of the cerebral endothelium [190–192]. Transporter systems with the same transport capabilities for neutral and cationic amino acids have been characterized in the BRB [193]. For example, LAT1 is expressed at the inner BRB, where it plays a key role in maintaining LNAA and neurotransmitters [194].

The high-affinity system L1 for LNAA is expressed symmetrically between luminal and abluminal membranes of the BBB, whereas the $Na^+$-dependent systems A and $B^{0,+}$ are apparently only expressed in the abluminal membrane. Transports of LNAA and small neutral amino acids were inhibited by plasma amino acids with extremely low uptake for L-cysteine, L-serine, L-alanine, L-proline, glycine, L-glutamate, and L-aspartate, possibly due to the low specific affinity or the limited expression of relevant transporters at the luminal surface of the brain endothelium [195].

Cationic amino acid influx is stereospecific, $Na^+$-independent, and can be inhibited by L-homoarginine, L-arginine, L-ornithine, and diaminobutyric acid but not BCH, McAIB, or L-glutamate. Saturable transport of L-arginine has also been detected in neuronal and astroglial cells, the choroidal epithelium, and the blood–choroid plexus barrier *in vivo* [196, 197].

The blood–brain transports of acidic amino acids such as L-aspartate and L-glutamate across the luminal membrane are limited [198]. The net efflux of L-glutamate from the brain is more than 20-fold higher than that of the influx, indicating that the low-capacity acidic transport system in BBB may be involved in mediating the efflux of excitatory neurotransmitters from the brain [188]. The high intracellular concentration of dietary taurine in the brain suggests that the sulfur-containing amino acid plays a role in volume regulation during osmotic stress [199]. Taurine transport across the luminal surface of the BBB was saturable and dependent on $Na^+$ and $Cl^-$ gradients [200]. The transport of β-alanine molecule was coupled to the transport of two $Na^+$ and one $Cl^-$. Although transport at both membranes was dependent on the $Na^+$ and $Cl^-$ gradients, influx was more efficient across the luminal membrane [190].

### 19.4.2 Transporters for Human Fetal Endothelium

Sufficient amino acid transport activity and placental nutrient uptake activity are indispensable for both maternal and fetal nutrient demands. Neutral amino acid transport has been shown to occur in human umbilical vein endothelial cell monolayers [201, 202]. For example, transport of homocysteine was pH insensitive and mediated by both $Na^+$-dependent and $Na^+$-independent saturable carriers, resembling systems ASC and L, respectively [203, 204]. Transport of L-leucine in fetal endothelial cells was saturable, $Na^+$-independent, and inhibited significantly by L-methionine, L-phenylalanine, BCH, and L-DOPA, but not by 2-(methylamino)-isobutyrate or MeAIB, L-cysteine, L-proline, L-hydroxyproline, β-alanine, L-aspartate, or glycine. These characteristics and the observed sensitivity of L-leucine transport to trans-stimulation are consistent with the influx (across the placenta) being mediated via system L, which was confirmed in human umbilical vein endothelial cells [205].

L-Arginine and L-ornithine, two cationic amino acids, are taken up by human umbilical vein endothelial cells [206]. Transport of L-arginine was pH insensitive and inhibited in a concentration-dependent manner by L-homoarginine and L-NMMA (N(G)-monomethyl-L-arginine), but not by MeAIB [207]. Furthermore, L-arginine transport was stereospecific and inhibited by 10-fold excess L-lysine and L-ornithine, but not by L-serine or N(G)-nitro-L-arginine or L-NNA or its methyl ester L-NAME (N(G)-nitro-L-arginine methyl ester) [207]. The influx of L-arginine and tetraphenylphosphonium (a membrane-potential-sensitive probe) is significantly reduced in umbilical vein endothelial cells depolarized with $K^+$ [205, 207]. System $y^+$ has been shown to possess system $y^+$L-like activity, whereas $y^+$LAT1 and $y^+$LAT2 mRNAs have been found in human umbilical vein endothelial cells [208, 209]. LAT1 is expressed in placental membranes supplying the developing fetus and the placenta with amino acids.

Aspartate and glutamate, two anionic amino acids, have been employed as tools to study the $Na^+$-dependence of anionic amino acid transport in human umbilical vein endothelial cells [210]. The results indicated that the entry of L-aspartate is predominantly $Na^+$-dependent and regulated by system $X^-AG$ [205].

### 19.4.3 Transporters for Aortic Endothelium

Significant uptake of large neutral, small neutral, cationic, and acidic amino acids was found in bovine and porcine aortic endothelial cells [211, 212]. L-Arginine transport kinetics in aortic endothelial cells was described by the Michaelis–Menten equation. The transport of L-arginine in aortic endothelial cells was sensitive to trans-stimulation and alterations in membrane potential. In addition, it was not inhibited by the extracellular pH, $Na^+$, or $Ca^{2+}$. On the other hand, the L-arginine transport was inhibited by L-lysine, L-ornithine, but not by MeAIB, BCH, 6-diazo-5-oxo-norleucine, L-glutamine, L-phenylalanine, L-cysteine, or L-glutamate [211, 213].

L-Arginine transport is regulated by system $y^+$ and coupled to the $Na^+$-dependent system $B^{0,+}$. Passive diffusion contributes only minimally to L-arginine influx into the cells [214]. System $b^{0,+}$ and $y^+L$ are not operative in bovine arotic endothelial cells [16].

A taurine transporter in bovine aortic endothelial cells shares a high degree of sequence homology with taurine transporters that have been cloned from mammalian brain, retina, and placenta [215, 216]. Influx of taurine is saturable and inhibited by β-alanine and homotaurine with a requirement of at least two $Na^+$ and one $Cl^-$ to transport taurine. The prolonged exposure of endothelial cells to elevated glucose may lead to depletion of intracellular taurine and apoptosis, resulting in a downregulation of the taurine transporter by glucose, which was similar to findings in retinal pigment epithelial cells [217, 218].

### 19.4.4 Transporters for Pulmonary Endothelium

The uptake of initial rapid tracer on L-alanine, glycine, L-methionine, L-phenylalanine, L-lysine, taurine, and L-glutamate has been determined to be in the range from 2% to 25%. The uptake was insensitive to inhibition by competitive amino acids, which suggested that endothelial cells in the intact lung did not rapidly take up amino acids. However, the transport of amino acids has been identified in cultured endothelial cells isolated from bovine and porcine pulmonary arteries. The uptake of L-glutamate in bovine pulmonary artery endothelial cell monolayers was not saturated up to 6 mM, was partially $Na^+$-dependent, and inhibited by L-aspartate and cysteate but not by MeAIB [219]. The inhibition of L-glutamine transport was insensitive to pH changes or regulators such as insulin, glucagon, or dexamethasone. L-Glutamine transport in pulmonary artery endothelial cells is regulated by the $Na^+$-dependent system ASC. In contrast, endothelial cells cultured from the rat lung microvasculature express system-N-like transport activity [220], which suggested that amino acid transporters are possibly expressed differentially in macro- and microvascular endothelium.

Approximately 70% of L-arginine transport in porcine pulmonary artery endothelial cells is mediated by the $Na^+$-independent systems $y^+$ and $b^{0,+}$. A high-affinity $Na^+$-dependent transport system resembling system $B^{0,+}$, is responsible for ~20–40% of total L-arginine uptake [221]. Cationic L-arginine analogs are potent inhibitors of NOS (NO synthase) and system $y^+$ in porcine pulmonary artery endothelial cells [222]. L-NMMA, *N*-amino-L-homoarginine, and *N*-iminoethyl-L-lysine inhibited system $y^+$ in endothelial cells. $Na^+$-independent L-arginine transport is regulated predominantly by the CAT family.

Transport of L-cystine and L-glutamate was reduced in bovine pulmonary artery endothelial cells exposed to diethyl maleate or DEM (an electrophilic agent to enzyme conjugate with GSH) or hyperoxia [223]. The increase in system $x_c^-$ transport activity was associated with an elevation in intracellular GSH, and both responses were dependent on protein and RNA synthesis. The induction of L-cystine transport by DEM or hyperoxia was not triggered by an initial depletion of GSH in pulmonary artery endothelial cells, which were exposed to arsenite, disulfiram, or the heavy metal cadmium [224]. The inhibition of GSH reductase with *N,N*-bis(2-chloroethyl)-*N*-nitrosourea (24h) is concentration-dependent and increases in L-cystine transport and GSH levels in pulmonary artery endothelial cells [224].

## 19.5 Regulation of Amino Acid Transport

### 19.5.1 Regulating Effects of Vasoactive Agonists

NO, an important vasoactive agonist, is produced from the essential amino acid L-arginine through the

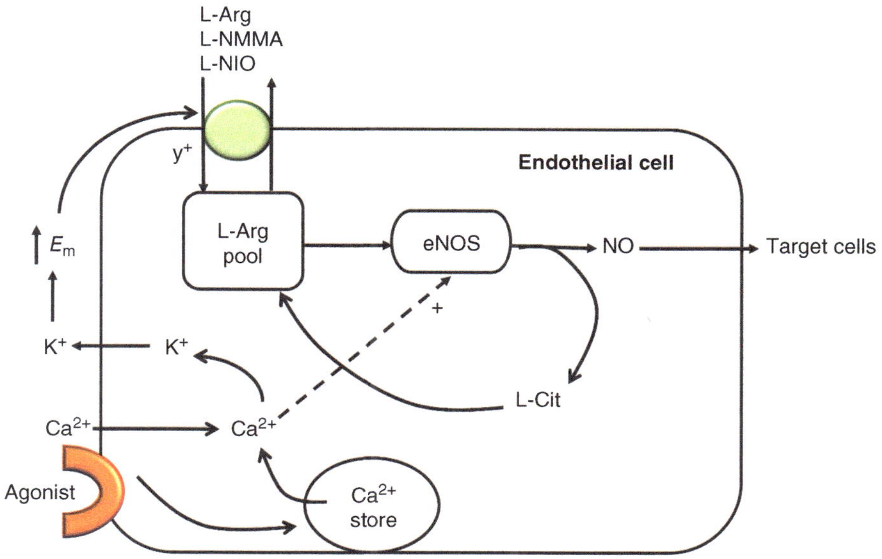


**Figure 19.4** The possible relationship between L-arginine transporter via system y+ and NO release in vascular endothelial cells. L-Arg, L-arginine; L-Cit, L-citrulline; $E_m$, hyperpolarization of the cell membrane. Source: Adapted from Bogle et al. [229].

catalytic action of NOS [225]. Transport of L-arginine is primarily dependent on the y+ transport system, which includes the CAT family of proteins [226]. Cellular L-arginine could be acutely activated by $Ca^{2+}$ and cause ionophore-induced NO release from L-arginine-deprived porcine aortic endothelial cells in 24 hours [227]. The endothelial cells release $^{15}NO$ when induced by bradykinin in the presence of [$^{15}N$] arginine, which showed that exogenous L-arginine is employed for NO synthesis. Brief exposure of L-arginine-deprived porcine aortic endothelial cells to bradykinin or ATP transiently activates the influx of L-arginine and triggers NO release [228]. Stimulation of endothelial cells with agonist elevates intracellular $Ca^{2+}$ resulting in activation of particulate eNOS (endothelial NO synthase), conversion of L-arginine into NO and L-citrulline and opening of $Ca^{2+}$-activated $K^+$ channels (Figure 19.4, [229]). In endothelial cells, constitutive activity of Kir-type $K^+$ channels and the activation of other $K^+$ channels by vasoactive agonists such as bradykinin and ATP induce hyperpolarization [230, 231]. The transport of L-arginine was reduced by 70 mM $K^+$ in cells, while agonist-induced NO and $PGI_2$ releases were still detected [229]. Inhibition of NO production by L-NNA decreased the uptake of L-arginine evoked by bradykinin [229], indicating that NO could modulate the transport of L-arginine.

Histamine does not have any effects on the transport of L-arginine [232, 233] in umbilical vein endothelial cells, but it markedly enhances intracellular accumulation of $Ca^{2+}$ and release of NO and $PGI_2$. Although a definitive relationship between L-arginine transport and NO production has not been determined in endothelial cells, system $y^+$ and eNOS are thought to control plasma membrane caveolae in porcine pulmonary artery endothelial cells [234]. Moreover, endothelial cells have a localized pool for L-arginine [105, 235], which is able to preferentially delivery the extracellular L-arginine to eNOS. Therefore, the acute rate of L-arginine utilization by cells could reflect in part the rate of refilling the localized L-arginine pool via CAT transporters.

NO production in vascular cells is important to the delivery of L-arginine. NO production by cytokine-activated macrophages is virtually silenced in $CAT^{2+}/^{-}$ knockout mice [236]. Moreover, both basal and ADP-stimulated NO production in platelets from chronic renal failure patients is dependent on enhancing the transport of L-arginine by the high-affinity amino acid transport system $y^+L$. Additionally, the roles of systems $y^+$ and $y^+L$ in regulating the delivery of L-arginine for NO production in different blood cells and vascular endothelial and smooth muscle cells have also been elucidated [16].

### 19.5.2 Regulation of Amino Acid Transport by Ischemia and Hypoxia

Since both CAT-1 and eNOS are located in the inner structure of endothelial cells, when the hypoxia environment affects the expression of cytoskeleton filaggrin, the quasi-transport of arginine and the production of NO are simultaneously blocked [237]. BBB transport of LNAA is mediated by LAT1. It has been found that the mRNA expression of LAT1 was finally downregulated due to hypoxia caused by an initial destabilization of mRNA [238]. The expression of LAT1 mRNA in T98G and A172 glioblastoma cells *in vitro* decreased significantly after hypoxia [239]. Under cerebral ischemia and hypoxia, a large amount of $Ca2^+$ influxes and activates phospholipase A2, and then releases a large amount of free arachidonic acid, which can significantly reduce the activity of glutamate transporter [240]. In certain brain areas, such as hippocampal CA1 area, the protein and mRNA levels of GLT-1 are significantly reduced after the ischemia and hypoxia environment [241, 242].

### 19.5.3 Regulation of Amino Acid Transport by Cytokines

Studies have shown that cytokines are also widely involved in the regulation of amino acid transporters. Cytokines can increase the expression of CAT2B and iNOS in rat aorta, and the expression of $y^+$ amino acid transport system on red blood cells or platelets [243–245]. In porcine vascular endothelial cells, LPS can activate the CAT by stimulating the autocrine function of TNF-α and IL-1. When TNF-α and IL-1 receptors are inhibited, the enhancement effect of LPS on the transport system begins to weaken [246]. At the same time, LPS or cytokines can also activate the cation transporter system to promote the transport of L-arginine on endothelial cells [247]. TNF-α can enhance the expression of L-cystine transporter, acidic amino acid transporter, and CAT2B transporter [54, 248]. Interleukin IL-1β and interferon-γ (IFN-γ) can co-stimulate rat cardiac microvascular endothelial cells, upregulate the expression of CAT family, and increase arginine transport [249].

## 19.6 Conclusion

Multiple transporter systems are responsible for the transport of amino acids across the apical and basolateral membrane of renal and intestinal epithelial cells and endothelial cells. Individual amino acids are absorbed by more than one transporter, which maintains adequate amino acid transport in case a transport system is inactivated or greatly downregulated. The transport of amino acids in kidney and intestine is of critical importance to the supply of neutral amino acids to all tissues and the homeostasis of plasma amino acids acid levels. The identification of most epithelial and endothelial amino acid transporters in the recent decades has provided an ever more clear picture, at the molecular level, of each amino acid transporter's functional activities on the apical and basolateral membranes of mammalian cells. It is now clear that amino acid transport in the apical membrane not only depends on metal ions such as $Na^+$ and $K^+$ but also employs the proton ($H^+$) gradient and gradient of other amino acids to efficiently transport amino acids across the luminal walls. Similarly, on the basolateral membrane, antiporters, in concert with other carriers, ensure that amino acids are released without depleting valuable nutrients in cells.

Currently, the molecular cloning and identification of all amino acid transporter in kidney, intestine, and other tissues have been demonstrated. However, a few amino acid transporters still need to be identified. For example, the $Na^+$-activated CAT, which may be related with rBAT in the proximal straight tubule. Compared with the luminal amino acid influx step, the efflux step across the basolateral membrane is not as well-known. The mechanism that determines how intracellular neutral amino acids activate efflux of cationic amino acids remains also to be resolved, and how glycine, proline, and the β-amino acids are transported across the basolateral membrane still remains to be clearly identified.

There is no doubt that whatever direction this field takes; further studies will be required to develop an entirely new or at least significantly improved understanding on the complex functions of each amino acid transporters on the epithelial membrane, the protein conformation changes associated with each amino acid transporters, expression of transporters and their physiological stimulation, and lastly tissue-, cell-, and developmental stage-specific expression. We expect to see more advancement in the amino acid transporter studies in the future.

## Abbreviations

| | |
|---|---|
| ABC | ATP-binding cassette |
| SLCs | the solute carriers |
| EAAT | excitatory amino acid transporter |
| TDM | trehalose dimycolate |
| BCH | 2-amino-2-norbornane-carboxylic acid |
| BBMV | brush-border membrane vesicles |
| CAT | cationic amino acid transporter |
| L-DOPA | levodopa |
| BBB | blood–brain barrier |
| GABA | γ-aminobutyric acid B-subtype receptor |

| | |
|---|---|
| OAT | organic anion transport |
| LNAA | large neutral amino acids |
| BCH | Cdc42GAP homology domain |
| L-NMMA | N(G)-monomethyl-L-arginine |
| L-NAME | N(G)-nitro-L-arginine methyl ester |

## References


1 Kaplan, J. (1993). Molecular biology of carrier proteins. *Cell* 72: 13–18.

2 Dassa, E. and Bouige, P. (2001). The ABC of ABCS: a phylogenetic and functional classification of ABC systems in living organisms. *Res. Microbiol.* 152: 211–229.

3 Vasiliou, V., Vasiliou, K., and Nebert, D.W. (2009). Human ATP-binding cassette (ABC) transporter family. *Hum. Genom.* 3: 281–290.

4 Schaller, L. and Lauschke, V.M. (2019). The genetic landscape of the human solute carrier (SLC) transporter superfamily. *Hum. Genet.* 138: 1359–1377.

5 Endou, H. (2000). Molecular mechanisms of drug transport. *Nippon Yakurigaku Zasshi* 116: 114–124.

6 Adibi, S., Gray, S., and Menden, E. (1967). The kinetics of amino acid absorption and alteration of plasma composition of free amino acids after intestinal perfusion of amino acid mixtures. *Am. J. Clin. Nutr.* 20: 24–33.

7 Christensen, H.N. (1984). Organic ion transport during seven decades. The amino acids. *Biochim. Biophys. Acta* 779: 255–269.

8 Christensen, H.N. (1990). Role of amino acid transport and counter-transport in nutrition and metabolism. *Physiol. Rev.* 70: 43–77.

9 Silbernagl, S., Foulkes, E., and Deetjen, P. (1975). Renal transport of amino acids. *Rev. Physiol., Biochem. Pharmacol.* 74: 105–167.

10 Stevens, B., Ross, H., and Wright, E. (1982). Multiple transport pathways for neutral amino acids in rabbit jejunal brush border vesicles. *J. Membr. Biol.* 66: 213–225.

11 Kim, D., Kanai, Y., Choi, H. et al. (2002). Characterization of the system L amino acid transporter in T24 human bladder carcinoma cell. *Biochim. Biophys. Acta* 1565: 112–122.

12 Fleck, C., Schwertfeger, M., and Taylor, P. (2003). Regulation of renal amino acid (AA) transport by hormones, drugs and xenobiotics-a review. *Amino Acids* 24: 347–374.

13 Bröer, S. and Fairweather, S. (2018). Amino acid transport across the mammalian intestine. *Compr. Physiol.* 9: 343–373.

14 Yahyaoui, R. and Pérez-Frías, J. (2019). Amino acid transport defects in human inherited metabolic disorders. *Int. J. Mol. Sci.* 21: 119.

15 Kandasamy, P., Gyimesi, G., Kanai, Y., and Hediger, M.A. (2018). Amino acid transporters revisited: new views in health and disease. *Trends Biochem. Sci* 43: 752–789.

16 Giovanni, E., Mann, D.L., and Yudilevich, L.S. (2003). Regulation of amino acid and glusoce transporters in endothelial and smooth muscle cells. *Physiol. Res.* 83: 183–252.

17 Bodoy, S., Fotiadis, D., Stoeger, C. et al. (2013). The small SLC43 family: facilitator system l amino acid transporters and the orphan EEG1. *Mol. Aspects Med.* 34: 638–645.

18 Bröer, S. (2008). Amino acid transport across mammalian intestinal and renal epithelia. *Physiol. Rev.* 88: 249–286.

19 Cherqui, S. and Courtoy, P.J. (2017). The renal Fanconi syndrome in cystinosis: pathogenic insights and therapeutic perspectives. *Nat. Rev. Nephrol.* 13: 115–131.

20 Fiermonte, G., Palmieri, L., Todisco, S. et al. (2002). Identification of the mitochondrial glutamate transporter. Bacterial expression, reconstitution, functional characterization, and tissue distribution of two human isoforms. *J. Biolumin. Chemilumin.* 277: 19289–19294.

21 Gasnier, B. (2004). The SLC32 transporter, a key protein for the synaptic release of inhibitory amino acids. *Pflugers Arch.* 447: 756–759.

22 Schiöth, H.B., Roshanbin, S., Hägglund, M.G., and Fredriksson, R. (2013). Evolutionary origin of amino acid transporter families SLC32, SLC36 and SLC38 and physiological, pathological and therapeutic aspects. *Mol. Aspects Med.* 34: 571–585.

23 Takanaga, H., Mackenzie, B., Suzuki, Y., and Hediger, M.A. (2005). Identification of mammalian proline transporter SIT1 (SLC6A20) with characteristics of classical system imino. *J. Biolumin. Chemilumin.* 280: 8974–8984.

24 Thwaites, D.T. and Anderson, C.M. (2011). The SLC36 family of proton-coupled amino acid transporters and their potential role in drug transport. *Br. J. Pharmacol.* 164: 1802–1816.

25 Oxender, D.L. and Christensen, H.N. (1963). Evidence for two types of mediation of neutral

amino acid transporter in Ehrlich cells. *Nature* 197: 765–767.

26 Barker, G.A. and Ellory, J.C. (1990). The identification of neutral amino acid transport system. *Exp. Phaysiol.* 75: 3–26.

27 Rosenberg, R., Young, J.D., and Ellory, J.C. (1980). L-Tryptophan transport in human red blood cells. *Biochin. Biophys. Acta* 598: 375–384.

28 MacLeod, C.L., Finley, K.D., and Kakuda, D.K. (1994). y+-Type cation amino acid transporter: expression and regulation of the mCAT genes. *J. Exp. Biol.* 196: 109–121.

29 Devés, R., Chavez, P., and Boyd, C. (1992). Identification of a new transport system (y+L) in human erythrocytes that recognizes lysine and leucine with high affinity. *J Physiol* 454: 491–501.

30 Van Winkle, L., Campione, A., and Gorman, J. (1988). Na+-independent transport of basic and zwitterionic amino acids in mouse blastocysts by a shared system and by processes which distinguish between these substrates. *J. Biolumin. Chemilumin.* 263: 3150–3163.

31 Bannai, S. (1984). Transport of cystine and cysteine in mammalian cells. *Biochim. Biophys. Acta* 779: 289–306.

32 Siow, R., Sato, H., Leake, D. et al. (1998). Vitamin C protects human arterial smooth muscle cells against atherogenic lipoproteins: effects of antioxidant vitamins C and E on oxidized LDL-induced increases in cystine transport and glutathione. *Arterioscler. Thromb. Vasc. Biol.* 18: 1662–1670.

33 Sweiry, J.H., Sastre, J., Viña, J. et al. (1995). A role for gamma-glutamyl-transpeptidase and the amino acid transport system $xc^-$ in cystine transport by a human pancreatic duct cell line. *J Physiol* 485: 167–177.

34 Verrey, F., Singer, D., Ramadan, T. et al. (2009). Kidney amino acid transport. *Pflugers Arch.* 458: 53–60.

35 Hammerman, M. (1982). $Na^+$-independent L-arginine transport in rabbit renal brush border membrane vesicles. *Biochim. Biophys. Acta* 685: 71–77.

36 Stieger, B., Stange, G., Biber, J., and Murer, H. (1983a). Transport of L-cysteine by rat renal brush border membrane vesicles. *J. Membr. Biol.* 73: 25–37.

37 Bhutia, Y.D., Babu, E., Ramachandran, S., and Ganapathy, V. (2015). Amino acid transporters in cancer and their relevance to "glutamine addiction": novel targets for the design of a new class of anticancer drugs. *Cancer Res.* 75: 1782–51788.

38 Kim, J.W., Closs, E.I., Albritton, L.M., and Cunningham, M. (1991). Transport of cationic amino acids by the mouse ecotropic retrovirus receptor. *Nature* 352: 725–728.

39 Wang, H., Kavanaugh, M.P., North, R.A., and Kabat, D. (1991). Cell-surface receptor for ecotropic murine retroviruses is a basic amino acid transporter. *Nature* 352: 729–731.

40 Kavanaugh, M.P. (1993). Voltage dependence of facilitated arginine flux mediated by the system y+ basic amino acid transporter. *Biochemistry* 32: 5781–5785.

41 Sperandeo, M.P., Borsani, G., Incerti, B. et al. (1998). The gene encoding a cationic amino acid transporter (SLC7A4) maps to the region deleted in the velocardiofacial syndrome. *Genomics* 49: 230–236.

42 Reade, M., Clark, M., Young, J., and Boyd, C. (2002). Increased cationic amino acid flux through a newly expressed transporter in cells overproducing nitric oxide from patients with septic shock. *Clin. Sci.* 102: 645–650.

43 Hosokawa, H., Sawamura, T., Kobayashi, S. et al. (1997). Cloning and characterization of a brain-specific cationic amino acid transporter. *J. Biolumin. Chemilumin.* 272: 8717–8722.

44 Ito, A., Tsao, P., Adimoolam, S. et al. (1999). Novel mechanism for endothelial dysfunction: dysregulation of dimethylarginine dimethylaminohydrolase. *Circulation* 99: 3092–3095.

45 Devés, R. and Boyd, C.A. (2000). Surface antigen CD98(4F2): not a single membrane protein, but a family of proteins with multiple function. *J. Membr. Biol.* 173: 165–177.

46 Wagner, C., Lang, F., and Bröer, S. (2001). Function and structure of heterodimeric amino acid transporters. *Am. J. Physiol. Cell Physiol.* 281: C1077–C1093.

47 Estévez, R., Camps, M., Rojas, A. et al. (1998). The amino acid transport system y+L/4F2hc is a heteromultimeric complex. *FASEB J* 12: 1319–1329.

48 Nakamura, E., Sato, M., Yang, H. et al. (1999). 4f2 (Cd98) heavy chain is associated covalently with an amino acid transporter and controls intracellular trafficking and membrane topology of 4F2 heterodimer. *J. Biolumin. Chemilumin.* 274: 3009–3016.

49 Pfeiffer, R., Rossier, G., Spindler, B. et al. (1999). Amino acid transport of y+L-type by heterodimers of 4F2hc/CD98 and members of the glycoprotein-associated amino acid transporter family. *EMBO J.* 18: 49–57.

50 Xu, D. and Hemler, M.E. (2006). Metabolic activation-related CD147-CD98 complex. *Mol. Cell. Proteomics* 4: 1061–1071.

51 Pineda, M., Fernandez, E., Torrents, D. et al. (1999). Identification of a membrane protein, LAT-2, that co-expresses with 4F2 heavy chain, an L-type amino acid transport activity with broad specificity for small and large zwitterionic amino acids. *J. Biolumin. Chemilumin.* 274: 19738–19744.

52 Segawa, H., Fukasawa, Y., Miyamoto, K. et al. (1999). Identification and functional characterization of a Na+-independent neutral amino acid transporter with broad substrate selectivity. *J. Biolumin. Chemilumin.* 274: 19745–19751.

53 Capbell, W. and Thompson, N. (2001). Overexpression LAT1/CD98 light chain is sufficient to increase system L amino acid transport activity in mouse hepatocytes but not fibroblasts. *J. Biolumin. Chemilumin.* 276: 16877–16884.

54 Bannai, S. and Tateishi, N. (1986). Role of membrane transport in metabolism and function of glutathione in mammals. *J. Membr. Biol.* 89: 1–8.

55 Ishii, T., Itoh, K., Takahashi, S. et al. (2000). Transcription factor Nrf2 coordinately regulates a group of oxidative stress-inducible genes in macrophages. *J. Biolumin. Chemilumin.* 275: 16023–16029.

56 Palacín, M., Estevez, R., Bertran, J., and Zorzana, A. (1998). Molecular biology of mammalian plasma membrane amino acid transporters. *Physiol. Rev.* 78: 969–1054.

57 Bertan, J., Magagnin, S., Werner, A. et al. (1992). Stimulation of system y(+)-like amino acid transport by the heavy chain of human 4F2 surface antigen in *Xenopus laevis* oocytes. *Proc. Natl. Acad. Sci. U.S.A.* 89: 5606–5610.

58 Fotiadis, D., Kanai, Y., and Palacín, M. (2013). The SLC3 and SLC7 families of amino acid transporters. *Mol. Aspects Med.* 34: 139–158.

59 Van Winkle, L., Campone, A., and Farrington, B. (1990). Development of system $B^{0,+}$ and a broad-scope Na+-dependent transporter of zwitterionic amino acids in preimplantation mouse conceptuses. *Biochim. Biophys. Acta* 1025: 225–233.

60 Sloan, J. and Mager, S. (1999). Cloning and functional expression of a human Na+ and Cl+-dependent neutral and cationic amino acid transporter $B^{0,+}$. *J. Biolumin. Chemilumin.* 274: 23740–23745.

61 Albers, A., Broer, A., Wagner, C. et al. (2001). Na+ transport by the neural glutamine transporter ATA1. *Pflugers Arch.* 443: 92–101.

62 Varoqui, H., Zhu, H., Yao, D. et al. (2000). Cloning and functional identification of a neuronal glutamine transporter. *J. Biolumin. Chemilumin.* 275: 4049–4054.

63 Wang, H., Huang, W., Sugawara, M. et al. (2000). Cloning and functional expression of ATA1, a subtype of amino acid transporter A, from human placenta. *Biochem. Biophys. Res. Commun.* 273: 1175–1179.

64 Sugawara, M., Nakanishi, T., Fei, Y. et al. (2000). Structure and function of ATA3, a new subtype of amino acid transport system A, primarily expressed in the liver and skeletal muscle. *Biochim. Biophys. Acta* 1509: 7–13.

65 Yao, D., Mackenzie, B., Ming, H. et al. (2000). A novel system A isoform mediating Na+/neutral amino acid cotransport. *J. Biolumin. Chemilumin.* 275: 22790–22797.

66 Hatanaka, T., Huang, W., Ling, R. et al. (2001). Evidence for the transport of neutral as well as cationic amino acids by ATA3, a novel and liver-specific subtype of amino acid transport system A. *Biochim. Biophys. Acta* 1510: 10–17.

67 Fukasawa, Y., Segawa, H., Kim, J. et al. (2000). Identification and characterization of a $Na^+$-independent neutral amino acid transporter that associates with the 4F2 heavy chain and exhibits substrate selectivity for small neutral D- and L-amino acids. *J. Biolumin. Chemilumin.* 275: 9690–9698.

68 Nakauchi, J., Matsuo, H., Kim, D. et al. (2000). Cloning and characterization of a human brain Na+-independent transporter for small neutral amino acids that transports D-serine with high affinity. *Neurosci. Lett.* 287: 231–235.

69 Matsuo, H., Kanai, Y., Kim, J. et al. (2002). Identification of a novel Na+-independent acidic amino acid transporter with structural similarity to the member of a heterodimeric amino acid transporter family associated with unknown heavy chains. *J. Biolumin. Chemilumin.* 277: 21017–21026.

70 Kilberg, M., Handlogten, M., and Christensen, H. (1981). Characteristics of system ASC for transport of neutral amino acids in the isolated rat hepatocyte. *J. Biolumin. Chemilumin.* 256: 3304–3312.

71 Malandro, M. and Kilberg, M. (1996). Molecular biology of mammalian amino acid transporters. *Annu. Rev. Biochem.* 65: 305–336.

72 Vadgama, J. and Christensen, H. (1984). Wide distribution of pH-dependent service of transport system ASC for both anionic and zwitterionic amino acids. *J. Biolumin. Chemilumin.* 259: 3648–3652.

73 Shafqat, S., Tamarappoo, B., Kilberg, M. et al. (1993). Cloning and expression of a novel Na+-dependent neutral amino acid transporter structurally related to mammalian Na+/glutamate cotransporters. *J. Biolumin. Chemilumin.* 268: 15351–15355.

74 Weiss, M., Derazi, S., Kilberg, M., and Anderson, K. (2001). Ontogeny and localization of the neutral amino acid transporter ASCT1 in rat brain. *Brain Res.* 130: 183–190.

75 Utsunomiya, N., Endou, H., and Kanai, Y. (1996). Cloning and functionalcharacterization of a system ASC-like Na-dependent neutral amino acid transporter. *J. Biolumin. Chemilumin.* 271: 14883–14890.

76 Zerangue, H. and Kavanaugh, M. (1996). Asct-1 is a neutral amino acid exchanger with chloride channel activity. *J. Biolumin. Chemilumin.* 271: 27991–27994.

77 Ahmed, A., Maxwell, D., Taylor, P., and Rennie, M. (1993). Glutamine transport in human skeletal muscle. *Am. J. Physiol. Endocrinol. Metab* 264: 993–1000.

78 Hundal, H., Rennie, M., and Watt, P. (1987). Characteristics of L-glutamine transport in perfused rat skeletal muscle. *J Physiol* 39: 283–305.

79 Nakanishi, T., Sugawara, M., Huang, W. et al. (2001). Structure, function and tissue expression pattern of human SN2, a subtype of the amino acid transport system N. *Biochem. Biophys. Res. Commun.* 281: 1343–1348.

80 Fei, Y., Sugawara, M., Nakanishi, T. et al. (2000). Primary structure, genomic organization, and functional and electrogenic characteristics of human system N1, a Na+- and H+-coupled glutamine transporter. *J. Biolumin. Chemilumin.* 275: 23707–23717.

81 Hägglund, M., Sreedharan, S., Nilsson, V. et al. (2011). Identification of SLC38A7 (SNAT7) protein as a glutamine transporter expressed in neurons. *J. Biolumin. Chemilumin.* 286: 20500–20511.

82 Marchesi, C., Asta, V., Rotoli, B. et al. (2006). Chlorpromazine, clozapine and olanzapine inhibit anionic amino acid transport in cultured human fibroblasts. *Amino Acids* 31: 93–99.

83 Kragh-Hansen, U., Roigaard-Petersen, H., Jacobsen, C., and Sheikh, M. (1984). Renal transport of neutral amino acids. Tubular localization of Na+-dependent phenylalanine- and glucose-transport systems. *Biochem. J* 220: 15–24.

84 Hoyer, J. and Gogelein, H. (1991). Sodium-alanine cotransport in renal proximal tubule cells investigated by whole-cell current recording. *J. Gen. Physiol.* 97: 1073–1094.

85 Young, J. and Edwards, K. (1964). Studies on the absorption, metabolism and excretion of methyldopa and other catechols and their influence on amino acid transport in rats. *J. Pharmacol. Exp. Ther.* 145: 102–112.

86 Webber, W., Brown, J., and Pitts, R. (1961). Interactions of amino acids in renal tubular transport. *Am. J. Phys* 200: 380–386.

87 Mircheff, A., Kippen, I., Hirayama, B., and Wright, E. (1982). Delineation of sodium-stimulated amino acid transport pathways in rabbit kidney brush border vesicles. *J. Membr. Biol.* 64: 113–122.

88 Fass, S., Hammerman, M., and Sacktor, B. (1977). Transport of amino acids in renal brush border membrane vesicles. Uptake of the neutral amino acid L-alanine. *J. Biolumin. Chemilumin.* 252: 583–590.

89 Kragh-Hansen, U. and Sheikh, M. (1984). Serine uptake by luminal and basolateral membrane vesicles from rabbit kidney. *J Physiol* 354: 55–67.

90 Jorgensen, K., Kragh-Hansen, U., and Sheikh, M. (1990). Transport of leucine, isoleucine and valine by luminal membrane vesicles from rabbit proximal tubule. *J Physiol* 422: 41–54.

91 Jessen, H. and Sheikh, M. (1991). Renal transport of taurine in luminal membrane vesicles from rabbit proximal tubule. *Biochim. Biophys. Acta* 1064: 189–198.

92 Pontoglio, M., Barra, J., Hadchouel, M. et al. (1996). Hepatocyte nuclear factor 1 inactivation results in hepatic dysfunction, phenylketonuria, renal Fanconi syndrome. *Cell* 84: 575–585.

93 Sepulveda, F. and Smith, M. (1978). Discrimination between different entry mechanisms for neutral amino acids in rabbit ileal mucosa. *J Physiol* 282: 73–90.

94 Paterson, J., Sepulveda, F., and Smith, M. (1980b). Stoichiometry versus coupling ratio in the cotransport of Na and different neutral amino acids. *Biochim. Biophys. Acta* 603: 288–297.

95 Munck, B. and Munck, L. (1994a). Phenylalanine transport in rabbit small intestine. *J Physiol* 480: 99–107.

96 Stevens, B.R., Kaunitz, J.D., and Wright, E.M. (1984). Intestinal transport of amino acids and sugars: advances using membrane vesicles. *Annu. Rev. Physiol.* 46: 417–433.

97 Windmueller, H. (1982). Glutamine utilization by the small intestine. *Adv. Enzymol. Relat. Areas Mol. Biol.* 53: 201–237.

98 Wu, G. (1998). Intestinal mucosal amino acid catabolism. *J. Nutr.* 128: 1249–1252.

99 Al-Mahroos, F., Abumrad, N., and Ghishan, F. (1990). Developmental changes in glutamine transport by rat jejunal basolateral membrane vesicles. *Proc. Soc. Exp. Biol. Med.* 194: 186–192.

100 Evers, J., Murer, H., and Kinne, R. (1976). Phenylalanine uptake in isolated renal brush border vesicles. *Biochim. Biophys. Acta* 426: 598–615.

101 Lash, L. and Jones, D. (1984). Characteristics of cysteine uptake in intestinal basolateral membrane vesicles. *Am. J. Physiol. Gastrointest. Liver Physiol.* 247: G394–G401.

102 Fincham, D., Mason, D., and Young, J. (1985). Characterization of a novel Na+-independent amino acid transporter in horse erythrocytes. *Biochem. J* 227: 13–20.

103 Kuhlmann, M. and Vadgama, J. (1999). Characterization of threonine transport into a kidney epithelial cell line (BSC-1). Evidence for the presence of Na+-independent system asc. *J. Biolumin. Chemilumin.* 266: 15042–15047.

104 Reynolds, R., Wald, H., McNamara, P., and Segal, S. (1980). An improved method for isolation of basolateral membranes from rat kidney. *Biochim. Biophys. Acta* 601: 92–100.

105 Closs, E.I., Scheld, J.S., Sharafi, M., and Forstermann, U. (2000). Substrate supply for nitric oxide synthase in macrophages and endothelial cells: role for cationic amino acid transporters. *Mol. Pharmacol.* 57: 68–74.

106 Bowring, M., Foreman, J., Lee, J., and Segal, S. (1987). Characteristics of lysine transport by isolated rat renal cortical tubule fragments. *Biochim. Biophys. Acta* 901: 23–29.

107 McNamara, P., Rea, C., and Segal, S. (1986). Lysine uptake by rat renal brush-border membrane vesicles. *Am. J. Physiol. Renal. Fluid Electrolyte Physiol.* 251: F734–F742.

108 Schwartzman, L., Blair, A., and Segal, S. (1967). Exchange diffusion of dibasic amino acids in rat-kidney cortex slices. *Biochim. Biophys. Acta* 135: 120–126.

109 Wilson, O. and Scriver, C. (1967). Specificity of transport of neutral and basic amino acids in rat kidney. *Am. J. Phys* 213: 185–190.

110 Ruszkowski, M., Arasimowicz, C., Knapowski, J. et al. (1962). Renal reabsorption of amino acids. *Am. J. Phys* 203: 891–896.

111 Rosenberg, L., Downing, S., and Segal, S. (1962). Competitive inhibition of dibasic amino acid transport in rat kidney. *J. Biolumin. Chemilumin.* 237: 2265–2270.

112 Crawhall, J. and Segal, S. (1967). The intracellular ratio of cysteine and cystine in various tissues. *Biochem. J* 105: 891–896.

113 Mcnamara, P., Pepe, L., and Segal, S. (1981). Cystine uptake by rat renal brush-border vesicles. *Biochem. J* 194: 443–449.

114 Stieger, B., Stange, G., Biber, J., and Murer, H. (1983b). Transport of L-lysine by rat renal brush border membrane vesicles. *Pflugers Arch.* 397: 106–113.

115 Munck, B.G. (1981). *Intestinal Absorption of Amino Acids*. New York: Raven Press.

116 Munck, L. and Munck, B. (1994b). Amino acid transport in the small intestine. *Physiol. Res.* 43: 335–345.

117 Munck, B. and Schultz, S. (1969). Lysine transport across isolated rabbit ileum. *J. Gen. Physiol.* 53: 157–182.

118 Cassano, G., Leszczynska, B., and Murer, H. (1983). Transport of L-lysine by rat intestinal brush border membrane vesicles. *Pflugers Arch.* 397: 114–120.

119 Wolfram, S., Giering, H., and Scharrer, E. (1984). $Na^+$-gradient dependence of basic amino acid transport into rat intestinal brush border membrane vesicles. *Comp. Biochem. Physiol. A* 78: 475–480.

120 Torras-Llort, M., Soriano-Garcia, J., Ferrer, R., and Moreto, M. (1998). Effect of a lysine-enriched diet on L-lysine transport by the brushborder membrane of the chicken jejunum. *Am. J. Physiol. Regul. Integr. Comp. Physiol.* 274: R69–R75.

121 Lawless, K., Maenz, D., and Cheeseman, C. (1987). Is leucine an allosteric modulator of the lysine transporter in the intestinal basolateral membrane? *Am. J. Physiol. Gastrointest. Liver Physiol.* 253: G637–G642.

122 Eleno, N., Deves, R., and Boyd, C. (1994). Membrane potential dependence of the kinetics of cationic amino acid transport systems in human placenta. *J Physiol* 479: 291–300.

123 Schneider, E., Hammerman, M., and Sacktor, B. (1980). Sodium gradient dependent L-glutamate transport in renal brush border membrane vesicles. Evidence for an electroneutral mechanism. *J. Biol. Chem.* 255: 7650–7656.

124 Weiss, S., McNamara, P., Pepe, L., and Segal, S. (1978). Glutamine and glutamic acid uptake by rat renal brushborder membrane vesicles. *J. Membr. Biol.* 43: 91–105.

125 Burckhardt, G., Kinne, R., Stange, G., and Murer, H. (1980). The effects of potassium and membrane potential on sodium-dependent glutamic acid uptake. *Biochim. Biophys. Acta* 599: 191–201.

126 Welbourne, T. and Matthews, J. (1999). Glutamate transport and renal function. *Am. J. Physiol. Renal. Physiol.* 277: F501–F505.

127 Kanner, B.I. (1983). Bioenergetics of neurotransmitter transport. *Biochim. Biophys. Acta* 726: 293–316.

128 Munck, B. and Munck, L. (1999). Effects of pH changes on systems ASC and B in rabbit ileum. *Am. J. Physiol. Gastrointest. Liver Physiol.* 276: G173–G184.

129 Maenz, D., Chenu, C., Breton, S., and Berteloot, A. (1992). pH-dependent heterogeneity of acidic amino acid transport in rabbit jejunal brush border membrane vesicles. *J. Biolumin. Chemilumin.* 267: 1510–1516.

130 Schultz, S. and Curran, P. (1970). Coupled transport of sodium and organic solutes. *Physiol. Rev.* 50: 637–718.

131 Schultz, S., Alvarez, O., Curran, P., and Yu-Tu, L. (1970). Dicarboxylic amino acid influx across brush border of rabbit ileum effects of amino acid charge on the sodium-amino acid interaction. *J. Gen. Physiol.* 56: 621–639.

132 Boyd, C. and Perring, V. (1981). Transamination and asymmetry in glutamate transport across the basolateral membrane of frog small intestine. *Biosci. Rep.* 1: 851–856.

133 Sacktor, B., Rosenbloom, I., Liang, C., and Cheng, L. (1981). Sodium gradient-and sodium plus potassium gradient-dependent L-glutamate uptake in renal basolateral membrane vesicles. *J. Membr. Biol.* 60: 63–71.

134 Scriver, C., Efron, M., and Schafer, I. (1964). Renal tubular transport of proline, hydroxyproline, glycine in health and in familial hyperprolinemia. *J. Clin. Invest.* 43: 374–385.

135 Scriver, C., Arthus, M., and Bergeron, M. (1982). Neonatal iminoglycinuria: evidence that the prolinuria originates in selective deficiency of transport activity in the proximal nephron. *Pediatr. Res.* 16: 684–687.

136 Baerlocher, K., Scriver, C., and Mohyuddin, F. (1971). The ontogeny of amino acid transport in rat kidney. II. Kinetics of uptake and effect of anoxia. *Biochim. Biophys. Acta* 249: 364–372.

137 Hillman, R. and Albrecht, I. (1968). Identification and analysis of multiple glycine transport systems in isolated mammalian renal tubules. *J. Biolumin. Chemilumin.* 243: 5566–5571.

138 McNamara, P., Ozegovic, B., Pepe, L., and Segal, S. (1976). Proline and glycine uptake by renal brush border membrane vesicles. *Proc. Natl. Acad. Sci. U.S.A.* 73: 4521–4525.

139 Medow, M., Roth, K., Goldmann, D. et al. (1986). Developmental aspects of proline transport in rat renal brush border membranes. *Proc. Natl. Acad. Sci. U.S.A.* 83: 7561–7564.

140 Wunz, T.M. and Wright, S.H. (1993). Betaine transport in rabbit renal brush-border membrane vesicles. *Am. Physiol. Jun.* 264 (6 Pt2): F948–F955.

141 Thwaites, D., McEwan, G., Brown, C. et al. (1993). Na+-independent, H+-coupled transepithelial beta-alanine absorption by human intestinal Caco-2 cell monolayers. *J. Biolumin. Chemilumin.* 268: 18438–18441.

142 Barfuss, D., Mays, J., and Schafer, J. (1980). Peritubular uptake and transepithelial transport of glycine in isolated proximal tubules. *Am. J. Physiol. Renal. Fluid Electrolyte Physiol.* 238: F324–F333.

143 Slack, E., Liang, T., and Sacktor, B. (1977). Transport of L-proline and D-glucose in luminal (brush border) and contraluminal (basal-lateral) membrane vesicles from the renal cortex. *Biochem. Biophys. Res. Commun.* 77: 891–897.

144 Davies, S., Maenz, D., and Cheeseman, C. (1987). A novel imino-acid carrier in the enterocyte basolateral membrane. *Biochim. Biophys. Acta* 896: 247–255.

145 Howard, A. and Hirst, B.H. (2011). The glycine transporter GLYT1 in human intestine: expression and function. *Biol. Pharm. Bull.* 34: 784–788.

146 Friedman, A., Jax, D., and Chesney, R. (1981). Developmental aspects of renal beta-amino acid transport. III. Ontogeny of transport in isolated renal tubule segments. *Pediatr. Res.* 15: 10–13.

147 Chesney, R., Gusowski, N., and Dabbagh, S. (1985). Renal cortex taurine content regulates renal adaptive response to altered dietary intake of sulfur amino acids. *J. Clin. Invest.* 76: 2213–2221.

148 Hammerman, M. and Sacktor, B. (1978). Transport of beta-alanine in renal brush border membrane vesicles. *Biochim. Biophys. Acta* 509: 338–347.

149 Turner, R.J. (1986). β-Amino acid transport across the renal brush-border membrane is coupled to both Na and Cl. *J. Biolumin. Chemilumin.* 261: 16060–16066.

150 Chesney, R., Gusowski, N., and Friedman, A. (1983). Renal adaptation to altered dietary sulfur amino acid intake occurs at luminal brushborder membrane. *Kidney Int.* 24: 588–594.

151 Rozen, R. and Scriver, C. (1982). Renal transport of taurine adapts to perturbed taurine homeostasis. *Proc. Natl. Acad. Sci. U.S.A.* 79: 2101–2105.

152 Miyamoto, Y., Tiruppathi, C., Ganapathy, V., and Leibach, F. (1989). Active transport of taurine in

rabbit jejunal brush-border membrane vesicle. *Am. J. Physiol. Gastrointest. Liver Physiol.* 257: G65–G72.

153 Moyer, M., Goodrich, A., Rolfes, M., and Suchy, F. (1988). Ontogenesis of intestinal taurine transport: evidence for a beta-carrier in developing rat jejunum. *Am. J. Physiol. Gastrointest. Liver Physiol.* 254: G870–G877.

154 Miyamoto, Y., Nakamura, H., Hoshi, T. et al. (1990). Uphill transport of beta-alanine in intestinal brush-border membrane vesicles. *Am. J. Physiol. Gastrointest. Liver Physiol.* 259: G372–G379.

155 Jones, D., Miller, L., and Chesney, R. (1993). Polarity of taurine transport in cultured renal epithelial cell lines: LLC-PK1 and MDCK. *Am. J. Physiol. Renal. Fluid Electrolyte Physiol.* 265: F137–F145.

156 Roig-Perez, S., Moreto, M., and Ferrer, R. (2005). Transepithelial taurine transport in Caco-2 cell monolayers. *J. Membr. Biol.* 204: 85–92.

157 Dantzler, W. and Silbernagl, S. (1976). Renal tubular reabsorption of taurine, gamma-aminobutyric acid (GABA) and beta-alanine studied by continuous microperfusion. *Pflugers Arch.* 367: 123–128.

158 del Amo, M., Urtti, A., and Yliperttula, M. (2008). Pharmacokinetc role of L-type amino acid transporters LAT1 and LAT2. *Eur. J. Pharm. Sci.* 35: 161–174.

159 Puris, E., Gynther, M., Auriola, S., and Huttunen, K.M. (2020). L-Type amino acid transporter 1 as a target for drug delivery. *Pharm. Res.* 37: 88.

160 Fraga, S., Pinho, M.J., and Soares-da-Silva, P. (2005). Expression of LAT1 and LAT2 amino acid transporters in human and rat intestinal epithelial cells. *Amino Acids* 29: 229–233.

161 Fraga, S., Sampaio-Maia, B., Serrao, M.P., and Soares-da-Silva, P. (2002a). Regulation of apical transporter of L-DOPA in human intestinal caco-2 cells. *Acta Physiol. Scand.* 175: 103–111.

162 Fraga, S., Serrao, M.P., and Soares-da-Silva, P. (2002b). The l-3,4-dihydroxyphenylalanine transporter in human and rat epithelial intestinal cells is a type 2 hetero amino acid exchanger. *Eur. J. Pharmacol.* 441: 127–135.

163 Fraga, S., Serrao, M.P., and Soares-da-Silva, P. (2002c). L-type amino acid transporters in two intestinal epithelial cell lines function as exchangers with neutral amino acids. *J. Nutr.* 132: 733–738.

164 Kido, Y., Tamai, I., Uchino, H. et al. (2001). A molecular and functional identification of large neutral amino acid transporters LAT1 and LAT2 and their pharmacological relevance at the blood–brain barrier. *J. Pharm. Pharmacol.* 53: 497–503.

165 Sampaio-Maia, B., Serrao, M.P., and Soares-da-Silva, P. (2001). Regulatory pathways and uptake of L-DOPA by capillary cerebral endothelial cells, astrocytes, and neuronal cells. *Am. J. Physiol. Cell Physiol.* 280: C333–C342.

166 Quiñones, H., Collazo, R., and Moe, O.W. (2004). The dopamine precursor L-dihydroxyphenylalanine is transported by the amino acid transporters rBAT and LAT2 in renal cortex. *Am. J. Physiol. Renal. Physiol.* 287: F74–F80.

167 Lennernäs, H., Nilsson, D., Aquilonius, S.M. et al. (1993). The effect of L-leucine on the absorption of levodopa, studied by regional jejunal perfusion in man. *Br. J. Clin. Pharmacol.* 35: 243–250.

168 Alexander, G.M., Schwartzman, R.J., Grothusen, J.R., and Gordon, S.W. (1994). Effect of plasma levels of large neutral amino acids and degree of parkinsonism on the blood-to-brain transport of levodopa in naive and MPTP parkinsonian monkeys. *Neurology* 44: 1491–1499.

169 Nilsson, D., Lennernas, H., Fasth, K.J. et al. (1999). Absorption of L-DOPA from the proximal small intestine studied in the rhesus monkey by positron emission tomography. *Eur. J. Pharm. Sci.* 7: 185–189.

170 Guo, L., Ren, J., and Jiang, X. (2012). Perspectives on brain-targeting drug delivery systems. *Curr. Pharm. Biotechnol.* 13: 2310–2318.

171 Ramadan, T., Camargo, S.M., Herzog, B. et al. (2007). Recycling of aromatic amino acids via TAT1 allows efflux of neutral amino acids via LAT2-4F2hc exchanger. *Pflugers Arch.* 454: 507–516.

172 Cornford, E.M., Young, D., Paxton, J.W. et al. (1992). Melphalan penetration of the blood–brain barrier via the neutral amino acid transporter in tumor-bearing brain. *Cancer Res.* 52: 138–143.

173 Goldenberg, G.J., Lam, H.Y., and Begleiter, A. (1979). Active carrier-mediated transport of melphalan by two separate amino acid transport systems in LPC-1 plasmacytoma cells in vitro. *J. Biolumin. Chemilumin.* 254: 1057–1064.

174 Greig, N., Momma, S., Sweeney, D.J. et al. (1987). Facilitated transport of melphalan at the rat blood–brain barrier by the large neutral amino acid carrier system. *Cancer Res.* 47: 1571–1576.

175 Bosanquet, A.G. and Gilby, E.D. (1984). Comparison of the fed and fasting states on the absorption of melphalan in multiple myeloma. *Cancer Chemother. Pharmacol.* 12: 183–186.

176 Kuhne, A., Kaiser, R., Schirmer, M. et al. (2007). Genetic polymorphisms in the amino acid transporters LAT1 and LAT2 in relation to the

pharmacokinetics and side effects of melphalan. *Pharmacogenet. Genom.* 17: 505–517.

177 Linnankoski, J., Makela, J.M., Ranta, V.P. et al. (2006). Computational prediction of oral drug absorption based on absorption rate constants in humans. *J. Med. Chem.* 49: 3674–3681.

178 Van Bree, J.B., Audus, K.L., and Borchardt, R.T. (1988). Carrier-mediated transport of baclofen across monolayers of bovine brain endothelial cells in primary culture. *Pharm. Res.* 5: 369–371.

179 Deguchi, Y. (2002). Application of in vivo brain microdialysis to the study of blood–brain barrier transport of drugs. *Drug Metab. Pharmacokinet.* 17: 395–407.

180 Deguchi, Y., Inabe, K., Tomiyasu, K. et al. (1995). Study on brain interstitial fluid distribution and blood–brain barrier transport of baclofen in rats by microdialysis. *Pharm. Res.* 12: 1838–1844.

181 Dickens, D., Webb, S.D., and Antonyuk, S. (2013). Transport of gabapentin by LAT1 (SLC7A5). *Biochem. Pharmacol.* 85: 1672–1683.

182 Stewart, B.H., Kugler, A.R., Thompson, P.R., and Bockbrader, H.N. (1993). A saturable transport mechanism in the intestinal absorption of gabapentin is the underlying cause of the lack of proportionality between increasing dose and drug levels in plasma. *Pharm. Res.* 10: 276–281.

183 Gidal, B.E., DeCerce, J., Bockbrader, H.N. et al. (1998). Gabapentin bioavailability: effect of dose and frequency of administration in adult patients with epilepsy. *Epilepsy. Res.* 31: 91–99.

184 Friesema, E.C., Jansen, J., Milici, C., and Visser, T.J. (2005). Thyroid hormone transporters. *Vitam. Horm.* 70: 137–167.

185 Friesema, E.C., Docter, R., Moerings, E.P. et al. (2001). Thyroid hormone transport by the heterodimeric human system L amino acid transporter. *Endocrinology* 142: 4339–4348.

186 Yan, R., Li, Y., Müller, J. et al. (2021). Mechanism of substrate transport and inhibition of the human LAT1-4F2hc amino acid transporter. *Cell Discovery* 7: 16.

187 Braun, D., Kinne, A., and Bräuer, A. (2011). Developmental and cell type-specific expression of thyroid hormone transporters in the mouse brain and in primary brain cells. *Glia* 59: 463–471.

188 Pardridge, W.M. (1983). Brain metabolism: a perspective from the bloodbrain barrier. *Physiol. Rev.* 63: 1481–1535.

189 Smith, Q. and Takasato, Y. (1986). Kinetics of amino acid transport at the blood-brain barrier studied using an in situ brain perfusion technique. *Ann. N.Y. Acad. Sci.* 481: 186–201.

190 Komura, J., Tamai, I., Senmaru, M. et al. (1996). A sodium and chloride ion-dependent transport of beta-alanine across the blood-brain barrier. *J. Neurochem.* 67: 330–335.

191 Lee, W., Hawkins, R., Peterson, D., and Vin, J. (1996). Role of oxoproline in the regulation of neutral amino acid transport across the blood-brain barrier. *J. Biolumin. Chemilumin.* 271: 19129–19133.

192 Mastroberardino, L., Spindler, B., Pfeiffe, R. et al. (1998). Amino-acid transport by heterodimers of 4F2hc/CD98 and members of a permease family. *Nature* 395: 288–291.

193 Tornquist, P., Alm, A., and Bill, A. (1990). Permeability of ocular vessels and transport across the blood-retinal-barrier. *Eye* 4: 303–309.

194 Barar, J., Rafi, M.A., PourseifK, M.M., and Omidi, Y. (2016). Blood-brain barrier transport machineries and targeted therapy of brain diseases. *Bioimpacts* 6: 225–248.

195 Sershen, H. and Lajtha, A. (1976). Capillary transport of amino acids in the developing brain. *Exp. Neurol.* 53: 465–474.

196 Stoffel, W., Sasse, J., Duker, M. et al. (1996). Human high-affinity, Na+-dependent L-glutamate/L-aspartate transporter GLAST-1 (EAAT-1): gene structure and localization to chromosome 5p11–p12. *FEBS Lett.* 386: 189–193.

197 Stuhlmiller, D. and Boje, K.M.K. (1995). Characterization of L-arginine and aminoguanidine uptake into isolated rat choroid plexus: differences in uptake mechanisms and inhibition by nitric oxide synthase inhibitors. *J. Neurochem.* 65: 68–74.

198 Hutchison, H., Eisenberg, H., and Haber, B. (1985). High-affinity transport of glutamate in rat brain microvessels. *Exp. Neurol.* 87: 260–269.

199 Huxtable, R.J. (1992). The physiological actions of taurine. *Physiol. Rev.* 72: 101–163.

200 Benrabh, H., Bourre, J., and Lefauconnier, J. (1995). Taurine transport at the blood-brain barrier: an *in vivo* brain perfusion study. *Brain Res.* 692: 57–65.

201 Lau, Y., Chen, J., Chen, B., and Hsieh, C. (1994). Transport of 2-aminoisobutyric acid in cultured endothelial cells. *Biochim. Biophys. Acta* 1194: 118–122.

202 Madge, L., Baydoun, A., Pearson, J., and Mann, G. (1992). Amino acid deprivation stimulates L-glutamine transport in cultured human venous endothelial cells (Abstract). *J Physiol* 452: P306.

203 Duell, P. and Malinow, M. (1997). Homocyst(e)ine: an important risk factor for atherosclerotic vascular disease. *Curr. Opin. Lipidol.* 8: 28–34.

204 Fryer, R., Wilson, B., Gubler, D. et al. (1993). Homocysteine, a risk factor for premature vascular disease and thrombosis, induces tissue factor activity in endothelial cells. *Arterioscl. Thromb.* 13: 1327–1333.

205 Bussolati, O., Sala, R., Astorri, A. et al. (1993). Characterization of amino acid transport in human endothelial cells. *Am. J. Physiol. Cell Physiol.* 265: C1006–C1014.

206 Mann, G., Pearson, J., Sheriff, C., and Toothill, V. (1989). Expression of amino acid transport systems in cultured human umbilical vein endothelial cells. *J Physiol* 410: 325–339.

207 Sobrevia, L., Cesare, P., Yudilevich, D., and Mann, G. (1995). Diabetes induced activation of system y+ and nitric oxide synthase in human endothelial cells: association with membrane hyperpolarization. *J Physiol* 489: 183–192.

208 Arancibia, C. and Sobrevia, L. (1999). Transport of L-arginine mediated by system y+L in human fetal endothelial cells. *J Physiol* 517: 47P.

209 Sala, R., Rotoli, B., Colla, E. et al. (2002). Two-way transport in human endothelial cells: TNF-alpha stimulation is restricted to system y(+). *Am. J. Physiol. Cell Physiol.* 282: C134–C143.

210 Pan, M., Wasa, M., and Souna, W. (1995). Tumor necrosis factor stimulates system xAG+ transport activity in human endothelium. *J. Surg. Res.* 58: 659–664.

211 Bogle, R., Coade, S., Moncada, S. et al. (1991). Bradykinin and ATP stimulate L-arginine uptake and nitric oxide release in vascular endothelial cells. *Biochem. Biophys. Res. Commun.* 180: 926–932.

212 Mann, G., Sheriff, C., and Pearson, J. (1990). NG-monomethyl-L-arginine and cationic amino acids inhibit L-arginine transport in perfused microcarrier cultures of aortic endothelial cells. In: *Nitric Oxide From L-Arginine: A Bioregulatory System* (ed. S. Moncada), 331–339. Amsterdam: *Elsevier Science.*

213 Bogle, R., Baydoun, A., Pearson, J. et al. (1992). L-Arginine transport is increased in macrophages generating nitricoxide. *Biochem. J* 284: 15–18.

214 Ogonowski, A., Kaesemeyer, W., Jin, L. et al. (2000). Effects of NO donors and synthase antagonists on endothelial cell uptake of L-arginine and superoxide production. *Am. J. Physiol. Cell Physiol.* 278: C136–C143.

215 Ramamoorthy, S., Leibach, F., Mahesh, V. et al. (1994). Functional characterization and chromosomal localization of a cloned taurine transporter from human placenta. *Biochem. J* 300: 893–900.

216 Vinnakota, S., Qian, X., Egal, H. et al. (1997). Molecular characterization and in situ localization of a mouse retinal taurine transporter. *J. Neurochem.* 69: 2238–2250.

217 Quinn, L.A. and McCumbee, W. (1998). Regulation of glucose transport by angiotensin II and glucose in cultured vascular smooth muscle cells. *J. Cell. Physiol.* 177: 94–102.

218 Stevens, M., Hosaka, Y., Masterson, J. et al. (1999). Downregulation of human taurine transporter by glucose in cultured retinal pigment epithelial cells. *Am. J. Physiol. Endocrinol. Metab* 277: E760–E771.

219 Herskowitz, K., Bode, B., Block, E., and Souba, W. (1991). The effects of endotoxin on glutamine transport by pulmonary artery endothelial cells. *J. Surg. Res.* 50: 356–361.

220 Sakurai, E., Sakurada, T., Ochiai, Y. et al. (2002). Stereoselective transport of histidine in rat lung microvascular endothelial cells. *Am. J. Physiol. Lung Cell Mol. Physiol.* 282: L1192–L1197.

221 Block, E., Herrera, H., and Couch, M. (1995). Hypoxia inhibits L-arginine uptake by pulmonary artery endothelial cells. *Am. J. Physiol. Lung Cell Mol. Physiol.* 269: L574–L580.

222 Jursky, F. and Nelson, N. (1996). Developmental expression of the glycine transporters GLYT1 and GLYT2 in mouse brain. *J. Neurochem.* 67: 336–344.

223 Deneke, S., Baxter, D., Phelps, D., and Fanburg, B. (1989). Increase in endothelial cell glutathione and precursor amino acid uptake by diethyl maleate and hyperoxia. *Am. J. Physiol. Lung Cell Mol. Physiol.* 257: L265–L271.

224 Susanto, I., Wright, S., Lawson, R., and Williams, C. (1998). Metallothionein, glutathione, and cystine transport in pulmonary artery endothelial cells and NIH/3T3 cells. *Am. J. Physiol. Lung Cell Mol. Physiol.* 274: L296–L300.

225 Rajapakse, N.W., Nanayakkara, S., and Kaye, D.M. (2015). Pathogenesis and treatment of the cardiorenal syndrome: Implications of L-arginine-nitric oxide pathway impairment. *Pharmacol. Ther.* 154: 1–12.

226 Singh, K., Al-Greene, N.T., Verriere, T.G. et al. (2016). The L-arginine transporter solute carrier family 7 member 2 mediates the immunopathogenesis of attaching and effacing bacteria. *PLoS Pathog.* 12: e1005984.

227 Palmer, R., Ashton, D., and Moncada, S. (1988). Vascular endothelial cells synthesize nitric oxide from L-arginine. *Nature* 333: 664–666.

228 Deneke, S., Lawrence, R., and Jenkinson, S. (1992). Endothelial cell cystine uptake and glutathione increase with *N*,*N*-bis(2-chloroethyl)-*N*-nitrosourea exposure. *Am. J. Physiol. Lung Cell Mol. Physiol.* 262: L301–L304.

229 Bogle, R., Baydoun, A., Pearson, J., and Mann, G. (1996). Regulation of L-arginine transport and nitric oxide release in superfused porcine aortic endothelial cells. *J Physiol* 490: 229–241.

230 Nilius, B. and Droogmans, G. (2001). Ion channels and their functional role in vascular endothelium. *Physiol. Rev.* 81: 1415–1459.

231 Takeda, K. and Klepper, M. (1990). Voltage-dependent and agonist-activated ionic currents in vascular endothelial cells: a review. *Blood Vessels* 27: 169–183.

232 Sobrevia, L., Nadal, A., Yudilevich, D., and Mann, G. (1996). Activation of L-arginine transport (system y+) and nitric oxide synthase by elevated glucose and insulin in human endothelial cells. *J Physiol* 490: 775–781.

233 Sobrevia, L., Yudilevich, D., and Mann, G. (1997). Activation of A2-purinoceptors by adenosine stimulates L-arginine transport (system y+) and nitric oxide synthesis in human fetal endothelial cells. *J Physiol* 499: 135–140.

234 Mcdonald, K., Zharikov, S., Block, E., and Kilberg, M. (1997). A caveolar complex between the cationic amino acid transporter 1 and endothelial nitric-oxide synthase may explain the "arginine paradox". *J. Biolumin. Chemilumin.* 272: 31213–33126.

235 Hardy, T. and May, J. (2002). Coordinate regulation of L-arginine uptake and nitric oxide synthase activity in cultured endothelial cells. *Free Radical Biol. Med.* 32: 121–132.

236 Nchlson, B., Manner, C., Kleeman, J., and Macleod, C. (2001). Sustained nitric oxide production in macrophages requires the arginine transporter CAT2. *J. Biolumin. Chemilumin.* 276: 15881–15885.

237 Zharikov, S.I., Sigova, A.A., Chen, S. et al. (2001). Cytoskeletal regulation of the L-arginine/NO pathway in pulmonary artery endothelial cells. *Am. J. Physiol. Lung Cell Mol. Physiol.* 280: L465–L473.

238 Boado, R.J., Li, J.Y., Tsukamoto, H., and Pardridge, W.M. (2003). Hypoxia induces de-stabilization of the LAT1 large neutral amino acid transporter mRNA in brain capillary endothelial cells. *J. Neurochem.* 85: 1037–1042.

239 Wada, Y., Hirose, K., Harada, T. et al. (2018). Impact of oxygen status on 10B-BPA uptake into human glioblastoma cells, referring to significance in boron neutron capture therapy. *J. Radiat. Res.* 59: 122–128.

240 Grewer, C., Gameiro, A., Zhang, Z. et al. (2008). Glutamate forward and reverse transport: from molecular mechanism to transporter-mediated release after ischemia. *IUBMB Life* 60: 609–619.

241 Pow, D.V., Naidoo, T., Lingwood, B.E. et al. (2004). Loss of glial glutamate transporters and induction of neuronal expression of GLT-1B in the hypoxic neonatal pig brain. *Brain. Res. Dev. Brain Res.* 153: 1–11.

242 Yeh, T.H., Hwang, H.M., Chen, J.J. et al. (2005). Glutamate transporter function of rat hippocampal astrocytes is impaired following the global ischemia. *Neurobiol. Dis.* 18: 476–483.

243 Brunini, T., Francischetti, E., Ellory, J. et al. (2001). Elevated nitric oxide in uraemic platelets is dependent on L-arginine transport. *J Physiol* 531: 33P.

244 Hanssen, H., Brunini, T.M., Conway, M. et al. (1998). Increased L-arginine transport in human erythrocytes in chronic heart failure. *Clin. Sci. (London)* 94: 43–48.

245 Shankar, R.R., Wu, Y., Shen, H.Q. et al. (2000). Mice with gene disruption of both endothelial and neuronal nitric oxide synthase exhibit insulin resistance. *Diabetes* 49: 684–687.

246 Cendan, J.C., Souba, W.W., Copeland, E.M. 3rd, and Lind, D.S. (1995). Cytokines regulate endotoxin stimulation of endothelial cell arginine transport. *Surgery* 117: 213–219.

247 Durante, W., Liao, L., Iftikhar, I. et al. (1996). Differential regulation of L-arginine transport and nitric oxide production by vascular smooth muscle and endothelium. *Circ. Res.* 78: 1075–1082.

248 Irie, K., Tsukahara, F., Fujii, E. et al. (1997). Cationic amino acid transporter-2 mRNA induction by tumor necrosis factor-alpha in vascular endothelial cells. *Eur. J. Pharmacol.* 339: 289–293.

249 Simmons, W.W., Ungureanu-Longrois, D., Smith, G.K. et al. (1996). Glucocorticoids regulate inducible nitric oxide synthase by inhibiting tetrahydrobiopterin synthesis and L-arginine transport. *J. Biolumin. Chemilumin.* 271: 23928–23937.

# 20

# Drug Transporters and Their Role in Absorption and Disposition of Peptides and Peptide-Based Pharmaceuticals

*David J. Lindley[1], Stephen M. Carl[1,2], Dea Herrera-Ruiz[3], Li F. Pan[1], Lori B. Ward[1], Jonathan M.E. Goole[4], Olafur S. Gudmundsson[2], Matthew Behymer[1], and Gregory T. Knipp[1]*

[1]Department of Industrial and Physical Pharmacy, College of Pharmacy, Purdue University, West Lafayette, IN, USA
[2]Bristol-Myers Squibb Research Institute, Discovery Pharmaceutics, Princeton, NJ, USA
[3]Universidad Autónoma del Estado de Morelos, Facultad de Farmacia, Cuernavaca, Mexico
[4]Pharmacotherapy and Galenic Pharmacy Research Department, Institute of Pharmacy, Universite Libre de Bruxelles, Brussels, Belgium

## 20.1 Introduction

The utility of peptide and protein-based therapeutics in the mitigation of human maladies has existed since 1922 when porcine insulin was first utilized in diabetic patients. Over the last few decades, our understanding of the importance of proteins and peptides being involved in a number of maladies has increasingly grown. Moreover, the literature is replete with investigations into protein–protein interactions and their role in diseases and has led to numerous efforts toward the discovery of therapeutic agents utilizing peptide and peptide-based molecules. While peptide based therapies could offer considerable promise in the mitigation of diseases, the continued efforts of the pharmaceutical industry to advance the field has had led to a limited return on investment with only about 2% of the marketed products in this chemical space [1]. Furthermore, this review discusses several strategies to address some of the ADME properties that may contribute to the poor development of peptide and peptide-based drugs, while offering some strategies to improve ADME and possibly enhance the chemical space covered. In this chapter, we will focus on developing a better understanding of the factors that control oral absorption of peptides.

The absorptive pathways for proteins and peptides across the gastrointestinal tract (GIT) have long been an investigational topic amongst pharmaceutical scientists, simply based upon the therapeutic potential that these molecules and related analogs possess. Initial discussions on the mechanisms of protein and peptide absorption began in the mid-19th century with the belief that whole proteins obtained in the diet must be absorbed intact into the peripheral circulation due to the constant presence of serum plasma proteins [2]. Currently, it is well established that the majority of proteins are degraded in the gastrointestinal lumen by both extra and intracellular proteases and peptidases to yield amino acids and oligopeptides, particularly di and tripeptides, for passive and active absorption. After absorption, these compounds serve as major constituents for the necessary synthesis of functional human proteins and metabolic fuel. The role of di and tripeptide transport in amino acid absorption was not initially apparent due to the absence of peptides found in the portal blood following a high protein meal, which suggested that luminal protein digestion had to proceed all the way to free amino acids before absorption could occur [3]. The apparent high capacity of the intestine to absorb amino acids and the difficulties in designing oral peptide and protein therapeutics further supported this theory [3].

It was in 1986 that Steinhardt et al. first demonstrated that the luminal absorption of amino acids from a solution of dipeptides was much greater than the absorption of mixtures of the free amino acids themselves, suggesting that smaller peptides might be of significant clinical and nutritional importance [4]. Supporting these findings was conflicting evidence that did not explain how certain hydrolysis-resistant dipeptides, such as carnosine (β-alanylhistidine) and anserine (β-alanyl 1-methylhistidine), could be intestinally absorbed intact [5]. Upon additional investigation, it was concluded that the end products of intestinal protein digestion were a mixture of free amino acids and small peptides that could traverse the physical, enzymatic, and immunologic barrier of the small intestinal epithelium [6]. These findings led to subsequent studies suggesting

*Oral Bioavailability and Drug Delivery: From Basics to Advanced Concepts and Applications*, First Edition. Edited by Ming Hu and Xiaoling Li.

the presence of saturable, carrier-mediated transporter system(s) capable of mediating peptide absorption. The results of these studies further illustrated the need for a better understanding of the physiological barriers and the pathways by which di and tripeptides as well as peptide-based compounds traversed enterocytes for absorption.

The luminal GI morphology varies significantly from the esophagus down through the intestine. The increase in folding from very little in the esophagus to the folds of Kerckring, villi, and microvilli in the small intestine where the significant increase in the surface area leads to enhanced absorption for xenobiotics and nutrients is important for formulation scientists to incorporate into dosage form design and tailoring drug release [7, 8]. The epithelial cells lining the GIT form a selective, polarized barrier to the free diffusion of compounds that may be harmful to the body, yet enhance the absorption of beneficial nutrients. Polarity across the enterocytes is imparted by the presence of tight junctional complexes formed between the cells, which enable a functionally distinct lumen-facing apical (lumen) membrane surface to form in opposition to a functionally distinct blood-facing basolateral (blood and lymphatic) membrane [7]. The GIT is a highly specialized organ that is regionally designed specifically for absorption of beneficial nutrients and protecting the body by resisting potentially harmful xenobiotic transport. Therefore, the extent of absorption of any compound depends on its specific physicochemical properties, the influence of the formulation governing regional dissolution, and the interactions between the compound and the physiological barriers presented by the epithelial interface of the GIT. The challenge for pharmaceutical scientists is to define molecular entities that have optimal activity, are formulated to provide regiospecific release for optimal absorption, and can sufficiently permeate the GIT to reach therapeutically beneficial plasma levels.

In general, the two major routes of permeation for the transfer of substances across physiological barriers are the transcellular and paracellular pathways [6]. The aforementioned tight junctional complexes, which exhibit a net negative charge, limit paracellular absorption by size, conformation, and charge, and are largely dependent on the cellular tight junctional barrier's porosity [3, 9, 10]. Conversely, transcellular absorption requires the transport of solutes across two functionally distinct and morphologically different cellular membranes (e.g. luminal versus abluminal facing membranes) [6, 10]. There are several potential mechanisms of transcellular drug transport including passive transcellular diffusion, ion channels, facilitated diffusion, active transport, and endocytosis [10, 11]. The net permeation of amino acid and peptide-based compounds through these routes is dependent on several factors that include: affinity and capacity for integral membrane transporter proteins, lipophilicity (resulting from hydrogen bonding potential and hydrophobicity in peptides and proteins), charge, molecular size and shape of the compound and the extent of endocytic activity (generally considered minimal in intestinal enterocytes). Taken together, transport through both passive paracellular and passive and transporter-mediated transcellular pathways contribute to the overall net permeation of any particular compound, where the precise permeation mechanisms involved can be extremely complex. In order to expedite the discovery and development of peptide and protein-based therapeutic agents, researchers currently try to delineate and consider the net contribution of individual transport mechanisms to the overall transport in efforts to accurately predict and assess the overall absorption, distribution, metabolism, excretion, and toxicity (ADMET). In this chapter, we will focus on the transporter proteins that can significantly increase ADME of peptide-based molecules in the GIT.

The importance of transporters in mediating the net permeation or absorption of compounds has been the subject of considerable debate in the last several years [12–15]. Kell and colleagues hypothesize that there are two primary schools of through regarding the pathways of diffusion: (i) investigators who believe the assertion that lipoidal phospholipid bilayer diffusion is important (BDII) versus (ii) investigators who subscribe to the assertion that in intact cells lipoidal phospholipid diffusion is negligible (PDIN) and can be largely ignored in most cases due to the significant numbers of carrier-mediated transporters (discussed below). In their reviews, Kell can colleagues rely on a lack of literature analysis demonstrating actual compounds in the phospholipid bilayer as a justification to question the relevance of BDII [13]. Instead, they attribute passive paracellular permeation and the potential of other unidentified or less well-characterized transporters that possess affinity and high capacity for substrates as the alternative reason for the traditional BDII theorists. In contrast, they maintain that the overwhelming amount of manuscripts on drug transporters that have been characterized combined with a significantly higher amount of uncharacterized transporters provide the evidence supports that PBIN is more feasible. While the possibility of the PBIN theory may exist, the question comes down to the overall ability to identify these transporters with a higher, non-saturated capacity for the diverse array of compound chemistries, including peptide-based substrates, currently believed to be passive, transcellular permeants. While BDII is not fully eliminated as a potential for permeation in the reviews, the goal of Kell and colleagues is to point out that the alternate hypothesis for a role of transporters and the potential for increased paracellular permeation is more likely to explain the difference between known transporter related observations and the remaining unidentified permeation (currently considered BDII). This is an interesting hypothesis, and if proven over the test of time, could dramatically change the way we view cellular permeability. However, there is still a significant amount of uncertainty in the discussion, particularly when the deviation between established transporter-mediated and the remaining extent of absorption is attributed to the role of unidentified transporters that have yet to be characterized or increased paracellular diffusion.

Alternatively, several investigators with an extensive background in the pharmaceutical ADME area have published reviews that support the potential for the passive lipoidal transcellular permeation (BDII) pathway as playing a role in the observed diffusion across cellular barriers [14–16]. They contend that it is the combination of both passive and carrier-mediated pathways that leads to the overall absorption of the chemicals. They cite numerous studies comparing black lipid membrane and Parallel Artificial Membrane Permeability Assays (PAMPA) with permeability studies across the Caco-2 and 2/4/A1 cell lines revealing strong evidence that the lipoidal pathway is important in absorption [15]. These papers add strength to the argument that the BDII mechanism cannot be ignored. Furthermore, they demonstrate that in most cases the truth is in the middle, refuting the absolute stance of BDII and PBIN put forth by Kell and colleagues.

For this chapter, we will discuss the data based on the assumption that both passive and carrier-mediated pathways are relevant. We will also discuss data that will provide insight into the evidence supporting the different routes of permeation available for these peptide- and protein-based therapeutics to be absorbed from the GIT.

### 20.1.1 Passive Permeability of Peptide and Peptide-Based Agents

Hydrogen bonding potential and the effect of conformation on smaller oligopeptides have been theorized to play an important role in controlling the route of passive permeation across cell barriers. Several studies have attempted to delineate the predominant route of passive permeation of conformationally flexible and cyclic peptides across *in vitro* and *in vivo* barriers, e.g. Caco-2 cell lines [17–20]. In earlier work, the effect of a linear versus β-turn structure forming tetra and pentapeptides containing hydrogen bonding side chains appeared to predominantly traverse the Caco-2 cell monolayer via the paracellular route [17]. However, upon replacement of side chain residues with similarly sized aliphatic amino acids had little effect on the paracellular diffusion of the linear analogs, but interestingly the presence of a β-turn structure in the peptides lead to an increase in passive transcellular diffusion. In subsequent studies across the bovine brain microvessel endothelial cell (BMEC) model, the theoretical paracellular permeability coefficients for the tetra and pentapeptides containing hydrogen bonding side chains and the linear aliphatic residues revealed that the paracellular route was predominant. It was demonstrated that replacement with the aliphatic side chains along with the ability of the β-turn structure formation significantly shifted permeation to approximately 90% passive transcellular route. The prevailing hypothesis was that upon forming a compact structure, intramolecular hydrogen bonding between the amides and carbonyls of the peptide backbone formed, thus reducing the desolvation energy (e.g. reducing the number of hydrogen bonds with water) and enabling a more favorable interaction with the lipid bilayer.

Additional studies performed in the laboratory revealed that utilizing cyclization strategies to form more compact oligopeptides led to an increase in the permeation of the peptides largely due to an alteration in their lipophilicity and reduced hydrogen bonding potential without altering the size [18–21]. However, the cyclization of some peptides did lead to some affinity for P-glycoprotein, which inhibited permeation as discussed below [21]. These studies combined with investigations by others have revealed some insight into the physicochemical properties that dictate oligopeptide permeation across cell barriers via a passive mechanism. This is particularly true for cyclic peptides, where a recent review highlighted some of the physicochemical properties of 125 cyclic peptides that were orally absorbed by mammals [22]. The summary of their findings where that cyclization can lead to reduced metabolism and degradation of peptides due to conformational locking. Furthermore, cyclization can lead to optimized conformations that may reduce hydrogen bonding, restrict conformational flexibility, and favor oral absorption by shielding polar atoms resulting in a reduction in the polar surface area [22]. Further research in the area is needed to help better understand the relationship between the key molecular properties of cyclic peptides and absorption to assist in driving down the dose by increasing exposure.

### 20.1.2 General Transporter Overview

Numerous classes of transporter proteins have been identified, each with different and sometimes overlapping substrate affinities and capacities, as well as specific tissue, cellular and temporal expression patterns [3, 23–25]. The transporters belong to one of two families, either the Solute Carrier Transporters (*SLC*) or the ATP Binding Cassette (*ABC*) Transporters. The SLC transporters represent a broad class of at least 43 separate subfamilies based upon their affinity and capacity for substrates, and in total is comprised of over 300 human isoform members [26–28]. SLC transporters generally participate in influx (transport a substrate into the cytoplasm) or efflux (transport a substrate out of the cytoplasm) of nutrients and ionic substrates across cell membranes. Several SLC members can act as facilitative transporters that bind substrates and function by transporting substrates across a membrane from a higher to lower concentration gradient [29]. Since a facilitative transporter is required and they often possess affinity for polar compounds, e.g. the Glucose Transporter family (GLUT or *SLC2A* members), to transport glucose and other carbohydrates into the body, the mechanism driving permeation is distinctly different than passive lipoidal diffusion. Many other SLC isoforms rely on ion gradients, typically through the cotransport or secondary transporters which change ion concentrations across a cell to provide the energy to actively transport substrates [29]. We will highlight some of these active SLC transporter isoforms below as they have been implicated in the transport of peptides and peptide-based therapeutics.

The ABC transporters are different from the SLC family members based on their containing at least one ATP-Binding domain [30, 31]. There are at least 49 members of the family and they have broad substrate affinity and capacity for nutrients (including peptides and small proteins) to hydrophobic compounds and their metabolites. The ATP binding cassette allows the conversion of ATP to ADP + Pi and the release of the energy to drive their function. We will largely focus on isoforms like P-glycoprotein (P-gp, *ABCB1*), the multidrug resistance-associated proteins (MRPs, *ABCC* isoforms), and the breast cancer resistance protein (BCRP, *ABCG2*) as they are the most widely studied transporters that affect **A**bsorption, **D**istribution, and play a role in concerted **M**etabolism and **E**xcretion (ADMET) of therapeutic agents including peptide-based drugs [31].

While the physicochemical nature of the drug and dosage form design will contribute to the solubility of the compound and therefore, its net permeability, absorption may also depend on the interactions with the aforementioned transporter families in the GIT. These compound-to-transporter interactions will also ultimately influence the drug's disposition in the body, as many of the SLC and ABC transporters are widely expressed in different tissues and in epithelial and endothelial barriers. Recently, it has become increasingly more evident that membrane transporter protein activity plays a critical role in determining the pharmacokinetic profiles of numerous compounds by actively transporting substrates between organs and tissues. The clinical importance of these transporters has been reviewed and prioritized by the International Transporter Consortium [23, 24, 32].

Table 20.1 outlines some of the main transporter families affecting peptide-based drug absorption. Of particular interest in this area of research is the

**Table 20.1** Some families of relevant transporters affecting peptide-based drug absorption.

| Family | Gene name | Family size | General affinity |
|---|---|---|---|
| POT | SLC15A | 4 | Di and tri-peptides, Peptide-like drugs |
| OAT | SLC22A | 5 | Organic anions, β-lactam antibiotics, antiviral drugs, and ACE inhibitors |
| OATP | SLCO/SLC21A | 52 | Thyroid hormones, bile salts, neural steroids and steroid conjugates |
| MCT | SLC16A | 14 | Short chain anionic compounds, β-lactam antibiotics, and NSAIDs |
| MRP | ABCC | 9 | Anionic drugs, leukotrienes, metabolite conjugates, bile salts etc. |
| P-gp | ABCB1 | 1 | Lipophilic and amphiphilic compounds such as antibiotics, anti-cancer drugs, HIV protease inhibitors and immunosuppressives |
| BCRP | ABCG2 | 1 | Chemotherapeutic agents, glucoronide or metabolic conjugates, fluorescent dyes |
| HPT-1 | Cadherin Member | 1 | Di and tri-peptides, Peptide-like drugs, Cephalosporins |

proton-dependent oligopeptide transporter (POT; *SLC15A*) family, which has been the most widely studied family of transporters that influence peptide and peptide-based drug absorption [33, 35]. Although members of this family are involved in the net influx of di and tripeptides in addition to many peptide-based compounds, they do have mixed affinities and capacities for other substrates. It is also reasonable to assume that every transporter family exhibits the potential to influence the ADMET of a particular substrate, albeit peptide-based or otherwise. This led to the realization that in order to fully determine how a particular peptide-based drug is absorbed in the body, one must consider all of the factors involved, which includes the potential interaction with other transporters expressed within a given tissue. The challenge for pharmaceutical scientists is to define molecular entities targeted to transporters that have optimal activity (e.g. affinity and high capacity) to enable sufficient absorption across the gastrointestinal epithelium to provide therapeutically beneficial plasma levels.

Of equal importance to the expression patterns of a transporter within a given tissue, is the specific membrane to which it is expressed due to the polarity conveyed by tight junctional complexes formed between the cells of polarized biological barriers. When considering the membrane localization, it is important to also determine the net function of the transporter. For instance, a transporter expressed on the apical membrane may function to increase absorption if it is an influx transporter or decrease absorption if it is an efflux transporter. On the other hand, a basolaterally expressed transporter could act to increase absorption in the case of an efflux transporter, or conversely increase the potential for accumulation, or net secretion of compounds in the case of an influx transporter. As such, it is important to not only keep in mind the transporter functionality (influx versus efflux) but its specific localization within a given cell type when considering its influence on ADMET.

Transporter functionality not only depends on its tissue and cellular expression patterns, but also on disease state, fasted versus fed state administration, drug–drug and drug–xenobiotic interactions, changes in expression (e.g. induction), and pharmacogenetic influences, to name a few. Therefore, a broad overview of these variables is necessary to understand and develop rational therapeutic dosage design and administration strategies. For example, drug delivery systems (DDSs) modifying the potential for interactions between drugs and transporters must be accounted for as they may have significant effects on the absorption process, as well as the disposition of certain compounds [34]. Based on this knowledge, different strategies have been applied to improve drug ADMET. This chapter aims to discuss the different transporters involved in the overall absorption and disposition of peptides and peptide-based compounds. This includes not only the influence of direct substrate transporters such as those in the POT family but also direct and indirect transport exhibited by coupling design strategies with the inhibition of efflux transporters affecting peptide-based drugs. This chapter will also discuss the GIT-specific transporter activity and expression, as well as some substrate–transporter interactions and *in vitro/in vivo* methodologies to test such functionalities. Finally, the effect of pharmacogenomic differences such as age, sex, and ethnicity, on transporter functionality and rational drug design will be discussed.

## 20.2 Transport Systems Mediating Peptide-based Pharmaceutical Absorption and Disposition: The Solute Carrier (SLC) Family

### 20.2.1 Proton-Coupled Oligopeptide Transporters (SLC15A)

The POT Superfamily consists of four distinct isoforms: Peptide Transporters 1 and 2 (PepT1 [*SLC15A1*] and PepT2 [*SLC15A2*]) and Peptide/Histidine Transporters 1 and 2 (PHT1 [*SLC15A4*] and PHT2 [*SLC15A3*]). Each of the four isoforms shares a common topological profile which consists of 12 putative α-helical transmembrane domains (TMDs), with intracellularly localized N- and C- termini, and numerous observed and hypothesized conserved consensus amino acid sequences [35–40]. Interestingly, while these similarities do exist, there is little homology between the PepT and PHT members [35].

POT isoform-mediated transport is hypothesized to be driven by an electrochemical proton gradient and the membrane potential as the primary driving forces to facilitate transport [35, 41, 42]. In mammalian cells the proton gradient is generally supplied by electroneutral proton–cation exchangers, e.g. $Na^+/H^+$ antiporters [43]. PepT substrates have been demonstrated to range from virtually any di and tripeptide combination of the 20 L-α-amino acids, to several peptidomimetic drugs, prodrugs, α-amino fatty acids, and non-peptidic compounds [35]. Pharmacologically active substrates of PepT transporters include: β-lactam antibiotics [35, 44].

ACE inhibitors such as captopril and fosinopril [45, 46], prodrugs of L-dopa [47–49], prodrugs of antiviral nucleosides, namely valacyclovir and valganciclovir [50, 51], and antibacterial peptide analogs, such as alafosfalin [52]. In contrast to the PepT transporters, relatively little is known concerning the transport activity of PHT1 and PHT2, although both rat isoforms have demonstrated dipeptide transport activity and affinity for the individual amino acid histidine [53, 54]. Additionally, results from our laboratory have also revealed PHT1 transport activity for the antiviral nucleoside prodrug valacyclovir [55].

In contrast, the functional role of PHT2 has yet to be studied, although preliminary investigations indicate that the rat isoform does exhibit affinity for histidine and histidyl-leucine in the kidney cell lines HEK293T and BHK [54]. Also of particular note is that while PepT1, PepT2, and PHT1 have each been demonstrated to be bound to the cellular membrane, the rat PHT2 isoform was determined to be predominantly intracellularly localized to the Golgi, lysosomes, and autophagosomes, although some cells did indicate localization on the outer nuclear membrane [54].

### 20.2.2 Organic Anion/Cation Transporters (SLC21A/SLCO/SLC22A)

SLC mediating the transport of organic anions are separated into two distinct families, the Organic Anion Transporter family (OAT; *SLC22A*) and the Organic Anion Transporting Polypeptide superfamily (OATP/SLCO; *SLC21A*) [56–59]. Five isoforms of the OAT family have been identified to date (OAT1-5) [58, 60]. OAT isoforms are characterized by the presence of twelve α-helix TMDs; one large hydrophilic extracellular loop between TMD1 and 2 carrying several potential glycosylation sites; and a large intracellular loop between TMD 6 and 7, containing multiple potential phosphorylation sites [60]. OATs have been demonstrated to transport a wide range of clinically significant organic anion drugs such as nonsteroidal anti-inflammatory drugs (NSAIDs), β-lactam antibiotics, antiviral drugs, diuretics, antitumor drugs, and angiotensin-converting enzyme inhibitors [59, 60].

The OATP superfamily consists of 52 members spanning 6 of the 13 *SLCO* subfamilies [56]. While members of the OATP superfamily also exhibit twelve TMDs, OATP isoforms differ dramatically from their OAT counterparts, exhibiting a large extracellular domain between TMD 9 and 10 (extracellular loop 5) and multiple glycosylation sites in extracellular loop 2 and 5 [56]. The sodium independent transport activity of various OATP isoforms has been demonstrated to be either glutathione [61, 62] or proton dependent [63], although it is unclear which mechanism predominates for each isoform. Substrates for OATP isoforms include organic anions, such as bile salts, bilirubin, prostaglandin and estrogen conjugates, neural steroids and steroid conjugates, lipophilic organic cations, organic dyes, thyroid hormones, and anionic oligopeptides [56, 59]. Some pharmacologically active substrates include digoxin, pravastatin, methotrexate, benzylpenicillin, tebipenem pivoxil, fexofenadine, and various NSAIDs [56, 59, 64]. Coupled with their widespread tissue localization, the biopharmaceutical relevance of this transporter family should be readily apparent.

The Organic Cation Transporter (OCT) and Carnitine and Organic Cation Transporter (OCTN) subfamilies are two such families within the *SLC22A* superfamily that mediate the absorption of endogenous monoamine neurotransmitters, thiamine, guanidine, and creatinine [59, 60, 65]. The OCT and OCTN isoforms are characterized by 12 α-helix TMDs, a large glycosylated extracellular loop between TMD1 and 2, and a large intracellular loop carrying phosphorylation sites between TMD 6 and 7 [60]. OCT transport is both sodium and proton concentration independent, having been demonstrated to be electrogenically dependent on the substrate concentration gradient and membrane potential, thus allowing for bidirectional substrate translocation [66–69]. In contrast, the mechanism of OCTN-mediated transport appears to be isoform and substrate specific [70–73].

### 20.2.3 Monocarboxylate Transporters (SLC16A)

The monocarboxylate transporter (MCT) family currently consists of 14 different isoforms, characterized by 12 TMDs and a large intracellular loop between TMDs 6 and 7 [74, 75]. In contrast to the other SLC families discussed above, a number of MCT isoforms appear to require the co-expression of an ancillary protein, such as CD147 or GP70 (both widely distributed cell surface glycoproteins), for proper intracellular trafficking, cell surface expression and function [76]. Short chain anionic compounds such as lactate, pyruvate, and butyrate are the primary MCT substrates, although a number of pharmaceutically relevant compounds, such as β-lactam antibiotics, penicillins, NSAIDs, valproic acid, atorvastatin, and nateglinide [(-)-*N*-(trans-4-Isopropylcyclohexanecarbonyl)-D-phenylalanine) have also been demonstrated to be MCT substrates [77–83]. MCT-mediated transport occurs via a proton-dependent symport mechanism, where prodrug targeting of MCT 1 has been hypothesized to increase oral absorption [84].

### 20.2.4 Non-SLC Member-Mediated Peptide Transport

Evidence for the role of the Human Peptide Transporter 1 (HPT-1) in the transport of peptide-based compounds was first reported by Danzig and colleagues in 1993 [85]. The HPT-1 cDNA is approximately 2.5 kb long, encoding a 120 kDa protein containing 832 amino acids. HPT-1 is related to proteins of the cadherin family and does not share significant homology with PepT1 (16% identity and 41% similarity in the amino acid sequence), although it has been demonstrated to facilitate transport of peptide-based drugs, predominantly of the β-lactam antibiotic class [85–87]. Chinese Hamster Ovary (CHO) cells transfected with HPT-1 demonstrated enhanced dipeptide and β-lactam transport without amino acid inhibitory effects, whereas the wild-type CHO did not exhibit appreciable oligopeptide transport [87].

## 20.3 ATP Binding Cassette (ABC) Transporters

To date, 7 subfamilies and 52 transporter members of the ABC transporters have been established, and collectively, they embrace a broad array of substrate specificities ranging from endogenous fatty acid metabolites to anticancer drugs [88]. These transporters are integral membrane proteins that impart multidrug resistance by effluxing xenobiotics from the cytoplasm across the cellular membrane in a polarized, energy-dependent manner [89]. Although these transporters differ in membrane topologies, they all share the presence of nucleotide-binding domains (NBDs) that enable them to bind and hydrolyze ATP to drive efflux [88].

The majority of the research pertaining to this family of transporters has focused on P-glycoprotein (P-gp, *ABCB1*), MRPs (*ABCC* isoforms), and the BCRP (*ABCG2*) [89]. Although there are many other efflux transporters potentially involved in peptide-based drug absorption and disposition, they will not be discussed here as, to date, they are not considered to be major players in mediating compound absorption and disposition. For a more comprehensive review of ABC transporters, refer to reviews written by Glavinas et al. [90] and Szakacs et al. [88].

### 20.3.1 P-glycoprotein (ABCB1) and Other ABCB Isoforms

P-Glycoprotein is probably the most widely characterized ABC transporter in terms of molecular structure, function, and regulation [91] P-gp is a 170–180 kDa transmembrane efflux protein encoded by the *ABCB1* gene in humans [92]. In addition to its major role in conferring drug resistance in tumors, P-gp also participates in excluding substances from a variety of healthy tissues including the intestine, kidney, liver, placenta, and brain [93]. P-gp exhibits affinity for a wide range of lipophilic and amphiphilic compounds with dissimilar molecular characteristics such as antibiotics, anticancer drugs, HIV protease inhibitors, and immunosuppressives [31].

Transporters associated with antigen presentation (TAP) are also members of the ABCB family, where they play a critical role in regulating immune response through class I major histocompatibility processing (MHC I) [94, 95]. Briefly, TAP 1 (*ABCB2*) and TAP 2 (*ABCB3*) intracellularly localize to the endoplasmic reticular membrane, forming a heterodimer whose function is to present oligopeptides from degraded endogenous proteins to the cell surface for MHC I processing. MHC I processing occurs in all nucleated cell types, where peptide fragments from the normal cell do not normally elicit an immune response by activation of T-cells [95]. If the cells become compromised by infection or through transformation to a cancerous form, other proteins are expressed and presented which leads to immune detection and potentially removal. Enterocytes can also respond to changes in the luminal contents that arise from foreign peptides and some have theorized that the TAP proteins can be utilized to stimulate an immune response [96]. Lastly, there is evidence that suggests transporters encoded by *ABCB8*, *ABCB9*, and *ABCB10* can also facilitate the intracellular trafficking of peptides in cells [97].

### 20.3.2 Multidrug Resistance-Associated Protein Family (MRP; ABCC)

MRPs are another class of relevant efflux transporters affecting drug absorption and disposition. To date, nine MRP isoforms have been identified (MRP1–MRP9) and expression varies in different tissues throughout the human body [98]. For instance, both MRP1 and MRP2 are expressed in the intestine, with MRP1 showing higher regional expression. However, MRP1 is localized to the basolateral membrane of enterocytes, whereas MRP2 is found at the apical membrane [99]. Due to the presence of multiple binding sites, MRP's bind and translocate a wide variety of substrates. For example, MRP1 is involved in the transport of anionic drugs, metabolite conjugates, leukotriene's, dianionic bile salts, glutathione-conjugates of peptidomimetics, and vinca alkaloids [100, 101]. MRP2 has overlapping substrate specificities with MRP1 such as leukotrienes and metabolite conjugates but also transports distinctly different compounds such as glucoronide conjugates of estradiol and acetaminophen [102].

### 20.3.3 Breast Cancer Resistance Protein (BCRP, ABCG2)

BCRP was initially cloned from a non-P-gp expressing line of cancerous breast cells resistant to treatment with doxorubicin and verapamil [103]. BCRP is present on the apical membrane of enterocytes, in the liver canalicular membrane, and at the luminal periphery of brain microvessels [104]. As observed for other efflux transporters, BCRP was found to be a conveyor of chemotherapeutic resistance, specifically in breast cancer. However, it has been demonstrated that BCRP has affinity for a broad range of other compounds. Therefore, this transporter seems to serve a protective role against xenobiotics, carcinogens, and endogenous substances [103]. Although BCRP shares overlapping substrate specificity with P-gp and MRP1, it also transports structurally diverse organic molecules including estrone-3-sulfate, 17B-estradiol and fluorophores [104].

## 20.4 Gastrointestinal Tract-Specific Transporter Activity

### 20.4.1 Solute Carrier (SLC) Family

The primary function of transporter proteins in the GIT is to mediate the absorption and bioavailability of both endogenous and exogenous nutrients and xenobiotics, and serve a protective role by modulating compound exposure. This could refer to apical and/or basolateral influx/efflux transporters mediating the vectorial transcellular flux of a compound to either the GI luminal or blood/lymphatic compartments. Since many therapeutic agents are designed based on functional peptide or protein epitopes, the current aim of many pharmaceutical scientists has been to identify those transporters that play a significant role in mediating their GI absorption and resultant bioavailability. Figure 20.1 is an oversimplified depiction of a representative human intestinal cell summarizing some of the families of transporters discussed thus far in this chapter and their known localization. Figure 20.1 illustrates the complexity of the system and the difficulties scientists face when designing peptide-based compounds to target specific transporters. Of particular interest is the potential to use amino acid prodrugs or peptidomimetic approaches to enhance the affinity for PepT1 (*SLC15A1*), which has been successfully employed for compounds including valacyclovir, valganciclovir, Phe-Pro-alendroate, rennin inhibitors, pulmonary disorders, and β-lactam antibiotics [35, 105, 106].

PepT1 is the primary POT member that mediates the GIT absorption of peptide-like therapeutic agents, although previous studies suggest that PHT1 (*SLC15A4*) may potentially play a significant role as well [53, 55]. Although widely expressed throughout the GIT, the functional role of PHT2 (*SLC15A3*) is relatively unknown [35, 107]. PHT2, as of yet, has not been found to significantly alter peptide-based compound absorption, but may participate in mediating the extent of permeation by affecting the intracellular disposition these agents based on its intracellular localization [35, 54]. More recent evidence has linked PHT1 and PHT2 to a role in inflammatory responses, and even functioning with PepT1 to mediate gastrointestinal inflammation [108–111]. Interestingly, while PepT2 (*SLC15A2*) transcripts have not been demonstrated to be expressed in the human GIT *in vivo*, data suggests potential expression in various *in vitro* cell models widely utilized to study peptide transporter function (*unpublished data from our laboratory*). If verified, such data would provide significant insights into mechanistically understanding much of the inherent variability with *in vitro* cell culture testing models.

Correlating with its role in mediating absorption at the apical surface, PepT1 expression has been shown to decrease along the GI tract, with ileal expression significantly lower than jejunal expression [107]. Interestingly, while PepT1 almost certainly mediates the apical influx of peptide-like compounds, the basolateral transporter mediating peptide efflux into the blood has still yet to be identified, although it has been functionally characterized [35].

There have been numerous literature reports detailing some of the structural determinants required for recognition by PepT1, as reviewed by several different laboratories [3, 6, 35, 44, 99, 105, 112, 141]. As reviewed previously by our laboratory [99], several laboratories have performed transportophoric investigations into the binding and transport characteristics of PepT1 and established of molecular descriptor models to predict some of critical functional group elements required of PepT-substrates [113–118]. Briefly, the studies revealed

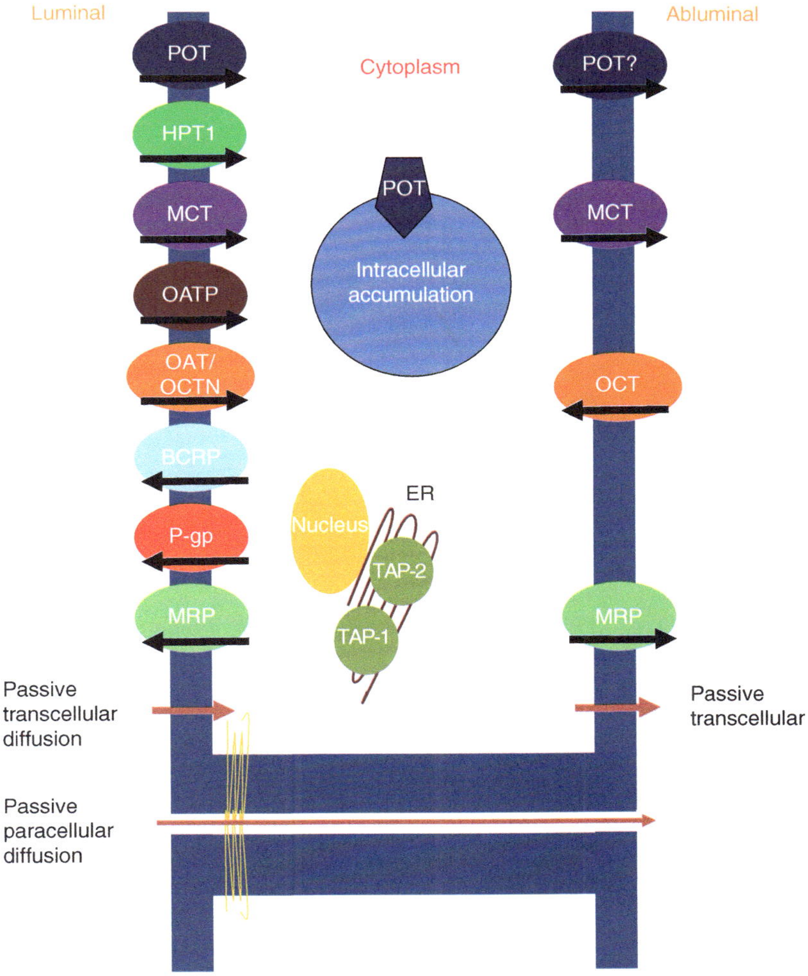


**Figure 20.1** A representative depiction of a number of transporters expressed in a human intestinal cell illustrating the complexity of the system. Source: Adapted from Carl et al. [2].

the following structural characteristics were required for high PepT1 affinity (<0.5 mM) which include: (i) L-amino acids were preferred, (ii) the presence of an acidic or hydrophobic function at the C-terminus, (iii) the presence of a weakly basic group in α-position at the N-terminus, (iv) the substrate should exhibit a ketomethylene moiety or an acid amide bond, and in the case of having a peptide bond, (v) to present it in a trans configuration. PepT1 has also been shown to mediate the transport of a variety of substrates possessing varying degrees of affinity and capacity, depending on their chemical structure [99]. Table 20.2 in this review highlights representative compounds that were determined to have been demonstrated to possess affinity and/or capacity in PepT1 studies [99]. More recently, Brandsh and colleagues have presented two extensive reviews that further expand molecular descriptors required for PepT1 recognition [6, 112]. The reviews by Brandsch and colleagues are an excellent resource for those interested in the area and are highly recommended for further reading.

Members of the OAT (*SLC22A6-11, 13,* and *20*) subfamily have not been identified in the GIT, although various OATP (*SLCO*), OCT (*SLC22A1* and *3*), and OCTN (SLC22A5) isoforms have been demonstrated to be expressed at varying levels [119]. For example, human OATP-B (*SLCO2B1*), OATP-D (*SLCO3A1*), OATP-E (*SLCO4A1*), OCT1 (*SLC22A1*), OCT3 (*SLC22A3*), and OCTN2 (*SLC22A5*) expression have each been demonstrated utilizing various methodologies [120, 121]. While there has been some debate over the direct cellular membrane localization OCT1, OCT3, and OCTN isoforms appear to be apically localized despite some functional studies indicating the basolateral expression of OCT1. It is believed that the OATP isoforms are apically localized. Taken together, these various subfamilies appear to function in concert to mediate the intestinal flux of various

**Table 20.2** A partial list of therapeutic compounds that have been demonstrated to interact (affinity and/or capacity) with PepT1.

| Substrate classes | Compounds | | Inhibitors |
|---|---|---|---|
| Cephalosporins | Cefaclor | Cefadroxil | Latamoxef |
| | Cefamandole | Cefatrizine | |
| | Cefepine | Cefixime | |
| | Cefodizime | Cefpirome | |
| | Cefroxadine | Ceftibuten | |
| | Cefotaxime | Ceftriaxone | |
| | Cephalexin | Cephradine | |
| | Cloxacillin [83] | Cyclacillin [83] | |
| | Loracarbef | Moxalactam | |
| Penicillins | Amoxicillin [83] | Benzylpenicillin | Carbenicillin |
| | Cloxacillin | Cyclacillin | |
| | Dicloxacillin | Metampicillin | |
| | Oxacillin | Phenoxymethyl-penicillin | |
| | Propicillin | | |
| ACE inhibitors[20] | Benazepril | Benzaprilat [83] | Benazeprilat |
| | Captopril | Ceronapril [83] | Enalaprilat |
| | Cilazapril [83] | 4-Cyclohexylcaptopril [83] | Fosinoprilat |
| | Enalapril | Enalaprilat [83] | Quinalaprilat |
| | Fosinopril | Fosinoprilat [83] | |
| | Lisonopril | Perindopril [83] | |
| | Quinalapril | Quinaprilat [83] | |
| | S-Benzylcaptopril [83] | | |
| Antivirals | Lopinavir Prodrugs [93] | Saquinavir Prodrugs [95–97] | |
| | Valaciclovir | Valganciclovir | |
| Others | Bestatin | | Glibenclamide |
| | α-methyldopa-phenylalanine | | Nateglinide |
| | Pro-Phe-alendronate | | Lys[Z($NO_2$)]-Pro |
| | Renin inhibitors | | |
| | Thrombin inhibitors | | |
| | Thyrotropin-releasing hormone | | |

Source: Adapted from Carl et al. [74].

bile conjugates and ionic compounds including peptides.

### 20.4.2 ATP Binding Cassette (ABC) Family

As discussed previously, oral absorption of peptide-based pharmaceuticals is highly dependent upon transport through intestinal enterocytes which express many of the efflux proteins mentioned above. Overcoming the effects of these proteins is of major concern for formulation scientists and synthetic chemists. Generally, prodrug strategies are utilized to enhance oral absorption of pharmaceutical entities by modifying their physicochemical properties including lipophilicity, permeability, and stability [122]. In addition, peptide prodrugs can also be formulated to target absorption via peptide transporters, circumvent efflux transport systems, or evade Cytochrome P450 (CYP) mediated metabolism as ways of improving drug delivery properties of peptide-based therapeutics [123]. For example, the HIV protease inhibitor saquinavir has been established to be a substrate for P-gp, MRP1, and MRP2 [124, 125]. Recent investigations have demonstrated that peptide prodrug derivatives of saquinavir (SQV) and lopinavir (LVR), two peptide-based protease inhibitors for the treatment of HIV, including valine-valine-saquinavir/lopinavir (Val-Val-SQV/LVR) and glycine-valine-saquinavir/lopinavir (Gly-Val-SQV/LVR) experience circumvention of various efflux transporter-mediated systems such as P-gp [124] and MRP2 [125]. In the case of saquinavir, an increase in the absorption of the Val-Val-SQV and Gly-Val-SQV prodrugs was observed in rat jejunum when contrasted with the unconjugated SQV. The same trend was observed for Val-Val-LVR and Gly-Val-LVR in both MDCK-MDR1 and MDCK-MRP2 transfected cells. Prodrug modification decreased the affinity for P-gp, which was confirmed by using specific inhibitors, as elaborated upon in the publications [123, 126]. Furthermore, the role of oligopeptide transporters (POTs) was further implicated when the increase in absorption for both prodrug derivatives was inhibited by concurrent Gly-Sar administration, a well-established peptide transporter substrate [126]. Thus the prodrug approach was doubly successful in that the prodrugs were found to be influxed by transport into cells, and had decreased affinity for the efflux transporter P-glycoprotein, two characteristics that will result in an increase in bioavailability.

Interestingly, MDCKII-MRP2 cells were also used to perform transport studies of SQV, Val-Val-SQV, and Gly-Val-SQV in the absence and presence of MK-571, a known MRP2 inhibitor [125]. In addition, single-pass intestinal perfusion studies in rats were conducted to assess the disposition of these prodrug derivatives through the evaluation of their absorption rate constants and intestinal permeabilities [125]. Results from these studies suggested that the modification to Val-Val-SQV eliminated its interaction with MRP2, whereas Gly-Val-SQV remained a substrate for MRP2 [125]. Moreover, peptide prodrug strategies targeted to evade efflux transporters hold a great potential in enhancing peroral delivery of peptide-based pharmaceutical agents.

There are also several other classes of peptide-based compounds that serve as substrates for ABC transporters. For example, cyclosporine and tacrolimus are potent immunosuppressive agents, cyclic peptides, and have been demonstrated to be substrates for P-gp, MRP-2, and/or BCRP [127, 128].

### 20.4.3 *In Vitro/In Vivo* Case Studies: Utilizing Transport Systems for Drug Delivery

#### 20.4.3.1 *In Vitro* Testing Methods

Though modern methods for predicting drug absorption have progressed far beyond the limited estimations of the pH-partition hypothesis, they remain at best a poor reflection of the incredible complexity and sophistication of human physiological conditions. Cell models cultured on Transwells$^{TM}$ are a commonly used *in vitro* approach that was developed to help predict drug absorption. For instance, Caco-2 cells, derived from a human colon adenocarcinoma, and BMECs [129] are useful for studying transport across the GIT and the blood–brain barrier, respectively [93]. Other cell lines typically used to specifically evaluate function of certain transporter proteins include the LLC-PK1, MDCK-II, COS-1, and COS-7 models, to name a few, as these cells are typically easy to culture and transfect. Cultured cells often grow into tight monolayers with a polarized character resembling the physiological environment, although this is not always the case with certain cell lines, e.g. COS cell lines [93].

Cellular *in vitro* models confer an additional advantage by enabling the identification and delineation of specific transport pathways in a physiologically relevant manner without the complexity of added endogenous compounds [93]. In addition, intracellular accumulation of substances as well as drug–drug interactions can also be evaluated. For example, a study conducted by Letrent et al. utilized BMEC's to study transport changes of morphine with the concurrent administration of the P-gp inhibitors cyclosporine and GF120918 [130]. Results of their studies demonstrated that incubation with P-gp inhibitors not only enhanced intracellular morphine and rhodamine (positive P-gp control) accumulation but also eliminated the observed net basolateral to apical efflux due to P-gp [93, 130].

Although cultured cell models prove their utility in evaluating many facets of cell transport, they inherit certain aspects that limit their capability to establish valuable correlations between *in vitro* and *in vivo* activity. For example, most wild-type cultured cell models do not provide information regarding single transporters but instead, give an overall assessment of the numerous transport mechanisms. In addition, the degree of transporter expression and the extent of transporter turnover *in vitro* still remains uncertainty [131]. Therefore, *in vitro* cellular models are incapable of resembling the dynamic yet complex physiological nature of the *in vivo* environment, as the net movement of compounds between tissues and blood may be dependent on factors such as membrane permeability, transporter activity, GI transit time at the site of permeation, changes in pH, dietary fluctuations, and protein binding events.

#### 20.4.3.2 Defining Absorption Windows

Drug absorption is a complex process that occurs through potential interaction with a number of physicochemical and physiological variables. Some drugs display regiospecific absorption that can be related to variations in drug solubility and stability as a result of changes in environmental pH along the GI tract, degradation by enzymes present in the intestinal lumen, interaction with endogenous components such as bile [132], as well as regiospecific changes in the expression of facilitated and active transporters [133]. Taking into account these factors, one can imagine that certain drugs may be limited to being absorbed in a specific segment of the GI tract. These particular segments are named "absorption windows." If the drug is a substrate for a transporter, potential passive diffusion will be supplemented by the local expression level of the protein(s), which may differ along the GIT (as discussed above). In general, for influx transporter substrates, regions with high transporter expression will have enhanced permeability, although the process depends also on the DDS transit time, solubility, etc. [33, 134] For efflux transporters, regions with high expression will have reduced permeability unless the transporters are saturated [89].

#### 20.4.3.3 Exploiting Influx Transporters

It has been defined that ideal influx drug transporters will be those with an intermediate substrate affinity, thereby allowing the transport of specific drugs sharing structural similarities. In addition, a high translocation capacity is required so that the protein allows the permeation of therapeutic compounds with a low risk of saturating the binding sites to reduce potential drug–nutrient interactions. Influx transporters (e.g. peptide transporters) provide a unique opportunity to specifically design and develop therapeutic agents with significantly higher molecular recognition with the end goal of increasing intestinal drug absorption and bioavailability. As discussed in Sections 20.2 and 20.4.1, this strategy has been utilized in the development of prodrugs exhibiting high affinity for certain transporters, for instance, PepT1 [135, 136]. Several authors have documented 3- to 5-fold increases in bioavailability for valacyclovir as compared to acyclovir, as mentioned previously [137–140]. Other drugs and prodrugs, such as cephradine, cefaclor, enalapril, and lisinopril among others are PepT substrates. The affinity for this transporter has shown important consequences in the pharmacokinetics of these compounds [6, 141].

Other transmembrane proteins have demonstrated an effect on compound transport. The transport of negatively charged drugs and metabolites by epithelial tissues, particularly the kidney, plays a significant role in controlling their distribution, concentration, and retention in the body. Thus, OATs impact both the therapeutic efficacy and potential toxicity of negatively charged compounds [142]. For instance, OATP-B has been shown to mediate the intestinal absorption of fexofenadine [143] and pravastatin [144]. Therefore, determinations of the intestinal expression patterns of other important OATP/Oatp family members could have critical clinical relevance. Unfortunately, limited *in vivo* data exists to ascertain the full functional significance of this transporter family with respect to oral drug absorption [99].

OCT family members are primarily responsible for the excretion and/or distribution of several drugs and also for serious tissue-specific side-effects, such as cisplatin-induced nephrotoxicity [145]. OCT1 has shown to be an influx transporter with affinity to some clinically used drugs such as oxaliplatin, metformin, and pindolol [146], however, the pharmacokinetic importance of OCT1 has not yet been fully elucidated. Targeting OCTs to increase oral bioavailability of drug substrates is not pharmaceutically striking considering that these substrates are transported across the sinusoidal membrane into the hepatocyte, and further across the canalicular membrane to the bile, thereby potentially increasing their elimination rate as well [99].

In contrast, OCTNs are localized on the apical membranes of intestinal and renal epithelial cells, so targeting OCTNs might be an efficient method to increase drug bioavailability through renal resorption (e.g. cephaloridine) [147]. Thus, these transporters could successfully increase drug bioavailability by both, increasing intestinal absorption and decreasing drug elimination, via a renal reabsorption process [99].

#### 20.4.3.4 Overcoming Efflux Transporter Effects

The inhibition of efflux transporters can be applied to basically two main fields: (i) improve the transport of efflux transporter substrate therapeutic agents into multidrug-resistant cells and (ii) drug delivery [148]. For instance, to circumvent multidrug resistance in cancer cells, researchers focus on the design of novel therapeutic non-substrate agents, the development of formulations which permit the drug to bypass efflux transporter effects [149, 150], and potentially inhibit their activity [151–153].

The low molecular weight efflux transporter inhibitors can be distinguished between three generations as shown in Table 20.3: (i) first-generation P-gp inhibitors are compounds that are in clinical use for other indications and exhibit additional inhibitory properties, (ii) second-generation inhibitors display little to no therapeutic activity and have enhanced inhibitory

**Table 20.3** Classification of P-gp Inhibitors.

| First generation | Second generation | Third generation | Polymeric inhibitors |
|---|---|---|---|
| Amiodarone quinidine verapamil itraconazole ketoconazole felodipine nifedipine nitrendipine chlorpromazine cyclosporine A lidocaine mibefradil progesterone RU 486 | Biricodar GF120918 KR30031 | LY335989 OC144093 (ONT 093) XR9676 | PEG-300 |
| | | | Pluronic P85 |
| | MS-209 | | D-alpha-tocopherol polyethylene glycol 1000 |
| | PSC 833 | | Succinate |
| tamoxifen derivates terfenadine | | | Tween 80 |
| | | | cremophor EL |
| | | | chitosan-4-thiobutylamidine |
| | | | dendrimers (PAMAM) thiomers |

Source: Adapted from Werle et al. [114].

properties, and (iii) third-generation compounds are the most potent and selective inhibitors [148]. First-generation inhibitors do not have clinical application to improve drug bioavailability because they also exhibit pharmacological activity. Second- and third-generation inhibitors have either been developed by pharmaceutical companies or possess extensive intellectual property protection, which limits their commercial availability and reduces publication of studies illustrating their utility [99, 148, 154].

#### 20.4.3.5 Active Pharmaceutical Ingredient and Excipient Effects

Another strategy to increase drug bioavailability via transporters is by modifying their functionality using coadministration of nontherapeutic substances. In this regard, several examples have been previously mentioned concerning efflux transporter inhibitors and an analogous approach has been developed for some influx transporters. For instance, Nozawa et al. [155] showed the stimulation of PepT-like transport by increasing the proton driving force using Eudragit L100-55 (a proton-releasing polymer). Utilizing such excipients and methodologies to enhance drug permeation is certainly becoming more popular for research scientists. Furthermore, identification of excipient interactions with transporters that lead to an increased bioavailability and impact on drug ADMET will be paramount for the future of small molecule drug development in the pharmaceutical industry. However, caution must be taken when trying to utilize excipients to overcome efflux- or metabolism-limited permeation though, as the direct interaction of an excipient with a transporter or enzyme may be construed as a pharmacological interaction. We recently reviewed purported excipient-transporter or enzyme interactions, and in many cases could not rule out alternative means (e.g. enhanced solubility saturating a transporter or enzyme and leading to higher passive permeation) [156]. We proposed that additional studies should be performed to delineate whether or not a direct excipient interaction is occurring or the result is arising from an indirect effect.

#### 20.4.3.6 Food–Drug Interactions

The majority of drugs commercially available worldwide are administered orally and the efficacy of these drugs is highly dependent on the extent of absorption to determine their oral bioavailability. In addition, drug disposition depends on many chemical and physiological variables that help determine the drug's pharmacokinetic/pharmacodynamic (PK/PD) profile. One major complication that affects the PK/PD profile is the effect of food on the rate and extent of nutrient-based drug absorption due to postprandial changes in the GIT [157].

Due to their confounding PK/PD effects, food–drug interactions are commonly associated with significantly altered clinical manifestations of drug therapy [158]. The effects of foods on HIV protease inhibitors have been well studied. For instance, grapefruit and orange juice have been demonstrated to alter SQV transport by inhibiting apically localized P-gp and MRP2 in Caco-2 and porcine kidney epithelial cell lines [159]. Similar effects due to grapefruit juice were demonstrated for indinavir in rat everted ileal sacs [160], although indinavir coadministration with either orange juice, or grapefruit juice in humans had no discernable effects on *in vivo* PK parameters [161].

Although the *in vitro* grapefruit juice effects from rats did not correlate to the human *in vivo* indinavir studies, the literature is replete with examples where these effects are evident. For instance, the OATP and

P-gp substrate fexofenadine, when coadministered with grapefruit, apple, and orange fruit juices decreased the fexofenadine AUC, $C_{max}$, and urinary excretion to 30% to 40% of those in a water control [143]. Interestingly, there were no changes in $T_{max}$, elimination half-life, renal clearance, or urinary volume, indicating that in the case of fexofenadine treatment, fruit juices are more potent inhibitors of OATP isoforms as opposed to P-gp, and have more discernable effects representing such [143]. Another recent case study illustrated the effects of pomelo juice, as opposed to cranberry juice on the PK of cyclosporine in humans [162]. These results, performed in 12 healthy male volunteers, illustrated significant variations in cyclosporine PK parameters suggesting that pomelo juice increased the bioavailability by potentially inhibiting one, or a number of efflux transporters. Clearly, co-ingested substrates can affect transporter proteins involved in intestinal absorption and can lead to either elevated blood plasma levels or virtually no drug absorption at all. The Food and Drug Administration (FDA) has recognized the clinical importance of food–drug interactions and has issued the Guidance for Industry entitled, "Food-Effect Bioavailability and Fed Bioequivalence Studies." [163]

For immediate-release solid oral dosage forms, bioequivalence studies may be waived if the compound meets certain requirements under the Biopharmaceutics Classification System (BCS). The BCS provides a scientific framework to classify compounds based on their solubility and intestinal permeability with the overall goal to account for the three major factors (including dissolution profile) governing intestinal absorption from an immediate-release dosage form. Wu and Benet amended the BCS, instead classifying compounds based on their relative solubility and degree of metabolism, resulting in the Biopharmaceutics Drug Disposition Classification System (BDDCS) [164]; a more comprehensive classification system which accounts for transporter and metabolic effects on drug absorption and disposition. As outlined above, transporter and metabolic effects, and perhaps more importantly, the effects of food–drug and drug–drug interactions on drug transporters and metabolic enzymes, cannot be ignored when attempting to understand the mechanisms by which a compound may traverse biological barriers. To this end, Custodio et al. [165] provided significant insights into the effects of food on drug absorption and disposition, by studying the interplay between compound absorption, elimination, and transport. Utilizing the BDDCS, they were able to accurately predict the potential effects of a high-fat meal on the PK parameters of BDDCS class 1, 2, and 3 compounds:

BDDCS Class 1 compounds, those that are extensively metabolized and have high solubility, would not be expected to exhibit significant transporter–drug interactions. As such, high-fat meals should have no significant effect on the extent of bioavailability. However, high-fat meals may delay stomach emptying and therefore cause an increase in $T_{max}$.

BDDCS Class 2 drugs, poorly soluble, highly metabolized compounds, may be subject to significant transporter effects, particularly efflux transporter effects, due to their limited solubility. As such, high-fat meals may increase bioavailability due to potential inhibition of efflux transporters, such a P-gp, in the intestine. Formulation changes that markedly increase the solubility of Class 2 compounds may decrease or eliminate high-fat meal effects and potentially reduce other transporter–drug interactions.

BDDCS Class 3 compounds are expected to be more susceptible to the influence of uptake transporters, due to their relative permeability. As such, high-fat meals will decrease the extent of bioavailability of Class 3 compounds due to intestinal uptake transporter inhibition. However, if a compound is a substrate for both influx and efflux transporters, the observed effects will depend on the extent of transporter inhibition, as well as the substrates relative affinity for those transporters. This can lead to either an unexpected increase in the extent of bioavailability or no observable meal effect. The $T_{max}$ may also be expected to be increased by a high-fat meal due to the combination of delayed stomach emptying and slower absorption.

It is difficult to predict what effect high-fat meals may have on BDDCS Class 4 compounds, as a combination of both Class 2 and Class 3 compound interactions are possible. However, when high-fat meal effects occur, they are usually manifested as an increase in the extent of bioavailability, resulting from the combination of increased solubilization of the drug in the intestine, as well as potential inhibition of efflux transporter(s).

#### 20.4.3.7 Drug–Drug Interactions

The issue of drug–drug interactions has generated significant concern within the pharmaceutical industry and amongst regulatory authorities in recent years [166]. Transporter-mediated drug–drug interactions are one of the most important potential side-effects of drug therapy and may result in adverse drug events. A number of reports summarize severe clinical incidents where drug–transporter interactions were hypothesized to be responsible for the inhibition of transporter-mediated uptake and efflux in clearance organs (liver and kidney) and at several physiological barriers [167, 168].

The FDA has recognized the importance of drug–drug interactions and has issued the *Draft Guidance for Industry: Drug Interaction Studies—Study Design, Data Analysis, and Implications for Dosing and Labeling* for public comment [169]. This guidance primarily addresses CYP450 and drug–transporter interactions [170], and provides recommendations for industrial evaluation of drug metabolism–transporter interactions during the drug discovery and development process. Briefly, the FDA proposes that an integrated approach, which includes both *in vitro* and *in vivo* studies, can help focus the design of drug interaction studies.

Increasingly, clinical evidence suggests that altering transporter function may be responsible for many of the observed drug–drug interactions, however, the potential effects of drug–metabolic enzyme interactions cannot be ruled out [169]. All of these concerns must be considered when coadministration of drugs is a possibility. Earlier identification for the potential for such interactions will help ensure drug efficacy and patient safety from potential unwanted and serious adverse events.

#### 20.4.3.8 Pharmacogenomics

Pharmacogenomics is defined as a branch of pharmacology which deals with the influence of genetic variation on drug response in patients by correlating gene expression or single-nucleotide polymorphisms (SNPs) with a drug's efficacy or toxicity [171]. A broader understanding of this area and identification of new technological advances is of increasing interest among pharmaceutical scientists in the evolving global market. Techniques such as microarray analysis and real-time polymerase chain reaction (RT–PCR, qPCR) have allowed the quick identification of a multitude of genetic differences between humans based on age, ethnicity, and gender. A number of these differences, such as SNPs, have been identified and can affect transporter functionality. For example, the influence of several commonly occurring PepT1cSNPs, SNPs that occur in the coding region of a transporter gene, have been studied [172]. In short, nine non-synonymous and four synonymous cSNPs were identified in the PepT1 gene in a population of 44 individuals with different ethnic backgrounds. Functional analysis revealed only one of the cSNPs, P586L resulted in significantly reduced Gly-Sar transport capacity ($V_{max}$) and decreased PepT1 protein expression levels.

At least 50 MDR1 gene SNPs MDR1 (P-glycoprotein) have been identified with several exhibiting the potential to affect patient PK/PD parameters [173]. Differences in MDR1 gene alleles have been found between Central Americans and Spaniards [174], as well as between European and African Americans [175]. Not only does the inherent genetic diversity seen between human populations complicate our understanding of SNP diversity and relevance, but also the variation in gene expression observed between humans and other species commonly utilized to study *in vivo* drug activity [176]. Given the gene expression variability seen in some commonly utilized *in vitro* cell models, such as Caco-2 cells [177], it is clear that a better understanding of the genetic differences across these pharmaceutical screening tools needs to be explored in order to treat specific patient populations.

## 20.5 Conclusions

Peptide and peptide-based drug absorption has long been a topic of considerable debate and interest among pharmaceutical scientists. Although there are exist mechanisms of protein absorption within the GIT, the bulk of nitrogen absorption occurs through the uptake and transport of amino acids and small peptides through the GIT. The restrictive, yet highly specialized transport properties exhibited by the GIT convey a protective functionality, resulting in increased difficulty in designing clinically effective agents. As such, there is high interest in identifying the molecular requirements a compound should exhibit to maximize its absorption characteristics. This is particularly important for peptides and peptide-based agents where passive permeation across the intestinal epithelium is limited due to the compound's hydrophilicity and therefore based primarily on the activity of transporter proteins. Understanding the individual contribution of these transporters can help target drug and/or formulation design to optimize the compound's overall absorption.

Of equal importance is the presence of these transporters in other organs such as the liver, kidney, and brain that affect the disposition of such compounds after absorption. In addition to the interactions of substrates with transporters, one must consider all of the potential interactions of all entities potentially coming into contact with the substrate and transporter of interest. Other drugs, excipients, fasted versus fed state, and the specific components of the meal can all affect a compound's bioavailability and/or potential toxicity. A lack of knowledge of such interactions with drugs such as those with a narrow therapeutic window, could significantly affect patient safety. Furthermore, a detailed knowledge of the pharmacogenomic differences within and between patients' needs to be further studied in order to properly treat specific patient populations. Also, knowledge of species differences is crucial when trying to make in *vitro/in vivo* correlations from such data.

In conclusion, knowledge in this field must move from a specific focus on well-established transporters, such as PepT1, to a broader understanding of all transporters and factors that affect peptide and peptide-based drug ADMET. It is growing increasingly more important to develop screening tools that will enable early detection of peptide-based drug affinities and capacities for these transporters to rapidly assess their developability. Future research in this field will no doubt utilize new technologies and techniques to make such evaluations with the goal to treat individuals and continue to advance the field.

## Acknowledgments

We would like to thank Drs. Stephen Carl, Olafur Gudmundsson, Dea Herrera-Ruiz, David J. Lindley, Li F. Pan, Lori B. Ward, and Jonathan M.E. Goole for their significant contributions to the initial version of this currently revised chapter. The authors would like to acknowledge financial support from the National Institute of General Medical Sciences (R01-GM65448) for the support of our earlier chapter version and work in this field.

## References

1 Di, L. (2015). Strategic approaches to optimizing peptide ADME properties. *AAPS J.* 17 (1): 134–143.

2 Matthews, D.M. (1991). *Protein Absorption: Development and Present State of the Subject*. New York, NY: Wiley-Liss, Inc.

3 Carl, S.M., Herrera-Ruiz, D., Bhardwaj, R.K. et al. (2007). Chapter 6: Mammalian oligopeptide transporters. In: *Drug Transporters: Molecular Characterization and Role in Drug Disposition* (ed. G. You and M.E. Morris), 105–145. Hoboken, NJ: John Wiley & Sons, Inc.

4 Steinhardt, H.J. and Adibi, S.A. (1986). Kinetics and characteristics of absorption from an equimolar mixture of 12 flycyl-dipeptides in human jejunum. *Gastroenterology* 90: 577–582.

5 Devlin, T.M. (2002). *Textbook of Biochemistry with Clinical Correlations*, 5e, 1097–1100. New York: John Wiley & Sons, Inc.

6 Brandsch, M., Knutter, I., and Bosse-Doenecke, E. (2008). Pharmaceutical and pharmacological importance of peptide transporters. *J. Pharm. Pharmacol.* 60: 543–585.

7 Washington, N., Washington, C., and Wilson, C.G. (2001). *Physiological Pharmaceutics: Barriers to Drug Absorption*, 2e, 109–142. New York: Taylor & Francis.

8 Tortora, G.J. and Grabowski, S.R. (1993). *Principles of Anatomy and Physiology*, 768. New York: Harper Collins College Publishers.

9 Knipp, G.T., Vander Velde, D., Siahaan, T.J., and Borchardt, R.T. (1997). The effect of β-turn structure on the passive diffusion of peptides across Caco-2 monolayers. *Pharm. Res.* 14: 1332–1340.

10 Knipp, G.T., Ho, N.F.H., Barsuhn, C.L., and Borchardt, R.T. (1997). Delineation of paracellular diffusion in Caco-2 monolayers effect of perturbation on the transport of model compounds which vary in charge and size. *J. Pharm. Sci.* 86: 1105–1110.

11 Oh, D.-M. and Amidon, G.L. (1999). Overview of membrane transport. In: *Membrane Transporters as Drug Targets* (ed. G.L. Amidon and W. Sadee), 1–27. New York, NY: Kluwer Academic/Plenum Publishers.

12 Dobson, P.D. and Kell, D.B. (2008). Carrier-mediated cellular uptake of pharmaceutical drugs: an exception or the rule? *Nat. Rev. Drug Discovery* 7 (3): 205–220.

13 Kell, D.B. and Oliver, S.G. (2014). How drugs get into cells: tested and testable predictions to help discriminate between transporter-mediated uptake and lipoidal bilayer diffusion. *Front. Pharmacol.* 5: 231.

14 Di, L., Artursson, P., Avdeef, A. et al. (2020). The critical role of passive permeability in designing successful drugs. *Chem. Med. Chem.* 15 (20): 1862–1874.

15 Sugano, K., Kansy, M., Artursson, P. et al. (2010). Coexistence of passive and carrier-mediated processes in drug transport. *Nat. Rev. Drug Discovery* 9 (8): 597–614.

16 Smith, D., Artursson, P., Avdeef, A. et al. (2014). Passive lipoidal diffusion and carrier-mediated cell uptake are both important mechanisms of membrane permeation in drug disposition. *Mol. Pharmaceutics* 11 (6): 1727–1738.

17 Knipp, G.T., Vander Velde, D.G., Siahaan, T.J., and Borchardt, R.T. (1997). The effect of beta-turn structure on the passive diffusion of peptides across Caco-2 cell monolayers. *Pharm. Res.* 14 (10): 1332–1340.

18 Gudmundsson, O.S., Vander Velde, D.G., Jois, S.D. et al. (1999). The effect of conformation of the acyloxyalkoxy-based cyclic prodrugs of opioid peptides on their membrane permeability. *J. Pept. Res.* 53 (4): 403–413.

19 Gudmundsson, O.S., Jois, S.D., Vander Velde, D.G. et al. (1999). The effect of conformation on the membrane permeation of coumarinic acid- and phenylpropionic acid-based cyclic prodrugs of opioid peptides. *J. Pept. Res.* 53 (4): 383–392.

20 Wang, B., Nimkar, K., Wang, W. et al. (1999). Synthesis and evaluation of the physicochemical properties of esterase-sensitive cyclic prodrugs of opioid peptides using coumarinic acid and phenylpropionic acid linkers. *J. Pept. Res.* 53 (4): 370–382.

21 Bak, A., Siahaan, T.J., Gudmundsson, O.S. et al. (1999). Synthesis and evaluation of the physicochemical properties of esterase-sensitive cyclic prodrugs of opioid peptides using an (acyloxy)alkoxy linker. *J. Pept. Res.* 53 (4): 393–402.

22 Nielsen, D.S., Shepherd, N.E., Xu, W. et al. (2017). Orally absorbed cyclic peptides. *Chem. Rev.* 117 (12): 8094–8128.

23 Zamek-Gliszczynski, M.J., Taub, M.E., Chothe, P.P. et al. (2018). Transporters in drug development: 2018 ITC recommendations for transporters of emerging clinical importance. *Clin. Pharmacol. Ther.* 104 (5): 890–899.

24 Tweedie, D., Polli, J.W., Berglund, E.G. et al. (2013). Transporter studies in drug development: experience to date and follow-up on decision trees from the International Transporter Consortium. *Clin. Pharmacol. Ther.* 94 (1): 113–125.

25 Bhardwaj, R.K., Herrera-Ruiz, D.R., Xu, Y. et al. (2008). Chapter 7: Intestinal transporters in drug absorption. In: *Biopharmaceutics Applications in Drug Development* (ed. R. Krishna and L. Yu), 175–261. New York, NY: Kluwer Press.

26 Lin, L., Yee, S.W., Kim, R.B., and Giacomini, K.M. (2015). SLC transporters as therapeutic targets: emerging opportunities. *Nat. Rev. Drug Discovery* 14 (8): 543–560.

27 Pizzagalli, M.D., Bensimon, A., and Superti-Furgam, G. (2020). A guide to plasma membrane solute carrier proteins. *FEBS J.* https://febs.onlinelibrary.wiley.com/doi/10.1111/febs.15531.

28 Rives, M.L., Javitch, J.A., and Wickenden, A.D. (2017). Potentiating SLC transporter activity: Emerging drug discovery opportunities. *Biochem. Pharmacol.* 135: 1–11.

29 Liu, X. (2019). SLC family transporters. *Adv. Exp. Med. Biol.* 1141: 101–202.

30 Liu, X. (2019). ABC family transporters. *Adv. Exp. Med. Biol.* 1141: 13–100.

31 Schinkel, A.H. and Jonker, J.W. (2003). Mammalian drug efflux transporters of the ATP binding cassette (ABC) family: an overview. *Adv. Drug Delivery Rev.* 55 (1): 3–29.

32 Hillgren, K.M., Keppler, D., Zur, A.A. et al. (2013). Emerging transporters of clinical importance: an update from the International Transporter Consortium. *Clin. Pharmacol. Ther.* 94 (1): 52–63.

33 Herrera-Ruiz, D., Wang, Q., Gudmundsson, O.S. et al. (2001). Spatial expression patterns of peptide transporters in the human and rat gastrointestinal tracts, Caco-2 in vitro cell culture model, and multiple human tissues. *AAPS Pharm. Sci.* 3 (1): E9.

34 Sissung, T.M., Gardner, E.R., Gao, R., and Figg, W.D. (2008). Pharmacogenetics of membrane transporters: a review of current approaches. *Methods Mol. Biol.* 448: 41–62.

35 Herrera-Ruiz, D. and Knipp, G.T. (2003). Current perspectives on established and putative mammalian oligopeptide transporters. *J. Pharm. Sci.* 92: 691–714.

36 Covitz, Y.K.-M., Amidon, G.L., and Sadée, W. (1998). Membrane topology of the human dipeptides transporter, hPepT1, determined by epitope insertions. *Biochemistry* 37: 15214–15221.

37 Lee, V.H.L. (2000). Membrane transporters. *Eur. J. Pharm. Sci.* 11: S41–S50.

38 Steiner, H.Y., Naider, F., and Becker, J.M. (1995). The PTR family: a new group of peptide transporters. *Mol. Microbiol.* 16: 825–834.

39 Fei, Y.J., Ganapathy, V., and Leibach, F.H. (1998). Molecular and structural features of the proton-coupled oligopeptide transporter superfamily. *Prog. Nucleic Acid Res. Mol. Biol.* 58: 239–261.

40 Saier, M.H. Jr., Eng, B.H., Fard, S. et al. (1999). Phylogenetic characteristics of novel transport protein families revealed by genome analyses. *Biochim. Biophys. Acta* 1422: 1–56.

41 Nussberger, S., Steel, A., Trott, D. et al. (1997). Symmetry of H+ binding to the intra- and extracellular side of the H+-coupled oligopeptide cotransporter PepT1. *J. Biolumin. Chemilumin.* 272: 7777–7785.

42 Kennedy, D.J., Leibach, F.H., Ganapathy, V., and Thwaites, D.T. (2002). Optimal absorptive transport

of the dipeptide glycylsarcosine is dependent on functional Na+/H+ exchange activity. *Pflugers. Arch. - Eur. J. Physiol.* 445: 139–146.

43 Thwaites, D.T., Kennedy, D.J., Raldua, D. et al. (2002). H/dipeptide absorption across the human intestinal epithelium is controlled indirectly via a functional Na/H exchanger. *Gastroenterology* 122: 1322–1333.

44 Rubio-Aliaga, I. and Daniel, H. (2002). Mammalian peptide transporters as targets for drug delivery. *Trends Pharmacol. Sci.* 23: 434–440.

45 Boll, M., Herget, M., Wagener, M. et al. (1996). Expression cloning and functional characterization of the kidney cortex high-affinity proton-coupled peptide transporter. *Proc. Natl. Acad. Sci. U.S.A.* 93: 284–289.

46 Shu, C., Shen, H., Hopfer, U., and Smith, D.E. (2001). Mechanism of intestinal absorption and renal reabsorption of an orally active ace inhibitor: uptake and transport of fosinopril in cell cultures. *Drug Metab. Dispos.* 29: 1307–1315.

47 Tamai, I., Nakanishi, T., Hayashi, K. et al. (1997). The predominant contribution of oligopeptide transporter PepT1 to intestinal absorption of beta-lactam antibiotics in the rat small intestine. *J. Pharm. Pharmacol.* 49: 796–801.

48 Tsuji, A., Tamai, I., Nakanishi, M., and Amidon, G.L. (1990). Mechanism of absorption of the dipeptide alpha-methyldopa-phe in intestinal brush-border membrane vesicles. *Pharm. Res.* 7: 308–309.

49 Wang, H.P., Lu, H.H., Lee, J.S. et al. (1996). Intestinal absorption studies on peptide mimetic alpha-methyldopa prodrugs. *J. Pharm. Pharmacol.* 48: 270–276.

50 Ganapathy, M.E., Huang, W., Wang, H. et al. (1998). Valacyclovir: a substrate for the intestinal and renal peptide transporters PepT1 and PepT2. *Biochem. Biophys. Res. Commun.* 246: 470–475.

51 Sugawara, M., Huang, W., Fei, Y.J. et al. (2000). Transport of valganciclovir, a ganciclovir prodrug, via peptide transporters PEPT1 and PEPT2. *J. Pharm. Sci.* 89: 781–789.

52 Neumann, J., Bruch, M., Gebauer, S., and Brandsch, M. (2004). Transport of the phosphonodipeptide alafosfalin by the H+/peptide cotransporters PEPT1 and PEPT2 in intestinal and renal epithelial cells. *Eur. J. Biochem.* 271: 2012–2017.

53 Yamashita, T., Shimada, S., Guo, W. et al. (1997). Cloning and functional expression of a brain peptide/histidine transporter. *J. Biolumin. Chemilumin.* 272: 10205–10211.

54 Sakata, K., Yamashita, T., Maeda, M. et al. (2001). Cloning of a lymphatic peptide/histidine transporter. *Biochem. J* 356: 53–60.

55 Bhardwaj, R.K., Herrera-Ruiz, D., Eltoukhy, N. et al. (2006). The functional evaluation of human peptide/histidine transporter 1 (hPHT1) in transiently transfected COS-7 cells. *Eur. J. Pharm. Sci.* 27: 533–542.

56 Hagenbuch, B. and Meier, P.J. (2003). The superfamily of organic anion transporting polypeptides. *Biochim. Biophys. Acta* 1609: 1–18.

57 van Montfoort, J.E., Hagenbuch, B., Groothuis, G.M. et al. (2003). Drug uptake systems in liver and kidney. *Curr. Drug Metab.* 4: 185–211.

58 Miyazaki, H., Sekine, T., and Endou, H. (2004). The multispecific organic anion transporter family: properties and pharmacological significance. *Trends Pharmacol. Sci.* 25: 654–662.

59 Koepsell, H. and Endou, H. (2004). The SLC22 drug transporter family. *Pflugers Arch.* 447: 666–676.

60 You, G. (2004). The role of organic ion transporters in drug disposition: an update. *Curr. Drug Metab.* 5: 55–62.

61 Li, L., Lee, T.K., Meier, P.J., and Ballatori, N. (1998). Identification of glutathione as a driving force and leukotriene C4 as a substrate for oatp1, the hepatic sinusoidal organic solute transporter. *J. Biolumin. Chemilumin.* 273: 16184–16191.

62 Li, L., Meier, P.J., and Ballatori, N. (2000). Oatp2 mediates bidirectional organic solute transport: a role for intracellular glutathione. *Mol. Pharmacol.* 58: 335–340.

63 Kobayashi, D., Nozawa, T., Imai, K. et al. (2003). Involvement of human organic anion transporting polypeptide OATP-B (SLC21A9) in pH-dependent transport across intestinal apical membrane. *J. Pharmacol. Exp. Ther.* 306: 703–708.

64 Kato, K., Shirasaka, Y., Kuraoka, E. et al. (2010). Intestinal absorption mechanism of tebipenem pivoxil, a novel oral carbapenem: involvement of human OATP family in apical membrane transport. *Mol. Pharmaceutics* 7 (5): 1747–1756.

65 Koepsell, H., Schmitt, B.M., and Gorboulev, V. (2003). Organic cation transporters. *Rev. Physiol., Biochem. Pharmacol.* 150: 36–90.

66 Busch, A.E., Quester, S., Ulzheimer, J.C. et al. (1996). Electrogenic properties and substrate specificity of the polyspecific rat cation transporter rOCT1. *J. Biolumin. Chemilumin.* 271: 32599–32604.

67 Busch, A.E., Karbach, U., Miska, D. et al. (1998). Human neurons express the polyspecific cation transporter hOCT2, which translocates monoamine

neurotransmitters, amantadine, and memantine. *Mol. Pharmacol.* 54: 342–352.
68 Gorboulev, V., Ulzheimer, J.C., Akhoundova, A. et al. (1997). Cloning and characterization of two human polyspecific organic cation transporters. *DNA Cell Biol.* 16: 871–881.
69 Kekuda, R., Prasad, P.D., Wu, X. et al. (1998). Cloning and functional characterization of a potential-sensitive, polyspecific organic cation transporter (OCT3) most abundantly expressed in placenta. *J. Biolumin. Chemilumin.* 273: 15971–15979.
70 Tamai, I., Ohashi, R., Nezu, J.I. et al. (2000). Molecular and functional characterization of organic cation/carnitine transporter family in mice. *J. Biolumin. Chemilumin.* 275: 40064–40072.
71 Tamai, I., China, K., Sai, Y. et al. (2001). Na(+)-coupled transport of L-carnitine via high-affinity carnitine transporter OCTN2 and its subcellular localization in kidney. *Biochim. Biophys. Acta* 1512: 273–284.
72 Tamai, I., Ohashi, R., Nezu, J.I. et al. (1998). Molecular and functional identification of sodium ion-dependent, high affinity human carnitine transporter OCTN2. *J. Biolumin. Chemilumin.* 273: 20378–20382.
73 Wagner, C.A., Lükewille, U., Kaltenbach, S. et al. (2000). Functional and pharmacological characterization of the human Na+/carnitine cotransporter hOCTN2. *Am. J. Physiol. Renal. Physiol.* 279: F584–F591.
74 Halestrap, A.P. and Price, N.T. (1999). The proton-linked monocarboxylate transporter (MCT) family: structure, function and regulation. *Biochem. J* 343: 281–299.
75 Halestrap, A.P. and Meredith, D. (2004). The SLC16 gene family-from monocarboxylate transporters (MCTs) to aromatic amino acid transporters and beyond. *Pflugers Arch. - Eur. J. Physiol.* 447: 619–628.
76 Philp, N.J., Ochrietor, J.D., Rudoy, C. et al. (2003). Loss of MCT1, MCT3 and MCT4 expression in the retinal pigment epithelium and neural retina of the 5A11/basigin-null mouse. *Invest. Ophthalmol. Vis. Sci.* 44: 1305–1311.
77 Li, Y.H., Ito, K., Tsuda, Y. et al. (1999). Mechanism of intestinal absorption of an orally active beta-lactam prodrug: uptake and transport of carindacillin in Caco-2 cells. *J. Pharmacol. Exp. Ther.* 290: 958–964.
78 Itoh, T., Tanno, M., Li, Y.-H., and Yamada, H. (1998). Transport of phenethicillin into rat intestinal brush border membrane vesicles: role of the monocarboxylic acid transport system. *Int. J. Pharm.* 172: 103–112.
79 Emoto, A., Ushigome, F., Koyabu, N. et al. (2002). H(+)-linked transport of salicylic acid, an NSAID, in the human trophoblast cell line BeWo. *Am. J. Phys* 282: C1064–C1075.
80 Utoguchi, N. and Audus, L.L. (2000). Carrier-mediated transport of valproic acid in BeWo cells, a human trophoblast cell line. *Int. J. Pharm.* 195: 115–124.
81 Hosoya, K., Kondo, T., Tomi, M. et al. (2001). MCT1-mediated transport of L-lactic acid at the inner blood-retinal barrier: a possible route for delivery of monocarboxylic acid drugs to the retina. *Pharm. Res.* 18: 1669–1676.
82 Wu, X.C., Whitfield, L.R., and Stewart, B.H. (2000). Atorvastatin transport in the Caco-2 cell model: contributions of P-glycoprotein and the proton-monocarboxylic acid co-transporter. *Pharm. Res.* 17: 209–215.
83 Okamura, A., Emoto, A., Koyabu, N. et al. (2002). Transport and uptake of nateglinide in Caco-2 cells and its inhibitory effect on human monocarboxylate transporter MCT1. *Br. J. Pharmacol.* 137: 391–399.
84 Varma, M.V., Ambler, C.M., Ullah, M. et al. (2010). Targeting intestinal transporters for optimizing oral drug absorption. *Curr. Drug Metab.* 11 (9): 730–742.
85 Dantzig, A.H., Hoskins, J.A. et al. (1994). Association of intestinal peptide transport with a protein related to the cadherin superfamily. *Science* 264: 430–433.
86 Hu, M., Chen, J. et al. (1994). Mechanism and kinetics of transcellular transport of a new beta-lactam antibiotic loracarbef across an intestinal epithelial membrane model system (Caco-2). *Pharm. Res.* 11: 1405–1413.
87 Yang CY. (1998) Studies on the human intestinal di-/tripeptide transporter HPT-1 as a potential carrier for orally administered drugs. Thesis, Purdue University.
88 Szakacs, G., Varadi, A., Ozvegy-Laczka, C., and Sarkadi, B. (2008). The role of ABC transporters in drug absorption, distribution, metabolism, excretion and toxicity (ADME-Tox). *Drug Discovery* 13: 379–393.
89 Murakami, T. and Takano, M. (2008). Intestinal efflux transporters and drug absorption. *Expert Opin. Drug Metab. Toxicol.* 4: 923–939.
90 Glavinas, H., Krajcsi, P., Cserepes, J., and Sarkadi, B. (2004). The role of ABC transporters in drug resistance, metabolism and toxicity. *Curr. Drug Delivery* 1: 27–42.

**91** Gottesman, M.M. and Ling, V. (2006). The molecular basis of multidrug resistance in cancer: the early years of P-glycoprotein research. *FEBS Lett.* 580: 998–1009.

**92** Ford, J.M. and Hait, W.N. (1990). Pharmacology of drugs that alter multidrug resistance in cancer. *Pharmcol. Rev.* 42: 155–199.

**93** Golden, P.L. and Pollack, G.M. (2003). Blood-brain barrier efflux transport. *J. Pharm. Sci.* 92: 1739–1753.

**94** Ritz, U. and Seliger, B. (2001). The transporter associated with antigen processing (TAP): structural integrity, expression, function, and its clinical relevance. *Mol. Med.* 7: 149–158.

**95** Lankat-Buttgereit, B. and Tampé, R. (2002). The transporter associated with antigen processing: function and implications in human diseases. *Physiol. Rev.* 82: 187–204.

**96** Lu, J., Wettstein, P.J., Higashimoto, Y. et al. (2001). TAP-independent presentation of CTL epitopes by Trojan antigens. *J. Immunol.* 166: 7063–7071.

**97** Herget, M. and Tampé, R. (2007). Intracellular peptide transporters in human -compartmentalization of the "peptidome". *Pflugers Arch.* 453: 591–600.

**98** Bera, T.K., Lee, S., Salvatore, G. et al. (2001). MRPS. a new member of ABC transporter superfamily, identified by EST database mining and gene prediction program, is highly expressed in breast cancer. *Mol. Med.* 7: 509–516.

**99** Bhardwaj, R.K., Herrera-Ruiz, D.R., Xu, Y. et al. (2008). Intestinal transporters in drug absorption. In: *Biopharmaceutics Applications in Drug Development* (ed. R. Krishna and L. Yu), 175–261. New York, NY: Kluwer Press.

**100** Burg, D., Wielinga, P., Zelcer, N. et al. (2002). Inhibition of the multidrug resistance protein 1 (MRP1) by peptidomimetic glutathione-conjugate analogs. *Mol. Pharmacol.* 62: 1160–1166.

**101** Kruh, G.D. and Belinsky, M.G. (2003). The MRP family of drug efflux pumps. *Oncogene* 22: 7537–7552.

**102** Suzuki, H. and Sugiyama, Y. (1998). Excretion of GSSG and glutathione conjugates mediated by MRP1 and cMOAT/MRP2. *Semin. Liver Dis.* 18: 359–376.

**103** Kusahara, H. and Sugiyama, Y. (2007). ATP-binding cassette, subfamily G (ABCG family). *Eur. J. Pharmacol.* 453: 735–744.

**104** Mao, Q.C. and Unadkat, J.D. (2005). Role of the breast cancer resistance protein (ABCG2) in drug transport. *AAPS J.* 7: E118–E133.

**105** Döring, F., Will, J., Amasheh, S. et al. (1998). Minimal molecular determinants of substrates for recognition by the intestinal peptide transporter. *J. Biolumin. Chemilumin.* 273: 23211–23218.

**106** Groneberg, D.A., Fischer, A., Chung, K.F., and Daniel, H. (2004). Molecular mechanisms of pulmonary peptidomimetic drug and peptide transport. *Am. J. Respir. Cell Mol. Biol.* 30: 251–260.

**107** Herrera-Ruiz, D., Wang, Q., Gudmundsson, O.S. et al. (2001). Spatial expression patterns of peptide transporters in the human and rat gastrointestinal tracts, Caco-2 in vitro cell culture, and multiple human tissues. *AAPS Pharm. Sci.* 3:Article 9.

**108** Kobayashi, T., Shimabukuro-Demoto, S., Yoshida-Sugitani, R. et al. (2014). The histidine transporter SLC15A4 coordinates mTOR-dependent inflammatory responses and pathogenic antibody production. *Immunity* 41 (3): 375–388.

**109** Song, F., Yi, Y., Li, C. et al. (2018). Regulation and biological role of the peptide/histidine transporter SLC15A3 in Toll-like receptor-mediated inflammatory responses in macrophage. *Cell Death Dis.* 9 (7): 770.

**110** Kobayashi, T., Tsutsui, H., Shimabukuro-Demoto, S. et al. (2017). Lysosome biogenesis regulated by the amino-acid transporter SLC15A4 is critical for functional integrity of mast cells. *Int. Immunol.* 29 (12): 551–566.

**111** Viennois, E., Pujada, A., Zen, J., and Merlin, D. (2018). Function, regulation, and pathophysiological relevance of the POT superfamily, specifically PepT1 in inflammatory bowel disease. *Compr. Physiol.* 8 (2): 731–760.

**112** Brandsch, M. (2009). Transport of drugs by proton-coupled peptide transporters: pearls and pitfalls. *Expert Opin. Drug Metab. Toxicol.* 5: 887–905.

**113** Brandsch, M., Knutter, I., and Leibach, F.H. (2004). The intestinal H+/peptide symporter PEPT1: structure-affinity relationships. *Eur. J. Pharm. Sci.* 21 (1): 53–60.

**114** Bailey, P.D., Boyd, C.A., Bronk, J.R. et al. (1999). How to make drugs orally active: a substrate model for the peptide transporter PepT1. *Angew. Chem. Int. Ed.* 39: 505–508.

**115** Swaan, P.W. and Tukker, J.J. (1997). Molecular determinants of recognition for the intestinal peptide carrier. *J. Pharm. Sci.* 86: 596–602.

**116** Zhang, E.Y., Knipp, G.T., Ekins, S., and Swaan, P.W. (2002a). Structural biology and function of solute transporters: implications for identifying and designing substrates. *Drug Metab. Rev.* 34: 709–750.

117 Biegel, A., Gebauer, S., Hartrodt, B. et al. (2005). Three-dimensional quantitative structure-activity relationship analyses of beta-lactam antibiotics and tripeptides as substrates of the mammalian H+/peptide cotransporter PEPT1. *J. Med. Chem.* 48 (13): 4410–4409.
118 Zhang, E.Y., Mitch, A.P., Cheng, C. et al. (2002). Modeling of active transport systems. *Adv. Drug Delivery Rev.* 54: 329–354.
119 Roth, M., Obaidat, A., and Hagenbuch, B. (2012). OATPs, OATs and OCTs: the organic anion and cation transporters of the SLCO and SLC22A gene superfamilies. *Br. J. Pharmacol.* 165 (5): 1260–1287.
120 Zhang, L., Dresser, M.J., Gray, A.T. et al. (1997). Cloning and functional expression of a human liver organic cation transporter. *Mol. Pharmacol.* 51: 913–921.
121 Tamai, I., Nezu, J., Uchino, H. et al. (2000). Molecular identification and characterization of novel members of the human organic anion transporter (OATP) family. *Biochem. Biophys. Res. Commun.* 273: 251–260.
122 Li, F., Maag, H., and Afredson, T. (2008). Prodrugs of nucleoside analogues for improved oral absorption and tissue targeting. *J. Pharm. Sci.* 97: 1109–1134.
123 Agarwal, S., Boddu, S.H.S., Jain, R. et al. (2008). Peptide prodrugs: Improved oral absorption of lopinavir, a HIV protease inhibitor. *Int. J. Pharm.* 359: 7–14.
124 Jain, R., Agarwal, S., Majumdar, S. et al. (2005). Evasion of P-gp mediated cellular efflux and permeability enhancement of HIV-protease inhibitor saquinavir by prodrug modification. *Int. J. Pharm.* 303: 8–19.
125 Jain, R., Agarwal, S., Mandava, N.K. et al. (2008). Interaction of dipeptide prodrugs of saquinavir with multidrug resistance protein-2 (MRP-2): evasion of MRP-2. *Int. J. Pharm.* 362: 44–51.
126 Jain, R., Duvvuri, S., Kansara, V. et al. (2007). Intestinal absorption of novel-dipeptide prodrugs of saquinavir in rats. *Int. J. Pharm.* 336: 233–240.
127 Mathis, A.S., Friedman, G.S., and Knipp, G.T. (2002). Do sex and ethnicity influence drug pharmacokinetics in solid organ transplantation? *Graft* 5: 294–302.
128 Hendrikse, N.H., Bart, J., de Vries, E.G. et al. (2008). P-glycoprotein at the blood-brain barrier and analysis of drug transport with positron-emission tomography. *J. Clin. Pharmacol.* 41: 48S–54S.
129 Takakura, Y., Audus, K.L., and Borchardt, R.T. (1992). Cultured brain microvessel endothelial cells as in vitro models of the blood-brain barrier. *NIDA Res. Monogr.* 120: 138–152.
130 Letrent, S.P., Polli, J.W., Humphreys, J.E. et al. (1999). P-glycoprotein-mediated transport of morphine in brain capillary endothelial cells. *Biochem. Pharmacol.* 58: 951–957.
131 Regina, A., Romero, I.A., Greenwood, J. et al. (1999). Dexamethasone regulation of P-glycoprotein activity in an immortalized rat brain endothelial cell line. *J. Neurochem.* 73: 1954–1963.
132 Charman, W.N., Porter, C.J., Mithani, S., and Dressman, J.B. (1997). Physiochemical and physiological mechanisms for the effects of food on drug absorption: the role of lipids and pH. *J. Pharm. Sci.* 86: 269–282.
133 Martinez, M.N. and Amidon, G.L. (2002). A mechanistic approach to understanding the factors affecting drug absorption: a review of fundamentals. *J. Clin. Pharmacol.* 42: 620–643.
134 Herrera-Ruiz, D., Faria, T.N., Bhardwaj, R.K. et al. (2004). A novel hPepT1 stably transfected cell line: establishing a correlation between expression and function. *Mol. Pharmaceutics* 1: 136–144.
135 Friedrichsen, G.M., Chen, W., Begtrup, M. et al. (2002). Synthesis of analogs of L-valacyclovir and determination of their substrate activity for the oligopeptide transporter in Caco-2 cells. *Eur. J. Pharm. Sci.* 16: 1–13.
136 Bhardwaj, R.K., Herrera-Ruiz, D., Sinko, P.J. et al. (2005). Delineation of HPEPT1 meditated uptake and transport of substrates with varying transporter affinities utilizing stably transfected HPEPT1/MDCK clones and Caco-2 cells. *J. Pharmacol. Exp. Ther.* 314: 1093–1100.
137 Weller, S., Blum, M., Doucette, M. et al. (1993). Pharmacokinetics of the aciclovir pro-drug valaciclovir after escalating single- and multiple-dose administration to normal volunteers. *Clin. Pharmacol. Ther.* 54: 595–605.
138 Soul-Lawton, J., Seaber, E., On, N. et al. (1995). Absolute bioavailability and metabolic disposition of valaciclovir, the L-valyl ester of aciclovir, following oral administration to humans. *Antimicrob. Agents Chemother.* 39: 2759–2764.
139 Steingrimsdottir, H., Gruber, A., Palm, C. et al. (2000). Bioavailability of aciclovir after oral administration of aciclovir and its prodrug valaciclovir to patients with leukopenia after chemotherapy. *Antimicrob. Agents Chemother.* 44: 207–209.
140 MacDougall, C. and Guglielmo, B.J. (2004). Pharmacokinetics of valaciclovir. *J. Antimicrob. Chemother.* 53: 899–901.

141 Rubio-Aliaga, I. and Daniel, H. (2008). Peptide transporters and their roles in physiological processes and drug disposition. *Xenobiotica* 38: 1022–1042.
142 Srimaroeng, C., Perry, J.L., and Pritchard, J.B. (2008). Physiology, structure, and regulation of the cloned organic anion transporters. *Xenobiotica* 2008 (38): 889–935.
143 Dresser, G.K., Bailey, D.G., Leake, B.F. et al. (2002). Fruit juices inhibit organic anion transporting polypeptide-mediated drug uptake to decrease the oral availability of fexofenadine. *Clin. Pharmacol. Ther.* 71: 11–20.
144 Triscari, J., O'Donnell, D., Zinny, M., and Pan, H.Y. (1995). Gastrointestinal absorption of pravastatin in healthy subjects. *J. Clin. Pharmacol.* 35: 142–144.
145 Ciarimboli, G. (2008). Organic cation transporters. *Xenobiotica* 38: 936–971.
146 Shu, Y., Brown, C., Castro, R.A. et al. (2008). Effect of genetic variation in the organic cation transporter 1, OCT1, on metformin pharmacokinetics. *Clin. Pharmacol. Ther.* 83: 273–280.
147 Kano, T., Kato, Y., Ito, K. et al. (2009). Carnitine/organic cation transporter OCTN2 (Slc22a5) is responsible for renal secretion of cephaloridine in mice. *Drug Metab. Dispos.* 37 (5): 1009–1016.
148 Werle, M. (2008). Polymeric and low molecular mass efflux transporter inhibitors for oral drug delivery. *J. Pharm. Sci.* 97: 60–70.
149 Mazel, M., Clair, P., Rousselle, C. et al. (2001). Doxorubicin–peptide conjugates overcome multidrug resistance. *Anti-Cancer Drug Des.* 12: 107–116.
150 Raub, T.J. (2006). P-glycoprotein recognition of substrates and circumvention through rational drug design. *Mol. Pharmacol.* 3: 3–25.
151 Varma, M.V., Ashokraj, Y., Dey, C.S., and Panchagnula, R. (2003). P-glycoprotein inhibitors and their screening: a perspective from bioavailability enhancement. *Pharmacol. Res.* 48: 347–359.
152 Woo, J.S., Lee, C.H., Shim, C.K., and Hwang, S.J. (2003). Enhanced oral bioavailability of paclitaxel by coadministration of the P-glycoprotein inhibitor KR30031. *Pharmacol. Res.* 20: 24–30.
153 Föger, F., Malaivijitnond, S., Wannaprasert, T. et al. (2008). Effect of a thiolated polymer on oral paclitaxel absorption and tumor growth in rats. *J. Drug Targeting* 16: 149–155.
154 Werle, M. (2008). Natural and synthetic polymers as inhibitors of drug efflux pumps. *Pharm. Res.* 25: 500–511.
155 Nozawa, T., Toyobuku, H., Kobayashi, D. et al. (2003). Enhanced intestinal absorption of drugs by activation of peptide transporter PEPT1 using proton-releasing polymer. *J. Pharm. Sci.* 92: 2208–2216.
156 Lavan, M. and Knipp, G. (2020). Considerations for determining direct versus indirect functional effects of solubilizing excipients on drug transporters for enhancing bioavailability. *J. Pharm. Sci.* 109 (6): 1833–1845.
157 Fagerholm, U. (2007). Prediction of human pharmacokinetics–gastrointestinal absorption. *J. Pharm. Pharmacol.* 59: 905–916.
158 Singh, B.N. (1999). Effects of food on clinical pharmacokinetics. *Clin. Pharmacokinet.* 37: 213–255.
159 Honda, Y., Ushigome, F., Koyabu, N. et al. (2004). Effects of grapefruit juice and orange juice components on P-glycoprotein- and MRP2-mediated drug efflux. *Br. J. Pharmacol.* 143: 856–864.
160 Panchagnula, R., Bansal, T., Varma, M.V., and Kaul, C.L. (2005). Co-treatment with grapefruit juice inhibits while chronic administration activates intestinal P-glycoprotein-mediated drug efflux. *Pharmazie* 60: 922–927.
161 Penzak, S.R., Acosta, E.P., Turner, M. et al. (2002). Effect of Seville orange juice and grapefruit juice on indinavir pharmacokinetics. *J. Clin. Pharmacol.* 42: 1165–1170.
162 Grenier, J., Fradette, C., Morelli, G. et al. (2006). Pomelo juice, but not cranberry juice, affects the pharmacokinetics of cyclosporine in humans. *Clin. Pharmacol. Ther.* 79: 255–262.
163 US Food and Drug Administration (2002). *Guidance for Industry: Food-Effect Bioavailability and Fed Bioequivalence Studies*. Rockville, MD: Food and Drug Administration http://www.fda.gov/cder/guidance/index.htm.
164 Wu, C.Y. and Benet, L.Z. (2005). Predicting drug disposition via application of BCS: transport/absorption/elimination interplay and development of a biopharmaceutics drug disposition classification system. *Pharm. Res.* 22: 11–23.
165 Custodio, J.M., Wu, C.Y., and Benet, L.Z. (2008). Predicting drug disposition, absorption/elimination/transporter interplay and the role of food on drug absorption. *Adv. Drug Delivery Rev.* 60: 717–733.
166 Tucker, G.T., Houston, J.B., and Huang, S.M. (2001). Optimizing drug development: strategies to assess drug metabolism/transporter interaction potential-towards a consensus. *Br. J. Clin. Pharmacol.* 52: 107–117.

167 Zhang, L., Strong, J.M., Qiu, W. et al. (2006). Scientific perspectives on drug transporters and their role in drug interactions. *Mol. Pharmaceutics* 3: 62–69.

168 Zhang, L., Zhang, Y., Strong, J.M. et al. (2008). A regulatory viewpoint on transporter-based drug interactions. *Xenobiotica* 38: 709–724.

169 Huang, S.M., Strong, J.M., Zhang, L. et al. (2008). New era in drug interaction evaluation: US Food and Drug Administration update on CYP enzymes, transporters, and the guidance process. *J. Clin. Pharmacol.* 48: 662–670.

170 US Food and Drug Administration (2006). Draft Guidance for Industry: Drug Interaction Studies—Study Design, Data Analysis and Implications for Dosing and Labeling. http://www.fda.gov/cder/guidance/6695dft.pdf; final guidance in the following Web site: http://www.fda.gov/cder/guidance/index.htm (under clinical pharmacology) (Accessed 4 August 2008).

171 Gumbo, T. (2008). Integrating pharmacokinetics, pharmacodynamics and pharmacogenomics to predict outcomes in antibacterial therapy. *Curr. Opin. Drug Discovery Dev.* 11: 32–42.

172 Zhang, E.Y., Fu, D.J., Pak, Y.A. et al. (2004). Genetic polymorphisms in human proton-dependent dipeptide transporter PEPT1: implications for the functional role of Pro586. *J Pharmacol Exp Ther* 310: 437–445.

173 Ishikawa, T., Hirano, H., Onishi, Y. et al. (2004). Functional evaluation of ABCB1 (P-glycoprotein) polymorphisms: high speed screening and structure-activity relationship analyses. *Drug Metab. Pharmacokinet.* 19: 1–14.

174 Vicente, J., Sinues, B., Fanlo, A. et al. (2008). Polymorphism C3435T of the MDR1 gene in Central Americans and Spaniards. *Mol. Biol .Rep.* 35: 473–478.

175 Kim, R.B., Leake, B., Choo, E.F. et al. (2001). Identification of functionally variant MDR1 alleles among European Americans and African Americans. *Clin. Pharmacol. Ther.* 70: 188–189.

176 Lu, H. and Klaassen, C. (2008). Gender differences in mRNA expression of ATP-binding cassette efflux and bile acid transporters in kidney, liver, and intestine of 5/6 nephrectomized rats. *Drug Metab. Dispos.* 36: 16–23.

177 Hayeshi, R., Hilgendorf, C., Artursson, P. et al. (2008). Comparison of drug transporter gene expression and functionality in Caco-2 cells from 10 different laboratories. *Eur. J. Pharm. Sci.* 35: 383–396.

# 21

# OATP Transporters in Hepatic and Intestinal Uptake of Orally Administered Drugs

*Wei Yue*[1], *Taleah Farasyn*[1], *Alexandra Crowe*[1], *Khondoker Alam*[1], *Lucila Garcia-Contreras*[1], *Yifan Tu*[2], *and Lu Wang*[3]

[1] *Department of Pharmaceutical Sciences, University of Oklahoma Health Sciences Center College of Pharmacy, Oklahoma, OK, USA*
[2] *Department of Pharmaceutical Science, Roche Pharma Research and Early Development, China Innovation Center of Roche, Pudong, Shanghai, China*
[3] *Department of Pharmacological and Pharmaceutical Sciences, College of Pharmacy, University of Houston, Houston, TX, USA*

## 21.1 Introduction

Membrane transport proteins play a key role in the absorption, distribution, and excretion of endogenous and xenobiotic substrates [1]. Over 400 transport proteins have been discovered in the human genome and have been organized into two major gene superfamilies known as the ATP-binding cassette (ABC) and solute carrier (SLC) [2–6]. Special attention has been paid to transporter proteins that are expressed on the epithelial barriers of the intestine, liver, kidney, and the blood–brain endothelial barrier for drug development, as these barriers are important determinants of drug disposition and pharmacokinetics. Organic anion-transporting polypeptides (OATPs) belong to the superfamily of SLC transport proteins. To date, 11 human OATPs have been identified that are categorized into six OATP families with amino acid sequence identities ≥40% within each family [7]. For human OATPs, the OATP1 family includes four human OATP members, OATP1A2, OATP1B1, OATP1B3, and OATP1C1. The OATP2 family includes human OATP2A1 and OATP2B1. The OATP4 family includes human OATP4A1 and OATP4C1. The OATP3, OATP5, and OATP6 families include human OATP3A1, OATP5A1, and OATP6A1, respectively [7].

The liver and intestine are two major tissues that play important roles in the oral bioavailability of drugs. This chapter focuses on OATPs in the liver and intestine that are involved in oral drug bioavailability and drug–drug interactions (DDIs) of orally administered drugs. Factors involved in regulation of the transporter function of OATPs in the liver and intestine are also discussed.

## 21.2 Hepatic OATP1B1 and OATP1B3

The genes encoding OATP1B1 (also called *SLC21A6*) and OATP1B3 (also called *SLC21A8*) are located on chromosome 12 (gene locus 12p12) [8, 9]. OATP1B1 shares ~80% amino acid homology with OATP1B3 [10]. OATP1B1 and OATP1B3 are primarily localized to the basolateral membrane of hepatocytes [10, 11]. While OATP1B1 and OATP1B3 are both highly expressed in normal human liver, they have different zonal expression patterns in the liver. OATP1B1 has a diffused expression pattern throughout the hepatic lobules, while OATP1B3 is primarily expressed around the central vein [10–12]. Hepatic uptake via OATP transporters, including OATP1B1 and OATP1B3, can be the rate-limiting step in systemic clearance of many drugs, including statins [13]. When hepatic uptake is the rate-limiting process in clearance of a drug, a reduction or depletion of OATP1B1 and OATP1B3 transport protein function due to genetic polymorphism or DDIs can result in increased systemic exposure to that substrate.

### 21.2.1 Substrate Transport Specificity and Transport Mechanism of OATP1B1 and OATP1B3

OATP1B1 and OATP1B3 play an important role in the uptake of a wide variety of drugs and endogenous compounds from the systemic blood circulation. They have a wide-ranging substrate selectivity that consists of anionic, neutral, and zwitterionic compounds. Many clinically important drugs are substrates and/or inhibitors of OATP1B1 and OATP1B3. They share

*Oral Bioavailability and Drug Delivery: From Basics to Advanced Concepts and Applications*, First Edition. Edited by Ming Hu and Xiaoling Li.

some common substrates. The most well-known drug substrates shared between OATP1B1 and OATP1B3 include the lipid-lowering HMG-CoA inhibitors known as statins [14, 15], estradiol-17β-beta-glucuronide [16], bromsulphthalein (BSP) [17], olmesartan [18], and bosentan [19]. In addition, many endogenous compounds, including bile acids, thyroid hormones, steroid sulfates, glucuronide conjugates, and peptides, can be transported by both OATP1B1 and OATP1B3 [8]. There are, however, some substrates that are transported preferentially by OATP1B1 (e.g. estrone-3-sulfate [14]), while others, such as cholecystokinin octapeptide (CCK-8) [20], telmisartan [21], the hepatotoxic cyclic peptide amanitin [22], and the cardiac glycoside ouabain [23], are transported primarily by OATP1B3.

OATP1B1 and OATP1B3 have 12 putative transmembrane (TM) domains with both N- and C-termini located intracellularly [24]. Some amino acids in TM domains 2, 6, 8, 9, and 10 and extracellular loop (ECL) 6 have been reported to be important for substrate transport by OATP1B1 and OATP1B3 [25–28]. Replacement of A45 in TM1, L545 in TM10, and T615 in ECL 6 of OATP1B1 with the respective amino acids in OATP1B3 enabled OATP1B1 to transport CCK-8, which is a specific substrate of OATP1B3 [25]. Estradiol-17β-glucuronide and estrone-3-sulfate are substrates of OATP1B1 commonly used for *in vitro* OATP1B1 transport assays [14]. Substituting the TM8 of OATP1B1 with that of OATP1B3 produces a chimeric transport protein that has 18-fold lower affinity toward estrone-3-sulfate than does wild-type OATP1B1, and abolishes the ability to transport estradiol-17β-glucuronide [28]. Replacing the TM9 of OATP1B1 with that of OATP1B3 decreases the affinity for estrone-3-sulfate without affecting the transport kinetics for estradiol-17β-glucuronide [28].

To date, the transport mechanism of OATPs has not been fully understood. Bicarbonate was first reported as the counter-ion in the transport of taurocholate in HeLa cells overexpressing rat Oatp [29]. Another study also reported that OATP-mediated substrate transport is accompanied by bicarbonate efflux [30]. Reduced glutathione (GSH) has been described as the driving force of rodent Oatp1-mediated transport [31]. One study demonstrated that uptake of bile acids by human OATP1B3 is cotransported by GSH [32]. However, this finding could not be reproduced by a later study [33].

It has been reported that transport activity of OATP1B1 and OATP1B3 is affected under circumstances of displacement of local pH [34]. Extracellular pH appears to differently affect OATP1B1- and OATP1B3-mediated substrate transport. Compared with a physiological pH of 7.4, an extracellular pH of 6.5 stimulates OATP1B3-mediated transport of taurocholate, estradiol-17β-glucuronide, and estrone-3-sulfate and OATP1B1-mediated transport of estrone-3-sulfate, while it inhibits OATP1B1-mediated transport of taurocholate and estradiol-17β-glucuronide [34].

### 21.2.2 Altered Disposition of OATP1B1 and 1B3 Substrates Due to Genetic Variation

More than 40 nonsynonymous single-nucleotide polymorphism (SNP) variants have been identified in OATP1B1 [35], among which the clinical relevance of c.521 T>C SNP (rs4149056), which results in an amino acid substitution (Val174Ala, V174A), has been supported by a compelling body of literature [36]. The c.521 T>C SNP has a high allele frequency of 8–20% in Europeans [37]. The c.521 T>C SNP of OATP1B1 has reduced transport activity *in vitro* [38]. Clinically, the c.521 T>C SNP is associated with an increased plasma area under the concentration-time profile (AUC) of multiple OATP1B1 substrates, including statins [37, 39–41]. For example, healthy individuals bearing the c.521 CC genotype had a three-fold increase in their AUC of simvastatin acid when compared with individuals with the c.521 TT genotype [41, 42]. As statins can cause the adverse drug reaction of myopathy in a concentration-dependent manner, the c.521 T>C is associated with increased risk of simvastatin acid-induced myopathy [42]. Based on these findings, the c.521 T>C is considered the most robust predictor of statin-induced myopathy [42]. Altering the recommended dosage of statins based on a patient's genotype has been proposed [42].

Two frequent coding, non-synonymous SNPs of OATP1B3, T334G and G699A, are in complete linkage disequilibrium [43]. *In vitro*, the 334G–699A variant of OATP1B3 has decreased substrate transport of mycophenolic acid glucuronide (MPAG) compared with the reference OATP1B3 [44]. In humans, the carriers of the 334G allele have significantly increased systemic exposure of mycophenolic acid (MPA), which is likely due to reduced MPAG hepatic uptake and the subsequent reduction in MPA biliary excretion and reabsorption through enterohepatic cycling [44].

Simultaneous genetic deficiencies of OATP1B1 and OATP1B3 have been linked to Rotor syndrome (RS), a benign and rare hereditary hyperbilirubinemia [45]. RS is associated with significantly reduced hepatic uptake of many diagnostic compounds that are OATP1B1 and OATP1B3 substrates, such as ($^{99}$m)Tc-mebrofenin [46, 47].

### 21.2.3 Altered Disposition of OATP1B1 and 1B3 Substrates Due to Drug–Drug Interactions

Multidrug therapy is becoming a common practice as more information on the benefits of various drug combinations is collected. Coadministration of two or more drugs may cause DDIs resulting in altered drug pharmacokinetics or pharmacodynamics. OATP1B1 and OATP1B3 are important determinants of DDIs. Many therapeutically important drugs (e.g. cyclosporine, gemfibrozil, antiretroviral agents, and antibiotics) are inhibitors of OATP1B1 and OATP1B3 [48, 49]. When these inhibitors are coadministered with lipid-lowering statins, they may increase the systemic exposure of statins and cause clinically significant adverse effects, such as myopathy [49, 50]. Pravastatin and rosuvastatin are OATP1B1 and OATP1B3 substrates that are subjected to minimal metabolism [51, 52]. Gemfibrozil has been reported to cause a ~2-fold increase in the AUC of pravastatin and rosuvastatin [52, 53]. The OATP1B1 and OATP1B3 inhibitor CsA [54] caused a 7.1- and 10.6-fold increase in the AUC and $C_{max}$ of rosuvastatin, respectively [55]. Such DDIs can potentially result in severe adverse effects, including life-threatening rhabdomyolysis [51].

Although some statins (e.g. atorvastatin) are substrates of both metabolic enzymes and OATP1B1 and OATP1B3, it has been reported that hepatic uptake via OATPs is the rate-determining process in the hepatic elimination of atorvastatin [13]. In a clinical micro-dosing study, eight healthy subjects were administered a cocktail consisting a microdose of atorvastatin along with a probe substrate for OATPs (pravastatin) and CYP3A4 (midazolam). The pharmacokinetics of this dose were observed for each patient at baseline and following a coadministered dose of a known potent OATP inhibitor (rifampicin) or CYP3A4 inhibitor (itraconazole). The atorvastatin dose-normalized AUC increased 12-fold when coadministered with rifampicin (OATP inhibitor), but was not affected by coadministration with itraconazole (CYP3A4 inhibitor), indicating that hepatic uptake via OATPs provides the major contribution to the hepatic elimination of atorvastatin. Clinical drugs involved in OATP1B1- and OATP1B3-mediated DDIs have been extensively reviewed in previous publications [56, 57].

### 21.2.4 Model-Based Prediction of DDIs Mediated by OATP1B1 and OATP1B3

The importance of assessing DDIs mediated by drug transporters, including OATP1B1 and OATP1B3, has been increasingly recognized in the past decade [58]. The US Food and Drug Administration (FDA) has published final guidance to help drug developers plan and evaluate studies to determine the DDI potential of an investigational drug product [59].

#### 21.2.4.1 Static Model for the Prediction of OATP1B1- and OATP1B3-Mediated DDIs

The US FDA has recommended calculating an $R$-value (Eq. (21.1)) to predict the ratio of the victim drug's AUC in the presence and absence of an investigational drug as the inhibitor. This process can be used to assess the OATP1B1- and OATP1B3-mediated DDI potential of inhibitors [59], with the following equations:

$$R = 1 + (f_{u,\,plasma} \times I_{in,\,max})/IC_{50} \qquad (21.1)$$

$$I_{in,\,max} = I_{max} + (F_a \times F_g \times k_a \times Dose)/Q_h/R_B \qquad (21.2)$$

where $I_{in,\,max}$ is maximum plasma inhibitor concentration at the inlet to the liver estimated based on Eq. (21.2). $I_{max}$ is the maximum plasma concentration of the inhibitor in systemic circulation, $IC_{50}$ is the *in vitro* inhibitor concentration resulting in 50% of maximal inhibition of OATP1B1 or OATP1B3 function, $f_{u,\,plasma}$ is the unbound fraction in the plasma, $F_a$ is the fraction absorbed, $F_g$ is the intestinal availability, $k_a$ is the absorption rate constant, $Q_h$ is the hepatic blood flow rate (1500 ml/min [60]), and $R_B$ is the blood-to-plasma concentration ratio.

Recently, time-dependent inhibition for some OATP1B1 and OATP1B3 inhibitors was discovered [61–65], where the inhibitory effects of some OATP1B1 and OATP1B3 inhibitors were potentiated after preincubation with the inhibitors. The US FDA guidance recommends adding a preincubation step as part of assay validation when determining $IC_{50}$ values for an investigational drug [59]. Currently, a cutoff $R$-value of 1.1 is recommended in the US FDA 2020 final guidance to determine the potential to cause OATP1B1- and OATP1B3-mediated DDIs [59]. The investigational drug has the potential to inhibit OATP1B1 and OATP1B3 *in vivo* if the R-value is >1.1. The cutoff $R$-values recommended by European Medicine Agency (EMA) and Japan's Pharmaceuticals and Medical Devices Agency (PMDA) are ≥1.04 and ≥1.25, respectively. The current US FDA cutoff R-value was established after a thorough comparison of multiple methods and cutoff values used for the prediction of *in vivo* DDI with *in vitro* inhibition potencies, including those recommended by three regulatory agencies (EMA, FDA, and PMDA), as shown in a recent review article [58].

#### 21.2.4.2 Physiologically Based Pharmacokinetic (PBPK) Models for the Prediction of OATP1B1- and OATP1B3-Mediated DDIs

PBPK models comprise multiple compartments that correspond to different organs or systems of the body [66]. The parameters that are integrated into PBPK models can be drug-dependent or system-dependent parameters [67]. Drug-dependent parameters include physicochemical properties (e.g. $pK_a$, molecular weight, and net charge), plasma protein and blood binding, permeability, solubility, transport protein contribution, and *in vitro* enzyme metabolism data. PBPK models also comprise species-specific systemic parameters that are independent of the compound being modeled. A PBPK modeling approach has been applied to predict OATP1B1- and OATP1B3-mediated DDIs [64, 65, 68, 69].

### 21.2.5 Factors Involved in Altered Transporter Function of OATP1B1 and OATP1B3

#### 21.2.5.1 Transcriptional Regulation of OATP1B1 and OATP1B3

Several transcription factors have been reported to regulate the expression of OATP1B1 and OATP1B3 [70]. The OATP1B1 promoter can be transactivated by liver X receptor (LXR) α [71] and farnesoid X receptor (FXR) [71], whereas the OATP1B3 promoter can be transactivated by FXR [72, 73], hepatic nuclear factor (HNF)1α [74], and growth hormone- and prolactin-activated transcription factor (STAT5) [73], and repressed by HNF 3β [75]. Although few are known for OATP1B1 and 1B3, epigenetic mechanisms can also regulate their expression. For example, 5-aza-2′-deoxycytidine, which inhibits DNA methylation, has been reported to increase the mRNA levels of OATP1B3 in various cancer cell lines [76].

#### 21.2.5.2 Posttranslational Regulation of OATP1B1 and OATP1B3

At the posttranslational level, both OATP1B1 and OATP1B3 are glycosylated proteins [10, 77]. For OATP1B1, abolishing all glycosylation sites reduces protein stability and reduced total protein levels [77]. The functional consequence of glycosylation on OATP1B3 transport function remains unclear.

OATP1B1 and OATP1B3 have been reported to be phosphorylated proteins. Powell et al. reported that OATP1B3 is a phosphorylated protein in primary human hepatocytes [78]. Increased phosphorylation of OATP1B3 is associated with the rapid downregulation of OATP1B3 transport function following protein kinase C (PKC) activation with phorbol 12-myristate 13-acetate (PMA) [78]. Farasyn et al. characterized phosphorylation sites of OATP1B1 using a phosphoproteomics approach [65]. The V174A-OATP1B1, which has reduced transport activity *in vitro* and *in vivo*, has similar plasma membrane levels to the wild-type OATP1B1; however, it has significantly increased total phosphorylation levels compared with the wild-type OATP1B1 [79]. Inhibition of Lck/Yes novel tyrosine kinase (LYN) was implicated in reduced transport function of OATP1B1 and OATP1B3 [80, 81].

Ubiquitin is a small 76-amino-acid protein. An isopeptide bond can form between a lysine residue on the protein and the carboxyl terminus of ubiquitin [82, 83]. Ubiquitination is an important posttranslational modification that can mediate proteasome degradation of proteins and the maintenance of protein homeostasis [84]. A recent study reported that both OATP1B1 and OATP1B3 can be ubiquitin-conjugated in HEK293 cells overexpressing OATP1B1 or OATP1B3 [85]. Proteasome inhibitor drug bortezomib down-regulates OATP1B3 transport function in a substrate-dependent manner, while it does not affect OATP1B1 transport function [85]. As OATP1B1 and OATP1B3 transport proteins degrade primarily via the lysosome degradation pathway [86], the effects of bortezomib on OATP1B3 transport function unlikely occurred via altered ubiquitination of the transporter.

## 21.3 OATP2B1 in the Intestine

OATP2B1 is the main intestinal OATP isoform that in humans is encoded by the *SLCO2B1* gene located on chromosome 11 (gene locus 11q13.4) [87]. Compared with the liver-specific expression of OATP1B1 and OATP1B3, OATP2B1 has a broader tissue expression profile, including expression in the liver, small intestine, heart, blood–brain barrier (BBB), and the placenta [88]. OATP2B1 is generally considered to be expressed on the apical (luminal) membrane of human enterocytes, mediates intestinal apical uptake of its substrate, and plays a role in drug absorption and distribution [89]. However, the apical versus basolateral membrane localization of OATP2B1 in enterocytes remains debated [89, 90]. Furthermore, the direction of transport of OATP2B1 on the basolateral membrane is under discussion (transport into or out of an enterocyte) [91]. OATP2B1 is highly expressed in endothelial cells of the BBB [92]. *In vivo* studies in mice suggested that OATP2B1 might play a role in brain drug delivery across the BBB [93].

### 21.3.1 Substrates of OATP2B1

Substrates of OATP2B1 include many drugs, natural products, and hormones [91]. OATP2B1 shares several common substrates with other OATPs. For example, substrates of OATP2B1, such as statins (e.g. atorvastatin [94] and rosuvastatin [15]) and BSP [17], overlap with those of OATP1B1 and/or OATP1B3. Interestingly, pH-dependent transport by OATP2B1 was observed on multiple substrates, including estrone-3-sulfate, taurocholate acid, and pravastatin [95, 96], where the transport efficiency is higher at a lower pH (e.g. 5.5) than at a neutral pH. This pH-dependency may affect the absorption and distribution of orally administered drugs, as the pH varies along the gastrointestinal tract. Further, it suggests a more important role of OATP2B1 in the intestine than in the liver, since the pH of intestinal lumen is in the acidic range, but portal blood is expected to be neutral.

### 21.3.2 Altered Disposition of OATP2B1 Substrates Due to Genetic Variation

A few SNPs have been discovered in the *SLCO2B1* gene. McFeely et al. summarized a list of polymorphisms with clinical relevance in a recent review article [91]. One of the most studied SNPs is *SLCO2B1*3* (rs2306168), with an allele frequency of 34.1% in the African–American population and 19.2% in the Asian population, but much lower in the Caucasian population (3.9%). An *in vitro* study compared uptake kinetics of estrone-3-sulfate in OATP2B1- and OATP2B1*3-expressed HEK293 cells. OATP2B1 protein content-adjusted $V_{max}$ values of *SLCO2B1*3* were only 42.5% of that in wild type, but no difference was observed in $K_m$, suggesting a decreased transport capacity and an unchanged affinity with the OATP2B1*3 variant [97]. Clinical studies confirmed affected pharmacokinetic parameters (e.g. decreased AUC and $C_{max}$) of drugs including fexofenadine (35.5% decrease in AUC) and celiprolol (29.1% decrease in AUC in heterozygotes and 50% decrease in AUC in homozygotes) of individuals with the SLCO2B1*3 compared with the wild type [98, 99]. While current research suggests a promising start to the knowledge of genetic variation of OATP2B1, more data are needed to recognize the polymorphisms and identify their clinical relevance.

### 21.3.3 Altered Disposition of OATP2B1 Substrates Due to Food–Drug and Drug–Drug Interactions

Many efforts have been made to describe the potential food–drug and DDIs mediated by OATP2B1. Since OATP2B1 is highly expressed in the intestine, most studies focused on the interaction during oral absorption. Among these, food–drug interactions are most studied *in vitro* and *in vivo* with studies involving fruit juices. Shirasaka et al. compared the inhibitory effects of apple juice, orange juice, and grapefruit juice and found that grapefruit juice had the smallest $IC_{50}$ in inhibiting the uptake of estrone-3-sulfate in OATP2B1-expressing *Xenopus* oocytes, followed by orange juice and apple juice, respectively [100]. *In vivo*, as reviewed by McFeely et al. [91], fruit juices reduce the AUC of several OATP2B1 substrates, including fexofenadine, aliskiren, and celiprolol. Apple juice and grapefruit juice decrease the AUC of fexofenadine by 85% and up to 52%, respectively. Apple juice, grapefruit juice, and orange juice decreased the AUC of aliskiren by 63%, 61%, and 62%, respectively. Grapefruit juice decreased the AUC of celiprolol by 84%. One example of an OATP2B1-mediated DDI is the altered drug disposition of rosuvastatin when it is coadministered with ronacaleret. Ronacaleret was reported to inhibit intestinal OATP2B1 and thus decreased the exposure of OATP2B1 substrate rosuvastatin to half in a clinical study [101].

Regardless of the preclinical and clinical findings on the potential drug–drug and drug–food interactions, current US FDA guidance has not recommended evaluating OATP2B1 in drug development due to the lack of sufficient *in vivo* and clinical evidence [59]. A white paper published by the International Transporter Consortium suggested exploring the role of OATP2B1 when *in vivo* substrate transport in the liver or intestine cannot be ascribed to other more common mechanisms (e.g. OATP1B1/1B3, intestinal BCRP/P-gp), and to consider OATP2B1 as a potential mechanism of DDIs or food–drug interactions when apparently decreased substrate absorption is observed [89].

### 21.3.4 Factors Involved in Altered Transporter Function of OATP2B1

Transcriptional factors have been reported to regulate the expression of OATP2B1 [70]. Specificity protein (Sp) 1 is involved in OATP2B1 expression in both liver and small intestine [102], while HNF 4α regulates hepatic expression of OATP2B1 [103]. A recent study showed a posttranscriptional regulation of OATP2B1 by miR-24. Overexpression of miR-24 down-regulates OATP2B1 promoter activity and mRNA levels *in vitro* [104, 105].

## 21.4 OATP1A2 in the intestine

SLCO1A2, the gene encoding OATP1A2, is located on chromosome 12p12 and is composed of 16 exons and 15 introns. OATP1A2 is a 670-amino-acid glycoprotein that shares 67~73% amino acid sequence identity with its rodent orthologs [106]. *In vitro*, OATP1A2 has been shown to transport many endogenous compounds (e.g. bile acid taurocholate [107]) and clinical drugs (e.g. fexofenadine [108]) [106]. OATP1A2 appears to express ubiquitously in the human body [106]. The highest expression of OATP1A2 was found in the brain [109, 110]. OATP1A2 mRNA was detected in the liver [107]. However, at the protein level, OATP1A2 was not detected in hepatocytes, but rather in cholangiocytes [111] of the liver. Currently, discrepant data exist regarding OATP1A2 expression in the small intestine. Earlier studies using immunohistochemical staining with a polyclonal OATP1A2 antibody showed positive staining in human duodenal tissue sections [112]. However, recent data using absolute protein quantification by a proteomics approach confirmed that OATP1A2 protein does not express in the small intestine [110, 113]. The latter study is in line with the lack of OATP1A2 mRNA expression in the small intestine reported by several other studies [114, 115].

OATP1A2 was assumed to mediate intestinal uptake of several drugs, including talinolol, fexofenadine, and aliskiren [116–118]. Based on the more recent data on the lack of OATP1A2 protein in the small intestine, the role of OATP1A2 in the small intestine is questionable. It was believed that most of the presumed OATP1A2-mediated uptake in the intestine could be attributed to OATP2B1, which has overlapping substrate specificity with OATP1A2 and is inhibited by the experimental inhibitors for OATP1A2 (e.g. grapefruits and other herbs) [117, 119]. Currently, the role of OATP1A2 in drug absorption in the gastrointestinal tract remains unclear.

## 21.5 Summary

OATP1B1 and OATP1B3 are localized to the basolateral membrane of human hepatocytes. They mediate hepatic uptake of many clinically important drugs and are determinants of transporter-mediated DDIs. Evaluation of investigational drugs as substrates and inhibitors of OATP1B1 and OATP1B3 has been included in the current US FDA final guidance for industry. OATP2B1 is the major intestinal OATP isoform. The pH-dependent substrate transport mediated by OATP2B1 suggests a more important role of OATP2B1 in intestinal absorption than in the liver. The role of OATP2B1 in drug–drug and food–drug interactions requires additional attention. The recent finding of the lack of OATP1A2 expression at protein levels in the small intestine triggered reevaluation of OATP2B1 in DDIs that were previously assumed to be mediated by OATP1A2. Future directions in this field include investigation of the *in vivo* function of OATP transporters in drug absorption and bioavailability using humanized animal models, as previously done for the OATP1B transporters [120]. Understanding factors involved in altered function of OATP transporter also merits further investigation in order to better predict altered drug disposition.

## Acknowledgement

W.Y. is supported, in part, by the National Institute of General Medical Sciences of the National Institutes of Health (NIH) under Award Number GM146956-01. Alexandra Crowe is an American Foundation of Pharmaceutical Education Fellow.

## References

1 International Transporter, C, Giacomini, K.M., Huang, S.M. et al. (2010). Membrane transporters in drug development. *Nat. Rev. Drug Discovery* 9 (3): 215–236.

2 Schinkel, A.H. and Jonker, J.W. (2003). Mammalian drug efflux transporters of the ATP binding cassette (ABC) family: an overview. *Adv. Drug Delivery Rev.* 55: 3–29.

3 Sai, Y. (2005). Biochemical and molecular pharmacological aspects of transporters as determinants of drug disposition. *Drug Metab. Pharmacokinet.* 20 (2): 91–99.

4 Cascorbi, I. (2006). Role of pharmacogenetics of ATP-binding cassette transporters in the pharmacokinetics of drugs. *Pharmacol. Therap.* 112 (2): 457–473.

5 Choudhuri, S. and Klaassen, C.D. (2006). Structure, function, expression, genomic organization, and single nucleotide polymorphisms of human ABCB1 (MDR1), ABCC (MRP), and ABCG2 (BCRP) efflux transporters. *Int. J. Toxicol.* 25 (4): 231–259.

6 Koepsell, H., Lips, K., and Volk, C. (2007). Polyspecific organic cation transporters: structure, function, physiological roles, and biopharmaceutical implications. *Pharm. Res.* 24 (7): 1227–1251.

7 Hagenbuch, B. and Meier, P.J. (2004). Organic anion transporting polypeptides of the OATP/SLC21 family: phylogenetic classification as OATP/SLCO

superfamily, new nomenclature and molecular/functional properties. *Pflügers Archiv.* 447 (5): 653–665.

8 Hagenbuch, B. and Meier, P.J. (2003). The superfamily of organic anion transporting polypeptides. *Biochim. Biophys. Acta* 1609 (1): 1–18.

9 Kullak-Ublick, G.A., Beuers, U., Meier, P.J. et al. (1996). Assignment of the human organic anion transporting polypeptide (OATP) gene to chromosome 12p12 by fluorescence in situ hybridization. *J. Hepatol.* 25 (6): 985–987.

10 Konig, J., Cui, Y., Nies, A.T., and Keppler, D. (2000). Localization and genomic organization of a new hepatocellular organic anion transporting polypeptide. *J. Biol. Chem.* 275 (30): 23161–23168.

11 König, J., Cui, Y., Nies, A.T., and Keppler, D. (2000). A novel human organic anion transporting polypeptide localized to the basolateral hepatocyte membrane. *Am. J. Physiol.* 278: G156–G164.

12 Ho, R., Tirona, R., Leake, B. et al. (2006). Drug and bile acid transporters in rosuvastatin hepatic uptake: function, expression, and pharmacogenetics. *Gastroenterology* 130 (6): 1793–1806.

13 Maeda, K., Ikeda, Y., Fujita, T. et al. (2011). Identification of the rate-determining process in the hepatic clearance of atorvastatin in a clinical cassette microdosing study. *Clin. Pharmacol. Therap.* 90 (4): 575–581.

14 Hirano, M., Maeda, K., Shitara, Y., and Sugiyama, Y. (2004). Contribution of OATP2 (OATP1B1) and OATP8 (OATP1B3) to the hepatic uptake of pitavastatin in humans. *J. Pharmacol. Exp. Therap.* 311 (1): 139–146.

15 Kitamura, S., Maeda, K., Wang, Y., and Sugiyama, Y. (2008). Involvement of multiple transporters in the hepatobiliary transport of rosuvastatin. *Drug Metab. Dispos.* 36 (10): 2014–2023.

16 Gui, C., Miao, Y., Thompson, L. et al. (2008). Effect of pregnane X receptor ligands on transport mediated by human OATP1B1 and OATP1B3. *Eur. J. Pharmacol.* 584 (1): 57–65.

17 Kullak-Ublick, G.A., Ismair, M.G., Stieger, B. et al. (2001). Organic anion-transporting polypeptide B (OATP-B) and its functional comparison with three other OATPs of human liver. *Gastroenterology* 120 (2): 525–533.

18 Yamada, A., Maeda, K., Kamiyama, E. et al. (2007). Multiple human isoforms of drug transporters contribute to the hepatic and renal transport of olmesartan, a selective antagonist of the angiotensin II AT1-receptor. *Drug Metab. Dispos.* 35 (12): 2166–2176.

19 Treiber, A., Schneiter, R., Hausler, S., and Stieger, B. (2007). Bosentan is a substrate of human OATP1B1 and OATP1B3: inhibition of hepatic uptake as the common mechanism of its interactions with cyclosporin A, rifampicin, and sildenafil. *Drug Metab. Dispos.* 35 (8): 1400–1407.

20 Ismair, M.G., Stieger, B., Cattori, V. et al. (2001). Hepatic uptake of cholecystokinin octapeptide by organic anion-transporting polypeptides OATP4 and OATP8 of rat and human liver. *Gastroenterology* 121 (5): 1185–1190.

21 Ishiguro, N., Maeda, K., Kishimoto, W. et al. (2006). Predominant contribution of OATP1B3 to the hepatic uptake of telmisartan, an angiotensin II receptor antagonist, in humans. *Drug Metab. Dispos.* 34 (7): 1109–1115.

22 Letschert, K., Faulstich, H., Keller, D., and Keppler, D. (2006). Molecular characterization and inhibition of amanitin uptake into human hepatocytes. *Toxicol. Sci.* 91 (1): 140–149.

23 Gozalpour, E., Greupink, R., Wortelboer, H.M. et al. (2014). Interaction of digitalis-like compounds with liver uptake transporters NTCP, OATP1B1, and OATP1B3. *Mol. Pharmaceutics* 11 (6): 1844–1855.

24 König, J. (2011). Uptake transporters of the human OATP family: molecular characteristics, substrates, their role in drug-drug interactions, and functional consequences of polymorphisms. *Handb. Exp. Pharmacol.* 201: 1–28.

25 DeGorter, M.K., Ho, R.H., Leake, B.F. et al. (2012). Interaction of three regiospecific amino acid residues is required for OATP1B1 gain of OATP1B3 substrate specificity. *Mol. Pharmaceutics* 9 (4): 986–995.

26 Gui, C. and Hagenbuch, B. (2008). Amino acid residues in transmembrane domain 10 of organic anion transporting polypeptide 1B3 are critical for cholecystokinin octapeptide transport. *Biochemistry* 47 (35): 9090–9097.

27 Gui, C. and Hagenbuch, B. (2009). Role of transmembrane domain 10 for the function of organic anion transporting polypeptide 1B1. *Protein Sci.* 18 (11): 2298–2306.

28 Miyagawa, M., Maeda, K., Aoyama, A., and Sugiyama, Y. (2009). The eighth and ninth transmembrane domains in organic anion transporting polypeptide 1B1 affect the transport kinetics of estrone-3-sulfate and estradiol-17beta-D-glucuronide. *J. Pharmacol. Exp. Therap.* 329 (2): 551–557.

29 Satlin, L.M., Amin, V., and Wolkoff, A.W. (1997). Organic anion transporting polypeptide mediates

organic anion/HCO3- exchange. *J. Biol. Chem.* 272 (42): 26340–26345.

30 Leuthold, S., Hagenbuch, B., Mohebbi, N. et al. (2009). Mechanisms of pH-gradient driven transport mediated by organic anion polypeptide transporters. *Am. J. Physiol. Cell Physiol.* 296 (3): C570–C582.

31 Li, L., Lee, T.K., Meier, P.J., and Ballatori, N. (1998). Identification of glutathione as a driving force and leukotriene C4 as a substrate for oatp1, the hepatic sinusoidal organic solute transporter. *J. Biol. Chem.* 273 (26): 16184–16191.

32 Briz, O., Romero, M.R., Martinez-Becerra, P. et al. (2006). OATP8/1B3-mediated cotransport of bile acids and glutathione: an export pathway for organic anions from hepatocytes? *J. Biol. Chem.* 281 (41): 30326–30335.

33 Mahagita, C., Grassl, S.M., Piyachaturawat, P., and Ballatori, N. (2007). Human organic anion transporter 1B1 and 1B3 function as bidirectional carriers and do not mediate GSH-bile acid cotransport. *Am. J. Physiol.* 293 (1): G271–G278.

34 Martinez-Becerra, P., Briz, O., Romero, M.R. et al. (2011). Further characterization of the electrogenicity and pH sensitivity of the human organic anion-transporting polypeptides OATP1B1 and OATP1B3. *Mol. Pharmacol.* 79 (3): 596–607.

35 Clarke, J.D. and Cherrington, N.J. (2012). Genetics or environment in drug transport: the case of organic anion transporting polypeptides and adverse drug reactions. *Expert Opin. Drug Metab. Toxicol.* 8 (3): 349–360.

36 Giacomini, K.M., Balimane, P.V., Cho, S.K. et al. (2013). International Transporter Consortium commentary on clinically important transporter polymorphisms. *Clin. Pharmacol. Therap.* 94 (1): 23–26.

37 Niemi, M., Pasanen, M.K., and Neuvonen, P.J. (2011). Organic anion transporting polypeptide 1B1: a genetically polymorphic transporter of major importance for hepatic drug uptake. *Pharmacol. Rev.* 63 (1): 157–181.

38 Tirona, R.G., Leake, B.F., Merino, G. et al. (2001). Identification of multiple allelic variants associated with altered transport activity among European and African-Americans. *J. Biol. Chem.* 276 (38): 35669–35675.

39 Pasanen, M.K., Fredrikson, H., Neuvonen, P.J., and Niemi, M. (2007). Different effects of SLCO1B1 polymorphism on the pharmacokinetics of atorvastatin and rosuvastatin. *Clin. Pharmacol. Therap.* 82 (6): 726–733.

40 Niemi, M., Pasanen, M.K., and Neuvonen, P.J. (2006). SLCO1B1 polymorphism and sex affect the pharmacokinetics of pravastatin but not fluvastatin. *Clin. Pharmacol. Therap.* 80 (4): 356–366.

41 Pasanen, M.K., Neuvonen, M., Neuvonen, P.J., and Niemi, M. (2006). SLCO1B1 polymorphism markedly affects the pharmacokinetics of simvastatin acid. *Pharmacogenet. Genomics* 16 (12): 873–879.

42 Group, S.C., Link, E., Parish, S. et al. (2008). SLCO1B1 variants and statin-induced myopathy--a genomewide study. *New Engl. J. Med.* 359 (8): 789–799.

43 Hamada, A., Sissung, T., Price, D.K. et al. (2008). Effect of SLCO1B3 haplotype on testosterone transport and clinical outcome in caucasian patients with androgen-independent prostatic cancer. *Clin. Cancer Res.* 14 (11): 3312–3318.

44 Picard, N., Yee, S.W., Woillard, J.B. et al. (2010). The role of organic anion-transporting polypeptides and their common genetic variants in mycophenolic acid pharmacokinetics. *Clin. Pharmacol. Therap.* 87 (1): 100–108.

45 van de Steeg, E., Stranecky, V., Hartmannova, H. et al. (2012). Complete OATP1B1 and OATP1B3 deficiency causes human Rotor syndrome by interrupting conjugated bilirubin reuptake into the liver. *J. Clin. Invest.* 122 (2): 519–528.

46 LeBouthillier, G., Morais, J., Picard, M. et al. (1992). Scintigraphic aspect of Rotor's disease with technetium-99m-mebrofenin. *J. Nucl. Med.* 33 (8): 1550–1551.

47 de Graaf, W., Hausler, S., Heger, M. et al. (2011). Transporters involved in the hepatic uptake of (99m)Tc-mebrofenin and indocyanine green. *J. Hepatol.* 54 (4): 738–745.

48 Annaert, P., Ye, Z.W., Stieger, B., and Augustijns, P. (2010). Interaction of HIV protease inhibitors with OATP1B1, 1B3, and 2B1. *Xenobiotica* 40 (3): 163–176.

49 Shitara, Y., Itoh, T., Sato, H. et al. (2003). Inhibition of transporter-mediated hepatic uptake as a mechanism for drug-drug interaction between cerivastatin and cyclosporin A. *J. Pharmacol. Exp. Therap.* 304 (2): 610–616.

50 Kiser, J.J., Gerber, J.G., Predhomme, J.A. et al. (2008). Drug/drug interaction between lopinavir/ritonavir and rosuvastatin in healthy volunteers. *J. Acquir. Immune Defic. Syndr.* 47: 570–578.

51 Chauvin, B., Drouot, S., Barrail-Tran, A., and Taburet, A.M. (2013). Drug-drug interactions

between HMG-CoA reductase inhibitors (statins) and antiviral protease inhibitors. *Clin. Pharmacokinet.* 52 (10): 815–831.

52 Schneck, D.W., Birmingham, B.K., Zalikowski, J.A. et al. (2004). The effect of gemfibrozil on the pharmacokinetics of rosuvastatin. *Clin. Pharmacol. Therap.* 75 (5): 455–463.

53 Kyrklund, C., Backman, J.T., Neuvonen, M., and Neuvonen, P.J. (2003). Gemfibrozil increases plasma pravastatin concentrations and reduces pravastatin renal clearance. *Clin. Pharmacol. Therap.* 73 (6): 538–544.

54 Shitara, Y., Takeuchi, K., Nagamatsu, Y. et al. (2012). Long-lasting inhibitory effects of cyclosporin A, but not tacrolimus, on OATP1B1- and OATP1B3-mediated uptake. *Drug Metab. Pharmacokinet.* 27 (4): 368–378.

55 Simonson, S.G., Raza, A., Martin, P.D. et al. (2004). Rosuvastatin pharmacokinetics in heart transplant recipients administered an antirejection regimen including cyclosporine. *Clin. Pharmacol. Therap.* 76 (2): 167–177.

56 Yoshida, K., Maeda, K., and Sugiyama, Y. (2012). Transporter-mediated drug--drug interactions involving OATP substrates: predictions based on in vitro inhibition studies. *Clin. Pharmacol. Therap.* 91 (6): 1053–1064.

57 Alam, K., Crowe, A., Wang, X. et al. (2018). Regulation of organic anion transporting polypeptides (OATP) 1B1- and OATP1B3-mediated transport: an updated review in the context of OATP-mediated drug-drug interactions. *Int. J. Mol. Sci.* 19 (3).

58 Vaidyanathan, J., Yoshida, K., Arya, V., and Zhang, L. (2016). Comparing various in vitro prediction criteria to assess the potential of a new molecular entity to inhibit organic anion transporting polypeptide 1B1. *J. Clin. Pharmacol.* 56 (Suppl 7): S59–S72.

59 US FDA in vitro metabolism- and transporter-mediated drug-drug interaction studies-guidance for industry. https://www.fda.gov/media/134582/download (accessed February 4).

60 Bradley, S.E., Ingelfinger, F.J. et al. (1945). The estimation of hepatic blood flow in man. *J. Clin. Invest.* 24: 890–897.

61 Amundsen, R., Christensen, H., Zabihyan, B., and Asberg, A. (2010). Cyclosporine A, but not tacrolimus, shows relevant inhibition of organic anion-transporting protein 1B1-mediated transport of atorvastatin. *Drug Metab. Dispos.* 38 (9): 1499–1504.

62 Gertz, M., Cartwright, C.M., Hobbs, M.J. et al. (2013). Cyclosporine inhibition of hepatic and intestinal CYP3A4, uptake and efflux transporters: application of PBPK modeling in the assessment of drug-drug interaction potential. *Pharm. Res.* 30 (3): 761–780.

63 Izumi, S., Nozaki, Y., Maeda, K. et al. (2015). Investigation of the impact of substrate selection on in vitro organic anion transporting polypeptide 1B1 inhibition profiles for the prediction of drug-drug interactions. *Drug Metab. Dispos.* 43 (2): 235–247.

64 Pahwa, S., Alam, K., Crowe, A. et al. (2017). Pretreatment with rifampicin and tyrosine kinase inhibitor dasatinib potentiates the inhibitory effects toward OATP1B1- and OATP1B3-mediated transport. *J. Pharm. Sci.* 106 (8): 2123–2135.

65 Farasyn, T., Crowe, A., Hatley, O. et al. (2019). Preincubation with everolimus and sirolimus reduces organic anion-transporting polypeptide (OATP)1B1- and 1B3-mediated transport independently of mTOR kinase inhibition: implication in assessing OATP1B1- and OATP1B3-mediated drug-drug interactions. *J. Pharm. Sci.* 108 (10): 3443–3456.

66 Zhao, P., Zhang, L., Grillo, J.A. et al. (2011). Applications of physiologically based pharmacokinetic (PBPK) modeling and simulation during regulatory review. *Clin. Pharmacol. Therap.* 89 (2): 259–267.

67 Jones, H.M., Gardner, I.B., and Watson, K.J. (2009). Modelling and PBPK simulation in drug discovery. *AAPS J.* 11 (1): 155–166.

68 Kayesh, R., Farasyn, T., Crowe, A. et al. (2020). Assessing OATP1B1- and OATP1B3-mediated drug-drug interaction potential of vemurafenib using R-value and physiologically-based pharmacokinetic models. *J. Pharm. Sci.* .

69 Shebley, M., Liu, J., Kavetskaia, O. et al. (2017). Mechanisms and predictions of drug-drug interactions of the hepatitis C virus three direct-acting antiviral regimen: paritaprevir/ritonavir, ombitasvir, and dasabuvir. *Drug Metab. Dispos.* 45 (7): 755–764.

70 Brouwer, K.L.R., Evers, R., Hayden, E. et al. (2022). Regulation of drug transport proteins-from mechanisms to clinical impact: a white paper on behalf of the international transporter consortium. *Clin. Pharmacol. Therap.* .

71 Meyer Zu Schwabedissen, H.E., Bottcher, K., Chaudhry, A. et al. (2010). Liver X receptor alpha and farnesoid X receptor are major transcriptional regulators of OATP1B1. *Hepatology* 52 (5): 1797–1807.

72 Jung, D., Podvinec, M., Meyer, U.A. et al. (2002). Human organic anion transporting polypeptide 8 promoter is transactivated by the farnesoid X receptor/bile acid receptor. *Gastroenterology* 122 (7): 1954–1966.

73 Wood, M., Ananthanarayanan, M., Jones, B. et al. (2005). Hormonal regulation of hepatic organic anion transporting polypeptides. *Mol. Pharmacol.* 68 (1): 218–225.

74 Jung, D., Hagenbuch, B., Gresh, L. et al. (2001). Characterization of the human OATP-C (SLC21A6) gene promoter and regulation of liver-specific OATP genes by hepatocyte nuclear factor 1alpha. *J. Biol. Chem.* 276 (40): 37206–37214.

75 Vavricka, S.R., Jung, D., Fried, M. et al. (2004). The human organic anion transporting polypeptide 8 (SLCO1B3) gene is transcriptionally repressed by hepatocyte nuclear factor 3beta in hepatocellular carcinoma. *J. Hepatol.* 40 (2): 212–218.

76 Ichihara, S., Kikuchi, R., Kusuhara, H. et al. (2010). DNA methylation profiles of organic anion transporting polypeptide 1B3 in cancer cell lines. *Pharm. Res.* 27 (3): 510–516.

77 Yao, J., Hong, W., Huang, J. et al. (2012). N-Glycosylation dictates proper processing of organic anion transporting polypeptide 1B1. *PLoS One* 7 (12): e52563.

78 Powell, J., Farasyn, T., Kock, K. et al. (2014). Novel mechanism of impaired function of organic anion-transporting polypeptide 1B3 in human hepatocytes: post-translational regulation of OATP1B3 by protein kinase C activation. *Drug Metab. Dispos.* 42 (11): 1964–1970.

79 Crowe, A., Zheng, W., Miller, J. et al. (2019). Characterization of plasma membrane localization and phosphorylation status of organic anion transporting polypeptide (OATP) 1B1 c.521 T>C nonsynonymous single-nucleotide polymorphism. *Pharm. Res.* 36 (7): 101.

80 Hayden, E.R., Chen, M., Pasquariello, K.Z. et al. (2021). Regulation of OATP1B1 function by tyrosine kinase-mediated phosphorylation. *Clin. Cancer Res.* 27 (15): 4301–4310.

81 Hove, V.N., Anderson, K., Hayden, E.R. et al. (2022). Influence of tyrosine kinase inhibition on organic anion transporting polypeptide 1B3-mediated uptake. *Mol. Pharmacol.* 101 (6): 381–389.

82 Pickart, C.M. and Fushman, D. (2004). Polyubiquitin chains: polymeric protein signals. *Curr. Opin. Chem. Biol.* 8 (6): 610–616.

83 Ashida, H., Kim, M., and Sasakawa, C. (2014). Exploitation of the host ubiquitin system by human bacterial pathogens. *Nat. Rev. Microbiol.* 12 (6): 399–413.

84 Hershko, A. and Ciechanover, A. (1992). The ubiquitin system for protein degradation. *Annu. Rev. Biochem.* 61 (1): 761–807.

85 Alam, K., Farasyn, T., Crowe, A. et al. (2017). Treatment with proteasome inhibitor bortezomib decreases organic anion transporting polypeptide (OATP) 1B3-mediated transport in a substrate-dependent manner. *PLoS One* 12 (11): e0186924.

86 Alam, K., Pahwa, S., Wang, X. et al. (2016). Downregulation of organic anion transporting polypeptide (OATP) 1B1 transport function by lysosomotropic drug chloroquine: implication in OATP-mediated drug-drug interactions. *Mol. Pharmaceutics* 13 (3): 839–851.

87 Kinzi, J., Grube, M., and Schwabedissen, M.Z., H.E. (2021). OATP2B1 – The underrated member of the organic anion transporting polypeptide family of drug transporters? *Biochem. Pharmacol.* 188: 114534.

88 Hagenbuch, B. (2010). Drug uptake systems in liver and kidney: a historic perspective. *Clin. Pharmacol. Therap.* 87 (1): 39–47.

89 Zamek-Gliszczynski, M.J., Taub, M.E., Chothe, P.P. et al. (2018). Transporters in Drug Development: 2018 ITC Recommendations for Transporters of Emerging Clinical Importance. *Clin. Pharmacol. Therap.* 104 (5): 890–899.

90 Muller, J., Keiser, M., Drozdzik, M., and Oswald, S. (2017). Expression, regulation and function of intestinal drug transporters: an update. *Biol. Chem.* 398 (2): 175–192.

91 McFeely, S.J., Wu, L., Ritchie, T.K., and Unadkat, J. (2019). Organic anion transporting polypeptide 2B1 - More than a glass-full of drug interactions. *Pharmacol. Therap.* 196: 204–215.

92 Schafer, A.M., Meyer Zu Schwabedissen, H.E., Bien-Moller, S. et al. (2020). OATP1A2 and OATP2B1 are interacting with dopamine-receptor agonists and antagonists. *Mol. Pharmaceutics* 17 (6): 1987–1995.

93 Yang, Z.Z., Li, L., Wang, L. et al. (2016). siRNA capsulated brain-targeted nanoparticles specifically knock down OATP2B1 in mice: a mechanism for acute morphine tolerance suppression. *Sci. Rep.* 6: 33338.

94 Grube, M., Kock, K., Oswald, S. et al. (2006). Organic anion transporting polypeptide 2B1 is a high-affinity transporter for atorvastatin and is

expressed in the human heart. *Clin. Pharmacol. Therap.* 80 (6): 607–620.

**95** Nozawa, T., Imai, K., Nezu, J. et al. (2004). Functional characterization of pH-sensitive organic anion transporting polypeptide OATP-B in human. *J. Pharmacol. Exp. Therap.* 308 (2): 438–445.

**96** Kobayashi, D., Nozawa, T., Imai, K. et al. (2003). Involvement of human organic anion transporting polypeptide OATP-B (SLC21A9) in pH-dependent transport across intestinal apical membrane. *J. Pharmacol. Exp. Therap.* 306 (2): 703–708.

**97** Nozawa, T., Nakajima, M., Tamai, I. et al. (2002). Genetic polymorphisms of human organic anion transporters OATP-C (SLC21A6) and OATP-B (SLC21A9): allele frequencies in the Japanese population and functional analysis. *J. Pharmacol. Exp. Therap.* 302 (2): 804–813.

**98** Imanaga, J., Kotegawa, T., Imai, H. et al. (2011). The effects of the SLCO2B1 c.1457C > T polymorphism and apple juice on the pharmacokinetics of fexofenadine and midazolam in humans. *Pharmacogenet. Genomics* 21 (2): 84–93.

**99** Ieiri, I., Doi, Y., Maeda, K. et al. (2012). Microdosing clinical study: pharmacokinetic, pharmacogenomic (SLCO2B1), and interaction (grapefruit juice) profiles of celiprolol following the oral microdose and therapeutic dose. *J. Clin. Pharmacol.* 52 (7): 1078–1089.

**100** Shirasaka, Y., Shichiri, M., Mori, T. et al. (2013). Major active components in grapefruit, orange, and apple juices responsible for OATP2B1-mediated drug interactions. *J. Pharm. Sci.* 102 (1): 280–288.

**101** Johnson, M., Patel, D., Matheny, C. et al. (2017). Inhibition of intestinal OATP2B1 by the calcium receptor antagonist ronacaleret results in a significant drug-drug interaction by causing a 2-fold decrease in exposure of rosuvastatin. *Drug Metab. Dispos.* 45 (1): 27–34.

**102** Maeda, T., Hirayama, M., Higashi, R. et al. (2006). Characterization of human OATP2B1 (SLCO2B1) gene promoter regulation. *Pharm. Res.* 23 (3): 513–520.

**103** Knauer, M.J., Girdwood, A.J., Kim, R.B., and Tirona, R.G. (2013). Transport function and transcriptional regulation of a liver-enriched human organic anion transporting polypeptide 2B1 transcriptional start site variant. *Mol. Pharmacol.* 83 (6): 1218–1228.

**104** Liu, W., Nakano, M., Nakanishi, T. et al. (2020). Post-transcriptional regulation of OATP2B1 transporter by a microRNA, miR-24. *Drug Metab. Pharmacokinet.* 35 (6): 515–521.

**105** Tajiri, A., Hirota, T., Kawano, S. et al. (2020). Regulation of organic anion transporting polypeptide 2B1 expression by MicroRNA in the human liver. *Mol. Pharmaceutics* 17 (8): 2821–2830.

**106** Franke, R.M., Scherkenbach, L.A., and Sparreboom, A. (2009). Pharmacogenetics of the organic anion transporting polypeptide 1A2. *Pharmacogenomics* 10 (3): 339–344.

**107** Kullak-Ublick, G.A., Hagenbuch, B., Stieger, B. et al. (1995). Molecular and functional characterization of an organic anion transporting polypeptide cloned from human liver. *Gastroenterology* 109 (4): 1274–1282.

**108** Cvetkovic, M., Leake, B., Fromm, M.F. et al. (1999). OATP and P-glycoprotein transporters mediate the cellular uptake and excretion of fexofenadine. *Drug Metab. Dispos.* 27 (8): 866–871.

**109** Kullak-Ublick, G.A., Fisch, T., Oswald, M. et al. (1998). Dehydroepiandrosterone sulfate (DHEAS): identification of a carrier protein in human liver and brain. *FEBS Lett.* 424 (3): 173–176.

**110** Drozdzik, M., Groer, C., Penski, J. et al. (2014). Protein abundance of clinically relevant multidrug transporters along the entire length of the human intestine. *Mol. Pharmaceutics* 11 (10): 3547–3555.

**111** Lee, W., Glaeser, H., Smith, L.H. et al. (2005). Polymorphisms in human organic anion-transporting polypeptide 1A2 (OATP1A2): implications for altered drug disposition and central nervous system drug entry. *J. Biol. Chem.* 280 (10): 9610–9617.

**112** Glaeser, H., Bailey, D.G., Dresser, G.K. et al. (2007). Intestinal drug transporter expression and the impact of grapefruit juice in humans. *Clin. Pharmacol. Therap.* 81 (3): 362–370.

**113** Groer, C., Bruck, S., Lai, Y. et al. (2013). LC-MS/MS-based quantification of clinically relevant intestinal uptake and efflux transporter proteins. *J. Pharm. Biomed. Anal.* 85: 253–261.

**114** Hilgendorf, C., Ahlin, G., Seithel, A. et al. (2007). Expression of thirty-six drug transporter genes in human intestine, liver, kidney, and organotypic cell lines. *Drug Metab. Dispos.* 35 (8): 1333–1340.

**115** Meier, Y., Eloranta, J.J., Darimont, J. et al. (2007). Regional distribution of solute carrier mRNA expression along the human intestinal tract. *Drug Metab. Dispos.* 35 (4): 590–594.

**116** Dresser, G.K., Bailey, D.G., Leake, B.F. et al. (2002). Fruit juices inhibit organic anion transporting polypeptide-mediated drug uptake to decrease the oral availability of fexofenadine. *Clin. Pharmacol. Therap.* 71 (1): 11–20.

117 Rebello, S., Zhao, S., Hariry, S. et al. (2012). Intestinal OATP1A2 inhibition as a potential mechanism for the effect of grapefruit juice on aliskiren pharmacokinetics in healthy subjects. *Eur. J. Clin. Pharmacol.* 68 (5): 697–708.

118 Schwarz, U.I., Seemann, D., Oertel, R. et al. (2005). Grapefruit juice ingestion significantly reduces talinolol bioavailability. *Clin. Pharmacol. Therap.* 77 (4): 291–301.

119 Tamai, I. (2012). Oral drug delivery utilizing intestinal OATP transporters. *Adv. Drug Delivery Rev.* 64 (6): 508–514.

120 van de Steeg, E., van Esch, A., Wagenaar, E. et al. (2013). Influence of human OATP1B1, OATP1B3, and OATP1A2 on the pharmacokinetics of methotrexate and paclitaxel in humanized transgenic mice. *Clin. Cancer Res.* 19 (4): 821–832.

# 22

## ABC Transporters in Intestinal and Liver Efflux

*Marilyn E. Morris and Tianjing Ren*

*University at Buffalo (State University of New York), School of Pharmacy and Pharmaceutical Sciences, Department of Pharmaceutical Sciences, Buffalo, NY, USA*

### 22.1 Introduction

The oral bioavailability of many drugs depends on their ability to cross intestinal membranes and avoid first-pass extraction either due to metabolism or efflux transporters. A lipophilic drug can cross membranes by passive diffusion, while hydrophilic drugs require active transport processes to enter cells. Whether a drug is lipophilic or hydrophilic, it can be extruded out of cells by a class of transporters known as the ATP binding cassette (ABC) family of transporters. These transporters play a pivotal role in protecting tissues such as brain, testis, and placenta, among others, from xenobiotics. The ABC superfamily of transporters consists of 51 human genes in 7 subfamilies [1], but not all transporters are involved in drug transport. The ABC transporters involved in drug transport are found in many tissues in the body including the liver, kidney, brain, and intestine. The liver and the intestine are the tissues most important in the oral bioavailability of drugs. The efflux transporters linked with drug transport in the liver and the intestine are ABCB1 (MDR1), ABCB4 (MDR2/3), ABCB11 (BSEP), ABCC1-6 (MRP1-6), and ABCG2 (BCRP).

The absorption of most nutrients and drugs occurs in the small intestine. Some efflux transporters are opportunistically located on the apical membrane of enterocytes that line the gut lumen (Figure 22.1). These transporters are responsible for the efflux of a wide range of substrates from enterocytes back into intestinal lumen, limiting the oral bioavailability of these compounds. The compounds that escape efflux in the small intestine enter the liver through the portal vein, where they can undergo hepatic first-pass extraction by metabolism and/or excretion into bile. Efflux transporters located on the canalicular membrane in hepatocytes facilitate the efflux of compounds against their concentration gradient into bile (Figure 22.2). Thus, efflux transporters of the ABC family can represent determinants of the oral bioavailability and disposition of xenobiotics.

ABC efflux transporters exhibit significant expression levels on the membranes of the intestine and the liver. The mRNA and protein expressions of ABCB1, ABCB11, ABCC1, ABCC2, ABCC3, ABCC4, and ABCG2 have been recently quantified in human livers and intestines from nine organ donors [2]. The following ranking of transporter protein abundance was observed in the duodenum: ABCC3 > ABCC4 > ABCC2 > ABCB1 > ABCG2; in the jejunum: ABCB1 > ABCC3 > ABCC2 = ABCG2 > ABCC4; in the ileum: ABCB1 > ABCG2 > ABCC3 > ABCC2 > ABCC4; and in the colon: ABCC3 > ABCC4 > ABCC2 > ABCB1 > ABCG2. The protein expression of ABCB11 and ABCC1 was not detected in the intestine. The protein expression of ABCB1 and ABCG2 increased from duodenum to distal jejunum while the ABCC2 and ABCC3 were maximal in jejunum. The transporter protein expression in the liver follows the order as ABCC2 > ABCC3 > ABCB1 = ABCB11 with undetectable ABCC1, ABCC4, and ABCG2. These findings are generally consistent with other studies, although Prasad et al. and others have noted low expression of ABCG2 in livers [3]. As well, considerable inter-species variability in expression has been reported [4]. Drozdzik et al. also demonstrated limited correlations between mRNA level and protein amounts in livers [2], which is in agreement with other reports, suggesting mRNA expression data may not be adequate in evaluating transporter protein abundances and functions [4, 5]. Due to the high expression of ABCB1 and ABCG2 in intestine and high expression of ABCB1 in the liver, most new molecular entities should be evaluated as ABCB1/ABCG2 substrates, as suggested in the FDA Guidance for Industry [6].

*Oral Bioavailability and Drug Delivery: From Basics to Advanced Concepts and Applications*, First Edition. Edited by Ming Hu and Xiaoling Li.

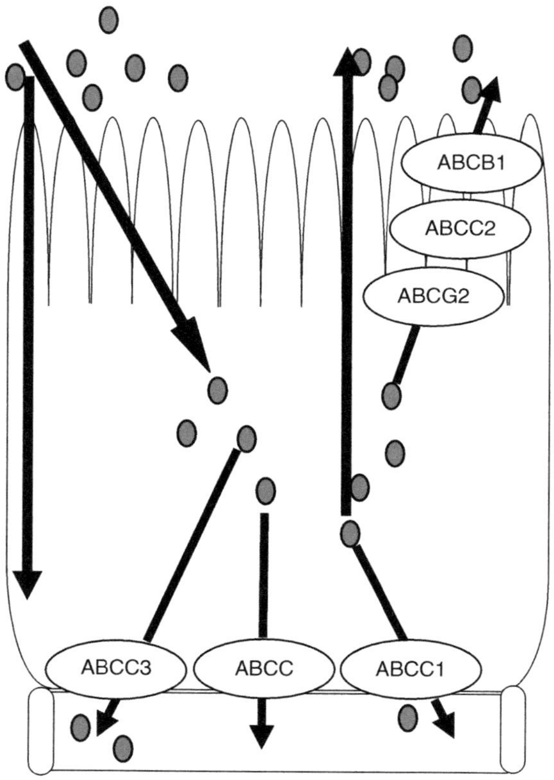


**Figure 22.1** ABC transporters in the intestine. Drug in the gut lumen can enter enterocytes by passive diffusion or an influx transporter. Once inside the cell, endogenous compounds, drugs, or metabolites (especially conjugated metabolites) can be effluxed back into the intestinal lumen by ABCB1, ABCC2, and ABCG2 transporters located at the apical membrane, or can be effluxed into blood by ABCC1, ABCC3, and ABCC4 at the basolateral membrane. mRNA expression of ABCC5 and ABCC6 has been reported in the intestine, but there is no indication of the protein levels of these transporters.

This chapter will focus on the ABC efflux transporters present in liver and intestine and their roles in the oral bioavailability and hepatobiliary excretion of xenobiotics. Factors that affect the expression and function of efflux transporters in the liver and intestine will also be discussed.

## 22.2 Apical Membrane Efflux Proteins

### 22.2.1 ABCB1 (P-glycoprotein)

ABCB1 was discovered by Juliano and Ling in 1976 [7] in Chinese hamster ovary cells selected for resistance to colchicines; these investigators named this protein P-glycoprotein (Pgp). In humans, ABCB1 is the only drug-transporting member while in rodents there are two members: Abcb1a and Abcb1b. Knockout mice deficient in both Abcb1 members have proved useful in uncovering the properties of this therapeutically important protein [8]. The most striking feature of human ABCB1 is its ability to transport a wide range of structurally diverse compounds, including many anticancer drugs, natural products, steroids, fluorescent dyes, and ionophores [9–11]. A list of ABCB1 substrates and inhibitors is included in Table 22.1.

A number of studies have emphasized the importance of ABCB1 as a determinant of the oral bioavailability and hepatobiliary excretion of xenobiotics. Plasma concentrations of paclitaxel, indinavir, saquinavir, and nelfinavir are higher in *Abcb1a/Abcb1b* knockout mice than in their wild-type counterparts after oral administration [17]. The bioavailability of the ABCB1 substrate digoxin was increased by ABCB1 inhibitors clarithromycin and quercetin and decreased by ABCB1 inducers St. John's wort and rifampin [45]. Shimizu et al. reported that plasma concentrations of fexofenadine, an ABCB1 substrate that is poorly metabolized, were significantly increased following the administration of itraconazole [46, 47]. In addition to digoxin and fexofenadine, dabigatran etexilate has also been listed as one of the clinical probes for ABCB1 for interaction studies [48]. As a solely intestinal ABCB1 substrate with high substrate affinity, dabigatran etexilate demonstrated higher selectivity and sensitivity than digoxin during a study of intestinal ABCB1-mediated drug interactions [49]. A recent study demonstrated increased oral bioavailability of ketamine in *Abcb1a* knockout and *Abcb1a/Abcg2* double knockout mice, indicating that ketamine is a dual Abcb1a/Abcg2 substrate [50]. Concomitant use of the ABCB1 inhibitors verapamil and lumefantrine in rats led to increased bioavailability of lumefantrine, suggesting a significant role of ABCB1 in the oral absorption of lumefantrine [51]. Based on a case report demonstrating the increased concentration/dose ratio of tacrolimus in patients cotreated with oral risperidone, Watanabe et al. further utilized Caco-2 cells to illustrate the involvement of ABCB1 inhibition by risperidone in this interaction [52].

The ability to predict potential ABCB1 substrates or inhibitors during the development phase of drug discovery would help in understanding the bioavailability and disposition of candidate drugs. Structure activity relationships (SARs) attempting to correlate physico-chemical or structural properties of the drug molecule important for ABCB1 transport have been derived, but a general model that can predict substrates and inhibitors is lacking. The recent focus has shifted to the classification models for large datasets to profile

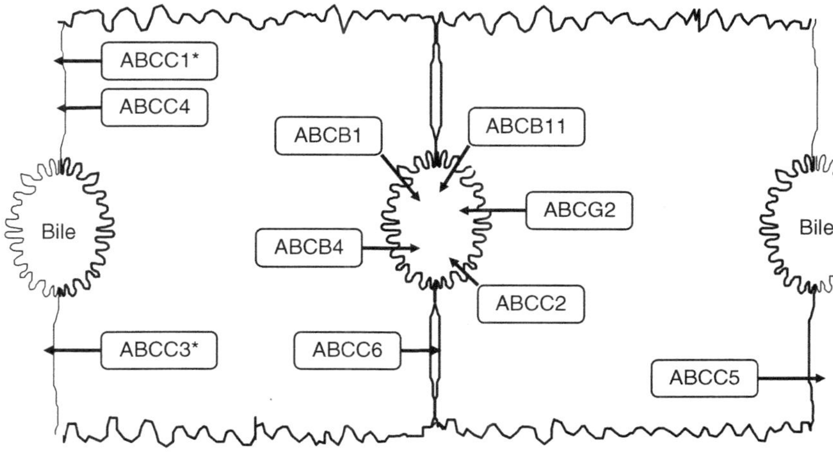


**Figure 22.2** ABC transporters in the hepatocyte. ABCB1, ABCB4, ABCB11, ABCC2, and ABCG2 are located at the canalicular membrane in hepatocytes and play a role in biliary excretion of xenobiotics and endogenous compounds. ABCC1 and ABCC3–6 transporters are located at the basolateral/lateral membranes in hepatocytes and play a role in the efflux of xenobiotics and endogenous compounds from the hepatocyte into sinusoidal blood.

the compound libraries by in silico machine learning [53]. Broccatelli et al. have established a workflow that combines pharmacophore and general physiochemical descriptors to predict ABCB1 inhibitors based on over 1200 compounds [54]. The final models indicated that flexibility, hydrophobic surface area, log $P$, and 3D structures are the crucial factors for predicting ABCB1 inhibitors. The classification model for ABCB1 substrates has been developed by Li et al. using 423 substrates and 300 non-substrates [55]. By analyzing the important structural fragments of compounds, the model suggested that H-bond acceptors arranged in distinct spatial patterns and flexibility are essential for ABCB1 substrate-likeness.

Disease states may also be associated with the changes in expression and function of ABCB1, leading to altered drug exposures. For example, the systemic exposure of nelfinavir, a ABCB1 substrate, was significantly higher in human immunodeficiency virus (HIV) and hepatitis C virus (HCV) coinfected patients compared to that in HIV infected patients without HCV [56]. The altered pharmacokinetics may be attributed to the downregulated ABCB1 in HCV patients.

### 22.2.2 ABCB4 (MDR2/3)

ABCB4 is a phospholipid transporter located on the canalicular membrane of hepatocytes. Even though bile acids form the major component of hepatic bile, phospholipids, and cholesterol are responsible for the formation of micelles that solubilize bile acids in the small intestine. ABCB4 transports cholesterol and phospholipids, particularly phosphatidylcholine [14]. The ABCB4-mediated efflux of phospholipids and cholesterol was remarkably increased in the presence of taurocholate while inhibited by verapamil in HEK/ABCB4 cells [57]. Taipalensuu et al. have reported the ABCB4 expression in human jejunum and Caco-2 cells [58], but little is known concerning the effects of ABCB4 on the oral bioavailability or the hepatobiliary excretion of drugs.

Studies with ABCB4 have shown that the phospholipid transporter is also involved in the transport of drugs. ABCB4 shares 82% similarity in amino acid sequence with ABCB1 and it has been proposed that ABCB4 shares substrate specificity with ABCB1, but exhibits a lower affinity for substrates [14]. In ABCB4-transfected cells, the apical to basolateral directional transport of digoxin, paclitaxel, daunorubicin, vinblastine, and ivermectin was increased significantly in comparison to the parental cell line [14]. Curcumin has been demonstrated to reverse doxorubicin resistance due to inhibition of ABCB4 efflux in breast cancer cells [59]. Inhibition of ABCB4 by azole antifungals such as itraconazole, posaconazole, and ketoconazole may lead to cholestatic liver injury due to the inhibition on ABCB4-mediated phosphatidylcholine secretion [13].

### 22.2.3 ABCB11 (BSEP)

ABCB11 is a monovalent bile salt efflux protein located predominantly on the canalicular membrane in liver hepatocytes. Excretion of bile salts from liver hepatocytes into bile canaliculi is a major driving force for bile flow and can be a rate-limiting step. ABCB11 is a high-affinity transporter involved in the transport of conjugated and unconjugated bile salts and serves two primary functions in the liver: to reduce the bile acid concentration in hepatocytes and to facilitate enterohepatic recirculation of bile salts [60]. Childs et al. discovered Abcb11 in the porcine liver in 1995 and named it the sister of Pgp, due to similarity in

**Table 22.1** Substrates and inhibitors of ABC transporters in liver and intestine.

| Transporter | Substrates | Inhibitors |
|---|---|---|
| **ABCB1** | Actinomycin D | Acacetin |
| (Clay and Sharom [9], Kadioglu and Efferth [10], Li et al. [11])[a] | Aldosterone | Agosterol |
| | Beauvericin | Amiodarone |
| | Biochanin A | Amorinin |
| | Calcein acetoxymethyl ester | Apigenin |
| | Cimetidine | Atorvastatin |
| | Colchicine | Atovaquone |
| | Daunorubicin | Azidopine |
| | Dexamethasone | Biochanin |
| | Digoxin | Biricodar |
| | Docetaxel | Catechin |
| | Doxorubicin | Catechin gallate (CG) |
| | Epirubicin | CBT-1 |
| | Etoposide | Cefoperazone |
| | Fexofenadine | Chloropromazine |
| | Gramicidin D | Chloroquine |
| | Hoechst 33342 | Chrysin |
| | Idarubicin | Cortisol |
| | Indinavir | Curcumin |
| | Irinotecan | Cyclosporine A |
| | Ivermectin | Dexniguldipine |
| | Kaempferol | 6′,7′-Epoxybergamottin |
| | Leupeptin | 6′,7′-Dihydroxybergamottin |
| | Loperamide | Diltiazem |
| | Mitomycin | Disulfiram |
| | Mitoxantrone | Elacridar |
| | Morphine | Epicatechin gallate (ECG) |
| | Narigenin | Epigallocatechin gallate (EGCG) |
| | Nelfinavir | Fangchinoline |
| | Nonactin | Genistein |
| | Nonylphenol ethoxylate | Ginsenoside |
| | Ondansetron | Hesperetin |
| | Paclitaxel | Laniquidar |
| | Pepstatin A | Loratidine |
| | Procyanidin B2 | Mibefradil |
| | Quercetin | Naringenin |
| | Rhodamine 123 | Naringin |
| | Ritonavir | Nifedipine |
| | Saquinavir | OC144-093 |
| | Teniposide | Pgp-4008 |
| | Tetramethylrosamine | Phloretin |
| | Topotecan | Progesterone |
| | Triton X-100 | Quercetin |
| | Vinblastine | Quinidine |
| | Vincristine | Quinine |

*(continued)*

**Table 22.1** (Continued)

| Transporter | Substrates | Inhibitors |
|---|---|---|
| | Vindesine | Reserpine |
| | Vinorelbine | Rotenone |
| | | Sakuranetin |
| | | Sertraline |
| | | Sinensetin |
| | | Stigmasterol |
| | | Syringaresinol |
| | | Tamoxifen |
| | | Tariquidar |
| | | Terfenadine |
| | | Tetrandine |
| | | *Trans*-flupenthixol |
| | | Trifluoperazine |
| | | Valspodar |
| | | Verapamil |
| | | Zosuquidar |
| **ABCB4** | Aureobasidin A | Cyclosporine A |
| (Kino et al. [12], Mahdi et al. [13][a), Smith et al. [14][a)) | Daunorubicin | Itraconazole |
| | Digoxin | Ketoconazole |
| | Ivermectine | Paclitaxel |
| | Paclitaxel | Posaconazole |
| | Vinblastine | Valspodar |
| | | Verapamil |
| | | Vinblastine |
| **ABCB11** | Calcein-acetoxymethyl ester | Bosentan |
| (Byrne et al. [15], Hirano et al. [16], Chan et al. [17][a), Stieger [18][a)) | Glycochenodeoxycholate | Chlorpromazine |
| | Glycocholate | CI-1034 |
| | Pravastatin | Ciglitazone |
| | Taurochenodeoxycholate | Cloxacillin |
| | Taurocholate | Colchicine |
| | Vinblastine | Cyclosporine A |
| | | Doxycycline |
| | | Erythromycin |
| | | Estolate |
| | | Fendiline |
| | | Fusidate |
| | | Glibenclamide |
| | | Lovastatin |
| | | Mideamycin |
| | | Nefadozone |
| | | Nicardipine |
| | | Nifedipine |
| | | PKI166 |
| | | Pravastatin |

*(continued)*

**Table 22.1** (Continued)

| Transporter | Substrates | Inhibitors |
|---|---|---|
| | | Prenylamine |
| | | Progesterone |
| | | Quinidine |
| | | Rifampicin |
| | | Ritonavir |
| | | Rosiglitazone |
| | | Saquinavir |
| | | Tamoxifen |
| | | Troglitazone |
| | | Troleandomycin |
| | | Verapamil |
| **ABCC1** | Anthracyclines | MK-571 |
| (Chan et al. [17][a], Li et al. [11][a], Schinkel and Jonker [8][a], Estudante et al. [19], Fahrmayr et al. [20]) | Calcein | Apigenin |
| | Daunorubicin | Benzbromarone |
| | Dehydroepiandrosterone sulfate (DHEAS) | Curcumin |
| | Dinitrophenyl *S*-glutathione (DNP-SG) | Genistein |
| | Doxorubicin | Kaempferol |
| | ECG | Naringenin |
| | Estradiol-17-β-glucuronide | Probenecid |
| | Estrone sulfate | Quercetin |
| | Etoposide | Schisandrin A |
| | Etoposide-glucuronide | Schisandrin B |
| | Ezetimibe-glucuronide | Schisandrol A |
| | Fluo 3 | Schisandrol B |
| | Grepafloxacin | *S*-Decylglutathione |
| | Leukotriene C4 | Sulfinpyrazone |
| | Methotrexate | |
| | Mitoxantrone | |
| | Rhodamine 123 | |
| | Teniposide | |
| | Vinblastine | |
| | Vinca alkaloids | |
| | Vincristine | |
| **ABCC2** | Acetaminophen | MK-571 |
| (Cui et al. [21], Huisman et al. [22], Fardel et al. [23][a], Li et al. [11][a]) | Ampicillin | Azithromycin |
| | Arsenic | Curcumin |
| | Azithromycin | Cyclosporine A |
| | Bilirubin | Glibenclamide |
| | BQ-123 | Indomethacin |
| | Camptothecin | Luteolin |
| | Ceftriaxone | Myricetin |
| | Cisplatin | Probenecid |
| | Docetaxel | Quercetin |
| | Doxorubicin | Rifampicin |
| | ECG | Robinetin |

*(continued)*

Table 22.1 (Continued)

| Transporter | Substrates | Inhibitors |
|---|---|---|
| | Epirubicin | |
| | Estradiol | |
| | Estradiol-17-β-glucuronide | |
| | Etoposide | |
| | Glutathione | |
| | Grepafloxacin | |
| | Indinavir | |
| | Irinotecan | |
| | Leukotriene C4 | |
| | Methotrexate | |
| | 4-*O*-methyl-EGCG | |
| | Paclitaxel | |
| | *p*-Aminohippurate | |
| | Pravastatin | |
| | Ritonavir | |
| | Saquinavir | |
| | SN-38 | |
| | Taurochenodeoxycholate sulfate | |
| | Taurolithocholate sulfate | |
| | Temocaprilat | |
| | Vincristine | |
| **ABCC3** | Acetaminophen glucuronide | Indomethacin |
| (Lee et al. [24], Zelcer et al. [25], Zeng et al. [26], Ghanem and Manautou [27][a)], Li et al. [11][a)]) | Acetaminophen-sulfate | Probenecid |
| | Baicalein-7-glucuronide | Sulfinpyrazone |
| | Bisglucuronosyl bilirubin | |
| | Bisphenol A-glucuronide | |
| | Cefadroxil | |
| | Daidzein-glucuronide | |
| | Dehydroepiandrosterone sulfate | |
| | DHEAS | |
| | DNP- SG | |
| | E217β-glucuronide | |
| | E3040-glucuronide | |
| | Enterodiol-glucuronide | |
| | Enterolactone-glucuronide | |
| | Epipodophyllotoxins | |
| | Equol-glucuronide | |
| | Estradiol-17-β-glucuronide | |
| | Ethinyl estradiol-glucuronide | |
| | Etoposide | |
| | Etoposide glucuronide | |
| | Ezetimibe-glucuronide | |
| | Fexofenadine | |
| | Folic acid | |
| | Gemfibrozil-glucuronide | |
| | Genistein-glucuronide | |

*(continued)*

**Table 22.1** (Continued)

| Transporter | Substrates | Inhibitors |
|---|---|---|
| | Glycitein-glucuronide | |
| | Glycocholate | |
| | Harmol-glucuronide | |
| | Harmol-sulfate | |
| | Hyocholate-glucuronide | |
| | Hyodeoxycholate-glucuronide | |
| | Leucovorin | |
| | Leukotriene C4 | |
| | Methotrexate | |
| | Methotrexate 7-hydroxymethotrexate | |
| | 4-Methylumbelliferone-glucuronide | |
| | 4-Methylumbelliferone-sulfate | |
| | Mono-/di-bilirrubin glucuronide | |
| | Monoglucuronosyl bilirubin | |
| | Morphine-3-glucuronide | |
| | Morphine-6-glucuronide | |
| | Resveratrol | |
| | Resveratrol-glucuronide | |
| | Secoisolariciresinol-glucuronide | |
| | Taurochenodeoxycholate-3-sulfate | |
| | Taurocholate | |
| | Taurolithocholate-3-sulfate | |
| | Teniposide | |
| | Toglitazone-glucuronide | |
| **ABCC4** | Adefovir | Benzbromarone |
| (Alemanno et al. [28], Ferslew et al. [29], Fukuda et al. [30], Furmanski et al. [31], Rius et al. [32], Berthier et al. [33][a], Haimeur et al. [34][a], Li et al. [11][a]) | Azidothymidine monophosphate (AZT) | Cefazoline |
| | Cefazoline | Cefmetazole |
| | Cefmetazole | Cefotaxime |
| | Cefotaxime | Ceftizoxime |
| | Ceftizoxime | Celecoxib |
| | Celecoxib | Cilostazol |
| | Chenodeoxycholylglycine (+ GSH) | DHEAS |
| | Chenodeoxycholyltaurine (+ GSH) | Dipyridamole |
| | Cholate (+ GSH) | Estradiol-17-β-glucuronide |
| | Cholylglycine (+ GSH) | Flurbiprofen |
| | Cyclic adenosine monophosphate (cAMP) | Furosemide |
| | Cyclic guanosine monophosphate (cGMP) | Hydrochlorothiazide |
| | Dasatinib | Ibuprofen |
| | Deoxycholylglycine (+ GSH) | Indomethacin |
| | DHEAS | Irbesartan |
| | Edaravone | Leucovorin |

(*continued*)

**Table 22.1** (Continued)

| Transporter | Substrates | Inhibitors |
|---|---|---|
| | Enalapril | Losartan |
| | Estradiol-17-β-glucuronide | Methotrexate |
| | Furosemide | Micafungin |
| | Ganciclovir | MK-571 |
| | Hydrochlorothiazide | Mycophenolic acid |
| | Indomethacin | Naproxen |
| | Lamivudine | Nelfinavir |
| | Leucovorin | Nitrofurantoin |
| | 6-Mercaptopurine | Olmesartan |
| | Methotrexate | Probenecid |
| | Mycophenolic acid | Quercetin, |
| | Nelfinavir | Sildenafil |
| | Olmesartan | Silymarin |
| | 9-(2-Phosphonyl methoxyethyl) adenine | Simvastatin |
| | Piperacillin | Sorafenib |
| | Probenecid | Sulfasalazine |
| | Sildenafil | Sulfinpyrazone |
| | Sulindac | Sulindac |
| | Tenofovir disoproxil fumarate | Tamoxifen |
| | 6-Thioguanine | Ticlopidine |
| | Topotecan | |
| | Ursodeoxycholylglycine (+ GSH) | |
| | Ursodeoxycholyltaurine (+ GSH) | |
| **ABCC5** | Adefovir | MK-571 |
| (Li et al. [11], Liu et al. [35], Pratt et al. [36], Wijnholds et al. [37])[a)] | cAMP | 4-Nitro-2-(3-phenylpropylamino) benzoate (NPPB) |
| | cGMP | Curcumin |
| | Cisplatin | Glybenclamide |
| | DNP-SG | Probenecid |
| | Doxorubicin | Quercetin |
| | 5-Flurouracil | Sildenafil |
| | 6-Mercaptopurine | Silymarin |
| | Methotrexate | Trequinsin |
| | Oxaliplatin | Vardenafil |
| | Pemetrexed | Zaprinast |
| | 9-(2-Phosphonylmeth-oxyethyl) adenine | |
| | *S*-(2,4-dinitrophenyl)glutathione | |
| | Thioguanine | |
| **ABCC6** | BQ-123 | Benzbromarone |
| (Belinsky et al. [38], Eadie et al. [39], Ilias et al. [40]) | Cisplatin | Indomethacin |
| | Dasatinib | Pantoprazole |
| | DNP-SG | Probenecid |
| | Doxorubicin | |
| | Etoposide | |
| | Leukotriene C4 | |

*(continued)*

**Table 22.1** (Continued)

| Transporter | Substrates | Inhibitors |
|---|---|---|
| | Nilotinib | |
| | *S*-Glutathionyl 2,4-dinitrobenzene | |
| | Teniposide | |
| **ABCG2** | Acetaminophen sulfate | Acacetin |
| (Basseville et al. [41][a), Lee et al. [42][a), Li et al. [11][a), Mao and Unadkat [43][a), Miyata et al. [44]) | Aconitine | Afatinib |
| | Aflatoxin B1 | Antifungal |
| | Albendazole sulfoxide | Apigenin |
| | 2-Amino-1-methyl-6-phenylimidazo (4,5-b)pyridine | Aripiprazole |
| | 2-Amino-3-methylimidazo[4,5-*f*] quinoline | Axitinib |
| | 3-Amino-1,4-dimethyl1-5*H*-pyrido [4,3-*b*]indole | Ayanin |
| | Anthrapyrazole (AZA) | Azoles |
| | Apixaban | Beclomethasone |
| | Axitinib | Benzoflavone |
| | AZT 5′-monophosphate | Biochanin A |
| | Barasertib hydroxy-QPA | Biricodar |
| | Benzo[*a*]pyrene-3-glucuronide | Boceprevir |
| | Benzo[*a*]pyrene-3-sulfate | Cannabinoids |
| | Bisantrene | Celastrol |
| | Boceprevir | Chrysin |
| | Bupropion | Chrysoeriol |
| | Canertinib | CI 1033 |
| | Cerivastatin | Ciclesonide |
| | Chlorine | Corticosterone |
| | Chlorothiazide | Curcumin |
| | CI 1033 | Cyanidin |
| | Cimetidine | Cyaniding-3,5-diglucoside |
| | Ciprofloxacin | Cyaniding-3-rutinoside |
| | Compound A | Cyclosporine A |
| | Cyanidin-3-galactoside | Delphinidin |
| | Cyanidin-3-glucoside | Delphinidin-3-glucoside |
| | Dantrolene | Dexamethasone |
| | Daunorubicin | DHP-014 |
| | Dehydroepiandrosterone sulfate | Diadzein |
| | Diflomotecan | Diethylstilbestrol |
| | Dihydrotestosterone | Digoxin |
| | Dimenhydrinate | Diosmetin |
| | Dipyridamole | Dipyridamole |
| | Dipyrrometheneboron Difluoride (BODIPY)-prazosin | Elacridar |
| | Dirithromycin | Ellagic acid |
| | Dolutegravir | Erberine |
| | Doxorubicin | Erlotinib |
| | E3040 | Estradiol-17-β-glucuronide |

*(continued)*

**Table 22.1** (Continued)

| Transporter | Substrates | Inhibitors |
|---|---|---|
| | Edaravone sulfate | Estrone |
| | Epirubicin | Febuxostat |
| | Erythromycin | Flavopiridol |
| | Estradiol-17-β-glucuronide | Fluoconazole |
| | Estradiol-17-β-sulfate | Fluvastatin |
| | Estrone-3-sulfate | Fumitremorgin C |
| | Ethinyl estradiol sulfate | Galangin |
| | Etoposide | Gefitinib |
| | Flavopiridol | Genistein |
| | Fluvastatin | Hesperetin |
| | Folic acid | 7-Hydroxystaurosporine |
| | Furosemide | Imatinib |
| | Gefitinib | Imatinib mesylate |
| | Gemifloxacin | Indolyl diketopiperazines |
| | Genistein | Itraconazole |
| | Glyburide | Ivermectin |
| | Grepafloxacin | Kaempferide |
| | GV 196771 | Kaempferol |
| | GW1843 | Ketoconazole |
| | Hematoporphyrin | Ko 143 |
| | Hoechst 33342 | Lapatinib |
| | 6-Hydroxy-5,7-dimethyl-2-methylamino-4- | Laricitrin |
| | 7-Hydroxystaurosporine (UCN-01) | Limonin |
| | Hypaconitine | Lopinavir |
| | Imatinib | Luteolin |
| | Indolocarbazole | Malvidin |
| | Irinotecan | Malvidin-3-galactoside |
| | J-107088 | Malvidin-3-glucoside |
| | Lamivudine | 6-α-Methylprednisolone |
| | Lapatinib | Mometasone |
| | Lysotracker Green | Myricetin |
| | Malvidin | Myricetin-3′,4′,5′-trimethylether |
| | Malvidin-3,5-diglucoside | Naringen-7-glucoside |
| | Malvidin-3-galactoside | Naringenin |
| | ME 3229 | Nelfinavir |
| | Mesaconitine | Nicardipine |
| | Mesylate | Niguldipine |
| | Methotrexate | Nilotinib |
| | Methotrexate diglutamate | Nimodipine |
| | Methotrexate triglutamate | Nitrendipine |
| | 4-Methylumbelliferone glucuronide | Novobiocin |
| | 4-Methylumbelliferone sulfate | Oleanolic acid |
| | Mitoxantrone | Omeprazole |
| | NB-506 | Ortataxel |
| | *N*-Desethyl sunitinib | Oxfendazole |
| | Nilotinib | Pantoprazole |

*(continued)*

**Table 22.1** (Continued)

| Transporter | Substrates | Inhibitors |
|---|---|---|
| | Nitrofurantoin | Pelargonidin |
| | Norfloxacin | Pelargonidin-3,5-diglucoside |
| | Ochratoxin A | Peonidin |
| | Ofloxacin | Petunidin |
| | Olomoucine II | Pinocembrin |
| | Oxfendazole | Pitavastatin |
| | Pantoprazole | Ponatinib |
| | Pemetrexed | 6-Prenylchrysin |
| | Peonidin-3-glucoside | Propyl gallate |
| | Petunidin | Quercetin |
| | Phenethyl isothiocynate | Quercitrin |
| | Pheophorbide A | Quizartinib |
| | Pitavastatin | Rabeprazole |
| | Prazosin | Regorafenib |
| | Protoporphyin IX | Reserpine |
| | Quercetin | Resveratrol |
| | Resveratrol | Retusin |
| | Resveratrol-3-glucuronide | Rilpivirine |
| | Resveratrol-3-sulfate | Silymarin |
| | Rhodamine 123 | Sinapic acid |
| | Rifampicin | Sirolimus |
| | Ritonavir | Stilbenoids |
| | Rivaroxaban | Sulfasalazine |
| | Rosuvastatin | Sunitinib |
| | Saquinavir | Tacrolimus |
| | Sildenafil | TAG-11 |
| | SN-38 | TAG-139 |
| | SN-38 glucuronide | Tamarixetin |
| | Sorafenib | Tamoxifen |
| | Sulfasalazine | Tariquidar (XR9576) |
| | Sunitinib | Tectochrysin |
| | Tandutinib | Telaprevir |
| | Temocapril | Teriflunomide |
| | Teniposide | Toremifene |
| | Teriflunomide | tRA96023 |
| | Tomudex | tRA98006 |
| | Topotecan | Trametinib |
| | Troglitazone sulfate | Triamcinolone |
| | Ulifloxacin | Tricetin |
| | Vardenafil | Tricetin 3′,4′,5′-trimethylether |
| | Zidovudine | Trifluoperazine |
| | | Tryprostatin A |
| | | Ursolic acid |
| | | Vismodegib |

a) Data taken from review papers.

mRNA sequence to ABCB1 [61]. Loss of Abcb11 in mice leads to reduced hepatobiliary bile acid transport and increased hydroxylation of bile acids in liver [62]. Mutations in ABCB11 result in progressive familial intrahepatic cholestasis type 2 (PFIC2), which presents with severe jaundice and liver dysfunction [63].

Studies with membrane vesicles expressing rat Abcb11 and human ABCB11 transfected Sf9 cells have demonstrated that taurochenodeoxycholate, tauroursodeoxycholate, taurocholate, glycocholate, and cholate are all transported by this protein. ABCB11 can transport some ABCB1 substrates, including vinblastine, calcein-acetoxymethyl ester, fexofenadine, and pravastatin, but not rhodamine 123, daunorubicin, digoxin, and paclitaxel [17, 64]. A list of ABCB11 substrates is found in Table 22.1. The role of ABCB11 in oral drug bioavailability and drug disposition appears to be limited, but inhibition of ABCB11 by drugs can lead to drug-induced liver injury [65]. Competitive inhibitors of ABCB11 such as cyclosporin, rifamycin, glibenclamide, and bosentan may cause increased plasma bile salt levels and acquired forms of intrahepatic cholestasis [18].

### 22.2.4 ABCC2 (MRP2)

ABCC2 is expressed in many tissues in the body and is located on the apical membrane in polarized cells. In the liver, ABCC2 is expressed on the canalicular membrane and facilitates the efflux of xenobiotics into bile; lack of functional ABCC2 at the canalicular membrane leads to Dubin-Johnson syndrome in humans. ABCC2 is also localized on the apical side of enterocytes with a higher concentration in the proximal jejunum and lower expression at the distal ileum demonstrated in rat intestine [19]. The transport-negative ($TR^-$ and Eisai hyperbilirubinemic rats (EHBR) lacking *Abcc2* have helped in understanding the effects of Abcc2 *in vivo* [66]. ABCC2 is structurally similar to ABCC1, ABCC3, ABCC6, and ABCC10 in that these transporters all have an additional membrane-spanning domain that precedes the core ABC structure; ABCC2 shares 49%, 58%, and 45% amino acid similarity with ABCC1, ABCC3, and ABCC6, respectively [17].

Endogenous compounds transported by ABCC2 include leukotriene $C_4$ ($LTC_4$), dinitrophenyl *S*-glutathione (DNP-SG), *S*-glutathionyl ethacrynate, mono and bis-glucuronosyl bilirubin, and estradiol-17-β-glucuronide [67, 68]. ABCC2 also transports sulfated bile salts such as taurochenodeoxycholate-3-sulfate, taurolithocholate-3-sulfate, and estrone-3-sulfate [17]. In Abcc2 deficient rats, the oral bioavailability of the antioxidant boldine is significantly reduced compared to wildtype rats [69] and plasma concentrations of methotrexate are higher [70]. There is evidence for the cotransport of glutathione (GSH) and xenobiotics by ABCC2 [71]. In addition, the glucuronide conjugates of xenobiotics could also be transported by ABCC2, including the glucuronides of ezetimibe and etoposide [20]. ABCC2 inhibitors include cyclosporine A, probenecid, rifampicin, luteolin, quercetin, and MK571 (Table 22.1). Some of these inhibitors also interact with ABCC1, ABCB1, ABCB11, SLCO (OATP), and other transporters; therefore, results of experiments using these inhibitors should be viewed with caution since they are not specific for ABCC2 [23]. 2-Indolylmethylenebenzofuranone derivatives have been identified as inhibitors with higher selectivity to ABCC2 than ABCC1 and ABCB1 [72].

### 22.2.5 ABCG2 (BCRP)

ABCG2 was isolated and characterized by three different groups working independently in the 1990s. ABCG2 was named BCRP because it was discovered in a breast cancer cell line co-selected for doxorubicin and verapamil. ABCG2 was cloned and named MXR for mitoxantrone resistance and named ABCP for ABC transporter expressed in placenta [34]. ABCG2 was fully characterized after its initial cloning and belongs to the G family of the ABC superfamily that also includes the *Drosophila* white, brown, and scarlet genes. ABCG2 is considered a "half" transporter because the secondary protein structure only contains one membrane-spanning domain (MSD) versus two or three for other ABC efflux transporters. Most scientists agree that ABC transporters need two MSDs and two nucleotide-binding domains (NBDs) and thus it is proposed that ABCG2 forms a homotetramer to function as a transporter [73].

Despite the differences in nucleotide sequences between ABCG2, ABCC1, and ABCB1, there is considerable substrate overlap. The first ABCG2 substrates discovered were chemotherapeutic agents such as mitoxantrone, topotecan, SN-38, irinotecan, and flavopiridol [74]. Unlike ABCB1, ABCG2 does not transport vincristine, vinblastine, paclitaxel, or cisplatin [8]. ABCG2-mediated transport is not dependent on intracellular GSH concentration, as seen for ABCC1 [75]. Sulfasalazine is also a substrate of intestinal ABCG2. Since sulfasalazine can be metabolized by intestinal bacteria, impairment of ABCG2 produces higher absorption of sulfasalazine and lower levels of microbial metabolites, which has been demonstrated in *Abcg2* knockout rats [76]. ABCG2 inhibitors include some ABCB1 inhibitors such as elacridar, cyclosporine

A, and tariquidar [77]. Gefitinib is also an ABCG2 inhibitor as it dramatically increased the oral bioavailability of irinotecan [78]. Febuxostat has been identified as a potent ABCG2 inhibitor during *in vitro* screening and *in vivo* studies [44]. It significantly increased the intestinal absorption of sulfasalazine in wild-type mice, but not in *Abcg2* knockout mice. A list of ABCG2 substrates and inhibitors is found in Table 22.1.

Multiple flavonoids were tested for their ability to inhibit BCRP-mediated transport. Many of the tested flavonoids were able to significantly increase mitoxantrone concentration in BCRP overexpressing cells with little effect in BCRP-negative cells. Chrysin had the lowest $EC_{50}$ value of $0.39 \pm 0.13\,\mu M$ [79]. Although the flavonoids chrysin and 7,8-benzoflavone did not alter the oral bioavailability of oral topotecan in Sprague Dawley rats or Abcb1a(−/−)/Abcb1b(−/−) mice [80], these flavonoids significantly increased the AUC of another BCRP substrate nitrofurantoin [81, 82]. Several structural requirements have been identified for potent flavonoid-BCRP interactions, including the presence of a 2,3-double bond in ring C, ring B attached at position 2, and hydroxylation at position 5, based on a QSAR model [83]. In addition to flavonoids, curcumin is another natural compound that inhibits ABCG2, increasing the AUC and $C_{max}$ of orally administered rosuvastatin in cynomolgus monkeys [84]. Other diet-drug interactions have been reported in the literature [81, 85].

The role of ABCG2 in biliary excretion was studied in Abcg2(−/−), Abcb1a(−/−)/Abcb1b(−/−), and Abcc2(−/−) mice [86]. Results showed that mouse Abcg2 contributed to the biliary excretion of all sulfate and glucuronide conjugates in mice, while Abcc2 mediated the biliary excretion of only glucuronide conjugates [86]. Hirano et al. studied the contribution of ABCG2 in the biliary excretion of pitavastatin [87]. *Abcc2* deficient rats showed similar biliary clearance rates of pitavastatin but the biliary clearance in *Abcg2(−/−)* mice was 10-fold lower than in wild-type mice, suggesting that ABCG2 can influence oral bioavailability and hepatobiliary transport of xenobiotics [87].

## 22.3 Basolateral/Lateral Membrane Efflux Proteins

### 22.3.1 ABCC1 (MRP1)

Cole et al. first identified ABCC1 in 1992 in a cell line resistant to doxorubicin and deficient in ABCB1 and named it multidrug resistance protein 1 (MRP1) [88]. ABCC1 is widely distributed in the body with the highest expression in the lung, testis, kidney, intestine, and brain. In polarized cells, ABCC1 is located at the basolateral membrane. In addition, ABCC1 expression could be detected in activated stellate cells in rats and humans but with lower expression levels compared to that in hepatocytes [89]. In the intestine, since ABCC1 is located on the basolateral membrane, it acts as an absorptive transporter to facilitate its substrates entering into the blood circulation [19]. Under certain conditions, ABCC1 can be upregulated in the liver and plays an important role in the efflux of endogenous as well as xenobiotic conjugates into the sinusoidal blood. *Abcc1* was upregulated at the basolateral/lateral membrane in rat hepatocytes after bile duct ligation, suggesting that Abcc1 plays a compensatory role in protecting the hepatocyte when there is little or no transport into bile [90].

ABCC1 has been shown to cotransport substrates along with reduced GSH. Initially, it was thought that ABCC1 was unable to transport vincristine and daunorubicin but following the addition of GSH, transport of these compounds was observed [34]. Other ABCC1 substrates include many GSH, glucuronide, and sulfate conjugates. Several anticancer drugs are shown to be transported by ABCC1, including vinca alkaloids, anthracyclines, etoposide, teniposide, mitoxantrone, and methotrexate [19]. A list of substrates is given in Table 22.1. However, no specific inhibitor for ABCC1 has been identified [19].

### 22.3.2 ABCC3 (MRP3)

ABCC3 is an organic anion transporter with broad substrate specificity similar to ABCC1 and ABCC2 and shares 58% and 48% amino acid identity with these transporters, respectively. ABCC3 is expressed in adrenal cortex, small intestine, and colon, with lower expression in liver, kidney, and prostate. ABCC3 is located on the basolateral membrane of hepatocytes, cholangiocytes, distal tubules in the kidney, and the enterocytes at the colon and ileum [34]. ABCC3 levels vary up to 80-fold in humans and its expression in the liver is induced in several disease states [91]. Increased hepatic levels of ABCC3 are observed with diseases such as Dubin-Johnson syndrome, progressive familial intrahepatic cholestasis (PFIC), and obstructive cholestasis [68]. ABCC3 plays a compensatory role when ABCC2 activity is impaired and serves to protect hepatocytes from toxicity due to increased concentrations of bilirubin glucuronides [92].

Human ABCC3 transports a wide range of endogenous and exogenous organic anions, including xenobiotic sulfate and glucuronide conjugates. However, unlike ABCC1 and ABCC2, ABCC3 showed limited transport of GSH [27]. The only GSH substrate of ABCC3 is DNP-SG, but ABCC3-mediated transport occurs with lower affinity than that for ABCC1 and ABCC2. An *in vitro* study showed that glucuronides of androsterone, epitestosterone, and etiocholanolone are transported by ABCC3, suggesting the role of ABCC3 in human androgen homeostasis [93]. In *Abcc3* knockout mice, the efflux transport of curcumin-*O*-glucuronide from hepatocyte to blood is found to be facilitated by Abcc3 [94]. ABCC3 also transports several clinically important drugs, such as etoposide, teniposide, methotrexate, and cefadroxil (Table 22.1) [27]. Indomethacin and probenecid represent known nonspecific inhibitors for ABCC3 (Table 22.1) [8]. The transport of ABCC3 substrates can also be influenced by disease states. By analyzing liver tissues, Canet et al. suggested that the induction of hepatic ABCC3 and the altered canalicular localization of ABCC2 in nonalcoholic steatohepatitis (NASH) patients may contribute to the observed change of acetaminophen pharmacokinetics [95].

### 22.3.3 ABCC4 (MRP4)

ABCC4 is an ATP-dependent organic anion transporter with a broad substrate specificity and wide tissue distribution. ABCC4 was first discovered by Lee et al. in 1998 and has been further characterized since then [96]. ABCC4 is located on chromosome 13q12.1 and the gene encodes for a protein of 1325 amino acids. Structurally, ABCC4 is more closely related to ABCB1 than some of the other ABCC proteins, in that ABCC4 lacks the initial MSD that is present in ABCC1–3 [34].

In the liver, ABCC4 is located on the basolateral membrane and can efflux substrates out of hepatocytes. ABCC4 may also be present in the basolateral membrane of enterocytes [97]. ABCC4 can play a role in the oral bioavailability of certain drugs but there are few studies. Ferslew et al., using human sandwich-cultured hepatocytes, showed that enalaprilat, the active metabolite of enalapril, is a ABCC4 substrate since the hepatic efflux of enalaprilat was significantly reduced by an ABCC4 inhibitor [29]. Methotrexate is also found to be transported by ABCC4 [33]. Endogenous ABCC4 substrates include folic acid, folinic acid, cyclic adenosine monophosphate (cAMP), cyclic guanosine monophosphate (cGMP), and bile acids [17]. The transport of cAMP and cGMP is influenced by GSH [98]. Nucleoside monophosphate analogs, used as antiretroviral drugs, were the first ABCC4 substrates to be identified. ABCC4 is involved in the efflux of purine analogs 6-mercaptopurine and 6-thioguanine along with the chemotherapeutic and immunotherapeutic agent ganciclovir [68]. There is evidence that ABCC4 mediates the transport of prostaglandins $PGE_1$, $PGE_2$, $PGE_3$, $PGF_{2\alpha}$, $PGF_{3\alpha}$, $PGD_2$, thromboxane $B_2$, and thromboxane $B_3$ [99–101].

Studies in mouse livers have shown that when Abcb11 is significantly down-regulated, up-regulation of Abcc4 occurs. This compensatory effect was studied in human ABCC4-enriched membrane vesicles and it was shown that ABCC4 can transport certain bile salts [34]. No specific inhibitors for ABCC4 have been found, but general organic anion transport inhibitors such as probenecid and indomethacin can inhibit ABCC4 activity. Nelfinavir, an HIV protease inhibitor, is suggested to be a substrate and inhibitor of ABCC4 [30]. See Table 22.1 for a list of ABCC4 substrates and inhibitors.

### 22.3.4 ABCC5 (MRP5)

ABCC5, like ABCC4, is primarily known for its transport of cyclic nucleotides cAMP and cGMP. ABCC4 shows a higher affinity to cAMP while ABCC5 has a higher affinity to cGMP [35]. The blockage of Abcc5 in zebrafish embryos led to significantly increased cellular cGMP levels [102]. ABCC5 is also capable of transporting purine derivatives, including 9-(2-phosphonylmethoxyethyl)adenine and 6-mercaptopurine [35]. Although most glutathione and glucuronide conjugates are not substrates of ABCC5, *S*-(2,4-dinitrophenyl)glutathione can be transported by ABCC5 [35]. Tissue distribution of ABCC5 is ubiquitous, with expression detected in the epithelial cells of urethra, smooth muscle cells of heart, basal membrane of syncytiotrophoblasts, blood–brain-barrier, and the basolateral membrane of hepatocytes. Whether ABCC5 plays a role in the oral bioavailability of drugs is unknown. The subcellular location of ABCC5 in polarized cells depends on the cell type, with both apical and basolateral expression. Belinsky et al. [103] showed that ABCC5 has 33–37% amino acid identity to ABCC1–4.

Organic anion transport inhibitors such as probenecid, sulfinpyrazone, and MK571 can inhibit ABCC5 [104]. *In vitro* studies in HEK293 and pancreatic cancer cells identified curcumin as an ABCC5 inhibitor [105]. Sildenafil, vardenafil, and their analogs inhibit ABCC5, which leads to inhibition of cGMP efflux [106, 107]. To fully understand the function and expression of ABCC5, more detailed characterization is needed.

### 22.3.5 ABCC6 (MRP6)

ABCC6 is located on chromosome 16p13.1 and encodes for a protein of 1503 amino acids and 180 kDa. ABCC6 is most closely related to ABCC1, ABCC2, and ABCC3 with 45%, 38%, and 43% amino acid identity [34]. In human and rat hepatocytes, ABCC6 is localized to the basolateral/lateral membrane [17].

Little is known about the substrate specificity of ABCC6. Belinsky et al. has shown that ABCC6 can transport several cytotoxic agents such as etoposide, doxorubicin, and cisplatin, but not vinblastine, vincristine, and paclitaxel [38]. Ilias et al. showed that human ABCC6 could transport the glutathione *S*-conjugate of $LTC_4$ and *N*-ethylmaleimide but not estradiol-17-β-glucuronide [40]. Based on the identified *in vitro* substrates, Hosen et al. generated the ABCC6 homology model to predict its substrate through molecular docking of potential binding sites and found many lipids and biliary secretion compounds as potential substrates, along with several compounds involved in vitamin D metabolism [108]. Indomethacin, probenecid, and pantoprazole have been used as ABCC6 inhibitors in identifying nilotinib and dasatinib as substrates of ABCC6 [39]. Hereditary deficiency of ABCC6 is the genetic basis of the disease pseudoxanthoma elasticum, and ABCC6 appears to play an important physiological role [109]. More information is needed to understand the substrate specificity and inhibitors of ABCC6 and clinical relevance in drug disposition.

## 22.4 Clinical Relevance of ABC Transporters in Oral Bioavailability of Drugs

The ABC family of efflux transporters are determinants of the absorption and disposition of substrates for these transporters. Drug–drug interactions between inhibitors and substrates of efflux transporters in the small intestine or liver can affect the oral bioavailability and disposition of substrates in the body. An unintentional increase in plasma concentrations for narrow therapeutic index drugs can lead to unwanted toxicity. On the other hand, for compounds with low oral bioavailability, drug–drug interactions involving inhibition of efflux transporters can enhance absorption and increase plasma concentrations, representing a beneficial drug–drug interaction.

ABCB1 inhibition was first used to reverse multidrug resistance conferred by ABCB1. Several studies using inhibitors to increase the oral bioavailability of drugs, including digoxin, paclitaxel, docetaxel, tacrolimus, saquinavir, fexofenadine, irinotecan, and cyclosporine A [110, 111]. Although digoxin as an ABCB1 substrate has been involved in many clinical interaction studies, the vast majority of reported studies (93 of 123) exhibited AUC change of less than 25%, which is generally not considered clinically significant [112]. The drugs requiring closer monitoring include valspodar, quinidine, and cyclosporin since more than a twofold AUC change of digoxin has been reported in interaction studies with these drugs [112]. ABCB1-associated clinical interactions have been reported for two oral anticoagulants, dabigatran etexilate, and rivaroxaban. The bioavailability of dabigatran etexilate was increased by 143% with single-dose verapamil treatment and decreased by 67% with 7-day rifampicin treatment [113, 114]. In the study of rivaroxaban, coadministered clarithromycin led to a 54% increase whereas coadministration of ketoconazole and ritonavir led to an increase of 153–158% in rivaroxaban exposure [115]. Other reports have indicated clinically relevant herb-drug interactions following the acute administration of St John's wort and the ABCB1 substrates talinolol, digoxin, and fexofenadine, leading to decreased bioavailability of the substrates due to ABCB1 induction by St John's wort [85, 116]. Inhibition on ABCB1 by herbal phytochemicals has also been reported, resulted in the increased oral bioavailability of the ABCB1 substrates. For example, short-term coadministration of quercetin significantly increased the oral bioavailability of fexofenadine in healthy volunteers [117].

ABCG2 is another ABC transporter of interest because of its ability to alter oral bioavailability and hepatobiliary excretion of a wide range of substrates. Clinically relevant drug–drug interactions involving ABCG2 have been reported for topotecan and irinotecan [77, 78]. The ABCG2 substrate rosuvastatin is primarily eliminated unchanged in human bile and associated with several clinical interactions. Rosuvastatin and rosuvastatin-lactone $AUC_{0-24}$ and $C_{max}$ were significantly increased and $t_{max}$ was significantly reduced when coadministered with atazanavir/ritonavir (ATV/RTV), which may be the result of inhibition of ABCG2 by ATV/RTV [118]. Increased rosuvastatin exposure was also found by coadministration with darolutamide, a novel androgen receptor antagonist for prostate cancer, and febuxostat, a common treatment for gout, suggesting their interaction risk with ABCG2 substrates [119, 120]. The oral bioavailability of sulphasalazine, which is limited by ABCG2, is significantly increased during coadministration with curcumin or the NK1 receptor antagonist rolapitant in healthy volunteers [121, 122]. Based on the clinically relevant ABCG2-mediated interactions, Lee et al. proposed the best available clinical substrates for interaction studies as oral sulfasalazine for intestinal ABCG2, oral rosuvastatin for both intestinal and hepatic ABCG2, and intravenous rosuvastatin for hepatic ABCG2 [42].

Clinically relevant drug–drug interactions can be anticipated for ABCC2 because ABCC2 can transport many naturally occurring anticancer agents, HIV protease inhibitors, and other drugs. Huisman et al. reported the increased transport of taxanes when coadministered with probenecid due to ABCC2 inhibition by probenecid [22]. Rifampin treatment in healthy subjects resulted in increased expression of ABCC2 in duodenal biopsy samples, suggesting the potential for increased intestinal ABCC2-mediated efflux [123].

Most of the clinically relevant drug interactions have been reported for ABCB1 and ABCG2. Uncovering the role of ABC transporters in clinical drug–drug interaction studies can be difficult because many of the substrates and inhibitors are also substrates for other transporters and metabolizing enzymes.

## 22.5 Pharmacogenomics of ABC Transporters

Genetic polymorphisms in ABC transporters can lead to changes in function and expression. Many single nucleotide polymorphisms (SNPs) have been identified in ABC transporters, although not all SNPs cause a change in phenotype. Genetic polymorphisms have been identified for ABCG2, ABCB1, ABCB11, ABCB4, ABCC2, ABCC3, and ABCC6. Since ABC transporters are involved in many physiological processes, certain polymorphisms in ABC transporters are also associated with inherited diseases.

With respect to drug therapy, polymorphisms in only ABCG2 have shown significant clinical associations with variability in drug concentrations and associated toxicities or responses in different populations [124]. Of the many identified SNPs in ABCG2 polymorphisms, 421C>A (Q141K) and 34G>A (V12M) occur most frequently, with allele frequencies of 30–60% in East Asians and 5–10% in Caucasians and African–American populations [43]. The amino acid substitution Q141K led to an increase in the oral bioavailability of topotecan [125]. Rosuvastatin showed significantly higher AUC and $C_{max}$ and lower clearance in healthy Chinese volunteers with the 421CA and 421AA variants [126]. Similar results were observed in healthy Finnish volunteers, as greater $AUC_{0-\infty}$ and $C_{max}$ values for rosuvastatin were reported for the 421AA genotype as compared to 421CC and 421CA genotypes [127]. The 421C>A polymorphism also led to increased oral bioavailability of atorvastatin in the Japanese population [128]. The mutation in 34G>A resulted in reduced bioavailability of teriflunomide sodium in 34A/A mutants compared to that in wild-type 34G/G in healthy Chinese subjects [129].

Further studies are needed to evaluate associations between genetic variants and associated clinically-relevant changes in drug concentration or response.

## 22.6 Regulation of Efflux Transporters

Short-term regulation of proteins often occurs by the trafficking of intracellular pools of proteins to the cell membrane, while long-term regulation occurs at the transcriptional and/or translational level through a superfamily of nuclear receptors. After ligand binding, most nuclear receptors bind to a heterodimeric partner such as retinoic acid X receptor (RXR), and the complex then binds to the regulatory regions of genes and modulates transcription. Many of the metabolizing enzymes and efflux transporters share the same nuclear receptors leading to the coordinated regulation of certain metabolizing enzymes and transport proteins [60]. Nuclear receptors that have been discovered include aryl hydrocarbon receptor (AhR), constitutive androstane receptor (CAR), pregnane X receptor (PXR), peroxisome proliferator-activated receptor α (PPARα), liver X receptor (LXR), farnesoid X receptor (FXR), estrogen receptor α (ERα) and others [130].

The induction of ABCB1 by rifampicin through PXR-dependent mechanisms has been extensively reported, leading to an average reduction of 20–67% in substrate exposure [131]. Several antiviral and anticancer agents, such as amprenavir, indinavir, saquinavir, lopinavir, vincristine, tamoxifen, vinblastine, docetaxel, cyclophosphamide, flutamide, ifosfamide, and paclitaxel demonstrated PXR-mediated ABCB1 induction in human colon adenocarcinoma LS180 cells [132, 133]. In addition to PXR, AhR is involved in ABCB1 regulation, and the cisplatin resistance mediated by ABCB1 is reversed by suppression of AhR nuclear translocator.

Kast et al. demonstrated that hepatic ABCC2 expression was induced by FXR, CAR, and PXR [134]. The induction of ABCC2 by carbamazepine was found to be PXR dependent in HepG2 cells and PXR residue Gln285 is critical in this interaction [135]. The antiviral agent saquinavir was found to increase mRNA of ABCB1, ABCC2, ABCC5, and ABCG2 in LS180 cells, likely through PXR induction [136]. Induced mRNA of ABCC2 was found in SW-480 cells with bortezomib treatment, which is suggested to be regulated by the nuclear factor-erythroid 2 related factor 2 (Nrf2) [137]. The induction of ABCC2 by pitavastatin may be regulated by LXR activation with the potential involvement of the sterol regulatory element (SRE) [138].

Nuclear receptors have been implicated in the regulation of ABCG2 by different authors [139–141]. Micro-RNAs, short noncoding RNAs, can regulate the expression of genes posttranscriptionally [142]. The involvement of micro-RNA 328 was assessed in MCF7 (parental) and MCF7/MX100 (ABCG2 over-expressing) cells using quantitative real time RT-PCR [143]. mRNA levels of ABCG2 and micro-RNA 328 were found to be inversely related to each other in the two cell lines and micro-RNA 328 was found to act on the 3′ untranslated region of ABCG2 [143]. Based on the increased expression of ABCG2 in human hepatocytes with phenobarbital treatment, Benoki et al. further confirmed that CAR transactivates ABCG2 through the DR5 motif located in its distal promoter [144]. The involvement of AhR in the regulation of ABCG2 is also evidenced in cells that respond to romidepsin with ABCG2 upregulation and in human placental cells, where ABCG2 mRNA induction was abrogated by an AhR-specific antagonist [145, 146]. The regulation of ABCG2 by PPARα was demonstrated in mice treated with PPARα agonist, while the regulation by PPARγ was shown in human placental cells following PPARγ agonist and antagonist treatments [147, 148].

Chen et al. implicated RXR in the regulation of ABCC3 in HepG2 cells [149]. The luciferase assays showed that the addition of retinoid x receptor α:retinoid acid receptor α (RXRα:RARα) abrogated the *ABCC3* gene activation, suggesting the RXRα:RARα as a repressor of ABCC3 expression [149]. This is further evidenced in RARα knockdown HepG2 cells, where endogenous ABCC3 expression was upregulated [149]. RXR is also involved in the regulation of ABCC6 in HepG2 cells. 9-Cis-retinoic acid, a RXR agonist, induced *ABCC6* promoter activity and mRNA expression in the cell line used [150]. In addition to RXR, TNFα, and ERα are involved in the induction of hepatic ABCC3 expression [151, 152] while HNF4α is involved in ABCC6 expression [153]. A recent study revealed the binding of miR-192-5p to the ABCC3 3′-UTR and miR-409-3p to the ABCB1 3′-UTR, further confirming the contribution of mucosal micro-RNA in the discrepancies of transporter expression and protein content in intestines [154].

ABCB11 is known to be regulated by FXR [155]. The FXRα2 regulation was predominately involved in transactivation of human ABCB11 through one of the FXR-responsive elements, IR1a [156]. In addition to FXR, small heterodimer partner (SHP), but not LXRα receptors, was found to be associated with non-alcoholic fatty liver disease (NAFLD) with upregulated ABCB11 [157].

## 22.7 Summary

Efflux transporters of the ABC super-family play an essential role in the oral bioavailability and hepatobiliary transport of xenobiotics. Efflux transporters located at the apical membrane include ABCB1, ABCB4, ABCB11, ABCC2, and ABCG2. Among these, ABCC2, ABCB1, and ABCG2 have been implicated in altering the oral bioavailability and biliary secretion of xenobiotics *in vivo*. ABCC1 and ABCC3–6 are efflux transporters of the ABC family located on the basolateral/lateral membranes of hepatocytes and intestinal cells; their physiological and pharmacological functions are not completely understood, including their role in the oral bioavailability of xenobiotics.

Characterization of the role of ABC-mediated efflux in the oral bioavailability of drugs will require the identification of specific inhibitors and substrates, as well as the use of knockout animals deficient in efflux transporters, in order to characterize the role of these transporters as determinants of drug oral bioavailability. The significance of ABC efflux transporters including the role of variations in their expression due to genetic polymorphisms, environment, diet, or disease in oral absorption and hepatobiliary elimination remains a current research area important in elucidating the determinants of drug–drug interactions and inter- and intra-subject variability.

## Abbreviations

| | |
|---|---|
| BCRP | breast cancer resistance protein |
| BSEP | bile salt export pump |
| cAMP | cyclic adenosine monophosphate |
| CAR | constitutive androstane receptor |
| cGMP | cyclic guanosine monophosphate |
| DNP-SG | dinitrophenyl *S*-glutathione |
| EHBR | Eisai hyperbilirubinemic rat |
| FXR | farnesoid X receptor |

GSH glutathione
$LTC_4$ leukotriene $C_4$
MDR multiple drug resistance
MRP multidrug resistance protein
MSD membrane-spanning domain
NBD nucleotide binding domains
OATP organic anion transporting polypeptide
PFIC progressive familial intrahepatic cholestasis
Pgp P-glycoprotein
PXR pregnane X receptor
RAR retinoid acid receptor
RXR retinoic acid X receptor
SNP single nucleotide polymorphism

## Acknowledgments

We thank Dr. Yash A. Gandhi who contributed to the first version of this chapter.

## References

1 HGNC (2020). HGNC database. https://www.genenames.org/data/genegroup/#!/group/417 (accessed 1 June 2020).
2 Drozdzik, M., Busch, D., Lapczuk, J. et al. (2019). Protein abundance of clinically relevant drug transporters in the human liver and intestine: a comparative analysis in paired tissue specimens. *Clin. Pharmacol. Ther.* 105: 1204–1212.
3 Prasad, B., Lai, Y., Lin, Y., and Unadkat, J.D. (2013). Interindividual variability in the hepatic expression of the human breast cancer resistance protein (BCRP/ABCG2): effect of age, sex, and genotype. *J. Pharm. Sci.* 102: 787–793.
4 Wang, L., Prasad, B., Salphati, L. et al. (2015). Interspecies variability in expression of hepatobiliary transporters across human, dog, monkey, and rat as determined by quantitative proteomics. *Drug Metab. Dispos.* 43: 367–374.
5 Kurzawski, M., Szelag-Pieniek, S., Łapczuk-Romańska, J. et al. (2019). The reference liver—ABC and SLC drug transporters in healthy donor and metastatic livers. *Pharmacol. Rep.* 71: 738–745.
6 FDA (2020a) In vitro drug interaction studies – cytochrome P450 enzyme – and transporter-mediated drug interactions: Guidance for industry. https://www.fda.gov/media/134582/download (accessed 1 June 2020).
7 Juliano, R.L. and Ling, V. (1976). A surface glycoprotein modulating drug permeability in Chinese hamster ovary cell mutants. *Biochim. Biophys. Acta* 455: 152–162.
8 Schinkel, A.H. and Jonker, J.W. (2003). Mammalian drug efflux transporters of the ATP binding cassette (ABC) family: an overview. *Adv. Drug Delivery Rev.* 55: 3–29.
9 Clay, A.T. and Sharom, F.J. (2014). Multidrug resistance protein: P-glycoprotein. In: *Drug Transporters: Molecular Characterization and Role in Drug Disposition* (ed. G. You and M.E. Morris), 141–160. Wiley.
10 Kadioglu, O. and Efferth, T. (2019). A machine learning-based prediction platform for P-glycoprotein modulators and its validation by molecular docking. *Cells* 8: 1286.
11 Li, Y., Revalde, J., and Paxton, J.W. (2017). The effects of dietary and herbal phytochemicals on drug transporters. *Adv. Drug Delivery Rev.* 116: 45–62.
12 Kino, K., Taguchi, Y., Yamada, K. et al. (1996). Aureobasidin A, an antifungal cyclic depsipeptide antibiotic, is a substrate for both human MDR1 and MDR2/P-glycoproteins. *FEBS Lett.* 399: 29–32.
13 Mahdi, Z.M., Synal-Hermanns, U., Yoker, A. et al. (2016). Role of multidrug resistance protein 3 in antifungal-induced cholestasis. *Mol. Pharmacol.* 90: 23–34.
14 Smith, A.J., van Helvoort, A., van Meer, G. et al. (2000). MDR3 P-glycoprotein, a phosphatidylcholine translocase, transports several cytotoxic drugs and directly interacts with drugs as judged by interference with nucleotide trapping. *J. Biol. Chem.* 275: 23530–23539.
15 Byrne, J.A., Strautnieks, S.S., Mieli-Vergani, G. et al. (2002). The human bile salt export pump: characterization of substrate specificity and identification of inhibitors. *Gastroenterology* 123: 1649–1658.
16 Hirano, M., Maeda, K., Hayashi, H. et al. (2005). Bile salt export pump (BSEP/ABCB11) can transport a nonbile acid substrate, pravastatin. *J. Pharmacol. Exp. Ther.* 314: 876–882.
17 Chan, L.M., Lowes, S., and Hirst, B.H. (2004). The ABCs of drug transport in intestine and liver: efflux proteins limiting drug absorption and bioavailability. *Eur. J. Pharm. Sci.* 21: 25–51.

18 Stieger, B. (2010). Role of the bile salt export pump, BSEP, in acquired forms of cholestasis. *Drug Metab. Rev.* 42: 437–445.

19 Estudante, M., Morais, J.G., Soveral, G., and Benet, L.Z. (2013). Intestinal drug transporters: an overview. *Adv. Drug Delivery Rev.* 65: 1340–1356.

20 Fahrmayr, C., König, J., Auge, D. et al. (2012). Identification of drugs and drug metabolites as substrates of multidrug resistance protein 2 (MRP2) using triple-transfected MDCK-OATP1B1-UGT1A1-MRP2 cells. *Br. J. Pharmacol.* 165: 1836–1847.

21 Cui, Y., Konig, J., Buchholz, J.K. et al. (1999). Drug resistance and ATP-dependent conjugate transport mediated by the apical multidrug resistance protein, MRP2, permanently expressed in human and canine cells. *Mol. Pharmacol.* 55: 929–937.

22 Huisman, M.T., Chhatta, A.A., van Tellingen, O. et al. (2005). MRP2 (ABCC2) transports taxanes and confers paclitaxel resistance and both processes are stimulated by probenecid. *Int. J. Cancer* 116: 824–829.

23 Fardel, O., Jigorel, E., Le Vee, M., and Payen, L. (2005). Physiological, pharmacological and clinical features of the multidrug resistance protein 2. *Biomed. Pharmacother.* 59: 104–114.

24 Lee, Y.M., Cui, Y., Konig, J. et al. (2004). Identification and functional characterization of the natural variant MRP3-Arg1297His of human multidrug resistance protein 3 (MRP3/ABCC3). *Pharmacogenetics* 14: 213–223.

25 Zelcer, N., Saeki, T., Reid, G. et al. (2001). Characterization of drug transport by the human multidrug resistance protein 3 (ABCC3). *J. Biol. Chem.* 276: 46400–46407.

26 Zeng, H., Liu, G., Rea, P.A., and Kruh, G.D. (2000). Transport of amphipathic anions by human multidrug resistance protein 3. *Cancer Res.* 60: 4779–4784.

27 Ghanem, C.I. and Manautou, J.E. (2019). Modulation of hepatic MRP3/ABCC3 by xenobiotics and pathophysiological conditions: role in drug pharmacokinetics. *Curr. Med. Chem.* 26: 1185–1223.

28 Alemanno, L., Massimi, I., Klaus, V. et al. (2018). Impact of multidrug resistance protein-4 inhibitors on modulating platelet function and high on-aspirin treatment platelet reactivity. *Thromb. Haemost.* 118: 490–501.

29 Ferslew, B.C., Köck, K., Bridges, A.S., and Brouwer, K.L.R. (2014). Role of multidrug resistance–associated protein 4 in the basolateral efflux of hepatically derived enalaprilat. *Drug Metab. Dispos.* 42: 1567–1574.

30 Fukuda, Y., Takenaka, K., Sparreboom, A. et al. (2013). Human immunodeficiency virus protease inhibitors interact with ATP binding cassette transporter 4/multidrug resistance protein 4: a basis for unanticipated enhanced cytotoxicity. *Mol. Pharmacol.* 84: 361–371.

31 Furmanski, B.D., Hu, S., Fujita, K.-i. et al. (2013). Contribution of ABCC4-mediated gastric transport to the absorption and efficacy of dasatinib. *Clin. Cancer Res.* 19: 4359–4370.

32 Rius, M., Hummel-Eisenbeiss, J., Hofmann, A.F., and Keppler, D. (2006). Substrate specificity of human ABCC4 (MRP4)-mediated cotransport of bile acids and reduced glutathione. *Am. J. Physiol. Gastrointest. Liver Physiol.* 290: G640–G649.

33 Berthier, J., Arnion, H., Saint-Marcoux, F., and Picard, N. (2019). Multidrug resistance-associated protein 4 in pharmacology: overview of its contribution to pharmacokinetics, pharmacodynamics and pharmacogenetics. *Life Sci.* 231: 116540.

34 Haimeur, A., Conseil, G., Deeley, R.G., and Cole, S.P. (2004). The MRP-related and BCRP/ABCG2 multidrug resistance proteins: biology, substrate specificity and regulation. *Curr. Drug Metab.* 5: 21–53.

35 Liu, Y.H., Di, Y.M., Zhou, Z.W. et al. (2010). Multidrug resistance-associated proteins and implications in drug development. *Clin. Exp. Pharmacol. Physiol.* 37: 115–120.

36 Pratt, S., Shepard, R.L., Kandasamy, R.A. et al. (2005). The multidrug resistance protein 5 (ABCC5) confers resistance to 5-fluorouracil and transports its monophosphorylated metabolites. *Mol. Cancer Ther.* 4: 855–863.

37 Wijnholds, J., Mol, C.A., van Deemter, L. et al. (2000). Multidrug-resistance protein 5 is a multispecific organic anion transporter able to transport nucleotide analogs. *Proc. Natl. Acad. Sci. U.S.A.* 97: 7476–7481.

38 Belinsky, M.G., Chen, Z.S., Shchaveleva, I. et al. (2002). Characterization of the drug resistance and transport properties of multidrug resistance protein 6 (MRP6, ABCC6). *Cancer Res.* 62: 6172–6177.

39 Eadie, L.N., Dang, P., Goyne, J.M. et al. (2018). ABCC6 plays a significant role in the transport of nilotinib and dasatinib, and contributes to TKI resistance in vitro, in both cell lines and primary patient mononuclear cells. *PLoS One* 13: e0192180.

40 Ilias, A., Urban, Z., Seidl, T.L. et al. (2002). Loss of ATP-dependent transport activity in

pseudoxanthoma elasticum-associated mutants of human ABCC6 (MRP6). *J. Biol. Chem.* 277: 16860–16867.

41 Basseville, A., Robey, R.W., Bahr, J.C., and Bates, S.E. (2014). Breast cancer resistance protein (BCRP) or ABCG2. In: *Drug Transporters: Molecular Characterization and Role in Drug Disposition* (ed. G. You and M.E. Morris), 187–221. Wiley.

42 Lee, C.A., O'Connor, M.A., Ritchie, T.K. et al. (2015). Breast Cancer Resistance Protein (ABCG2) in clinical pharmacokinetics and drug interactions: practical recommendations for clinical victim and perpetrator drug-drug interaction study design. *Drug Metab. Dispos.* 43: 490–509.

43 Mao, Q. and Unadkat, J.D. (2015). Role of the breast cancer resistance protein (BCRP/ABCG2) in drug transport – an update. *AAPS J.* 17: 65–82.

44 Miyata, H., Takada, T., Toyoda, Y. et al. (2016). Identification of febuxostat as a new strong ABCG2 inhibitor: potential applications and risks in clinical situations. *Front. Pharmacol.* 7: 518.

45 Wang, Y.H., Chao, P.D., Hsiu, S.L. et al. (2004). Lethal quercetin-digoxin interaction in pigs. *Life Sci.* 74: 1191–1197.

46 Gurley, B.J., Swain, A., Williams, D.K. et al. (2008). Gauging the clinical significance of P-glycoprotein-mediated herb-drug interactions: comparative effects of St. John's wort, Echinacea, clarithromycin, and rifampin on digoxin pharmacokinetics. *Mol. Nutr. Food Res.* 52: 772–779.

47 Shimizu, M., Uno, T., Sugawara, K., and Tateishi, T. (2006). Effects of single and multiple doses of itraconazole on the pharmacokinetics of fexofenadine, a substrate of P-glycoprotein. *Br. J. Clin. Pharmacol.* 62: 372–376.

48 FDA (2020b) Drug development and drug interactions: table of substrates, inhibitors and inducers. https://www.fda.gov/drugs/drug-interactions-labeling/drug-development-and-drug-interactions-table-substrates-inhibitors-and-inducers#table4-1 (accessed 1 June 2020).

49 Chu, X., Galetin, A., Zamek-Gliszczynski, M.J. et al. (2018). Dabigatran etexilate and digoxin: comparison as clinical probe substrates for evaluation of P-gp inhibition. *Clin. Pharmacol. Ther.* 104: 788–792.

50 Ganguly, S., Panetta, J.C., Roberts, J.K., and Schuetz, E.G. (2018). Ketamine pharmacokinetics and pharmacodynamics are altered by P-glycoprotein and breast cancer resistance protein efflux transporters in mice. *Drug Metab. Dispos.* 46: 1014–1022.

51 Wahajuddin, Raju, K.S.R., Singh, S.P., and Taneja, I. (2014). Investigation of the functional role of P-glycoprotein in limiting the oral bioavailability of lumefantrine. *Antimicrob. Agents Chemother.* 58: 489–494.

52 Watanabe, N., Higashi, H., Nakamura, S. et al. (2018). The possible clinical impact of risperidone on P-glycoprotein-mediated transport of tacrolimus: a case report and in vitro study. *Biopharm. Drug Dispos.* 39: 30–37.

53 Montanari, F. and Ecker, G.F. (2015). Prediction of drug-ABC-transporter interaction – recent advances and future challenges. *Adv. Drug Delivery Rev.* 86: 17–26.

54 Broccatelli, F., Carosati, E., Neri, A. et al. (2011). A novel approach for predicting P-glycoprotein (ABCB1) inhibition using molecular interaction fields. *J. Med. Chem.* 54: 1740–1751.

55 Li, D., Chen, L., Li, Y. et al. (2014). ADMET evaluation in drug discovery. 13. Development of in silico prediction models for P-glycoprotein substrates. *Mol. Pharmaceutics* 11: 716–726.

56 Regazzi, M., Maserati, R., Villani, P. et al. (2005). Clinical pharmacokinetics of nelfinavir and its metabolite M8 in human immunodeficiency virus (HIV)-positive and HIV-hepatitis C virus-coinfected subjects. *Antimicrob. Agents Chemother.* 49: 643–649.

57 Morita, S., Kobayashi, A., Takanezawa, Y. et al. (2007). Bile salt–dependent efflux of cellular phospholipids mediated by ATP binding cassette protein B4. *Hepatology* 46: 188–199.

58 Taipalensuu, J., Tornblom, H., Lindberg, G. et al. (2001). Correlation of gene expression of ten drug efflux proteins of the ATP-binding cassette transporter family in normal human jejunum and in human intestinal epithelial Caco-2 cell monolayers. *J. Pharmacol. Exp. Ther.* 299: 164–170.

59 Wen, C., Fu, L., Huang, J. et al. (2019). Curcumin reverses doxorubicin resistance via inhibition the efflux function of ABCB4 in doxorubicin-resistant breast cancer cells. *Mol. Med. Rep.* 19: 5162–5168.

60 Ferslew, B.C., Köck, K., and Brouwer, K.L.R. (2014). Drug transport in the liver. In: *Drug Transporters: Molecular Characterization and Role in Drug Disposition* (ed. G. You and M.E. Morris), 245–271. Wiley.

61 Childs, S., Yeh, R.L., Georges, E., and Ling, V. (1995). Identification of a sister gene to P-glycoprotein. *Cancer Res.* 55: 2029–2034.

62 Fuchs, C.D., Paumgartner, G., Wahlström, A. et al. (2017). Metabolic preconditioning protects

BSEP/ABCB11−/− mice against cholestatic liver injury. *J. Hepatol.* 66: 95–101.
63 Telbisz, Á. and Homolya, L. (2016). Recent advances in the exploration of the bile salt export pump (BSEP/ABCB11) function. *Expert Opin. Ther. Targets* 20: 501–514.
64 Cheng, Y., Woolf, T.F., Gan, J., and He, K. (2016). In vitro model systems to investigate bile salt export pump (BSEP) activity and drug interactions: a review. *Chem. Biol. Interact.* 255: 23–30.
65 Funk, C., Pantze, M., Jehle, L. et al. (2001). Troglitazone-induced intrahepatic cholestasis by an interference with the hepatobiliary export of bile acids in male and female rats. Correlation with the gender difference in troglitazone sulfate formation and the inhibition of the canalicular bile salt export pump (Bsep) by troglitazone and troglitazone sulfate. *Toxicology* 167: 83–98.
66 Yue, W., Lee, J.K., Abe, K. et al. (2011). Decreased hepatic breast cancer resistance protein expression and function in multidrug resistance-associated protein 2-deficient ($TR^-$) rats. *Drug Metab. Dispos.* 39: 441–447.
67 Gotoh, Y., Suzuki, H., Kinoshita, S. et al. (2000). Involvement of an organic anion transporter (canalicular multispecific organic anion transporter/multidrug resistance-associated protein 2) in gastrointestinal secretion of glutathione conjugates in rats. *J. Pharmacol. Exp. Ther.* 292: 433–439.
68 Nies, A.T. and Lang, T. (2014). Multidrug resistance proteins of the ABCC subfamily. In: *Drug Transporters: Molecular Characterization and Role in Drug Disposition* (ed. G. You and M.E. Morris), 161–185. Wiley.
69 Cermanova, J., Prasnicka, A., Dolezelova, E. et al. (2016). Pharmacokinetics of boldine in control and Mrp2-deficient rats. *Physiol. Res.* 65: S489–S497.
70 Masuda, M., I'Izuka, Y., Yamazaki, M. et al. (1997). Methotrexate is excreted into the bile by canalicular multispecific organic anion transporter in rats. *Cancer Res.* 57: 3506–3510.
71 Evers, R., de Haas, M., Sparidans, R. et al. (2000). Vinblastine and sulfinpyrazone export by the multidrug resistance protein MRP2 is associated with glutathione export. *Br. J. Cancer* 83: 375–383.
72 Baiceanu, E., Nguyen, K.-A., Gonzalez-Lobato, L. et al. (2016). 2-Indolylmethylenebenzofuranones as first effective inhibitors of ABCC2. *Eur. J. Med. Chem.* 122: 408–418.
73 Graf, G.A., Yu, L., Li, W.P. et al. (2003). ABCG5 and ABCG8 are obligate heterodimers for protein trafficking and biliary cholesterol excretion. *J. Biol. Chem.* 278: 48275–48282.
74 Jonker, J.W., Smit, J.W., Brinkhuis, R.F. et al. (2000). Role of breast cancer resistance protein in the bioavailability and fetal penetration of topotecan. *J. Natl. Cancer Inst.* 92: 1651–1656.
75 Gauthier, C., Ozvegy-Laczka, C., Szakacs, G. et al. (2013). ABCG2 is not able to catalyze glutathione efflux and does not contribute to GSH-dependent collateral sensitivity. *Front. Pharmacol.* 4: 138–138.
76 Zamek-Gliszczynski, M.J., Bedwell, D.W., Bao, J.Q., and Higgins, J.W. (2012). Characterization of SAGE Mdr1a (P-gp), Bcrp, and Mrp2 knockout rats using loperamide, paclitaxel, sulfasalazine, and carboxydichlorofluorescein pharmacokinetics. *Drug Metab. Dispos.* 40: 1825–1833.
77 Kruijtzer, C.M., Beijnen, J.H., Rosing, H. et al. (2002). Increased oral bioavailability of topotecan in combination with the breast cancer resistance protein and P-glycoprotein inhibitor GF120918. *J. Clin. Oncol.* 20: 2943–2950.
78 Stewart, C.F., Leggas, M., Schuetz, J.D. et al. (2004). Gefitinib enhances the antitumor activity and oral bioavailability of irinotecan in mice. *Cancer Res.* 64: 7491–7499.
79 Zhang, S., Yang, X., and Morris, M.E. (2004). Combined effects of multiple flavonoids on breast cancer resistance protein (ABCG2)-mediated transport. *Pharm. Res.* 21: 1263–1273.
80 Zhang, S., Wang, X., Sagawa, K., and Morris, M.E. (2005). Flavonoids chrysin and benzoflavone, potent breast cancer resistance protein inhibitors, have no significant effect on topotecan pharmacokinetics in rats or mdr1a/1b (−/−) mice. *Drug Metab. Dispos.* 33: 341–348.
81 Wang, X. and Morris, M.E. (2007). Effects of the flavonoid chrysin on nitrofurantoin pharmacokinetics in rats: potential involvement of ABCG2. *Drug Metab. Dispos.* 35: 268–274.
82 Wang, X. and Morris, M.E. (2008). Pharmacokinetics and bioavailability of the flavonoid 7,8-benzoflavone in rats. *J. Pharm. Sci.* 97: 4546–4556.
83 Zhang, S., Yang, X., Coburn, R.A., and Morris, M.E. (2005). Structure activity relationships and quantitative structure activity relationships for the flavonoid-mediated inhibition of breast cancer resistance protein. *Biochem. Pharmacol.* 70: 627–639.
84 Karibe, T., Imaoka, T., Abe, K., and Ando, O. (2018). Curcumin as an in vivo selective intestinal breast cancer resistance protein inhibitor in

Cynomolgus monkeys. *Drug Metab. Dispos.* 46: 667–679.

85 Wang, X. and Morris, M.E. (2014). Diet/Nutrient interactions with drug transporters. In: *Drug Transporters: Molecular Characterization and Role in Drug Disposition* (ed. G. You and M.E. Morris), 409–432. Wiley.

86 Zamek-Gliszczynski, M.J., Nezasa, K., Tian, X. et al. (2006). The important role of Bcrp (Abcg2) in the biliary excretion of sulfate and glucuronide metabolites of acetaminophen, 4-methylumbelliferone, and harmol in mice. *Mol. Pharmacol.* 70: 2127–2133.

87 Hirano, M., Maeda, K., Matsushima, S. et al. (2005). Involvement of BCRP (ABCG2) in the biliary excretion of pitavastatin. *Mol. Pharmacol.* 68: 800–807.

88 Cole, S.P., Bhardwaj, G., Gerlach, J.H. et al. (1992). Overexpression of a transporter gene in a multidrug-resistant human lung cancer cell line. *Science* 258: 1650–1654.

89 Wlcek, K. and Stieger, B. (2014). ATP-binding cassette transporters in liver. *Biofactors* 40: 188–198.

90 Pei, Q.L., Kobayashi, Y., Tanaka, Y. et al. (2002). Increased expression of multidrug resistance-associated protein 1 (mrp1) in hepatocyte basolateral membrane and renal tubular epithelia after bile duct ligation in rats. *Hepatol. Res.* 22: 58–64.

91 Lang, T., Hitzl, M., Burk, O. et al. (2004). Genetic polymorphisms in the multidrug resistance-associated protein 3 (ABCC3, MRP3) gene and relationship to its mRNA and protein expression in human liver. *Pharmacogenet. Genomics* 14: 155–164.

92 Belinsky, M.G., Dawson, P.A., Shchaveleva, I. et al. (2005). Analysis of the in vivo functions of Mrp3. *Mol. Pharmacol.* 68: 160–168.

93 Järvinen, E., Kidron, H., and Finel, M. (2020). Human efflux transport of testosterone, epitestosterone and other androgen glucuronides. *J. Steroid Biochem. Mol. Biol.* 197: 105518.

94 Jia, Y.M., Zhu, T., Zhou, H. et al. (2020). MRP3 is responsible for the efflux transport of curcumin glucuronide from hepatocytes to the blood. *Drug Metab. Dispos.*

95 Canet, M.J., Merrell, M.D., Hardwick, R.N. et al. (2015). Altered regulation of hepatic efflux transporters disrupts acetaminophen disposition in pediatric nonalcoholic steatohepatitis. *Drug Metab. Dispos.* 43: 829–835.

96 Lee, K., Belinsky, M.G., Bell, D.W. et al. (1998). Isolation of MOAT-B, a widely expressed multidrug resistance-associated protein/canalicular multispecific organic anion transporter-related transporter. *Cancer Res.* 58: 2741–2747.

97 Proctor, W.R., Ming, X., Bourdet, D. et al. (2016). Why does the intestine lack basolateral efflux transporters for cationic compounds? A provocative hypothesis. *J. Pharm. Sci.* 105: 484–496.

98 Russel, F.G., Koenderink, J.B., and Masereeuw, R. (2008). Multidrug resistance protein 4 (MRP4/ABCC4): a versatile efflux transporter for drugs and signalling molecules. *Trends Pharmacol. Sci.* 29: 200–207.

99 Reid, G., Wielinga, P., Zelcer, N. et al. (2003). The human multidrug resistance protein MRP4 functions as a prostaglandin efflux transporter and is inhibited by nonsteroidal antiinflammatory drugs. *Proc. Natl. Acad. Sci. U.S.A.* 100: 9244–9249.

100 Tanaka, N., Kawai, J., Hirasawa, N. et al. (2020). ATP-binding cassette transporter C4 is a prostaglandin D2 exporter in HMC-1 cells. *Prostaglandins Leukot. Essent. Fatty Acids* 159: 102139.

101 Tanaka, N., Yamaguchi, H., and Mano, N. (2014). Transport of eicosapentaenoic acid-derived PGE3, PGF3α, and TXB3 by ABCC4. *PLoS One* 9: e109270.

102 Long, Y., Li, Q., Li, J., and Cui, Z. (2011). Molecular analysis, developmental function and heavy metal-induced expression of ABCC5 in zebrafish. *Comp. Biochem. Physiol. B: Biochem. Mol. Biol.* 158: 46–55.

103 Belinsky, M.G., Bain, L.J., Balsara, B.B. et al. (1998). Characterization of MOAT-C and MOAT-D, new members of the MRP/cMOAT subfamily of transporter proteins. *J. Natl. Cancer Inst.* 90: 1735–1741.

104 Reid, G., Wielinga, P., Zelcer, N. et al. (2003). Characterization of the transport of nucleoside analog drugs by the human multidrug resistance proteins MRP4 and MRP5. *Mol. Pharmacol.* 63: 1094–1103.

105 Li, Y., Revalde, J.L., Reid, G., and Paxton, J.W. (2011). Modulatory effects of curcumin on multi-drug resistance-associated protein 5 in pancreatic cancer cells. *Cancer Chemother. Pharmacol.* 68: 603–610.

106 Kashgari, F.K., Ravna, A., Sager, G. et al. (2020). Identification and experimental confirmation of novel cGMP efflux inhibitors by virtual ligand screening of vardenafil-analogues. *Biomed. Pharmacother.* 126: 110109.

107 Sager, G., Ørvoll, E.Ø., Lysaa, R.A. et al. (2012). Novel cGMP efflux inhibitors identified by virtual ligand screening (VLS) and confirmed by experimental studies. *J. Med. Chem.* 55: 3049–3057.

108 Hosen, M.J., Zubaer, A., Thapa, S. et al. (2014). Molecular docking simulations provide insights in the substrate binding sites and possible substrates of the ABCC6 transporter. *PLoS One* 9: e102779.

109 Le Saux, O., Martin, L., Aherrahrou, Z. et al. (2012). The molecular and physiological roles of ABCC6: more than meets the eye. *Front. Genet.* 3: 289.

110 Huisman, M.T., Smit, J.W., and Schinkel, A.H. (2000). Significance of P-glycoprotein for the pharmacology and clinical use of HIV protease inhibitors. *AIDS* 14: 237–242.

111 Kruijtzer, C.M., Beijnen, J.H., and Schellens, J.H. (2002). Improvement of oral drug treatment by temporary inhibition of drug transporters and/or cytochrome P450 in the gastrointestinal tract and liver: an overview. *Oncologist* 7: 516–530.

112 Fenner, K., Troutman, M., Kempshall, S. et al. (2009). Drug–drug interactions mediated through P-glycoprotein: clinical relevance and in vitro–in vivo correlation using digoxin as a probe drug. *Clin. Pharmacol. Ther.* 85: 173–181.

113 Härtter, S., Koenen-Bergmann, M., Sharma, A. et al. (2012). Decrease in the oral bioavailability of dabigatran etexilate after co-medication with rifampicin. *Br. J. Clin. Pharmacol.* 74: 490–500.

114 Härtter, S., Sennewald, R., Nehmiz, G., and Reilly, P. (2013). Oral bioavailability of dabigatran etexilate (Pradaxa®) after co-medication with verapamil in healthy subjects. *Br. J. Clin. Pharmacol.* 75: 1053–1062.

115 Mueck, W., Kubitza, D., and Becka, M. (2013). Co-administration of rivaroxaban with drugs that share its elimination pathways: pharmacokinetic effects in healthy subjects. *Br. J. Clin. Pharmacol.* 76: 455–466.

116 Schwarz, U.I., Hanso, H., Oertel, R. et al. (2007). Induction of intestinal P-glycoprotein by St John's wort reduces the oral bioavailability of talinolol. *Clin. Pharmacol. Ther.* 81: 669–678.

117 Kim, K.A., Park, P.W., and Park, J.Y. (2009). Short-term effect of quercetin on the pharmacokinetics of fexofenadine, a substrate of P-glycoprotein, in healthy volunteers. *Eur. J. Clin. Pharmacol.* 65: 609–614.

118 Busti, A.J., Bain, A.M., Hall, R.G. 2nd, et al. (2008). Effects of atazanavir/ritonavir or fosamprenavir/ritonavir on the pharmacokinetics of rosuvastatin. *J. Cardiovasc. Pharmacol.* 51: 605–610.

119 Lehtisalo, M., Keskitalo, J.E., Tornio, A. et al. (2020). Febuxostat, but not allopurinol, markedly raises the plasma concentrations of the breast cancer resistance protein substrate rosuvastatin. *Clin. Transl. Sci.* 0: 1–8.

120 Zurth, C., Koskinen, M., Fricke, R. et al. (2019). Drug–drug interaction potential of darolutamide: in vitro and clinical studies. *Eur. J. Drug Metab. Pharmacokinet.* 44: 747–759.

121 Kusuhara, H., Furuie, H., Inano, A. et al. (2012). Pharmacokinetic interaction study of sulphasalazine in healthy subjects and the impact of curcumin as an in vivo inhibitor of BCRP. *Br. J. Pharmacol.* 166: 1793–1803.

122 Wang, X., Zhang, Z., Arora, S. et al. (2018). Effects of rolapitant administered intravenously or orally on the pharmacokinetics of digoxin (P-glycoprotein substrate) and sulfasalazine (breast cancer resistance protein substrate) in healthy volunteers. *J. Clin. Pharmacol.* 58: 202–211.

123 Fromm, M.F., Kauffmann, H.M., Fritz, P. et al. (2000). The effect of rifampin treatment on intestinal expression of human MRP transporters. *Am. J. Pathol.* 157: 1575–1580.

124 Giacomini, K., Balimane, P., Cho, S. et al. (2013). International Transporter Consortium commentary on clinically important transporter polymorphisms. *Clin. Pharmacol. Ther.* 94: 23–26.

125 Sparreboom, A., Loos, W.J., Burger, H. et al. (2005). Effect of ABCG2 genotype on the oral bioavailability of topotecan. *Cancer Biol. Ther.* 4: 650–658.

126 Zhang, W., Yu, B.N., He, Y.J. et al. (2006). Role of BCRP 421C>A polymorphism on rosuvastatin pharmacokinetics in healthy Chinese males. *Clin. Chim. Acta* 373: 99–103.

127 Keskitalo, J.E., Zolk, O., Fromm, M.F. et al. (2009). ABCG2 polymorphism markedly affects the pharmacokinetics of atorvastatin and rosuvastatin. *Clin. Pharmacol. Ther.* 86: 197–203.

128 Tsamandouras, N., Guo, Y., Wendling, T. et al. (2017). Modelling of atorvastatin pharmacokinetics and the identification of the effect of a BCRP polymorphism in the Japanese population. *Pharmacogenet. Genomics* 27: 27–38.

129 Yao, X., Wu, Y., Jiang, J. et al. (2019). Safety, pharmacokinetics, and pharmacogenetics of single-dose teriflunomide sodium and leflunomide in healthy Chinese subjects. *Clin. Drug Investig.* 39: 643–651.

130 Catania, V.A., Sanchez Pozzi, E.J., Luquita, M.G. et al. (2004). Co-regulation of expression of phase II metabolizing enzymes and multidrug resistance-associated protein 2. *Ann. Hepatol.* 3: 11–17.

131 Elmeliegy, M., Vourvahis, M., Guo, C., and Wang, D.D. (2020). Effect of P-glycoprotein (P-gp) inducers

on exposure of P-gp substrates: review of clinical drug–drug interaction studies. *Clin. Pharmacokinet.* 59: 699–714.

132 Harmsen, S., Meijerman, I., Febus, C. et al. (2010). PXR-mediated induction of P-glycoprotein by anticancer drugs in a human colon adenocarcinoma-derived cell line. *Cancer Chemother. Pharmacol.* 66: 765–771.

133 Kwatra, D., Vadlapudi, A.D., Vadlapatla, R.K. et al. (2014). Binary and ternary combinations of anti-HIV protease inhibitors: effect on gene expression and functional activity of CYP3A4 and efflux transporters. *Drug Metab. Drug Interact.* 29: 101–110.

134 Kast, H.R., Goodwin, B., Tarr, P.T. et al. (2002). Regulation of multidrug resistance-associated protein 2 (ABCC2) by the nuclear receptors pregnane X receptor, farnesoid X-activated receptor, and constitutive androstane receptor. *J. Biol. Chem.* 277: 2908–2915.

135 Grewal, G.K., Singh, K.D., Kanojia, N. et al. (2017). Exploring the carbamazepine interaction with human Pregnane X Receptor and effect on ABCC2 *in vitro* and *in silico* approach. *Pharm. Res.* 34: 1444–1458.

136 König, S.K., Herzog, M., Theile, D. et al. (2010). Impact of drug transporters on cellular resistance towards saquinavir and darunavir. *J. Antimicrob. Chemother.* 65: 2319–2328.

137 Ebert, B., Kisiela, M., Wsól, V., and Maser, E. (2011). Proteasome inhibitors MG-132 and bortezomib induce AKR1C1, AKR1C3, AKR1B1, and AKR1B10 in human colon cancer cell lines SW-480 and HT-29. *Chem. Biol. Interact.* 191: 239–249.

138 Kobayashi, M., Gouda, K., Chisaki, I. et al. (2013). Regulation of multidrug resistance protein 2 (MRP2, ABCC2) expression by statins: involvement of SREBP-mediated gene regulation. *Int. J. Pharm.* 452: 36–41.

139 Gorczyca, L. and Aleksunes, L.M. (2020). Transcription factor-mediated regulation of the BCRP/ABCG2 efflux transporter: a review across tissues and species. *Expert Opin. Drug Metab. Toxicol.* 16: 239–253.

140 To, K.K., Polgar, O., Huff, L.M. et al. (2008). Histone modifications at the ABCG2 promoter following treatment with histone deacetylase inhibitor mirror those in multidrug-resistant cells. *Mol. Cancer Res.* 6: 151–164.

141 Vore, M. and Leggas, M. (2008). Progesterone acts via progesterone receptors A and B to regulate breast cancer resistance protein expression. *Mol. Pharmacol.* 73: 613–615.

142 Zhu, H., Wu, H., Liu, X. et al. (2008). Role of MicroRNA miR-27a and miR-451 in the regulation of MDR1/P-glycoprotein expression in human cancer cells. *Biochem. Pharmacol.* 76: 582–588.

143 Pan, Y.Z., Morris, M.E., and Yu, A.M. (2009). MicroRNA-328 negatively regulates the expression of breast cancer resistance protein (BCRP/ABCG2) in human cancer cells. *Mol. Pharmacol.* 75: 1374–1379.

144 Benoki, S., Yoshinari, K., Chikada, T. et al. (2012). Transactivation of ABCG2 through a novel cis-element in the distal promoter by constitutive androstane receptor but not pregnane X receptor in human hepatocytes. *Arch. Biochem. Biophys.* 517: 123–130.

145 Neradugomma, N.K., Liao, M.Z., and Mao, Q. (2017). Buprenorphine, norbuprenorphine, R-methadone, and S-methadone upregulate BCRP/ABCG2 expression by activating aryl hydrocarbon receptor in human placental trophoblasts. *Mol. Pharmacol.* 91: 237–249.

146 To, K.K.W., Robey, R., Zhan, Z. et al. (2011). Upregulation of ABCG2 by romidepsin via the aryl hydrocarbon receptor pathway. *Mol. Cancer Res.* 9: 516–527.

147 Eldasher, L.M., Wen, X., Little, M.S. et al. (2013). Hepatic and renal Bcrp transporter expression in mice treated with perfluorooctanoic acid. *Toxicology* 306: 108–113.

148 Lin, Y., Bircsak, K.M., Gorczyca, L. et al. (2017). Regulation of the placental BCRP transporter by PPAR gamma. *J. Biochem. Mol. Toxicol.* 31: e21880.

149 Chen, W., Cai, S.Y., Xu, S. et al. (2007). Nuclear receptors RXRalpha:RARalpha are repressors for human MRP3 expression. *Am. J. Physiol. Gastrointest. Liver Physiol.* 292: G1221–G1227.

150 Ratajewski, M., Bartosz, G., and Pulaski, L. (2006). Expression of the human ABCC6 gene is induced by retinoids through the retinoid X receptor. *Biochem. Biophys. Res. Commun.* 350: 1082–1087.

151 Chai, J., He, Y., Cai, S.Y. et al. (2012). Elevated hepatic multidrug resistance-associated protein 3/ATP-binding cassette subfamily $C_3$ expression in human obstructive cholestasis is mediated through tumor necrosis factor alpha and c-Jun $NH_2$-terminal kinase/stress-activated protein kinase–signaling pathway. *Hepatology* 55: 1485–1494.

152 Ruiz, M.L., Rigalli, J.P., Arias, A. et al. (2013). Estrogen receptor-α mediates human multidrug resistance associated protein 3 induction by

17α-ethynylestradiol. Role of activator protein-1. *Biochem. Pharmacol.* 86: 401–409.

**153** de Boussac, H., Ratajewski, M., Sachrajda, I. et al. (2010). The ERK1/2-hepatocyte nuclear factor 4alpha axis regulates human ABCC6 gene expression in hepatocytes. *J. Biol. Chem.* 285: 22800–22808.

**154** Bruckmueller, H., Martin, P., Kähler, M. et al. (2017). Clinically relevant multidrug transporters are regulated by microRNAs along the human intestine. *Mol. Pharmaceutics* 14: 2245–2253.

**155** Plass, J.R., Mol, O., Heegsma, J. et al. (2002). Farnesoid X receptor and bile salts are involved in transcriptional regulation of the gene encoding the human bile salt export pump. *Hepatology* 35: 589–596.

**156** Song, X., Chen, Y., Valanejad, L. et al. (2013). Mechanistic insights into isoform-dependent and species-specific regulation of bile salt export pump by farnesoid X receptor. *J. Lipid Res.* 54: 3030–3044.

**157** Aguilar-Olivos, N.E., Carrillo-Córdova, D., Oria-Hernández, J. et al. (2015). The nuclear receptor FXR, but not LXR, up-regulates bile acid transporter expression in non-alcoholic fatty liver disease. *Ann. Hepatol.* 14: 487–493.

# 23

# Interplay Between Metabolic Enzymes and Transporters

*Zuoxu Xie[1], Lu Wang[1], Zicong Zheng[1], Yifan Tu[1], Yi Rong[1], Ming Hu[1], and Stephen Wang[2]*

[1] *Department of Pharmacological and Pharmaceutical Sciences, College of Pharmacy, University of Houston, Houston, TX, USA*
[2] *Novartis Institutes for Biomedical Research, Cambridge, MA, USA*

## 23.1 Pathways and Functions of Drug Metabolic Enzymes and Transporters

### 23.1.1 Xenobiotics Metabolism Pathways

Xenobiotic metabolism is usually associated with a variety of enzymes, which typically help to increase the efficiency of the reaction to biotransform a substrate into a product. With different enzymes involved, two categories of xenobiotic metabolism are well recognized today: Phase I and Phase II metabolism.

#### 23.1.1.1 Phase I Xenobiotic Metabolism Pathways (see Also Chapters 11 and 12)

Phase I enzymes can catalyze a wide array of biochemical reactions. However, a common theme apparent for all Phase I enzyme reactions is that final products are usually modified to contain a functional group (i.e. hydroxyl, amine, sulfa, and carboxylic acid) that could be chemically bioreactive [1]. This gives support to the well-documented concept that Phase II metabolism (i.e. glucuronidation, glycosidation, sulfation, methylation, acetylation, glutathione conjugation, amino acid conjugation, fatty acid conjugation, and condensation) is the "true" detoxification step for xenobiotic elimination from the body (discussed in detail below). Because Phase I reactions add functional groups to a substrate so that Phase II metabolism might occur, Phase I metabolism might therefore serve as a necessary prerequisite step that allows the detoxification of xenobiotics. The bulk of research in the area of Phase I metabolism has traditionally been focused on cytochrome P450 (CYP) proteins (450 stands for the maximal absorption of the enzyme at 450 nm using carbon monoxide binding spectrum). As a superfamily of enzyme proteins, CYPs can play a very important role in Phase I oxidative metabolism of xenobiotics.

***23.1.1.1.1 Importance of CYPs*** It is reported that nearly 3/4 of all marketed drugs undergo CYP-mediated Phase I metabolism. Among them, CYP3A4 contributes to about 40% of all P450s [2]. Because CYPs are responsible for the primary metabolism of a large variety of xenobiotics and their metabolism *in vivo* [3], great emphasis has been placed on the determination of the mechanistic details of CYP reactions. The importance of CYPs can be observed from various perspectives. From a basic research perspective, a current opinion for CYPs with regards to its mechanistic purposes in the elimination of xenobiotics might be to prepare a substrate for Phase II metabolism via the addition of reactive functional groups. However, this is not to say that Phase I metabolites cannot be excreted/eliminated from the body via both active and passive efflux. From a drug design point of view, a thorough understanding of CYPs' behavior in itself, as well as its activity with xenobiotics, will help predict possible adverse reactions to drugs as well as provide solutions to improve drug efficacy and administration [4]. Therefore, the Food and Drug Administration (FDA) as well as other regulatory agencies have also paid much attention to this subject. In the guidance for the industry with respect to drug interaction studies, a clear emphasis is placed on the Phase I metabolism of an investigational new drug to be defined during early drug development.

***23.1.1.1.2 Structure of CYPs*** CYPs are located in the endoplasmic reticulum and have the capacity for oxidation of various xenobiotics. Because CYPs have a prosthetic group known as iron protoporphyrin IX, they are categorized as heme-containing enzymes. This heme iron is bonded to the rest of the enzyme by four pyrrole nitrogen atoms of protoporphyrin IX and two axial ligands to the protoporphyrin plane. These two

*Oral Bioavailability and Drug Delivery: From Basics to Advanced Concepts and Applications,* First Edition. Edited by Ming Hu and Xiaoling Li.

axial ligands (known as the fifth and sixth ligands) and their complex configuration states essentially allow CYPs to hydroxylate a diverse variety of substrates. Therefore, the overarching catalytic functions of CYPs are to transfer electrons (oxidation) from NADH to a large number of different substrates.

***23.1.1.1.3 Nomenclature of CYPs*** Because of the enormous varieties of CYPs that exist in nature, the classification of these enzymes has also been a rather complex issue due to newly found isoforms that are consistently published in literature. Amidst this confusion, Dr. David R. Nelson has a very well-updated website that reports all of the known CYPs genes across species. In addition to this, gene sequence alignments and information about known structures are also available: (https://drnelson.uthsc.edu/CytochromeP450.html). The nomenclature for CYPs is essentially based on gene sequence similarity. Through divergent evolution, many subclassifications of CYPs have been adopted into gene families and subfamilies. In general, all CYPs share 40% in identical sequences to each other. Take CYP2D6 as an example, nomenclature dictates that a P450 isoform belongs to the 2D subfamily and is the sixth gene to have been completely sequenced in this subfamily. With respect to individual isoforms in a particular subfamily for example CYP3A, the isoforms CYP3A4 and CYP3A5, while both in the same subfamily, would have to share at least 70% of their respective gene sequence as they belong to the same subfamily.

***23.1.1.1.4 Polymorphism and Complications*** Although the amino acid composition is very diverse in CYPs and enzyme mutations occur quite frequently in nature, the protein structures of CYPs are quite well conserved. Therefore, as mentioned earlier, complicated pathologies resulting from mutated CYPs, such as the Antley–Bixler syndrome [1], are seen in subjects with mutations in the genetic makeup of their CYPs. Another example in which polymorphism can have a great impact on patient reaction to drugs is seen in the case of warfarin (*S* enantiomer form has higher activity than *R* form). CYP2C9 is the main isoform responsible for *S*-warfarin metabolism. Patients with polymorphisms primarily CYP2C9*2 and CYP2C9*3 (both hetero and homozygous) have major deficiencies in the metabolism of *S*-warfarin and thus experience higher incidences of serious life-threatening bleeding [5].

CYP3A4*22 is a recently identified polymorphism with a genotyped frequency of 5.2–6.1% in Caucasians [6, 7]. The polymorphism has been reported to associate with reduced CYP3A4 mRNA expression and enzyme activity [8]. Clinical consequences such as increased blood levels or decreased clearance have been reported for its substrates including statins, cyclosporin A, and tacrolimus [9]. For example, systemic exposure (as measured via area under concentration-time curve) of ticagrelor in blood samples of *CYP3A4*22* carriers is 89% higher than control group in a clinical pharmacokinetic study [10].

**23.1.1.2 Phase II Xenobiotic Metabolism Pathways**

Phase I metabolism (mainly CYPs) has received far more notoriety in comparison to Phase II conjugating enzymes [11]. However, we now know that many xenobiotics and their metabolites undergo Phase II conjugation reactions. With respect to isoflavones, Phase II metabolism is the major biotransformation pathway responsible for their metabolism. The predominant family of Phase II enzyme responsible for metabolism of isoflavones is UDP-glycosyltransferases (UGTs), which mediate conjugation reaction at multiple functional groups by adding glucose residues from a sugar donor (the cofactor). The fact that isoflavones are metabolized mainly via UGTs is not uncommon since UGT-catalyzed glucuronidation reactions are responsible for 35% of all drugs metabolized by Phase II enzymes [4].

***23.1.1.2.1 Importance of UDP-glycosyltransferase*** The human UGTs are a superfamily of enzymes that metabolize via conjugation a variety of endogenous substrates (e.g. bilirubin, steroid hormones, thyroid hormones, bile acids, and fat-soluble vitamins) and exogenous compounds (e.g. xenobiotics/drugs, chemical carcinogens, environmental pollutants, and dietary substances) [5, 12]. As the most involved Phase II enzyme, UGTs contribute to the metabolism of about 15% of all marketed drugs [2]. UGTs are expressed intracellularly on the luminal side of the endoplasmic reticulum, a location that confers both an advantage and a disadvantage [12]. This specific configuration layout allows these enzymes to have direct access to Phase I biotransformation reaction metabolites. However, the topology of UGTs has caused a specific restriction with regards to access of drug substrates, cofactors, and glucuronidated products to and out from the active sites of UGT enzymes. This translates into a recognized phenomenon in which we see a decrease in UGT activity from studies using organelle microsomes. This phenomenon termed "latency" can cause erroneous predictions of *in vivo* effects of UGT enzymes when it is based on data obtained from *in vitro* microsome experiments without accounting for this effect [13, 14]. Therefore, utilization of surfactants has been incorporated into reaction mixtures to overcome this latency effect.

***23.1.1.2.2 Nomenclature of UGTs*** The UGT superfamily is comprised of 4 families: UGT1, UGT2, UGT3, and UGT8 [15]. Very much like that of CYPs, each family of UGTs must share at least 50% homology in their DNA sequences, whereas UGTs in each subfamily must share at least 60% homology in their DNA sequences[16]. The nomenclature system uses Arabic numerals to represent the family (e.g. UGT1) whereas a letter designates the subfamily (e.g. UGT1A) and followed again by Arabic numeral denoting the individual gene (e.g. UGT1A1) [17]. Historically, UGT was also used as the abbreviation for UDP-glucuronosyltransferase to represent the two most studied enzymes, UGT1 and 2, when they were first identified in the superfamily. These two families of enzymes got their names from the fact that they preferably utilize UDP-glucuronic acid (UDPGA, or UDPGlcA) as the co-factor in the conjugation reaction. When UGT3 and 8 were characterized in later years, it was revealed that these two other members of the superfamily do not use UDPGA as the sugar donor. Rather, UGT3A1 uses UDP-*N*-acetylglucosamine, UGT3A2 utilizes UDP-glucose and UDP-xylose, and UGT8 prefers UDP-galactose as their cofactors, respectively.

With respect to their expression in mammals, UGT1As come from a single gene locus (located on chromosome 2q37) that is currently known to encode for nine functional proteins of UGT1A1, 3, 4, 5, 6, 7, 8, 9, and 10. The mechanism by which the UGT1A mRNA transcript is spliced has been documented with the existence of four common downstream exons in conjunction with an upstream variable exon. Therefore, UGT1A nomenclature, for example, is based on the UGT1A gene and the relative position of the first variable exon with respect to the four common exons on the un-spliced mRNA. From this organizational layout, the UGT1A1 isoform is assumed to be the closest whilst UGT1A12 is the most distant with respect to gene location between the variable and common exons [17]. In addition to this, it is worthwhile to note that the gene sequences of the four common exons are conserved between all the mRNA transcripts within the UGT1A subfamily and are believed to make up the C-terminal portion of UGT1As, which is responsible for cofactor binding. However, the variable exon is selected from a subset of 13 exon-1 sequences of the gene locus and transcripts for the N-terminal domain of the protein responsible for substrate binding [18].

With respect to the UGT2 family, multiple gene loci located on chromosome 4q13 comprise the following functional proteins: UGT2A1, 2, 3, UGT2B4, 7, 10, 11, 15, 17, and 28 [15]. The UGT2 family also shares common properties with UGT1A subfamily of enzymes in that they also have a conserved protein sequence at the C-terminal portion, whereas the N-terminal portion tends to be substrate-specific. However, the mRNA transcript in this family is spliced from six exons whereas only five exons are spliced in the UGT1A subfamily [17].

Two members are identified in UGT3 family: UGT3A1 and UGT3A2. The two enzymes utilize sugar donors other than UDP-glucuronic acid, as is required in UGT1 and UGT2 enzyme reactions. Instead, UGT3A1 uses UDP-N-acetylglucosamine and UGT3A2 uses UDP-glucose and UDP-xylose as a cofactor, respectively [19, 20]. Both UGT3A1 and 3A2 genes are located on chromosome 5p13.2, where UGT3A1 gene is located downstream of UGT3A2 in the direction of transcription. UGT3A1 and 3A2 genes comprise seven exons and encode 523 amino acids in the mature protein, sharing 78% sequence identity [21].

UGT8, or previously named ceramide galactosyltransferase (CGT), uses exclusively UDP-galactose as its sugar donor. The only member of UGT8 family, UGT8A1 gene, or simply UGT8, is located on chromosome 4q26 and encoded by five exons [22].

***23.1.1.2.3 Localization and Expression of UGTs (Table 23.1)*** Liver is considered the major organ for Phase II metabolism in the body as it is directly exposed to the influx of both endogenous and exogenous compounds from the hepatic portal vein during oral absorption. The mRNA of UGT1A1, 3, 4, 6, 9, UGT2A3, 2B4, 7, 10, 15, 17,28, and UGT3A1 are found highly expressed in the liver, as well as UGT 1A5, 2A1, 2A2, and 2B17, which are moderately expressed in liver [23–26].

On the other hand, extrahepatic UGTs also exists in tissues that are responsible for absorption and/or excretion of drugs such as gastrointestinal (GI) tract (especially small intestine), kidney, lung and proximal digestive tract. Small intestine is the most important organ for drug absorption in oral administration route. Compounds with low bioavailability (such as raloxifene, some flavonoids) are considered to undergo extensive glucuronidation in intestine [27]. UGT1A1, 5, 8, 10, 2A3, 2B15, and 17 are detected highly in GI tract at mRNA level, with UGT1A1 and 2B7 at a lower level, while kidney expresses UGT1A6, 9, 2B7, 3A1 abundantly, along with UGT2A1 and UGT8 [20, 22, 23, 26].

Some of the UGT isoforms play a role in the conjugation of endogenous compounds such as hormones therefore these isoforms are distributed differently. For

**Table 23.1** Distribution of UGT isoforms at mRNA expression level.

| | Liver | GI tract | Kidney |
|---|---|---|---|
| High expression | UGT1A1, 3, 4, 6, 9 | UGT1A5, 8, 10 | UGT1A6, 9 |
| | UGT2A3, 2B4, 7, 10, 15, 17, 28 | UGT2A3, 2B15, 17, | UGT2B7 |
| | UGT3A1 | | UGT3A1 |
| Moderate to low expression | UGT1A5, | UGT1A1 | UGT2A1, |
| | UGT2A1, 2 | UGT2B7 | UGT8 |

example, UGT8 biosynthesizes galactosyl-ceramide which is crucial in central and peripheral nervous system, and the mRNA of UGT8 is found to be expressed at a high level in the brain [28].

***23.1.1.2.4 Polymorphism and Complications*** Genetic polymorphisms for UGT enzymes have been documented for virtually all isoforms of UGT1A1, 3, 4, 6, 7, 8, 9, UGT2A1, UGT2B4, 7, 15, and 28 [29]. The potential danger for patients with genetic polymorphisms is that subjects with allelic variants can have enhanced or reduced glucuronidation activity. Therefore, complications can arise with regard to drug disposition, which could lead to severe or fatal drug-drug interactions.

For instance, UGT1A1 is a major conjugating enzyme responsible for the glucuronidation of a wide selection of xenobiotics. In addition to this, UGT1A1 is vital for the homeostasis of bilirubin. A genetic polymorphism in the promoter region of the UGT1A1 gene (denoted as UGT1A1*28) have shown to cause UGT1A1 expression reduction and is associated with Gilbert's syndrome [16, 30]. Typically observed in the Caucasian population (approximately 7–19% are homozygous for this allele), Gilbert's syndrome causes a mild form of unconjugated hyperbilirubinemia, with no apparent pathological consequences [31].

While polymorphisms in the promoter region seem relatively inconsequential, polymorphisms in the coding regions of the UGT1A subfamily gene can lead to severe diseases. A rare (yet similar 1% frequency) but severe unconjugated hyperbilirubinemia, known as the Crigler–Najjar (CN) type I and II syndromes [31] is seen for subjects with polymorphisms in the coding regions of UGT1A gene. Severe hyperbilirubinemia leads to jaundice or kernicterus, which is due to the passage of bilirubin into the brain, resulting in necrosis of neurons and glial tissues [32, 33].

***23.1.1.2.5 Animal Model of Crigler–Najjar Type I Syndrome*** CN syndrome is a very rare disease with an incidence of around 1/1 000 000 live births [34]. Considered as a form of familial hyperbilirubinemia (and thus described under the name of hereditary unconjugated hyperbilirubinemia), this disease is associated with a complete hepatic deficit of bilirubin glucuronosyltransferase (UGT1As) activity [35]. The distinction between type I and II is that type I is more severe in terms of the high levels of serum bilirubin when compared to type II.

The animal model for this disease was originally discovered in 1938 by C.H. Gunn [36]. The breeding stock of the Connaught Laboratories (University of Toronto, Canada) at that time had mutant animals that were jaundiced. The original stocks were of Wistar origin. Further studies revealed that liver slices from these animals could not conjugate bilirubin, and this was later confirmed to be due to the absence of UGT1As [37].

Experiments in Gunn rats have provided important insights into physiological, biochemical, and molecular mechanisms of bilirubin conjugation and disposition [38]. Presently, the Gunn rat model is being used to develop methods for gene therapy for inherited bilirubin-UGT deficiency. Although Gunn rat has been utilized as a natural model for bilirubin encephalopathy, some studies have actually explored the distribution/expression and functional capacity of other enzymes or transporter proteins involved in the disposition of various endogenous or xenobiotics.

### 23.1.2 Pathways of Drug Elimination

Although the glucuronidation of isoflavones is very important to the overall disposition process, various other factors and mechanisms must also be accounted for. One important area is the transport of compounds with respect to drug transporter(s) and their impact. Traditional dogma indicates that upon the ingestion of foods, drugs, or xenobiotics, the transport of molecules from these sources across the intestinal lumen dictates the bioavailability of these compounds. Three general routes of transport are passive diffusion, facilitated diffusion, and active transport. The route of transport for molecules typically depends on their size and respective chemical polarity in that small hydrophobic molecules are usually passively transported across the lumen. With regards to flavonoids and isoflavones, it is well documented that isoflavones in glycoside form are poorly absorbed. However, upon rapid conversion to the aglycone (no sugar attached) form via enzymes such as lactase-phlorizin hydrolase, aglycones are shown to be rapidly absorbed in a passive manner [39, 40].

However, current research in oral absorption places major emphasis on active transport, by which molecules can be either actively absorbed (by which the molecule travels across the intestinal lumen via uptake transporters such as sodium-dependent glucose transporter [SGLT1]) or actively pumped out of the intestinal epithelial cells via efflux transporters such as P-glycoprotein. Recent data from our laboratory using Caco-2 model indicated the possibility of isoflavones (extracted from Red Clover supplement products) requiring active or facilitated transport. Upon crossing the apical side of the intestinal lumen, isoflavones as well as other molecules of interest can be metabolized (Phase I or Phase II; described in detail in the above sections of this chapter) into conjugates. These conjugates, as well as the unmodified molecule, can be actively transported across the basal membrane of intestinal epithelial cells into the blood or efflux back into the intestinal lumen. Therefore, these processes of active transport are facilitated by many types of major active transport proteins such as P-glycoprotein (which belongs to the multidrug resistance-type), multidrug resistance-associated proteins (MRPs) and other organic anion transporters which together play a major role in governing the overall bioavailability of foods, drugs and other xenobiotics.

#### 23.1.2.1 P-glycoprotein (see Also Chapter 22)

The history of P-gp goes hand in hand with human cancer and subsequent chemotherapy treatments. Upon discovery of cytotoxic drugs that can destroy cancer cells, researchers also discovered that drug resistance (for multiple drugs) can impede the effects of these drugs. P-glycoprotein was the first active transporter protein to have been discovered; a little over 50 years ago. Two research groups must be acknowledged with great contributions to the discovery of P-gp and determining this transporter as a major protein responsible for multidrug resistance (MDR). In 1968, the Ling lab initially used somatic cell genetics to investigate colchicine resistance in Chinese hamster ovary cells linking P-gp to MDR [41]. In the meantime, Dr. Gottesman's lab was able to clone amplified DNA sequences related to MDR phenotype in cell lines [42]. From both of their work we now know that MDR gene and P-gp gene were of the same. Today, we know that P-gp, and other more recently identified protein(s) act as efflux pumps with broad specificity for a variety of substrates.

***23.1.2.1.1 Structure of P-gp*** As a member of ATP-binding cassette (ABC) efflux proteins, P-gp consists of 1280 amino acids with a molecular mass of approximately 170 kDa. P-gp structure shares extensive sequence homology and domain organization compared to other members of superfamily. The most conserved regions are the ATP-binding cassette domain ranging approximately between 200 and 250 amino acids consisting of two nucleotide-binding domains (NBDs) located towards the cytoplasmic face of the membrane [43]. More specifically, P-gp contains 12 transmembrane regions split into two "halves" each contains an NBD.

***23.1.2.1.2 Substrates of P-gp*** Substrates for P-gp are typically lipophilic and cationic. As previously mentioned, a large number of compounds are substrates/inhibitors for P-gp and include anticancer agents, antibiotics, antivirals, calcium channel blockers, immunosuppressive agents, and plant chemicals usually found in normal diet [44]. With respect to isoflavones, typical Phase II conjugates exist as organic anions and thus do not interact with P-gp for their elimination/excretion from the cell.

***23.1.2.1.3 Modulation of P-gp activity*** The nature of P-gp is to limit the absorption of certain compounds by driving efflux back into the lumen and thus limits their bioavailability. Data from both *in vitro* and *in vivo* studies using human intestinal epithelial cell lines [45] and P-gp-knockout mice [46], respectively, show that the disruption of P-gp activity would lead to some potentially hazardous problems with regards to drug disposition. Essentially, the disruption of P-gp simply means that elimination via excretion of a compound in the body will be substantially decreased (if the compound is predominantly eliminated by P-gp) and thus leading to increased exposure that could lead to toxicity due to net fluctuations in the concentration of compound(s).

#### 23.1.2.2 Multidrug-Resistance Associated Proteins (see Also Chapter 22)

As another ABC transporter family, MRPs were discovered much later than P-gp. For instance, MRP1 was first cloned in 1992 from a human small cell lung cancer cell line that did not express P-gp [47]. MRP2 was later cloned in 1996 from rat liver designated as cMRP [48] and from human liver and named canalicular multispecific organic anion transporter (cMOAT) [49]. MRP3 was discovered later and shown to be closest related to MRP1 sharing 58% amino acid identity [48]. In general, all currently know MRPs (MRP1 to MRP9) were all discovered later when compared to P-gp and relatively little information exists currently with regards to both function and expression patterns of the most recently discovered MRPs (MRP7-9).

***23.1.2.2.1 Structure of MRPs*** As previously mentioned, the general structure of P-gp is that they comprise of 12 transmembrane regions split into two "halves" each containing an NBD. While MRPs do share this common structure with P-gp, MRPs structures are just a little bit more complex than P-gp. MRP1-3 and MRP6-7 are seen to have a total of 17 transmembrane regions with an additional five transmembrane regions at the N-terminus. However, these additional five transmembrane regions do not have an additional NBD. More interestingly, MRP4 and MRP5 do not have these additional regions and thus resemble that of P-gp with only 12 transmembrane regions.

***23.1.2.2.2 Substrates of MRPs*** The substrates for MRPs are different with regards to each individual isoform as well as their respective localization. However, the overall mechanistic functions of MRPs are to eliminate compounds from the cell via efflux. Studies have shown that these class of transporters can confer resistance to cytotoxic drugs such as vincristine [50] and peptides [51], heavy metal anions [52] as well as endogenous metabolites such as bilirubin glucuronides [53]. Specifically, MRP2 has shown to be centrally involved in the detoxification process via its ability to secrete metabolites into bile. The importance of MRP2 is clearly evidenced in patients who suffer from Dubin-Johnson syndrome with defective MRP2 and thus impaired ability to secrete glutathione and conjugates as well as bilirubin glucuronides [54]. With respect to flavonoids/isoflavones disposition, previous studies in our laboratory using *in vitro* Caco-2 transport model have shown MRP2 to be a central efflux transporter that is responsible for the excretion of metabolites from the cell [55, 56]. MRP3 has shown an affinity for bile salts with an emphasis in mediating bile salt reabsorption into the systemic circulation. While it is clear that there are some major overlaps with regards to substrate specificity for these MRPs, they do seem to have some specific roles with regards to the disposition of xenobiotics as well as endogenous compounds. In addition to this, much is yet to be discovered with regards to the substrate specificities as well as their functions for the more recently discovered transporters (MRP7-9).

***23.1.2.2.3 Modulation of MRPs activity*** As previously mentioned, substrate specificity for MRPs can differ with regards to each individual isoform as well as their localization in the cell. Therefore, the disruption of MRP activity can differ greatly with regards to the impact on the disposition process of xenobiotics or endogenous compounds. In general, various MRPs have significant overlap with regards to substrate specificity, and thus a disruption of a specific MRP isoform such as MRP3 will probably not result in major physiological dysfunction simply because MRP1 (if expressed and functional) may also excrete substrates and thus compensate for the MRP3 dysfunction. However, this is not the case for all MRPs in that certain isoforms (i.e. MRP2) seem to be more critically important for the elimination of certain substrates (i.e. bilirubin). As mentioned in the previous section, the dysfunction of MRP2 in humans can lead to major consequences resulting in Dubin–Johnson syndrome [54].

***23.1.2.2.4 Animal Models of Dubin–Johnson Syndrome*** First discovered in 1985, the TR-rat model is an animal model of the human Dubin–Johnson syndrome [49]. Both rats in this TR-model as well as patients who suffer from Dubin–Johnson syndrome are characterized by autosomal recessive conjugated hyperbilirubinemia. TR-rats are defective in the canalicular multispecific organic anion transporter (cMOAT2 or MRP2), which mediates hepatobiliary excretion of numerous organic anions [54]. Furthermore, the complementary DNA for rat cMOAT, a homolog of the human multidrug resistance gene (hMRP1), was isolated and shown to be expressed in the canalicular membrane of hepatocytes. Due to this, the transport of compounds such as conjugated bilirubin from the liver to bile is severely impaired, whereas uptake of organic anions from plasma to liver is normal [57]. Furthermore, the bile flow in these mutant rats is reduced to about 50%, while liver marker enzyme activities in mutant rat serum are normal [57]. Liver morphology is also normal.

While the TR-rats provide a unique model to study mechanisms of bile formation and cholestasis, we can utilize this model to assess the potential impact as well as the possible compensatory mechanisms elicited due to the absence of MRP2s [57–59]. In the absence of this efflux transporter, we can use this rat model to reveal major deficiencies and/or compensations with respect to the metabolism and overall disposition process of compounds such as apigenin and genistein; both of which could be heavily dependent on MRP2-mediated efflux as their main pathway for the excretion of Phase II glucuronidated metabolites.

#### 23.1.2.3 Organic Anion Transporters (see Also Chapter 21)

The origin of the discovery of organic anion transporters began almost 70 years ago [60]. At that time, the anionic dye phenolsulphonphthalein was observed to be in very high concentrations in renal convoluted tubules and thus suggested a tubular secretion process. This essentially gave rise to the discovery of organic anion transport (organic cation transport was categorized as well) [60].

***23.1.2.3.1 Structure of OATs*** Organic anion transporter 1 (OAT1) was first cloned in 1997 from rat kidneys and functionally identified as an organic anion transporter [61, 62]. From a structural perspective, organic anion and cation transporters share a predicted 12-transmembrane domain (TMD) structure with a large extracellular loop between TMD 1 and 2, carrying potential *N*-glycosylation sites. Conserved amino acid motifs revealed a relationship to the sugar transporter family within the major facilitator superfamily.

***23.1.2.3.2 Substrates for OATs*** Substrates for OATs typically include weak acids that have a net negative charge on carboxylate or sulfonyl residues at physiological pH. Although OATs have the capacity for endogenous transport via active secretion by the proximal tubules for compounds such as riboflavin, their principal function of OATs is to eliminate xenobiotic agents [63]. This includes many of the products of Phase I and Phase II hepatic biotransformation, as well as anionic drugs [64].

***23.1.2.3.3 Modulation of OATs activity*** The disruption or alteration of OAT activity due to functional disruptions or underexpression of OATs is reflected in pathophysiological states. Essentially, the accumulation of uremic toxins (of which many are organic anions) in the kidney due to OAT dysfunction will lead to general renal disease [65].

#### 23.1.2.4 Organic Anion Transporting Polypeptides

As members of solute carrier (SLC) transporters, organic anion transporting polypeptides (OATPs) are a group of carriers that mediate cellular uptake of a great number of unrelated amphipathic organic substrates. OATPs originally belonged to SLC21A superfamily. In 2004, when the nomenclature of this superfamily was updated, SLC21A was renamed SLCO [66].

***23.1.2.4.1 Structure of OATPs*** However, investigators have not elucidated any of the OATPs' crystal structures so far. Hence, they used homology modeling as a tool to construct OATPs' molecular structure, which can improve the theoretical prediction. Researchers have predicted that all OATPs contain 10 or 12 transmembrane domains based on hydropathy analysis [67]. Human OATPs are usually composed of 643 to 724 amino acids, except for OATP5A1, which consists of 848 amino acids. In 2010, Waver and Hagenbuch found that several conserved positively charged amino acids R57, K361, and R580 in OATP1B1 might be part of its substrate-binding sites based on a putative model [68].

***23.1.2.4.2 Substrates of OATPs*** OATPs are a group of membrane SLCs that have a wide spectrum of amphipathic organic substrates [69]. Although OATPs usually transport amphipathic compounds, they can also mediate the disposition and elimination of neutral and cationic compounds [70]. Therefore, a very large number of compounds, which are mainly amphipathic molecules with molecular weights greater than 350 Daltons, are OATPs' substrates. The latter includes bile acids, steroid conjugates, thyroid hormones, anionic peptides, other xenobiotics, and different types of drugs, such as statins, antivirus, antibiotics, and anticancer drugs [71]. And it is noteworthy that different OATPs have partially overlapping and partially distinct substrate specificities [72].

***23.1.2.4.3 Modulation of OATPs Activity*** Transcriptional regulation can significantly control the expression of OATPs, which will alter OATPs' activity. These transcription signal-related factors include HNF1a, HNF3b, Stat5, IFN-g, and IL-1b [71]. The OATPs expression in different tissue types appears to have various mechanisms for regulation. Kullak-Ublick et al. found that increased levels of taurocholate, a bile acid, can up-regulate OATP1A2 expression, but thyroid hormone decreased the expression of OATP1A2 in the liver [73]. However, Miki et al. figured out that OATP1A2 expression is closely related to steroid and xenobiotic receptor expression in breast carcinoma tissues and cell lines [74]. In addition, disease states are also associated with altered OATP expression levels [75]. For instance, inflammatory bowel disease is associated with altered OATP expression levels, which may be linked to the inflammation of the tissues. Researchers found higher OATP2B1 and OATP4A1 levels in terminal ileum and colon in IBD patients [76]. And for OATP1B1, its polymorphism has a significant effect on physician's decision about how to use statins. Because most statins are known as substrates of OATP1B1 [77]. For example, Yohei Nishizato et al. found that, in their clinical study of pravastatin, individuals with OATP1B1*15 allele had reduced total and nonrenal clearance compared to those with OATP1B1*1b allele [78]. To avoid unexpected drug disposition and potential adverse drug–drug interaction, it is essential to confirm the mechanisms that modulate OATPs' activities.

## 23.2 Interplay Between Metabolic Enzymes and Transporters

### 23.2.1 What is Interplay or Coupling?

In early research focusing on cytochrome P450s, it was thought that the rate and the overall extent of metabolism were the predominant steps that would rate-limit the overall disposition of compounds that were subject to metabolism [79]. As research progressed, we began to understand that an intricate relationship exists between these Phase I metabolic enzymes and efflux transporters such as P-gp [4]. This interplay, in which not only a single metabolic enzyme and/or transporter but a relationship/communication between at least two proteins (i.e. one metabolic enzyme and one transporter) is involved, which would ignite a new wave of excitement in what is now known as drug transporter research, leading to subsequent findings and trends that still guides the field today [80].

Similar to this trend, until recently, Phase II conjugating enzymes such as UGTs and their ability to metabolize compounds (Phase II metabolism) were believed to be the most important step in the disposition of a xenobiotic substrate. The ability for UGTs (as well as other Phase I and Phase II metabolic enzymes) to biotransform a lipophilic compound into a more hydrophilic metabolite (via the addition of a glucuronic acid) was initially believed to be the major biological barrier to the entry of xenobiotics/drugs into systemic circulation and represent one of the main pathways for their elimination [81]. However, this belief, which places great emphasis on UGT activity, does not comprehensively cover the overall disposition of xenobiotic/drugs.

While metabolism is certainly important as it serves as the initiation step of a long chain of reactions ultimately leading to elimination, we also understand that many conjugates are hydrophilic and cannot permeate the cell membrane once they are biotransformed by UGTs. Take, for example, the case of flavonoids, their metabolites are classified as organic anions and thus cannot be taken up or excreted without the aid of transporters. While UGT metabolism is a very necessary component for the disposition process of xenobiotics/drugs, we have now begun to pay more attention to the various transporters capable of transporting negatively charged hydrophilic conjugates across the cellular membrane. Although much work has been done on the study of transporters, we have barely begun to understand the interplay or coupling of these two steps.

### 23.2.2 Type of Interaction Among Metabolic Enzyme(s) and Transporter(s)

#### 23.2.2.1 Interaction Between Metabolic Enzyme(s) and Metabolic Enzyme(s)

Metabolic enzymes, such as CYP450s and UGTs, are known to have coordinated metabolic steps in that, typically, a CYP will be involved in an oxidative reaction with our compound of interest. Following this, the newly formed metabolite (depending on the specific pathway of metabolism in Phase I) is now a suitable subject ready for Phase II (e.g. UGTs) metabolism. This type of interplay between metabolic enzymes has been well established and documented.

#### 23.2.2.2 Interaction Between Metabolic Enzyme(s) and Transporter(s)

Upon metabolism via either Phase I and/or Phase II mechanisms, the metabolites are loosely transferred (the mechanism of transfer is not completely clear) to efflux transporters. Following this, the metabolite finds itself outside of the biological structure (e.g. enterocyte) and is subject to various mechanisms such as reentry into the enterocyte upon additional transformational processes (e.g. hydrolysis) that might or might not be enzymatic and/or eventual elimination (typically fecal or renal).

The various mechanisms/ways of interplay between metabolic enzyme(s) and efflux transporters will be described in more detail in the following sections. But nevertheless, this way of interacting has been a hot topic of research for many scientists.

#### 23.2.2.3 Interaction Between Transporter(s) and Metabolic Enzyme(s)

From a brief overview of this section title, the reader will first think that this is a typo in which the topic of discussion is exactly the same as that of Section 23.2.2.2.

However, this is not the same. The interaction between efflux transporters and metabolic enzyme(s) comes from observations that efflux transporters can provide feedback to the metabolic enzyme(s). Therefore, this feedback signaling can allow efflux transporters to relay messages to metabolic enzymes to either increase (positive) or decrease (negative) their functional capacity. This change in functional capacity can usually be observed as either or both in terms of expression and functionality (i.e. metabolism rate increase) for the metabolic enzyme in question. In addition, uptake transporters can also affect the metabolism of the compound in the body. As a rate-limiting step, the uptake of compounds into the hepatocytes determines the extent of metabolism that leads to the change of pharmacokinetic profiles.

#### 23.2.2.4 Interaction Between Transporter(s) and Transporter(s)

Transporters can also completely bypass the involvement of metabolic enzymes. Typically, in this case, we see that compounds are not extensively metabolized (or at all) but still utilizes an uptake transporter for entry into the cell and an efflux transporter for a subsequent exit out of the cell. This interaction can allow the efflux of compounds out of the apical side of the cell (e.g. enterocyte) or it can allow the facilitation of a compound from entry into the cell and eventual passage through the basolateral side as well.

### 23.2.3 Complexities of Coupling

Based on the different interaction types we discussed above, one would imagine a scenario of the traditional one-two punch in which a compound would enter the cell of interest (in this case, let us focus on oral absorption, and thus this would be an enterocyte), then quickly get metabolized (Phase I or II or both), and then be immediately taken up by efflux transporters and pumped back across the intestinal lumen. However, this is rather a simplified/unified type of mechanism that was detected. Current findings and literature data suggest that the way metabolic enzymes interact with transporters and the way in which metabolic enzymes interact with each other as well as transporters and their connection to each other are highly complex and are not as simple as a one-two step. As of today, several theories exist that explain the mechanistic details and complexities of this coupling process which will be detailed below.

### 23.2.4 Double Jeopardy Theory

The "double jeopardy" theory was purposed by Benet and coworkers [79, 82]. This theory was originally proposed to explain the mechanistic details for the

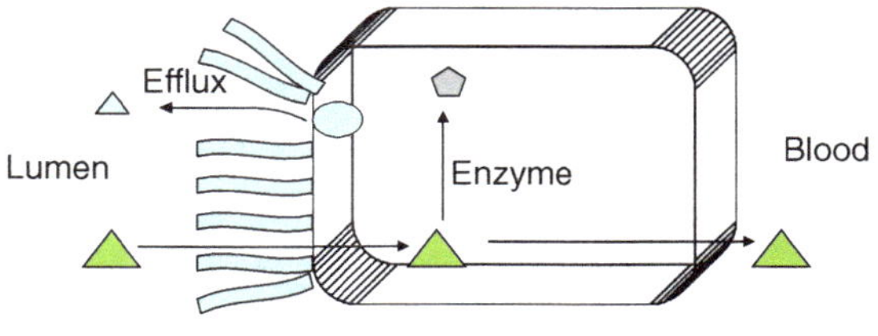


**Figure 23.1** Schematic representation of the double jeopardy theory. Substrates are represented by triangles whereas products of enzymes or metabolites are represented by pentagons.

coupling of CYPs and the efflux transporter P-gp. The mechanism (see Figure 23.1) of action for this coupling theory is a little less complex than the revolving doors theory (below). Essentially, the substrate is assumed to be absorbed (at the apical side of the membrane) and is metabolized by the enzyme CYPs (typically CYP3A4). This substrate is subjected to metabolism and thus is termed as "prosecuted" for the first time by CYP3A4. With respect to efflux, P-gp is assumed to take the parent compound and transport it via efflux back into the intestinal lumen. Meanwhile, it is important to note that the substrate having encountered the CYP one-time already but has not been metabolized, can pass through the other side (basolateral) of the membrane and thus has reached the systemic circulation. But if the intact parent compound makes it back into the intestinal lumen via efflux, it can be absorbed one more time at the apical side of the membrane (further down the intestinal tract) and thus is met again by the enzyme CYPs and is subjected to metabolism again. This repeated "prosecution" thus gives the term "Double Jeopardy" to this theory.

#### 23.2.4.1 Consequences of Disruption

In general, the consequences of the disruption of this coupling mechanism as well as any other coupling mechanism can vary depending on the substrate in question with regards to toxicity and interactions with other drug mechanisms as well as compensatory mechanisms or effects by other enzymes or transporters in the system. With regards to double jeopardy theory and thus more specifically with CYPS and P-gp, the disruption of P-gp activity would result in a less repeated chance of exposure time for the substrate in question at the apical membrane. This would lead to more substrate inside the intestine (i.e. increase in absorption) and thus could potentially increase the bioavailability of both the substrate and/or its metabolite in the systemic circulation. On the other hand, if CYPs activity was disrupted, the amount of metabolite would decrease

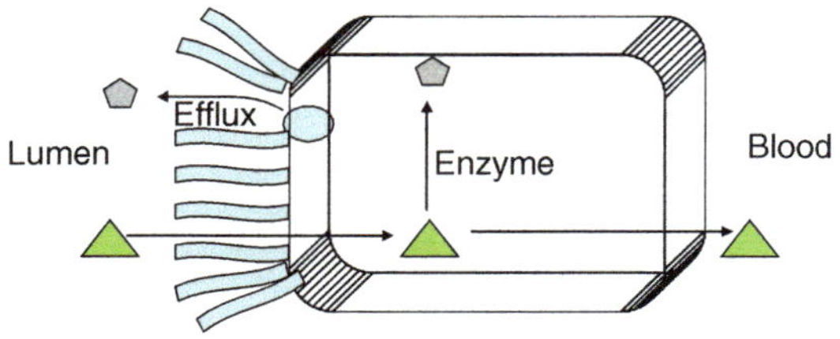


**Figure 23.2** Schematic representation of the revolving door theory. Substrates are represented by triangles whereas products of enzymes or metabolites are represented by pentagons.

severely and thus perhaps allowing more substrate to reach systemic circulation intact.

### 23.2.5 Revolving Doors Theory

The revolving doors theory was proposed more recently and hypothesize that efficient coupling of the conjugating enzymes (UGTs) and efflux transporters enables both the enterohepatic recycling and enteric recycling processes [81]. Therefore, unlike the previous theory, this "revolving door" theory emphasizes the importance of efflux transporters and their ability to modulate the cellular excretion of hydrophilic metabolites. In addition to this, this theory also recognizes the transport of hydrophilic metabolites to both apical and basolateral sides of the cells. This latter point adds complexity to this theory and its mechanism (see Figure 23.2).

#### 23.2.5.1 Mechanism Description

The mechanism of the revolving door theory is best illustrated step by step for a model compound such as the soy isoflavone genistein (aglycone). When genistein is absorbed into the intestinal cell, it has two immediate fates. One is that it will pass through the basolateral side of the membrane and reach the systemic circulation. Research has found that this is not the case as bioavailability is extremely low for this compound upon ingestion [83]. The fate of genistein in the intestinal cell is more likely that it will be metabolized by UGTs [83, 84]. After biotransformation by UGTs, the genistein metabolite again has two fates. One, the genistein metabolite, as an organic anion, can be transported by efflux transporters (MRP1 and/or MRP3) located at the basolateral side of the lumen and eventually reaching systemic circulation. The other choice is that they will be transported back across the apical side of the intestinal lumen by efflux transporters such as MRP2 or perhaps BCRP. According to our research [85], the efflux step is most likely the rate-limiting step for this coupling process. In other words, the amount of metabolites made by UGTs may not significantly alter or influence the disposition of genistein or its metabolites because it is dependent on the efflux step by transporters. However, in the event that no metabolites have been made due to a deficiency in UGT activity, the efflux cannot occur simply because there will be no metabolites during the efflux period. Therefore, the efflux step is dependent on the premise that UGT metabolism can produce sufficient amounts of metabolites. This is ultimately the theme of the revolving door theory, which places emphasis on the rate of efflux (i.e. the speed at which the revolving door is spinning) as the determining factor for the disposition of genistein and its metabolite.

Upon efflux into the intestinal lumen, genistein metabolites (e.g. glucuronides or sulfates) are converted back to the genistein aglycone through bacterial hydrolysis and thus can re-enter the intestinal cell and this whole process occurring at the intestinal level is known as enteric recycling. This extensive metabolism at the enteric level is also known as a part of the first-pass metabolism. With regards to the liver, the genistein that passes through the UGTs in the intestine will be subjected to liver UGTs. These metabolites, as well as any metabolites that made it through the intestinal level, will be again secreted in bile into the intestinal lumen, which must be hydrolyzed by microflora back to its parent compound and again ready for absorption. This process is known as enterohepatic recycling and again the revolving door theory is very much involved in this scheme.

#### 23.2.5.2 Consequences of Disruption

The consequences of disruption for this coupling theory are again dependent on the substrate and its metabolite(s) in question. The reason for this is because the body has an elaborate system of various metabolic enzymes (i.e., UGTs, SULTs) as well as transporters (i.e. MRPs, BCRP, and OATs) that have overlapping substrate specificities. Because of the fact that detoxification is an essential function for the survival of the organism, the body has developed through evolution a very complex system that has multiple enzymes and transporters that can compensate for each other should any one particular member is made dysfunctional. Therefore, in order to severely disrupt coupling, a large number and wide variety of enzymes (i.e. both UGTs and SULTs) will have to be made dysfunctional.

### 23.2.6 Enterohepatic Recycling and Hepatoenteric Recycling

As a classic recycling process, enterohepatic recycling (EHR) was first proposed in the 1950s for bile acid recirculation [86]. Subsequently, its importance has been appreciated and mentioned in almost all medical and pharmacological textbooks due to its vital role in the disposition of various endogenous and exogenous compounds [87]. During this process, the interaction and coupling between transporters and metabolic enzymes enable the recycling of compounds from the liver to intestine and their transport back to the liver.

Generally, the compounds absorbed by enterocytes may be excreted into the mesenteric vein as a parent or metabolic form, which may be taken up by hepatocytes through passive diffusion and/or assistance of uptake transporters. After that, the parent compounds or their Phase II metabolites (major form) will be excreted into the bile via the efflux transporter on the apical side of hepatocytes, and then deposited back into the intestine. In the intestine, the parent drug excreted or biotransformed by the microflora is available for re-absorption.

Traditionally, investigators hold the notion that Phase II metabolites participating in EHR are mainly from the liver since it is considered as the major metabolic organ in the body [88]. That means the compounds undergoing EHR are principally transformed into the Phase II metabolites by the enzyme, such as UGTs and SULTs, in hepatocytes. Following this, the relatively hydrophilic Phase II metabolites are transferred into the bile by the efflux transporters for reabsorption. Therefore, the hepatic enzyme(s)-transporter(s) coupling is recognized as the most important in EHR.

Recently, our lab found that certain phenolic glucuronides could be taken up hepatocytes rapidly and efficiently, whereas others were not [89]. These phenomena suggest that the Phase II metabolites of some phenolic compounds may be sourced from the intestine instead of the liver.

Meanwhile, due to the high hydrophilicity of the Phase II metabolites, the transport of them from enterocytes to bile needs the coordinated actions of the basolateral enteric efflux transporters, basolateral hepatic uptake transporters, and apical hepatic efflux transporters [90]. In this case, the intestinal enzyme(s)-transporter(s) coupling occupies a predominant position in the recycling of compounds. Therefore, a new disposition mechanism called “Hepatoenteric Recycling” (HER) is proposed to distinguish it from EHR. Facilitated by the new concept, HER, the complex coupling between metabolic enzyme(s) and transporter(s) in the recycling of certain phenolic compounds can be illuminated more clearly.

#### 23.2.6.1 Mechanistic Description of EHR and HER

Like the revolving doors theory, the complex interplay between metabolic enzyme(s) and transporter(s) in EHR and HER is best illustrated step by step using model compounds. Here, raloxifene and icaritin (ICT) are taken as examples to introduce two coordinate patterns, parallel mode and tandem mode, between EHR and HER respectively.

Raloxifene is a selective estrogen receptor modulator for the treatment of osteoporosis. The pharmacokinetic studies show that only 2% raloxifene can reach systemic circulation while more than 60% raloxifene can be absorbed [92]. This low absolute bioavailability is attributed to the extensive hepatic and extrahepatic glucuronidation, through which the raloxifene will be transformed to raloxifene-4′-β-glucuronide (Ral-4′-G) and raloxifene-6-β-glucuronide (Ral-6-G), with a plasma Ral-4′-G: Ral-6-G ratio of about 8 : 1 [27, 93, 94]. Although researchers have noticed the importance of intestinal glucuronidation in the presystemic metabolism of raloxifene, the fate of metabolites in the body has not been illuminated. Recently, our lab showed the role of OATPs in the transport of raloxifene metabolites and proposed that the aforementioned new concept, HER, could help explain the coupling of metabolic enzymes and transporters in the disposition of raloxifene and its metabolites.

Based on the previous studies and our recent work, raloxifene undergoes EHR and HER in the parallel mode, which means some raloxifene involved in the EHR, while the others go through the HER (see Figure 23.3). After absorption, raloxifene has two fates: one is to pass through the metabolic barrier and reach the mesenteric vein; another is to be metabolized by the UGTs. Research has found that the UGT1A8 and 1A10, which are absent in the liver and favor Ral-4-G formation, are the primary contributors in the formation of Ral-4′-G, which is the primary metabolite, in the intestine [93, 95, 96]. Later, this primary metabolite will undergo enteric cycling (via MRP2) or HER to go back to the intestine. For the part involved in HER, Ral-4′-G will be transported by efflux transporters, such as MRP1 and MRP3, at the basolateral side to the mesenteric vein [95, 97, 98]. After that, it was assumed that most soluble Phase II metabolites will go into the systemic circulation without extensive uptake by hepatocytes. However, according to our recent studies, the OATPs located at the basolateral side of hepatocytes can efficiently uptake Ral-4′-G from blood to hepatocytes. Furthermore, we found that there is a strong correlation between the uptake of OATP overexpressing cells and recycle ratios of many phenolic glucuronides. This

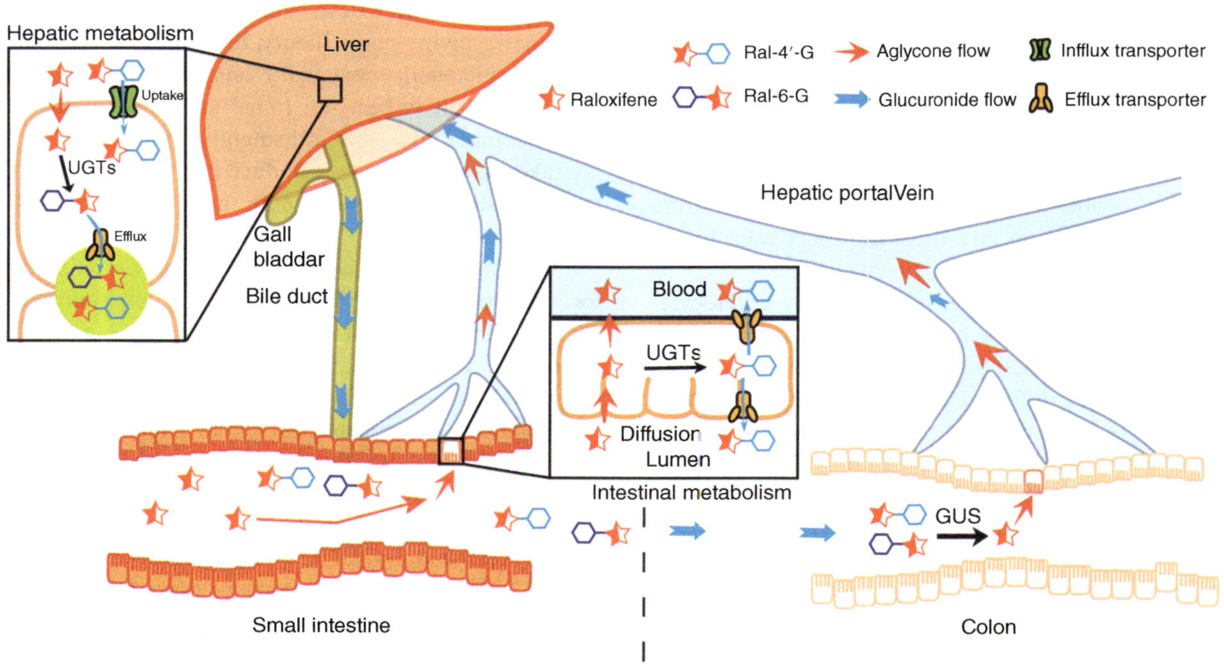


**Figure 23.3** The pictorial representation of recycling of raloxifene in which enterohepatic recycling and HER work in parallel. Source: Adapted from Tu et al. [91].

indicates that the hepatic uptake step is most likely the rate-limiting step for this coupling process. Upon taken up by the hepatocytes, Ral-4′-G will be excreted into the bile via the efflux transporter, such as MRP2, on the apical side. Following the bacterial hydrolysis, by which the Ral-4′-G can be converted back to raloxifene, the Ral-4′-G efflux into the intestinal lumen can reenter the enterocytes as an aglycone.

As to the fate of raloxifene mentioned previously, raloxifene molecules that penetrate the intestinal metabolic barrier will reach hepatocytes via passive diffusion. In hepatocytes, the large amount of UGT1A1 becomes the primary catalyst of the raloxifene metabolic clearance via the 6-glucuronide pathway [93]. Like Ral-4′-G, Ral-6-G will be transported into the bile duct with the help of MRP2. At last, after being transformed back to aglycone, raloxifene have completed one round of EHR, and then start a new journey, which could be enteric recycling, EHR, or HER.

Raloxifene is such a good example that enteric recycling schemes, EHR and HER, are all involved in its disposition. During this process, the couplings of conjugating enzymes and transporters in different types of recycling work separately and together in the parallel mode to enable the recycling of raloxifene in the body.

In contrast to the parallel mode, the tandem mode represents a pattern in which the EHR and HER are lined up one behind another. In other words, the coupling of enzymes and transporters in one type of recycling will act on the products of the other type of recycling to achieve the recycling of compounds. Take ICT as an example, our recent studies indicated that the generation of diglucuronide is stepwise. The recycling starts from the ICT in the intestinal lumen, followed by the generation of monoglucuronides in enterocytes, and then further converted into the diglucuronide in the liver, secreted into bile, and then finally gets hydrolyzed back to ICT by microflora, and reabsorb into the colon to complete tandem recycling.

ICT is a prenylflavonoid derivative, recognized as a candidate for the treatment of hepatocellular carcinoma. Previous studies have demonstrated that ICT-3, 7-diglucuronide (ICT-DIG) is the major metabolite, and icaritin-3-glucuronide (ICT-3-G) and icaritin-7-glucuronide (ICT-7-G) are also detected in the systemic circulation [99, 100]. It provides us a good opportunity to study the coupling between conjugating enzymes and transporters since the glucuronidation predominates in the metabolic pathway (see Figure 23.4).

After taken orally, ICT will be absorbed by enterocytes. Subsequently, abundant intestinal UGTs, including UGT1A1, 1A3, 1A8, 1A9, and 2B7, can mainly convert ICT into ICT-3-G or ICT-7-G but not ICT-DIG [99, 101].

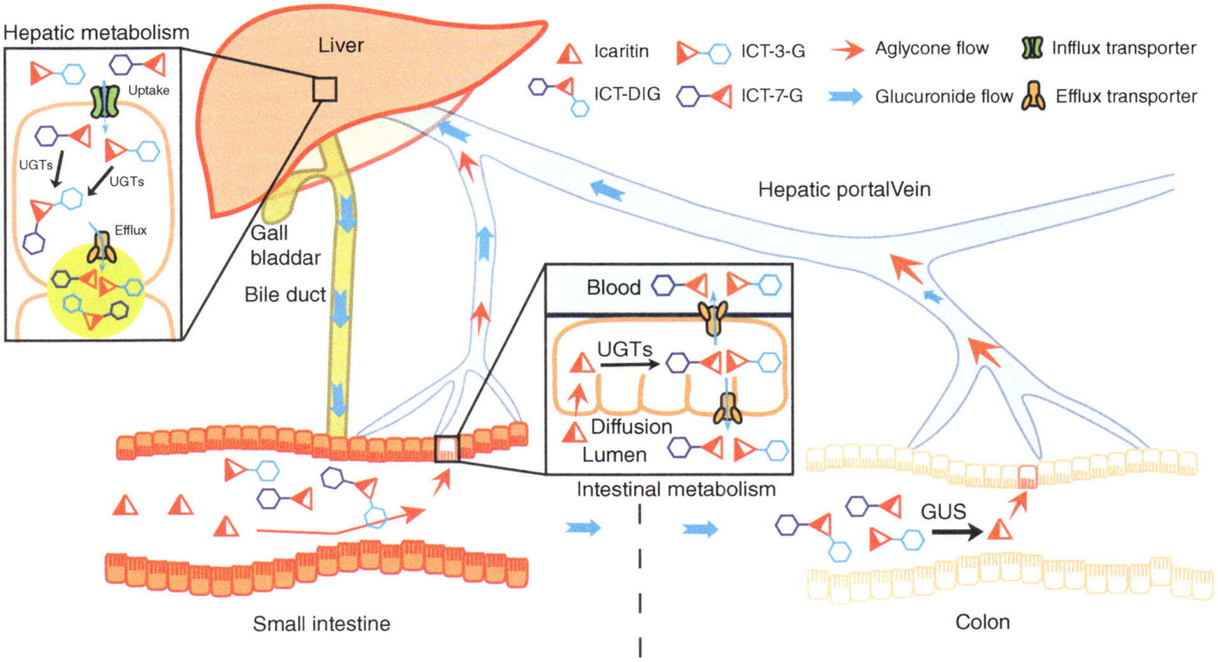


**Figure 23.4** The pictorial representation of recycling of icaritin in which enterohepatic recycling and HER work tandemly. Source: Adapted from Tu et al. [91].

This has been demonstrated by our rat jejunal microsomes incubation assay and Caco-2 cell assay, in which we saw a large amount of monoglucuronides instead of diglucuronide in the reaction system. The ICT-3-G and ICT-7-G formed in the enterocytes will be transported into the intestinal lumen or mesenteric vein. Although the transporter participated in these transportations have not been figured out, the research on a similar compound, wushanicaritin, reveals that BCRP and MRPs may be involved in the excretion of ICT-3-G and ICT-7-G [102]. For the monoglucuronides reaching the portal vein, according to our studies, they will be taken up by hepatocytes via OATPs. In hepatocytes, the ICT-3-G and ICT-7-G generated in the intestine will be subjected to hepatic UGTs that lead to the formation of ICT-DIG. According to our studies, the newly formed ICT-DIG will be transported into the bile duct immediately, because the high concentration of ICT-DIG was found in the bile but not in the liver. The ICT-DIG in the bile will be excreted into the intestinal lumen, followed by the hydrolysis of microflora to get ready for reabsorption. During this process, the conjugating enzymes and transporters in EHR and HER work stepwise, known as tandem mode, to achieve the recycling of ICT.

#### 23.2.6.2 Consequences of Disruption

Similar to the discussion in the Double Jeopardy Theory and Revolving Doors Theory, the consequences of disruption depend on the enzymes, transporters, and substrates in question. The dysfunction of any enzymes or/and transporters can cause a pharmacokinetic change of parent compound and its metabolite(s) that may lead to toxicity or less of efficacy. On the other hand, dysfunctional enzymes or transporters can be compensated by others since our body is a very complex system that involves multiple enzymes and transporters. Further discussion on the disruption and utilization of EHR and HER has been conducted in Chapter 32 by Tu et al.

## 23.3 Conclusion

The overall point of this chapter is to give the readers a time scope of the evolving scientific landscape with respect to our understanding and appreciation of the complexities in teasing apart the relationship between enzymes and transporters. We have observed and researched only the basic levels of interactions through the cumulative efforts of basic research over the decades.

We began this chapter with an introduction to metabolic enzymes (both Phase I and Phase II) as well as transporters. We detailed the nomenclatures that exist within and between the families of proteins. We then described the defining characteristics of coupling and, from there, covered the four basic fundamental interactions in which coupling exists: (i) enzyme to enzyme, (ii) enzyme-to-transporter, (iii) transporter to enzyme, and (iv) transporter to transporter.

As we proceed further into this chapter, we discuss several predominant theories that explain the above interactions: (i) Double Jeopardy Theory, (ii) Revolving Doors Theory, and (iii) Enterohepatic Recycling and HER. While these theories are fundamentally strong in their explanation of the overall coupling process, we notice that it is far more complex than the scope of these theories. It is rare to find a compound whose disposition in the body can be simply described by one theory. In all likelihood, enzymes and transporters might form thousands of coupling types, and when combined with others, these couplings can further generate various interplay patterns/permutations. Moreover, these different combinations of coupling work together in the disposition of a compound in our body.

While we appreciate the many layers of complexities that are likely to be faced and hopefully better characterized/understood in the future, much work is needed in this field to get us there, with the key finding/focus being the ultimate set of controller(s) that can determine which transporter and/or enzyme are expressed and what their respective functional capacity is like from animals to humans.

## References


1 Masters, G.A., Argiris, A.E., Hahn, E.A. et al. (2006). A randomized phase II trial using two different treatment schedules of gemcitabine and carboplatin in patients with advanced non-small-cell lung cancer. *J. Thorac. Oncol.* 1: 19–24.

2 Guengerich, F.P. (2006). Cytochrome P450s and other enzymes in drug metabolism and toxicity. *AAPS J.* 8: E101–E111.

3 Fisher, M.B., Campanale, K., Ackermann, B.L. et al. (2000). In vitro glucuronidation using human liver microsomes and the pore-forming peptide alamethicin. *Drug Metab. Dispos.* 28: 560–566.

4 Lamb, D.C., Guengerich, F.P., Kelly, S.L., and Waterman, M.R. (2006). Exploiting Streptomyces coelicolor A3(2) P450s as a model for application in drug discovery. *Expert Opin. Drug Metab. Toxicol.* 2: 27–40.

5 Dervieux, T., Meshkin, B., and Neri, B. (2005). Pharmacogenetic testing: proofs of principle and pharmacoeconomic implications. *Mutat. Res.* 573: 180–194.

6 Elens, L., Van Gelder, T., Hesselink, D.A. et al. (2013). CYP3A4* 22: promising newly identified CYP3A4 variant allele for personalizing pharmacotherapy. *Pharmacogenomics* 14: 47–62.

7 Okubo, M., Murayama, N., Shimizu, M. et al. (2013). The CYP3A4 intron 6 C> T polymorphism (CYP3A4* 22) is associated with reduced CYP3A4 protein level and function in human liver microsomes. *J. Toxicol. Sci.* 38: 349–354.

8 Wang, D., Guo, Y., Wrighton, S. et al. (2011). Intronic polymorphism in CYP3A4 affects hepatic expression and response to statin drugs. *Pharmacogenomics J.* 11: 274–286.

9 Werk, A.N. and Cascorbi, I. (2014). Functional gene variants of CYP3A4. *Clin. Pharmacol. Ther.* 96: 340–348.

10 Holmberg, M.T., Tornio, A., Paile-Hyvärinen, M. et al. (2019). CYP3A4* 22 impairs the elimination of ticagrelor, but has no significant effect on the bioactivation of clopidogrel or prasugrel. *Clin. Pharmacol. Ther.* 105: 448–457.

11 Cerny, M.A. (2016). Prevalence of non-cytochrome P450-mediated metabolism in Food and Drug Administration-approved oral and intravenous drugs: 2006-2015. *Drug Metab. Dispos.* 44: 1246–1252.

12 Evans, W.E. and Relling, M.V. (1999). Pharmacogenomics: translating functional genomics into rational therapeutics. *Science* 286: 487–491.

13 Ritter, J.K. (2000). Roles of glucuronidation and UDP-glucuronosyltransferases in xenobiotic bioactivation reactions. *Chem. Biol. Interact.* 129: 171–193.

14 Tukey, R.H. and Strassburg, C.P. (2000). Human UDP-glucuronosyltransferases: metabolism, expression, and disease. *Annu. Rev. Pharmacol. Toxicol.* 40: 581–616.

15 Meech, R., Hu, D.G., McKinnon, R.A. et al. (2019). The UDP-glycosyltransferase (UGT) superfamily: new members, new functions, and novel paradigms. *Physiol. Rev.* 99: 1153–1222.

16 Lin, J.H. and Wong, B.K. (2002). Complexities of glucuronidation affecting in vitro in vivo extrapolation. *Curr Drug Metab* 3: 623–646.

17 Mackenzie, P.I., Owens, I.S., Burchell, B. et al. (1997). The UDP glycosyltransferase gene superfamily: recommended nomenclature update based on evolutionary divergence. *Pharmacogenetics* 7: 255–269.

18 Burchell, B. (2003). Genetic variation of human UDP-glucuronosyltransferase: implications in disease and drug glucuronidation. *Am. J. Pharmacogenom.* 3: 37–52.

19 MacKenzie, P.I., Rogers, A., Elliot, D.J. et al. (2011). The novel UDP glycosyltransferase 3A2: cloning, catalytic properties, and tissue distribution. *Mol. Pharmacol.* 79: 472–478.

20 Mackenzie, P.I., Rogers, A., Treloar, J. et al. (2008). Identification of UDP glycosyltransferase 3A1 as a UDP N-acetylglucosaminyltransferase. *J. Biol. Chem.* 283: 36205–36210.

21 Meech, R. and Mackenzie, P.I. (2010). UGT3A: novel UDP-glycosyltransferases of the UGT superfamily. *Drug Metab. Rev.* 42: 45–54.

22 Meech, R., Mubarokah, N., Shivasami, A. et al. (2015). A novel function for UDP glycosyltransferase 8: galactosidation of bile acids. *Mol. Pharmacol.* 87: 442–450.

23 Ohno, S. and Nakajin, S. (2009). Determination of mRNA expression of human UDP-glucuronosyltransferases and application for localization in various human tissues by real-time reverse transcriptase-polymerase chain reaction. *Drug Metab. Dispos.* 37: 32–40.

24 Strassburg, C.P., Oldhafer, K., Manns, M.P., and Tukey, R.H. (1997). Differential expression of the UGT1A locus in human liver, biliary, and gastric tissue: identification ofUGT1A7 and UGT1A10 transcripts in extrahepatic tissue. *Mol. Pharmacol.* 52: 212–220.

25 Turgeon, D., J-Sb, C., Er, L. et al. (2001). Relative enzymatic activity, protein stability, and tissue distribution of human steroid-metabolizing UGT2B subfamily members. *Endocrinology* 142: 778–787.

26 Uhlon, M., Fagerberg, L., and Hallstrom, B. (2015). Proteomics. Tissue-based map of the human proteome. *Science* 347: 1260419.

27 Mizuma, T. (2009). Intestinal glucuronidation metabolism may have a greater impact on oral bioavailability than hepatic glucuronidation metabolism in humans: a study with raloxifene, substrate for UGT1A1, 1A8, 1A9, and 1A10. *Int. J. Pharm.* 378: 140–141.

28 Bosio, A., Binczek, E., Le Beau, M.M. et al. (1996). The human gene CGT encoding the UDP-galactose ceramide galactosyl transferase (cerebroside synthase): cloning, characterization, and assignment to human chromosome 4, band q26. *Genomics* 34: 69–75.

29 Mackenzie, P.I., Gregory, P.A., Gardner-Stephen, D.A. et al. (2003). Regulation of UDP glucuronosyltransferase genes. *Curr Drug Metab* 4: 249–257.

30 Beaulieu, M., Levesque, E., Hum, D.W., and Belanger, A. (1996). Isolation and characterization of a novel cDNA encoding a human UDP-glucuronosyltransferase active on C19 steroids. *J. Biolumin. Chemilumin.* 271: 22855–22862.

31 Levesque, E., Turgeon, D., Carrier, J.S. et al. (2001). Isolation and characterization of the UGT2B28 cDNA encoding a novel human steroid conjugating UDP-glucuronosyltransferase. *Biochemistry* 40: 3869–3881.

32 Bosma, P.J. (2003). Inherited disorders of bilirubin metabolism. *J. Hepatol.* 38: 107–117.

33 Miners, J.O., McKinnon, R.A., and Mackenzie, P.I. (2002). Genetic polymorphisms of UDP-glucuronosyltransferases and their functional significance. *Toxicology* 181–182: 453–456.

34 Muslu, N., Turhan, A.B., Eskandari, G. et al. (2007). The frequency of UDP-glucuronosyltransferase 1A1 promoter region (TA)7 polymorphism in newborns and it's relation with jaundice. *J. Trop. Pediatr.* 53: 64–68.

35 Rastogi, A., Krishnani, N., and Pandey, R. (2006). Dubin-Johnson syndrome – a clinicopathologic study of twenty cases. *Indian J. Pathol. Microbiol.* 49: 500–504.

36 Jansen, P.L. (1999). Diagnosis and management of Crigler–Najjar syndrome. *Eur. J. Pediatr.* 158 (Suppl 2): S89–S94.

37 Seppen, J., Bosma, P.J., Goldhoorn, B.G. et al. (1994). Discrimination between Crigler-Najjar type I and II by expression of mutant bilirubin uridine diphosphate-glucuronosyltransferase. *J. Clin. Invest* 94: 2385–2391.

38 Arias, I.M. (1959). A defect in microsomal function in nonhemolytic acholuric jaundice. *J. Histochem. Cytochem.* 7: 250–252.

39 Chowdhury, J.R., Kondapalli, R., and Chowdhury, N.R. (1993). Gunn rat: a model for inherited deficiency of bilirubin glucuronidation. *Adv. Vet. Sci. Comp. Med.* 37: 149–173.

40 Schmid, R., Axelrod, J., Hammaker, L., and Swarm, R.L. (1958). Congenital jaundice in rats, due to a defect in glucuronide formation. *J. Clin. Invest* 37: 1123–1130.

41 Liu, Y. and Hu, M. (2002). Absorption and metabolism of flavonoids in the Caco-2 cell culture model and a perused rat intestinal model. *Drug Metab. Dispos.* 30: 370–377.

42 Liu, Y., Liu, Y., Dai, Y. et al. (2003). Enteric disposition and recycling of flavonoids and ginkgo flavonoids. *J. Altern. Complement Med.* 9: 631–640.
43 Kessel, D., Botterill, V., and Wodinsky, I. (1968). Uptake and retention of daunomycin by mouse leukemic cells as factors in drug response. *Cancer Res.* 28: 938–941.
44 Gottesman, M.M. and Ling, V. (2006). The molecular basis of multidrug resistance in cancer: the early years of P-glycoprotein research. *FEBS Lett.* 580: 998–1009.
45 Hyde, S.C., Emsley, P., Hartshorn, M.J. et al. (1990). Structural model of ATP-binding proteins associated with cystic fibrosis, multidrug resistance and bacterial transport. *Nature* 346: 362–365.
46 Evans, A.M. (2000). Influence of dietary components on the gastrointestinal metabolism and transport of drugs. *Ther. Drug Monit.* 22: 131–136.
47 Hunter, J., Hirst, B.H., and Simmons, N.L. (1991). Epithelial secretion of vinblastine by human intestinal adenocarcinoma cell (HCT-8 and T84) layers expressing P-glycoprotein. *Br. J. Cancer* 64: 437–444.
48 Buchler, M., Konig, J., Brom, M. et al. (1996). cDNA cloning of the hepatocyte canalicular isoform of the multidrug resistance protein, cMrp, reveals a novel conjugate export pump deficient in hyperbilirubinemic mutant rats. *J. Biolumin. Chemilumin.* 271: 15091–15098.
49 Cole, S.P., Bhardwaj, G., Gerlach, J.H. et al. (1992). Overexpression of a transporter gene in a multidrug-resistant human lung cancer cell line. *Science* 258: 1650–1654.
50 Paulusma, C.C., Bosma, P.J., Zaman, G.J. et al. (1996). Congenital jaundice in rats with a mutation in a multidrug resistance-associated protein gene. *Science* 271: 1126–1128.
51 Belinsky, M.G. and Kruh, G.D. (1999). MOAT-E (ARA) is a full-length MRP/cMOAT subfamily transporter expressed in kidney and liver. *Br. J. Cancer* 80: 1342–1349.
52 Stride, B.D., Grant, C.E., Loe, D.W. et al. (1997). Pharmacological characterization of the murine and human orthologs of multidrug-resistance protein in transfected human embryonic kidney cells. *Mol. Pharmacol.* 52: 344–353.
53 de Jong, M.C., Slootstra, J.W., Scheffer, G.L. et al. (2001). Peptide transport by the multidrug resistance protein MRP1. *Cancer Res.* 61: 2552–2557.
54 Cole, S.P., Sparks, K.E., Fraser, K. et al. (1994). Pharmacological characterization of multidrug resistant MRP-transfected human tumor cells. *Cancer Res.* 54: 5902–5910.
55 Kawabe, T., Chen, Z.S., Wada, M. et al. (1999). Enhanced transport of anticancer agents and leukotriene C4 by the human canalicular multispecific organic anion transporter (cMOAT/MRP2). *FEBS Lett.* 456: 327–331.
56 Paulusma, C.C. and Oude Elferink, R.P. (1997). The canalicular multispecific organic anion transporter and conjugated hyperbilirubinemia in rat and man. *J. Mol. Med.* 75: 420–428.
57 Chen, J., Lin, H., and Hu, M. (2005). Absorption and metabolism of genistein and its five isoflavone analogs in the human intestinal Caco-2 model. *Cancer Chemother. Pharmacol.* 55: 159–169.
58 Jansen, P.L., Peters, W.H., and Lamers, W.H. (1985). Hereditary chronic conjugated hyperbilirubinemia in mutant rats caused by defective hepatic anion transport. *Hepatology* 5: 573–579.
59 Johnson, B.M., Zhang, P., Schuetz, J.D., and Brouwer, K.L. (2006). Characterization of transport protein expression in multidrug resistance-associated protein (Mrp) 2-deficient rats. *Drug Metab. Dispos.* 34: 556–562.
60 Zamek-Gliszczynski, M.J., Xiong, H., Patel, N.J. et al. (2003). Pharmacokinetics of 5 (and 6)-carboxy-2',7'-dichlorofluorescein and its diacetate promoiety in the liver. *J. Pharmacol. Exp. Ther.* 304: 801–809.
61 Anzai, N., Kanai, Y., and Endou, H. (2006). Organic anion transporter family: current knowledge. *J. Pharm. Sci.* 100: 411–426.
62 Zamek-Gliszczynski, M.J., Nezasa, K., Tian, X. et al. (2006b). The important role of Bcrp (Abcg2) in the biliary excretion of sulfate and glucuronide metabolites of acetaminophen, 4-methylumbelliferone, and harmol in mice. *Mol. Pharmacol.* 70: 2127–2133.
63 Sekine, T., Watanabe, N., Hosoyamada, M. et al. (1997). Expression cloning and characterization of a novel multispecific organic anion transporter. *J. Biolumin. Chemilumin.* 272: 18526–18529.
64 Sweet, D.H., Wolff, N.A., and Pritchard, J.B. (1997). Expression cloning and characterization of ROAT1. The basolateral organic anion transporter in rat kidney. *J. Biolumin. Chemilumin.* 272: 30088–30095.
65 Moller, J.V. and Sheikh, M.I. (1982). Renal organic anion transport system: pharmacological, physiological, and biochemical aspects. *Pharmacol. Rev.* 34: 315–358.
66 Hagenbuch, B. and Meier, P.J. (2004). Organic anion transporting polypeptides of the OATP/SLC21 family: phylogenetic classification as OATP/SLCO superfamily, new nomenclature and

molecular/functional properties. *Pflügers Arch.* 447: 653–665.

67 Wang, P., Hata, S., Xiao, Y. et al. (2008). Topological assessment of oatp1a1: a 12-transmembrane domain integral membrane protein with three N-linked carbohydrate chains. *Am. J. Physiol. Gastrointest. Liver Physiol.* 294: G1052–G1059.

68 Weaver, Y. and Hagenbuch, B. (2010). Several conserved positively charged amino acids in OATP1B1 are involved in binding or translocation of different substrates. *J. Membr. Biol.* 236: 279–290.

69 Kullak-Ublick, G., Ismair, M., Stieger, B. et al. (2001). Organic anion-transporting polypeptide B (OATP-B) and its functional comparison with three other OATPs of human liver. *Gastroenterology* 120: 525–533.

70 Bossuyt, X., Muller, M., and Meier, P. (1996). Multispecific amphipathic substrate transport by an organic anion transporter of human liver. *J. Hepatol.* 25: 733–738.

71 Roth, M., Obaidat, A., and Hagenbuch, B. (2012). OATPs, OATs and OCTs: the organic anion and cation transporters of the SLCO and SLC22A gene superfamilies. *Br. J. Pharmacol.* 165: 1260–1287.

72 Hagenbuch, B. and Meier, P. (2003). The superfamily of organic anion transporting polypeptides. *Biochim. Biophys. Acta* 1609: 1–18.

73 Kullak-Ublick, G., Beuers, U., Fahney, C. et al. (1997). Identification and functional characterization of the promoter region of the human organic anion transporting polypeptide gene. *Hepatology (Baltimore, Md)* 26: 991–997.

74 Miki, Y., Suzuki, T., Kitada, K. et al. (2006). Expression of the steroid and xenobiotic receptor and its possible target gene, organic anion transporting polypeptide-A, in human breast carcinoma. *Cancer Res.* 66: 535.

75 Clarke, J.D. and Cherrington, N.J. (2012). Genetics or environment in drug transport: the case of organic anion transporting polypeptides and adverse drug reactions. *Expert Opin. Drug Metab. Toxicol.* 8: 349–360.

76 Wojtal, K., Eloranta, J., Hruz, P. et al. (2009). Changes in mRNA expression levels of solute carrier transporters in inflammatory bowel disease patients. *Drug Metab. Dispos.* 37: 1871–1877.

77 Romaine, S.P.R., Bailey, K.M., Hall, A.S., and Balmforth, A.J. (2010). The influence of SLCO1B1 (OATP1B1) gene polymorphisms on response to statin therapy. *Pharmacogenomics J.* 10: 1–11.

78 Nishizato, Y., Ieiri, I., Suzuki, H. et al. (2003). Polymorphisms of OATP-C (SLC21A6) and OAT3 (SLC22A8) genes: consequences for pravastatin pharmacokinetics. *Clin. Pharmacol. Ther.* 73: 554–565.

79 Deguchi, T., Kusuhara, H., Takadate, A. et al. (2004). Characterization of uremic toxin transport by organic anion transporters in the kidney. *Kidney Int.* 65: 162–174.

80 Shugarts, S. and Benet, L.Z. (2009). The role of transporters in the pharmacokinetics of orally administered drugs. *Pharm. Res.* 26: 2039–2054.

81 Sweet, D.H., Bush, K.T., and Nigam, S.K. (2001). The organic anion transporter family: from physiology to ontogeny and the clinic. *Am. J. Physiol. Renal. Physiol.* 281: F197–F205.

82 Jeong, E.J., Liu, X., Jia, X. et al. (2005a). Coupling of conjugating enzymes and efflux transporters: impact on bioavailability and drug interactions. *Curr Drug Metab* 6: 455–468.

83 Benet, L.Z., Cummins, C.L., and Wu, C.Y. (2004). Unmasking the dynamic interplay between efflux transporters and metabolic enzymes. *Int. J. Pharm.* 277: 3–9.

84 Cummins, C.L., Jacobsen, W., and Benet, L.Z. (2002). Unmasking the dynamic interplay between intestinal P-glycoprotein and CYP3A4. *J. Pharmacol. Exp. Ther.* 300: 1036–1045.

85 Coldham, N.G., Zhang, A.Q., Key, P., and Sauer, M.J. (2002). Absolute bioavailability of [14C] genistein in the rat; plasma pharmacokinetics of parent compound, genistein glucuronide and total radioactivity. *Eur. J. Drug Metab. Pharmacokinet.* 27: 249–258.

86 Norman, A. and Sjövall, J. (1958). On the transformation and enterohepatic circulation of cholic acid in the rat bile acids and steroids 68. *J. Biol. Chem.* 233: 872–885.

87 Brunton, L., Knollman, B., and Hilal-Dandan, R. (2017). *Goodman and Gilman's The Pharmacological Basis of Therapeutics*, *13*e. McGraw-Hill Education.

88 Erlinger, S. (1996). New insights into the mechanisms of hepatic transport and bile secretion. *J. Gastroenterol. Hepatol.* 11: 575–579.

89 Zeng, M., Sun, R., Basu, S. et al. (2016). Disposition of flavonoids via recycling: direct biliary excretion of enterically or extrahepatically derived flavonoid glucuronides. *Mol. Nutr. Food Res.* 60: 1006–1019.

90 Zamek-Gliszczynski, M.J., Hoffmaster, K.A., K-i, N. et al. (2006a). Integration of hepatic drug transporters and phase II metabolizing enzymes: mechanisms of hepatic excretion of sulfate, glucuronide, and glutathione metabolites. *Eur. J. Pharm. Sci.* 27: 447–486.

91 Tu, Y., Wang, L., Rong, Y. et al. (2021). Hepatoenteric recycling is a new disposition mechanism for orally administered phenolic drugs and phytochemicals in rats. *Elife* 10: e58820.

92 Hochner-Celnikier, D. (1999). Pharmacokinetics of raloxifene and its clinical application. *Eur. J. Obstet. Gynecol. Reprod. Biol.* 85: 23–29.

93 Kemp, D.C., Fan, P.W., and Stevens, J.C. (2002). Characterization of raloxifene glucuronidation in vitro: contribution of intestinal metabolism to presystemic clearance. *Drug Metab. Dispos.* 30: 694–700.

94 Sun, D., Jones, N.R., Manni, A., and Lazarus, P. (2013). Characterization of raloxifene glucuronidation: potential role of UGT1A8 genotype on raloxifene metabolism in vivo. *Cancer Prev. Res. (Phila)* 6: 719–730.

95 Jeong, E.J., Liu, Y., Lin, H., and Hu, M. (2005b). Species- and disposition model-dependent metabolism of raloxifene in gut and liver: role of UGT1A10. *Drug Metab. Dispos.* 33: 785–794.

96 Kishi, N., Takasuka, A., Kokawa, Y. et al. (2016). Raloxifene glucuronidation in liver and intestinal microsomes of humans and monkeys: contribution of UGT1A1, UGT1A8 and UGT1A9. *Xenobiotica* 46: 289–295.

97 Jeong, E.J., Lin, H., and Hu, M. (2004). Disposition mechanisms of raloxifene in the human intestinal Caco-2 model. *J. Pharmacol. Exp. Ther.* 310: 376–385.

98 Trdan Lusin, T., Mrhar, A., Stieger, B. et al. (2012). Influence of hepatic and intestinal efflux transporters and their genetic variants on the pharmacokinetics and pharmacodynamics of raloxifene in osteoporosis treatment. *Transl. Res.* 160: 298–308.

99 Rong, Y., Meng, Z., Li, J. et al. (2018). Application of ultra high-performance liquid chromatography tandem mass spectrometry to investigate the regioselective glucuronidation of icaritin in vitro. *J. Pharm. Biomed. Anal.* 154: 444–453.

100 Zhang, B., Chen, X., Zhang, R. et al. (2017). Metabolite profiling, pharmacokinetics, and in vitro glucuronidation of icaritin in rats by ultra-performance liquid chromatography coupled with mass spectrometry. *J. Anal. Methods Chem.* 2017: 1073607.

101 Wang, L., Hong, X., Yao, Z. et al. (2018). Glucuronidation of icaritin by human liver microsomes, human intestine microsomes and expressed UDP-glucuronosyltransferase enzymes: identification of UGT1A3, 1A9 and 2B7 as the main contributing enzymes. *Xenobiotica* 48: 357–367.

102 Qin, Z., Li, S., Yao, Z. et al. (2018). Chemical inhibition and stable knock-down of efflux transporters leads to reduced glucuronidation of wushanicaritin in UGT1A1-overexpressing HeLa cells: the role of breast cancer resistance protein (BCRP) and multidrug resistance-associated proteins (MRPs) in the excretion of glucuronides. *Food Funct.* 9: 1410–1423.

# 24

# Systemic Versus Local Bioavailability Enabled by Recycling

*Yifan Tu, Lu Wang, and Ming Hu*

*College of Pharmacy, University of Houston, Houston, TX, USA*

## 24.1 Introduction

Bioavailability is a parameter used to describe the rate and the extent of administrated drug that reaches the systemic circulation. Bioavailability also determines the *in vivo* exposure of a drug, which is often directly associated with the efficacy of an administrated drug. The value of bioavailability varies from 0 to 1 (100%). For intravenous (IV) injected drugs, the bioavailability is set at 100%. For other routes of administration, such as oral, ocular, rectal, transdermal, subcutaneous, or sublingual administration, the bioavailability values are generally lower than that of IV. Mathematically, absolute bioavailability is calculated by using the area under drug blood concentration (AUC) versus time of extravascular administration divided by AUC of the IV administration. Other important factors that determine a drug's exposure are the maximal systemic drug concentration ($C_{max}$) and time ($T_{max}$) when concentration reaches $C_{max}$. The calculation of bioavailability is summarized in Table 24.1.

**Table 24.1** Calculation of absolute bioavailability and relative bioavailability.

| Absolute bioavailability | Relative bioavailability |
|---|---|
| $F_{abs} = 100\% * \frac{AUC_{ref}}{AUC_{I.V.}} * \frac{Dose_{I.V.}}{Dose_{ref}}$ | $F_{rel} = 100\% * \frac{AUC_{ref\ 1}}{AUC_{ref\ 2}} * \frac{Dose_{ref\ 2}}{Dose_{ref\ 1}}$ |

There are many factors governing absorption, distribution, metabolism and elimination, and they may influence the bioavailability of a drug. These factors could be classified as drug's physicochemical properties (e.g. solubility and p$K_a$ values) and biological and physiological parameters (e.g. epithelial permeability) that impact a drug's bioavailability. In addition, there are several important physiological processes that facilitate local recycling and enhance local gut bioavailability. As a result, they may also influence systemic bioavailability. One of them is enterohepatic recycling (EHR). Recycling is an important drug disposition process that enables recycling in the digestive (intestine–liver) system.

In this chapter, we will discuss how two types of EHR could influence bioavailability. First, we will introduce the definition of bioavailability. We will give a brief introduction on the differences between systemic bioavailability and local gut bioavailability. Next, we will discuss three different types of digestive system recycling processes and focus on the discussion of how recycling could influence bioavailability. Lastly, we will discuss the clinical significance of the local gut bioavailability enabled by recycling.

## 24.2 Systemic Bioavailability

Systemic bioavailability, or bioavailability, is defined as the rate and extent of an administrated drug that enters the systemic circulation [1]. The value ranges from 0 to 1 (100%). By definition, drugs administrated from intravascular injection have 100% bioavailability. Other dosing routes, such as oral, ocular, rectal, transdermal, subcutaneous, and sublingual administration, usually have less than 100% bioavailability. Low bioavailability is usually caused by poor absorption in gastrointestinal tract and/or extensive first-pass metabolism in small intestine and/or liver.

There are two types of systemic bioavailability: absolute bioavailability and relative bioavailability. Absolute bioavailability is calculated by comparing the AUC between extravascular administrated drug and the same drug administrated by intravascular injection [2, 3]. The comparison must be normalized by dose. However, absolute bioavailability is not quite commonly used because the calculation of absolute bioavailability

*Oral Bioavailability and Drug Delivery: From Basics to Advanced Concepts and Applications,* First Edition. Edited by Ming Hu and Xiaoling Li.

requires the AUC of IV-injected drugs, which is not applicable for all drugs due to safety, cost, or practical considerations. For example, drugs with high toxicity or low solubility are not able to be directly administrated by IV injection. Therefore, another parameter called relative bioavailability is usually applied for the evaluation of a drug's *in vivo* exposure.

Relative bioavailability is calculated by comparing the bioavailability (AUC is commonly used in calculation) of certain administration route (or formulation) with a reference administration route (or formulation) of the same drug (Research, 2014). Normally, the compared route (or formulation) is a well-established standard. Relative bioavailability is one of the factors used to evaluate the bioequivalence of two or more different drug products.

## 24.3 Local Bioavailability

Systemic bioavailability describes the rates and extent of drugs that entered the systemic circulation. To reach the systemic circulation, the drug molecules must penetrate the small intestinal barrier and escape from small intestinal and hepatic metabolism to finally become "bioavailable" in the systemic circulation. They could be further sent to their targeting organs via the systemic circulation. Different from systemic bioavailability, local bioavailability is a term that describes drugs that are only bioavailable in specific organs or tissues. These drugs may come from small intestine absorption or systemic circulation, but they are "trapped" or "concentrated" in local organs or tissues, which is usually not correlated with their presence in the systemic circulation. With an unexpected higher concentration in certain organ(s) and tissue(s), the local bioavailability is usually limited and not much attention is paid to their occurrence. Unlike systemic bioavailability, the calculation of local bioavailability could not be done by comparing the blood AUC. It is nearly impossible to measure local AUC as blood samples normally are taken from the systemic circulation. A more practical approach to evaluate local bioavailability is to measure the amount of drug in a tissue or a biological matrix that represents the local tissue (e.g. saliva drug concentration for buccal local bioavailability).

For this particular chapter, the local bioavailability we refer to is the local gut bioavailability, where the gut segment in question is likely the terminal ileum and colon. Intestine–liver system is such a classical local system that "traps" certain drugs between intestine and liver. These orally administrated drugs are absorbed in small intestine and recycled within intestine and between intestine and liver in the form of original compounds and their phase II metabolites. To evaluate the local gut bioavailability of the drug in the "intestine–liver" system, the drug concentration in bile could be a good biomarker.

In order to "trap" drugs in local system instead of letting them enter the systemic circulation, there exists a special mechanism that facilitates the process. Recycling is such an important mechanism that enables the recycling between liver and intestine. Recycling not only "traps" absorbed compounds in gut or within the intestine–liver system, but it also "steals" drugs from the systemic circulation and keeps them in local intestine–liver system.

## 24.4 Factors Affecting Bioavailability

Bioavailability could be affected by many factors [4–7]. Generally, all the factors could be classified into two types: drug's physicochemical properties or biological/physiological properties. Factors that may influence drug bioavailability are summarized in Table 24.2. Drug properties include physical and chemical properties like solubility, stability, hydrophilicity, drug formulation, and potential interaction with other drugs/food (cause enzyme induction or inhibition). These properties will influence drug absorption, distribution, metabolism, and elimination, which affect drug bioavailability. Physiological conditions are even more complicated. Simple factors like fasted/fed conditions and gastric emptying rate are relatively straightforward. Other factors like circadian rhyme differences, individual variation, and disease state may change the enzyme expression level or enzyme activity level. These enzymes include metabolic enzymes (like CYPs, 5′-diphospho-glucuronosyltransferase (UGTs), and sulfotransferases (SULTs)) and transporters (uptake and efflux transporters like organic anion transporting polypeptides (OATPs), P-gp, and multidrug resistance-associated proteins (MRPs)). The changes in their activity or expression level would also affect the drug absorption, distribution, metabolism and elimination, which change drug bioavailability accordingly.

In addition to those factors mentioned previously, several important physiological or biological processes are also affecting bioavailability. One of the important processes is called EHR [8]. The efficiency of EHR could greatly change the fraction of orally administrated drugs entering systemic circulation. We recently proposed hepatoenteric recycling (HER) as an alternative to

**Table 24.2** Factors that influence drug bioavailability.

| Drug physicochemical properties | Impact |
|---|---|
| Stability | Drug absorption |
| Solubility | Drug absorption |
| Hydrophilicity | Drug absorption |
| Drug formulations | Drug administration routine/absorption |
| Potential interaction (with food/other drugs) | Absorption or metabolism enzyme inhibition/induction |
| Physiological conditions | Impact |
| Fasted/fed condition | Oral drug absorption difference |
| Gastric emptying rate | Oral drug absorption difference |
| Circadian rhyme | Absorption or metabolism enzyme activity |
| Individual variation | Enzyme expression level and activity |
| Disease state | Enzyme expression level and activity |

EHR, and Figures 24.1 and 24.2 summarize two major recycling processes. Briefly, compounds with higher recycling efficiency are more likely to stay in the digestive system (i.e. intestine and liver) instead of entering systemic circulation, primarily due to the action of efficient hepatic uptake and bile excretion. Metabolized in either liver or intestine, quite large amounts of compounds recycle between liver and intestine in the form of their phase II metabolites. Even those entered systemic circulation could also be recaptured when they are taken up by the liver and reintroduced into the recycling process. Thus, higher recycling efficiency would reduce the fraction of drugs entering systemic circulation, even "steal" drugs from systemic circulation. Therefore, drugs with high her/HER efficiency are prone to have lower systemic bioavailability.

Local bioavailability (i.e. colon and terminal ileum) could also be greatly affected by all the factors mentioned previously. The change in drug absorption, distribution, metabolism, and eliminiation, would influence both systemic bioavailability and local bioavailability. More importantly, her/HER plays a predominant role in affecting local bioavailability, because the local recycling of drugs in intestine–liver system completely depends on her/HER. Therefore, the efficiency of her/HER directly decides drug's local bioavailability in the terminal ileum and colon.

## 24.5 Enterohepatic Recycling (EHR)

EHR, known since 1958 for bile acid recycling [9], allows a compound to be absorbed in small intestine and return to small intestine via bile secretion [10]. It is a recycling of biliary acids, bilirubin, drugs, or other substances from liver to bile, followed by reentry into small intestine. It starts from small intestine absorption and ended up in reabsorption at the lower part of the intestine (i.e. colon). The study of EHR could be traced back to 1950s, initially conceptualized for bile acid recirculation [9].

In the 1960s, this concept was subsequently used for drugs and xenobiotics that were recycled in addition to the recycling of bile acids [11, 12]. These drugs and xenobiotics would be absorbed in small intestine and metabolized via phase II metabolism (a detoxification mechanism). Finally, they would end as phase II metabolites, mostly bio-transformed via phase II enzymes such as uridine UGTs and SULTs. These phase II metabolites would be further excreted into bile and returned to small intestine. Although most bile acid conjugates are absorbed intact from the intestine and then transported into the hepatocytes as conjugates, there is no evidence indicating that glucuronides and sulfates, two of the most common phase II metabolites, can be efficiently absorbed intact from the intestine. Hence, we typically called this process (involving glucuronides) recycling not recirculation. With the participation of gut flora, these metabolites would be hydrolyzed and "regenerated" into their aglycones forms. These regenerated aglycone could be further reabsorbed in intestine and complete a loop of recycle. The word "recirculation" was changed to "recycling" since the aglycone must be regenerated from its metabolite by microflora hydrolysis in terminal ilium and colon.

Despite the involvement of phase II enzymes, the participation of transporters is essential to facilitate recycling. As shown in Fig 24.3 and Fig. 24.4, transporters expressed in enterocytes and hepatocytes

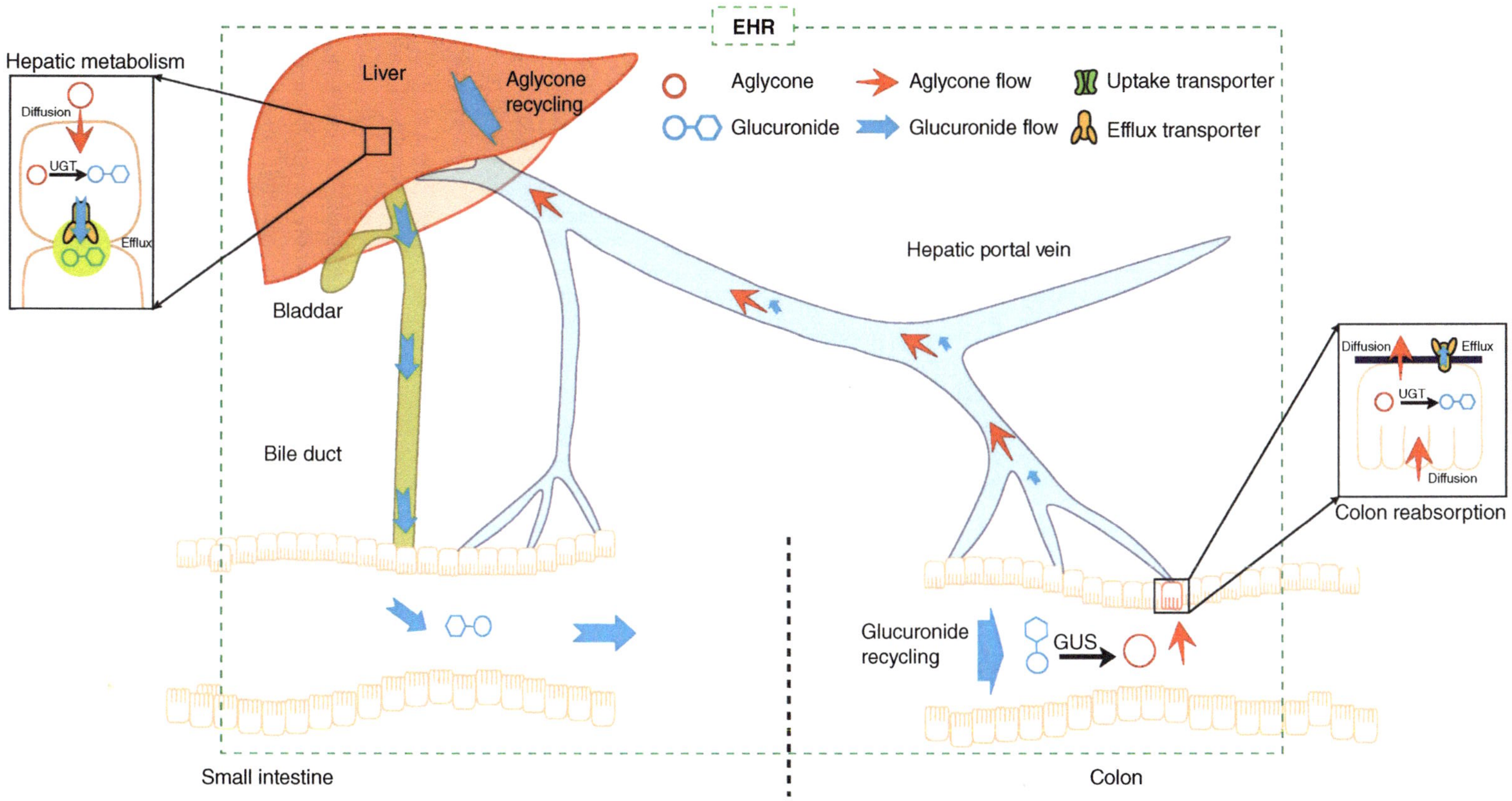


**Figure 24.1** Enterohepatic recycling (EHR) starts from hepatic metabolism, where the aglycones (parent compounds) are converted to their phase II conjugates (mostly, glucuronides). This figure indicates the process of HER. Red arrows shows the flow of aglycones and blue arrow shows the flow of glucuronides. These glucuronides will further be excreted into bile and return to intestine via bile secretion. In downstream intestine (terminal ileum and colon), glucuronides will be converted to their aglycone form by β-glucuronidases (GUS) of intestinal bacteria. These converted aglycones will be reabsorbed in intestine to complete the recycling loop.

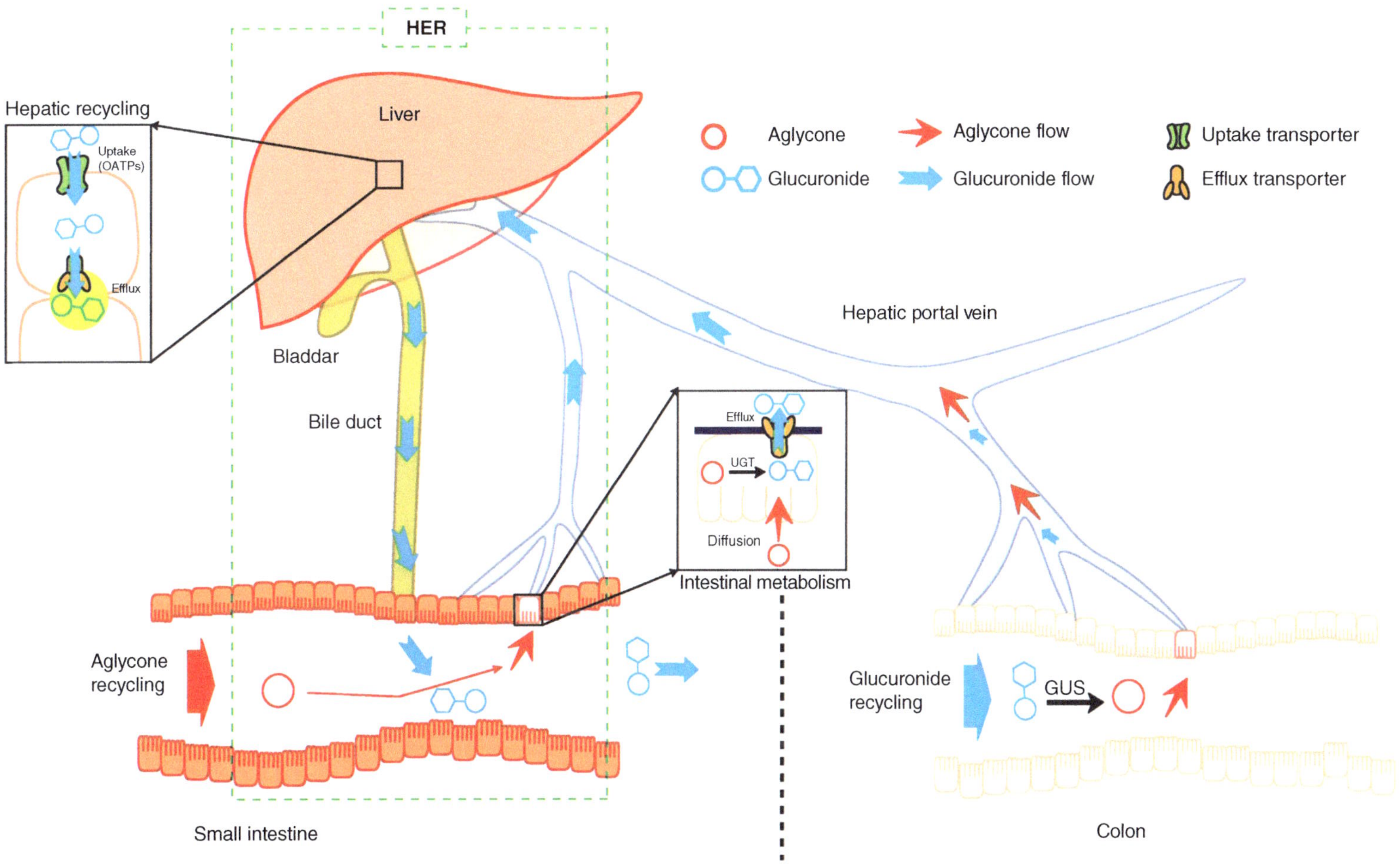


**Figure 24.2** Hepatoenteric recycling (HER) starts from intestinal metabolism, where the aglycones (parent compounds) are converted to their phase II conjugates (mostly, glucuronides). This figure indicates the process of HER. Red arrows shows the flow of aglycones and blue arrow shows the flow of glucuronides. These glucuronides will enter liver via hepatic portal vein and are taken up by hepatocytes in liver. Without further hepatic metabolism, the glucuronides would be excreted into bile and return to intestine via bile secretion. In downstream intestine (terminal ileum and colon), glucuronides will be converted to their aglycone form by β-glucuronidases (GUS) of intestinal bacteria. These converted aglycones will be reabsorbed in intestine to complete the recycling loop.

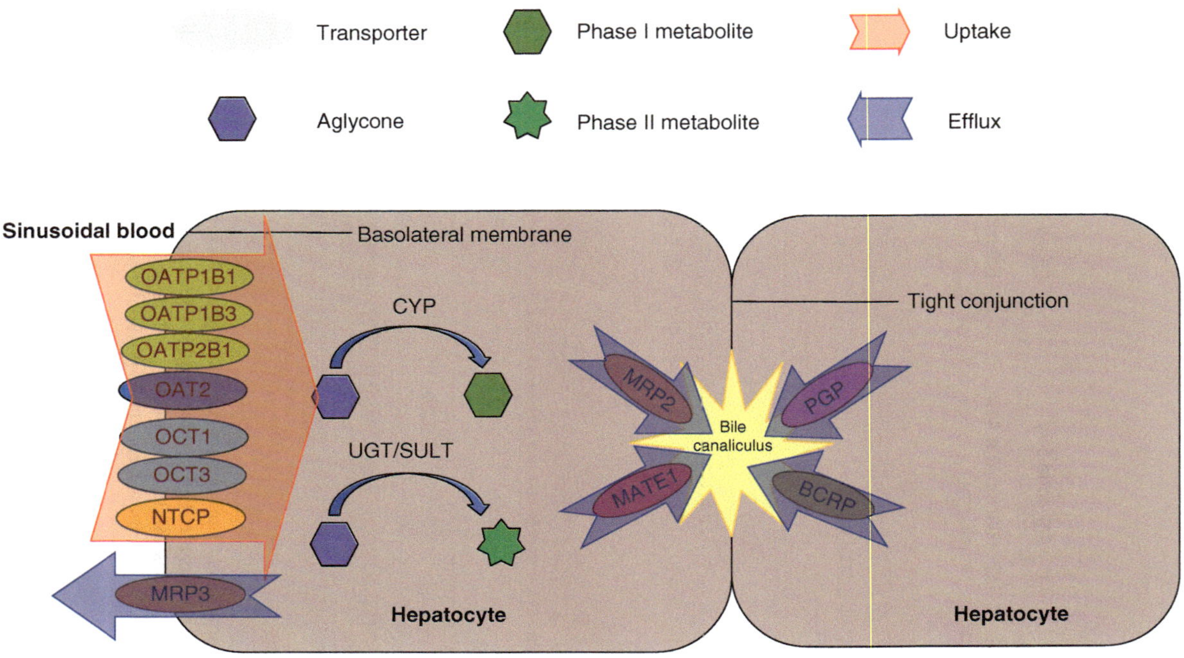


**Figure 24.3** Uptake and efflux transporters expressed in hepatocytes. This figure indicates the localization of uptake and efflux transporters in hepatocytes. There are many uptake transporters expressed on the basolateral side of hepatocytes, which mediate the active uptake from hepatic blood. There are also efflux transporters expressed on both apical and basolateral sites of hepatocytes, which excrete their substrates into hepatic blood flow or bile.

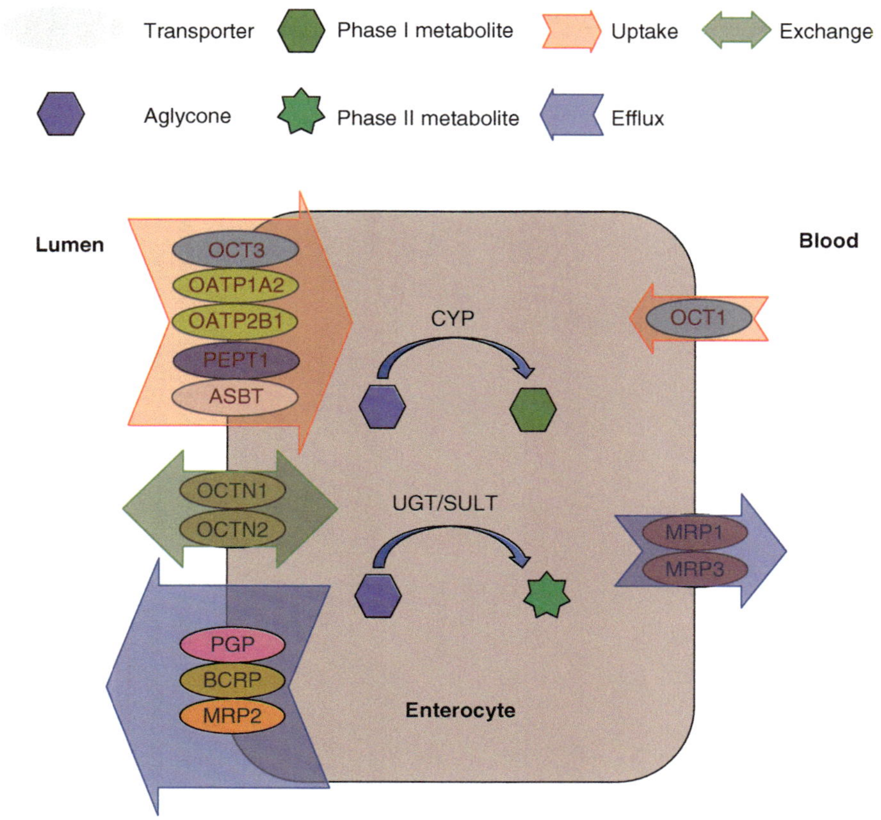


**Figure 24.4** Uptake and efflux transporters are expressed in enterocytes. This figure indicates the localization of uptake and efflux transporters in enterocytes. There are many uptake transporters expressed on both apical and basolateral sides of enterocytes, which mediate the active uptake from both sides. There are also efflux transporters expressed on both apical and basolateral sites of hepatocytes, which excrete their substrates into lumen or blood flow. There are also exchangers expressed on the basolateral side, which mediate the exchange of ions and chemicals.

**Table 24.3** Drugs that undergo enterohepatic recycling.

| Anticancer | Anticholesterol | Antiosteoporosis | Analgesics | Antihypertension | NSAIDs | Others |
|---|---|---|---|---|---|---|
| Sorafenib | Ezetimibe | Raloxifene | Morphine | Metolazone | Diclofenac | Mycophenolic acid |
| SN-38 | Cholestyramine | | Methadone | | | Progesterone |
| | Atorvastatin | | | | | Vitamin A |
| | | | | | | Estradiol acetate |

mediated the recycling process. These transporters are special membrane proteins that are expressed on the apical side (bile duct side) of hepatocytes [13, 14]. They are responsible for the active transporting of substrates from hepatocytes to bile duct canaliculus by consuming ATP. There are MRP 2, breast cancer resistant protein (BCRP), multidrug resistance protein 1 (MDR 1, commonly called P-glycoprotein), and multidrug resistance protein 3 (MDR 3) existing in hepatocytes, and they are responsible for the efflux of both endogenous and exogenous compounds. There is one more efflux transporter called bile salt export pump (BSEP), which mostly mediates the efflux of bile salts and bile acids. Currently, a variety of drugs, including anticancer (e.g. sorafenib, SN-38) [15, 16], anticholesterol (e.g. ezetimibe) [17], anti-osteoporosis (e.g. raloxifene) [18], analgesics (e.g. morphine) [19] and others (e.g. mycophenolic acid for organ transplant) [20], are known to undergo EHR. Some examples are summarized in Table 24.3.

## 24.6 Hepatoenteric Recycling (HER)

The definition of EHR does not specify where these metabolites are formed before they are excreted into bile. It is postulated that the glucuronides found in bile are mainly formed in liver, since bile acids are known to be synthesized in liver. Many microsomes studies have shown that the glucuronidation rate in liver was much faster than intestine. Therefore, liver is considered as the most important metabolism organ in EHR. Most of the drug conjugates found in bile are presumed to be formed in liver, just like bile acid conjugates, and then further excreted into bile.

However, certain phenolic compounds are rapidly metabolized in the small intestine before they reach liver. Some of these metabolites (mostly glucuronides) are further efficiently taken up by hepatocytes and then rapidly excreted into the bile, whereas other phenolic metabolites are not taken up by the hepatocytes. Considering the hydrophilicity of these glucuronides, basolateral hepatic uptake transporters like OATPs are required for them to enter hepatocytes. Thus, the small intestine could be another metabolites generation organ with function classically assigned to liver in the EHR. Therefore, HER is another recycling scheme within the "intestine–liver" system and can be used to describe recycling of phase II metabolites that are circulating in the blood, which are likely to be generated extrahepatically (e.g. in the intestine and kidney).

Metabolism enzymes like UGTs and SULTs are still required in HER, just like they were required in EHR. Different from EHR, compounds were metabolized mainly in small intestine instead of in liver. Phase II metabolites instead of aglycones further exit the intestinal epithelial via basolateral efflux transporters. When some of these metabolites entered liver via hepatic portal vein, they are taken up by hepatocytes and excreted into bile, whereas other phase II metabolites are not taken up by the hepatocytes. Even for compounds that are not absorbed in small intestine but only available in the systemic circulation, they are also capable of being taken up by hepatocytes and entering HER. Clearly, the uptake transporters existing on the basolateral side of hepatocytes played a significant role in HER.

Same as the efflux transporters, the uptake transporters are a class of special membrane proteins that express on the basolateral side of hepatocytes. They are responsible for taking up compounds from blood and transport them into the hepatocytes. The major basolateral uptake transporters expressed in liver are OATPs 1B1/1B3, organic anion transporters (OATS) 2, organic cation transporters (OCTs) 1/3, and $Na^+$-taurocholate co-transporting polypeptides (NTCPs, also known as liver bile acid transporter, LBAT). OATPs are the most important uptake transporters mediating the uptake of xenobiotics. Several reports have shown that polymorphisms in OATPs will influence the pharmacokinetics of drugs and their glucuronides [1, 3, 11, 15]. One study showed that fecal excretion of ezetimibe was decreased in patients with lower functional activity of OATP1B1 [11], suggestive of its likely role in ezetimibe's recycling. Therefore, the active transporting mediated by uptake transporters is a critical step in HER.

## 24.7 Enteroenteric Recycling (EER)

For many orally administered phenolic compounds that are extensively metabolized to phase II conjugates (i.e. glucuronides or sulfates), the metabolites are often excreted directly from enterocytes into intestinal lumen. In intestine, they will join bile-secreted phase II conjugates as described above to move to the terminal ileum and colon. These metabolites will further be exposed to microbial hydrolases that can reconvert conjugates into aglycones. The aglycones could be reabsorbed to complete the loop of recycling within intestine, which we used to call this process enteroenteric recycling (EER) [21, 22]. Although initially described for the intestinal recycling of phenolics, it can also be used to describe any recycling within a particular organ, including oral cavity [23].

## 24.8 Summary

Bioavailability indicates drug *in vivo* exposure in the systemic circulation. Local bioavailability is another parameter that reflects drug *in vivo* exposure available in local tissues or organs (like intestine–liver system). These two parameters are greatly affected by a drug's physicochemical properties and biological/physiological conditions of the subjects administered with the drug.

In addition to these factors, there are several important biological/physiological processes that facilitate the local recycling and impact the systemic bioavailability. EHR is one of the processes that influences systemic bioavailability and directly impacts the local bioavailability. Generally speaking, hepatic metabolism is the critical step in her, for it generates phase II metabolites. However, some phenolic glucuronides generated extrahepatically are directly taken up by hepatocytes. These glucuronides may be generated before reaching hepatocytes and participate in EHR. This recycling process is termed as HER. In this route, drugs are metabolized in small intestine, taken up, and recycled by the liver.

In EHR and HER, the roles of small intestine and liver are switched. Liver serves as metabolism organ, and small intestine serves as recycling organ in EHR, while they work the other way around in HER. The delineation of the disposition site in EHR/HER helps understand the accurate metabolism site of a drug. For drugs disposed by HER, patients with intestine diseases should take special caution. For drugs disposed by EHR, patients with liver diseases should take special caution. The disease states in liver or intestine would change the metabolism of drug and influence its bioavailability.

The understanding of EHR and HER helps provide an accurate evaluation of systemic and local bioavailability. It gives a more accurate drug dosing management in clinical situation for patients with hepatic or intestinal diseases. Furthermore, it provides a better understanding of drug *in vivo* efficacy for those drugs that have active metabolites.

## References


1 Price, G. and Patel, D.A. (2020). Drug bioavailability. In: *StatPearls*, Treasure Island (FL).

2 Hinderling, P.H. (1997). Commentary 'Prediction of absolute bioavailability for drugs using oral and renal clearance following a single oral dose: a critical review' (I. Mahmood). *Biopharm. Drug Dispos.* 18: 821–824.

3 Research CfDEa (2014). Bioavailability and Bioequivalence Studies Submitted in NDAs or INDs---General Considerations.

4 Lach, J.L. (1972). Some factors affecting bioavailability of drug products. *J. Iowa Med. Soc.* 62: 441–442. passim.

5 McGilveray, I.J. (1975). Some factors affecting bioavailability. *Int. J. Clin. Pharmacol. Biopharm.* 11: 340–341.

6 Nielsen, I.L. and Williamson, G. (2007). Review of the factors affecting bioavailability of soy isoflavones in humans. *Nutr. Cancer* 57: 1–10.

7 Walker, W.A. (2002). Developmental issues: physiologic factors affecting the safety and bioavailability. *Nutr. Today* 37: 174–175.

8 Adlercreutz, H., Martin, F., Jarvenpaa, P., and Fotsis, T. (1979). Steroid absorption and enterohepatic recycling. *Contraception* 20: 201–223.

9 Norman, A. and Sjovall, J. (1958). On the transformation and enterohepatic circulation of cholic acid in the rat: bile acids and steroids 68. *J. Biol. Chem.* 233: 872–885.

10 Roberts, M.S., Magnusson, B.M., Burczynski, F.J., and Weiss, M. (2002). Enterohepatic circulation: physiological, pharmacokinetic and clinical implications. *Clin. Pharmacokinet.* 41: 751–790.

11 Hartwig, Q.L. (1961). Interrupted enterohepatic circulation of bile and its effect on urinary and biliary excretion of I-131. *Proc. Soc. Exp. Biol. Med. Soc. Exp. Biol. Med.* 107: 502–505.

12 Lester, R., Ostrow, J.D., and Schmid, R. (1961). Enterohepatic circulation of bilirubin. *Nature* 192: 372.

13 Huang, S.M., Strong, J.M., Zhang, L. et al. (2008). New era in drug interaction evaluation: US Food and Drug Administration update on CYP enzymes, transporters, and the guidance process. *J. Clin. Pharmacol.* 48: 662–670.

14 Nakanishi, T. and Tamai, I. (2015). Interaction of drug or food with drug transporters in intestine and liver. *Curr. Drug Metab.* 16: 753–764.

15 Edginton, A.N., Zimmerman, E.I., Vasilyeva, A. et al. (2016). Sorafenib metabolism, transport, and enterohepatic recycling: physiologically based modeling and simulation in mice. *Cancer Chemother. Pharmacol.* 77: 1039–1052.

16 Younis, I.R., Malone, S., Friedman, H.S. et al. (2009). Enterohepatic recirculation model of irinotecan (CPT-11) and metabolite pharmacokinetics in patients with glioma. *Cancer Chemother. Pharmacol.* 63: 517–524.

17 Soulele, K. and Karalis, V. (2019). On the population pharmacokinetics and the enterohepatic recirculation of total ezetimibe. *Xenobiotica* 49: 446–456.

18 Hochner-Celnikier, D. (1999). Pharmacokinetics of raloxifene and its clinical application. *Eur. J. Obstet. Gynecol. Reprod. Biol.* 85: 23–29.

19 Horton, T.L. and Pollack, G.M. (1991). Enterohepatic recirculation and renal metabolism of morphine in the rat. *J. Pharm. Sci.* 80: 1147–1152.

20 Colom, H., Lloberas, N., Andreu, F. et al. (2014). Pharmacokinetic modeling of enterohepatic circulation of mycophenolic acid in renal transplant recipients. *Kidney Int.* 85: 1434–1443.

21 Chen, J., Lin, H., and Hu, M. (2003). Metabolism of flavonoids via enteric recycling: role of intestinal disposition. *J. Pharmacol. Exp. Therap.* 304: 1228–1235.

22 Hu, M., Chen, J., and Lin, H. (2003). Metabolism of flavonoids via enteric recycling: mechanistic studies of disposition of apigenin in the Caco-2 cell culture model. *J. Pharmacol. Exp. Therap.* 307: 314–321.

23 Mallery, S.R., Budendorf, D.E., Larsen, M.P. et al. (2011). Effects of human oral mucosal tissue, saliva, and oral microflora on intraoral metabolism and bioactivation of black raspberry anthocyanins. *Cancer Prev. Res.* 4: 1209–1221.

# 25

# Intestinal Microbiome and Its Impact on Metabolism and Safety of Drugs

*Xin Y. Chu*[1] *and Paul C.L. Ho*[1,2]

[1]*Department of Pharmacy, Faculty of Science, National University of Singapore, Singapore, Singapore*
[2]*School of Pharmacy, Monash University Malaysia, Kuala Lumpur, Malaysia*

## 25.1 Introduction

The human intestinal microbiome is made up of a large population of microbes, with an estimate of over 100 trillion microbes located in the gastrointestinal tract [1, 2]. The quantity and variety of microbes increase along the stomach, duodenum, jejunum, ileum, and colon, with an estimated number of $10^1$, $10^3$, $10^4$, $10^7$, and $10^{12}$ respectively [3, 4]. Therefore, drugs that reach the lower GI tract or undergo enterohepatic circulation have the greatest tendency to interact with the microbes in the gut [5]. Given that many new drug candidates tend to have low solubility and/or permeability, which allows more interaction time in the GI tract [6], it is essential to examine the influence of intestinal microbiome on drugs. Microbes colonize the gut since birth and its composition generally becomes stable once adulthood is reached [7, 8]. Nonetheless, our bacteria makeup can be altered by diet, use of drugs (especially antibiotics), and genetics [9, 10].

Intestinal bacteria is essential for catabolizing nondigestible carbohydrates into short-chain fatty acids (SCFAs) [11, 12], metabolism of endogenous compounds including bile acids and cholesterol [13, 14], synthesis of vitamins (B and K) [15], defense against pathogens, and regulation of immune system [16, 17]. The importance of intestinal microbiome on drug metabolism and safety has only been recognized in recent decades and is rapidly gaining much interest. Indeed, intestinal microbiome can directly transform drugs to bioactive, inactive, or toxic metabolites via microbial enzymes in the intestinal tract [18]. Drug metabolism in the liver and intestinal bacteria differs in that intestinal bacteria typically carry out reductive and hydrolytic reactions producing hydrophobic metabolites whereas metabolism in the liver mainly involves oxidation and conjugation producing hydrophilic metabolites [5]. It can also affect the disposition of a drug via indirect mechanisms, for instance, by altering gene expression that modulates host metabolizing enzymes or producing microbial metabolites that interfere with metabolism of drugs [19–21]. The modulation of gut bacteria can have significant clinical implications on the safety, activity, and bioavailability of drugs.

The diversity of microbes, and the fact that the interplay between the intestinal microbiome is extremely complex, emphasize the importance of extensive studies encompassing any new advances and developments in this area. Hence, there is a need for continuous effort to elucidate the impact on other drugs. This review aims to summarize and evaluate the impact of intestinal microbiome on drug metabolism and its therapeutic significance.

## 25.2 Direct Metabolism by Intestinal Microbiome

### 25.2.1 Conversion to Bioactive Metabolites

One of the earliest antimicrobial drugs, Prontosil and Neoprontosil, exert their antibacterial activity via conversion to the active sulphanilamide metabolite by bacteria azoreductases [22, 23]. Previously, it had been assumed that the liver was the main site of reduction of these drugs by hepatic azoreductases but it was later demonstrated that intestinal microbiome also contributed significantly to this conversion. Administration of antibiotics also suppressed the conversion to their active metabolite in rats. Prontosil is a lipid-soluble drug that is readily absorbed in the intestine after oral dosing and can be metabolized by either the gut bacteria before absorption or liver after absorption. On the

*Oral Bioavailability and Drug Delivery: From Basics to Advanced Concepts and Applications,* First Edition. Edited by Ming Hu and Xiaoling Li.

other hand, Neoprontosil is a water-soluble drug that is hardly absorbed through the intestine. When administered orally, most of the drug would be metabolized by gut bacteria to sulphanilamide. When administered intraperitoneally, it was found that nearly 70% of the drug was excreted unchanged into the bile, which could be converted to the active sulphanilamide in the gut. This means that Neoprontosil is mainly metabolized to the active form by the gut flora regardless of oral or intraperitoneal administration [23].

Sulfasalazine, which has been used for ulcerative colitis, is poorly absorbed from the small intestine [24, 25]. Instead, it acts as a carrier molecule, which reaches the colon and undergoes azoreduction to form sulphapyridine and the active moiety, 5-aminosalicyclic acid (5-ASA), which has a local anti-inflammatory effect on the inflamed mucosa [24, 26, 27]. Majority of sulphapyridine was found to be systemically absorbed [27, 28], and accounted for most of the adverse effects [28]. To circumvent this problem, a newer prodrug Balsalazide was synthesized with an azo linkage of inactive 4-aminobenzoyl-*b*-alanine and 5-ASA [29]. Among the bacterial species, the *Clostridium* species possessed the highest amount of azoreductase activity. The rate of metabolism for sulfasalazine was also higher than that of balsalazide and olsalazine. A higher amount of 5-ASA was also retrieved from Olsalazine as it consists of an azo linkage of two 5-ASA molecules [30]. Therefore, this will eventually have an effect on the pharmacokinetics and dosing regimens of the drug. In addition, the intestinal microflora is capable of acetylating 5-ASA to form the inactive acetyl-5-ASA [31, 32]. *N*-Acetyltransferase activity was detected mainly in obligate anaerobes, including *Bacteroides and Clostridium difficile*, and the enzyme activity could vary up to 10-fold between individuals [33]. This indicates that the considerable interindividual variability in gut bacteria metabolism can potentially contribute to variations in drug efficacy. Figure 25.1 shows the metabolism of sulfasalazine, olsalazine, and balsalazide to active 5-ASA and their side products by intestinal azoreductases [34].

Lovastatin is another prodrug where its lactone ring undergoes hydrolysis, generating the active β-hydroxy acid metabolite by gut bacteria, as shown *in vitro* and

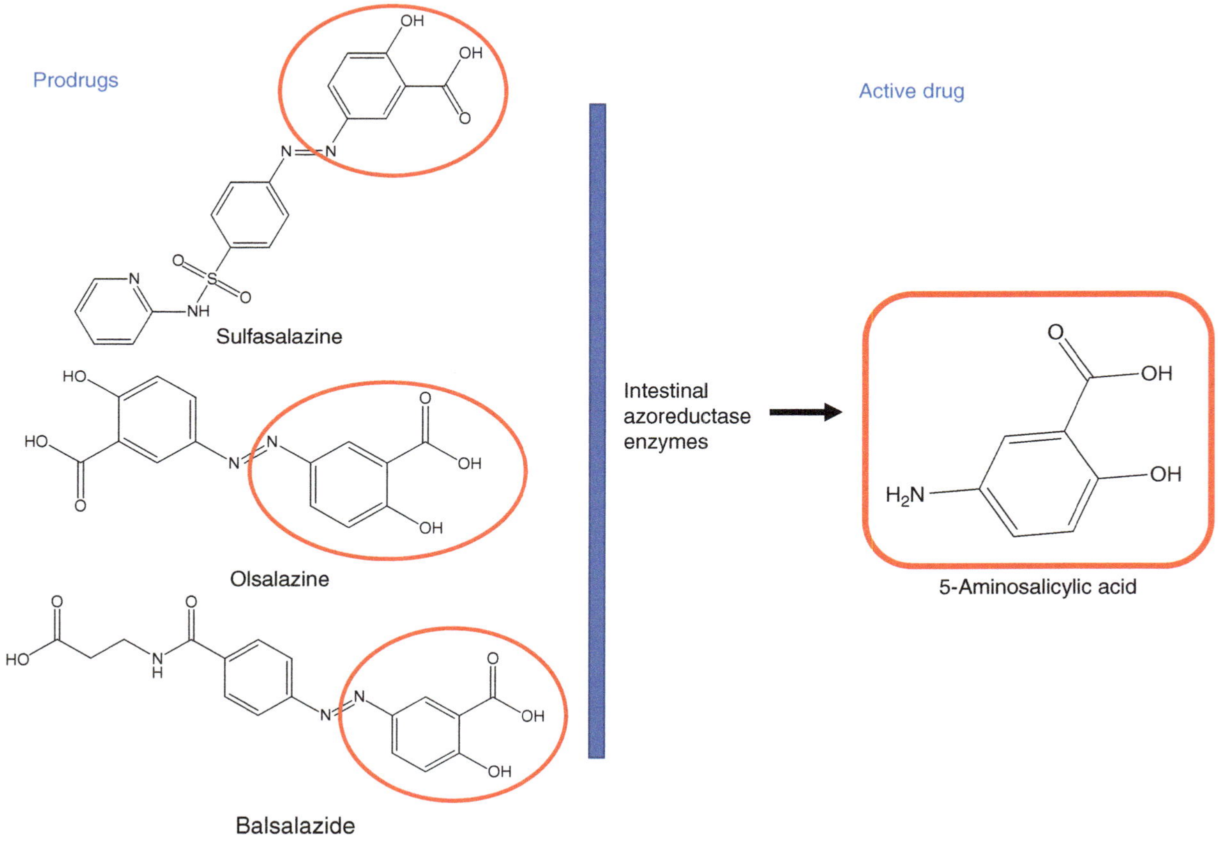


**Figure 25.1** Metabolism of the prodrugs sulfasalazine, olsalazine, and balsalazide to their active drug 5-ASA by the intestinal azoreductase enzymes.

*in vivo*. A significantly lower area-under-curve (AUC) of the active hydroxy metabolite was also observed in antibiotics-treated rats. Hence, apart from liver microsomal enzymes, gut bacteria can also produce a considerable amount of the active metabolite. While certain antibiotics can inactivate CYP450 enzymes, resulting in rise in lovastatin plasma levels, the effect of antibiotics on gut bacteria can lower the efficacy of lovastatin [35]. Therefore, gut bacteria interaction with antibiotics as a potential mediator of drug–drug interaction certainly warrants more attention.

Sulindac is converted to the active sulfide metabolite via bacterial sulphoxide reduction. It was estimated that about half of the active metabolite recovered was produced by gut bacteria [36]. While considerable reduction occurs in the liver, the enzymatic activity in caecum contents was significantly higher than that of liver [37]. Similarly, sulphinpyrazone, a drug used for gout and thromboembolic disorders, is metabolized to the active sulfide metabolite by gut bacteria. In contrast to sulindac, it was found that gut bacteria is the major, if not the only site of conversion to its active metabolite [38, 39].

In murine models, thioguanine when given orally improved ulcerative colitis even with the knockout of host hypoxanthine-guanine phosphoribosyltransferase (Hprt), a key purine salvage enzyme mediating the conversion of thioguanine to active thioguanine nucleotides (TGN) and colonic bacteria was shown to be capable of this conversion. Therefore, it is hypothesized that rectal thioguanine can reduce intestinal inflammation locally through gut bacterial conversion without causing liver toxicity associated with oral thioguanine. In addition, unlike mercaptopurine, conversion of thioguanine to TGN by the purine enzyme, inosine monophosphate dehydrogenase (IMPDH) is not rate limited by the host or bacteria, hence allowing a rapid therapeutic action [40, 41].

Lactulose, which is used for hepatic encephalopathy depends on gut bacterial conversion. In hepatic encephalopathy, the ability for ammonia to be detoxified by hepatic enzymes and subsequently excreted by the kidneys is reduced, leading to rise in blood ammonia levels [42–45]. Lactulose is converted by gut flora into fructose and galactose, which is then further converted to lactic and acetic acid in the gut. An acidic environment is created and the ammonia is readily converted to its cation, which promotes its excretion in the feces [46].

Lastly, sodium picosulfate, a laxative, is also converted into active 4,4′-dihydroxydiphenyl-(2 pyridyl)-methane by sulfotransferase-producing *Eubacterium* in the colon. The bioactivation was found to be activated in the presence of endogenous or exogenous phenolic compounds such as tannic acid and flavonoids [47].

### 25.2.2 Conversion to Inactive Metabolites

Digoxin, a cardiac glycoside drug is transformed by intestinal bacteria to inactive digoxin reduction products (DRP) such as dihydrodigoxin via reduction of its lactone ring. It was estimated that approximately 10% of individuals excrete a substantial amount of DRP (more than 40%) with significantly reduced cardiac activity. Administration of broad-spectrum antibiotics also reduced the level of DRPs excreted in humans [48, 49]. The use of highly bioavailable capsules also showed fewer conversion to inactive metabolites [50]. It was later discovered that the conversion to DRP is associated with the bacterium *Eggerthella lenta* [51–53]. Using RNA-sequencing, a two-gene operon that is highly upregulated in the presence of digoxin was identified, known as the cardiac glycoside reductase (*cgr*) operon, which is predictive of digoxin inactivation. [54]. Digoxin reduction by *E. lenta* is also inhibited by arginine via blocking of *cgr* expression, which increases digoxin levels [53, 54]. Therefore, host diet modification presents a potential avenue to manipulate microbial drug metabolism, and nutritional guidelines that can improve drug outcomes are a valuable area for further investigation. Differences in gut metabolic activity are also a potential source for interindividual variability. For instance, it was reported that New Yorkers have a higher capability to inactivate digoxin via conversion to DRP products as compared to South Indians [55]. Another study comparing against Bangladeshis in Dhaka also demonstrated significant interethnic variability [56]. Therefore, digoxin efficacy can vary significantly across different populations, leading to much unpredictability. Identifying patient populations that have a higher possibility to undergo significant metabolism by the intestinal microbiota can potentially help to manage clinical risk and improve therapeutic outcomes. While antibiotics can potentially raise digoxin concentrations due to their effect on intestinal microflora, certain antibiotics such as levofloxacin [57] and garenoxacin [58] were found to have no effect on digoxin metabolism. Therefore, there might be a need for assessment of potential interactions with antibiotic therapies or other factors that may influence the intestinal microbiota and affect digoxin levels. Figure 25.2 shows the metabolism of active digoxin to inactive dihydrodigoxin by *cgr* in *E. lenta*, leading to lower bioavailability, while inhibition of the breakdown of

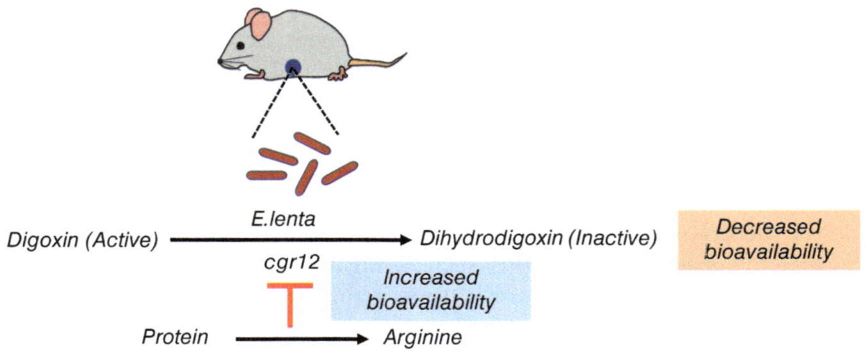


**Figure 25.2** Metabolism of active digoxin to inactive dihydrodigoxin by *cgr* in *E. lenta* and inhibition of the breakdown of digoxin by arginine derived from protein.

digoxin by arginine derived from protein causes higher bioavailability [59].

### 25.2.3 Increase Bioavailability

Valproic acid, a drug commonly used for seizures has seen several case reports of significantly reduced valproic acid concentrations to a sub-therapeutic level in patients taking carbapenem antibiotics concomitantly [60–63]. It is postulated that the markedly reduced drug levels are the result of alterations in absorption, distribution, and metabolism processes [64, 65]. Data from rats indicated that in the presence of panipenem, valproic acid concentration decreased significantly in rats without bile duct cannulation. In rats with bile duct cannulation, no difference in plasma valproic acid concentration was seen with or without panipenem. Panipenem also does not affect extent of biliary excretion of valproic acid and its metabolites. The authors suggested that one possible mechanism was the inhibition of enterohepatic circulation by reducing bacterial glucuronidase responsible for deconjugation of the glucuronide metabolite, and hence less parent drug is recirculated [66]. Caution ought to be exercised in concomitant administration of valproic acid and carbapenems and if doses are raised when carbapenem is used, one can expect to step down the doses when carbapenem is withdrawn.

Similarly, bacterial glucuronidase also contributes to the bioavailability of mycophenolate mofetil (MMF), an immunosuppressant. During short bowel decontamination (SBD) in liver transplant patients, an initial peak representing the initial absorption of MMF was followed by an absence of the secondary peak representing enterohepatic recirculation of the drug, which was present when administered after discontinuation of SBD. The reduced bioavailability of MMF during SBD was likely due to decreased bacterial deconjugation of the glucuronide metabolite and it was estimated that enterohepatic recycling contributes to 7–54% of the bioavailability of MMF [67]. Another study noted that vancomycin was able to eliminate β-glucuronidase producing bacteria and thereby reduced its gastrointestinal adverse effects including weight loss and colonic inflammation [68].

### 25.2.4 Decrease Bioavailability

Levodopa crosses the blood–brain-barrier, which is then converted to dopamine to exert its antiparkinson effects; it was discovered that gut bacteria is capable of transformation reactions resulting in decreased bioavailability of levodopa. Experiments demonstrated that a levodopa metabolite, *m*-hydroxyphenylacetic acid, was found in urine of conventional rats but not germ-free rats [69] and also in anaerobic incubations with rat caecal contents [70]. It was suggested that the dehydroxylation of dopamine to *m*-tyramine was solely attributed to intestinal bacteria, which could be converted to *m*-hydroxyphenylacetic acid by the host and bacteria [69]. Urinary excretion of *m*-tyramine [71] and *m*-hydroxyphenylacetic acid [72] also decreased in patients treated with levodopa and neomycin. Recent studies identified tyrosine decarboxylase in *Enterococcus faecalis* capable of decarboxylating levodopa to dopamine [73, 74] and molybdenum cofactor-dependent dehydroxylase (Dadh) in *E. lenta* capable of dehydroxylating dopamine to m-tyramine [74]. The decarboxylase inhibitor carbidopa was found to be ineffective against bacterial decarboxylases [73, 74]. Such conversions can potentially lead to variability in response, but how significantly gut bacterial conversion contributes to resistance to levodopa therapy and increased dosing regimens require further investigation. Figure 25.3 shows the metabolism of levodopa by gut bacterial enzymes [74].

The incidence of *Helicobacter pylori* infection is higher in Parkinson's disease (PD) patients as opposed to healthy subjects, although the reason behind this observation is unclear [75, 76]. Evidence demonstrated

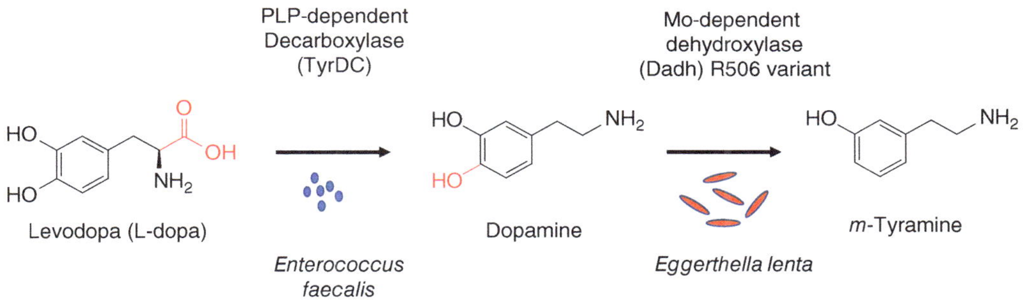


**Figure 25.3** Schematic conversion pathway of levodopa by gut bacterial enzymes: Tyrosine decarboxylase in *Enterococcus faecalis* decarboxylates levodopa to dopamine and molybdenum cofactor-dependent dehydroxylase (Dadh) in *Eggerthella lenta* dehydroxylates dopamine to *m*-tyramine.

that PD patients infected with *H. pylori* are associated with negative motor performances, most likely due to *H. pylori* interfering with levodopa action. Eradication of *H. pylori* improved plasma levodopa levels, [77], led to improvement in time to onset, increased "ON" time, and better motor symptoms [75, 78]. There are several postulations on how *H. pylori* can affect the absorption of levodopa. It was suggested that gastritis resulting from *H. pylori* can induce changes in gastric acid secretion hence affecting gut pH [79–81]. *Helicobacter pylori* may also disrupt gastric myoelectric function, leading to alterations in gastric motility and delay in gastric emptying [82, 83]. It was also reported levodopa adsorbs bacterial surface adhesins of *H. pylori* [84]. Taken together, this can result in erratic absorption and fluctuations. Therefore, it has been suggested that PD patients ought to be screened and treated for *H. pylori*, especially for patients with unpredictable and inconsistent responses to levodopa therapy [75].

Zonisamide, an anticonvulsant, undergoes reduction of the benzisoxazole ring to form 2-sulphamoylacetylphenol, with highest reductase activity found in *Clostridium sporogenes*. Excretion of 2-sulphamoylacetylphenol in urine and feces in rats markedly decreased after treatment with antibiotics, which was later restored after recontamination of gut flora. Therefore, other than the liver enzymes, intestinal bacteria also contributes to biotransformation of zonisamide [85].

Yoo et al. reported that amlodipine was metabolized to a major pyridine metabolite via oxidation in human and rat fecal incubations, and amlodipine level decreases [86]. When administered ampicillin, rats also showed a higher systemic exposure to amlodipine. Although the pyridine metabolite is mainly formed from metabolism by hepatic CYP3A4 enzymes [87], these findings showed that gut bacterial conversion also mediates drug–drug interaction with antibiotics and it may be crucial to consider both the effect of antibiotics on CP450 and gut bacteria metabolism [86].

The *in vitro* stability of H2 receptor antagonist ranitidine was accessed and human fecal inoculations showed that ranitidine concentration decreased over time, suggesting that it was degraded by colonic bacteria via N-oxide bond cleavage [88]. It was demonstrated that colonic-release formulations of ranitidine had the lowest bioavailability, followed by enteric formulations and then immediate-release formulations [89]. This conversion partly contributed to its poor colonic bioavailability. A later study showed that nizatidine undergoes a similar N-oxide bond cleavage by colonic bacteria while no such metabolism occurred with cimetidine and famotidine [90]. Therefore, bacterial metabolism has important implications on drug delivery and absorption.

### 25.2.5 Production of Toxic Metabolites

In rats caecal suspensions, nitrazepam undergoes nitroreduction to form teratogenic 7-aminonitrazepam. After administration of antibiotics to pregnant rats, the urinary and fecal excretion of teratogenic metabolites, and nitrazepam-induced teratogenicity decreased significantly. Although the liver is also capable of reductive metabolism, reductive activity was significantly higher in caecal contents than in liver homogenates [91]. *Clostridium leptum* was later identified to have highly specific nitroreductase activity [92].

5-Fluorocytosine toxicity exhibits similar characteristics to 5-fluorouracil toxicity and it has been demonstrated that 5-FC can be deaminated to 5-FU by intestinal microbiome [93, 94]. 5-FC metabolism to 5-FU is catalyzed by fungal cytosine deaminase, which human cells lacked, resulting in selective toxicity [95]. However, an *in vitro* semicontinuous culture with human feces demonstrated the metabolism of 5-FC to 5-FU was also mediated by bacteria [93]. *Escherichia coli*

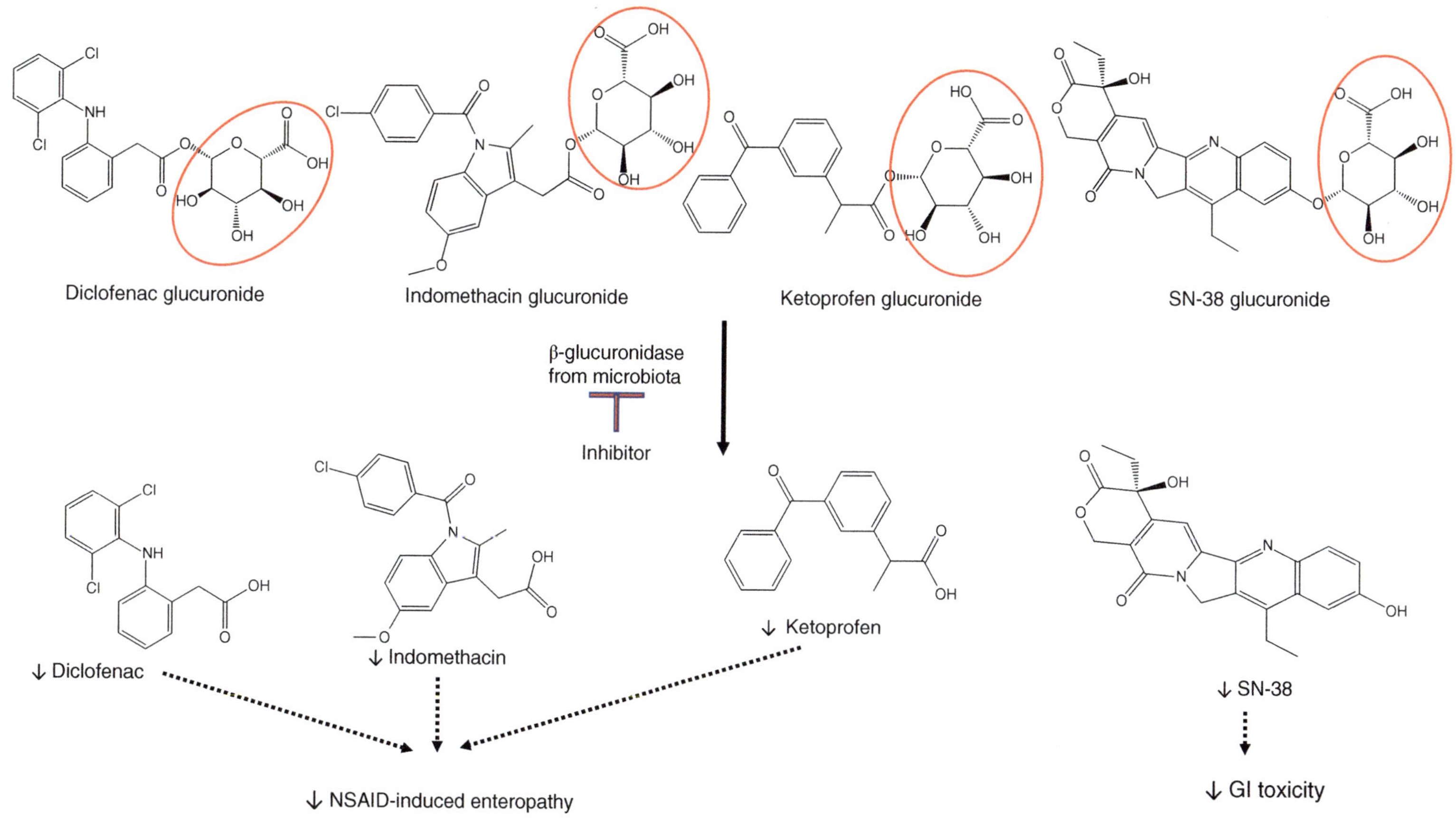


**Figure 25.4** Decrease in intestinal toxic effects of diclofenac, indomethacin, ketoprofen, and active SN-38 via inhibition of β-glucuronidase, preventing the conversion of the glucuronides back to the active/toxic drugs in the intestinal lumen.

was found to contain substantial cytosine deaminase activity, which continued to be active even in *E. coli* suspensions that were nonviable [94].

The active SN-38 metabolite of irinotecan is detoxified by hepatic enzymes to a glucuronide conjugate, which is subsequently biliary excreted into the intestines. The deconjugation of inactive SN-38 glucuronide in the intestine frequently causes gastrointestinal toxicity. Furthermore, the degree of mucosal damage correlates well with bacterial activity. [96, 97]. *Escherichia coli* especially was shown to exacerbate irinotecan-induced toxicity [98]. Administration of antibiotics conferred protective effects by eliminating β-glucuronidase producing bacteria, reducing level of SN-38 in the intestine, thus alleviating irinotecan-induced diarrhea [99–102]. An alternative targeted approach would be small molecule bacterial β-glucuronidase inhibitors that do not kill the commensal bacteria and mammalian cells [103, 104]. Other strategies proposed to prevent the reactivation of SN-38 include Kampo medicine baicalin [102] and repurposing known drugs such as amoxapine as β-glucuronidase inhibitors [105, 106]. Furthermore, it was proposed in a recent study that intestinal β-glucuronidase may serve as a biomarker for irinotecan-associated diarrhea severity, thereby facilitating toxicity risks stratification and treatment adaptations [107]. Similarly, high concentrations of NSAID aglycone in the intestine that resulted in NSAID-induced enteropathy were alleviated by treatment with β-glucuronidase inhibitors, as shown in studies with diclofenac [108, 109], indomethacin and ketoprofen [110]. Inactive regorafenib-glucuronide that is reactivated to regorafenib by bacterial glucuronidases also frequently leads to diarrhea, which is alleviated by β-glucuronidase inhibitors [111]. Figure 25.4 shows the decrease in toxic effects of diclofenac, indomethacin, ketoprofen, and active SN-38 via inhibition of β-glucuronidase [112].

*Clostridium scindens* ATCC 35704, a species of gut bacteria with desA and desB gene encoding for steroid-17,20-desmolase is a rare bacterial enzyme capable of converting glucocorticoids including prednisone, prednisolone, dexamethasone, and 9-fluorocortisol to 11-oxy-androgens via side chain cleavage, with highest activity against prednisone. 1,4-Androstanediene-3,11, 17-trione, the side chain cleavage product of prednisone, has been demonstrated to cause proliferation of androgen-dependent prostate cancer cells at a level similar to 5α-dihydrotestosterone (DHT) in cell culture experiments [113]. Therefore, metabolism by gut bacteria may potentially affect host pathophysiology and its physiological consequences ought to be taken into consideration. This rare enzymatic pathway can also cause important interindividual differences with respect to disease risk.

## 25.3 Indirect Mechanisms Affecting Drug Metabolism

### 25.3.1 Alterations of Host Gene Expression Modulating Drug Metabolizing Enzymes and Transporters

Livers from germ-free mice have a marked downregulation of CYP3A genes and an upregulation of CYP4A genes at both mRNA and protein levels in contrast to conventional mice. The decrease in CYP3A and increase in CYP4A expressions correspond to decrease in the pregnane X receptor (PXR) and increase in peroxisome proliferator-activated receptor a (PPARa) signaling respectively, possibly attributable to differential amount of microbial metabolites in germ-free versus conventional states. These observations were consistent across many studies [114–117]. Apart from CYP enzymes, gut microbiota has the capability to affect non-CYP phase I and II enzymes. The upregulation of Aldh3a2, downregulation of Aldh1b1, gstp1, gstm3, and sult5a1 in germ-free mice were also consistent across studies [115, 116]. A marked upregulation of organic anion transporting polypeptide 1b2 transporter was also observed [115]. However, contradictory results regarding the regulation of nuclear receptors have been observed. For instance, Selwin et al. reported no change in gene expression of PXR in livers of germ-free and conventional mice [115], while another study observed a downregulation of mRNA of PXR in germ-free mice [117]. Studies also observed a decrease [117], increase [118], or no change in constitutive androstane receptor (CAR) mRNA expression in germ-free mice [115]. Changes in mRNA levels of certain drug metabolizing enzymes in germ-free and conventional mice were also conflicting [115–118]. It was noted that this inconsistency could be due to strain differences [115], different quantification techniques (RNA-Sequencing versus reverse transcription qPCR), vehicle effects, and/or statistical methods [116]. Hence, ascertaining the impact of intestinal microbiome on gene expression of some nuclear receptors and metabolic enzymes is challenging as it is likely a result of complex and dynamic interplay of many factors, as seen in a study where intestinal microbiome significantly affects the ontogeny of many liver drug-metabolizing genes in an age and gender-dependent fashion [114]. The modulation of hepatic enzymes has important pharmacological

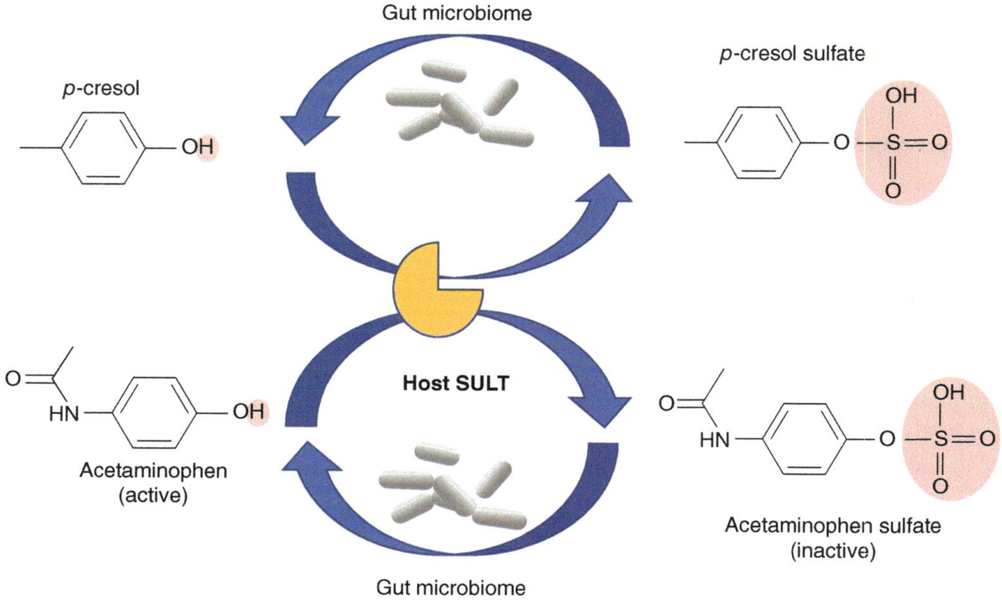


**Figure 25.5** Microbial metabolite, *p*-cresol, competes with acetaminophen for host metabolizing enzymes.

consequences. For instance, phenobarbital showed a more efficient metabolism in germ-free mice than colonized ones in which germ-free animals showed a higher expression of important drug-metabolizing genes [118]. Therefore, gut bacteria undoubtedly may affect expression of hepatic enzymes and transporters. However, more research is needed regarding the extent in which normal variations in the intestinal microbiome regulate drug metabolizing genes and in turn, affect the metabolic activity and its pharmacological consequences.

### 25.3.2 Competition for Host Enzymes with Microbial Metabolites

Acetaminophen is mainly deactivated by two major phases II conjugation processes, namely glucuronidation and sulfonation [119] while a minor portion undergoes the toxic pathway producing a reactive metabolite, *N*-acetyl-*p*-benzoquinone imine (NAPQI), whereby glutathione (GSH) detoxifies it [120, 121]. Individuals with high *p*-cresol sulfate level in urine before acetaminophen administration correlated with a low proportion of acetaminophen sulfate after acetaminophen administration. The bacterial metabolite *p*-cresol is subjected to competitive *o*-sulfonation forming *p*-cresol sulfate and hence depletes the liver's capacity for acetaminophen sulfonation, which predisposes some individuals to acetaminophen-induced hepatotoxicity [122]. *p*-Cresol is generated from tyrosine by gut bacteria, *C. difficile*, with the enzyme identified as *p*-hydroxyphenylacetate decarboxylase [123]. It is well accepted that acetaminophen-induced hepatotoxicity displays diurnal variability, where administration at night results in more serious liver injury [124, 125]. A recent study using metabolomic analysis indicated that the gut microbial metabolite 1-phenyl-1,2-propanedione (PPD) displayed a similar diurnal variation and had a synergistic effect on hepatotoxicity by directly reducing liver GSH pool, resulting in less detoxification of NAPQI. In addition, reduction of PPD by *Saccharomyces cerevisiae* decreased its intestinal levels and resulted in markedly reduced hepatotoxicity [126]. This indicates that probiotics present exciting opportunities in improving drug treatment outcomes. Figure 25.5 shows that microbial metabolite, *p*-cresol, competes with acetaminophen for host metabolizing enzymes [59].

### 25.3.3 Enzyme Induction/Inhibition by Microbial Metabolites

Sorivudine is metabolized by gut bacteria to (*E*)-5-(2-bromovinyl)uracil (BVU), a suicide inhibitor that is activated by hepatic dihydropyrimidine dehydrogenase (DPD), in which the activated BVU then irreversibly inactivates DPD, an enzyme responsible for 5-fluorouracil degradation. Inhibition of 5-FU degradation enhances its anti-tumor effects and toxicity caused by concomitant administration of sorivudine and 5-FU or its prodrugs such as tegafur [127–131]. Similarly, brivudine also potentiates the toxicity of chemotherapeutic pyrimidine analogs with a similar mechanism, and the accumulation of BVU in serum was significantly higher in conventional mice in contrast to germ-free mice. The highest activity for the bacterial conversion was found for *Bacteroides thetaiotaomicron* and *Bacteroides ovatus*, with *bt4554* gene encoding a predicted purine nucleoside phosphorylase essential for the conversion [132].

CD6168, an alkene reduction product of deleobuvir, is a major metabolite almost exclusively formed by gut bacteria [133, 134], with CD6168 representing about 15% of total drug-related material and 27% of the parent in plasma [133]. According to European Medicines Agency (EMA) and US Food and Drug Administration (FDA) guidelines, metabolites exceeding 25% of the parent drug's area-under-curve or exceeding 10% of total drug-related material should be evaluated *in vitro* for drug–drug interaction (DDI). Indeed, CD6168 is a potent inducer of CYP3A4, which counteracts the inhibition by deleobuvir. When metabolites were considered, the overall net effect on CYPs was closer to the observed DDI observed clinically, which emphasizes the significance of incorporating these metabolites in *in-vitro* DDI studies [135]. Table 25.1 summarizes the impact of intestinal

**Table 25.1** Summary of impact of intestinal microbiome on drug pharmacokinetics.

| Drug | Pharmacological consequences | Type of reaction | Metabolite |
|---|---|---|---|
| Prontosil | Prodrug activation | Azoreduction | Sulphanilamide |
| Neoprontosil | Prodrug activation | Azoreduction | Sulphanilamide |
| Sulfasalazine | Prodrug activation | Azoreduction | 5-ASA, sulphapyridine |
| Balsalazide | Prodrug activation | Azoreduction | 5-ASA, 4-aminobenzoyl-*b*-alanine |
| Olsalazine | Prodrug activation | Azoreduction | 5-ASA |
| Lovastatin | Prodrug activation | Hydrolysis | Active β-hydroxy acid metabolite |
| Sulindac | Prodrug activation | Sulphoxide reduction | Active sulfide metabolite |
| Sulphinpyrazone | Prodrug activation | Sulphoxide reduction | Active sulfide metabolite |
| Thioguanine | Prodrug activation | — | Thioguanine nucleotides |
| Lactulose | Prodrug activation | Hydrolysis | Organic acids (acetic acid, lactic acid) |
| Sodium picosulfate | Prodrug activation | Desulfation | 4,4′-dihydroxydiphenyl-(2 pyridyl)-methane |
| Digoxin | Drug inactivation | C=C reduction of lactone ring | Dihydrodigoxin |
| Valproic acid | Increase bioavailability | Deconjugation | Release of parent drug from glucuronides |
| Mycophenolate mofetil | Increase bioavailability | Deconjugation | Release of parent drug from glucuronides |
| Levodopa | Decrease bioavailability | Dehydroxylation, Decarboxylation | *m*-Tyramine, dopamine |
| Zonisamide | Decrease bioavailability | N-oxide bond reduction in benzisoxazole ring | 2-Sulphamoylacetylphenol |
| Amlodipine | Decrease bioavailability | Oxidation | Pyridine metabolite |
| Ranitidine, Nizatidine | Decrease bioavailability | N-oxide bond cleavage | — |
| Nitrazepam | Toxic metabolites | Nitroreduction | 7-Aminonitrazepam |
| 5-Fluorocytosine | Toxic metabolites | Deamination | 5-Fluorouracil |
| Irinotecan | Toxic metabolites | Deconjugation | Release of parent drug from glucuronides |
| Diclofenac, Ketoprofen, Indomethacin | Toxic metabolites | Deconjugation | Release of parent drug from glucuronides |
| Regorafenib | Toxic metabolites | Deconjugation | Release of parent drug from glucuronides |
| Prednisone | Toxic metabolites | Side chain cleavage | 1,4-androstanediene-3,11,17-trione |
| Sorivudine, Brivudine | Enzyme inhibition by bacteria-generated metabolite | — | (*E*)-5-(2-bromovinyl)uracil (BVU) |
| Deleobuvir | Enzyme induction by bacteria-generated metabolite | Alkene reduction | CD6168 |

microbiome on drug pharmacokinetics discussed in this review.

## 25.4 Impact of Intestinal Microbiome on Drug Treatment in Clinical Practice

### 25.4.1 Aminosalicylates

Aminosalicylates such as sulfasalazine, balsalazide, olsalazine, and ipsalazine are prodrugs used for the treatment of inflammatory disease (IBD) [136]. These prodrugs are reductively cleaved by colonic bacteria at the azo bond to liberate the active drugs in the large intestine. It was found that oral treatment with 5-aminosalicylate (5-ASA) was able to change gut microbial community composition [137]. The anti-inflammatory SCFA-producing bacteria, such as *Firmicutes* was found to decrease and the pro-inflammatory *Escherichia-Shigella* increased in the inflamed mucosae of ulcerative colitis (UC) patients, inferring to be the main attributes to the mucosal inflammation of UC patients. These situations were reversed after 5-ASA treatment, leading to its treatment efficacy.

### 25.4.2 Tacrolimus

Tacrolimus is an immunosuppressant with narrow therapeutic index used to suppress organ rejection in solid organ transplant patients [138]. Treatment with antibiotics in the first month following kidney transplant was found to increase tacrolimus trough variability in certain clinical population [139]. In another study, patients requiring an escalation in tacrolimus dose over time were found to have much higher levels of fecal *Faecalibacterium prausnitzii* (11.8%) in the first week of posttransplantation when compared to patients who did not require dose escalation (0.8%) [140]. In a subsequent study, incubation of *F. prausnitzii* with tacrolimus led to formation of two metabolites [141]. The major one (M1) was found 15-fold less potent than tacrolimus as an immunosuppressant. Screening of 22 gut microflora revealed that *Clostridiales* are extensive tacrolimus metabolizers. Finding in the study suggested that intestinal microflora might contribute to the low and variable oral absorption of tacrolimus.

### 25.4.3 Digoxin

Digoxin, a cardiac drug commonly used for the treatment of heart failure and atrial fibrillation was found to be metabolized extensively by gut microflora to form cardio-inactive, reduced metabolites [21]. *Eggerthella lenta* is a gut microflora responsible for metabolizing digoxin into the inactive dihydrodigoxin [54]. Treatment of antibiotic, erythromycin, or tetracycline before or with digoxin administration reduced the urinary excretion of the digoxin-reduced metabolites and increased the serum digoxin concentrations as much as twofold. Thus, changes in the intestinal microflora can markedly alter the state of digitalization [48].

### 25.4.4 Levodopa

Levodopa is a first-line drug used for the treatment of Parkinson's disease (PD) [142]. Levodopa is metabolized by gut microbiota to dopamine in the intestine. Clinical observation indicates that PD patients require significantly different doses of levodopa. This highly interindividual variability in drug response has been inferred to be caused by different gut microbial population and communities. Studies also indicated that concurrent *Helicobacter pylori* (*H. pylori*) infection might contribute to erratic levodopa absorption [75, 77]. Eradication of *H. pylori* infection in PD patients may increase levodopa absorption and improve treatment response. A recent study indicated that intestinal bacterial tyrosine decarboxylases can efficiently convert levodopa to dopamine, even in the presence of tyrosine, a competitive substrate, or inhibitors of human decarboxylase [73]. High abundance of gut bacterial tyrosine decarboxylase in patients with Parkinson's disease can reduce levodopa bioavailability. That is particularly important as Parkinson's disease patients often have impaired intestinal motility [143] that is also associated with small intestinal bacterial overgrowth [144].

## 25.5 Conclusion and Future Perspectives

Other than liver, gut bacteria is also pivotal in the transformation of many drugs with obvious implications on efficacy, bioavailability, and safety. The impact of intestinal microbiome on drug pharmacokinetics has a large variety of mechanisms and can even affect host gene expression of metabolizing enzymes. The metabolic function of the intestinal microbiome can also complement that of the liver. More importantly, some metabolites produced by the intestinal microbiome are unique metabolites not produced in the liver. As such, any differential activity or toxicity exerted by these metabolites should not be overlooked. Drug therapy has always been subjected to interindividual variability, which could be due to differences in genetic makeup. However, there is also increasing evidence that our unique microbial composition can partly contribute to different therapeutic outcomes on top of host metabolism. Unraveling the influence of intestinal microbiome can lead to advancement in personalized medicine whereby an individual's microbial composition and its predictive biomarkers are taken into consideration for clinical management. As mentioned above, fecal β-glucuronidase levels and abundance of *cgr* gene in fecal samples could potentially serve as biomarkers to predict drug response. Therefore, identification of genetic screening tools and accurate biomarkers that help to individualize optimal drug therapy is an area with much room for development. In addition, the intestinal microbiome undoubtedly represents a drug target with vast potential to be subjected to alteration and modulation of its composition and functional activity. Strategies to modulate the intestinal microbiota present many opportunities to improve therapeutic outcomes, such as diet, probiotics, and drugs targeting the intestinal microbiota such as bacterial enzyme inhibitors. Thus, more research is required to identify viable targets and the efficacy and downstream consequences of such targeted intervention. Besides, other future works may include how normal variations in gut bacteria due to external factors such as environmental changes, stress, diet, drugs, or medical conditions could affect the metabolism of drugs and drug metabolizing genes and the significance of their impact on drug safety, efficacy, or bioavailability.

Antibiotics, which can modulate the intestinal microbiome are also frequently coadministered with other drugs in clinical settings. Hence, the impact of antibiotics on drug pharmacokinetics and pharmacodynamics should also be investigated. Furthermore, a mechanistic understanding in this area would give us additional insights for rational drug design and hence lead to improvement in drug discovery and development. One potential future application is the design of better drug formulations and dosage forms to improve drug delivery while taking account potential interactions with gut bacteria. For instance, such an understanding will aid in the decision for the targeted location of release in the gastrointestinal tract. Another application would be the design of prodrugs based on metabolism by gut bacteria and design of drugs that would not be subjected to unfavorable metabolism by gut bacteria. Therefore, there is always a need to elucidate its impact on drug metabolism. Furthermore, given the complex nature of the intestinal microbiome, it is important to further investigate the specific bacteria species involved, specific enzymes, and gene encoding the bacterial enzyme involved in the metabolism of those drugs. Moving forward, the impact of intestinal microbiome shall not be overlooked during the development phase of a drug and consideration ought to be given to the intestinal microbiome and its effect on drug metabolism.

## References


1 Ley, R.E., Peterson, D.A., and Gordon, J.I. (2006). Ecological and evolutionary forces shaping microbial diversity in the human intestine. *Cell* 124 (4): 837–848.

2 Qin, J., Li, R., Raes, J. et al. (2010). A human gut microbial gene catalogue established by metagenomic sequencing. *Nature* 464 (7285): 59–65.

3 Dieterich, W., Schink, M., and Zopf, Y. (2018). Microbiota in the gastrointestinal tract. *Med. Sci.* 6 (4): 116.

4 Sekirov, I., Russell, S.L., Antunes, L.C., and Finlay, B.B. (2010). Gut microbiota in health and disease. *Physiol. Rev.* 90 (3): 859–904.

5 Sousa, T., Paterson, R., Moore, V. et al. (2008). The gastrointestinal microbiota as a site for the biotransformation of drugs. *Int. J. Pharm.* 363 (1–2): 1–25.

6 Davis, S.S. (2005). Formulation strategies for absorption windows. *Drug Discovery Today* 10 (4): 249–257.

7 Faith, J.J., Guruge, J.L., Charbonneau, M. et al. (2013). The long-term stability of the human gut microbiota. *Science* 341 (6141): 1237439.

8 Rajilic-Stojanovic, M., Heilig, H.G., Tims, S. et al. (2012). Long-term monitoring of the human intestinal microbiota composition. *Environ. Microbiol.* 15 (4): 1146–1159.

9 Hasan, N. and Yang, H. (2019). Factors affecting the composition of the gut microbiota, and its modulation. *PeerJ* 7: e7502.

10 Wen, L. and Duffy, A. (2017). Factors influencing the gut microbiota, inflammation, and type 2 diabetes. *J. Nutr.* 147 (7): 1468S–1475S.

11 den Besten, G., van Eunen, K., Groen, A.K. et al. (2013). The role of short-chain fatty acids in the interplay between diet, gut microbiota, and host energy metabolism. *J. Lipid Res.* 54 (9): 2325–2340.

12 Topping, D.L. (1996). Short-chain fatty acids produced by intestinal bacteria. *Asia Pac. J. Clin. Nutr.* 5 (1): 15–19.

13 Gérard, P. (2013). Metabolism of cholesterol and bile acids by the gut microbiota. *Pathogens* 3 (1): 14–24.

14 Kriaa, A., Bourgin, M., Potiron, A. et al. (2019). Microbial impact on cholesterol and bile acid metabolism: current status and future prospects. *J. Lipid Res.* 60 (2): 323–332.

15 Ramakrishna, B.S. (2013). Role of the gut microbiota in human nutrition and metabolism. *J. Gastroenterol. Hepatol.* 28 (Suppl 4): 9–17.

16 Kinnebrew, M.A. and Pamer, E.G. (2012). Innate immune signaling in defense against intestinal microbes. *Immunol. Rev.* 245 (1): 113–131.

17 Pickard, J.M., Zeng, M.Y., Caruso, R., and Núñez, G. (2017). Gut microbiota: role in pathogen colonization, immune responses, and inflammatory disease. *Immunol. Rev.* 279 (1): 70–89.

18 Saad, R., Rizkallah, M.R., and Aziz, R.K. (2012). Gut pharmacomicrobiomics: the tip of an iceberg of complex interactions between drugs and gut-associated microbes. *Gut Pathogens* 4 (1): 16.

19 Haiser, H.J. and Turnbaugh, P.J. (2013). Developing a metagenomic view of xenobiotic metabolism. *Pharmacol. Res.* 69 (1): 21–31.

20 Swanson, H.I. (2015). Drug metabolism by the host and gut microbiota: a partnership or rivalry? *Drug Metab. Dispos.* 43 (10): 1499–1504.

21 Wilson, I.D. and Nicholson, J.K. (2017). Gut microbiome interactions with drug metabolism, efficacy, and toxicity. *Transl. Res.* 179: 204–222.

22 Gingell, R. and Bridges, J.W. (1971). Intestinal azo reduction and glucuronide conjugation of prontosil. *Biochem. J.* 125 (2): 24P.

23 Gingell, R., Bridges, J.W., and Williams, R.T. (1971). The role of the gut flora in the metabolism of prontosil and neoprontosil in the rat. *Xenobiotica* 1 (2): 143–156.

24 Peppercorn, M.A. and Goldman, P. (1972). The role of intestinal bacteria in the metabolism of salicylazosulfapyridine. *J. Pharmacol. Exp. Ther.* 181 (3): 555–562.

25 Schröder, H. and Campbell, D.E. (1972). Absorption, metabolism, and excretion of salicylazosulfapyridine in man. *Clin. Pharmacol. Ther.* 13 (4): 539–551.

26 Das, K.M., Eastwood, M.A., McManus, J.P., and Sircus, W. (1973). The metabolism of salicylazosulphapyridine in ulcerative colitis. I. the relationship between metabolites and the response to treatment in inpatients. *Gut* 14 (8): 631–641.

27 Peppercorn, M.A. and Goldman, P. (1973). Distribution studies of salicylazosulfapyridine and its metabolites. *Gastroenterology* 64 (2): 240–245.

28 Klotz, U. (1985). Clinical pharmacokinetics of sulphasalazine, its metabolites and other prodrugs of 5-aminosalicylic acid. *Clin. Pharmacokin.* 10 (4): 285–302.

29 Chan, R.P., Pope, D.J., Gilbert, A.P. et al. (1983). Studies of two novel sulfasalazine analogs, ipsalazide and balsalazide. *Digest. Dis. Sci.* 28 (7): 609–615.

30 Sousa, T., Yadav, V., Zann, V. et al. (2014). On the colonic bacterial metabolism of azo-bonded prodrugs of 5-aminosalicylic acid. *J. Pharm. Sci.* 103 (10): 3171–3175.

31 Dull, B.J., Salata, K., and Goldman, P. (1987). Role of the intestinal flora in the acetylation of sulfasalazine metabolites. *Biochem. Pharmacol.* 36 (21): 3772–3774.

32 van Hogezand, R.A., Kennis, H.M., van Schaik, A. et al. (1992). Bacterial acetylation of 5-aminosalicylic acid in faecal suspensions cultured under aerobic and anaerobic conditions. *Eur. J. Clin. Pharmacol.* 43 (2): 189–192.

33 Deloménie, C., Fouix, S., Longuemaux, S. et al. (2001). Identification and functional characterization of arylamine *N*-acetyltransferases in eubacteria: evidence for highly selective acetylation of 5-aminosalicylic acid. *J. Bacteriol.* 183 (11): 3417–3427.

34 Zhang, J., Zhang, J., and Wang, R. (2018). Gut microbiota modulates drug pharmacokinetics. *Drug Metab. Rev.* 50 (3): 357–368.

35 Yoo, D.H., Kim, I.S., Van Le, T.K. et al. (2014). Gut microbiota-mediated drug interactions between lovastatin and antibiotics. *Drug Metab. Dispos.* 42 (9): 1508–1513.

36 Strong, H.A., Warner, N.J., Renwick, A.G., and George, C.F. (1985). Sulindac metabolism: the importance of an intact colon. *Clin. Pharmacol. Ther.* 38 (4): 387–393.

37 Lee, S.C. and Renwick, A.G. (1995). Sulphoxide reduction by rat intestinal flora and by *Escherichia coli* in vitro. *Biochem. Pharmacol.* 49 (11): 1567–1576.

38 Renwick, A.G., Evans, S.P., Sweatman, T.W. et al. (1982). The role of the gut flora in the reduction of sulphinpyrazone in the rat. *Biochem. Pharmacol.* 31 (16): 2649–2656.

39 Strong, H.A., Oates, J., Sembi, J. et al. (1984). Role of the gut flora in the reduction of sulfinpyrazone in humans. *J. Pharmacol. Exp. Ther.* 230 (3): 726–732.

40 Florin, T., Movva, R., Begun, J. et al. (2018). Colonic thioguanine pro-drug: investigation of

microbiome and novel host metabolism. *Gut Microbes* 9 (2): 175–178.

41 Oancea, I., Movva, R., Das, I. et al. (2017). Colonic microbiota can promote rapid local improvement of murine colitis by thioguanine independently of T lymphocytes and host metabolism. *Gut* 66 (1): 59–69.

42 Cordoba, J. (2014). Hepatic encephalopathy: from the pathogenesis to the new treatments. *ISRN Hepatol.* 2014: 236268–236268.

43 Elwir, S. and Rahimi, R.S. (2017). Hepatic encephalopathy: an update on the pathophysiology and therapeutic options. *J. Clin. Transl. Hepatol.* 5 (2): 142–151.

44 Frederick, R.T. (2011). Current concepts in the pathophysiology and management of hepatic encephalopathy. *Gastroenterol. Hepatol.* 7 (4): 222–233.

45 Tranah, T.H., Paolino, A., and Shawcross, D.L. (2015). Pathophysiological mechanisms of hepatic encephalopathy. *Clin. Liver Dis.* 5 (3): 59–63.

46 Elkington, S.G., Floch, M.H., and Conn, H.O. (1969). Lactulose in the treatment of chronic portal-systemic encephalopathy. A double-blind clinical trial. *N. Engl. J. Med.* 281 (8): 408–412.

47 Kim, D.H., Hyun, S.H., Shim, S.B., and Kobashi, K. (1992). The role of intestinal bacteria in the transformation of sodium picosulfate. *Jpn. J. Pharmacol.* 59 (1): 1–5.

48 Lindenbaum, J., Rund, D.G., Butler, V.P. Jr., et al. (1981a). Inactivation of digoxin by the gut flora: reversal by antibiotic therapy. *N. Engl. J. Med.* 305 (14): 789–794.

49 Lindenbaum, J., Tse-Eng, D., Butler, V.P. Jr., and Rund, D.G. (1981b). Urinary excretion of reduced metabolites of digoxin. *Am. J. Med.* 71 (1): 67–74.

50 Rund, D.G., Lindenbaum, J., Dobkin, J.F. et al. (1983). Decreased digoxin cardioinactive-reduced metabolites after administration as an encapsulated liquid concentrate. *Clin. Pharmacol. Ther.* 34 (6): 738–743.

51 Dobkin, J.F., Saha, J.R., Butler, V.P. Jr., et al. (1982). Inactivation of digoxin by *Eubacterium lentum*, an anaerobe of the human gut flora. *Trans. Assoc. Am. Physicians* 95: 22–29.

52 Robertson, L.W., Chandrasekaran, A., Reuning, R.H. et al. (1986). Reduction of digoxin to 20R-dihydrodigoxin by cultures of *Eubacterium lentum*. *Appl. Environ. Microbiol.* 51 (6): 1300–1303.

53 Saha, J.R., Butler, V.P. Jr., Neu, H.C., and Lindenbaum, J. (1983). Digoxin-inactivating bacteria: identification in human gut flora. *Science* 220 (4594): 325–327.

54 Haiser, H.J., Gootenberg, D.B., Chatman, K. et al. (2013). Predicting and manipulating cardiac drug inactivation by the human gut bacterium *Eggerthella lenta*. *Science* 341 (6143): 295–298.

55 Mathan, V.I., Wiederman, J., Dobkin, J.F., and Lindenbaum, J. (1989). Geographic differences in digoxin inactivation, a metabolic activity of the human anaerobic gut flora. *Gut* 30 (7): 971–977.

56 Alam, A.N., Saha, J.R., Dobkin, J.F., and Lindenbaum, J. (1988). Interethnic variation in the metabolic inactivation of digoxin by the gut flora. *Gastroenterology* 95 (1): 117–123.

57 Chien, S.C., Rogge, M.C., Williams, R.R. et al. (2002). Absence of a pharmacokinetic interaction between digoxin and levofloxacin. *J. Clin. Pharm. Ther.* 27 (1): 7–12.

58 Nord, C.E., Meurling, L., Russo, R.L. et al. (2003). Effect of garenoxacin on eubacteria in the normal intestinal microflora when administered concomitantly with digoxin. *J. Chemother.* 15 (3): 244–247.

59 Spanogiannopoulos, P., Bess, E.N., Carmody, R.N., and Turnbaugh, P.J. (2016). The microbial pharmacists within us: a metagenomic view of xenobiotic metabolism. *Nat. Rev. Microbiol.* 14 (5): 273–287.

60 Fudio, S., Carcas, A., Piñana, E., and Ortega, R. (2006). Epileptic seizures caused by low valproic acid levels from an interaction with meropenem. *J. Clin. Pharm. Ther.* 31 (4): 393–396.

61 Llinares Tello, F., Bosacoma Ros, N., Hernández Prats, C. et al. (2003). Pharmacokinetic interaction between valproic acid and carbapenem-like antibiotics: a discussion of three cases. [Interacción farmacocinética entre ácido valproico y antibióticos carbapenémicos: descripción de tres casos]. *Farm Hosp.* 27 (4): 258–263.

62 Nacarkucuk, E., Saglam, H., and Okan, M. (2004). Meropenem decreases serum level of valproic acid. *Pediatr. Neurol.* 31 (3): 232–234.

63 Nagai, K., Shimizu, T., Togo, A. et al. (1997). Decrease in serum levels of valproic acid during treatment with a new carbapenem, panipenem/betamipron. *J. Antimicrob. Chemother.* 39 (2): 295–296.

64 Mancl, E.E. and Gidal, B.E. (2009). The effect of carbapenem antibiotics on plasma concentrations of valproic acid. *Ann. Pharmacother.* 43 (12): 2082–2087.

65 Mori, H., Takahashi, K., and Mizutani, T. (2007). Interaction between valproic acid and carbapenem antibiotics. *Drug Metab. Rev.* 39 (4): 647–657.

66 Kojima, S., Nadai, M., Kitaichi, K. et al. (1998). Possible mechanism by which the carbapenem antibiotic panipenem decreases the concentration of valproic acid in plasma in rats. *Antimicrob. Agents Chemother.* 42 (12): 3136–3140.

67 Schmidt, L.E., Rasmussen, A., Norrelykke, M.R. et al. (2001). The effect of selective bowel decontamination on the pharmacokinetics of mycophenolate mofetil in liver transplant recipients. *Liver Transpl.* 7 (8): 739–742.

68 Taylor, M.R., Flannigan, K.L., Rahim, H. et al. (2019). Vancomycin relieves mycophenolate mofetil-induced gastrointestinal toxicity by eliminating gut bacterial β-glucuronidase activity. *Sci. Adv.* 5 (8): eaax2358.

69 Goldin, B.R., Peppercorn, M.A., and Goldman, P. (1973). Contributions of host and intestinal microflora in the metabolism of L-dopa by the rat. *J. Pharmacol. Exp. Ther.* 186 (1): 160–166.

70 Bakke, O.M. (1971). Degradation of DOPA by intestinal microorganisms in vitro. *Acta Pharmacol. Toxicol.* 30 (1): 115–121.

71 Sandler, M., Goodwin, B.L., and Ruthven, C.R. (1971). Therapeutic implications in Parkinsonism of m-tyramine formation from L-dopa in man. *Nature* 229 (5284): 414–415.

72 Sandler, M., Karoum, F., Ruthven, C.R., and Calne, D.B. (1969). *m*-Hydroxyphenylacetic acid formation from L-dopa in man: suppression by neomycin. *Science* 166 (3911): 1417–1418.

73 van Kessel, S.P., Frye, A.K., El-Gendy, A.O. et al. (2019). Gut bacterial tyrosine decarboxylases restrict levels of levodopa in the treatment of Parkinson's disease. *Nat. Commun.* 10 (1): 310.

74 Maini Rekdal, V., Bess, E.N., Bisanz, J.E. et al. (2019). Discovery and inhibition of an interspecies gut bacterial pathway for Levodopa metabolism. *Science* 364: 6445.

75 Hashim, H., Azmin, S., Razlan, H. et al. (2014). Eradication of *Helicobacter pylori* infection improves levodopa action, clinical symptoms and quality of life in patients with Parkinson's disease. *PLoS One* 9 (11): e112330.

76 Wan Yahya, N., Mohamed Ibrahim, N., Achok, H. et al. (2013). High prevalence of *Helicobacter pylori* infection in Malaysian Parkinson's disease patients. *Res. Rev. Parkinsonism* 2013, 3: 63.

77 Pierantozzi, M., Pietroiusti, A., Galante, A. et al. (2001). *Helicobacter pylori*-induced reduction of acute levodopa absorption in Parkinson's disease patients. *Ann. Neurol.* 50 (5): 686–687.

78 Lee, W.Y., Yoon, W.T., Shin, H.Y. et al. (2008). *Helicobacter pylori* infection and motor fluctuations in patients with Parkinson's disease. *Mov. Disord.* 23 (12): 1696–1700.

79 Annibale, B., Di Giulio, E., Caruana, P. et al. (2002). The long-term effects of cure of *Helicobacter pylori* infection on patients with atrophic body gastritis. *Aliment. Pharmacol. Ther.* 16 (10): 1723–1731.

80 Calam, J. (1999). *Helicobacter pylori* modulation of gastric acid. *Yale J. Biol. Med.* 72 (2–3): 195–202.

81 Feldman, M., Cryer, B., and Lee, E. (1998). Effects of *Helicobacter pylori* gastritis on gastric secretion in healthy human beings. *Am. J. Phys.* 274 (6): G1011–G1017.

82 Miyaji, H., Azuma, T., Ito, S. et al. (1999). The effect of helicobacter pylori eradication therapy on gastric antral myoelectrical activity and gastric emptying in patients with non-ulcer dyspepsia. *Aliment. Pharmacol. Ther.* 13 (11): 1473–1480.

83 Thor, P., Lorens, K., Tabor, S. et al. (1996). Dysfunction in gastric myoelectric and motor activity in *Helicobacter pylori* positive gastritis patients with non-ulcer dyspesia. *J. Physiol. Pharmacol* 47 (3): 469–476.

84 Niehues, M. and Hensel, A. (2009). In-vitro interaction of L-dopa with bacterial adhesins of *Helicobacter pylori*: an explanation for clinical differences in bioavailability? *J. Pharm. Pharmacol.* 61 (10): 1303–1307.

85 Kitamura, S., Sugihara, K., Kuwasako, M., and Tatsumi, K. (1997). The role of mammalian intestinal bacteria in the reductive metabolism of zonisamide. *J. Pharm. Pharmacol.* 49 (3): 253–256.

86 Yoo, H.H., Kim, I.S., Yoo, D.H., and Kim, D.H. (2016). Effects of orally administered antibiotics on the bioavailability of amlodipine: gut microbiota-mediated drug interaction. *J. Hypertens.* 34 (1): 156–162.

87 Zhu, Y., Wang, F., Li, Q. et al. (2014). Amlodipine metabolism in human liver microsomes and roles of CYP3A4/5 in the dihydropyridine dehydrogenation. *Drug Metab. Dispos.* 42 (2): 245–249.

88 Basit, A.W. and Lacey, L.F. (2001). Colonic metabolism of ranitidine: implications for its delivery and absorption. *Int. J. Pharm.* 227 (1–2): 157–165.

89 Basit, A.W., Podczeck, F., Newton, J.M. et al. (2004). The use of formulation technology to assess regional gastrointestinal drug absorption in humans. *Eur. J. Pharm. Sci.* 21 (2–3): 179–189.

90 Basit, A.W., Newton, J.M., and Lacey, L.F. (2002). Susceptibility of the H2-receptor antagonists

cimetidine, famotidine and nizatidine, to metabolism by the gastrointestinal microflora. *Int. J. Pharm.* 237 (1–2): 23–33.

91 Takeno, S. and Sakai, T. (1991). Involvement of the intestinal microflora in nitrazepam-induced teratogenicity in rats and its relationship to nitroreduction. *Teratology* 44 (2): 209–214.

92 Rafii, F., Sutherland, J.B., Hansen, E.B. Jr., and Cerniglia, C.E. (1997). Reduction of nitrazepam by *Clostridium leptum*, a nitroreductase-producing bacterium isolated from the human intestinal tract. *Clin. Infect. Dis.* 25 (Suppl 2): S121–S122.

93 Harris, B.E., Manning, B.W., Federle, T.W., and Diasio, R.B. (1986). Conversion of 5-fluorocytosine to 5-fluorouracil by human intestinal microflora. *Antimicrob. Agents Chemother.* 29 (1): 44–48.

94 Vermes, A., Kuijper, E.J., Guchelaar, H.-J., and Dankert, J. (2003). An in vitro study on the active conversion of flucytosine to fluorouracil by microorganisms in the human intestinal microflora. *Chemotherapy* 49 (1–2): 17–23.

95 Polak, A. and Scholer, H.J. (1975). Mode of action of 5-fluorocytosine and mechanisms of resistance. *Chemotherapy* 21 (3–4): 113–130.

96 Brandi, G., Dabard, J., Raibaud, P. et al. (2006). Intestinal microflora and digestive toxicity of irinotecan in mice. *Clin. Cancer Res.* 12 (4): 1299–1307.

97 Takasuna, K., Hagiwara, T., Hirohashi, M. et al. (1996). Involvement of beta-glucuronidase in intestinal microflora in the intestinal toxicity of the antitumor camptothecin derivative irinotecan hydrochloride (CPT-11) in rats. *Cancer Res.* 56 (16): 3752–3757.

98 Stringer, A.M., Gibson, R.J., Bowen, J.M. et al. (2009). Irinotecan-induced mucositis manifesting as diarrhoea corresponds with an amended intestinal flora and mucin profile. *Int. J. Exp. Pathol.* 90 (5): 489–499.

99 Kehrer, D.F., Sparreboom, A., Verweij, J. et al. (2001). Modulation of irinotecan-induced diarrhea by cotreatment with neomycin in cancer patients. *Clin. Cancer Res.* 7 (5): 1136–1141.

100 Kodawara, T., Higashi, T., Negoro, Y. et al. (2016). The inhibitory effect of ciprofloxacin on the β-glucuronidase-mediated deconjugation of the irinotecan metabolite SN-38-G. *Basic Clin. Pharmacol. Toxicol.* 118 (5): 333–337.

101 Takasuna, K., Hagiwara, T., Hirohashi, M. et al. (1998). Inhibition of intestinal microflora beta-glucuronidase modifies the distribution of the active metabolite of the antitumor agent, irinotecan hydrochloride (CPT-11) in rats. *Cancer Chemother. Pharmacol.* 42 (4): 280–286.

102 Takasuna, K., Hagiwara, T., Watanabe, K. et al. (2006). Optimal antidiarrhea treatment for antitumor agent irinotecan hydrochloride (CPT-11)-induced delayed diarrhea. *Cancer Chemother. Pharmacol.* 58 (4): 494–503.

103 Roberts, A.B., Wallace, B.D., Venkatesh, M.K. et al. (2013). Molecular insights into microbial beta-glucuronidase inhibition to abrogate CPT-11 toxicity. *Mol. Pharmacol.* 84 (2): 208–217.

104 Wallace, B.D., Wang, H., Lane, K.T. et al. (2010). Alleviating cancer drug toxicity by inhibiting a bacterial enzyme. *Science* 330 (6005): 831–835.

105 Ahmad, S., Hughes, M.A., Yeh, L.-A., and Scott, J.E. (2012). Potential repurposing of known drugs as potent bacterial β-glucuronidase inhibitors. *J. Biomol. Screening* 17 (7): 957–965.

106 Kong, R., Liu, T., Zhu, X. et al. (2014). Old drug new use – amoxapine and its metabolites as potent bacterial beta-glucuronidase inhibitors for alleviating cancer drug toxicity. *Clin. Cancer Res.* 20 (13): 3521–3530.

107 Chamseddine, A.N., Ducreux, M., Armand, J.-P. et al. (2019). Intestinal bacterial β-glucuronidase as a possible predictive biomarker of irinotecan-induced diarrhea severity. *Pharmacol. Ther.* 199: 1–15.

108 LoGuidice, A., Wallace, B.D., Bendel, L. et al. (2012). Pharmacologic targeting of bacterial β-glucuronidase alleviates nonsteroidal anti-inflammatory drug-induced enteropathy in mice. *J. Pharmacol. Exp. Ther.* 341 (2): 447–454.

109 Yauw, S.T.K., Arron, M., Lomme, R.M.L.M. et al. (2018). Microbial glucuronidase inhibition reduces severity of diclofenac-induced anastomotic leak in rats. *Surg. Infect.* 19 (4): 417–423.

110 Saitta, K.S., Zhang, C., Lee, K.K. et al. (2014). Bacterial β-glucuronidase inhibition protects mice against enteropathy induced by indomethacin, ketoprofen or diclofenac: mode of action and pharmacokinetics. *Xenobiotica* 44 (1): 28–35.

111 Ervin, S.M., Hanley, R.P., Lim, L. et al. (2019). Targeting regorafenib-induced toxicity through inhibition of gut microbial β-glucuronidases. *ACS Chem. Biol.* 14 (12): 2737–2744.

112 Noh, K., Kang, Y.R., Nepal, M.R. et al. (2017). Impact of gut microbiota on drug metabolism: an update for safe and effective use of drugs. *Arch. Pharmacal Res.* 40 (12): 1345–1355.

113 Ly, L.K., Rowles, J.L. 3rd, Paul, H.M. et al. (2019). Bacterial steroid-17,20-desmolase is a taxonomically

rare enzymatic pathway that converts prednisone to 1,4-androstanediene-3,11,17-trione, a metabolite that causes proliferation of prostate cancer cells. *J. Steroid Biochem. Mol. Biol.* 199: 105567.

114 Selwyn, F.P., Cheng, S.L., Bammler, T.K. et al. (2015a). Developmental regulation of drug-processing genes in livers of germ-free mice. *Toxicol. Sci.* 147 (1): 84–103.

115 Selwyn, F.P., Cui, J.Y., and Klaassen, C.D. (2015b). RNA-seq quantification of hepatic drug processing genes in germ-free mice. *Drug Metab. Dispos.* 43 (10): 1572–1580.

116 Selwyn, F.P., Cheng, S.L., Klaassen, C.D., and Cui, J.Y. (2016). Regulation of hepatic drug-metabolizing enzymes in germ-free mice by conventionalization and probiotics. *Drug Metab. Dispos.* 44 (2): 262–274.

117 Toda, T., Saito, N., Ikarashi, N. et al. (2009). Intestinal flora induces the expression of Cyp3a in the mouse liver. *Xenobiotica* 39 (4): 323–334.

118 Bjorkholm, B., Bok, C.M., Lundin, A. et al. (2009). Intestinal microbiota regulate xenobiotic metabolism in the liver. *PLoS One* 4 (9): e6958.

119 Lau, G.S. and Critchley, J.A. (1994). The estimation of paracetamol and its major metabolites in both plasma and urine by a single high-performance liquid chromatography assay. *J. Pharm. Biomed. Anal.* 12 (12): 1563–1572.

120 Dahlin, D.C., Miwa, G.T., Lu, A.Y., and Nelson, S.D. (1984). *N*-Acetyl-*p*-benzoquinone imine: a cytochrome P-450-mediated oxidation product of acetaminophen. *Proc. Natl. Acad. Sci. U.S.A.* 81 (5): 1327–1331.

121 Holme, J.A., Dahlin, D.C., Nelson, S.D., and Dybing, E. (1984). Cytotoxic effects of *N*-acetyl-*p*-benzoquinone imine, a common arylating intermediate of paracetamol and *N*-hydroxyparacetamol. *Biochem. Pharmacol.* 33 (3): 401–406.

122 Clayton, T.A., Baker, D., Lindon, J.C. et al. (2009). Pharmacometabonomic identification of a significant host-microbiome metabolic interaction affecting human drug metabolism. *Proc. Natl. Acad. Sci. U.S.A.* 106 (34): 14728–14733.

123 Selmer, T. and Andrei, P.I. (2001). *p*-Hydroxyphenylacetate decarboxylase from *Clostridium difficile*. A novel glycyl radical enzyme catalysing the formation of *p*-cresol. *Eur. J. Biochem.* 268 (5): 1363–1372.

124 Kakan, X., Chen, P., and Zhang, J. (2011). Clock gene mPer2 functions in diurnal variation of acetaminophen induced hepatotoxicity in mice. *Exp. Toxicol. Pathol.* 63 (6): 581–585.

125 Kim, Y.C. and Lee, S.J. (1998). Temporal variation in hepatotoxicity and metabolism of acetaminophen in mice. *Toxicology* 128 (1): 53–61.

126 Gong, S., Lan, T., Zeng, L. et al. (2018). Gut microbiota mediates diurnal variation of acetaminophen induced acute liver injury in mice. *J. Hepatol.* 69 (1): 51–59.

127 Desgranges, C., Razaka, G., De Clercq, E. et al. (1986). Effect of (*E*)-5-(2-bromovinyl)uracil on the catabolism and antitumor activity of 5-fluorouracil in rats and leukemic mice. *Cancer Res.* 46 (3): 1094–1101.

128 Kanamitsu, S.I., Ito, K., Okuda, H. et al. (2000). Prediction of in vivo drug-drug interactions based on mechanism-based inhibition from in vitro data: inhibition of 5-fluorouracil metabolism by (*E*)-5-(2-bromovinyl)uracil. *Drug Metab. Dispos.* 28 (4): 467–474.

129 Ogura, K., Nishiyama, T., Takubo, H. et al. (1998). Suicidal inactivation of human dihydropyrimidine dehydrogenase by (*E*)-5-(2-bromovinyl)uracil derived from the antiviral, sorivudine. *Cancer Lett.* 122 (1–2): 107–113.

130 Okuda, H., Nishiyama, T., Ogura, K. et al. (1997). Lethal drug interactions of sorivudine, a new antiviral drug, with oral 5-fluorouracil prodrugs. *Drug Metab. Dispos.* 25 (5): 270–273.

131 Okuda, H., Ogura, K., Kato, A. et al. (1998). A possible mechanism of eighteen patient deaths caused by interactions of sorivudine, a new antiviral drug, with oral 5-fluorouracil prodrugs. *J. Pharmacol. Exp. Ther.* 287 (2): 791–799.

132 Zimmermann, M., Zimmermann-Kogadeeva, M., Wegmann, R., and Goodman, A.L. (2019). Separating host and microbiome contributions to drug pharmacokinetics and toxicity. *Science* 363 (6427): 600.

133 Chen, L.Z., Sabo, J.P., Philip, E. et al. (2015). Mass balance, metabolite profile, and in vitro-in vivo comparison of clearance pathways of deleobuvir, a hepatitis C virus polymerase inhibitor. *Antimicrob. Agents Chemother.* 59 (1): 25–37.

134 McCabe, M., Sane, R.S., Keith-Luzzi, M. et al. (2015). Defining the role of gut bacteria in the metabolism of deleobuvir: in vitro and in vivo studies. *Drug Metab. Dispos.* 43 (10): 1612–1618.

135 Sane, R.S., Ramsden, D., Sabo, J.P. et al. (2016). Contribution of major metabolites toward complex drug-drug interactions of deleobuvir: in vitro predictions and in vivo outcomes. *Drug Metab. Dispos.* 44 (3): 466–475.

**136** Dhaneshwar, S.S. (2014). Colon-specific prodrugs of 4-aminosalicylic acid for inflammatory bowel disease. *World J. Gastroenterol.* 20 (13): 3564–3571.

**137** Xu, J., Chen, N., Wu, Z. et al. (2018). 5-Aminosalicylic acid alters the gut bacterial microbiota in patients with ulcerative colits. *Front. Microbiol.* 9: 1274.

**138** Scalea, J.R., Levi, S.T., Ally, W., and Brayman, K.L. (2016). Tacrolimus for the prevention and treatment of rejection of solid organ transplants. *Expert Rev. Clin. Immunol.* 12 (3): 333–342.

**139** Zheng, Y.P., Masand, A., Wagner, M. et al. (2019). Identification of antibiotic administration as a potentially novel factor associated with tacrolimus trough variability in kidney transplant recipients: a preliminary study. *Transplant. Direct* 5 (9): e485.

**140** Lee, J.R., Muthukumar, T., Dadhania, D. et al. (2015). Gut microbiota and tacrolimus dosing in kidney transplantation. *PLoS One* 10 (3): e0122399.

**141** Guo, Y.K., Crnkovic, C.M., Won, K.J. et al. (2019). Commensal gut bacteria convert the immunosuppressant tacrolimus to less potent metabolites. *Drug Metab. Dispos.* 47 (3): 194–202.

**142** Brüssow, H. (2020). Parkinson disease, levodopa and the gut microbiota – when microbiology meets pharmacology. *Environ. Microbiol.* 22 (3): 808–812.

**143** Pellegrini, C., Antonioli, L., Colucci, R. et al. (2015). Gastric motor dysfunctions in Parkinson's disease: current pre-clinical evidence. *Parkinsonism Relat. Disord.* 21 (12): 1407–1414.

**144** Gabrielli, M., Bonazzi, P., Scarpellini, E. et al. (2011). Prevalence of small intestinal bacterial overgrowth in Parkinson's disease. *Mov. Disord.* 26 (5): 889–892.

# 26

# Drug–Drug Interactions and Drug–Dietary Chemical Interactions

*Mengbi Yang, Yuanfeng Lyu, and Zhong Zuo*

*School of Pharmacy, Faculty of Medicine, The Chinese University of Hong Kong, Hong Kong SAR*

## 26.1 Introduction

Nowadays, patients frequently use more than one medication at a time. The incidence of potential drug–drug interactions (DDIs) could be as high as 80% based on the specific clinical settings and the number of drugs prescribed [1]. DDIs can be mainly classified into pharmacokinetics-based interactions and pharmacodynamics-based interactions, leading to either favorable or toxic effects. Among our recent literature search of clinically evidenced DDIs reported since 2007, about 113 clinical studies are found to be related to pharmacodynamics-based DDIs, while 36 clinical studies are on pharmacokinetics-based DDIs. Most of the pharmacodynamics-based DDIs are related to disease conditions, which require complex combination therapies, including cancers, AIDS, and series of psychological disorders, and their clinical outcomes are usually additive effects that could be predicted based on the mechanism of actions for the drugs involved. Similarly, drug–dietary chemical interactions (DDCI) will occur via these pathways leading to either beneficial or adverse effects. The current chapter mainly focuses on understanding the clinical occurrence and related mechanisms of DDIs and DDCIs during the past three decades with the major interaction pathways covered in this chapter illustrated in Figure 26.1. In addition, updates on drug interactions with therapeutic biologics and interactions mediated by the gut microbiome will be introduced. Based on the amount of clinical and mechanistic evidence available, DDIs along with major mechanisms will be discussed, with emphasis on pharmacokinetics-based interactions during drug oral absorption, metabolism, and distribution, while clinical-evidenced DDCIs of commonly used dietary supplements will be highlighted.

## 26.2 Drug–Drug Interactions (DDIs)

Major pharmacokinetics-based DDIs in human is summarized in Table 26.1, in which the specific mechanism of interactions, victim drug, and perpetrator drug together with their pharmacokinetics outcomes are listed. The major pharmacokinetics-based DDI happened during the oral absorption, hepatic metabolism and tissue distribution, and kidney elimination processes. The related mechanisms include formation of insoluble complex in gastrointestinaal (GI) tract, alteration in gastric pH, GI motility, and activity of transporters and metabolic enzymes, which are further elaborated as follows:

### 26.2.1 Interactions Involved in Oral Absorption of Small Molecules

Since GI tract is the primary absorption site for orally administrated drugs, its physiological properties such as pH and motility will unavoidably affect the absorption of drugs. Despite a large number of drugs are absorbed via passive diffusion, alterations of intestinal influx and/or efflux transporters also play an important role in the oral absorption of certain drugs with transporters involved in their absorption process. In addition, induction or inhibition of the intestinal metabolizing enzymes can also modulate the first-pass metabolism in the GI tract, leading to significant alteration of drug bioavailability.

#### 26.2.1.1 Alteration of pH in the GI Tract

pH of the GI tract can be influenced by drug, food, drink, or other xenobiotics. The most common perpetrator drugs, including antacids, proton pump inhibitors, and histamine $H_2$ receptor antagonists, can increase gastric pH and lead to a pH-dependent dissolution and subsequent changes in the bioavailability/systemic

*Oral Bioavailability and Drug Delivery: From Basics to Advanced Concepts and Applications*, First Edition. Edited by Ming Hu and Xiaoling Li.

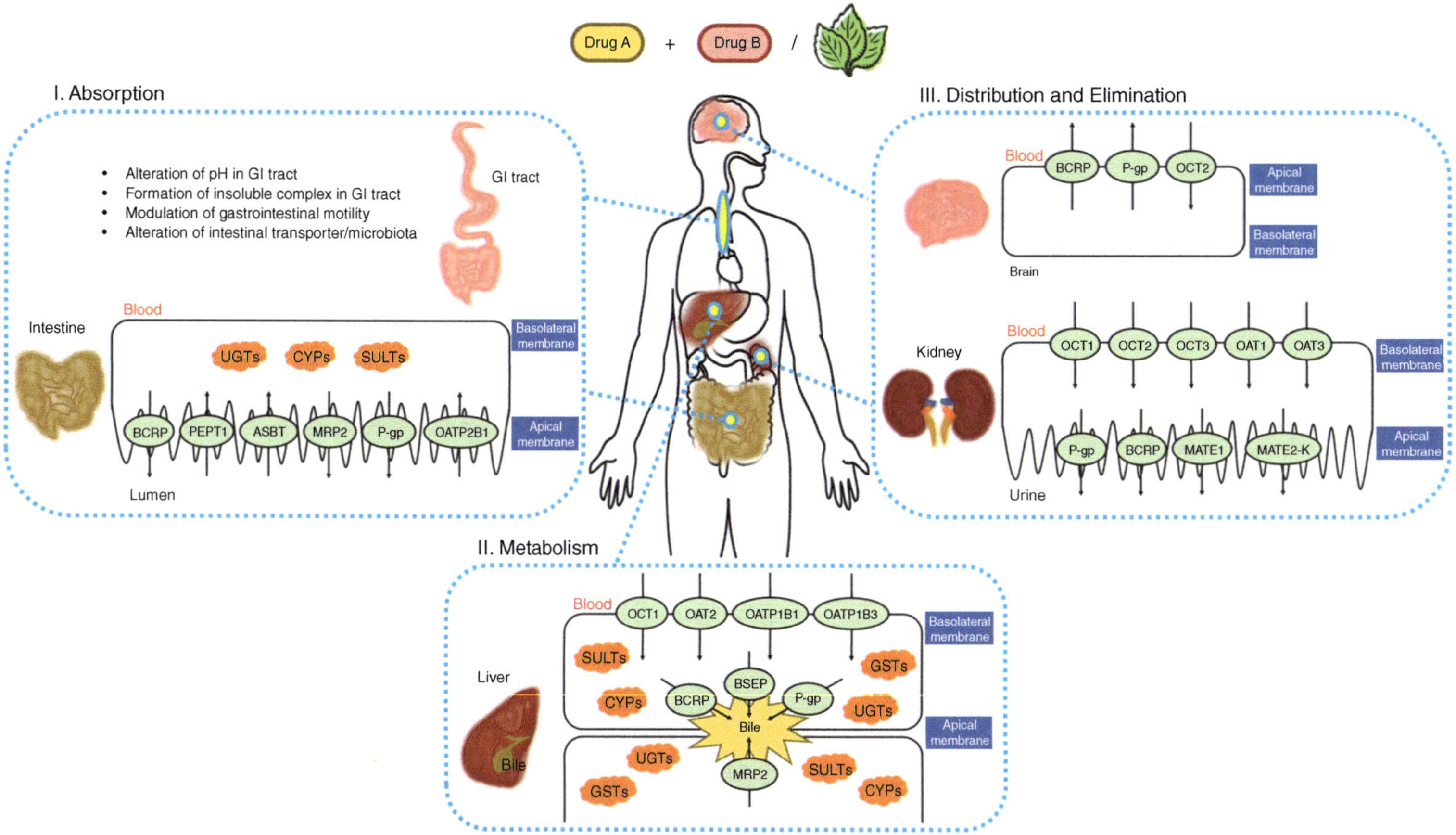


**Figure 26.1** Major pharmacokinetics based DDI and HDI pathways illustrated in this chapter.

**Table 26.1** Summary of major pharmacokinetics drug–drug interactions in human based on their mechanisms.

| Interaction types | Interaction mechanisms | Victim drug | Perpetrator drug | Outcomes of DDI | References |
|---|---|---|---|---|---|
| **Alteration of pH and formation of insoluble complex in the GI tract** | **↑ Gastric pH** | Cefpodoxime Proxetil, Cefuroxime Axetil, Cefditoren Pivoxil | $H_2$ Receptor Antagonists | ↓ 30 – 40% in F | [2] |
| | | Cefdinir, Cefpodoxime, Cefditoren | Antacids | ↓ 20 – 40% in F | [2] |
| | | Atorvastatin | Antacids | ↓ 38% in $C_{max}$ | [3] |
| | | Cefpodoxime Proxetil | Famotidine, Maalox | ↓ 40% in F | [4] |
| | | Indinavir | Omeprazole | ↓ 50% in AUC | [5] |
| | | Atazanavir | Omeprazole | ↓ 79% in F | [5] |
| | | | Lansoprazole | ↓ 94% in F | [6] |
| | | Ketoconazole | Omeprazole | ↓ 80% in AUC | [7] |
| | | Fluvastatin | Omeprazole, Cimetidine, Ranitidine | ↑ 43 – 70% in $C_{max}$ and 24–33% in AUC | [8] |
| | | Capecitabine | Proton Pump Inhibitors | ↓ Plasma concentration | [9] |
| | **↑ Gastric pH and Complex formation** | Enoxacin | Maalox | ↓ 73% in F | [10] |
| | | Pefloxacin | Aluminium, Magnesium | ↓ 56% in F | [11] |
| | **Complex formation** | Ciprofloxacin | Magnesium, Aluminum, Zinc, Calcium, | ↓ 50 – 90% in $C_{max}$ and AUC | [2] |
| | | | Iron | ↓ 57% in AUC | [12] |
| | | Moxifloxacin | Iron | ↓ 39% in AUC | |
| | | Sotalol | Maalox | ↓ 26% in $C_{max}$ and 21% in AUC | [13] |
| | | Mycophenolate Mofetil | Iron Ion Preparations | ↓ 90% in F | [14] |
| | | Lomefloxacin | Aluminium, Magnesium | ↓ 40% and 50% in AUC and $C_{max}$ | [15] |
| | | Cimetidine | Mylanta | ↓ 23% in AUC | [16] |
| | | Ranitidine | Mylanta | ↓ 26% in AUC | [16] |
| | | Famotidine | Mylanta | ↓ 19% in AUC | [16] |
| | | Nizatidine | Mylanta | ↓ 12% in AUC | [16] |
| **Modulation of GI motility** | **↓ GI motility** | Metformin | Propantheline | ↑ 19% in AUC | [17] |
| | | Desmopressin | Loperamide | ↑ 3.1-fold in AUC and 2.3-fold in $C_{max}$ | [18] |
| **Alteration of intestinal transporter** | **Competition for P-gp** | Digoxin | Talinolol | ↑ 23% in AUC and 45% in $C_{max}$ | [19] |
| | | | Atorvastatin | ↑ 15% AUC and 20% in $C_{max}$ | [20] |
| | **↓ P-gp** | Fexofenadine | Ketoconazole, Erythromycin | ↑ 2.5-fold in $C_{max}$ | [21] |
| | | Digoxin | Quinidine, Verapamil | ↑ $C_{max}$ | [21] |
| | | Ranitidine | Ketoconazole | ↑ 78% and 74% in $C_{max}$ and AUC | [22] |

*(continued)*

**Table 26.1** (Continued)

| Interaction types | Interaction mechanisms | Victim drug | Perpetrator drug | Outcomes of DDI | References |
|---|---|---|---|---|---|
| | **↑ P-gp** | Ranitidine | Rifampicin | ↓ 53% and 52% in $C_{max}$ and AUC | [22] |
| | | Digoxin | Rifampicin | ↓ 30.3% and 58% in AUC and $C_{max}$ | [23] |
| | | Talinolol | Rifampicin | ↓ 35% in AUC | [19] |
| | ↓ BCRP | Topotecan | Elacridar | ↑ 1.8-fold in F | [24] |
| **Alteration of first-pass metabolism via interplay between CYP3A4 and P-gp** | **↑ CYP3A and P-gp** | Itraconazole, Ketoconazole, Voriconazole, Diltiazem, Verapamil, Metoprolol, Propranolol, Carvedilol, Doxycycline, Fluconazole | Rifampin | ↓ $C_{max}$ up to 70% | [2] |
| | | Cyclosporine | Rifampin | ↓ 3.7 in F and 2.8-fold in $C_{max}$ | [25] |
| | | Nifedipine | Rifampicin | ↓ eightfold in F | [26] |
| | **↓ CYP3A and P-gp** | Lovastatin | Cyclosporine, Itraconazole | ↑ 20-fold in AUC | [26] |
| | | Atorvastatin | Erythromycin | ↑ 38% in $C_{max}$ and 33% in AUC | [16] |
| | | Cyclosporine | Ketoconazole | ↑ 5.3 in F and 2.7-fold in $C_{max}$ | [25] |
| **Interactions involving liver metabolic enzymes** | **Competition on CYP3A4** | Nevirapine | Mefloquine | ↓ 47% in median plasma concentration | [27] |
| | **↑ CYP3A4** | Bedaquiline | Rifampin | ↓ 45% in AUC | [28] |
| | **↓ CYP3A4** | Midazolam | Erythromycin, Clarithromycin, Telitheromycin | ↑ 2–6 fold in AUC | [2] |
| | | Simvastatin | Erythromycin, Clarithromycin, Telitheromycin | ↑ 4–8 fold in AUC | [2] |
| | | Cyclosporine, Tacrolimus | Erythromycin, Clarithromycin | ↑ 2–6 fold in $C_{max}$ | [2] |
| | | Clopidogrel | Ritonavir and Cobicistat | ↓ 69% in AUC of the active metabolite | [29] |
| | | Prasugrel | Ritonavir and Cobicistat | ↓ 52% and 41% in AUC and $C_{max}$ of active metabolite | [29] |
| | | Docetaxel | Ritonavir | ↑ 590% in AUC | [30] |
| | | | Ketoconazole | ↑ 210% in AUC | |
| | | Atorvastatin | Itraconazole | ↑ 63% in AUC | [31] |
| | **↓ CYP3A5** | Cyclosporine | Amlodipine | ↑ the plasma concentration | [32] |
| | **↓ CYP2D6** | Metoprolol | Diphenhydramine | ↓ 60% metabolic clearance in extensive metabolizers | [33] |
| | | Ecstasy | Paroxetine | ↑ 30% in $C_{max}$ | [34] |

*(continued)*

**Table 26.1** (Continued)

| Interaction types | Interaction mechanisms | Victim drug | Perpetrator drug | Outcomes of DDI | References |
|---|---|---|---|---|---|
| | | Nebivolol | Paroxetine | ↑ 510% in AUC | [35] |
| | | Atomoxetine | Paroxetine | ↑ 550% in AUC | [36] |
| | | | Fluvoxamine | ↑ 133% in AUC | [37] |
| | **↓ CYP2D6 and UGT2B10** | Atomoxetine | Bupropion | ↑ 71% and 410% in $C_{max}$ and AUC | [38] |
| | **↑ CYP2C9** | Phenytoin | Lopinavir and Ritonavir | ↓ 23% in AUC | [39] |
| | **↑ CYP2C9 and CYP2C19** | Voriconazole | Ritonavir | ↓ 66% and 82% in $C_{max}$ and AUC | [40] |
| **Interactions involving transporters in liver, kidney and other organs** | **↓ OATP1B1** | Ambrisentan | Clarithromycin | ↑ 27 % in $C_{max}$ | [41] |
| | **↓ OATP1B1** | Pravastatin | Ritonavir | ↑ 113 % in AUC in slow elimination population | [42] |
| | **↓ OCT2** | Metformin | Cimetidine | ↑ 20 % in AUC in slow elimination population | [43] |
| | **↓ OCT2 and MATE1** | Metformin | Ranitidine | ↓ 15 % in renal clearance in slow elimination population | [44] |
| | **↓ OATP2** | Rosuvastatin | Gemfibrozil | ↑ 121% and 88% in $C_{max}$ and AUC | [45] |

exposure of the victim drugs. Various clinical reports on oral drug absorption influenced by the alteration of pH in the GI tract are listed in Table 26.1. Oral bioavailabilities of cephalosporin antibacterial agents, such as cefpodoxime proxetil, cefuroxime axetil, and cefditoren pivoxil, could be significantly reduced by 30–40% when coadministered with perpetrator drug such as histamine $H_2$ receptor antagonist famotidine due to their decreased solubility at the increased gastric pH [2, 4]. Similarly, presence of antacid during the treatment with antibiotics, such as cefaclor, cefdinir, cefpodoxime, and cefditoren, could lead to 20–40% reduction in oral bioavailabilities of these victim antibiotics in human subjects [2]. Besides antibiotics, the gastric solubilities of victim drugs such as viral protease inhibitors are also pH dependent with decreased oral absorption observed when gastric pH increased. Drugs such as antacids, $H_2$ receptor antagonists, and proton pump inhibitors, which increase gastric pH may reduce the solubilities of the protease inhibitors in the GI tract. For example, coadministered omeprazole could decrease the aqueous solubility of indinavir from more than 100 mg/ml at pH 3 to less than 0.1 mg/ml at pH 5, leading to the subsequent 50% reduction in its oral bioavailability in healthy volunteers [5]. Coadministration with omeprazole or lansoprazole could lead to a potential virologic failure of atazanavir due to up to 79–94% reductions in its oral bioavailability [5, 6]. In addition to antibiotics and protease inhibitors, oral absorptions of several statins including atorvastatin and fluvastatin were found to be influenced by gastric pH. Coadministered oral antacid suspension containing magnesium and aluminum hydroxides could significantly decrease the plasma concentrations of both atorvastatin and its active metabolites by approximately 38% due to the reduced dissolution of atorvastatin in GI tract in presence of these antacids [3]. In contrast, coadministered histamine $H_2$-receptor antagonist, including omeprazole, cimetidine, or ranitidine, could increase gastric pH and reduce the GI tract degradation of fluvastatin, leading to its significantly increased $C_{max}$ (from 43% to 70%) and AUC (from 24% to 33%) [8, 46]. A more recent study found coadministered proton pump inhibitors could also reduce the systemic exposure of orally administered chemotherapy drug capecitabine [9].

#### 26.2.1.2 Formation of Insoluble Complex in GI Tract

Since antacids and sucralfates contain multivalent cations such as aluminum, magnesium, zinc, calcium, and iron, they would chelate the victim drugs to form insoluble complexes in the upper GI tract, resulting in their reduced oral absorptions. As listed in Table 26.1, the commonly reported victim drugs include $H_2$ receptor antagonists and various antibiotics, in particular quinolones and tetracyclines. Area under the curve (AUC) for plasma concentrations versus time profiles of $H_2$-receptor antagonists, cimetidine, ranitidine, nizatidine, and famotidine decreased by 23%, 26%, 12%, and 19%, respectively, when coadministrated with antacid Mylanta [16]. Tetracyclines are also prone to react with antacids, forming insoluble or poorly soluble chelates that lead to a significant reduction in their oral absorption in the GI tract [47]. Besides, fluoroquinolone antibacterial drugs interact with multivalent cation-containing agents to form insoluble complexes, leading to up to 90% reduction in their plasma $C_{max}$ and AUC [2, 14].

#### 26.2.1.3 Modulation of Gastrointestinal Motility

The rate and extent of oral absorption of many drugs can be significantly affected by the alteration of GI motility, such as the increase in the rate of gastric emptying and upper intestinal motility by GI prokinetic agents. Half an hour of pretreatment with propantheline in healthy volunteers prolonged gastric emptying and intestinal transit time, resulting in 19% increase in the plasma AUC of the coadministered metformin [17]. Similarly, pretreatment with loperamide could also slow the GI motility, leading to the 3.1- and 2.3-folds increase in plasma AUC and peak/maximum concentration ($C_{max}$) of co-treated desmopressin, respectively, but the time to reach the $T_{max}$ was markedly longer indicating the decreased rate of absorption [18].

#### 26.2.1.4 Alteration of Intestinal Transporters

During drug absorption process, lipophilic drugs are absorbed mainly by passive diffusion, whereas absorption of a number of aqueous soluble drugs are generally mediated by active influx and efflux transporters in the small intestine. Inevitably, DDI will take place when coadministered multiple drugs are substrates of the same transporters. In addition, certain drugs could inhibit or induce activity of the specific transporters, which may also lead to interactions with other drugs that are substrates of the transporters involved. Among all the transporters involved in the absorption processes, P-glycoprotein (P-gp) has been investigated most extensively. Many P-gp substrates drugs including digoxin, talinolol, clarithromycin, dipyridamole, and hydroxy-methylglutaryl coenzyme A reductase inhibitors and atorvastatin, may cause DDIs when they are administered together. Competition for the intestinal P-gp transporter, an efflux transporter located at the apical side of the intestinal epithelium membrane, during their absorption in the small intestine could lead to significantly increased oral bioavailability of coadministered digoxin and talinolol [19] or 15% and 20% rise in the AUC and $C_{max}$ of digoxin in presence of atorvastatin [20]. As also indicated in Table 26.1, clinically significant DDIs have been observed with the involvement of perpetrator drugs that are P-gp inhibitors, such as ketoconazole, erythromycin, verapamil, quinidine, probenecid, and cimetidine [21]. Besides, P-gp has been shown to be inducible by perpetrator drugs including rifampicin, phenobarbital, dexamethasone, isosafrole, clotrimazole, and reserpine [21]. Pretreatment with rifampicin resulted in a 3.5-fold increase in intestinal P-gp levels, leading to significantly decreased AUC (30.3%) and $C_{max}$ (58%) of coadministered digoxin in healthy volunteers [23]. Moreover, nine consecutive days pre-treatment with rifampicin (600 mg/day) could induce intestinal P-gp level for 4.2-fold and thus decrease the serum AUC (35%) of coadministered talinolol, a P-gp substrate [19].

In addition to P-gp, BCRP also has the potential to impact the oral bioavailability of its substrates. For example, Elacridar could inhibit the intestinal BCRP, leading to the increased bioavailability of coadministered topotecan to 2.8-fold [24]. Besides, other transporters expressed in the intestine include PEPT1, MRP2, OCT1, and OATPs and potential DDIs related to these transporters are also worth of investigation, especially for drugs with low membrane permeability. According to the FDA guidance, a drug candidate, which is a major P-gp or BCRP substrate with poor membrane permeability or solubility, is recommended to perform clinical drug–drug interaction study based on its safety margin, therapeutic index, and likely concomitant medications [48].

#### 26.2.1.5 Alteration of Intestinal First-Pass Metabolism Via Interplay Between Metabolic Enzymes and Transporters (see also Chapter 23)

It has been extensively reported that the metabolizing enzymes such as CYP3A4 and P-gp expressed in the small intestinal epithelial cells could possibly interplay during oral absorption process of a drug. The spatial relationship of P-gp across the plasma membrane and CYP3A4 inside the cell on the endoplasmic reticulum suggests that P-gp may act to regulate the exposure of substrate drugs to the CYP3A4-mediating metabolism in small intestinal epithelial cells. P-gp and CYP3A4 share many substrates and inhibitors and their common substrates can be absorbed via passive diffusion or possibly through an influx transporter into the enterocytes from lumen of the small intestinal tract, and then be moved directly into the systemic circulation or metabolized by CYP3A4 in the enterocytes or secreted back into the intestinal lumen via P-gp. Drug that is pumped back into the lumen can be reabsorbed at a distal site of the small intestine and enter the systemic circulation again, which creates a cycling effect. If one such drug is given concurrently with perpetrator drug that inhibits both P-gp and CYP3A4, such as ketoconazole, increased oral absorption and bioavailability of the victim drug will be observed. P-gp is considered as a "gate keeper" for the drug exposure to the later CYP3A4-mediating reactions in the cycling process.

The common substrates shared by CYP3A4 and P-gp include colchicine, cortisol, cyclosporine, diltiazem, etoposide, nicardipine, paclitaxel, rapamycin, saquinavir, tacrolimus, verapamil, and vinca alkaloids, while dexamethasone, phenobarbital, phenytoin, rifampin are the examples of their common inducers [25]. Although concurrent administration of either two of such substrates has a potential to cause interactions, the clinically significant DDIs are mainly resulted from the preparator drugs as the inducers or inhibitors of CYP3A4 and/or P-gp systems rather than their substrates only. Coadministration with rifampin, an inducer for both CYP3A4 and P-gp, could reduce the oral bioavailability and the $C_{max}$ of cyclosporine by about 3.7-fold and 2.8-folds, respectively, due to the increased apical efflux and first-pass metabolism in the GI tract [25]. Due to significant induction of the P-gp -mediated transport and CYP3A4-mediated first-pass metabolism in both intestine and liver, rifampin could significantly decrease the oral bioavailability of the coadministered nifedipine by almost eightfold, [26]. On the other hand, concomitant administration of CYP3A4 and P-gp inhibitors could enhance the oral bioavailability of victim drugs, which are the substrates of both P-gp and CYP3A4. Ketoconazole, the inhibitor for CYP3A4 and P-gp, significantly increased the oral bioavailability and $C_{max}$ of coadministered cyclosporine by about 5.3-fold and 2.7-fold, respectively [25]. Tacrolimus is a substrate of both CYP3A4 and P-gp and its oral bioavailability could be markedly increased in the presence of preparator drug ketoconazole, which could significantly inhibit its P-gp-mediated transport and/or the CYP3A4-catalyzed metabolism in the small intestine [49].

In addition to CYPs, intestinal phase II metabolism enzymes such as UDP-glucuronosyltransferases (UGTs) and sulfotransferases (SULTs) also contribution to the metabolic barrier in the intestine. The UGT1 and UGT2 families are primarily enzymes involved in drug glucuronidation. Among them, UGT1A1, UGT1A6, UGT1A10, UGT2A3, UGT2B7, and UGT2B17 are the most abundant within small intestine [50]. UGTs are sensitive to induction and inhibition by xenobiotics. SULT1B1 and SULT1A3 are the two predominant SULTs in the small intestine, accounting for 36 and 31% of the total, respectively [50]. SULTs are less responsive to induction than CYPs and UGTs [51]. Although many inhibitors/inducers on UGTs and SULTs have been reported based on their in vitro activity [52, 53], clinical evidences supporting the intestinal DDIs mediated by UGTs or SULTs are rare. Further studies on potential DDIs related to these enzymes are warranted.

#### 26.2.1.6 Alteration of Intestinal Microbiota (see also Chapter 25)

It is well known that intestinal microbiota can mediate drug metabolism including reduction, oxidation, dehydroxylation, and decarboxylation. In recent decade, intestinal microbiota has been identified as a vital drug target in many disease treatments and its changes in the GI tract may influence the metabolism of administered drugs, leading to their altered pharmacokinetics [52]. Unlike inhibition of metabolizing enzymes or transporters, alteration of intestinal microbiota may sustain even after the elimination of administered antibiotics from the intestine. Among the intestinal microbiota-mediated DDIs, interactions with antibiotics are most reported. Many antibiotics can disturb the pharmacokinetics of a coadministered drug via alterations on the enzymatic activities and composition of gut microbiota [54], leading to potential therapeutic effect changes. Research has demonstrated that two-thirds (176 out of 271 tested) of the assayed orally administered drugs are metabolized by at least one bacterial strain among 76 human gut bacteria [55]. Some of the metabolism have shown specificity to chemical structures. For example, Bacteroidetes strain specifically hydrolyze drugs that contain ester or amide groups. Compounds that are metabolized by most other bacteria contain a nitro or azo group, which is prone to reduction in anaerobic metabolism [55]. Metabolism of sulfinpyrazone to sulfinpyrazone sulfide in the gut contents and their plasma pharmacokinetic profiles were found to be altered in patients with the antibiotic metronidazole treatment [56]. In addition, altered microbiota composition may also lead to pharmacodynamics changes in the concomitant drugs. It was noticed that the presence of a certain type of bacteria may have an impact on chemotherapy and immunotherapy [57, 58]. Clinical trials are currently conducted on microbiota interventions, such as probiotics and fecal microbiota transplant, to explore their influence on the efficacy and toxicity of coadministered chemotherapeutic agents, immunotherapeutic agents, and anti-inflammatory drugs [59].

### 26.2.2 Pharmacokinetics Interactions Involving Liver Metabolic Enzymes

Since metabolic enzymes, including Phase I and Phase II, play an extremely important role in pharmacokinetics of drugs, understanding variation in their activity is crucial for better understanding related DDIs and improving the clinical use of drugs. Advances are being made in understanding the role of specific Phase I and Phase II enzymes in the pharmacokinetic-based DDIs as listed in Table 26.1 and illustrated below.

Among all the isoforms of enzymes involved in DDIs, cytochrome P450s (CYPs) are the ones being investigated the most as shown in Figure 26.1. As the major Phase I enzyme, CYPs mediates the oxidization of foreign substances to enhance their aqueous solubility for elimination from the body with CYP1 to CYP4 families focusing on oxidation of both exogenous and endogenous substrates [31, 52, 60–64]. CYP1A2, CYP2C9, CYP2C19, CYP2D6, CYP3A4, and CYP2B6 covered 75.9% of the hepatic CYPs enzymes in abundance and accounted for the metabolism of 97% of marketed drug. Among these isoenzymes, CYP3A plays the most important role in drug hepatic metabolism and is involved in many reported DDIs. Table 26.1 summarized the human studies with significant DDIs induced by the substrates and inducers of CYP3A4. For example, the plasma concentrations of docetaxel, a substrate of CYP3A4, can be significantly increased by three and sixfold by CYP3A inhibitors ritonavir and ketoconazole, respectively [30]; while the systemic exposure of another CYP3A4 substrate, bedaquiline can also be reduced by 45% when coadministrated with CYP3A4 inducer rifampin [28]. Besides CYP3A4, CYP2D6 has also been found mediating many DDIs. The CYP2D6 inhibitor paroxetine can significantly induce the systemic exposure of coadministrated drugs (including ecstasy, nebivolol, and atomoxetine) to up to sixfold, leading to undesired adverse effects [34–36]. Since large variations in protein content, mRNA levels, and intrinsic activities of 10 P450s (CYP3A4, 1A2, etc.) have been revealed and certain single nucleotide polymorphisms had a significant impact on P450 expression, clinical studies on DDIs mediated by such CYPs often take such polymorphisms into considerations to identify the potential extensive or poor metabolizer populations for safe use of medication.

Besides CYPs, conjugative enzymes including UDP-glucuronyltransferases (UGTs), sulfonyl transferases (SULTs), and glutathione S-transferases (GSTs) mainly mediate the phase II reactions. Among the UGTs, UGT1, and UGT2 families are glucuronidation enzymes involved in many reported DDIs [65]. For example, the systemic exposure of atomoxetine is increased 410% by co-treatment of Bupropion, which is believed a combined result of the inhibition of glucuronidation as well as CYP2D6 [38].

### 26.2.3 Pharmacokinetics Interactions Involving Transporters in Liver, Kidney, and Other Organs

Since transporters control the uptake and efflux of drugs and serve as the gatekeepers for cells, alteration of their function or expression may significantly change the systemic and tissue exposure of drugs, leading to significant changes in pharmacokinetics. As shown in Table 26.1, induction or inhibition of transporters by coadministered drugs can induce pharmacokinetics and pharmacodynamics DDI. Unlike the alteration of metabolic enzymes, which can be directly observed from clinical studies via the changes of pharmacokinetic profiles of the metabolites, modulation of transporters-mediated DDI are often evidenced by pre-clinical studies using cell lines or transgenic animals based on their strong indications of transporters' involvement from the clinical DDIs studies.

Figure 26.1 also summarizes the transporters in major organs that could be affected by drugs, leading to selective distribution of drugs to specific tissues. Efflux transporters like P-gp and BCRP on the blood–brain barrier can suppress the distribution of drugs into the brain, whereas the uptake transporters such as OCT2 could play key roles in the uptake of neuroactive drugs such as serotonin, norepinephrine, dopamine, choline, and histamine into the brain. In addition, transporters located in liver and kidney played primary roles in drug elimination. Since P-gp is responsible for the biliary excretion of drugs such as amphiphilic cationic drugs including doxorubicin, digoxin, and vinblastine, its inducer/inhibitor can not only alter the intestinal absorption but also the biliary excretion of these drugs. Compared to the DDI reports on P-gp, clinical evidence on DDIs mediated by other efflux transporters such as BCRP, MRP2, and MRP4 remain limited.

In addition to above-mentioned efflux transporters, the SLC superfamily (including OAT, OCT, and OATP families) transporters are responsible for drug uptake into hepatocytes and active secretion/reabsorption of drugs in the proximal tubule region of kidney. Among these transporters, OATP1B1/1B3 highly expressed in liver are considered to be of particular importance for hepatic drug elimination and related DDIs. Gemfibrozil and clarithromycin were found to significantly increase the $C_{max}$ and AUC of coadministered rosuvastatin and ambrisentan via inhibiting their OATP2 and OATP1B1 mediated hepatic eliminations, respectively [41, 45]. The OCTs and MATE families expressed in kidney are responsible for renal secretion of a variety of drugs. Ranitidine could inhibit the OCT2 and MATE1-mediated kidney elimination of metformin leading to reduction in its renal clearance [44].

It is known that groups of transporters and metabolizing enzymes including phase I enzymes (e.g. CYPs), phase II enzymes (e.g. UGTs and GST), and transporters (e.g. P-gp, BCRP, OATPs, and MRP2) are regulated by the nuclear receptors PXR and CAR [66]. These nuclear receptors can be activated by administered drugs, leading to the induction of related proteins. The expression of CAR and PXR is tissue specific. Both PXR and CAR are expressed in the liver. CARs have little extrahepatic expression. PXRs are also expressed in the intestine, while its expression in the kidney is significantly lower than that in liver. Therefore, DDIs involving transporter or enzyme inducers highly depend on the induction mechanism as well as the tissue concentrations of the inducers. Generally, inducer activate CARs may have major effect on the liver, while activator of PXRs may alter the expression levels of transporters and enzymes in both liver and intestine [66].

It was also noticed that some transporter-mediated DDIs can only be observed when genetic polymorphisms of transporters are taken into consideration. For example, an 808G > $T$ genetic variation of OCT2 has been found to be associated with a reduced metformin renal clearance. Coadministration of cimetidine in patients carrying this genetic variation can lead to 30% decrease in renal clearance and 20% increase in the AUC of metformin [43]. In other situation, reduced function variants do not show any significant pharmacokinetic change until after the addition of inhibitors. The subjects with MATE1 variant rs2289669 ($G > A$) only had significantly lower metformin renal clearance compared with carriers of GG genotype after treatment with the OCT2 and MATE1 inhibitor ranitidine [44].

### 26.2.4 Pharmacokinetics and Pharmacodynamics Interactions Between Drugs and Biologics

With the growing development of therapeutic biologics including therapeutic proteins, monoclonal antibodies (mAbs), vaccines, and peptide and nucleic acid derivatives, the risk of DDIs with biologics is increasing in clinical practice. The pharmacokinetics of biologics is generally quite different from those of small molecules. To avoid rapid degradation in the gastrointestinal tract after oral administration, most of the therapeutic biologics are administered via alternative routes including intravenous, intramuscular, and subcutaneous injections. Since the distribution of therapeutic biologics is mainly mediated by interstitial penetration, lymphatic drainage, transcytosis, and receptor-mediated cell uptake, therapeutic proteins usually have a limited volume of distribution and do not bind to plasma proteins with negligible biliary and renal excretion. Besides catabolism via proteolytic degradation, target antigen-mediated disposition also plays a role in the elimination of therapeutic proteins [52, 67]. Fragment crystallizable receptor-mediated antibody recycling by monocytes, macrophages, and dendritic cells is a salvage pathway that could prolong the half-lives of many mAbs. Since immune responses participate in both the catabolism and the antibody recycling process, immunogenicity can significantly influence the clearance of therapeutic proteins [68]. Compared to therapeutic proteins, nucleic acid and peptide drugs with high plasma binding undergo much slower renal excretion with elimination mainly by peptidases and nucleases [67, 69].

Although direct competition between therapeutic biologics and small molecules is not likely due to their distinct pharmacokinetic properties, DDIs may still occur via indirect pathyways. It was reported that certain biologics with immunomodulatory effects could alter CYP activities via modulating the cytokine levels and result in pharmacokinetics changes for victim drugs that are substrates of the related CYPs [70]. According to the newest FDA guideline on DDIs of therapeutic proteins, such CYP-mediated DDIs are based on three types of mechanisms: (i) the therapeutic protein is a proinflammatory cytokine, (ii) the therapeutic protein can cause an increase in proinflammatory cytokine levels, and (iii) the therapeutic protein can modulate proinflammtory cytokines in conditions with elevated cytokine levels [64]. Since tocilizumab could induce CYP3A4 activity via decreasing interleukin 6 levels, it could reduce the systemic exposure of coadministered CYP substrate drug simvastatin [71]. Similarly, influenza vaccination has been reported to decrease CYP activity by triggering inflammation, such that it could affect the systemic exposure of CYP substrate drug clozapine [72].

Besides affecting cytokine levels, therapeutic proteins may also alter pharmacokinetic profiles of coadministered small molecules via changing the physiological processes in human. GLP-1 receptor agonists such as dulaglutide and albiglutide could delay the gastric emptying and immunosuppressive agents may decrease the immunogenicity of the therapeutic protein [64]. Immunosuppressant methotrexate could decrease the clearance of coadministered mAbs including infliximab [73], golimumab [74], and adalimumab [75].

Similar to DDIs for small molecule drugs, pharmacodynamics-based DDIs of therapeutics biologics are more commonly reported than their pharmacokinetics-based DDIs. Owing to their complex signal pathways, insulin could interact with numerous drugs, including hormones, antidiabetics, antibiotics, and antipsychotics, and recombinant growth hormones could interact with small molecule hormones such as glucocorticoids, estrogens, and thyroxin [52, 68]. Although DDIs among biologics with similar indications usually result in additive or synergistic efficacy, they may also induce toxicity. Combination use of anakinra and etanercept could lead to severe adverse effects including increased risk of infection and neutropenia without significant improvement in the rheumatoid arthritis treatment [76].

## 26.3 Drug–Dietary Chemical Interactions in Oral Bioavailability

Dietary supplements, containing dietary chemicals with identified structures as their major ingredients, were defined by FDA as the products such as herbals, vitamins, minerals, and other products that people use in addition to the foods they eat. During the past two decades, supported by more and more clinical evidence, drug–dietary interactions, regardless of the mechanism involved, are frequently reported as a potential contributor to the altered therapeutics of drugs. Nowadays, it is common to see cautions for the potential drug–dietary chemical interactions appeared in the package labeling and/or inserts of a number of medications. Despite some of the interactions are beneficial, most cases reported interactions that would lead to various undesirable effects. The uniqueness of dietary supplement is that they quite often exist as a mixture of different dietary chemicals or an extract from various herbal plants with dosage forms like tablet or capsule. However, the identities of the active components and/or the quantities of each active component in the mixture or the extract are not always fully specified and the wide variation between different batches of herbal products also leads to poor reproducibility of the tests, which became the biggest barrier for the mechanistic studies of DDCIs. Unlike the DDIs, due to the complexity of the components in the dietary supplements and limited studies on the detailed pharmacokinetic parameters of specific dietary chemicals, very limited studies have reported the related pharmacokinetic changes of the specified dietary chemicals. Regardless, the level of evidence on DDCIs is still insufficient. Although the clinical evidenced DDCIs studies are getting more and more, majority of their clinical studies were conducted in healthy populations, where the impact of the herbs on the pharmacodynamics effects of the concurrent drug may not be determined. DDCIs basically share the same mechanisms as DDIs. Alteration of pH in the GI tract and gastrointestinal motility was barely reported. To avoid physio-chemical interactions between herbal components and drugs, it is usually recommended that herbs should be taken two hours before or after the drugs.

As summarized in Table 26.2, majority of the clinical evidenced DDCIs are related to the alteration of liver metabolic enzymes and transporters. Regarding the victim drugs involved in DDCIs, effect of herbs/herbal components on the pharmacokinetics and pharmacodynamics of drugs with narrow therapeutic indexes, especially drugs for CNS and cardiovascular diseases, are often reported [116]. Among all the clinically evidenced DDCIs, garlic, pepper, St John's Wort, ginkgo biloba, ginger, ginseng, curcumin, grapefruit juice, and vitamins are the most commonly reported ones with possible mechanisms investigated and highlighted below, whereas other commonly used dietary supplements such as milk thistle, cranberry juice, and black cohosh extract seem to have minimal impact on their coadministered drugs.

### 26.3.1 Garlic

Garlic is one of the most popular dietary substances with various beneficial effects. Coadministrations with garlic have been reported to alter pharmacokinetics of

**Table 26.2** Summary of major dietary product with Drug–dietary chemical interactions in human.

| Dietary supplements | Substance (Content)/Component | Dose and Dosing frequency | Victim drug (dose and dosing frequency) | Effect of interaction[a] | Potential mechanism[b] | References |
|---|---|---|---|---|---|---|
| Garlic | Extract (NA) | 1 caplet, BID, day 5~25 | Saquinavir (1200 mg, TID, day 1~4, 22~25, and 36~39) | ↓ 51% in AUC and 54% in $C_{max}$ | ↑ CYP3A4 | [77] |
| | Enteric-coated garlic tablets (2000 mg of fresh garlic bulb; 3.71 mg of allicin per tablet) | 1 tablet, BID, 2-week | *Rac*-warfarin (25 mg) | = PK | NA | [78] |
| Pepper | Piperine (NA) | 20 mg, QD, 7-day | Propranolol (40 mg) | ↓ $T_{max}$, $C_{max}$ and AUC | ↓ CYP1A1 | [79] |
| | | | Theophylline (150 mg) | ↑ $C_{max}$ and AUC | ↓ CYP1A2 | |
| | | 1 g | Phenytoin (300 mg) | ↑ AUC | ↓ P-gp | [80] |
| | Kava kava rhizome extract (75 mg kavalactones per capsule) | 1227 mg, TID | Midazolam (8 mg) | = PK | Not a potent CYP3A inhibitor *in vivo* | [60] |
| | | 136.3 mg, TID, 14-day | Debrisoquine (5 mg) | = Debrisoquine urinary recovery ratios | Not a potent CYP2D6 inhibitor *in vivo* | [61] |
| St John's Wort | Extract (NA) | 300 mg, TID | Cyclosporine (NA) | ↓ 47% in steady state concentration | ↑ CYPs | [81] |
| | | 300 mg, TID, 14-day | Indinavir (800 mg) | ↓ 57% in AUC | ↑ CYPs | [82] |
| | | | Alprazolam (2 mg) | ↓ twofold in AUC | ↑ CYPs | [62] |
| | | | Digoxin (0.5 mg) | ↓ 18% in AUC | ↑ P-gp | [83] |
| | | 300 mg, TID, 30-day | Fentanyl (NA) | = PK | Not known | [84] |
| | Extract (3% hyperforin) | 300 mg, TID, 14-day | Debrisoquine (5 mg) | = Debrisoquine urinary recovery ratios | Not a potent CYP2D6 inhibitor *in vivo* | [61] |
| | Jarsin 300® (NA) | 300 mg, TID, 12-day | Talinolol (50 mg, po) | ↓ 25% in AUC | ↑ P-gp | [85] |
| | | | Talinolol (30 mg, iv) | ↑ 35% in $CL_{nonrenal}$ | | |

*(continued)*

**Table 26.2** (Continued)

| Dietary supplements | Substance (Content)/Component | Dose and Dosing frequency | Victim drug (dose and dosing frequency) | Effect of interaction[a] | Potential mechanism[b] | References |
|---|---|---|---|---|---|---|
| Ginkgo Biloba | Extract (NA) | 140 mg, BID, 12-day | Omeprazole (40 mg) | ↓ AUC ratio of omeprazole/5-OH omeprazole | ↑ CYP2C19 | [86] |
| | | 120 mg, BID, 2-week | Midazolam (8 mg) | ↓ 34% in AUC and 31% in $C_{max}$ | ↑ CYP3A | [87] |
| | | | Fexofenadine (120 mg) | = PK | NA | |
| | | 360 mg, QD, 14-day | Talinolol (100 mg) | ↑ 22% in AUC and 36% in $C_{max}$ | ↓ P-gp | [88] |
| Curcumin | Nanoparticles | 3 g, QD, 12-week | Acitretin (0.4 mg/kg, QD, 12-week) | = Cholesterol serum levels | NA | [89] |
| | ZhiKePing,<br>50 mg capsule contained demethoxycurcumin and bisdemethyoxycur-cumin | 300 mg, QD, 6-day | Talinolol (50 mg) | ↓ AUC and $C_{max}$; ↑ CL/F | May relate to upregulation of further ATP-binding cassette transporters | [90] |
| *Curcuma rhizomes* | Extract (NA) | 1.5 g, BID, 7-day | Dextromethorphan (30 mg) | ↑ Urinary metabolic ratio of dextromethorphan /dextrorphan | ↓ CYP2D6 | [91] |
| Grapefruit juice | | NA | Cyclosporine (5 mg/kg) | ↑ 45% in AUC | ↓ CYP3A4 | [92] |
| | | 240 ml, 7-day | Terfenadine (60 mg, BID, 14-day) | ↑ AUC and $C_{max}$ | ↓ CYP3A4 | [93] |
| | | NA | Midazolam (NA) | ↑ AUC | ↓ CYP3A4 | [94] |
| | | 250 ml, 13~38 days | Diltiazem (120 mg) | ↑ 20% in AUC | ↓ P-gp or CYPs | [95] |
| | | 200~300 ml | Felodipine (10 mg) | ↑ 47% in AUC and 50% in $C_{max}$ | ↓ CYP3A4 | [96] |
| Caffeine | | 1.5 g/m$^2$/day, 3-day continuously, 2~5 courses, 2-week intervals from June 1989 to Dec 1991 | Cisplatin (120 mg/m$^2$, intra-arterial, 1-day, 2~5 courses, 2-week intervals) | Radiographic responders at stages II and III were significantly improved compared to the non-responders ($p = 0.004$ and $p = 0.034$) | NA | [97] |

| | | | | | | |
|---|---|---|---|---|---|---|
| | | 1.5 g/$m^2$/day, 3-day continuously, 2~5 courses, 3-week intervals from Jan 1992 to Dec 1995 | Cisplatin (120 mg/$m^2$, 1-day); Doxorubicin (30 mg/$m^2$/day, 2-day); 2~5 courses, 3-week intervals | | | |
| | | 1.5 g/$m^2$/day, 3-day continuously, 3~5 courses, 3-week intervals from Jan 1996 to Dec 1998 | Cisplatin (120 mg/$m^2$, 1-day); Doxorubicin (30 mg/$m^2$/day, 2-days) after radiation therapy (2 Gy/day, 5-day); 3~5 courses, 3-week intervals | | | |
| Vitamin | Vitamin D | 100000 U bolus intramuscularly + 50000 U orally weekly, 24-week | Budesonide or budesonide + formoterol (NA) | ↑ Forced expiratory volume in 1 second [$FEV_1$] | NA | [98] |
| | Vitamin C | 500 mg, BID, 90-day | Metformin (500 mg, BID, 90-day) | - Improved fasting blood sugar, HbA1c, lipid profile, insulin, homeostasis model assessment of insulin resistance;<br>- Reduced glutathione and Quantitative Insulin Sensitivity Check Index | NA | [99] |
| | Vitamin E | 400 mg, BID, 90-day | | | | |
| | Vitamin C + Vitamin E | Vitamin C 500 mg + Vitamin E 400 mg, BID, 90-day | | | | |
| Vitamin | Vitamin B12 + Folic acid | Vitamin B12 1000 μg, every 9 weeks, starting 1 week before chemotherapy, finishing at least 3 weeks after the last treatment dose;<br>Folic acid 450 μg, starting at least 1 week prior to chemotherapy, finishing at least 3 weeks after the last treatment dose | Gemcitabine (1250 mg/$m^2$, iv, day-1 and 8);<br>Cisplatin (80 mg/$m^2$, iv, day-1 in a 3-weekly cycle) | = PK | NA | [100] |

*(continued)*

**Table 26.2** (Continued)

| Dietary supplements | Substance (Content)/Component | Dose and Dosing frequency | Victim drug (dose and dosing frequency) | Effect of interaction[a] | Potential mechanism[b] | References |
|---|---|---|---|---|---|---|
| | Folic acid | 0.8 mg, median 4.4-year | Enalapril (10 mg, median 4.4-year) | ↓ Development of proteinuria in diabetic patients with hypertension | NA | [101] |
| Milk thistle | Silymarin 140 mg; 173.0~186.7 mg dry extract of milk thistle per capsule | 4 capsules | Nifedipine (10 mg) | = Absorption or metabolism of nifedipine | Not a potent CYP3A4 inhibitor *in vivo* | [102] |
| | 80% silymarin per capsule | 300 mg, TID,14-day | Debrisoquine (5 mg) | = Debrisoquine urinary recovery ratios | Not a potent CYP2D6 inhibitor *in vivo* | [61] |
| Black cohosh | Extract (2.5% triterpene glycosides per tablet) | 40 mg, BID, 14-day | | | | |
| *Echinacea purpurea* | Extract (2.2 mg isobutylamides per capsule) | 267 mg, TID, 14-day | | | | |
| | Extract (NA) | 20 drops, TID, day-7 to day-21) | Docetaxel (135 mg, iv, day-1 and 22) | = PK | NA | [103] |
| Echinacea | 600 mg of E. angustifolia roots and 675 mg of E. purpurea root | 1275 mg, QID, 7-day | Warfarin (25 mg) | ↓ AUC of (S)-warfarin | NA | [104] |
| Cranberry juice | 500 mg cranberry juice concentrate per tablet | 2 capsules, TID, 2-week | *Rac*-warfarin (25 mg) | = PK | NA | [78] |
| | 27% cranberry | 180 ml, BID, 5-day | Diclofenac (25 mg) | | Not a potent CYP2C9 modulator *in vivo* | [105] |
| Resveratrol | NA | 500 mg, QD, 10-day | Chlorzoxazone (250 mg) | ↑ AUC and $C_{max}$; ↓ CL/F | ↓ CYP2E1 | [106] |
| | | | Diclofenac (100 mg) | ↑ AUC and $C_{max}$; ↓ CL/F | ↓ CYP2C9 | [106] |
| Shiitake mushroom | NA | 250 g, TID, 1-day | Gabapentin (600 mg) | ↓ $C_{max}$ | NA | [107] |
| Goldenseal (*Hydrastis canadensis*) | Root extract (24.1 mg isoquinoline alkaloids per capsule) | 1070 mg, TID, 14-day | Debrisoquine (5 mg) | ↓ Debrisoquine urinary recovery ratios | ↓ CYP2D6 | [61] |
| | | 1323 mg, TID, 14-day | Midazolam (8 mg) | ↑ AUC and $C_{max}$; ↓ CL/F | ↓ CYP3A | [60] |

| | | | | | | |
|---|---|---|---|---|---|---|
| *Radix astragali* | Extract granules (NA) | 4 g, BID, 7-day | Fexofenadine (120 mg) | = Systematic exposure of fexofenadine | Not a potent modulator of P-gp *in vivo* | [108] |
| *Radix/rhizoma notoginseng* | Extract (Sanchitongshu capsule: 100 mg panaxatriol saponin contained 50% Ginsenoside Rg1, 6% Ginsenoside Re, 11% Notoginsenoside R1 per capsule) | 1 capsule (200 mg), QD, 28-day | Aspirin (50 mg, QD, 28-day) | Ameliorated neurological deficit (increased score of ESS: $t = -5.02$, $p < 0.0001$) and activities of daily living (increased score of BI: $t = -2.4$, $p = 0.0178$) | NA | [109] |
| *Schisandra sphenanthera* | Extract (11.25 mg deoxyschizandrin per capsule) | 1 capsule, BID, 7-day | Midazolam (15 mg) | ↑ AUC and $C_{max}$; ↓ 52% in CL/F | ↓ CYP3A | [110] |
| *Hibiscus sabdariffa* | Aqueous extract (NA) | 300 ml | Simvastatin (40 mg) | ↑ 44.6% in CL/F; ↓ 18% in $C_{max}$ and 30.8% in AUC | NA | [111] |
| | | 300 ml, QD, 2-day | Diclofenac (25 mg) | ↓ Excretion | | [112] |
| | | 1 000 ml Zobo drink | Acetaminophen (1 g) | ↑ CL/F | | [113] |
| | | 300 ml Karkadi (Sudanese beverage) | Chloroquine (600 mg) | ↓ AUC and $C_{max}$ | | [114] |
| *Moringa oleifera* | Leaf suspension (NA) | 3 g, QD, 7-day | Amodiaquine (10 mg/kg) | ↓ $C_{max}$ | May relate to CYP2C8 | [115] |

*Notes*: [a]↑, ↓, =: "increase," "decrease," and "no significant effect," respectively; [b]↑ and ↓: "induce" and "inhibit," respectively; NA: not available.

several therapeutic drugs. Due to the potential induction of CYP3A4 activity by garlic, the plasma AUC and $C_{max}$ of coadministered saquinavir decreased by 51 and 54%, respectively [77]. However, no alteration of warfarin pharmacokinetics by its coadministered garlic tablet was observed [78].

### 26.3.2 Pepper

Piperine is an alkaloid naturally occurred in plants such as black pepper and long pepper. Black pepper has been used in the traditional Chinese medicinal practice to treat seizure disorders, and its active ingredient, piperine, exhibited various beneficial effects including anticarcinogenic and anti-inflammatory activities. Several studies demonstrated that concomitant ingestion of piperine conferred potential alterations of bioavailabilities of various drugs. In a crossover clinical study, six healthy subjects in each group received a single oral dose of propranolol or theophylline alone or in combination with piperine daily for seven days. The resultant shorter $T_{max}$ as well as higher $C_{max}$ and AUC were observed in the subjects who received both piperine and propranolol. Similarly, higher $C_{max}$ and AUC values and a longer elimination half-life of theophylline were found with coadministration of piperine with theophylline. Since propranolol and theophylline are mainly metabolized by CYP1A1 and CYP1A2, the enhanced bioavailability of both propranolol and theophylline were suggested to be due to the potential inhibition of CYP1A1/2 by piperine [79]. In addition, the influence of piperine on P-gp transporter was also suggested in some studies. Treatment with piperine in rodents and human resulted in increased plasma concentrations of phenytoin, the substrates of P-gp, indicating the inhibition of P-gp by piperine [80]. Series of clinical HDI studies using probe drug cocktails from Gurley et al. demonstrated no inhibition on CYP3A or 2D6 by pepper [60, 61].

### 26.3.3 St John's Wort

As a popular herbal dietary supplement for the treatment of depression, St John's Wort was also a well-known inducer for both CYP and P-gp. Its concomitant administration with cyclosporine was reported to lead to a potential risk of transplant rejection in kidney transplant patients due to a 47% reduction in systemic cyclosporine concentration [81]. In addition, St John's Wort was also found to reduce the AUC of HIV-1 protease inhibitor indinavir by approximately 57% [82]. Such decreased oral bioavailabilites of various drugs by St John's Wort were found to be mainly related to its potential induction of CYP3A4. A 14-day course of St John's Wort administration caused twofold decrease in the plasma concentrations and AUC of alprazolam (substrate of CYP3A4) and also twofold increase in alprazolam clearance [62]. Since CYP3A4 substrates account for at least 50% of all currently marketed drugs, caution should be made with the concomitant intake of St John's Wort with therapeutic drugs that are substrates of CYP3A4.

In addition to the effects on CYP3A4, the influence of St John's Wort on P-gp was also suggested to be responsible for the altered oral bioavailability of various drugs. Two-week treatment with St John's Wort extract to healthy volunteers resulted in 1.5-fold increased expressions of duodenal P-gp and CYP3A4 in both intestine and liver and a subsequent 18% decrease in oral bioavailability of coadministered digoxin, the substrate of both P-gp and CYP3A4 [83]. Similarly, increased P-gp expression in human duodenal mucosa was found after 12-day treatment of St John's Wort, which could lead to 25% reduction on the oral bioavailability of talinolol, a non-CYP3A4 metabolizable P-gp substrate [85].

### 26.3.4 Ginkgo Biloba

Ginkgo biloba, originating from ancient China, has been widely used for memory improvement in the world. Several clinical studies demonstrated the induction of cytochrome P450 enzymes including CYP2C19 and 3A by ginkgo biloba. Pretreatment of ginkgo biloba extract for 12 days would increase the CYP2C19-mediated metabolism of omeprazole to its metabolite 5-hydroxyomeprazole, which was more significant in CYP2C19 poor metabolizers [86]. Similarly, a four-week treatment with ginkgo biloba decreased the AUC and $C_{max}$ of midazolam (probe of CYP3A) by 34% and 31%, respectively in 14 healthy volunteers [87]. Besides inducing the CYPs, Ginkgo Biloba was found to be able to increase AUC and $C_{max}$ of Talinolol via inhibition of P-gp [88].

### 26.3.5 Curcumin and Curcuma Rhizomes Extract

Curcuma rhizomes extract with curcumin as the major component is a mixture predominantly composed of turmeric root extract and other spices such as coriander and fenugreek. It has been shown to possess activity in the treatment and prevention of cancer, multiple sclerosis, and Alzheimer's disease. A study in healthy volunteers suggested an upregulation of P-gp by curcumin based on the observed evidence of the decreased

$C_{max}$ and AUC values of coadministered talinolol, a P-gp substrate [90].

### 26.3.6 Grapefruit Juice

Since 1990s, extensive reports on the potential interactions between flavonoids containing grapefruit juice and various coadministered drugs have stimulated remarkable interest in the studies of dietary flavonoids. Grapefruit juice was reported to increase oral bioavailability of cyclosporine by 45% and decrease the $C_{max}$ and AUC of its metabolite by 21% and 15%, respectively [92]. Further studies on grapefruit juice consistently indicated that it could markedly improve oral bioavailability of CYP3A4 substrates including felodipine [96], terfenadine [93], and midazolam [94]. It was speculated that such inhibitory effects on CYP3A4 and/or P-gp by grapefruit juice might be responsible for the reported increased oral bioavailability of various drugs, which are CYP3A4 and/or P-gp substrates. In grapefruit juice takers, there was about 20% increase in the AUC of diltiazem, the substrate of both CYP3A4 and P-gp [95]. It was suggested that flavonoids, including naringin, naringenin, quercetin, and kaempferol, were the active ingredients in grapefruit juice responsible for the inhibition of intestinal CYP3A4 activity and could lead to the increased oral bioavailabilities of coadministered drugs, which are CYP3A4 substrates [117].

### 26.3.7 Caffeine

Caffeine could be coadministered with Cisplatin and Doxorubicin for the chemotherapy medication. The response rates of radiographic and histological indicated that caffeine-participated chemotherapy could be beneficial to the survival of cancer patients [97].

### 26.3.8 Vitamins

Unlike the often-reported pharmacokinetics-based DDCIs for above-mentioned dietary supplements, pharmacodynamics-based interactions are commonly reported for vitamins. Coadministered vitamin C and vitamin E could improve fasting blood sugar and decrease the production of glutathione [99], whereas coadministration of folic acid and enalapril could lead to the development of proteinuria in diabetic patients with hypertension [101].

## 26.4 Summary

Knowledges on both DDIs and DDCIs are highly important to ensure the safety and efficacy of multiple drug therapy. DDIs mainly occurred during intestinal absorption, hepatic metabolism, kidney-mediated elimination, and tissue distribution with intestinal pH, mobility, CYPs, and transporters in both intestine and liver as the key factors. With the recent research and development, intestinal microbiota-mediated DDIs and drug–biologics interactions set the future research directions for DDI. Comparing with DDIs, clinical evidenced studies on DDCIs and related mechanisms remain insufficient, which are mainly due to the complexity of the compositions of dietary chemicals, inconsistent quality of the dietary products, and the lack of well-designed clinical trials. Further mechanistic studies on dietary chemicals and dietary supplements with known active ingredients and compositions are needed for better understanding of the risk, incidence, and mechanisms of drug–dietary chemical interactions so as to accurately predicate their potential outcomes in the drug therapy.

## Abbreviations

| | |
|---|---|
| DDIs | drug–drug interactions |
| DDCIs | drug–dietary chemical interactions |
| GI | gastrointestinal |
| AUC | area under the curve |
| CYP | cytochrome P450 |
| UGT | uridine diphosphate glucuronosyl transferases |
| SULT | sulfotransferases |
| MRP2 | multidrug resistance protein 2 |
| P-gp | P-glycoprotein |
| OCT | organic cation transporter |
| OAT | organic anion transporter |
| BCRP | breast cancer resistance protein |
| BESEP | bile salt export pump |
| MATE | multidrug and toxic compound extrusion |
| mAbs | monoclonal antibodies |

## References

1 Kannan, B., Nagella, A.B., Prabhu, A.S. et al. (2016). Incidence of potential drug-drug interactions in a limited and stereotyped prescription setting-comparison of two free online Pharmacopoeias. *Cureus* 8, e886.
2 Pai, M.P., Momary, K.M., and Rodvold, K.A. (2006). Antibiotic drug interactions. *Med. Clin. North Am.* 90: 1223–1255.
3 Williams, D. and Feely, J. (2002). Pharmacokinetic-pharmacodynamic drug interactions with HMG-CoA reductase inhibitors. *Clini. Pharmacokinet.* 41: 343–370.
4 Saathoff, N., Lode, H., Neider, K. et al. (1992). Pharmacokinetics of cefpodoxime proxetil and interactions with an antacid and an $H_2$ receptor antagonist. *Antimicrob. Agents Chemother.* 36: 796–800.
5 Winston, A. and Boffito, M. (2005). The management of HIV-1 protease inhibitor pharmacokinetic interactions. *J. Antimicrob. Chemother.* 56: 1–5.
6 Tomilo, D.L., Smith, P.F., Ogundele, A.B. et al. (2006). Inhibition of atazanavir oral absorption by lansoprazole gastric acid suppression in healthy volunteers. *Pharmacotherapy* 26: 341–346.
7 Blume, H., Donath, F., Warnke, A., and Schug, B.S. (2006). Pharmacokinetic drug interaction profiles of proton pump inhibitors. *Drug Safety* 29: 769–784.
8 Bays, H. and Dujovne, C. (1998). Drug interactions of lipid-altering drugs. *Drug Safety* 5: 355–371.
9 Chu, M.P., Hecht, J.R., Slamon, D. et al. (2017). Association of proton pump inhibitors and capecitabine efficacy in advanced gastroesophageal cancer: secondary analysis of the TRIO-013/LOGiC randomized clinical trial. *JAMA Oncol.* 3: 767–773.
10 Grasela, T.H., Schentag, J.J., Sedman, A.J. et al. (1989). Inhibition of enoxacin absorption by antacids or ranitidine. *Antimicrob. Agents Chemother.* 33: 615–617.
11 Jaehde, U., Sörgel, F., Stephan, U., and Schunack, W. (1994). Effect of an antacid containingmagnesium and aluminumon absorption,metabolism, and mechanism of renal elimination of pefloxacin in humans. *Antimicrob. Agents Chemother.* 38: 1129–1133.
12 Aminimanizani, A., Beringer, P., and Jelliffe, R. (2001). Comparative pharmacokinetics and pharmacodynamics of the newer fluoroquinolone antibacterials. *Clini. Pharmacokinet.* 43: 169–187.
13 Laer, S., Neumann, J., and Scholz, H. (1997). Interaction between sotalol and an antacid preparation. *Br. J. Clin. Pharmacol.* 43: 269–272.
14 Morii, M., Ueno, K., Ogawa, A. et al. (2000). Impairment of mycophenolate mofetil absorption by iron ion. *Clin. Pharmacol. Ther.* 68: 613–616.
15 Shimada, J., Shiba, K., Oguma, T. et al. (1992). Effect of antacid on absorption of the quinolone lomefloxacin. *Antimicrob. Agents Chemother.* 36: 1219–1224.
16 Sullivan, T.J., Reese, J.H., Jauregui, L. et al. (1994). Short report: a comparative study of the interaction between antacid and $H_2$ receptor antagonists. *Aliment. Pharmacol. Ther.* 8: 123–126.
17 Marathe, P.H., Wen, Y., Norton, J. et al. (2000). Effect of altered gastric emptying and gastrointestinal motility on metformin absorption. *Br. J. Clin. Pharmacol.* 50: 325–332.
18 Callréus, T., Lundahl, J., Höglund, P., and Bengtsson, P. (1999). Changes in gastrointestinal motility influence the absorption of desmopressin. *Eur. J. Clin. Pharmacol.* 55: 305–309.
19 Takano, M., Yumoto, R., and Murakami, T. (2006). Expression and function of efflux drug transporters in the intestine. *Pharmacol. Ther.* 109: 137–161.
20 Boyd, R.A., Stern, R.H., Stewart, B.H. et al. (2000). Atorvastatin coadministration may increase digoxin concentrations by inhibition of intestinal P-glycoprotein-mediated secretion. *J. Clin. Pharmacol.* 40: 91–98.
21 Ayrton, A. and Morgan, P. (2001). Role of transport proteins in drug absorption, distribution and excretion. *Xenobiotica* 31: 469–497.
22 Machavaram, K.K., Gundu, J., and Yamsani, M.R. (2006). Effect of ketoconazole anf rifampicin on the pharmacokinetics of ranitidine in healthy human volunteers: a possible role of P-glycoprotein. *Drug Metabol. Drug Interact.* 22: 47–65.
23 Greiner, B., Eichelbaum, M., Fritz, P. et al. (1999). The role of intestinal P-glycoprotein in the interaction of digoxin and rifampin. *J. Clin. Invest.* 104: 147–153.
24 Kuppens, I.E., Witteveen, E.O., Jewell, R.C. et al. (2007). A phase I, randomized, open-label, parallel-cohort, dose-finding study of elacridar (GF120918) and oral topotecan in cancer patients. *Clin. Cancer Res.* 13: 3276–3285.
25 Wacher, V.J., Silverman, J.A., Zhang, Y., and Benet, L.Z. (1998). Role of P-glycoprotein and cytochrome

P450 3A in limiting oral absorption of peptides and peptidomimetics. *J. Pharm. Sci.* 87: 1322–1330.

26 Doherty, M.M. and Charman, W.N. (2002). The mucosa of the small intestine how clinically relevant as an organ of drug metabolism? *Clini. Pharmacokinet.* 41: 235–253.

27 Haaland, R.E., Otieno, K., Martin, A. et al. (2018). Reduced nevirapine concentrations among HIV-positive women receiving mefloquine for intermittent preventive treatment for malaria control during pregnancy. *AIDS Res. Human Retrov.* 34: 912–915.

28 Healan, A.M., Griffiss, J.M., Proskin, H.M. et al. (2018). Impact of rifabutin or rifampin on bedaquiline safety, tolerability, and pharmacokinetics assessed in a randomized clinical trial with healthy adult volunteers. *Antimicrob. Agents Chemother.* 62: e00855–e00817.

29 Marsousi, N., Daali, Y., Fontana, P. et al. (2018). Impact of boosted antiretroviral therapy on the pharmacokinetics and efficacy of clopidogrel and prasugrel active metabolites. *Clini. Pharmacokinet.* 57: 1347–1354.

30 Rudek, M.A., Chang, C.Y., Steadman, K. et al. (2014). Combination antiretroviral therapy (cART) component ritonavir significantly alters docetaxel exposure. *Cancer Chemother. Pharmacol.* 73: 729–736.

31 Lennernas, H. (2003). Clinical pharmacokinetics of atorvastatin. *Clin. Pharmacokinet.* 42: 1141–1160.

32 Cai, J., Huang, Z., Yang, G. et al. (2011). Comparing antihypertensive effect and plasma ciclosporin concentration between amlodipine and valsartan regimens in hypertensive renal transplant patients receiving ciclosporin therapy. *Am. J. Cardiovasc. Drugs* 11: 401–409.

33 Hamelin, B.A., Bouayad, A., Méthot, J. et al. (2000). Significant interaction between the nonprescription antihistamine diphenhydramine and the CYP2D6 substrate metoprolol in healthy men with high or low CYP2D6 activity. *Clin. Pharmacol. Ther.* 67: 466–477.

34 Farré, M., Abanades, S., Roset, P.N. et al. (2007). Pharmacological interaction between 3, 4-methylenedioxymethamphetamine (ecstasy) and paroxetine: pharmacological effects and pharmacokinetics. *J. Pharmacol. Exp. Ther.* 323: 954–962.

35 Briciu, C., Neag, M., Muntean, D. et al. (2014). A pharmacokinetic drug interaction study between nebivolol and paroxetine in healthy volunteers. *J. Clin. Pharm. Ther.* 39: 535–540.

36 Belle, D.J., Ernest, C.S., Sauer, J.-M. et al. (2002). Effect of potent CYP2D6 inhibition by paroxetine on atomoxetine pharmacokinetics. *J. Clin. Pharmacol.* 42: 1219–1227.

37 Todor, I., Popa, A., Neag, M. et al. (2017). Evaluation of the potential pharmacokinetic interaction between atomoxetine and fluvoxamine in healthy volunteers. *Pharmacol. Ther. Part A* 99: 84–88.

38 Todor, I., Popa, A., Neag, M. et al. (2016). Evaluation of a potential metabolism-mediated drug-drug interaction between atomoxetine and bupropion in healthy volunteers. *J. Pharm. Pharm. Sci.* 19: 198–207.

39 Lim, M.L., Min, S.S., Eron, J.J. et al. (2004). Coadministration of lopinavir/ritonavir and phenytoin results in two-way drug interaction through cytochrome P-450 induction. *JAIDS - J Acquir. Immune Defic. Syndr.* 36: 1034–1040.

40 Liu, P., Foster, G., Gandelman, K. et al. (2007). Steady-state pharmacokinetic and safety profiles of voriconazole and ritonavir in healthy male subjects. *Antimicrob. Agents Chemother.* 51: 3617–3626.

41 Markert, C., Hellwig, R., Burhenne, J. et al. (2013). Interaction of ambrisentan with clarithromycin and its modulation by polymorphic SLCO1B1. *Eur. J. Clin. Pharmacol.* 69: 1785–1793.

42 Aquilante, C.L., Kiser, J.J., Anderson, P.L. et al. (2012). Influence of SLCO1B1 polymorphisms on the drug-drug interaction between darunavir/ritonavir and pravastatin. *J. Clin. Pharmacol.* 52: 1725–1738.

43 Wang, Z.-J., Yin, O.Q.P., Tomlinson, B., and Chow, M.S.S. (2008). OCT2 polymorphisms and in-vivo renal functional consequence: studies with metformin and cimetidine. *Pharmacogenet. Genomics* 18: 637–645.

44 Cho, S.K. and Jae-Yong, C. (2016). The MATE1 rs2289669 polymorphism affects the renal clearance of metformin following ranitidine treatment. *Int. J. Clin. Pharmacol. Ther.* 54: 253.

45 Schneck, D.W., Birmingham, B.K., Zalikowski, J.A. et al. (2004). The effect of gemfibrozil on the pharmacokinetics of rosuvastatin. *Clin. Pharmacol. Ther.* 75: 455–463.

46 Scripture, C.D. and Pieper, J.A. (2001). Clinical pharmcokinetics of fluvastatin. *Clini. Pharmacokinet.* 40: 263–281.

47 Sadowski, D.C. (1994). Drug interactions with antacids. Mechanism and clinical significance. *Drug Safety* 11: 395–407.

48 U.S. Food and Drug Administration. In Vitro Drug Interaction Studies — Cytochrome P450 Enzyme-

and Transporter-Mediated Drug Interactions (2020) https://www.fda.gov/regulatory-information/search-fda-guidance-documents/vitro-drug-interaction-studies-cytochrome-p450-enzyme-and-transporter-mediated-drug-interactions [Access on 20 Nov 2020]

49 Floren, L.C., Bekersky, I., Benet, L.Z. et al. (1997). Tacrolimus oral bioavailability doubles with coadmininstration of ketoconazole. *Clin. Pharmacol. Ther.* 62: 41–49.

50 Wolf, K.K., Paine, M.F., and Watkins, P.B. (2018). *Metabolic Barrier of the Gastrointestinal Tract*, Comprehensive Toxicology, 3e, vol. 3, 74–98.

51 Coleman, M.D. (2010). *Human drug metabolism: an introduction*, 2nde. West Sussex: Wiley-Blackwell.

52 Li, Y., Meng, Q., Yang, M. et al. (2019). Current trends in drug metabolism and pharmacokinetics. *Acta Pharm. Sin. B* 9: 1113–1144.

53 Liu, Y., Ramírez, J., House, L., and Ratain, M.J. (2010). Comparison of the drug-drug interactions potential of erlotinib and gefitinib via inhibition of UDP-glucuronosyltransferases. *Drug Metab. Dispos.* 38: 32–39.

54 Kim, D.-H. (2015). Gut microbiota-mediated drug-antibiotic interactions. *Drug Metab. Dispos.* 43: 1581–1589.

55 Zimmermann, M., Zimmermann-Kogadeeva, M., Wegmann, R., and Goodman, A.L. (2019). Mapping human microbiome drug metabolism by gut bacteria and their genes. *Nature* 570: 462–467.

56 Strong, H.A., Renwick, A.G., George, C.F. et al. (1987). The reduction of sulphinpyrazone and sulindac by intestinal bacteria. *Xenobiotica* 17: 685–696.

57 Lehouritis, P., Cummins, J., Stanton, M. et al. (2015). Local bacteria affect the efficacy of chemotherapeutic drugs. *Sci. Rep.* 5: 14554.

58 Routy, B., Le Chatelier, E., Derosa, L. et al. (2018). Gut microbiome influences efficacy of PD-1--based immunotherapy against epithelial tumors. *Science* 359: 91–97.

59 Wilkinson, E.M., Ilhan, Z.E., and Herbst-Kralovetz, M.M. (2018). Microbiota--drug interactions: impact on metabolism and efficacy of therapeutics. *Maturitas* 112: 53–63.

60 Gurley, B.J., Swain, A., Hubbard, M.A. et al. (2008). Supplementation with goldenseal (Hydrastis canadensis), but not kava kava (Piper methysticum), inhibits human CYP3A activity in vivo. *Clin. Pharmacol. Ther.* 83: 61–69.

61 Gurley, B.J., Swain, A., Hubbard, M.A. et al. (2008). Clinical assessment of CYP2D6-mediated herb-drug interactions in humans: effects of milk thistle, black cohosh, goldenseal, kava kava, St. John's wort, and Echinacea. *Mol. Nutr. Food Res.* 52: 755–763.

62 Markowitz, J.S., Donovan, J.L., DeVane, C.L. et al. (2003). Effect of St John's wort on drug metabolism by induction of cytochrome P450 3A4 enzyme. *JAMA* 290: 1500–1504.

63 Nebert, D.W. and Dalton, T.P. (2006). The role of cytochrome P450 enzymes in endogenous signalling pathways and environmental carcinogenesis. *Nat. Rev. Cancer* 6 (12): 947–960.

64 US Food and Drug Administration (2020). Drug-drug interaction assessment for therapeutic proteins guidance for industry.

65 Nair, P.C., Meech, R., Mackenzie, P.I. et al. (2015). Insights into the UDP-sugar selectivities of human UDP-glycosyltransferases (UGT): a molecular modeling perspective. *Drug Metabol. Rev.* 47: 335–345.

66 Elmeliegy, M., Vourvahis, M., Guo, C., and Wang, D.D. (2020). Effect of P-glycoprotein (P-gp) inducers on exposure of P-gp substrates: review of clinical drug–drug interaction studies. *Clini. Pharmacokinet.* 59: 699–714.

67 Dietrich, U., Dürr, R., and Koch, J. (2013). Peptides as drugs: from screening to application. *Curr. Pharm. Biotechnol.* 14: 501–512.

68 Zhou, H. and Meibohm, B. (2013). *Drug-drug interactions for therapeutic biologics*. John Wiley & Sons, Inc.

69 Vugmeyster, Y. (2012). Pharmacokinetics and toxicology of therapeutic proteins: advances and challenges. *World J. Biol. Chem.* 3: 73.

70 U.S. Food and Drug Administration. Drug-Drug Interaction Assessment for Therapeutic Proteins Guidance for Industry (2020) https://www.fda.gov/regulatory-information/search-fda-guidance-documents/drug-drug-interaction-assessment-therapeutic-proteins-guidance-industry [Access on 25 Oct 2020]

71 Schmitt, C., Kuhn, B., Zhang, X. et al. (2011). Disease-drug-drug interaction involving tocilizumab and simvastatin in patients with rheumatoid arthritis. *Clin. Pharmacol. Ther.* 89: 735–740.

72 Raaska, K., Raitasuo, V., and Neuvonen, P. (2001). Effect of influenza vaccination on serum clozapine and its main metabolite concentrations in patients with schizophrenia. *Eur. J. Clin. Pharmacol.* 57: 705–870.

73 Baert, F., Noman, M., Vermeire, S. et al. (2003). Influence of immunogenicity on the long-term efficacy of infliximab in Crohn's disease. *New Engl. J. Med.* 348: 601–608.

74 Zhuang, Y., Xu, Z., Frederick, B. et al. (2012). Golimumab pharmacokinetics after repeated subcutaneous and intravenous administrations in patients with rheumatoid arthritis and the effect of concomitant methotrexate: an open-label, randomized study. *Clin. Ther.* 34: 77–90.
75 Ferri, N., Bellosta, S., Baldessin, L. et al. (2016). Pharmacokinetics interactions of monoclonal antibodies. *Pharmacol. Res.* 111: 592.
76 Genovese, M.C., Cohen, S., Moreland, L. et al. (2004). Combination therapy with etanercept and anakinra in the treatment of patients with rheumatoid arthritis who have been treated unsuccessfully with methotrexate. *Arthritis Rheumatism* 50: 1412–1419.
77 Piscitelli, S.C., Burstein, A.H., Welden, N. et al. (2002). The effect of garlic supplements on the pharmacokinetics of saquinavir. *Clin. Infect. Dis.* 34: 234–238.
78 Mohammed Abdul, M.I., Jiang, X., Williams, K.M. et al. (2008). Pharmacodynamic interaction of warfarin with cranberry but not with garlic in healthy subjects. *Br. J. Pharmacol.* 154: 1691–1700.
79 Bano, G., Raina, R.K., Zutshi, U. et al. (1991). Effect of piperine on bioavailability and pharmacokinetics of propranolol and theophylline in healthy volunteers. *Eur. J. Clin. Pharmacol.* 41: 615–617.
80 Velpandian, T., Jasuja, R., Bhardwaj, R.K. et al. (2001). Piperine in food: interference in the pharmacokinetics of phenytoin. *Eur. J. Drug Metabol. Pharmacokinet.* 26: 241–247.
81 Breidenbach, T., Kliem, V., Burg, M. et al. (2000). Profound drop of cyclosporin a whole blood trough levels caused by St. John's wort (Hypericum perforatum). *Transplantation* 69: 2229–2230.
82 Piscitelli, S.C., Burstein, A.H., Chaitt, D. et al. (2000). Indinavir concentrations and St John's wort. *Lancet* 355: 547–548.
83 Dürr, D., Stieger, B., Kullak-Ublick, G.A. et al. (2000). St John's Wort induces intestinal P-glycoprotein/MDR1 and intestinal and hepatic CYP3A4. *Clin. Pharmacol. Ther.* 68: 598–604.
84 Loughren, M.J., Kharasch, E.D., Kelton-Rehkopf, M.C. et al. (2020). Influence of St. John's Wort on intravenous Fentanyl Pharmacokinetics, Pharmacodynamics, and clinical effects: a Randomized clinical trial. *Anesthesiology* 132: 491–503.
85 Schwarz, U.I., Hanso, H., Oertel, R. et al. (2007). Induction of intestinal P-glycoprotein by St John's wort reduces the oral bioavailability of talinolol. *Clin. Pharmacol. Ther.* 81: 669–678.
86 Yin, O.Q., Tomlinson, B., Waye, M.M. et al. (2004). Pharmacogenetics and herb-drug interactions: experience with Ginkgo biloba and omeprazole. *Pharmacogenetics* 14: 841–850.
87 Robertson, S.M., Davey, R.T., Voell, J. et al. (2008). Effect of Ginkgo biloba extract on lopinavir, midazolam and fexofenadine pharmacokinetics in healthy subjects. *Curr. Med. Res. Opinion* 24: 591–599.
88 Fan, L., Tao, G.Y., Wang, G. et al. (2009). Effects of Ginkgo biloba extract ingestion on the pharmacokinetics of talinolol in healthy Chinese volunteers. *Ann. Pharmacother.* 43: 944–949.
89 Bilia, A.R., Bergonzi, M.C., Isacchi, B. et al. (2018). Curcumin nanoparticles potentiate therapeutic effectiveness of acitrein in moderate-to-severe psoriasis patients and control serum cholesterol levels. *J. Pharm. Pharmacol.* 70: 919–928.
90 Juan, H., Terhaag, B., Cong, Z. et al. (2007). Unexpected effect of concomitantly administered curcumin on the pharmacokinetics of talinolol in healthy Chinese volunteers. *Eur. J. Clin. Pharmacol.* 63: 663–668.
91 Al-Jenoobi, F.I., Al-Thukair, A.A., Alam, M.A. et al. (2015). Effect of Curcuma longa on CYP2D6- and CYP3A4-mediated metabolism of dextromethorphan in human liver microsomes and healthy human subjects. *Eur. J. Drug Metabol. Pharmacokinet.* 40: 61–66.
92 Ku, Y.M., Min, D.I., and Flanigan, M. (1998). Effect of grapefruit juice on the pharmacokinetics of microemulsion cyclosporine and its metabolite in healthy volunteers: does the formulation difference matter? *J. Clin. Pharmacol.* 38: 959–965.
93 Benton, R.E., Honig, P.K., Zamani, K. et al. (1996). Grapefruit juice alters terfenadine pharmacokinetics, resulting in prolongation of repolarization on the electrocardiogram. *Clin. Pharmacol. Ther.* 59: 383–388.
94 Goho, C. (2001). Oral midazolam-grapefruit juice drug interaction. *Pediatric Dentistry* 23: 365–366.
95 Christensen, H., Asberg, A., Holmboe, A.B., and Berg, K.J. (2002). Coadministration of grapefruit juice increases systemic exposure of diltiazem in healthy volunteers. *Eur. J. Clin. Pharmacol.* 58: 515–520.
96 Bailey, D.G. and Dresser, G.K. (2004). Interactions between grapefruit juice and cardiovascular drugs. *Am. J. Cardiovasc. Drugs* 4: 281–297.
97 Takeuchi, A., Tsuchiya, H., Yamamoto, N. et al. (2007). Caffeine-potentiated chemotherapy for patients with high-grade soft tissue sarcoma:

long-term clinical outcome. *Anticancer Res.* 27: 3489–3495.

98 Arshi, S., Fallahpour, M., Nabavi, M. et al. (2014). The effects of vitamin D supplementation on airway functions in mild to moderate persistent asthma. *Ann. Allergy Asthma Immunol.* 113: 404–409.

99 El-Aal, A.A., El-Ghffar, E.A.A., Ghali, A.A. et al. (2018). The effect of vitamin C and/or E supplementations on type 2 diabetic adult males under metformin treatment: a single-blinded randomized controlled clinical trial. *Diabetes Metab. Syndr. Clin. Res. Rev.* 12: 483–489.

100 van Zweeden, A.A., van Groeningen, C.J., Honeywell, R.J. et al. (2018). Randomized phase 2 study of gemcitabine and cisplatin with or without vitamin supplementation in patients with advanced esophagogastric cancer. *Cancer Chemother. Pharmacol.* 82: 39–48.

101 Li, Y., Liang, M., Wang, G. et al. (2017). Effects of folic acid therapy on the new-onset proteinuria in chinese hypertensive patients: a post Hoc analysis of the Renal substudy of CSPPT (China Stroke Primary Prevention Trial). *Hypertension* 70: 300–306.

102 Fuhr, U., Beckmann-Knopp, S., Jetter, A. et al. (2007). The effect of silymarin on oral nifedipine pharmacokinetics. *Planta Med.* 73: 1429–1435.

103 Goey, A.K., Meijerman, I., Rosing, H. et al. (2013). The effect of Echinacea purpurea on the pharmacokinetics of docetaxel. *Br. J. Clin. Pharmacol.* 76: 467–474.

104 Abdul, M.I., Jiang, X., Williams, K.M. et al. (2010). Pharmacokinetic and pharmacodynamic interactions of echinacea and policosanol with warfarin in healthy subjects. *Br. J. Clin. Pharmacol.* 69: 508–515.

105 Ushijima, K., Tsuruoka, S., Tsuda, H. et al. (2009). Cranberry juice suppressed the diclofenac metabolism by human liver microsomes, but not in healthy human subjects. *Br. J. Clin. Pharmacol.* 68: 194–200.

106 Bedada, S.K. and Neerati, P. (2016). Resveratrol pretreatment affects CYP2E1 activity of chlorzoxazone in healthy human volunteers. *Phytother. Res.* 30: 463–468.

107 Toh, D.S., Limenta, L.M., Yee, J.Y. et al. (2014). Effect of mushroom diet on pharmacokinetics of gabapentin in healthy Chinese subjects. *Br. J. Clin. Pharmacol.* 78: 129–134.

108 Zhou, Q., Ye, Z., Ruan, Z., and Zeng, S. (2013). Investigation on modulation of human P-gp by multiple doses of Radix Astragali extract granules using fexofenadine as a phenotyping probe. *J. Ethnopharmacol.* 146: 744–749.

109 He, L., Chen, X., Zhou, M. et al. (2011). Radix/rhizoma notoginseng extract (sanchitongtshu) for ischemic stroke: a randomized controlled study. *Phytomedicine* 18: 437–442.

110 Xin, H.W., Wu, X.C., Li, Q. et al. (2009). Effects of Schisandra sphenanthera extract on the pharmacokinetics of midazolam in healthy volunteers. *Br. J. Clin. Pharmacol.* 67: 541–546.

111 Showande, S.J., Adegbolagun, O.M., Igbinoba, S.I., and Fakeye, T.O. (2017). In vivo pharmacodynamic and pharmacokinetic interactions of Hibiscus sabdariffa calyces extracts with simvastatin. *J. Clin. Pharm. Ther.* 42: 695–703.

112 Fakeye, T.O., Adegoke, A.O., Omoyeni, O.C., and Famakinde, A.A. (2007). Effects of water extract of Hibiscus sabdariffa, Linn (Malvaceae) 'Roselle' on excretion of a diclofenac formulation. *Phytother. Res.* 21: 96–98.

113 Kolawole, J.A. and Maduenyi, A. (2004). Effect of zobo drink (Hibiscus sabdariffa water extract) on the pharmacokinetics of acetaminophen in human volunteers. *Eur. J. Drug Metabol. Pharmacokinet.* 29: 25–29.

114 Mahmoud, B.M., Ali, H.M., Homeida, M.M., and Bennett, J.L. (1994). Significant reduction in chloroquine bioavailability following coadministration with the Sudanese beverages Aradaib, Karkadi and Lemon. *J. Antimicrob. Chemother.* 33: 1005–1009.

115 Olawoye, O.S., Adeagbo, B.A., and Bolaji, O.O. (2018). Moringa oleifera leaf powder alters the pharmacokinetics of amodiaquine in healthy human volunteers. *J. Clin. Pharm. Ther.* 43: 626–632.

116 Tsai, H.-H., Lin, H.-W., Simon Pickard, A. et al. (2012). Evaluation of documented drug interactions and contraindications associated with herbs and dietary supplements: a systematic literature review. *Int. J. Clin. Pract.* 66: 1056–1078.

117 Ameer, B. and Weintraub, R.A. (1997). Drug interactions with grapefruit juice. *Clini. Pharmacokinet.* 33: 103–121.

# 27

# Regulatory Considerations in Metabolism- and Transport-Based Drug Interactions

*Xinning Yang[1], Sue-Chih Lee[2], Xinyuan Zhang[1], and Lei Zhang[2]*

[1] *Food and Drug Administration, Offices of Clinical Pharmacology, Office of Translational Sciences, Center for Drug Evaluation and Research, Silver Spring, MD, USA*
[2] *Food and Drug Administration, Offices of Research and Standards, Office of Generic Drugs, Center for Drug Evaluation and Research, Silver Spring, MD, USA*

## Disclaimer

The views expressed are those of authors and do not necessarily reflect FDA's views or policy.

## 27.1 Overview of Drug–Drug Interactions

The desirable and undesirable effects of a drug are generally related to its concentration at the sites of action, which is related to the amount administered (dose) and to the drug's absorption, distribution, metabolism, and/or excretion (ADME). All these processes can be influenced by both intrinsic and extrinsic factors such as age, race, gender, disease states, drug interactions, and food effects. Observed changes arising from pharmacokinetic drug–drug interactions (DDI) can be substantial such as an order of magnitude or more decrease or increase in the blood concentrations of a drug or metabolites. These large changes in exposure can alter the safety and efficacy profile of a drug in important ways (Figure 27.1). If these potential DDIs were not properly studied during drug development or could not be managed by proper instructions in the labeling, they could cause serious safety concerns after the drugs are on the market leading to drug being withdrawn from the market. In fact, five of 11 drugs withdrawn from the US market from 1998 to 2002 were prone to serious pharmacokinetic DDIs [1]. Many interactions involved inhibition of metabolizing enzymes and transporters, resulting in increased systemic exposure and subsequent adverse drug reactions. In other cases, induction of metabolizing enzymes and transporters resulted in reduced systemic exposure leading to a risk of loss of efficacy of coadministered drugs. In addition, inhibition of metabolizing enzymes may decrease the formation of active metabolites that are primarily responsible for efficacy of some drugs and lead to impaired efficacy. Conversely, induction of metabolizing enzymes may increase the formation of toxic metabolites leading to safety concerns.

Our understanding of pharmacokinetic DDI has improved dramatically over the years because of the scientific progress in research areas of metabolizing enzymes and drug transporters. Drug metabolizing enzymes and their role in DDI have been intensively investigated. Cytochrome P450 (CYP)-mediated DDIs have attracted most attention because majority of drugs are cleared via CYP-mediated pathways. The major CYPs recommended by regulatory agencies to evaluate during drug development include CYP1A2, 2B6, 2C8, 2C9, 2C19, 2D6, and 3A. They can be inhibited or induced (except CYP2D6) by concomitant drug treatment (Table 27.1). In addition to CYPs, various other drug metabolizing enzymes (some Phase 2 enzymes are listed in Table 27.2) have been recognized to play important roles in pharmacokinetics of some drugs [2]. Historically, most of these enzymes have attracted less attention than CYP enzymes mainly due to the lack of tools (e.g. specific substrates or inhibitors) to study them and/or a lower incidence of observed adverse DDIs. There has been interest in *N*-acetyltransferases (NATs) and associated polymorphisms resulting in fast and slow acetylators [3] and acyl glucuronyl transferases [4], both leading to formation of toxic metabolites [5].

For non-CYP enzymes, there has been an increased interest in DDIs involving UDP glucuronyl transferases (UGT). For example, polymorphism of UGT1A1 was shown to affect exposure of SN-38, an active metabolite of irinotecan, and has efficacy and safety implications [6, 7]. Corona, et al. [8] reported that lopinavir–ritonavir

*Oral Bioavailability and Drug Delivery: From Basics to Advanced Concepts and Applications*, First Edition. Edited by Ming Hu and Xiaoling Li.

**Table 27.1** Examples of *in vivo* substrate, inhibitor, and inducer for specific CYP enzymes recommended for study.[b)c)]

| CYP | Substrate | Inhibitor | Inducer |
|---|---|---|---|
| 1A2 | Caffeine[a)], tizanidine,[a)] alosetron, duloxetine, ramelteon, tasimelteon, theophylline, melatonin | Fluvoxamine,[a)] enoxacin | Smokers versus nonsmokers[d], rifampin, phenytoin |
| 2B6 | Bupropion, efavirenz | Ticlopidine, clopidogrel | Rifampin,[a)] carbamazepine, efavirenz |
| 2C8 | Repaglinide,[a)] dasabuvir, montelukast, pioglitazone, rosiglitazone | Gemfibrozil,[a)]clopidogrel[a)] | Rifampin[a)] carbamazepine |
| 2C9 | Tolbutamide,[a)] *S*-warfarin,[a)] celecoxib, flurbiprofen, phenytoin | Fluconazole,[a)] amiodarone, miconazole, (use of PM versus EM subjects)[e], sulfaphenazole, voriconazole | Rifampin,[a)] carbamazepine, enzalutamide, ritonavir |
| 2C19 | Omeprazole,[a)] *S*-lansoprazole,[a)] diazepam, esomeprazole, *S*-mephenytoin, voriconazole | Fluvoxamine,[a)] fluconazole,[a)]fluoxetine, ticlopidine, (use of PM versus EM subjects)[e] | Rifampin,[a)] apalutamide, enzalutamide, phenytoin, ritonavir |
| 2D6 | Desipramine,[a)] dextromethorphan,[a)] atomoxetine, nebivolol,[a)] venlafaxine, metoprolol | Paroxetine,[a)] fluoxetine,[a)] bupropion, quinidine, terbinafine, (use of PM versus EM subjects)[e] | None identified |
| 3A4/5 | Midazolam,[a)] triazolam,[a)] buspirone, everolimus, eletriptan, felodipine, ibrutinib, naloxegol, sildenafil, simvastatin, tacrolimus, sirolimus | Itraconazole,[a)] clarithromycin,[a)] atazanavir, cobicistat, indinavir/ritonavir, ketoconazole, nefazodone, nelfinavir, posaconazole, saquinavir/ritonavir, telithromycin, voriconazole | Rifampin,[a)] carbamazepine,[a)] phenytoin,[a)] apalutamide, efavirenz, enzalutamide, phenobarbital, lumacaftor |

a) Index substrates, inhibitors, or inducers of corresponding CYP enzymes.
b) This is not an exhaustive list, for an updated list, see the following: https://www.fda.gov/drugs/drug-interactions-labeling/drug-development-and-drug-interactions-table-substrates-inhibitors-and-inducers.
c) Substrates for any particular CYP enzyme listed in this table are those with plasma AUC values increased by twofold or higher when coadministered with inhibitors of that CYP enzyme; for CYP3A, only those with plasma AUC increased by fivefold or higher are listed. Inhibitors listed are those that increase plasma AUC values of substrates for that CYP enzyme by twofold or higher. For CYP3A inhibitors, only those that increase AUC of CYP3A substrates by fivefold or higher are listed. Inducers listed are those that decrease plasma AUC values of substrates for that CYP enzyme by 20% or more.
d) A clinical study can be conducted in smokers as compared to nonsmokers, when appropriate.
e) A clinical study can be conducted in poor metabolizers (PM) as compared to extensive metabolizers (EM) for the specific CYP enzyme (in lieu of an interaction study with an inhibitor), when appropriate.

**Table 27.2** Phase 2 enzymes.

| Reaction | Enzyme | Functional group |
|---|---|---|
| Glucuronidation | UDP-Glucuronyltransferase | –OH, –COOH, $-NH_2$, –SH |
| Glycosidation | UDP-Glycosyltransferase | –OH, –COOH, –SH |
| Sulfation | Sulfotransferase | $-NH_2$, $-SO_2NH_2$, –OH |
| Methylation | Methyltransferase | –OH, $-NH_2$ |
| Acetylation | Acetyltransferase | $-NH_2$, $-SO_2NH_2$, –OH |
| Amino acid conjugation | | –COOH |
| Glutathione conjugation | Glutathione-*S*-transferase | Epoxide, organic halide |
| Fatty acid conjugation | | –OH |
| Condensation | | Various |

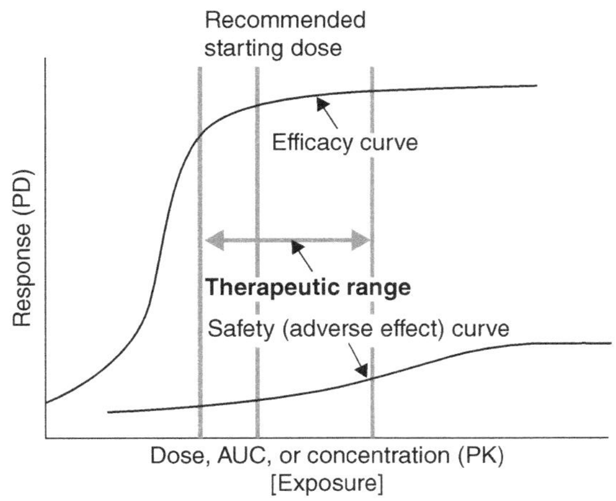


**Figure 27.1** Hypothetical exposure–response relationship. determination of a therapeutic range based on exposure (dose, AUC, $C_{max}$, etc.) and response data.

and irinotecan coadministration increased the SN-38 AUC by 204% due to a dual effect of CYP3A inhibition leading to more diversion of irinotecan metabolism to SN-38 and UGT1A1 inhibition resulting in reduced inactivation of SN-38 to SN-38 glucuronide. In addition, certain glucuronide metabolites may cause DDIs. Ogilvie, et al. [9] determined that gemfibrozil glucuronide was a metabolism-based inhibitor for CYP2C8 ($IC_{50}$ of 1.8 μM after preincubation) which may have implications for the mechanism of the clinical interaction reported between gemfibrozil and CYP2C8 substrates, such as cerivastatin, repaglinide, rosiglitazone, and pioglitazone. In addition, gemfibrozil glucuronide is an inhibitor of OATP1B1, and cerivastatin and repaglinide are substrates of OATP1B1 [10].

Similar to the CYP enzymes, UGT are encoded by a UGT gene "superfamily" with 22 human UGT proteins identified to date [11]. Unlike CYP enzymes, there is no clear consensus with respect to the tools, i.e. experiment systems and conditions, selective substrates, inhibitors, and inducers for studying the UGT enzymes. Recombinant human UGTs, many available from commercial sources, have often been used to investigate the individual UGT isoforms responsible for the formation of a glucuronide metabolite. Most but not all of the isoforms are expressed in the liver. UGT1A7, UGT1A8, and UGT1A10 are only expressed in the gastrointestinal tract, and UGT1A1, 1A3, 1A4, 1A6, 1A9, 2B4, 2B7, 2B10, 2B15, and 2B17 are considered to be the isoforms of importance in hepatic drug elimination based on data available [11].

*In vitro–in vivo* correlation for drugs that are eliminated by glucuronidation is less mature compared to CYP enzymes [12]. For example, the use of microsomes to determine the intrinsic clearance of drugs that are eliminated by glucuronidation is challenging since UGTs are integral proteins of the endoplasmic reticulum and are dependent on lipid for catalytic activity. An article by Miners provides excellent discussion of the "pitfalls" and "promises" for predicting *in vivo* drug glucuronidation in humans [13]. A review written by Kiang, et al. provides additional information on clinical DDIs [14].

In addition to the effects of drug metabolizing enzymes on the pharmacokinetics of drugs, increasing attention is being given to transporters where emerging evidence indicates an important role in modulating drug ADME as well as the historical importance of transporters in the development of drug-resistant tumors. Transporters, acting alone or in concert with drug metabolizing enzymes, can affect the pharmacokinetics and/or pharmacodynamics of a drug. Examples include the inhibition or induction of transport proteins, such as P-glycoprotein (P-gp, ABCB1), breast cancer resistance protein (BCRP, ABCG2), organic anion transporting polypeptides (OATPs, SLCO family), organic anion transporters (OATs, SCL22A family), organic cation transporters (OCTs, SCL22A family), and multidrug and toxin extrusion transporters (MATEs, SLC47As) (Table 27.3). Examples of clinically relevant transporter-based DDIs include the interactions between digoxin and quinidine [16, 17], rosuvastatin and cyclosporine interaction [18–20], penicillin and probenecid [21], and metformin and cimetidine [22, 23]. Additional transporters such as multidrug resistance-associated proteins (MRPs, ABCC) and bile salt export pump (BSEP, ABCB11) could play a role in drug-induced liver toxicity [24].

The increased clinical use of therapeutic proteins raises concerns regarding their impact on drug metabolism or transporters and the potential for drug interactions. These interactions may involve plasma proteins (specific binding proteins and/or major plasma carriers such as albumins) [25] or may affect the CYP enzyme system at the transcriptional or posttranslational levels (e.g. cytokines) [26–28]. Regulation of Phase 2 enzymes and transporters by cytokines or cytokine antagonists may share similarities with CYP regulation [27]. For example, treatment of rheumatoid arthritis patients with tocilizumab, a monoclonal anti-IL-6 receptor antibody, reversed the enzyme suppression caused by IL-6 and normalized CYP3A4-mediated clearance of simvastatin [29]. Specific drug interaction studies were conducted during drug development and results were included in the

**Table 27.3** Selected human drug transporters, their locations, function, and examples of their substrates and inhibitors.

| Gene | Aliases | Tissue | Function | Substrate | Inhibitor |
|---|---|---|---|---|---|
| *ABC family* | | | | | |
| *ABCB1* | P-gp, MDR1 | Intestine, liver, kidney, brain, placenta | efflux | Digoxin, dabigatran etexilate, fexofenadine, talinolol, edoxaban | Cyclosporine, clarithromycin, itraconazole, quinidine, cobicistat, verapamil |
| *ABCG2* | BCRP | Intestine, liver, breast, brain, placenta | efflux | Rosuvastatin, sulfasalazine, topotecan, methotrexate | Cyclosporine, darolutamide, eltrombopag, regorafenib, curcumin |
| *SLC family* | | | | | |
| *SLCO1B1/SLCO1B3* | OATP1B1/OATP 1B3 | Liver | uptake | Atorvastatin, rosuvastatin, pitavastatin, pravastatin, simvastatin, fexofenadine, fluvastatin, rifampin, bosentan, elagolix, lovastatin, repaglinide | Cyclosporine, rifampin, gemfibrozil, regorafenib |
| *SLC22A2* | OCT2 | Kidney, brain | uptake | Metformin, creatinine, amphetamine, methamphetamine | Dolutegravir, isavuconazole |
| *SLC22A6* | OAT1 | Kidney, brain | uptake | Cidofovir, adefovir, furosemide, tenofovir | Probenecid |
| *SLC22A8* | OAT3 | Kidney, brain | uptake | Bumetanide, baricitinib, cefaclor, furosemide, methotrexate, benzylpenicillin (penicillin G) | Probenecid , gemfibrozil, teriflunomide |
| *SLC47A1* | MATE1 | Kidney, liver, intestine | Efflux | Metformin, creatinine, amantadine, memantine, dofetilide | Cimetidine, pyrimethamine, trimethoprim, vandetanib |
| *SLC47A2* | MATE2-K | Kidney | Efflux | Metformin, creatinine, dofetilide | Cimetidine, pyrimethamine, trimethoprim, vandetanib |

Source: Adapted from FDA [15].

**Figure 27.2** FDA DDI guidance history.

labeling [30]. Generally, these interactions between therapeutic proteins and small molecule drugs may not be adequately detected by *in vitro* assessments [31, 32]. Therefore, *in vivo* drug interaction studies may be indicated when a therapeutic protein is known to modulate enzymes or transporters (e.g. proinflammatory cytokines), and is designed to be administered with other drug product(s) as a combination regimen, especially if the coadministered drug has a narrow therapeutic range. Evaluation of DDIs for therapeutic proteins is still an evolving area. The 2020 FDA guidances for industry on *in vitro* and clinical drug interaction studies did not include a section on DDIs for therapeutic proteins [33, 34], but a separate draft guidance on DDI assessment for therapeutic proteins was published in 2020 and finalized in 2023 [35]). Overview of clinically reported therapeutic protein–drug interactions and the information included in the product package insert has recently been summarized elsewhere [36].

Since the publication of the first DDI guidance in 1997, FDA has updated DDI guidances multiple times over the past two decades to incorporate updated knowledge and recommendations on DDI evaluation during drug development so that DDIs can be managed properly (Figure 27.2). Drug withdrawal due to DDIs has rarely occurred since 2002 which may be attributed to potential DDIs being evaluated early so that proper mitigation strategies can be implemented prior to drug approval [1].

## 27.2 Regulatory Recommendations of DDI Studies

Drug interaction potential is recognized as important consideration in the evaluation of a new molecular entity (NME) and is an integral part of drug development and regulatory review prior to its market approval [37–42]. Metabolism, transport, and drug interaction information is important to benefit/risk assessment of drug products. Several FDA guidance documents developed since the late-1990s reflect the Agency's view that the metabolism and transport of an NME should be defined (Figure 27.2). Potential DDIs resulting from effects of selected drugs on the NME and effects of the NME on other selected drugs should be explored during drug development to ensure an adequate assessment of an NME's safety and effectiveness. Two final DDI guidances, one focusing on *in vitro* assessment and the other on clinical evaluation for DDIs, were released by the FDA in January 2020 [33, 34]. Compared to previous drafts, the guidances include updated information about recommendations on evaluation of DDIs mediated by CYP inhibition and induction, certain drug transporters, and metabolites of drugs, which are described in detail in later sessions of this chapter.

It is important that DDI studies explore whether an NME is likely to significantly affect the metabolic elimination and transport of drugs already in the marketplace and likely in medical practice to be taken concomitantly and, conversely, whether drugs in the marketplace are likely to affect the metabolic elimination and transport of the investigational drug. It is also important to evaluate drugs that are not substantially metabolized or transported but can have important effects on the metabolism or transport of concomitant drugs. For example, quinidine is a potent inhibitor of CYP2D6 but is metabolized by CYP3A4. Thus, the metabolic and transporter DDI liability should be explored, even for an investigational compound that is not eliminated significantly by metabolic enzymes or transporters. An integrated approach that includes *in vitro* and *in vivo*

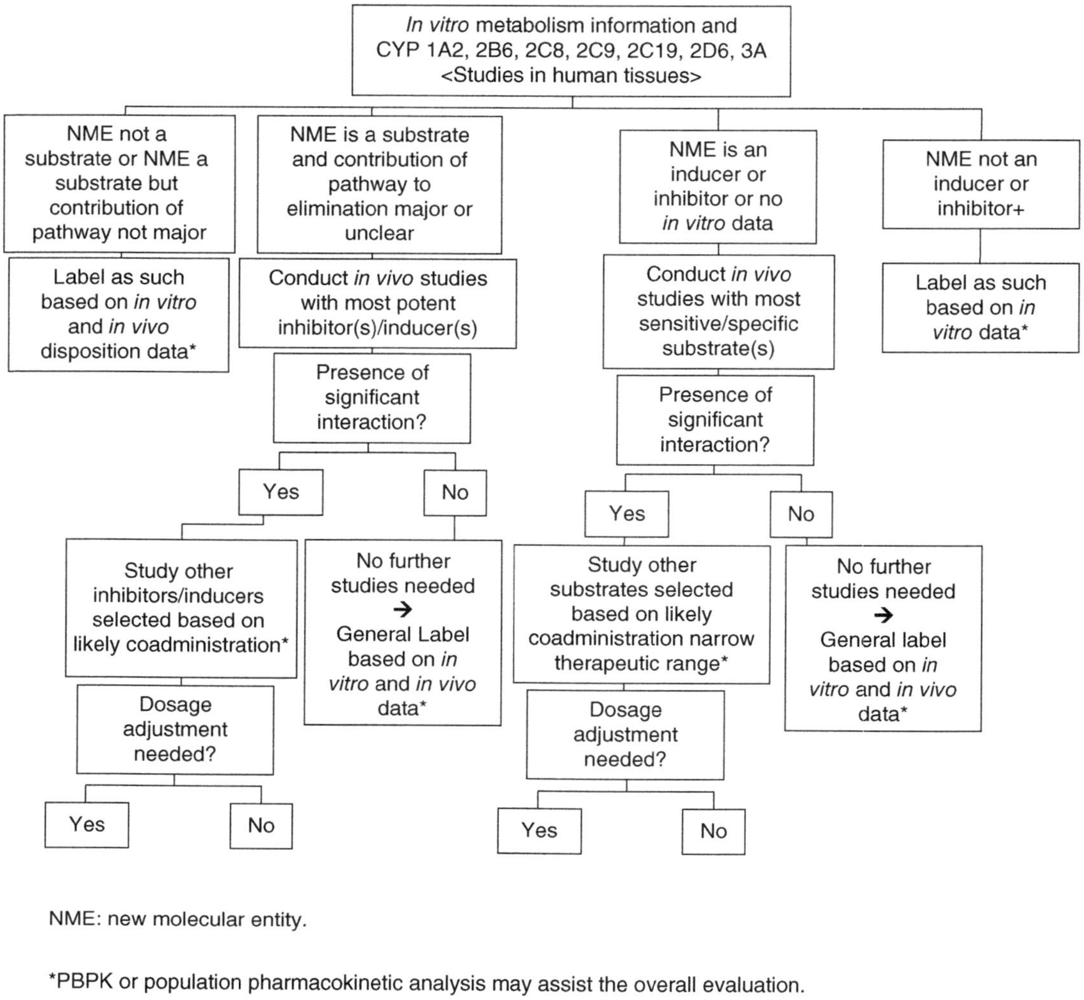


**Figure 27.3** CYP-based drug–drug interaction studies decision framework.

studies and *in silico* tools (e.g. PBPK modeling) can help determine the need of *in vivo* DDI studies, study design, and inform labeling recommendations (Figure 27.3).

## 27.3 Highlights of the Final Guidances for Industry: *In Vitro* and Clinical Drug Interaction Studies – Cytochrome P450 Enzyme- and Transporter-Mediated Drug Interactions

The final DDI guidance documents provide recommendations for industry regarding the evaluation of drug metabolism and transporter interactions during drug development [33, 34]. It recommends that *in vitro* drug interaction studies be conducted early in the development process to help determine whether further *in vivo* DDI investigations are necessary.

### 27.3.1 Highlights Concerning Drug Metabolizing Enzyme-Based DDIs

The following CYP enzymes are recommended for routine assessment to identify potential metabolism-mediated interactions: CYP1A2, CYP2B6, CYP2C8, CYP2C9, CYP2C19, CYP2D6, and CYP3A.

#### 27.3.1.1 Establishment of the NME's Metabolic Profile and Determine Whether NME is a Substrate of a Particular CYP In Vitro

Initially, the metabolic profile of the NME is investigated using human liver tissues such as microsomes or hepatocytes, since liver is the most important metabolic organ. If the results suggest that a drug undergoes substantial metabolism and the above CYP enzymes do not play a major role, further experiments may be considered to explore the involvement of other CYP enzymes (e.g. CYP2A6, CYP2J2, CYP4F2, and CYP2E1), non-CYP Phase 1 enzymes (e.g. alcohol/aldehyde dehydrogenase, aldehyde oxidase, carboxylesterase, flavin monooxygenase, monoamine oxidase, and xanthine oxidase), or Phase 2 enzymes (e.g. UGTs, sulfotransferases, glutathione *S*-transferase, glycine-*N*-acyltransferase, and NATs).

If human *in vivo* data indicate that CYP enzymes may contribute $\geq$25% to the total clearance of the NME, studies should be conducted using human liver microsomes (HLMs) or recombinant enzymes to determine the individual CYP enzymes responsible for the drug's metabolism. If a particular CYP plays a major role in the metabolism of an NME, an *in vivo* study with a strong inhibitor or inducer for that CYP (refer to the later section about classification of CYP inhibitors and inducers) is recommended to determine whether inhibition or induction of this particular pathway may lead to change in the NME's pharmacokinetics. Negative results would alleviate further *in vivo* studies with less strong inhibitors or inducers. If results are positive, further assessment with less potent inhibitors or inducers would generally be needed to give advice on dosage adjustment.

#### 27.3.1.2 Criteria for Identifying an NME as a CYP Enzyme Inhibitor *In Vitro* and Determining Whether an *In Vivo* Study Should be Conducted

*In vitro* experiments are routinely conducted to investigate whether a drug may be a reversible inhibitor or a time-dependent inhibitor (TDI) for CYP1A2, 2B6, 2C8, 2C9, 2C19, 2D6, and 3A, and to determine inhibition potency ($IC_{50}$ or Ki for reversible inhibitor and $k_{inact}$ and $K_I$ for TDI) when initial screening indicates inhibition potential. Such experiments are often conducted using HLMs and sometimes human hepatocytes. The *in vivo* DDI potential for a reversible inhibitor cannot be ruled out if *in vitro* data demonstrate an $R_1$ value $\geq$1.02 ($R_1 = 1 + [I]/K_i$) where $[I]$ is the unbound mean maximum plasma concentration ($C_{max,u}$) for a drug at steady state of the highest therapeutic dose and $K_i$ is the unbound inhibition constant for the NME toward a particular CYP [43]. When evaluating the potential of the NME to inhibit CYP3A, at least 2 structurally different CYP3A substrates should be used and the lower value of Ki is usually used to calculate $R_1$ value. Because CYP3A has significant presence in intestine and some CYP3A substrates undergo extensive first-pass intestinal metabolism, the contribution of intestinal CYP3A inhibition needs to be considered when evaluating the *in vivo* DDI potential of a CYP3A inhibitor. Thus, for an orally administered NME that reversibly inhibits CYP3A, a $R_{1,gut}$ is also calculated as $(1 + (\text{Dose}/250\,\text{ml})/K_i)$. If the $R_{1,gut}$ value is equal to or greater than 11, the *in vivo* DDI potential of an NME to inhibit CYP3A cannot be excluded.

For TDIs, a $R_2$ value is calculated as $(1 + k_{obs}/k_{deg})$, where $k_{obs}$ is the observed apparent first-order inactivation rate of the affected enzyme and $k_{deg}$ is the apparent first-order degradation rate constant of the affected enzyme. The parameter $k_{obs}$ is defined as $(k_{inact} \times 50 \times C_{max,u})/(50 \times C_{max,u} + K_I)$, where $K_I$ is the unbound inhibitor concentration causing half-maximal inactivation and $k_{inact}$ is the maximal inactivation rate constant. If an $R_2$ value is equal to or larger than 1.25, the TDI is considered likely to inhibit the affected CYP *in vivo*. The $R_2$ value varies depending on $k_{deg}$. While the guidance does not provide $k_{deg}$, there are literature-reported values for hepatic CYPs that are commonly used, e.g. 0.018/hour (corresponding $t_{1/2}$ is 38.6 hours) for CYP1A2 [44], 0.0217/hour (corresponding $t_{1/2}$ is 32 hours) for CYP2B6 [45], 0.0315/hour (corresponding $t_{1/2}$ is 22 hours) for CYP2C8 [46], 0.0067/hour (corresponding $t_{1/2}$ is 104 hours) for CYP2C9 [46], 0.0267/hour (corresponding $t_{1/2}$ is 26 hours) for CYP2C19 [46], and 0.0137/hour (corresponding $t_{1/2}$ is 51 hours) for CYP2D6 [47, 48]. As to hepatic CYP3A, there is some controversy. Initially, a $k_{deg}$ of 0.0077/hour corresponding to a $t_{1/2}$ around 90 hours was proposed [49]. However, a literature analysis based on 54 *in vivo* DDI studies involving TDIs suggested that this $k_{deg}$ often resulted in overprediction and proposed a different $k_{deg}$ value of 0.03/hour corresponding to a $t_{1/2}$ of 23 hours [50]. A later analysis suggested a $k_{deg}$ of 0.0193/hour (i.e. $t_{1/2}$ of 36 hours) [51]. As to intestinal CYPs, a universal $k_{deg}$ of 0.029/hour (corresponding to $t_{1/2}$ of 24 hours) is used because the enterocytes on intestine villi are replaced daily [52, 53].

These *R* values also referred to as basic models are used as initial DDI evaluation to provide qualitative prediction of the likelihood of *in vivo* DDI. When an *R* value for an inhibitor ($R_1$ and $R_{1,gut}$ for reversible inhibitors, $R_2$ for TDIs) is equal to or greater than the corresponding threshold as mentioned above, further assessment of *in vivo* DDI potential of an NME is needed by either conducting an *in vivo* study with a sensitive substrate of the inhibited CYP or applying mechanistic static or dynamic (e.g. PBPK) models to make a quantitative prediction of *in vivo* DDI and then based on that determining the need of performing an *in vivo* study.

Though different regulatory agencies have generally similar recommendations for the basic models, there is a difference in the evaluation of DDIs due to TDIs of CYP3A [54]. The guidelines from European Medicines Agency (EMA) and Japan's Pharmaceuticals and Medical Devices Agency (PMDA) recommend the $R$ value using $50 \times C_{max,u}$ and also 1/10 of Dose/250 ml [55, 56]. The latter is introduced to take into account inhibition of intestinal CYP3A. In contrast, the FDA guidance does not require calculation of $R_2$ value using $0.1 \times$ Dose/250 ml. This is not because the FDA guidance ignores the contribution of intestinal CYP3A inhibition to *in vivo* DDIs caused by TDIs of CYP3A. The decision is based on data analysis which showed that even just using $R_2$ value calculated with $50 \times C_{max,u}$, there was no false negative prediction and the *in vivo* DDI was often over-predicted [43]. Thus, introducing an additional calculation using $0.1 \times$ Dose/250 ml does not provide extra value for qualitative prediction. In fact, it has been recognized for quite a while that the DDI potential of TDI is generally over-predicted. Thus, there is still a need to conduct further research to explore approaches to improve the *in vitro* to *in vivo* extrapolation for TDIs. One potential solution is using a lower inhibitory concentration to calculate $R$ value. The current scaling factor (50×) applied to $C_{max,u}$ is arbitrary. The aim of using a scaling factor is to cover situations where intracellular free drug concentrations in liver cells may be higher than the free plasma concentrations for certain compounds, e.g. drugs that are substrates of hepatic uptake transporters. However, there is no systematic research to examine which scaling factor is optimal. Based on preliminary data analysis to correlate $R_2$ value-predicted DDI potential to *in vivo* DDI results (a study is considered as positive DDI if the AUC ratio is 1.25 or more for a sensitive substrate in the presence of an inhibitor versus the substrate given alone), it appears that a lower scaling factor than 50 could be appropriate [43]. This was supported by two recent publications, (i) one study evaluating TDIs of CYP3A suggested that, if HLM is used to determine $K_{inact}$ and $K_I$, no scaling factor is needed, while if hepatocyte is used, a scaling factor of 3 can be used [57]; (ii) the other study focusing on TDIs of non-CYP3A enzymes (i.e. CYP1A2, CYP2C8, CYP2C9, CYP2C19, and CYP2D6) showed that a scaling factor of 3 led to two false negative predictions when using hepatocytes and no false negatives using HLM as *in vitro* system [58].

It is noted that $R_{1,gut}$ value calculated using Dose/250 ml for reversible inhibitors of CYP3A also leads to substantial false positive prediction. An analysis showed that the positive predictive error (i.e. proportion of clinical studies conducted that actually did not need to be (no DDI risk)) was 21% [43]. This may be due to overestimation of inhibitor concentration. The relevant concentration is supposed to be free intracellular drug concentration in enterocytes, while Dose/250 ml is a conservative estimation of drug concentration in the gut lumen. For Biopharmaceutical Classification Class (BCS) 2 or 4 drugs that have limited solubility, the dose administered may not be dissolved in 250 ml of gastrointestinal fluid. Thus, further research is encouraged to explore whether a different inhibitor concentration used to calculate $R_{1,gut}$ may provide a better prediction while maintaining minimal false negative prediction.

#### 27.3.1.3 Evaluation of an NME as a Potential CYP Enzyme Inducer *In Vitro* and Determining Whether an *In Vivo* Study Should be Conducted

The FDA *in vitro* DDI guidance suggests that induction studies be carried out using at least three lots of freshly isolated or cryopreserved human hepatocytes in culture or immortalized cell lines including a positive control, and the results from each hepatocyte lot (or batch of cell line) be evaluated individually, i.e. if one hepatocyte lot shows positive signal then an NME is determined carrying DDI liability as an inducer. Initially, enzyme activity was recommended as the endpoint (in the 2006 draft FDA DDI guidance), and if the increase in enzyme activity for NME-treated cells is ≥40% of a positive control, the NME is considered to be an enzyme inducer and *in vivo* induction DDI studies are recommended. However, a later analysis based on data generated from three hepatocyte lots for 20 inducers and 15 non-inducers of CYP3A showed that there were substantial false negative predictions with this approach [61]. The false negative prediction level was still at a level unacceptable even when a threshold of 20% was used instead of 40% relative to response to positive control. In addition, a limitation with enzyme activity is that some inducer drugs are also inhibitors for the investigated CYP and in such situations, the activity results are interfered by the inhibition effect. Another limitation is the dynamic range of certain CYP activity changes is generally less than that of mRNA and thus it is more challenging to derive induction kinetic parameters (e.g. $E_{max}$: the maximum induction effect; and $EC_{50}$: the concentration causing half-maximal effect) from activity data. Thus, since the 2012 draft FDA DDI guidance, mRNA measurement was recommended as a preferred endpoint to assess induction potential for an NME because of its larger dynamic range and lack of interference with concurrent inhibition of enzyme activity. While induction potential evaluated with enzyme activity is still acceptable if concurrent inhibition can be ruled out, recommendation on the criterion to assess CYP activity data from induction experiment has not been provided in the current guidance, since further research is still needed to identify the appropriate threshold (e.g. response relative to positive control).

There are several ways to evaluate the mRNA data, as elaborated in the 2017 draft FDA *in vitro* DDI guidance, i.e. mRNA fold-change method, correlation method, basic model, and mechanistic models. Among these, the mRNA fold-change method is used most often. In the 2017 version, it was recommended that an NME is considered as an inducer if the CYP mRNA level in NME-treated cells is twofold or more than that in the vehicle-treated cells and also the CYP mRNA increase by an NME is ≥20% of the response to the positive control. Recent research by the induction working group of the International Consortium for Innovation and Quality (IQ Consortium, an organization of pharmaceutical companies) indicated general consensus that CYP induction is evaluated with measuring mRNA expression in separate human hepatocytes from three donors and supported twofold induction as the threshold for positive findings of CYP3A4 mRNA induction when there is a concentration-dependent mRNA increase [60]. But the group raised a question about normalizing the data as percentage of the response to positive control, due to the empirically divergent responses in rifampin (as positive control) and most test inducers [60]. Considering this and also the findings of an earlier study which showed certain false negative predictions when applying the 20% threshold for response relative to rifampin to a dataset of 20 inducers and 15 non-inducers of CYP3A4 [61], the 20% threshold was removed from the 2020 final version of FDA *in vitro* DDI guidance. However, there may still be utility of applying a threshold to the percentage of positive control. For example, it is often observed that omeprazole (the most commonly used positive control for CYP1A2 induction) leads to a large fold of induction of CYP1A2 mRNA (sometimes even more than 100-fold). It is questionable that an NME which induces CYP1A2 mRNA slightly more than twofold of vehicle control would result in significant induction of CYP1A2 *in vivo*. The analysis conducted by the IQ induction working group also showed a high false positive rate for CYP3A *in vivo* induction potential prediction [60], indicating the need to refine the criteria. Besides the twofold cutoff for mRNA induction relative to the vehicle group, applying a threshold to the percentage of response to positive controls has potential utility to reduce false positive prediction. Further analysis is desirable to examine whether there may be an optimal threshold that can decrease false positive prediction while maintaining minimal false negatives when using mRNA fold-change method.

Another update in the final FDA *in vitro* DDI guidance is providing recommendations on the relevant concentration of a drug for induction potential evaluation when using the mRNA fold-change method. Usually, CYP mRNA levels are measured at a series of concentrations of a drug with the higher end often exceeding *in vivo* concentrations many folds, for the purpose to characterize the induction–concentration profile and $E_{max}$. However, an mRNA fold of increase ≥2 at such high concentrations may not be clinically relevant. Based on free drug theory, it is the free intracellular drug concentration in liver cells that binds with nuclear receptor and then regulates transcription of CYP gene. In the EMA guideline, a scaling factor of 50 is applied to unbound plasma concentration ($C_{max,u}$) assuming that this may be conservative enough to cover free drug concentration in liver cells [55]. As mentioned earlier, this is an arbitrary factor. The analysis conducted by the IQ induction working group showed that a lower scaling factor of 30 led to just one more false negative prediction than using $50 \times C_{max,u}$ for mRNA fold-change method (i.e. ≥2-fold or more increase compared to vehicle group at a drug concentration around $30 \times C_{max,u}$ or $50 \times C_{max,u}$ is considered as positive findings) [60], and this false negative prediction can be avoided by capping the $f_{u,p}$ (free fraction of drug in plasma) at 0.01 for drugs that have measured human plasma protein binding more than 99% (note that the analysis by the IQ induction working group did not cap $f_{u,p}$). Capping $f_{u,p}$ at 0.01 is recommended by regulatory guidances from FDA, EMA, and PMDA and is also applied for CYP inhibition assessment, due to concerns with the ability to accurately measure $f_{u,p}$ for very highly protein-bound drugs and the subsequent impact of that uncertainty on $R$ value calculation. Thus, in the 2020 final FDA guidance, it is recommended to examine mRNA fold change of CYPs at concentration(s) of a drug around $30 \times C_{max,u}$ with $f_{u,p}$ capped at 0.01 when there is a concentration-dependent increase of CYP mRNA level. As mentioned earlier, there are substantial false positive predictions with mRNA fold-change method and other approaches based on the dataset generated by the IQ induction working group [60]. Therefore, it is valuable to conduct further research to refine the criteria to decrease false positive predictions without loosening the control of false negatives. One potential solution is to use a lower scaling factor than 30 or 50.

Another approach to evaluate the mRNA data is the correlation method. This method seems having the best prediction accuracy [62], but needs extensive experimental work. It requires establishment of a calibration curve which correlates *in vivo* DDI magnitudes (i.e. AUC ratios of CYP substrates) to relative induction scores (RIS) which is calculated from *in vitro* data for a set of known inducers with different induction strength of a CYP enzyme. Using this curve, the *in vivo* induction effect of an NME can be predicted from its RIS, and if the projected DDI magnitude is considered clinically significant, a follow-up *in vivo* DDI study is warranted. The RIS is defined as $(E_{max} \times C_{max,u})/(EC_{50} + C_{max,u})$ [63]. Thus, the correlation method requires measuring CYP mRNA at a serial of concentrations for not only the NME investigated but also a group of reference inducers in order to characterize their $E_{max}$ and $EC_{50}$. The calibration may be established once for one batch of hepatocytes and does not need to be determined for each experiment using this batch to reduce the burden of excessive experiment work. The EMA guideline recommends that at least two inducers at specified concentrations need to be included as positive controls, to get an idea about whether the induction response of the hepatocyte in the experiment evaluating an NME is comparable to the response in the earlier experiment setting up the calibration curve. Sometimes, $E_{max}$ or $EC_{50}$ cannot be estimated due to an incomplete *in vitro* induction profile (e.g. limited by solubility or cytotoxicity of the tested drug). Alternative correlation approaches may be used if validated, such as using $C_{max,u}/EC_{50}$, $AUC/EC_{50}$, or $AUC/F_2$ ($F_2$ = *in vitro* test concentration leading to a twofold increase of baseline mRNA) to replace RIS [64, 65].

Besides the mRNA fold-change method and the correlation method, a basic model can also be used to evaluate induction potential based on mRNA data. To use this method, an $R$ value ($R_3$) will be calculated as $1/(1 + d \times (E_{max} \times 10 \times C_{max,u})/(EC_{50} + 10 \times C_{max,u}))$, where $d$ represents an empirical calibration factor enabling *in vitro* to *in vivo* induction scaling and is determined for each hepatocyte batch. Its value is estimated by correlating the predicted and observed induction effects (i.e. AUC ratios of substrates) of a set of known inducers and performing a linear regression to identify the $d$ value that can minimize the prediction error (e.g. geometric mean fold error (GMFE), root mean squared error (RMSE)) [66]. Thus, to estimate the $d$ factor, extensive experiment work is also needed to characterize the $E_{max}$ and $EC_{50}$ of a group of inducers. There has been research exploring alternative approach by estimating $d$ factor only based on one inducer, which suggested that the prediction accuracy of using $d$ factor derived from rifampin alone was comparable to that using $d$ factor from a set of 15 known CYP3A inducers [67]. Sometimes, the $d$ factor is just assumed as 1, when the induction profile of positive control(s) has not been characterized and thus $E_{max}$ and $EC_{50}$ values are unavailable.

The EMA guideline does not have $R_3$ value as a stand-alone method. Instead, it is considered as part of a mechanistic static model (also called as net-effect model) which takes into consideration of reversible inhibition, inactivation (i.e. TDI), and induction of hepatic CYPs and also intestinal CYP (usually CYP3A), as well as the contribution of a CYP to the metabolic clearance of a substrate drug (i.e. $f_m$) [55, 64]. The mechanistic static model predicts the AUC ratio of a substrate of the affected CYP when administered along with an NME versus the substrate given by itself. This model is usually used when the initial evaluation (using basic models or simple criteria as described above for inhibition and induction) cannot rule out the *in vivo* DDI potential of an NME. It also has an advantage over simple equations or criteria in that it can provide concurrent prediction for a drug that is an inhibitor and also an inducer for a CYP. With the rapid progress of PBPK in the past decade, this dynamic modeling approach has also been widely used to make a more quantitative prediction after the initial assessment. Its advantage over the mechanistic static model is considering the change of inhibitor or inducer drug concentration over time. A later session of this chapter will discuss more about the application of PBPK in DDI evaluation. It should be noted, while a mechanistic static or PBPK model can be used to make concurrent predictions for inhibition and induction, it is recommended to also provide predictions considering inhibition mechanism alone and induction effect alone. This is due to a concern that there may be overprediction for inhibition effect (e.g. for TDI or intestinal CYP3A inhibition) which could mask the induction effect leading to a false negative prediction of the overall induction effect, and similarly overprediction of induction effect that may mask the overall inhibition effect.

Studies indicate that activation of the pregnane X receptor (PXR) nuclear receptor results in co-induction of CYP3A4 and CYP2Cs (i.e. CYP2C8, CYP2C9, and CYP2C19), and generally CYP2Cs are less inducible than CYP3A4 [61]. Thus, CYP induction potential of an NME is usually evaluated for CYP1A2, CYP2B6, and CYP3A4, which are regulated by nuclear receptors aryl hydrocarbon receptor (AhR), constitutive androstane receptor (CAR), and PXR, respectively. If *in vitro* result for CYP3A4 induction is determined not clinically relevant, then there is no need for additional induction studies for CYP2C enzymes. When *in vitro* result shows positive finding for CYP3A4 induction, the potential of an NME to induce CYP2Cs is needed. While this can be done with *in vitro* induction experiment, sometimes it is challenging to interpret the data since CYP2Cs are generally less sensitive to inducer effect and even the positive control (e.g. rifampin) may only show limited fold of increase of CYP2Cs mRNA [59]. Thus, it may be inconclusive to determine whether a drug may induce CYP2Cs *in vivo* from *in vitro* data. One approach to bypass this challenge is to leverage *in vivo* DDI study results with CYP3A substrates. When a drug significantly induces CYP3A4 *in vitro*, this may trigger an *in vivo* study with a sensitive substrate of CYP3A. If the result showed no significant alteration of the substrate AUC, and if it is known that the drug does not significantly inhibit CYP3A *in vitro*, it can be concluded that the drug does not induce CYP3A *in vivo* and thus it is not expected that the drug will significantly induce CYP2Cs, either. For an NME that is an inducer and also a reversible inhibitor of CYP3A, the Dose/250 ml may be a large value and results in $R_{1,\mathrm{gut}}$ exceeding the threshold of 11. Thus, *in vivo* inhibitory potential on CYP3A cannot be readily ruled out, and no significant change of CYP3A substrate exposure with the presence of the NME could be a net effect of induction and inhibition which counteracts each other. An approach to delineate this is to add an additional period in the *in vivo* study, so that there are three treatment periods with single doses of a sensitive CYP3A substrate that undergoes substantial first-pass intestinal metabolism (e.g. midazolam) – the substrate given alone, the substrate concomitantly administered with a single dose of the NME, and the substrate taken along with the NME after multiple dosing of NME [68]. If the concentrations of the substrate are not significantly changed after a single dose of the NME, it can be concluded that the $R_{1,\mathrm{gut}}$ value leads to false positive prediction, and lack of significant change of the substrate exposure after steady-state of the NME indicates no inhibition and no induction of CYP3A *in vivo*, which also eliminate the need to evaluate the induction potential for CYP2Cs. It should be noted this approach does not work for inducer drugs that are also TDIs of CYP3A, since TDI effect becomes larger after multiple dosing and thus *in vivo* observations cannot separate the contribution of inhibition and induction. This approach will not work for a reversible inhibitor (or its metabolite) that has a long half-life and substantial accumulation at steady state, unless a high single dose of such an inhibitor can be given and achieve the exposure comparable to that at steady-state. Besides being applied to CYP2C induction, this strategy of leveraging *in vivo* results with CYP3A substrate may also be adopted for evaluating induction potential of a drug toward certain UGTs which are also co-regulated by PXR but less inducible than CYP3A4.

#### 27.3.1.4 Metabolites as Substrates or Inhibitors for CYP Enzymes

A number of NMEs undergo metabolism and the metabolites generated may also carry DDI liability. When a metabolite is associated with significant adverse events or a metabolite considerably contributes to the efficacy of a drug, it is important to also characterize how the metabolite is eliminated, since another medication that affects the metabolic enzymes responsible for clearance of the metabolite may result in clinically significant change of the metabolite exposure and thus affects the safety or efficacy of a drug. On the other hand, metabolites may inhibit or induce CYPs and cause or contribute to inhibitory or induction effect of a drug on the metabolism of other medications [69]. Thus, in the 2012 FDA draft DDI guidance, it was recommended that, when a parent drug does not significantly inhibit CYPs, there is still a need to conduct *in vitro* experiments to assess whether a metabolite may inhibit major CYPs if the metabolite has systemic exposure (i.e. AUC) ≥25% of the AUC of its parent drug. Since then, several analyses were performed with different flowcharts proposed on how to determine when DDI potential of metabolites needs to be evaluated [70–72]. One of the frameworks takes into consideration the polarity of metabolites relative to parent drugs, based on an assumption that lipophilic compounds are more likely to bind with and inhibit CYPs. Because majority of metabolites are more polar than parent drugs, evaluating the inhibition potential of CYPs is only warranted when those metabolites at least have the same level of exposure (AUC) as parent drugs. For the metabolites that are less polar than parent drugs, because they may be more potent inhibitors, there is a need to evaluate their inhibitory potential *in vitro* as long as those metabolites have AUCm/AUCp ≥25% [71]. In addition, if a metabolite has structure alert indicating that it may act as a TDI, this metabolite needs to be evaluated even though its AUCm/AUCp < 100%, because TDIs generally carry more risk of *in vivo* DDI. This approach stratifying metabolites based on the exposure and polarity was adopted in the 2017 FDA

**Table 27.4** Classification of strong, moderate, and weak perpetrators for CYP enzymes [33].

| Variable | Strong inhibitor[a)] | Moderate inhibitor | Weak inhibitor |
|---|---|---|---|
| Change in AUC of a sensitive index substrate | ↑ ≥5-fold | ↑ ≥2- to <5-fold | ↑ ≥1.25- to <2-fold |
| | **Strong inducer**[a)] | **Moderate inducer** | **Weak inducer** |
| | ↓ ≥80% | ↓ ≥50% to <80% | ↓ ≥20% to <50% |

a) For strong inhibitors or inducers, data may be from nonsensitive index substrates.
AUC, area under the concentration–time curve.
Source: Adapted from FDA [33].

**Table 27.5** Classification of sensitive and moderate sensitive substrate for CYP enzymes [33].

| Variable | Sensitive substrate[a)] | Moderate sensitive substrate |
|---|---|---|
| Change in AUC with a strong index inhibitor | ↑ ≥5-fold | ↑ ≥2- to <5-fold |

a) For sensitive substrate, data may be from any known index substrates for a particular pathway or whose AUC ratio in poor metabolizers for a specific enzyme is greater than or equal to fivefold compared to extensive metabolizers.
AUC, area under the concentration–time curve.
Source: Adapted from FDA [33].

draft DDI guidance and also the 2020 FDA final version. However, there are differences from the recommendation of EMA guideline that only relies on the exposure of metabolite and requires *in vitro* evaluation when (i) a metabolite is considered as a major metabolite (i.e. the AUC of this metabolite is >10% of the AUC of drug-related material based on radioactive moieties from mass balance study) and also (ii) the metabolite also has AUCm/AUCp > 25%. For calculation of AUCm/AUCp, the EMA guideline recommends using unbound concentrations when protein binding information is available for parent drug and metabolites. Total drug concentrations may be used in the absence of such information. In addition, EMA only asks for exanimating Phase 1 metabolites, while gemfibrozil glucuronide is known to be a TDI of CYP2C8 and may contribute more to the inhibitory effect of gemfibrozil on CYP2C8 than the parent drug [9]. Apparently, further research is needed to compare the two criteria from FDA and EMA and explore an alternative approach to harmonize the recommendations from different regulatory agencies. It should also be noted that when a mechanistic static or PBPK model is used to justify lack of *in vivo* CYP inhibitory effect for a drug, the inhibitory potential of metabolite(s) of the drug needs to be taken into account.

#### 27.3.1.5 Classification of CYP Enzyme Inhibitors, Inducers, and Substrates

The *in vivo* DDI guidance includes a recommendation for classifying CYP1A2, CYP2B6, CYP2C8, CYP2C9, CYP2C19, CYP2D6, and CYP3A4 inhibitors as strong (≥5-fold increase in the AUC of sensitive index substrates for specific CYP enzymes), moderate (2-fold to 5-fold increase in AUC), and weak (<2-fold increase in AUC) inhibitors (Table 27.4). A strong inducer decreases the AUC of a sensitive index CYP substrate by ≥80 percent. A moderate inducer decreases the AUC of a sensitive index CYP substrate by ≥50 to <80% and a weak inducer decreases the AUC of a sensitive index CYP substrate by ≥20 to <50% (Table 27.4). Similarly, substrate is classified as sensitive substrates for a particular enzyme (≥5-fold increase in its AUC in the presence of index CYP inhibitors), or moderately sensitive substrates (2-fold to 5-fold increase in AUC of strong index CYP inhibitors) (Table 27.5). Index perpetrators are drugs that inhibit or induce a given metabolic pathway by a defined magnitude when administered with a sensitive substrate and are commonly used in prospective DDI studies. Index substrates have defined changes in systemic exposure when administered with a strong inhibitor for a specific drug elimination pathway.

This classification information helps to determine whether other drugs that have not been investigated in a DDI study have clinically significant DDIs with the investigational drug and therefore need to be mentioned in labeling. For example, if an NME is a sensitive CYP3A substrate as demonstrated by an *in vivo* study with a strong CYP3A inhibitor (e.g. itraconazole), there may be a warning in the labeling regarding a potential drug interaction with other strong or even moderate CYP3A inhibitors [37, 39, 73].

Currently, there is no standardized classification system for Phase 2 metabolizing enzymes or transporters.

#### 27.3.1.6 Pharmacogenetic Considerations

If a drug is metabolized by a polymorphic enzyme (such as CYP2D6, CYP2C9, or CYP2C19), the comparison of pharmacokinetic parameters of this drug in poor metabolizers and extensive metabolizers would be expected to indicate the extent of interaction of this drug with strong inhibitors of these enzymes and may make interaction studies with strong inhibitors unnecessary (for CYP2C9, this may also apply to moderate inhibitors, since there is very few strong inhibitor of CYP2C9) [33]. When such studies demonstrate a significant interaction, further evaluation with weaker inhibitors may be necessary. It is noted that DDI and drug–gene interactions could be gene-dependent. The magnitude of DDI as compared to polymorphism effect can depend on the potency of various inhibitors and the allele function [74–76].

#### 27.3.1.7 Evaluation of Multiple CYP Inhibitors or Elimination Pathways may be Appropriate

In order to define drug safety at the highest anticipated exposure, a study using an inhibitor affecting multiple CYPs may be appropriate to evaluate the maximum impact of CYP inhibition, as it represents the worst-case scenario. It is useful if (i) there is a concentration-dependent safety concern of an NME; (ii) the NME is metabolized via multiple CYP enzymes; and (iii) residual (i.e. non-inhibitable) drug clearance is low. If such a study reveals a safety concern, then further studies with specific inhibitors that affect individual CYP pathways may be warranted [35, 37]. When an NME is metabolized by multiple CYPs among which there is a polymorphic CYP, the effect of an inhibitor of an alternate pathway of that polymorphic CYP is likely to be larger in the poor metabolizers of that CYP than in extensive metabolizers because both metabolic pathways are impaired. Thus, a study (either an *in vivo* study or PBPK simulation) may be considered to assess such interactions involving CYP inhibition and also genetic polymorphism [77, 78]. Similarly, for a drug undergoing metabolism and renal elimination, drug exposure may be significantly increased in renally impaired patients who take inhibitors of a CYP and such an increase may need clinical intervention, although the change of drug exposure may not be clinically significant in patients with normal renal function taking CYP inhibitors or patients with renal impairment but not taking inhibitors [79, 80].

#### 27.3.1.8 Use of Cocktails for *In Vivo* Evaluation of an NME's Drug Interaction Potential

The simultaneous use of a combination of CYP drug substrates, or a CYP "cocktail," to evaluate interaction potential of an NME can be efficient. Several cocktail combinations have been reported in the literature. To provide valid results, the CYP substrates in the cocktail should be selective for the corresponding CYP enzyme and must not interact with each other [81]. The sample size of these studies should be adequate to detect metabolic differences. If results of the cocktail study are negative (no interaction), additional *in vivo* studies would be unnecessary. Positive findings may necessitate further *in vivo* investigations with less sensitive substrates of the affected CYP enzyme [33, 37]. Results from a well-conducted cocktail study can be interpreted and presented in drug labeling the same way as results from any other well-conducted DDI study.

### 27.3.2 Highlights Concerning Drug Transporter-Based DDIs

Membrane transporters in various tissues of the human body govern the transport of endogenous and exogenous substances in and out of cells in the body. With more than 400 transporters identified, approximately 30 transporters are known to play a role in drug transport [82]. Transporter-based interactions have increasingly been documented in recent years. Analogous to drug metabolizing enzyme-based DDIs, coadministration of a drug that is an inhibitor or an inducer of a drug transporter may affect the pharmacokinetics of another drug that is a substrate for that transporter. Some observed drug interactions that could not be explained adequately early on by CYP-mediated interactions were later found to be at least in part mediated by transporters. Examples include the digoxin-quinidine interaction [16, 17], fexofenadine-ketoconazole interaction [83, 84], and rosuvastatin–cyclosporine interaction [18, 19]. The first two interactions are mediated by inhibition of P-gp and the last one is likely mediated by OATP1B1, OATP1B3, and BCRP. More transporter-based interactions have since been reported, which also see interactions involving other transporters such as OATs, OCTs, and MATEs [85, 86]. These observations have helped to establish the now-recognized view that metabolic drug interactions alone may not adequately account for the individual variation in absorption, distribution, and elimination of drugs and that transporters, either alone or in concert with drug metabolizing enzymes, can also play a significant role in these processes. Table 27.3 lists some of the major human transporters and reported substrates and inhibitors.

#### 27.3.2.1 Clinically Important Transporters for Routine Evaluations During Drug Development

Proliferations in transporter research in the past decades have yielded substantial amount of data allowing systematic evaluations to provide regulatory recommendations on transporter studies [82, 87, 88]. To date, nine clinically important transporters are recommended in FDA DDI guidances for routine evaluations of DDI potential for investigational drugs, which include four efflux transporters (P-gp, BCRP, MATE1, and MATE2-K) and five uptake transporters (OATP1B1, OATP1B3, OAT1, OAT3, and OCT2) [33, 34]. With respect to tissue localization, important transporters in the intestine include P-gp and BCRP, in the liver include OATP1B1, OATP1B3, P-gp, and BCRP, and in kidneys include OAT1, OAT3, OCT2, MATE1, MATE2-K, and P-gp (Table 27.3). As these transporters are present in the organs for oral drug absorption and/or drug elimination, DDIs involving these transporters can alter the systemic exposure to the drug, which can be readily assessed. Depending on the specific transporters present in tissues, transporter-mediated drug interactions may have effects on tissue concentrations impacting the safety or efficacy profiles of the drug that is not reflected at the systemic level, which may be investigated on a case-by-case basis.

#### 27.3.2.2 Mechanistic Approach in Transporter Studies

As with the evaluation of metabolizing enzyme-based interactions, transporter studies as recommended in the FDA guidance also adopt a risk-based approach to evaluate an investigational drug as a transporter substrate and as an inhibitor first through *in vitro* studies then followed by *in vivo* interaction studies as appropriate. For example, if the *in vitro* findings identify an investigational drug as a substrate for a particular transporter, *in vivo* interaction studies may be conducted depending on the importance of that transporter in the pharmacokinetic of the drug, clinical considerations such as safety margin of the drug, and the likelihood of concomitant medications that is a modulator of that transporter. Acquiring mechanistic understanding using this approach helps to predict qualitatively the interaction potential with other untested drugs.

#### 27.3.2.3 *In Vitro* Transporter Assay Systems and Assay Validation

Three main types of *in vitro* assay systems are used to evaluate an investigational drug as a substrate or inhibitor of specific transporters: membrane vesicles, cell-based polarized bidirectional assay systems for efflux transporters, and cell-based assay systems for uptake transporters. The FDA *in vitro* DDI guidance provides examples of assay systems for the nine transporters, which include membrane vesicles, primary cells (e.g. hepatocytes), transporter-expressing cell lines (e.g. Caco-2 cells), and transfected cells (e.g. HEK293, CHO, MDCK, and LLC-$PK_1$) [34]. Appropriate systems should be selected based on the specific transporter to be studied and the purpose of the study.

Variability in study results can arise from various sources [89, 90]. It is important that the assays used to classify a drug as a substrate or inhibitor of specific transporters are optimized and validated. The FDA guidance also provides some considerations about the best practice for conducting *in vitro* transporter studies [34]). A few key points are described here. When a transfected cell line is used for transporter assay, the corresponding empty vector should serve as a control system to correct for endogenous transporter activity and nonspecific binding. The assay system used should be characterized with *in vitro* positive and negative controls to verify that the system functions appropriately. Examples of *in vitro* substrates and inhibitors for various transporters that may be used for this purpose can be found in the FDA website [73]. Other substrates or inhibitors can also be used as long as scientifically justified. When cell systems used express multiple transporters, efforts should be made to determine the contribution of specific transporter(s). For example, when Caco-2 cells are used to evaluate a drug as a P-gp substrate, more than one P-gp inhibitor should be used in the evaluation, because Caco-2 expresses multiple efflux transporters and some P-gp inhibitors are not selective for P-gp and thus using multiple inhibitors may help differentiate effects of P-gp and other transporters.

#### 27.3.2.4 *In Vitro* Evaluation of an Investigational Drug as a Transporter Substrate

To determine which transporters to be studied *in vitro* for a drug, its ADME characteristics also need to be considered. For example, evaluation of an investigational drug as a substrate for the liver uptake transporters (OATP1B1 and OATP1B3) should be conducted when there are *in vitro* or *in vivo* data suggesting that hepatic clearance (hepatic metabolism or biliary secretion) is significant (i.e. $\geq$25% of systemic clearance) and there is information to support active hepatic uptake by OATP1B (e.g. limited passive permeability; high drug concentrations in the liver relative to other tissues). Similarly, *in vitro* evaluation of an investigational drug as a substrate for renal uptake transporters (OAT1, OAT3, and OCT2) and renal efflux transporters (MATE1 and

MATE2-K) should be conducted when there are data to suggest that active renal secretion is significant (i.e. $\geq$ 25% of systemic clearance). Active renal secretion can be estimated by deducting glomeruli filtration part ($GFR \times f_{u,p}$) from renal clearance by assuming no active reabsorption. On the other hand, because both P-gp and BCRP are present in the organs of drug absorption (intestine) and elimination (liver and kidney), investigational drugs generally need to be evaluated as substrates for these transporters except (i) BCS class 1 drugs that are highly soluble and highly permeable or (ii) drugs whose administration route does not involve oral absorption and not undergoing significant biliary excretion or active renal secretion. It should be noted inhibition of P-gp or BCRP may still have an impact on the pharmacokinetics of an orally administered BCS class 1 drug if it has substantial first-pass intestinal metabolism and thus P-gp or BCRP may interplay with CYP3A in enterocytes. In addition, for central nervous system (CNS) drugs, it may be important to characterize whether the drug is transported by P-gp or BCRP since these transporters are present on the blood–brain barriers (BBB) and may limit penetration of the drug into brain and potentially affect its efficacy.

Generally, a drug is deemed an *in vitro* substrate for a transporter if the transport of the drug observed in an appropriate study system is ≥2-fold of that with a corresponding empty vector system and a known inhibitor (at a concentration at least 10 times of its $K_i$ or $IC_{50}$) can decrease the drug transport by 50% or more [34]. To evaluate a drug as a substrate for P-gp (or BCRP), a bidirectional transport assay is often used. In this case, the decision criteria include an efflux ratio (ER; i.e. the flux of basolateral to apical divided by the flux of apical to basolateral) of ≥2 and an ER decrease by ≥50% by at least one known P-gp (or BCRP) inhibitor. Alternative criteria can be applied if validated for the experimental system used. The decision framework for determining an investigational drug as a transporter substrate is presented in Figure 27.4. Although MATEs are efflux transporters, the transporter experiments are typically

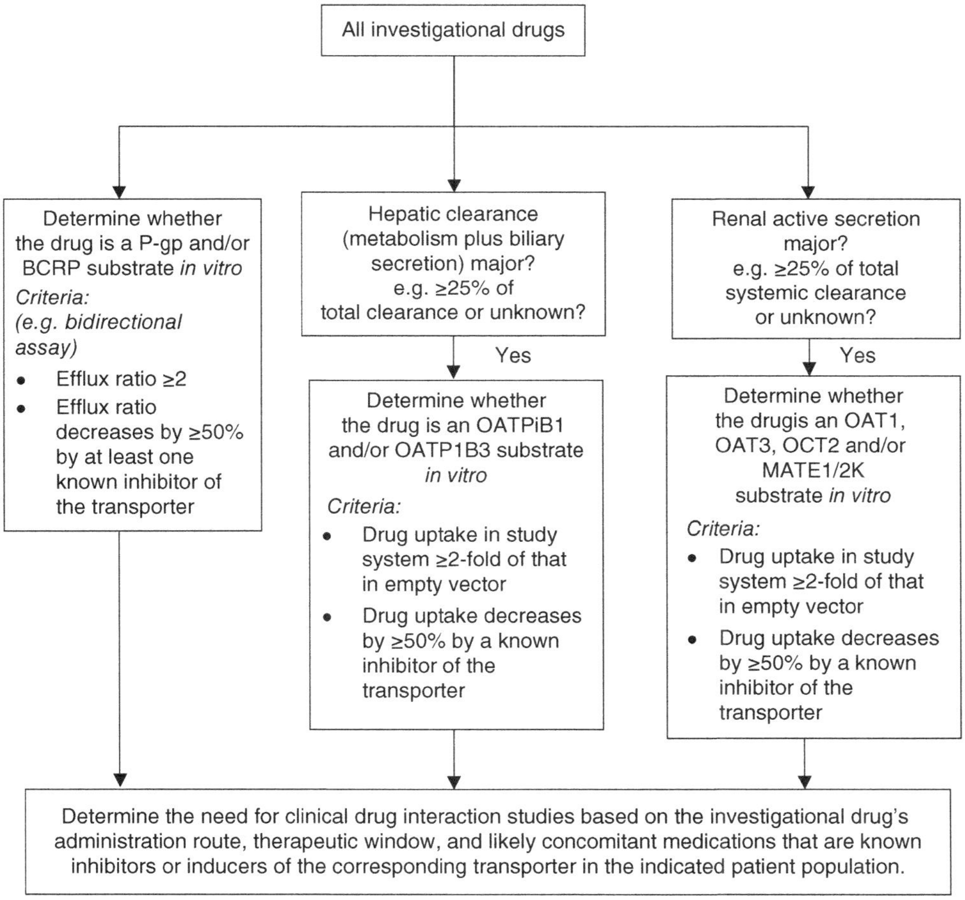


**Figure 27.4** Evaluation of an Investigational Drug as a Substrate of Transporters (P-gp, BCRP, OAT1, OAT3, OATP1B1, OATP1B3, OCT2, and MATE1/MATE2-K). Source: Adapted from Refs. [15, 86].

conducted in an uptake manner by reversing the pH gradient via internal acidification or external alkalization, and thus, the study results are expressed as drug uptake.

#### 27.3.2.5 *In Vitro* Evaluation of an Investigational Drug as a Transporter Modulator

Transporter modulation by a drug may result in the inhibition or induction of the transporter. At this time, however, *in vitro* evaluation of transporter induction is not well established, and *in vivo* transporter induction studies have been mainly reported for P-gp and there are some controversial data with OATP1B1 [91–93]. Because P-gp and CYP3A share similar mechanisms in induction, evaluation of P-gp induction is generally not needed when the potential for *in vivo* CYP3A induction is ruled out [33]. Similar strategy described earlier for evaluation of CYP2C induction is also applicable for assessing P-gp induction since P-gp is co-induced with CYP3A by PXR agonists but less inducible than CYP3A. On the other hand, evaluation of an investigational drugs as an inhibitor of the nine transporters mentioned earlier is recommended. The current FDA *in vitro* DDI guidance [33] describes three separate decision criteria (Table 27.6), i.e. one for P-gp and BCRP, another for liver transporters OATP1B1 and OATP1B3, and the other for renal transporters OAT1/OAT3, OCT2, and MATE1/MATE2-K. Of note, an inhibitor of a transporter is not necessarily a substrate for that transporter [94, 95]. As such, negative findings in the substrate study do not negate the need for evaluating the drug as an inhibitor of that transporter.

Generally, in each criterion, the ratio (*i.e.* the decision metric) of the clinically relevant concentration of the investigational drug to the *in vitro* transporter inhibition potency of that drug (e.g. $IC_{50}$, which is the concentration of the drug to achieve 50% of maximal inhibition of the transporter activity) is compared to a predefined cutoff value to determine whether there is a potential for the drug to inhibit the transporter *in vivo* [34] (Table 27.6). The cutoff value associated with each transporter was determined based on analyses of available *in vitro* and clinical drug interaction data for various substrate-inhibitor drug pairs pertaining to that transporter. The need for conducting a follow-up *in vivo* DDI study depends on clinical considerations such as the likelihood of concomitant medication that is a substrate for the transporter and the safety margin of that substrate.

*27.3.2.5.1 P-gp and BCRP* The potential for an orally administered investigational drug to inhibit these two efflux transporters is assessed at both the gut level and systemic level. Drug concentration in the gut lumen ($I_{gut}$) is roughly estimated as Dose/250 ml while the systemic level ($I_1$) is the maximum total plasma concentration ($C_{max}$) following administration. Initially, an investigational drug was considered to have the potential for drug interactions *in vivo* if $I_{gut}/(IC_{50}$ or $K_i) \geq 10$ or $I_1/(IC_{50}$ or $K_i) \geq 0.1$. However, later research suggested that considering $I_{gut}/(IC_{50}$ or $K_i) \geq 10$ alone seems sufficient to predict *in vivo* DDI risk [96, 97]. Hence, in the current FDA *in vitro* DDI guidance, it is recommended only using $I_{gut}$ to assess DDI potential for inhibitors that are orally administered. The inhibition decision framework for P-gp and BCRP transporters is illustrated in Figure 27.5. As mentioned earlier, $I_{gut}$ may be overestimated by Dose/250 ml for BCS Class 2 or 4 drugs. Moreover, it is the intracellular drug that interacts with P-gp or BCRP and thus the relevant inhibitor concentration is actually free drug inside enterocytes

**Table 27.6** Decision criteria recommended in the FDA final *in vitro* DDI guidance (2020).

| Transporters | Relevant inhibitor concentration | Recommended decision criteria indicating potential DDI *in vivo* |
|---|---|---|
| P-gp or BCRP | Gut concentration<br>$[I]_2$ ($[I]_{gut}$) = Dose/250 ml | $I_2/IC_{50} \geq 10$ (for oral drugs) |
| OATP1B1/ OATP1B3 | Free hepatic inlet concentration<br>$I_{in,max,u} = f_{u,p} \times (C_{max} + (F_a F_g \times k_a \times Dose)/Q_h/R_B)$ | $I_{in,\ max,u}/IC_{50} \geq 0.1$ |
| OAT1/OAT3 | Free systemic concentration<br>$I_u = C_{max,u}$ | $I_{max,u}/IC_{50} \geq 0.1$ |
| OCT2/MATE1/ MATE2-K | Free systemic concentration<br>$I_u = C_{max,u}$<br>(as a "surrogate" for MATEs) | $I_{max,u}/IC_{50} \geq 0.1$ |

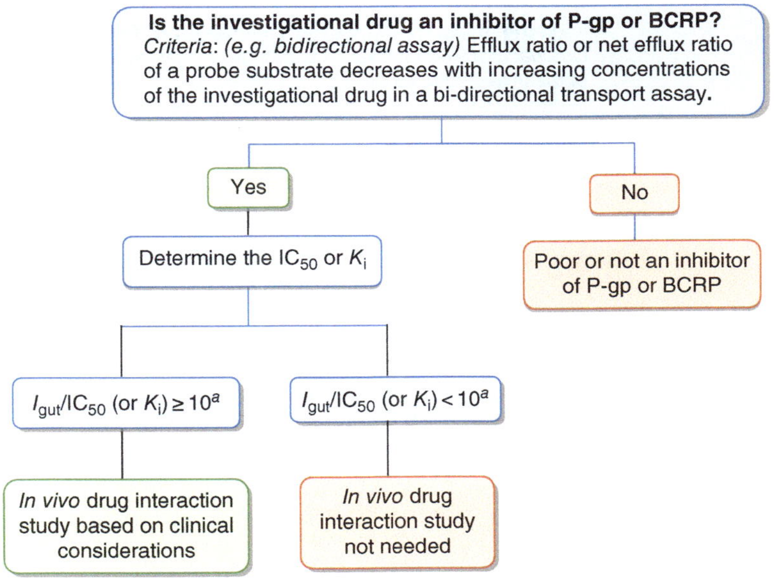


**Figure 27.5** Evaluation of an investigational drug as an inhibitor of P-gp or BCRP. Source: Adapted from FDA [15].

instead of $I_{gut}$. In the previous analysis, majority of the false positive predictions occur with BCS Class 2 or 4 drugs [97], suggesting the need to conduct further research to explore potential solution to increase the prediction performance for such compounds. It should be noted that $I_{gut}$ is not applicable to inhibitors that are not orally administered or drugs themselves not inhibiting P-gp or BCRP but having metabolites that are P-gp or BCRP inhibitors. In such situations, possible *in vivo* DDI is driven by inhibition of P-gp or BCRP in the liver and/or kidney unless the inhibitory metabolites are primarily formed in enterocytes. Tentatively, the FDA guidance recommends using total $C_{max}/IC_{50}$ or $K_i$ compared to a cutoff of 0.1, while the EMA guideline considers unbound $C_{max,u}/IC_{50}$ or $K_i$ and a lower cutoff of 0.02. Recently, an analysis was conducted comparing these two criteria, showing that applying $C_{max,u}/IC_{50}$ or $K_i \geq 0.02$ generated a more balanced sensitivity (76%) and specificity (78%). While it led to more false negatives, three of the six false negatives were associated with verapamil. When considering a major metabolite, norverapamil that has similar AUC to its parent drug and seems to have more potent inhibitory effect, those false negative predictions would be corrected as true positives [98].

***27.3.2.5.2 OATP1B1 and OATP1B3*** It is noteworthy that the clinically relevant concentration of a drug considered as an inhibitor for these hepatic uptake transporters is the maximum unbound plasma concentration at the inlet to the liver, $I_{u,inlet,max}$, which is the maximum unbound concentration derived from both the hepatic artery and portal vein. A drug is considered to have the potential for DDI *in vivo* if $I_{u,in,max}/(IC_{50}$ or $K_i) \geq 0.1$ [99]. Because several drugs were found to be TDIs of OATP1B transporters [100–103], preincubation of tested drugs with OATPB1-expressing cells should be considered. The value of $I_{u,inlet,max}$ can be estimated using the question $I_{u,inlet,max} = f_{u,p} \times (I_{max} + k_a \times \text{Dose} \times F_a \times F_g/Q_h/R_B)$, where $f_{u,p}$ is the unbound fraction in plasma, while $k_a$, Dose, $F_a$, $F_g$, $Q_h$, and $R_B$ represent absorption rate constant, oral dose of an inhibitor, fraction of the dose of the inhibitor absorbed, fraction of the dose of the inhibitor escaping first-pass intestinal metabolism, estimated hepatic blood flow and blood-to-plasma concentration ratio, respectively. If unknown, the values for $F_a$ and $F_g$ are taken as 1 while ka is assumed to be 0.1/min to obtain the worst-case estimates. The inhibition decision framework for OATP1B1 and OATP1B3 transporters is illustrated in Figure 27.6.

While the EMA guideline also recommends using $I_{u,inlet,max}$, a lower cutoff of 0.04 instead of 0.1 is proposed. Either one results in substantial false positive predictions as shown by positive predictive error of 35% (FDA) and 43% (EMA) [99], which means the proportion of studies that were conducted that actually did not need to be (no DDI risk). Thus, it is valuable to find a way to improve the *in vitro* to *in vivo* extrapolation. In the past several years, there is rapid growth in research to identify and characterize endogenous biomarkers for transporters [104]. This may provide a solution to better translate *in vitro* findings to *in vivo*. Once validated, a biomarker can be measured during early clinical studies such as single-dose escalation and multiple-dose escalation studies of an investigational

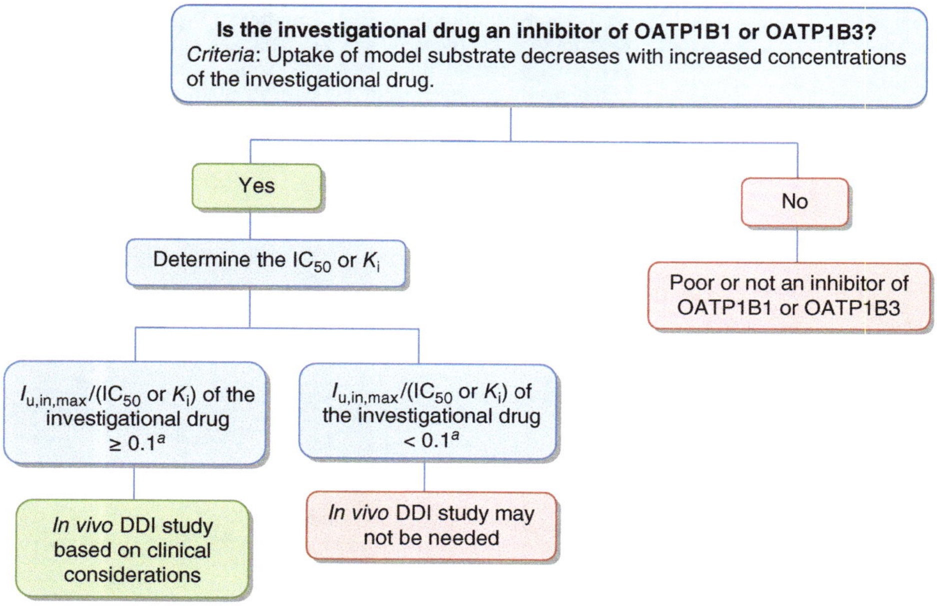


**Figure 27.6** Evaluation of an Investigational Drug as an Inhibitor of OATP1B1 or OATP1B3. Source: Adapted from FDA [15].

drug, and changes of the biomarker level from baseline can inform the *in vivo* effect of the drug on a transporter that is primarily responsible for the disposition of that biomarker [105]. Coproporphyrin I (CP-I), a derivative of heme synthesis, seems to be a promising biomarker for OATP1B. A few examples have been recently published to illustrate the utility of CP-I in drug development to de-risk OATP1B-related DDIs and understand the mechanism of complex DDIs for GDC-0810, glecaprevir, JNJ-A, probenecid, and fenebrutinib [106–110]. Further research is encouraged to gain more experience with the application of CP-I and other biomarkers in DDI risk evaluation for OATP1B inhibitors.

*OAT1/OAT3, OCT2, and MATE1/MATE2-K*: A drug is considered to have the potential to inhibit these transporters *in vivo* if $I_{u,max}/(IC_{50}$ or $K_i) \geq 0.1$ [111, 112] (Figure 27.7). Inhibiting OCT2 and/or MATEs has implications not only for DDI but also for creatinine-drug interaction. While creatinine clearance is mainly mediated by glomeruli filtration, it undergoes active secretion which involves OCT2 and MATEs, respectively, mediating the uptake of creatinine into kidney tubular cells and the transport of creatinine out of tubular cells into urine [113]. Thus, an inhibitor of OCT2 and/or MATEs that does not have nephrotoxicity can still decrease creatinine renal clearance and hence increase serum creatinine level (typically with an average increase of 10–30%) without affecting actual glomeruli filtration rate (as measured by iohexol, iothalamate, inulin, etc.) or other renal function biomarkers (e.g. cystatin C, blood urea nitrogen, β-microglobulin) [114, 115]. Therefore, evaluating inhibition potential of a drug on OCT2/MATEs provides a mechanistic understanding for serum creatinine increase if observed and helps interpret the safety profile of a drug in terms of renal toxicity [115, 116].

It should be noted that these decision frameworks for transporter drug interaction prediction are meant to offer a qualitative (yes/no) prediction primarily to rule out the potential of a drug to inhibit transporters *in vivo* based on *in vitro* inhibition data and does not provide a quantitative prediction on the extent of drug interaction effect occurring in humans. In addition, the decision frameworks only consider one transporter at a time while in clinical settings, a drug may interact with multiple enzymes and/or transporter pathways with another drug. Therefore, as with most studies of intrinsic/extrinsic factors, *in vitro* transporter assessments, in many cases, are reductionist and do not consider multiple mechanisms for drug interaction prediction. Development of more sophisticated *in vitro* models coupled with advancement in PBPK modeling that incorporates multiple drug interaction mechanisms may help with better drug interaction predictions.

***27.3.2.5.3 Evaluation of Metabolites as Transporter Substrates or Inhibitors*** Besides parent drugs, potential drug interactions caused by metabolites either as substrates or modulators of transporters may need to be considered. For example, published data have shown that some Phase 2 metabolites can be better inhibitors of various transporters leading to a higher chance of DDIs than the parent drug [117]. If only parent drugs

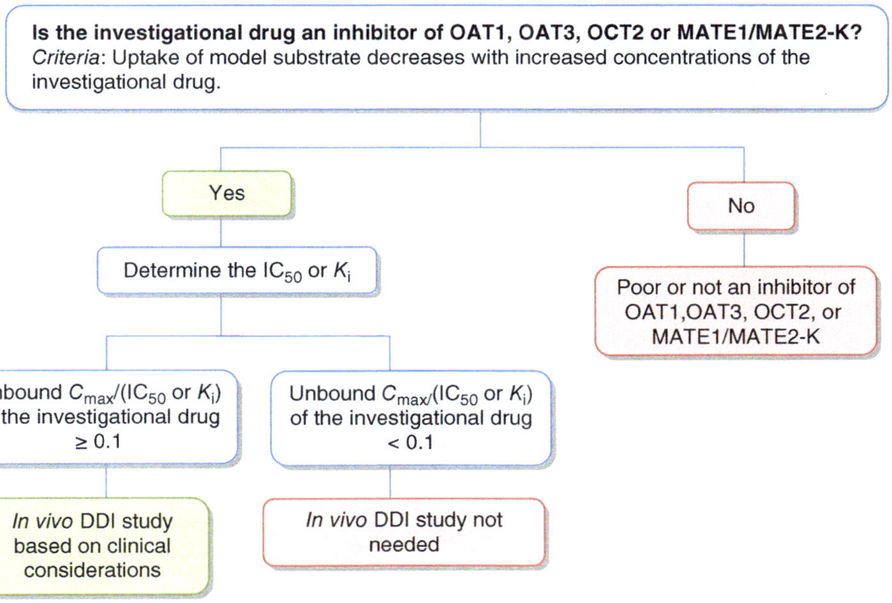


**Figure 27.7** Evaluation of an Investigational Drug as an Inhibitor of OAT1, OAT3, OCT2 or MATE1/MATE2-K. Source: Adapted from FDA [15].

were considered, prediction on drug interaction might not be accurate. The general principles and strategies for evaluation of parent drugs could be applied for metabolites where applicable, for a better drug interaction prediction based on *in vitro* assessment. For example, evaluation of a metabolite as a transporter substrate should be considered when the metabolite contributes significantly to the pharmacological activity (i.e. ≥50% of overall activity of a drug). Evaluation of a metabolite as a transporter inhibitor should also be considered, while it is not clear whether the polarity of a metabolite relative to parent drug is appropriate for stratifying metabolites. Currently, there is less experience with metabolites as transporter inhibitors than CYP inhibitors and further research will be helpful to develop recommendations on when *in vitro* assessment of inhibitory effect of metabolites toward transporter is needed.

### 27.3.3 Points to Consider for *In Vivo* Drug Interaction Study Design

Clinical DDI studies are prospectively designed to determine both the presence and the magnitude of suspected DDIs. The guidance [33] describes various types of clinical DDI studies that are used in drug development and their purposes including stand-alone DDI studies (with index substrates or perpetrators or with concomitant use drugs) in healthy subjects or patients and nested DDI studies as part of patient studies where DDI is not the main objective. Additionally, *in silico* DDI studies may be conducted with PBPK modeling (see more discussion later in this chapter).

Stand-alone clinical DDI studies are conducted using appropriate index substrates/perpetrators or likely coadministered drugs. Clinical probe is expected to show appropriate sensitivity, i.e. that the extent of change in its exposure due to inhibition correlates with the potency of coadministered inhibitor and can be measured with robust and validated bioanalytical methods. For CYP enzymes, DDI results from index substrates or perpetrators can generally be extrapolated to other substrates or perpetrators sharing the same pathway with similar sensitivity or potency (Table 27.1). However, for most transporters, general lack of index substrates or perpetrators makes such extrapolation more challenging. The doses of probe and perpetrator drugs should be clinically relevant for appropriate characterization of a DDI risk. For certain probes, potential saturation of drug transporters and the impact on DDI needs to be considered. For example, the inhibitory effects of several P-gp inhibitors are larger with a microdose of dabigatran etexilate (DE) than therapeutic dose of DE. This was speculated to be due to saturation of intestinal P-gp transport of DE at high dose levels [118].

Lists of recommended clinical probes and perpetrators for clinically relevant enzymes and transporters are available on the FDA website https://www.fda.gov/drugs/drug-interactions-labeling/drug-development-and-drug-interactions-table-substrates-inhibitors-and-inducers [73]. A majority of the listed transporter substrates, inducers, or inhibitors cannot be considered as index drugs for DDI studies because they lack specificity for one transporter and are often substrates of multiple transporters and/or enzymes. However, clinical interaction studies conducted with these drugs can provide useful information about potential DDIs with concomitant drugs.

Depending on the outcome of the *in vitro* evaluation, and other information, clinical studies may be conducted to answer specific questions about drug interactions. Whether a particular study should or should not be conducted will vary and depend on the drug in development and its intended clinical use. The selection of a particular study design also depends on a number of factors for both the substrate and interacting drug, including (i) whether use of the substrate and/or interacting drug is acute or chronic; (ii) safety considerations, including whether a drug is likely to have narrow therapeutic range; (iii) pharmacokinetic and pharmacodynamic characteristics of the substrate and interacting drugs; and (iv) assessment of induction as well as inhibition.

Considering the "intent" of a study includes (i) whether use of a probe drug is to describe the drug interaction potential so as to improve understanding of the mechanistic basis of DDIs; (ii) evaluating specific DDI studies for clinical reasons where coadministration is likely or the clinical consequences of an interaction are of concern (i.e. how to adjust the dose when two specific drugs are coadministered); and (iii) designing a study to maximize seeing an interaction by using the highest therapeutic dose and shortest dosing interval of the inhibitor. In some instances, doses smaller than those to be used clinically may be recommended for substrates on safety grounds. If the pharmacokinetics of a substrate is dose proportional, the DDI effect observed with a lower dose is considered extrapolatable to higher dose.

Results of DDI studies should be reported as 90% confidence intervals (90% CI) about the geometric mean ratio (GMR) of the observed pharmacokinetic measures with and without the interacting drug. In order to make specific labeling recommendations, it would be helpful to recommend specific "no effect" boundaries, or clinical equivalence intervals, for a DDI. The "no effect" boundaries represent the interval within which a change in a systemic exposure measure is considered not clinically meaningful. The selection of the number of subjects for a given DDI study will depend on how small an effect is clinically important to detect or rule out, the inter- and intra-subject variability in pharmacokinetic measurements, and possibly other factors or sources of variability not well recognized.

#### 27.3.3.1 Evaluation of a Drug as an Inhibitor or Inducer for CYPs

In studying an investigational drug as the interacting (inhibiting or inducing) drug, the choice of substrates for initial *in vivo* studies depends on the CYP enzymes affected by the interacting drug. When testing inhibition, the substrate selected should generally be one whose pharmacokinetics are markedly altered by coadministration of known specific inhibitors of the enzyme, i.e. sensitive substrates are preferably used, and if not available, moderately sensitive substrates can be used. Some examples of substrates include (i) midazolam or triazolam for CYP3A inhibition, (ii) buproprion for CYP2B6 (with the evaluation of hydroxybuproprion), (iii) caffeine or tizanidine for CYP1A2 inhibition, (iv) repaglinide for CYP2C8, (v) tolbutamide or warfarin for CYP2C9 (with the evaluation of S-warfarin), (vi) omeprazole or lansoprazole (with the evaluation of *S*-lansoprazole) for CYP2C19, and (vii) desipramine, dextromethorphan, or nebivolol for CYP2D6 inhibition (Table 27.1). Additional examples of substrates, along with inhibitors and inducers of specific CYP enzymes, are provided on the FDA's drug interaction website [73]. If the initial study shows that an investigation drug either inhibits or induces metabolism, further studies using less sensitive substrates, also considering the likelihood of coadministration, may be useful. If the initial study is negative with the most sensitive substrate, it can be presumed that less sensitive substrates will be unaffected.

It should be noted that some of the substrates mentioned above are metabolized by multiple enzymes. In some cases, measuring the concentrations of metabolites besides parent drugs may be needed or valuable. For example, buproprion is metabolized to three major metabolites: hydroxybuproprion, erythrohydrobupropion, and threohydrobupropion [119]. The latter two metabolites are generated by non-CYP enzymes. Only hydroxybupropion is primarily formed by CYP2B6. Thus, when evaluating the effect of a CYP2B6 inhibitor, hydroxybupropion has to be measured and the effect of the inhibitor should be evaluated based on the change of hydroxybupropion exposure, since buproprion exposure is less affected due to compensation of other metabolic pathways, as observed for two CYP2B6 inhibitors – ticlopidine and clopidogrel [120]. Another example is omeprazole that is mainly metabolized by CYP2C19 responsible for the formation of 5-hydroxyomeprazole and to a lesser extent by CYP3A mediating the formation of omeprazole sulfone. In a situation where an inhibitor of CYP2C19 is also a CYP3A inducer, the increase of omeprazole concentration due to CYP2C19 inhibition will be counteracted by the CYP3A induction effect that decreases omeprazole exposure, and thus only measuring omeprazole concentration change may under-estimate the CYP2C19 inhibition effect of the interacting drug [121]. In such case, the metabolic ratio between 5-hydroxyomeprazole and omeprazole would be a better indicator.

#### 27.3.3.2 Evaluation of a Drug as a Substrate for CYPs

In studying an investigational drug as a substrate of a CYP, usually, a strong inhibitor is chosen for the initial *in vivo* DDI study to reflect worst-case scenario. Some examples of inhibitors for major CYPs are listed in Table 27.1 and more examples are available on the DDI website of FDA. Ketoconazole used to be the prototype CYP3A inhibitor selected for *in vivo* studies. However, due to concern with potential liver injury risk, it is recommended to replace ketoconazole with other strong CYP3A inhibitors, among which itraconazole is often used and sometimes clarithromycin is chosen. While clarithromycin is a strong inhibitor of CYP3A (as demonstrated by its effect on midazolam and triazolam exposure) [39], there are PBPK analyses suggesting that clarithromycin may result in less DDI inhibitory effect than itraconazole for some drugs [122, 123]. There are also some considerations when itraconazole is used. The exposure of itraconazole depends on its formulation and food intake status. When a capsule formulation of itraconazole was taken under fasted state, the concentrations of itraconazole were lower than those of capsule administered with food or oral solution formulation regardless of food intake status [124]. Thus, if a capsule formulation of itraconazole is used, it is preferably administered with food. In addition, the oral solution formulation contains hydroxypropyl-β-cyclodextrin which in some cases may form complex with certain substrates and interfere with the effect of CYP3A inhibition. An example is fenebrutinib whose AUC was increased 2.3-fold by itraconazole solution, however, its $C_{max}$ was slightly decreased by 11% [125]. When *in vitro* and/or *in vivo* results indicate that complex formation may occur, further study or PBPK analysis is warranted to delineate the effect of CYP3A inhibition [126]. In addition, there is also suggestion that ritonavir may be an appropriate alternative for ketoconazole [127]. But, a low dose of ritonavir (e.g. 100 mg twice daily) needs to be used, because high doses of ritonavir may induce other CYPs (e.g. CYP1A2, CYP2C9, and CYP2C19) [128]. Thus for investigational drugs that are metabolized by CYP3A and also other CYP enzymes which can be induced by ritonavir, the DDI effect due to CYP3A inhibition may be underestimated when a high dose of ritonavir is used, since induction of other CYP may counteract with the impact of CYP3A inhibition on the pharmacokinetic of an investigational drug.

Besides evaluating the impact of metabolic inhibitors, it is also important to characterize the effect of enzyme inducers which may reduce the efficacy of substrates. Overall, there are less inducers than inhibitors (Table 27.1). Among these, rifampin is most often used. It should be noted that rifampin is also a potent inhibitor of OATP1B and inhibits P-gp. Thus, when an investigational drug is also a substrate of these transporters, simultaneous administration of the drug with rifampin may not accurately capture the effect of enzyme induction. In such cases, staggered administration of the substrate (e.g. 24 hours after rifampin dosing) is recommended [93], and an extra period can be included in the study to evaluate the pharmacokinetic change of the substrate when given along with a single dose of rifampin which will characterize the effect of transporter inhibition.

When studying an investigational drug as a substrate, it is usually administered as a single dose given alone and after multiple doses of perpetrators. However, when the drug is an auto-inhibitor or autoinducer of its own metabolism, its clearance will change after multiple-dose administration compared to single dose. Thus, the DDI magnitude characterized after single-dose administration of an auto-inhibitor or autoinducer with perpetrators does not reflect the DDI effect after chronic treatment of the auto-inhibitor or inducer. For example, while crizotinib is predominantly metabolized by CYP3A, it is also a moderate inhibitor of CYP3A and increased midazolam AUC 3.7-fold after multiple dosing. Ketoconazole increased crizotinib AUC 3.16-fold after a single dose of crizotinib, however, the steady-state AUC of crizotinib was only increased by 57% (i.e. 1.57-fold) by itraconazole [129, 130]. This is probably because CYP3A has been inhibited by crizotinib after multiple dosing. Similar phenomenon was also observed for autoinducers. Efavirenz is a moderate inducer of CYP3A and CYP2B6, and it is metabolized by these two enzymes. While rifampin reduced efavirenz AUC by 51–61% after single-dose administration of efavirenz, the rifampin effect diminished to 2–22% reduction when multiple doses of efavirenz were administered [131–135]. Therefore, it is preferred conducting *in vivo* DDI studies with multiple-dosing of an auto-inhibitor or autoinducer, since this is more clinically relevant for majority of drugs which are used as chronic treatment. Alternatively, PBPK modeling approach may be applied to bridge the gap if the *in vivo* study with a perpetrator is conducted with single doses of an auto-inhibitor or autoinducer.

It should also be noted that an auto-inhibitor or autoinducer may be able to affect perpetrator concentrations. For example, itraconazole, a strong CYP3A inhibitor, is also primarily metabolized by CYP3A, and its concentration was decreased by 65–88% with rifampin and by 93% with phenytoin [136, 137]. Posaconazole, another strong CYP3A inhibitor, is mainly metabolized by UGT1A4, and its AUC was reduced by about 50% with several moderate or strong CYP3A inducers – efavirenz, phenytoin, and rifabutin (UGTs may be co-induced with CYP3A by PXR agonists) [138–140]. Thus, after multiple dose administration, a substrate that is an autoinducer of CYP3A may decrease the concentration of those inhibitors in *in vivo* studies. When PBPK modeling is utilized, this should also be taken into account [141, 142]. When extrapolating the DDI effect of those inhibitors on an autoinducer substrate to other strong CYP3A inhibitors, it needs to be considered whether the pharmacokinetics of other CYP3A inhibitors can also be affected by the autoinducer and the impact on DDI magnitude.

### 27.3.4 Issues to Consider in the Interpretation of Drug Interaction Study Results

A specific objective of DDI studies is to determine whether the interaction is sufficiently large to necessitate a dosage adjustment of the drug itself or other drugs with which it might be used, or whether the interaction would require additional therapeutic monitoring or warrant a contraindication. In some instances, understanding how to adjust dose or dosage regimen in the presence of an interacting drug, or how to avoid interactions, may allow marketing of a drug that would otherwise have been associated with an unacceptable level of toxicity. Sometimes a drug interaction can be used intentionally to increase levels or reduce elimination of another drug (e.g. ritonavir and lopinavir). In a few cases, the degree of interaction caused by a drug, or the degree to which other drugs alter its metabolism, can be such that it cannot be marketed safely (e.g. terfenadine, cisapride, and mibefradil).

The clinical significance of increases or decreases in the systemic exposure of coadministered drugs found in clinical studies conducted in healthy subjects or patient populations should be determined based on the known relationships between dose and/or plasma concentrations and clinical endpoints or biomarkers of efficacy or safety [143]. When evaluating the significance of a pharmacokinetic interaction, it is usually assumed that the pharmacokinetic–pharmacodynamic (PK-PD) relationship of a drug is unchanged in the presence of other drugs. Recommendations based on this relationship can then be made with respect to changing the dose, adjusting the dosing interval, or including suitable language in the labeling, such as "close monitoring," "concomitant use not recommended," or "contraindications." Figure 27.1 illustrates that, via evaluation of the relationship between exposure (dose, AUC, $C_{max}$, etc.) and response (key efficacy and safety measures, either clinical endpoints, surrogate endpoints, or validated biomarkers), a range of systemic exposures corresponding to effective and safe administration can be determined. With this information, doses or dosing regimens not tested clinically may be recommended if these adjusted doses or dosage regimens in patients in certain conditions (e.g. presence of an enzyme inhibitor or hepatic dysfunction) would provide similar systemic exposure as that in patients without these conditions. In some cases, it may not be necessary to adjust the dose or dosing regimen as the changes in exposure would still be within the therapeutic range of exposure defined in other clinical trials. For example, raltegravir, the first-in-class integrase inhibitors for the treatment of HIV-1 infection, is mainly metabolized by UGT1A1 [144]. Raltegravir plasma levels were markedly increased (with coadministration with atazanavir alone and in combination with ritonavir, which is consistent with inhibition of UGT1A1 (Table 27.7). However, concomitant use of raltegravir and atazanavir was well tolerated in the Phase 2 and Phase 3 studies. Based on these data, atazanavir may be coadministered with raltegravir without dose adjustment of raltegravir as suggested in the labeling. The effect of ritonavir (100 mg twice-daily) alone on the pharmacokinetics of raltegravir is not significant [145].

## 27.4 Role of Physiologically Based Pharmacokinetic (PBPK) Modeling in DDI Assessment

PBPK modeling and simulation is a tool that can predict the pharmacokinetics of a drug in humans. It consists of multiple mathematical equations that mechanistically describe the ADME of a drug product. The PBPK modeling and simulation tool can be used to evaluate the effect of intrinsic factors (such as age, race, organ dysfunction, diseases, pregnancy/lactation, gender, generics, and others), and extrinsic factors (such as DDIs, food effect, alcohol use, smoking, environment,

**Table 27.7** Effect of atazanavir on raltegravir [144].

| Coadministered drug and dose | *N* | Reltegravie dose/schedule | Ratio (90% CI) of raltegravir pharmacokinetic parameters with/without coadministered drug (no effect = 1.00) | | |
|---|---|---|---|---|---|
| | | | $C_{min}$ | $AUC_{tau}$ | $C_{max}$ |
| Atazanavir 400 mg QD | 10 | 100 mg SD | 1.95 (1.30, 2.92) | 1.72 (1.47, 2.02) | 1.53 (1.11, 2.12) |
| Atazanavir 400 mg QD | 14 | 1200 mg SD | 1.26 (1.08, 1.46) | 1.67 (1.34, 2.10) | 1.16 (1.01, 1.33) |
| Atazanavir/ritonavir 300/100 mg QD | 10 | 400 mg BID | 1.77 (1.39, 2.25) | 1.41 (1.12, 1.78) | 1.24 (0.87, 1.77) |

SD = Single dose administration of raltegravir
BID: twice daily
QD: once daily
Source: Adapted from ISENTRESS (Raltegravir) Labeling [144].

medical practices, and others) on the pharmacokinetic of a drug [146].

Over the past decade, the application of PBPK modeling and simulation in drug development and regulatory submissions had made a significant progress. Between June 2008 and July 2010, the Office of Clinical Pharmacology (OCP) at the FDA received 13 submissions that contained PBPK analyses including 7 for investigational new drugs (INDs) and 6 for new drug applications (NDAs) [146]. Between July 2010 and December 2011, OCP received additional 12 submissions containing PBPK analyses [147]. In 2013, approximately 50 submissions either requested scientific advice on using the PBPK approach or included such an approach in regulatory filings [148]. Between 2008 and 2017, the OCP received 130 INDs and 94 NDAs that contained PBPK analyses [149]. The number continues to grow. In 2018 and 2019 only, OCP received 56 INDs, 57 NDAs, and 3 biologics license applications (BLAs) that contained PBPK analyses [150]. A variety of PBPK applications had been submitted to the FDA including predicting absorption and the effects of drug interactions, food intake, organ impairment, diseases, pharmacogenetic polymorphisms, and age. Among all the areas of applications, the PBPK analyses for DDI evaluations accounted for approximately 60% of the total number of applications [149, 150].

To date, two dedicated PBPK guidances were issued by the US FDA: (i) the PBPK format and content guidance entitled "Physiologically Based Pharmacokinetic Analyses – Format and Content" [151], and (ii) the PBPK analyses for biopharmaceutics and chemistry manufacturing and controls (CMC) applications for oral products guidance entitled "The Use of Physiologically Based Pharmacokinetic Analyses – Biopharmaceutics Applications for Oral Drug Product Development, Manufacturing Changes, and Controls" [152].

General considerations of using PBPK modeling to predict enzyme- or transporter- mediated DDIs are also laid out in the FDA *in vitro* DDI guidance [34]. When using PBPK modeling to evaluate the effect of other drugs on an investigational drug as an enzyme substrate, the applicant or sponsor should minimally address the following questions:

- whether the PBPK model can describe all the available clinical pharmacokinetic data of the investigational drug using different dosing regimens (e.g. single-dose, and/or multiple-dose administration), and dosing routes (e.g. intravenous or oral administration);
- whether the elimination pathways are quantitatively assigned in the substrate's model based on available *in vitro* and *in vivo* data;
- whether the index perpetrator models are validated regarding their modulating effect on enzyme activity in humans;
- whether there are sensitivity analyses being performed for parameters having a high level of uncertainty; and
- whether the substrate and modulator models include the major disposition and interaction mechanisms and are validated if complex metabolic and transporter mechanisms are expected.

When using PBPK modeling to evaluate the effect of an investigational drug as an enzyme perpetrator, the applicant or sponsor should minimally address the following questions:

- whether the PBPK model can describe all the available clinical pharmacokinetic data of the investigational drug using different dosing regimens (e.g. single-dose, and/or multiple-dose administration), and dosing routes (e.g. intravenous or oral administration);

- whether the index substrate models are validated regarding the effect of altered enzyme activity on their pharmacokinetics in humans;
- whether the inhibition and induction mechanisms are separately considered in addition to the combination of inhibition and induction to ensure a conservative prediction of *in vivo* enzyme inhibition or induction;
- whether the simulation uses the highest clinical dose of the investigational drug to ensure the highest modulational potential; and
- whether there are sensitivity analyses being performed for parameters exhibiting high levels of uncertainty.

In principle, the general considerations of using PBPK modeling to predict transporter- mediated DDIs are similar to enzyme-mediated DDIs [153, 154]. Compared to the enzyme-mediated DDI perdition, large knowledge gaps exist in transporter biology, limited experience in determining and modeling the kinetics of transporters, and challenges in validating the local effects of transporters. To evaluate the interplay between transporter and enzymes for a substrate of transporters and enzymes, the PBPK model needs to be validated with clinical data from a wide range of relevant DDI and/or pharmacogenetic studies. For drugs that are potential transporter inhibitors, the transporter substrate models should be established and validated with proper human studies.

PBPK models can be used in lieu of some *in vivo* DDI studies as indicated in the Clinical DDI guidance [33], such as to evaluate the effects of moderate inhibitors or inducers on the pharmacokinetics of an investigational drug. The guidance also indicates that new uses of PBPK modeling to replace *in vivo* studies are continuously being considered by the FDA because of evolving science.

Majority of the enzyme-mediated DDI evaluations involves CYP enzymes with limited number of PBPK analyses evaluating UGT-mediated DDIs [149]. A survey conducted by FDA scientists for the PBPK submissions between 2009 and 2013 suggested that PBPK models for CYP3A-mediated inhibition predicted the observed clinical DDIs (15 substrates and 26 DDI cases) within a 2-fold threshold, and approximately 80% predicted the observed DDI magnitude within a 1.25-fold threshold [155]. A similar survey of the PBPK submissions between 2009 and 2014 suggested that PBPK models for CYP3A-mediated induction predictions were within the 1.25-fold of the observed data in 77% (10/13) and 73% (10/12) cases for AUC ratio and $C_{max}$ ratio, respectively [156]. In recent submissions, PBPK models have been used to evaluate complex CYP3A-mediated DDI evaluations [157]. For example, PBPK has been used to evaluate the DDI potentials for investigational drugs that are substrates and TDIs or inducers of CYP3A (e.g. duvelisib, fedratinib, and lorlatinib) [158–160]. In some cases of complex DDIs, there are mutual interactions between substrates and modulators, e.g. some CYP3A modulators (such as itraconazole) are also CYP3A substrates whose exposure may potentially be impacted by substrates that can inhibit or induce CYP3A [142]. These interactions may not be observed in a dedicated DDI study where the investigational drug (the substrate) is generally given as a single dose administration. Simulations may help estimate the interactions at steady state.

PBPK analyses to evaluate transporter-mediated DDI can be scientifically challenging yet still play a role in regulatory submission. For an investigational drug as a transporter substrate, we have seen the PBPK approach being used to estimate the contribution from P-gp interaction to the observed DDI for dual substrates of P-gp and CYP3A including rivaroxaban, prucalopride, upadacitinib, and naloxegol [161–164]. PBPK analyses were also conducted for simeprevir and letermovir to evaluate the roles of OATP1B1/3 in their disposition [165, 166]. It is often challenging to quantitatively predict a transporter-mediated DDI using a PBPK approach where extensive *in vivo* studies are usually needed for model development and validation. For example, PBPK was used to evaluate the effect of OAT3 inhibitors on the pharmacokinetics of baricitinib which is an OAT3 substrate [167]. The PBPK model was validated with its ability to reproduce the *in vivo* DDI effect of probenecid (an OAT3 inhibitor) on baricitinib exposure and then was used to predict the effect of ibuprofen, another OAT3 inhibitor, whose PBPK model was previously verified with interaction of ibuprofen and an OAT3 substrate, pemetrexed [168]. Finally, PBPK was used to predict the effect of diclofenac on baricitinib.

When using PBPK approach to evaluate the effect of an investigation drug on transporters in the absence of an *in vivo* DDI study or endogenous biomarker data, the model is considered as a completely bottom-up approach, and therefore, *in vitro* studies become very critical. To evaluate a drug as a transporter inhibitor, the $IC_{50}$ or Ki value is one of the important parameters in predicting its effect *in vivo*. Literature reports indicated that, in general, *in vitro* Ki values tend to underpredict the transporter-mediated DDI magnitude [154]. Additional research is warranted to close the gap of *in vitro* to *in vivo* extrapolation for transporter modulators. Sensitivity analysis has been used to explore the range of DDI when there is uncertainty with some parameters in the model including inhibition potency parameter. This is also used to identify the major transporter contributing to observed *in vivo* DDIs for drugs inhibiting multiple transporters *in vitro*.

Overall, in the past decade, PBPK modeling and simulation for DDI evaluation have made tremendous progress impacting drug development and regulatory decision-making. PBPK is still evolving in addressing other clinical pharmacology and biopharmaceutics questions.

## 27.5 A Labeling Example to Illustrate the Translation of Complicated Drug Interaction Results to Labeling: Tipranavir

Tipranavir is a non-peptidic HIV protease inhibitor and the approved dosage regimen is tipranavir 500 mg capsules coadministered with 200 mg of ritonavir twice daily (bid).

### 27.5.1 Drug Interaction Potential Assessment during NDA Review

*In vitro* and *in vivo* data suggest that tipranavir is a substrate and inducer of CYP3A and P-gp [169, 170]. Ritonavir, in this combination, acts as a pharmacokinetic booster to increase the bioavailability of tipranavir by inhibiting CYP3A and/or P-gp, which results in increased tipranavir concentrations. The *in vivo* effect of tipranavir/ritonavir (500/200 mg bid) at steady-state on CYP3A is confirmed to be inhibition. Review of several *in vivo* studies includes mass balance study, studies with loperamide, clarithromycin, and protease inhibitors (amprenavir, lopinavir, and saquinavir) suggest that tipranavir-/ritonavir-induced P-gp at steady-state [170, 171].

*In vitro* drug metabolism data also suggest that tipranavir is a potential inhibitor of CYP1A2, CYP2C9, CYP2C19, and CYP2D6 with a respective $R_1$ value ($C_{max,u}/K_i$) greater than 1.02 ($f_{u,p}$ was capped at 0.01). The potential *in vivo* net effect of tipranavir/ritonavir on CYP2D6 is inhibition because ritonavir also inhibits CYP2D6. However, the net effects of tipranavir/ritonavir on CYP1A2, CYP2C9, and CYP2C19 are unknown because ritonavir may induce these enzymes, and whether tipranavir is also an inducer to CYP1A2, 2C9, and 2C19 is not known.

Based on these findings, the sponsor conducted a cocktail drug interaction study [171] to evaluate tipranavir's inhibition or induction potential on P-gp and CYPs using digoxin (as a P-gp substrate) and various probe substrates for CYPs, respectively, as a post-marketing commitment. A single oral dose of digoxin was given on Days 1, 8, and 18. Tipranavir/ritonavir 500/200 mg was given twice daily from Day 7 through Day 22. Thus, effects of both short (two doses) and long treatments of tipranavir/ritonavir on digoxin were evaluated. The results confirmed that tipranavir/ritonavir induced P-gp after multiple doses. Namely, P-gp activity, as measured by orally administered digoxin, was inhibited after the initial doses of tipranavir/ritonavir (75% increase in digoxin AUC) but induction of P-gp occurred over time [172]. At steady state of tipranavir/ritonavir, digoxin exposure was similar to baseline conditions (digoxin alone) due to the counteracting concurrent induction and inhibition effects of tipranavir/ritonavir on P-gp. The P-gp inhibitory effect could be mainly due to ritonavir because tipranavir is a weak P-gp inhibitor as shown in Caco-2 cells using digoxin as a substrate. The reason for different effects of tipranavir/ritonavir on digoxin versus loperamide remains unclear, though P-gp induction occurred in both cases. It is noted that digoxin is minimally metabolized, while loperamide undergoes metabolism mediated by CYP3A and CYP2C8 [173].

A single oral dose of caffeine (CYP1A2), dextromethorphan (CYP2D6), warfarin (CYP2C9), omeprazole (CYP2C19), and midazolam (CYP3A) as a cocktail was given on Days 1, 7, and 17. Tipranavir/ritonavir 500/200 mg was given twice daily from Day 7 through Day 22. Thus, effects of both single dose and multiple doses of tipranavir/ritonavir on respective probe substrates of CYP1A2, CYP2D6, CYP2C9, and CYP2C19 were evaluated. The results demonstrated that: there was no effect on CYP1A2 after the first dose of tipranavir/ritonavir, but there was induction at steady state (caffeine AUC decreased by 34%), which was probably due to ritonavir as another study conducted with ritonavir alone (400 mg bid) reduced caffeine AUC by 65% [174]; there was potent inhibition of CYP2D6 after the first dose and at steady state (dextromethorphan AUC was increased 3.95-fold and 5.46-fold, respectively); there was slight inhibitory effect on CYP2C9 after the first dose indicated by 22% increase in S-warfarin AUC but modest induction effect at steady state of tipranavir/ritonavir counteracting the inhibitory effect (S-warfarin AUC was 89% of that at baseline), which is consistent with weak induction effect of ritonavir as observed in another study of ritonavir alone (100 mg once daily or twice daily) [175]; and there was modest inhibition of CYP2C19 after the first dose of tipranavir/ritonavir that increased omeprazole AUC by 76%, but there was marked induction at steady state where omeprazole AUC was reduced by 70%. Lastly, tipranavir/ritonavir after single dose dramatically increased midazolam (orally administered) AUC 27-fold, in line with strong CYP3A inhibition effect by ritonavir. Interestingly, the effect became smaller at steady state of tipranavir/ritonavir during which midazolam AUC was increased 10-fold of that when midazolam was given alone. Similar trend was also observed with intravenously administered midazolam (AUC was increased 5-fold and 2.9-fold, respectively, after single dose and multiple doses of tipranavir/ritonavir). These results indicated that there was induction of CYP3A after continuous dosing of tipranavir/ritonavir, though the effect was overridden by the inhibitory effects. The occurrence of CYP3A was also consistent with the expectation that CYP2C9 and CYP2C19 are often co-induced with CYP3A since these enzymes are all regulated by PXR nuclear receptor.

### 27.5.2 Labeling Impact

The data obtained concerning potential tipranavir/ritonavir drug interactions provided information for better prediction and management of known and potential drug interactions of this drug. Based on the facts that the net *in vivo* effect of tipranavir/ritonavir (500/200 bid) at steady-state on CYP3A is inhibition but the net effect on P-gp may be induction or mixed, it was concluded that it is difficult to predict the net effect of tipranavir/ritonavir on oral bioavailability and plasma concentrations of drugs that are dual substrates of CYP3A and P-gp. The net effect will vary depending on the relative affinity of the coadministered drugs for CYP3A and P-gp, and the extent of intestinal first-pass metabolism/efflux. Tipranavir labeling has extensive drug interaction information including P-gp. For example, precautionary wordings regarding the complexity of CYP3A and P-gp effects of tipranavir/ritonavir are provided in "Drug Interactions" in Section 7 of the labeling (Table 27.8 as extracted from the labeling) [171]. The phenotypic cocktail study results with regard to CYP1A2, CYP2D6, CYP2C9, and CYP2C19 drug

**Table 27.8** Drug interaction information in the APTIVUS labeling.

| Concomitant drug class: drug name | Effect on concentration of tipranavir or concomitant drug | Clinical comment |
|---|---|---|
| **Calcium Channel Blockers:**<br>Diltiazem<br>Felodipine<br>Nicardipine<br>Nisoldipine<br>Verapamil | Combination with APTIVUS/ritonavir not studied. Cannot predict effect of TPV/ritonavir on calcium channel blockers that are dual substrates of CYP 3A and P-gp due to conflicting effect of TPV/ritonavir on CYP 3A and P-gp.<br>↕ Diltiazem<br>↑ Felodipine (CYP3A substrate but not Pgp substrate)<br>↕ Nicardipine<br>↕ Nisoldipine (CYP3A substrate but not clear whether it is a P-gp substrate)<br>↕ Verapamil | Caution is warranted and clinical monitoring of patients is recommended |
| **Immunosuppressants:**<br>Cyclosporine<br>Sirolimus<br>Tacrolimus | Combination with APTIVUS/ritonavir not studied. Cannot predict effect of TPV/ritonavir on immunosuppressants due to conflicting effect of TPV/ritonavir on CYP 3A and P-gp.<br>↕ Cyclosporine<br>↕ Sirolimus<br>↕ Tacrolimus | Increased frequency of monitoring of plasma levels of immunosuppressant drugs is recommended |

↑:increase; ↕: unable to predict.
Extracted from Table 27.4 in the labeling: Established and Other Potentially Significant Drug Interactions: Alterations in Dose or Regimen May be Recommended Based on Drug Interaction Studies or Predicted Interaction [171].
Source: Adapted from APTIVUS (Tipranavir) Labeling [171].

**Table 27.9** Example of PBPK application in DDI assessment for CYP3A/P-gp dual substrates.

| Substrate | Perpetrator | AUCR | $C_{max}R$ | DDI mechanisms (inhibition) | US Prescribing Information | References |
|---|---|---|---|---|---|---|
| Prucalopride | Ketoconazole 200 mg | 1.4 | 1.4 | CYP3A (strong inhibition), P-gp, BCRP | No dose adjustment | [161] |
| Upadacitinib | Ketoconazole 400 mg | 1.75 | 1.7 | CYP3A (strong inhibition), P-gp | Use with caution | [162, 176] |
| Naloxegol | Ketoconazole 400 mg | 12.85 | 9.58 | CYP3A (strong inhibition), P-gp | Contraindicate | [164, 177, 178] |
| | Diltiazem | 3.41 | 2.86 | CYP3A (moderate inhibition), P-gp (no or weak inhibition) | Avoid use | |
| | Quinidine | 1.39 | 2.47 | P-gp | No dose adjustment | |
| Neratinib | Ketoconazole | 4.8 | 3.2 | CYP3A (strong inhibition), P-gp | Avoid use | [179] |
| | Verapamil | 4.0 (p) | 3.9 (p) | CYP3A (moderate inhibition), P-gp | Avoid use | |
| | Fluconazole | 1.7 (p) | 1.3 (p) | CYP3A (moderate inhibition) | No dose adjustment | |

AUCR: the ratio of a substrate's AUC in the presence and absence of a perpetrator; $C_{max\,R}$: the ratios of a substrate's $C_{max}$ in the presence and absence of a perpetrator; (p) not studies in *in vivo* study but predicted by PBPK.

interaction potential are also communicated in Clinical Pharmacology – Drug Interactions section of the labeling of APTIVUS.

## 27.6 Examples to Illustrate the Use of PBPK in Supporting Labeling for Drugs that are Dual CYP3A/P-GP Substrates

Both CYP3A and P-gp exist in the gastrointestinal tract and can potentially affect a drug's absorption and bioavailability. CYP3A and P-gp pathways sometimes are coupled. In this section, a few examples where the investigational drugs are dual CYP3A/P-gp substrates are provided to illustrate how the PBPK modeling approach was explored trying to separate the effect from each pathway (Table 27.9).

Prucalopride and upadacitinib are CYP3A and P-gp substrates based on *in vitro* studies. DDI studies with ketoconazole, a P-gp, and strong CYP3A inhibitor, showed that the AUC of these substrates was increased less than 2-fold. PBPK analyses were conducted to explore the contribution from P-gp interaction to the overall observed interaction (Table 27.9). Naloxegol appears to be a sensitive CYP3A substrate based on the DDI study with ketoconazole which increased AUC of naloxegol more than 10-fold. It is also a P-gp substrate and quinidine, a P-gp inhibitor with no significant CYP3A inhibition effect, increased naloxegol AUC by about 40% and $C_{max}$ by 150%. PBPK analyses were conducted to evaluate the potential effect of P-gp on the disposition of naloxegol and DDIs related to P-gp (Table 27.9) [164]. Neratinib is also a P-gp and CYP3A dual substrate. Ketoconazole increased its AUC by almost fivefold, and $C_{max}$ by about threefold. PBPK analyses were conducted and used to predict the effect of moderate CYP3A inhibitors with or without P-gp inhibition potential. The model predicted that verapamil, a P-gp and moderate CYP3A inhibitor, increases neratinib AUC and $C_{max}$ about fourfold, while fluconazole, a moderate CYP3A inhibitor increases AUC by 70% and $C_{max}$ by 30%. The results are reflected in the recent US prescribing information (Table 27.9).

As incorporating transporters in a PBPK model is not a trivial task when conducting PBPK analyses, additional questions should be asked, such as whether the interaction with CYP3A/P-gp dual inhibitors can be clinically significant based on the known information; whether P-gp contributes significantly to ADME of the drug based on results of *in vitro*, mass balance, and *in vivo* pharmacokinetic and DDI studies; whether there may be an identifiability issue if a middle-out model is developed based on the available studies; and whether the potential DDI risks can be addressed by appropriate labeling language.

## 27.7 Summary

The understanding of drug interactions based on alterations in drug metabolism and transport has matured to the point that the FDA includes considerable detail on the evaluation of metabolic enzymes and transporters-mediated DDIs *in vitro*, *in silico*, and *in vivo* in the recently published guidances [34, 35]. In addition to the new guidances, the FDA maintained a Drug Development and Drug Interactions Website as a venue to provide additional information supplemental to the guidances for evaluation of drug interactions [73]. The Agency's current view in the evaluation of DDIs during drug development includes the following principles:

- Understanding of metabolism, transport, and DDI liability is important to the benefit/risk assessment of a drug.
- Using an integrated approach incorporating *in vitro, in silico,* and *in vivo* metabolism and transport studies to elucidate the underlying mechanisms and to evaluate the potential for DDIs can reduce the number of *in vivo* studies needed and optimize our knowledge.
- Classification of CYP inhibitors, inducers, and substrates can aid in the study design and cross-labeling of drugs. FDA periodically updated the Drug Interactions and Drug Development website to recommend index substrates, inhibitors, or inducers that may be used to study enzyme-based DDIs or clinical substrate, inhibitors, or inducers that may be used to inform labeling for both enzyme- and transporter-based DDIs (note: those lists are not comprehensive).
- Appropriately designed studies are critical to providing important information for proper labeling.
- The clinical significance of DDIs based on clinical studies or *in silico* prediction should be interpreted in the context of well-defined exposure–response data and analyses.
- Drug interaction information should be appropriately placed in the labeling. With improved understanding of the molecular bases of DDIs and the interplay of various intrinsic and extrinsic factors affecting these interactions, risks associated with DDIs can be assessed and managed to minimize untoward effects.

DDI guidance documents published by the FDA over the past two decades have enabled proper drug interaction evaluation, mitigation, and management during drug development and regulatory assessment for personalized dosing in the era of polypharmacy and have ensured safe and effective use of medicines.

## Acknowledgement

The authors thank Dr. E. Dennis Bashaw from FDA for his valuable comments and are grateful to Dr. Donna Volpe from FDA for her help with creating Figure 27.2.

## References

1 Giacomini, K.M., Krauss, R.M., Roden, D.M. et al. (2007). When good drugs go bad. *Nature* 446: 975–977.
2 Saravanakumar, A., Sadighi, A., Ryu, R., and Akhlaghi, F. (2019). Physicochemical properties, biotransformation, and transport pathways of established and newly approved medications: a systematic review of the top 200 most prescribed drugs vs. the FDA-approved drugs between 2005 and 2016. *Clin. Pharmacokinet.* 58: 1281–1294.
3 Hein, D.W., Doll, M.A., Fretland, A.J. et al. (2000). Molecular genetics and epidemiology of the NAT1 and NAT2 acetylation polymorphisms. *Cancer Epidemiol. Biomarkers Prev.* 9: 29–42.
4 Aithal, G.P. and Day, C.P. (2007). Nonsteroidal anti-inflammatory drug-induced hepatotoxicity. *Clin. Liver Dis.* 11: 563–vii.
5 Spielberg, S.P. (1996). N-acetyltransferases: pharmacogenetics and clinical consequences of polymorphic drug metabolism. *J. Pharmacokinet. Biopharm.* 24: 509–519.
6 CAMPTOSAR (Irinotecan) Injection Labeling (2022). http://www.fda.gov/cder/foi/label/2006/020571s030lbl.pdf; https://www.accessdata.fda.gov/drugsatfda_docs/label/2020/020571s051lbl.pdf.
7 Hahn, K.K., Wolff, J.J., and Kolesar, J.M. (2006). Pharmacogenetics and irinotecan therapy. *Am. J. Health Syst. Pharm.* 63: 2211–2217.
8 Corona, G., Vaccher, E., Sandron, S. et al. (2008). Lopinavir-ritonavir dramatically affects the pharmacokinetics of irinotecan in HIV patients with Kaposi's sarcoma. *Clin. Pharmacol. Ther.* 83: 601–606.
9 Ogilvie, B.W., Zhang, D., Li, W. et al. (2006). Glucuronidation converts gemfibrozil to a potent, metabolism-dependent inhibitor of CYP2C8: implications for drug-drug interactions. *Drug Metab. Dispos.* 34: 191–197.
10 Tornio, A., Neuvonen, P.J., Niemi, M., and Backman, J.T. (2017). Role of gemfibrozil as an inhibitor of CYP2C8 and membrane transporters. *Expert Opin. Drug Metab. Toxicol.* 13: 83–95.

11 Miners, J.O., Rowland, A., Novak, J.J. et al. (2020). Evidence-based strategies for the characterisation of human drug and chemical glucuronidation in vitro and UDP-glucuronosyltransferase reaction phenotyping. *Pharmacol. Ther.* 107689.

12 Docci, L., Klammers, F., Ekiciler, A. et al. (2020). In vitro to in vivo extrapolation of metabolic clearance for UGT substrates using short-term suspension and long-term co-cultured human hepatocytes. *AAPS J.* 22: 1–12.

13 Miners, J.O., Knights, K.M., Houston, J.B., and Mackenzie, P.I. (2006). In vitro-in vivo correlation for drugs and other compounds eliminated by glucuronidation in humans: pitfalls and promises. *Biochem. Pharmacol.* 71: 1531–1539.

14 Kiang, T.K., Ensom, M.H., and Chang, T.K. (2005). UDP-glucuronosyltransferases and clinical drug-drug interactions. *Pharmacol. Ther.* 106: 97–132.

15 FDA Guidance for Industry (Draft): Drug Interaction Studies — Study Design, Data Analysis, Implications for Dosing, and Labeling Recommendations (2012). https://downloads.regulations.gov/FDA-2006-D-0036-0032/content.pdf.

16 Fromm, M.F., Kim, R.B., Stein, C.M. et al. (1999). Inhibition of P-glycoprotein-mediated drug transport: a unifying mechanism to explain the interaction between digoxin and quinidine. *Circulation* 99: 552–557.

17 Hager, W.D., Fenster, P., Mayersohn, M. et al. (1979). Digoxin-quinidine interaction pharmacokinetic evaluation. *N. Engl. J. Med.* 300: 1238–1241.

18 Billington, S., Shoner, S., Lee, S. et al. (2019). Positron emission tomography imaging of [(11)C]rosuvastatin hepatic concentrations and hepatobiliary transport in humans in the absence and presence of cyclosporin A. *Clin. Pharmacol. Ther.* 106: 1056–1066.

19 Simonson, S.G., Raza, A., Martin, P.D. et al. (2004). Rosuvastatin pharmacokinetics in heart transplant recipients administered an antirejection regimen including cyclosporine. *Clin. Pharmacol. Ther.* 76: 167–177.

20 Mochizuki, T., Zamek-Gliszczynski, M.J., Yoshida, K. et al. (2022). Effect of cyclosporin A and impact of dose staggering on OATP1B1/1B3 endogenous substrates and drug probes for assessing clinical drug interactions. *Clin. Pharmacol. Ther.* 111 (6): 1315–1323.

21 Brown, G.R. (1993). Cephalosporin-probenecid drug interactions. *Clin. Pharmacokinet.* 24: 289–300.

22 Elsby, R., Chidlaw, S., Outteridge, S. et al. (2017). Mechanistic in vitro studies confirm that inhibition of the renal apical efflux transporter multidrug and toxin extrusion (MATE) 1, and not altered absorption, underlies the increased metformin exposure observed in clinical interactions with cimetidine, trimethoprim or pyrimethamine. *Pharmacol. Res. Perspect.* 5.

23 Muller, F., Weitz, D., Mertsch, K. et al. (2018). Importance of OCT2 and MATE1 for the cimetidine-metformin interaction: insights from investigations of polarized transport in single- and double-transfected MDCK cells with a focus on perpetrator disposition. *Mol. Pharmaceutics* 15: 3425–3433.

24 Hillgren, K.M., Keppler, D., Zur, A.A. et al. (2013). Emerging transporters of clinical importance: an update from the international transporter consortium. *Clin. Pharmacol. Ther.* 94: 52–63.

25 Bertucci, C. and Domenici, E. (2002). Reversible and covalent binding of drugs to human serum albumin: methodological approaches and physiological relevance. *Curr. Med. Chem.* 9: 1463–1481.

26 Abdel-Razzak, Z., Loyer, P., Fautrel, A. et al. (1993). Cytokines down-regulate expression of major cytochrome P-450 enzymes in adult human hepatocytes in primary culture. *Mol. Pharmacol.* 44: 707–715.

27 Aitken, A.E., Richardson, T.A., and Morgan, E.T. (2006). Regulation of drug-metabolizing enzymes and transporters in inflammation. *Annu. Rev. Pharmacol. Toxicol.* 46: 123–149.

28 Elkahwaji, J., Robin, M.A., Berson, A. et al. (1999). Decrease in hepatic cytochrome P450 after interleukin-2 immunotherapy. *Biochem. Pharmacol.* 57: 951–954.

29 Schmitt, C., Kuhn, B., Zhang, X. et al. (2011). Disease-drug-drug interaction involving tocilizumab and simvastatin in patients with rheumatoid arthritis. *Clin. Pharmacol. Ther.* 89: 735–740.

30 ACTEMRA (Tocilizumab) Injection Labeling (2022). https://www.accessdata.fda.gov/drugsatfda_docs/label/2020/125276s129,125472s042lbl.pdf.

31 Evers, R., Dallas, S., Dickmann, L.J. et al. (2013). Critical review of preclinical approaches to investigate cytochrome p450-mediated therapeutic protein drug-drug interactions and recommendations for best practices: a white paper. *Drug Metab. Dispos.* 41: 1598–1609.

32 Kenny, J.R., Liu, M.M., Chow, A.T. et al. (2013). Therapeutic protein drug-drug interactions: navigating the knowledge gaps-highlights from the 2012

AAPS NBC roundtable and IQ consortium/FDA workshop. *AAPS J.* 15: 933–940.

33 FDA Guidance for Industry: Clinical Drug Interaction Studies — Cytochrome P450 Enzyme- and Transporter-Mediated Drug Interactions (2020). https://www.fda.gov/media/134581/download.

34 FDA Guidance for Industry: In Vitro Drug Interaction Studies — Cytochrome P450 Enzyme- and Transporter-Mediated Drug Interactions (2020). https://www.fda.gov/media/134582/download.

35 FDA Guidance for Industry (Final): Drug-Drug Interaction Assessment for Therapeutic Proteins (2023). https://www.fda.gov/media/140909/download.

36 Jing, X., Ji, P., Schrieber, S.J. et al. (2020). Update on therapeutic protein-drug interaction: information in labeling. *Clin. Pharmacokinet.* 59: 25–36.

37 Huang, S.M., Temple, R., Throckmorton, D.C., and Lesko, L.J. (2007). Drug interaction studies: study design, data analysis, and implications for dosing and labeling. *Clin. Pharmacol. Ther.* 81: 298–304.

38 Huang, S.M., Strong, J.M., Zhang, L. et al. (2008). New era in drug interaction evaluation: US Food and Drug Administration update on CYP enzymes, transporters, and the guidance process. *J. Clin. Pharmacol.* 48: 662–670.

39 Rekic, D., Reynolds, K.S., Zhao, P. et al. (2017). Clinical drug-drug interaction evaluations to inform drug use and enable drug access. *J. Pharm. Sci.* 106: 2214–2218.

40 Yoshida, K., Zhao, P., Zhang, L. et al. (2017). In vitro-in vivo extrapolation of metabolism- and transporter-mediated drug-drug interactions-overview of basic prediction methods. *J. Pharm. Sci.* 106: 2209–2213.

41 Zhang, L., Zhang, Y.D., and Huang, S.M. (2009). Scientific and regulatory perspectives on metabolizing enzyme-transporter interplay and its role in drug interactions - challenges in predicting drug interactions. *Mol. Pharm.* 6: 1766–1774.

42 Zhang, L., Zhang, Y.D., Zhao, P., and Huang, S.M. (2009). Predicting drug-drug interactions: an FDA perspective. *AAPS J.* 11: 300–306.

43 Vieira, M.L., Kirby, B., Ragueneau-Majlessi, I. et al. (2014). Evaluation of various static in vitro-in vivo extrapolation models for risk assessment of the CYP3A inhibition potential of an investigational drug. *Clin. Pharmacol. Ther.* 95: 189–198.

44 Faber, M.S. and Fuhr, U. (2004). Time response of cytochrome P450 1A2 activity on cessation of heavy smoking. *Clin. Pharmacol. Ther.* 76: 178–184.

45 Renwick, A.B., Watts, P.S., Edwards, R.J. et al. (2000). Differential maintenance of cytochrome P450 enzymes in cultured precision-cut human liver slices. *Drug Metab. Dispos.* 28: 1202–1209.

46 Backman, J.T., Honkalammi, J., Neuvonen, M. et al. (2009). CYP2C8 activity recovers within 96 hours after gemfibrozil dosing: estimation of CYP2C8 half-life using repaglinide as an in vivo probe. *Drug Metab. Dispos.* 37: 2359–2366.

47 Liston, H.L., DeVane, C.L., Boulton, D.W. et al. (2002). Differential time course of cytochrome P450 2D6 enzyme inhibition by fluoxetine, sertraline, and paroxetine in healthy volunteers. *J. Clin. Psychopharmacol.* 22: 169–173.

48 Venkatakrishnan, K. and Obach, R.S. (2005). In vitro-in vivo extrapolation of CYP2D6 inactivation by paroxetine: prediction of nonstationary pharmacokinetics and drug interaction magnitude. *Drug Metab. Dispos.* 33: 845–852.

49 Yang, J., Liao, M., Shou, M. et al. (2008). Cytochrome p450 turnover: regulation of synthesis and degradation, methods for determining rates, and implications for the prediction of drug interactions. *Curr. Drug Metab.* 9: 384–394.

50 Wang, Y.H. (2010). Confidence assessment of the Simcyp time-based approach and a static mathematical model in predicting clinical drug-drug interactions for mechanism-based CYP3A inhibitors. *Drug Metab. Dispos.* 38: 1094–1104.

51 Rowland Yeo, K., Walsky, R.L., Jamei, M. et al. (2011). Prediction of time-dependent CYP3A4 drug-drug interactions by physiologically based pharmacokinetic modelling: impact of inactivation parameters and enzyme turnover. *Eur. J. Pharm. Sci.* 43: 160–173.

52 Darwich, A.S., Aslam, U., Ashcroft, D.M., and Rostami-Hodjegan, A. (2014). Meta-analysis of the turnover of intestinal epithelia in preclinical animal species and humans. *Drug Metab. Dispos.* 42: 2016–2022.

53 Greenblatt, D.J., von Moltke, L.L., Harmatz, J.S. et al. (2003). Time course of recovery of cytochrome p450 3A function after single doses of grapefruit juice. *Clin. Pharmacol. Ther.* 74: 121–129.

54 Sudsakorn, S., Bahadduri, P., Fretland, J., and Lu, C. (2020). 2020 FDA drug-drug interaction guidance: a comparison analysis and action plan by pharmaceutical industrial scientists. *Curr. Drug Metab.* 21: 403–426.

55 EMA Guideline on the Investigation of Drug Interactions (2013). https://www.ema.europa

.eu/en/documents/scientific-guideline/guideline-investigation-drug-interactions-revision-1_en.pdf.
56 Japanese MHLW (PMDA) Guidance on Drug Interaction for Drug Development and Appropriate Provision of Information. (2019). https://www.pmda.go.jp/files/000228122.pdf.
57 Tseng, E., Eng, H., Lin, J. et al. (2021). Static and dynamic projections of drug-drug interactions caused by cytochrome P450 3A time-dependent inhibitors measured in human liver microsomes and hepatocytes. *Drug Metab. Dispos.* 49 (10): 947–960.
58 Ramsden, D., Perloff, E.S., Whitcher-Johnstone, A. et al. (2022). Predictive in vitro-in vivo extrapolation for time dependent inhibition of CYP1A2, CYP2C8, CYP2C9, CYP2C19, and CYP2D6 using pooled human hepatocytes, human liver microsomes, and a simple mechanistic static model. *Drug Metab. Dispos.* 50 (2): 114–127.
59 Hariparsad, N., Ramsden, D., Palamanda, J. et al. (2017). Considerations from the IQ induction working group in response to drug-drug interaction guidance from regulatory agencies: focus on down-regulation, CYP2C induction, and CYP2B6 positive control. *Drug Metab. Dispos.* 45: 1049–1059.
60 Kenny, J.R., Ramsden, D., Buckley, D.B. et al. (2018). Considerations from the innovation and quality induction working group in response to drug-drug interaction guidances from regulatory agencies: focus on CYP3A4 mRNA in vitro response thresholds, variability, and clinical relevance. *Drug Metab. Dispos.* 46: 1285–1303.
61 Fahmi, O.A., Kish, M., Boldt, S., and Obach, R.S. (2010). Cytochrome P450 3A4 mRNA is a more reliable marker than CYP3A4 activity for detecting pregnane X receptor-activated induction of drug-metabolizing enzymes. *Drug Metab. Dispos.* 38: 1605–1611.
62 Einolf, H.J., Chen, L., Fahmi, O.A. et al. (2014). Evaluation of various static and dynamic modeling methods to predict clinical CYP3A induction using in vitro CYP3A4 mRNA induction data. *Clin. Pharmacol. Ther.* 95: 179–188.
63 Fahmi, O.A., Boldt, S., Kish, M. et al. (2008). Prediction of drug-drug interactions from in vitro induction data: application of the relative induction score approach using cryopreserved human hepatocytes. *Drug Metab. Dispos.* 36: 1971–1974.
64 Fahmi, O.A. and Ripp, S.L. (2010). Evaluation of models for predicting drug-drug interactions due to induction. *Expert Opin. Drug Metab. Toxicol.* 6: 1399–1416.
65 Zhang, J.G., Ho, T., Callendrello, A.L. et al. (2014). Evaluation of calibration curve-based approaches to predict clinical inducers and noninducers of CYP3A4 with plated human hepatocytes. *Drug Metab. Dispos.* 42: 1379–1391.
66 Fahmi, O.A., Maurer, T.S., Kish, M. et al. (2008). A combined model for predicting CYP3A4 clinical net drug-drug interaction based on CYP3A4 inhibition, inactivation, and induction determined in vitro. *Drug Metab. Dispos.* 36: 1698–1708.
67 Vermet, H., Raoust, N., Ngo, R. et al. (2016). Evaluation of normalization methods to predict CYP3A4 induction in six fully characterized cryopreserved human hepatocyte preparations and HepaRG cells. *Drug Metab. Dispos.* 44: 50–60.
68 Ramsden, D., Fung, C., Hariparsad, N. et al. (2019). Perspectives from the innovation and quality consortium induction working group on factors impacting clinical drug-drug interactions resulting from induction: focus on cytochrome 3A substrates. *Drug Metab. Dispos.* 47: 1206–1221.
69 Yeung, C.K., Fujioka, Y., Hachad, H. et al. (2011). Are circulating metabolites important in drug-drug interactions?: quantitative analysis of risk prediction and inhibitory potency. *Clin. Pharmacol. Ther.* 89: 105–113.
70 Callegari, E., Kalgutkar, A.S., Leung, L. et al. (2013). Drug metabolites as cytochrome p450 inhibitors: a retrospective analysis and proposed algorithm for evaluation of the pharmacokinetic interaction potential of metabolites in drug discovery and development. *Drug Metab. Dispos.* 41: 2047–2055.
71 Yu, H. and Tweedie, D. (2013). A perspective on the contribution of metabolites to drug-drug interaction potential: the need to consider both circulating levels and inhibition potency. *Drug Metab. Dispos.* 41: 536–540.
72 Yu, H., Balani, S.K., Chen, W. et al. (2015). Contribution of metabolites to P450 inhibition-based drug-drug interactions: scholarship from the drug metabolism leadership group of the innovation and quality consortium metabolite group. *Drug Metab. Dispos.* 43: 620–630.
73 FDA Drug Development and Drug Interactions Website (2023). https://www.fda.gov/drugs/drug-interactions-labeling/drug-development-and-drug-interactions-table-substrates-inhibitors-and-inducers (accessed 8 November 2023).
74 Bahar, M.A., Setiawan, D., Hak, E., and Wilffert, B. (2017). Pharmacogenetics of drug-drug interaction and drug-drug-gene interaction: a systematic review

on CYP2C9, CYP2C19 and CYP2D6. *Pharmacogenomics* 18: 701–739.

75 Conrado, D.J., Rogers, H.L., Zineh, I., and Pacanowski, M.A. (2013). Consistency of drug-drug and gene-drug interaction information in US FDA-approved drug labels. *Pharmacogenomics* 14: 215–223.

76 Lagishetty, C.V., Deng, J., Lesko, L.J. et al. (2016). How informative are drug-drug interactions of gene-drug interactions? *J. Clin. Pharmacol.* 56: 1221–1231.

77 Gardin, A., Ufer, M., Legangneux, E. et al. (2019). Effect of fluconazole coadministration and CYP2C9 genetic polymorphism on siponimod pharmacokinetics in healthy subjects. *Clin. Pharmacokinet.* 58: 349–361.

78 Huth, F., Gardin, A., Umehara, K., and He, H. (2019). Prediction of the impact of cytochrome P450 2C9 genotypes on the drug-drug interaction potential of siponimod with physiologically-based pharmacokinetic modeling: a comprehensive approach for drug label recommendations. *Clin. Pharmacol. Ther.* 106: 1113–1124.

79 Moore, K.T., Vaidyanathan, S., Natarajan, J. et al. (2014). An open-label study to estimate the effect of steady-state erythromycin on the pharmacokinetics, pharmacodynamics, and safety of a single dose of rivaroxaban in subjects with renal impairment and normal renal function. *J. Clin. Pharmacol.* 54: 1407–1420.

80 XARELTO (Rivaroxaban) Labeling (2023). https://www.accessdata.fda.gov/drugsatfda_docs/label/2020/202439s031,022406s035lbl.pdf.

81 Fuhr, U., Jetter, A., and Kirchheiner, J. (2007). Appropriate phenotyping procedures for drug metabolizing enzymes and transporters in humans and their simultaneous use in the "cocktail" approach. *Clin. Pharmacol. Ther.* 81: 270–283.

82 International Transporter, Consortium, Giacomini, K.M., Huang, S.M. et al. (2010). Membrane transporters in drug development. *Nat. Rev. Drug Discovery* 9: 215–236.

83 Cvetkovic, M., Leake, B., Fromm, M.F. et al. (1999). OATP and P-glycoprotein transporters mediate the cellular uptake and excretion of fexofenadine. *Drug Metab. Dispos.* 27: 866–871.

84 ALLEGRA (Fexofenadine) Labeling. (2008). http://www.accessdata.fda.gov/drugsatfda_docs/label/2008/020872s018,021963s002lbl.pdf.

85 Gessner, A., Konig, J., and Fromm, M.F. (2019). Clinical aspects of transporter-mediated drug-drug interactions. *Clin. Pharmacol. Ther.* 105: 1386–1394.

86 Lee, S.C., Arya, V., Yang, X. et al. (2017). Evaluation of transporters in drug development: current status and contemporary issues. *Adv. Drug Delivery Rev.* 116: 100–118.

87 Ellens, H., Deng, S., Coleman, J. et al. (2013). Application of receiver operating characteristic analysis to refine the prediction of potential digoxin drug interactions. *Drug Metab. Dispos.* 41: 1367–1374.

88 Tweedie, D., Polli, J.W., Berglund, E.G. et al. (2013). Transporter studies in drug development: experience to date and follow-up on decision trees from the international transporter consortium. *Clin. Pharmacol. Ther.* 94: 113–125.

89 Brouwer, K.L., Keppler, D., Hoffmaster, K.A. et al. (2013). In vitro methods to support transporter evaluation in drug discovery and development. *Clin. Pharmacol. Ther.* 94: 95–112.

90 Volpe, D.A. (2016). Transporter assays as useful in vitro tools in drug discovery and development. *Expert Opin. Drug Discovery* 11: 91–103.

91 Elmeliegy, M., Vourvahis, M., Guo, C., and Wang, D.D. (2020). Effect of P-glycoprotein (P-gp) inducers on exposure of P-gp substrates: review of clinical drug-drug interaction studies. *Clin. Pharmacokinet.* 59: 699–714.

92 Rodrigues, A.D., Lai, Y., Shen, H. et al. (2020). Induction of human intestinal and hepatic organic anion transporting polypeptides: where is the evidence for its relevance in drug-drug interactions? *Drug Metab. Dispos.* 48: 205–216.

93 Zamek-Gliszczynski, M.J., Patel, M., Yang, X. et al. (2020). Intestinal P-gp and putative hepatic OATP1B induction: international transporter consortium perspective on drug development implications. *Clin. Pharmacol. Ther.* .

94 Iwaki, K., Sakaeda, T., Kakumoto, M. et al. (2006). Haloperidol is an inhibitor but not substrate for MDR1/P-glycoprotein. *J. Pharm. Pharmacol.* 58: 1617–1622.

95 Lomitapide (NDA 203858) Clinical Pharmacology and Biopharmaceutics review (2012). https://www.accessdata.fda.gov/drugsatfda_docs/nda/2012/203858Orig1s000ClinpharmR.pdf.

96 Agarwal, S., Arya, V., and Zhang, L. (2013). Review of P-gp inhibition data in recently approved new drug applications: utility of the proposed [I(1)]/IC(50) and [I(2) ]/IC(50) criteria in the P-gp decision tree. *J. Clin. Pharmacol.* 53: 228–233.

97 Zhou, T., Arya, V., and Zhang, L. (2019). Comparing various in vitro prediction methods to assess the potential of a drug to inhibit P-glycoprotein

(P-gp) transporter in vivo. *J. Clin. Pharmacol.* 59: 1049–1060.

98 Taskar, K.S., Yang, X., Neuhoff, S. et al. (2022). Clinical relevance of hepatic and renal P-gp/BCRP inhibition of drugs: An international transporter consortium perspective. *Clin. Pharmacol. Ther.* 112 (3): 573–592.

99 Vaidyanathan, J., Yoshida, K., Arya, V., and Zhang, L. (2016). Comparing various in vitro prediction criteria to assess the potential of a new molecular entity to inhibit organic anion transporting polypeptide 1B1. *J. Clin. Pharmacol.* 56 (Suppl 7): S59–S72.

100 Amundsen, R., Christensen, H., Zabihyan, B., and Asberg, A. (2010). Cyclosporine A, but not tacrolimus, shows relevant inhibition of organic anion-transporting protein 1B1-mediated transport of atorvastatin. *Drug Metab. Dispos.* 38: 1499–1504.

101 Pahwa, S., Alam, K., Crowe, A. et al. (2017). Pretreatment with rifampicin and tyrosine kinase inhibitor dasatinib potentiates the inhibitory effects toward OATP1B1- and OATP1B3-mediated transport. *J. Pharm. Sci.* 106: 2123–2135.

102 Shitara, Y., Nagamatsu, Y., Wada, S. et al. (2009). Long-lasting inhibition of the transporter-mediated hepatic uptake of sulfobromophthalein by cyclosporin a in rats. *Drug Metab. Dispos.* 37: 1172–1178.

103 Tatrai, P., Schweigler, P., Poller, B. et al. (2019). A systematic in vitro investigation of the inhibitor preincubation effect on multiple classes of clinically relevant transporters. *Drug Metab. Dispos.* 47: 768–778.

104 Chu, X., Liao, M., Shen, H. et al. (2018). Clinical probes and endogenous biomarkers as substrates for transporter drug-drug interaction evaluation: perspectives from the international transporter consortium. *Clin. Pharmacol. Ther.* 104: 836–864.

105 Arya, V., Reynolds, K.S., and Yang, X. (2022). Using endogenous biomarkers to derisk assessment of transporter-mediated drug-drug interactions: a scientific perspective. *J. Clin. Pharmacol.* 62 (12): 1501–1506.

106 Jones, N.S., Yoshida, K., Salphati, L. et al. (2020). Complex DDI by fenebrutinib and the use of transporter endogenous biomarkers to elucidate the mechanism of DDI. *Clin. Pharmacol. Ther.* 107: 269–277.

107 Kalluri, H.V., Kikuchi, R., Coppola, S. et al. (2020). Coproporphyrin I can serve as an endogenous biomarker for OATP1B1 inhibition: assessment using a glecaprevir/pibrentasvir clinical study. *Clin. Transl. Sci.* .

108 Kunze, A., Ediage, E.N., Dillen, L. et al. (2018). Clinical investigation of coproporphyrins as sensitive biomarkers to predict mild to strong OATP1B-mediated drug-drug interactions. *Clin. Pharmacokinet.* 57: 1559–1570.

109 Liu, L., Cheeti, S., Yoshida, K. et al. (2018). Effect of OATP1B1/1B3 inhibitor GDC-0810 on the pharmacokinetics of pravastatin and coproporphyrin I/III in healthy female subjects. *J. Clin. Pharmacol.* 58: 1427–1435.

110 Zhang, Y., Holenarsipur, V.K., Kandoussi, H. et al. (2020). Detection of weak organic anion-transporting polypeptide 1B inhibition by probenecid with plasma-based coproporphyrin in humans. *Drug Metab. Dispos.* 48: 841–848.

111 Dong, Z., Yang, Y., Arya, V., and Zhang, L. (2016). Comparing various in vitro prediction criteria to assess the potential of a new molecular entity (NME) to inhibit organic anion transporter 1 and 3 (OAT1 and OAT3) in vivo. *Clin. Pharmacol. Ther.* 99.

112 Dong, Z., Yang, Y., Arya, V., and Zhang, L. (2016). Comparing various in vitro prediction criteria to assess the potential of a new molecular entity (NME) to inhibit OCT2 and MATE transporters in vivo. *Clin. Pharmacol. Ther.* 99.

113 Mathialagan, S., Rodrigues, A.D., and Feng, B. (2017). Evaluation of renal transporter inhibition using creatinine as a substrate in vitro to assess the clinical risk of elevated serum creatinine. *J. Pharm. Sci.* 106: 2535–2541.

114 Arya, V., Yang, X., Balimane, P. et al. (2014). Creatinine as an endogenous marker for renal function—emerging role of transporters in the overall assessment of renal toxicity. *Clin. Pharmacol. Therap.* 95.

115 Chu, X., Bleasby, K., Chan, G.H. et al. (2016). Transporters affecting biochemical test results: creatinine-drug interactions. *Clin. Pharmacol. Ther.* 100: 437–440.

116 Chu, X., Bleasby, K., Chan, G.H. et al. (2016). The complexities of interpreting reversible elevated serum creatinine levels in drug development: does a correlation with inhibition of renal transporters exist? *Drug Metab. Dispos.* 44: 1498–1509.

117 Zamek-Gliszczynski, M.J., Chu, X., Polli, J.W. et al. (2014). Understanding the transport properties of metabolites: case studies and considerations for drug development. *Drug Metab. Dispos.* 42: 650–664.

118 Chu, X., Galetin, A., Zamek-Gliszczynski, M.J. et al. (2018). Dabigatran etexilate and digoxin: comparison as clinical probe substrates for evaluation of P-gp inhibition. *Clin. Pharmacol. Ther.* 104: 788–792.

119 Park, J., Vousden, M., Brittain, C. et al. (2010). Dose-related reduction in bupropion plasma concentrations by ritonavir. *J. Clin. Pharmacol.* 50: 1180–1187.

120 Turpeinen, M., Tolonen, A., Uusitalo, J. et al. (2005). Effect of clopidogrel and ticlopidine on cytochrome P450 2B6 activity as measured by bupropion hydroxylation. *Clin. Pharmacol. Ther.* 77: 553–559.

121 Cenobamate (NDA 212839) Clinical Pharmacology and Biopharmaceutics Review (2022). https://www.accessdata.fda.gov/drugsatfda_docs/nda/2019/212839Orig1s000ClinPharmR.pdf.

122 Abemaciclib (NDA 208716) Multidisciplinary Review (2017). https://www.accessdata.fda.gov/drugsatfda_docs/nda/2017/208716Orig1s000MultidisciplineR.pdf.

123 Zhou, D., Podoll, T., Xu, Y. et al. (2019). Evaluation of the drug-drug interaction potential of acalabrutinib and its active metabolite, ACP-5862, using a physiologically-based pharmacokinetic modeling approach. *CPT Pharm. Syst. Pharmacol.* 8: 489–499.

124 Liu, L., Bello, A., Dresser, M.J. et al. (2016). Best practices for the use of itraconazole as a replacement for ketoconazole in drug-drug interaction studies. *J. Clin. Pharmacol.* 56: 143–151.

125 Durk, M.R., Jones, N.S., Liu, J. et al. (2020). Understanding the effect of hydroxypropyl-beta-cyclodextrin on Fenebrutinib absorption in an itraconazole-Fenebrutinib drug-drug interaction study. *Clin. Pharmacol. Ther.* 108: 1224–1232.

126 Chen, Y., Ma, F., Jones, N.S. et al. (2020). Physiologically-based pharmacokinetic model-informed drug development for fenebrutinib: understanding complex drug-drug interactions. *CPT Pharm. Syst. Pharmacol.* 9: 332–341.

127 Greenblatt, D.J. and Harmatz, J.S. (2015). Ritonavir is the best alternative to ketoconazole as an index inhibitor of cytochrome P450-3A in drug-drug interaction studies. *Br. J. Clin. Pharmacol.* 80: 342–350.

128 Greenblatt, D.J. (2016). Evidence-based choice of ritonavir as index CYP3A inhibitor in drug-drug interaction studies. *J. Clin. Pharmacol.* 56: 152–156.

129 XALKORI (Crizotinib) Capsules Labeling (2022). https://www.accessdata.fda.gov/drugsatfda_docs/label/2019/202570s028lbl.pdf.

130 Xu, H., O'Gorman, M., Tan, W. et al. (2015). The effects of ketoconazole and rifampin on the single-dose pharmacokinetics of crizotinib in healthy subjects. *Eur. J. Clin. Pharmacol.* 71: 1441–1449.

131 Cho, D.Y., Ogburn, E.T., Jones, D., and Desta, Z. (2011). Contribution of N-glucuronidation to efavirenz elimination in vivo in the basal and rifampin-induced metabolism of efavirenz. *Antimicrob. Agents Chemother.* 55: 1504–1509.

132 Cho, D.Y., Shen, J.H., Lemler, S.M. et al. (2016). Rifampin enhances cytochrome P450 (CYP) 2B6-mediated efavirenz 8-hydroxylation in healthy volunteers. *Drug Metab. Pharmacokinet.* 31: 107–116.

133 Habtewold, A., Aklillu, E., Makonnen, E. et al. (2016). Long-term effect of rifampicin-based anti-TB regimen coadministration on the pharmacokinetic parameters of efavirenz and 8-hydroxy-efavirenz in Ethiopian patients. *J. Clin. Pharmacol.* 56: 1538–1549.

134 Kwara, A., Tashima, K.T., Dumond, J.B. et al. (2011). Modest but variable effect of rifampin on steady-state plasma pharmacokinetics of efavirenz in healthy African-American and Caucasian volunteers. *Antimicrob. Agents Chemother.* 55: 3527–3533.

135 Lopez-Cortes, L.F., Ruiz-Valderas, R., Viciana, P. et al. (2002). Pharmacokinetic interactions between efavirenz and rifampicin in HIV-infected patients with tuberculosis. *Clin. Pharmacokinet.* 41: 681–690.

136 Ducharme, M.P., Slaughter, R.L., Warbasse, L.H. et al. (1995). Itraconazole and hydroxyitraconazole serum concentrations are reduced more than tenfold by phenytoin. *Clin. Pharmacol. Ther.* 58: 617–624.

137 Jaruratanasirikul, S. and Sriwiriyajan, S. (1998). Effect of rifampicin on the pharmacokinetics of itraconazole in normal volunteers and AIDS patients. *Eur. J. Clin. Pharmacol.* 54: 155–158.

138 Krishna, G., Parsons, A., Kantesaria, B., and Mant, T. (2007). Evaluation of the pharmacokinetics of posaconazole and rifabutin following co-administration to healthy men. *Curr. Med. Res. Opin.* 23: 545–552.

139 Krishna, G., Sansone-Parsons, A., and Kantesaria, B. (2007). Drug interaction assessment following concomitant administration of posaconazole and phenytoin in healthy men. *Curr. Med. Res. Opin.* 23: 1415–1422.

140 Krishna, G., Moton, A., Ma, L. et al. (2009). Effects of oral posaconazole on the pharmacokinetics of atazanavir alone and with ritonavir or with

efavirenz in healthy adult volunteers. *J. Acquir. Immune Defic. Syndr.* 51: 437–444.

141 Dai, D., Yang, H., Nabhan, S. et al. (2019). Effect of itraconazole, food, and ethnic origin on the pharmacokinetics of ivosidenib in healthy subjects. *Eur. J. Clin. Pharmacol.* 75: 1099–1108.

142 Prakash, C., Fan, B., Ke, A. et al. (2020). Physiologically based pharmacokinetic modeling and simulation to predict drug-drug interactions of ivosidenib with CYP3A perpetrators in patients with acute myeloid leukemia. *Cancer Chemother. Pharmacol.* 86: 619–632.

143 FDA Guidance for Industry: Exposure-Response Relationships - Study Design, Data Analysis, and Regulatory Applications. (2003). https://www.fda.gov/media/71277/download.

144 ISENTRESS (Raltegravir) Labeling (2021). https://www.accessdata.fda.gov/drugsatfda_docs/label/2020/022145s042,203045s016,205786s008lblrpl.pdf.

145 Iwamoto, M., Wenning, L.A., Petry, A.S. et al. (2008). Minimal effects of ritonavir and efavirenz on the pharmacokinetics of raltegravir. *Antimicrob. Agents Chemother.* 52: 4338–4343.

146 Zhao, P., Zhang, L., Grillo, J.A. et al. (2011). Applications of physiologically based pharmacokinetic (PBPK) modeling and simulation during regulatory review. *Clin. Pharmacol. Ther.* 89: 259–267.

147 Zhao, P., Rowland, M., and Huang, S.M. (2012). Best practice in the use of physiologically based pharmacokinetic modeling and simulation to address clinical pharmacology regulatory questions. *Clin. Pharmacol. Ther.* 92: 17–20.

148 Sinha, V., Zhao, P., Huang, S.M., and Zineh, I. (2014). Physiologically based pharmacokinetic modeling: from regulatory science to regulatory policy. *Clin. Pharmacol. Ther.* 95: 478–480.

149 Grimstein, M., Yang, Y., Zhang, X. et al. (2019). Physiologically based pharmacokinetic modeling in regulatory science: an update from the U.S. Food and Drug Administration's Office of Clinical Pharmacology. *J. Pharm. Sci.* 108: 21–25.

150 Zhang, X., Yang, Y., Grimstein, M. et al. (2020). Application of PBPK modeling and simulation for regulatory decision making and its impact on US prescribing information: an update on the 2018-2019 submissions to the US FDA's Office of Clinical Pharmacology. *J. Clin. Pharmacol.* 60 (Suppl 1): S160–S178.

151 FDA Guidance for Industry: Physiologically Based Pharmacokinetic Analyses — Format and Content (2018). https://www.fda.gov/media/101469/download.

152 FDA Guidance for Industry (Draft): The Use of Physiologically Based Pharmacokinetic Analyses — Biopharmaceutics Applications for Oral Drug Product Development, Manufacturing Changes, and Controls. (2020). https://www.fda.gov/media/142500/download.

153 Pan, Y., Hsu, V., Grimstein, M. et al. (2016). The application of physiologically based pharmacokinetic Modeling to predict the role of drug transporters: scientific and regulatory perspectives. *J. Clin. Pharmacol.* 56 (Suppl 7): S122–S131.

154 Taskar, K.S., Pilla Reddy, V., Burt, H. et al. (2020). Physiologically-based pharmacokinetic models for evaluating membrane transporter mediated drug-drug interactions: current capabilities, case studies, future opportunities, and recommendations. *Clin. Pharmacol. Ther.* 107: 1082–1115.

155 Wagner, C., Pan, Y., Hsu, V. et al. (2015). Predicting the effect of cytochrome P450 inhibitors on substrate drugs: analysis of physiologically based pharmacokinetic modeling submissions to the US Food and Drug Administration. *Clin. Pharmacokinet.* 54: 117–127.

156 Wagner, C., Pan, Y., Hsu, V. et al. (2016). Predicting the effect of CYP3A inducers on the pharmacokinetics of substrate drugs using physiologically based pharmacokinetic (PBPK) modeling: an analysis of PBPK submissions to the US FDA. *Clin. Pharmacokinet.* 55: 475–483.

157 Zhang, X. (2019). Physiologically based pharmacokinetic modeling & simulation in OCP submissions: case studies. In: *Development of Best Practices in Physiologically Based Pharmacokinetic Modeling to Support Clinical Pharmacology Regulatory Decision-Making*. MD: Silver Spring.

158 Duvelisib (NDA 11155) Multi-Discipline Review (2018). https://www.accessdata.fda.gov/drugsatfda_docs/nda/2018/211155Orig1Orig2s000MultidisciplineR.pdf.

159 Fedratinib (NDA 212327) Multi-Discipline Review (2019). https://www.accessdata.fda.gov/drugsatfda_docs/nda/2019/212327Orig1s000MultidisciplineR.pdf.

160 Lorlatinib (NDA 210868) Multi-Discipline Review (2018). https://www.accessdata.fda.gov/drugsatfda_docs/nda/2018/210868Orig1s000MultidisciplineR.pdf.

161 Prucalopride (NDA 210166) Multi-Discipline Review (2018). https://www.accessdata.fda.gov/drugsatfda_docs/nda/2018/210166Orig1s000MultidisciplineR.pdf.

**162** Upadacitinib (NDA 211675) Clinical Pharmacology and Biopharmaceutics Review (2019). https://www.accessdata.fda.gov/drugsatfda_docs/nda/2019/211675Orig1s000ClinPharmR.pdf.

**163** Willmann, S., Coboeken, K., Kapsa, S. et al. (2020). Applications of physiologically based pharmacokinetic modeling of rivaroxaban-renal and hepatic impairment and drug-drug interaction potential. *J. Clin. Pharmacol.* .

**164** Zhou, D., Bui, K., Sostek, M., and Al-Huniti, N. (2016). Simulation and prediction of the drug-drug interaction potential of naloxegol by physiologically based pharmacokinetic modeling. *CPT Pharm. Syst. Pharmacol.* 5: 250–257.

**165** Snoeys, J., Beumont, M., Monshouwer, M., and Ouwerkerk-Mahadevan, S. (2016). Mechanistic understanding of the nonlinear pharmacokinetics and intersubject variability of simeprevir: a PBPK-guided drug development approach. *Clin. Pharmacol. Ther.* 99: 224–234.

**166** Wang, Y.H., Chen, D., Hartmann, G. et al. (2019). PBPK modeling strategy for predicting complex drug interactions of Letermovir as a perpetrator in support of product labeling. *Clin. Pharmacol. Ther.* 105: 515–523.

**167** Posada, M.M., Cannady, E.A., Payne, C.D. et al. (2017). Prediction of transporter-mediated drug-drug interactions for baricitinib. *Clin. Transl. Sci.* 10: 509–519.

**168** Posada, M.M., Bacon, J.A., Schneck, K.B. et al. (2015). Prediction of renal transporter mediated drug-drug interactions for pemetrexed using physiologically based pharmacokinetic modeling. *Drug Metab. Dispos.* 43: 325–334.

**169** Antiviral Drugs Advisory Committee Meeting, May 19, 2005 (FDA Presentations, Drug Interactions). (2005). http://www.fda.gov/ohrms/dockets/ac/05/slides/2005-4139S1_10_FDA-Zhang.ppt.

**170** APTIVUS Clinical Pharmacology Biopharmaceutics Review (2005). http://www.fda.gov/cder/foi/nda/2005/21814_000_AptivusTOC.htm.

**171** APTIVUS (Tipranavir) Labeling (2020). https://www.accessdata.fda.gov/drugsatfda_docs/label/2020/021814s021,022292s014lbl.pdf.

**172** Dumond, J.B., Vourvahis, M., Rezk, N.L. et al. (2010). A phenotype-genotype approach to predicting CYP450 and P-glycoprotein drug interactions with the mixed inhibitor/inducer tipranavir/ritonavir. *Clin. Pharmacol. Ther.* 87: 735–742.

**173** IMODIUM (Loperamide Hydrochloride) Capsules Labeling (2016). https://www.accessdata.fda.gov/drugsatfda_docs/label/2016/017690s005lbl.pdf.

**174** Kirby, B.J., Collier, A.C., Kharasch, E.D. et al. (2011). Complex drug interactions of HIV protease inhibitors 2: in vivo induction and in vitro to in vivo correlation of induction of cytochrome P450 1A2, 2B6, and 2C9 by ritonavir or nelfinavir. *Drug Metab. Dispos.* 39: 2329–2337.

**175** Morcos, P.N., Chang, L., Kulkarni, R. et al. (2013). A randomised study of the effect of danoprevir/ritonavir or ritonavir on substrates of cytochrome P450 (CYP) 3A and 2C9 in chronic hepatitis C patients using a drug cocktail. *Eur. J. Clin. Pharmacol.* 69: 1939–1949.

**176** RINVOQ (Upadacitinib) Extended-Release Tablets Labeling (2023). https://www.accessdata.fda.gov/drugsatfda_docs/label/2020/211675s001lbl.pdf.

**177** MOVANTIK (Naloxegol) Tablets Labeling (2016). https://www.accessdata.fda.gov/drugsatfda_docs/label/2016/204760s001lbl.pdf.

**178** Naloxegol (NDA 204760) Clinical Pharmacology and Biopharmaceutics Review (2014). https://www.accessdata.fda.gov/drugsatfda_docs/nda/2014/204760Orig1s000ClinPharm.pdf.

**179** NERLYNX (Neratinib) Tablets Labeling (2021). https://www.accessdata.fda.gov/drugsatfda_docs/label/2020/208051s005s006lbl.pdf

# 28

# Formulation Approaches to Improve Oral Bioavailability of Drugs

*Zeren Wang[1], Chandan Bhugra[2], and Shun Chen[1]*

[1] Shenzhen Pharmacin Co., Ltd, Shenzhen, PR China
[2] GlaxoSmithKline, Collegeville, PA, USA

## 28.1 Introduction

The barriers to the bioavailability of an oral drug have been discussed in Chapter 1 of this book. Certain barriers may be reduced and improved by developing an appropriate formulation, while other barriers may only be removed or improved by structure modification of active moieties.

Among these barriers, poor bioavailability of poorly water-soluble compounds may be improved by formulation technologies, which is the main topic of this chapter. In addition, poor bioavailability of polar molecules such as peptides due to their poor permeabilities may also be manipulated through formulation, but the topic will be discussed in Chapter 31.

### 28.1.1 Why NCEs in the Current Pipelines are Poorly Water Soluble

Historically, biologically active molecules were mostly derived from botanical extracts. Aspirin is the first synthetic molecule modified from biologically active molecules in the extract of willow bark. It is as early as in 1960s that the scientists found association of lipophilicity of a drug with its hydrophobic binding to proteins [1] and up until the 1990s most molecules were relatively water soluble in comparison to the water solubilities of the molecules currently in development. This change in water solubility of the molecules paralleled the introduction of combinatorial chemistry and high throughput screening in the 1990s, when newly discovered chemical entities became more poorly water soluble [2]. The properties of the molecules identified from high throughput screening shifted toward higher molecular weight and increasing lipophilicity, resulting in decreased aqueous solubility [3, 4]. In last 20 years of the twenty-first century, this trend is getting even more significant perhaps due to addition of more hydrophobic substituents to the lead molecules with the aim to further optimize potency and selectivity potentially improving chances of increased efficacy of drug candidates (often with an unfavorable outcome from solubility perspective). Furthermore, newer therapeutic targets such as kinases with ion channels, nuclear receptors, and protein–protein interactions often require planar templates with high lipophilicity (wide aromatic regions) as well as intermolecular hydrogen bonds to yield strong intermolecular interactions, thus resulting in molecules that are highly lipophilic and/or with high crystal-lattice energies [5, 6].

This trend of poor solubility is more prominent for oncology NCEs. Many current oncology molecules inhibit protein kinases. These molecules tend to have planar templates with multiple conjugated aromatic rings and a network of hydrogen bond donors/acceptors [7]. These type of molecules not only have high lipophilicity, but also high crystalline lattice energy due to having strong $\pi$–$\pi$ interactions between molecules that could create tight crystal packing, resulting in extremely poor water solubility.

### 28.1.2 Issues Related to Poorly Water-Soluble Drugs

Active ingredients of an oral pharmaceutical drug product will need to be dissolved in gastric and intestinal fluid in order to be absorbed systemically before it can take pharmacological effect. Poorly water-soluble active ingredients tend to slowly/incompletely dissolve and thus be absorbed partially during their transition time in the gastrointestinal (GI) tract, leading to poor bioavailability. In order to maintain adequate

*Oral Bioavailability and Drug Delivery: From Basics to Advanced Concepts and Applications,* First Edition. Edited by Ming Hu and Xiaoling Li.

plasma concentration of "poorly water soluble" active ingredients, the corresponding drug products often have issues with high dose requirement, multiple dosing regimen or more number of tablets, and significant food effect, also resulting in high inter-subject and intra-subject variations in *in vivo* exposure, etc. These issues could potentially translate to variable efficacy or unacceptable side effects profile.

Oral administration of drug products, manufactured using poorly water-soluble drug substances, with or after meal often could increase *in vivo* exposure, or bioavailability, sometimes as significantly as more than 10-fold increase [8], in comparison to the exposure with drug administration under fasted condition. Foods with different levels of fat and calorie, i.e. low, moderate, or high-fat diet, could translate to variable bioavailability due to factors related to concomitant food type. Some products take advantage of the food effect by requiring administration of the products with food, while some other products require administration of the products on an empty stomach, i.e. no food should be consumed for at least two hours before dose is taken and for at least one hour after the dose is taken. For some drugs, this requirement of administration on an empty stomach is even listed in the black box warning of the product label, to avoid potentially toxic exposure level caused by administration after a meal [9]. This food effect is often particularly more significant with highly hydrophobic compounds, causing inconveniences to patients and high inter- and intra-subject variabilities of efficacy and adverse events.

In addition to food effect, for weakly basic poorly water-soluble compounds, coadministration with antiacid drugs could also affect their *in vivo* exposure. Acidic environment of the stomach could help to dissolve these weakly basic poorly water-soluble compounds and thus provide better bioavailability. However, antiacid drugs will increase the pH of the stomach and thus inhibit dissolution and *in vivo* exposure of these compounds.

## 28.2 Theoretical Considerations for Formulation Development of Poorly Water-Soluble Drugs

Solubility is an intrinsic material property that is defined by the structures of the molecule and solvent. For orally administered pharmaceutical products, active pharmaceutical ingredients need to be dissolved first before they could be absorbed through gastric intestinal tract into systemic circulation for desired efficacy. However, the time that an administered product can reside in the GI tract is limited and is usually up to 24 hours. That means, the solid particles of the API need to be dissolved within this time period as a worst case but more ideally within one to two hours after oral administration.

As it is discussed in Chapter 3, dissolution of API particles *in vivo* is highly related to the amount of dissolved API and kinetic solubility values of these API particles. In Sections 28.2.1 and 28.2.2, theoretical considerations applied during formulation development of poorly soluble compounds are discussed.

### 28.2.1 Factors Affecting Solubility

#### 28.2.1.1 Crystal Lattice Energy (High Melting Temperature)

Many solid chemicals, including most of the pharmaceutical active ingredients, are crystalline solids and a few others are amorphous. Crystalline solids are the solids having definite geometrical arrangements of ions and molecules. Thermodynamically, solids in crystalline state are more stable in comparison to their amorphous counterpart, liquid, or molten state.

In crystalline solids, ions, and molecules are bound together by ionic or molecular forces. Among different forces, ionic force in salt crystalline solids or hydrogen bonding force in molecular solids are the strongest forces. Crystalline solids bound with high forces will therefore have stronger lattice energy, higher melting temperature, and thus, lower solubility.

Dannenfelser and Yalkowsky [10] and Jain et al. [11] have proposed the relationship of solubility with the melting temperature of nonionic crystalline solids as the following equation:

$$\log X \approx -0.01(\mathrm{MP} - 25) \tag{28.1}$$

where MP is the melting point temperature of the solute expressed in °C, and $X$ represents solubility of the crystal expressed in molar ratio.

According to Eq. (28.1), for each 100 °C increase in melting point temperature, solubility of the solute decreases roughly by 10-fold. Usually, for a nonionic free form (free base or free acid) pharmaceutical solid melting point temperature of more than 200 °C is considered as high.

#### 28.2.1.2 Hydrophobic Effect (High Log *P* Values)

Hydrophobic effect describes the phenomenon by which nonpolar molecules avoid contact with water molecules when exposed to an aqueous environment. Hydrophobic molecules and water molecules will tend to separate and form oily layer and water layer due to entropy effect [12]. When a nonpolar molecule is mixed with water the hydrophobic effect will result in poor solubility of the molecule in water. This poor water solubility of nonpolar chemicals is due to the hydrophobic effect.

Log *P*, or octanol–water partition coefficient in logarithm, is a physicochemical property used to quantitatively define the hydrophobicity of a chemical. It is calculated by the following equation:

$$\text{Log}\,P = -\text{Log}\left(\frac{S_o}{S_W}\right) \tag{28.2}$$

where $S_o$ is the solubility of the chemical in octanol phase and $S_w$ is the solubility of the chemical in water phase.

From Eq. (28.2), every one log increase in the Log *P* value, results in the decrease in solubility of the chemical in water by 10-fold.

An ideal range of Log *P* values for a pharmaceutical compound or drug candidate that is orally administered is 1–3 [13]. In this range, a compound is neither polar enough that transport of molecule across membrane of GI tract happens passively nor is hydrophobic enough for it to have low water solubility to not dissolve prior to absorption.

Based on Sections 28.2.1.1 and 28.2.1.2, high crystallinity represented by both high melting point temperature and high hydrophobic effect represented by high Log *P* value are the two major parameters that affect the solubility of pharmaceutical solids.

## 28.2.2 Definition of Poorly Water-Soluble Pharmaceutical Compounds

### 28.2.2.1 US Pharmacopeia

In order to define solubility behavior of an active pharmaceutical ingredient in different solvents, US Pharmacopeia has a section (Description and Relative Solubility) to define different categories of solubility in a solvent. Note that the solubility categories of pharmacopeia are very generic. They are typically used for assessing solubility in different solvents and in different dosage forms. For oral products, it is important to assess the impact of aqueous solubility of API for *in vivo* absorption. Following three methods for characterization of the API aqueous solubility of oral products have been used to correlate *in vivo* absorption and the aqueous solubility of an active ingredients.

### 28.2.2.2 Biopharmaceutics Classification System

The concept of Biopharmaceutics Classification System (or simply called BCS) is discussed in detail in Chapter 9 [14–16]). An API is considered highly soluble when the highest single dose is soluble in 250 ml or less of aqueous media within the pH range of 1–6.8 at 37 ± 1 °C. If API of an oral drug product is classified as highly soluble based on this BCS method (class 1 or class 3), particularly classified as highly dissolving or very highly dissolving, the API is likely dissolved in the stomach and therefore *in vivo* absorption is likely to be not affected by the solubility and or dissolution of the immediate release dosage form.

It is necessary to note that this classification provides the extreme scenario, that is, *in vivo* absorption of highly soluble active ingredients is highly likely complete and is not affected by *in vivo* environment at the absorption site, i.e. pH of stomach or the levels of bile salts in the small intestinal. On the other hand, although complete *in vivo* absorption of a lower solubility molecule can happen, their lower solubility values make the *in vivo* absorption of the low soluble molecules more susceptible to *in vivo* environment at the absorption site. In this case, formulation of the drug product may need to be optimized in order to overcome potential absorption impact from its *in vivo* environment.

### 28.2.2.3 Time to Complete Dissolution of API Particles

After an oral drug product is ingested, API particles take time to dissolve. Dissolution rate is determined by the solubility of the API, based on Noyes–Whitney equation [17] as shown in Eq. (28.3),

$$\frac{dm}{dt} = -A\frac{D}{h}(C_s - C_t) \tag{28.3}$$

where *m* is the mass of an API particle, *t* is time of dissolution, *A* is its surface area, $C_s$ is the solubility of API, $C_t$ is the concentration of API at a specific dissolution time, *D* is diffusivity of API molecule in aqueous solution, and *h* is the thickness of the diffusion layer at the particle surface during dissolution.

This Noyes–Whitney equation was developed based on a solid slab. Wang and Flanagan proposed Eq. (28.4) to describe the dissolution profile of a spherical particle [18, 19].

$$\frac{dm}{dt} = -AD(C_s - C_t)\left(\frac{1}{r} + \frac{1}{h}\right) \tag{28.4}$$

For simplicity, Eq. (28.4) may be simplified to Eq. (28.5) to describe the dissolution profile in sink condition.

$$\frac{dm}{dt} = -ADC_s\left(\frac{1}{r} + \frac{1}{h}\right) \tag{28.5}$$

Equation (28.5) can be further simplified to the following:

$$\frac{dr}{dt} = -\frac{D}{\rho}C_s\left(\frac{1}{r} + \frac{1}{h}\right) \tag{28.6}$$

Hintz and Johnson [20], and Lu et al. [21] proposed that for dissolution of a spherical particle (with particle radius less than 30 μm) in well-suspended environment thickness of the diffusion layer, *h*, is approximately the same as the particle radius. When the particle radius is 30 μm or larger, the thickness of the diffusion layer is almost constant, about 30 μm.

For particles with radius less than 30 μm, following Eqs. (28.7) and (28.8) can be derived from above Eq. (28.6).

$$\frac{dr}{dt} = -\frac{DC_s}{\rho} * \frac{2}{r} \tag{28.7}$$

$$r * dr = -\frac{2DC_s}{\rho} * dt \tag{28.8}$$

By integrating both sides of Eq. (28.8), Eq. (28.9) can be achieved.

$$\frac{1}{2}r^2 = \frac{2DC_s}{\rho} * t \tag{28.9}$$

By rearranging Eq. (28.9), Eq. (28.10) can be obtained.

$$t = \frac{\rho}{4DC_s} * r^2 \tag{28.10}$$

In Eq. (28.10), *t* represents the time to complete dissolution of a spherical particle, in well-suspended environment, with solubility $C_s$ and particle radius *r* in sink condition.

In pharmaceutical industry, particle size of API particles is often expressed in diameter, *d*. Thus, Eq. (28.11) can be used to estimate the time to complete dissolution of a spherical particle with a certain particle size expressed in diameter.

$$t = \frac{\rho}{16DC_s} * d^2 \tag{28.11}$$

For most organic pharmaceutical solids, solid density is about 1.5 g/cm$^3$, and diffusivity of pharmaceutical small molecules in aqueous solutions is in the range of $5 \times 10^{-6}$ cm$^2$/s. With these parameters, Table 28.1 shows some examples of dissolution time as the function of particle size and solubility.

It is necessary to note that there are several assumptions applied in the model proposed by Hintz and Johnson, and Eq. (28.11). In reality, the particles in GI tract are not well suspended and the dissolution is unlikely in sink conditions. Thus, the actual *in vivo* dissolution is likely slower than the values listed in this table. However, Eq. (28.11) and Table 28.1 provide some estimates of the time necessary to completely dissolve pharmaceutical particles in terms of solubility and particle size.

Sheng et al [22] performed experiments in USP dissolution apparatus and showed that their results were not significantly different from the model proposed by Hintz and Johnson and Eq. (28.11).

**Table 28.1** Dissolution time of particles with different particle size and solubility.

| $C_s$ (mg/ml) | d (μm) | *t* (s) | *t* (m) | *t* (h) |
|---|---|---|---|---|
| 1 | 50 | 469 | 7.8 | 0.13 |
| | 25 | 117 | 2.0 | 0.033 |
| | 10 | 18.8 | 0.31 | 0.0052 |
| 0.1 | 50 | 4688 | 78 | 1.3 |
| | 25 | 1172 | 20 | 0.33 |
| | 10 | 188 | 3.1 | 0.052 |
| 0.01 | 50 | $4.7 \times 10^4$ | 781 | 13 |
| | 25 | $1.2 \times 10^4$ | 195 | 3.3 |
| | 10 | $1.9 \times 10^3$ | 31 | 0.52 |
| 0.001 | 50 | $4.7 \times 10^5$ | 7813 | 130 |
| | 25 | $1.2 \times 10^5$ | 1953 | 33 |
| | 10 | $1.9 \times 10^4$ | 313 | 5.2 |

Data in Table 28.1 could predict whether an API with a certain solubility and particle size could possibly be solubilized completely in stomach (residence time less than 60 minutes), in small intestinal (residence time about 4–6 hours), or in colon (residence time about 20 hours).

#### 28.2.2.4 MAD Theory

In addition to solubility considerations as discussed above, rate of permeation of API molecules through intestinal membrane cells into systemic circulation also takes time. For most of pharmaceutically active molecules, permeation across intestinal membrane cells is driven by passive diffusion, which is dependent on the concentration gradient of the API across the intestinal membrane. Based on the diffusion theory, rate of permeation is defined by Eq. (28.12):

$$\frac{dm}{dt} = A_m D_m \frac{1}{h_m} * (C_l - C_p) \tag{28.12}$$

where *m* is the amount of API permeated through the intestinal membrane, $A_m$ is the surface of intestinal membrane that the API solution is in contact with, $D_m$ is the diffusivity of the API molecules in the intestinal membrane, $C_l$ is the concentration of the API in intestinal lumen, $C_p$ is the concentration of the API in plasma, and $h_m$ is the thickness of the intestinal membrane.

Equation (28.12) can be further simplified to Eq. (28.13), by assuming that $C_p$, the concentration of API in plasma is zero and lumping three parameters $A_m$, $D_m$, and $h_m$ together to be $k_m$, the coefficient of permeation.

$$\frac{dm}{dt} = k_m C_l \tag{28.13}$$

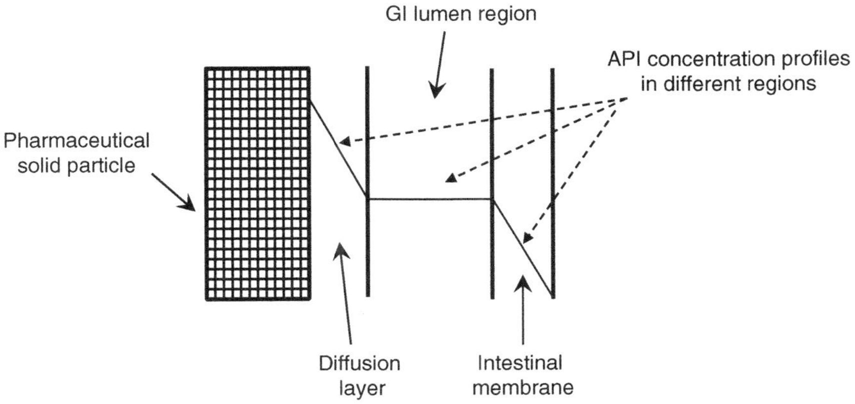


**Figure 28.1** Illustration of concentration profiles of API during its dissolution and permeation.

Based on Eq. (28.13), the rate of permeation, which ultimately determines the rate of absorption, is determined by $k_m$, the permeation coefficient, and $C_l$, the API concentration in intestinal lumen.

Figure 28.1 illustrates the concentration gradients of a pharmaceutical solid during dissolution and permeation in the GI lumen and plasma. For simplicity, the simple slab model is used to represent the solid particle, instead of the spherical particle model.

According to Figure 28.1 and Eq. (28.13), rate of drug absorption is ultimately determined by the concentration gradient of the API across the intestinal membrane, and $C_l$, the concentration of API in lumen, is further determined by the solubility and dissolution rate of the solid drug molecule (i.e. API).

Johnson and Swindell [23] studied the rate of absorption by combining both the dissolution model and permeation model. Their model provides estimation of the maximum absorbable dose with different solubility values and particle size distributions of APIs and proposed the concept of Maximum Absorbable Dose (MAD). Curatolo [24] further simplified the model by assuming that the rate of dissolution is significantly higher than the rate of permeation. In this case, the concentration of API in lumen is almost the same as its solubility. Based on this assumption, the maximum absorbable dose of an API can be estimated by Eq. (28.14), as shown below:

$$\text{MAD} = S \times K_a \times V_i \times T_i \qquad (28.14)$$

where $S$ is the solubility of the API at pH 6.5 (mg/ml), $K_a$ is the transintestinal absorption rate constant (min$^{-1}$), $V_i$ is the small intestinal water volume (ml), and $T_i$ is small intestinal transit time (min). By further assuming $V_i$ is 250 ml and $T_i$ is 4.5 hours as the small intestinal transit time, and $K_a$ is 0.003 min$^{-1}$ for low permeable molecules and 0.03 min$^{-1}$ for high permeable molecules, the author listed the maximum absorbable doses with different solubility values. The MAD values assuming highly permeable molecules are shown in Table 28.2.

**Table 28.2** Estimation of maximum absorbable dose based on API solubility.

| Solubility (mg/ml) | Solubility (μm/ml) | MAD (mg) |
|---|---|---|
| 1 | 1000 | 2025 |
| 0.1 | 100 | 202.5 |
| 0.01 | 10 | 20.25 |
| 0.001 | 1 | 2.025 |
| 0.0001 | 0.1 | 0.2025 |

Note that this MAD estimation provides the maximum doses possibly absorbed. In reality, the actual doses that may be absorbed are likely less than the values estimated by this MAD model.

## 28.3 Formulation Considerations for the Development of Poorly Water-Soluble Drugs

### 28.3.1 Formulation Development Strategy to Reduce Oral Absorption Barrier

There are many formulation approaches that can be used to reduce or remove absorption barrier due to poor solubility of an API. However, the science-based formulation development should start with scientific understanding of the mechanism of the absorption barrier and the scientist should be equipped with the right formulation approaches to tackle the barrier. Based on the illustration of Figure 28.1, a drug molecule needs to first dissociate from its solid particle and diffuse through its diffusion layer into the lumen of the GI tract, and then permeate through the GI tract membrane into systemic circulation. This process involves two steps, the dissolution step, and the permeation step. The rate of dissolution is controlled by the concentration gradient at the diffusion layer, defined by Eqs. (28.3) and (28.4), while the rate of permeation is controlled by the concentration gradient at the intestinal membrane, defined by Eq. (28.12).

Considering both the estimates in Tables 28.1 and 28.2, a formulator should first determine whether the rate-limiting step of *in vivo* absorption of active ingredients is dissolution-limited or solubility-limited. The solubility values in the Fasted Simulated Small Intestinal Fluid (FaSSIF) and Fed Simulated Small Intestinal Fluid (FeSSIF) should be determined during pre-formulation activities and are used to derive estimates according to Eqs. (28.11) and (28.14), similar to Tables 28.1 and 28.2. If based on Table 28.1, the time to dissolve API particles is much longer than the intestinal residence time, more particularly longer than the small intestinal residence time, formulation approaches to enhance dissolution should be considered. If based on Table 28.2 the Maximum Absorbable Dose is higher than the dose to be administered, formulation approaches to enhance the solubility (or apparent solubility) in the intestinal environment should be considered. Table 28.3 lists formulation approaches based on cases with different solubility values of the API (in FaSSIF) and the doses required. The actual formulation approaches listed in the table are discussed in detail in Sections 28.3.2 and 28.3.3.

### 28.3.2 Formulation Approaches to Enhance the Rate of Dissolution

Based on the discussion in Section 28.2.2, for the cases that the API dose is lower than MAD, but the dissolution time of API particles is longer than one hour, *in vivo* absorption of the drug product may be limited by dissolution of the API particles. In order to reduce the absorption barrier, technologies used for particle size reduction are often adopted as part of formulation development. Various milling technologies have been widely used to reduce and control the particle size distribution of API. Hammer milling, micronization (or jet milling), and nano-milling are the milling technologies that have been used as part of formulation development and therefore are utilized in the marketed drug products. Since nano-milling is discussed in detail in Chapter 30, hammer milling and micronization are discussed in Sections 28.3.2.1 and 28.3.2.2.

#### 28.3.2.1 Hammer Milling

Hammer milling is one of the conventional milling technologies and the most commonly applied process to reduce and control particle size distribution of API particles. During the milling process, the API particles are constantly impacted (or hammered) by metal bars (sometimes also called hammer bars) rotating at a high speed in an enclosed chamber. These metal bars acting as hammers strike the API particles in mid-air with such an impact that the particles are shattered. The particles are repeatedly stricken by the hammer impacts, collisions with the walls of the milling chamber as well as particle-on-particles impacts. In addition, the particles are further size reduced in the gap between the tip of the metal bars and the chamber wall. At the bottom of the milling chamber, there is a screen that allows the final material with the properly sized particles to pass. The mesh size of the screen can be changed depending on the target for particle size distribution. The schematic diagram of a hammer mill is shown in Figure 28.2.

The materials suitable for hammer milling are typically brittle materials, which is the case for most active pharmaceutical ingredients. Elastic or plastic materials are not suitable for hammer milling. In addition, low melting materials are also not suitable. Temperatures up to 90 °C could be generated during milling and therefore thermolabile molecules are not good candidates for hammer milling process. It is noted that some recent adaptations to hammer milling equipment such as cooling accessories may alleviate degradation due to thermal sensitivity. The processing parameters that may be controlled during the milling process are the rotating speed of the shaft and the screen mesh size. The milled

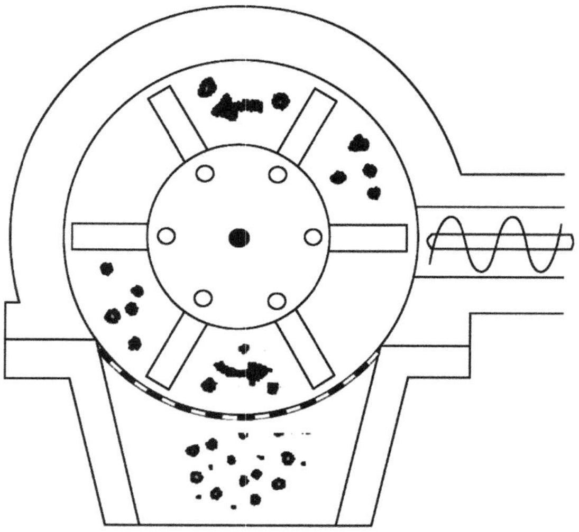

**Figure 28.2** Schematic diagram of a hammer mill.

**Table 28.3** Formulation development strategy to reduce the absorption barrier of APIs.

| Solubility (μg/mL) | Dose (mg) | Formulation approaches |
|---|---|---|
| 1000 | 100 | • highly soluble based on BCS<br>• significantly below MAD based on Table 28.2<br>• particle size has minimum impact on dissolution in GI tract based on Table 28.1<br>• thus, conventional formulation with low-cost excipients and low-cost processing technology should be considered |
| 100 | 100 | • not highly soluble based on BCS<br>• slightly below MAD based on Table 28.2<br>• particle size has some impact on dissolution in GI tract based on Table 28.1<br>• thus, some simple solubility enhancement techniques such as selection of a salt form should be considered<br>• in addition, conventional milling techniques such as hammer milling of the API may be considered |
| 100 | 10 | • highly soluble based on BCS<br>• significantly below MAD based on Table 28.2<br>• particle size has some impact on dissolution in GI tract based on Table 28.1<br>• thus, conventional milling techniques such as hammer milling of the API should be considered |
| 10 | 100 | • not highly soluble based on BCS<br>• significantly higher than MAD based on Table 28.2<br>• particle size has significant impact on dissolution in GI tract based on Table 28.1<br>• in this case, particle size reduction will not enough to remove absorption barrier<br>• solubility enhancement approaches such as solid dispersion or self-emulsifying drug delivery system should be considered |
| 10 | 10 | • not highly soluble based on BCS<br>• slightly higher than MAD based on Table 28.2<br>• particle size has significant impact on dissolution in GI tract based on Table 28.1<br>• in this case, significant particle size reduction may be enough to remove absorption barrier<br>• thus, dissolution enhancement approaches such as micronization or jet milling of API should be considered |
| 1 | 100 | • not highly soluble based on BCS<br>• significantly higher than MAD based on Table 28.2<br>• particle size reduction will not be enough to improve absorption based on Table 28.1<br>• solubility enhancement approaches such as solid dispersion or self-emulsifying drug delivery system should be considered |
| 1 | 10 | • not highly soluble based on BCS<br>• significantly higher than MAD based on Table 28.2<br>• particle size reduction will not be enough to improve absorption based on Table 28.1<br>• solubility enhancement approaches such as solid dispersion or self-emulsifying drug delivery system should be considered |

material may be milled again to further reduce the size of particles. The range of particle size distribution may be controlled by hammer milling at D90 from 20 to 100 μm.

Hammer milling could be applied as the last step of drug substance manufacturing, or as the first step of drug product manufacturing. However, since hammer milling process is relatively low cost and widely used, almost all drug substance manufacturing facilities are equipped with hammer mills.

Hammer mill often comes with bulky equipment, and the smallest equipment will still need hundreds of grams of drug substance. Therefore, for early formulation development of poorly soluble new chemical entities, manual grinding with mortar and pestle may be applied to assess the effect of particle size reduction on improvement in dissolution and by extension to potential bioavailability improvement.

#### 28.3.2.2 Micronization

The term "micronization" refers to the milling process that could produce the particles down to the size at single-digit microns. The most commonly used micronization technique today is air jet milling. Pressurized air gas used in this process forces the particles to move at a high speed and then to collide with each other and break into smaller, uniform-size particles. Since the pressured air expands significantly at the nozzle when the air stream is sprayed out, the air temperature decreases during the milling process and thus this design could prevent temperature from increasing. Therefore, it is ideal for heat-sensitive products because the temperature remains relatively constant throughout the process. Cryogenic micronization is a relatively new process and can be more effective for some low-melting or more heat-sensitive materials. In addition to suitability to heat-sensitive materials, there are additional advantages for jet milling. There are little mechanical moving parts in the jet milling region of the equipment. Therefore, less equipment wear and tear could happen. The jet milling equipment comes in different sizes. The small version could work with just as less as one gram of drug substance. Milling parameters that can be adjusted include air pressure and material feed rate. Multiple milling cycles could also be applied. By adjusting and controlling these processing parameters, the desired particle size distribution of API may be produced.

There are many publications that show bioavailability improvement after micronization of API. A typical example of a marketed product using micronized API is Tricor (fenofibrate). At the dose of 200 mg, this product was approved based on its bioequivalence with nonmicronized fenofibrate product at the dose of 300 mg [25]. In the patent publication [26] of this product, the inventors found that under the pressurized air jet stream, the API particles could be milled to achieve a mean particle size of less than 15 μm, preferably less than 5 μm. In addition, the inventors also found that co-micronization with a solid surfactant such as sodium lauryl sulfate (SLS) could further improve dissolution significantly. For hydrophobic molecules such as fenofibrate, as surface area of its solid particles significantly increased, more hydrophobic functional groups are exposed at the surface of the solid particles, thus causing poor wetting of the solid particles and further causing poor dissolution. This micronized formulation was further improved by spraying the micronized API onto excipients in a fluid bed dryer [27] to avoid agglomeration of micronized API particles. This process can not only improve dissolution, but also improve oral bioavailability. Therefore, the fenofibrate dose of this new product is further reduced to 160 mg but is bioequivalent to 200 mg of the comicronized but not spray-dried drug product.

### 28.3.3 Formulation Approaches to Enhance the Solubility in the GI Lumen

With majority of new chemical entities belonging to BCS Class 2 category, ensuring acceptable bioavailability for poorly water-soluble drug molecules has become a routine challenge for many formulation scientists. Typical approaches from drug substance perspective were discussed in Sections 28.3.2.1 and 28.3.2.2 wherein particle size reduction methodologies to increase drug particle surface area, leading to increase in kinetic concentrations via milling. However, milling alone may not be a feasible option for many drug candidates especially when bioavailability is solubility limited and additional approaches may need to be considered. Routinely, scientists employ increasingly complex solutions based on the level of challenge presented and typically initiate with drug substance form selection which is often followed by addition of pH modifiers and/or complexing agents with and without drug substance particle size reduction. Subsequently, more complex solutions are evaluated such as the use of amorphous dispersion and or self-emulsifying dispersion technology. Sections 28.3.3.1–28.3.3.4 will discuss routine practices succinctly while discussing details on more emerging technologies.

#### 28.3.3.1 Salt Form Selection

For drug candidates that are ionizable, salt formation to increase kinetic solubility is commonplace. Salt form physical and chemical stability, crystallinity, hygroscopicity, thermal behavior, manufacturability, and polymorphic landscape are aspects often considered during nomination of the salt form of the drug candidate [28].

For salt formation, and for ensuring physiological performance of the selected salt form, selection of suitable counter ion is critical. Typically, successful salt formation is driven by the difference in the p$K_a$ of the drug molecule and that of the counterion, and a difference of two pH units is considered adequate (i.e. for basic drugs p$K_a$ of the counterion is two pH units below and for acidic drugs p$K_a$ of the counterion is two pH unit higher) [29]. The difference in the p$K_a$'s by two pH units ensures the solubility of the salt form prevails over the solubility of the corresponding free acid/base. Additionally, the difference in p$K_a$'s also ensures stability from salt disproportionation perspective.

#### 28.3.3.2 Addition of Complexing Agents

Use of complexing agents in oral drug products to improve bioavailability through enhanced solubility is also considered. The benefit of using complexing agents such as cyclodextrins, more specifically the water-soluble cyclodextrins, is dependent upon the ability of cyclodextrin to form a complex with drug molecules, the stoichiometry of the complex formed, and the value of the equilibrium constant that determines the strength of complex formed and ease of release of drug molecule from the complex. Key considerations include intended dose, solubility enhancement provided by inclusion complex, and method of preparation. A review of literature evidence where cyclodextrins have been used to enhance dissolution and potentially bioavailability is presented in a recent literature article [30]. Use of cyclodextrins in multiple dosage forms such as oral solution, suspension, and solid formulations has been explored and successfully commercialized.

#### 28.3.3.3 Addition of pH Modifiers

For BCS class 2 compounds where salt form selection is either not feasible and or not independently expected to overcome the challenge to the solubility limited bioavailability of drug candidate, addition of pH modifiers may be considered.

Many drug molecules are either weakly acidic or weakly basic and have pH-dependent solubility, therefore dissolution is expected to be dependent on physiological pH, GI transit time, and solubility in the local environment.

Often pH modifiers, i.e. acidic and basic excipients, are added to alter the microenvironmental pH via faster dissolution of the excipients thereby boosting drug dissolution. Eq. (28.3) (Noyes–Whitney Equation) in Section 28.2.2.3 describes the driving force for drug dissolution, i.e. difference in the saturated concentration of drug at the surface of the drug crystal and the concentration in the bulk solution. By utilizing pH modifiers, the microenvironmental pH is altered, favoring drug concentration at the surface thereby driving increased drug in solution with a potential increase in drug absorption.

For weakly acidic drugs solubility is high in neutral and basic pH environments and excipient choices may include calcium phosphate, magnesium hydroxide, magnesium oxide, meglumine, and sodium carbonate; for weakly basic drugs, acidic pH offers the maximum advantage and excipient choices may include citric, fumaric, and tartaric acid [31].

Selection of pH modifiers can be guided by Henderson-Hasselbalch equation which describes the concentration dependence on pH. For example, for weakly monobasic drugs, drug solubility $C_s$ is described by Eq. (28.15) below where $C_0$ is intrinsic solubility of the drug molecule. Additional considerations for selecting excipients include, but are not limited to, use in other marketed products, acceptability from safety perspective, manufacturability, and stability criterion.

$$C_s = C_0 \left[1 + 10^{(pK_a - pH)}\right] \quad (28.15)$$

#### 28.3.3.4 Solid Dispersion Systems

Solid dispersion systems are the mixtures of two or more solids that are intimately mixed or dispersed but physically inseparable. In pharmaceutical field, solid dispersion systems were defined, per Chiou and Riegelman [32], as "the dispersion of one or more active ingredients in an inert carrier or matrix at solid state prepared by the melting [fusion], solvent, or melting-solvent method". From solid state point of view, solid dispersion systems can be classified as crystalline (such as eutectic or monotectic), partially crystalline, or amorphous dispersions. From miscibility point of view, solid dispersion systems can also be classified as continuous or complete miscible solid solutions or partially continuous or incomplete miscible solid suspensions. For poorly water-soluble active ingredients in BCS Class II and Class IV category, the most commonly used solid dispersion systems are amorphous solid dispersions.

In such situations, the amorphous form of the drug substance which is higher energy and thereby higher solubility offers an advantage. High energy also means higher propensity for instability and conversion to less soluble crystalline form during manufacturing, storage, or even dissolution of the amorphous solid dispersions, resulting in the lost advantage. Therefore, scientists have often resorted to the approach of using solid dispersions by molecularly dispersing drug molecules in high molecular weight polymers. Use of polymers raises the glass transition temperature of the solid dispersion thereby reducing crystallization tendency.

Use of amorphous solid dispersions in comparison to amorphous drug substance alone is also preferred so as to ensure that the drug substance remains in solution for longer durations of time due to the presence of polymeric excipient which may also act as a supersaturating agent. This is well depicted by the parachute effect imparted by the polymer (Figure 28.3) while amorphous drug substance alone may only be able to offer the spring effect on drug concentration.

Sections 28.3.3.4.1 and 28.3.3.4.2 describe the development strategies of amorphous solid dispersion systems in terms of theoretical as well as practical considerations.

***28.3.3.4.1 Theoretical Considerations*** Selection of amorphous formulation over its crystalline counterpart is guided by the dose number and the ratio of the solubility of the amorphous and crystalline form, i.e. the advantage offered is sufficient to consider amorphous route as the solution to the solubility problem. Therefore, it is pertinent to know the true solubility advantage of the amorphous form. An experimental procedure was

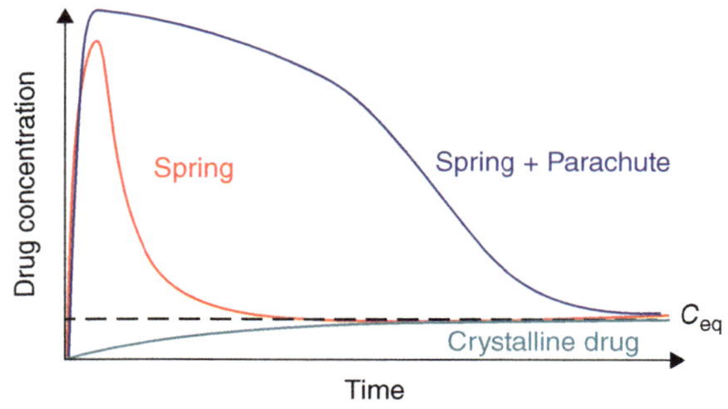


**Figure 28.3** Spring and Parachute effect showing the effect of polymer on drug solubility. Source: Adapted as-is from Brouwers et al. [33].

published [34] to determine the solubility advantage via utilization of thermal properties of the crystalline form and that of the amorphous form. Murdande et al. [35] published a more refined methodology of calculating the solubility advantage. The authors from this study calculated the solubility advantage of the amorphous state for indomethacin as an example, i.e. ratio of amorphous to crystalline solubility of 7.0, and showed it to be very close to the observed experimental value of 4.9. Calculations such as these, may help formulation scientist to ascertain if going to the amorphous route offers sufficient advantage from theoretical perspective.

The amorphous solid dispersion systems may be considered as frozen supersaturated solutions which are thermodynamically unstable, but may provide benefits due to theoretical solubility advantage. For a multicomponent amorphous solid dispersion system, the thermodynamic stability (or instability) is defined by the difference between the thermodynamic driving force for crystallization of the active ingredient (drug substance) and the thermodynamic driving force for the miscibility of the active ingredient with the polymer matrix. The thermodynamic driving force for crystallization of the active ingredient is the major driving force for instability of the amorphous solid dispersion system and is correlated with the physicochemical properties of the crystalline active ingredient, such as the melting temperature and heat of melting of the crystalline material, which cannot be manipulated by formulation development. The thermodynamic driving force for miscibility, however, is correlated with the specific interaction such as hydrogen bonding between the active ingredient and the polymer matrix, and thus could compensate for the thermodynamic driving force for crystallization and reduce the overall thermodynamic driving force for instability.

The lattice-based Flory-Huggins theory has been the most frequently discussed theory for describing the interactions between the drug and the polymer. According to FH theory, free energy of mixing for an API-polymer system [36] is mostly correlated with the strength of molecular interactions between the API and the polymer matrix. While negative interaction parameter suggests strong attractive interactions favoring miscibility, positive interaction parameter suggests cohesive interactions suggestive of de-mixing. To this end of consideration, polymers with strong specific interaction with the active ingredients are often considered as the matrix of choice for the amorphous solid dispersion systems. Interaction parameters are more typically estimated via (i) using solubility parameters wherein differences in the solubility parameters in drug and polymers can be utilized to estimate interaction parameters, or, (ii) using melting point depression methodology wherein interaction is determined via degree of change in melting point of API in different polymer systems. In a more recent work [37], the authors described a more fundamental means to describe interactions between drug-polymer combinations for a given drug wherein interactions can be ranked ordered via understanding of the hydrogen bonding potential between the drug and polymer choices. The authors correlated the rank-ordering of interactions with the amorphicity index (AI%) determined by monitoring the crystallization behavior of the dispersions.

It is necessary to point out that the drug loading in solid dispersion systems is conceptually similar to drug concentration in a supersaturated solution. The higher the drug loading, the higher is supersaturation of the solid solution, and consequently, more instability of the amorphous solid dispersion system can be expected. The effect of drug loading to the instability of an amorphous solid dispersion system is also reflected in the Flory-Huggins theory [36]. However, higher the drug loading, the smaller size of a dosage form. Therefore, as a formulation scientist, appropriate drug loading in an amorphous solid dispersion system needs to be balanced between its physical instability (during storage and dissolution) and the size of the dosage form.

In addition to the considerations of the thermodynamic driving force to the physical instability of amorphous solid dispersions, kinetic considerations are also critical. If an amorphous solid dispersion system is truly a frozen supersaturated solid solution, even though the system is physically unstable, due to its low molecular mobility, the amorphous solid dispersion system may be kinetically stable for the shelf-life of the system. The molecules in the system are rigid and difficult to move causing phase separation. However, the molecular mobility of actual amorphous systems depends on the difference between its storage temperature and its glass transition temperature ($T_g$). As the storage temperature goes beyond its $T_g$, the amorphous system becomes a rubbery liquid, while the system becomes glassy solid when the storage temperature goes below its $T_g$. The glass transition temperature of an amorphous dispersion is therefore often the kinetic parameter determined (either by measuring the $T_g$ of the prepared dispersion or calculated from $T_g$'s of individual components) to ascertain the risk toward "physical instability" of amorphous dispersions. A rule of $T_g$-50 °C has often been applied to guide pharmaceutical scientists to rank order relative stability of amorphous dispersions from kinetic point of view [38]. To this end of consideration, polymers with high glass temperatures are often considered as the matrix for amorphous solid dispersion systems.

It is clear from literature review that factors more than one affect tendency to crystallize from the amorphous state. It is intuitive that drug molecules need to move to find the next drug molecule, in an amorphous dispersion, and orient themselves to allow stable crystal nuclei to form subsequently resulting in crystal growth. However, molecular mobility is impacted by molecular interactions and stronger interactions have shown to result in longer relaxation times in amorphous states, i.e. lower molecular mobility, and a reduced tendency to crystallize [39, 40]. Therefore, an understanding of molecular interactions driving the polymer selection for amorphous dispersion, storage temperature relative to its glass transition temperature, adequate protection from humidity during shelf life can pave the way for a stable amorphous dispersion.

##### 28.3.3.4.2 Practical Considerations

**Choice of Polymers** Although numerous polymer choices are evaluated in literature to understand the impact of polymer interactions, molecular mobility, and the overall impact on stability of the amorphous dispersion [40], the selection pool of polymers for product development, scale-up, and subsequent commercialization of drug products is often limited. A recent survey of commercially available amorphous dispersions reveals that out of nine FDA approved dispersions since 2000, two products utilized polyvinylpyrollidone-vinyl acetate (PVP-VA), three products utilized hydroxypropyl methylcellulose (HPMC), and the more recent four products utilized hydroxypropyl methylcellulose acetyl succinate (HPMCAS) [41]. Since HPMCAS has carboxylic acid functional groups, it will form strong hydrogen bonds as specific interactions with drug molecules having alkaline groups. Thus, it was found that it could provide the best stability when used as the amorphous solid dispersion polymer matrix.

Selection of polymers is often driven by recent successes, acceptability of a polymer in the technology order of preference for each organization, and available toxicology coverage thereby increasing the hurdle to evaluate new polymers. From a technical perspective, solubility of polymer in solvents, glass transition temperature of amorphous dispersion, and miscibility of the drug substance in the chosen polymer driven by the functional groups of the polymer and resulting interaction with drug molecules, supersaturation potential, and hygroscopicity of the polymer are often considered. Low polymer solubility in aqueous and organic solvents may limit the use in spray drying process technology, and similarly, a very high glass transition may limit the use in hot-melt extrusion. While stability of the amorphous dispersion is critical to maintain the supersaturation advantage, supersaturation itself may drive the selection of the polymer. The two above aspects, in consideration of the ease of processability in each technology order of preference and other practical aspects, may dictate the final polymer choice.

Typically, a 1 : 1 to a 1 : 3 ratio of drug and polymer is utilized in amorphous dispersions. The choice of drug loading in a polymer is based on a balance of (i) desired bioavailability advantage as determined from a human pharmacokinetic study and its relation to drug loading/supersaturation, (ii) glass transition temperature of drug substance and that of resulting amorphous dispersion relative to the measured risk of phase separation on stability (physical and dissolution), (iii) clinical dose and acceptability on number of dosage units, and (iv) lastly on the downstream processability.

**Processing Technologies** Two processing technologies are more frequently utilized from product development and commercialization perspective, i.e. spray drying and hot-melt extrusion. Other technologies such as solvent-controlled precipitation and kinetisol have been infrequently used as well (note: kinetisol technology has not been used in commercial products to date).

For experiments early in development solvent evaporation via rotovap or small-scale extruders are routinely utilized to optimize drug loading, selection of polymer, and de-risk dispersion from stability perspective.

Spray dried dispersion process is typically developed on lab-scale spray dryers with subsequent scale-up. First key aspect for spray drying process development is selection of solvent system. For this either aqueous, organic, or an appropriate ratio of two solvent types is selected based on drug/polymer solubility in the solvent system selected, drying efficiency, residual solvent content and its allowed limits in the resultant dispersion, and environmental factors such as solvent disposal considerations.

From a technical perspective solvent selection may be guided by the interactions of solvent with the molecules of the drug substance and with that of the polymer, i.e. using Flory-Huggins theory. This becomes particularly important considering that during the spray drying process as solvent evaporates the diminishing levels of solvent may show affinity toward one or the other component depending on the interaction parameter of solvent with individual components. A greater difference in solvent affinity with the two components may drive phase separation in the drug-polymer system during/after the spray drying process. Therefore, a system wherein the Flory-Huggins interaction parameter of solvent with the drug molecule is similar to the interaction parameter of solvent with the polymer is a preferred choice [41].

From a processing perspective, the inlet temperature is an important parameter guiding the stability aspects of the spray-dried dispersion powder, i.e. when considering phase stability, and the residual solvent content in the spray-dried dispersion. Inlet temperature is typically guided by the boiling point of the selected solvent and is selected to maintain a desired outlet temperature.

In addition to considerations above, performance (i.e. supersaturation potential) and physical stability of spray-dried dispersion, powder properties are equally pertinent to ensuring consistent downstream processing for solid oral product development (Figure 28.4). The properties include particle size distribution of the spray dried powder, shape/morphology of the particles, and its bulk density/flow properties. Understanding of desirable atomization patterns (and therefore aspects such as feed rate and atomization air flow rate), as well as chamber volume and chamber height, are important considerations when scaling up the spray drying process.

While solubility of drug-polymer in solvent is a key first criterion for spray drying process development, in hot-melt extrusion process it is the melting point of API, the glass transition temperature of the polymer in comparison to the degradation temperature, and the rheological properties that plays an important role.

From a processing perspective, barrel temperature (considering the critical material properties mentioned above), screw design, feed rate of polymer-drug blend and screw speed play an important role. A combination of these parameters helps to define the level of mixing needed between the polymer and drug substance to ensure phase stability of the resultant dispersion. Figure 28.5 below describes the various parameters that may be important to describe the hot-melt extruded dispersion.

For downstream processing of either spray-dried powder and/or hot-melt extruded granulation typically a direct compression and or dry granulation process is chosen to avoid any direct solvent process preventing any physical stability concerns associated with a wet granulation type process. While direct compression may be suitable for hot-melt extruded dispersion due to high bulk density of the milled extrudates, roller compaction of the spray-dried dispersion may provide the required densification needed to achieve good flow properties, a pre-requisite for a tablet compression/capsule filling process.

Standard excipients including brittle or ductile fillers, a disintegrant, and lubricants are selected based on blend flow/tableting requirements and are routine.

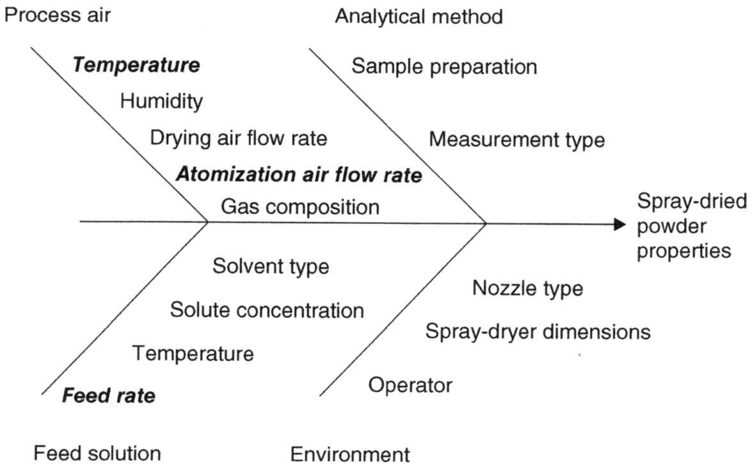


**Figure 28.4** Ishikawa diagram showing factors impacting the quality of spray-dried powder. Source: Adapted as-is from Baldinger et al. [42].

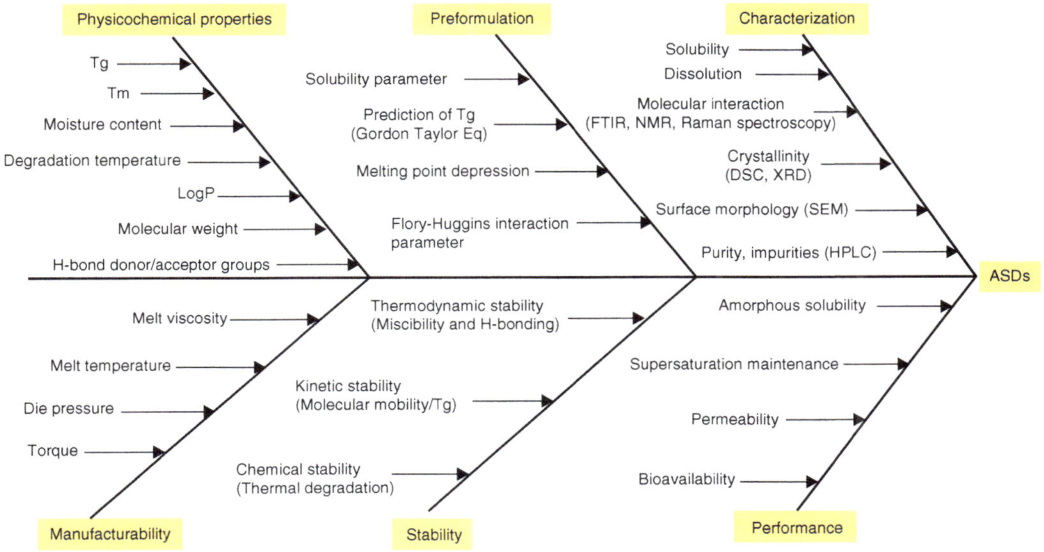


**Figure 28.5** Ishikawa diagram showing factors impacting the hot-melt extruded dispersion product. Source: Adapted as-is from Butreddy et al. [43].

## 28.4 Other Formulation Approaches

Apart from solid dispersion technology, as discussed above, there are other formulation approaches that could be explored to improve oral bioavailability of poorly water-soluble drug molecules. These approaches include self-emulsifying drug delivery systems (SEDDS) which are more discussed in Chapter 29, gastro-retentive drug delivery systems as discussed in Chapter 33, utilizing a prodrug approach as discussed in Chapter 32, and utilizing permeability enhancement approaches for poorly permeable drug molecules as discussed more in Chapter 31.

## References


1 Hansch, C. and Fujita, T. (1964). p-σ-π Analysis. A method for the correlation of biological activity and chemical structure. *J. Am. Chem. Soc.* 86: 1616–1626.

2 Broach, J.R. and Thorner, J. (1996). High-throughput screening for drug discovery. *Nature* 384 (Suppl 6604): 14–16.

3 Lipinski, C.A., Lombardo, F., Dominy, B.W., and Feeney, P.J. (1997). Experimental and computational approaches to estimate solubility and permeability in drug discovery and development settings. *Adv. Drug Delivery Rev.* 23: 3–25.

4 Lipinski, C.A. (2000). Drug-like properties and the causes of poor solubility and poor permeability. *J. Pharmacol. Toxicol. Methods* 44: 235–249.

5 Keseru, M.G. and Makara, G.M. (2006). Hit discovery and hit-to-lead approaches. *Drug Discovery Today* 11: 741–748.

6 Wells, J.A. and McClendon, C.L. (2007). Reaching for high-hanging fruit in drug discovery at protein–protein interfaces. *Nature* 450: 1001–1009.

7 Di, L., Fish, P.V., and Mano, T. (2012). Bridging solubility between drug discovery and development. *Drug Discovery Today* 17: 486–495.

8 FULL PRESCRIBING INFORMATION FOR ZYTIGA® (August 2021). https://www.accessdata.fda.gov/drugsatfda_docs/label/2021/202379s035lbl.pdf (accessed 10 July 2023).

9 FULL PRESCRIBING INFORMATION FOR TASIGNA® (December 2020). https://www.accessdata.fda.gov/drugsatfda_docs/label/2020/022068s033lbl.pdf (accessed 10 July 2023).

10 Dannenfelser, R.M. and Yalkowsky, S.H. (1996). Estimation of entropy of melting from molecular structure: a non-group contribution method. *Ind. Eng. Chem. Res.* 35: 1483–1486.

11 Jain, A., Yang, G., and Yalkowsky, S.H. (2004). Estimation of total entropy of melting of organic compounds. *Ind. Eng. Chem. Res.* 43: 4376–4379.

12 Chandler, D. (2005). Interfaces and the driving force of hydrophobic assembly. *Nature* 437: 640–647.

13 Papich, M.G. and Martinez, M.N. (2015). Applying biopharmaceutical classification system (BCS) criteria to predict oral absorption of drugs in dogs: challenges and pitfalls. *AAPS J.* 17: 948–964.
14 Amidon, G.L., Lennernäs, H., Shah, V.P., and Crison, J.R. (1995). A theoretical basis for a biopharmaceutic drug classification: the correlation of in vitro drug product dissolution and in vivo bioavailability. *Pharm. Res.* 12: 413–420.
15 Yu, L.X., Amidon, G.L., Polli, J.E. et al. (2002). Biopharmaceutics classification system: the scientific basis for biowaiver extensions. *Pharm. Res.* 19: 921–925.
16 CDER/FDA (August 2000). Guidance for industry, Waiver of in vivo bioavailability and bioequivalence studies for immediate release solid oral dosage forms based on a biopharmaceutics classification system.
17 Noyes, A.A. and Whitney, W.R. (1897). The rate of solution of solid substances in their own solutions. *J. Am. Chem. Soc.* 19: 930–934.
18 Wang, J. and Flanagan, D.R. (1999). General solution for diffusion-controlled dissolution of spherical particles. 1. Theory. *J. Pharm. Sci.* 88: 731–738.
19 Wang, J. and Flanagan, D.R. (2002). General solution for diffusion-controlled dissolution of spherical particles. 2. Evaluation of experimental data. *J. Pharm. Sci.* 91: 534–542.
20 Hintz, R.J. and Johnson, K.C. (1989). The effect of particle size distribution on dissolution rate and oral absorption. *Int. J. Pharm.* 51: 9–17.
21 Lu, A.T., Frisella, M.E., and Johnson, K.C. (1993). Dissolution modeling: factors affecting the dissolution rates of polydisperse powders. *Pharm. Res.* 10: 1308–1314.
22 Sheng, J.J., Sirois, P.J., Dressman, J.B., and Amidon, G.L. (2008). Particle diffusional layer thickness in a USP dissolution apparatus II: a combined function of particle size and paddle speed. *J. Pharm. Sci.* 97: 4815–4829.
23 Johnson, K. and Swindell, A. (1996). Guidance in the setting of drug particle size specifications to minimize variability in absorption. *Pharm. Res.* 13: 1795–1798.
24 Curatolo, W. (1998). Physical chemical properties of oral drug candidates in the discovery and exploratory development settings. *Pharm. Sci. Technol. Today* 1: 387–393.
25 Munoz, A., Guichard, J.P., and Reginault, P. (1994). Micronised fenofibrate. *Atherosclerosis* 110: S45–S48.
26 Curtet, B., Teillaud, E., and Reginault, Ph. (1990). Novel dosage form of fenofibrate. US Patent 4,895,726.
27 Stamm, A. and Seth, P. (2000). Fenofibrate pharmaceutical composition having high bioavailability and method for preparing it. US Patent 6,074,670.
28 Morris, K.R., Fakes, M.G., Thakur, A.B. et al. (1994). An integrated approach to the selection of optimal salt form for a new drug candidate. *Int. J. Pharm.* 105: 209–217.
29 Tong, W.Q. and Whitesell, G. (1998). In situ salt screening – a useful technique for discovery support and preformulation studies. *Pharm. Dev. Technol.* 3: 215–223.
30 Conceição, J., Adeoye, O., Cabral-Marques, H.M., and Lobo, J.M.S. (2018). Cyclodextrins as excipients in tablet formulations. *Drug Discovery Today* 23: 1274–1284.
31 Taniguchi, C., Kawabata, Y., Wada, K. et al. (2014). Microenvironmental pH-modification to improve dissolution behavior and oral absorption for drugs with pH-dependent solubility. *Expert. Opin. Drug Deliv.* 11: 505–516.
32 Chiou, W.L. and Riegelman, S. (1971). Pharmaceutical applications of solid dispersion systems. *J. Pharm. Sci.* 60: 1281–1302.
33 Brouwers, J., Brewster, M.E., and Augustijns, P. (2009). Supersaturating drug delivery systems: the answer to solubility-limited oral bioavailability? *J. Pharm. Sci.* 98: 2549–2572.
34 Hancock, B.C. and Parks, M. (2000). What is the true solubility advantage for amorphous pharmaceuticals? *Pharm. Res.* 17: 397–404.
35 Murdande, S.B., Pikal, M.J., Shanker, R.M., and Bogner, R.H. (2010). Solubility advantage of amorphous pharmaceuticals: I. A thermodynamic analysis. *J. Pharm. Sci.* 99: 1254–1264.
36 Marsac, P.J., Shamblin, S.L., and Taylor, L.S. (2006). Theoretical and practical approaches for prediction of drug-polymer miscibility and solubility. *Pharm. Res.* 23: 2417–2426.
37 Eerdenbrugh, B.V. and Taylor, L.S. (2011). An ab initio polymer selection methodology to prevent crystallization in amorphous solid dispersions by application of crystal engineering principles. *CrystEngComm* 13: 6171–6178.
38 Hancock, B.C., Shamblin, S.L., and Zografi, G. (1995). Molecular mobility of amorphous pharmaceutical solids below their glass transition temperatures. *Pharm. Res.* 12: 799–806.
39 Bhugra, C. and Pikal, M. (2008). Role of thermodynamic, molecular, and kinetic factors in crystallization from the amorphous state. *J. Pharm. Sci.* 97: 1329–1349.

**40** Kothari, K., Ragoonanan, V., and Suryanarayanan, R. (2014). Influence of molecular mobility on the physical stability of amorphous pharmaceuticals in the supercooled and glassy States. *Mol. Pharm.* 11: 3048–3055.

**41** Li, J., Patel, D., and Wang, G. (2017). Use of spray-dried dispersions in early pharmaceutical development: theoretical and practical challenges. *AAPS J.* 19: 321–333.

**42** Baldinger, A., Clerdent, L., Rantanen, J. et al. (2012). Quality by design approach in the optimization of the spray-drying process. *Pharm. Dev. Technol* 17: 389–397.

**43** Butreddy, A., Bandari, S., and Repka, M.A. (2021). Quality-by-design in hot melt extrusion based amorphous solid dispersions: An industrial perspective on product development. *Eur. J. Pharm. Sci.* 158: Article 105655.

# 29

# Lipid-Based and Self-Emulsifying Oral Drug Delivery Systems

*Pushkaraj Wagh, Jonathan Moreno, Christopher Nayar, and Jeffrey Wang*

*Department of Pharmaceutical Sciences, College of Pharmacy, Western University of Health Sciences, Pomona, CA, USA*

## 29.1 Introduction

Oral dosing is the preferred route of drug delivery for the treatment of many diseases because it offers a convenient, cost-effective, noninvasive mode of drug administration. However, formulating a successful oral dosage form is often a challenging task because many drugs exhibit low oral bioavailability, mainly because of their poor aqueous solubility properties. In recent years, drug discovery strategies, based on the technologies of combinatorial chemistry and high-throughput screening, have often produced compounds with poor water solubility. It is estimated that oral delivery of 50% of drug candidates is hampered because of high lipophilicity of the drug itself, and about 40% of new drug candidates exhibit low solubility in water, which leads to poor oral bioavailability, high intra- and inter-subject variability, and lack of dose proportionality. To be effective as an oral dosage form, drugs need to overcome many limiting factors to be well absorbed across the gastrointestinal (GI) tract [1]. These include chemical stability and solubility in the GI fluid, reasonable intestinal permeability, and resistance to metabolism within the gut lumen, epithelium, and liver [2]. Since the absorption rate from the GI lumen is largely controlled by dissolution, pharmaceutical scientists have used many approaches to increase the solubility of the drug, such as salt formation and particle size reduction. However, these approaches have their drawbacks; for example, salt formation of neutral compounds is not feasible and the salts formed may convert back to their original acid or base forms and lead to aggregation in the GI tract [3].

Over the past decade, the use of lipid-based drug delivery systems (DDSs) to enhance the *in vivo* efficiency of both conventional small molecules and biotechnological products such as peptides and proteins, has been well established. These lipid-based DDSs are mostly composed of materials obtained from biological sources such as phospholipids, cholesterol, cholesterol esters, and triglycerides, which are compatible with the biological membrane of the human body. In a lipid-based DDS, the drug can be delivered using two main strategies: (i) lipid-based carrier – the drug is incorporated or encapsulated into a lipid carrier; and (ii) lipidization – the drug is modified or conjugated with a lipid.

Lipid-based carriers are lipid formulations where the drug is incorporated into or encapsulated inside the lipid membrane [4]. Encapsulation of the drug avoids enzymatic degradation and helps the drug to stay for longer time in the body. However, when encapsulation of the drug (especially with a peptide and protein drug) is difficult or does not yield satisfactory results, lipidization is considered. Lipidization is the process of increasing the lipophilicity of a drug molecule with one or several lipid moieties. Drugs that have been investigated using this approach include desmopressin, tetragastrin, insulin, salmon calcitonin, leu-enkephalin, and interferon-gamma. Favorable changes in pharmacokinetics and pharmacodynamics of the modified drugs have been observed, including the prolongation of the drug effect and enhanced bioavailability. Owing to space limitation, we only discuss lipid-based carriers (with an emphasis on self-emulsifying DDS) in this chapter.

## 29.2 Lipid-based Drug Delivery Systems

On the basis of their structure, size, and physiochemical properties, lipid-based carriers can be further classified into liposomes, proliposomes, solid lipid

*Oral Bioavailability and Drug Delivery: From Basics to Advanced Concepts and Applications*, First Edition. Edited by Ming Hu and Xiaoling Li.

**Table 29.1** Lipid formulation classification system.

| Characteristics | | Excipients in formulation<br>Content of formulation (% w/w)<br>Oils: triglycerides or mixed mono and diglycerides | Water-insoluble surfactants (HLB <12) | Water-soluble surfactants (HLB >12) | Hydrophilic cosolvents (e.g. PEG, propylene glycol, Transcutol) |
|---|---|---|---|---|---|
| Type I | Pure oils<br>Limited or no dispersion<br>Digestion required | 100 | — | — | — |
| Type II | SEDDS<br>Moderate dispersion needed to form emulsion<br>Likely to require digestion | 40–80 | 20–60 | — | — |
| Type IIIA | SMEDDS<br>Rapid dispersion to form micro- or nano-emulsion<br>May need digestion | 40–80 | — | 20–40 | 0–40 |
| Type IIIB | SMEDDS<br>Rapid dispersion to form micro- or nano-emulsion<br>Digestion likely not needed | <20 | — | 20–50 | 20–50 |
| Type IV | Oil-free<br>Rapid dispersion results in micellar solution<br>No digestion needed | — | 0–20 | 30–80 | 0–50 |

nanoparticles, lipid/oil suspensions, emulsions, and self-emulsifying drug delivery systems (SEDDSs). Some of the commercial products are listed in Tables 29.1 and 29.2.

### 29.2.1 Liposomes

Liposomes are spherical, self-enclosed structures composed of lipid bilayer enclosing aqueous compartments. Since their discovery by Bangham and coworkers in the 1960s [5], there has been a tremendous interest among the scientific community to use liposomes as carriers for different molecules including small molecule drugs, proteins, peptides, DNA, siRNA, and vaccines [6].

In general, liposomes can be divided into two major classes on the basis of their size and number of bilayers: multilamellar vesicles (MLVs) and unilamellar vesicles (ULVs). The size of MLVs ranges from 100 nm to more than 1000 nm, and they are the most commonly formed vesicles. On the other hand, unilamellar liposomes can be further divided into small unilamellar vesicles (SUVs), with a size range of <100 nm, and large unilamellar vesicles (LUVs), with a size range of 100–1000 nm. A large number of methods have been proposed for liposome preparation, including thin-film hydration, lipid hydration in organic solvents, freeze drying, microfluidization, solvent injection, detergent dialysis, extrusion, and sonification [5, 7]. MLVs are relatively easy to prepare and are the most commonly formed liposomes [7]. MLVs can be processed into SUVs or LUVs by sonication or extrusion.

Commonly used lipids include natural phospholipids such as egg lecithin and soy lecithin, and semisynthetic phospholipids such as dimyristoyl phosphatidylcholine (DMPC), dipalmitoyl phosphatidylcholine (DPPC), distearoyl phosphatidylcholine (DSPC), dimyristoyl phosphatidylethanolamine (DSPE), and dipalmitoyl phosphatidylglycerol (DPPG). Generally, lipids undergo gel–liquid crystalline transition at a temperature known as the phase transition temperature ($T_c$). As the vesicle exists as solid (gel) below and as a liquid above this temperature, $T_c$ can be considered a very important factor when choosing the right lipid for the preparation of liposomes. Other bilayer additives such as cholesterol and alpha-tocopherol are often added to the lipid composition to increase the rigidity and encapsulation efficiency of the drug [7]. Most of the commonly used lipids such as DMPC, DSPC, and DPPC have

**Table 29.2** Examples of commercial liposomal, emulsion, oil based suspension products.

| Drug name | Use | Strength (mg/ml) | Dosage form | Inactive ingredients | MFG by/for |
|---|---|---|---|---|---|
| Doxil® (Doxorubin) | Breast cancer, ovarian cancer, AIDs related Kaposi sarcoma | 2 | IV | Sodium *N*-(carbonyl-methoxypolyethylene glycol 2000)-1,2-distearoyl-*sn*-glycero-3-phosphoethanolamine, hydrogenated soybean lecithin, cholesterol, ammonium sulfate, histidine, HCl, NaOH, sucrose | Sequus Pharmaceuticals |
| Abelcet® (Amphotericin B) | Fungal Infections | 5 | IV | Dimyristoylphosphatidylcholine, dimyristoylphosphatidylgylcerol, NaCl, water | Sigma-Tau Pharmaceuticals |
| Ambisome® (Amphotericin B) | Fungal Infections | 50 | IV | Alpha tocopherol, cholesterol, sistearoyl phosphatidylglycerol, NaCl, hydrogenated soy phosphatidylcholine, disodium succinate hexahydrate, sucrose | Astellas Pharma |
| Amphotec® (Amphotericin B) | Fungal Infections | 5 | IV | Sodium cholesteryl sulfate, tromethamine, disodium edetate dihydrate, lactose monohydrate, HCl | Ben Venue Laboratories |
| Vyxeos® (Daunorubicin and Cytarabine) | AML | 0.44 | IV | Distearoylphosphatidylcholine, sistearoylphosphatidylglycerol, cholesterol, copper gluconate, trolamine, sucrose | Jazz Pharmaceuticals |
| Depodur® (Morphine Sulfate) | Severe pain | 10 | Epidural | Cholesterol, distearoylphosphatidylglycerol, ticapryline, triolein, 1,2-dioleoyl-*sn*-glycero-3-phosphocholine | SkyPharma Inc. |
| Onivyde® (Irinotecan) | Metastatic pancreatic adenocarcinoma | 4.3 | IV | 1,2-Distearoyl-*sn*-glycero-3-phosphocholine, cholesterol, sodium *N*-(carbonyl-methoxypolyethylene glycol 2000)-1,2-distearoyl-*sn*-glycero-3-phosphoethanolamine, hydroxyethylpiperazine ethane sulfonic acid, NaCl | Ipsen Bioscience Inc. |
| Ikervis® (cyclosporine A) | Ocular keratitis | 1 | Ocular drops | Medium-chain triglycerides, cetalkonium chloride, glycerol, Tyloxapol, Poloxamer 188, sodium hydroxide, water | Santen Pharmaceuticals |
| Diprivan® (propofol) | Anaesthesia sedation | 10 | IV | Soy bean oil, glycerol, egg lecithin, sodium edetate, NaOH | Fresenius Kabi |
| CuroSurf® (poractant alfa) | Respiratory distress syndrome | 80 | Intra-tracheal | NaCl, $NaHCO_3$ | Chiesi Farmaceutici |
| Eradia® (metronidazole) | Giardiasis in dogs | 125 | Oral | Butylhydroxytoluene | Virbac |
| Onpattro® (patisiran) | polyneuropathy of hereditary transthyretin-mediated amyloidosis | 2 | IV | Mc-3, 1,2-dimyristoyl-*sn*-glycero-3-carboxaminopropylpolyethylene glycol 2000 methyl ether, 1,2-distearoyl-*sn*-glycero-3-phosphocholine, cholesterol, dibasic sodium phosphate heptahydrate, monobasic potassium phosphate, NaCl | Alnylam Pharmaceuticals, Inc. |

neutral charge; however, both cationic lipids such as 1,2-dioleoyl-3-trimethylammonium-propane (DOTAP) and *N*-[1-(2,3-dioleyloxy)propyl]-*N*,*N*,*N*,-trimethylammonium chloride (DOTMA) and anionic lipids such as 1,2-dimyristoyl-*sn*-glycero-3-[phosphor-*rac*-(1-glycerol)] (DMPG) and 1,2-dimyristoyl-*sn*-glycero-3-[phosphor-L-serine] (DMPS) are also used to make liposomes. Charged liposomes can also be obtained by the addition of additives such as stearylamine (positive charge) or dicetylphosphate (negative charge) [7]. In general, cationic liposomes are often used in the delivery of DNA, RNA, and vaccines.

Liposomes were mostly used as carriers for parenteral delivery of therapeutic drugs until the "insulin rush" during the 1970s. Insulin was one of the first drugs to be investigated for oral administration using liposomes. However, oral liposomal delivery of insulin was a subject of controversy as two different laboratories reported contradicting results during the mid-1970s. Ryman et al. and Gregoriadis et al. investigated oral administration of insulin-entrapped liposomes in diabetic rats and reported a reduction in blood glucose levels between 30% and 60%. Furthermore, Gregoriadis et al. observed a 20–30% reduction in blood glucose levels in normal rats. However, Ryman et al. did not observe such an effect and concluded that the results might depend on the diabetic state of the animal, the insulin sensitivity, and the rate of destruction of circulating insulin [8, 9]. This variation could also be attributed to the method of preparation and composition of phospholipids in these liposomes, as different researchers employed their own methods of making liposomes [10].

Other biotechnological products that have been investigated for oral administration as liposomes include recombinant human erythropoietin, streptokinase, factor VIII, factor IX, and heparin [11]. It was reported that approximately 0.74–31% of erythropoietin was pharmacologically available when administered as an oral liposomal preparation in comparison to intravenous (IV) administration in rats [12]. Encapsulation of biotechnological products, especially proteins and peptides, into liposomes is difficult because of the nature, size, molecular weight, and the charge of these molecules. Heparin, a high-molecular-weight polysaccharide with a high negative charge density, was orally administered in a liposomal formulation to beagle dogs, resulting in an increased heparin activity [13]. Hemkar et al. investigated liposomal entrapped factor VIII and concluded that it had a prolonged activity when compared to factor IV therapy [14]. Similarly, it was reported that liposome-entrapped streptokinase enzyme was more effective than the free streptokinase itself in improving thrombolysis in an experimental rabbit model [15].

Small molecules such as cefotaxime and tacrolimus have also yielded satisfactory results when entrapped in liposomes. The anticancer drug $N_3$-*O*-toluyl-5-fluorouracil, a prodrug of 5-fluorouracil, showed an increased GI absorption when administered as liposomes [16]. McAlister et al. reported that tacrolimus-entrapped liposomes not only showed an increased 25% better GI absorption than free tacrolimus but also reduced toxicity [17].

Liposomes have a great potential in delivering antigens and vaccine adjuvants to stimulate mucosal immune response, mainly because of their biocompatibility, protection of antigen, and targeting capabilities [11]. Liposomes have the capacity to deliver multiple components simultaneously, which generally increases the targeting specificity. Jackson et al. observed significant reduction in dental caries and increased salivary IgA anti-*Streptococcus mutans* antibody when liposome-entrapped, antiidiotic antibodies were orally administered in rats [18]. Lipopolysaccharide and outer membrane protein preparations (OMP) of Bordetella pertussis incorporated into multilamellar liposomes elicited a fourfold increase in secretory antibodies to OMP [11]. An oral enterotoxigenic *Escherichia coli* vaccine containing colonization factor antigen (CFA/I) associated with activated whole cell vibrio cholerae vaccine incorporated into liposomes increased titers of specific antibodies to CFA/I antigen in about 83% of animals [19]. Similarly, a significant immune response was found against *Herpes simplex* virus, *Streptococcus sobrinus*, cholera toxin, diptheria toxin, tetanus toxoid, and many others [11]. Cong et al. conducted a study in which they formulated sialic acid-conjugate liposomes with doxorubicin for targeting neutrophils in tumor therapy. They found that sialic acid-modified liposomal doxorubicin showed increased cellular uptake by peripheral neutrophil cells and *in vivo* circulation studies in Kunming mice showed improved $t_{1/2}$ and AUC [20]. Singh et al. targeted HER2 receptor for inhibiting tumor growth by capecitabine-loaded liposomes with THP for targeted therapy. *In vivo,* study showed significant tumor growth regression of capecitabine-loaded liposomes with THP when compared to capecitabine-loaded liposomes without THP in tumor-bearing mice [21]. In another study, doxorubicin-loaded corosolic acid liposomes showed higher *in vitro* cellular uptake along with stronger cytotoxicity against tumor cells than liposomes without corosolic acid. *In vivo* studies in 4T1-bearing mice indicate that doxorubicin-loaded corosolic acid liposomes showed significant tumor growth inhibition and longer life span in comparison to liposomes without corosolic acid [20].

Several parenteral liposome products have been approved by the Food and Drug Administration (FDA) for clinical use. These include the anticancer drugs daunorubicin and doxorubicin, and the antibiotic amphotericin B. Although liposomes have the potential to be developed as oral dosage forms, issues with toxicity and particularly formulation stability need to be further assessed before commercialization.

### 29.2.2 Proliposomes

In the recent past, proliposomes have gained considerable attention owing to their higher efficiency in manufacturing and formulation stability compared to liposomes. Proliposomes are free-flowing dry powders that can form liposomes upon hydration. Potluri and Betageri reported that proliposomes of progesterone, a poorly water-soluble hormone, showed enhanced dissolution and transport across the intestinal membrane *in vitro* [22]. In a separate study, Song et al. showed increased bioavailability of the peptide, salmon calcitonin, when administered as proliposomes to rats [23]. Another poorly bioavailable hydrophilic drug, cefotaxime, also showed enhanced oral bioavailability and intestinal lymphatic transport when given as proliposomes [24]. In another study, proliposomes of bioflavonoids from *Selaginella doederleinii*, showed significant increase in the oral bioavailability which enhanced *in vivo* cytotoxic effect without systemic toxicity which was evaluated using the HT-29 xenogarft bearing mice models in rats [25]. Dronedarone hydrochloride proliposomes, prepared by using solvent evaporation method, showed about 148% of relative bioavailability compared to pure drug. The optimized proliposomal formulation resulted in 2.6-fold increase in apparent permeability when tested in Caco-2 cells [26]. Similarly, *In vitro* and *in vivo* evaluation of isoliquiritigenin-loaded TPGS-modified proliposomes showed higher drug release profiles and increase in relative bioavailability as compared to isoliquiritigenin suspension [27]. Compared to liposomes, proliposomes formulations are easier to prepare, more stable, and less expensive. Therefore, proliposomes have great potential to be developed as oral dosage forms.

### 29.2.3 Solid Lipid Nanoparticles

Solid lipid nanoparticles (SLNs) or lipospheres or nanospheres are submicron colloidal particles made of solid lipids with the size ranging from 50 to 1000 nm. SLNs consist of a hydrophobic core encapsulated by a phospholipid monolayer. Since lipids are generally biocompatible, solid lipid nanoparticles offer an alternative DDS to synthetic polymeric nanoparticles by eliminating the risk of acute and chronic toxicity.

A variety of methods have been proposed for the preparation of SLNs, including high-pressure homogenization, microemulsion formation, and solvent emulsification [28]. A variety of drugs including proteins and peptides can be potentially delivered as SLNs through various routes of administration [29]. Although SLNs can be administered intravenously, burst release of the active ingredients poses a practical challenge. SLNs can incorporate both lipophilic and hydrophilic drugs. Furthermore, SLNs can be easily scaled up to industrial production.

Many drugs have been extensively investigated using SLNs for oral administration. Insulin-incorporated SLNs showed a considerable hypoglycemic effect when administered orally to diabetic rats [30]. Zhang et al. studied insulin incorporated in lecitin-modified SLNs and concluded that sufficient insulin was absorbed through the intestine and that it was enough to reduce the blood glucose levels [31]. In another study, salmon calcitonin incorporated into chitosan-coated SLNs provided a continuous and slow release of the peptide [32]. Pandey et al. investigated the feasibility of antitubercular chemotherapy using oral SLNs by incorporating rifampicin, isoniazid, and pyrazinamide, and concluded that only five oral doses on every 10th day of these SLNs were required to obtain an equivalent therapeutic effect as that of 46 conventional daily doses [33]. Therefore, SLN-based chemotherapy could reduce the dosing frequency and increase patient compliance [34]. Lerata successfully extracted variabilin from sponge *Ircinia* sp. And formulated SLNPs which showed improved biological activity against PC-3, HT-26, and MCF-7 cell lines [35]. In another study, it was demonstrated that levosulpiride-loaded SLNs enhanced solubilization and bioavailability. They found that prepared SLNs showed about 7 times increased bioavailability as compared to levosulpiride powder [36]. Al-Heibshy et al. prepared SLNs of poorly soluble drug rosuvastatin calcium and their pharmacokinetic evaluation demonstrated that $C_{max}$ increased by 1.4-fold and $AUC_{last}$ by 8.5-fold in comparison with the pure drug [37]. Father Abd-Ellatef et al. showed that curcumin-loaded SLNs were 5–10-folds more effective than free curcumin in increasing the intracellular retention and cytotoxicity of doxorubicin in Pgp-expressing triple-negative breast cancer [38]. However, the use of SLNs as oral DDSs is still at an infant stage, and so far no commercial product has been approved by the FDA.

### 29.2.4 Lipid/Oil Suspensions

Oral administration of oils has been in use since ancient times. Besides dietary consumption, many oils such as peppermint oil and lemongrass oil are used for their medicinal properties. Oil-based suspension can be a promising system for the delivery of a variety of drugs. Lipid or oil suspension is a heterogeneous fluid containing dispersed particles. Generally, drugs used for sustained release and with short half-lives are administered as lipid suspensions. Oil suspensions are highly viscous in nature and their viscosity can be further increased by the addition of gelling agents such as aluminum monostearate [39]. Piao et al. investigated the oral administration of solid-in-oil suspension of diclofenac sodium, a nonsteroidal anti-inflammatory drug (NSAID), and reported reduced gastric ulcerogenicity as well as high blood drug levels in rats [40]. Ampicillin trihydrate produced prolonged plasma concentrations when administered orally as an oily suspension in rabbits [41]. Orally administered suspension of toltrazuril in a soybean oil carrier demonstrated improved stability, increased bioavailability, decreased clearance rate, and improved efficacy against coccidiosis in rabbits [42]. In another case, improved bioavailability of the anti-osteoporotic drug ipriflavone was observed when administered as an oral corn oil suspension [43].

Although oil suspensions have gained considerable popularity for their use in parenteral, topical, and oral drug delivery during past couple of decades, not many of them have been investigated on commercial scales. Numerous advantages exist for the use of oil-based suspensions, including the ability to formulate drug products at concentrations above the aqueous solubility and the improved stability seen with water-labile drugs. Currently, however, few commercially available products of this type have been marketed, likely due to the investment cost of establishing and validating new manufacturing systems. Stability issues continue to remain a concern in commercializing lipid-based suspensions, as the mobility of the drug in liquid vehicles increases the opportunity for drug oxidation following interactions with excipient impurities [44]. In preclinical applications, the advantages of oil-based suspensions may be utilized within toxicology research where administration of high drug doses is needed for dose-response assays [45].

Despite demonstrating stable oil-based suspensions of clorazepate dipotassium stabilized with polyoxyethylated vegetable oil, and subsequently filing an international patent (Global Patent No. WO19910-08734A1, 1990) for an oil-based oral suspension delivery vehicle in 1991, Abbott laboratories did not commercialize the technology. Similarly, Virbac SA filed a 2018 patent application (United States of America Patent No. US20150335620A1, 2018) for development of an oral suspension of metronidazole in an oil carrier for treatment of giardia in livestock and pets; this product has since been brought to the European market under the brand name Eradia. Oil-based suspensions of omeprazole have also been investigated, with results showing improved absorption relative to other pediatric omeprazole formulations [46]. The advantages of oil-based oral suspensions can also be obtained using compounding pharmacies, which are able to create patient-specific oral suspensions using commercially available oil suspension vehicles including Fixed Oil Suspension Vehicle (PCCA, Houston Tx) and Fixed Oil Suspension (Humco, Austin Tx). These products are commonly used to make liquid medications more palatable for patients with difficulty swallowing tablets, as well as for the creation of veterinary compounds and oil-filled soft capsules.

### 29.2.5 Emulsions

An emulsion can be defined as a mixture of two immiscible liquids, an aqueous phase, and an oil phase. Either of these phases exists as a continuous or a dispersed phase in an emulsion. Emulsions can be of oil-in-water type (o/w) or water-in-oil (w/o) type, depending on the nature of the continuous and dispersed phase. On the basis of the size of the dispersed particles, emulsions can be further designated as microemulsions or nanoemulsions. Emulsions provide an excellent method for the delivery of both hydrophilic and lipophilic drugs since the hydrophilic drug can be added to the aqueous phase and the lipophilic to the oil/lipid phase. Sometimes, a cosolvent or a surfactant may be added to increase drug solubilization and formulation stability. Methods for the preparation of emulsions include high-pressure homogenization, microfluidization, ultrasonification, and solvent diffusion [47].

Microemulsions and nanoemulsions can also be termed as transparent emulsions and micellar emulsions, which are two of the most extensively investigated systems among emulsions in oral delivery. Muller et al. investigated the pharmacokinetics of valspodar, a

P-glycoprotein modulator and an analog of cyclosporine D, as an oral microemulsion and reported that the formulation produced an improved and less variable oral delivery system when studied over a clinically relevant dose range [48]. Methotrexate, an anticancer drug, was studied in an oral microemulsion drug carrier system on breast cancer cell lines and GI cell lines. The microemulsion exerted low cytotoxicity on normal cells but was effective in inducing apoptosis in cancer cells [49]. Microemulsions of various poorly soluble drugs such as ibuprofen, ketoprofen, tamoxifen, testosterone, tolbutamide, and cyclosporine have also been investigated [50, 51].

The intestinal lymphatic absorption of these microemulsions depends on the long-chain or medium-chain glyceride lipids present in the lipid phase. [52] Khoo et al. studied the intestinal lymphatic transport of halofantrine, a model lymphatically transported drug, as a microemulsion and concluded that enhanced lymphatic absorption was observed with the long-chain lipids and also when the subject was in fasted state. In contrast, another group investigated microemulsions containing medium-chain glycerides and reported an increased intestinal absorption [53]. This discrepancy may result from several factors including the type of emulsion formed (o/w or w/o), the nature of the drug including solubility, the surfactants used, and also the other additives in the formulations.

The delivery of proteins and peptides with the aid of microemulsions has also gained considerable interest in the recent past [54]. Kim et al. investigated deoxycholic acid-conjugated low-molecular-weight heparin microemulsion in monkeys and reported that the microemulsion had an enhancing effect on drug absorption [55].

### 29.2.6 Self-Emulsifying Drug Delivery Systems (SEDDS) and Self-Microemulsifying Drug Delivery Systems (SMEDDS)

An emulsion is often considered unstable owing to the very nature of its formation. Moreover, the use of high-pressure homogenization and other high-shear-exerting methods for its preparation may further impart a detrimental effect on drug stability. To overcome these obstacles, dry emulsions or SEDDSs/ SMEDDS have been developed.

SEDDS and SMEDDS are self-emulsifying isotropic, thermodynamically stable mixtures of oils, surfactants, and hydrophilic solvents and cosolvents [56]. The main driving force of emulsification is the low interfacial tension, which is achieved by using two or more emulsifiers. One emulsifier is predominantly water soluble (surfactant); while the other is oil soluble (cosurfactant) [57]. On mild agitation followed by dilution in aqueous media, such as the GI fluid, these systems can form fine oil-in-water (o/w) emulsions or microemulsions as shown in Figure 29.1. SEDDS and SMEDDS, with emulsion-like properties, can be administered as a dispersed oil solution. In addition, they do not show physical instability in terms of agglomeration or separation of the dispersed phase. The apparent difference between SEDDS and SMEDDS is that SEDDS produce opaque emulsion with a larger droplet size (>300 nm), while SMEDDS produce clear emulsions with a smaller droplet size (<250 nm) [56]. These dosage forms also allow the drug to be formulated as both ready-to-use aqueous solutions and nonaqueous concentrates. These concentrates, encapsulated in soft gelatin capsules, are expected to form solubilized drug-containing microemulsions *in vivo* upon dilution in the stomach [59]. With the incorporation of oils, surfactants, hydrophilic solvents, and cosolvents, SEDDS

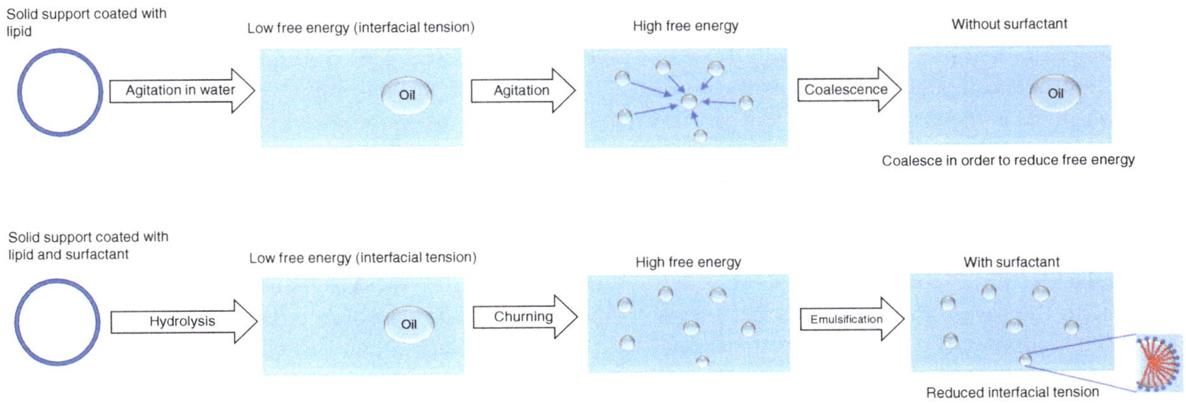


**Figure 29.1** Illustration of emulsion formation in aqueous media. Source: Modified from Betageri (2019) [58].

and SMEDDS can enhance the rate and extend of oral absorption of lipophilic drugs while yielding more reproducible blood–concentration–time profiles [60].

#### 29.2.6.1 Components of SEDDS and SMEDDS

An efficient self-emulsifying system (SES) requires very specific combinations of pharmaceutical excipients. The outcome of SEDDS and SMEDDS has been sown to be determined by the nature of the oil/surfactant pair, the surfactant concentration and oil/surfactant ratio, and the temperature at which self-emulsification occurs [61]. Other factors affecting SES are the polarity of the emulsion, droplet size, and charge [58]. Drug solubility is a critical quality parameter in the selection of excipients in SES formulation. Typically, the lipids used consist of a fatty acid ester or a medium/long-chain saturated, partially unsaturated, or unsaturated hydrocarbon chain. Some examples include mineral oil, lanolin, refined animal oil, fatty acids, fatty alcohols, and mono-/di-/tri-glycerides [62]. One of the most important excipients in the SEDDS/SMEDDS formulation is the oil. Not only can oil solubilize the lipophilic drug or facilitate self-emulsification but it can also increase the fraction of lipophilic drug transported via the intestinal lymphatic system, leading to an increase in absorption from the GI tract [61].

The next important component of self-emulsifying formulations is the surfactant. Although there are many compounds that exhibit surfactant capability, a high hydrophilic-lipophilic balance (HLB) is preferred as it ensures efficient self-dispersibility and stability of the emulsion. Examples of widely used surfactants include Span 80, Tween 80, Tween 20, and Cremphor RH40 [61]. The property of the surfactant determines the stabilizing interaction of the hydrophilic end of the surfactant with the aqueous base [59]. Since toxicity is of major concern, emulsifiers of natural origin are preferred as they are considered safer than their synthetic counterparts. The use of ionic surfactants is also possible, but not preferred, due to their potential toxicity [59].

Another important component of SESs is the cosolvent. A cosolvent aids in drug solubility and is often included in self-emulsifying formulations to decrease the GI distress and lowering the interfacial tension to a small or negative value. These cosolvents can also serve as cosurfactants in SMEDDS formulations. In microemulsions, these cosurfactants substantially accumulate with the surfactant at the interfacial layer. Cosurfactants also improve permeation of the dispersion media and decrease the shear required to disperse globules [63]. Some of the common cosolvents are ethanol, propylene glycol (PG), and polyethylene glycol (PEG). They are suitable for oral delivery and are capable of dissolving large quantities of the hydrophilic surfactant or the drug in the lipid base.

The Lipid Formulation Classification System (LFCS) was introduced by Pouton in 2000 and updated in 2006 (Table 29.1) [64]. The LFCS classifies lipid-based formulations into four different classes according to their composition. The LFCS is designed to interpret *in vivo* studies and to enable identification of thermostable formulations for specific drugs in relation to their physicochemical properties [65]. SEDDS is classified as a type II emulsion or water-insoluble lipid formulation. These formulations contain isotropic mixtures of lipids and lipophilic surfactants (HLB < 12). Self-emulsification is usually obtained at surfactant levels above 25% (w/w). At higher surfactant levels (>50–60%), the emulsification may be compromised by the formation of viscous liquid crystalline gels at the oil/water interface [66]. It is important to determine the surfactant concentration correctly because too much surfactant can also lead to GI irritation [59]. These systems provide a large interfacial area, offering enhanced partitioning of the drug between oil droplets which are responsible for absorption.

SMEDDS, on the other hand, is classified as Type III lipid-based formulation. These formulations have hydrophilic surfactants (HLB > 12) in addition to cosolvents such as ethanol, PG, and PEG [66]. Hydrophilic surfactants are necessary components that give SMEDDS formulations their smaller particle size. Depending on the oil and surfactant concentrations, Type III SMEDDS can be further divided into Type IIIA (oil content ~60%, cosurfactants 0–40%) and Type IIIB (oil content <20%, surfactant 20–50%). Overall, the major advantage of Type III lipid formulations includes the formation of a clear dispersion on aqueous dilution and drug absorption without digestion [58].

#### 29.2.6.2 Characterization of SEDDS and SMEDDS

The use of SEDDS or SMEDDS is one of the most effective approaches to improve solubility, dissolution, and oral absorption for poorly water-soluble drugs [67]. The preparation of SESs is a simple process yielding a mixture of drugs, lipids, surfactants, and cosurfactants, which form fine oil-in-water (o/w) emulsion or microemulsion, it is necessary to have a system that can characterize these self-emulsifying formulations. Several methods have been proposed in the literature to characterize the self-emulsifying performance.

The simplest and perhaps also the best way to assess self-emulsification performance is by visual inspection. Visual evaluation can provide important information

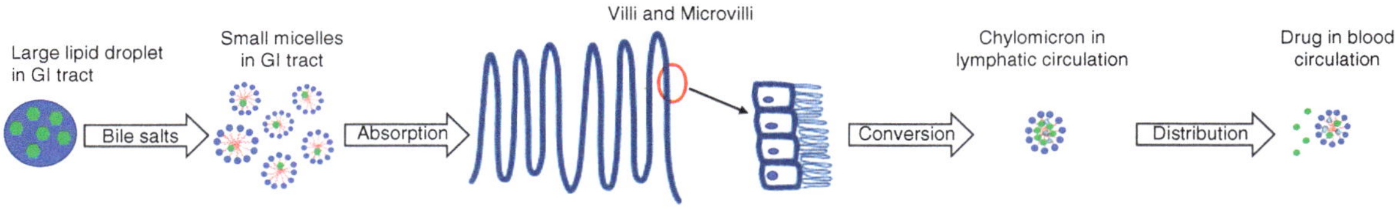


**Figure 29.2** Illustration of absorption of lipid-based drug formulations.

about the self-emulsifying properties of the mixture and about the resulting dispersion system. Following aqueous dilution of SEDDS, a clear isotropic transparent solution indicates microemulsion formation and an opaque, milky white appearance indicates a macroemulsion [68]. The absence of phase separation and/or precipitation indicates the formulation's stability.

Several tools are available to characterize and estimate the efficiency of self-emulsification of SEDDS/SMEDDS to determine the rate of equilibrium reached by the dispersion and the reproducibility of this process. Droplet size analysis evaluates the rate of emulsification and the particle size distribution by using dynamic light scattering techniques [69]. Another technique used for the determination of emulsion droplet size is photon correlation spectroscopy. The properties of the surfactant determine droplet size distribution, which is a critical quality attribute and plays a role in effective drug release, *in vivo* absorption and formulation stability.

In addition to emulsion droplet size, another crucial factor is emulsion droplet polarity. Polarity represents the affinity of the drug for oil and/or water. Droplet polarity also promotes the rapid release of the drug into the aqueous phase. Polarity is affected by the HLB, chain length, degree of unsaturation of the fatty acid, molecular weight of the hydrophilic portion, and concentration of the emulsifier [61]. The charge of the oil droplets in conventional SEDDS is negative owing to the presence of ionized free fatty acids. Incorporation of a cationic lipid, such as oleylamine, yields cationic SEDDS. Zeta potential indicates the formulation stability after appropriate dilution. A stable formulation exhibits a higher zeta potential. However, a zwitterion charge is shown to have better biocompatibility and a higher residence time in the blood as compared to particles displaying either surface charge [70].

Other physical characterization tests include the evaluation of emulsification time using a basket dissolution apparatus. The formation of a clear solution under agitation is observed after adding the formulation in a dropwise manner into the basket containing water. The time needed for emulsification depends on the oil phase and oil/surfactant ratio [71]. The surfactant's micelle formation capacity is evaluated by measuring the cloud point, or the temperature above which the homogeneous solution loses its transparency. It is important for the formulations to exhibit a cloud point greater than 37.5 °C to retain the self-emulsification qualities [72]. The structure and morphology of the SES are investigated by refractive index and transmission electron microscopy. The emulsions are investigated at different temperatures (4 and 25 °C); insignificant changes at these temperatures indicate constant microemulsion structure as well as thermodynamic stability [73].

*In vivo* lipid digestion of DDS is very complex but similar to dietary lipid digestion (Figure 29.2). Lipid digestion in the stomach is initiated by the emulsification and hydrolysis of triglycerides to free fatty acids and diglycerides by gastric lipase. In general, gastric lipase is active only at lower pH (i.e. in the stomach) [74]. The lipolysis products, combined with the movement of the stomach, further help in the emulsification of the lipids [75]. The lipid emulsion is then transferred into the duodenum, where the fatty acids become ionized and accumulate on the emulsion particle, while triglycerides and diglycerides partition into the core of the emulsion particle. On the surface of the emulsion particle, both free fatty acids and bile salts form a lamellar layer surrounding it. Lipids composed mainly of monoacylglycerides, lyso-phospholipids, and ionized fatty acids desorb from the emulsion surface to form MLV and ULVs [76]. These vesicles either disintegrate in the unstirred water lining the epithelium or get absorbed. The pancreas secretes several enzymes including lipase and colipase, which aid in the digestion of triglycerides. Pancreatic lipase breaks the triglycerides into 2-monoacylglyceride and two free fatty acid molecules [76], which might further undergo a similar process. However, the interaction between the drug and the lipid is still not well understood. Lipid particles are absorbed through the lymphatic system and eventually enter the blood circulation (Figure 29.2). It is reported that the absorption of the drugs through SEDDS/SMEDDS is dependent on the globule size of the formulation. An emulsion with a lower globule size (<250 nm) exhibits enhanced bioavailability as compared to an emulsion with a higher globule size (>300 nm) [77]. Absorption and bioavailability need further study as a function of globule size while maintaining constant concentrations of lipids and surfactants.

#### 29.2.6.3 Examples of SEDDS and SMEDDS

The discovery of self-emulsifying, lipid-based systems has allowed for improved oral bioavailability of poorly water-soluble, lipophilic drugs. However, despite the proven utility of self-emulsifying, lipid-based formulations, relatively few lipid-based products have been commercialized [66]. This is because the use of SEDDS and SMEDDS is limited by their drug-loading capacity and the usage level of excipients. Surfactants and cosolvents can be toxic at high doses and may be limited in their daily and per-dose uptake levels [59]. In addition, SEDDS and SMEDDS formulations also cause an unpredictable improvement in oral bioavailability owing to the possibility of drug precipitation upon aqueous dilution *in vivo*. Thus, it is important to be able to understand how the different components of SEDDS and SMEDDS play their roles in enhancing the oral bioavailability of lipophilic drugs. The following section discusses the different designs used in formulating SEDDS and SMEDDS. By understanding the advantages and disadvantages of these formulations, we can possibly overcome the unpredictability of aqueous dilution *in vivo*, leading to suitable SEDDS and SMEDDS formulations for oral drug delivery.

Owing to the inability to predict *in vivo* performance, successful development of lipid-based formulations is significantly hindered. Therefore, the choice of formulation components in the design of SEDDS and SMEDDS has been largely driven by the solubility of a candidate drug in the dosage form, the ease of dispersion of the formulation, and the particle size of the resulting emulsion droplets [66]. Although drug solubility and the ease of dispersion remain important in the design criteria, many studies have suggested that the ability to successfully assess the impact of lipid digestion on the solubilization capacity of a lipid-based formulation is also necessary to accurately explain *in vivo* performance [78–80].

One of the first compounds to be formulated using SEDDS was WIN 54954, an antiviral agent. When this formulation was tested in dogs in 1992, it was found that there was no difference in bioavailability but improved reproducibility of the plasma profile and increased Cmax were observed [81]. Another published SEDDS formulation in 1998 was the formulation of halofantrine, a drug to treat malaria. With an unknown mechanism of action and an erratic absorption profile of the drug, experiments were conducted to determine if the bioavailability could be improved using self-emulsifying formulations [82]. Of the different formulations of SEDDS and SMEDDS prepared, the LCT SMEDDS formulation consisting of 29% soybean oil, 29% Maisine 35-1, 30% Cremophor EL, and 7% ethanol resulted in higher bioavailability.

Self-emulsifying formulations were used not only to improve bioavailability and reduce variability but also to enhance the stability of the drug. 1,3-Bis(2-chloethyl)-1-nitrosurea (BCNU) is one of the chemotherapeutic agents for treating malignant brain tumors and is known to cross the blood–brain barrier owing to its low molecular weight and lipophilicity. However, the doe-limiting side effects and its short half-life in the plasma render BCNU very ineffective. When BCNU was prepared as in a SEDDS formulation, its stability was greatly enhanced. Self-emulsifying BCNU resulted in a slower degradation and a significant fourfold increase in the half-life *in vivo* in comparison to the free drug [83]. Both SEDDS and SMEDDS formulations are better than their ready-to-use emulsions and microemulsions counterparts. Self-emulsifying formulations have improved physical stability profiles upon long-term storage and can be filled directly into soft or hard gelatin capsules for convenient oral delivery [84].

Another successful example of bioavailability improvement using self-emulsifying formulations is a SMEDDS containing fenofribrate. Fenofibrate is a lipid-regulating agent that has a chemical, pharmacological, and clinical similarities to the other fibrate drugs such as clofibrate and gemfibrozil. The low oral bioavailability of fenofibrate is due to its low solubility and dissolution limitations [85]. A formulation of SMEDDS with fenofibrate was prepared with 31.5% Labrafac CM10, 47.3% Tween 80, and 12.7% PEG 400. This optimized formulation was discovered after a number of combinations were evaluated for the micro-emulsification existence area using pseudoternary phase diagrams and the release rate was investigated using an *in vitro* dissolution test. This formulation was then investigated in terms of the lipid-lowering efficacy using a Triton-induced hypercholesterolemia model in rats. It was found that the SMEDDS formulation significantly reduced the serum lipid level as compared with free fenofibrate. This provided evidence that SMEDDS can improve the bioavailability of fenofibrate over their traditional formulations [85].

One of the most recent SMEDDS formulations was developed for VIP, a semisynthetic derivative of vincamine. VIP is widely used in the treatment of chronic cerebral vascular ischemia, acute stroke, senile cerebral dysfunction, and Alzheimer's disease. VIP can also be

**Table 29.3** Examples of commercial SEDDS and SMEDDS products.

| Drug product name | Use | BCS class | Strength (mg) | Dosage form | Inactive ingredients | Manufactured by/for |
|---|---|---|---|---|---|---|
| Sandimmune® (cyclosporine A/I) | Indicated for the organ rejection prophylaxis in allogenic transplants of kidney, liver, and heart | IV | 25/100 | Soft gelatin capsule | Corn oil, linoleoylmacrogol glycerides, and sorbitol | Novartis pharmaceuticals corporation |
| Neora® (cyclosporine) | Systemic immunosuppressant | IV | 10/25/50/100 | Soft gelatin capsule | Corn oil-mono-di-triglycerides, polyoxyl 40 hydrogenated castor oil NF, DL-α-tocopherol USP | Novartis Pharmaceuticals Corporation |
| Gengraf® (Cyclosporine A/III) | Systemic immunosuppressant | IV | 25/100 | Hard gelatin capsule | Polyethylene glycol NF, polyoxyl 35 castor oil NF, polysorbate 80 NF, propylene glycol USP, sorbitan monooleate NF, titanium dioxide | AbbVie Inc. |
| Norvir® (Ritonavir) | Combination with other antiretroviral agents for the treatment of HIV-1 infection | II | 100 | Soft gelatin capsule | Butylated hydroxytoluene, ethanol, oleic acid, polyoxyl 35, and castor oil | AbbVie Inc. |
| Fortovase® (Saquinavir) | Inhibitor of the human immunodeficiency virus (HIV) protease | IV | 200 | Soft gelatin capsule | Medium chain mono-and diglycerides, povidone, and dl-alpha-tocopherol | Roche Laboratories Inc. |
| Agenerase® (Amprenavir) | Inhibitor of the human immunodeficiency virus (HIV) protease | II | 50 | Soft gelatin capsule | d-alpha tocopheryl PEG 1000 succinate (TPGS), PEG 400, and propylene glycol | GlaxoSmithKline |
| Depakene® (Valproic acid) | Monotherapy and adjunctive therapy in the treatment of patients with complex partial seizures that occur either in isolation or in association with other types of seizures | II | 250 | Soft gelatin capsule | Corn oil, glycerin, methylparaben, and propylparaben | AbbVie Inc. |
| Rocaltrol® (Calcitriol) | Management of secondary hyperparathyroidism and management of hypocalcemia | II | 0.25/0.5 | Soft gelatin capsule | Triglyceride of coconut oil | Roche Products Limited |
| Targretin® (Bexarotene) | Treatment of cutaneous manifestations of cutaneous T-cell lymphoma in patients who are refractory to at least one prior systemic therapy | II | 75 | Soft gelatin capsule | Polyethylene glycol 400, NF, Polysorbate 20, NF, povidone, USP, and butylated hydroxyanisole, NF | Ligand Pharmaceuticals/ Eisai Ltd. |
| Vesanoid® (Tretinoin) | Retinoid that induces maturation of acute promyelocytic leukemia (APL) | II | 10 | Soft gelatin capsule | Beeswax, butylated hydroxyanisole, edetate disodium, hydrogenated soybean oil flakes, hydrogenated vegetable oils, and soybean oil | Roche Laboratories Inc. |
| Accutane® (Isotretinoin) | Severe recalcitrant nodular acne | II | 10/20/40 | Soft gelatin capsule | Beeswax, butylated hydroxyanisole, edetate disodium, hydrogenated soybean oil flakes, hydrogenated vegetable oil, and soybean oil | Roche Laboratories Inc. |
| Aptivus® (Tipranavir) | Combination antiretroviral treatment of HIV-1 | II | 250 | Soft gelatin capsule | Dehydrated alcohol (7% w/w or 0.1 g per capsule), polyoxyl 35 castor oil, propylene glycol, mono/diglycerides of caprylic/capric acid | Boehringer Ingelheim Pharmaceuticals, Inc. |

used for anticonvulsion, cognition enhancement, neuroprotection, and antioxidation. VIP is mainly absorbed from the small intestine with a low oral bioavailability of 7% [86]. To develop an optimized formulation of VIP-SMEDDS for increasing the solubility and bioavailability, Chen et al. used solubility assay, compatibility tests, and pseudoternary phase diagram analysis for formulation optimization [86]. Eventually, a formulation with a ratio of 40 : 10 : 40 : 10 (w/w) of Labrafac, oleic acid, Cremophor EL, and Transcutol P, respectively, was prepared, and this formulation allowed for an increase in *in vitro* dissolution, and an increase in absorption from *in situ* intestine of rats. The VIP-SMEDDS formulation also improved the oral bioavailability in rats by twofold over the VIP crude powder suspension.

Most commercially approved SES dosage forms are available as soft gelatin capsules and a few are as hard gelatin capsules. These formulations can be prepared as dry powders by adsorption onto adsorbent materials such as microcrystalline cellulose [87]. Up to date, there are no self-emulsifying tablets commercially available.

#### 29.2.6.4 Commercial SEDDS and SMEDDS products

Several SES drug products have been filed as new drug application (NDA) and abbreviated new drug application (ANDA). Among the commercially available products, some NDAs such as Agenerase, Depakene, Rocaltrol, Targretin, Accutane, and Aptivus are filed under Type I – new molecular entity. Other products are filed under subclassification Type 3 – new dosage form; examples are Sandimmune, Neoral, Norvir, Fortovase, and Vesanoid. A list of commercially available drug products is presented in Table 29.3 [58].

## 29.3 Advantages and Limitations of Lipid-Based and Self-Emulsifying Drug Delivery Systems

A lipid-based and self-emulsifying DDS has many advantages over conventional DDS, as it comprises lipids derived from biological origin, which makes them highly biocompatible when administered into human body [39]. Poorly soluble and lipophilic drugs can be administered through these systems. Furthermore, protection against enzymatic degradation is vital when delivering sensitive molecules such as proteins and peptides. Interestingly, lipid-based carriers such as stable liposomes are absorbed intact by pinocytosis across gut wall, and sometimes through the lymphatic system [10]. However, obtaining overall consistency including in particle size, drug encapsulation, drug release characteristics, and particle shape is a difficult task in lipid-based DDS, especially with liposomes and SLNs. Furthermore, lipids have different capacities to incorporate different drugs based on their crystallization and polymorphism, which usually leads to inconsistency [39]. Although liposomes have the potential as lipid-based carriers in oral administration of drugs including proteins, peptides, DNA, and vaccines, issues with toxicity and stability need to be addressed before an oral product emerges from the market.

## 29.4 Summary

The boom in drug discovery and development facilitates the development of lipid-based and self-emulsifying DDS. In particular, a significant number of drug candidates are poorly water soluble. Furthermore, the past decade has witnessed a tremendous growth in the field of biotechnological products. Lipid-based and self-emulsifying DDSs provide an excellent method for the delivery of hydrophobic small molecule drugs as well as hydrophilic biotechnology-derived products. These systems can provide biological stability for drugs by preventing them from chemical and enzymatic degradation and enhance drug absorption. For a particular drug candidate, choosing the best DDS among all available DDSs would depend on many factors including the physiochemical nature of the drug, chemical and biological stability of the drug in the DDS, preparation method, and stability of the DDS during both storage and administration and the *in vivo* processing of the lipids.

## References

1 Reinholz, J., Landfester, K., and Mailänder, V. (2018). The challenges of oral drug delivery via nanocarriers. *Drug Delivery* 25 (1): 1694–1705.

2 Rezhdo, O., Speciner, L., and Carrier, R.L. (2016). Lipid-associated oral delivery: mechanisms and analysis of oral absorption enhancement. *J. Controlled Release* 240: 544–560.

3 Date, A.A., Hanes, J., and Ensign, M.L. (2016). Nanoparticles for oral delivery: design, evaluation

and state-of-the-art. *J. Controlled Release* 240: 504–526.

4 Irby, D., Du, C., and Li, F. (2017). Lipid-drug conjugate for enhancing drug delivery. *Mol. Pharm.* 14 (5): 1325–1338.

5 Johnson, S.M. and Bangham, A.D. (1969). Potassium permeability of single compartment liposomes with and without valinomycin. *Biochim. Biophys. Acta (BBA)* 193 (1): 82–91.

6 Baca-Estrada, M.E., Foldvari, M., Babiuk, S.L., and Babiuk, L.A. (2000). Vaccine delivery: lipid-based delivery systems. *J. Biotechnol.* 83 (1-2): 91–104.

7 Betageri, G.V. and Kulkarni, S.B. (1997). Preparation of liposomes. In: *Microspheres, Microcapsules & Liposomes, VI: Preparation & Chemical Applications* (ed. R. Arshady), 489–521. London: Citus Books.

8 Dapergola, G. and Gregoriadis, G. (1976). Hypoglycemic effect of liposome-entrapped insulin administered intragastrically into rats. *Lancet* 2 (7990): 824–827.

9 Patel, H.M. and Ryman, B.E. (1976). Oral administration of insulin by encapsulation within liposomes. *FEBS Lett.* 62: 60–63.

10 Woodley, J.K. (1985). Liposomes for oral administration of drugs. *Crit. Rev. Therap. Drug Carrier Syst.* 2 (1): 1–18.

11 Rogers, J.A. and Anderson, K.E. (1998). The potential of liposomes in oral drug delivery. *Crit. Rev. Therap. Drug Carrier Syst.* 15 (5): 421–480.

12 Maitani, Y., Hazama, M., Tojo, Y. et al. (1996). Oral administration of recombinant human erythropoietin in liposomes in rats: influence of lipid composition and size of liposomes on bioavailability. *J. Pharm. Sci.* 85 (4): 440–445.

13 Ueno, M., Nagasaki, T., Horikoshi, I., and Sakuragawa, N. (1982). Oral administration of liposomally-entrapped heparin to beagle dogs. *Chem. Pharm. Bull.* 30: 2245–2247.

14 Hemkar, H.C., Hermens, W.T., Muller, A.D., and Zwaal, R. (1980). Oral treatment of haemophilia A by gastrointestinal absorption of factor VIII entrapped in liposomes. *Lancet* 1 (8159): 70–71.

15 Perkins, W.R., Vaughan, D.E., Plavin, S.R. et al. (1997). Streptokinase entrapment in interdigitation-fusion liposomes improves thrombolysis in an experimental rabbit model. *Thromb. Haemost.* 77 (6): 1174–1178.

16 Sun, W., Zhang, N., Li, A. et al. (2007). Preparation and evaluation of N(3)-O-toluyl-fluorouracil-loaded liposomes. *Int. J. Pharm.* 353 (1-2): 243–250.

17 McAlister, V.C., Keshavamurthy, M., and Lee, T.D. (1999). Oral delivery of liposomal tacrolimus: increased efficacy and reduced toxicity. *Transplant. Proc.* 31 (1-2): 1110.

18 Jackson, S., Mestecky, J., Childers, N.K., and Michalek, S.M. (1990). Liposomes containing anti-idiotypic antibodies: an oral vaccine to induce protective secretory immune responses specific for pathogens of mucosal surfaces. *Infect. Immun.* 58 (6): 1932–1936.

19 Dima, V.F., Ionescu, M.D., Palade, R. et al. (2001). Stimulation of mucosal immune response following oral administration of enterotoxigenic *Escherichia coli* fimbriae (CFA/I) entrapped in liposomes in conjunction with inactivated whole-cell vibrio cholerae vaccine. *Rouman. Archiv. Microbiol. Immunol.* 60 (1): 27–54.

20 Li, C., Qiu, Q., Liu, M. et al. (2020). Sialic acid-conjugate modified liposomes targeting neutrophils for improved tumour therapy. *Biomater. Sci.* 8: 2189–2201.

21 Singh, M.K., Kiran, S.S.P., Kumar Reddy Sanapalli, B. et al. (2020). HER2 targeted biological macromolecule modified liposomes for improved efficacy of capecitabine in breast cancer. *Int. J. Biol. Macromol.* 150: 631–636.

22 Potluri, P. and Betageri, G.V. (2006). Mixed-micellar proliposomal systems for enhanced oral delivery of progesterone. *Drug Delivery* 13 (3): 227–232.

23 Song, K.-H., Chung, S.-J., and Shim, C.-K. (2005). Enhanced intestinal absorption of salmon calcitonin (sCT) from proliposomes containing bile salts. *J. Controlled Release* 106 (3): 298–308.

24 Ling, S.S., Magosso, E., Khan, N.A. et al. (2006). Enhanced oral bioavailability and intestinal lymphatic transport of a hydrophilic drug using liposomes. *Drug Dev. Ind. Pharm.* 32 (3): 335–345.

25 Chen, B., Wang, X., Lin, D. et al. (2019). Proliposomes for oral delivery of total biflavonoids extract from Selaginella doederleinii: formulation development, optimization, and in vitro-in vivo characterization. *Int. J. Nanomed.* 14: 6691–6706.

26 Kovvasu, S., Kunamaneni, P., Yeung, S. et al. (2019). Formulation of dronedarone hydrochloride-loaded proliposomes: in vitro and in vivo evaluation using Caco-2 and rat model. *AAPS PharmSciTech* 20 (6).

27 Liu, J., Wang, Q., Adu-Frimpong, M. et al. (2019). Preparation, in vitro and in vivo evaluation of isoliquiritigenin-loaded TPGS modified proliposomes. *Int. J. Pharm.* 563: 53–62.

28 Battagalia, L., Trotta, M., Gallarate, M. et al. (2007). Solid lipid nanoparticles formed by solvent-in-water emulsion-diffusion technique: development and

influence on insulin stability. *J. Microencapsulation* 24 (7): 672–684.

29 Almeida, A.J. and Souto, E. (2007). Solid lipid nanoparticles as a drug delivery system for peptides and proteins. *Adv. Drug Delivery Rev.* 59 (6): 478–490.

30 Sarmento, B., Martins, S., Ferreira, D., and Souto, E.B. (2007). Oral insulin delivery by means of solid lipid nanoparticles. *Int. J. Nanomed.* 2 (4): 743–749.

31 Zhang, N., Ping, Q., Huang, G. et al. (2006). Lectin-modified solid lipid nanoparticles as carriers for oral administration of insulin. *Int. J. Pharm.* 327 (1-2): 153–159.

32 Garcia-Fuentes, M., Torres, D., and Alonsa, M.J. (2005). New surface-modified lipid nanoparticles as delivery vehicles for salmon calcitonin. *Int. J. Pharm.* 296 (1-2, 132): 122.

33 Pandey, R., Sharma, S., and Khuller, G.K. (2005). Oral solid lipid nanoparticle-based antitubercular chemotherapy. *Tuberculosis* 85 (5-6): 415–420.

34 Yang, S., Zhu, J., Lu, Y. et al. (1999). Body distribution of camptothecin solid lipid nanoparticles after oral administration. *Pharm. Res.* 16: 751–757.

35 Lerata, M.S., D'Souza, S., Sibuyi, N.R. et al. (2020). Encapsulation of variabilin in stearic acid solid lipid nanoparticles enhances its anticancer activity in vitro. *Molecules* 25 (4): 830.

36 Khaleeq, N., Din, F.-U., Khan, A.S. et al. (2020). Development of levosulpiride-loaded solid lipid nanoparticles and their in vitro and in vivo comparison with commercial product. *J. Microencapsulation* 37 (2): 160–169.

37 Al-Heibshy, F., Basaran, E., Arslan, R. et al. (2020). Physicochemical characterization and pharmacokinetic evaluation of rosuvastatin calcium incorporated solid lipid nanoparticles. *Int. J. Pharm.* 578: 119106.

38 Father Abd-Ellatef, G.-E., Gazzano, E., Chirio, D. et al. (2020). Curcumin-loaded solid lipid nanoparticles bypass p-glycoprotein mediated doxorubicin resistance in triple negative breast cancer cells. *Pharmaceutics* 12 (2): 96.

39 Rawat, M., Singh, D., Saraf, S., and Saraf, S. (2008). Lipid carriers: a versatile delivery vehicle for proteins and peptides. *Yakugaku Zasshi* 128: 269–280.

40 Piao, H., Hirata, A., Yokoyama, H. et al. (2007). Reduction in gastric ulcerogenicity during multiple administration of diclofenac sodium by a novel solid-in-oil suspension. *Pharm. Dev. Technol.* 12 (3): 321–325.

41 Alhamami, O.M. (2003). Effects of oils and pharmaceutical excipients on the bioavailability of ampicillin orally administered, different oily and suspensions in rabbit. *Drug Dev. Ind. Pharm.* 29 (1): 51–60.

42 Xu, X., Zhang, S., Wang, C. et al. (2021). Pharmacokinetics of toltrazuril and its metabolites after oral and parenteral administration of novel oil-based suspension based on micro-environmental pH-modifying solid dispersion in rabbits. *Vet. Parasitol.* 299: 109580. https://doi.org/10.1016/j.vetpar.2021.109580.

43 Rohatagi, S., Barrett, J.S., Sawyers, W. et al. (1997). Pharmacokinetics of ipriflavone and metabolites after oral administration of corn oil suspension relative to the Osteofix tablet. *Am. J. Therapeut.* 4 (7-8): 229–238.

44 Surendra Prasad Peravali (2017). Lipid-based suspension can unlock new horizons in oral bioavailability https://www.ns-healthcare.com/analysis/potential-in-lipid-based-suspension-research-5963540/.

45 Shah, A.K. and Agnihotri, S.A. (2011). Recent advances and novel strategies in pre-clinical formulation development: an overview. *J. Control. Release* 156 (3): 281–296. https://doi.org/10.1016/j.jconrel.2011.07.003.

46 Cruz-Sanmartin, M.R., Perez-Martinez, J.I., and Rabasco, A.M. (2015). Preparation and characterization of an oily suspension of omeprazole for administration in pediatrics. *Int. J. Pharm. Sci. Res.* 6 (10): 4216–4225. https://doi.org/10.13040/IJPSR.0975-8232.6(10).4216-25.

47 Constantinides, P.P., Chaubal, M.V., and Shorr, R. (2008). Advances in lipid nanodispersions for parenteral drug delivery and targeting. *Adv. Drug Delivery Rev.* 60 (6): 757–767.

48 Muller, E.A., Kovarik, J., Uresin, Y. et al. (1997). Optimizing the absorption of valspodar, a P-glycoprotein modulator, Part I: selecting an oral formulation and exploring its clinical pharmacokinetics/dynamics. *J. Clin. Pharmacol.* 37 (11): 1001–1008.

49 Karasula, H.Y., Karabulat, B., Göker, E. et al. (2007). Controlled release of methotrexate from w/o microemulsion and its in vitro antitumor activity. *Drug Delivery* 14 (4): 225–233.

50 Araya, H., Tomita, M., and Hayashi, M. (2005). The novel formulation design of o/w microemulsion for improving the gastrointestinal absorption of poorly water soluble compounds. *Int. J. Pharm.* 305 (1-2): 61–74.

51 Gao, Z.G., Choi, H.G., Shin, H.J. et al. (1998). Physicochemical characterization and evaluation of a microemulsion system for oral delivery of cyclosporine A. *Int. J. Pharm.* 161 (1): 75–86.

52 Khoo, S.-M., Shackleford, D.M., Porter, C.J. et al. (2003). Intestinal lymphatic transport of halofantrine

occurs after oral. administration of a unit-dose lipid-based formulation to fasted dogs. *Pharm. Res.* 20: 1460–1465.

53 Constantinides, P.P., Welzel, G., Ellens, H. et al. (1996). Water-in-oil microemulsions containing medium-chain fatty acids/salts: formulation and intestinal absorption enhancement evaluation. *Pharm. Res.* 13: 210–215.

54 Zheng, J.Y. and Fulu, M.-Y. (2006). Decrease of genital organ weights and plasma testosterone levels on rats following oral administration of leuprolide microemulsion. *Int. J. Pharm.* 307 (2): 209–215.

55 Kim, S.K., Lee, E.H., Vaishali, B. et al. (2005). Tricaprylin microemulsion for oral delivery of low molecular weight haparin conjugates. *J. Controlled Release* 105 (1-2): 32–42.

56 Dokania, S. and Joshi, A.K. (2014). Self-microemulsifying drug delivery system (SMEDDS) – challenges and road ahead. *Drug Delivery* 22 (6): 675–690.

57 Rahman, A., Hussain, A., and Hussain, S. (2012). Role of ecipients in successful development of self-emulsifying/,icroemulsifying drug delivery system (SEDDS/SMEDDS). *Drug Delivery* 39 (1): 1–19.

58 Betageri, G.V. (2019). Self-emulsifying drug delivery systems and their marketed products: a review. *Asian J. Pharm. (AJP)* 13 (02). https://doi.org/10.22377/ajp.v13i02.3102.

59 Narang, A.S., Delmare, D., and Gao, D. (2007). Stable drug encapsulation in micelles and microemulsions. *Int. J. Pharm.* 345 (1-2): 9–25.

60 Gershanik, T. and Benita, S. (2000). Self-dispersing lipid formulations for improving oral absorption of lipophilic drugs. *Eur. J. Pharm. Biopharm.* 50 (1): 179–188.

61 Gursoy, R.N. and Benita, S. (2004). Self-emulsifying drug delivery systems(SEDDS) for improved oral delivery of lipophilic drugs. *Biomed. Pharmacother.* 58 (3): 173–182.

62 Tang, B., Cheng, G., Gu, J.C., and Xu, C.H. (2008, July). Development of solid self-emulsifying drug delivery systems: preparation techniques and dosage forms. *Drug Discovery Today* 13 (13-14): 606–612.

63 Khedekar, K. and Mittal, S. (2013). Self-emulsifying drug delivery system: a review. *Int. J. Pharm. Sci. Res.* 4 (12): 4494–4507.

64 Pouton, W.C. (2006). Formulation of poorly water-soluble drugs for oral administration: physicochemical and physiological issues and the lipid formulation classification system. *Eur. J. Pharm. Sci.* 29 (3-4): 278–287.

65 Sharma, G., Wilson, K., Van der Walle, C.F. et al. (2010). Microemulsions for oral delivery in streptozotocin induced diabetic rats. *Eur. J. Pharm. Biopharm.* 76 (2): 159–169.

66 Porter, C.J., Pouton, C.W., Cuine, J.F., and Charman, W.N. (2008). Enhancing intestinal drug solubilisation using lipid-based delivery systems. *Adv. Drug Delivery Rev.* 60 (6): 673–691.

67 Zhang, P., Liu, Y., Feng, N., and Xu, J. (2008). Preparation and evaluation os self-emulsifying drug delivery system of oridonin. *Int. J. Pharm.* 355 (1-2): 269–276.

68 Kohli, K., Chopra, S., Dhar, D. et al. (2010). Self-emulsifying drug delivery systems: an approach to enhance oral bioavailability. *Drug Discovery Today* 15 (21-22): 958–965.

69 Constantinides. (1995). Lipid microemulsions for improving drug dissolution and oral absorption: physical and biopharmaceutical aspects. *Pharm. Res.* 12 (11): 1561–1572.

70 Joshi, R.R. and Devarajan, V.P. (2015). Anionic self micro-emulsifying drug delivery system (SMEDDS) of docetaxel for circulation longevity. *J. Pharm. Drug Delivery Res.* 4 (3).

71 Balakrishnan, P., Lee, B.-J., Oh, D.H. et al. (2009). Enhanced oral bioavailability of dexibuprofen by a novel solid self-emulsifying drug delivery system (SEDDS). *Eur. J. Pharm. Biopharm.* 72 (3): 539–545.

72 Elnaggar, Y.S., El-Massik, M.A., and Abdallah, O.Y. (2009). Self-nanoemulsifying drug delivery systems of tamoxifen citrate: design and optimization. *Int. J. Pharm.* 380 (1-2): 133–141.

73 Karamustafa, F. and Celebi, N. (2008). Development of an oral microemulsion formulation of alendronate: effects of oil and co-surfactant type on phase behaviour. *J. Microencapsulation* 25 (5): 315–323.

74 Carriere, F., Barrowman, J.A., Verger, R., and Laugier, R. (1993). Secretion and contribution to lipolysis of gastric and pancreatic lipases during a test meal in humans. *Gastroenterology* 105 (3): 876–888.

75 Mullertz, A. (2007). Lipid-based drug delivery systems: choosing the right in vitro tools. *Am. Pharm. Rev.* 10 (3): 102–107.

76 Staggers, J.E., Hernell, O., Stafford, R.J., and Carey, M.C. (1990). Physical-chemical behavior of dietary and biliary lipids during intestinal digestion and absorption. 1. Phase behavior and aggregation states of model lipid systems patterned after aqueous duodenal contents of healthy adult human beings. *Biochemistry* 29: 2028–2040.

77 Tarr, B.D. and Yalkowsky, S.H. (1989). Enhances instestinal absorption of clyclosporine in rats through the reduction of emulsion droplet size. *Pharm. Res.* 6 (1): 40–43.
78 Christensen, J.Ø., Schultz, K., Mollgaard, B. et al. (2004). Solubilisation of poorly water-soluble drugs during in vitro lipolysis of medium and long chain triacyglycerols. *Eur. J. Pharm. Sci.* 23 (3): 287–296.
79 MacGregor, K.J., Embleton, J.K., Lacy, J.E. et al. (1997). Influence of lipolysis on drug absorption from the gastro-intestinal tract. *Adv. Drug Delivery Rev.* 25: 33–46.
80 Porter, C.J. and Charman, W.N. (2001). In vitro assessment of oral lipid based formulations. *Adv. Drug Delivery Rev.* 50 Suppl 1, 50: S127–S147.
81 Charman, S.A., Charman, W.N., Rogge, M.C. et al. (1992). Self-emulsifying drug delivery systems: formulation and biopharmaceutic evaluation of an investigational lipophilic compound. *Pharm. Res.* 9 (1): 87–93.
82 Khoo, S.M., Humberstone, A.J., Porter, C.J. et al. (1998). Formulation design and bioavailability assessment of lipophilic self-emulsifying formulations of Halofantrine. *Int. J. Pharm.* 167: 155–164.
83 Chae, G.S., Lee, J.S., Kim, S.H. et al. (2005). Enhancement of solubility of BCNU using self-emulsifying drug delivery system (SEDDS) and in vitro antitumor activity of self-emulsified BCNU-loaded PLGA wafer. *Int. J. Pharm.* 301 (1-2): 6–14.
84 Li, P., Ghosh, A., Wagner, R. et al. (2005). Effect of combined use of nonionic surfactant on formation of oil-in-water microemulsions. *Int. J. Pharm.* 288: 27–34.
85 Patel, A.R. and Vavia, P.R. (2007). Preparation and in vivo evaluation of SMEDDS (self-microemulsifying drug delivery systems). *AAPS J.* 9 (3): E344–E352.
86 Chen, Y., Li, G., Wu, X. et al. (2008). Self-microemulsifying drug delivery system (SMEDDS) of vinpocetine: formulation development and in vivo assessment. *Biol. Pharm. Bull.* 31 (1): 118–125.
87 Nikolakakis, I. and Partheniadis, I. (2017). Self-emulsifying granules and pellets: Composition and fomration mechanisms for instant or controlled release. *Pharmaceutics* 9 (4): 50.

# 30

# Oral Delivery of Nanoparticles: Challenges and Opportunities

*Qing Lin, Ling Zhang, and Zhirong Zhang*

*Department of Pharmaceutics, West China School of Pharmacy, College of Polymer Science and Engineering, Sichuan University, Chengdu, China*

## 30.1 Introduction

Oral administration appears as a preferred route for drug delivery, largely owing to the greater level of convenience and more patient compliance when self-administration is allowed and dosage can be easily adjusted. Therefore, oral administration is most suitable for the treatment of chronic diseases. The convenience also exists in the manufacturing processes where it does not require strict sterile conditions, leading to much lower manufacturing costs. The oral drug delivery market is projected to reach US$150.063 billion by 2024 from US$100.050 billion in 2018, growing at a compound annual growth rate of 6.99% [1]. Therefore, it is anticipated that development of oral dosage forms will continue to dominate the global pharmaceutical market. Rising health expenditure and geriatric population will further boost the market growth in the forecast years.

Physiological features of the gastrointestinal (GI) tract facilitate oral drug delivery. Absorptive enterocytes and the villi formed along the GI tract provide extensive luminal surface area – it has been long considered 180–300 $m^2$ [2], though recently reported that the total mucosal surface of the digestive tract interior might average 30–40 $m^2$, which is still more than enough for uptake of nutrients and drugs [3, 4]. In addition, various types of cells such as goblet cells, Paneth cells, and specialized M cells may also participate in drug absorption [5]. However, properties of drug molecules themselves and physiological barriers formed based on those cells and tissues would also be barriers to absorption of some active pharmaceutical ingredients (APIs) [6] (Figure 30.1). To cross the mucus barrier, it requires API of appropriate aqueous solubility and membrane permeability. Only molecules equipped with both robust chemical stability and suitable metabolic stability would be able to survive the conditions of diverse pH in the GI tract ranging from 1.2–7.6 and the variable GI layer thickness. Chemical design and drug discovery screening still dominate the research and development of the small molecular drugs yet the final success rate from Phase I initiated to FDA approved is generally around 7.9% [6]. Oral delivery is even more challenging for delivering macro-biopharmaceuticals resulting from their high molecular weight and low permeability probably due to high hydrophilicity, as well as unfavorable physicochemical and biological stability. Development of biologics such as protein-/nuclear-based molecules would need considering their manufacture and storage costs.

Away from remolding drug molecules, platforms based on nanoparticles have provided broader possibilities for development in oral drug delivery. Massive research reported oral delivery conducted by nanoparticles, however, to date a few of nanoparticle preparations have successfully appeared on the market, such as albumin-bounded nanoparticles (e.g. Abraxane) and liposomal preparations (e.g. Doxil, DaunoXome, and Marqibo). Relatively easily uptaken but difficultly transported, nanoparticle pharmaceuticals seem quite limited to preferably apply in product translation. To help better understand the current situation, several factors possibly affecting absorption and transport of orally administrated nanoparticles have been discussed in the limited length; *in vitro* and *in vivo* methods to characterize oral nanoparticle systems also make sense in preclinical investigations. Furthermore, parts of state-of-the-art carriers and delivery systems have been introduced, and finally, suspending challenges have been given. The readers might be left a first impression – challenges seem presenting along almost every stage, and meanwhile, opportunities coexist.

*Oral Bioavailability and Drug Delivery: From Basics to Advanced Concepts and Applications*, First Edition. Edited by Ming Hu and Xiaoling Li.

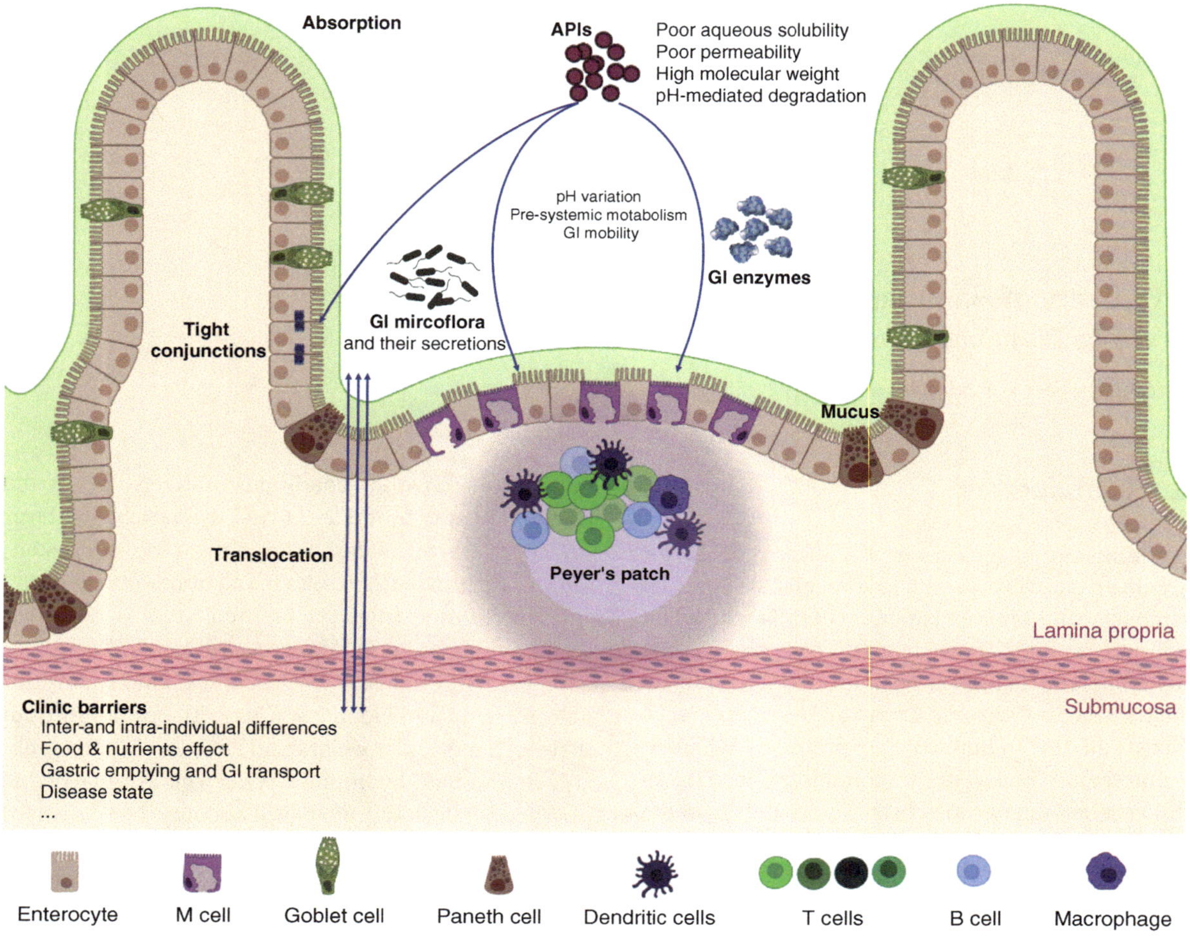


**Figure 30.1** Challenges and barriers affecting oral drug delivery.

However, all those given texts in the chapter could be regarded as a reference that triggers readers' interest in digging a wider and deeper range of literature learning as well as exploring any occurring thoughts in this research field.

## 30.2 Role of Nanoparticle Shape, Size, and Surface in Oral Delivery of Nanoparticles

Up-to-date studies have agreed that the uptake and transport of particles is strongly dependent on the particle size, shape, and physicochemical properties themselves.

Nanoparticles of disparate geometries could have diverse flow properties in the body and could accumulate in organs and tissues differently as well as interact with targeted cells dissimilarly [7]. Compared to spherical nanoparticles, worm-like nanoparticles appeared more applicable in inhibiting phagocytosis by macrophages [8]. Still compared with spherical particles, rod-shaped polystyrene ones displayed stronger cellular uptake and transport across intestinal cells. Cellular uptake of rods and disks was 2-fold higher than spherical ones by Caco-2 cells, where the difference was 3-folds between rods and spheres in Caco-2/HT-29 coculture [9].

Utilizing different-sized nanoparticles for oral protein delivery, 300 nm in size or less seemed ideal for oral delivery due to better internalization by intestinal epithelial cells and M cells, thus possibly to conduct enhanced intestinal transport, compared to larger-sized particles whose transport across Caco-2/HT-29 monolayers was negligible [10]. Similar results appeared in studies adopting nanoparticles of diverse materials for oral delivery of various APIs.

Nanoparticles with either positive or negative zeta potential may process different affinity with types of cells. Generally, as the potential of cell membrane presents slightly negative, positively surficial properties might favor the uptake; however, possible cytotoxicity induced by cationic materials needs taking into consideration. Nanoparticles modified with corresponding targeting ligands could realize more efficient uptake types of cancer cells as well as enhanced accumulation at metastatic tumor sites [11].

The diversity of cells uptaking these particles might possibly be explained through a view of mechanics, including both theoretical analysis and molecular simulations, though it might appear more ideal. It was reported a maximum of five distinct wrapping phases based on the stability of full wrapping, partial wrapping, and no wrapping states. Conclusively, it was easier for stiffer particles to achieve full wrapping than softer ones, which experienced smaller energy changes during wrapping and might be more favorable in case full wrapping is not necessary [12]. For shaped nanoparticles, the uptake speed also depended on its contact angle when interacting with cell membrane. For instance, rod-like particles would get through cell membrane more easily when in the radial than in the lateral [12].

Although it seems easy for these particles in the process of uptake, transcytosis might be far harder. Intact solid lipid nanoparticles or PLGA ones when administrated orally seemed not to travel across Caco-2 monolayers after quick uptake by intestinal cells [13, 14]. The transport of nanoparticles across Caco-2 monolayers or Caco-2/HT-29 cells was negligible for all particles investigated regardless of shape and ligand conjugation [9]. Contrarily, M cells have high endocytic ability and were reported to demonstrate robust capability for transcytosis of both small molecules and macro-biologics as well as nanoparticles.

## 30.3 Characterization Methods of Nanoparticles for Oral Delivery

Proper choice of characterization techniques for evaluating nanoparticle formulations *in vitro* and *in vivo* is crucial in preclinic stage. Methods commonly adopted to assess the orally administrated preparations have been detailly reviewed [15, 16].

*In vitro* models are relatively easy to acquire and of higher efficiency in preliminary screening and of lower costs. Among the massive *in vitro* models, Caco-2 monolayer might be used most widely in simulating intestinal mucus epithelial barriers in Transwells [17]. However, in the case of nanoparticles, one major limitation of Caco-2 monolayers is that the cells are literally enterocytes while lacking mucus-secreting goblet cells located among those epithelial cells. In addition, it is also implied that M cells probably function importantly in nanoparticle uptake and absorption, leading researchers to develop Caco-2 cells cocultured with M-cell-like ones. It was reported that nanoparticle translocation in Caco-2/Raji or Caco-2/HT29-MTX/M cell cocultures was increased compared to cultures with Caco-2 alone [18–20]. Therefore, to choose and develop a proper Caco-2-cell-based model would also be wise when assessing and predicting a nanoplatform.

A new *in vitro* model named as "organ-on-a-chip" is a microfluidic cell culture device combining 3D cell culture with microfabrication techniques, aiming to better simulate tissue–tissue interfaces, mechanical environment, and vascular perfusion [21]. Besides mimicking intestinal epithelium or stomach permeability, this system also incorporates local conditions inside the tract, such as flow of GI fluids, which could be a crucial limiting factor participating in uptake and retention of nanoparticles [22, 23]. Utilizing artificial microenvironment and microfluidics, a coculture of Caco-2 and HT29-MTX cells to engineer "intestine," and HepG2/C3A cells to act as "liver" were created as well as multiple chambers simulating body fluid from other related organs [24]. Than 50 nm polystyrene nanoparticles were detected to have crossed the "intestine" by flowing them through the apical side of the "intestine" and also detected to have translocated through the "liver" [24].

However, an indispensable constitution of preclinical development for a new-formulated oral-delivered nanoparticle preparation is animal experimentations. *In vivo* studies are of complex nature, which demands utmost careful designs and tests, as well as ethical records; and to properly select animal models and methodology is associated with final reliable results. Therefore, an overview to illustrate rationale and operations of animal models and *in vivo* techniques for oral nanoparticle delivery is for your reference [16].

To which degree to connect the relationship of the tested *in vitro* features between *in vivo* properties had aroused concerns over the years. In 1950s, a concept of *in vitro* – *in vivo* correlation (IVIVC) was introduced initially for oral solid dosage forms [25]. IVIVC models and computing have kept advancing and perfecting [26]. Besides forecasting absorption of nanoparticles after oral administration, predicting *in vivo* properties for more complicated systems would be more challenging, such as nanoparticles eliminated rapidly or remaining intact when uptaken or translocated through intestinal epithelium. In addition, behaviors of whatever drug preparations including nanoparticle formulations would definitely differ among animal species, where another challenge exists – how to precisely converse data acquired from animal studies to human beings are still in progress.

## 30.4 State-of-the-Art Carriers Designed and Applied in Oral Delivery of Nanoparticles

Recent developments in oral nanoparticles targeting to specific sites in the GI tract have kept progressing. Most of the preparations formulated to stay in the stomach consist of materials that are mucoadhesive, thus to extend gastric retention time [27–29]. Besides, site-specific release systems have also been developed to target gastric tumor cells, focusing on efficient treatment of gastric carcinogenesis [30]. Nanoparticles targeting to small intestinal tracts have been widely investigated, possibly due to the fact that this part of the GI tract undertakes major responsibility in systemic absorption of large-ranged APIs no matter small-molecular drugs or biologics with macro molecular weights. In the past two decades, the formulations of oral nanoparticles were mostly based on enteric polymeric materials or those combatively constructed with other polymers [31–34]. More recently, articles reported fictionized nanoparticles with specific targeting ligands in order to enhance uptake or transmembrane transport specially aiming at intestinal epithelial cells, mucus-secreting goblet cells, or M cells [35–37]. Along the digestive tract, colon-located nanoplatform might be investigated most extensively. Colon could be a local therapeutic site for treatment of inflammatory bowel disease. In this stage, enteric-coated nanoparticles with various modifications and functionalization boosted [38, 39].

Apart from specifically delivering agents to specific sites of the GI tract, targeting to intestinal lymphatic system (ILS) also help to improve oral bioavailability of the APIs, especially those would experience heavy enzymatic degradation and ineluctable hepatic first-pass effect [40]. ILS circumvents all over the liver and drugs are allowed to directly enter systemic circulation via thoracic duct. Nanoparticles fabricated from lipids or polymers would be commonly transported into ILS through lacteal and Peyer's patches, so strategies like surface modification of particles with ligands special for Peyer's patches would obviously improve the drug captured in ILS.

As noticed earlier, overcoming barriers in oral absorption could always be a challenging in endocytosis and transcytosis of nanoparticles even after they are already at the desired GI site. To get through the mucosal barrier, modulating nanoparticle surface properties would contribute. Generally, the strategies mainly conclude as application of mucoadhesion in mucosal drug delivery, mucus-disrupting approach to enhance mucus-diffusion, and engineer mucus-penetrating nanoparticles inspired from nature [41]. In addition, different shaped nanoparticle may also induce different endocytosis when interacting with cells as described before. Furthermore, it was experimentally proven that nanoparticle transcytosis could be improved by tailoring particles' elasticity combined with biomimetic surface [42].

## 30.5 Challenges and Coexisting Opportunities

Oral drug delivery, owing to its noninvasiveness and high patient compliance, is a highly sought-after route for drug administration. To take the advantages, nanoparticles have been widely investigated for oral delivery of small molecules and some biopharmaceuticals. However, except that barriers to nanoparticle drugs remain a problem, challenges still exist elsewhere.

Recently, high-efficient delivery of biopharmaceuticals seems to have attracted unprecedented research focuses, and thus, extension of platforms robust enough in efficiently delivering wider ranges of molecules would be a spot. Due to the features of those macromolecules themselves such as structure complexity and multifaceted instability, it would require the constructed nanoplatform could not only precisely deliver the macromolecules to specific sites but also efficiently help them penetrate through cell layers and barriers to enhance the APIs' absorption and translocation.

Another concern might be undisclosed toxicity that nanoparticles may induce when interacting with cells and remaining in tissues. What would happen as long as nanoparticles contact with and enter cells has been caught a glimpse among hundreds of reports. Those phenomena may differ in cellular and molecular levels – the majority of the research progress has focused on unfavorable apoptosis, pyroptosis, and ferroptosis [43–45], which also possibly happen among normal tissues. A list of methods has been reviewed regarding assessments for *in vitro* and *in vivo* toxicity of nanoparticles yet they still need to keep updating [46].

In recent decades, there exists ubiquitous interest in oral nanoparticle delivery, witnessing a mountainous pile of publications and billions of dollars contributed in this field. But unfortunately, as of today the first oral nanoparticle preparation that is FDA-approved and intends to apply in the clinics has still been on the way. To address this gap, limitations in clinical translation must be recognized. For design of nanoparticle formulations, one important challenge practically that is often not considered in multiple kinds of research is maintaining simplicity in preparation. Complexity could often introduce variability both in manufacture and storage, even more importantly from the perspective of clinical translation, complexity can limit the nanoparticle preparation from helping patients. Caution is supposed to be taken that a number of preclinical studies have been conducted at a basic science level; by comparison, a disproportionately low number of studies would further proceed to clinical trials. Clinical trials might suffer from the same concerns as unsolved – safety remains the superior important obstacle to clinical translation, where those related to nanoparticle formulations' toxicity and safety cannot be neglected since this alone could stop the drug from being introduced into human body.

The GI tract undertakes multiple tasks and presents a wide array of functionalities, and oftentimes, to design and construct oral nanoparticle formulations that take advantage of or work in relation with physiological and biological processes might lead to more favorable outcomes [47]. In addition, proper selection of *in vitro* and *in vivo* experimental methods is quite critical for preclinical evaluations. As methodology continues to advance, more clear pathways to clinical studies are projected to emerge. However, with no doubt will the tremendous potential for nanomedicine to positively impact the effectiveness of oral drug delivery.

## References


1 Knowledge Sourcing Intelligence LLP Oral Drug Delivery Market - Forecasts from 2019 to 2024. 2019.

2 Duthie, G. and Gardner, A. (2006). *Physiology of the Gastrointestinal Tract*. Wiley.

3 Helander, H.F. and Fändriks, L. (2014). Surface area of the digestive tract–revisited. *Scand. J. Gastroenterol.* 49 (6): 681–689.

4 https://medicalxpress.com/news/2014-04-surface-area-digestive-tract-large.html (accessed on 03 October 2020).

5 Sehgal, A., Donaldson, D.S., Pridans, C. et al. (2018). The role of CSF1R-dependent macrophages in control of the intestinal stem-cell niche. *Nat. Commun.* 9 (1): 1–17.

6 Informa Pharma Intelligence New Clinical Development Success Rates 2011-2020 Report. 2021.

7 Blanco, E., Shen, H., and Ferrari, M. (2015). Principles of nanoparticle design for overcoming biological barriers to drug delivery. *Nat. Biotechnol.* 33 (9): 941.

8 Barua, S., Yoo, J.-W., Kolhar, P. et al. (2013). Particle shape enhances specificity of antibody-displaying nanoparticles. *Proc. Natl. Acad. Sci.* 110 (9): 3270–3275.

9 Banerjee, A., Qi, J., Gogoi, R. et al. (2016). Role of nanoparticle size, shape and surface chemistry in oral drug delivery. *J. Controlled Release* 238: 176–185.

10 He, C., Yin, L., Tang, C., and Yin, C. (2012). Size-dependent absorption mechanism of polymeric nanoparticles for oral delivery of protein drugs. *Biomaterials* 33 (33): 8569–8578.

11 Kolhar, P., Anselmo, A.C., Gupta, V. et al. (2013). Using shape effects to target antibody-coated nanoparticles to lung and brain endothelium. *Proc. Natl. Acad. Sci.* 110 (26): 10753–10758.

12 Yi, X., Shi, X., and Gao, H. (2011). Cellular uptake of elastic nanoparticles. *Phys. Rev. Lett.* 107 (9): 098101.

13 Hu, X., Fan, W., Yu, Z. et al. (2016). Evidence does not support absorption of intact solid lipid nanoparticles via oral delivery. *Nanoscale* 8 (13): 7024–7035.

14 He, B., Lin, P., Jia, Z. et al. (2013). The transport mechanisms of polymer nanoparticles in Caco-2 epithelial cells. *Biomaterials* 34 (25): 6082–6098.

15 Wong, C.Y., Al-Salami, H., and Dass, C.R. (2020). Cellular assays and applied technologies for characterisation of orally administered protein nanoparticles: a systematic review. *J. Drug Targeting* 1–15.

16 Shrestha, N. and Préat, V. (2020). In vivo testing of orally delivered nanoparticles. In: *Nanotechnology for Oral Drug Delivery*, 459–480. Elsevier.

17 Tan, H.-Y., Trier, S., Rahbek, U.L. et al. (2018). A multi-chamber microfluidic intestinal barrier model using Caco-2 cells for drug transport studies. *PLoS One* 13 (5): e0197101.

18 Beloqui, A., Brayden, D.J., Artursson, P. et al. (2017). A human intestinal M-cell-like model for investigating particle, antigen and microorganism translocation. *Nat. Protoc.* 12 (7): 1387.

19 Schimpel, C., Teubl, B., Absenger, M. et al. (2014). Development of an advanced intestinal in vitro triple culture permeability model to study transport of nanoparticles. *Mol. Pharmaceutics* 11 (3): 808–818.

20 Damge, C., Maincent, P., and Ubrich, N. (2007). Oral delivery of insulin associated to polymeric

nanoparticles in diabetic rats. *J. Controlled Release* 117 (2): 163–170.

21 Bhatia, S.N. and Ingber, D.E. (2014). Microfluidic organs-on-chips. *Nat. Biotechnol.* 32 (8): 760–772.

22 Delon, L.C., Guo, Z., Oszmiana, A. et al. (2019). A systematic investigation of the effect of the fluid shear stress on Caco-2 cells towards the optimization of epithelial organ-on-chip models. *Biomaterials* 225: 119521.

23 Lourenço, B.N., Dos Santos, T., Oliveira, C. et al. (2018). Bioengineering a novel 3D in vitro model of gastric mucosa for stomach permeability studies. *Acta Biomater.* 82: 68–78.

24 Esch, M.B., Mahler, G.J., Stokol, T., and Shuler, M.L. (2014). Body-on-a-chip simulation with gastrointestinal tract and liver tissues suggests that ingested nanoparticles have the potential to cause liver injury. *Lab Chip* 14 (16): 3081–3092.

25 Jain, P., Pawar, R.S., Pandey, R.S. et al. (2017). In-vitro in-vivo correlation (IVIVC) in nanomedicine: is protein corona the missing link? *Biotechnol. Adv.* 35 (7): 889–904.

26 Suarez-Sharp, S., Li, M., Duan, J. et al. (2016). Regulatory experience with in vivo in vitro correlations (IVIVC) in new drug applications. *AAPS J.* 18 (6): 1379–1390.

27 Zhou, Y., Zhu, X., Lin, S. et al. (2020). A novel nanoparticle preparation to enhance the gastric adhesion and bioavailability of xanthatin. *Int. J. Nanomed.* 15: 5073.

28 Angsantikul, P., Thamphiwatana, S., Zhang, Q. et al. (2018). Coating nanoparticles with gastric epithelial cell membrane for targeted antibiotic delivery against *Helicobacter pylori* infection. *Adv. Ther.* 1 (2): 1800016.

29 Cao, S.-J., Xu, S., Wang, H.-M. et al. (2019). Nanoparticles: oral delivery for protein and peptide drugs. *AAPS PharmSciTech* 20 (5): 190.

30 Lin, Y.-H., Chen, Z.-R., Lai, C.-H. et al. (2015). Active targeted nanoparticles for oral administration of gastric cancer therapy. *Biomacromolecules* 16 (9): 3021–3032.

31 Sampathkumar, K., Riyajan, S., Tan, C.K. et al. (2019). Small-intestine-specific delivery of antidiabetic extracts from *Withania coagulans* using polysaccharide-based enteric-coated nanoparticles. *ACS Omega* 4 (7): 12049–12057.

32 Gou, J., Liang, Y., Miao, L. et al. (2018). The promoting effect of enteric materials on the oral absorption of larotaxel-loaded polymer-lipid hybrid nanoparticles. *Eur. J. Pharm. Sci.* 124: 288–294.

33 Li, C., Zhou, K., Chen, D. et al. (2019). Solid lipid nanoparticles with enteric coating for improving stability, palatability, and oral bioavailability of enrofloxacin. *Int. J. Nanomed.* 14: 1619.

34 Salah, E., Abouelfetouh, M.M., Pan, Y. et al. (2020). Solid lipid nanoparticles for enhanced oral absorption: a review. *Colloids Surf., B* 111305.

35 Verma, A., Sharma, S., Gupta, P.K. et al. (2016). Vitamin B12 functionalized layer by layer calcium phosphate nanoparticles: a mucoadhesive and pH responsive carrier for improved oral delivery of insulin. *Acta Biomater.* 31: 288–300.

36 García-Díaz, M., Birch, D., Wan, F., and Nielsen, H.M. (2018). The role of mucus as an invisible cloak to transepithelial drug delivery by nanoparticles. *Adv. Drug Delivery Rev.* 124: 107–124.

37 Xia, D., Tao, J., He, Y. et al. (2015). Enhanced transport of nanocage stabilized pure nanodrug across intestinal epithelial barrier mimicking *Listeria monocytogenes*. *Biomaterials* 37: 320–332.

38 Gugulothu, D., Kulkarni, A., Patravale, V., and Dandekar, P. (2014). pH-sensitive nanoparticles of curcumin–celecoxib combination: evaluating drug synergy in ulcerative colitis model. *J. Pharm. Sci.* 103 (2): 687–696.

39 Vong, L.B., Mo, J., Abrahamsson, B., and Nagasaki, Y. (2015). Specific accumulation of orally administered redox nanotherapeutics in the inflamed colon reducing inflammation with dose–response efficacy. *J. Controlled Release* 210: 19–25.

40 Managuli, R.S., Raut, S.Y., Reddy, M.S., and Mutalik, S. (2018). Targeting the intestinal lymphatic system: a versatile path for enhanced oral bioavailability of drugs. *Expert Opin. Drug Delivery* 15 (8): 787–804.

41 Wu, L., Shan, W., Zhang, Z., and Huang, Y. (2018). Engineering nanomaterials to overcome the mucosal barrier by modulating surface properties. *Adv. Drug Delivery Rev.* 124: 150–163.

42 Zheng, Y., Xing, L., Chen, L. et al. (2020). Tailored elasticity combined with biomimetic surface promotes nanoparticle transcytosis to overcome mucosal epithelial barrier. *Biomaterials* 262: 120323.

43 Li, L., Sun, S., Tan, L. et al. (2019). Polystyrene nanoparticles reduced ROS and inhibited ferroptosis by triggering lysosome stress and TFEB nucleus translocation in a size-dependent manner. *Nano Lett.* 19 (11): 7781–7792.

44 Kim, S.E., Zhang, L., Ma, K. et al. (2016). Ultrasmall nanoparticles induce ferroptosis in nutrient-deprived cancer cells and suppress tumour growth. *Nat. Nanotechnol.* 11 (11): 977.

**45** Ploetz, E., Zimpel, A., Cauda, V. et al. (2020). Metal–organic framework nanoparticles induce pyroptosis in cells controlled by the extracellular pH. *Adv. Mater.* 1907267.

**46** Kumar, V., Sharma, N., and Maitra, S.S. (2017). In vitro and in vivo toxicity assessment of nanoparticles. *Int. Nano Lett.* 7 (4): 243–256.

**47** Date, A.A., Hanes, J., and Ensign, L.M. (2016). Nanoparticles for oral delivery: design, evaluation and state-of-the-art. *J. Controlled Release* 240: 504–526.

# 31

# Oral Delivery of Therapeutic Peptides: Strategies for Product Development

*Puchun Liu*

*PL Consulting – Pharmaceutical Leverage, New York, USA*

## 31.1 Introduction

According to the US Food and Drug Administration (FDA), "peptide" is generally considered as any α-amino acid polymer with specific defined sequence of 40 amino acids or less in size. Comparing to proteins (polypeptides), peptides have fewer functions and less 3-D structure, are less likely to be posttranslationally modified, and are generally more easily characterized [1]. Differentiating from traditional small molecule drugs, "therapeutic peptides" are revolutionizing the biotech and pharmaceutical market, spurring the creation of next-generation products with reduced immunogenicity, improved safety, and greater effectiveness.

One less-attractive quality of therapeutic peptides is their delivery method, which for the vast majority of peptide products on the market is by "invasive" injection, often requiring frequent injections. There is consequently significant driving force in developing "noninvasive" routes of administration for therapeutic peptides. The per oral (PO) route of therapeutic peptides would be simpler and more convenient for systemic delivery, and additionally, safer and more effective for pre-systemic local actions when compared with "invasive" injectable routes. For many decades, if not a century [2], most often developed as injectable formulations, a few "oral peptide" drugs have been approved. Advances in peptide delivery technologies have contributed to growing interest, and many more are in development. Extra and continuous efforts have been taken to enable oral peptide delivery technologies by overcoming major challenges associated with no or poor permeability across diffusion barriers and no or poor stability against degradation barriers for the peptides in gastrointestinal tract (GIT) [3]. Better scientific understanding of breakthrough technologies to improve the product developability justifies continued efforts to identify and optimize opportunities for innovation. Successful oral drug delivery technologies will bring more peptide therapeutics to fruition and be available to serve the needs of the patients.

Table 31.1 summaries the commercially available oral administrated products of 8 peptides that the US FDA has approved as New Drug Application (NDA) starting from cyclosporine (Sandimmune® solution and soft gelatin capsule) in 1983 and 1990, and (Neoral® solution and soft gelatin capsule) in 1995. Cyclosporine is a natural peptide calcineurin inhibitor known for its immunomodulatory properties that prevent organ transplant rejection and treat various inflammatory and autoimmune conditions. The success achieved with cyclosporine formulations remains one clear example of what can be achieved. However, for nearly 30 years since then, only 3 additional oral peptides have been approved by the FDA. It has been realized that effective oral peptide products remain highly compound-specific.

Nevertheless, the FDA's approval of 5 more oral peptides and 1 formulation (Firvanq® solution of previously approved vancomycin) during the last 10 years represents a great progress in oral peptide products after our last update in the first edition of this book [4].

As NDA 505(b)(1) regulatory pathway [Type 1 New Molecular Entity (NME)], the FDA approved 2 intestinal locally acting guanylin hormones: linaclotide (Linzess® capsule) in 2012 [5] and plecanatide (Trulance® tablet) in 2017 [6] for treatment of irritable bowel syndrome with constipation (IBS-C) and chronic idiopathic constipation (CIC). Both synthetic peptides are potent agonists of guanylate cyclase C (GC-C) receptor expressed on the luminal aspect of the intestinal epithelium [7–9]. At the beginning of 2021, the FDA approved 1 systemic action NME voclosporin (Lupkynis™ capsule) in combination

*Oral Bioavailability and Drug Delivery: From Basics to Advanced Concepts and Applications*, First Edition. Edited by Ming Hu and Xiaoling Li.

**Table 31.1** Commercially Available Oral Products of Therapeutic Peptides (NDA Approved by the US FDA).

| Product (NDA approval year, number, type, company) | Indication (action site, systemic BA) | Peptide [molecular weight (MW), isoelectric point (pI)] | Peptide structure and sequence |
|---|---|---|---|
| *Sandimmune® solution*<br>(**1983,** NDA050574, Type 3 New Dosage Form, Novartis Pharmaceuticals Corp.)<br>*Sandimmune® soft gelatin capsule*<br>(**1990,** NDA050625, Type 3 New Dosage Form, Novartis Pharmaceuticals Corp.)<br>*Neoral® solution, Neoral® soft gelatincapsule* (**1995,** NDA050715, Type 3 New Dosage Form, Novartis Pharmaceuticals Corp.) | Immunosuppression<br>(***systemic delivery,*** *absolute BA 20-40%*) | **Cyclosporine**<br>(MW 1203, no pI, acidic pKa 11.8) | 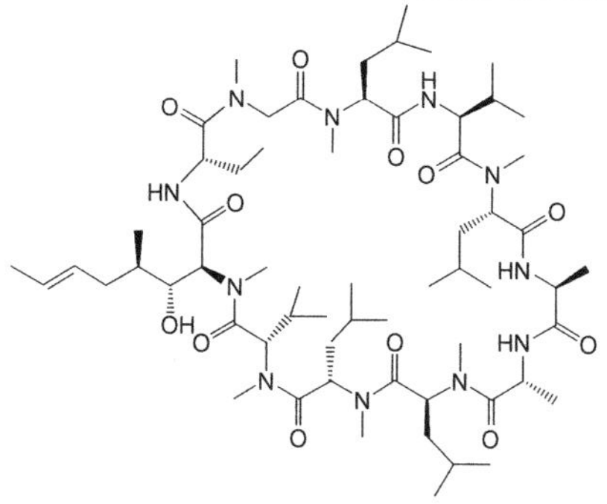<br>[R-[R*,R*-(*E*)]]-cyclic-(L-Ala-D-Ala-*N*-methyl-L-Leu-*N*-methyl-L-Leu-*N*-methyl-L-Val-3-hydroxy-*N*,4-dimethyl-L-2-amino-6-octenoyl-L-α-aminobutyryl-*N*-methyl-Gly-*N*-methyl-L-Leu-L-Val-*N*-methyl-L-Leu) |
| *Vancocin® HCl capsule*<br>(**1986,** NDA050606, Type 3 New Dosage Form, Ani Pharmaceuticals, Inc.)<br>*Firvanq® solution*<br>(**2018,** NDA208910, Type 5 New Formulation or New Manufacturer, Azurity Pharmaceuticals Inc.) | Antibiotic (Clostridium difficile-associated diarrhea; Enterocolitis caused by Staphylococcus aureus including methicillin-resistant strains)<br>(***local action - small intestines***) | **Vancomycin**<br>(MW 1449, pI 8.5)<br>(tricyclic glycopeptide) | 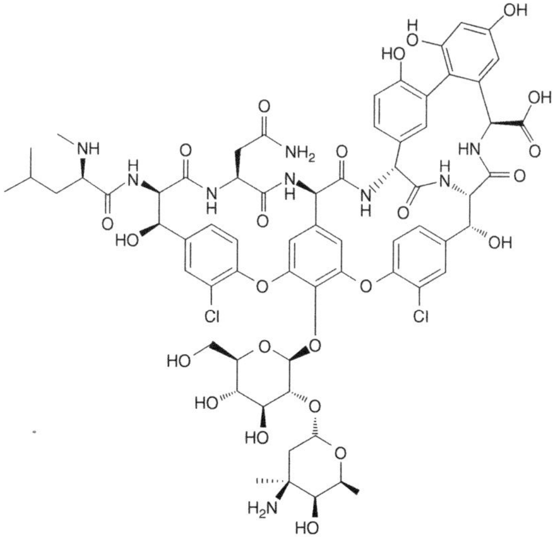<br>22H-8,11:18,21-Dietheno-23,36-(iminomethano)-13,16:31,35-dimetheno-1H,16H-[1,6,9]oxadiazacyclohexadecino[4,5-*m*][10,2,16]benzoxadiazacyclotetracosine-26-carboxylic acid, 3-(2-amino-2-oxoethyl)-44-[[2-*O*-(3-amino-2,3,6-trideoxy-3-C-methyl-α-L-lyxo-hexopyranosyl)-β-D-glucopyranosyl]oxy]-10,19-dichloro-2,3,4,5,6,7,23,24,25,26,36,37,38,38a-tetradecahydro-7,22,28,30,32-pentahydroxy-6-[[4-methyl-2-(methylamino)-1-oxopentyl]amino]-2,5,24,38,39-pentaoxo-, [3S-[3R*,6S*(S*),7S*,22S*,23R*,26R*,36S*,38aS*]] |

| | | | |
|---|---|---|---|
| *Ddavp*® *tablet*<br>(**1995,** NDA019955, Type 3 New Dosage Form, Ferring Pharmaceuticals Inc.) | Central diabetes insipidus, primary nocturnal enuresis<br>(***systemic delivery***, *absolute BA 0.08-0.16%*) | **Desmopressin**<br>(MW 1069, pI 10.2) | 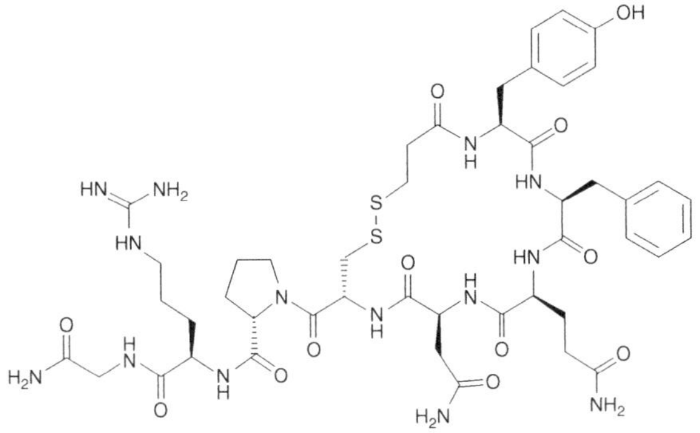<br>Gly-amide, *N*-(3-mercapto-1-oxopropyl)-L-Tyr-L-Phe-L-Gln-L-Asn-L-Cys-L-Pro-D-Arg-, cyclic (1→5)-disulfide |
| *Linzess*® *capsule*<br>(**2012,** NDA202811, Type 1 NME, Allergan Sales LLC) | Irritable bowel syndrome with constipation (IBS-C); chronic idiopathic constipation (CIC)<br>(***local action - small intestines***) | **Linaclotide**<br>(MW 1527, pI 3.6) | 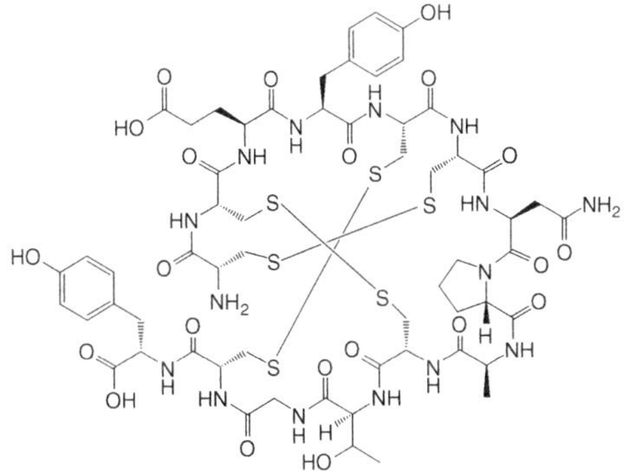<br>L-Cys-L-Cys-L-Glu-L-Tyr-L-Cys-L-Cys-L-Asn-L-Pro-L-Ala-L-Cys-L-Thr-Gly-L-Cys-L-Tyr, cyclic (1-6), (2-10), (5-13)-tris (disulfide) |
| *Trulance*® *tablet*<br>(**2017,** NDA208745, Type 1 NME, Salix PharmaceuticalsInc.) | Irritable bowel syndrome with constipation (IBS-C); chronic idiopathic constipation (CIC)<br>(***local action - small intestines***) | **Plecanatide**<br>(MW 1682, pI 3.1) | 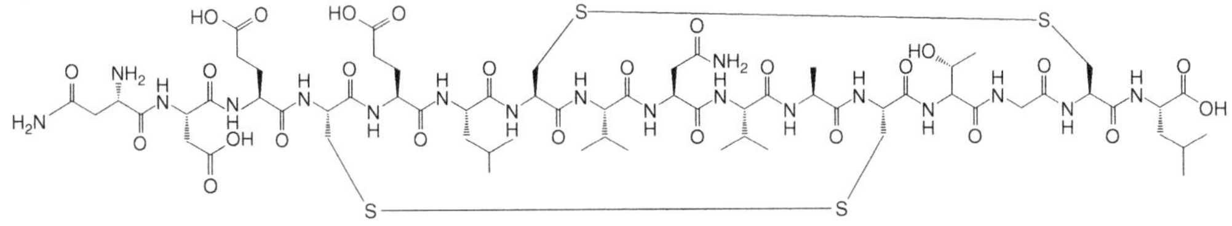<br>L-Leu, L-Asn-L-α-Asp-L-α-Glu-L-Cys-L-α-Glu-L-Leu-L-Cys-L-Val-L-Asn-L-Val-L-Ala-L-Cys-L-Thr-Gly-L-Cys-, cyclic (4→12), (7→15)-bis(disulfide) |

Table 31.1 (Continued)

| Product (NDA approval year, number, type, company) | Indication (action site, systemic BA) | Peptide [molecular weight (MW), isoelectric point (pI)] | Peptide structure and sequence |
|---|---|---|---|
| *Rybelsus*® *tablet* (**2019/2020,** NDA213051/ 213182, Type 3 New Dosage Form, Novo Nordisk Inc.) | Type 2 diabetes mellitus (T2DM) (***systemic delivery***, *absolute BA 0.4-1%*) | **Semaglutide** (MW 4114, pI 5.4) | 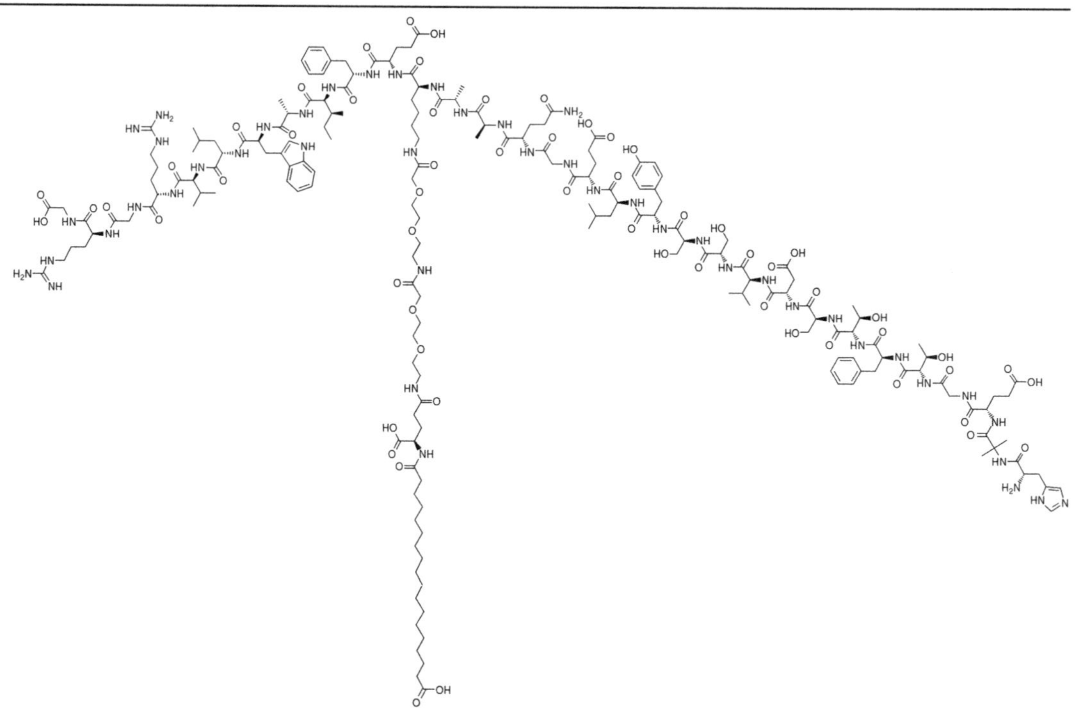 H-His-Aib-Glu-Gly-Thr-Phe-Thr-Ser-Asp-Val-Ser-Ser-Tyr-Leu-Glu-Gly-Gln-Ala-Ala-Lys(C18diacid-γ-Glu-OEG-OEG)]-Glu-Phe-Ile-Ala-Trp-Leu-Val-Arg-Gly-Arg-Gly-OH |

| | | | |
|---|---|---|---|
| *Mycapssa® delayed release capsule*<br>(**2020**, NDA208232, Type 5 New Formulation or New Manufacturer, Chiasma Inc.) | Acromegaly<br>(***systemic delivery***, *absolute BA ~0.5%*) | **Octreotide**<br>(MW 1029, pI 11.7) | 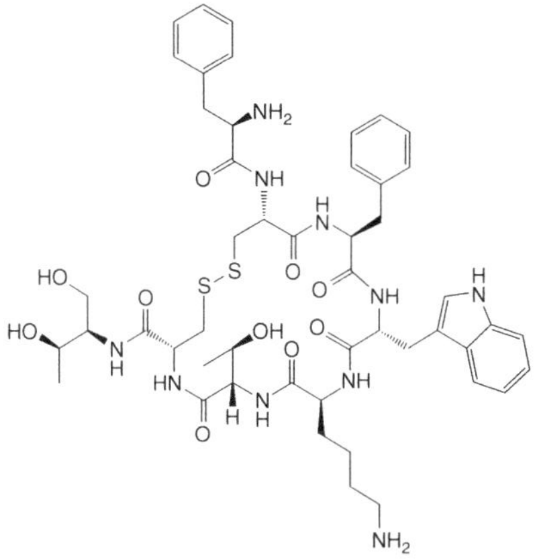<br>H2N-D-Phe-Cys-Phe-D-Trp-Lys-Thr-Cys-Thr-OH, cyclic (2→7)-disulfide |
| *Lupkynis® capsule*<br>(**2021**, NDA213716, Type 1 NME, Aurinia Pharmaceuticals, Inc.) | Lupus nephritis (immunosuppression)<br>(***systemic delivery,*** *absolute BA ?%*) | **Voclosporin** (MW 1215, no pI, acidic pKa 11.8 | 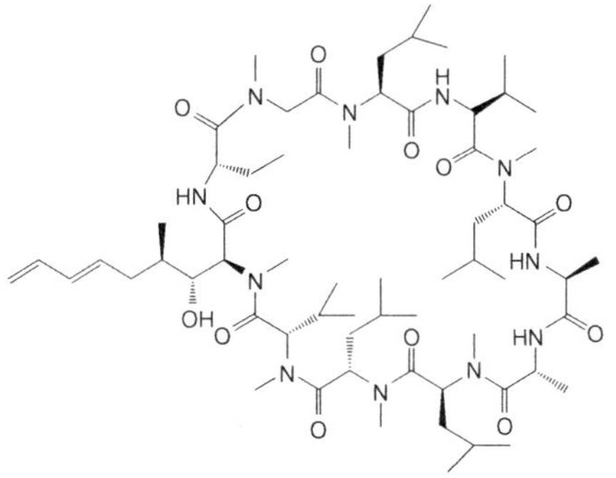<br>Cyclo{{(6E)-(2S,3R,4R)-3-hydroxy-4-methyl-2-(methylamino)-6,8-nonadienoyl}-L-2- aminobutyryl-*N*-methyl-glycyl-*N*-methyl-L-leucyl-L-valyl-*N*-methyl-L-leucyl-L-alanyl-D-alanyl-*N*-methyl-L-leucyl-*N*-methyl-L-leucyl-*N*-methyl-L-valyl} |

with a background immunosuppressive therapy regimen to treat adult patients with active lupus nephritis. Voclosporin (an analog of cyclosporine) works by blocking interleukin-2 expression and T-cell-mediated immune responses and stabilizing the podocyte in the kidney. As an NME first approved (as a product) by the FDA, the 505(b)(1) process is an application that contains full reports of investigations of safety and effectiveness.

As NDA 505(b)(2) regulatory pathway, the FDA approved 2 systemic action products (Rebelsus® in 2019, Type 3 New Dosage Form; Mycapssa® in 2020, Type 5 New Formulation or New Manufacturer) converting injectable therapeutics into oral formulations. Semaglutide (Rebelsus® tablet) is a glucagon-like peptide-1 (GLP-1) receptor agonist indicated as an adjunct to diet and exercise to improve glycemic control in adults with type 2 diabetes mellitus (T2DM) [10–12]. Octreotide (Mycapssa® delayed-release capsule) is a somatostatin analog indicated for long-term maintenance treatment in acromegaly patients [13–16]. The 505(b)(2) pathway also contains full safety and effectiveness reports but allows at least some of the information required for approval to come from previous studies not conducted by or for the applicant. The 505(b)(2) process can take an existing drug molecule and makes some modifications in the product, often significantly advancing the medication for the patients' benefit, such as changes in dosage form, strength, formulation, dosing regimen or route of administration, new combination products, and new therapeutic indications.

This chapter begins with an overview of approaches to enabling oral peptide delivery focusing on key issues of protecting chemical stability and enhancing physical permeability of therapeutic peptides in GIT after PO administration. An observation of low and variable bioavailability (BA) data with pharmacokinetics (PK) analysis reveals the peptides' instability problem being the main cause of the large inter/intra-individual variations. Knowing the current limitations of the most advanced state-of-the-art technologies, the author provides several recommendations on the strategies for oral peptide product development for readers to consider.

## 31.2 Overview of Approaches to Enabling Oral Peptide Delivery

Imparting good oral delivery of peptides while maintaining their pharmacological efficacy is the first key challenge that has hindered the development of oral peptides into useful therapeutic products. Technological feasibility to enabling oral peptide delivery depends on simultaneous protection of peptides' stability against degradation barriers and enhancement of peptides' permeability across diffusion barriers in the GIT [17].

The peptides' (acidic, enzymatic, and microbial) degradation barriers in the human GIT (stomach, small intestine, and colon) lumen are well-written in Kimball's biology textbook [18]. The stomach wall is lined with millions of gastric glands. Parietal cells secrete hydrochloric acid and intrinsic factors. Chief cells synthesize and secrete pepsinogen, the precursor to an endopeptidase pepsin. Pepsin breaks down proteins into small peptides, favoring peptide bonds on the C-terminal side of tyrosine, phenylalanine, and tryptophan. The liquefied contents of the stomach pass into the duodenum where most of ingested vitamins and minerals are absorbed. Two ducts enter the duodenum: the gall bladder (and hence the liver) and the exocrine portion of the pancreas. Bile contains bile acids (emulsify ingested fat), sodium bicarbonate (neutralizes the acidic contents from stomach), amylase (hydrolyzes starch), lipase (hydrolyzes ingested fats), and nucleases (hydrolyze ingested nucleic acids). Most important to the peptides, pancreatic fluid has four "zymogens" that are immediately converted into 4 active proteolytic enzymes: (i) trypsin cleaves peptide bonds on the C-termini of arginine and lysine, (ii) chymotrypsin cuts on the C-termini of tyrosine, phenylalanine, and tryptophan, (iii) elastase cuts peptide bonds next to alanine and serine, and (iv) carboxypeptidase removes, one by one, the amino acids at the C-termini of peptides. The epithelial cells in small intestine have villi and microvilli on their surface which provide a large area of absorption for the nutrients and water, and also contain various other enzymes and transporters. The large intestine or colon contains an enormous population of microorganisms (mainly Firmicutes and Bacteroidetes). Most of the species live there perfectly harmlessly, and some are actually beneficial as they synthesize vitamins and digest polysaccharides for which we have no enzymes. Reabsorption of water is the chief function of the colon; the large amounts of water secreted into the GIT by the various digestive glands must be reclaimed to avoid dehydration.

Figure 31.1 is a diagram showing human GIT's diffusion barriers (mucus layers, mucosal/epithelium), in addition to (acidic, enzymatic, and microbial) degradation barriers for oral peptide delivery. Continuously secreted by goblet cells along the GIT, mucus acts as a lubricant to facilitate the passage of digestive matter and to protect the underlying epithelium from pathogens and mechanical stress.

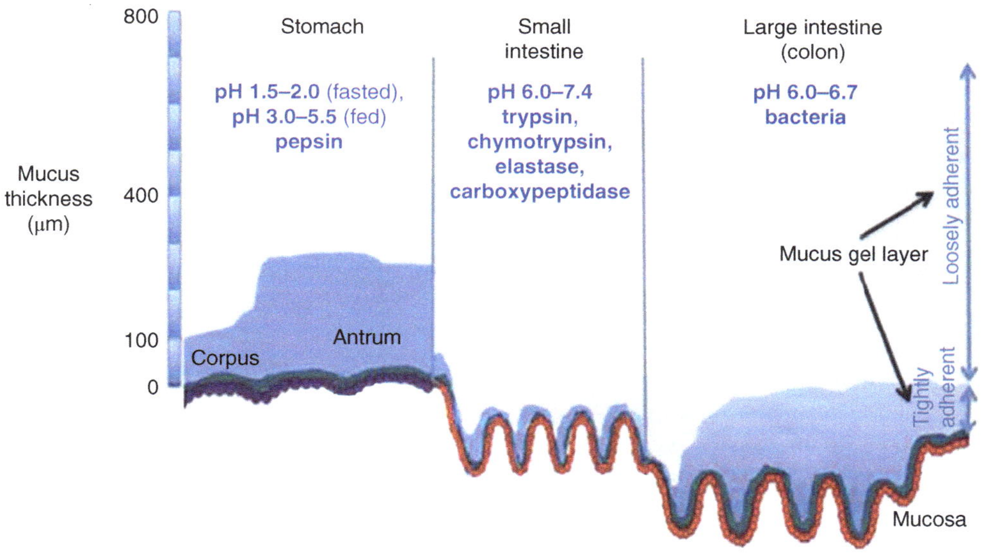


**Figure 31.1** A diagram of human GIT (stomach, small intestine, and colon), labeling degradation barriers (lumen pH, peptidases, and bacteria) and diffusion barriers (mucus, mucosal/epithelium) and showing the thickness of mucus layers: (i) stomach: inner, tightly adherent layer 100–300 μm; outer, loosely adherent layer 200 μm. (ii) small intestine: inner, tightly adherent layer negligible; outer, loosely adherent layer 100–200 μm. (iii) colon: inner, tightly adherent layer 200–300 μm; outer, loosely adherent layer 150–700 μm. Source: Adapted from Moran et al. [19].

Mucus is composed of water and mucin protein molecules coated with proteoglycans, which gives the mucus a negative charge. The mucus layer is not only a physical barrier limiting particles' movements based on their sizes but also selectively binds with particles through both hydrophilic and lipophilic interactions. The heterogeneity of mucus components offers a wide range of potential chemical interaction sites for peptides, while the mesh-like architecture given to mucus by the intermolecular cross-linking of mucin molecules results in a dense network that physically, and in a size-dependent manner, hinders the diffusion of nanoparticles through mucus [20]. While the small intestine has only one thin layer of loosely attached mucus, the stomach (protects the mucosal tissue from gastric acid) and colon (provides a stable environment for the enteric microflora) have a thick two-layered mucus system comprising of an inner, tightly adherent layer and an outer, loosely attached layer [19, 21].

Attempts have been carried out to enable oral peptide delivery by protecting peptides' chemical stability against (acidic, enzymatic, and microbial) degradation barriers and enhancing peptides' physical permeability across (mucus, epithelium) diffusion barriers in the GIT. Very recently, Brayden et al. [22] published a comprehensive review on the formulation and medicinal chemistry approaches for systemic delivery of peptides by oral route. Readers are also referred to additional new review articles for the details [23–32].

Oral delivery for pre-systemic liver targeting provides a unique advantage for certain peptides. First, oral delivery mimics the physiological path of endogenous secretion of some peptide hormones. For example, oral insulin is expected to have a distinct effect on the hepatic aspect of carbohydrate metabolism, hepatic insulin resistance, and at the same time, avoid hyperinsulinemia and minimize the risk of hypoglycemia [33]. Second application is for peptide drugs of hepato-portal destination. The effects of various antiviral and chemotherapy peptides could be improved if high concentration in the liver is required through the mesenteric veins and the portal vein.

For local-acting peptides on GIT (stomach, small intestine, and colon) disorders, high local concentration in the lumen, epithelial, or subepithelial tissue optimize the therapy with reduced systemic exposure and toxicity [34, 35]. The oral route for new peptides targeting the receptors that are situated locally within the GIT lumen provides a unique opportunity for peptide therapeutics acting as topical agents with low systemic exposures and favorable safety profiles [7]. In particular, several authors have addressed the unmet medical need for oral peptides in colon to treat colon-associated diseases [8, 9, 24, 36].

### 31.2.1 Protecting Peptides' Stability Against GIT Degradation Barriers

For the acidic, enzymatic, and microbial degradation barriers in the GIT (Figure 31.1), enteric coatings can enable passage of peptides through the low pH stomach conditions, and little is known about bacteria degradation, which may be very important for the colon local action. Modulation of local pH and use of safe enzyme inhibitors are the two approaches to help minimize the peptides' breakdown by the GIT peptidases. Some stability–permeability combined approaches are described in Section 31.2.3 (Combining techniques for both stability and permeability of peptides) with not much effective breakthroughs in the technology development.

Instead of "direct inhibiting GIT enzymes," a more pragmatic approach is to protect the drugs from GIT metabolism without altering the natural flora of enzymes in the GIT. Pepsin in stomach is active at low pH, but it can be rapidly inactivated above pH 5. Used in Rebelsus® tablet (semaglutide), sodium *N*-(8-[2-hydroxybenzoyl]amino) caprylate (SNAC) – actually its sodium salt – increases the local stomach pH, transiently inactivating pepsin in the stomach. Intestinal and pancreatic enzymes trypsin, chymotrypsin, and elastase also have an optimal pH of neutral to basic to be active. Inclusion of acids in the formulation results in a local, transient decrease in pH, inactivating the resident peptidases.

The most clinically advanced example of direct enzyme inhibition as a component of an oral peptide formulation may be an oral insulin formulation developed by Oramed. This formulation includes soybean trypsin inhibitor as well as a chelating agent that scavenges calcium, which is a cofactor for many proteases [33]. However, side effects such as systemic intoxication and disturbed digestion of food proteins are often associated with the use of enzyme inhibitors [37].

### 31.2.2 Enhancing Peptides' Permeability Across GIT Diffusion Barriers

For most of peptides which are hydrophilic, mucus and epithelial layers are the diffusion barriers for systemic delivery (including pre-systemic liver targeting), while the mucus is the main barrier for GIT locally acting peptides (Figure 31.1). The peptide's "permeability" (defined as permeation rate or flux here for the simplicity) across the epithelium transcellularly (lipid-like membrane) is a product of its solubility and diffusivity, normalized by the barrier membrane thickness. The peptide's molecular size determines the diffusivity and its lipophilicity determines the solubility. When the peptide goes through epithelium paracellularly (aqueous pore), the peptide's permeability equals its diffusivity depending upon the molecular size relative to the pore size. For the transcellular pathway, the hydrophilic peptide molecule can be lipidized by transcellular permeation enhancers (or drug carriers) to increase the peptide solubility in the epithelium membrane. A paracellular permeation enhancer can transiently open the tight junction between cells to increase the peptide's diffusivity.

Used in Rebelsus® tablet (semaglutide), SNAC as the first of Eligen® drug carriers facilitate transcellular absorption through the gastric membrane enabling semaglutide to reach systemic circulation [11, 38, 39]. SNAC was previously approved by the FDA as a medical food (Eligen®-Vitamin B12). Other two Eligen® carriers, sodium 8-(*N*-2-hydroxy-5-chlorobenzoyl)-amino-caprylate (5-CNAC) and monosodium *N*-(4-chlorosalicyloyl)-4-aminobutyrate (4-CNAB) used for salmon calcitonin (sCT) and insulin, respectively, in clinical studies [40]. Unlike SNAC, sodium caprylate (C8) used in octreotide (Mycapssa® delayed-release capsule) [16] acts by transient alteration of epithelial tight junctions and/or membrane perturbation. The permeation mechanism that involves modulating the tight junction is not without safety concerns for treatment of chronic conditions. Once tight junctions have been opened, transport is enhanced not only for drugs but also for potentially toxic or unwanted molecules present in the GIT [41, 42].

Use of lipid-based formulation is another way to lipidize the peptide, enhancing the peptide's permeability across the epithelium. The lipid-based self-emulsifying formulations include bile salt stabilized vesicles [43], hydrophobic ion pairs [44–46], Zeta potential-changing system [47], and oily droplets which can be made mucoadhesive or highly mucus permeating [48]. Encapsulating the drug within colloidal particles (microemulsions, emulsions, solid lipid particles, liposomes, and microgels) is reported [28].

For GIT locally acting peptides, the overall hydrophilic and negatively charged mucus with hydrophobic domains presents an interactive barrier limiting the peptides to reach the epithelial target beneath the "tightly adherent" mucus layer [49–51]. The pore sizes of mesh-like mucus architecture mucus were reported to be around 100–200 nm [52]. Ionic liquids interact with the mucus layer resulting in its thinning [53]. Taking an example from nature, virus-mimicking nanoparticles with highly densely charged surfaces were designed to improve their mucus permeability [54].

Mucolytic enzyme-decorated systems decorated with mucin-cleaving enzymes such as papain or bromelain [55]. In terms of safety, a key property of a mucus-modulating delivery system is that it must not damage the protective function of the mucus layer [56].

### 31.2.3 Combining Techniques for both Stability and Permeability of Peptides

Peptide-entrapped nanoparticles were thought to combine techniques for both stability and permeability for systemic delivery of peptides [23, 31]. Many laboratory work and animal studies have been done with self-emulsified nano-carrier [57, 58], negatively charged nanoparticles [59], chemically modified nanocarriers with absorption enhancers and mucous adhesion [60], lipid-based nanocarriers [61], lipid-based nanosuspensions [62], hybrid lipid-mucoadhesive polymer nanoparticles [26], nanocarriers including polymeric nanoparticles, solid lipid nanoparticles, liposomes and micelles [63], Zeta potential-changing polyphosphate nanoparticles [64], cell-penetrating peptides with nanoparticles [65], conjugation of cell-penetrating peptides with nanoparticles [66], and nanotechnology-based formulations [67, 68]. However, administration in nanoparticle formats has largely disappointed with insufficient particle loading, lack of controlled release, low epithelial particle uptake, and lack of scalable synthesis [22, 69, 70].

Perhaps, the most interesting development is an ingestible self-orienting milliposts inspired by the leopard tortoise's ability to passively reorient [71]. Other devices are luminal unfolding microneedle [72], RaniPill™ microneedle [73], unidirectional mucoadhesive patch [74], microcontainer [75], ink-jet printed enteric device [76], and hand-held ultrasound device [77]. The disruptive technologies based on engineered devices promise improvements, but scale-up and toxicology aspects are issues to be addressed [22].

### 31.2.4 Optimizing Peptides' Structures for both Stability and Permeability

Cyclosporine (Table 31.1) provides an example where the drug is relatively stable from peptidase degradation in the GIT (rather, some oxidation by CYP450 enzymes). Understanding of how cyclosporin cross-cell membranes may unlock wider opportunities in drug discovery [78]. In human gastric fluid, the relatively large peptides (somatostatin, calcitonin, secretin, glucagon, and insulin) were metabolized rapidly, while the smaller peptides showed good stability. In human small intestinal fluid, however, both small and large peptides degraded rapidly with the exception of the cyclic peptide cyclosporin and the disulfide-bridge containing peptides octreotide and desmopressin (Table 31.1), which showed good stability [79].

Chemically modified peptides range from endogenous transport mechanisms to favorable interactions with the absorptive membrane at the cell surface [80]. Natural peptides typically have poor absorption, distribution, metabolism, and elimination (ADME) properties. Strategies have been developed to improve peptide drugability through enhancing permeability, reducing proteolysis and renal clearance, and prolonging half-life [3]. In vivo, in vitro, and in silico tools are available to evaluate ADME properties, and structural modification strategies are in place to improve peptide developability [81, 82]. Covalent methods include conjugating fatty acids via amidation, esterification, reversible aqueous lipidization, and cyclization [83].

Cyclization of peptides minimizes degradation in the GIT, blood, and tissues by removing cleavable N- and C-termini and by shielding components from metabolic enzymes also folds peptides into bioactive conformations that determine exposure of polar atoms to solvation by water and lipids, and therefore can influence permeation [84]. Lipophilic cyclic peptide natural products such as antimicrobials or human peptide hormones [85] introduce favorable physicochemical properties to peptidic macrocycles that maintain activity and oral BA [86]. In parallel, medicinal chemists are synthesizing stable hydrophobic macrocyclic candidate peptides of lower molecular weight (MW), permeability correlated well with the extent of intramolecular hydrogen bonding, the incorporation of γ-amino acids offers a route to increase backbone diversity and improve ADME properties [87].

By utilizing solution phase chemistry, peptides with less than 30 amino acids in length are significantly more economical to manufacture than longer peptides [88], reducing the cost of active pharmaceutical ingredient (API) for oral peptide products.

## 31.3 Observation and Data Analysis of Low BA with Large Variabilities

In addition to animal (rat, dog, and monkey) PK results previously published [4], we have reviewed those from human studies of several peptides using Eligen® technology (SNAC and its analogs such as 4-CNAB, 5-CNAC, and another), including sCT [89], insulin [90], human growth hormone (hGH) [91], and semaglutide [11, 39].

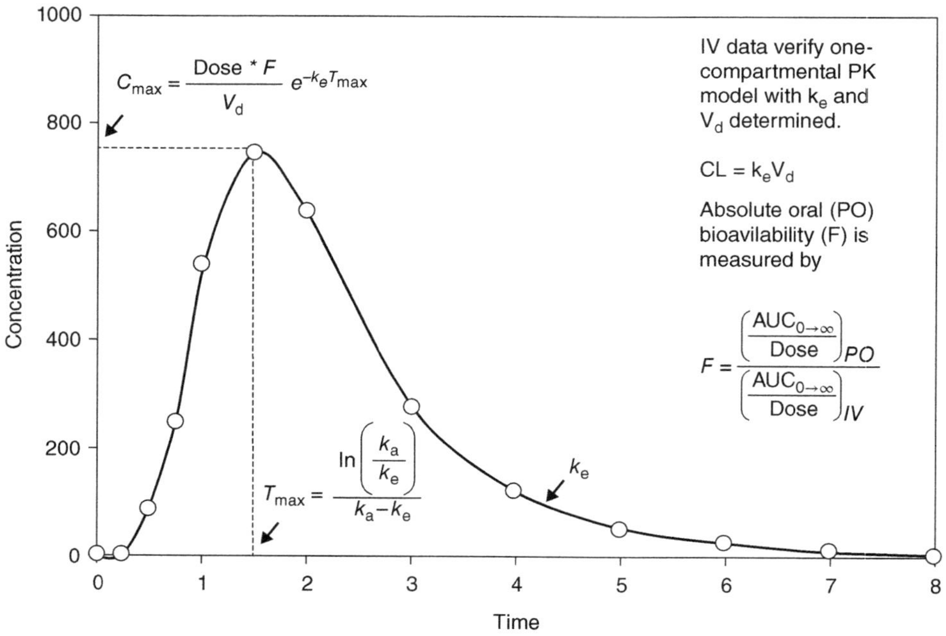


**Figure 31.2** A representative PK profile of oral peptide products in human and animals, showing all PK parameters to analyze the data observation of small variability in $T_{max}$ and large variability in $C_{max}$ and *AUC* (one-compartment PK model mathematics used for simplicity of the same concept as multi-compartment models).

Again, the important observation is a small variability in $T_{max}$, but a large variability in $C_{max}$ even from the same subject (intrasubject).

Figure 31.2 represents a typical oral peptide concentration–time PK profile, showing the peak concentration ($C_{max}$) and the time to achieve the peak concentration ($T_{max}$). For the purpose of simplicity, one-compartment model mathematics leads to the following expressions for $T_{max}$ and $C_{max}$ (the same concept applies to other compartment or non-compartment models):

$$T_{max} = \frac{\ln\left(\frac{k_a}{k_e}\right)}{k_a - k_e} \tag{31.1}$$

$$C_{max} = \frac{Dose^*F}{V_d} e^{-k_e T_{max}} \tag{31.2}$$

where $k_e$ and $k_a$ are the first-order elimination and absorption rate constants, respectively, $V_d$ is the volume distribution, Dose is the oral dose, and $F$ is the absolute oral systemic BA (expressed in fraction, comparing to equal-dose intravenous administration).

By assuming systemic delivery of typical GIT - portal vein - liver - systemic pathway (no enteric and enterohepatic recycling), the absolute oral systemic BA $F$ is given by

$$F = F_{abs}(1 - f_{GI})(1 - f_H) \tag{31.3}$$

where $F_{abs}$ is the fraction of drug absorbed into the GIT wall, $f_{GI}$ is the fraction metabolized in the GIT (lumen and intracellular), and $f_H$ is the fraction extracted during the first-pass hepatic metabolism. We measured these parameters in a rat model [92]. By comparing area under the curve (AUC) from intravenous (IV) versus intraportal injections of drug at equal doses, approximately 30% of insulin is extracted by the liver (that is, insulin in rat $f_H = 0.3$). From in situ stomach perfusion (after washout of pepsin) and portal sampling, an $F_{abs}$-related parameter, insulin permeability coefficient increases from $2 \times 10^{-7}$ cm/s (no SNAC) to $4 \times 10^{-6}$ cm/s (at 30 mg/ml SNAC), showing a very significant SNAC-enhancement of insulin "permeability" across rat stomach epithelium. From in vitro incubation with fresh rat stomach fluid, an $f_{GI}$-related parameter, insulin degradation half-life ($t_{1/2}$) increases from 0.5 min (no SNAC) to 2 minutes (at 25 mg/ml SNAC), showing a moderate SNAC (pH effect)-protection of insulin "stability" against stomach pepsin.

Based on Eqs. (31.1)–(31.3), an analysis was conducted and the results are summarized in Table 31.2. First, the small variation in $T_{max}$ suggests that variability associated with drug absorption ($k_a$) is reasonably small, and the contribution to $T_{max}$ from drug elimination ($k_e$) is independent of absorption. Hence, the

**Table 31.2** Data analysis of observed small variability on $T_{max}$ and large variability on $C_{max}$ and *AUC* of oral peptide products.

| PK or related biopharmaceutical parameters | | Drug dose-dependent[a)] | Carrier dose-dependent[a)] | Remarks |
|---|---|---|---|---|
| $t_{1/2}$ | $k_e$ | No | No | |
| $T_{max}$ | $k_e$ | No | No | |
| | $k_a$ | No | ***Yes*** | Small variability in $T_{max}$ suggests small variability in $k_a$ at the fixed carrier dose. |
| $C_{max}$, $AUC_{0\to\infty}$ | $k_e$ | No | No | |
| | $k_a$ | No | ***Yes*** | (see above) |
| | $V_d$ | No | No | |
| | Dose | Yes | No | |
| | $F$ | No | ***Yes*** | Large variability in $C_{max}$ (and *AUC*) is due to large variability in *F* at the fixed doses of both the drug and the carrier. |
| $F$ | $F_{abs}$ | No | ***Yes*** | With small variability in $k_a$ at the fixed carrier dose (see $T_{max}$), a large variability in drug concentration at the absorption site can cause large variability in $F_{abs}$, and then *F*. |
| | $f_{GI}$ | No | ***Yes*** | The weak enzyme inhibition by the carrier at the site of absorption suggests that the value of *F* is small and more sensitive to variability. |
| | $f_H$ | No | No | |

a) Assumed no saturation kinetics with relatively low concentrations of the drug and the carrier at absorption sites.

large variability in $C_{max}$ is then due to a large variability in *F*, since other parameters ($V_d$, *Dose*) are not related. It can be concluded that a large variability in drug concentration at the absorption site can lead to the large variation in $F_{abs}$ and *F*. Due to the enzymatic degradation, only a small fraction of drug (with large inter- and intra-subject variability) is available for absorption even if permeability is enhanced by SNAC or analogs. The transit, weak enzyme inhibition by SNAC or analogs (pH increase effect at local stomach by sodium salt) at the site of absorption suggests that the value of *F* is small and more sensitive to variability.

This conclusion is consistent with that by other investigators. Sugihara et al. [93] suggested that the factors affecting intra and intersubject variabilities include the relative expression of digestive enzymes, peptidases, and relevant transporters, potentially contributing to the timing and extent of GIT peptide delivery. Tyagi et al. [94] further concluded that physiological barriers within the GIT are the root causes of the low BA and high variability.

## 31.4 Recommended Strategies for Oral Peptide Product Development

Now the question is not "can the oral peptide products be developed?" but "what and how to do making it happen?" There have been many expert opinions in the literature, with some excerpts listed here. "The low bioavailability accompanied by large intra and intersubject variability leads to translational issues, preventing the development of successful peptide therapeutics" [95]. "Oral bioavailability remains low and variable therefore high, potentially toxic doses of peptide drugs are needed to elicit a therapeutic effect leading to high cost of the final product" [96]. "Bioavailability is still low though, even with the most advanced state-of-the-art technologies, limiting their application to high potency peptides with large therapeutic windows" [97]. "Criteria of peptide selection, formulation design, system components and excipients, intestinal mechanism of action, efficacy in man, and safety issues are only currently suitable for a limited range of peptides" [98]. "While promising formulation approaches have been studied, issues including reproducibility of response, intersubject variability, and duration of response require further optimization before commercially viable products are possible" [67]. "Oral delivery to local GIT targets is currently more promising because of their accessibility and the lacking requirement for intestinal permeability enhancement" [2]. "For those that meet the necessary criteria, advances in formulation technologies coupled with favorable market dynamics will continue to drive interest across the entire prescription drug spectrum for safe and effective orally administered peptide therapeutics" [99]. "Small and potent peptides that are resistant to degradation and that present a large therapeutic window certainly represent the best candidates for systemic absorption" [11]. "Desirable features of candidates should include a large therapeutic index, some stability in the GIT, a long elimination half-life, and a relatively low clearance rate" [22].

Each oral peptide product is unique and presents its own set of feasibility assessments and development challenges. A comprehensive understanding of the properties of each peptide is necessary to design the most effective dosage form and develop an optimal oral peptide product. It is critical to focus on the effort by selecting an appropriate peptide candidate from achieving its full potential, the merits of each application have to be examined individually in terms of clinical benefit, market potential, and technical feasibility [4, 100]. The positive and negative effects need to be weighed carefully before large expenditures for developmental work are committed. Outside the scope of this chapter, the important issue of peptide stability in the manufacturing process and during the shelf-life of the drug product needs to be evaluated. In the following, the author shares some thoughts related to the strategies for oral peptide product development.

### 31.4.1 Where is the Action Site? (distinguish systemic versus pre-systemic liver versus local GIT)

It is important to understand the peptide's therapeutic target or PD action site, distinguishing among systemic delivery, pre-systemic liver targeting, or local GIT targeting. Different approaches to enabling oral delivery, and even more different strategies for oral product development will be applied as to create the target product profile.

1) *Systemic delivery*: GI absorption needs both chemical stability (against acid + peptidase + microbial) and physical permeability (across mucus + epithelium); Systemic exposure determines both efficacy and safety.
2) *Pre-systemic portal vein/liver targeting*: GI absorption needs both chemical stability (against acid + peptidase + microbial) and physical permeability (across mucus + epithelium); Pre-systemic (mesenteric-portal vein to liver) exposure determines efficacy, systemic exposure determines (systemic) safety. Examples: some endogenous peptides targeting the liver (e.g. insulin for T2DM) or those peptides targeting hepato-portal destination (e.g. antiviral, chemotherapy).
3) *Local GIT targeting*: Local disorder site (stomach, small intestine, or colon) release (not absorption) needs chemical stability against local (acid, peptidases, and/or microbials if apply), but physical permeability across mucus; local disorder site exposure determines efficacy, systemic exposure determines (systemic) safety.

### 31.4.2 How Wide is the Therapeutic Window? (Ensure "Variable BA" not Efficacy/Safety Issue)

It is important to understand the therapeutic window of this peptide drug for this indication, ensuring that the inter and intrasubject variability on the oral BA is not an efficacy/safety issue, and is predictable. Except for the GIT local targeting, currently oral peptides may apply only to the clinical indications with *a wide therapeutic window*. Along with this strategy, the following are two special ("not to do" and "to do") situations for the oral peptide product development:

1. *Not apply to the indications of acute condition*, only consider those of chronic condition. Example: Insulin has a wide therapeutic window for T2DM, but has a very narrow therapeutic window for T1DM, so oral insulin currently is not ready to replace the injectable insulin for T1DM!
2. *Apply to the indications of repeat-dose based efficacy* which usually have a wide therapeutic window. Example: hGH needs daily exposure which repeats and cumulates for several days to raise the level of insulin-like growth factor I (IGF-1) for an efficacy response.

### 31.4.3 What is the Peptide's Structure? (Hope Cyclic-, Oligo- with Stable/Permeable Amino Acids)

It is important to understand the peptide's molecular structure at the amino acid level (size, shape, and amino acid residues), identifying the positive and negative elements contributing to the peptide's ultimate "stability" and "permeability" in the GIT [101]. While the amino acid evaluation below is based on the 20 most common ones (Table 31.3), the real-world peptides may be those with various attachments to the common amino acids (Table 31.1). For example, hydrophobic N-methylated amides in cyclosporine make it more permeable.

1. *Cyclic peptides:* are "more stable" and "more permeable" than their linear counterparts.
2. *Oligo peptides:* 2–20 amino acid residues are generally "more permeable" than the larger peptides with similar amino acids.
3. *"Less stable" amino acids*
   i) by pepsine in gastric fluids: phenylalanine (Phe), tyrosine (Tyr), tryptophan (Trp), or leucine (Leu); especially C-terminal bonding with Phe, Trp, or Tyr.
   ii) by trypsin in small intestinal fluids: arginine (Arg) or lysine (Lys).
   iii) by α-chymotrypsin in small intestinal fluids: tyrosine (Tyr), tryptophan (Trp), phenylalanine (Phe), leucine (Leu), or Methionine (Met).
   iv) by carboxypeptidase A in small intestinal fluids: all amino acids except those bonding with aspartate (Asp), glutamate (Glu), arginine (Arg), lysine (Lys), or Proline (Pro).
   v) by pancreatic elastase in small intestinal fluids: alanine (Ala), serine (Ser), glycine (Gly), or valine (Val).
4. *"More permeable" amino acids*: isoleucine (Ile), leucine (Leu), valine (Val); tyrosine (Tyr), phenylalanine (Phe), tryptophan (Trp).

### 31.4.4 Is the Technology Toolbox Good Enough? (Tackle Stability and Permeability Challenges)

It is important to understand the current status and future trends on the development of the most advanced state-of-the-art technologies, prioritizing to tackle the technical challenge on both (either) "stability" and (or) "permeability" with the knowledge of action site, therapeutic window, and peptide's structure. There are a couple of additional comments:

1. The "low BA" issue is a combined result of poor "stability" and poor "permeability" problems in the GIT. Interplay of the "stability" protection and "permeability" enhancement must be optimized and maximized. Long way to go, but it is critical that this combined feature as a function of time at the absorption site be studied at the cellular/molecular levels of peptide (amino acid), epithelium membrane and permeation enhancers, and peptidases and enzyme inhibitors.
2. The "large BA variability" issue is believed to be caused mainly by the variable "stability" problem (inter and intrasubject variability on the peptide concentration at the GIT absorption site). It is the author's opinion to tackle the "stability" challenge (not much breakthrough) more than the "permeability" (some success such as SNAC). Even a small improvement on the "stability" issue can result in a significant reduction of the BA variability.

### 31.4.5 Does it Worth? (Analyze "low BA" Cost in a Big Picture of Pharmacoeconomics)

It is important to understand not only additional cost due to the "low BA," but also a big picture of total cost, effectiveness, and benefit of an oral peptide product, either as new molecular entity [505(b)(1)] or converting from an existing injection product [505(b)(2)]. A cost-effectiveness analysis of oral semaglutide (Rybelsus®) has concluded that the additional costs (a dose on the order of 100 times) incurred to produce extra semaglutide for the oral product are cost-effective, given the greater quality of life experience when taking a daily oral over weekly injection [72].

Assuming that "oral" and "injectable" products have the same efficacy/safety profiles as the FDA approves, we are doing an exercise of "oral" versus "injection" cost-benefit analysis:

(1) *Cost (direct + indirect + intangible)*: "oral" increase/decrease from "injection:"
   i) Direct – "oral" increases: cost of larger quantity of API due to its "low BA; "oral" decreases: (a) API's unit price ("oral" versus "injection" quality grades), (b) cost of the product manufacturing ("oral" – no need of aseptic equipment/process), (c) cost of the product transportation and storage ("oral" solid dosage – no need of refrigerator temperature).
   ii) Indirect – "oral" decreases: cost of hospital/clinic visit and time for patient (personally administer or schedule appointments) and caregiver if "injection" is intravenous (IV), intramuscular (IM), or other nonself-injection.
   iii) Intangible – "oral" decreases: cost associated with "injection" pain, fear, inconvenience, associated risks, and higher incidence of adverse events.

(2) *Benefits (clinical + commercial)*: converting "injection" to "oral;"

**Table 31.3** 20 Most Common Amino Acids Commonly Found in Peptides.

| | | | | | |
|---|---|---|---|---|---|
| Amino Acids with *aliphatic-hydrophobic* side chains | 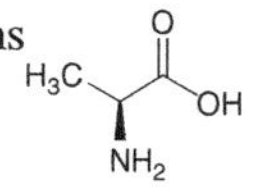<br>Alanine (Ala)<br>MW = 89, pI = 6.0 | 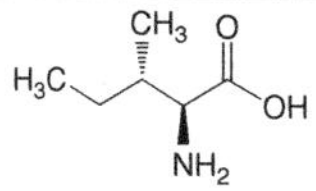<br>Isoleucine (Ile)<br>MW = 131, pI = 6.2 | 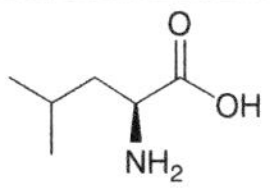<br>Leucine (Leu)<br>MW = 131, pI = 6.2 | 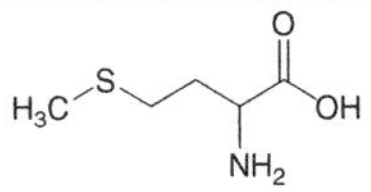<br>Methionine (Met)<br>MW = 149, pI = 6.0 | 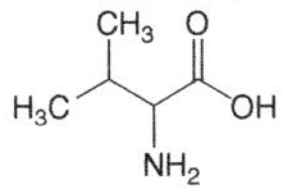<br>Valine (Val)<br>MW = 117, pI = 6.2 |
| Amino Acidswith *aromatic-hydrophobic* side chains | 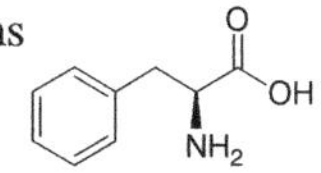<br>Phenylalanine (Phe)<br>MW = 165, pI = 6.0 | 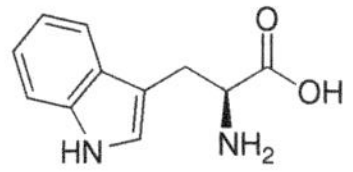<br>Tryptophan (Trp)<br>MW = 204, pI = 6.0 | 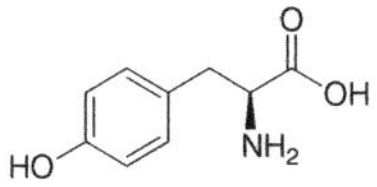<br>Tyrosine (Tyr)<br>MW = 181, pI = 5.5 | | |
| Amino Acidswith *polar neutral*side chains | 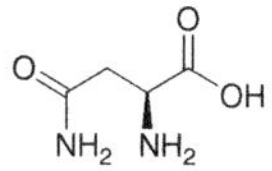<br>Asparagine (Asn)<br>MW = 132, pI = 5.2 | 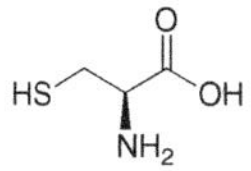<br>Cysteine (Cys)<br>MW = 121, pI = 5.7 | 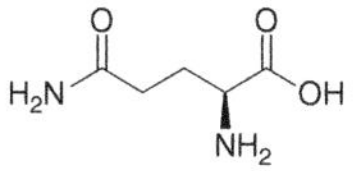<br>Glutamine (Gln)<br>MW = 146, pI = 5.7 | 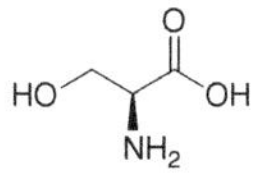<br>Serine (Ser)<br>MW = 105, pI = 5.7 | 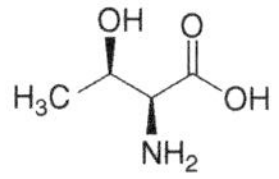<br>Threonine (Thr)<br>MW = 119, pI = 5.6 |
| Amino Acids with *negatively charged* side chains | 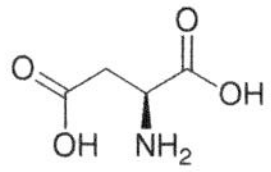<br>Aspartate (Asp)<br>MW = 133, pI = 3.4 | 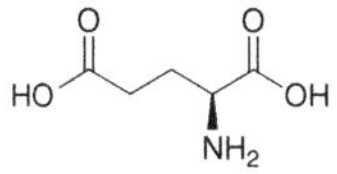<br>Glutamate (Glu)<br>MW = 146, pI = 2.8 | | | |
| Amino Acids with *positively charged* side chains | 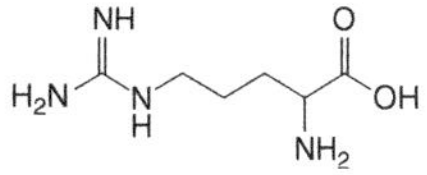<br>Arginine (Arg)<br>MW = 174, pI = 10.8 | 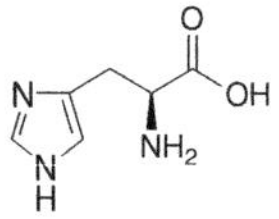<br>Histidine (His)<br>MW = 155, pI = 8.0 | 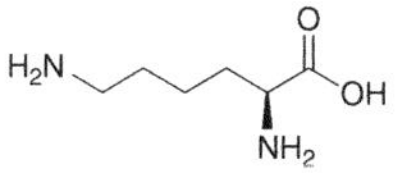<br>Lysine (Lys)<br>MW = 146, pI = 9.8 | | |
| *Unique* Amino Acids | 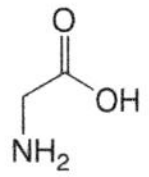<br>Glycine (Gly)<br>MW = 75, pI = 5.8 | 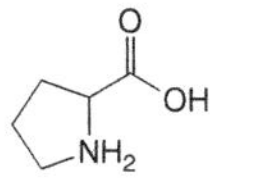<br>Proline (Pro)<br>MW = 115, pI = 7.1 | | | |

i) Improved patient compliance and convenience – "oral" is far more patient-centric and managed care friendly (no direct involvement of the health care provider), more favorable which results in improved clinical outcomes, and greater quality of life.
ii) Physiology mimetics and hepato-portal destination – "oral" provides certain endogenous peptides a unique advantage to mimic the physiological path of GIT to portal vein to liver and various antiviral and chemotherapy peptides to improve if high concentration in the liver is required.
iii) GIT local targeting therapy – "oral" lowers and optimizes the efficacy-related dose, no systemic exposure, and good safety profiles.
iv) Lifecycle management – "oral" facilitates the extension of patent life and overall value for priority portfolio assets by developing more advantageous and differentiated products (expand many drug market opportunities – a variety of new indications, diseases preventions, other population (e.g. children) that currently impossible due to "injection"), FDA's encouragement on "injection"-to-"oral" conversion (quicker approval [505(b)(2)] pathway).
v) Health-care systems and society – "oral" generates credible estimates of the total healthcare delivery costs for its products and the likelihood that these products will find favorable recommendations from healthcare payers.

(3) *Cost-benefit analysis*. By comparing "oral" with its "injectable" counterpart, the benefit-to-cost ratio measures both costs and benefits of the two. If "oral" is superior to "injectable" and costs less, then "oral" is said to be dominant. If "oral" has greater benefit but also costs more than "injectable," then the ratio is used in determining whether the benefits gained from "oral" are worth the additional cost.

## Abbreviations

ADME absorption, distribution, metabolism, and elimination
API active pharmaceutical ingredient
BA bioavailability
CIC chronic idiopathic constipation
4-CNAB *N*-(4-chlorosalicyloyl)-4-aminobutyrate (monosodium)
5-CNAC 8-(*N*-2-hydroxy-5-chlorobenzoyl)-amino-caprylate (sodium)
FDA the Food and Drug Administration
GC-C guanylate cyclase C
GI gastrointestinal
GIT gastrointestinal tract
GLP-1 glucagon-like peptide-1
GRAS generally regard as safe
hGH human growth hormone
IBS-C irritable bowel syndrome with constipation
IGF-1 insulin like growth factor-1
IM intramuscular injection
IV intravenous injection
MW molecular weight
NDA New Drug Application
NME new molecular entity
PD pharmacodynamics
pI isoelectric point
PK pharmacokinetics
PO per oral
sCT salmon calcitonin
SNAC sodium *N*-(8-[2-hydroxybenzoyl]amino) caprylate
T1DM type 1 diabetes mellitus
T2DM type 2 diabetes mellitus

## References

1 Wu, L. (2019). Regulatory considerations for peptide therapeutics. In: *Peptide Therapeutics: Strategy and Tactics for Chemistry, Manufacturing, and Controls* (ed. V. Srivastava). London: Royal Society of Chemistry.
2 Moroz, E., Matoori, S., and Leroux, J.-C. (2016). Oral delivery of macromolecular drugs: where we are after almost 100 years of attempts. *Adv. Drug Delivery Rev.* 101: 108–121.
3 Fosgerau, K. and Hoffmann, T. (2015). Peptide therapeutics: current status and future directions. *Drug Discovery Today* 21: 122–128.
4 Liu, P. and Dinh, S. (2011). Oral delivery of protein/peptide therapeutics. In: *Oral Bioavailability: Basic Principles, Advanced Concepts, and Applications* (ed. M. Hu and X. Li). New Jersey: Wiley.
5 Currie MG, Mahajan-Miklos S, Norman T and Milne GT (2007) Methods and compositions for the

treatment of gastrointestinal disorders. *US Patent* 7,304,036.

6 Comiskey S, Feng R, Foss J and Shailubhai K (2017) Formulations of guanylate cyclase C agonists and methods of use. *US Patent 9,616,097*.

7 Fretzen, A. (2018). Peptide therapeutics for the treatment of gastrointestinal disorders. *Bioorg. Med. Chem.* 26: 2863–2872.

8 Islam, B.N., Sharman, S.K., and Browning, D.D. (2018). Clinical utility of plecanatide in the treatment of chronic idiopathic constipation. *Int. J. Gen. Med.* 11: 323–330.

9 Weinberg, D.S., Lin, J.E., Foster, N.R. et al. (2017). Bioactivity of oral linaclotide in human colorectum for cancer chemoprevention. *Cancer Prev. Res. (Philadelphia, Pa.)* 10: 345–354.

10 Anderson, S.L., Beutel, T.R., and Trujillo, T.M. (2020). Oral semaglutide in type 2 diabetes. *J. Diabetes Complications* 34: 107520.

11 Buckley, S.T., Bækdal, T.A., Vegge, A. et al. (2019). Transcellular stomach absorption of a derivatized glucagon-like peptide-1 receptor agonist. *Sci. Transl. Med.* 10: eaar7047.

12 Sauerberg P, Bjerregaard S and Nielsen FS (2016) Solid compositions comprising a GLP-1 agonist and a salt of N-(8-(2-hydroxybenzoyl)amino)caprylic acid. *US Patent 9,278,123*.

13 Fattah, S. and Brayden, D.J. (2017). Progress in the formulation and delivery of somatostatin analogs for acromegaly. *Ther. Delivery* 8: 867–878.

14 Mamluk R, Tzabari M, Marom K, Salama P and Weinstein I (2016) Pharmaceutical compositions and related methods of delivery. *US Patent 9,265,812*.

15 Melmed, S., Popovic, V., Bidlingmaier, M. et al. (2015). Safety and efficacy of oral octreotide in acromegaly: results of a multicenter Phase III trial. *J. Clin. Endocrinol. Metabol.* 100: 1699–1708.

16 Tuvia, S., Pelled, D., Karen Marom, K. et al. (2014). A novel suspension formulation enhances intestinal absorption of macromolecules via transient and reversible transport mechanisms. *Pharm. Res.* 31: 2010–2021.

17 Varamini, P. and Toth, I. (2016). Recent advances in oral delivery of peptide hormones. *Expert Opin. Drug Delivery* 13: 507–522.

18 Kimball, J.W. (2020). The human gastrointestinal tract. In: *Kimball's Biology Pages* (ed. J.W. Kimball). LibreTexts Tufts University & Harvard University.

19 Moran, A.P., Gupta, A., and Joshi, L. (2011). Sweet-talk: role of host glycosylation in bacterial pathogenesis of the gastrointestinal tract. *Gut* 60: 1412–1425.

20 Murgia, X., Loretz, B., Hartwig, O. et al. (2018). The role of mucus on drug transport and its potential to affect therapeutic outcomes. *Adv. Drug Delivery Rev.* 124: 82–97.

21 Hua S (2020) Advances in oral drug delivery for regional targeting in the gastrointestinal tract - influence of physiological, pathophysiological and pharmaceutical factors. *Front. Pharmacol.* 11:524. doi: https://doi.org/10.3389/fphar.2020.00524

22 Brayden, D.J., Hill, T.A., Fairlie, D.P. et al. (2020). Systemic delivery of peptides by the oral route: formulation and medicinal chemistry approaches. *Adv. Drug Delivery Rev.* 157: 2–36. https://doi.org/10.1016/j.addr.2020.05.007.

23 Drucker, D.J. (2020). Advances in oral peptide therapeutics. *Nat. Rev. Drug Discovery* 19: 277–289.

24 Durán-Lobato, M., Niu, Z., and Alonso, M.J. (2020). Oral delivery of biologics for precision medicine. *Adv. Mater.* 32: e1901935.

25 He, S., Liu, Z., and Xu, D. (2019). Advance in oral delivery systems for therapeutic protein. *J. Drug Targeting* 27: 283–291.

26 Ibrahim, Y.H.-E.Y., Regdon, G. Jr., Hamedelniel, E.I., and Sovány, T. (2020). Recently used techniques and materials to improve the efficiency of orally administered proteins/peptides. *DARU J. Pharm. Sci.* 28: 403–416.

27 Liu, C., Kou, Y., Zhang, X. et al. (2018). Strategies and industrial perspectives to improve oral absorption of biological macromolecules. *Expert Opin. Drug Delivery* 15: 223–233.

28 Perry, S.L. and McClements, D.J. (2020). Recent advances in encapsulation, protection, and oral delivery of bioactive proteins and peptides using colloidal systems. *Molecules* 25: 1161.

29 Richard, J. (2017). Challenges in oral peptide delivery: lessons learnt from the clinic and future prospects. *Ther. Delivery* 8: 663–684.

30 Wong, C.Y., Martine, J., and Dass, C.R. (2016). Oral delivery of insulin for treatment of diabetes: status quo, challenges and opportunities. *J. Pharm. Pharmacol.* 68: 1093–1108.

31 Wright, L., Barnes, T.J., and Prestidge, C.A. (2020). Oral delivery of protein-based therapeutics: gastroprotective strategies, physiological barriers and in vitro permeability prediction. *Int. J. Pharm.* 585: 119488.

32 Xu, Y., Shrestha, N., Préat, V., and Beloqui, A. (2020). Overcoming the intestinal barrier: a look

into targeting approaches for improved oral drug delivery systems. *J. Controlled Release* 322: 486–508.
33 Arbit, E. and Kidron, M. (2017). Oral insulin delivery in a physiologic context: review. *J. Diabetes Sci. Technol.* 11: 825–832.
34 Bak, A., Ashford, M., and Brayden, D.J. (2018). Local delivery of macromolecules to treat diseases associated with the colon. *Adv. Drug Delivery Rev.* 136-137: 2–27.
35 Vass, P., Démuth, B., Hirsch, E., and Nagy, B. (2019). Drying technology strategies for colon-targeted oral delivery of biopharmaceuticals. *J. Controlled Release* 296: 162–178.
36 Odenwald, M.A. and Turner, J.R. (2017). The intestinal epithelial barrier: a therapeutic target? *Nat. Rev. Gastroenterol. Hepatol.* 14: 9–21.
37 de Limaa, V.C.O., Piuvezamb, G., Macielc, B.L.L., and de Araujo Morais, A.H. (2019). Trypsin inhibitors: promising candidate satietogenic proteins as complementary treatment for obesity and metabolic disorders? *J. Enzyme Inhib. Med. Chem.* 34: 405–419.
38 Bucheit, J.D., Pamulapati, L.G., Carter, N. et al. (2020). Oral semaglutide: a review of the first oral Glucagon-like peptide 1 receptor agonist. *Diabet. Technol. Ther.* 22: 10–18.
39 Granhall, C., Donsmark, M., Blicher, T.M. et al. (2019). Safety and pharmacokinetics of single and multiple ascending doses of the novel oral human GLP-1 analogue, oral semaglutide, in healthy subjects and subjects with Type 2 diabetes. *Clin. Pharmacokinet.* 58: 781–791.
40 Karsdal, M.A., Riis, B.J., Mehta, N. et al. (2014). Lessons learned from the clinical development of oral peptides. *Br. J. Clin. Pharmacol.* 79: 720–732.
41 Maher, S., Brayden, D.J., Casettari, L., and Illum, L. (2019). Application of permeation enhancers in oral delivery of macromolecules: an update. *Pharmaceutics* 11: 41.
42 Twarog, C., Fattah, S., Heade, J. et al. (2019). Intestinal permeation enhancers for oral delivery of macromolecules: a comparison between salcaprozate sodium (SNAC) and sodium caprate (C10). *Pharmaceutics* 11: 78.
43 Ahmad, J., Singhal, M., Amin, S. et al. (2017). Bile salt stabilized vesicles (bilosomes): a novel nano-pharmaceutical design for oral delivery of proteins and peptides. *Curr. Pharm. Des.* 23: 1575–1588.
44 Bonengel, S., Jelkmann, M., Abdulkarim, M. et al. (2018). Impact of different hydrophobic ion pairs of octreotide on its oral bioavailability in pigs. *J. Controlled Release* 273: 21–29.
45 Dumont, C., Jannin, V., Miolane, C. et al. (2019). A proof-of-concept for developing oral lipidized peptide nanostructured lipid carrier formulations. *J. Drug Delivery Sci. Technol.* 54: 101394.
46 Griesser, J., Hetényi, G., Moser, M. et al. (2017). Hydrophobic ion pairing: key to highly pay-loaded self-emulsifying peptide drug delivery systems. *Int. J. Pharm.* 520: 267–274.
47 Nazir, I., Fürst, A., Lupo, N. et al. (2019). Zeta potential changing self-emulsifying drug delivery systems: a promising strategy to sequentially overcome mucus and epithelial barrier. *Eur. J. Pharm. Biopharm.* 144: 40–49.
48 Mahmood, A. and Bernkop-Schnürch, A. (2019). SEDDS: a game changing approach for the oral administration of hydrophilic macromolecular drugs. *Adv. Drug Delivery Rev.* 142: 91–101.
49 Boegh, M. and Nielsen, H.M. (2015). Mucus as a barrier to drug delivery - Understanding and mimicking the barrier properties. *Basic Clin. Pharm. Toxicol.* 116: 179–186.
50 Pawar, V.K., Meher, J.G., Singh, Y. et al. (2014). Targeting of gastrointestinal tract for amended delivery of protein/peptide therapeutics: strategies and industrial perspectives. *J. Controlled Release* 196: 168–183.
51 Shan, W., Zhu, X., Liu, M. et al. (2015). Overcoming the diffusion barrier of mucus and absorption barrier of epithelium by self-assembled nanoparticles for oral delivery of insulin. *ACS Nano* 9: 2345–5236.
52 Pearson, J.P., Chater, P.I., and Wilcox, M.D. (2016). The properties of the mucus barrier, a unique gel -How can nanoparticles cross it? *Ther. Delivery* 7: 229–244.
53 Banerjee, A., Ibsen, K., Brown, T. et al. (2018). Ionic liquids for oral insulin delivery. *Proc. Natl. Acad. Sci.* 115: 7296–7301.
54 de Sousa, I.P., Steiner, C., Schmutzler, M. et al. (2015). Mucus permeating carriers: formulation and characterization of highly densely charged nanoparticles. *Eur. J. Pharm. Biopharm.* 97: 273–279.
55 Menzel, C. and Bernkop-Schnürch, A. (2018). Enzyme decorated drug carriers: targeted swords to cleave and overcome the mucus barrier. *Adv. Drug Delivery Rev.* 124: 164–174.
56 Chater, P.J., Wilcox, M.D., and Pearson, J.P. (2018). Efficacy and safety concerns over the use of mucus modulating agents for drug delivery using nanoscale systems. *Adv. Drug Delivery Rev.* 124: 184–192.
57 Abdulkarim, M., Sharma, P.K., and Gumbleton, M. (2019). Self-emulsifying drug delivery system:

Mucus permeation and innovative quantification technologies. *Adv. Drug Delivery Rev.* 142: 62–74.

58 Leonaviciute, G. and Bernkop-Schnürch, A. (2015). Self-emulsifying drug delivery systems in oral (poly)peptide drug delivery. *Expert Opin. Drug Delivery* 12: 1703–1716.

59 Artursson, P. and Lundquist, P. (2020). A new opening for orally taken peptide drugs. *Nat. Biomed. Eng.* 4: 12–13.

60 Cao, S.-J., Xu, S., Wang, H.-M. et al. (2019). Nanoparticles: oral delivery for protein and peptide drugs. *AAPS PharmSciTech* 20: 190. https://doi.org/10.1208/s12249-019-1325-z.

61 Niu, Z., Conejos-Sánchez, I., Griffin, B.T. et al. (2016). Lipid-based nanocarriers for oral peptide delivery. *Adv. Drug Delivery Rev.* 106: 337–354.

62 Dumont, C., Bourgeois, S., Fessi, H., and Janninet, V. (2018). Lipid-based nanosuspensions for oral delivery of peptides: a critical review. *Int. J. Pharm.* 541: 117–135.

63 Ismail, R. and Csóka, I. (2017). Novel strategies in the oral delivery of antidiabetic peptide drugs – Insulin, GLP-1 and its analogs. *Eur. J. Pharm. Biopharm.* 115: 257–267.

64 Akkus, Z.B., Nazir, I., Jalil, A. et al. (2019). Zeta potential-changing polyphosphate nanoparticles: a promising approach to overcome the mucus and epithelial barrier. *Mol. Pharmaceutics* 16: 2817–2825.

65 Bouttefeux, O., Beloqui, A., and Préat, V. (2016). Delivery of peptides via the oral route: diabetes treatment by peptide-loaded nanoparticles. *Curr. Pharm. Des.* 22: 1161–1176.

66 Silva, S., Almeida, A.J., and Valeet, N. (2019). Combination of cell-penetrating peptides with nanoparticles for therapeutic application: a review. *Biomolecules* 9: 22.

67 Griffin, B.T., Guo, J., Presas, E. et al. (2016). Pharmacokinetic, pharmacodynamic and biodistribution following oral administration of nanocarriers containing peptide and protein drugs. *Adv. Drug Delivery Rev.* 106: 367–380.

68 Lakkireddy, H.R., Urmann, M., Besenius, M. et al. (2016). Oral delivery of diabetes peptides - Comparing standard formulations incorporating functional excipients and nanotechnologies in the translational context. *Adv. Drug Delivery Rev.* 106: 196–222.

69 Han, Y., Gao, Z., Chen, L. et al. (2019). Multifunctional oral delivery systems for enhanced bioavailability of therapeutic peptides/proteins. *Acta Pharm. Sin. B* 9: 902–922.

70 Lundquist, P. and Artursson, P. (2016). Oral absorption of peptides and nanoparticles across the human intestine: opportunities, limitations and studies in human tissues. *Adv. Drug Delivery Rev.* 106: 256–276.

71 Abramson, A., Caffarel-Salvador, E., Khang, M. et al. (2019a). An ingestible self-orienting system for oral delivery of macromolecules. *Science* 363: 611–615.

72 Abramson, A., Halperin, F., Kim, J., and Traverso, G. (2019b). Quantifying the value of orally delivered biologic therapies: a cost-effectiveness analysis of oral semaglutide. *J. Pharm. Sci.* 108: 3138–3145.

73 Hashim, M., Korupolu, R., Syed, B. et al. (2019). Jejunal wall delivery of insulin via an ingestible capsule in anesthetized swine-A pharmacokinetic and pharmacodynamic study. *Pharmacol. Res. Perspect.* 7: e00522.

74 Banerjee, A., Wong, J., Gogoi, R. et al. (2017). Intestinal micropatches for oral insulin delivery. *J. Drug Targeting* 25: 608–615.

75 Jørgensen JR, Yu F, Venkatasubramanian R, Nielsen LH, Nielsen HM, Boisen A, Rades T and Müllertz A (2020) In vitro, ex vivo and in vivo evaluation of microcontainers for oral delivery of insulin. *Pharmaceutics* 12: pii: E48

76 Nemeth, C.L., Lykins, W.R., Tran, H. et al. (2019). Bottom-up fabrication of multilayer enteric devices for the oral delivery of peptides. *Pharm. Res.* 36: 89.

77 France, M.M., Del Rio, T., Travers, H. et al. (2019). Ultra-rapid drug delivery in the oral cavity using ultrasound. *J. Controlled Release* 304: 1–6.

78 Giordanetto, F. and Kihlberg, J. (2014). Macrocyclic drugs and clinical candidates: what can medicinal chemists learn from their properties? *J. Med. Chem.* 57: 278–295.

79 Wang, J., Yadav, V., Smart, A.L. et al. (2015). Towards oral delivery of biopharmaceuticals: an assessment of the gastrointestinal stability of 17 peptide drugs. *Mol. Pharmaceutics* 12: 966–973.

80 Buckley, S.T., Hubálek, F., and Rahbek, U.L. (2016). Chemically modified peptides and proteins - Critical considerations for oral delivery. *Tissue Barriers* 4 (2): e1156805.

81 Di, L. (2015). Strategic approaches to optimizing peptide ADME properties. *AAPS J.* 17: 134–143.

82 Räder, A.F.B., Weinmüller, M., Reichart, F. et al. (2018). Orally active peptides: is there a magic bullet? *Angew. Chem.* 57: 14414–14438.

83 Zupančič, O. and Bernkop-Schnürch, A. (2017). Lipophilic peptide character - What oral barriers fear the most. *J. Controlled Release* 255: 242–257.

84 Nielsen, D.S., Shepherd, N.E., Xu, W. et al. (2017). Orally absorbed cyclic peptides. *Chem. Rev.* 117: 8094–8128.

85 Zorzi, A., Deyle, K., and Heinis, C. (2017). Cyclic peptide therapeutics: past, present and future. *Curr. Opin. Chem. Biol.* 38: 24–29.

86 Naylor, M.R., Bockus, A.T., Blanco, M.-J., and Lokey, R.S. (2017). Cyclic peptide natural products chart the frontier of oral bioavailability in the pursuit of undruggable targets. *Curr. Opin. Chem. Biol.* 38: 141–147.

87 Bockus, A.T., Lexa, K.W., Pye, C.R. et al. (2015). Probing the physicochemical boundaries of cell permeability and oral bioavailability in lipophilic macrocycles inspired by natural products. *J. Med. Chem.* 58: 4581–4589.

88 Badiani, K. (2012). Peptides as drugs. *Inter. Pharm. Ind.* 4: 84–90.

89 Buclin, T., Rochat, C., Burckhardt, M. et al. (2002). Bioavailability and biological efficacy of a new oral formulation of salmon calcitonin in healthy volunteers. *J. Bone Miner. Res.* 17: 1478–1485.

90 Kidron, M., Dinh, S., Menachem, Y. et al. (2004). A novel per-oral insulin formulation: Proof of concept study in non-diabetic subjects. *Diabet. Med.* 21: 354–357.

91 Dinh S, Liu P, Arbit E, Shepard T and Trapnell C (2004) A Phase I, double-blind, randomized, dose escalating crossover study of an oral formulation of human growth hormone in healthy male volunteers. *31st CRS Annual Meeting, Honolulu, HI.*

92 Liu, P., Zhang, H., Lee, J., and Dinh, S. (2003). Oral delivery of insulin. In: *In situ portal perfusion in a rat model*. San Francisco, CA: *3rd Annual Diabetes Technology Meeting*.

93 Sugihara, M., Takeuchi, S., Sugita, M. et al. (2015). Analysis of intra- and inter-subject variability in oral drug absorption in human bioequivalence studies of 113 generic products. *Mol. Pharmaceutics* 12: 4405–4413.

94 Tyagi, P., Pechenov, S., and Subramony, J. (2018). Oral peptide delivery: translational challenges due to physiological effects. *J. Controlled Release* 287: 167–176.

95 Maher, S., Ryan, B., Duffy, A., and Brayden, D.J. (2014). Formulation strategies to improve oral peptide delivery. *Pharm. Pat. Anal.* 3: 313–336.

96 Smart, A.L., Gaisford, S., and Basit, A.W. (2014). Oral peptide and protein delivery: intestinal obstacles and commercial prospects. *Expert Opin. Drug Delivery* 11: 1323–1335.

97 Hassani, L.N., Lewis, A., and Richard, J. (2015). Oral peptide delivery: technology landscape & current status. *ONdrugDelivery* 59: 12–17.

98 Aguirre, T.A.S., Teijeiro-Osorio, Z., Rosa, M. et al. (2016). Current status of selected oral peptide technologies in advanced preclinical development and in clinical trials. *Adv. Drug Delivery Rev.* 106: 223–241.

99 Shields, P. (2017). Oral peptide therapeutics–a holy grail or quixotic quest? *Drug Discov. World* .

100 Liu, P. and Dinh, S. (2010). Oral delivery of protein and peptide therapeutics - PK/PD and metabolism considerations. In: *Proteins and Peptides: Pharmacokinetic, Pharmacodynamic, and Metabolic Outcomes* (ed. R. Mrsny and A. Daugherty). New York: Informa Healthcare.

101 Vlieghe, P., Lisowski, V., Martinez, J., and Khrestchatisky, M. (2010). Synthetic therapeutic peptides: science and market. *Drug Discovery Today* 15: 40–56.

# 32

# Prodrugs to Improve Oral Delivery

*Arjun D. Patel, Shuchi Gupta, and Mamoun Alhamadsheh*

*Department of Pharmaceutics & Medicinal Chemistry, Thomas J. Long School of Pharmacy, University of the Pacific, Stockton, CA, USA*

## 32.1 Introduction

The oral route of administration is the most preferred route of drug delivery due to many advantages including safety, ease of ingestion, higher patient compliance, lack of pain, and lower cost of care [1]. Unfortunately, oral drug administration comes with numerous challenges due to the complex absorption mechanisms involved. Orally administered drugs must first pass through the gastrointestinal tract (GIT) wall mucosa consisting of polarized enterocytes linked together through tight junctions. These enterocytes express several drug transporters which are responsible for both drug absorption and expulsion. The systemic bioavailability of oral drugs is significantly altered by these transporters as well as by intestinal enzymes which play a role in first-pass metabolism.

First-pass metabolism is a phenomenon which occurs first in intestinal enterocytes and then in liver hepatocytes. In this process, drugs are metabolized before reaching the systemic circulation and thus require an escalation of doses for patients [2]. The cytochrome P450 enzyme family (CYPs) is the main source of metabolizing enzymes accounting for about 75% of the first-pass effect on oral drugs [3]. Inside the CYP family, CYP3A4 accounts for 70% of all CYP enzymes in small intestinal enterocytes [4]. Several approaches are being developed to improve oral drug absorption by co-administering the parent drug with CYP inhibitors. For example, the calcium channel blocker felodipine undergoes extensive first-pass metabolism by CYP3A4 in the small intestine and liver, resulting in low bioavailability. Coadministration of felodipine with grapefruit juice resulted in fourfold increase in the blood levels of felodipine. The primary mechanism for grapefruit juice to reduce the metabolism of felodipine is by inhibition of CYP3A4 enzyme [5].

Two of the main drug transporters in enterocytes are the ATP-binding cassette (ABC) superfamily and the solute carrier (SLC) family. The ABC superfamily includes well-known efflux transporters such as P-glycoprotein (P-gp), multidrug resistance-associated protein 2 (MRP2), and breast cancer resistance protein (BCRP) [6, 7]. The SLC family includes transporters such as the organic anion transporter-1 (OAT1 or SLC22A6) [8]. A main difference between the two families is that the ABC family hydrolyzes adenosine-5′-triphosphate (ATP) for efflux whereas the SLC family has transporters which use either facilitated diffusion or active transport for influx of small molecules. Because of their abundance at the apical (luminal) membrane of enterocytes, the ABC transporters limit the intestinal absorption of many important drugs such as statins, antibiotics, HIV protease inhibitors, immunosuppressants, and cardiac drugs. The high expression levels of P-gp in tumor cells are responsible for the active efflux of many anticancer agents. For example, the low oral bioavailability of paclitaxel, due to efflux through P-gp, can be enhanced by coadministration with the P-gp inhibitor elacridar [9]. The SLC uptake transporters (e.g. peptide transporter 1, PEPT1) present an opportunity for a prodrug approach in which the parent drug is endowed with a moiety that allows it to become a substrate for transporters present on apical side of the enterocytes in the GIT. For example, prodrugs targeting the SLC intestinal transporters increase oral bioavailability of the parent drug, not only by helping uptake of drugs into enterocytes but also by increasing water solubility, promoting stability of the drug, and evading efflux transport and intestinal first-pass metabolism [8].

The prodrug approach has been proven effective to overcome issues related to lower oral bioavailability. Currently, there are two treatments for influenza,

*Oral Bioavailability and Drug Delivery: From Basics to Advanced Concepts and Applications,* First Edition. Edited by Ming Hu and Xiaoling Li.

oseltamivir (Tamiflu) and zanamivir. Oseltamivir is an orally available ethyl ester prodrug that is converted to the active oseltamivir carboxylate by hepatic esterases. The prodrug masks the polarity of the active oseltamivir carboxylate, allowing for enhanced oral bioavailability (~80%) [10]. Several reports indicate that resistance against oseltamivir is growing [11]. In comparison, zanamivir has no reports of influenza resistance but it has a low bioavailability (~2%) [12]. Gupta et al. [13] utilized a carrier-mediated intestinal transport of an L-valyl prodrug by PEPT1 to improve the poor bioavailability of zanamivir. As a result, the L-valyl-zanamivir prodrug conjugated with an acyloxy ester had a three-fold increase in uptake by PEPT1 overexpressing cells. Aside from uptake transporters, efflux transporters have also been targeted to improve oral absorption. This chapter will discuss various prodrug approaches utilized to enhance the oral bioavailability and reduce side effects of drugs.

## 32.2 Factors Associated With Oral Drug Absorption

The oral bioavailability of a drug is an outcome of a complex interplay of different factors such as pH, GI secretions, enterocyte tight junctions, and solubility. The collection of these different factors can be categorized into two main topics: (i) physiological barriers and (ii) physiochemical and formulation characteristics (Figure 32.1) [14]. To achieve optimal drug penetration, drugs should have certain characteristics. The following section will discuss the optimal characteristics to maximize oral absorption of drug molecules.

## 32.3 Intestinal Physiology and Background

The GIT is a ~30 ft long tube consisting of two main subdivisions: the upper and lower GIT. The upper portion consists of the mouth, esophagus, stomach, and duodenum. The lower portion accounts for most of the small intestine (jejunum and ileum) as well as the large intestine up to the anus. The exact location where the upper and lower GI tract separate is the suspensory muscle in the duodenum [15]. There are four main layers of the GIT: mucosa; submucosa; muscularis propria; and the serosa (in order from surface of the GI lumen towards mesentery). Of these four, the mucosa is the main barrier to consider regarding absorption of drugs, food, vitamins, etc. Furthermore, exploration of the mucosa yields three subdivisions: mucosal epithelium (single layer of specialized epithelial cells; villi), lamina propria (mucous glands, lymph nodules, and capillaries), and muscularis mucosae (smooth muscle that produces local movement) [16].

A pronounced physical barrier facing drug absorption is the phospholipid bilayer of the mucosal epithelial cells. This barrier is mainly responsible for blocking drug molecules from freely diffusing into blood capillaries through a process known as passive diffusion.

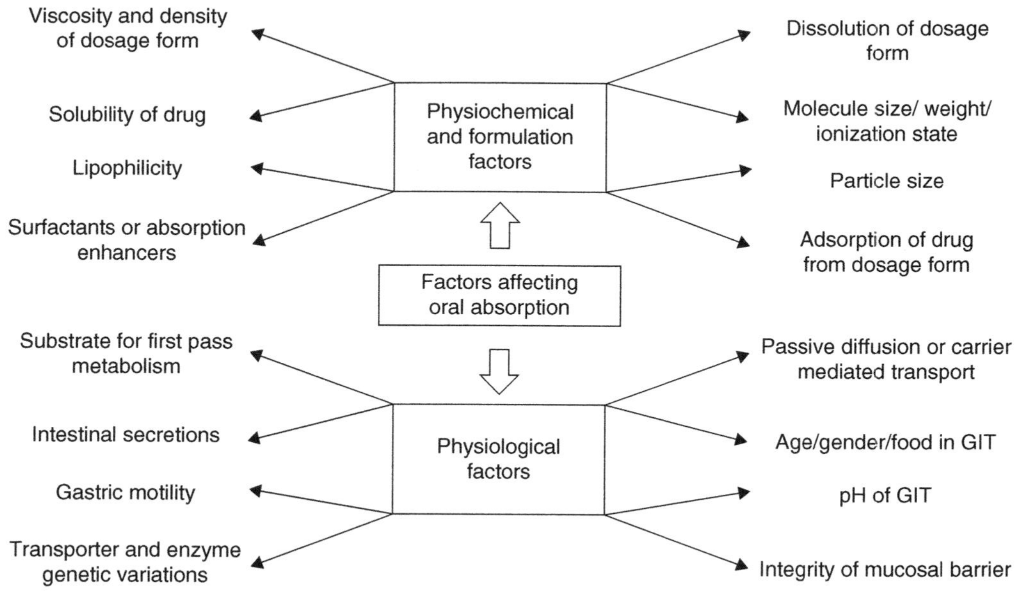


**Figure 32.1** Factors associated with oral drug absorption.

This is a process where a concentration gradient forces movement of molecules from areas of high solute levels to low solute levels [17]. The rate of drug passing (flux) through a membrane down a concentration gradient follows Fick's law of diffusion (Eq. (32.1)).

Fick's law of diffusion:

$$\frac{dQ}{dt} = DAK\frac{(C_x - C_y)}{h} \tag{32.1}$$

| | | |
|---|---|---|
| $dQ/dt$ | = | Rate of drug diffusion |
| $D$ | = | Diffusion coefficient of the drug through the membrane |
| $A$ | = | Surface area of the membrane through which molecule can pass |
| $K$ | = | Lipid/water partition coefficient of molecule |
| $h$ | = | Thickness of the permeated membrane |
| $C_x$; $C_y$ | = | Drug concentrations in areas $x$ and $y$, respectively |

Penetration of this barrier through passive diffusion is highly influenced by a number of factors such as the lipophilicity, molecular size, and ionization state of drug molecules. The ionization state depends on the surrounding pH which varies throughout the GIT, ranging from pH 1–3 in the stomach; 6.5–7.5 in the small intestine; and 5.5–7 in the large intestine [18]. Unionized drugs are best for passive diffusion as they are relatively less polar compared to ionized forms.

Passive drug absorption can further be classified into three main pathways, paracellular, transcellular, and carrier-mediated transport: (i) paracellular transport is the movement of drug molecule through channels/pores found between the cells. The main barrier facing this pathway is the tight junctions between the cells. These are formed by complexes of proteins such as occludin, claudins, and junctional adhesion molecules (JAMs). Tight junction complexes further associate with cell surface membrane proteins such as zonula occludens, thus forming a barrier against paracellular transport. As a result, only low molecular weight molecules of hydrophilic nature cross through the paracellular pathway; (ii) transcellular transport is where drug molecules diffuse directly through enterocytes. This transport method is mostly associated with small lipophilic drugs because of their ability to associate with the phospholipid bilayers of cell membrane; (iii) carrier-mediated transport (also known as facilitated diffusion) involves transporter proteins on the apical side of enterocytes facilitating the simple diffusion of drug without using ATP. Here, specific transporter proteins recognize specific ligands (e.g. vitamin B12, transferrin, and lectins) and facilitate their diffusion through enterocytes into circulation [17].

In comparison to passive transport, the uptake of drug molecules by active transport involves the transfer of molecules against concentration gradient through the expenditure of ATP. This transport mechanism is prone to saturation at high drug concentrations. Primary active transport involves a transmembrane protein directly catabolizing ATP to internalize a drug molecule. Secondary active transport involves using ATP to establish an electrochemical gradient (i.e. $Na^+/K^+$ ATPase effluxes 3 $Na^+$ ions per 2 $K^+$ ions influxed). This electrochemical gradient is then used by another transporter to uptake drug molecules. Common active transporters include $Na^+/K^+$ ATPase, $H^+/K^+$ ATPase, $Ca^{+2}$ ATPase, and ABC transporters [17, 19]. Substrates of this pathway of drug absorption include amino acids, short peptides, sugars, and vitamins. Various examples of prodrug approaches utilizing the activity of these transporters in their designs will be discussed in subsequent sections.

## 32.4 Strategies to Improve the Bioavailability of Orally Administered Drugs

Improvements in synthesis technology and combinatorial chemistry have led to the design of potential drug molecules using target-based approach. However, the majority of these molecules are unsuccessful as drugs due to poor physicochemical and biopharmaceutical properties. There are several factors which affect oral drug delivery including poor drug solubility, poor drug stability, and limited permeability across the mucosal barriers of the GIT. These biopharmaceutical properties can be altered through modification of fundamental physicochemical properties such as hydrophobicity, hydrogen bonding, chemical reactivity, and steric and electronic factors (which include ionization constants). Different approaches have been employed to alter the physicochemical properties, which will ultimately improve the bioavailability of orally administered drugs. These approaches involve formulation [20] and prodrug strategies [21].

To overcome the challenge of poor absorption, oral drugs can be formulated in three different forms: amorphous formulations, crystalline solid formulations, and lipid formulations [20]. Amorphous formulations are solid dispersions which are formed using a variety of technologies like spray drying and may include surfactants and polymers that will provide more surface area during dispersion. The amorphous drug form enhances the solubility of poorly water-soluble drugs by creating a supersaturated solution of the drug thus increasing its absorption [22]. In crystalline solid formulations, physicochemical properties of the drug molecule are altered by different processes including salt formation [23] and micronization [24] to increase surface area and thus increase absorption. Drug absorption is enhanced using lipid formulations by dispersing the drug molecule into inactive lipid vehicles like surfactant dispersions [25], emulsions, nanoparticle emulsions, and many other forms, thus increasing the permeability through the intestinal membrane. Lipid-based dispersions can be solid or liquid formulations. Neoral® is an oral formulation of cyclosporine in which microemulsion was used to increase dissolution and hence absorption [26]. The following sections will discuss prodrug approaches for optimizing various physicochemical properties important for enhancing oral bioavailability.

## 32.5 Prodrug Overview and Classification

Prodrugs are pharmacologically inactive molecules that are converted, through chemical or enzymatic processes or a combination of the two, to an active form within the body. Adrian Albert [27] was the first to recognize the strategy of prodrug, though few prodrugs like methenamine, phenacetin, and prontosil were already being used [28]. The aim of designing prodrugs is to optimize the pharmaceutical, pharmacokinetic, and pharmacodynamic properties of compounds by improving their chemical stability, aqueous solubility, sensory (taste and smell), ADME properties (absorption, distribution, metabolism, excretion) and decreasing their toxicity [29]. During the period of 2008 to 2017 over 12% of all small molecule new chemical entities approved by FDA were prodrugs [21] (Table 32.1). Prodrug strategies have been successfully used to develop targeted drug delivery systems, especially in chemotherapy, leading to increased therapeutic efficacy.

Prodrugs can be divided into two categories: bioprecursors and carrier-linked prodrugs (Figure 32.2). Bioprecursors are prodrugs that do not have a carrier and are modified into new molecules that will be transformed *in vivo* into active metabolites through a chemical or metabolic transformation [29]. Nabumetone is a nosteroidal anti-inflammatory bioprecursor prodrug that is rapidly metabolized in the liver, through a series of oxidative decarboxylations, to the active metabolite 6-methoxy-2-naphthyl acetic acid [30].

Carrier-linked prodrugs are obtained by modification of a functional group (e.g. ester, amide, oxime, carbamate, or *N*-Mannich) of an active drug molecule by covalent conjugation to a carrier (also called promoiety). The promoiety is conjugated to the drug through a linker system that is bio-reversible. In the body, the carrier-linked prodrug undergoes either enzymatic or chemical transformation to release the active drug. Ideally, the promoiety should be safe, stable, and pharmacologically inactive during administration. The promoiety should also undergo biodegradation to non-toxic metabolites [28]. Carrier-linked prodrugs can be further subdivided into three categories: (i) bipartite; (ii) tripartite; and (iii) mutual prodrugs [31]. In bipartite prodrugs, the promoiety is directly linked to the drug

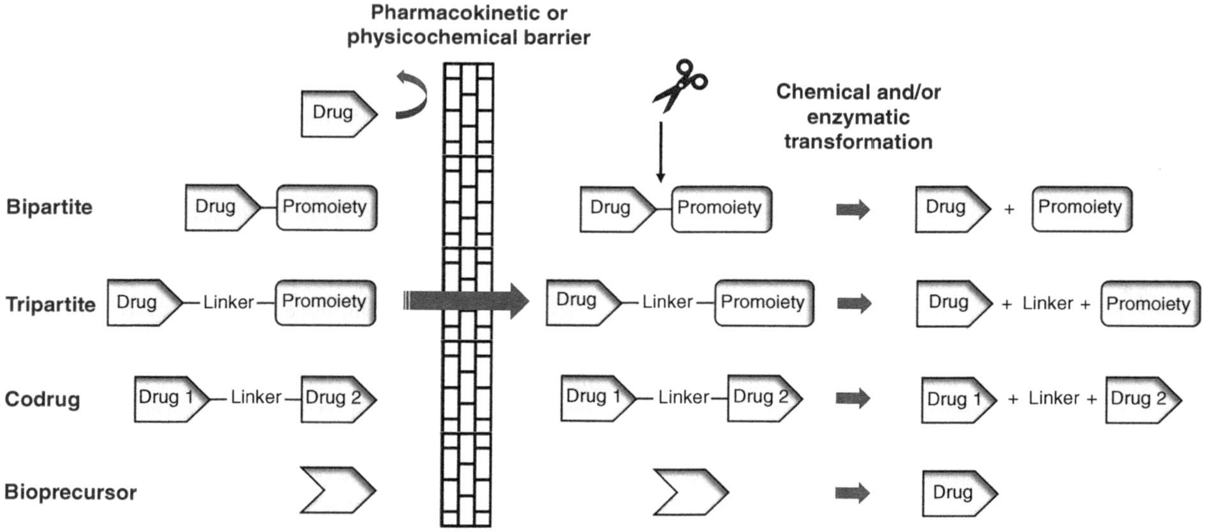


**Figure 32.2** Classification of prodrugs and concept for their *in vivo* bioactivation.

**Table 32.1** Chemical structures and properties of approved orally administered prodrugs.

| Prodrug structure (promoiety highlighted)/Active form | Prodrug approach/ activation mechanism | Improved property | Therapeutic indication |
|---|---|---|---|
| 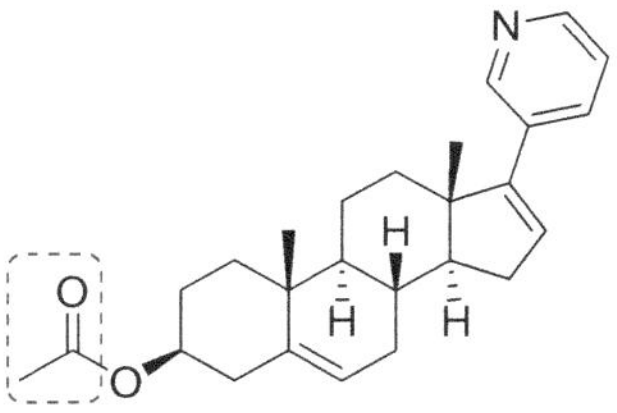 Abiraterone acetate/Abiraterone | Acetic acid ester/Hydrolysis by esterases | Lipophilicity and permeability | Metastatic castration-resistant prostate cancer |
| 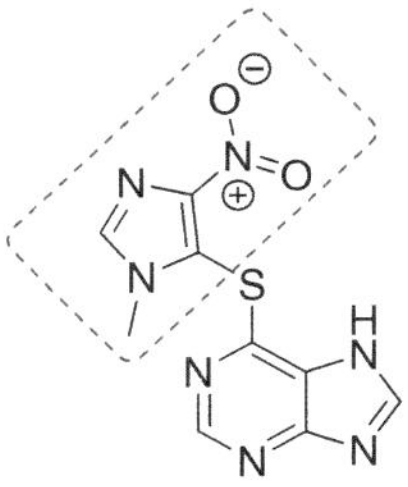 Azathioprine/6-Mercaptopurine | Imidazole derivative/Non enzymatic reductive cleavage of thioether | Bioavailability | Immunosuppressant |
| 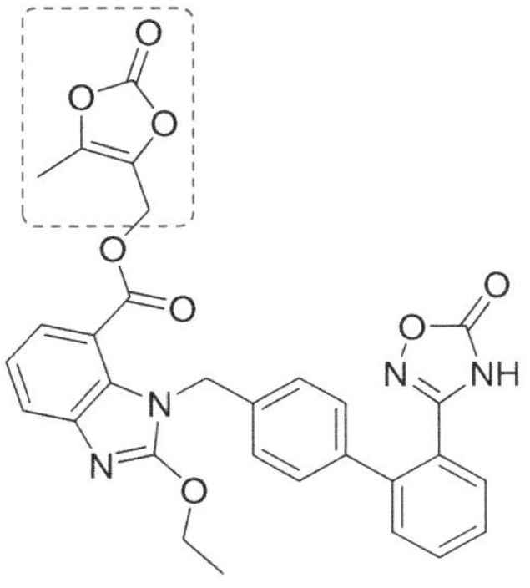 Azilsartan Medoxomil/Azilsartan | Oxodioxolyl methyl ester hydrolyzed in GI tract during absorption by esterases | Lipophilicity and permeability | Hypertension |
| 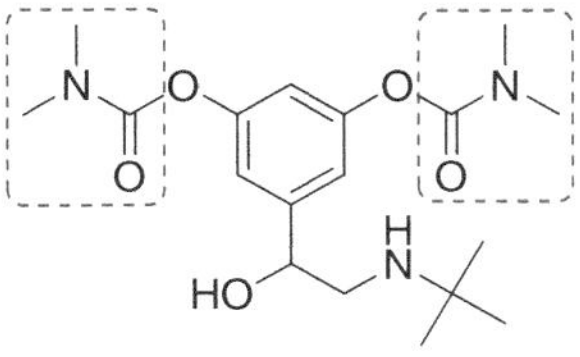 Bambuterol/Tertbutaline | Bis-dimethylcarbamate hydrolysis by esterases | Prolong duration of action | Asthma |
| 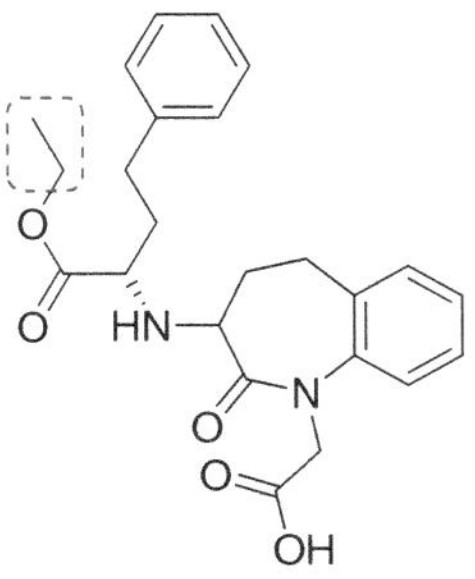 Benazepril/Benazeprilat | Ethyl ester hydrolysis by esterases | Lipophilicity and permeability | Hypertension |

**Table 32.1** (Continued)

| **Prodrug structure (promoiety highlighted)/Active form** | **Prodrug approach/ activation mechanism** | **Improved property** | **Therapeutic indication** |
|---|---|---|---|
| 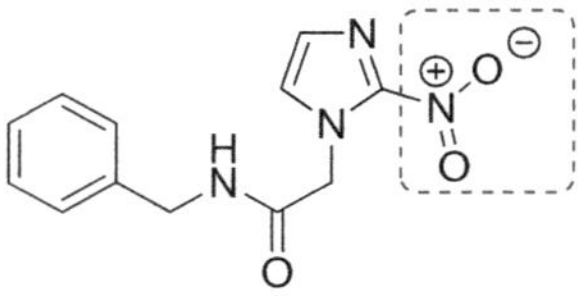 Benznidazole/Reactive metabolites | 2-Nitroimidazole prodrug/Reduction by nitroreductase | Bioavailability | Chagas disease |
| 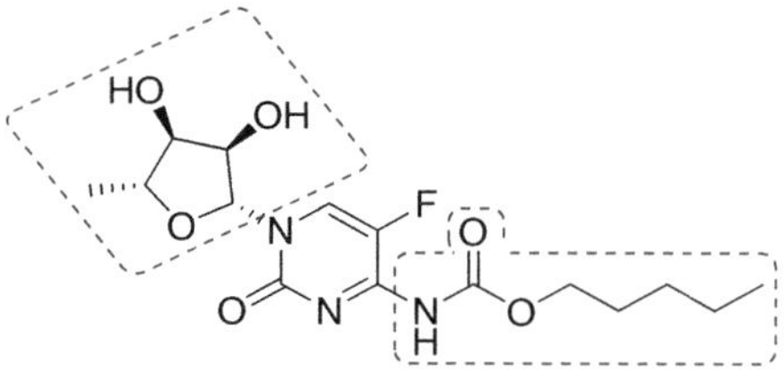 Capecitabine/5-Florouracil | Esterification and glycosylation/ Carboxylesterase, Cytidine deaminase, and Thymidine phosphorylase | Increased tumor selectivity and lower GI toxicity | Used to treat various cancers (i.e. breast cancer, gastric cancer, colorectal cancer) |
| 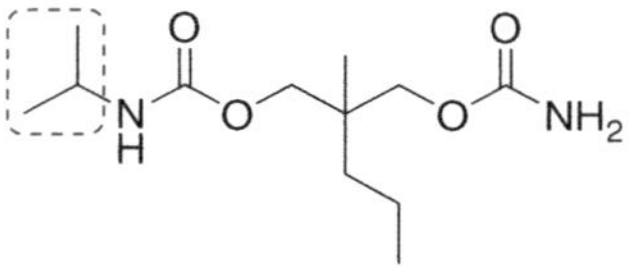 Carisoprodol/Meprobamate | Alkylation of carbamate/N-deal-kylation of amide by CYP450 enzymes | Lipophilicity and permeability | Central acting skeletal muscle relaxant |
| 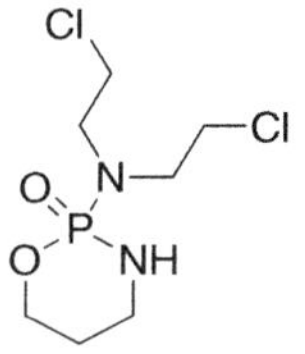 Cyclophosphamide/4-OH-cyclophosphamide | CYP3A4 mediated hydroxylation of aryl-substituted cyclic phosphodiester | Enhanced hepatic selectivity | Anticancer agent |
| 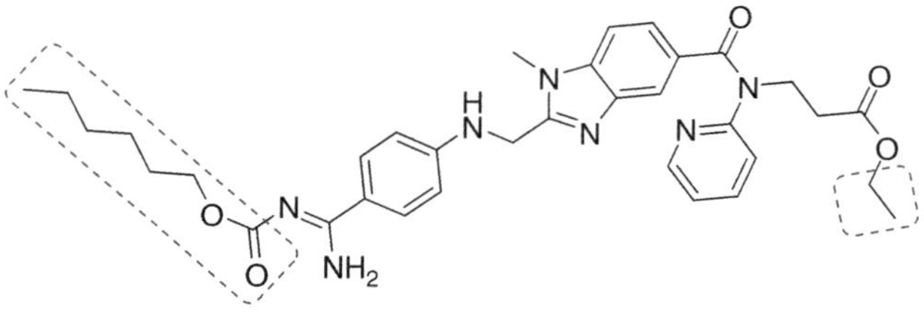 Dabigatran Etexilate/Dabigatran | Hydrolysis of ethyl ester and etexilate groups by esterases | Lipophilicity and permeability | Anticoagulant |
| 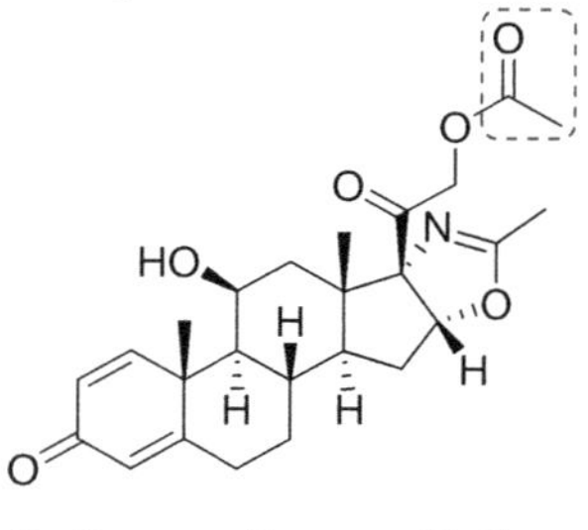 Deflazacort/Desacetyldeflazacort | Hydrolysis of acetyl group by esterases | Lipophilicity and permeability | Muscular dystrophy Immunosup-pressant |
| 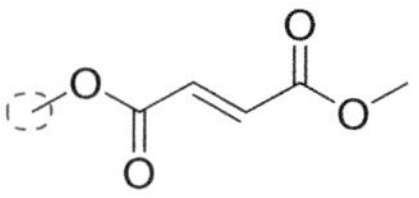 Dimethyl fumarate/Monomethyl fumarate | Hydrolysis of methyl ester by esterases | Lipophilicity and permeability | Multiple sclerosis |

**Table 32.1** (Continued)

| Prodrug structure (promoiety highlighted)/Active form | Prodrug approach/ activation mechanism | Improved property | Therapeutic indication |
|---|---|---|---|
| 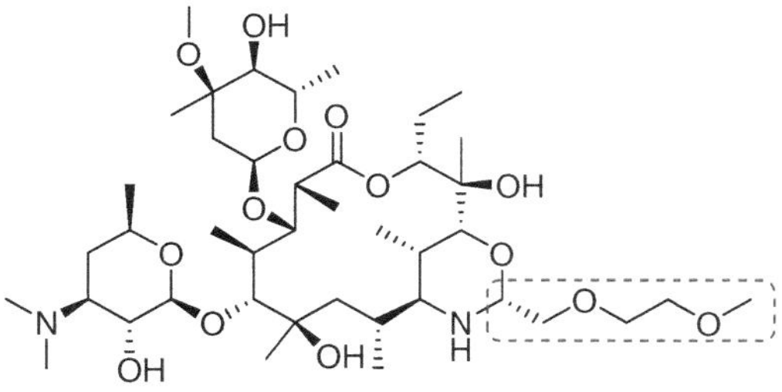 Dirithromycin/Erythromycyclamine | Spontaneous hydrolysis of oxazine ring in acidic or alkaline environments | Lipophilicity and permeability | Antibiotic |
| 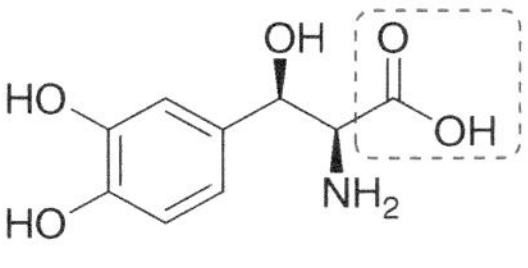 Droxidopa/Norepinephrine | Enzymatic decarboxylation | Improved brain permeability | Neurogenic orthostatic hypotension |
| 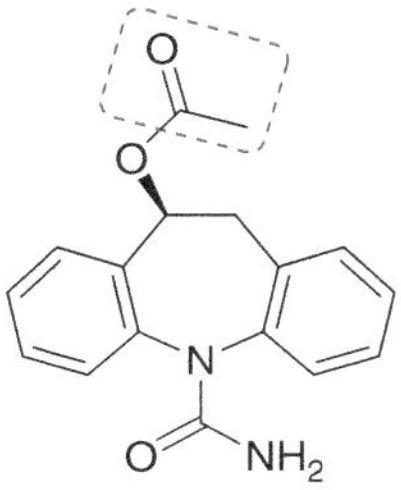 Eslicarbazepine acetate/Eslicarbazepine | Acetate ester hydrolysis | Lipophilicity and permeability | Anticonvulsant |
| 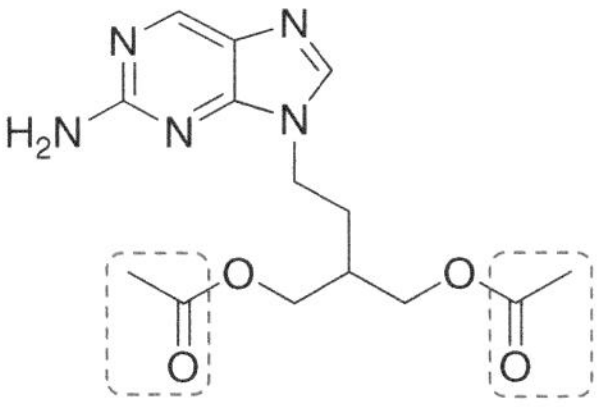 Famciclovir/Penciclovir | Acetate ester hydrolysis | Lipophilicity and permeability | Antiviral |
| 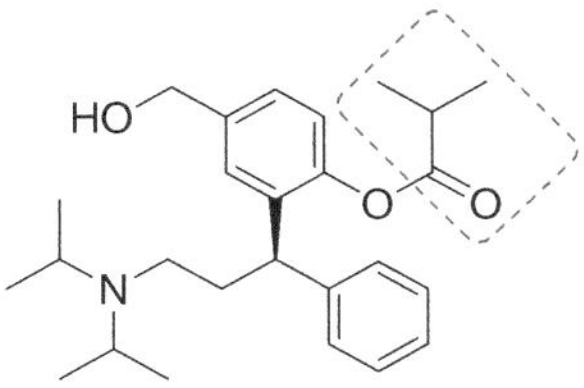 Fesoterodine/Desfesoterodine | Isobutyryl ester hydrolysis by esterases | Less pharmacokinetic variation across patients | Overactive bladder syndrome |
| 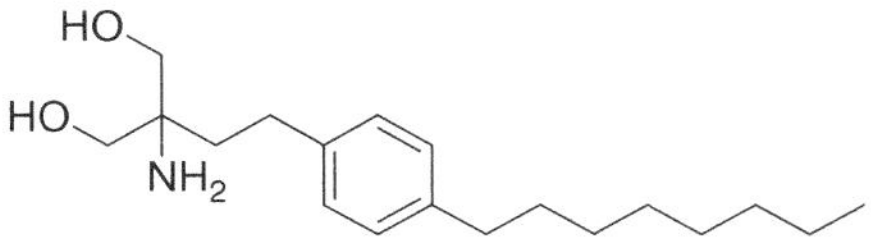 Fingolimod/Fingolimod Phosphate | Phosphorylation | Lipophilicity and permeability | Multiple Sclerosis |

**Table 32.1** (Continued)

| Prodrug structure (promoiety highlighted)/Active form | Prodrug approach/ activation mechanism | Improved property | Therapeutic indication |
|---|---|---|---|
| 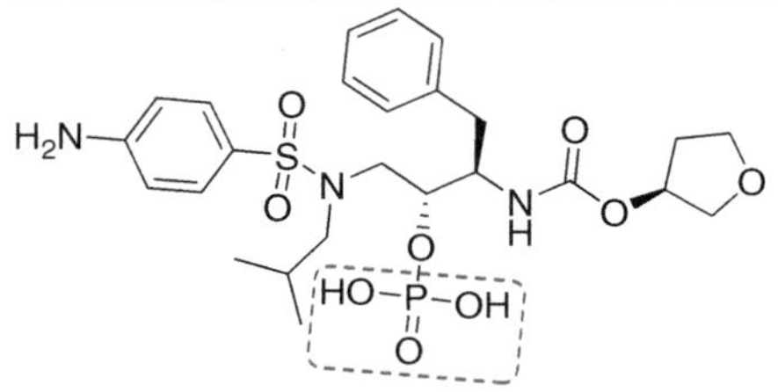 Fosamprenavir/Amprenavir | Hydrolysis by alkaline phosphatase | Improved aqueous solubility | HIV infections |
| 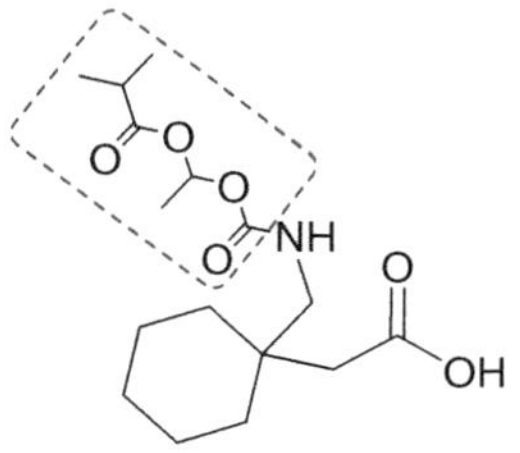 Gabapentin Enacarbil/Gabapentin | Hydrolysis of Isobutanoyloxy-ethoxy carbamate by esterases | Lipophilicity and permeability | Anticonvulsant and alleviates neuropathic pain |
| 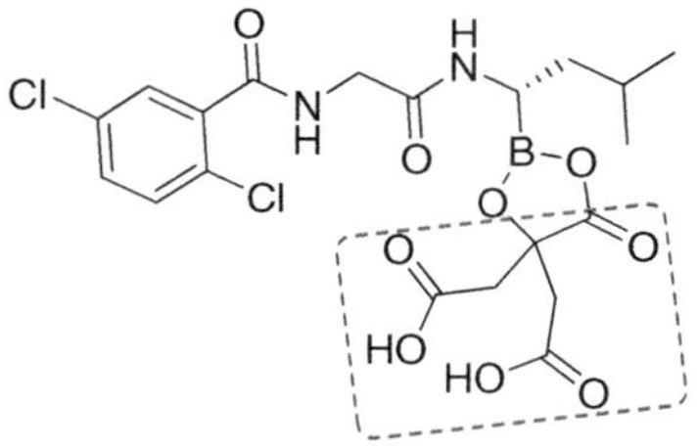 Ixazomib Citrate/Ixazomib | Hydrolysis of citrate ester in aqueous environment | Improved stability and dissolution rate in aqueous environment | Multiple Myeloma |
| 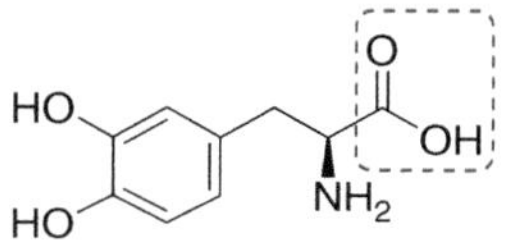 Levodopa/Dopamine | Decarboxylation by AADC (aromatic amino acid decarboxylase) | Improved uptake into brain by LAT1 | Parkinson's disease |
| 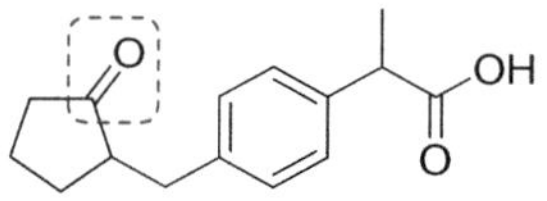 Loxoprofen/*trans*-OH Metabolite | Reduction of carbonyl group into hydroxyl | Lipophilicity and permeability | Anti-inflammatory |
| 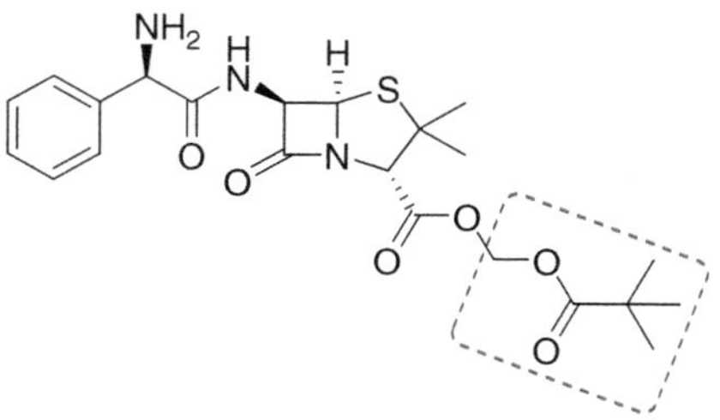 Pivampicillin/Ampicillin | Pivaloxymethylester hydrolysis by esterases | Lipophilicity and permeability | β-lactam antibiotic |

**Table 32.1** (Continued)

| **Prodrug structure (promoiety highlighted)/Active form** | **Prodrug approach/ activation mechanism** | **Improved property** | **Therapeutic indication** |
|---|---|---|---|
| 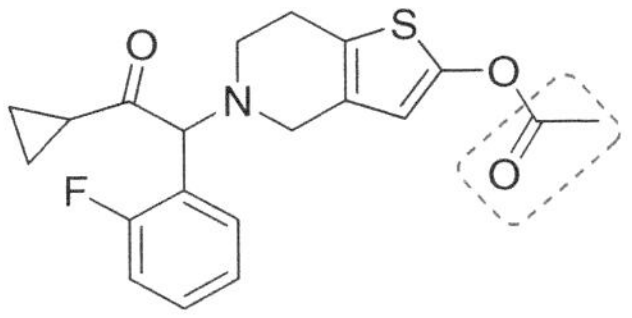 Prasugrel/R-138727 | Hydrolysis of acetyl ester to release thiolactone ring | Faster conversion to active metabolite | Antiplatelet |
| 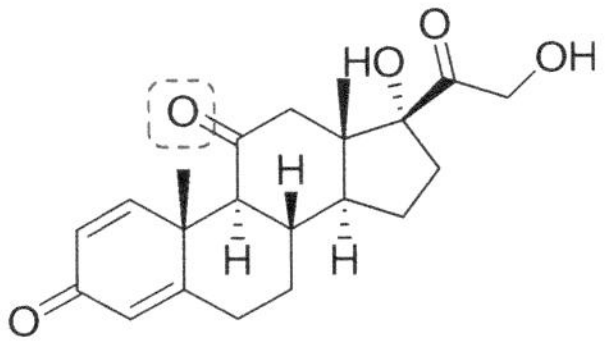 Prednisone/Prednisolone | Reduction of 11-dione to hydroxyl | Lipophilicity and permeability | Immunosuppressant |
| 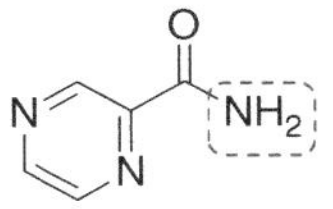 Pyrazinamide/Pyrazinoic acid | Hydrolysis of amide group | Lipophilicity and permeability | Anti-tuberculosis agent |
| 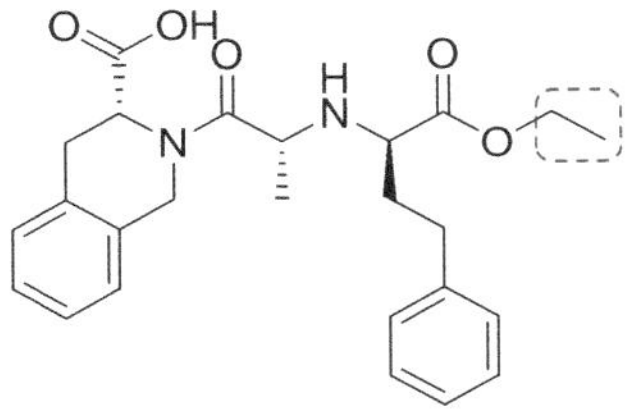 Quinapril/Quinaprilat | Hydrolysis of ethyl ester | Lipophilicity and permeability | Hypertension, congestive heart failure, and chronic renal failure |
| 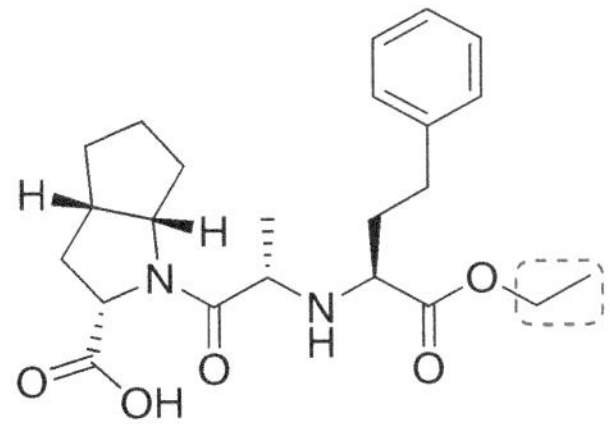 Ramipril/Ramiprilat | Hydrolysis of ethyl ester | Lipophilicity and permeability | Hypertension |
| 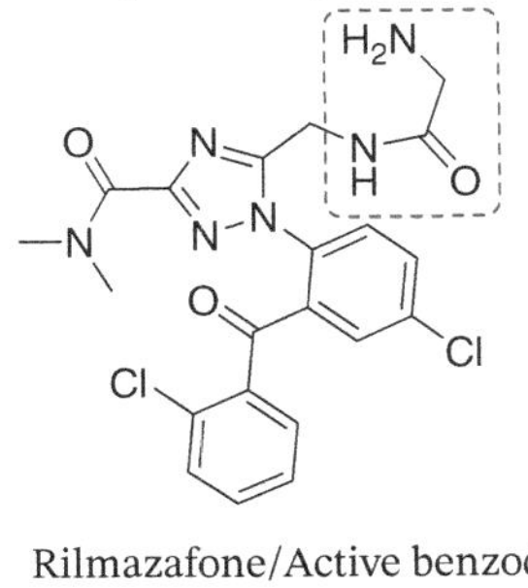 Rilmazafone/Active benzodiazapine form | Amide hydrolysis by aminopeptidase in intestine | Improved aqueous solubility | Sedative and hypnotic |

**Table 32.1** (Continued)

| Prodrug structure (promoiety highlighted)/Active form | Prodrug approach/ activation mechanism | Improved property | Therapeutic indication |
|---|---|---|---|
| 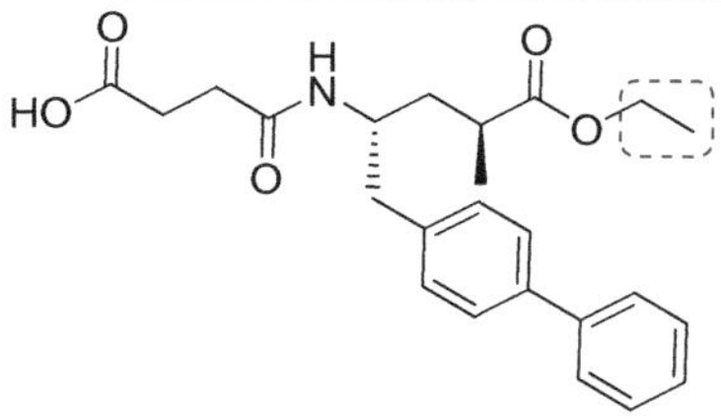 Sacubitril/Sacubitrilat | Ethyl ester hydrolysis by esterases | Lipophilicity and permeability | Antihypertensive drug |
| 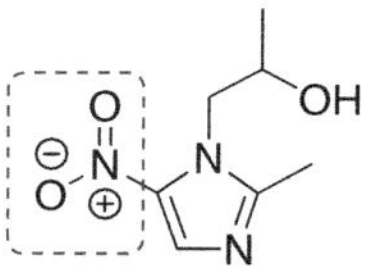 Secnidazole | Bacterial enzyme reduction of nitro group | Bioavailability | Bacterial vaginosis |
| 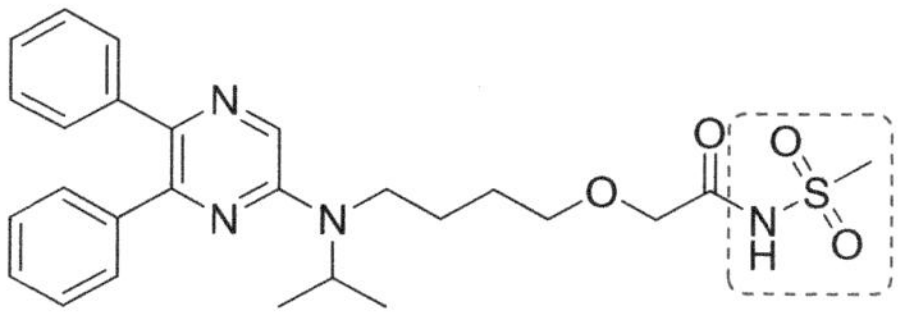 Selexipag/Radical anions | Hydrolysis of *N*-acyl sulfonamide | Prolonged duration of action | Pulmonary hypertension |
| 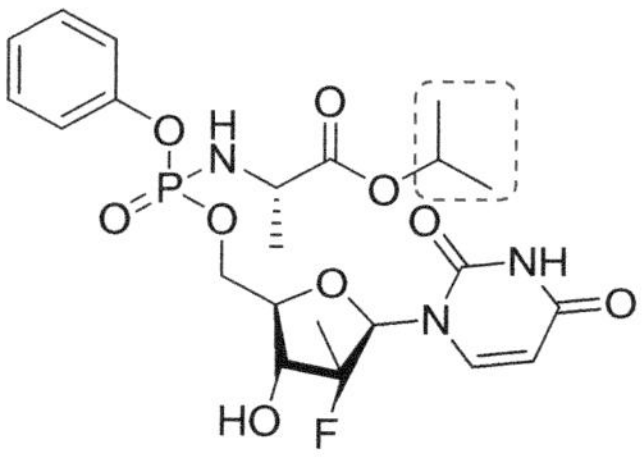 Sofosbuvir/2′-deoxy-2′-α-fluoro-β-*C*-methyluridine-5′-triphosphate | Aryloxy phosphoramidate pronucleotide/Ester hydrolysis of isopropyl ester and kinase action to phosphorylate into active form | Permeability and targeted delivery to liver | Hepatitis C infection |
| 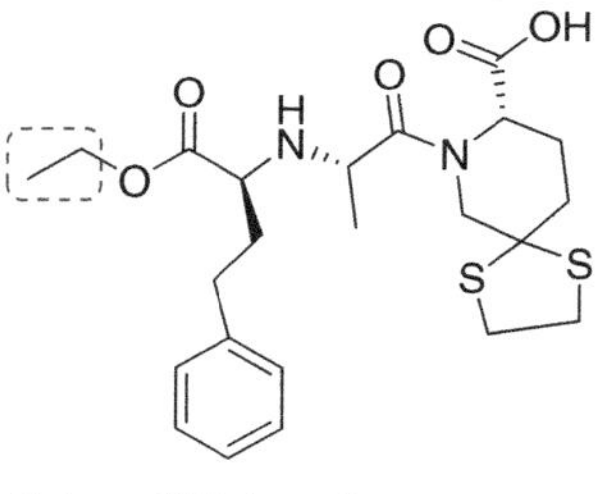 Spirapril/Spiraprilat | Ethyl ester hydrolysis by esterases | Lipophilicity and permeability | Hypertension |
| 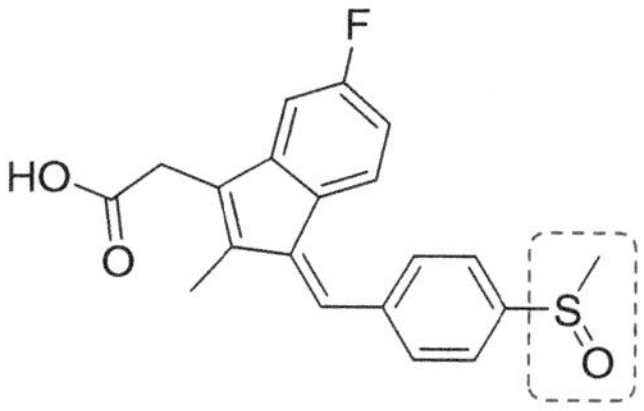 Sulindac/Sulindac sulfide and sulfone | Hepatic metabolism to active sulfide form | Improved aqueous solubility | Anti-inflammatory |

**Table 32.1** (Continued)

| Prodrug structure (promoiety highlighted)/Active form | Prodrug approach/ activation mechanism | Improved property | Therapeutic indication |
|---|---|---|---|
| 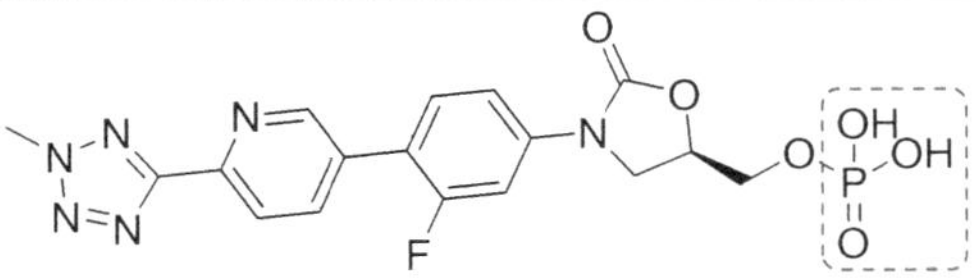 Tedizolid Phosphate/Tedizolid | Phosphate ester hydrolysis | Improved aqueous solubility | Topical antibiotic |
| 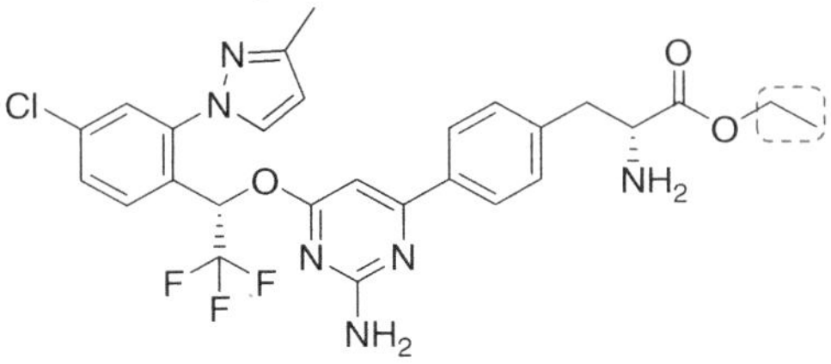 Telotristat Etiprate/Telotristat | Ethyl ester hydrolysis | Improved aqueous solubility | Diarrhea associated with carcinoid syndrome |
| 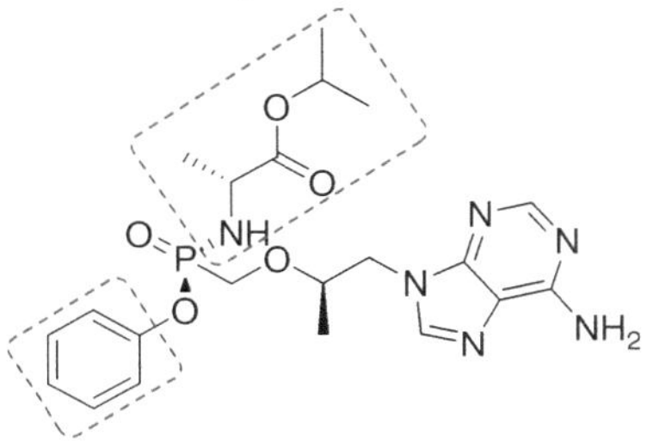 Tenofovir alafenamide/Tenofovir | Aryloxy phosphoramidate pronucleotide/Ester hydrolysis of isopropyl ester and kinase action to phosphorylate into active form | Lipophilicity and permeability | HIV infections |
| 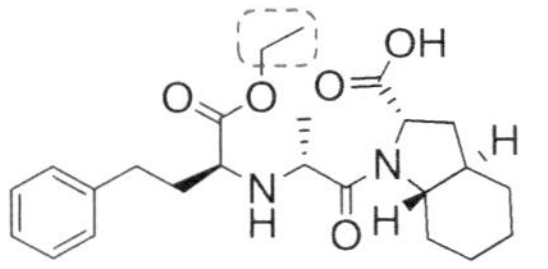 Trandolapril/Trandolaprilat | Ethyl ester hydrolysis | Lipophilicity and permeability | Hypertension |
| 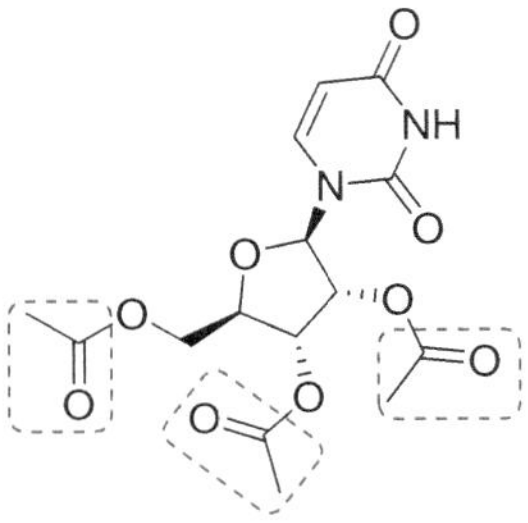 Uridine triacetate/Uridine | Hydrolysis of acetate groups by esterases | Lipophilicity and permeability | HIV infections |
| 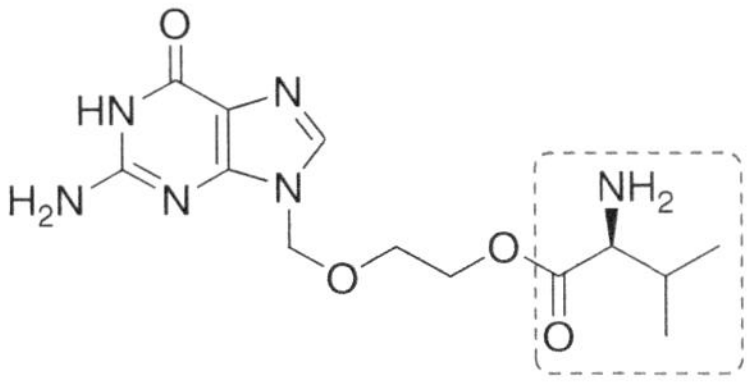 Valacyclovir/Acyclovir | Hydrolysis of valine ester by valacyclovir hydrolase | Increase bioavailability and uptake by hPEPT1 | Antiviral |

**Table 32.1** (Continued)

| Prodrug structure (promoiety highlighted)/Active form | Prodrug approach/ activation mechanism | Improved property | Therapeutic indication |
|---|---|---|---|
| 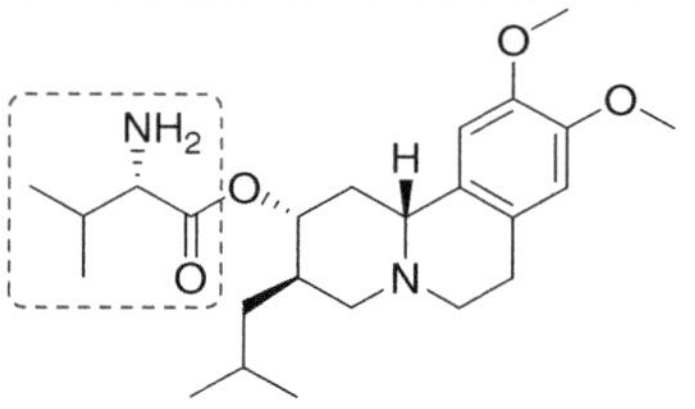 Valbenazine/Dihydrotetrabenazine | Hydrolysis of valine ester by valacyclovir hydrolase | Enhanced oral bioavailability | Tardive dyskinesia |
| 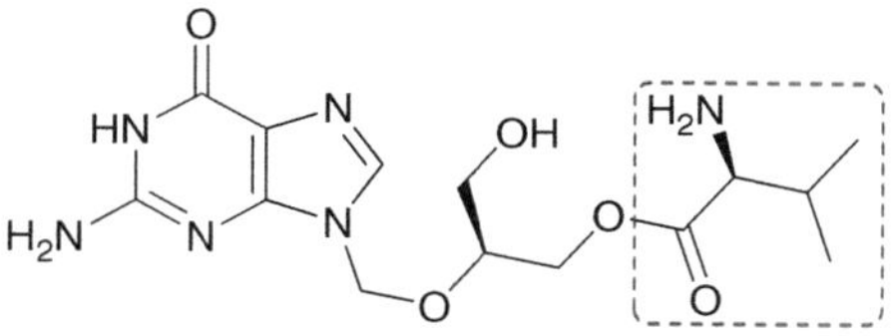 Valganciclovir/Ganciclovir | Hydrolysis of valine ester by valacyclovir hydrolase. | Increase uptake by hPEPT1 | Antiviral |
| 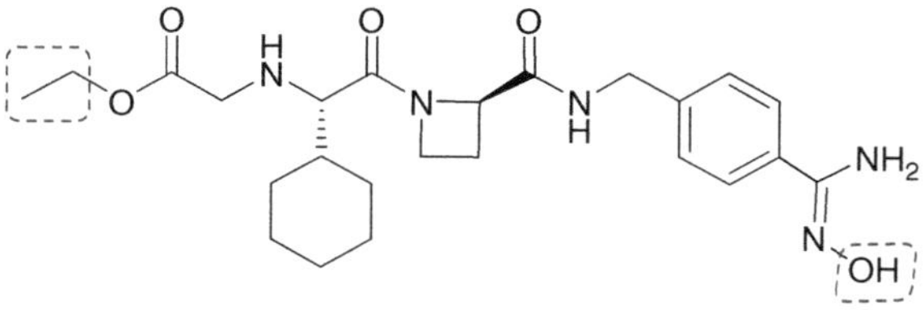 Ximelagatran/Melagatran | Ethyl ester hydrolysis by esterase and reduction of hydroxylamidine | Lipophilicity and permeability | Anticoagulant |

molecule (Figure 32.2). For example, the phosphate promoiety in the antibacterial drug tedizolid phosphate is directly attached to the hydroxyl group of the parent drug, tedizolid. The phosphate group is hydrolyzed by phosphatase enzymes in the GIT to give the biologically active tedizolid [32]. In the tripartite prodrugs, the promoiety is attached to the drug via a self-immolative linker (Figure 32.2). The prodrug pivampicillin is the pivaloyloxymethyl ester of ampicillin, which uses methylene spacer to link ampicillin and the pivalic acid promoiety. The incorporation of the pivalic ester promoiety increases the lipophilicity of the molecule resulting in higher oral bioavailability [33].

Mutual prodrugs or codrugs are a type of carrier-linked prodrug in which two pharmacologically active drugs are linked together. The two drugs act as carriers to each other thereby increasing the efficacy synergistically. Codrugs liberate both active molecules in the same target tissue after their bioconversion [21]. The antibacterial agent sultamicillin is a codrug which upon hydrolysis releases both the β-lactam antibiotic ampicillin and the β-lactamase inhibitor sulbactam [34].

## 32.6 Prodrug Strategies to Improve Aqueous Solubility

Aqueous solubility is an important factor in determining the oral bioavailability of drug molecules. Formulation strategies such as salt formation, micronization, complexation (e.g. cyclodextrins), or surfactants have been used to improve solubility [35]. When these formulation strategies fail to result in adequate enhancement in solubility or cause toxicity, the prodrug approach provides an alternative method to improve the solubility [36]. In the prodrug approach, the aqueous solubility can be increased by several folds utilizing polar non-ionizable promoieties (e.g. polyethylene glycol or sugars). Solubility can be even enhanced to a greater degree by using ionizable promoieties (e.g. succinic acid, amino, or phosphate groups) [21]. Phosphate esters are the most extensively used functional group to increase the oral bioavailability of many drugs with poor aqueous solubility. Phosphate prodrugs are stable in solution and readily undergo hydrolysis to the parent drug by alkaline phosphatases present in the brush border of the enterocytes in the GIT [37]. The release of the ionizable promoiety is typically fast and happens before or during absorption and therefore, the permeability of the parent drug is not affected. Fosamprenavir is a phosphate ester prodrug of the HIV protease inhibitor amprenavir [38]. Amprenavir had poor aqueous solubility and is therefore formulated using a combination of solubilizing excipients. To achieve the desired therapeutic concentration, patients had to take eight large capsules of amprenavir twice a day. On the other hand, the phosphate monoester fosamprenavir has higher aqueous solubility which decreased the daily dose to four tablets [39, 40]. The main advantage of fosamprenavir is providing a patient-compliant dosage regimen for patients with HIV-1 infection. Orally administered fosamprenavir is hydrolyzed by alkaline phosphatase in the apical membrane of the intestinal epithelium to liberate amprenavir. Tedizolid phosphate, prednisolone phosphate, estramustine phosphate, and fludarabine phosphate are other examples of orally administered phosphate prodrugs. Isavuconazonium sulfate represents a novel prodrug strategy to enhance solubility. Isavuconazonium sulfate is an acyloxyalkyl triazolium salt-based prodrug of the antifungal agent isavuconazole [41]. The introduction of sarcosine (N-methylglycine) in the promoiety of isavuconazole allows the generation of isavuconazonium sulfate salt, which has much higher aqueous solubility than the parent drug [42]. Isavuconazonium sulfate is hydrolyzed by plasma esterases and subsequent chemical reactions to release isavuconazole, thus enhancing the oral bioavailability [43].

## 32.7 Prodrug Approaches for Enhancing Absorption

The oral bioavailability of a drug is dependent upon its ability to permeate through the GIT mucosa and then survive the first-pass metabolism effect of enterocyte/liver enzymes. Many prodrug approaches have been developed to improve the oral absorption of drug molecules. The prodrug approach allows for molecules to cross the cell membrane barrier better than parent drug and then convert back to their active form once absorbed into the systemic circulation [44]. One example is the antiviral penciclovir, used to treat various herpesvirus infections. Penciclovir has a poor oral absorption (<5%) stemming from its polar dihydroxy alkyl chain. However, famciclovir, a prodrug of penciclovir, has acylated versions of those polar hydroxyl groups. This modification increased the oral bioavailability of famciclovir to 77% [45]. After ingestion of famciclovir, the molecule undergoes hepatic deacetylation and oxidation, thus giving the active penciclovir. Another example is propranolol, which has low oral bioavailability because of its high first-pass metabolism through O-glucuronidation. Garceau et al. [46] created a hemisuccinate version of propranolol to prevent glucuronidation of the active molecule. As a result, the prodrug was able to enhance the oral bioavailability of propranolol by fivefold over orally administered propranolol hydrochloride. The pyrimidine analogs 5-FU (5-Fluorouracil) and its prodrug capecitabine are used in the treatment of solid tumors. Unfortunately, 5-FU has highly variable GIT absorption and must be administered intravenously. To address this problem, multiple modifications are used to ensure a stable pharmacokinetic profile and good oral bioavailability for the prodrug, capecitabine. The first modification is the addition of a pentofuranose moiety to 5-FU followed by converting the 4-carbonyl moiety into a carbamoyl pentane. As a result, capecitabine is readily absorbed from the GIT and activated by liver metabolism with a predictable pharmacokinetic absorption profile. After passing the intestinal barrier, activation occurs by hydrolysis of the carbamate ester through liver carboxylesterase into 5′-deoxy-5-fluorouridine (5′-DFUR) which is converted to 5-FU through deamination and removal of the pentofuranose [47].

## 32.8 Prodrug Approaches for Targeting Enzymes

Certain prodrugs are designed to only activate at the target site of action. This activation is designed to be dependent upon a specific enzyme that is unique to a specific location. Inactivated prodrugs are designed to be otherwise inert resulting in less off-target side effects. Sulfasalazine is one such prodrug used for treatment of inflammatory bowel disease. Bacterial azoreductase found in the colon reduces the azo linkage between 5-aminosalicylic acid (5-ASA) and sulfapyridine. After activation, the released 5-ASA exerts an anti-inflammatory effect with minimal overall systemic absorption [48].

In addition to using naturally present site-specific enzymes, artificial enzymes have been introduced to target sites to activate prodrugs as well. This approach, known as antibody-directed enzyme prodrug therapy (ADEPT), was first demonstrated by Blakey et al. [49]. Their approach used a carboxypeptidase G2 enzyme (CPG2) conjugated to the fab fragment of an antitumor monoclonal antibody (A5B7) to selectively activate a nitrogen mustard prodrug (di-iodophenyl mustard). Francis et al. [50] used the A5CP antibody-CPG2 enzyme conjugate pretreatment to activate ZD2767P in colorectal carcinoma tumors for a phase I trial. ZD2767P is nitrogen mustard inactivated through attachment of a L-glutamic acid promoiety which is selectively removed by CPG2. In future work, the authors noted the need for a more specific and efficient delivery system to introduce enzyme into target sites. Gene directed enzyme prodrug therapy (GDEPT) is the other form of ADEPT which has been thoroughly explored. Rainov et al. [51] used the prodrug ganciclovir and herpes simplex virus-1 thymidine kinase (HSV-TK) for the treatment of glioblastoma solid tumors. In their approach, a retroviral vector transfected glioblastoma cells to express HSV-TK. Upon subsequent ganciclovir administration, transfected glioblastoma cells selectively formed phosphorylated ganciclovir resulting in the inhibition of mitosis. The authors noted in their phase III trial the need for improved transgene delivery vectors.

## 32.9 Prodrug Approaches for Targeting Membrane Transporters

Membrane transporters play an important role in controlling uptake and efflux of sugars, peptides, amino acids, bile acids, and drugs across cell membranes. Specific membrane transporters are expressed in the apical and/or basolateral membranes of enterocytes in the intestine. These transporters can be targeted using prodrug approach for increasing the oral bioavailability of drugs. The approach would be especially useful for ionized and polar drugs, that are largely transported across cell membranes using carrier-mediated transport. Prodrugs that target specific membrane transporters are designed to resemble physiological substrates, which drives the uptake of these drug molecules by the transporters present in the intestinal epithelium thus altering the pharmacokinetics of the drug. This prodrug strategy offers exciting new targets in drug design. The following section will discuss examples of prodrugs targeting these transporters.

### 32.9.1 Bile Acid Transporters

Bile acids are steroidal molecules synthesized from the catabolism of cholesterol in the liver and are secreted into the intestine to act as biological surfactants facilitating absorption of fat-soluble vitamins. Bile salts undergo enterohepatic circulation in which the majority of bile acids are reabsorbed into the liver. The main bile acid transporters are the apical sodium-dependent bile acid transporter (ASBT), the sodium taurocholate cotransporting polypeptide (NTCP), the bile salt export pump (BSEP), and the organic solute transporter alpha–beta (OSTα-OSTβ) located mainly in hepatocytes and enterocytes of the ileal epithelium [52]. The specificity and efficiency of these transporters for bile acids make them a great target for designing prodrugs to increase intestinal absorption [53]. For example, to enhance the oral bioavailability of the antiviral Acyclovir, four bile acid prodrugs (acyclovir chenodeoxycholate, deoxycholate, cholate, and ursodeoxycholate) were designed to target the human ASBT [54]. The prodrug analog acyclovir valylchenodeoxycholate showed 12-fold increase in permeability and 2-fold increase in bioavailability compared to acyclovir alone. In a more recent study, liver-targeted drug delivery of the anticancer camptothecin was evaluated through conjugation with deoxycholic acid [55]. The study in mice showed that the deoxycholic acid conjugate had increased stability and higher biodistribution and liver targeting than camptothecin.

### 32.9.2 Monocarboxylic Acid Transporters

Monocarboxylate transporters (MCTs) belong to solute carrier family (SLC16) and are responsible for the transport of nutrients, monocarboxylates like pyruvate and lactate, and ketone bodies like acetoacetate [56]. This function results in maintaining important cellular processes such as glycolysis and intracellular pH. There are 14 isoforms of monocarboxylate transporters which are involved in transport of drugs such as probenecid, bumetanide and nateglinide, thyroid hormones, carnitine, creatinine, and certain amino acids. Since MCTs are widely distributed in different tissues and have broad range of substrates they can affect the pharmacokinetics of many drug molecules. MCTs are also overexpressed in colon, breast, bone, and renal cancers, making them the target of many chemotherapy strategies. Therefore, intestinal MCTs represent promising target for designing oral prodrugs [57].

### 32.9.3 Nucleoside Transporters

Human nucleoside transporters (hNTs) are responsible for the transport of purines and pyrimidines across cell membranes. They can be divided into two categories: concentrative nucleoside transporters (CNTs) and equilibrative nucleoside transporters (ENTs). CNTs are sodium or proton-dependent transporters which are responsible for the uptake of nucleosides while ENTs are sodium-independent and regulate both efflux and uptake of nucleobases and nucleosides. The nucleoside transporters play an important role in the transport of various nucleoside-derived prodrugs such as the anticancer agents gemcitabine and fludarabine and the antivirals ribavirin and zidovudine. There has been significant increase in our understanding of the structure and function of nucleoside transporters [58]. Therefore, it is conceivable that substrates for these transporters can be used for the delivery of prodrugs conjugates into target cells.

### 32.9.4 Amino Acid Transporters

Amino acid transporters are membrane-bound transport proteins responsible for the transport of amino acids. These transporters belong to solute carrier family. L-type (LAT)-1 amino acid transporters are one of the most widely studied amino acid transporters and are expressed at the blood–brain barrier (BBB), bone marrow, placenta, blood–retinal barrier (BRB) and some tumors [59]. LAT-1 is responsible for the influx of several amino acids including tryptophan, histidine, alanine, asparagine, cysteine, glutamine, glycine, proline, serine, tyrosine, phenylalanine, threonine, valine, and glutamine. Due to high expression in brain and some tumors LAT-1 has already been targeted in the drug delivery of CNS drugs such as L-dopa, drug for Parkinson's disease, baclofen, gabapentin, and anticancer prodrug melphalan. Several prodrugs, with structural similarities to endogenous substrates of LAT-1, have been synthesized and studied. Different amino acid derivatives of several drugs including ketoprofen and gabapentin have been designed and synthesized as prodrugs and showed improved pharmacokinetic profiles by targeting LAT-1. The LAT1-mediated drug delivery is a promising approach that could potentially be used in prodrug strategy for the targeting the brain or cancer tissue [59].

### 32.9.5 Peptide Transporters

Peptide transporters belong to the solute carrier 15 (SLC15) family and are responsible for the proton-coupled transport of dipeptides and tripeptides [60]. PEPT1 is mainly expressed at the apical membrane of enterocytes in the small intestine, while PEPT2 is expressed in the kidney and brain [61]. PEPT1 is responsible for the absorption of dietary nitrogen in the small intestine while PEPT2 mediates renal uptake of di/tripeptides from the glomerular filtrate in the renal proximal tubule [62]. Peptide/histidine transporters PHT1 and PHT2 allow transport of both di/tripeptides and histidine. Various prodrugs such as beta-lactam antibiotics (e.g. cefadroxil), angiotensin-converting enzyme inhibitors (e.g. enalapril), L-Dopa, antiviral agents (e.g. valacyclovir, oseltamivir, and valganciclovir), doxorubicin and didanosine prodrugs interact with the SLC15 family of transporters [63].

Incorporation of a valine promoiety in the antiviral prodrug valacyclovir resulted in a fivefold higher oral bioavailability compared to the parent drug. The valine promoiety allowed acyclovir to be transported through intestinal PEPT1. Various dipeptide ester prodrugs of acyclovir and val-val-acyclovir conjugates have been shown to display higher affinity towards PEPT1 [64].

Oseltamivir carboxylate is an oral prodrug of oseltamivir, a neuraminidase inhibitor used in the treatment of influenza. A more potent analog of oseltamivir carboxylate was developed by incorporating a guanidine moiety in the molecule. However, incorporating the guanidine moiety resulted in low membrane permeability. A valine ester prodrug of this molecule targeting PEPT1 exhibited higher intestinal permeability than the parent oseltamivir carboxylate [64]. PEPT1 and PHT1 are also linked to inflammatory bowel disease and therefore prodrugs targeting these transporters could be useful in the treatment of this disease [60].

## 32.10 Conclusion

The prodrug approach has recently gained interest in drug design. Over the past decade, prodrugs have accounted for more than 10% of the new drug molecules approved by Food and Drug Administration (FDA). Prodrugs provide the ability to improve the pharmacokinetic profile of molecules, enhance drug targeting to specific tissues, and overcome physiological barriers facing these drug molecules (e.g. first-pass metabolism, poor penetration of GIT barriers, and low solubility). In addition, prodrug approaches have allowed the targeting of drugs to specific tissues in cancers, hepatic diseases, and viral infections. The design of a prodrug takes into consideration the properties of the promoiety and the parent molecule in order to optimize the final compound. Most of the prodrug strategies have focused on enhancing lipophilicity and hence the permeability of the parent drug molecule. Enhancing aqueous solubility is another advantage of the prodrug design, which has recently been exploited in few of the approved drugs using a phosphate group. Intestinal transporters targeted prodrug derivatization has recently gained momentum in site-selective delivery especially for anticancer agents.

In the past, the prodrug approach was only considered after exhausting all other medicinal chemistry options. Today, various prodrug strategies continue to be an integral part in the early stages of the drug discovery process. The prodrug approach is anticipated to continue being harnessed more frequently to tackle complex drug targets or to optimize existing therapeutics.

## Abbreviations

| | |
|---|---|
| 5′-DFUR | 5′-deoxy-5-fluorouridine |
| 5-ASA | 5-aminosalicylic acid |
| 5-FU | 5-florouracil |
| ABC | ATP-binding cassette |
| ADEPT | antibody-directed enzyme prodrug therapy |
| ADME | absorption distribution metabolism excretion |
| ASBT | apical sodium-dependent bile acid transporter |
| ATP | adenosine-5′-triphosphate |
| BCRP | breast cancer resistance protein |
| BSEP | bile salt export pump |
| CNT | concentrative nucleoside transporter |
| CPG2 | carboxypeptidase G2 enzyme |
| CYP | cytochrome P450 |
| CYP3A4 | cytochrome P450 3A4 |
| ENT | equilibrative nucleoside transporter |
| FDA | Food and Drug Administration |
| GDEPT | gene directed enzyme prodrug therapy |
| GIT | gastrointestinal tract |
| HIV | human immunodeficiency virus |
| Hnt | human nucleoside transporter |
| HSV-TK | herpes simplex virus-1 thymidine kinase |
| JAM | junctional adhesion molecules |
| LAT1 | L-type amino acid transporter |
| MCT | monocarboxylate transporters |
| MRP2 | multidrug resistance-associated protein 2 |
| NTCP | sodium taurocholate co-transporting polypeptide |
| OAT-1 | organic anion transporter 1 |
| OST | organic solute transporter |
| PEPT | peptide transporter |
| P-Gp | P-glycoprotein |
| PHT | peptide/histidine transporter |
| SLC | solute carrier |

## References


1 Sastry, S.V., Nyshadham, J.R., and Fix, J.A. (2000). Recent technological advances in oral drug delivery – a review. *Pharm. Sci. Technol. Today* 3: 138–145.

2 Chordiya, M., Gangurde, H., and Borkar, V. (2017). Technologies, optimization and analytical parameters in gastroretentive drug delivery systems. *Curr. Sci.* 112: 946–953.

3 Guengerich, F. (2008). Cytochrome P450 and chemical toxicology. *Chem. Res. Toxicol.* 21 (1): 70–83.

4 Watkins, P.B., Wrighton, S.A., Schuetz, E.G. et al. (1987). Identification of glucocorticoid-inducible cytochromes P-450 in the intestinal mucosa of rats and man. *J. Clin. Invest.* 80: 1029–1036.

5 Bailey, D.G., Spence, J.D., Munoz, C., and Arnold, J.M.O. (1991). Interaction of citrus juices with felodipine and nifedipine. *Lancet* 337: 268–269.

6 Hediger, M.A., Romero, M.F., Peng, J.B. et al. (2004). The ABCs of solute carriers: physiological, pathological and therapeutic implications of human membrane transport proteins. *Pflugers Archiv.* 447: 465–468.

7 Van Waterschoot, R.A., ter Heine, R., Wagenaar, E. et al. (2010). Effects of cytochrome P450 3A (CYP3A) and the drug transporters P-glycoprotein (MDR1/ABCB1) and MRP2 (ABCC2) on the pharmacokinetics of lopinavir. *Br. J. Pharmacol.* 160 (5): 1224–1233.

8 Lin, L., Yee, S.W., Kim, R.B., and Giacomini, K.M. (2015). SLC transporters as therapeutic targets: emerging opportunities. *Nat. Rev. Drug Discovery* 14: 543–560.

9 Hendrikx, J.J.M.A., Lagas, J.S., Wagenaar, E. et al. (2014). Oral co-administration of elacridar and ritonavir enhances plasma levels of oral paclitaxel and docetaxel without affecting relative brain accumulation. *Br. J. Cancer* 110: 2669–2676.

10 Aoki, F.Y. and Doucette, K.E. (2001). Oseltamivir: a clinical and pharmacological perspective. *Expert Opin. Pharmacother.* 2: 1671–1683.

11 Le, Q.M., Kiso, M., Someya, K. et al. (2005). Avian flu: Isolation of drug-resistant H5N1 virus. *Nature* 437 (7062): 1108.

12 Cass, L., Efthymiopoulos, C., and Bye, A. (1999). Pharmacokinetics of zanamivir after intravenous, oral inhaled or intranasal administration to healthy volunteers. *Clin. Pharmacokinet.* 36: 1–11.

13 Gupta, S.V., Gupta, D., Sun, J. et al. (2011). Enhancing the intestinal membrane permeability of zanamivir: a carrier mediated prodrug approach. *Mol. Pharmaceutics* 8: 2358–2367.

14 Abuhelwa, A.Y., Williams, D.B., Upton, R.N., and Foster, D.J. (2017). Food, gastrointestinal pH, and models of oral drug absorption. *Eur. J. Pharm. Biopharm.* 112: 234–248.

15 Mudie, D.M., Amidon, G.L., and Amidon, G.E. (2010). Physiological parameters for oral delivery and in vitro testing. *Mol. Pharmaceutics* 7: 1388–1405.

16 Turner, J.R. (2009). Intestinal mucosal barrier function in health and disease. *Nat. Rev. Immunol.* 9: 799–809.

17 Yang, N.J. and Hinner, M.J. (2015). Getting across the cell membrane: an overview for small molecules, peptides, and proteins. *Methods Mol. Biol.* 1266: 29–53.

18 Philip, A. and Philip, B. (2010). Colon targeted drug delivery systems: a review on primary and novel approaches. *Oman Med. J.* 25: 70–78.

19 Rees, D.C., Johnson, E., and Lewinson, O. (2009). ABC transporters: the power to change. *Nat. Rev. Mol. Cell Biol.* 10: 218–227.

20 Gupta, S., Kesarla, R., and Omri, A. (2013). Formulation strategies to improve the bioavailability of poorly absorbed drugs with special emphasis on self-emulsifying systems. *ISRN Pharm.* 2013: 1–16.

21 Rautio, J., Meanwell, N.A., Di, L., and Hageman, M.J. (2018). The expanding role of prodrugs in contemporary drug design and development. *Nat. Rev. Drug Discovery* 17: 559–587.

22 Huang, Y. and Daib, W.G. (2014). Fundamental aspects of solid dispersion technology for poorly soluble drugs. *Acta Pharm. Sin. B* 4 (1): 18–25.

23 Serajuddin, A.T.M. (2007). Salt formation to improve drug solubility. *Adv. Drug Deliv.* 59: 603–616.

24 Brittain, H.G. (2002). Effects of mechanical processing on phase composition. *J. Pharm. Sci.* 91: 1573–1580.

25 Chaudhari, S.P. and Dugar, R.P. (2017). Application of surfactants in solid dispersion technology for improving solubility of poorly water soluble drugs. *J. Drug Delivery Sci. Technol.* 41: 68–77.

26 Parquet, N., Reigneau, O., Humbert, H. et al. (2000). New oral formulation of cyclosporin A (Neoral) pharmacokinetics in allogeneic bone marrow transplant recipients. *Bone Marrow Transplant.* 25: 965–968.

27 Albert, A. (1958). Chemical aspects of selective toxicity. *Nature* 182: 421–423.

28 Huttunen, K.M., Raunio, H., and Rautio, J. (2011). Prodrugs-from serendipity to rational design. *Pharmacol. Rev.* 63: 750–771.

29 Rautio, J., Kumpulainen, H., Heimbach, T. et al. (2008). Prodrugs: design and clinical applications. *Nat. Rev. Drug Discovery* 7: 255–270.

30 Hedner, T., Samulesson, O., Wahrborg, P. et al. (2004). Nabumetone: therapeutic use and safety profile in the management of osteoarthritis and rheumatoid arthritis. *Drugs* 64: 2315–2343.

31 Karaman, R. (2013). Prodrugs design based on inter- and intramolecular chemical processes. *Chem. Biol. Drug Des.* 82: 643–668.

32 Wong, E. and Rab, S. (2014). Tedizolid phosphate (sivextro): a second-generation oxazolidinone to treat acute bacterial skin and skin structure infections. *Pharm. Ther.* 39 (8): 555–579.

33 Von Daehne, W., Godtfredsen, W.O., Roholt, K., and Tybring, L. (1970). Pivampicillin, a new orally active ampicillin ester. *Antimicrob. Agents Chemother.* 10: 431–437.

34 Baltzer, B., Binderup, E., von Daehne, W. et al. (1980). Mutual pro-drugs of beta-lactam antibiotics and beta-lactamase inhibitors. *J. Antibiot.* 33: 1183–1192.

35 Williams, H.D., Trevaskis, N.L., Charman, S.A. et al. (2013). Strategies to address low drug solubility in discovery and development. *Pharmacol. Rev.* 65 (1): 315–499.

36 Stella, V.J. and Nti-Addae, K.W. (2007). Prodrug strategies to overcome poor water solubility. *Adv. Drug Delivery Rev.* 59 (7): 677–694.

37 Heimbach, T., Fleisher, D., and Kaddoumi, A. (2007). Overcoming poor aqueous solubility of drugs for oral delivery. In: *Prodrugs: Challenges and Rewards*, vol. 1 (ed. V.J. Stella, R.T. Borchardt, M.J. Hageman, et al.), 157–215. Springer.

38 Sorbera, L.A., Martín, L., Castañer, J., and Castañer, R.M. (2001). Fosamprenavir. Anti-HIV, HIV protease inhibitor. *Drugs Future* 26: 224–231.

39 Falcoz, C., Jenkins, J.M., Bye, C. et al. (2002). Pharmacokinetics of GW433908, a prodrug of amprenavir, in healthy male volunteers. *J. Clin. Pharmacol.* 42: 887–898.

40 Wire, M.B., Shelton, M.J., and Studenberg, S. (2006). Fosamprenavir: clinical pharmacokinetics and drug interactions of the amprenavir prodrug. *Clin. Pharmacokinet.* 45: 137–168.

41 Walker, R.C., Zeuli, J.D., and Temesgen, Z. (2016). Isavuconazonium sulfate for the treatment of fungal infection. *Drugs Today* 52: 7–16.

42 Ohwada, J., Tsukazaki, M., Hayase, T. et al. (2003). Design, synthesis and antifungal activity of a novel water soluble prodrug of antifungal triazole. *Bioorg. Med. Chem. Lett.* 13: 191–196.
43 Schmitt-Hoffmann, A., Roos, B., Heep, M. et al. (2006). Single-ascending-dose pharmacokinetics and safety of the novel broad-spectrum antifungal triazole BAL4815 after intravenous infusions (50, 100, and 200 milligrams) and oral administrations (100, 200, and 400 milligrams) of its prodrug, BAL8557, in healthy volunteers. *Antimicrob. Agents Chemother.* 50: 279–285.
44 Beaumont, K., Webster, R., Gardner, I., and Dack, K. (2003). Design of ester prodrugs to enhance oral absorption of poorly permeable compounds: challenges to the discovery scientist. *Curr. Drug Metab.* 4 (6): 461–485.
45 Vere Hodge, R.A. (1993). Famciclovir and penciclovir. The mode of action of famciclovir including its conversion to penciclovir. *Antiviral Chem. Chemother.* 4: 67–84.
46 Garceau, Y., Davis, I., and Hasegawa, J. (1978). Plasma propranolol levels in beagle dogs after administration of propranolol hemisuccinate ester. *J. Pharm. Sci.* 67 (10): 1360–1363.
47 Li, F., Maag, H., and Alfredson, T. (2008). Prodrugs of nucleoside analogues for improved oral absorption and tissue targeting. *J. Pharm. Sci.* 97 (3): 1109–1134.
48 Hoult, J.R. (1986). Pharmacological and biochemical actions of sulphasalazine. *Drugs* 32: 18–26.
49 Blakey, D.C., Valcaccia, B.E., East, S. et al. (1993). Antitumor effects of an antibody-carboxypeptidase g2 conjugate in combination with a benzoic acid mustard prodrug. *Cell Biophys.* 22 (1-3): 1–8.
50 Francis, R.J., Sharma, S.K., Springer, C. et al. (2002). A phase I trial of antibody directed enzyme prodrug therapy (ADEPT) in patients with advanced colorectal carcinoma or other CEA producing tumours. *Br. J. Cancer* 87 (6): 600–607.
51 Rainov, N.G., Kramm, C.M., Banning, U. et al. (2000). Immune response induced by retrovirus-mediated HSV-tk/GCV pharmacogene therapy in patients with glioblastoma multiforme. *Gene Ther.* 7 (21): 1853–1858.
52 Dawson, P.A., Lan, T., and Rao, A. (2009). Bile acid transporters. *J. Lipid Res.* 50: 2340–2357.
53 Sievänen, E. (2007). Exploitation of bile acid transport systems in prodrug design. *Molecules* 12: 1859–1889.
54 Tolle-Sander, S., Lentz, K.A., Maeda, D.Y. et al. (2004). Increased acyclovir oral bioavailability via a bile acid conjugate. *Mol. Pharmaceutics* 1 (1): 40–48.
55 Xiao, L., Yu, E., Yue, H., and Li, Q. (2019). Enhanced liver targeting of camptothecin via conjugation with deoxycholic acid. *Molecules* 24: 1179.
56 Felmlee, M., Jones, R.S., Rodriguez-Cruz, V. et al. (2020). Monocarboxylate transporters (SLC16): function, regulation, and role in health and disease. *Pharmacol. Rev.* 72: 466–485.
57 Sun, Y., Zhao, D., Wang, G. et al. (2019). A novel oral prodrug-targeting transporter MCT 1: 5-fluorouracil-dicarboxylate monoester conjugates. *Asian J. Pharm. Sci.* 14: 631–639.
58 Gray, J.H., Owen, R.P., and Giacomini, K.M. (2004). The concentrative nucleoside transporter family, SLC28. *Eur. J. Phys.* 447: 728–734.
59 Puris, E., Gynther, M., Auriola, S., and Huttunen, K.M. (2020). L-Type amino acid transporter 1 as a target for drug delivery. *Pharm. Res.* 37: 88.
60 Smith, D.E., Clémençon, B., and Hediger, M.A. (2013). Proton-coupled oligopeptide transporter family SLC15: Physiological, pharmacological and pathological implications. *Mol. Aspects Med.* 34: 323–336.
61 Ganapathy, M.E., Brandsch, M., Prasad, P.D. et al. (1995). Differential recognition of b-lactam antibiotics by intestinal and renal peptide transporters, PEPT 1 and PEPT 2. *J. Biol. Chem.* 270: 25672–25677.
62 Newstead, S. (2017). Recent advances in understanding proton coupled peptide transport via the POT family. *Curr. Opin. Struct. Biol.* 45: 17–24.
63 Alexander, S.P.H., Kelly, E., Mathie, A. et al. (2019). The concise guide to pharmacology 2019/20: transporters. *Br. J. Pharmacol.* 176: S397–S493.
64 Murakami, T. (2016). A minireview: usefulness of transporter-targeted prodrugs in enhancing membrane permeability. *J. Pharm. Sci.* 105: 2515–2526.

# 33

# Gastroretentive Drug Delivery Systems

*Vrushali Waknis[1] and Ajit S. Narang[2]*

[1] AbbVie, Inc., Irvine, CA, USA
[2] Department of Pharmaceutical Sciences, ORIC Pharmaceuticals, Inc., South San Francisco, CA, USA

## 33.1 Introduction

Gastroretentive, or gastric retentive, indicates retention of a dosage form in the stomach for an extended period compared to normal gastric emptying time. Thus, a gastroretentive drug delivery system (GRDDS) is one of the approaches of modulating the delivery of a drug administered orally with a desire to improve patient outcomes by enhancing and/or modulating the biopharmaceutical presentation of the drug at the site of drug absorption [1]. For example,

- A drug that undergoes passive absorption in the intestine and is more soluble in the acidic gastric environment compared to the basic intestinal milieu may be presented at its absorption site in the solution state – thus increasing the total amount of the drug absorbed and potentially reducing variability in oral drug absorption.
  - The improved bioavailability and lower variability could be particularly valuable for drugs with a small window of drug absorption, i.e. for drugs whose absorption is limited to the upper portions of the gastrointestinal (GI) tract.
  - A GRDDS effectively removes gastric emptying and site of drug release in the GI tract as a variable in drug absorption.
- A drug that is retained longer and released slowly in the stomach has the potential to increase the duration of drug absorption from a single dosage form, thus reducing the needed frequency of drug administration.
- A drug that is retained for a longer duration in the stomach for an intended local action in the stomach, such as propantheline bromide and benactyzine hydrochloride.

For a drug to be suitable candidate for a GRDDS, it should have certain preferred attributes such as,

- Stability in the acidic gastric environment
- Relatively low dose
- Solubility at the low stomach pH

GRDDS have been studied extensively over the years and several informative and in-depth review articles [2–7] are available in the literature that elaborates on the select aspects of these devices. Minimizing repetition, this chapter will focus on high-level overview and building a pragmatic understanding of the technology platforms, their cross-functional development considerations, and products in development and market that utilize the principles of gastroretentive drug delivery.

## 33.2 Oral Drug Delivery – Challenges and Opportunities

Oral administration is the most common and the most preferred route of drug delivery. It affords a natural (e.g. same as food) route of administration, improved patient compliance (vs. injectables), and also simplifies development and manufacturing for the new drug candidates.

Oral drug delivery can be divided into two main categories:

1. Immediate release (IR), which includes most conventional dosage forms that provide pharmacologic activity, including potential relief from ailment, typically in 0.5–1 hour after ingestion. Technically, these dosage forms could be defined as the ones that achieve about 80% of drug release within 45 minutes of when in contact with an aqueous solution.
2. Modified release (MR), the dosage forms that modulate drug release with either timed or spatial control

*Oral Bioavailability and Drug Delivery: From Basics to Advanced Concepts and Applications*, First Edition. Edited by Ming Hu and Xiaoling Li.

to allow for improved patient compliance and disease outcomes. MR could be divided further into:

a. Programmed release that can be subdivided into
   i. Sustained or extended (prolonged over a specific time interval)
   ii. Controlled (rate of drug release over a period of time is predefined and controlled),
   iii. Delayed (drug release starts after a lag period upon contact with aqueous medium).
b. Local release, i.e. at a particular site within the GI tract (focusing on oral route of administration). These could be, for example,
   i. Sublingual or buccal
   ii. Colonic
   iii. Gastroretentive or gastric retentive

GRDDS are both, spatial and programmed release drug delivery systems (DDS). In other words, a DDS that is designed to retain in the stomach for an extended period must also sustain/extend or control drug release (i.e. MR), which would allow the drug to be released slowly over that period of time or after a lag period (instead of, for example, an IR system).

To develop an effective gastroretentive system, it is crucial to understand gastric physiology and its impact on drug delivery [8, 9].

## 33.3 Human Gastric Physiology Relevant to GRDDS Design

GI tract starts at the mouth and pharynx followed by the esophagus, stomach, small intestine made up of duodenum, jejunum, and ileum, and large intestine that consists of cecum, colon, rectum, and anus. The three major sections that are relevant for dosage form dissolution and drug absorption are the stomach, small intestine, and large intestine. Muscular contractions of both the stomach and small intestine move food and other materials down the GI tract [2].

### 33.3.1 Gastrointestinal Transit Times

Normal motility of the GI tract is characterized in terms of transit time through different "compartments" of the GI tract, which are utilized in modeling drug absorption. The transit time is defined as the time taken for a dosage form or its components to pass through a compartment. For example, the following parameters are utilized in the GastroPlus software (Simulations Plus, Inc., Lancaster, CA) for simulation of human drug absorption after oral administration [2].

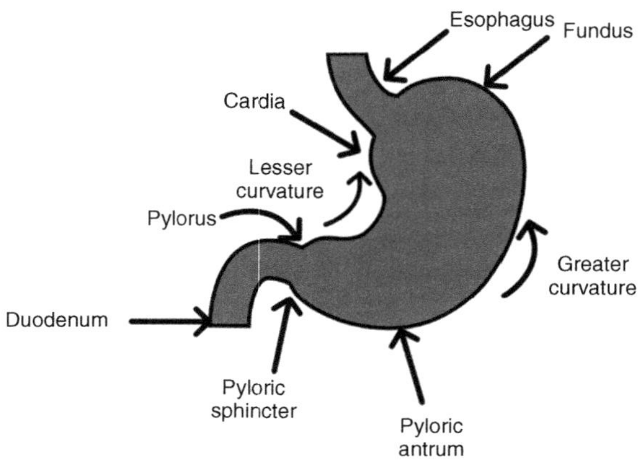


**Figure 33.1** Anatomical compartments of the human stomach. Source: Modified from Tripathi et al. [10].

- **Stomach**: Gastric emptying time in the fasted transit time is generally less than half an hour, while high fat breakfast can increase the gastric emptying time to several hours.
- **Small intestine**: Transit time through different intestinal segments is estimated based on the volume of fluid in each segment. The average small intestinal transit time is considered about 3 hours.
- **Caecum**: Transit time for human caecum is 4.5 hours.
- **Colon**: Human colon transit time is generally considered to be 13.5 hours.

Of all the stages of GI transit, gastric emptying provides greatest influence on the rate of oral drug absorption since an orally administered dosage form encounters the stomach first [2]. The different parts of the human stomach are provided in Figure 33.1.

### 33.3.2 Physiology of Gastric Emptying

In the stomach, the pyloric valve controls the passage of food and fluids to the small intestine, and in the small intestine, the ileocaecal valves modulate the passage of food and other materials from the small intestine to the large intestine [1]. The rate of transfer of drug product from one segment of the GI tract to the next can influence the time available for drug dissolution or absorption available in one component. The mechanism by which gastric emptying in fasted state occurs is by migrating myoelectric/motor complex (MMC). MMC consists of four phases as shown in Figure 33.2.

Phase III is referred to as the housekeeper wave which is made of intense contractions that propel undigested food out. This wave is a series of contractions that begin at the apex of the stomach and moves down the length of the GI tract – sweeping out the tract as it moves

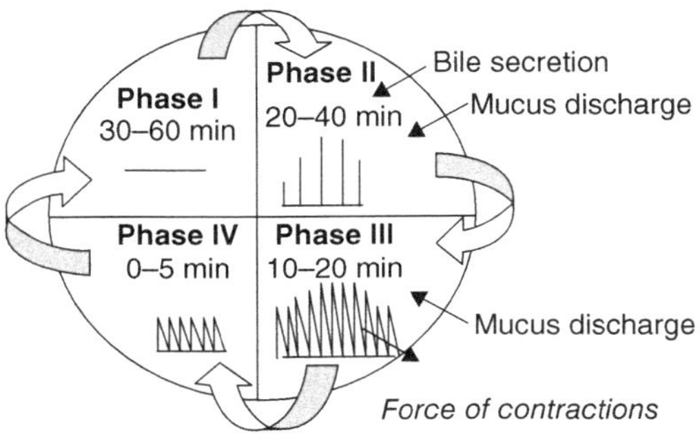


**Figure 33.2** Four phases of MMC; phase III is considered to be the most intense housekeeper wave. Source: Chawla et al. [11] / Pharmaceutical Technology.

along. These contractions occur over a period of about five to fifteen minutes and are repeated on a regular basis (every two hours) throughout the period that the stomach remains in the fasted state [5]. Hence the dosage form must be designed to resist these contractile motions of the stomach [11, 12].

The stomach with food/fed state goes into postprandial motility characterized by mixing and peristalsis. In addition to the emptying of stomach contents, the gastric muscles exert mechanical pressure on the dosage form. In addition to the timing of dosing, e.g. fed vs fasted state (and the phase of fasted state), the motility of the stomach is an important consideration for successful dosage form design [11–13].

### 33.3.3 Factors Impacting Gastric Emptying

Transit time from the stomach to the small intestine depends upon:

a. ***Fed or Fasted state of the stomach***: A fasted gastric transit time has been observed to be typically less than two hours while a fed gastric transit time can range from two to six hours [14]. As illustrated in Figure 33.3, the influence of calories in a meal can have a direct correlation with gastric retention time thus influencing the dosage form retention time in stomach significantly [1].
b. ***Type of meal***: Indigestible polymers or a high protein or fat meal can increase the gastric retention times by upto 4–10 hours.
c. ***Concomitant administration of other drugs or herbs***: certain drugs and herbs e.g. anticholinergics can decrease GI motility thereby increasing gastric retention time [3].
d. ***Age and gender***: It has been shown that gastric retention in adults over 70 years of age is longer. Similarly, healthy adult males have shorter gastric retention times (3.4 ± 0.6 hours) than females of the same age (4.6 ± 1.2 hours) regardless of weight, height, and body surface area [14].
e. ***Size of the dosage form***: The critical size of a GRDDS for retention in the stomach can be inferred from the average size of the pylorus in adult humans (about 12.7 mm ± 7 mm in diameter [1]). Hence, non-disintegrating 7 mm tablets pass the pyloric valve in the fed state, but 13 mm tablets are retained in the fasted state [1]. It has also been found that multiparticulate systems avoid the "all or none" phenomenon of gastric emptying compared to monolithic systems and may offer an advantage for gastroretention.
f. ***Subject posture***: Posture of subject ingesting a GRDDS will impact the performance of the dosage form. Floating dosage forms tend to remain buoyant above the gastric contents in an upright position and are protected from gastric postprandial emptying. However, this buoyancy of dosage form is not similar in a supine subject position since when the dosage form is in a distal position it may be swept away by the gastric emptying motions. Hence, dosing a buoyancy-based GRDDS in a supine position is discouraged [11].

### 33.3.4 Interaction of Drug Bioavailability with Gastric Physiology

Basic drugs that are administered as solid particles or tablets, the most common pharmaceutical scenario, must first dissolve in the acidic gastric environment before being transported as drug solutions to the upper intestinal tract, where most of the drug absorption takes place. In some cases, rapid gastric emptying can lead to incomplete drug dissolution in the stomach, leading to transfer of partially undissolved drug particles in the duodenum. This can not only lead to incomplete drug dissolution and absorption, but the undissolved drug particles can serve as nucleation sites for precipitation of dissolved drug in the duodenum. This can further reduce the extent of drug absorption and also introduce inter-individual variability. This phenomenon is the main reason for variability in drug absorption in monkey models for many drugs that exhibit pH-based solubility leading to supersaturation in the duodenum [4].

## 33.4 Technologies

To enhance physiological gastric retention for drug dissolution and absorption, formulation technologies

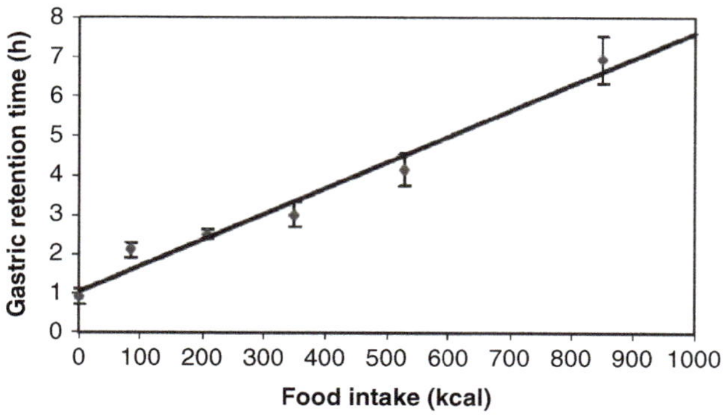


**Figure 33.3** Correlation between gastric retention time and calorie content of a meal. Source: Waterman [1, 2] / Taylor & Francis.

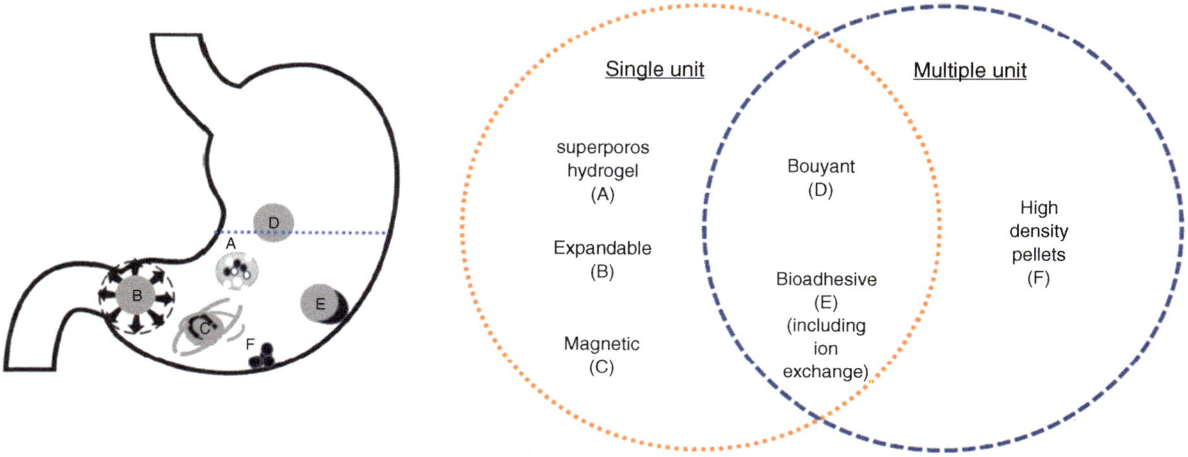


**Figure 33.4** Different technologies available for development of GRDDS and their mechanism of action in the stomach. Source: Adapted from Refs. [10, 15].

have been investigated. These technologies are focused on physical principles such as density and adhesion that help with gastric retention. These principles may vary by the biology of the target species. For example, higher-density devices may be suitable for retention in the rumen of grazing animals, while this principle can reduce gastric retention in "vertical" species such as humans. All these technologies for improved gastric retention are discussed in the following section. The technologies are visually represented by Figure 33.4 that also depicts the common systems that can be designed for both monolithic and multiparticulate systems.

### 33.4.1 Buoyant Systems

#### 33.4.1.1 Buoyant Tablets

If one imagines a liquid or semisolid filled gastric milieu that is moving with peristalsis and assisted by gravity to the small intestine through a valve at the bottom, a floating dosage form could be expected to remain in the gastric compartment for a prolonged duration. Buoyant GRDDS has been one of the most commonly evaluated approaches [16]. Ways to design a buoyant system involve -

- Layering an enteric coat onto an empty gelatin capsule followed by a drug-containing layer that also incorporates sustained release properties [17]. The entrapped air in the capsule provides the buoyancy, enabling a system that floats on gastric fluids. A key criterion here would be the extent of buoyancy – taking into consideration the drug load and weight of the coating on the capsule. Hence, only low-dose drugs, such as propantheline bromide and benactyzine hydrochloride, might be amenable for such application.
- Coating the dosage form with a cross-linked polymeric (polymethyl vinyl ether/ maleic anhydride) film that enables them to swell in gastric media in the presence of gastric fluid, allowing the tablet or capsule to balloon and float [18]. The film also provides a rate-limiting barrier to drug transport. When evaluated in human volunteers, the system extended the gastric residence time of tablets to about

six hours compared to three hours for uncoated non-disintegrating tablets. There was no effect in the fasted state [18].

- Effervescent systems, like effervescent tablets that contain a basic compound such as sodium bicarbonate, which, in the presence of gastric acid, creates carbon dioxide within the system. Such systems can be prepared by incorporating a large molecular weight insoluble hydrocolloid polymer, such as Methocel K4M, along with sodium bicarbonate and other excipients to create a dosage form that swells in the presence of gastric fluid. Such a dosage form can be made into either tablets or capsules or even bilayer tablets [13]. By their very nature, these systems tend to be large swelling tablets or capsules.

An interesting question that follows is whether the resulting gastric retention arises from the size of the dosage form or from the buoyancy effect [2]. Timmermans et al. addressed this specific question by preparing floating and non-floating dosage forms of three different sizes (5, 0, and 000 capsules) and then determined their relative retention times in human subjects in both upright and supine conditions [19]. The authors observed that the floating dosage forms are retained longer than the non-floating ones in the upright position, but no significant difference existed in the supine condition. The smaller-sized dosage forms had greater differences between the floating and the non-floating capsules, compared to the larger-sized capsules. This could be attributed to the combined effect of size and floatation that would be more evident in larger-size dosage forms. The systems described by these authors were used to develop several commercial products marked by Roche including Madopar® and Valrelease® but both products were ultimately withdrawn from the market. Another variation of the buoyant system due to effervescence is the raft system. In the raft system, due to effervescence and gel-forming polymer, a floating viscous gel is formed, which floats on top of the gastric fluid. Such a system is useful in treatment of conditions such as gastroesophageal reflux diseases (GERD) [20].

#### 33.4.1.2 Buoyant Multiparticulates

Several buoyant multiparticulate and/or minitablets have been investigated [21]. For example, Ichikawa et al. coated a conventional multiparticulate granulation with an effervescent layer composed of sodium bicarbonate and tartaric acid, which was further coated with a swelling membrane layer of ethyl cellulose and shellac [22]. This system was tested both *in vitro* and in beagle dogs. A significant increase in drug exposure was observed for p-aminobenzoic acid but not for isosorbide dinitrate. These ambivalent results are fairly typical of those found with the multiparticulate systems with the overriding conclusion that there may be some small benefit of buoyancy compared to conventional systems, but the advantages gained are less than that seen with large non-disintegrating dosage forms [15, 21].

### 33.4.2 Expandable Devices

#### 33.4.2.1 Expandable Devices in Veterinary Medicine

An expandable GRDDS incorporates features that allow the device to expand in one or more dimensions following administration. The Laby device [6, 7] used for long-term delivery of various drugs to grazing sheep is composed of a plastic housing that has two “wings” on either side of the capsule that are held in place during administration by a water dispersible tape. Following administration, the tape dissolves and the wings expand thereby ensuring that the device is neither regurgitated nor transported down the GI tract. Within the plastic housing is a series of erodible matrix tablets that are stacked and held in place against an end cap of prescribed diameter by a metal spring. Numerous drugs including ivermectin, albendazole, selenium, and iodine have been formulated into this device with delivery periods ranging from weeks to years. Numerous products are commercially available through the Captec Company in Australia and elsewhere using this technology [5].

Another example is an expandable sheet device, the Paratect Flex Bolus [23], which is a trilaminate sheet that is about $4'' \times 8'' \times 0.07''$, wherein a central polymeric matrix containing a drug load of 40% by weight morantel tartrate is sandwiched between two outer polymeric layers that are impermeable to drug transit. A series of holes are punched through the trilaminate in a symmetrical manner through which drug release occurs. The release rate is relatively constant with time and the device delivers the drug for about 90 days. For administration and storage, the device is rolled into a cylinder, restrained in the rolled configuration by a piece of tape held in place by a water-soluble adhesive. Upon administration, the adhesive dissolves and the device unrolls to a size sufficient to prevent regurgitation. This device was commercialized in the 1990's and remains on the market today for treatment of GI worms in grazing cattle [1].

Even as the veterinary systems have been shown to work for up to 135 days in grazing animals (taking advantage of the rumen, a stomach compartment sac that holds on to the feed), a human GRDDS is usually intended for shorter periods of retention in the stomach due to biology and safety considerations.

#### 33.4.2.2 Expandable Devices in Human Medicine

Expandable systems expand in the stomach (after oral administration) to a size that can neither be regurgitated nor pass the pyloric sphincter – under both fed and fasted conditions and also under the influence of compressive gastric motility waves. Such a system, of necessity, would need to meet several characteristics:

- Being small enough that it can be administered orally, such as the shape and size of a typical tablet or capsule.
- Expansion in the stomach to a size substantially larger than the largest open size of the pyloric sphincter, which would be more than about 2 in. in diameter.
- Triggering mechanism in the presence of high water activity, such as liquid water in the stomach, but not under varying and sometimes high humidity conditions during storage and use. Ideally, such a triggering mechanism would be an all-or-none kind of phenomenon with a sharp transition over a very small range of changes in environmental conditions (e.g. humidity).
- The expanded state of the device maintains sufficient integrity (without excessive stiffness or rigidity), resistance to gastric acid and enzymes, food components (including fats and surfactants in certain foods), and compressive motility motions, while maintaining flexibility (e.g. bends) so that it does not damage gastric lining.

Several configurations of the design of the dosage form have been studied. For example, Curatolo and Lo designed a system with wings that open to dimensions of ~3.5–4 cm when fully expanded [24]. The retention arms are made of an enteric polymer. A controlled-release tablet is inserted and retained in the central cylinder. For administration, the device is placed into a hard gelatin capsule [1].

When tested in beagle dogs, the device was administered in the fasted state and the animals were fed seven hours post administration. The GI transit time was followed by radiography at regular intervals for 24 hours. All five devices were still present in the stomach at 20 hours, while at 24 hours one device had moved from the stomach. In a pharmacokinetic study using glipizide as a model drug, the controlled-release tablet was administered with and without the retention system. With the retention system, the exposure doubled and the $T_{max}$ was delayed from ~3 to ~12 hours [1].

While this study demonstrates the utility and proof of principle of such expandable systems, meeting the criteria outlined at the beginning of this section, such devices still face the limitations with respect to application to low-dose drugs, relatively large size of dosage form for administration, and the challenge of robust and reproducible commercial manufacturing and supply.

The importance of the shape and size of the expandable systems has been discussed in other studies [8–11, 16, 23, 24], Cargill et al. evaluated molded polymeric devices, such as a tetrahedron shaped, that were designed to be placed in a gelatin capsule for administration and then unfold in the stomach following ingestion. They evaluated the gastric transit times in dogs for numerous sizes, geometric shapes, and polymer physical properties. In a study of human volunteers, it was found that the devices were not well retained when administered to fasted subjects (median residence time of about 3 hours) and had a mean residence time of about 6.5 hours in the fed state. Hence, despite showing some initial promise in animals, the devices were not particularly promising in man. Additional drawbacks to this approach were that it is difficult to see how a drug delivery device could be added to the system and the system is difficult to manufacture and assemble into hard shell gelatin capsules [25].

#### 33.4.2.3 Expandable Buoyant Systems

Elan and Depomed [3, 12, 13] developed an Endogastric Therapeutic System that employed both expansion and buoyancy to achieve retention [4, 18, 26]. This system was composed of a sealed pouch that incorporates the drug and sodium bicarbonate. The pouch is folded and placed within a hard gelatin capsule for administration. Upon oral ingestion, the pouch unfolds upon dissolution of the capsule gelatin shell. When gastric acid penetrates the pouch, the reaction with bicarbonate creates effervescence leading to an inflation of the pouch to a volume of about 20 ml. The pouch has sufficient resiliency to withstand gastric contractions and sufficient size and/or buoyancy to resist passage through the pyloric valve. Drug is released via diffusion through the pouch, just as the carbon dioxide generated by the reaction of sodium bicarbonate with gastric acid. Eventually, the device deflates and passes down the GI tract.

Pouches of various sizes (20 × 20 mm to 25 × 25 mm) were evaluated in humans using gamma scintigraphy to ascertain pouch location in the GI tract over time. For individuals fed a high-fat meal the pouches were retained in the stomach for at least 16 hours post administration. Retention was much less under fasting or following a light breakfast. The significant dependence of the gastric retention on meals can become a significant limitation in clinical practice, leading to variable and potentially subtherapeutic exposures for drugs that, by selection, require longer duration exposure.

Hoffman and colleagues [14–18] developed an expanding sheet-shaped system, about 2.5 cm × 5 cm, that can be folded along the shorter axis and placed within a hard gelatin capsule for administration. This device, known as the "Accordion Pill," is a trilaminate sheet composed of two outer layers of biodegradable drug-release rate-controlling polymer films that are attached to a frame prepared from ethylcellulose and polylactic acid. Drug and suitable excipients are sandwiched between the biodegradable films. Upon oral administration the capsule dissolves, the film unfolds, and the drug diffuses through the membrane at controlled rate (see, https://www.intecpharma.com/technology/, last accessed December 2020). This device has been tested in humans with findings of increased gastric retention and drug exposure [21, 22, 27, 28].

### 33.4.3 Swelling Systems

Polymeric tablets that swell considerably in the stomach upon oral administration can be retained in the stomach and provide the advantages of GRDDS with controlled drug release. These systems are distinguished from the expandable systems in that the dosage form retains its shape upon swelling in gastric fluids.

These systems have been pioneered by several investigators [19–23, 29–33]. An example of such a system is controlled release tablets made from high molecular weight hydrophilic polymers hydroxypropyl methylcellulose (HPMC) and polyethylene oxide (PEO) – that not only provide a mechanism for controlled drug release but also swell rapidly when placed in aqueous fluids. For example, tablets that are about 13 mm × 18 mm in the unswollen state can swell to over 200% by volume and are retained for an extended duration of time in the stomach (~12 hours) in the fed state with a high-calorie meal [24, 34]. These tablets then disintegrate after about 15 hours. These works have contributed to the development of two commercial products (Glumetza® and Proquin® XR) [25].

The swelling mechanism and trigger for these systems become critically important in patient outcomes since premature rapid swelling in the mouth or esophagus during the process of ingestion can lead to significant patient distress or an emergency situation. In addition, these systems seem to require a fed condition and in certain cases high-fat meal, which does put a label restriction and can impact patient outcomes with compliance disparities. As much as commercialization of products that require concomittant administration with food has been shown viable, it is interesting to note that another metformin product (Fortamet®) has a similar PK profile when given with food [25]. Fortamet is not marketed as a GI retention system but is a large (12 mm in diameter) nondisintegrating osmotic tablet [4]. Both products have been shown to be clinically equivalent to 2× daily administration of an immediate-release metformin tablet (Glucophage®) when given once daily with the evening meal [25].

A key limitation of these swelling systems could be the size of the pill. For the swollen size of a dosage form to be larger than that of the pyloric sphincter and given that swelling of the dosage form would be in the range of ~2×, the dosage form needs to be large. This becomes a key undesirable attribute in the clinical setting.

### 33.4.4 Bioadhesive Systems

Bioadhesion has been considered a promising strategy for enhanced localized retention of a dosage form, thereby promising not only prolonged duration of drug absorption through a site but also potentially local drug action. Bioadhesive systems for oral drug delivery are designed as mucoadhesive systems, with polymers that have the potential to interact and entangle with the mucus lining of the gastric epithelial cells. Bioadhesion is molecularly a result of several additive nonspecific (van der Waals, hydrogen bonding, and/or hydrophobic interactions) and specific interactions between the complementary structures in the mucus and the polymer (entanglement).

Adhesion of the bioadhesive system to the gastric mucosa is a complex event and is regulated to a great extent by the intrinsic properties of the bioadhesive polymer, the biological substrate, and the gastric environment. The gastric mucus layer is created biologically to play several important functions of protecting the underlying tissues from various diffusing/corrosive elements such as enzymes, acids, and other toxic molecules. As a viscoelastic gel, the gastric mucosa helps in the transport of food over the epithelium, thereby minimizing potential erosive damage. The mucus layer, in addition to providing protection, also imparts a barrier to drug absorption [35].

#### 33.4.4.1 Criteria and Challenges to Mucoadhesion

The usefulness of a bioadhesive or mucoadhesive system as a GRDDS has several prerequisites and inherent challenges. These include:

- Spontaneous, strong, and prolonged adhesion of the system to the gastric mucosa.

- Hydrophilic polymers that show strong bioadhesive characteristics often also show loss of bioadhesion when fully hydrated. The rate and state of hydration of these polymers, hence, compete against their bioadhesion characteristics.
- ***Mucosal recycling***: The continuous biological recycling of the gastric mucus to maintain the mucus layer against digestive gastric environment leads to a dynamic environment that requires continuous refresh of molecular bonds and interactions that contribute to mucoadhesion [36].

- Reproducible performance and translation from animal models to humans.
  - In contrast to bioadhesive systems that may be designed for prolonged contact and retention in the buccal mucosa or skin, for example, an orally administered dosage form has uncertain probability in the extent, timing, and pressure of contact with the mucosal surface – especially with the variability that comes with concomitant food and fluid ingestion.
  - The translation from animal studies to humans faces challenges of significant differences in the geometry (horizontal versus vertical orientation of the GI tract), size, peristalsis and propulsion forces of the stomach wall, food and fluid components, and a myriad of other variables – that have often resulted in a proof of principle demonstrated in an animal species, but not quite reproduced in a human clinical study.
- Optimal bioadhesion in the acidic gastric pH conditions.
  - Bioadhesive polymers, such as poly(acrylic acid) (PAA), usually have ionizable functional groups that provide hydrophilicity and bonds with mucus components. Such ionizable functional groups have pH-dependent ionization and the mucoadhesion is lower in the unionized state at the acidic gastric pH.
- An ideal bioadhesive DDS should have the bioadhesion and drug delivery as mechanistically independent.
  - This could be conceived in a polymeric system wherein the bioadhesive polymer is coated onto either a controlled release tablet or multiparticulate. Alternatively, a polymer composition with a hydrophilic surface (for bioadhesion) and a hydrophobic core (for drug entrapment and release) could serve such a role. Such a polymeric system design, however, is difficult to achieve.
  - Polymer design influences mucoadhesion. For example, tethered polymers [27] serve as bioadhesion promoters that help form a rigid gel network [37]. Bioadhesion is generally a result of surface interactions between hydrophilic polymer surfaces and mucins. Adding tethers of linear and block copolymers containing neutral or ionizable structures can provide increased interdigitation and interpenetration with the mucus – leading to increased mucoadhesion [37].
- The bioadhesive DDS should not cause local irritation or long-term toxicity, nor should it exacerbate any preexisting conditions.
- The strength of the mucoadhesive interaction should be able to withstand the high turbulence of the stomach environment, including that from food and fluids; the peristaltic and compressive motions of the gastric valve, and potential variation in the surface and location of adhesion.

#### 33.4.4.2 Bioadhesive Tablets

Several studies have been conducted with mucoadhesive dosage forms, such as tablets.

When a matrix system containing drug and a hydrophilic cross-linked acrylic acid polymer was used as the bioadhesive material and griseofulvin as a model drug, 3–4 fold higher exposure was observed in rabbits compared to conventional dispersions. Whether this was the result of increased transit time or simply some controlled release mechanism is not clear. The authors speculated that the mechanism of improvement is probably due to the increase in GI transit time and the intimacy of the drug with the absorbing membrane brought about by the bioadhesive polymer [38].

In a radioscintigraphic study, two different tablet formulations, based on the hydrophilic bioadhesive polymers polycarbophil and carbopol, were evaluated for gastric emptying and small intestinal transit time. Results indicated that both bioadhesive formulations were not dramatically different from the nonadhesive controlled formulation in terms of stomach emptying, small intestinal transit time, and time to reach the colon [39].

Similar results were obtained with an oral bioadhesive controlled release formulation of furosemide [29, 40]. A bioadhesive tablet system based on Carbopol 934 and containing furosemide, a model drug with erratic absorption and a narrow absorption window, was evaluated in a radio scintigraphy transit time study in humans. The extent of absorption from the bioadhesive formulation was less than the control non-bioadhesive formulation. The reasons suggested for this poor performance were that in vivo adhesive properties were insufficient to overcome the powerful gastric contractions of the housekeeping waves in the fasted state.

In a study with another bioadhesive polymer, Säkkinen et al. evaluated the gastric retention of chitosan granules in the stomach by gamma-scintigraphy in fasted human volunteers. Although chitosan showed marked bioadhesive capabilities *in vitro*, retention of the chitosan formulation in the upper small intestine was not reproducible and the duration of retention was similar to the lactose granules used as a control [25].

In developing a bioadhesive system for furosemide, a model drug with a narrow absorption window in the upper GI tract, the administration of furosemide in chitosan granules resulted in lower bioavailability than that from a conventional immediate-release formulation, again indicating that these formulations were not achieving target behavior as polymers in man [31, 41].

These and other [1] studies indicate that hydrophilic polymers may not be the ideal candidates for systems; efforts have been made to evaluate other types of polymers that may have utility for use as gastric retention DDS [2].

#### 33.4.4.3 Bioadhesive Multiparticulates

Bioadhesive systems based on multiparticulates, microparticles, and nanoparticles have inherent advantages to mucoadhesive performance in comparison to monolithic matrix tablets. Less weight reduces gravitational pull out of the gastric cavity, increases diffusion within the gastric fluids, and provides greater surface area per unit mass for mucoadhesion – being able to get entrapped in the mucus – exhibiting prolonged gastric residence [32, 33, 42, 43].

For example, ion exchange resin, such as cholestyramine, displays prolonged gastric residence and uniform distribution over the gastric mucosa when given in a small volume of water to fasted subjects [34, 44]. In a systematic study of dose size, food, and surface coating effects on the gastric residence time of the resin in healthy human volunteers by gamma scintigraphy, Thairs et al. found that the prolonged gastric residence and uniform distribution of ionic resins are not influenced by the dose size and that the binding of the dose to the mucosa is sufficiently strong to retain the dose during feeding four hour after administration [34]. Contrary to general perceptions, charge-based interactions associated with ion exchange resins played a minor role in mucoadhesion [34, 44].

In an *in vivo* efficacy study with drug-loaded bioadhesive nanoparticles, the bioadhesive system resulted in the complete clearance of *Helicobacter pylori* infection in the stomach compared to amoxicillin suspension [35, 45].

#### 33.4.4.4 Bioadhesive Microdevices

The principles of bioadhesion have been explored with small drug reservoirs (microreservoirs), in the shape of thin flat discs, modified on one side with a bioadhesive agent while the other side provides a controlled drug release membrane surface [36, 46]. These devices are designed for adhesion to the gastric mucosa on one side while the other side is expected to release the drug in the gastric milieu. In addition to unidirectional drug release, the mucoadhesion of the bioadhesive surface can be further enhanced with lectins [37, 47].

It is debatable whether the asymmetric coating is sufficient to orient this microdevice delivery system differentially toward the target stomach lining.

## 33.5 New Drug Development Considerations

At the interface of drug discovery and development, which is roughly aligned with preclinical to clinical transition of new drugs, an oral product is designed to be as simple a dosage form as possible, such as an IR capsule or tablet. This strategy is aligned with the primary locus of uncertainty, innovation, and risk in the process of drug discovery being the molecule itself. A dosage form design is driven primarily by the indication for the drug, its physicochemical properties, biorelevant assessments, preclinical data, and projected patient population.

The place for dosage form design to utilize alternate (to conventional immediate-release products) technological options is establishing the need (e.g. molecule properties, developability constraints, and timelines) and the opportunity (e.g. disease area, patient population, and competitive needs). Thus, knowing the drug characteristics, e.g. solubility, stability, and dissolution rate in acidic pH, including SGF (simulated gastric fluid), versus intestinal pH, including FeSSIF or FaSSIF (fed or fasted-state simulated small intestinal fluid); permeability characteristics; the projected site of drug absorption; and information from preclinical studies regarding potential risks (such as variability in drug absorption) can all help inform the need for an extended-release gastroretentive system. In addition, an understanding of the projected impact of a long-acting and GRDDS on human PK prior to the clinical studies, such as using biopharmaceutics modeling software such as GastroPlus (Simulations Plus) or Simsyp, can further inform the potential clinical utility of these systems.

Additional consideration of patient compliance, based on business needs (e.g. competitive products for the same indication) and input from patients or patient representatives could design a patient-centric drug delivery system. Indications such as Parkinson's or peripheral neuropathy can be debilitating without consistent medication routines. A drug delivery system that may reduce the dosing frequency can be considered highly impactful in such cases. For example, patients in weekly dosing groups have been shown to be almost twice as likely to adhere to the dosing regimen than the daily dosing group [48]. Additionally, a patient-compliant delivery system, not limited to GRDDS, can help overcome challenges of therapy in cases such as AIDS and malaria, where supervised mass drug administration at a reduced frequency can help achieve full remission/cure [48].

In new drug development, GRDDS therefore, lies in establishing a niche where it can provide an order of potential patient benefit and/or avoid a tangible risk. Such a position is understandably easier to establish for known drugs, such as generic compounds. Hence, applications of GRDDS technology that have transitioned into the clinical and commercial realm have predominantly been of established drugs.

### 33.5.1 *In vitro* Development

Conventional basket and paddle dissolution testing systems are insufficient to mimic GI milieu and obtain reproducible drug release data for a GRDDS in various fluids. Some complexity in dissolution testing includes aspects such as the use of sinkers during drug release testing. For example, the use of sinkers during dissolution may impede with the mechanism of drug release for swellable and expanding dosage forms; while without sinkers, it may be challenging to obtain reproducible dissolution data for buoyant systems. Additionally, since these systems are so diverse in their mechanism to achieve gastroretention, one common method of measurement of *in vitro* performance is inadequate. To address such issues, alternate testing systems with limited exposure to fluid have been developed for in vitro evaluation of GRDDS, e.g. a shaker water bath maintained at 37 °C and 100 rpm to mimic stomach fluid dynamics [20]. For a hydrodynamically balanced system (HBS), a simple visual observation, as reported by Sheth and Tossounian could be an effective tool for initial assessment of this system. The ability of the HBS system to maintain its shape and size for upto eight hours was easily noted by such a visual observation [49]. For multiparticulate systems, on the other hand, an exact measurement of the count of sedimented versus floating particles is needed to ascertain the utility of a floating multiparticulate system.

In addition to conventional solid oral dosage forms attributes such as weight variation, uniformity of dosage units, dosage form disintegration, and eventual drug release, the following attributes (as detailed in Table 33.1) should be studied to ensure in vitro gastroretention of the dosage form. Demonstration of acceptable excipient compatibility for the drug substance for formulation into GRDDS is imperative. Eventually, commercial success of GRDDS technology will depend on ensuring consistent performance of the drug product from batch to batch.

Additional considerations that are critical to the success of GRDDS are the gastric motility and the stress of this motility on the dosage forms. Innovative methods such as an inflated balloon to add a realistic physiological pressure to the dissolution testing have been used to mimic the gastric milieu [20, 50]. Other testing innovations include paddle (USP-2) dissolution testing with polystyrene beads that mimic the intense pressure due to gastric motility. Biorelevant systems that may be utilized for testing include the TNO's TIM-1, the human gastric simulator (HGS), and dynamic gastric model (DGM) – in addition to various buffered aqueous solutions, FaSSIF, FeSSIF, and SGF [20].

### 33.5.2 Preclinical Development

As much as the design of GRDDS is intended for humans, these systems and their principles must be tested and established, including the demonstration of safety and reproducibility, in preclinical animal models before human testing. The primary challenges in the development of a GRDDS are the *in vitro in vivo* correlation (IVIVC) and interspecies scaling [51]. The latter is inherently difficult due to the anatomy (e.g. size of gastric valve, vertical versus horizontal orientation of the GI tract) and physiology (e.g. pH and volume of aqueous buffers, and dosage form residence times, in various GI compartments) differences among animal species versus human. For example, rabbit gastric pH is similar to humans and may be an acceptable species for early dosage form evaluation. Additional species that are considered in early development are beagle dogs and pigs [6, 52, 53]. The choice of an animal model for a small monolithic or multiparticulate dosage form is simpler, and a rodent species may be suitable.

**Table 33.1** Type of GRDDS and its *in vitro* assessment technique for gastroretention.

| Type of GRDDS | Parameter studied | Method of measurement |
|---|---|---|
| Bouyant/Low Density Systems: Monolithic and Multiparticulate | Floating Lag Time (FLT)<br>Total Floating Time (TFT)<br>Floating Strength<br>Specific Gravity | Test is conducted at 37 °C in 900 ml of simulated gastric or intestinal fluid in USP dissolution apparatus;<br>FLT is time taken for the dosage form to rise to the surface;<br>TFT is the total time the dosage form stays afloat in the media Floating strength is the force measured to determine floating of bouyant dosage forms over time.<br>For multiparticulate systems, the number of bouyant particles also needs to be measured in addition to FLT and TFT.<br>Specific gravity is measured by benzene displacement method. |
| Expandable Systems | Unfolding Study | *In vitro* unfolding of the dosage form study. |
| Swellable Systems | Swelling Study | Calculation of swellability index (SI) or extent of erosion (E) helps understand uptake of fluid and mass loss in comparison to initial weight, respectively, when the dosage form is exposed to dissolution media<br>Exposed surface parameter (ESP), which is measured 2D by X-ray imaging could be useful to assess the unfolding and percent area exposed to gastric media [27] |
| Mucoadhesive Monolithic and Multiparticulate Systems | Viscosity and Rheology<br>Muco or Bioadhesive Strength | Viscosity measurement upon contact with simulated gastric fluids.<br>Measuring the force required to separate the polymer from artificial or animal gastric tissue membrane.<br>For mucoadhesive multiparticulate systems, a test of adhesion before and after contact for a specific period of time either in vitro or in a perfused animal model followed by counting of particles adhered to the tissue is suggested as a representative in vitro evaluation for adhesion of such systems. |
| Ion Exchange Resin GRDDS | Particle Size, Ion Exchange Capacity, and Moisture Content | Conventional particle size measurement techniques e.g. sieve analysis or laser diffraction; Karl Fischer for moisture content and ion exchange capacity tests will be based on type of ion exchange resin used |

Source: Adapted from Chawla et al., and Schneider F. et al. [11, 20].

Following are the key considerations for the selection of species for evaluation of an animal model for GRDDS

1. Smaller species such as rabbit or rodent may not be suitable for expandable systems, primarily due to GI tract size [6].
2. Although beagle dogs are one of the most studied animal species in preclinical drug development, factors such as timing of the gastric MMC wave after a meal may be significantly longer than humans and may not reflect human residence time of the GRDDS [52].
3. Beagle dogs may also exert different pressures on the dosage form during gastric emptying compared to humans [16, 52].
4. Pigs have similar GIT composition, colon microflora, and peristaltic waves as humans, but may have long gastric emptying times. Fasted studies in pigs have demonstrated residence times ranging from 1.5 hours for pellets and liquids to 6 hours for large tablets (22.0 × 8.7 × 5.1 mm) [53]. There are other studies, which demonstrate a gastric residence time of > 5 days, which has not been replicated in a clinical setting [20, 53].

Hence, an understanding of gastric and GI physiology, and drug absorption dynamics, across different species is imperative to the design and development of GRDDS for human use.

### 33.5.3 Clinical Development

As much as early animal studies are important to drug development and transition to clinical, the resultant data for GRDDS must be treated with caution that it may not be entirely representative. Since the dosage form depends on several anatomical features and physiological parameters, eventually, even with adequate *in vitro* and preclinical studies, a clinical study is essential to determine the relative merit and value of a GRDDS. Such a clinical study may require comparison of a GRDDS with an IR product, including that of pharmacokinetic parameters and potential adverse effects.

This adds complexity to the clinical trial design as well as data interpretation, especially for new drugs. For example, a typical phase I clinical study starts with a single ascending dose evaluation of drug safety and pharmacokinetics. It is the first time a drug is given to a human, with the primary objective of drug safety evaluation. Unless the need for a GRDDS is established preclinically (such as a drug whose target is in the gastric lumen, such as *H. pylori* infection), clinical SAD study is typically preferred with an IR product so that the clinical results enable understanding of the molecule itself. It is possible to have GRDDS as a study arm in one of the cohorts of the SAD study. Given the time and resources needed for the development of a GRDDS, such an arm would be logistically challenging to incorporate in a timely manner.

These considerations lead toward the development of GRDDS for established drugs whose clinical use has already identified potential opportunities in drug safety, pharmacokinetics, and efficacy that can be unlocked with prolonged gastroretention.

### 33.5.4 Diagnostic Tools for Preclinical and Clinical Development

In preclinical and clinical studies for GRDDS, in addition to establishing successful drug pharmacokinetic (PK) levels, an ability to visualize the delivery system is a critical attribute for clinical understanding. Some common diagnostic techniques in preclinical and clinical studies are gamma ($\gamma$) scintigraphy, magnetic resonance imaging (MRI), radiology, gastroscopy, ultrasonography, and magnetic marker monitoring.

#### 33.5.4.1 Gamma Scintigraphy

Gamma ($\gamma$) scintigraphy is based on incorporating a stable radionuclide isotope, e.g. $^{153}$Samarium (half-life ($t_{1/2}$): 46.2 hours, photon energy 103 keV), $^{111}$Indium ($t_{1/2}$: 2.8 days, photon energy: 247 keV), and $^{99m}$Technetium ($t_{1/2}$: 6 hours, photon energy: 140 keV), in the dosage form. The radionuclide is irradiated *in vivo* with a neutron source causing gamma radiation to be released that can be tracked by a gamma camera. Due to the low radiation exposure, this technique offers a safe way to observe the dosage form *in vivo*. This technique,

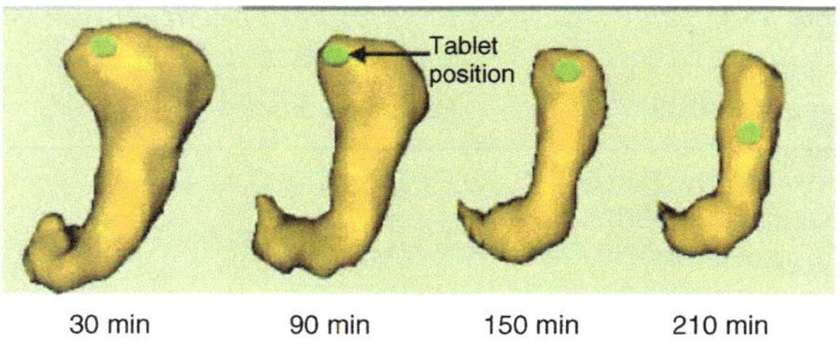


**Figure 33.5** Location of the tablet in the stomach of a healthy volunteer. In this volunteer, the tablet was always floating on the surface of the stomach contents. Source: Steingotter et al. [55] / Springer Nature.

however, cannot provide adequate topographic information and may suffer in resolution [14]. Additionally, the use of radionuclide should be protected from gastric contents and have its activity last for the duration of formulation manufacture followed by dosing and the duration of study [18, 54].

#### 33.5.4.2 Radiology

Utilization of X-rays to visualize the GI transit of the dosage form, in addition to disintegration of the dosage form and dissolution of the drug, can be performed by incorporating a radio-opaque material in the dosage form and taking radiographs after ingestion that would help locate the position of the dosage form. The major advantage of this system is its easy application and low cost. The high exposure to X-rays, however, is a serious drawback of this methodology [14, 20].

#### 33.5.4.3 Magnetic Resonance Imaging (MRI)

MRI has a distinct advantage over radiology or $\gamma$ scintigraphy in that it not only provides information about the dosage form but also provides topographical measurements for the location of the ingested dosage form, as shown in Figure 33.5 [55].

Application of supra or paramagnetic compounds to enhance signals or diminish interference from tissues is another advantage of MRI, in addition to lack of ionizing radiation.

MRI utilizes strong magnetic field gradients and radio waves to generate images of the body organs, including GI tract. This technique has a distinct advantage of being safe for the subject, avoiding unnecessary radiation exposure, and superior in imaging for the ingested GRDDS [14, 20, 55].

#### 33.5.4.4 Ultrasound

Ultrasound is a safe, noninvasive, and effective technique for visualization of GRDDS inside the GIT. Ultrasound waves are generated due to mechanical

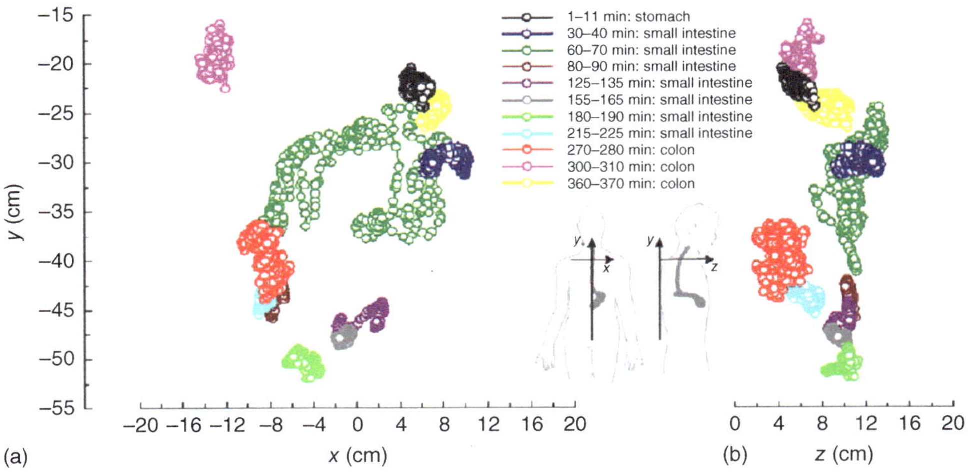


**Figure 33.6** GI tract transit in a subject as per magnetic moment monitoring. The locations of the ER tablets are shown in frontal view (a) and in a view from the side (b). Each circle represents the mean location of the tablet during one second. Source: Weitschies et al. [56] / with permission of Elsevier.

oscillations generated by many crystals inside a transducer. As these ultrasound waves travel through the body, they detect density differences within tissues and return to the transducer as echoes. These echoes are then converted back to electrical waves followed by its conversion into images by the transducer. Presence of gases in the intestine will lead to loss of signal and may not be useful for the transit testing of a dosage form through the GIT, however, will be useful for GRDDS localized in the stomach. This technique can also useful for non-disintegrating or slowly disintegrating dosage forms in the gastric environment. This tool appears to be underutilized in GRDDS research but may have significant benefits being a noninvasive technique [14, 20].

#### 33.5.4.5 Gastroscopy

Gastroscopy is an invasive visualization technique utilizing a camera/scope to view contents inside the GI tract. This technique is straightforward, however, the presence of food or blood in the stomach may interfere with its viewing capability [14]. Secondly, there may be limitations to the scope to view the entire GI tract. Additionally, the scope will need to be inserted and re-inserted in order to assess the entire lifetime of the GRDDS which may not be always feasible for human studies. Other methods would be a capsule camera or a PillCam to overcome the intrusive nature of the endoscopy. However, the PillCam emptying might be earlier than the GRDDS and it also may move freely in the GIT. Even if administered with a significant delay after the GRDDS, it will still involve challenges due to the ability to be carried across GIT earlier than the GRDDS thus negating the utility of such a system [20, 22, 27].

#### 33.5.4.6 Magnetic Marker Monitoring (MMM)

Also known as sometimes also called Magnetic Moment Imaging (MMI) or Gastrointestinal Magnetomarkergraphy (GIMG), MMM is the monitoring of the drug product in the intestinal tract by monitoring a magnetic marker (MM), such as $Fe_3O_4$ (added to the drug product), which generates a magnetic signal.

$Fe_3O_4$ is commonly added to drug products as a coloring and opacifying agent, does not exhibit pharmacological activity, and is considered inactive. The passage of the dosage form is monitored by superconducting quantum interference devices (SQUIDs). It is a technique for spatial monitoring of the drug product retention and its transit, as shown in Figure 33.6. The only drawback for this technique would be, if not done correctly in a special room, it may be affected by interference from the environment [20, 56].

As mentioned earlier, clinical trials designed to study a GRDDS almost invariably involve one or more diagnostic tools as a part of subject monitoring to collect data regarding spatial positioning of the GRDDS.

## 33.6 Commercial GRDDS Products and Investigational New Products

Commonly marketed GRDDS are based on floating technology including raft formation and expanding/swellable technology. Mucoadhesive systems although being developed remain in development primarily due to challenges of adhesion and impact of gastric motility on adhesion. GRDDS-based marketed formulations are listed in Table 33.2.

Clinical trials (active, enrolling, complete, inactive, or unknown status) studying GRDDS for multiple indications are listed in Table 33.3.

It is important to notice that most commercial and clinical study products are for chronic conditions that benefit from lower frequency of dosing and most products utilize either the buoyancy or swelling principle for achieving gastroretention. Out of the multiple investigational new products being tested, some advanced clinical-stage products and products utilizing novel methods of GRDDS are discussed in the next section.

Following early efforts of prolonging gastric retention by Cargill et al., the original concept of layers of polymer has been modified to currently atleast three systems in clinical trials that have evidence of substantial of increase in gastric retention time – Acuform by Assertio (previously Depomed), Accordion Pill by Intec Pharma, Lyndra's stellate dosage form, hydrogel-based Plenity device, and the Geoclock tablet and Soctec device.

### 33.6.1 Acuform®

Acuform® by Assertio Therapeutics (originally Depomed) utilizes its swellable polymer technology in multiple marketed products as well as has a next-generation technology in development. This technology is supposed to offer 8–10 hours of upper GI retention time. The only drawbacks noted with these systems are the overall soft nature of the capsule and the need to be dosed in a fed state for longer gastric retention [48].

### 33.6.2 Accordion Pill® (AP)

Accordion Pill® (AP) by Intec Pharma is based on drug release based on folds of polymer offering the extended retention and release of the drug as shown in Figure 33.7. This system has been shown to extend release up to 12 hours. The other polymer is enteric which offers resistance to the gastric environment. It is emphasized that AP's drug release mechanism is independent of its gastric retention mechanism. This system can also be tailored to offer both IR and GR release profiles. However, if the enteric coating fails in the GI, similar to other enteric release tablets, there could be a risk of dose dumping and lack of efficacy. Similar to Acuform, this system is expected to be dosed with meals to enhance its gastroretentive period e.g. in the phase II study of carbidopa/levodopa, fasted retention time was four to five hours compared to seven to eight hours in a fed state) [48].

### 33.6.3 Lyndra®

Lyndra® delivery system is a stellate system composed of a central elastomeric core, drug-loaded "arms" and disintegrating matrix that joins the arms to the core as shown in Figure 33.8.

This design is folded in a size 00EL capsule for oral dosing. Once the capsule opens up in the stomach, the stellate design unfolds and due to its size resists passage through pylorus until complete disintegration of the dosage form occurs thereby allowing passage of the core and remnant polymer. Achieving once-weekly administration with such a sustained oral drug delivery system has been proposed as a greater patient compliance application for oral contraceptives.

Notably, the application of such a system would be restricted to low-dose drugs where drug release is the biopharmaceutical rate-limiting factor in drug absorption. The application could be extended to conditions that require chronic therapy with potential compliance issues including psychiatric disorders, Parkinsons, Alzheimers, HIV, and opioids therapy. The application of combinations of several commonly used pharmaceutical excipients to modulate drug release behavior of these systems has been studied by Lyndra [59, 60].

### 33.6.4 Plenity®

Plenity® recently received FDA approval in 2019 to market its hydrogel-based weight loss reduction device. Plenity marketed by Gelesis, Inc., is a cellulose-based oral capsules, which simulates fullness in the stomach and thus reduces the urge to eat more. The clinical study tested the efficacy of these capsules in the clinical setting with or without diabetes, and was proven to be effective and caused greater weight loss over placebo without any increase in safety risks.

This cellulosic device offers no caloric value and is modified by cross-linking with citric acid creating a three-dimensional (3D) matrix made of individual gel fragments by absorption of water in the stomach enhancing satiety in patients. Plenity maintains its 3D structure and mechanical properties during transit through the small intestine. Once it arrives in the large intestine, the hydrogel is partially broken down by enzymes and loses its 3D structure along with most of its

**Table 33.2** Commercial GRDDS-based products and investigational new products.

| S. No. | Technology/ delivery system | Brand name | Active | Indication | Manufacturer |
|---|---|---|---|---|---|
| 1. | Bilayer Floating Capsule | Cytotec® | Misoprostol | Gastroprotection | Pfizer, UK |
| 2. | Floating Capsule | Madopar HBS® | Levodopa and Benserzide | Parkinson's disease | Roche, UK |
| 3. | Floating Capsule | Prolopa HBS® | Levodopa and Benserzide Hydrochloride | Parkinson's disease | Roche, UK |
| 4. | Floating Capsule | Valrelease® | Diazepam | Anxiety | Roche, UK |
| 5. | Floating, swelling system (Tablet) | Inon Ace Tablets® | Simethicone,<br>aluminum–magnesium salts | Antifoaming agent, antacid | Sato Pharma, Japan |
| 6. | Minextab Floating®<br>floating and swelling system | Metformin HCl | Metformin | Hyperglycemia | Galanix, France |
| 7. | Minextab Floating® – floating and swelling system | Ceflacor LP | Ceflacor | Antibiotic | Galanix, France |
| 8. | Minextab Floating®<br>floating and swelling system | Tramadol LP | Tramadol | Synthetic opioid analgesic | Galanix, France |
| 9. | Effervescent floating system (Tablets) | Prazopress XL® | Prazosin HCl | Alpha-1 antagonist | Sun Pharma, Japan |
| 10. | Effervescent floating system (Film-coated tablet) | Zanocin OD® | Ofloxacin | Fluoroquinolone antibiotics | Ranbaxy, India |
| 11. | Effervescent floating system (Film-coated tablet) | Cifran OD® | Ciprofloxacin | Fluoroquinolone antibiotics | Ranbaxy, India |
| 12. | Effervescent floating system (Film-coated tablet) | Riomet OD® | Metformin HCl | Hyperglycemia | Ranbaxy, India |
| 13. | Colloidal gel forming floating system | Conviron® | Ferrous sulfate | Iron Deficiency Anemia | Ranbaxy, India |
| 14. | Raft forming system | Topalkan® | Alginic acid;<br>Aluminum–Magnesium salts | Antacid | Pierre Fabre Medicament, France |
| 15. | Raft forming system | Almagate FlatCoat® | Aluminum–Magnesium salts | Antacid | Pierre Fabre Medicament, France |
| 16. | Raft forming system | Liquid Gaviscon® | Sodium Alginate, Sodium<br>Bicarbonate, Calcium Carbonate | Antacid<br>(in reflux esophagitis) | Reckitt Benckiser Healthcare, UK |
| 17. | Polymer-based swelling technology | Gabapentin GR® | Gabapentin | Seizure and Pain | Assertio/Depomed Inc., USA |
| 18. | Polymer-based swelling technology | Proquin XR® | Ciprofloxacin | Antibiotic | Assertio/Depomed Inc., USA |
| 19. | Polymer-based swelling technology | Glumetza® | Metformin HCl | Hyperglycemia | Assertio/Depomed Inc., USA |
| 20. | Polymer-based swelling technology | Proquin XR | Ciprofloxacin | Antibiotoic | Assertio/Depomed Inc., USA |
| 21. | Polymer-based swelling technology | Plenity | Hydrogel based | Obesity | Gelesis Inc., USA,<br>Approved: 2019 |

**Table 33.3** Clinical trials (as of December 2020) utilizing gastroretentive method of drug delivery.

| S. No. | Title | Sponsor/Collaborators | Conditions | Interventions |
|---|---|---|---|---|
| 1 | Phase 1b Study Evaluating OPomD in Relapsed or Refractory Multiple Myeloma | Amgen | Relapsed or Refractory Multiple Myeloma | Drug: Immediate Release (IR) and Gastroretentive (GR) Oprozomib with Dexamethasone and Pomalidomide |
| 2 | Magnetic Marker Monitoring of Furosemide-containing GR Formulation in Healthy Male Subjects (Fasting and Fed Conditions) | LTS Lohmann Therapie-Systeme AG | Edema | Drug: GRDF furosemide |
| 3 | Healthy Volunteer Study Using 3 Different Formulations of Firategrast | GlaxoSmithKline | Multiple Sclerosis, Relapsing–Remitting | Firategrast IR tablet; Firategrast MR tablet<br>Firategrast GR solution |
| 4 | Study Determining Gastric-Retentive and Modified Release Properties of Prototype Capsules in Healthy Subjects | Lyndra Inc. | Gastric Retention\|Healthy | Memantine Hydrochloride MR Prototype Capsule Formulation<br>Procedure: Magnetic Resonance Imaging |
| 5 | Trial of Novel Oral Zinc Cysteine Preparation in Alzheimer's Disease | Adeona Pharmaceuticals | Alzheimer's Disease; Mild Cognitive Impairment | GR zinc cysteine tablet; and comparatory physically to active comparator containing |
| 6 | The Gastric Residence Time (GRT) of Soctec Capsule After a Standardized Breakfast | SkyePharma AG\|University of Zurich | Healthy | Device: Soctec |
| 7 | Once Daily Gabapentin in the Treatment of Post Amputation Pain | Chicago Anesthesia Pain Specialists | Pain; Quality of Life | Drug: Gralise |
| 8 | Phase 2 studies for Chronic Idiopathic Constipation | Gelesis Inc. | Chronic Idiopathic Constipation, Irritable Bowel | Device: Hydrogel capsule |
| 9 | Mild to Moderate Crohn's Disease | Soligenix Inc. | Mild to Moderate Pediatric Crohn's Disease | Drug: Beclamethasone Dipropionate (SGX 203) |
| 10 | Crohn's Disease | Alphasigma S.p.A | Crohn's Disease | Rifaximin |

absorption capacity. The released water is reabsorbed, and the remaining cellulosic material is expelled in the feces [61].

*In vitro* assessment of the properties of Plenity that would be relevant to its functionality in vivo were reported in a previously published abstract that compared the elastic modulus and viscosity of Plenity with common processed functional fibers (psyllium, guar gum, and glucomannan) and masticated mixed green salad after digestion in SGF small intestine fluid (SIF), and small colonic fluid (SCF) over a period of 30—180 minutes. During the in vitro incubation, the elastic modulus of Plenity approached those of mixed salad remnants, both of which were orders of magnitude higher than glucomannan [62].

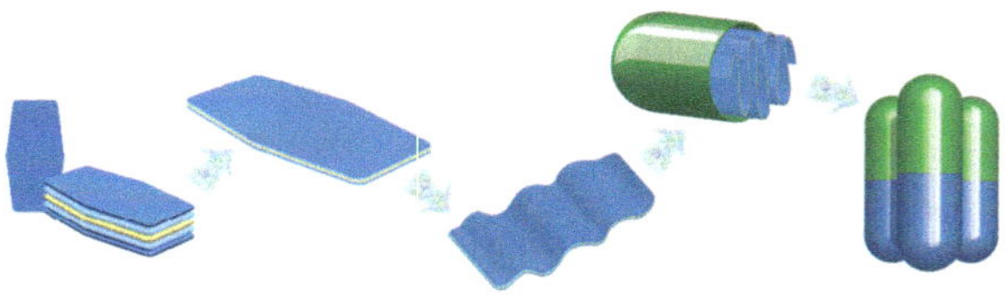

**Figure 33.7** Accordion Pill technology. Source: Adapted from http://www.intecpharma.sk/ [57].

The preparation and use of polymer hydrogels with high elastic modulus have been covered by Gelesis

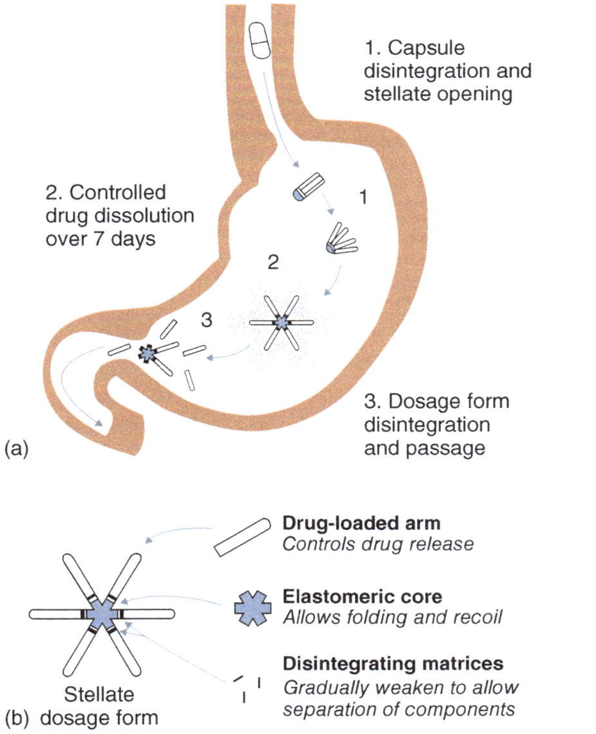


**Figure 33.8** Lyndra's stellate dosage form offering gastroretention by floatation, its size, and erosion of polymer matrix. (a) Shows capsule opening as it enters and transitions through the stomach into (b) a stellate dosage form, showing its various parts. Source: Kanasty et al. [58] / with permission of Elsevier.

Inc. in a series of patents. In US Patent 10,584,183B2, Sannino et al. study cross-linking of high viscosity carboxymethyl cellulose (CMC) to provide desired absorption and rheological properties [61]. Such cross-linking forms superabsorbent polymer hydrogels (SAPs) that can absorb and retain large quantities of water per unit mass of the dry polymer. Pressure, pH, and other environmental conditions impact the absorption and retention of water by these cross-linked polymers, which are retained by molecular interactions with the polymeric and cross-link's backbone.

Cellulose-based hydrogels are widely used in the industry and can be obtained by both physical and chemical means. Modified cellulosic polymers, such as methylcellulose (MC) and HPMC form thermally reversible gels in aqueous solutions at certain concentrations. The gelation mechanism of these systems involves hydrophobic association of polymeric chains bearing the functional groups with hydrogen bonding capability and electronegative atoms, such as hydroxyl groups. The sol–gel transition of these thermally reversible gels is attributed to water penetration, swelling, and entanglement of polymers chains at lower temperatures (sol) and the loss of part of the water of hydration at the higher temperatures (gel). In contrast to the physically cross-linked hydrogels, the entanglement structure of the chemically cross-linked hydrogels is more robust to disruption by environmental conditions such as temperature, salt concentration, or mechanical pressure. Such are the hydrogels prepared and used in the Gelesis polymers. Different grades of polymers reflect differences in the elastic modulus and hydration rate [61–63].

### 33.6.5 Geoclock® and Soctec®

SkyePharma has Geoclock and Soctec technologies available for development of gastroretentive dosage forms. Geoclock is based on tablet in a tablet design with an outer wax-based coated tablet and the inner tablet offering a controlled or IR as per product requirements. First, the outer waxy tablet disintegrates and allows for a lag time that can be defined to suit the product's needs. Soctec is a system that can offer controlled release including gastroretention. Soctec® is a bouyant capsule that is designed to withstand the peristaltic movements of the stomach due to a denser counterweight inside the capsule [39, 64].

## 33.7 Future Outlook

Gastric retention of a drug delivery system is a function of both the gastric physiology, fed state, and the dosage form characteristics. Fed state stomach, for example, has a higher residence time of the dosage form in the stomach compared to the fasted state; and the residence time increases with the caloric content of food. The impact of food and gastric physiology alone can have the standard, nondisintegrating dosage forms display significant gastric retention times. With these data and benefits factored in, the benefit of dosage form technologies to promote gastric retention seems variable and unclear. Nonetheless, several technology options have been developed to different degrees of refinement in promoting gastroretention of dosage forms upon oral administration. Systems have been designed for both animal health and for human health applications. While significant progress and commercial application have been shown for animal devices, clinical demonstration and commercialization in human clinical medicine are limited. Important contributors to these are enhanced safety concerns and rigor in clinical drug development, focus on simpler and more reliable strategies for new drug development, and the limitations in the dose and indications for which these technologies can provide significant benefit. The progress and the application, however, continue – mostly with low-dose established drugs and in niche indications where gastric retention can help treat local disorders or provide significant tangible benefit to the patient.

## Acknowledgments

We thank Dr. John R. Cardinal (J. R. Cardinal Consulting LLC, Wilmington, NC) and Dr. Avinash Nangia (Vaunnex Inc., Sharon, Massachusetts), whose chapter in the first edition [4] of this book was updated and modified to the current state.

## References

1 Cardinal, J.R. and Nangia, A. (2011). *Oral Bioavailability*, 329–341. Wiley Online Library.
2 Waterman, K.C. (2007). A critical review of gastric retentive controlled drug delivery. *Pharm. Dev. Technol.* 12: 1–10, https://doi.org/10.1080/10837450601168680.
3 Arora, S., Ali, J., Ahuja, A. et al. (2005). Floating drug delivery systems: a review. *AAPS PharmSciTech* 6: E372–E390, https://doi.org/10.1208/pt060347.
4 Hou, S.Y., Cowles, V.E., and Berner, B. (2003). Gastric retentive dosage forms: a review. *Crit. Rev. Ther. Drug Carrier Syst.* 20: 459–497, https://doi.org/10.1615/critrevtherdrugcarriersyst.v20.i6.30.
5 Hwang, S.J., Park, H., and Park, K. (1998). Gastric retentive drug-delivery systems. *Crit. Rev. Ther. Drug Carrier Syst.* 15: 243–284.
6 Mandal, U.K., Chatterjee, B., and Senjoti, F.G. (2016). Gastro-retentive drug delivery systems and their in vivo success: a recent update. *Asian J. Pharm. Sci.* 11: 575–584, https://doi.org/10.1016/j.ajps.2016.04.007.
7 Setia, M., Kumar, K., and Teotia, D. (2018). Gastro-retentive floating beads a new trend of drug delivery system. *J. Drug Delivery Ther.* 8, https://doi.org/10.22270/jddt.v8i3.1717.
8 Qiu, Y. and Lee, P.I. (2017). *Developing Solid Oral Dosage Form- Pharmaceutical Theory and Practice* (ed. Y. Qiu et al.), Ch. 19, 519–550. Elsevier.
9 Qiu, Y., Li, X., and Duan, J.Z. (2014). Influence of drug property and product design on in vitro-in vivo correlation of complex modified-release dosage forms. *J. Pharm. Sci.* 103: 507–516, https://doi.org/10.1002/jps.23804.
10 Tripathi, J., Thapa, P., Maharjan, R., and Jeong, S.H. (2019). Current state and future perspectives on gastroretentive drug delivery systems. *Pharmaceutics* 11, https://doi.org/10.3390/pharmaceutics11040193.
11 Chawla, G., Gupta, P., Koradia, V., and Bansal, A. (2009). Gastroretention: a means to address regional variability in intestinal drug absorption. *Pharm. Technol.* 50–68.
12 Van Den Abeele, J., Rubbens, J., Brouwers, J., and Augustijns, P. (2017). The dynamic gastric environment and its impact on drug and formulation behaviour. *Eur. J. Pharm. Sci.* 96: 207–231, https://doi.org/10.1016/j.ejps.2016.08.060.
13 Awasthi, R. and Kulkarni, G.T. (2012). Development of novel gastroretentive floating particulate drug delivery system of gliclazide. *Curr. Drug Delivery* 9: 437–451, https://doi.org/10.2174/156720112802650716.
14 Awasthi, R. and Kulkarni, G.T. (2016). Decades of research in drug targeting to the upper gastrointestinal tract using gastroretention technologies: where do we stand? *Drug Delivery* 23: 378–394, https://doi.org/10.3109/10717544.2014.936535.
15 Lopes, C.M., Bettencourt, C., Rossi, A. et al. (2016). Overview on gastroretentive drug delivery systems for improving drug bioavailability. *Int. J. Pharm.* 510: 144–158, https://doi.org/10.1016/j.ijpharm.2016.05.016.
16 Fix, J.A., Cargill, R., and Engle, K. (1993). Controlled gastric emptying. III. Gastric residence time of a nondisintegrating geometric shape in human volunteers. *Pharm. Res.* 10: 1087–1089, https://doi.org/10.1023/a:1018939512213.
17 Prinderre, P., Sauzet, C., and Fuxen, C. (2011). Advances in gastro retentive drug-delivery systems. *Expert Opin. Drug Delivery* 8: 1189–1203, https://doi.org/10.1517/17425247.2011.592828.
18 Wilding, I.R., Coupe, A.J., and Davis, S.S. (2001). The role of gamma-scintigraphy in oral drug delivery. *Adv. Drug Delivery Rev.* 46: 103–124, https://doi.org/10.1016/s0169-409x(00)00135-6.
19 Timmermans, J. and Moës, A.J. (1994). Factors controlling the buoyancy and gastric retention capabilities of floating matrix capsules: new data for reconsidering the controversy. *J. Pharm. Sci.* 83: 18–24, https://doi.org/10.1002/jps.2600830106.
20 Schneider, F., Koziolek, M., and Weitschies, W. (2019). In vitro and in vivo test methods for the evaluation of gastroretentive dosage

forms. *Pharmaceutics* 11, https://doi.org/10.3390/pharmaceutics11080416.

21 Kagan, L., Lapidot, N., and Afargan, M. (2006). Gastroretentive Accordion Pill: enhancement of riboflavin bioavailability in humans. *J. Controlled Release* 113: 208–215, https://doi.org/10.1016/j.jconrel.2006.03.022.

22 Klausner, E.A., Eyal, S., Lavy, E. et al. (2003). Novel levodopa gastroretentive dosage form: in-vivo evaluation in dogs. *J. Controlled Release* 88: 117–126, https://doi.org/10.1016/s0168-3659(02)00487-x.

23 Cargill, R. et al. (1988). Controlled gastric emptying. 1. Effects of physical properties on gastric residence times of nondisintegrating geometric shapes in beagle dogs. *Pharm. Res.* 5: 533–536, https://doi.org/10.1023/a:1015981627525.

24 Cargill, R. et al. (1989). Controlled gastric emptying. II. In vitro erosion and gastric residence times of an erodible device in beagle dogs. *Pharm. Res.* 6: 506–509, https://doi.org/10.1023/a:1015976709043.

25 Säkkinen, M. et al. (2004). Gamma scintigraphic evaluation of the fate of microcrystalline chitosan granules in human stomach. *Eur. J. Pharm. Biopharm.* 57: 133–143, https://doi.org/10.1016/s0939-6411(03)00097-3.

26 Wilson, A. et al. (2002). Pharmacoscintigraphic evaluation of a novel controlled release (Endogastric Therapeutic System) for gastric retention in healthy volunteers. *AAPS Pharm. Sci. Tech.* 4: T3274.

27 Klausner, E.A. et al. (2003). Novel gastroretentive dosage forms: evaluation of gastroretentivity and its effect on levodopa absorption in humans. *Pharm. Res.* 20: 1466–1473, https://doi.org/10.1023/a:1025770530084.

28 Klausner, E. *et al.* in *Proceed Int Symp Control Rel Bioact Mater*.

29 Berner, B. and Louie-Helm, J. (2000). Tablet shapes to enhance gastric retention of swellable controlled-release oral dosage forms. USOO6488962B1, filed 20 June 2000 and issued 3 December 2002.

30 Shell, J.W., Louie-Helm, J., and Markey, M. (2001). Extending the Duration of Drug Release Within the Stomach during the Fed Mode. US 6,635,280 B2, filed 17 June 2008 and issued 5 June 2012.

31 Gusler, G. et al. (2001). Pharmacokinetics of metformin gastric-retentive tablets in healthy volunteers. *J. Clin. Pharmacol.* 41: 655–661, https://doi.org/10.1177/00912700122010546.

32 Urquhart, J. and Theeuwes, F. (1984). Drug delivery system comprising a reservoir containing a plurality of tiny pills. US patent 360,477.

33 Mamajek, R.C. and Moyer, E.S. (1980). Drug Dispensing Device and Method. US patent 4,207,890.

34 Erni, W. and Held, K. (1987). The hydrodynamically balanced system: a novel principle of controlled drug release. *Eur. Neurol.* 27 (Suppl 1): 21–27, https://doi.org/10.1159/000116171.

35 Sigurdsson, H.H., Loftsson, T., and Lehr, C.-M. (2006). Assessment of mucoadhesion by a resonant mirror biosensor. *Int. J. Pharm.* 325: 75–81, https://doi.org/10.1016/j.ijpharm.2006.06.027.

36 Rubinstein, A. and Tirosh, B. (1994). Mucus gel thickness and turnover in the gastrointestinal tract of the rat: response to cholinergic stimulus and implication for mucoadhesion. *Pharm. Res.* 11: 794–799, https://doi.org/10.1023/a:1018961204325.

37 Peppas, N.A. and Huang, Y. (2004). Nanoscale technology of mucoadhesive interactions. *Adv. Drug Delivery Rev.* 56: 1675–1687, https://doi.org/10.1016/j.addr.2004.03.001.

38 Tur, K.M., Ch'ng, H.-S., and Baie, S. (1997). Use of bioadhesive polymer to improve the bioavailability of griseofulvin. *Int. J. Pharm.* 148: 63–71, https://doi.org/10.1016/S0378-5173(96)04827-2.

39 Harris, D., Fell, J.T., Sharma, H.L., and Taylor, D.C. (1990). GI transit of potential bioadhesive formulations in man: a scintigraphic study. *J. Controlled Release* 12: 45–53, https://doi.org/10.1016/0168-3659(90)90182-S.

40 Santus, G. et al. (1997). An in vitro-in vivo investigation of oral bioadhesive controlled release furosemide formulations. *Eur. J. Pharm. Biopharm.* 44: 39–52, https://doi.org/10.1016/S0939-6411(97)00100-8.

41 Säkkinen, M., Tuononen, T., Jürjenson, H. et al. (2003). Evaluation of microcrystalline chitosans for gastro-retentive drug delivery. *Eur. J. Pharm. Sci.* 19: 345–353, https://doi.org/10.1016/s0928-0987(03)00121-0.

42 Ponchel, G., Montisci, M.-J., Dembri, A. et al. (1997). Mucoadhesion of colloidal particulate systems in the gastro-intestinal tract. *Eur. J. Pharm. Biopharm.* 44: 25–31, https://doi.org/10.1016/S0939-6411(97)00098-2.

43 Vasir, J.K., Tambwekar, K., and Garg, S. (2003). Bioadhesive microspheres as a controlled drug delivery system. *Int. J. Pharm.* 255: 13–32, https://doi.org/10.1016/s0378-5173(03)00087-5.

44 Thairs, S. et al. (1998). Effect of dose size, food and surface coating on the gastric residence and distribution of an ion exchange resin. *Int. J. Pharm.* 176: 47–53, https://doi.org/10.1016/S0378-5173(98)00305-6.

45 Umamaheshwari, R.B., Ramteke, S., and Jain, N.K. (2004). Anti-*Helicobacter pylori* effect of mucoadhesive nanoparticles bearing amoxicillin in experimental gerbils model. *AAPS PharmSciTech* 5: 60–68, https://doi.org/10.1208/pt050232.

46 Ahmed, A., Bonner, C., and Desai, T.A. (2002). Bioadhesive microdevices with multiple reservoirs: a new platform for oral drug delivery. *J. Controlled Release* 81: 291–306, https://doi.org/10.1016/s0168-3659(02)00074-3.

47 Tao, S.L., Lubeley, M.W., and Desai, T.A. (2003). Bioadhesive poly(methyl methacrylate) microdevices for controlled drug delivery. *J. Controlled Release* 88: 215–228, https://doi.org/10.1016/s0168-3659(03)00005-1.

48 Altreuter, D.H. et al. (2018). Changing the pill: developments toward the promise of an ultra-long-acting gastroretentive dosage form. *Expert Opin. Drug Delivery* 15: 1189–1198, https://doi.org/10.1080/17425247.2018.1544615.

49 Sheth, P.R. and Tossounian, J. (1984). The hydrodynamically balanced system (Hbs™): a novel drug delivery system for oral use. *Drug Dev. Ind. Pharm.* 10: 313–339, https://doi.org/10.3109/03639048409064653.

50 Garbacz, G., Klein, S., and Weitschies, W. (2010). A biorelevant dissolution stress test device – background and experiences. *Expert Opin. Drug Delivery* 7: 1251–1261, https://doi.org/10.1517/17425247.2010.527943.

51 Davis, S.S. and Wilding, I.R. (2001). Oral drug absorption studies: the best model for man is man! *Drug Discovery Today* 6: 127–130, https://doi.org/10.1016/s1359-6446(00)01653-6.

52 Lin, J.H. (1995). Species similarities and differences in pharmacokinetics. *Drug Metab. Dispos.* 23: 1008–1021.

53 Davis, S.S., Illum, L., and Hinchcliffe, M. (2001). Gastrointestinal transit of dosage forms in the pig. *J. Pharm. Pharmacol.* 53: 33–39, https://doi.org/10.1211/0022357011775163.

54 Burke, M.D., Staton, J.S., Vickers, A.W. et al. (2007). A novel method to radiolabel gastric retentive formulations for gamma scintigraphy assessment. *Pharm. Res.* 24: 695–704, https://doi.org/10.1007/s11095-006-9189-9.

55 Steingoetter, A. et al. (2003). Magnetic resonance imaging for the in vivo evaluation of gastric-retentive tablets. *Pharm. Res.* 20: 2001–2007, https://doi.org/10.1023/b:pham.0000008049.40370.5a.

56 Weitschies, W. et al. (2005). Impact of the intragastric location of extended release tablets on food interactions. *J. Controlled Release* 108: 375–385, https://doi.org/10.1016/j.jconrel.2005.08.018.

57 Harro, H. (2023). Accordion Pill (R): Laminated, folded, encapsulated. (10), year 2020. https://www.harro-magazine.com/en/10-2020/accordion-pill-laminated-folded-encapsulated/.

58 Kanasty, R. et al. (2019). A pharmaceutical answer to nonadherence: once weekly oral memantine for Alzheimer's disease. *J. Controlled Release* 303: 34–41, https://doi.org/10.1016/j.jconrel.2019.03.022.

59 Bellinger, A., Kanasty, R., Yang, J.H. et al. (2017). Gastric residence systems for sustained delivery of adamantane-class drugs, US Patent Application US20190231697A1.

60 Janczura, M., Sip, S., and Cielecka-Piontek, J. (2022). The development of innovative dosage forms of the fixed-dose combination of active pharmaceutical ingredients. *Pharmaceutics* 14: 834–856.

61 Sannino, A., Demitri, C., Zohar, Y. et al. (2018). Method for producing hydrogels coupling high elastic modulus and absorbance, US Patent 10,584,183.

62 Greenway, F.L., Aronne, L.J., Raben, A. et al. (2019). A randomized, double-blind, placebo-controlled study of gelesis 100: a novel nonsystemic oral hydrogel for weight loss. *Obesity* 27: 205–216, https://doi.org/10.1002/oby.22347.

63 Demitri, C., Zohar, Y., Heshmati, H.M. et al. (2018). Satiety, weight loss, and glycemic control-enhancing properties vary between functional fibers, mixed vegetables, and a novel hydrogel (Gelesis 100). In: *78th Scientific Sessions of the American Diabetes Association,* LB38.

64 Skyepharma. (2023). Extended release gastroretentive platform technology.

# 34

# Enhancing Oral Bioavailability Using 3D Printing Technology

*Timothy Tracy*[1,2], *Senping Cheng*[1], *Lei Wu*[1], *Xin Liu*[1], *and Xiaoling Li*[1,3]

[1] *Triastek, Inc., Nanjing, Jiangsu, China*
[2] *Tracy Consultants, LLC, Huntsville, AL, USA*
[3] *Department of Pharmaceutics and Medicinal Chemistry, Thomas J. Long School of Pharmacy, University of the Pacific, Stockton, CA, USA*

## 34.1 Introduction

For orally administered drugs, bioavailability must be sufficient so that adequate drug concentrations and exposure are achieved to generate the desired pharmacodynamic response. Factors that affect oral bioavailability include first-pass metabolism, drug solubility, dissolution rate, absorption window (related to gastric transit time), and permeability [1, 2]. In addition, for biomacromolecules, such as peptides, stability within the gastric milieu will also impact oral bioavailability [3]. As briefly discussed below, both physicochemical properties of the drug molecule, as well as gastrointestinal (GI) tract processes can serve to affect the degree of bioavailability, and subsequently the drug's pharmacokinetics (PKs). However, these challenges have been further amplified in recent years as several of the recently approved drugs and those currently in development are increasingly lipophilic and of higher molecular weight (e.g. 600–700 Da) to better interact with allosteric targets and achieve increased potency. These molecules, frequently classified as Biopharmaceutics Classification System (BCS) Class IV (low solubility and low permeability), provide additional challenges to formulators in their attempts to achieve adequate oral bioavailability.

The oral bioavailability of a drug will impact both the PKs and pharmacodynamics that are observed. Relatedly, the rate of absorption will impact both the $C_{max}$ and the $T_{max}$, the drug's overall concentration–time profile, and subsequently the effect (pharmacodynamics) observed. Likewise, the extent of a drug's absorption will not only impact the $C_{max}$ but the overall area under the curve (AUC) (exposure) of the drug, with concomitant effects on the pharmacodynamics.

The rate and extent of drug absorption can be impacted by both the physicochemical characteristics of the drug, as well as the GI tract processes that result in either degradation, metabolism, or active transport of the drug [4]. Drugs that are poorly soluble can exhibit both a slow rate of absorption and reduced extent of absorption. Even if a drug exhibits reasonable solubility characteristics, if it is poorly permeable it may not cross the intestinal membrane well and thus, exhibit a diminished extent of absorption and thus, a reduced AUC. The presence of efflux transporters in the gut, such as P-glycoprotein (ATP-binding cassette subfamily B member 1, ABCB1), actively pump foreign compounds, such as drugs, that have permeated into the intestinal cells back into the intestinal lumen resulting in decreased absorption [5]. Furthermore, drug-metabolizing enzymes in the gut, such as CYP3A4/5 and UDP glucuronosyl transferases, can metabolize drugs thus reducing the amount reaching the systemic circulation.

Though these characteristics, poor solubility and permeability, and transporter proteins or metabolizing enzymes, can reduce oral bioavailability, strategies can be employed to overcome these issues and thus, increase drug absorption.

## 34.2 3D Printing in Pharmaceutical Applications

The first application of 3D printing technology to the development of pharmaceuticals occurred in 1996 in the laboratory of Dr. Michael Cima at the Massachusetts Institute of Technology (MIT) [6]. Since that time, a variety of 3D printing techniques, discussed below, have

*Oral Bioavailability and Drug Delivery: From Basics to Advanced Concepts and Applications,* First Edition. Edited by Ming Hu and Xiaoling Li.

been implemented in the production of pharmaceuticals resulting in numerous scientific publications [7]. By the end of 2021, more than 600 scientific manuscripts had been published on the application of 3D printing to the development of pharmaceuticals, with almost 80% of the manuscripts published between 2018 and 2021. The largest number of publications has been around the extrusion-based 3D printing methods (fused deposition modeling (FDM), semisolid extrusion, and melt extrusion deposition), comprising approximately one-half of the total publications though the number of articles describing other 3D printing techniques continues to grow. Thus, there has recently been a rapid expansion of research in the area with the number of publications per year expanding substantially.

A standardized classification system related to 3D printing technologies has been introduced by ASTM International (formerly the American Society for Testing and Materials) and the International Organization for Standardization (ISO) [8]. These standards categorize the various 3D printing techniques into seven main groups: material extrusion, powder bed fusion, vat photopolymerization, direct energy deposition, binder jetting, material jetting, and sheet lamination. The techniques most frequently used in the pharmaceutical area can be loosely divided into three primary categories: extrusion-based systems, powder-based systems, and liquid-based systems (Figure 34.1) [9], although other techniques such as screen printing technology are now being applied to the development of pharmaceuticals. Below, we briefly introduce each of these systems and their various embodiments, as well as the advantages and disadvantages of each system.

### 34.2.1 Extrusion-Based Systems

Material extrusion, according to the ISO/ASTM categorization, is also referred to as extrusion-based 3D printing, which has been the most widely investigated and utilized 3D printing technology applied to the production of pharmaceutical dosage forms [10]. Because the tablet is formed layer-by-layer without the constraint of a powder bed or resin solution, this technique allows the most free-forming of both internal and external tablet structures. Additionally, multiple print heads can be utilized to easily combine multiple active pharmaceutical ingredients (APIs) even within different compartments or layers within the tablet, allowing additional flexibility as compared to the other 3D printing techniques described below.

Two widely known applications of material extrusion methods are FDM and semisolid extrusion. In recent years, additional novel applications of material extrusion technology have been developed, including melt extrusion deposition (first patent in 2017), direct powder extrusion (first reported in 2019), and direct powder printing (first reported in 2020) [11–13]. In each of these methods, the drug-containing material is extruded through a nozzle (printhead) whereupon the extrudate is deposited on a plate and the structure is formed layer by layer by altering the position of the nozzle in the $X$, $Y$, and $Z$ directions. This methodology allows truly free-form structures to be created that can contain compartments or unique internal geometries and be of virtually any shape and size. Utilizing multiple nozzle heads, one can deposit different drug compositions throughout the tablet to allow for controls over disintegration and release kinetics.

In FDM (Figure 34.1c), a drug-infused filament is moved by rollers, heated, and once in a molten state deposited onto the build plate. The filament must possess the required mechanical properties to be bent and compressed through the rollers, as well as the proper melting temperature to allow extrusion [14]. Because of these mechanical property requirements, the choice of excipients that can be included in the final tablet

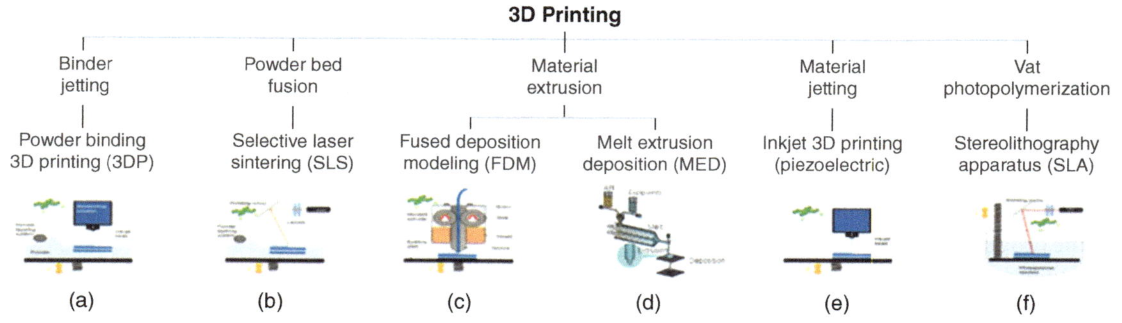


**Figure 34.1** Illustration of various types of 3D printing technology, including (a) powder binding, (b) selective laser sintering, (c) fused deposition modeling, (d) melt extrusion deposition, (e) inkjet 3D printing, and (f) stereolithography. Source: Alhnan et al. [9]/with permission of Springer Nature.

is limited, as compared to other methods. For some drugs, attaining sufficient drug-loading of the filament may limit the amount of drug that can be incorporated into a tablet, thus limiting its utility. Finally, because the system is under increased temperature due to the mechanical requirements of the filaments, the thermostability of the drug must be considered and may be a limiting factor that prevents the use of certain drugs with lower thermostability.

For semisolid extrusion, the drug is incorporated into a semisolid matrix, either at room temperature or elevated temperature, and then the semisolid material is extruded through a nozzle onto the build plate [15]. The use of a semisolid drug matrix overcomes the mechanical property requirements of FDM. However, because of the semisolid state of the material, post-extrusion drying is required, which can make the process time-consuming. In addition, because volatile solvents are often used in forming the semisolid matrix and to aid in drying, one must consider their potential interaction with the drug of interest.

Melt extrusion deposition (Figure 34.1d), a proprietary technology of Triastek was developed to overcome some of the disadvantages of FDM and semisolid extrusion (i.e. does not require any drying or post-printing processing, is not limited in the excipients that can be used and is not constrained by the amount of drug that can be loaded into a filament) [16]. Melt extrusion deposition begins with dry powders of the drug(s) of interest and excipients, mixes them inline, and melts them into an extrudable form, which is then deposited through one or more nozzles onto the build plate to create the 3D structure. This process has the advantage of not requiring any post-processing (e.g. no drying) and allows for a "continuous" manufacturing process and multiple mixing chambers and printing nozzles allow for easy incorporation of multiple drug substances, as well as different excipient formulations to control drug release. Like the other extrusion-based methods, the amount of drug that can be incorporated can be a factor though most standard tablet drug loads, e.g. below 400 mg, can be accomplished. Additionally, because the process does involve a slightly elevated temperature one must be cognizant of the drug's thermostability, although low temperature processes can also be incorporated into the printing process.

### 34.2.2 Powder-Based Systems

Two primary powder-based 3D printing techniques have been applied to the formulation of drugs, to date. These include binder jetting, also known as drop on powder, and powder bed fusion. Both techniques use a bed of powder containing the drug of interest and once a layer has been "bonded," another layer of powder is spread and the process repeated. However, the two processes differ substantially in how this "bonding" process is accomplished.

The binder jetting technique (Figure 34.1) is similar to the familiar inkjet printing process in that a liquid binding agent is sprayed from a printhead onto the surface, much like printing on paper. Traditionally, the drug is incorporated with the excipients in powder form to allow for greater loading. If drug is loaded in the print fluids, the drug must be soluble in the combination aqueous/organic solvent of the print fluid and not cause the print fluid to become too viscous so as to not be effectively ejected from the print heads [17]. With each successive layer of powder that is "bound" by the printing fluid, another layer of powder is spread, and the process repeated. Because no compression is involved, unlike traditional tableting processes, very porous structures can be created making this technology particularly amenable to creating oral dispersible tablets. In addition, multiple compartments or internal structures can be created to modify or control drug release kinetics. The primary disadvantage of drop on powder technique is the requirement of substantial post-printing processes (removal of excess unbound powder and tablet drying) and compromising of tablet hardness, which can affect storage, shipping and handling.

Within the pharmaceutical field, selective laser sintering is the most widely used method of powder bed fusion technology [18]. Though selective laser sintering (Figure 34.1) uses the same bed of drug-containing powder as drop on powder methods, this technique differs substantially in how the powder is bound to form the tablet. A beam of high energy is applied by laser to solidify the drug-excipient powders, as compared to applying a binding agent. Because a high-resolution laser beam is employed, high resolution tablet structures can be achieved [19]. Thus, selective laser sintering can be used to fabricate complex and precise structures, including lattice designs or dual miniprintlets for controlled release [20]. Selective laser sintering also requires removal of excess unbound powder, like binder jetting, but in addition, the powder bed must be heated before printing to the optimum process temperature and then slowly cooled down to avoid stress on the product. This results in a lengthy post-printing process. Finally, one must be cognizant of the potential for degradation of the drug due to the high energy input from the laser beam, in addition the heating of the powder bed.

### 34.2.3 Liquid-Based Systems

The liquid-based 3D printing systems for pharmaceuticals include vat photopolymerization and material jetting. In each of these cases, the drug is dissolved in liquid and then either an additive or a subtractive process employed. These two techniques are the least employed 3D printing techniques for pharmaceuticals for reasons discussed below.

Stereolithography (Figure 34.1), the most widely used vat photopolymerization method, involves the drug being dissolved in a resin solution, which is subjected to a high energy laser beam, like selective laser sintering but using a liquid as the substrate. In the process of photopolymerization, free radicals are released after the interaction between the photo-initiator and UV light [21]. By varying the composition of the monomers in the resin, one can alter the release characteristics of the drug. This method can also be used to produce various shapes and sizes of tablets and some of the most interesting applications have been in the production of a multi-drug polypill. Because the process has been customized to allow pausing of printing and switching of the resin tray to include a different drug-containing resin, one can create multi-layer tablets with different drug components in each layer. However, post-printing curing of the tablet product is required, the potential for production of carcinogenic oligomers from the activation of the resin and the expense of the equipment are downsides to this method. Other liquid-based printing methods have also been investigated, such as digital light processing (DLP), liquid crystal display (LCD) and volumetric printing to enhance printing speed [22, 23].

In contrast, drop on demand (Figure 34.1) applications of material jetting methods use an inkjet printing head, analogous to that of drop on powder, but instead the drug is typically contained in the printing fluid. Because of the precise nature of the printhead openings, high spatial resolution structures can be produced. The droplets are deposited in thin layers and solidified by either evaporation of solvent or cooling. One can also incorporate photosensitive polymers into the solution and solidification achieved by exposure to ultraviolet light. By altering the thickness and surface area, different release kinetics can be achieved. Researchers have even used beeswax in the printing fluid and achieved a honeycomb architecture, which by varying the opening sizes in the honeycombs the release kinetics can also be controlled [24]. This method is especially useful for extremely low drug doses (e.g. low milligram to microgram doses) since the drug is dissolved in the print fluid. However, the amount of material that can be ejected from the printhead is limited and nozzle blockage can also pose challenges.

## 34.3 Novel Tablet Structures Possible with 3D Printing

### 34.3.1 Unique Tablet External Geometries

Three-dimensional (3D) printing allows the formation of an almost unlimited number of external tablet structures not possible with traditional tableting or encapsulation equipment. Furthermore, because one does not have to worry about the "resolution" of the tableting die or whether the final product will eject from the die without compromising the tablet structure, the design possibilities are more numerous. These features allow the production of not only unique geometric shapes but also of honeycomb/weave designs (Figure 34.2a–c) [25–27]. In most cases, the primary purpose of these designs is to control the amount of surface area to which the solvent (e.g. GI fluid) has access. For example, by using a weave or honeycomb design, one can substantially increase the tablet's external surface area and thus, greatly influence the rate of dissolution or disintegration [30]. By changing the size of the infill or "weave" or the number of weaves, one can precisely control the rate of disintegration and thus, dissolution that is not possible with a conventional round or convex tablet design. For example, for a drug that is best dissolved in lower pH values, e.g. pH 1–2 of the stomach, one might create a design with a substantial amount of external surface area so that the tablet disintegrates quickly in the stomach exposing drug to this acidic pH, which could serve to enhance bioavailability. Conversely, if the drug is more completely dissolved in a more basic environment, e.g. pH 5–6 of the small intestine, decreasing the external surface area can slow the dissolution or disintegration until the tablet is past the stomach and reaches the small intestine, again as a way to potentially increase absorption. These two examples illustrate how 3D printing can be utilized to effectuate the release of drugs at the optimal absorption window.

### 34.3.2 Unique Tablet Internal Geometries

The potential internal geometry designs possible with 3D printing techniques are equally as varied as external geometries and because these designs can combine the unique internal geometries with different combinations of excipients in different areas of a single tablet (Figure 34.2d,e) [28, 29], something not possible with traditional compression tableting, provides almost

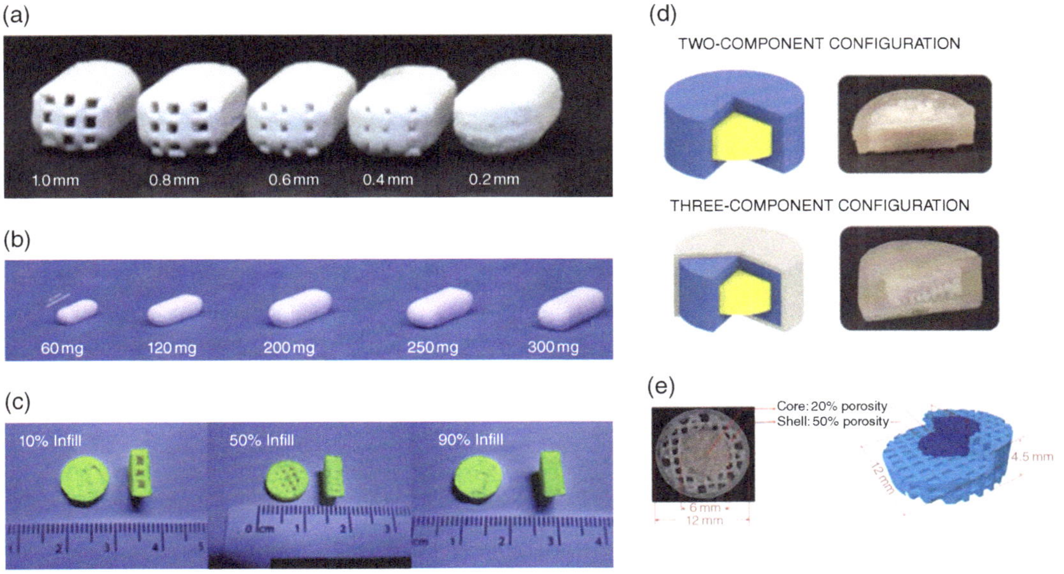


**Figure 34.2** (a) Rendered images of caplet designs with decreasing channel size with nine long channels. (b) The manipulation of printing scale allows the fabrication of Eudragit RL-based tablets with increasing dose. (c) Cross-sectional views of 3DP fabricated tablets containing fluorescein under UV light (top 10%, middle 50%, and bottom 90%). (d) Chronotopic™ system with core-shell structure for pulsatile-release. (e) Core-shell structure with different porosity ratios. Core: 20% porosity; Shell: 50% porosity. Source: Goyanes et al. [25]/from Elsevier; Pietrzak et al. [26]/from Elsevier; Sadia et al. [27]/from Elsevier; Zhang et al. [28]/from Elsevier; Melocchi et al. [29]/from Oxford University Press.

unlimited combinations of possibilities for different delivery and clinical needs. With respect to the internal structures that are possible, we will bin the internal geometry types into: layered designs, release control caps, multiple compartments, and buoyancy-enhancing structures.

#### 34.3.2.1 Layered Designs

Layering of tablet constituents can be customized via 3D printing to control release rates and thus achieve desired PK profiles. Zheng et al. [16] have employed a non-dissolving perimeter structure which forces dissolution from only the top and/or bottom of the tablet, to allow for customization of release characteristics. Examples of these kinds of structures are presented in Figure 34.3. By varying the diameter of the drug-containing layer one can control the amount of drug exposed to GI fluid and thus, the amount of drug released. For example (Figure 34.3a), for a drug where one desires higher concentrations earlier successively smaller perimeter layers can be utilized to provide a greater amount of drug for dissolution early in the time course to achieve higher drug concentrations but successively smaller perimeter layers below to maintain the concentrations but not continue to increase them. This could be used to control how much drug is released in particular locations in the GI tract, where absorption might be enhanced. Vice-versa, one can use successively larger perimeter layers wherein the amount of drug released is more gradually increased resulting in more drug being released lower in the GI tract where the drug might be more soluble, beyond the highest concentrations of drug-metabolizing enzymes in the gut or to avoid unwanted high drug concentrations early in the PK profile. This strategy can also be used to compensate for the gradually decreasing drug absorption along the GI tract by increasing the concentration of drug present. Resulting release profiles from these tablet designs are presented in Figure 34.3b, wherein more immediate release or slower release profiles are observed depending on the tablet's internal structure. As a reminder, in the examples of Figure 34.3a, both the sides and the bottom of the tablet consist of a non-dissolvable layer to allow this degree of control. However, one could imagine allowing drug release from both top and bottom of the tablet and simply producing a "mirror-image" profile of the layers. This degree of release control offers substantial advantages over conventional tablet formulation methods both in terms of release rate but also time to an effective formulation. Because tablets can be made individually or in very small batches (e.g. 10–20) one can readily design and test numerous tablet designs in a very short period of time.

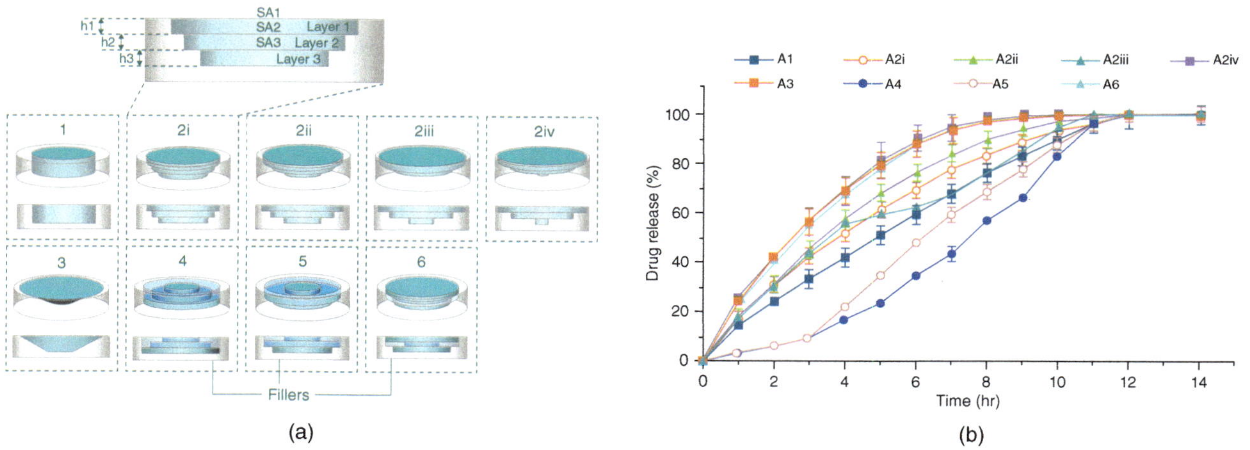


**Figure 34.3** (a) Design A: Multilayered drug compartment in core–shell structure tablets with constant or varied surface area (SA). Drug compartment (teal), shell (pale white), fillers (bluish). [1] Drug compartment with constant SA. [2] Drug compartment with stepwise decreasing SA. [3] Drug compartment with continuous decreasing SA. [4] Drug compartment with stepwise increasing SA. [5] Drug compartment with increasing–decreasing SA. [6] Drug compartment with decreasing–increasing SA. (b) Experimental metoprolol release profiles of 3D printed A1–A6 tablets. Source: Zheng et al. [16]/with permission of Elsevier.

#### 34.3.2.2 Release Control Caps

Three-dimensional printing allows additional levels of release control by employing release control caps. These "caps" are layers of excipient material that can be designed to dissolve or disintegrate at various rates, such that the time before the GI fluids reach the drug-containing layer is prolonged, resulting in a delayed release of drug. By altering the chemical makeup of this control cap, as well as cap thickness, one can achieve a substantial delay in drug release. Examples of the implementation of release control caps are shown in Figure 34.4 [16]. In this example, a tablet with either no release control cap, a 0.2 mm release control cap, or a 0.4 mm release control cap was used to delay the release of drug (Figure 34.4a,b). Whereas the tablet with no release control cap began releasing drug immediately, release of drug from the tablet with the 0.2 mm release control cap was delayed approximately 1.5 hours with a similar release profile to the no cap tablet, thereafter (Figure 34.4c). Likewise, a doubling of the size of the release control cap (0.2 mm increased to 0.4 mm) resulted in a doubling of time (~3 hours) until release of drug was initiated, again with a similar release profile to the other two formulations thereafter

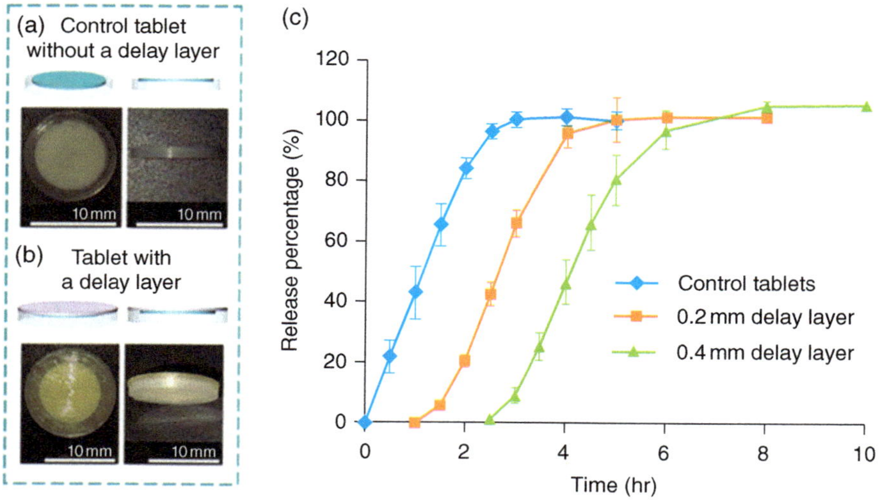


**Figure 34.4** *In vitro* and *in vivo* studies of MED™ 3D printed Design B tofacitinib tablets. Printed tofacitinib tablets without a delay layer (a) and with a delay layer (b): shell (transparent), drug core (white), delay layer (transparent yellow). (c) Release profiles of tofacitinib tablets with a delay layer with increasing layer thickness (0, 0.2, and 0.4 mm). Data are presented as mean ± SD ($n = 6$). Source: Zheng et al. [16]/from Elsevier.

(Figure 34.4c). In this manner, the release profile of the drug can be maintained but delayed for a period. One can envision this type of formulation as being advantageous in several instances. For example, when the peak physiological event (e.g. the peak level of blood pressure) occurs in the early morning hours before the patient arises, and thus, with conventional morning dosing the peak drug concentrations would not occur until after the peak physiological event. However, using a release control cap to delay absorption, for example, for six hours would allow bedtime dosing but peak drug concentrations occurring in the early morning hours before awakening and during the physiological event. Similarly, one could use this type of formulation to delay release within the GI tract to optimize absorption or to deliver drug to a specific location in the GI tract for a "local" effect. With respect to this latter concept of GI-targeted delivery, one must take care to consider the limited volume of fluid in the latter portions of the GI tract within which drug can dissolve.

#### 34.3.2.3 Multiple Compartments

Multiple compartments can also be used to deliver pulsatile release of drug for a variety of purposes. For example, one could mimic twice or three times a day dosing from a single tablet dosed once a day using pulsatile dosing. The different compartments can be designed with different release times (e.g. different time control layers) from which drug is released at the desired times. This type of design can be used to deliver, for example, two peaks ($C_{max}$) of drug over the dosing interval from only a single tablet. This design can also allow for release of drug at predetermined intervals, analogous to additional doses, to maintain a desired concentration of the drug. In another embodiment, the use of multiple compartments and pulsed release can be used in conjunction with traditional solubility enhancers to increase a drug's bioavailability. Through the ability to create multiple compartments, each compartment with its own unique release characteristics and "sub-compartments" containing functional excipients, provides formulators a multitude of options to enhance bioavailability.

## 34.4 Application of 3D Printing in Oral Bioavailability Enhancement

### 34.4.1 3D Printing Combined with Solubility Enhancement Techniques (SEDDS, ASD, etc.)

The ability to create unique external and internal geometries, multiple internal compartments, control of release rate, and inclusion of excipients in a nonuniform manner across the tablet geometry made possible by 3D printing have been used to demonstrate several examples of oral bioavailability enhancement. For example, the direct orally acting anticoagulant dabigatran etexilate is very polar, and permanently charged with no in vivo bioavailability. However, the marketed commercial formulation is a double prodrug to mask the charged group and once absorbed the ester is removed by hydrolytic cleavage, but the bioavailability remains very low [31]. Liu et al. developed a novel 3D printed formulation of dabigatran etexilate utilizing multiple compartments, pulsatile release, and a solubility enhancer to greatly improve the bioavailability of dabigatran [32]. In this formulation, dabigatran is loaded into three separate compartments programmed using time control layers of varying thickness to release drug in pulses over time (Figure 34.5a). Below each of these layers is incorporated a self-emulsifying drug delivery system (SEDDS) comprised of Capmul MCM C8: Cremophor EL 35: Transcutol P: Soluplus: API = 8.3%: 49.3%: 24.6%: 5.3%: 12.5% (wt%) to increase the solubility of the active drug. A mixture of dabigatran and the SEDDS is loaded into each compartment, where it is released. This formulation was administered orally to rats and the PKs compared to a 3D printed SEDDS-containing formulation with one compartment. In Figure 34.5c, one can see that dabigatran formulated with SEDDS in the tablet containing multiple compartments with pulsatile release characteristics exhibited a threefold increase in AUC and $C_{max}$ and a twofold increase in half-life as compared to a 3D printed immediate release formulation with only one compartment that contained dabigatran mixed with the SEDDS. These results demonstrate the true power of 3D printing since it is the unique tablet geometry and release characteristics that produce the enhancement in oral bioavailability beyond any improvements due to the inclusion of a solubility enhancer. These kinds of formulations provide numerous possibilities to enhance the exposure of poorly bioavailable drugs, both those already on the market and those in development. One can also envision the use of other types of solubility enhancers, permeability enhancers, etc. that can be incorporated into the tablet design for programmed release to optimize absorption of the drug in GI tract.

Three-dimensional printing techniques have also been used to develop a fixed-dose combination product (efavirenz, tenofovir, and emtricitabine) with enhanced bioavailability for the treatment of HIV. Since these three products represent a range of

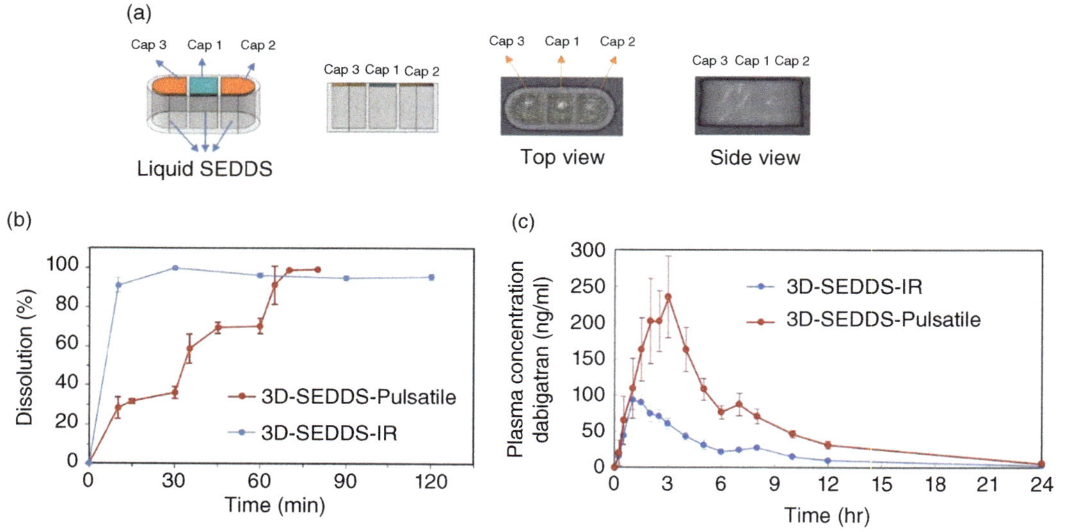


**Figure 34.5** (a) Schematic drawing and photographs of SEDDS-loaded pulsatile released tablet with multiple compartments. (b) The dissolution of dabigatran from SEDDS-loaded immediate-released ($n = 3$) and pulsatile-released tablets ($n = 5$) at pH 2.0. Data: mean ± SD. (c) Plasma concentration of dabigatran after oral administration of pulsatile released tablets or immediate release tablets. Data: mean ± SE ($n = 3$). Source: Liu [32]/with permission of American Association of Pharmaceutical Scientists.

physicochemical properties and a range of BCS classes (efavirenz – BCS II, tenofovir – BCS III, and emtricitabine – BCS Class I) these researchers investigated a humic acid–polyquaternium 10 complex (HA-PQ10) to enhance the solubility and permeability of these drugs and to control the release. This matrix is particularly beneficial since it can load both hydrophilic and lipophilic drugs. An extrusion technique was employed for printing of the "sludge" containing drug material incorporated into the HP-PQ10 matrix [33]. A 3D-Bioplotter® was used to print the material, which allowed printing of the material at room temperature. This is an advantage, since the HA-PQ10 is not compatible with higher temperatures, and one does not have to worry about drug thermo-instability. A rectangular prism structure with a weave design was found to be optimal (Figure 34.6). The size, shape, and infill (weave) of the tablet were optimized to obtain the desired drug release profiles. The PKs of the drugs from the 3D printed formulation were compared to those from a commercially available tablet. For the efavirenz component, the $T_{max}$ was increased ~5-fold, the $C_{max}$ was essentially unchanged, and the bioavailability was increased 109% in the 3D printed tablet as compared to the commercial formulation. For tenofovir, the $T_{max}$ and $C_{max}$ were roughly unchanged, but the bioavailability was increased ~220% in the 3D printed tablet as compared to the commercial formulation. Finally, the $T_{max}$ of emtricitabine was increased ~3-fold, the $C_{max}$ was essentially unchanged, and the bioavailability was increased ~135% in the 3D printed tablet as compared to the commercial formulation. These results demonstrate the utility of using 3D printing to not only enhance the bioavailability but also improve the PKs of fixed-dose combination products containing drugs with markedly different physicochemical properties.

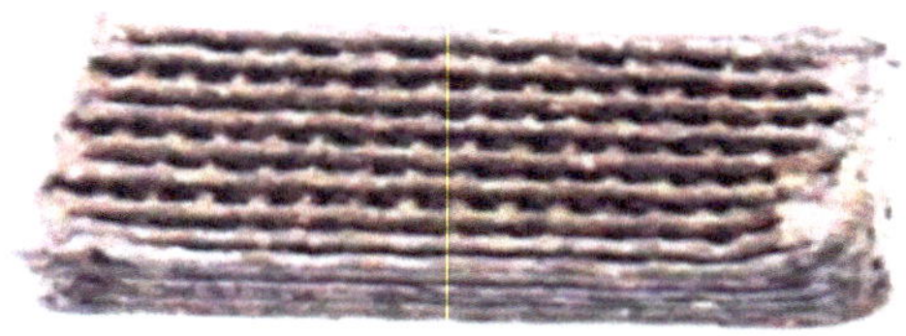

EFV, TDF, and FTC combined in one sludge

**Figure 34.6** Digital images of results from the tablet design attempt for 3DP FDC. Source: Siyawamwaya et al. [33]/from Elsevier.

Ahmed et al. used extrusion-based 3D printing to create multi-compartment tablets to increase the bioavailability of rosuvastatin and glimepiride in an attempt to create a formulation that would address both dyslipidemia and diabetes [34]. These tablets were created using a self-nanoemulsifying drug delivery system (SNEDDS) using cumin oil as the emulsifier. Glimepiride is hydrophobic in nature with a relatively short half-life in humans [35] and rosuvastatin exhibits low solubility [36], which limits its bioavailability. Formulations of each agent alone or in combination and incorporated into the cumin oil SNEDDS were then optimized with respect to mechanical characteristics

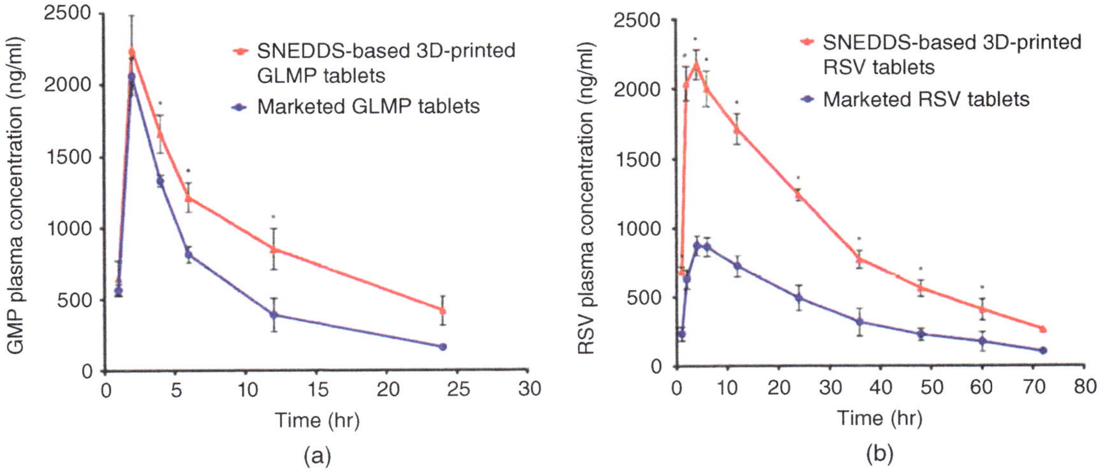


**Figure 34.7** Plasma level–time curves for glimepiride (a) and rosuvastatin (b) after oral administration of the 3DP and marketed tablets to male Wistar rats ($n = 6$). Note: * Indicates significant difference, at $p < 0.05$, between groups. Source: Ahmed et al. [34]/MDPI/CC BY 4.0.

(e.g. friability) and release rate to achieve the desired kinetics. Either glimepiride or rosuvastatin 3D printed tablets or a marketed formulation were then administered to rats and the plasma levels of glimepiride (Figure 34.7a) and rosuvastatin (Figure 34.7b) were monitored. The half-life of glimepiride was increased by approximately 64%, while the bioavailability was increased by approximately 88% in the 3D printed tablets containing SNEDDS versus the marketed formulation. In the case of rosuvastatin, the half-life remained relatively unchanged whereas the bioavailability was increased by approximately 147% as compared to the marketed formulation. These results demonstrate that combining solubility-enhancing agents with 3D printing can result in substantial increases in bioavailability. It is of note, that the distinct contribution of the 3D printed tablets was not assessed in the absence of SNEDDS, however, the 3D printing process did facilitate the production of the SNEDDS tablets as compared to traditional tableting methods.

These same investigators evaluated glimepiride PKs across three different formulations: 3D printed tablets of glimepiride that were formulated with a black seed oil-based SNEDDS system, a liquisolid formulation glimepiride tablet formulated with the black seed oil SNEDDS system and a traditional compressed tablet (no SNEDDS) [37]. Again, using an extrusion-based method, the 3D printed tablets were optimized for their mechanical properties as well as glimepiride release rate. In comparison to the studie described above, however, a liquisolid formulation was also developed using the same SNEDDS system to allow a direct comparison of the benefits of the 3D printed formulation as well as to a non-SNEDDS system traditional tablet formulation. This liquisolid technique allows for the conversion of liquid drugs into an apparently dry, non-adherent, free-flowing, and compressible powder [38]. Evaluating the in vitro release rate, SNEDDS-containing 3D printed tablets released glimepiride at approximately the same rate but to a different extent than non-SNEDDS-containing 3D printed tablets, with the 3D printed formulation releasing 100% of the glimepiride by 12 hours, whereas the non-SNEDDS-containing 3D printed tablets had only released 60% of the glimepiride by 12 hours (Figure 34.8a). In contrast, both the liquisolid (SNEDDS-containing) tablets and an extemporaneously prepared compression tablet had completely released the glimepiride contents in vitro within two hours and the commercial formulation had only released about 60% in two hours (Figure 34.8b). When the three formulations were administered orally to rats, the 3D printed SNEDDS-containing formulation and the liquisolid SNEDDS-containing formulation achieved virtually identical PK profiles, while the relative bioavailability of the compressed tablet was approximately 50% of the other two formulations (Figure 34.8c). Thus, both SNEDDS-containing formulations produced a substantial increase in bioavailability, but the 3D printed formulation was not different from the liquisolid formulation. Although 3D printing techniques were used in this study, the benefits of using 3D printing for bioavailability enhancement were not demonstrated.

Choudury et al., [39] evaluated using a 3D printed capsule design to enhance berberine bioavailability.

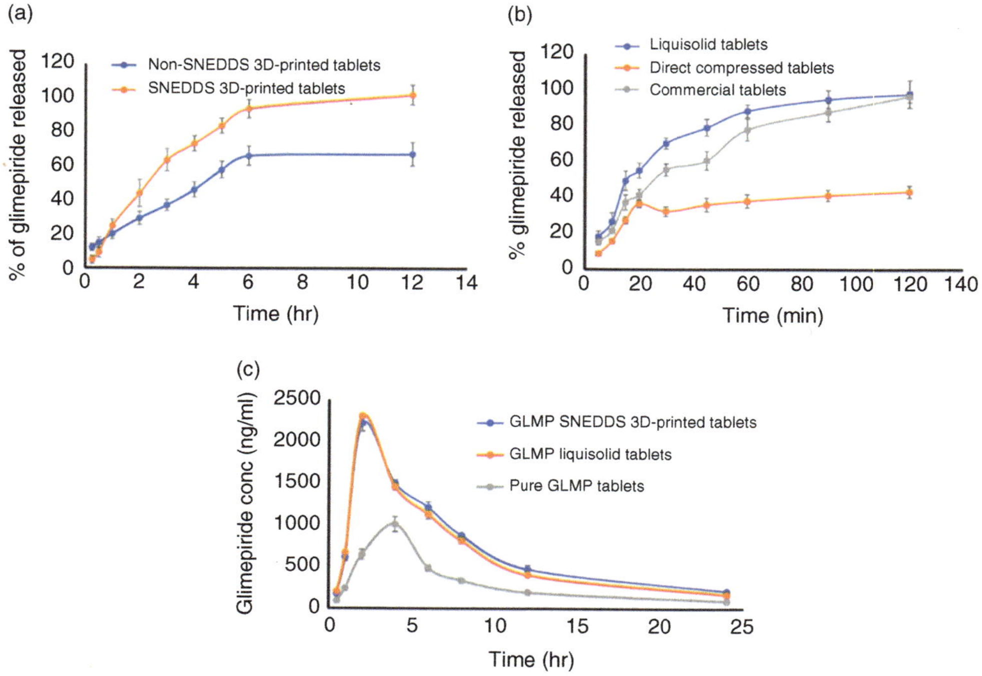


**Figure 34.8** In vitro release of glimepiride from (a) the prepared 3D-printed tablets, (b) the liquisolid, direct compressed, and commercial tablets. (c) The plasma concentration versus time profile of glimepiride following the oral administration of the prepared tablets to Male Wistar rats (Mean ± SD, $n = 6$). Source: Ahmed et al. [37]/MDPI/CC BY 4.0.

Berberine is an alkaloid with an extremely low bioavailability [40], which has limited its clinical usefulness. These researchers developed sustained-release microparticles containing berberine using a pH-independent release polymer. These microspheres were then loaded into 3D printed capsules (cap and body) and the bioavailability was evaluated as compared to 3D printed capsules containing non-microsphere berberine. The berberine AUC was increased almost fourfold when capsules containing the berberine microspheres were administered, as compared to the non-microsphere capsules (Figure 34.9). However, it is difficult to assess the impact of the 3D printed tablets since a non-3D printed comparison capsules were not studied. Yet, since the microspheres were liquid in nature, the 3D printing allows for development of a capsule matrix that is not dissolved by the microspheres, as compared to a traditional gelatin capsule.

In addition to the in vivo examples discussed above, a number of researchers have conducted in vitro evaluations of 3D printing technology in improving oral bioavailability [28, 41–49]. These studies also combined 3D printing with traditional solubility-enhancing technologies and through dissolution studies, validated the applicability of 3D printing to potentially enhance oral bioavailability. From the in vitro and in vivo studies, it is clear that 3D printing is a promising tool that can be utilized to enhance the bioavailability of poorly soluble drugs, particularly through the ability to create complex internal and external tablet geometries.

### 34.4.2 Enhancing Bioavailability Through Controlling the Location of Drug Release

#### 34.4.2.1 Gastro-Retention

For some drugs, dissolution and/or absorption or stability is optimal in the acidic pH of the stomach, as opposed to the more neutral/basic pH of the small intestine and greater GI tract. Typically, as a tablet enters the stomach it sinks to the bottom of the stomach near the pylorus, where gastric emptying causes the tablet to move into the small intestine. Few, if any, options exist to promote effective retention of the tablet within the stomach so that disintegration, dissolution, and even absorption can occur in the stomach and prior to the tablet moving into the small intestine [50]. The ability to control internal tablet geometry and the contents of any internal structures (or lack of contents) provided by 3D printing, makes formulation of a gastro-retentive (GR) tablet design possible. For example, overall tablet density can be controlled by creating a chamber or multiple chambers within the tablet structure that are hollow, thus resulting in a low-density tablet and

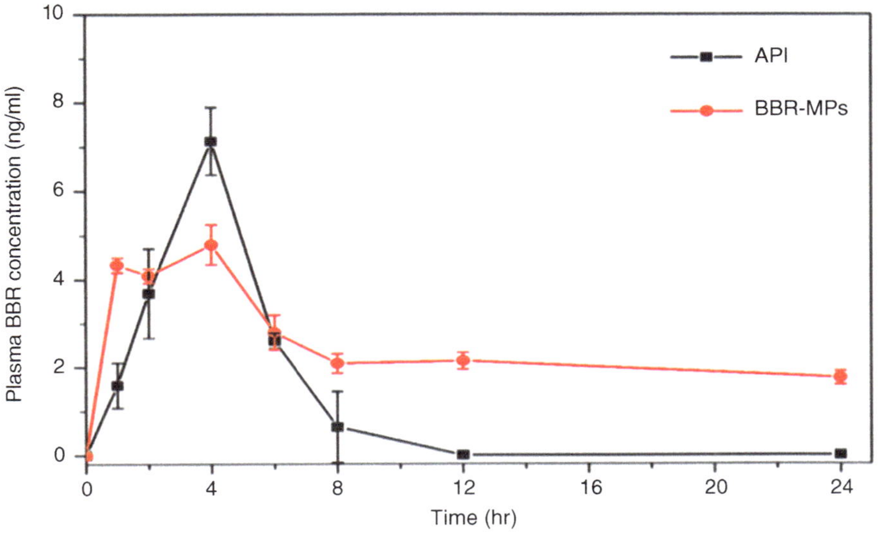


**Figure 34.9** Mean plasma concentration versus time profiles for BBR after oral administration of pristine BBR and BBR-loaded MPs filled in 3D printed hollow capsules at 5 mg/kg in rabbits ($n = 3$). BBR: berberine, MPs: microparticles. Source: Choudhury et al. [39]/with permission of Springer Nature.

creating buoyancy. Furthermore, these chambers can be created from non-dissolvable materials such that the chamber remains intact while drug is contained in and subsequently released from an adjacent chamber (Figure 34.10a). Using a tablet containing levodopa and carbidopa with a hollow chamber produced by the MED® extrusion-based process, an imaging agent was incorporated into the tablets, and the formulation was administered to dogs (both with and without a hollow chamber). Following administration, radiographic

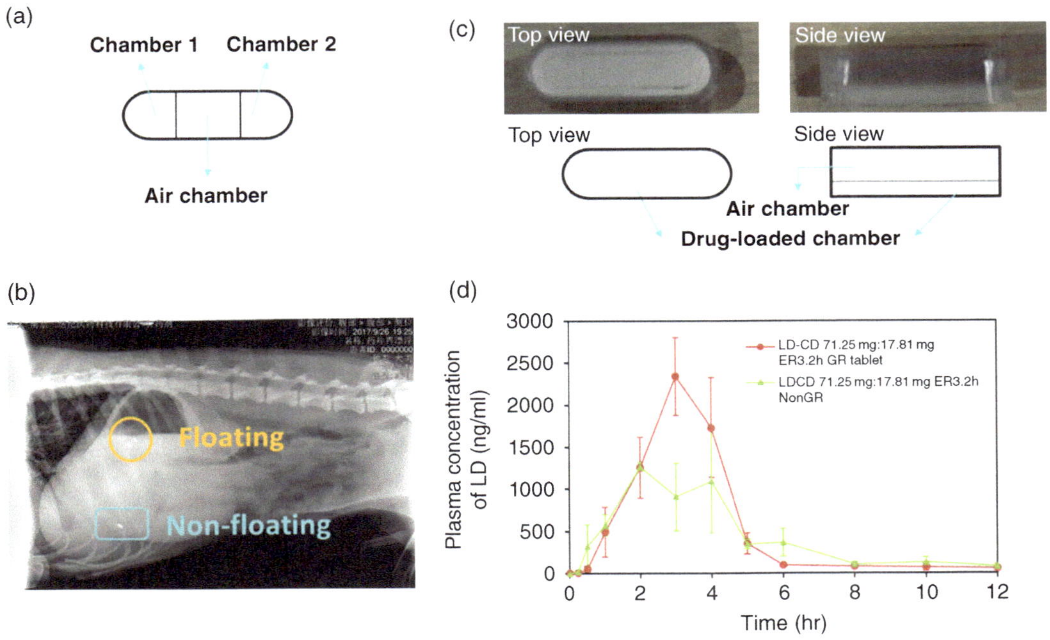


**Figure 34.10** The schematic drawing of gastro-retentive tablet with an air chamber (a) and the X-ray imaging result of the floating and non-floating tablets after oral administration to Beagle dogs (b). The imaging agent loaded in Chamber 1 and Chamber 2 is barium sulfate. (c) The photographs and schematic drawing of LD and CD co-loaded gastro-retentive tablet with an air chamber. (d) The plasma concentration of LD after oral administration of LD and CD-loaded tablets ($n = 3$, mean ± SE). LD: levodopa, CD: carbidopa.

imaging was carried out over time, as well measurement of the levodopa drug concentration–time profile. In the first experiment, tablets without drugs were administered simultaneously and the imaging was carried out. It is clear from Figure 34.10b that the non-floating tablet immediately sinks to the bottom of the stomach near the pylorus whereas the GR tablet with the hollow chamber remains floating at the surface of the gastric contents near the top of the stomach. Following separate administration of the hollow chamber formulation (Figure 34.10c) and the non-buoyant formulation, drug concentrations were measured after the two tablets were given separately. One observes a substantially higher $C_{max}$ and AUC for levodopa following administration of the GR tablet as compared to the tablet without the hollow chamber (Figure 34.10d). For drugs with an absorption window in the upper GI tract, GR technology enables continuous and adequate exposure of drugs within the absorption window, thus facilitating the drug dissolution and absorption.

This ability to increase drug bioavailability using a GR formulation has also been demonstrated by Shin et al., [51] using the antiviral drug acyclovir. Employing FDM techniques for drug formulation, these investigators created a three-compartment two-part capsule design wherein the two ends of the capsule contain air-pocket chambers to provide buoyancy and a middle compartment with "windows" for drug release where a sustained release acyclovir tablet was inserted (Figure 34.11a). The size of the drug-releasing windows was adjusted to achieve optimal dissolution rate of the acyclovir tablet contained therein to coincide with the dissolution of the capsule as a whole. Too large of windows resulted in the drug-containing tablet being released from the GR capsule too early and thus, not achieving the full effects of the buoyancy. Likewise, too small of windows resulted in the drug dissolving too slowly and the outer capsule dissolving and thus, the buoyancy being lost. Once optimized, the capsules were administered to Beagle dogs and the PKs of acyclovir were measured. When compared to either an immediate-release formulation of acyclovir or the sustained-release tablet without the buoyant capsule, the combination of the sustained-release tablet contained within a buoyant capsule resulted in a 22% increase in acyclovir bioavailability (Figure 34.11b).

Chai et al. [52] have also employed FDM to create a GR tablet design for the delivery of domperidone. The bioavailability of domperidone, which is used in the treatment of nausea and vomiting, is particularly amenable to GR designs since it is very soluble in acidic conditions but much less soluble under basic conditions [53]. Thus, formulations that promote retention of the tablet and subsequent drug release in the more acidic environment of the stomach should enhance bioavailability as compared to more traditional formulations that may release part or all the domperidone post-stomach. These investigators created tablets with the drug contained in the outer shell structure and the inner compartment of the tablet being hollow, thus creating a reduction in overall tablet density. Furthermore, the researchers were able to control the rate of domperidone release from the tablet by optimizing the formulation composition, resulting in a substantially slower in vitro release rate as compared to the commercially available tablets (Figure 34.12a). Shell numbers and infill percentages were proven not to influence the release rates of tablets. To allow for imaging of the tablet location in the GI tract over time, barium sulfate was added to the tablet formulations. The final formulation

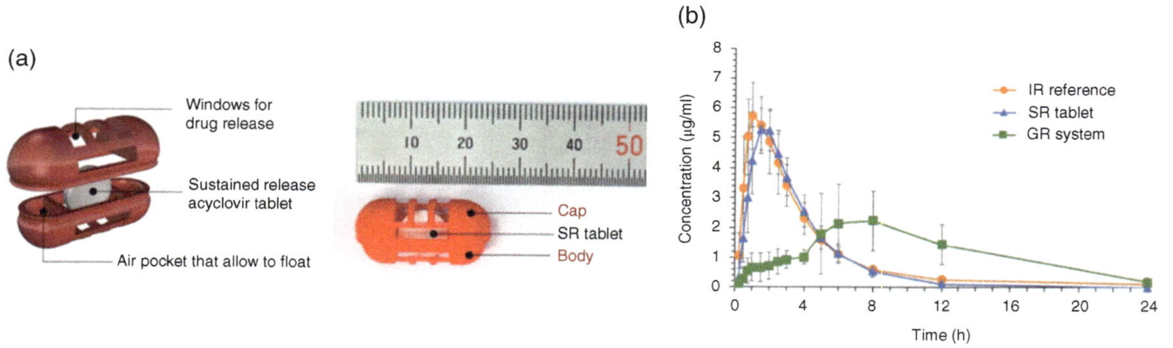


**Figure 34.11** (a) The structure of the gastro-retentive (GR) system. The novel GR system is composed of the 3D printed gastro-floating device and acyclovir sustained-release (SR) tablet inside the device. (b) *In vivo* pharmacokinetics of the GR system in Beagle dogs. Plasma concentration versus time profiles of acyclovir following oral administration of the immediate-release (IR), sustained-release (SR), or GR system of acyclovir in Beagle dogs. Data are presented mean ± SD ($n = 5$). Source: Shin et al. [51]/from PLOS ONE.

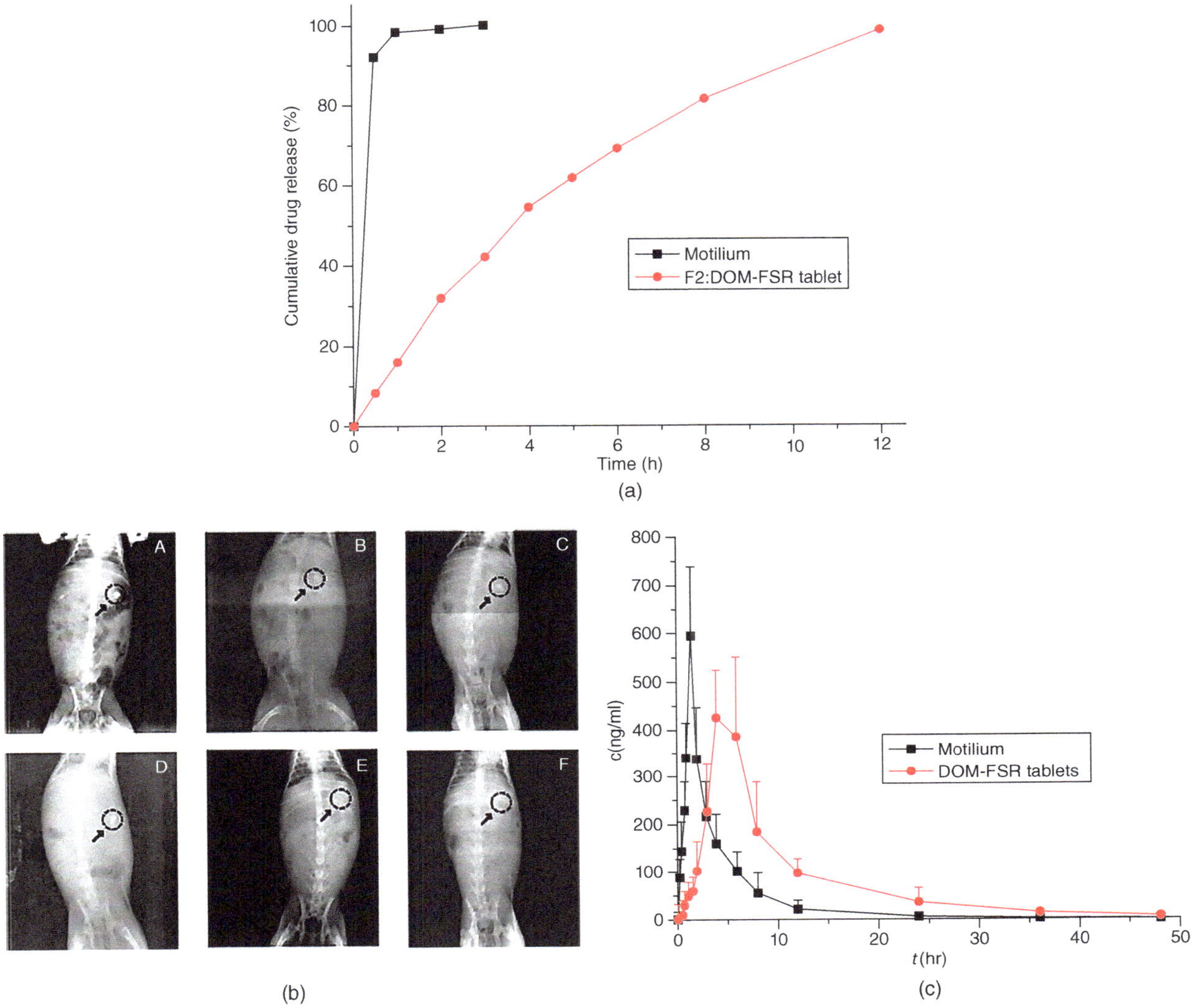


**Figure 34.12** Drug release profiles of the formulations used in the experiments ($n = 6$): (a) comparison of commercial fast-release tablets and DOM-FSR tablets. (b) A-F X-rays indicating the positions of the $BaSO_4$–labeled DOM-FSR tablets in the gastrointestinal tract of New Zealand rabbits at different time points. X-rays taken at (A) 2 hours, (B) 4 hours, (C) 6 hours, (D) 8 hours, (E) 10 hours, and (F) 12 hours. (c) Plasma pharmacokinetic profile after administering either commercial fast release tablets or DOM-FSR tablets to rabbits. Source: Chai et al. [52]/from Springer Nature/CC BY 4.0.

was then administered to rabbits with tablet location being monitored radiographically over 12 hours and the PKs of domperidone measured over 48 hours. The 3D printed formulation remained within the stomach of the rabbits for up to eight hours, becoming less distinct at ten hours and virtually undetectable at twelve hours (Figure 34.12b). As compared to the commercially available formulation (Figure 34.12c), the 3D printed GR formulation exhibited a longer $T_{max}$ (~5 versus ~1.5 hours), a slightly lower $C_{max}$ (mean of ~500 ng/ml versus mean of ~600 ng/ml), a longer half-life (mean of ~2.4 hours versus mean of ~1.3 hours) and substantially prolonged plasma concentrations. It is particularly noteworthy that the goal of enhanced bioavailability was achieved as the AUC was increased 222% with the 3D printed GR tablet as compared to the commercial formulation.

It is noted that many researchers have fabricated GR systems using various 3D printing technologies and conducted mechanistic studies. Recently, Pan at el. demonstrated that 3D printing (paste extrusion printing) can be used to affect the floating time by adjusting the internal structure of tablets without changing the drug release mechanism from first-order kinetics [54]. The utility of 3D printing to produce GR tablets, demonstrated in both in vitro and in vivo studies, has been widely proven [46, 55–64]. In comparison, the use of 3D printing to enhance oral bioavailability has been less investigated but has substantial potential to produce clinically significant benefits.

#### 34.4.2.2 Colon Targeting

Targeting drug release to specific locations in the GI tract can be used to enhance bioavailability or to deliver drug locally to treat bowel disorders, such as Crohn's disease or ulcerative colitis [65]. With respect to enhancing bioavailability, there are several strategies that can be employed. If a drug is best dissolved at higher pH then formulations that do not release drug until the high pH or lower enzyme activity of the intestine may be advantageous. If a drug is highly metabolized in the intestine, for example, by CYP3A4, then formulations that delay release until lower in the intestine where the CYP3A4 content is lower may increase bioavailability [66]. Conversely, if the drug is a substrate of the efflux transporter P-glycoprotein then formulations that release drug higher in the intestine where P-glycoprotein content is lower may enhance bioavailability.

Almeida et al. [67] explored the use of 3D printing to create camptothecin tablets with enhanced bioavailability and targeting release in the colon. Camptothecin is an anticancer drug derived from natural product sources and has demonstrated efficacy in a number of cancers. However, it suffers from poor oral bioavailability and has been classified as a poor solubility and poor permeability BCS Class IV drug and is also subject to extensive first-pass metabolism, as well as efflux by P-glycoprotein [68]. Thus, the challenge has been how to increase solubility and permeability and protect camptothecin from intestinal metabolism, all of which influence bioavailability. To address these issues, these investigators created chitosan-based polymeric micelles of camptothecin, which were incorporated into 3D printed printfills sealed with an enteric layer to protect the nanosystem from the GI milieu. The enteric layer was designed to prevent dissolution in the stomach and not release drug until reaching colonic pH. Using a dissolution media intended to mimic the intestinal absorption conditions, the permeability of the camptothecin-containing printfills was measured in both Caco-2 cells and a HT29-MTX mucus-producing cell line. The permeability of camptothecin was increased by approximately 30% in Caco-2 cells and approximately 70% in the HT29-MTX cells, as compared to free camptothecin (Figure 34.13). Lack of release at lower pH values and release at higher pH values was also demonstrated (data not shown). Though in vivo studies were not conducted, these

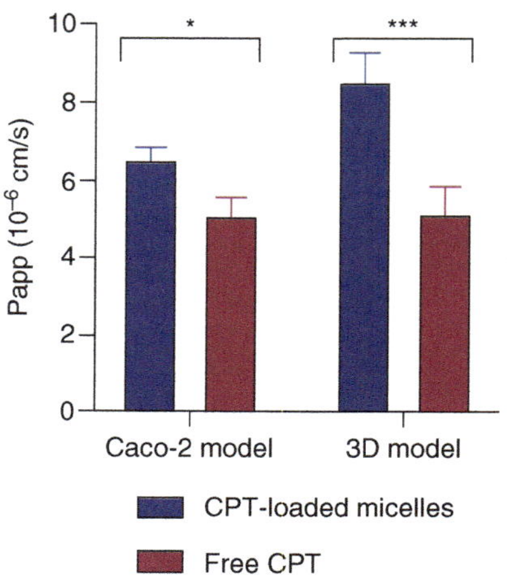


**Figure 34.13** Apparent permeability coefficients (Papp) of CPT-loaded micelles and free CPT across Caco-2 monoculture model and across the 3D model. All experiments were conducted from the apical to basolateral side in HBSS at 37 °C. Error bars represent mean ± SD ($n = 3$). $^{*}p < 0.05$, $^{***}p < 0.001$, and $^{****}p < 0.0001$ denotes a significant difference by two-way analysis of variance (ANOVA) Sidak's multiple comparison test. Source: Almeida et al. [67]/with permission of Elsevier.

results demonstrate in an in vitro system that release can be delayed until reaching the higher pH values of the colon and that permeability can be enhanced by the use of polymeric micelles incorporated into 3D printing produced printfills. In vitro, studies have also demonstrated the ability to produce a pH-sensitive release profile via 3D printed tablets.

### 34.4.3 Optimizing Drug Pharmacokinetics to Enhance Bioavailability

One can imagine situations whereby efforts to optimize the PKs of a drug could also have an impact on drug bioavailability. For example, in an effort to achieve steady and prolonged drug concentrations, 3D printing could be used to formulate a tablet with zero-order release kinetics. If the drug is actively taken up by gastric drug transporters (e.g. OATs, organic anion transporter proteins) that can be saturated, this slower and more constant release of drug could avoid the transporter saturation and provide a greater absorption of the drug. It is also known that some drugs show variability in drug absorption and Pks at different times of the day (due to diurnal rhythms). By controlling time of release and the temporal window of drug release, one could potentially modulate drug absorption according to the time of administration and time of day to enhance a drug's bioavailability. Kadry et al. [69] used diltiazem-containing filaments for FDM to

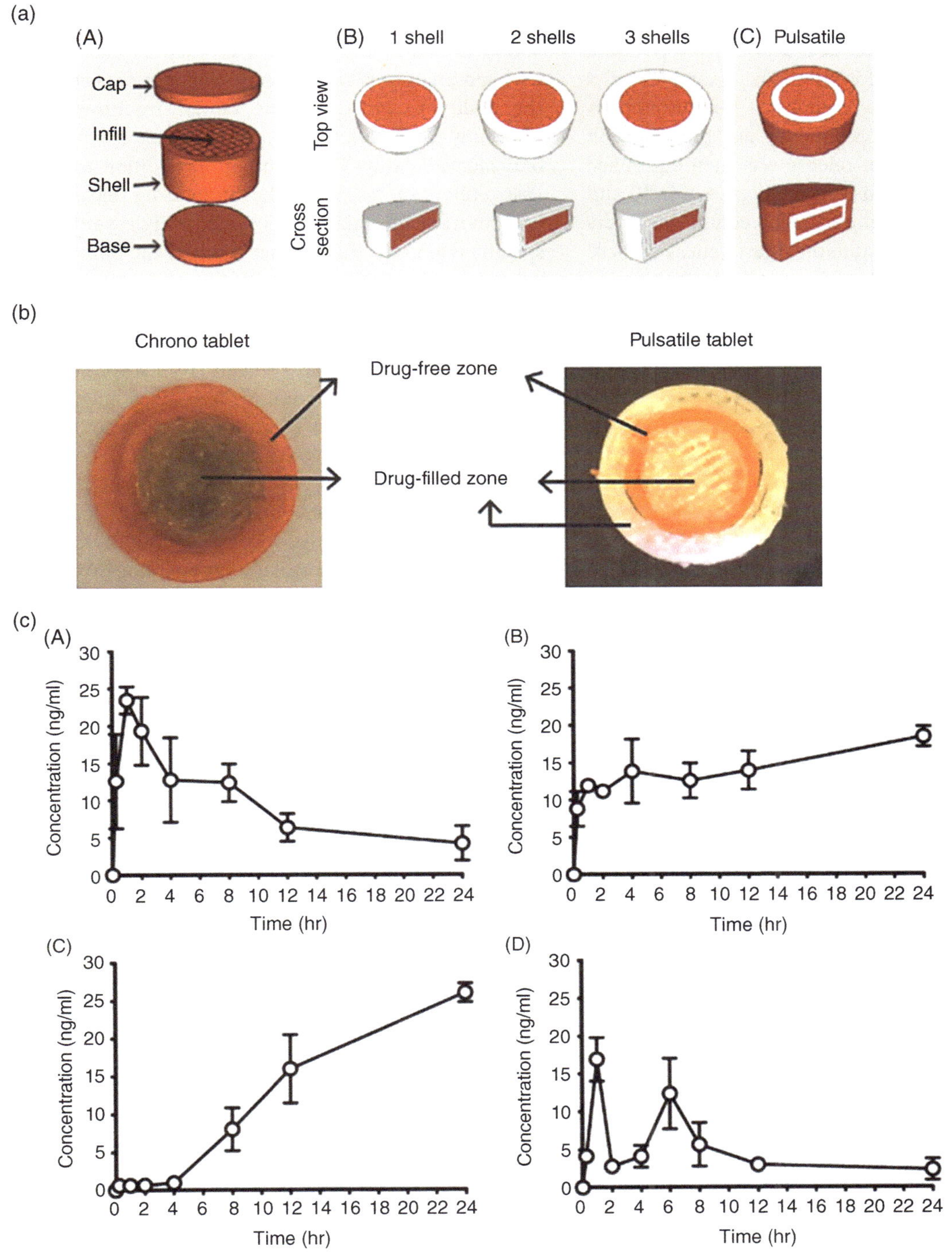


**Figure 34.14** (a) Design of various tablets: (A) Caps, base, shell, and infill patterns of a hypothetical tablet. (B) Chrono controlled-release tablets with drug-free outer shells with 1, 2, or 3 shells that are 0.3, 0.6, 0.9 mm thick. Drug-free layers are in gray and drug-loaded layers in red. (C) Pulsatile-release tablets, whereby one tablet is printed within another tablet, separated by a drug-free layer (gray). (b) Photographs of chrono and pulsatile tablets with drug-free (orange) and drug-loaded layers (no color). Orange filaments were used to distinguish drug-free filament from drug-laden filament. Colored tablets were fabricated only for taking pictures, and tablets used in dissolution and absorption studies were printed without any dyes. (c) Oral absorption profiles of (A) immediate-, (B) extended-, (C) chrono-, and (D) pulsatile-release tablets. The data represents [mean ± SD ($n = 3$)]. Source: Kadry et al. [69]/from Elsevier.

produce tablets containing varying thicknesses/layers of diltiazem-containing and diltiazem-free components to produce immediate, extended, delayed, and pulsatile release (Figure 34.14a) formulations. Examples of a tablet with delayed release intended for chronotherapy and a pulsatile release tablet are presented in Figure 34.14b. The rate of release was then optimized to the desired rate by adjusting the degree of infill and the number/thickness of the nondrug-containing layers of the tablets. To demonstrate the effectiveness of these formulations, diltiazem-containing tablets formulated to provide immediate release, extended release, delayed release for chronotherapy, and pulsatile release were administered to rats and diltiazem PKs monitored over 24 hours (Figure 34.14c). All four desired release profiles were achieved, demonstrating a good *in vitro–in vivo* correlation, with the extended-release formulation almost mimicking zero order release, the delayed release formulation exhibiting at least an hour delay before diltiazem release was initiated and the pulsatile release formulation exhibiting two distinct peaks at ~1 and ~6 hours after administration. Though they did not specifically study the effect of these release characteristics and resulting PKs on diltiazem bioavailability, one can imagine how these formulations could help achieve a goal of avoiding saturation of drug transporters (zero-order release profile) or enhanced absorption during a specific time of day (chronotherapy-based release profile). These kinds of profiles can be difficult to achieve with conventional formulation technology and further demonstrates the potential of 3D printing technology in the development of pharmaceuticals.

## 34.5 Future Outlook for 3D Printing and Bioavailability Enhancement

The use of 3D printing techniques to enhance bioavailability is still in the nascent stages as to date. Most of the efforts have been focused on refining and optimizing the 3D printing process. Now that this has been accomplished, investigators are turning their attention to how to utilize the myriad of design possibilities to enhance oral bioavailability and optimize a drug's PKs. As demonstrated earlier in this chapter, 3D printing allows for an expansion of the use of solubilizing and permeability-enhancing agents in enhancing bioavailability through compartmentalization, pulsatile release, etc. to better enhance drug absorption. We have likely only scratched the surface of possibilities for designs that maximize the impact of solubility and permeability enhancement.

Some of the most important work in oral bioavailability enhancement may come from leveraging the ability to create unique tablet geometries themselves. One can imagine many different embodiments wherein the ability to create unique tablet structures, both internally and externally, is leveraged to enhance oral bioavailability. For example, in addition to using gastro-retention to control the site of drug release, one can envision formulations designed to release drug in specific regions of the GI tract wherein the optimal pH exists for solubility and absorption of a drug. This concept of pH-sensitive drug release has been demonstrated in vitro using MED-produced tablets wherein tablets were formulated to release (via a pH-sensitive layer) at either pH 1.2 (stomach pH), pH 5.8 (duodenal pH), or pH 6.8 (jejunal pH) [16]. Controlling location of release could also be used to avoid areas of the GI tract where drug-metabolizing enzymes or efflux transporters are highest in concentration. Similarly, peptidases are highest in concentration in the upper regions of the GI tract and thus, developing formulations of peptide drugs that do not release until later in the GI tract could avoid intestinal breakdown of the peptide and thus, facilitate absorption. Another embodiment could involve making 3D printed tablets designed to embed in the intestinal mucosa in a directional manner such that drug is only released from the tablet surface that is embedded in the mucosa. This has the potential to minimize exposure of peptides to the luminal fluids and accompanying peptidases to enhance oral bioavailability. These latter formulations for peptide delivery would be a major advancement as they could make possible the oral administration of peptides with adequate absorption, which is not feasible today.

In addition to controlling tablet internal geometry to control the site or rate of drug release, the ability to produce unique tablet external geometries can also be leveraged to enhance bioavailability. For drugs with substantial first-pass metabolism via the gut, delivery of the drug by the oral route can be challenging. However, sublingual and buccal administration can avoid first-pass metabolism. Tablets administered buccally or sublingually have typically been fast dissolving and thus, may result in rapid and high drug concentrations. Three-dimensional printing could be used to develop a formulation shaped, for example, to fit neatly in the buccal cavity with potentially mucoadhesive properties that could deliver controlled release of the drug to achieve the desired PK profile and avoid first-pass metabolism. In addition to controlling drug release, this ability to control external geometry also could be used to create a delivery vehicle more acceptable to the patient than a conventional-shaped tablet.

Finally, one can imagine a 3D printed tablet with some mechanism (e.g. spring, hinged arms, or similar mechanism) wherein the tablet, once exposed to the stomach milieu expands the spring or hinged arms due to swelling, such that the tablet cannot pass through the pylorus. This would maintain the tablet (gastro-retention) within the stomach cavity regardless of fluid volume or emptying time. The tablet spring or arms would be formulated so that they dissolve at a time after the drug has been completely released, allowing the tablet contents to pass on through the GI tract.

The above are just a few examples of how 3D printing could be used to enhance oral bioavailability and we are confident that researchers will imagine many more.

## 34.6 Summary

Though still relatively new, 3D printing technologies have the potential to revolutionize drug development and delivery. Each of the three technologies (extrusion-based, powder-based, and liquid-based) have their own unique advantages and caveats, suggesting that they are complementary, and each may serve a unique niche in drug development and delivery. To date, the primary application of the 3D printing methods has been to control drug release and delivery. Now that several novel applications have been developed and validated, researchers are turning to leveraging 3D printing technology to enhancing oral bioavailability, long a goal of oral drug development but constrained by traditional tableting technologies. 3D printing has allowed a true leveraging of solubility and permeability-enhancing agents by better incorporating them with the drug of interest but also controlling how the drug is released into the intestinal milieu. Novel internal and external geometries, including multiple compartments, unique structures to increase surface area, and air-filled chambers to enhance buoyancy are now being utilized to control the rate, location, and mode of drug release to enhance oral bioavailability.

The past 20 years have seen a boom in the development and application of 3D printing technologies to enhance drug delivery, bioavailability, and PKs. The next 20 years will likely see even more advances, many not even imagined today.

## References


1 El-Kattan, A. and Varma, M. (2012). Oral absorption, intestinal metabolism and human oral bioavailability. In: *Topics on Drug Metabolism* (ed. J. Paxton). InTechOpen.

2 van der Merwe, J., Steenekamp, J., Steyn, D., and Hamman, J. (2020). The role of functional excipients in solid oral dosage forms to overcome poor drug dissolution and bioavailability. *Pharmaceutics* 12 (5): E393.

3 Zizzari, A.T., Pliatsika, D., Gall, F.M. et al. (2021). New perspectives in oral peptide delivery. *Drug Discovery Today* 26 (4): 1097–1105.

4 Li, Y., Meng, Q., Yang, M. et al. (2019). Current trends in drug metabolism and pharmacokinetics. *Acta Pharm. Sin. B.* 9 (6): 1113–1144.

5 Pangeni, R., Kang, S., Jha, S.K. et al. (2021). Intestinal membrane transporter-mediated approaches to improve oral drug delivery. *J. Pharm. Investig.* 51 (2): 137–158.

6 Wu, B.M., Borland, S.W., Giordano, R.A. et al. (1996). Solid free-form fabrication of drug delivery devices. *J. Controlled Release* 40 (1): 77–87.

7 Gao, G., Ahn, M., Cho, W.W. et al. (2021). 3D printing of pharmaceutical application: drug screening and drug delivery. *Pharmaceutics* 13 (9): 1373.

8 Mohammed, A., Elshaer, A., Sareh, P. et al. (2020). Additive manufacturing technologies for drug delivery applications. *Int. J. Pharm.* (580): 119245.

9 Alhnan, M.A., Okwuosa, T.C., Sadia, M. et al. (2016). Emergence of 3D printed dosage forms: opportunities and challenges. *Pharm. Res.* 33 (8): 1817–1832.

10 Seoane-Viaño, I., Januskaite, P., Alvarez-Lorenzo, C. et al. (2021). Semi-solid extrusion 3D printing in drug delivery and biomedicine: personalised solutions for healthcare challenges. *J. Controlled Release* (332): 367–389.

11 Cheng S, Li X, Deng F, CN, Lu H, et al. (2022). 3D printing device and method. United States Patent: 11364674. https://image-ppubs.uspto.gov/dirsearch-public/print/downloadPdf/11364674.

12 Goyanes, A., Allahham, N., Trenfield, S.J. et al. (2019). Direct powder extrusion 3D printing: Fabrication of drug products using a novel single-step process. *Int. J. Pharm.* (567): 118471.

13 Fanous, M., Gold, S., Muller, S. et al. (2020). Simplification of fused deposition modeling 3D-printing paradigm: feasibility of 1-step direct powder printing for immediate release dosage form production. *Int. J. Pharm.* (578): 119124.

14 Quodbach, J., Bogdahn, M., Breitkreutz, J. et al. (2021). Quality of FDM 3D printed medicines for pediatrics: considerations for formulation development, filament extrusion, printing process and printer design. *Ther. Innov. Regul. Sci.*

15 Cui M, Pan H, Fang D, Qiao S, Wang S, Pan W. *Fabrication of high drug loading levetiracetam tablets using semi-solid extrusion 3D printing*. 2020; *J. Drug Delivery Sci. Technol.*, 57, 101683.

16 Zheng, Y., Deng, F., Wang, B. et al. (2021). Melt extrusion deposition (MED™) 3D printing technology – A paradigm shift in design and development of modified release drug products. *Int. J. Pharm.* (602): 120639.

17 Sen, K., Mehta, T., Sansare, S. et al. (2021). Pharmaceutical applications of powder-based binder jet 3D printing process – a review. *Adv. Drug Delivery Rev.* 177: 113943.

18 Awad, A., Fina, F., Goyanes, A. et al. (2020). 3D printing: Principles and pharmaceutical applications of selective laser sintering. *Int. J. Pharm.* (586): 119594.

19 Fina, F., Goyanes, A., Madla, C.M. et al. (2018). 3D printing of drug-loaded gyroid lattices using selective laser sintering. *Int. J. Pharm.* 547 (1–2): 44–52.

20 Awad, A., Fina, F., Trenfield, S.J. et al. (2019). 3D printed pellets (Miniprintlets): a novel, multi-drug, controlled release platform technology. *Pharmaceutics* 11 (4): E148.

21 Kyle, S., Jessop, Z.M., Al-Sabah, A., and Whitaker, I.S. (2017). "Printability" of candidate biomaterials for extrusion based 3D printing: state-of-the-art. *Adv. Healthcare Mater.* 6 (16).

22 Madžarević, M. and Ibrić, S. (2021). Evaluation of exposure time and visible light irradiation in LCD 3D printing of ibuprofen extended release tablets. *Eur. J. Pharm. Sci.* (158): 105688.

23 Rodriguez-Pombo, L., Xu, X., Seijo-Rabina, A. et al. (2022). Volumetric 3D printing for rapid production of medicines. *Addit. Manuf.* 52. https://doi.org/10.1016/j.addma.2022.102673.

24 Kyobula, M., Adedeji, A., Alexander, M.R. et al. (2017). 3D inkjet printing of tablets exploiting bespoke complex geometries for controlled and tuneable drug release. *J. Controlled Release* (261): 207–215.

25 Goyanes, A., Buanz, A.B.M., Basit, A.W., and Gaisford, S. (2014). Fused-filament 3D printing (3DP) for fabrication of tablets. *Int. J. Pharm.* 476 (1–2): 88–92.

26 Pietrzak, K., Isreb, A., and Alhnan, M.A. (2015). A flexible-dose dispenser for immediate and extended release 3D printed tablets. *Eur. J. Pharm. Biopharm.* 96: 380–387.

27 Sadia, M., Arafat, B., Ahmed, W. et al. (2018). Channelled tablets: an innovative approach to accelerating drug release from 3D printed tablets. *J. Controlled Release* (269): 355–363.

28 Zhang, J., Xu, P., Vo, A.Q., and Repka, M.A. (2021). Oral drug delivery systems using core-shell structure additive manufacturing technologies: a proof-of-concept study. *J. Pharm. Pharmacol.* 73 (2): 152–160.

29 Melocchi, A., Uboldi, M., Briatico-Vangosa, F. et al. (2021). The chronotopic™ system for pulsatile and colonic delivery of active molecules in the era of precision medicine: feasibility by 3D printing via fused deposition modeling (FDM). *Pharmaceutics* 13 (5): 759.

30 Fanous, M., Gold, S., Hirsch, S. et al. (2020). Development of immediate release (IR) 3D-printed oral dosage forms with focus on industrial relevance. *Eur. J. Pharm. Sci.* (155): 105558.

31 Eisert, W.G., Hauel, N., Stangier, J. et al. (2010). Dabigatran: an oral novel potent reversible nonpeptide inhibitor of thrombin. *Arterioscler. Thromb. Vasc. Biol.* 30 (10): 1885–1889.

32 Liu, X. (2021). *Advances in 3D Printing Technology for Modified Release Products Development*. American Association of Pharmaceutical Scientists Meeting, Pittsburgh, PA.

33 Siyawamwaya, M., du Toit, L.C., Kumar, P. et al. (2019). 3D printed, controlled release, tritherapeutic tablet matrix for advanced anti-HIV-1 drug delivery. *Eur. J. Pharm. Biopharm.* 138: 99–110.

34 Ahmed, T.A., Felimban, R.I., Tayeb, H.H. et al. (2021). Development of multi-compartment 3D-printed tablets loaded with self-nanoemulsified formulations of various drugs: a new strategy for personalized medicine. *Pharmaceutics* 13 (10): 1733.

35 Badian, M., Korn, A., Lehr, K.H. et al. (1994). Absolute bioavailability of glimepiride (Amaryl) after oral administration. *Drug Metab. Drug Interact.* 11 (4): 331–339.

36 Schachter, M. (2005). Chemical, pharmacokinetic and pharmacodynamic properties of statins: an update. *Fundam. Clin. Pharmacol.* 19 (1): 117–125.

37 Ahmed, T.A., Alotaibi, H.A., Alharbi, W.S. et al. (2022). Development of 3D-Printed, liquisolid and directly compressed glimepiride tablets, loaded with black seed oil self-nanoemulsifying drug delivery system: in vitro and in vivo characterization. *Pharm. Basel Switz.* 15 (1): 68.

38 Lu, M., Xing, H., Jiang, J. et al. (2017). Liquisolid technique and its applications in pharmaceutics. *Asian J. Pharm. Sci.* 12 (2): 115–123.

39 Choudhury, D., Jala, A., Murty, U.S. et al. (2022). In vitro and in vivo evaluations of berberine-loaded microparticles filled in-house 3D printed hollow capsular device for improved oral bioavailability. *AAPS PharmSciTech.* 23 (4): 89.

40 Chen, W., Miao, Y.Q., Fan, D.J. et al. (2011). Bioavailability study of berberine and the enhancing effects of TPGS on intestinal absorption in rats. *AAPS PharmSciTech.* 12 (2): 705–711.

41 Zhang, J., Thakkar, R., Kulkarni, V.R. et al. (2021). Investigation of the fused deposition modeling additive manufacturing I: influence of process temperature on the quality and crystallinity of the dosage forms. *AAPS PharmSciTech.* 22 (8): 258.

42 Saydam, M. and Takka, S. (2020). Improving the dissolution of a water-insoluble orphan drug through a fused deposition modelling 3-Dimensional printing technology approach. *Eur. J. Pharm. Sci.* (152): 105426.

43 Pistone, M., Racaniello, G.F., Arduino, I. et al. (2022). Direct cyclodextrin-based powder extrusion 3D printing for one-step production of the BCS class II model drug niclosamide. *Drug Del. Transl. Res.* 12 (8).

44 Li, P., Jia, H., Zhang, S. et al. (2019). Thermal extrusion 3D printing for the fabrication of puerarin immediate-release tablets. *AAPS PharmSciTech.* 21 (1): 20.

45 Kulkarni, V.R., Kazi, M., Shahba, A.A.W. et al. (2022). Three-dimensional printing of a container tablet: a new paradigm for multi-drug-containing bioactive self-nanoemulsifying drug-delivery systems (bio-SNEDDSs). *Pharmaceutics* 14 (5): 1082.

46 Kimura, S.I., Ishikawa, T., Iwao, Y. et al. (2019). Fabrication of zero-order sustained-release floating tablets via fused depositing modeling 3D printer. *Chem. Pharm. Bull.* 67 (9): 992–999.

47 Clark, E.A., Alexander, M.R., Irvine, D.J. et al. (2020). Making tablets for delivery of poorly soluble drugs using photoinitiated 3D inkjet printing. *Int. J. Pharm.* (578): 118805.

48 Bhatt, U., Malakar, T.K., Murty, U.S., and Banerjee, S. (2021). 3D printing of immediate-release tablets containing olanzapine by filaments extrusion. *Drug Dev. Ind. Pharm.* 47 (8): 1200–1208.

49 Alhijjaj, M., Belton, P., and Qi, S. (2016). An investigation into the use of polymer blends to improve the printability of and regulate drug release from pharmaceutical solid dispersions prepared via fused deposition modeling (FDM) 3D printing. *Eur. J. Pharm. Biopharm.* 108: 111–125.

50 Das, S., Kaur, S., and Rai, V.K. (2021). Gastro-retentive drug delivery systems: a recent update on clinical pertinence and drug delivery. *Drug Del. Transl. Res.* 11 (5): 1849–1877.

51 Shin, S., Kim, T.H., Jeong, S.W. et al. (2019). Development of a gastroretentive delivery system for acyclovir by 3D printing technology and its in vivo pharmacokinetic evaluation in Beagle dogs. *PLoS One* 14 (5): e0216875.

52 Chai, X., Chai, H., Wang, X. et al. (2017). Fused deposition modeling (FDM) 3D printed tablets for intragastric floating delivery of domperidone. *Sci. Rep.* 7 (1): 2829.

53 Khan, A., Iqbal, Z., Khan, A. et al. (2016). Modulation of pH-independent release of a class II drug (domperidone) from a polymeric matrix using acidic excipients. *Dissolution Technol.* 23 (01): 32–40.

54 Fang, D., Pan, H., Cui, M. et al. (2022). Fabrication of three-dimensional-printed ofloxacin gastric floating sustained-release tablets with different structures. *J. Drug Delivery Sci. Technol.* 67.

55 Lin, X., Fu, H., Hou, Z. et al. (2021). Three-dimensional printing of gastro-floating tablets using polyethylene glycol diacrylate-based photocurable printing material. *Int. J. Pharm.* (603): 120674.

56 Melocchi, A., Uboldi, M., Inverardi, N. et al. (2019). Expandable drug delivery system for gastric retention based on shape memory polymers: development via 4D printing and extrusion. *Int. J. Pharm.* (571): 118700.

57 Li, P., Zhang, S., Sun, W. et al. (2019). Flexibility of 3D extruded printing for a novel controlled-release puerarin gastric floating tablet: design of internal structure. *AAPS PharmSciTech.* 20 (6): 236.

58 Lamichhane, S., Park, J.B., Sohn, D.H., and Lee, S. (2019). Customized novel design of 3D printed pregabalin tablets for intra-gastric floating and controlled release using fused deposition modeling. *Pharmaceutics* 11 (11): E564.

59 Jeong, H.M., Weon, K.Y., Shin, B.S., and Shin, S. (2020). 3D-printed gastroretentive sustained release drug delivery system by applying design of experiment approach. *Mol. Basel Switz.* 25 (10): E2330.

60 Huanbutta, K. and Sangnim, T. (2019). Design and development of zero-order drug release gastroretentive floating tablets fabricated by 3D printing technology. *J. Drug Delivery Sci. Technol.* 52: 831–837.

61 Giri, B.R., Song, E.S., Kwon, J. et al. (2020). Fabrication of intragastric floating, controlled release 3D

printed theophylline tablets using hot-melt extrusion and fused deposition modeling. *Pharmaceutics* 12 (1): E77.
62 Reddy Dumpa, N., Bandari, S., Repka, A., and M. (2020). Novel gastroretentive floating pulsatile drug delivery system produced via hot-melt extrusion and fused deposition modeling 3D printing. *Pharmaceutics* 12 (1): E52.
63 Charoenying, T., Patrojanasophon, P., Ngawhirunpat, T. et al. (2020). Three-dimensional (3D)-printed devices composed of hydrophilic cap and hydrophobic body for improving buoyancy and gastric retention of domperidone tablets. *Eur. J. Pharm. Sci.* (155): 105555.
64 Charoenying, T., Patrojanasophon, P., Ngawhirunpat, T. et al. (2020). Fabrication of floating capsule-in-3D-printed devices as gastro-retentive delivery systems of amoxicillin. *J. Drug Delivery Sci. Technol.* (55): 101393.
65 Li, X., Lu, C., Yang, Y. et al. (2020 Sep). Site-specific targeted drug delivery systems for the treatment of inflammatory bowel disease. *Biomed. Pharmacother. Biomed. Pharmacother.* 129: 110486.
66 Kato, M. (2008). Intestinal first-pass metabolism of CYP3A4 substrates. *Drug Metab. Pharmacokinet.* 23 (2): 87–94.
67 Almeida, A., Linares, V., Mora-Castaño, G. et al. (2021 Oct). 3D printed systems for colon-specific delivery of camptothecin-loaded chitosan micelles. *Eur. J. Pharm. Biopharm.* 167: 48–56.
68 Gupta, E., Vyas, V., Ahmed, F. et al. (2000). Pharmacokinetics of orally administered camptothecins. *Ann. N.Y. Acad. Sci.* 922: 195–204.
69 Kadry, H., Al-Hilal, T.A., Keshavarz, A. et al. (2018). Multi-purposable filaments of HPMC for 3D printing of medications with tailored drug release and timed-absorption. *Int. J. Pharm.* 544 (1): 285–296.

# 35

# Anatomical and Physiological Factors Affecting Oral Drug Bioavailability in Rats, Dogs, Monkeys, and Humans

*Ayman El-Kattan*

*IFM Therapeutics, Boston, MA, USA*

## 35.1 Introduction

The discovery and development of therapeutically efficacious drugs are a time-consuming, resource-intensive, and complicated process. Recent estimates of three billion dollars show that developing a new drug in the United States requires 10–15 years to reach the marketplace [1]. Based on the data from Tufts Center for the Study of Drug Development, the average capitalized cost to bring one new biopharmaceutical product to market, including the cost of failures, is $1.24 billion in 2005 dollars [2]. Most of the small molecular drugs that are available in the marketplace are administered via the oral route, which is a convenient and cost-effective route of administration [3]. Thus, oral bioavailability is one of the key parameters frequently evaluated in drug discovery when considering a new chemical entity (NCE) for further development, as poor oral bioavailability is one of the major contributing factors to therapeutic variability [4]. Hellriegel et al. reported a significant inverse relationship between the oral bioavailability of drugs from several therapeutic classes and the coefficient of interindividual variability in their oral bioavailability [5]. The increased variability associated with poor oral bioavailability may predispose the patient to a higher risk of exposure to toxic or subtherapeutic plasma drug concentration. This is particularly important for drugs with narrow therapeutic windows or potential for resistance development such as cytotoxic drugs and antibiotics [4].

Evaluation of a NCE's pharmacokinetic profile including oral bioavailability determination is routinely performed in animal species such as rats and dogs during drug discovery and early preclinical development. Typically, the *in vivo* pharmacokinetic evaluations are performed in rodents during early discovery since NCEs are synthesized in small quantities, which are insufficient to perform such studies in larger animal species [6]. This data, in conjunction with that obtained from *in silico* and *in vitro* studies, are commonly used to project the pharmacokinetic profile in humans in order to provide guidance for chemical synthesis to optimize the absorption, distribution, metabolism, and elimination (ADME) properties of a chemical series. When more promising candidates are identified, synthesis of larger quantity is typically made for pharmacokinetic assessment in a non-rodent species such as dog or monkey to refine the projection of human pharmacokinetic profiles in support of further development of the NCE. However, the approach of predicting oral bioavailability in humans from animal and *in vitro* data still present many challenges due to species-dependent physiological and metabolic differences.

The purpose of this chapter is to provide a critical review of the species similarities and differences in the absorption, distribution, and metabolism mechanisms that affect oral bioavailability in humans and in the three commonly used preclinical species, namely rats, dogs, and monkeys.

## 35.2 Determinants of Oral Bioavailability

### 35.2.1 Absorption

### 35.2.2 Physiological Factors that Impact Oral Drug Absorption

An extensive review of the anatomical and physiological differences in gastrointestinal tract between preclinical animal species and humans has been described previously and the following is a brief overview [7–9].

***35.2.2.1.1 Gastrointestinal Anatomy and Physiology*** In rats, dogs, monkeys, and humans, the gastrointestinal

*Oral Bioavailability and Drug Delivery: From Basics to Advanced Concepts and Applications*, First Edition. Edited by Ming Hu and Xiaoling Li.

tract consists mainly of the stomach, small intestine (duodenum, jejunum, and ileum), and large intestine (cecum, colon, and rectum). The dimensions and surface areas of these components differ among these species. Table 35.1 illustrates a comparison among the radius, anatomical lengths, and relative contributions of the major subdivisions of the gastrointestinal tract in rats, dogs, monkeys, and humans, [7–10].

Although the body weight of human (70 kg) is larger than that of dog (10 kg) and rat (0.25 kg), the length of the human gastrointestinal tract (8.35 m) is only 2.0-fold and 5.6-fold longer than the dog and rat gastrointestinal tract, respectively (Tables 35.1 and 35.2). However, the relative length of the small intestine, which is considered the primary site of drug absorption, to the total length of the gastrointestinal tract is 83% in rats, 86% in dogs, and 81% in humans. As for the large intestine, its relative length in rats, dogs, and humans is 17%, 14%, and 19%, respectively [11, 12].

The radius of the small intestine in human is 1.5–2 cm, while in rats, dogs, and monkeys it is 0.2, 0.5, and 0.4 cm, respectively (Table 35.1). Interestingly, the radius of the human small intestine decreases as one progresses distally from the upper to the end of the small intestine (1.7 to 1.0 cm, respectively).

A comparison of the absolute surface areas of the gastrointestinal tracts in rats, dogs, monkeys, and humans is shown in Table 35.2. The human small intestine

**Table 35.1** Comparison of the anatomical lengths of the intestinal tract and its major subdivision in rats, dogs, monkeys, and humans [7].

| Species | Region of Intestine | Radius (cm) | Length (m) | % of Total |
|---|---|---|---|---|
| Rat | Small Intestine | 0.2 | 1.25 | 83 |
| | Large Intestine | | 0.25 | 17 |
| | Total Intestinal Tract | | 1.50 | |
| Dog | Small Intestine | 0.5 | 4.14 | 86 |
| | Large Intestine | | 0.68 | 14 |
| | Total Intestinal Tract | | 4.82 | |
| Monkey | Small Intestine | 0.4[a)]/0.8[b)] | ND | ND |
| | Large Intestine | | ND | ND |
| | Total Intestinal Tract | | ND | ND |
| Human | Small Intestine | 1.5–2 | 6.80 | 81 |
| | Large Intestine | | 1.55 | 19 |
| | Total Intestinal Tract | | 8.35 | |

a) Rhesus monkey.
b) Cynomolgus monkey.
ND: Not determined.
Source: Adapted from DeSesso et al. [7].

**Table 35.2** Comparison of the absolute surface areas and surface area of the gastrointestinal tract relative to total body surface area in rats, dogs, monkeys, and humans [7, 10].

| | Human | | Rat | | Dog | | Monkey | |
|---|---|---|---|---|---|---|---|---|
| Body Weight (kg) | 70 | | 0.3 | | 10.7–12.6 | | 2.7–3.0 | |
| Body surface area (BSA; $m^2$) | 1.8 | | 0.03–0.06 | | 0.39–0.78 | | 0.3 | |
| Region of Intestinal Tract | Absolute SA ($m^2$)[a)] | SA Relative to BSA | Absolute SA ($m^2$)[a)] | SA Relative to BSA | Absolute SA ($m^2$)[a)] | SA Relative to BSA | Absolute SA ($m^2$)[a)] | SA Relative to BSA |
| Stomach | 0.053 | 0.029 | 0.00062 | 0.01–0.02 | 0.0344–0.0426 | 0.04–0.11 | 0.0143–0.0306 | 0.048–0.1 |
| Small Intestine | 252 | 140 | 1.6 | 26.67–53.33 | 24.75–35 | 31.73–89.74 | 54.6–66 | 18220 |
| Large Intestine | 0.35 | 0.19 | 0.034 | 0.57–1.13 | 0.023–0.245 | 0.029–0.63 | 0.23–0.31 | 0.23–0.31 |

a) Surface area.

surface area is 158 $m^2$ and 11-fold larger than the rat and dog small intestine surface area, respectively. Yet, the human small intestine is only 4 fold larger than the monkey small intestine surface area. The large difference in the surface area between rat and dog versus human and monkey small intestine is attributed to the fact that human and monkey small intestine has three anatomical modifications that significantly increase the surface area of the small intestine [13, 14]. Indeed, rat and dog small intestine have only two of these. The human and monkey small intestine has an additional grossly observable fold of mucosa (plicae circularis or folds of kerckring) that increase the surface area by 3-fold. From the plicae circularis project microscopic finger-like pieces of tissue called villi that increase the surface area by 5 to 10-fold depending on the species. Each villus is covered in microvilli which increases the surface area by 20 to 25-folds. When the small intestine is normalized to the total body surface area, the relative surface area of the human small intestine is 4 and 2.3-fold larger than that of the rat and dog, respectively. However, it is only 0.7-fold the relative surface area of the monkey small intestine [7].

Unlike the small intestine, the large intestine surface area does not have villi and is divided into geographical areas by transverse furrows. In addition, the large intestine enterocytes differ slightly from that of the small intestine and its microvilli are less packed [9]. Overall, this contributes to the smaller surface area of the large intestine in all species investigated (Table 35.2) and is consistent with the fact that small intestine is the major site of drug absorption in these species.

***35.2.2.1.2 Unstirred Water Layer*** Adjacent to the absorptive membrane of the enterocyte is an unstirred water layer (UWL) that functions as a potential barrier for the permeation of various drug molecules across the intestinal membrane. The UWL is comprised of strings of water lamella that become more stirred toward the lumen bulk and away from the apical surface of the enterocyte. The thickness of this UWL in rat and dog is comparable to that in human and about 300 μm [15]. In the literature, monkey UWL was not determined. However, based on antipyrine effective intestinal permeability calculated in monkey and considering that monkey plicate expansion is smaller compared to human, it is suggested that the thickness of monkey UWL is similar to that in human [16].

Chiou evaluated the impact of UWL adjacent to the intestinal membrane on the rate and extent of absorption for passively absorbed drugs with different membrane absorption half-lives (10 to 300 minutes) in humans, dogs, rabbits, rats, and mice. Results suggested that the presence of the UWL is generally expected to have a relatively minor if any, effect on the rate and extent of absorption of drugs across the gastrointestinal tract [17].

***35.2.2.1.3 Gastrointestinal Transit Times*** The absorption rate of a drug molecule is generally a function of drug absorption through the gastrointestinal tract. Absorption rate is determined by the residence time and absorption in each gastrointestinal segment [18]. Gastric emptying time, also known as gastric transit time, depends on the content of the stomach. The gastric transit time describes the time required for the stomach to empty its content in the small intestine [19]. In human, the half-life of the stomach content for small particles and liquids is calculated to be 10 minutes and 1 hour in fasted and fed conditions, respectively. Higher caloric density and smaller meal volume are associated with longer gastric transit time [19]. This mainly impacts the rate of absorption of rapidly dissolved and well-absorbed drugs (BCS Class 1) and leads to a decrease in drug $C_{max}$ and a corresponding longer $T_{max}$ and/or lag time in the fed conditions. However, the extent of absorption and oral bioavailability is typically not affected [19–21]. It should be emphasized that in the fasted conditions, the gastric emptying rate of a solution for dogs, monkeys, and humans is similar [22]. While in the fed conditions, dogs gastric emptying time is longer than that reported in human [8, 23].

Intestinal transit time influences the absorption of drugs with limited mucosal permeability, substrates for carrier-mediated uptake, subject to intestinal degradation, or products whose dissolution is the rate-limiting step for systemic absorption [24]. In contrast to gastric transit time, intestinal transit time is independent of the feeding conditions and the physical composition of the intestinal contents [25]. The rat, dog, monkey, and human intestinal transit time is around three to four hours [8, 22, 26]. In rats, the transit time in the large intestine is around 15 hours and the transit time in human and dog large intestine can vary from 8 to 72 hours [9, 27]. Overall, the intestinal content and formulation size have minimal impact on the transit time in the human large intestine [28].

***35.2.2.1.4 The Gastrointestinal pH*** Most of the drug molecules are either acidic or basic in nature. The extent of ionization plays a pivotal role in determining their dissolution rate and passive permeability across the gastrointestinal tract. Therefore, pH at the absorption site is a critical factor in facilitating or decreasing the

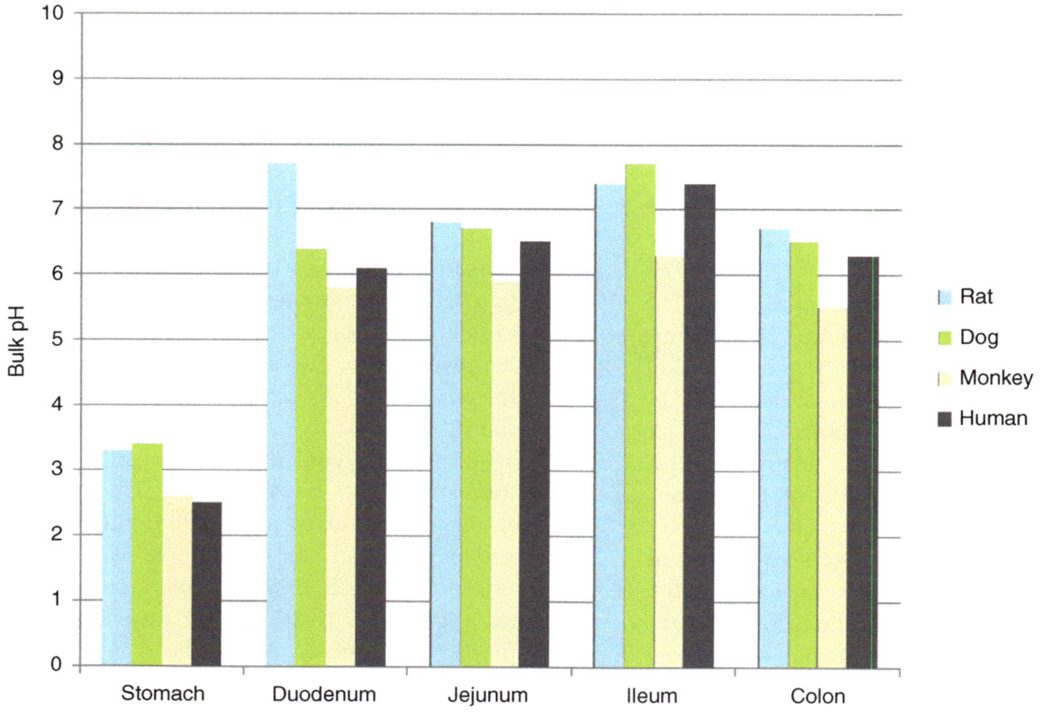


**Figure 35.1** Comparative bulk pH across the gastrointestinal tract of rat, dog, monkey, and human [24, 29].

dissolution rate and absorption of various ionizable drug molecules [1]. Figure 35.1 illustrates the pH values of the gastrointestinal tract of rats, dogs, monkeys, and humans. In rats and dogs, the pH of luminal content (chyme) in the stomach is around 3.3; while in humans and monkeys, the pH of chyme is more acidic and around 2.5 [24]. When the chyme arrives in the duodenum, it is quickly neutralized by the secretion of the pancreatic bicarbonate and bile. The pH values of chyme become progressively more alkaline in the distal portion of the small intestine in rats, dogs, monkeys, and humans. However, the pH of chyme in the large intestine is generally more acidic than the pH observed in the small intestine in these species, possibly due to fermentation mediated by the microbial flora [30, 31].

***35.2.2.1.5 Bile Fluid*** Bile is produced by hepatocytes and is drained through the bile ducts that penetrate the liver [7]. During this process, the epithelial cells add a watery solution that is rich in bicarbonates, which increases the alkalinity of the solution. In humans and dogs, bile is stored and concentrated up to five times its original potency in the gall bladder by water reabsorption. Gall bladder content is emptied following food intake. Emptying process is mediated by secretin and cholecystokinin, which stimulates the gallbladder to contract and secrete bile as the acidic chyme enters the duodenum. Human and dog bile secretion rate is 22 and 19–36 ml/kg/day, respectively. However, rats lack a gall bladder and bile is secreted continuously at an estimated rate of 48 ml/kg as a dilute solution into the duodenum each day [9, 24]. In monkey, the bile secretion rate is 25 ml/kg [29].

Bile acts as a detergent that emulsifies fats by increasing the surface area to help digestive enzymes action, and thus, aid in their absorption in the small intestine. Bile is a complex mixture of bicarbonate solution, cholesterol, phospholipids such as lecithin and bilirubin, and bile salts such as those of taurocholic acid and deoxycholic acid derivatives. Bile salts are also amphipathic water-soluble molecules that are derived from cholesterol and ionized at physiological conditions [32]. There are major differences in the chemical structure of various bile salts identified. These are shown in stereochemistry of the hydroxyl groups and their numbers. Overall, the bile salt is around 20 Å long, with an average radius of about 3.5 Å [33]. Bile salts are combined with phospholipids to break down fat globules via the process of emulsification. Emulsified droplets then are organized into many micelles, which increase their absorption. Since bile increases the absorption of fats, it also plays a pivotal role in the absorption of the fat-soluble vitamins, steroids, and lipophilic xenobiotics [34]. Indeed, bile increases these molecules' dissolution and solubility through improving the wetting properties of solid particles and enhancing the solubility of lipophilic drugs in micellar solutions, respectively. However, bile tends to reduce the dissolution rate of hydrophilic compounds and their absorption rate and extent [34].

***35.2.2.1.6 Bacterial Microflora*** In rats, dogs, monkeys, and humans, bacterial microflora exist in most of the gastrointestinal tract and is an important component of the luminal content. In rats and dogs, there is significant number of bacterial microflora in the stomach as well as small and large intestines. In humans, there is no bacterial microflora in the stomach and upper small intestine. This lack of bacterial microflora is mainly attributed to the low pH of the gastric content of human stomach. However, large number of bacterial microflora populates the human's distal small and large intestines [35].

Intestinal bacterial microflora plays a crucial role in the metabolism of various chemicals, endogenous substrates, and xenobiotics through hydrolysis of glycosides, sulfate esters, amide, esters, dehydroxylation, deamidation, decarboxylation, dealkylation, acetylation, esterification, and reduction of double bonds, nitro, and diazo groups [36].

Indeed, hydrolysis of glucuronide conjugates is a pivotal metabolic reaction, which is mediated by the β-glucuronidase enzyme that is produced by the intestinal bacterial microflora. The abundance of this enzyme appears to be higher in both rat and dog gastrointestinal tracts compared to human [9]. However, the β-glucuronidase activity in the human gastrointestinal tract is still substantial and generates 0.02 and 0.9 μmol of product formed/h/g content for the proximal and distal regions of small intestine, respectively [37]. Interestingly, this process sometimes contributes to the enterohepatic circulation of several drugs. Where following biliary excretion, drug conjugates are broken down by β-glucuronidase, which is then followed by intestinal reabsorption of parent molecule. Enterohepatic circulation typically leads to multiple-peak phenomenon and a longer terminal half-life.

In regard to *in vivo* models for human intestinal microflora metabolism, conventional laboratory animals are limited. A germ-free rat model with associated human colonic bacteria has been used as an *in vivo* model for human projections [36]. Since there are significant species differences in the distribution, amount, strains, and metabolic activities of microflora that populate the gastrointestinal tract of preclinical species and human, more studies are needed to enable a meaningful interpretation of the data observed in preclinical models and further emphasize the need to exercise caution in the projection of compounds profiles in human based on preclinical data [37, 38].

***35.2.2.1.7 Lymphatic Absorption*** The intestinal lymphatic route plays a key role in the absorption of drugs that are highly lipophilic and may provide an opportunity for development of controlled release formulation via this route of delivery [39]. Similarities and differences in drug lymphatic uptake are determined by the lymph flow to various absorption sites, mechanisms of drug lymphatic absorption, mechanism of lipid digestion, and bile composition and secretion pattern [24]. However, little is known about species differences in lymphatic uptake of drugs and their impact on species differences in drug absorption and bioavailability [40].

#### 35.2.2.2 Mechanism of Oral Absorption

Following oral dosing, drug molecules can cross the luminal membrane through various mechanisms such as passive diffusion or active transport. Passive diffusion is comprised of two pathways: (i) the paracellular pathway in which a drug diffuses through the aqueous pores at the tight junctions between the intestinal enterocytes; and (ii) the transcellular (lipophilic) pathway, which requires drug diffusion across the lipid cell membrane of the enterocyte. Active transport pathway is mediated by transporters and is functionally divided into active drug influx and efflux. The relevance of each route to the absorption of a compound is determined by the compound's physicochemical properties and its affinity for various transport proteins.

***35.2.2.2.1 Passive Diffusion*** In paracellular diffusion, drug molecules are absorbed by diffusion and convective volume flow through the intercellular spaces governed by the tight junctions or Zonula Occludens, which formulate the major rate-limiting barrier [41]. Drug molecules absorbed through this pathway are typically small in size (e.g. molecular weight [MW] < 350 Da) and hydrophilic (Log $P < 0$). Because the junctional complex has a net negative charge, positively charged molecules pass more readily, whereas negatively charged molecules are repelled. There are significant species differences in drug absorption via this mechanism among rat, dog, and human. For example, He et al. conducted cross-species comparison of the absolute bioavailability of several hydrophilic and metabolically stable drugs (acyclovir, atenolol, nedocromil, dazmegrel, sumatriptan, TRH, and nadolol) (Table 35.3). The authors reported a good correlation between the bioavailability of these drugs in humans and rats,

**Table 35.3** The absorption of a set of hydrophilic drugs in rat, dog, and human [42].

| Compound | Acyclovir | Atenolol | Nedocromil | Dazmegrel | Sumatriptan | TRH | Nadolol |
|---|---|---|---|---|---|---|---|
| Mwt (Dalton) | 225 | 266 | 371 | 283 | 295 | 362 | 309 |
| Log D | −1.5 | −1.9 | <−4 | −1 | −1.5 | −2.4 | −1.4 |
| Rat $F_a$ | 0.23 | 0.56 | 0.02 | 0.64 | 0.54 | 0.02 | 0.19 |
| Dog $F_a$ | 0.75 | 1 | 0.8 | 0.68 | 0.92 | 0.13 | 0.9 |
| Human $F_a$ | 0.28 | 0.55 | 0.02 | 0.98 | 0.57 | 0.015 | 0.3 |

Source: Adapted from He et al. [42].

with rats slightly underpredicting the bioavailability in humans. In contrast, there was no apparent correlation in bioavailability between dog and human, with dog overpredicting the bioavailability and oral absorption in human. For example, acyclovir and nadolol oral bioavailability was 75% and 90% in dog, but was only 28% and 30%, respectively, in human (Table 35.3). The authors concluded that rat is a better model than dog for predicting human bioavailability of paracellularly absorbed compounds [43]. Furthermore, the paracellular pathway offers a limited window for absorption since it accounts for ~ 0.01–0.1% of the total surface area of intestinal membrane. In addition, the tight junctions between cells become tighter traveling from the jejunum toward the colon. Therefore, compounds that are absorbed via paracellular pathway are not optimum for traditional controlled release formulation that target delivery through the colon. For example, metformin is a compound that is predominantly absorbed via paracellular pathway and its absorption is confined to the small intestine. As a result, metformin is not a good candidate for controlled-release formulation that targets the colon. Instead, it is formulated using sustained-release gastric retentive technology. Where the tablet swells into a gel-like mass. This would limit its passage through the stomach pylorus and extends its gastric residence. Once the gel-like mass is formed, metformin slowly dissolves and diffuses from the polymer-containing tablet matrix, leading to a longer half-life [44].

The transcellular pathway is the main route of absorption for most drug molecules. In general, the rate of passive transcellular permeability is mainly determined by the rate of transport across the apical cell membrane. This is controlled by the physicochemical properties of the absorbed molecules. Unlike the paracellular pathway, compounds that are absorbed through the transcellular pathway are unionized, with Log $P > 0$ and MW > 350 Da. In addition, the hydrogen-bonding capacity as determined by the number of hydrogen bond donors and hydrogen bond acceptors is <10 and 5, respectively [45].

##### *35.2.2.2.2 Active Transport (see also Chapters 18–22)*

Intestinal transporters have a significant impact on the absorption of extended clearance classification system namely class 3a, 3b, and 4 [46, 47]. These are membrane-bound proteins that are either located on the apical or basolateral side of the enterocyte. Depending on the direction in which they translocate substrates across the cell membrane, intestinal transporters are classified as efflux or influx transporters. Intestinal efflux transporters are part of the ATP-binding cassette superfamily (ABC superfamily). These are primary transporters that use ATP hydrolysis to mediate the active export of xenobiotics from the cytoplasm to intestinal lumen or blood depending on their location at the enterocyte surface. Intestinal influx transporters are part of the solute carrier (SLC) superfamilies and they mainly employ a proton and $Na^+$ gradient created by primary active carriers ($Na^+/K^+$-ATPase, $Na^+/H^+$-ATPase) to drive the transport of endogenous substrates and xenobiotics into the enterocyte and called secondary transporters.

Recent studies suggest that a plethora of transporter genes are expressed in the rat, dog, monkey, and human gastrointestinal tracts [48–50]. Below is an overview of the widely studied intestinal efflux and influx transporters and their reported species differences.

**Efflux Transporters (see also Chapters 22–23)** The most investigated efflux transporters are p-glycoprotein (P-gp), breast cancer resistance protein (BCRP), and multidrug resistance-associated proteins 2 (MRP2) [51]. They are located at the apical surface of the enterocyte and limit the oral absorption of widely prescribed drugs such as statins, antibiotics, HIV protease inhibitors, immunosuppressants, anticancer, and cardiac drugs. These substrates are usually transported against their concentration gradient using ATP hydrolysis as a source of energy. Expression of some of these transporters in the gastrointestinal tract has been quantified, which displays regional distribution patterns [52–54].

*P-gp (ABCB1)* P-gp (*ABCB1*) is considered the best-characterized member of the apical efflux transporters [51]. It is a product of the multidrug resistance (MDR1) gene and consists of two homologous halves with a total of twelve transmembrane helices and one ABC. This membrane-bound protein plays a critical role in limiting the intestinal absorption of a large number of lipophilic drugs that are either neutral or basic such as anthracyclines (doxorubicin and daunorubicin), alkaloids (reserpine, vincristine, and vinblastine), specific peptides (valinomycin and cyclosporine), steroid hormones (aldosterone and hydrocortisone), and local anesthetics (dibucaine) [55–58]. This membrane-bound protein also mediates the active secretion of substrates e.g. digoxin from the systemic circulation into the intestinal lumen [59]. Dog, monkey, and human, only have one MDR1 gene, while mouse and rat contain two paralogous genes (Mdr1a and Mdr1b) that encode for P-gp [60]. Separate research groups studied the relative mRNA expression of MDR1 in rat, dog, monkey, and human gastrointestinal tract using semiquantitative reverse-transcription polymerase chain reaction (RTC-PCR). Interestingly, dog and monkey small intestine appear to have higher relative mRNA expression of MDR1 compared to that of human [61, 62]. However, rat duodenum had lower mRNA expression of MDR1 compared to that of human [52].

At the protein level, rat, dog, and rhesus monkey P-gp are 86.6%, 90.9%, and 96.3% identical to human P-gp amino acid sequence, respectively [63]. In humans, relative protein expression of P-gp is not homogenous but rather has a regional dependence that progressively increases from the proximal to distal region of the small intestine [54, 64]. Therefore, the absorption and oral bioavailability of a P-gp substrate may depend on the site of absorption and delivery relative to this regional distribution pattern [65, 66]. Therefore, potential species differences in regional distribution and function of P-gp should be taken into consideration when predicting absorption of a NCE in humans from preclinical species [36].

*In vitro*, species differences were reported in the P-gp-mediated transport activity, which was also shown to be substrate dependent [67]. For example, Xia et al. used human MDR1, rhesus monkey MDR1, and beagle dog Mdr1 transfected insect cell membranes and a set of 21 compounds with different physicochemical properties to investigate species difference in the P-gp binding affinity (αKa) via P-gp ATPase assays. The compounds list included ritonavir, saquanivir, vinblastine, quercetin, verapamil, digoxin, progesterone, rhodamine 123, propranolol, cyclosporine, etoposodie, amiodarone, tamoxifene, thioridazone, physostigamine, paclitaxel, erythromycin, chloroquine, daunomycin, hoechst33342, miconazole, terfenadine, diltiazem, and dexamethasone. Consistent with the closer similarity in the gene and amino acid sequence between human and monkey MDR1, the authors reported a good correlation coefficient ($R^2 = 0.83$) of $\alpha Ka$ between the two species. However, the correlation coefficient ($R^2$) of $\alpha Ka$ between human and beagle dog P-gp was only 0.36 [68]. Another research group compared efflux ratio for 12 compounds generated in seven multidrug resistance 1 (MDR1) transfected LLCK-PK1 cell lines, in which the cDNA was from human, monkey, dog, rat (Mdr1aandMdr1b), and mouse (Mdr1a and Mdr1b). The group conducted a correlation analysis of the efflux ratio between two animal species and reported only a good correlation between human and monkey MDR1 ($R^2 = 0.954$). Other combinations had poor correlation [69].

*BCRP (ABCG2)* BCRP (*ABCG2*) is a 72 kDa membrane protein that is composed of 655 amino acids and arranged in six transmembrane segments (TM) and has one ABC at the amino terminus [70]. BCRP is also believed to be a half-transporter that may function as a homodimer or homotetramer. BCRP limits the oral absorption of substrate compounds and/or facilitates their active secretion back into gastrointestinal tract. BCRP exhibits broad substrate specificity and diverse chemical space. Substrates to BCRP include chemotherapy agents (mitoxantrone, camptothecins, and tyrosine kinase inhibitors), antivirals (zidovudine and lamivudine), HMG-CoA reductase inhibitors (statins), benzimidazoles, and antibiotics (ciprofloxacin and rifampicin) [58]. Research groups studied the relative intestinal mRNA distribution of BCRP in various preclinical species and human small intestine. Similar to MDR1, BCRP relative mRNA expression in small intestine of dog and monkey is higher compared to that of human [62, 71, 72]. It is important to realize that BCRP intestinal relative mRNA expression tends to be lower compared to overall ABCB1 (MDR1) expression along the dog intestine [71]. This expression is dissimilar to the findings observed in human jejunum where ABCG2 mRNA expression is higher than ABCB1 [72].

*MRP2 (ABCC2)* MRP2 (*ABCC2*) is another member of ABC transporter superfamily, which is colocated with BCRP and P-gp transporters at the apical membrane of the enterocytes. MRP2 has an apparent molecular mass of 190–200 kDa and is composed of 1545 amino acids [73]. MRP2 is composed of two TM and two ABCs with an N-terminal TM domain that is composed of five TM segments [73]. Rat, dog, and monkey Mrp2 share 78%, 84%, and 96% identity with the amino acid sequence of human MRP2, respectively [73–76]. Known MRP2 substrates are lipophilic and usually acidic in nature. They are either therapeutic drugs (e.g. rosuvastatin, bosentan, methotrexate, cyclosporine, cabamazepine, rifamipin, pravastatin, cisplatin, etoposide, doxorubicin, epirubicin, and mitoxantrone) or phase II metabolites (anionic drug conjugates of glucuronides, sulfates, or glutathiones), and endogenous compounds [77–80]. Different research groups investigated the relative mRNA expression of MRP2 in rat, dog, monkey, and human small intestine and colon using RTC-PCR. Overall, MRP2 mRNA expression in human small intestine and colon is lower compared to that shown in preclinical species tested [52, 62, 71, 72]. Ninomiya et al. studied the intrinsic transport activity of leukotriene C4, 17β-estradiol 17-(β-D-glucuronide), bromosulfophthalein, and cholecystokinin octapeptide. Vesicles studies were conducted using recombinant mouse, rat, dog, and monkey Mrp2 cDNA transfected insect cells. It is interesting to note that the substrate affinity ($K_m$) for the compounds investigated was comparable. However, intrinsic transport activity ($V_{max}/K_m$) was only different for leukotriene C4 and bromsulphthalein. The group attributed the findings to differences in the $K_m$ or $V_{max}$ values as well as to major differences in substrate and modulator recognition among species studied. Furthermore, preclinical studies demonstrated the central role of MRP2 in limiting the oral absorption and active intestinal secretion of its substrates following intravenous dosing ([81, 82]).

Taken together, compared to rat and dog, monkey major efflux transporters (MDR1, BCRP, and MRP2) have close similarity to those expressed in human gastrointestinal tract in term of gene and amino acid sequence, function, and regional distribution. However, significantly higher expression of efflux transporters along the monkey gastrointestinal tract relative to human was reported. This higher expression may lead to a considerably lower intestinal membrane permeability and fraction absorbed in monkey relative to human. Indeed, this factor has been postulated for rationalizing furosamide (a known BCRP and MRP2 substrate) lower intestinal permeability and fraction absorbed in monkey ($F_a$ = 0.38) relative to human ($F_a$ = 0.73) [62, 83]. Further studies to understand the impact of higher efflux transporter expression along the monkey gastrointestinal tract on intestinal permeability and $F_a$ using a larger set of efflux transporters substrates are warranted. These studies should also be accompanied with both a quantification of efflux transporter protein amount along the monkey gastrointestinal tract, similar to the work recently published by Oswald and coworkers. In addition, integration of these findings in physiological-based pharmacokinetics models (PBPK) to better characterize absorption in preclinical models in comparison with that of humans is pivotal [54]. Overall, this modeling allows better understanding of the potential factors that lead to species differences in oral bioavailability determination for NCE that are substrates for efflux transporters. Also, these models help to increase confidence in the decision-making process to advance an NCE into first-in-human studies based on preclinical pharmacokinetic studies.

**Influx Transporters (see also Chapters 19–21)** The role of influx transporters is critical for enabling the intestinal absorption of BDDCS class III and IV drug molecules [84]. These compounds have low passive permeability and tend to be hydrophilic in nature [58]. Examples of intestinal influx transporters include PEPT1 transporter, organic cation transporter (OCT), organic anion transporter (OAT), plasma membrane monoamine transporter (PMAT), large neutral amino acids (system L), bile acids, nucleosides, monocarboxylic acid transporters, and organic anion-transporting polypeptides (OATP) [85]. This section focuses on the most investigated intestinal influx transporters with a wealth of clinical data, which includes PEPT1 and OATP2B1 transporters.

*PEPT1 (SLC15A1)* Human PEPT1 (*SLC15A1*) is a low-affinity and high-capacity transporter protein that is electrogenic and pH gradient dependent. The cloned human PEPT1 cDNA sequence encodes a 708 amino acid protein (MW 79 kDa) with several putative glycosylation and phosphorylation sites. There are twelve TM domains and a large extracellular loop between the IX and X TM domains, which possess intracellularly localized N- and C- termini [58]. PEPT1 is a well-characterized influx transporter that plays a pivotal role in the oral absorption of hydrophilic peptide-based drugs that include beta-lactam antibiotics such as cephalosporins and penicillins, angiotensin-converting enzyme inhibitors such as zofenopril, fosinopril, benazepril, quinapril, trandolapril, spirapril, cilazapril, ramipril, moexipril, quinaprilat, enalapril maleate, and perindopril, selected rennin inhibitors, antitumor agents such as bestatin, and dopamine receptor antagonists such as sulpiride [86–90].

The gene sequences of PEPT1 transporter appear to be highly conserved in rat, dog, monkey, and human, with the rat, dog, and monkey gene sequences being

82, 85, and 94%, respectively, identical to that of humans [91–93]. Regional intestinal distribution pattern of PEPT1 also appears to be comparable among rat, dog, and human where higher expression was reported in small intestine relative to colon in the three species tested [54, 71, 94, 95]. Interestingly, RTC-PCR studies indicated PEPT1 mRNA expression in monkey gastrointestinal tract is limited to the small intestine [93].

Based on reported absorption fraction ($F_a$) calculated in rat, dog, monkey, and human following oral absorption, PEPT1 substrates tend to have comparable absorption profiles. This is consistent with their close similarity in their gene sequence and mRNA regional distribution in the gastrointestinal tract in species tested [96–98]. Overall, preclinical species such as rat, dog, and monkey are viable models to project human intestinal absorption of compounds absorbed via active uptake by PEPT1 transporter.

*OATP2B1 (SLCO2B1)* Human OATP2B1 (*SLCO2B1*) is part of the OATP superfamily, which is another important class of influx transporters expressed in the small intestine [99–101]. Indeed, human OATP2B1 protein is immunolocalized at the apical surface of the enterocytes [102] and Caco-2 cell monolayers [100]. It is worth noting that mRNA expression and protein quantification studies indicated minimal or no expression of other members of OATP superfamily (e.g. OATP1A1, OATP1B1, and OATP1B3) in human gastrointestinal tract [54]. Heterologous expression of OATP2B1 produces a sodium-independent and pH gradient-dependent transporter [103]. The isolated human OATP2B1 cDNA comprises 4108 base pair (bp) that encodes a protein with 709 amino acids. It consists of twelve TM domains and its structure showed relatively shorter helices X and XII compared to the OATP1 family. While it has narrow substrate specificity relative to other OATPs, OATP2B1 plays a key role in the intestinal absorption of certain xenobiotics. For example, it transports sulfate-conjugated steroids such as estrone-3-sulfate, dehydroepiandrosterone-sulfate, 3-hydroxy-3-methylglutaryl-coenzyme A (HMG-CoA) reductase inhibitors, which include rosuvastatin, fluvastatin, pravastatin, atorvastatin, cerivastatin, pitavastatin, lovastatin, and simvastatin, the antidiabetic agent glibenclamide, bosentan, and fexofenadine [58, 104–106].

Human OATP2B1 direct rat ortholog transporter, rOatp2b1 also known as Oatp9 (Slc21a9), has 77% identical amino acid sequence [107]. Notwithstanding the similarity in the function, rat Oatp2b1 has wider substrate specificity relative to human OATP2B1 [108].

Cynomolgus monkey OATP2B1 amino acid sequence is 96.6% identical to human OATP2B1. Shen et al. studied the uptake of 3 known OATP2B1 substrates (estradiol-17β-D-glucuronide, cholecystokinin octapeptide, and estrone-3-sulfate) in HEK-293 cells stably expressing cynomolgus monkey and human OATP2B1. Consistent with the close similarity in amino acid sequence, the group reported comparable uptake profiles in monkey and human OATP2B1 expressing cells for the three model drugs tested under similar assay conditions.

#### 35.2.2.3 Summary of Intestinal Transporters that Impacts Oral Drug Absorption

Marked species differences were reported in the anatomical and physiological factors that influence drug absorption in rats, dogs, monkeys, and humans. In a series of papers, Chiou *et al.* reported that the fraction of oral dose absorbed in rats and monkeys for a set of drugs with diverse physicochemical properties correlate well with that observed in human ($R^2 = 0.971$ and 0.974, respectively). However, a less robust correlation was observed, with several drugs showing much better absorption in dogs than in humans ($R^2 = 0.5123$). Thus, caution should be exercised when extrapolating bioavailability data from dog to human [98]. It should be stressed that an additional factor should be considered and involves the role that intestinal efflux and infux transporters play in oral drug absorption. Little is known regarding the expression, regional distribution, substrate specificity, and functional characterization of efflux and influx transporters particularly in upper preclinical species such as dog and monkey gastrointestinal tract relative to that in human. Indeed, investigating these areas will be pivotal in improving our overall understanding of the impact of potential species differences in tissue localization and functions of intestinal transporters on drug absorption and oral bioavailability among rat, dog, monkey, and human. This knowledge will also improve the quality of prediction of human absorption based on available preclinical data.

### 35.2.3 Metabolism

Extrapolation of preclinical data to project human clearance and bioavailability for NCE predominantly eliminated by metabolizing enzymes requires a thorough understanding of the metabolic fate and identification of rate-determining step for the compound clearance [109]. Appropriate *in vitro* and *in vivo* models of these species are essential to establish *in vitro* and *in vivo* correlation (IVIVC) between the NCE intrinsic metabolic clearance and its clearance observed *in vivo*. It is also a prerequisite for the successful prediction of human pharmacokinetics [110]. The following section will focus on cross-species comparison of the major drug-metabolizing enzyme families that are involved in the hepatic and intestinal first-pass metabolism, which are cytochrome P450, uridine 5′-diphosphate (UDP)-glucuronosyltransferase, and aldehyde oxidase metabolizing enzymes.

#### 35.2.3.1 Cytochrome P450s

The cytochrome P450 (CYP) superfamily of enzymes is divided into multiple families and subfamilies, with individual isoform identified and grouped into each family or subfamily based on similarity in amino acid sequences [111, 112]. In mammals, CYPs are membrane-bound enzymes that are located in the endoplasmic reticulum of the cells. As a result, CYP enzyme substrates need to have high cellular permeability to be metabolized. Compounds with low permeability would have limited access to these metabolizing enzymes unless they are substrates to active uptake transporters [113, 114].

CYP enzymes have been found in almost all human body organs; however, liver and intestine are the major sites of CYP-mediated drug metabolism [115, 116]. The main human CYP enzymes involved in drug metabolism are CYPs 1A2, 2C9, 2C19, 2D6, and 3A4 with CYPs 2B6, 2C8, and 3A5 playing a role in the metabolism of certain therapeutic agents [117–124]. It is interesting to note that CYPs 3A4, 2C, and 2D6, comprise nearly 70% of the total hepatic CYP protein with around 80% of drugs available in the market are metabolized predominantly by these enzymes [116, 125].

Overall, CYP enzymes have broad substrate specificity, which renders their substrates more hydrophilic and facilitates their rapid elimination through kidney and/or bile. The rapid elimination is usually mediated by transporters lining the kidney proximal tubules and canalicular side of liver hepatocytes. These transporters also make some molecules more amenable for phase II conjugation by enzymes such as uridine diphosphate glucuronosyltransferases (UGTs) and sulfonyl transferase enzymes. Profound species differences in CYP enzyme's amino acid sequence, expression, substrate specificity, and catalytic activity are well documented in the literature [126]. This lead to major species differences in the metabolism and pharmacokinetic profiles for CYPs substrates.

***35.2.3.1.1 CYP3A*** In humans, CYP3A is the major drug metabolizing enzymes subfamily. While its liver content is only 30% of the CYPs in human liver, CYP3A is responsible for the metabolism of around 50% of the drugs available on the market [127]. It consists of four isozymes (CYP3A4, CYP3A5, CYP3A7, and CYP3A43). The most abundant liver CYP protein level is CYP3A4 with a median value of 68.1 pmol/mg microsomal protein [128]. CYP3A5 is polymorphically expressed enzyme with >50% of the total CYP3A in some Individuals [116, 121, 129, 130]. CYP3A enzymes are expressed in the intestine and located at the apex of enterocytes with the greatest enzymatic activity in the duodenum, followed by jejunum and ileum [131]. CYP3A7 is considered a fetus-specific form of CYP3A; however, the CYP3A7*1C polymorphism causes the persistence of enzymatic activity of CYP3A7 during adult life [132]. CYP3A43 is minimally expressed in liver with low catalytic activity compared to CYP3A4 [133]. Although gender differences in CYP3A4 activity utilizing the probe substrate midazolam have been observed, the overall impact on the systemic exposure to midazolam was of negligible clinical importance [134].

In rats, CYP3A consists of several isoforms including CYP3A1, CYP3A2, CYP3A9, CYP3A18, CYP3A23, and CYP3A62. It is interesting to note that the expression of these isoforms in rats appears to be gender-specific. For example, CYP3A2 and CYP3A18 are male-specific isoforms, whereas CYP3A9 is a female-dominant isoform [135, 136]. CYP3A1 and CYP3A2 were only detected in the liver. Whereas, CYP3A62 is the predominant form in the intestine and its expression is similar to that of human CYP3A4. CYP3A9 and CYP3A18 were detected in both liver and intestines [137]. GeneChip analysis of both rat and human intestinal samples showed no correlation between the enzyme expression level of CYP3A4 and CYP3A9 in either duodenum or colon [52]. As for the enzymatic activity of CYP3A enzymes, nifedipine, and sildenafil are known probe substrates for human CYP3A4 that are not metabolized by rat CYP3A1, but by rat CYP2C [138, 139]. However, CYP3A is involved in the metabolism of midazolam in both rats and humans [140]. Species differences have also been reported in CYP3A enzymes induction, where rifampin is a stronger inducer for human than rat [141], whereas dexamethasone is a stronger inducer in rats than humans CYP3A [141].

In dogs, CYP3A consists of two isoforms (CYP3A12 and CYP3A26) [142]. These enzymes showed marked differences in the rates of metabolism for several steroids (hydroxylation reactions) [142]. 6-β-Hydroxytestosterone activity (human CYP3A probe metabolic pathway) has been observed in dog intestines [143]. Ketoconazole is a potent inhibitor of 6-β-hydroxytestosterone activity in both dogs and humans human liver microsomes [144]. The effect of ketoconazole in rat liver microsomes is dependent on sex [144]. Ketoconazole is also a moderate inhibitor of 1′-hydroxy bufuralol metabolism in dog (CYP2D) but not in human liver microsomes. While Troleandomycin is a potent *in vitro* inhibitor of 6-β-hydroxytestosterone activity in both dogs and humans; its specificity against other CYP enzymes has not been fully evaluated [145].

In cynomolgus monkey, there is one CYP3A isoform, which is CYP3A8. CYP3A8 comprises around 20% of the total CYPs expressed in the liver [6]. While CYP3A8 shares 93% identity with the amino acid sequence of the human CYP3A4, there are major species differences in terms of protein levels and catalytic activity between the two isoforms. For example, CYP3A8 has around 4–5 folds higher protein levels compared with CYP3A4. Furthermore, typical CYP3A4 probe substrates such as midazolam and erythromycin have significantly higher enzymatic rate in monkey than in human liver microsomes. The higher enzymatic rate is partially attributed to the higher levels of CYP3A protein in monkey relative to human liver.

In summary, there are major species differences in the CYP3A isoforms expression: inducibility; substrate specificity; and catalytic activity. This renders extrapolation of clearance for CYP3A substrates from animals to humans quite perilous.

***35.2.3.1.2 CYP2C*** In humans, the CYP2C superfamily consists of CYP2C8, CYP2C9, CYP2C18, and CYP2C19, which display race-related differences and polymorphic expression [112, 146, 147]. Around 16% of drugs available in the market are partially/completely metabolized by this superfamily [6]. CYP2C superfamily makes ≈ 20% of the total CYP located in the liver. The abundance of CYP2C protein is greater in human liver compared to small intestine [148]. The order of relative abundance of CYP2C protein was CYP2C9 > CYP2C8 > CYP2C19.

It should be emphasized that CYP2C18 is not detected in either human liver or small intestine and only expressed in human skin epidermis [6]. Sex differences in CYP2C metabolism have not been observed in humans [149, 150].

In rats, the CYP2C family consists of several isoforms, including CYP2C6, CYP2C7, CYP2C11, CYP2C12, CYP2C13, CYP2C22, and CYP2C23. There are sex and developmental stage-dependent differences in the expression of the CYP2C family in rats [151, 152]. For example, CYP2C11 is a male-specific enzyme that is expressed in adult rat liver and only induced at puberty, where it comprises 50% of the total CYP content in adult male rat liver. Similarly, CYP2C13 is a male-specific enzyme that is expressed in rat liver and brain. Furthermore, CYP2C12 has higher protein expression in the liver of adult female rat than male. Interestingly, these differences do not exist in either immature or adult rats. Unlike these CYP2C isoforms, CYP2C7 enzyme is female-specific, while CYP2C6 expression is gender independent.

In dogs, the CYP2C family consists of several isoforms, including CYP2C21 and CYP2C41[6]. They share 70% identity with amino acid and gene sequences. Canine CYP2C21 and CYP2C41 display 74–83% gene and 67–76% amino acid similarity with human CYP2Cs. However, canine CYP2C21 is less homologous to the human CYP2Cs compared to CYP2C41. CYP2C21 is expressed in all dogs, while CYP2C41 expression is rather polymorphic. The *in vitro* metabolism of several human CYP2C9 probes (tolbutamide, 7-hydroxylation of S-warfarin) is impaired in dogs. In contrast, the *in vitro* metabolism of the human CYP2C9 probe R-mephenytoin (4-hydroxylation) was much greater in dog than human microsomes [153]. These examples illustrate important species differences between human and dog CYP2C.

The CYP2C family consists of two isoforms (CYP2C20 and 2C43) that are expressed in the monkey liver. They share 83% and 77% identity with amino acid and gene sequences, respectively. Consistent with the high similarity in gene and amino acid sequences of CYP2C43 with human CYP2C19, this enzyme metabolized (*S*)-mephenytoin, a known probe substrate of human CYP2C19. However, CYP2C43 did not metabolize tolbutamide, a positive probe substrate of human CYP2C9. Therefore, it is widely accepted that CYP2C43 is functionally related to human CYP2C19 rather CYP2C9 [6].

***35.2.3.1.3 CYP2D*** In humans, CYP2D subfamily consists of one isoform (CYP2D6), which is expressed in various tissues including liver and small intestine [116, 154, 155]. Although CYP2D6 only accounts for 4% of the total CYP content, it is involved in the metabolism of around 30% of the drugs available in the market at present [156]. CYP2D6 exhibits polymorphism with more than 80 identified alleles within the human population [155]. This leads to significant interindividual variability in CYP2D6 substrates disposition and pharmacokinetics.

In rats, there are six CYP2D isoforms (CYP2D1, CYP2D2, CYP2D3, CYP2D4, CYP2D5, and CYP2D18) [112]. These isoforms share >70% identity with human CYP2D6 gene sequence [157]. *In vitro* evaluation showed that expressed CYP2D isozymes including rat CYP2D1, CYP2D2, CYP2D3, and CYP2D4 and human recombinant CYP2D6 isoforms, possessed catalytic activity for the 1′-hydroxylation of bufuralol [158]. However, the 1′2′-ethenylation of bufuralol was catalyzed only by rat CYP2D4 and human CYP2D6, indicating a difference between the rat and human CYP2D catalytic properties. The inhibition profile is also different for rat CYP2D as compared to human as evidenced by the fact that human (but not rat) CYP2D is inhibited by quinidine, while rat CYP2D is weakly inhibited by quinine [159, 160].

In dogs, CYP2D15 is the major CYP2D enzyme with several identified variants namely CYP2D15 WT2, CYP2D15 V1 and CYP2D15 V2 [161]. *In vitro,* bufuralol 1′-hydroxylation enzyme kinetics were similar and quinidine inhibited its metabolic reaction to a similar degree in both dogs and humans. However, in contrast to human microsomal metabolism, quinine also showed moderate inhibition in dog. In addition, antihuman CYP2D did not inhibit 1′-hydroxylation of bufuralol in dog but did moderately inhibit 1′-hydroxylation of bufuralol in rat and human liver microsomes [144].

In monkeys, significant strain differences in the expression of various CYP2D isoforms have been reported [6]. For example, CYP2D17 is expressed in the liver of cynomolgus monkey. It shares 93% identity with human CYP2D6, while CYP2D42 is expressed in Rhesus monkey. Similar trends have been observed with other monkey strains [6].

***35.2.3.1.4 CYP1A*** The CYP1A subfamily consists of CYP1A1 and CYP1A2 isoforms that are expressed in rats, dogs, monkeys, and humans. CYP1A2 is constitutively expressed in the livers of both humans [162] and rats [163], but is expressed at low levels in the livers of untreated dogs and monkeys [164]. CYP1A1 is constitutively expressed at very low levels in rat, dog [165], monkey, and human liver and present in the small intestine [154, 166, 167], lung [168], and placenta ([169, 170]) of rats and humans. The presence of extrahepatic CYP1A1 for dog has not been established. CYP1A1 expression levels are variable in human small intestine and have been shown to be inducible by cigarette smoking and ingestion of chargrilled meats [171, 172]. Furafylline inhibits both rat and human CYP1A2 but the potency is 1000-fold less for the rat than human. This species difference in furafylline inhibition between rat and human suggests a major difference in the active site topography between rat and human orthologs of CYP1A2 [173, 174]. Cross-species differences are also apparent between dog and human orthologs of CYP1A. In a cross-species comparison study; antihuman 1A antibody and human CYP1A inhibitors (α-napthoflavone and furafylline) inhibited the metabolism of chlorzoxazone (human CYP2E1 probe substrate) in dog microsomes in addition to the human CYP1A2 probe substrate (7-ethoxy-resorufin *O*-dealkylation) [144].

#### 35.2.3.2 UGT

UGT superfamily of enzymes consists of two families (UGT1 and UGT2). These enzymes primarily catalyze the linkage of glucuronic acid component of UDP-glucuronic acid, to a substrate with acceptor functional group such as carboxylic acid moiety, hydroxyl, and amine groups [175]. This glucuronidation process is effective in rendering lipophilic xenobiotics and toxins water soluble, which facilitates their renal and/or biliary excretion. Known substrates of UGT enzymes are amitriptyline, acetominophen, carvedilol, clofibric acid, fenofibrate, furosemide, naloxone, naproxen, olanzapine, propofol, raloxifene, and valporic acid [156].

In rats and humans, UGTs are located in various tissues with tissue specificity dependent on the isoform. In humans, UGT1A1, 1A3, 1A4, 1A6, 1A9, 2B4, 2B7, 2B15, and 2B17 have been identified in the liver based on mRNA and protein levels [113, 176]. UGT2B7 and UGT2B15 are the most abundant UGT enzymes in the liver with average protein levels of $82.9 \pm 36.1$ and $62.1 \pm 31.5$ pmol/mg protein, respectively [128]. UGT1A1, 1A3, 1A6, 1A8, and 2B7 have been identified in human small intestine based on protein levels determined using proteomic approaches. In rats, the UGTs identified in liver-based on mRNA levels include UGT1A1, 1A5, 1A6, 2B1, 2B2, 2B3, 2B6, and 2B12 [177]. UGT1A1, 1A2, 1A3, 1A6, 1A7, 2B3, 2B8, and 2B12 have been identified in rat intestines based on mRNA levels [177].

In a recent publication, Heikkinen et al. quantified protein levels of UGT and CYP metabolizing enzymes in beagle dog intestine and liver using mass spectrometry-based quantitative proteomics [178]. The group quantified seven UGT enzymes in both liver and intestine. Interestingly, UGT1A11 and UGT1A9 were only present in the intestine, while UGT1A7 and UGT2B31 were only reported in dog liver. The most abundant UGTs in dog intestine were UGT1A11 and UGT1A2. Whereas, UGT2B31 was the most abundant UGT enzyme in the liver. When compared to the overall abundance of CYP enzymes, the UGT enzymes had similar abundance to the sum of CYP enzymes in the dog liver. However, UGT enzymes' overall abundance were four times higher than that reported with CYP enzymes in dog intestine.

The mRNA expression levels of eight UGT isoforms in large intestine, liver, and small intestine of cynomolgus monkey were assessed using real-time reverse transcription PCR by Nishimura et al. [179]. It is interesting to note that UGT enzymes expression was tissue specific. The highest expression of six UGT enzymes isoforms namely UGT1A6, UGT1A9, UGT2B9, UGT1B18, UGT1B20, and UGT1B3 were reported in the monkey liver. Whereas, two isoforms (UG1A1 and UGT1A8) had the highest expression in monkey large intestine. High expression was also noted in small intestine. It should be emphasized that UGT1A8 mRNA has a selective expression in cynomolgus monkey large and small intestines.

UGT-mediated intestinal metabolism has a significant impact on the disposition and pharmacokinetics of UGT substrates. Furukawa et al. reported species differences in the intestinal glucuronidation activities between humans, rats, dogs, and monkeys [180]. While correlated, human in vitro intrinsic clearance ($CL_{int,u,UGT}$) calculated using intestinal micorsomes for known UGT substrates was typically lower than that reported in other species investigated. This result is not unexpected and consistent with a general trend of lower clearance in human relative to other species that is partially ascribed to the lower human hepatic blood flow in comparison with preclinical species [29]. An inverse relationship between $CL_{int,u,UGT}$ and $F_aF_g$ of UGT substrates for human, rat, dog, and monkey was observed. In dogs, the $CL_{int,u,UGT}$ values relating to $F_aF_g$ values were three to four times higher than other species investigated. These species differences are ascribed to different UGT isoforms mediating the metabolism of these compounds in each species.

#### 35.2.3.3 Aldehyde Oxidase

Human aldehyde oxidase (AO) is a soluble complex molybdoflavoprotein that unlike CYP and UGT enzymes is located in the cytosolic compartment of cells. Molybdenum is an integral part of the enzyme and is required along with flavin adenine dinucleotide (FAD) and iron–sulfur center for AO enzyme activity [181]. AO catalyzes the oxidation of organic aldehyde into carboxylic acid and nucleophilic oxidation of nitrogenous heterocyclic compounds [113]. Examples on xenobiotics that are substrates for AO include 6-deoxypenciclovir, 6-*O*-benzylguanine, acyclovir, BIBX1382, carbazeran, famciclovir, methotrexate, phenanthridine, phthalazine, zaleplon, ziprasidone, zoniporide, PF-4217903, and PF-945863 [182–186].

Moriwaki and coworkers demonstrated the widespread distribution of AO in various excretory organs of human body, using immunohistochemistry staining [187]. For example, it was detected in lung (epithelial cells and alveolar cells), gastrointestinal tract (small intestine and colon), and kidney with its highest expression in the liver [181]. Similar trends were also observed in preclinical species [188].

There are significant species and strain (rat and mice) differences of AO. For example, both humans and monkeys have one single active AO gene (*AOX1*) that codes one functional enzyme. On the other hand, mice and rats have four separate *Aox* genes, which are *Aox1, Aox3, Aox4, and Aox3l*. They encode an equivalent number of structurally conserved enzymes. Interestingly, *Aox3* is the most expressed *Aox* form in rats and mice strains and contributes to the majority of the metabolic activity in the liver cytosol of these preclinical species. This leads to species differences in the metabolic profiles for AO substrates assessed in humans and rats or mice [189].

From enzymatic activity perspective, marked species differences were observed. In general, AO enzymatic activity tends to be high in both monkey and human, this is followed by rat [190]. Interestingly, no AO enzymatic activity is reported in dogs. This is attributed to the inactivation of the *AOX1* gene and suppression/pseudogenization of *AOX3* counterpart [188].

In summary, caution should be exercised when projecting the human clearance of AO substrate. Instead of using allometric scaling from preclinical species with different AO genetic profiles e.g. mice, rats, and dogs, discovery DMPK scientists are recommended to scale human clearance using *in vitro* tools that are based on human hepatocytes and S9 fraction [191].

Failing to adhere to these guidelines may lead to clinical trial attrition due to compound poor human oral pharmacokinetic profile, which has devastating economic consequences. More importantly, a poor human oral pharmacokinetic profile impedes our ability to develop novel therapeutic agents for the treatment of devastating diseases in a timely manner. For example, FK3453 is a novel drug that was developed by Astellas for the treatment of Parkinson. The compound showed promising pharmacokinetic profiles in rats and dogs with high metabolic stability in liver microsomes. This result suggests minor contribution of CYP enzymes to FK3453 overall metabolism in species tested. Allometric scaling from rat and dog projected a favorable human intrinsic clearance of 1.3 ml/min/kg. When tested in human, the compound had poor oral plasma exposure with the oxidative metabolite of the aminopyrimidine moiety (M4) identified as a major metabolite. Furthermore, FK3453 observed human intrinsic clearance was

3000 ml/min/kg. Follow-up *in vitro* studies confirmed AO critical role in the metabolism of FK3453 and contribution to its high hepatic extraction ratio and poor oral bioavailability in humans. Overall, the results from this study led to FK3453 clinical trial termination. Similar trends were also reported with the development of a selective c-MET inhibitor, SGX523, and a novel p38 MAP kinase inhibitor. Their clinical trials were discontinued since their observed clearance in humans was significantly higher than that projected based on allometric scaling from rats and dogs leading to high hepatic extraction and poor oral pharmacokinetics profile in humans [192, 193].

#### 35.2.3.4 Species Differences in the First Pass Metabolism and its Impact on Oral Bioavailability

In addition to characterizing the species differences in the metabolic profile of an NCE (including the specific enzymatic pathways), an understanding of the species differences in the rate and extent of first-pass metabolism is crucial for the appropriate projection of human bioavailability from preclinical data. The importance of this understanding is illustrated using the examples of atomoxetine and indinavir below.

Atomoxetine is a CYP2D6 substrate with an absolute human oral bioavailability of 94% and 63% in poor and extensive metabolizers of CYP2D6, respectively [194]. The moderate to high human oral bioavailability suggests nearly complete oral absorption of atomoxetine. However, preclinical evaluations indicated that the absolute oral bioavailability of atomoxetine in rat was only 4% [195], but was 74% in dog [195]. Overall, the disposition of atomoxetine is similar in rats, dogs, and humans with a primary oxidative metabolite of 4-hydroxyatomoxetine that is subsequently conjugated to form 4-hydroxyatomoxetine-*O*-glucuronide. In a radiolabeled study in rats administered $^{14}$C-atomoxetine, atomoxetine AUC following oral administration accounted for only 2% of the total $^{14}$C AUC as compared to 30% of the $^{14}$C AUC following intravenous administration, indicating extensive first-pass metabolism in rats [195]. In a corresponding radiolabeled study in dogs, atomoxetine AUC following oral administration accounted for 33% of the total $^{14}$C AUC as compared to 39% of the $^{14}$C AUC following intravenous administration, indicating considerably less pronounced first-pass metabolism [195]. This example clearly illustrates the importance of understanding not only the species differences in a drug's metabolic fate but also the extent of species differences in the first-pass metabolism when utilizing preclinical data to project human oral bioavailability.

Indinavir, a CYP3A4 substrate, is an HIV protease inhibitor for which variable oral bioavailability has been observed in preclinical species, ranging from 72% in dogs to 19% in monkeys, and 24% in rats [196]. This variability was mainly attributed to species differences in the extent of hepatic first-pass metabolism. Chemical and immunochemical inhibition studies indicated the potential involvement of CYP3A isoforms in the metabolism of indinavir in rats, dogs, and monkeys [196], which is consistent with the observation that CYP3A4 is the main isoforms responsible for the oxidative metabolism of indinavir in human liver microsomes [197]. The *in vitro* profile of indinavir metabolism was qualitatively similar across species [196]. In addition, an *in vitro–in vivo* correlation was established in rats and dogs using the *in vivo* hepatic clearance and hepatic first-pass extraction ratio obtained from *in vitro* rat and dog metabolic data, respectively. Based on the *in vitro–in vivo* correlation established in rats and dogs, the *in vitro* intrinsic clearance of indinavir in human liver microsomes projected a small first-pass metabolism in humans ($E_h = 0.25$), which was consistent with indinavir' high oral bioavailability (60–65%) observed in humans at clinically relevant doses [196, 198]. This example depicts the importance of establishing an *in vitro–in vivo* correlation in tested preclinical species so as to use it as a basis to project human clearance and oral bioavailability.

#### 35.2.3.5 Summary of Physiological Factors that Impact Drug Metabolism

Species differences in metabolic enzymes among rats, dogs, monkeys, and humans have been well characterized with the CYP enzymes and to a lesser extent with the UGT superfamily of enzymes. An important implication of these species differences is that the projection of hepatic and intestinal extraction as well as oral bioavailability from rat, dog, and monkey data to human needs to be made with caution, taking into account potential species differences in first-pass metabolism. Despite this limitation, there are approaches that can be utilized to improve the robustness of the projection of human oral bioavailability as illustrated by the examples from atomoxetine and indinavir. Knowledge of the validity (or lack thereof) regarding the *in vitro–in vivo* relationship for metabolic clearance in the preclinical species, along with an understanding of the similarities and differences in the metabolic profile and clearance rate between rat/dog and human *in vitro*, will provide important insight into the appropriateness of utilizing preclinical and/or *in vitro* data to predict human oral bioavailability.

## 35.3 Summary

Oral bioavailability is a pivotal factor in the successful discovery and development of NCEs. As discussed in this chapter, the key determinants of oral bioavailability are driven by complex interplay of the physicochemical property of a compound, its absorption (solubility and permeability including passive permeability and active transport), and first-pass intestinal and hepatic metabolism characteristics. Many of the physiological factors that influence these processes show distinct species differences among rats, dogs, monkeys, and humans. Considerable advances have been made to understand these critical factors, which in turn has led to the development of various *in silico* and *in vitro* models to assess a drug candidate's potential to be orally bioavailable early in the discovery program. While passive permeability can be reasonably predicted based on the physicochemical properties of a compound such as lipophilicity and polarity, key gaps in our understanding of the species differences in active transport and metabolism still exist. The lack of a complete or comprehensive understanding renders extrapolation of the bioavailability data from preclinical models such as rat, dog, and monkey *in vivo* studies to humans challenging. Future advances in these areas undoubtedly will improve our ability to predict oral bioavailability in humans and further increase the success and expedite the process of developing successful drugs.

## References

1 Thomas, V.H., Bhattachar, S., Hitchingham, L. et al. (2006). The road map to oral bioavailability: an industrial perspective. *Expert Opinion on Drug Metabolism & Toxicology* 2 (4): 591–608.
2 Kaitin, K.I. (2008). Obstacles and opportunities in new drug development. *Clinical Pharmacology and Therapeutics* 83 (2): 210–212.
3 Lipinski, C.A., Lombardo, F., Dominy, B.W., and Feeney, P.J. (2001). Experimental and computational approaches to estimate solubility and permeability in drug discovery and development settings. *Advanced Drug Delivery Reviews* 46 (1–3): 3–26.
4 Bardelmeijer, H.A., van Tellingen, O., Schellens, J.H., and Beijnen, J.H. (2000). The oral route for the administration of cytotoxic drugs: strategies to increase the efficiency and consistency of drug delivery. *Investigational New Drugs* 18 (3): 231–241.
5 Hellriegel, E.T., Bjornsson, T.D., and Hauck, W.W. (1996). Interpatient variability in bioavailability is related to the extent of absorption: implications for bioavailability and bioequivalence studies. *Clinical Pharmacology and Therapeutics* 60 (6): 601–607.
6 Martignoni, M., Groothuis, G.M., and de Kanter, R. (2006). Species differences between mouse, rat, dog, monkey and human CYP-mediated drug metabolism, inhibition and induction. *Expert Opinion on Drug Metabolism & Toxicology* 2 (6): 875–894.
7 DeSesso, J.M. and Jacobson, C.F. (2001). Anatomical and physiological parameters affecting gastrointestinal absorption in humans and rats. *Food and Chemical Toxicology* 39 (3): 209–228.
8 Dressman, J.B. (1986). Comparison of canine and human gastrointestinal physiology. *Pharmaceutical Research* 3 (3): 123–131.
9 Kararli, T.T. (1995). Comparison of the gastrointestinal anatomy, physiology, and biochemistry of humans and commonly used laboratory animals. *Biopharmaceutics & Drug Disposition* 16 (5): 351–380.
10 DeSesso, J.M. and Williams, A.L. (2008). Contrasting the gastrointestinal tracts of Mammals: factors that influence absorption. *Annual Reports in Medicinal Chemistry* 43: 353–371.
11 Hebel, R. and Stromberg, M.W. (1986). *Anatomy and Embryology of the Laboratory Rat*, 271. Worthsee, Germany: BioMed Verlag.
12 Snyder, W.S., Cook, M.J., Nasset, E.S. et al. (1975). *Report of the Task Group on Reference Man*. New York.
13 Rowland, M. and Tozer, T.N. (1995). *Clinical Pharmacokinetics Concepts and Application*. Philadelphia: Lippincott Williams & Wilkins.
14 Shargel, L. and Yu, A. (1999). *Applied Biopharmaceutics and Pharmacokinetics*. Appleton and Lange: Stamford.
15 Pappenheimer, J.R. (1998). Scaling of dimensions of small intestines in non-ruminant eutherian mammals and its significance for absorptive mechanisms. *Comparative Biochemistry and Physiology. Part A, Molecular & Integrative Physiology* 121 (1): 45–58.
16 Takahashi, M., Washio, T., Suzuki, N. et al. (2010). Investigation of the intestinal permeability and first-pass metabolism of drugs in cynomolgus

monkeys using single-pass intestinal perfusion. *Biological & Pharmaceutical Bulletin* 33 (1): 111–116.

17 Chiou, W.L. (1994). Effect of 'unstirred' water layer in the intestine on the rate and extent of absorption after oral administration. *Biopharmaceutics & Drug Disposition* 15 (8): 709–717.

18 Kimura, T. and Higaki, K. (2002). Gastrointestinal transit and drug absorption. *Biological & Pharmaceutical Bulletin* 25 (2): 149–164.

19 Mudie, D.M., Amidon, G.L., and Amidon, G.E. (2010). Physiological parameters for oral delivery and in vitro testing. *Molecular Pharmaceutics* 7 (5): 1388–1405.

20 Custodio, J.M., Wu, C.Y., and Benet, L.Z. (2008). Predicting drug disposition, absorption/elimination/transporter interplay and the role of food on drug absorption. *Advanced Drug Delivery Reviews* 60 (6): 717–733.

21 Singh, B.N. (1999). Effects of food on clinical pharmacokinetics. *Clinical Pharmacokinetics* 37 (3): 213–255.

22 Ikegami, K., Tagawa, K., Narisawa, S., and Osawa, T. (2003). Suitability of the cynomolgus monkey as an animal model for drug absorption studies of oral dosage forms from the viewpoint of gastrointestinal physiology. *Biological & Pharmaceutical Bulletin* 26 (10): 1442–1447.

23 Lui, C.Y., Amidon, G.L., Berardi, R.R. et al. (1986). Comparison of gastrointestinal pH in dogs and humans: implications on the use of the beagle dog as a model for oral absorption in humans. *Journal of Pharmaceutical Sciences* 75 (3): 271–274.

24 Martinez, M., Amidon, G., Clarke, L. et al. (2002). Applying the biopharmaceutics classification system to veterinary pharmaceutical products. Part II. Physiological considerations. *Advanced Drug Delivery Reviews* 54 (6): 825–850.

25 Garanero, G., Ramanathan, S., and Amidon, G. (2005). Gastrointestinal dissolution and absorption of drugs. In: *Drug Bioavailability: Estimation of Solubility, Permeability, Absorption and Bioavailability*, vol. 18 (ed. H.V.D. Waterbeemd, H. Lennernas, and Artursson), 191–210. Zurich: Wiley-VCH.

26 Stenberg, P., Bergstrom, C.A.S., Luthman, K., and Artursson, P. (2002). Theoretical predictions of drug absorption in drug discovery and development. *Clinical Pharmacokinetics* 41 (11): 877–899.

27 Enck, P., Merlin, V., Erckenbrecht, J.F., and Wienbeck, M. (1989). Stress effects on gastrointestinal transit in the rat. *Gut* 30 (4): 455–459.

28 Barrow, L., Steed, K.P., Spiller, R.C. et al. (1992). Scintigraphic demonstration of lactulose-induced accelerated proximal colon transit. *Gastroenterology* 103 (4): 1167–1173.

29 Davies, B. and Morris, T. (1993). Physiological parameters in laboratory animals and humans. *Pharmaceutical Research* 10 (7): 1093–1095.

30 Lentz, K.A. (2008). Current methods for predicting human food effect. *The AAPS Journal* 10 (2): 282–288.

31 Lentz, K.A., Quitko, M., Morgan, D.G. et al. (2007). Development and validation of a preclinical food effect model. *Journal of Pharmaceutical Sciences* 96 (2): 459–472.

32 Hofmann, A.F. and Hagey, L.R. (2008). Bile acids: chemistry, pathochemistry, biology, pathobiology, and therapeutics. *Cellular and Molecular Life Sciences* 65 (16): 2461–2483.

33 Monte, M.J., Marin, J.J., Antelo, A., and Vazquez-Tato, J. (2009). Bile acids: chemistry, physiology, and pathophysiology. *World Journal of Gastroenterology* 15 (7): 804–816.

34 Charman, W.N., Porter, C.J., Mithani, S., and Dressman, J.B. (1997). Physiochemical and physiological mechanisms for the effects of food on drug absorption: the role of lipids and pH. *Journal of Pharmaceutical Sciences* 86 (3): 269–282.

35 Borriello, S.P. (1984). Bacteria and gastrointestinal secretion and motility. *Scandinavian Journal of Gastroenterology. Supplement* 93: 115–121.

36 Hurst, S., Loi, C.M., Brodfuehrer, J., and El-Kattan, A. (2007). Impact of physiological, physicochemical and biopharmaceutical factors in absorption and metabolism mechanisms on the drug oral bioavailability of rats and humans. *Expert Opinion on Drug Metabolism & Toxicology* 3 (4): 469–489.

37 Roberts, M.S., Magnusson, B.M., Burczynski, F.J., and Weiss, M. (2002). Enterohepatic circulation: physiological, pharmacokinetic and clinical implications. *Clinical Pharmacokinetics* 41 (10): 751–790.

38 Drasar, B.S., Montgomery, F., and Tomkins, A.M. (1986). Diet and faecal flora in three dietary groups in rural northern Nigeria. *J Hyg (Lond)* 96 (1): 59–65.

39 Trevaskis, N.L., Porter, C.J., and Charman, W.N. (2006). The lymph lipid precursor pool is a key determinant of intestinal lymphatic drug transport. *The Journal of Pharmacology and Experimental Therapeutics* 316 (2): 881–891.

40 Edwards, G.A., Porter, C.J., Caliph, S.M. et al. (2001). Animal models for the study of intestinal

lymphatic drug transport. *Advanced Drug Delivery Reviews* 50 (1–2): 45–60.

41 Lennernas, H. (1995). Does fluid flow across the intestinal mucosa affect quantitative oral drug absorption? is it time for a reevaluation? *Pharmaceutical Research* 12 (11): 1573–1582.

42 He, Y.L., Murby, S., Warhurst, G. et al. (1998). Species differences in size discrimination in the paracellular pathway reflected by oral bioavailability of poly(ethylene glycol) and D-peptides. *Journal of Pharmaceutical Sciences* 87 (5): 626–633.

43 Bader, T.R., Herneth, A.M., Blaicher, W. et al. (1998). Hepatic perfusion after liver transplantation: noninvasive measurement with dynamic single-section CT. *Radiology* 209 (1): 129–134.

44 Graham, G.G., Punt, J., Arora, M. et al. (2011). Clinical pharmacokinetics of metformin. *Clinical Pharmacokinetics* 50 (2): 81–98.

45 Avdeef, A. (2001). Physicochemical profiling (solubility, permeability and charge state). *Current Topics in Medicinal Chemistry* 1 (4): 277–351.

46 El-Kattan, A. (2017). Extended clearance classification system (ECCS) and its utility in predicting clearance rate determing step. In: *Oral Bioavailability Assessment: Basics and Strategies for Drug Discovery and Development* (ed. M. Lee), 309–329. Wiley.

47 El-Kattan, A.F., Varma, M.V., Steyn, S.J. et al. (2016). Projecting ADME behavior and drug-drug interactions in early discovery and development: application of the extended clearance classification system. *Pharmaceutical Research* 33 (12): 3021–3030.

48 Bleasby, K., Castle, J.C., Roberts, C.J. et al. (2006). Expression profiles of 50 xenobiotic transporter genes in humans and pre-clinical species: a resource for investigations into drug disposition. *Xenobiotica* 36 (10-11): 963–988.

49 Hilgendorf, C., Ahlin, G., Seithel, A. et al. (2007). Expression of thirty-six drug transporter genes in human intestine, liver, kidney, and organotypic cell lines. *Drug Metabolism and Disposition* 35 (8): 1333–1340.

50 Shimoyama, Y., Kirat, D., Akihara, Y. et al. (2007). Expression of monocarboxylate transporter 1 (MCT1) in the dog intestine. *The Journal of Veterinary Medical Science* 69 (6): 599–604.

51 International Transporter Consortium, Giacomini, K.M., Huang, S.M. et al. (2010). Membrane transporters in drug development. *Nature Reviews. Drug Discovery* 9 (3): 215–236.

52 Cao, X., Gibbs, S.T., Fang, L. et al. (2006). Why is it challenging to predict intestinal drug absorption and oral bioavailability in human using rat model. *Pharmaceutical Research* 23 (8): 1675–1686.

53 Groer, C., Bruck, S., Lai, Y. et al. (2013). LC-MS/MS-based quantification of clinically relevant intestinal uptake and efflux transporter proteins. *Journal of Pharmaceutical and Biomedical Analysis* 85: 253–261.

54 Oswald, S., Groer, C., Drozdzik, M., and Siegmund, W. (2013). Mass spectrometry-based targeted proteomics as a tool to elucidate the expression and function of intestinal drug transporters. *The AAPS Journal* 15 (4): 1128–1140.

55 Abel, S., Beaumont, K.C., Crespi, C.L. et al. (2001). Potential role for P-glycoprotein in the non-proportional pharmacokinetics of UK-343,664 in man. *Xenobiotica* 31 (8-9): 665–676.

56 El Ela, A.A., Hartter, S., Schmitt, U. et al. (2004). Identification of P-glycoprotein substrates and inhibitors among psychoactive compounds – implications for pharmacokinetics of selected substrates. *The Journal of Pharmacy and Pharmacology* 56 (8): 967–975.

57 Salphati, L. and Benet, L.Z. (1998). Effects of ketoconazole on digoxin absorption and disposition in rat. *Pharmacology* 56 (6): 308–313.

58 Varma, M.V., Ambler, C.M., Ullah, M. et al. (2010). Targeting intestinal transporters for optimizing oral drug absorption. *Current Drug Metabolism* 11 (9): 730–742.

59 Drescher, S., Glaeser, H., Murdter, T. et al. (2003). P-glycoprotein-mediated intestinal and biliary digoxin transport in humans. *Clinical Pharmacology and Therapeutics* 73 (3): 223–231.

60 Devault, A. and Gros, P. (1990). Two members of the mouse MDR gene family confer multidrug resistance with overlapping but distinct drug specificities. *Molecular and Cellular Biology* 10 (4): 1652–1663.

61 Nishimura, M. and Naito, S. (2005). Tissue-specific mRNA expression profiles of human ATP-binding cassette and solute carrier transporter superfamilies. *Drug Metabolism and Pharmacokinetics* 20 (6): 452–477.

62 Takahashi, M., Washio, T., Suzuki, N. et al. (2008). Characterization of gastrointestinal drug absorption in cynomolgus monkeys. *Molecular Pharmaceutics* 5 (2): 340–348.

63 Kim, I.W., Booth-Genthe, C., and Ambudkar, S.V. (2008). Relationship between drugs and functional activity of various mammalian P-glycoproteins (ABCB1). *Mini Reviews in Medicinal Chemistry* 8 (3): 193–200.

64 Mouly, S. and Paine, M.F. (2003). P-glycoprotein increases from proximal to distal regions of human small intestine. *Pharmaceutical Research* 20 (10): 1595–1599.

65 Lown, K.S., Mayo, R.R., Leichtman, A.B. et al. (1997). Role of intestinal P-glycoprotein (mdr1) in interpatient variation in the oral bioavailability of cyclosporine. *Clinical Pharmacology and Therapeutics* 62 (3): 248–260.

66 Tamura, S., Ohike, A., Ibuki, R. et al. (2002). Tacrolimus is a class II low-solubility high-permeability drug: the effect of P-glycoprotein efflux on regional permeability of tacrolimus in rats. *Journal of Pharmaceutical Sciences* 91 (3): 719–729.

67 Katoh, M., Suzuyama, N., Takeuchi, T. et al. (2006). Kinetic analyses for species differences in P-glycoprotein-mediated drug transport. *Journal of Pharmaceutical Sciences* 95 (12): 2673–2683.

68 Xia, C.Q., Xiao, G., Liu, N. et al. (2006). Comparison of species differences of P-glycoproteins in beagle dog, rhesus monkey, and human using Atpase activity assays. *Molecular Pharmaceutics* 3 (1): 78–86.

69 Takeuchi, T., Yoshitomi, S., Higuchi, T. et al. (2006). Establishment and characterization of the transformants stably-expressing MDR1 derived from various animal species in LLC-PK1. *Pharmaceutical Research* 23 (7): 1460–1472.

70 Han, B. and Zhang, J.T. (2004). Multidrug resistance in cancer chemotherapy and xenobiotic protection mediated by the half ATP-binding cassette transporter ABCG2. *Current Medicinal Chemistry. Anti-Cancer Agents* 4 (1): 31–42.

71 Haller, S., Schuler, F., Lazic, S.E. et al. (2012). Expression profiles of metabolic enzymes and drug transporters in the liver and along the intestine of beagle dogs. *Drug Metabolism and Disposition* 40 (8): 1603–1610.

72 Seithel, A., Karlsson, J., Hilgendorf, C. et al. (2006). Variability in mRNA expression of ABC- and SLC-transporters in human intestinal cells: comparison between human segments and Caco-$_2$ cells. *European Journal of Pharmaceutical Sciences* 28 (4): 291–299.

73 Nies, A.T. and Keppler, D. (2007). The apical conjugate efflux pump ABCC2 (MRP2). *Pflügers Archiv* 453 (5): 643–659.

74 Buchler, M., Konig, J., Brom, M. et al. (1996). cDNA cloning of the hepatocyte canalicular isoform of the multidrug resistance protein, cMrp, reveals a novel conjugate export pump deficient in hyperbilirubinemic mutant rats. *The Journal of Biological Chemistry* 271 (25): 15091–15098.

75 Conrad, S., Viertelhaus, A., Orzechowski, A. et al. (2001). Sequencing and tissue distribution of the canine MRP2 gene compared with MRP1 and MDR1. *Toxicology* 156 (2–3): 81–91.

76 Yasunaga, M., Takemura, M., Fujita, K. et al. (2008). Molecular cloning and functional characterization of cynomolgus monkey multidrug resistance-associated protein 2 (MRP2). *European Journal of Pharmaceutical Sciences* 35 (4): 326–334.

77 Borst, P., Evers, R., Kool, M., and Wijnholds, J. (2000). A family of drug transporters: the multidrug resistance-associated proteins. *Journal of the National Cancer Institute* 92 (16): 1295–1302.

78 Kato, R., Nishide, M., Kozu, C. et al. (2010). Is cyclosporine A transport inhibited by pravastatin via multidrug resistant protein 2? *European Journal of Clinical Pharmacology* 66 (2): 153–158.

79 Misaka, S., Muller, F., and Fromm, M.F. (2013). Clinical relevance of drug efflux pumps in the gut. *Current Opinion in Pharmacology* 13 (6): 847–852.

80 Pfeifer, N.D., Yang, K., and Brouwer, K.L. (2013). Hepatic basolateral efflux contributes significantly to rosuvastatin disposition: I. Characterization of basolateral vs. biliary clearance using a novel protocol in sandwich-cultured hepatocytes. *The Journal of Pharmacology and Experimental Therapeutics* 347 (3): 727–736.

81 Dietrich, C.G., de Waart, D.R., Ottenhoff, R. et al. (2001a). Mrp2-deficiency in the rat impairs biliary and intestinal excretion and influences metabolism and disposition of the food-derived carcinogen 2-amino-1-methyl-6-phenylimidazo. *Carcinogenesis* 22 (5): 805–811.

82 Dietrich, C.G., de Waart, D.R., Ottenhoff, R. et al. (2001b). Increased bioavailability of the food-derived carcinogen 2-amino-1-methyl-6-phenylimidazo [4,5-*b*]pyridine in MRP2-deficient rats. *Molecular Pharmacology* 59 (5): 974–980.

83 Lin, X., Skolnik, S., Chen, X., and Wang, J. (2011). Attenuation of intestinal absorption by major efflux transporters: quantitative tools and strategies using a Caco-$_2$ model. *Drug Metabolism and Disposition* 39 (2): 265–274.

84 Benet, L.Z., Amidon, G.L., Barends, D.M. et al. (2008). The use of BDDCS in classifying the permeability of marketed drugs. *Pharmaceutical Research* 25 (3): 483–488.

85 Estudante, M., Morais, J.G., Soveral, G., and Benet, L.Z. (2013). Intestinal drug transporters:

an overview. *Advanced Drug Delivery Reviews* 65 (10): 1340–1356.

86 Bretschneider, B., Brandsch, M., and Neubert, R. (1999). Intestinal transport of beta-lactam antibiotics: analysis of the affinity at the H+/peptide symporter (PEPT1), the uptake into Caco-2 cell monolayers and the transepithelial flux. *Pharmaceutical Research* 16 (1): 55–61.

87 Knutter, I., Wollesky, C., Kottra, G. et al. (2008). Transport of angiotensin-converting enzyme inhibitors by H+/peptide transporters revisited. *The Journal of Pharmacology and Experimental Therapeutics* 327 (2): 432–441.

88 Terada, T., Saito, H., Mukai, M., and Inui, K. (1997). Characterization of stably transfected kidney epithelial cell line expressing rat H+/peptide cotransporter PEPT1: localization of PEPT1 and transport of beta-lactam antibiotics. *The Journal of Pharmacology and Experimental Therapeutics* 281 (3): 1415–1421.

89 Watanabe, K., Sawano, T., Endo, T. et al. (2002). Studies on intestinal absorption of sulpiride (2): transepithelial transport of sulpiride across the human intestinal cell line Caco-$_2$. *Biological & Pharmaceutical Bulletin* 25 (10): 1345–1350.

90 Watanabe, K., Sawano, T., Jinriki, T., and Sato, J. (2004). Studies on intestinal absorption of sulpiride (3): intestinal absorption of sulpiride in rats. *Biological & Pharmaceutical Bulletin* 27 (1): 77–81.

91 Miyamoto, K., Shiraga, T., Morita, K. et al. (1996). Sequence, tissue distribution and developmental changes in rat intestinal oligopeptide transporter. *Biochimica et Biophysica Acta* 1305 (1–2): 34–38.

92 Perkins, E.J. and Abraham, T. (2007). Pharmacokinetics, metabolism, and excretion of the intestinal peptide transporter 1 (SLC15A1)-targeted prodrug (1*S*,2*S*,5*R*,6*S*)-2-[(2'*S*)-(2-amino)propionyl] aminobicyclo[3.1.0.]hexen-2,6-di carboxylic acid (LY544344) in rats and dogs: assessment of first-pass bioactivation and dose linearity. *Drug Metabolism and Disposition* 35 (10): 1903–1909.

93 Zhang, E.Y., Emerick, R.M., Pak, Y.A. et al. (2004). Comparison of human and monkey peptide transporters: PEPT1 and PEPT2. *Molecular Pharmaceutics* 1 (3): 201–210.

94 Englund, G., Rorsman, F., Ronnblom, A. et al. (2006). Regional levels of drug transporters along the human intestinal tract: co-expression of ABC and SLC transporters and comparison with Caco-2 cells. *European Journal of Pharmaceutical Sciences* 29 (3-4): 269–277.

95 Lu, H. and Klaassen, C. (2006). Tissue distribution and thyroid hormone regulation of Pept1 and Pept2 mRNA in rodents. *Peptides* 27 (4): 850–857.

96 Chiou, W.L. and Barve, A. (1998). Linear correlation of the fraction of oral dose absorbed of 64 drugs between humans and rats. *Pharmaceutical Research* 15 (11): 1792–1795.

97 Chiou, W.L. and Buehler, P.W. (2002). Comparison of oral absorption and bioavailablity of drugs between monkey and human. *Pharmaceutical Research* 19 (6): 868–874.

98 Chiou, W.L., Jeong, H.Y., Chung, S.M., and Wu, T.C. (2000). Evaluation of using dog as an animal model to study the fraction of oral dose absorbed of 43 drugs in humans. *Pharmaceutical Research* 17 (2): 135–140.

99 Kullak-Ublick, G.A., Ismair, M.G., Stieger, B. et al. (2001). Organic anion-transporting polypeptide B (OATP-B) and its functional comparison with three other OATPs of human liver. *Gastroenterology* 120 (2): 525–533.

100 Sai, Y., Kaneko, Y., Ito, S. et al. (2006). Predominant contribution of organic anion transporting polypeptide OATP-B (OATP2B1) to apical uptake of estrone-3-sulfate by human intestinal Caco-$_2$ cells. *Drug Metabolism and Disposition* 34 (8): 1423–1431.

101 Tamai, I., Nezu, J., Uchino, H. et al. (2000). Molecular identification and characterization of novel members of the human organic anion transporter (OATP) family. *Biochemical and Biophysical Research Communications* 273 (1): 251–260.

102 Kobayashi, D., Nozawa, T., Imai, K. et al. (2003). Involvement of human organic anion transporting polypeptide OATP-B (SLC21A9) in pH-dependent transport across intestinal apical membrane. *The Journal of Pharmacology and Experimental Therapeutics* 306 (2): 703–708.

103 Nozawa, T., Imai, K., Nezu, J. et al. (2004). Functional characterization of pH-sensitive organic anion transporting polypeptide OATP-B in human. *The Journal of Pharmacology and Experimental Therapeutics* 308 (2): 438–445.

104 Imanaga, J., Kotegawa, T., Imai, H. et al. (2011). The effects of the SLCO2B1 c.1457C >T polymorphism and apple juice on the pharmacokinetics of fexofenadine and midazolam in humans. *Pharmacogenetics and Genomics* 21 (2): 84–93.

105 Treiber, A., Schneiter, R., Hausler, S., and Stieger, B. (2007). Bosentan is a substrate of human OATP1B1 and OATP1B3: inhibition of hepatic

uptake as the common mechanism of its interactions with cyclosporin A, rifampicin, and sildenafil. *Drug Metabolism and Disposition* 35 (8): 1400–1407.

**106** Varma, M.V., Rotter, C.J., Chupka, J. et al. (2011). pH-sensitive interaction of HMG-CoA reductase inhibitors (statins) with organic anion transporting polypeptide 2B1. *Molecular Pharmaceutics* 8 (4): 1303–1313.

**107** Chu, X., Bleasby, K., and Evers, R. (2013). Species differences in drug transporters and implications for translating preclinical findings to humans. *Expert Opinion on Drug Metabolism & Toxicology* 9 (3): 237–252.

**108** Hagenbuch, B. and Meier, P.J. (2003). The superfamily of organic anion transporting polypeptides. *Biochimica et Biophysica Acta* 1609 (1): 1–18.

**109** Yoshida, K., Maeda, K., and Sugiyama, Y. (2013). Hepatic and intestinal drug transporters: prediction of pharmacokinetic effects caused by drug-drug interactions and genetic polymorphisms. *Annual Review of Pharmacology and Toxicology* 53: 581–612.

**110** Hosea, N.A., Collard, W.T., Cole, S. et al. (2009). Prediction of human pharmacokinetics from preclinical information: comparative accuracy of quantitative prediction approaches. *Journal of Clinical Pharmacology* 49 (5): 513–533.

**111** Gonzalez, F.J. (1988). The molecular biology of cytochrome P450s. *Pharmacological Reviews* 40 (4): 243–288.

**112** Nelson, D.R., Koymans, L., Kamataki, T. et al. (1996). P450 superfamily: update on new sequences, gene mapping, accession numbers and nomenclature. *Pharmacogenetics* 6 (1): 1–42.

**113** Di, L. (2014). The role of drug metabolizing enzymes in clearance. *Expert Opinion on Drug Metabolism & Toxicology* 10 (3): 379–393.

**114** Varma, M.V., Steyn, S.J., Allerton, C., and El-Kattan, A.F. (2015). Predicting clearance mechanism in drug discovery: extended clearance classification system (ECCS). *Pharmaceutical Research* 32: 3785–3802.

**115** Peters, W. and Kremers, P. (1989). Cytochromes P-450 in the intestinal mucosa of man. *Biochemical Pharmacology* 38 (9): 1535–1538.

**116** Shimada, T., Yamazaki, H., Mimura, M. et al. (1994). Interindividual variations in human liver cytochrome P-450 enzymes involved in the oxidation of drugs, carcinogens and toxic chemicals: studies with liver microsomes of 30 Japanese and 30 Caucasians. *The Journal of Pharmacology and Experimental Therapeutics* 270 (1): 414–423.

**117** Casabar, R.C.T., Wallace, A.D., Hodgson, E., and Rose, R.L. (2006). Metabolism of endosulfan-alpha by human liver microsomes and its utility as a simultaneous in vitro probe for CYP2B6 and CYP3A4. *Drug Metabolism and Disposition* 34 (10): 1779–1785.

**118** Dai, D., Zeldin, D.C., Blaisdell, J.A. et al. (2001). Polymorphisms in human CYP2C8 decrease metabolism of the anticancer drug paclitaxel and arachidonic acid. *Pharmacogenetics* 11 (7): 597–607.

**119** Desai, P.B., Duan, J.Z., Zhu, Y.W., and Kouzi, S. (1998). Human liver microsomal metabolism of paclitaxel and drug interactions. *European Journal of Drug Metabolism and Pharmacokinetics* 23 (3): 417–424.

**120** McGinnity, D.F., Parker, A.J., Soars, M., and Riley, R.J. (2000). Automated definition of the enzymology of drug oxidation by the major human drug metabolizing cytochrome P450s. *Drug Metabolism and Disposition* 28 (11): 1327–1334.

**121** Mirghani, R.A., Sayi, J., Aklillu, E. et al. (2006). CYP3A5 genotype has significant effect on quinine 3-hydroxylation in Tanzanians, who have lower total CYP3A activity than a Swedish population. *Pharmacogenetics and Genomics* 16 (9): 637–645.

**122** O'Donnell, C.J., Grime, K., Courtney, P. et al. (2007). The Development of a Cocktail CYP2B6, CYP2C8, and CYP3A5 inhibition assay and a preliminary assessment of utility in a drug discovery setting. *Drug Metabolism and Disposition* 35 (3): 381–385.

**123** Spatzenegger, M. and Jaeger, W. (1995). Clinical importance of hepatic cytochrome P450 in drug metabolism. *Drug Metabolism Reviews* 27 (3): 397–417.

**124** Turpeinen, M., Raunio, H., and Pelkonen, O. (2006). The functional role of CYP2B6 in human drug metabolism: substrates and inhibitors in vitro, in vivo and in silico. *Current Drug Metabolism* 7 (7): 705–714.

**125** Smith, D.A., Abel, S.M., Hyland, R., and Jones, B.C. (1998). Human cytochrome P450s: selectivity and measurement in vivo. *Xenobiotica* 28 (12): 1095–1128.

**126** Stevens, J.C., Shipley, L.A., Cashman, J.R. et al. (1993). Comparison of human and rhesus monkey in vitro phase I and phase II hepatic drug metabolism activities. *Drug Metabolism and Disposition* 21 (5): 753–760.

**127** Williams, J.A., Hurst, S.I., Bauman, J. et al. (2003). Reaction phenotyping in drug discovery: moving forward with confidence? *Current Drug Metabolism* 4 (6): 527–534.

128 Achour, B., Russell, M.R., Barber, J., and Rostami-Hodjegan, A. (2014). Simultaneous quantification of the abundance of several cytochrome P450 and uridine 5′-diphospho-glucuronosyltransferase enzymes in human liver microsomes using multiplexed targeted proteomics. *Drug Metabolism and Disposition* 42 (4): 500–510.

129 Kuehl, P., Zhang, J., Lin, Y. et al. (2001). Sequence diversity in CYP3A promoters and characterization of the genetic basis of polymorphic CYP3A5 expression. *Nature Genetics* 27 (4): 383–391.

130 Wrighton, S.A., Brian, W.R., Sari, M.A. et al. (1990). Studies on the expression and metabolic capabilities of human liver cytochrome P450IIIA5 (HLp3). *Molecular Pharmacology* 38 (2): 207–213.

131 Paine, M.F., Khalighi, M., Fisher, J.M. et al. (1997). Characterization of interintestinal and intraintestinal variations in human CYP3A-dependent metabolism. *The Journal of Pharmacology and Experimental Therapeutics* 283 (3): 1552–1562.

132 Smit, P., van Schaik, R.H., van der Werf, M. et al. (2005). A common polymorphism in the CYP3A7 gene is associated with a nearly 50% reduction in serum dehydroepiandrosterone sulfate levels. *The Journal of Clinical Endocrinology and Metabolism* 90 (9): 5313–5316.

133 Gellner, K., Eiselt, R., Hustert, E. et al. (2001). Genomic organization of the human CYP3A locus: identification of a new, inducible CYP3A gene. *Pharmacogenetics* 11 (2): 111–121.

134 Chen, M., Ma, L., Drusano, G.L. et al. (2006). Sex differences in CYP3A activity using intravenous and oral midazolam. *Clinical Pharmacology and Therapeutics* 80 (5): 531–538.

135 Robertson, G.R., Farrell, G.C., and Liddle, C. (1998). Sexually dimorphic expression of rat CYP3A9 and CYP3A18 genes is regulated by growth hormone. *Biochemical and Biophysical Research Communications* 242 (1): 57–60.

136 Yamazoe, Y., Murayama, N., Shimada, M. et al. (1988). A sex-specific form of cytochrome P-450 catalyzing propoxycoumarin O-depropylation and its identity with testosterone 6 beta-hydroxylase in untreated rat livers: reconstitution of the activity with microsomal lipids. *J. Biochem. (Tokyo)* 104 (5): 785–790.

137 Matsubara, T., Kim, H.J., Miyata, M. et al. (2004). Isolation and characterization of a new major intestinal CYP3A form, CYP3A62, in the rat. *The Journal of Pharmacology and Experimental Therapeutics* 309 (3): 1282–1290.

138 Smith, D.A. (1991). Species differences in metabolism and pharmacokinetics: are we close to an understanding? *Drug Metabolism Reviews* 23 (3–4): 355–373.

139 Warrington, J.S., Von Moltke, L.L., Shader, R.I., and Greenblatt, D.J. (2002). In vitro biotransformation of sildenafil (Viagra) in the male rat: the role of CYP2C11. *Drug Metabolism and Disposition* 30 (6): 655–657.

140 Ghosal, A., Satoh, H., Thomas, P.E. et al. (1996). Inhibition and kinetics of cytochrome P4503A activity in microsomes from rat, human, and cDNA-expressed human cytochrome P450. *Drug Metabolism and Disposition* 24 (9): 940–947.

141 Lu, C. and Li, A.P. (2001). Species comparison in P450 induction: effects of dexamethasone, omeprazole, and rifampin on P450 isoforms 1A and 3A in primary cultured hepatocytes from man, Sprague-Dawley rat, minipig, and beagle dog. *Chemico-Biological Interactions* 134 (3): 271–281.

142 Fraser, D.J., Feyereisen, R., Harlow, G.R., and Halpert, J.R. (1997). Isolation, heterologous expression and functional characterization of a novel cytochrome P450 3A enzyme from a canine liver cDNA library. *The Journal of Pharmacology and Experimental Therapeutics* 283 (3): 1425–1432.

143 Prueksaritanont, T., Gorham, L.M., Hochman, J.H. et al. (1996). Comparative studies of drug-metabolizing enzymes in dog, monkey, and human small intestines, and in Caco-$_2$ cells. *Drug Metabolism and Disposition* 24 (6): 634–642.

144 Bogaards, J.J., Bertrand, M., Jackson, P. et al. (2000). Determining the best animal model for human cytochrome P450 activities: a comparison of mouse, rat, rabbit, dog, micropig, monkey and man. *Xenobiotica* 30 (12): 1131–1152.

145 Chauret, N., Gauthier, A., Martin, J., and Nicoll-Griffith, D.A. (1997). In vitro comparison of cytochrome P450-mediated metabolic activities in human, dog, cat, and horse. *Drug Metabolism and Disposition* 25 (10): 1130–1136.

146 Goldstein, J.A. (2001). Clinical relevance of genetic polymorphisms in the human CYP2C subfamily. *British Journal of Clinical Pharmacology* 52 (4): 349–355.

147 Lapple, F., von Richter, O., Fromm, M.F. et al. (2003). Differential expression and function of $CYP_2C$ isoforms in human intestine and liver. *Pharmacogenetics* 13 (9): 565–575.

**148** Blaisdell, J., Goldstein, J.A., and Bai, S.A. (1998). Isolation of a new canine cytochrome P450 CDNA from the cytochrome P450 2C subfamily (CYP2C41) and evidence for polymorphic differences in its expression. *Drug Metabolism and Disposition* 26 (3): 278–283.

**149** Mugford, C.A. and Kedderis, G.L. (1998). Sex-dependent metabolism of xenobiotics. *Drug Metabolism Reviews* 30 (3): 441–498.

**150** Ohhira, S., Enomoto, M., and Matsui, H. (2006). In vitro metabolism of tributyltin and triphenyltin by human cytochrome P-450 isoforms. *Toxicology* 228 (2-3): 171–177.

**151** Mode, A., Ahlgren, R., Lahuna, O., and Gustafsson, J.A. (1998). Gender differences in rat hepatic CYP2C gene expression–regulation by growth hormone. *Growth Hormone & IGF Research* 8 Suppl B: 61–67.

**152** Rich, K.J. and Boobis, A.R. (1997). Expression and inducibility of P450 enzymes during liver ontogeny. *Microscopy Research and Technique* 39 (5): 424–435.

**153** Graham, M.J., Bell, A.R., Crewe, H.K. et al. (2003). mRNA and protein expression of dog liver cytochromes P450 in relation to the metabolism of human CYP2C substrates. *Xenobiotica* 33 (3): 225–237.

**154** Paine, M.F., Hart, H.L., Ludington, S.S. et al. (2006). The human intestinal cytochrome P450 “pie”. *Drug Metabolism and Disposition* 34 (5): 880–886.

**155** Zanger, U.M., Raimundo, S., and Eichelbaum, M. (2004). Cytochrome P450 2D6: overview and update on pharmacology, genetics, biochemistry. *Naunyn-Schmiedeberg's Archives of Pharmacology* 369 (1): 23–37.

**156** Williams, J.A., Hyland, R., Jones, B.C. et al. (2004). Drug-drug interactions for UDP-glucuronosyltransferase substrates: a pharmacokinetic explanation for typically observed low exposure (AUCi/AUC) ratios. *Drug Metabolism and Disposition* 32 (11): 1201–1208.

**157** Venhorst, J., ter Laak, A.M., Commandeur, J.N. et al. (2003). Homology modeling of rat and human cytochrome P450 2D (CYP2D) isoforms and computational rationalization of experimental ligand-binding specificities. *Journal of Medicinal Chemistry* 46 (1): 74–86.

**158** Hiroi, T., Chow, T., Imaoka, S., and Funae, Y. (2002). Catalytic specificity of CYP2D isoforms in rat and human. *Drug Metabolism and Disposition* 30 (9): 970–976.

**159** Muralidharan, G., Hawes, E.M., McKay, G. et al. (1991a). Quinidine but not quinine inhibits in man the oxidative metabolic routes of methoxyphenamine which involve debrisoquine 4-hydroxylase. *European Journal of Clinical Pharmacology* 41 (5): 471–474.

**160** Muralidharan, G., Hawes, E.M., McKay, G., and Midha, K.K. (1991b). Quinine is a more potent inhibitor than quinidine in rat of the oxidative metabolic routes of methoxyphenamine which involve debrisoquine 4-hydroxylase. *Xenobiotica* 21 (11): 1441–1450.

**161** Paulson, S.K., Engel, L., Reitz, B. et al. (1999). Evidence for polymorphism in the canine metabolism of the cyclooxygenase 2 inhibitor, celecoxib. *Drug Metabolism and Disposition* 27 (10): 1133–1142.

**162** Shimada, T., Martin, M.V., Pruess-Schwartz, D. et al. (1989). Roles of individual human cytochrome P-450 enzymes in the bioactivation of benzo(a)pyrene, 7,8-dihydroxy-7,8-dihydrobenzo(a)pyrene, and other dihydrodiol derivatives of polycyclic aromatic hydrocarbons. *Cancer Research* 49 (22): 6304–6312.

**163** Iba, M.M., Storch, A., Ghosal, A. et al. (2003). Constitutive and inducible levels of CYP1A1 and CYP1A2 in rat cerebral cortex and cerebellum. *Archives of Toxicology* 77 (10): 547–554.

**164** Graham, R.A., Downey, A., Mudra, D. et al. (2002). In vivo and in vitro induction of cytochrome P450 enzymes in beagle dogs. *Drug Metabolism and Disposition* 30 (11): 1206–1213.

**165** Uchida, T., Komori, M., Kitada, M., and Kamataki, T. (1990). Isolation of cDNAs coding for three different forms of liver microsomal cytochrome P-450 from polychlorinated biphenyl-treated beagle dogs. *Molecular Pharmacology* 38 (5): 644–651.

**166** Guengerich, F.P. (1997). Comparisons of catalytic selectivity of cytochrome P450 subfamily enzymes from different species. *Chemico-Biological Interactions* 106 (3): 161–182.

**167** Lindell, M., Lang, M., and Lennernas, H. (2003). Expression of genes encoding for drug metabolising cytochrome P450 enzymes and P-glycoprotein in the rat small intestine; comparison to the liver. *European Journal of Drug Metabolism and Pharmacokinetics* 28 (1): 41–48.

**168** Shimada, T., Yamazaki, H., Mimura, M. et al. (1996). Characterization of microsomal cytochrome P450 enzymes involved in the oxidation of xenobiotic chemicals in human fetal liver and adult lungs. *Drug Metabolism and Disposition* 24 (5): 515–522.

169 Hakkola, J., Pasanen, M., Hukkanen, J. et al. (1996a). Expression of xenobiotic-metabolizing cytochrome P450 forms in human full-term placenta. *Biochemical Pharmacology* 51 (4): 403–411.

170 Hakkola, J., Raunio, H., Purkunen, R. et al. (1996b). Detection of cytochrome P450 gene expression in human placenta in first trimester of pregnancy. *Biochemical Pharmacology* 52 (2): 379–383.

171 Benet, L.Z., Izumi, T., Zhang, Y. et al. (1999). Intestinal MDR transport proteins and P-450 enzymes as barriers to oral drug delivery. *Journal of Controlled Release* 62 (1–2): 25–31.

172 Fisher, J.M., Wrighton, S.A., Watkins, P.B. et al. (1999). First-pass midazolam metabolism catalyzed by 1alpha,25-dihydroxy vitamin D3-modified Caco-2 cell monolayers. *The Journal of Pharmacology and Experimental Therapeutics* 289 (2): 1134–1142.

173 Sesardic, D., Boobis, A.R., Murray, B.P. et al. (1990a). Furafylline is a potent and selective inhibitor of cytochrome P450IA2 in man. *British Journal of Clinical Pharmacology* 29 (6): 651–663.

174 Sesardic, D., Edwards, R.J., Davies, D.S. et al. (1990b). High affinity phenacetin O-deethylase is catalysed specifically by cytochrome P450d (P450IA2) in the liver of the rat. *Biochemical Pharmacology* 39 (3): 489–498.

175 Wu, B., Kulkarni, K., Basu, S. et al. (2011). First-pass metabolism via UDP-glucuronosyltransferase: a barrier to oral bioavailability of phenolics. *Journal of Pharmaceutical Sciences* 100 (9): 3655–3681.

176 King, C.D., Rios, G.R., Green, M.D., and Tephly, T.R. (2000). UDP-glucuronosyltransferases. *Current Drug Metabolism* 1 (2): 143–161.

177 Shelby, M.K., Cherrington, N.J., Vansell, N.R., and Klaassen, C.D. (2003). Tissue mRNA expression of the rat UDP-glucuronosyltransferase gene family. *Drug Metabolism and Disposition* 31 (3): 326–333.

178 Heikkinen, A.T., Friedlein, A., Matondo, M. et al. (2014, 2015). Quantitative ADME proteomics - CYP and UGT enzymes in the beagle dog liver and intestine. *Pharmaceutical Research* 32: 74–90.

179 Nishimura, M., Koeda, A., Morikawa, H. et al. (2009). Tissue-specific mRNA expression profiles of drug-metabolizing enzymes and transporters in the cynomolgus monkey. *Drug Metabolism and Pharmacokinetics* 24 (2): 139–144.

180 Furukawa, T., Naritomi, Y., Tetsuka, K. et al. (2014). Species differences in intestinal glucuronidation activities between humans, rats, dogs and monkeys. *Xenobiotica* 44 (3): 205–216.

181 Kitamura, S., Sugihara, K., Nakatani, K. et al. (1999). Variation of hepatic methotrexate 7-hydroxylase activity in animals and humans. *IUBMB Life* 48 (6): 607–611.

182 Beedham, C., Miceli, J.J., and Obach, R.S. (2003). Ziprasidone metabolism, aldehyde oxidase, and clinical implications. *Journal of Clinical Psychopharmacology* 23 (3): 229–232.

183 Clarke, S.E., Harrell, A.W., and Chenery, R.J. (1995). Role of aldehyde oxidase in the in vitro conversion of famciclovir to penciclovir in human liver. *Drug Metabolism and Disposition* 23 (2): 251–254.

184 Jordan, C.G., Rashidi, M.R., Laljee, H. et al. (1999). Aldehyde oxidase-catalysed oxidation of methotrexate in the liver of guinea-pig, rabbit and man. *The Journal of Pharmacy and Pharmacology* 51 (4): 411–418.

185 Rashidi, M.R., Smith, J.A., Clarke, S.E., and Beedham, C. (1997). In vitro oxidation of famciclovir and 6-deoxypenciclovir by aldehyde oxidase from human, guinea pig, rabbit, and rat liver. *Drug Metabolism and Disposition* 25 (7): 805–813.

186 Zientek, M., Jiang, Y., Youdim, K., and Obach, R.S. (2010). In vitro-in vivo correlation for intrinsic clearance for drugs metabolized by human aldehyde oxidase. *Drug Metabolism and Disposition* 38 (8): 1322–1327.

187 Moriwaki, Y., Yamamoto, T., Takahashi, S. et al. (2001). Widespread cellular distribution of aldehyde oxidase in human tissues found by immunohistochemistry staining. *Histology and Histopathology* 16 (3): 745–753.

188 Garattini, E. and Terao, M. (2012). The role of aldehyde oxidase in drug metabolism. *Expert Opinion on Drug Metabolism & Toxicology* 8 (4): 487–503.

189 Garattini, E. and Terao, M. (2013). Aldehyde oxidase and its importance in novel drug discovery: present and future challenges. *Expert Opinion on Drug Discovery* 8 (6): 641–654.

190 Beedham, C., Peet, C.F., Panoutsopoulos, G.I. et al. (1995). Role of aldehyde oxidase in biogenic amine metabolism. *Progress in Brain Research* 106: 345–353.

191 Akabane, T., Tanaka, K., Irie, M. et al. (2011). Case report of extensive metabolism by aldehyde oxidase in humans: pharmacokinetics and metabolite profile

of FK3453 in rats, dogs, and humans. *Xenobiotica* 41 (5): 372–384.

192 Diamond, S., Boer, J., Maduskuie, T.P. Jr., et al. (2010). Species-specific metabolism of SGX523 by aldehyde oxidase and the toxicological implications. *Drug Metabolism and Disposition* 38 (8): 1277–1285.

193 Zhang, X., Liu, H.H., Weller, P. et al. (2011). In silico and in vitro pharmacogenetics: aldehyde oxidase rapidly metabolizes a p38 kinase inhibitor. *The Pharmacogenomics Journal* 11 (1): 15–24.

194 Sauer, J.M., Ring, B.J., and Witcher, J.W. (2005). Clinical pharmacokinetics of atomoxetine. *Clinical Pharmacokinetics* 44 (6): 571–590.

195 Mattiuz, E.L., Ponsler, G.D., Barbuch, R.J. et al. (2003). Disposition and metabolic fate of atomoxetine hydrochloride: pharmacokinetics, metabolism, and excretion in the Fischer 344 rat and beagle dog. *Drug Metabolism and Disposition* 31 (1): 88–97.

196 Lin, J.H., Chiba, M., Balani, S.K. et al. (1996). Species differences in the pharmacokinetics and metabolism of indinavir, a potent human immunodeficiency virus protease inhibitor. *Drug Metabolism and Disposition* 24 (10): 1111–1120.

197 Chiba, M., Hensleigh, M., Nishime, J.A. et al. (1996). Role of cytochrome P450 3A4 in human metabolism of MK-639, a potent human immunodeficiency virus protease inhibitor. *Drug Metabolism and Disposition* 24 (3): 307–314.

198 Yeh, K.C., Stone, J.A., Carides, A.D. et al. (1999). Simultaneous investigation of indinavir nonlinear pharmacokinetics and bioavailability in healthy volunteers using stable isotope labeling technique: study design and model-independent data analysis. *Journal of Pharmaceutical Sciences* 88 (5): 568–573.

# 36

# *In Vivo* Methods for Oral Bioavailability Studies

*Ana Ruiz-Garcia*[1] *and Marival Bermejo*[2]

[1] *Metrum Research Group, San Diego, CA, USA*
[2] *Department of Engineering: Pharmacy and Pharmaceutical Technology Section, School of Pharmacy, Universidad Miguel Hernández de Elche, Elche, Alicante, Spain*

## 36.1 Introduction

Oral administration is the preferred and most convenient method of administration for patients as well as caregivers. It does not require any healthcare support from the administration itself; it can be done anywhere (home, work, restaurant, etc.), which assures compliance of the prescribed treatment. Until the appearance of most biopharmaceuticals, which rarely can be orally absorbed, best-selling medicines were oral (compare Table 36.1A from 2006 with Table 36.1B from 2018). Nevertheless, when considering the number of prescriptions, most of the prescribed drugs are administered by the oral route to treat conditions such as depression, high cholesterol, high blood pressure, and pain/inflammation [1].

When a drug is administered by an extra-vascular route, the amount of drug or active moiety that reaches the systemic circulation is not necessarily the amount of drug present in the dosage form. Due to this fact, it became necessary in drug development to introduce the concept of bioavailability (F).

According to the Food and Drug Administration (FDA), bioavailability is defined as "the rate and extent to which the active ingredient or active moiety is absorbed from a drug product and becomes available at the site of action. For drug products that are not intended to be absorbed into the bloodstream, bioavailability may be assessed by measurements intended to reflect the rate and extent to which the active ingredient or active moiety becomes available at the site of action" [2].

Efficacy and safety are without no doubt relevant aspects for any drug but there cannot be any efficacy if the drug does not reach its target. For any extra-vascular administration route that implies as required condition being bioavailable. For a new promising oral candidate if its plasma concentrations are low at a given dose, due to low oral bioavailability, the solution cannot be as simple as increasing the dose. This approach has many shortcuts from the economic impact associated with the cost of the wasted active molecule, to the potential adverse effects of the unabsorbed drug in the lower parts of the gastrointestinal (GI) site, to the pharmaceutics considerations as the dosage form size in case of bulky drugs that would lead to swallowing difficulties and impairment of patient compliance. Optimizing drug bioavailability in the preclinical stages is a key factor in clinical development success [3].

## 36.2 Factors that Affect Oral Availability

Upon administration, the active ingredient dissolves from the pharmaceutical form and travels across membranes. The rate of dissolution depends on many factors as well as the permeability across membranes [4, 5]. Absorption and thus systemic availability are not always complete for drugs given orally; the reasons for this may be poor solubility, low permeability, instability in luminal fluids, formulation issues, and/or high metabolism (in lumen or after absorption). Factors affecting bioavailability can be classified as those affecting absorbability and oral fraction absorbed, mainly solubility, permeability, and dissolution rate, and those affecting first-pass metabolism. The Biopharmaceutic classification system (BCS) [4] marked a milestone in our understanding of the determinants of oral fraction absorbed (see also Chapter 9). BCS classifies drugs

*Oral Bioavailability and Drug Delivery: From Basics to Advanced Concepts and Applications*, First Edition. Edited by Ming Hu and Xiaoling Li.

**Table 36.1A** Adapted from FORBES: The best selling drugs in America (02.27.06).

| Drug | Generic name | Sales (billion) | Route | Indication |
|---|---|---|---|---|
| Lipitor | Atorvastatin | 8.4 | Oral | High cholesterol |
| Zocor | Simvastatin | 4.4 | Oral | High cholesterol |
| Nexium | Esomeprazole | 4.4 | Oral | Heartburn |
| Prevacid | Lansoprazole | 3.8 | Oral | Heartburn |
| Advair Diskus | fluticasone/salmeterol | 3.6 | Inhaled | Asthma |
| Plavix | Clopidogrel | 3.5 | Oral | Heart disease |
| Zoloft | Sertraline | 3.1 | Oral | Depression |
| Epogen | Epoetin Alfa | 3 | SC | Anemia |
| Procrit | Epoetin Alfa | 3 | SC | Anemia |
| Aranesp | Darbepoetin Alfa | 2.8 | SC | Anemia |
| Enbrel | Etanercept | 2.7 | SC | Rheumatoid arthritis |
| Norvasc | Amlodipine | 2.6 | Oral | High blood pressure |
| Seroquel | Quetiapine | 2.6 | Oral | Schizophrenia |
| Effexor XR | Venlafaxine | 2.6 | Oral | Depression |
| Zyprexa | Olanzapine | 2.5 | Oral | Schizophrenia |
| Singulair | Montelukast | 2.5 | Oral | Asthma and allergies |
| Protonix | Pantoprazole | 2.4 | Oral | Heartburn |
| Risperdal | Risperidone | 2.3 | Oral | Schizophrenia |
| Neulasta | Pegfilgrastim | 2.2 | SC | Chemotherapy side effects |
| Remicade | Infliximab | 2.2 | SC | Rheumatoid arthritis |

**Table 36.1B** Top selling drugs in 2018.

| Drug | Trade name | 2018 sales (million USD) | Administration (route) | Main indication |
|---|---|---|---|---|
| Adalimumab | Humira | 19 936 | Subcutaneous | Rheumatoid arthritis |
| Apixaban | Eliquis | 9872 | Oral | Anticoagulant |
| Lenalidomide | Revlimid | 9685 | Oral | Multiple myeloma |
| Nivolumab | Opdivo | 7570 | Intravenous | Oncology |
| Pembrolizumab | Keytruda | 7171 | Intravenous | Oncology |
| Etanercept | Enbrel | 7126 | Subcutaneous | Rheumatoid arthritis |
| Trastuzumab | Herceptin | 6981 | Subcutaneous and intravenous | Breast cancer |
| Bevacizumab | Avastin | 6847 | Intravenous | Colon cancer |
| Rituximab | Rituxan, MabThera | 6750 | Intravenous | Non-Hodgkin's lymphoma |
| Rivaroxaban | Xarelto | 6589 | Oral | Anticoagulant |
| Aflibercept | Eylea | 6551 | Intravenous and intravitreal | Ophthalmology |
| Infliximab | Remicade | 5908 | Intravenous | Immunosuppressive |
| Pneumococcal conjugate vaccine | Prevnar 13<br>Prevenar 13 | 5802 | Intramuscular | Infectious diseased |
| Ustekinumab | Stelara | 5156 | Subcutaneous | Immunosuppressive |
| Pregabalin | Lyrica | 4970 | Oral | Epilepsy |

according to their permeability P: (High-HP or Low-LP) and solubility S: (HS or LS) to ascertain the limiting steps in their absorption. Briefly, Class I drugs (HP, HS), do not have solubility issues and when formulated in rapidly dissolving immediate-release dosage forms their fraction absorbed is high (higher than 85%). Class II drugs (LS, HP) are limited by solubility and dissolution rate. Class III compounds (HS, LP) are those permeability-limited as their solubility is high and Class IV (LS, LP) have low oral fraction absorbed due to both unfavorable dissolution and permeation.

### 36.2.1 Solubility and Permeability

Low solubility and/or permeability can lead to a low fraction of oral absorption, limiting oral bioavailability due to an insufficient time at the GI tract. The drug is exposed to the entire GI tract for no more than one to two days and to the small intestine for only two to four hours, unless gastric emptying is considerably delayed. If the drug does not dissolve readily or if the drug is incapable of penetrating the epithelial membrane (highly ionized and polar), there may be insufficient time at the absorption site. Not only is the bioavailability low in this case, but it tends to be highly variable. In addition, individual variations in age, sex, activity, genetic phenotype, stress, disease (achlorhydria, malabsorption syndromes), and previous GI surgery can alter and further increase variability in drug bioavailability.

Carrier-mediated systems located in the gut wall or hepatocytes may play a key role in determining exposure. Intestinal absorption is often in times mainly due to passive diffusion, where lipophilicity and molecular weight are the main players that will determine rate and extent of exposure. However, many molecules travel through lipoid membranes by specialized carrier-mediated mechanisms, both in the absorptive and the secretory directions. These carriers have been classified according to their energy consumption; passive or facilitated and active transporters [6]. ATP-binding Cassette (ABC) transporter family, ion and water channels, ion pumps, and the solute carriers (SLC) are the major families involved in the carrier-mediated transport of xenobiotics through lipoidal membranes (see Figure 36.1).

### 36.2.2 Formulation

Most active pharmaceutical ingredients (API) are active at low doses/amounts that make necessary a physical

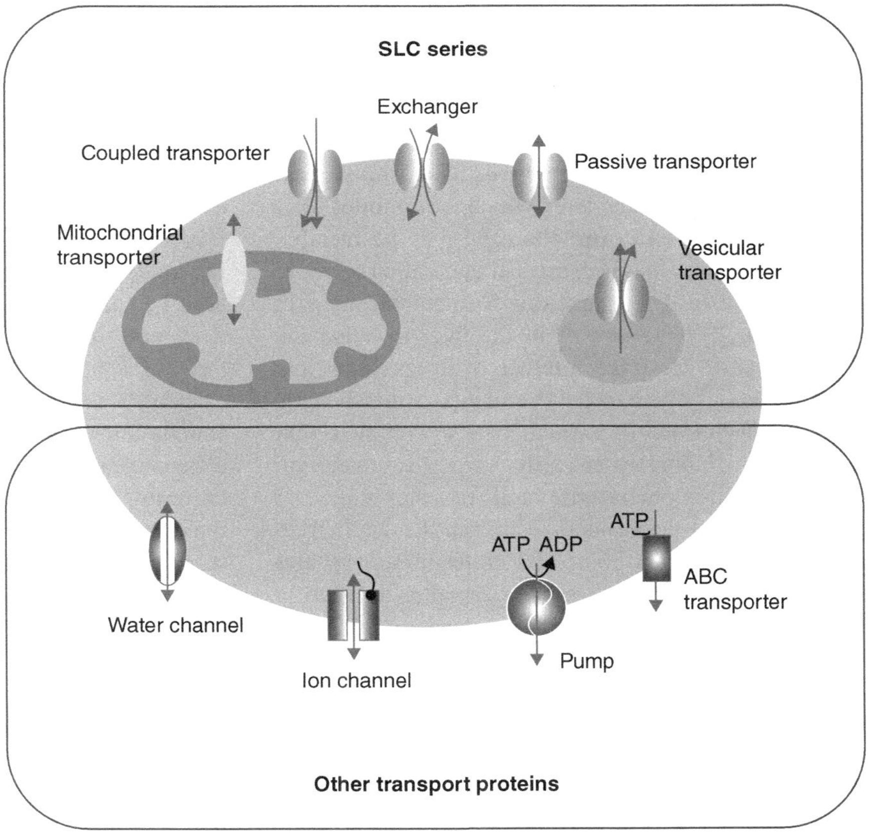


**Figure 36.1** Different carrier-mediated transporters are expressed in the lipoidal membrane or intra-cellular compartment membranes. Source: Hediger et al. [6] / Springer Nature.

vehicle for handling the drug and ensuring compound stability, content uniformity, and ease of administration by a given route. Nevertheless, in the last century, because of several clinical failures [7, 8], it became obvious that pharmaceutical formulations affect oral bioavailability and the physical characteristics of the API, the technological variables and the excipients may affect the mentioned factors in Section 36.2.1, which determine fraction absorbed, and first-pass loses.

Formulation factors that affect bioavailability may be classified into two categories: factors that affect drug dissolution or release from the pharmaceutical form and factors related to excipients or inactive ingredients that may affect physiological processes as luminal degradation (drug stability), membrane permeability, metabolic processes, and GI transit. Excipients effects not only have been extensively reviewed evidencing effects in release rate, permeability, and transit [9, 10], but also in metabolism [11]. In general, those effects are desired and constitute a part of the formulation design as in the so-called "bio-enabling formulations" (lipid-based delivery, amorphous solid dispersions, mesoporous silica, nano-suspensions, and cyclodextrin formulations) [12] devoted to the enhancement of solubility and dissolution rate, while in some cases, the effects are undesired and evidenced after bioequivalence failures [13–15].

### 36.2.3 Metabolism

Upon oral administration, a pharmaceutical active ingredient before reaching bloodstream, must pass through the gut wall and liver, which are common sites of drug metabolism. Thus, the drug may be metabolized before it can be measured in general circulation. The decrease in drug input due to metabolization in the intestinal epithelium or in the liver is called the first-pass effect. A large number of drugs show low bioavailabilities due to extensive first-pass metabolism. In many instances, the extraction is so complete that the bioavailability is virtually zero (isoproterenol, norepinephrine, phenacetin, and testosterone). The concept of first-pass metabolism can be applied to other administration routes as transpulmonary and transdermal as the drug can be metabolized during its passage through the barrier before reaching the systemic circulation.

Thus, oral bioavailability is the product of the fraction of the oral dose entering the GI tract enterocytes ($F_{abs}$), the fraction of $F_{abs}$ that passes through the intestinal wall into the portal vein without being metabolized ($F_{GI}$), and the fraction of unchanged drug that reaches the portal vein and hence, not metabolized by the liver, $F_H$.

$$F_{oral} = F_{abs} \cdot F_{GI} \cdot F_H \tag{36.1}$$

where

$$F_{GI} = 1 - E_{GI} \tag{36.2}$$

$$F_H = 1 - E_H \tag{36.3}$$

Being $E_{GI}$ and $E_H$ the intestinal and hepatic extraction ratios, respectively.

The hepatic extraction ratio can be calculated from intravenous clearance data as follows:

$$E_H = {}^{Cl_{blood}}/_{Q_H} \tag{36.4}$$

where $Cl_{blood}$ is the drug blood clearance—assuming no other routes for drug elimination—(often in times we obtain from the pharmacokinetic analysis plasma or serum clearance, in these cases, the blood/plasma or blood/serum ratio needs to be calculated) and $Q_H$ is the liver blood flow. For high hepatic extraction drugs (i.e. with high intrinsic clearance), $F_H$ could be affected if the unbound drug fraction, intrinsic clearance, or splanchnic blood flow change. For low hepatic extraction drugs (i.e. low intrinsic clearance), $F_H$ is not affected by changes in those factors.

To assess human intestinal wall metabolism, both direct and indirect approaches can be used. As a direct approach, the administration of the drug directly in both, the intestinal lumen and the intestinal wall vein is required. The difference in AUCs between both routes of administration will correspond to the drug fraction extracted by intestinal wall metabolism, $F_{GI}$.

As indirect approach, some assumptions need to be made: Oral bioavailability is the result of the absorbed drug fraction into the intestinal wall that does not undergo intestinal or hepatic metabolism (see Eq. (36.3)). $F_{oral}$ and $E_H$ can be experimentally calculated whereas $E_{GI}$ is indirectly derived. Intravenous administration will allow the calculation of total plasma clearance. Hepatic clearance is assumed to be the result of subtracting from the total clearance value, the renal clearance value. Assuming that the liver is the sole first-pass metabolism site after IV administration $F_H = 1 - E_H$ (see Eqs. (36.2) and (36.4)). $F_{abs}$ needs to be obtained in previous studies usually carried out with radiolabeled drugs [16].

Another common cause of low bioavailability is the presence of competing reactions. Reactions that compete with absorption and can reduce bioavailability include complex formation; hydrolysis by gastric pH or digestive enzymes; conjugation in gut wall;

adsorption to other drugs and metabolism by luminal microflora.

Thus, there is a difference between absorption and bioavailability as a drug can be absorbed almost completely but its bioavailability may be relatively low as a consequence of one or many of the above-mentioned processes that take place until the active ingredient reaches the blood circulation.

Several studies fall under the classification of bioavailability studies including:

- Bioavailability studies to define the rate and extent of absorption relative to a reference dosage form (e.g. intravenous injection, true solution, and suspension).
- Bioequivalence studies comparing pharmaceutical equivalents for the purpose of establishing equivalent extents and/or rates of absorption.
- Dosage strength equivalence studies show that equivalent doses of different dosage forms deliver the same amount of drug (e.g. 10 × 1 mg versus 1 × 10 mg pills).

The conventional way of estimating extravascular bioavailability (BA) is through sequential administration of the systemic (intravenous, IV) and extra-vascular doses with a washout period between administrations. A first administration is given and plasma/serum samples are collected and then the second administration occurs sometime later and plasma/samples are collected. The order of the intravenous and oral doses may be randomized.

By definition, administration by IV gives 100% bioavailability. To mathematically define bioavailability, the Area Under the Curve (AUC) of the drug concentration versus time plot must be measured. The absolute bioavailability of a drug is defined by the AUC of a dose delivered by an extra-vascular route divided by the AUC of an IV bolus. The underlying assumption in this setting is that clearance remains constant between the two dose administrations. However, some drugs are able to modify some of the processes that take place during their absorption, distribution, and elimination stages by induction or inhibition, which will affect the time course of the second dose.

For an IV administration, the amount of drug eliminated at a time equal to infinity is the same that the amount of active substance reaching the systemic circulation (dose administered):

$$\frac{dQ_{el}}{dt} = k_{el} \cdot Q = k_{el} \cdot V \cdot C \quad (36.5)$$

$$dQ_{el} = k_{el} \cdot V \cdot C \cdot dt \quad (36.6)$$

$$\int dQ_{el} = \int k_{el} \cdot V \cdot C \cdot dt \quad (36.7)$$

$$\int dQ_{el} = k_{el} \cdot V \int C \cdot dt \quad (36.8)$$

$$\int_0^\infty dQ_{el} = \mathrm{Cl} \int_0^\infty C \cdot dt \quad (36.9)$$

$$\int_0^\infty dQ_{el} = \mathrm{Dose}_{IV} = \mathrm{Cl} \cdot \mathrm{AUC}_{IV} \quad (36.10)$$

For an extra-vascular administration, $F$ corresponds to the dose fraction that reaches the systemic circulation. When time equals infinity, the active ingredient has been completely eliminated from the body.

$$\frac{dQ_{el}}{dt} = k_{el} \cdot Q = k_{el} \cdot V \cdot C \quad (36.11)$$

$$dQ_{el} = k_{el} \cdot V \cdot C \cdot dt \quad (36.12)$$

$$\int dQ_{el} = \int k_{el} \cdot V \cdot C \cdot dt \quad (36.13)$$

$$\int dQ_{el} = k_{el} \cdot V \int C \cdot dt \quad (36.14)$$

$$\int_0^\infty dQ_{el} = \mathrm{Cl} \int_0^\infty C \cdot dt \quad (36.15)$$

$$\int_0^\infty dQ_{el} = \mathrm{Dose}_{ex} \cdot F = \mathrm{Cl} \cdot \mathrm{AUC}_{ex} \quad (36.16)$$

Hence $F$ is defined as follows:

$$\frac{Dose_{iv}}{Dose_{ex} \cdot F} = \frac{AUC_{iv}}{AUC_{ex}} \quad (36.17)$$

$$F = \frac{Dose_{iv} \cdot AUC_{ex}}{Dose_{ex} \cdot AUC_{iv}} \quad (36.18)$$

Usually, the bioavailability studies are performed by administering the same intra- and extra-vascular dose, where the previous equation is reduced to:

$$F = \frac{AUC_{ex}}{AUC_{iv}} \quad (36.19)$$

If different doses are used for the extra-vascular and IV administrations and the drug has a nonlinear profile, then clearance is not constant and hence it cannot be taken out of the integration, therefore F cannot be estimated using the above formulas.

In many instances, intravenous administration is not feasible for various reasons. In such cases, the relative bioavailability is calculated by comparing the pharmaceutical form to a secondary standard (oral solution):

$$F_{rel} = \frac{AUC_{oral}}{AUC_{oral\,solution(standard)}} \quad (36.20)$$

The assumption here is that clearance is similar for both formulations at a given dose.

Many drugs have a low and variable bioavailability and the reasons are often not very well known. Incomplete bioavailability as mentioned above could be the result of several mechanisms such as first-pass metabolism (GI as well as hepatic) and carrier-mediated system, which their expression and affinity may contribute to high variability.

Bioavailability studies are, in general, performed during drug discovery in several animal species, with the goal of having an approximate idea of what the bioavailability is going to be in humans. For small species, it is very common to perform a cannulation of a major vein prior to drug administration in order to have easy access to the bloodstream and withdraw blood samples frequently enough to build drug concentration–time profiles and hence calculate AUC as a measure of exposure. With large species (including humans) cannulation is not necessary since the main veins are easily accessible for blood withdrawal.

In Section 36.3, some general recommendations for performing these types of studies in animals are provided along with some useful literature and URL resources. Some examples of bioavailability estimations in rats by a permanent jugular vein cannulation technique are presented to illustrate the steps and procedures described in this chapter. Finally, intestinal perfusion studies are described in a human setting and what are the mathematical considerations in order to correctly interpret the data.

## 36.3 *In Vivo* Animal Techniques

### 36.3.1 Animal Use and The Three Rs Policy. Ethical Guidelines

In the last decade, there has been an increase in awareness of the care and well-being of experimental animals, which has directed more attention to the so-called "Three Rs" policy (Reduce, refine, replace). The "Three Rs" originated after a proposal that was made in 1954 by Charles Hume, founder of the Universities Federation for Animal Welfare (UFAW), where the UFAW should undertake a scientific study of humane techniques in laboratory animal experiments. The project was managed by a committee under the chairmanship of Nobel prize winner and immunologist, Sir Peter Medawar. Included among its members was also William Lane-Petter, Secretary of the Research Defence Society of Great Britain. Christine Stevens, founder of the Animal Welfare Institute (AWI) in the United States provided financial support for the project. W.M.S. Russell, a zoologist, and R. L. Burch, a microbiologist, were appointed to carry out the work, which led to the publication of the book *The Principles of Humane Experimental Technique* in 1959 [17, 18]. The "Three Rs" principle of Russell and Burch is based on concepts of refinement of the experimental methods to eliminate or reduce animal pain and distress; reduction in the number of animals consistent with sound experimental design; and the replacement of animal methods with nonanimal methods wherever feasible.

Nowadays, these principles are reflected in regulations and laws. For instance, the European Union Directive 86/609EEC (EU, 1986), which indicates the obligation to refine animal experiments in order to cause the minimum amount of stress. Similar regulations have become mandatory around the world. National and supranational research centers and institutes have been created to promote the "Three R" policy in research. The ECVAM (European Center for the Validation of Alternative Methods https://ec.europa.eu/jrc/en/eurl/ecvam) and the ICCVAM (Interagency Coordinating Committee on the Validation of Alternative Methods-USA https://ntp.niehs.nih.gov/whatwestudy/niceatm/iccvam/index.html) facilitate the validation, review, and adoption of new scientifically-sound testing methods, including alternatives that can refine, reduce, and replace animal use. Their web pages constitute a good resource for standards, protocols, and surgical methods, which ensure that the animals are handled correctly and compassionately, with as little suffering as possible. In Europe, other institutions like the National center for the replacement, refinement, and reduction of animals in research, (https://www.nc3rs.org.uk/) provide useful information for researchers and technicians. Essential updated references can be found on their web pages [19–22].

In the United States, the Animal Welfare Act requires that all procedures involving animals, be reviewed by an Institutional Animal Care and Use Committee (IACUC). The review by an IACUC is required to ensure that the protocols are worthwhile and that there is a minimal number of animals used. It is also required that the investigators document an adequate consideration of alternatives to any procedure that can cause more than momentary pain or distress (either with or without the use of anesthetics). Guidelines in searching for alternative procedures have been prepared to assist investigators and IACUC members in these considerations [23].

## 36.4 Animals Used in Bioavailability Studies

In order to have good predictability in animal models of the human outcome, considering that systemic availability is the composite of intestinal absorption and first-pass metabolism, the animal model selected has to present similar characteristics in both aspects. Ideally, a similar GI anatomy and physiology are desirable or, at least, one that is easy to scale up. A comparable metabolic ability in terms of type of metabolic enzymes and expression levels is the target. Therefore, similarities in intestinal fluid composition, intestinal pH, transit times, para-cellular barrier characteristics profile, and expression level of transporters. The CYP profile expression and levels could, on the other hand, determine differences in first-pass extraction. For instance, rats and humans show similar drug intestinal absorption profiles and similar transporter expression patterns in the small intestine, while the two species exhibit distinct expression levels and patterns for metabolizing enzymes in the intestine (CYP3A4/CYP3A9 and UDPG) of 12 to 193-fold. Therefore, a rat model can be used to estimate the absorption rate and intestinal permeability, but not to predict drug metabolism or oral bioavailability in humans [24].

On the other hand, the animal size will affect several aspects such as the blood volume that can be withdrawn; the difficulty of the surgical procedures, drug administration, animal handling, and husbandry costs. These factors need to be considered in order to select the most adequate animal model.

While small animals such as rats, mice, guinea pigs, and rabbits, are most suitable for determining the mechanism of drug absorption and bioavailability values for powder or solution formulations, larger animals (dogs, minipigs, and monkeys) are used to assess absorption of more complex pharmaceutical formulations. The understanding of the physiological, anatomical, and biochemical differences between the GI tracts of different animal species is necessary to select the correct animal model to mimic the bioavailability of compounds in humans and/or to interpret the data coming from them when some discrepancy arises [25].

The animal species that are commonly used could be divided into two groups according to size/body weight: small animals such as mice, rats, and guinea pigs and large species such as dogs, rabbits, mini-pigs, sheep, and monkeys. Different species respond differently to the same dose per weight (mg/kg) due to varying physiological and biological conditions [26]. Moreover, the metabolic profile changes among species, which affects the PK profile of a given drug; thus, a drug might have a good exposure in one species (e.g. mice) while having poor one in the other (e.g. monkeys). Due to the animal size and their blood volumes, there is a limit as to how much one can dose without significantly affecting the animal's well-being, which will directly influence the maximum exposure in animals (see Table 36.2).

Dogs and primates are often the large species of choice. The intestinal tract of the dog is relatively short and simple compared to other animals [27]. For most nonruminants animals, especially dogs, the intestinal transit time constitutes mainly the time requires for small-intestinal transit as the small intestine represents more than 80% of the total intestinal length (see Table 36.3).

**Table 36.3** Comparative lengths of small intestine.

| Animal | Relative length of small intestine (%) | Average absolute length (m) | Ratio of body length to intestinal length |
|---|---|---|---|
| Dog | 85 | 4.14 | 1 : 6 |
| Cat | 83 | 7.72 | 1 : 4 |
| Sheep | 80 | 26.2 | 1 : 27 |
| Pig | 78 | 18.3 | 1 : 14 |
| Horse | 75 | 22.4 | 1 : 12 |
| Rabbit | 61 | 3.56 | 1 : 10 |

Source: Adapted from Von Richter et al. [16].

**Table 36.2** Animal size, blood volumes, and recommended dosing volumes.

| Animal species | Weight (Kg) | Blood volume (ml) | Blood flor (ml/min) | | Total surface area ($m^2$) | Life-span (year) | Normal dosing volumes (ml/kg) | |
|---|---|---|---|---|---|---|---|---|
| | | | Liver | Kidney | | | Oral | IV |
| Mouse | 0.02 | 1.7 | 1.8 | 1.3 | 0.008 | 2.7 | 10 | 5 |
| Rat | 0.25 | 13.5 | 13.8 | 9.2 | 0.023 | 4.7 | 5–10 | 2.5 |
| Rabbit | 2.5 | 165 | 177 | 80 | 0.17 | 8.0 | 5–10 | 1–2 |
| Rhesus Monkeys | 5 | 367 | 218 | 138 | 0.32 | 22 | 5–8 | 0.5–1 |
| Dog | 10 | 900 | 309 | 216 | 0.51 | 20 | 5–8 | 0.5–1 |

Source: Adapted from Ruiz et al. [15].

Over the past two decades, there has been increased interest in monoclonal antibodies. The most common administration route for large molecules is extra-vascular, most exactly subcutaneously (SC). Several factors have made primates the preclinical species of choice for testing not only toxicity but also pharmacokinetics and hence bioavailabilty [28]. The similarities between humans and primates at the level of their immune system, cross-reactivity, pharmacology, and antibody kinetics are strong arguments to advocate in favor of the use of this controlled species. Although surrogates are often in times prescreened in mice, they are still just one more screening tool that will not prevent the use of primates.

## 36.5 General Considerations for Blood Sampling

The estimation of the AUC value for serum/plasma drug concentration–time profiles can be performed under different sampling designs (e.g. intensive versus sparse or destructive sampling), which will determine the type of statistical analysis required. Pharmacokinetic sampling schedule can be optimized in order to get a better estimation of pharmacokinetic parameters [29].

In general, for more than one sample per animal, a limit of less than 10% of total blood volume TBV (i.e. 2.56 ml in rat) on any single occasion and less than 25% TBV (= 6.4 ml) in 28 days is recommended. For repeat bleeds at short intervals, suggested limit <1% TBV (= 0.25 ml) in 24 hours—cannulation should be considered for small species (see Tables 36.4–36.6).

Jugular vein and carotid artery cannulations are the most widely used surgeries in research labs around the world. The main goals are to confirm intravenous dosage of the test article that is in study and allow either vein or arterial blood collection. Cannulation considerably reduces the stress of multiple sampling observed in tail vein or orbital sinus techniques. Several authors have described in detail these worldwide surgery techniques listed in Table 36.5 [30–37].

In summary, cannulation should be an option to consider when repeated samples are required as it avoids multiple needle entries at any one site. Since surgery is required, the appropriate anesthesia and analgesia should be provided to minimize any caused pain. The inserted catheter is exteriorized at the nape of the neck or the base of the tail. Post-surgery time should be allowed prior to drug dosage and blood sampling.

**Table 36.4** Limit volumes and recovery periods.

| Single sampling | | Multiple sampling | |
|---|---|---|---|
| % Circulatory Blood Volume Removed | Approximate recovery period | % Circulatory Blood Volume Removed in 24 hours | Approximate recovery period |
| 7.5% | 1 week | 7.5% | 1 week |
| 10% | 2 weeks | 10–15% | 2 weeks |
| 15% | 4 weeks | 20% | 3 weeks |

Source: Adapted from Karl-Heinz et al. [21] / John Wiley & Sons.

**Table 36.5** Total blood volumes and recommended maximum blood sample volumes for species of given body weights.

| Species (weight) | Blood volume (ml) | 7.5% | 10% | 15% | 20% |
|---|---|---|---|---|---|
| Mouse (25 g) | 1.8 | 0.1 | 0.2 | 0.3 | 0.4 |
| Rat (250 g) | 16 | 1.2 | 1.6 | 2.4 | 3.2 |
| Rabbit (4 kg) | 224 | 17 | 22 | 34 | 45 |
| Dog (10 kg) | 850 | 64 | 85 | 127 | 170 |
| Macaque (Rhesus) (5 kg) | 280 | 21 | 28 | 42 | 56 |
| Macaque (Cynomolgus) (5 kg) | 325 | 24 | 32 | 49 | 65 |
| Marmoset (350 g) | 25 | 2.0 | 2.5 | 3.5 | 5 |
| Minipig (15 kg) | 975 | 73 | 98 | 146 | 195 |

Source: Adapted from Karl-Heinz et al. [21] / John Wiley & Sons.

## 36.6 Statistical Considerations for Data Handling. (AUC Calculations in Sparse Sampling Designs)

Estimation of AUC and its variance in complete sampling design is simple, but it is not straightforward in serial sacrifice or sparse (destructive) sampling designs (where each subject is sampled once or more than once but an incomplete concentration–time profile is obtained).

The AUC variance estimation is made in different ways, depending on the available data for the AUC calculation. Thus, for rich designs in which the complete plasma level curves can be obtained from each individual, the individual AUC is calculated by compartmental or non-compartmental methods, and then the population AUC is calculated as the average of all individual values. In this case, it is simple to calculate the AUC variance. Nevertheless, there are situations in which the complete design is not possible because of

**Table 36.6** Table 36.5 Summary of advantages and disadvantages of the various methods of blood sampling.

| Route/vein | General anesthesia | Tissue damage | Repeat bleeds | Volume | Species |
|---|---|---|---|---|---|
| Jugular | No | Low | Yes | +++ | Rat, Dog, Rabbit |
| Cephalic | No | Low | Yes | +++ | Macaque, Dog |
| Saphenous/lateral tarsal | No | Low | Yes | ++(+) | Mouse/Rat, Marmoset/Macaque, Dog |
| Marginal ear | No (local) | Low | Yes | +++ | Rabbit, Minipig |
| Femoral | No | Low | Yes | +++ | Marmoset/Macaque |
| Sublingual | Yes | Low | Yes | +++ | Rat |
| Lateral tail | No | Low | Yes | +++(+) | Rat, mouse, Marmoset |
| Central ear artery | No (local) | Low | Yes | +++ | Rabbit |
| Cranial vena cava | No | Low | Yes | +++ | Minipig |
| Tail tip amputation (<1–3 mm) | Yes | Mod | Limited | + | Mouse/Rat |
| Retrobulbar plexus | Yes | Mod/High | Yes | +++ | Mouse/Rat |
| Cardiac[a)] | Yes | Mod | No | +++ | Mouse/Rat/Rabbit |

a) Only carried out as a terminal procedure under general anesthesia.
Adapted from Karl-Heinz et al. [21] / John Wiley & Sons.

ethical, experimental o simply economical reasons. In these cases, the incomplete design could be:

***'Serial sacrifice' design***: only one sample per subject is available.

***'Batch' design***: more than one sample (but not a complete profile) is taken from each subject. Thus, the design consists of different batches where the samples are taken at the same times in all subjects in the same batch. There are no common sample times between batches. Therefore, a design is obtained by excluding batches that cover the complete time profile. In these situations, it is not possible to calculate individual AUC, and the AUC variance calculation is more complicated. There are different methods for estimating the AUC and its variability such us Bailer's approach [38, 39], Bailer model [40], Bootstrap [41, 42], Jackknife method [29], Bayesian estimation [43], partial areas under the curve [44], and Satterthwaithe's approximation [45].

## 36.7 Practical Examples in Rat Model

The terms catheter/catheterization and cannula/cannulation are used interchangeably, although the term catheter connotes flexible tubing whereas the term cannula connotes metal tubing. Several techniques for cannulation of the rat jugular vein are described in the literature. Thrivikraman et al. [30], and Burvin et al. [31], are useful references in particular because of their schemes and photos of the different stages during the procedure. The crossover design for the oral and iv administrations seems to be inadequate [46] as the cannulation produces hemodynamic changes that alter the clearance.

When adapting any of the previous procedures for experimental work the basic steps of the protocol that the researcher has to standardize are the following:

- Selection of the catheter-cannula material, sterilization, and preparation before surgery—different materials may change the degree of biocompatibility and thus, determine the length of the experiment depending on which material is used.
- Preparation of the rat for the surgery i.e. selection of the anesthesia mixture.
- Isolation of the vein/artery by blunt dissection.
- The length of the catheter has to be adapted to the animal type-size-weight.
- Volume of drug formulation, administration, and administration rate (i.e. iv bolus versus iv infusion).
- Volume and timing of the samples (see Sections 36.5 and 36.6).

A particular example of these aspects and some pharmacokinetic studies with this technique are listed below.

### 36.7.1 Jugular Vein Cannulation

Depending on the body weight of the rats, the length of the medical grade silicon tubing introduced varies (for rats weighing between 250 and 360 g, a length of

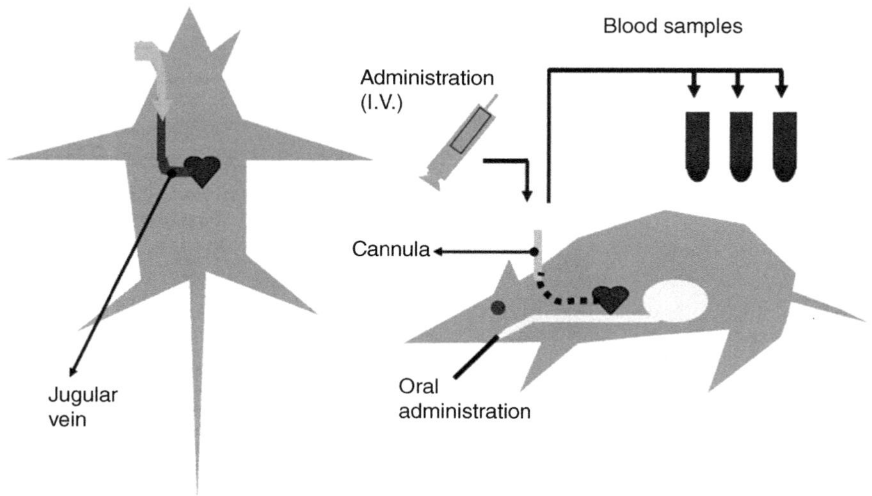


**Figure 36.2** Cartoon representing a ported animal dosed and undergoing sequential blood withdraws.

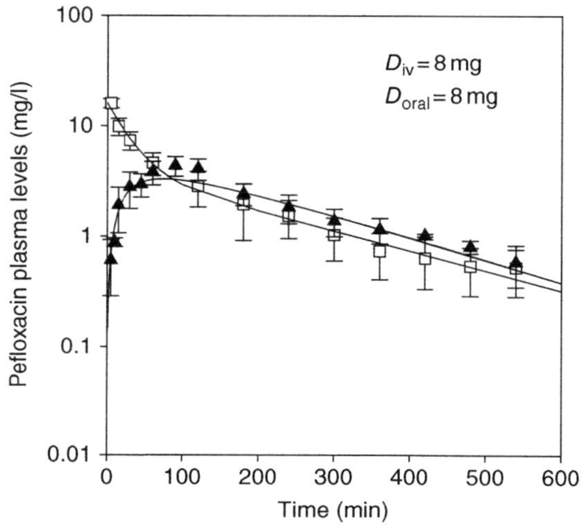


**Figure 36.3** Example of average plasma time profiles in rat obtained after oral and iv administration of Pefloxacin to a group of eight rats. The iv administration and the blood sampling from both routes were performed after the cannulation of the jugular vein of the rat. Source: Sanchez-castaño et al. [47] / with permission of Elsevier.

3.2–3.4 cm, internal diameter 0.5 mm; outer diameter 0.94 mm). The cannula is introduced into the jugular vein toward the heart and externalized at the neck under an anesthesia mixture. The catheter is fixed with ligatures at the muscles near the insertion point. The catheter remains filled with heparinized normal saline solution and it is occluded to prevent backflow. After recovery (overnight), an additional silicon tube is connected to the free end of the neck cannula in order to allow, if necessary, drug solution administration and blood sampling. Figure 36.2 summarizes the step process for blood collection in a ported animal after drug administration (see the results in Figure 36.3).

## 36.8 Intestinal Perfusion (see also Chapter 42)

As oral fraction absorbed is one of the main determinants of oral bioavailability (see Eq. (36.3)), predicting $F_{abs}$ is a key step in drug candidate selection for oral route. As $F_{abs}$ direct measurement requires mass balance studies with radiolabeled drug, which add a considerable burden not always achievable, there was in the past decades a lot of efforts to develop alternative experimental methods to estimate indirectly $F_{abs}$. In the absence of dissolution limitations, i.e. with drug in solution, intestinal permeability is the main determinant of $F_{abs}$. The experimental demonstration of this concept was achieved; thanks to a technique of intestinal perfusion in man [4, 48].

Single-pass intestinal perfusion technique (SPIP) [16] is been used to study intestinal drug permeability. This technique is widely used due to its proximity to the in vivo conditions, lower sensitivity to pH variations, maintenance of blood flow to the intestine, and good correlation with human absorption data [49–53].

This technique allows quantitative and mechanistic evaluation of drug absorption and first-pass-liver disposition during the absorption phase [16, 51]. Briefly, the perfusion tube, orally introduced under fluoroscopic guidance, consists of a multi-channel tube with two inflatable balloons placed at a given distance apart isolating a portion of the small intestine [48, 54] (see Figure 36.4).

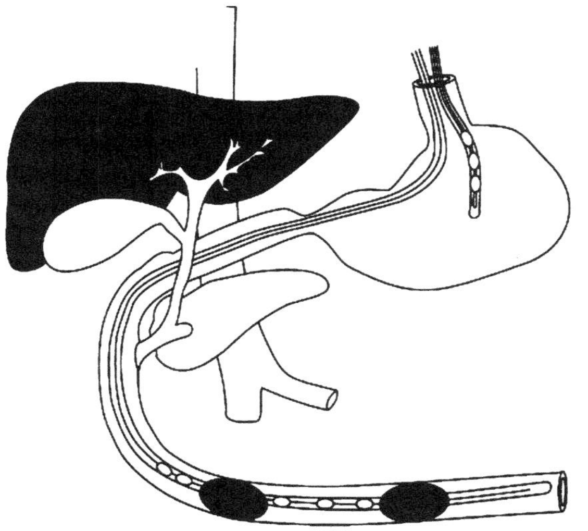

**Figure 36.4** Scheme of the intestinal perfusion technique. Source: Adapted from Tannergren et al. [39] / with permission of Elsevier.

A drug solution and a nonabsorbable marker are infused into the intestinal segment. The nonabsorbable marker is used to correct changes in the outlet drug concentration due to fluid absorption and/or secretion. Polyethylene glycol 4000 has been widely used as reference marker, especially radioactive polyethylene glycol[$^{14}$C]-PEG [55–58]. Calculating the recovery of the marker is important because it allows us to verify that there was no obstruction of the portion of the intestine isolated during the study and that none of the markers was lost or absorbed.

After reaching a steady state (30–60 minutes) [45, 59], collections of the perfusate leaving the intestinal segment are obtained. The difference between the inlet and outlet drug amount is the assumed drug fraction absorbed.

Despite the relevant information that human perfusion studies can provide, it is clear that it is not a good screening method in the preclinical setting, thus consequently several perfusion methods in other species such as rats were developed and validated against human intestinal permeability data. Among those methods, single-pass intestinal perfusion (SPIP) in rat and closed-loop perfusion method have shown comparable results and good predictability of human permeability and human oral fraction absorbed [60, 61].

In closed-loop perfusion method, the intestinal segment is isolated with the aid of two syringes (proximal and distal). An adequate volume (adapted to the segment length) of the drug solution is introduced in the segment and syringes are used alternatively to get a sample of the luminal content. Thus, absorption is characterized also from drug luminal disappearance.

## 36.9 Mathematical Considerations

In SPIP in man, sample collection is performed for short time intervals (10–20 minutes). The net water flux (NWF) in the isolated segment will be calculated as follows:

$$NWF = \left(1 - \frac{RM_{out}}{RM_{in}}\right) \cdot \frac{Q_{in}}{L} \tag{36.21}$$

where $Q_{\text{in}}$ is the flow rate of the perfusate, $L$ is the length of the isolated intestinal segment, and $\text{RM}_{\text{out}}$ and $\text{RM}_{\text{in}}$ are the reference marker concentrations entering and leaving the isolated intestinal segment, respectively.

Prior to any calculation of the amount of drug absorbed, the drug concentration in the collected perfusate has to be corrected for water flux using Eq. (36.21), then the fraction of drug being absorbed, $f_{\text{abs,}}$ and the effective permeability, $P_{\text{eff}}$ can be calculated (see Eqs. (36.22) and (36.23)).

$$f_{abs} = 1 - \left(\frac{C_{out} \cdot RM_{in}}{C_{in} \cdot RM_{out}}\right) \tag{36.22}$$

$$P_{eff} = \frac{C_{in} - C_{out}}{C_{out}} \cdot \frac{Q_{in}}{2\pi rL} \tag{36.23}$$

where $C_{\text{in}}$ and $C_{\text{out}}$ are the drug concentrations entering and leaving the intestinal segment, respectively. The surface of the intestinal segment corresponds to a rectangle of length equals the length of the intestinal segment, $L$ and width is the diameter of the intestinal cylinder, hence $r$ corresponds to the intestinal radius.

In rats, the experimental setting is slightly different in the SPIP method versus man. The intestinal segment in the rat is isolated and the fluid is continuously perfused and collected at the end of the segment. Drug inlet and outlet concentrations are also measured and corrected with the aid of a nonabsorbable marker. There are several mass transport methods to estimate the rat permeability, but the simplest one is the parallel tube method in which the $P_{\text{eff}}$ is estimated as [62].

$$P_{\text{eff}} - Q \cdot \frac{\text{Ln}\left({}^{C_{\text{out}}}/{}_{C_{\text{in}}}\right)}{2\pi rL} \tag{36.24}$$

where the meaning of the terms is the same as previously defined.

In the closed loop method, absorption rate coefficient ($k_{\text{a}}$) is estimated from the concentration profile decrease using a first-order equation [60]:

$$C_t = C_0 \cdot e^{(-ka \cdot t)} \tag{36.25}$$

where $C_t$ is the drug concentration at time $t$, $C_0$ is the intercept at time = 0.

Absorption rate coefficients retransformed into permeability values as:

$$P_{\text{eff}} = \frac{ka \cdot r}{2} \tag{36.26}$$

For the measurement of the intestinal radius, several approaches can be taken [52]. A simplistic approach could be that knowing the length of the intestinal segment and the volume of fluid that inside the segment producing the intestinal distension ($V$), $r$ can be calculated as follows:

$$V = \pi \cdot r^2 \cdot L \tag{36.27}$$

$$r = \sqrt{V/\pi \cdot L} \tag{36.28}$$

## References


1 Hales, C.M., Servais, J., Martin, C.B., and Kohen, D. (2019). NCHS Data Brief No. 347 August 2019. Prescription Drug Use Among Adults Aged 40–79 in the United States and Canada. https://www.cdc.gov/nchs/data/databriefs/db347-h.pdf.

2 Bioavailability and Bioequivalence Studies Submitted in NDAs or INDs — General Considerations. Authors: U.S. Department of Health and Human Services Food and Drug Administration Center for Drug Evaluation and Research (CDER) March 2014 Biopharmaceutics.

3 Ruiz-Garcia, A., Bermejo, M., Moss, A., and Casabó, V.G. (2008). Pharmacokinetics in drug discovery. *J. Pharm. Sci.* 97 (2): 654–690. https://doi.org/10.1002/jps.21009.

4 Amidon, G.L., Lennernas, H., Shah, V.P., and Crison, J.R. (1995). A theoretical basis for a biopharmaceutic drug classification: the correlation of in vitro drug product dissolution and in vivo bioavailability. *Pharm. Res.* 12: 413–420.

5 Waiver of In Vivo Bioavailability and Bioequivalence Studies for Inmediate-Release Solid Oral Dosages Forms Based on a Biopharmaceutics Classification System. U.S. Department of Health and Human Services Food and Drug Administration Center for Drug Evaluation and Research (CDER) August 2000.

6 Hediger, M.A., Romero, M.F., Peng, J.B. et al. (2004). The ABCs of solute carriers: physiological, pathological and therapeutic implications of human membrane transport proteins: introduction. *Pflugers Arch.* 447 (5): 465–468.

7 Greenblatt, D.J., Smith, T.W., and Koch-Weser, J. (1976). Bioavailability of drugs: the digoxin dilemma. *Clin. Pharmacokinet.* 1 (1): 36–51. https://doi.org/10.2165/00003088-197601010-00004.

8 Tyrer, J.H., Eadie, M.J., Sutherland, J.M., and Hooper, W.D. (1970). Outbreak of anticonvulsant intoxication in an Australian city. *Br. Med. J.* 4 (5730): 271–273. https://doi.org/10.1136/bmj.4.5730.271.

9 Flanagan, T. (2019). Potential for pharmaceutical excipients to impact absorption: a mechanistic review for BCS Class 1 and 3 drugs. *Eur. J. Pharm. Biopharm.* 141: 130–138. https://doi.org/10.1016/j.ejpb.2019.05.020. Epub 2019 May 22.

10 Ruiz-Picazo, A., Lozoya-Agullo, I., González-Álvarez, I. et al. (2020). Effect of excipients on oral absorption process according to the different gastrointestinal segments. *Expert Opin. Drug Delivery* 17: 1–20. https://doi.org/10.1080/17425247.2020.1813108.

11 Patel, R., Barker, J., and ElShaer, A. (2020). Pharmaceutical excipients and drug metabolism: a mini-review. *Int. J. Mol. Sci.* 21 (21): 8224. https://doi.org/10.3390/ijms21218224.

12 Ditzinger, F., Price, D.J., Ilie, A. et al. (2019). Lipophilicity and hydrophobicity considerations in bio-enabling oral formulations approaches - a PEARRL review. *J. Pharm. Pharmacol.* 71 (4): 464–482. https://doi.org/10.1111/jphp.12984. Epub 2018 Aug 2.

13 Bermejo, M., Kuminek, G., Al-Gousous, J. et al. (2019). Exploring bioequivalence of dexketoprofen trometamol drug products with the gastrointestinal simulator (GIS) and precipitation pathways analyses. *Pharmaceutics* 11 (3): 122. https://doi.org/10.3390/pharmaceutics11030122.

14 Colón-Useche, S., González-Álvarez, I., Mangas-Sanjuan, V. et al. (2015). Investigating the discriminatory power of BCS-biowaiver in vitro methodology to detect bioavailability differences between immediate release products containing a class I drug. *Mol. Pharmaceutics* 12 (9): 3167–3174. https://doi.org/10.1021/acs.molpharmaceut.5b00076.

15 Ruiz-Picazo, A., Colón-Useche, S., Perez-Amorós, B. et al. (2019). Investigation to explain bioequivalence failure in pravastatin immediate-release products. *Pharmaceutics* 11 (12): 663. https://doi.org/10.3390/pharmaceutics11120663.

16 Von Richter, O., Greiner, B., Fromm, M.F. et al. (2001). Determination of in vivo absorption,

metabolism, and transport of drugs by the human intestinal wall and liver with a novel perfusion technique. *Pharm. Drug Dispos.* 70 (3): 217–227.

17 Russell, W.M.S. and Burch, R.L. (1959). *The Principles of Humane Experimental Technique*. London: Methuen.

18 Zurlo, J., Rudacille, D., and Alan, M. (1996). Goldberg the three rs: the way forward. *Environ. Health Perspect.* 104 (8): 878–880.

19 Remie, R. (2000). Experimental surgery, Chapter 27. In: *The Laboratory Rat* (ed. G.J. Krinke), 523–568. San Diego, CA: Academic Press.

20 NRC (1996). *Guide for the Care and Use of Laboratory Animals*. Washington, DC: National Academy Press.

21 (2001). EFPIA/ECVAM paper on good practice in administration of substances and removal of blood. *J. Appl. Toxicol.* 21: 15–23.

22 (1993). Removal of blood from laboratory mammals and birds. First report of the BVA/FRAME/RSPCA/UFAW joint working group on refinement. *Lab. Anim.* 27: 1–22.

23 Stokes, W.S. and Jensen, D.J.B. (1995). Guidelines for institutional animal care and use committees: consideration of alternatives. *Contemp. Top. Lab. Anim. Sci.* 34 (3): 51–60.

24 Cao, X., Gibbs, S.T., Fang, L. et al. (2006). Why is it challenging to predict intestinal drug absorption and oral bioavailability in human using rat model. *Pharm. Res.* 23 (8): 1675–1686.

25 Kararli, T.T. (1995). Comparison of the gastrointestinal anatomy, physiology, and biochemistry of humans and commonly used laboratory animals. *Biopharm. Drug Dispos.* 16 (5): 351–380.

26 Li, P. and Zhao, L. (2007 Aug 16). Developing early formulations: practice and perspective. *J. Pharm.* 341 (1–2): 1–19.

27 Sabnis, S. (1999). Factors influencing the bioavailability of peroral formulations of drugs for dogs. *Vet. Res. Commun.* 23: 425–447.

28 Chapman, K., Pullen, N., Graham, M., and Ragan, I. (2007). Preclinical safety testing of monoclonal antibodies: the significance of species relevance. *Nat. Rev. Drug Discovery* 6: 120–126.

29 Aarons L, Ogungbenro K. Optimal design of pharmacokinetic studies. *Basic Clin. Pharmacol. Toxicol.* 2010;106(3):250-5. doi: https://doi.org/10.1111/j.1742-7843.2009.00533.x.

30 Thrivikraman, K.V., Huot, R.L., and Plotsky, P.M. (2002). Jugular vein catheterization for repeated blood sampling in the unrestrained conscious rat. *Brain Res. Protoc.* 10 (2): 84–94.

31 Ram Burvin, A., Moshe Zloczower, A., and Eddy Karnieli, A. (1998). Double-vein jugular/inferior vena cava clamp technique for long-term in vivo studies in rats. *Physiol. Behav.* 63 (4): 511–515.

32 van Herck, H., Baumans, V., Brandt, C.J. et al. (2001). Blood sampling from the retro-orbital plexus, the saphenous vein and the tail vein in rats: comparative effects on selected behavioural and blood variables. *Lab. Anim.* 35 (2): 131–139.

33 Luzzi, M., Skoumbourdis, E., Baumans, V. et al. (2005). Collecting blood from rodents: a discussion by the Laboratory Animal Refinement and Enrichment Forum. *Anim. Technol. Welfare* 4 (2): 99–102.

34 Fluttert, M., Dalm, S., and Oitzl, M.S. (2000). A refined method for sequential blood sampling by tail incision in rats. *Lab. Anim.* 34 (4): 372–378.

35 Toft, M.F., Petersen, M.H., Dragstead, N., and Hansen, A.K. (2005). The impact of different blood sampling methods on laboratory rats under different types of anaesthesia. *Lab. Anim.* 40: 261–274.

36 *Rat Jugular Vain and Carotid Artery Catheterization for Acute Survival Studies: A Practical Guide (Paperback)* by Angela Heiser, J.H.K. Liu Springer; 1 (2007).

37 *Experimental and Surgical Techniques in the Rat*, 2 by H. B. Waynforth, Paul A. Flecknell Academic Press (1992).

38 Yuan, J. (1993). Estimation of variance for AUC in animal studies. *J. Pharm. Sci.* 82: 761–763.

39 Gagnon, R.C. and Peterson, J.J. (1998). Estimation of confidence intervals for area under the curve from destructively obtained pharmacokinetic data. *J. Pharmacokinet. Biopharm.* 26: 87–102.

40 Nedelman, J.R., Gibiansky, E., and Lau, D.T. (1995). Applying Bailer's method for AUC confidence intervals to sparse sampling. *Pharm. Res.* 12: 124–128.

41 Bonate, P.L. (1998). Coverage and precision of confidence intervals for area under the curve using parametric and non-parametric methods in a toxicokinetic experimental design. *Pharm. Res.* 15: 405–410.

42 Pai, S.M., Fettner, S.H., Hajian, G. et al. (1996). Characterization of AUCs from sparsely sampled populations in toxicology studies. *Pharm. Res.* 13: 1283–1290.

43 Magni, P., Bellazzi, R., De Nicolao, G. et al. (2002). Nonparametric AUC estimation in population studies with incomplete sampling: a Bayesian approach. *J. Pharmacokinet. Pharmacodyn.* 29: 445–471.

44 Holder, D.J., Hsuan, F., Dixit, R., and Soper, K. (1999). A method for estimating and testing area under the curve in serial sacrifice, batch, and complete data designs. *J. Biopharm. Stat.* 9: 451–464.

45 Nedelman, J.R. and Jia, X. (1998). An extension of Satterthwaite's approximation applied to pharmacokinetics. *J. Biopharm. Stat.* 8: 317–328.

46 Torres-Molina, F., Aristorena, J.C., Garcia-Carbonell, C. et al. (1992). Influence of permanent cannulation of the jugular vein on pharmacokinetics of amoxycillin and antipyrine in the rat. *Pharm. Res.* 9 (12): 1587–1591.

47 Sanchez-Castaño, G., Ruiz-Garcia, A., Banon, N. et al. (2000). Intrinsic absolute bioavailability prediction in rats based on in situ absorption rate constants and/or in vitro partition coefficients: 6-fluoroquinolones. *J. Pharm. Sci.* 89 (11): 1395–1403.

48 Lennernäs H, Ahrenstedt O, Hällgren R, Knutson L, Ryde M, Paalzow LK (1992). Regional jejunal perfusion, a new in vivo approach to study oral drug absorption in man. *Pharm. Res.* 9 (10): 1243–51. doi: https://doi.org/10.1023/a:1015888813741.

49 Lennernäs, H., Ahrenstedt, O., and Ungell, A.L. (1994). Intestinal drug absorption during induced net water absorption in man; a mechanistic study using antipyrine, atenolol and enalaprilat. *Br. J. Clin. Pharmacol.* 37 (6): 589–596.

50 Fagerholm, U., Johansson, M., and Lennernäs, H. (1996). Comparison between permeability coefficients in rat and human jejunum. *Pharm. Res.* 13 (9): 1336–1342.

51 Lennernas, H., Knutson, L., Knutson, T. et al. (2002). The effect of amiloride on the in vivo effective permeability of amoxicillin in human jejunum: experience from a regional perfusion technique. *Eur. J. Pharm. Sci.* 15: 271–277.

52 Lennernäs, H., Nylander, S., and Ungell, A.L. (1997). Jejunal permeability: a comparison between the using chamber technique and the single-pass perfusion in humans. *Pharm. Res.* 14 (5): 667–671.

53 Tannergreen, C., Petri, N., Knutson, L. et al. (2003). Multiple transport mechanisms involved in the intestinal absorption and first-pass extraction of fexofenadine. *Clin. Pharmacol. Ther.* 423–436.

54 Knutson, L., Odlind, B., and Hallgren, R. (1989). A new technique for segmental jejunal perfusion in man. *Am. J. Grastroenterol.* 84: 1278–1284.

55 Jacobson, E.D., Bondy, D.C., Broitman, S.A., and Fordtran, J.S. (1963). Validity of polyethylene glycol in estimating intestinal water volume. *Gastroenterology* 44: 761–767.

56 Soergel, K.H. and Hogan, W.J. (1967). On the suitability of poorly absorbed markers as dilution indicators in the gastrointestinal tract. *Gastroenterology* 52: 1056–1057.

57 Schiller, L.R., Santa Ana, C.A., Porter, J., and Fordtran, J.S. (1997). Validation of polyethylene glycol 3350 as a poorly absorbable marker for intestinal perfusion studies. *Dig. Dis. Sci.* 42: 1–5.

58 Mourad, F.H. (2004). Animal and human models for studying effects of drugs on intestinal fluid transport in vivo. *J. Pharmacol. Toxicol. Methods* 50: 3–12.

59 Modigliani, R., Rambaud, J.C., and Bernier, J.J. (1973). The method of intraluminal perfusion of the human small intestine: I Principle and technique. *Digestion* 9: 176–192.

60 Lozoya-Agullo I, Zur M, Beig A, Fine N, Cohen Y, González-Álvarez M, Merino-Sanjuán M, González-Álvarez I, Bermejo M, Dahan A (2016). Segmental-dependent permeability throughout the small intestine following oral drug administration: Single-pass vs. Doluisio approach to in-situ rat perfusion. *Int. J. Pharm.* 515 (1–2): 201–208. doi: https://doi.org/10.1016/j.ijpharm.2016.09.061

61 Lozoya-Agullo I, Zur M, Wolk O, Beig A, González-Álvarez I, González-Álvarez M, Merino-Sanjuán M, Bermejo M, Dahan A (2015). In-situ intestinal rat perfusions for human Fabs prediction and BCS permeability class determination: Investigation of the single-pass vs. the Doluisio experimental approaches. *Int. J. Pharm.* 480 (1–2): 1–7. doi: https://doi.org/10.1016/j.ijpharm.2015.01.014

62 Amidon, G.L., Kou, J., Elliott, R.L., and Lightfoot, E.N. (1980). Analysis of models for determining intestinal wall permeabilities. *J. Pharm. Sci.* 69: 1369–1373.

# 37

# Caco-2 Cell Culture Model for Oral Drug Absorption

*Kaustubh Kulkarni*[1], *Lu Wang*[2], *and Ming Hu*[2]

[1] *Novartis Institutes for BioMedical Research, Cambridge, MA, USA*
[2] *Department of Pharmacological and Pharmaceutical Sciences, College of Pharmacy, University of Houston, Houston, TX, USA*

## 37.1 Introduction

For oral drug absorption, intestine is the organ where the action occurs in humans and animals. The small intestine is the longest section of the digestive tract and possesses vast areas available for drug absorption and possible metabolism. Small intestine is divided into three segments duodenum, jejunum, and ileum. Ileum, the end part of small intestine, forms a passage to the large intestine or colon. For the past 20 years, the Caco-2 monolayer cell culture model (Figure 37.1) is becoming a routine, yet, the most important model to study intestinal absorption and metabolism of drugs *in vitro* in both pharmaceutical industries and academia. This chapter will describe details about what Caco-2 cell culture model is and why it is used to test the oral absorption and metabolism of different drug candidates, nutrients, and other commonly orally ingested chemicals.

The chapter will be divided into areas that cover its brief history, pharmaceutical applications, and recent progress of the Caco-2 cell culture model. We will begin by describing the anatomical structure of human intestine in relation to drug metabolism and transport, and along the way, we will point out some of its similarities to the Caco-2 cell culture model. Then we will discuss the different mechanisms of transport involved in drug absorption and metabolite excretion along with the rationale for the use of Caco-2 cell culture model to study them. Later, we will discuss different types of Caco-2 cells and how to grow them in laboratory environment while dealing with some practical challenges. We will also discuss the traditional utilities of Caco-2 cell model along with its limitations and correlation of the data obtained to other models. Furthermore, we will talk about some recent progress (siRNA, gene overexpression technique, and intestinal chips) that contributes to the increased flexibility and utility of this model to determine the transporter and metabolic enzymes involved. These recent approaches give us more mechanistic insight into the drug metabolism and disposition without the use of humans/animals. At the end of the book chapter, we will discuss the result analysis in the Caco-2 cell culture model using an example of apical to basolateral transport of model compound naringenin.

### 37.1.1 Intestinal Anatomic Structures and Their Relationship to Caco-2 Cells

The Caco-2 cells are human colon adeno-carcinoma cells that differentiate spontaneously, forming a tight monolayer when they confluence. To better understand the Caco-2 cell culture model and its imperative use, we will first describe the anatomy of the human intestinal epithelial cells and then understand possible underlying mechanisms of drug absorption in the intestinal cells.

#### 37.1.1.1 Anatomy of the Intestinal Cell

In rodents and humans, the small intestine is approximately 3 to 4 times longer than their respective body lengths. To give an example, small intestine of a human is about 6 m in length. Now, if the small intestine is viewed as a simple pipe, its luminal surface area would be of the order of one-half of a square meter. But in reality, the absorptive surface area of small intestine is much larger, almost the size of a tennis court, which is about 100 $m^2$. Although the morphology of the small intestine is similar to the rest of the digestive tract, to account for this huge absorptive area, we need to consider the three important features of small intestine, which are mucosal folds, villi, and microvilli.

*Oral Bioavailability and Drug Delivery: From Basics to Advanced Concepts and Applications,* First Edition. Edited by Ming Hu and Xiaoling Li.

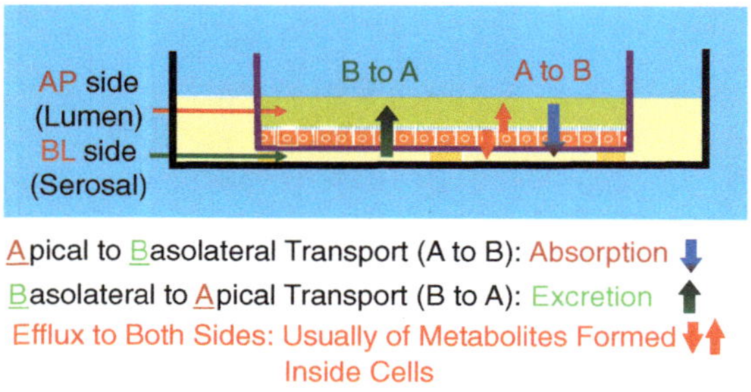


**Figure 37.1** A schematic of the Caco-2 cell culture model.

#### 37.1.1.2 Mucosal Folds

The inner surface of the small intestine is not flat but thrown into circular folds. This increases the surface area for absorption of the drug while mixing intestinal contents by acting as baffles.

#### 37.1.1.3 Villi

The mucus membrane, also called mucosa, consists of large amounts of projections, which actually protrude into the lumen and are lined with epithelial cells.

#### 37.1.1.4 Microvilli

The luminal plasma membrane of absorptive epithelial cells is studded with densely-packed brush-like microvilli, which increases the area available for drug absorption tremendously (>10 folds). The brush border membrane of the intestine is where the actual drug absorption and metabolism take place.

### 37.1.2 Intestinal Absorption of Drugs

Absorption of drug molecules across the intestinal epithelium is a complex process involving multiple pathways (Figure 37.2). The absorption of the drug is determined by two major factors, uptake, and efflux of the drug molecules. The intestinal uptake takes place via following host of mechanisms alone or in combinations.

#### 37.1.2.1 Transcellular/Paracellular Passive Diffusion

It is concentration gradient driven and does not require energy. It occurs mostly through the cell membrane of enterocytes (transcellular route) or through the tight junctions between the enterocytes (paracellular route).

#### 37.1.2.2 Carrier-Mediated Transport

Carrier-mediated absorption occurs via an active process or by a facilitated transport mechanism. The major differences between these two transport processes are that facilitated transport only occurs along the concentration gradient and does not require energy. On the other hand, active transport can function even against the concentration gradient but it requires the expenditure of metabolic energy (e.g. ATP).

Functionally, there are two types of transporters such as uptake and efflux transporters, present on the apical (luminal) and/or basolateral (serosal) side of the mature enterocyte. The transporter proteins responsible for the absorption are called uptake transporters and are located on the apical membrane. Additional uptake transporters are responsible for the uptake of the drug molecule into the cell and may also be present on the basolateral side of the cell. Uptake transporters are of different kinds such as solute carrier (SLC), peptide transporters, organic anion transporter peptides (OATP), and so on. The efflux transporters exert an action that is exactly opposite of the uptake transporters. Efflux transporters such as ATP binding cassette (ABC) transporters are responsible for the secretory transport of the drug and/or its metabolite out of the cell either on the apical side or on the basolateral side. Therefore, efflux transporters are expressed

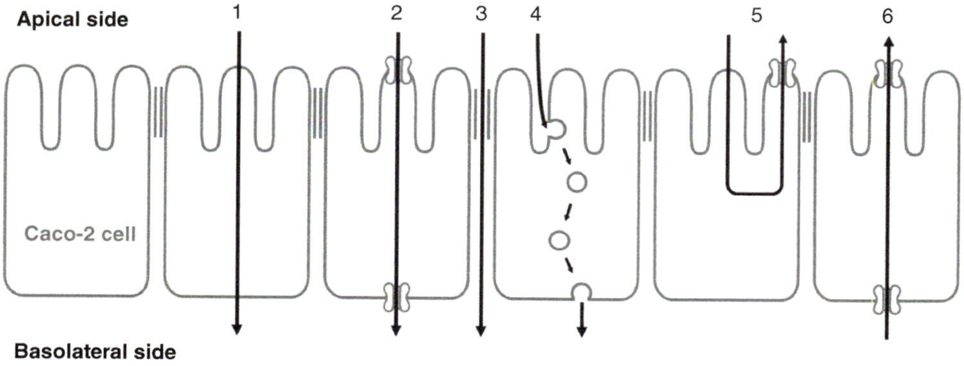


**Figure 37.2** Multiple pathways for intestinal absorption of a compound: (1) passive; (2) transporter-mediated uptake; (3) paracellular; (4) receptor-mediated endocytosis; (5) transporter-mediated efflux; (6) transporters-mediated intestinal excretion from basolateral side.

on the apical as well as basolateral side of the cell. Intestinal cells have different efflux transporters present on the cell surface, and they are divided into different subclasses including multidrug resistance (MDR) transporters such as *p*-glycoprotein (P-gp/Mdr1), multidrug resistance-related proteins (MRP 1-9), and breast cancer resistance protein (BCRP), which limit the absorption and in turn the *in vivo* bioavailability of drugs.

#### 37.1.2.3 Endocytosis

This interesting transport process uses receptors present on the cell membrane. In this process, the compound binds to the receptor on the surface of the cell, and the entire complex is then engulfed by the cell. The complex pushes the compound molecule onto the other side of the cell membrane and then brings the receptor back onto the cell surface. This process is very selective for the substrates and is very slow as compared to above-described absorption processes. The slow speed is because of the time taken by receptor proteins for internalization during the transport process.

It is well known that, at any given time, drug molecules may use more than one type of uptake mechanism for their intestinal absorption. More recently, it has become clear that oral drug absorption is affected by apical uptake and efflux transporters that usually transport drug in the opposite ways, and that intestinal metabolism coupling with efflux function to apical or basolateral side, limiting the oral bioavailability of the drugs. Because a drug (i.e. the parent or one of its metabolites) can be a substrate of multiple efflux transporters at the same time, these transporter may compensate for each other's deficiencies. For example, a drug molecule can be a substrate of Mrp2, 3, and 4, and inhibition of Mrp2 efflux transporter expression levels in the cells may increase the Mrp-3 and 4 transporter expression levels at the same time, resulting in no apparent change in intracellular concentrations. This compensation phenomenon makes it difficult to use an approach of inhibition of efflux transporters to accurately predict the *in vivo* permeability characteristics of any dietary or pharmaceutical compound.

### 37.1.3 Intestinal Metabolism of Drugs

Once a drug enters the cell by passive diffusion or carrier-mediated transport processes, it may undergo phase I and/or phase II metabolism inside the cell. A detailed description of intestinal metabolism is present elsewhere in this book. Briefly, phase I metabolism primarily includes hydrolysis, oxidation, or reduction reactions. A large portion of the phase I reaction is catalyzed by cytochrome P450 (CYP) enzymes present inside the cell. On the other hand, phase II metabolism is via conjugation reaction and mainly catalyzed by different transferases such as UDP-glucuronosyltransferases (UGTs), sulfotransferases (SULTs), and glutathione transferases (GTs). The objective of phase I and II metabolisms is to make the drug molecule more polar, so they can be eliminated from the body through urine or bile. For phase II metabolites, once the hydrophilic metabolite is formed inside the cells, most of it is usually excreted out via apically located efflux transporters (e.g. Mrp2 or BCRP), while some of it is excreted via the basolaterally located Mrp3, 4, and 6. Recent studies in our lab have shown an interesting phenomenon of coupling efflux transporters with the metabolism by UGT and SULT enzymes [1, 2]. In this coupling phenomenon, the metabolite is a substrate of one efflux transporter and the saturation of the particular transporter may inhibit further metabolism of the drug molecule by a negative feedback mechanism either partially or completely. In another situation, metabolite excretion may be compensated by some other transporter thereby increasing drug metabolism inside the cell. Furthermore, there are reports suggesting upregulation of metabolizing enzymes and transporters due to high concentration of drugs or dietary components [3, 4].

### 37.1.4 Rationale for Caco-2

Recent advances in the field of biomedicine, molecular biology, and high throughput screening have put forth significant numbers of potential drug molecules that may be viable medicinal cures for known diseases. To study the systemic bioavailability of all these compounds and their chemical analogs in humans/rodents is practically impossible within limited time frame and is cost prohibitive. Therefore, a cell culture model is preferred because that can achieve faster speed, needs only a smaller quantity of compounds, and present less analytical challenges for sample measurement.

## 37.2 Description

Caco-2 cell lines have come a long way as a model system to provide a detailed and mechanistic insight into the metabolism and transport of various drugs *in vitro*. Since there are various types of Caco-2 cell strains or clones available, in this section, we will explain the sources and properties of different strains/clones and their usefulness in understanding the mechanism of metabolism and transport of drugs in the Caco-2 cell culture model.

### 37.2.1 Caco-2 Cell

As mentioned previously, Caco-2 cell line is a cancer cell line derived from human colon adeno-carcinoma cells. The initial parental Caco-2 cell line was obtained from American Tissue Culture Collection (ATCC). In 1983, Pinto and coworkers demonstrated that Caco-2 cells upon differentiation express several morphological and biochemical characteristics of small intestinal enterocytes [5]. Caco-2 cells, unlike other cells, explicitly grow in a monolayer and show a cylindrically polarized morphology with microvilli on the apical side. Caco-2 cells also have tight junctions between adjacent cells and express many small intestinal enzymes and transporters. Enzyme expressions in Caco-2 cell lines suggest a closer resemblance of these cells to fetal rather than to adult ileal enterocytes. Taken together, Caco-2 cells exhibit a better morphological and functional enterocytic differentiation compared to other colon carcinoma cell lines [2, 6].

### 37.2.2 The Development of Caco-2 Model

Hidalgo and coworkers pioneered the work of using Caco-2 cells as an *in vitro* model to determine the intestinal absorption of various compounds in humans [7]. The work done in Borchardt's lab in establishing Caco-2 cell monolayer as a model to test drug's intestinal epithelial permeability was important for the establishment of the Caco-2 cells as the cells of choice for drug absorption mechanism studies and for absorption screening [8]. Since 1989, the use of Caco-2 cell model has grown phenomenally and it is now the gold standard of intestinal models, well received by the academic and pharmaceutical industry world.

In earlier days, Caco-2 cell culture, even in stationary phase of growth showed heterogeneity in subpopulations with different morphologies. The Caco-2 cells may exhibit different morphological as well as functional properties depending upon the cell culture maintenance, clonal origin, and various other factors. This often creates problems while comparing results between laboratories and should be considered very seriously. Therefore, to improve the homogeneity and the stability of the cell population, clonal cell lines were isolated and characterized. Additionally, it is recommended that the cell lines are used within 10 passages because even cloned cells have tendency to change over time.

### 37.2.3 Caco-2 Cell TC-7 Clone

Earlier, it was hard to characterize non-differentiated Caco-2 cells depending upon different parameters such as transport rate, brush border assembly, and other differentiated characteristics. Hence, the isolation of clones from the parental Caco-2 cell line was attempted to obtain a more homogeneous population suitable for specific studies. Since 1990s, Caco-2/TC-7 cells have been widely used to study phase I/II metabolism and transport studies because of their unique characteristics.

Table 37.1 shows origin and characteristics of some of the commonly used Caco-2 clones. From the table, different Caco-2 cell lines can be used for very different functional experiments because of their special characteristics. Caco-2/TC7 is a common choice of cell line for performing the metabolism and transport experiments because of its stable expression levels of multiple transporters and enzymes especially the phase II conjugating enzymes such as UGT and SULT. Recently, O'Sullivan lab has isolated a couple of copper-resistant Caco-2 clones, which can be utilized in studying dysfunction of copper homeostasis [9]. Subclones with a low expression level of carboxylesterase 1 were isolated in Imai lab, which closely mimics the human small intestine enzyme expression and may be used for prodrug screening [10].

**Table 37.1** Origin and characteristics of different Caco-2 cell clones.

| Clones | Origin | Characteristics |
|---|---|---|
| Caco-2/40 | Imperial Royal College London | Better taurocholic transport; morphological similarity |
| Caco-2/TC7 | Caco-2/WT | Better taurocholic transport; High inducibility for CYP1A1, UDP-GST; Stable CYP3A4, 3A5 expression |
| Caco-2/BBE1, 2 | Caco-2/ATCC | Brush border expression comparable to human enterocytes |
| Caco-2/15 | Caco-2/ATCC | High sucrose-isomaltase expression |
| Caco-2/AQ | Caco-2/15 | Low confluence time |

### 37.2.4 Shorter Growth Time

A significant impediment to the use of Caco-2 cell monolayers is the need to grow the cells for approximately 19–21 days before they can be used. This creates a significant challenge for the operators of the cell monolayers as a single contamination often results in a loss of productivity for one to two months. One possible approach to reduce the cost is to develop alternative cell culture procedures, media, or even clones that can give you a quality monolayer within a week. With a short growth period, much pressure can be alleviated from occasional contamination. Many investigators have published papers by trying to modify the culture media or protocols so that the cell are ready in five days or less [11–13]. However, shorter growth time has not afforded these Caco-2 cell monolayers the same properties as the longer growth Caco-2 cells (e.g. less expression of certain enzymes and transporters) [11], and to this date, Caco-2 cells with standard growing conditions remain the cell membrane of choice for researchers.

## 37.3 Utility

The recent advancement in the pharmaceutical industry and medicinal research has extended the use of Caco-2 model in many different ways. In this section, we are presenting some of the ways that Caco-2 cell culture model can be used to test and mechanistically study the enzyme(s) and transporter(s) responsible for the absorption and disposition of the test compounds.

### 37.3.1 Use in Absorption Screening

The Caco-2 cell monolayers, a model of human intestinal epithelium, are extensively used for permeability screening purposes because experiments can be completed in a relatively short period of time and only a small amount of chemicals is needed for each screening experiment. Currently, the expression levels of the many uptake transporters (e.g. NPC1, NPC1L1, and PEPT1) and efflux transporters (e.g. P-gp, Mrp, and BCRP) are known in Caco-2 cells [14, 15]. It has also been shown that the expression levels of certain uptake and efflux transporters are almost similar to the human intestinal epithelial cells [14, 15]. The available knowledge and extensively published literature make this model very useful in performing the absorption screening.

Despite the wide acceptance of this model as the most relevant absorption screening model, the limitations of the model should also be mentioned. Caco-2 cells grow slowly as compared to other cell lines making it time consuming and contamination prone. The expression levels of the enzymes and transporter proteins change with cell passages and should be accounted for when correlating permeability (across the cell monolayer) data with *in situ* and *in vivo* data. Also, it has been shown that the expression levels of these proteins change from lab to lab and even from operator to operator in the same lab due to the variation in the growing and handling of the cells [4, 16–20]. Furthermore, certain enzyme isoforms or transporters may not be present in the Caco-2 cells, which may be responsible for the metabolism and thereby transport of metabolite of test compound [17, 21, 22]. This may give a very different absorption and metabolism profile for the test compound as compared to the other models. Caco-2 cells show minimal nuclear receptor effects as compared to *in vivo* experiments on intestinal cells. Because nuclear receptors (CAR, PXR, FXR) may upregulate the enzyme and/or transporter protein expression levels, the Caco-2 cell culture model is not commonly used for regulation studies.

### 37.3.2 Use in Absorption Mechanism Determination

The Caco-2 cell culture is the best model to study the primary absorption mechanism of test compounds *in vitro* because of at least three possible reasons: the ease of use, the access to both sides of Caco-2 cell monolayer grown on filters, and the variety of experimental controls at hands to modify the cell lines, in order to study different aspects of drug transport (efflux or uptake or both). The apical to basolateral (A-B) transport study of test compounds with more than four different concentrations can provide information about the mechanism of transport (passive or transporter dependent). The basolateral to apical (B-A) transport study of the compound provides insight into the possible involvement of the efflux transporter in the oral bioavailability of the test compounds. Higher B-A than A-B transport across Caco-2 cell monolayer usually means the involvement of efflux transporters in the transport of the test compounds. If the transport of a compound in the B-A direction is 10–20-fold higher as compared to A-B transport, it usually means that an efflux transporter plays a predominant role in limiting its oral absorption.

Earlier work done by Seithel and coworkers have shown that P-gp, BCRP, and Mrp2 transporter proteins showed intermediate or higher expression levels in Caco-2 cells as compared to human intestinal epithelial cells [18]. For uptake transporters, PEPT1/2, NPC1L1, NPC1, OAT, OATP, GLUT1/2, and MCT1, are shown to be present in the Caco-2 cells [15, 17, 18, 23, 24]. PEPT1/2 transporters are used by the small peptides and certain drugs (Valciclovir®) for transport across the intestinal membrane (see chapter dedicated to this topic) [25]. NPC1L1, NPC1, and NPC2 transporter proteins are confirmed as the uptake transporter proteins for cholesterol [23, 24]. NPC1L1 transporter protein expression has shown high levels in small intestine in humans [1, 26]. Recent publications have shown the expression levels of NPC1L1 transporter protein to be similar in Caco-2 cells and human jejunum [15]. OATP2B1 is expressed at significant levels in Caco-2 cells, and hence Caco-2 cells can be used to study the transport of drugs that are substrates of OATP2B1 [17].

Similar to the above-mentioned limitations in absorption screening, expression levels of the uptake and efflux transporters present in Caco-2 cell monolayers should be taken into account in data analysis and in understanding the mechanisms of transport. Care should be taken to consider the expression levels of the uptake and efflux transporters as well as the metabolizing enzymes involved in the disposition of the test compounds in order to extrapolate data obtained from the Caco-2 cell model to humans.

To confirm the involvement of uptake and efflux transporters, selective chemical inhibitors can be used. Caco-2 cells provide a stable platform to conduct the chemical inhibitor studies, and $IC_{50}$ values could be used to confirm selectivity of the chemical inhibitor for a particular transporter protein. The limitations of use of chemical inhibitors in Caco-2 cell model are as follows. Chemical inhibitor is not always highly selective for any particular protein at a given concentration. Chemical inhibitors usually do not produce 100% inhibition of a particular transporter protein, which may explain less than 100% inhibition of uptake or efflux transport of test compound. Results obtained from chemical inhibitor studies in Caco-2 cells sometimes do not correlate well with *in situ* or *in vivo* studies because inhibitors *in vivo* may be absorbed or somehow exhausted first before absorption of the target compound occurs. Hence, it is not always possible to predict the *in vivo* effects of a particular chemical inhibitor without actually studying it *in vivo*.

### 37.3.3 Use in Metabolism Studies

Caco-2 (e.g. Caco-2/TC7 cell lines express different types of metabolizing enzymes) (phase I and II) present inside the cell with varying degrees of expression levels [19, 20, 22]. Phase I metabolizing enzymes consist of various oxidative enzymes including cytochrome P450 isozymes, whereas phase II metabolizing enzymes include various conjugate enzymes including different UGT and SULT isoforms. In case of different CYP isoforms, Caco-2 cells have shown equivalent or many folds higher expression levels of CYP1A1, 1A2, 2C8-19, 2D6, 2E1, and 3A5 as compared to human intestine and liver [27]. The major intestinal and liver CYP isoform CYP3A4 was shown to have very low expression levels in the Caco-2 cells [27]. In addition to studies by Meinl et al. and Siissalo et al., expression levels of different UGT isoforms present in the Caco-2 cells have been published in our own lab as well [20, 22, 28]. To provide more details, UGT1A1, 1A3, 1A4, and 1A6 had high expression levels in Caco-2/TC-7 cell lines whereas levels of UGT1A8 and UGT1A9 were poor and UGT1A7 and 1A10 were almost undetectable. The UGT2B4, 2B7, 2B10, 2B11, 2B15, 2B17, and 2B28 expression levels are also sufficiently high in Caco-2 cells. In case of sulfotransferases, SULT1A1, 1A2, 1A3, 1B1, 1C1, 1C2, 2A1, and 1E1 isoforms have shown high expression levels whereas SULT2B1a, 2B1b, and 4A1 are absent in Caco-2 cells [22].

### 37.3.4 Use in Intestinal Excretion Studies

It has become clear recently that intestine may act as an elimination route for an increasing number of drugs (Zhang et al., 2021). Caco-2 cell culture model can be used to determine the mechanism by which intestinal excretion of a drug occurs by using the B-A transport studies as shown in Figure 37.1. An example of drug excreted significantly by the intestine is apixaban, which replaced warfarin as the major "blood thinner."

### 37.3.5 Other Uses

Caco-2 cells can also be used to confirm the enzymes and transporters involved in the metabolism and transport of the test compound using overexpression/deletion of a particular protein. Addition and or deletion of genes techniques can also be successfully implemented in Caco-2 cell model to access the mechanistic pathways of test compound's absorption and metabolism in the cells. This is discussed in detail later in this chapter.

## 37.4 Recent Progress

As mentioned before, a lot of progress has been made in Caco-2 cell culture model since 1989 when this model was first implemented for studying drug absorption.

New techniques have been successfully implemented to add or delete/silence genes to confirm the role of a particular uptake/efflux transporter and/or a metabolizing enzyme in the absorption and metabolism of drugs.

### 37.4.1 Addition or Deletion of Genes

#### 37.4.1.1 Addition of Genes

With the available information of enzyme and transporter expression, we are able to study the metabolism and transport of a parent compound and its metabolites in Caco-2 cells. For those enzymes or transporters of interest that have no or low expression in Caco-2 cells, overexpression of such proteins in Caco-2 cells may be useful to study the metabolism and transport of test compounds. For example, Hu et al. have shown that the permeability of various drugs such as propranolol and nifedipine did not change with overexpression of CYP3A4 in transfected and wild-type Caco-2/TC7 cells [25]. Yamanashi et al. showed successful overexpression of NPC1L1 uptake transporter protein to study transport of cholesterol in Caco-2 cells [24].

To study the effect of UGT1A6 isoforms, Liu et al. used siRNA technique to lower UGT1A6 enzyme expression up to 60% to 80% in Caco-2 cells [28]. Again, using the same siRNA technique, Yang et al. silenced the gene regulating Mrp2 efflux transporter protein in Caco-2 cells [29]. Watanabe et al showed successful use of siRNA technique in knockout of P-gp in Caco-2 cells [30]. To understand the advantages, disadvantages, and mechanism of action, siRNA gene silencing technique will be discussed below in more detail.

#### 37.4.1.2 siRNA Technique

The application of small interfering RNA (siRNA) is a very novel concept in understanding the transport of the compound in *in vivo* and *in vitro* studies. Even though the applications of siRNA technique *in vivo* are difficult to do, it is a well-established and well-documented *in vitro* procedure. As mentioned before, Dr. Hu and Dr. Sugiyama's labs have already successfully transfected the siRNA to knockout the UGT1A6 enzyme and Mrp2 transporter protein in Caco-2 cells [29, 31].

The advantages of this technique over genetically mutated cells and chemical inhibitors are overwhelming. In genetically mutated cells, the phenomenon of compensation for the overexpression or removal of certain transporter is very much a possibility. In case of chemical inhibitors, it is known that many inhibitors are not transporter/enzyme specific, thus leading to compound interactions that are unexpected or results that are difficult to understand and explain. In using siRNA technique, we can be much more precise in silencing a particular enzyme/transporter to gain mechanistic insight into the drug-enzyme or drug-transporter interactions that influence the absorption and/or disposition of the test compound. Interested readers should consult specialized book and review articles about the advantages and disadvantages of siRNA silencing technique.

#### 37.4.1.3 CRISPR/CAS9 Technique

Dr. Negoro Ryosuke at Ritsumeikan University (Japan) used the Precision Integration into Target Chromosomes system to insert UGT1A1 and CYP3A4 into the Caco-2 cells [13]. Unlike previous variants, however, these Caco-2 cells appear to have similar light microscopy characteristics, normal TEER values, and stably express these two active enzymes over more than five-month period. Earlier work on expression has focused on CYP3A4 and the results appeared to be promising as well [32, 33]. Another Caco-2 cell variant has a tetracycline-inducible system allowing the users to have different levels of CYP3A4 expression [34]. In general, not enough characterization studies have been run using these cells to examine the expression of other transporters and enzymes, but these cells may represent significant improvement of the original Caco-2 cells.

### 37.4.2 Intestine-On-a-Chip

Intestine-on-a-chip, or intestine chip, is a novel *in vitro* model that provides cells with a dynamic microenvironment. Although there are different designs of organ-chip models, the basic concepts are similar. Generally, two chambers are horizontally separated by a porous membrane in a chip, where cells are seeded in either or both chambers. Liquid flow is allowed into both channels to apply shear stress on cell layers. In some models, vertical mechanical force is available to mimic the intestinal peristalsis.

Intestine chip models have been used to assess intestinal metabolism, transport, and immune functions. It has several advantages compared to the traditional Transwell model. It is reported that intestine-on-a-chip model using Caco-2 cells takes less time (usually within a week) to form barrier function compared to Transwell model (about three weeks) [35]. In addition, coculturing with other cell types in both channels allows cell-cell communication. For example, Caco-2 cells may be cocultured with U937 cells (immune responsive model) to assess the immune response and inflammation in intestine [36]. Scientists have also cocultured Caco-2 cells with intestinal microbes or artificial microbiota to study the interactions between intestinal cells and microbiome and microbiome-related inflammations [35, 37, 38].

One significant drawback of the intestine-on-a-chip model is the cost of the microfluidic equipment and consumables. Many laboratories are designing their own chips based on their use, and also a few chip models are commercially available to scientists. Recently, a tape-based intestine chip using Caco-2/BBe1 has been developed, providing a possibility for low-cost microfluidic models [39].

## 37.5 Significance of Caco-2 Cell Culture Model in Drug Discovery and Development

Caco-2 cell culture model is a very useful and diagnostic tool to characterize and determine drug permeation/absorption in human intestine. The strong mechanistic data obtained from these studies in very early stages of drug discovery and development can be of utmost importance in making intelligent decisions about the absorption and disposition profile of the compound in rodents and humans. The ease of manipulation of this model by overexpression of a target protein, chemical inhibition, and protein downregulation by siRNA technique to understand and reconfirm the enzymes and/or transporters involved shows the great potential of this model in drug development. At present, there is no other model that can match the broad capabilities of the Caco-2 cell culture model to study the processes involved in intestinal drug absorption. Despite some noted deficiencies in expression of certain transporters and metabolic enzymes, the Caco-2 cell culture model remains the gold standard for studying drug transport and metabolism in the intestine.

## 37.6 Example

Below presented is an example of apical-to-basolateral transport study for naringenin (10 μM) with some illustrations (Figures 37.3–37.6, Table 37.2) showing the permeability of naringenin and its metabolite in Caco-2 cell monolayer.

### 37.6.1 Drug Transport Study

The drug transport studies in Caco-2 cells are done using the 6, 24, and 96 wells plate. Here we will discuss the widely used six-well plate transport study. As mentioned before, the experimental window for the transport study is from 19th to 22nd day post seeding. The media in Caco-2 cell plate is changed one day before the experiment to get rid of all the waste and the dead cells and provide fresh nutrients to the cells. On the day of the study, the cells in the transport plate are washed thoroughly with 3 ml of freshly prepared Hank's Balanced Salt Solution (HBSS, adjusted to pH 7.4) from inside and outside the insert containing the Caco-2 cells, respectively. This wash cleans out all the media content from the plate. After the initial wash with HBSS buffer, another 3 ml of HBSS buffer solution is added to both sides of the insert in each well. To confirm the monolayer integrity, transepithelial electrical resistance (TEER) value should be measured. This TEER value for

**Date:** 2009/5/17
**Cell info: Caco-2 Cells: TC7 P38, 19days**
**Expt Info: Temperature 37°. 10 μM Naringenin (A to B) Caco-2 cell transport**
**Analytical Instrument: UPLC**
**Detection wavelength: 286 nm**
**Internal Standard: 100 μM Testesterone in 4% Acetic Acid + 96% ACN**

| μM | Naringenin | IS | Ratio |
|---|---|---|---|
| 0.78 | 7115 | 209458 | 0.03 |
| 1.56 | 14940 | 192081 | 0.08 |
| 3.13 | 29144 | 211418 | 0.14 |
| 6.25 | 59475 | 206978 | 0.29 |
| 12.5 | 117584 | 198308 | 0.59 |
| 25 | 236165 | 206225 | 1.15 |
| 50 | 469115 | 206173 | 2.28 |
| 100 | 927895 | 213676 | 4.34 |

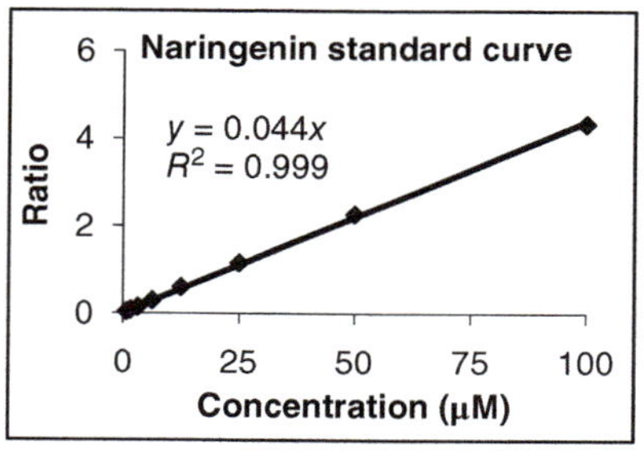


| TEER values for Caco-2 cell plate | | | |
|---|---|---|---|
| Blank | 120 | 125 | 126 |
| A | 225 | 226 | 244 |
| B | 232 | 227 | 242 |
| C | 236 | 237 | 242 |

**Slope** 0.044

**Figure 37.3** Caco-2 cell transport: Experimental, cell, and analytical instrumental information.

| | Slope | a = | Donor (ml) | Take (ml) | $R^2$ | Dilute (ml) | Initial amt (nmol) |
|---|---|---|---|---|---|---|---|
| **Donor side A** | 0.044 | 0 | 2.5 | 0.4 | 1 | 0.4 | 28.95 |

| RT min | 1.134 | Naringenin glucuronide (1.134 min) | | | | | 1.499 | | Naringenin (1.499 min) | | | | (1.8 min) |
|---|---|---|---|---|---|---|---|---|---|---|---|---|---|
| hr | M1 | Conc (μM) | Accu Conc (μM) | Accu amount (nmol) | Avg | SD | Naringenin | Conc (μM) | Accu Conc (μM) | Accu amount (nmol) | Avg | SD | IS |
| 0 | 0 | 0.00 | | | | | 103845 | 11.27 | | | | | 209458 |
| | 0 | 0.00 | | | | | 93800 | 11.10 | | | | | 192081 |
| | 0 | 0.00 | | | | | 107734 | 11.58 | | | | | 211418 |
| 1 | 3432 | 0.38 | 0.38 | 0.94 | 1.04 | 0.09 | 62608 | 6.87 | 7.58 | 18.94 | 18.85 | 0.22 | 206978 |
| | 3704 | 0.42 | 0.42 | 1.06 | | | 58837 | 6.74 | 7.44 | 18.60 | | | 198308 |
| | 4027 | 0.44 | 0.44 | 1.11 | | | 62135 | 6.85 | 7.61 | 19.01 | | | 206225 |
| 2.0 | 6264 | 0.69 | 0.75 | 1.88 | 2.16 | 0.25 | 43478 | 4.79 | 5.83 | 14.57 | 14.64 | 0.11 | 206173 |
| | 8285 | 0.88 | 0.95 | 2.37 | | | 45410 | 4.83 | 5.83 | 14.58 | | | 213676 |
| | 7678 | 0.82 | 0.89 | 2.22 | | | 45365 | 4.83 | 5.91 | 14.77 | | | 213566 |
| 3 | 9038 | 0.98 | 1.10 | 2.75 | 3.03 | 0.25 | 36411 | 3.95 | 5.12 | 12.79 | 12.60 | 0.17 | 209754 |
| | 10491 | 1.14 | 1.30 | 3.24 | | | 35183 | 3.84 | 5.00 | 12.50 | | | 208368 |
| | 10240 | 1.09 | 1.24 | 3.09 | | | 35079 | 3.75 | 5.00 | 12.50 | | | 212733 |
| 4.0 | 10255 | 1.09 | 1.27 | 3.17 | 3.68 | 0.51 | 31900 | 3.40 | 4.66 | 11.65 | 11.57 | 0.30 | 213110 |
| | 11885 | 1.47 | 1.68 | 4.20 | | | 26107 | 3.24 | 4.49 | 11.24 | | | 183312 |
| | 11373 | 1.27 | 1.46 | 3.66 | | | 30709 | 3.42 | 4.73 | 11.81 | | | 204107 |

**Receiver side B**

| RT min | 1.134 | Naringenin glucuronide (1.134 min) | | | | | 1.499 | | Naringenin (1.499 min) | | | | (1.8 min) |
|---|---|---|---|---|---|---|---|---|---|---|---|---|---|
| hr | M1 | Conc (μM) | Accu Conc (μM) | Accu amount (nmol) | Avg | SD | Naringenin | Conc (μM) | Accu Conc (μM) | Accu amount (nmol) | Avg | SD | IS |
| 1 | 8871 | 0.88 | 0.88 | 2.21 | 2.28 | 0.30 | 21490 | 2.14 | 2.14 | 5.36 | 5.43 | 0.08 | 228011 |
| | 9683 | 1.04 | 1.04 | 2.61 | | | 20421 | 2.20 | 2.20 | 5.51 | | | 210727 |
| | 9060 | 0.81 | 0.81 | 2.02 | | | 24287 | 2.17 | 2.17 | 5.42 | | | 254819 |
| 2.0 | 15732 | 1.71 | 1.85 | 4.63 | 4.85 | 0.21 | 25115 | 2.73 | 3.08 | 7.69 | 7.53 | 0.14 | 208901 |
| | 16301 | 1.79 | 1.95 | 4.89 | | | 24095 | 2.64 | 2.99 | 7.49 | | | 207259 |
| | 17209 | 1.89 | 2.02 | 5.04 | | | 23884 | 2.62 | 2.97 | 7.41 | | | 207292 |
| 3 | 19877 | 2.66 | 2.95 | 7.38 | 6.65 | 0.68 | 25190 | 3.37 | 3.86 | 9.65 | 8.34 | 1.14 | 169998 |
| | 20094 | 2.10 | 2.42 | 6.04 | | | 25291 | 2.65 | 3.13 | 7.81 | | | 217165 |
| | 20718 | 2.29 | 2.62 | 6.54 | | | 22984 | 2.54 | 3.02 | 7.55 | | | 205287 |
| 4.0 | 21982 | 2.39 | 2.86 | 7.16 | 6.99 | 0.22 | 22707 | 2.47 | 3.09 | 7.72 | 7.33 | 0.34 | 209028 |
| | 22718 | 2.44 | 2.83 | 7.06 | | | 22245 | 2.39 | 2.89 | 7.22 | | | 211709 |
| | 22388 | 2.28 | 2.70 | 6.74 | | | 22986 | 2.34 | 2.82 | 7.06 | | | 223301 |

**Figure 37.4** Summary of Naringenin (10 μM) transport Experiment in Caco-2 cell Model.

**Figure 37.5** Apical and Basolateral Transport of Naringenin and Naringenin Glucuronide in Caco-2 cell.

**10 μM naringenin (A to B) Caco-2 cell transport**

| | AP-M1 | | AP-Nar | | | BL-M1 | | BL-Nar | |
|---|---|---|---|---|---|---|---|---|---|
| hrs | nmol | SD | nmol | SD | | nmol | SD | nmol | SD |
| 1 | 1.04 | 0.09 | 18.9 | 0.22 | | 2.28 | 0.30 | 5.43 | 0.08 |
| 2 | 2.16 | 0.25 | 14.6 | 0.11 | | 4.85 | 0.21 | 7.53 | 0.14 |
| 3 | 3.03 | 0.25 | 12.6 | 0.17 | | 6.65 | 0.68 | 8.3 | 1.14 |
| 4 | 3.68 | 0.51 | 11.6 | 0.30 | | 7.0 | 0.2 | 7.3 | 0.34 |

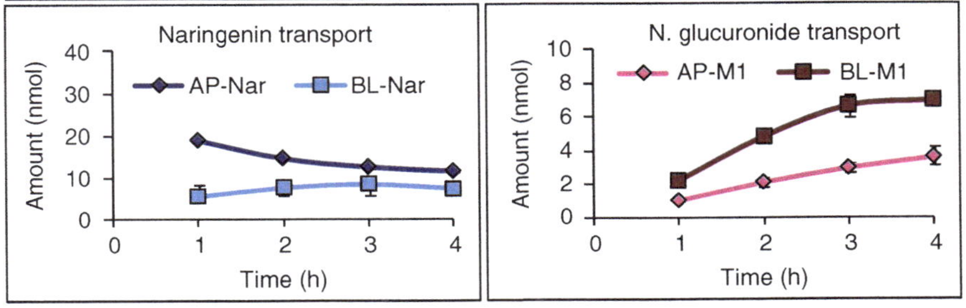

| Donor-naringenin | | | | | |
|---|---|---|---|---|---|
| **Hrs** | **Well-1** | **Well-2** | **Well-3** | **Average** | **SD** |
| 1 | 26.02 | 25.74 | 25.96 | 25.91 | 0.15 |
| 2 | 21.64 | 21.72 | 21.72 | 21.70 | 0.04 |
| 3 | 19.86 | 19.64 | 19.45 | 19.65 | 0.21 |
| 4 | 18.72 | 18.38 | 18.76 | 18.62 | 0.21 |

| Donor-naringenin glucuronide | | | | | |
|---|---|---|---|---|---|
| **Hrs** | **Well-1** | **Well-2** | **Well-3** | **Average** | **SD** |
| 1 | 0.94 | 1.06 | 1.11 | 1.04 | 0.09 |
| 2 | 1.88 | 2.37 | 2.22 | 2.16 | 0.25 |
| 3 | 2.75 | 3.24 | 3.09 | 3.03 | 0.25 |
| 4 | 3.17 | 4.20 | 3.66 | 3.68 | 0.51 |

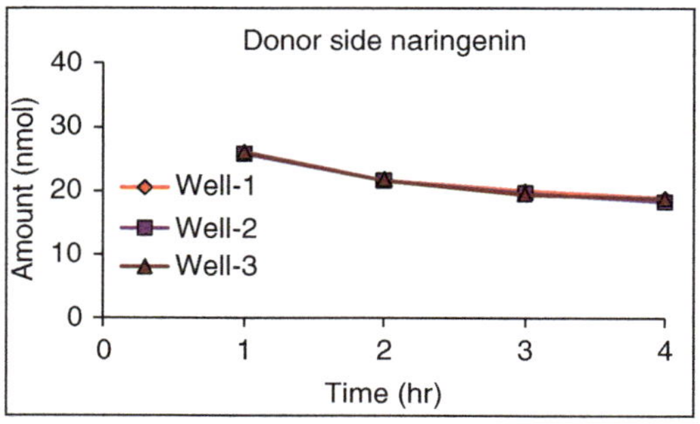


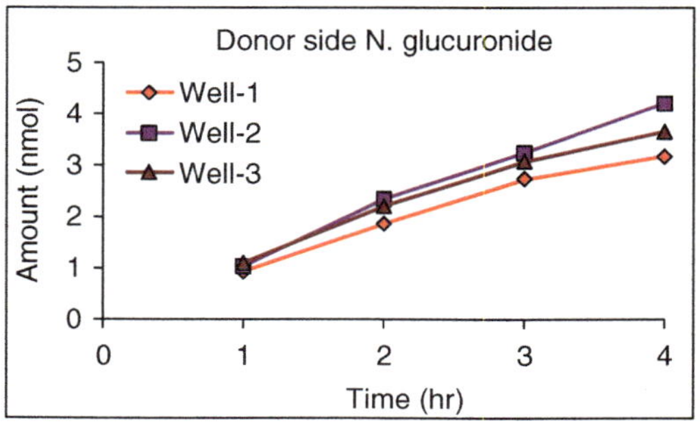


| Receiver-naringenin | | | | | |
|---|---|---|---|---|---|
| Hrs | **Well-1** | **Well-2** | **Well-3** | **Average** | **SD** |
| 1 | 5.36 | 5.51 | 5.42 | 5.43 | 0.08 |
| 2 | 7.69 | 7.49 | 7.41 | 7.53 | 0.14 |
| 3 | 8.14 | 7.81 | 7.55 | 7.83 | 0.29 |
| 4 | 7.47 | 7.22 | 7.06 | 7.25 | 0.21 |

| Receiver-naringenin glucuronide | | | | | |
|---|---|---|---|---|---|
| **Hrs** | **Well-1** | **Well-2** | **Well-3** | **Average** | **SD** |
| 1 | 2.21 | 2.61 | 2.02 | 2.28 | 0.30 |
| 2 | 4.63 | 4.89 | 5.04 | 4.85 | 0.21 |
| 3 | 6.19 | 6.04 | 6.54 | 6.26 | 0.26 |
| 4 | 6.97 | 7.06 | 6.74 | 6.92 | 0.16 |

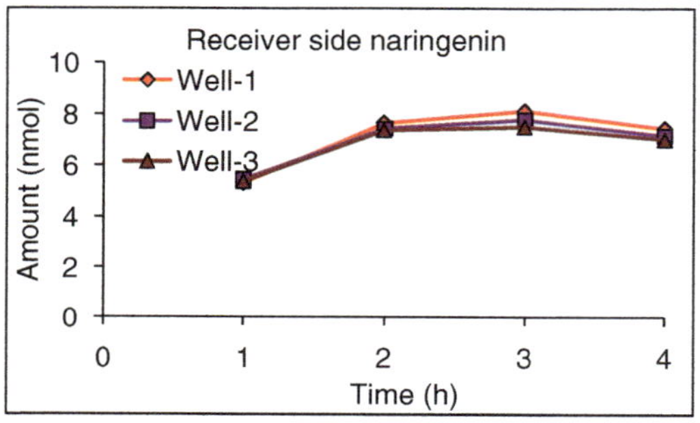


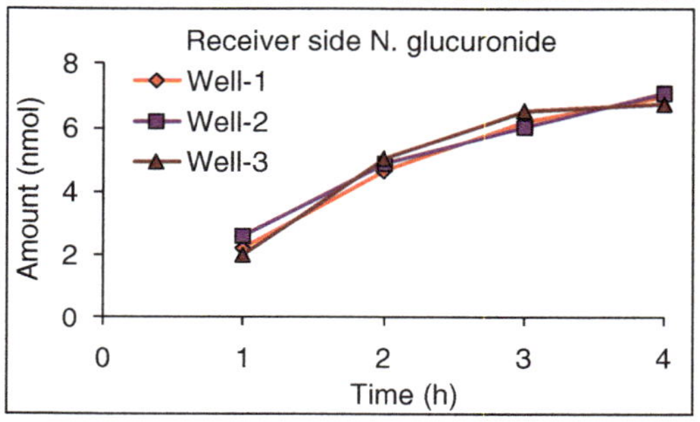


**Figure 37.6** Difference in Transport of Naringenin and Naringenin Glucuronide in Individual Well of Caco-2 cell Transport plate.

the insert (4.2 $cm^2$) containing the Caco-2 cells is usually at least 100 ohms (or >420 ohm $cm^2$) higher than the control insert with no cell monolayer, confirming the presence of the tight junctions and formation of intact Caco-2 monolayer in the test insert. After this, another 2 ml of HBSS buffer (pH 7.4) is added to inside and outside of the insert in each of the 6 inserts in the plate, and the plate is kept for incubation at 37 °C on an incubating shaker (50 rpm) for an hour.

After an hour of incubation, the transport experiment is started. Depending on the study, 2.5 ml of solution containing test compound in HBSS buffer (pH 7.4) is added inside (for A-B transport) or outside (for B-A transport) of the insert, and 2.5 ml of HBSS buffer solution is added on the other side of the insert. Usually, the Caco-2 transport experiment is carried out for four hours with sample points taken at 0, 1, 2, 3, and 4 hours, however, this can be adjusted to shorter or longer time based on the rate of absorption and metabolism of the test compounds. The sample volume collected from each side is usually 400–600 μl and should be replaced each time with same volume of buffer solution to maintain the steady volume in the donor and receiver side. Throughout the experiment, the plate is maintained at 37 °C with continuous circular movements.

At the end of the experiment, drug and buffer solutions are removed from each well and 3 ml of ice-cold HBSS buffer (pH 7.4) is added on inside and outside of the insert to remove residual solution left over from the experiments. The insert membrane with Caco-2 cells attached is then cut, removed and put in a microcentrifuge tube placed in ice. After an appropriate solution (e.g. HBSS) is added, the membrane is then sonicated for 30 minutes at 4 °C and centrifuged for 10 minutes at 15 000 rpm preferably at 4 °C. This cell lysate/supernatant is then collected and stored at −20 °C until analyzed. The cell lysate is analyzed for the protein concentration and the amount

**Table 37.2** Difference Percent Recovery of Naringenin and Percent Excretion of Naringenin Glucuronide in Caco-2 cell monolayer.

| Naringenin Aglycone Recovery in Caco-2 cell Transport Experiment | | | | | |
|---|---|---|---|---|---|
| **Donor** | **Nar(D) (nmol)** | **Nar(R) (nmol)** | **M1(D) (nmol)** | **M1(R) (nmol)** | **Recovery (%)** |
| **Well-1** | | | | | **A** |
| Start | 28.95 | 0.00 | 0.00 | 0.00 | 102.58 |
| End | 11.65 | 7.72 | 3.17 | 7.16 | |
| **Well-2** | | | | | **B** |
| Start | 28.95 | 0.00 | 0.00 | 0.00 | 102.67 |
| End | 11.24 | 7.22 | 4.20 | 7.06 | |
| **Well-3** | | | | | **C** |
| Start | 28.95 | 0.00 | 0.00 | 0.00 | 101.12 |
| End | 11.81 | 7.06 | 3.66 | 6.74 | |

| Naringenin Metabolite Excretion in Caco-2 cell Transport Experiment | | | | | | | |
|---|---|---|---|---|---|---|---|
| **Donor** | **Nar(D) (nmol)** | **M1(D) (nmol)** | **Excreted (%)** | **Receiver** | **Nar(D) (nmol)** | **M1(R) (nmol)** | **Excreted (%)** |
| **Well-1** | | | **A** | **Well-1** | | | **A** |
| Start | 28.95 | 0.00 | 10.96 | 0.00 | 28.95 | 0.00 | 24.72 |
| End | 11.65 | 3.17 | | End | 11.65 | 7.16 | |
| **Well-2** | | | **B** | **WeII-2** | | | **B** |
| Start | 28.95 | 0.00 | 14.52 | 0.00 | 28.95 | 0.00 | 24.40 |
| End | 11.24 | 4.20 | | End | 11.24 | 7.06 | |
| **Well-3** | | | **C** | **WeII-3** | | | **C** |
| Start | 28.95 | 0.00 | 12.64 | 0.00 | 28.95 | 0.00 | 23.29 |
| End | 11.81 | 3.66 | | End | 11.81 | 6.74 | |

of drug accumulated inside the cell during the transport study.

### 37.6.2 Data Analysis

To explain the Caco-2 cell transport study model a little more in detail, we have shown an example of 10 μM naringenin apical to basolateral transport in Caco-2 studies along with all the calculations using Microsoft excel software.

In Figure 37.3, we have shown the in detail experimental, Caco-2 cell and HPLC analysis information about naringenin transport experiment. Caco-2 cell plate information such as type of Caco-2 cells, passage number, number of days post seeding, and TEER values is important and should be well-documented. The Caco-2 cell plate usually has a narrow window of usage, 19 through 21 days post seeding. During this window of time, the Caco-2 cells are mature and maintain an intact monolayer to perform a transport study. The TEER value difference of 100 ohms between test cell wells and blank well is considered appropriate for the intactness of the cell monolayer. A log of experimental conditions such as 37 °C incubation temperature of Caco-2 cell plate at time of experiment and concentration of the test compound used is also important and should be well-documented for reproducibility of the results. The analytical instruments with the methods being used and the preparation of internal standard are also mentioned. The standard curve for naringenin (100 μM to 0.8 μM) was prepared in the same pH HBSS buffer used for the transport study of naringenin 10 μM (test compound concentration used in Caco-2 cell experiment). The slope of the standard curve line is calculated and used for future calculations of concentration of the naringenin on the donor and receiver side of the plate in the experiment (Figure 37.3).

In Figure 37.4, a summary of the entire naringenin transport experimental calculations is shown. To calculate the concentrations of naringenin aglycone and metabolite ($C_P$), respective HPLC peak area ratio (Parent or metabolite/Internal Standard) is multiplied by slope of the standard curve. To calculate the accumulated concentration of naringenin and naringenin glucuronide on each side of the Caco-2 cell monolayer, following equations are used.

$$\sum C_{\text{Pd}}^{t=n} = C_{\text{Pd}}^{t=n} - 0.16\, C_{\text{Pd}}^{t=n} + 0.16\, C_{\text{Pd}}^{t=n-1}$$

$$\sum C_{\text{Md}}^{t=n} = C_{\text{Md}}^{t=n} + 0.16\, C_{\text{Md}}^{t=n-1}$$

$$\sum C_{\text{Pr}}^{t=n} = C_{\text{Pr}}^{t=n} + 0.16\, C_{\text{Pr}}^{t=n-1}$$

$$\sum C_{\text{Mr}}^{t=n} = C_{\text{Mr}}^{t=n} + 0.16\, C_{\text{Mr}}^{t=n-1}$$

where

$\sum C_{\text{Pd}}^{t=n}$ and $\sum C_{\text{Pr}}^{t=n}$ are accumulated concentrations of naringenin on donor and receiver side, respectively, at time $n = 1, 2, 3$, and 4 hours.

$\sum C_{\text{Md}}^{t=n}$; $\sum C_{\text{Mr}}^{t=n}$ are the concentrations of naringenin glucuronide on donor and receiver side, respectively, at time $n = 1, 2, 3$, and 4 hours.

$C_{\text{Pd}}^{t=n}$; $C_{\text{Md}}^{t=n} C_{\text{Pr}}^{t=n}$; $C_{\text{Md}}^{t=n}$; $C_{\text{Mr}}^{t=n}$ are the concentrations of naringenin and naringenin glucuronide on donor and receiver side, respectively, at time "$n$" hours.

A constant 0.16 = [(2.5–2.1)/2.5], which is withdrawal of 0.4 ml solution from 2.5 ml donor/receiver side solution.

$C_{\text{Pd}}^{t=n-1}$; $C_{\text{Md}}^{t=n-1}$; $C_{\text{Pr}}^{t=n-1}$; $C_{\text{Mr}}^{t=n-1}$ are the concentrations of naringenin and naringenin glucuronide on donor and receiver side, respectively, at time "$n$–1" hours.

While calculating the accumulated concentrations of naringenin on donor side for time $t = 1, 2, \ldots, n$, the equation takes into account the sample volume of 0.4 ml taken out and equivalent test sample volume added to the donor side. However, for the calculations of accumulated concentrations of naringenin on receiver side as well as naringenin glucuronide on donor and receiver side, only volume of 0.4 ml lost for each sample point is taken into account since no test compound is added to the receiver side at any time point.

To calculate the actual accumulated amount from concentration of naringenin and/or its metabolite on each side of the Caco-2 cell monolayer, accumulated concentration value is multiplied by the volume of the solution (2.5 ml) used. Using these values, statistical summaries such as mean and standard deviation for each sample point for $n = 3$ samples were also calculated.

To understand the transport of naringenin parent and metabolite across the Caco-2 cell monolayer, graph of accumulated amount (nmol) against time (hours) was plotted for donor and receiver side (Figure 37.5). This graph summarizes the entire naringenin transport experiment, which in turn explains the absorption of naringenin and production and excretion of naringenin metabolite in Caco-2 cells (Figure 37.4). The graph shows that the rate of transport of naringenin across the Caco-2 cell monolayer is high at 10 μM concentration reaching equilibrium concentrations on apical and basolateral side in only four hours of experimental study time. Naringenin glucuronide graph showed different profiles for donor (apical) and receiver (basolateral) side of the Caco-2 cell monolayer. Higher excretion of naringenin glucuronide was observed on the basolateral side as compared to apical side over the period of four hours. In the graph, plateauing lines of the naringenin glucuronide profiles for apical and basolateral side might be due to either saturation of metabolizing enzymes and/or efflux transporters.

In Figure 37.6, we have shown accumulated amount of naringenin and naringenin glucuronide for each of $n = 3$ wells on donor and receiver side of the Caco-2 cell monolayer. This information shows significant differences observed for each of the $n = 3$ wells used and thereby avoid possible manual calculation mistakes.

Finally, to calculate the recovery of naringenin aglycone and excretion of naringenin glucuronide on apical and basolateral side of the Caco-2 cell monolayer as shown in Table 37.2, the following set of equations were used.

Now, to calculate the recovery of naringenin on the donor and receiver sides, the equations are as follows.

$$C_{\text{Pr}} = \frac{\left[\sum C_{\text{Pd}}^{t=4} + \sum C_{\text{Pr}}^{t=n} + \sum C_{\text{Md}}^{t=4} + \sum M_{\text{Mr}}^{t=4}\right]}{M_{\text{Pd}}^{t=0}} \times 100$$

The percent recovery in the range of 90–110 is considered good. The accumulated amount is taken into consideration to calculate the recovery to take into account the amount lost and the amount added for each time point of $t = 1, 2, 3$, and 4 hours.

To calculate the total amount of naringenin glucuronide excreted on apical and basolateral sides,

$$\%M_{\text{Md}} = \sum C_{\text{Md}}^{t=4} / M_{\text{Pd}}^{t=0} \times 100$$

$$\%M_{\text{Mr}} = \sum M_{\text{Mr}}^{t=4} / M_{\text{Pd}}^{t=0} \times 100$$

where $\%M_{\text{Md}}$ and $\%M_{\text{Mr}}$ are the percent excretion of naringenin glucuronide on donor and receiver side, respectively, for each of $n = 3$ wells.

The percent excretion of naringenin metabolite on the apical and basolateral side may help in making an informative decision about the major route of elimination of the compound as well as the possible efflux transporters involved in the transport of naringenin metabolites in enterocytes.

## 37.7 Concluding Remarks

Caco-2 cell culture model is a great tool to determine oral absorption and metabolism of various drug candidates, dietary nutrients, and other orally ingested chemicals without the use of animals or humans. Caco-2 cell culture model is an obvious choice for scientists in academia industry and regulatory agencies for its advantages such as faster screening of drug molecules as well as the easy manipulation of the monolayer to efficiently elucidate the mechanism of action of absorption and metabolism. Overexpression, silencing of genes as well as the use of chemical inhibitors/activators for certain enzymes/transporters have been successfully achieved and well-documented. The ease of use and multiple controls at hand to confirm the role of certain enzyme(s) and/or transporter(s) make this model a great tool.

## References


1 Altmann, S.W. et al. (2004). Niemann-pick C1 like 1 protein is critical for intestinal cholesterol absorption. *Science* 303 (5661): 1201–1204.

2 Chantret, I. et al. (1988). Monensin inhibits the expression of sucrase-isomaltase in Caco-2 cells at the mRNA level. *FEBS Lett.* 235 (1-2): 125–128.

3 Chen, Y. et al. (2008). Genistein induction of human sulfotransferases in HepG2 and Caco-2 cells. *Basic Clin. Pharmacol. Toxicol.* 103 (6): 553–559.

4 Richert, L. et al. (2009). Use of mRNA expression to detect the induction of drug metabolising enzymes in rat and human hepatocytes. *Toxicol. Appl. Pharmacol.* 235 (1): 86–96.

5 Grasset, E. et al. (1985). Epithelial properties of human colonic carcinoma cell line Caco-2: effect of secretagogues. *Am. J. Phys* 248 (5 Pt 1): C410–C418.

6 Matsumoto, H. et al. (1990). Biosynthesis of alkaline phosphatase during differentiation of the human colon cancer cell line Caco-2. *Gastroenterology* 98 (5 Pt 1): 1199–1207.

7 Hidalgo, I.J. et al. (1989). Characterization of the human colon carcinoma cell line (Caco-2) as a model system for intestinal epithelial permeability. *Gastroenterology* 96 (3): 736–749.

8 Hidalgo, I.J. and Borchardt, R.T. (1990). Transport of a large neutral amino acid (phenylalanine) in a human intestinal epithelial cell line: Caco-2. *Biochim. Biophys. Acta* 1028 (1): 25–30.

9 O'Doherty, C. et al. (2020). Gene expression profiling of copper-resistant Caco-2 clones. *Metallomics* 12 (10): 1521–1529.

10 Ohura, K. et al. (2016). Establishment and characterization of a novel Caco-2 subclone with a similar low expression level of human carboxylesterase 1 to human small intestine. *Drug Metab. Dispos.* 44 (12): 1890–1898.

11 Caldwell, G.W. et al. (2014). Permeability assessment using 5-day cultured Caco-2 cell monolayers. In: *Optimization in Drug Discovery: In Vitro Methods* (ed. G.W. Caldwell and Z. Yan), 49–76. Totowa, NJ: Humana Press.

12 Da Violante, G. et al. (2004). Short term Caco-2/TC7 cell culture: comparison between conventional 21-d and a commercially available 3-d system. *Biol. Pharm. Bull.* 27 (12): 1986–1992.

13 Uchida, M. et al. (2009). A modified fast (4 day) 96-well plate Caco-2 permeability assay. *J. Pharmacol. Toxicol. Methods* 59 (1): 39–43.

14 Maubon, N. et al. (2007). Analysis of drug transporter expression in human intestinal Caco-2 cells by real-time PCR. *Fundam. Clin. Pharmacol.* 21 (6): 659–663.

15 Sane, A.T. et al. (2006). Localization and role of NPC1L1 in cholesterol absorption in human intestine. *J. Lipid Res.* 47 (10): 2112–2120.

16 Englund, G. et al. (2006). Regional levels of drug transporters along the human intestinal tract: co-expression of ABC and SLC transporters and comparison with Caco-2 cells. *Eur. J. Pharm. Sci.* 29 (3-4): 269–277.

17 Hayeshi, R. et al. (2008). Comparison of drug transporter gene expression and functionality in Caco-2 cells from 10 different laboratories. *Eur. J. Pharm. Sci.* 35 (5): 383–396.

18 Seithel, A. et al. (2006). Variability in mRNA expression of ABC- and SLC-transporters in human intestinal cells: comparison between human segments and Caco-2 cells. *Eur. J. Pharm. Sci.* 28 (4): 291–299.

19 Siissalo, S. et al. (2007). Effect of cell differentiation and passage number on the expression of efflux proteins in wild type and vinblastine-induced Caco-2 cell lines. *Eur. J. Pharm. Biopharm.* 67 (2): 548–554.

20 Siissalo, S. et al. (2008). The expression of most UDP-glucuronosyltransferases (UGTs) is increased significantly during Caco-2 cell differentiation, whereas UGT1A6 is highly expressed also in

undifferentiated cells. *Drug Metab. Dispos.* 36 (11): 2331–2336.

21 Engman, H.A. et al. (2001). CYP3A4, CYP3A5, and MDR1 in human small and large intestinal cell lines suitable for drug transport studies. *J. Pharm. Sci.* 90 (11): 1736–1751.

22 Meinl, W. et al. (2008). Sulfotransferase forms expressed in human intestinal Caco-2 and TC7 cells at varying stages of differentiation and role in benzo[*a*]pyrene metabolism. *Drug Metab. Dispos.* 36 (2): 276–283.

23 Liscum, L. (2007). A role for NPC1 and NPC2 in intestinal cholesterol absorption–the hypothesis gutted. *Biochem. J* 408 (1): e1–e3.

24 Yamanashi, Y. et al. (2007). Niemann-Pick C1-like 1 overexpression facilitates ezetimibe-sensitive cholesterol and beta-sitosterol uptake in CaCo-2 cells. *J. Pharmacol. Exp. Ther.* 320 (2): 559–564.

25 Hu, M. et al. (1999). Transport and metabolic characterization of Caco-2 cells expressing CYP3A4 and CYP3A4 plus oxidoreductase. *Pharm. Res.* 16 (9): 1352–1359.

26 Iyer, S.P. et al. (2005). Characterization of the putative native and recombinant rat sterol transporter Niemann-Pick C1 Like 1 (NPC1L1) protein. *Biochim. Biophys. Acta* 1722 (3): 282–292.

27 Lampen, A. et al. (1998). Catalytic activities, protein- and mRNA-expression of cytochrome P450 isoenzymes in intestinal cell lines. *Xenobiotica* 28 (5): 429–441.

28 Liu, Y. and Hu, M. (2002). Absorption and metabolism of flavonoids in the caco-2 cell culture model and a perused rat intestinal model. *Drug Metab. Dispos.* 30 (4): 370–377.

29 Yang, Q. et al. (2007). Ezrin and radixin both regulate the apical membrane localization of ABCC2 (MRP2) in human intestinal epithelial Caco-2 cells. *Exp. Cell. Res.* 313 (16): 3517–3525.

30 Watanabe, T. et al. (2005). Construction of a functional transporter analysis system using MDR1 knockdown Caco-2 cells. *Pharm. Res.* 22 (8): 1287–1293.

31 Liu, X. et al. (2007). Disposition of flavonoids via enteric recycling: determination of the UDP-glucuronosyltransferase isoforms responsible for the metabolism of flavonoids in intact Caco-2 TC7 cells using siRNA. *Mol. Pharmaceutics* 4 (6): 873–882.

32 Ohta, Y. et al. (2020). Development of Caco-2 cells expressing four CYPs via a mammalian artificial chromosome. *BMC Biotech.* 20 (1): 44.

33 Takenaka, T. et al. (2017). Development of Caco-2 cells co-expressing CYP3A4 and NADPH-cytochrome P450 reductase using a human artificial chromosome for the prediction of intestinal extraction ratio of CYP3A4 substrates. *Drug Metab. Pharmacokinet.* 32 (1): 61–68.

34 Ichikawa, M. et al. (2021). Generation of tetracycline-controllable CYP3A4-expressing Caco-2 cells by the piggyBac transposon system. *Sci. Rep.* 11 (1): 11670.

35 Kim, H.J. et al. (2012). Human gut-on-a-chip inhabited by microbial flora that experiences intestinal peristalsis-like motions and flow. *Lab Chip* 12 (12): 2165–2174.

36 Ramadan, Q. and Jing, L. (2016). Characterization of tight junction disruption and immune response modulation in a miniaturized Caco-2/U937 coculture-based in vitro model of the human intestinal barrier. *Biomed. Microdevices* 18 (1): 11.

37 Maurer, M. et al. (2019). A three-dimensional immunocompetent intestine-on-chip model as in vitro platform for functional and microbial interaction studies. *Biomaterials* 220: 119396.

38 Shah, P. et al. (2016). A microfluidics-based in vitro model of the gastrointestinal human-microbe interface. *Nat. Commun.* 7: 11535.

39 Winkler, T.E. et al. (2020). Low-cost microphysiological systems: feasibility study of a tape-based barrier-on-chip for small intestine modeling. *Lab Chip* 20 (7): 1212–1226.

# 38

# OATP Overexpressed Cells and Their Use in Drug Uptake Studies

*Lu Wang, Zuoxu Xie, Yifan Tu, and Ming Hu*

*Department of Pharmacological and Pharmaceutical Sciences, College of Pharmacy, University of Houston, Houston, TX, USA*

## 38.1 Introduction to OATP Cell Assay

Organic anion transporting polypeptides (OATPs) are an important family of uptake transporters in the distribution of many endogenous compounds such as bile acids and hormones, and xenobiotics like statins [1]. The involvement of OATPs may lead to the unexpected phenomenon *in vivo*, including nonlinear pharmacokinetics and intra and intersubject variability in the pharmacokinetics, pharmacodynamics, and toxicity [2, 3]. The significance of OATPs in pharmacokinetics and drug–drug interactions (DDI) studies has seized increasing attention in research and clinical practice [4]. By knowing whether and which OATPs are involved in the disposition process and further, the role of a compound as a substrate or inhibitor in the process, we will be able to explain and predict pharmacokinetic and pharmacodynamic properties of a drug and potential drug interactions. In research practice, both *in vitro* cell assay and *in vivo* animal models are commonly used to assess the involvement of OATP. Although the information obtained from the *in vitro* cell assay may not be sufficient to represent all that happens in human, the results of *in vitro* cell studies are able to provide support for hypothetical mechanisms and help to make a decision for further studies *in vivo* or in human [5].

### 38.1.1 Importance and Application

*In vitro* assay using OATP overexpressed cell lines is a powerful tool for screening both transporter substrates (compound as a substrate) and inhibitors (compound as an inhibitor toward a probe substrate). It can be widely used in every stage of drug discovery, from preclinical development to post-market surveillance.

One major use of OATP cell assay is to determine the DDI potential of a compound of interest. The evidence from the assay can help to predict OATP-mediated DDIs and pharmacokinetic change for the compound or any combination in question, and therefore, allow the identification of patients at risk of developing adverse events caused by DDIs or genetic polymorphisms [6]. In addition, the OATP cell assay can provide solid information to determine the necessity of conducting clinical DDI study for a compound of interest and optimize the regimen for reducing the adverse events and improving therapeutic effects for a commercially available drug.

Another important use of the OATP cell assay is to understand the organ tissue distribution of a compound in the body. For example, OATP2B1 is becoming a target to study brain penetration. Bednarczyk and Sanghvi compared the *in vitro* OATP2B1 uptake properties of various drugs with *in vivo* results, looking for the impact of OATP2B1 on intestinal absorption and blood–brain barrier permeability [7]. Yang *et al.* determined uptake kinetic parameters of morphine and its glucuronides in HEK293-OATP2B1 cells to describe their accumulation in brain along with *in vivo* studies [8].

The OATP cell assay can further be utilized in drug design to determine the OATP-targeting potency of a compound. Many promising compounds or drugs fail in development or are withdrawn from the market due to adverse effects derived from nontargeted exposure. To achieve tissue-selective delivery and reduce the nontargeted effect, investigators adopt a transporter-mediated tissue targeting strategy, in which the structural features recognized by the transporters are integrated into a compound of interest to make it become a substrate of transporters [9]. For example, Pfefferkorn *et al.* transformed a systemically acting glucokinase activator (an antidiabetic agent) to a liver selective activator by reducing its passive permeability and making it a substrate of liver-specific OATP1B1 and OATP1B3. The modification in its structure aimed at avoiding

*Oral Bioavailability and Drug Delivery: From Basics to Advanced Concepts and Applications*, First Edition. Edited by Ming Hu and Xiaoling Li.

side effects of hypoglycemia due to excessive pancreas distribution, and at the same time providing effective glycemic control [10].

*In vitro*, cell assays are usually composed of early steps of a study. Compared to *in vivo* study, *in vitro* cell assay provide a cost-efficient and time-saving method to investigate the uptake and inhibition property, and is specifically suitable for large-scale screening.

### 38.1.2 Guidance from Regulatory Agencies on *in vitro* Studies

DDI is one of the key issues in regulatory review of New Drug Application in the scope of Food and Drug Administration (FDA) in the United States [5]. Similarly, European Medicines Agency (EMA) and Pharmaceuticals and Medical Devices Agency of Japan (PMDA) also published guidelines on how to evaluate OATP-mediated DDI with *in vitro* systems [11, 12]. Currently, only OATP1B1 and 1B3 are listed in the above-mentioned recommendations.

OATP1B1 and OATP1B3 are key uptake transporters that are specifically expressed on the sinusoidal membrane of hepatocytes, *i.e.* transporting compounds from blood into hepatocytes. A relatively low systemic exposure may be observed for compounds that are substrates of these transporters. In the United States, FDA has recommendations on evaluating whether a compound is a substrate of OATP1B1 and OATP1B3. When a compound of interest is estimated to have significant hepatic uptake or elimination, namely, when hepatic metabolism and/or biliary excretion of the compound contributes to over 25% of the total clearance of the compound, conducting an *in vitro* uptake study is recommended. In addition, when the hepatic uptake is clinically significant, e.g., when the compound undergoes massive hepatic metabolism or is expected to bring pharmacological effect in the liver, it is also recommended to investigate the contribution of the transporters [5]. Accordingly, in EU and Japanese guidance on DDI, the investigation of the involvement of OATP1B1 and 1B3 is recommended for compounds with ≥25% hepatic elimination [11, 12]. The cutoff values to determine whether a compound is a substrate of a transporter may be justified and accommodated in each lab. An often-used criterion is that a known inhibitor can decrease over 50% uptake of the compound of interest; another is when uptake of the compound in transporter overexpressed cells is over two-fold higher than its uptake in control cells.

For inhibition properties, FDA guidance recommends conducting *in vitro* studies to determine the inhibition effect of the compound of interest on the uptake of a known substrate of OATP1B1 and 1B3 for all new drug entities. Usually, a concentration of the compound at 10-fold of clinically relevant concentration is used in cell assay as an inhibitor. The cutoff criterion is calculated differently in FDA and EMA, but both include maximum hepatic inlet concentration of the compound estimated from *in vivo* data and $IC_{50}/K_i$ of the compound determined from cell assay. Recent studies reveal that preincubation of an inhibitor may significantly decrease the $IC_{50}$ value of the compound in OATP1B1/1B3 cell system [13, 14]; therefore, adding a preincubation step may be considered when performing inhibition studies.

In addition, substrate and inhibition properties of the major metabolites and/or active metabolites of the compound may also be considered.

When identified as a substrate or inhibitor, further, *in vivo* studies with known inhibitors or probe substrates and likely concomitant medications are recommended by these agencies.

### 38.1.3 Commonly Used Cell Lines

Commonly used cell lines in OATP uptake study include primary human hepatocytes and other cell lines (over)expressing OATPs. The transport activity of cells should be validated prior to the studies, with at least one probe substrate and the uptake is inhibitable by known inhibitors.

Stable HEK-293 overexpressing 1B1/1B3 cell lines were initially developed by Dietrich Keppler lab in 2000, along with stably transfected Madin-Darby canine kidney (MDCK) cells of both transporters [15, 16]. Later, as a result of emerging science, more laboratories were able to develop their own stable transfected cell lines.

HEK-293 are widely used because they are relatively easy to maintain and transfect. Other cell lines (over)expressing OATPs frequently used include Chinese hamster ovary (CHO) cell, MDCK cell, etc. With the increasing realization of the importance in drug interactions, many OATP overexpressed cell lines are currently commercially available.

Human primary hepatocytes are also used for liver-specific OATP1B1 and 1B3. As compounds may be substrates or inhibitors of multiple transporters, hepatocytes are often used to assess the general uptake or inhibition property in liver. Moreover, hepatocytes can also be used together with probe substrates or inhibitors to identify the compound's interaction with a specific transporter. Cell lines without specific transporter expression by the gene knockout or using

**Table 38.1** Examples of *in vitro* substrates and inhibitors of OATPs.

| | Substrate | Inhibitor |
|---|---|---|
| OATP1A2 | Estrone-3-sulfate, fexofenadine | Rifamycin, rifampicin, verapamil, hesperidin, naringin |
| OATP1B1 | Estrone-3-sulfate, bromsulphthalein (BSP), estradiol-17-β-glucuronide ($E_2G$), statins (e.g. pitavastatin, pravastatin, and rosuvastatin) | Cyclosporine, $E_2G$, estrone-3-sulfate, rifampicin, rifamycin SV |
| OATP1B3 | Cholecystokinin octapeptide (CCK-8), telmisartan, $E_2G$, BSP, statins (e.g. pitavastatin, pravastatin, and rosuvastatin) | Cyclosporine, $E_2G$, estrone-3-sulfate, rifampicin, rifamycin SV |
| OATP2B1 | Estrone-3-sulfate, BSP, statins (e.g. atorvastatin and fluvastatin) | Rifampin, cyclosporine, BSP, rifamycin SV |

silencing mRNA may also be used, according to EMA recommendation [11].

Regardless of the choice of the cell line, the *in vitro* system should be justified on the basis of current scientific literature.

### 38.1.4 Commonly Used *In Vitro* Substrates and Inhibitors

Table 38.1 is a list of examples that can be used as *in vitro* substrates and inhibitors. As scientific knowledge evolves, the choice of substrates and inhibitors should be based on current scientific research.

## 38.2 Materials

### 38.2.1 Chemicals and Reagents

Cell culture medium: DMEM supplemented with 10% fetal bovine serum, 1% antibiotic antimycotic solution, and 600 μg/ml geneticin

Working medium: Hanks' Balanced Salt Solution (HBSS). 9.7 g Hanks' Balanced Salt solution powder is supplemented with 0.372 g sodium bicarbonate, 3.502 g D-glucose, 5.963 g HEPES, 185 mg calcium chloride, and 98 mg magnesium sulfate, and dissolved in 1 l water. The pH of the solution is adjusted to 7.4.

Stock solutions of test compound, *in vitro* substrate, and inhibitor prepared in dimethyl sulfoxide (DMSO)

### 38.2.2 Supplies and Equipment

Twenty-four-well cell culture plates
Incubating shaker
Vortex mixer
Sonicator
Centrifuge
Detection system (e.g. LC–MS)

## 38.3 Methods

### 38.3.1 Preoperative Preparations

#### 38.3.1.1 Cell Preparation

OATP1B1 overexpressed HEK-293 cells are generously provided by Dr. Wei Yue [17, 18]. Methods of cell culture and uptake studies in HEK-293 are designed similarly to those published previously [18] and are adapted in the current laboratory settings.

HEK-293-OATP1B1 cells are maintained in cell culture medium in 38.2.1. Generally, passage numbers 3–15 after reviving are used to study uptake properties. Cells are seeded in 24-well plates for each assay, and the assay is performed when cell confluency reaches around 80% in the wells.

#### 38.3.1.2 Solution Preparation

Compounds of interest are diluted in HBSS from stock solutions at desired concentration (usually 1–10 μM, based on physiological conditions and the LLOQ of the detection method). For minimum cytotoxicity and potential effect from DMSO, the final concentration of DMSO should not exceed 1%.

In substrate studies for calculating $K_m$ and $V_{max}$, a range of concentrations of the test compound should be prepared. In inhibition study, multiple concentrations (usually no less than 4) of a test compound (as an inhibitor) are needed against a known substrate at a certain concentration.

### 38.3.2 Uptake Study

a) Warm up 50 ml HBSS at 37 °C before experiment. Likewise, all compound solutions should be prepared and warmed at 37 °C before starting the experiment. Chill another 80 ml HBSS on ice.
b) Start aspirating the media out carefully without touching the cells.

c) Gently wash the cells with 400 μl pre-warmed HBSS twice.
d) Pipet 400 ml pre-warmed HBSS into each well and incubate at 37 °C for 15 minutes so that potential interfering compounds from the cell culture media can be released from the cells.
e) Aspirate HBSS out and load 400 μl pre-warmed compound solution into the wells. Triplicates are recommended for more accurate results.
f) Incubate at 37 °C for desired duration of time (see Section 38.3.3.b).
g) To terminate the incubation, aspirate compound solution out immediately and wash the cells with 400 ml ice-cold HBSS for three times.
h) Add 200 ml HBSS into each well. Dissociate all adherent cells from the bottom of the wells by gentle pipetting and harvest the cell suspension. Use a cell scraper if necessary. Inspect the base of the plate to ensure all cells are removed.
i) Sonicate the cell suspension for 30 min to break down the cells.
j) Centrifuge the cell lysate for at 15 000 rpm for 15 minutes.
k) Transfer 150 μl supernatant into a fresh tube. Keep the remaining solution for protein concentration determination using BCA assay.
l) Add 150 μl acetonitrile with internal standard to mix with the 150 μl supernatants, followed by vortex-mixing.
m) Centrifuge the mixture at 15 000 rpm for 15 minutes to precipitate aggregated proteins.
n) Transfer the supernatant into fresh tubes for further analysis. Samples are kept at −80 °C if not analyzed immediately.

### 38.3.3 Preliminary Studies

a) The cells are in the best condition on the third day of seeding on 24-well plates in our lab's settings. It may be helpful to run a test study from one to five days after seeding to evaluate the cell condition.
b) A time-dependent study should be performed prior to obtaining kinetics data. The incubation time for kinetic study should be within the linear increasing range in the time course. In our experiment settings, uptake of most compounds is linear up to 10 minutes, therefore subsequent experiments are performed at 5 minutes.
c) A set of preliminary experiments may be performed to get an approximate range of $K_m$ and $IC_{50}$ to better estimate the concentration range of the compound chosen in the substrate and inhibition study so that these kinetic parameters can be included amid the concentrations.

## 38.4 Data Analysis

An example of the worksheet of an uptake study is shown below. In this example, we are trying to determine the kinetic parameters of $E_2G$, an OATP1B1 substrate, and the inhibition properties of rifampicin, a known strong inhibitor of OATP1B1.

Quantification analysis from LC/MS was performed with Analyst 1.7.1. Data analysis was performed with GraphPad Prism 8.4.2. The kinetic parameters were calculated with preset formula in the software.

### 38.4.1 Calculation of $K_m$ and $V_{max}$ in Substrate Study

The *in vitro* uptake property of a substrate in a cell system follows Michaelis–Menten model. In the ideal situation, there is a steep increase in the uptake rate at low concentrations and a saturated plateau at high concentrations in the rate of uptake versus substrate concentration plot. $V_{max}$, i.e. the maximum rate of uptake, is observed when the transporter is saturated, and $K_m$ is the substrate concentration at half of the maximum rate of uptake. A larger $V_{max}$ indicates better capacity of the transporter for its substrate, and a smaller $K_m$ shows better affinity of the transporter for the substrate.

Concentration in cell lysates is normalized by internal standard. Rate of uptake ($R$) is calculated with the following equation:

$$R = \frac{A_s}{t \times A_p} = \frac{C_s \times V_s}{t \times C_p \times V_p} \tag{38.1}$$

where $A_s$ is the total amount of compound of interest (substrate) that is taken up into the cells, $A_p$ is the amount of proteins extracted from the cells, and $t$ is the incubation time (five minutes in the current experiment setting). $A_s$ is calculated as the product of $C_s$ and $V_s$ and $A_p$ is the product of $C_p$ and $V_p$, where $C_s$ is the concentration determined in the cell lysate samples (e.g. from LC/MS results), $V_s$ is the volume of the cell lysate collected from each well (200 μl in the current setting), and $C_p$ is the soluble protein concentration determined with a Bicinchoninic Acid (BCA) test using bovine serum albumin as standard. $V_p$ is the volume of protein which in the current setting, is the same as $V_s$.

$K_m$ and $V_{max}$ values are calculated with nonlinear regression – Michaelis–Menten function.

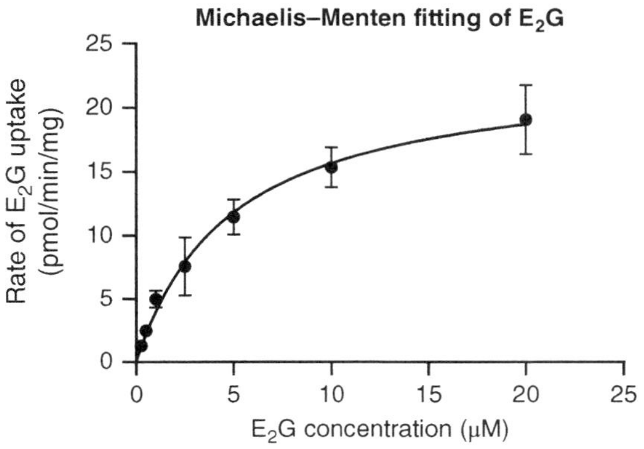


**Figure 38.1** Michaelis–Menten Fitting curve of $E_2G$. Substrate ($E_2G$) concentration range from 0.25 to 20 μM. Data are expressed as mean ± SD ($n = 3$).

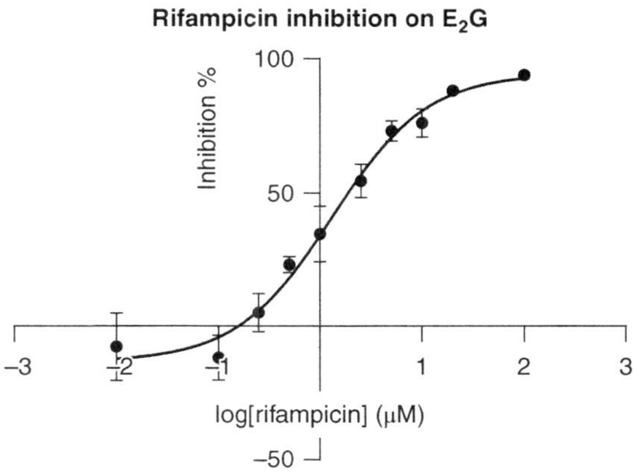


**Figure 38.2** Rifampicin inhibition curve on $E_2G$. Substrate ($E_2G$) concentration is 2 μM and inhibitor (rifampicin) concentrations range from 10 nM to 100 μM. Data are expressed as mean ± SD ($n = 3$).

Table 38.2 shows an example of uptake study of $E_2G$ in OATP1B1 overexpressed cells. The best-fit $V_{max}$ calculated from the model is 23.27 (95% confidence interval: 20.29–27.12) pmol/min/mg protein and the $K_m$ is 4.844 (95% confidence interval: 3.311–7.166) μM with $R^2$ of 0.9524. The curve in Figure 38.1 represents the best-fit line of Michaelis–Menten model for calculating $K_m$ and $V_{max}$.

### 38.4.2 Calculation IC50 in Inhibition Study

$IC_{50}$ is the inhibitor concentration that causes a 50% inhibition in the uptake process. Smaller $IC_{50}$ denotes stronger inhibition potency of the inhibitor. In the current study, we compare the intracellular concentration with different concentrations of the inhibitor. The following equation is used:

$$\text{Inhibition percentage} = \left(1 - \frac{\text{sample conc.}}{\text{control conc.}}\right) * 100\% \tag{38.2}$$

where control concentration is the calculated intracellular concentration of the substrate without inhibitors, and sample concentration is the intracellular concentration of the substrate with different inhibitor concentrations at the same incubation condition (incubation time, substrate concentration, protein concentration, …).

The $IC_{50}$ is determined by fitting dose-response inhibition model to the data with nonlinear regression analysis, where inhibitor concentration is represented in log 10 scale.

In the example below, the $IC_{50}$ value is determined as 1.283 μM (95% confidence interval: 0.9161–1.837, $R^2$ = 0.9685). Figure 38.2 shows the best-fitted $IC_{50}$ curve. Raw data for this figure is listed in Table 38.3.

## 38.5 Notes

a) Cell passage should be recorded for tracking purpose, as the cell may lose its transporting function at a high passage number thus affect experimental results.
b) Concentrations of the test compound should be chosen with rationale. Physiologically/clinically relevant concentrations or concentrations acquired from *in vivo* data should be included in the experiment. In addition, the compound should be soluble in all testing concentrations in HBSS. The solution should be (nearly) isosmotic and no cytotoxicity should be observed.
c) A positive and a negative control is usually needed in each plate to confirm the cell status.
d) A preincubation step may be considered as some time-dependent inhibition has been observed recently for OATP 1B1 and 1B3 (references see FDA guidance). A lower $IC_{50}$ is observed for the test compound when using a preincubation step. In this case, cells are incubated with inhibitors only and followed by the co-incubation of the compound and the inhibitor.
e) The addition and removal of ice-cold HBSS in Section 38.3.2.g terminate the uptake process and remove compounds from the well. The timing should be critically followed for short incubation durations (especially when shorter than 10 minutes).
f) It is recommended to label the tubes before starting the experiment.

g) Relative standard deviation (RSD, RSD = standard deviation/mean * 100%) of intracellular concentration is calculated to evaluate the precision of the results. In the current setting, the RSD among triplicates usually falls within 10–20%. If the RSD is larger than 30%, the results may not be reliable.
h) Generally, a HEK-293 vector cell line (HEK-293-mock) is needed as a control. The net rate of uptake is calculated with the concentration difference between overexpress cell line and mock cell line. In the current experiment setting, we tested uptake properties HEK-293 overexpressing OATP2B1 cells. No significant uptake was observed for $E_2G$ and the intracellular concentration for $E_2G$ is negligible (data not shown); therefore, we consider the OATP2B1 overexpressed cells as a control and did not include it in the calculation. However, for unknown compounds especially with high lipophilicity, passive diffusion across cell membrane is not negligible, and the control cells are absolutely necessary. Supporting information Representative raw data for a substrate kinetics and inhibition study is listed as below.

| **Date of experiment:** | **MM/DD/YY** |
|---|---|
| **Cell info:** | Strain, passage number, days in culture |
| **Substrate/inhibitor:** | $E_2G$/rifampicin |
| **Incubation time:** | 5 min |

**Table 38.2** Data calculation for rate of $E_2G$ uptake. Intracellular concentration is determined from LC/MS.

| Sample # | Substrate Conc. (μM) | Intracellular concentration (nM) | | | Rate of $E_2G$ uptake (pmol/min/mg) | | |
|---|---|---|---|---|---|---|---|
| E1 | 0 | <LLOQ | <LLOQ | <LLOQ | – | – | – |
| E2 | 0.25 | 7.07 | 5.82 | 5.28 | 1.50 | 1.24 | 1.12 |
| E3 | 0.5 | 12.5 | 13.1 | 9.09 | 2.66 | 2.79 | 1.93 |
| E4 | 1 | 20.3 | 23.6 | 26.5 | 4.32 | 5.02 | 5.64 |
| E5 | 2.5 | 35.8 | 46.4 | 24.7 | 7.62 | 9.87 | 5.26 |
| E6 | 5 | 46.6 | 56.4 | 58.7 | 9.91 | 12.00 | 12.49 |
| E7 | 10 | 63.9 | 77.9 | 74.8 | 13.60 | 16.57 | 15.91 |
| E8 | 20 | 85.7 | 79.6 | 104 | 18.23 | 16.94 | 22.13 |

Rate of $E_2G$ uptake is calculated with Eq. (38.1). Protein concentration is determined to be 0.94 mg/ml.

**Table 38.3** Data calculation for rifampicin inhibition on $E_2G$.

| Sample # | Inhibitor Conc. (μM) | $Log_{10}$ (Inhibitor Conc.) | Inhibition (%) | | |
|---|---|---|---|---|---|
| R1 | 0.01 | −2 | −16.81 | −13.02 | 6.70 |
| R2 | 0.1 | −1 | −10.75 | −3.92 | −20.61 |
| R3 | 0.25 | −0.6 | 9.73 | −3.16 | 8.98 |
| R4 | 0.5 | −0.3 | 26.32 | 20.11 | 23.33 |
| R5 | 1 | 0 | 34.61 | 24.48 | 45.20 |
| R6 | 2.5 | 0.4 | 61.09 | 53.95 | 48.66 |
| R7 | 5 | 0.7 | 73.98 | 76.52 | 69.15 |
| R8 | 10 | 1 | 81.30 | 70.76 | 76.52 |
| R9 | 20 | 1.3 | 86.42 | 89.13 | 88.95 |
| R10 | 100 | 2 | 92.96 | 94.39 | 94.55 |

The inhibition percentage is calculated with Eq. (38.2).

## References


1 Stieger, B. and Hagenbuch, B. (2014). Organic anion-transporting polypeptides. *Curr. Top. Membr.* 73: 205–232.

2 Ali, Y., Shams, T., Wang, K. et al. (2020). The involvement of human organic anion transporting polypeptides (OATPs) in drug-herb/food interactions. *Chin. Med.* 15: 71.

3 Kalliokoski, A. and Niemi, M. (2009). Impact of OATP transporters on pharmacokinetics. *Br. J. Pharmacol.* 158: 693–705.

4 Alam, K., Crowe, A., Wang, X. et al. (2018). Regulation of organic anion transporting polypeptides (OATP) 1B1- and OATP1B3-mediated transport: an updated review in the context of OATP-mediated drug-drug interactions. *Int. J. Mol. Sci.* 19.

5 FDA (2020). In vitro drug interaction studies — cytochrome P450 enzyme- and transporter-mediated drug interactions: guidance for industry.

6 Nakanishi, T. and Tamai, I. (2012). Genetic polymorphisms of OATP transporters and their impact on intestinal absorption and hepatic disposition of drugs. *Drug Metab. Pharmacokinet.* 27: 106–121.

7 Bednarczyk, D. and Sanghvi, M.V. (2020). Organic anion transporting polypeptide 2B1 (OATP2B1), an expanded substrate profile, does it align with OATP2B1's hypothesized function? *Xenobiotica* 50: 1128–1137.

8 Yang, Z.Z., Li, L., Wang, L. et al. (2016). siRNA capsulated brain-targeted nanoparticles specifically knock down OATP2B1 in mice: a mechanism for acute morphine tolerance suppression. *Sci. Rep.* 6: 33338.

9 Zhou, J., Xu, J., Huang, Z., and Wang, M. (2015). Transporter-mediated tissue targeting of therapeutic molecules in drug discovery. *Bioorg. Med. Chem. Lett.* 25: 993–997.

10 Pfefferkorn, J.A., Guzman-Perez, A., Litchfield, J. et al. (2012). Discovery of (S)-6-(3-cyclopentyl-2-(4-(trifluoromethyl)-1H-imidazol-1-yl)propanamido) nicotinic acid as a hepatoselective glucokinase activator clinical candidate for treating type 2 diabetes mellitus. *J. Med. Chem.* 55: 1318–1333.

11 EMA (2012). Guideline on the investigation of drug interactions.

12 PMDA (2019). Guideline on drug interaction for drug development and appropriate provision of information (English version).

13 Amundsen, R., Christensen, H., Zabihyan, B., and Åsberg, A. (2010). Cyclosporine A, but not tacrolimus, shows relevant inhibition of organic anion-transporting protein 1B1-mediated transport of atorvastatin. *Drug Metab. Dispos.* 38: 1499.

14 Pahwa, S., Alam, K., Crowe, A. et al. (2017). Pretreatment with rifampicin and tyrosine kinase inhibitor dasatinib potentiates the inhibitory effects toward OATP1B1- and OATP1B3-mediated transport. *J. Pharm. Sci.* 106: 2123–2135.

15 Konig, J., Cui, Y., Nies, A.T., and Keppler, D. (2000a). Localization and genomic organization of a new hepatocellular organic anion transporting polypeptide. *J. Biol. Chem.* 275: 23161–23168.

16 Konig, J., Cui, Y., Nies, A.T., and Keppler, D. (2000b). A novel human organic anion transporting polypeptide localized to the basolateral hepatocyte membrane. *Am. J. Physiol. Gastrointest. Liver Physiol.* 278: G156–G164.

17 Alam, K., Farasyn, T., Crowe, A. et al. (2017). Treatment with proteasome inhibitor bortezomib decreases organic anion transporting polypeptide (OATP) 1B3-mediated transport in a substrate-dependent manner. *PLoS One* 12: e0186924.

18 Alam, K., Pahwa, S., Wang, X. et al. (2016). Downregulation of organic anion transporting polypeptide (OATP) 1B1 transport function by lysosomotropic drug chloroquine: implication in OATP-mediated drug-drug interactions. *Mol. Pharm.* 13: 839–851.

# 39

# Use of Human Intestinal and Hepatic Tissue Fractions and Microbiome as Models in Assessment of Drug Metabolism and its Impact on Oral Bioavailability

*Hani Zaher*[1] *and George Zhang*[1,2]

[1] *Corning Gentest*[SM] *Contract Research Services, Corning Life Sciences, Woburn, MA, USA*
[2] *Arvinas Inc. New Haven, CT, USA*

## 39.1 Introduction

Bioavailability describes the rate and the concentration at which the active moiety (drug or metabolite) enters systemic circulation, thereby accessing the site of action. It is expressed as a percentage of the dose that is initially administered. For orally administered drugs, their intestinal absorption and hepatic metabolism are key players in determining a drug's systemic bioavailability and thus therapeutic effect. Drug absorption and metabolism are both complicated processes, with many physicochemical, biochemical, and physiological factors involved. In addition, metabolism and transporter-mediated drug interaction can also impact bioavailability of victim drugs. Therefore, drug-drug interaction studies are required during drug discovery and development [1]. Determining the contribution of each of these processes is essential in understanding a drug's level in the bloodstream and maintaining its optimum therapeutic outcome and safety [2].

Drugs administered by the intravascular route exhibit the bioavailability ($F$) = 1. However, most drugs are taken orally with $F < 1$ due to drug lost between oral administration and first appearance in the systemic circulation, e.g. via first-pass metabolism and efflux transporters such as P-glycoprotein (P-gp) in the intestine. The drug must survive the milieu in the gastrointestinal (GI) tract, cross the gut wall, and then pass through the portal vein to the liver. If a drug molecule survives that gauntlet, it will reach the systemic circulation. Very often and for a variety of reasons, a reduced fraction of the active drug reaches its site of action or blood circulation [3].

After oral administration, bioavailability may be estimated by comparison with either the intravascular route, yielding the absolute bioavailability, or another pharmaceutical dosage form of the drug by the same route, yielding the relative bioavailability [4].

In this chapter, the role of conventional human intestinal and hepatic tissue fraction models in the assessment of oral drug bioavailability will be discussed. Next, we will discuss *in vitro* assays used to estimate the clearance of low-turnover compounds. Lastly, we will cover the role that microbial flora plays in oral drug bioavailability.

## 39.2 Gastrointestinal Tract and Absorption (see Also Chapter 5)

Gastrointestinal absorption is the first step in a sequence of events that lead an orally administered drug to its intended site of action. The absorption process itself is influenced by the surface area of the GIT mucosal membrane, which varies from the stomach to the rectum. It is particularly useful to understand where along the GIT a drug is absorbed since absorption can vary substantially from site to site. For example, drugs that are weak acids tend to absorb readily within the stomach, while drugs that are weak bases are poorly absorbed there and will have most of the dose passed into the small intestine, which consists of the duodenum, jejunum, and ileum [5]. The small intestine itself is composed of three layers: circular folds (valves of Kerckring, plicae circulkares), intestinal villi (fingerlike extensions of intestinal mucosa), and microvilli (projections found on the apical surface of each intestinal cell: enterocytes). Each villus consists of individual microvilli approximately 0.1 μm in diameter and 1 μm in height; each epithelial cell may have as many as 1000 microvilli. The microvilli play a major role in the absorption of drugs by enlarging the absorbing surface approximately 25 times [6]. In contrast, absorption of drugs from the

*Oral Bioavailability and Drug Delivery: From Basics to Advanced Concepts and Applications*, First Edition. Edited by Ming Hu and Xiaoling Li.

large intestine differs significantly from absorption in the small intestine because of several physiological, physicochemical, and biopharmaceutical factors. The permeability through passive transport is thought to be lower in colon than in other intestinal tissue because of the smaller surface area and tighter junctions in the epithelial cell layer, which support the fact that drug absorption in humans is primarily mediated from the small intestine [7].

Several physiochemical factors impact on drug absorption. pH partition plays a critical role in the absorption mechanism of drugs for which absorption is governed by the equilibrium between ionized and non-ionized forms at the different physiological pH values along the GIT. Other physical properties of the drug such as molecular weight, lipophilicity, polar surface area (PSA), and number of free rotatable bonds (RB) collectively impact on the absorption of small molecules into the bloodstream. For example, reducing the number of RB and PSA was critical to improve oral drug bioavailability [8]. On average, both the number of RB and PSA increased with molecular weight up to a cutoff molecular weight of 500 g/mol. Molecular weight trends suggest that increasing the size of molecules above 400–500 g/mol would lead to a steady decline in bioavailability, mainly due to the effect on absorption ($F_a$) by the decrease in solubility and permeability. The study also suggested that compounds with 10 or fewer RB and PSA equal to or less than 140 Å showed good oral bioavailability [8]. The above findings were consistent with the most widely used concepts to qualitatively predict oral drug absorption called Lipinski's rule of five [9, 10]. Overall, parameters examined in the study showed an impact only on $F_a$ but not on $F_g$ and $F_h$, suggesting that the relationship between rule-of-five and bioavailability is associated mainly with intestinal absorption.

After oral administration, the active moiety leaves the dosage formulation and dissolves in the aqueous digestive fluids of GIT mucosal membrane before passing into the blood. Solubility and permeability are two fundamental properties determining the absorption of an orally active drug. Based on these properties, drugs are classified according to the biopharmaceutical classification system (BCS) into four categories: class I—high solubility and high permeability, class II—low solubility and high permeability, class III—high solubility and low permeability, and class IV—low solubility and low permeability [11]. The most frequent causes of low oral absorption are attributed to poor solubility and/or low permeability [12]. A prerequisite for the absorption of oral drugs is that they must be in aqueous solution, which depends on drugs' dissolution rate and aqueous solubility. On the other hand, hydrophilic compounds have low permeability across the biological membranes and permeability becomes the rate-limiting step in achieving adequate absorption [11, 12]. There are numerous approaches available to enhance aqueous solubility of poorly water-soluble drugs. Solubility improvement techniques can be categorized into physical modification by particle size reduction like micronization [13, 14], nanoparticles [15], modification of the crystal habit polymorphs and amorphous forms [16], and chemical modifications of the drug substance salt formation [17] or complexation [18].

## 39.3 Mechanisms of Drug Absorption and Concept of Oral Bioavailability (see also Chapters 4–6)

Drugs are primarily absorbed in the small intestinal regions and several mechanisms of oral drug absorption have been shown including passive transcellular diffusion, paracellular transport, and carrier-mediated active transport. Of these, passive diffusion is recognized as the main mechanism for absorption for most lipophilic compounds [19].

Passive diffusion is comprised of two pathways: the paracellular pathway, in which drug diffuses through the aqueous pores at the tight junctions between the intestinal enterocytes, and the transcellular (lipophilic) pathway, which requires drug diffusion across the lipid cell membrane of the enterocytes. The active transport pathway is mediated by transporters and is divided into active drug influx and efflux. It is important to note that the relevance of each route is determined by the compound's physicochemical properties and its potential affinity for various transport proteins [20]. The most referenced drug interactions are via the efflux transporter P-glycoprotein (P-gp) and breast cancer resistance protein (BCRP). A summary of intestinal transport processes involved in the movement of small molecules from the intestinal lumen into the bloodstream is described in detail [21]. Taken together, intestinal absorption is a complex process and is influenced by various factors such as anatomical, physicochemical (lipophilicity, solubility, p$K_a$ value, molecular weight), biochemical (uptake/efflux transport processes, metabolism), and biopharmaceutical properties (drug release and dissolution from the drug product).

Absolute bioavailability is usually determined by calculating the respective plasma drug exposure assessed as the total area under the drug plasma concentration versus time curve (AUC) after oral and intravenous administration [4] as described in Eq. (39.1).

$$\text{Absolute Bioavailability}(\%) = \frac{\text{AUC}_{\text{oral}}}{\text{AUC}_{\text{iv}}} \times \frac{\text{Dose}_{\text{iv}}}{\text{Dose}_{\text{oral}}} \times 100\% \tag{39.1}$$

Absolute oral bioavailability can also be derived from clearance as described in Eqs. (39.2)–(39.5)

$$\text{CL}_{\text{h}} = \text{CL}_{\text{tot}} - \text{CL}_{\text{r}} \tag{39.2}$$

$$F_{\text{h}} = 1 - \text{CL}_{\text{h}}/Q_{\text{h}} \tag{39.3}$$

$$F_{\text{a}}\,F_{\text{g}} = F_{\text{oral}}/F_{\text{h}} \tag{39.4}$$

where

$\text{CL}_{\text{tot}}$: total plasma clearance
$\text{CL}_{\text{h}}$: hepatic clearance
$\text{CL}_{\text{r}}$: renal clearance
$Q_{\text{h}}$: hepatic blood flow rate

Rearrangement of Eq. (39.4)

$$F_{\text{oral}} = F_{\text{a}} \cdot F_{\text{g}} \cdot F_{\text{h}} \tag{39.5}$$

where

$F_{\text{oral}}$: absolute oral bioavailability.
$F_{\text{a}}$: fraction absorbed: the fraction of the dose entering the cellular space of the enterocytes from the intestinal lumen.
$F_{\text{g}}$: bioavailability in the gut: the fraction of the drug entering the enterocytes that escapes first-pass gut metabolism.
$F_{\text{h}}$: bioavailability in liver: the fraction of the drug that escapes first-pass hepatic metabolism and biliary secretion.

Experimentally, it is difficult to evaluate $F_{\text{a}}$ and $F_{\text{g}}$ separately, these parameters are often treated as the product of $F_{\text{a}}$ and $F_{\text{g}}$ ($F_{\text{a}}\,F_{\text{g}}$) (Eq. (39.5)), which is the fraction of the administered drugs that reach the portal blood [22]. The product $F_{\text{a}}\,F_{\text{g}}$ is determined by the drug's physical–chemical (solubility), and biochemical properties (membrane permeability, transporters) as well as its susceptibility to intestinal metabolism [22].

## 39.4 Intestinal Metabolism and Oral Bioavailability (see Also Chapter 11)

Like absorption, both intestinal and hepatic metabolism impact on oral bioavailability. Therefore, understanding drug metabolic clearance (or stability) and metabolism pathways are important steps for drug discovery and development and the latter studies are typically called reaction phenotyping studies [23]. Throughout the years studies from industry and academia have demonstrated the impact of intestinal metabolism on small-molecule bioavailability [3, 24]. CYP3A4 has been shown to be a major phase I drug-metabolizing enzyme responsible for the metabolic clearance of approximately 30% of clinically important drugs [25, 26]. This enzyme is abundantly expressed in the intestine and liver, and the resultant first-pass metabolism (FPM) would significantly reduce drug's bioavailability. CYP3A4 metabolizes drugs from different therapeutic areas, such as benzodiazepines (e.g. alprazolam and midazolam), macrolide antibiotics (e.g. clarithromycin and erythromycin), calcium channel blockers (e.g. diltiazem, verapamil), carbamazepine, human immunodeficiency virus-1 protease inhibitors, (e.g. indinavir and saquinavir) and kinase inhibitors (e.g. cobimetinib). Furthermore, clinically significant drug-drug or drug-food interactions may also alter bioavailability. For example, inhibition of CYP3A4 with grapefruit juice in the intestine and liver gave rise to altered systemic exposure of drugs *in vivo*, resulting in labeling changes for several medications [27].

Our knowledge of intestinal metabolism has increased in recent years. Evidence points to lower protein abundance and catalytic activity for the majority of phase 1 and phase II drug-metabolizing enzymes (DMEs) within the gut wall, compared to the liver. However, the gut is the first organ exposed to a drug following oral dosing. Therefore, the resulting high concentration of drug in the enterocytes during the absorption phase may lead to substantial metabolic extraction before the drug enters the liver [28]. Indeed, a growing body of evidence demonstrates that the GIT not only contributes to low $F_{\text{oral}}$, through restricting the fraction absorbed, but also by metabolism that can occur as a drug transit through the gut wall. Most of the principal DMEs located in the liver are also localized in the gut [29]. For phase I enzymes, CYP3A4 is assumed to be the predominant CYP expressed in the small intestine followed by CYP2C9 and 2C19 [30]. Intestinal CYP3A4 expression in human represents ~1% of the hepatic estimate [30]. Regional differences in CYP450 activities were also observed with the jejunum compared with the duodenum and ileum [31]. Regarding phase II metabolism, conjugation enzymes have been detected in the intestine, including uridine diphosphate-glucuronosyltransferases (UGTs), sulfotransferases (SULTs), and glutathione *S*-transferases (GSTs). The distribution of these enzymes along the

GIT is not even and the expression levels are not assumed to be like hepatic levels. A recent report showed regional differences in human intestinal drug metabolism and indicated that phase I metabolism predominantly occurred in the ileum, while phase II metabolism mostly took place in the colon [32]. Based on the literature conjugation reactions, glucuronidation and sulfation in particular, are also important for the metabolism of drugs in the intestine (e.g. terbutaline, afimoxifene, endoxifen, fulvestrant, raloxifene, acetaminophen, phenylephrine, and others). As a result of enterocyte uptake and metabolic processes located there, a large percentage of drugs undergo some degree of metabolism before they reach the portal blood [33].

## 39.5 *In Vitro* Systems Applied to Assess Intestinal Metabolism

In drug discovery, *in vitro* systems are used to assess the role that intestinal metabolism plays in mediating poor $F_{\text{oral}}$ in animal PK models. One such application is to facilitate troubleshooting of "compound series" focusing on issues such as the underlying causes and consequences of poor oral exposure in the rat [34]. Secondly, as shown in Eq. (39.1), $F_{\text{oral}}$ is a multiparametric endpoint. Each independent parameter $F_a$ $F_g$ and $F_h$ can be measured across species using *in vitro* systems with scaling factors to predict the corresponding value in humans. Reasonable extrapolation of human intestinal bioavailability data generated from *in vitro* human intestinal metabolism has been reported [35, 36]. In addition, the human intestinal model is particularly important given that species differences in absorption, distribution, metabolism, and excretion (ADME) can greatly affect the correlation of $F_{\text{oral}}$ [30, 37].

A large variety of *in vitro* experimental models are available to assess the impact of intestinal metabolism on drug bioavailability. In this section, we will focus on the application of human intestinal microsomes (HIM), human intestinal S9 fraction (HIS9), cryopreserved human intestinal mucosa (CHIM), and cryopreserved human enterocytes (CHE).

### 39.5.1 Human Intestinal Microsomes

Microsomes are vesicles obtained from the processing of endoplasmic reticulum membranes by differential centrifugation of various tissues. They are frequently used in drug discovery because they are convenient in terms of costs, commercial availability, and ease of use. Microsomes generally contain membrane-bound phase I enzymes namely CYPs, flavine-containing monooxygenases (FMO), esterases, amidases, epoxide hydrolases, and the phase II enzymes such as UGTs. Certain enzyme reactions such as CYPs and UGTs need to be incubated in reaction mixture supplemented with the corresponding essential cofactors such as a nicotinamide adenine dinucleotide phosphate (NADPH) regenerating system for CYPs and uridine diphosphoglucuronic acid (UGPDA) for UGTs.

HIM are used in early drug discovery to determine the fraction metabolized by the gut mediated by CYP and/or UGT. HIM has been used to assess oral bioavailability of saquinavir, testosterone, and niclosamide. It is well documented that HIV protease inhibitors have poor oral bioavailability. Early work conducted on saquinavir (Invirase®) showed poor oral bioavailability and patients took up to 4–5 tablets/day to maintain therapeutic efficacy. The poor oral bioavailability may be associated with intestinal clearance. The potential of HIM to metabolize saquinavir in 17 different HIM preparations was assessed by Fitzsimmons and Collins [38], demonstrating that saquinavir was metabolized by HIM to multiple mono- and dihydroxylated species. Ketoconazole and troleandomycin, selective inhibitors of CYP3A4, potently inhibited the formation of all oxidative metabolites of saquinavir while selective inhibitors for other CYPs had little inhibitory effect. In addition, all saquinavir metabolites highly correlated with testosterone 6β-hydroxylation and recombinant CYP3A4 oxidized saquinavir to the same metabolic profile observed with HIM. Indinavir, another potent HIV protease inhibitor and a substrate for human hepatic CYP3A4, was a comparatively poor substrate for HIM and inhibited the oxidative metabolism of saquinavir to all metabolites. These data showed that saquinavir was metabolized by human intestinal CYP3A4, which in turn contributed to its poor oral bioavailability [39].

It is widely recognized that liver is a major site of glucuronidation for orally administered small molecules. It is now clear that human intestinal glucuronidation is a common cause of incomplete and variable absolute bioavailability for an orally dosed drug. Recently, a study conducted by Zhang et al. [40] showed that UGT2B17 protein abundance and activity compared to UGT2B15 was much higher in enterocytes than hepatocytes and that UGT2B17 but not UGT2B15 was shown to play a significant role in testosterone first-pass metabolism [40]. Imatinib, a selective inhibitor of UGT2B17 [41] and HIM obtained from UGT2B17 deleted genes [42] significantly decreased testosterone glucuronide formation. Together, the data suggested a critical role of intestinal

UGT2B17 in first-pass metabolism of testosterone and other UGT2B17 oral substrates.

Niclosamide, a drug with various pharmacological properties showed poor *in vivo* efficacy due to low bioavailability. The impact of *in vitro* matrices HIM and human liver microsomes (HLM) on niclosamide bioavailability was examined by Fan et al. [43]. In HIM reactions, hydroxy-niclosamide formation was not detected, but rates of niclosamide-glucuronide formation were much greater than in liver. Michaelis Menten parameters $K_m$ and $V_{max}$ for microsomal niclosamide-glucuronide formation showed large differences between the two tissues, with higher $K_m$ and $V_{max}$ in the intestine than the liver. These results suggest that the observed poor oral bioavailability of niclosamide was partially due to intestinal glucuronidation [43]. These studies suggest a need for the incorporation of intestinal glucuronidation into clearance prediction, especially for compounds metabolized by intestine specific UGTs.

Throughout the years, progress on *in vitro-in vivo* extrapolation (IVIVE) of intestinal metabolism has been made but not like the extent of established liver models. This is due in part to lack of standardized methodology, consensus for the optimal isolation of subcellular fractions in the intestine, regional difference in enzymes expression along the GIT, and preparation of human enterocytes, which in turn make data comparison between laboratories difficult. Unlike liver, little is known about the physiological scalars necessary for extrapolation of data generated from the various *in vitro* intestinal systems. However, the limited number of studies and frequent failure to correct for losses during sub-cellular fraction preparation preclude confidence in IVIVE using microsomal scaling factors typified for the liver [44]. As a result, different strategies have been utilized to scale intestinal $CL_{int}$, for example, based on CYP3A abundance. It is noteworthy that these values come from samples prepared by mucosal scraping, which can bias the estimate due to the highly mechanical nature of the procedure which is known to dilute or deteriorate the CYP450s [2, 45].

### 39.5.2 Human Intestinal S9 Fraction

The S9 fraction contains both cytosol and microsomes and represents a nearly complete collection of all drug-metabolizing enzymes. The main components of human intestinal S9 (HIS9) fractions are CYPs, UGTs, aldehyde oxidase (AO), xanthine oxidase (XO), SULT, methyltransferase, monoamine oxidases (MAO), *N*-acetyltransferase (NAT), and GST, which are considered to be major contributors for the metabolism of a variety of drugs in the market. The S9 dataset is therefore richer in content and provides an opportunity for medicinal chemists in discovery to stabilize compounds against both Phase I and II simultaneously. Similar to microsomes, S9 fraction assays need cofactors for certain enzymes such as CYPs and UGTs. There have been reports on an important role of the human intestinal SULTs, (cytosolic enzymes), in the metabolism of a wide variety of drugs such as acetaminophen [46], ractopamine and salbutamol [47], phenylephrine [48, 49], and opioid drugs (morphine, hydromorphone, oxymorphone, butorphanol, nalbuphine, levorphanol, nalorphine, and naltrexone) [50]. For example, using HIS9 and HLS9 fractions, opicapone 3-*O*-sulfate formation from opicapone, a newly approved drug for treatment of Parkinson's disease, was higher in HIS9 fractions compared to HLS9 fractions. The results obtained in the study support a major role of intestinal conjugation of opicapone to opicapone 3-O-sulfation in the low bioavailability of the drug [51].

Unlike oxidation by CYP, the role of reductive enzymes is not well established, and their metabolism in the intestine is not well understood. The significance of reductive metabolism in human intestine and establishment of a quantitative prediction method of intestinal first-pass metabolism by cytosolic reductive enzymes, using haloperidol, mebendazole, and ziprasidone were investigated by Nishimuta et al. [52]). Human intestinal cytosol (HIC) showed intestinal FPM of haloperidol, mebendazole, and ziprasidone to be mediated by non-CYP450 enzyme carbonyl reductase. It has been shown that carbonyl reductases have higher metabolic activities in human intestine than in human liver [53]. The metabolic activities for all these three compounds were also largely inhibited by menadione, a carbonyl reductase-selective chemical inhibitor [52]. In addition, a quantitative prediction of intestinal bioavailability ($F_g$) for these compounds was established using an artificial membrane permeability assay and metabolic activities in HIS9 fraction [52].

### 39.5.3 Human Intestinal Mucosa and Enterocytes

CHIM has been recently developed for potential use as *in vitro* surrogate model for the small intestine system to assess oral bioavailability of drugs [31, 54]. Briefly, the small intestine is dissected into three segments (duodenum, jejunum, and ileum), digested by collagenase, homogenized, and subsequently intestinal mucosa is stored in liquid nitrogen for later use. Overall, the system showed retention of common CYP450 and non-CYP450 activities in all three segments with higher CYP3A4 and MAO enzyme activities compared with other DMEs. A recent study using CHIM system demonstrated that grapefruit juice, a classic CYP3A4 inhibitor, produced the expected dose dependent CYP3A inhibition thereby substantiating that CHIM may be a useful *in vitro* experimental model for the evaluation of enteric CYP3A inhibition [55].

An important step in the development of new drugs is to evaluate the extent of drug bioavailability associated with metabolism and transporters during absorption in the small intestine. Reliable *in vitro* systems to do this can expedite the development process, but the current systems are often unsuitable because they lack the appropriate metabolic enzymes (e.g. Caco-2 cell monolayers) or are not representative of the physiological conditions present in the intact intestinal cells (e.g. isolated microsomes). Recently isolation and cryopreservation of human intestinal enterocytes (CHE) have been reported [56]. The CHE in suspension showed similar profile of enzymes to those of mucosa [56]. However, the use of enterocytes in culture to study transporters and metabolism or transporter-metabolism interplay might be difficult due to their short survival time in suspension and their inability to attach in culture plates for longer-term cultures [57]. In the course of improving the use of CHE, cofactor-supplemented and permeabilized cryopreserved human enterocytes were developed [58]. The model has recently been used for clearance studies [59]. The results showed that metabolic clearance of CYP3A4/5 substrates such as midazolam, amprenavir, and loperamide was lower in enterocytes compared to that of hepatocytes, which agrees with previously published studies [30]. In contrast, UGT and carboxylesterase (CES) activities were higher in enterocytes compared to hepatocytes. In addition, conjugation of salbutamol by SULT1A3 was detected in enterocytes and not in hepatocytes, indicating higher SULT activity in enterocytes than hepatocytes. As expected, the permeabilized enterocytes did not show active P-gp efflux functions [59]. Taken together, the data suggested that this model may be useful for assessing first-pass clearance in human intestine.

### 39.5.4 EpiIntestinal Microtissues

A novel *in vitro* system for the estimation of fraction absorbed in human intestine was published by Takeneka et al. [60]. This *in vitro* system is called EpiIntestinal microtissues and is made of Caco-2 cells coexpressing recombinant human CYP3A4 and NADPH-CYP reductase. The advantage of EpiIntestinal microtissues is the presence of both intact barrier function and co-expressed DME activities. The combined barrier function and DME activities in EpiIntestinal microtissues make it possible to evaluate intestinal first-pass availability ($F_a$ $F_g$) in humans in a single experiment. The activities of relevant drug transporters and drug-metabolizing enzymes, including P-gp, BCRP, CYP3A4, CYP2J2, UGT, and CES were detected in functional assays with selective substrates and inhibitors. Compared to Caco-2, EpiIntestinal microtissues proved to be a more acceptable model for the investigation of drug absorption and metabolism in human GIT [61]. Some limitations of the EpiIntestinal system are microtissue availability and lack of microtissue donor information (medical, ethnicity, sex, and other relevant information). In addition, pharmacogenetic data obtained from a single donor expressing polymorphic metabolizing enzymes are certainly not representative for the patient population.

A major limitation in investigating intestinal metabolism is the lack of reproducibility in available *in vitro* systems [28, 62]. This is in contrast to *in vitro* tools available for studying hepatic metabolism, including microsomes, subcellular fractions, and cryopreserved human hepatocyte (CHH), which are standard assays used in drug discovery. The region of the intestine used for *in vitro* preparations can differ widely, as can the preparation techniques for tissue fractionation and isolation of intestinal subcellular fractions, leading to inconsistent quality of preparations, which could impact on scaling factors [3] and limit their use in intestinal metabolism assessment [63]. Primary human enterocytes show promising activity [56], but isolation may also be more sensitive to the method employed, due to the heterogeneous nature of the small intestine compared to the liver [3, 54, 58]. Regardless of enterocytes' metabolic system, knowledge deficits in enzyme abundance and resulting lack of scaling factors are major limitations of intestinal *in vitro* tools, which are further hampered in preclinical animal models due to interspecies differences [28]. Therefore, modeling of events on first pass through the intestine requires understanding of the complex interplay between passive permeability, active transport, binding, relevant blood flows, and the intrinsic activity and capacity of enzyme systems. *In vitro* scaling and physiologically based pharmacokinetic modeling methods with varying levels of mechanistic complexity were developed to predict *in vivo* human $F_g$ [64–66].

## 39.6 *In Vitro* Systems Applied to Assess Human Hepatic First-Pass Metabolism (see Also Chapter 12)

Extensive hepatic metabolism is often associated with variable systemic bioavailability and leads to lower efficacy of small molecule drug. The prediction of the fraction of drug escaping hepatic metabolism ($F_h$), as well as transporter-mediated secretion, has been challenged by the lack of appropriate preclinical models. Application of *in vitro* hepatic systems to determine the clearance of low-turnover metabolic compounds has also been a challenge. In addition, hepatic transporters such as organic anion transporting polypeptides (OATP) 1B1 and 1B3 have been recognized as important determinants in drug hepatic clearance, in particular for statins, sartans, and glinides drugs [67, 68]. Nevertheless, over the past decade, significant progress has been made to predict hepatic metabolic clearance using *in vitro* models, which include recently developed tools [69]. This section will review the utility of human HLM, human liver S9 fractions (HLS9), human recombinant cytochromes P450, and CHHs in the assessment of bioavailability.

### 39.6.1 Human Liver Microsomes

HLM are a conventional model to assess the metabolic stability of drug candidates that are eliminated by metabolism in the liver and to predict the pharmacokinetic (PK) parameters involved in the dosing and dosage frequency. HLM were used in the early 90s [70] to predict the oral bioavailability of derivatives of progestogenic 19-nortestoterones where a comparison of the scaled *in vivo* $CL_{int}$ values with the corresponding *in vivo* $CL_{int}$ values showed a reasonable correlation. HLM has been routinely used to predict hepatic clearance in early stage of drug discovery to help developing lower-clearance drug candidates. Well-stirred model is commonly used to predict *in vivo* clearance. With increasing knowledge of IVIVE for metabolic clearance in the past decades, it is generally believed that use of HLM underpredicts the *in vivo* clearance due to the absence of certain phase II and transporter-related properties [71–74]. However, it offers the ability for enhanced-throughput clearance screening at relatively low cost. HLM are also routinely used for assessment of human CYP and UGT-mediated metabolism and for understanding how metabolic pathway impacts bioavailability [1, 23, 75]. In the reaction phenotyping studies, HLM are incubated with a single or multiple concentration of test compound supplemented with respective cofactors such as NADPH or UGPDA. After incubation, loss of test compound and/or metabolite formation are quantitated, and this will indicate the involvement of CYP or UGT metabolism. Furthermore, use of HLM in conjunction with selective chemical inhibitors for each CYP isoform can estimate $f_m$, the fraction of hepatic clearance of the substrate mediated by these CYP enzymes [1, 23]. For example, the antiplatelet agent clopidogrel is a pro-drug known to undergo intestinal FPM by CYP2C9 to 2-oxo-clopidrogel (active biological entity) [76, 77]. Recently, a study by Ford [78] using HLM showed that the conversion of clopidogrel to its active metabolite was mediated by CYP3A oxidation, suggesting that CYP3A in enterocyte may play a role in clopidogrel bioavailability and efficacy. This was further supported by a clinical drug interaction study with a CYP3A inhibitor grapefruit juice, which caused a sixfold reduction in the AUC of the active metabolite. Therefore, the active metabolite of clopidogrel is now considered to be primarily a CYP3A4/5-mediated with CYP2C19 playing a minor role [78, 79].

It should be noted that the use of HLM in metabolism studies may produce confounding results. For example, 4-oxidation metabolism of carbazeran in HLM incubation with NADPH as cofactor was thought to be mediated by CYPs [80]. Recently, using a complementary experimental approach demonstrated that AO was responsible for the biotransformation of carbazeran to carbazeran 4-oxidation. The significant metabolism measured in HLM with NADPH could be the outcome of cytosolic contamination with HLM [81]. Therefore, it is recommended to quantify the extent of cytosolic contamination of each lot of HLM and to use proper controls such as microsomal incubations without NADPH for HLM metabolism experiments.

### 39.6.2 Human Liver S9 Fractions

In general, HLS9 fractions as a test system are comparable to that of hepatocytes to screen compounds for metabolic stability assays [82]. For example, a comparative study on two different test systems, HLH and HLS9 fractions, with 7-ethoxy coumarin showed that phase-I metabolite 7-hyroxycoumarin was observed in both systems. In addition, 7-hyroxycoumarin glucuronide and sulfate conjugates were observed in hepatocyte and HLS9 [83]. In addition, it has been reported that

HLS9 and hepatocyte stability assays are binned into the same category 70–84% of the time. The *in vivo* rat plasma clearance has been reasonably predicted from data generated from S9 fraction and hepatocyte stability assays [84]. In another study, the rate for stability versus plasma clearance was 45% for hepatocytes and 43% for HLS9 [82]. Together, data suggest using HLS9 assay for high throughput metabolic screening purposes would provide the combined benefit of comprehensive and high-quality data at a reasonable expense for drug discovery programs.

Typical screening assays are performed in HLM as they contain the CYP enzymes, and it is possible to overlook the contribution of non-microsomal clearance until later stages of the drug discovery process. There has been an increase in efforts to improve the metabolic stability of new drug entities by reducing their susceptibility to CYP-mediated metabolism [85]. However, such strategies often lead to an increase in metabolic clearance by non-CYP enzymes, such as AO, XO, SULT, MAO, and other reductases and hydrolases, [23]. Therefore, HLS9 has been routinely used for non-CYP metabolism to screen drug candidates. AO-mediated metabolism was found responsible for the low exposure of FK3453 in humans [86], which highlights the importance of understanding the metabolism of new drug candidates in systems that contain the full array of metabolizing enzymes. In contrast, XO is not a commonly screened enzyme in drug discovery. In general, to assess XO metabolic activity for a new chemical entity *in vitro*, allopurinol was used to selectively inhibit XO activities [87]. These inhibition experiments are most often performed with human liver cytosol. Since the process of liver harvest involves liver perfusion with solutions containing allopurinol, human cytosol fractions could be devoid of any XO activity [88]. Therefore, it is recommended that human liver cytosol /HLS9 batches be screened for oxypurinol (metabolite of allopurinol) and/or XO activity prior to testing for XO-mediated metabolism of a new chemical entity.

### 39.6.3 Recombinant Enzymes

Recombinant enzymes (e. g. Supersomes™, Bactosomes™) are a technology designed for studying individual human DMEs. They are normally generated from the recombinant expression of a single DME in insect cells or *Escherichia coli* as a host via the use of a baculovirus vector. The simplicity, commercial availability, and ability to study a single enzyme create the appeal of this system. Comparison of the metabolic activities of recombinant enzymes and microsomes illustrates that recombinant enzymes accurately reflect the *in vivo* situation with respect to $K_m$ values, however, Vmax values are not always reproducible [89]. Recombinant enzymes are an excellent system to confirm hypotheses generated from other *in vitro* systems (i.e. HIM, HIS9, enterocytes, HLM, HLS9, hepatocytes, and others) regarding the contribution of individual enzymes in the clearance of a drug [1, 23]. The quantitative extrapolation of human CYP-mediated metabolic activity in the recombinant CYP isoform to that in the liver microsomes, so-called relative activity factor (RAF) can also be used to estimate $f_m$ [72]. As discussed in previous sections, a systemic analysis of suspected enzymes from *in vitro* systems combined with the determination of enzyme kinetic parameters can identify the major responsible enzyme(s) involved in the clearance of a drug [23]. For example, orally administered astemizole (Hismanal®) is an antihistaminic drug with great absorption but extensive first-pass metabolism to *O*-desmethylastemizole and desmethylastemizole in HIM and HLM systems suggested a role of CYP in the first-pass metabolism of this molecule [90]. Human CYPs involved in the O-demethylation of astemizole could not be identified early on in discovery. While searching for the identity of CYPs responsible for astemizole extraction, higher activities of the astemizole O-demethylation were observed in the rabbit small intestine than in the liver (about threefold). The data suggested the possible involvement of CYP2J2, since this enzyme is dominantly expressed in the small intestine of rabbits [91]. Indeed, higher activity of astemizole O-demethylation was detected in recombinant CYP2J2. Expression levels of the immunoreactive protein for CYP2J2 in the small intestine were also well correlated with the activities of the astemizole O-demethylation. The CYP2J2 substrates arachidonic acid and ebastine strongly inhibited the microsomal astemizole O-demethylation in human small intestines and recombinant CYP2J2. These results indicated the involvement of CYP2J2 in the pre-systemic elimination of astemizole in the human small intestine [92].

### 39.6.4 Human Hepatocytes

CHHs are commercially available and have been implemented in various fields of research, including induction, inhibition, metabolism, clearance, cytotoxicity, and transporter studies. Due to interindividual variability of donors, hepatocyte metabolism studies are often performed with a pool of multiple donors. A major advantage of cryopreserved hepatocytes is the ability to reproduce an experiment under identical conditions (e.g. long-term storage of the same lot or batch of cells). Like other primary cells, hepatocytes are a more physiological relevant model system, compared with microsomes, because they contain a full complement of enzymes, cofactors and relevant uptake transports [93–96]. The presence of uptake transporters and cofactors is considered to be the missing link between microsomes and CHH, explaining discrepancies between the two systems. Therefore, prediction of human hepatic clearance for complex mechanisms of enzyme-transporter interplay (transporter-mediated hepatic uptake and canalicular excretion) has been a major challenge in drug discovery and development due to the limitation of compound- and transporter-dependent scaling factors, inability to detect transporter activity for highly lipophilic permeable compounds, and the need to know the rate-determining step prior to clearance prediction [68, 97]. Nevertheless, hepatocytes are still a conventional model for clearance IVIVE studies. Like HLM, systematic underprediction of *in vivo* $CL_{int}$ from hepatocyte suspension model appears to be universally apparent [71, 72, 74]. A recent study with a large set of data (>140 compounds) showed that early half of test compounds exhibited equivalent scaled intrinsic clearance for both HLM and hepatocytes [98]. But significant difference between HLM and hepatocyte clearances were observed. The reasons for the disconnection were unknown but appeared not correlated with human *in vivo* CL, human volume distribution, Log $D$, molecular weight, or compound p$K_a$. The disconnection may also not be related to uptake transporters [98]. Interestingly, CYP3A substrates displayed a significantly higher $CL_{int}$ in HLM versus hepatocytes and overlapping substrate specificities between CYP3A and Pgp may contribute to the observed disconnection [98]. In another report, the average underprediction of *in vivo* $CL_{int}$ was less in HLM than that in hepatocytes and this occurred more frequently with high turnover compounds, most likely due to an increasing incidence of permeation rate limitation of the high turnover drugs and cofactor depletion in hepatocyte during incubation [74]. Analysis of the relation between plasma protein binding and transporter-mediated hepatic uptake also provided possible explanations for the underprediction of hepatic clearance and use of hepatocytes provided better prediction of hepatic clearance compared to HLM, for example, 73% of the drugs studied, scaled clearances were within twofold of the actual clearance after consideration of plasma binding and uptake transporters [72]. Indeed, recent studies showed that use of human hepatocytes in suspension in conjunction with measuring liver-to-plasma partitioning (Kpuu) resulted in good IVIVE of hepatic clearance for a number of drugs [68].

## 39.7 Long-Term Hepatocyte Culture and Slow Metabolizing Drug Candidate

In drug discovery programs during lead optimization, one of the common parameters that scientists focus on is the optimization of metabolic stability to identify a lead drug candidate with acceptable clearance and therapeutically feasible dosage regimen. Metabolic soft-spot analysis in simple *in vitro* systems such as hepatocytes and S9 fractions is well suited for this objective [82]. On the other hand, once the lead compound(s) become metabolically stable, scientists have to manage the challenge to accurately characterize and differentiate slowly metabolized drug candidates from both a clearance rate and metabolite production standpoint. Slowly metabolized drugs require longer incubation times to enable a robust signal that can be measured with confidence [99, 100], which is not practical using conventional *in vitro* systems. In addition, as the number of low-clearance compounds in pharmaceutical programs increases due to the application of highly effective metabolic clearance reduction strategies in lead optimization to allow once-daily administration of moderate drug quantities, more sophisticated *in vitro* liver models are required [99, 100].

Several investigators have tested new methodologies to extend incubation times and more sufficiently measure metabolism of low-turnover drugs. These methods include a relay of sequential incubations with cryopreserved human hepatocytes in suspension, Human HepatoPac® (micropatterned coculture hepatocyte system), HμREL® hepatocyte coculture, and hepatocyte spheroid culture, among others. In this section, the advantages, and disadvantages of each of the listed *in vitro* methodologies as it relates to the prediction of clearance of low-turnover metabolite identification are described below.

### 39.7.1 Human Hepatocytes Relay Method

Primary hepatocytes in suspension are commonly used to predict *in vivo* clearance. However, incubation times typically limit their use to four to six hours, which is not long enough to accurately evaluate the metabolic stability of slowly metabolized compounds. The hepatocyte relay method was designed to repeatedly transfer the supernatant from the previous hepatocyte incubations to the next freshly thawed hepatocyte suspension at the end of the 4-hour incubation to prolong the exposure time to active enzymes in hepatocytes [101]. A cumulative incubation time of 20 hours (five relays) or longer in hepatocytes can be achieved using this method [99]. The relay method was tested using drugs with known low *in vivo* intrinsic clearance (diazepam, disopyramide, theophylline, timolol, tolbutamide, S-warfarin, and zolmitriptan) that were metabolized by various CYPs with known low human *in vivo* intrinsic clearance. The results showed that the relay method provided a good IVIVE in predicting human *in vivo* clearance. The difference between *in vitro* and *in vivo* intrinsic clearance was within twofold for most compounds, which is similar to the standard prediction accuracy for moderate to high clearance compounds using hepatocytes. The need to replace hepatocytes multiple times during the course of a relay assay makes the assay more labor intensive (four to five consecutive days) and increases material costs. Also, data manipulation can prove challenging. As the method represents a mathematical approach to integrate several separate four-hour incubations into one algorithm, there are potential issues with error propagation and statistical significance of the defined $CL_{int}$ [102]. Although the relay method is effective in predicting *in vivo* clearance for compounds that are cleared via hepatic metabolism, there could be a notable under prediction for drugs that are substrates of uptake and efflux transporters [99].

### 39.7.2 Human HepatoPac and HμREL Hepatocyte Coculture

HepatoPac is a coculture of hepatocytes, supported by mouse embryonic 3T3 fibroblasts with retention of *in vivo*–like hepatocyte functions in culture (e.g. albumin secretion, urea synthesis, drug-metabolizing activities, and canalicular efflux transports) [103–105]. Using this system, 17 compounds known to be primarily metabolized by the liver (10 compounds with low clearance and 7 compounds with intermediate clearance) were incubated for 7 days without media change. The $CL_h$ values were calculated based on the parent depletion profiles in HepatoPac and were compared with the reported *in vivo* $CL_h$ values for all compounds. Overall, HepatoPac cultures with the well-stirred model predicted within twofold of the *in vivo* $CL_h$ for 11 of 17 compounds; 7 of 10 compounds with low clearance, and 4 of 7 compounds with intermediate clearance. Using threefold as the criterion for accuracy of prediction, 14 of 17 compounds were predicted, and among them 10 of 10 for low clearance compounds, and 4 of 7 for intermediate clearance compounds. The plasma protein binding correction significantly improved prediction using this model [106]. The extended functionality of hepatocytes in human HepatoPac suggests that additional important factors such as drug transporters, which can play a crucial role in drug clearance may also be appropriately represented *in vitro* [107]. Limitations of using this method in a discovery setting include the higher cost and increased labor intensiveness relative to cryopreserved hepatocytes in suspension due to extended incubation time. However, for later-staged chemical series where metabolism is difficult to measure this model may be an option. The longevity of drug-metabolizing enzyme activity and general health of cells have also enabled numerous ADME applications, including the ability to study the interplay between drug metabolism and transport.

Regarding the interplay between DMEs and transporters, the system was tested using faldaprevir, a selective inhibitor of the hepatitis C virus (HCV) NS3/NS4 protease [108]. During preclinical evaluations, rats dosed with the drug showed a significant amount of faldaprevir present in the liver suggesting a possible role of uptake transporters [109]. Faldaprevir is a low-clearance drug in humans and undergoes oxidation to two hydroxylated metabolites detected in feces (40% of the administered dose) without exhibiting appreciable levels of circulating metabolites [110]. In human HepatoPac cultures, two predominant hydroxylated metabolites were formed comparable to *in vivo* and faldaprevir was concentrated in the liver with active uptake by multiple transporters. The data suggested that human HepatoPac cultures provided valuable insights into the metabolism and disposition of faldaprevir in humans and demonstrated the importance of enzyme and transporter interplay in the clearance of the drug [111].

HμREL is a cell-based hepatic model comprised of primary cryopreserved hepatocytes cultured together with cells of a non-parenchymal (stromal type) used for *in vivo* clearance prediction and metabolite identification of slowly metabolized drugs [112]. The validity of this system to predict metabolites of slowly cleared drugs was tested using four low-turnover drugs, over seven days incubation. Several relevant metabolites that were identified in HμREL system were also reported in human *in vivo* studies. Overall, the HμREL model maintained metabolic activity over a diverse array of metabolic enzymes and was able to predict *in vivo* metabolism for low-turnover compounds or compounds that undergo multistep biotransformation [113]. In general, the HμREL model is less labor intensive than the relay method. Another advantage over other coculture models is that pooled hepatocytes can be plated in the HμREL model, Nonetheless, just like any *in vitro* hepatocyte model, the metabolite data produced in the HμREL model are dependent on the enzyme activity in the hepatocyte pools selected. In addition, the suitability of the HμREL model for enzyme phenotyping remains a major challenge for drugs that are slowly metabolized and require longer incubation time [113].

### 39.7.3 Human Hepatocytes Spheroid Cultures

Primary human hepatocyte (PHH) spheroid culture can maintain mature hepatic phenotypes for three to five weeks and is a useful tool to study metabolism of slowly metabolizing compounds. Multiple methods for the generation of spheroids have been developed including stirring bioreactors, aggregation in hanging drops, and culture on ultralow attachment (ULA) surfaces, namely Corning® spheroid microplates with the latter seemingly becoming the predominant spheroid-formation technique in recent years [114–118].

Spheroid formed functional bile canaliculi and exhibited hepatocyte function such as secretion of albumin and urea synthesis over the course of at least two weeks in culture [118]. In addition, the spheroids retained their various phase I and phase II enzymes, and transporter functions in the extended three to five weeks of culture time [114, 116–119]. Recently, Kanebratt et al. [120] reported that hepatocyte spheroids cultured for 21 days maintained the expression and activities of CYPs enzymes (CYP1A2, 2B6, 2C9, 2D6, and 3A4). Interestingly, the activities of CYP2B6 and 3A4 decreased, while the activities of CYP2C9 and 2D6 increased over time. The intrinsic clearance values and metabolite formation for compounds that undergo multiple metabolic pathways were comparable to standard *in vitro* 2D hepatic models and the model successfully predicted *in vivo* $CL_{int}$ within threefold from observed values for low clearance compounds [120].

Moreover, spheroids were successfully used to evaluate the impact of genetic polymorphisms in drug-metabolizing enzymes on the metabolic fate of their substrates. Dextromethorphan is primarily metabolized by CYP2D6 to dextrorphan; however, when cells from a donor were used who was phenotypically and genotypically categorized as a poor CYP2D6 metabolizer (CYP2D6*4/*10), the metabolic flux was shunted toward CYP3A4-dependent metabolism, resulting in the predominant formation of 3-methoxymorphinan [119]. These findings encourage the use of such systems to evaluate the effects of genetic polymorphisms, many of which are common in the general population on the metabolism and metabolic fingerprints of drugs of interest [121]. Moreover, spheroids have been suggested as promising tools for the identification of drug-drug interactions by systematically evaluating the effects of co-exposures on metabolic profiles [122].

## 39.8 Microbiome and Absorption: A New Perspective

Like enterocytes and hepatocytes, gut microorganisms have been shown to have an impact on the bioavailability of orally administered drugs. In recent years, many studies have shown that bacteria secrete microbial enzymes that metabolize drug molecules via hydrolysis, hydroxylation, deamidation, decarboxylation, and reduction of azide groups [123, 124]. Examples of drugs that have been shown to be affected by microbial enzymes are omeprazole [125], digoxin [126], irinotecan, [127] glycyrrhizin [128], and among others.

Interplay between host cytochrome P450 and gut bacterial enzymes was also involved in drug metabolism in GIT [126, 129]. Both human and anaerobic gut bacteria contributed to the metabolism of fostamatinib, a tyrosine kinase inhibitor and a major fecal metabolite observed in humans was believed to be the result of hepatic CYP-mediated O-demethylation and subsequent O-demethylations and dehydroxylation by gut bacteria [130].

Use of an *in vitro* colon model has revealed that simvastatin was extensively metabolized by the colon microbiota [131]. Simvastatin was hydroxylated and subjected to β-oxidation, followed by glutathione conjugation and glucuronidation in the liver. The demethylation, carbon-carbon bond cleavage, α or β oxidation, dihydroxylation, and cyclization of simvastatin were thought to arise by the colon bacteria [131]. Other drugs determined to be significantly metabolized by colonic bacteria using *in vitro* cultures include ranitidine, an H2 antagonist [132].

Gut microorganisms may also promote enterohepatic recirculation of drugs. For example, SN-38 active form of the pro-anticancer drug irinotecan (CPT-11) can be conjugated by hepatic UGT1A1 to form the inactive SN-38-glucouronide. The later underwent biliary secretion into the intestines where it was deconjugated back to SN-38 by β-glucuronidase produced by bacteria including *E. coli*, *Bacteroides vulgatus*, and *Clostridium ramosum* [133].

The potential impact of the microbiota on drug pharmacokinetics has also been examined using rats. For example, the administration of a live probiotic (*E. coli Nissle 1917*) to rats increased the bioavailability of amiodarone [134]; however, it is yet to be determined whether these changes in the pharmacokinetics of amiodarone are due to alterations in drug transport or bacterial metabolism. A study performed in human patients also indicated that the microbiome may alter the pharmacokinetics of tacrolimus, a drug with a narrow therapeutic index, and may affect the therapeutic drug levels [135], since analysis of the fecal microbiota in 19 patients involved in this study indicated that those requiring higher doses of tacrolimus also harbor an abundance of fecal *Faecal ibacterium prausnitzii*. It remains to be determined whether an abundance of commensal bacteria, like *Faecal ibacterium prausnitzii*, which are often associated with a "healthy gut" [136] corresponds to an "optimal" drug-metabolizing capacity in the gut.

Intestinal microbiome may also modulate the expression levels of DMEs and transporters in the intestine and the liver of wild and germ-free (GF) mice [137]. Overall, the study showed that genes encoding DMEs were downregulated in GF mice compared to wild-type mice. In contrast, levels of certain uptake and efflux transporters and many phase II enzymes were increased in GF mice. Interestingly, this differential expression was not specific to the intestinal tissue but was also seen in the liver [137]. In a different study, where GF mice and antibiotic-treated mice were used as surrogate models for dysbiosis, a change in the level of expression of DMEs and transporters was observed in the liver and kidneys in both models. The study suggested a potential role of dysbiosis in drug pharmacokinetics and bioavailability [138].

As far as studying the impact of intestinal flora on oral bioavailability is concerned, *in vitro* systems such as Fecal Slurry, Isolated Pure Bacterial Cultures, and Complex Intestinal Models have been used. The most common approach to estimate metabolism by microflora is by direct incubation with bacterial preparations, which can be either from gut or fecal content or pure cultures of microorganisms characteristic of the GIT [129]. Such studies allow a ranking of the degradation rate of test compounds against compounds with characterized or known degradation rates to draw conclusions to the extent of *in vivo* microflora metabolism. [139]. A more advanced model reflects not only the colon conditions but also the preceding steps in the stomach and small intestine. The rationale for this elaborate *in vitro* model design is that *in vivo*, the matrix presented to the colon has undergone complex interactions with transporters, bile, and enzymes at a range of pH levels. Part of the matrix will thus be modified and slowly presented to the microbiota. The Complex Intestinal Model was developed to simulate more accurate conditions in the large intestine (colon), including pH, anaerobiosis, and gradual intake of predigested meal compounds from the small intestine [140]. The compartments contain a high-density metabolically active microflora of human origin (healthy subjects or patients). Selected beneficial or pathogenic bacteria can be added to this microflora. Single or repeated addition of the compounds to the model is possible. Samples can be taken in real time for chemical and microbial analysis and functional cell-based assays. The results can provide information about bioaccessibility, biofunctionality, produced metabolites, and changes in microbiota.

## 39.9 Summary

This chapter described the application of intestinal and hepatic tissue fractions as conventional *in vitro* models to assess oral absorption and metabolism on the bioavailability of drug molecules in pre-clinical studies. Understanding the contribution of intestinal and hepatic metabolism to oral bioavailability aids in estimating clinical pharmacokinetic parameters including oral doses. Over the years, emphasis on reducing metabolic clearance of drugs has increased with the goal of improving patient compliance by decreasing the frequency of drug regimen. Currently, efforts are aimed at developing newer *in vitro* hepatic models (i.e. relay method, hepatocyte coculture, and spheroid methods) to improve the prediction of oral bioavailability of such drug molecules. As we continue to understand how the gut microbiota modulates metabolism of drugs, advanced pre-clinical models have been developed to understand its overall impact on patient response, factors that contribute to interindividual differences, and the mechanisms that underlie the host-microbiome interplay. Certainly, there will be additional, newer, and better-modified models developed in the future to support an efficient drug discovery and development process.

## Acknowledgments

Authors would like to thank Dr. Elke Perloff for her valuable feedback during the preparation of this chapter.

## Abbreviations

| | |
|---|---|
| ADME | absorption distribution metabolism and excretion |
| AO | aldehyde oxidase |
| AUC | area under curve |
| BCRP | breast cancer resistance protein |
| BCS | Biopharmaceutical Classification System |
| CHHs | cryopreserved human hepatocytes |
| CHEs | cryopreserved human enterocytes |
| CES | carboxylesterase |
| CHIM | cryopreserved human intestinal mucosa |
| $C_{max}$ | maximum concentration |
| CYP | Cytochrome P450 |
| DMEs | drug-metabolizing enzymes |
| F | Bioavailability |
| $F_a$ | the fraction absorbed |
| $F_g$ | the intestinal availability |
| $F_h$ | the fraction transferring to systemic circulation |
| FMO | flavin-monooxygenase |
| FPM | First-pass metabolism |
| GIT | Gastrointestinal tract |
| HIC | Human intestinal cytosol |
| HIS9 | Human intestinal S9 fraction |
| HLM | Human liver microsomes |
| HIM | Human intestine microsomes |
| IVIVE | *in vitro -in vivo* extrapolation |
| MAO | monoamine oxidase |
| NAT | *N*-acetyltransferase |
| PK | Pharmacokinetic |
| PHH | primary human hepatocytes |
| RB | free rotatable bonds |
| PSA | polar surface area |
| SULT | Sulfotransferases |
| $T_{1/2}$ | half-life |
| $T_{max}$ | time of maximum concentration |
| UGT | Uridine diphosphate glucuronosyl transferases |

## References


1 Food and Drug Administration Center for Drug Evaluation and Research (CDER) (2020). *In vitro* drug interaction studies - cytochrome P450 enzyme- and transporter-mediated drug interactions guidance for industry.

2 Jones, C.R., Hatley, O.J.D., Ungell, A. et al. (2016). Gut wall metabolism. Application of pre-clinical models for the prediction of human drug absorption and first-pass elimination. *AAPS J.* 18: 589–604.

3 Hatley, O.J.D., Jones, C.R., Galetine, A., and Rostami-Hodjegana, A. (2017). Optimization of intestinal microsomal preparation in the rat: a systematic approach to assess the influence of various methodologies on metabolic activity and scaling factors. *Biopharm. Drug Dispos.* 38: 187–208.

4 Rowland, M. and Tozer, T.N. (1995). *Clinical Pharmacokinetics Concepts and Applications*, 2e, 33–48. London: Lea and Febiger Philadelphia.

5 Kiela, P.R. and Ghishan, F.K. (2016). Physiology of intestinal absorption and secretion. *Best Pract. Res. Clin. Gastroenterol.* 30: 145–159.

6 Vertzonia, M., Augustijnsb, P., Grimmc, M. et al. (2019). Impact of regional differences along the gastrointestinal tract of healthy adults on oral drug absorption: an UNGAP review. *Eur. J. Pharm. Sci.* 134: 153–175.

7 Hua, S. (2020). Advances in oral drug delivery for regional targeting in the gastrointestinal tract-influence of physiological, pathophysiological and pharmaceutical factors. *Front. Pharmacol.* 11: 1–22.

8 Veber, D.F., Johnson, S.R., Cheng, H.Y. et al. (2002). Molecular properties that influence the oral bioavailability of drug candidate. *J. Med. Chem.* 45: 2615–2623.

9 Lipinski, C.A. (1995). Computational alerts for potential absorption problems: profiles of clinically tested drugs. In: *Tools for Oral Absorption. Part Two. Predicting Human Absorption. AAPS Meeting Miami, AAPS.*

10 Lipinski, C.A. (2000). Drug-like properties and the causes of poor solubility and poor permeability. *J. Pharmacol. Toxicol. Methods* 44: 235–249.

11 Dahan, A., Miller, J.M., and Amidon, G.L. (2009). Prediction of solubility and permeability class membership: provisional BCS classification of the world's top oral drugs. *AAPS J.* 11: 740–746.

12 Benet, L.Z. (2013). The role of BCS (biopharmaceutics classification system) and BDDCS (biopharmaceutics drug disposition classification system). *Drug Development J. Pharm. Sci.* 102: 34–42.

13 Chu, K.R., Lee, E., Jeong, S.H., and Park, E.S. (2012). Effect of particle size on the dissolution behaviors of poorly water-soluble drugs. *Arch. Pharmacal Res.* 35: 1187–1195.

14 Sun, J., Wang, F., Sui, Y. et al. (2012). Effect of particle size on solubility, dissolution rate, and oral bioavailability: evaluation using coenzyme $Q_{10}$ as naked nanocrystals. *Int. J. Nanomed.* 7: 5733–5744.

15 Mokarram, R.A., Zadeh, K.A., Keshavarz, M. et al. (2010). Preparation and *in-vitro* evaluation of indomethacin nanoparticles. *DARU* 18: 185–192.

16 Chadha, R., Kuhad, A., Arora, P., and Kishor, S. (2012). Characterization and evaluation of pharmaceutical solvates of atorvastatin calcium by thermoanalytical and spectroscopic studies. *Chem. Cent. J.* 6: 114–129.

17 Serajuddin, A.T. (2007). Salt formation to improve drug solubility. *Adv. Drug Delivery Rev.* 59: 603–616.

18 Shimpi, S., Chauhan, B., and Shimpi, P. (2005). Cyclodextrins: application in different routes of drug administration. *Acta Pharm.* 55: 139–156.

19 Cao, X., Gibbs, S.T., Fang, L. et al. (2006). Why is it challenging to predict intestinal drug absorption and oral bioavailability in human using rat model? *Pharm. Res.* 23: 1675–1860.

20 Hurst, S., Loi, C.M., Brodfuehrer, J., and El-Kattan, A. (2007). Impact of physiological, physicochemical and biopharmaceutical factors in absorption and metabolism mechanisms on the drug oral bioavailability of rats and humans. *Expert Opin. Drug Metab. Toxicol.* 3: 469–489.

21 Lodish, H., Berk, A., Zipursky, S.L. et al. (2000). *Molecular Cell Biology*, 4e, 524–546. New York: W. H. Freeman.

22 Tanaka, Y., Kitamura, Y., Maeda, K., and Sugiyama, Y. (2016). Explication of definitional description and empirical use of fraction of orally administered drugs absorbed from the intestine ($F_a$) and intestinal availability ($F_g$): effect of P-glycoprotein and CYP3A on $F_a$ and $F_g$. *J. Pharm. Sci.* 105: 431–442.

23 Bohnert, T., Patel, A., Templeton, I. et al. (2016). Evaluation of a new molecular entity as a victim of metabolic drug-drug interactions—an industry perspective. *Drug Metab. Dispos.* 44: 1399–1423.

24 Davies, M., Peramuhendige, P., Golding, L.K.M. et al. (2020). Evaluation of *in vitro* models for assessment of human intestinal metabolism in drug discovery. *Drug Metab. Dispos.* 48: 1169–1182.

25 Rendic, S. and Guengerich, F.P. (2015). Survey of human oxidoreductases and cytochrome P450 enzymes involved in the metabolism of xenobiotic and natural chemicals. *Chem. Res. Toxicol.* 28: 38–42.

26 Zanger, U.M. and Schwab, M. (2013). Cytochrome P450 enzymes in drug metabolism: regulation of gene expression, enzyme activities, and impact of genetic variation. *Pharmacol. Ther.* 138: 103–141.

27 Bailey, D.G., Dresser, G., Malcolm, J., and Arnold, O. (2013). Grapefruit–medication interactions: forbidden fruit or avoidable consequences. *CMAJ* 5: 309–316.

28 Peters, S.A., Jones, C.R., Ungell, A.L., and Hatley, O.J.D. (2016). Predicting drug extraction in the human gut wall: assessing contributions from drug metabolizing enzymes and transporter proteins using preclinical models. *Clin. Pharmacokinet.* 55: 673–696.

29 Speer, J.E., Wang, Y., Fallon, J.K. et al. (2019). Evaluation of human primary intestinal monolayers for drug metabolizing capabilities. *J. Biol. Eng.* 13: 82–97.

30 Xie, F., Ding, X., and Zhang, Q.Y. (2016). An update on the role of intestinal cytochrome P450 enzymes in drug disposition. *Acta Pharm. Sin. B* 6: 374–383.

31 Li, A.P., Ho, M.C.D., Alam, N. et al. (2020). Inter-individual and inter-regional variations in enteric drug metabolizing enzyme activities: Results with cryopreserved human intestinal mucosal epithelia (CHIM) from the small intestines of 14 donors. *Pharmacol. Res. Perspect.* 8: 1–17.

32 Iswandana, R., Irianti, M.I., Oosterhuis, D. et al. (2018). Regional differences in human intestinal drug metabolism. *Drug Metab. Dispos.* 46: 1879–1885.

33 Gavhane, Y.N. and Yadav, A.V. (2012). Loss of orally administered drugs in GI tract. *Saudi Pharm. J.* 20: 331–344.

34 Li, Y., Meng, Q., Yang, M. et al. (2019). Current trends in drug metabolism and pharmacokinetics. *Acta Pharm. Sin.* 9: 1113–1144.

35 Furukawa, T., Yamano, K., Naritomi, Y. et al. (2012). Method for predicting human intestinal first-pass metabolism of UGT substrate compounds. *Xenobiotica* 42: 980–988.

36 Kadono, K., Akabane, T., Tabata, K. et al. (2010). Quantitative prediction of intestinal metabolism in humans from a simplified intestinal availability model and empirical scaling factor. *Drug Metab. Dispos.* 38: 1230–1237.

37 Wu, B., Kulkarni, K., Basu, S. et al. (2011). First-Pass metabolism via UDP-glucuronosyltransferase: a barrier to oral

bioavailability of phenolics. *J. Pharm. Sci.* 100: 3655–3682.

**38** Fitzsimmons, M.F. and Collins, M.J. (1997). Selective biotransformation of the human immunodeficiency virus protease inhibitor saquinavir by human small-intestinal cytochrome P4503A4. *Drug Metab. Dispos.* 25: 256–266.

**39** Eagling, V.A., Wiltshire, H., Whitcombe, I.W., and Back, D.J. (2002). CYP3A4-mediated hepatic metabolism of the HIV-1 protease inhibitor saquinavir *in vitro*. *Xenobiotica* 32: 1–17.

**40** Zhang, H., Basit, A., Busch, D. et al. (2018). Quantitative characterization of UDP-glucuronosyltransferase 2B17 in human liver and intestine and its role in testosterone first-pass metabolism. *Biochem. Pharmcol.* 156: 32–42.

**41** Zhang, N., Liu, Y., and Jeong, H. (2015). Drug-Drug interaction potentials of tyrosine kinase inhibitors via inhibition of UDP-glucuronosyltransferases. *Sci. Rep.* 5: 17778. https://doi.org/10.1038/srep17778.

**42** Bhatt, D.K., Basit, A., Zhang, H. et al. (2018). Hepatic abundance and activity of androgen- and drug-Metabolizing enzyme UGT2B17 are associated with genotype, age, and sex. *Drug Metab. Dispos.* 46: 888–896.

**43** Fan, X., Li, H., Ding, X., and Zhang, Q.Y. (2019). Contributions of hepatic and intestinal metabolism to the disposition of niclosamide a repurposed drug with poor bioavailability. *Drug Metab. Dispos.* 47: 756–763.

**44** Choi, G.W., Lee, Y.B., and Cho, H.Y. (2009). Interpretation of non-clinical data for prediction of human pharmacokinetic parameters: *in vitro-in vivo* extrapolation and allometric Scaling. *Pharmaceutics* 11: 2–32.

**45** Cho, H.J., Kim, J.E., Kim, D.D., and Yoon, I.S. (2014). *In vitro–in vivo* extrapolation (IVIVE) for predicting human intestinal absorption and first-pass elimination of drugs: principles and applications. *Drug Dev. Ind. Pharm.* 40: 989–998.

**46** Bairam, A.F., Rasool, M.I., Alherz, F.A. et al. (2018). Effects of human *SULT1A3/SULT1A4* genetic polymorphisms on the sulfation of acetaminophen and opioid drugs by the cytosolic sulfotransferase SULT1A3. *Arch. Biochem. Biophys.* 648: 44–52.

**47** Ko, A., Kurogi, K., Davidson, G. et al. (2012). Sulfation of ractopamine and salbutamol by the human cytosolic sulfotransferases. *J. Biochem.* 152: 275–283.

**48** Bairam, A.F., Rasool MI, M.I., Alherz FA, F.A. et al. (2019). Impact of *SULT1A3/SULT1A4* genetic polymorphisms on the sulfation of phenylephrine and salbutamol by human SULT1A3 allozymes. *Pharmacogenet. Genomics* 29: 99–105.

**49** Yamamoto, A., Kim, J., Liu, M.Y. et al. (2014). Sulfation of phenylephrine by the human cytosolic sulfotransferases. *Drug Metab. Lett.* 8: 96–100.

**50** Kurogi, K., Chepak, A., Hanrahan, M.T. et al. (2014). Sulfation of opioid drugs by human cytosolic sulfotransferases: Metabolic labeling study and enzymatic analysis. *Eur. J. Pharm. Sci.* 62: 40–48.

**51** Lopes, C.M. (2013). Characterization of Opicapone Sulfation: Sulfotransferase Isoforms Ph.D. Thesis, Universidad Do Porto.

**52** Nishimuta, H., Sato, K., Yabuki, M., and Komuro, S. (2011). Prediction of the intestinal first-pass metabolism of CYP3A and UGT substrates in humans from *in vitro* data. *Drug Metab. Pharmacokinet.* 26: 592–601.

**53** Malátková, P. and Wsól, V. (2014). Carbonyl reduction pathways in drug metabolism. *Drug Metab. Rev.* 46: 96–123.

**54** Li, A.P., Alam, N., Amaral, K. et al. (2018). Cryopreserved human intestinal mucosal epithelium: A novel *in vitro* experimental system for the evaluation of enteric drug metabolism, cytochrome P450 induction, and enterotoxicity. *Drug Metab. Dispos.* 46: 1562–1571.

**55** Loretz, C., MCD, H., Alam, N. et al. (2020). Application of cryopreserved human intestinal mucosa and cryopreserved human enterocytes in the evaluation of herb-drug interactions: evaluation of CYP3A inhibitory potential of grapefruit juice and commercial formulations of twenty-nine herbal supplements. *Drug Metab. Dispos.* 48: 1084–1091.

**56** Ho, M.C.D., Ring, N., Amaral, K. et al. (2017). Human enterocytes as an *in vitro* model for the evaluation of intestinal drug metabolism: characterization of drug-metabolizing enzyme activities of cryopreserved human enterocytes from twenty-four donors. *Drug Metab. Dispos.* 45: 686–691.

**57** Li, A.P. (2020). *In vitro* human cell-based experimental models for the evaluation of enteric metabolism and drug interaction potential of drugs and natural products. *Drug Metab. Dispos.* 48: 980–992.

**58** Li, A.P., Amaral, K., and Ho, M.D. (2018). A novel *in vitro* experimental system for the evaluation of enteric drug metabolism: cofactor-supplemented permeabilized cryopreserved human enterocytes (MetMax Cryopreserved Human Enterocytes). *Drug Metab. Lett.* 12: 132–137.

**59** Wong, S., Doshi, U., Vuong, P. et al. (2018). Utility of pooled cryopreserved human enterocytes as an

*in vitro* model for assessing intestinal clearance and drug-drug interactions. *Drug Metab. Lett.* 12: 3–13.
60 Takenaka, T., Kazuki, K., Harada, N. et al. (2017). Development of Caco-2 cells co-expressing CYP3A4 and NADPH-cytochrome P450 reductase using a human artificial chromosome for the prediction of intestinal extraction ratio of CYP3A4 substrates. *Drug Metab. Pharmacokinet.* 32: 61–68.
61 Cui, Y., Claus, S., Schnell, D. et al. (2020). In-depth characterization of EpiIntestinal microtissue as a model for intestinal drug absorption and metabolism in human. *Pharmaceutics* 12: 405–417.
62 Rostami-Hodjegan, A., Tamai, I., and Pang, K.S. (2017). Revisiting the role of gut wall in the fate of orally administered drugs: why now and to what effect? *Biopharm. Drug Dispos.* 38: 87–93.
63 Küblbeck, J., Hakkarainen, J.J., Petsalo, A. et al. (2016). Genetically modified Caco-2 cells with improved cytochrome P450 metabolic capacity. *J. Pharm. Sci.* 105: 941–949.
64 Gertz, M., Houston, J.B., and Galetin, A. (2011). Physiologically based pharmacokinetic modeling of intestinal first-pass metabolism of CYP3A substrates with high intestinal extraction. *Drug Metab. Dispos.* 39: 1633–1642.
65 Huang, S.M. and Rowland, M. (2012). The role of physiologically based pharmacokinetic modeling in regulatory review. *Clin. Pharmacol. Ther.* 91: 542–549.
66 Kostewicz, E.S., Aarons, L., Bergstrand, M. et al. (2014). PBPK models for the prediction of *in vivo* performance of oral dosage forms. *Eur. J. Pharm. Sci.* 57: 300–321.
67 Kalliokoski, A. and Niemi, M. (2009). Impact of OATP transporters on pharmacokinetics. *Br. J. Pharmacol.* 158: 693–705.
68 Riccardi, K.A., Tess, D.A., Lin, J. et al. (2019). A novel unified approach to predict human hepatic clearance for both enzyme- and transporter-mediated mechanisms using suspended human hepatocytes. *Drug Metab. Dispos.* 47: 484–492.
69 Bowman, C.M. and Benet, L.Z. (2019). *In vitro-in vivo* extrapolation and hepatic clearance dependent underprediction. *J. Pharm. Sci.* 108: 2500–2504.
70 Kuhnz, W. and Gieschen, H. (1998). Predicting the oral bioavailability of 19- nortestosterone progestins *in vivo* from their metabolic stability in human liver microsomal preparations *in vitro*. *Drug Metab. Dispos.* 26: 1120–1127.
71 Brown, H.S., Griffin, M., and Houston, J.B. (2007). Evaluation of cryopreserved human hepatocytes as an alternative *in vitro* system to microsomes for the prediction of metabolic clearance. *Drug Metab. Dispos.* 35: 293–301.
72 Chiba, M., Ishii, Y., and Sugiyama, Y. (2009). Prediction of hepatic clearance in human from *in vitro* data for successful drug development. *AAPS J.* 11: 262–276.
73 Clarke, S.E. and Jeffrey, P. (2008). Utility of metabolic stability screening: comparison of *in vitro* and *in vivo* clearance. *Xenobiotica* 31: 591–598.
74 Wood, F.L., Houston, J.B., and Hallifax, D. (2017). Clearance prediction methodology needs fundamental improvement: Trends common to rat and human hepatocytes/microsomes and implications for experimental methodology. *Drug Metab. Dispos.* 45: 1178–1188.
75 Jia, L. and Liu, X. (2007). The conduct of drug metabolism studies considered good practice (II): *in vitro* experiments. *Curr. Drug Metab.* 8: 822–829.
76 Conrado, D.J., Gonzalez, C.D., and Derendorf, H. (2010). Role of drug absorption in the pharmacokinetics of therapeutic interventions for stroke. *Ann. N.Y. Acad. Sci.* 1207: 134–142.
77 Laine, L. and Hennekens, C. (2010). Proton pump inhibitor and clopidogrel interaction: fact or fiction? *Am. J. Gastroenterol.* 105: 34–41.
78 Ford, N.F. (2016). The metabolism of clopidogrel: CYP2C19 is a minor pathway. *J. Clin. Pharmacol.* 56: 1474–1483.
79 Ford, N.F. and Taubert, D. (2013). Clopidogrel, CYP2C19, and a black box. *J. Clin. Pharmacol.* 53: 241–248.
80 Kaye, B., Offerman, J.L., Reid, J.L. et al. (1984). A species difference in the presystemic metabolism of carbazeran in dog and man. *Xenobiotica* 14: 935–945.
81 Xie, J., Saburulla, N.F., Chen, S. et al. (2019). Evaluation of carbazeran 4-oxidation and $O_6$-benzylguanine 8-oxidation as catalytic markers of human aldehyde oxidase: impact of cytosolic contamination of liver microsomes. *Drug Metab. Dispos.* 47: 26–37.
82 Richardson, S.J., Bai, A., Kulkarni, A.A., and Moghaddam, M.F. (2016). Efficiency in drug discovery: liver S9 fraction assay as a screen for metabolic stability. *Drug Metab. Lett.* 10: 83–90.
83 Sahi, J., Grepper, S., and Smith, C. (2010). Hepatocytes as a tool in drug metabolism, transport and safety evaluations in drug discovery. *Curr. Drug Discov. Technol.* 7: 188–198.
84 Kulkarni, A., Riggs, J., Phan, C. et al. (2014). Proposing advancement criteria for efficient DMPK

triage of new chemical entities. *Future Med. Chem.* 6: 131–139.

85 Rashidi, M.R. and Soltani, S. (2017). An overview of aldehyde oxidase: an enzyme of emerging importance in novel drug discovery. *Expert Opin. Drug Discovery* 12: 305–316.

86 Akabane, T., Tanaka, K., Irie, M. et al. (2011). Case report of extensive metabolism by aldehyde oxidase in humans: pharmacokinetics and metabolite profile of FK3453 in rats, dogs, and humans. *Xenobiotica* 41: 372–384.

87 Panoutsopoulos, G.I., Kouretas, D., and Beedham, C. (2004). Contribution of aldehyde oxidase, xanthine oxidase, and aldehyde dehydrogenase on the oxidation of aromatic aldehydes. *Chem. Res. Toxicol.* 17: 1368–1376.

88 Barr, J.T., Choughule, K.V., Nepal, S. et al. (2014). Why do most human liver cytosol preparations lack xanthine oxidase activity? *Drug Metab. Dispos.* 42: 695–699.

89 Plant, N. (2004). Strategies for using *in vitro* screens in drug metabolism. *Drug Discovery Today* 9: 328–336.

90 Matsumoto, S., Hirama, T., Matsubara, T. et al. (2002). Involvement of CYP2J2 on the intestinal first-pass metabolism of antihistamine drug astemizole. *Drug Metab. Dispos.* 30: 1240–1245.

91 Matsumoto, S., Hirama, T., Kim, H.J. et al. (2003). *In vitro* inhibition of human small intestinal and liver microsomal asemizole O - demethylation: different contribution of CYP2J2 in the small intestine and liver. *Xenobiotica* 33: 615–623.

92 Das, A., Weigle, T., Arnold, W.R. et al. (2020). CYP2J2 molecular recognition: a new axis for therapeutic design. *Pharmacol. Ther.* 215: 1–74.

93 Badolo, L., Trancart, M.M., Gustavsson, L., and Chesné, C. (2011). Effect of cryopreservation on the activity of OATP1B1/3 and OCT1 in isolated human hepatocytes. *Chem. Biol. Interact.* 190: 165–170.

94 De Bruyn, T., Ye, Z.W., Peeters, A. et al. (2011). Determination of OATP-, NTCP- and OCT- mediated substrate uptake activities in individual and pooled batches of cryopreserved human hepatocytes. *Eur. J. Pharm. Sci.* 43: 297–307.

95 Shitara, Y., Li, A.P., Kato, Y. et al. (2003). Function of uptake transporters for taurocholate and estradiol 17beta-D-glucuronide in cryopreserved human hepatocytes. *Drug Metab. Pharmacokinet.* 18: 33–41.

96 Zhang, D., Luo, G., Ding, X., and Lu, C. (2012). Preclinical experimental models of drug metabolism and disposition in drug discovery and development. *Acta Pharm. Sin. B* 2: 549–561.

97 Kusuhara, I. and Sugiyama, Y. (2009). *In vitro-in vivo* extrapolation of transporter-mediated clearance in the liver and kidney. *Drug Metab. Pharmacokinet.* 24: 37–52.

98 Williamson, B., Harlfinger, S., and McGinnity, D.F. (2020). Evaluation of the disconnect between hepatocyte and microsome intrinsic clearance and *in vitro in vivo* extrapolation performance. *Drug Metab. Dispos.* 48: 1137–1146.

99 Di, L. and Obach, R.S. (2015). Addressing the challenges of low clearance in drug research. *AAPS J.* 17: 352–357.

100 Hutzler, J.M., Ring, B.J., and Anderson, S.R. (2015). Low-turnover drug molecules: a current challenge for drug metabolism scientists. *Drug Metab. Dispos.* 43: 1917–1928.

101 Di, L., Trapa, P., Obach, R.S. et al. (2012). A novel relay method for determining low-clearance values. *Drug Metab. Dispos.* 40: 1860–1865.

102 Peng, C.C., Doshi, U., Prakash, C., and Li, A.P. (2016). A novel plated hepatocyte relay assay (PHRA) for *In vitro* evaluation of hepatic metabolic clearance of slowly metabolized compounds. *Drug Metab. Lett.* 10: 3–15.

103 Ballard, T.E., Wang, S., Cox, L.M. et al. (2016). Application of a micropatterned cocultured hepatocyte system to predict preclinical and human specific drug metabolism. *Drug Metab. Dispos.* 44: 172–179.

104 Chan, T.S., Yu, H., Moore, A. et al. (2013). Meeting the challenge of predicting hepatic clearance of compounds slowly metabolized by cytochrome P450 using a novel hepatocyte model HepatoPac. *Drug Metab. Dispos.* 41: 2024–2032.

105 Khetani, S.R. and Bhatia, S.N. (2008). Microscale culture of human liver cells for drug development. *Nat. Biotechnol.* 26: 120–126.

106 Lin, C., Shi, J., Moore, A., and Khetani, S.R. (2016). Prediction of drug clearance and drug-drug interactions in microscale cultures of human hepatocytes. *Drug Metab. Dispos.* 44: 127–1376.

107 Shitara, Y., Horie, T., and Sugiyama, Y. (2006). Transporters as a determinant of drug clearance and tissue distribution. *Eur. J. Pharm. Sci.* 27: 425–446.

108 Sulkowski, M.S. (2013). HCV therapy in HIV-infected patients. *Liver Int.* 33 (Suppl 1): 63–67.

109 White, P.W., Llinàs-Brunet, M., Amad, M. et al. (2010). Preclinical characterization of BI 201335, a C-terminal carboxylic acid inhibitor of the hepatitis C virus NS3-NS4A protease. *Antimicrob. Agents Chemother.* 54: 4611–4618.

110 Chen, L.Z., Rose, P., Mao, Y. et al. (2014). Mass balance and metabolite profiling of steady-state faldaprevir, a hepatitis C virus NS3/4 protease inhibitor, in healthy male subjects. *Antimicrob. Agents Chemother.* 58: 2369–2376.

111 Ramsden, D., Tweedie, D.J., Chan, T.S. et al. (2014). *In vitro* and *in vivo* metabolism and transport of faldaprevir in human using a novel cocultured human hepatocyte system HepatoPac. *Drug Metab. Dispos.* 42: 394–406.

112 Hultman, I., Vedin, C., Abrahamsson, A. et al. (2016). Use of HμREL human coculture system for prediction of intrinsic clearance and metabolite formation for slowly metabolized compounds. *Mol. Pharmaceutics* 13: 2796–2807.

113 Burton, R.D., Hieronymus, T., Chamem, T. et al. (2018). Assessment of the biotransformation of low turnover drugs in the H$\mu$REL human hepatocyte coculture model. *Drug Metab. Dispos.* 46: 1617–1625.

114 Bell, C.C., Hendriks, D.F., Moro, S.M. et al. (2016). Characterization of primary human hepatocyte spheroids as a model system for drug-induced liver injury, liver function and disease. *Sci. Rep.* 6: 25187. https://doi.org/10.1038/srep25187: 1–13.

115 Messner, S., Agarkova, I., Moritz, W., and Kelm, J.M. (2013). Multi-cell type human liver microtissues for hepatotoxicity testing. *Arch. Toxicol.* 87: 209–213.

116 Messner, S., Fredriksson, L., Lauschke, V.M. et al. (2018). Transcriptomic, proteomic, and functional long-term characterization of multicellular three-dimensional human liver microtissues. *Appl. In Vitro Toxicol.* 4: 1–13.

117 Rebelo, S.P., Costa, R., Silva, M.M. et al. (2017). Three-dimensional co-culture of human hepatocytes and mesenchymal stem cells: improved functionality in long-term bioreactor cultures Tissue. *J. Tissue Eng. Regen. Med.* 11: 2034–2045.

118 Tostões, R.M., Leite, S.B., Serra, M. et al. (2012). Human liver cell spheroids in extended perfusion bioreactor culture for repeated-dose drug testing. *Hepatology* 55: 1227–1236.

119 Vorrink, S.U., Ullah, S., Schmidt, S. et al. (2017). Endogenous and xenobiotic metabolic stability of primary human hepatocytes in long-term 3D spheroid cultures revealed by a combination of targeted and untargeted metabolomics. *FASEB J.* 31: 2696–2708.

120 Kanebratt, K.P., Janefeldt, A., Vilen, L. et al. (2020). Primary human hepatocyte spheroid model as a 3D *in vitro* platform for metabolism studies. *J. Pharm. Sci.* 110: 422–431.

121 Zhou, Y., Ingelman-Sundberg, M., and Lauschke, V. (2017). Worldwide distribution of cytochrome P450 alleles: a meta-analysis of population-scale sequencing projects. *Clin. Pharmacol. Ther.* 102: 688–700.

122 Ingelman-Sundberg, M. and Lauschke, V.M. (2018). Human liver spheroids in chemically defined conditions for studies of gene-drug, drug-drug and disease-drug interactions. *Pharmacogenomics* 19: 1133–1138.

123 Neelis, E., Koning, B.D., Rings, E. et al. (2019). The gut microbiome in patients with intestinal failure: current evidence and implications for clinical practice. *J. Parenter. Enteral. Nutr.* 43: 194–205.

124 Roeselers, G., Ponomarenko, M., Lukovac, S., and Wortelboer, H.M. (2013). *Ex vivo* systems to study host-microbiota interactions in the gastrointestinal tract. *Best Pract. Res. Clin. Gastroenterol.* 27: 101–113.

125 Watanabe, K., Yamashita, S., Furuno, K. et al. (1995). Metabolism of omeprazole by gut flora in rats. *J. Pharm. Sci.* 84: 516–517.

126 Li, H. and Jia, W. (2013). Co-metabolism of microbes and host: implications for drug metabolism and drug-induced toxicity. *Clin. Pharmacol. Ther.* 94: 574–581.

127 Vanhoefer, U., Harstrick, A., Achterrath, W. et al. (2001). Irinotecan in the treatment of colorectal cancer: clinical overview. *J. Clin. Oncol.* 19: 1501–1518.

128 Kim, D.H., Hong, S.W., Kim, B.T. et al. (2000). Biotransformation of glycyrrhizin by human intestinal bacteria and its relation to biological activities. *Arch. Pharmacal Res.* 23: 172–177.

129 Swanson, H.I. (2015). Drug metabolism by the host and gut microbiota: a partnership or rivalry? *Drug Metab. Dispos.* 43: 1499–1504.

130 Sweeny, D.J., Li, W., Clough, J. et al. (2010). Metabolism of fostamatinib, the oral methylene phosphate prodrug of the spleen tyrosine kinase inhibitor R406 in humans: contribution of hepatic and gut bacterial processes to the overall biotransformation. *Drug Metab. Dispos.* 38: 1166–1176.

131 Aura, A.M., Mattila, I., Hyötyläinen, T. et al. (2011). Drug metabolome of the simvastatin formed by human intestinal microbiota *in vitro*. *Mol. Biosyst.* 7: 437–446.

132 Basit, A.W. and Lacey, L.F. (2001). Colonic metabolism of ranitidine: implications for its delivery and absorption. *Int. J. Pharm.* 227: 157–165.

**133** Kajinami, K., Akao, H., Polisecki, E., and Schaefer, E.J. (2005). Pharmacogenomics of statin responsiveness. *Am. J. Cardiol.* 96: 65–70.

**134** Matuskova, Z., Anzenbacherova, E., Vecera, R. et al. (2014). Administration of a probiotic can change drug pharmacokinetics: effect of *E. coli* Nissle 1917 on amidarone absorption in rats. *PLoS One* 9: e87150. https://doi.org/10.1371/journal.pone.0087150.

**135** Lee, J.R., Muthukumar, T., Dadhania, D. et al. (2015). Gut microbiota and tacrolimus dosing in kidney transplantation. *PLoS One* 10: 1–17.

**136** Scott, K.P., Antoine, J.M., Midtvedt, T., and van Hemert, S. (2015). Manipulating the gut microbiota to maintain health and treat disease. *Microb. Ecol. Health Dis.* 26: 25877. https://doi.org/10.3402/mehd.v26.25877).

**137** Fu, Z.D., Selwyn, F.P., Cui, J.Y., and Klaassen, C.D. (2017). RNA-seq profiling of intestinal expression of xenobiotic processing genes in germ-free mice. *Drug Metab. Dispos.* 45: 1225–1238.

**138** Kuno, T., Hirayama-Kurogi, M., Ito, S., and Ohtsuki, S. (2016). Effect of intestinal flora on protein expression of drug-metabolizing enzymes and transporters in the liver and kidney of germ-free and antibiotics-treated mice. *Mol. Pharmaceutics* 13: 2691–2701.

**139** Papanicolas, L.E., Choo, J.M., Wang, Y. et al. (2019). Bacterial viability in faecal transplants: which bacteria survive? *EBioMedicine* 41: 509–516.

**140** Venema, K. (2015). The TNO *in vitro* model of the colon (TIM-2). In: *The Impact of Food Bioactives on Health* (ed. K. Verhoeckx et al.). Cham: Springer https://doi.org/10.1007/978-3-319-16104-4_26.

# 40

# Liver Perfusion and Primary Hepatocytes for Studying Drug Metabolism and Metabolite Excretion

*Paresh P. Chothe*[1], *Sean Xiaochun Zhu*[1], *Sandeepraj Pusalkar*[1], *Chuang Lu*[2], *and Cindy Xia*[1]

[1] *Global Drug Metabolism and Pharmacokinetics, Takeda Development Center Americas, Inc. (TDCA), Cambridge, MA, USA*
[2] *Drug Metabolism and Pharmacokinetics, Sanofi-Genzyme, Waltham, MA, USA*

## 40.1 Introduction

The liver is the largest internal organ in the human body and is strategically located between the digestive tract and the rest of the body [1]. It is the primary organ for metabolism and excretion of most endogenous and exogenous chemicals, as regulated by both drug-related transporters (Phase 0 and III) and drug-metabolizing enzymes (Phase I and II). Phase 0 involves uptake of chemicals into hepatocytes via either passive diffusion or active transport processes, which are mediated by sinusoidal uptake transporters such as organic anion transporter (OAT), organic cation transporter (OCT), organic anion transporting polypeptide (OATP), and $Na^+$-taurocholate cotransporting polypeptide (NTCP). Inside hepatocytes, molecules may undergo oxidative biotransformations by Phase I enzymes, mediated predominantly by cytochrome P450 (CYP), or direct conjugations by Phase II enzymes, after which the oxidized products may be further conjugated. Phase III involves the elimination of the intact parent molecule and its metabolites from hepatocytes into the bile duct via passive diffusion or canalicular efflux transporters, or from hepatocytes back into the systemic circulation via sinusoidal efflux or bidirectional uptake transporters. The scope of this chapter is limited to a discussion of metabolism and excretion within the hepatic systems. The localization of hepatobiliary transporters is illustrated in Figure 40.1.

Because the liver has a large capacity for extraction of drugs from the blood circulation, hepatocytes are generally exposed to high concentrations of chemicals. Metabolism by hepatocytes can produce active, inactive, or reactive metabolites, with consequences for pharmacologic activity and/or toxicity, sometimes to the extent of liver dysfunction or failure. Hepatic metabolism can have a huge impact on oral bioavailability via the first-pass effect, which can be affected by enzyme and transporter inhibition and induction.

Several in vitro and ex vivo liver models, including supersomes, microsomes, cytosols, S9 fractions, transfected cell lines, primary hepatocytes (suspension, plated, and sandwich culture), liver slices, and liver perfusions, have become norms for studying hepatic clearance of drugs. Of these models, liver perfusion and primary hepatocytes provide two very important tools used for understanding the role of the liver in the overall disposition and potential toxicity of compounds. Since these models provide different levels of cellular structure and architecture, the information derived about a single compound could vary greatly between the two models; liver perfusion would be expected to provide results more closely mimicking what would occur in vivo. The level of information derived from liver perfusion studies is more refined than that from hepatocyte studies, in which there are multiple additional barriers to uptake, mechanisms for biliary secretion, cell-to-cell interactions, and hepatocyte zonal differences. For both systems, a wide variety of conditions can be tailored to study specific aspects of drug metabolism, distribution, and elimination in liver – for example, enzyme inhibition and induction, biliary excretion, and transporter-based uptake and efflux showing the varied and distinctive flexibilities of these systems. This chapter describes liver perfusion and hepatocyte studies and discusses, with examples, the utility of each in terms of their ability to elucidate pathways of drug metabolism, drug interactions, and excretion. As with any in vitro or isolated tissue approach, the limitations of these techniques should be borne in mind when relating results to those from the whole body.

*Oral Bioavailability and Drug Delivery: From Basics to Advanced Concepts and Applications*, First Edition. Edited by Ming Hu and Xiaoling Li.

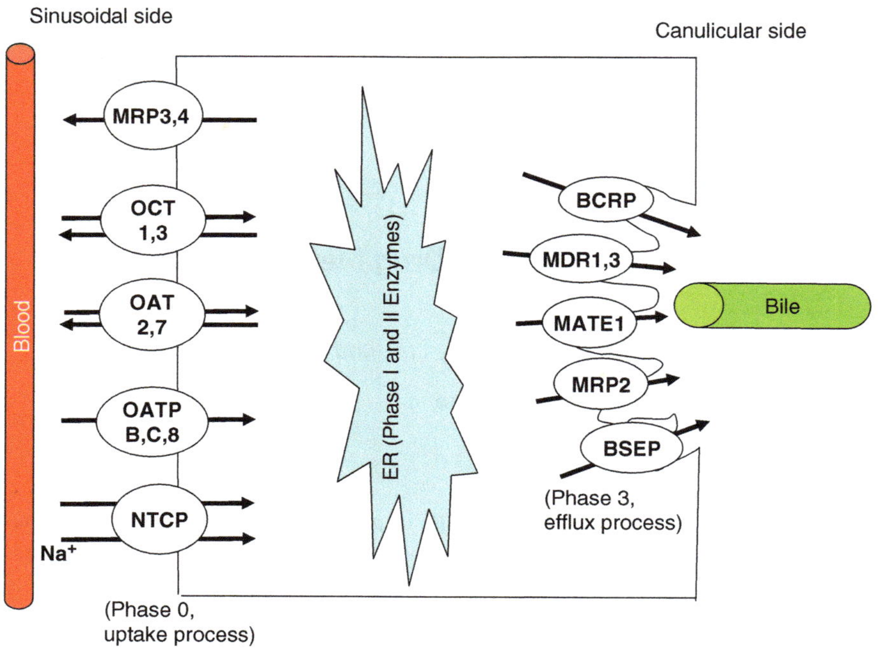


**Figure 40.1 Localization of hepatobiliary transporters in hepatocytes. ABC transporter in liver**: BCRP, breast cancer resistance protein (ABCG2); BSEP, bile salt export pump (ABCB11); MDR, multidrug resistance; P-glycoprotein (P-gp) encoded by MDR1 (ABCB1) and 3 (ABCB4); MRP2, multidrug resistance protein 2 (ABCC2), 3 (ABCC3), and 4 (ABCC4); SLC transporters in liver: NTCP, sodium-dependent bile acid transporter (SLC10A1); MATE1, multidrug and toxic compound extrusion (SLC47A1); OAT, organic anion transporter (SLC22A); OATP, organic anion transporting polypeptide (SLC21A); OCT, organic cation transporter (SLC22A). The human transporter genes are designated by all capital letters and the animal transporter genes are denoted by all lower-case letters or a capital letter followed by lower-case letters.

## 40.2 Liver Perfusion

### 40.2.1 The Technique

Organ perfusion is a useful tool for determining the mechanistic aspects of drug clearance and interaction within a particular organ. Most major organs have been used in this way; however, the predominantly used organ is the liver, since it plays a major role in the elimination of drugs from the body. Livers from small animals — including rats [2], fish [3], and sheep [4] — and lobes from humans have been perfused [5]. The technique is more or less standardized in terms of surgery and setup, with multiple variations in the constituents of the perfusate. The technique has also been used to understand aspects of the toxicology and pharmacology of molecules.

Since the details of liver perfusion technique are well described elsewhere [6, 7], it has been discussed briefly in this chapter but with adequate references for readers. Also, this chapter focuses on the use of the more abundant rat liver. Livers from higher nonclinical species and humans could be used similarly, but their lack of accessibility and viability would be restrictive to frequent use.

Liver perfusion can be performed in situ or in an isolated liver perfusion system; these techniques can be used interchangeably. The advantages of the former include less organ preparation, less organ damage, and less interruption of perfusion. However, with the latter, liver-specific mechanistic studies can be conducted to provide definitive answers, with less interference from other organs. Since isolated liver perfusion is more generally utilized, the utility of this technique will be discussed in more detail.

Isolated liver perfusion systems are commercially available (e.g. Harvard Apparatus, Holliston, MA). Options for perfusates are numerous, allowing multiple variations in albumin and red blood cell (RBC) binding, and other factors that may affect the metabolism and hence the bioavailability of a compound. Though the best perfusate is heparinized whole blood from the same species, it is not always readily available. Bovine blood, bovine serum albumin (BSA), and bovine or human erythrocytes have also been used [8] to mimic in vivo conditions because they provide hemoglobin as the oxygen carrier and ideal protein binding conditions.

The portal vein and hepatic artery, delivering 80% and 20% of the hepatic blood flow, respectively, can be used for perfusion using peristaltic pumps. In human liver perfusions for transplantation, such conditions can cause ischemic-type biliary lesions, decreasing the viability of the liver. In general, for rats only the portal vein perfusion is used. However, for intravenously administered compounds, perfusion through the hepatic artery would be more appropriate. For isolated rat liver perfusions (the most widely used model), prewarmed, diluted rat blood has also been used, with proper in-line oxygenation (5% carbon dioxide in oxygen). The perfusate exits, and is collected from, the liver through the abdominal vena cava.

There are three modes for perfusion: single-pass, recycling, and retrograde [9]. As the name signifies, in the single-pass method the perfusate passes through the liver only once, entering through portal vein and exiting from hepatic vein, i.e. the normal perfusion. In the recycling system, the perfusate is recycled continuously. In the retrograde method, the perfusate enters liver through the hepatic vein and exits at the portal vein. The normal and retrograde perfusions help to study zonal differences in metabolizing enzymes, engaging first the upstream and then the down stream enzymes, and vice-versa, respectively (St-Pierre MV, Schwab AJ, Goresky CA, Lee WF and Pang KS [10]. The volume of blood that would be required for single-pass perfusion would be impractical, therefore, 30% blood in Krebs-Ringer bicarbonate-buffer with low-viscosity at physiologic pH is a better alternative. A typical flow rate for portal-plus-hepatic artery perfusion is 60 ml/min at a hydrostatic pressure of 15–20 cm water.

Bile flow is typically used as an indication of the viability of the system. A rat bile flow rate of 0.4–0.6 ml/h/10 g liver is considered normal for a viable liver [11]. This is lower than the rate in vivo, largely because bile acids are depleted during perfusion. Since bile salts are usually recycled in vivo, adding sodium taurocholate to the perfusate can extend the viability of the liver and thus the duration of the study. This technique has also been used during prolonged bile collection studies in vivo [12]. Oxygen consumption can also be used as a measure of liver viability [6]; other viability parameters were covered by Thurman et al [13]. As with any experiment, it is important to run a control experiment, especially to eliminate effects due to nonspecific binding to the perfusion apparatus.

### 40.2.2 Scope

Liver perfusion provides a closed system, allowing accurate determination of the mass balance of a compound as it relates to disposition by the liver. Additionally, the perfusate, bile, and liver tissue can be analyzed for the parent compound and its metabolites, whether free or protein-bound [14]. Results from these studies can provide an idea of the extent of the hepatic first-pass effect, the role of transporters involved in the process of uptake from the perfusate and elimination via canalicular transporters, the potential for toxicity from reactive metabolites, and the net hepatic effect on oral bioavailability *in vivo*. The effect of CYP isoform- or transporter-inducing or inhibiting agents on metabolic and excretion profiles can be studied using livers from animals treated with such agents. Additionally, the effect of protein binding can be readily assessed by modulating the albumin concentration in the perfusate. Greater amounts of metabolites can be generated using liver perfusion in a recirculation circuit for compounds that cannot be administered at high doses in vivo because of toxicity concerns. Since a metabolite profile would be representative of that in vivo, particularly for the formation of secondary metabolites, liver perfusion can aid in the positive identification of minor metabolites formed in animals and humans that would otherwise be more tedious to characterize. However, recirculation of perfusate could also lead to artificial buildup of polar metabolites that would, under in vivo conditions, be excreted renally.

The uses of the liver perfusion technique also extend to toxicity testing, for example, when animals cannot be subjected to high exposures because they would cause multi-organ toxicity or general intolerability. Because the homeostasis achieved with liver perfusion is similar to that *in vivo*, the toxicity assessment will be equivalent to that possible using a hepatocyte or tissue slice system. One limitation is that liver perfusion is an acute study, and hence will mainly be reflective of acute toxicities. Liver perfusion has also been used in pharmacologic assessments [15].

### 40.2.3 Utility of Liver Perfusion

#### 40.2.3.1 Enzyme Systems and Metabolic Interactions

The liver perfusion model, which leaves all the metabolizing enzymes intact, offers unique advantages for studying metabolism and acute toxicity, such as disposition kinetics. One example of this is the use of a recirculated perfusion system to investigate the metabolism of tramadol, an analgesic agent, in the isolated perfused rat liver [16]. Tramadol was added to the perfusion medium at a concentration of 300 ng/ml, and the perfusate samples were collected for 180 min. The concentrations of tramadol and its three main metabolites, O-desmethyltramadol (M1), N-desmethyltramadol

(M2), and N,O-didesmethyltramadol (M5), were detected in perfusate samples by a rapid high-performance liquid chromatography (HPLC) method. Throughout the study, even at early sampling time points, the Phase I metabolism of tramadol led to the formation of the M1 metabolite. The M5 metabolite was detectable after 50 minutes in 6 out of 10 perfused livers and the M2 metabolite was not detectable in any experiments. The kinetic parameters of tramadol and the two detectable metabolites (M1 and M5) were then calculated for the perfusate samples. The tramadol concentration decreased from 298 to 160 ng/ml, with a mean half-life of 232 min and hepatic clearance of 0.73 ml/min. After 180 min, the mean concentration of M1 reached 59.5 ng/ml, resulting in a metabolic ratio of 16%, while the mean concentration of the M5 metabolite reached 14.6 ng/ml, resulting in a metabolic ratio of 2% using the area under the concentration versus time curve from 0 to 180 minutes ($AUC_{0180min}$). Thus, the system shows utility for studying time profile of metabolism.

The liver perfusion model can also be used to elucidate the extent and mechanisms of the first-pass metabolism of drugs in the liver after oral administration. For example, perfused rat liver was used to investigate the first-pass metabolism of the peptide drug metkephamid and of thyroid-stimulating hormone (TSH)-releasing hormone (TRH) [17]. After single-pass constant perfusion through rat liver, the fraction of intact metkephamid recovered reached steady-state very quickly. Metkephamid was hydrolyzed enzymatically at the surface of hepatocytes and the endothelial cells of microvessels, rather than being taken up by hepatocytes. The fraction of metkephamid recovered intact was approximately 40% under protein-free conditions but increased to 70–75% after addition of BSA to the perfusate. The fraction of metkephamid bound to BSA was approximately 50%, suggesting that only the free fraction of metkephamid in the plasma was metabolized in the liver. Calculations based on the tube model showed that approximately 30–35% of the metkephamid absorbed from the intestine undergoes first-pass metabolism before entering the systemic circulation in vivo. In contrast, the fraction of TRH metabolized in the liver was less than 10%, indicating that first-pass metabolism contributes remarkably little to the bioavailability of TRH. These results demonstrate that hepatic first-pass metabolism of metkephamid contributes to its low systemic bioavailability. This information has important implications in the oral delivery of many kinds of peptides.

The liver perfusion model has been used to investigate reactive metabolites generated in the liver. This type of study was useful in investigating the metabolites of the nonsteroidal anti-inflammatory drug (NSAID) diflunisal, which can cause potentially life-threatening hypersensitivity reactions. Using an isolated rat liver recirculating perfusion, Wang and Dickinson were able to show the formation of reactive acyl glucuronides and their covalent binding to hepatic proteins [14]. The study showed a differential disposition of a sulfate conjugate (14% of the dose), eliminated exclusively via perfusate, whereas the acyl glucuronide (8%), the phenolic glucuronide (11%), and a novel diglucuronide (iso-acyl and phenolic, 2%) were excreted mainly via bile (Figure 40.2). Covalent protein adducts of diflunisal were detected in both the liver and the perfusate (<1% each). The covalent adduct formation, measured in perfusate and liver, increased considerably when acyl glucuronide (or its isomerized mixture) was added to the perfusate, confirming the role of acyl glucuronide in the formation of covalent adducts, which could play a role in the hypersensitivity reaction reported for this drug.

Metabolic interactions can also be evaluated in the liver perfusion model. The effects of acute and short-term (7-day) parenteral nutrition (PN, consisting of 24.2% dextrose and 5.2% amino acids) on hepatic lidocaine elimination were studied using an isolated perfused rat liver [18]. In the acute study, lidocaine was co-infused with PN solution or normal saline at 0.123 ml/min for 80 minutes. There was no substantial difference in hepatic lidocaine elimination between control and PN-treated groups. In the post-PN study, only lidocaine was infused for 80 minutes. The hepatic clearance of lidocaine and the excretion ratio were reduced by 29.8% and 31.1%, respectively, compared to what occurred in livers isolated from control animals with free access to standard chow and water. Mass balance at steady state showed that recovery was higher for lidocaine (46% versus 26%) and lower for *N*-(*N*,*N*-diethylglycyl)-2-hydroxymethyl-6-methylaniline, a metabolite of lidocaine (0.2% versus 2.8%) in the PN group compared to chow-fed controls. Recoveries of other lidocaine metabolites — *N*-(*N*-ethylglycyl)-2,6-xylidene, 3-hydroxylidocaine, 3-hydroxy-*N*-(*N*-ethylglycyl)-2,6-xylidene, and *N*-(*N*-ethylglycyl)-2-hydroxymethyl-6-methylaniline — were similar in the PN- and chow-fed groups. These findings suggested that acute PN infusion has no effect on lidocaine metabolism and that the impaired hepatic lidocaine elimination after 7 days of PN occurs as a result of selective inhibition of

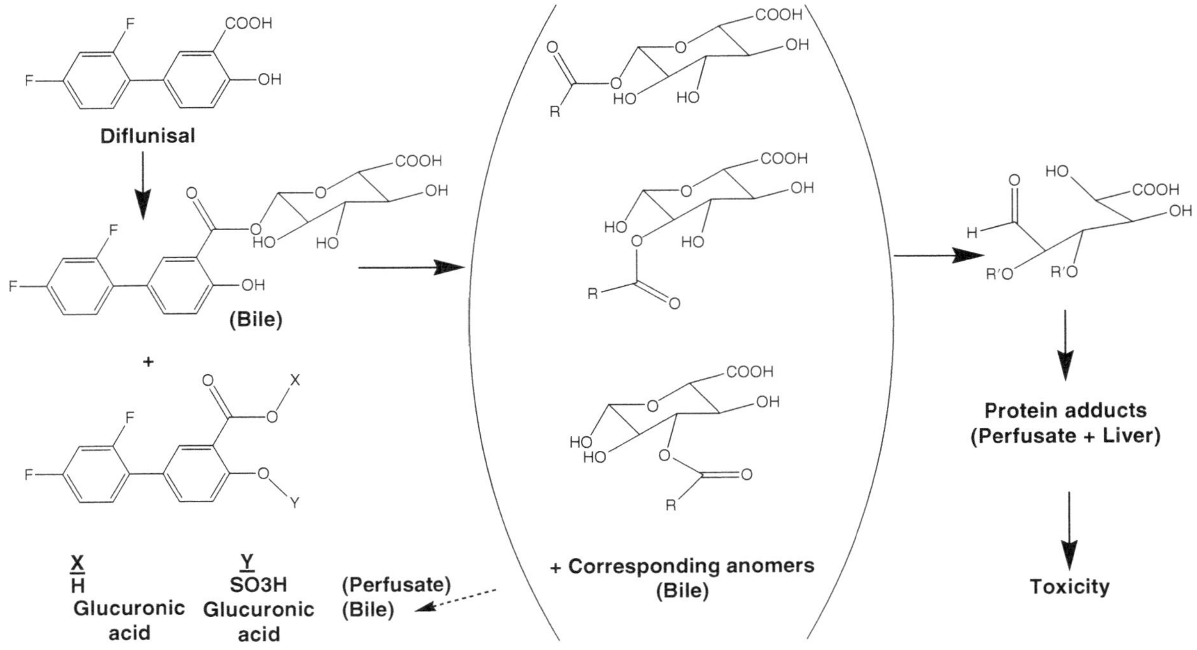


Figure 40.2 Metabolism of diflunisal using rat liver perfusion

hepatic enzyme activities, such as those involved in aryl methyl hydroxylation.

Because of the short-term nature of liver perfusion studies, their utility in CYP isoform induction studies is limited. However, the livers from CYP isoform-induced rats could be utilized for such studies. It is widely recognized that hepatocytes are a more appropriate system for induction studies because they can be maintained for several days to allow the onset of induction (see Section 3.3.2.2, CYP Isoform Induction).

#### 40.2.3.2 Effect of Protein Binding

Liver perfusion studies can also be used to examine the effects of protein binding on the metabolism of a drug. For example, a murine in situ single-pass liver perfusion study nicely showed that albumin in the perfusate affected the uptake of the pesticide parathion. Parathion toxicity is dependent on the extent of its metabolism. At the typical BSA concentration of 4% in the perfusate, the hepatic extraction of parathion was 0.19, and the extent of protein binding was 96% [19]. No reduction in CYP levels was observed when the BSA concentration was varied from 0% to 4%. However, when the albumin concentration was lowered to 1%, the extraction ratio increased to 0.49, and the extent of protein binding dropped to 88%, suggesting that albumin binding in the perfusate confined parathion to the perfusate.

The isolated rat perfusion study has also been used to assess the effect of BSA on the metabolic interactions between naproxen and the well-known pollutant bisphenol A [20] in the absence and presence of physiologically relevant concentration of BSA. The hepatic clearance of both naproxen and bisphenol A either alone or in combination was significantly impacted in the presence of BSA. Given high protein binding of both naproxen (~95%) and bisphenol A (~97%) the free drug concentration is significantly reduced in the presence of BSA with subsequent reduction in their metabolism as indicated by decrease in apparent $V_{max}$ of naproxen and bisphenol A by a factor of 1.3 and 6, respectively. In the same perfusion study, full competitive metabolic inhibition was also captured by concomitant exposure of both naproxen and bisphenol A.

#### 40.2.3.3 Transporters

Since the isolated perfused liver retains the intact hepatocyte morphology and proteins, it is a useful model to study transporter-mediated biliary drug elimination and drug–drug interactions (DDIs). This technique was used to compare excretion of fexofenadine in multiple mouse phenotypes. Comparison of perfusion studies in adenosine triphosphate (ATP)-binding cassette, subfamily C (Abcc)2/- (multidrug resistant protein [Mrp]2-/-), Abcc3-/- (Mrp3-/-), Abcc4-/- (Mrp4-/-), and wild-type mouse livers showed that basolateral excretion of fexofenadine was impaired in Mrp3-/- mice, resulting in an approximately 50% decrease in perfusate recovery; hepatobiliary disposition of fexofenadine was unaltered in

Mrp4-/- mice; and fexofenadine was redirected from the canalicular to the basolateral membrane for excretion in Mrp2-/- mice. In Mrp2-/- / Mrp3-/- (double-knockout) mice, fexofenadine biliary excretion was impaired, but perfusate recovery was similar to that in wild-type mice and more than 2-fold higher than that in Mrp3-/- mice, presumably because of compensatory basolateral transport mechanisms. These results demonstrate that multiple transport proteins, including Mrp2 and Mrp3 but not Mrp4, are involved in the hepatobiliary disposition of fexofenadine, a zwitterionic drug [21].

Liver perfusion studies have shown that transporter and metabolism enzymes can act in concert to govern drug metabolism in hepatocytes and elimination in bile [22–26]. For example, rat liver perfusion studies showed that the area under the concentration versus time curve (AUC) of digoxin (a substrate of OATP2, P-glycoprotein [P-gp], and CYP3A4) in perfusate was increased by treatment with rifampicin (an OATP2 inhibitor) and decreased by treatment with quinidine (a P-gp inhibitor) [23]. It was therefore concluded that rifampicin limits hepatic enzyme exposure to digoxin by inhibiting sinusoidal Oatp2 uptake transport in rats, whereas quinidine increases hepatic enzyme exposure to digoxin by inhibiting canalicular P-gp efflux. Separately, the rat liver perfusion model was used to study the differences in the metabolism and elimination of octylphenols, which are used in variety of detergents and plastics [27]. Comparing 4-*n*-octylphenol and 4-*tert*-octylphenol, which have different side chains and different 3-dimensional structures, it was found that the former was directly conjugated by uridine diphosphate-glucuronosyltransferase (UGT) 2B1 and the latter was hydrolyzed before being conjugated by UGT2B1. In a liver perfusion study using normal rats, the metabolites of 4-*tert*-octylphenol were readily excreted into bile, while the metabolites of 4-*n*-octylphenol accumulated in the liver. However, when the study was done in Eisai hyperbilirubinemic rats (EHBR), metabolites from both compounds had little excretion into the bile, suggesting that the clearance of the glucuronides of both compounds was mediated by Mrp2 [27]. These data emphasize the importance of uptake and efflux transporters for hepatic drug metabolism [22].

The rat liver perfusion model has also been successfully used to elucidate DDIs. Cyclosporin A is a pan inhibitor of liver transporters and known to affect the PK and hepatobiliary disposition of many drugs. Parasrampuria R et al. showed the effect of cyclosporin A on hepatobiliary disposition of methotrexate [28] using isolated perfused rat liver model. At low doses, cyclosporin A inhibited hepatic extraction ratio and clearance but did not affect bile/liver concentration ratio indicating inhibition of OATP-mediated uptake of methotrexate. However, at higher dose, cyclosporin A inhibited both OATP-mediated uptake and MRP2-mediated biliary excretion of methotrexate. In a similar study by Khezrian M et al. the effect of cyclosporin A on hepatobiliary disposition of etoposide was investigated using a recirculating isolated perfused rat liver model [29]. The study demonstrated the dose-dependent nonspecific effect of cyclosporin A on multiple transporters, OATPs, P-gp, MRP2, and BSEP that are known to be involved in the hepatobiliary disposition of etoposide.

## 40.3 Primary Hepatocytes

### 40.3.1 Techniques

Primary hepatocytes are a popular and practical in vitro system for drug biotransformation research. Hepatocytes can be isolated from the liver of various animal species by the traditional collagenase perfusion operation developed by Howard and Pesch [30]. This method requires the whole liver and therefore cannot be used for human livers, which are mainly obtained from patients who undergo partial liver resection. Therefore, a modified perfusion method has been developed for human liver parts [31]. With this method, it is preferable that perfusion take place immediately after resection. When this is not possible, the tissue can be stored at 4 °C for up to 48 hours in University of Wisconsin (UW) solution, without any meaningful loss of viability [32].

Fresh isolated or cryopreserved hepatocytes can be kept viable in suspension for only a few hours, but they can be maintained for a maximum of 4 days in monolayer culture. Both cultured hepatocytes [33–35] and hepatocyte suspensions [33, 34, 36] have been demonstrated to be powerful tools to analyze the specific metabolic profiles of a variety of drugs, with good in vitro–in vivo correlations. However, it is widely recognized that cultured hepatocytes are subject to a gradual loss of liver-specific functions, particularly decreased CYP isoform expression. The extent of this loss varies for the specific CYP isoforms; for some isoforms, such as CYP3A4 and 2E1, it becomes evident after a few days of culture, while others, such as CYP1A2 and 2C9, can remain nearly unaffected by the isolation and culturing processes [37]. Various methods of maintaining the liver-specific characteristics of hepatocytes during prolonged culture have been explored. These include the application of culturing matrices; the addition

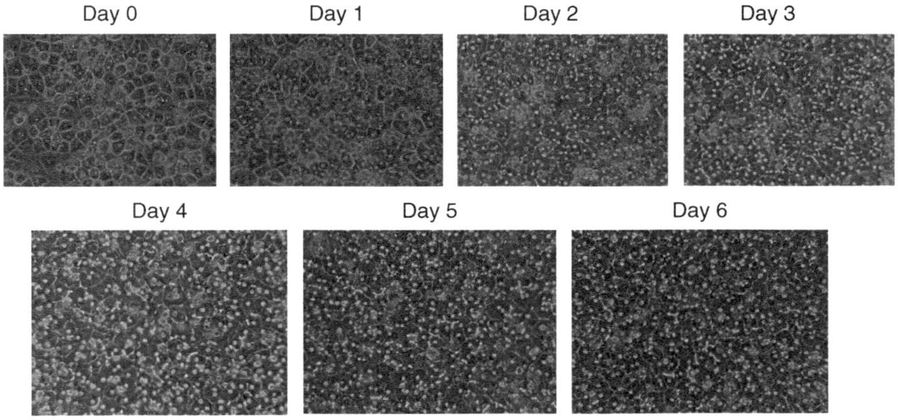


**Figure 40.3** Morphologic changes in sandwich-cultured cryopreserved human hepatocytes (overlaid with Matrigel)

of specific nutrients, hormones, and inducers to the culture medium; and the coculturing of hepatocytes with other cell types (e.g. the hepatic Kupffer cells) [38].

Culturing hepatocytes in a sandwich configuration between two layers of gelled extracellular matrix proteins, with collagen I and Matrigel™ being the most commonly used, dramatically prolongs the longevity of cultures displaying hepatocyte-specific functions and bile duct canalicular network formation [39, 40]. This culture method can be used to study not only biotransformation but also transporter-mediated biliary excretion. In addition, media formulation (e.g. the addition or omission of serum or specific hormone mixtures) has a substantial influence on the morphologic development and cell survival of hepatocytes in culture [41–43]. Cells are generally seeded in media containing fetal calf serum since this enhances the surface attachment ability of hepatocytes [44]. The morphologic distinction between hepatocytes seeded onto collagen-coated plates without a collagen gel overlay (conventional monolayers) and those seeded onto collagen gel with a subsequent collagen gel overlay (sandwich) is visible only a few hours after seeding: conventional monolayer hepatocytes have already adopted their polygonal shape and established extensive cell–cell contacts, whereas in sandwich cultures this takes markedly longer, with cells remaining mostly spherical and singular at this time point. In general, conventional monolayers appear more flattened than sandwich-cultured cells, a result of the lack of a three-dimensional extracellular matrix environment. All cells display clear cytoplasm and well-delineated plasma membranes. After overnight incubation, sandwich culture hepatocytes form aggregates with a typical cuboidal shape. For sandwich-cultured hepatocytes, the time of overlay is crucial to the phenotypic outcome: the second layer should be overlaid within the first 24 hours, and the best results are obtained if the monolayer/sandwich hybrid is prepared at the initial medium change, about 4 hours after seeding. In this configuration, the morphology at the time of overlay can be preserved for over one week. In the sandwich culture model, structures resembling bile canaliculi form during the 24 hours after overlay. By overlaying monolayer cultures of primary hepatocytes with collagen or Matrigel™, it is possible to obtain cultures displaying some features of collagen sandwich cultures over extended periods of time (e.g. bile canaliculi-like structures) (Figure 40.3), although the overall cell morphology is still different [45].

The cryopreserved hepatocyte model has been widely accepted for evaluation of hepatic clearance, metabolic pathways, hepatobiliary transporters, and biliary elimination since cryopreserved hepatocytes have been shown to retain the activity of most Phase I and II enzymes and Phase 0 and III transporters [46–48]. Thanks to successful cryopreservation techniques, human and animal hepatocytes are now commercially available and widely used for ADME/Toc studies [49].

Although 2D cultured hepatocytes ares still a widely used tool for assessing drug metabolism and transporter-related functions, there are still some limitations to the model including structural and functional instability. In particular, the limitations related to the loss of hepatic function and altered gene expression profile of drug-metabolizing enzymes and drug transporters make this model only suitable for short-term cultures. To overcome this limitation several next-generation liver models have been developed such as spheroid systems [50, 51], hepatocytes coculture with stromal cells as either micropatterned coculture [52] or a random coculture [53], and microfluidic liver-on-a-chip model [54]. Unlike the 2D hepatocytes culture which can last for a few days, these new models have been shown to be viable over four-weeks and proven to be useful for assessment of liver disease [55, 56], and long-term drug-induced liver toxicity (DILI) [51]. These models have also been used to assess the hepatic clearance of metabolically stable compounds [55, 57]. The spheroids are typically 200–300 micron in size, contain approximately 1000–1400 hepatocytes, and have been shown to be useful for assessment of CYP induction assessment [58].

### 40.3.2 Scope

Primary hepatocytes are now recognized as one of the most relevant and practical model to study drug metabolism and transporter interactions. When isolated and handled appropriately, they contain a broad complement of metabolizing enzymes and transport proteins, organized in a physiologically relevant context and regulated via cellular processes that occur within the liver in vivo (e.g., nuclear hormone-mediated xenosensors). Microsomes continue to be the first-line screening model for high-throughput drug metabolism assays. However, studies undertaken with microsomes are increasingly being replaced or complemented by the use of hepatocytes. The practical and technical difficulties posed by hepatocyte isolation have been largely circumvented by the use of cryopreserved cells and therefore cryopreserved hepatocytes are routinely used in absorption, distribution, metabolism, excretion, and toxicology (ADMET) studies. As a result of the increasing demand for a more relevant model, the application of hepatocytes to higher-throughput assays has been improved by more sophisticated automation instruments and miniaturization methods [45].

Sandwich cultures are the preferred culture format of the Food and Drug Administration (FDA) (to maintain the viability and morphology, not necessarily to increase the induction response) for CYP isoform induction studies. The sandwich culture format allows hepatocytes to remain relatively differentiated compared with cultures with no overlay [59]. In 2004, the FDA issued a draft guidance suggesting that, instead of freshly isolated hepatocytes, cryopreserved hepatocytes could be used if they were shown to attach sufficiently in cultures. As with fresh hepatocytes, before each experiment, cryopreserved hepatocytes should be shown to be responsive to positive control compounds and the fold induction should be similar to that seen in fresh cells.

Sandwich-cultured hepatocytes with biliary pocket formation can also be used to evaluate canalicular efflux transporters and the biliary excretion of a compound and its metabolites [40, 60, 61]. This model could potentially be used for transporter induction studies.

### 40.3.3 Utility of Primary Hepatocytes (see Also Chapter 39)

#### 40.3.3.1 Transporter Studies

Freshly isolated or cryopreserved hepatocyte suspension has been considered the gold standard to assess the hepatic uptake of drugs [48, 60, 62], however, its application in efflux transporter studies is limited as efflux transporters have been shown to be internalized after the isolation process of hepatocytes [60, 63]. Shitara et al. [48] examined NTCP-mediated taurocholate and OATP-mediated estradiol-17β-glucuronide (E17βG) uptake in human hepatocytes before and after cryopreservation. The uptake of compounds into suspended hepatocytes was determined by a centrifugation filtration method (Figure 40.4). It was determined that NTCP and OATP were only partially active after cryopreservation. Compared to the uptake in freshly isolated human hepatocytes from the same donor, $Na^+$-dependent and independent taurocholate uptake were reduced 59% and 29%, respectively, and the mean uptake of E17βG was reduced 68%, in cryopreserved hepatocytes, indicating that the cryopreservation process can change the activities of uptake transporters. Kinetic analysis revealed that the Michaelis constant ($K_m$) for taurocholate and E17βG uptake into cryopreserved hepatocytes were approximately 2–8 μM and approximately 3–18 μM, respectively, which are similar to the $K_m$ value of taurocholate in human NTCP-transfected cell line (6 μM) and $K_m$ value of E17βG in OATP2- (8 μM) and OATP8- (5 μM) transfected cell lines [48]. Similarly, Badolo L et al. [64] demonstrated the effect of cryopreservation on the activity of OATP1B1/OATP1B3 and OCT1 using model substrates E17βG and 1-methyl-4-phenylpyridinium (MPP+), respectively, in isolated human hepatocytes. The study indicated that $CL_{uptake}$ of E17βG was decreased on average by 47% in cryopreserved human hepatocytes which is attributed to reduced $V_{max}$ as $K_m$ remained unaffected. Interestingly, the passive diffusion of E17βG was also significantly decreased after cryopreservation. On the contrary, cryopreservation did not affect $K_m$, $V_{max}$, and passive diffusion of MPP+. Although these studies indicated partial reduction in the functionality of some uptake transporters due to cryopreservation cryopreserved hepatocytes are routinely used for *in vitro* ADME investigations in drug discovery as they are more convenient and flexible tool compared to freshly isolated hepatocytes.

Hepatocytes in suspension are not an ideal model to study the canalicular efflux transporters as they lose their plasma membrane expression due to internalization, however, when cultured in sandwich culture

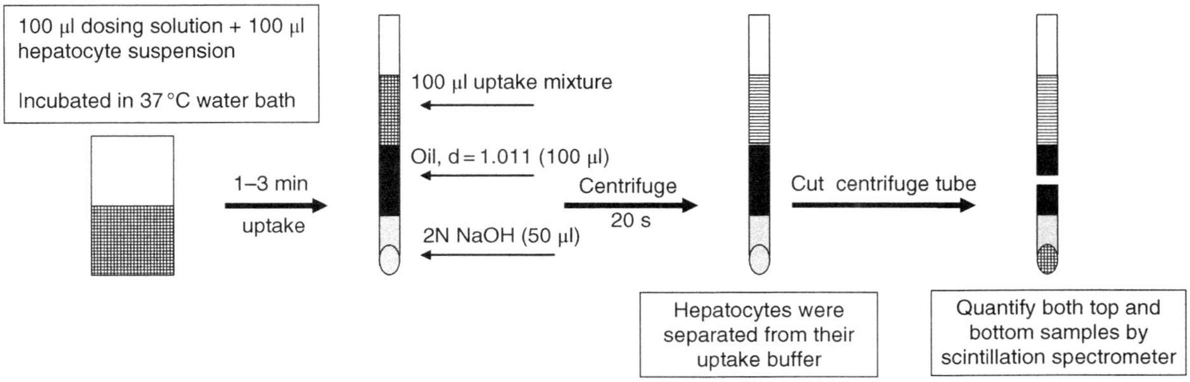


**Figure 40.4 Flowchart of the hepatocyte uptake study.** Prior to the uptake studies, 100 μl of hepatocytes ($3.0 \times 10^6$ cells/ml) in Krebs-Henseleit buffer (KHB) is pre-warmed at 37°C for 3 min. The uptake assay is initiated by adding 100 μl of test compound solution. After incubation for 3 min at 37°C or 4°C, the reaction is terminated by separating the cells from the assay buffer by centrifugal filtration. An aliquot of 100 μl incubation mixture is added in a centrifuge tube (450 μl) containing 50 μl 2N NaOH under a layer of 200 μl of oil (density = 1.015, a mixture of silicone oil and mineral oil). The sample tubes are then centrifuged for 20 seconds using a bench. top microfuge (10,000 × g) allowing the hepatocytes to pass through the oil layer into the alkaline solution. The centrifuge tubes have three layers after centrifugation: the top layer is the assay buffer (100 μl), the middle layer is the oil (200 μl), and the bottom layer is the cells in alkaline solution (50 μl). The centrifuge tubes are cut at the oil layer, and the cell and assay buffer compartments are put in two separate scintillation vials. After the hepatocytes are dissolved in the alkaline solution, the samples of cell or assay buffer are mixed with 5 ml of scintillation cocktail and the radioactivity in each sample is measured by a liquid scintillation spectrometer. Source: Adapted from Shitara et al. [48].

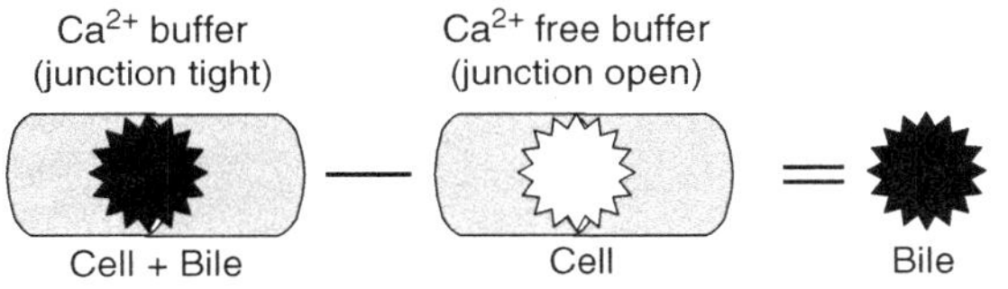


**Figure 40.5 Working model of sandwich-cultured hepatocytes.** Because depletion of $Ca^{2+}$ opens tight junctions, accumulation of compounds in hepatocytes and bile ducts is determined in both $Ca^{2+}$ buffer and $Ca^{2+}$-free buffer. Biliary elimination is then calculated as the difference between these two measurements. Source: [61] / American Society for Pharmacology and Experimental Therapeutics.

format, hepatocytes have been shown to maintain the characteristic hepatocytic architecture including tight junctions, canalicular biliary network, and the functional transporters with proper localization. As a result of these features, sandwich-cultured hepatocytes have become a valuable tool to evaluate uptake clearance, basolateral efflux clearance, and biliary clearance of drugs. The standard protocol involving incubation of a compound in the presence and absence of $Ca^{2+}$ is followed to measure the biliary clearance of drugs. The depletion of $Ca^{2+}$ in incubation medium opens up tight junctions by which compound excreted into bile is released into the incubation medium. Thus the accumulation of a compound in hepatocytes and bile ducts is determined [61]. Biliary clearance of a compound is then calculated as the difference between the measurement in the presence and absence of $Ca^{2+}$ (Figure 40.5). The successful utilization of human hepatocytes in sandwich culture in biliary clearance measurement is attributed to the polarized location of canalicular efflux transporters including P-gp, BCRP, MRP2, and BSEP which are involved in the excretion of drugs into bile. The polarized localization of P-gp and MRP2 in canalicular membranes was demonstrated in sandwich-cultured rat and human hepatocytes by determining the colocalization of these transporters with dipeptidyl peptidase IV (DPPIV), a marker protein of canalicular membranes [60]. The function of P-gp in sandwich-cultured hepatocytes was demonstrated using active biliary elimination of rhodamine 123, $^3$H-digoxin [65], and $^3$H-DPDPE [60]. The elimination of CDF, a fluorescent MRP2 substrate, and a hydrolyzed metabolite of carboxy-dichlorofluorescein diacetate (CDFDA), into the bile duct network was correlated with the expression of MRP2 on the canalicular membrane of sandwich-cultured hepatocytes [60]. Oatp1 (OATP2 for human hepatocytes), Oatp2 (OATP8), and Ntcp (NTCP) were expressed in sandwich-cultured rat or human hepatocytes, and their functions were maintained throughout the culture period [60, 66]. Taurocholate, an endogenous bile salt, is taken up into the sandwich-cultured hepatocytes predominantly by NTCP in a $Na^+$-dependent manner [66] and actively secreted into the bile canalicular network by BSEP [67]. Recent study demonstrated that sandwich-cultured rat and human hepatocytes can be used to identify compounds that inhibit the hepatic uptake and/or excretion of bile acids, which may cause DILI [68].

Sandwich-cultured hepatocytes were also used to determine the effects of various types of P-gp modulators on the biliary excretion index (BEI; a relative measure of the extent of biliary excretion, which can be calculated from Eq. (40.1)) and the *in vitro* biliary clearance ($CL_{bile}$, which is calculated from Eq. (40.2)) of rhodamine 123. Considerable reductions in rhodamine 123 BEI and $CL_{bile}$ were noted in the presence of the P-gp inhibitors verapamil (30–100 μM) and progesterone (100 μM). The P-gp activator quercetin (10–100 μM) enhanced rhodamine 123 $CL_{bile}$ by approximately fourfold, with only a minor effect on BEI, suggesting that quercetin had a more pronounced effect on uptake at the basolateral membrane than on excretion across the canalicular membrane. However, it is not clear about the mechanism of the quercetin inhibitory effect on rhodamine 123 uptake in hepatocyte. Treatment of hepatocytes for 48 h with dexamethasone (10 μM) resulted in large enhancement of $CL_{bile}$, whereas rifampin (5–50 μM) increased both BEI and $CL_{bile}$, indicating that the inductive effects of dexamethasone and rifampin occurred at the basolateral and canalicular membranes, respectively. Total rhodamine 123 uptake in sandwich-cultured rat hepatocytes was partly saturable and was affected by the presence of typical Oatp1a4 substrates (digoxin, quinine, *d*-verapamil, and estradiol-17ß-glucuronide) [69].

$$BEI = \frac{(Accumulation_{cells+bile} - Accumulation_{cells})}{Accumulation_{cells+bile}} \times 100\% \quad (40.1)$$

$$CL_{bile} = \frac{(Accumulation_{cells+bile} - Accumulation_{cells})}{AUC\ media} \quad (40.2)$$

Yang K et al. [70] demonstrated that sandwich cultured rat hepatocytes can be useful tool for determining the impact of impaired function of canalicular efflux transporter/s on the biliary excretion of drugs by RNA interference (RNAi)-mediated transporter knockdown strategy. In this study, Bcrp knockdown was successfully achieved by RNAi in Mrp2-deficient (TR(-)) rat hepatocytes and the absence of both Bcrp and Mrp2 showed almost complete abolition of rosuvastatin excretion into bile confirming the primary role of these transporters in rosuvastatin biliary clearance.

Sandwich-cultured hepatocytes can be used to evaluate hepatobiliary transporter-mediated DDIs (including inhibition and induction) and liver toxicities [67]. For example, troglitazone (a Type II antidiabetic drug) and bosentan (an antihypertensive drug), both of which caused liver cholestasis and were therefore withdrawn from market, were shown to inhibit taurocholate biliary elimination by inhibiting both Ntcp and Bsep in sandwich-cultured rat hepatocytes [67]. In our own laboratory, we have used an inverted fluorescent microscope to visualize the hepatobiliary transporter-mediated DDIs between carboxy-dichlorofluorescein (CDF) (a fluorescent MRP2 and OATP substrate) and MK571 (an MRP2 inhibitor) in sandwich-cultured rat hepatocytes. CDFDA, a nonfluorescent compound, passively diffuses into cells and hydrolyzes to CDF, thereby bypassing the active uptake process at the sinusoidal membrane of hepatocytes and enabling the investigation of efflux transporter-mediated processes. As shown in Figure 40.6, the fluorescent signal of CDF was concentrated in the canalicular bile duct network after a 10-minutes incubation of CDFDA in sandwich-cultured rat hepatocytes in the presence of $Ca^{2+}$. Upon co-incubation with MK571 (a MRP2 inhibitor), the fluorescent signal was completely abolished in the bile duct network (shown in the picture taken by phase contrast microscope), suggesting that MRP2 regulated excretion of CDF into the bile [71].

Interestingly sandwich cultured hepatocyte model has also been shown to investigate the species difference in hepatobiliary disposition of bile salts. Yang K et al. [72] examined the hepatocellular excretion of taurocholate in human and rat sandwich cultured hepatocytes. The study demonstrated that biliary excretion predominates while both biliary excretion and basolateral efflux contributes equally to the hepatocellular excretion of taurocholate in human and rat sandwich cultured hepatocytes, respectively. Although cultured hepatocytes can provide information about the overall outcome of transporter-related DDIs and hepatotoxicity, their utility is limited by the lack of detailed results regarding the individual interactions between the xenobiotic and uptake or efflux transporters. This limitation could be overcome by using a transporter-specific substrate or inhibitor or by using cell lines or membrane vesicles expressing a specific transporter. In addition, the results obtained from a single donor of hepatocytes may not be reflective of the entire human population because of possible polymorphisms in transporter genes and other physiological factors such as age, sex, medication history, and disease condition.

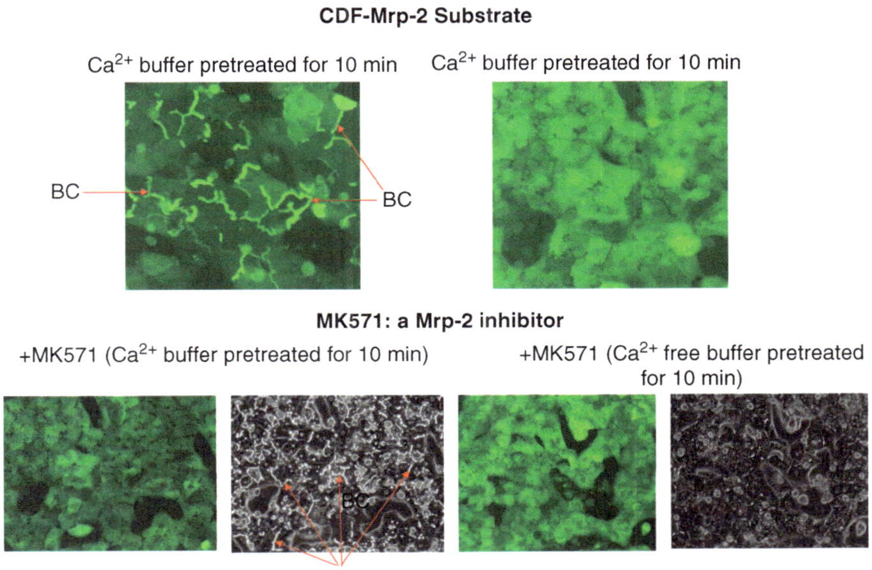


**Figure 40.6 Fluorescence and phase-contrast micrographs of hepatocytes treated with CDFDA in the presence and absence of MK571.** These results demonstrate that depletion of $Ca^{2+}$ opens the tight junction and enables compounds to be released from bile canaliculi (BC). MK571, which did not disrupt BC, blocked the excretion of CDF into BC via its inhibitory effect on Mrp2.

#### 40.3.3.2 Enzymatic Studies

Hepatocytes are widely used in metabolism studies for both determination of intrinsic clearance and metabolite profiling. The determination of in vitro intrinsic clearance ($CL_{int}$) for drug candidates in the early discovery stage is a common practice in the pharmaceutical industry [73–75]. The $CL_{int}$ values of drug candidates can help to confirm whether metabolism is the main clearance pathway when it is compared to the total body clearance in vivo. It is also helpful, in selection of a lead candidate, to rank order drug candidates on the basis of their metabolic stabilities. In addition, when the species and sex differences in metabolic clearance are considered, $CL_{int}$ can help to project the metabolic clearance of drug candidates in humans.

The in vitro $CL_{int}$ may be derived from enzyme kinetic data such as maximum velocity $(V_{max})/K_m$ [76–78] or from the rate of parent compound disappearance when sub-$K_m$ drug substrate concentrations are used [74, 79–81]. In the early discovery stage, the metabolite standards are generally not available, thus the parent disappearance method is often the method of choice. This method also captures the total clearance of a drug via all pathways, instead of focusing on major pathways by methods where metabolite formation is measured: in the hepatocyte system, the membrane barrier and all enzymes and cofactors are present at their physiologically relevant concentrations, allowing sequential metabolism by Phase I enzymes followed by Phase II conjugation enzymes. The disadvantage is that hepatocytes tend to have a low turnover rate and the drug incubation concentration may be limiting if it causes cytotoxicity at high concentrations [46].

***40.3.3.2.1 CYP Isoform Inhibition*** Traditionally, CYP isoform inhibition studies are conducted using human liver microsomes, providing accurate values for enzyme kinetic parameters, such as $V_{max}$ and $K_m$. However, the use of human hepatocytes to evaluate enzyme inhibition is drawing more attention these days because it can generate results closer to those observed in vivo. The advantage of using hepatocytes, again, is that they are intact living cells with full complements of enzymes and cofactors at physiologic concentrations [46, 82]. Hepatocytes can also better mimic in vivo hepatic nonspecific binding. For compounds that are substrates of transporters, the intracellular concentration may greatly differ from that generally used in human liver microsomal incubations. For example, if Phase II conjugations are involved in the compound's clearance, human liver microsomal incubation could fail to predict the compound's true stability and therefore fail to predict the potential for human DDIs or cytotoxicity. This could happen with either substrates or inhibitors. Zhao and colleagues [83] have demonstrated a discrepancy in time-dependent inhibition between hepatocyte and microsomal systems with amprenavir, diltiazem, erythromycin, raloxifene, and troleandomycin. The authors attribute the difference to factors such as metabolic stability, nonspecific binding, and active transporters. In another study, Brown and colleagues [84] studied the interaction between four substrates — tolbutamide, dextromethorphan, midazolam, and phenytoin — and six inhibitors — miconazole, fluconazole, ketoconazole, quinidine, fluoxetine, and fluvoxamine. Although most of these compounds have high nonspecific binding in hepatocytes, up to thousand-fold range, the $K^i$ values observed in microsomal and hepatocyte incubations were comparable. This suggests that the free concentration plays a role in the inhibition of drug-metabolizing enzymes and that the higher concentration of drug needed to cause, in hepatocytes, the extent of inhibition seen in microsomes, is due to nonspecific binding and membrane permeability [84].

***40.3.3.2.2 CYP Isoform Induction*** The correlations between CYP isoform induction in vitro and in vivo are well documented in the literature. It is known that an in vivo inducer almost always shows induction in vitro. However, looking at the reverse, the finding of in vitro induction may not necessarily translate into in vivo induction. For example, omeprazole is a known CYP1A2 inducer that gives maximum induction at around 10–30 μM in human hepatocytes [85]. In healthy humans, omeprazole is quickly metabolized by CYP2C19, and therefore does not reach the concentration necessary to cause CYP1A2 induction. CYP1A2 induction by omeprazole is only observed in people with poor CYP2C19 metabolism, known as the PM population. These individuals cannot metabolize omeprazole effectively and therefore have high plasma concentrations of the drug [86].

It is well known that the extent of CYP isoform induction varies by species [85, 87]. Besides the genetic differences between species (e.g. the difference in the nuclear receptors that mediate CYP isoform induction), there are also differences in metabolism pathways that could contribute to species differences in CYP isoform induction. Therefore, human hepatocytes are the most reliable in vitro model to predict human CYP isoform induction. It should be noted that in vitro methods are only tools to understand mechanisms or provide a level of understanding of processes. In vivo, the observations could be vastly different because of the presence of a multitude of other factors and the dynamic nature of the whole-organism system. This was exemplified by studying diclofenac in vitro in monkey hepatocytes and in vivo in monkeys under enzyme-inducing conditions [88]. While in monkey hepatocytes rifampin caused sizeable induction of CYP2C9-mediated diclofenac metabolism, pretreatment of monkeys with rifampin before diclofenac administration did not alter the pharmacokinetics of diclofenac. Rifampin is known to induce PXR and CAR, nuclear receptors that activate multiple enzymes, such as CYP2B6, CYP2C8/9/19, CYP3A4, UGT1A1, GSTs, and P-gp [89]. CYP3A was induced as expected, as measured by reduction in midazolam exposure, in a separate study in monkeys [89]. Examination of metabolism by monkey hepatocytes indicated that diclofenac was metabolized largely via acyl glucuronidation, with <10% of the metabolism accounted for by hydroxylation. This explained the lack of decrease in diclofenac exposure in vivo upon pretreatment of monkeys with rifampin and suggested that diclofenac may not be a good probe for induction of CYP2C9 in vivo. The degree of induction of different CYP isozymes regulated by the same receptor may also be different. Sahi et al. studied coregulation of CYP2C9 with CYP3A and 2B6 by PXR and CAR and demonstrate that hepatic CYP2C9 is differentially regulated by agonists of CAR and PXR ([90]). The induction profile of CYP2C9 was similar to that of CYP3A4.

The utility and superiority of the hepatocyte spheroid culture for CYP induction assessment was demonstrated by Berger et al. [58]. Berger investigated the suitability of in vivo Basel phenotyping cocktail (caffeine [CYP1A2], efavirenz [CYP2B6], losartan [CYP2C9], omeprazole [CYP2C19], metoprolol [CYP2D6], and midazolam [CYP3A4]) for *in vitro* basal metabolism and CYP induction assessment using four different hepatocyte cell systems (HepG2 cells, HepaRG cells, and primary cryopreserved human hepatocytes in 2-dimensional [2D] culture or in 3D-spheroid coculture). Their findings indicated the 3D-PHH spheroid culture and HepaRG had the highest non-induced levels of CYP activities although CYP1A2 and 2C9 activities could not be determined in HepaRG and 2D-PHH. In HepG2 cells, only CYP2C19 and CYP3A4 activities could be determined under non-induced conditions. Additionally, mRNA expression was at least fourfold higher for all CYPs in 3D-PHH compared to the other cell systems. Similarly, the 3D-PHH spheroid systems were also found to be best suited for CYP induction assessment with 3- to 50-fold increase in mRNA and at least a 2-fold increase in CYP activity for all inducible CYPs after treatment with 20 μM rifampicin. The CYP1A2 and CYP2C9 activity did not increase in 2D-PHH and HepaRG with HepG2 cells the least suitable for CYP induction assessment. The Basel phenotyping cocktail was shown to be suitable for the *in vitro* assessment of CYP activity and induction.

***40.3.3.2.3 Enzyme-Transporter Interplay (see also Chapter 23)*** The hepatobiliary disposition of xenobiotics may involve passive and/or active uptake, metabolism by CYP isoforms, and excretion of the parent compound and/or metabolite(s) into bile. Although *in vitro* systems have been used to evaluate these individual processes discretely, mechanistic *in vitro* studies of the sequential processes of uptake, metabolism, and biliary or basolateral excretion are limited. Sandwich-cultured hepatocytes may provide an opportunity to do such studies, since they have intact cell morphology, form bile duct canalicular networks, maintain appropriate enzyme levels, and keep the natural localization of hepatobiliary transporters. To test their utility, the hepatobiliary disposition of terfenadine and fexofenadine, a model drug/metabolite pair, was investigated in this cell model [91]. The metabolism of terfenadine and the biliary excretion of terfenadine and fexofenadine were determined in control and dexamethasone (DEX)-treated sandwich-cultured rat hepatocytes. DEX treatment increased the rates of formation of terfenadine metabolites azacyclonol and fexofenadine about 20- and 2-fold, respectively. The BEI of fexofenadine, when generated by terfenadine metabolism, was not very different from the BEI of preformed fexofenadine ($15\% \pm 2\%$ versus $19\% \pm 2\%$, respectively). Pharmacokinetic modeling revealed that the rate constant for hepatocyte uptake was faster for terfenadine than for preformed fexofenadine (2.5 versus $0.08\,h^{-1}$, respectively), whereas the biliary excretion rate constant for preformed fexofenadine exceeded that of terfenadine (0.44 versus $0.039\,h^{-1}$, respectively). Of interest, the rate constants for basolateral excretion of terfenadine and fexofenadine were comparable (3.2 versus $1.9\,h^{-1}$, respectively) and increased only slightly with DEX treatment. These studies demonstrate the utility of the sandwich-cultured hepatocyte model, coupled with pharmacokinetic modeling, to evaluate the hepatobiliary disposition of generated metabolites.

The prediction of clinically relevant DDI involving enzyme-transporter interplay using sandwich cultured hepatocytes model was evident in a study by Fukuda H et al. [92]. Candesartan cilexetil (CCX) is a prodrug which is metabolized to candesartan (CAN) by esterases in the liver. Based on membrane vesicle study, CCX indicated higher potency towards BSEP inhibition with $IC_{50}$ of 6.2 μM compared to CAN which showed $IC_{50}$ of 70.5 μM. As a result, CCX did not show inhibition of BSEP in sandwich cultured hepatocytes as it is metabolized to less potent CAN. However, when diisopropyl fluorophosphate (an irreversible cholinesterase inhibitor) inhibited the metabolism of CCX to CAN in human sandwich cultured hepatocytes it led to significant inhibition of BSEP-mediated transport of taurocholate due to increased intracellular concentration of CCX.

***40.3.3.2.4 Protein Binding (see Also Chapter 7)*** When studying drug efficacy or DDI potential, the free drug concentration, theoretically, represents the amount of drug available for interaction with the target or inhibition of drug-metabolizing enzymes or transporters. However, it is often dependent on the difference in drug affinity to nonspecific binding proteins and the targeted binding protein, as well as the kinetics of these proteins, such as the dissociation rate constant ($k_{off}$). With some drugs, a better in vitro–in vivo correlation is achieved with the total concentration [93], whereas with others, the free concentration is more relevant [94].

Making clinical DDI predictions from early in vitro studies is an important effort in the pharmaceutical industry. The traditional approaches are to use either a physiologically based pharmacokinetic model (PBPK) or a simple prediction model, such as the one shown in Eq. (40.3). PBPK models consider the metabolic fate of both substrate and inhibitor, but many other parameters also need to be taken into consideration, such as the limited information on the drug plasma concentration, absorption and excretion rates in different compartments, and the rate of synthesis of drug-metabolizing enzymes. Currently, the Drug Metabolism and Pharmacokinetics (DMPK) community has not reached a consensus on CYP isoform synthesis rates, because of the difficulty of making direct measurements. For most compounds, the liver concentration is often higher than the plasma concentration. For example, Chen and colleagues [89] have shown that ketoconazole, a widely used CYP3A4 inhibitor, has a totally different PK profile in plasma than in liver. The plasma concentration of ketoconazole declines steadily after each dose but the liver concentration stays above the effective concentration for CYP3A4 inhibition. Many groups use simple DDI models, such as that shown in Eq. (40.3), to predict clinical outcomes [94–97].

$$\frac{AUC_I}{AUC} = \frac{CL}{CL_I} = \frac{F_{g,I}}{F_g} \frac{1}{\left[\frac{fm_{3A4}}{1+\frac{I}{K_i}}\right] + [1 - fm_{3A4}]} \tag{40.3}$$

In this approach, it is critical to obtain an accurate ratio of the enzyme-site concentration of the inhibitor $[I]$ to its inhibition constant ($Ki$), the $[I]/Ki$. The $Ki$ value usually can be determined from in vitro microsomal incubation assays, although the extent of protein binding and the choice of substrate and enzyme systems may influence the outcome. However, the value of $[I]$ cannot be determined directly and must be estimated. Vast efforts have been focused on finding alternative values to mimic $[I]$ [94, 95, 98–100]. To overcome this problem, we applied an indirect approach of incubating human hepatocytes in human plasma to mimic in vivo plasma protein binding and hepatocyte nonspecific binding. The enzyme velocity measured from the hepatocyte incubation was then used to predict the effect of an inhibitor on in vivo clearance [96, 101–103]. Eqs. (40.4, 40.5), and (40.6) illustrate the link between a measurable velocity ratio ($v_I/v$) and the unmeasurable $[I]/Ki$ ratio.

Under a competitive inhibition scenario [104]:

$$fA = \frac{v_I}{v} = \frac{\frac{V_{max}[S]}{K_m\left(1+\frac{I}{Ki}\right)+[S]}}{\frac{V_{max}[S]}{K_m+[S]}} = \frac{K_m + [S]}{K_m\left(1+\frac{I}{Ki}\right)+[S]} \tag{40.4}$$

where fA is the fractional activity remaining in the presence of the inhibitor.

When $[S] \ll K_m$:

$$fA = \frac{v_I}{v} = \frac{1}{1+\frac{I}{Ki}} \tag{40.5}$$

On the other hand, under a noncompetitive inhibition scenario [104]:

$$fA = \frac{v_I}{v} = \frac{\frac{V_{max}[S]}{K_m\left(1+\frac{I}{Ki}\right)+[S]\left(1+\frac{I}{Ki}\right)}}{\frac{V_{max}[S]}{K_m+[S]}} = \frac{1}{1+\frac{I}{Ki}} \tag{40.6}$$

Applying this approach, good quantitative predictions of the potential for DDIs between a set of drugs and the potent CYP3A4 inhibitor ketoconazole, and the moderate CYP3A4 inhibitor fluconazole, were obtained [96].

***40.3.3.2.5 Uptake-Dependent Hepatic $CL_{int}$*** Unlike cell-free systems, such as microsomes and S9 fractions, hepatocytes have an intact cell membrane. Any drug must penetrate the membrane before reaching the drug-metabolizing enzymes. This process could be mediated via passive diffusion or active transporters, such as OAT, OATP, and NTCP. For a long time, it was assumed that drugs, at least those that are not substrates of transporters, can cross the cell membrane instantly. Therefore, the intracellular free concentration was estimated using the extracellular buffer concentration. In one of our earlier studies [82], we observed a discrepancy in intrinsic clearance between hepatocyte incubations and microsomes. This difference was greater when clearance of the compound was higher. This observation was attributed to the effect of permeability on the compound's clearance. In general, for compounds that are not substrates of transporters and are subject to metabolism, the intracellular free concentration is always lower than the extracellular free concentration, because of the involvement of metabolism [82]. Therefore, the apparent intrinsic clearance follows a hyperbolic relationship, with the rate of uptake (permeability rate) being the upper limit (Eq. (40.4)). If the uptake rate was ultrafast (uptake velocity $[V_{up}] \gg$ metabolism velocity $[V_{met}]$), then the intracellular concentration would be equal to the extracellular concentration, and the metabolism would reach its maximum rate ($V_{met,max}$). Otherwise, the uptake rate will have an impact on the apparent metabolism rate.

$$v_{met,app} = \frac{v_{met,max} \times v_{up}}{v_{met,max} + v_{up}} \tag{40.7}$$

A recent study by Parker and Houston [105] in rats using saquinavir, ritonavir, and nelfinavir supported our hypothesis. The hepatic clearance rates were close to the hepatocyte uptake rates, whereas the microsomal clearance rates were much higher. This is in line with our determination that many compounds do not cross hepatocyte membranes instantly, and that clearance in hepatocytes is affected by both the metabolism rate and the uptake or efflux rate.

#### 40.3.3.3 Drug Metabolism Studies

Drug metabolism is an essential support function for small-molecule drug discovery and development. In drug discovery stage, there are two major roles for metabolite profiling and identification. First, it can inform the medicinal chemists of the metabolic soft-spot of a drug candidate so that the molecule can be modified to optimize the metabolic stability to attain a desired pharmacokinetic and drug metabolism properties. Second, the metabolite profile across different species can help select appropriate animal species to evaluate the nonclinical safety of a molecule. The early drug metabolism study is typically conducted using liver subcellular fractions including liver microsomes, liver S9, cytosol, and hepatocytes [106]. A limitation of liver subcellular fractions is that each of them only contains a subset of drug metabolism enzymes and may miss the metabolism information if the absent enzyme is liable for the biotransformation. In comparison, hepatocytes contain full sets of metabolism enzymes including oxidative, conjugative, and hydrolytic enzymes [84, 107].

Hepatocyte suspensions have been widely used in drug metabolism study in pharmaceutical industry [106, 108]. However, hepatocyte suspensions can only maintain the enzyme activity for around 4 h [109]. Therefore, the system is suitable to investigate the metabolism of a drug candidate which has moderate to high intrinsic clearance. For the compound with a low intrinsic clearance, minimum metabolite information can be obtained from the hepatocyte suspension incubation for the metabolite identification study because of the low yield of metabolites. New methods and technologies were thus developed to prolong the exposure time of a drug candidate to metabolic enzymes using primary hepatocytes.

Hepatocyte relay method was developed to extend the metabolism time for a drug candidate [110]. The method involved transferring the supernatant from hepatocyte incubation after 4 h to another freshly thawed hepatocyte suspensions and continuing the same cycle again to extend the exposure of the drug substance to hepatic enzymes. The method has been utilized to generate human excretory and circulating metabolites for slowly metabolized drugs and demonstrated an improved performance compared to that from previous methods [111].

Micropatterned hepatocyte coculture system is a miniaturized culture system for hepatocytes with optimized microscale architecture which can maintain the drug metabolism enzyme expression levels for several weeks [112]. The system has been used to generate major human excretory and circulating drug metabolites for slowly metabolized compounds and exceeded the performance of hepatocyte suspensions and other in vitro systems with this regard [113]. The same system was also applied to predict preclinical and human-specific drug metabolism, which showed good correlation between in vitro and in vivo data [55]. HμREL system is another in vitro hepatocyte coculture model with ease for use. The system can be incubated for up to 7 days without changing the incubation media while maintaining the enzyme activity and ideal for the metabolism study of compounds with low intrinsic clearance. It has been evaluated to produce human in vivo metabolites for low-turnover compounds and demonstrated superior performance over the hepatocyte suspensions [53].

## 40.4 Organ Perfusion Versus Hepatocyte Studies

The advantages of liver perfusion over techniques like liver microsomes, S9, homogenates, slices, and hepatocytes are obvious in the sense that liver perfusion provides a system more similar to the environment in vivo, with the organ architecture maintained and most of the enzymes and transporter systems intact and functioning. The micro systems, such as hepatocytes, nonetheless provide certain utilities and conveniences, so that any researcher with minimal training can use them. These include dissecting out the involvement and contribution of individual enzymes or transporters and producing larger quantities of primary and/or secondary metabolites. In contrast, organ perfusion requires specific skills in surgery and instrumentation, making this technique more challenging.

The main intrinsic advantage of liver perfusion over the micro techniques is the entry of compounds through the vascular endothelium, as happens in vivo, instead of via direct access to parenchymal cells, as in hepatocytes and tissue slice systems. The maintenance of tight junctions between hepatocytes in the liver restricts the entry of certain compounds into the organ. This makes for better in vitro–in vivo correlation of clearance since both the transporter and metabolic components are maintained. Moreover, compared to isolated hepatocytes, the liver perfusion model preserves the variety of cells that are critical to liver toxicity, especially metabolism-based toxicity. In humans, for example, Kupffer cells are important to the regulation of the production of proinflammatory cytokines, such as interleukin (IL)-1 and IL-6: It is known that patients with cancer and human immunodeficiency virus (HIV) who are receiving IL-2 treatment may have suppressed CYP3A4/5 activity, which could lead to differential metabolic clearance and sequential toxic effects [114, 115]. However, this toxicity was not initially observed in a study using human hepatocyte culture alone [116]. A later study using a hepatocyte-Kupffer cell coculture system demonstrated that IL-2 suppresses CYP3A4/5 activity in hepatocytes only in the presence of Kupffer cells. Therefore, in this case, cultured human hepatocytes alone failed to predict the adverse effects/cytotoxicity of IL-2. Freshly isolated hepatocytes, in turn, are reported to contain glutathione levels at a concentration much lower than that in the intact liver [117]. This could clearly have implications for the development of oxidative stress, depending on the in vitro system used.

Another limitation is that isolated liver perfusion can be done for a maximum duration of about 3-4 h, which is long enough to study acute effects on metabolism and toxicity. But to identify chronic effects, e.g. enzyme induction, the use of this system would require additional perfusion studies with livers from rats pretreated to induce enzyme levels that would allow measurement of the effects of induction on and by the test compound. The use of different livers could increase the variability of the results, and it may be necessary to replicate the studies in order to get a handle on the extent or significance of the variability. Therefore, to identify chronic effects, hepatocytes are generally used, as they are more amenable to longer incubations (4 weeks or more) and are relatively less complicated to use and maintain.

## 40.5 Perspectives

The utility of organ perfusion is apparent considering the preserved structure of the organ and cellular interactions. Smaller systems perhaps cannot fully replace the benefits that this technique offers, for short-term studies. However, the utilization of this technique is limited by the unavailability of human livers, poor reproducibility, maximum study duration (no longer than 4 h), and labor intensity. In this high-throughput era, the benefits of micro systems outshine those of the macro in vitro/in situ system when it comes to mechanistic studies. The major reasons for this are the ease, cost, lack of time limitations, and wide availability of the system components. Advances made in the hepatocyte suspension and sandwich-culture techniques can now allow biliary secretion and transporter-based interaction studies to be conducted with ease, given the commercial availability of plateable hepatocytes. We believe that hepatocytes will continue to play a greater role in the conduct of metabolism, transporter, and DDI studies.

## Acknowledgements

We would like to acknowledge the excellent editorial help of Ms. Alexis Khalil of the Millennium NCDS Management, Strategy, and Operations department.

## Abbreviations

| | |
|---|---|
| ABC transporter family | ATP-binding cassette transporters |
| ADMET | absorption, distribution, metabolism, excretion, and toxicology |
| AUC | area under concentration versus time curve |
| BCRP | breast cancer resistance protein |
| BSA | bovine serum albumin |
| BSEP | bile salt export pump |
| CYP | cytochrome P450 |
| DDI | drug–drug interaction |
| HPLC | high-performance liquid chromatography |
| MATE1 | multidrug and toxin extrusion1 |
| MDR | multidrug resistance |
| MRP2 | multidrug resistance protein 2 |
| NR | Nuclear receptor |
| NTCP | sodium-dependent bile acid transporter |
| OAT | organic anion transporter |
| OATP | organic anion transporting polypeptide |
| OCT | organic cation transporter |
| P-gp | P-glycoprotein |
| PN | parenteral nutrition (consisting of 24.2% dextrose and 5.2% amino acids) |
| RBC | red blood cell |
| SLC transporters | solute carrier transporters |
| UGT | Uridine diphosphate glucuronosyl transferases. |

## References

1 Moffett, D.F., Moffett, S.B., and Schauf, C.L. (1993). *Human Physiology*. Louis, MO: Mosby-Year Book St.

2 Chen, J. and Pang, K.S. (1997). Effect of flow on first-pass metabolism of drugs: single pass studies on 4-methylumbelliferone conjugation in the serially perfused rat intestine and liver preparations. *J. Pharm. Exp. Ther.* 280: 24–31.

3 Andersson, T., Foerlin, L., and Hansson, T. (1983). Biotransformation of 7-ethoxycoumarin in isolated perfused rainbow trout liver. *Drug Metab. Dispos.* 11: 494–498.

4 Ring, J.A., Ghabrial, H., Ching, M.S. et al. (1998). Propranolol elimination by right and left fetal liver: studies in the intact isolated perfused fetal sheep liver. *J. Pharm. Exp. Ther.* 284: 535–541.

5 Melgert, B.N., Olinga, P., Weert, B. et al. (2001). Cellular distribution and handling of liver-targeting preparations in human livers studied by a liver lobe perfusion. *Drug Metab. Dispos.* 29: 361–367.

6 Mehendale, H.M. (1976). Uptake and disposition of chlorinated biphenyls by isolated perfused rat liver. *Drug Metab. Dispos.* 4: 124–132.

7 Meijer, D.K.F., Groothuis, G.M.M., Mulder, G.J., and Swart, P.J. (1994). The isolated perfused Rat liver as a tool to study drug transport, drug metabolism and cell specific drug delivery. In: *In Vitro and Ex Vivo*

*Test Systems to rationalize Drug Design and Delivery* (ed. D. Crommelin, P. Couvreur, and D. Duchene), 17–31. Paris: Editions de Sante.

8 Schimmel, R.J. and Knobil, E. (1969). Role of free fatty acids in stimulation of gluconeogenesis during fasting. *Am. J. Physiol.* 217: 1803–1808.

9 Rothbarth, J., Sparidans, R.W., Beijnen, J.H. et al. (2002). Reduced liver uptake of arterially infused melphalan during retrograde rat liver perfusion with unaffected liver tumor uptake. *J. Pharm. Exp. Ther.* 303: 736–740.

10 St-Pierre, M.V., Schwab, A.J., Goresky, C.A. et al. (1989). The multiple-indicator dilution technique for characterization of normal and retrograde flow in once-through rat liver perfusions. *Hepatology* 9 (2): 285–296.

11 Mitzkat, H.J. and Meyer, U. (1973). Metabolic state of isolated perfused rat liver and model-induced metabolism modifications. In: *Isolated Liver Perfusion and Its Applications* (ed. I. Bartosek, A. Guaitani, and L.L. Miller), 79–86. New York: Raven Press.

12 Balani, S.K., Kauffman, L.R., DeLuna, F.A., and Lin, J.H. (1999). Nonlinear pharmacokinetics of efavirenz (DMP-266), a potent HIV-1 reverse transcriptase inhibitor, in rats and monkeys. *Drug Metab. Dispos.* 27: 41–45.

13 Thurman, R.G., Reinke, L.A., and Kauffman, F.C. (1979). The isolated perfused liver: a model to define biochemical mechanisms of chemical toxicity. *Rev. Biochem, Toxicol.* 1: 249–285.

14 Wang, M. and Dickinson, R.G. (1998). Disposition and covalent binding of diflunisal and diflunisal acyl glucuronide in the isolated perfused rat liver. *Drug Metab. Dispos.* 26: 98–104.

15 Compagnon, P., Clement, B., Campion, J.P., and Boudjema, K. (2001). Effects of hypothermic machine perfusion on rat liver function depending on the route of perfusion. *Transplantation* 72 (4): 606–614.

16 Rouini, M.-R., Ghazi-Khansari, M., Ardakani, Y.H. et al. (2008). A disposition kinetic study of tramadol in rat perfused liver. *Biopharm. Drug Dispos.* 29: 231–235.

17 Taki, Y., Sakane, T., Nadai, T. et al. (1998). First-pass metabolism of peptide drugs in rat perfused liver. *J. Pharm. Pharmacol.* 50: 1013–1018.

18 Ke, J., Tam, Y.K., Koo, W.W.K. et al. (1990). Effects of parenteral nutrition on hepatic elimination of lidocaine: a study using the isolated perfused rat liver. *J. Pharm. Exp. Ther.* 255: 351–356.

19 Sultatos, L.G. and Minor, L.D. (1986). Factors affecting the biotransformation of the pesticide parathion by the isolated perfused mouse liver. *Drug Metab. Dispos.* 14: 214–220.

20 Bounakta, S., Bteich, M., Mantha, M. et al. (2018). Predictions of bisphenol A hepatic clearance in the isolated rat liver (IPRL): impact of albumin binding and co-administration with naproxen. *Xenobiotica* 48 (2): 135–147.

21 Tian, X., Swift, B., Zamek-Gliszczynski, M.J. et al. (2008). Impact of basolateral multidrug resistance-associated protein (Mrp) 3 and Mrp4 on the hepatobiliary disposition of fexofenadine in perfused mouse livers. *Drug Metab. Dispos.* 36: 911–915.

22 Benet, L.Z., Cummins, C.L., and Wu, C.Y. (2004). Unmasking the dynamic interplay between efflux transporters and metabolic enzymes. *Int. J. Pharm.* 277: 3–9.

23 Lau, Y.Y., Wu, C.-Y., Okochi, H., and Benet, L.Z. (2004). Ex situ inhibition of hepatic uptake and efflux significantly changes metabolism: hepatic enzyme-transporter interplay. *J. Pharmacol. Exp. Ther.* 308: 1040–1045.

24 Lau, Y.Y., Okochi, H., Huang, Y., and Benet, L.Z. (2006). Multiple transporters affect the disposition of atorvastatin and its two active hydroxy metabolites: application of in vitro and ex situ systems. *J. Pharmacol. Exp. Ther.* 316: 762–771.

25 Liu, L. and Pang, K.S. (2005). The roles of transporters and enzymes in hepatic drug processing. *Drug Metab. Dispos.* 33: 1–9.

26 Xu, H., Kulkarni, K.H., Singh, R. et al. (2009). Disposition of Naringenin via Glucuronidation Pathway is affected by Compensating Efflux transporters of Hydrophilic Glucuronides. *Mol. Pharmaceutics* 6 (6): 1703–1715.

27 Nomura, S., Daidoji, T., Inoue, H., and Yokota, H. (2008). Differential metabolism of 4-n- and 4-tert-octylphenols in perfused rat liver. *Life Sci.* 83: 223–228.

28 Parasrampuria, S. and Mehvar, R. (2010). Dose-dependent inhibition of transporter-mediated hepatic uptake and biliary excretion of methotrexate by cyclosporine A in an isolated perfused rat liver model. *J. Pharm. Sci.* 99 (12): 5060–5069.

29 Khezrian, M., Sheikholeslami, B., Dadashzadeh, S. et al. (2015). Effects of cyclosporine A on the hepatobiliary disposition and hepatic uptake of etoposide in an isolated perfused rat liver model. *Cancer Chemother. Pharmacol.* 75 (5): 961–968.

30 Howard, R.B. and Pesch, L.A. (1968). Respiratory activity of intact, isolated parenchymal cells from rat liver. *J. Biol. Chem.* 243: 3105–3109.

31 Puviani, A.C., Ottolenghi, C., Tassinari, B. et al. (1998). An update on high-yield hepatocyte isolation methods and on the potential clinical use of isolated liver cells. *Comp. Biochem. Physiol. A: Mol. Integr. Physiol.* 121A: 99–109.

32 Guyomard, C., Chesne, C., Meunier, B. et al. (1990). Primary culture of adult rat hepatocytes after 48-hour preservation of the liver with cold UW solution. *Hepatology (Baltimore)* 12: 1329–1336.

33 Chenery, R.J., Ayrton, A., Oldham, H.G. et al. (1987). Diazepam metabolism in cultured hepatocytes from rat, rabbit, dog, guinea pig, and man. *Drug Metab. Dispos.* 15: 312–317.

34 Cross, D.M. and Bayliss, M.K. (2000). A commentary on the use of hepatocytes in drug metabolism studies during drug discovery and development. *Drug Metabol. Rev.* 32: 219–240.

35 Le Bigot, J.F., Begue, J.M., Kiechel, J.R., and Guillouzo, A. (1987). Species differences in metabolism of ketotifen in rat, rabbit and man: demonstration of similar pathways in vivo and in cultured hepatocytes. *Life Sci.* 40: 883–890.

36 Bayliss, M.K., Bell, J.A., Jenner, W.N. et al. (1999). Utility of hepatocytes to model species differences in the metabolism of loxtidine and to predict pharmacokinetic parameters in rat, dog and man. *Xenobiotica* 29: 253–268.

37 George, J., Goodwin, B., Liddle, C. et al. (1997). Time-dependent expression of cytochrome P450 genes in primary cultures of well-differentiated human hepatocytes. *J. Lab. Clin. Med.* 129: 638–648.

38 LeCluyse, E.L., Bullock, P.L., and Parkinson, A. (1996). Strategies for restoration and maintenance of normal hepatic structure and function in long-term cultures of rat hepatocytes. *Adv. Drug Delivery Rev.* 22: 133–186.

39 Dunn, J.C.Y., Tompkins, R.G., and Yarmush, M.L. (1991). Long-term in vitro function of adult hepatocytes in a collagen sandwich configuration. *Biotechnol. Progr.* 7: 237–245.

40 Liu, X., LeCluyse, E.L., Brouwer, K.R. et al. (1999a). Biliary excretion in primary rat hepatocytes cultured in a collagen-sandwich configuration. *Am. J. Phys* 277: G12–G21.

41 Pascussi, J.-M., Drocourt, L., Fabre, J.-M. et al. (2000a). Dexamethasone induces pregnane X receptor and retinoid X receptor-alpha expression in human hepatocytes: synergistic increase of CYP3A4 induction by pregnane X receptor activators. *Mol. Pharmacol.* 58: 361–372.

42 Pascussi, J.-M., Gerbal-Chaloin, S., Fabre, J.-M. et al. (2000b). Dexamethasone enhances constitutive androstane receptor expression in human hepatocytes: consequences on cytochrome P450 gene regulation. *Mol. Pharmacol.* 58: 1441–1450.

43 Sidhu, J.S., Liu, F., and Omiecinski, C.J. (2004). Phenobarbital responsiveness as a uniquely sensitive indicator of hepatocyte differentiation status: requirement of dexamethasone and extracellular matrix in establishing the functional integrity of cultured primary rat hepatocytes. *Exp. Cell. Res.* 292: 252–264.

44 Williams, G.M., Bermudez, E., and Scaramuzzino, D. (1977). Rat hepatocyte primary cell cultures. III. improved dissociation and attachment techniques and the enhancement of survival by culture medium. *In Vitro* 13: 809–817.

45 Hewitt, N.J., Lechon, M.J.G., Houston, J.B. et al. (2007). Primary hepatocytes: current understanding of the regulation of metabolic enzymes and transporter proteins, and pharmaceutical practice for the use of hepatocytes in metabolism, enzyme induction, transporter, clearance, and hepatotoxicity studies. *Drug Metabol. Rev.* 39: 159–234.

46 Li, A.P., Lu, C., Brent, J.A. et al. (1999). Cryopreserved human hepatocytes: characterization of drug-metabolizing enzyme activities and applications in higher throughput screening assays for hepatotoxicity, metabolic stability, and drug-drug interaction potential. *Chem. Biol. Interact.* 121: 17–35.

47 Ruegg, C.E., Silber, P.M., Mughal, R.A. et al. (1997). Cytochrome-P450 induction and conjugated metabolism in primary human hepatocytes after cryopreservation. *In Vitro Toxicol.* 10: 217–222.

48 Shitara, Y., Li, A.P., Kato, Y. et al. (2003). Function of uptake transporters for taurocholate and estradiol 17b-D-glucuronide in cryopreserved human hepatocytes. *Drug Metabol. Pharmacokinet.* 18: 33–41.

49 Li, A.P. and Lu, C. (2009). *Evaluation of Inhibitors of Drug Metabolism in Human Hepatocytes, in Enzyme Inhibition in Drug Discovery and Development: The Good and The Bad (Lu C and Li AP eds).* New York: John Wiley & Sons.

50 Foster, A.J., Chouhan, B., Regan, S.L. et al. (2019). Integrated in vitro models for hepatic safety and metabolism evaluation of a human liver-chip and liver spheroid. *Arch. Toxicol.* 93: 1021–1037.

51 Lauschke, V.M., Shafagh, R.Z., Hendriks, D.F.G., and Ingelman-Sundberg, M. (2019). 3D primary

hepatocyte culture systems for analyses of liver diseases, drug metabolism, and toxicity: emerging culture paradigms and applications. *Biotechnol. J.* 14: 1800347.

52 Lin, C. and Khetani, S. (2018). Micropatterned co-cultures of human hepatocytes and stromal cells for the assessment of drug clearance and drug-drug interactions. *Curr. Protocol. Toxicol.* 72: 14.17.1-14.17.23. doi:https://doi.org/10.1002/cptx.23.

53 Burton, R.D., Hieronymus, T., Chamem, T. et al. (2018). Assessment of the biotransformation of low-turnover drugs in the HμREL human hepatocyte coculture model. *Drug Metab. Dispos.* 46: 1617–1625.

54 Chao, P., Maguire, T., Novik, E. et al. (2009). Evaluation of microfluidic based cell culture platform with primary human hepatocytes for the prediction of hepatic clearance in human. *Biochem. Pharmacol.* 78: 625–632.

55 Ballard, T.E., Wang, S., Cox, L.M. et al. (2016). Application of a micropatterned cocultured hepatocyte system to predict preclinical and human specific human specific drug metabolism. *Drug Metab. Dispos.* 44: 172–179.

56 Britta, B., Svanberg, P., Janefeldt, A. et al. (2016). Determination of human hepatocyte intrinsic clearance for slowly metabolized compounds: comparison of a primary hepatocyte/stromal cell co-culture with plated primary hepatocytes and HepaRG. *Drug Metab. Dispos.* 44 (4): 527–533.

57 Bonn, B., Svanberg, P., Janefeldt, A. et al. (2016). Determination of human hepatocyte intrinsic clearance for slowly metabolized compounds: comparison of a primary hepatocyte/stromal cell co-culture with plated primary hepatocytes and HepaRG. *Drug Metab. Dispos.* 44 (4): 527–533.

58 Berger B, Donzelli M, Maseneni S, Boess F, Roth A, Krähenbühl S and Haschke M (2016) Comparison of liver cell models using the basel Phenotyping Cocktail. *Front. Pharmacol.* 7:443. doi: https://doi.org/10.3389/fphar.2016.00443

59 Swift, B., Pfeifer, N.D., and Kim, L.R.B. (2010). Sandwich-cultured hepatocytes: an in vitro model to evaluate hepatobiliary transporter-based drug interactions and hepatotoxicity. *Drug Metab. Rev.* 42 (3): 446–471.

60 Hoffmaster, K.A., Turncliff, R.Z., LeCluyse, E.L. et al. (2004). P-glycoprotein expression, localization, and function in Sandwich-Cultured primary Rat and human Hepatocytes: relevance to the Hepatobiliary Disposition of a model Opioid Peptide. *Pharm. Res.* 21: 1294–1302.

61 Liu, X., Lecluyse, E.L., Brouwer, K.R. et al. (1999b). Use of Ca2+ modulation to evaluate biliary excretion in sandwich-cultured rat hepatocytes. *J. Pharmacol. Exp. Ther.* 289: 1592–1599.

62 Hirano, M., Maeda, K., Shitara, Y., and Sugiyama, Y. (2004). Contribution of OATP2 (OATP1B1) and OATP8 (OATP1B3) to the hepatic uptake of pitavastatin in humans. *J. Pharmacol. Exp. Ther.* 311: 139–146.

63 Bow, D.A., Perry, J.L., Miller, D.S. et al. (2008). Localization of P-gp (Abcb1) and Mrp2 (Abcc2) in freshly isolated rat hepatocytes. *Drug Metab. Dispos.* 36 (1): 198–202.

64 Badolo, L., Trancart, M.M., Gustavsson, L., and Chesne, C. (2011). Effect of cryopreservation on the activity of OATP1B1/3 and OCT1 in isolated human hepatocytes. *Chem. Biol. Interact.* 190 (2-3): 165–170.

65 Annaert, P.P., Turncliff, R.Z., Booth, C.L. et al. (2001). P-glycoprotein-mediated in vitro biliary excretion in sandwich-cultured rat hepatocytes. *Drug Metab. Dispos.* 29: 1277–1283.

66 Liu, X., Brouwer, K.L.R., Gan, L.-S.L. et al. (1998). Partial maintenance of taurocholate uptake by adult rat hepatocytes cultured in a collagen sandwich configuration. *Pharm. Res.* 15: 1533–1539.

67 Kemp, D.C., Zamek-Gliszczynski, M.J., and Brouwer, K.L.R. (2005). Xenobiotics Inhibit Hepatic Uptake and Biliary Excretion of Taurocholate in Rat Hepatocytes. *Toxicol. Sci.* 83: 207–214.

68 Wolf, K.K., Vora, S., Webster, L.O. et al. (2009, 2010). Use of cassette dosing in sandwich-cultured rat and human hepatocytes to identify drugs that inhibit bile acid transport. *Toxicol. in Vitro* 24 (1): 297–309.

69 Annaert, P.P. and Brouwer, K.L.R. (2005). Assessment of drug interactions in hepatobiliary transport using rhodamine 123 in sandwich-cultured rat hepatocytes. *Drug Metab. Dispos.* 33: 388–394.

70 Yang, K., Pfeifer, N.D., Hardwick, R.N. et al. (2014). An experimental approach to evaluate the impact of impaired transport function on hepatobiliary drug disposition using Mrp2-deficient TR- rat sandwich-cultured hepatocytes in combination with Bcrp knockdown. *Mol. Pharmaceutics* 11 (3): 766–775.

71 Xia, C.Q., Milton, M.N., and Gan, L.-S. (2007). Evaluation of drug-transporter interactions using in vitro and in vivo models. *Curr. Drug Metabol.* 8: 341–363.

72 Yang, K., Pfeifer, N.D., Kock, K., and Brouwer, K.L. (2015). Species differences in hepatobiliary disposition of taurocholic acid in human and rat

sandwich-cultured hepatocytes: implications for drug-induced liver injury. *J. Pharm. Exp. Ther.* 353 (2): 415–423.

73 Houston, J.B. (1994). Utility of in vitro drug metabolism data in predicting in vivo metabolic clearance. *Biochem. Pharmacol.* 47: 1469–1479.

74 Lave, T., Dupin, S., Schmitt, C. et al. (1997). Integration of in vitro data into algometric scaling to predict hepatic metabolic clearance in man: application to 10 extensively metabolized drugs. *J. Pharm. Sci.* 86: 584–590.

75 Obach, R.S., Baxter, J.G., Liston, T.E. et al. (1997). The prediction of human pharmacokinetic parameters from preclinical and in vivo metabolism. *J. Pharm. Exp. Ther.* 283: 46–58.

76 Griffin, S.J. and Houston, J.B. (2004). Comparison of fresh and cryopreserved rat hepatocytes suspensions for the prediction of in vitro intrinsic clearance. *Drug Metab. Dispos.* 32: 552–558.

77 Lin, J.H., Chiba, M., Balani, S.K. et al. (1996). Species differences in the pharmacokinetics and metabolism of indinavir, a potent human immunodeficiency virus protease inhibitor. *Drug Metab. Dispos.* 24: 1111–1120.

78 Tan, E. and Pang, S. (2001). Sulfation is the rate limiting in the futile cycling between estrone and estrone sulfate in enriched periportal and perivenous rat hepatocytes. *Drug Metab. Dispos.* 29: 335–346.

79 Jones, H.M. and Houston, J.B. (2004). Substrate depletion approach for determining in vitro metabolic clearance: time dependencies in hepatocyte and microsomal incubations. *Drug Metab. Dispos.* 32: 973–982.

80 Lau, Y.Y., Sapidou, E., Gui, X. et al. (2002). Development of a novel in vitro model to predict hepatic clearance using fresh, cryopreserved, and sandwich-cultured hepatocytes. *Drug Metab. Dispos.* 30: 1446–1454.

81 Obach, R.S. (1999). Prediction of human clearance of twenty-nine drugs from hepatic microsomal intrinsic clearance data: an examination of in vitro half-life approach and nonspecific binding to microsomes. *Drug Metab. Dispos.* 27: 1350–1359.

82 Lu, C., Li, P., Gallegos, R. et al. (2006). Comparison of intrinsic clearance in liver Microsomes and Hepatocytes from Rats and humans, evaluation of free fraction and uptake in hepatocytes. *Drug Metab. Dispos.* 34: 1600–1650.

83 Zhao, P., Kunze, K.L., and Lee, C.A. (2005). Evaluation of time-dependent inactivation of CYP3A in cryopreserved human hepatocytes. *Drug Metab. Dispos.* 32: 335–346.

84 Brown, H.S., Chadwick, A., and Houston, J.B. (2007). Use of isolated hepatocyte preparations for cytochrome P450 inhibition studies: comparison with microsomes for ki determination. *Drug Metab. Dispos.* 35: 2119–2126.

85 Lu, C. and Li, A.P. (2001). Species comparison in P450 induction: effects of dexamethasone, omeprazole, and rifampin of P450 isoforms 1A and 3A in primary cultured hepatocytes from man, Sprague-Dawley rat, minipig, and beagle dog. *Chem. Biol. Interact.* 16: 271–281.

86 Rost, K.I., Brosicke, H., Brockmoller, J. et al. (1992). Increase of cytochrome P4501A2 activity by omeprazole: evidence by the 13C-[N-3methyl]-caffeine breath test in poor and extensive metabolizers of S-mephenytoin. *Clin. Pharmacol. Ther.* 52: 170–180.

87 Graham, M.J. and Lake, B.G. (2008). Induction of drug metabolism: species differences and toxicological relevance. *Toxicology* 254 (3): 184–191.

88 Prueksaritanont, T., Li, C., Tang, C. et al. (2006). Rifampin induces the in vitro oxidative metabolism, but not the in vivo clearance of diclofenac in rhesus monkeys. *Drug Metab. Dispos.* 34: 1806–1810.

89 Chen J and Raymond K (2006) Roles of rifampicin in drug-drug interactions: underlying molecular mechanisms involving the nuclear pregnane X receptor. *Ann. Clin. Microbiol. Antimicrob.* 5: No pp given.

90 Sahi, J., Shord, S.S., Lindley, C. et al. (2009). Regulation of cytochrome P450 2C9 expression in primary cultures of human hepatocytes. *J. Biochem. Mol. Toxicol.* 23 (1): 43–58.

91 Turncliff, R.Z., Hoffmaster, K.A., Kalvass, J.C. et al. (2006). Hepatobiliary disposition of a drug/metabolite pair: comprehensive pharmacokinetic modeling in sandwich-cultured rat hepatocytes. *J. Pharm. Exp. Ther.* 318: 881–889.

92 Fukuda, H., Nakanishi, T., and Tamai, I. (2014). More relevant prediction for in vivo drug interaction of candesartan cilexetil on hepatic bile acid transporter BSEP using sandwich-cultured hepatocytes. *Drug Metabol. Pharmacokinet.* 29 (1): 94–96.

93 Von Moltke, L.L., Greenblatt, D.J., Duan, S.X. et al. (1998). Inhibition of desipramine hydroxylation (cytochrome P450-2D6) in vitro by quinidine and by viral protease inhibitors: relation to drug interactions in vivo. *J. Pharm. Sci.* 87: 1184–1189.

94 Obach, R.S., Walsky, R.L., Venkatakrishnan, K. et al. (2006). The utility of in vitro cytochrome P450 inhibitioon data in the prediction of drug-drug interactions. *J. Pharm. Exp. Ther.* 316: 336–348.

95 Ito, K., Hallifax, D., Obach, R.S., and Houston, J.B. (2005). Impact of parallel pathways of drug elimination and multiple cytochrome P450 involvement on drug-drug interactions: CYP2D6 paradigm. *Drug Metab. Dispos.* 33: 837–844.

96 Lu, C., Miwa, G.T., Prakash, S.R. et al. (2007). A novel model for the prediction of drug-drug interactions in humans based on in vitro CYP phenotyping. *Drug Metab. Dispos.* 35: 79–85.

97 Rostami-Hodjegan, A. and Tucker, G.T. (2004). "In Silico" simulation to assess the "in vivo" consequences of "in vitro" metabolic drug-drug interactions. *Drug Discovery Today: Technol.* 1: 441–448.

98 Bachmann, K.A. (2006). Inhibition constants, inhibitor concentrations and the prediction of inhibitory drug drug interaction: pitfalls, progress and promise. *Curr. Drug Metabol.* 7: 1–14.

99 Blanchard, N., Richert, L., Coassolo, P., and Lave, T. (2004). Qualitative and Quantitative assessment of drug-drug interaction potential in man, based on Ki, IC50 and inhibitor concentration. *Curr. Drug Metabol.* 5: 147–156.

100 Cook, C.S., Berry, L.M., and Burton, E. (2004). Prediction of in vivo drug interaction with eplerenone in man from in vitro metabolic inhibition data. *Xenobiotica* 34: 215–228.

101 Lu, C., Balani, S.K., Qian, M.G. et al. (2010). Quantitative prediction and clinical observation of a CYP3A inhibitor-based drug – drug interactions with MLN3897, a potent CCR1 antagonist. *J. Pharm. Exp. Ther., in press* 332 (2): 562–568.

102 Lu, C., Berg, C., Prakash, S.R. et al. (2008a). Prediction of pharmacokinetic drug-drug interactions using human hepatocyte suspension in plasma – part II. In vitro – in vivo correlation with fluconazole. *Drug Metab. Dispos.* 36: 1255–1260.

103 Lu, C., Hatsis, P., Berg, C. et al. (2008b). Prediction of pharmacokinetic drug-drug interactions using human hepatocyte suspension in plasma – part III. in vitro – in vivo correlation with ketoconazole. *Drug Metab. Dispos.* 36: 1261–1266.

104 Segel, H.S. (1974). *Enzyme kinetics*. New York: John Wiley & Son, Inc.

105 Parker, A.J. and Houston, J.B. (2008). Rate-limiting steps in hepatic srug clearance: comparison of hepatocellular uptake and metabolism with microsomal metabolism of saquinavir, nelfinavir, and ritonavir. *Drug Metab. Dispos.* 36: 1375–1384.

106 Dalvie, D., Obach, R.S., Kang, P. et al. (2009). Assessment of three human in vitro systems in the generation of major human excretory and circulating metabolites. *Chem. Res. Toxicol.* 22: 357–368.

107 McGinnity, D.F., Soars, M.G., Urbanowicz, R.A., and Riley, R.J. (2004). Evaluation of fresh and cryopreserved hepatocytes as in vitro drug metabolism tools for the prediction of metabolic clearance. *Drug Metab. Dispos.* 32: 1247–1253.

108 Xu, L., Woodward, C., Khan, S., and Prakash, C. (2012). In vitro metabolism of BIIB021, an inhibitor of heat shock protein 90, in liver microsomes and hepatocytes of rats, dogs, and humans and recombinant human cytochrome P450 isoforms. *Drug Metab. Dispos.* 40: 680–693.

109 Di, L. and Obach, R.S. (2015). Addressing the challenges of low clearance in drug research. *AAPS J.* 17: 352–357.

110 Di, L., Trapa, P., Obach, R.S. et al. (2012). A novel relay method for determining low-clearance values. *Drug Metab. Dispos.* 40: 1860–1865.

111 Ballard, T.E., Orozco, C.C., and Obach, R.S. (2014). Generation of major human excretory and circulating drug metabolites using a hepatocyte relay method. *Drug Metab. Dispos.* 42: 899–902.

112 Khetani, S.R. and Bhatia, S.N. (2008). Microscale culture of human liver cells for drug development. *Nat. Biotechnol.* 26: 120–126.

113 Wang, W.W., Khetani, S.R., Krzyzewski, S. et al. (2010). Assessment of a micropatterned hepatocyte coculture system to generate major human excretory and circulating drug metabolites. *Drug Metab. Dispos.* 38: 1900–1905.

114 De Boer, A.W., Markowitz, N., Lane, H.C. et al. (2003). A randomized controlled trial evaluating the efficacy and safty of intermittent 3-,4- and 5-day cycles of intravenous recombinant human interleukin-2 combined with antiretroviral therapy (ART) versus ART alone in HIV-seropositive patients with 100-300 CD4+ T cells. *Clin. Immunol.* 106: 188–196.

115 Elias, D., France, F., Triebel, F. et al. (1995). Phase I-II randomized study on prehepatectomy recombinant interleukin-2 immunotherapy in patients with metastatic of the colon and rectum. *J. Am. Coll. Surgeon* 181: 303–310.

**116** Kashuba, A.D.M., Hawke, R.L., Tonkin, J., and LeCluyes, E.L. (2000). Direct exposure to interleukins (IL)-2,-10 and -12 does not significantly suppress CYP3A activity in human hepatocytes. *Clin. Pharmacol. Ther.* 67: 100.

**117** Kera, Y., Penttila, K.E., and Lindros, K.O. (1988). Glutathione replenishment capacity is lower in isolated perivenous than in periportal hepatocytes. *Biochemist* 254: 411–417.

# 41

# Determination of Regulation of Drug Metabolizing Enzymes and Transporters

*Sergio C. Chai[1], Taosheng Chen[1], and Wen Xie[2]*

[1] *Department of Chemical Biology and Therapeutics, St. Jude Children's Research Hospital, Memphis, TN, USA*
[2] *Center for Pharmacogenetics and Department of Pharmaceutical Sciences, University of Pittsburgh, Pittsburgh, PA, USA*

## 41.1 Introduction

Drug metabolizing enzymes, including the Phase I and Phase II enzymes, and drug transporters, play central roles in the absorption, distribution, metabolism, excretion, and detoxification of xenobiotics, including clinical drugs [1–3]. Many drugs are taken orally. For drugs intended to act systemically, a significant fraction of the dose can be eliminated during its first passage through a sequence of organs before entry into the general circulation. For some drugs, the degree of first-pass elimination can be large enough such that oral bioavailability is significantly reduced, resulting in a reduced clinical response. Of these first-pass eliminating organs, the small intestine and liver are the most commonly implicated, in part because they express the highest levels of drug-metabolizing enzymes and transporters. Alterations in enteric and hepatic drug-metabolizing enzymes and transporters expression can significantly influence the oral bioavailability of drugs.

In the past two decades, orphan nuclear receptors such as pregnane X receptor (PXR) and constitutive androstane receptor (CAR) have been established as species-specific xenobiotics sensors that regulate the expression of many drug-metabolizing enzymes and transporters [4–12]. The identification of PXR and CAR as xenobiotic receptors represents a major step forward in understanding the pharmacological and genetic control of the expression of drug-metabolizing enzymes and the implications of this regulation in drug metabolism, drug–drug interactions, and human diseases.

PXR and CAR belong to the nuclear receptor (NR) family of transcriptional factors [13–15]. All NR proteins exhibit a characteristic modular structure that consists of five to six domains of homology (designated A to F, from the N-terminal to the C-terminal end) on the basis of regions of conserved sequence and function (Figure 41.1). The DNA-binding domain (DBD, region C) and the ligand-binding domain (LBD, region E) are the most highly conserved domains. Two key motifs are required for transcriptional activation of NR activity: activation function (AF)-1 in the N-terminal A/B region and AF-2 in the C-terminal LBD region [16, 17]. Like many other NRs, upon ligand binding, PXR or CAR translocates to nucleus and heterodimerizes with RXR. The PXR–RXR or CAR–RXR heterodimers bind to the specific xenobiotic response elements (XREs) present in the promoter regions of drug-metabolizing enzyme and transporter genes [18, 19].

It is now established that PXR and CAR are the master regulators of a broad spectrum of drug-metabolizing enzymes, as well as drug transporters. Examples of PXR and CAR target genes include the Phase I enzymes CYP2B6, CYP2B9, CYP2C8, CYP2C9, CYP2C19, CYP3A4, and CYP3A7; Phase II enzymes glutathione *S*-transferases (GSTs), UDP-glucuronosyltransferases (UGTs) and sulfotransferases (SULTs); and the transporters multidrug resistance protein 1 (MDR1), MDR2, multidrug resistance-associated protein 2 (MRP2), and the organic anion transporter polypeptide 2 (OATP2) [15, 20–22]. This chapter focuses on the in vitro and in vivo methodologies to study NR-mediated regulation of drug-metabolizing enzymes and transporters.

## 41.2 In vivo Methods

### 41.2.1 Creation of Xenobiotic Receptor Mouse Models

Mouse models are useful tools to study NR-mediated gene regulation and associated pathophysiological events. Homologous recombination (knockout) and

*Oral Bioavailability and Drug Delivery: From Basics to Advanced Concepts and Applications*, First Edition. Edited by Ming Hu and Xiaoling Li.

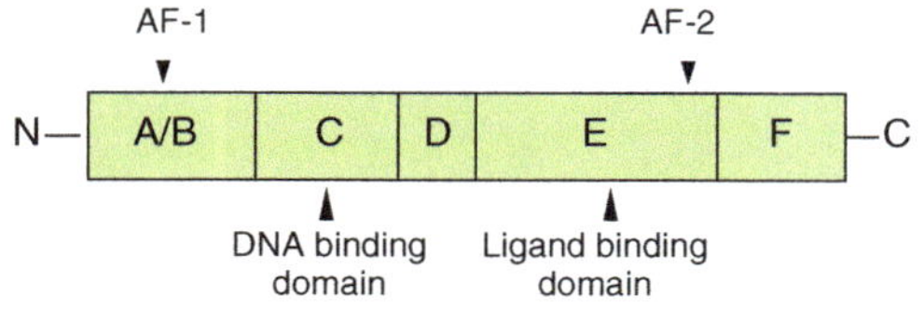


**Figure 41.1** Domain structure of nuclear receptors. Nuclear receptors are composed of several functional domains. The variable N-terminal (A/B) region contains ligand-independent AF-1 transactivation domain. Region C is conversed and contains the DNA binding domain responsible for sequence-specific recognition of DNA. The hinge domain (D) links the DNA to the ligand binding domain (LBD). The LBD (E/F) contains the ligand binding pocket, the receptor dimerization domain, and the ligand-dependent AF-2 core transactivation domain at the COOH-terminal portion.

transgenic techniques have been extensively used to study the function of xenobiotic receptors. Using PXR as an example, Figure 41.2 summarizes strategies to create the loss-of-function knockout model, gain-of-function transgenic models, and the combined "humanized" function models [23].

#### 41.2.1.1 Creation of PXR Knockout Mice

***41.2.1.1.1 Construction of PXR Targeting Vector*** Mouse PXR genomic DNA was isolated by screening a 129/Sv library (Stratagene, La Jolla, CA) using a PXR cDNA probe. A targeting vector was generated by replacing the second and third exons of PXR with a PGK-Neo selection marker, in conjunction with a negative selection marker (PGK-TK) (Figure 41.3a). The resulting mutant allele has a deletion of two exons spanning nucleotides 339–660, which include amino acids 63–170 of the PXR DBD.

***41.2.1.1.2 Embryonic Stem (ES) Cell Transfection and Selection*** The targeting vector was linearized by Not I digestion and transfected into J1 ES cells by electroporation. G418 (200 mg/ml)- and ganciclovir (0.2 mM)-resistant ES clones were screened for designated homologous recombination by Southern blot analysis.

***41.2.1.1.3 Generation of Chimeric Mice, Germ Line Transmission, and Confirmation of Loss of PXR Expression*** $PXR^{+/-}$ ES cells were microinjected into C57BL6/J blastocysts, which were subsequently transplanted into the uteri of pseudopregnant ICR mice. Chimeric male progeny were crossed with C57BL6/J females. Germline transmission of the mutant allele was detected in agouti progeny by Southern blot analysis (Figure 41.3b). The absence of PXR mRNA expression in the liver and small intestine of PXR null mice was confirmed by Northern blot analysis (Figure 41.3c) [24].

#### 41.2.1.2 Creation of hPXR and CAR Transgenic Mice

***41.2.1.2.1 Construction of Alb-hPXR Transgene*** To generate the transgenic mice with expression of the wild-type hPXR, the hPXR cDNA was cloned downstream of the mouse albumin (Alb) promoter/enhancer [25]. A SV40 intron/poly (A) sequence was placed downstream of hPXR cDNA (Figure 41.4a).

***41.2.1.2.2 Generation of Alb-hPXR Transgenic Mice and Confirmation of Transgene Expression*** Transgene was excised and purified from the plasmid vector and microinjected into the pronuclei of fertilized mouse eggs. The injected eggs were then implanted into the oviduct of pseudopregnant female mice. Transgene-positive mice were screened by PCR on tail DNA [24]. The transgene expression was confirmed by Northern blot analysis (Figure 41.4b).

***41.2.1.2.3 Creation of Tetracycline-Inducible CAR Transgenic Mice*** To create a transgenic mouse system that allowed conditional expression of a constitutively activated CAR (VP-CAR) in the liver. Two lineages of transgenic mice were used as diagramed in Figure 41.5a. First, we created the TetRE-VP-CAR transgene that encodes VP-CAR under the control of a minimal cytomegalovirus promoter and the TetRE [26]. The TetRE-VPCAR mice were then bred with the Lap-tTA activator line from the Jackson Laboratory (Bar Harbor, ME) to generate bi-transgenic animals. Driven by the liver-specific Lap (CCAAT/enhancer-binding protein-β) promoter, the Lap-tTA transgene directs the expression of the tetracycline-responsive transcriptional activator (tTA) exclusively in the hepatocytes [27]. We anticipated that tTA bound to TetRE and consequently induced the expression of VP-CAR only in the absence of doxycycline (Dox). Addition of Dox, which will be supplied in drinking at a concentration of 2 mg/ml, will result in the displacement of tTA from TetRE and will silence VP-CAR expression (Tet-Off). Northern blot analysis demonstrating the expected transgene expression and Dox regulation is shown in Figure 41.5b.

#### 41.2.1.3 Creation of hPXR "Humanized" Mice

***41.2.1.3.1 Generation of hPXR Humanized Mice*** As diagramed in Figure 41.2, the Alb-hPXR transgene was bred into a PXR-null background, and the resulting PXR-null/hPXR-transgenic mice lacked mPXR but had hPXR transgene expressed in the liver [24, 28]. Although

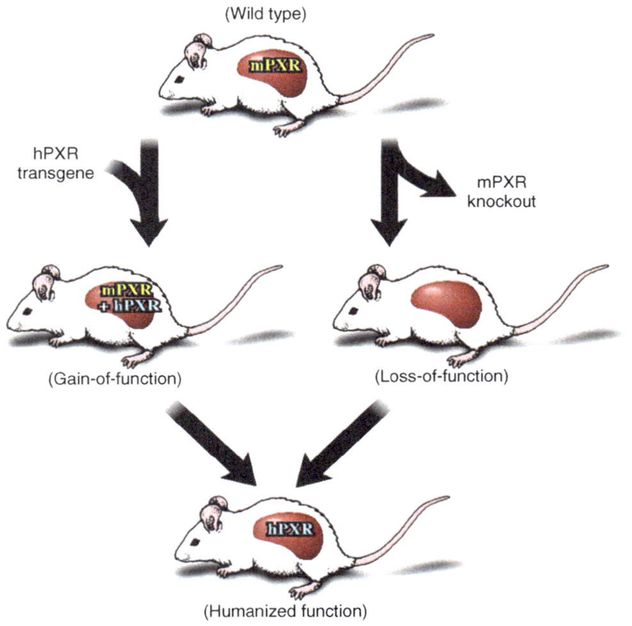


**Figure 41.2** Strategies to create the loss-of-function knockout, gain-of-function transgenic, and combined "humanized" function models. Source: Gong et al. [23] / with permission of Elsevier.

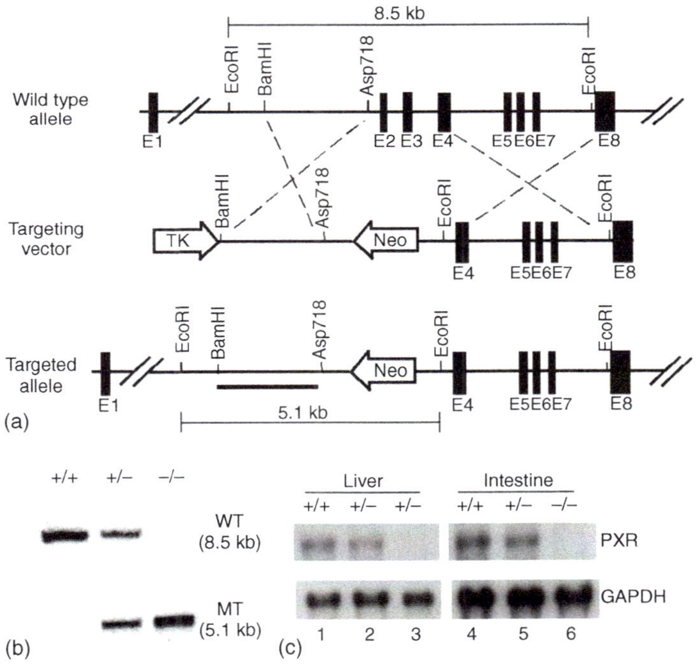


**Figure 41.3** Creation of PXR null mice. (a) Restriction map of the PXR gene and strategy to generate PXR mutant allele. The PXR probes used for Southern blot and expected fragment sizes after *Eco*R I digestion are indicated. E, exon; Neo, neomycin resistance gene; TK, thymidine kinase promoter. (b) Southern blot of *Eco*R I-digested genomic DNA. WT, wild type; MT, Mutant. C. Loss of PXR expression in the liver and small intestine of PXR−/− mice. Source: Gong et al. [23] / with permission of Elsevier.

it is not the primary focus of this book chapter, it should be noted that a similar strategy has been used to create the hCAR humanized mice [29]. Moreover, alternative models of hPXR and hCAR humanized mice have been developed and characterized and some of them have become commercially available [30, 31].

***41.2.1.3.2 Humanized Drug Response Profile in the "Humanized" Mice*** The drug induction of Cyp3a11 was compared between the wild type and humanized mice by Northern blot analysis (Figure 41.6). In humanized mice, Cyp3a11 was no longer induced by rodent PXR activator pregnenolone-16α-carbonitrile (PCN) but was efficiently induced by the human-specific PXR inducer rifampicin (RIF). In contrast, no induction occurred in wild-type mice by RIF. Expressing hPXR and responding to human-specific PXR inducer exclusively, the "humanized" mice represent a major step

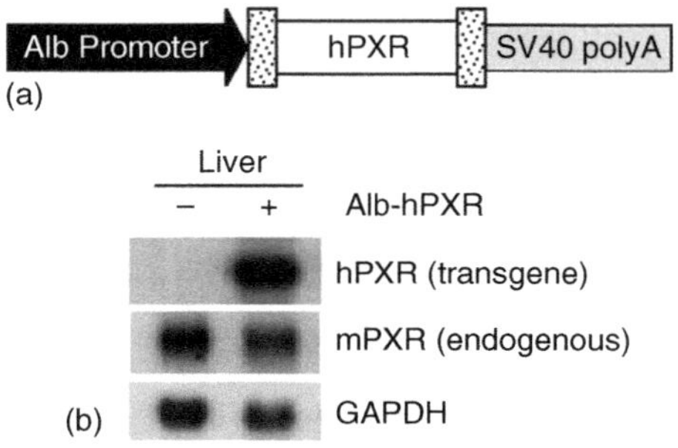


**Figure 41.4** Creation of PXR transgenic mice. (a). Schematic representation of the Alb-hPXR transgene construct. (b). Northern blot analysis of mouse and human PXR gene expression in the liver of wild type (−) and transgenic (+) mice. Source: Gong et al. [23] / with permission of Elsevier.

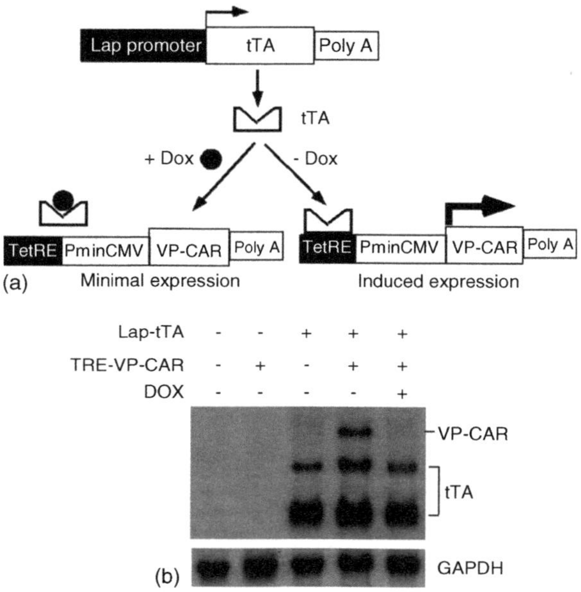


**Figure 41.5** Creation of transgenic mice that harbor conditional expression of the activated CAR in the liver. (a). Schematic outline of the Lap-tTA/TetRE-VP-CAR two-component Tet-Off transgenic system. The Lap-tTA transgene directs the expression of the tTA activator to the liver. The binding of tTA to the TetRE and the induction of the transgene VP-CAR should only occur in the absence of Dox. (b). Liver-specific conditional expression of VP-CAR. Liver RNAs of mice with indicated genotypes were subjected to Northern blot analysis. The mouse in the right-most lane was subjected to five days of Dox treatment. Source: Gong et al. [23] / with permission of Elsevier.

toward generating humanized toxicological models to predict xenobiotic enzyme inducibility and drug-drug interactions.

### 41.2.2 Measurements of Expression of Drug-Metabolizing Enzymes in Animal Models

#### 41.2.2.1 Detection of Enzyme Expression at the mRNA Level

***41.2.2.1.1 Northern Blot Analysis*** Total RNA was extracted using the Trizol reagent (Invitrogen, Calsbad, CA). An equal amount of total RNA (20 μg) was separated on 1.25% agarose–6% formaldehyde gel and transferred to the nylon membrane (Amersham) by capillary action in 10x SSC (0.75 M NaCl, 0.075 M Na citrate). After UV crosslinking, the membrane was pre-hybridized at 42°C for 2 h in pre-hybridization buffer containing 50% formamide, 5x SSC, 10 mM $Na_3PO_4$, 1 mM EDTA, 0.5% SDS, 10x Denhardts, and 20 μg/ml of denatured salmon sperm DNA (Sigma). The membrane was then hybridized with specific probe overnight at 42 °C in hybridization solution (pre-hybridization buffer plus 200 μg/ml of denatured salmon sperm DNA, 5% Dextran sulfate, and $1 \times 10^6$ cpm/ml radiolabeled cDNA probe labeled by the random primed labeling kit (Roche). Membrane was washed twice (15 minutes each) in 2x SSC and 0.1% SDS at 42 °C, twice (15 minutes each) in 1x SSC and 0.1% SDS at 65 °C, and twice (15 minutes each) in 0.1XSSC and 0.1% SDS at 65 °C. The hybridization signal was detected by exposing the membrane to X-ray film at −80 °C. The membranes were stripped and rehybridized with a GAPDH probe for loading control [32]. An example result is shown in Figure 41.6.

***41.2.2.1.2 Semi-Quantitative RT-PCR*** Total RNA was treated with RNase-free DNase I to avoid DNA contamination. One μg of total RNA was reverse transcribed into cDNA by AMV reverse transcriptase (Roche) using

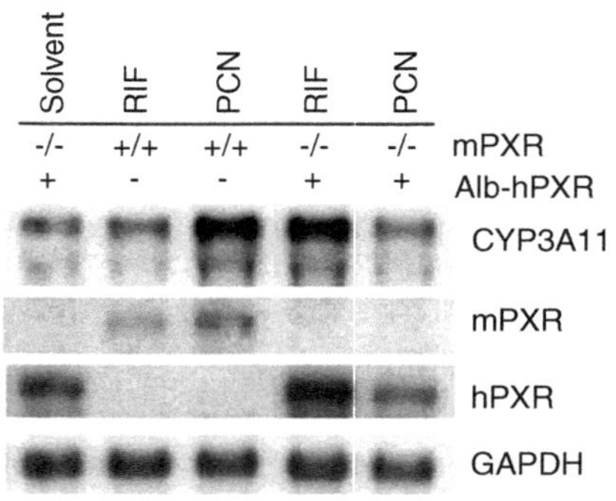


**Figure 41.6** Drug response profile in the humanized mice. Mice with indicated genotypes were treated with a single dose of RIF (5 mg $kg^{-1}$) or PCN (40 mg $kg^{-1}$) 24 hour prior to liver harvesting. Liver RNA was isolated and subjected to Northern blot analysis with indicated probes. Source: Gong et al. [23] / with permission of Elsevier.

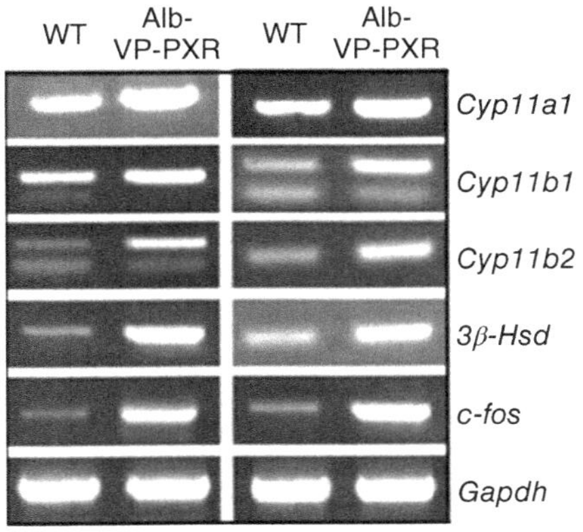


**Figure 41.7** Semiquantitative RT-PCR analysis on the expression of steroidogenic enzyme and c-fos mRNA in the wild-type (WT) and Alb-VP-PXR mice. Source: Reproduced with permission from Zhai et al. [33] / Oxford University Press.

oligo-dT as primer. PCR amplification of cDNA was performed in a 50-μl reaction mixture containing 10 mM Tris-HCl buffer (pH 8), 20 mM KCl, 0.1% Triton X-100; 1.5 mM $MgCl_2$, 0.2 mM of each dNTP, 50 pmol of each primer, and 2U of Taq DNA polymerase (Promega). A typical PCR cycle consists of 94 °C for 30 seconds, 55 °C for 40 seconds, and 72 °C for 60 seconds. The amplified products were separated by electrophoresis in 2% agarose gel and visualized with ethidium bromide under UV light. Amplification of the ubiquitously expressed β-actin or GAPDH was performed under the same conditions as a loading control. An example result of the semi-quantitative method used to measure drug-metabolizing enzyme expression is shown in Figure 41.7.

***41.2.2.1.3 Real-time PCR*** Total RNA isolation, DNase I treatment, and first-strand cDNA synthesis were performed as described in Section 41.3.1.2. Taqman or SYBR green real-time PCR was performed using the Applied Biosystems 7300 real-time PCR system (Applied Biosystems, Inc., Foster City, CA) [33]. An example result of the real-time method used to measure drug-metabolizing enzyme expression is shown in Figure 41.8.

#### 41.2.2.2 Detection of Enzyme Expression at the Protein Level by Western Blot Analysis

Equal amounts of protein were denatured by boiling the protein samples for 5 minutes and loaded onto 10% Tris-glycine SDS-polyacrylamide Gel. Proteins were transferred to nitrocellulose membranes by electroblotting using Tris-glycine buffer with 20% (v/v) methanol. Membranes were blocked for 2 h with 5% nonfat dried milk in PBS containing 0.1% Tween 20 (PBS-T), and then incubated for 1 h with specific antibody diluted in PBS-T plus 5% nonfat dried milk. After washing with PBS-T three times, the membranes were incubated for 1 hour with horseradish peroxidase-labeled secondary antibody in PBS-T plus 5% nonfat dried milk. The blots were then washed three times and proteins were visualized by Enhanced Chemiluminescence (ECL) method (Amersham) [11]. An example result of Western blot used to measure drug- metabolizing enzyme expression is shown in Figure 41.9.

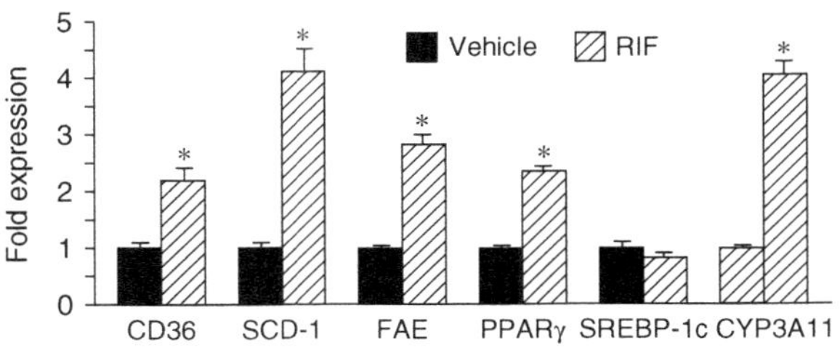


**Figure 41.8** Real-time PCR analysis on liver RNA derived from the vehicle- and RIF-treated hPXR humanized mice. FAE, long chain free fatty acid elongase; PPAR, peroxisome proliferator-activated receptor; SCD-1, stearoyl-CoA desaturase-1; SREBP, sterol regulatory element-binding protein. Source: Zhou et al. [28] / with permission of Elsevier.

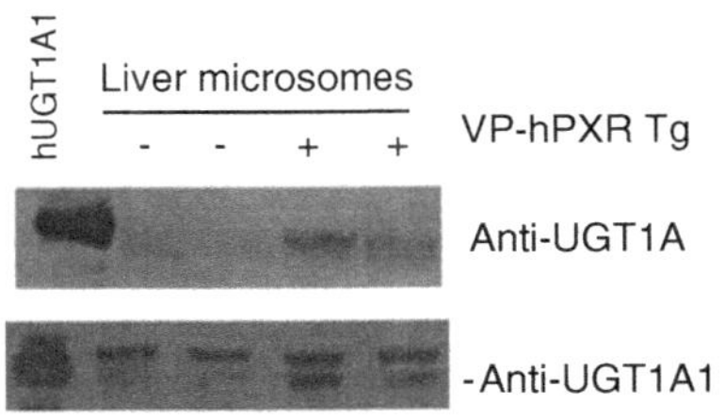


**Figure 41.9** Induction of UGT1A expression by PXR activation. Liver microsomes were prepared from WT and VP-hPXR transgenic mice and subjected to Western blot profiling by using a pan anti-UGT1A antibody and a specific anti-UGT1A1 antibody. Source: Reproduced with permission from Rosenfeld et al. [34] / Proceedings of the National Academy of Sciences PNAS.

#### 41.2.2.3 Use of Xenobiotic Receptor Mouse Models for In vivo Drug Metabolism Studies

In addition to their use in dissecting the transcriptional control of drug-metabolizing enzymes and transporters, the use of humanized or knockout mice offers a unique opportunity to study the in vivo relationship between drug concentrations and the corresponding physiological effects. For example, the effects of a drug discovery candidate drugs could be evaluated in such a model by administering several doses of the drug candidate and measuring the effects on the RNA expression or activity of various enzymes. It was demonstrated that increasing concentrations of rifampicin (1–10 mg/kg) lead to corresponding increases in CYP3A11 mRNA levels in hPXR mice by this approach [24]. During the drug development process, this type of study could be employed with new drug candidates to assess the upregulation of drug- metabolizing enzymes or transporters. By administering the drug candidates to animals for several days at doses yielding plasma exposures similar to those predicted to be efficacious in humans, the effect of the candidate drugs on hPXR-mediated enzyme and transporter expression can be assessed in this in vivo setting.

The study of drug dose (or plasma concentration) and their corresponding pharmacological effects by mathematical modeling is often referred to as pharmacokinetic–pharmacodynamic analysis. This type of modeling has shown utility in comparing the drug concentration–effect relationship in vivo with in vitro assays or in quantitating the time course of drug effects in relation to plasma/organ concentrations. Humanized PXR or CAR mice offer a platform to study the relationship among dose, plasma, and liver concentrations of new and existing drugs to the induction effects on CYP3A and 2B genes, respectively, in a more complex in vivo system as compared to current in vitro models. In a similar manner, knockout models can be used to assess the involvement of certain nuclear hormone receptors in the expression of enzymes or transporters, changes in physiology, and the toxicity of certain drugs [9, 24, 29, 35].

These newer in vivo models offer a dynamic system incorporating drug absorption, distribution, metabolism, and elimination, albeit in mouse, in contrast to the static system of cell culture where cells are exposed continually to drug and perhaps metabolites. In addition, the concentration–effect relationship can be expanded to compare in vivo liver/plasma concentrations with concentrations used in the in vitro human hepatocyte system and their corresponding in vivo and in vitro effects on gene transcription, enzyme induction, or toxicity. This type of comparison may actually improve the extrapolation of in vitro human hepatocyte concentration-induction effects to the prediction of induction and toxicological responses in patients.

## 41.3 In vitro Methods

### 41.3.1 Reporter Gene

Functional cell-based assays are amenable for identifying ligands under a more physiologically relevant condition compared to biochemical assays, but the data can be more convoluted for interpretation due to the complexities inherent to cellular systems (e.g. off-target events can indirectly affect the observed signal) [36]. The most widely utilized assays rely on the luciferase-based transactivation approaches because of their high sensitivity and dynamic range. They can be performed by transiently or stably expressing the NR along with a luciferase reporter gene driven by the corresponding target gene promoter [37–44]. Establishing stable cell lines is more time consuming than using transient transfections, but they display lower experiment-to-experiment variations [45].

The reporter gene system is a highly sensitive, easily detectable, and quantitative method designed to evaluate NR-mediated gene expression. In the following example, cells are transiently transfected with PXR or CAR, together with luciferase reporter genes that contain either a PXR or CAR responsive element upstream of a minimal thymidine kinase (tk) promoter or the natural promoters of PXR/CAR target genes [46].

#### 41.3.1.1 Plasmids

The protein-coding region of hPXR was PCR amplified and sub-cloned into NcoI and BamH1 sites of the vector pCDG1. The CYP3A23 cellular promoter-reporter, PGL3–CYP3A23, was cloned by inserting the PCR-amplified 5′ regulatory sequence of rat CYP3A23 gene (nucleotides −1,360 to +82) into the PGL3 vector (Promega). The tk-CYP3A23-Luc was generated by insertion of three copies of the DR-3 (TGAACTtcaTGAACT) of PXR response element from the rat CYP3A23 gene into the tk-Luc vector [5, 24]. The pSV-β-galactosidase control vector is used as an internal control in all transfections to monitor transfection efficiencies and normalize luciferase activities.

#### 41.3.1.2 Transient Transfection of HepG2 Cells

The following procedure is used to transfect plasmid DNA into HepG2 cells in a 48-well format [47, 48]. All amounts and volumes are given on a per-well basis.

(1) The day before transfection, trypsinize and count the cells. Plate $3 \times 10^4$ cells per well in 0.2 ml of complete growth medium. Cell density should be 50–80% confluent on the day of transfection.
(2) For each well of cells to be transfected, dilute 0.4 μg of DNA (including 0.2 μg luciferase reporter, 0.1 μg receptor, and 0.1 μg β-galactosidase control vector) into 50 μl of DMEM without serum and phenol red.
(3) For a 48-well plate, dilute 11 μl of polyethylenimine (PEI) polymer to 2.7 ml of DMEM without serum

and phenol red. For each well of cells to be transfected, combine 50 μl diluted PEI solution with the above 50 μl DNA mixture. Mix gently and incubate for 20 min at room temperature to form DNA-PEI complexes.

(4) Dilute the DNA-PEI mixture with 100 μl DMEM without serum and phenol red for each well transfection.
(5) Aspirate medium from cells and add 200 μl of diluted transfection mix to each well. Mix gently by rocking the plate back and forth. Incubate the cells at 37 °C in a $CO_2$ incubator for 18–24 hours post-transfection before ligand treatment.

#### 41.3.1.3 Ligand Treatment

(1) Supplement phenol red-free DMEM medium with 10% heat-inactivated, delipidated, Dextran-coated charcoal-treated fetal bovine serum, 0.1 mM MEM nonessential amino acids solution, 2 mM L-glutamine, 100 units/ml penicillin, and 100 μg/ml streptomycin.
(2) Prepare dilutions of drug stock solutions and vehicle (1 : 1000) in above medium. Aspirate transfection mix from cells and dispense 200 μl of diluted drug or vehicle to each well.
(3) Return cells to an incubator at 37 °C for an additional 18 hours prior to the determination of reporter luciferase activity.

#### 41.3.1.4 Luciferase and β-galactosidase Assays

(1) Aspirate culture medium from each well after 24 hours incubation with the test compound and freeze the plate at −80 °C for 30 minutes. Add 150 μl of cell lysis buffer (20 mM tricine, 8 mM $MgCl_2$, 4 mM EGTA PH 8.0, 1 mM DTT, 1% Triton X100, 10% glycerol) and shake at 4° for 30 minutes.
(2) Transfer 50 μl of medium from each well of treated plates to daughter plates to assay for β-galactosidase activity as an internal control for transfection efficiency using substrate *O*-nitrophenyl-β-D-galactoside (ONPG). Add 100 μl of ONPG (2 mg/ml) in β-galactosidase assay buffer (60 mM $Na_2HPO_4$, 40 mM $NaH_2PO_4$, 10 mM KCl, 1.0 mM $MgSO_4$, 50 mM β-mercaptoethanol) to each well and incubate at 37 °C for 10–20 minutes until the yellow color appears and read absorbance at 420 nm.
(3) Transfer 50 μl of cell lysate to a luminometer plate and add 50 μl of luciferase assay reagent (20 mM tricine, 1.07 mM $(MgCO_3)_4{\cdot}Mg(OH)_2{\cdot}5H_2O$, 10 mM $MgSO_4$, 0.1 mM EDTA, 1.0 mM ATP, 10 mM DTT, 0.2 mM Coenzyme A, 0.5 mM luciferin) and measure firefly luciferase activity immediately.

#### 41.3.1.5 Data Analysis

(1) Calculate the normalized reporter activity of each sample by dividing the luciferase activity (reporter activity) by the β-galactosidase activity (internal standard).
(2) Calculate the PXR activity of each sample from the ratio of normalized reporter activity for PXR test ligand-treated cells relative to that of vehicle-treated (control; the control value is arbitrarily set at 1) cells. Data are presented as mean activities ± SD from triplicate analyses. An example result of the reporter gene assay used to examine the regulation of drug-metabolizing enzyme or transporter is shown in Figure 41.10.

### 41.3.2 Electrophoretic Mobility Shift Assay

Electrophoretic mobility shift assay (EMSA), also referred to as a gel shift assay, gel mobility shift assay, band shift assay, or gel retardation assay, is a common technique used to study nuclear receptor-DNA interactions. DNA-protein complexes will migrate more slowly than unbound radioactive probe and are thus visualized by discrete bands of radioactivity on an acrylamide gel image [46, 49]. An example result of EMSA method is shown in Figure 41.11a–c.

#### 41.3.2.1 Primer Design

Design complementary nucleotide primers containing binding site of interest. Labeling with $^{32}$P-dCTP will ideally occur at the end of primer. The incorporation of a CGA sequence at the 5-prime end will allow for the incorporation of $^{32}$P-dCTP during radiolabeling.

#### 41.3.2.2 Primer Annealing

(1) Dissolve two complementary nucleotides in TE buffer to a final concentration of 1 mg/ml.
(2) Mix 0.5 μl each of forward and reverse oligos, 5 μl of 10X restriction buffer 2 (NEB), and 44 μl of MQ water in a 1.5 ml Eppendorf tube.
(3) Bring 1 liter of water to boil in a beaker and place the Eppendorf(s) in a floating rack and into the water. Take beaker off the heat and leave it at room temperature to cool slowly for successful annealing. Annealed duplexes can be stored at −20 °C for several months.

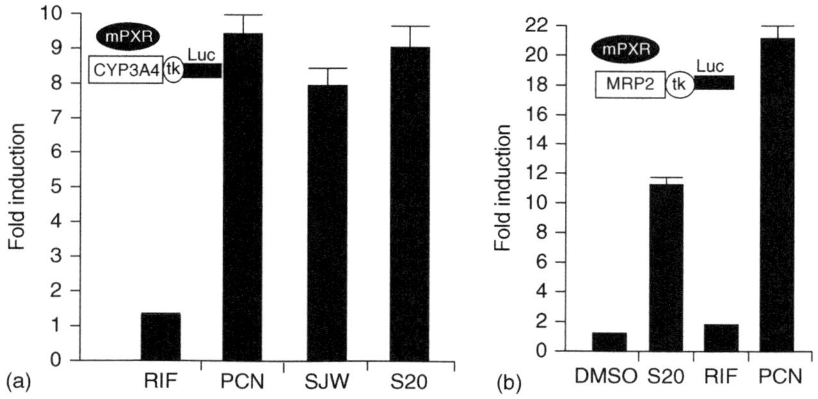


**Figure 41.10** S20 induces the expression of drug-metabolizing enzymes and transporters in reporter gene assays. (a), PXR-mediated activation of CYP3A4 reporter gene by S20. CV-1 cells were cotransfected with the PXR-responsive tk-CYP3A4-Luc reporter and mPXR expression vector. Transfected cells were treated with RIF, PCN, SJW (St John's wort), and S20. Results are shown as -fold induction over vehicle controls and represent the average from triplicate assays. (b), PXR-mediated activation of MRP2 reporter gene by S20. CV-1 cells were cotransfected with the PXR-responsive tk-MRP2-Luc reporter and expression vector for mPXR before ligand treatment. Source: Mu et al. [47] / with permission of American Society for Pharmacology and Experimental Therapeutics.

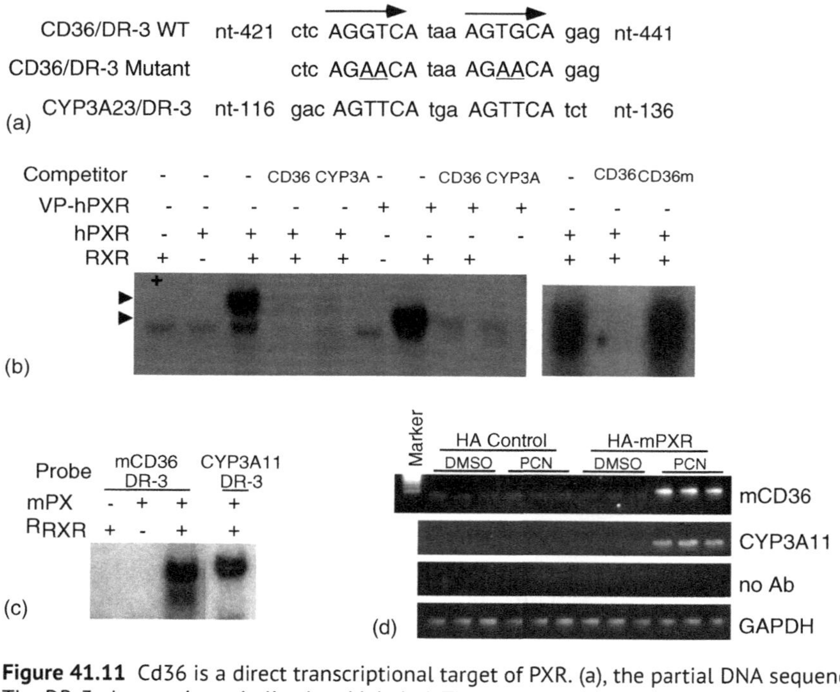


**Figure 41.11** Cd36 is a direct transcriptional target of PXR. (a), the partial DNA sequences of the mouse Cd36 gene promoter. The DR-3 element is capitalized and labeled. The mutant variant is shown with the mutated nucleotides (nt) underlined. (b), hPXR/RXR or VP-hPXR/RXR heterodimers bound to mCd36/DR-3 as revealed by EMSA. Arrows indicate specific shift bands. (c), binding of the mPXR/ RXR heterodimers to mCd36/DR-3 as revealed by EMSA. The Cyp3a11/DR-3 binding was included as the positive control. (d), ChIP analysis to show the recruitment of mPXR onto the mCD36 promoter. HA-tagged mPXR or the HA vector control was used to transfect wild-type mouse livers by a hydrodynamic gene delivery method. Mice were treated with DMSO or PCN for 8 hour before sacrificing. ChIP assay was performed with the use of an anti-HA antibody (Ab). Each lane represents one mouse. Source: Zhou et al. [28] / with permission of Elsevier.

#### 41.3.2.3 Primer Labeling

(1) Pipette 2.5 μl of each probe into each tube.
(2) Make mixes as follows (for 1 tube):
    2.5 μl of 10X restriction enzyme buffer 2 (NEB)
    1 μl "-dCTP" (5 mM each of dATP, dGTP, dTTP)
    16 μl of MQ water
    1 μl Klenow enzyme
    Add enzyme last and keep the mix on ice until used.
(3) Bring the mixes into the radioactive working area. Add 2 μl of $^{32}$P-dCTP to tube. Flick mix and add 22.5 μl of mix to each tube. Keep at 37 °C for 20 minutes.

#### 41.3.2.4 Probe Purification

(1) Redistribute Sephadex in a spin column by flicking column to disperse contents throughout.
(2) Place column into a tube. Spin down columns at 3000 rpm for 5 minutes to remove the storage buffer. Discard tube with storage buffer. Place column into a new tube.
(3) After probe incubation period is up, dilute probe with 75 μl MQ water.
(4) Place the entire contents of diluted probe tube (~100 μl) into centre of column. Spin columns for 5 minutes at 3000 rpm. Discard column into a radioactive waste bin.
(5) Remove 1 μl of purified probe into 5 ml Scinti Verse SX1-4 scintillation cocktail (Fisher Scientific Products) to be used for post-purification cpm (count per minute) determination. A successful labeling should result in a cpm value of more than 10 000.
(6) Store probes in radiation safety box in a −20 freezer until use.

#### 41.3.2.5 Preparation of an Acrylamide Gel

(1) Clean a 20 × 20 cm plain glass plate and a 20 × 20 notched glass plate with 70% ethanol. Add spacers at sides and bottom and clip together at equidistant spacing using large paper clips.
(2) Mix 6.7 ml of 30% acrylamide stock solution, 5 ml of 5X TBE, 1.25 ml of 50% glycerol, 500 μl of 10% ammonium persulphate, and 36.5 ml MQ water in 50 ml Conical tube.
(3) When ready to pour, add 50 μl of TEMED (gel will begin polymerizing at this point). Invert several times and carefully pour between plates, avoiding the formation of bubbles. Insert a 14-well comb at top between plates. Let sit at room temperature for about an hour before using. The gel can also be wrapped in Saran wrap and stored in cold room for later use.

#### 41.3.2.6 Preparation of In Vitro Synthesized NR Proteins

In vitro transcribed and translated nuclear receptors PXR and RXR proteins are prepared by using TNT® Coupled Reticulocyte Lysate Systems (Promega, Madison, WI).

#### 41.3.2.7 Binding Reaction

(1) To make 5X binding buffer:
    10 ml: 500 μl of 1 M Tris-HCl, pH 7.5 (50 mM)
    500 μl 5 M NaCl (250 mM)
    60 μl 0.5 M EDTA (3 mM)
    2 ml glycerol (20%)

    Add 50 μl 1 M DTT (5 mM) and 1 ml 10 mg/ml BSA (1 mg/ml) before use.
(2) Binding reaction:
    4 μl of 5X binding buffer
    1 μl dI-dC polymer (1 μg/μl)
    4 μl TNT protein (2 μl of PXR and RXR each)
    2 μl probe
    9 μl of MQ water

    Let tube stand at room temperature for 20 min before continuing.

#### 41.3.2.8 Gel Loading and Running

(1) Have gel(s) ready the night before, storing in cold room. Using fine line marker, mark bottoms of wells of gel(s) and number to assist in loading. Remove end spacer with forceps.
(2) Using a 1000 μl pipette set at 500 μl, pipette buffer up and down in each well to rinse out. Pre-run the gel at 150V for at least 0.5 hour.
(3) After gel has pre-run and incubation period is finished, carry materials needed from hot room into cold room.
(4) Using fine-point 20 μl gel tips, carefully load gel wells with incubation mixture. Run the gel at 300V for approximately 1.5 hours.

#### 41.3.2.9 Gel Drying

(1) Remove gel plate(s) from apparatus.
(2) Remove side spacers with forceps. Using handle of forceps, slowly separate glass plates to expose one side of gel (make note of which side it is relative to where you started loading).

(3) Place a sheet of Whatman paper on the center of gel surface and carefully smooth outward toward edges. Place a second sheet on top of first and carefully invert gel and paper so that second glass plate is now on top. Carefully remove second glass plate.
(4) Cover gel surface with Saran wrap, avoiding wrinkling as much as possible. Label on the Whatman paper so that directionality of gel is clearly labeled.
(5) Place one sheet of blue roll flat on gel dryer mat. Lay gel and Whatman assembly on top, and cover carefully with another sheet of blue roll. Cover all this with gel dryer sheet and cover. Turn on the vacuum pump, check for wrinkling of top blue roll, and immediately remedy if needed.
(6) Set up cycle, temp (80 °C), and time (about 1 hour) and start gel drying. When paper looks completely flat (no bulge from gel), gel is completely dry.

#### 41.3.2.10 Autoradiograph

(1) When gel is dry, lift up one corner of gel dryer sheet and then turn off the vacuum and gel dryer. Discard blue roll and an extra sheet of Whatman paper.
(2) Tape Whatman containing gel inside an autoradiograph cassette. Label outside of cassette with gel number and date.
(3) Carry cassette(s) into dark room with box of blue-sensitive film. Turn out the white light and turn on photo-safe red light. Place film on gel and close the cassette. Make sure cassette is securely closed before turning off the red light and leaving dark room.
(4) Carry cassette(s) to −80 °C freezer and incubate in freezer overnight.
(5) The next day, bring the cassette into dark room, turning off the white light and turning on photo-safe red light. Remove film from cassette and feed it into the developer.

### 41.3.3 Chromatin Immunoprecipitation (ChIP) Assay

Chromatin immunoprecipitation (ChIP) is a widely used method to identify binding region of putative target gene associated with specific nuclear receptors. Both cell lines and animal tissues can be used to study the regulation of drug-metabolizing enzymes and drug transporters mediated by nuclear receptor. The following is a method using a mouse liver to determine the binding of PXR with putative target gene promoter [28]. An example result of ChIP method is shown in Figure 41.11d.

#### 41.3.3.1 Chromatin Crosslinking and Cell Lysis

(1) Eight-weeks-old wild type female mice were pretreated with i.p. injection of DMSO or PCN (40 mg/kg) for 30 mins before being transfected with the pCMX-HA-mPXR or pCMX-HA control plasmid by a hydrodynamic gene delivery method. 5 μg of plasmid was used for each of the mouse of an average body weight of 20 g. The mice were sacrificed eight hours after transfection and the liver tissues were harvested.
(2) 50 mg of reach tissue was minced and added with formaldehyde to a final concentration of 1%. After rocking at room temperature for 15 minutes, glycine was added to a final concentration of 0.125 M to stop the crosslinking.
(3) The tissues were collected by centrifuge and homogenized by a Dounce homogenizer in cold 1X PBS. The resultant cell pellets were then re-suspended in lysis buffer (5 mM PIPES pH 8.0, 85 mM KCL, 0.5% NP40) supplemented with proteinase inhibitor cocktail. Nuclei pellets were collected by centrifugation at 5000 rpm for 5 minutes at 4 °C and re-suspended in 100 μl of nuclei lysis buffer (50 mM Tris-Cl pH 8.1, 10 mM EDTA, 1% SDS) supplemented with proteinase inhibitor cocktail.

#### 41.3.3.2 Sonication to Shear DNA

(1) The lysates are sonicated five times at setting 10 with the sonicator (model Virsonic 60, Virtis, NY) for 7–8 seconds each time. This step needs to be optimized according to the sonicator and the cell line used. At this point, it is important to verify the length of the chromatin fragments resulting from the sonication step. An aliquot of the sonicated lysates is removed, incubated at 65 °C for at least 6 hours to reverse the formaldehyde crosslinking, purified using QIAquick spin PCR purification kit (Qiagen, CA), and analyzed on a 2% agarose gel. The average size of the DNA fragments should be approximately 200–1000 bp. While this verification step is in progress, the remaining sonicated lysates can be frozen at −80 °C. Further sonication steps can be carried out to obtain the desired fragment length.
(2) The sonicated lysates are centrifuged at 12 000 × g at 4 °C for 10 minutes; the supernatants collected and diluted 10× in ChIP dilution buffer (0.5% Triton X-100, 2 mM EDTA, 150 mM NaCl, 20 mM HEPES, pH 8) to achieve a final SDS concentration of 0.1%.

#### 41.3.3.3 Chromatin Immunoprecipitation and PCR Analysis

(1) Immunoclearing of the diluted chromatin is achieved by incubating for 1 hours at 4 °C with 60 μl of salmon sperm DNA–protein A agarose provided in 50% slurry in 10 mM Tris–HCl, pH 8, 1 mM EDTA, 0.05% sodium azide (Upstate Biotechnology, Inc., Lake Placid, NY). Pellet agarose by brief centrifugation and remove 10 μl (1%) of the supernatant as Input.
(2) Immunoprecipitation (IP) is performed using 2 μg of HA antibody (Santa Cruz, Cat No: sc-805) or normal IgG overnight at 4 °C. The precipitated complex was collected by incubating for 1 hour at 4 °C with 60 μl of salmon sperm DNA–protein A agarose.
(3) The precipitates are washed sequentially for 10 minutes each with low-salt wash buffer (0.1% SDS, 1% Triton X-100, 2 mM EDTA, 20 mM Tris–HCl, pH 8.1, 150 mM NaCl), high-salt wash buffer (0.1% SDS, 1% Triton X-100, 2 mM EDTA, 20 mM Tris–HCl, pH 8.1, 500 mM NaCl), and LiCl wash buffer (0.25 mM LiCl, 1% NP-40, 1% deoxycholate, 1 mM EDTA, 10 mM Tris–HCl, pH 8.1). Precipitates are then washed three times with TE buffer and eluted twice for 15–30 minutes each time on vortex set at 3 with 75 μl of elution buffer (1% SDS, 0.1 M $NaHCO_3$).
(4) The pooled elutes and the input are incubated at 65 °C for at least 6 hour to reverse the formaldehyde crosslinking. The isolated DNA fragments are then purified according to the QIAquick Spin Kit protocol (Qiagen, CA) and used for semi-quantitative PCR or quantitative real-time PCR analysis. PCR was carried out with specific primer of target gene promoter binding site. Primer of Cyp3a11 promoter that also contains the DR-3 element was used as the positive control. GAPDH was used as the loading control.

### 41.3.4 Measurement of Transporter Activity

Transporter activity can be monitored by measuring the efflux or retention of a fluorescent probe from cells. Enhancement of the uptake of the fluorophore Calcein has been applied in the development of P-gp inhibitors [50] Similarly, efflux of the probe Rhodamine 123 has been used in conjunction with flow cytometry to determine P-gp modulation [51].

### 41.3.5 Modulation of Cytochrome p450 Enzymes

The modulation of various cytochrome p450 enzymes can be determined by measuring product formation in liver microsomes by liquid chromatography coupled with mass spectrometry [52]. Particular substrates are employed to monitor the activity of a certain enzyme: tacrine for CYP1A2, bupropion for CYP2B6,

amodiaquine for CYP2C8, tolbutamide for CYP2C9, mephenytoin for CYP2C19, dextromethorphan for CYP2D6, and midazolam for CYP3A4. The metabolites detected are hydroxytacrine for CYP1A2, hydroxybupropion for CYP2B6, desethylamodiaquine for CYP2C8, α-hydroxytolbutamide for CYP2C9, 4-hydroxymephenytoin for CYP2C19, dextromethorphan for CYP2D6, and 1-hydroxymidazolam for CYP3A4 [52].

A more amenable assay for large-scale screening involves the use of a luminogenic substrate, which is metabolized by CYP enzymes to generate D-luciferin [53]. In turn, D-luciferin serves as a substrate for firefly luciferase to produce luminescence. This approach can be implemented in cell-based and biochemical formats, where enzyme-selective luminogenic substrates are incubated with membrane preparations or cells overexpressing the particular cytochrome p450 under investigation. Because the luminogenic substrate and the metabolite D-luciferin are cell-permeable, cells can remain intact for additional experiments.

## 41.4 Biochemical, Biophysical and Structural Analysis of NRs Using Purified Proteins

Biochemical and biophysical analysis of purified proteins complement experiments conducted on cell-based systems, providing unique perspectives and information arising from conditions with well-defined biological components. Notably, they allow for the determination of direct interactions of ligands to their target proteins, binding affinity, and mechanistic insights at the molecular level. These studies are valuable in the discovery, design, and development of PXR and CAR ligands, and in the understanding of the function of these NRs. This section focuses on the expression and purification of proteins, which are subsequently employed in biochemical, biophysical, and structural studies of ligand interactions to PXR and CAR (Figure 41.12).

### 41.4.1 Protein Purification Strategies

There are several reported PXR and CAR purification protocols, depending on the intended use for the purified protein preparations (e.g. biochemical assays and x-ray crystallography). With very few exceptions, the successful purification of the full length of these NRs

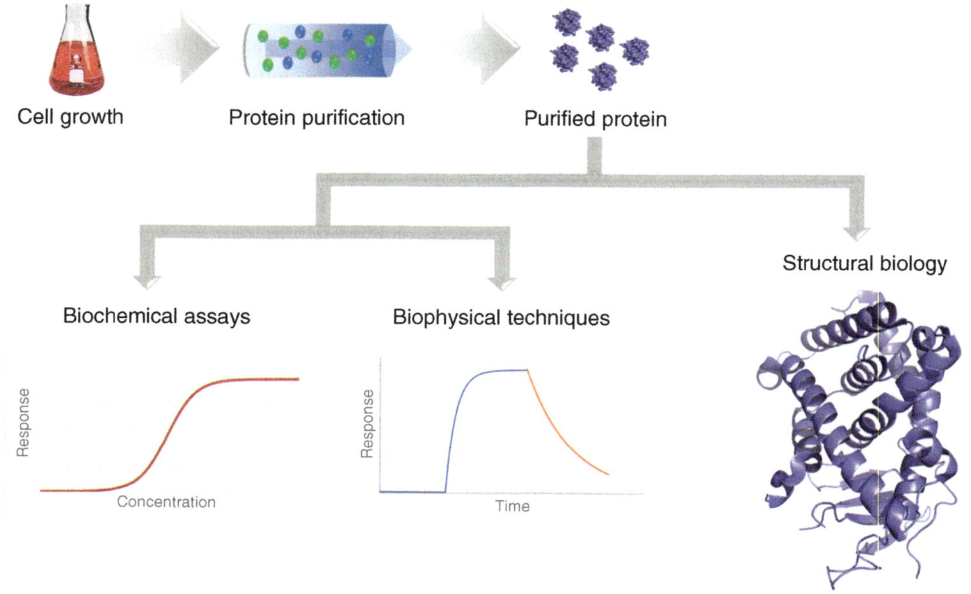


**Figure 41.12** Overview of a typical pipeline needed to characterize purified NR proteins using biochemical, biophysical, and structural methods.

has been hampered by protein instability and aggregation. Therefore, most studies focus on the purification of the LBD, which is the most relevant domain for the study of ligand-induced PXR or CAR transactivation. The following is a protocol used to obtain hPXR LBD for crystallization [41].

#### 41.4.1.1 Expression Vector

Because the presence of a coactivator peptide improves PXR solubility and stability, the cDNA fragments of the hPXR LBD (residues 130–434) and mouse steroid receptor coactivator 1 (SRC-1; residues 623–710) are expressed from the same vector. The *Escherichia coli* codon-optimized cDNA fragments are inserted into pETDuet1 plasmid at NcoI/HindIII and NdeI/KpnI sites, respectively. Alternatively, hPXR LBD can be expressed at lower amounts without a coactivator peptide by subcloning the hPXR LBD cDNA fragment into pETDuet1 at NcoI/HindIII sites.

#### 41.4.1.2 Protein Expression

*Escherichia coli* BL21(DE3) cells transformed with the plasmid are grown in LB media containing 100 mg/l ampicillin at 37 °C. Cells are induced to express hPXR LBD by the addition of 0.8 mM IPTG when the optical density value of the media (measured at 600 nm) reaches 0.6, after which cells are grown overnight at 18 °C. Cell pellets are obtained by centrifugation at 4000 g at 4 °C, which are then resuspended in buffer A (50 mM Tris-HCl, pH 7.5, 150 mM NaCl, 10% glycerol) containing EDTA-free protein inhibitor cocktail (10 ml buffer per gram of cells). Cells are disrupted by sonication on ice, followed by centrifugation of the lysate.

#### 41.4.1.3 Protein Purification

The clear portion of the lysate is loaded onto a Ni-NTA column followed by a sequential column washing with buffer A containing 0, 50, 100, or 200 mM imidazole. The flow-through fractions from the column are collected and analyzed by SDS-PAGE. The fraction obtained from the elution at 200 mM imidazole is diluted (1 : 2) in buffer A. The partially purified PXR is incubated overnight at 4 °C with a SRC-1 peptide (CPSSHSSLTERHKILHRLLQEGSPS) at 1.5-fold concentration. The sample is filtered with a 0.45-μm membrane and loaded onto a MonoS column pre-equilibrated with buffer B (20 mM Tris-HCl, pH 7.5, 50 mM NaCl, 5% glycerol, 5 mM DTT). A gradient of 0 to 50% buffer C (20 mM Tris-HCl, pH 7.5, 1 M NaCl, 5% glycerol, 5 mM DTT) is initially applied in 25 column volumes (CV), followed by a second gradient of 50 to 100% buffer C in 10 CV. The elution fractions containing hPXR LBD are combined and loaded onto a desalting column for buffer exchange to 20 mM Tris-HCl, pH 7.8, 250 mM NaCl, 5% glycerol, 5 mM DTT, and 2.5 mM EDTA. The protein can be concentrated using centrifugal filter units with a molecular weight cut-off of 10 kDa.

### 41.4.2 X-ray Crystallography

X-ray crystallography is one of the principal methods to obtain the three-dimensional structures of proteins. Following the first reported crystal structure of hPXR LBD [54], a number of hPXR LBD structures have been elucidated in complex with agonists [55]. On the other hand, only a handful of CAR LBD structures have been reported [56–58]. These structural studies have contributed significantly to the understanding of PXR's ligand promiscuity, and they layout the foundation for the discovery, design, and development of PXR and CAR ligands [55]. In addition, crystallography can explain the molecular basis of agonism, antagonism, and inverse agonism. For instance, it was important in the understanding of the super-agonism by TCPOBOP and the inverse agonism by androstenol on CAR [56, 57].

#### 41.4.2.1 Crystallization of hPXR LBD

The primary method to obtain hPXR LBD crystals is through hanging drop vapor diffusion, although the sitting drop vapor diffusion approach has also been applied [59]. Most of the structures of hPXR LBD in complex with ligands have been obtained by co-crystallization, while only a handful of complexes have been attained by soaking hPXR LBD crystals with the ligand.

Most of the crystals of hPXR LBD have been obtained in the presence of a short coactivator SRC-1 peptide, and some when the peptide is tethered to the C-terminus of the LBD. The most common crystallization conditions include 8–16% 2-propanol in 50–100 mM imidazole, with pH values in the range of 7 and 8 [59]. For instance, crystals of hPXR LBD in complex with the agonist SJB7 were obtained by hanging drop vapor diffusion using 50 mM imidazole, pH 7.8, and 10% 2-propanol [41]. The crystals were cryoprotected in 15, 25, and 35% ethylene glycol in a sequential manner prior to freezing in liquid nitrogen and data collection in a synchrotron.

### 41.4.3 Time-Resolved Fluorescence Energy Transfer Assay

Time-resolved fluorescence energy transfer (TR-FRET) measures the time-resolved energy transfer between a donor fluorophore and an acceptor fluorophore as fluorescence intensities [60]. TR-FRET is a robust and sensitive assay because it is a ratiometric technique, measuring the ratio of the fluorescence emitted by the donor and acceptor fluorophores. A donor fluorophore is typically a chemical complex containing a lanthanide, such as terbium or europium. Because lanthanides have a long fluorescence lifetime, delayed detection can be exploited to reduce the spectral interference by compounds with intrinsic fluorescence. Although the target protein can be labeled directly with the donor fluorophore by covalent attachment, it is more efficient to use a labeled antibody that recognizes a portion of the target protein, such as a poly-histidine or glutathione-*S*-transferase tag. Similar approaches have been applied using the streptavidin-biotin binary complex. A wider choice of acceptor fluorophores is available, which include the commonly utilized compounds fluorescein, allophycocyanin, BODIPY dyes, Alexa Fluor dyes, and their derivatives.

Two main TR-FRET strategies have been applied to study ligand interactions with PXR (Figure 41.13) [61]. The first one monitors the competitive displacement of a labeled ligand by unlabeled test compounds [42, 62, 63]. The second approach is to investigate the ligand-induced (agonist or antagonist) enhancement in the recruitment of a labeled coactivator or corepressor peptide to PXR or CAR [41]. The coactivator peptides contain the conserved LXXLL motif (where L = leucine, X = any amino acid), whereas the corepressor peptides have the I/L-X-X-I/V-I motif (I = isoleucine, L = leucine, V = valine, X = any amino acid).

#### 41.4.3.1 TR-FRET Assay to Monitor the Displacement of a Labeled Ligand from PXR

A mixture containing 5 nM hPXR LBD tagged with GST, 5 nM of anti-GST antibody labeled with terbium, and 40 nM of the fluorescence-labeled hPXR agonist Fluormore PXR Green is dispensed (20 μl per well) into 384-well low-volume solid black microtiter plates using a liquid dispenser or multi-channel pipettor. Test compounds (usually dissolved in DMSO) are added at various concentrations. In addition, positive (a known PXR agonist such as SR-12813) and negative (DMSO) controls are included. The plate is incubated at 25 °C for 30 minutes followed by fluorescence measurement using a plate reader. The excitation wavelength is set at 340 nm, and the emission intensities are collected at 490 and 520 nm, with a 100 μs delay time and 200 μs integration time. The 520 nm/490 nm fluorescence signal ratio is used as the raw data to be normalized using the positive and negative controls. A sigmoidal equation is used to fit the dose-response curves to obtain $IC_{50}$ values.

The above-mentioned protocol requires a kit from Invitrogen (Carlsbad, CA, USA), which provides a proprietary labeled PXR ligand and buffer with undisclosed components. An alternative reagent mix can be prepared using BODIPY FL vindoline (100 nM) as the labeled probe, with the assay buffer composed of 50 mM Tris, pH 7.5, 20 mM $MgCl_2$, 0.1 mg/ml BSA and 0.05 mM DTT [63].

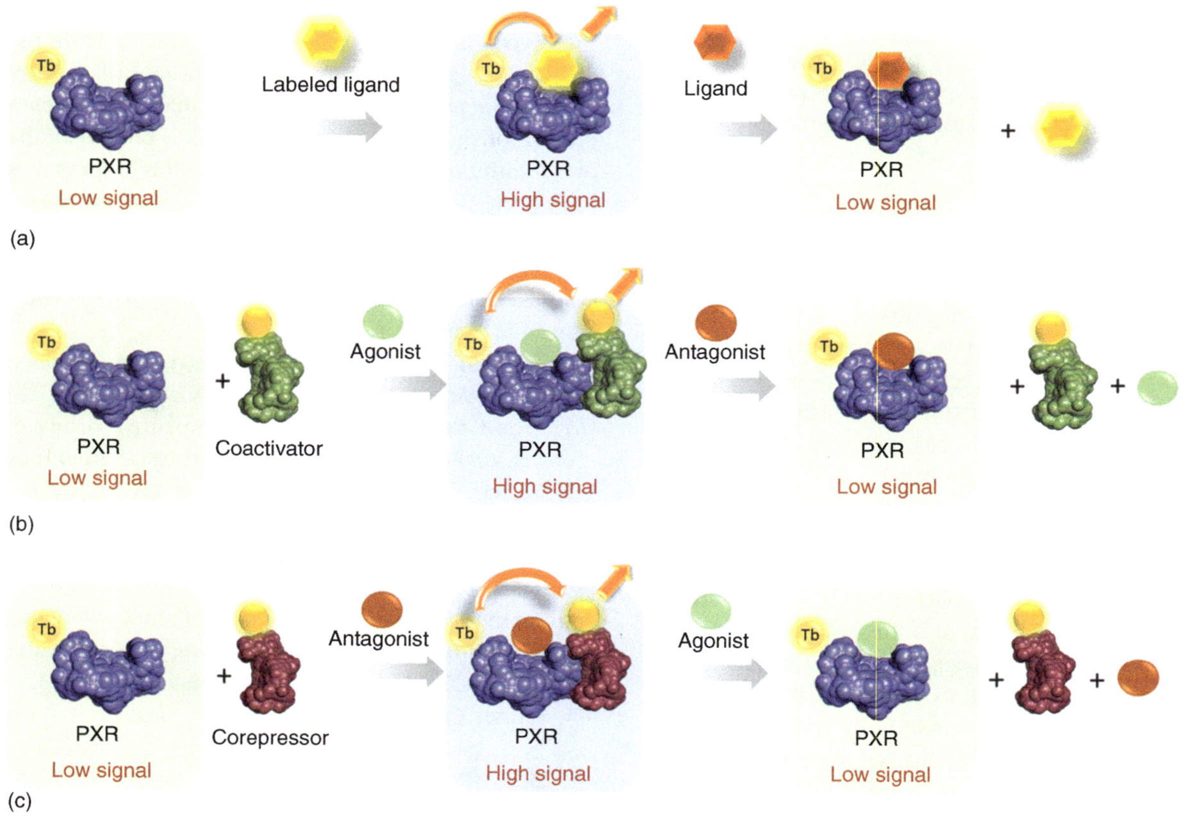


**Figure 41.13** Illustration of time-resolved fluorescence energy transfer (TR-FRET) assays to investigate interactions of ligands to PXR and CAR. (a). Displacement of the fluorescent-labeled ligand by an unlabeled ligand. (b). Coactivator recruitment enhancement by agonist with subsequent dissociation due to antagonist. (c) Corepressor recruitment enhancement by antagonist with subsequent dissociation due to agonist. Tb, terbium.

#### 41.4.3.2 TR-FRET Assay to Monitor the Interaction of PXR with Coactivator or Corepressor Peptide

The biological reagent is comprised of 5 nM hPXR LBD tagged with GST and 5 nM terbium-labeled anti-GST antibody in a suitable TR-FRET buffer (50 mM Tris, pH 7.5, 20 mM $MgCl_2$, 0.1 mg/ml BSA, 0.05 mM DTT) [41]. The solution also includes 500 nM fluorescein-labeled NCoR corepressor peptide (DPASNLGLEDIIRKALMGSFDDK) for assays monitoring corepressor recruitment to PXR or 100 nM fluorescein labeled SRC-1 coactivator peptide [FAM-$(PEG)_5$-CPSSHSSLTERHKILHRLLQEGSPS] for experiments measuring coactivator association. The mixture (20 μl per well) is dispensed into 384-well low-volume solid black plates. Test compounds at various concentrations and suitable internal controls (positive and negative) are added to the plate and incubated for 120 minutes at RT. A plate reader is used to obtain the TR-FRET signal by monitoring the fluorescence emission ratio (520 nm/490 nm) with an excitation wavelength of 340 nm, 100 μs delay time, and 200 μs integration time. The normalized dose-response curves are fitted with a sigmoidal equation to calculate $IC_{50}$ values.

### 41.4.4 Protein Thermal Shift Assay

Protein thermal shift assay (TSA) monitors the thermal stability of a target protein by applying a temperature gradient (Figure 41.14a). A fluorescent dye is added to the solution, which emits a fluorescence signal upon binding to exposed hydrophobic regions of the temperature-induced denatured protein. Plotting fluorescence intensity as a function of temperature results in a melting curve (melt profile), where the melting temperature ($T_m$) at the inflection point of the curve can be determined. Ligand binding to the target protein can increase its thermal stability, which is observed as an increase in $T_m$ values. TSA is a technique that investigates direct interactions of ligands to the target protein, but the calculated $T_m$ values do not necessarily correlate with binding affinity.

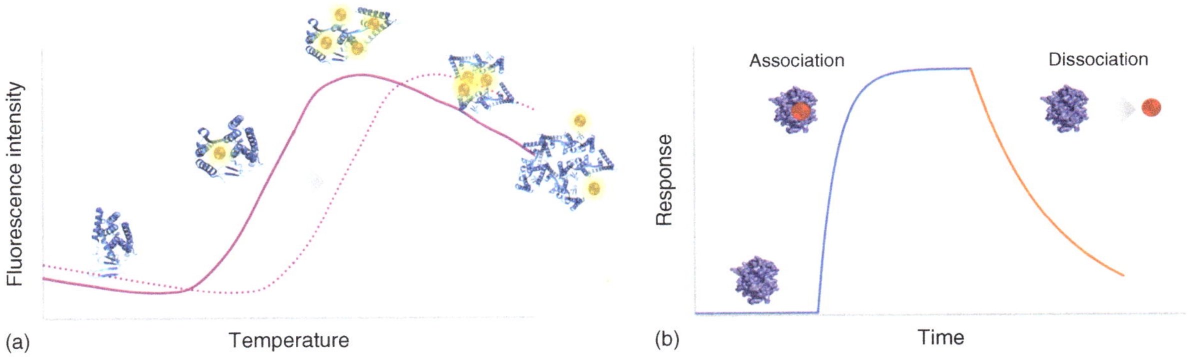


**Figure 41.14** Illustration of the principles of protein thermal shift assay and surface plasmon resonance assay. (a) In the protein thermal shift assay, the native protein starts to unfold as the temperature increases, exposing hydrophobic residues that interact with a fluorescent dye. The increase in fluorescence intensity generates the prototypical melt curve. The apex of the curve represents the fully denatured protein with maximal dye binding, after which the signal decreases due to protein aggregation. (b) In surface plasmon resonance assay, the experiment begins with a protein immobilized on the surface of channels within a biosensor chip. During the association phase, binding is monitored while the compound flows over the captured protein, causing an increase in the instrument's response. During the dissociation phase, the flow of the compound over the capture protein is stopped, leading to a decrease in response. The sensogram provides kinetic parameters such as association and dissociation constants and binding affinity.

#### 41.4.4.1 Monitoring Ligand-CAR Interactions by Protein Thermal Shift Assay

A volume of 10 μl of 10 μM hCAR LBD in assay buffer (50 mM HEPES, pH 8.0, 200 mM NaCl) is dispensed into a standard 384-well real-time PCR assay plate [64]. Test compounds at various concentrations are transferred to the plate together with suitable positive and negative controls. The plate is incubated for 30 minutes at RT prior to the addition (10 μl per well) of a 1 : 1000 dilution of a SYPRO Orange dye stock solution (Sigma-Aldrich, St. Louis, MO). The plate is sealed with a clear film and centrifuged at 1000 g to force the solution to the bottom of the wells. Fluorescence signal of the protein-bound SYPRO orange dye can be monitored using a real-time PCR system, with the excitation and emission wavelengths set at 492 nm and 610 nm, respectively. The temperature applied to the plate is increased from 25 to 95 °C at a rate of 1 °C/min. The resulting melting curves can be fitted with the Boltzmann equation to determine $T_m$ values at the inflection point of the curves.

### 41.4.5 Tryptophan Fluorescence Quenching Assay

Tryptophan residues within proteins have intrinsic fluorescence that can be altered by changes in the local microenvironment upon ligand binding [65]. This assay measures direct interactions of the ligand with the target protein, and the ligand dose-dependent behavior allows for the estimation of dissociation constants [38]. However, there can be interference of the tryptophan-derived fluorescence signal by compounds with intrinsic fluorescence.

#### 41.4.5.1 Ligand-PXR Interactions Monitored by Tryptophan Fluorescence Quenching Assay

A solution comprised of 0.3 mM hPXR LBD in assay buffer (20 mM phosphate buffer, pH 7.4, 1 mM DTT) and the tested compound is dispensed into a quartz cuvette of 10 mm pathlength maintained at 25 °C [38]. The fluorescence spectra are recorded by exciting the sample at 280 nm and scanning the emission signal from 300 to 400 nm, with an emission maximum at 340 nm. The resulting spectra are baseline corrected by subtracting the spectra of solutions without hPXR LBD. The presence of a ligand in close proximity to tryptophan residues causes a decrease in fluorescence intensity, which can be expressed as $(F_0-F_c)/(F_0-F_{min})$, where $F_0$ is the signal from the ligand-free hPXR LBD, $F_c$ is the background corrected signal at a particular ligand concentration, and $F_{min}$ is the signal arising at the highest ligand concentration. The plot of $(F_0-F_c)/(F_0-F_{min})$ as a function of ligand concentration can be fitted with a nonlinear single binding site model.

### 41.4.6 Surface Plasmon Resonance

Surface Plasmon Resonance (SPR) is a label-free biophysical technique that allows for the determination of direct binding of a ligand to an immobilized protein, which is captured on surfaces in a biosensor chip (Figure 41.14b). It can be used to obtain real-time

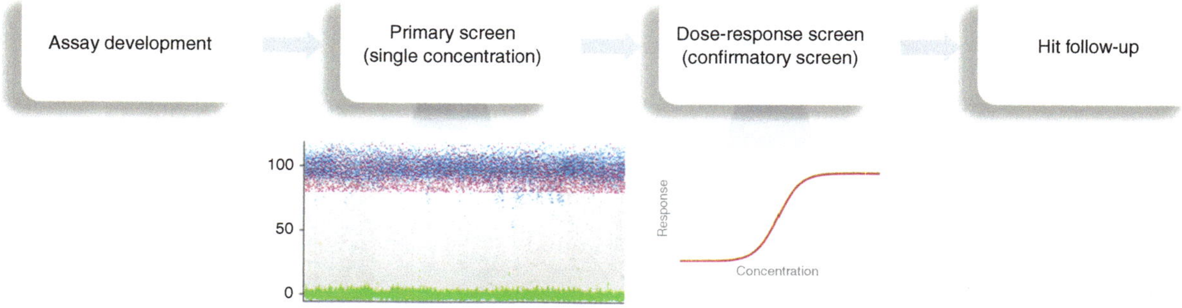


**Figure 41.15** Schematic representation of a typical HTS process. Following assay development, compounds are initially screened at single concentration (primary screen). The chart reflects a hypothetical screening of 500 000 compounds (grey dots), with the signal normalized to positive (blue) and negative (green) controls. Hits above the cutoff of 80% are indicated in pink. Hits are retested in a confirmatory screen in a dose-dependent manner, generating dose-response curves for the determination of potency values.

kinetic parameters such as association and dissociation constants and the binding affinity $K_d$ [66].

#### 41.4.6.1 Probing Ligand Interactions to hCAR LBD by SPR

SPR was used to investigate direct interactions of various ligands (CINPA1, PK11195, and CITCO) to hCAR LBD, which was tethered to a Ni-NTA biosensor using primary amines [64]. This particular protocol is for experiments conducted using a SensiQ Pioneer instrument (Pall FortèBio, Fremont, CA). In order to exclude immobilization through residues lining the ligand binding pocket, 10 μM hCAR LBD is pre-incubated with 30 μM PK11195 for 30 minutes at RT in running buffer (50 mM Hepes, pH 8.0, 200 mM NaCl, 0.01% Triton X-100 and 2% DMSO) prior to the capturing step. The reference and ligand surfaces (SPR terminology) are conditioned with 0.5 M EDTA at a flow rate of 25 μl/minutes for 5 minutes, followed by a 5 minutes application of 50 μM $NiCl_2$ to the ligand surface only. A 1 : 1 mixture of 0.1 M N-hydroxysuccinimide (NHS) and 0.4 M N-(3-dimethylaminopropyl)-N′-ethylcarbodiimide hydrochloride (EDC) is applied for 5 minutes to activate the surface of the biosensor. The pretreated hCAR LBD is injected at a flow rate of 10 μl/min until reaching a signal of 4000 response units. Buffer is allowed to flow into the channels until the baseline signal is stabilized at which point hCAR LBD is in the apo form. Compounds to be tested are diluted in buffer, ensuring that the final DMSO concentration matches that of the running buffer. The compounds are injected using a flow rate of 150 μl/min to the ligand and reference surfaces of the biosensor chip using the OneStep dynamic module. The data is fitted with a 1:1 kinetic model to obtain kinetic parameters and binding affinities.

### 41.4.7 Absorbance-Based Assay to Monitor Ligand Binding to Cytochrome p450

Cytochrome p450s contain a heme group at the active site, which is essential for the catalytic process. A water molecule coordinates the iron at the heme, and the displacement of the axial water by a ligand can be monitored by UV-vis spectroscopy [67]. Type I inhibitors generate an absorbance shift to shorter wavelengths (blue shift). For example, clobetasol dipropionate causes a spectral shift from 416 to 386 nm in CYP3A5 [68]. On the other hand, the axial water molecule can be exchanged by direct coordination of the ligand to the heme iron, resulting in an absorbance shift to longer wavelengths (red shift): ketoconazole induces a shift from 416 to 424 nm when bound to CYP3A5 [52].

### 41.4.8 High Throughput Screening Strategies to Identify Agonistic or Antagonistic PXR and CAR Ligands

#### 41.4.8.1 Overview of HTS Campaigns

High throughput screening (HTS) has become essential in drug discovery, whose objective is to identify biologically active compounds that can be used as starting points for drug design and the development of chemical tools to probe biological processes. It is based on the concerted application of specialized instrumentation and data analysis strategies to reliably screen large collections of compounds in a relatively fast and efficient manner.

There are several major steps in a typical HTS campaign (Figure 41.15). An appropriate assay is developed in a biologically relevant context and amenable for miniaturization; experiments are conducted in high-density microtiter plates in order to increase throughput, while minimizing the consumption of reagents and chemicals. The assay is scrutinized for robustness and reliability, and stringent quality control criteria are normally imposed. The initial screening of compounds is often carried out at a single concentration (primary screen). Hits obtained from the primary screen that meet certain activity cutoff are confirmed in a dose-dependent manner (dose-response screen), where they are tested at various compound concentrations in order to determine potency (commonly expressed as $EC_{50}$ or $IC_{50}$ values). Compounds of interest can serve as the basis for lead optimization through chemical modifications in order to improve potency, selectivity, and PK/PD profiles.

#### 41.4.8.2 Chemical Library

One of the most valuable and highly regarded assets in HTS is the chemical library, which is a collection of compounds properly formatted for screening. It is comprised of chemicals mostly from commercial sources, but it can also be enriched by molecules synthesized in-house or extracted from natural sources. The proper record keeping of the identity, purity, and concentration of these chemicals is of paramount importance. Compounds are reliably traced by barcodes and the position within a microtiter plate (well position).

#### 41.4.8.3 Instrumentation in HTS

A wide array of specialized instrumentation has been developed for HTS with the purpose of being able to handle microtiter plates and operate small solution volumes. Among the most common devices include liquid dispensers, such as those needed to dispense biological reagents in bulk (e.g. cells or proteins), transfer compounds (e.g. pipetting systems, acoustic, and pintool-based transfers), and exchange solutions (e.g. plate washers) [69]. Plate readers monitor signals from biological or chemical events, with conventional detection modes, which include absorbance, fluorescence intensity, fluorescence polarization, TR-FRET, and luminescence. In addition, dedicated incubators are capable to organize large numbers of plates and labware while providing all the features of a stand-alone incubator for cell growth. Plates are shuffled between devices using robotic arms, which are controlled by scheduling software in fully automated robotic platforms. On the other hand, similar screening throughput can be accomplished in semiautomated mode for simple assays, which is the case in most academic settings.

#### 41.4.8.4 HTS Assay

There are several types of assays that can be implemented for HTS, which include most biochemical and cell-based assays that monitor common electromagnetic radiation modes as endpoints. In particular, TR-FRET has been used in assays involving homogeneous PXR and CAR preparations, and luminescence has been a prevalent detection mode in reporter cell-based assays.

Novel hPXR antagonists were discovered during the screening of a chemical library of 160,000 compounds. The HepG2 human liver carcinoma cell line was designed to stably express hPXR and a luciferase reporter (luminescence detection), which was subjected under the control of the CYP3A4 promoter [41]. Similarly, a hCAR selective inhibitor was identified from screening a focus library of 1000 compounds using HepG2 cells that were transiently transfected with pcDNA-FLAG-hCAR1 and CYP2B6-luciferase plasmids [39]. Secondary and orthogonal assays allowed for further confirmation and characterization of these lead compounds. TR-FRET-based cell-free assays were used to determine the competitive displacement of the labeled probe by tested compounds [41], and to investigate the enhancement in coregulatory peptide recruitment [39].

#### 41.4.8.5 Data Analysis

Internal positive and negative controls are included in every plate along with tested compounds. DMSO is normally utilized as the negative control because it is used to dissolve tested compounds. Ideally, positive controls are chemicals that elicit the desired biological effect in the assay. In the absence of such compounds, "artificial" positive controls have been used by altering components in the reagents, such as omitting a protein or cells in an assay designed to identify inhibitors.

The raw data obtained from screening is typically normalized using the internal negative and positive controls as the lower bound (0%) and upper bound (100%) of the assay activity or inhibition, respectively. In the primary screen, a certain value is used as a cutoff to select hits for follow-up experiments. In the dose-response phase, the normalized data is fitted with a sigmoidal model (e.g. Hill equation) that allows for the determination of potency.

Internal controls are also employed to monitor assay performance. Variability in the raw data across plates is expected due to instrumental and biological factors. The quality of an assay can be quantified with statistical parameters such as Z′-factor, which takes into consideration the variations in the signals corresponding to the positive and negative controls [70].

## 41.5 Conclusions

Induction of drug-metabolizing enzymes and transporters is one of the key factors influencing drug bioavailability. Xenobiotic nuclear receptors PXR and CAR have been shown to mediate the transcriptional regulation of drug-metabolizing enzymes and transporters. The PXR or CAR knockout and transgenic mice have helped to identify the target genes involved in drug metabolism and disposition. The creation of hPXR- and hCAR-humanized mice has provided a solution to the problem of species differences in ligand specificity of PXR and CAR. These identified target genes are further analyzed and confirmed by in vitro methods, such as reporter gene assay, EMSA, and ChIP. However, careful interpretation of data obtained from animal models is vital when knockout or the expression of a constitutive active NR can affect the expression level of other NRs with shared target genes [34]. Biochemical, biophysical, and structural analyses of NRs using purified proteins are also important in order to understand the ligand-NR interactions and the dynamic recruitment of NR co-activators and co-repressors at the molecular and structural level. Last but not least, high throughput screening has been proved to be a powerful tool to identify agonistic or antagonistic PXR and CAR ligands.

Increased knowledge of the nuclear receptor-mediated regulation of drug- metabolizing enzymes and transporters has improved our understanding of the drug absorption, disposition, metabolism, and excretion (ADME). Additionally, this information is likely to provide us with novel tools for improving the drug oral bioavailability in drug discovery and development. Moreover, emerging evidence suggests that in addition to functioning as xenobiotic receptors, PXR and CAR have numerous endobiotic functions by regulating the homeostasis of endobiotics and thus the expression and activities of PXR and CAR are implicated in many diseases, ranging from metabolic disease to cholestasis, gallstone disease, inflammatory bowel disease, cardiovascular disease, and cancer [71–73]. As such, PXR and CAR have been proposed as therapeutic targets. The collective application of animal models and cell-based and cell-free systems enhances the discovery of PXR and CAR modulators that can be used as starting points for the design of potential therapeutics, and in the development of chemical probes that can be employed to understand biological processes related to the function of these NRs.

## Acknowledgments

The original results from our labs that are described in this chapter were generated with the support of NIH grants ES012479, ES014626, CA107011 (to W.X.); and GM118041, GM086415, GM110034 (to T.C.). W.X. is currently supported in part by NIH grants ES030429 and DK135538 and the Joseph Koslow Endowed Professorship from the University of Pittsburgh School of Pharmacy.

## References

1 Dobson, P.D. and Kell, D.B. (2008). Carrier-mediated cellular uptake of pharmaceutical drugs: an exception or the rule? *Nat. Rev. Drug Discovery* 7: 205–220.
2 Mizuno, N., Niwa, T., Yotsumoto, Y., and Sugiyama, Y. (2003). Impact of drug transporter studies on drug discovery and development. *Pharmacol. Rev.* 55: 425–461.
3 Nebert, D.W. and Russell, D.W. (2002). Clinical importance of the cytochromes P450. *Lancet* 360: 1155–1162.
4 Bertilsson, G., Heidrich, J., Svensson, K. et al. (1998). Identification of a human nuclear receptor defines a new signaling pathway for CYP3A induction. *Proc. Natl. Acad. Sci. U.S.A.* 95: 12208–12213.
5 Blumberg, B., Sabbagh, W. Jr., Juguilon, H. et al. (1998). SXR, a novel steroid and xenobiotic-sensing nuclear receptor. *Genes Dev.* 12: 3195–3205.
6 Chai, X.J., Zeng, S., and Xie, W. (2013b). Nuclear receptors PXR and CAR: implications for drug metabolism regulation, pharmacogenomics and beyond. *Expert Opin. Drug Metab. Toxicol.* 9: 253–266.
7 Kliewer, S.A., Moore, J.T., Wade, L. et al. (1998). An orphan nuclear receptor activated by pregnanes defines a novel steroid signaling pathway. *Cell* 92: 73–82.
8 Swales, K. and Negishi, M. (2004). CAR, driving into the future. *Mol. Endocrinol.* 18: 1589–1598.
9 Wei, P., Zhang, J., Egan-Hafley, M. et al. (2000). The nuclear receptor CAR mediates specific xenobiotic induction of drug metabolism. *Nature* 407: 920–923.
10 Xie, W., Radominska-Pandya, A., Shi, Y. et al. (2001). An essential role for nuclear receptors

SXR/PXR in detoxification of cholestatic bile acids. *Proc. Natl. Acad. Sci. U.S.A.* 98: 3375–3380.

11 Xie, W., Yeuh, M.F., Radominska-Pandya, A. et al. (2003). Control of steroid, heme, and carcinogen metabolism by nuclear pregnane X receptor and constitutive androstane receptor. *Proc. Natl. Acad. Sci. U.S.A.* 100: 4150–4155.

12 Yan, J. and Xie, W. (2016). A brief history of the discovery of PXR and CAR as xenobiotic receptors. *Acta Pharm. Sin. B* 6: 450–452.

13 Gronemeyer, H., Gustafsson, J.A., and Laudet, V. (2004). Principles for modulation of the nuclear receptor superfamily. *Nat. Rev. Drug Discovery* 3: 950–964.

14 Mangelsdorf, D.J., Thummel, C., Beato, M. et al. (1995). The nuclear receptor superfamily: the second decade. *Cell* 83: 835–839.

15 Xie, W., Uppal, H., Saini, S.P. et al. (2004). Orphan nuclear receptor-mediated xenobiotic regulation in drug metabolism. *Drug Discovery Today* 9: 442–449.

16 Aranda, A. and Pascual, A. (2001). Nuclear hormone receptors and gene expression. *Physiol. Rev.* 81: 1269–1304.

17 Bain, D.L., Heneghan, A.F., Connaghan-Jones, K.D., and Miura, M.T. (2007). Nuclear receptor structure: implications for function. *Annu. Rev. Physiol.* 69: 201–220.

18 Blumberg, B. and Evans, R.M. (1998). Orphan nuclear receptors--new ligands and new possibilities. *Genes Dev.* 12: 3149–3155.

19 Xie, W. and Evans, R.M. (2001). Orphan nuclear receptors: the exotics of xenobiotics. *J. Biol. Chem.* 276: 37739–37742.

20 Klaassen, C.D. and Slitt, A.L. (2005). Regulation of hepatic transporters by xenobiotic receptors. *Curr. Drug Metabol.* 6: 309–328.

21 Runge-Morris, M. and Kocarek, T.A. (2005). Regulation of sulfotransferases by xenobiotic receptors. *Curr. Drug Metabol.* 6: 299–307.

22 Zhou, J., Zhang, J., and Xie, W. (2005). Xenobiotic nuclear receptor-mediated regulation of UDP-glucuronosyl-transferases. *Curr. Drug Metabol.* 6: 289–298.

23 Gong, H., Sinz, M.W., Feng, Y. et al. (2005). Animal models of xenobiotic receptors in drug metabolism and diseases. *Methods Enzymol.* 400: 598–618.

24 Xie, W., Barwick, J.L., Downes, M. et al. (2000a). Humanized xenobiotic response in mice expressing nuclear receptor SXR. *Nature* 406: 435–439.

25 Pinkert, C.A., Ornitz, D.M., Brinster, R.L., and Palmiter, R.D. (1987). An albumin enhancer located 10 kb upstream functions along with its promoter to direct efficient, liver-specific expression in transgenic mice. *Genes Dev.* 1: 268–276.

26 Xie, W., Chow, L.T., Paterson, A.J. et al. (1999). Conditional expression of the $ErbB_2$ oncogene elicits reversible hyperplasia in stratified epithelia and up-regulation of TGFalpha expression in transgenic mice. *Oncogene* 18: 3593–3607.

27 Kistner, A., Gossen, M., Zimmermann, F. et al. (1996). Doxycycline-mediated quantitative and tissue-specific control of gene expression in transgenic mice. *Proc. Natl. Acad. Sci. U.S.A.* 93: 10933–10938.

28 Zhou, J., Zhai, Y., Mu, Y. et al. (2006). A novel pregnane X receptor-mediated and sterol regulatory element-binding protein-independent lipogenic pathway. *J. Biol. Chem.* 281: 15013–15020.

29 Zhang, J., Huang, W., Chua, S.S. et al. (2002). Modulation of acetaminophen-induced hepatotoxicity by the xenobiotic receptor CAR. *Science* 298: 422–424.

30 Cheng, J., Ma, X.C., and Gonzalez, F.J. (2011). Pregnane X receptor- and CYP3A4-humanized mouse models and their applications. *Br. J. Pharmacol.* 163: 461–468.

31 Scheer, N., Snaith, M., Wolf, C.R., and Seibler, J. (2013). Generation and utility of genetically humanized mouse models. *Drug Discovery Today* 18: 1200–1211.

32 Gong, H., Singh, S.V., Singh, S.P. et al. (2006). Orphan nuclear receptor pregnane X receptor sensitizes oxidative stress responses in transgenic mice and cancerous cells. *Mol. Endocrinol.* 20: 279–290.

33 Zhai, Y., Pai, H.V., Zhou, J. et al. (2007). Activation of pregnane X receptor disrupts glucocorticoid and mineralocorticoid homeostasis. *Mol. Endocrinol.* 21: 138–147.

34 Rosenfeld, J.M., Vargas, R. Jr., Xie, W., and Evans, R.M. (2003). Genetic profiling defines the xenobiotic gene network controlled by the nuclear receptor pregnane X receptor. *Mol. Endocrinol.* 17: 1268–1282.

35 Qatanani, M., Wei, P., and Moore, D.D. (2004). Alterations in the distribution and orexigenic effects of dexamethasone in CAR-null mice. *Pharmacol. Biochem. Behav.* 78: 285–291.

36 Zaman, G.J.R., de Roos, J.A.D.M., Blomenrohr, M. et al. (2007). Cryopreserved cell facilitate cell-based drug discovery. *Drug Discovery Today* 12: 521–526.

37 Banerjee, M., Chai, S.C., Wu, J. et al. (2016). Tryptophan 299 is a conserved residue of human pregnane X receptor critical for the functional consequence of ligand binding. *Biochem. Pharmacol.* 104: 131–138.

38 Banerjee, M. and Chen, T.S. (2013). Differential regulation of CYP3A4 promoter activity by a new class

of natural product derivatives binding to pregnane X receptor. *Biochem. Pharmacol.* 86: 824–835.

39 Cherian, M.T., Lin, W.W., Wu, J., and Chen, T.S. (2015). CINPA1 is an inhibitor of constitutive androstane receptor that does not activate pregnane X receptor. *Mol. Pharmacol.* 87: 878–889.

40 Dong, H., Lin, W., Wu, J., and Chen, T. (2010). Flavonoids activate pregnane x receptor-mediated $CYP_3A_4$ gene expression by inhibiting cyclin-dependent kinases in $HepG_2$ liver carcinoma cells. *BMC Biochem.* 11: 23.

41 Lin, W.W., Wang, Y.M., Chai, S.C. et al. (2017). SPA70 is a potent antagonist of human pregnane X receptor. *Nat. Commun.* 8: 741.

42 Lin, W.W., Wu, J., Dong, H.Q. et al. (2008). Cyclin-dependent kinase 2 negatively regulates human pregnane X receptor-mediated $CYP_3A_4$ gene expression in $HepG_2$ liver carcinoma cells. *J. Biol. Chem.* 283: 30650–30657.

43 Sinz, M., Kim, S., Zhu, Z. et al. (2006). Evaluation of 170 xenobiotics as transactivators of human pregnane X receptor (hPXR) and correlation to known CYP3A4 drug interactions. *Curr. Drug Metabol.* 7: 375–388.

44 Wang, Y.M., Lin, W., Chai, S.C. et al. (2013). Piperine activates human pregnane X receptor to induce the expression of cytochrome P450 3A4 and multidrug resistance protein 1. *Toxicol. Appl. Pharmacol.* 272: 96–107.

45 Raucy, J.L. and Lasker, J.M. (2010). Current in vitro high throughput screening approaches to assess nuclear receptor activation. *Curr. Drug Metabol.* 11: 806–814.

46 Xie, W., Barwick, J.L., Simon, C.M. et al. (2000b). Reciprocal activation of xenobiotic response genes by nuclear receptors SXR/PXR and CAR. *Genes Dev.* 14: 3014–3023.

47 Mu, Y., Stephenson, C.R., Kendall, C. et al. (2005). A pregnane X receptor agonist with unique species-dependent stereoselectivity and its implications in drug development. *Mol. Pharmacol.* 68: 403–413.

48 Mu, Y., Zhang, J., Zhang, S. et al. (2006). Traditional Chinese medicines Wu Wei Zi (Schisandra chinensis Baill) and Gan Cao (Glycyrrhiza uralensis Fisch) activate pregnane X receptor and increase warfarin clearance in rats. *J. Pharmacol. Exp. Ther.* 316: 1369–1377.

49 Saini, S.P., Sonoda, J., Xu, L. et al. (2004). A novel constitutive androstane receptor-mediated and CYP3A-independent pathway of bile acid detoxification. *Mol. Pharmacol.* 65: 292–300.

50 Chen, C.Y., Liu, N.Y., Lin, H.C. et al. (2016). Synthesis and bioevaluation of novel benzodipyranone derivatives as P-glycoprotein inhibitors for multidrug resistance reversal agents. *Eur. J. Med. Chem.* 118: 219–229.

51 Shekari, F., Sadeghpour, H., Javidnia, K. et al. (2015). Cytotoxic and multidrug resistance reversal activities of novel 1,4-dihydropyridines against human cancer cells. *Eur. J. Pharmacol.* 746: 233–244.

52 Wright, W.C., Chenge, J., Wang, J. et al. (2020). Clobetasol propionate is a heme-mediated selective inhibitor of human Cytochrome P450 3A5. *J. Med. Chem.* 63: 1415–1433.

53 Cali, J.J., Ma, D., Sobol, M. et al. (2006). Luminogenic cytochrome P450 assays. *Expert Opin. Drug Metab. Toxicol.* 2: 629–645.

54 Watkins, R.E., Wisely, G.B., Moore, L.B. et al. (2001). The human nuclear xenobiotic receptor PXR: Structural determinants of directed promiscuity. *Science* 292: 2329–2333.

55 Chai, S.C., Cherian, M.T., Wang, Y.M., and Chen, T.S. (2016). Small-molecule modulators of PXR and CAR. *BBA-Gene Regul. Mech.* 1859: 1141–1154.

56 Shan, L., Vincent, J., Brunzelle, J.S. et al. (2004). Structure of the murine constitutive androstane receptor complexed to androstenol: a molecular basis for inverse agonism. *Mol. Cell* 16: 907–917.

57 Suino, K., Peng, L., Reynolds, R. et al. (2004). The nuclear xenobiotic receptor CAR: Structural determinants of constitutive activation and heterodimerization. *Mol. Cell* 16: 893–905.

58 Xu, R.X., Lambert, M.H., Wisely, B.B. et al. (2004). A structural basis for constitutive activity in the human CAR/RXR alpha heterodimer. *Mol. Cell* 16: 919–928.

59 Buchman, C.D., Chai, S.C., and Chen, T.S. (2018). A current structural perspective on PXR and CAR in drug metabolism. *Expert Opin. Drug Metab. Toxicol.* 14: 635–647.

60 Sittampalam, G.S., Kahl, S.D., and Janzen, W.P. (1997). High-throughput screening: advances in assay technologies. *Curr. Opin. Chem. Biol.* 1: 384–391.

61 Chai, S.C., Lin, W.W., Li, Y.T., and Chen, T.S. (2019). Drug discovery technologies to identify and characterize modulators of the pregnane X receptor and the constitutive androstane receptor. *Drug Discovery Today* 24: 906–915.

62 Lin, W.W. and Chen, T.S. (2013). A vinblastine fluorescent probe for pregnane X receptor in a time-resolved fluorescence resonance energy transfer assay. *Anal. Biochem.* 443: 252–260.

63 Lin, W.W., Liu, J.Y., Jeffries, C. et al. (2014). Development of BODIPY FL Vindoline as a novel and high-affinity pregnane X receptor fluorescent probe. *Bioconjugate Chem.* 25: 1664–1677.

64 Cherian, M.T., Chai, S.C., Wright, W.C. et al. (2018). CINPA1 binds directly to constitutive androstane receptor and inhibits its activity. *Biochem. Pharmacol.* 152: 211–223.

65 Akbar, S.M., Sreeramulu, K., and Sharma, H.C. (2016). Tryptophan fluorescence quenching as a binding assay to monitor protein conformation changes in the membrane of intact mitochondria. *J. Bioenerg. Biomembr.* 48: 241–247.

66 Renaud, J.P., Chung, C.W., Danielson, U.H. et al. (2016). Biophysics in drug discovery: impact, challenges and opportunities. *Nat. Rev. Drug Discovery* 15: 679–698.

67 Mak, P.J. and Denisov, I.G. (2018). Spectroscopic studies of the cytochrome P450 reaction mechanisms. *Biochim. Biophys. Acta, Proteins Proteomics* 1866: 178–204.

68 Wang, J., Buchman, C.D., Seetharaman, J. et al. (2021). Unraveling the structural basis of selective inhibition of human cytochrome P450 3A5. *J. Am. Chem. Soc.* 44: 18467–18480.

69 Chai, S.C., Goktug, A.N., Cui, J. et al. (2013a). Practical Considerations of Liquid Handling Devices in Drug Discovery. In: *Drug Discovery*. London, UK: IntechOpen.

70 Goktug, A.N., Chai, S.C., and Chen, T. (2013). Data Analysis Approaches in High Throughput Screening. In: *Drug Discovery*. London, UK: IntechOpen.

71 Banerjee, M., Robbins, D., and Chen, T.S. (2015). Targeting xenobiotic receptors PXR and CAR in human diseases. *Drug Discovery Today* 20: 618–628.

72 Gao, J. and Xie, W. (2012). Targeting xenobiotic receptors PXR and CAR for metabolic diseases. *Trends Pharmacol. Sci.* 33: 552–558.

73 Oladimeji, P.O. and Chen, T. (2018). PXR: more than just a master xenobiotic receptor. *Mol. Pharmacol.* 93: 119–127.

# 42

# Intestinal Perfusion Methods for Oral Drug Absorptions

*Li Li*[1], *Dinh Bui*[1], *Wei Zhu*[2], *and Eun-Jung (Zenobia) Jeong*[3]

[1] *Department of Pharmacological and Pharmaceutical Sciences, College of Pharmacy, University of Houston College of Pharmacy, Houston, TX, USA*
[2] *Janssen Research & Development, Raritan, NJ, USA*
[3] *Korean Institute of Toxicology, Daejeon, South Korea*

## 42.1 Introduction

The intestinal perfusion method is an in situ model widely used to characterize and determine the absorption and bioavailability of drugs, nutrients, or other related compounds for the purpose of preclinical evaluation of drugs absorption and/or metabolism. The intestinal perfusion model in rodents is an FDA-recognized model to study drug absorption for its high predictive value of drug absorption in humans [1–3]. Rat in situ intestinal perfusion model is the most popular model among the intestinal perfusion models.

The intestinal perfusion technique in rodents was developed based on an *in situ* experiment to study metabolism, excretion, and tissue distribution but not the dynamic absorption process in rats [4–8]. Over the years, the model has been modified and improved to fulfill the research requirement. The models introduced by Schanker in 1958 and Kakemi in 1967 have exhibited the ability to study the dynamics of absorption process by a single pass or circulated solution [7, 9]. In 1969, Deluisio et al reported an in situ method to determine drug absorption rate from rat intestinal segments [10]. The model was developed in mid-1980s [11, 12], which was more sophisticated than the classic method of "through-and-through" perfusion method of Deluisio. Originally, the model utilizes one segment of the intestine per experiment and perfuses a compound with or without an inhibitor at a particular pH to determine the absorption characteristics of a compound [11, 12]. It was shown shortly thereafter that there is a correlation between jejunal wall permeability, $P_w$, and percent absorption in humans [13]. Ming Hu and his group integrated the portal vein and bile duct cannulation into the model in order to study biliary excretion and disposition of flavonoids [14, 15]. Compatibility with emerging trends in pharmaceutical profiling such as knockout or transgenic animals, this model generates unparalleled possibilities to gain mechanistic insight into specific absorption processes [2].

The perfused rat intestinal model is a good transition model between *in vitro* models (such as Caco-2 model and intestinal strips) and *in vivo* pharmacokinetic models. The special feature of this model is that the blood supply to the perfused segment of intestine is maintained throughout the experimental period, and it is expected to have appropriate or physiological expression levels of various nutrient transporters and enzymes, which are normally underexpressed in the Caco-2 model. Because of this special feature, the model is especially suited for the study of active transport and intestinal metabolism. Perfused rat intestinal model and its variants could be used to determine the absorption mechanisms of a drug or drug candidate, absorption and metabolism in different regions of the intestine, and contribution of intestinal metabolism to the overall metabolism of a particular compound. This information significantly contributes to the estimation of the efficacy and toxicity potential and prediction of the first-in-human dose. Drug exposure data derived from animal intestinal perfusion models can be used for PBPK modeling and extrapolated to humans through allometric scaling.

## 42.2 Application and Recent Development of the Intestinal Perfusion Method

### 42.2.1 Intestinal Regional Absorption Studies

In situ rat intestinal perfusion model is widely used to determine absorption of drugs in different regions

*Oral Bioavailability and Drug Delivery: From Basics to Advanced Concepts and Applications*, First Edition. Edited by Ming Hu and Xiaoling Li.

of the intestine [16, 17]. The regional differences in composition of intestinal membrane, location of transporters, and distribution of metabolic enzymes have been well documented. Membrane composition and transporters of small intestinal mucosal cells are different from those of large intestinal cells (enterocytes and colonocytes). And the anatomical structure of small intestinal membrane is different from large intestine, such as the villi structures of small intestine that greatly increase its surface area for absorption [18]. Taken together, studies of these regional differences are often the determinant of differences in drug transport at various segments of the intestine [19].

Regional perfusion studies are also important for determining if a drug candidate could be used in sustained-release formulations that are expected to release its content in regions where absorption is slow or nonexistent. These studies may serve as a prelude to extensive formulation studies that may take a much longer time and larger resources to complete. An important extension of the original perfusion model was made in late 1990's when simultaneous perfusion of all four segments of the intestine that includes duodenum, jejunum, ileum, and colon was enabled. Initially, these studies were performed to study absorption and metabolism of flavonoids, the results of which were published in 2002 [20]. This modification significantly reduced the number of animals necessary for obtaining similar results using single-segment perfusion. Because of the smaller number of animals used, it also decreased the method variability, potentially further reducing the need to use a large number of animals.

### 42.2.2 Intestinal Metabolism Studies

Intestinal perfusion can be modified to measure segment-dependent production of metabolites, which is particularly useful for determining the excretion of metabolites in different regions of the intestine. The regional small intestinal metabolism and permeability of captopril, didanosine, mannitol, ofloxacin, and zidovudine were investigated in rats using single-pass intestinal perfusion procedure or intestinal homogenates to demonstrate regional differences in intestinal bioavailability [21]. More recently, we have been using it to determine the intestinal excretion of hydrophilic phase II metabolites (i.e. glucuronides and sulfates) of flavones and isoflavones, which was found to be dependent on both the enzyme activities and capability of the efflux transporters [22]. The in situ rat intestinal perfusion model, together with rat intestinal and liver microsome model, has been used to determine the importance of coupling of efflux transporters and metabolic enzymes in intestinal disposition of isoflavones, and investigate effect of structural difference of isoflavones on their intestinal disposition [23].

### 42.2.3 Bile Duct Cannulation: Contribution of Intestine vs. Liver

Intestinal perfusion can be modified to determine the contribution of intestinal wall versus liver in the excretion of phase II metabolite by introducing bile duct cannulation during the intestinal perfusion experiment [22]. To verify the relative metabolic contribution of intestine versus liver, intestinal and biliary excretion of phase II conjugates of flavonoids were measured in a rat perfusion model [22, 24]. The whole small intestine from the upper duodenum to the end of the ileum or four sites of intestine were perfused, and bile and jugular vein were also cannulated for the collection of bile and blood samples. Perfusion studies suggest that intestine is the main organ for genistein glucuronide formation and excretion in rats and may serve as its main first-pass metabolism organ.

In mice intestinal perfusion model, the bile sample was obtained from the gallbladder at the end of the experiment due to the very small size of bile duct [25]. Bile duct was tied before the intestine and bile was allowed to accumulate inside the gallbladder before and during the perfusion. In contrary to intestinal perfusate, sulfate was not detected although glucuronide was detected in mouse bile.

### 42.2.4 Use of Inhibitors or Gene Knockout Animals

By adding inhibitors of specific transporter into the perfusate and co-perfuse with the compound of interest or using the specific gene knockout animals, we can investigate the involvement of certain transporter in the efflux of drugs, and its influence on drug permeability in intestine. For example, it was reported by Verma and Panchagnula using single-pass rat perfusion model, that the inhibition of P-gp significantly increased the permeability ($p < 0.05$) of digoxin, which is a known P-gp substrate [26]. Sesink et al. perfused quercetin with and without the presence of fumitremorgin C, a specific BCRP inhibitor, in MRP2-deficient rats to determine the role of these two transporters [27]. In MRP2-deficient rats, intestinal uptake of quercetin from quercetin-3-glucoside, efflux of quercetin glucuronides to the gut lumen, and plasma concentration of quercetin were similar to that in control rats. When intestinal BCRP was inhibited by FTC in MRP2-deficient rats, total plasma concentrations of quercetin and its methylated metabolite isorhamnetin after 30 min of perfusion were more than twice that of controls, whereas uptake of free quercetin from the intestinal lumen was not affected. Instead, inhibition of BCRP lowered the efflux of quercetin glucuronides into the perfusion fluid by approximately fourfold. Based on the results, the author concluded that BCRP limits net intestinal absorption of quercetin by pumping quercetin glucuronides back into the lumen.

### 42.2.5 Intestinal Perfusion in Mice and Other Animals

Although rat intestinal perfusion is the most popular model, this perfusion method maybe applied to other species such as mice. Use of alternative species is sometimes motivated by the availability of particular species or physiological similarity between that species and humans. Although intestinal mucosal membrane structure is comparable across species, the activities and patterns of intestinal metabolic enzymes, transporters, microorganisms could differ among species [19].

Differences in subpopulation of intestinal enzymes (or enzyme isoforms) and transporters are often involved in the species-specific pharmacokinetics of xenobiotics. Species differences in oral bioavailability can be investigated and underlying mechanism could be understood by intestinal perfusion using different animal species.

Recently, we investigated the intestinal absorption and metabolism of formononetin, a flavonoid, using single-pass mouse intestinal perfusion model along with Caco-2 cell culture model [25]. Although sulfated conjugates represent important metabolites in humans, very little or no sulfated flavonoid conjugates were found in the rat intestinal perfusate [25]. In contrary to the rat intestinal perfusion, where four sites of intestine can be simultaneously perfused, two segments of small intestine were perfused simultaneously (a "two-site model") because of the shorter length of mouse small intestine (total ~25 cm) than rats (total ~80 cm). Bile duct was tied before the intestine and bile was allowed to accumulate inside the gallbladder before and during the perfusion. At the end of perfusion, bile samples were collected by removing gallbladder. The results showed that mouse intestinal perfusion model is superior in understanding the disposition of flavonoids since mouse intestine metabolized formononetin into sulfated conjugates, as expected from human intestine. In addition, the excretion of conjugates appeared to be mediated by efflux transporters, allowing the mouse model to be used to investigate the coupling of sulfotransferase and organic anion transporters such as MRPs. Because of the availability of genetically modified mice, this technique maybe especially suited for understanding the coupling of the conjugating enzymes and efflux transporters responsible for removing hydrophilic conjugated metabolites from the intestine.

Besides rats and mice, intestinal perfusion model has been used in other animals such as rabbits. The intestinal transport parameters of sildenafil citrate were investigated using an in situ intestinal perfusion technique in rabbits. The absorptive clearance did not correlate with the net water flux in the four anatomical regions studied, duodenum, ileum, ascending colon, and rectum, indicating a mainly passive diffusion mechanism through a transcellular pathway. Only 34% of sildenafil that disappeared from the intestinal segment appeared in the systemic circulation and confirmed that the incomplete bioavailability of sildenafil is mainly due to presystemic elimination [28].

## 42.3 Data Interpretation and Method Comparison

While studying the mechanisms of drug metabolism especially those related to phase II metabolism, it is often necessary to use additional models to determine which step is the rate-limiting step in the overall disposition process, which often aids to interpret the final data. Here, we also described a few additional *in vitro* models that maybe used to determine intestinal absorption and metabolism of drugs and compared them to the intestinal perfusion model.

## 42.4 Common *In Vitro* Methods Studying Intestinal Permeability and Metabolism

### 42.4.1 Rat Subcellular Fractions (S9 Fractions and Microsomes)

With this method, different segments of rat intestine (duodenum, jejunum, ileum, and colon) are excised and each segment is cut open lengthwise on an ice-cold plate. The intestinal strips are then blot-dried, and the mucosal layer is carefully scraped off and homogenized in an ice-cold buffer containing EDTA, dithiothreitol, and PMSF. After centrifugation at 9000 g for 15 minutes, the supernatant is collected as S9 fractions. Since this fraction contains both cytosol and microsomes, it is generally incubated with compounds of interest and cofactors, such as NADPH, UDPGA, and PAPS for the measurement of metabolic enzyme activities [29].

After intestinal mucosal scraping and epithelial cell homogenization, microsomes can be precipitated by differential centrifuge at 100 000 x g for 60 minutes. The pellets were then resuspended in 250 mM sucrose solution and stored at −80 °C until use. Microsomes are commonly used for the metabolism study of mitochondria enzymes, such as P450 and UGTs [30–32]. Both S9 fractions and microsomes are very useful because the studies are narrowly focused on the activities of intestinal enzymes. The experiment conditions can be more easily controlled than *in situ* or *in vivo* models. Moreover, these two methods are theoretically applicable to all species and can be used to study phase I and phase II drug metabolism or to get the enzyme isoform's activity profiles. Some of the subcellular preparations are also often commercially available.

These two subcellular preparations can also be used to study regional differences in enzyme activities along the rat intestinal tract [33], and to study the inhibition mechanisms of metabolic enzymes. However, these two methods cannot provide information on transporter activities along the intestine, which is critical for absorption and permeability of a compound. The cofactors of reaction system are not presented at physiological concentrations so the metabolic activities maybe lower than *in vivo*. Additionally, the preparation of intestinal S9 fractions and microsomes is more prone to artifacts compared to the preparation of liver fractions. For example, the high expression of intestinal proteases has been proven to reduce P450 activities during mucosa scraping, so that protease inhibitors are needed in the entire preparation procedure. The freeze/thaw cycle is also considered to cause activity loss for P450 enzymes [34]. In our own experience, intestinal UGT activities present in microsomes are particularly heat sensitive [35].

### 42.4.2 Cultured Intestinal Cell Model

#### 42.4.2.1 Caco-2 Cells and Caco-2 TC-7 Clone (see Also Chapter 37)

Caco-2 cells are derived from human adenocarcinoma cells of colon. They undergo spontaneous differentiation while reaching confluency, and after they are cultured for about 20 days, the cells will form quality monolayers with well-established polarity and tight junctions. Unlike normal colonocytes, Caco-2 cells behave like ileum cells and express drug transporters (e.g. bile acid transporters) similar to the human small intestine [36, 37]. Caco-2 cells have been the most extensively characterized and useful cell model for studying drug permeability and are recognized by the US Food and Drug Administration as a useful tool to evaluate drug absorption. Moreover, Caco-2 cell monolayers are easy to handle and require less amount of compound to determine absorption, so they are used as a rapid *in vitro* screening tool in pharmaceutical industry to rank order a large number of compounds in terms of absorption potential [38].

Caco-2 TC-7 cells are one of the subclones isolated from Caco-2 cells [39]. It was reported to have very similar cell morphology as in Caco-2 cells, such as the gene expression, the polarity of membrane, and the formation of tight junctions [40]. The permeability values of passively absorbed drugs in TC-7 clones correlated well as in parental Caco-2 cells to the extent of absorption in humans [41]. However, the enzyme and transporter expressions differ in TC-7 clones compared to Caco-2 cells, indicating that the metabolic and transport profiles might be different for certain drugs [40].

However, since the drug metabolic enzymes and transporters are often underexpressed in Caco-2 cells than *in vivo*, such as CYP3A4, CYP2C9, MRP1, and OAT [42–44], the quantitative prediction of drug metabolism and transport for carrier-mediated compounds *in vivo* becomes more difficult. The expression of enzymes and transporters of Caco-2 cells varies between different labs during cultivation and between different passage numbers [45, 46]. The differences in the composition of cell membrane and paracellular radius also imply that the transport characteristics of Caco-2 cells are different from that of small intestine *in vivo* [47, 48]. Additionally, the long culture time (20 days) presents another drawback for the utilization of this method.

2/4/A1 cells. The 2/4/A1 cells were firstly reported by Tavelin et al. that originated from fetal rat small intestine [49]. This cell line also forms differentiated monolayer with tight junctions. Since the tight junction in Caco-2 cells is too tight compared to normal human intestinal epithelial cells, 2/4/A1 cells were shown to better mimic the permeability of the human small intestine, especially for paracellular and passive transcellular drug transport [49, 50]. Using six compounds of interest, it was also reported that this cell line predicts the percentage of absorption ($f_a$) better than Caco-2 cells according to BCS classification [51]. However, the lack of expression for several important drug transporter proteins, such as PepT1, MDR1, MRP2, and BCRP in 2/4/A1 cells poses a major limitation to its use as a model to study carrier-mediated drug transport [50]. Additionally, the variability of cells in small temperature changes (from 37 to 39 °C) is another drawback of this method. Currently, the data from this cell line is limited and further experiments are needed to evaluate this model. In addition to this model, other cell culture models, such as LLCPK-1 cells, have been proposed to replace Caco-2 cells.

### 42.4.3 Animal Tissue-based *In Vitro* Method

#### 42.4.3.1 Everted Sac

The everted sac was firstly introduced by Wilson and Wiseman in 1954 in the transport study of amino acids from mucosal to serosal side of intestine [52]. Over the years, modifications of the method have been applied to increase the viability of the tissues. In brief, the intestine is excised, washed, everted over a glass rod, filled with fresh oxygenated culture media, and divided into sacs [53]. According to the literature, the length of the sac can vary between 1 and 10 cm [54]. The drug is then perfused along the mucosal side and the samples on serosal side were collected at various time points.

This model is fast and inexpensive, and it is ideal for studying the absorption mechanism of drugs for both passive and active transport. It can be used to study the regional differences along the intestine [53]. This model has also been reported to be applicable to both rats and mice, and to study phase I, and phase II metabolism as well as transport. Inhibition studies of enzymes and transporters, such as CYP3A4 and MDR1, have also been reported [54, 55]. However, the major disadvantage of this model is that the lack of blood and nerve supply can lead to a rapid loss of tissue viability. The tissue morphology may also be damaged by everting the intestine and generate potentially misleading results. Finally, operator's errors can be quite significant as there is inadequate method to assess tissue damage due to handling variability. Further, the operator must be highly efficient as the period of time between intestinal excision, eversion, and sac cannulation should not exceed 8 min. Because of this, the method is challenging to execute and not suitable for studying slow absorption of drugs.

#### 42.4.3.2 Ussing Chamber

In 1951, Ussing and Zehren invented the Ussing chamber for the transport study of sodium in frog skin [56]. In the Ussing chamber preparation, sheets of the intestinal mucosal tissue are mounted in between two chambers containing buffers. Both mucosal and serosal sides of the chambers are usually supplied continuously with oxygenated buffers, such as KRB. Compounds of interest can be applied to either the luminal or serosal sides and the transport samples to either side can be measured at different time points for the calculation of permeability.

The Ussing chamber is ideal for the bidirectional transport and metabolism in different regions of intestine. It is also possible to perform the study using human intestinal tissue so that the comparison of permeability values across the species can be investigated [57]. The amount of drug required is relatively small and the samples are analytically clean for further processing. However, like everted sac, the Ussing chamber does not have blood and nerve supply so that the viability of the tissues may decrease rapidly during the experiment. The changes in morphology and functionality of transporter proteins during the process of surgery and mounting of the tissues may also generate misleading data. Furthermore, the enzymes and transporters mounted tissues are not under *in vivo* physiological conditions, resulting in the potential underestimation of metabolism and permeability values.

### 42.4.4 Comparison and Relationship of Perfusion Model and Other Models for Studying Drug Metabolism and Absorption in Intestine

#### 42.4.4.1 Intestinal Perfusion and Subcellular Fractions

The major difference between perfusion model and subcellular fraction models is that subcellular fraction studies provide specific information of metabolic pathways and enzyme activities, whereas in perfusion studies we can measure metabolism and absorption at the same time. The experimental procedures of subcellular fractions studies, including S9 fractions and microsomes, are well-developed and relatively easy to operate by the laboratory personnel. Since the effect of transporter is excluded from consideration in these subcellular models, it allows us to focus on the mechanism of enzyme activities. It is more probing and provides straightforward information of metabolic pathways and enzyme isoform information. However, since the reaction system is not occurring in a living organism, it can only provide information on relative enzyme activities, which may not predict the actual activities *in vivo*. The latter is especially true for the metabolism and excretion of hydrophilic phase II metabolites since these metabolites cannot be eliminated via passive diffusion. Using both perfusion and subcellular fraction models, we are able to measure and compare the metabolic rate and the metabolite excretion rate. Recently, Wang et al. investigated the glucuronidation of six isoflavones in rat intestinal microsomes and their glucuronide conjugates excretion rates in single-pass rat intestinal perfusion study. The excretion of glucuronides was not predicted properly by any of the kinetics parameters from the microsome metabolic study. This observation elucidated the need for the incorporation of involved efflux transporters in determination of the excretion of phase II metabolites, since those metabolites are very hydrophilic and require certain transporters to get effluxed to the lumen [23].

#### 42.4.4.2 Intestinal Perfusion and *In Vitro* Cell Line Model

In Caco-2 cell transport study, by measuring the drug transport from apical to basolateral or basolateral to apical side, we can obtain the information on enzyme presence, enzyme activities as well as transporter presence and activities in both apical and basolateral sides. Additionally, the use of Caco-2 cells in the prediction of drug permeability is a high-throughput screening method compared to intestinal perfusion model. However, the $P_{eff}$ data obtained from Caco-2 cells could not predict the percentage of absorption ($f_a$) in humans for compounds with low $P_{eff}$ such as creatinine, atenolol, and terbutaline [58], even though these compounds had no solubility/dissolution problem. The predictability of human $f_a$ is particularly weak when active transport is involved, probably because of the lack of or low expression of certain important transporters in Caco-2 cells than in *in vivo* [59]. On the other hand, it was shown by Cao et al. that rats and humans have similar patterns of expression of small intestinal transporters and drug absorption profiles for compounds undergoing active transport [60]. A reasonable correlation obtained from in situ perfusion between rat and human $P_{eff}$ has been reported ($r^2 = 0.7$).

#### 42.4.4.3 Intestinal Perfusion and Ussing Chamber

While comparing human jejunal $P_{eff}$ with corresponding rat $P_{eff}$ value generated by Using chamber model, it has been reported that they have the same rank order for drugs with both passive diffusion and carrier-mediated mechanisms. The *in vitro* rat $P_{eff}$ was highly correlated but about 4 to 5 times lower than corresponding human $P_{eff}$ data, probably due to lack of blood flow and a lower supply of cofactor *in vitro*, which is important for the optimal function of transport proteins [61]. The regional rat permeability data acquired by Ussing chamber model were able to show that low permeability drugs have an even lower permeability in the colon, whereas high permeability drugs showed a slightly higher permeability in colon when passive diffusion is the dominant transport mechanism [62]. The utilization of Ussing chamber is a valuable predictive tool for human in vivo drug absorption, especially for passively diffused drugs. The permeability data from single-pass perfusions correlated highly with data from the Ussing chamber in rats ($r^2 = 0.90$). However, there were deviations between these two models for high permeability drugs vs. the low permeability drugs, possibly due to the blood supply for the maintenance of membrane characteristics and the sink condition in in situ perfusion model for high permeability drugs [63].

### 42.4.5 Implication and Extrapolation of Animal Data in Predicting Human Intestinal Permeability

#### 42.4.5.1 Human Intestinal Perfusion Method

In 1992, Lennernas et al. developed a method for performing regional single-pass perfusion of the proximal jejunum *in vivo* in humans. In this new "Loc-I-Gut" method, a multichannel tube is introduced with 175 cm long and 5.3 mm external diameter [64]. It contains six channels and is provided distally with two 40 mm long, elongated latex balloons, placed 10 cm apart, and each of them connected to one of the smaller channels. The two wider channels in the center of the tube are for infusion and aspiration of perfusate. The balloons are filled with air when the proximal balloon has passed the ligament of Treitz. Compare to the older open or semi-open perfusion in humans, the main advantage is that the occlusion of the segment between two balloons minimizes contamination with luminal fluids both proximally and distally into the perfusate. The leakage from the segment is also small. Using this method, human *in vivo* jejunal permeability values of 42 compounds have been determined [65]. However, ethnic issues and the limited numbers of subjects are the major limitations of this method, especially for the development of new chemical identity in pharmaceutical industry.

#### 42.4.5.2 Prediction of Human Intestinal Drug Absorption and Bioavailability with Animal Data

Although it is believed that intestinal mucosal membrane structure is comparable across species, the intertwined expression of metabolic enzymes and transporters in the gut wall and the interspecies physiological and metabolic differences between human and experimental animals complicate the prediction of intestinal absorption in human from animal data. Many transporters (such as pepT1, SGLT-1, GLUTS, MRP2, NT2, and high-affinity glutamate transporter) share similar expression levels in both humans and rats with regional dependent expression patterns, which have high expression in the small intestine and low expression in the colon. However, discrepancy was observed for several other transporters (such as MDR1, MRP3, GLUT1, and GLUT3) in both duodenum and colon of humans and rats. On the contrary, the expressions of metabolizing enzymes (CYP3A4/CYP3A9 and UDPG) showed 12 to 193-fold difference between human and rat intestines with distinct regional dependent expression patterns [60].

Small intestinal permeability values for drugs with passive transport mechanisms in rats and humans are reported to correlate well, at least when the intestinal permeability is moderate or high [58, 60], and rats and humans have similar patterns of expression of small intestinal transporters and drug absorption profiles for compounds that undergo active transport. It appears that the rat perfusion model is mainly useful for predicting the active uptake potential [18]. For passively absorbed compounds, the experimental rat and literature human jejunal effective permeability values were highly correlated [66]. In another study, a significant correlation was obtained between rat ileum permeability and the literature on human intestinal absorption ($r = 0.891$; $p < 0.01$) [67]. Following the recommended procedure for assessing the suitability of permeability determination by FDA, permeability of 20 test drugs was determined using the in situ single-pass perfusion model in rats. The test compounds were co-perfused through jejunal intestinal segments with an internal permeability reference standard, metoprolol, and the rat single-pass perfusion method accurately categorized the selected test drugs into the proper permeability class as determined in human experiments [68].

While comparing the drug intestinal permeability between rats and humans using single-pass perfusion model, a reasonable correlation ($r^2 = 0.8$) was observed with both carrier-mediated absorption and passive diffusion mechanisms between these two species, although the rat permeability values are 5 to 10 times lower than human data. A moderate correlation with $r^2 > 0.56$ was also found for the expression levels of transporters in the duodenum of humans and rats, which indicated the molecular mechanisms of the correlation of drug absorption between these two species [60]. However, no correlation was found for the expressions of metabolizing enzymes between rat and human intestine, providing the underlying mechanism for the observation that no correlation ($r^2 = 0.29$) was found in oral drug bioavailability. These results indicated that the rat and human show similar transporter expression patterns and intestinal drug absorption profiles in small intestine, whereas the metabolic enzyme levels and oral availability values are distinct between these two species. Therefore, it is plausible to use rat perfusion data to predict oral drug absorption and permeability in human small intestine, but the oral bioavailability is not easily predicted based on the rat data.

## 42.5 Summary

In the first four sections of this chapter, we discussed the development of the perfusion model and how it had impacted drug development. We also discussed the recent development and applications of this model, including the study of regional differences along the intestine, the measurement and analysis of drug metabolism, the contribution of intestinal versus liver metabolism and transport, the use of inhibitors and gene knockout animals in understanding the underlying mechanisms in intestine. The comparison and relationship with various other *in vitro* models of intestinal drug absorption, such as subcellular fraction models, intestinal cell line models and tissue-based *in vitro* models have also been discussed. With continuous blood and nerve supply during the experiment, which mimics *in vivo* environment and other conditions, the single-pass animal perfusion method seems to be more reliable in terms of the prediction of drug permeability and absorption. The use of single-pass perfusion has been expanded to different species, such as mice, rabbits, and humans. The drug permeability values from rat and human perfusion data are correlated well, whereas the metabolic enzyme expression levels and oral bioavailabilities between these two species are not correlated, making it challenging for the prediction of human bioavailability in drug discovery area. In the last section of this chapter, we will provide a detailed description of the single-pass perfusion model.

## 42.6 Methodologies and Experimental Data Analysis

In this section, a typical method to perform intestinal perfusion will be provided using rats as an example. For other species, such as mice, the method can be applied after some modification.

### 42.6.1 Materials and Preparation Before Experiments

#### 42.6.1.1 Chemicals

a. Hank's balanced salt (modified) HBSS (Sigma)
b. D-glucose (JRH Bioscience)
c. HEPES (JRH Bioscience)

d. Sodium bicarbonate NaHCO3 (Sigma)
e. Sodium chloride (NaCl) (J.T. Baker)
f. [$^{14}$C]-PEG4000 (Perkin Elmer Life Sciences)

#### 42.6.1.2 Reagents

a. Anesthetizing agent: Isoflurane (preferred) or Urethane * at 2 g/kg
b. Saline (0.9% NaCl) (autoclaved or USP grade).
c. HBSS buffer (pH 6.5 or 7.4): 9.801 g/l HBSS powders, 0.372 g/l $NaHCO_3$, 3.502 g/l glucose, 5.963 g/l HEPES, and 1.164 g/l NaCl.

Preparation note: Dissolve all powders in 1 l previously autoclaved double distilled water, stir, and adjust the pH to 6.5 or 7.4 using the NaOH solution. Store at 4°C in an autoclaved water after filtration through a 0.22 μm filter (Millipore).

* Urethane will only be used as anesthetizing agent if isoflurane inhalation is unable to maintain the proper anesthetic plane for the entire length of perfusion experiment.

#### 42.6.1.3 Equipment

a. PHD 2000 infuse/withdraw pump (Harvard Apparatus)
b. Water bath with cooling capability
c. Heating lamp
d. Electric heating blanket
e. Scissors (one coarse for cutting skin and muscle fine scissors to cut skin or muscle), flat head and pointed stainless steel twisters (not too sharp), and liquid scintillation counter

#### 42.6.1.4 Other Materials

a. Tubing
   i. Silicon tubing with O.D. 0.5 inch or 0.25 inch and wall thickness of 0.0625 inch
   ii. PE10 tubing "0.157 × 0.125" (VWR)
   iii. PE20 tubing "0.157 × 0.125" (VWR)
   iv. PE50 tubing "0.157 × 0.125" (VWR)
   Note: Teflon or other tubing may be used to further decrease adsorption.
b. Syringe. Plastic, glass and stainless steel are available for different compounds. We recommend plastic syringes since they are disposable and inexpensive. However, some chemicals may stick to plastic. Therefore, it is prudent to measure drug concentration in the perfusate before it is loaded into the syringe, and after it has flowed out of the cannulation tubing, to check for binding to various perfusion apparatus.

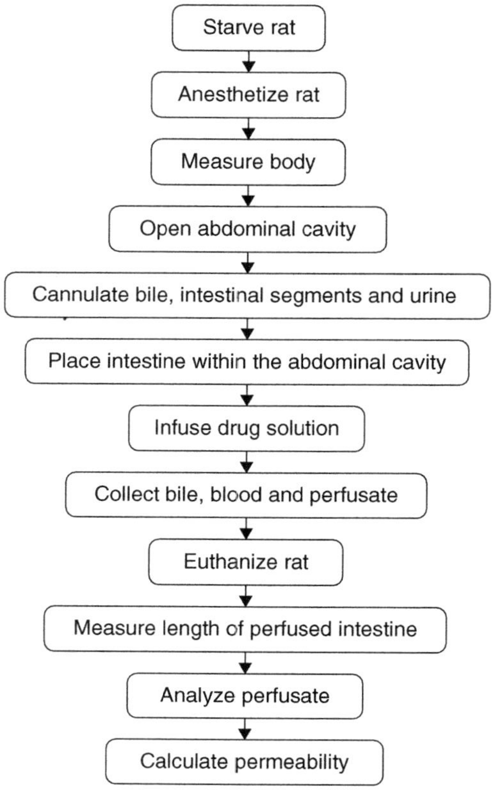


**Figure 42.1** Typical flow chart of intestinal perfusion steps.

### 42.6.2 Animals

Male Sprague Dawley rats, 250–350 g, 70–90 days old, are suitable. For female rats, we typically use a body weight of 230–280 g, 80–90 days old, since female rats are smaller than male rats of the same age. Rats at the recommended body weight and age have the best surgery success rate, and survival rate and are able to collect required volume of blood at sampling time. Also, it could be used as the ideal weight to set up the validation. In addition to Sprague–Dawley rats, we have used various strains of Wistar rats, Pirc, and F344 rats, and they are quite similar to Sprague–Dawley rats.

### 42.6.3 Experiment Procedure

#### 42.6.3.1 Intestinal Perfusion Steps and Comments

Intestinal perfusion is described below and demonstrated using Figures 42.1 and 42.2.

The following are detailed steps for the perfusion procedure:

1. Restrict access to food for rats on or before 5 pm (but no more than 16 h) on the preceding day while maintaining freshwater for rats overnight. Rats should be kept in the cage with toxicology

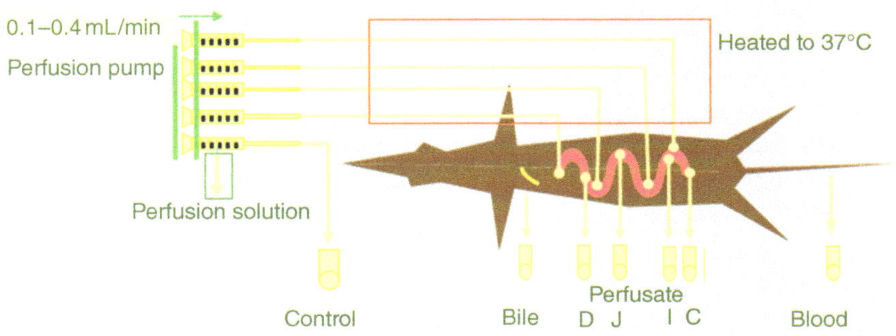


**Figure 42.2** Four-site rat intestinal perfusion.

floor (regular cage with toxicology floor over the bedding). This is done to fast the animal before the surgery to keep the rat intestine clean for the experiment. The wire mesh is used to prevent the animal from eating its own feces.

2. Turn on the water bath and a heating blanket in the morning. Keep the circulating water bath at 37 °C to maintain the temperature of the perfusate constant.
3. Anesthetize rats with gaseous anesthetic agent (e.g. isoflurane) or urethane. Rat is first placed in induction box and anesthesia is initiated using isoflurane vapors at 3–5%. Once the rat reaches the desired plane of anesthesia, urethane (50% w/v) is given to the rat at the doses of 2 g/kg.
   Note: The dose of urethane should be divided. Usually, half of the dose (2 g/kg) required based on body weight is given via i.p. injection. Wait for 30–60 minutes to ensure that the first dose of urethane is absorbed. Then, the second dose, 25% of total dose, is given and animal is monitored carefully. Wait for another 30–60 minutes, the final dose is around 25% of total dose as well. However, it can be adjusted according to the anesthetic status of the rats. Occasionally, urethane is freshly prepared on the experimental day. The use of needles larger than 25 G may cause unnecessary stress to the rats and bleeding.
4. Weigh rats and record total body weight. Rats should not react to tail pinch before cutting into the abdominal cavity to ensure compliance with IACUC protocol. Be aware that rats tend to die prematurely if they are in too much stress (e.g. excessive bleeding during surgery and experiment).
5. While the rat is reaching the desired plane of anesthesia under isoflurane or by urethane injection, do the following preparations:
   a. Prepare perfusate buffer: known concentration of the test compound is prepared in HBSS buffer (pH 7.4)
   b. Filled up five 60 cc plastic syringes with 50 ml of perfusate solution and placed on the perfusion pump (Harvard Apparatus)
   c. Flush five perfusion tubings with double distilled water. Connect four of them to the perfusing tubing, the last one is left as a control which is connected to small tubing and kept at one end of the perfusion pump
   d. Once connected, turn on the Harvard apparatus pump with the speed of 3 ml/min to allow the liquid to pass through the five tubings, and all the air bubbles are taken out from the system
   e. Once the bubbles are removed and the system is ready up to this step, the flow rate of the pump is reduced to a targeted speed (e.g. 0.191 ml/min)
6. During and after this setup process, the rat is carefully monitored for its vitals such as heartbeat and breathing pattern to ensure that it stays in the proper anesthetic plane. Rat is also pinched interdigitally (a pinch of the webbing between the [hind] toes) to measure the anesthetic depth. Once the proper plane is achieved, additional dose of urethane anesthetic will not be required.
7. The anesthetic plane is reconfirmed once more by pinching the webbing between the hind toes of the rat (interdigital pinch) before the surgery. Once the rat shows no sign of twitching while pinched, its tongue is moved to the side of the mouth with a hemostat clamp to free the airway from obstruction. Observe for the presence of liquid in the airway throughout the experiments. Clear any liquid from the mouth/nose with gauze or a cotton-tipped swab to keep the airway open.
8. Once the animal is completely under the desired anesthetic plane, it is placed on a circulating water heating pad covered with animal diaper maintained at around 28–30 °C. The body temperature is checked every 30 minutes using KSC infrared pocket thermometer with laser sighting

and recorded. Optimal desired temperature is 25–32 °C.

9. The rat is put on its back the abdominal portion is shaved with rodent electric trimmer.
10. Cut the skin and then open the abdominal cavity by a middle incision of 3–4 cm along the visible white line (to prevent excessive bleeding) on the rat abdominal muscle. Extra precautions must be taken and try not to harm the organs such as intestine, and liver and not to cause excessive bleeding.
11. Once the abdominal cavity is open, warm saline is sprayed on the entire intestine. Locate the duodenum and bile duct which could be easily found on the right side of the rat under the stomach.
12. ***Bile duct cannulation***. Carefully separate the bile duct by removing the fatty tissues from the surrounding surface. Make a small cut with a microvascular scissor and insert a tubing (PE10) and secure the tubing with a surgical suture.
13. ***Duodenum cannulation***. After bile duct cannulation, an incision is made on duodenum, and inlet perfusion tubing is placed inside the duodenum (approx. 1 cm inside) and secured tight with surgical suture. This is the inlet cannula for the duodenum. At approximately 10 cm down (away from the stomach) from the inlet cannula, another incision is made to cannulate the duodenum. This is the outlet tubing for duodenum.
14. ***Jejunum cannulation***. Approximately 5 cm down from the duodenum outlet cannula, an incision is made for the jejunum inlet cannula, which is again secured using surgical suture. At approximately 10 cm down from the jejunum inlet, another incision is made to secure outlet perfusion tubing for jejunum. Outlet cannula will be inserted after a brief wash of the jejunal segment with 5 ml saline to remove any remaining food content. Saline wash can be skipped if it is assured that no food/luminal content is present.
15. ***Ileum cannulation***. To cannulate terminal ileum, the first step is to locate the cecum. At approximately 1 cm from the ileo-colonic junction, one incision is made for outlet tubing of ileum, which is secured by silk suture as well. At approximately 10 cm up from the outlet cannula, another incision is made to secure inlet perfusion tubing for terminal ileum. Inlet cannula will be inserted after a brief wash of the terminal ileal segment with 5–10 ml saline to remove any remaining food content. Saline wash can be skipped if the operator is assured that no food content is present.
16. ***Colon cannulation***. One more incision is made on the other side of the cecum for colon inlet tubing, which is secured using suture. Anal opening is taken as an outlet for the colon perfusion. Outlet cannula is inserted after a thorough wash of the colonic segment with 20–40 ml saline to remove any remaining fecal content. If required, colon is washed from both anal to inlet and inlet to anal to remove all the fecal content.

    Note: The intestinal segment should not be overhandled or grabbed with sharp objects since intestine can be bruised extremely easily. Handle the intestine gently using hand and insert the cannula, after an opening equivalent to one-third to one-half of the intestinal radius is cut using a fine scissors. All cannulation process is optional in the rat intestinal perfusion experiments based on different experimental purposes.
17. After the cannulation, carefully transfer the rat to the perfusion table where it is placed on a circulating water heating pad covered with animal diaper maintained at around 28–30 °C. Lay the rest of the intestinal segment flat on the abdominal surface of the rats. Cover the whole area with paper towel wetted with normal saline. Put a piece of plastic wrap on the towel to keep the intestinal segments moist. The use of 37 °C normal saline to wet the towels is recommended before putting towels on the intestine. The goal here is to keep the bleeding to the minimal, obstruction-free perfusate flow, and warm and moist intestine. Set the infuse pump to a desired flow rate (e.g. 0.191 ml/min). Clear the intestine for about 30 minutes. by using perfusate containing the compound of interest and nonabsorbable water flux marker (commonly [$^{14}$C]-PEG-4000 and 0.1 mM cold PEG4000 are used) in a desired buffer (e.g. HBSS). This step serves to clean the intestine and lets the absorption proceed to a steady state. If there is insufficient amount of drug, one can use buffer to wash out the foodstuff and then start with a perfusion of the compound (empty the intestine first with an air push). If the flow rate is less than 0.1 ml/min, the wash perfusion time may have to be longer (e.g. 45 min).
18. Collect the perfusate at 30-minutes interval for next 120 min, which is for a flow rate of 0.191 ml/min. Different sample intervals maybe used when different flow rates are used. In general, shorter intervals for faster flow rate and vice versa. We normally collect four samples, at least 1 ml of perfusate, and calculate the permeability individually.

19. The bile samples were collected every 30 minutes as well. The first 30 min sample was treated as blank bile.
20. A small volume of rat blood was collected every 30 minutes from the tail vein. Clean the tail using an acceptable antiseptic scrub (e.g. betadine). Cut the tip (2–3 mm) of the tail off with a sharp surgical scalpel. Use gentle massage and slight pressure to force approximately 100 μL of blood into the collection vial. Apply direct pressure to the cut for 1–2 minutes to facilitate blood clogging and formation of a clot.
21. During the whole perfusion, rat is continuously and carefully monitored for its vitals such as body temperature, heart rate, and breathing patterns. Slow heart rate with no jerking motion and consistent breathing is a sign of good surgery and appropriate anesthesia effect, which is expected under the desired anesthetic plane.
22. After the perfusion experiments finish, rat is sacrificed using overdose of isoflurane followed by cervical dislocation. The remaining urethane is also injected into the animal's body for euthanasia and environmental protection.
23. The perfused segments are carefully removed from the body. Measure the length of perfused intestine by wetting them with normal saline (4 °C) and carefully laying them flat without stretching. It is necessary to develop a standard way of measuring the length since it is a step where an operator's error may be significant. Finally, half of the intestinal segment (by weight) may be homogenized and/or solubilized to determine the amount bound to the enterocytes. The other half may be fixed using a histological procedure to observe any damage to the mucus.
24. Prepare the collected perfusate for further measurements. We commonly centrifuge the samples at >8000 rpm for 30 minutes at room temperature to pellet any proteins or cells or cellular debris, before the further process and analysis. Normally, no additional cleanup procedure is required. It is often necessary to prepare the samples against chemical and microbiological instability. These samples have bacteria that can consume the tested compound at room temperature (most autosamplers) in less than a day. We commonly acidify the samples to pH 2 or less. Sometimes, antioxidants are added too.
25. In addition to perfusate samples, one could cannulate jugular vein and portal vein, and harvest the organs of tested animals at the end of each experiment if so desired. After rats are euthanized, organs are harvested at the end of each study. Weights of heart, lung, liver, spleen, kidney, intestine, colon, and cecum are measured, and the amount of test compound in each organ is determined by an appropriate method.

#### 42.6.3.2 Alternative Cannulation

Most common cannulation we usually make is the bile cannulation and single-pass or four-site intestinal perfusion. Jugular vein, portal vein, and urine cannulation are optional cannulation for rats undergoing perfusion experiments based on different experimental purposes. The cannulation operation is described as below:

1. ***Jugular vein cannulation***. Make a 1.5 cm incision on the left neck. Locate and identify the jugular vein via removing the fatty tissue from the surface after dissecting different layers. Insert a tubing (PE20) through a small incision made by a microvascular scissor. Secure the tubing with suture, which will allow drug administration to or blood sampling from the jugular vein.
2. ***Portal vein cannulation***. Locate the superior mesenteric vein in the mesentery of jejunum, which is directly connected to the portal vein. Puncture the superior mesenteric vein with a venous indwelling needle (no tie-up with suture to allow free blood flow), which will allow drug administration or blood sampling from the portal vein.
3. ***Urinary bladder cannulation***. Locate the top (most cranial) portion of bladder. Insert a soft tubing (PE50) and secure the tubing with a surgical suture, which will allow urine to flow into the collection vials.

### 42.6.4 Data Processing

#### 42.6.4.1 Calculation of Drug Concentration in Blood, Perfusate, and Bile

Sample processing method and quantification method are developed and validated for each compound via UPLC or LC-MS/MS. Samples are processed and analyzed using the appropriate methods. Drug concentrations measured from UPLC or LC-MS/MS will be directly used in relevant data analysis. Amounts of drugs can be calculated by multiplying the concentrations with corresponding perfusate or bile volumes, which are measured experimentally. Statistical analyses will be conducted if the differences or trend identification are needed in the experiments. Tables 42.1 and 42.2 are typical spreadsheets for bile and blood drug concentration analyses in rat perfusion study.

#### 42.6.4.2 Calculation of Permeability

This method measures the steady-state uptake of a compound from the perfusate by determining the rate

**Table 42.1** A typical bile sample analyst spreadsheet in rat intestinal perfusion study.

| **Compound X concentration in bile samples** | | | | | | |
|---|---|---|---|---|---|---|
| Experiment date: mm/dd/yy | | | | | | |
| Sample analyzed date: mm/dd/yy | | | | | | |
| **Standard curve for bile** | | | | | | |
| **Analyte conc. (nM)** | **Analyte peak area (counts)** | **IS peak area (counts)** | | **Calculated conc. (nM)** | | **Accuracy (%)** |
| 1.95 | 1.27E+04 | 2.58E+05 | | 1.73 | | 88.7 |
| 3.91 | 2.51E+04 | 2.57E+05 | | 3.49 | | 89.4 |
| 7.81 | 5.16E+04 | 2.80E+05 | | 7.94 | | 102 |
| 15.63 | 9.84E+04 | 2.60E+05 | | 15.5 | | 99.3 |
| 31.25 | 2.00E+05 | 2.64E+05 | | 32.1 | | 103 |
| 62.5 | 4.21E+05 | 2.70E+05 | | 62.8 | | 101 |
| 125 | 8.22E+05 | 2.96E+05 | | 134 | | 107 |
| 250 | 1.64E+06 | 2.99E+05 | | 267 | | 107 |
| 500 | 2.93E+06 | 2.68E+05 | | 520 | | 104 |
| **Bile samples** | | | | | | |
| **Rat #1** | **Analyte peak area (counts)** | **IS peak area (counts)** | **Calculated conc (nM)** | **Actual conc (μM)** | **Sample volume (μl)** | **Bile amount (nmol)** |
| 0 min | 0.00E+00 | 2.66E+05 | 0 | 0 | 0 | 0 |
| 30 min | 1.11E+06 | 2.62E+05 | 192 | 1.92 | 346 | 0.66 |
| 60 min | 1.00E+06 | 2.55E+05 | 178 | 1.78 | 369 | 0.66 |
| 90 min | 9.79E+05 | 2.56E+05 | 173 | 1.73 | 423 | 0.73 |
| 120 min | 9.42E+05 | 2.58E+05 | 165 | 1.65 | 418 | 0.69 |
| 150 min | 1.24E+06 | 2.63E+05 | 214 | 2.14 | 387 | 0.83 |

of disappearance from the perfusate. And it is used to calculate an unbiased intestinal wall permeability ($P_w$). At steady state, $P_w$ of a compound is calculated using the following equations:

$$P_W = \frac{P_{eff}}{1 - P_{eff}/P_{AQ}} \tag{42.1}$$

where

$$P_{eff} = \frac{1 - C_m/C_o}{4Gz} \tag{42.2}$$

$$P_{aq} = (A(Gz)^{1/3})^{-1} \tag{42.3}$$

$$A = 10.0\,Gz + 1.01\ 0.004 \leq Gz \leq 0.01$$
$$A = 4.5\,Gz + 1.065\ 0.01 \leq Gz \leq 0.03$$
$$A = 2.5\,Gz + 1.125\ 0.03 \leq Gz \tag{42.4}$$

In Eqs. (42.2) to (42.4), $C_o$ and $C_m$ are inlet and outlet concentrations, respectively; while Gz or Graetz number $\left(Gz = \frac{\pi DL}{2Q}\right)$, is a scaling factor that incorporates flow rate ($Q$) and intestinal length ($L$), and diffusion coefficients ($D$) to make the permeability dimensionless; A is a correction factor for aqueous resistance of the intestine. $C_m$ was adjusted for water flux as indicated by the concentration of [$^{14}$C]-PEG4000, a nonabsorbable marker compound. For experiments performed in the small intestine, the data points are discarded if the water flux exceeds 0.55%/cm of intestine. For experiments in colon, this limit maybe relaxed to 1.5%/cm due to the slow flow rate (sometimes necessary due to slower absorption) and rapid water absorption. Permeability parameters can be calculated using simple software such as Microsoft Excel program. Table 42.3 is a typical spreadsheet for permeability calculation in rat perfusion study.

**Table 42.2** A typical blood sample analyst spreadsheet in rat intestinal perfusion study.

**Compound X concentration in blood samples**

Experiment date:
mm/dd/yy

Sample analyzed date:
mm/dd/yy

| Standard curve for blood | | | | |
|---|---|---|---|---|
| **Analyte conc. (nM)** | **Analyte peak area (counts)** | **IS peak area (counts)** | **Calculated conc. (nM)** | **Accuracy (%)** |
| 0.98 | 1.69E+04 | 2.73E+05 | 9.08 | 92.7 |
| 1.95 | 3.38E+04 | 2.63E+05 | 20.8 | 106 |
| 3.91 | 6.94E+04 | 2.74E+05 | 43.2 | 111 |
| 7.81 | 1.32E+05 | 2.77E+05 | 84.5 | 108 |
| 15.63 | 2.55E+05 | 2.84E+05 | 166 | 106 |
| 31.25 | 4.97E+05 | 2.85E+05 | 326 | 104 |
| 62.5 | 9.29E+05 | 2.89E+05 | 627 | 100 |
| 125 | 1.87E+06 | 2.90E+05 | 1280 | 103 |
| 250 | 3.02E+06 | 2.83E+05 | 2150 | 86.1 |
| 500 | 5.33E+06 | 2.97E+05 | 4120 | 82.3 |

| Blood samples | | | |
|---|---|---|---|
| **Rat #1** | **Analyte peak area (counts)** | **IS peak area (counts)** | **Calculated conc. (nM)** |
| 0 min | 0.00E+00 | 2.97E+05 | 0 |
| 30 min | 4.64E+05 | 2.91E+05 | 29.1 |
| 60 min | 6.47E+05 | 2.91E+05 | 46.3 |
| 90 min | 9.47E+05 | 3.01E+05 | 71.5 |
| 120 min | 7.13E+05 | 2.52E+05 | 62.8 |
| 150 min | 3.18E+05 | 2.82E+05 | 16.3 |

#### 42.6.4.3 Interpretation of Drug Concentration in Blood, Perfusate, and Bile

Bile drug concentration and amount can be used to calculate the percentage of the dose that is excreted into bile, providing more crude understanding of biliary route. Urine drug analysis can indicate the percentage of the drug that is eliminated via urine. Tail vein blood levels inform on the systemic drug concentration during the whole perfusion experiments. If the drug metabolites are analyzed, such extended data analysis provides more comprehensive drug distribution information.

#### 42.6.4.4 Interpretation of Permeability Data

$P_w$ measures the permeability of the compound across the biological membrane, while $P_{aq}$ measures the permeability across the unstirred water layer. The $P_w$ values are much better representation of the intestinal membrane permeability than $P_{eff}$, because the permeability contribution of unstirred water layer ($P_{aq}$) is factored out. Since permeability across the unstirred water layer is not highly dependent on compound structure, it is more beneficial to use $P_w$ which is highly dependent on compound structure. For example, compounds that are generally well absorbed (>75%) have a $P_w$ and could approach infinity as $P_{aq}$ dominated the $P_{eff}$. Under such circumstances, a change in $P_{eff}$ does not significantly impact the overall permeation, whereas a change in $P_{aq}$ does. However, when a compound is highly metabolized in the cells, metabolism may drive or significantly impact $P_w$, and in that case, $P_w$ value might not be particularly relevant as a permeability measure.

## Acknowledgment

This Chapter was originally written by Wei Zhu and Eu-Jung Jeong. Updates for this Chapter are provided by Li Li and Dinh Bui.

**Table 42.3** A typical permeability calculation spreadsheet in rat intestinal perfusion study.

**Compound X Perfusion in Intestine**

| | |
|---|---|
| Experiment date: mm/dd/yy | Inhibitor: |
| Sample analyzed date | Concentration: xx |
| Site of perfusion: 4-sites (Duo, Jej, Ileu, Colon) | pH = xx |
| Birthday of the Rats: mm/dd/yy | Drug: Compound X |

**Rat characteristics**

| Rat#1 | *W* (gm) | *L* (cm) | *Q* (ml/min) | Gz | *A* | $P_a^*$ |
|---|---|---|---|---|---|---|
| Duo | 260 | 4 | 0.2 | 0.0197 | 1.1539 | 3.2063 |
| Jej | 260 | 13 | 0.2 | 0.0642 | 1.2854 | 1.943 |
| Ileu | 260 | 8.5 | 0.2 | 0.042 | 1.2299 | 2.3397 |
| Colon | 260 | 5.5 | 0.2 | 0.0272 | 1.1872 | 2.8025 |

| | CPMA | | | | | | $\%H_2O$ change | | | |
|---|---|---|---|---|---|---|---|---|---|---|
| *T* (min) | Duo | Jej | Ileu | Colon | Original | BLK | Duo | Jej | Ileu | Colon |
| 30–60 | 2795 | 2778 | 2924 | 2867 | 2862 | 28 | −2.43 | −3.08 | 2.13 | 0.17 |
| 60–90 | 2795 | 2842 | 2910 | 2783 | 2862 | 28 | −2.43 | −0.74 | 1.65 | −2.87 |
| 90–120 | 2863 | 2898 | 2818 | 2699 | 2862 | 28 | 0.04 | 1.25 | −1.6 | −6.13 |
| 120–150 | 2778 | 2886 | 2829 | 2782 | 2862 | 28 | −3.08 | 0.84 | −1.18 | −2.9 |
| Avg. | 2808 | 2851 | 2870 | 2783 | 2862 | 28 | −1.97 | −0.43 | 0.25 | −2.93 |

**Check volume change with $^{14}$C-PEG 4000**

**Sample's peak area ratio from HPLC (*T* = xx min)**

| | Peak area | | | | Corrected peak height | | | | |
|---|---|---|---|---|---|---|---|---|---|
| T(min) | Duo | Jej | Ileu | Colon | Original | Duo | Jej | Ileu | Colon |
| 30–60 | 1089 | 817 | 760 | 644 | 1372 | 1116 | 843 | 744 | 643 |
| 60–90 | 1049 | 948 | 831 | 668 | 1372 | 1075 | 955 | 818 | 688 |
| 90–120 | 1002 | 856 | 789 | 698 | 1372 | 1001 | 846 | 802 | 744 |
| 120–150 | 1034 | 998 | 793 | 711 | 1372 | 1066 | 989 | 802 | 732 |
| Avg. | 1287 | 905 | 793 | 680 | 1372 | 1065 | 908 | 792 | 702 |
| | | | | | %ABS | 22 | 34 | 42 | 49 |

**Calculate permeability**

| | $P_{eff}^* = (1\text{-}C_m/C_0)/4\,Gz$ | | | | $P_w^* = P_{eff}^*/(1\text{-}P_{eff}^*/P_a^*)$ | | | |
|---|---|---|---|---|---|---|---|---|
| *T* (min) | Duo | Jej | Ileu | Colon | Duo | Jej | Ileu | Colon |
| 30–60 | 2.36 | 4.88 | 5.79 | 6.7 | 8.89 | −3.23 | −3.93 | −4.8 |
| 60–90 | 2.74 | 3.85 | 5.11 | 6.3 | 18.68 | −3.92 | −4.31 | −5.04 |
| 90–120 | 3.42 | 4.86 | 5.26 | 5.8 | −51.76 | −3.24 | −4.22 | −5.42 |
| 120–150 | 2.82 | 3.53 | 5.25 | 5.9 | 23.33 | −4.32 | −4.22 | −5.34 |
| Avg. | 2.83 | 4.28 | 5.35 | 6.2 | −0.22 | −3.68 | −4.17 | −5.15 |
| S.D. | 0.44 | 0.69 | 0.3 | 0.4 | 34.89 | 0.54 | 0.17 | 0.28 |

## References


1 Gao, Y., Shao, J., Jiang, Z. et al. (2014). Drug enterohepatic circulation and disposition: constituents of systems pharmacokinetics. *Drug Discovery Today* 19: 326–340.

2 Stappaerts, J., Brouwers, J., Annaert, P., and Augustijns, P. (2015). In situ perfusion in rodents to explore intestinal drug absorption: challenges and opportunities. *Int. J. Pharm.* 478: 665–681.

3 Zheng, L., Chen, J., Zhu, Y. et al. (1994). Comparison of the transport characteristics of D- and L-methionine in a human intestinal epithelial model (Caco-2) and in a perfused rat intestinal model. *Pharm. Res.* 11: 1771–1776.

4 Hogben, C.A., Schanker, L.S., Tocco, D.J., and Brodie, B.B. (1957). Absorption of drugs from the stomach. II. the human. *J. Pharmacol. Exp. Ther.* 120: 540–545.

5 Hogben, C.A., Tocco, D.J., Brodie, B.B., and Schanker, L.S. (1959). On the mechanism of intestinal absorption of drugs. *J. Pharmacol. Exp. Ther.* 125: 275–282.

6 Schanker, L.S., Shore, P.A., Brodie, B.B., and Hogben, C.A. (1957). Absorption of drugs from the stomach. I. The rat. *J. Pharmacol. Exp. Ther.* 120: 528–539.

7 Schanker, L.S., Tocco, D.J., Brodie, B.B., and Hogben, C.A. (1958). Absorption of drugs from the rat small intestine. *J. Pharmacol. Exp. Ther.* 123: 81–88.

8 Shore, P.A., Brodie, B.B., and Hogben, C.A. (1957). The gastric secretion of drugs: a pH partition hypothesis. *J. Pharmacol. Exp. Ther.* 119: 361–369.

9 Kakemi, K., Arita, T., Hori, R., and Konishi, R. (1967). Absorption and excretion of drugs. XXXII. Absorption of barbituric acid derivatives from rat small intestine. *Chem. Pharm. Bull.* 15: 1883–1887.

10 Doluisio, J.T., Billups, N.F., Dittert, L.W. et al. (1969). Drug absorption. I. an in situ rat gut technique yielding realistic absorption rates. *J. Pharm. Sci.* 58: 1196–1200.

11 Hu, M., Sinko, P., DeMeere, A. et al. (1988). Membrane permeability parameters for some amino acids and β-lactam antibiotics: application of the boundary layer approach. *J. Theor. Biol.* 131: 107–114.

12 Sinko, P.J. and Amidon, G.L. (1988). Characterization of the oral absorption of beta-lactam antibiotics. I. Cephalosporins: determination of intrinsic membrane absorption parameters in the rat intestine in situ. *Pharm. Res.* 5: 645–650.

13 Amidon, G.L., Sinko, P.J., and Fleisher, D. (1988). Estimating human oral fraction dose absorbed: a correlation using rat intestinal membrane permeability for passive and carrier-mediated compounds. *Pharm. Res.* 5: 651–654.

14 Xu, H., Kulkarni, K.H., Singh, R. et al. (2009). Disposition of naringenin via glucuronidation pathway is affected by compensating efflux transporters of hydrophilic glucuronides. *Mol. Pharmaceutics* 6: 1703–1715.

15 Zeng, M., Sun, R., Basu, S. et al. (2016). Disposition of flavonoids via recycling: direct biliary excretion of enterically or extrahepatically derived flavonoid glucuronides. *Mol. Nutr. Food Res.* 60: 1006–1019.

16 Hu, M. and Amidon, G.L. (1988). Passive and carrier-mediated intestinal absorption components of captopril. *J. Pharm. Sci.* 77: 1007–1011.

17 Hu, M., Roland, K., Ge, L. et al. (1998). Determination of absorption characteristics of AG337, a novel thymidylate synthase inhibitor, using a perfused rat intestinal model. *J. Pharm. Sci.* 87: 886–890.

18 Fagerholm, U. (2007). Prediction of human pharmacokinetics--gastrointestinal absorption. *J. Pharm. Pharmacol.* 59: 905–916.

19 Jeong, E.J., Jia, X., and Hu, M. (2005). Disposition of formononetin via enteric recycling: metabolism and excretion in mouse intestinal perfusion and Caco-2 cell models. *Mol. Pharmaceutics* 2: 319–328.

20 Liu, Y. and Hu, M. (2002). Absorption and metabolism of flavonoids in the caco-2 cell culture model and a perused rat intestinal model. *Drug Metab. Dispos.* 30: 370–377.

21 Sinko, P.J. and Hu, P. (1996). Determining intestinal metabolism and permeability for several compounds in rats. Implications on regional bioavailability in humans. *Pharm. Res.* 13: 108–113.

22 Chen, J., Lin, H., and Hu, M. (2003). Metabolism of flavonoids via enteric recycling: role of intestinal disposition. *J. Pharmacol. Exp. Ther.* 304: 1228–1235.

23 Wang, S.W., Chen, J., Jia, X. et al. (2006). Disposition of flavonoids via enteric recycling: structural effects and lack of correlations between in vitro and in situ metabolic properties. *Drug Metab. Dispos.* 34: 1837–1848.

24 Jia, X., Chen, J., Lin, H., and Hu, M. (2004). Disposition of flavonoids via enteric recycling: enzyme-transporter coupling affects metabolism of biochanin A and formononetin and excretion of their phase II conjugates. *J. Pharmacol. Exp. Ther.* 310: 1103–1113.

25 Jeong, E.J., Liu, X., Jia, X. et al. (2005). Coupling of conjugating enzymes and efflux transporters: impact

on bioavailability and drug interactions. *Curr. Drug Metabol.* 6: 455–468.

26 Varma, M.V., Kapoor, N., Sarkar, M., and Panchagnula, R. (2004). Simultaneous determination of digoxin and permeability markers in rat in situ intestinal perfusion samples by RP-HPLC. *J. Chromatogr. B* 813: 347–352.

27 Sesink, A.L., Arts, I.C., de Boer, V.C. et al. (2005). Breast cancer resistance protein (Bcrp1/Abcg2) limits net intestinal uptake of quercetin in rats by facilitating apical efflux of glucuronides. *Mol. Pharmacol.* 67: 1999–2006.

28 Osman, M.A., El Maghraby, G.M., and Hedaya, M.A. (2006). Intestinal absorption and presystemic disposition of sildenafil citrate in the rabbit: evidence for site-dependent absorptive clearance. *Biopharm. Drug Dispos.* 27: 93–102.

29 Svensson, U.S. and Ashton, M. (1999). Identification of the human cytochrome P450 enzymes involved in the in vitro metabolism of artemisinin. *Br. J. Clin. Pharmacol.* 48: 528–535.

30 Mae, T., Inaba, T., Konishi, E. et al. (2000). Identification of enzymes responsible for rifalazil metabolism in human liver microsomes. *Xenobiotica* 30: 565–574.

31 Mandagere, A.K., Thompson, T.N., and Hwang, K.K. (2002). Graphical model for estimating oral bioavailability of drugs in humans and other species from their Caco-2 permeability and in vitro liver enzyme metabolic stability rates. *J. Med. Chem.* 45: 304–311.

32 Sumida, K., Ooe, N., Nagahori, H. et al. (2001). An in vitro reporter gene assay method incorporating metabolic activation with human and rat S9 or liver microsomes. *Biochem. Biophys. Res. Commun.* 280: 85–91.

33 Sohlenius-Sternbeck, A.K. (2006). Determination of the hepatocellularity number for human, dog, rabbit, rat and mouse livers from protein concentration measurements. *Toxicol. in Vitro* 20: 1582–1586.

34 Takemoto, K., Yamazaki, H., Tanaka, Y. et al. (2003). Catalytic activities of cytochrome P450 enzymes and UDP-glucuronosyltransferases involved in drug metabolism in rat everted sacs and intestinal microsomes. *Xenobiotica* 33: 43–55.

35 Joseph, T.B., Wang, S.W., Liu, X. et al. (2007). Disposition of flavonoids via enteric recycling: enzyme stability affects characterization of prunetin glucuronidation across species, organs, and UGT isoforms. *Mol. Pharmaceutics* 4: 883–894.

36 Englund, G., Rorsman, F., Rönnblom, A. et al. (2006). Regional levels of drug transporters along the human intestinal tract: co-expression of ABC and SLC transporters and comparison with Caco-2 cells. *Eur. J. Pharm. Sci.* 29: 269–277.

37 Seithel, A., Karlsson, J., Hilgendorf, C. et al. (2006). Variability in mRNA expression of ABC-and SLC-transporters in human intestinal cells: comparison between human segments and Caco-2 cells. *Eur. J. Pharm. Sci.* 28: 291–299.

38 Artursson, P. and Karlsson, J. (1991). Correlation between oral drug absorption in humans and apparent drug permeability coefficients in human intestinal epithelial (Caco-2) cells. *Biochem. Biophys. Res. Commun.* 175: 880–885.

39 Meunier, V., Bourrie, M., Berger, Y., and Fabre, G. (1995). The human intestinal epithelial cell line Caco-2; pharmacological and pharmacokinetic applications. *Cell Biol. Toxicol.* 11: 187–194.

40 Pfrunder, A., Gutmann, H., Beglinger, C., and Drewe, J. (2003). Gene expression of CYP3A4, ABC-transporters (MDR1 and MRP1-MRP5) and hPXR in three different human colon carcinoma cell lines. *J. Pharm. Pharmacol.* 55: 59–66.

41 Grès, M.-C., Julian, B., Bourrié, M. et al. (1998). Correlation between oral drug absorption in humans, and apparent drug permeability in TC-7 cells, a human epithelial intestinal cell line: comparison with the parental Caco-2 cell line. *Pharm. Res.* 15: 726–733.

42 Läpple, F., von Richter, O., Fromm, M.F. et al. (2003). Differential expression and function of CYP2C isoforms in human intestine and liver. *Pharm. Genom.* 13: 565–575.

43 Sun, D., Lennernas, H., Welage, L.S. et al. (2002). Comparison of human duodenum and Caco-2 gene expression profiles for 12,000 gene sequences tags and correlation with permeability of 26 drugs. *Pharm. Res.* 19: 1400–1416.

44 Taipalensuu, J., Törnblom, H., Lindberg, G. et al. (2001). Correlation of gene expression of ten drug efflux proteins of the ATP-binding cassette transporter family in normal human jejunum and in human intestinal epithelial Caco-2 cell monolayers. *J. Pharmacol. Exp. Ther.* 299: 164–170.

45 Engman, H.A., Lennernäs, H., Taipalensuu, J. et al. (2001). CYP3A4, CYP3A5, and MDR1 in human small and large intestinal cell llines suitable for drug transport studies. *J. Pharm. Sci.* 90: 1736–1751.

46 Ingels, F.M. and Augustijns, P.F. (2003). Biological, pharmaceutical, and analytical considerations with respect to the transport media used in the absorption screening system, Caco-2. *J. Pharm. Sci.* 92: 1545–1558.

47 Saitoh, R., Sugano, K., Takata, N. et al. (2004). Correction of permeability with pore radius of tight junctions in Caco-2 monolayers improves the prediction of the dose fraction of hydrophilic drugs absorbed by humans. *Pharm. Res.* 21: 749–755.

48 Tanaka, Y., Taki, Y., Sakane, T. et al. (1995). Characterization of drug transport through tight-junctional pathway in Caco-2 monolayer: comparison with isolated rat jejunum and colon. *Pharm. Res.* 12: 523–528.

49 Tavelin, S., Milovic, V., Ocklind, G. et al. (1999). A conditionally immortalized epithelial cell line for studies of intestinal drug transport. *J. Pharmacol. Exp. Ther.* 290: 1212–1221.

50 Tavelin, S., Taipalensuu, J., Söderberg, L. et al. (2003). Prediction of the oral absorption of low-permeability drugs using small intestine-like 2/4/A1 cell monolayers. *Pharm. Res.* 20: 397–405.

51 Matsson, P., Bergström, C.A., Nagahara, N. et al. (2005). Exploring the role of different drug transport routes in permeability screening. *J. Med. Chem.* 48: 604–613.

52 Wilson, T.H. and Wiseman, G. (1954). The use of sacs of everted small intestine for the study of the transference of substances from the mucosal to the serosal surface. *J. Physiol.* 123: 116–125.

53 Bohets, H., Annaert, P., Mannens, G. et al. (2001). Strategies for absorption screening in drug discovery and development. *Curr. Top. Med. Chem.* 1: 367–383.

54 Hashimoto, Y., Sasa, H., Shimomura, M., and Inui, K.-i. (1998). Effects of intestinal and hepatic metabolism on the bioavailability of tacrolimus in rats. *Pharm. Res.* 15: 1609–1613.

55 Ballent, M., Lifschitz, A., Virkel, G. et al. (2006). Modulation of the P-glycoprotein-mediated intestinal secretion of ivermectin: in vitro and in vivo assessments. *Drug Metab. Dispos.* 34: 457–463.

56 Ussing, H.H. and Zerahn, K. (1951). Active transport of sodium as the source of electric current in the short-circuited isolated frog skin. *Acta Physiol. Scand.* 23: 110–127.

57 Le Ferrec, E., Chesne, C., Artusson, P. et al. (2001). In vitro models of the intestinal barrier: the report and recommendations of ECVAM workshop 46. *Altern. Lab. Anim.* 29: 649–668.

58 Fagerholm, U., Johansson, M., and Lennernäs, H. (1996). Comparison between permeability coefficients in rat and human jejunum. *Pharm. Res.* 13: 1336–1342.

59 Hu, M. and Borchardt, R.T. (1990). Mechanism of L-α-methyldopa transport through a monolayer of polarized human intestinal epithelial cells (Caco-2). *Pharm. Res.* 7: 1313–1319.

60 Cao, X., Gibbs, S.T., Fang, L. et al. (2006). Why is it challenging to predict intestinal drug absorption and oral bioavailability in human using rat model. *Pharm. Res.* 23: 1675–1686.

61 Lennernäs, H. (1997). Human jejunal effective permeability and its correlation with preclinical drug absorption models. *J. Pharm. Pharmacol.* 49: 627–638.

62 Ungell, A.-L., Nylander, S., Bergstrand, S. et al. (1998). Membrane transport of drugs in different regions of the intestinal tract of the rat. *J. Pharm. Sci.* 87: 360–366.

63 Lennernäs, H. (2007). Animal data: the contributions of the Ussing Chamber and perfusion systems to predicting human oral drug delivery in vivo. *Adv. Drug Delivery Rev.* 59: 1103–1120.

64 Lennernäs, H., Ahrenstedt, Ö., Hällgren, R. et al. (1992). Regional jejunal perfusion, a new in vivo approach to study oral drug absorption in man. *Pharm. Res.* 9: 1243–1251.

65 Lennernäs, H. (2007). Intestinal permeability and its relevance for absorption and elimination. *Xenobiotica* 37: 1015–1051.

66 Zakeri-Milani, P., Valizadeh, H., Tajerzadeh, H. et al. (2007). Predicting human intestinal permeability using single-pass intestinal perfusion in rat. *J. Pharm. Pharm. Sci.* 10: 368–379.

67 Varma, M.V. and Panchagnula, R. (2005). Enhanced oral paclitaxel absorption with vitamin E-TPGS: effect on solubility and permeability in vitro, in situ and in vivo. *Eur. J. Pharm. Sci.* 25: 445–453.

68 Kim, J.-S., Mitchell, S., Kijek, P. et al. (2006). The suitability of an in situ perfusion model for permeability determinations: utility for BCS class I biowaiver requests. *Mol. Pharmaceutics* 3: 686–694.

# 43

# In Silico Prediction of Oral Drug Absorption

*Jin Dong[1], Zhu Zhou[2], Yujuan Zheng[3], and Miki Susanto Park[4]*

[1] *Clinical Pharmacology and Quantitative Pharmacology, Clinical Pharmacology and Safety Sciences, R&D, AstraZeneca, Gaithersburg, Maryland, USA*
[2] *Department of Chemistry, York College, City University of New York, Jamaica, New York, USA*
[3] *Life Sciences Department, Simulations Plus, Inc., Lancaster, California, USA. Current affiliation: Amador Bioscience, Quantitative Clinical Pharmacology Department, Germantown, Maryland, USA*
[4] *Department of Pharmaceutics and Medicinal Chemistry, Thomas J. Long School of Pharmacy, University of the Pacific, Stockton, CA, USA*

## 43.1 Introduction

As discussed in the preceding chapters, there are many processes involved in an active pharmaceutical ingredient (API) to be absorbed systemically following oral delivery. The API needs to dissolve, stay in solution and be able to permeate across gastrointestinal (GI) membranes and escape chemical and enzymatic degradations in the stomach, gut, and liver. Therefore, oral bioavailability is a complex process that is influenced by many factors (e.g. solubility, permeability and stability of the API, excipients, particle size, pH in the GI tract, fed or fasting state, GI transit time, other concomitant drug and herb administration, age, gender, disease status, etc.). The rate and extent of oral drug absorption are governed by the interplay of a drug's physicochemical properties, its formulation, and the anatomy and physiological characteristics of the user. Since the oral route is still the preferred route of drug administration, partly due to ease of administration and patient compliance, there is a need to be able to accurately predict the oral bioavailability of an API during early drug discovery and development stages to reduce drug attrition rates. *In vitro* and *in vivo* studies to estimate **h**uman **o**ral **b**io**a**vailability (**HOBA**) are costly and time consuming, so many computational models have been developed to predict HOBA [1]. However, even though significant progress has been made in this field over the last few decades, due to the complexities and interplay of many processes involved in oral drug absorption, the accuracy of currently available in silico HOBA prediction is still too low to completely replace experimental *in vitro* and *in vivo* studies and more improvement in the knowledge database is needed to improve confidence in computational HOBA prediction [2]. Nevertheless, biorelevant dissolution coupled with PBPK modeling has emerged as a reliable tool to minimize the use of *in vivo* studies, and in silico models are being increasingly used by the regulatory agency and pharmaceutical companies to support decisions regarding formulation development, bioequivalence, and other bridging studies [2].

The purpose of this chapter is to provide an overview of the currently available in silico HOBA prediction models and provide select case studies to illustrate how HOBA modeling and simulations are used and carried out. The currently available in silico oral drug absorption prediction can be broadly divided into nonphysiological and physiological in silico models [3]. Examples of nonphysiological models are quantitative structure–property relationship (QSPR) and rule of thumbs (RoT) which use chemical descriptors to predict properties related to oral drug absorption, such as solubility, permeability, human intestinal absorption (HIA), and metabolism. An example of a physiological in silico model is physiologically based pharmacokinetics (PBPK) modeling [3]. In this chapter, we will briefly introduce QSPR modeling and more focus will be given to PBPK modeling.

## 43.2 QSPR Modeling

The QSPR approach uses a statistical method to correlate a set of molecular structure descriptors to correlate with experimental values to develop a computational predictive model for either HIA or oral bioavailability [3]. The predictive capability of the

*Oral Bioavailability and Drug Delivery: From Basics to Advanced Concepts and Applications,* First Edition. Edited by Ming Hu and Xiaoling Li.

**Table 43.1** A summary of available software programs to calculate molecular descriptors.

| Software | Website | Free or Fee |
|---|---|---|
| Chemistry Development Kit (CDK) | http://cdk.sourceforge.net | Free |
| Mold2 | http://www.fda.gov/ScienceResearch/BioinformaticsTools/Mold2 | Free |
| MOLE db | http://michem.disat.unimib.it/mole_db/ | Free |
| MOLGEN | http://molgen.de | Free |
| PaDEL Descriptors | http://www.yapcwsoft.com/dd/padeldescriptor/ | Free |
| VolSurf+ | http://www.moldiscovery.com/software/vsplus/ | Free |
| ADMEWORKS | https://www.fujitsu.com/global/solutions/business-technology/tc/sol/admeworks/ | Fee |
| CODESSA PRO | http://www.codessa-pro.com | Fee |
| Molconn-Z | http://www.edusoft-lc.com/molconn/ | Fee |
| Molecular Operating Environment (MOE) | http://www.chemcomp.com/software-chem.htm | Fee |

Source: Adapted from Cabrera et al. [3].

model depends on the quality and quantity of the dataset and the biological complexity employed. Currently, the availability of a good quality dataset of HOBA is probably the greatest challenge to improving the performance of existing QSPR computational models [3]. The largest public domain database now has more than 1000 drugs [4] but they were collected using literature data, unlike the dataset used by Veber et al. [5] and Andrew et al. [6], which are in-house datasets owned by GSK (over 1100 compounds) and Glaxo Wellcome (591 structures), respectively. In-house databases have the advantage over the public domain databases in that they are consistent and have less variability [1]. Furthermore, the public databases also have publication bias toward high or moderate oral bioavailability drugs that could limit the structural diversity and influence the predictive capability of in silico models that were developed using the public databases as the training set [3].

To predict ADME properties of interest, the chemical information from a compound is transformed into a useful number by mathematical or logical procedures, and the transformed product is called a molecular descriptor [3]. There are free and commercial software programs available to calculate molecular descriptors, as listed in Table 43.1. For example, Mold2 is a descriptors generator software that can quickly calculate descriptors from a two-dimensional chemical structure for small and large datasets. It is a free software that is available for download via the US Food and Drug Administration website (http://www.fda.gov/ScienceResearch/BioinformaticsTools/Mold2). Mold2 is created by the Center for Bioinformatics, led by Dr. Weida Tong, at the National Center for Toxicological Research (NCTR). The chemistry development kit (CDK) is "a collection of modular Java libraries for processing chemical information (Cheminformatics). The modules are free and open-source and are easy to integrate with other open-source or in-house projects." It can be downloaded at https://cdk.github.io/. Mole db is a free online molecular descriptors database, developed by the Milano Chemometrics and QSAR Research Group. Queries can be made at http://michem.disat.unimib.it/mole_db/.

There are thousands of molecular descriptors available (1D, 2D, and 3D) and appropriate molecular descriptors that can successfully model HOBA need to be selected based on the understanding of factors that affect HOBA and redundant molecular descriptors need to be eliminated using covariance analysis [1]. HIA defines the upper limit of HOBA and HIA data from *in vitro*/*in vivo* analysis, as well as molecular descriptors of HIA, such as log *P* and log *D* can be good descriptors of HOBA [1]. Due to the first-pass gut and liver effect, molecular descriptors that identify chemical structures involved in metabolism reactions should be good descriptors of HOBA as well [1]. In general, 1D descriptors allow easy mechanistic interpretation of bioavailability models but have limited predictive capacity while 3D molecular descriptors, such as WHIM and Volsurf, provide more structural information but are more complex due to the need to select the right 3D conformation and the calculation takes much longer compared to 1D descriptors. Taking into account the speed and potential mechanistic interpretation, 2D descriptors, such as the electrotopological state indices (E-state) and the TOPS-MODE descriptors, could be

**Table 43.2** A summary of available QSPR software programs to calculate ADME property related to oral drug absorption.

| Software | Website | Free or Fee |
|---|---|---|
| AdmetSAR | http://lmmd.ecust.edu.cn/admetsar2 | Free |
| Chembench | http://chembench.mml.unc.edu | Free |
| FAF-Drugs 4 | http://fafdrugs4.mti.univ-paris-diderot.fr | Free |
| SwissADME | http://www.swissadme.ch/ | Free |
| ADMET Predictor | http://www.simulations-plus.com/software/admetpredictor/ | Fee |
| ACD/ADME Suite | http://acd-adme-suite.software.informer.com/ | Fee |
| ACD/percepta predictors | http://www.acdlabs.com/products/percepta/predictors.php | Fee |
| Impact-F | http://www.pharmainformatic.com/html/impact-f.html | Fee |
| StarDrop | http://www.optibrium.com | Fee |

Source: Adapted from Cabrera et al. [3].

a good compromise molecular descriptors to use for HOBA models [3].

The last step in the computational HOBA model is to construct a QSPR model using the calculated molecular descriptors. There are many statistical tools available and the most commonly used are multiple linear regressions (MLR), partial least square fitting (PLS), genetic algorithms (GA), artificial neural network (ANN), support vector machine-regression (SVMR), Gaussian processes (GP), and many of these have been incorporated into specific software [1], as listed in Table 43.2. Some of the software are free and some are commercially available. For example, SwissADME is a free Web tool that is available since 2017. It is a part of the SwissDrugDesign project, developed by the Swiss Institute of Bioinformatics (SIB) molecular modeling group. Using a compound's structure or SMILES code as input, SwissADME "allows you to compute physicochemical descriptors as well as to predict ADME parameters, pharmacokinetic properties, druglike nature and medicinal chemistry friendliness of one or multiple small molecules to support drug discovery" (http://www.swissadme.ch).

## 43.3 PBPK Modeling

PBPK modeling is a mathematical model that uses differential equations to describe biological processes in a human body to predict drug blood/plasma concentration profile as a function of time. In a PBPK model, major organs/tissues are represented by compartments and are connected to one another by blood flows. A compartment can be categorized as either eliminating or non-eliminating tissues. PBPK models are mechanistic in nature and incorporate information about a drug's physicochemical properties, biochemical process, formulation properties, and administration protocol as well as the body's anatomical and physiological structure [7]. Figure 43.1 shows a schematic of a general whole body PBPK model [8].

The first PBPK model was introduced in 1937 by Torsten Teorell, known as the father of pharmacokinetics [9] but it was not until the last decade or so that its usage has gained in popularity. The uses of PBPK models in drug development include pharmacokinetic interspecies and interindividual scaling, extrapolation of indications, dose optimization, prediction of DDIs, predictions of PK in pediatrics, geriatrics, disease states, and special population extrapolations, etc.

OrBiTo (Innovative Tools for Oral Biopharmaceutics) is a European project initiated by IMI (Innovative Medicines Initiative). As stated in Lennernäs et al. [10], the overall objective of OrBiTo is to "streamline and optimize the development of orally administered drug products through the creation and efficient application of biopharmaceutics tools. This toolkit will include both experimental and computational models developed on improved understanding of the highly dynamic GI physiology relevant to the GI absorption of drug products in both fasted and fed states." One of OrBiTo's work packages (WP4) is to optimize absorption related processes in PBPK models and "a key objective for OrBiTo is to evaluate the predictive accuracy of available in silico and *in vitro* biopharmaceutical models and to provide validation data in different dosing scenarios to increase their use in regulatory submissions and decision-making throughout the drug development process." [10] As part of their objectives, OrBiTo project tested and compared three PBPK software packages (GI-Sim, Simcyp Simulator, and GastroPlus®) using a blinded "bottom-up" study of human pharmacokinetics

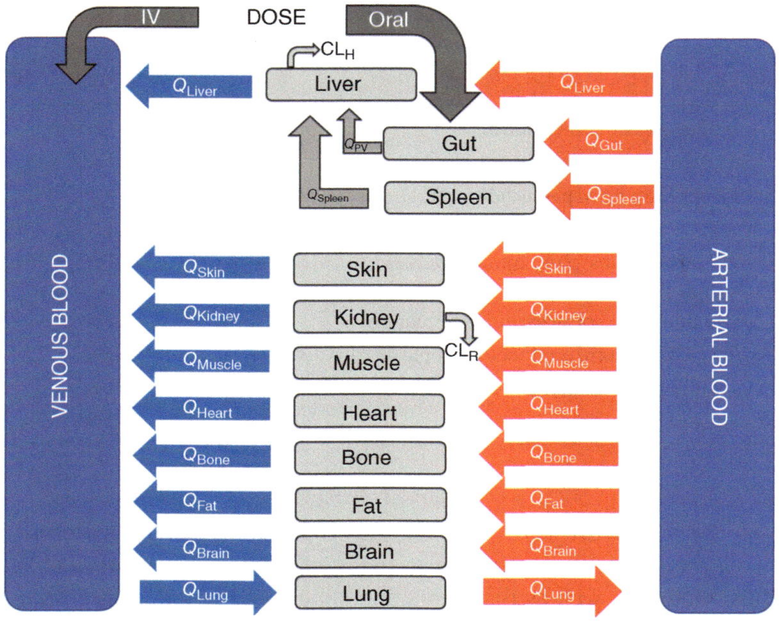


**Figure 43.1** A general schematic of a whole body PBPK model. Source: Adapted from Bravo-Goméz et al. [8].

and reported that in general: (i) the bioavailability of orally administered APIs that permeated the intestine poorly was underpredicted; (ii) the bioavailability of APIs with acidic p$K_a$ was underpredicted; (iii) the bioavailability of weak bases was overpredicted; and (iv) the relative bioavailability of both highly hydrophobic compounds and poorly aqueous-soluble APIs was underpredicted [2].

As part of the OrBiTo project, a survey was sent out in May 2015 to 13 large pharmaceutical companies in EFPIA (European Federation of Pharmaceutical Industry Association) and the response rate was 85% [11]. The survey found that biorelevant dissolution-absorption PBPK modeling and simulation were used by 88% of the responding companies in the early drug development process, especially in learning the impact of API particle size and different pharmaceutical dosage forms on intestinal drug absorption. The survey showed that PBPK-based modeling and simulation were used to support salt selection, impact of drug substance polymorphism, and support for low oral bioavailability of a drug substance but not much of these are included in regulatory files [2, 10]. There are interests that with improved in silico tools to predict drug absorption, there will be evolution in the area of biowaivers that will help expedite the development of high-quality products for patients [2, 10].

Li et al. [12] assessed the predictive performance of PBPK modeling in prospectively predicting food effect. Based on literature search and published US FDA reviews (2006–2016) of 48 food effect predictions of 27 compounds (22 bases, 2 acids, 2 ampholytes, and 1 neutral) using PBPK modeling, they concluded that approximately 50% of food effect predictions were predicted within 1.25-fold of the observed effect while 75% were predicted within 2-fold. The authors stated that although 75% of predictions were within twofold of observed data, the ability to draw any conclusion is still limited and there is publication bias in the knowledge base, as only "good" results tend to be published. The data also implied that food effect prediction using PBPK model may be difficult for compounds with a relatively low intestinal absorption ($f_a < 30\%$).

Table 43.3 lists some available software programs to perform PBPK modeling and simulation [3]. Case studies using GastroPlus and Simcyp Simulator will be presented later in the chapter. A PBPK model is composed of three parts: absorption, disposition, and physiology models. The absorption model is developed using *in vitro* experimental data such as intrinsic/pH dependent (biorelevant) solubility, dissolution/release profile in biorelevant media, precipitation time, super-saturation ratio, permeability, carrier-mediated transport, and first-pass metabolism in the gut and liver.

There are various mechanistic oral absorption models used in PBPK modeling, as shown in Figure 43.2. ACAT (advanced compartmental and transit) model is used by GastroPlus, ADAM (advanced dissolution, absorption, and metabolism) model is used by Simcyp

**Table 43.3** A summary of software programs to perform PBPK modeling and simulation.

| Software | Website |
|---|---|
| ADAPT5 | https://bmsr.usc.edu/software/adapt/ |
| ADMEWORKS DDI Simulator | http://www.fqs.pl/en/chemistry/products/ddi-simulator |
| Berkeley-Madonna | http://www.berkeleymadonna.com |
| Cloe PK™ | http://www.cyprotex.com/cloepredict/ |
| GastroPlus | http://www.simulations-plus.com |
| GI-SIM™ | In-house software of AstraZeneca |
| Intellipharm® PK | www.intellipharm.com |
| PK-SIM™ | https://github.com/Open-Systems-Pharmacology/Suite/releases/tag/v9.1 |
| SimBiology/MATLAB® | https://www.mathworks.com/products/simbiology.html |
| Simcyp Simulator | http://www.certara.com/software/simcyp-pbpk/ |
| Stella™ | http://www.iseesystems.com/ |

Source: Adapted from Adapted from Cabrera et al. [3].

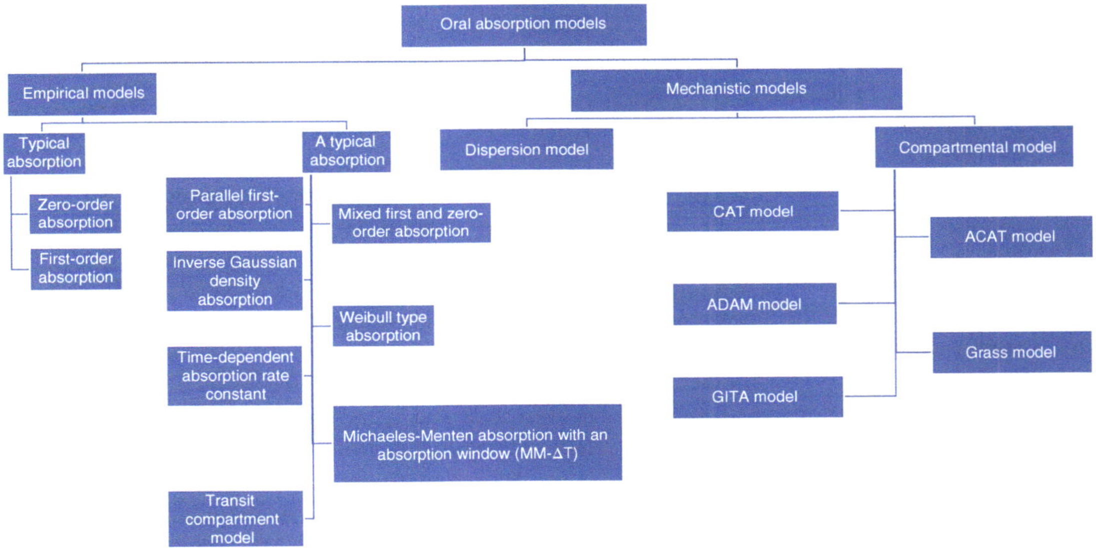


**Figure 43.2** Various oral absorption models. Source: Abuhelwa et al. [13] / with permission of Elsevier.

Simulator, PK-SIM uses dispersion model while Stella uses a self-developed absorption model [14]. We will expand on ACAT and ADAM models in later sections.

## 43.4 PBBM Modeling as a Subset of PBPK Modeling

First-order oral absorption has a long history in PK modeling for describing the process of oral absorption. However, over the years, mechanistic oral absorption models have been increasingly used to replace the oversimplified first-order absorption model to incorporate the potential nonlinear oral absorption processes in dissolution, precipitation, passive permeation, intestinal metabolism, and/or transport. Also, mechanistic absorption models are promising since they enable the assessment of the influence of key GI tract physiological features such as transit time, fluid pH and volume, and their interactions with drug properties.

It has been proposed that the specific part of "mechanistic oral absorption modeling" in PBPK modeling to be brought under the term "physiologically based biopharmaceutics modeling (PBBM)" [15]. Thus, PBBM may be considered as a subset of PBPK. According to Suarez-Sharp et al. [16], PBBM is defined as mechanistic models to mimic physiological conditions and incorporate dissolution information while accounting for relevant physicochemical and physiological factors leading to a prediction of systemic exposure versus time. The main aim of PBBM is to establish an *in vitro–in vivo* correlation (IVIVC) or relationship to support formulation-related applications such as developing a bio-predictive dissolution method to predict clinical performance or support biowaivers, constructing model-informed formulation selection, or defining a safe space to support biowaiver or regulatory flexibility [15, 16].

## 43.5 Applications of PBPK/PBBM Modeling

### 43.5.1 Establishing Safe Space Using Mechanistic Absorption Model

One of the most important applications of mechanistic absorption model such as the ACAT and the ADAM model is to establish safe space of *in vitro* dissolution for ensuring drug product bioequivalence by incorporating drug substance characteristics and mechanistic elements of drug release/dissolution relevant to interactions of the pharmaceutical product with physiological conditions and events [15, 16]. Specifically, the safe space is defined as "boundaries defined by *in vitro* specifications (i.e. dissolution or other relevant drug product quality attributes), within which drug product batches are anticipated to be bioequivalent to one another, or less optimally, but still possible, bioequivalent to the pivotal clinical batch(es)" [16]. It is also expected that the mechanistic absorption models could be used to predict the impact of variations in the critical material attributes (CMAs), critical formulation variables (CFVs), and critical process parameters (CPPs) [15]. However, the typical inputs for the current mechanistic absorption model and PBPK models only account for limited properties of the formulation such as drug substance particle size. It is still challenging to predict the impact of chemistry, manufacturing, and control (CMC) changes on formulation performances by modeling and simulation [15–18].

Benefits of establishing safe space include (i) assuring consistent performance throughout drug product life cycle via product specification setting; (ii) enabling regulatory flexibility; (iii) waiving/reducing the number of clinical studies needed in support of major manufacturing changes; and (iv) providing an opportunity for taking a major step in accelerating drug product development/lowering the cost of drug development [16, 19].

In general, there are four typical approaches to establish safe space (Figure 43.3): (i) bracketing approach; (ii) exposure–response analysis; (ii) *in vitro–in vivo* correlation (IVIVC); and (4) Physiologically based biopharmaceutics model (PBBM) based *in vitro–in vivo* relationship (IVIVR) [16, 19, 20]. Below we will only briefly introduce bracketing and exposure–response analysis approaches while expanding more on IVIVC and PBBM-based IVIVR.

#### 43.5.1.1 Bracketing Approach

If bioequivalence can be demonstrated between strengths or drug product variants with the extremes in dissolution profiles, all *in vitro* dissolution profiles falling within the extremes bioequivalence dissolution bounds are considered to be bioequivalence [21].

#### 43.5.1.2 Exposure–Response Analysis

If the difference in efficacy and safety between strengths or drug product variants with the extremes in dissolution profiles can be demonstrated to be not clinically different even though their bioequivalence studies fail to meet the standard bioequivalence acceptance intervals of 80–125%. All *in vitro* dissolution profiles falling within the extremes dissolution bounds are considered not to cause clinically significant differences [22].

#### 43.5.1.3 *In Vitro–In Vitro* Correlation (IVIVC) (see also Chapter 16)

Given that IVIVC is covered by a separate chapter in this book and is not the main topic of the present one, we will only briefly introduce level A IVIVC here. According to the United States Pharmacopeia (USP), conventional level A IVIVCs are defined as point-to-point relationships between *in vitro* dissolution and the *in vivo* input rate of the drug from the dosage form [17, 23]. A two-step conventional IVIVC approach that includes deconvolution and then convolution has been widely used. First, deconvolution methods such as Wagner–Nelson, Loo-Riegelman, or general numerical methods are used to estimate the *in vivo* systemic input–time profile based on a plasma concentration–time profile after intravenous or immediate release (IR) formulation administration. Then, a correlation between the *in vitro* release profile and the estimated *in vivo* systemic input profile is formed to build IVIVC [17, 18]. The developed IVIVC could be used to predict oral exposure based on the *in vitro* dissolution profile using convolution. The disadvantage of the conventional IVIVC is that they rely on the assumption of linear pharmacokinetics. However, very often, the absorption is not a linear process [17, 18].

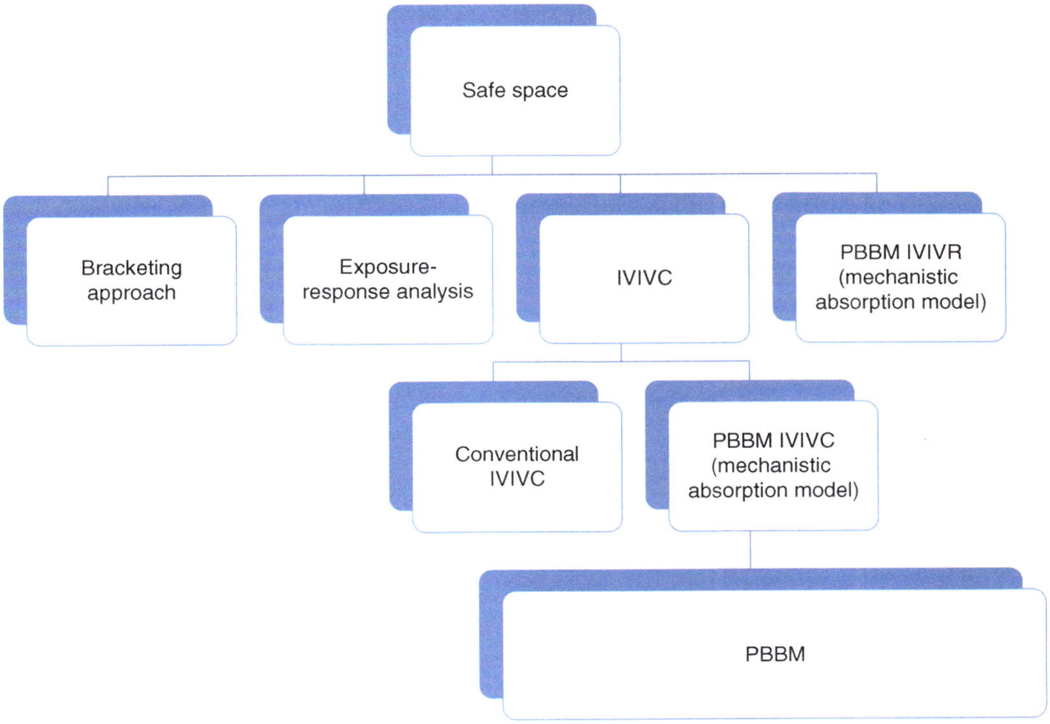


**Figure 43.3** Approaches for building "safe space."

Thus, an advanced absorption model using mechanistic modeling of drug release, dissolution, and permeation across the intestinal membranes is preferred for estimating the *in vivo* absorption from observed *in vivo* plasma–time profiles [17, 18].

#### 43.5.1.4 PBBM Based IVIVR

IVIVR is also an effort to link *in vitro* dissolution and *in vivo* dissolution/absorption. In many publications, IVIVC and IVIVR are used interchangeably, especially PBBM-based IVIVC and PBBM-based IVIVR. However, according to Sandra Suarez-Sharp et al. [16], IVIVR is a less quantitative approach than IVIVC, which has no regulatory application unless a safe space is established (e.g. when combined with bracketing approach).

Currently, compared with PBBM-based IVIVR, IVIVC typically gives higher regulatory flexibility and wider safe space [16]. However, PBBM-based IVIVR holds the promise to predict *in vivo* dissolution directly using *in vitro* dissolution-derived parameters. A case study of PBBM-based IVIVR successfully supporting regulatory submission of lesinurad will be shown in the later section. When developing PBBM-based IVIVR, tabulated *in vitro* dissolution profile, Weibull function or Z-factor fitted to the *in vitro* dissolution profile, effective particle size, etc. could be used as dissolution-derived parameters inputs [15, 17, 19, 20].

***43.5.1.4.1 Tabulated In vitro Dissolution Profile or Weibull Function Fitted to the In Vitro Dissolution Profile*** Tabulated *in vitro* dissolution profile could be directly incorporated as the *in vivo* dissolution profile. Or the *in vitro* dissolution profile could be fitted by using the Weibull function and the resulting Weibull function could be used to simulate *in vivo* dissolution.

$$\%\ \text{Dose Released} = \text{Max} \times \left(1 - f_1\left[\frac{-(t-T)^{b_1}}{A_1}\right]\right) \tag{43.1}$$

$$\%\ \text{Dose Released} = \text{Max} \times \left(1 - f_1\left[\frac{-(t-T)^{b_1}}{A_1}\right] - f_2\left[\frac{-(t-T)^{b_2}}{A_2}\right]\right) \tag{43.2}$$

$$\%\ \text{Dose Released} = \text{Max} \times \left(1 - f_1\left[\frac{-(t-T)^{b_1}}{A_1}\right] - f_2\left[\frac{-(t-T)^{b_2}}{A_2}\right] - f_3\left[\frac{-(t-T)^{b_3}}{A_3}\right]\right) \tag{43.3}$$

where *% Dose Released* is the percentage of dose released; Max is the maximum amount to release; $T$ is the lag time; $f_i$ is the fraction of the maximum amount to release; $A_i$ is the time scaling factor (SF); and $b_i$ is the shape factor. In GastroPlus, there are three options to fit Weibull function: single-Weibull, double-Weibull, or triple-Weibull function.

Both approaches (tabulated *in vitro* dissolution profile and Weibull function) are essentially the same for integrating *in vitro* dissolution data to simulate *in vivo* dissolution. The assumption behind the two approaches is that the rate and extent of *in vivo* dissolution is the same as the *in vitro* dissolution. Both approaches are not mechanistic because the *in vitro* and *in vivo* dissolution conditions such as pH, media volume, temperature, and surfactants are typically very different considering currently available *in vitro* dissolution assays and cannot mimic *in vivo* dissolution conditions well [24, 25]. The difference between the two approaches is that the Weibull function provides a more continuous description of the *in vitro* dissolution than the tabulated *in vitro* dissolution profile.

***43.5.1.4.2 Z-factor Model*** The *Z*-factor model is a semi-mechanistic dissolution model based on variations of the dissolution equation derived by Noyes and Whitney [26]. It was described by Nicolaides et al. in 2001 [27] and Takano et al. in 2006 [28]. In the *Z*-factor model, the term $\frac{3D_w}{\rho r T}$ in the Nernst–Brunner model is replaced by a *Z* parameter fitted to the *in vitro* dissolution profile [28, 29]:

$$\frac{dM_D}{dt} = zM_{u,0}\left(\frac{M_{u,t}}{M_{u,0}}\right)^{2/3}(C_s - C_l) \tag{43.4}$$

where $D_w$ is the diffusion coefficient; $C_s$ is the drug solubility in the lumen compartment (could be affected by pH and bile salt concentration); $C_l$ is the drug concentration in the lumen compartment; $\rho$ is the density of the dissolving drug particles (after disintegration); $r$ is the spherical particle radius of particles in the compartment (initial or current depending on user selection); $T$ is the temperature; and $M_u$ and $M_D$ are the undissolved and dissolved amount of drug, respectively; and $M_{u,0}$ and $M_{u,t}$ are the undissolved amount of drug at time point of 0 and $t$, respectively.

Because *Z*-factor represents the $\frac{3D_w}{\rho r T}$ term in the Nernst–Brunner model, the *Z*-factor model assumes a spherical particle shape and uniform particle size. Since both $\rho$ and $r$ are features of a specific formulation, the *Z*-factor is a formulation-dependent parameter [29]. When obtaining the value by fitting across multiple dissolution profiles, only the profiles obtained from exactly the same formulation (i.e. the same formulation compositions and the same particle size of the API and excipients) can be used. Theoretically, the dose amount, dissolution volume, and type of dissolution media (affecting the solubility of API) do not affect the *Z*-factor value. However, if the solubility is pH dependent, *Z*-factors at various pH across the physiological range should be experimentally measured and the *Z*-factor-pH profile is incorporated into the model.

Complete immediate disintegration of the formulation is another assumption behind the *Z*-factor model because $M_{u,0}$ in the *Z*-factor model represents the complete dose of an API used in the *in vitro* dissolution experiment. This assumption may be reasonable for drugs formulated as powders, suspensions, and possibly even granules which have no significant disintegration time and the complete dose is immediately available for dissolution. But it should be used cautiously for solid oral dosage forms such as tablets and capsules when disintegration could significantly impact the mass available for initial dissolution [26]. Many solid oral dosage forms, including tablets and capsules, may exhibit a sigmoidal dissolution profile shape and the *Z*-factor model usually does not adequately capture the measured profiles. This could be due to the inappropriate use of the *Z*-factor model without taking into account the initial disintegration process [26]. In these cases, it is not appropriate to fit the *in vitro* dissolution data using the traditional *Z*-factor model. Instead, Hofsäss and Dressman [26] proposed a *Z*-factor estimation approach considering both the dosage form disintegration and particle dissolution as two processes, independently. This novel approach may yield a more realistic estimation of the *Z*-factor for dosages with a significant disintegration process.

***43.5.1.4.3 Effective Particle Size*** API particle size is the only formulation property that could be directly incorporated in the current PBPK/PBBM models. It is impossible to assess the influence of CMC changes and the resulting formulation performance with the current model [15].

The more recent proposed product particle size distribution (P-PSD) approach is promising to improve the prediction of *in vivo* dissolution since it incorporates effects such as product disintegration, particles wettability, and actual surface area available for dissolution rather than using API particle size [17, 24, 25, 30, 31]. P-PSD is the effective particle size fitted to the *in vitro* dissolution data [24]. The fitted P-PSD must be verified to be able to explain the *in vitro* dissolution at various tested *in vitro* dissolution conditions, including the ones throughout the physiological pH range and/or in the presence of natural or synthetic surfactants, to verify the fitted particle size values [24]. The use of P-PSD, instead of the API particle size, that incorporates formulation effects such as product disintegration, particles wettability, and actual surface area available for dissolution is believed to be more mechanistic and allows more accurate *in vivo* dissolution prediction [17, 24, 25].

Once a PBBM-based IVIVR model is developed, it may be used to predict the impact of variations in the CMAs, CFVs, CPPs, and the establishment of a safe space (i.e. the formulation or manufacturing space) [15]. Typically, virtual bioequivalence simulation is performed to take into account the influence of intersubject variabilities.

## 43.6 PBPK Software

### 43.6.1 Unique Features of GastroPlus

#### 43.6.1.1 ACAT Model

Yu and Amidon developed the original compartmental absorption and transit (CAT) model for estimating the fraction of dose absorbed and the rate of drug absorption for passively transported drugs in immediate-release products in the late 1990s [32]. The CAT model describes the drug transit in the GI tract by dividing the GI tract into a set of seven small intestinal compartments. The first version of GastroPlus was based on the work of Yu and Amidon and was established in August 1998 [33]. However, the original CAT model does not consider the influence of the finite rate of dissolution, the pH dependence of the solubility of drugs, controlled release (CR), absorption in the colon, gut metabolism, carrier-mediated transport, and many other factors that could significantly affect drug absorption within the intestinal tract. To improve the prediction of oral absorption, an advanced compartmental absorption and transit (ACAT) model was developed by Simulations Plus, Inc. in 2001. The ACAT model describes the drug transit and absorption in the GI tract by a series of nine compartments as shown in Figure 43.4.

Each of the nine GI tract compartments in the ACAT model is divided into three sub compartments: "undissolved," "dissolved," and "enterocyte" to describe undissolved, dissolved, and absorbed drug molecules in the GI tract [29]. One additional sub compartment of "unreleased" is only used for CR dosage forms. Herein, we take the CR dosage form as an example to further illustrate how the ACAT model describes drug oral absorption. After oral administration, the drug in a CR formulation can be released into solution (dissolved) or as solid particles (undissolved). Undissolved drug molecules may undergo passive permeation, carrier-mediated transport, and lumenal degradation in the gut lumen. Both undissolved and dissolved drugs undergo first-order transit through the GI tract. Unabsorbed drug in the colon before the end of the transit process is eventually excreted. Once drug molecules are absorbed into enterocytes, they may undergo metabolism, passive/active transport into the portal vein, apical efflux transport back to the gut lumen, and lysosomal trapping inside the enterocytes. All the abovementioned processes could be incorporated into the ACAT model.

The original ACAT model was developed with constant volumes for each intestinal compartment. Dynamic fluid volume ACAT model was developed and added in GastroPlus recently to simulate changes in intestinal volume based on fluid intake, absorption, and secretion [29, 33]. The model assumes constant fluid secretion, first-order water absorption, and first-order emptying rates between compartments. However, dynamic fluid volume ACAT model is only available for fasted state simulation since no reliable dynamic fluid data obtained under fed state has been published.

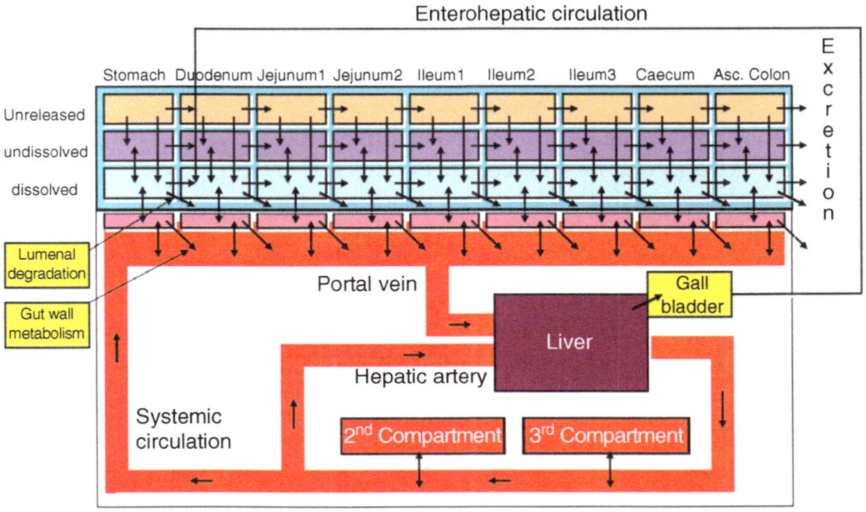


**Figure 43.4** A schematic graph of the advanced compartmental absorption and transit (ACAT) model [13, 29]. Source: Adapted from GastroPlus user manual by Simulations Plus, Inc.

Given that variations in surface-to-volume ratio, pH, paracellular absorption, and other absorption-rate-determining effects differ from one GI tract region to another, the drug $P_{eff}$ (effective permeability) value for each GI tract compartment varies from one region to another. Ideally, regional $P_{eff}$ value for each compartment should be used in the simulation. However, regional $P_{eff}$ is not available in most cases. In the ACAT model, the effective absorption rate coefficient in each compartment is the product of a primary $P_{eff}$ and the absorption scale factor (ASF) corresponding to that compartment. Typically, the primary effective permeability is converted from *in vitro* Caco-2 or MDCK apparent permeability or measured jejunum permeability from animal studies. Then, the primary $P_{eff}$ is scaled to the regional $P_{eff}$ to account for the variations in GI tract regions.

GastroPlus has eight options for ASF model: User-Defined, Gradient Multiplier, Gradient Additive log *D* Model, Opt log *D* Model, Opt log *D* Model SA/V, Opt log *D* Model SA/V 6.1, Theoretical, and Theoretical SA/V. ASFs are calculated automatically from the C1–C4 coefficients generated by GastroPlus when the selected ASF Model is Opt log *D* Model, Opt log *D* Model SA/V, Opt log *D* Model SA/V 6.1, or log *D* Model [29]. Herein, we only introduce the Opt log *D* Model SA/V 6.1 model as it is the default option.

The Opt log *D* Model SA/V 6.1 model assumes that increased ionization results in decreased permeability [29, 33]. It incorporates a log *D*-dependent calibration of ASFs based on the pH-partition theory. When using the Opt log *D* Model SA/V 6.1 model, the ASF for each GI tract regional compartment (except the colon compartment) is calculated by

$$\mathrm{ASF} = A \times C_2 \times 10^{\left(C_1 \times \frac{\Delta \log D_{pH} - C}{\Delta \log D_{6.5} - C}\right)}. \tag{43.5}$$

where $C_1$ and $C_2$ are two fitted constants that regulate the degree to which log *D* can influence ASF for the small intestine; *C* is another fitted constant to avoid the singular condition; $\Delta \log D_{\mathrm{pH}}$ is the difference between log *D* at the corresponding regional compartment pH and log *P*; $\Delta \log D_{6.5}$ is the difference between log *D* at pH 6.5 and log *P*; parameter *A* is equal to $2/R$*SEF for all intestinal compartments. Parameter A represents the surface area-to-volume ratio for each compartment and is based on the known radius of each intestinal compartment. SEF is the surface area enhancement factor. It accounts for the higher density of microvilli in the upper small intestine compared to the distal sections.

The ASF for the colon compartment is calculated by

$$\mathrm{ASF} = A \times C_3 \times 10^{C_4 \times \log D} \tag{43.6}$$

where $C_3$ and $C_4$ are fitted constants that regulate the degree to which log *D* can influence colonic ASF. $C_1$, $C_2$, $C_3$, and $C_4$ are four constants optimized against fraction absorbed in human for a diverse set of compounds of a GastroPlus internal library for which experimental parameters of the most relevant inputs (solubility, log *P*, and permeability) were available. However, GastroPlus users may need to calibrate the ASFs for their own drugs given that $C_1{\sim}C_4$ are drug-specific parameters and the "best compromise" to compounds in the GastroPlus internal library may not accurately describe the absorption of users' particular compounds. To modify $C_1{\sim}C_4$ individually the users need to set the ASF Model to "User-Defined."

#### 43.6.1.2 IVIVCPlus Module

In GastroPlus, IVIVC could be developed using the IVIVCPlus module. The IVIVCPlus module includes conventional deconvolution methods such as Wagner–Nelson, Loo-Riegelman, and numerical methods [29]. However, different from traditionally defined numerical methods allowing a wide range of input functions such as polyexponential and polynomial functions, Weibull function with one, two, or three terms is specifically applied for deconvolution when the numerical method is used.

In addition, a state-of-the-art mechanistic deconvolution method using the full GastroPlus ACAT model could be used when building IVIVC for oral formulations [29]. Different from conventional IVIVC based on a correlation between *in vitro* release profile (fraction *in vitro* release) and estimated *in vivo* systemic input (fraction absolute bioavailability), the IVIVC using the mechanistic deconvolution method is built based on *in vitro* release profile (fraction *in vitro* release) and estimated *in vivo* release profile (fraction *in vivo* release).

Moreover, while the IVIVC is traditionally referred to and applied in oral dosage formulations only, GastroPlus with mechanistic deconvolution could accommodate the development of IVIVC using intramuscular model for intramuscular CR dosage form and the full GastroPlus TCAT model for subcutaneous CR and transdermal reservoir patch formulations [29].

#### 43.6.1.3 Lysosomal Trapping in Absorption and Drug Distribution

Lysosomes are single-membrane intracellular organelles containing at least 60 hydrolases within an acidic environment with a low internal pH of 4–5 [34]. Lysosomes play a key role in various metabolic processes including recycling defective organelles and breaking down of macromolecules. The lower lysosomal pH compared

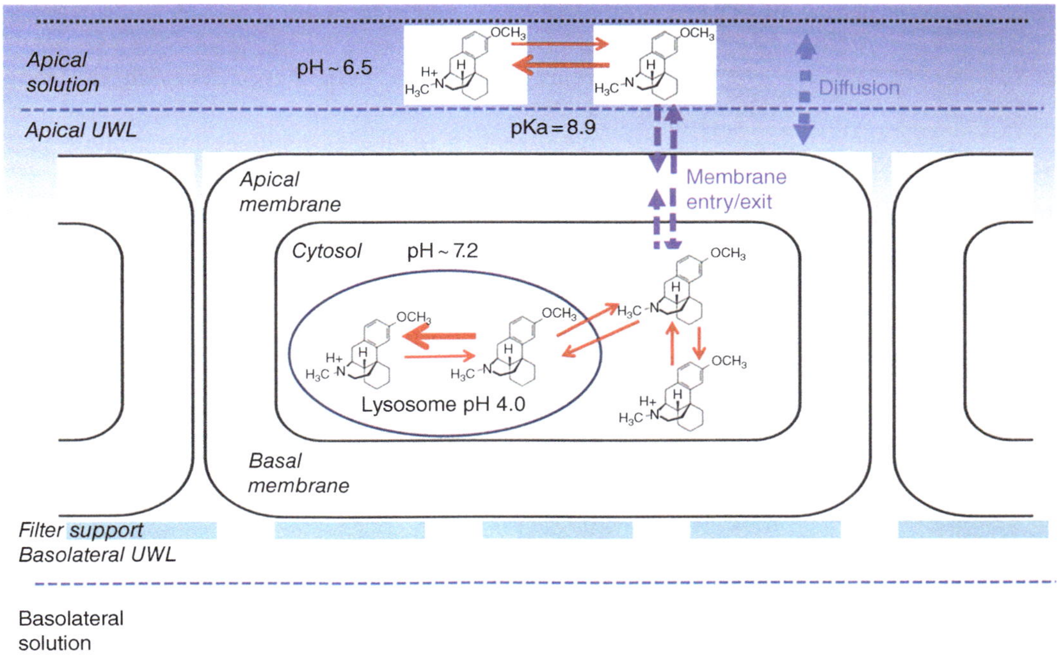


**Figure 43.5** The schematic graph shows the mechanism of lysosomal trapping of dextromethorphan in the enterocytes. Source: Bolger et al. [35] / with permission of Elsevier.

with the cytosol pH creates a pH gradient responsible for the sequestration of certain types of amine-containing molecules, a process known as lysosomal trapping (Figure 43.5) [34–38]. It is believed that lipophilic weak bases with a log $P > 2$ and $pK_a > 6.5$ are the ones likely to undergo lysosomal trapping [35].

Lysosomal trapping could significantly affect drugs' intestinal absorption and distribution [35]. Lipophilic weak bases could readily diffuse across the GI lumen (pH ~ 6.5) to the enterocyte cytosol (pH ~ 7.2). However, after diffusion into the acidic lysosomes in the cytosol, the lower pH in the lysosomes results in protonation of the bases to increase their polarity and reduce the back-diffusion of entrapped molecules from the lysosomes to the cytosol. This results in a longer dwell time and an overall slower rate of entry of the drug from the enterocytes into the portal vein. Thus, lysosomal trapping could prolong the intestinal absorption of lipophilic weak bases and cause longer $T_{max}$ than expected. In GastroPlus, the lysosomal trapping could be incorporated by lowering the value of the parameter of unbound fraction in the enterocyte ($f_{u,ent}$). This parameter may be predicted by MembanePlus™. However, very often, it has to be fitted to observed *in vivo* concentration–time profiles due to limited quantitative understanding of the process.

Similarly, the lipophilic weak bases could suffer from lysosomal trapping in the organs with an abundance of lysosomes (e.g. liver, kidney, spleen, and lung). It is anticipated that these lipophilic weak bases have higher tissue-to-plasma partition coefficient due to the higher extent of lysosomal trapping and so forth resulting in a higher volume of distribution.

#### 43.6.1.4 Case Studies

***43.6.1.4.1 Case Study 1: Establishing Safe Space of* In Vitro *Dissolution Using IVIVR for Lesinurad [24]*** Lesinurad (Zurampic®) (Figure 43.6) is a selective uric acid reabsorption inhibitor for treating high blood uric acid levels associated with gout. It is administered orally as an immediate-release tablet. Lesinurad is a weak acid classified as a BCS class II drug with high passive

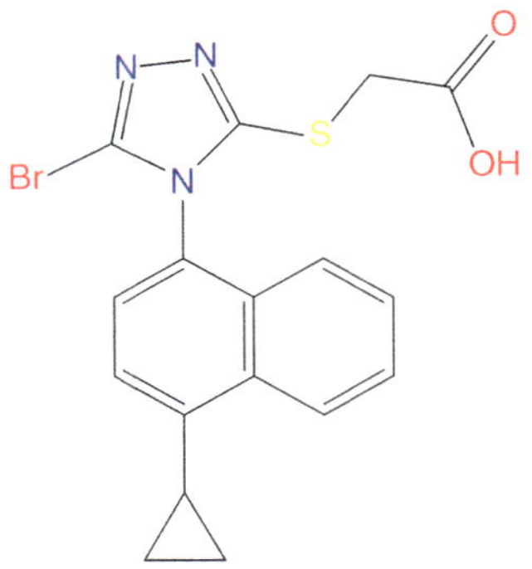


**Figure 43.6** Structure of lesinurad.

**Table 43.4** Key physicochemical and pharmacokinetic parameters for lesinurad.

| Parameter | Value |
|---|---|
| Molecular weight | 404.3 |
| log $P$ | 2.85 |
| p$K_a$ | 3.2 (acidic) |
| pH 4 buffer (citrate 25 mM, $I = 0.1$ M), final pH 4.0 | 0.045 mg/ml |
| pH 5 buffer (acetate 25 mM, $I = 0.1$ M), final pH 5.1 | 0.60 mg/ml |
| pH 6 buffer (citrate 22 mM, $I = 0.1$ M), final pH 5.9 | 5.2 mg/ml |
| pH 6.5 buffer (phosphate 25 mM) + NaOH, final pH 6.0 | 17 mg/ml |
| pH 6 buffer (citrate 22 mM) + NaOH, final pH 5.9 | 43 mg/ml |
| FaSSIF pH 6.5, final pH 5.6 | 3.2 mg/ml |
| FeSSIF pH 5.0, final pH 5.0 | 1.7 mg/ml |
| SGF pH 1.6 (HCl 30 mM, $I = 0.1$ M), final pH 1.5 | 0.0061 mg/ml |
| FaSSIF pH 6.5 + NaOH, final pH 5.9 | 43 mg/ml |
| calc. jejunal $P_{eff}$ | $3 \times 10^{-4}$ cm/s |
| $f_{u,p}$ (unbound fraction in plasma) | <2% |
| $R_{bp}$ (blood to plasma ratio) | 0.55 |
| $F$ (oral bioavailability) | 101% |
| Total CL (clearance) | 5.98 l/h |
| $CL_h$ (0.687 of total CL) | 4.11 l/h |
| $CL_r$ (0.313 of total CL) | 1.87 l/h |
| $V_{ss}$ | 20.3 l |

permeability and pH-dependent solubility. It has limited solubility in the stomach under acidic conditions but rapid and complete dissolution in intestinal conditions at higher pH (close to 7). Clinical studies showed that it is quickly and practically completely absorbed from the gut. Table 43.4 listed key physicochemical and pharmacokinetic parameters for lesinurad.

During the review of the marketing applications, it was proposed that drug product batches with a dissolution specification of $Q$ = 80% in 30 minutes are anticipated to be bioequivalent to the clinical reference batch. A PBPK model developed using GastroPlus version 9.0.0007 (Simulations Plus, Inc. Lancaster, CA) was submitted to US FDA to support the proposed batch quality control strategy [24].

A three compartment with ACAT model was developed for simulating and predicting lesinurad PK in human. Data from the absolute bioavailability study (Study RDEA-594-131) and human ADME study (Study RDEA-594-112) were used for model development (Table 43.5). Because of high intersubject variability in the pharmacokinetics of RDEA-594-131, individual models for each subject enrolled in Study RDEA-594-131 were constructed by fitting their corresponding plasma concentration–time profiles after intravenous and oral administration (Figure 43.7).

The three compartmental model was obtained by fitting the individual pharmacokinetic data from Study RDEA-594-131 after intravenous administration to describe the disposition of lesinurad in human. Additionally, the fractions of hepatic and renal elimination were obtained from Study RDEA-594-112. Plasma concentration–time profiles from Study RDEA-594-131 after oral administration were used to fit individual jejunal $P_{eff}$ and gastric emptying patterns after incorporating the corresponding individual three compartment model fitted to the IV data from the same individual.

Four options were used to integrate dissolution when developing the ACAT model:

(A) particle size distribution fitted to the *in vitro* dissolution data of batch 12A015. The fitted particle size distribution with 10 bins was verified with the *in vitro* dissolution data of batch ELAB. The calculation and parameter estimation method was shown in the Supplementary Material of the reference [24]. The results showed that it can predict observed *in vitro* dissolution rate in dissolution media with various pH (3–5) and with or without surfactant (1% SDS) (Figure 43.7B).
(B) single-Weibull function fitted to the *in vitro* dissolution data of batch 12A015. The dosage form is switched from IR to controlled release dissolved (CRD) assuming released drug particles are dissolved.
(C) $Z$-factor fitted to the *in vitro* dissolution data of batch 12A015. The results showed that the fitted $Z$-factors did not adequately capture the measured profiles.
(D) single-Weibull function fitted to the *in vitro* dissolution data of batch 12A015. The dosage form is switched from IR to controlled release undissolved (CRU) assuming released drug particles are undissolved.

All four options for integrating the *in vitro* dissolution data were explored. Model options A, B, and C all seem to predict reasonably well the observed $C_{max}$ in Study RDEA-594-131. Model A was selected as the best model because of the following reasons: first, most importantly, among all four options, option A recovered the observed bioinequivalence in RDEA-594-129 the best. Second, the fitted $Z$-factors in model C did not adequately capture

**Table 43.5** Clinical studies used for model development and validation for lesinurad.

| Study no. | Study type | Batch no. for oral administration | Study details including dose and regimen |
|---|---|---|---|
| RDEA-594-131 | Absolute bioavailability study | 12A015, 400 mg | An open-label study in 10 healthy adult volunteers.<br>Oral administration with a concomitant intravenous micro dose of $^{14}$C-lesinurad. |
| RDEA-594-112 | Human ADME study | — | Oral solution administration of 600 mg dose of $^{14}$C-lesinurad in 6 healthy adult male volunteers under fasted conditions |
| RDEA-594-129 | Relative bioavailability study | 12A015, 400 mg;<br>ELAB, 400 mg. | A randomized, open-label crossover study in 18 healthy adult volunteers.<br>Batch ELAB was compared to 12A015<br>Oral administration under fasted condition |

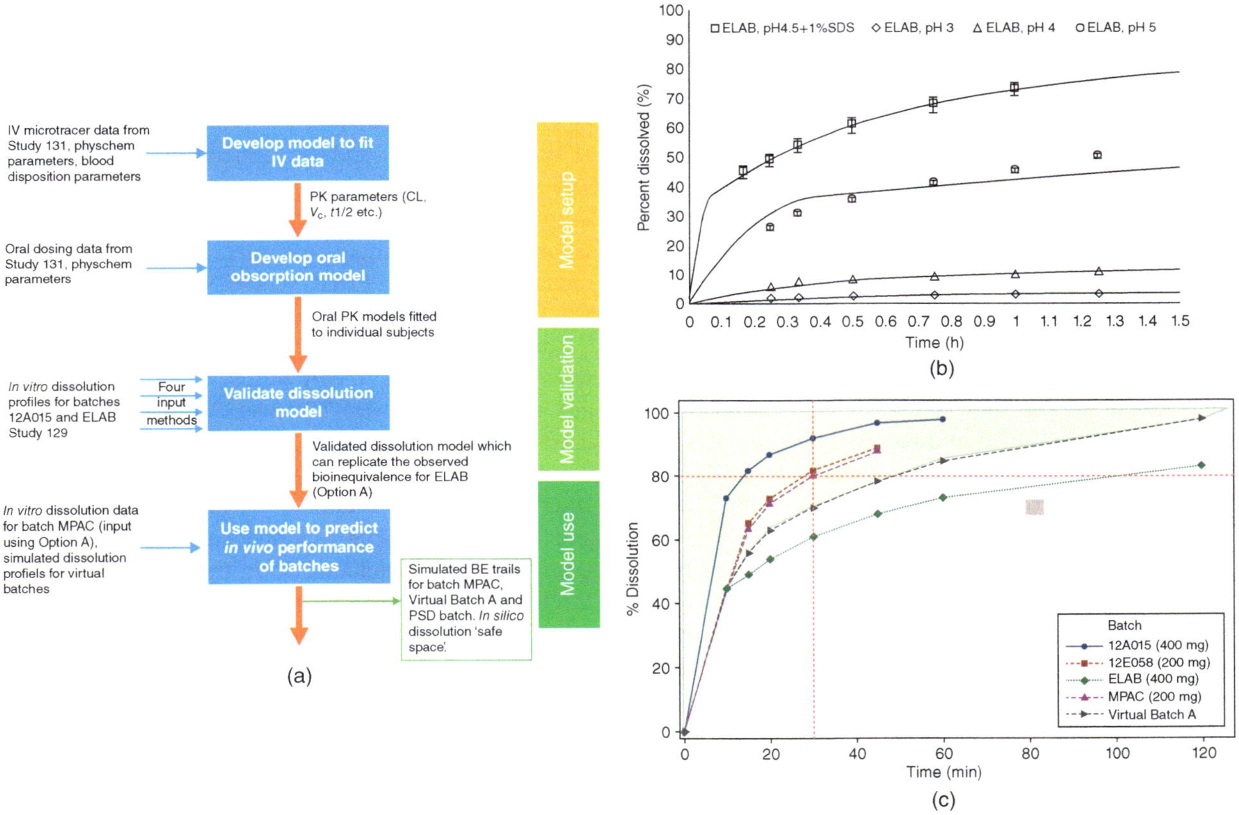


**Figure 43.7** (a) Modeling strategy. (b) verification of fitted particle size distribution derived from QC dissolution method profile (pH 4.5 + 1% SLS) using *in vitro* dissolution data obtained at pH 3, pH 4, and pH 5. (c) "safe space" for anticipated bioequivalence to lesinurad IR tablets [24]. Source: Pepin et al. [24] / American Chemical Society.

the measured profiles. Third, models B and D directly use *in vitro* dissolution data, assuming the same rate and extent of dissolution in both *in vitro* and *in vivo* situations, and is not mechanistic.

Then, the developed model with option A was used to justify the proposed dissolution specification of $Q = 80\%$ in 30 minutes for establishing bioequivalency to the clinical batch of 12A015 and explore the "safe space." Batch MPAC has a slower dissolution profile than batch 12A015 and its percent of dissolution in 30 minutes is right about 80%. This batch was used to show bioequivalence is anticipated if the percent of dissolution in 30 minutes is higher than 80%. A virtual crossover trial simulation was performed for batch MPAC and 12A015 with twenty-five virtual subjects. The virtual population simulation showed that the geometric mean ratios for both AUC and $C_{max}$ are very close to one. Thus, batch MPAC is anticipated to be bioequivalent to batch 12A015.

Furthermore, a virtual batch A was generated to search for an *in vitro* dissolution profile that would keep the $C_{max}$ exposure ratio to 12A015 between approximately 1 and 0.9 (which would be expected to pass an adequately powered bioequivalence study) to find the "safe space." Another virtual crossover trial simulation was performed for batch A and 12A015. Again, the geometric mean ratios for $C_{max}$ and AUC are close to one. Although it is not exactly the edge of the "safe space" for defining "pass" or "failure" of bioequivalence using *in vitro* dissolution data, it provides additional confidence in the proposed specification of $Q = 80\%$ in 30 minutes since this specification is well within the "safe space" shown by the virtual batch A (Figure 43.7c).

***43.6.1.4.2 Case Study 2: A PBPK Model of Dextromethorphan with Lysosomal Trapping to Explain the long $T_{max}$ and Investigate the Sensitivity of Dissolution Rates on its PK [35]*** Dextromethorphan (DEX) is a synthetic noncompetitive *N*-methyl-D-aspartate receptor antagonist widely used as an over-the-counter medicine for cold and cough. It has high solubility and permeability throughout the GI tract and has complete oral absorption. However, DEX has an unusually long $T_{max}$ of ~2.5 hours after oral administration as a BCS class I drug.

DEX has the identified common properties of lysosomotropic agents mentioned earlier, such as a log $P > 2$ (log $P = 3.97$) and a basic p$K_a$ between 6.5 and 11 (basic p$K_a = 8.91$); therefore, it could be sequestered in the lysosomes during the absorption process resulting in delayed $T_{max}$ (Figure 43.5). In addition, its lysosomal trapping potential has been confirmed by *in vitro* cell-based assays [38]. It was proposed that the high lysosomal trapping in the enterocytes caused the long $T_{max}$ of DEX. A PBPK model using GastroPlus version 9.0 (Simulations Plus, Inc. Lancaster, CA) was constructed to demonstrate how lysosomal trapping may explain this unusual phenomenon [35].

Bolger et al. [35] showed that the *in vitro* and *in vivo* dissolution rates, which is a common sensitive factor in affecting bioequivalence, do not have significant impacts on the bioequivalence of DEX by developing the PBPK model for DEX. Before developing the PBPK model, Bolger et al. [35] evaluated whether lysosomal trapping could have a significant influence on the intracellular distribution of DEX by using the program MembanePlus (Simulations Plus, Inc., Lancaster, CA) with the simulations of the *in vitro* Caco-2 permeability assay. The simulated accumulation of DEX with lysosomal pH = 4.0 (physiological pH) was then compared to the simulated accumulation of DEX with lysosomal pH = 6.5. The simulation results showed the simulated lysosome concentration is ~4 orders of magnitude higher than the cytoplasm concentration at a lysosomal pH = 4.0. On the other hand, the concentration in the lysosomal compartment is reduced to ~5-fold of that in the cytoplasm with the lysosomal pH = 6.5. The simulation results supported the hypothesis of significant lysosomal trapping of DEX.

The primary metabolic enzyme of DEX is CYP2D6 enzyme, which is highly polymorphic. Its rate and extent of metabolism are very different in extensive (EM) and poor (PM) metabolizers. The metabolite of DEX, dextrorphan (DXO), undergoes rapid phase II metabolism to yield the glucuronide metabolite. The PBPK model was developed for both the parent drug DEX and its metabolite DXO. The clearance, volume of distribution, and oral absorption components of the DEX PBPK model were developed as described below.

The metabolism of DEX by CYP2D6 and CYP3A4 was accounted by incorporating both enzymes in gut and liver. *In vitro* CYP3A4 $K_m$ and $V_{max}$ were used for CYP3A4 while the $V_{max}$ for CYP2D6 was fitted to match the literature reported DEX and total DXO plasma concentration–time profiles for both EM and PM subjects. The reported concentration of the DXO metabolite includes both free and conjugated DXO in most PK studies of DEX and the observed PK of the total metabolites corresponds best to the physicochemical properties of the glucuronide rather than the free DXO. Thus, the clearance of total DXO metabolites (DXO and DXO-O-glucuronide) was estimated from the renal clearance of DXO-*O*-glucuronide conjugate.

Besides delayed oral absorption, lysosomal trapping could affect the distribution of DEX. Using the GastroPlus default mechanistic Lukacova method with the experimental blood-to-plasma ratio ($R_{bp}$) usually underpredicts the volume of distribution for lysosomotropic agents due to lysosomal trapping in tissues, for example, in liver and lung. A novel tissue partition coefficients ($K_{ps}$) calculation method of Lukacova method with lysosomal trapping was introduced, starting from GastroPlus version 9.7. Unfortunately, this option was not available to be used for the DEX model. Instead, the tissue partition coefficients ($K_{ps}$) for DEX were predicted using the GastroPlus default mechanistic Lukacova method with adjustment of the blood-to-plasma concentration ratio for DEX to 1.65 to match for the observed increased volume of distribution at steady state ($V_{d,ss}$). $R_{bp}$ was adjusted to scale $K_{ps}$ to take into account of lysosomal trapping because $R_{bp}$ is commonly used as a substitute to account for the unknown interactions between drug molecules and acidic components of tissue cells when predicting $K_{ps}$. Once $K_{ps}$ were scaled up, $R_{bp}$ was changed back to the experimental value since the experimental $R_{bp}$ value is needed for the correct calculation of blood clearance.

The Poulin and Theil extracellular method for $K_p$ calculation was applied to get $V_{d,ss}$ of the total DXO metabolite. To incorporate the effects of lysosomal trapping, the unbound drug in the enterocytes ($f_{u,ent}$) was reduced to 1.3% from the default 100%. This reduction effectively decreased the rate of mass transfer from inside the enterocytes across the basolateral membrane into the portal vein.

The simulated result of the developed model not only matched the observed data of parent compound DEX but also its primary metabolite, DXO. The simulated result matches the observed plasma concentration–time profile after a single dose of 30 mg in both EM and PM populations very well with a reduced $f_{u,ent} = 1.3\%$ (Figure 43.8). However, the model overpredicted the $C_{max}$ of DEX and DXO and yielded a much shorter $T_{max}$ when simulated with the default $f_{u,ent}$ of 100% (assuming no lysosomal trapping in the enterocytes).

The developed model was used to test the sensitivity of dissolution rate on bioequivalence for DEX in EM subjects. Starting with the 60 mg solution formulation and then changing to tablet formulations with artificial monodisperse particle sizes of 25, 50, and 100 μm, the bioequivalence of each formulation was tested to the clinical data from Silvasti et al. [39] The simulated PK parameters including $C_{max}$ and AUC of DEX fell within 0.8–1.25 for all formulations tested.

The model suggested that the dissolution and enterocyte absorption of DEX are rapid, but it enters portal vein and plasma slowly due to lysosomal trapping. The authors suggested that the *in vitro* dissolution of DEX tablets is irrelevant in setting product specifications for drugs like DEX that are subject to a high extent of lysosomal trapping.

### 43.6.2 Unique Features of Simcyp Simulator

ADAM model is implemented in Simcyp population-based ADME simulator. Simcyp Ltd was formed as a spinout company from the University of Sheffield in 2001. ADAM model is structured on the CAT and ACAT model and uses the modified Wang and Flanagan method to model dissolution instead of the Noyes–Whitney equation [40, 41]. Different from the CAT model but like the ACAT model, the ADAM models consider more types of formulations such as solid dosage IR and solid dosage CR; the impact of pharmaceutical and physicochemical issues in each GI tract section can be considered in the ADAM models, for example, disintegration, de-aggregation, dissolution, solubility, and supersaturation/precipitation; in the ADAM models, the user can specify the particle size characteristics. It is noted that in the enterohepatic recirculation model, the ADAM model can only be used in PKPD profiles mode and for compounds loaded in the substrate, substrate primary metabolite, and inhibitor position [42].

Simcyp was first designed for mechanistic prediction of drug metabolism which then expanded to predict drug absorption. The Simcyp population-based ADME Simulator can be used for mechanistic modeling and simulation for drug disposition in healthy and special populations (pediatrics, obese, renally impaired, etc.) and for prediction of the extent of drug–drug interactions (DDIs). The simulator combines the relevant physicochemical properties of drug and formulation with physiological, demographic, genetic information, and experimental data collected from *in vitro* cellular system to predict *in vivo* pharmacokinetics profiles and parameters. The simulator can predict outcomes across populations rather than in an average individual so the characteristics of individuals at the extremes of risk can be evaluated. Unlike GastroPlus, Simcyp incorporates *a priori* estimates of interindividual variability based on physiological parameters. For example, a population mean and interindividual variability of bile acid concentrations and GI pH in the fasted and fed states are considered, and how these factors affect the drug's dissolution rate and solubility are modeled [43].

In the ADAM model, GI tract is divided into nine sections and they are stomach, duodenum, jejunum I, jejunum II, ileum I, ileum II, ileum III, ileum IV, and colon (Figure 43.9). This model can also incorporate physiological factors such as gastric emptying time, intestinal and colonic transit times, blood flow, physiologically relevant regional luminal pH, permeability, and distribution of CYP enzymes that can affect absorption of a drug [13, 41]. It also includes an enterohepatic recirculation model. In the ADAM model, the absorption of drug from the stomach is considered

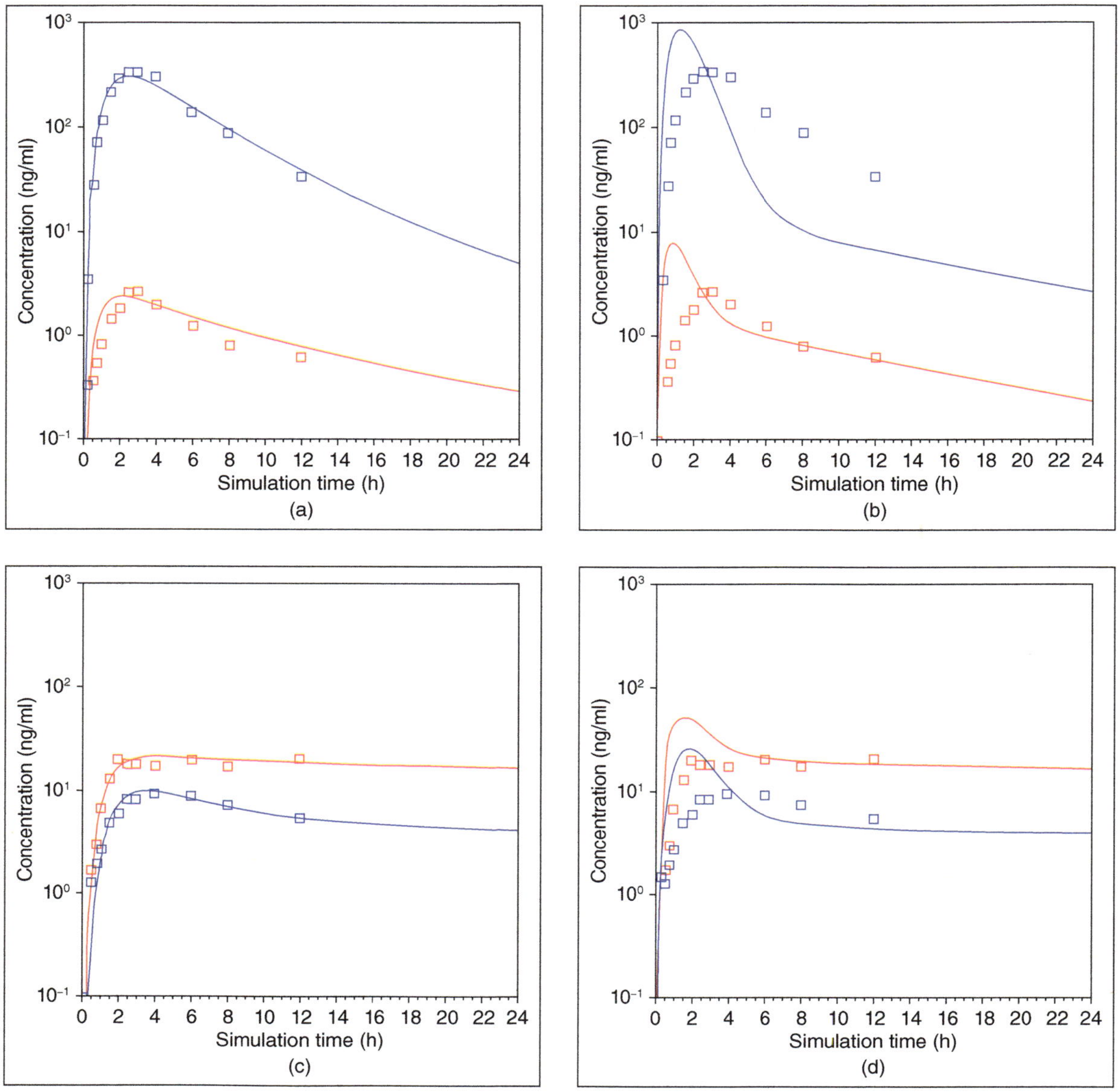


**Figure 43.8** Reported (squares) and simulated (lines) mean plasma concentration–time profiles of DEX after a single oral dose of 30 mg immediate-release tablet in healthy volunteers in EM (A and B) and PM (C and D). (A and C) $f_{u,ent}$ = 1.3% and (B and D) $f_{u,ent}$ = 100%. DEX (red) and total DXO (blue). Source: Bolger et al. [35] / with permission of Elsevier.

insignificant compared to the small intestine. Also, the drug metabolism in colon is assumed insignificant in this model. The movement of drug through each segment of the GI tract is assumed to be first-order kinetics. Transit times in the colon are defined by colonic mean resident times while transit times in the stomach are assigned by gastric mean resident times. The transit times in seven sections of the small intestine are defined as a fraction of the total small intestine transit time in proportion to the length of each segment. There are several physiological factors affecting the residence time of drugs in the stomach such as small intestine transit time.

Considering interindividual variability in intestinal transit, ADAM incorporates a Weibull probability distribution equation to the observed data. For drug-related factors such as solubility and dissolution, the ADAM model provides different options for the input information which can be measured in biorelevant media or aqueous buffer. Additionally, dissolution rate from solid dosage forms can be calculated by diffusion layer models if *in vitro* dissolution rates profiles are not

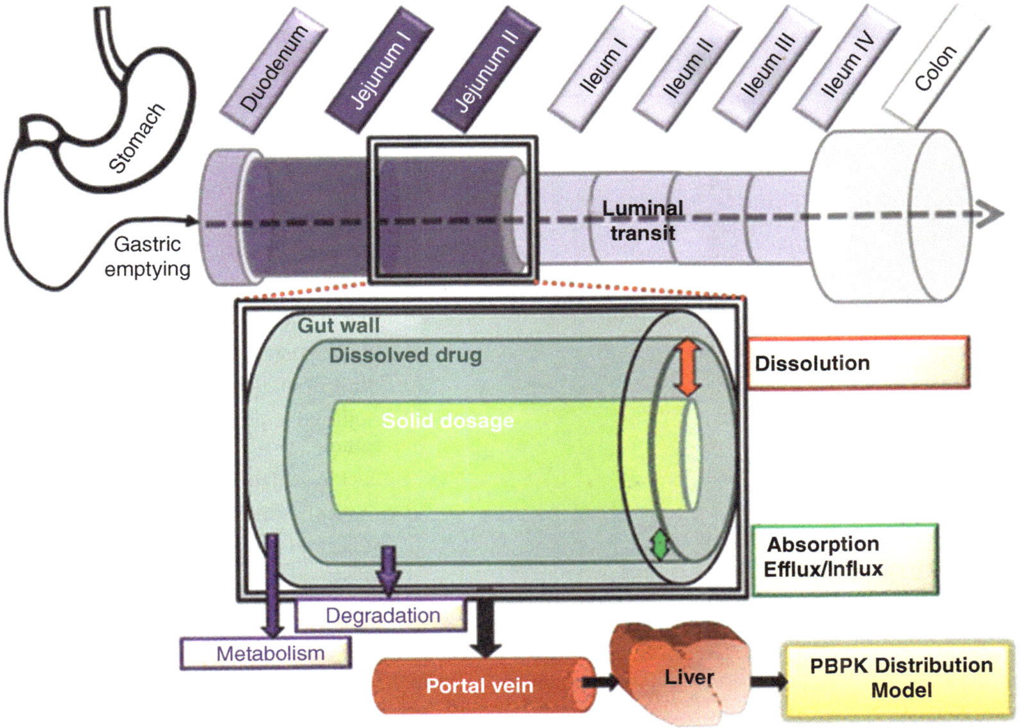


**Figure 43.9** Schematic representation of the ADAM model. Source: Reproduced from Pathak et al. [44] / with permission of American Chemical Society.

available [45]. The ADAM sub-model considers the interplay between transporters and enzymes in the gut and links to a full PBPK model. Transporter-mediated intestinal absorption of drugs can be described by the Michaelis–Menten equation which uses the related *in vitro* Michaelis constant or drug–transporter affinity constant $K_m$ and maximal transporter-mediated drug efflux uptake ($J_{max}$) [45]. The case study of bosutinib listed below is an example of such application of the simulator.

In an optional extension model named multilayer gut wall ADAM model (M-ADAM), each segment of the GI tract (corresponding to duodenum, jejunum I/II, ileum I/II/III/IV, and colon segments in the ADAM model) are expanded into a multilayer gut wall. It incorporates an unstirred boundary layer for absorption from luminal fluid to enterocyte. It also includes a permeability-limited basolateral membrane separating the intestinal interstitial fluid and the enterocyte. Additionally, it incorporates a lymphatic absorption route from the intestinal interstitial fluid to the systemic circulation. The M-ADAM model can only be used when in PKPD profiles mode and for compounds loaded in the substrate position [42].

#### 43.6.2.1 Case Study: The Impact of P-gp and CYP3A4 on drug Absorption DDIs for Bosutinib [46]

***43.6.2.1.1 Introduction*** This is a case study using PBPK model by Simcyp to predict how transporters (P-glycoprotein, P-gp) and enzymes-(CYP3A4) mediated DDIs impact oral absorption. There is overlap in substrate specificity between CYP3A4 and P-gp as they are both expressed in many same tissues and organs and both are regulated by nuclear receptors (e.g. Pregnane X Receptor) [47]. Therefore, understanding the contribution of CYP3A4 and P-gp on drug pharmacokinetics is very important. A PBPK model was developed to include both CYP3A4 and P-gp kinetics to describe bosutinib's plasma concentration–time profiles. After model verification, it was applied to predict other DDIs that have not been studied. All data, methods, table, and figures in the following discussions are cited from Yamazaki et al. [46] unless other references are provided.

Bosutinib (4-anilino-3-quinolinecarbonitrile monohydrate, $C_{26}H_{29}Cl_2N_5O_3 \cdot H_2O$), an orally administered tyrosine kinase inhibitor, has been widely approved for the treatment of adult patients with chronic, accelerated, or blast-phase Philadelphia chromosome-positive chronic myelogenous leukemia with intolerance or

**Table 43.6** Physicochemical and pharmacokinetic parameters of bosutinib used for PBPK models.

| Parameter (*U*) | Value | Sources |
|---|---|---|
| Molecular weight | 530 | Calculated |
| log *P* | 3.1 | Measured |
| p$K_a$ (monobase) | 7.9 | Measured |
| $f_{u,plasma}$ | 0.063 | Measured *in vitro* |
| B/P | 1.2 | Measured *in vitro* |
| Solubility (mg/ml) | 0.02–11 (at pH 1–8) | Measured *in vitro* |
| $P_{eff,man}$ ($10^{-4}$ cm/s) | 1.8 | Calculated from *in vitro* data |
| $f_{u,gut}$ | 0.063 | Predicted by sensitive analysis |
| $V_{ss}$ (l/kg) | 28 | Clinical study results |
| $K_p$ scalar | 3.7 | Adjusted to predict the observed $V_{ss}$ value |
| $CL_{int,CYP3A4}$ (μl/min/mg protein) | 560 | Back-calculated in Simcyp |
| $CL_{renal}$ (l/hr) | 1.2 | Clinical study results |
| P-gp $K_m$ (μM) | 0.38 | Measured in Caco-2 cells |
| P-gp $J_{max}$ (pmol/min) | 15 | Measured in Caco-2 cells |
| P-gp $J_{max}$ SF | 1–25 | Optimized to recover the observed results |

Source: Reproduced from Yamazaki et al. [46] / with permission of American Society for Pharmacology and Experimental Therapeutics.

resistance to previous tyrosine kinase inhibitor therapy in more than 30 countries including Japan, European Union countries, and the United states [48]. Bosutinib is primarily metabolized by CYP3A4. The predominantly circulating metabolites identified in plasma are oxydechlorinated (M2) bosutinib (19% of parent exposure) and N-desmethylated (M5) bosutinib (25% of parent exposure), with bosutinib N-oxide (M6) as a minor circulating inactive metabolite. M2 and M5 were deemed inactive and are associated with little (<5%) of the bosutinib activity observed *in vitro* [49]. Bosutinib is an ABCB1 (P-gp, the most studied transporter among the adenosine triphosphate (ATP)-binding cassette-transporters) substrate and inhibitor *in vitro*. It has moderate *in vitro* passive permeability and exhibits pH-dependent aqueous solubility over the pH range of 1–8 [46]. Bosutinib exhibits linear PK in the dose range of 200–800 mg but is approximately supra-proportional at doses of 50–200 mg. A saturation of intestinal P-gp-mediated efflux on absorption is considered as the primary reason for nonlinear PK at 50–200 mg. Therefore, understanding the contribution of intestinal P-gp-mediated efflux to bosutinib pharmacokinetics is very important. Additionally, as bosutinib is a CYP3A4 substrate, coadministration of a strong CYP3A4 inhibitor or CYP3A4 inducer with bosutinib would lead to DDIs, thus quantitatively rationalizing bosutinib DDI is very crucial.

##### *43.6.2.1.2 Methodology*

**Model Development and Simulation** Initially, Simcyp population-based simulator version 13.1 (Simcyp Limited, Sheffield, UK), was used to develop the PBPK models to predict bosutinib oral exposures in patients after single and multiple doses. Clinical study results of single-dose bosutinib DDI with ketoconazole and rifampin including the DDI data with renal and hepatic impairment patients are used to refine and verify the model. However, this previously developed model could not adequately predict the newly conducted absolute bioavailability study's data, thus a further refinement of bosutinib PBPK model is needed. As P-gp kinetics can only be incorporated into the PBPK-ADAM models in Simcyp [45], the ADAM model implemented in Simcyp (version 17.1) was used to incorporate intestinal P-gp kinetic parameters into the refined PBPK model. Physicochemical and biopharmaceutical parameters are either back-calculated, estimated, or measured from the *in vitro* or clinical study results as shown in Table 43.6. For example, the ADAM model integrates the biopharmaceutical and physicochemical properties including the intestinal effective permeability and pH-dependent solubility to predict bosutinib $F_a$. Moreover, the ADAM model defined the disintegration profile based on first-order kinetics with 100% maximal disintegration, a lag time of 0.25 hours, and a disintegration constant of 0.01/hour. Under the assumption of steady state condition, the estimated $K_m$ (concentration required to achieve half of the maximal transport rate) and $J_{max}$ (maximal transport rate) were inputted as 0.38 mM and 15 pmol/min, respectively, based on the kinetic model and $K_m$ corrected for *in vitro* nonspecific binding. $K_m$ was assumed to be intrinsic and was fixed for the PBPK model, an *in vitro*-to-*in vivo* SF for P-gp $J_{max}$ was optimized to adequately recover the observed results.

36 virtual healthy participants, age between 20 and 50 years with a female/male ratio of 0.5, in six trials (6 subjects each) whose CYP3A4 degradation rate constant ($k_{deg}$) was 0.0193/hour and 0.03/hour in liver and intestine, respectively, were chosen for all clinical trial simulations with sampling interval set to 0.2 hours in Simcyp. The gastric emptying time was updated from the default values to maximal values which are two hours in fasted and four hours in fed states based on the clinical observation data ($t_{max}$; four to six hours).

**Model Refinement and Verification** The bioavailability study was used for model refinement while DDI studies with ketoconazole and rifampin were used for both model refinement and verification. The bioavailability clinical study was conducted in 13 to 14 healthy male subjects in fed conditions. It is a two-way crossover study and the subjects either received a single dose of bosutinib intravenously (120 mg in a one-hour infusion) or orally (500 mg) [50].

The vendor-verified compound files in Simcyp library were used for bosutinib DDI predictions, such as ketoconazole (sim-ketoconazole 400 mg QD) and rifampin (sv-rifampicin-md). For CYP3A4-mediated inhibition, ketoconazole competitive $K_i$ was set as 0.015 μM ($f_{u,mic}$ = 0.97), while for P-gp-mediated efflux, ketoconazole was assumed to inhibit in a competitive manner. As ketoconazole *in vitro* $IC_{50}$ values against P-gp varied differently in different studies, sensitivity analysis and optimization (SAO) for ketoconazole inhibition constant ($K_i$) were performed. For CYP3A4-mediated induction, the input parameters listed as: rifampin induction $E_{max}$ = 16, $EC_{50}$ = 0.32 μM, and competitive $K_i$ = 15 μM ($f_{u,mic}$ = 1). As repeated dose of rifampin was reported to increase intestinal P-gp expression, thus rifampin was assumed to increase SFs for P-gp $J_{max}$ and proportionally increases in P-gp-mediated efflux activities in PBPK-ADAM models. For model refinement and verification, SAO for the model input parameters such as P-gp $J_{max}$ SFs, ketoconazole $K_i$ for intestinal P-gp, and rifampin-mediated increases in intestinal P-gp abundance were performed to adequately recover observed results.

**Model Application** Bosutinib (500 mg single dose) was administered on day 5 with and without oral administration of dual CYP3A/P-glycoprotein inhibitors, itraconazole (200 mg once daily for 16 days), and verapamil (80 mg three times a day for 16 days) in a virtual population of healthy participants for model application. The vendor-verified compound files in Simcyp library were used, such as itraconazole (sv-itraconazole_fed capsule), itraconazole metabolite (sv-OH-itraconazole), verapamil (sv-verapamil), and verapamil metabolite (sv-norverapamil). For CYP3A4-mediated inhibition, itraconazole competitive $K_i$ and metabolite competitive $K_i$ were set as 0.0013 μM ($f_{u,mic}$ = 1) and 0.0023 μM ($f_{u,mic}$ = 1), respectively. For P-gp-mediated inhibition, as no P-gp inhibition parameters were available in the default compound file, the lower end of itraconazole P-gp $K_i$ values was used based on the reported values found in the Metabolism and Transport Drug Interaction Database ((DIDB®; School of Pharmacy, University of Washington, Seattle, WA). For verapamil, the input parameters for CYP3A4 mediated inhibition were listed in the compound files as follows: mechanism-based inhibition $K_i$ = 2.21 μM ($f_{u,mic}$ = 1), maximal inactivation rate ($k_{inact}$) = 2/hour, verapamil metabolite mechanism-based inhibition $K_i$ = 10.3 μM ($f_{u,mic}$ = 1) and $k_{inact}$ = 18/hour. For P-kinetics, the input parameters were as follows: verapamil $K_m$ = 0.734 μM ($f_{u,mic}$ = 1), $J_{max}$ = 2.814 pmol/min, and $K_i$ = 0.16 μM ($f_{u,mic}$ = 1); verapamil metabolite $K_i$ = 0.04 μM ($f_{u,mic}$ = 1).

***43.6.2.1.3 Results and Discussion*** The predicted bosutinib's (100 mg) plasma concentration–time profiles in the control group were considerably overpredicted by the PBPK-ADAM model with P-gp inputs. The ratios of predicted-to-observed pharmacokinetic parameters (P/O) were used to evaluate predictive model performance. It was calculated as predicted parameters divided by observed parameters and the values within 50% of the observed results were provisionally considered to be acceptable. SAO can evaluate the effects on the overall outputs of uncertainty around input parameters, which might improve model fitting and provide additional mechanistic insights [51]. After the SAO for $J_{max}$ SFs in PBPK-ADAM models (25 was used), the P/O ratios for $C_{max}$ and AUC values were 1.15 and 0.85, respectively, in the control group. For the 500 mg bosutinib group, the PBPK-ADAM model with P-gp input ($J_{max}$ SF set as 2 and 4 for bioavailability and DDI studies, respectively) adequately predicted the plasma concentration–time profiles. The P/O ratios for models are 0.95–1.0 for $C_{max}$ and 0.94–1.2 for AUC. Therefore, $J_{max}$ SF of 4 was used for the following model

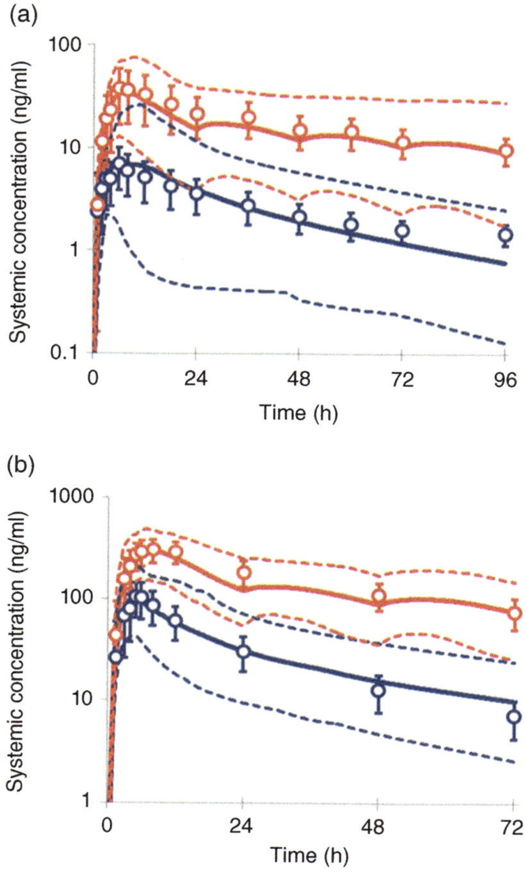


**Figure 43.10** Observed and PBPK model-predicted bosutinib's plasma concentrations after bosutinib coadministered with (red color) and without (blue color) ketoconazole. (a) Bosutinib (100 mg) with and without ketoconazole, ketoconazole $K_i$ set as 0.2 μM. (b) Bosutinib 5100 mg) with and without ketoconazole, ketoconazole $K_i$ set as 0.2 μM. The observed (showed as O) are expressed as mean ± SD while the predicted plasma concentrations are shown as mean (–) with 5th and 95th percentiles (- -), respectively. Source: Reproduced from Yamazaki et al. [46] / with permission of American Society for Pharmacology and Experimental Therapeutics.

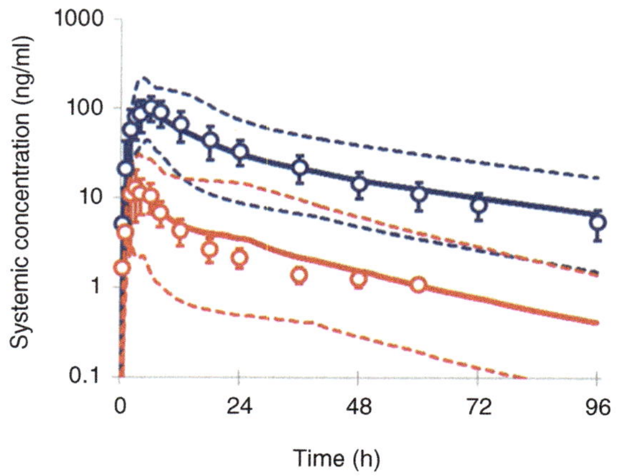


**Figure 43.11** Observed and PBPK model-predicted bosutinib's plasma concentrations after bosutinib (500 mg) coadministered with (red color) and without (blue color) rifampin. Input parameters of P-gp kinetic parameters in bosutinib PBPK models were $J_{max}$ SFs of 40. The observed (showed as O) are expressed as mean ± SD while the predicted plasma concentrations are shown as mean (–) with 5th and 95th percentiles (- -), respectively. Source: Reproduced from Yamazaki et al. [46] / with permission of American Society for Pharmacology and Experimental Therapeutics.

verification based on the DDI results. Compared to $F_a$ value (0.7) estimated from the mass-balance study, the predicted $F_a$ in these studies was 0.5–0.7.

DDIs prediction of bosutinib (100 and 500 mg) with ketoconazole was performed by the PBPK-ADAM models with the optimized $J_{max}$ SFs. If ketoconazole-mediated P-gp inhibition was not included (not include ketoconazole $K_i$ for P-gp), P/O ratios for $C_{max}$ and AUC were 0.56 and 0.67, respectively, at 100 mg, whereas those were 0.69 and 0.70, respectively, at 500 mg (Figures 43.10a,c). Thus, $K_i$ for P-gp was incorporated. Following the SAO for ketoconazole $K_i$, the predicted $C_{max}$ and AUC were 0.96 and 0.80, respectively, at 100 mg, while those were 0.83, and 0.80, respectively, at 500 mg (Figures 43.10a,b). The median value of $F_a$ was predicted to increase from 0.5 to 0.7 caused by P-gp inhibition by ketoconazole.

Similarly, DDI prediction of bosutinib (500 mg) with rifampin was performed by the PBPK-ADAM model with the optimized $J_{max}$ SFs, which consistently under predicted the effect of rifampin on bosutinib exposure. Following the SAO for $J_{max}$ SFs as intestinal P-gp abundances increased by rifampin-mediated P-gp induction, P/O ratios for $C_{max}$ and AUC of 1.01 and 0.26, respectively, with $J_{max}$ and induction fold set as 4 and 10, respectively (Figure 43.11). $F_a$ value was predicted from 0.6 to 0.2 caused by P-gp induction. DDIs between bosutinib and dual CYP3A/P-gp inhibitors itraconazole and verapamil were predicted by the PBPK-ADAM model for model application.

#### 43.6.2.2 Conclusion

The PBPK model for bosutinib has adequately been developed, refined, and verified based on the clinically observed results which include not only CYP3A4-mediated metabolism but also P-gp-mediated efflux on absorption. Though incorporating P-gp kinetics in the PBPK models are very important, modeling for P-gp substrate drugs is still very challenging.

## 43.7 Summary

A lot of progress has been made in computational modeling of human oral absorption over the last few decades. Modeling and simulation have been increasingly being accepted by the regulatory agencies for supporting regulatory submission. However, due to the complexities of numerous processes that are involved in oral drug absorption, accurate prediction of oral bioavailability by computational modeling is still a challenge. There is a need to improve the prediction accuracy by increasing our understanding of the dissolution and absorption process such as amorphous API dissolution, API precipitation, the impact of formulation manufacturing process parameters like disintegration, wettability, granulation performance, dynamics of gut physiology, and the involvement of intestinal enzymes and transporters. Additionally, perhaps, the use of a hybrid QSPR-PBPK model, multiscale absorption and transit model (MAT) [52], and machine learning approaches such as KNIME automated workflow [53] could further improve the prediction accuracy.

## References


1 Wang, J. and Hou, T. (2015). Advances in computationally modeling human oral bioavailability. *Adv. Drug Delivery Rev.* https://doi.org/10.1016/j.addr.2015.01.001.

2 Dahlgren, D. and Lennernäs, H. (2019). Intestinal permeability and drug absorption: predictive experimental, computational and in vivo approaches. *Pharmaceutics* https://doi.org/10.3390/pharmaceutics11080411.

3 Cabrera-Pérez, M.Á. and Pham-The, H. (2018). Computational modeling of human oral bioavailability: what will be next? *Expert Opin. Drug Discovery* https://doi.org/10.1080/17460441.2018.1463988.

4 Tian, S., Li, Y., Wang, J. et al. (2011). ADME evaluation in drug discovery. 9. Prediction of oral bioavailability in humans based on molecular properties and structural fingerprints. *Mol. Pharma.* 8 (3): 841–851. https://doi.org/10.1021/mp100444g.

5 Veber, D.F., Johnson, S.R., Cheng, H.Y. et al. (2002). Molecular properties that influence the oral bioavailability of drug candidates. *J. Med. Chem.* 45 (12): 2615–2623. https://doi.org/10.1021/jm020017n.

6 Andrews, C.W., Bennett, L., and Yu, L.X. (2000). Predicting human oral bioavailability of a compound: development of a novel quantitative structure-bioavailability relationship. *Pharm. Res.* 17 (6): 639–644. https://doi.org/10.1023/A:1007556711109.

7 Kuepfer, L., Niederalt, C., Wendl, T. et al. (2016). Applied concepts in PBPK modeling: how to build a PBPK/PD model. *CPT Pharm. Syst. Pharmacol.* 5 (10): 516–531. https://doi.org/10.1002/psp4.12134.

8 Bravo-Gómez, M.E., Camacho-García, L.N., Castillo-Alanís, L.A. et al. (2019). Revisiting a physiologically based pharmacokinetic model for cocaine with a forensic scope. *Toxicol. Res. (Camb).* 8 (3): 432–446. https://doi.org/10.1039/c8tx00309b.

9 Paalzow, L.K. (1995). Torsten Teorell, the father of pharmacokinetics. *Ups. J. Med. Sci.* 100 (1): 41–46. https://doi.org/10.3109/03009739509178895.

10 Lennernäs, H., Lindahl, A., Van Peer, A. et al. (2017). In vivo predictive dissolution (IPD) and biopharmaceutical modeling and simulation: future use of modern approaches and methodologies in a regulatory context. *Mol. Pharm.* 14 (4): 1307–1314. https://doi.org/10.1021/acs.molpharmaceut.6b00824.

11 Flanagan, T., Van Peer, A., and Lindahl, A. (2016). Use of physiologically relevant biopharmaceutics tools within the pharmaceutical industry and in regulatory sciences: where are we now and what are the gaps? *Eur. J. Pharm. Sci.* 91: 84–90. https://doi.org/10.1016/j.ejps.2016.06.006.

12 Li, M., Zhao, P., Pan, Y., and Wagner, C. (2018). Predictive performance of physiologically based pharmacokinetic models for the effect of food on oral drug absorption: current status. *CPT Pharm. Syst. Pharmacol.* 7 (2): 82–89. https://doi.org/10.1002/psp4.12260.

13 Abuhelwa, A.Y., Williams, D.B., Upton, R.N., and Foster, D.J.R. (2017). Food, gastrointestinal pH, and models of oral drug absorption. *Eur. J. Pharm. Biopharm.* 112: 234–248.

14 Kaur, N., Narang, A., and Bansal, A.K. (2018). Use of biorelevant dissolution and PBPK modeling to predict oral drug absorption. *Eur. J. Pharm. Biopharm.* 129: 222–246. https://doi.org/10.1016/j.ejpb.2018.05.024.

15 Pepin, X.J.H., Parrott, N., Dressman, J. et al. (2020). Current state and future expectations of translational modeling strategies to support drug product development, manufacturing changes and controls: a workshop summary report. *J. Pharm. Sci.*

16 Suarez-Sharp, S., Cohen, M., Kesisoglou, F. et al. (2018). Applications of clinically relevant dissolution

testing: workshop summary report. *AAPS J.* 20 (6): https://doi.org/10.1208/s12248-018-0252-3.

17 Stillhart, C., Pepin, X., Tistaert, C. et al. (2019). PBPK absorption modeling: establishing the in vitro–in vivo link—industry perspective. *AAPS J.* 21 (2): 19.

18 Suarez-Sharp, S., Li, M., Duan, J. et al. (2016). Regulatory experience with in vivo in vitro correlations (IVIVC) in new drug applications. *AAPS J.* 18 (6): 1379–1390.

19 Yang Zhao, Sandra Suarez (2019). FDA expectations in building a safe space to gain regulatory flexibility based on Physiologically Based Biopharmaceutics Modeling (PBBM). https://cersi.umd.edu/sites/cersi.umd.edu/files/Day%203-1%20Zhao%20Suarez%20LM.pdf (accessed 12 June 2023).

20 Delvadia PR (2018). Biowaiver Approaches for Solid Oral Dosage Forms in New Drug Applications. https://pqri.org/wp-content/uploads/2018/12/Biowaiver-Approaches-for-Solid-Oral-Dosage-Forms-in-New-Drug-Applications_V8_Final.pdf (accessed 12 June 2023).

21 Suarez-Sharp, S., Delvadia, P.R., Dorantes, A. et al. (2016). Regulatory perspectives on strength-dependent dissolution profiles and biowaiver approaches for immediate release (IR) oral tablets in new drug applications. *AAPS J.* 18 (3): 578–588.

22 FDA guidance. U.S. Food and Drug Administration (2003). *Exposure-Response Relationship-Study Design, Data Analysis, and Regulatory Applications.*

23 FDA Guidance (1997). Extended Release Oral Dosage Forms: Development, Evaluation, and Application of In Vitro/In Vivo Correlations.

24 Pepin, X.J.H., Flanagan, T.R., Holt, D.J. et al. (2016). Justification of drug product dissolution rate and drug substance particle size specifications based on absorption PBPK modeling for lesinurad immediate release tablets. *Mol. Pharm.* 13 (9): 3256–3269.

25 Pepin, X.J.H., Moir, A.J., Mann, J.C. et al. (2019). Bridging in vitro dissolution and in vivo exposure for acalabrutinib. Part II. A mechanistic PBPK model for IR formulation comparison, proton pump inhibitor drug interactions, and administration with acidic juices. *Eur. J. Pharm. Biopharm.* 142: 435–448.

26 Hofsäss, M.A. (2020). Suitability of the z-factor for dissolution simulation of solid oral dosage forms: potential pitfalls and refinements. *J. Pharm. Sci.*

27 Nicolaides, E., Symillides, M., Dressman, J.B., and Reppas, C. (2001). Biorelevant dissolution testing to predict the plasma profile of lipophilic drugs after oral administration. *Pharm. Res.* 18 (3): 380–388. https://doi.org/10.1023/A:1011071401306.

28 Takano, R., Sugano, K., Higashida, A. et al. (2006). Oral absorption of poorly water-soluble drugs: computer simulation of fraction absorbed in humans from a miniscale dissolution test. *Pharm. Res.* 23 (6): 1144–1156.

29 SimulationsPlus (2019). GastroPlus(TM) User Manual for version 9.7.

30 Pepin, X.J.H., Sanderson, N.J., Blanazs, A. et al. (2019). Bridging in vitro dissolution and in vivo exposure for acalabrutinib. Part I. Mechanistic modelling of drug product dissolution to derive a P-PSD for PBPK model input. *Eur. J. Pharm. Biopharm.* 142: 421–434.

31 Cristofoletti, R., Hens, B., Patel, N. et al. (2019). Integrating drug-and formulation-related properties with gastrointestinal tract variability using a product-specific particle size approach: case example ibuprofen. *J. Pharm. Sci.* 108 (12): 3842–3847.

32 Yu, L.X. and Amidon, G.L. (1999). Estimating oral drug absorption. *Int. J. Pharm.* 186: 119–125.

33 Hens, B. and Bolger, M.B. (2019). Application of a dynamic fluid and pH model to simulate intraluminal and systemic concentrations of a weak base in GastroPlus®. *J. Pharm. Sci.* 108 (1): 305–315.

34 Dielschneider, R.F., Henson, E.S., and Gibson, S.B. (2017). Lysosomes as oxidative targets for cancer therapy. *Oxid. Med. Cell. Longev.* 2017.

35 Bolger, M.B., Macwan, J.S., Sarfraz, M. et al. (2019). The irrelevance of in vitro dissolution in setting product specifications for drugs like dextromethorphan that are subject to lysosomal trapping. *J. Pharm. Sci.* 108 (1): 268–278.

36 Borghardt, J.M., Weber, B., Staab, A. et al. (2016). Investigating pulmonary and systemic pharmacokinetics of inhaled olodaterol in healthy volunteers using a population pharmacokinetic approach. *Br. J. Clin. Pharmacol.* 81 (3): 538–552.

37 Heikkinen, A.T., Mönkkönen, J., and Korjamo, T. (2009). Kinetics of cellular retention during Caco-2 permeation experiments: role of lysosomal sequestration and impact on permeability estimates. *J. Pharmacol. Exp. Ther.* 328 (3): 882–892.

38 Kazmi, F., Hensley, T., Pope, C. et al. (2013). Lysosomal sequestration (trapping) of lipophilic amine (cationic amphiphilic) drugs in immortalized human hepatocytes (Fa2N-4 cells). *Drug Metab. Dispos.* 41 (4): 897–905.

39 Silvasti, M., Karttunen, P., Tukiainen, H. et al. (1987). Pharmacokinetics of dextromethorphan and dextrorphan: a single dose comparison of three

preparations in human volunteers. *Int. J. Clin. Pharmacol. Ther. Toxicol.* 25 (9): 493–497.

40 Agoram, B., Woltosz, W.S., and Bolger, M.B. (2001). Predicting the impact of physiological and biochemical processes on oral drug bioavailability. *Adv. Drug Delivery Rev.* 50: https://doi.org/10.1016/S0169-409X(01)00179-X.

41 Lin, L. and Wong, H. (2017). Predicting oral drug absorption: mini review on physiologically-based pharmacokinetic models. *Pharmaceutics* 9 (4): https://doi.org/10.3390/pharmaceutics9040041.

42 Limited CU (2019). *Help file*. In: *Simcyp Population Based Simulator V19R1 [Computer Software]*.

43 Upton, R.N., Foster, D.J.R., and Abuhelwa, A.Y. (2016). An introduction to physiologically-based pharmacokinetic models. *Paediatr. Anaesth.* 26 (11): 1036–1046. https://doi.org/10.1111/pan.12995.

44 Pathak, S.M., Ruff, A., Kostewicz, E.S. et al. (2017). Model-based analysis of biopharmaceutic experiments to improve mechanistic oral absorption modeling: an integrated in vitro in vivo extrapolation perspective using ketoconazole as a model drug. *Mol. Pharm.* 14 (12): 4305–4320. https://doi.org/10.1021/acs.molpharmaceut.7b00406.

45 Jamei, M., Marciniak, S., Feng, K. et al. (2009). The Simcyp® population-based ADME simulator. *Expert Opin. Drug Metab. Toxicol.* 5 (2): 211–223. https://doi.org/10.1517/17425250802691074.

46 Yamazaki, S., Loi, C.M., Kimoto, E. et al. (2018). Application of physiologically based pharmacokinetic modeling in understanding bosutinib drug-drug interactions: importance of intestinal P-glycoprotein. *Drug Metab. Dispos.* 46 (8): 1200–1211. https://doi.org/10.1124/dmd.118.080424.

47 Yamazaki, S., Costales, C., Lazzaro, S. et al. (2019). Physiologically-based pharmacokinetic modeling approach to predict rifampin-mediated intestinal P-glycoprotein induction. *CPT Pharm. Syst. Pharmacol.* 8 (9): 634–642. https://doi.org/10.1002/psp4.12458.

48 Syed, Y.Y., McCormack, P.L., and Plosker, G.L. (2014). Bosutinib: a review of its use in patients with Philadelphia chromosome-positive chronic myelogenous leukemia. *BioDrugs* 28 (1): 107–120. https://doi.org/10.1007/s40259-013-0082-x.

49 Abbas, R., Boni, J., and Sonnichsen, D. (2015). Effect of rifampin on the pharmacokinetics of bosutinib, a dual Src/Abl tyrosine kinase inhibitor, when administered concomitantly to healthy subjects. *Drug Metab. Pers. Ther.* 30 (1): 57–63. https://doi.org/10.1515/dmdi-2014-0026.

50 Hsyu, P.-H., Pignataro, D.S., and Kyle, M. (2018). Absolute bioavailability of bosutinib in healthy subjects from an open-label, randomized, 2-period crossover study. *Clin. Pharmacol. Drug Dev.* 7 (4): 373–381.

51 Zhao, P., Rowland, M., and Huang, S.M. (2012). Best practice in the use of physiologically based pharmacokinetic modeling and simulation to address clinical pharmacology regulatory questions. *Clin. Pharmacol. Ther.* 92 (1): 17–20. https://doi.org/10.1038/clpt.2012.68.

52 Kannan, R. and Przekwas, A. (2020). A multiscale absorption and transit model for oral drug delivery: formulation and applications during fasting conditions. *Int. J. Numer. Methods Biomed. Eng.* 36 (3): 1–20. https://doi.org/10.1002/cnm.3317.

53 Falcón-Cano, G., Molina, C., and Cabrera-Pérez, M.Á. (2020). ADME prediction with KNIME: development and validation of a publicly available workflow for the prediction of human oral bioavailability. *J. Chem. Inf. Model.* https://doi.org/10.1021/acs.jcim.0c00019.

# 44

## Computational Modeling of Drug Oral Bioavailability

*Lon W.R. Fong, Beibei Huang, Rajan Chaudhari, and Shuxing Zhang*

*Intelligent Molecular Discovery Laboratory, Department of Experimental Therapeutics, The University of Texas MD Anderson Cancer Center, Houston, TX, USA*

### 44.1 Introduction

The processes by which a drug is taken from the point of administration to systemic circulation can be roughly divided into three steps: dissolution, absorption, and metabolism. Dissolution is the dissolving of a drug into the gastrointestinal fluid. Once dissolved, the drug is often absorbed into the bloodstream in the small intestine by permeating the cells of the intestinal epithelium, although absorption can occur elsewhere. Finally, as with most ingested xenobiotics, the absorbed drugs are subject to metabolism, or chemical modification, primarily by enzymes in the liver but possibly as well in other organs (e.g. small intestine) or even other organisms (e.g. gut flora) before entering systemic circulation. The first two steps of drug disposition, i.e. dissolution and absorption, can be controlled to some extent by changing the physicochemical properties such as polarity/hydrophobicity and/or acidity/alkalinity of the drug. These properties can be adjusted either by changing the chemical structure of the drug itself or by using different formulations (e.g. syrups versus coated tablets). The third step a drug undergoes before entering systemic circulation, first-pass (pre-systemic) metabolism, can be controlled by adding functional groups to or removing groups from the core scaffold of the drug. These changes are usually intended to change the drug's affinity to the hepatic enzymes responsible for first-pass metabolism, rather than change the physicochemical properties per se.

For drugs with systemic effects, bioavailability, signified by $F$, is the fraction of an administered drug that enters systemic circulation in its active form. The main metric of bioavailability is maximal plasma concentration ($C_{max}$), the highest concentration an administered drug reaches in the blood. A related metric that is often considered in pharmacokinetics (PK) studies is time-to-peak-concentration ($T_{max}$), the time it takes for an administered drug to reach its maximal plasma concentration. Other specialized metrics, such as first-pass intestinal extraction ratios or regional dose-absorbed fractions, may be used as well when the situation demands. It is important to note that bioavailability is not the same as absorption; for orally administered drugs, absorption is how readily a drug transfers from the intestines into the bloodstream. Although absorption is a major factor in bioavailability, a well-absorbed drug will not necessarily have good bioavailability as first-pass (pre-systemic) metabolism may reduce the amount of drug in its original form that enters systemic circulation [1].

Drugs can be administered in a number of different ways depending on several factors. The basic categories of drug administration routes include parenteral, oral, nasal, ocular, transmucosal, and transdermal. Parenteral routes, which are those besides the mouth and alimentary canal, include intravenous (directly into systemic circulation via a vein), intramuscular (into a muscle), and subcutaneous (into the tissue directly under the skin) [2]. For drugs with intended systemic effects, the most common methods of administration are intravenous and oral.

Since drugs administered intravenously directly enter the systemic circulation, bioavailability for this administration route is usually considered to be 100%, and we will not further address it in this book chapter. Orally administered drugs, however, must undergo a series of processes to take them from the point of administration to systemic circulation as described above, and the efficiencies of these processes determine the final concentration of the drug entering systemic circulation. Predicting the fraction of the drug administered that actually enters systemic circulation is, therefore, not a straightforward process. In this book chapter, we

*Oral Bioavailability and Drug Delivery: From Basics to Advanced Concepts and Applications*, First Edition. Edited by Ming Hu and Xiaoling Li.

provide an overview of the numerous computational techniques that have been developed to address the challenges.

## 44.2 Computational Modeling of Bioavailability

Bioavailability can either be modeled qualitatively or quantitatively. Qualitative, or rule-based, approaches model bioavailability as a binary variable (bioavailable or not) and are often used to screen large sets of compounds or to make "go/no-go" decisions in the drug-discovery phase of development. Qualitative models, while relying on quantitative physicochemical properties to make a prediction, are usually rules of thumb and do not follow a specified algorithm, leaving the decision to the individual using the model. Quantitative models, on the other hand, may also model bioavailability as a binary variable but follow a particular formula in forming a prediction. Of course, the output variable of quantitative models may also be numeric as well; for bioavailability, this is most often $F$, $C_{max,}$ or $T_{max}$. Some models may predict other PK measurements as well, such as the efficiency of the disposition pathway (e.g. biliary drug excretion and CYP metabolism) or accumulation of the drug in the target organ or tissue. The two main categories of quantitative models are physiologically based pharmacokinetic (PBPK) models and statistical models. PBPK models, which are usually represented as systems of differential equations, are based on existing knowledge, such as known physiological principles or established parameters; in contrast, statistical models, which include those generated through machine learning approaches, are based on observed data and may model the relationship between the input and output variables in a way different from or even contrary to the established body of knowledge.

### 44.2.1 Qualitative Modeling

The simplest models of oral bioavailability are qualitative models, which, as their name suggests, are used to classify compounds as either orally bioavailable or not, rather than predict a measurable quantity. These "rules of thumb," which use the compounds' physiochemical properties, are greatly simplified compared to quantitative models to facilitate use by non-computational researchers. The most widely used of these models is perhaps Lipinski's "Rule of Five," [3] which classifies compounds as orally bioavailable when they meet the following criteria: (i) a molecular mass under 500 daltons, (ii) an octanol-water partition coefficient under 5, (iii) fewer than five hydrogen-bond donors, and (iv) fewer than 10 hydrogen-bond acceptors. Other models include Veber et al.'s rules and Martin's scoring system. Veber et al.'s model classify as orally bioavailable compounds that have (i) 10 or fewer rotatable bonds and (ii) polar surface areas (PSAs) less than or equal to 140 Å [2] (alternatively, 12 or fewer hydrogen-bond donors and acceptors) [4]. Martin's system, named ABS, is based on the observation that the physicochemical properties affecting solubility and permeability change for a given compound based on the compound's ionization state. As such, it incorporates ionization state, Lipinski's Rule-of-Five properties, and PSA in predicting bioavailability [5].

The simplicity of these qualitative models comes at the cost of accuracy. A study by Hou and colleagues demonstrated that Veber et al.'s rules cannot predict oral bioavailability with high confidence, limiting their use in human applications [6]. Martin's system likewise is of uncertain usefulness in human settings, as it was developed using data from rats. Other factors may also need to be taken into consideration when using qualitative models: Zhu et al. compared the performance of 35 such models of oral bioavailability and discovered that while the best performing of these rules showed satisfactory sensitivity and specificity in classifying lowly metabolized compounds, the specificity was much worse when classifying highly metabolized compounds. The usefulness of such rules may be limited then to culling large sets of compounds at the beginning of drug discovery, using the rules to determine which compounds to ignore, as false positivity is of greater concern than false negativity at that point [7].

### 44.2.2 Quantitative Modeling

#### 44.2.2.1 Physiologically Based Pharmacokinetic (PBPK) Modeling

PBPK modeling represents physiological processes as mathematical equations with the aim of predicting pharmacokinetic quantities (e.g. $C_{max}$) or simulating pharmacokinetics profiles (e.g. plasma concentration–time curves) of a drug. Differential equations for the quantity of the drug in its different states or physical locations are most often used for this type of modeling, and the parameters of these equations correspond to physiological quantities such as blood flow rate or organ volumes. In their simplest forms, these equations model the entire gastrointestinal tract (GIT) as a single, well-mixed tank, while more complex models use systems of differential equations for the quantity of drug in each of several compartments, where each compartment corresponds to a particular organ or tissue (or segment thereof) involved in drug disposition, such as the intestines or the liver. Compared to statistical modeling techniques, PBPK modeling is more "bottom-up" than "top-down" in its approach as it is more mechanistic and driven by basic physiological knowledge rather than observed data [8–10]. Nonetheless, modern PBPK models are not completely based on theory and do incorporate experimental data. PBPK models are most often used in the later phases of drug discovery and development, where they are mainly used to predict in-human doses. However, they have been used in a number of other applications including prediction of PK profiles, drug metabolism, and drug–drug interactions, as well as development of drug formulations [11–22].

PBPK models tend to model oral absorption [23], although more recent models have begun to address metabolism as well. Dissolution is usually described using a form of the Noyes-Whitney equation, which models the process as a first-order reaction with the rate of dissolution a function of the following quantities: a drug-specific dissolution constant, the concentration of drug at time $t$, the drug's solubility, and the volume of the intestinal fluid [23, 24]. Permeation, the transport of the dissolved drug across the intestinal wall into the hepatic portal vein, is also modeled as a first-order process, with the rate of drug permeation a function of an absorption constant and the concentration of dissolved drug. We will discuss different types of models that have been developed over the years.

***44.2.2.1.1 Mixing-Tank Model*** The mixing-tank model is one of the most basic types of PBPK model and represents the entire GIT as a single, well-mixed tank. Dressman and Fleisher pioneered this model, which was one of the first to integrate the dissolution and permeation processes [23, 25]. In the mixing-tank model, the drug enters the GIT as a solid (whose dose is given by $X_{\text{solid}}$), which dissolves at a rate given by a Noyes–Whitney expression [24], giving a dose of dissolved drug ($X_{\text{diss}}$), which then permeates across the intestinal wall into the portal vein at a rate given by $k_a$. As shown in Figure 44.1, this model divides the quantity of the drug into three states—solid ($X_{\text{solid}}$), dissolved ($X_{\text{diss}}$), and absorbed ($X_a$)—each with a corresponding differential equation (44.1)–(44.3); the model is described mathematically as the system of these three equations. The amount of solid drug $X_{\text{solid}}$ in the GIT is given by

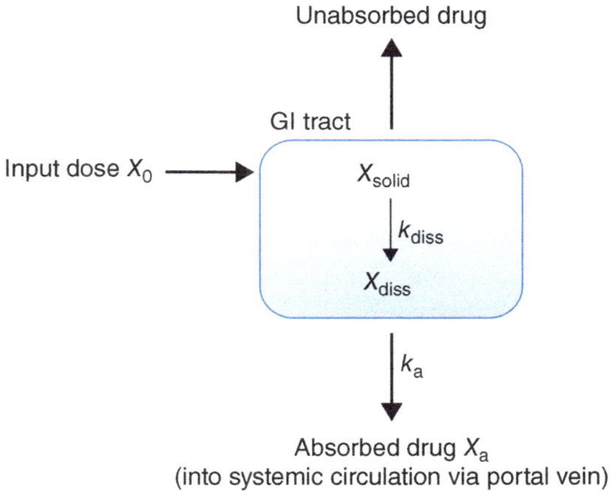


**Figure 44.1** A scheme of Dressman and Fleisher's mixing-tank model, which models the GIT as a single, well-mixed tank into which the drug enters with input dose $X_0$. The model incorporates dissolution and permeation, giving a total of three states of the drug (solid, dissolved, and absorbed), with the quantity of drug in each state given by Eqs. (44.1)–(44.3), respectively.

$$-\frac{dX_{\text{solid}}}{dt} = \frac{DSA_t}{h}\left(C_s - \frac{X_{\text{diss}}}{V}\right) + \frac{QX_{\text{solid}}}{V} \quad (44.1)$$

The first term is a Noyes–Whitney expression for dissolution [24], in which $D$ is drug diffusivity, $SA_t$ is the surface area of $N_t$ drug particles in the GIT at time $t$, $h$ is the boundary-layer thickness, $C_s$ is the aqueous solubility, $X_{\text{diss}}$ is the amount of drug in solution, and $V$ is the volume of fluid in the GIT. The second term accounts for the removal of solid drug by outflow from the GIT; $Q$ is the flow rate. The amount of dissolved drug $X_{\text{diss}}$ is given by

$$-\frac{dX_{\text{diss}}}{dt} = \frac{DSA_t}{h}\left(C_s - \frac{X_{\text{diss}}}{V}\right) + \frac{QX_{\text{diss}}}{V} + k_a X_{\text{diss}} \quad (44.2)$$

where, as with the expression for the amount of solid drug $X_{\text{solid}}$, the first term is a Noyes–Whitney expression for dissolution, and the second term accounts for the removal of dissolved drug by outflow from the GIT. The third term accounts for the absorption of the dissolved drug into the portal vein; $k_a$ is the first-order absorption rate constant. Finally, the amount of absorbed drug $X_a$ is given by

$$\frac{dX_a}{dt} = k_a X_{\text{diss}} \quad (44.3)$$

The initial conditions for this system are $X_{\text{solid}}(t = 0) = X_0$, $X_{\text{diss}}(t = 0) = 0$, and $X_a(t = 0) = 0$.

Of course, the simplicity of this model does not allow for the modeling of other processes that affect bioavailability, such as drug metabolism, and it masks the heterogeneity in absorption-rate constants present along the GIT. Nonetheless, the mixing-tank model serves as the basis for more-complicated models used now, such as advanced compartmental absorption and transit (ACAT) and advanced dissolution absorption metabolism (ADAM) as discussed in subsequent sections. The remaining PBPK models discussed in this chapter are principally similar, dividing the quantity of the drug into different states or physical locations and using a differential equation to represent the quantity of the drug in a given state, location, or combination thereof; the main differences are in the number of divisions (compartments) of the drug quantity, whether to account for certain physiological processes such as metabolism in the equations, and the parameters used in the equations.

#### 44.2.2.2 Compartmental Absorption and Transit Model

Building on the mixing-tank model, Yu et al. introduced the compartmental absorption and transit (CAT) model, which simulated the GIT not as a single anatomical compartment but as a series of compartments [26]. CAT models represent the GIT as seven compartments (the number that best fits small-intestine transit time), with the first compartment representing the duodenum, the next two the jejunum, and the last four the ileum. Yu later updated the CAT model by dividing each of the seven compartments into two compartments, one representing dissolved drug and other undissolved drug, allowing for the modeling of dissolution rate-limited absorption [27] (Figure 44.2).

#### 44.2.2.3 Extensions of and Variations on the Compartmental Absorption and Transit Model

Despite its improvements over the mixing-tank model, CAT still did not account for other factors affecting drug absorption, such as rate of dissolution or drug metabolism by the liver and/or intestines. Several multi-compartmental models have been developed to provide a more fine-grained approximation of the process of intestinal drug absorption. Among the most popular are the ACAT model, developed by Agoram et al. and implemented in the GastroPlus software program [28], and ADAM model, developed by Jamei et al. and implemented in the SimCYP program [23, 29]. Table 44.1 provides a list of PBPK models with currently available software implementations.

Instead of the seven compartments in CAT models, ACAT represents the GIT with nine compartments, each accounting for three states of a drug: unreleased, undissolved, and dissolved. ACAT also incorporates factors not accounted for by CAT, including physicochemical factors (e.g. solubility), physiological factors (e.g. hepatic metabolism), and dosage factors. In an early usage of ACAT, Parrott et al. used the model to predict oral bioavailability in rats, although the results were less than ideal due to ACAT's inability to accurately account for solubility [36, 37]. Later variations of the model, however, were able to address some of these and other issues. By incorporating *in vitro*, *in vivo*, and *in silico* data, other researchers were able to produce more accurate predictions or use the model to provide mechanistic explanations for certain observations on the bioavailability of drugs. Paixão and colleagues incorporated *in vitro* and *in silico* permeability and intrinsic clearance data, which led to promising results [38]. Similarly, Daga et al. observed that the use of inputs calculated based on universal rules alone in building global PBPK models could not produce the accuracy needed for lead optimization; in response, they used their own experimental intrinsic clearance data to build a QSPR model for a particular medicinal chemistry series. By combining the intrinsic clearance calculated by their local QSPR models and other inputs calculated from structures alone, the researchers were able to obtain adequate bioavailability predictions in rats for their particular series [39]. Olivares-Morales and colleagues, meanwhile, incorporated *in vivo*, *in vitro*, and *in silico* data to explain the differences in bioavailability between two different formulations of the drug Oxybutynin [40].

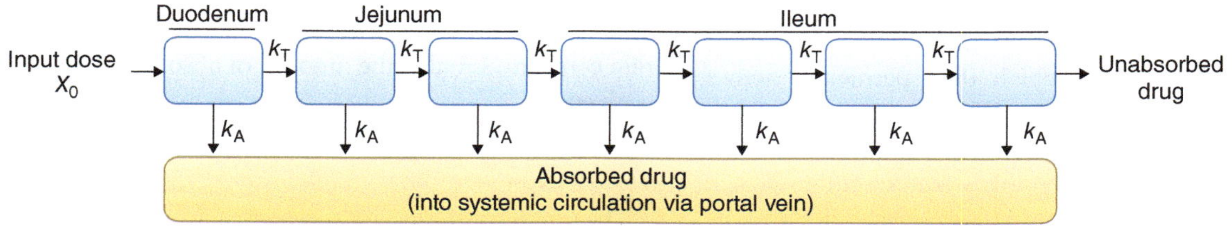


**Figure 44.2** A simplified CAT model. Compared to the mixing-tank model, which represented the GIT as a single well-mixed tank, CAT models divide the GIT into several compartments, with one or more compartments corresponding to the duodenum, jejunum, and ileum. Later updates of the CAT model also accounted for dissolution within each compartment.

**Table 44.1** A list of PBPK models with available software implementations.

| Name | Publisher | License type | Physiological model structure | References |
|---|---|---|---|---|
| GastroPlus | SimulationsPlus | Proprietary | ACAT; GIT divided into nine anatomically defined segments, with one for the stomach, seven for the small intestine, and one for the colon. These segments are further compartmentalized to account for different drug states (unreleased, undissolved, and dissolved) and different regions of the intestine (lumen and enterocytic regions.) | [28] |
| SimCYP | Certara | Proprietary | ADAM; GIT divided into nine anatomically defined segments representing the stomach, small intestine, and colon. These segments are further compartmentalized to account for four drug states: unreleased, undissolved, dissolved, and absorbed. | [29] |
| PK-Sim | Open Systems Pharmacology | GNU General Public License | GIT divided into five anatomically defined segments (stomach, small intestine, large intestine, pancreas, and spleen), each divided further into three sub-compartments (vascular, interstitial, and cellular space.) | [30–32] |
| Cloe HIA | Cyprotex | Proprietary | GIT divided into nine anatomically defined segments representing the stomach, duodenum, jejunum, ileum, and colon. | [33] |
| Maxsim2 | Maxsim2 | Proprietary | GIT treated as a single anatomical segment, which may be further subdivided into a plasma compartment and a tissue compartment. | [34] |
| PKQuest | PKQuest (David G. Levitt) | Not specified; free to use | GIT (defined as organs drained by the portal vein [stomach, small intestine, large intestine, pancreas, and spleen]) treated as a single anatomical segment. | [35] |

A multi-compartmental model developed by Grass and later implemented in the software IDEA, accounts for physiological parameters affecting oral absorption (e.g. pH and transit time) as well as drug compound parameters (e.g. permeability and dissolution rate) [41, 42]. A comparison of IDEA and GastroPlus on 28 drugs showed similar performance in predicting absorption classes and fractions absorbed [43]. However, a review by Huang et al. noted that IDEA is best suited for earlier development phases because of its ease of use, but for late-phase development as well as for formulation evaluation, GastroPlus is ideal as it is more powerful for integrating data and takes into account factors IDEA does not, including dosage form, first-pass metabolism, and transport process [44].

ADAM also divides the GIT into nine sections, with blood flow segregated among the different sections. This approach has also enjoyed widespread use and generated accurate predictions in plasma concentration profiles, first-pass intestinal extraction ratios, and regional dose-absorbed fractions [29]. Yau et al. compared ADAM, $Q_{gut}$ (covered in a subsequent section), and Competing Rates model in predicting $F_g$, the fraction of drug having escaped both intestinal first-pass metabolism and transporter-mediated secretion to become available in the hepatic portal blood. They showed ADAM has a slight advantage in accuracy over the competing models [45]. Finally, though structurally similar to ACAT, ADAM also allows for the simulation of populations and their variabilities [37].

Several other multi-compartmental models have also been developed, each focusing on a particular area of the absorption process. The segregated intestinal flow model (SFM) and the $Q_{gut}$ model were developed to address differences in absorption by tissue region; absorption only occurs in a particular region of the intestine—the enterocytes at the tips of the villi—while the lumen does not participate in absorption. SFM, first conceived of by Cong et al. and further developed into the segmental segregated flow model (SSFM) by Pang's group [46–49], and $Q_{gut}$, developed by Yang et al. and based on Wilkinson and Shand's "well-stirred" model of hepatic drug clearance [50], are principally similar, compartmentalizing the blood flow perfusing the enterocytic region from the rest of the blood flow perfusing the small intestine; however, as Ando et al. and Hisaka et al. point out, $Q_{gut}$ does not consider drug concentration in the enterocytic region where nonlinear processes such as metabolism occur, making the model unable to analyze nonlinear pharmacokinetics [51, 52]. SFM, by contrast, considers intestinal metabolism, a significant factor in bioavailability for orally dosed drugs.

Indeed, SFM was able to explain differences in intestinal metabolism between intravenous and oral dosing [47]. The gastrointestinal transit absorption (GITA) model, developed by Sawamoto and colleagues [53], and the translocation model (TLM), developed by Ando and colleagues [51], were developed to address time- and location-dependent heterogeneity in intestinal drug absorption. GITA divides the GIT into eight segments, and for each segment, the absorption rate constant was determined by an *in situ* closed-loop method while the GI transit rate constant was determined using data from *in vivo* studies. Because it considers inter-segmental heterogeneity in drug absorption and transit through the GIT, GITA was able to predict site-specific oral drug absorption. GITA has also been employed to study the effects on oral absorption of drug interactions with food or other drugs [44, 53]. TLM, meanwhile, was developed using the free Napp program as a free alternative to commercial PBPK models and simplified relative to other models. As such, it models the GIT with minimal anatomical segmentation, allowing the underlying structure to be easily accessible to users [51].

### 44.2.3 Statistical Modeling

Statistical modeling refers to a broad category of statistics-based approaches that generate a model from data consisting of known inputs and outputs. In modeling bioavailability, the most prominent are quantitative structure-activity relationship (QSAR) or quantitative structure-property relationship (QSPR) models. We will use *QSPR* as a blanket term herein. In these models, the input data are molecular descriptors or other representations of the physicochemical properties of compounds and the output data are their corresponding activities or properties (e.g. metrics of bioavailability). After a model is constructed, it can be used to predict outputs given new inputs. The general workflow of statistical modeling of bioavailability can be expressed in the following steps:

1. curation of experimental data, split into training and test sets,
2. calculation of molecular descriptors or other quantities used as inputs,
3. construction, or "training," of statistical models,
4. and validation of models.

Experimental data is used to "train" models. This process involves identifying a mathematical relationship between the input variables (also called *predictor variables* or *features*; molecular descriptors; or other representations of the physicochemical properties of compounds in the case of QSPR) and output variables (also called *target variables*, e.g. metrics of bioavailability). Usually, not all the available experimental data is used to train the model; in a division of the experimental data often called a *train-test split*. A small subset, called the *test set*, is held back with which to validate the model later on, and the remaining data, the train set, is actually used to train the model. Validation involves using the model built with the train set to predict the output variables of the test set and comparing the predicted values to the actual values. The better the agreement between the predicted and actual values, the more accurate the model is said to be.

The algorithms that build these models fall under the broad category of machine learning (ML) approaches that discover the "rules" for a particular predictive model on their own (artificial intelligence) from the training data, rather than having those rules hard-coded into the algorithm, as compared to PBPK models in which the rules for the predictive model are decided beforehand by human intelligence and usually based on established knowledge found in the literature. Early ML QSPR models were trained using algorithms such as support vector machines (SVM) and regression analysis. Improvements in computing technology and advances in algorithm development in recent years have led to the popularity of artificial neural networks, a subset of machine learning somewhat based on the structure of biological neural networks in animal brains and in which the inputs are progressively passed through multiple layers, each layer extracting higher level features from the output of the previous layer. These types of algorithms perform best when using many layers, an approach called *deep learning*, which has only recently become practical due to the intensive computing costs associated with additional layers. A list of examples of the major categories of ML approaches covered here can be found in Table 44.2.

Because of the many possible ways a molecule can be represented in computational models (e.g. molecular descriptors, fragments, or fingerprints), the selection of the optimal set of input variables for models is another challenge that has received attention in the field of computational bioavailability modeling. Genetic algorithms, heuristics for solving optimization problems and based on evolutionary principles, were among the popular methods used for this task early on [66–68]. Later efforts that proved successful used correlation-based feature selection or even QSPR-specific feature-selection methods [56, 59].

**Table 44.2** A list of examples of the types of machine learning models covered in this chapter, and their respective performances. Where applicable, such as from studies that compared multiple machine learning models, only the best-performing model from each study are included in this table. Subscripts on performance metric variables indicate the type of cross-validation used.

| References | Algorithm | Target variable(s) (type) | Predictor variable(s) (type) (number) | Performance metrics | Dataset size |
|---|---|---|---|---|---|
| Andrews et al. [54] | Stepwise multiple linear regression | Oral bioavailability (percentage) | Substructure counts or combinations thereof (integer) (85) | $Q^2_{LOO} = 0.63$<br>$Q^2_{80/20} = 0.58$<br>$R^2 = 0.71$ | 591 |
| Moda et al. [55] | Partial least-squares regression | Oral bioavailability (percentage) | Weighted 2D molecular fragments | $Q^2 = 0.70$<br>$R^2 = 0.93$ | 302 |
| Ahmed and Ramakrishnan [56] | Logistic regression | HIA (binary);<br>Caco-2 permeability (binary) | Molecular descriptors (numeric) (47) | $\text{Accuracy}_{90/10} =$ 85.09% (HIA)<br>$\text{Accuracy}_{90/10} =$ 78.30% (Caco-2 permeability) | 605 (HIA)<br>705 (Caco-2 permeability) |
| Xu et al. [57] | SVM regression | Log oral bioavailability (numeric) | Molecular descriptors (numeric) (1536) | $Q^2_{80/20} = 0.73$<br>$R^2 = 0.84$ | 805 |
| Kumar et al. [58] | SVM classifier | Oral bioavailability (binary) | Molecular descriptors (numeric) (12) | $\text{Accuracy}_{90/10} =$ 96.85% | 511 |
| Ponzoni et al. [59] | Random-forest classifier | HIA (binary) | Molecular descriptors (numeric) | $\text{Accuracy}_{67/33} =$ 86.96% | 202 |
| Wang et al. [60] | Random-forest classifier | HIA (binary) | Molecular descriptors (numeric) (11) | $\text{Accuracy}_{80/20} =$ 88.5% | 970 |
| Suenderhauf et al. [61] | CHAID | HIA (binary) | Molecular descriptors (numeric) | CCR = 0.883<br>MCC = 0.751 | 458 |
| Ponzoni et al. [59] | Neural-network regression | HIA (percentage) | Molecular descriptors | $\rho_{75/25} = 0.76$ | 202 |
| Suenderhauf et al. [61] | Multi-layer-perceptron regression | HIA (percentage) | Molecular descriptors | $R^2 = 0.6$ | 458 |
| Du-Cuny et al. [62] | *k*-nearest neighbors | Caco-2 permeability (numeric) | Molecular descriptors<br>(5) | $Q^2 = 0.95$<br>$R^2_{test} = 0.64$<br>$R^2_{external} = 0.81$ | 109 |
| Yoshida and Topliss [63] | ORMUCS | Oral bioavailability (categorical);<br>oral bioavailability (percentage) | Molecular descriptors (numeric) (18) | $\text{Accuracy}_{LOO} = 67\%$<br>$\rho_{LOO} = 0.812$ | 232 |
| Falcón-Cano et al. [64] | Consensus | Oral bioavailability (binary) | Molecular descriptors (numeric) (>3800) | $\text{Accuracy}_{90/10} = 76\%$ | 1448 |
| Kim et al. [65] | Consensus | Log*K* (oral bioavailability); oral bioavailability (binary) | Molecular descriptors | $R^2 = 0.40$<br>Accuracy = 76% | 362 |

LOO = leave-one-out cross-validation. A subscript of form [number]/[number] indicates the train-test split of the dataset used for cross-validation; e.g. $Q^2_{80/20}$ indicates an 80/20 train-test split; i.e. 80% of the dataset was used to train the model and 20% was used to evaluate the trained model. Lack of a subscript indicates an unspecified cross-validation method or that no cross-validation was performed. HIA = intestinal absorption. $\rho$ = Spearman's rank correlation coefficient. CCR = corrected classification rate. MCC = Matthews correlation coefficient. CHAID = chi-squared automatic interaction detector. ORMUCS = ordered multicategorical classification method using the simplex technique. LogK(oral bioavailability) is an "equilibrium-like" term: $\log K(\%\text{F}) = \log\left(\frac{\%F}{100-\%F}\right)$, in which $\%F$ is oral bioavailability.

#### 44.2.3.1 Regression Analysis

As one of the simplest classes of ML algorithms, regression analysis involves estimating the relationship between a set of input variables and their corresponding outputs. One of the earliest forerunners of a computational bioavailability model was built in the 1970s by Cammarata and Rogers, who used multiple regression analysis to predict the solubility of drugs in biological systems [69]. Nearly 30 years later, Andrews et al. introduced one of the first true QSPR models of bioavailability, a stepwise multiple linear regression (MLR) model [54]. Since then, several other regression models have been developed and have been used to predict the bioavailability of both drugs and environmental toxins. Despite their relative simplicity, regression models have performed well in multiple studies. Among these models was one developed by Moda et al. using a hologram-QSAR approach, which uses partial least squares (PLS) regression [55, 70]. The model produced impressive results on the researchers' dataset, with their most significant model achieving a $Q^2$ of 0.70 and an $R^2$ of 0.93, but performance of this model was not directly compared to other ML algorithms [55]. Ahmed and Ramakrishnan did directly compare the performance of 21 ML algorithms in predicting intestinal absorption (HIA), showing logistic regression as the most accurate, outperforming even tree-based and neural-network algorithms, approaches that have generally performed quite well [56]. (See the following sections for an overview of these algorithms.)

#### 44.2.3.2 Support Vector Machines

Support vector machines (SVMs) are another class of machine learning algorithms that have been successfully used to predict bioavailability. Classification is the more typical use for SVMs, although they have been employed for regression as well in predicting bioavailability. In the use of SVMs for classification, the input variables of each sample in the training dataset are organized as an input vector and mapped into a higher-dimension space where the data is linearly separated into their different classes (i.e. the output variable). In the new higher-dimension space, each sample is represented as a point, and the training involves finding the hyperplane of maximum distance to a data point belonging to any class.

The performance of SVMs in predicting bioavailability compares favorably to that of other ML algorithms. A study by Xu et al. comparing SVMs, partial least squares regression (PLS), and MLS showed SVMs to be the most accurate, with both the highest $R^2$ (0.84) and $Q^2_{\text{test}}$ (0.73) of the three classes of algorithms [57]. Similarly, a study by Kumar comparing SVMs, ANNs, and Bayesian classifiers in predicting the oral bioavailability of a set of 511 chemically diverse drugs showed SVMs have superior performance in five of the six performance metrics (Youden's index, area under the curve, accuracy, sensitivity, specificity, and precision) used to evaluate the algorithms [71]. Interestingly, in Suenderhauf et al.'s evaluation of different ML approaches in predicting HIA, SVMs were among the best in numerical prediction but had only mediocre results in classification [61].

#### 44.2.3.3 Decision Trees

Decision tree-based algorithms are another popular class of machine learning techniques used for both classification and regression problems. At its most basic, a decision tree is a tree-like or flowchart-like model in which the leaves (terminal nodes) represent the output variables—class labels or probability distribution over the classes in the case of classification models or numeric values in the case of regression models—and the branches represent conjunctions of the input variables pertaining to a given output (class or numeric value). Each internal node is labeled with an input feature, and the branches coming off from the node represent possible values of that input feature. Training a decision tree involves finding the partitions on the attributes of a set of training data such that at each partition, the resulting nodes are as pure as possible, i.e. containing only or mostly data points from a single class.

Decision trees are not often used alone; popular decision tree-based algorithms known as *ensemble* methods build multiple decision trees and aggregate the output from each individual tree, e.g. by consensus, to produce a final prediction. Ensemble models such as random forests and gradient-boosted trees have enjoyed widespread success in a variety of applications, and they have demonstrated their capacity for modeling oral bioavailability as well. Although not as popular in the early days of computational bioavailability models, tree-based methods, and random forests in particular, have been used in several more recent studies and performed comparably to or even outperformed techniques with a proven track record such as SVM or linear regression [56, 59, 60, 65].

#### 44.2.3.4 Artificial Neural Networks

Artificial neural networks (ANNs, or simply *neural networks*, NNs) are algorithms roughly based on the structure of biological neural networks. The basic unit of an NN is an artificial neuron (or simply *neuron*), a mathematical function meant to model the action of a biological neuron. In a typical setup, a neuron takes one or more input signals in the form of real-numbered values, in a process representing the reception of a signal via its dendrites by a neuron from a previous neuron; weights these inputs, and sums them together; then passes this sum through a function called a *transfer* or *activation function*. The output of this function is then passed on to the next neuron in the network, in a process representing the transfer of a neuron's action potential along its axon. In a typical NN, neurons are arranged in parallel into groups called *layers*; in deep learning, NNs consist of multiple sequential layers, allowing an input signal to be propagated through and processed by large numbers of neurons to generate the final output. Training an NN involves finding the set of input weights for each neuron such that for a given set of inputs, the discrepancy between the output produced by the NN and the actual output corresponding to those inputs is minimized.

NNs are not a new concept, having been around in a primitive conception since at least the 1940s [72, 73] and implemented since at least the late 1950s when Rosenblatt created the perceptron [74, 75], an algorithm for supervised learning of binary classifiers and considered to be the first modern neural network. However, as Minsky and Papert discovered in the late 1960s [76], NNs require multiple layers to produce results accurate enough for practical purposes, but until recently, computers lacked the power to process these deeper NNs [76]. In recent years, with great advances being made in computer processing power, NNs have become a feasible option for machine learning applications and have seen a dramatic rise in popularity. Algorithmic development has likewise been rapid, with improvements addressing fundamental issues such as efficiency and the problem of overfitting. The use of multiple processing layers in deep learning allows these algorithms to learn representations of data with multiple levels of abstraction, leading to the success of deep neural networks across different biomedical fields as diverse as image-based diagnosis and protein-structure prediction [77–82]. Although at least one NN-based model was developed to predict bioavailability in the early 2000s [83], it wasn't until the past decade that NNs have seen more widespread use. Their performances generally compare favorably to other methods; Ahmed and Ramakrishnan's comparison of ML methods in predicting HIA showed multilayer perceptrons to be among the most accurate models, with performances similar to those of random forests [56]. Suenderhauf and colleagues' comparison of ML methods in predicting HIA, meanwhile, showed multilayer perceptrons to be the best of the tested algorithms in terms of numerical predictions but with only middling results in classification [61].

#### 44.2.3.5 Other Algorithms

Another one of the first ML algorithms applied to the modeling of bioavailability used simplices, an approach to solving linear programming models. The algorithm, named ORMUCS (ordered multicategorical classification method using the simplex technique), was a classifier developed by Yoshida and Topliss around the same time Andrews et al. developed their MLR-based model [63]. Other algorithms that have been used, with varying degrees of success, include *k*-nearest neighbors (kNN), adaptive fuzzy partitioning (AFP), and Bayesian approaches [61, 64].

## 44.3 Conclusions

Computational modeling of oral bioavailability has seen major improvements in the past few decades, to the point where *in silico* models are commonly used in conjunction with bench experiments in commercial drug design and development. Statistical and physiologically based models remain the two major classes of computational models. While both are widely used, few direct comparisons of these two types of models have been performed, making it difficult to say with certainty in what situations each class would have a comparative advantage. However, the strengths and weaknesses of each type is well defined: statistical models, as they rely only on observational data and not on established physiological rules, are both freer to identify patterns that physiologically based models would miss and more subject to inaccuracies due to errors in the data used to build the models. Also, statistical models are limited by the amount of available experimental data, which in some cases may be scarce due to the financial and time costs associated with performing certain experiments. Physiologically based models, on the other hand, have the opposite challenges and advantages: they rely less on large amounts of experimental data and are based on established biological principles, making them less susceptible to inaccuracies due to errors in the observed data and more easily interpretable from a mechanistic standpoint [37]. However, the accuracy of physiologically based models is constrained by the complexity of the models. The more closely the model takes into accounts all the intricacies of actual physiological processes, the more computationally expensive the model becomes. Moreover, as new discoveries are made, the theories underpinning these models may become outdated.

One obvious solution to these shortcomings would be to use hybrid models that incorporate both statistical and physiologically based modeling techniques. The most common method of integrating these two types of models would be to use QSPR models to predict parameters for the PBPK models, such as partition coefficients or metabolic constants. Many of the published studies using such hybrid approaches are in the field of toxicology, with few applications in drug discovery or design [39, 84–88]. Although hybrid models would in theory use the strengths of each type of model to make up for the weaknesses of the other, it is difficult to say whether these integrated approaches significantly improve upon the performance of statistical or physiologically based models used alone, as direct comparisons of the different types of models are scarce and, again, published studies of hybrid models themselves are not widely available.

As this is the case with all statistical models, QSPR models of bioavailability would most benefit from more training data. Improvements in and more widespread use of high-throughput screening technology will almost certainly assist in this area. PBPK models, meanwhile, would most benefit from advances in technology that would allow for the quicker computation of more complex models, or, alternatively, from innovations that simplify the mathematics behind the models while maintaining acceptable accuracy. Finally, more direct comparisons of statistical, physiologically based, and hybrid models are needed to identify the areas to which each is best suited. The use of predictive modeling in all areas of biomedical research has expanded tremendously over the past few decades, and *in silico* prediction of bioavailability is one of the areas that has seen technical improvements that have made it an indispensable part of the drug-design process. Nevertheless, computational models of bioavailability still rely heavily on experimental data; continued advances in the state of the art and increased data sharing among researchers is necessary for *in silico* methods to be used in more independent capacities.

## Acknowledgment

This work was partially supported by CPRIT RP170333, NIH 1R01CA225955-01, Koch Foundation Awards, and the University of Texas MD Anderson Cancer Center Institutional Research Grant (IRG) Program. We also thank Luis Hernandez for the initial discussion of the book chapter.

## References

1 Sim, D.S.M. (2015). Drug absorption and bioavailability. In: *Pharmacological Basis of Acute Care* (ed. Y.K. Chan, K.P. Ng, and D.S.M. Sim), 17–26. (Springer International Publishing https://doi.org/10.1007/978-3-319-10386-0_3.

2 Tandel, H., Florence, K., and Misra, A. (2011). Chapter 9 - Protein and peptide delivery through respiratory pathway. In: *Challenges in Delivery of Therapeutic Genomics and Proteomics* (ed. A. Misra), 429–479. Elsevier https://doi.org/10.1016/B978-0-12-384964-9.00009-8.

3 Lipinski, C.A., Lombardo, F., Dominy, B.W., and Feeney, P.J. (1997). Experimental and computational approaches to estimate solubility and permeability in drug discovery and development settings. *Adv. Drug Delivery Rev.* 23: 3–25.

4 Veber, D.F. et al. (2002). Molecular properties that influence the oral bioavailability of drug candidates. *J. Med. Chem.* 45: 2615–2623.

5 Martin, Y.C. (2005). A bioavailability score. *J. Med. Chem.* 48: 3164–3170.

6 Hou, T., Wang, J., Zhang, W., and Xu, X. (2007). ADME evaluation in drug discovery. 7. Prediction of oral absorption by correlation and classification. *J. Chem. Inf. Model.* 47: 208–218.

7 Zhu, J., Wang, J., Yu, H. et al. (2011). Recent developments of in silico predictions of oral bioavailability. *Comb. Chem. High Throughput Screening* 14: 362–374.

8 Bolger, M.B., Agoram, B., Fraczkiewicz, R., and Steere, B. (2004). Simulation of absorption, metabolism, and bioavailability. In: *Drug Bioavailability*, 420–443. John Wiley & Sons, Ltd https://doi.org/10.1002/3527601473.ch18.

9 Bolger, M.B., Fraczkiewicz, R., and Lukacova, V. (2009). Simulations of absorption, metabolism, and bioavailability. In: *Drug Bioavailability*, 453–495. John Wiley & Sons, Ltd https://doi.org/10.1002/9783527623860.ch17.

10 Korzekwa, K. and Nagar, S. (2017). On the nature of physiologically-based pharmacokinetic models-a priori or a posteriori? mechanistic or empirical? *Pharm. Res.* 34: 529–534.

11 Zhao, P. et al. (2011). Applications of physiologically based pharmacokinetic (PBPK) modeling and simulation during regulatory review. *Clin. Pharmacol. Ther.* 89: 259–267.

12 Sager, J.E., Yu, J., Ragueneau-Majlessi, I., and Isoherranen, N. (2015). Physiologically based pharmacokinetic (PBPK) modeling and simulation approaches: a systematic review of published models, applications, and model verification. *Drug Metab. Dispos.* 43: 1823–1837.

13 Obach, R.S. et al. (1997). The prediction of human pharmacokinetic parameters from preclinical and in vitro metabolism data. *J. Pharmacol. Exp. Ther.* 283: 46–58.

14 Wu, B., Zhang, S., and Hu, M. (2011). Evaluation of 3,3′,4′-trihydroxyflavone and 3,6,4′-trihydroxyflavone (4′-O-glucuronidation) as the in vitro functional markers for hepatic UGT1A1. *Mol. Pharmaceutics* 8: 2379–2389.

15 Wu, B., Kulkarni, K., Basu, S. et al. (2011). First-pass metabolism via UDP-glucuronosyltransferase: a barrier to oral bioavailability of phenolics. *J. Pharm. Sci.* 100: 3655–3681.

16 Wu, B., Wang, X., Zhang, S., and Hu, M. (2012). Accurate prediction of Glucuronidation of structurally diverse phenolics by human UGT1A9 using combined experimental and in silico approaches. *Pharm. Res.* 29: 1544–1561.

17 Wu, B., Dong, D., Hu, M., and Zhang, S. (2013). Quantitative prediction of glucuronidation in humans using the in vitro- in vivo extrapolation approach. *Curr. Top. Med. Chem.* 13: 1343–1352.

18 Ji, B. et al. (2019). Prediction of drug-drug interactions between opioids and overdosed benzodiazepines using physiologically based pharmacokinetic (PBPK) modeling and simulation. *Drugs R&D* 19: 297–305.

19 Li, S. et al. (2019). Prediction of pharmacokinetic drug-drug interactions causing atorvastatin-induced rhabdomyolysis using physiologically based pharmacokinetic modelling. *Biomed. Pharmacother.* 119: 109416.

20 Chetty, M. et al. (2014). Applications of linking PBPK and PD models to predict the impact of genotypic variability, formulation differences, differences in target binding capacity and target site drug concentrations on drug responses and variability. *Front. Pharmacol.* 5: 258.

21 Kesisoglou, F., Balakrishnan, A., and Manser, K. (2016). Utility of PBPK absorption modeling to guide modified release formulation development of Gaboxadol, a highly soluble compound with region-dependent absorption. *J. Pharm. Sci.* 105: 722–728.

22 Nothnagel, L. et al. (2019). Predictive PBPK modeling as a tool in the formulation of the drug candidate TMP-001. *Eur. J. Pharm. Biopharm.* 134: 144–152.

23 Lin, L. and Wong, H. (2017). Predicting oral drug absorption: mini review on physiologically-based pharmacokinetic models. *Pharmaceutics* 9: 41.

24 Noyes, A.A. and Whitney, W.R. (1897). The rate of solution of solid substances in their own solutions. *J. Am. Chem. Soc.* 19: 930–934.

25 Dressman, J.B. and Fleisher, D. (1986). Mixing-tank model for predicting dissolution rate control or oral absorption. *J. Pharm. Sci.* 75: 109–116.

26 Yu, L.X., Crison, J.R., and Amidon, G.L. (1996). Compartmental transit and dispersion model analysis of small intestinal transit flow in humans. *Int. J. Pharm.* 140: 111–118.

27 Yu, L.X. (1999). An integrated model for determining causes of poor oral drug absorption. *Pharm. Res.* 16: 1883–1887.

28 Agoram, B., Woltosz, W.S., and Bolger, M.B. (2001). Predicting the impact of physiological and biochemical processes on oral drug bioavailability. *Adv. Drug Delivery Rev.* 50 (Suppl 1): S41–S67.

29 Jamei, M. et al. (2009). Population-based mechanistic prediction of oral drug absorption. *AAPS J.* 11: 225–237.

30 Willmann, S. et al. (2003). PK-Sim®: a physiologically based pharmacokinetic 'whole-body' model. *Biosilico* 1: 121–124.

31 Willmann, S., Schmitt, W., Keldenich, J. et al. (2004). A physiological model for the estimation of the fraction dose Absorbed in humans. *J. Med. Chem.* 47: 4022–4031.

32 Eissing, T. et al. (2011). A computational systems biology software platform for multiscale modeling and simulation: integrating whole-body physiology, disease biology, and molecular reaction networks. *Front. Physiol.* 2: 4.

33 Thomas, S., Brightman, F., Gill, H. et al. (2008). Simulation modelling of human intestinal absorption using Caco-2 permeability and kinetic solubility data for early drug discovery. *J. Pharm. Sci.* 97: 4557–4574.

34 Gabrielsson, J., Andersson, K., Tobin, G. et al. (2014). Maxsim2—Real-time interactive simulations for computer-assisted teaching of pharmacokinetics and pharmacodynamics. *Comput. Meth. Programs Biomed.* 113: 815–829.

35 Levitt, D.G. (2002). PKQuest: a general physiologically based pharmacokinetic model. introduction and

application to propranolol. *BMC Clin. Pharmacol.* 2: 5.

36 Parrott, N., Paquereau, N., Coassolo, P., and Lavé, T. (2005). An evaluation of the utility of physiologically based models of pharmacokinetics in early drug discovery. *J. Pharm. Sci.* 94: 2327–2343.

37 Cabrera-Pérez, M.Á. and Pham-The, H. (2018). Computational modeling of human oral bioavailability: what will be next? *Expert Opin. Drug Discovery* 13: 509–521.

38 Paixão, P., Gouveia, L.F., and Morais, J.A.G. (2012). Prediction of the human oral bioavailability by using in vitro and in silico drug related parameters in a physiologically based absorption model. *Int. J. Pharm.* 429: 84–98.

39 Daga, P.R., Bolger, M.B., Haworth, I.S. et al. (2018). Physiologically based pharmacokinetic modeling in lead optimization. 1. evaluation and adaptation of GastroPlus to predict bioavailability of medchem series. *Mol. Pharmaceutics* 15: 821–830.

40 Olivares-Morales, A., Ghosh, A., Aarons, L., and Rostami-Hodjegan, A. (2016). Development of a novel simplified PBPK absorption model to explain the higher relative bioavailability of the OROS® formulation of oxybutynin. *AAPS J.* 18: 1532–1549.

41 Grass, G.M. (1997). Simulation models to predict oral drug absorption from in vitro data. *Adv. Drug Delivery Rev.* 23: 199–219.

42 Norris, D.A., Leesman, G.D., Sinko, P.J., and Grass, G.M. (2000). Development of predictive pharmacokinetic simulation models for drug discovery. *J. Controlled Release* 65: 55–62.

43 Parrott, N. and Lavé, T. (2002). Prediction of intestinal absorption: comparative assessment of GASTROPLUS and IDEA. *Eur. J. Pharm. Sci.* 17: 51–61.

44 Huang, W., Lee, S.L., and Yu, L.X. (2009). Mechanistic approaches to predicting oral drug absorption. *AAPS J.* 11: 217–224.

45 Yau, E., Petersson, C., Dolgos, H., and Peters, S.A. (2017). A comparative evaluation of models to predict human intestinal metabolism from nonclinical data. *Biopharm. Drug Dispos.* 38: 163–186.

46 Cong, D., Doherty, M., and Pang, K.S. (2000). A new physiologically based, segregated-flow model to explain route-dependent intestinal metabolism. *Drug Metab. Dispos.* 28: 224–235.

47 Pang, K.S., Yang, Q.J., and Noh, K. (2017). Unequivocal evidence supporting the segregated flow intestinal model that discriminates intestine versus liver first-pass removal with PBPK modeling. *Biopharm. Drug Dispos.* 38: 231–250.

48 Noh, K. and Pang, K.S. (2019). Theoretical consideration of the properties of intestinal flow models on route-dependent drug removal: Segregated Flow (SFM) vs. Traditional (TM). *Biopharm. Drug Dispos.* 40: 195–213.

49 Pang, K.S., Peng, H.B., and Noh, K. (2020). The segregated intestinal flow model (SFM) for drug absorption and drug metabolism: implications on intestinal and liver metabolism and drug-drug interactions. *Pharmaceutics* 12: 312.

50 Yang, J., Jamei, M., Yeo, K.R. et al. (2007). Prediction of intestinal first-pass drug metabolism. *Curr. Drug Metabol.* 8: 676–684.

51 Ando, H., Hisaka, A., and Suzuki, H. (2015). A new physiologically based pharmacokinetic model for the prediction of gastrointestinal drug absorption: translocation model. *Drug Metab. Dispos.* 43: 590–602.

52 Hisaka, A., Nakamura, M., Tsukihashi, A. et al. (2014). Assessment of intestinal availability (FG) of substrate drugs of Cytochrome P450s by analyzing changes in pharmacokinetic properties caused by drug–drug interactions. *Drug Metab. Dispos.* 42: 1640–1645.

53 Sawamoto, T., Haruta, S., Kurosaki, Y. et al. (1997). Prediction of the plasma concentration profiles of orally administered drugs in rats on the basis of gastrointestinal transit kinetics and absorbability. *J. Pharm. Pharmacol.* 49: 450–457.

54 Andrews, C.W., Bennett, L., and Yu, L.X. (2000). Predicting human oral bioavailability of a compound: development of a novel quantitative structure-bioavailability relationship. *Pharm. Res.* 17: 639–644.

55 Moda, T.L., Montanari, C.A., and Andricopulo, A.D. (2007). Hologram QSAR model for the prediction of human oral bioavailability. *Bioorg. Med. Chem.* 15: 7738–7745.

56 Ahmed, S.S.S.J. and Ramakrishnan, V. (2012). Systems biological approach of molecular descriptors connectivity: optimal descriptors for oral bioavailability prediction. *PLoS One* 7: e40654.

57 Xu, X. et al. (2012). A novel chemometric method for the prediction of human oral bioavailability. *Int. J. Mol. Sci.* 13: 6964–6982.

58 Kumar, R., Sharma, A., and Varadwaj, P.K. (2011). A prediction model for oral bioavailability of drugs using physicochemical properties by support vector machine. *J. Nat. Sci, Biol. Med.* 2: 168–173.

59 Ponzoni, I. et al. (2017). Hybridizing feature selection and feature learning approaches in QSAR modeling for drug discovery. *Sci. Rep.* 7: 1–9.

60 Wang, N.-N. et al. (2017). Predicting human intestinal absorption with modified random forest approach: a comprehensive evaluation of molecular representation, unbalanced data, and applicability domain issues. *RSC Adv.* 7: 19007–19018.

61 Suenderhauf, C., Hammann, F., Maunz, A. et al. (2011). Combinatorial QSAR modeling of human intestinal absorption. *Mol. Pharmaceutics* 8: 213–224.

62 Du-Cuny, L. et al. (2009). Computational modeling of novel inhibitors targeting the Akt pleckstrin homology domain. *Bioorg. Med. Chem.* 17: 6983–6992.

63 Yoshida, F. and Topliss, J.G. (2000). QSAR model for drug human oral bioavailability. *J. Med. Chem.* 43: 2575–2585.

64 Falcón-Cano, G., Molina, C., and Cabrera-Pérez, M.Á. (2020). ADME prediction with KNIME: development and validation of a publicly available workflow for the prediction of human oral bioavailability. *J. Chem. Inf. Model.* 60: 2660–2667.

65 Kim, M.T., Sedykh, A., Chakravarti, S.K. et al. (2014). Critical evaluation of human oral bioavailability for pharmaceutical drugs by using various cheminformatics approaches. *Pharm. Res.* 31: 1002–1014.

66 Wang, J., Krudy, G., Xie, X.-Q. et al. (2006). Genetic algorithm-optimized QSPR models for bioavailability, protein binding, and urinary excretion. *J. Chem. Inf. Model.* 46: 2674–2683.

67 Ma, C.Y. et al. (2008). Prediction models of human plasma protein binding rate and oral bioavailability derived by using GA-CG-SVM method. *J. Pharm. Biomed. Anal.* 47: 677–682.

68 Tian, S., Li, Y., Wang, J. et al. (2011). ADME evaluation in drug discovery. 9. Prediction of oral bioavailability in humans based on molecular properties and structural fingerprints. *Mol. Pharmaceutics* 8: 841–851.

69 Cammarata, A. and Rogers, K.S. (1971). Electronic representation of lipophilic parameter Pi. *J. Med. Chem.* 14: 269.

70 Heritage, T.W. and Lowis, D.R. (1999). Molecular Hologram QSAR. In: *Rational Drug Design*, vol. 719, 212–225. (American Chemical Society).

71 Kumar, R., Sharma, A., Varadwaj, P., Ahmad, A. & Ashraf, G. Classification of oral bioavailability of drugs by machine learning approaches: a comparative study. (2012). https://doi.org/10.6062/JCIS.2011.02.03.0045.

72 McCulloch, W.S. and Pitts, W. (1943). A logical calculus of the ideas immanent in nervous activity. *Bull. Math. Biophys.* 5: 115–133.

73 Chakraverty, S., Sahoo, D.M., and Mahato, N.R. (2019). McCulloch–Pitts Neural Network Model. In: *Concepts of Soft Computing: Fuzzy and ANN with Programming* (ed. S. Chakraverty, D.M. Sahoo, and N.R. Mahato), 167–173. (Springer https://doi.org/10.1007/978-981-13-7430-2_11.

74 Rosenblatt, F. (1958). The perceptron: a probabilistic model for information storage and organization in the brain. *Psychol. Rev.* 65: 386–408.

75 Russell, S.J., Russell, S., and Norvig, P. (2019). *Artificial Intelligence: A Modern Approach*. Upper Saddle River, NJ: Prentice Hall.

76 Minsky, M. and Papert, S. (1969). *Perceptrons; an Introduction to Computational Geometry*. Cambridge, MA: MIT Press.

77 Hosny, A., Parmar, C., Quackenbush, J. et al. (2018). Artificial intelligence in radiology. *Nat. Rev. Cancer* 18: 500–510.

78 Ngiam, K.Y. and Khor, I.W. (2019). Big data and machine learning algorithms for health-care delivery. *Lancet Oncol.* 20: e262–e273.

79 Torrisi, M., Pollastri, G., and Le, Q. (2020). Deep learning methods in protein structure prediction. *Comput. Struct. Biotechnol. J.* 18: 1301–1310.

80 Wardah, W., Khan, M.G.M., Sharma, A., and Rashid, M.A. (2019). Protein secondary structure prediction using neural networks and deep learning: a review. *Comput. Biol. Chem.* 81: 1–8.

81 Noé, F., De Fabritiis, G., and Clementi, C. (2020). Machine learning for protein folding and dynamics. *Curr. Opin. Struct. Biol.* 60: 77–84.

82 Senior, A.W. et al. (2020). Improved protein structure prediction using potentials from deep learning. *Nature* 577: 706–710.

83 Turner, J.V., Maddalena, D.J., and Agatonovic-Kustrin, S. (2004). Bioavailability prediction based on molecular structure for a diverse series of drugs. *Pharm. Res.* 21: 68–82.

84 Price, K. and Krishnan, K. (2011). An integrated QSAR-PBPK modelling approach for predicting the inhalation toxicokinetics of mixtures of volatile organic chemicals in the rat. *SAR QSAR Environ. Res.* 22: 107–128.

85 Knaak, J.B. et al. (2012). Parameters for pyrethroid insecticide QSAR and PBPK/PD models for human risk assessment. *Rev. Environ. Contam. Toxicol.* 219: 1–114.

86 Yang, R.S. et al. (1998). Approaches to developing alternative and predictive toxicology based on PBPK/PD and QSAR modeling. *Environ. Health Perspect.* 106 (Suppl 6): 1385–1393.

**87** Verhaar, H.J. et al. (1997). A proposed approach to study the toxicology of complex mixtures of petroleum products: the integrated use of QSAR, lumping analysis and PBPK/PD modeling. *Environ. Health Perspect.* 105 (Suppl 1): 179–195.

**88** Vlot, A.H.C. et al. (2017). Target and tissue selectivity prediction by integrated mechanistic pharmacokinetic-target binding and quantitative structure activity modeling. *AAPS J.* 20: 11.

# 45

# Blood–Brain Barrier Permeability Assessment for Small-Molecule Drug Discovery Using Computational Techniques

*Yankang Jing*[1,2,3,4,5] *and Xiang-Qun Xie*[1,2,3,4,5]

[1]*Department of Pharmaceutical Sciences, Computational Chemical Genomics Screening Center, University of Pittsburgh, Pittsburgh, PA, USA*
[2]*Pharmacometrics and System Pharmacology (PSP) PharmacoAnalytics Program, School of Pharmacy, University of Pittsburgh, Pittsburgh, PA, USA*
[3]*National Center of Excellence for Computational Drug Abuse Research, University of Pittsburgh, Pittsburgh, PA, USA*
[4]*Drug Discovery Institute, University of Pittsburgh, Pittsburgh, PA, USA*
[5]*Departments of Computational Biology and Structural Biology, School of Medicine, University of Pittsburgh, Pittsburgh, PA, USA*

## 45.1 Introduction

The blood–brain barrier (BBB) is a physiological and biochemical barrier between the CNS and the peripheral tissues that serves as the primary active interface [1] to prevent the neurotoxic plasma components, blood cells, and pathogens from entering the brain [2]. The BBB only permits the movement of selective molecules that are essential for keeping the brain function to maintain the homeostasis of the CNS [3]. For drugs administered orally, the blood-brain barrier (BBB) poses a more substantial obstacle as it represents the primary driving force for permeation across the BBB. Free drug concentrations in the blood are typically low due to the high degree of protein binding most drugs exhibit (see also Chapter 7).

Understanding the BBB permeability of small-molecule drugs is vitally important in not only CNS but also non-CNS drug discovery. In CNS drug development, it is crucial for small-molecule drugs to cross the BBB to achieve their therapeutic effect. However, the failure rate for an effective drug targeting CNS diseases is very high compared to most other non-CNS areas of drug discovery, and there is an urgent need for drugs against many CNS diseases that lack effective treatment [4]. For example, CNS disorders such as Alzheimer's disease (AD) are becoming one of the major burdensome disease areas [5]. Also, the deaths from major CNS disorders were ranked third (10.8%) in the United States among all causes of death and fifth (5.5%) globally ([6]). Meanwhile, the global sales of CNS disease drugs and related therapeutical products totaled more than $80 billion in the year 2019 and are forecasted to be more than $100 billion in the year 2022 [7]. The BBB is the major hurdle for CNS drug delivery; therefore, the role of the BBB must be considered to develop a successful treatment for CNS diseases [8]. On the other hand, it is also important for non-CNS drug development to avoid CNS penetration as it may cause potential CNS toxicities.

## 45.2 Basic Principle of the BBB Permeation

The BBB is a physiological and biochemical barrier between the CNS and the peripheral tissues. The BBB serves as the primary active interface between the changeable blood environment of CNS and the extracellular fluid [1] and prevents the neurotoxic plasma components, blood cells, and pathogens from entering the brain [2]. By regulating the transport of small molecules or macromolecules into and out of the brain, the BBB only permits the movement of selective molecules that are essential for keeping the brain function to maintain the homeostasis of the CNS [3].

Generally, the BBB is a capillary wall composed of brain endothelial cells, basement membrane, pericytes, vascular smooth muscle cells, astrocytes, and others [9]. The main physiological structures responsible for the BBB transport mainly include the tight cell–cell junctions, endothelial and pericyte transporters, and perivascular transport [2]. There are several types of transport routes for molecules to cross the BBB: paracellular and transcellular diffusion, carrier-mediated transport, and transcytosis [10].

*Oral Bioavailability and Drug Delivery: From Basics to Advanced Concepts and Applications*, First Edition. Edited by Ming Hu and Xiaoling Li.

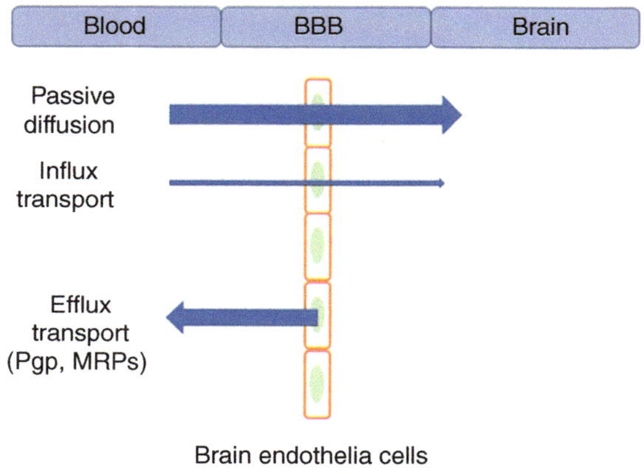


**Figure 45.1** General pathway across the BBB. The transport of small molecules across the BBB mainly through passive diffusion and influx transport. Efflux transport proteins can also help with the elimination of endogenous and exogenous compounds.

The transport of most of the molecules between the brain and vascular system is mainly through transcellular transport (Figure 45.1); therefore, a few small molecules can enter the brain by paracellular and transcellular diffusion due to the tight junctions between capillary endothelial cells [8]. The brain endothelial cells also provide alternate transport pathways such as through active efflux transport proteins, including P-glycoprotein (P-gp) and breast cancer resistance protein (BCRP); other substances including large molecules may access the brain through transcytosis [11]. In addition, the BBB also helps with the clearance of metabolites and toxins in the brain and regulates the composition and volume of the cerebrospinal fluid [3].

## 45.3 Role of the BBB in Drug Delivery

Because of these characteristics, the BBB becomes one of the most complicated microenvironments in drug discovery and limits the development of novel drugs targeting CNS diseases. For example, more than 98% of small-molecule drug candidates have limited delivery to the brain [12], with water-soluble molecule drugs in the blood being prevented from entering the CNS and lipophilic molecule drugs being excluded by efflux transporters [13]. Thus, it is vitally important to optimize the BBB impermeability of the drug so as to offer an effective treatment for CNS diseases. More researchers have made efforts on this topic (Figure 45.2).

## 45.4 Experimental Methods for Assessing BBB Permeability

The scientific interests in the pathology and physiology of the BBB accelerated the development of various experimental methods for BBB permeability assessment. To speed up the discovery of novel CNS drugs, different types of experimental BBB models have been established, including both in vitro and in vivo methods [14, 15].

In vitro methods for evaluating the BBB permeability of small molecules involve two types. The first one aims to test the physiochemical properties of the chemical. For example, the parallel artificial membrane permeability assay (PAMPA) methods were designed to evaluate the passive permeability of the compound across BBB by measuring the kinetic properties using porcine brain lipid in dodecane as the artificial permeability membrane [16]. In addition, BBB permeability can also be evaluated using log $P$ and log $D$ from an *in vitro* assay. Generally, the optimal log $D$ for brain penetration is 1 to 3 [17].

Another type of *in vitro* method to assess BBB permeability adopts cell-based assays. Commonly used cell-based assays include Caco-2 methods and MDR1

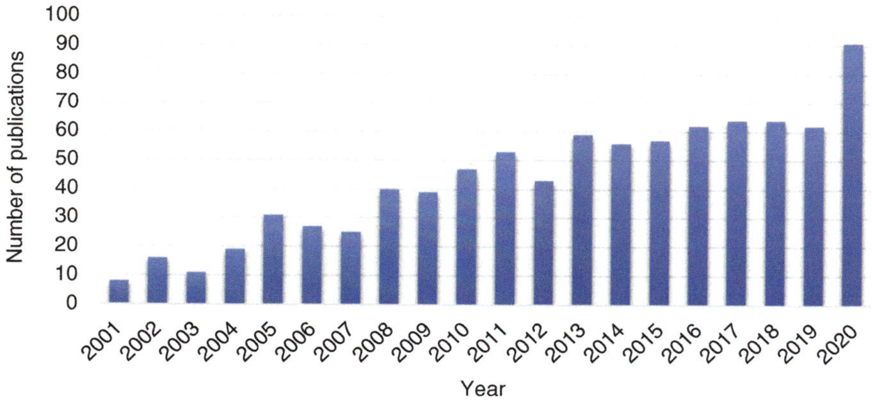


**Figure 45.2** Survey on BBB permeability modeling-related publications by SciFinder to show the general trend.

methods. Caco-2 methods were initially developed as a permeation model for assessing gastrointestinal absorption and then were extended to BBB permeability. It incorporates lipid bilayer membranes, P-gp efflux, and some other transporters [18]. MDCK/MDR1 methods are widely used for BBB permeability prediction rather than brain distribution [19]. The MDCK cells have low transporter expression levels and are also commonly used as a model for passive diffusion permeability prediction. In summary, cell-based BBB assays have a superiority compared to physicochemical methods because they are living and closer to the actual situation. However, those methods may still not closely resemble the complex conditions at the BBB.

*In vivo* methods are also widely used in the industry for BBB permeability prediction. For example, the *in vivo* B/P experiment is considered one of the standard approaches to evaluate BBB permeability [20]. The B/P experiment can provide brain distribution data over time. Typically, animal samples are firstly dosed at time zero and sacrificed at designated time points; then the compound concentrations in the plasma and brain are measured over time; and finally, the B/P ratio can be calculated (log BB value is the log of B/P). The advantage of this approach is that the B/P values are continuous data that is easy to use for ranking candidates. However, those methods also have limitations. One is that the experiment is expensive and time-consuming. Second, the B/P ratio is a brain distribution value that is largely determined by the nonspecific binding of drug molecules against brain tissue rather than brain permeability. Lastly, the experiment does not provide data for the unbound drug concentration in the extracellular fluid of the brain.

In summary, diverse types of experimental methods are available for assessing brain penetration. However, the cost of the experiments increases when choosing more advanced experiments and getting more complex data. Therefore, it should be careful to select proper methods. One common way is to first assess the brain penetration of a compound using higher throughput methods such as using physicochemical properties or PAMPA–BBB methods. These methods provide an initial prediction of passive BBB permeability, which can be useful for early screening. Then *in vivo* studies can be used next for further evaluation. Those results from *in vivo* experiments provide a more in-depth assessment of the BBB penetration as well as validate the results from the early assessment. If the *in vivo* and *in vitro* brain penetration data are significantly different, other methods such as P-gp efflux, hepatic clearance, plasma protein binding, nonspecific brain tissue binding, and others can be tested using other *in vitro* assays.

## 45.5 Computational Method to Predict BBB Permeability

As mentioned, there is an urgent need to identify a new therapeutic strategy for CNS disease; however, the survival rate of drug candidates, as well as the cost of drug discovery, limits the cycle time. Computational methods can be applied as a great alternative to the costly and labor-intensive screening experiments carried out in laboratories in the early drug discovery, which helps to increase the survival of drug candidates and to move the medicinal chemistry design to a higher probability space for success. Many review articles have already summarized different computational methods for BBB permeability prediction [21–26]. In the following sections, a more general discussion on the available tools as well as different types of computational methods is presented.

### 45.5.1 Commercial Software and Free Computational Tools for BBB Permeability Prediction

Currently, there have been multiple commercial software and online tools predicting the BBB penetration. Several commonly used commercial software and free online tools are listed in Tables 45.1 and 45.2, respectively. These tools can be used for the preliminary prediction of the BBB permeability of a certain small molecule in the early stage drug discovery. However, users need to evaluate the predictability of different software or tools before deciding, which one to purchase or use. That is because the user's query compounds may have different chemical scaffolds with the training data for certain tools or software; therefore, the validation of those tools using the user's data is necessary. Most of the commercial software includes not only BBB prediction but the entire ADMET prediction module, while some free online tools may be specific to BBB permeability prediction. In addition, as the source code of most of the online tools was publicly available, another option for users is to develop their own predictive models using the published computational methods and their data. In the next few sections, we will briefly introduce different types of computational methods for BBB prediction.

**Table 45.1** Commercial software for predicting BBB permeability.

| Name | Company | Website |
|---|---|---|
| ADMET Predictor | Simulations-Plus | https://www.simulations-plus.com/software/admetpredictor/ |
| VolSurf+ | Molecular Discovery | https://www.moldiscovery.com/software/vsplus/ |
| QikProp | Schrodinger | https://www.schrodinger.com/ |
| Percepta | ACD/Labs | https://www.acdlabs.com/products/percepta/ |
| Pipeline Pilot | BIOVIA | https://www.3ds.com/products-services/biovia/products/data-science/pipeline-pilot/ |
| MolCode ToolBox | COMPUDRUG | https://www.compudrug.com/molcode_toolbox |
| StarDrop | Optibrium | https://www.optibrium.com/project/adme-qsar/ |
| ADMET Plugin | ChemAxon | https://chemaxon.com/products/calculators-and-predictors |
| SeeSAR | Biosolveit | https://www.biosolveit.de/SeeSAR |

**Table 45.2** Free online tools for predicting BBB permeability.

| Name | Publications | Website |
|---|---|---|
| BBB Predictor | [27] | https://www.cbligand.org/BBB/ |
| BBPpred | [28] | http://bbppred.xialab.info/ |
| LightBBB | [29] | http://bioanalysis.cau.ac.kr:7030/ |
| PreADMET | [30] | https://preadmet.bmdrc.kr/ |
| ADMETlab | [31] | http://admet.scbdd.com/ |
| SwissADME | [32] | http://www.swissadme.ch/ |
| admetSAR | [33] | http://lmmd.ecust.edu.cn/admetsar1 |
| ALOGPS | [34] | http://www.vcclab.org/lab/alogps/ |
| XLOGP | [35] | https://www.ics.uci.edu/~dock/manuals/xlogp2.1/intro.html |

### 45.5.2 Rule-of-Thumb Methods for BBB Prediction

The association between drug-like molecules and their physicochemical properties has been reported using rule-of-thumb methods, such as the rule of five (RO5) that defines the association between four physicochemical parameters of drug-like compounds and their aqueous solubility and intestinal permeability [36]. This type of method is supposed to summarize the property space for a collection of molecules and provide simple but effective rules for selecting CNS drugs [37]. Several studies were reported to develop this analysis of methods for CNS discovery (Table 45.3).

In an early analysis of comparative properties for BBB penetration by Fischer et al., an estimated threshold of $pK_a$ for BBB penetration was identified between 4 and 10 [39]. Early studies using rule-based methods for BBB are usually combined with Quantitative structure–activity relationship (QSAR) analysis. In those studies, the BBB permeability was correlated with physicochemical properties such as the polar surface area (PSA) as well as hydrogen-bond information. For example, Ghose et. al. systematically explored the chemical space of CNS drugs and summarized a six-property-rule of high-quality CNS molecules (Figure 45.3): topological polar surface area (TPSA) $< 76$ Å$^2$, number of nitrogen $>= 1$, number of linear chains (outside rings) $< 7$, number of polar hydrogens $< 3$, volume of 740–970 Å$^3$, and solvent assessable surface area of 460–580 Å$^2$ [44]. Based on the selected properties, they developed a tree-based QSAR model using recursive partition algorithms to differentiate CNS drugs from non-CNS drugs.

Similar work was conducted by Wager et al. from Pfizer reporting a different chemical space for optimal CNS molecules using another six molecular properties: *C*log *P*, *C*log *D*, molecular weight (MW), TPSA, hydrogen-bond donor (HBD), $pK_a$ [42]. Furthermore, they further upgraded their rule-based method by developing a scoring system called central nervous system multiparameter optimization (CNS MPO) [43]. After evaluating the satisfaction of the six fundamental physicochemical properties, a composite score ranging from 0 to 6 was calculated by adding the transformed

**Table 45.3** Summary of rule-based methods for predicting BBB permeability.

| Year | Authors | Link | Citation | References |
|---|---|---|---|---|
| 1980 | Victor A. Levin | https://doi.org/10.1021/jm00180a022 | 594 | [38] |
| 1998 | Fischer et al. | https://doi.org/10.1007/s002329900434 | 268 | [39] |
| 1998 | Waterbeemd et al. | https://doi.org/10.3109/10611869808997889 | 625 | [40] |
| 2000 | Österberg et al. | https://doi.org/10.1021/ci000065l | 617 | [41] |
| 2010 | Wager et. al. | https://doi.org/10.1021/cn100007x | 298 | [42] |
| 2010 | Wager et. al. | https://doi.org/10.1021/cn100008c | 582 | [43] |
| 2012 | Ghose et. al. | https://doi.org/10.1021/cn200100h | 256 | [44] |
| 2013 | Pham-The et.al. | https://doi.org/10.1002/minf.201200166 | 27 | [37] |

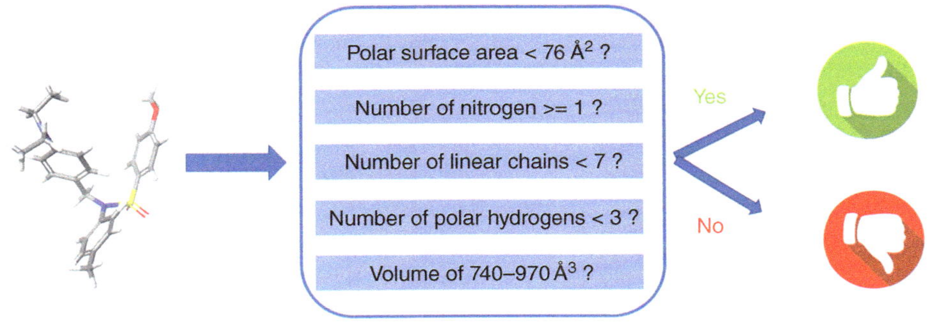


**Figure 45.3** Overview of rule-based methods for predicting BBB permeability.

value (range from 0 to 1) for the satisfaction of each of the properties.

In summary, this kind of method is simple and straightforward, but it clarifies and provides guidelines for developing compounds with good BBB permeability so that medicinal chemists can develop and optimize a candidate with an increased probability of success and favorable safety profiles.

### 45.5.3 Traditional QSAR/QSPR to Predict log BB

QSAR or quantitative structure–property relationship (QSPR) analysis is very important computational methods for a small molecular study, which can be used to predict either the biological effect against a certain protein (in QSAR) or the physicochemical properties based on the chemical structure (in QSPR) [45]. Generally, a mathematical and/or statistical relationship (equation) will be established between the compound descriptors and the target affinity or physicochemical property to provide reliable predictions on the biological activity of target molecules (Figure 45.4). For example, a simple QSAR/QSPR model to predict BBB permeability could be a linear combination of $A\log P$ and PSA as shown in Eq. (45.1) [46]:

$$\log \text{BB} = 0.241 * A \log P - 0.0143 * \text{PSA} - 0.0213 \tag{45.1}$$

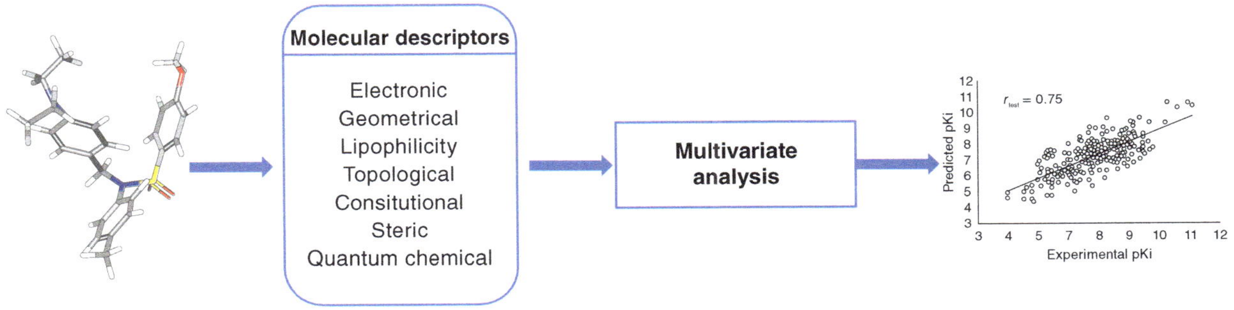


**Figure 45.4** Overview of traditional QSAR/QSPR methods for predicting BBB permeability.

**Table 45.4** Summary of traditional QSAR/QSPR methods for predicting BBB permeability.

| Year | Authors | Link | Citations | References |
|---|---|---|---|---|
| 1980 | Victor A. Levin | https://doi.org/10.1021/jm00180a022 | 594 | [38] |
| 1998 | Waterbeemd et al. | https://doi.org/10.3109/10611869808997889 | 625 | [40] |
| 1999 | Kelder et al. | https://doi.org/10.1023/a:1015040217741 | 619 | [47] |
| 2000 | Österberg et al. | https://doi.org/10.1021/ci000065l | 617 | [41] |
| 2001 | Platts et al. | https://doi.org/10.1016/s0223-5234(01)01269-7 | 230 | [48] |
| 2002 | Iyer et al. | https://doi.org/10.1023/a:1020792909928 | 117 | [49] |
| 2003 | Subramanian et al. | https://doi.org/10.1023/b:jcam.0000017372.32162.37 | 67 | [46] |
| 2003 | Michael C. Hutter | https://doi.org/10.1023/a:1027359714663 | 36 | [50] |
| 2004 | Pan et al. | https://doi.org/10.1021/ci0498057 | 39 | [51] |
| 2005 | Ma et al. | https://doi.org/10.1111/j.1745-7254.2005.00068.x | 103 | [52] |
| 2008 | Zhang et al. | https://doi.org/10.1007/s11095-008-9609-0 | 128 | [53] |
| 2009 | Lanevskij et al. | https://doi.org/10.1002/jps.21405 | 64 | [54] |
| 2010 | Fan et al. | https://doi.org/10.1021/ci900384c | 50 | [55] |
| 2011 | Muehlbacher et al. | https://doi.org/10.1007/s10822-011-9478-1 | 59 | [56] |
| 2013 | Pham-The et al. | https://doi.org/10.1002/minf.201200166 | 27 | [37] |
| 2015 | Wang et al. | https://doi.org/10.1007/s11095-015-1687-1 | 43 | [57] |
| 2015 | Varadharjajan et al. | https://doi.org/10.1002/jps.24301 | 22 | [58] |
| 2016 | Zhang et al. | https://doi.org/10.1155/2015/292683 | 24 | [59] |
| 2016 | Dixon et al. | https://doi.org/10.4155/fmc-2016-0093 | 73 | [60] |
| 2017 | Ciura et al. | https://doi.org/10.1016/j.jpba.2017.05.041 | 21 | [61] |

A list of the representative works using traditional QSAR/QSPR to predict BBB permeability is summarized in Table 45.4.

Early exploration of QSAR/QSPR analyses for CNS drugs can be traced back to the early 1980s by Levin [38], finding that there was a linear relationship between the rat brain permeability of a compound and its lipid solubility when its MW was under a threshold of 400–600. Later QSAR/QSPR studies mainly adopted the B/P ratio or log BB value as the representation of BBB permeability to simulate, which were collected from radio-labeled drug determinations in the blood and brain of anesthetized rats.

Kelder et al. developed a QSAR model to predict BBB permeability using a relatively large dataset containing 2366 compounds (776 CNS drugs, 1590 non-CNS drugs) [47], with a conclusion that general CNS drugs had a much less PSA compared to non-CNS drugs. Similar rules were discussed in other QSAR studies. For example, Waterbeemd et al. reported that the optimal cutoff for CNS penetration is PSA less than 90 Å$^2$ and MW less than 450 [40]. Another QSAR study discussed that the sum of nitrogen (N) and oxygen (O) atom numbers can be used to assess BBB penetration [41].

Besides using general physicochemical properties such as TPSA and log *P* as the variable to describe the small-molecule characteristics, other molecular descriptors were also explored. Platts et al. developed a QSAR model based on linear free energy relation (LFER) methods, which involved LFER descriptors in the QSAR equation [48]. Those descriptors included molar refraction, polarizability, hydrogen-bond acidity and basicity, and solute McGowan volume. A wide range of solvation and transport processes were modeled for calculating those solute descriptors.

Other studies are using special types of descriptors to build QSAR/QSPR models, such as descriptors generated using molecular dynamics [49], molecular mechanism [17, 30, 50, 62], and experiments [61]. More QSAR studies can be found in other publications [52, 54, 58, 60, 63].

Furthermore, a QSAR/QSPR study was conducted by combining different methods. For example, Pan et al. developed QSAR models to assess BBB penetration by combining traditional QSAR methods with clustering methods. They first split a collection of 150 diverse compounds into different subclasses based on similarity using the clustering method, and then built QSAR equations for each subclass, finally achieving an accuracy of $r^2 = 0.69$ and $q^2 = 0.60$. Although the data set was small and the model performance was not very high, their attempt to generalize the model for wide chemical diversity was valuable [51, 52]. Besides, different machine learning algorithms can be combined with genetic algorithms (GA) to achieve better performance [59].

In summary, traditional QSAR/QSPR methods to predict BBB permeability are widely used in modern drug discovery since QSAR/QSPR helps with the understanding of the effect of structure on activity/property, leading to the synthesis of novel analogs. However, there are also disadvantages of QSAR/QSPR. Firstly, the experimental error and noise may heavily increase the false correlation in such regression models. Secondly, the descriptors used for generating QSAR/QSPR models may cross-correlate with each other, which makes it harder to build a robust model. Thirdly, different types of variables are not suggested to be mixed when building regression models, which limits the selection of descriptors to represent the compounds.

### 45.5.4 Machine-Learning-Based Regression and Classification

Traditional QSAR/QSPR studies generally applied linear methods to analyze the BBB penetration of small molecules [64]. However, the correlation between BBB permeability and compound descriptors can be more complex. Machine learning's popularity and pharmacoanalytics surged in recent decades, marking the beginning of significant changes. The general steps of machine-learning-based modeling are depicted in Figure 45.5. Machine learning algorithms have the capability to do nonlinear fitting between dependent variables and independent variables, and thus usually provide more accurate results when compared to general linear methods. Secondly, machine learning models can handle larger datasets with more independent variables using feature selection methods, which allow scientists to incorporate more chemical descriptors for building models. The representative studies of machine-learning-based QSAR/QSPR and BBB classification models are listed in Table 45.5.

For example, Zhang et al. published their BBB predictive regression models using combined machine learning methods [53]. In their methods, the k-nearest neighbors (k-NN) method was used for selecting important features from a total of 346 Dragon, MOE, and MolConnZ descriptors, and the support vector machine (SVM) regression algorithms were used for model construction. More recently, Wang et al. developed machine-learning-based BBB permeability models using a relatively large dataset containing 439 compounds and their log BB values [57]. Multiple machine learning algorithms including Random Forest (RF), SVM, and kNN were adopted to build models. A total of 192 descriptors calculated by Molecular Operating Environment (MOE) software were used as the molecular representations, which included physical properties, atom and bond counts, connectivity and shape information, pharmacophore features, and others.

Machine learning can not only handle regression tasks but also deal with classification problems. Besides using machine learning in regression tasks for predicting the exact value of BBB permeability parameters such as log BB, many computational models were developed attempting to discriminate BBB permeable

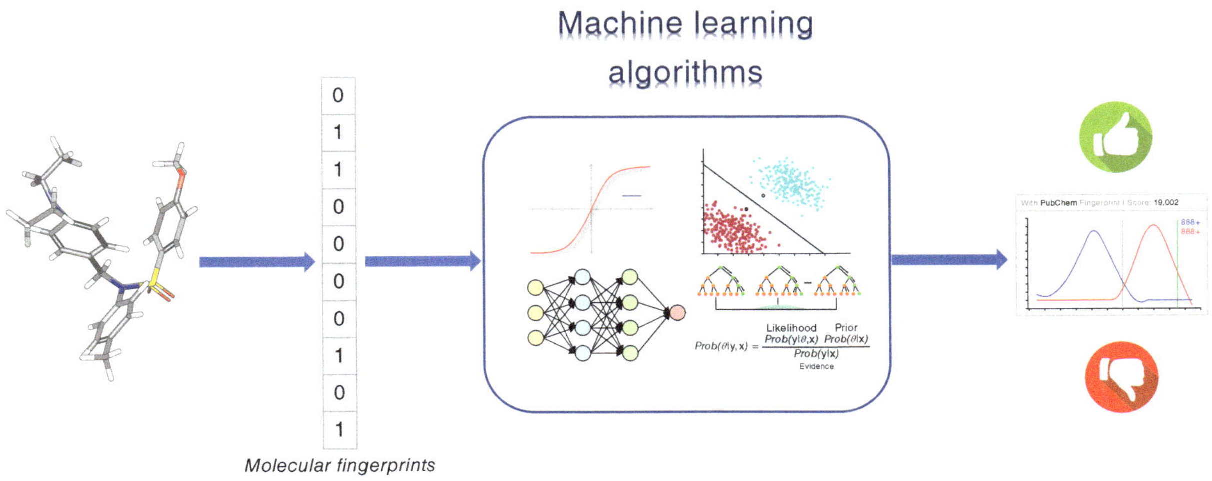


**Figure 45.5** Overview of machine-learning-based methods for predicting BBB permeability.

**Table 45.5** Summary of machine learning methods for predicting BBB permeability.

| Year | Authors | Link | Citations | References |
|---|---|---|---|---|
| 2002 | Doniger et al. | https://doi.org/10.1089/10665270260518317 | 162 | [65] |
| 2004 | Winkler et al. | https://doi.org/10.1016/j.jmgm.2004.03.010 | 98 | [66] |
| 2005 | Li et al. | https://doi.org/10.1021/ci050135u | 107 | [67] |
| 2008 | Kortagere et al. | https://doi.org/10.1007/s11095-008-9584-5 | 75 | [68] |
| 2012 | Martins et al. | https://doi.org/10.1021/ci300124c | 66 | [69] |
| 2012 | Suenderhauf et al. | https://doi.org/10.3390/molecules170910429 | 68 | [70] |
| 2013 | Yan et al. | https://doi.org/10.1080/1062936X.2012.729224 | 29 | [71] |
| 2015 | Brito-S' anchez et al. | https://doi.org/10.1002/minf.201400118 | 21 | [72] |
| 2016 | Jiang et al. | https://doi.org/10.1142/S0219720016500050 | 7 | [73] |
| 2017 | Gao et al. | https://doi.org/10.1093/bioinformatics/btw713 | 40 | [74] |
| 2017 | Castillo-Garit et al. | https://doi.org/10.2174/1573406413666170209124302 | 27 | [75] |
| 2017 | Lingineni et al. | https://doi.org/10.1007/s11030-016-9715-6 | 17 | [76] |
| 2018 | Yuan et al. | https://doi.org/10.1208/s12248-018-0215-8 | 22 | [77] |
| 2018 | Wang et al. | https://doi.org/10.1002/cmdc.201800533 | 45 | [78] |
| 2019 | Plisson et al. | https://doi.org/10.3390/md17020081 | 15 | [64] |
| 2019 | Roy et al. | https://doi.org/10.1021/acsomega.9b01512 | 15 | [79] |
| 2020 | Singh et al. | https://doi.org/10.1016/j.jmgm.2019.107516 | 10 | [80] |

compounds from non-permeable compounds [76, 80]. Generally, in those classification models, compounds were labeled as CNS+ (or CNS−) based on the *in vivo* or *in vitro* experimental results, such as B/P ratio or log BB, indicating that a compound can penetrate the brain (or not). One of the advantages for generating classification models is that labeling compounds into binary classes (CNS+/CNS−) may be considered a relatively reasonable way to merge several small datasets from multiple resources into a large one since their experimental standards for validating BBB permeability were various.

Early attempts of building classification models for BBB permeability prediction appeared in the early 2000s. For example, Kortagere et al. developed classification models using the SVM classification technique with a combined BBB dataset from seven resources [68]. Those individual datasets were all quite small and used different experimental standards to get the log BB value. By using different thresholds of the log BB value suggested by the authors to specify BBB+ and BBB-compounds, it would be hard to merge those small datasets into a large dataset [81]. Besides SVM methods [65, 73], different machine learning classification algorithms, such as linear method [72], decision tree [67, 70, 75], and Bayesian approaches [66, 69], were also discussed and compared. Similar to traditional QSAR/QSPR study, different types of molecular descriptors were adopted in the classification study; for example, solvation energy descriptors [79], and binary molecular fingerprints descriptors [78] were used. Recent studies also tried to combine different types of data in their BBB classification modeling [77, 82]. In addition to those molecular descriptors, which describe the internal molecular characteristics, other descriptors such as drug side effects were used for predicting BBB permeability [74].

In summary, the use of machine learning methods significantly promoted the development of computational prediction of BBB permeability in early drug discovery and preclinical research. With well-established feature selection and model fitting techniques, scientists can quickly build their machine-learning-based model using a large dataset and achieve relatively good performance. However, there are still disadvantages to those machine learning models. One is that most of the machine learning models lack interpretability due to the complex nonlinear fitting; therefore, those "black box" models help a little with the explanation of the mechanism and further optimization of the target compound. Secondly, although machine learning algorithms can deal with large datasets, redundancy, and over-fitting problems often make the prediction and classification unreliable [83].

### 45.5.5 Advance Deep Neural Network and Deep Learning for BBB Permeability

Among machine learning algorithms, artificial neural networks (ANN) can be considered as a special one that is related to deep-learning methods in modern artificial intelligence (AI) [84]. The architecture of Artificial Neural Networks (ANN) was inspired by the structure and function of the human brain, which composed of a collection of connected units, known as artificial neurons that are loosely modeled after biological neurons. Neurons that receive signals can process them and pass signals to other connected neurons. The connections between these neurons are characterized by weights, representing the strength or influence of one neuron's signal on another. Those weights can be adjusted and optimized based on real-world data through computational learning processes.

An early attempt to use ANN to predict BBB permeability was conducted by Garg and Verma [85]. They used physicochemical properties such as log *P* and MW as the descriptors and used them to predict the log BB value in the neural network regression models. Similar work was reported by Hemmateenejad et al. combining GAs for feature selection [86]. Classification models using ANN were also reported in other studies [71, 76, 87].

Deep learning is the rebranding of the traditional ANN method. In deep learning, neural networks are "deep," which means that it is composed of more than one hidden layer between the input layer and output layer (Figure 45.6). Deep-learning methods have the general advantages of not only providing complex nonlinear fitting but also automatically dealing with the feature selection from the raw data. Also, deep learning using the graphic processing units (GPU) computing technique can speed up the simulation. Therefore, deep-learning methods have been extremely successful and widely used to develop various kinds of AI applications in multiple domains including small molecular drug discovery [88]. There have been several studies using deep learning for BBB permeability prediction (Table 45.6).

One early deep-learning work related to BBB permeability prediction was reported by Ma et al. in

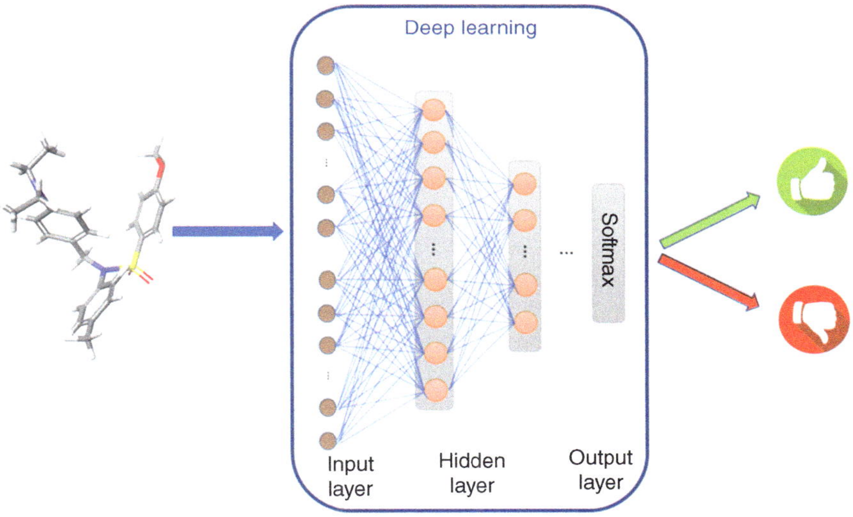


**Figure 45.6** Overview of deep-learning methods for predicting BBB permeability.

**Table 45.6** Summary of deep-learning methods for predicting BBB permeability.

| Year | Authors | Link | Citations | References |
|---|---|---|---|---|
| 2006 | Garg and Verma | https://doi.org/10.1021/ci050303i | 92 | [85] |
| 2006 | Hemmateenejad et al. | https://doi.org/10.1002/jcc.20437 | 33 | [86] |
| 2008 | Guerra et al. | https://doi.org/10.1002/qsar.200710019 | 35 | [87] |
| 2013 | Yan et al. | https://doi.org/10.1080/1062936X.2012.729224 | 29 | [71] |
| 2015 | Ma et al. | https://doi.org/10.1021/ci500747n | 543 | [89] |
| 2019 | Miao et al. | https://doi.org/10.1038/s41598-019-44773-4 | 24 | [90] |
| 2020 | Alsenan et al. | https://doi.org/10.1016/j.compbiolchem.2020.107377 | 14 | [91] |

2015 [89]. In the study, deep neural network models were generated for predicting different tasks including p-glycoprotein substrate prediction, which was commonly considered when evaluating the BBB permeability of small molecules. And specifically, a total of 5135 atom pair descriptors that describe the detailed molecular chemical information were adopted as the input data, and other traditional machine learning methods may not be capable of handling so many input descriptors.

In 2019, Miao et al. developed their BBB classification models using deep neural networks [90]. They compared it to the published models built using SVM methods [74] and found that the deep-learning-based model had a better performance than traditional SVM models. Recently, a deep-learning model using a recurrent neural network (RNN) was reported by Alsenan et al. [91]. RNN is a special type of deep-learning architecture for dealing with sequential data, and it is widely used in natural language processing and signaling processing [88]. In the study, they used molecular fingerprints as the molecular sequence to be fitted into the RNN model and attempted to classify molecules into high penetration group (BBB+) and low penetration group (BBB−).

In summary, although deep learning is a powerful new technique for developing predictive models, only limited numbers of BBB permeability studies were reported using deep learning, and more deep-learning-based works should be encouraged. There have been several well-developed free deep-learning toolkits for small molecular drug discovery published [92, 93]. These toolkits were validated using BBB permeability data and could be used directly for BBB permeability prediction. However, there are still limitations of deep-learning-based models. One limitation is that like most traditional machine-learning-based models, deep-learning-based models are hard to interpret. Another limitation is that those deep-learning methods are commonly used in classification tasks only specifying "yes" or "no," and how to apply deep learning for successful regression as well as ranking a task needs to be explored.

### 45.5.6 Summary and Discussion

Building predictive models using computational methods has become a vitally important step for recent decades in the development of therapeutical small molecules to speed up the drug discovery speed and improve the success rate, especially the preclinical ADMET evaluation. Different from the target identification, BBB permeation is a process related to not only one certain target. As discussed in Section 45.2, there are multiple mechanisms involved in this process, making it a challenge to predict the BBB penetration using mechanism-based methods. Attempts of those predictions include physiologically based pharmacokinetic (PBPK) modeling to predict drug concentration in cerebrospinal fluid (CSF) [94], however, only passive transfer was considered in the model. The simulation of more complex processes such as efflux transport and exocytosis, as well as the systems' biological effect, is still a challenge. Therefore, the ligand-based method becomes the mainstream method for BBB prediction.

In addition, the current available dataset for developing the BBB permeability model can be divided into two categories: the quantitative data and the binary data. The quantitative data involves the log BB and the log PS data. The log BB data gives the log ratio of the total steady-state concentration in the brain to the total steady-state concentration in the blood, while the log PS is the logarithm of the permeable surface area product of a compound at a given time. The binary data only gives the suggestion of whether the small molecule can go across the BBB (BBB+/BBB−). The major available public resources of the BBB permeability data are published papers, most of which are binary data. One of the recent published articles covered more than 7000 compounds in their dataset [29], which can be considered as one of the largest public available datasets. Pharmaceutical companies usually own their own large database for internal modeling; however, those data are hard for academic institutions to get for free. Moreover, the issue of those BBB permeability datasets is not only the lack of data. Those merged datasets come from different assays with different standards across years, and most of them are binary data. One more concern is that the complex properties that define the BBB are often altered in disease states [95]; however, current data resources mainly come from experiments that may not consider such situations. All those factors bring more challenges for computational models to predict the BBB permeability accurately.

It is encouraging that the computational methods for BBB permeability prediction have become more and more powerful and accurate. However, those methods were supposed to be easily explainable in order to be trusted by scientists. For the future vision of the development on BBB permeability prediction, one vision could be that by combining with the causal inference, deep-learning methods can be developed to do that; and, the AI technique using those deep-learning methods is supposed to give suggestions for compound optimization. However, another doubt about those state-of-art deep-learning techniques is that deep-learning methods usually require abundant data to train the model recurrently, but the volume of the data in modern drug discovery, especially BBB permeability prediction, is quite small.

## Abbreviations

| | |
|---|---|
| AD | Alzheimer's disease |
| ADMET | absorption distribution metabolism excretion and toxicity |
| AI | artificial intelligence |
| ANN | artificial neural networks |
| BBB | blood–brain barrier |
| BCRP | breast cancer resistance protein |
| CNS | central nervous system |
| FDA | U.S. Food and Drug Administration |
| GA | genetic algorithms |
| HBD | hydrogen-bond donor |
| kNN | k-Nearest Neighbor |
| MW | molecular weight |
| PAMPA | parallel artificial membrane permeability assay |
| P-gp | p-glycoprotein |
| QSAR | quantitative structure–activity relationship |
| QSPR | quantitative structure–property relationship |
| RF | random forest |
| RNN | recurrent neural network |
| SVM | support vector machine |
| TPSA | topological polar surface area |

## References


1 Redzic, Z. (2011). Molecular biology of the blood-brain and the blood-cerebrospinal fluid barriers: similarities and differences. *Fluids Barriers CNS* 8: 3.

2 Sweeney, M.D., Zhao, Z., Montagne, A. et al. (2019). Blood-brain barrier: from physiology to disease and back. *Physiol. Rev.* 99: 21–78.

3 Małkiewicz, M.A., Szarmach, A., Sabisz, A. et al. (2019). Blood-brain barrier permeability and physical exercise. *J. Neuroinf.* 16: 15.

4 Gribkoff, V.K. and Kaczmarek, L.K. (2017). The need for new approaches in CNS drug discovery: Why drugs have failed, and what can be done to improve outcomes. *Neuropharmacology* 120: 11–19.

5 Collaborators, G.U.N.D., Feigin, V.L., Vos, T. et al. (2021). Burden of neurological disorders across the US From 1990-2017: a global burden of disease study. *JAMA Neurol.* 78: 165–176.

6 (IHME) IfHMaE (2019) GBD Compare data visualization.

7 Dealmakers, B. (2020). *A view into the central nervous system disorders market*. Nature NEWS FEATURE: Nature.

8 Wong, K.H., Riaz, M.K., Xie, Y. et al. (2019). Review of current strategies for delivering Alzheimer's disease drugs across the blood-brain barrier. *Int. J. Mol. Sci.* 20.

9 Sharif, Y., Jumah, F., Coplan, L. et al. (2018). Blood brain barrier: a review of its anatomy and physiology in health and disease. *Clin. Anat.* 31: 812–823.

10 Dong, X. (2018). Current strategies for brain drug delivery. *Theranostics* 8: 1481–1493.

11 Bors, L.A. and Erdő, F. (2019). Overcoming the blood–brain barrier. Challenges and tricks for CNS drug delivery. *Sci. Pharm.* 87: 6.

12 Pardridge, W.M. (2005). The blood-brain barrier: Bottleneck in brain drug development. *NeuroRx* 2: 3–14.

13 Banks, W.A. (2009). Characteristics of compounds that cross the blood-brain barrier. *BMC Neurol.* 9: S3.

14 Bagchi, S., Chhibber, T., Lahooti, B. et al. (2019). In-vitro blood-brain barrier models for drug screening and permeation studies: an overview. *Drug Des.Dev. Ther.* 13: 3591–3605.

15 Vastag, M. and Keseru, G.M. (2009). Current in vitro and in silico models of blood-brain barrier penetration: a practical view. *Curr. Opin. Drug Discovery Dev.* 12: 115–124.

16 Ottaviani, G., Martel, S., and Carrupt, P.-A. (2006). Parallel artificial membrane permeability assay: a new membrane for the fast prediction of passive human skin permeability. *J. Med. Chem.* 49: 3948–3954.

17 Leo, A., Hansch, C., and Elkins, D. (1971). Partition coefficients and their uses. *Chem. Rev.* 71: 525–616.

18 Hubatsch, I., Ragnarsson, E.G.E., and Artursson, P. (2007). Determination of drug permeability and prediction of drug absorption in Caco-2 monolayers. *Nat. Protoc.* 2: 2111–2119.

19 Tang, F., Horie, K., and Borchardt, R.T. (2002). Are MDCK cells transfected with the human MDR1 gene a good model of the human intestinal mucosa? *Pharm. Res.* 19: 765–772.

20 Kerns, E.H. and Di, L. (2008). Chapter 28 – Blood-brain barrier methods. In: *Drug-like Properties: Concepts, Structure Design and Methods* (ed. E.H. Kerns and L. Di), 311–328. San Diego: Academic Press.

21 Bradbury, M.W. (1993). The blood-brain barrier. *Exp. Physiol.* 78: 453–472.

22 Clark, D.E. (2003). In silico prediction of blood-brain barrier permeation. *Drug Discovery Today* 8: 927–933.

23 Gupta, S. (1989). QSAR studies on drugs acting at the central nervous system. *Chem. Rev.* 89: 1765–1800.

24 Pajouhesh, H. and Lenz, G.R. (2005). Medicinal chemical properties of successful central nervous system drugs. *NeuroRx* 2: 541–553.

25 Pardridge, W.M. (1998). CNS drug design based on principles of blood-brain barrier transport. *J. Neurochem.* 70: 1781–1792.

26 Saxena, D., Sharma, A., Siddiqui, M.H., and Kumar, R. (2019). Blood brain barrier permeability prediction using machine learning techniques: an update. *Curr. Pharmaceut. Biotechnol.* 20: 1163–1171.

27 Liu, H., Wang, L., Lv, M. et al. (2014). AlzPlatform: an Alzheimer's disease domain-specific chemogenomics knowledgebase for polypharmacology and target identification research. *J. Chem. Inf. Model.* 54: 1050–1060.

28 Dai, R., Zhang, W., Tang, W. et al. (2021). BBPpred: sequence-based prediction of blood-brain barrier peptides with feature representation learning and logistic regression. *J. Chem. Inf. Model.* 61: 525–534.

29 Shaker, B., Yu, M.-S., Song, J.S. et al. (2020). LightBBB: computational prediction model of blood–brain-barrier penetration based on LightGBM. *Bioinformatics.*

30 Lee, S., Park, S., Lee, I., and No, K. (2007). PreAD-MET Ver. v2. 0. *Seoul, Korea: BMDRC.*

31 Dong, J., Wang, N.-N., Yao, Z.-J. et al. (2018). ADMETlab: a platform for systematic ADMET evaluation based on a comprehensively collected ADMET database. *J. Cheminf.* 10: 29.

32 Daina, A., Michielin, O., and Zoete, V. (2017). SwissADME: a free web tool to evaluate pharmacokinetics, drug-likeness and medicinal chemistry friendliness of small molecules. *Sci. Rep.* 7: 42717.

33 Cheng, F., Li, W., Zhou, Y. et al. (2012). admetSAR: a comprehensive source and free tool for assessment of chemical ADMET properties. *J. Chem. Inf. Model.* 52: 3099–3105.

34 Tetko, I.V. and Poda, G.I. (2004). Application of ALOGPS 2.1 to predict log D distribution coefficient for pfizer proprietary compounds. *J. Med. Chem.* 47: 5601–5604.

35 Wang, R., Fu, Y., and Lai, L. (1997). A new atom-additive method for calculating partition coefficients. *J. Chem. Inf. Comput. Sci.* 37: 615–621.

36 Lipinski, C.A., Lombardo, F., Dominy, B.W., and Feeney, P.J. (1997). Experimental and computational approaches to estimate solubility and permeability in drug discovery and development settings. *Adv. Drug Delivery Rev.* 23: 3–25.

37 Pham-The, H., González-Álvarez, I., Bermejo, M. et al. (2013). The use of rule-based and QSPR approaches in ADME profiling: a case study on caco-2 permeability. *Mol. Inf.* 32: 459–479.

38 Levin, V.A. (1980). Relationship of octanol/water partition coefficient and molecular weight to rat brain capillary permeability. *J. Med. Chem.* 23: 682–684.

39 Fischer, H., Gottschlich, R., and Seelig, A. (1998). Blood-brain barrier permeation: molecular parameters governing passive diffusion. *J. Membr. Biol.* 165: 201–211.

40 van de Waterbeemd, H., Camenisch, G., Folkers, G. et al. (1998). Estimation of blood-brain barrier crossing of drugs using molecular size and shape, and H-bonding descriptors. *J. Drug Targeting* 6: 151–165.

41 Österberg, T. and Norinder, U. (2000). Prediction of polar surface area and drug transport processes using simple parameters and PLS statistics. *J. Chem. Inf. Comput. Sci.* 40: 1408–1411.

42 Wager, T.T., Chandrasekaran, R.Y., Hou, X. et al. (2010a). Defining desirable central nervous system drug space through the alignment of molecular properties, in vitro ADME, and safety attributes. *ACS Chem. Neurosci.* 1: 420–434.

43 Wager, T.T., Hou, X., Verhoest, P.R., and Villalobos, A. (2010b). Moving beyond rules: the development of a central nervous system multiparameter optimization (CNS MPO) approach to enable alignment of druglike properties. *ACS Chem. Neurosci.* 1: 435–449.

44 Ghose, A.K., Herbertz, T., Hudkins, R.L. et al. (2012). Knowledge-based, central nervous system (CNS) lead selection and lead optimization for CNS drug discovery. *ACS Chem. Neurosci.* 3: 50–68.

45 Peter, S.C., Dhanja, J.K., Malik, V. et al. (2019). Quantitative structure-activity relationship (QSAR): modeling approaches to biological applications. In: *Encyclopedia of Bioinformatics and Computational Biology*, 661–676.

46 Subramanian, G. and Kitchen, D.B. (2003). Computational models to predict blood–brain barrier permeation and CNS activity. *J. Comput.-Aided Mol. Des.* 17: 643–664.

47 Kelder, J., Grootenhuis, P.D., Bayada, D.M. et al. (1999). Polar molecular surface as a dominating determinant for oral absorption and brain penetration of drugs. *Pharm. Res.* 16: 1514–1519.

48 Platts, J.A., Abraham, M.H., Zhao, Y.H. et al. (2001). Correlation and prediction of a large blood–brain distribution data set—an LFER study. *Eur. J. Med. Chem.* 36: 719–730.

49 Iyer, M., Mishra, R., Han, Y., and Hopfinger, A.J. (2002). Predicting blood-brain barrier partitioning of organic molecules using membrane-interaction QSAR analysis. *Pharm. Res.* 19: 1611–1621.

50 Hutter, M.C. (2003). Prediction of blood–brain barrier permeation using quantum chemically derived information. *J. Comput.-Aided Mol. Des.* 17: 415–443.

51 Pan, D., Iyer, M., Liu, J. et al. (2004). Constructing optimum blood brain barrier QSAR models using a combination of 4D-molecular similarity measures and cluster analysis. *J. Chem. Inf. Comput. Sci.* 44: 2083–2098.

52 Ma, X.L., Chen, C., and Yang, J. (2005). Predictive model of blood-brain barrier penetration of organic compounds. *Acta Pharmacol. Sin.* 26: 500–512.

53 Zhang, L., Zhu, H., Oprea, T.I. et al. (2008). QSAR modeling of the blood-brain barrier permeability for diverse organic compounds. *Pharm. Res.* 25: 1902–1914.

54 Lanevskij, K., Japertas, P., Didziapetris, R., and Petrauskas, A. (2009). Ionization-specific prediction of blood-brain permeability. *J. Pharm. Sci.* 98: 122–134.

55 Fan, Y., Unwalla, R., Denny, R.A. et al. (2010). Insights for predicting blood-brain barrier penetration of CNS targeted molecules using QSPR approaches. *J. Chem. Inf. Model.* 50: 1123–1133.

56 Muehlbacher, M., Spitzer, G.M., Liedl, K.R., and Kornhuber, J. (2011). Qualitative prediction of blood-brain barrier permeability on a large and refined dataset. *J. Comput.-Aided Mol. Des.* 25: 1095–1106.

57 Wang, W., Kim, M.T., Sedykh, A., and Zhu, H. (2015). Developing enhanced blood-brain barrier permeability models: integrating external bio-assay data in QSAR modeling. *Pharm. Res.* 32: 3055–3065.

58 Varadharajan, S., Winiwarter, S., Carlsson, L. et al. (2015). Exploring in silico prediction of the unbound brain-to-plasma drug concentration ratio: model validation, renewal, and interpretation. *J. Pharm. Sci.* 104: 1197–1206.

59 Zhang, D., Xiao, J., Zhou, N. et al. (2015). A genetic algorithm based support vector machine model for blood-brain barrier penetration prediction. *BioMed Res. Int.* 2015: 292683.

60 Dixon, S.L., Duan, J., Smith, E. et al. (2016). AutoQSAR: an automated machine learning tool for best-practice quantitative structure–activity relationship modeling. *Future Med. Chem.* 8: 1825–1839.

61 Ciura, K., Belka, M., Kawczak, P. et al. (2017). Combined computational-experimental approach to predict blood-brain barrier (BBB) permeation based on "green" salting-out thin layer chromatography supported by simple molecular descriptors. *J. Pharm. Biomed. Anal.* 143: 214–221.

62 Man, V.H., Li, M.S., Derreumaux, P. et al. (2020). Molecular mechanism of ultrasound interaction with a blood brain barrier model. *J. Chem. Phys.* 153: 045104.

63 van de Waterbeemd, H. and Kansy, M. (1992). Hydrogen-bonding capacity and brain penetration. *Chim. Int. J. Chem.* 46: 299–303.

64 Plisson, F. and Piggott, A.M. (2019). Predicting blood–brain barrier permeability of marine-derived kinase inhibitors using ensemble classifiers reveals potential hits for neurodegenerative disorders. *Mar. Drugs* 17: 81.

65 Doniger, S., Hofmann, T., and Yeh, J. (2002). Predicting CNS permeability of drug molecules: comparison of neural network and support vector machine algorithms. *J. Comput. Biol.* 9: 849–864.

66 Winkler, D.A. and Burden, F.R. (2004). Modelling blood–brain barrier partitioning using Bayesian neural nets. *J. Mol. Graphics Modell.* 22: 499–505.

67 Li, H., Yap, C.W., Ung, C.Y. et al. (2005). Effect of selection of molecular descriptors on the prediction of blood–brain barrier penetrating and nonpenetrating agents by statistical learning methods. *J. Chem. Inf. Model.* 45: 1376–1384.

68 Kortagere, S., Chekmarev, D., Welsh, W.J., and Ekins, S. (2008). New predictive models for blood–brain barrier permeability of drug-like molecules. *Pharm. Res.* 25: 1836–1845.

69 Martins, I.F., Teixeira, A.L., Pinheiro, L., and Falcao, A.O. (2012). A bayesian approach to in silico blood-brain barrier penetration modeling. *J. Chem. Inf. Model.* 52: 1686–1697.

70 Suenderhauf, C., Hammann, F., and Huwyler, J. (2012). Computational prediction of blood-brain barrier permeability using decision tree induction. *Molecules* 17: 10429–10445.

71 Yan, A., Liang, H., Chong, Y. et al. (2013). In-silico prediction of blood–brain barrier permeability. *SAR QSAR Environ. Res.* 24: 61–74.

72 Brito-Sánchez, Y., Marrero-Ponce, Y., Barigye, S.J. et al. (2015). Towards better BBB passage prediction using an extensive and curated data set. *Mol. Inf.* 34: 308–330.

73 Jiang, L., Chen, J., He, Y. et al. (2016). A method to predict different mechanisms for blood-brain barrier permeability of CNS activity compounds in Chinese herbs using support vector machine. *J. Bioinf. Comput. Biol.* 14: 1650005.

**74** Gao, Z., Chen, Y., Cai, X., and Xu, R. (2017). Predict drug permeability to blood-brain-barrier from clinical phenotypes: drug side effects and drug indications. *Bioinformatics* 33: 901–908.

**75** Castillo-Garit, J.A., Casanola-Martin, G.M., Le-Thi-Thu, H. et al. (2017). A simple method to predict blood-brain barrier permeability of drug- like compounds using classification trees. *Med. Chem.* 13: 664–669.

**76** Lingineni, K., Belekar, V., Tangadpalliwar, S.R., and Garg, P. (2017). The role of multidrug resistance protein (MRP-1) as an active efflux transporter on blood-brain barrier (BBB) permeability. *Mol. Diversity* 21: 355–365.

**77** Yuan, Y., Zheng, F., and Zhan, C.G. (2018). Improved prediction of blood-brain barrier permeability through machine learning with combined use of molecular property-based descriptors and fingerprints. *AAPS J.* 20: 54.

**78** Wang, Z., Yang, H., Wu, Z. et al. (2018). In silico prediction of blood-brain barrier permeability of compounds by machine learning and resampling methods. *ChemMedChem* 13: 2189–2201.

**79** Roy, D., Hinge, V.K., and Kovalenko, A. (2019). To pass or not to pass: predicting the blood–brain barrier permeability with the 3D-RISM-KH molecular solvation theory. *ACS Omega* 4: 16774–16780.

**80** Singh, M., Divakaran, R., Konda, L.S.K., and Kristam, R. (2020). A classification model for blood brain barrier penetration. *J. Mol. Graphics Modell.* 96: 107516.

**81** Kunwittaya, S., Nantasenamat, C., Treeratanapiboon, L. et al. (2013). Influence of logBB cut-off on the prediction of blood-brain barrier permeability. *Biomed. Appl. Technol.J.* 1: 16–34.

**82** Zhang, Y.Y., Liu, H., Summerfield, S.G. et al. (2016). Integrating in silico and in vitro approaches to predict drug accessibility to the central nervous system. *Mol. Pharm.* 13: 1540–1550.

**83** Ghasemi, F., Mehridehnavi, A., Pérez-Garrido, A., and Pérez-Sánchez, H. (2018). Neural network and deep-learning algorithms used in QSAR studies: merits and drawbacks. *Drug Discovery Today* 23: 1784–1790.

**84** Krogh, A. (2008). What are artificial neural networks? *Nat. Biotechnol.* 26: 195–197.

**85** Garg, P. and Verma, J. (2006). In silico prediction of blood brain barrier permeability: an Artificial Neural Network model. *J. Chem. Inf. Model.* 46: 289–297.

**86** Hemmateenejad, B., Miri, R., Safarpour, M.A., and Mehdipour, A.R. (2006). Accurate prediction of the blood–brain partitioning of a large set of solutes using ab initio calculations and genetic neural network modeling. *J. Comput. Chem.* 27: 1125–1135.

**87** Guerra, A., Páez, J.A., and Campillo, N.E. (2008). Artificial neural networks in ADMET modeling: prediction of blood–brain barrier permeation. *QSAR Comb. Sci.* 27: 586–594.

**88** Jing, Y., Bian, Y., Hu, Z. et al. (2018). Correction to: deep learning for drug design: an artificial intelligence paradigm for drug discovery in the big data era. *AAPS J.* 20: 79.

**89** Ma, J., Sheridan, R.P., Liaw, A. et al. (2015). Deep neural nets as a method for quantitative structure–activity relationships. *J. Chem. Inf. Model.* 55: 263–274.

**90** Miao, R., Xia, L.-Y., Chen, H.-H. et al. (2019). Improved classification of blood-brain-barrier drugs using deep learning. *Sci. Rep.* 9: 8802.

**91** Alsenan, S., Al-Turaiki, I., and Hafez, A. (2020). A Recurrent Neural Network model to predict blood–brain barrier permeability. *Comput. Biol. Chem.* 89: 107377.

**92** Korshunova, M., Ginsburg, B., Tropsha, A., and Isayev, O. (2021). OpenChem: a deep learning toolkit for computational chemistry and drug design. *J. Chem. Inf. Model.* 61: 7–13.

**93** Wu, Z., Ramsundar, B., Feinberg, E.N. et al. (2018). MoleculeNet: a benchmark for molecular machine learning. *Chem. Sci.* 9: 513–530.

**94** Verscheijden, L.F.M., Koenderink, J.B., de Wildt, S.N., and Russel, F.G.M. (2019). Development of a physiologically-based pharmacokinetic pediatric brain model for prediction of cerebrospinal fluid drug concentrations and the influence of meningitis. *PLoS Comput. Biol.* 15: e1007117.

**95** Profaci, C.P., Munji, R.N., Pulido, R.S., and Daneman, R. (2020). The blood–brain barrier in health and disease: Important unanswered questions. *J. Exp. Med.* 217: e20190062.

# Index

Note: Page numbers in *italics* refer to figures; those in **bold** to tables.

## *b*

## q